2016 IEEE Applied Power Electronics Conference and Exposition (APEC 2016)

Long Beach, California, USA
20-24 March 2016

Pages 715-1467

IEEE Catalog Number: CFP16APE-POD
ISBN: 978-1-4673-9551-9

**Copyright © 2016 by the Institute of Electrical and Electronic Engineers, Inc
All Rights Reserved**

Copyright and Reprint Permissions: Abstracting is permitted with credit to the source. Libraries are permitted to photocopy beyond the limit of U.S. copyright law for private use of patrons those articles in this volume that carry a code at the bottom of the first page, provided the per-copy fee indicated in the code is paid through Copyright Clearance Center, 222 Rosewood Drive, Danvers, MA 01923.

For other copying, reprint or republication permission, write to IEEE Copyrights Manager, IEEE Service Center, 445 Hoes Lane, Piscataway, NJ 08854. All rights reserved.

***This publication is a representation of what appears in the IEEE Digital Libraries. Some format issues inherent in the e-media version may also appear in this print version.**

IEEE Catalog Number: CFP16APE-POD
ISBN (Print-On-Demand):   978-1-4673-9551-9
ISBN (Online):   978-1-4673-9550-2
ISSN: 1048-2334

Additional Copies of This Publication Are Available From:

Curran Associates, Inc
57 Morehouse Lane
Red Hook, NY 12571 USA
Phone: (845) 758-0400
Fax: (845) 758-2633
E-mail: curran@proceedings.com
Web: www.proceedings.com

TECHNICAL PAPERS

Session T01: Three-Phase AC-DC Converters
Location: 101A
March 22, 2016 8:30 - 12:00
Session Chairs: Gerry Moschopoulos, *Western University, Canada*
Patrick Wheeler, *University of Nottingham*

Hardware Implementation and Characterization of SiC-Based Hybrid Three-Phase Rectifier Employing Third Harmonic Injection 1
M. Makoschitz, *Technische Universität Wien, Austria*
M. Hartmann, *Schneider Electric SE, Austria*
H. Ertl, *Technische Universität Wien, Austria*

Voltage Oriented Control of the Three-Level Vienna Rectifier using Vector Control Method 9
Jeevan Adhikari, *National University of Singapore, Singapore*
Prasanna IV, *National University of Singapore, Singapore*
S.K. Panda, *National University of Singapore, Singapore*

Compensation of Neutral Point Deviation in 3-Level NPC Converter Under Unbalanced Grid Conditions 17
Kyungsub Jung, *Chungbuk National University, Korea, South*
Yongsug Suh, *Chungbuk National University, Korea, South*

High Power Factor Modular Polyphase AC/DC Converters with Galvanic Isolation based on Resistor Emulators 25
Javier Sebastián, *Universidad de Oviedo, Spain*
Ignacio Castro, *Universidad de Oviedo, Spain*
Diego G. Lamar, *Universidad de Oviedo, Spain*
Aitor Vázquez, *Universidad de Oviedo, Spain*
Kevin Martín, *Universidad de Oviedo, Spain*

Reduced Duty-Cycle Loss and Output Inductor Current Ripple in a ZVS Switched Three-Phase Isolated PWM Rectifier 33
Jahangir Afsharian, *Ryerson University, Canada*
Dewei David Xu, *Ryerson University, Canada*
Tao Zhao, *Ryerson University, Canada*
Bing Gong, *Murata Power Solution, Canada*
Zhihua Yang, *Murata Power Solution, Canada*

Analysis, Design, and Evaluation of Three-Phase Three-Wire Isolated AC-DC Converter Implemented with Three Single-Phase Converter Modules 38
Laszlo Huber, *Delta Products Corporation, United States*
Misha Kumar, *Delta Products Corporation, United States*
Milan M. Jovanović, *Delta Products Corporation, United States*
Dinggang Ping, *Delta Electronics Shanghai Co., Ltd., China*
Gang Liu, *Delta Electronics Shanghai Co., Ltd., China*

Startup Procedure for Three-Phase Three-Wire Isolated AC-DC Converter Implemented with Three Single-Phase Converter Modules 46
Misha Kumar, *Delta Products Corporation, United States*
Laszlo Huber, *Delta Products Corporation, United States*
Milan M. Jovanović, *Delta Products Corporation, United States*
Dinggang Ping, *Delta Electronics Shanghai Co., Ltd., China*
Gang Liu, *Delta Electronics Shanghai Co., Ltd., China*

Control of a Single-Stage Three-Phase Boost Power Factor Correction Rectifier 54
Ayan Mallik, *University of Maryland, United States*
Bryan Faulkner, *Virginia Polytechnic Institute and State University, United States*
Alireza Khaligh, *University of Maryland, United States*

A Bidirectional Single-Stage Three-Phase Rectifier with High-Frequency Isolation and Power Factor Correction 60
Bruno Ricardo de Almeida, *Universidade Federal do Ceará, Brazil*
Demercil de Souza Oliveira Jr., *Universidade Federal do Ceará, Brazil*
Paulo P. Praça, *Universidade Federal do Ceará, Brazil*

Session T02: High Frequency and Fast-Response DC-DC Converters
Location: 104A
March 22, 2016 8:30 - 12:00
Session Chairs: Olivier Trescases, *University of Toronto*
Jeff Nilles, *Texas Instruments*

A 5 MHz, 12 V, 10 A, Monolithically Integrated Two-Phase Series Capacitor Buck Converter 66
Pradeep S. Shenoy, *Texas Instruments Inc., United States*
Orlando Lazaro, *Texas Instruments Inc., United States*
Ramanathan Ramani, *Texas Instruments Inc., United States*
Mike Amaro, *Texas Instruments Inc., United States*
Wlodek Wiktor, *Texas Instruments Inc., United States*
Joseph Khayat, *Texas Instruments Inc., United States*
Brian Lynch, *Texas Instruments Inc., United States*

A 10-MHz Isolated Class-Φ_2 Synchronous Resonant DC-DC Converter 73
Yuan Zhou, *Nanjing University of Aeronautics and Astronautics, China*
Zhiliang Zhang, *Nanjing University of Aeronautics and Astronautics, China*
Xue-Wen Zou, *Nanjing University of Aeronautics and Astronautics, China*
Zhou Dong, *Nanjing University of Aeronautics and Astronautics, China*
Xiaoyong Ren, *Nanjing University of Aeronautics and Astronautics, China*

865 MHz Switching-Speed Step-Down DC-DC Power Converter for Envelope Tracking 79
Vivek Mehrotra, *Teledyne Scientific Company, United States*
Andrea Arias, *Teledyne Scientific Company, United States*
Joshua Bergman, *Teledyne Scientific Company, United States*
Charles Neft, *Teledyne Scientific Company, United States*
Miguel Urteaga, *Teledyne Scientific Company, United States*
Berinder Brar, *Teledyne Scientific Company, United States*

Current Parking Regulator for Zero Droop/Overshoot Load Transient Response 86
Sudhir S. Kudva, *Nvidia Corporation, United States*
William J. Dally, *Nvidia Corporation, United States*
Thomas H. Greer III, *Nvidia Corporation, United States*
C. Thomas Gray, *Nvidia Corporation, United States*

A 5MHz, 24V-to-1.2V, AO^2T Current Mode Buck Converter with One-Cycle Transient Response and Sensorless Current Detection for Medical Meters 94

Xugang Ke, *University of Texas at Dallas, United States*
Joseph Sankman, *Texas Instruments Inc., United States*
Dongsheng Ma, *University of Texas at Dallas, United States*

Capacitively-Aided Switching Technique for High-Frequency Isolated Bus Converters 98

Seungbum Lim, *Massachusetts Institute of Technology, United States*
Alex J. Hanson, *Massachusetts Institute of Technology, United States*
Juan A. Santiago-González, *Massachusetts Institute of Technology, United States*
David J. Perreault, *Massachusetts Institute of Technology, United States*

A 10 MHz, 48-to-5V Synchronous Converter with Dead Time Enabled 125 ps Resolution Zero-Voltage Switching 106

Alexander Barner, *Robert Bosch GmbH, Germany*
Jürgen Wittmann, *Hochschule Reutlingen, Germany*
Thoralf Rosahl, *Robert Bosch GmbH, Germany*
Bernhard Wicht, *Hochschule Reutlingen, Germany*

Plug-and-Play Electronic Capacitor for VRM Applications 111

Or Kirshenboim, *Ben-Gurion University of the Negev, Israel*
Alon Cervera, *Ben-Gurion University of the Negev, Israel*
Bar Halivni, *Ben-Gurion University of the Negev, Israel*
Eli Abramov, *Ben-Gurion University of the Negev, Israel*
Mor Mordechai Peretz, *Ben-Gurion University of the Negev, Israel*

Adaptive Voltage Positioning (AVP) Design of Multi-Phase Constant On-Time I^2 Control for Voltage Regulators with Ramp Compensations 118

Kuang-Yao Cheng, *Texas Instruments Inc., United States*
Yipeng Su, *Texas Instruments Inc., United States*

Session T03: Microgrids and Hybrid Systems
Location: 104B
March 22, 2016 8:30 - 12:00
Session Chairs: Yunwei Li, *University of Alberta*
Joesep Guerrero, *Aalborg University*

Reactive Power Support Capabilities of Nonsynchronous Interconnection Systems in Microgrid Applications 125

Yong-Duk Lee, *University of Connecticut, United States*
Sung-Yeul Park, *University of Connecticut, United States*

Zero Standby Power High Efficiency Hot Plugging Outlet for 380VDC Power Delivery System 132

Kai Tan, *North Carolina State University, United States*
Chang Peng, *North Carolina State University, United States*
Pengkun Liu, *North Carolina State University, United States*
Xiaoqing Song, *North Carolina State University, United States*
Alex Q. Huang, *North Carolina State University, United States*

Design of Control System for Smooth Mode Transfer in Smart Microgrid Application 138
Mingzhi Gao, *Zhejiang University, China*
Canhui Zhang, *Zhejiang University, China*
Maohang Qiu, *Zhejiang University, China*
Min Chen, *Zhejiang University, China*
Aron Levy, *Technology Dynamics Inc., United States*

Resonance Propagation Modeling and Analysis of AC Filters in a Large-Scale Microgrid ... 143
Yusi Liu, *University of Arkansas, United States*
Chris Farnell, *University of Arkansas, United States*
H. Alan Mantooth, *University of Arkansas, United States*
Juan Carlos Balda, *University of Arkansas, United States*
Roy A. McCann, *University of Arkansas, United States*
Cheng Deng, *University of Arkansas, United States*

A New Bidirectional DC-DC Converter for Fuel Cell, Solar Cell and Battery Systems 150
Ankur Patel, *Vicor Corporation, United States*

A Multiport Isolated DC-DC Converter 156
Yan-Kim Tran, *École Polytechnique Fédérale de Lausanne, Switzerland*
Drazen Dujic, *École Polytechnique Fédérale de Lausanne, Switzerland*

A Seamless Transfer Control Method with High Load Sharing Performance for Modular ESS ... 163
Jung-Hoon Ahn, *Sungkyunkwan University, Korea, South*
Won-Yong Sung, *Sungkyunkwan University, Korea, South*
Chang-Yeol Oh, *Sungkyunkwan University, Korea, South*
Byoung-Kuk Lee, *Sungkyunkwan University, Korea, South*
Yun-Sung Kim, *Dongahelecomm Corporation, Korea, South*

A Plug-and-Play Ripple Mitigation Approach for DC-Links in Hybrid Systems 169
Sinan Li, *University of Hong Kong, Hong Kong*
Albert T.L. Lee, *University of Hong Kong, Hong Kong*
Siew-Chong Tan, *University of Hong Kong, Hong Kong*
S.Y. Ron Hui, *University of Hong Kong, Hong Kong*

**Active Control of Low Frequency Common Mode Voltage to Connect AC Utility and 380
V DC Grid** 177
Fang Chen, *Virginia Polytechnic Institute and State University, United States*
Rolando Burgos, *Virginia Polytechnic Institute and State University, United States*
Dushan Boroyevich, *Virginia Polytechnic Institute and State University, United States*
Xuning Zhang, *Virginia Polytechnic Institute and State University, United States*

Session T04: Control Strategies for Inverters and Motor Drives
Location: 103C
March 22, 2016 8:30 - 12:00
Session Chairs: Bilal Akin, *Univeristy of Texas, Dallas*
Babak Nahid-Mobarakeh, *University of Lorraine*

**A Three-Level Space Vector Modulation Scheme for Paralleled Two Converters to
Reduce Zero-Sequence Circulating Current and Common Mode Voltage** 185
Zhongyi Quan, *University of Alberta, Canada*
Yunwei Li, *University of Alberta, Canada*

Nonlinearity Analysis and Linear Modulation Method for Two Level Voltage Source Inverter with Low Switching to Operating Frequency Ratio 193
Yongjae Lee, *Seoul National University, Korea, South*
Jung-Ik Ha, *Seoul National University, Korea, South*

Synchronization Strategies in Cascaded H-Bridge Multi Level Inverters for Carrier based Sinusoidal PWM Techniques 199
Saroj Kumar Sahoo, *Indian Institute of Technology Kharagpur, India*
Tanmoy Bhattacharya, *Indian Institute of Technology Kharagpur, India*

Design and Implementation of a Sinusoidal Flux Controller for Core Loss Measurements 207
Burak Tekgun, *University of Akron, United States*
Ali R. Boynuegri, *University of Akron, United States*
Md Asif Mahmood Chowdhury, *University of Akron, United States*
Yilmaz Sozer, *University of Akron, United States*

Implementation of Deadbeat-Direct Torque and Flux Control for Synchronous Reluctance Machines to Minimize Loss Each Switching Period 215
Michael Saur, *Universität der Bundeswehr München, Germany*
Francisco Ramos, *Universität der Bundeswehr München, Germany*
Aday Perez, *Universität der Bundeswehr München, Germany*
Dieter Gerling, *Universität der Bundeswehr München, Germany*
Robert D. Lorenz, *University of Wisconsin at Madison, United States*

Addressing the Unbalance Loading Issue in Multi-Drive Systems with a DC-Link Modulation Scheme for Harmonic Reduction 221
Yongheng Yang, *Aalborg University, Denmark*
Pooya Davari, *Aalborg University, Denmark*
Firuz Zare, *Danfoss Power Electronics A/S, Denmark*
Frede Blaabjerg, *Aalborg University, Denmark*

Input Current Interharmonics in Adjustable Speed Drives Caused by Fixed-Frequency Modulation Techniques 229
Hamid Soltani, *Aalborg University, Denmark*
Pooya Davari, *Aalborg University, Denmark*
Poh Chiang Loh, *Aalborg University, Denmark*
Frede Blaabjerg, *Aalborg University, Denmark*
Firuz Zare, *Danfoss Power Electronics A/S, Denmark*

Low-Frequency Voltage Ripples in the Flying Capacitors of the Nested Neutral-Point-Clamped Converter 236
Amer M.Y.M. Ghias, *University of Sharjah, U.A.E.*
Josep Pou, *University of New South Wales, Australia*
Salvador Ceballos, *TECNALIA, Spain*
Vassilios G. Agelidis, *University of New South Wales, Australia*

DC Bus Capacitor Discharge of Permanent Magnet Synchronous Machine Drive Systems for Hybrid Electric Vehicles 241
Ziwei Ke, *Oregon State University, United States*
Julia Zhang, *Oregon State University, United States*
Michael W. Degner, *Ford Motor Company, United States*

Session T05: Si Devices and Power Module Packaging

Location: 101B

March 22, 2016 8:30 - 12:00

Session Chairs: Iulian Nistor, *Corporate Research, ABB Inc.*

Brian Rowden,

C_{OSS} Hysteresis in Advanced Superjunction MOSFETs 247

J.B. Fedison, *Enphase Energy, Inc., United States*

M.J. Harrison, *Enphase Energy, Inc., United States*

Compact Electrothermal Models for Unbalanced Parallel Conducting Si-IGBTs 253

Roozbeh Bonyadi, *University of Warwick, United Kingdom*

Olayiwola Alatise, *University of Warwick, United Kingdom*

Ji Hu, *University of Warwick, United Kingdom*

Zarina Davletzhanova, *University of Warwick, United Kingdom*

Yeganeh Bonyadi, *University of Warwick, United Kingdom*

Jose Ortiz-Gonzalez, *University of Warwick, United Kingdom*

Li Ran, *University of Warwick, United Kingdom*

Philip Mawby, *University of Warwick, United Kingdom*

General 3D Lumped Thermal Model with Various Boundary Conditions for High Power IGBT Modules 261

Amir Sajjad Bahman, *Aalborg University, Denmark*

Ke Ma, *Aalborg University, Denmark*

Frede Blaabjerg, *Aalborg University, Denmark*

Improved 6.5kV FREEMD-Pair based on SiC JFET and Si IGBT 269

Xiaoqing Song, *North Carolina State University, United States*

Alex Q. Huang, *North Carolina State University, United States*

Chang Peng, *North Carolina State University, United States*

Liqi Zhang, *North Carolina State University, United States*

On the Comparative Assessment of 1.7 kV, 300 a Full SiC-MOSFET and Si-IGBT Power Modules 276

Muhammad Nawaz, *ABB Corporate Research, Sweden*

Kalle Ilves, *ABB Corporate Research, Sweden*

Suppression of Reverse Recovery Ringing 3.3kV/450A Si/SiC Hybrid in Low Internal Inductance Package Next High Power Density Dual; nHPD^2 283

Katsuaki Saito, *Hitachi Europe Ltd., United Kingdom*

Daisuke Kawase, *Hitachi Power Semiconductor, Ltd., Japan*

Masamitsu Inaba, *Hitachi Power Semiconductor, Ltd., Japan*

Keiichi Yamamoto, *Hitachi Power Semiconductor, Ltd., Japan*

Katsunori Azuma, *Hitachi Power Semiconductor, Ltd., Japan*

Seiichi Hayakawa, *Hitachi Power Semiconductor, Ltd., Japan*

New Layout Concepts in MW-Scale IGBT Modules for Higher Robustness during Normal and Abnormal Operations 288

Paula Diaz Reigosa, *Aalborg University, Denmark*

Francesco Iannuzzo, *Aalborg University, Denmark*

Stig Munk-Nielsen, *Aalborg University, Denmark*

Frede Blaabjerg, *Aalborg University, Denmark*

Design, Package, and Hardware Verification of a High Voltage Current Switch 295

Ankan De, *North Carolina State University, United States*
Adam Morgan, *North Carolina State University, United States*
Vishnu Mahadeva Iyer, *North Carolina State University, United States*
Haotao Ke, *North Carolina State University, United States*
Xin Zhao, *North Carolina State University, United States*
Kasunaidu Vechalapu, *North Carolina State University, United States*
Subhashish Bhattacharya, *North Carolina State University, United States*
Douglas C. Hopkins, *North Carolina State University, United States*

Investigation of Short Circuit in a IGBT Power Module with Three-Level Neutral Point Clamped Type 2 (NPC2, T-NPC, Mixed Voltage) Topology 303

Kevin Lenz, *Danfoss Silicon Power, Germany*
Vladan Jerinic, *Danfoss Silicon Power, Germany*
Reiner Hinken, *Danfoss Silicon Power, Germany*

Session T06: DC-DC Converter Control
Location: 102AB
March 22, 2016 8:30 - 12:00
Session Chairs: Sombuddha Chakraborty, *Texas Instruments*
Rafael Pena Alzola, *University of British Columbia*

Closed-Loop Design and Time-Optimal Control for a Series-Capacitor Buck Converter 308

Timur Vekslender, *Ben-Gurion University of the Negev, Israel*
Ofer Ezra, *Ben-Gurion University of the Negev, Israel*
Yevgeny Bezdenezhnykh, *Ben-Gurion University of the Negev, Israel*
Mor Mordechai Peretz, *Ben-Gurion University of the Negev, Israel*

Unified Constant On/Off-Time Hybrid Compensation for Fast Recovery in Digitally Current-Mode Controlled Point-of-Load Converters 315

K. Hariharan, *Indian Institute of Technology Kharagpur, India*
Santanu Kapat, *Indian Institute of Technology Kharagpur, India*
Siddhartha Mukhopadhyay, *Indian Institute of Technology Kharagpur, India*

Digital Implementation of Adaptive Synchronous Rectifier (SR) Driving Scheme for LLC Resonant Converters 322

Chao Fei, *Virginia Polytechnic Institute and State University, United States*
Fred C. Lee, *Virginia Polytechnic Institute and State University, United States*
Qiang Li, *Virginia Polytechnic Institute and State University, United States*

Digital Synchronous Rectification Controller for LLC Resonant Converters 329

Maryam S. Amouzandeh, *University of Toronto, Canada*
Behzad Mahdavikhah, *University of Toronto, Canada*
Aleksandar Prodić, *University of Toronto, Canada*
Brent McDonald, *Texas Instruments Inc., United States*

A Novel Adaptive Synchronous Rectification Method for Digitally Controlled LLC Converters ... 334

Fan Wang, *Texas Instruments Inc., United States*
Brent A. McDonald, *Texas Instruments Inc., United States*
Jeff Langham, *Texas Instruments Inc., United States*
Bo Fan, *Texas Instruments Inc., China*

Influence of the ADC Zero Bin on the Performance of an Integrated DC-DC Converter 339
S. Vesti, *Infineon Technologies Austria AG, Austria*
M. Agostinelli, *Infineon Technologies Austria AG, Austria*
H. Koltsov, *Infineon Technologies Austria AG, Austria*
S. Marsili, *Infineon Technologies Austria AG, Austria*

Improved Current-Mode Control with Single-Cycle Load Transient 343
Virginia Li, *Virginia Polytechnic Institute and State University, United States*
Pei-Hsin Liu, *Virginia Polytechnic Institute and State University, United States*
Qiang Li, *Virginia Polytechnic Institute and State University, United States*
Fred C. Lee, *Virginia Polytechnic Institute and State University, United States*

A Mixed-Signal Ripple-Based Controller for a 16 V, 10 MHz Integrated Buck Converter 350
Sergii Tkachov, *Infineon Technologies Austria AG, Austria*
Matteo Agostinelli, *Infineon Technologies Austria AG, Austria*

**New Control Concept for Soft-Switching Flyback Converters with Very High
Switching Frequency** 355
A.M. Connaughton, *Technische Universität Graz, Austria*
K. Krischan, *Technische Universität Graz, Austria*
K.K. Leong, *Infineon Technologies AG, Austria*
A. Muetze, *Technische Universität Graz, Austria*

Session T07: Solar Energy Systems
Location: 104C
March 22, 2016 8:30 - 12:00
Session Chairs: Babak Fahimi, *UT- Dallas*
Morgan Kiani, *Texas Christian University*

**Analysis, Modeling and Control of an Interleaved Isolated Boost Series Resonant
Converter for Microinverter Applications** 362
Luciano A. Garcia-Rodriguez, *University of Arkansas, United States*
Cheng Deng, *University of Arkansas, United States*
Juan Carlos Balda, *University of Arkansas, United States*
Andrés Escobar-Mejía, *Universidad Tecnologica de Pereira, Colombia*

**Benchmarking of Constant Power Generation Strategies for Single-Phase Grid-
Connected Photovoltaic Systems** 370
Ariya Sangwongwanich, *Aalborg University, Denmark*
Yongheng Yang, *Aalborg University, Denmark*
Frede Blaabjerg, *Aalborg University, Denmark*
Huai Wang, *Aalborg University, Denmark*

**Advanced Slip Mode Frequency Shift Islanding Detection Method for Single Phase Grid
Connected PV Inverters** 378
Bahador Mohammadpour, *Queen's University, Canada*
Majid Pahlevani, *Queen's University, Canada*
Sajjad Makhdoomi Kaviri, *Queen's University, Canada*
Praveen Jain, *Queen's University, Canada*

Direct MPPT Control of PWM Converters for Extreme Transient PV Applications 386
Ignacio Galiano Zurbriggen, *University of British Columbia, Canada*
Francisco Paz, *University of British Columbia, Canada*
Martin Ordonez, *University of British Columbia, Canada*

Feeding Partial Power into Line Capacitors for Low Cost and Efficient MPPT of Photovoltaic Strings 392
Ali Elrayyah, *Qatar Environment and Energy Research Institute, Qatar*
Mohammed Badawey, *University of Akron, United States*
Yilmaz Sozer, *University of Akron, United States*

Single Phase Cascaded H5 Inverter with Leakage Current Elimination for Transformerless Photovoltaic System 398
Xiaoqiang Guo, *Yanshan University, China*
Xiaoyu Jia, *Yanshan University, China*
Zhigang Lu, *Yanshan University, China*
Josep M. Guerrero, *Aalborg University, Denmark*

Optimal Low Switching Frequency Pulse Width Modulation of Current-Fed Three-Level Inverter for Solar Integration 402
Gnana Sambandam Kulothungan, *National University of Singapore, Singapore*
Akshay K. Rathore, *National University of Singapore, Singapore*
Amarendra Edpuganti, *National University of Singapore, Singapore*
Dipti Srinivasan, *National University of Singapore, Singapore*

Low Leakage Current Single-Phase PV Inverters with Universal Neutral-Point-Clamping Method 410
Liwei Zhou, *Shandong University, China*
Feng Gao, *Shandong University, China*

Modular Subpanel Photovoltaic Converter System: Analysis and Control 417
Yuan Li, *Sichuan University / Northeastern University, China*
Yue Zheng, *Northeastern University, United States*
Su Sheng, *Northeastern University, United States*
Brad Scandrett, *PowerFilm, Inc., United States*
Brad Lehman, *Northeastern University, United States*

Session T08: Advanced Converter for Power Systems used in Transportation
Location: 103AB
March 22, 2016 8:30 - 12:00
Session Chairs: Omer Onar, *Oak Ridge National Laboratory*
Khurram Afridi, *University of Colorado, Boulder*

Integrated DC-DC Converter Design for Electric Vehicle Powertrains 424
Saeed Anwar, *University of Tennessee, United States*
Weimin Zhang, *University of Tennessee, United States*
Fred Wang, *University of Tennessee, United States*
Daniel J. Costinett, *University of Tennessee, United States*

A 1 MHz Bi-Directional Soft-Switching DC-DC Converter with Planar Coupled Inductor for Dual Voltage Automotive Systems 432
Chenhao Nan, *Arizona State University, United States*
Raja Ayyanar, *Arizona State University, United States*

A Bridgeless Totem-Pole Interleaved PFC Converter for Plug-In Electric Vehicles 440
Yichao Tang, *University of Maryland, United States*
Weisheng Ding, *University of Maryland, United States*
Alireza Khaligh, *University of Maryland, United States*

Stability Analysis of Hybrid AC/DC Power Systems for More Electric Aircraft 446
Mehdi Karbalaye Zadeh, *Norwegian University of Science and Technology, Norway*
Roghayeh Gavagsaz-Ghoachani, *Université de Lorraine, France*
Babak Nahid-Mobarakeh, *Université de Lorraine, France*
Serge Pierfederici, *Université de Lorraine, France*
Marta Molinas, *Norwegian University of Science and Technology, Norway*

On the Concept of the Multi-Source Inverter 453
Lea Dorn-Gomba, *McMaster University, Canada*
Pierre Magne, *McMaster University, Canada*
Clement Barthelmebs, *McMaster University, Canada*
Ali Emadi, *McMaster University, Canada*

Time-Domain Analysis of a Wide-DC-Range Series Resonant Dual-Active-Bridge Bidirectional Converter with a New Passive Auxilliary Circuit 460
Alireza Safaee, *Queen's University, Canada*
Praveen Jain, *Queen's University, Canada*
Alireza Bakhshai, *Queen's University, Canada*

A New High Capacity Compact Power Modules for High Power EV/HEV Inverters 468
Seiichiro Inokuchi, *Mitsubishi Electric Corporation, Japan*
Shoji Saito, *Mitsubishi Electric Corporation, Japan*
Arata Izuka, *Mitsubishi Electric Corporation, Japan*
Yuki Hata, *Mitsubishi Electric Corporation, Japan*
Shinji Hatae, *Mitsubishi Electric Corporation, Japan*
Toshiya Nakano, *Powerex, Inc., United States*
Eric R. Motto, *Powerex, Inc., United States*

Modular Pet, Two-Phase Air-Cooled Converter Cell Design and Performance Evaluation with 1.7kV IGBTs for MV Applications 472
Frederick Kieferndorf, *ABB Switzerland Ltd, Switzerland*
Uwe Drofenik, *ABB Switzerland Ltd, Switzerland*
Francesco Agostini, *ABB Switzerland Ltd, Switzerland*
Francisco Canales, *ABB Switzerland Ltd, Switzerland*

A Phase Shift Full Bridge based Reconfigurable PEV Onboard Charger with Extended ZVS Range and Zero Duty Cycle Loss 480
Haoyu Wang, *ShanghaiTech University, China*

Session T09: Gate Drives, Failure Analysis, and Protection
Location: 102C
March 22, 2016 8:30 - 12:00
Session Chairs: Zhiliang Zhang, *Nanjing University of Aeronautics and Astronautics*
Indumini Ranmuthu, *Texas Instruments*

Series Arc Fault Detection Method based on Statistical Analysis for DC Microgrids 487
Gab-Su Seo, *Seoul National University, Korea, South*
Jung-Ik Ha, *Seoul National University, Korea, South*
Bo-Hyung Cho, *Seoul National University, Korea, South*
Kyu-Chan Lee, *Smart Power Supply Co., Ltd., Korea, South*

Arc Welding Inverter with Embedded Digital Active EMI Controller 493
Junpeng Ji, *Xi'an Jiaotong University, China*
Wenjie Chen, *Xi'an Jiaotong University, China*
Xu Yang, *Xi'an Jiaotong University, China*

A Thermo-Sensitive Electrical Parameter with Maximum dI_C/dt during Turn-Off for High Power Trench/Field-Stop IGBT Modules 499
Yuxiang Chen, *Zhejiang University, China*
Haoze Luo, *Zhejiang University, China*
Wuhua Li, *Zhejiang University, China*
Xiangning He, *Zhejiang University, China*
Jun Ma, *Shanghai Electric, China*
Guodong Chen, *Shanghai Electric, China*
Ye Tian, *Shanghai Electric, China*
Enxing Yang, *Shanghai Electric, China*

A Software Frequency Response Analysis Method to Monitor Degradation of Power MOSFETs in Basic Single-Switch Converters 505
Serkan Dusmez, *University of Texas at Dallas, United States*
Manish Bhardwaj, *Texas Instruments Inc., United States*
Lei Sun, *University of Texas at Dallas, United States*
Bilal Akin, *University of Texas at Dallas, United States*

A New Capacitance Estimation Method of Supercapacitor Bank using a Bank Impedance and Current Injection 511
Junwon Lee, *Chungnam National University, Korea, South*
Hyunsik Jo, *Chungnam National University, Korea, South*
Hanju Cha, *Chungnam National University, Korea, South*

Gate Driver Design for 1.7kV SiC MOSFET Module with Rogowski Current Sensor for Shortcircuit Protection 516
Jun Wang, *Virginia Polytechnic Institute and State University, United States*
Zhiyu Shen, *Virginia Polytechnic Institute and State University, United States*
Christina Dimarino, *Virginia Polytechnic Institute and State University, United States*
Rolando Burgos, *Virginia Polytechnic Institute and State University, United States*
Dushan Boroyevich, *Virginia Polytechnic Institute and State University, United States*

2 MHz High-Density Integrated Power Supply for Gate Driver in High-Temperature Applications 524

Remi Perrin, *Université Claude Bernard Lyon 1, France*
Bruno Allard, *Université Claude Bernard Lyon 1, France*
Cyril Buttay, *Université Claude Bernard Lyon 1, France*
Nicolas Quentin, *Université Claude Bernard Lyon 1, France*
Wenli Zhang, *Virginia Polytechnic Institute and State University, United States*
Rolando Burgos, *Virginia Polytechnic Institute and State University, United States*
Dushan Boroyevich, *Virginia Polytechnic Institute and State University, United States*
Philippe Preciat, *Labinal Power Systems, France*
Donatien Martineau, *Labinal Power Systems, France*

Design Consideration of Gate Driver Circuits and PCB Parasitic Parameters of Paralleled E-Mode GaN HEMTs in Zero-Voltage-Switching Applications 529

Juncheng Lu, *Kettering University, United States*
Hua Bai, *Kettering University, United States*
Alan Brown, *Hella Corporate Center USA Inc., United States*
Matt McAmmond, *Hella Corporate Center USA Inc., United States*
Di Chen, *GaN Systems Inc., Canada*
Julian Styles, *GaN Systems Inc., Canada*

A Gate Driver of SiC MOSFET for Suppressing the Negative Voltage Spikes in a Bridge Circuit 536

Qi Zhou, *Shandong University, China*
Feng Gao, *Shandong University, China*

Session T10: Control of AC-DC Converters
Location: 102AB
March 23, 2016 8:30 - 10:10
Session Chairs: Tsorng-Juu Liang, *National Cheng-Kung University (Taiwan)*
Laszlo Balogh, *Fairchild Semiconductor*

Interleaved Boost based AC/DC Bidirectional Converter with Four Quadrant Power Control based on One-Cycle Controller (OCC) 544

Snehal Bagawade, *Queen's University, Canada*
Praveen Jain, *Queen's University, Canada*

A New Control Scheme to Improve Load Transient Response of Single Phase PWM Rectifier with Auxiliary Current Injection Circuit 552

Naga Brahmendra Yadav Gorla, *National University of Singapore, Singapore*
Sandeep Kolluri, *National University of Singapore, Singapore*
Pritam Das, *National University of Singapore, Singapore*
Sanjib Kumar Panda, *National University of Singapore, Singapore*

Active Capacitor with Ripple-Based Duty Cycle Modulation for AC-DC Applications 558

Ching-Chieh Yang, *National Taiwan University, Taiwan*
Yang-Lin Chen, *National Taiwan University, Taiwan*
Yaow-Ming Chen, *National Taiwan University, Taiwan*

Novel Approach to Current-Mode Control in DCM/CCM Boundary Boost PFC 564

Giovanni Gritti, *STMicroelectronics, Italy*
Claudio Adragna, *STMicroelectronics, Italy*

Reducing the Switching Frequency Variation Range for CRM Buck PFC Converter by Variable On-Time Control 572

Xiaoping Wang, *Nanjing University of Science and Technology, China*
Kai Yao, *Nanjing University of Science and Technology, China*
Junfang Zhang, *Nanjing University of Science and Technology, China*

Session T11: GaN-Based DC-DC Converters
Location: 104A
March 23, 2016 8:30 - 10:10
Session Chairs: Alexis Kwasinski, *University of Pittsburgh*
Regan Zane, *Utah State*

High Efficiency 20-400 MHz PWM Converters using Air-Core Inductors and Monolithic Power Stages in a Normally-Off GaN Process 580

Alihossein Sepahvand, *University of Colorado at Boulder, United States*
Yuanzhe Zhang, *University of Colorado at Boulder, United States*
Dragan Maksimović, *University of Colorado at Boulder, United States*

Thermal Evaluation of Chip-Scale Packaged Gallium Nitride Transistors 587

David Reusch, *Efficient Power Conversion Corporation, United States*
Johan Strydom, *Efficient Power Conversion Corporation, United States*
Alex Lidow, *Efficient Power Conversion Corporation, United States*

Over 300kHz GaN Device based Resonant Bidirectional DCDC Converter with Integrated Magnetics 595

Gang Liu, *Fudan University, China*
Dan Li, *Fudan University, China*
Yungtaek Jang, *Delta Products Corporation, United States*
Jianqiu Zhang, *Fudan University, China*

Effective Control & Software Techniques for High Efficiency GaN FET based Flexible Electrical Power System for Cube-Satellites 601

Ashish Shrivastav, *North Carolina State University, United States*
Shikhar Singh, *IBM, United States*
Anirudh Mahajan, *North Carolina State University, United States*
Subhashish Bhattacharya, *North Carolina State University, United States*

A 98.8% Efficient Bidirectional Full-Bridge Isolated DC-DC GaN Converter 609

Rakesh Ramachandran, *University of Southern Denmark, Denmark*
Morten Nymand, *University of Southern Denmark, Denmark*

Session T12: Electric Machines
Location: 101A
March 23, 2016 8:30 - 10:10
Session Chairs: Bilal Akin, *Univeristy of Texas, Dallas*
Bulent Sarlioglu, *University of Wisconsin - Madison*

Comparison of Lateral- and Cylindrical-Stator Electrical Machines for High-Speed Direct-Drive Applications in Confined Spaces 615

Arda Tüysüz, *ETH Zürich, Switzerland*
Johann W. Kolar, *ETH Zürich, Switzerland*

Novel Contactless Axial-Flux Permanent-Magnet Electromechanical Energy Harvester 623
Michael Flankl, *ETH Zürich, Switzerland*
Arda Tüysüz, *ETH Zürich, Switzerland*
Ivan Subotic, *Liverpool John Moores University, United Kingdom*
Johann W. Kolar, *ETH Zürich, Switzerland*

Design of Rare-Earth Free Five-Phase Outer-Rotor IPM Motor Drive for Electric Bicycle 631
Md. Zakirul Islam, *University of Akron, United States*
Seungdeog Choi, *University of Akron, United States*

Transverse Flux Machines with Rotary Transformer Concept for Wide Speed Operations without using Permanent Magnet Material 638
Iftekhar Hasan, *University of Akron, United States*
Md Wasi Uddin, *University of Akron, United States*
Yilmaz Sozer, *University of Akron, United States*

Field Oriented Modeling and Control of Six Phase, Open-Delta Winding, Interior Permanent Magnet Synchronous Machines Considering Current Unbalance and Zero Sequence Currents 643
Murat Senol, *RWTH Aachen University, Germany*
Michael Schubert, *RWTH Aachen University, Germany*
Georges Engelmann, *RWTH Aachen University, Germany*
Rik W. De Doncker, *RWTH Aachen University, Germany*
Thorben Grosse, *RWTH Aachen University, Germany*
Kay Hameyer, *RWTH Aachen University, Germany*

Session T13: Advances in Magnetics
Location: 101B
March 23, 2016 8:30 - 10:10
Session Chairs: Matthew Wilkowski, *Enpirion*
Charles Sullivan, *Dartmouth*

Passive Integration using FMLF Technique for Integrated Boost Resonant Converters 651
Cheng Deng, *University of Arkansas, United States*
Luciano Andres Garcia Rodriguez, *University of Arkansas, United States*
Juan Zou, *Xiangtan University, China*
Juan Carlos Balda, *University of Arkansas, United States*

Magnetic Characterization Technique and Materials Comparison for Very High Frequency IVR 657
Dongbin Hou, *Virginia Polytechnic Institute and State University, United States*
Fred C. Lee, *Virginia Polytechnic Institute and State University, United States*
Qiang Li, *Virginia Polytechnic Institute and State University, United States*

Large-Signal Power Circuit Characterization of On-Silicon Coupled Inductors for High Frequency Integrated Voltage Regulation 663
S. Kulkarni, *Tyndall National Institute, Ireland*
Z. Pavlovic, *Tyndall National Institute, Ireland*
S. Kubendran, *Tyndall National Institute, Ireland*
C. Carretero, *Universidad de Zaragoza, Spain*
N. Wang, *Tyndall National Institute, Ireland*
C. O'Mathuna, *Tyndall National Institute / University College Cork, Ireland*

Point-of-Load Inductor with High Swinging and Low Loss at Light Load 668

Ting Ge, *Virginia Polytechnic Institute and State University, United States*
Khai Ngo, *Virginia Polytechnic Institute and State University, United States*
Jim Moss, *Texas Instruments Inc., United States*

Iron Loss Evaluation of Three-Phase Inductor for Three-Phase PWM Inverter 676
Hiroaki Matsumori, *Tokyo Metropolitan University, Japan*
Toshihisa Shimizu, *Tokyo Metropolitan University, Japan*
Koushi Takano, *Iwatsu Test Instrument Corporation, Japan*
Ishii Hitoshi, *Iwatsu Test Instrument Corporation, Japan*

Session T14: System Design and Layout for Improved Performance
Location: 102C
March 23, 2016 8:30 - 10:10
Session Chairs: Jeff Nilles, *Texas Instruments*
Ernie Parker, *Crane Aerospace & Electronics*

CMOS Gate Drive IC with Embedded Cross Talk Suppression Circuitry for SiC Devices 684
Jeffery Dix, *University of Tennessee, United States*
Zheyu Zhang, *University of Tennessee, United States*
Benjamin J. Blalock, *University of Tennessee, United States*

Optimal Design of a Voltage Regulator based Resonant Switched-Capacitor Converter IC 692
Eli Abramov, *Ben-Gurion University of the Negev, Israel*
Alon Cervera, *Ben-Gurion University of the Negev, Israel*
Mor Mordechai Peretz, *Ben-Gurion University of the Negev, Israel*

Novel Highly Integrated Current Measurement Method for Drive Inverters 700
N. Langmaack, *Technische Universität Braunschweig, Germany*
G. Tareilus, *Technische Universität Braunschweig, Germany*
M. Henke, *Technische Universität Braunschweig, Germany*

**A Novel DBC Layout for Current Imbalance Mitigation in SiC MOSFET Multichip
Power Modules** 704
Helong Li, *Aalborg University, Denmark*
Stig Munk-Nielsen, *Aalborg University, Denmark*
Szymon Bęczkowski, *Aalborg University, Denmark*
Xiongfei Wang, *Aalborg University, Denmark*

**A Double-End Sourced Multi-Chip Improved Wire-Bonded SiC MOSFET Power
Module Design** 709
Miao Wang, *Ohio State University, United States*
Fang Luo, *Ohio State University, United States*
Longya Xu, *Ohio State University, United States*

Session T15: Modeling of AC Energy Converters and Systems
Location: 104B
March 23, 2016 8:30 - 10:10
Session Chairs: Jaber Abu Qahouq, *The University of Alabama*
Xiongfei Wang, *Aalborg University*

Comparing Extended Kalman Filter and Particle Filter for Estimating Field and Damper Bar Currents in Brushless Wound Field Synchronous Generator for Stator Winding Fault Detection and Diagnosis 715
Sivakumar Nadarajan, *National University of Singapore, Singapore*
S.K. Panda, *National University of Singapore, Singapore*
Bicky Bhangu, *Rolls-Royce Singapore Pte. Ltd., Singapore*
Amit Kumar Gupta, *Rolls-Royce Singapore Pte. Ltd., Singapore*

Analytical Determination of Conduction Power Losses for Active Neutral-Point-Clamped Multilevel Converter 720
Vahid Dargahi, *Clemson University, United States*
Arash Khoshkbar Sadigh, *Extron Electronics, United States*
Keith Corzine, *Clemson University, United States*

Multifrequency Small-Signal Model of Voltage Source Converters Connected to a Weak Grid for Stability Analysis 728
Xing Li, *Huazhong University of Science and Technology, China*
Hua Lin, *Huazhong University of Science and Technology, China*

A New Approach to Control the Modified LinVerter for High Frequency Applications 733
Peyman Farhang, *University of Southern Denmark, Denmark*
Stefan Mátéfi-Tempfli, *University of Southern Denmark, Denmark*

Small-Signal Terminal Characteristics Modeling of Three-Phase Boost Rectifier with Variable Fundamental Frequency 739
Zeng Liu, *Xi'an Jiaotong University, China*
Jinjun Liu, *Xi'an Jiaotong University, China*
Dushan Boroyevich, *Virginia Polytechnic Institute and State University, United States*

Session T16: Manufacturing, Test, and Reliability
Location: 103C
March 23, 2016 8:30 - 10:10
Session Chairs: Jim Marinos, *Payton Group*
Brian Narveson, *Narveson Innovative Consulting*

Reliability Analysis of a High-Efficiency SiC Three-Phase Inverter for Motor Drive Applications 746
Juan Colmenares, *KTH Royal Institute of Technology, Sweden*
Diane-Perle Sadik, *KTH Royal Institute of Technology, Sweden*
Patrik Hilber, *KTH Royal Institute of Technology, Sweden*
Hans-Peter Nee, *KTH Royal Institute of Technology, Sweden*

RCP Evaluation of Electrolytic Capacitor Degradation for SMPS Failure Prediction 754
Hiroshi Nakao, *Fujitsu Laboratories Ltd., Japan*
Yu Yonezawa, *Fujitsu Laboratories Ltd., Japan*
Yoshiyasu Nakashima, *Fujitsu Laboratories Ltd., Japan*
Fujio Kurokawa, *Nagasaki University, Japan*

Modular Test System Architecture for Device, Circuit and System Level Reliability Testing 759
Roland Sleik, *Kompetenzzentrum Automobil- und Industrieelektronik GmbH, Austria*
Michael Glavanovics, *Kompetenzzentrum Automobil- und Industrieelektronik GmbH, Austria*
Sascha Einspieler, *Kompetenzzentrum Automobil- und Industrieelektronik GmbH, Austria*
Annette Muetze, *Technische Universität Graz, Austria*
Klaus Krischan, *Technische Universität Graz, Austria*

**EMI Noise Cancelation by Optimizing Transformer Design without Need for the
Traditional Y-Capacitor** 766
Yongjiang Bai, *Xi'an Jiaotong University, China*
Wenjie Chen, *Xi'an Jiaotong University, China*
Ruirui He, *Xi'an Jiaotong University, China*
Dan Zhang, *Silergy Corp., China*
Xu Yang, *Xi'an Jiaotong University, China*

**Manufacturing, Assembly and Production Qualifications of High Density, High
Reliability POL DC-DC Converters** 772
Fariborz Musavi, *CUI Inc., United States*

Session T17: Soft-Switching Converters in Renewable Energy Systems
Location: 104C
March 23, 2016 8:30 - 10:10
Session Chairs: Khurram Afridi, *University of Colorado at Boulder*
Katherine Kim, *Ulsan NIST*

**Power Flow Control and ZVS Analysis of Three Limb High Frequency Transformer
based Three-Port DAB** 778
Ritwik Chattopadhyay, *North Carolina State University, United States*
Subhashish Bhattacharya, *North Carolina State University, United States*

**A Novel Multi-Input Converter using Soft-Switched Single-Switch Input Modules with
Integrated Power Factor Correction Capability for Hybrid Renewable Energy Systems** 786
Sanjida Moury, *York University, Canada*
John Lam, *York University, Canada*
Vineet Srivastava, *Cistel Technology Inc., Canada*
Ron Church, *Cistel Technology Inc., Canada*

**Analysis and Design of Impulse Commutated ZCS Three-Phase Current-Fed Push-Pull
DC/DC Converter** 794
Radha Sree Krishna Moorthy, *National University of Singapore, Singapore*
Akshay Kumar Rathore, *National University of Singapore, Singapore*

ZCS Resonant Converter based Parallel Balancing of Serially Connected Batteries String 802
Ilya Zeltser, *Rafael Advanced Defense Systems Ltd., Israel*
Or Kirshenboim, *Ben-Gurion University of the Negev, Israel*
Nadav Dahan, *Ben-Gurion University of the Negev, Israel*
Mor Mordechai Peretz, *Ben-Gurion University of the Negev, Israel*

A Novel Topology of High Voltage and High Power Bidirectional ZCS DC-DC Converter based on Serial Capacitors 810

Lejia Sun, *Xi'an Jiaotong University, China*
Fang Zhuo, *Xi'an Jiaotong University, China*
Feng Wang, *Xi'an Jiaotong University, China*
Tianhua Zhu, *Xi'an Jiaotong University, China*

Session T18: Solid State Lighting
Location: 103AB
March 23, 2016 8:30 - 10:10
Session Chairs: Jim Spangler, *Spangler Prototype Inc*
Nan Chen, *ABB*

Control Scheme for TRIAC Dimming High PF Single-Stage LED Driver with Adaptive Bleeder Circuit and Non-Linear Current Reference 816

Weizhong Ma, *Hangzhou Dianzi University, China*
Xiaogao Xie, *Hangzhou Dianzi University, China*
Yang Han, *Hangzhou Dianzi University, China*
Hao Deng, *Hangzhou Dianzi University, China*

Three Phase Converter with Galvanic Isolation based on Loss-Free Resistors for HB-LED Lighting Applications 822

Ignacio Castro, *Universidad de Oviedo, Spain*
Diego G. Lamar, *Universidad de Oviedo, Spain*
Manuel Arias, *Universidad de Oviedo, Spain*
Javier Sebastián, *Universidad de Oviedo, Spain*
Marta M. Hernando, *Universidad de Oviedo, Spain*

A ZV-ZCS Electrolytic Capacitor-Less AC/DC Isolated LED Driver with Continous Energy Regulation 830

John Lam, *York University, Canada*
Nader A. El-Taweel, *York University, Canada*

High Efficiency and Power Density GaN-Based LED Driver 838

Eric Faraci, *Texas Instruments Inc., United States*
Michael Seeman, *Texas Instruments Inc., United States*
Bin Gu, *Texas Instruments Inc., United States*
Yogesh Ramadass, *Texas Instruments Inc., United States*
Paul Brohlin, *Texas Instruments Inc., United States*

A Novel LED Drive System based on Matrix Rectifier 843

Baoping Shi, *Nanjing University of Aeronautics and Astronautics, China*
Bo Zhou, *Nanjing University of Aeronautics and Astronautics, China*
Jiadan Wei, *Nanjing University of Aeronautics and Astronautics, China*
Xianhui Qin, *Nanjing University of Aeronautics and Astronautics, China*
Yuanyu Yang, *Nanjing University of Aeronautics and Astronautics, China*
Bing Liu, *Nanjing University of Aeronautics and Astronautics, China*

Session T19: Resonant and Soft Switching DC-DC Converters
Location: 101A
March 23, 2016 14:00 - 17:30
Session Chairs: Mahshid Amirabadi, *Northeastern University*
Ray Orr, *Solantro*

LLC Synchronous Rectification using Coordinate Modulation 848
Mehdi Mohammadi, *University of British Columbia, Canada*
Navid Shafiei, *University of British Columbia, Canada*
Martin Ordonez, *University of British Columbia, Canada*

Low Parasitics Planar Transformer for LLC Resonant Battery Chargers 854
Mohammad Ali Saket, *University of British Columbia, Canada*
Navid Shafiei, *University of British Columbia, Canada*
Martin Ordonez, *University of British Columbia, Canada*
Marian Craciun, *Delta-Q Technologies Corporation, Canada*
Chris Botting, *Delta-Q Technologies Corporation, Canada*

New Symmetrical Bidirectional L3C Resonant DC-DC Converter with Wide Voltage Range 859
Minjae Kim, *Seoul National University of Science and Technology, Korea, South*
Shinyoung Noh, *Seoul National University of Science and Technology, Korea, South*
Sewan Choi, *Seoul National University of Science and Technology, Korea, South*

Influence of the Junction Capacitance of the Secondary Rectifier Diodes on Output Characteristics in Multi-Resonant Converters 864
Stefan Ditze, *Fraunhofer Institute for Integrated Systems and Device Technology, Germany*
Thomas Heckel, *Fraunhofer Institute for Integrated Systems and Device Technology, Germany*
Martin März, *Fraunhofer Institute for Integrated Systems and Device Technology, Germany*

A Triple Active Bridge DC-DC Converter Capable of Achieving Full-Range ZVS 872
Ling Jiang, *University of Tennessee, United States*
Daniel Costinett, *University of Tennessee, United States*

A Novel High Gain Step-Up Resonant DC-DC Converter for Automotive Application 880
Fei Shang, *Illinois Institute of Technology, United States*
Mahesh Krishnamurthy, *Illinois Institute of Technology, United States*
Alexander Isurin, *Vanner Inc., United States*

Series Injection Enabled Full ZVS Light Load Operation of a 15kV SiC IGBT based Dual Active Half Bridge Converter 886
Awneesh Tripathi, *North Carolina State University, United States*
Sachin Madhusoodhanan, *North Carolina State University, United States*
Krishna Mainali, *North Carolina State University, United States*
Kasunaidu Vechalapu, *North Carolina State University, United States*
Subhashish Bhattacharya, *North Carolina State University, United States*

Soft Switching for Half Bridge Current Doubler for High Voltage Point of Load Converter in Data Center Power Supplies 893
Yutian Cui, *University of Tennessee, United States*
Weimin Zhang, *University of Tennessee, United States*
Leon M. Tolbert, *University of Tennessee, United States*
Daniel J. Costinett, *University of Tennessee, United States*
Fred Wang, *University of Tennessee, United States*
Benjamin J. Blalock, *University of Tennessee, United States*

An Algorithm to Analyze Circulating Current for Multi-Phase Resonant Converter 899
Hongliang Wang, *Queen's University, Canada*
Yang Chen, *Queen's University, Canada*
Zhiyuan Hu, *Queen's University, Canada*
Laili Wang, *Queen's University, Canada*
Tianshu Liu, *Queen's University, Canada*
Wenbo Liu, *Queen's University, Canada*
Yan-Fei Liu, *Queen's University, Canada*
Jahangir Afsharian, *Murata Power Solutions, Canada*
Zhihua Yang, *Murata Power Solutions, Canada*

Session T20: Control Applications and Modulation Schemes
Location: 102C
March 23, 2016 14:00 - 17:30
Session Chairs: Masoud Karimi Ghartemani, *Mississippi state University*
Paul Bauer, *University of Lorraine*

A Simple Active Damping Method for Active Power Filters 907
Huawei Yuan, *Tsinghua University, China*
Xinjian Jiang, *Tsinghua University, China*

Simultaneous Voltage and Current Compensation of the 3-Phase Electric Spring with Decomposed Voltage Control 913
Shuo Yan, *University of Hong Kong, Hong Kong*
Tianbo Yang, *University of Hong Kong, Hong Kong*
C.K. Lee, *University of Hong Kong, Hong Kong*
Siew-Chong Tan, *University of Hong Kong, Hong Kong*
S.Y. Ron Hui, *University of Hong Kong / Imperial College London, Hong Kong*

Self-Synchronization Operation of Global Synchronous Pulsewidth Modulation with Communication Fault Tolerant and Simplified Calculation Capabilities 921
Tao Xu, *Shandong University, China*
Feng Gao, *Shandong University, China*

Design Considerations and Predictive Direct Current Control of Active Regenerative Rectifiers for Harmonic and Current Ripple Reduction 928
Alberto Berzoy, *Florida International University, United States*
A.A.S. Mohamed, *Florida International University, United States*
Osama Mohammed, *Florida International University, United States*

A Robust Controller for Medium Voltage AC Collection Grid for Large Scale Photovoltaic Plants based on Medium Frequency Transformers 936
Bahaa Hafez, *Texas A&M University, United States*
Prasad Enjeti, *Texas A&M University, United States*
Shehab Ahmed, *Texas A&M University at Qatar, Qatar*

Optimal Low Switching Frequency Pulse Width Modulation of Current-Fed Five-Level Inverter for Solar Integration 943
Gnana Sambandam Kulothungan, *National University of Singapore, Singapore*
Akshay K. Rathore, *National University of Singapore, Singapore*
Amarendra Edpuganti, *National University of Singapore, Singapore*
Dipti Srinivasan, *National University of Singapore, Singapore*

Design and Implementation of D-Σ Digital Controlled Multi-function Inverter to Achieve APF, Active Power Injection and Rectification 951
T.-F. Wu, *National Tsing Hua University, Taiwan*
H.-C. Hsieh, *National Chung Cheng University, Taiwan*
L.-C. Lin, *National Tsing Hua University, Taiwan*
C.-H. Chang, *National Tsing Hua University, Taiwan*

Operation and Analysis of an Improved Transformerless Unified Power Flow Controller 959
Yang Liu, *Michigan State University, United States*
Shuitao Yang, *Michigan State University / Ford Motor Company, United States*
Fang Zheng Peng, *Michigan State University, United States*

Design Consideration of Converter based Transmission Line Emulation 966
Bo Liu, *University of Tennessee, United States*
Shuoting Zhang, *University of Tennessee, United States*
Sheng Zheng, *University of Tennessee, United States*
Yiwei Ma, *University of Tennessee, United States*
Fred Wang, *University of Tennessee, United States*
Leon M. Tolbert, *University of Tennessee, United States*

Session T21: Advances in Wide BandGap Devices
Location: 104A
March 23, 2016 14:00 - 17:30
Session Chairs: Doug Hopkins, *North Carolina State University*
Alex Huang, *North Carolina State University*

Short-Circuit Characterization of 10 kV 10A 4H-SiC MOSFET 974
Emanuel-Petre Eni, *Aalborg University, Denmark*
Szymon Bęczkowski, *Aalborg University, Denmark*
Stig Munk-Nielsen, *Aalborg University, Denmark*
Tamas Kerekes, *Aalborg University, Denmark*
Remus Teodorescu, *Aalborg University, Denmark*

Record-Low 10mΩ SiC MOSFETs in TO-247, Rated at 900V 979
Vipindas Pala, *Wolfspeed, A Cree Company, United States*
Gangyao Wang, *Wolfspeed, A Cree Company, United States*
Brett Hull, *Wolfspeed, A Cree Company, United States*
Scott Allen, *Wolfspeed, A Cree Company, United States*
Jeffrey Casady, *Wolfspeed, A Cree Company, United States*
John Palmour, *Wolfspeed, A Cree Company, United States*

Performance Evaluation of Multiple Si and SiC Solid State Devices for Circuit Breaker Application in 380VDC Delivery System 983
Kai Tan, *North Carolina State University, United States*
Pengkun Liu, *North Carolina State University, United States*
Xijun Ni, *North Carolina State University, United States*
Chang Peng, *North Carolina State University, United States*
Xiaoqing Song, *North Carolina State University, United States*
Alex Q. Huang, *North Carolina State University, United States*

Evaluation of High Voltage Cascode GaN HEMTs in Parallel Operation 990
He Li, *Ohio State University, United States*
Xuan Zhang, *Ohio State University, United States*
Lucheng Wen, *Ohio State University, United States*
John Alex Brothers, *Ohio State University, United States*
Chengcheng Yao, *Ohio State University, United States*
Ke Zhu, *Ohio State University, United States*
Jin Wang, *Ohio State University, United States*
Liming Liu, *ABB Inc., United States*
Jing Xu, *ABB Inc., United States*
Joonas Puukko, *ABB Inc., United States*

A New Driving Concept for Normally-On GaN Switches in Cascode Configuration 996
Bernhard Zojer, *Infineon Technologies Austria AG, Austria*

Avoiding Divergent Oscillation of Cascode GaN Device Under High Current Turn-Off Condition 1002
Weijing Du, *Virginia Polytechnic Institute and State University, United States*
Xiucheng Huang, *Virginia Polytechnic Institute and State University, United States*
Fred C. Lee, *Virginia Polytechnic Institute and State University, United States*
Qiang Li, *Virginia Polytechnic Institute and State University, United States*
Wenli Zhang, *Virginia Polytechnic Institute and State University, United States*

Temperature-Dependent Turn-On Loss Analysis for GaN HFETs 1010
Edward A. Jones, *University of Tennessee, United States*
Fred Wang, *University of Tennessee, United States*
Daniel Costinett, *University of Tennessee, United States*
Zheyu Zhang, *University of Tennessee, United States*
Ben Guo, *United Technologies Research Center, United States*

Analysis of Parasitic Elements of SiC Power Modules with Special Emphasis on Reliability Issues 1018
Diane-Perle Sadik, *KTH Royal Institute of Technology, Sweden*
Juan Colmenares, *KTH Royal Institute of Technology, Sweden*
Hans-Peter Nee, *KTH Royal Institute of Technology, Sweden*
Konstantin Kostov, *Acreo Swedish ICT AB, Sweden*
Florian Giezendanner, *Alstom Power Sweden AB, Sweden*
Per Ranstad, *Alstom Power Sweden AB, Sweden*

Static and Dynamic Characterization of GaN HEMT with Low Inductance Vertical Phase Leg Design for High Frequency High Power Applications 1024
Nidhi Haryani, *Virginia Polytechnic Institute and State University, United States*
Xuning Zhang, *Virginia Polytechnic Institute and State University, United States*
Rolando Burgos, *Virginia Polytechnic Institute and State University, United States*
Dushan Boroyevich, *Virginia Polytechnic Institute and State University, United States*

Session T22: Motor Drive Design and Inverter Topologies
Location: 101B
March 23, 2016 14:00 - 17:30
Session Chairs: Yingying Kuai, *Caterpillar Inc.*
Jin Wang, *The Ohio State University*

A Family of Single-Phase Current Source Converters with Double Outputs 1032
Louelson A. Costa, *Universidade Federal de Campina Grande, Brazil*
Maurício B.R. Corrêa, *Universidade Federal de Campina Grande, Brazil*
Montiê A. Vitorino, *Universidade Federal de Campina Grande, Brazil*
Gutemberg G. Dos Santos, *Universidade Federal de Campina Grande, Brazil*
Darlan A. Fernandes, *Universidade Federal da Paraíba, Brazil*

**Multiple-Output Boost Resonant Inverter for High Efficiency and Cost-Effective
Induction Heating Applications** 1040
Hector Sarnago, *Universidad de Zaragoza, Spain*
Oscar Lucia, *Universidad de Zaragoza, Spain*
José M. Burdío, *Universidad de Zaragoza, Spain*

Development of 2-kW Interleaved DC-Capacitor-Less Single-Phase Inverter System 1045
Runruo Chen, *Michigan State University, United States*
Hulong Zeng, *Michigan State University, United States*
Deepak Gunasekaran, *Michigan State University, United States*
Yunting Liu, *Michigan State University, United States*
Fang Z. Peng, *Michigan State University, United States*

Single Stage Transformer Isolated High Frequency AC Link based Open End Drive 1051
Srikant Gandikota, *University of Minnesota, United States*
Ned Mohan, *University of Minnesota, United States*

**A Quasi-Z-Source Integrated Multi-Port Power Converter with Reduced Capacitance for
Switched Reluctance Motor Drives** 1057
Fan Yi, *University of Texas at Dallas, United States*
Wen Cai, *University of Texas at Dallas, United States*

**A Fault-Tolerant Topology of T-Type NPC Inverter with Increased Thermal
Overload Capability** 1065
Jiangbiao He, *Marquette University, United States*
Nathan Weise, *Marquette University, United States*
Lixiang Wei, *Rockwell Automation, United States*
Nabeel A.O. Demerdash, *Marquette University, United States*

**A Novel Analysis and Design Method of Phase Lead Filters in Repetitive Controllers for
Pulse-Width Modulated Inverters** 1071
Shunfeng Yang, *Nanyang Technological University, Singapore*
Peng Wang, *Nanyang Technological University, Singapore*
Yi Tang, *Nanyang Technological University, Singapore*
Michael Zagrodnik, *Rolls-Royce Singapore Pte. Ltd., Singapore*
Xiaolei Hu, *Nanyang Technological University, Singapore*
King Jet Tseng, *Nanyang Technological University, Singapore*

Research on the Filter of Load Side Converter in BDFG based Ship Shaft Power Generation System 1078

Meilin Wang, *Huazhong University of Science and Technology, China*
Hua Lin, *Huazhong University of Science and Technology, China*
Hongbin Yang, *Huazhong University of Science and Technology, China*
Xingwei Wang, *Huazhong University of Science and Technology, China*

Investigation of Common Mode Current Related DC-Bus Overvoltage in Multiple Converter Systems 1084

Jiangbiao He, *Rockwell Automation, United States*
Zoran Vrankovic, *Rockwell Automation, United States*
Patrick E. Ozimek, *Rockwell Automation, United States*
Craig Winterhalter, *Rockwell Automation, United States*

Session T23: Modeling of Magnetic Circuits and Systems
Location: 102AB
March 23, 2016 14:00 - 17:30
Session Chairs: Ed Herbert,
Jin Ye, *San Francisco State University*

High Frequency AC Inductor Analysis and Design for Dual Active Bridge (DAB) Converters .. 1090

Zhe Zhang, *Technical University of Denmark, Denmark*
Michael A.E. Andersen, *Technical University of Denmark, Denmark*

A Comprehensive Assessment of PM Motor Topology Impact on Magnet Defect Fault Signatures 1096

Mohsen Zafarani, *University of Texas at Dallas, United States*
Taner Goktas, *University of Texas at Dallas, United States*
Bilal Akin, *University of Texas at Dallas, United States*

High Frequency Modeling for Transformer Common Mode Noise Coupling Path based on Multiconductor Transmission Line Theory 1102

Peipei Meng, *Wuhan University of Technology, China*
Xiangming Zhang, *Naval University of Engineering, China*

Leakage Flux Modelling of Multi-Winding Transformer using Permeance Magnetic Circuit ... 1108

Min Luo, *École Polytechnique Fédérale de Lausanne, Switzerland*
Drazen Dujic, *École Polytechnique Fédérale de Lausanne, Switzerland*
Jost Allmeling, *Plexim GmbH, Switzerland*

Modeling Magnetic Devices using SPICE: Application to Variable Inductors 1115

J. Marcos Alonso, *Universidad de Oviedo, Spain*
Gilberto Martínez, *Continental Automotive R&D, Mexico*
Marina Perdigão, *Universidade de Coimbra, Portugal*
Marcelo Cosetin, *Universidade Federal de Santa Maria, Brazil*
Ricardo N. do Prado, *Universidade Federal de Santa Maria, Brazil*

Investigation of a Thermal Model for a Permanent Magnet Assisted Synchronous Reluctance Motor 1123

Joseph Herbert, *University of Akron, United States*
A.K.M. Arafat, *University of Akron, United States*
Guo-Xiang Wang, *University of Akron, United States*
Seungdeog Choi, *University of Akron, United States*

Design Procedure for Multi-Phase External Rotor Permanent Magnet Assisted Synchronous Reluctance Machines 1131
Sai Sudheer Reddy Bonthu, *University of Akron, United States*
Seungdeog Choi, *University of Akron, United States*

Applicability and Limitations of an M2Spice-Assisted "Planar-Magnetics-in-the-Circuit" Simulation Approach 1138
Samantha J. Gunter, *Massachusetts Institute of Technology, United States*
Minjie Chen, *Massachusetts Institute of Technology, United States*
Stephanie A. Pavlick, *Massachusetts Institute of Technology, United States*
Rose A. Abramson, *Massachusetts Institute of Technology, United States*
Khurram K. Afridi, *University of Colorado at Boulder, United States*
David J. Perreault, *Massachusetts Institute of Technology, United States*

Session T24: Inverter/Converter Control
Location: 103C
March 23, 2016 14:00 - 17:30
Session Chairs: Siavash Pakdelian, *UMass Lowell*
Behrooz Mirafzal, *Kansas State University*

Solution of Input Double-Line Frequency Ripple Rejection for High-Efficiency High-Power Density String Inverter in Photovoltaic Application 1148
Xiaonan Zhao, *Virginia Polytechnic Institute and State University, United States*
Lanhua Zhang, *Virginia Polytechnic Institute and State University, United States*
Rachael Born, *Virginia Polytechnic Institute and State University, United States*
Jih-Sheng Lai, *Virginia Polytechnic Institute and State University, United States*

Fractional-Order Phase Lead Compensation for Multi-Rate Repetitive Control on Three-Phase PWM DC/AC Inverter 1155
Zhichao Liu, *University of South Carolina, United States*
Bin Zhang, *University of South Carolina, United States*
Keliang Zhou, *University of Glasgow, United Kingdom*

A Robust Modified Model Predictive Control (MMPC) based on Lyapunov Function for Three-Phase Active-Front-End (AFE) Rectifier 1163
M. Parvez, *University of Malaya, Malaysia*
S. Mekhilef, *University of Malaya, Malaysia*
Nadia M.L. Tan, *Universiti Tenega Nasional, Malaysia*
Hirofumi Akagi, *Tokyo Institute of Technology, Japan*

Adaptive Reference Model Predictive Control for Power Electronics 1169
Yun Yang, *University of Hong Kong, Hong Kong*
Siew-Chong Tan, *University of Hong Kong, Hong Kong*
Shu-Yuen Ron Hui, *Imperial College London, United Kingdom*

Power Switch Lifetime Extension Strategies for Three-Phase Converters 1176
Serkan Dusmez, *University of Texas at Dallas, United States*
Enes Ugur, *University of Texas at Dallas, United States*
Bilal Akin, *University of Texas at Dallas, United States*

Current Controller Modeling for an Interleaved Boost with Voltage Multiplier Cells for PV Applications 1183
Alessandro Pevere, *Katholieke Universiteit Leuven, Belgium*
Urmimala Chatterjee, *Katholieke Universiteit Leuven, Belgium*
Johan Driesen, *Katholieke Universiteit Leuven, Belgium*

New Active Capacitor Voltage Balancing Method for Five-Level Stacked Multicell Converter 1191
Arash Khoshkbar Sadigh, *Extron Electronics, United States*
Vahid Dargahi, *Clemson University, United States*
Keith Corzine, *Clemson University, United States*

Gate Signal Jitter Elimination and Noise Shaping Modulation for High-SNR Class-D Power Amplifiers 1198
M. Mauerer, *ETH Zürich, Switzerland*
A. Tüysüz, *ETH Zürich, Switzerland*
J.W. Kolar, *ETH Zürich, Switzerland*

Analysis and Compensation of Inverter Nonlinearity for Three-Level T-Type Inverters 1206
Hyeon-Sik Kim, *Seoul National University, Korea, South*
Yong-Cheol Kwon, *Seoul National University, Korea, South*
Seung-Jun Chee, *Seoul National University, Korea, South*
Seung-Ki Sul, *Seoul National University, Korea, South*

Session T25: Topics in Renewable Energy Systems I
Location: 104B
March 23, 2016 14:00 - 17:30
Session Chairs: Fei Gao, *University of Technology of Belfort-Montbéliard*
Kent Wanner, *John Deere*

Front-End Isolated Quasi-Z-Source DC-DC Converter Modules in Series for Photovoltaic High-Voltage DC Applications 1214
Yushan Liu, *Texas A&M University at Qatar, Qatar*
Haitham Abu-Rub, *Texas A&M University at Qatar, Qatar*
Baoming Ge, *Texas A&M University, United States*

Analysis of Non Detection Zone for Multiple Distributed PCS based on Equivalent Single PCS using Reactive Power Approach 1220
Byeong-Heon Kim, *Seoul National University, Korea, South*
Seung-Ki Sul, *Seoul National University, Korea, South*

Optimal Power Scheduling for a Grid-Connected Hybrid PV-Wind-Battery Microgrid System .. 1227
Adriana Luna, *Aalborg University, Denmark*
Nelson Diaz, *Aalborg University, Denmark*
Mehdi Savaghebi, *Aalborg University, Denmark*
Juan C. Vásquez, *Aalborg University, Denmark*
Josep M. Guerrero, *Aalborg University, Denmark*
Kai Sun, *Tsinghua University, China*
Guoliang Chen, *Shanghai Solar Energy & Technology Co., Ltd., China*
Libing Sun, *Shanghai Solar Energy & Technology Co., Ltd., China*

High Efficiency Power Converter for a Doubly-Fed SOEC/SOFC System 1235
Kevin Tomas-Manez, *Technical University of Denmark, Denmark*
Alexander Anthon, *Technical University of Denmark, Denmark*
Zhe Zhang, *Technical University of Denmark, Denmark*

A Hierarchical Active Balancing Architecture for Li-Ion Batteries 1243
Han-Dong Gui, *Nanjing University of Aeronautics and Astronautics, China*
Zhiliang Zhang, *Nanjing University of Aeronautics and Astronautics, China*
Dong-Jie Gu, *Nanjing University of Aeronautics and Astronautics, China*
Yang Yang, *Nanjing University of Aeronautics and Astronautics, China*
Zhouyu Lu, *Nanjing University of Aeronautics and Astronautics, China*
Yan-Fei Liu, *Queen's University, Canada*

A Series-DG based Autonomous Islanding Microgrid 1249
Beihua Liang, *Tianjin University, China*
Yun Wei Li, *University of Alberta, Canada*
Jinwei He, *Tianjin University, China*
Chengshan Wang, *Tianjin University, China*

An Enhanced Droop Control Scheme for Resilient Active Power Sharing in Paralleled
Two-Stage PV Inverter Systems 1253
Hongpeng Liu, *Harbin Institute of Technology, China*
Yongheng Yang, *Aalborg University, Denmark*
Xiongfei Wang, *Aalborg University, Denmark*
Poh Chiang Loh, *Aalborg University, Denmark*
Frede Blaabjerg, *Aalborg University, Denmark*
Wei Wang, *Harbin Institute of Technology, China*
Dianguo Xu, *Harbin Institute of Technology, China*

Voltage Closed-Loop Virtual Synchronous Generator Control of Full Converter Wind
Turbine for Grid-Connected and Stand-Alone Operation 1261
Yiwei Ma, *University of Tennessee, United States*
Liu Yang, *University of Tennessee, United States*
Fred Wang, *University of Tennessee, United States*
Leon M. Tolbert, *University of Tennessee, United States*

DC Voltage Ripple Quantification for a Flywheel-Battery based Hybrid Energy
Storage System 1267
Christopher R. Lashway, *Florida International University, United States*
Ahmed T. Elsayed, *Florida International University, United States*
Osama A. Mohammed, *Florida International University, United States*

Session T26: Electric Vehicle Charging Systems
Location: 104C
March 23, 2016 14:00 - 17:30
Session Chairs: Jim Spangler, *Spangler Prototype Inc*
Hadi Malek, *Ford*

Adaptive Loss Reduction Charging Strategy Considering Variation of Internal
Impedance of Lithium-Ion Polymer Batteries in Electric Vehicle Charging Systems 1273
Nari Kim, *Sungkyunkwan University, Korea, South*
Jung-Hoon Ahn, *Sungkyunkwan University, Korea, South*
Dong-Hee Kim, *Sungkyunkwan University, Korea, South*
Byoung-Kuk Lee, *Sungkyunkwan University, Korea, South*

A Pulse Width Modulated LLC Type Resonant Topology Adpated to Wide Output Voltage Range 1280
Haoyu Wang, *ShanghaiTech University, China*

A Series Resonant Circuit for Voltage Equalization of Series Connected Energy Storage Devices 1286
Yanqi Yu, *University of British Columbia, Canada*
Raed Saasaa, *University of British Columbia, Canada*
Wilson Eberle, *University of British Columbia, Canada*

Implementation of 3.3-kW GaN-Based DC-DC Converter for EV On-Board Charger with Series-Resonant Converter that Employs Combination of Variable-Frequency and Delay-Time Control 1292
Yungtaek Jang, *Delta Products Corporation, United States*
Milan M. Jovanović, *Delta Products Corporation, United States*
Juan M. Ruiz, *Delta Products Corporation, United States*
Misha Kumar, *Delta Products Corporation, United States*
Gang Liu, *Delta Electronics Shanghai Co., Ltd., China*

Dual Active Bridge-Based Full-Integrated Active Filter Auxiliary Power Module for Electrified Vehicle Applications with Single-Phase Onboard Chargers 1300
Ruoyu Hou, *McMaster University, Canada*
Ali Emadi, *McMaster University, Canada*

All-SiC Inductively Coupled Charger with Integrated Plug-In and Boost Functionalities for PEV Applications 1307
M. Chinthavali, *Oak Ridge National Laboratory, United States*
O.C. Onar, *Oak Ridge National Laboratory, United States*
S.L. Campbell, *Oak Ridge National Laboratory, United States*
L.M. Tolbert, *Oak Ridge National Laboratory, United States*

Switching Condition and Loss Modeling of GaN-Based Dual Active Bridge Converter for PHEV Charger 1315
Lingxiao Xue, *Virginia Polytechnic Institute and State University, United States*
Dushan Boroyevich, *Virginia Polytechnic Institute and State University, United States*
Paolo Mattavelli, *Università degli Studi di Padova, Italy*

Analysis of Cascaded Multi-Output-Port Converter for Wireless Plug-In Hybrid/On-Board EV Chargers 1323
Erdem Asa, *Hevo Power Inc. / New York University, United States*
Kerim Colak, *Istanbul Ulasim A.S., Turkey*
Dariusz Czarkowski, *New York University, United States*

Comparative Analysis of High Step-Down Ratio Isolated DC/DC Topologies in PEV Applications 1329
Zhiqing Li, *ShanghaiTech University, China*
Haoyu Wang, *ShanghaiTech University, China*

Session T27: Utility Interface and Inverter Applications
Location: 103AB
March 23, 2016 14:00 - 17:30
Session Chairs: Akshay Kumar Rathore, *Concordia University*
Yichao Tang, *Texas Instruments*

DC to Single-Phase AC Voltage Source Inverter with Power Decoupling Circuit based on Flying Capacitor Topology for PV System 1336
Hiroki Watanabe, *Nagaoka University of Technology, Japan*
Keisuke Kusaka, *Nagaoka University of Technology, Japan*
Keita Furukawa, *Nagaoka University of Technology, Japan*
Koji Orikawa, *Nagaoka University of Technology, Japan*
Jun-Ichi Itoh, *Nagaoka University of Technology, Japan*

GaN FET and Hybrid Modulation based Differential-Mode Inverter 1344
Sudip K. Mazumder, *NextWatt LLC, United States*
Ankit Gupta, *University of Illinois at Chicago, United States*
Shirish Raizada, *University of Illinois at Chicago, United States*
Harshit Soni, *University of Illinois at Chicago, United States*
Nikhil Kumar, *University of Illinois at Chicago, United States*
Paromita Mazumder, *NextWatt LLC, United States*
Parijat Bhattachaarjee, *NextWatt LLC, United States*

Thermal and Electrical Co-Design of a Modular High-Density Single-Phase Inverter using Wide-Bandgap Devices 1350
Steven Chung, *University of Toronto, Canada*
Miad Nasr, *University of Toronto, Canada*
David Guirguis, *University of Toronto, Canada*
Masafumi Otsuka, *University of Toronto, Canada*
Shahab Poshtkouhi, *University of Toronto, Canada*
David K.W. Li, *University of Toronto, Canada*
Vishal Palaniappan, *University of Toronto, Canada*
David Romero, *University of Toronto, Canada*
Cristina Amon, *University of Toronto, Canada*
Ray Orr, *Solantro Semiconductor, Canada*
Olivier Trescases, *University of Toronto, Canada*

Reactive Power Compensation with Improvement of Current Waveform Quality for Single-Phase Buck-Type Dynamic Capacitor 1358
Xinwen Chen, *Huazhong University of Science and Technology, China*
Ke Dai, *Huazhong University of Science and Technology, China*
Chen Xu, *Huazhong University of Science and Technology, China*
Ziwei Dai, *Huazhong University of Science and Technology, China*
Li Peng, *Huazhong University of Science and Technology, China*

Circulating Current Reduction for a D-Σ Digital Controlled Transformerless UPS 1364
T.-F. Wu, *National Tsing Hua University, Taiwan*
T.-H. Shiu, *National Tsing Hua University, Taiwan*
P.-H. Lin, *National Tsing Hua University, Taiwan*
L.-C. Lin, *National Tsing Hua University, Taiwan*
J.-W. Huang, *Industrial Technology Research Institute, Taiwan*

A Multi-Function Three-Level Dynamic Voltage Corrector with Wide Correction Range and Short Circuit Fault Isolation 1371

Jiankun Cao, *Nanjing University of Aeronautics and Astronautics, China*
Pengling Ding, *Nanjing University of Aeronautics and Astronautics, China*
Haichun Liu, *Nanjing University of Aeronautics and Astronautics, China*
Shaojun Xie, *Nanjing University of Aeronautics and Astronautics, China*

Effects and Analysis of Minimum Pulse Width Limitation on Adaptive DC Voltage Control of Grid Converters 1376

Bo Sun, *Aalborg University, Denmark*
Ionut Trintis, *Aalborg University, Denmark*
Stig Munk-Nielsen, *Aalborg University, Denmark*
Josep M. Guerrero, *Aalborg University, Denmark*

Improved Three-Phase Micro-Inverter using Dynamic Dead Time Optimization and Phase-Skipping Control Techniques 1381

S. Milad Tayebi, *University of Central Florida, United States*
Xianmin Mu, *University of Central Florida, United States*
Issa Batarseh, *University of Central Florida, United States*

Correcting Current Imbalances in Three-Phase Four-Wire Distribution Systems 1387

Vinson Jones, *University of Arkansas, United States*
Juan Carlos Balda, *University of Arkansas, United States*

Session T28: Isolated DC-DC Converters
Location: 104A
March 24, 2016 8:30 - 11:20
Session Chairs: Dragan Maksimovic, *UC Boulder*
Zhong Ye, *Texas Instruments*

New Design Methdology for Megahertz-Frequency Resonant DC-DC Converters using Impedance Control Network Architecture 1392

Yushi Liu, *University of Colorado at Boulder, United States*
Ashish Kumar, *University of Colorado at Boulder, United States*
Jie Lu, *University of Colorado at Boulder, United States*
Dragan Maksimovic, *University of Colorado at Boulder, United States*
Khurram K. Afridi, *University of Colorado at Boulder, United States*

Dual Voltage Regulations of Single Switch Flyback Converter using Variable Switching Frequency 1398

Jin-Woong Kim, *Seoul National University, Korea, South*
Jung-Ik Ha, *Seoul National University, Korea, South*

On-Chip PLL-Based Methods for Synchronizing Active Switches Across the Isolation Boundary in DC-DC Converters 1403

Shahab Poshtkouhi, *University of Toronto, Canada*
Miad Fard, *University of Toronto, Canada*
Olivier Trescases, *University of Toronto, Canada*

An Isolated Soft-Switching Buck-Boost Converter Utilizing Two Transformers and Embedded Bidirectional Switches on Secondary-Side for Wide Voltage Applications 1410

Tingting Liu, *Nanjing University of Aeronautics and Astronautics, China*
Hongfei Wu, *Nanjing University of Aeronautics and Astronautics, China*
Yan Xing, *Nanjing University of Aeronautics and Astronautics, China*
Kai Sun, *Tsinghua University, China*

Effect of Transformer Design on Operation of Fundamental Duty Modulation for Dual-Active-Bridge Converter 1416

Wooin Choi, *Seoul National University, Korea, South*
Moonhyun Lee, *Seoul National University, Korea, South*
Bo-Hyung Cho, *Seoul National University, Korea, South*

A High Step-Up Bidirectional Isolated Dual-Active-Bridge Converter with Three-Level Voltage-Doubler Rectifier for Energy Storage Applications 1424

Xiaohai Zhan, *Nanjing University of Aeronautics and Astronautics, China*
Hongfei Wu, *Nanjing University of Aeronautics and Astronautics, China*
Yan Xing, *Nanjing University of Aeronautics and Astronautics, China*
Hongjuan Ge, *Nanjing University of Aeronautics and Astronautics, China*
Xi Xiao, *Tsinghua University, China*

Digitized Self-Oscillating Loop for Piezoelectric Transformer-Based Power Converters 1430

Marzieh Ekhtiari, *Technical University of Denmark, Denmark*
Thomas Andersen, *Technical University of Denmark, Denmark*
Zhe Zhang, *Technical University of Denmark, Denmark*
Michael A.E. Andersen, *Technical University of Denmark, Denmark*

Session T29: Multilevel Converters
Location: 101A
March 24, 2016 8:30 - 11:20
Session Chairs: Maryam Saeedifard, *Georgia Tech*
Julia Zhang, *Oregon State University*

An Isolated Topology for Reactive Power Compensation with a Modularized Dynamic-Current Building-Block 1437

Hao Chen, *Georgia Institute of Technology, United States*
Anish Prasai, *Varentec, Inc., United States*
Deepak Divan, *Georgia Institute of Technology, United States*

Design and Control of a Compact MMC Submodule Structure with Reduced Capacitor Size using the Stacked Switched Capacitor Architecture 1443

Yuan Tang, *University of Warwick, United Kingdom*
Minjie Chen, *Massachusetts Institute of Technology, United States*
Li Ran, *University of Warwick, United Kingdom*

Fundamental Frequency Sorting Strategy for Capacitor Voltage Balance of Modular Multilevel Converters with Phase Disposition PWM 1450

Kun Wang, *Zhejiang University, China*
Yan Deng, *Zhejiang University, China*
Wenyu Li, *Zhejiang University, China*
Hao Peng, *Zhejiang University, China*
Guipeng Chen, *Zhejiang University, China*
Xiangning He, *Zhejiang University, China*

Active Voltage Balancing Control for 10kV Three-Level Converter using Series-Connected HV-IGBTs 1456

Shiqi Ji, *Tsinghua University, China*
Ting Lu, *Tsinghua University, China*
Zhengming Zhao, *Tsinghua University, China*
Hualong Yu, *Tsinghua University, China*
Fred Wang, *University of Tennessee, United States*

Average-Value Model of Modular Multilevel Converters Considering Capacitor Voltage 1462

Heya Yang, *Zhejiang University, China*
Yuxiang Chen, *Zhejiang University, China*
Wuhua Li, *Zhejiang University, China*
Xiangning He, *Zhejiang University, China*
Wei Sun, *China Electric Power Research Institute, China*
Yongning Chi, *China Electric Power Research Institute, China*
Yan Li, *China Electric Power Research Institute, China*

New Submodule Circuits for Modular Multilevel Current Source Converters with DC Fault Ride through Capability 1468

Xinyu Yu, *Tsinghua University, China*
Yingdong Wei, *Tsinghua University, China*
Qirong Jiang, *Tsinghua University, China*

Voltage and Power Balance Control Strategy for Three-Phase Modular Cascaded Solid Stated Transformer 1475

Zhiyu Zhang, *Zhejiang University, China*
Hengyang Zhao, *Zhejiang University, China*
Shihang Fu, *Zhejiang University, China*
Jianjiang Shi, *Zhejiang University, China*
Xiangning He, *Zhejiang University, China*

Session T30: Multilevel and Matrix Converters for Motor Drives
Location: 102C
March 24, 2016 8:30 - 11:20
Session Chairs: SeonHwan Hwang, *Kyungnam University, Korea*
Xiaohu Liu, *GE*

New Flying-Capacitor-Based Multilevel Converter with Optimized Number of Switches and Capacitors Controlled with a New Logic-Form-Equation based Active Voltage Balancing Technique 1481

Vahid Dargahi, *Clemson University, United States*
Arash Khoshkbar Sadigh, *Extron Electronics, United States*
Keith Corzine, *Clemson University, United States*

New Low-Cost Five-Level Active Neutral-Point Clamped Converter 1489

Hongliang Wang, *Queen's University, Canada*
Lei Kou, *Queen's University, Canada*
Yan-Fei Liu, *Queen's University, Canada*
Paresh C. Sen, *Queen's University, Canada*
Sucheng Liu, *Anhui University of Technology, China*

Medium Voltage (≥ 2.3 kV) High Frequency Three-Phase Two-Level Converter Design and Demonstration using 10 kV SiC MOSFETs for High Speed Motor Drive Applications .. 1497

Sachin Madhusoodhanan, *North Carolina State University, United States*
Krishna Mainali, *North Carolina State University, United States*
Awneesh Tripathi, *North Carolina State University, United States*
Kasunaidu Vechalapu, *North Carolina State University, United States*
Subhashish Bhattacharya, *North Carolina State University, United States*

Novel Three Phase Multi-Level Inverter Topology with Symmetrical DC-Voltage Sources . 1505

Ahmed Salem, *Aswan University, Egypt*
Emad M. Ahmed, *Aswan University, Egypt*
Mahrous Ahmed, *Aswan University, Egypt*
Mohamed Orabi, *Aswan University, Egypt*

A 2 kW, Single-Phase, 7-Level, GaN Inverter with an Active Energy Buffer Achieving 216 W/in^3 Power Density and 97.6% Peak Efficiency 1512

Yutian Lei, *University of Illinois at Urbana-Champaign, United States*
Christopher Barth, *University of Illinois at Urbana-Champaign, United States*
Shibin Qin, *University of Illinois at Urbana-Champaign, United States*
Wen-Chuen Liu, *University of Illinois at Urbana-Champaign, United States*
Intae Moon, *University of Illinois at Urbana-Champaign, United States*
Andrew Stillwell, *University of Illinois at Urbana-Champaign, United States*
Derek Chou, *University of Illinois at Urbana-Champaign, United States*
Thomas Foulkes, *University of Illinois at Urbana-Champaign, United States*
Zichao Ye, *University of Illinois at Urbana-Champaign, United States*
Zitao Liao, *University of Illinois at Urbana-Champaign, United States*
Robert C.N. Pilawa-Podgurski, *University of Illinois at Urbana-Champaign, United States*

Indirect Matrix Converter based Open-End Winding AC Drives with Zero Common-Mode Voltage 1520

Saurabh Tewari, *MTS Systems Corporation, United States*
Ranjan K. Gupta, *First Solar, Inc., United States*
Apurva Somani, *Dynapower Company LLC, United States*
Ned Mohan, *University of Minnesota, United States*

Precharging Strategy for Soft Startup Process of Modular Multilevel Converters based on Various SM Circuits 1528

Jiangchao Qin, *Arizona State University, United States*
Suman Debnath, *Oak Ridge National Laboratory, United States*
Maryam Saeedifard, *Georgia Institute of Technology, United States*

Session T31: System Design Techniques for Reduced EMI
Location: 101B
March 24, 2016 8:30 - 11:20
Session Chairs: John Vigars, *Allegro Microsystems*
Doug Hopkins, *North Carolina State University*

Conducted EMI Analysis and Filter Design for MHz Active Clamp Flyback Front-End Converter 1534

Xiucheng Huang, *Virginia Polytechnic Institute and State University, United States*
Junjie Feng, *Virginia Polytechnic Institute and State University, United States*
Fred C. Lee, *Virginia Polytechnic Institute and State University, United States*
Qiang Li, *Virginia Polytechnic Institute and State University, United States*
Yuchen Yang, *Virginia Polytechnic Institute and State University, United States*

EMC Investigation of a Very High Frequency Self-Oscillating Resonant Power Converter 1541
Jeppe A. Pedersen, *Technical University of Denmark, Denmark*
Arnold Knott, *Technical University of Denmark, Denmark*
Michael A.E. Andersen, *Technical University of Denmark, Denmark*

Numerical Optimization of Passive Line Filter Components for Suppression of Electromagnetic Interference (EMI) 1547
Carsten Henkenius, *Universität Paderborn, Germany*
Norbert Fröhleke, *Universität Paderborn, Germany*
Joachim Böcker, *Universität Paderborn, Germany*
Heiko Figge, *Delta Energy Systems GmbH, Germany*

Electromagnetic Noise Coupling and Mitigation for Fast Response On-Die Temperature Sensing in High Power Modules 1554
Chengcheng Yao, *Ohio State University, United States*
Pengzhi Yang, *Ohio State University, United States*
Mingzhi Leng, *Ohio State University, United States*
He Li, *Ohio State University, United States*
Lixing Fu, *Ohio State University, United States*
Jin Wang, *Ohio State University, United States*
Ke Zou, *Ford Motor Company, United States*
Chingchi Chen, *Ford Motor Company, United States*

Ultra-Low Inductance Vertical Phase Leg Design with EMI Noise Propagation Control for Enhancement Mode GaN Transistors 1561
Xuning Zhang, *Virginia Polytechnic Institute and State University, United States*
Zhiyu Shen, *Virginia Polytechnic Institute and State University, United States*
Nidhi Haryani, *Virginia Polytechnic Institute and State University, United States*
Dushan Boroyevich, *Virginia Polytechnic Institute and State University, United States*
Rolando Burgos, *Virginia Polytechnic Institute and State University, United States*

Decoupling of Interaction between WBG Converter and Motor Load for Switching Performance Improvement 1569
Zheyu Zhang, *University of Tennessee, United States*
Fred Wang, *University of Tennessee, United States*
Leon M. Tolbert, *University of Tennessee, United States*
Benjamin J. Blalock, *University of Tennessee, United States*
Daniel J. Costinett, *University of Tennessee, United States*

Control and Characterization of Electromagnetic Emissions in Wide Band Gap based Converter Modules for Ungrounded Grid-Forming Applications 1577
Robert Cuzner, *University of Wisconsin at Milwaukee, United States*
Rasoul Hosseini, *University of Wisconsin at Milwaukee, United States*
Andrew Lemmon, *University of Alabama, United States*
James Gafford, *Mississippi State University, United States*
Michael Mazzola, *Mississippi State University, United States*

Session T32: Modeling of DC Energy Converters and Systems
Location: 102AB
March 24, 2016 8:30 - 11:20
Session Chairs: Santanu Kapat, *IIT Kharagpur*
Sombuddha Chakraborty, *Texas Instruments*

A Practical Switching Time Model for Synchronous Buck Converters 1585
Yuan Rao, *Texas Instruments Inc., United States*
Surinder P. Singh, *Texas Instruments Inc., United States*
Taisuke Kazama, *Texas Instruments Inc., United States*

Off-Line Identification of Digitally Controlled Power Converters using an Analog Frequency Response Analyzer 1591
Marco Meola, *Zentrum Mikroelektronik Dresden AG, Germany*
Anthony Kelly, *Altera Corporation, Ireland*

Extended Wide-Load Range Model for Multi-Level DC-DC Converters and a Practical Dual-Mode Digital Controller 1597
Nenad Vukadinović, *University of Toronto, Canada*
Aleksandar Prodić, *University of Toronto, Canada*
Brett A. Miwa, *Maxim Integrated, United States*
Cory B. Arnold, *Maxim Integrated, United States*
Michael W. Baker, *Maxim Integrated, United States*

Burst Mode Control and Switched-Capacitor Converters Losses 1603
Michael Evzelman, *Utah State University, United States*
Regan Zane, *Utah State University, United States*

Equivalent Circuit Modeling of LLC Resonant Converter 1608
Shuilin Tian, *Virginia Polytechnic Institute and State University, United States*
Fred C. Lee, *Virginia Polytechnic Institute and State University, United States*
Qiang Li, *Virginia Polytechnic Institute and State University, United States*

Small Signal Modeling of the Hysteretic Modulator with a Current Ripple Synthesizer 1616
Yi Huang, *Intersil Corporation, United States*
Chun Cheung, *Intersil Corporation, United States*

A Black-Box Modeling Approach for DC Nanogrids 1624
A. Francés, *Universidad Politécnica de Madrid, Spain*
R. Asensi, *Universidad Politécnica de Madrid, Spain*
O. García, *Universidad Politécnica de Madrid, Spain*
R. Prieto, *Universidad Politécnica de Madrid, Spain*
J. Uceda, *Universidad Politécnica de Madrid, Spain*

Session T33: Gate Drive Techniques
Location: 103C
March 24, 2016 8:30 - 11:20
Session Chairs: Christopher Bridge, *SIMPLIS Technologies*
Martin Ordonez, *University of British Columbia*

Design and Evaluation of Isolated Gate Driver Power Supply for Medium Voltage Converter Applications 1632
Krishna Mainali, *North Carolina State University, United States*
Sachin Madhusoodhanan, *North Carolina State University, United States*
Awneesh Tripathi, *North Carolina State University, United States*
Kasunaidu Vechalapu, *North Carolina State University, United States*
Ankan De, *North Carolina State University, United States*
Subhashish Bhattacharya, *North Carolina State University, United States*

General-Purpose Clocked Gate Driver (CGD) IC with Programmable 63-Level Drivability to Reduce IC Overshoot and Switching Loss of Various Power Transistors 1640
Koutarou Miyazaki, *University of Tokyo, Japan*
Seiya Abe, *Kyushu Institute of Technology, Japan*
Masanori Tsukuda, *Kyushu Institute of Technology, Japan*
Ichiro Omura, *Kyushu Institute of Technology, Japan*
Keiji Wada, *Tokyo Metropolitan University, Japan*
Makoto Takamiya, *University of Tokyo, Japan*
Takayasu Sakurai, *University of Tokyo, Japan*

An Integrated SiC CMOS Gate Driver 1646
Matthew Barlow, *University of Arkansas, United States*
Shamim Ahmed, *University of Arkansas, United States*
H. Alan Mantooth, *University of Arkansas, United States*
A. Matt Francis, *Ozark Integrated Circuits, Inc., United States*

Digital Active Gate Drives using Sequential Optimization 1650
Daniel J. Rogers, *University of Oxford, United Kingdom*
Boris Murmann, *Stanford University, United States*

One Adaptive Turn-Off Method for PFC Converter with Voltage Spike Limitation 1657
Qunfang Wu, *Nanjing University of Aeronautics and Astronautics, China*
Qin Wang, *Nanjing University of Aeronautics and Astronautics, China*
Lan Xiao, *Nanjing University of Aeronautics and Astronautics, China*
Jialin Xu, *Nanjing University of Aeronautics and Astronautics, China*
Hongxu Li, *Nanjing University of Aeronautics and Astronautics, China*

A Digital Implementation for PWM Phase-Frequency Synchronization in SMPS Systems 1663
Luca Bizjak, *Infineon Technologies Austria AG, Austria*
Emanuele Bodano, *Infineon Technologies Austria AG, Austria*
Ante Gotovac, *Infineon Technologies Austria AG, Austria*
Sergii Tkachov, *Infineon Technologies Austria AG, Austria*

A High Accuracy and High Bandwidth Current Sense Circuit for Digitally Controlled DC-DC Buck Converters 1670
David Stack, *Altera Corporation, Ireland*
Anthony Kelly, *Altera Corporation, Ireland*
Thomas Conway, *University of Limerick, Ireland*

Session T34: Energy Storage Systems
Location: 104B
March 24, 2016 8:30 - 11:20
Session Chairs: Wei Qiao, *University of Nebraska Lincoln*
Yilmaz Sozer, *University of Akron*

Modular Multilevel Dual Active Bridge DC-DC Converter with ZVS and Fast DC Fault Recovery for Battery Energy Storage Systems 1675
Yuxiang Shi, *Florida State University, United States*
Rui Li, *Florida State University, United States*
Hui Li, *Florida State University, United States*

An Analytical Framework to Design a Dynamic Frequency Control Scheme for Microgrids using Energy Storage 1682
Ajit A. Renjit, *Ohio State University, United States*
Feng Guo, *NEC Laboratories America, Inc., United States*
Ratnesh Sharma, *NEC Laboratories America, Inc., United States*

Comparative Evaluation of LiFePO$_4$ Cell SOC Estimation Performance with ECM Structure and Noise Model/Data Rejection in the EKF for Transportation Application 1690
Hyun-jun Lee, *Soongsil University, Korea, South*
Joung-hu Park, *Soongsil University, Korea, South*
Jonghoon Kim, *Chosun University, Korea, South*

A Power Sharing Scheme for Series Connected Offshore Wind Turbines in a Medium Voltage DC Collection Grid 1695
Michael T. Daniel, *Texas A&M University, United States*
Prasad N. Enjeti, *Texas A&M University, United States*

Fault Ride-Through Performance Evaluation of an Interleaved Grid-Connected Converter Employing Low Switching Frequency 1702
Lorand Bede, *Aalborg University, Denmark*
Ghanshyamsinh Gohil, *Aalborg University, Denmark*
Mihai Ciobotaru, *University of New South Wales, Australia*
Tamas Kerekes, *Aalborg University, Denmark*
Remus Teodorescu, *Aalborg University, Denmark*
Vassilios G. Agelidis, *University of New South Wales, Australia*

Analysis of Two Charging Modes of Battery Energy Storage System for a Stand-Alone Microgrid 1708
Jongmin Jo, *Chungnam National University, Korea, South*
Hanju Cha, *Chungnam National University, Korea, South*

Proposition and Experimental Verification of a Bi-Directional Isolated DC/DC Converter for Battery Charger-Discharger of Electric Vehicle 1713
Ryota Kondo, *Mitsubishi Electric Corporation, Japan*
Yusuke Higaki, *Mitsubishi Electric Corporation, Japan*
Masaki Yamada, *Mitsubishi Electric Corporation, Japan*

Session T35: Topics on Inductive and Capacitive Wireless Power Transfer
Location: 104C
March 24, 2016 8:30 - 11:20
Session Chairs: Chris Mi, *San Diego State University*
Omer Onar, *Oak Ridge National Laboratory*

A CLLC-Compensated High Power and Large Air-Gap Capacitive Power Transfer System for Electric Vehicle Charging Applications 1721
Fei Lu, *University of Michigan at Ann Arbor, United States*
Hua Zhang, *Northeastern Polytechnical University, China*
Heath Hofmann, *University of Michigan at Ann Arbor, United States*
Chris Mi, *San Diego State University, United States*

A Large Air-Gap Capacitive Power Transfer System with a 4-Plate Capacitive Coupler Structure for Electric Vehicle Charging Applications 1726
Hua Zhang, *Northwestern Polytechnical University, China*
Fei Lu, *University of Michigan at Ann Arbor, United States*
Heath Hofmann, *University of Michigan at Ann Arbor, United States*
Weiguo Liu, *Northwestern Polytechnical University, China*
Chris Mi, *San Diego State University, United States*

Dynamic Wireless Power Transfer System for Electric Vehicles to Simplify Ground Facilities – Power Control and Efficiency Maximization on the Secondary Side – 1731
Katsuhiro Hata, *University of Tokyo, Japan*
Takehiro Imura, *University of Tokyo, Japan*
Yoichi Hori, *University of Tokyo, Japan*

Uniform-Gain Frequency Tracking of Wireless EV Charging for Improving Alignment Flexibility 1737
Yabiao Gao, *University of Georgia, United States*
Antonio Ginart, *University of Georgia / Sonnenbatterie GmbH, United States*
Kathleen Blair Farley, *Southern Company Services, Inc., United States*
Zion Tsz Ho Tse, *University of Georgia, United States*

Design and Optimization of a Multi-Coil System for Inductive Charging with Small Air Gap .. 1741
Christopher Joffe, *Fraunhofer Institute for Integrated Systems and Device Technology, Germany*
Andreas Roßkopf, *Fraunhofer Institute for Integrated Systems and Device Technology, Germany*
Stefan Ehrlich, *Fraunhofer Institute for Integrated Systems and Device Technology, Germany*
Christian Dobmeier, *Fraunhofer Institute for Integrated Systems and Device Technology, Germany*
Martin März, *Fraunhofer Institute for Integrated Systems and Device Technology, Germany*

Core Design for Better Misalignment Tolerance and Higher Range of Wireless Charging for HEV 1748
Mostak Mohammad, *University of Akron, United States*
Sangshin Kwak, *Chung-ang University, Korea, South*
Seungdeog Choi, *University of Akron, United States*

A 25 kW Industrial Prototype Wireless Electric Vehicle Charger 1756
Mariusz Bojarski, *Hevo Power Inc., United States*
Erdem Asa, *Hevo Power Inc. / New York University, United States*
Kerim Colak, *Istanbul Ulasim A.S., Turkey*
Dariusz Czarkowski, *New York University, United States*

Session T36: Wireless Power Transfer
Location: 103AB
March 24, 2016 8:30 - 11:20
Session Chairs: Sriram Jala Reddy, *Ford Motors*
Michael Masquelier, *WAVE*

Full-Bridge Series Resonant Multi-Inverter Featuring New 900-V SiC Devices for Improved Induction Heating Appliances 1762
Mario Pérez-Tarragona, *Universidad de Zaragoza, Spain*
Héctor Sarnago, *Universidad de Zaragoza, Spain*
Óscar Lucía, *Universidad de Zaragoza, Spain*
José M. Burdío, *Universidad de Zaragoza, Spain*

A Novel Phase Control of Single Switch Active Rectifier for Inductive Power Transfer Applications 1767
Kerim Colak, *Istanbul Ulasim A.S., Turkey*
Erdem Asa, *Hevo Power Inc. / New York University, United States*
Dariusz Czarkowski, *New York University, United States*

Optimal Shaped Dipole-Coil Design and Experimental Verification of Inductive Power Transfer System for Home Applications 1773
Duy T. Nguyen, *Korea Advanced Institute of Science and Technology, Korea, South*
Eun S. Lee, *Korea Advanced Institute of Science and Technology, Korea, South*
Byeung G. Choi, *Korea Advanced Institute of Science and Technology, Korea, South*
Chun T. Rim, *Korea Advanced Institute of Science and Technology, Korea, South*

A Novel Time-Sharing Current-Fed ZCS High Frequency Inverter-Applied Resonant DC-DC Converter for Inductive Power Transfer 1780
Kyohei Konishi, *Kobe University, Japan*
Tomokazu Mishima, *Kobe University, Japan*
Mutsuo Nakaoka, *University of Malaya, Malaysia*

Optimization of Coils for Magnetically Coupled Resonant Wireless Power Transfer System based on Maximum Output Power 1788
Dan Jiang, *Nanjing University of Aeronautics and Astronautics, China*
Yong Yang, *Nanjing University of Aeronautics and Astronautics, China*
Fuxin Liu, *Nanjing University of Aeronautics and Astronautics, China*
Xinbo Ruan, *Nanjing University of Aeronautics and Astronautics, China*
Xuling Chen, *Nanjing University of Aeronautics and Astronautics, China*

Online Regulation of Receiver-Side Power and Estimation of Mutual Inductance in Wireless Inductive Link based on Transmitter-Side Electrical Information 1795
Jeff Po Wa Chow, *City University of Hong Kong, Hong Kong*
Henry Shu-Hung Chung, *City University of Hong Kong, Hong Kong*
Chun Sing Cheng, *City University of Hong Kong, Hong Kong*

Dynamic Period Switching of PRS-PWM with Run-Length Limiting Technique for Spurious and Ripple Reduction in Fast Response Wireless Power Transmission 1802
Takahiro Moroto, *Keio University, Japan*
Toru Kawajiri, *Keio University, Japan*
Hiroki Ishikuro, *Keio University, Japan*

Session T37: Single-Phase AC-DC Converters
Location: 102AB
March 24, 2016 14:00 - 17:30
Session Chairs: Dusty Becker, *Emerson Network Power*
Pritam Das, *National University of Singapore*

A Flyback AC/DC Converter using Power Semiconductor Filter for Input Power
Factor Correction 1807
Chung-Pui Tung, *City University of Hong Kong, Hong Kong*
Henry Shu-Hung Chung, *City University of Hong Kong, Hong Kong*

Reducing the Variation Range of the Switching Frequency for CRM Boost PFC
Converter by Injecting 3rd Harmonic into the Input Current 1815
Yi Wang, *Nanjing University of Science and Technology, China*
Kai Yao, *Nanjing University of Science and Technology, China*

A Sustained Increase of Input Current Distortion in Active Input Current Shapers to
Eliminate Electrolytic Capacitor for Designing AC to DC HB-LED Drivers for Retrofit
Lamps Applications 1823
D.G. Lamar, *Universidad de Oviedo, Spain*
M. Arias, *Universidad de Oviedo, Spain*
A. Rodriguez, *Universidad de Oviedo, Spain*
J. Sebastian, *Universidad de Oviedo, Spain*
A. Fernandez, *European Space Agency, Netherlands*
J.A. Villarejo, *Universidad de Cartagena, Spain*

Reduced Current Stress Bridgeless Cuk PFC Converter with New Voltage Multiplier Circuit ... 1831
Yi-Hung Liao, *National Penghu University of Science and Technology, Taiwan*

Implementation of Multi-Level Bridgeless PFC Rectifiers for Mid-Power Single
Phase Applications 1835
Trong Tue Vu, *Eisergy Ltd., Ireland*
George Young, *Eisergy Ltd., Ireland*

US Mains Stacked Very High Frequency Self-Oscillating Resonant Power Converter with
Unified Rectifier 1842
Jeppe A. Pedersen, *Technical University of Denmark, Denmark*
Mickey P. Madsen, *Technical University of Denmark, Denmark*
Jakob D. Mønster, *Technical University of Denmark, Denmark*
Thomas Andersen, *Technical University of Denmark, Denmark*
Arnold Knott, *Technical University of Denmark, Denmark*
Michael A.E. Andersen, *Technical University of Denmark, Denmark*

Digital-Based Interleaving Control for GaN-Based MHz CRM Totem-Pole PFC 1847
Zhengyang Liu, *Virginia Polytechnic Institute and State University, United States*
Zhengrong Huang, *Virginia Polytechnic Institute and State University, United States*
Fred C. Lee, *Virginia Polytechnic Institute and State University, United States*
Qiang Li, *Virginia Polytechnic Institute and State University, United States*

A Novel AC-to-DC Adaptor with Ultra-High Power Density and Efficiency 1853
Yan-Cun Li, *Virginia Polytechnic Institute and State University, United States*
Fred C. Lee, *Virginia Polytechnic Institute and State University, United States*
Qiang Li, *Virginia Polytechnic Institute and State University, United States*
Xiucheng Huang, *Virginia Polytechnic Institute and State University, United States*
Zhengyang Liu, *Virginia Polytechnic Institute and State University, United States*

A Single-Stage Single-Phase Isolated AC-DC Converter based on LLC Resonant Unit and T-Type Three-Level Unit for Battery Charging Applications 1861
Yikai Gao, *University of Texas at Dallas, United States*
Wen Cai, *University of Texas at Dallas, United States*
Fan Yi, *University of Texas at Dallas, United States*

Session T38: Non-Isolated DC-DC Converters
Location: 101A
March 24, 2016 14:00 - 17:30
Session Chairs: Pradeep Shenoy, *Texas Instruments*
Juan Rivas-Davila, *Stanford*

DC-DC Power Converter Controller for SOC Balancing of Paralleled Battery System 1868
Jaber A. Abu Qahouq, *University of Alabama, United States*
Lin Zhang, *University of Alabama, United States*
Yuan Cao, *University of Alabama, United States*
Bharat Balasubramanian, *University of Alabama, United States*

Ultra-Step-Up DC-DC Converter with Integrated Autotransformer and Coupled Inductor ... 1872
Yam P. Siwakoti, *Aalborg University, Denmark*
Frede Blaabjerg, *Aalborg University, Denmark*
Poh Chiang Loh, *Aalborg University, Denmark*

Optimal Dynamic Phase Add/Drop Mechanism in Multiphase DC-DC Buck Converters 1878
Anandha Ruban T T, *Texas Instruments India Pvt. Ltd., India*
Preetam Tadeparthy, *Texas Instruments India Pvt. Ltd., India*
Sankaran Aniruddhan, *Indian Institute of Technology Madras, India*
Vikram Gakhar, *Texas Instruments India Pvt. Ltd., India*
Muthusubramanian Venkateswaran, *Texas Instruments India Pvt. Ltd., India*

A Universal Self-Calibrating Dynamic Voltage and Frequency Scaling (DVFS) Scheme with Thermal Compensation for Energy Savings in FPGAs 1882
Shuze Zhao, *University of Toronto, Canada*
Ibrahim Ahmed, *University of Toronto, Canada*
Carl Lamoureux, *University of Toronto, Canada*
Ashraf Lotfi, *Altera Corporation, United States*
Vaughn Betz, *University of Toronto, Canada*
Olivier Trescases, *University of Toronto, Canada*

Morphing Switched-Capacitor Step-Down DC-DC Converters with Variable Conversion Ratio 1888
Song Xiong, *University of Hong Kong, Hong Kong*
Ying Huang, *University of Hong Kong, Hong Kong*
Siew-Chong Tan, *University of Hong Kong, Hong Kong*
Shu-Yuen Ron Hui, *University of Hong Kong, Hong Kong*

Compact Modular Switched-Capacitor DC/DC Converters with Exponential Voltage Gain 1894
Ying Huang, *University of Hong Kong, Hong Kong*
Song Xiong, *University of Hong Kong, Hong Kong*
Siew-Chong Tan, *University of Hong Kong, Hong Kong*
Shu-Yuen Ron Hui, *University of Hong Kong, Hong Kong*

Study and Implementation of a High Step-Up Voltage DC-DC Converter using Coupled-Inductor and Cascode Techniques 1900
Tsorng-Juu Liang, *National Cheng Kung University, Taiwan*
Yung-Ting Huang, *National Cheng Kung University, Taiwan*
Jian-Hsing Lee, *National Cheng Kung University, Taiwan*
Lo Pang-Yen Ting, *National Cheng Kung University, Taiwan*

20 mV Input, 4.2 V Output Boost Converter with Methodology of Maximum Output Power for Thermoelectric Energy Harvesting 1907
Taichi Ogawa, *Toshiba Corporation, Japan*
Takeshi Ueno, *Toshiba Corporation, Japan*
Takayuki Miyazaki, *Toshiba Corporation, Japan*
Tetsuro Itakura, *Toshiba Corporation, Japan*

Clarification of Relationship between Current Ripple and Power Density in Bidirectional DC-DC Converter 1911
Hoai Nam Le, *Nagaoka University of Technology, Japan*
Koji Orikawa, *Nagaoka University of Technology, Japan*
Jun-Ichi Itoh, *Nagaoka University of Technology, Japan*

Session T39: Inverter Applications and Technologies
Location: 101B
March 24, 2016 14:00 - 17:30
Session Chairs: Ali Khajehoddin, *University of Alberta*
Wen Cai, *University of Texas, Dallas*

Grid-Voltage Feedforward based Control for Grid-Connected LCL-Filtered Inverter with High Robustness and Low Grid Current Distortion in Weak Grid 1919
Jinming Xu, *Nanjing University of Aeronautics and Astronautics, China*
Qiang Qian, *Nanjing University of Aeronautics and Astronautics, China*
Shaojun Xie, *Nanjing University of Aeronautics and Astronautics, China*
Binfeng Zhang, *Nanjing University of Aeronautics and Astronautics, China*

Evaluation of PV Frequency-Watt Function for Fast Frequency Reserves 1926
J. Neely, *Sandia National Laboratories, United States*
J. Johnson, *Sandia National Laboratories, United States*
J. Delhotal, *Sandia National Laboratories, United States*
S. Gonzalez, *Sandia National Laboratories, United States*
M. Lave, *Sandia National Laboratories, United States*

A Systematic Design Method and Verification for a Zero-Ripple Interface for PV/Battery-to-Grid Applications 1934
Suvankar Biswas, *University of Minnesota, United States*
Ned Mohan, *University of Minnesota, United States*
William Robbins, *University of Minnesota, United States*

Grid-Voltage-Feedforward Active Damping for Grid-Connected Inverter with LCL Filter 1941

Minghui Lu, *Aalborg University, Denmark*
Xiongfei Wang, *Aalborg University, Denmark*
Frede Blaabjerg, *Aalborg University, Denmark*
S.M. Muyeen, *Petroleum Institute, U.A.E.*
Ahmed Al-Durra, *Petroleum Institute, U.A.E.*
Siyu Leng, *Petroleum Institute, U.A.E.*

A High Power Density Single-Phase Inverter using Stacked Switched Capacitor Energy Buffer 1947

Colin McHugh, *University of Colorado at Boulder, United States*
Sreyam Sinha, *University of Colorado at Boulder, United States*
Jeffrey Meyer, *University of Colorado at Boulder, United States*
Saad Pervaiz, *University of Colorado at Boulder, United States*
Jie Lu, *University of Colorado at Boulder, United States*
Fan Zhang, *University of Colorado at Boulder, United States*
Hua Chen, *University of Colorado at Boulder, United States*
Hyeokjin Kim, *University of Colorado at Boulder, United States*
Usama Anwar, *University of Colorado at Boulder, United States*
Ashish Kumar, *University of Colorado at Boulder, United States*
Alihossein Sepahvand, *University of Colorado at Boulder, United States*
Scott Jensen, *University of Colorado at Boulder, United States*
Beomseok Choi, *University of Colorado at Boulder, United States*
Daniel Seltzer, *University of Colorado at Boulder, United States*
Robert Erickson, *University of Colorado at Boulder, United States*
Dragan Maksimovic, *University of Colorado at Boulder, United States*
Khurram K. Afridi, *University of Colorado at Boulder, United States*

A Novel Single-Stage Dual-Active Bridge based Isolated DC-AC Converter 1954

Shiladri Chakraborty, *Indian Institute of Technology Kharagpur, India*
Souvik Chattopadhyay, *Indian Institute of Technology Kharagpur, India*

Ultra-Low Ripple Inverters for Distributed Generation Applications 1962

Ang Shen, *Missouri University of Science and Technology, United States*
Pourya Shamsi, *Missouri University of Science and Technology, United States*
Mehdi Ferdowsi, *Missouri University of Science and Technology, United States*

A 15 kV SiC MOSFET Gate Drive with Power Over Fiber based Isolated Power Supply and Comprehensive Protection Functions 1967

Xuan Zhang, *Ohio State University, United States*
He Li, *Ohio State University, United States*
John A. Brothers, *Ohio State University, United States*
Jin Wang, *Ohio State University, United States*
Lixing Fu, *Texas Instruments Inc., United States*
Mico Perales, *MH GoPower Co., Ltd., Taiwan*
John Wu, *MH GoPower Co., Ltd., Taiwan*

A 15-kV Class Intelligent Universal Transformer for Utility Applications 1974

Jih-Sheng Lai, *Virginia Polytechnic Institute and State University, United States*
Wei-Han Lai, *Enertronics, Inc., United States*
Seung-Ryul Moon, *Virginia Polytechnic Institute and State University, United States*
Lanhua Zhang, *Virginia Polytechnic Institute and State University, United States*
Arindam Maitra, *Electric Power Research Institute, United States*

Session T40: Modeling, Modulation and Control of Motor Drive
Location: 102C
March 24, 2016 14:00 - 17:30
Session Chairs: Jin Wang, *The Ohio State University*
River-TinHo Li, *ABB*

Modulation Technique for Common Mode Voltage Reduction in a Matrix Converter Drive Operating with High Voltage Transfer Ratio 1982
Varsha Padhee, *Rockwell Automation, United States*
Ashish Kumar Sahoo, *University of Minnesota, United States*
Ned Mohan, *University of Minnesota, United States*

Soft-Switched Discontinuous Pulse-Width Pulse-Density Modulation Scheme 1989
Arash Rahnamaee, *University of Illinois at Chicago, United States*
Alireza Mojab, *University of Illinois at Chicago, United States*
Hossein Riazmontazer, *University of Illinois at Chicago, United States*
Sudip K. Mazumder, *University of Illinois at Chicago, United States*
Milos Zefran, *University of Illinois at Chicago, United States*

A Novel Flux Estimator based on SOGI with FLL for Induction Machine Drives 1995
Rende Zhao, *China University of Petroleum, China*
Zhen Xin, *Aalborg University, Denmark*
Poh Chiang Loh, *Aalborg University, Denmark*
Frede Blaabjerg, *Aalborg University, Denmark*

Performance Characterization of Random Pulse Width Modulation Algorithms in Industrial and Commercial Adjustable Speed Drives 2003
Kevin Lee, *Eaton Corporation, United States*
Guangtong Shen, *Purdue University, United States*
Wenxi Yao, *Zhejiang University, China*
Zhengyu Lu, *Zhejiang University, China*

Stability Analysis and Controller Synthesis for Digital Single-Loop Voltage-Controlled Inverters 2011
Xiongfei Wang, *Aalborg University, Denmark*
Poh Chiang Loh, *Aalborg University, Denmark*
Frede Blaabjerg, *Aalborg University, Denmark*

High Efficiency, Hybrid Selective Harmonic Elimination Phase-Shift PWM Technique for Cascaded H-Bridge Inverters to Improve Dynamic Response and Operate in Complete Normal Modulation Indices 2019
Amirhossein Moeini, *University of Florida, United States*
Zhao Hui, *University of Florida, United States*
Shuo Wang, *University of Florida, United States*

Implementation and Experimental Validation of Efficiency Improvement in PMSM Drives through Switching Frequency Reduction 2027
Parag Kshirsagar, *United Technologies Research Center, United States*
Krishnan Ramu, *Virginia Polytechnic Institute and State University, United States*

Sensorless Speed Control of Symmetrical Triple-Star Nine-Phase Interior Permanent Magnet Machines 2035
Olorunfemi Ojo, *Tennessee Technological University, United States*
Medhi Ramezani, *Tennessee Technological University, United States*

Mitigation of Common-Mode Noise in Wide Band Gap Device based Motor Drives 2043

Sneha Narasimhan, *Rockwell Automation, United States*
Saurabh Tewari, *MTS Systems Corporation, United States*
Eric Severson, *University of Minnesota, United States*
Rohit Baranwal, *University of Minnesota, United States*
Ned Mohan, *University of Minnesota, United States*

Session T41: Gate Drivers and Integrated Packaging
Location: 103C
March 24, 2016 14:00 - 17:30
Session Chairs: Qiang Li, *Virginia Tech*
Jean-Luc Schanen, *Ecole Nationale Supérieure de l'Energie*

A High-Efficient Driving Isolated Drive-by-Microwave Half-Bridge Gate Driver for a GaN Inverter 2051

Shuichi Nagai, *Panasonic Corporation, Japan*
Yasufumi Kawai, *Panasonic Corporation, Japan*
Osamu Tabata, *Panasonic Corporation, Japan*
Songbaek Choe, *Panasonic Corporation, Japan*
Noboru Negoro, *Panasonic Corporation, Japan*
Tesuzo Ueda, *Panasonic Corporation, Japan*

Sensing Gallium Nitride HEMT Junction Temperature using Gate Drive Output Transient Properties 2055

He Niu, *University of Wisconsin at Madison, United States*
Robert D. Lorenz, *University of Wisconsin at Madison, United States*

Design and Application of a 1200V Ultra-Fast Integrated Silicon Carbide MOSFET Module ... 2063

Suxuan Guo, *North Carolina State University, United States*
Liqi Zhang, *North Carolina State University, United States*
Yang Lei, *North Carolina State University, United States*
Xuan Li, *North Carolina State University, United States*
Wensong Yu, *North Carolina State University, United States*
Alex Q. Huang, *North Carolina State University, United States*

Active Gate Charge Control Strategy for Series-Connected IGBTs 2071

Fan Zhang, *Xi'an Jiaotong University, China*
Xu Yang, *Xi'an Jiaotong University, China*
Yu Ren, *Xi'an Jiaotong University, China*
Ying Chen, *Xi'an Jiaotong University, China*
Ruifeng Gou, *Xi'an XD Power Systems Co., LTD, China*

A MV Intelligent Gate Driver for 15kV SiC IGBT and 10kV SiC MOSFET 2076

Awneesh Tripathi, *North Carolina State University, United States*
Krishna Mainali, *North Carolina State University, United States*
Sachin Madhusoodhanan, *North Carolina State University, United States*
Akshat Yadav, *North Carolina State University, United States*
Kasunaidu Vechalapu, *North Carolina State University, United States*
Subhashish Bhattacharya, *North Carolina State University, United States*

Linear Temperature Sensors in High-Voltage GaN-HEMT Power Devices 2083
Richard Reiner, *Fraunhofer Institute for Applied Solid State Physics, Germany*
Patrick Waltereit, *Fraunhofer Institute for Applied Solid State Physics, Germany*
Beatrix Weiss, *Fraunhofer Institute for Applied Solid State Physics, Germany*
Matthias Wespel, *Fraunhofer Institute for Applied Solid State Physics, Germany*
Dirk Meder, *Fraunhofer Institute for Applied Solid State Physics, Germany*
Michael Mikulla, *Fraunhofer Institute for Applied Solid State Physics, Germany*
Rüdiger Quay, *Fraunhofer Institute for Applied Solid State Physics, Germany*
Oliver Ambacher, *Fraunhofer Institute for Applied Solid State Physics, Germany*

An Innovative Power Module with Power-System-in-Inductor Structure 2087
Laili Wang, *Sumida Corporation, Canada*
Doug Malcolm, *Sumida Corporation, Canada*
Yan-Fei Liu, *Queen's University, Canada*

Thermal Analysis of a Magnetic Packaged Power Module 2095
Laili Wang, *Sumida Corporation, Canada*
Doug Malcolm, *Sumida Corporation, Canada*
Wenbo Liu, *Queen's University, Canada*
Yan-Fei Liu, *Queen's University, Canada*

**Analysis of a Low-Inductance Packaging Layout for Full-SiC Power Module Embedding
Split Damping** 2102
Yu Ren, *Xi'an Jiaotong University, China*
Xu Yang, *Xi'an Jiaotong University, China*
Fan Zhang, *Xi'an Jiaotong University, China*
Linlin Tan, *Xi'an Jiaotong University, China*
Xiangjun Zeng, *Xi'an Jiaotong University, China*

Session T42: Component Modeling
Location: 103AB
March 24, 2016 14:00 - 17:30
Session Chairs: Sheldon Williamson, *University of Ontario Institute of Technology*
Abhijit Pathak, *Infineon/IR*

**Comprehensive Parametric Analyses of Thermally Aged Power MOSFETs for Failure
Precursor Identification and Lifetime Estimation based on Gate Threshold Voltage** 2108
Serkan Dusmez, *University of Texas at Dallas, United States*
Bilal Akin, *University of Texas at Dallas, United States*

Modeling and Design Guidelines of High Density Power Inductor for Battery Power Unit 2114
Zhigang Dang, *University of Alabama, United States*
Jaber A. Abu Qahouq, *University of Alabama, United States*

Degradation of Low Voltage Metal Oxide Varistors in Power Supplies 2122
Dawood Talebi Khanmiri, *Northeastern University, United States*
Roy Ball, *Mersen USA, United States*
Jerry Mosesian, *Mersen USA, United States*
Brad Lehman, *Northeastern University, United States*

Characterization and Modeling of SiC MOSFET Body Diode 2127
Kang Peng, *University of South Carolina, United States*
Soheila Eskandari, *University of South Carolina, United States*
Enrico Santi, *University of South Carolina, United States*

A Simple Behavioral Electro-Thermal Model of GaN FETs for SPICE Circuit Simulation 2136
Liyao Wu, *Georgia Institute of Technology, United States*
Maryam Saeedifard, *Georgia Institute of Technology, United States*

Decomposition and Electro-Physical Model Creation of the CREE 1200V, 50A 3-Ph SiC Module 2141
Adam J. Morgan, *North Carolina State University, United States*
Yang Xu, *North Carolina State University, United States*
Douglas C. Hopkins, *North Carolina State University, United States*
Iqbal Husain, *North Carolina State University, United States*
Wensong Yu, *North Carolina State University, United States*

A Three-Legged MATLAB/Simulink Transformer Model using a Fictitious Delta Winding 2147
Thomas A. Nondahl, *Rockwell Automation, United States*
Jingbo Liu, *Rockwell Automation, United States*
Peter B. Schmidt, *Rockwell Automation, United States*

A Lifetime Prediction Method for LEDs Considering Mission Profiles 2154
Xiaohui Qu, *Southeast University, China*
Huai Wang, *Aalborg University, Denmark*
Xiaoqing Zhan, *City University of Hong Kong, Hong Kong*
Frede Blaabjerg, *Aalborg University, Denmark*
Henry Shu-Hung Chung, *City University of Hong Kong, Hong Kong*

Enhanced Li-Ion Battery Modeling using Recursive Parameters Correction 2161
Jae-Gu Kim, *Sungkyunkwan University, Korea, South*
Jung-Hoon Ahn, *Sungkyunkwan University, Korea, South*
Byoung-Kuk Lee, *Sungkyunkwan University, Korea, South*

Session T43: Grid and Utility Interface
Location: 104A
March 24, 2016 14:00 - 17:30
Session Chairs: Manish Bhardwaj, *Texas Instruments*
Nan Chen, *ABB*

Robust Sensorless Control of Grid Connected Converters with LCL Line Filters using Frequency Adaptive Observers as AC Voltage Estimators 2167
Vlatko Miskovic, *Danfoss Drives, United States*
Vladimir Blasko, *United Technologies Research Center, United States*
Thomas Jahns, *University of Wisconsin at Madison, United States*
Robert Lorenz, *University of Wisconsin at Madison, United States*
Haojiong Zhang, *Danfoss Drives, United States*

Active Stabilization of Direct Matrix Converter Input Side Filter through Grid Current Control 2175
Martin Leubner, *Technische Universität Dresden, Germany*
Nico Remus, *Technische Universität Dresden, Germany*
Marc Stübig, *Technische Universität Dresden, Germany*
Wilfried Hofmann, *Technische Universität Dresden, Germany*

Impedance-Based Stability Analysis of Single-Phase Inverter Connected to Weak Grid with Voltage Feed-Forward Control 2182
Jiangfeng Wang, *Nanjing University of Aeronautics and Astronautics, China*
Jianhui Yao, *Nanjing University of Aeronautics and Astronautics, China*
Haibing Hu, *Nanjing University of Aeronautics and Astronautics, China*
Yan Xing, *Nanjing University of Aeronautics and Astronautics, China*
Xiaobin He, *Shanghai Institute of Space Power-Sources, China*
Kai Sun, *Tsinghua University, China*

New Configuration of Dynamic Voltage Restorer for Medium Voltage Application 2187
Arash Khoshkbar Sadigh, *Extron Electronics, United States*
Vahid Dargahi, *Clemson University, United States*
Keith Corzine, *Clemson University, United States*

Studies on the Clustered Voltage Balancing Mechanism for Cascaded H-Bridge STATCOM ... 2194
Daorong Lu, *Nanjing University of Aeronautics and Astronautics, China*
Haibing Hu, *Nanjing University of Aeronautics and Astronautics, China*
Yan Xing, *Nanjing University of Aeronautics and Astronautics, China*
Xiaobin He, *Shanghai Institute of Space Power-Sources, China*
Kai Sun, *Tsinghua University, China*
Jianhui Yao, *Nanjing University of Aeronautics and Astronautics, China*

Design of a Fast Response Time Single-Phase PLL with DC Offset Rejection Capability ... 2200
Abhijit Kulkarni, *Indian Institute of Science, India*
Vinod John, *Indian Institute of Science, India*

Four New Applications of Second-Order Generalized Integrator Quadrature Signal Generator 2207
Zhen Xin, *Aalborg University, Denmark*
Rende Zhao, *China University of Petroleum, China*
Xiongfei Wang, *Aalborg University, Denmark*
Poh Chiang Loh, *Aalborg University, Denmark*
Frede Blaabjerg, *Aalborg University, Denmark*

Three-Phase Multiple Harmonic Sequence Detection based on Generalized Delayed Signal Superposition 2215
Yong Lu, *Xi'an Jiaotong University, China*
Guochun Xiao, *Xi'an Jiaotong University, China*
Xiongfei Wang, *Aalborg University, Denmark*
Frede Blaabjerg, *Aalborg University, Denmark*

Hybrid Modelling and Control of Single-Phase Grid-Connected NPC Inverters 2223
Xingda Yan, *University of Southampton, United Kingdom*
Zhan Shu, *University of Southampton, United Kingdom*
Suleiman M. Sharkh, *University of Southampton, United Kingdom*

Session T44: Topics in Renewable Energy Systems II
Location: 104B
March 24, 2016 14:00 - 17:30
Session Chairs: Akshay Kumar Rathore, *Concordia University*
Yichao Tang, *Texas Instruments*

Stability Criterion and Controller Parameter Design of Radial-Line Renewable Systems with Multiple Inverters 2229
Wenchao Cao, *University of Tennessee, United States*
Xuan Zhang, *University of Tennessee, United States*
Yiwei Ma, *University of Tennessee, United States*
Fred Wang, *University of Tennessee, United States*

Stability Analysis and Improvement of Solid State Transformer (SST)-Paralleled Inverters System using Negative Impedance Feedback Control 2237
Qing Ye, *Florida State University, United States*
Hui Li, *Florida State University, United States*

Compensator-Less Structures for Droop Control of Single Phase Inverters in a Flexible Microgrid 2245
Onkar Vitthal Kulkarni, *Indian Institute of Technology Bombay, India*
Suryanarayana Doolla, *Indian Institute of Technology Bombay, India*
B.G. Fernandes, *Indian Institute of Technology Bombay, India*

Comparative Evaluation of the Loss and Thermal Performance of Advanced Three Level Inverter Topologies 2252
Alexander Anthon, *Technical University of Denmark, Denmark*
Zhe Zhang, *Technical University of Denmark, Denmark*
Michael A.E. Andersen, *Technical University of Denmark, Denmark*
Grahame Holmes, *RMIT University, Australia*
Brendan McGrath, *RMIT University, Australia*
Carlos Teixeira, *RMIT University, Australia*

Dual Buck Inverter with Series Connected Diodes and Single Inductor 2259
Liwei Zhou, *Shandong University, China*
Feng Gao, *Shandong University, China*

Magnetic Integration of the Harmonic Filter Inductor for Dual-Converter Fed Open-End Transformer Topology 2264
Ghanshyamsinh Gohil, *Aalborg University, Denmark*
Lorand Bede, *Aalborg University, Denmark*
Remus Teodorescu, *Aalborg University, Denmark*
Tamas Kerekes, *Aalborg University, Denmark*
Frede Blaabjerg, *Aalborg University, Denmark*

Mechanism Analysis and Mitigation of Instability in Grid-Connected Voltage Source Inverter with LCL Filters based on Terminal Impedance 2272
Teng Liu, *Xi'an Jiaotong University, China*
Zeng Liu, *Xi'an Jiaotong University, China*
Jinjun Liu, *Xi'an Jiaotong University, China*
Qingyun Dou, *Xi'an Jiaotong University, China*

Seven-Switch Five-Level Active Neutral-Point Clamped Converter and Optimal Modulation Strategy 2278
Hongliang Wang, *Queen's University, Canada*
Lei Kou, *Queen's University, Canada*
Yan-Fei Liu, *Queen's University, Canada*
Paresh C. Sen, *Queen's University, Canada*
Sucheng Liu, *Anhui University of Technology, China*

A Simple Variable Step Size Method for Maximum Power Point Tracking using Commercial Current Mode Control DC-DC Regulators 2286
Su Sheng, *Northeastern University, United States*
Brad Lehman, *Northeastern University, United States*

Session T45: Envelope Tracking and Resonant Conversion
Location: 104C
March 24, 2016 14:00 - 17:30
Session Chairs: Brian Zahnstecher, *PowerRox*
Davide Giacomini, *Infineon*

Envelope Tracking GaN Power Supply for 4G Cell Phone Base Stations 2292
Yuanzhe Zhang, *University of Colorado at Boulder, United States*
Johan Strydom, *Efficient Power Conversion Corporation, United States*
Michael de Rooij, *Efficient Power Conversion Corporation, United States*
Dragan Maksimović, *University of Colorado at Boulder, United States*

Envelope Tracking Power Supply for Volume-Sensitive Low-Power Applications based on a Resonant Switched-Capacitor Converter 2298
Alon Cervera, *Ben-Gurion University of the Negev, Israel*
Mor Mordechai Peretz, *Ben-Gurion University of the Negev, Israel*

A Passive-Impedance-Matching Concept for Multi-Phase Resonant Converter 2304
Hongliang Wang, *Queen's University, Canada*
Yang Chen, *Queen's University, Canada*
Yan-Fei Liu, *Queen's University, Canada*

LLC Converter with Auxiliary Switch for Hold Up Mode Operation 2312
Yang Chen, *Queen's University, Canada*
Hongliang Wang, *Queen's University, Canada*
Yan-Fei Liu, *Queen's University, Canada*
Jahangir Afsharian, *Murata Power Solutions, Canada*
Zhihua Yang, *Queen's University, Canada*

A Common Capacitor Multi-Phase LLC Resonant Converter 2320
Hongliang Wang, *Queen's University, Canada*
Yang Chen, *Queen's University, Canada*
Zhiyuan Hu, *Queen's University, Canada*
Laili Wang, *Queen's University, Canada*
Yajie Qiu, *Queen's University, Canada*
Wenbo Liu, *Queen's University, Canada*
Yan-Fei Liu, *Queen's University, Canada*
Jahangir Afsharian, *Murata Power Solutions, Canada*
Zhihua Yang, *Murata Power Solutions, Canada*

LLC Resonant Converter Design for Bendable Power Converter 2328
Kwun Yuan Godwin Ho, *University of Hong Kong, Hong Kong*
M.H. Bryan Pong, *University of Hong Kong, Hong Kong*
Shu-Yuen Ron Hui, *University of Hong Kong, Hong Kong*

Design Consideration of MHz Active Clamp Flyback Converter with GaN Devices for Low Power Adapter Application 2334
Xiucheng Huang, *Virginia Polytechnic Institute and State University, United States*
Junjie Feng, *Virginia Polytechnic Institute and State University, United States*
Weijing Du, *Virginia Polytechnic Institute and State University, United States*
Fred C. Lee, *Virginia Polytechnic Institute and State University, United States*
Qiang Li, *Virginia Polytechnic Institute and State University, United States*

A New Capacitor Voltage Balancing Control for Hybrid Modular Multilevel Converter with Cascaded Full Bridge 2342
Mahendra B. Ghat, *Indian Institute of Technology Bombay, India*
Anshuman Shukla, *Indian Institute of Technology Bombay, India*
Richa Mishra, *Indian Institute of Technology Bombay, India*

Sensorless Scheduling of the Modular Multilevel Series-Parallel Converter: Enabling a Flexible, Efficient, Modular Battery 2349
Stefan M. Goetz, *Duke University, United States*
Zhongxi Li, *Duke University, United States*
Angel V. Peterchev, *Duke University, United States*
Xinyu Liang, *North Carolina State University, United States*
Chengduo Zhang, *North Carolina State University, United States*
Srdjan M. Lukic, *North Carolina State University, United States*

Session D01: AC-DC Converters
Location: Poster Area
March 24, 2016 11:30 - 14:00
Session Chairs: Nathan Weise, *Marquette*
Daniel Costinett, *University of Tennessee-Knoxville*

An Input Current Calculation Switching Driver for High Power-Factor and Phase-Cut Dimmer Compatibility 2355
Hyunchul Eum, *Fairchild Semiconductor International, Inc., Korea, South*
Youngjong Kim, *Fairchild Semiconductor International, Inc., Korea, South*
Kuohsien Huang, *Fairchild Semiconductor International, Inc., Taiwan*

High Frequency Range Conducted Common-Mode Noise Suppression in SMPS 2360
Jinping Zhou, *Delta Electronics Shanghai Co., Ltd., China*
Yicong Xie, *Delta Electronics Shanghai Co., Ltd., China*
Min Zhou, *Delta Electronics Shanghai Co., Ltd., China*

Improved Medium Voltage AC-DC Rectifier based on 10kV SiC MOSFET for Solid State Transformer (SST) Application 2365
Qianlai Zhu, *North Carolina State University, United States*
Li Wang, *North Carolina State University, United States*
Liqi Zhang, *North Carolina State University, United States*
Wensong Yu, *North Carolina State University, United States*
Alex Q. Huang, *North Carolina State University, United States*

Suppression of Circulating Current in Parallel Operation of Three-Level Converters 2370
Young-Kwang Son, *Seoul National University, Korea, South*
Seung-Jun Chee, *Seoul National University, Korea, South*
Younggi Lee, *Seoul National University, Korea, South*
Seung-Ki Sul, *Seoul National University, Korea, South*
Changjin Lim, *LG Electronics, Korea, South*
Sungjae Huh, *LG Electronics, Korea, South*
Jaeyoon Oh, *LG Electronics, Korea, South*

Hybrid Bridgeless DCM SEPIC Rectifier Integrated with a Modified Switched Capacitor Cell ... 2376
Paulo Junior Silva Costa, *Universidade Federal de Santa Catarina, Brazil*
Telles Brunelli Lazzarin, *Universidade Federal de Santa Catarina, Brazil*
Carlos Henrique Illa Font, *Universidade Tecnológica Federal do Paraná, Brazil*

**LCL Filter Design for Three-Phase Two-Level Power Factor Correction using Line
Impedance Stabilization Network** 2382
Alireza Kouchaki, *University of Southern Denmark, Denmark*
Morten Nymand, *University of Southern Denmark, Denmark*

Sensorless Current Rebuilding Strategy in a Single Phase Bridgeless PFC 2389
Felipe López, *Universidad de Cantabria, Spain*
Paula Lamo, *Universidad de Cantabria, Spain*
Alberto Pigazo, *Universidad de Cantabria, Spain*
F.J. Azcondo, *Universidad de Cantabria, Spain*

A Compact Electrolytic-Free Two-Stage Universal Input Offline LED Driver 2395
Saad Pervaiz, *University of Colorado at Boulder, United States*
Ashish Kumar, *University of Colorado at Boulder, United States*
Khurram K. Afridi, *University of Colorado at Boulder, United States*

Session D02: DC-DC Converters I
Location: Poster Area
March 24, 2016 11:30 - 14:00
Session Chairs: Charles Sullivan, *Dartmouth*
Mahshid Amirabadi, *Northeastern University*

**Design Methodology for a High Insulation Voltage Power Transmission Function for
IGBT Gate Driver** 2401
Sokchea Am, *Grenoble Institute of Technology, France*
Pierre Lefranc, *Grenoble Institute of Technology, France*
David Frey, *Grenoble Institute of Technology, France*
Mahmoud Ibrahim, *Grenoble Institute of Technology, France*

**Optimized Design of GaN Switching Capacitor based Envelope Tracking Power Supply
for Satellite Applications** 2409
Qian Jin, *Nanjing University of Aeronautics and Astronautics, China*
M. Vasić, *Universidad Politécnica de Madrid, Spain*
O. Garcia, *Universidad Politécnica de Madrid, Spain*
P. Alou, *Universidad Politécnica de Madrid, Spain*
J.A. Oliver, *Universidad Politécnica de Madrid, Spain*
J.A. Cobos, *Universidad Politécnica de Madrid, Spain*

An Isolated High Step-Up Converter with Continuous Input Current and LC Snubber 2415

K.I. Hwu, *National Taipei University of Technology, Taiwan*
W.Z. Jiang, *National Taipei University of Technology, Taiwan*
Y.T. Yau, *National Taipei University of Technology, Taiwan*

Output-Inductor-Less Full-Bridge Converter with SiC-MOSFETs for Low Noise and ZVS Operation 2422

Kazuhide Domoto, *Nagasaki University, Japan*
Yoichi Ishizuka, *Nagasaki University, Japan*
Seiya Abe, *Kyushu Institute of Technology, Japan*
Tamotsu Ninomiya, *Green Electronics Research Institute, Kitakyushu, Japan*

Reduction Technique of Leakage Flux Effects on GaN-HEMTs in 5 MHz / 100 W Isolated DC-DC Converters 2430

Akinori Hariya, *Nagasaki University, Japan*
Tomoya Koga, *Nagasaki University, Japan*
Ken Matsuura, *TDK Corporation, Japan*
Hiroshige Yanagi, *TDK-Lambda Corporation, Japan*
Satoshi Tomioka, *TDK-Lambda Corporation, Japan*
Yoichi Ishizuka, *Nagasaki University, Japan*
Tamotsu Ninomiya, *City of Kitakyushu, Japan*

A High-Voltage Level Shifter with Sub-Nano-Second Propagation Delay for Switching Power Converters 2437

Ahmed Abdelmoaty, *Ohio State University, United States*
Mohammad Al-Shyoukh, *TSMC Inc., United States*
Ayman Fayed, *Ohio State University, United States*

Dual-Output, Three-Level GaN-Based DC-DC Converter for Battery Charger Applications 2441

Ren Ren, *Nanjing University of Aeronautics and Astronautics, China*
Bo Liu, *University of Tennessee, United States*
Edward A. Jones, *University of Tennessee, United States*
Fred Wang, *University of Tennessee, United States*
Zheyu Zhang, *University of Tennessee, United States*
Daniel Costinett, *University of Tennessee, United States*

Quadruple Active Bridge DC-DC Converter as the Basic Cell of a Modular Smart Transformer 2449

Levy F. Costa, *Christian-Albrechts-Universität zu Kiel, Germany*
Giampaolo Buticchi, *Christian-Albrechts-Universität zu Kiel, Germany*
Marco Liserre, *Christian-Albrechts-Universität zu Kiel, Germany*

Analytical Model of a Phase-Shift Controlled Three-Level Zero-Voltage Switching Converter 2457

Cas Bakker, *Prodrive Technologies, Netherlands*
Bas Vermulst, *Technische Universiteit Eindhoven, Netherlands*
Anton Driessen, *Prodrive Technologies, Netherlands*

High Efficiency Design for ISOP Converter System with Dual Active Bridge DC-DC Converter 2465

Masaki Sato, *Nagasaki University, Japan*
Kazuhide Domoto, *Nagasaki University, Japan*
Yoichi Ishizuka, *Nagasaki University, Japan*
Masahiro Yamaguchi, *Tohoku University, Japan*
Shinya Manabe, *RICOH Electronic Devices Co., Ltd., Japan*
Hiizu Okubo, *RICOH Electronic Devices Co., Ltd., Japan*
Atsushi Itagaki, *Ryowa Electronics Co., Ltd., Japan*

Wide Input Range Power Converters using a Variable Turns Ratio Transformer 2473

Ziwei Ouyang, *Technical University of Denmark, Denmark*
Michael A.E. Andersen, *Technical University of Denmark, Denmark*

Design Approaches for Fast Supercapacitor Chargers for Applications like SCATMA, SRUPS 2479

Nicoloy Gurusinghe, *University of Waikato, New Zealand*
Nihal Kularatna, *University of Waikato, New Zealand*
W. Howell Round, *University of Waikato, New Zealand*
D. Alistair Steyn-Ross, *University of Waikato, New Zealand*

Stack Multiphase Asymmetrical Half-Bridge Topology Offering Advance Performance and Efficiency 2485

Trong Tue Vu, *Eisergy Ltd., Ireland*
George Young, *Eisergy Ltd., Ireland*

Session D03: DC-DC Converters II
Location: Poster Area
March 24, 2016 11:30 - 14:00
Session Chairs: Jason Stauth, *Dartmouth*
Yan-Fei Liu, *Queens*

Design of a Novel APWM Half-Bridge DC-DC Resonant Converter with Load-Independent Soft-Switching and Reduced Circulating Current 2491

Kawsar Ali, *National University of Singapore, Singapore*
Sandeep Kolluri, *National University of Singapore, Singapore*
Naga Brahmendra Yadav Gorla, *National University of Singapore, Singapore*
Pritam Das, *National University of Singapore, Singapore*
Sanjib Kumar Panda, *National University of Singapore, Singapore*

A Low-Volume Hybrid Step-Down DC-DC Converter based on the Dual use of Flying Capacitor 2497

S.M. Ahsanuzzaman, *University of Toronto, Canada*
Yingxian Ma, *University of Toronto, Canada*
Abrar Ahmed Pathan, *University of Toronto, Canada*
Aleksandar Prodić, *University of Toronto, Canada*

Fractional Pulse Skipping in Digitally Controlled DC-DC Converters for Improved Light-Load Efficiency and Power Spectrum 2504

Bipin Chandra Mandi, *Indian Institute of Technology Kharagpur, India*
Santanu Kapat, *Indian Institute of Technology Kharagpur, India*
Amit Patra, *Indian Institute of Technology Kharagpur, India*

A New Compact and High Efficiency Resonant Converter 2511
Sheng-Yang Yu, *Texas Instruments Inc., United States*

A 10-MHz eGaN FETs based Isolated Class-Φ_2 DCX 2518
Xuewen Zou, *Nanjing University of Aeronautics and Astronautics, China*
Zhiliang Zhang, *Nanjing University of Aeronautics and Astronautics, China*
Zhou Dong, *Nanjing University of Aeronautics and Astronautics, China*
Yuan Zhou, *Nanjing University of Aeronautics and Astronautics, China*
Xiaoyong Ren, *Nanjing University of Aeronautics and Astronautics, China*
Qianhong Chen, *Nanjing University of Aeronautics and Astronautics, China*

Multi-Level Capacitor Clamped DC-DC Multiplier/Divider with Variable and Fractional Voltage Gain – An (n/m)X DC-DC Converter 2525
Deepak Gunasekaran, *Michigan State University, United States*
Liang Qin, *Wuhan University, China*
Ujjwal Karki, *Michigan State University, United States*
Yuan Li, *Sichuan University, China*
Fang Z. Peng, *Michigan State University, United States*

Multi-Mode Quasi-Z-Source Series Resonant DC/DC Converter for Wide Input Voltage Range Applications 2533
Dmitri Vinnikov, *Ubik Solutions LLC, Estonia*
Andrii Chub, *Tallinn University of Technology, Estonia*
Indrek Roasto, *Ubik Solutions LLC, Estonia*
Liisa Liivik, *Tallinn University of Technology, Estonia*

Hybrid Serial-Output Converter for Integrated LED Lighting Applications 2540
T. McRae, *University of Toronto, Canada*
A. Prodić, *University of Toronto, Canada*
G. Lisi, *Texas Instruments Inc., United States*
W. McIntrye, *Texas Instruments Inc., United States*
A. Aguilar, *Texas Instruments Inc., United States*

Analysis and Modeling of a Modular ISOP Full Bridge based Converter with Input Filter ... 2545
P. Zumel, *Universidad Carlos III de Madrid, Spain*
E. Oña, *Universidad Carlos III de Madrid, Spain*
C. Fernandez, *Universidad Carlos III de Madrid, Spain*
M. Sanz, *Universidad Carlos III de Madrid, Spain*
A. Lazaro, *Universidad Carlos III de Madrid, Spain*
A. Barrado, *Universidad Carlos III de Madrid, Spain*
A. Vazquez, *Universidad de Oviedo, Spain*
D.G. Lamar, *Universidad de Oviedo, Spain*

Wide-Input High Power Density Flexible Converter Topology for DC-DC Applications 2553
Parth Jain, *University of Toronto, Canada*
Aleksandar Prodić, *University of Toronto, Canada*
Alexander Gerfer, *Würth Elektronik eiSos GmbH & Co. KG, Germany*

High Efficiency LLC Converter Design for Universal Battery Chargers 2561
Navid Shafiei, *University of British Columbia, Canada*
Ali Arefifar, *University of British Columbia, Canada*
Mohammad Ali Saket, *University of British Columbia, Canada*
Martin Ordonez, *University of British Columbia, Canada*

A New High Power Density Modular Multilevel DC-DC Converter with Localized Voltage Balancing Control for Arbitrary Number of Levels 2567

Ahmed Morsy, *Texas A&M University, United States*
Yong Zhou, *Texas A&M University, United States*
Prasad Enjeti, *Texas A&M University, United States*

Design and Control of a Fault Tolerant Soft Switching DC-DC Converter for High Power High Voltage Applications 2573

Tao Li, *Rensselaer Polytechnic Institute, United States*
Leila Parsa, *Rensselaer Polytechnic Institute, United States*

Accurate Parametric Steady State Analysis and Design Tool for DC-DC Power Converters 2579

Mohammad Daryaei, *University of Alberta, Canada*
Mohammad Ebrahimi, *University of Alberta, Canada*
S. Ali Khajehoddin, *University of Alberta, Canada*

Analysis of Multi-Output Half-Wave Semi-Synchronous Rectifier with a Uniform Magnetic Field Transmitter 2587

Erdem Asa, *Hevo Power Inc. / New York University, United States*
Kerim Colak, *Istanbul Ulasim A.S., Turkey*
Dariusz Czarkowski, *New York University, United States*

High Gain QZS DC/DC Converter with Coupled Inductor 2592

Rafael V. Silva, *Universidade Federal do Ceará, Brazil*
Antônio A.A. Freitas, *Universidade Federal do Ceará, Brazil*
Marcus R. Castro, *Universidade Federal do Ceará, Brazil*
Fernando L.M. Antunes, *Universidade Federal Rural do Semi-Árido, Brazil*
Edilson M. Sá Jr., *Universidade Federal do Ceará, Brazil*

Session D04: Utility Interface
Location: Poster Area
March 24, 2016 11:30 - 14:00
Session Chairs: Ali Khajehoddin, *University of Alberta*
Julia Zhang, *Oregon State University*

A Power Decoupling Method with Small Capacitance Requirement based on Single-Phase Quasi-Z-Source Inverter for DC Microgrid Applications 2599

Dingyi He, *University of Texas at Dallas, United States*
Wen Cai, *University of Texas at Dallas, United States*
Fan Yi, *University of Texas at Dallas, United States*

Operation Analysis of High Efficiency Grid Connected Bi-Directional Power Conversion System for Various Storage Battery Systems with Bi-Directional Switch Circuit Topology 2607

Go Yamada, *Panasonic Corporation, Japan*
Takaaki Norisada, *Panasonic Corporation, Japan*
Fumito Kusama, *Panasonic Corporation, Japan*
Keiji Akamatsu, *Panasonic Corporation, Japan*
Masakazu Michihira, *Kobe City College of Technology, Japan*

Fault Tolerant Control of MMC with Redundant Sub-Modules based on Carrier Phase Shift Modulation 2613

Kai Li, *Tsinghua University, China*
Zhengming Zhao, *Tsinghua University, China*
Liqiang Yuan, *Tsinghua University, China*
Sizhao Lu, *Tsinghua University, China*
Bing Pan, *State Grid Smart Grid Research Institute, China*
Zhengang Lu, *State Grid Smart Grid Research Institute, China*

A New Topology of Multilevel VSC Converter for Hybrid HVDC Transmission System 2620

Jae-Jung Jung, *Seoul National University, Korea, South*
Shenghui Cui, *RWTH Aachen University, Germany*
Seung-Ki Sul, *Seoul National University, Korea, South*

Performance of Solid State Transformers Under Imbalanced Loads in Distribution Systems 2629

Tao Yang, *University College Dublin, Ireland*
Ronan Meere, *University College Dublin, Ireland*
Cathal O'Loughlin, *University College Dublin, Ireland*
Terence O'Donnell, *University College Dublin, Ireland*

Steady-State Analysis of Modular Multilevel Converter (MMC) Under Unbalanced Grid Conditions 2637

Xiaojie Shi, *University of Tennessee, United States*
Yalong Li, *University of Tennessee, United States*
Zhiqiang Wang, *University of Tennessee, United States*
Bo Liu, *University of Tennessee, United States*
Leon M. Tolbert, *University of Tennessee, United States*
Fred Wang, *University of Tennessee, United States*

Design and Control of a Compensated Submodule Testing Scheme for Modular Multilevel Converter 2645

Yuan Tang, *University of Warwick, United Kingdom*
Li Ran, *University of Warwick, United Kingdom*
Olayiwola Alatise, *University of Warwick, United Kingdom*
Philip Mawby, *University of Warwick, United Kingdom*

A Voltage Independent Islanding Detection Method and Low Voltage Ride through of a Two-Stage PV Inverter 2652

Partha Pratim Das, *Indian Institute of Technology Kharagpur, India*
Souvik Chattopadhyay, *Indian Institute of Technology Kharagpur, India*
Shiladri Chakraborty, *Indian Institute of Technology Kharagpur, India*

Low Cost and High Efficiency Topology for Flexible Integration of Multi-PV and Batteries in Resonant-Based Converters 2660

Ali Elrayyah, *Qatar Environment and Energy Research Institute, Qatar*

Real-Time Integrated Model of a Micro-Grid with Distributed Clean Energy Generators and their Power Electronics 2666

Weiqiang Chen, *University of Connecticut, United States*
Ali M. Bazzi, *University of Connecticut, United States*
James Hare, *University of Connecticut, United States*
Shalabh Gupta, *University of Connecticut, United States*

Minimization of Inter-Module Leakage Current in Cascaded H-Bridge Multilevel Inverters for Grid Connected Solar PV Applications 2673
V.V.S. Pradeep Kumar, *Indian Institute of Technology Bombay, India*
B.G. Fernandes, *Indian Institute of Technology Bombay, India*

Effect of Grid Inductance on Grid Current Quality of Parallel Grid-Connected Inverter System with Output LCL Filter and Closed-Loop Control 2679
Wooyoung Choi, *University of Wisconsin at Madison, United States*
Woongkul Lee, *University of Wisconsin at Madison, United States*
Bulent Sarlioglu, *University of Wisconsin at Madison, United States*

Small Signal Modeling and Control of a Grid Tied Converter without a Syncronization Unit 2687
Subhajyoti Mukherjee, *Missouri University of Science and Technology, United States*
Pourya Shamsi, *Missouri University of Science and Technology, United States*
Mehdi Ferdowsi, *Missouri University of Science and Technology, United States*

Bridgeless SEPIC PFC Converter for Low Total Harmonic Distortion and High Power Factor ... 2693
Yasemin Onal, *Bilecik Seyh Edebali University, Turkey*
Yilmaz Sozer, *University of Akron, United States*

Effectiveness of Pareto-Front Analysis Applied to the Design of a Single-Phase PFC Rectifier 2700
Mahmoud Ibrahim, *Eaton Corporation, France*
Luc Gonnet, *Eaton Corporation, France*
Pierre Lefranc, *Grenoble Institute of Technology, France*
David Frey, *Grenoble Institute of Technology, France*
Jean-Paul Ferrieux, *Grenoble Institute of Technology, France*
Sokchea Am, *Grenoble Institute of Technology, France*

State Space Analysis and Duty Cycle Control of a Switched Reactance based Center-Point-Clamped Reactive Power Compensator 2706
Pankaj Kumar Bhowmik, *University of North Carolina at Charlotte, United States*
Somasundaram Essakiappan, *University of North Carolina at Charlotte, United States*
Madhav Manjrekar, *University of North Carolina at Charlotte, United States*

A SiC-Based Power Converter Module for Medium-Voltage Fast Charger for Plug-In Electric Vehicles 2714
Srdjan Srdic, *North Carolina State University, United States*
Chi Zhang, *North Carolina State University, United States*
Xinyu Liang, *North Carolina State University, United States*
Wensong Yu, *North Carolina State University, United States*
Srdjan Lukic, *North Carolina State University, United States*

Shunt Active Power Filter based on Cascaded Transformers Coupled with Three-Phase Bridge Converters 2720
Gregory A. de Almeida Carlos, *Universidade Federal de Campina Grande, Brazil*
Cursino B. Jacobina, *Universidade Federal de Campina Grande, Brazil*
João Paulo R. Méllo, *Universidade Federal de Campina Grande, Brazil*
Euzeli C. dos Santos Jr., *Indiana University - Purdue University, United States*

Independent DC Link Voltage Control of Cascaded Multilevel PV Inverter 2727
Qingyun Huang, *North Carolina State University, United States*
Wensong Yu, *North Carolina State University, United States*
Alex Q. Huang, *North Carolina State University, United States*

New Active Damping Method for LCL Filter Resonance based on Two Feedback System 2735

Mahmoud A. Gaafar, *Kyushu University, Japan*
Gamal M. Dousoky, *Minia University, Egypt*
Masahito Shoyama, *Kyushu University, Japan*

Static Synchronous Generator Model for Investigating Dynamic Behaviors and Stability Issues of Grid-Tied Inverters 2742

Liansong Xiong, *Xi'an Jiaotong University, China*
Xiaokang Liu, *Xi'an Jiaotong University, China*
Feng Wang, *Xi'an Jiaotong University, China*
Fang Zhuo, *Xi'an Jiaotong University, China*

Session D05: Motor Drives and Inverters: Modeling and Control I
Location: Poster Area
March 24, 2016 11:30 - 14:00
Session Chairs: Liming Liu, *ABB Inc.*
Thomas Gietzold, *United Technologies Aerospace Systems*

Initial Orientation and Sensorless Starting Strategy of Wound-Rotor Synchronous Starter/Generator 2748

Jichang Peng, *Northwestern Polytechnical University, China*
Weiguo Liu, *Northwestern Polytechnical University, China*
Jinhao Meng, *Northwestern Polytechnical University, China*
Tao Meng, *Northwestern Polytechnical University, China*
Guangzhao Luo, *Northwestern Polytechnical University, China*

A Novel Method for Polarity Detection of Non-Salient PMSMs in Initial Position Estimation 2754

Bing Liu, *Nanjing University of Aeronautics and Astronautics, China*
Bo Zhou, *Nanjing University of Aeronautics and Astronautics, China*
Jiadan Wei, *Nanjing University of Aeronautics and Astronautics, China*
Long Wang, *Nanjing University of Aeronautics and Astronautics, China*
Tianheng Ni, *Nanjing University of Aeronautics and Astronautics, China*

A Speed Adaptive Sensorless Flux Observer for the Induction Motor Drive using Sylvester Criterion Design 2759

Mihai Comanescu, *Penn State Altoona, United States*

Discontinuous PWM for Low Switching Losses in Indirect Matrix Converter Drives 2764

Yeongsu Bak, *Ajou University, Korea, South*
Kyo-Beum Lee, *Ajou University, Korea, South*

Model Predictive Control for Extended Kalman Filter based Speed Sensorless Induction Motor Drives 2770

Jie Li, *Xi'an University of Technology, China*
Li-Heng Zhang, *Xi'an University of Technology, China*
Ying Niu, *Xi'an University of Technology, China*
Hai-Peng Ren, *Xi'an University of Technology, China*

Research on Excitation Control Methods for the Two-Phase Brushless Exciter of Wound-Rotor Synchronous Starter/Generators in the Starting Mode 2776
Ningfei Jiao, *Northwestern Polytechnical University, China*
Weiguo Liu, *Northwestern Polytechnical University, China*
Tao Meng, *Northwestern Polytechnical University, China*
Jichang Peng, *Northwestern Polytechnical University, China*
Shuai Mao, *Northwestern Polytechnical University, China*

A High Performance Speed Regulator Design for AC Machines 2782
Adil Khurram, *American University of Sharjah, U.A.E.*
Habibur Rehman, *American University of Sharjah, U.A.E.*
Shayok Mukhopadhyay, *American University of Sharjah, U.A.E.*

Zero-Sequence Current Suppression for Open-End Winding Induction Motor Drive with Resonant Controller 2788
Hajime Kubo, *Meidensha Corporation, Japan*
Yasuhiro Yamamoto, *Meidensha Corporation, Japan*
Takeshi Kondo, *Meidensha Corporation, Japan*
Kaushik Rajashekara, *University of Texas at Dallas, United States*
Bohang Zhu, *University of Texas at Dallas, United States*

Optimized Control of High-Performance Servo-Motor Drives in the Field-Weakening Region 2794
Jack Bermingham, *Moog Ireland Ltd, Ireland*
Gerard O'Donovan, *Moog Ireland Ltd, Ireland*
Ray Walsh, *Moog Ireland Ltd, Ireland*
Michael Egan, *University College Cork, Ireland*
Gordon Lightbody, *University College Cork, Ireland*
John G. Hayes, *University College Cork, Ireland*

Motor Current Reference Generation for Reducing Motor Currents in Drive Systems with Single-Phase Diode Rectifier and Small DC-Link Capacitor 2801
Young-Ho Chae, *Seoul National University, Korea, South*
Jung-Ik Ha, *Seoul National University, Korea, South*

A Simple Double Mapping based SVPWM Method for Balancing DC-Link Capacitor Voltages of Five-Level Diode-Clamped Converters 2806
Aparna Saha, *University of Akron, United States*
Ali Elrayyah, *Qatar Environment and Energy Research Institute, Qatar*
Yilmaz Sozer, *University of Akron, United States*

Session D06: Motor Drives and Inverters: Modeling and Control II
Location: Poster Area
March 24, 2016 11:30 - 14:00
Session Chairs: Bulent Sarlioglu, *University of Wisconsin - Madison*
                Yichao Tang, *Texas Instruments*

Capacitor-Clamped Inverter based Transient Suppression Method for Azimuth Thruster Drives 2813
Shantha Gamini Jayasinghe, *Australian Maritime College, University of Tasmania, Australia*
Viknash Shagar, *Australian Maritime College, University of Tasmania, Australia*
Hossein Enshaei, *Australian Maritime College, University of Tasmania, Australia*
Danyal Mohammadi, *Boise State University, United States*
Mahinda Vilathgamuwa, *Queensland University of Technology, Australia*

Active Common-Mode Voltage Reduction in a Fault-Tolerant Three-Phase Inverter 2821
Danyal Mohammadi, *Boise State University, United States*
Said Ahmed-Zaid, *Boise State University, United States*

Power Cycling Lifetime Improvement of Three-Level NPC Inverters with an Improved DPWM Method 2826
Jiangbiao He, *Marquette University, United States*
Lixiang Wei, *Rockwell Automation, United States*
Nabeel A.O. Demerdash, *Marquette University, United States*

Synchronous Optimal Pulsewidth Modulation Digital Implementation Concept for Multilevel Converters 2833
Jackson Lago, *Universidade Federal de Santa Catarina, Brazil*
Marcelo Lobo Heldwein, *Universidade Federal de Santa Catarina, Brazil*

Analytical Determination of Conduction Losses for Modified Flying Capacitor Multicell Converters 2840
Vahid Dargahi, *Clemson University, United States*
Arash Khoshkbar Sadigh, *Extron Electronics, United States*
Keith Corzine, *Clemson University, United States*

Comparison of Electrical Losses in an Inverter-Fed Five-Phase and Three-Phase Permanent Magnet Assisted Synchronous Reluctance Motor 2847
Akm Arafat, *University of Akron, United States*
Seungdeog Choi, *University of Akron, United States*

A Hybrid Adaptive Observer for the Speed and Flux Estimation of Induction Motors 2855
Mihai Comanescu, *Penn State Altoona, United States*

Determination of CM Choke Parameters for SiC MOSFET Motor Drive based on Simple Measurements and Frequency Domain Modeling 2861
Di Han, *University of Wisconsin at Madison, United States*
Casey Morris, *University of Wisconsin at Madison, United States*
Woongkul Lee, *University of Wisconsin at Madison, United States*
Bulent Sarlioglu, *University of Wisconsin at Madison, United States*

An Improved Model Predictive Current Control of Permanent Magnet Synchronous Motor Drives 2868
Yongchang Zhang, *North China University of Technology, China*
Sugu Gao, *North China University of Technology, China*
Wei Xu, *Huazhong University of Science and Technology, China*

Analysis of Magnet Defect Faults in Permanent Magnet Synchronous Motors through Fluxgate Sensors 2875
Taner Goktas, *University of Texas at Dallas, United States*
Kun Wang Lee, *University of Texas at Dallas, United States*
Mohsen Zafarani, *University of Texas at Dallas, United States*
Bilal Akin, *University of Texas at Dallas, United States*

Session D07: Motor Drives and Inverters: Topologies
Location: Poster Area
March 24, 2016 11:30 - 14:00
Session Chairs: Amirnaser Yazdani, *Ryerson University*
Babak Nahid-Mobarakeh, *University of Lorraine*

Performance Comparison of Transfer Switch Topologies in Switched-Doubly-Fed Machine Drives 2881
Arijit Banerjee, *Massachusetts Institute of Technology, United States*
Steven B. Leeb, *Massachusetts Institute of Technology, United States*
James L. Kirtley, *Massachusetts Institute of Technology, United States*

Multilevel Converter Topologies for High-Power High-Speed Switched Reluctance Motor: Performance Comparison 2889
Devendra Patil, *University of Texas at Dallas, United States*
Shiliang Wang, *University of Texas at Dallas, United States*
Lei Gu, *University of Texas at Dallas, United States*

Bidirectional Magnetically Coupled T-Source Inverter for Extra Low Voltage Application 2897
Thomas Baier, *Friedrich-Alexander-Universität Erlangen-Nürnberg, Germany*
Bernhard Piepenbreier, *Friedrich-Alexander-Universität Erlangen-Nürnberg, Germany*

Active Virtual Ground: Single Phase Grid-Connected Voltage Source Inverter Topology ... 2905
River Tin-Ho Li, *ABB China Ltd., China*
Carl Ngai-Man Ho, *University of Manitoba, Canada*

Design and Evaluation of 30kVA Inverter using SiC MOSFET for 180°C Ambient Temperature Operation 2912
Feng Qi, *Ohio State University, United States*
Miao Wang, *Ohio State University, United States*
Longya Xu, *Ohio State University, United States*
Bo Zhao, *State Grid Corporation of China, China*
Zhe Zhou, *State Grid Corporation of China, China*
Xizhou Ren, *State Grid Corporation of China, China*

A DC to Three-Phase Boost-Buck Inverter with Stored Energy Modulation and a Tiny DC Link Capacitor 2919
Mahima Gupta, *University of Wisconsin at Madison, United States*
Giri Venkataramanan, *University of Wisconsin at Madison, United States*

Drive Circuits for Ultra-Fast and Reliable Actuation of Thomson Coil Actuators used in Hybrid AC and DC Circuit Breakers 2927
Chang Peng, *North Carolina State University, United States*
Alex Huang, *North Carolina State University, United States*
Iqbal Husain, *North Carolina State University, United States*
Bruno Lequesne, *E-Motors Consulting, LLC, United States*
Roger Briggs, *Energy Efficiency Research, LLC, United States*

Improved Transformerless Dual Buck Inverters with Buffer Inductors 2935
Liwei Zhou, *Shandong University, China*
Feng Gao, *Shandong University, China*

A 99% Efficiency SiC Three-Phase Inverter using Synchronous Rectification 2942
Shan Yin, *Nanyang Technological University, Singapore*
K.J. Tseng, *Nanyang Technological University, Singapore*
C.F. Tong, *Nanyang Technological University, Singapore*
Rejeki Simanjorang, *Rolls-Royce Singapore Pte. Ltd., Singapore*
C.J. Gajanayake, *Rolls-Royce Singapore Pte. Ltd., Singapore*
Amit K. Gupta, *Rolls-Royce Singapore Pte. Ltd., Singapore*

**Comparison and Evaluation of Common Mode EMI Filter Topologies for GaN-Based
Motor Drive Systems** 2950
Casey T. Morris, *University of Wisconsin at Madison, United States*
Di Han, *University of Wisconsin at Madison, United States*
Bulent Sarlioglu, *University of Wisconsin at Madison, United States*

**Analysis of Thermal Cycling Stress on Semiconductor Devices of the Modular Multilevel
Converter for Drive Applications** 2957
Xiangyu Han, *Georgia Institute of Technology, United States*
Qichen Yang, *Georgia Institute of Technology, United States*
Liyao Wu, *Georgia Institute of Technology, United States*
Maryam Saeedifard, *Georgia Institute of Technology, United States*

Fault Tolerant Topologies of Five-Level Active Neutral-Point-Clamped Converters 2963
Jun Li, *ABB Inc., United States*

Session D08: Advanced Components and Devices
Location: Poster Area
March 24, 2016 11:30 - 14:00
Session Chairs: Abhijit Pathak, *Infineon/IR*
Doug Hopkins, *North Carolina State University*

**Dynamic Characterization of the Input and Reverse Transfer Capacitances in Power
MOSFETs under High Current Conduction** 2969
Cristino Salcines, *Universität Stuttgart, Germany*
Ingmar Kallfass, *Universität Stuttgart, Germany*
Hisao Kakitani, *Keysight Technologies International, Japan*
Atsushi Mikata, *Keysight Technologies International, Japan*

Medium Voltage Power Switch based on SiC JFETs 2973
Xueqing Li, *United Silicon Carbide, Inc., United States*
Hao Zhang, *United Silicon Carbide, Inc., United States*
Peter Alexandrov, *United Silicon Carbide, Inc., United States*
Anup Bhalla, *United Silicon Carbide, Inc., United States*

**Numerical Model and Experimental Study on Comparison of Semiconductor Pulsed
Power Devices** 2981
Lin Liang, *Huazhong University of Science and Technology, China*
Changdong Chen, *Huazhong University of Science and Technology, China*
Fang Luo, *Ohio State University, United States*

A Normalization Procedure of DC-Side Stray Inductance for High-Speed Switching Circuit 2986
Masato Ando, *Tokyo Metropolitan University, Japan*
Keiji Wada, *Tokyo Metropolitan University, Japan*

Thermal Network Parameter Identification of IGBT Module based on the Cooling Curve of Junction Temperature 2992

Xiong Du, *Chongqing University, China*
Tengfei Li, *Chongqing University, China*
Jun Zhang, *Chongqing University, China*
Heng-Ming Tai, *University of Tulsa, United States*
Pengju Sun, *Chongqing University, China*
Luowei Zhou, *Chongqing University, China*

Design and Evaluation of High Current PCB Embedded Inductor for High Frequency Inverters 2998

Mehrdad Biglarbegian, *University of North Carolina at Charlotte, United States*
Neel Shah, *University of North Carolina at Charlotte, United States*
Iman Mazhari, *University of North Carolina at Charlotte, United States*
Johan Enslin, *University of North Carolina at Charlotte, United States*
Babak Parkhideh, *University of North Carolina at Charlotte, United States*

Prognosis of Wire Bond Lift-Off Fault of an IGBT based on Multisensory Approach 3004

Moinul Shahidul Haque, *University of Akron, United States*
Jeihoon Baek, *Korean Rail Research Institute, Korea, South*
Joseph Herbert, *University of Akron, United States*
Seungdeog Choi, *University of Akron, United States*

Electrical Parasitics and Thermal Modeling for Optimized Layout Design of High Power SiC Modules 3012

Amir Sajjad Bahman, *Aalborg University, Denmark*
Frede Blaabjerg, *Aalborg University, Denmark*
Atanu Dutta, *University of Arkansas, United States*
Alan Mantooth, *University of Arkansas, United States*

Calculation of Losses in PCB Windings for Multi-Coil Contactless Charging Systems 3020

J. Serrano, *Universidad de Zaragoza, Spain*
J. Acero, *Universidad de Zaragoza, Spain*
I. Lope, *BSH Home Appliances Group, Spain*
C. Carretero, *Universidad de Zaragoza, Spain*
J.M. Burdío, *Universidad de Zaragoza, Spain*
R. Alonso, *Universidad de Zaragoza, Spain*

Design of Efficient Loads for Domestic Induction Heating Applications by Means of Non-Magnetic Thin Metallic Layers 3026

Jesús Acero, *Universidad de Zaragoza, Spain*
Claudio Carretero, *Universidad de Zaragoza, Spain*
Rafael Alonso, *Universidad de Zaragoza, Spain*
José Miguel Burdío, *Universidad de Zaragoza, Spain*

A New Evaluation Circuit with a Low-Voltage Inverter Intended for Capacitors used in a High-Power Three-Phase Inverter 3032

Kazunori Hasegawa, *Kyushu Institute of Technology, Japan*
Ichiro Omura, *Kyushu Institute of Technology, Japan*
Shin-Ichi Nishizawa, *Kyushu Institute of Technology / National Institute of Advanced Industrial Science and Technology, Japan*

Energy Absorption Capability of Low Voltage Metal Oxide Varistors in AC and Impulse Currents 3038

Dawood Talebi Khanmiri, *Northeastern University, United States*
Roy Ball, *Mersen USA, United States*
Craig McKenzie, *Mersen USA, United States*
Brad Lehman, *Northeastern University, United States*

Optimization and Experimental Validation of Medium-Frequency High Power Transformers in Solid-State Transformer Applications 3043

M.A. Bahmani, *Chalmers University of Technology, Sweden*
T. Thiringer, *Chalmers University of Technology, Sweden*
M. Kharezy, *SP Technical Research Institute of Sweden, Sweden*

Evaluation of Core Loss in Magnetic Materials Employed in Utility Grid AC Filters 3051

Remus Beres, *Aalborg University, Denmark*
Xiongfei Wang, *Aalborg University, Denmark*
Frede Blaabjerg, *Aalborg University, Denmark*
Claus Leth Bak, *Aalborg University, Denmark*
Hiroaki Matsumori, *Tokyo Metropolitan University, Japan*
Toshihisa Shimizu, *Tokyo Metropolitan University, Japan*

A Novel Gate Assisted Circuit to Reduce Switching Loss and Eliminate Shoot-Through in SiC Half Bridge Configuration 3058

Shan Yin, *Nanyang Technological University, Singapore*
K.J. Tseng, *Nanyang Technological University, Singapore*
C.F. Tong, *Nanyang Technological University, Singapore*
Rejeki Simanjorang, *Rolls-Royce Singapore Pte. Ltd., Singapore*
C.J. Gajanayake, *Rolls-Royce Singapore Pte. Ltd., Singapore*
Amit K. Gupta, *Rolls-Royce Singapore Pte. Ltd., Singapore*

Session D09: System Design Considerations for Power Electronics
Location: Poster Area
March 24, 2016 11:30 - 14:00
Session Chairs: John Vigars, *Allegro Microsystems*
Ernie Parker, *Crane Aerospace & Electronics*

Methods to Enhance the Thermal Performance of a 3D Power Package 3065

Jonathan Noquil, *Texas Instruments Inc., United States*
Ozzie Lopez, *Texas Instruments Inc., United States*
Tianyi Luo, *Lehigh University, United States*

Highly Reliable and Cost Effective Thick Film Substrates for Power LEDs 3069

Paul Gundel, *Heraeus Deutschland GmbH & Co. KG, Germany*
Ryan Persons, *Heraeus Deutschland GmbH & Co. KG, Germany*
Melanie Bawohl, *Heraeus Deutschland GmbH & Co. KG, Germany*
Mark Challingsworth, *Heraeus Deutschland GmbH & Co. KG, Germany*
Christoph Czwickla, *Heraeus Deutschland GmbH & Co. KG, Germany*
Virginia Garcia, *Heraeus Deutschland GmbH & Co. KG, Germany*
Christina Modes, *Heraeus Deutschland GmbH & Co. KG, Germany*
Ilias Nikolaidis, *Heraeus Deutschland GmbH & Co. KG, Germany*
Jessica Reitz, *Heraeus Deutschland GmbH & Co. KG, Germany*
Caitlin Shahbazi, *Heraeus Deutschland GmbH & Co. KG, Germany*
Torsten Nowak, *Fraunhofer-Institut für Zuverlässigkeit und Mikrointegration, Germany*

Design and Evaluation of SiC-Based High Power Density Inverter, 70kW/Liter, 50kW/kg ... 3075
Koji Yamaguchi, *IHI Corporation, Japan*

An Improved Automatic Layout Method for Planar Power Module ... 3080
Puqi Ning, *Chinese Academy of Sciences, China*
Xuhui Wen, *Chinese Academy of Sciences, China*
Yaohua Li, *Chinese Academy of Sciences, China*
Xiongxuan Ge, *Chinese Academy of Sciences, China*

Practical Implementation Schemes of Motor Speed Measurement by Magnetic Encoder on Electric Power Steering Applications ... 3086
Jae-Hyun Lee, *Hyundai Mobis, Korea, South*

Low-Cost Input Impedance Estimator of DC-to-DC Converters for Designing the Control Loop in Cascaded Converters ... 3090
M. Sanz, *Universidad Carlos III de Madrid, Spain*
A. Lázaro, *Universidad Carlos III de Madrid, Spain*
M. Bermejo, *Universidad Carlos III de Madrid, Spain*
D. López del Moral, *Universidad Carlos III de Madrid, Spain*
P. Zumel, *Universidad Carlos III de Madrid, Spain*
C. Fernández, *Universidad Carlos III de Madrid, Spain*
A. Barrado, *Universidad Carlos III de Madrid, Spain*

On-Chip High Performance Magnetics for Point-of-Load High-Frequency DC-DC Converters ... 3097
Dragan Dinulovic, *Würth Elektronik eiSos GmbH & Co. KG, Germany*
Mahmoud Shousha, *Würth Elektronik eiSos GmbH & Co. KG, Germany*
Martin Haug, *Würth Elektronik eiSos GmbH & Co. KG, Germany*
Alexander Gerfer, *Würth Elektronik eiSos GmbH & Co. KG, Germany*
Mike Wens, *MinDCet NV, Belgium*
Jef Thone, *MinDCet NV, Belgium*

Effects of Auxiliary Source Connections in Multichip Power Module ... 3101
Helong Li, *Aalborg University, Denmark*
Stig Munk-Nielsen, *Aalborg University, Denmark*
Szymon Bęczkowski, *Aalborg University, Denmark*
Xiongfei Wang, *Aalborg University, Denmark*
Emanuel-Petre Eni, *Aalborg University, Denmark*

Session D10: Modeling and Simulation
Location: Poster Area
March 24, 2016 11:30 - 14:00
Session Chairs: Marco Meola, *ZMD AG*
Mehdi Ferdowsi, *Missouri University of Science & Technology*

Modelling Technique Utilizing Modified Sigmoid Functions for Describing Power Transistor Device Capacitances Applied on GaN HEMT and Silicon MOSFET ... 3107
H.L. Yeo, *Nanyang Technological University, Singapore*
K.J. Tseng, *Nanyang Technological University, Singapore*

Design and Precise Modeling of a Novel Digital Active EMI Filter ... 3115
Junpeng Ji, *Xi'an Jiaotong University, China*
Wenjie Chen, *Xi'an Jiaotong University, China*
Xu Yang, *Xi'an Jiaotong University, China*

Development of a Hybrid Emulation Platform based on RTDS and Reconfigurable Power Converter-Based Testbed 3121

Shuoting Zhang, *University of Tennessee, United States*
Yiwei Ma, *University of Tennessee, United States*
Liu Yang, *University of Tennessee, United States*
Fred Wang, *University of Tennessee, United States*
Leon M. Tolbert, *University of Tennessee, United States*

Online Temperature Estimation for Phase Change Composite – 18650 Lithium Ion Cells based Battery Pack 3128

Mohamad Salameh, *Illinois Institute of Technology, United States*
Ben Schweitzer, *AllCell Technologies, United States*
Peter Sveum, *AllCell Technologies, United States*
Said Al-Hallaj, *AllCell Technologies, United States*
Mahesh Krishnamurthy, *Illinois Institute of Technology, United States*

Modeling and Fault Diagnosis of Inter-Turn Short Circuit for Five-Phase PMSM based on Particle Swarm Optimization 3134

Jianwei Yang, *Northwestern Polytechnical University, China*
Manfeng Dou, *Northwestern Polytechnical University, China*
Zhiyong Dai, *Northwestern Polytechnical University, China*
Dongdong Zhao, *Northwestern Polytechnical University, China*
Zhen Zhang, *Northwestern Polytechnical University, China*

Comprehensive Modeling, Testing, and Experimental Validation of Ultracapacitor Open Circuit Voltage Characteristics 3140

Amandeep Singh, *University of Ontario Institute of Technology, Canada*
Najath Abdul Azeez, *University of Ontario Institute of Technology, Canada*
Sheldon S. Williamson, *University of Ontario Institute of Technology, Canada*

Novel SPICE Model for Common Mode Choke Including Complex Permeability 3146

Katsuya Nomura, *Toyota Central R&D Labs., Inc., Japan*
Naoto Kikuchi, *Toyota Central R&D Labs., Inc., Japan*
Yoshitoshi Watanabe, *Toyota Central R&D Labs., Inc., Japan*
Shuntaro Inoue, *Toyota Central R&D Labs., Inc., Japan*
Yoshiyuki Hattori, *Toyota Central R&D Labs., Inc., Japan*

Session D11: Control I
Location: Poster Area
March 24, 2016 11:30 - 14:00
Session Chairs: Bilal Akin, *Univeristy of Texas, Dallas*
Brian Zahnstecher, *PowerRox LLC*

Analysis and Design of Capacitive Power Transmission System Employing Out-of-Band Wireless Feedback Link 3153

Sung-Jin Choi, *University of Ulsan, Korea, South*
Hee-Su Choi, *University of Ulsan, Korea, South*

Introducing Fourier-Based Modeling and Control of Active-Bridge Converters 3158

B.J.D. Vermulst, *Technische Universiteit Eindhoven, Netherlands*
J.L. Duarte, *Technische Universiteit Eindhoven, Netherlands*
C.G.E. Wijnands, *Technische Universiteit Eindhoven, Netherlands*
E.A. Lomonova, *Technische Universiteit Eindhoven, Netherlands*

A Stability Analysis and Efficiency Improvement of Synchronverter 3165
Prasanna Piya, *Mississippi State University, United States*
Masoud Karimi-Ghartemani, *Mississippi State University, United States*

Compensation of Switching Dead-Time Effects in Voltage-Fed PWM Inverters using FPGA-Based Current Oversampling 3172
Bastian Weber, *Leibniz Universität Hannover, Germany*
Tobias Brandt, *Leibniz Universität Hannover, Germany*
Axel Mertens, *Leibniz Universität Hannover, Germany*

Control Strategy of High Power Converters with Synchronous Generator Characteristics for PMSG-Based Wind Power Application 3180
Yuzhi Zhang, *University of Arkansas, United States*
Haoyan Liu, *University of Arkansas, United States*
H. Alan Mantooth, *University of Arkansas, United States*

Phase Compensation, ZVS Operation of Wireless Power Transfer System based on SOGI-PLL 3185
Pingan Tan, *Xiangtan University, China*
Haibing He, *Xiangtan University, China*
Xieping Gao, *Xiangtan University, China*

A Novel Low-Cost Online State of Charge Estimation Method for Reconfigurable Battery Pack 3189
Ni Lin, *University of Nebraska at Lincoln, United States*
Song Ci, *University of Nebraska at Lincoln, United States*
Dalei Wu, *University of Tennessee at Chattanooga, United States*

Effect of Decoupling Terms on the Performance of PR Current Controllers Implemented in Stationary Reference Frame 3193
Sizhan Zhou, *Xi'an Jiaotong University, China*
Jinjun Liu, *Xi'an Jiaotong University, China*

Fuzzy Predictive DTC of Induction Machines with Reduced Torque Ripple and High Performance Operation 3200
Alberto Berzoy, *Florida International University, United States*
Osama Mohammed, *Florida International University, United States*
Johnny Rengifo, *Universidad Simon Bolivar, Venezuela*

Session D12: Control II
Location: Poster Area
March 24, 2016 11:30 - 14:00
Session Chairs: Martin Ordonez, *University of British Columbia*
Jiangbiao He, *GE Global Research*

Fixed-Frequency Generalized Peak Current Control (GPCC) for Inverters 3207
Mohammad Ebrahimi, *University of Alberta, Canada*
S. Ali Khajehoddin, *University of Alberta, Canada*

Improved Control Strategy of 1 MHz LLC Converter for High Frequency Resolution 3213
Hwa-Pyeong Park, *Ulsan National Institute of Science and Technology, Korea, South*
Jee-Hoon Jung, *Ulsan National Institute of Science and Technology, Korea, South*

Bumpless Control for Reduced THD in Power Factor Correction Circuits 3219
Joel Steenis, *Microchip Technology, United States*
Alex Dumais, *Microchip Technology, United States*

Mixed-Signal Hysteretic Internal Model Control of Buck Converters for Ultra-Fast Envelope Tracking 3224
V. Inder Kumar, *Indian Institute of Technology Kharagpur, India*
Santanu Kapat, *Indian Institute of Technology Kharagpur, India*

A Continuous Actor-Critic Maximum Power Point Tracker Applied to Low Power Wind Turbine Systems 3231
J.L. Wattes, *Universidade Federal do Ceará, Brazil*
A.J.S. Dias Jr., *Universidade Federal do Ceará, Brazil*
A.P.S. Braga, *Universidade Federal do Ceará, Brazil*
P.P. Praça, *Universidade Federal do Ceará, Brazil*
A.U. Barbosa, *Universidade Federal do Ceará, Brazil*
D.S. de Souza Oliveira Jr., *Universidade Federal do Ceará, Brazil*

Multi-Band Mixed-Signal Hysteresis Current Control for EMI Reduction in Switch-Mode Power Supplies 3237
Arindam Mandal, *Indian Institute of Technology Kharagpur, India*
V. Inder Kumar, *Indian Institute of Technology Kharagpur, India*
Santanu Kapat, *Indian Institute of Technology Kharagpur, India*

A Parabolic Current Control based Digital Current Control Strategy for High Switching Frequency Voltage Source Inverters 3243
Lanhua Zhang, *Virginia Polytechnic Institute and State University, United States*
Rachael Born, *Virginia Polytechnic Institute and State University, United States*
Xiaonan Zhao, *Virginia Polytechnic Institute and State University, United States*
Jih-Sheng Jason Lai, *Virginia Polytechnic Institute and State University, United States*
Hongbo Ma, *Southwest Jiaotong University, China*

Finite Control Set Model Predictive Control of Dual-Output Four-Leg Indirect Matrix Converter Under Unbalanced Load and Supply Conditions 3248
Ozan Gulbudak, *University of South Carolina, United States*
Enrico Santi, *University of South Carolina, United States*

A Silicon Carbide Integrated Circuit Implementing Nonlinear-Carrier Control for Boost Converter Applications 3255
Richard Kyle Harris, *University of Tennessee, United States*
Benjamin M. McCue, *University of Tennessee, United States*
Benjamin D. Roehrs, *University of Tennessee, United States*
Charles Roberts II, *University of Tennessee, United States*
Benjamin J. Blalock, *University of Tennessee, United States*
Daniel J. Costinett, *University of Tennessee, United States*
Kouros Sariri, *Frequency Management International, United States*
George Megyei, *Frequency Management International, United States*
Cheng-Po Chen, *GE Global Research, United States*
Avinash Kashyap, *GE Global Research, United States*
Reza Ghandi, *GE Global Research, United States*

A New Current Mode Constant on Time Control with Ultrafast Load Transient Response 3259
Syed Bari, *Virginia Polytechnic Institute and State University, United States*
Qiang Li, *Virginia Polytechnic Institute THD and State Universtiy, United States*
Fred C. Lee, *Virginia Polytechnic Institute and State University, United States*

A Web-Based Tool for Compensation Design of Power Converters using Hybrid Optimization 3266

Srikanth Pam, *Texas Instruments Inc., India*
Yudhister Satija, *Texas Instruments Inc., India*
Pradeep Chawda, *Texas Instruments Inc., United States*
Makram Mansour, *Texas Instruments Inc., United States*
Robert Hanrahan, *Texas Instruments Inc., United States*
Jeff Perry, *Texas Instruments Inc., United States*

Second Order Sliding Mode Controlled Point of Load Power Supply 3273

Prasanta K. Achanta, *University of Colorado at Boulder, United States*
David C. Jones, *University of Colorado at Boulder, United States*
Dragan Maksimovic, *University of Colorado at Boulder, United States*
Serhii M. Zhak, *Linear Technology Corporation, United States*
Brett Miwa, *Maxim Integrated, United States*
Cory Arnold, *Maxim Integrated, United States*

Vibration and Torque Ripple Reduction of Switched Reluctance Motors through Current Profile Optimization 3279

Cong Ma, *University of Nebraska at Lincoln, United States*
Liyan Qu, *University of Nebraska at Lincoln, United States*
Rakesh Mitra, *Nexteer Automotive, United States*
Prerit Pramod, *Nexteer Automotive, United States*
Rakib Islam, *Nexteer Automotive, United States*

Modified Predictive Current Control of Neutral-Point Clamped Converter with Reduced Switching Frequency 3286

Dinto Mathew, *Indian Institute of Technology Bombay, India*
Anshuman Shukla, *Indian Institute of Technology Bombay, India*
Santanu Bandyopadhyay, *Indian Institute of Technology Bombay, India*

Implicit Finite Control Set Model Predictive Current Control for Modular Multilevel Converter based on IPA-SQP Algorithm 3291

Hamed Nademi, *ABB AS, Norway*
Lars Einar Norum, *Norwegian University of Science and Technology, Norway*

Resolution Requirements to Avoid Limit Cycling in LLC Resonant Converter 3297

Shadi Dashmiz, *University of Toronto, Canada*
Behzad Mahdavikhah, *University of Toronto, Canada*
Aleksandar Prodić, *University of Toronto, Canada*
Brent McDonald, *Texas Instruments Inc., United States*

Session D13: Renewable Energy Systems I
Location: Poster Area
March 24, 2016 11:30 - 14:00
Session Chairs: Akshay Kumar Rathore, *Concordia University*
Xiaoqiang Guo, *Yanshan University, China*

Reduction of Storage Capacity in DC Microgrids using PV-Embedded Series DC Electric Springs 3302

Ming-Hao Wang, *University of Hong Kong, Hong Kong*
Siew-Chong Tan, *University of Hong Kong, Hong Kong*
Shu-Yuen Ron Hui, *University of Hong Kong, Hong Kong*

A Vector Control Strategy of Grid-Connected Brushless Doubly Fed Induction Generator based on the Vector Control of Doubly Fed Induction Generator 3310
Sheng Hu, *Wuhan University of Technology, China*
Guorong Zhu, *Wuhan University of Technology, China*

An Energy Router based on Multi-Winding High-Frequency Transformer 3317
Xianzhuo Liu, *Tsinghua University, China*
Zedong Zheng, *Tsinghua University, China*
Kui Wang, *Tsinghua University, China*
Yongdong Li, *Tsinghua University, China*

Noise Suppression of the DWT-Based MRA on Mother Wavelet and Decomposition Level Optimization for a Robust Adaptive SOC Estimator in Multi-Cell Battery String 3322
Jonghoon Kim, *Chosun University, Korea, South*
Chang Yoon Chun, *Seoul National University, Korea, South*
Woonki Na, *California State University, Fresno, United States*

A Feedforward Control based Power Decoupling Scheme for Voltage-Controlled Grid-Tied Inverters 3328
Baojin Liu, *Xi'an Jiaotong University, China*
Zeng Liu, *Xi'an Jiaotong University, China*
Jinjun Liu, *Xi'an Jiaotong University, China*
Teng Wu, *Xi'an Jiaotong University, China*
Shike Wang, *Xi'an Jiaotong University, China*

Light Load Efficiency Improvement of Solar Farms Three-Phase Two-Stage Module Integrated Converter 3333
Ahmadreza Amirahmadi, *University of Central Florida, United States*
Utsav Somani, *University of Central Florida, United States*
Mahmood Alharbi, *University of Central Florida, United States*
Charlie Jourdan, *University of Central Florida, United States*
Issa Batarseh, *University of Central Florida, United States*

Switching System Stability Analysis of DC Microgrids with DBS Control 3338
Na Zhi, *Xi'an University of Technology, China*
Hui Zhang, *Xi'an University of Technology, China*
Xi Xiao, *Tsinghua University, China*

A Grid-Connected WECS with Power Limiting Control 3346
Jéssica Santos Guimarães, *Universidade Federal do Ceará, Brazil*
Demercil de Souza Oliveira Jr., *Universidade Federal do Ceará, Brazil*
Juliano de Oliveira Pacheco, *Universidade Federal do Ceará, Brazil*
Paulo P. Peixoto, *Universidade Federal do Ceará, Brazil*

Overshoot Control of the Electromagnetic Torque during Fault Recovery for an SCIG with a STATCOM 3353
Zahra Mahmoodzadeh, *Washington State University, United States*
Mehrdad Yazdanian, *Washington State University, United States*
Hooman Ghaffarzadeh, *Washington State University, United States*
Ali Mehrizi-Sani, *Washington State University, United States*

A Self-Adaptive Power Balance Control Strategy for PV Inverters in Islanded Microgrids 3358
Zhenxiong Wang, *Xi'an Jiaotong University, China*
Hao Yi, *Xi'an Jiaotong University, China*
Fang Zhuo, *Xi'an Jiaotong University, China*
Zhigang Zhang, *Xi'an Jiaotong University, China*

High Performance ZVT with Bus Clamping Modulation Technique for Single Phase Full Bridge Inverters 3364
Yinglai Xia, *Arizona State University, United States*
Raja Ayyanar, *Arizona State University, United States*

Small AC Signal Droop based Secondary Control for Microgrids 3370
Teng Wu, *Xi'an Jiaotong University, China*
Zeng Liu, *Xi'an Jiaotong University, China*
Jinjun Liu, *Xi'an Jiaotong University, China*
Baojin Liu, *Xi'an Jiaotong University, China*
Shike Wang, *Xi'an Jiaotong University, China*

Mode Transition Control Strategy for Multiple Inverter based Distributed Generators Operating in Grid-Connected and Stand-Alone Mode 3376
Onkar Vitthal Kulkarni, *Indian Institute of Technology Bombay, India*
Suryanarayana Doolla, *Indian Institute of Technology Bombay, India*
B.G. Fernandes, *Indian Institute of Technology Bombay, India*

An Autonomous Power Management Strategy based on DC Bus Signaling for Solid-State Transformer Interfaced PMSG Wind Energy Conversion System 3383
Rui Gao, *North Carolina State University, United States*
Iqbal Husain, *North Carolina State University, United States*
Alex Q. Huang, *North Carolina State University, United States*

An Isolated Buck-Boost Type High-Frequency Link Photovoltaic Microinverter 3389
Shiladri Chakraborty, *Indian Institute of Technology Kharagpur, India*
Souvik Chattopadhyay, *Indian Institute of Technology Kharagpur, India*

Energy Management and Stabilization of a Hybrid DC Microgrid for Transportation Applications 3397
Mehdi Karbalaye Zadeh, *Norwegian University of Science and Technology, Norway*
Louis-Marie Saublet, *Université de Lorraine, France*
Roghayeh Gavagsaz-Ghoachani, *Université de Lorraine, France*
Babak Nahid-Mobarakeh, *Université de Lorraine, France*
Serge Pierfederici, *Université de Lorraine, France*
Marta Molinas, *Norwegian University of Science and Technology, Norway*

A Low-Cost Solar Micro-Inverter with Soft-Switching Capability Utilizing Circulating Current 3403
Xiaohu Liu, *GE Global Research, United States*
Mohammed Agamy, *GE Global Research, United States*
Dong Dong, *GE Global Research, United States*
Maja Harfman-Todorovic, *GE Global Research, United States*
Luis Garces, *GE Global Research, United States*

Session D14: Renewable Energy Systems II
Location: Poster Area
March 24, 2016 11:30 - 14:00
Session Chairs: Haoyu Wang, *Shanghai Tech University*
Robert Pilawa-Podgurski, *University of Illinois at Urbana-Champaign*

Design and Stability Analysis for an Autonomous DC Microgrid with Constant Power Load ... 3409
Qianwen Xu, *Nanyang Technological University, Singapore*
Xiaolei Hu, *Nanyang Technological University, Singapore*
Peng Wang, *Nanyang Technological University, Singapore*
Jianfang Xiao, *Nanyang Technological University, Singapore*
Leonardy Setyawan, *Nanyang Technological University, Singapore*
Changyun Wen, *Nanyang Technological University, Singapore*
Lee Meng Yeong, *Rolls-Royce Singapore Pte. Ltd., Singapore*

MPC-SVM Method for Vienna Rectifier with PMSG used in Wind Turbine Systems 3416
June-Seok Lee, *Korea Railroad Research Institute, Korea, South*
Yeongsu Bak, *Ajou University, Korea, South*
Kyo-Beum Lee, *Ajou University, Korea, South*
Frede Blaabjerg, *Aalborg University, Denmark*

An Equivalent Circuit Model for State of Energy Estimation of Lithium-Ion Battery 3422
Kaiyuan Li, *Nanyang Technological University, Singapore*
King Jet Tseng, *Nanyang Technological University, Singapore*

Distributed Optimal Control of Reactive Power and Voltage in Islanded Microgrids 3431
Yanbo Wang, *Aalborg University, Denmark*
Xiongfei Wang, *Aalborg University, Denmark*
Zhe Chen, *Aalborg University, Denmark*
Frede Blaabjerg, *Aalborg University, Denmark*

New Start-Up Scheme for HF Transformer Link Photovoltaic Inverter 3439
Abhijit Kulkarni, *Indian Institute of Science, India*
Vinod John, *Indian Institute of Science, India*

Analysis and Improvement of Harmonic Quasi Resonant Control for LCL-Filtered Grid-Connected Inverters in Weak Grid 3446
Qiang Qian, *Nanjing University of Aeronautics and Astronautics, China*
Jinming Xu, *Nanjing University of Aeronautics and Astronautics, China*
Shaojun Xie, *Nanjing University of Aeronautics and Astronautics, China*
Lin Ji, *Nanjing University of Aeronautics and Astronautics, China*

Model Predictive Control Method to Reduce Common-Mode Voltage and Balance the Neutral-Point Voltage in Three-Level T-Type Inverter 3453
Xiangyang Xing, *Shandong University, China*
Alian Chen, *Shandong University, China*
Zicheng Zhang, *Shandong University, China*
Jie Chen, *Shandong University, China*
Chenghui Zhang, *Shandong University, China*

Convergence Analysis of Distributed Control for Operation Cost Minimization of Droop Controlled DC Microgrid based on Multiagent 3459
Chendan Li, *Aalborg University, Denmark*
Juan C. Vásquez, *Aalborg University, Denmark*
Josep M. Guerrero, *Aalborg University, Denmark*

A Novel Model Predictive Control Algorithm to Suppress the Zero-Sequence Circulating Currents for Parallel Three-Phase Voltage Source Inverters 3465
Zicheng Zhang, *Shandong University, China*
Alian Chen, *Shandong University, China*
Xiangyang Xing, *Shandong University, China*
Chenghui Zhang, *Shandong University, China*

Design of Dynamic Voltage Restorer and Active Power Filter for Wind Power Systems Subject to Unbalanced and Harmonic Distorted Grid 3471
Woei-Luen Chen, *Chang Gung University, Taiwan*
Meng-Jie Wang, *Chang Gung University, Taiwan*

Dynamic Variable Coupling Analysis and Modeling of Proton Exchange Membrane Fuel Cells for Water and Thermal Management 3476
Daming Zhou, *Université de Technologie de Belfort-Montbéliard, France*
Elena Breaz, *Université de Technologie de Belfort-Montbéliard, France*
Alexandre Ravey, *Université de Technologie de Belfort-Montbéliard, France*
Fei Gao, *Université de Technologie de Belfort-Montbéliard, France*
Abdellatif Miraoui, *Université de Technologie de Belfort-Montbéliard, France*
Ke Zhang, *Northwestern Polytechnical University, China*

Voltage and Frequency Control of Electric Spring based Smart Loads 3481
Yun Yang, *University of Hong Kong, Hong Kong*
Siew-Chong Tan, *University of Hong Kong, Hong Kong*
Shu-Yuen Ron Hui, *University of Hong Kong, Hong Kong*

Second Harmonic Current Compensator with Improved One-Cycle-Control 3488
Li Zhang, *Nanjing University of Aeronautics and Astronautics, China*
Xinbo Ruan, *Nanjing University of Aeronautics and Astronautics, China*
Xiaoyong Ren, *Nanjing University of Aeronautics and Astronautics, China*

Frequency Adaptive Control of a Smart Transformer-Fed Distribution Grid 3493
Zhi-Xiang Zou, *Christian-Albrechts-Universität zu Kiel, Germany*
Giovanni De Carne, *Christian-Albrechts-Universität zu Kiel, Germany*
Giampaolo Buticchi, *Christian-Albrechts-Universität zu Kiel, Germany*
Marco Liserre, *Christian-Albrechts-Universität zu Kiel, Germany*

A Synchronization Scheme for Single-Phase Grid-Tied Inverters under Harmonic Distortion and Grid Disturbances 3500
Lenos Hadjidemetriou, *University of Cyprus, Cyprus*
Elias Kyriakides, *University of Cyprus, Cyprus*
Yongheng Yang, *Aalborg University, Denmark*
Frede Blaabjerg, *Aalborg University, Denmark*

Series-Parallel Connection of Low-Voltage Sources for Integration of Galvanically Isolated Energy Storage Systems 3508
Ramy Georgious, *Universidad de Oviedo, Spain*
Jorge Garcia, *Universidad de Oviedo, Spain*
Angel Navarro, *Universidad de Oviedo, Spain*
Sarah Saeed, *Universidad de Oviedo, Spain*
Pablo Garcia, *Universidad de Oviedo, Spain*

Saturation Controller-Based Direct Power Control for Doubly-Fed Induction Generator 3514
Chun Wei, *University of Nebraska at Lincoln, United States*
Zhe Zhang, *Nexteer Automotive, United States*
Wei Qiao, *University of Nebraska at Lincoln, United States*
Liyan Qu, *University of Nebraska at Lincoln, United States*

Inductance-Simulating Control for DFIG-Based Wind Turbine to Ride-Through Grid Faults 3521
Donghai Zhu, *Huazhong University of Science and Technology, China*
Xudong Zou, *Huazhong University of Science and Technology, China*
Yong Kang, *Huazhong University of Science and Technology, China*
Lu Deng, *Wuhan NARI Limited Company of State Grid Electric Power Research Institute, China*
Qingjun Huang, *State Key Laboratory of Disaster Prevention & Reduction for Power Grid Transmission and Distribution Equipment, China*

Session D15: Transportation Power Electronics
Location: Poster Area
March 24, 2016 11:30 - 14:00
Session Chairs: Ted Bohn, *Argonne National Labs*
                    Khurram Afridi, *University of Colorado, Boulder*

Misalignment Effect on Efficiency of Wireless Power Transfer for Electric Vehicles 3526
Yabiao Gao, *University of Georgia, United States*
Antonio Ginart, *University of Georgia / Sonnenbatterie GmbH, United States*
Kathleen Blair Farley, *Southern Company Services, Inc., United States*
Zion Tsz Ho Tse, *University of Georgia, United States*

Genetic Algorithm Design of a 3D Printed Heat Sink 3529
Tong Wu, *University of Tennessee, United States*
Burak Ozpineci, *Oak Ridge National Laboratory, United States*
Curtis Ayers, *Oak Ridge National Laboratory, United States*

Evaluation of Power Flow Control for an All-Electric Warship Power System with Pulsed Load Applications 3537
J. Neely, *Sandia National Laboratories, United States*
L. Rashkin, *Sandia National Laboratories, United States*
M. Cook, *Sandia National Laboratories, United States*
D. Wilson, *Sandia National Laboratories, United States*
S. Glover, *Sandia National Laboratories, United States*

Reduced Active Switch AC to DC Rectifier with High Frequency Isolation for Electric Vehicle Chargers 3545
José Juan Sandoval, *Texas A&M University, United States*
Taeyong Kang, *Texas A&M University, United States*
Prasad Enjeti, *Texas A&M University, United States*

A Wide Bandgap Device based Multilevel Switched-Capacitor Converter 3553
Diogo Cesar Santos de Moura, *North Dakota State University, United States*
Boris Curuvija, *North Dakota State University, United States*
Dong Cao, *North Dakota State University, United States*

Session D16: Power Topologies, Distribution, and Control
Location: Poster Area
March 24, 2016 11:30 - 14:00
Session Chairs: Tiefu Zhao, *Eaton*
Xiaonan Lu, *Argonne National Laboratory*

Novel Circulating Current Suppression Strategy for MMC based on Quasi-PR Controller 3560
Shengbao Geng, *Shanghai Jiao Tong University, China*
Yiliang Gan, *Shanghai Jiao Tong University, China*
Yungui Li, *Shanghai Jiao Tong University, China*
Lijun Hang, *Shanghai Jiao Tong University, China*
Guojie Li, *Shanghai Jiao Tong University, China*

Assymmetric Duty-Cycle Phase-Shift Modulation for Power Management in Double Half-Bridge Inverter with Partly Coupled Inductive Loads 3566
C. Carretero, *Universidad de Zaragoza, Spain*
H. Sarnago, *Universidad de Zaragoza, Spain*
O. Lucia, *Universidad de Zaragoza, Spain*
J. Acero, *Universidad de Zaragoza, Spain*
J.M. Burdío, *Universidad de Zaragoza, Spain*

Control Implementation for a Wide Voltage Range High Efficiency Power Supply Utilizing Low Voltage MOSFETs 3570
Werner Konrad, *Technische Universität Graz, Austria*
Gerald Deboy, *Infineon Technologies AG, Austria*
Annette Muetze, *Technische Universität Graz, Austria*

A Single-Phase Dual Frequency Inverter based on Multi-Frequency Selective Harmonic Elimination 3577
Chongwen Zhao, *University of Tennessee, United States*
Daniel Costinett, *University of Tennessee, United States*
Brad Trento, *University of Tennessee, United States*
Daniel Friedrichs, *Medtronic, United States*

Grid Connected DC Distribution Network Deploying High Power Density Rectifier for DC Voltage Stabilization 3585
Danillo B. Rodrigues, *Universidade Federal do Triângulo Mineiro, Brazil*
Paulo R. Silva, *Universidade Federal de Uberlândia, Brazil*
Gustavo B. Lima, *Universidade Federal do Triângulo Mineiro, Brazil*
Ernane A.A. Coelho, *Universidade Federal de Uberlândia, Brazil*
Luiz C.G. Freitas, *Universidade Federal de Uberlândia, Brazil*

Even-Harmonic Repetitive Control for Circulating Current Suppression in Modular Multilevel Converters 3591

Shunfeng Yang, *Nanyang Technological University, Singapore*
Peng Wang, *Nanyang Technological University, Singapore*
Yi Tang, *Nanyang Technological University, Singapore*
Michael Zagrodnik, *Rolls-Royce Singapore Pte. Ltd., Singapore*
Xiaolei Hu, *Nanyang Technological University, Singapore*
King Jet Tseng, *Nanyang Technological University, Singapore*

A New DSC-PLL using Recursive Discrete Fourier Transform for Robustness to Frequency Variation 3598

Jaedo Lee, *Korea Institute of Nuclear Safety, Korea, South*
Hanju Cha, *Chungnam National University, Korea, South*

A Four-Quadrant Modulation Technique for Cascaded Multilevel Inverters to Extend Solution Range for Selective Harmonic Elimination/Compensation 3603

Hui Zhao, *University of Florida, United States*
Shuo Wang, *University of Florida, United States*

Online Battery Impedance Spectrum Measurement Method 3611

Jaber A. Abu Qahouq, *University of Alabama, United States*

Analysis and Control of a Reduced Switch Converter for Active Magnetic Bearings 3616

Dong Jiang, *Huazhong University of Science and Technology, China*
Parag Kshirsagar, *United Technologies Research Center, United States*

A Novel Balanced Winding Topology to Mitigate EMI without the Need for a Y-Capacitor 3623

Yongjiang Bai, *Xi'an Jiaotong University, China*
Xu Yang, *Xi'an Jiaotong University, China*
Xinlei Li, *Silergy Corp., China*
Dan Zhang, *Silergy Corp., China*
Wenjie Chen, *Xi'an Jiaotong University, China*

Topology and Control Strategy for Accelerated Lifetime Test Setup of DC-Link Capacitor of Wind Turbine Converter 3629

Youngjong Ko, *Christian-Albrechts-Universität zu Kiel, Germany*
Holger Jedtberg, *Christian-Albrechts-Universität zu Kiel, Germany*
Giampaolo Buticchi, *Christian-Albrechts-Universität zu Kiel, Germany*
Marco Liserre, *Christian-Albrechts-Universität zu Kiel, Germany*

Voltage Droop Compensation based on Resonant Circuit for Generalized High Voltage Solid-State Marx Modulator 3637

Hiren Canacsinh, *Instituto Superior de Engenharia de Lisboa, Portugal*
Luís M. Redondo, *Instituto Superior de Engenharia de Lisboa, Portugal*
J. Fernando Silva, *Instituto Superior Técnico, Portugal*
Beatriz Borges, *Instituto Superior Técnico, Portugal*

Four H-Bridge based Shunt Active Power Filter for Three-Phase Four Wire System 3641

Edgard L.L. Fabricio, *Universidade Federal da Paraíba, Brazil*
Cursino B. Jacobina, *Universidade Federal de Campina Grande, Brazil*
Gregory A.A. Carlos, *Universidade Federal de Campina Grande, Brazil*
Maurício B.R. Correa, *Universidade Federal de Campina Grande, Brazil*

High-Frequency AC Distributed Power Delivery System 3648
Mengqi Wang, *University of Michigan at Dearborn, United States*
Qingyun Huang, *North Carolina State University, United States*
Wensong Yu, *North Carolina State University, United States*
Alex Q. Huang, *North Carolina State University, United States*

**Effect of the Capacitance Distribution on the Output Impedance of the Half-Wave
Cockcroft-Walton Voltage Multiplier** 3655
Liran Katzir, *Tel Aviv University, Israel*
Doron Shmilovitz, *Tel Aviv University, Israel*

Session D17: Emerging and Renewable Power
Location: Poster Area
March 24, 2016 11:30 - 14:00
Session Chairs: Katherine Kim, *Ulsan NIST*
Dimitri Torregrossa, *EPFL*

A Cost Effective High Performance LED Driver Powered by Electronic Ballasts 3659
Jianwen Shao, *STMicroelectronics, United States*
Thomas Stamm, *STMicroelectronics, United States*

**Model Predictive Control of Z-Source Four-Leg Inverter for Standalone Photovoltaic
System with Unbalanced Load** 3663
Sertac Bayhan, *Gazi University, Turkey*
Mohamed Trabelsi, *Texas A&M University at Qatar, Qatar*
Haitham Abu-Rub, *Texas A&M University at Qatar, Qatar*

**Efficiency Optimization of an Integrated Wireless Power Transfer System by a
Genetic Algorithm** 3669
Rosario Pagano, *Integrated Device Technology Inc., United States*
Siamak Abedinpour, *Integrated Device Technology Inc., United States*
Angelo Raciti, *Università degli Studi di Catania, Italy*
Salvatore Musumeci, *Università degli Studi di Catania, Italy*

**Loss Analysis of a High Efficiency GaN and Si Device Mixed Isolated Bidirectional
DC-DC Converter** 3677
Fei Xue, *North Carolina State University, United States*
Ruiyang Yu, *North Carolina State University, United States*
Alex Q. Huang, *North Carolina State University, United States*

**Dynamic Efficiency Tracking Controller for Reconfigurable Four-Coil Wireless Power
Transfer System** 3684
Yuan Cao, *University of Alabama, United States*
Zhigang Dang, *University of Alabama, United States*
Jaber A. Abu Qahouq, *University of Alabama, United States*
Evan Phillips, *University of Alabama, United States*

Wireless Power and Data Transfer System for Smart Bridge Sensors 3690
Yujin Jang, *Korea Advanced Institute of Science and Technology, Korea, South*
Jung Kyu Han, *Korea Advanced Institute of Science and Technology, Korea, South*
Shin Young Cho, *Korea Advanced Institute of Science and Technology, Korea, South*
Gun-Woo Moon, *Korea Advanced Institute of Science and Technology, Korea, South*
Ji-Min Kim, *Korea Advanced Institute of Science and Technology, Korea, South*
Hoon Sohn, *Korea Advanced Institute of Science and Technology, Korea, South*

Inrush Transient Current Analysis and Suppression of Photovoltaic Grid-Connected Inverters during Voltage Sag 3697
Zhongyu Li, *China University of Petroleum, China*
Rende Zhao, *China University of Petroleum, China*
Zhen Xin, *Aalborg University, Denmark*
Josep M. Guerrero, *Aalborg University, Denmark*
Mehdi Savaghebi, *Aalborg University, Denmark*
Peide Li, *Shandong Jinan Power Equipment Factory Co., LTD, China*

A Highly Reliable Single-Stage Converter for Electric Vehicle Applications 3704
S.A.Kh. Mozaffari Niapour, *Northeastern University, United States*
Mahshid Amirabadi, *Northeastern University, United States*

Simple and Efficient Low Power Photovoltaic Emulator for Evaluation of Power Conditioning Systems 3712
Jesus Gonzalez-Llorente, *Universidad Sergio Arboleda, Colombia*
Andres Rambal-Vecino, *Universidad Sergio Arboleda, Colombia*
Luciano A. Garcia-Rodriguez, *University of Arkansas, United States*
Juan C. Balda, *University of Arkansas, United States*
Eduardo I. Ortiz-Rivera, *University of Puerto Rico at Mayaguez, Puerto Rico*

Data Transmission Method without Additional Circuits in Bidirectional Wireless Power Transfer System 3717
Yeongrack Son, *Seoul National University, Korea, South*
Jung-Ik Ha, *Seoul National University, Korea, South*

Improved Impedance Source Inverter for Hybrid/Electric Vehicle Application with Continuous Conduction Operation 3722
Thilak Senanayake, *University of Tsukuba, Japan*
Ryuji Iijima, *University of Tsukuba, Japan*
Takanori Isobe, *University of Tsukuba, Japan*
Hiroshi Tadano, *University of Tsukuba, Japan*

Comparing Extended Kalman Filter and Particle Filter for Estimating Field and Damper Bar Currents in Brushless Wound Field Synchronous Generator for Stator Winding Fault Detection and Diagnosis

Sivakumar Nadarajan, *Student member IEEE*

Department of Electrical and Computer Engineering,
National University of Singapore.

Advanced Technology Centre,
Rolls-Royce Singapore Pte. Ltd.,
Singapore.

Bicky Bhangu

Rolls-Royce Singapore Pte. Ltd.,
Singapore.

S.K. Panda, Senior *member IEEE*

Department of Electrical and Computer Engineering,
National University of Singapore,
Singapore.

Amit Kumar Gupta, Senior *Member IEEE*

Advanced Technology Centre,
Rolls-Royce Singapore Pte. Ltd.,
Singapore.

Abstract— Condition monitoring of the Brushless Wound Field Synchronous Generator (BWFSG) is important as it is widely used in mission and safety critical applications such as marine vessels. The frequency signatures in rotor field current are well known indicators to detect and diagnose stator winding short circuits in synchronous generators, in addition the damper bars current could also be used for fault detection and diagnosis. However, BWFSG these currents are not accessible. Hence, it is important to use the mathematical model and state estimation techniques to estimate these parameters. This paper compares the performance of state estimation techniques such as the Extended Kalman Filter (EKF) and Particle Filter (PF) in estimating field current and damper bars current for stator winding fault detection and diagnosis. The experimental validation results confirmed that the performance of the EKF is better than that of the PF in terms of number of stator winding fault signatures extracted from estimating the field current and damper bars currents. Thus, the future work proposed to use the EKF for developing Model-Based Condition Monitoring (MBCM) system for the BWFSG.

Keywords—Synchronous generator, Condition monitoring, fault detection and diagnosis, marine generator, aerospace generator, winding fault, brushless synchronous generator

NOMENCLATURE

v	Instantaneous voltage
i	Instantaneous current
L	Inductance
r	Resistance
ω_r	Angular velocity of the rotor in rad/s

Subscripts

l	Leakage quantity
m	Magnetizing quantity
q	*q-axis* quantity

d	*d-axis* quantity
0	Zero sequence quantity
s	Stator quantity
r	Rotor quantity
kq	Damper bar *q-axis* leakage quantity
fd	Rotor field *d-axis* leakage quantity
kd	Damper bar *d-axis* leakage quantity

Superscript

$'$	Rotor quantity referred to the stator

I. INTRODUCTION

THE Brushless Wound Field Synchronous generator (BWFSG) is widely used in marine vessels for on power generation, any unforeseen failures in the BWFSG may lead to complete shutdown of the critical loads and cause revenue loss. The winding fault is one of the most common faults in the synchronous generators [1] and if this fault is undetected at early stage i.e. at turn-to-turn short circuit (TTSC), then it may leads to catastrophic failures such as phase-ground or phase-phase faults. Hence, it is important to develop condition monitoring techniques to detect and diagnose the TTSC faults. The condition monitoring approaches are generally classified as signal-based, model-based and data-driven approaches. The research work on conventional slip-ring based synchronous generator concluded that the rotor field current carries useful information about the TTSC faults [2]. In addition, the damper bars currents could also potentially carry the fault signatures as the negative sequence flux due to TTSC fault can induce current in damper bars. However, these currents are not accessible in the BWFSG. The Model-Based Condition Monitoring (MBCM) approach for the BWFSG can utilize the

978-1-4673-9551-9/16 $31.00 © 2016 IEEE

mathematical model and state estimation techniques to estimate the rotor quantities such as field current and damper bar current and these use the fault signatures in estimated rotor quantities to detect and diagnose the fault. Hence, MBCM approach is more suitable for rotor current based condition monitoring of the BWFSG. The state estimator is a key sub-system of MBCM approach. In general, the physics based electrical machine models are nonlinear in nature, hence the Extend Kalman Filter (EKF) is widely used as state estimation or parameter identification technique for electrical machines [3]-[5]. The Particle Filter (PF) is another state estimation technique popular in recent years [6][7] and researchers claim that it can outperform EKF [7][8]. The performance characteristics of the EKF and PF for condition monitoring of the BWFSG are not discussed in previous works. Hence, this paper compares performance of these two state estimation techniques for a given BWFSG model in detecting stator winding TTSC faults from the estimated rotor field and damper bar currents.

II. STATE SPACE MODEL OF BWFSG

Since the purpose of this study is to compare accuracy of the EKF and PF in estimating the rotor field current and damper bar currents, the healthy portion of the model in [9] is considered in this paper. The BWFSG presented in [9] contains three damper bars, two in q-axis and one in d-axis. To reduce the computation, this paper assumes the BWFSG has two damper bars, one in q-axis and other in d-axis. The state space representation of the developed healthy BWFSG model is given in (1) and (2).

$$X(t) = A.X(t) + B.U(t) \qquad (1)$$
$$Y(t) = C.X(t) \qquad (2)$$

where,

$$A = \frac{A_s}{L},$$

$$B = \frac{1}{L} diag\{[1,1,1,1,1,1]\},$$

$$C = \begin{bmatrix} 1 & 0 & 0 & 0 & 0 & 0 \\ 0 & 1 & 0 & 0 & 0 & 0 \\ 0 & 0 & 1 & 0 & 0 & 0 \end{bmatrix}$$

$A_s =$

$$\begin{bmatrix} R_s & \omega_r(L_{ls}+L_{md}) & 0 & 0 & -L_{md}\omega_r & -L_{md}\omega_r \\ -\omega_r(L_{ls}+L_{mq}) & R_s & 0 & L_{mq}\omega_r & 0 & 0 \\ 0 & 0 & R_s & 0 & 0 & 0 \\ 0 & 0 & 0 & -R_{kq} & 0 & 0 \\ 0 & 0 & 0 & 0 & -R_{fd} & 0 \\ 0 & 0 & 0 & 0 & 0 & -R_{kd} \end{bmatrix}$$

$L=$

$$\begin{bmatrix} -(L_{ls}+L_{mq}) & 0 & 0 & L_{mq} & 0 & 0 \\ 0 & -(L_{ls}+L_{md}) & 0 & 0 & L_{md} & L_{md} \\ 0 & 0 & -L_{ls} & 0 & 0 & 0 \\ -L_{mq} & 0 & 0 & L_{lkq1}+L_{mq} & 0 & 0 \\ 0 & -L_{md} & 0 & 0 & L_{lfd}+L_{md} & L_{md} \\ 0 & -L_{md} & 0 & 0 & L_{md} & L_{lfd}+L_{md} \end{bmatrix}$$

the state vector, $X(t) = [i_{qs}\ i_{ds}\ i_{0s}\ i_{kq}\ i_{fd}\ i_{kd}]^T$,
the output vector, $Y(t) = [i_{qs}\ i_{ds}\ i_{0s}]^T$, and
the input vector, $U(t) = [v_{qs}\ v_{ds}\ v_{0s}\ v_{kq}\ v_{fd}\ v_{kd}]^T$.

To apply the state estimation techniques such as EKF and Particle filter, the model must be discretize. The discretized state space model in (1) and (2) is given in (3) and (4)

$$X(k + 1) = A_d.X(k) + B_d.U(k) \qquad (3)$$
$$Y(k) = C_d.X(k) + D_d.U(k) \qquad (4)$$

Where,

$$A_d = e^{|AT_s|} \approx I + AT_s; \quad B_d \approx BT_s; \quad C_d = C; \quad D_d = 0$$
$$T_s - \text{is the sampling period}$$

III. STATE ESTIMATION - EXTENDED KALMAN FILTER

The EKF is a derivative of the Kalman filter for non-linear systems. The EKF based state estimation provides optimal solution for the system with Gaussian noise. It also has excellent capability to maintain accuracy of the state estimation even under parameter variation, measurement noise and inaccurate model [4] and [5]. The non-linear state-space system in (3)-(4) can be rewritten as in (5)-(6)

$$X(k + 1) = A_d.X(k) + B_d.U(k) + w(k) \quad (14)$$
$$Y(k) = C_d.X(k) + D_d.U(k) + v(k) \qquad (15)$$

Where,
$w(k)$ – Zero-means Gaussian noise disturbance in state vector with covariance matrix Q
$v(k)$ – Zero-means Gaussian noise disturbance in measurement vector with covariance matrix R

The EKF has two key steps which include step to predict the state and a setup to correct the predicted state called update state. Both these steps are performed for each iteration, refer [3]-[5] and [10] for the details of the equations. The key matrices involved in these two steps are system gradient matrix (F) and measurement matrix (H). Since, the parameters of interest (rotor field and damper bar currents) are already the states in the model (14) and (15). The F and H matrix are equal to A_d and C_d matrixes, respectively. The steps involved in implementing EKF of the developed the BWFSG model are as follow,

Step 1: Initialize state vector $X_e(0)$, covariance error $P(0)$, state noise covariance Q and measurement noise covariance R.

Step 2: Prediction phase
 2a) State prediction
  $\widehat{X}(k + 1) = A_d.\widehat{X}(k) + B_d.U(k)$
 2c) Calculate covariance matrix of prediction error
  $M(k + 1) = F(k).P(k).F^T(k) + Q$

Step 3: Correction phase
 3a) Calculate Kalman gain
  $K(k + 1) = M(k + 1).H^T.(H.M(k + 1).H^T + R)^{-1}$
 3b) Update covariance error matrix
  $P(k + 1) = (1 - K(k + 1).H).M(k + 1)$
 3c) Update state estimation
  $\widehat{X}(k + 1) = \widehat{X}(k + 1) + K(k + 1)\left(Y(k + 1) - H.\widehat{X}(k + 1)\right)$

Step 4: $\hat{X}(k) = \hat{X}(k+1)$, and go to step 2.

The field current and damper bars current are obtained from the estimated states at step 4.

IV. STATE ESTIMATION – PARTICLE FILTER

The PF also has same two steps as the EKF, predict state and update state [11]. The PF uses Monte Carlo numerical approximation methods to recursively calculate the predicted and updated states. The PF can be applicable to a system with non-linear and non-Gauassian noise. In the PF, the set of random particles are initialized, each particles contains a sample (value) and corresponding weights $\{x_k^i, w_k^i\}$. The initialized particles are used to predict the states and then the weight of the particles are then adjusted based on the prediction error and posterior probability density function. A re-sampling technique is used to bring more copies of the higher weight particles for the next iteration. The description of the implemented PF in this study can be found in an author of this paper previous work [12].

V. EXPERIMENTAL SETUP

To compare the performance of the EKF and PF state estimation techniques, an experimental setup is designed and built. The experimental setup has 14kVA BWFSG generator with provisions to introduce various degrees and severity of stator winding short circuit faults. The experimental setup also equipped with various sensors to measure three phase stator currents and terminal voltages. Fig. 1 shows the complete experimental setup.

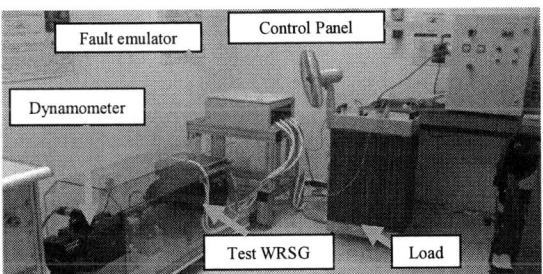

Fig. 1. Complete Experimental setup

VI. EXPERIMENTAL VALIDATION RESULTS

The performance of the EKF and PF state estimators are evaluated by using the three phase voltages and three phase currents measured from the BWFSG in experimental setup under healthy, 5% stator winding TTSC, and 20% stator winding TTSC conditions. Fig. 2 (a) and (b) show the estimated stator currents, q-axis damper bar current (i_{kq}), field current (i_{fd}) and d-axis damper bar current (i_{kd}) by using the EKF and PF state estimation techniques, respectively under healthy condition. The estimated stator currents from the EKF closely match with the measured currents as shown in Fig. 2(a), in addition the estimated damper bars currents are closer to zero which is the desired behavior of the damper bars in the

BWFSG under steady state. Hence it can be confirmed that the developed EKF estimation technique able to estimate the states in the BWFSG model accurately.

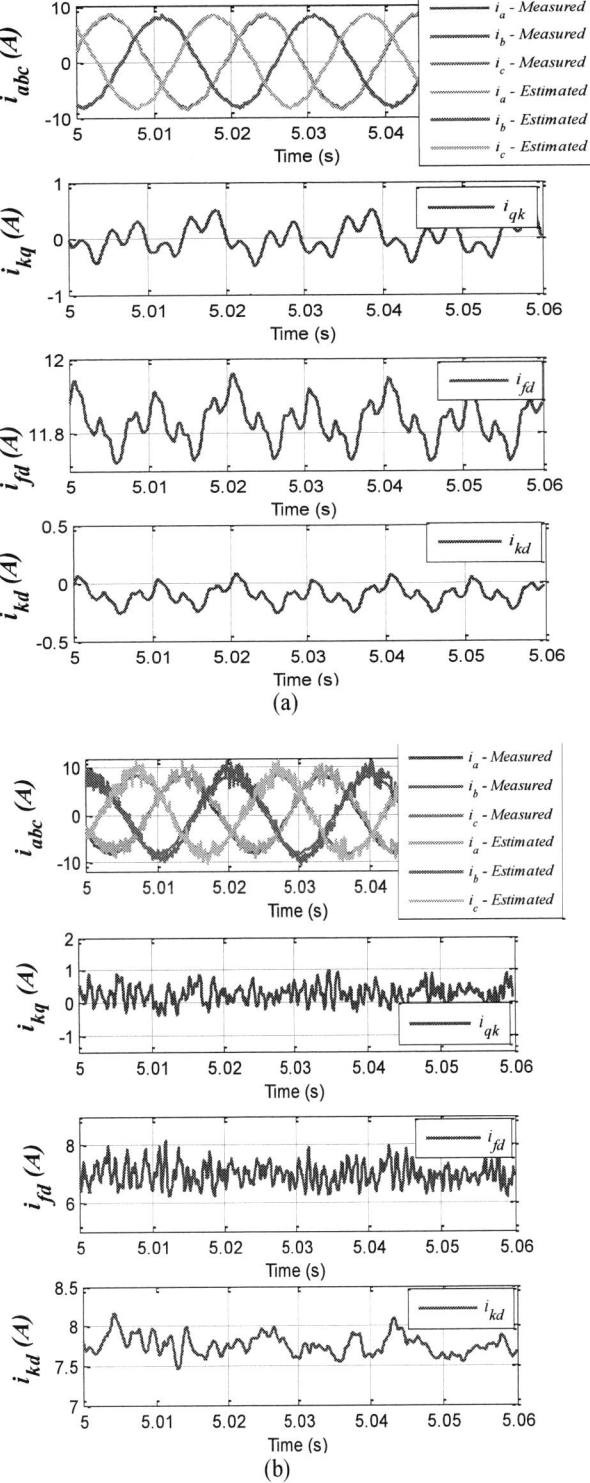

Fig. 2. Estimated stator currents, i_{kq}, i_{fd}, and i_{kd} under healthy condition by using (a) EKF (b) PF state estimation techniques

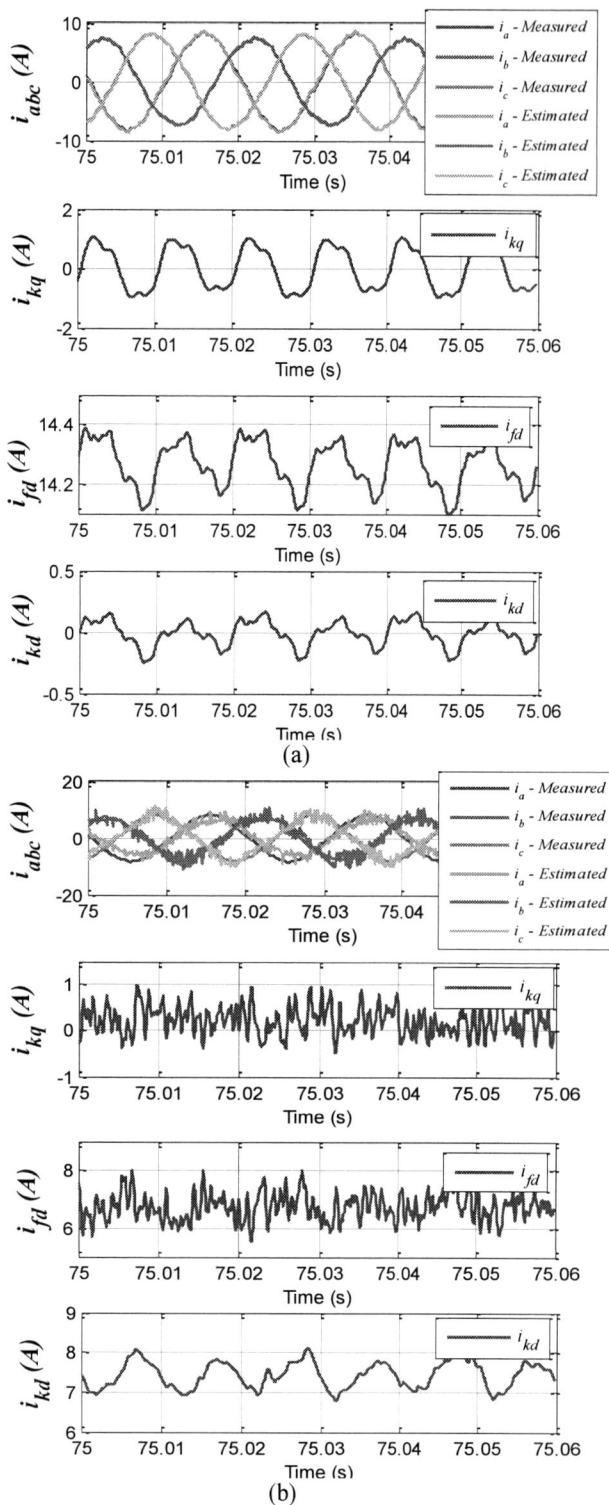

Fig. 3. Estimated stator currents, i_{kq}, i_{fd}, and i_{kd} under 20% TTSC fault by using (a) EKF (b) PF state estimation techniques

From Fig. 2(b) it clear that the performance of PF is not satisfactory when compared to the EKF, because the estimated stator current is noisy and the d-axis damper bar current

supposed to be zero but it is showing some finite value. Figs. 3(a) and 3(b), show the estimated stator and rotor currents under 20% TTSC by using the EKF and the PF, respectively. The 2nd harmonic oscillation in the estimated the field current and the damper bars currents are more pronounced when EKF estimator is used, this the expected behavior of the field and damper bar currents presented in [1]. Whereas when PF is used there is no significant visible 2nd harmonic oscillation in estimated field and damper bar currents waveforms. However, it is important to analyse frequency spectrum of estimated rotor currents by using both the EKF and PF to identify fault signatures as well as to identify suitable state estimation technique for the BWFSG condition monitoring.

Fig. 4 shows the frequency spectrum of estimated q-axis damper bar current, field current and d-axis damper bar current from EKF state estimator under healthy and faulty conditions. Fig. 5 shows the frequency spectrum of the estimated rotor currents from the PF under healthy and faulty conditions. From the frequency spectrums it can be noted that the 2nd harmonic component increased significantly under TTSC fault for all three rotor currents when EKF is used, whereas for PF only field and d-axis damper bar currents have significant change and no change in q-axis damper bar currents. From Fig. 4 and 5, it can be concluded that if EKF state estimation technique is used for developing the MBCM for the BWFSG, then it could provide more fault indicators (i.e. 3 indicators one from each current) to detect and diagnose TTSC fault when compared to the PF which provide only two indicators. The robustness of fault detection increases with increasing numbers of fault indicators, hence EKF is proposed as suitable state estimation technique for developing MBCM for the BWFSG in this research.

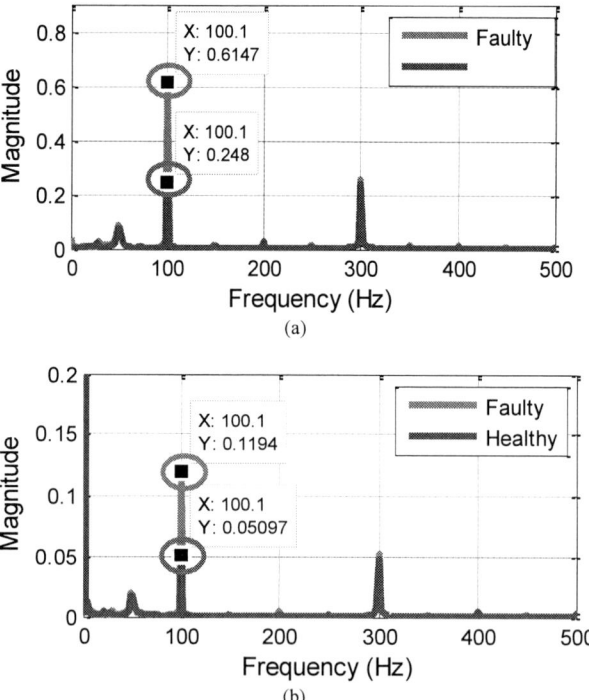

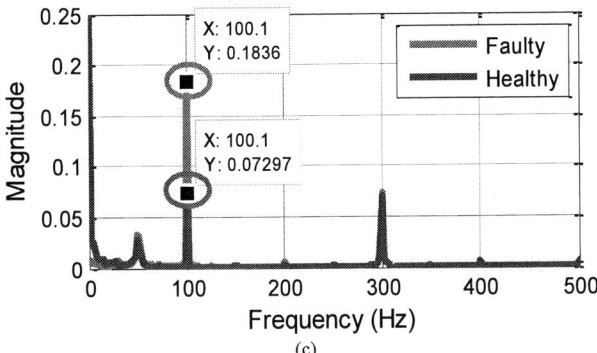

Fig. 4. Frequency spectrum of (a) i_{kq}, (a) i_{fd}, and (a) i_{kd} under healthy and 20% TTSC fault using EKF

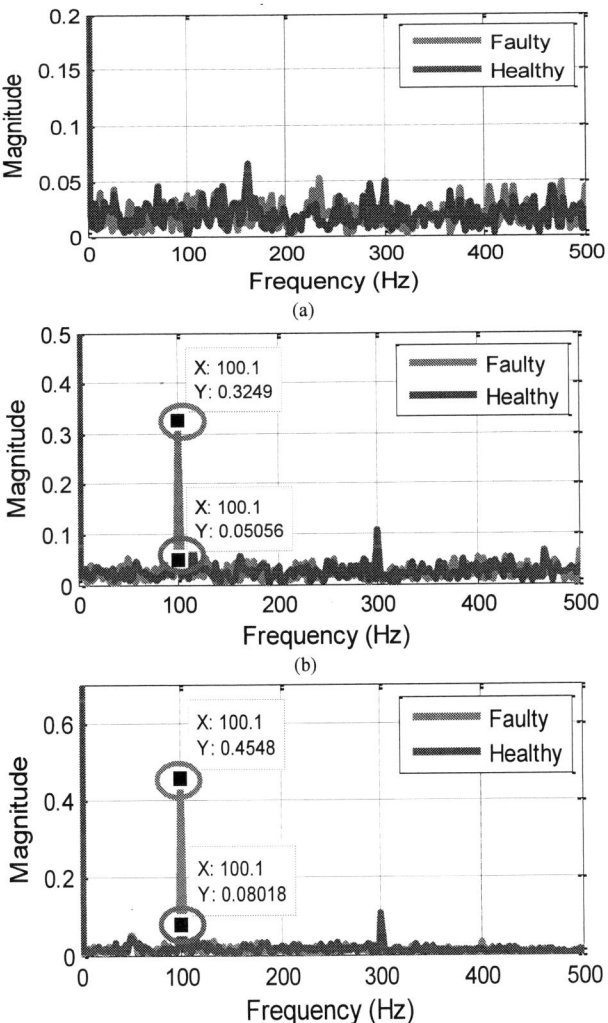

Fig. 5. Frequency spectrum of (a) i_{kq}, (a) i_{fd}, and (a) i_{kd} under healthy and 20% TTSC fault using PF

VII. CONCLUSIONS AND FUTURE WORK

This paper implemented the EKF and PF state estimation techniques for the developed BWFSG model. The performances of the EKF and PF have been evaluated under healthy and stator winding fault conditions by using the experimental data. The stator winding TTSC fault signature (i.e. 2nd harmonic oscillation) in estimated field and damper bars currents is used as one of the key performance indices to validate the suitability of EKF and PF in developing MBCM system for the BWFSG. The validation results show that the EKF state estimation could provide more fault indicators for fault detection and diagnosis compared to that of the PF. Hence the EKF may increase the robustness of fault detection, thus future work will focus on developing complete MBCM system for the BWFSG by using the EKF state estimation.

REFERENCES

[1] IEEE Recommended Practice for the Design of Reliable Industrial and Commercial Power Systems," IEEE Std 493-2007, 2007

[2] Hemmati, S.; Shokri, Sh; Saied, S.A, "Modeling and simulation of internal short circuit faults in large hydro generators with wave windings," in Int. Conf. POWERENG, 2011, pp.1,6, 11-13

[3] Fernandez Gomez, A.J.; Jaramillo, V.H.; Ottewill, J.R., "Fault detection in electric motors by means of the extended Kalman Filter as disturbance estimator," Control (CONTROL), 2014 UKACC International Conference on , vol., no., pp.432,437, 9-11 July 2014

[4] Xinan Zhang; Foo, G.; Don Vilathgamuwa, M.; Tseng, K.J.; Bhangu, B.S.; Gajanayake, C., "Sensor fault detection, isolation and system reconfiguration based on extended Kalman filter for induction motor drives," Electric Power Applications, IET , vol.7, no.7, pp.607,617, Aug. 2013

[5] Foo, G.H.B.; Xinan Zhang; Vilathgamuwa, D.M., "A Sensor Fault Detection and Isolation Method in Interior Permanent-Magnet Synchronous Motor Drives Based on an Extended Kalman Filter," Industrial Electronics, IEEE Transactions on , vol.60, no.8, pp.3485,3495, Aug. 2013Mansouri, M.; Nounou, H.;

[6] Climente-Alarcon, V.; Antonino-Daviu, J.A.; Haavisto, A.; Arkkio, A., "Particle Filter-Based Estimation of Instantaneous Frequency for the Diagnosis of Electrical Asymmetries in Induction Machines," Instrumentation and Measurement, IEEE Transactions on , vol.63, no.10, pp.2454,2463, Oct. 2014

[7] Nounou, M., "State estimation and application to induction machines - A comparative study," Multi-Conference on Systems, Signals & Devices (SSD), 2014 11th International , vol., no., pp.1,6, 11-14 Feb. 2014

[8] Restaino, R.; Zamboni, W., "Comparing particle filter and extended kalman filter for battery State-Of-Charge estimation," IECON 2012 - 38th Annual Conference on IEEE Industrial Electronics Society , vol., no., pp.4018,4023, 25-28 Oct. 2012

[9] Nadarajan, S.; Panda, S.K.; Bhangu, B.; Gupta, A.K., "Hybrid Model for Wound-Rotor Synchronous Generator to Detect and Diagnose Turn-to-Turn Short-Circuit Fault in Stator Windings,"Industrial Electronics, IEEE Transactions on , vol.62, no.3, pp.1888,1900, March 2015

[10] Fernandez Gomez, A.J.; Jaramillo, V.H.; Ottewill, J.R., "Fault detection in electric motors by means of the extended Kalman Filter as disturbance estimator," Control (CONTROL), 2014 UKACC International Conference on , vol., no., pp.432,437, 9-11 July 2014

[11] Movaghati, S.; Moghaddamjoo, A.; Tavakoli, A., "Road Extraction From Satellite Images Using Particle Filtering and Extended Kalman Filtering," Geoscience and Remote Sensing, IEEE Transactions on , vol.48, no.7, pp.2807,2817, July 2010

[12] Ming Yu; Danwei Wang; Ukil, A.; Vaiyapuri, V.; Sivakumar, N.; Jayampathi, C.; Gupta, A.K.; VietHung Nguyen, "Model-based failure prediction for electric machines using particle filter," in Control Automation Robotics & Vision (ICARCV), 2014 13th International Conference on , vol., no., pp.1811-1816, 10-12 Dec. 2014

Analytical Determination of Conduction Power Losses for Active Neutral-Point-Clamped Multilevel Converter

Vahid Dargahi*, *Student Member, IEEE*, Arash Khoshkbar Sadigh†, *Member, IEEE*, and Keith Corzine*, *Senior Member, IEEE*

*Microgrid and Power Electronics Laboratory, Holcombe Department of Electrical and Computer Engineering, Clemson University, Clemson, SC 29634, USA
†Extron Electronics, Anaheim, CA 92805, USA

Abstract— **Active neutral-point-clamped (ANPC) converter is a well-known type of multilevel converter which is commercially available in high-power medium-voltage drive market. Since conduction loss investigation is advantageous in design phase of converters, this paper presents an analytical approach to calculate and investigate conduction power loss in ANPC converter. First, rms and average currents of IGBTs and anti-parallel diodes are analytically calculated by considering the associated duty cycle of each IGBT and diode, converter modulation index, load current, and load power factor. Numerical results of the derived analytical equations to calculate rms and average current of IGBT/diode are compared against simulation results and experimental measurements. All simulation, analytic, and experimental results agree well with each other which substantiate derived closed-form equations. Afterwards, obtained equations for rms and average current computations are utilized to calculate conduction power losses in a 12.4MVA 3.3kV 9-level (line-to-line) ANPC converter analytically. For this purpose, a 2.5kV 1.5kA IGBT module from ABB is considered as a power switch.**

Index Terms- **Active Neutral-Point-Clamped Multilevel Converter, Conduction Loss, IGBT Average and RMS Current Ratings.**

I. INTRODUCTION

One of the recently commercialized breeds of multilevel converters is active neutral-point-clamped (ANPC) converter introduced by ABB [1]. The ANPC converter is a hybrid multilevel based on combination of neutral-point-clamped (NPC) and flying-capacitor-multicell (FCM) converters which features advantages of both converters. This power converter is a well-established topology in medium-voltage (MV) drive market. For example, ANPC technology has now been matured into ACS2000 MV drive constructed by ABB [2].

Power conversion through power electronic converters/inverters employing semiconductor devices exhibiting internal inherent characteristics like on-state resistance and forward voltage drop account for conduction loss, hence slightly dropped efficiency and some energy loss inside these converters. The conduction losses are calculated with obtaining the required characteristics of on-state resistance and forward voltage drop from datasheet of insulated-gate bipolar transistor (IGBT) and diode through linearization method and corresponding average and rms current of phase current waveform [3]–[6]. For appropriate design of converters, it is a concern of utmost importance and practical interest to calculate the rms and average currents ratings of IGBTs/diodes in order to: 1) select IGBTs/diodes properly 2) calculate conduction power losses of converter. It is noteworthy that analytical calculation of rms and average currents of IGBTs/diodes also results in analytical investigation of conduction power losses which is advantageous in design procedure. All of the available related articles in this topic have investigated switching and conduction losses of multilevel converters through simulation/experimental results [7]–[11].

Using simulation/experimental results for investigation of the conduction power losses is time consuming, cumbersome, and arduous task because the converter circuit needs to be run for numerous cases considering different values of modulation index and load power factor. Taking into account advantages of ANPC converters and their escalating utilization in MV drive market necessitates the investigation of the ANPC converter efficiency with any number of voltage levels as a function of converter output power, load power factor, and modulation index. Hence, this paper presents an analytical approach to calculate and investigate conduction losses in ANPC converters.

II. CALCULATION OF CONDUCTION POWER LOSSES IN IGBTS AND DIODES

The IGBT collector-emitter voltage drop v_{CE} when it is conducting can be approximated very well as follows [12]:

$$v_{CE} = V_{CE0} + R_C \cdot i_C \tag{1}$$

where V_{CE0} represents IGBT on-state zero-current collector-emitter forward voltage drop and R_C is collector-emitter on-state resistance. The same approximation can be used for anti-parallel diode, giving:

$$v_F = V_{F0} + R_F \cdot i_F \tag{2}$$

where V_{F0} represents anti-parallel diode on-state zero-current forward voltage drop and R_F is anti-parallel diode on-state resistance. These important parameters can be obtained directly

978-1-4673-9551-9/16 $31.00 © 2016 IEEE

from IGBT datasheet. The instantaneous values of IGBT conduction loss ($p_{CT}(t)$) and average loss (P_{CT}) are:

$$p_{CT}(t) = v_{CE}(t) \cdot i_C(t) = V_{CE0} \cdot i_C(t) + R_C \cdot i_C^2(t) \quad (3)$$

$$P_{CT} = \frac{1}{2\pi} \int_0^{2\pi} [V_{CE0} \cdot i_C(t) + R_C \cdot i_C^2(t)] d(\omega t) \quad (4)$$

$$P_{CT} = V_{CE0} \cdot I_{C,avg} + R_C \cdot I_{C,rms}^2 \quad (5)$$

where $I_{C,avg}$ and $I_{C,rms}$ are the average and rms currents of IGBT, respectively. Similar to IGBT, average value of diode conduction loss P_{CD} is:

$$P_{CD} = V_{F0} \cdot I_{D,avg} + R_F \cdot I_{D,rms}^2 \quad (6)$$

where $I_{D,avg}$ and $I_{D,rms}$ are average and rms currents of anti-parallel diode, respectively. As it can be seen in Eq. 5 and Eq. 6, it is required to obtain average and rms current flowing through IGBT and diode in order to calculate their conduction power losses. The main contribution of this paper is to derive analytical equations expressing average and rms currents of semiconductors, i.e., IGBT and diode, in terms of load power factor, load peak current and modulation index.

III. ANALYTICAL CALCULATION OF AVERAGE AND RMS CURRENTS OF IGBTs/DIODES FOR ANPC CONVERTER

In this section, analytical approach is presented to calculate average and rms currents flowing through IGBT/diodes in terms of load power factor, load peak current and modulation index in ANPC converter which is shown in Fig. 1. Semiconductors in ANPC converter can be categorized as follows:

- High-frequency (HF) IGBTs/diodes (S_1 to S_n, D_1 to D_n).
- Top/bottom Low-frequency (LF) IGBTs/diodes (S_{J1}, S_{J4}, D_{J1}, D_{J4}).
- Middle Low-frequency (LF) IGBT/diodes (S_{J2}, S_{J3}, D_{J2}, D_{J3}).

Phase current and duty cycle of LF and HF IGBTs/diodes can be expressed according to the following equations.

$$i_\phi(t) = I_P \sin(\omega t) \quad (7)$$

$$D_{HF}(t) = \begin{cases} M\sin(\omega t + \varphi) & -\varphi < \omega t < \pi - \varphi \\ 1 + M\sin(\omega t + \varphi) & \pi - \varphi < \omega t < 2\pi - \varphi \end{cases} \quad (8)$$

$$D_{TB,LF}(t) = \begin{cases} M\sin(\omega t + \varphi) & -\varphi < \omega t < \pi - \varphi \\ 0 & \pi - \varphi < \omega t < 2\pi - \varphi \end{cases} \quad (9)$$

$$D_{M,LF}(t) = \begin{cases} 1 - M\sin(\omega t + \varphi) & -\varphi < \omega t < \pi - \varphi \\ 0 & \pi - \varphi < \omega t < 2\pi - \varphi \end{cases} \quad (10)$$

where I_P is peak current, M is modulation index, $\omega = 2\pi f$ is angular frequency and f is output voltage frequency.

Average current of HF IGBT (S_x) can be calculated as follows:

$$\begin{aligned} I_{CHF,avg} &= \frac{1}{2\pi} \int_0^\pi [i_\phi(t) \cdot D_{HF}(t)] d(\omega t) \\ &= \frac{MI_P}{4}\cos(\varphi) + \frac{I_P}{2\pi}(1 - \cos(\varphi)) \end{aligned} \quad (11)$$

By following the same procedure, average current of HF anti-parallel diode (D_x) can be obtained as follows:

$$\begin{aligned} I_{DHF,avg} &= \frac{1}{2\pi} \int_\pi^{2\pi} [-i_\phi(t) \cdot D_{HF}(t)] d(\omega t) \\ &= \frac{I_P}{2\pi}(1 + \cos(\varphi)) - \frac{MI_P}{4}\cos(\varphi) \end{aligned} \quad (12)$$

Similarly, average current of top/bottom LF IGBTs and diodes can be obtained as follows:

$$\begin{aligned} I_{CTB,avg} &= \frac{1}{2\pi} \int_0^\pi [i_\phi(t) \cdot D_{TB,LF}(t)] d(\omega t) \\ &= \frac{MI_P}{4\pi}((\pi - \varphi)\cos(\varphi) + \sin(\varphi)) \end{aligned} \quad (13)$$

$$\begin{aligned} I_{DTB,avg} &= \frac{1}{2\pi} \int_\pi^{2\pi} [-i_\phi(t) \cdot D_{TB,LF}(t)] d(\omega t) \\ &= \frac{MI_P}{4\pi}(\sin(\varphi) - \varphi\cos(\varphi)) \end{aligned} \quad (14)$$

The same way, average current of middle LF IGBTs and diodes can be obtained as follows:

$$\begin{aligned} I_{CM,avg} &= \frac{1}{2\pi} \int_0^\pi [i_\phi(t) \cdot D_{M,LF}(t)] d(\omega t) \\ &= \frac{I_P}{2\pi}(1 + \cos(\varphi)) - \frac{MI_P}{4\pi}(\sin(\varphi) + (\pi - \varphi)\cos(\varphi)) \end{aligned} \quad (15)$$

$$\begin{aligned} I_{DM,avg} &= \frac{1}{2\pi} \int_\pi^{2\pi} [-i_\phi(t) \cdot D_{M,LF}(t)] d(\omega t) \\ &= \frac{I_P}{2\pi}(\cos(\varphi) - 1) - \frac{MI_P}{4\pi}(\sin(\varphi) - \varphi\cos(\varphi)) \end{aligned} \quad (16)$$

As the next step, rms current of HF IGBTs can be calculated as follows:

$$\begin{aligned} I_{CHF,rms} &= \sqrt{\frac{1}{2\pi} \int_0^\pi [i_\phi^2(t) \cdot D_{HF}(t)] d(\omega t)} \\ &= I_P \sqrt{\frac{2M}{3\pi}\cos(\varphi) + \frac{1}{8\pi}(2\varphi - \sin(2\varphi))} \end{aligned} \quad (17)$$

By following the same procedure, rms current of HF diode can be obtained as follows:

$$\begin{aligned} I_{DHF,rms} &= \sqrt{\frac{1}{2\pi} \int_\pi^{2\pi} [(-i_\phi(t))^2 \cdot D_{HF}(t)] d(\omega t)} \\ &= I_P \sqrt{\frac{1}{4} - \frac{1}{8\pi}(2\varphi - \sin(2\varphi)) - \frac{2M}{3\pi}\cos(\varphi)} \end{aligned} \quad (18)$$

In the same manner, rms current of top/bottom LF IGBTs and diodes can be obtained as follows:

$$\begin{aligned} I_{CTB,rms} &= \sqrt{\frac{1}{2\pi} \int_0^\pi [i_\phi^2(t) \cdot D_{TB}(t)] d(\omega t)} \\ &= I_P \sqrt{\frac{M}{12\pi}(\cos(2\varphi) + 4\cos(\varphi) + 3)} \end{aligned} \quad (19)$$

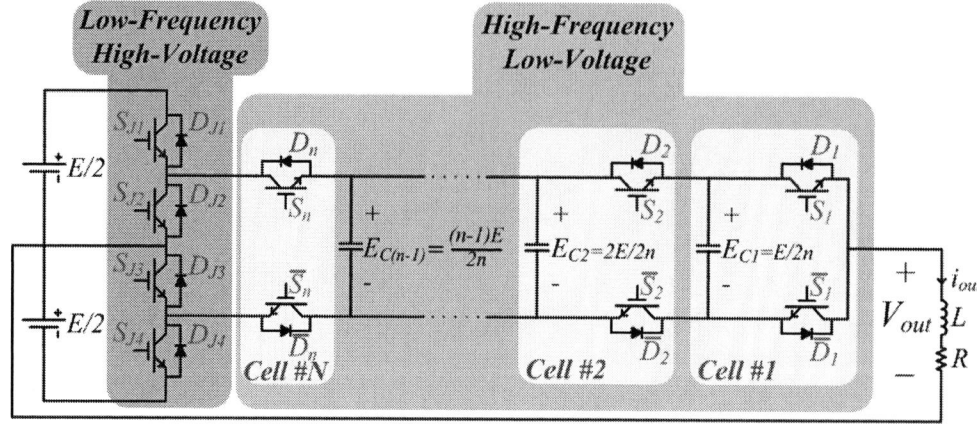

S_{J1}, D_{J1}, S_{J4}, D_{J4} are referred as Top-Bottom IGBT/Diodes (CTB/DTB)

S_{J2}, D_{J2}, S_{J3}, D_{J3} are referred as Middle IGBT/Diodes (CM/DM)

Fig. 1. Generalized topology of an n-cell $n+1$-level ANPC converter.

$$
\begin{aligned}
I_{DTB,rms} &= \sqrt{\frac{1}{2\pi}\int_{\pi}^{2\pi}[(-i_{\phi}(t))^2 \cdot D_{TB}(t)]d(\omega t)} \\
&= I_P\sqrt{\frac{M}{12\pi}(\cos(2\varphi) - 4\cos(\varphi) + 3)}
\end{aligned} \tag{20}
$$

Likewise, rms current of middle LF IGBTs and diodes can be obtained as follows:

$$
\begin{aligned}
I_{CM,rms} &= \sqrt{\frac{1}{2\pi}\int_{0}^{\pi}[i_{\phi}^2(t) \cdot D_M(t)]d(\omega t)} \\
&= \sqrt{\begin{array}{c}\dfrac{I_P^2}{4} - \dfrac{I_P^2}{8\pi}(2\varphi - \sin(2\varphi)) - \\[2mm] \dfrac{MI_P^2}{12\pi}(\cos(2\varphi) + 4\cos(\varphi) + 3)\end{array}}
\end{aligned} \tag{21}
$$

$$
\begin{aligned}
I_{DM,rms} &= \sqrt{\frac{1}{2\pi}\int_{\pi}^{2\pi}[(-i_{\phi}(t))^2 \cdot D_M(t)]d(\omega t)} \\
&= I_P\sqrt{\frac{1}{8\pi}(2\varphi - \sin(2\varphi)) - \frac{M}{12\pi}(\cos(2\varphi) - 4\cos(\varphi) + 3)}
\end{aligned} \tag{22}
$$

Finally, the obtained analytical equations for average and rms current flowing through IGBTs and diodes can be substituted in Eq. 5 and Eq. 6 to calculate conduction loss in ANPC converter.

IV. VALIDATION OF DERIVED CLOSED-FORM EQUATIONS

In order to verify the derived equations for calculation of average and rms currents flowing through IGBTs and diodes in ANPC converters, numerical computation results of the derived equations are compared against simulation and experimental results in the following two sub-sections.

A. Validation through Simulation Results

The numerical computation and simulation studies are done for 12.4MVA 3.3kV three-phase 9-level (line-to-line) ANPC converter (comprising 2 switching-power-cells per phase). In this study, a dc link of 5.4kV which is common in all three phases is used for ANPC converter. The flying capacitor voltage in the considered case study is assumed to be stabilized at 1.35kV under proper operation. Therefore, HF IGBTs in should withstand 1.35kV while LF IGBTs should withstand 2.7kV. Since the utilization factor regarding the voltage of the high-power medium-voltage switches is practically around 50% to 60%, the 2.5kV switches are required to withstand 1.35kV. For this purpose, ABB 5SNA 1500E250300 HiPak 2.5kV 1.5kA IGBT module with parameters of $V_{CE0} = 1.2$V, $R_C = 1m\Omega$, $V_{F0} = 1.1$V, $R_F = 0.4m\Omega$ is considered for all HF IGBTs/didoes. Since voltage rating of LF IGBTs is 2.7kV, two of the aforementioned IGBT modules which are connected in series are utilized for LF IGBTs. The maximum of line peak current I_P is around 3.062 kA in all studies to avoid overcurrent situation in all IGBTs and diodes in the considered case for the ANPC converter. In order to verify the derived equations, numerical computation results are compared against simulation results. This comparison is shown in Figs. 2- 4 illustrating almost a zero error between numerical and simulation results which validates derived analytical equations. According to derived analytical equations as well as simulation results in Figs. 2- 4, it is worth mentioning that by increasing the modulation index, the average and rms currents of the HF IGBTs, top/bottom LF IGBTs/diodes increase whereas the average and rms currents of HF anti-parallel diodes, middle LF IGBTs/diodes decrease. In addition, average and rms currents of all LF anti-parallel diodes are negligible as compared to HF anti-parallel diodes, and this fact is significant for high

978-1-4673-9551-9/16 $31.00 © 2016 IEEE

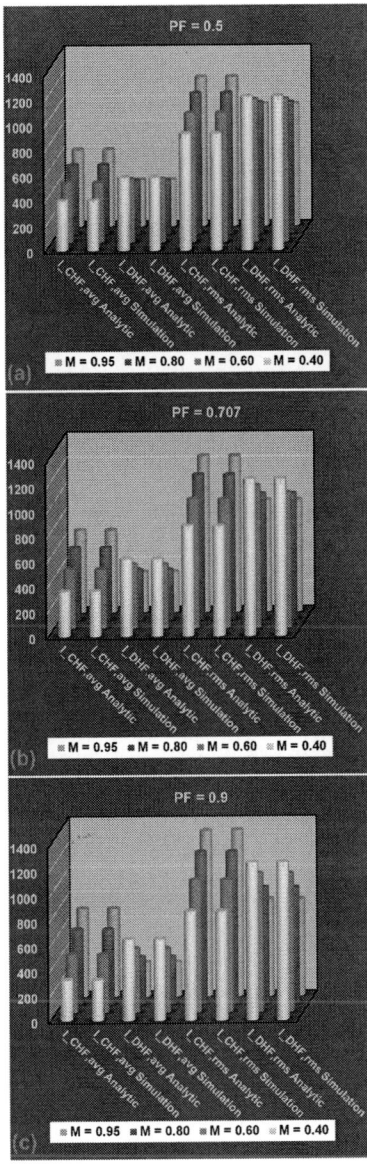

Fig. 2. Comparison between numerical computation and simulation results of the average and rms currents of the HF IGBTs and diodes for ANPC converter for power factors (PF) of: (a) 0.5 (b) 0.707 (c) 0.9.

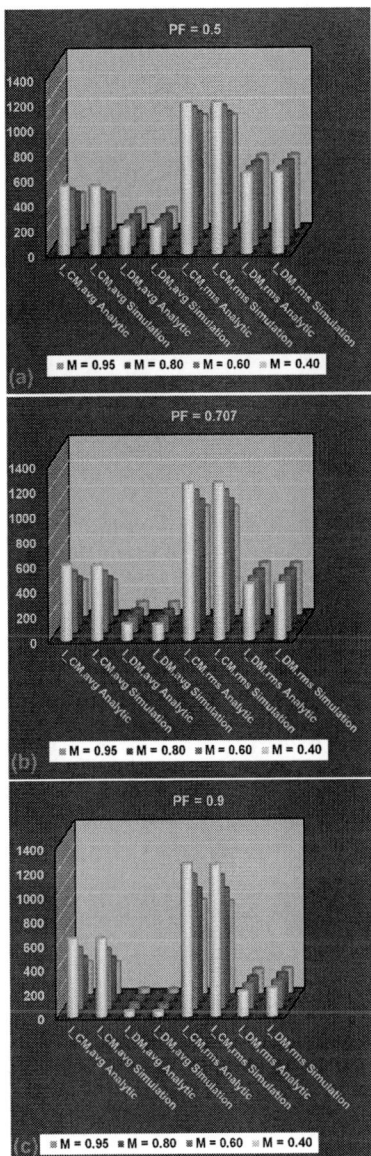

Fig. 3. Comparison between numerical computation and simulation results of the average and rms currents of the middle LF IGBTs and diodes for ANPC converter for power factors (PF) of: (a) 0.5 (b) 0.707 (c) 0.9.

power factors. It is noteworthy of mentioning that middle LF anti-parallel diodes have higher values of average and rms currents in comparison with top/bottom LF anti-parallel diodes. Additionally, average and rms current of all IGBTs are higher than average and rms current of anti-parallel diodes.

B. Validation through Experimental Measurements

In this subsection, a laboratory-scale converter of a 1.5kW five-level single-phase ANPC converter is built in order to make a comparative and verifying analysis between numerical computation results and measured experimental results. The DC link voltage for the ANPC converter is 600V and utilized IGBTs and ultrafast diodes are IXGH48N60B3 and RURP3060, respectively. The TMS320F28335 DSP manufactured by Texas Instruments (TI) has been used to modulate the ANPC converter. The gate driver ICs for IGBTs are IR2184 with gate resistance of $R_{Gate} = 10\Omega$. The DCP021515 IC is used to provide required isolated dc voltages for gate drive circuitry of IGBT modules. To validate the derived closed-form solutions for analytic computation of rms and average currents flowing

978-1-4673-9551-9/16 $31.00 © 2016 IEEE

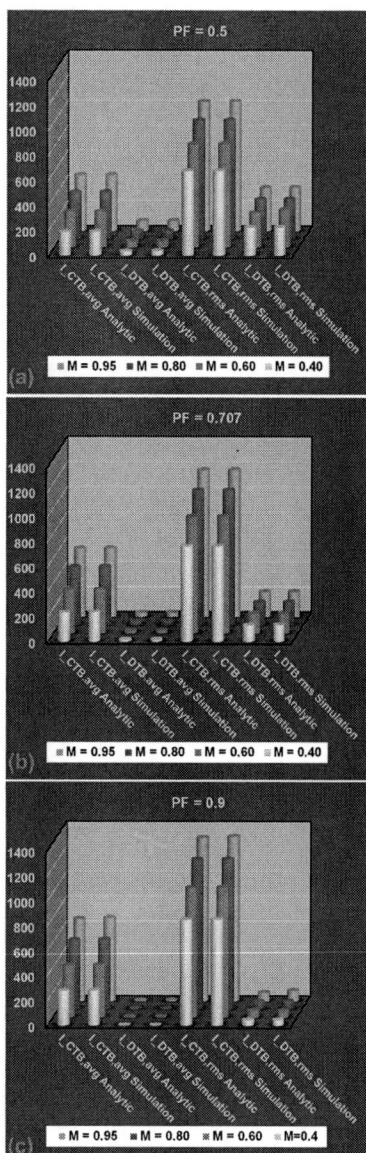

Fig. 4. Comparison between numerical computation and simulation results of the average and rms currents of the top/bottom LF IGBTs and diodes for ANPC converter for power factors (PF) of: (a) 0.5 (b) 0.707 (c) 0.9.

TABLE I
Error (%) between the numerical computations and experimental measurements of the average and rms currents of IGBTs and diodes in the 1.5 kW five-level single-phase ANPC converter

PF	0.26	0.48	0.63	0.73	0.82	0.9
$I_{CHF,avg}$	2.52 (1.67)	2.47 (1.69)	2.29 (1.72)	2.07 (1.75)	1.79 (1.81)	1.40 (1.91)
$I_{DHF,avg}$	2.02 (1.24)	1.65 (1.28)	1.34 (1.24)	1.10 (1.85)	0.88 (1.75)	0.63 (1.71)
$I_{CTB,avg}$	1.48 (1.27)	1.74 (1.14)	1.80 (1.76)	1.75 (1.49)	1.60 (2.07)	1.31 (1.24)
$I_{DTB,avg}$	0.63 (2.39)	0.33 (2.23)	0.17 (2.68)	0.09 (2.69)	0.04 (2.35)	0.01 (2.75)
$I_{CM,avg}$	1.39 (1.34)	1.31 (1.56)	1.16 (1.34)	1.01 (1.18)	0.84 (1.85)	0.62 (1.61)
$I_{DM,avg}$	1.03 (1.17)	0.73 (1.38)	0.49 (2.16)	0.32 (1.95)	0.19 (1.25)	0.08 (1.56)
$I_{CHF,rms}$	5.22 (1.29)	4.91 (1.23)	4.46 (1.88)	3.99 (1.45)	3.45 (1.38)	2.69 (1.92)
$I_{DHF,rms}$	4.86 (1.96)	4.24 (1.57)	3.57 (1.53)	2.98 (1.86)	2.41 (1.52)	1.73 (1.94)
$I_{CTB,rms}$	3.94 (1.67)	4.20 (2.18)	4.08 (1.95)	3.79 (2.14)	3.35 (2.35)	2.66 (2.53)
$I_{DTB,rms}$	2.29 (2.27)	1.47 (2.48)	0.91 (2.34)	0.56 (2.25)	0.32 (2.83)	0.13 (2.77)
$I_{CM,rms}$	4.29 (2.14)	3.97 (1.55)	3.45 (1.41)	2.93 (1.15)	2.39 (1.41)	1.72 (1.26)
$I_{DM,rms}$	3.42 (2.91)	2.55 (2.64)	1.80 (2.22)	1.25 (2.25)	0.82 (2.54)	0.41 (2.12)

of the experiments, the load inductance is kept constant while the resistance is changed to attain the required power factors mentioned in the test. In Table I, all notations of the switches and diodes are based on Fig. 1. As it is obvious, the maximum error is roughly 3% which sounds reasonable and acceptable for experimental verification. The obtained errors can be due to the miscalculation in the oscilloscope computation, variation of the load impedance under test, and/or interference of the noise with current probe measured values.

V. Conduction Power Loss Investigation

After verifying the derived equations to calculate the average and rms current of IGBTs and diodes in ANPC converter through simulation and experimental results, these closed-from equations are utilized to calculate the conduction power losses of IGBTs, diodes, and switching-power-cells as well as the whole three-phase ANPC converter. The obtained results as a function of converter modulation index and load power factor for a 12.4MVA 3.3kV 9-level (line-to-line) ANPC converter are shown in various cases in Figs. 5- 9.

In Figs. 5- 6, load impedance is considered constant; hence, its current varies linearly with converter modulation index while load peak current in Figs. 7- 9 is considered constant. It can be pointed out from Fig. 5 that by increasing the converter modulation index the conduction power loss of all IGBTs and anti-parallel diodes increase. Moreover, it can be concluded that by increasing load power factor, conduction power loss of high-frequency and top/bottom low-frequency IGBTs increase whereas increasing load power factor decreases

through IGBTs and anti-parallel diodes, experimental tests are conducted for different load power factors. The analytic results and the associated errors expressed in percentage (numbers in parenthesis) between analytically calculated values and experimental measurements for each case are listed in Table I while demonstrating a good accuracy for derived closed-form equations. For experimental measurements, the converter modulation index (M) is set at 0.9 whereas load power factor varies from 0.26 to 0.90 (6 different cases). At power factor of 0.9, the resistiveinductive load is 38Ω-58mH. For the rest

978-1-4673-9551-9/16 $31.00 © 2016 IEEE

conduction power losses in high-frequency and top/bottom low-frequency anti-parallel diodes. The same phenomenon occurs in the middle low-frequency IGBTs and anti-parallel diodes. According to Fig. 6, it is worth mentioning that increasing load power factor does not have a significant impact on the conduction power loss in both of the high and low-frequency switching-power-cells. Herein, one high-frequency switching-power-cell contains two complementary high-frequency IGBTs and their anti-parallel diodes, and the low-frequency switching-power-cell (which is only one) contains all low-frequency IGBTs and their anti-parallel diodes.

Cell components conduction power loss versus modulation index and power factor in a 12.4MVA 3.3kV ANPC converter considering constant load current are depicted in Figs. 7- 9. Based on these figures, the low-frequency IGBTs and switching-power-cells possess more conduction losses as compared to the high-frequency IGBTs and power-cells. According to Fig. 9(c), the maximum conduction power loss of the 12.4MVA 3.3kV three-phase 9-level ANPC occurs for almost purely resistive load and its value is roughly 81 kW. Even though this amount of power loss produces a lot of heat and its dissipation needs significant heat sink with advanced cooling system (i.e., water cooling system), this is negligible as compared to the converter output power which is 12.4 MVA. In other words, total conduction power loss of the converter is 0.653% of the output power.

VI. CONCLUSION

This paper presented closed-form solutions to calculate rms and average currents of IGBTs/diodes in terms of converter modulation index, load current, and load power factor for ANPC converters. Numerical results of the derived analytical equations for calculation of rms and average currents were compared with simulation and experimental results. The analytic, simulation, and experimental results matched perfectly which substantiated derived closed-form solutions. Afterwards, obtained equations for calculation of rms and average currents were utilized to calculate conduction power losses.

REFERENCES

[1] D. Andler, R. Alvarez, S. Bernet, and J. Rodriguez, "Switching loss analysis of 4.5-kv-5.5-ka igcts within a 3l-anpc phase leg prototype," *Industry Applications, IEEE Transactions on*, vol. 50, no. 1, pp. 584–592, Jan 2014.

[2] ——, "Experimental investigation of the commutations of a 3l-anpc phase leg using 4.5-kv-5.5-ka igcts," *Industrial Electronics, IEEE Transactions on*, vol. 60, no. 11, pp. 4820–4830, Nov 2013.

[3] P. Alemi and D.-C. Lee, "Power loss comparison in two- and three-level pwm converters," in *Power Electronics and ECCE Asia (ICPE ECCE), 2011 IEEE 8th International Conference on*, May 2011, pp. 1452–1457.

[4] Q. Tu and Z. Xu, "Power losses evaluation for modular multilevel converter with junction temperature feedback," in *Power and Energy Society General Meeting, 2011 IEEE*, July 2011, pp. 1–7.

[5] O. Senturk, L. Helle, S. Munk-Nielsen, P. Rodriguez, and R. Teodorescu, "Power capability investigation based on electrothermal models of press-pack igbt three-level npc and anpc vscs for multimegawatt wind turbines," *Power Electronics, IEEE Transactions on*, vol. 27, no. 7, pp. 3195–3206, July 2012.

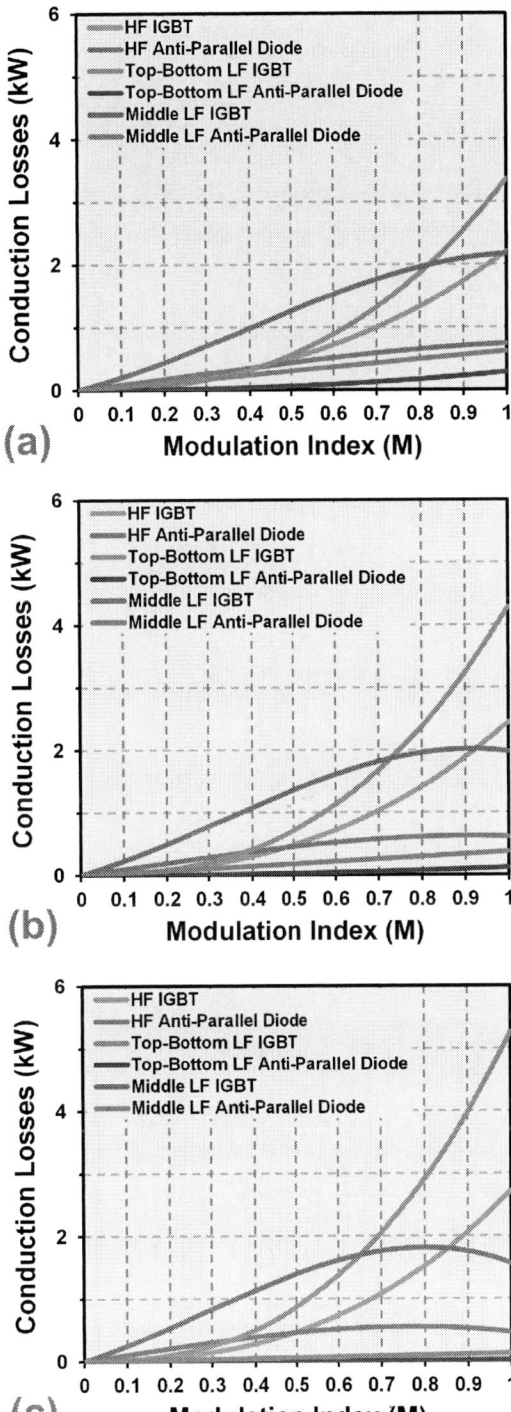

(a)

(b)

(c)

Fig. 5. High-frequency and low-frequency IGBTs/diodes power loss versus modulation index in a 12.4MVA 3.3kV 9-level (line-to-line) ANPC converter considering constant load impedance for load power factors of: (a) 0.5; (b) 0.707; (c) 0.9.

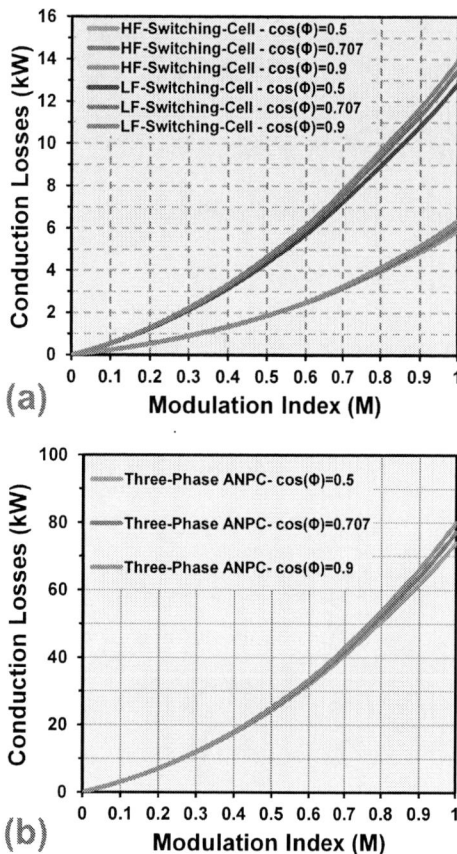

(a)

(b)

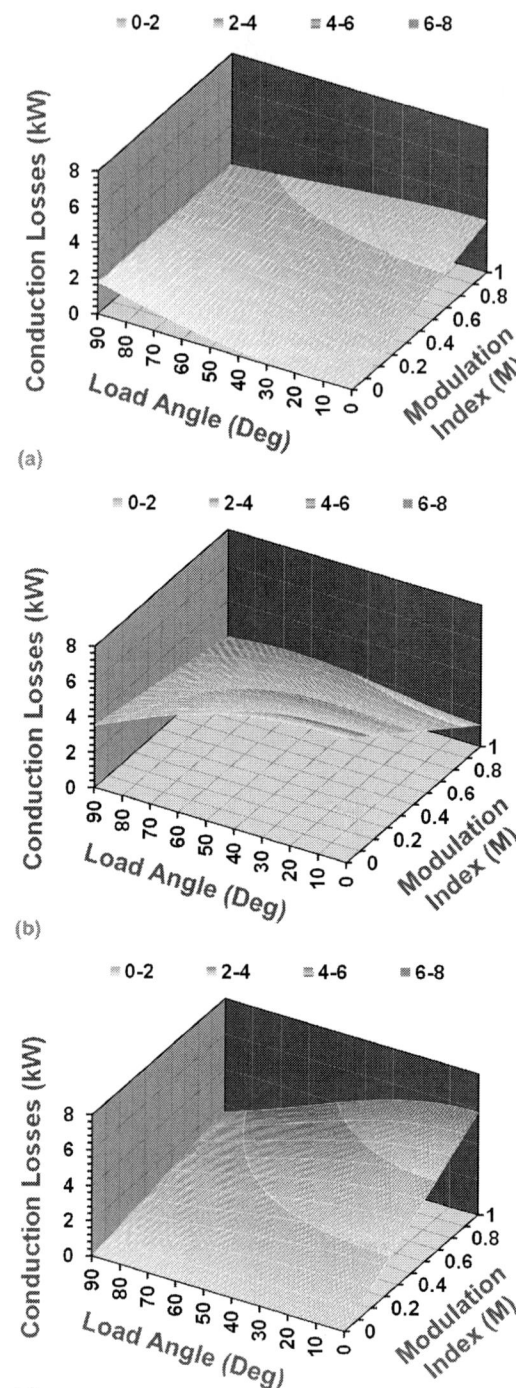

(a)

(b)

(c)

Fig. 6. Conduction power loss versus modulation index in a 12.4MVA 3.3kV 9-level (line-to-line) ANPC converter considering constant load impedance: (a) one high-frequency and one low-frequency switching-power-cells conduction losses; (b) total conduction power loss of the three-phase ANPC converter.

Fig. 7. Cell components conduction power loss versus modulation index and power factor in a 12.4MVA 3.3kV 9-level (line-to-line) ANPC converter considering constant load current (Ip=3062A): (a) high-frequency IGBT conduction power loss (b) middle low-frequency IGBT conduction power loss (c) top/bottom low-frequency IGBT conduction power loss.

[6] S. Rohner, S. Bernet, M. Hiller, and R. Sommer, "Modulation, losses, and semiconductor requirements of modular multilevel converters," *Industrial Electronics, IEEE Transactions on*, vol. 57, no. 8, pp. 2633–2642, Aug 2010.

[7] M. Schweizer and J. Kolar, "Design and implementation of a highly efficient three-level t-type converter for low-voltage applications," *Power Electronics, IEEE Transactions on*, vol. 28, no. 2, pp. 899–907, Feb 2013.

[8] S. Rohner, S. Bernet, M. Hiller, and R. Sommer, "Modulation, losses, and semiconductor requirements of modular multilevel converters," *Industrial Electronics, IEEE Transactions on*, vol. 57, no. 8, pp. 2633–2642, Aug 2010.

[9] A. Bazzi, P. Krein, J. Kimball, and K. Kepley, "Igbt and diode loss estimation under hysteresis switching," *Power Electronics, IEEE Transactions on*, vol. 27, no. 3, pp. 1044–1048, March 2012.

[10] F. Cazakevicius, R. Krug, H. Figueira, R. Beltrame, and H. Hey, "Loss and thermal analysis of semiconductor devices applied to an electric circuit simulator," in *Power Electronics Conference (COBEP), 2011 Brazilian*, Sept 2011, pp. 1050–1055.

[11] Z. Zhang, Z. Xu, and Y. Xue, "Valve losses evaluation based on piecewise analytical method for mmc-hvdc links," *Power Delivery, IEEE Transactions on*, vol. 29, no. 3, pp. 1354–1362, June 2014.

[12] D. Graovac, M. Pursche, and A. Kniep, "Mosfet power losses calculation using the data-sheet parameters," *Infineon Technologies*, 2008.

978-1-4673-9551-9/16 $31.00 © 2016 IEEE

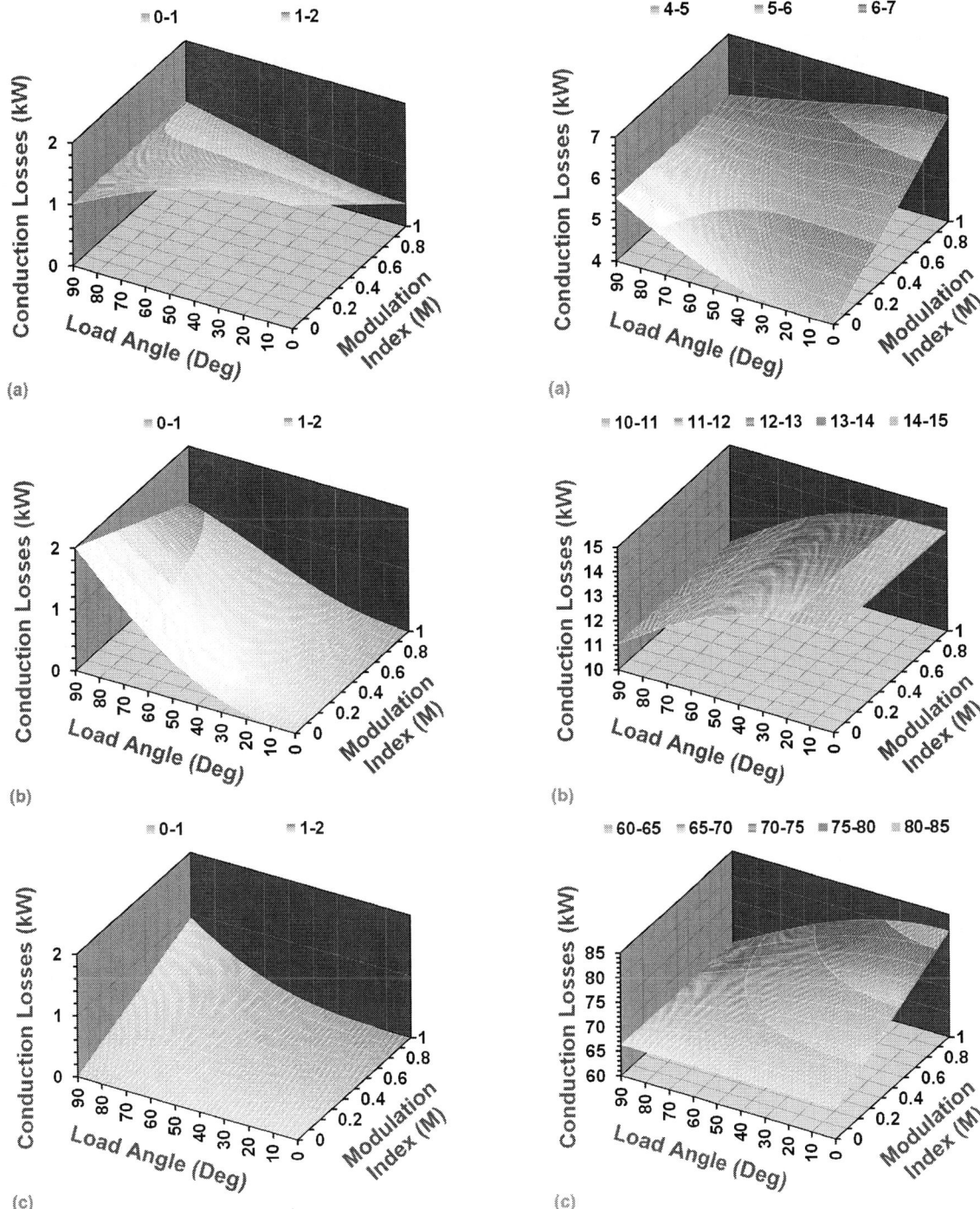

Fig. 8. Cell components conduction power loss versus modulation index and power factor in a 12.4MVA 3.3kV 9-level (line-to-line) ANPC converter considering constant load current (Ip=3062A): (a) high-frequency anti-parallel diode conduction power loss (b) middle low-frequency anti-parallel diode conduction power loss (c) top/bottom low-frequency anti-parallel diode conduction power loss.

Fig. 9. Cell components conduction power loss versus modulation index and power factor in a 12.4MVA 3.3kV 9-level (line-to-line) ANPC converter considering constant load current (Ip=3062A): (a) one high-frequency switching-power-cell conduction loss; (b) one low-frequency switching-power-cell conduction loss; (c) total conduction power loss of a three-phase ANPC converter.

Multifrequency Small-Signal Model of Voltage Source Converters Connected to a Weak Grid for Stability Analysis

Xing Li, Hua Lin

State Key Laboratory of Advanced Electromagnetic Engineering and Technology
Huazhong University of Science and Technology
Wuhan, China
hust_lx@hust.edu.cn, lhua@mail.hust.edu.cn

Abstract—**To investigate the instability issues of three-phase voltage source converters (VSCs) connected to a weak grid, this paper develops a modeling method based on the harmonic-balance approach. The nonlinear coordinate transformation in the control system would generate two sideband components. If the grid is not strong while the bandwidth of phase-locked loop (PLL) is high enough, the sideband components would be coupled, and both of them have influence on system stability, neither of them could be neglected. Hence, there would be two feedback loops of grid-current representing two sideband frequencies. Taking this effect into account, the multifrequency small-signal model is proposed. It is further demonstrated that the sum of the two grid-current loop gains determines the stability and transient performance. Simulations and experimental results were conducted to validate the analysis.**

Keywords—LCL filter; grid-connected converter; system stability; weak grid; PLL

I. INTRODUCTION

With the increasing demand of renewable energy sources in distributed power generation systems (DPGSs), as the interface between DPGSs and a power grid, voltage source converters (VSCs) are playing an increasingly important role in injecting high-quality power into the grid [1]. The design of the control loop of VSCs normally assumes an ideal AC voltage sources. However, a significant grid impedance exists in actual AC grid, especially in weak systems [2]. The actual grid impedance would affect the operation of the VSC and even destabilize the converter-grid system [3].

Much research efforts have been devoted to deal with the stability issues of the grid-connected system. One method is by directly applying the Nyquist stability criterion [4] to the loop gain of the system. In [5], [6] the grid impedance is regarded as part of the LCL filter of the VSC, and the averaged switch model is derived to obtain the loop gain. However, the impact of the phase-locked loop (PLL) on the stability of the VSC is not considered. The PLL was discovered to have a negative impact on system stability that a PLL with high bandwidth could destabilize the VSC connected to a weak grid [7]-[9].

The most popular method is the impedance-based method [10], [11] which takes the ratio between the grid impedance and the VSC impedance as the loop gain of the converter-grid system. The methods to model VSC impedance can be classified into two. One is modeling the converter in the *dq*-reference frame to obtain the VSC impedance matrix [12]-[15]. However, it is difficult to interpret and measure impedance responses in the *dq*-coordinate system [16]. The other is harmonic linearization method [17], [18] which models the VSC impedance directly in the phase domain and the obtained VSC impedance has clear physical interpretations and can be directly measured [16].

In this paper, we attempt to address the stability issues of the weak grid connected VSCs using the impedance model in [17]. However, it is found out that though the impedance model is rather precise, it failed to determine the system stability using the impedance ratios, especially when the bandwidth of the system is high. Instead, this paper applies the harmonic linearization technique to develop a multi-frequency model to obtain the system loop gain for stability analysis of the three-phase LCL-type grid-connected converters. The rest of this paper is organized as follows: Section II describes the system. Section III introduces the multifrequency model to predict the system behavior. Section IV includes verifications of the proposed model from its application in analysis of harmonic resonance. Section V concludes this paper.

II. SYSTEM DESCRIPTION

Fig. 1 shows the configuration of a voltage source converter feeding into the grid through an LCL filter. The converter is assumed to be supplied by a stiff dc source, hence the dynamic property of the dc input is neglected. L_1 is the converter-side inductor, C is the filter capacitor, L_2 is the grid-side inductor, and L_g is the grid inductor. The parasitic resistance of the circuit is neglected to represent the worst case.

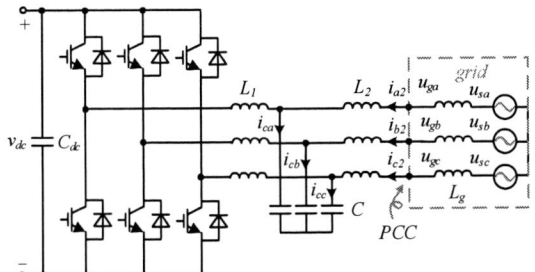

Fig. 1. Three-phase voltage source converter connecting to a nonideal grid.

This paper investigates the capacitor-current-feedback active damping with a $\alpha\beta$-domain current controller for the digitally controlled LCL-type grid-connected converter, as depicted in Fig. 2 where the influence of PLL is taken into consideration. By assuming a balanced three phase grid voltage, the current control scheme can be applied at a per-phase basis. In part (a), the grid-current references in $\alpha\beta$-frame($i_{\alpha,ref}$, $i_{\beta,ref}$) is obtained from the references in dq-frame (I_{dr}, I_{qr}), as depicted in equation (1). Part (b) is the grid-current loop, and controller $G_i(s)$ used is the proportional-resonant (PR) controller, K is the capacitor-current-feedback coefficient, $G_d(s)$ depicts the effect of the digital computation delay and the pulse width modulation (PWM) delay [19]. The close loop gain of the grid-current is given by (3). The PLL strategy in synchronous reference (d-q) frame (SRF) [20] is considered, and $G_{pll}(s)$ is a PI controller, as depicted in part (c). The grid impedance $Z_g(s)=L_g s$, and $G_v(s)$ models the delay caused by the Hall voltage sensor. Due to the high cutoff frequency considered in this paper, the delay caused by the ADC prefilter is neglected.

$$\begin{bmatrix} i_{\alpha,ref}(t) \\ i_{\beta,ref}(t) \end{bmatrix} = T(\theta(t)) \begin{bmatrix} I_{dr} \\ I_{qr} \end{bmatrix} \tag{1}$$

where the transformation from dq-frame to the stationary $\alpha\beta$-frame is defined as follows:

$$T(\theta(t)) = \begin{bmatrix} \cos\theta(t) & -\sin\theta(t) \\ \sin\theta(t) & \cos\theta(t) \end{bmatrix} \tag{2}$$

$$T_i(s) = \frac{i_{2\alpha}}{i_{\alpha,ref}} = \frac{G_i(s)G_d(s)}{L_t L_1 C s^3 + (L_1 + L_t)s + (KL_t C s^2 + G_i(s))G_d(s)} \tag{3}$$

where the grid inductor is considered to be part of the grid side inductor of the LCL filter, and $L_t = L_2 + L_g$.

III. MULTIFREQUENCY SMALL-SIGNAL MODEL

A small signal perturbation is added to the input angular angle of the transformation (1) as depicted in Fig. 2, and its time domain expression can be written as

$$\theta(t) = \theta_1(t) + \Delta\theta(t) = \theta_1(t) + \Delta\theta\cos(\omega_p t + \varphi) \tag{4}$$

where $\Delta\theta$, φ are the magnitude and phase of the perturbation at angular frequency ω_p. Under the steady state condition, $\theta(t)=\theta_1(t)=\omega_1 t$, ω_1 is the angular frequency of the grid. In the small-signal analysis, it is assumed that the perturbation is small enough that will not change the operating point of the converter [21], hence the trigonometric functions of $\Delta\theta(t)$ can be simplified as follows:

$$\begin{cases} \sin\Delta\theta(t) \approx \Delta\theta(t) \\ \cos\Delta\theta(t) \approx 1 \end{cases} \tag{5}$$

When analyzing the small-signal stability and the transient performance of a converter, only the consequence of the perturbation, i.e. the components at the perturbation frequency and the sideband frequencies, needs to be considered in the models [21]. In the frequency domain, the perturbation part $\theta(t)$ can be written as follows:

$$\Delta\theta[\pm\omega_p] = \frac{\Delta\theta}{2}e^{j\varphi} \tag{6}$$

It is in fact the Fourier coefficient of $\Delta\theta(t)$. The negative angular frequency component has no physical meaning and is neglected in the paper.

A. Model of the grid-current reference in $\alpha\beta$-domain

In order to deal with the nonlinearity in coordinate transformation, $T(\theta(t))$ can be broken into two parts as follows:

$$T(\theta(t)) = T(\theta_1(t)) \begin{bmatrix} 1 & -\Delta\theta(t) \\ \Delta\theta(t) & 1 \end{bmatrix} \tag{7}$$

where a small angle approximation of trigonometric functions in (5) is applied. Substituting (7) into (1), and neglecting second order perturbation, the perturbation part of grid-current reference in $\alpha\beta$-frame in the time domain are calculated as follows:

$$\begin{bmatrix} \Delta i_{\alpha,ref}(t) \\ \Delta i_{\beta,ref}(t) \end{bmatrix} = T(\theta_1(t)) \begin{bmatrix} -I_{qr} \\ I_{dr} \end{bmatrix} \Delta\theta(t) \tag{8}$$

Applying convolution to (8), the expression of $\Delta i_{\alpha,ref}$ and $\Delta i_{\beta,ref}$ in the frequency domain can be written as:

$$\Delta i_{\alpha,ref}[\omega_p \pm \omega_1] = \pm \frac{jI_{dr} \mp I_{qr}}{2}\Delta\theta[\omega_p] \tag{9}$$

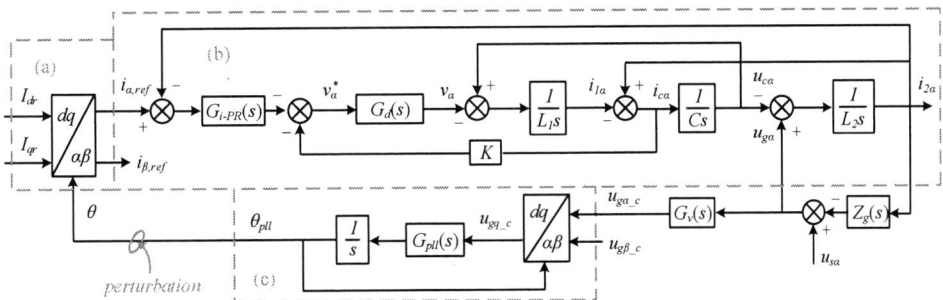

Fig. 2. Controller schematic of the three-phase VSC in a stationary $\alpha\beta$ reference frame.

978-1-4673-9551-9/16 $31.00 © 2016 IEEE

$$\Delta i_{\beta,ref}[\omega_p \pm \omega_1] = \frac{I_{dr} \pm jI_{qr}}{2}\Delta\theta[\omega_p] \qquad (10)$$

It can be obtained from (9) and (10) that the relationship between $\Delta i_{\alpha,ref}$ and $\Delta i_{\beta,ref}$ is

$$\Delta i_{\alpha,ref}[\omega_p \pm \omega_1] = \pm j\Delta i_{\beta,ref}[\omega_p \pm \omega_1] \qquad (11)$$

B. Frequency model of grid-current loop

The current feedback loop is shown in the dashed box (b) of Fig. 2. Based on the control block, the grid current can be obtained:

$$\Delta i_{2\alpha}[\omega_p \pm \omega_1] = T_i(j\omega_p \pm j\omega_1)\Delta i_{\alpha,ref}[\omega_p \pm \omega_1] \qquad (12)$$

And the PCC voltage is

$$\Delta u_{g\alpha}[\omega_p \pm \omega_1] = -Z_g(j\omega_p \pm j\omega_1)\Delta i_{2\alpha}[\omega_p \pm \omega_1] \qquad (13)$$

By taking the time delay caused by the Hall voltage sensor into consideration, the PCC voltage sent to the PLL is

$$\Delta u_{g\alpha\_c}[\omega_p \pm \omega_1] = \Delta u_{g\alpha}[\omega_p \pm \omega_1]G_v(j\omega_p \pm j\omega_1) \qquad (14)$$

The current loop is linear, hence the input voltage of the PLL have the similar relationship as shown in (11), that is

$$\Delta u_{g\alpha\_c}[\omega_p \pm \omega_1] = \pm ju_{g\beta\_c}[\omega_p \pm \omega_1] \qquad (15)$$

C. Frequency model of PLL

The control strategy of the PLL is shown in the dashed box (c) of Fig. 2. The outputs of the coordinate transformation in the PLL is

$$\begin{bmatrix} u_{gd\_c}(t) \\ u_{gq\_c}(t) \end{bmatrix} = T^{-1}(\theta_{PLL}(t))\begin{bmatrix} u_{g\alpha\_c}(t) \\ u_{g\beta\_c}(t) \end{bmatrix} \qquad (16)$$

where $\theta_{PLL}(t)$ is the output angle of the PLL and $T^{-1}(\theta(t))$ is the inverse matrix of $T(\theta(t))$ which can be separated into two part as follows:

$$T^{-1}(\theta(t)) = T^{-1}(\theta_1(t))\begin{bmatrix} 1 & \Delta\theta(t) \\ -\Delta\theta(t) & 1 \end{bmatrix} \qquad (17)$$

where a small angle approximation of trigonometric functions in (5) is applied. Substituting (17) into (16), and neglecting second and higher order perturbation, it can be obtained that:

TABLE I
PARAMETERS OF GRID-TIED CONVERTER PROTOTYPE

Symbol	Description	Value
U_{gd}	Grid phase-neutral peak voltage	$30\sqrt{2}$V
ω_1	Grid angular frequency	$2\pi\times50$rad/s
f_s	Switching frequency	10kHz
v_{dc}	Bus-voltage	120V
L_2	grid-side inductor of LCL	0.8mH
L_1	converter-side inductor of LCL	1.2mH
C	Capacitor of the LCL	18uF
K	Coefficient of capacitor current feedback	6
L_g	Grid inductor	0.6mH
ω_v	cutoff frequency of Hall transducer	20000rad/s
k_{ip}	Proportional gain of ac/dc current controller	9.6
k_{ir}	R parameter of ac/dc current controller	210
I_{dr}	D channel current reference of VSC	-7.3A
I_{qr}	Q channel current reference of VSC	0A
k_{ppll}	Proportional gain of PLL controller	48
k_{ipll}	Integrator gain of PLL controller	51776

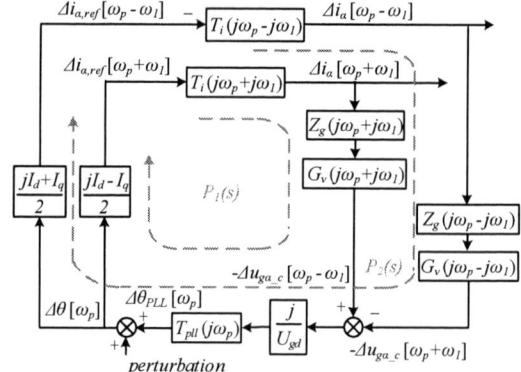

Fig. 3. The multi-frequency model of the grid-connected converter.

$$\begin{bmatrix} \Delta u_{gd\_c}(t) \\ \Delta u_{gq\_c}(t) \end{bmatrix} = T^{-1}(\theta_{PLL}(t))\begin{bmatrix} \Delta u_{g\alpha\_c}(t) \\ \Delta u_{g\beta\_c}(t) \end{bmatrix} + \begin{bmatrix} U_{gq} \\ -U_{gd} \end{bmatrix}\Delta\theta_{PLL}(t) \qquad (18)$$

where U_{gd}, $U_{gq}=0$ are the steady state values of PCC voltage in dq reference frame, respectively.

Applying convolution to (18), and from (15) the expression of Δu_{gq_c} in the frequency domain can be written as:

$$\Delta u_{gq\_c}[\omega_p] = j\Delta u_{g\alpha\_c}[\omega_p - \omega_1] - j\Delta u_{g\alpha\_c}[\omega_p + \omega_1] \\ - U_{gd}\Delta\theta_{PLL}[\omega_p] \qquad (19)$$

Note that the perturbation part of the output of PLL

$$\Delta\theta_{PLL}[\omega_p] = \Delta u_{gq\_c}[\omega_p]H_{pll}(j\omega_p) \qquad (20)$$

where the controller of the PLL $H_{pll}(j\omega_p)=G_{pll}(j\omega_p)/j\omega_p$; then, from (19) and (20), we can obtain that:

$$\Delta\theta_{PLL}[\omega_p] = \frac{jT_{pll}(j\omega_p)}{U_{gd}}(\Delta u_{g\alpha\_c}[\omega_p - \omega_1] - \Delta u_{g\alpha\_c}[\omega_p + \omega_1]) \qquad (21)$$

where $T_{pll}(j\omega_p)$ is the closed-loop gain of the PLL defined as

$$T_{pll}(j\omega_p) = \frac{U_{gd}H_{pll}(j\omega_p)}{1 + U_{gd}H_{pll}(j\omega_p)} \qquad (22)$$

With (9), (12), (13), (14), and (21), the proposed multi-frequency model of the grid-connected converter is illustrated in Fig. 3. In this model, there are two feedback loops representing the two sideband frequencies. It is calculated from the model that the sum of the loop gain of the two feedback loops is

$$P(s) = -\frac{\Delta\theta_{PLL}[\omega_p]}{\Delta\theta[\omega_p]} = P_1(s) + P_2(s) = \frac{L_gT_{pll}(s)}{2U_{gd}} \cdot$$

$$\sum_{k=-1,1}(I_{dr} + jkI_{qr})(s + jk\omega_1)T_i(s + jk\omega_1)G_v(s + jk\omega_1) \qquad (23)$$

that determines the stability and transient performance of the converter-grid system, where $s=j\omega_p$.

If the grid is strong, the grid impedance could be neglected, thus the two feedback loops are both cutoff. On the other hand, if the grid is weak, these two feedback loops are coupled with each other through the PLL, hence the stability of the system cannot be determined by only one of the two loops. $T_{pll}(s)$ is

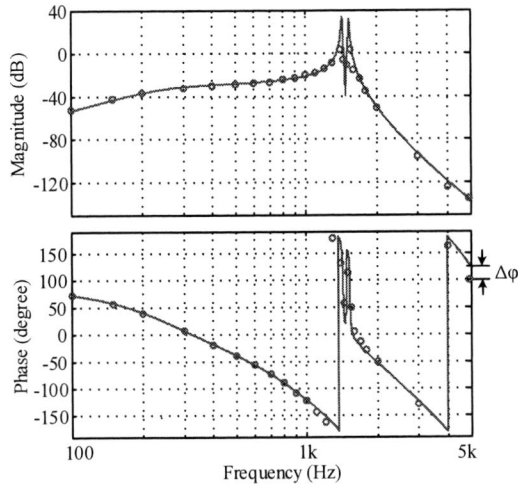

Fig. 4. Bode plot of the frequency response of the loop gain. Solid line: model prediction; Dots: Numerical simulation results.

very small for the frequency range higher than the PLL's bandwidth, hence if the bandwidth is low, $T_{pll}(s)$ could be neglected which means the two feedback loops no longer exist. Under these circumstances, the converter-grid system is stable as long as the current loop and the PLL are stable.

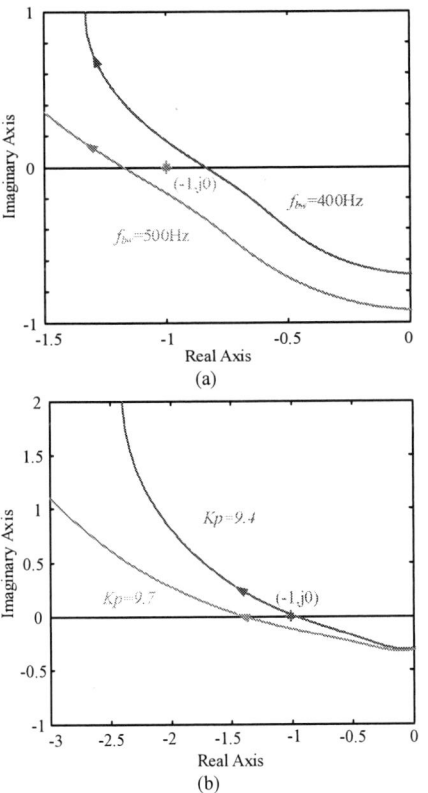

Fig. 5. Nyquist curve of $P(s)$ around (-1,j0) with: (a) different bandwidth of PLL; (b) changing proportional gain of the grid current controller.

Up to half of the switching frequency, Fig. 4 compares the loop gain in the proposed model and in the numerical simulation. The values of the circuit components used in the simulation are shown in Table I, with the bandwidth of the PLL f_{bw}=500Hz. The power flows from the converter to the grid. As depicted in Fig. 4, the proposed model successfully predicts two peaks of magnitude of the loop gain in the switching model simulation. At around half of the switching frequency, there is a $\Delta\varphi$-gap between the phase of $P(s)$ and the numerical simulation results. It is caused by the nonlinear pulse-width modulator. The nonlinearity of pulse-width modulator only affects the frequency region around or higher than half of the switching frequency [21]. However, harmonic resonance rarely happens at this frequency region because the magnitude of $P(s)$ here is much smaller than one. Thus, the nonlinearity of pulse-width modulator is not considered in this paper.

IV. MODEL VERIFICATION

A three-phase grid-connected converter with $\alpha\beta$-domain controller has been built and tested to verify the proposed multi-frequency model. The current controller was implemented in a Texas Instrument TMS320F28335 DSP board. Parameters for this experimental setup are provided in Table I, and the converter works as a rectifier with K=4.5. If the bandwidth of the PLL f_{bw}=400Hz, k_{ppll}=38, k_{ipll}=33136.

The Nyquist curves of $P(s)$ around (-1,j0) with different bandwidth of PLL are plotted in Fig. 5(a), as an example. The corresponding time-domain waveform is shown in Fig. 6(a). At

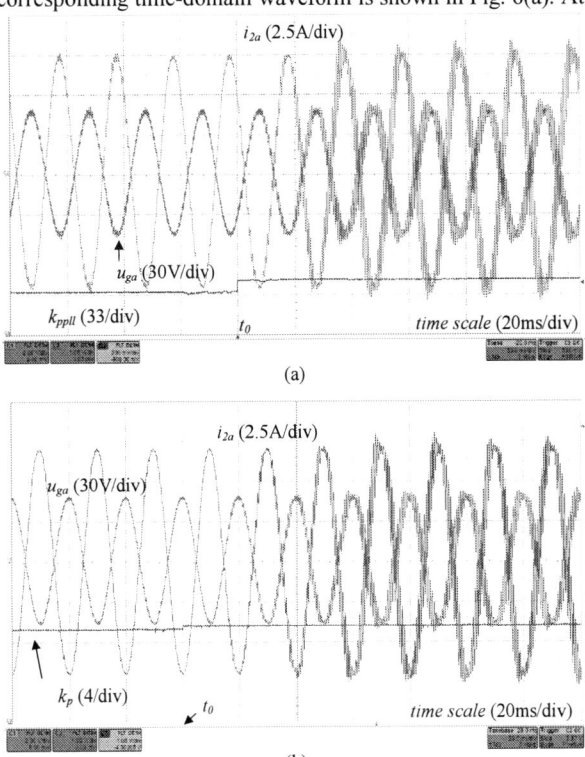

Fig. 6. Waveform of A-phase grid current and PCC voltage with: (a) f_{bw} jump from 400Hz to 500Hz at time t_0; (b) K_p jumps from 9.4 to 9.7 with fbw=500Hz.

time t_0, the f_{bw} jumps from 400Hz to 500 Hz and the grid current is no longer stable. When f_{bw} is set to be 500Hz, the Nyquist curve of $P(s)$ encloses (-1,j0) indicating that harmonic resonance would occur in the system. This explains the harmonic resonance in Fig. 6(a). Reducing the bandwidth of PLL is an effective way to eliminate harmonic resonance.

The Nyquist curves of $P(s)$ around (-1,j0) with different proportional gain of the grid current controller are plotted in Fig. 5(b). At time t_0, as shown in Fig. 6(b), the K_p jumps from 9.4 to 9.7 and the grid current is no longer stable. When K_p is set to be 9.7, the Nyquist curve of $P(s)$ encloses (-1,j0), which would cause harmonic resonance in the weak grid connected system. Reducing the bandwidth of grid current loop is also an effective way to eliminate harmonic resonance.

V. CONCLUSION

Three-phase LCL-type VSCs connected to a weak power grid are studied in this paper. The significant impedance present in the grid could drive the converter to enter an instability region. This paper points out that the overall system stability should be reconsidered in the light of a more complete model that takes into account sideband effect caused by the nonlinear coordinate transformation when the effect of the PLL is considered. To describe this effect and improve the design, this paper introduces a multi-frequency mode using harmonic linearization method. Key parameters of the grid-connected converter including the PLL parameters that would affect the system's stability significantly could be discussed and identified based on the proposed model in the near future.

REFERENCES

[1] F. Blaabjerg, R. Teodorescu, M. Liserre, and A. V. Timbus, "Overview of control and grid synchronization for distributed power generation systems," *IEEE Trans. Ind. Electron.*, vol. 53, no. 5, pp. 1398–1409, Oct. 2006.

[2] Y. A.-R. I. Mohamed, "Mitigation of converter-grid resonance, grid-induced distortion, and parametric instabilities in converter-based distributed," *IEEE Trans. Power Electron.*, vol. 26, no. 3, pp. 983–996, Mar. 2011.

[3] M. Liserre, R. Teodorescu, and F. Blaabjerg, "Stability of photovoltaic and wind turbine grid-connected inverters for a large set of grid impedance values," *IEEE Trans. Power Electron.*, vol. 21, pp. 263–272, Jan. 2006.

[4] R. D. Middlebrook, "Input filter considerations in design and application of switching regulators," in *Proc. IEEE Ind. Applicat. Soc. Conf.*, Oct. 1976, pp. 94–107.

[5] D. Pan, X. Ruan, C. Bao, "Capacitor-current-feedback active damping with reduced computation delay for improving robustness of LCL-type grid-connected inverter," *IEEE Trans. Power Electron.*, vol. 29, pp. 3414–3427, July 2014.

[6] M. Liserre, R. Teodorescu, and F. Blaabjerg, "Stability of photovoltaic and wind turbine grid-connected inverters for a large set of grid impedance values," *IEEE Trans. Power Electron.*, vol. 21, pp. 263-272, 2006.

[7] K. M. Alawasa, Y. A. R. I. Mohamed, and W. Xu, "Active Mitigation of Subsynchronous Interactions Between PWM Voltage-Source Converters and Power Networks," *IEEE Trans. Power Electron.*, vol. 29, pp. 121-134, 2014.

[8] L. Harnefors, M. Bongiorno, and S. Lundberg, "Input-Admittance Calculation and Shaping for Controlled Voltage-Source Converters," *IEEE Trans. Ind. Electron.*, vol. 54, pp. 3323-3334, 2007.

[9] J. Z. Zhou, D. Hui, F. Shengtao, Z. Yi, and A. M. Gole, "Impact of Short-Circuit Ratio and Phase-Locked-Loop Parameters on the Small-Signal Behavior of a VSC-HVDC Converter," *IEEE Trans. Power Delivery*, vol. 29, pp. 2287-2296, 2014.

[10] W. Xiongfei, F. Blaabjerg, M. Liserre, C. Zhe, H. Jinwei, and L. Yunwei, "An Active Damper for Stabilizing Power-Electronics-Based AC Systems," *IEEE Trans. Power Electron.*, vol. 29, pp. 3318-3329, 2014.

[11] S. Jian, "Impedance-Based Stability Criterion for Grid-Connected Inverters," *IEEE Trans. Power Electron.*, vol. 26, pp. 3075-3078, 2011.

[12] W. Cheng, H. Meng, C. K. Tse, and R. Xinbo, "Effects of Interaction of Power Converters Coupled via Power Grid: A Design-Oriented Study," *IEEE Trans. Power Electron.*, vol. 30, pp. 3589-3600, 2015.

[13] M. Belkhayat, *Stability criteria for AC power systems with regulated loads*, Ph.D. dissertation, Purdue University, West Lafayette, IN, USA, Dec. 1997.

[14] W. Bo, D. Boroyevich, R. Burgos, P. Mattavelli, and S. Zhiyu, "Analysis of D-Q Small-Signal Impedance of Grid-Tied Inverters," *IEEE Trans. Power Electron.*, vol. 31, pp. 675-687, 2016.

[15] W. Bo, D. Dong, D. Boroyevich, R. Burgos, P. Mattavelli, and S. Zhiyu, "Impedance-Based Analysis of Grid-Synchronization Stability for Three-Phase Paralleled Converters," *IEEE Trans. Power Electron.*, vol. 31, pp. 26-38, 2016.

[16] J. Sun, "Small-signal methods for ac distributed power systems—A Review," *IEEE Trans. Power Electron.*, vol. 24, no. 11, pp. 2545–2554, Nov. 2009.

[17] M. Cespedes and S. Jian, "Impedance Modeling and Analysis of Grid-Connected Voltage-Source Converters," *IEEE Trans. Power Electron.*, vol. 29, pp. 1254-1261, 2014.

[18] S. Jian, B. Zhonghui, and K. J. Karimi, "Input Impedance Modeling of Multipulse Rectifiers by Harmonic Linearization," *IEEE Trans. Power Electron.*, vol. 24, pp. 2812-2820, 2009.

[19] P. Mattavelli, F. Polo, F. Dal Lago, and S. Saggini, "Analysis of Control-Delay Reduction for the Improvement of UPS Voltage-Loop Bandwidth," *IEEE Trans. Ind. Electron.*, vol. 55, pp. 2903-2911, Aug. 2008.

[20] C. Se-Kyo, "A phase tracking system for three phase utility interface inverters," *IEEE Trans. Power Electron.*, vol. 15, pp. 431–438, May 2000.

[21] Y. Qiu, M. Xu, K. Y, J. S, F.C. Lee, "Multifrequency Small-Signal Model for Buck and Multiphase Buck Converters," *IEEE Trans. Power Electron.*, vol. 21, no. 5, pp 1185-1192, Sep. 2006.

A New Approach to Control the Modified LinVerter for High Frequency Applications

Peyman Farhang and Stefan Mátéfi-Tempfli
Mads Clausen Institute, Faculty of Engineering, University of Southern Denmark
Sønderborg, Denmark
Email: farhang@mci.sdu.dk

Abstract—In this paper not only a modified multi-device LinVerter is proposed but also a novel approach to control this topology is introduced. This topology is able to improve the conventional LinVerter performance using parallel power devices and designing a sequential switching scheme. In addition, from control perspective, a novel control approach based on bidirectional interface between LTspice and MATLAB is created. In this case, the circuit is modeled in LTspice environment and a Chaos Optimization Algorithm (COA) is coded in MATLAB in order to find out the optimal solution in control process. In fact, this new approach combines the advantages of LTspice for simulation of different circuit configurations using actual component obtained from manufactures' models with advanced intelligent techniques capabilities from COA in MATLAB. First, the performance of proposed multi-device LinVerter along with new control technique is evaluated in different conditions, and then effectiveness and robustness of the whole suggested designs are presented.

Keywords—linVerter; bidirectional interface; LTspice; MATLAB; chaos optimization

I. INTRODUCTION

While inverters are widely used in industrial applications, new and effective designs for improving inverters performance are still constantly expanded in different literatures. One of the most useful structures which is based on combination of two well-known buck and boost converters termed LinVerter [1]. Despite the fact that this new circuit has some advantages [2], the main problem is the large size of components which leads to high cost and weight of this type of converter. In addition, due to power dissipation restrictions and also from thermal design point of view, increasing the switching frequency to have higher dynamics and effective performance is not feasible. As an alternative, in [1-3] paralleling structure using a kind of interleaved method to overcome this problem, reduce the losses, and operating in high switching frequency has been addressed. Interleaved switching represents a concept supported by phase shifting between different power devices which are put in parallel. So, regarding the duty cycle value, overlapping mode is possible to occur [4]. In this condition, a power device in a certain parallel leg is turned on while the power switch in another one is still in on-state. Turning on the power device only as long as all other parallel devices stay at off-state would be the desired approach. Therefore, the interleaved method is not a good candidate for this. On the other hand, using the parallel topologies causes problems with respect to all diodes which are connected in parallel to each

other. In more details, once the power devices are off, all corresponding parallel diodes are forward biased and current is allowed to flow through these parallel diodes. Because of statistical tolerance, these diodes have no identical characteristics and the major part of the current is just passing through a specific diode which has the lowest internal resistance [5]. Also, owning to negative temperature coefficient, the conductivity of this specific diode will improve and eventually the diode will be destroyed [6]. To overcome these drawbacks, a modified LinVerter based on parallel legs is proposed. First of all, a sequential switching approach is applied to design a switching scheme for switching the power devices in a sequential manner in such a way that parallel power switches are never operated at the same time instant. Consequently, this technique not only can decrease the switching losses by decreasing the switching frequency of each parallel device but it also is able to reduce the conduction losses by limiting the maximum turn-on time of the each paralleled power switches. Secondly, to avoid the problem discussed above about the behavior of parallel diodes, paralleled thyristors are used in this modified version. Based on designed sequential switching scheme, all thyristors are switched actively in order to control the current path in different conditions. It is noticeable that even though in [4-6] magnetic freewheeling control is employed to switch freewheeling diodes of the parallel IGBTs in a sequential scheme, it has still problems like design complexity and large size. Another drawback is the fact that magnetic freewheeling control approach can be affected by load especially when the inductance of the load is higher than the inductance of the magnetic coils. So, in this case, it cannot guide the current through a specific diode.

In addition to proposing a new multi-device structure for LinVerter, a novel approach to control this modified LinVerter is introduced based on creation of bidirectional interface between LTspice and MATLAB. Some studies have been done to design different controller concepts and simulate the converters based on the approximate dynamic model which is acquired from linearization of state-space model as the common approach [7, 8]. Because of considering some assumptions and simplifications to derive small signal model, it cannot adequately describe the system real behavior. Furthermore, other studies tried to simulate their circuits and controllers by MATLAB/Simulink. Since typically generic elements especially for transistors and diodes have been used in MATLAB, the results are not so close to reality. By contrast, using LTspice allows users to have realistic results because it

978-1-4673-9551-9/16 $31.00 © 2016 IEEE   733

provides a wide component database with manufacturer specifications. Besides, this software is able to use various component models which are downloadable from manufactures' websites [9, 10]. On the other hand, LTspice itself has some limitations especially in utilizing advanced intelligent techniques to model and simulate while employing this bidirectional interface enables LTspice to solve these obstacles and brings some benefits. From the control point of view, a Modified PID controller (MPID) based on the Chaos Optimization Algorithm (COA) is designed to achieve desired current in the output of multi-device LinVerter which is even able to operate in high frequency.

II. PROPOSED TECHNIQUE AND MULTI-DEVICE LINVERTER

In this section, different designed blocks of the case study consist of modified Linverter and its new control approach are discussed in details.

A. Proposed LinVerter

Fig. 1 demonstrates the proposed multi-device LinVerter structure combined with novel control approach for performance improvement. In this case, one switch is used in each leg. In addition, thyristors are employed in another part. Based on the proposed topology, parallel legs are connected to each other and are switched one after another with respect to PWM signals created by intelligent control and phase shifting process. Also the thyristors are controlled in such a way that in each time only one of them is allowed to flow the current to the load. In fact, by making a balance between buck and boost sides, this topology is able to control the current for flowing in both directions. Therefore, compared to conventional LinVerter, by utilizing parallel switches and thyristors based on sequential switching technique, losses can be limited to reasonable amount which it causes the switching frequency to increase. Another salient feature of this topology is the fact that designed sequential switching scheme for parallel devices tries to deal with mentioned problems about diodes behavior to conduct the load current.

B. Proposed Sequential Switching Technique

Sequential switching technique [11] makes an effort to transfer gate signals to parallel power devices sequentially, not simultaneously. To get deeper understanding of how this concept provides pulse patterns, an example for sequential switching of a specific PWM signal is illustrated in Fig. 2. As shown in this figure, each pulse of effective PWM transferred

to each power device in a specific sequence. Using this technique, the power losses shared among different parallel legs by reduction of switching frequency of each device. Actually, the effective switching frequency and the maximum turn-on time of each paralleled device can be determined as follow:

$$F_{effectiveSW} = n \times F_{deviceSW} \tag{1}$$

$$T_{max-on-time\ powerdevice} = T_{effectiveSW} = \frac{1}{F_{effectiveSW}} \tag{2}$$

where n is the number of power devices in parallel, $F_{effectiveSW}$ is total switching frequency of LinVerter, $F_{deviceSW}$ is switching frequency of each parallel device, $T_{max-on-time}$ and $T_{effectiveSW}$ are the maximum turn-on time of each paralleled device and total switching time period of LinVerter, respectively.

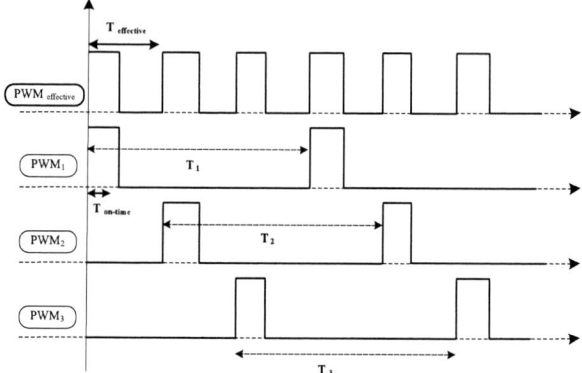

Fig. 2. Sequential switching for three parallel switches

C. Chaos Optimization Algorithm

Chaos as one of the most strong optimization algorithm among others not only tries to keep its solution diversity by producing some neighborhoods around the optimal solutions but it also helps to prevent the search process from becoming premature. In addition, it avoids to be trapped in local optimal solution [12].

COA variables are often generated using well-known logistic map which can be defined by:

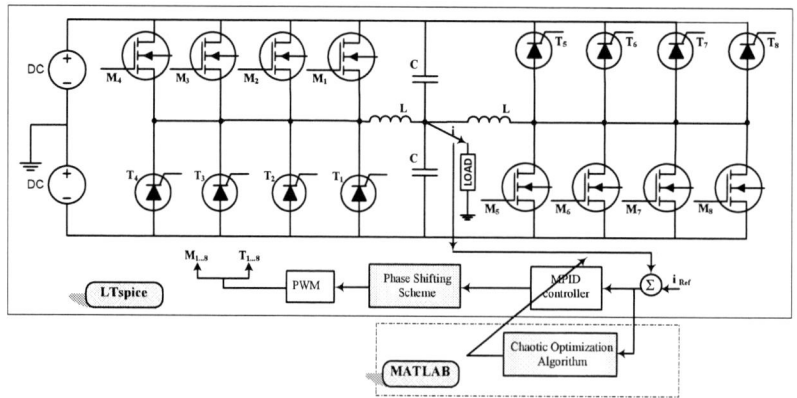

Fig. 1. Modified Multi-device LinVerter

978-1-4673-9551-9/16 $31.00 © 2016 IEEE

$$\gamma_i(k+1) = \beta\,\gamma_i(k)\big(1-\gamma_i(k)\big) \qquad (3)$$

where β is defined as a control parameter and $\gamma_i(0)$ is set between 0 and 1. Although at first glance, this equation seems to be simple, the solutions display a rich variety of behaviors. Equation (3) creates chaotic evolutions which its output is like a stochastic output. Besides, it should be noticed that value of $\gamma_i(k)$ is never repeated and actually the deterministic equation is really depends on initial conditions. $\gamma_i(0)$ is mapped into the variance ranges of optimization variables by the following equations [13]:

$$x_i(k) = x_i^* + \alpha_i\big(2\gamma_i(k)-1\big) \qquad (4)$$

$$\alpha_i = 0.01\big(b_i - a_i\big),\, x_i = [a_i, b_i] \qquad (5)$$

where x_i is the optimization variable, x_i^* is the best found solution at that iteration, and α is the feasible region. The flowchart of Chaos algorithm is shown in Fig. 3.

D. Proposed Bidirectional Interface

Fig. 4 demonstrates how the COA (in MATLAB) interacts with the model in the LTspice environment during the optimization process. This interface allows data to transfer bi-directionally between COA in MATLAB and the simulated circuit in LTspice. Based on this new technique, the bidirectional interfaces performed in a closed loop so that the optimization process will be continued automatically and without the need to open the LTspice file manually [14]. The following steps are considered in design procedure to create the mentioned interaction:

1) Open the simulated model in LTspice by using MATLAB; 2) Set the given initial parameters by COA in LTspice model directly from MATLAB; 3) Access to LTspice by MATLAB and simulate the circuit modeled in LTspice; 4) Read the output results from LTspice [15, 16] and calculate the fitness function in MATLAB for COA; 5) Perform COA to set new values of controller parameters in LTspice; 6) Repeat the step 3 in order to obtain the optimal parameters.

As shown in Fig. 4, first of all the initial parameters of the

COA are set. Then, the simulation procedure is conducted based on interaction between simulated model and COA, the results in data are allowed to flow towards MATLAB. Afterwards, the output results started to be analyzed by COA so as to find the optimal parameters for controller. The whole process repeats until a satisfactory consequence appears.

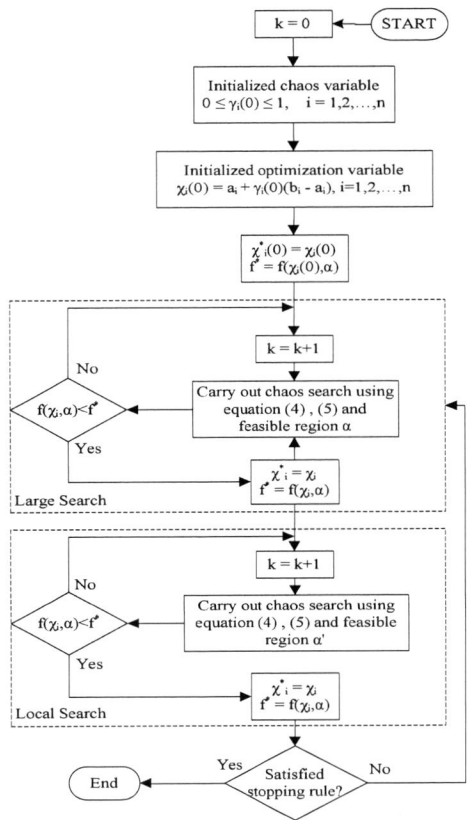

Fig. 3. Flowchart of the chaos algorithm

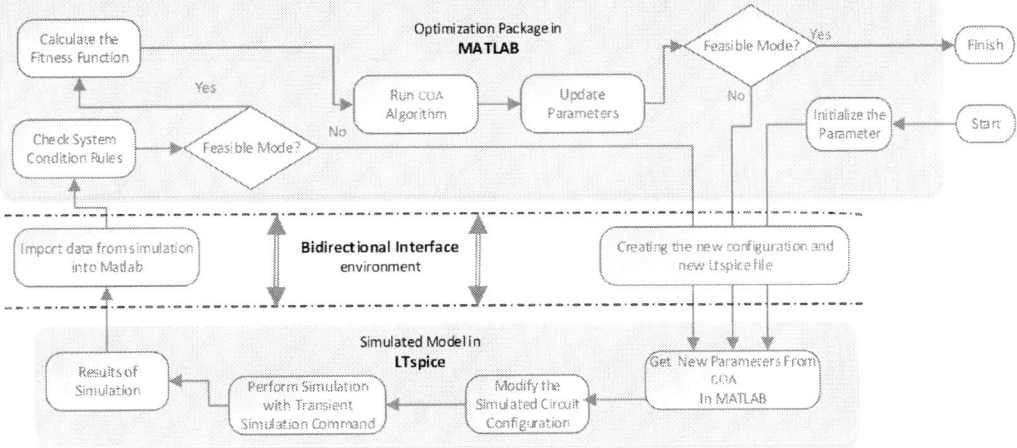

Fig. 4. Interaction between COA optimization algorithm and simulated circuit in LTspice

III. SIMULATION RESULTS

In this study, the performance of the proposed modified LinVerter along with suggested novel control approach is evaluated. To have better understanding and evaluation, these studies are categorized and presented in the following sections:

A. COA-based MPID for Proposed Modified LinVerter

While new and effective designs and theories are continually extended in the subject of control, PID controllers as a three-term controllers are still very popular in industry due to their simplicity in design and low price, together with of showing satisfactory performances [17,18].

Because of advantages of this type of controller, there were a lot of continuous attempts in order to enhance its robustness and effectiveness. Hence, the PID controller is often modified as follow [17]:

$$G(s) = K_p(1 + \frac{1}{T_i S} + \frac{T_d S}{\frac{T_d}{N} S + 1}) \qquad (6)$$

where K_P is the proportional gain, T_i is the integral time constant and T_d is the derivative time constant. In addition, N generally assumes a value between 1 and 33 [17, 19].

In this case, COA is employed to find out the optimal parameters for MPID (K_P, T_i, T_d, N) by minimizing the fitness function given in following equation:

$$Fitness\ Function = \int_{0}^{T_{sim}} t. \left| e(t) \right| dt \qquad (7)$$

where T_{sim} is the simulation time and $e(t)$ means the difference between desired and resulted current. In the first step, COA starts to evaluate the fitness function based on the initial population for control parameters. These initial values are obtained from specific search space between minimum and maximum values for each parameter. Then, it tries to update these values and go toward the optimal position. In other words, this algorithm calculates the fitness function in each iteration and makes an effort to find appropriate controller parameters so as to have minimum possible fitness function. In this study, COA executes the process continuously for finding optimal parameters with respect to number of iterations. Here the iteration and number of population are set 100 and 10, respectively. Fig. 5 depicts how the fitness function is moving toward its minimum value during this process according to the number of iterations. Finally, the optimal parameters found by COA are listed in Table I.

TABLE I. OPTIMIZED PARAMETERS

Controllers	Parameters
MPID	K_p=200.1, T_i=0. 21, T_d=0.34, N=25

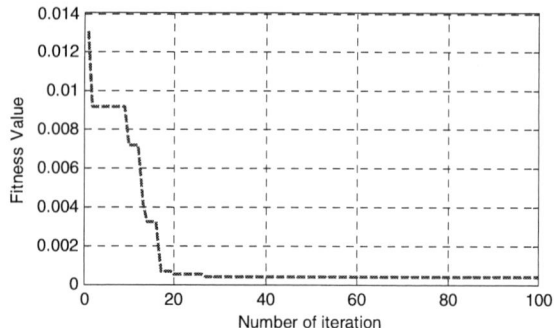

Fig. 5. Convergence of COA

B. Proposed Phase Shifting Scheme for Sequential Switching

The Logic circuits are designed to split main switching signals so that paralleled power devices and thyristors could operate sequentially [11]. The phase shifting scheme block generates different switching signals in such a way that in each time just one device is conducting. To clarify more, Fig. 6 demonstrates the results of the phase shift switching logic block in the controller part. In this figure, performance of designed circuit in order to switch different paralleled devices in a specific sequence is illustrated.

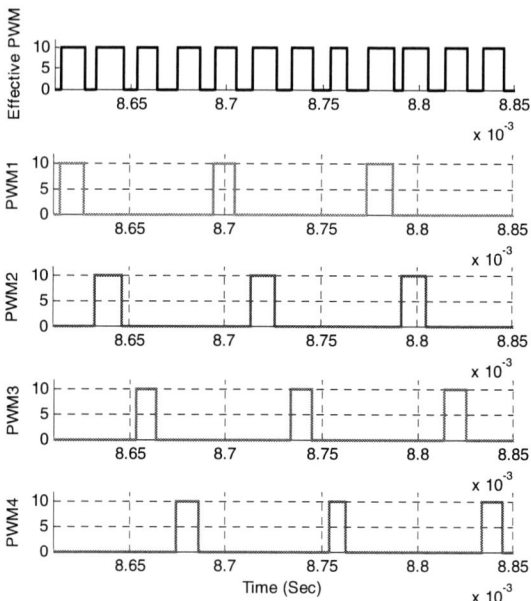

Fig. 6. Performance of proposed sequential switching technique

C. Proposed Modified LinVerter Results

In this section, the performance of proposed technique for modified Linverter is evaluated by doing simulation between LTspice and MATLAB. The DC voltage is set 10 V and the values for capacitors and inductors for tracking a 500 Hz signal are 20 µF and 1 mH, respectively. It is clear that increase of switching frequency causes inductor and capacitors values to reduce [20]. The responses of the modified LinVerter for tracking the different signals are illustrated in Figs. 7 and 8. As can be seen in these figures, in this topology, the output current can follow the desired current not only in low frequency (500 Hz) but it also tracks the reference in high frequency (30 kHz) as fast as possible. For more investigation and assessment of the robustness of the proposed technique, the tracking of a specific reference (which is changing by time) are shown in Fig. 9. Although the amplitude and frequency of the reference current is changing, it is evident that the output is still tracking the reference.

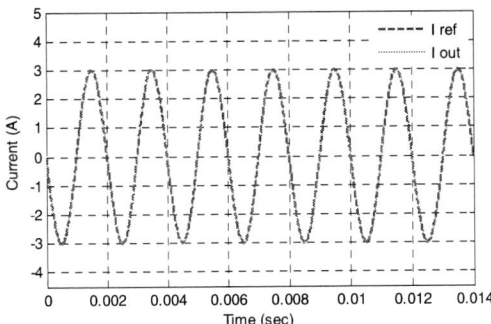

Fig. 7. Current tracking of 500 Hz

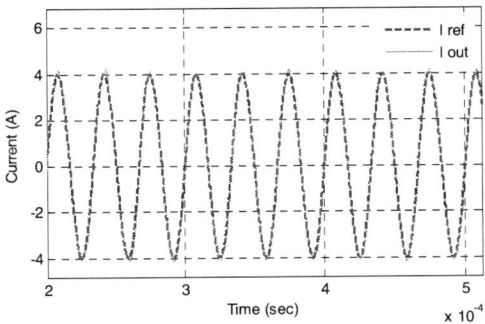

Fig. 8. Current tracking of 30 kHz signal

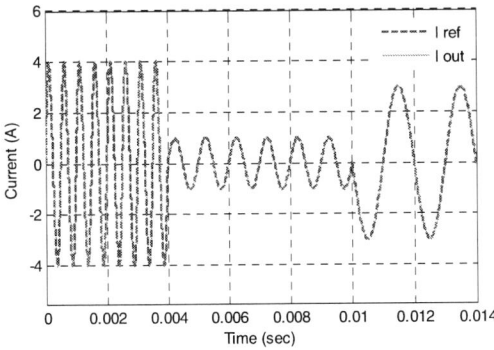

Fig. 9. Performance of modified LinVerter

IV. CONCLUSION

First of all, a new multi-device LinVerter based on the proposed sequential switching concept has been developed in this paper. Then, a new approach to control this LinVerter by creating a bidirectional interface between LTspice and MATLAB has been proposed. The COA is utilized to solve the optimization problem in order to find the optimal parameters of MPID controller which is a part of designed sequential switching scheme in LTspice. The performance of the system has been evaluated in different conditions and results demonstrate the effectiveness and robustness of the proposed method.

ACKNOWLEDGMENT

This work was supported by the Green PET Lab PhD project funded by the Region of Southern Denmark, Syddansk Vækstforum and European Regional Development Fund.

REFERENCES

[1] S.L. Baciu, S. Trabelsi, B Amlang, and W. Schumacher, "LINVERTER a low-harmonic and high-bandwidth inverter based on a parallel multilevel structure," 35th Annual IEEE Power Electronics Specialists Conference (PESC), 2004, pp. 3927-3931.

[2] S. Grubic, B. Amlang, W. Schumacher, and Andree Wenzel, "A high-performance electronic hardware-in-the-loop drive–load simulation using a linear inverter (LinVerter)," IEEE Transactions on Industrial Electronics, vol. 57, no .4, pp. 1208-1216, Apr. 2010.

[3] T. Boller, R. M. Kennel, and J. Holtz, "Increased power capability of standard drive inverters by sequential switching," IEEE International Conference on Industrial Technology (ICIT), , 2010, pp. 769-774.

[4] A. C. Ferreira, and R. M. Kennel, "Interleaved or sequential switching-for increasing the switching frequency," 7th International Conference on Power Electronics, 2007, pp. 738-741.

[5] R. M. Kennel, T. Boller, and J. Holtz, "Replacement of electrical (load) drives by a hardware-in-the-loop system," International Aegean Conference on Electrical Machines and Power Electronics (ACEMP)-Electromotion Joint Conference, 2011, pp. 17-25..

[6] R. Kennel, "Power electronics, hardware-in-the-loop systems, the example of the Virtual machine," Technical university of Munich, Electrical drive systems and power electronics, presentation & lectures, 2013.[Online].Available:www.eal.ei.tum.de/fileadmin/tueieal/www/courses/LGUS/lecture/2013-S/Hardware_in_the_Loop_www.pptx

[7] A. Amirahmadi, M. Rafiei, K. Tehrani, G. Griva, and I.Batarseh, " Optimum design of integer and fractional-order PID controllers for boost converter using SPEA Look-up Tables," Journal of Power Electronics, vol. 15, no. 1, pp. 160-176, Jan. 2015.

[8] K. Arab Tehrani, A. Amirahmadi, S. M. R. Rafiei, G. Griva, L. Barrandon, M. Hamzaoui, I. Rasoanarivo, and F.M.Sargos, "Design of fractional order PID controller for boost converter based on multi-objective optimization," 14th International Power Electronics and Motion Control Conference (EPE-PEMC), 2010, pp. 179-185.

[9] S. S. Raghuwanshi, A. Singh, and Y. mokhariwale, "A comparison & performance of simulation tools MATLAB/SIMULINK, PSIM & PSPICE for power electronics circuits," International Journal of Advanced Research in Computer Science and Software Engineering, vol. 2, no. 3, pp. 187-191, Mar. 2012.

[10] G. Brocard, The LTSpice IV Simulator: Manual, Methods and Applications, Würth Elektronik eiSos Gmbh &Co. KG, Waldenburg, 2011.

[11] Y-J. Lee, and A.Emadi, "Phase shift switching scheme for DC/DC boost converter with switches in parallel," IEEE Vehicle Power and Propulsion Conference (VPPC), Sept. 2008, pp. 1-6.

978-1-4673-9551-9/16 $31.00 © 2016 IEEE

[12] S. Wang, and B. Meng, "Chaos particle swarm optimization for resource allocation problem," Proceedings of the IEEE International Conference on Automation and Logistics August. 2007, pp. 464-467.

[13] L. Shengsong, W. Min and H. Zhijian, "Hybrid algorithm of chaos optimisation and SLP for optimal power flow problems with multimodal characteristic," IEE Proceedings- Generation, Transmission and Distribution, vol. 150, no. 5, pp. 543-547, Sept. 2003.

[14] P. Farhang, A. Drimus, S. Mátéfi-Tempfli, "New technique for voltage tracking control of a boost converter based on the PSO algorithm and LTspice," 56th International Scientific Conference on Power and Electrical Engineering of Riga Technical University (RTUCON), Oct 2015.

[15] P. Wagner, Fast Import of Compressed Binary .RAW Files Created with LTspice Circuit Simulator, 22 Mar 2009 (Updated 26 Apr 2009). [Online]. Available: www.mathworks.com/matlabcentral/fileexchange

[16] D. Dorran, runLTspice, October 29, 2014. [Online]. Available: https://dadorran.wordpress.com/

[17] A. Visioli, Practical PID control, Springer-Verlag London. 2006.

[18] D. Maiti, A. Acharya, M. Chakraborty, A. Konar, and R. Janarthanan, "Tuning PID and $PI^\lambda D^\delta$ controllers using the integral time absolute error criterion," 4th International Conference on Information and Automation for Sustainability, pp. 457-462, 2008.

[19] K. Heong Ang, G. Chong, and Y. Li. , "PID control systems analysis, design, and technology," IEEE Transactions on Control Systems Technology, vol. 13, no. 4, pp. 559-576. July 2005.

[20] O. Hegazy, J. Van Mierlo, P. Lataire, "Analysis, modeling, and implementation of a multidevice interleaved DC/DC converter for fuel cell hybrid electric vehicles," IEEE Transactions on Power Electronics, , vol. 27, no. 11. pp. 4445-4458. Nov. 2012

Small-Signal Terminal Characteristics Modeling of Three-Phase Boost Rectifier with Variable Fundamental Frequency

Zeng Liu[*†], Jinjun Liu[*], and Dushan Boroyevich[†]

[*]State Key Lab of Electrical Insulation and Power Equipment, Xi'an Jiaotong University, Xi'an, Shaanxi, China
[†]Center for Power Electronics System, Virginia Polytechnic Institute and State University, Blacksburg, VA, United States
Email: zeng.liu@ieee.org

Abstract — AC power electronics system are prone to instability due to the interaction between the power converters, and terminal characteristics of individual power converter based criteria are very attractive for analyzing stability of whole system. However, more and more AC power electronics systems emerge with the feature of variable fundamental frequency, while existing approaches for stability analysis are just suitable for systems with constant fundamental frequency. To overcome this problem, terminal characteristics of the three-phase boost rectifier are modeled considering the dynamic behavior of the fundamental frequency in this paper, which will be beneficial for terminal characteristics based stability analysis of three-phase AC power electronics system with variable fundamental frequency. Finally, the proposed model is verified in frequency domain.

Keywords — *modeling; stability analysis; terminal characteristics; variable fundamental frequency.*

I. INTRODUCTION

Stability of power electronics system is generally analyzed by dividing it into a source subsystem and a load subsystem at the bus, and then applying stability criterion of cascaded system to the small-signal terminal characteristics of them. For conventional three-phase AC system, these small-signal terminal characteristics can be represented by source output impedance and the load input admittance in synchronous reference frame (SRF) [1-3], and the overall system stability is determined by applying Generalized Nyquist Criterion (GNC) to the product of them [4-5]. Moreover, several GNC based stability criteria have been presented in order to reduce the computation complexity [6-9]. The physical meaning behind this is that the source and load are just coupled by the bus voltage and the transmitted current from the source to the load, and therefore small-signal source output impedance and load input admittance can fully represents the terminal characteristics of them.

In several emerging applications of three-phase AC power electronics system, such as micro-grids, the droop control is attractive for power sharing among the parallel power sources [10], where the droop between active power and fundamental frequency, and the droop between the reactive power and the

voltage magnitude are generally adopted [11-13]. It means that the fundamental frequency of such system is varied with the load power, which will no more be fixed, and the coupling between the source and the load not only includes the voltage and the current in SRF, but also covers the fundamental frequency. Consequently, the condition that small-signal terminal characteristics of the source and the load can be fully represented by the output impedance or the input admittance is not valid in the emerging system with variable fundamental frequency. In order to analyze the stability for the system with variable fundamental frequency by stability criterion, it is very urgent to explorer the small-signal terminal characteristics of the source and the load, which is seldom presented in existing publications [2-3].

To tackle the aforementioned problem, this paper proposes a full small-signal model of terminal characteristics for three-phase boost rectifier with variable fundamental frequency, which is the most common load in AC power electronics system. The contribution of this paper can be highlighted as below. At beginning, the small-signal model of the power stage is built in SRF, considering the dynamic of fundamental frequency, and it is found that the input current in SRF is affected by the fundamental frequency. Then, the dynamics of the fundamental frequency in the control system is identified in this paper at the first time. Finally, the full small-signal model of the three-phase boost rectifier covering the action of the fundamental frequency in both power stage and control system is present, and the analytical expressions of the terminal characteristics is derived.

II. DESCRIPTION OF SYSTEM STUDIED

The terminal characteristics of three-phase boost rectifier with the variable fundamental frequency are studied in this paper. The power stage is shown in Fig. 1, where the fundamental angular frequency ω_1 of the three-phase source voltage, expressed by (1), is variable.

$$\begin{cases} v_{Sa} = V_m \cos(\omega_1 t) \\ v_{Sb} = V_m \cos(\omega_1 t - 2\pi/3) \\ v_{Sc} = V_m \cos(\omega_1 t + 2\pi/3) \end{cases} \tag{1}$$

This work was supported by the National Natural Science Foundation of China under Grant 51407141, and the Delta Environmental & Educational Foundation under Grant DREG2014004.

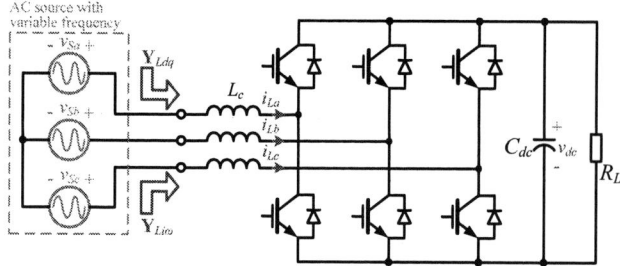

Fig. 1. Power stage of the three-phase boost rectifier studied, where the fundamental frequency of the input voltage is variable.

The system is regulated in a SRF in Fig. 2, the phase angle of which θ^c is obtained by the phase-locked loop (PLL), composed by a proportional-plus-integral (PI) compensator G_{PLL} and an integrator, from the three-phase input voltage v_{Sabc}. The output DC voltage v_{dc} is regulated by the external voltage loop, which is composed by a PI compensator G_V, and the three-phase input current i_{Labc} is regulated by the inner current loop with a PI compensator G_{IL} in D- and Q-axis respectively.

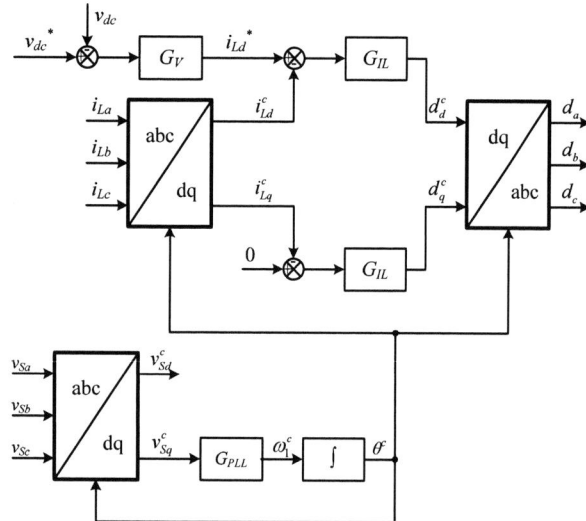

Fig. 2. Control block diagram of the three-phase boost rectifier studied.

III. TERMINAL CHARACTERISTICS MODELING

The terminal characteristics of the boost rectifier depend on both power stage and the control system, which will be modeled respectively considering the dynamic behavior of fundamental frequency.

A. Modeling of Power Stage

With the moving average and the coordinate transformation [14], the average model of the power stage in SRF can be presented by (2), where d_d and d_q are the average duty cycle of three-phase leg in SRF, and i_{Ld} and i_{Lq} denote the three-phase current of the inductor L_c in SRF, and v_{Sd} and v_{Sq} denote the three-phase source voltage v_{Sabc} in SRF.

Then small-signal linearization around operation point for each variable of the power stage is performed, including the angular frequency $\omega_1 = \Omega_1 + \hat{\omega}_1$, and the small-signal circuit of

the power stage can be presented by Fig. 3. It can be found that there are two additional controlled voltage sources in the D- and Q-axis, whose voltage are determined by the small-signal of fundamental frequency $\hat{\omega}_1$.

$$\begin{cases} \begin{bmatrix} v_{Sd} \\ v_{Sq} \end{bmatrix} = L_c \dfrac{d}{dt}\begin{bmatrix} i_{Ld} \\ i_{Lq} \end{bmatrix} + \begin{bmatrix} 0 & -\omega_1 L_c \\ \omega_1 L_c & 0 \end{bmatrix}\begin{bmatrix} i_{Ld} \\ i_{Lq} \end{bmatrix} + \dfrac{v_{dc}}{2}\begin{bmatrix} d_d \\ d_q \end{bmatrix} \\ \dfrac{3}{4}\left(i_{Ld}d_d + i_{Lq}d_q\right) = C_{dc}\dfrac{dv_{dc}}{dt} + \dfrac{v_{dc}}{R_L} \end{cases} \quad (2)$$

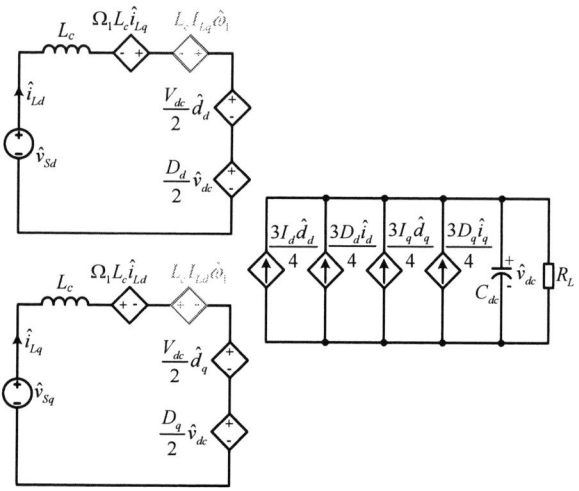

Fig. 3. Small-signal circuit of the power stage in SRF.

Furthermore, the small-signal model of the power stage can be treated as a system with the excitation of the source voltage $\hat{v}_{Sdq}$, the duty cycle $\hat{d}_{dq}$ and the angular frequency $\hat{\omega}_1$, and the response of the inductor current $\hat{i}_{Ldq}$ and DC voltage $\hat{v}_{dc}$. The block diagram of the small signal model is shown in Fig. 4, where the vector $\hat{v}_{odq}$ is defined as $\begin{bmatrix} \hat{v}_{dc} & 0 \end{bmatrix}^T$ and the vector $\hat{\omega}_{Sdq}$ is defined as $\begin{bmatrix} \hat{\omega}_1 & 0 \end{bmatrix}^T$. And the transfer function matrices in Fig. 4 have been derived out in this paper, which are expressed by (3)–(8) respectively.

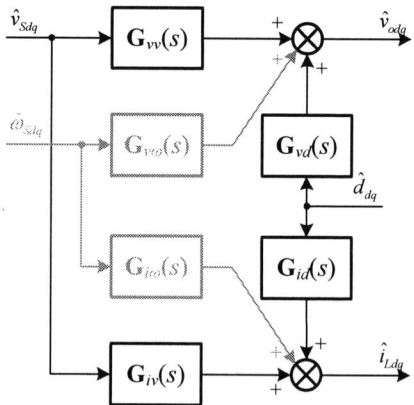

Fig. 4. Block diagram of the small-signal model of power stage, where all the small-signal variables are two dimensional vectors.

$$\mathbf{G}_{vv} = \frac{1}{\Delta_1}\begin{bmatrix} s\left(3R_L D_d\right) - 3\Omega_1 R_L D_q & s\left(3R_L D_q\right) + 3\Omega_1 R_L D_d \\ 0 & 0 \end{bmatrix} \tag{3}$$

$$\mathbf{G}_{iv} = \frac{1}{\Delta_1}\begin{bmatrix} s^2\left(4R_L C_{dc}\right) + 4s + \dfrac{3R_L D_q^2}{2L_c} & s\left(4\Omega_1 R_L C_{dc}\right) + 4\omega_1 - \dfrac{3R_L D_d D_q}{2L_c} \\[3mm] s\left(-4\Omega_1 R_L C_{dc}\right) - 4\omega_1 - \dfrac{3R_L D_d D_q}{2L_c} & s^2\left(4R_L C_{dc}\right) + 4s + \dfrac{3R_L D_d^2}{2L_c} \end{bmatrix} \tag{4}$$

$$\mathbf{G}_{vd} = \frac{1}{\Delta_1}\begin{bmatrix} s^2\left(3L_c R_L I_{Ld}\right) + s\left(-\dfrac{3R_L D_d V_{dc}}{2}\right) & s^2\left(3L_c R_L I_{Lq}\right) + s\left(-\dfrac{3R_L D_q V_{dc}}{2}\right) \\[2mm] +3\Omega_1^2 L_c R_L I_{Ld} + \dfrac{3\Omega_1 R_L D_q V_{dc}}{2} & +3\Omega_1^2 L_c R_L I_{Lq} - \dfrac{3\Omega_1 R_L D_d V_{dc}}{2} \\[3mm] 0 & 0 \end{bmatrix} \tag{5}$$

$$\mathbf{G}_{id} = \frac{1}{\Delta_1}\begin{bmatrix} -s^2\left(2R_L C_{dc} V_{dc}\right) - s\left(\dfrac{3R_L D_d I_{Ld}}{2} + 2V_{dc}\right) & -s\left(2\Omega_1 R_L C_{dc} V_{dc} + \dfrac{3R_L D_d I_{Lq}}{2}\right) \\[2mm] -\dfrac{3\Omega_1 R_L D_q I_{Ld}}{2} - \dfrac{3R_L D_q^2 V_{dc}}{4L_c} & -\dfrac{3\Omega_1 R_L D_q I_{Lq}}{2} - 2\Omega_1 V_{dc} + \dfrac{3R_L D_d D_q V_{dc}}{4L_c} \\[3mm] s\left(2\Omega_1 R_L C_{dc} V_{dc} - \dfrac{3R_L D_q I_{Ld}}{2}\right) & -s^2\left(2R_L C_{dc} V_{dc}\right) - s\left(\dfrac{3R_L D_q I_{Lq}}{2} + 2V_{dc}\right) \\[2mm] +\dfrac{3\Omega_1 R_L D_d I_{Ld}}{2} + 2\Omega_1 V_{dc} + \dfrac{3R_L D_d D_q V_{dc}}{4L_c} & +\dfrac{3\Omega_1 R_L D_d I_{Lq}}{2} - \dfrac{3R_L D_d^2 V_{dc}}{4L_c} \end{bmatrix} \tag{6}$$

$$\mathbf{G}_{v\omega} = \frac{1}{\Delta_1}\begin{bmatrix} s\left(3L_c R_L D_d I_{Lq} - 3L_c R_L D_q I_{Ld}\right) - 3L_c R_L I_{Ld} D_d \Omega_1 - 3L_c R_L I_{Lq} D_q \Omega_1 & 0 \\ 0 & 0 \end{bmatrix} \tag{7}$$

$$\mathbf{G}_{i\omega} = \frac{1}{\Delta_1}\begin{bmatrix} s^2\left(4L_c R_L C_{dc} I_{Lq}\right) + s\left(4L_c I_{Lq} - 4L_c R_L C_{dc} I_{Ld}\Omega_1\right) + \dfrac{3D_q^2 I_{Lq} R_L}{2} + \dfrac{3D_d D_q I_{Ld} R_L}{2} - 4L_c I_{Ld}\Omega_1 & 0 \\[2mm] -s^2\left(4L_c R_L C_{dc} I_{Lq}\right) - s\left(4L_c I_{Ld} + 4L_c R_L C_{dc} I_{Lq}\Omega_1\right) - \dfrac{3D_d^2 I_{Ld} R_L}{2} - \dfrac{3D_d D_q I_{Lq} R_L}{2} - 4L_c I_{Lq}\Omega_1 & 0 \end{bmatrix} \tag{8}$$

where

$$\Delta_1 = s^3\left(4L_c R_L C_{dc}\right) + s^2\left(4L_c\right) + s\left(4\Omega_1^2 L_c R_L C_{dc} + \frac{3R_L D_d^2}{2} + \frac{3R_L D_q^2}{2}\right) + 4\Omega_1^2 L_c$$

B. Modeling of Control System

In the control system, the input current of the boost rectifier is controlled in the SRF with the rotation angle θ^c. Therefore it should be noted that there are two SRFs in the modeling of the boost rectifier. The one is system SRF, called as SRF for short in this paper, which is used to define the source output impedance and the load input admittance, and its rotating angle θ is aligned with the phase angle of the source voltage. The other one appears in the control system of the boost rectifier, which will be called as converter SRF in this paper, and its rotation angle θ^c is obtained by the PLL. Since the input admittance defined in SRF depends on the current regulators G_{IL} in the control system, which performs in the converter SRF, it is necessary to gain the quantitative relationship between

these two SRFs for building the small-signal model of the control system.

The three-phase source voltage can be emulated with inverse Park transformation with given amplitude v_{Sd} in D-axis and v_{Sq} in Q-axis, and the angular frequency ω_1, and the angle θ generated by integrating ω_1. Meanwhile, the PLL in the control system tracks the angle θ, and generates the angle θ^c for the converter SRF. The difference between this two phase angles is represented by $\Delta\theta$ in Fig. 5(a). In the steady state, the $\Delta\theta$ equals zero since the PI compensator is adopted for G_{PLL}.

The block diagram in Fig. 5(a) can be simplified to Fig. 5(b) by combining the inverse Park transformation and Park transformation together, as well as combining these two integrators together. Therefore, according to Fig. 5(b) the relationship between the bus voltage in the SRF and the converter SRF can be expressed by (9), and the $\Delta\theta$ is represented by (10) in the time domain.

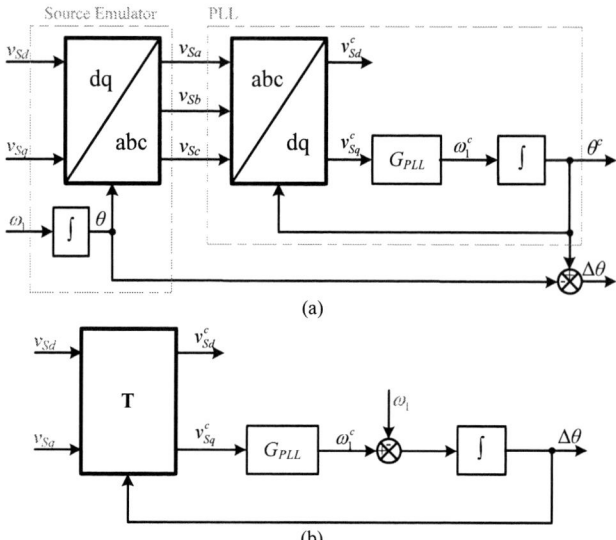

(a)

(b)

Fig. 5. Impaction of the fundamental frequency on the control system of the boost rectifier: (a) original block diagram; (b) simplified block diagarm.

$$\begin{bmatrix} v_{Sd}^c \\ v_{Sq}^c \end{bmatrix} = \mathbf{T}(\Delta\theta)\begin{bmatrix} v_{Sd} \\ v_{Sq} \end{bmatrix} = \begin{bmatrix} \cos(\Delta\theta) & \sin(\Delta\theta) \\ -\sin(\Delta\theta) & \cos(\Delta\theta) \end{bmatrix}\begin{bmatrix} v_{Sd} \\ v_{Sq} \end{bmatrix} \quad (9)$$

$$\Delta\theta = \int\left[G_{PLL}v_{Sq}^c - \omega_1\right]dt \quad (10)$$

Then the small-signal linearization can be performed for (9) and (10) since the operation points in these two SRFs are identical, and the small-signal model can be expressed by (11).

$$\Delta\hat{\theta}(s) = G_{\theta v}(s)\hat{v}_{Sd}(s) + G_{\theta\omega}(s)\hat{\omega}_1(s) \quad (11)$$

where

$$G_{\theta v}(s) = \frac{G_{PLL}(s)}{s + V_{Sd}G_{PLL}(s)} \qquad G_{\theta\omega}(s) = -\frac{1}{s + V_{Sd}G_{PLL}(s)}$$

It can be found that the small-signal difference between rotating angles $\Delta\hat{\theta}$ depends on the source voltage in Q-axis of the SRF $\hat{v}_{Sq}$ and the angular frequency $\hat{\omega}_1$.

Since the small-signal dynamic of the control system is modeled in SRF, while the input current is controlled in converter SRF, it is required to derive the input and output of the current regulator in SRF, which can be expressed by (12) and (13) respectively.

$$\hat{i}_{Ldq}^c(s) = \hat{i}_{Ldq}(s) + \mathbf{G}_{piv}(s)\hat{v}_{Sdq}(s) + \mathbf{G}_{pi\omega}(s)\hat{\omega}_{Sdq}(s) \quad (12)$$

$$\hat{d}_{dq} = \hat{d}_{dq}^c + \mathbf{G}_{pdv}(s)\hat{v}_{Sdq} + \mathbf{G}_{pd\omega}(s)\hat{\omega}_{Sdq} \quad (13)$$

where

$$\mathbf{G}_{piv}(s) = \begin{bmatrix} 0 & I_{Ld}G_{\theta v}(s) \\ 0 & -I_{Lq}G_{\theta v}(s) \end{bmatrix}$$

$$\mathbf{G}_{pi\omega}(s) = \begin{bmatrix} I_{Ld}G_{\theta\omega}(s) & 0 \\ -I_{Lq}G_{\theta\omega}(s) & 0 \end{bmatrix}$$

$$\mathbf{G}_{pdv}(s) = \begin{bmatrix} 0 & -D_d G_{\theta v}(s) \\ 0 & D_q G_{\theta v}(s) \end{bmatrix}$$

$$\mathbf{G}_{pd\omega}(s) = \begin{bmatrix} -D_d G_{\theta\omega}(s) & 0 \\ D_q G_{\theta\omega}(s) & 0 \end{bmatrix}$$

C. Deviation of terminal characteristics with variable fundamental frequency

Combining the small-signal models of the power stage and the control system together, the full small-signal model of the whole system can be gained. Since there are two small-signal excitations in the whole system, i.e. the source voltage in SRF $\hat{v}_{Sdq}$ and the fundamental frequency $\hat{\omega}_1$, the model of the whole system can be grouped into two parts by the excitations. This first part, shown in Fig. 6(a), is excited by the source voltage in SRF, while the second part, shown in Fig. 6(b), is excited by the fundamental frequency. In Fig. 6, the voltage regulator and current regulator are represented by two dimensional matrixes $\mathbf{H}_v(s)$ and $\mathbf{H}_i(s)$, which are expressed by (14) and (15).

$$\mathbf{H}_v(s) = \begin{bmatrix} G_V(s) & 0 \\ 0 & 0 \end{bmatrix} \quad (14)$$

$$\mathbf{H}_i(s) = \begin{bmatrix} G_{IL}(s) & 0 \\ 0 & G_{IL}(s) \end{bmatrix} \quad (15)$$

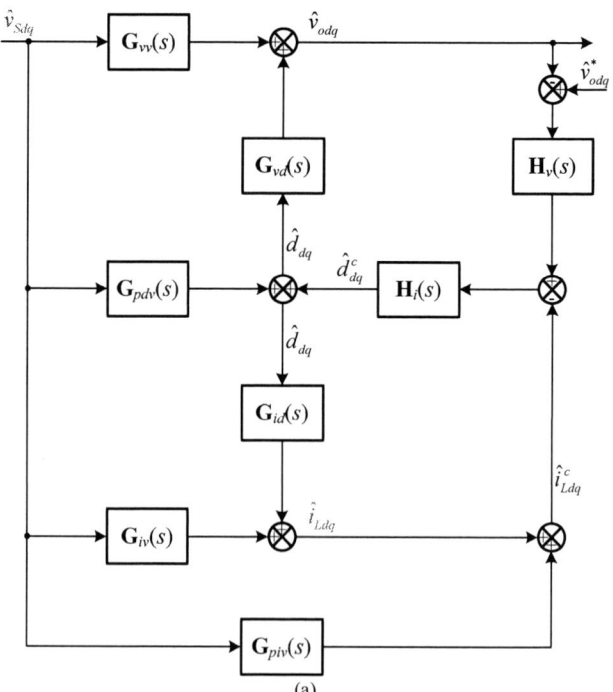

(a)

978-1-4673-9551-9/16 $31.00 © 2016 IEEE

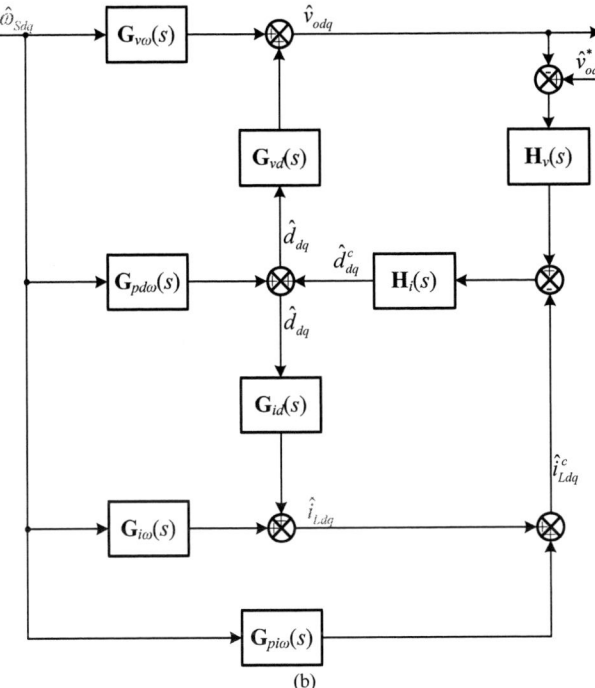

(b)

Fig. 6. Full small-signal model of the three-phase boost rectifier with variable fundamental frequency in SRF, excited by (a) input voltage and (b) fundamental frequency, where all the transfer functions are two dimensional matrices.

Then the terminal characteristics of the three-phase boost rectifier can be divided into two parts. The first one represents the relationship from the input voltage to the input current in SRF, i.e. conventional input admittance, which is expressed by (16). The second part, expressed by (17), represents the relationship from the fundamental frequency to the input current in SRF, which is called as input current-frequency admittance in this paper.

It can be found that the input current-frequency admittance is with the same structure as the conventional input admittance, which is introduced by the fundamental voltage drop of the filter inductor in the power stage and the PLL in the control system.

IV. MODEL VERIFICATION

In order to verify the proposed terminal characteristics model of the three-phase boost rectifier with variable fundamental frequency, the simulation validation has been done, and the parameters of the system are given in TABLE I.

The verification process is divided into two steps. Firstly, the terminal characteristics are calculated based on the proposed model (16) and (17) respectively in MATLAB, and the results are shown in Fig. 7(a) and Fig. 8(a). Secondly, the

terminal characteristics is measured with the approach of frequency sweep in SABER, and the small-signal perturbation is injected in the input voltage in SRF and the fundamental frequency respectively, and then the measurement results are presented in Fig. 7(b) and Fig. 8(b).

TABLE I PARAMETERS OF SYSTEM IN VERIFICATION

Parameters	Value
Source voltage amplitude V_m	80 V
Fundamental frequency Ω_1	100π rad/s
Filter inductance L_c	3.5 mH
DC capacitance C_{dc}	1500 μF
Load resistance R_L	90 Ω
DC voltage reference v_{dc}^*	200 V
Proportional gain of PLL regulator G_{PLL}	55
Integral gain of PLL regulator G_{PLL}	200
Proportional gain of current regulator G_{IL}	0.45
Integral gain of current regulator G_{IL}	3.5
Proportional gain of DC voltage regulator G_V	5
Integral gain of DC voltage regulator G_V	100

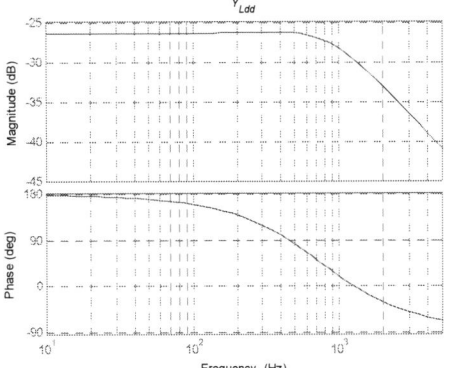

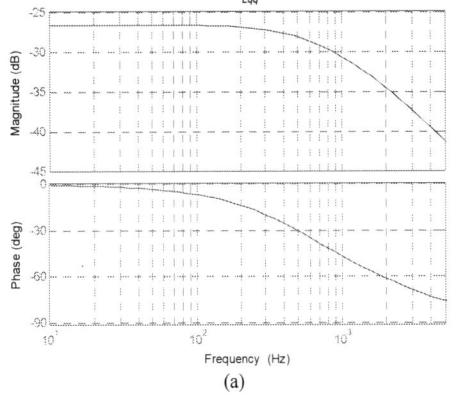

(a)

$$\mathbf{Y}_{Ldq} = \left[\mathbf{I} + \mathbf{G}_{id}\left(\mathbf{I} + \mathbf{H}_i\mathbf{H}_v\mathbf{G}_{vd}\right)^{-1}\mathbf{H}_i\right]^{-1} \cdot \left[\mathbf{G}_{iv} + \mathbf{G}_{id}\left(\mathbf{I} + \mathbf{H}_i\mathbf{H}_v\mathbf{G}_{vd}\right)^{-1}\left(-\mathbf{H}_i\mathbf{H}_v\mathbf{G}_{vv} + \mathbf{G}_{pdv} - \mathbf{H}_i\mathbf{G}_{piv}\right)\right] \quad (16)$$

$$\mathbf{Y}_{Li\omega} = \left[\mathbf{I} + \mathbf{G}_{id}\left(\mathbf{I} + \mathbf{H}_i\mathbf{H}_v\mathbf{G}_{vd}\right)^{-1}\mathbf{H}_i\right]^{-1} \cdot \left[\mathbf{G}_{i\omega} + \mathbf{G}_{id}\left(\mathbf{I} + \mathbf{H}_i\mathbf{H}_v\mathbf{G}_{vd}\right)^{-1}\left(-\mathbf{H}_i\mathbf{H}_v\mathbf{G}_{v\omega} + \mathbf{G}_{pd\omega} - \mathbf{H}_i\mathbf{G}_{pi\omega}\right)\right] \quad (17)$$

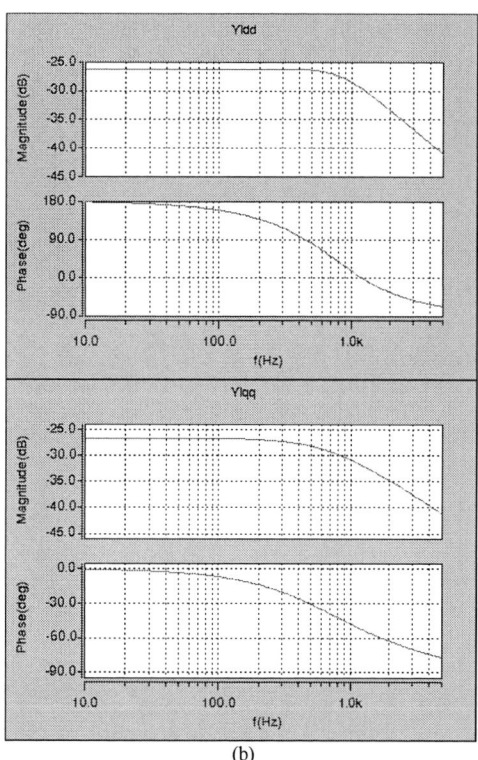

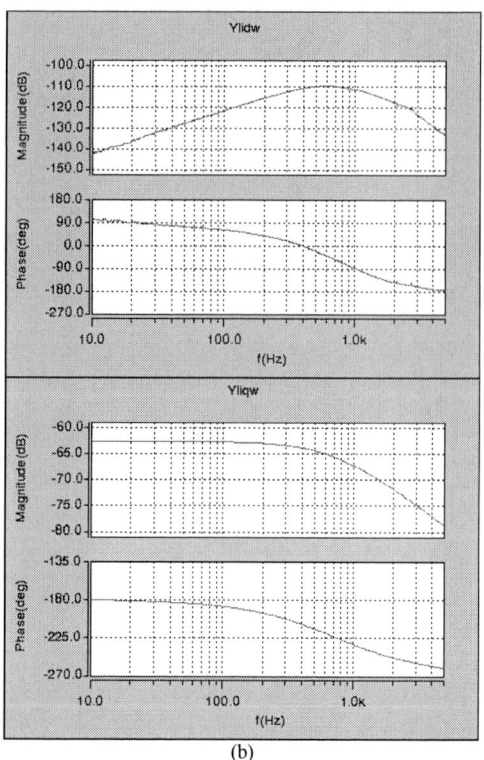

Fig. 7. Bode plots of the conventional input admittance, i.e. the transfer functions from input voltage to the input current in SRF, obtained by (a) calculation based on the proposed model, and (b) measurement based on frequency sweep.

Fig. 8. Bode plots of the input current-frequency admittance, i.e. the transfer functions from fundamental angular frequency to the input current in SRF, obtained by (a) calculation based on the proposed model, and (b) measurement based on frequency sweep.

It can be found that the calculation results based on the proposed model coincides to the measurement results very well, and thus the accuracy of the proposed model is verified.

V. CONCLUSIONS

Aiming at stability analysis of power electronics system with variable fundamental frequency, this paper proposes a small-signal model for terminal characteristics of three-phase boost rectifier considering the dynamic behavior of fundamental frequency, which is composed by the transfer function from the input voltage to the input current in SRF, and the transfer function from the fundamental angular frequency to the input current in SFR. It is revealed that the dynamic of fundamental frequency is introduced by both the fundamental voltage drop of the filter inductor and the PLL. The proposed model can be applied for the stability analysis of system with variable fundamental frequency, which will be explored in future publications.

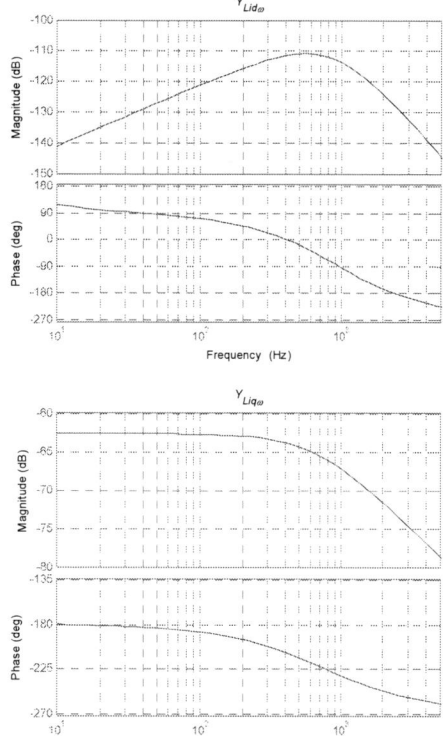

(a)

REFERENCES

[1] J. Sun, "Small-Signal Methods for AC Distributed Power Systems-A Review," *IEEE Transactions on Power Electronics*, vol. 24, pp. 2545-2554, 2009.

[2] B. Wen, D. Boroyevich, P. Mattavelli, Z. Shen, and R. Burgos, "Influence of phase-locked loop on input admittance of three-phase voltage-source converters," in *Twenty-Eighth Annual IEEE Applied Power Electronics Conference and Exposition*, 2013, pp. 897-904.

[3] L. Harnefors, M. Bongiorno, and S. Lundberg, "Input-Admittance Calculation and Shaping for Controlled Voltage-Source Converters,"

IEEE Transactions on Industrial Electronics, vol. 54, pp. 3323-3334, 2007.

[4] M. Belkhayat, "Stability Criterion for AC Power Systems with Regulated Loads," Doctor of Philosophy, Purdue University, 1997.

[5] B. Wen, D. Boroyevich, R. Burgos, P. Mattavelli, and Z. Shen, "Small-Signal Stability Analysis of Three-Phase AC Systems in the Presence of Constant Power Loads Based on Measured D-Q Frame Impedances," *IEEE Transactions on Power Electronics*, vol. 30, pp. 5952-5963, 2015.

[6] H. Mao, D. Boroyevich, and F. C. Lee, "Novel reduced-order small-signal model of a three-phase PWM rectifier and its application in control design and system analysis," *IEEE Transactions on Power Electronics*, vol. 13, pp. 511-521, 1998.

[7] S. Chandrasekaran, D. Borojevic, and D. K. Lindner, "Input filter interaction in three phase AC-DC converters," in *30th Annual IEEE Power Electronics Specialists Conference*, 1999, pp. 987-992 vol.2.

[8] R. Burgos, D. Boroyevich, F. Wang, K. Karimi, and G. Francis, "On the Ac stability of high power factor three-phase rectifiers," in *2nd IEEE Energy Conversion Congress and Exposition*, 2010, pp. 2047-2054.

[9] Z. Liu, J. Liu, W. Bao, and Y. Zhao, "Infinity-Norm of Impedance-Based Stability Criterion for Three-Phase AC Distributed Power Systems With Constant Power Loads," *IEEE Transactions on Power Electronics*, vol. 30, pp. 3030-3043, 2015.

[10] J. M. Guerrero, L. Hang, and J. Uceda, "Control of Distributed Uninterruptible Power Supply Systems," *IEEE Transactions on Industrial Electronics*, vol. 55, pp. 2845-2859, 2008.

[11] M. C. Chandorkar, D. M. Divan, and R. Adapa, "Control of parallel connected inverters in standalone AC supply systems," *IEEE Transactions on Industry Applications*, vol. 29, pp. 136-143, 1993.

[12] J. Rocabert, A. Luna, F. Blaabjerg and P. Rodríguez, "Control of power converters in ac microgrids," *IEEE Transactions on Power Electronics*, vol. 27, pp. 4734–4749, November 2012.

[13] J. M. Guerrero, M. Chandorkar, T. Lee, and P. C. Loh, "Advanced control architectures for intelligent microgrids—part I: decentralized and hierarchical control," *IEEE Transactions on Industrial Electronics*, vol. 60, pp. 1254–1262, April 2013.

[14] S. Hiti, D. Boroyevich, and C. Cuadros, "Small-signal modeling and control of three-phase PWM converters," in *IEEE Industrial Application Society Annual Meeting*, 1994, pp. 1143-1150.

Reliability Analysis of a High-Efficiency SiC Three-Phase Inverter for Motor Drive Applications

Juan Colmenares[1], Diane-Perle Sadik[1], Patrik Hilber[2], Hans-Peter Nee[1]

[1]Department of Electrical Energy Conversion
[2]Department of Electromagnetic Engineering
KTH Royal Institute of Technology
Stockholm, Sweden

Abstract— Silicon Carbide as an emerging technology offers potential benefits compared to the currently used Silicon. One of these advantages is higher efficiency. If this is targeted, reducing the on-state losses is a possibility to achieve it. Parallel-connecting devices decrease the on-state resistance and therefore reducing the losses. Furthermore, increasing the amount of components introduces an undesired tradeoff between efficiency and reliability. A reliability analysis has been performed on a three-phase inverter for motor drive applications rated at 312 kVA. This analysis has shown that the gate voltage stress determines the reliability of the complete system. Nevertheless, decreasing the positive gate-source voltage could increase the reliability of the system approximately 8 times without affecting the efficiency significantly. Moreover, adding redundancy in the system could also increase the mean time to failure approximately 5 times.

Keywords— Inveter, Markov Chain, Motor Drive Application, Power Modules, Reliability, SiC, Silicon Carbide, Voltage Source Conveter.

I. INTRODUCTION

Silicon Carbide (SiC), as a wide band gap material, offers three main potential benefits compare to the currently used Silicon (Si). These benefits can be listed as higher efficiency, higher switching frequency and higher temperature of operation [1]. SiC technology with its benefits has been shown in many applications, such as, power inverters [2], modular multilevel converters [3], hybrid electric vehicles [4], resonant converters [5], and dc-dc converters [6].

If high efficiency is targeted, the most important parameter that has to be controlled is the power losses. These power losses can be divided into two components, the switching losses and the conduction losses. In order to reduce the switching losses it is important to ensure a low inductive path so as a fast switching performance of the device is achieved. One possibility to reduce the conduction losses is to parallel connect either discrete components or to build power modules with several chips connected in parallel [6]-[9]. Another possibility, which could result in not only reduction of the conduction losses but also in a fast switching performance, is the parallel connection of lower current rated power modules. This will result in low losses and higher power ratings. This has been proposed and tested in [10] [11], where and efficiency higher than 99 % has been reported for a power rating higher than 300 kVA.

Nevertheless, increasing the amount of components has a negative impact on the reliability performance of the system. This will introduce an undesired tradeoff between efficiency and reliability. Several studies have been done regarding the reliability performance of SiC devices. These studies have analyzed different modes of failure of the devices, such as; short-circuit behavior and protection [12]-[14], long term reliability [15]-[23] and gate-oxide stability and threshold voltage instability [24]-[29], as well as, high temperature conditions [30]-[32].

However, these studies have been focused on the variation of the electrical internal parameters, such as the threshold voltage, of the devices. Therefore, it is the main proposal of this paper to analyze the reliability aspect of a SiC power electronic system using the information derived from the reliability tests. An estimation of the life expectancy of a three-phase two-level voltage source converter (VSC) for motor drive applications, as well as, deriving important information about what parameters govern the reliability of the system will be done. Section II, will give a description and experimental results of the system analyzed in this paper. In Section III, a reliability analysis is described and the results are discussed in Section IV. Finally, the conclusions are presented in Section V.

II. VOLTAGE SOURCE CONVERTER

The three-phase two-level VSC for motor drive applications analyzed in this paper has a switching frequency of 20 kHz and, an output current of 450 A RMS, a dc-link voltage of 650 V. Assuming a typical line-to-line output voltage of 400 V, the rated output power will be 312 kVA.

It makes use of parallel-connection of power modules. This will reduce the on-state resistance, which will result in a reduction of the conduction losses. This is important when high efficiency is targeted. The power module used in this system is the Cree Inc. CAS100H12AM1 (see Fig. 1(a)). This is an all-SiC power module rated at 1200 V and 168 A. It has an on-state resistance of 16 mΩ at room temperature (RT). Each switch position is built with parallel-connected SiC metal-oxide silicon field-effect transistors (MOSFETs) and antiparallel SiC Schottky diodes. Fig. 2 shows a photograph of the partially built VSC analyzed in this paper.

978-1-4673-9551-9/16 $31.00 © 2016 IEEE

The requirements for high efficiency and current density are met when ten power modules are connected in parallel. This keeps the switching losses unchanged, while reducing the conduction losses by a factor of ten. Furthermore, the ten parallel-connected power modules for each phase were chosen without any sorting whatsoever. Additionally, a proper current density means that at rated load current, the current flowing through each of the power modules is sufficiently high to increase the junction temperature to the level where the on-state resistance is well within the positive-temperature-coefficient range. Being in this range is an essential condition for the auto-balancing mechanism of the current. This guarantees a uniform sharing of the current among the parallel-connected power modules. Moreover, an even number of power modules reduce the system complexity and is an important factor if a symmetrical placement is targeted [10].

Figure 1: Photograph of the SiC power module.

Figure 2: Photograph of the partially built VSC.

In order to reach even higher efficiency it was decided to use the SiC MOSFETs also in the reverse direction. A blanking time of 600 ns was used. Also a switching frequency of 20 kHz was chosen in order to reduce the size of the passive components, such as the dc-link capacitance. For this application it was decided to use MKP capacitors (Metallized Polypropylene Film Capacitors), in particular MKP1848C66012JY5, which meet the requirement of voltage ratings and capacitance. Finally, a total dc capacitance of 720 μF was found for the desired ripple of the dc-link voltage. Additionally, distributed gate-drive units were connected directly to the gate pins of the power modules such that a reduced stray inductance of the gate leads is achieved. This is important for minimizing the Miller effect and achieving a high switching speed, necessary for the high efficiency

approach. Table I presents the electrical parameters of the VSC. A more detailed explanation of the methodology and construction process of this converter can be found in [11].

TABLE I.   ELECTRICAL PARAMETERS OF THE THREE-PHASE INVERTER

Power Rating, Sn	312 kVA
Input Voltage	650 V DC
Output Line-to-Line Voltage	400 V RMS
Output Phase Current	450 A RMS
Switching Frequency, f_s	20 kHz
DC Capacitance, C	720 μF
Blanking Time, T_D	600 ns

Fig. 3(a) and Fig. 3(b), show the experimental results of the inverter operating at 4 kHz and 20 kHz respectively, supplying a motor drive. These results were obtained at more than 200 kVA. Symmetrical phase current with a motor drive as a load and closed-loop control are shown. The line-to-line voltage is also illustrated showing typical characteristics for pulse-width modulation (PWM) as expected. Table II shows the experimental results driving an electrical machine with efficiency higher than 99% was achieved.

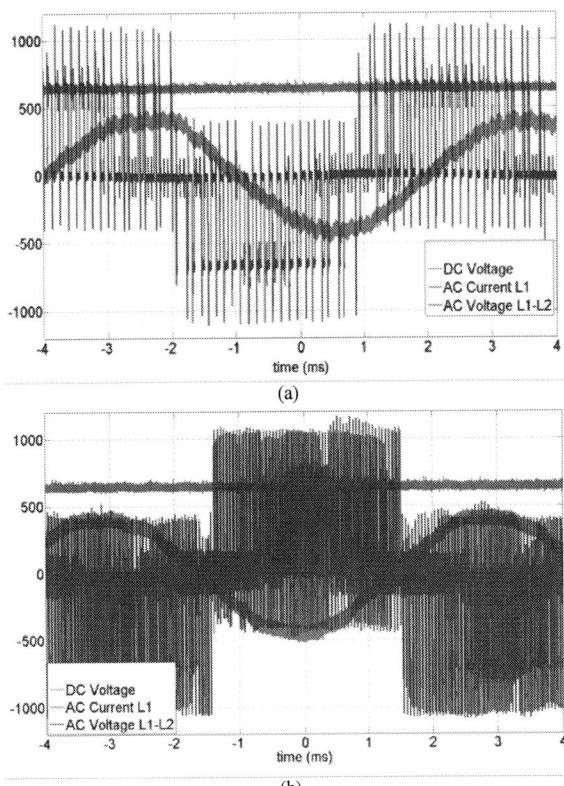

Figure 3: Inverter waveforms during operation at nominal power of 207 kVA and switching frequency of (a) 4 kHz and, (b) 20 kHz.

TABLE II. EXPERIMENTAL RESULTS OF THE THREE-PHASE INVERTER AT 20 kHz

Output Line-to-Line Voltage	397.88 V RMS
Output Phase Current	293.45 A RMS
Input Voltage	649.83 V DC
Input Current	153.44 A DC
Speed	2000 rpm
Torque	406.07 N/m
Efficiency, η	99.07 %

Furthermore, in order to perform a reliability analysis it is important to extrapolate the mean time to failure (MTTF) from the accelerated tests of each component. It is important to note that the power module used to build the converter consists of parallel connection of single chips from the first generation of Cree devices. These components have evolved into a second generation and even a third where the reliability issues as well as performance have been investigated and improved [18] [19]. Therefore, in this paper, the reliability calculations have been performed using the data from the second generation. A corresponding design, targeting high efficiency using parallel connection of power modules using the Cree Inc. CAS300M12BM2 (see Fig. 4(a)) power module rated at 1200 V and 300 A, with an on-state resistance of 5 mΩ at RT, has been proposed for the calculations, as shown in Fig. 4(b).

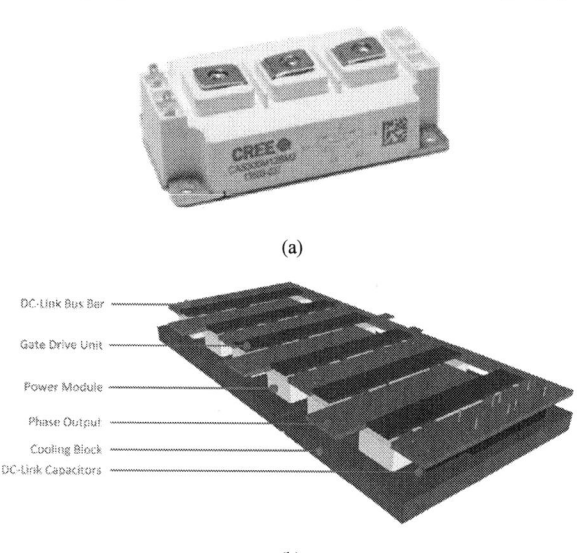

(a)

(b)

Figure 4: (a) Photograph of the SiC power module, (b) Layout of the proposed VSC.

Using the information from the manufacturer, available in [15] [16] and considering the values for the rated application of the converter, Fig. 5 for the gate-source voltage (V_{GS}) stress of the MOSFETs, Fig. 6 for the drain-source voltage (V_{DS}) stress of the MOSFETs and, Fig. 7 for the temperature stress of the diodes, were plotted. The MTTF values have been

derived for each failure mode in the case of the MOSFETs. Table III summarizes the MTTF and failure rate of the components at the rated condition, including the DC-link capacitors used for the construction of the inverter.

TABLE III. MTTF AND FAILURE RATE OF THE COMPONENTS

Component	Rated Condition	MTTF [hours]	Failure Rate, λ
MOSFETs, V_{GS} Stress	20 V @150 °C	10^7	9.03×10^{-4}
MOSFETs, V_{DS} Stress	1000 V @150 °C	6.5×10^7	1.35×10^{-4}
Power Diode	150 °C	6×10^7	1.46×10^{-4}
DC-Link Capacitor	1000 V	10^8	8.76×10^{-5}

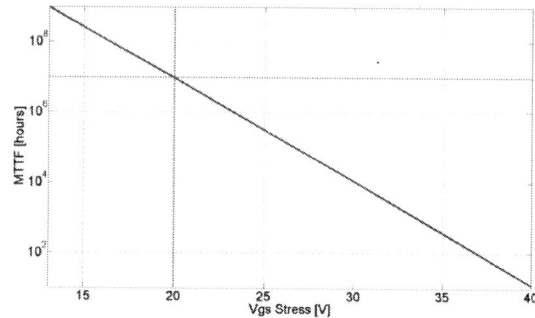

Figure 5: Extrapolated MTTF of 10^7 hours at V_{GS} = 20 V. TDDB of Gate Oxide on 20 A for the Gen2 of SiC MOSFETs [16].

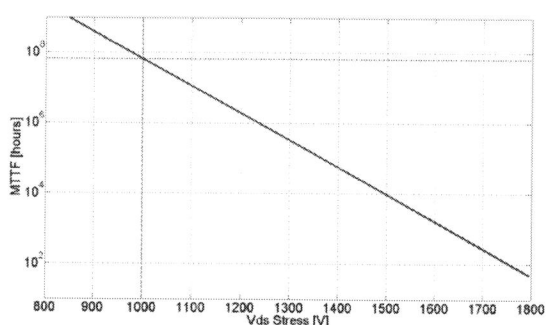

Figure 6: Extrapolated MTTF of 6.5×10^7 hours at V_{DS} = 1000 V. Accelerated field testing at 150 °C for the Gen2 of SiC MOSFETs [16].

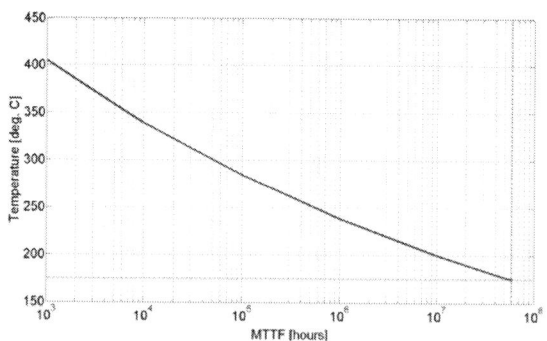

Figure 7: Extrapolated MTTF of 6×10^7 hours at 175 °C for the SiC Power Diode [15].

The extrapolated values are related to the single chip. Therefore a reliability calculation must be done in order to obtain the MTTF and failure rate of the power module (see Table IV). This calculation basically consists in the series connection of all the single chips used in order to build the power modules, as shown in Fig. 8. The failure rate for the power module considering both failure modes, the V_{GS} stress and the V_{DS} stress, were calculated using Eq. 1. This is mainly due to the fact that all the chips must be on the safe state so as the module can operate properly.

$$\lambda_{Module} = \sum_{i=1}^{10} \lambda_{Diode_i} + \sum_{i=1}^{10} \lambda_{MOSFET_i} \quad (1)$$

TABLE IV. MTTF AND FAILURE RATE OF THE SiC MOSFET POWER MODULE

Failure Mode	MTTF [years]	Failure Rate, λ
V_{GS} Stress	95.32	0.0105
V_{DS} Stress	356.16	0.0028

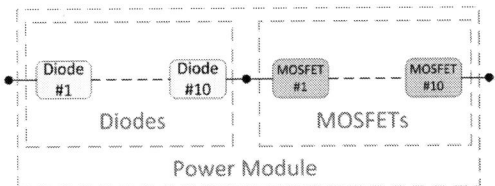

Figure 8: Reliability block diagram for the SiC power module.

III. RELIABILITY ANALYSIS

In this paper, as shown in Table II and Table III, two failure modes have been analyzed. These are: V_{DS} stress and V_{GS} stress. Moreover, two cases of analysis have been performed. The first case, no redundancy, while in the other hand; the second case, takes advantage of the parallel connection of power modules and introduces the so-called, active redundancy.

A. No Redundancy

When no redundancy is considered, all the components of the system are connected in series for the reliability analysis as shown in Fig. 9. This means, that all the components must be working properly to consider that the system is working. Table V shows the calculated MTTF, using Eq. 2, for each failure mode as well as the failure rate for the rated condition shown above in Table I. It is possible to note that the life expectancy is dominated by the gate voltage stress, and it is approximately 8 years.

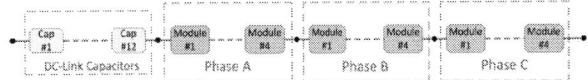

Figure 9: Reliability block diagram for the high-efficiency SiC three-phase inverter without redundancy.

$$\lambda_{Inv} = \sum_{i=1}^{12} \lambda_{Cap_i} + \sum_{i=1}^{12} \lambda_{Module_i} \quad (2)$$

TABLE V. MTTF AND FAILURE RATE OF THE HIGH-EFFICIENCY SiC THREE-PHASE INVERTER WITHOUT REDUNDANCY

Failure Mode	MTTF [years]	Failure Rate, λ
V_{GS} Stress	7.88	0.1269
V_{DS} Stress	28.78	0.0347

B. Active Redundancy

For this case, active redundancy, the VSC required only half of the power modules per phase in order to operate in the safe at the rated conditions (see Fig. 10). Therefore, a certain degree of redundancy could be achieved. In order to do that, additional components should be included in the switching loop, such as disconnectors. These components will add parasitic inductances that modify the switching performance and consequently the efficiency of the system. However, for this paper, all these additional components are considered to be more reliable than the power modules and capacitors.

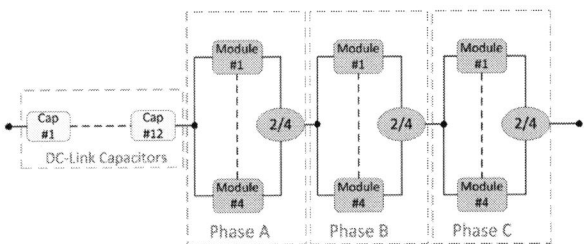

Figure 10: Reliability block diagram for the high-efficiency SiC three-phase inverter with active redundancy.

In order to calculate the MTTF of the system, several calculations must be performed, considering the fact that the failure rate is not constant. For redundant systems, a higher reliability is expected for shorter mission times.

For each phase of the inverter four different states are considered. These states are: (a) fully functional (all power modules working), (b) one failed (one power module failed), (c) two failed (two power modules failed) and, (d) phase failed (more than two power modules failed). Finally, the failure rate for each state must be calculated using the methodology for series and parallel system. For instance, for the first state (a), Fig. 11 shows the equivalent system of a single phase, and using Eq. 3, the failure rate was calculated. Similarly, calculations were performed for each of the other states.

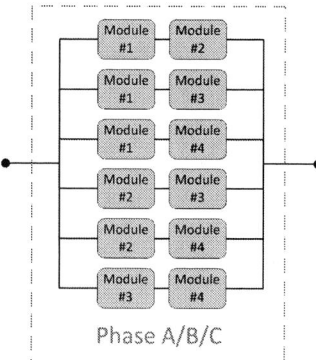

Figure 11: Reliability block diagram for one phase of the high-efficiency SiC three-phase inverter with active redundancy.

$$\lambda_{Phase} = \frac{2\lambda_{Module}}{\Sigma_{i=1}^{6} \frac{1}{i}} \qquad (3)$$

Consequently, Fig. 12 shows the transition between the considered states using the so called Markov Models. The probabilities of failure with respect to the mission time were calculated for each state of the system and for each failure mode as seen in Fig. 13(a) and Fig. 13(b). From these probabilities, the failure rate of the system could be calculated, as well as the probability density function, which contains the MTTF (see Fig. 13(c) and Fig. 13(d)). It must be noted that the plots for V_{DS} stress and V_{GS} stress, are extended to 400 years and 100 years respectively in order to illustrate how the failure rate changes regarding the mission times. Finally, active redundancy increases the MTTF several times, approximately 41 years, as shown in Table VI.

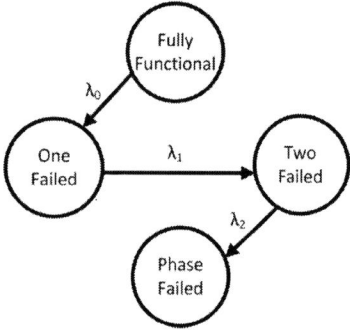

Figure 12: Markov diagram for one phase of the high-efficiency SiC three-phase inverter with active redundancy.

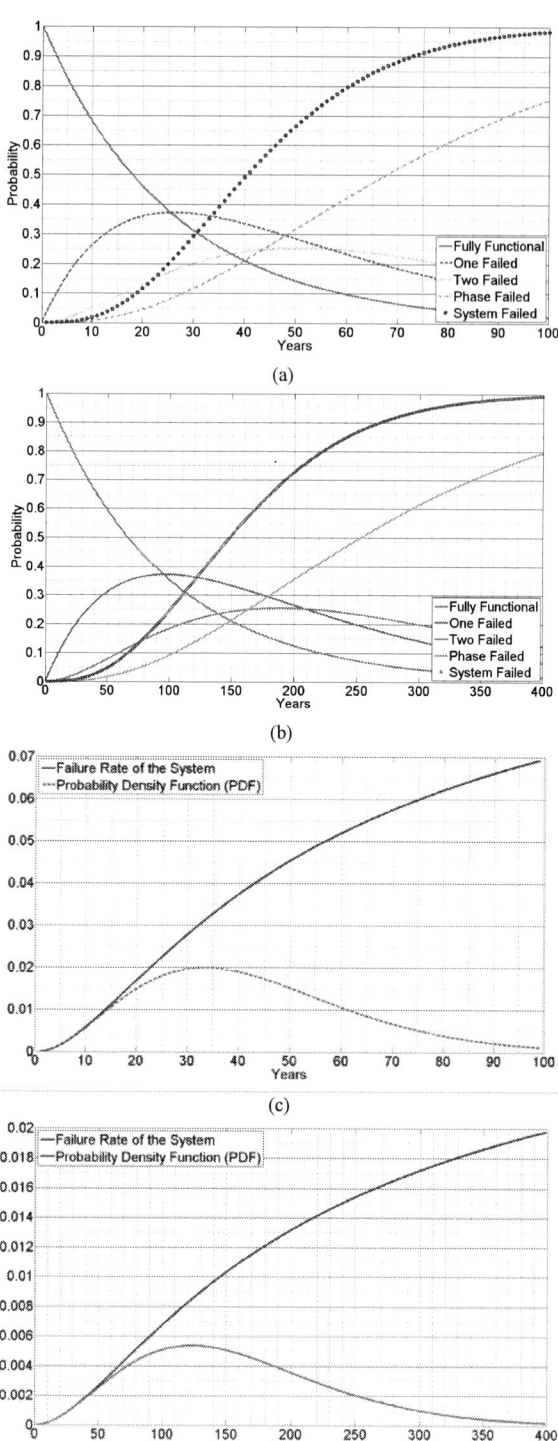

Figure 13: Probability of each state of Markov Models and Probability of system failure regarding the mission time for the (a) gate-source voltage stress, (b) drain-source voltage stress. Failure Rate and Probability density function regarding the mission time for the (c) gate-source voltage stress, (d) drain-source voltage stress.

TABLE VI.   MTTF AND FAILURE RATE OF THE HIGH-EFFICIENCY SiC THREE-PHASE INVERTER WITH ACTIVE REDUNDANCY

Failure Mode	MTTF [years]
V_{GS} Stress	41.29
V_{DS} Stress	155.34

IV. DISCUSSION

As shown in the previous section, the gate-source voltage stress determines the life expectancy of the system. A more detailed description on how the MTTF varies regarding the gate voltage is shown in Fig. 14. Higher life expectancy is obtained by reducing V_{GS}. However, if active redundancy is included, a similar value of MTTF could be achieved as when the V_{GS} is reduced without reducing it.

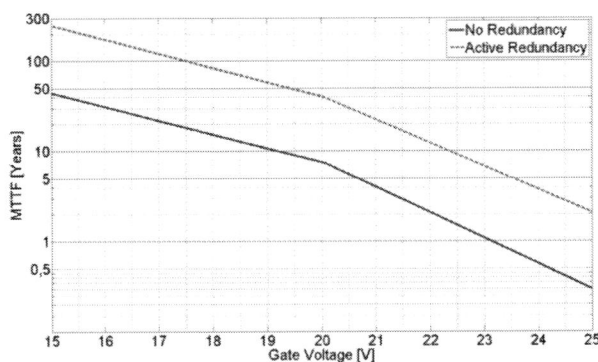

Figure 14: MTTF as function of Gate Voltage. No Redundancy, (blue), Active Redundancy (dash green).

Furthermore, reducing the gate-source voltage, will impact different parameters of the system such as the on-state resistance and the switching speeds, thus, affecting the total losses of the system. Using a simulation model of the power module, it is possible to estimate how these losses depend on the gate voltage. The simulation model of the investigated power module (CAS300M12BM2) was developed using ANSYS Simplorer and Q3D. Finally, the power module was simulated with LTSPICE including the parasitic elements derived from ANSYS. Fig. 15(a) and Fig. 15(b) show the experimental results and the simulation results, respectively, for the turn-ON transitions. Similarly, Fig. 15(c) and Fig. 15(d) show the results for the turn-OFF transition. It can be noted that the simulation results fit the experimental results appropriately. Therefore, an estimation of how the total losses change regarding V_{GS} could be performed.

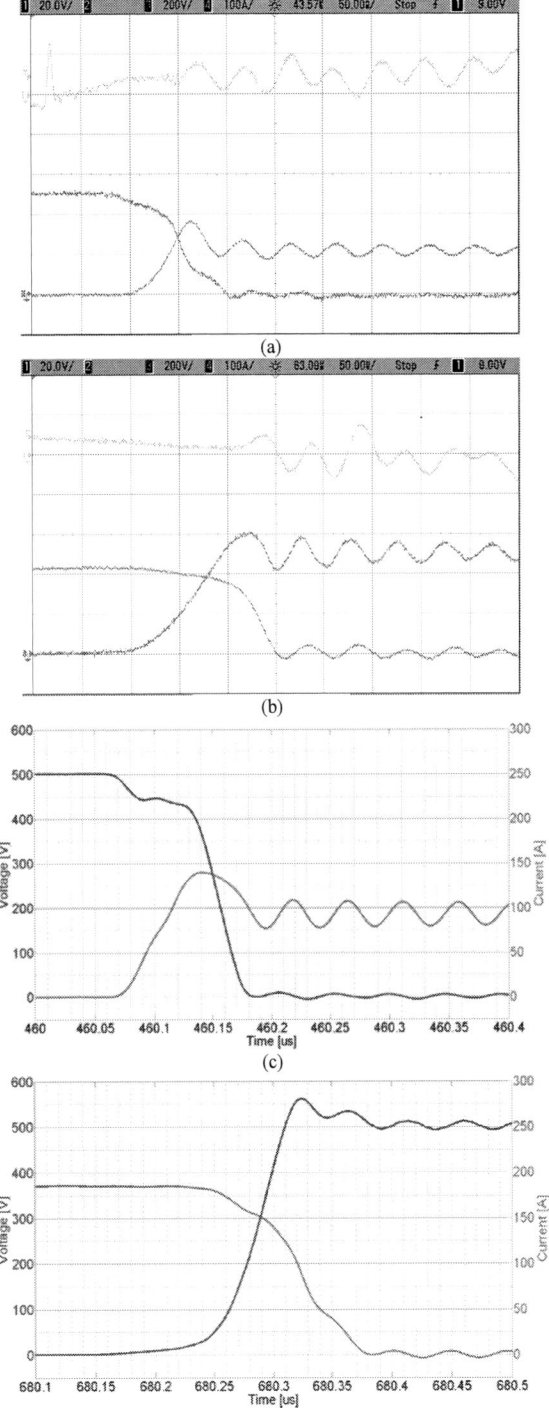

Figure 15: (a) Turn-ON, (b) Turn-OFF switching waveforms of the power module. Measured drain-source voltage of the SiC MOSFET, (purple, 200 V/div), drain-source current of the SiC MOSFET, (pink, 100 A/div), gate-source voltage of the SiC MOSFET, (yellow, 20 V/div), (time-base 50 ns/div), (c) Turn-ON, (d) Turn-OFF transients of the simulated SiC MOSFET power module, drain-source voltage of the SiC MOSFET, (blue, 100 V/div), drain-source current of the SiC MOSFET, (green, 50 A/div).

978-1-4673-9551-9/16 $31.00 © 2016 IEEE

Using the developed simulation model, the transient performance at different gate voltages were analyzed and plotted in Fig. 16. It is possible to estimate how the switching losses as well as the conduction losses vary depending on the gate voltage. As expected, the lower V_{GS}, the higher the losses.

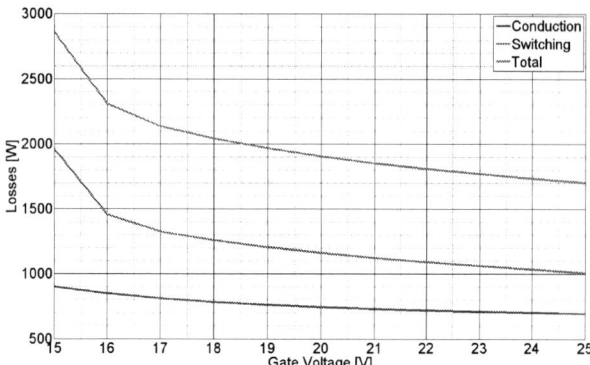

Figure 16: Conduction losses (blue), Switching losses (green), Total losses (red) regarding of the Gate Voltage.

Therefore, reducing the gate voltage will affect the efficiency of the system as shown in Fig. 17. By decreasing the positive gate-source voltage, 5 V, the reliability of the system is increased approximately 8 times and the efficiency is reduced by approximately 0.4 %. Nevertheless, this reduction of the efficiency is not significant compared to the MTTF improvement.

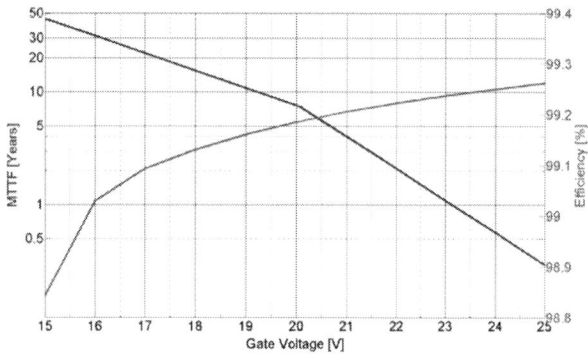

Figure 17: MTTF (blue) and Efficiency (green) regarding of the Gate Voltage.

Several assumptions have been made in order to perform the reliability study. First, when active redundancy is introduced in the study implies that the power modules that fail during operation are disconnected. As soon as this occurs, the remaining power modules will conduct higher current in order to maintain the rated output power. By conducting higher current the device temperatures will increase affecting

the reliability of the system. Nevertheless, the reliability calculations have been done with extrapolated values at 150 °C, i.e. that the study performed is a worst-case analysis.

Also, during the blanking time, part of the current flows through the intrinsic body diodes of the MOSFETs. Several studies have been dealing with the reliability of the body diode [20]-[22]. These studies show a stable body diode performance under a 1000 hour DC body diode stress of 22 A.

Additionally, the negative gate-source voltage, used in many applications, might apply more stress and therefore determine the reliability of the complete system. Several studies have been performed already regarding this aspect [21]-[23]. These studies show a stable threshold voltages for a $V_{GS} = -15$ V of stress for 1000 hours at 150°C. The average threshold voltage shift for the devices under this stress was approximately -50 mV.

However, it is the hypothesis of the authors that the use of the body diode of the device as well as the negative bias stress might impact the reliability of the system and could determine the life of the inverter. Therefore, this must be investigated in order to determine the failure rate and MTTF in respect to these conditions.

Moreover, the packaging itself could be the factor that determines the reliability of the system. However, for the power module analyzed in this paper, the package is similar to the one used for silicon IGBT power modules, which has been proven for several applications under different reliability standards. This is not considered in this work as a dominant factor of the reliability.

Finally, different strategies could be used so as to increase the reliability of a system, such as reducing the temperature. In order to do this, a lower current must flow through the power modules, i.e. more devices connected in parallel, which will not necessarily increase the reliability of the system. Instead adding additional components might have a negative effect on the operating life of the system. Another possibility is to decrease the voltage level which could also increase the reliability of the system. However, in this case, an additional transformer is needed so as to satisfy the rated conditions of the system. Nevertheless, the drain-source voltage stress does not determine the reliability of the system. Lastly, higher quality components, i.e. better SiC chips will also increase the final reliability.

V. CONCLUSIONS

A possible solution for higher efficiencies using SiC has been presented, using parallel connection of power modules. Experimental results of the proposed VSC driving a motor verify an efficiency of 99%. A reliability analysis has been performed on a 312 kVA VSC for motor drive applications. Two different failure modes have been studied, the V_{GS} stress and the V_{DS} stress. Additionally, two possible cases were analyzed: no redundancy and active redundancy. This analysis has shown that the gate-source voltage stress determines the reliability and MTTF of the complete system. Nevertheless, decreasing the positive gate-source voltage could increase the reliability of the system approximately 8 times without affecting the efficiency significantly. Moreover, adding redundancy in the system could also increase the MTTF approximately 5 times.

REFERENCES

[1] J. Rabkowski, D. Peftitsis, H.-P. Nee, "Silicon Carbide Power Transistors: A New Era in Power Electronics Is Initiated," *IEEE Ind. Electron. Mag.*, vol. 6, no. 2, pp. 17–26, Jun. 2012.

[2] J. Rabkowski, D. Peftitsis, H.-P. Nee, "Design steps toward a 40-kVA SiC JFET inverter with natural-convection cooling and an efficiency exceeding 99.5%," *IEEE Trans. Ind. Appl.*, vol. 49, no. 4, pp. 1589–1598, Jul./Aug. 2013.

[3] Peftitsis, G. Tolstoy, A. Antonopoulos, J. Rabkowski, J.-K. Lim, M. Bakowski, L. Ängquist, H.-P. Nee, "High-Power Modular Multilevel Converters With SiC JFETs," *IEEE Trans. Power Electron.*, vol. 27, no. 1, pp. 28–36, Jan. 2012.

[4] Wrzecionko, J. Biela, J.W. Kolar, "SiC power semiconductors in HEVs: Influence of junction temperature on power density, chip utilization and efficiency ", *35th Annual Conf. of IEEE Industrial Electron. IECON '09*, 2009, pp. 3834 – 3841.

[5] Tolstoy, P. Ranstad, J. Colmenares, D. Peftitsis, F. Giezendanner, J. Rabkowski, H.-P. Nee, "An experimental analysis of how the dead-time of SiC BJT and SiC MOSFET impacts the losses in a high-frequency resonant converter", *2014 16th Eur. Conf. on Power Electron. and Appl. (EPE)*, Sept. 2014.

[6] J. Rabkowski, D. Peftitsis, M. Zdanowski, H.-P. Nee, "A 6 kW, 200 kHz boost converter with parallel-connected SiC bipolar transistors," *IEEE Trans. Power Electron.*, vol. 29, no. 5, pp. 2482-2491, May 2014.

[7] J. B. Casady, "SiC Power Devices and Modules Maturing Rapidly," Power Electronics Europe, vol. 1/2013, no. 1, pp. 16–19, Jan-2013.

[8] L. Stevanovic, A. Bolotnikov, E. Kaminski, R. Beaupre, S. Kennerly, D. Lilienfeld, "Advanced SiC MOSFETs for High Power Applications," *CFES 2012-2013 Annual Conference*, 25-Jan-2013.

[9] A. Dutta, W. Shijie, Z. Jinchang, S.S. Ang, C. June-Chien, C. Chang-Sheng, "The design and fabrication of a 50KVA 450A silicon carbide power electronic module," *2013 4th IEEE International Symposium on Power Electronics for Distributed Generation Systems (PEDG)*, July 2013, pp. 1-5.

[10] J. Colmenares, D. Peftitsis, J. Rabkowski, H.-P. Nee, "Switching Performance of Parallel-Connected Power Modules with SiC MOSFETs", *2014 International Power Electron. Conf. (IPEC)*, May 2014.

[11] J. Colmenares, D. Peftitsis, D. Sadik , G. Tolstoy ,J. Rabkowski, H.-P. Nee, "High-Efficiency 312-kVA Three-Phase Inverter using Parallel Connection of Silicon Carbide MOSFET Power Modules", *IEEE Trans. on Ind. Appl*, Dec. 2015.

[12] D.-P. Sadik, J. Colmenares, G. Tolstoy, D. Peftitsis, J. Rabkowski, H.-P. Nee, "Analysis of Short-Circuit Conditions for Silicon Carbide Power Transistors and Suggestions for Protection", *2014 16th Eur. Conf. on Power Electron. and Appl. (EPE)*, Sept. 2014.

[13] X. Huang, G. Wang, Y. Li, A. Q. Huang, B. J. Baliga, "Short-circuit capability of 1200V SiC MOSFET and JFET for fault protection," in *2013 Twenty-Eighth Annual IEEE Applied Power Electronics Conference and Exposition (APEC)*, 2013, pp. 197–200.

[14] A. Castellazzi, T. Funaki, T. Kimoto, T. Hikihara, "Short-circuit tests on SiC power MOSFETs", *2013 IEEE 10th International Conference on Power Electronics and Drive Systems (PEDS)*, 2013, pp. 1297–1300.

[15] A. Ward, "SiC Power Diode Reliability ",*CREE InC.*, Oct. 2008.

[16] S. Allen, "Silicon Carbide MOSFETs for High Powered Modules", *CREE InC.*, March 2013.

[17] D. A. Gajewski, S. H. Ryu, M. Das, B. Hull, J. Young, J. W. Palmour, "Reliability Performance of 1200 V and 1700 V 4H-SiC DMOSFETs for Next Generation Power Conversion Applications," *Mater. Sci. Forum*, vol. 778–780, Feb. 2014, pp. 967–970.

[18] R.A. Wood, T.E. Salem, "Long-term operation and reliability study of a 1200-V, 880-A all-SiC dual module," *2012 International Symposium on Power Electronics, Electrical Drives, Automation and Motion (SPEEDAM)*, June 2012, pp. 1520-1525.

[19] T. Ueda, "Reliability issues in GaN and SiC power devices," *2014 IEEE International Reliability Physics Symposium(IRPS)*, June 2014.

[20] A. Agarwal, H. Fatima, S. Haney, S. H. Ryu, "A New Degradation Mechanism in High-Voltage SiC Power MOSFETs," *IEEE Electron. Device Letters*, vol. 28, no. 7, pp. 587-589, July 2007.

[21] B. Hull, S. Allen, Q. Zhang, D. Gajewski, V. Pala, J. Richmond, S. Ryu, M. O'Loughlin, E. Van Brunt, L. Cheng, A. Burk, J. Casady, D. Grider, J. Palmour, "Reliability and stability of SiC power mosfets and next-generation SiC MOSFETs," *2014 IEEE Workshop on Wide Bandgap Power Devices and Applications (WiPDA)*, pp.139-142, Oct. 2014.

[22] D.-P. Sadik, J. K. Lim, P. Ranstad, H.-P. Nee, "Investigation of Long-term Parameter Variations of SiC Power MOSFETs", *2015 17th Eur. Conf. on Power Electron. and Appl. (EPE)*, Sept. 2015.

[23] R. Green, A. Lelis, D. Habersat, "Application of reliability test standards to SiC Power MOSFETs," *2011 IEEE International Reliability Physics Symposium (IRPS)*, April 2011.

[24] N. Thanh-That, A. Ahmed, T.V. Thang, P. Joung-Hu, "Gate Oxide Reliability Issues of SiC MOSFETs Under Short-Circuit Operation," *IEEE Trans. Power Electron.*, vol.30, no.5, pp. 2445-2455, May 2015.

[25] M. A. Anders, P. M. Lenahan, A. J. Lelis, "Negative bias instability in 4H-SiC MOSFETS: Evidence for structural changes in the SiC," *2015 IEEE International Reliability Physics Symposium (IRPS)*, April 2015.

[26] A.J. Lelis, R. Green, D.B. Habersat, M. El, "Basic Mechanisms of Threshold-Voltage Instability and Implications for Reliability Testing of SiC MOSFETs," , *IEEE Trans. on Electron. Devices*, vol. 62, no. 2, pp. 316-323, Feb. 2015.

[27] T. Santini, M. Sebastien, M. Florent, L. V. Phung, B. Allard, "Gate oxide reliability assessment of a SiC MOSFET for high temperature aeronautic applications," *2013 IEEE ECCE Asia Downunder (ECCE Asia)*, pp. 385-391, June 2013.

[28] A. J. Lelis, D. Habersat, R. Green, A. Ogunniyi, M. Gurfinkel, J. Suehle, N. Goldsman, "Time Dependence of Bias-Stress-Induced SiC MOSFET Threshold-Voltage Instability Measurements,*" IEEE Trans. on Electron Devices*, vol. 55, no. 8, pp. 1835-1840, Aug. 2008.

[29] A. J. Lelis, R. Green, D. Habersat, N. Goldsman, "Effect of Threshold-Voltage Instability on SiC DMOSFET Reliability," *IEEE International Integrated Reliability Workshop*, pp.72-76, Oct. 2008.

[30] D. Watt, "Cree SiC Diodes High Temperature Reliability Trials." Aerospace Lab, TT Electronics Semelab Limited, 2012.

[31] S. Tanimoto, H. Ohashi, "Reliability issues of SiC power MOSFETs toward high junction temperature operation," *Phys. Status Solidi A*, vol. 206, no. 10, pp. 2417–2430, 2009.

[32] L. C. Yu, G. T. Dunne, K. S. Matocha, K. P. Cheung, J. S. Suehle, K. Sheng, "Reliability Issues of SiC MOSFETs: A Technology for High-Temperature Environments," *IEEE Trans. Device Mater. Reliab.*, vol. 10, no. 4, pp. 418–426, Dec. 2010.

RCP Evaluation of Electrolytic Capacitor Degradation for SMPS Failure Prediction

Hiroshi Nakao*†, Yu Yonezawa*,and Yoshiyasu Nakashima*
Computer Systems Laboratory
Fujitsu Laboratories LTD. *
Kawasaki-Shi, Japan
E-mail: Nakao.h@jp.fujitsu.com

Fujio Kurokawa†
Dept. of Advanced Technology and Science for Sustainable Development
Nagasaki University†
Bunkyo-Machi, Japan

Abstract— **This paper presents a new cost effective method for a failure prediction of a digitally controlled Switching Mode Power Supply (SMPS). Electrolytic capacitor is known as one of the highest failure rate components in SMPS, so we tried to detect an Equivalent Series Resistance (ESR) degradation of the electrolytic capacitor directly from the data fetched to a digital controller of SMPS. With a SPICE simulation and a Rapid Control Prototyping (RCP) evaluation, we confirm the degradation can be detected by the data at a transient response under a normal operation. Even only 10 percent of a load step change, which commonly occurs in SMPS for servers, causes detectable transient response degradation.**

Keywords— *Rapid Control Prototyping, RCP, Electrolytic Capacitor, Failure Prediction, Server power supply*

I. INTRODUCTION

In recent years, the scale of data centers has become larger and larger, and more and more servers are installed in them. Therefore, cost issues in data centers such as energy, equipment, operational, and maintenance costs have become more significant [1]. From the point of view of power electronics, energy cost has been reduced by developing high efficiency Switching Mode Power Supply (SMPS) such as 80 plus [2], cooling optimization [1], higher voltage supplying electricity [3], and so on. On the other hand, we believe cost effective failure prediction can be a key technology for reducing equipment, operational, and maintenance costs. If reliable and cost effective failure prediction is available, we can cut back redundancy of power supply units or reduce maintenance cost with planned exchange of the expendable units.

An Electrolytic Capacitor is known as one of the highest failure rate components in SMPS. There are many studies on failure prediction of electrolytic capacitors [4]-[11]. The degradation of Electrolytic Capacitors occurs with consumption of encapsulated electrolyte, and Equivalent Series Resistance (ESR) is more sensitive than capacitance in orders of magnitude for electrolyte consumption [4]. So, almost all studies evaluate Electrolytic Capacitor degradation as ESR degradation using switching ripple. However, for server SMPS, very small ripple with higher frequency, such as less than 120 mVpp with higher frequency than several tens kHz [12], must be evaluated, which requires high speed and high sensitivity

detection circuit. It is not so easy to evaluate switching ripple in operation with a cost effective method.

On the other hand, in digitally controlled SMPS, input/output and voltage/current data are AD converted and fetched to a digital controller to control switching timing. If capacitor degradation can be detected using only these data, cost effective failure prediction will be available. In this paper, capacitor degradation evaluation of digitally controlled SMPS using a Rapid Control Prototyping (RCP) system is reported.

II. RAPID CONTROL PROTOTYPING

Figure 1 shows the RCP system used for this study. RCP is one of the important processes in a Model Based Development (MBD) scheme [13], [14], [15]. The RCP system is connected to the SMPS power line instead of the digital controller, and quickly tests the control algorithm evaluated by the Model in a Loop Simulation (MILS), without a coding process for a specific digital controller. Very recently, we reported an RCP system designed for a server SMPS that supports an over 100-kHz control frequency and less than sub-nanosecond time resolution power [16]. An RCP system acquires IV signals from the convertor circuit, calculates PWM values with high performance CPU and sends PWM signals to the gate drivers.

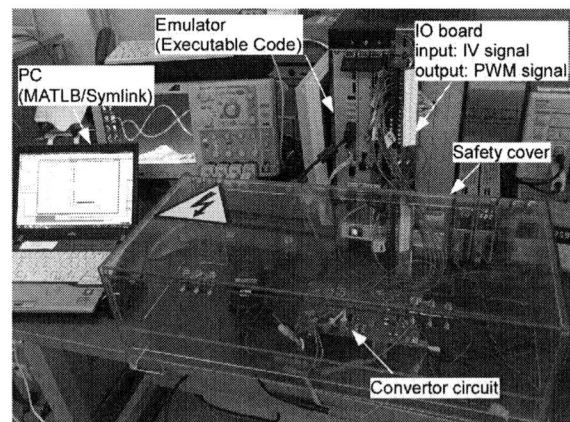

Figure 1-a. Setup image of the RCP system designed for SMPS.

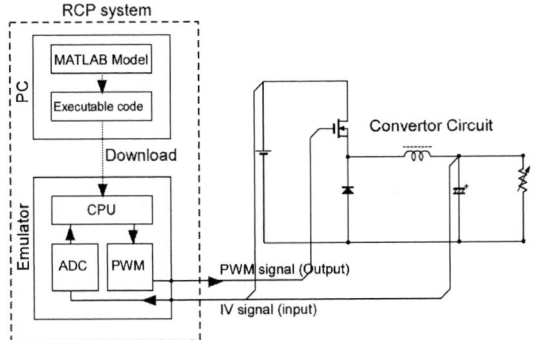

Figure 1-b. Schematic image of the RCP system. The RCP system is connected to the Converter circuit instead of the digital controller.

To develop high functional SMPS, the RCP has two great advantages for conventional development processes, programming directly to an SMPS digital controller. The first advantage is separating debugging processes in hardware and firmware. Using sufficient high performance control emulator, we can debug the control algorithm without worrying about side effects such as delay in switching timing, code size overflow, incorporation of hand coding bugs, and so on. They can easily happen in the hand coding process for the new algorithm. By selecting after algorithm and hardware debugging, we can select the most cost effective digital controller with sufficient performance. The second advantage is direct monitoring of IO fetched data. The RCP system can evaluate the same IO data sent to the digital controller for controlling SMPS, so there is no distortion such as probing problems. We can evaluate sensitivity for circuit component degradation and noise immunity of the control circuit directly from the controller fetched data.

III. SPICE SIMULATION

Firstly, SPICE simulation was performed for sensitivity analysis. Fig. 2a is schematic diagram of a 500 W Full-bridge DC-DC converter. SPICE parameters are summarized in Table 1. Step response and switching ripple were estimated using LTSPICE [17]. Fig. 2b shows waveforms for +30% load step response for a non-degraded Capacitor. Vout ripple is ±50 mV with 200 kHz frequency and is detectable enough using a suitable band-pass filter, high speed amplifier and high speed AD converter (ADC). However, this requires additional cost. Usually a digital controller for SMPS does not have high speed

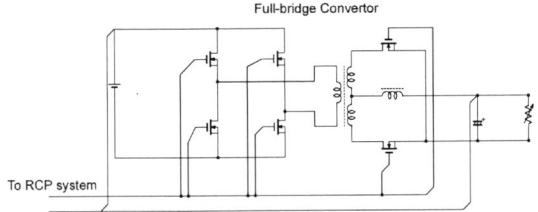

Figure 2-a. A schematic diagram of a 500W Full-bridge DC-DC converter with synchronous rectifier.

Table 1: SPICE Parameters for Full-bridge DC-DC Converter

Item	Specification and parameters
Circuit topology	Full bridge converter with synchronous rectifiers
Input voltage	400 V
Output voltage	12 V
Input Capacitor	330 uF with ESR 800 mΩ
Output Capacitor	1500 uF with ESR 25 mΩ x 5
Output Inductor	2 uH
Transformer	21:1 turn
Chopper FET	Infineon SPP20N60CFD
Rectifier FET	Fairchild FDP032N08
Switching frequency	100 kHz
AAF cutoff frequency	20 kHz
Compensator type	Simple integrator

ADC for over MHz sampling.

Ripple after the Anti-Aliasing Filter (AAF), which is the same value to be fetched to the digital controller, is dumped at ±1mV. This fact is consistent with AAF design, for dump switching ripple to be smaller than the Least Significant Bit (LSB) of ADC. For example, the LSB of 3.3V full-range 10 bit ADC is 3 mV.

On the other hand, the voltage drop, which depends on the product between the current step height and ESR, is detectable. This voltage drop is dominated by the compensator and

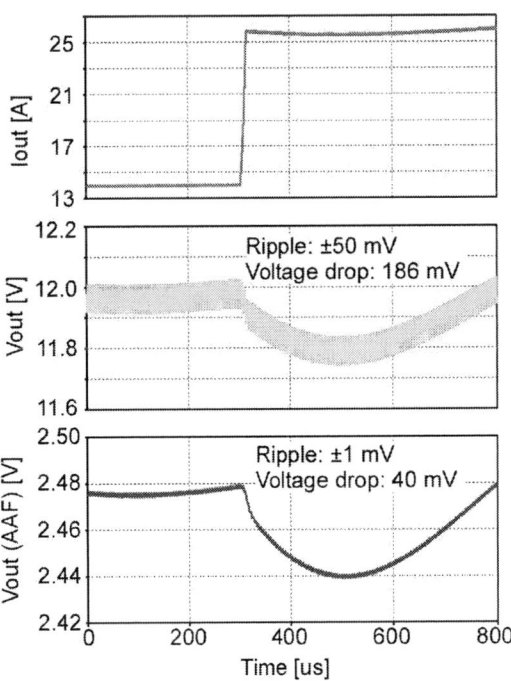

Figure 2-b. SPICE simulated waveforms at step load change.

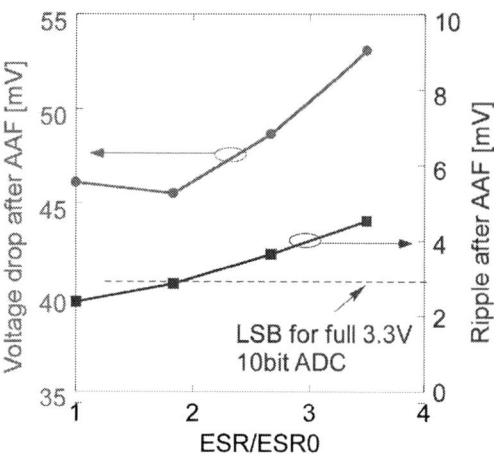

Figure 2-c. Voltage drop and ripple after AAF evaluated with various ESR.

sufficiently slower than AAF cutoff.

Fig. 2c shows the voltage drop and the ripple after AAF for degraded ESR. Even if ESR is degraded to 3 times greater than the initial value, ripple after AAF is smaller than the twice value of LSB. However, according to the sampling theorem, the fetched deviation to the digital controller must be much smaller than this simulated ripple. This means that capacitor degradation detection with switching ripple only is difficult.

On the other hand, voltage drop distortion is several times greater than LSB with ESR degradation, so we try to detect the capacitor degradation using this step response.

IV. EXPERIMENTAL SETUP

A 500W Full-bridge DC-DC converter is used as a test bed [16]. Five electrolytic capacitors (Nippon Chemi-Con Co. EKY-160ELL152MJ30S 16 V, 1500 uF, ESR 31 mΩ) are installed as output filters. C and ESR of this converter are varied as parameters: 1) reducing the number of parallel capacitors from 5 to 3, which means that C and ESR varied from 7.5 mF and 6.5 mΩ to 4.5 mF and 10.3 mΩ, respectively; and 2) adding serial resistance to ESR from 6.5 mΩ to 26.2 mΩ. AD sampling for converter control is 100 kHz, the same as the control frequency, and recorded to RCP memory as thinned data of 10 kHz. Electric load is used as load and transient response for step load change of half load (250 W) ±10, 20, 30 % with 160 mA/us setting.

V. RESUTS AND DISCUSSIONS

Figure 3 shows waveforms for output voltage measured by an oscilloscope (a, b) and the RCP system (c, d). ESRs of output capacitors are 6.2 mΩ (a, c) and 21.2 mΩ (b, d), respectively. In responding to ESR degradation, both broadening of envelop of waveform and increase of degradation of transient response are observed in the RCP fetched data, which is the same as the oscilloscope waveforms. Unlike the result of SPICE simulation, envelope broadening was observed in RCP fetched data. We think this difference is

Table 2: Specifications of 500 W Full-bridge DC-DC Converter

Item	Specification and designed value
Circuit topology	Full bridge converter with synchronous rectifiers
Input voltage	400 V
Output voltage	12 V
Output Current	0 to 41.67 A
Output power	0 to 500 W
Input Capacitor	330 uF ESR 800 mΩ
Output Capacitor	1500 uF ESR 25 mΩ x 5
Output Inductor	2 uH
Transformer	21:1 turn
Chopper FET	Infineon SPP20N60CFD
Rectifier FET	Fairchild FDP032N08
Switching frequency	100 kHz
Sampling period (control)	10 us (100 kHz)
Sampling period (failure prediction)	100 us (10 kHz)
Compensator type	3 pole 3 zero (Type 3 equivalent)
Crossover frequency	2 kHz
Gain margin	10 dB
Phase margin	45 degrees

depend on the switching noise caused by stray inductance and capacitance which is not considered in the SPICE simulation. There is some dispersion in peak-to-peak values of transient responses, too. We suppose that these dispersions are jitters affected by quantization error and switching noise. So some statistical work is required for failure prediction.

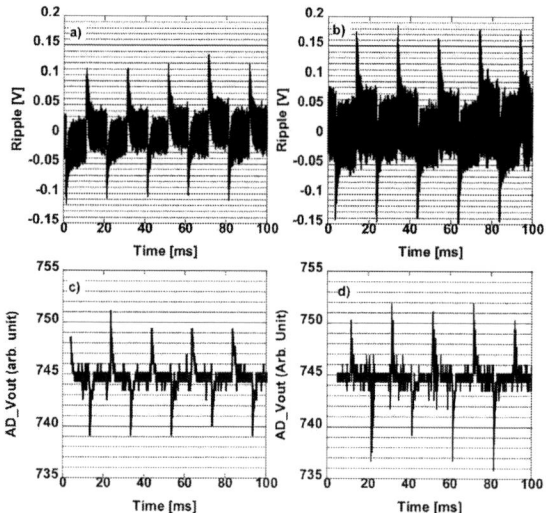

Figure 3. Waveforms of ±30% step response of output voltage. a) Raw value with initial ESR of 6.2 mΩ, b) with ESR degraded to 21.2 mΩ. c) AD converted value with initial ESR, d) with ESR degraded to 21.2 mΩ.

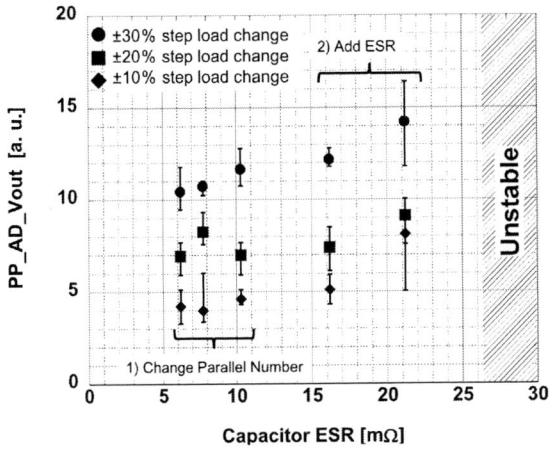

Figure 4. AD converted peak-to-peak value of step response for various Capacitor ESRs. ± 10% load step change seems to be detectable using the statistical method.

Figure 4 plots RCP fetched peak-to-peak values for step response vs. capacitor ESRs. Both conditions 1) and 2) are plotted together. Each condition shows almost the same trend and ESR is dominant in peak-to-peak value of step response. Even ± 10% step load change, which is commonly observed in server power supplies, causes a detectable degradation of the transient response. Usually, the same height step loads exist daily in the same server, such as step load from standby mode to a certain process, daily back up, and so on. By filtering step response used for failure prediction by a certain height step load, we can evaluate the trend in capacitance degradation.

Sampling periods for failure prediction is thinned to 10 kHz, however, degradation of transient response is clearly observed. That is, this method is light-load for low-cost digital controller. Memory and storage available in low-cost digital controller are limited and we can use the host server for complicated statistical work.

When the ESR is increased to 26 mΩ, which is four times greater than the initial value, the SMPS become unstable. This also means this evaluation method has enough sensitivity for failure prediction of SMPS.

VI. CONCLUSION AND FUTURE WORK

We proposed the new failure prediction method suitable for digitally controlled SMPS. Using SPICE simulation and RCP system, we evaluated the IV signals fetched to digital controller under normal operation. Because of the dumping with the AAF, the switching ripples, commonly evaluated for failure prediction is not detectable from digital controller without adding a high sensitivity and high speed ADC. However, transient response corresponding to the step load change is detectable. And we confirmed ESR degradation of electrolytic capacitors of SMPS is predictable with some statistical work. Even only ± 10% step load change causes detectable degradation of transient response. This method needs no additional circuit and cost effective. We are now trying to implement this RCP verified failure prediction algorithm into a digital controller using the Production Code Generation

process [18] and verify the prediction algorithm using a real server system.

ACKNOWLEDGMENT

The authors extend their thanks to Dr. Tomotake Sasaki of Fujitsu Laboratories LTD., Mr. Hisato Hosoyama and Mr. Atsusi Manabe of Fujitsu Advanced Technology LTD. for their useful comments about RCP and PCG. The authors also wish to say thanks for the support from Dr. Takeshi Horie Head of Computer systems laboratory at Fujitsu Laboratories Ltd. who gave us the chance to perform this research. Special thanks to Mr. Kouichi Kumon Member of the Board at Fujitsu Laboratories Ltd. who gave us a lot of support and useful advice.

REFERENCES

[1] L. A. Barroso, J. Clidaras, and U. Hölzle, "The datacenter as a computer: an introduction to the design of warehouse-scale machines, second edition", Morgan&Claypool, pp. 67-98, July (2013).

[2] 80 PLUS [Online], http://www.plugloadsolutions.com/80PlusPowerSupplies.aspx

[3] T. Babasaki, T. Tanaka, Y. Nozaki, T. Tanaka, T. Aoki, and F. Kurokawa, "Developing of higher voltage direct-current power-feeding prototype system", in Proc. INTELEC 2009: 31st International telecommunications energy conference, Incheon, Korea, pp. 1-5, Oct. (2009).

[4] K. Harada, A. Katsuki, and M. Fujiwara, "Use of ESR for deterioration diagnosis of electrolytic capacitor", IEEE trans. on power electronics, vol. 8, pp. 355-361, Oct. (1993).

[5] A. Lahyani, P. Venet, G. Grellet, and P.-J. Viverge "Failure prediction of electrolytic capacitors during operation of a switchmode power supply", IEEE trans. on power electronics, vol. 13, pp.1199-1207, Nov. (1998).

[6] A. M.R. Amaral and A. J. M. Cardoso, "Use of ESR to predict failure of output filtering capacitors in boost converters" in Proc. ISIE2004: 2004 IEEE international symposium on industrial electronics, vol.2, pp. 1309 – 1314, Ajaccio, France, May (2004).

[7] Y.-M. Chen, H.-C. Wu, M.-W. Chou, and K.-Y. Lee, "Online failure prediction of the electrolytic capacitor for LC filter of switching-mode power converters", IEEE trans. on industrial electronics, vol.55, pp.400-406, Jan. (2008).

[8] K. Abdennadher, P. Venet, G. Rojat, J. M. R´etif and C. Rosset "Online monitoring method and electrical parameter ageing laws of aluminium electrolytic capacitors used in UPS", in Proc PE '09: 13th European conference on power electronics and applications, pp. 1-9, Barcelona, Spain, Sep. (2009).

[9] K. Abdennadher, G. Rojat, J. M. Rétif, and C. Rosset, "A real-time predictive-maintenance system of aluminum electrolytic capacitors used in uninterrupted power supplies", IEEE trans. on industry applications, vol. 46, pp. 1644-1652, July/Aug. (2010).

[10] H. M. Pang. and P. M. H. Bryan, "A life prediction scheme for electrolytic capacitors in power converters without current sensor", in Proc. APEC2010: 2010 twenty-fifth annual IEEE applied power electronics conference and exposition, pp. 973-979, Palm Springs, USA, Mar. (2010).

[11] L. Liu, Y. Guan, M. Wu, and L. Wu, "Failure prediction of electrolytic capacitors in switching-mode power converters", in Proc. PHM2012: 2012 IEEE conference on prognostics and system health management, pp. 1-5, Beijing, Chaina, Sep. (2012).

[12] "ATX12V power supply design guide", Intel (2004).

[13] J. Sch¨auffele and T. Zurawka, "Automotive software engineering: principles, processes, methods and tools." Warrendale, Pa. : SAE International, (2005).

[14] R. K. Jurgen, Ed., "Automotive Software: PT-127". Warrendale, Pa. : SAE International, (2006).

[15] J. Krasner, "Comparring embedded design outcomes with and without model-based design," American Technology International," White Paper, Oct (2010).

[16] Y. Yonezawa, T. Sasaki, H. Hosoyama, H. Nakao, A. Manabe, J. Kaneko, Y. Nakashima and T. Maruyama, "Rapid control prototyping for server power supply with high-resolution PWM", in Proc. APEC2015: 2015 the 30th annual IEEE applied power electronics conference and exposition, pp. 2635-2641, Charlotte, USA , Mar. (2015).

[17] LTSPICE [Online] http://www.linear.com/designtools/software/

[18] T. Sasaki, H. Hosoyama, Y. Yonezawa, A. Manabe, K. Huang, X. Liu, J. Chen, J. Kaneko, and Y. Nakashima, "Production code generation for server power supply controller", in Proc. APEC2015: 2015 the 30th annual IEEE applied power electronics conference and exposition, pp. 2656-2663, Charlotte, USA, Mar. (2015).

Modular Test System Architecture for Device, Circuit and System Level Reliability Testing

Roland Sleik[*†], Michael Glavanovics[†], Sascha Einspieler[†‡],
Annette Muetze[*], Klaus Krischan[*]

[*]Electric Drives and Machines Institute, Graz University of Technology, Graz, Austria
[†]KAI Kompetenzzentrum Automobil- und Industrie-Elektronik GmbH, Villach, Austria
[‡]Smart Grids Group/Lakeside Labs, Alpen-Adria-Universität Klagenfurt, Austria

Email: roland.sleik@k-ai.at

Abstract—Reliability stress testing of power semiconductors requires significant development effort for a test apparatus to provide the required functionality. This paper presents a modular test system architecture which focuses on flexibility, reusability and adaptability to future test requirements. Different types of tests for different devices in application circuit configuration can be implemented based on the same modular test system concept. Vital parameters of the device under test (DUT) can be acquired in situ during the running stress test. This enables to collect drift data of this parameters. The control and data acquisition parts of the test system are separated from the actual test circuit. With this physical separation, the same control part can be used for different types of tests. Experimental results of a prototype test system are provided.

Keywords—aging, DC-to-DC converter, drift data, in situ measurement, life time, power semiconductors, reliability testing, stress conditions, test apparatus, wear-out.

I. INTRODUCTION

A. Aim of work

Power semiconductors are exposed to thermo-mechanical stress during their functional use in application. The stress occurs due to power cycling of the device, which results in its aging, and eventually the device will fatigue and may fail. Therefore, reliability characterization of power semiconductor devices is of paramount importance during development of new devices [1]. To improve the reliability and operating safety of power electronic systems, knowledge about aging mechanisms and failure causes is crucial. Conducting life stress tests is necessary to provide results which can be compared to simulation data of device behavior in stress condition. This paper focuses on in situ life-cycle analysis and stress testing of DUTs in application circuit configuration. It presents a modular stress test system which is designed to meet several different test requirements and can be relatively easily adjusted to meet demands of such analyses in the future.

Most reports on reliability testing focus on the behavior of the DUT. Usually, the test systems are designed to analyze the wear-out of *one specific* device under certain stress conditions. This paper focuses on the design of the test system itself: A system is proposed (Fig. 1) that can be easily adapted to changing test requirements and DUTs such as discrete transistors, integrated

Figure 1: Prototype modular test system for active thermal cycling.

circuits like a half-bridge configuration including driver, or a semiconductor power switch with additional built-in protection functions. A modular architecture is chosen for operating DUTs in a test setup equivalent to their real application circuit.

B. Novelty of the proposed modular test system

A well established test method consists in testing the bare DUT in a socket. This method has the advantage that characteristic device parameters, such as leakage current or transistor capacitances, can be well determined. Unfortunately, application-specific high load currents cannot be fed to the DUT when mounted in such a test socket.

Another conventional approach on the other side of production process is to submit the completely assembled field-application to a test. Here the whole system is put in an environmental chamber and tested in an exact application configuration. The drawback to this method is that only little life-relevant data is gained as only few parameters can be monitored in such a configuration.

The modular test system described in this paper shows a solution in-between these two methods. As the devices are soldered on DUT boards, no device characteristic parameters can be measured which may be seen as a drawback to the proposed structure. In return, the devices can be committed to an application-equivalent high load condition. Through this condition drift data of voltages, currents, on-state resistance (R_{dson}) and device temperatures are acquired. This is the key novelty of this proposed new test system: gaining life time drift data out of an application-equivalent test setup.

The second advance compared to standard industry practice (e.g. HTOL [2]) is that it is possible to measure vital parameters of the DUT in situ during a running stress test. Thus, no interim measurements outside the test system are necessary.

The acquired life time data gained from various stress tests can be modeled by a Coffin-Manson approach which is covered in detail in [3].

This paper is organized as follows: Section I-C gives an overview on the status quo of other available test systems found in publications. In Section II the different types of stress tests are covered which can all be conducted with the modular test system proposed here. Section III describes the modular test system in detail and Section IV shows the first prototype implementation of such a system. Section V concludes this paper with a summary and an outlook to future work.

C. Discussion of state-of-art

There are several examples of existing test system implementations. All works cited in this paragraph focus on the device to be tested and the test system is designed to fit for this *one* purpose. Only the DUT and life test results of this DUT are focused on. The literature on these systems do not report on any attempt to use the available test hardware in a systematic way for different DUTs or types of tests, or both, as it is in the focus of the system proposed and discussed in this paper.

The authors of [4] introduce a test setup for high power IGBT modules with a common centralized test system architecture. The measured collector-emitter voltage V_{ce} of an IGBT transistor is cited to be the best indicator for the wear-out condition (bond-wire lift-off) of the DUT. The data acquisition during on-line stress testing will also be available by the modular stress test system proposed here. The implementation of the V_{ce} measurement circuit is described in detail in [5].

In [6], a test bench for IGBT modules similar to the one proposed in [4] is introduced. In addition to V_{ce} the junction temperature T_j is also monitored on-line. Again, this system has a centralized architecture which will make adaptations to future requirements quite difficult.

A test system for stress testing automotive smart power switches is proposed in [7]. It provides a high level of automation. A large number of devices can be submitted to stress in parallel. Three parameters (V_{ds}, I_{load}, R_{dson}) are

acquired in situ during a stress test run for each DUT. A similar centralized architecture has also been used for the test system described in [8] in which the inductive clamping behavior of power MOSFETs after turn off is analyzed.

In [9] and [10] life stress test investigations of automotive smart power switches are presented. Both papers focus on the findings from the reliability stress investigations. Unfortunately little information on the test system is provided. The papers explain defects and failures on the DUTs, that occurred at the power semiconductors through repetitive short circuit operation.

II. Types of Stress Tests

Three common types of stress tests are discussed below. In all cases, the DUTs are placed in an environmental chamber for operation at elevated ambient temperatures. Interim device condition determination is performed during stress tests. Many reliability test systems do not support device condition determination, thus demanding regular user interaction. The proposed modular test system is designed to perform this condition determination in situ without user interaction. Moreover, the device condition can be acquired more frequently as the measurements are initiated automatically by the control software.

A. Active thermal cycling

This type of test either applies to discrete power transistors comprised in DC-DC topologies (such as buck, boost or Čuk converter) or to ASICs with integrated power transistors in a DC-DC topology, including their drivers. For operating life stress tests on solid state devices, the JEDEC 22-A108D standard [2] is available. The instructions for the High Temperature Forward Bias test (HTFB) [2] explicitly specify the necessary conditions for this type of test.

The stress test duration should reflect the time frame of the corresponding application. Therefore, the aging and thus the test time are accelerated by submitting the device to higher stresses than during nominal operation. The ambient temperature and self-heating of the power device are taken into account and the resulting junction temperature T_j is kept below or equal to the maximum specified operating temperature (typical values: 85 °C to 125 °C for consumer devices, 150 °C for automotive devices). The bias input supply voltage is set to the maximum level specified for the DUT to still comply with the data sheet specification. The DUTs are connected to adjustable electronic loads outside the environmental chamber. The load is set to stress the DUTs at, or near the maximum rated current [2]. The DUTs are submitted to intermittent loads which toggle between high load of nearly 100 % until T_j reaches the maximum operating temperature and low load of 10 % until T_j reaches the ambient temperature. In practical reliability tests, the timing is chosen to reach the appropriate T_j temperature. This thermal toggling increases the stress on the DUT and accelerates its aging.

B. Automotive repetitive short circuit testing

This test is designed to stress smart power switches (SPSs), used in automotive applications. SPSs are equipped with

978-1-4673-9551-9/16 $31.00 © 2016 IEEE

integrated driving, protection and diagnostic functions. In automotive applications they have taken the place of electromechanical relays for switching all kinds of electric loads, such as incandescent bulbs, solenoids and motors [11]. Protection against electric and thermal overload, a common feature of SPSs, is usually based on over-current detection and limitation, junction temperature sensing and protective shutdown, if electrical or thermal maximum ratings, or both, are exceeded [12].

Standard AEC-Q100-012 [13] defines the test procedures for repetitive short circuit characterization, including test circuit configuration, ambient temperature and impedances. Test equipment designed to perform stress on SPSs is proposed in [7], but it again has a centralized and less flexible architecture. The source and load side impedances are selectable by passive components (i.e. air coils and power resistors). These impedances emulate the cable harness in a car. The DUTs are turned on in short circuit condition. The SPS will limit the DUT's load current and eventually shut down when the DUT's junction temperature T_j reaches the temperature limit (typical value: $150\,°C$). The SPS is repeatedly put in short circuit until malfunction occurs and life time data is recorded. The proposed modular test system is able to perform these types of tests, as the application hardware modules are exchangeable.

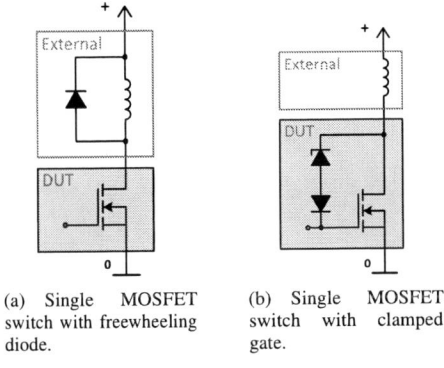

(a) Single MOSFET switch with freewheeling diode.

(b) Single MOSFET switch with clamped gate.

Figure 2: Circuits for switching inductive loads.

C. Inductive load clamping

Integrated power switches for automotive applications must be capable of switching inductive loads, while employing a minimum of additional components. The common approach would be to use a freewheeling path as in Fig. 2(a) to demagnetize the inductance after transistor turn-off. In automotive applications the cost efficient solution is an integrated gate clamp, as shown in Fig. 2(b), which limits the drain-source voltage of the power MOSFET during turn-off. In fact the MOSFET turns on again until the load inductance is fully demagnetized. The main drawback is that the inductive energy which dissipates in the power switch during turn-off causes a significant rise in junction temperature, posing severe stress on the DUT.

Therefore, repetitive inductive clamping stress tests are of interest, too. In [8], a test system designed for such purpose is introduced which again has a centralized architecture. For the definition of stress conditions, passive inductive loads are replaced by active driving circuits which provide the DUTs with current pulses of arbitrary shape. These driving circuits emulate the triangular switch-off current ramp which an integrated power switch would see with a real inductive load. Also here, the DUTs are placed in an environmental chamber to submit them to the worst ambient temperatures to accelerate the ageing process. The proposed modular test system is able to perform these types of tests as well.

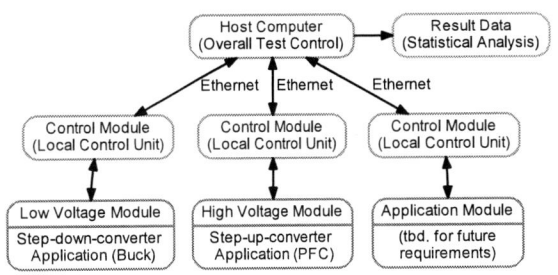

Figure 3: Test system architecture.

III. MODULAR TEST SYSTEM DESIGN

The proposed modular test system architecture is shown in Fig. 3. The test control is split into two instances, namely the host computer and the control module. The host computer is *one* unit which controls the overall test flow and communicates with the control modules. It also controls external periphery, such as power supplies and electronic loads, and stores the measured data to the file system. The control modules may be *many* units (typically in multiples of 8) and are connected to the host computer via Ethernet network. The host computer forwards the stress pattern to the control modules and receives preprocessed (digitized and filtered) measurement data and status information. Each application module is connected to one control module which controls the application test, performs measurement data acquisition and logs device status information. This information is sent back to the host computer which stores the data. The application modules are tailored to individual types of test.

The essential advantage of this test system architecture is the separation of the control and data acquisition parts from the actual test circuit. With this architecture the same control and data acquisition parts can be used for different types of tests. Only the test circuits (blue boxes in Fig. 3) need to be redesigned. In the following, the test circuit will be referred to as application module.

A. Control module

The control module (Fig. 4) is the core control element for the test. It consists of the following circuit blocks:

978-1-4673-9551-9/16 $31.00 © 2016 IEEE

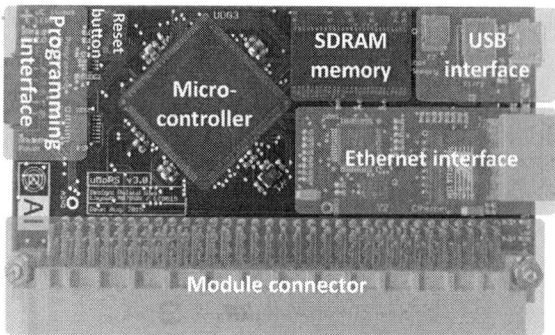

Figure 4: Control module.

- Infineon XMC4500 microcontroller [14].
- 8 MB SDRAM memory..
- 13 digital input / output channels.
- 2 SPI interfaces, 6 PWM output channels.
- 4 digital-to-analog converter channels.
- 8 differential and 12 single-ended analog-to-digital converter channels.
- Over-voltage protection on all signals on the module connector.
- Available interfaces: 100 Mbit Ethernet, USB, programming.

This module is plugged onto an application module and controls the running test sequence; on a synchronous boost converter application, it provides the PWM-signal via a driver circuit to the half-bridge switching transistors. It senses the output voltage and closes the control loop by a digitally implemented PI-controller. During the test, it collects the measured data which usually are input and output voltages and currents (V_{in}, V_{out}, I_{in}, I_{out}) of the converter as well as the case temperature T_c. All data are digitized on the control module and sent to the host computer via Ethernet network.

The major advantage of this concept is that the measurements are directly performed on the application module. No long and complicated sense wiring to a data logger outside of the environmental chamber is required. Altogether, 20 analog measurement channels are available on the control module. This ensures the expandability for future applications or extensions to existing ones. Only a small change in the script code is needed for logging further signals, provided the signal is routed to the control module.

On the control module the implemented differential operational amplifiers are provided with different input voltage dividers to offer several voltage ranges. For the whole system the voltage attenuation of analog measurements is implemented as a stacked concept. It is possible to directly use the provided input ranges implemented on the control module. Another possibility is (referring to Fig. 5) to change the series resistor values on the application board in order to adapt the input voltage range. With this stacked topology it is possible to flexibly adapt the input voltage range to the specific application where it is needed.

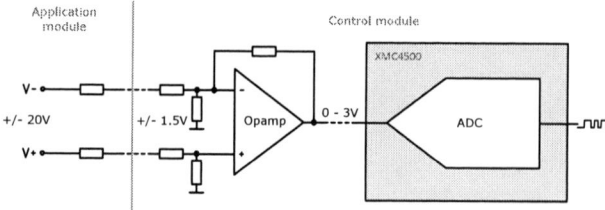

Figure 5: Analog measurement - stacked concept.

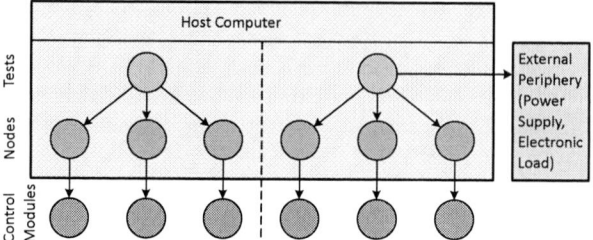

Figure 6: Software architecture: FSM control levels.

1) Software Architecture: The firmware implemented on the control module, together with a dedicated software environment on the host computer, allows an arbitrary stress pattern description. The test can be entered in the high level script language Lua [15] and no low-level microcontroller-specific code is needed for programming new test scenarios.

The software architecture, depicted in Fig. 6, follows a top-down approach. The test procedure on the host computer and the control modules is described by using multiple finite state machines (FSMs), in which each depicted circle represents a single FSM. The test procedure is described by the FSM-structure in which each single state can hold a chunk of Lua-code. A state transition can be triggered by internal or external events as well as by a special command invoked from the upper FSM.

The upper most circular, named test-FSM, controls the execution of multiple node-FSMs where each node subsequently controls a single control module FSM. Additionally, the test-FSM is capable of controlling external periphery, e.g. power supply or electronic load, to set voltage levels, current limits and cover automated power-up procedures. Based on this system approach, various test scenarios can be driven at the same time by simply modifying the Lua-script or FSM-structure, or both. A detailed description of the software architecture used with the proposed test system can be found in [16].

B. Application modules

An application module (blue boxes in Fig. 3) implements the main test circuit and is designed for one specific type of test. (E.g., for "active thermal cycling" the main test circuit may be a step-down or step-up converter.) Although they have different functions, some circuit blocks are common for all application modules. The most relevant ones are:

1) Guard block: An essential function for a stress test system is to ensure that the test setup is protected from catastrophic failure. When performing long-term tests on power semiconductors, the DUT may fail in short circuit. Under any circumstances the test system must be preserved in case of DUT damage. Preserving the DUT from further destruction immediately after failure also supports later physical analysis of failure root causes. The input current to the power circuit is monitored and an over-current limit is defined for each DUT. In case of short-circuit this limit is exceeded and the application module shuts down the power input immediately. This feature is implemented in hardware with a current sensor and an analog comparator to guarantee a fast response time (typical value: less than $2\,\mu s$). The failure event is then stored in an error latch and can only be reset by a test engineer intervention. Depending on the test circuit it may be necessary to monitor additional parameters. In this case a second parameter, such as the input voltage, is monitored which can trigger a power shut-down as well.

2) Device monitoring: All application modules incorporate voltage and current measurements (V_{in}, V_{out}, I_{in}, I_{out}) for input and output of the power circuit. Further voltage measurements can be analog status signals from the DUT and temperature signals (e.g., DUT case temperature T_c). Moreover, the control module provides several digital I/O channels to log digital status signals from the DUT or send digital stimuli to the DUT.

3) DUT board: In the modular stress test system the DUT is separated from the application module. This simplifies the task of replacing a failed DUT, while the application module can be reused. The DUT board is connected to the application module by a special connector, which may differ on various application boards. The type of this connector depends on the voltage and current ratings and the number of needed sense signals.

4) Module connector: As the control module is used on several application modules, all the application modules share the same type of connector [17].

5) Board identification: All application modules and DUT boards are equipped with a unique identity IC. This enables auto-detection of the connected boards by the control module.

6) Analog signal conditioning: As stated in Section III-A, operational amplifiers are provided on the control module to attenuate analog differential voltage signals to the single-ended voltage range of the analog-to-digital converters of the microcontroller. Additionally, all application modules have an analog signal conditioning block for performing further signal translations. E. g., a precision current source is implemented on an application module supporting high accuracy measurements of resistive temperature sensors. The voltage signal is amplified on the application module and then fed to the control module.

IV. RESULTS

Exemplarily, the realization of the application module named "low voltage module" (Fig. 8) is discussed in the following:

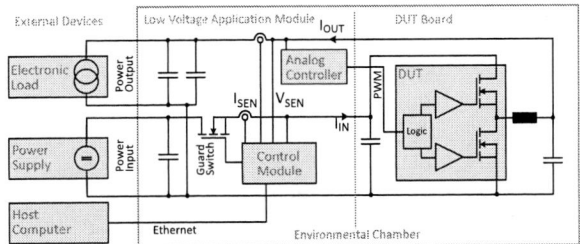

Figure 7: Low voltage module - simplified circuit.

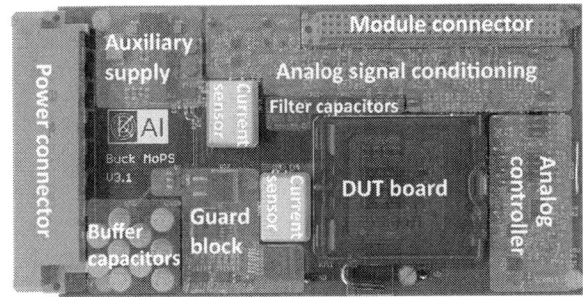

Figure 8: Low voltage application module and DUT board (red). DUT was removed for failure analysis.

A. Stress test example application

This application module submits the DUT to active thermal cycling. It comprises a step-down converter circuit designed for integrated power devices. The target device integrates two power MOSFET transistors in synchronous buck converter configuration, shown as the blue box "DUT" in Fig. 7. It has built-in gate drivers and sensing circuitry for current and temperature monitoring. It implements a temperature protection feature to shut down itself in improper operating conditions. It requires merely one PWM-signal input for operation. Interlocking and generation of correct turn-on signals for upper and lower transistor are generated internally.

The control loop of the DUT is closed by a dedicated analog controller from Linear Technologies [18]. This analog controller senses the output voltage and provides an appropriate PWM-signal to the DUT. In this case of application board, the connected control module operates only the test sequence and recording measurement data. The control loop of the step-down converter is closed by the aforementioned analog controller. The reason therefore is, that the converter runs at high switching frequencies (500 kHz to 2 MHz) and the microcontroller on the control module is not powerful enough to handle both the control loop and test sequence control, plus measurement data acquisition.

The DUT is available on a DUT board, shown in red in Fig. 8. To achieve an application-equivalent circuit behavior, the passive components (such as input and output capacitors and the filter inductor) must also be placed on the DUT board. This especially holds true for the input capacitance, as it influences the commutation loop inductance.

The application module holds the DUT board in a specialized socket with a high current carrying capability and low insertion impedances. It provides additional input and output capacitors (referred to as buffer and filter capacitors in Fig. 8) to support the power supply. It is equipped with two current transducers (yellow boxes in Fig. 8) for input and output current measurement. Conditioning for the logged current signals is achieved by operational amplifiers placed on the application module (block "analog signal conditioning" in Fig. 8) and fed directly to analog-to-digital converters of the control module. The operational amplifiers for the input and output voltage measurement are placed on the control module and by choosing the a reasonable series resistor, as shown in Fig. 5, the input voltage range is set. Signal conditioning for analog monitoring signals (T_{MON}, I_{MON}) implemented in the IC is done in a similar manner. For a case temperature measurement T_c the resistive temperature sensor is supplied from a constant current source and a voltage signal amplification is provided. The control module and the DUT board are plugged onto the application module and then placed in an environmental chamber. Fig. 10 shows all three boards plugged together. A guard block (Fig. 8) is likewise implemented on the application module, and will shut down in case of DUT failure.

The required connections to devices outside the chamber are:

- Power input: connected to a high power supply.
- Load output: connected to an electronic multi-load for emulation of arbitrary load conditions.
- The control module is connected via Ethernet connection to the host computer.
- An auxiliary supply (5 V) for the application functions and the control module is necessary.

B. Test result data

A prototype test system with a low voltage application board in active thermal cycling configuration has been completed a year ago and has proven a valuable tool for chip development. The following results are gained from this system. Fig. 9(a) gives typical waveforms of the step-down converter application. The input voltage of 16 V is converted down to 1.5 V at an average load current of 10 A. The figure shows the control PWM input signal, which is set to approximately 10 % duty cycle. Furthermore, the voltage at the switching node of the power transistors and the inductor current are given. Fig. 9(b) presents life time results of two stress tests. First, "Design 1" was tested and DUT failures where monitored. After that, the design of the DUT was improved and the same test was repeated. The test results of "Design 2" show that all DUTs now survive the stress test. The blue triangles of this test all lie in one line, because the stress test was stopped after reaching the full test time, at which time all DUTs were still operable. As with the proposed test system measurements are performed in situ during the running test, the exact failure times (diamonds and triangles) can be determined. Moreover, drift data of the voltages and currents as in Fig. 9(a) are continuously recorded and possible deviations are unveiled.

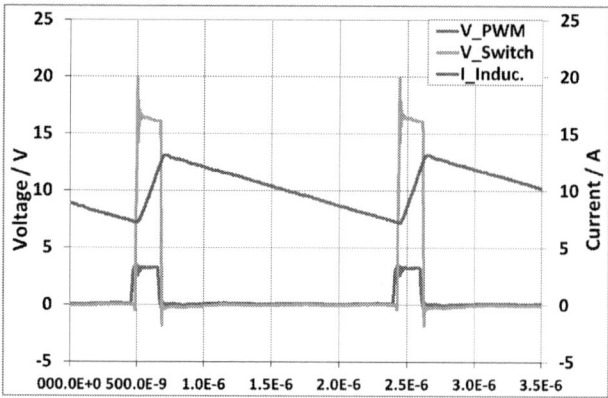

(a) Measured waveforms.

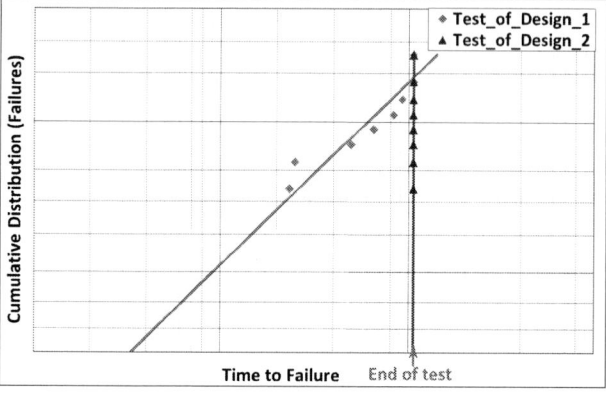

(b) Life time results.

Figure 9: Measurement results of life time tests conducted with prototype modular test system.

V. CONCLUSION AND OUTLOOK

A modular architecture for a fully automated stress test system has been presented. It can easily be adapted to future test requirements and different types of tests. The system is suitable for stressing a broad range of power transistors used in inverters. The test setup is built up as an application-equivalent circuit and can be designed for transistors of different voltage classes. Reliability tests for different power transistors can easily be realized, as the test circuit is clearly separated from the control and data acquisition part. Vital parameters of the DUTs are continuously logged during the running stress test. Drift data of voltages, currents and temperatures are collected to provide a base for subsequent statistical analysis. An example of a prototype test system realization for active thermal cycling was shown.

The next step is to develop a high voltage application module. This will incorporate a step-up converter circuit built up with wide band-gap transistors in the 500 V regime. The challenge with this module is that discrete transistors are used and the control loop, including interlock times, will be operated by microcontroller of the control module. Further development will focus on the modularization of circuit blocks at the level

978-1-4673-9551-9/16 $31.00 © 2016 IEEE

Figure 10: Low voltage application module, DUT board and control module plugged together.

of the application modules. This will reduce the development time of new test circuitry as readily designed circuit blocks are available for implementation.

VI. ACKNOWLEDGMENT

The authors would like to thank Sybille Ofner and Benjamin Steinwender for their support in the setup and programming of the control module.

This work was jointly funded by the Austrian Research Promotion Agency (FFG, Project No. 846579) and the Carinthian Economic Promotion Fund (KWF, contract KWF-1521/26876/38867).

REFERENCES

[1] F. Blaabjerg, K. Ma, and D. Zhou, "Power electronics and reliability in renewable energy systems," in *Industrial Electronics (ISIE), 2012 IEEE International Symposium on.* IEEE, 2012, pp. 19–30.

[2] *JEDEC Standard No. 22-A108D - Temperature, Bias, and Operating Life*, JEDEC Solid State Technology Association Std., Rev. A108D, November 2010. [Online]. Available: http://www.jedec.org/

[3] O. Bluder, J. Pilz, M. Glavanovics, and K. Plankensteiner, "A Bayesian Mixture Coffin-Manson Approach to Predict Semiconductor Lifetime," in *Proceedings of SMTDA 2012: Stochastic modeling Techniques and Data Analysis*, 2012.

[4] A. R. de Vega, P. Ghimire, K. B. Pedersen, I. Trintis, S. Beczckowski, S. Munk-Nielsen, B. Rannestad, and P. Thogersen, "Test setup for accelerated test of high power IGBT modules with online monitoring of V_{ce} and V_f voltage during converter operation," in *Power Electronics Conference (IPEC-Hiroshima 2014-ECCE-ASIA), 2014 International.* IEEE, 2014, pp. 2547–2553.

[5] S. Beczkowski, P. Ghimre, A. de Vega, S. Munk-Nielsen, B. Rannestad, and P. Thogersen, "Online V_{ce} measurement method for wear-out monitoring of high power IGBT modules," in *Power Electronics and Applications (EPE), 2013 15th European Conference on*, 2013, pp. 1–7. [Online]. Available: http://ieeexplore.ieee.org/stamp/stamp.jsp?arnumber=6634390

[6] F. Forest, A. Rashed, J.-J. Huselstein, T. Martiré, and P. Enrici, "Fast power cycling protocols implemented in an automated test bench dedicated to IGBT module ageing," *Microelectronics Reliability*, vol. 55, no. 1, pp. 81–92, 2015.

[7] M. Glavanovics, H.-P. Kreuter, R. Sleik, and C. Schreiber, "Cycle stress test equipment for automated short circuit testing of smart power switches according to the AEC Q100-012 standard," in *Power Electronics and Applications, 2009. EPE '09. 13th European Conference on*, 2009, pp. 1–7. [Online]. Available: http://ieeexplore.ieee.org/stamp/stamp.jsp?arnumber=5278683

[8] M. Glavanovics, H. Kock, H. Eder, V. Kosel, and T. Smorodin, "A new cycle test system emulating inductive switching waveforms," in *Power Electronics and Applications, 2007 European Conference on*, 2007, pp. 1–9. [Online]. Available: http://ieeexplore.ieee.org/stamp/stamp.jsp?arnumber=4417742

[9] R. Letor, S. Russo, and R. Crisafulli, "Life time prediction and design for reliability of Smart Power devices for automotive exterior lighting," in *Integrated Power Systems (CIPS), 2008 5th International Conference on.* VDE, 2008, pp. 1–5.

[10] S. Russo, R. Letor, O. Viscuso, L. Torrisi, and G. Vitali, "Fast thermal fatigue on top metal layer of power devices," *Microelectronics Reliability*, vol. 42, no. 9-11, pp. 1617–1622, 2002.

[11] M. Glavanovics, H. Estl, and A. Bachofner, "Reliable smart power system ICs for automotive and industrial applications-the Infineon smart multichannel switch family," in *Proceedings of PCIM Europe*, 2001.

[12] M. Glavanovics and M. D., "42V: The Dynamic Reliability Challenge." Munich, Germany: 42V PowerNet Conference, 2002.

[13] A. E. Council, "AEC-Q100-012: Short Circuit Reliability Characterization of Smart Power Devices for 12 V Systems," 2006.

[14] 32-Bit Industrial Microcontroller based on ARM® Cortex™-M. Infineon Technologies. Accessed 2015-02. [Online]. Available: http://www.infineon.com/xmc

[15] P.-R. Lua.org. Lua. Accessed 2015-10. [Online]. Available: http://lua.org

[16] B. Steinwender, M. Glavanovics, and W. Elmenreich, "Executable test definition for a state machine driven embedded test controller module," in *Industrial Informatics (INDIN), 2015 IEEE 13th International Conference on.* IEEE, 2015, pp. 168–173.

[17] ERNI. (2015, 02) Connectors DIN 41612 type C. Website. ERNI Electronics GmbH & Co. KG. [Online]. Available: http://www.erni.com/en/products/show/category/type-c-2/

[18] T. Linear. LTC3861 - Dual, Multiphase Step-Down Voltage Mode DC/DC Controller. Website. Linear Technology Corporation. Accessed 2015-10. [Online]. Available: http://www.linear.com/product/LTC3861

EMI Noise Cancelation by Optimizing Transformer Design without Need for the Traditional Y-Capacitor

Yongjiang Bai[1], Wenjie Chen[1], Ruirui He[1], Dan Zhang[2], Xu Yang[1]

Email: baiyj@mail.xjtu.edu.cn   cwj@mail.xjtu.edu.cn   yangxu@mail.xjtu.edu.cn

[1]Institute of Electrical Engineering, Xi'an JiaoTong University, Xi'an, China

[2]Silergy Corp. A206, Ascends Innovation Hub, XHTZ, Xi'an, China

Abstract—At present, Y capacitor is widely adopted inside an isolated switched mode power supply to pass EMI tests. However, to meet certain safety standard, engineers are trying to reduce the Y-cap in such equipment as cell phone and pad chargers, medical instruments, handheld and portable equipment. As we know, the removal of Y-cap offers a big challenge for the EMI design. To solve the problem, this paper proposed an optimized transformer design method. By means of this method, the EMI noise can be reduced without the need of the traditional Y-cap. Analysis of the transformer architecture as well as the auxiliary winding has been carried out. The experiment results demonstrated the effectiveness and feasibility of the proposed method.

Keywords—*EMI; noise cancelation; transformer design; Y-capacitor*

I. INTRODUCTION

Electromagnetic interference (EMI) generated by switched mode power supply (SMPS) is an important problem in power electronics products. Usually, the attenuation of common mode (CM) current in 2-wires AC input SMPS is more difficult than those 3-wires SMPS [5],[10]. So, for these 2-wire isolated SMPS, a big Y-capacitor is an essential component to reduce the common mode current [7-8].

On the other hand, standards on safety leakage current limit the value of this Y-cap. Generally, the maximum safety leakage current allowed is based on the specific classification of applications. And it is much more stringent in medical and handheld equipment [5]. For example, person who uses a charging cell phone with the metal case will have the risk of getting an electric shock owing to the existence of Y-capacitor. And this leakage current will affect the performance of the touch screen too. So for the cell phone charger, the leakage current is mainly caused by the Y-cap. It is required to get rid of Y-capacitor in these applications. However, which in turn, will generate big challenge in EMI design for these equipments.

Up to now, few studies have been done to reduce CM noise without the help of Y-cap [5]. In [1], [2], [5], [8], an additional balanced winding is added to cancel the CM current goes through parasitic capacitance between transformer primary side and secondary side. In [3], [5], a shield winding is added in series of a compensation capacitor to cancel the CM current goes through inter-winding capacitors. However,

This work was supported by the National Natural Science Foundation of China under Project 51277145.

these methods don't mention the auxiliary winding which is commonly configured at all and can only reduce the CM noise that goes through one propagation path. And it requires accurately control of parasitic capacitance, which is not only very difficult but also time consuming in mass production. In [5-7], the Y-cap is used to bypass the CM noise from the propagation path. But it is forbidden in those no-Y-cap applications. In [8-9], new topologies are proposed to reduce the CM noise at the expense of cost and complexity. In general, there is a lack of manuscript that investigating the no-Y-cap EMI design method from system viewpoint.

On considering the above mentioned problems, this paper proposed a transformer optimization method as well as the design rules to reduce the EMI noise without the help of tradition Y-cap. The impact of Y-cap on EMI noise and system stabilization is analyzed in detail. A more precise transformer model is also proposed. Based upon that, the impact of both the shielding winding and the winding structure on the CM noise are evaluated. Design guidelines for a flyback transformer without Y-cap are presented here. The effectiveness and feasibility of the proposed method is verified by the experimental results in 10W cell phone charger.

II. THE IMPACT OF Y-CAP ON EMI NOISE AND STABILIZATION

Conducted EMI noise is usually decoupled to differential mode (DM) and common mode (CM) noise. DM current affects the low frequency of CE below 0.1MHz. And it can be effectively attenuated by balancing noise current path with a DM filter and a high frequency capacitor in parallel with the bulk capacitor. The CM current affects the high frequency of CE above 1MHz. CM noises due to the transients change of switching voltages are largely coupled to the earth ground or chassis from primary winding of the transformer. Both common and differential mode currents affect the EMI between 0.1MHz and 1MHz.As a very popular topology, the flyback converter shown in Fig. 1 is used as an example to illustrate the CM noise propagation path.

To solve the problem, the secondary side CM noise propagation path is also illustrated in Fig.1.When a Y-cap is added between the primary return terminal GND and secondary return terminal GNDS, the CM noise current will change original path from Cse to Y2. By means of this method, the CM noise flowing to the secondary side will be shunted and better EMI performance will be obtained. On the contrary, if the Y-cap is removed for the leakage current standard requirements, all the CM noise has only one path by the Cse and it must cause worse EMI results.

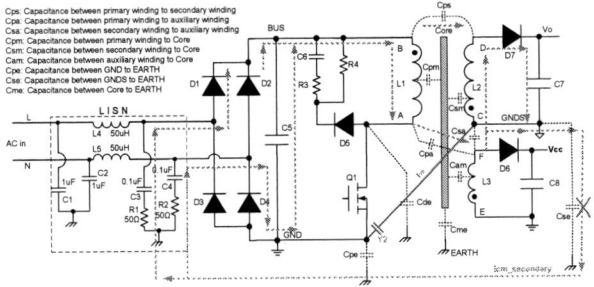

Fig.1 Shunting effects of Y2 for Common mode noise when Q1 is being turned off

III. OPTIMIZING TRANSFORMER STRUCTURE TO REDUCE CM NOISE WITHOUT Y-CAPACITOR

In a general way, the transformer of the flyback has three windings such as primary winding, auxiliary winding and secondary winding. Sandwich structure transformer is very popular in those applications for its smaller leakage inductance as well as higher efficiency. However, for the design of a flyback converter without Y-cap, it means much bigger parasitic capacitance which is harm to reducing CM noise. On the contrary, the non-sandwich has smaller parasitic capacitance and it means better EMI. So the following analysis is based on non-sandwich transformer structure.

To get rid of the Y-cap, the transformer model should be established. Fig. 2(a) shows the half window of transformer 1# structure implemented with EE core. The primary winding is evenly divided into three layer windings. The auxiliary winding and secondary winding is distributed symmetrically in a dedicated layer. The terminal B, E and C are corresponding terminals. According to ref [4], the winding to core capacitors can be neglected because its capacitance is too small and the primary winding inner layers capacitors can also be neglected because they do not contribute to CM noise.

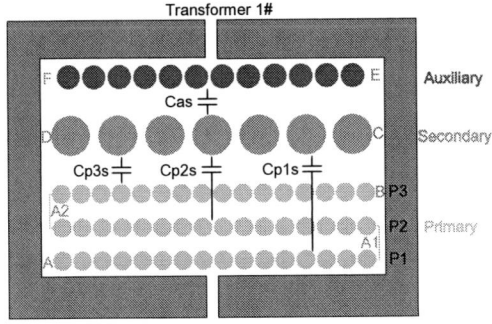

Fig. 2 (a) Inter-winding of transformer 1#

According to the current formula of capacitor, the displacement current between primary side and secondary side only depends on the voltage jump slope and the parasitic capacitance.

When the Q1 is in switching mode, the point A can be considered as a pulse voltage source which amplitude is V_A from turn on state to turn off state. And it can be expressed as following:

$$V_A = V_{in} + V_O * \frac{N_p}{N_s} \qquad (1)$$

Similarly, the point D can be considered as a pulse voltage source which amplitude is V_D from turn on state to turn off state. And it can be expressed as following:

$$V_D = V_O + V_{in} * \frac{N_s}{N_p} = V_A * \frac{N_s}{N_p} \qquad (2)$$

Similar way, the point F can be considered as a pulse voltage source which amplitude is V_F from turn on state to turn off state. And it can be expressed as following:

$$V_F = V_O * \frac{N_a}{N_s} + V_{in} * \frac{N_a}{N_p} = V_A * \frac{N_a}{N_p} \qquad (3)$$

The voltage distribution of the three windings is shown in Fig. 2 (b). When the Q1 is turned off, the voltage of point A, F and D is rising up relative to point B, E and C respectively. By means of the above equations, the displacement current propagation path is shown by the arrow in Fig. 2 (c).

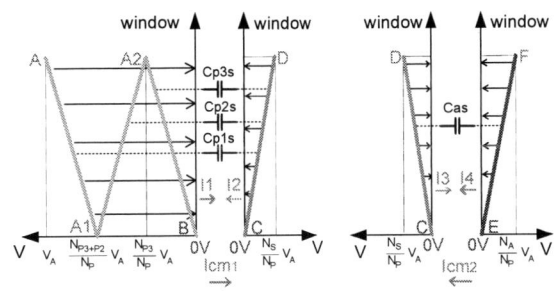

Fig. 2 (b) Voltage and current distribution of transformer 1#

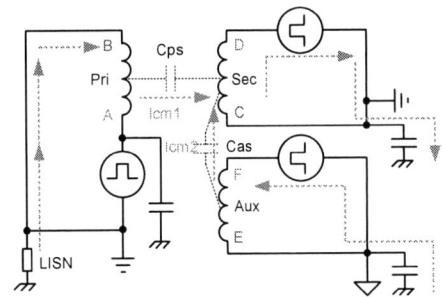

Fig. 2 (c) CM current propagation of transformer 1#

978-1-4673-9551-9/16 $31.00 © 2016 IEEE

Assuming that the positive current orient direction is from primary side to secondary side, the displacement current between primary winding and the secondary winding is given as:

$$
\begin{aligned}
I_{CM1} = {} & \frac{C_{p3s}}{2}\left(\frac{N_{p3}}{N_p}\frac{V_A}{\Delta t} - \frac{N_s}{N_p}\frac{V_A}{\Delta t}\right) + \\
& \frac{C_{p2s}}{2}\left(\frac{N_{p3}+N_{p2}+N_{p3}}{N_p}\frac{V_A}{\Delta t} - \frac{N_s}{N_p}\frac{V_A}{\Delta t}\right) + \\
& \frac{C_{p1s}}{2}\left(\left(\frac{N_{p3}+N_{p2}}{N_p}+1\right)\frac{V_A}{\Delta t} - \frac{N_s}{N_p}\frac{V_A}{\Delta t}\right)
\end{aligned} \quad (4)
$$

In the same way, the displacement current between auxiliary winding and the secondary winding is given as:

$$
I_{CM2} = \frac{C_{as}}{2}\left(\frac{N_a}{N_p}\frac{V_A}{\Delta t} - \frac{N_s}{N_p}\frac{V_A}{\Delta t}\right) \quad (5)
$$

According the physical distance between different layers, some assumptions are made to simplify the equations as following:

$$
C_{p3s} = C_{as},\ C_{p2s} = \frac{C_{as}}{2},\ C_{p2s} = \frac{C_{as}}{3},\ N_{p1} = N_{p2} = N_{p3} = \frac{N_p}{3} \quad (6)
$$

From (4), (5) and (6), the total displacement current from primary to secondary can be expressed by:

$$
I_{CM\_1\#} = I_{CM1} + I_{CM2} = C_{as}\left(\frac{25}{36} + \frac{1}{2}\frac{N_a}{N_p} - \frac{17}{12}\frac{N_s}{N_p}\right)\frac{V_A}{\Delta t} \quad (7)
$$

In order to reduce the displacement current, the parasitic capacitance of the primary and secondary windings has to be decreased by increasing the physical distance from the primary side windings to secondary winding, although the voltage pulsation between the primary and secondary side is very large. While, adding several turns of tapes may not be feasible considering the limited window area. To solve this problem, the auxiliary winding is inserted in middle layer as shown in Fig. 3(a).

By means of this method, some assumptions are made to simplify the equations as following:

$$
\begin{aligned}
C_{p3s} &= \frac{1}{2}C_{as},\ C_{p2s} = \frac{1}{3}C_{as},\ C_{p1s} = \frac{1}{4}C_{as}, \\
N_{p1} &= N_{p2} = N_{p3} = \frac{1}{3}N_P
\end{aligned} \quad (8)
$$

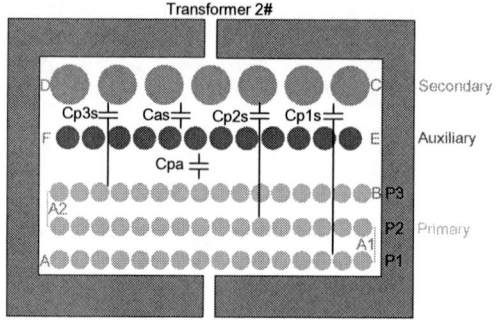

Fig. 3 (a) Inter-winding of transformer 2#

Therefore, the total displacement current formula from primary and secondary can be expressed by:

$$
I_{CM\_2\#} = I_{CM1} + I_{CM2} = C_{as}\left(\frac{11}{24} + \frac{1}{2}\frac{N_a}{N_p} - \frac{25}{24}\frac{N_s}{N_p}\right)\frac{V_A}{\Delta t} \quad (9)
$$

In the general low voltage output application, $I_{CM\text{-}2\#} < I_{CM\text{-}1\#}$ can be obtained in case of $N_s/N_p < 17/27$. The displacement current from primary side to secondary side has been reduced by placing the auxiliary winding in the middle of primary winding and secondary winding. In a similar way, the voltage pulsation and displacement current are analyzed in Fig. 3(b). And the displacement current propagation path is shown by the arrow in Fig. 3. (c).

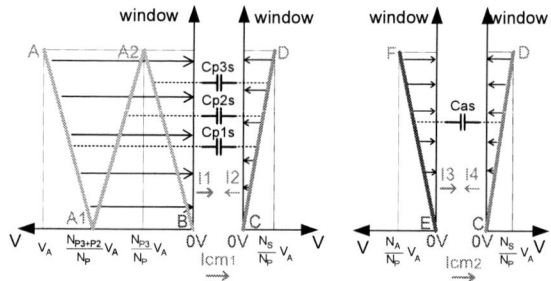

Fig. 3 (b) Voltage and current distribution of transformer 2#

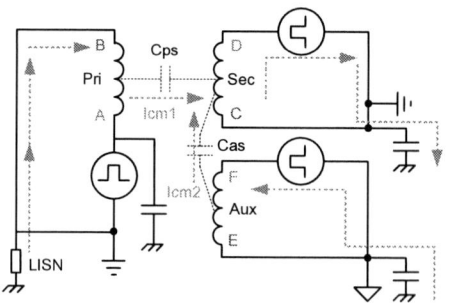

Fig. 3 (c) CM current propagation of transformer 2#

In order to reduce the displacement current further, the shield winding is added between primary winding and auxiliary winding as shown in Fig. 4(a). The starting point of the shielding winding is connected to point B, and the end point is not connected to any electrical node. The voltage pulsation and displacement current are analyzed in Fig. 4(b). And the displacement current propagation path is shown by the arrow in Fig. 4(c).

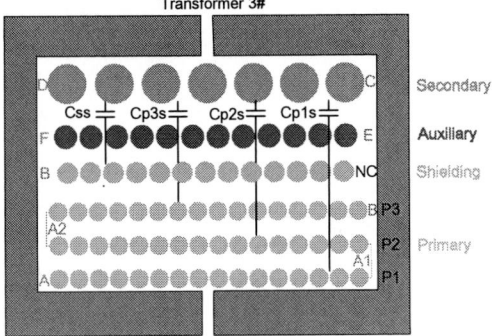

Fig. 4 (a) Inter-winding of transformer 3#

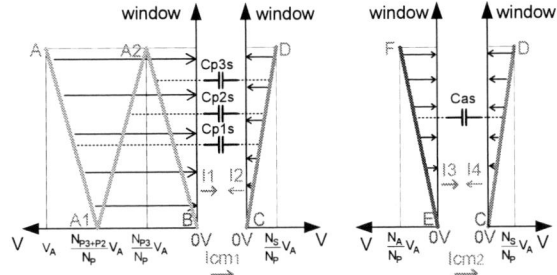

Fig. 4 (b) Voltage and current distribution of transformer 3#

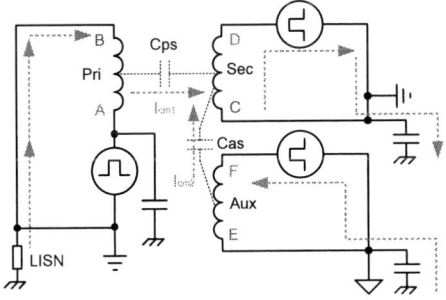

Fig. 4 (c) CM current propagation of transformer 3#

Based on above method, the displacement current between shielding winding and the secondary winding is given as:

$$I_{CM\_3\#} = \frac{C_{ss}}{2} \left(\frac{N_{shielding}}{N_p} + \frac{N_s}{N_p} \right) \frac{V_A}{\Delta t} \qquad (10)$$

For simplify, some assumptions can be made as following:

$$C_{ss} = \frac{1}{2} C_{as}, C_{p3s} = \frac{1}{3} C_{as}, C_{p2s} = \frac{1}{4} C_{as}, C_{p1s} = \frac{1}{5} C_{as}$$

$$N_{p1} = N_{p2} = N_{p3} = \frac{1}{3} N_P \qquad (11)$$

So I_{CM3} flows from secondary side to primary side, so it can cancel part of the total current. Therefore, the total displacement current formula from primary and secondary can be expressed by:

$$
\begin{aligned}
I_{CM\_3\#} &= I_{CM1} + I_{CM2} - I_{CM3} \\
&= C_{as} \left(\frac{25}{72} + \frac{1}{2} \frac{N_a}{N_p} - \frac{19}{15} \frac{N_s}{N_p} - \frac{1}{4} \frac{N_{shielding}}{N_p} \right) \frac{V_A}{\Delta t}
\end{aligned} \qquad (12)
$$

Based upon above analysis, the shielding winding not only increases the physical distance between primary winding and secondary winding, but also cancels part of the displacement current from primary side to secondary side. If a proper Nshielding is selected, the total displacement current could be reduced to zero even more.

In short, to get rid of the Y-cap, the parasitic capacitors of the power transformer are analyzed. And accurate relationship between different windings and different layers is built. The distribution of voltage along layer is analyzed to get the CM current between different layers. Based on above transformer model, the CM displacement current from the primary side to the secondary side is calculated in equations. So three different optimized transformers based on the design rules are analyzed and validated in following section.

IV. EXPERIMENTAL RESULTS

In order to verify the proposed CM noise reduction technique to get rid of the Y-cap, a flyback converter prototype is built. It is a CCCV charger operating in quasi-resonant mode for smart phone or touch Pads. The basic specification of the prototype is shown in Tab.1.

TABLE I.   PROTOTYPE SPECIFICATION

Input	L and N 2-wire, 85Vac~264Vac 47Hz~63Hz,
Output	5V, 2.1A, V_{ripple}<100mV and I_{OCP}<2.5A with USB port
Efficiency	Energy star Level 6 and No load loss <70mW
EMC and leakage current	EN555022 Class B limits; Leakage current <0.1mA
Size	W*L*H=31.5mm*40mm*19.6mm
Switching frequency	Quasi-Resonant flyback with 70kHz at full load in 115Vac
Transformer	EE16 core: Np=102Ts, Na=16Ts,Ns=6Ts

The prototype is shown in Fig. 5(a). The flyack converter operation waveforms are shown in Fig. 5(b). The voltage of pulsating points V_A, V_{ISEN}, V_{OUT} and V_D are shown as CH1, CH2, CH3 and CH4 respectively. Fig. 5(c) shows the test environment for the conducted EMI of home information technology device, as governed by European Union. Normally we test quasi-peak level limits as well as average level limits. All the experiments are test based on this platform.

Fig. 5 (a) Inter-winding of transformer 3#

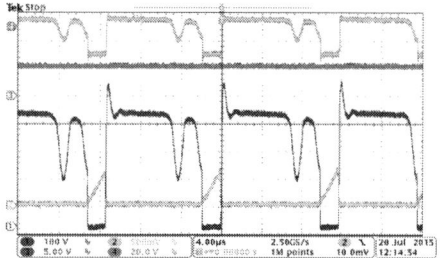

Fig. 5 (b) CH1: V_A, CH2: V_{ISEN}, CH3: V_{OUT}, CH4: V_D

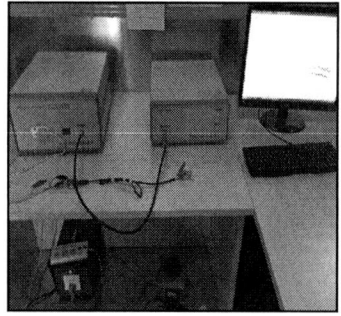

Fig. 5 (c) EMC test platform

Fig. 6 (a) shows EMI CE spectrum at 220Vac@50Hz by a traditional sandwich transformer with 1nF Y-cap between the primary to the secondary side. With the help of Y-cap, the EMI can meet the EN55022 EMC standard easily. When the Y-cap is removed for benefit of lower leakage current, the test EMI spectrum is shown in Fig. 6 (b). It turns to be worse and can't meet the EN55022 EMC standard. It is over the limitation line about 10dB. Therefore the Y-cap can't be removed easily and it will change the EMC by the propagation of CM displacement current.

Based on the analysis in section III, in order to get rid of the Y-cap, three types of different transformer structure are proposed. By means of the transformer optimization method, the CM noise can be significantly reduced. Fig. 7(a) shows the

EMI test result by the first transformer 1# which structure is shown in Fig. 2(a) without the Y-cap. Regard as the design rules, the second transformer 2# is designed by exchanging the secondary side winding and auxiliary winding position. According above analysis, the EMI will be better than the transformer 1#. As expected, the EMI spectrum shown in Fig. 7 (b) is about 3dB better than Fig. 7 (a).

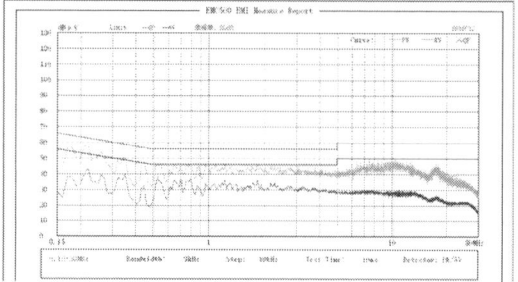

Fig. 6 (a) 5V2.1A EMI spectrum of sandwich transformer with Y-cap

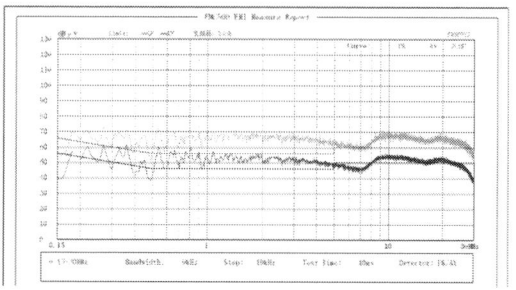

Fig. 6 (b) 5V2.1A EMI spectrum of sandwich transformer without Y-cap

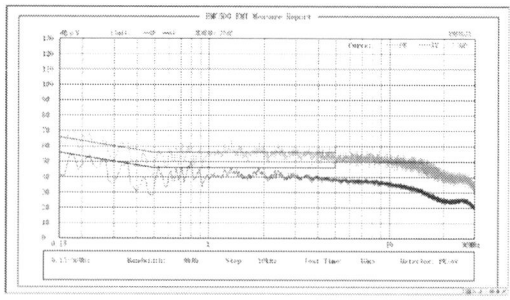

Fig. 7 (a) 5V2.1A EMI spectrum of 1# non-sandwich transformer

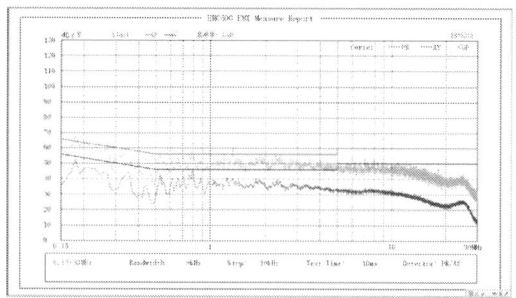

Fig. 7 (b) 5V2.1A EMI spectrum of 2# non-sandwich transformer

For better EMI performance, a shielding layer is added to the transformer 2#, and it is called transformer 3#. And the EMI test result is much better than the above two as expected. The CM noise is significantly reduced and it can meet the EN55022 EMI standard. The EMI spectrum is shown in Fig. 8 (a) and Fig. 8 (b).

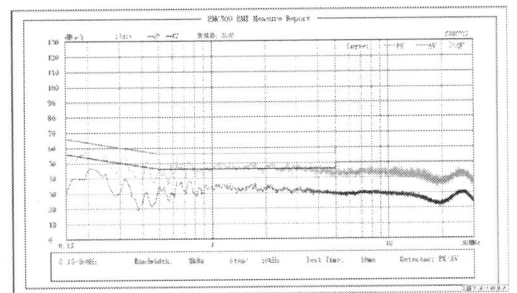

Fig. 8 (a) 5V2.1A EMI spectrum of 3# non-sandwich transformer, L-line @220V

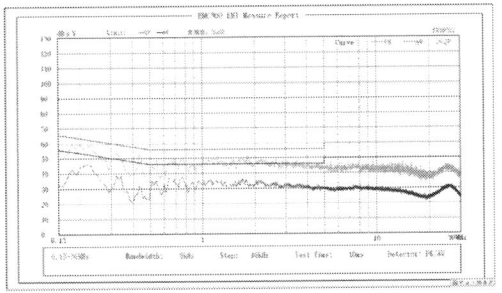

Fig. 8 (b) 5V2.1A EMI spectrum of 3# non-sandwich transformer, N-line @220V

According to the above experiments, the optimization method and design rules for converter without need of Y-cap are verified by the prototype test results.

V. CONCLUSION

In this paper, the impact of Y-cap on EMI noise reduction of isolated power converter is analyzed. Characteristics of inter-winding capacitances are discussed. And design rules for transformer structure of an isolated power converter without the need of Y-cap are present here. Analysis of the transformer architecture as well as auxiliary winding has been carried out. The CM current related to transformer winding structure is discussed theoretically and CM noise reduction by optimizing the winding structure is proposed. Experimental results validate the above analysis and the proposed noise reduction methods for converter without Y-cap.

REFERENCES

[1] Hung-I Hsieh, Sheng-Fang Shih, "Effects of Transformer Structures on the Noise Balancing and Cancellation Mechanisms of Switching Power Converters" in IPEC-Hiroshima 2014 - ECCE-ASIA, pp: 2380-2384.

[2] Yuchen Yang, Daocheng Huang, Fred C. Lee and Qiang Li, "Transformer shielding technique for common mode noise reduction in isolated converters," IEEE 2013 ECCE, pp. 4149 - 4153.

[3] Zengyi Lu, Wei Chen, "Common mode EMI noise reduction technique by noise path configuration of high frequency power transformer," in Proc. IEEE Conf. on Power Electronics and Motion Control, 2009, pp. 954-956.

[4] Pengju Kong and Fred C. Lee, "Transformer Structure and Its Effects on Common Mode EMI Noise in Isolated Power Converters" in IEEE APEC 2010, VOL. 2, pp: 1424-1429.

[5] Jin-ho Choi, Majid Madafshar, Kevin Parmenter, "Designing common-mode (CM) EMI noise cancellation without Y-Capacitor" in IEEE APEC 2007, pp. 936 – 940.

[6] Pingping Chen, Honghao Zhong, Zhaoming Qian, Zhengyu Lu, "The Passive EM1 Cancellation Effects of Y Capacitor and CM Model of Transformers Used in Switching Mode Power Supplies (SMPS) " in PESC 04. 2004 IEEE 35th Annual pp: 1076-1079.

[7] Milind M. Jha, Kunj Behari Naik and Shyama P. Das, "Types of Electro Magnetic Interferences in SMPS and Using Y-Capacitor for Mitigation of Mixed Mode Noise" in ICPCES,2010 International Conference on,pp1-6.

[8] Yick Po Chan, Bryan Man Hay Pong, Ngai Kit Poon, and Joe Chui Pong Liu, "Common-Mode noise cancellation in Switching-Mode Power Supplies using an equipotential transformer modeling technique" in IEEE Transactions on Electronmagnetic Compatibility Vol.54,No.3,June 2012.

[9] Mohammad Rouhollah Yazdani, , Hosein Farzanehfard, and Jawad Faiz, "EMI Analysis and Evaluation of an Improved ZCT Flyback Converter" in IEEE Transactions on Electronmagnetics,VOL.26, NO.8, AUGUST 2011.

[10] Wenjie Chen, Xu Yang, Jing Xue and Fred Wang. "A novel filter topology with active motor CM impedance regulator in PWM ASD system", IEEE Transactions on Industrial Electronics.vol.61, no.12, Dec.2014,pp.6938-3946.

Manufacturing, Assembly and Production Qualifications of High Density, High Reliability POL DC-DC Converters

Fariborz Musavi

Director of Engineering
CUI Inc., Tualatin, OR, USA
fmusavi@cui.com

Abstract—In this paper, a set of consistent methods of design for manufacturing and assembly techniques, which are used to minimize product cost and increase reliability through design and process improvements for high density, high reliability Point-of-Load DC-DC converters are discussed. The Pin-in-Paste process, also called through-hole reflow technology, along with consideration for component packaging in tape and reel, the pick-up cap design for automatic handling, PCB hole size selections and solder heating profile are also discussed in details as the preferred assembly techniques.

Furthermore, a detailed production qualification procedure is introduced to ensure methods and process used in production are validated, and the process consistently produces a product that meets its specifications, reduces production costs, and meets quality and regulatory requirements. Finally, selection and design of automated testing equipment system to complete the manufacturing of these high reliability products are discussed. Several examples of POL modules manufactured in different current level and module orientation and pin style, i.e. vertical, horizontal through-hole and surface mount are given to verify the proof of concept and improvements reported.

I. INTRODUCTION

Over the past two decades, the electronics industry has seen the supply voltage for high-density logic and processors drop from an average of 5V to 1V or less. At the same time, power envelopes for server-class equipment, which are generally thermally limited, are still potentially 100W or more. The result is a current demand that is beginning to exceed the 100A level at the point of load (POL), a shift that not only challenges conventional power-conversion architectures [1-3], but also introduces new challenges for manufacturing these products. In many of these advanced systems, the available PCB space is not increasing but reducing, consequently places much higher pressure on designers to utilize sophisticated PCB design with multi-layers and several ounces copper thickness. Also it is known that the design of a product determines the assembly process and in turn, the assembly process prescribes the assembly equipment [4]. Hence, if the product is designed without this knowledge there is a risk that the company invests in unnecessary equipment. The choice of assembly equipment constrains the assembly process and in turn also limits the product design. It is not only the direct cost of manufacturing that must be considered. The implications on quality and subsequently, reliability must be considered. The introduction of a manual process or the selection of a process that a manufacturer is not familiar with can have a devastating effect on yield [5, 6].

II. DESIGN FOR MANUFACTURE

Design for Manufacturing (DFM) and design for assembly (DFA) or DFMA are the most crucial part of product design. There is a balance to be struck between material and manufacturing cost. Too often the focus is placed on material and not the total product cost. It is not only the direct cost of manufacturing that must be considered. The implications on quality and subsequently, reliability must be considered. The introduction of a manual process or the selection of a process that a manufacturer is not familiar with can have a devastating effect on yield [5]. This section gives an overview of SMT manufacturing process along with assembly technologies available today.

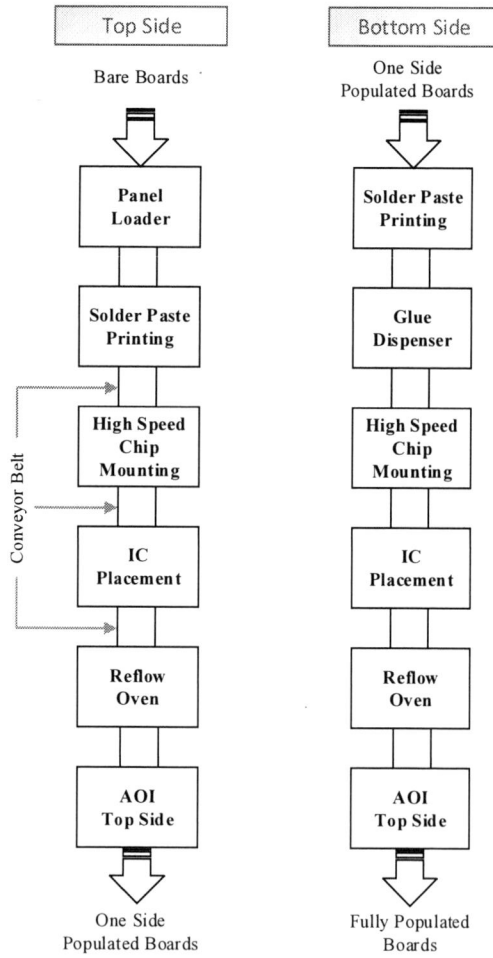

Figure 1. Work phases in an SMT assembly line.

A. SMT Manufacturing Process:

A basic overview of the generic high volume surface mount process available today is demonstrated in Fig. 1. All the machines shown in Fig. 1 would be linked together with panel conveyors, preventing any manual handling of the PCBs. For double sided assembly the whole process must be repeated for the second side. The process starts with panel loader. The objective is to load the panel on to the conveyor without contamination. Contaminants will inhibit the soldering process causing poor quality and consequent malfunction of the unit. In order to attach large SMT inductors to PCB, dispensing machines are used after printing paste on panels. The high-speed chip placement machines are designed to place uncomplicated devices fast. At this point, all small components are placed first on the panel. Once larger components start to be placed, the line must slow down to accommodate the increase in component mass. For fine pitch components, IC placement machine is used. After all components are placed, the reflow process starts where the components get solder to panel. There are many types of reflow ovens but the most common today are forced air convection models. Generally ICT (in circuit testing) would not be performed due to space constraints on these boards. Only AOI (Automated optical inspection) is performed at the end of assembly line.

B. Assembly Technologies:

Table I demonstrates the primary methods currently used for through-hole connector assembly: hand soldering, selective wave solder, pin-in-paste (PIP) reflow and solder preforms. A key obstacle to successful design and manufacture of mixed technology boards (surface mount technology, SMT and plated through hole, PTH components) is achieving an acceptable through-hole component solder fillet with reasonable yields, costs, and process development. The debate is more complex when evaluating available processing equipment versus purchasing and installing new equipment.

In the PIP method, the additional solder required to fill the though-hole barrel is deposited by overprinting the pad in the area of each connector pin, using standard SMT equipment. During reflow, the solder wicks to each pin forming the solder fillet.

III. PIN IN PASTE RELOW PROCESS

The design engineer's task to determine the best board configuration for assembly is made increasingly difficult by tighter component densities, changing board thicknesses, more fine-pitch devices, changing surface finishes, greater power demands, increasing reliability concerns, Pb-Free, mixed technology and much more. The Pin in Paste method, also called through-hole reflow technology [7-13], has become the ultimate solution to almost all challenges mentioned above. In addition, reaching a higher degree of automation with existing manufacturing equipment is one primary advantage. Further advantages are elimination of the soldering by hand or wave methods, reduction in manufacturing floor space, reduction in manufacturing equipment consequently in investment cost, compatible with the existing processes, use of no-clean soldering process possible, higher reliability at PCB level due

to fewer soldering processes and lower heating stress at the component level. Fig. 2 elaborates the pin-in-paste process steps. The greatest challenge in the PIP process is to design a stencil that delivers an adequate amount of solder paste to the through-hole component. Step-stencils are used to address this challenge. When overprinting does not deliver adequate volume, a step stencil from >10mil thick to <6mil thick can be achieved. In order to avoid the solder paste being pushed out of the hole during component placement, a proper pin and connector design should be considered along with an adequate hope design.

TABLE I.   ASSEMBLY OF THROUGH-HOLE COMPONENTS FOR MIXED TECHNOLOGY BOARDS

Process	Technique	Advantages	Disadvantages	Application
Single Point	Hand Soldering	Flexible solution, Low material costs	Labor cost, Through-put cost, Manual, Difficult quality control	High mix, low volume
Single Point	Robotic Soldering	Flexible solution, Low material costs	Throughput is diminished, Programming, sart-up time, Additional Process and equipment to maintain	High mix, low volume with all pins accessible to robots heat source and wire
Wave Process	Selective Wave Fountain	Low material cost	Unproven High start-up cost Start-up time and programming Additional process and equipment to maintain	Older technologies
Wave Process	Selective Wave Pallets	Low material cost Utilize existing equipment High volume of solder for thick boards and mechanical connections	Difficult to achieve high yields Not flexible- board specific pallet Cost and maintenance of pallets (cleaning) Additional process and equipment to maintain	Two sided SMT boards with edge through-hole devices
Wave Process	Wave Solder	Low material cost Established process in many plants	Consistent yields are difficult to achieve in wave solder process Additional process and equipment to maintain	Large, single sided SMT with numerous connectors
Perform Process	With Solder Paste	Exact amount of solder No additional flux Utilizes existing equipment Minimizes the extent of overprint Eliminates the need for step stencils	Adds cost to the BOM May add time to placement operation	Excellent solution for high-reliability mixed technology boards with many fine pitch devices
Perform Process	with a Flux Coating	Exact amount of solder	Adds cost to the BOM May require a fixture to retain perform in correct location	When an exact amount of solder with minimum flux residue is required
Paste Process	Pin-In-Paste (PIP) Overprint plus dispense	Additions of solder available to solder joints after component placement	Larger quantity of flux residue expected Additional process and equipment to maintain	Niche applications where board space is not available to permit sufficient overprint or hold-out area for a step stencil
Paste Process	Pin-In-Paste (PIP) Overprint (Combined with a step stencil)	Proven process High yield Utilizes existing equipment No additional process or equipment to maintain	Large quantity of flux residue Difficult to design with increasing component density Increases maintenance of placement equipment	Thin PCBs that have minimum barrel fill requirements and low density boards with sufficient overprint space

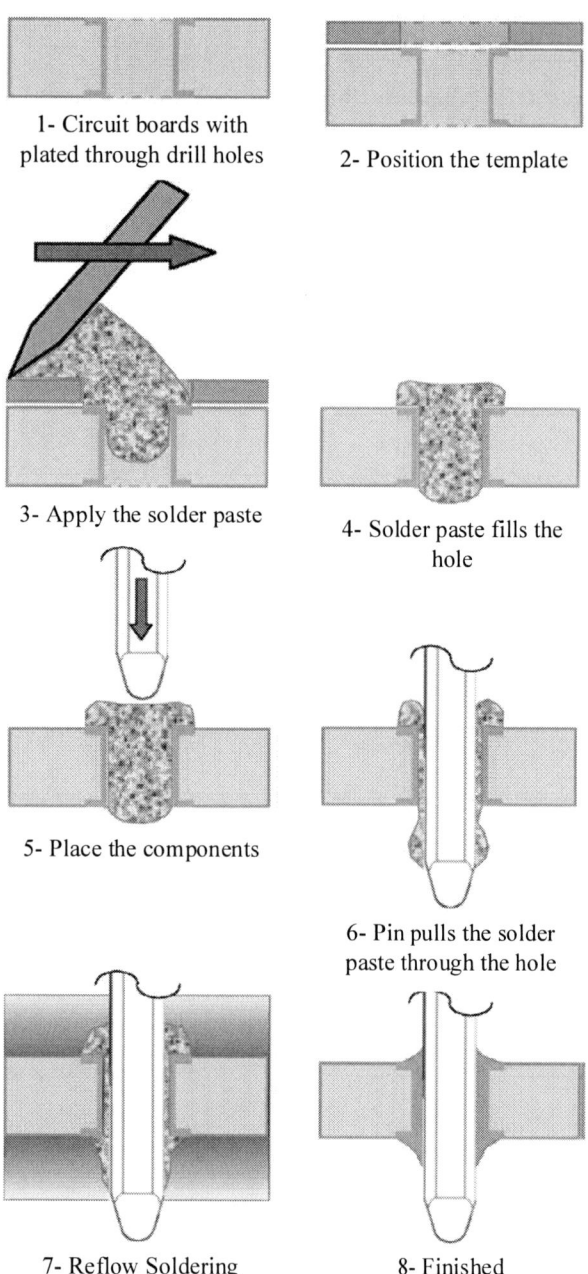

1- Circuit boards with plated through drill holes

2- Position the template

3- Apply the solder paste

4- Solder paste fills the hole

5- Place the components

6- Pin pulls the solder paste through the hole

7- Reflow Soldering

8- Finished

Figure 2. The pin-in-paste process steps.

IV. CONNECTOR AND PIN DESIGN CONSIDERATIONS

The major factors affecting the design engineer are compatibility of components with reflow temperatures and solder volume. Often the method used to specify component compatibility is by its material characteristics. The plastic used in connector and power pin design must comply with reflow process temperature. Clearance under the component should be available at the base with some form of stand-off pip or foot. A minimum of 0.015" prevents the part contacting the paste. The component packaging options should also be evaluated for automation [14-16]. The solderable lead

terminations need some attention. Ideally a tin/lead coating is preferred for soldering. If brass pins are used as the base material the pin must be first plated with copper to a minimum of 2.00um before the tin/lead is applied [11]. Lead length has been mentioned as it can affect solder paste push out. As a guide the pin length protruding from the board should be 1-1.5mm in length. The required solder paste volume is given by (1):

$$V_{Solder\ Paste} = 2 \times (\pi \frac{D_{Hole}^2 - D_{Lead}^2}{4} h) \qquad (1)$$

Where D_{Hole} is diameter of Hole in PCB, D_{Lead} is diameter of Pin/Connector Lead and h is the PCB thickness.

V. PRODUCTION QUALIFICATION

As the need for ultra-reliable electronic equipment increases, especially in life-critical applications, there will be an increased need to implement rigorous qualification procedures [17, 18].

Production qualification testing or life testing is designed to make sure the product not only has no design errors, but also has a reliable manufacturing process. These tests comprises of the followings:

A. Temperature, Humidity and Bias (THB):

The Temperature, Humidity and Bias test is used to test for moisture induced failures such as conductive anodic filaments (CAF), corrosion and dendrites. In addition, it will expose any mechanical or stress to the product during manufacturing process. The THB test starts with a pre-conditioning soak of 30 units for 72 hours unpowered, at an ambient temperature of 85° C and relative humidity of 85%. After the initial soak, test will start at rated maximum input voltage, minimum output load and at an ambient temperature of 85° C and relative humidity of 85%. The typical THB test profile, with all cycles and test conditions are given in Fig. 3.

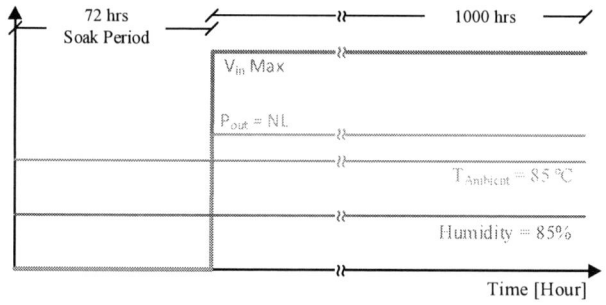

Figure 3. Typical THB test profile with its limits.

A LabVIEW test software was developed to interface and control DC power supplies for input voltage, the environmental chamber for humidity and temperature control, the test boards and fixtures for monitoring and capturing data. A snap shot of start page for control software interface in THB test is given in Fig. 4.

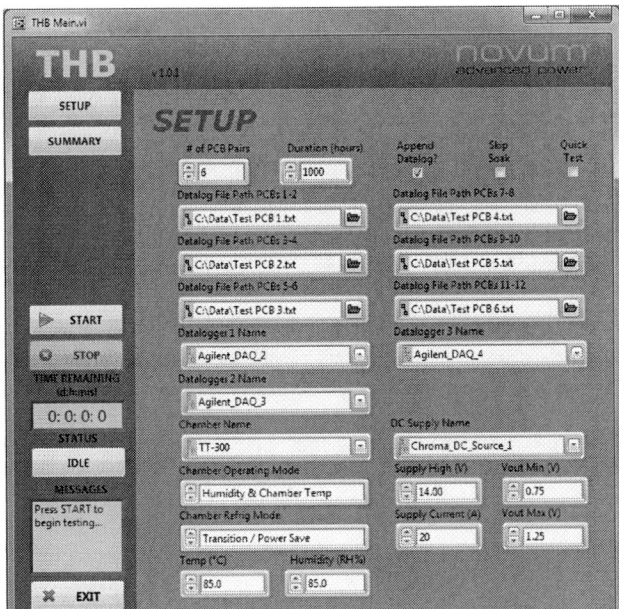

Figure 4. A snap shot of start page for control software interface.

B. High Tempereture Operating Bias (HTOB)

This test is to find weaknesses that appear only after long operation times (not infant mortality). This test is not intended to test or prove product life. The HTOB test has two distinct cycles. The only difference between these two cycles are the applied input voltage to UUT. It starts with maximum input voltage, maximum output load, and maximum rated temperature. The following cycle has exactly the same condition, but the applied voltage to the UUT is minimum. The typical HTOB test profile, with all cycles and test conditions are given in Fig. 5.

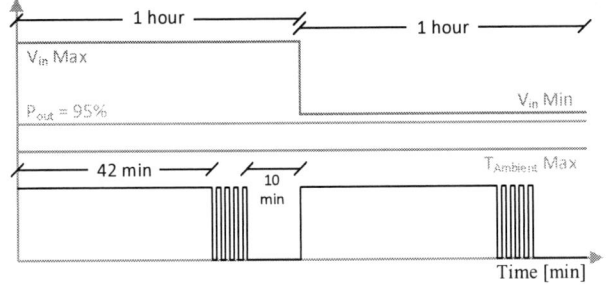

Figure 5. Typical HTOB test profile with its limits.

C. Temperature Cycling (TC)

The purpose of this test is to find weaknesses in solder joints, glued interfaces and other parts that are stressed by differences in the coefficient of thermal expansion (CTE). The TC test stars with all units unpowered. The minimum rate of change of ambient temperature is 20 °C/min. So in order to get to 125° C from -40° C, the chamber should be able to reach to that point in at least 8.25 minutes. Also the dwell time at either temperature limit is 30 minutes. Fig. 6 illustrates the typical TC test profile with its limits.

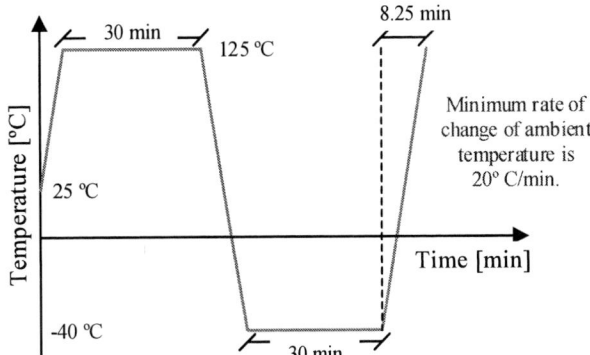

Figure 6. Typical TC test profile with its limits.

D. Power and Temperature Cycling (PTC)

The purpose of this test is to detect product weaknesses that are exposed at the corners of the operating conditions. The PTC test starts with a constant maximum input voltage. The output power is at 50% of rated power at the start of this test. The temperature is ramped from its minimum to the maximum rated temperature. It dwells at this point for 30 minutes. The units are power cycled once and then turned off. The tests enters the second identical cycle but the power is off during this period. Fig. 7 illustrates the typical PTC test profile with its limits.

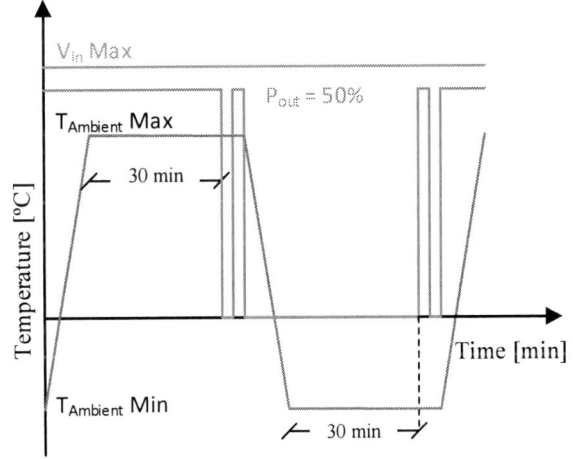

Figure 7. Typical PTC test profile with its limits.

E. SMT Attachement relibility (SMT)

The purpose of this test is to provide assurance that a surface mount mounted device will not detach or disconnect from the board to which it is mounted during its service life. The solderability and wettability of the device's SMT leads are tested per IPC/ECA J-STD-002, as referred in IPC-9701.

VI. AUTOMATED TEST EQUIPMENT SYSTEM (ATE)

ATE is a system that performs an automated production test on each POL unit with minimum human interaction.

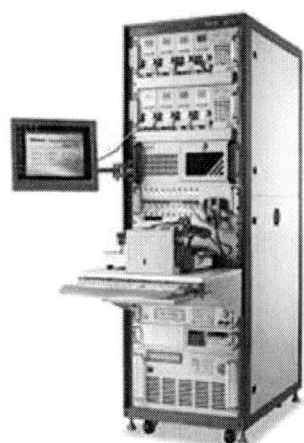

Figure 8.   ATE test system.

Fig. 8 illustrates an example of ATE test system. Each production POL unit shall go through the conformation test flow, as shown in Fig. 9. ATE test is the most important part in this flow. IPC-9592A [17] specifies that ATE test shall be performed to 100% production units.

ATE1 consists of ATE programming and a subset sequence testing and ATE2 is a full sequence testing. In ATE1, the test system scans 2D barcode on the label of unit to get the serial number. This serial number contains all manufacture information. The serial number is also written into MFR_SERIAL register of the unit. The digital measurement of the output current is calibrated.   The efficiency at half load is measured. The results of every step will be recorded in ATE test record form.

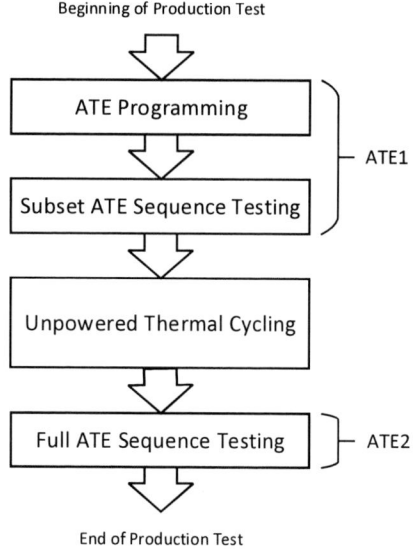

Figure 9.   ATE test in manufacture conformance test.

From the manufacture conformance test flowchart in Fig. 9, we can see that the unit goes through Unpowered Thermal

Cycling (UTC) test after ATE1 then it will go to ATE2. In ATE2, a full ATE sequence test is performed.

During ATE2, the test system scans 2D barcode one more time then measures startup input current at high input voltage; measures and calculates the efficiency at high input voltage and full load; measures and calculates load regulation; measures and calculates line regulation; verifies output voltage is able to adjust to high and low values; measures output voltage ripple; verifies output over current protection and record the protect point and finally verifies output short circuit protection.

If the test unit is successfully programmed, calibrated and the voltage/efficiency are verified to be in the right range, the center controller should give a visual/sound signal to show PASS. The operator could then prepare the system for the next unit.

VII.   Conclusions

A set of consistent methods of design for manufacturing and assembly techniques, which are used to minimize product cost and increase reliability through design and process improvements for high density, high reliability Point-of-Load DC-DC converters were discussed. The techniques addressed included connector and pin design, reliability, electrical and thermal conductivity, as well as manufacturing considerations such as pick and place compatibility, inspection and rework. The Pin-in-Paste process, also called through-hole reflow technology, along with consideration for component packaging in tape and reel, the pick-up cap design for automatic handling, PCB hole size selections and solder heating profile were also discussed in details as the preferred assembly techniques.

Furthermore, a detailed production qualification procedure was introduced to ensure methods and process used in production are validated, and the process consistently produces a product that meets its specifications, reduces production costs, and meets quality and regulatory requirements. Finally, selection and design of automated testing equipment system to complete the manufacturing of these high reliability products were discussed.   Several examples of POL modules manufactured in different current level and module orientation and pin style, i.e. vertical, horizontal through-hole and surface mount were given to verify the proof of concept and improvements reported.

References

[1]   F. Musavi. (April 8, 2014) Breaking the 100A Barrier in POL Converters Requires Fresh Thinking. *EE Times Magazine*. 60 - 66.

[2]   F. Musavi. (March 08, 2014) The path to 100A at the point of load, and beyond. *EDN Network Magazine*.

[3]   F. Musavi. (January/February 2015) Advanced ICs drive demand for "perfect" power. *Power System Design Magazine*.

[4]   S. Eskilander, "Design For Automatic Assembly - A Method For Product Design: DFA2," *Doctorate Thesis, Production Engineering, KTH Royal Institute of Technology*, Stockholm, Sweden, 2001.

[5]   K. McMillan, "Design For Manufacture," presented at the *IEE Colloquium on The Design of Digital Cellular Handsets*, London.

[6] S. J. Smed, M.; Johtela, T.; Olli Nevalainen, O., "Techniques and Applications of Production Planning in Electronics Manufacturing Systems," Turku Centre for Computer Science, *TUCS Technical Report* No 320, December 1999.

[7] G. P. S. K. Subbarayan, P. ; Lewin, S. ; Raut, R. ; Sethuraman, S., "Investigation for Use of Pin in Paste Reflow Process with Combination of Solder Preforms to Eliminate Wave Soldering," in Proceedings of *2011 IPC APEX Conference,*, Las Vegas, NV, pp. 318 - 333.

[8] "Paste in Hole Printing," Application NoteJanuary 4, 1999.

[9] Pin In Paste Application Note [Online]. Available: *www.littelfuse.com*

[10] J. B. S. Hinerman, K. ; Westby, G.R. . The Pin-in-Paste (or AART) Process for Odd Form and Through Hole Printed Circuit Boards [Online]. Available: available at: *www4.uic.com/wcms/WCMS2.nsf/index/Resources_42.html*

[11] B. Willis. A Practical Guide to Design Assembly and Reflow of Through Hole Components [Online]. Available: *www.smartgroup.org*

[12] Surface Mount Technology SMT – A Modern Production Process [Online]. Available: *www.phoenixcontact.com*

[13] R. C. P. Lasky, K. ; Berntson, R.B. Through-Hole Assembly Options for Mixed Technology Boards [Online]. Available: *www.techni-tool.com*

[14] R. E. B. Truesdale, D.W. ; Hicks, T.G., "Qualification of the tape-and-reel packing system ensures quality to the customer," presented at the *IEEE/CPMT Electronics Manufacturing Technology Symposium*, Austin, TX.

[15] C. O. D. Troxtell, B. ; Purdom, R. ; Zuniga, E. Semiconductor Packing Methodology [Online].

[16] Tape and Reel Packaging Standards [Online]. Available: *http://onsemi.com*

[17] IPC, "IPC-9592A: Requirements for Power Conversion Devices for the Computer and Telecommunications Industries," ed, 2010.

[18] M. G. Pecht, "Design For Qualification," in *Annual Reliability and Maintainability Symposium*, Atlanta, GA.

Power Flow Control and ZVS Analysis of Three Limb High Frequency Transformer Based Three-Port DAB

Ritwik Chattopadhyay
FREEDM Systems Centre, Department of ECE
North Carolina State University
Raleigh, NC, USA
email: rchatto@ncsu.edu

Subhashish Bhattacharya
FREEDM Systems Centre, Department of ECE
North Carolina State University
Raleigh, NC, USA
email: sbhatta4@ncsu.edu

Abstract— **The work presented in this paper focuses on the power flow control and study of three-limb high frequency transformer enabled three port Dual Active Bridge(DAB) converter, using 1200V and 1700V SiC devices. The advantage of using three-limb transformer is given by the low inter-winding parasitic capacitances due to placement of windings on different limbs and elimination of inter-winding insulations unlike concentric windings. The focus of the paper is based on power characterization for independent power flow of three port DAB using modulation control, and ZVS scenario study under the modulation control. Detailed analysis for power flow control and turn-on/turn-off ZVS conditions have been realized. A laboratory scale prototype using SiC Mosfets has been made and experimental results have been obtained for 500V dc input voltages and 1000V dc output voltage at 8.2kW of power.**

Keywords — *RES; DAB; Three Limb; High Frequency Transformer; ZVS; SiC Mosfet.*

I. INTRODUCTION

Renewable Energy Sources(RES) integration into medium voltage grids(4.16kV or 13.8kV) are quite challenging as the grid integrated inverter require a medium voltage dc bus voltage of 7kV-8kV dc for 4.16kV ac grid and 22kV dc or higher for 13.8kV ac grid[1][2]. Generating a high dc bus voltage from low voltage PV/RES of voltage level 400V-800V require the use of very high voltage(>10kV) switching devices or series connected lower voltage dc buses from multiple converters. The very high voltage fast switching devices of the range 10kV or higher have much limitation in terms of reliability and robustness, which encourages the use of series connected multiple dc buses of the type shown in [3] to achieve a high voltage dc link. The work presented in this paper involves the use of 1200V and 1700V SiC Mosfet based multiport dc-dc converter suitable for RES integration. 1700V SiC Mosfet from [4], have the capability of switching at high frequency(around 10kHz and higher) and at a switching voltage of 1kV or higher.

Researchers have realized several converter architecture for multiple RES based renewable energy integration involving high frequency magnetic cores for isolation and integration[3][4]. The type of high frequency transformer based power electronic converters, reported in literature, have mostly use the transformer geometry type having all the windings on same limb or on the same flux path[6]-[9]. The arrangement of different windings on same flux path or in concentric manner results in lower leakage but high parasitic inter winding capacitances. These type of transformers typically involve use of low switching speed devices like Si-IGBT and much lower switching frequency(around 10kHz). Positioning of the windings in a concentric manner, requires thick inter-winding insulation to be placed in between windings, causing high inter-winding capacitances. Use of low switching frequency using Si devices, for voltages greater than 1kV, causes the transformer size and weight to be high due to limitation in switching frequency. 1200V and 1700V SiC - Mosfet devices are capable of switching at 50kHz or higher at a high switching speed with dv/dt varying from 30-50kV/µS[10]. Converters using high speed SiC devices, require the high frequency transformer inter-winding/coupling capacitance to be low enough for minimal EMI effect due to high dv/dt and high switching frequency of SiC devices. Henceforth, it is necessary to reduce inter-winding capacitance by increasing the physical distance between the windings. For a multi-port transformer, the low inter-winding parasitic capacitance is achievable by placing different windings on separate limbs. The Dual Active Bridge(DAB) converters explained in [11]-[13], have presented the preliminary concept of multi limb transformer(fig.1) based DAB structure for integration of RES. This paper work analyzes into the details of the operation of the multi-limb transformer based converter.

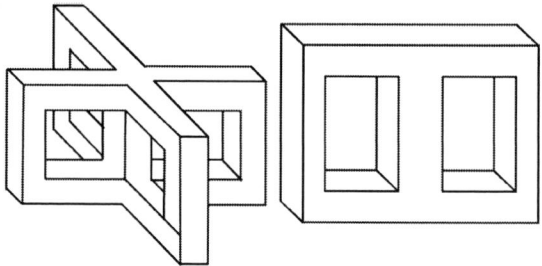

Fig. 1. Multi-Limb Transformer (Five Limb and Three Limb)

II. MULTI-LIMB TRANSFORMER BASED DUAL ACTIVE BRIDGE(DAB) CONVERTER OPERATION

A. Converter Equivalent Circuit

Multi-limb transformer DAB converter analysis is done with equivalent circuit analysis with the three limb transformer based three port DAB. Fig. 2 shows the three port Dual active bridge structure using a three limb transformer having three windings on its three limbs. The two side limb

978-1-4673-9551-9/16 $31.00 © 2016 IEEE

converters(using 1200V SiC Mosfts) are fed from two PV sources and the middle limb converter(using 1700V SiC Mosfets) creates a higher voltage dc bus.

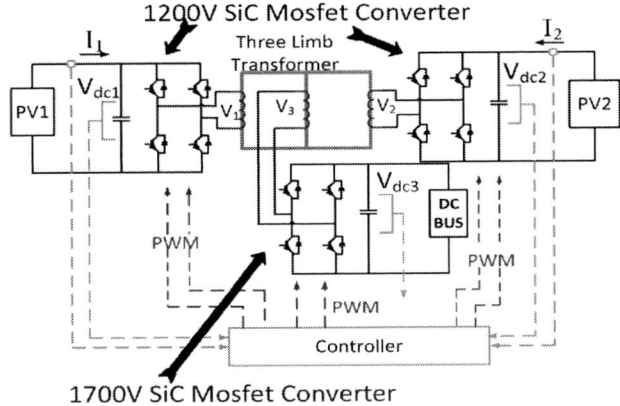

Fig. 2. Three port DAB Converter using Three Limb Transformer

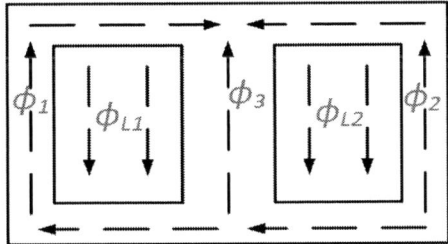

Fig. 3. Flux Flow Path for Three Limb Transformer

Figure 3 shows the flux path flow for three limb transformer, where ϕ_1, ϕ_2 and ϕ_3 are the induced fluxes in the three limbs created by their respective winding voltages applied by the three H-bridge converters. Fluxes ϕ_{L1} and ϕ_{L2} are the leakage fluxes that flows through the window area of the transformer. From the transformer geometry, equation (1) is derived by equating fluxes at the central limb. Differentiating equation (1), the equivalent electrical circuit equation (2) can be derived, where V_1, V_2, V_3 are the applied voltages and V_{L1}, V_{L2} are the voltage drops across the leakage inductances, where N_1, N_2, N_3 are the respective winding turns of the three limbs.

$$\phi_3 = (\phi_1 - \phi_{L1}) + (\phi_2 - \phi_{L2}) \tag{1}$$

$$\frac{V_3}{N_3} = \frac{(V_1 - V_{L1})}{N_1} + \frac{(V_2 - V_{L2})}{N_2} \tag{2}$$

Leakage inductance drop of a transformer depends on the transformer load current and is expressed as $V_L = \omega L I$, where ω is the frequency of the applied voltages, L is the equivalent leakage and I is the RMS value of load current. Referring all the voltages to the middle limb winding 3, the equation (2) is derived, where L_{13}' and L_{23}' are the equivalent leakage inductances between windings 1-3 and 2-3.

$$V_3 = (V_1' - \omega L_{13}' I_1') + (V_2' - \omega L_{23}' I_2') \tag{3}$$

Equation (3) represents the KVL for three voltage sources connected in series, and neglecting magnetizing current of the transformer, the MMF of the three windings are same ($N_1 I_1 = N_2 I_2 = N_3 I_3$). Hence, equation (3) reduces to (4), where $L_{eq} = L_{12} + L_{23}$ and I' is the winding current referred to middle limb winding. The equivalent electrical circuit is shown in figure (4). The equivalent series resistance R_s is neglected for analysis purpose.

$$V_3 = V_1' + V_2' - \omega L_{eq} I' \tag{4}$$

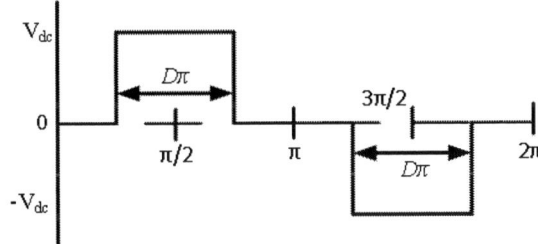

Fig. 4. Equivalent circuit of Three Limb DAB Converter

B. Decoupled Power Flow Control for Multi-port DAB

The power flow from one side limb H-bridge to another side limb H-bridge of the three port DAB converter of figure 2 can be realized by circuit analysis of equivalent circuit of fig.4. The possible voltage levels that can be applied to the winding voltages by an H-bridge are V_{dc}, 0, $-V_{dc}$, as shown in figure 5. The fourier representation of three level voltage of kth port, shown in figure 5 is given by equation (5).

Fig. 5. Three Level Voltage Output of an H-Bridge

$$V_k = \sum_{n=odd} \frac{4V_{dc}}{n\pi} \sin\frac{nD\pi}{2} \sin n\omega t \tag{5}$$

$$i_n = \frac{V_{1n}\angle n\delta_1 + V_{2n}\angle n\delta_2 - V_{3n}\angle 0}{Z_n \angle \theta_n} \tag{6}$$

$$P_k = \sum_{n=1,3,5}^{\infty} P_{kn} = \sum_{n=1,3,5}^{\infty} \text{Real}(V_{kn} \times i_n^*) \tag{7}$$

$$P_{kn} = \frac{V_{kn}^2 \cos\theta_n}{Z_n} + \frac{V_{kn}V_{jn}\cos(n\delta_k - n\delta_j + \theta_n)}{Z_n}$$
$$- \frac{V_{kn}V_{3n}\cos(n\delta_k + \theta_n)}{Z_n} \tag{8}$$

The power flowing from one side limb winding of the three limb transformer is given by equation (7), which is sum of real powers due to each harmonic components of the voltage and

978-1-4673-9551-9/16 $31.00 © 2016 IEEE

current, where V_{kn} is the RMS voltage of nth harmonic component of kth winding. Real power due to n^{th} harmonic components of the voltage and current for k^{th} winding is given by equation (8), where V_{jn} is the RMS voltage of the other side limb winding. The expressions for Z_n and θ_n are given in equation (9) and (10).

$$Z_n = \sqrt{R_s^2 + (n\omega L_{eq})^2} \quad (9) \qquad \theta_n = \tan^{-1}(\frac{n\omega L_{eq}}{R_s}) \quad (10)$$

For a high frequency transformer $n\omega L_{eq} >> R_s$, hence $\theta \approx \pi/2$, and if the phase leading angles δ_k and δ_j are kept at a same angle δ, then equation (8) can be expressed as given in equation (11), and equation (7) is reduces to equation (12). If the voltage waveform V_3 for middle limb voltage winding is kept at a fixed duty cycle D_3, then power output of a single side limb source can be controlled by controlling its own duty cycle (D_1 or D_2) of its output voltage, as given by equation (13) and (14).

$$P_{kn} = \frac{V_{kn} V_{3n} \sin(n\,\delta)}{Z_n} \quad (11) \quad P_k = \sum_{n=1,3,5}^{\infty} \frac{V_{kn} V_{3n} \sin(n\,\delta)}{Z_n} \quad (12)$$

$$P_1 = f(D_1) \quad (13), \quad P_2 = f(D_2) \quad (14)$$

C. Switching Technique

The typical winding voltage waveforms and transformer winding current waveform during a half cycle period for the three port Dual active bridge is shown in figure 6. Two different kind of switching techniques are possible for getting the three level voltage waveforms of figure 6. First technique is to switch every H-bridge leg at 50% duty cycle and shift the two switching legs of the same H-bridge by some angle to create a three level voltage output. Figure 7 shows the switching technique for this method, where S_1, S_2 are switches of one leg and S_3, S_4 are switches for another leg. For this switching technique with 50% duty cycle, one switching leg of the H-bridge connected to middle limb winding is taken as reference and the phase leading/lagging angles of the switching legs of other H-bridge are calculated using δ, D_1, D_2, D_3.

Switching technique 2, as shown in figure 8, has unequal duty cycle for switches, where voltage output of H-bridge of side limb windings are kept at same reference angle by using the same carrier wave for both the bridges, which causes the output voltages to be at sync with each other. The middle limb winding voltage is made to lag the side limb winding voltages by lagging the carrier wave of middle limb H-bridge by δ. The difference between the two switching techniques is unequal switch conduction. For technique 1, neglecting the deadband, the current flows through the Mosfet channel for the whole switching period, while for technique 2 the current flows through body diode of mosfets during some portion of the switching cycle. As a consequence, in switching technique 1, all the mosfets have equal conduction losses while in technique 2, all the mosfets do not have equal conduction loss, mosfets remaining on for lower time has higher conduction loss as conduction through body diode of SiC-Mosfets creates more losses than conduction through channel.

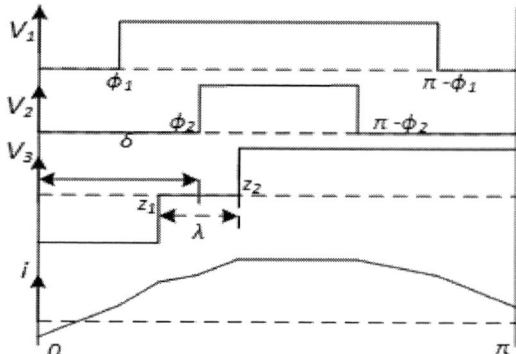

Fig. 6. Transformer Winding Voltages and Current

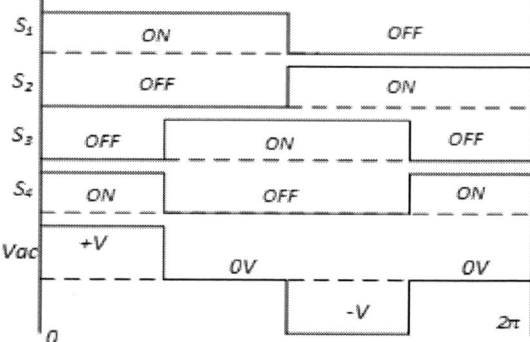

Fig. 7. Switching Technique with 50% duty Cycle and Phase Shift

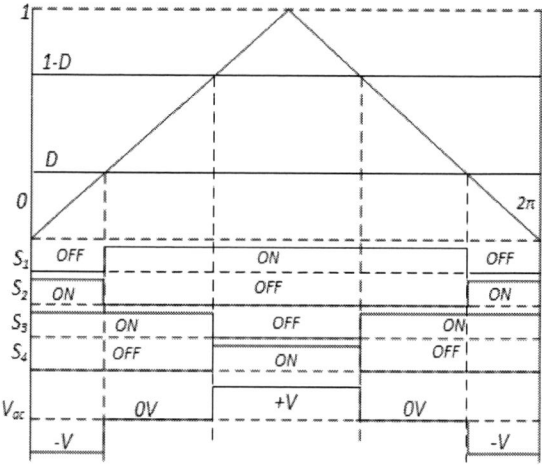

Fig. 8. Switching Technique with Asymmetrical Duty Cycle

D. Modes of operation and ZVS

Depending on different values of phase shift angle and zero voltage duration of winding voltages, six possible cases of power flow scenarios, along with transformer line current expressions at switching instants of three port DAB and corresponding DAB winding voltages and current diagrams are shown in figures 9(a)-(f).

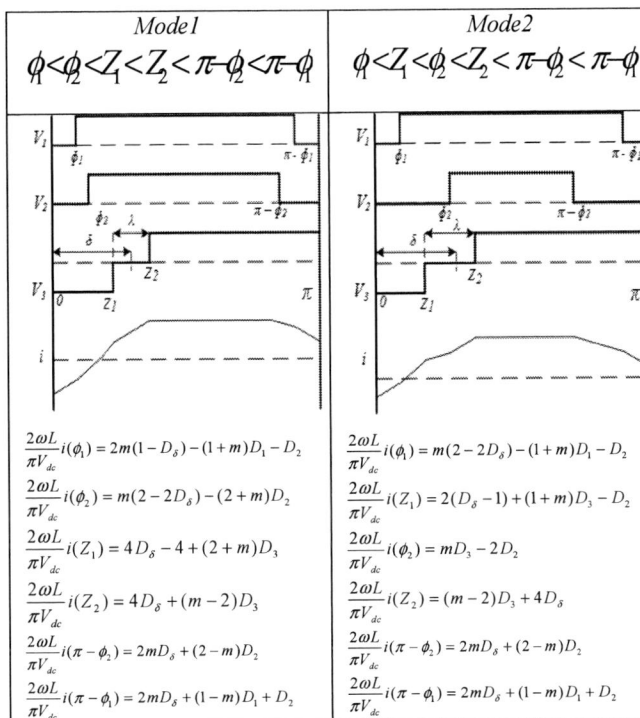

Mode1	Mode2
$\phi_1 < \phi_2 < Z_1 < Z_2 < \pi-\phi_2 < \pi-\phi_1$	$\phi_1 < Z_1 < \phi_2 < Z_2 < \pi-\phi_2 < \pi-\phi_1$

$$\frac{2\omega L}{\pi V_{dc}} i(\phi_1) = 2m(1-D_\delta) - (1+m)D_1 - D_2$$
$$\frac{2\omega L}{\pi V_{dc}} i(\phi_2) = m(2-2D_\delta) - (2+m)D_2$$
$$\frac{2\omega L}{\pi V_{dc}} i(Z_1) = 4D_\delta - 4 + (2+m)D_3$$
$$\frac{2\omega L}{\pi V_{dc}} i(Z_2) = 4D_\delta + (m-2)D_3$$
$$\frac{2\omega L}{\pi V_{dc}} i(\pi-\phi_2) = 2mD_\delta + (2-m)D_2$$
$$\frac{2\omega L}{\pi V_{dc}} i(\pi-\phi_1) = 2mD_\delta + (1-m)D_1 + D_2$$

Fig. 9.(a) Mode 1

$$\frac{2\omega L}{\pi V_{dc}} i(\phi_1) = m(2-2D_\delta) - (1+m)D_1 - D_2$$
$$\frac{2\omega L}{\pi V_{dc}} i(Z_1) = 2(D_\delta-1) + (1+m)D_3 - D_2$$
$$\frac{2\omega L}{\pi V_{dc}} i(\phi_2) = mD_3 - 2D_2$$
$$\frac{2\omega L}{\pi V_{dc}} i(Z_2) = (m-2)D_3 + 4D_\delta$$
$$\frac{2\omega L}{\pi V_{dc}} i(\pi-\phi_2) = 2mD_\delta + (2-m)D_2$$
$$\frac{2\omega L}{\pi V_{dc}} i(\pi-\phi_1) = 2mD_\delta + (1-m)D_1 + D_2$$

Fig. 9.(b) Mode 2

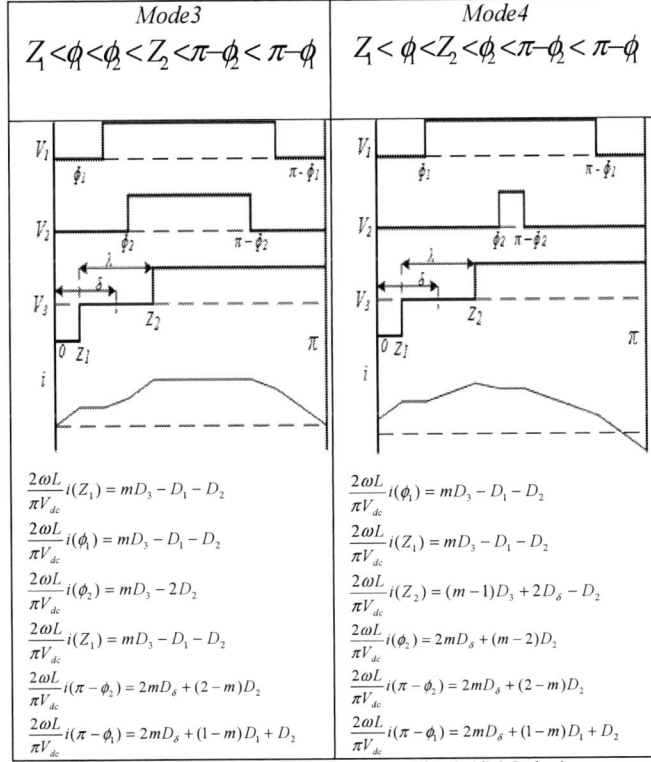

Mode3	Mode4
$Z_1 < \phi_1 < \phi_2 < Z_2 < \pi-\phi_2 < \pi-\phi_1$	$Z_1 < \phi_1 < Z_2 < \phi_2 < \pi-\phi_2 < \pi-\phi_1$

$$\frac{2\omega L}{\pi V_{dc}} i(Z_1) = mD_3 - D_1 - D_2$$
$$\frac{2\omega L}{\pi V_{dc}} i(\phi_1) = mD_3 - D_1 - D_2$$
$$\frac{2\omega L}{\pi V_{dc}} i(\phi_2) = mD_3 - 2D_2$$
$$\frac{2\omega L}{\pi V_{dc}} i(Z_1) = mD_3 - D_1 - D_2$$
$$\frac{2\omega L}{\pi V_{dc}} i(\pi-\phi_2) = 2mD_\delta + (2-m)D_2$$
$$\frac{2\omega L}{\pi V_{dc}} i(\pi-\phi_1) = 2mD_\delta + (1-m)D_1 + D_2$$

Fig. 9.(c) Mode 3

$$\frac{2\omega L}{\pi V_{dc}} i(\phi_1) = mD_3 - D_1 - D_2$$
$$\frac{2\omega L}{\pi V_{dc}} i(Z_1) = mD_3 - D_1 - D_2$$
$$\frac{2\omega L}{\pi V_{dc}} i(Z_2) = (m-1)D_3 + 2D_\delta - D_2$$
$$\frac{2\omega L}{\pi V_{dc}} i(\phi_2) = 2mD_\delta + (m-2)D_2$$
$$\frac{2\omega L}{\pi V_{dc}} i(\pi-\phi_2) = 2mD_\delta + (2-m)D_2$$
$$\frac{2\omega L}{\pi V_{dc}} i(\pi-\phi_1) = 2mD_\delta + (1-m)D_1 + D_2$$

Fig. 9.(d) Mode 4

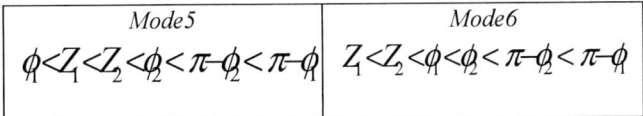

Mode5	Mode6
$\phi_1 < Z_1 < Z_2 < \phi_2 < \pi-\phi_2 < \pi-\phi_1$	$Z_1 < Z_2 < \phi_1 < \phi_2 < \pi-\phi_2 < \pi-\phi_1$

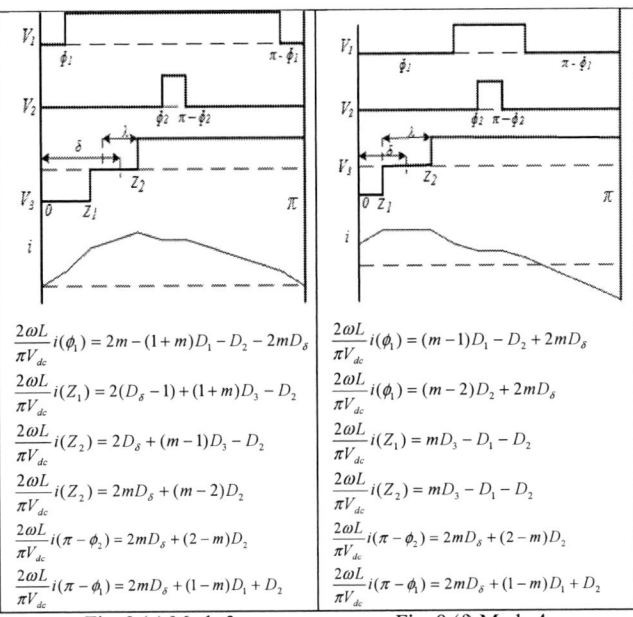

$$\frac{2\omega L}{\pi V_{dc}} i(\phi_1) = 2m - (1+m)D_1 - D_2 - 2mD_\delta$$
$$\frac{2\omega L}{\pi V_{dc}} i(Z_1) = 2(D_\delta-1) + (1+m)D_3 - D_2$$
$$\frac{2\omega L}{\pi V_{dc}} i(Z_2) = 2D_\delta + (m-1)D_3 - D_2$$
$$\frac{2\omega L}{\pi V_{dc}} i(Z_2) = 2mD_\delta + (m-2)D_2$$
$$\frac{2\omega L}{\pi V_{dc}} i(\pi-\phi_2) = 2mD_\delta + (2-m)D_2$$
$$\frac{2\omega L}{\pi V_{dc}} i(\pi-\phi_1) = 2mD_\delta + (1-m)D_1 + D_2$$

Fig. 9.(e) Mode 3

$$\frac{2\omega L}{\pi V_{dc}} i(\phi_1) = (m-1)D_1 - D_2 + 2mD_\delta$$
$$\frac{2\omega L}{\pi V_{dc}} i(\phi_1) = (m-2)D_2 + 2mD_\delta$$
$$\frac{2\omega L}{\pi V_{dc}} i(Z_1) = mD_3 - D_1 - D_2$$
$$\frac{2\omega L}{\pi V_{dc}} i(Z_2) = mD_3 - D_1 - D_2$$
$$\frac{2\omega L}{\pi V_{dc}} i(\pi-\phi_2) = 2mD_\delta + (2-m)D_2$$
$$\frac{2\omega L}{\pi V_{dc}} i(\pi-\phi_1) = 2mD_\delta + (1-m)D_1 + D_2$$

Fig. 9.(f) Mode 4

In figures 9(a)-(f), D_1, D_2, D_3 are the duty cycles for the three H-bridges, $D_\phi = \delta/\pi$ and $(V_{dc1} : V_{dc2} : V_{dc3}) = (V_{dc} : V_{dc} : mV_{dc})$. The natural ZVS turn-on condition of a device in a switching leg would depend on the direction of the current flowing through the switching node. The natural ZVS turn-on condition of the devices for the aforementioned three port DAB does not change depending on switching techniques. Considering the current going into middle limb bridge and current coming out of side limb H-bridges as positive current direction, the ZVS turn-on conditions for the all the switching transitions are given below in (15).

$$i(\phi_1) <= 0, \; i(\phi_2) <= 0, \; i(Z_1) >= 0, \; i(Z_2) >= 0, \qquad (15)$$
$$i(\pi-\phi_1) >= 0, \; i(\pi-\phi_2) >= 0$$

It can be clearly observed from equation (15) and the current values from fig. 9(a)-(f), natural ZVS turn-on is not possible for the whole power range. For the H-bridges connected to side limb windings, the ZVS turn-on is mainly lost during $i(\phi_1), i(\phi_2)$. At $i(\pi-\phi_1)$ and $i(\pi-\phi_2)$, ZVS turn-on is present for whole range of power for all values of D_1, D_2, D_3 and δ. The turn-off ZVS occurs when the turn-on ZVS is lost. However, for 1200V and 1700V SiC-Mosfets, the turn-on loss is dominant than turn-off loss, hence achieving a ZVS during turn-on of SiC Mosfets is mostly important. To observe the range of turn-on ZVS for 1200V SiC-Mosfet based side limb H-bridges, 3-D plots are shown in figures 10-11 for different values of D_3 and δ, which depict the combined power flow from the side limb bridges to the middle limb bridge for turn-on ZVS at $i(\phi_1)$ or $i(\phi_2)$. It can be clearly observed that the turn-on ZVS is present near full power range for any constant values of D_3 and δ. With increasing D_3 and δ, the power range and ZVS range increases. The range of turn-on ZVS for 1700V SiC-Mosfet based middle limb H-bridge in 3-D plot,s are shown in figures 12-13 for different values of D_3 and δ.

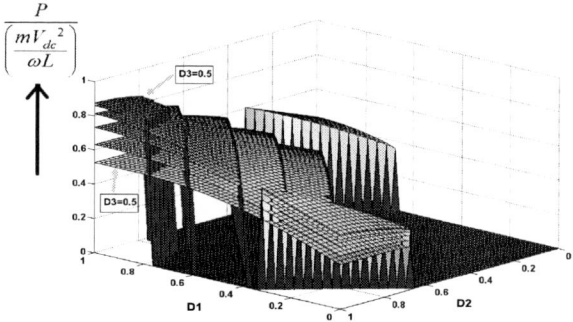

Fig. 10. Output Power for Full Turn-On ZVS of One Side-Limb Bridge($\delta=\pi/6$)

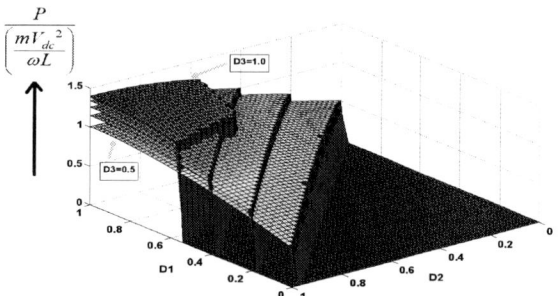

Fig. 11. Output Power for Full Turn-On ZVS of One Side-Limb Bridge($\delta=\pi/3$)

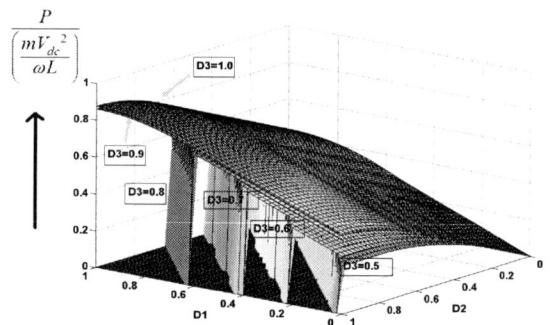

Fig. 12. Output Power for Full Turn-On ZVS of Middle-Limb Bridges($\delta=\pi/6$)

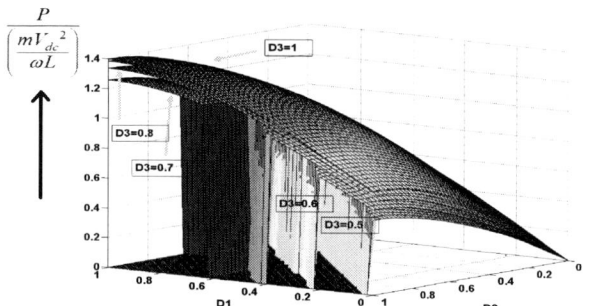

Fig. 13. Output Power for Full Turn-On ZVS of Middle-Limb Bridges($\delta=\pi/3$)

E. Three Limb Transformer Parameters

To demonstrate the operation and effectiveness of the aforementioned three port Dual Active bridge converter, a 10kW three limb high frequency transformer has been built using Ferrite E cores 0R49928EC with 1:1:1 turns ratio on each limb. Four E cores, two are joined at the limb cross section and two are joined at the back surfaces, to form the three limb transformer as shown in figure 14. Number of turns on each of the limbs are 20 with the middle limb has cross section area twice that of the side limbs. Operating frequency of the transformer is 50kHz with maximum switching voltages $V_{dc1}=V_{dc2}=700V$ on the side limbs and $V_{dc3}=1400V$. The transformer winding is made of 10AWG equivalent litz wire.

Fig. 14. Transformer Prototype for Laboratory Experiment

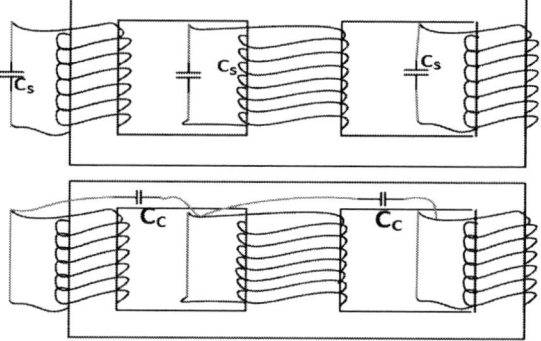

Fig. 15. Transformer Winding Self(C_S) and Coupling(C_C) Capacitance Model

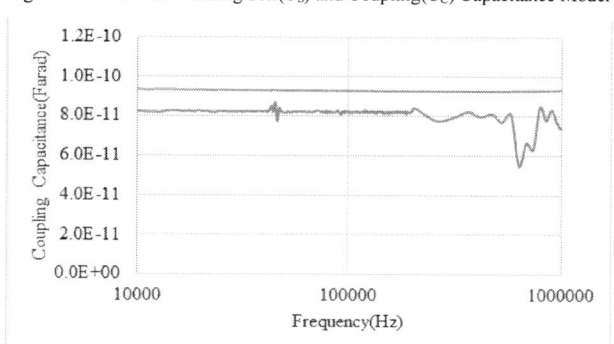

Fig. 16. Coupling Capacitance(C_C) Measurement Between Middle Limb Winding and Side Limb Windings

For the aforementioned three limb transformer operation, assessment of the parasitic capacitance model of the transformer is necessary. Figure 15 show the self and coupling capacitance model of the three limb transformer model, where C_C is the coupling capacitance and C_S is the self-capacitance of the windings. Measurement of C_C and C_S is done using impedance analyzer with frequency variation from 10kHz to 1MHz. Figure 16 show the two coupling capacitance measurement plot over the frequency range. Both the coupling capacitances are nearly equal around 80pF and 90pF. The self-capacitances of the windings have been found to vary around 200pF to 350pF. Figures 17 and 18 show the short circuit or leakage inductance and open circuit or magnetizing inductance of the transformer, all referred to middle limb winding. Table 1 shows the results of the impedance analyzer measurement for the three limb transformer.

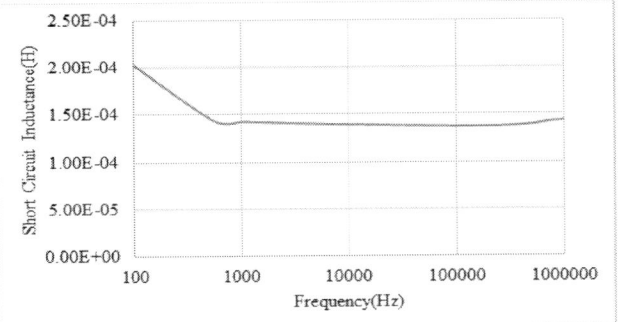

Fig. 17. Short Circuit Inductance Measurement From Middle Limb Winding

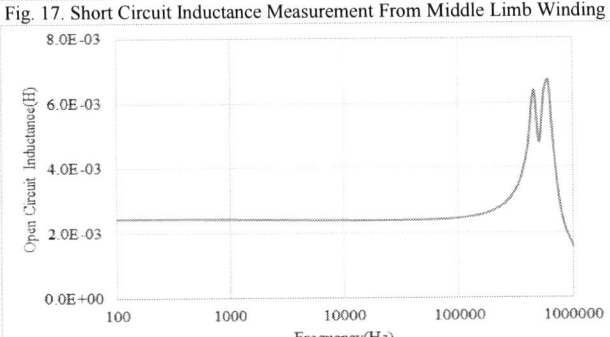

Fig. 18. Open Circuit Inductance Measurement From Middle Limb Winding

TABLE 1: Transformer Parameters	
Side limb Rated Voltage	700V
Middle Limb Rated Voltage	1400V
Operating Frequency	50 kHz
Winding AC Current(max)	20A (RMS)
Leakage Inductance(short circuit inductance referred to middle limb winding)	144uH(around 10kHz-100kHz)
Magnetizing Inductance(open circuit inductance referred to middle limb winding)	2.4 mH(around 10kHz-100kHz)
Coupling Capacitance(C_C)	80pF-90pF
Self Capacitance (C_S)	200pF-350pF
Measured Winding AC Resistance at 50kHz	0.16Ω

It can be observed from table 1 that the open circuit measured inductance or magnetizing inductance is nearly 20 times higher than short circuit series inductance or the leakage inductance, which solidifies the assumption made in subsection A, that the magnetizing current is negligible compared to load current. For a 700V dc voltage, the magnetizing current is around 0.9A while the full load current is upto 20A.

For analyzing the three limb transformer core loss, knowledge on the flux flow path and volume is necessary. Taking the flux path shown in figure 3, it can be deduced that the flux induced by the applied voltage flows through the volume of the limbs and in the connecting portion between the limbs the flux varies depending on the leakage flux which is depends on load current. Transformer core loss for non-sinusoidal waveform has been explained by i^2GSe method in [15]. Transformer core loss as per i^2GSe method, varies with peak flux density Bm and dB/dt, both of which are dependent on applied voltage. Considering only the induced flux due to applied voltage and neglecting the losses in connecting portion between limbs, core loss of the transformer is predicted for the transformer shown in figure 16. Clearly, the core loss varies with the duty cycles D_1, D_2 and D_3 only, higher the duty cycles, higher are the losses. Figure 19 shows the core loss for V_{dc}=700V due to variations in D_1, D_2 and D_3.

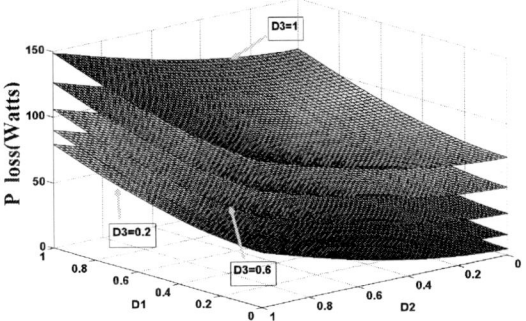

Fig. 19. Transformer Core Loss Variation for $V_{dc1}=V_{dc2}$=700V, V_{dc3}=1400V

III. RESULTS AND DISCUSSIONS

A laboratory prototype of the three port DAB converter of figure 2 has been built and experimental studies have been performed. The side limb H-bridge converters use 1200V/31A SiC Mosfet C2M0080120 and the middle limb H-bridge uses 1700V/50A(TO-247) SiC Mosfet from Cree. Gate resistance used for 1200V SiC Mosfet is 2.5Ω, and zero gate resistance is used for the 1700V SiC Mosfet. High speed switching isolated gate driver, capable of operating at high dv/dt has been developed and used. For validating the operation of the prototype, experiments have been performed with 500V dc voltage source connected to the dc bus of side limb H-bridges and resistive load bank is connected to the dc bus of middle limb H-bridge converter. For experimental study of the prototype, the dc bus of 1700V SiC Mosfet based H-bridge is boosted upto 1kV. The side limb H-bridge with voltage V_{dc1}, controls its own input dc current I_1 and the other side limb H-bridge of V_{dc2} controls its output power to control the output dc bus V_{dc3} at 1kV. The prototype has been tested upto 8.2 kW of dc power.

Figures 20-24 show the transformer waveforms at different values of I_1, δ and D_3 for 8.2kW. It can be observed that with increasing δ, D_3 value can be smaller to provide same output power. In figure 20, both the input H-bridges, delivering almost equal power have partial ZVS turn-on, while in figure 21, one bridge having higher power has full ZVS turn-on while the other providing lower power has partial ZVS turn-on. In figure 22 and 23, the middle limb H-bridge has partial ZVS. Hence higher value of δ can give lower transformer loss, as per figure 19, but can cause loss in ZVS. At 8.2 kW of power, for δ = 30°, efficiency is 97.3% and at δ = 54°, efficiency is 97%.

978-1-4673-9551-9/16 $31.00 © 2016 IEEE

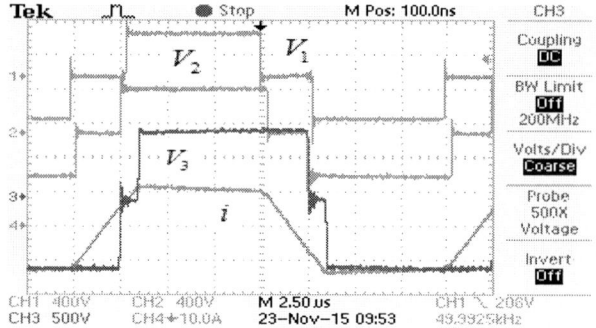

Fig. 20. Transformer Waveforms for P3=8.2kW at δ = 30°, I₁=8A, D₃=0.9

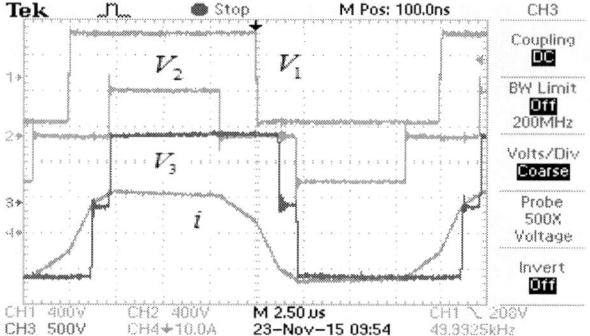

Fig. 21. Transformer Waveforms for P3=8.2kW at δ = 30°, I₁=10A, D₃=0.9

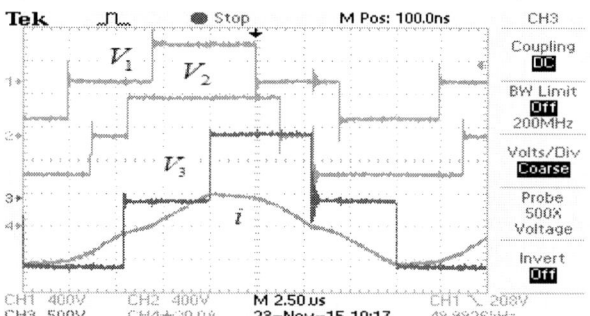

Fig. 22. Transformer Waveforms for P3=8.2kW at δ = 54°, I₁=8A, D₃=0.6

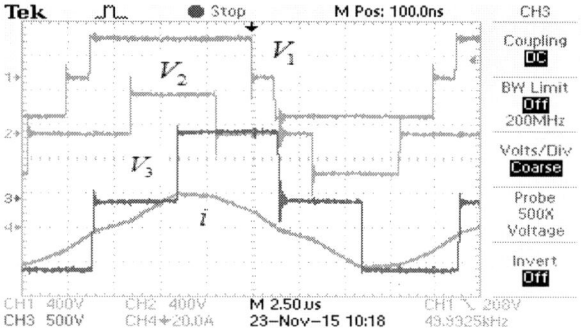

Fig. 23. Transformer Waveforms for P3=8.2kW at δ = 54°, I₁=11A, D₃=0.6

For lower power levels as 6.3kW and 2.9kW, partial ZVS turn-on cases are shown in figures 24 and 25. It can be evidently inferred from the waveforms that full ZVS turn-on is present when the corresponding converter is delivering major portion of the total power, like source Vdc1 in figure 24 and source Vdc2 in figure 25, where both the source have full ZVS

turn-on when the other side limb H-bridge has partial ZVS turn-on, while the middle limb H-bridge has full ZVS turn-on for both the cases.

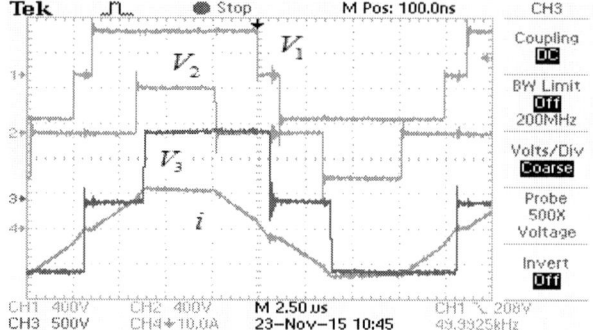

Fig. 24. Transformer Waveforms for P3=6.3kW at δ = 30°, I₁=8A, D₃=0.7

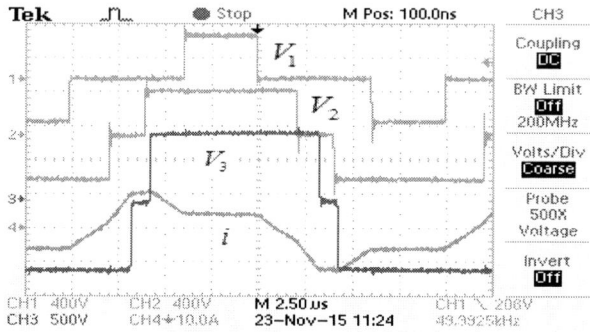

Fig. 25. Transformer Waveforms for P3=2.9kW at δ = 15°, I₁=2A, D₃=0.7

The independent power control as explained in section 2, is verified at a lower voltage of 300V input for V_{dc1}-V_{dc2}, and 500V-600V for V_{dc3}, and at a lower power of 1.6kW. In figure 26, the output voltage reference is changed from 600V to 500V and the output voltage changes but the current I_1 is regulated at reference dc value of 4A. In figure 27, the V_{dc3} reference is changed back to 600V and the voltage V_{dc3} goes transient while I_1 remains constant.

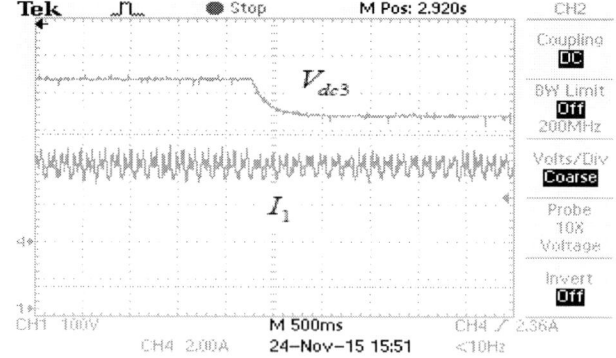

Fig. 26. V_{dc3} Change to 500V and Current I_1

In figure 28, V_{dc3} reference is kept at 500V, as the reference current of I_1 is changed from 4A to 3A, the current I_1 goes to 3A while voltage V_{dc3} has a small transient and remains at 500V. In figure 29, V_{dc3} reference is kept at 500V, as the reference current of I_1 is changed from 3A to 4A, the current I_1

978-1-4673-9551-9/16 $31.00 © 2016 IEEE

goes to 4A while voltage V_{dc3} has a small transient and remains at 500V.

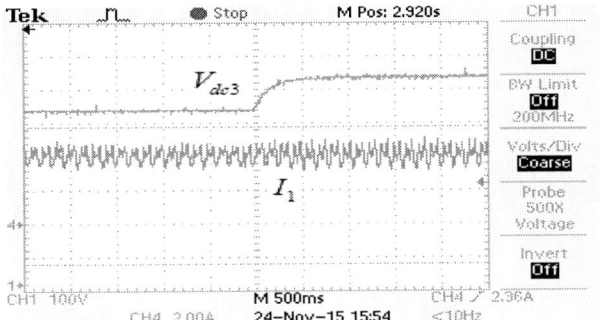

Fig. 27. V_{dc3} Change to 600V and Current I_1

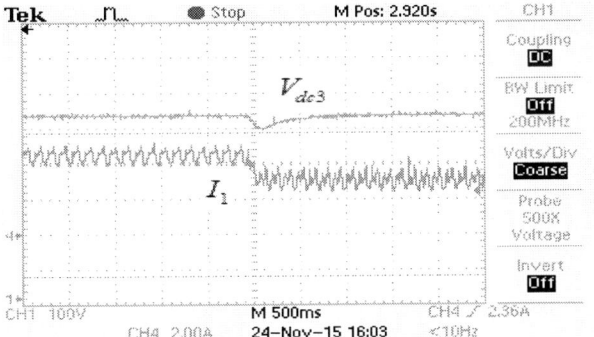

Fig. 28. Current I_1 Change to 3A and V_{dc3}

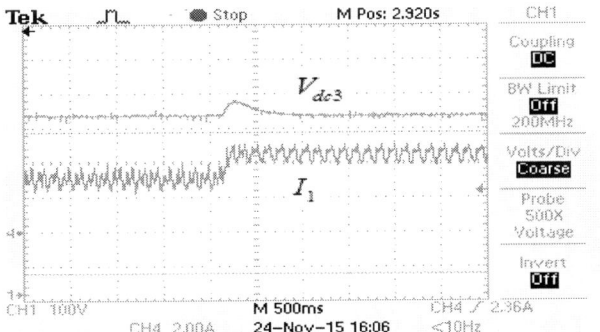

Fig. 29. Current I_1 Change to 4.5A and V_{dc3V}

IV. CONCLUSIONS

From the experimental results, the independent power control proposed in section 2 is clearly proven and the converter operation under different ZVS conditions for same operating point, as depicted in section2, is observed successfully.

ACKNOWLEDGMENT

This work made use of FREEDM ERC shared facilities supported by NSF under award no. EEC-0812121. Special acknowledgement to Akshat Yadav, Sayan Acharya and Samir Hazra for helping in prototype setup and experimental validation.

REFERENCES

[1] K. Hatua, S. Dutta, A. Tripathi, S. Baek, G. Karimi, S. Bhattacharya, "Transformer less Intelligent Power Substation design with 15kV SiC IGBT for grid interconnection," *IEEE Energy Conversion Congress and Exposition (ECCE), 2011,* pp. 4225 – 4232.

[2] S. Madhusoodhanan, A. Tripathi, K. Mainali, A. Kadavelugu, D. Patel, "Three-phase 4.16 kV medium voltage grid tied AC-DC converter based on 15 kV/40 a SiC IGBTs," *IEEE Energy Conversion Congress and Exposition (ECCE), 2015,* pp. 6675 - 6682.

[3] A. Azidehak, R. Chattopadhyay, S. Acharya, A.K. Tripathi, M.G. Kashani, G. Chavan, S. Bhattacharya, "Control of modular dual active bridge DC/DC converter for photovoltaic integration," *IEEE Energy Conversion Congress and Exposition (ECCE), 2015,* pp. 3400 - 3406.

[4] S. Hazra, A. De, L. Cheng, J. Palmour, M. Schupbach, B. Hull, S. Allen, S. Bhattacharya, "High Switching Performance of 1700V, 50A SiC Power MOSFET over Si IGBT/BiMOSFET for Advanced Power Conversion Applications," *IEEE Trans. Power Elec., Issue: 99, Volume: PP, Year: 2015.*

[5] Qiang Mei, Xu Zhen-lin, Wei-Yang Wu, "A novel multi-port DC-DC converter for hybrid renewable energy distributed generation systems connected to power grid," *IEEE International Conference on Industrial Technology(ICIT), 2008,* pp. 1 - 5.

[6] D. Gunasekaran, L. Umanand, "Integrated magnetics based multi-port bidirectional DC-DC converter topology for discontinuous-mode operation," *Power Electronics, IET, 2012, vol. 5, Issue 7, 2012,* pp. 935-944.

[7] Zhao Chuanhong, J.W. Kolar, "A novel three-phase three-port UPS employing a single high-frequency isolation transformer," *IEEE 35th Annual Power Electronics Specialists Conference(PESC), 2004,* pp. 4135 - 4141.

[8] M. Jafari, Z. Malekjamshidi, M.R. Islam, Zhu Jianguo, "Modeling of magnetic flux in multi-winding toroidal core high frequency transformers using 3D reluctance network model," *IEEE 11th International Conference on Power Electronics and Drive Systems (PEDS), 2015,* pp. 413 - 418.

[9] Jun Zhou, Qianhong Chen, Xinbo Ruan, Siu Chung Wong, Tse, C.K., "Experimental measurement and modeling of multi-winding high-voltage transformer," *International Conference on Electrical Machines and Systems(ICEMS), 2008,* pp. 4411 - 4415.

[10] Jian Wang, A.F. Witulski, J.L. Vollin, T.K. Phelps, G.I. Cardwell, "Derivation, calculation and measurement of parameters for a multi-winding transformer electrical model," *Fourteenth Annual Applied Power Electronics Conference and Exposition(APEC), 1999,* vol.1, pp. 220 - 226.

[11] D. Aggeler, J. Biela, J.W. Kolar, "Controllable dυ/dt behaviour of the SiC MOSFET/JFET cascode an alternative hard commutated switch for telecom applications," *Twenty-Fifth Annual IEEE Applied Power Electronics Conference and Exposition (APEC), 2010,* pp. 1584 – 1590.

[12] F. Jauch, J. Biela, "An innovative bidirectional isolated multi-port converter with multi-phase AC ports and DC ports," *15th European Conference on Power Electronics and Applications(EPE), 2013,* pp. 1-7.

[13] S. Dutta, S. Roy, S. Bhattacharya, "A mode switching, multiterminal converter topology with integrated fluctuating renewable energy source without energy storage," *in IEEE 2014 Twenty-Ninth Annual Applied Power Electronics Conference and Exposition(APEC), 2014,* pp. 419-426.

[14] R. Chattopadhyay, S. Bhattacharya, "Modular Isolated DC-DC Converter with Multi-Limb Transformer for Interfacing of Renewable Energy Sources", *in IEEE 2015 Applied Power Electronics Conference and Exposition(APEC), 2015,* pp. 3039- 3046.

[15] J. Muhlethaler, J. Biela, J.W. Kolar, A. Ecklebe, "Improved Core-Loss Calculation for Magnetic Components Employed in Power Electronic Systems," *IEEE Trans. Power Elec., 2012, vol. 27, Issue. 2.,*pp.964-973.

978-1-4673-9551-9/16 $31.00 © 2016 IEEE

A Novel Multi-Input Converter Using Soft-Switched Single-Switch Input Modules with Integrated Power Factor Correction Capability for Hybrid Renewable Energy Systems

Sanjida Moury[1], *IEEE Student Member*, John Lam [2], *IEEE Senior Member,* Vineet Srivastava [3], Ron Church [4]

[1][2] Dept. of Electrical Engineering and Computer Science

Lassonde School of Engineering, York University

4700 Keele Street, Toronto, M3J 1P3, Canada

[3][4] Cistel Technology Inc.

30 Concourse Gate, Nepean, K2E 7V7, Canada

e-mail address: [1]sanjida@cse.yorku.ca, [2]johnlam@cse.yorku.ca, [3]vineet@cistel.com, [4]ron.church@cistel.com

Abstract- **The energy consumption of the information technology (IT) data centers and servers have been increasing rapidly for the past decade. In order to reduce global warming and conserve depleting sources of fossil fuel, hybrid wind-solar energy systems can be an attractive alternative energy source to supply clean and sustainable energy for powering the IT data centers and servers. To minimize the cost and power losses in individual power converter of the conventional hybrid renewable energy system, multi-input converters (MICs) are an attractive solution. This paper proposed a new quasi-resonant soft-switched MIC with integrated power factor correction (PFC) for hybrid wind-solar energy systems. The proposed MIC topology requires only one switch in each input module and all the switches are able to achieve zero voltage switching (ZVS) turn-on and zero current switching (ZCS) turn-off for individual and simultaneous operation for different operating conditions. The operating principles of the proposed circuit are provided. Results are given on a hybrid wind-solar energy system with 48V-output to highlight the performance of the proposed circuit.**

Keyword- multi input converter; hybrid renewable energy system; power factor correction

I. INTRODUCTION

The energy consumption of Information Technology (IT) data centers and servers has grown rapidly with the growth of online applications and services including email, blogging, social networking, and cloud computing. As of 2013, the industrial and IT sectors account for 38% of the nation's energy use and 34% in related greenhouse gas emissions [1]. Due to environmental concern and the growing energy demand, it is essential to search for green and energy-efficient solutions for powering the IT data centers and servers [2]-[4]. To obtain clean and sustainable uninterrupted energy, hybrid renewable energy system is the way of the future due to the volatile nature of renewable energy resources. In hybrid renewable energy systems, the combination of wind and solar

energy is the primary choice as they are the most promising sources among the renewable energy resources and inherently, the electric power generation of PV array and wind turbine is complementary. In a conventional hybrid renewable energy system as shown in Fig. 1, individual power converter for each energy source is used to extract and convert the renewable energy source to useable electric power. In order to increase the converter efficiency and to reduce the cost and size of the power electronics interface, multi-input converter (MIC) configuration [4]-[5] by sharing the output filtering circuit components has been introduced for hybrid renewable energy systems.

Research on MIC for hybrid renewable energy system has been presented in [4]-[17] to replace the individual power converters. The idea is to integrate multiple independent converters into a single converter with multiple inputs and share the common components such as power devices and/or magnetic components of the original converters. In [4][6][17], non-isolated MICs resulted from the series connection of the output stages of two switch-mode DC-DC converters have been presented. Isolated MICs with full-bridge configuration has been presented in [11]. To reduce the switching losses, MICs with soft-switching operations have also been presented in [13]-[16]. However, the presented methods require either multiple switches in each input module [13] or additional auxiliary switch [14] in the converter to achieve soft-switching function.

Each of the presented MIC circuit in the literature has its own merits. However, most of the approaches presented do not consider the power factor correction (PFC) requirement when the MIC circuit consists of an AC-powered input module, such as working with wind turbine system. In this case, a high quality balanced 3-phase current is required to be drawn from the input to optimize the lifetime of the turbine generator. To meet the aforementioned requirement, an additional PFC stage is usually required in the input module.

978-1-4673-9551-9/16 $31.00 © 2016 IEEE

As a result, the size and cost of the power electronic interface will be increased substantially.

In this paper, a new MIC with integrated PFC function and soft-switched single-switch input modules is proposed for hybrid wind-solar systems. The proposed converter is capable of supporting individual as well as simultaneous operation with multiple energy sources. The proposed MIC is also able to achieve ZVS turn-on and ZCS turn-off for individual and simultaneous operation. The derivation and operating principles of the proposed circuit will be discussed in the next section. Results will be provided on a 48V-output hybrid wind-solar energy system to highlight the features of the proposed circuit.

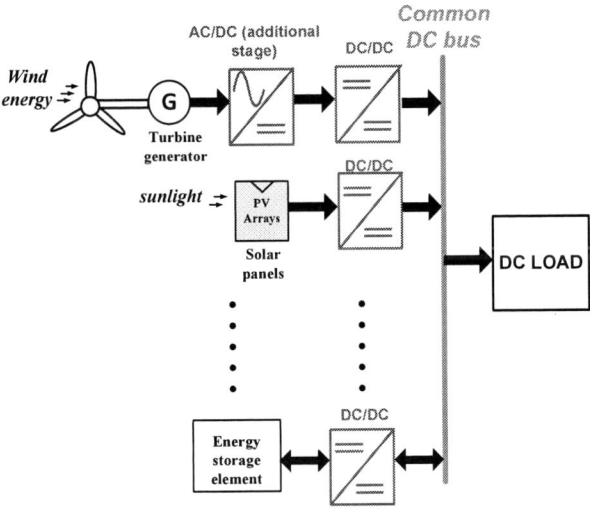

Figure 1: Conventional hybrid energy system power electronics interface

II. CIRCUIT DESCRIPTION

Based on several different single-switch soft-switched circuits presented in literature, which include the Class-E converter [18], ZVS and ZCS quasi-resonant converter (QRC) [19], 4 different soft-switched MICs are first derived as shown in Fig. 2. All of these MICs are capable of supporting individual as well as simultaneous operations. When a MIC is required to work with both solar and wind energy, a balanced three-phase close to sinusoidal output current of the turbine generator is essential. An additional power factor corrector circuit will increase the overall size and cost of the system. Hence, in this paper, a step is taken further to combine a 3-phase boost PFC circuit with the MIC presented in Fig. 2(d) as shown in Fig. 3(a). A new soft-switched MIC with integrated boost PFC in the AC-powered input module is then derived from Fig. 3(a) and the final circuit is shown in Fig. 3(b).

The proposed converter consists of two ZVS QRC modules- one for working with solar energy and the other one is for working with wind energy. The switch S_1 with tank circuit L_{r1} and C_{r1} and the switch S_2 with tank circuit L_{r2} and C_{r2} form the resonant switches for the ZVS QRC. As a result, when the power switches (S_1 and S_2) conduct, the resonant capacitors (C_{r1} and C_{r2}) are shorted and no resonance occurs for that time portion. Resonance occurs when power switches

(S_1 and S_2) are OFF. Diode D_{22} creates the unidirectional conduction path to the shared power switch S_2 for the PFC circuit. Both ZVS QRC modules share the same output filter components (L_f and C_f) and the output diodes (D_1 and D_2). The individual operation mode of the proposed converter associated to each energy source is shown in Fig. 4. For simultaneous operation mode, the key operating waveforms are shown in Fig. 5, where d_1 and d_2 represents the duty cycle of the input module that works with solar and wind energy respectively, with $d_1 > d_2$ and T_s is the switching period. Note that the switching frequencies for both input modules are assumed to be the same in this analysis to simplify the operating stages illustration. The operation of the converter can be analyzed by 7 operating stages within a switching period. It should be emphasized that depending on the availability of the input energy sources, the proposed circuit in Fig. 3(b) can be extended to consists of i input modules to meet the power demand.

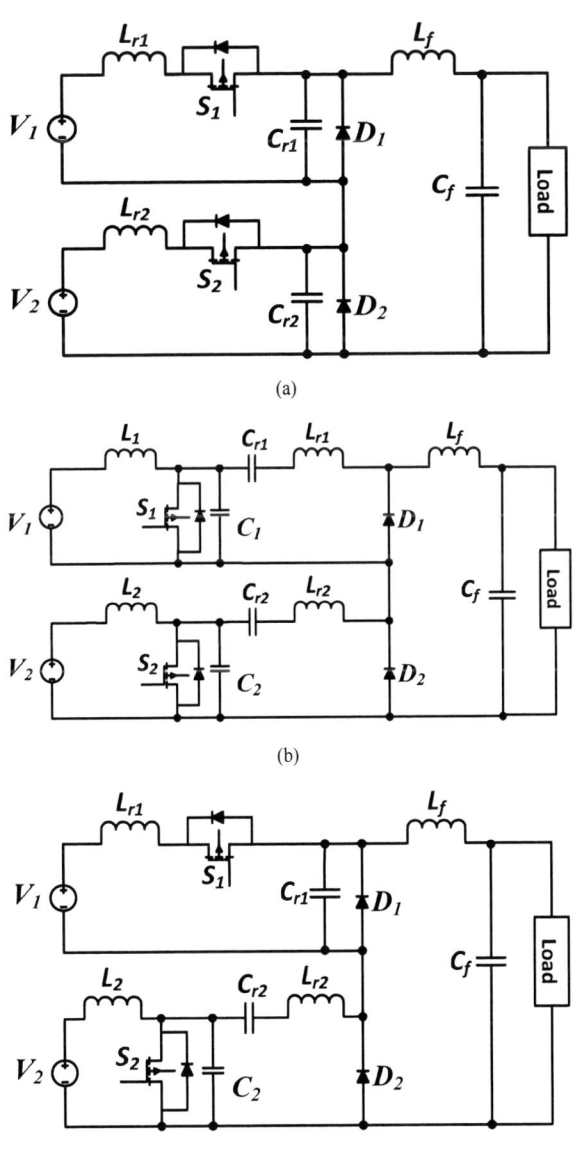

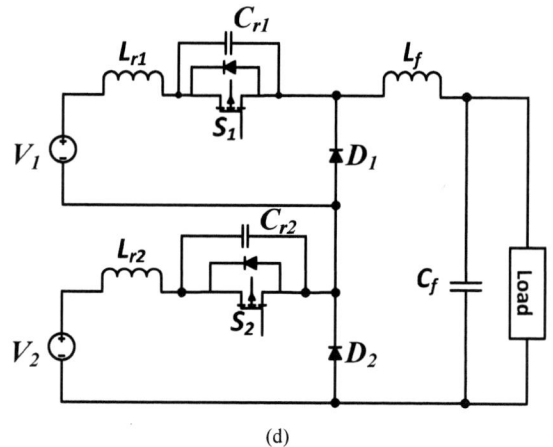

(d)

Figure 2: Soft-switched MICs: (a) ZCS QRC-ZCS QRC (b) class-E-class-E (c)ZCS QRC - class-E (d)ZVS QRC -ZVS QRC

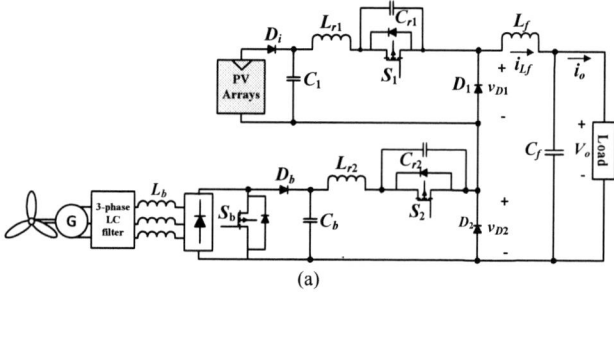

(a)

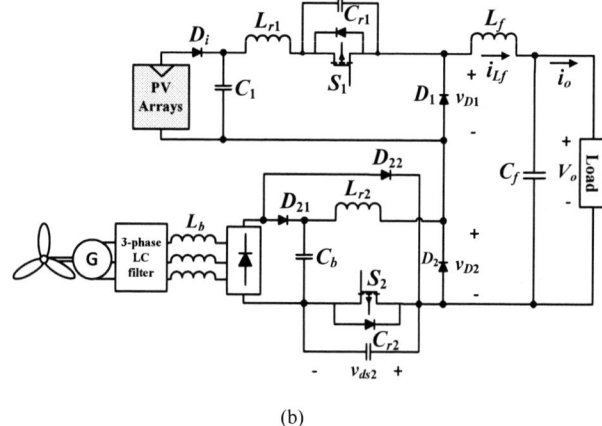

(b)

Figure 3: Proposed MIC (a) ZVS- QRC with additional PFC circuit (b) ZVS-QRC with integrated PFC function (in module 2)

[$t_0 < t < t_1$]: Prior to this stage, S_1 and S_2 are off. At t_0, the gate signal is applied to S_1 and S_2. The negative i_{r2} flows through the anti-parallel diode of S_2, which forces D_{22} to conduct and the resulting current flows through D_2. Since i_{ds1} and i_{ds2} are negative previously and so S_1 and S_2 turn on under ZVS. i_{D22} rises linearly throughout this stage. Since D_1 and D_2 are on, i_{Lf} continues to discharge linearly through D_1 and D_2.

[$t_1 < t < t_2$]: i_{ds2} forces D_1 and D_2 to turn off and energy starts

transferring from L_{r2} to L_f. So, i_{Lf} starts increasing.

[$t_2 < t < t_3$]: i_{ds1} increases to be equal to i_{Lf} and i_{Lf} continues rising linearly. Both L_{r1} and L_{r2} are now transferring energy to L_f

[$t_3 < t < t_4$]: The gate signal is removed from S_2, v_{ds2} rises slowly from zero, achieving ZCS turn-off. D_2 and D_{21} turn on and i_{Lf} continues increasing through D_2. C_{r2} starts charging and when V_{cr2} reaches to V_2, resonance starts in module 2.

[$t_4 < t < t_5$]: The gate signal of S_1 is removed, v_{ds1} starts rising slowly from zero and S_1 turns off under ZCS. D_1 turns on and i_{Lf} starts discharging linearly through D_1 and D_2. Resonance for module 1 happens in this time period.

[$t_5 < t < t_6$]: v_{ds2} drops to zero at the beginning of this stage. The negative i_{r2} starts flowing through the anti-parallel diode of S_2. D_{22} turns on and i_{D22} rises linearly. The resulting current of negative i_{r2} and i_{d22} flows through D_2.

[$t_6 < t < t_7$]: v_{ds1} drops to zero. The negative i_{r1} flows through the anti-parallel diode of S_1.

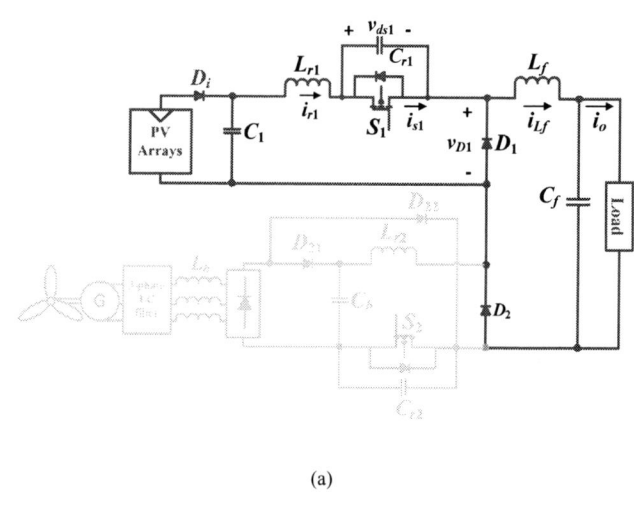

(a)

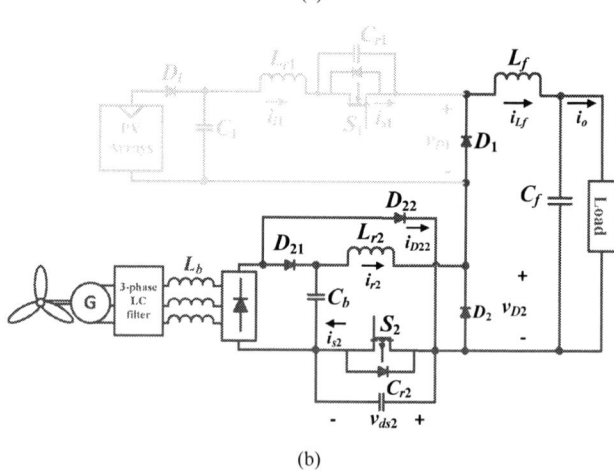

(b)

Figure 4: Individual operation mode (a) with only PV source (b) with only wind energy source

978-1-4673-9551-9/16 $31.00 © 2016 IEEE

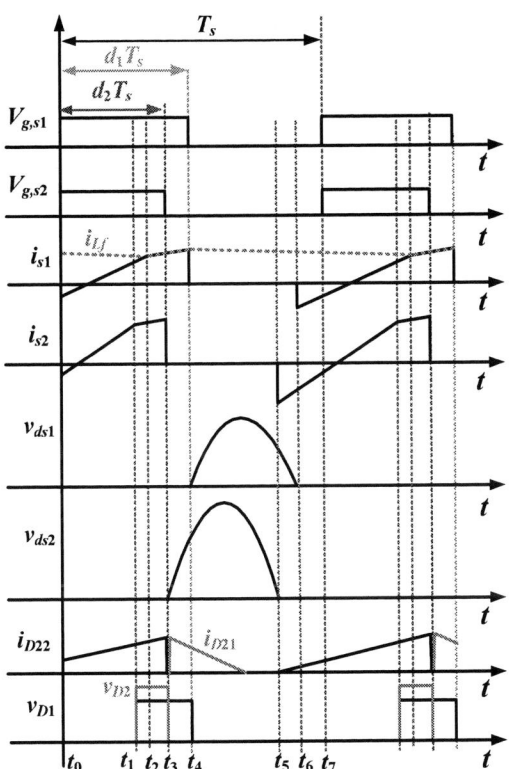

Figure 5: Waveforms in simultaneous operation (assuming $d_1 > d_2$)

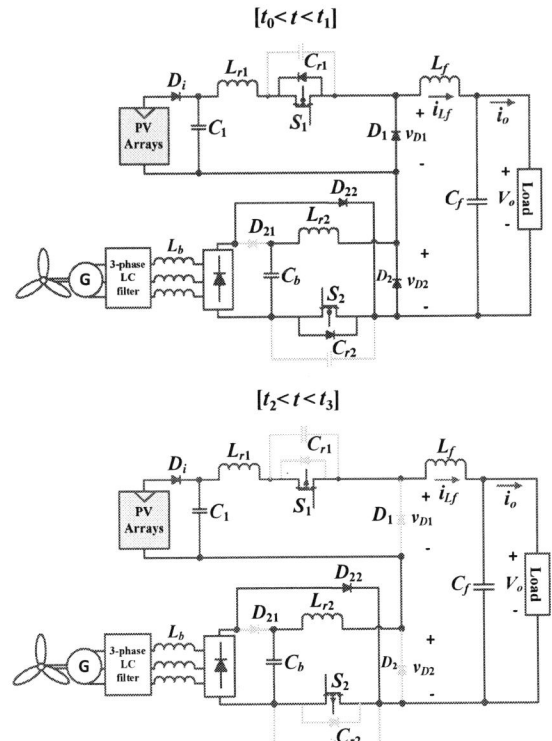

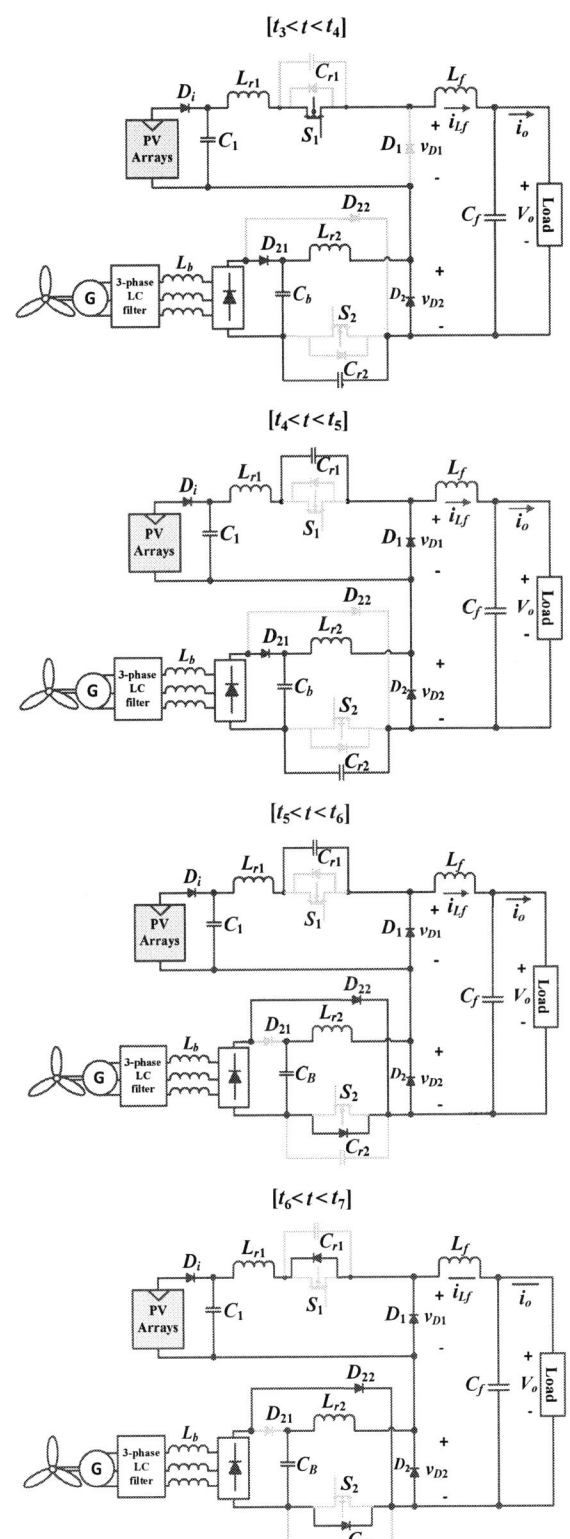

Figure 6: Operating stages of the proposed converter in simultaneous operation mode

The output voltage of the converter is governed by (1), where T_s is the switching period and $v_{Di}(t)$ is given by (2), where V_i is the input voltage of the ith module. The current and voltage of both switches are expressed for different time interval in (3) - (6) where, ω_{sw1} and ω_{sw2} are the angular switching frequency of solar module and wind module respectively. The initial switch currents are represented by $i_{ds1}(0)$ and $i_{ds2}(0)$.

$$V_0 = \frac{1}{T_S} \int_0^{T_S} \left(v_{D1}(t) + v_{D2}(t) + \ldots + v_{Di}(t) \right) dt \tag{1}$$

$$v_{Di}(t)\big|_{i=1,2\ldots} = V_i - L_{ri} C_{ri} \frac{d^2 v_{dsi}}{dt^2} - v_{dsi} \tag{2}$$

$$i_{ds1} = \begin{cases} \dfrac{1}{\omega_{sw1} L_{r1}} \displaystyle\int_0^{\omega_{sw1}t} V_1 d(\omega_{sw1}t) + i_{ds1}(0) & t_0 \leq t \leq t_2 \\[6pt] I_0 & t_2 \leq t \leq t_4 \end{cases} \tag{3}$$

$$i_{ds2} = \begin{cases} \dfrac{1}{\omega_{sw2} L_{r2}} \displaystyle\int_0^{\omega_{sw2}t} V_2 d(\omega_{sw2}t) + i_{ds2}(0) + i_{D22} & t_0 \leq t \leq t_1 \\[6pt] I_0 + i_{D22} & t_1 \leq t \leq t_3 \end{cases} \tag{4}$$

$$v_{ds1} = \begin{cases} \dfrac{1}{\omega_{sw1} C_{r1}} \displaystyle\int_{\omega_{sw1}t_4}^{\omega_{sw1}t} i_{cr1} d(\omega_{sw1}t) + v_{cr1}(\omega_{sw1}t_4) & until\ v_{ds1}=V_1 \\[6pt] \dfrac{1}{\omega_{sw1} C_{r1}} \displaystyle\int_{\omega_{sw1}t_4'}^{\omega_{sw1}t} i_{cr1} d(\omega_{sw1}t) + v_{cr1}(\omega_{sw1}t_4') & at\ resonance \end{cases} \quad t_4 \leq t \leq t_6 \tag{5}$$

$$v_{ds2} = \begin{cases} \dfrac{1}{\omega_{sw2} C_{r2}} \displaystyle\int_{\omega_{sw2}t_3}^{\omega_{sw2}t} i_{cr2} d(\omega_{sw2}t) + v_{cr2}(\omega_{sw2}t_3) & until\ v_{ds2}=V_2 \\[6pt] \dfrac{1}{\omega_{sw2} C_{r2}} \displaystyle\int_{\omega_{sw2}t_3'}^{\omega_{sw2}t} i_{cr2} d(\omega_{sw2}t) + v_{cr2}(\omega_{sw2}t_3') & at\ resonance \end{cases} \quad t_3 \leq t \leq t_5 \tag{6}$$

The resonant components, L_{ri} and C_{ri} of the QRC circuit determine the characteristics of the converter, which can be defined by (7) and (8), where f_{Ri} is the resonant frequency of the ith module, Z_{Ni} is the characteristic impedance of the ith module and can be expressed as in (9), with R_L represents the load resistance and M_{Vi} represents the voltage gain of the ith module. The resonant frequency is related with the switching frequency, f_{swi} and the voltage gain, M_{Vi} according to (10).

$$f_{Ri} = \frac{1}{2\pi \sqrt{L_{ri} C_{ri}}} \tag{7}$$

$$Z_{Ni} = \sqrt{\frac{L_{ri}}{C_{ri}}} \tag{8}$$

$$Z_{Ni} = \frac{R_L}{M_{Vi}} \tag{9}$$

$$f_{Ri} = \frac{6.21 \times f_{swi}}{2\pi (1 - M_{Vi})} \tag{10}$$

The input power factor of the module where a 3-phase AC supply is used as the input can be obtained by using the 3-phase boost PFC circuit as given by (11) [20], where $I_{in2,1}$ is the fundamental input current and $I_{in2,n}$ is the nth harmonic component input current of wind module.

$$PF = \left[\frac{I_{in2,1}/\sqrt{2}}{\sqrt{\displaystyle\sum_n^\infty \left(\frac{I_{in2,n}}{\sqrt{2}} \right)^2}} \right] \tag{11}$$

The boost inductor, L_b and capacitor, C_b can be expressed by (12) and (13) respectively, where h is the distortion factor of the input current, f_b is the frequency of the dominant harmonic component, $i_{b2,fb}$ is the dominant harmonic component current and d_2 is the duty cycle of the wind module [20].

$$L_b = 2 \times \frac{v_2}{I_{in2,1}} \times \frac{1}{f_{sw2}} \times h \tag{12}$$

$$C_b = \frac{i_{b2,fb}}{v_2} \times (1 - d_2) \times \frac{100}{\%ripple} \times \frac{1}{f_b} \tag{13}$$

III. RESULTS AND PERFORMANCE

A double-input MIC of the proposed circuit is designed to give an output voltage of 48V with 14A. In this design, input source 1 is connected to $140V_{dc}$ that emulates 4 series connected 35V PV panels. Input source 2 is connected to the output of a $100V_{rms}$ 60Hz wind turbine permanent magnet 3-phase generator. It should be noted that the switching frequency in each module is controlled to maintain the same output voltage. In this design, the range of the operating switching frequency is: 255 - 320 kHz. The components parameters are calculated by (7)-(10) and are determined as follows: L_{r1}=6.2μH, L_{r2}=16.3μH, C_{r1}=19.3nF and C_{r2}=10.9nF.

TABLE I. Circuit component of the designed converter in PSIM

	Component values
L_{r1}	6.2 μH
C_{r1}	20 nF
L_{r2}	16 μH
C_{r2}	11 nF
L_b	15 μH
C_b	47 μF
L_f	650 μH
C_f	3.3 μF

To find the parameter for boost converter, h, f_b and $i_{b2,fb}$ are determined from the Fast Fourier Transform (FFT) of the input current. Using (12) and (13), L_b and C_b are calculated. TABLE I summarizes the circuit components used in the design example. Fig. 7 and Fig. 8 show the switching waveforms of the individual operation with solar module and wind module respectively at full load condition.

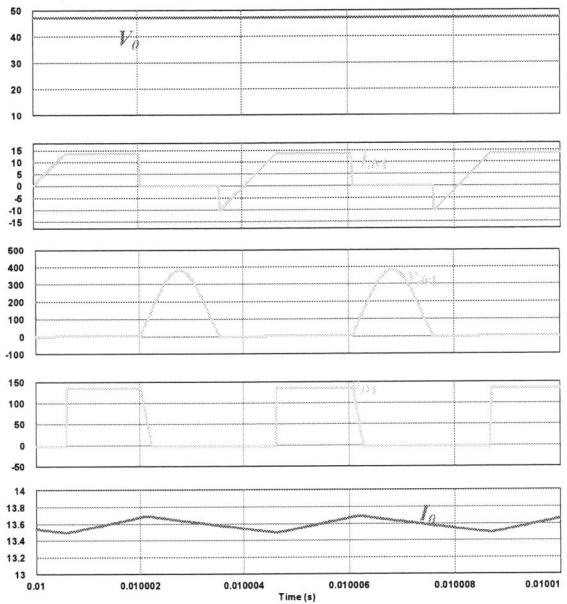

Figure 7: Individual operation for solar energy (f_{sw1} = 255kHz)

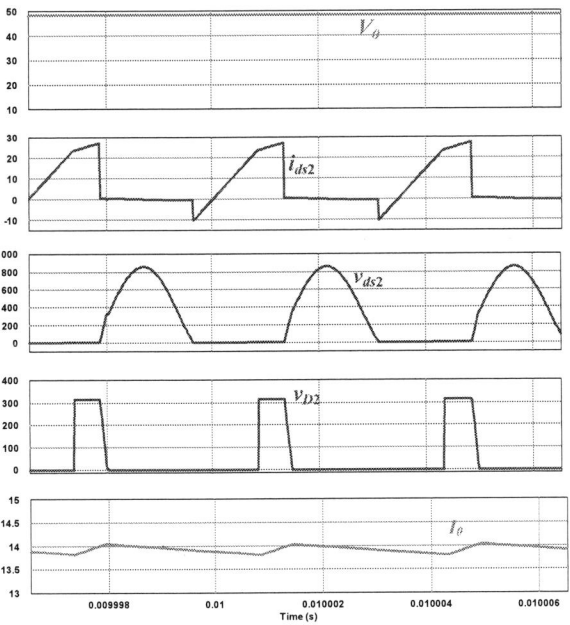

Figure 8: Individual operation for wind energy (f_{sw2} = 275kHz)

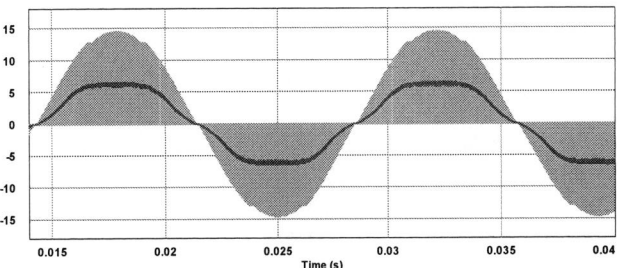

Figure 9: Per-phase input current of the input module that works with wind energy (individual operation)

Fig. 10 shows the simultaneous operation of the two input at full load condition when both of the input voltages operate at their rated values. In all three cases, d_1 and d_2 remain same (d_1= 0.5 and d_2= 0.4). The switching frequency of both input modules, f_{sw1} and f_{sw2} are varied to maintain the same output voltage. It can be observed that ZVS turn on and ZCS turn off is achieved for all cases. Fig. 9 and Fig. 11 show the per-phase input current waveform of the input module that works with the 3-phase generator for individual and simultaneous operation respectively.

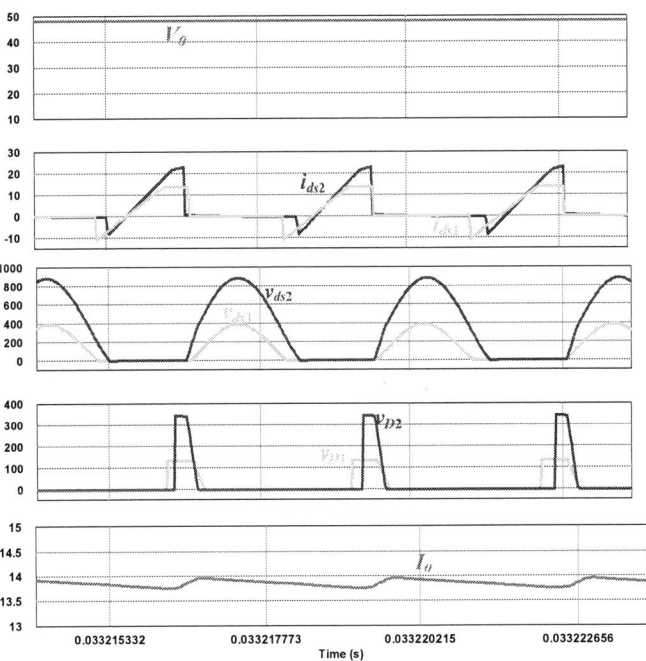

Figure 10: Simultaneous operation of solar and wind energy (f_{sw1} = 310kHz, f_{sw2} = 320kHz)

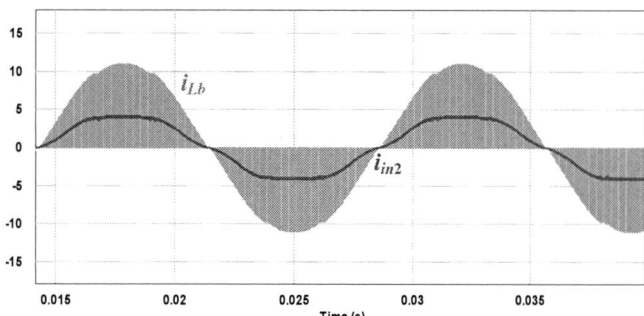

Figure 11: Per-phase input current of the input module that works with wind energy (simultaneous operation)

The efficiency of individual solar and wind module is 91.4% and 93.2% respectively and for the simultaneous operation at full load is 90.6%. The input power factor achieved for both individual wind and simultaneous operation are at least 0.97.

A 48V 4.8A-output experimental prototype of the proposed circuit with 2 inputs as shown in Fig. 12 has been developed in the laboratory and is currently under testing. The control logic circuit for each input module is mounted at the bottom of the prototype. Input 1 is connected to a 125V DC supply and input 2 is connected to an $80V_{rms}$ 3-phase AC source. The components parameters are given in TABLE II. The performance of the prototype has first been verified through results obtained from SIMETRIX 7.20 with the components listed in TABLE II.

Figure 12: Experimental prototype of the proposed MIC

Fig. 13 shows the full load performance when both inputs operate at the rated input voltages. The switching frequency is 255kHz for module 1 and 275kHz for module 2. Soft-switching operation is observed from Fig. 13 and the

efficiency is measured to be 91.8%. The input PF of the 3-phase current is 0.98.

TABLE II. Circuit components of the experimental prototype

	Component parameters/ part number
$L_{r1} L_{r2}$	SER2918H-103 (10 µH)
C_{r1}	BFC233860682 (6.8 nF, 1kV)
C_{r2}	R413I21000000M (10nF, 1kV)
L_b	SER2915H-153 (15 µH)
C_b	EZP-E50256LTA (25 µF, 500V)
L_f	GA3199-AL (650 µH)
C_f	B32522C1335J (3.3 µF, 100V)
S_1	STP17NK40ZFP (400V)
S_2	STP25N60M2-EP (600V)
D_1, D_2, D_{21}, D_{22}	MUR1560 (600V, 15A)

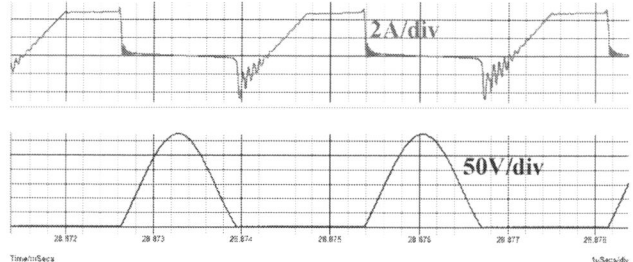

(a) Switching waveforms in S_1

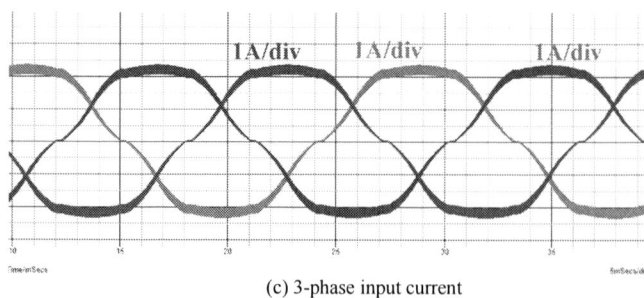

(b) Switching waveforms in S_2

(c) 3-phase input current

Figure 13 Results obtained from SIMETRIX at full load condition

978-1-4673-9551-9/16 $31.00 © 2016 IEEE

IV. CONCLUSION

With the IT electricity consumption increasing at an alarming rate and environmental concerns, it is essential to look into green IT power solutions, such as using hybrid renewable energy systems to power data centers and server rooms. This paper presented a new soft-switched MIC with integrated PFC function for hybrid renewable energy systems. The operating principle and the characteristics of the proposed converter have been described in this paper. The converter is capable of supporting individual as well as simultaneous operations, depends on the availability of the input sources. Finally, results have been given on a double input converter system with 48V-output to verify the theoretical analysis and to highlight the merits of the proposed circuit.

REFERENCES

[1] Natural Resources Canada, "Energy Efficiency - Industry" June. 2015

[2] J. Li; Z. Bao and Z. Li, "Modeling Demand Response Capability by Internet Data Centers Processing Batch Computing Jobs" *IEEE Trans on. Smart Grid*, vol. 6, no. 2, Mar. 2015, pp. 737–747.

[3] A. Kwasinski and P.T. Krein, "A Microgrid-based Telecom Power System using Modular Multiple- Input DC-DC Converters" *in Proceedings of the IEEE 2005 International Telecommunications Energy Conference, INTELEC 2005*, pp. 515 - 520.

[4] Yuan-Chuan Liu; Yaow-Ming Chen, "A Systematic Approach to Synthesizing Multi-Input DC–DC Converters," *IEEE Trans. on Power Electronics*, vol.24, no.1, Jan. 2009, pp.116-127.

[5] Kwasinski, A., "Identification of Feasible Topologies for Multiple-Input DC-DC Converters" *IEEE Trans. on Power Electronics*, vol. 24, no. 3, March 2009, pp. 856 – 861.

[6] Satpathy, A.S.; Kishore, N.K.; Kastha, D. and Sahoo, N.C., "Control Scheme for a Stand-Alone Wind Energy Conversion System" *IEEE Trans. on Energy Con*version, vol. 29, no. 2, June 2014, pp.418 - 425.

[7] Y.-M. Chen; C.-S. Cheng and H.-C. Wu, "Grid-connected hybrid PV/wind power generation system with improved DC bus voltage regulation strategy" *in Proceedings of the IEEE 2006 Applied Power Electronics Conference and Exposition*, pp. 1088 – 1094.

[8] Dusmez, S.; Xiong Li and Akin, B. , "A new multi-input three-level integrated DC/DC converter for renewable energy systems" *in Proceedings of the IEEE 2015 Applied Power Electronics Conference and Exposition (APEC)*, pp. 641 - 646.

[9] Amirabadi, M.; Toliyat, H.A. and Alexander, W.C., "A multi-input AC link PV inverter with reduced size and weight" *in Proceedings of the IEEE 2012 Applied Power Electronics Conference and Exposition (APEC)*, pp. 389- 396.

[10] Qin Wang; Jie Zhang; Xinbo Ruan and Ke Jin, "Isolated Single Primary Winding Multiple-Input Converters" *IEEE Trans. on Power Electronics*, vol. 26, no. 12, 2011, pp. 3435 – 3442.

[11] Yaow-Ming Chen; Yuan-Chuan Liu and Feng-Yu Wu, "Multi-input DC/DC converter based on the multi winding transformer for renewable energy applications" *IEEE Trans. on Industry Applications*, vol. 38 , no. 4, Jul/August 2002, pp. 1096 – 1104.

[12] Cheng-Wei Chen; Chien-Yao Liao; Kun-Hung Chen and Yaow-Ming Chen, "Modeling and Controller Design of a Semiisolated Multi-input Converter for a Hybrid PV/Wind Power Charger System" *IEEE Trans. on Power Electronics*, vol. 30, no. 9, Sept. 2015, pp. 4843 – 4853

[13] Fuxin Liu; Zhicheng Wang; Yunyu Mao and Xinbo Ruan, "Asymmetrical Half-Bridge Double-Input DC/DC Converters Adopting Pulsating Voltage Source Cells for Low Power

Applications" *IEEE Trans. on Power Electronics*, vol. 29, no. 9, Sept. 2014, pp. 4741 – 4751.

[14] R.-J. Wai; C.-Y. Lin; J.-J. Liaw and Y.-R. Chang, "Newly Designed ZVS Multi-Input Converter" *IEEE Trans. on Industrial Electronics*, vol. 58, no. 2, Feb. 2011, pp. 555–566.

[15] J. Lam and P. K. Jain, "A novel electrolytic capacitor-less multi-input DC/DC converter with soft- switching capability for hybrid renewable energy system" *in Proceedings of the IEEE 2014 European Conference on Power Electronics and Applications (EPE'14-ECCE Europe)*, pp. 1 – 10.

[16] S.-Y. Yu and A. Kwasinski, "Analysis of Soft-Switching Isolated Time-Sharing Multiple-Input Converters for DC Distribution Systems" *IEEE Trans. on Power Electronics*, vol. 28, no. 4, Apr. 2013, pp. 1783 – 1794.

[17] Y.-M. Chen; Y.-C. Liu; S.-C. Hung; and C.-S. Cheng, "Multi-Input Inverter for Grid-Connected Hybrid PV/Wind Power System" *IEEE Trans. on Power Electronics*, vol. 22, no. 3, May 2007, pp. 1070 – 1077.

[18] M. K. Kazimierczuk and X. T. Bui, "Class-E DC/DC converters with a capacitive impedance inverter" *IEEE Trans. on Industrial Electronics*, vol. 36, no. 3, August 1989, pp. 425 – 43.

[19] K.-H. Liu and F. C. Lee, "Zero-voltage switching technique in DC/DC converters" *IEEE Trans. on Power Electronics*, vol. 5, no. 3, July 1990, pp. 293 – 304.

[20] Prasad, A. R., Phoivos D. Ziogas, and Stefanos Manias, "An active power factor correction technique for three-phase diode rectifiers" *IEEE Trans. on Power Electronics*, vol. 6, no. 1, Jan. 1991, pp. 83 – 92.

978-1-4673-9551-9/16 $31.00 © 2016 IEEE

Analysis and Design of Impulse Commutated ZCS Three-phase Current-fed Push-pull DC/DC Converter

Radha Sree Krishna Moorthy
Department of Electrical and Computer Engineering
National University of Singapore
Singapore
radha_k@u.nus.edu

Akshay Kumar Rathore
Department of Electrical and Computer Engineering
National University of Singapore
Singapore
akshayrathore@nus.edu.sg

Abstract—The paper proposes and analyzes an impulse commutation based zero current switching (ZCS) current-fed three-phase push-pull dc/dc converter. The classic problem of turn-off spike in current-fed converters has been addressed via impulse commutation. A small high frequency capacitor is used to solve the problem. The resonance between the leakage inductances and the parallel capacitors facilitates zero current operation of the devices via impulse commutation and confines the maximum switch voltage to V_o/n naturally. Variable frequency modulation aids in regulating the output voltage. Impulse commutation offers reduction in the component count (snubber or active-clamp) and superior performance with low circulating currents. Simulation results using PSIM 9.3 and experimental results on 1kW laboratory prototype have been demonstrated to corroborate the aforementioned claims.

Keywords—Current-fed converter; impulse commutation; natural voltage clamping; zero current switching.

I. INTRODUCTION

Researchers around the world focus on building highly efficient dc/dc power converters be it voltage-fed or current-fed, though the later has been far less researched [1]. The buck type voltage-fed converters require high transformer turns ratio to boost the low dc voltage. Additionally, these voltage-fed converters suffer from high rectifier diode ringing, duty cycle loss etc. In recent years, the current-fed topology with its innate boost nature, inherent short circuit protection, no duty cycle loss and negligible rectifier diode ringing [2] has gained a lot of attention among researchers. These step up converters significantly reduce the transformer turns ratio when compared to the voltage-fed counterparts, thereby improving the converter's efficiency.

The historic problem with the current-fed converter is the severe spike seen across the devices during turn-off [3]. Use of passive snubber circuits [4] or active clamp circuits [5-7] has been reported in the literature to overcome this issue. Though they deal with the spike effectively, they derogate the converters' performance. Passive circuits dissipate the energy in the leakage inductances degrading the efficiency while active clamp circuits require additional components including a high side switch and a high frequency capacitor [5-7]. An alternate commutation technique utilizing the resonance between leakage inductance and a parallel capacitor was proposed in [8, 9]. A short resonance impulse during the commutation results in zero current switching (ZCS) of the devices eliminating the turn-off spike utilizing the substantial transformer parasitics.

In high power applications, three phase systems offer enormous potential in terms of power density and the current stress on switches [10, 11]. Considering the benefits of a three-phase current-fed system, an impulse commutated three inductor based current-fed converter has been proposed in [12]. In this paper, the idea of impulse commutation has been extended for three-phase current-fed push-pull converter. The push-pull converter with the low side switches reduces the gate drive requirements compared to the full bridge topologies. The role of magnetizing inductance becomes prominent in the proposed topology with differences in the transformer utilization compared to the three inductor based topology in [12] and this will be evident from the steady-state analysis.

Organization of this paper is as follows. The operation and the steady-state analysis of the proposed converter has been explained in Section II, followed by the illustration of the converter design using an example in Section III. Finally, Section IV puts forth the simulation and experimental results required to validate the concept of impulse commutation in the three-phase push-pull current-fed converter and the converter's performance.

II. STEADY-STATE OPERATION AND ANALYSIS OF THE CONVERTER

For simplifying the analysis of the converter shown in Fig. 1, the following assumptions have been made. a) The input boost inductor is large to maintain constant current through it. b) All semiconductor devices are ideal and lossless. c) Leakage inductances of the transformer and the parasitic capacitance between the phases are equal i.e., $L_{s1} = L_{s2} = L_{s3}$ and $C_{p1} = C_{p2} = C_{p3}$.

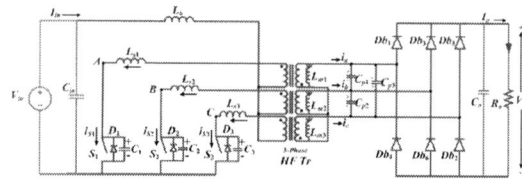

Fig. 1. Impulse commutated three phase current-fed push-pull converter

Variable frequency modulation exercises control over the power transfer and load voltage regulation under all operating conditions. The gating signals for the devices are phase shifted by 120° with an overlap decided by the constant duty cycle D. The chosen D is greater than 33 % with the maximum possible value being 66 %. The analysis has been carried out for one-third of the high frequency (HF) cycle and for the next cycles these intervals repeat in the same sequence with other symmetrical devices conducting. The equivalent circuits illustrating the different modes of operation of the converter are laid out in Fig. 2 followed by the steady-state operating waveforms in Fig. 3.

A. Interval 1 (Fig. 2(a): $t_0 < t < t_1$) – Power transfer mode

In this power transfer mode, primary switch S_1 carries the input current I_{in} with the secondary diodes D_{b1}, D_{b2} and D_{b6} feeding power to the load. Thereby, clamping the voltage across the parallel capacitor C_{p1} is clamped to load voltage V_o. Constant current $I_{in}/3n$ flows through the magnetizing inductances. Constant current $2I_{in}/3n$, $-I_{in}/3n$ and $-I_{in}/3n$ flows through phase A, B and C in the secondary. Final values: i_{Ls1} $(t_1) = i_{S1}(t_1) = I_{in}$; $v_{Cp1}(t_1) = V_o$; $v_{Cp2}(t_1) = 0$; $v_{Cp3}(t_1) = -V_o$; i_a $(t_1) = 2I_{in}/3n$; $i_b(t_1) = i_c(t_1) = -I_{in}/3n$.

B. Interval 2 (Fig. 2(b): $t_1 < t < t_2$) – Discharge of device capacitor

At $t = t_1$, switch S_2 is turned ON. The device capacitance C_2 discharges via the switch in a very short duration of time.

C. Interval 3 (Fig. 2(c): $t_2 < t < t_3$)

With the primary switches S_1 and S_2 conducting, S_2 commences to take over the input current I_{in} from S_1. Thereby, the current through the series inductors L_{s1} and L_{s2} and hence the respective switches starts decreasing and increasing respectively a slope of $V_o/2nL_s$. These currents can be given as

$$i_{Ls1} = i_{S1} = I_{in} - \frac{V_o}{2nL_s}(t - t_2) \tag{1}$$

$$i_{Ls2} = i_{S2} = \frac{V_o}{2nL_s}(t - t_2) \tag{2}$$

With the rectifier diodes transferring power to the load, the secondary phase currents can be given as,

$$i_a(t) = \frac{i_{Ls1}(t)}{n} - \frac{I_{in}}{3n} \tag{3}$$

$$i_b(t) = \frac{i_{Ls2}(t)}{n} - \frac{I_{in}}{3n} \tag{4}$$

Constant current $-I_{in}/3n$ flows through phase C on the secondary side. At the end of the interval, one of the rectifier diodes D_{b6} gets reverse biased as the secondary current in phase B (i_b) reaches zero. Final values: $i_{Ls1}(t_3) = i_{S1}(t_3) = 2I_{in}/3$; $i_{Ls2}(t_3) = i_{S2}(t_3) = I_{in}/3$; $v_{Cp1}(t_3) = V_o$; $v_{Cp3}(t_3) = -V_o$; $v_{Cp2}(t_3) = 0$; $i_a(t_3) = I_{in}/3n$; $i_b(t_3) = 0$ and $i_c(t_3) = -I_{in}/3n$. The duration of this interval can be given as

$$T_{32} = (t_3 - t_2) = \frac{2nI_{in}L_s}{3V_o} \tag{5}$$

D. Interval 4 (Fig. 2(d): $t_3 < t < t_4$) - Resonance

The primary switches S_1 and S_2 continue to conduct during this interval. The resonance between the series inductors L_{s1} and L_{s2} of the phase A and B and the parallel capacitor C_{p1} commences during this interval. The switch currents continue to increase and decrease in a resonant fashion. The frequency of resonance and the characteristic impedance at resonance can be given as

$$f_r = 1 / 2\pi \sqrt{L_{eq}C_p'} \tag{6}$$

$$Z_r = \sqrt{\frac{L_{eq}}{C_p'}} \tag{7}$$

Where L_{eq} can be given as $L_{s1} + L_{s2}$ and C_p' is the parallel capacitance reflected to the primary. The switch currents can be given as

$$i_{Ls1} = i_{S1} = \frac{2I_{in}}{3} - \frac{V_o}{nZ_r}\sin(2\pi f_r(t - t_3)) \tag{8}$$

$$i_{Ls2} = i_{S2} = \frac{I_{in}}{3} + \frac{V_o}{nZ_r}\sin(2\pi f_r(t - t_3)) \tag{9}$$

The current i_b starts to increase in positive direction during this interval. This current discharges and charges the parallel capacitors C_{p1} and C_{p2} respectively. The secondary phase currents i_a and i_b can be given by equations (3) and (4) respectively.

At the end of the interval, the input current gets shared equally between the two switches. The rectifier diodes D_{b1} and D_{b2} remain forward biased. The voltage across C_{p3} remains clamped at $-V_o$. Final values: $i_{Ls1}(t_4) = i_{Ls2}(t_4) = i_{S1}(t_4) = i_{S2}(t_4) = I_{in}/2$; $v_{Cp3}(t_4) = -V_o$; $i_a(t_4) = I_{in}/6n$; $i_b(t_4) = I_{in}/6n$ and $i_c(t_4) = -I_{in}/3n$. The duration of this interval can be given as

$$T_{43} = (t_4 - t_3) = \frac{1}{2\pi f_r}\sin^{-1}\left(\frac{nI_{in}Z_r}{6V_o}\right) \tag{10}$$

E. Interval 5 (Fig. 2(d): $t_4 < t < t_5$)

The primary switches S_1 and S_2 continue to conduct during this interval. The current through S_1 increases beyond I_{in} while current through S_2 decreases below $I_{in}/2$. The currents can be given as

$$i_{Ls1} = i_{S1} = \frac{I_{in}}{2} - \frac{V_o}{nZ_r}\sin(2\pi f_r(t - t_4)) \tag{11}$$

$$i_{Ls2} = i_{S2} = \frac{I_{in}}{2} + \frac{V_o}{nZ_r}\sin(2\pi f_r(t - t_4)) \tag{12}$$

The rectifier diodes D_{b1} and D_{b2} remain forward biased and the voltage across C_{p3} remains clamped at V_o. At the end of the interval, the secondary current i_a reaches zero and later increases in the negative direction. Final values: $i_{Ls1}(t_5) = i_{S1}(t_5) = I_{in}/3$; $i_{Ls2} = i_{S2} = 2I_{in}/3$; $v_{Cp3}(t_5) = -V_o$; $i_a(t_5) = 0$; $i_b(t_5) = I_{in}/3n$ and $i_c(t_5) = -I_{in}/3n$. The duration of this interval can be given as

$$T_{54} = (t_5 - t_4) = \frac{1}{2\pi f_r}\sin^{-1}\left(\frac{nI_{in}Z_r}{6V_o}\right) \tag{13}$$

F. Interval 6 (Fig. 2(e): $t_5 < t < t_6$)

During this interval, the series inductor currents and thereby the switch currents i_{S1} and i_{S2} continue to decrease and increase in a resonant fashion. The currents during this interval of operation can be given as

$$i_{Ls1} = i_{S1} = \frac{I_{in}}{3} - \frac{V_o}{nZ_r}\sin(2\pi f_r(t - t_4)) \tag{14}$$

$$i_{Ls2} = i_{S2} = \frac{2I_{in}}{3} + \frac{V_o}{nZ_r}\sin(2\pi f_r(t - t_4)) \tag{15}$$

The charging and discharging of the parallel capacitors C_{p1} and C_{p2} continues with C_{p3}'s voltage clamped at V_o. At the end of the interval, the diode D_{b1} and D_{b2} get reverse biased as i_{S1} reaches zero. The secondary current in the phase A becomes equal to the magnetizing component ($-I_{in}/3n$). Final values: $i_{Ls1}(t_6) = i_{S1}(t_6) = 0$; $i_{S2}(t_6) = i_{Ls2}(t_6) = I_{in}$; $v_{Cp3}(t_6) = -V_o$; $i_a(t_6) = i_c(t_6) = -I_{in}/3n$ and $i_b(t_6) = 2I_{in}/3n$. The duration of this interval can be given as

$$T_{65} = (t_6 - t_5) = \frac{1}{2\pi f_r}\sin^{-1}\left(\frac{nI_{in}Z_r}{3V_o}\right) \tag{16}$$

G. Interval 7 (Fig. 2(f): $t_6 < t < t_7$) - ZCS

The conduction of the anti-parallel body-diode of switch S_1 can be witnessed in this interval as i_{S1} increases in the negative direction. The switch S_1 can be turned-off with ZCS during body-diode conduction. The current through S_2 increases above I_{in} and the peak I_p and thereby the condition for ZCS can be formulated as:

$$I_p = |i_{Ls}(t)|_{max} = \frac{2I_{in}}{3} + \frac{V_o}{nZ_r} \geq I_{in} \tag{17}$$

The parallel capacitor C_{p3} also starts charging during this interval. At the end of the interval, i_{S1} reaches its peak value I_p and the rectifier diodes D_{b3} and D_{b2} go forward biased. Final values: $i_{Ls2}(t_7) = i_{S2}(t_7) = I_p$; $i_a(t_7) = (2I_{in}/3 - I_p/n)$; $i_b(t_7) = (I_p/n - I_{in}/3n)$ and $i_c(t_7) = -I_{in}/3n$.

H. Interval 8 (Fig. 2(g): $t_7 < t < t_8$) - ZCS

The body-diode of the switch S_1 continues to conduct during this interval. The C_{p2} discharges completely thus having zero voltage across it during this interval. The switch S_1 has been turned off with ZCS. The voltage across C_{p2} gets clamped to $-V_o$ during this interval. At the end of the interval i_{Ls1} reaches zero ceasing the body-diode conduction and i_{S2} reaches I_{in}. Final values: $i_{Ls1}(t_8) = i_{S1}(t_8) = 0$; $i_{Ls2} = i_{S2} = I_{in}$; $v_{Cp2} = -V_o$; $i_a(t_8) = i_c(t_8) = -I_{in}/3n$ and $i_b(t_8) = 2I_{in}/3n$. The duration of this interval can be given as

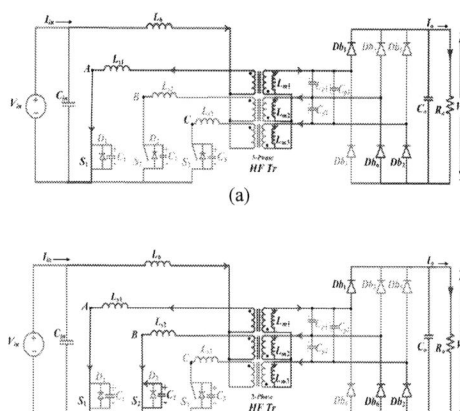

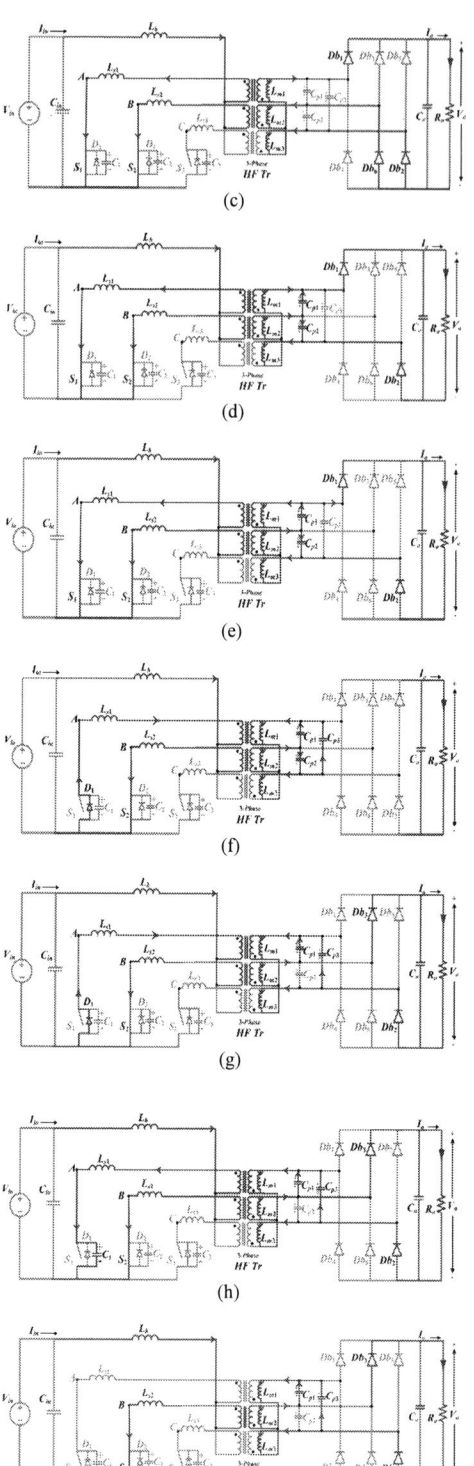

Fig. 2. Equivalent circuits depicting the different intervals of operation of the proposed converter

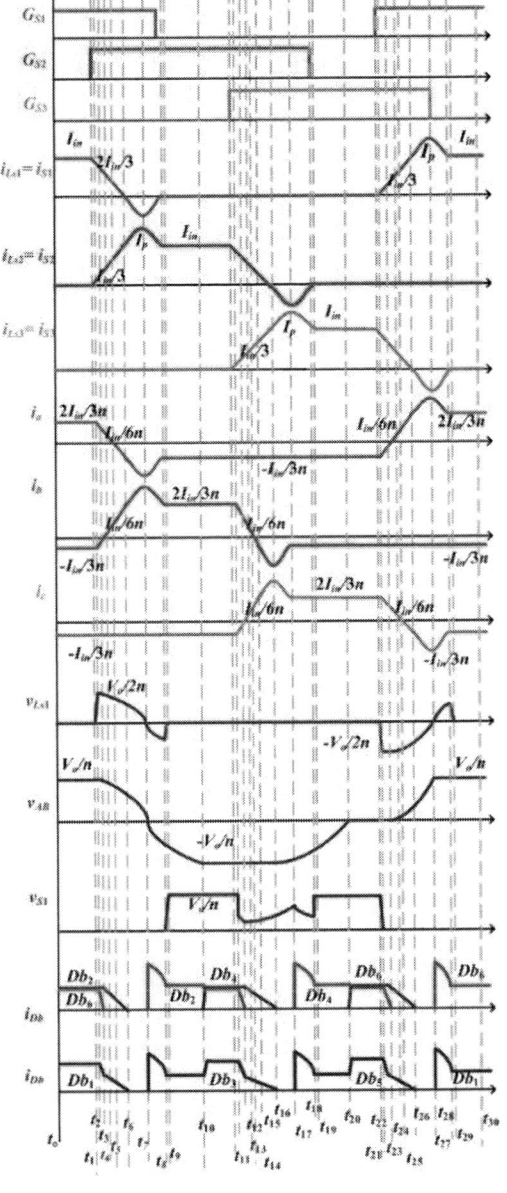

$$T_{86} = (t_8 - t_6) = \frac{\pi - 2\pi f_r (T_{65} + T_{53})}{2\pi f_r} \qquad (18)$$

I. Interval 9 (Fig. 2(h): $t_8 < t < t_9$) – device capacitor charging

During this interval, the device capacitance C_1 of the switch S_1 charges to a maximum voltage of V_o/n from 0.

J. Interval 10 (Fig. 2(i): $t_9 < t < t_{10}$)

During this interval, constant current I_{in} flows through the switch S_2 and the series inductor L_{s2}. The charging and discharging of the parallel capacitors C_{p1} and C_{p3} continues. Constant current $-I_{in}/3n$, $2I_{in}/3n$ and $-I_{in}/3n$ flows through the

secondary in phase A, B and C respectively. At the end of the interval, the voltage across C_{p3} is clamped to V_o while voltage across C_{p1} becomes zero. Thereby, the rectifier diode D_{b4} gets forward biased. Final values: $i_{Ls2}(t_{10}) = i_{S2}(t_{10}) = I_{in}$; $v_{Cp1}(t_{10}) = 0$; $v_{Cp2}(t_{10}) = -V_o$; $v_{Cp3}(t_{10}) = V_o$; $i_a(t_{10}) = i_c(t_{10}) = -I_{in}/3n$ and $i_b(t_{10}) = 2I_{in}/3n$. The duration of this interval can be given as

$$T_{109} = (t_{10} - t_9) = \frac{C_p' R_L}{n^2} \qquad (19)$$

III. CONVERTER DESIGN

This Section introduces the converter design using a design example with the following specifications: Input voltage $V_{in} = 42 – 48$ V, Output voltage $V_o = 380$ V, Output power $P_o = 1$ kW and range of frequency of operation $f_s = 75 – 88$ kHz. The equations for DC voltage gain, the necessary condition for ZCS etc. have been provided in this Section.

A. DC voltage gain

The DC voltage gain of the proposed converter can be approximated as

$$M = \frac{V_o}{V_{in}} = \frac{n\eta}{\frac{2}{3} - 0.88 f_n} \qquad (20)$$

Where f_n is the normalized frequency which is the ratio of the switching and the resonant frequency ($f_n = f_s/f_r$). The variation in gain with the normalized frequency for different values of turns ratio has been shown in Fig. 4.

B. Voltage and Current ratings of the devices

The voltage stress across the primary switches, the rectifier diodes and the parallel capacitors are

$$V_{S1} \sim V_{S3} = \frac{V_o}{n} \qquad (21)$$

$$V_{Db1} \sim V_{Db6} = V_o \qquad (22)$$

$$V_{Cp1} \sim V_{Cp3} = V_o \qquad (23)$$

The approximate expressions for the rms and the average currents through the switches and the rectifier diodes which can be employed for the estimation of conduction losses in the converter have been tabulated in Table-I.

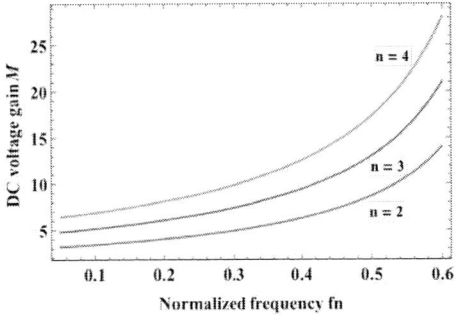

Fig. 4. Plot of gain vs. the normalized frequency for different turns ratio

TABLE I.   DEVICE CURRENTS

Currents	Equations
$I_{S1,rms} \sim I_{S2,rms}$	$I_{in} \cdot \sqrt{\dfrac{2f_n}{9} + \left(D - \dfrac{2f_n}{3}\right)}$
$I_{Ls1,rms} \sim I_{Ls2,rms}$	$I_{in} \cdot \sqrt{\dfrac{2f_n}{9} + \left(D - \dfrac{2f_n}{3}\right)}$
$I_{Db1,avg} \sim I_{Db6,avg}$	$I_o/3$

C. Duty ratio selection

The minimum and the maximum value of duty ratio required for ZCS operation is given by (24) and (25).

$$D_{min} = 0.33 + \frac{2nI_{in}L_s f_s}{3V_o} + \frac{f_n}{\pi}\sin^{-1}\left(\frac{nZ_r I_{in}}{6V_o}\right) + \frac{f_n}{2\pi}\sin^{-1}\left(\frac{nZ_r I_{in}}{3V_o}\right) \quad (24)$$

$$D_{max} = 0.33 + \frac{2nI_{in}L_s f_s}{3V_o} + \frac{f_n}{\pi}\sin^{-1}\left(\frac{nZ_r I_{in}}{6V_o}\right) + \frac{f_n}{2\pi}\sin^{-1}\left(\frac{nZ_r I_{in}}{3V_o}\right) + \frac{f_n}{2} \quad (25)$$

The duty ratio D is selected such that $D_{min} < D < D_{max}$ for all operating conditions to perpetuate ZCS for $V_{in} = 42 - 48$ V and from full load to light load. The duty cycle was fixed at 0.48 in this case.

D. Turns ratio

The transformer turns ratio determines the range of frequency variation as evident from (20) and also determines the voltage appearing across the primary switches. With lower value of turns ratio, the device voltages shoot up and the use of high voltage devices becomes mandatory. High voltage devices with higher on-state resistances increase the conduction losses in the high current side. On the other had a higher turns ratio, results in increased conduction losses in the transformer and also impacts the switching losses as the range of operating frequency stretches. As a trade-off between the two scenarios, the turns ratio was chosen as 3.

E. Resonant tank parameters

The resonant tank parameters can be computed using the resonant frequency and the characteristic impedance at resonance expressed in (6) and (7) respectively. In this example, the resonant frequency and the characteristic impedance are 0.28 MHz and 8.606 respectively. Thereby, the computed L and C values are 3 μH and 9 nF respectively.

F. ZCS condition

The stored energy in the parallel capacitors brings about ZCS of the primary switches under all operating conditions. Thereby, it is necessary to ensure that sufficient energy is stored in them to allow the anti-parallel body-diode conduction without considerably increasing the circulating current in the converter. This imposes a constraint on the characteristic impedance at resonance and the condition can be given as

$$Z_r < \frac{3V_{in,min}R_{FL}}{nV_o} \quad (26)$$

Where $V_{in,\,min}$ is the minimum input voltage and R_{FL} is the full load resistance. Z_r is selected such that ZCS is perpetuated for input from $42 - 48$ V and load variations while keeping the circulations as low as possible.

G. Boost inductor design

The boost inductance L_b can be given as

$$L_b = \frac{V_{Lb} - (T_{109} + T_{10})}{\Delta i} \quad (27)$$

Where $V_{Lb} = V_{S1} - V_{Lm1} - V_{in}$ and $T_{10} = T_s/3 - (T_{32} + T_{63} + T_{86} + T_{109})$. The boost inductance was computed to be 132 μH for $\Delta i = 0.05$ A.

H. Body-diode conduction time

The period for which the anti-parallel body-diode conducts giving rise to ZCS is given by (18). The body-diode conduction dominates at light load conditions as evident from (18) as the ratio of I_p/I_{in} increases at light loads.

IV.   SIMULATION AND EXPERIMENTAL RESULTS

The proposed converter was simulated in the software platform PSIM 9. 3. 3. This section puts forth the simulation results followed by the results acquired from a laboratory prototype rated at 1 kW.

The current through the primary switches S_1 and S_3 and their corresponding voltages for $V_{in} = 42$ V at 1 kW are depicted in Fig. 5(a). The switch currents are phase shifted by 120° from each other and they carry the input current I_{in} in the non-overlap region. Resonance occurs when during the overlap period when the current has to be transferred from one phase to another. At any point of time maximum of two switches are ON.

Clearly, the switches operate with ZCS and the voltage across them shoots up only after the body-diode conduction ceases. The gating signals for the switches are removed during the body-diode conduction. The secondary currents and the voltage across the diodes in a leg are highlighted in Fig. 5(b). When i_a is positive diode D_{b1} is forward biased transferring power to the load while D_{b4} remains reverse biased.

Further, the switch current and the voltage across S_1 are depicted for $V_{in} = 48$ V at 200 W in Fig. 6. Even with varying voltage and load, ZCS operation is maintained. The parallel capacitor voltages v_{Cp1}, v_{Cp2} and v_{Cp3} clamped either at V_o, 0 or $-V_o$ once the commutation completes. Series inductor currents i_{Ls1}, i_{Ls2} and i_{Ls3} are identical to the switch currents i_{S1}, i_{S2} and i_{S3}, respectively shown in Fig. 6(a) and 6(b) respectively.

A 1kW prototype developed in laboratory is shown in Fig. 7. The details of the prototype are laid out in Table II. The gating signals for the primary switches were generated using Altera cyclone IV DE0 Nano. The experimental results for the various operating condition such as $V_{in} = 48$ V, 1 kW, $V_{in} = 42$ V, 1 kW and $V_{in} = 48$ V, 200 W are illustrated in Figs. 8 to 10, respectively. Fig. 8(a) highlights the ZCS operation of the primary switch S_3. It is evident that the gating signal for the switch is removed once the body-diode begins to conduct and the voltage across the switch rises after the body-diode conduction ends. Thereby, the input current gets transferred from one phase to another whilst the switch in the outgoing phase is turned-off with ZCS. From Fig. 8(b), it can be seen that the transformer primary currents are identical to that of switch currents and phase shifted from one another by 120°.

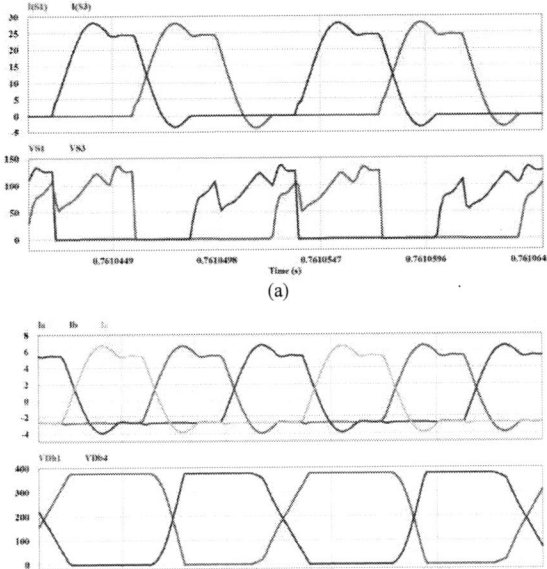

(a)

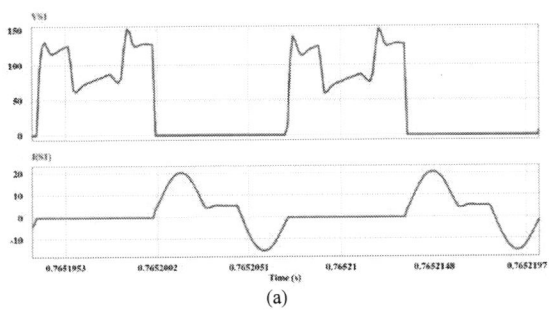

(b)

Fig. 5. Simulation results for V_{in} = 42 V, 1 kW. (a) Switch currents i_{S1} and i_{S3} and the corresponding switch voltages and (b) The secondary currents i_a, i_b and i_c and the voltage across diodes in one leg.

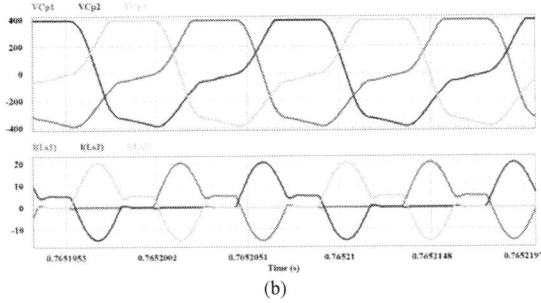

(a)

Fig. 6. Simulation results for V_{in} = 48 V, 200 W. (a) current through and voltage across switch S_1 and (b) Parallel capacitor voltages and series inductor currents.

It is to be noted that the required leakage inductance has been incorporated inside the transformer itself eliminating the need for additional series inductors. The secondary current i_c in phase C together with the voltages across the rectifier diodes D_{b2} and D_{b5} are highlighted in Fig. 8(c). Positive current in phase C forward biases the rectifier diode D_{b5} thereby transferring power from source to the load. The three single phase transformers connected in star-star configuration behave like a fly back transformer storing energy in the magnetizing inductance. The negative portion of the diode current is the transformer magnetizing current.

TABLE II.   PARAMETERS OF THE LABORATORY PROTOTYPE

Boost inductor L_b	MS250090 with 12 turns, L_b = 52.2 μH
Primary switches $S_1 \sim S_3$	IPP110N20N3; 200V; 88A; Rds,on = 10.7mΩ
Three HF Single-phase Transformers	3 cores of N97 material ETD 59 geometry, Primary turns, N_1 = 16; secondary turns N_2 = 48; Leakage inductances referred to the primary: L_{s1} = 2.8 μH, L_{s2} = 3 μH and L_{s3} = 3.1 μH. Magnetizing inductance in primary L_m = 290 μH
External parallel capacitance C_p	9 nF, 1000V ceramic capacitor
Output Capacitors C_o	100 μF, 400V electrolytic capacitor & 2.2μF, 400V HF film capacitor
Rectifier Diodes $D_{b1} \sim D_{b6}$	STTH30R04; 400 V, 30 A VF = 0.97 V

The magnetizing current ripple should be kept as low as possible to reduce the core losses in the transformer. Attention is to be paid to the transformer design to minimize the losses and to prevent saturation of the core also.

Finally, the voltage across the parallel capacitors phase shifted from one another by 120° is shown in Fig. 8(d). At any given instant one of the capacitor's voltage is clamped to either V_o, 0 or $-V_o$. The charging and discharging of the other two capacitors continues leading to the body-diode conduction of the outgoing switch. The capacitor voltage and the rectifier diode voltages serve as proof for output voltage regulation.

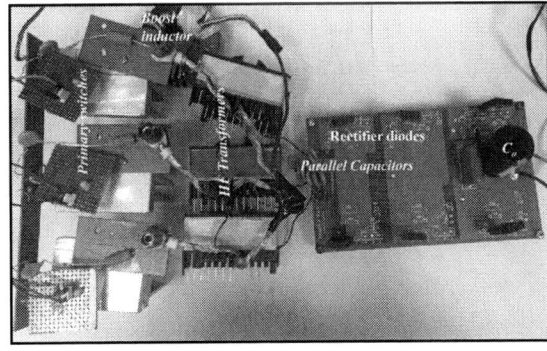

Fig. 7. Laboratory prototype of the proposed converter

The same has been observed for other operating conditions i.e., V_{in} = 42 V, 1 kW and V_{in} = 48 V, 200 W in Figs. 9 and 10, respectively. Frequency modulation ensures load voltage regulation and ZCS with natural voltage clamping. Maximum efficiency obtained was 93 % at V_{in} = 48 V at full load.

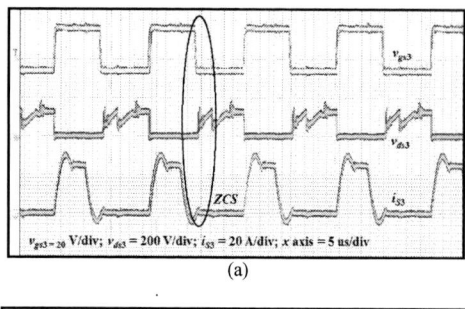

(a)

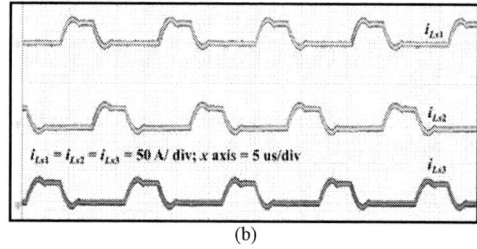

(b)

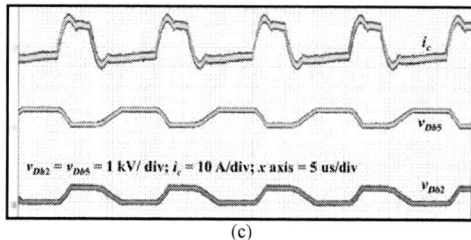

(c)

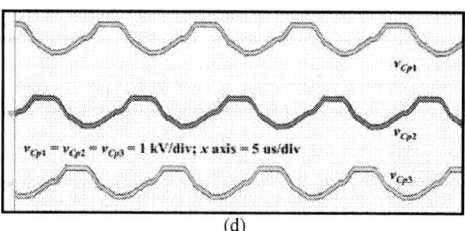

(d)

Fig. 8. Experimental results for V_{in} = 48 V, 1 kW. (a) gate to source voltage v_{gs1}, drain to source voltage v_{ds1} and current through switch S_1, (b) current through the primary of the HF transformers, (c) secondary current i_c and voltage across the rectifier diodes in one leg v_{Db2} and v_{Db5} and (d) voltage across the parallel capacitors.

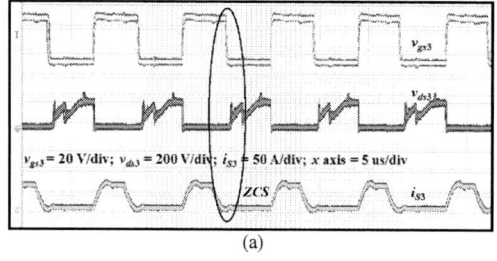

(a)

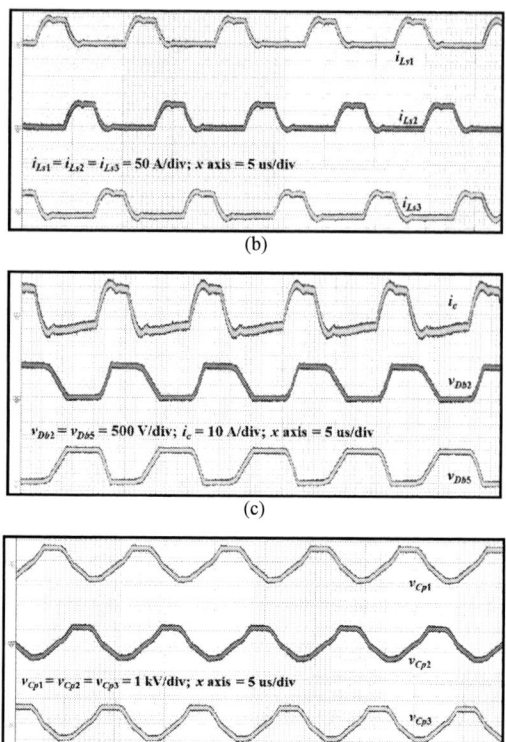

(b)

(c)

(d)

Fig. 9. Experimental results for V_{in} = 42 V, 1 kW. (a) gate to source voltage v_{gs1}, drain to source voltage v_{ds1} and current through switch S_1, (b) current through the primary of the HF transformers, (c) secondary current i_c and voltage across the rectifier diodes in one leg v_{Db2} and v_{Db5} and (d) voltage across the parallel capacitors.

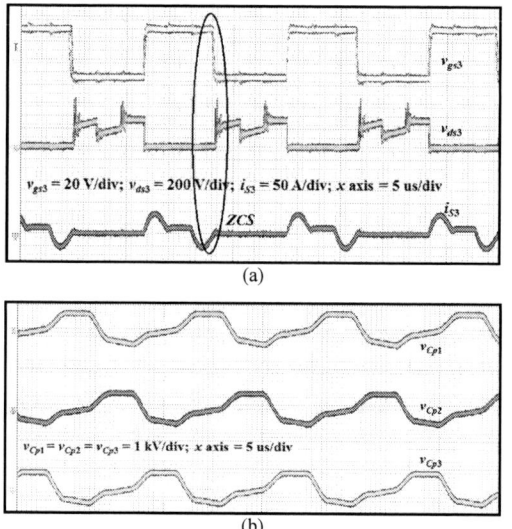

(a)

(b)

Fig. 10. Experimental results for V_{in} = 48 V, 200 W. (a) gate to source voltage v_{gs1}, drain to source voltage v_{ds1} and current through switch S_1 and (b) voltage across the parallel capacitors.

V. CONCLUSION

The paper elaborates the use of impulse commutation in a current-fed three-phase push-pull converter. Impulse commutation facilitates zero current operation of semiconductor devices and clamps the voltage across devices naturally to V_o/n making it immune to variations in frequency or duty cycle of operation. Both ZCS and device voltage clamping is load adaptive and can be sustained with input voltage variations. Load voltage regulation is made viable via variable frequency modulation. Experimental results validate the claims of ZCS and natural voltage clamping through impulse commutation. It is suitable for low voltage high current applications.

REFERENCES

[1] M. Ishibashi, M. Nakaoka, and Y. Konishi, "Performance evaluations of three-phase current-fed soft switching PWM converter with switched capacitor-type quasi-resonant snubber," in *Proc. Inst. Elect. Eng.—Electr. Power Appl.*, vol. 148, no. 5, pp. 431–437, Sep. 2001.

[2] A. K. Rathore , A. K. S. Bhat and R. Oruganti "Wide range ZVS active-clamped L-L type current-fed dc-dc converter for fuel cells to utility interface: Analysis, design and experimental results", *IEEE Trans. Ind. Electron.*, vol. 59, no. 1, pp. 473-485, 2012.

[3] H. R. E. Larico and I. Barbi, "Three-phase Weinberg isolated dc–dc converter: Analysis, design and experimentation," *IEEE Trans. Ind. Electron.*, vol. 59, no. 2, pp. 888–896, Feb. 2012.

[4] S. V. G. Oliveira and I. Barbi, "A three-phase step-up dc–dc converter with a three-phase high-frequency transformer for dc renewable power

sources applications," *IEEE Trans. Ind. Electron.*, vol. 58, no. 8, pp. 3567–3580, Aug. 2011.

[5] C. M. C. Duarte and I. Barbi, "An improved family of ZVS-PWM activeclamping dc-to-dc converters," *IEEE Trans. Power Electron.*, vol. 17, no. 1, pp. 1–7, Jan. 2002.

[6] H. Cha, J. Choi and P. N. Enjeti, "A three-phase current-fed DC/DC converter with active clamp for low-DC renewable energy sources," *IEEE Trans. Power Electron.*, vol. 23, no. 6, pp. 2784-2793, Nov. 2008.

[7] H.Cha, J.Choi and P.Enjeti, "A three-phase current-fed dc–dc converter with active clamp for low-dc renewable energy sources," *IEEE Trans. Power Electron.*, vol. 23, no. 6, pp. 2784–2793, Nov. 2008.

[8] B. Yuan, X. Yang, X. Zeng, J. Duan, J. Zhai and L. Donghao, "Analysis and design of a high step-up current-fed multiresonant DC-DC converter with low circulating energy and zero-current switching for all active switches," *IEEE Trans. Ind. Electron.*, vol. 59, no. 2, pp. 964-978, 2012.

[9] R.-Y. Chen, T.-J. Liang, J.-F. Chen, R.-L. Lin and K.-C. Tseng, "Study and implementation of a current-fed full-bridge boost DC/DC converter with zero-current switching for high-voltage applications," *IEEE Trans. Ind. Appl.*, vol. 44, no. 4, pp. 1218-1226, Jul./Aug. 2008.

[10] C. F. Chuang, C. T. Pan and H. C. Cheng, "A novel transformer-less interleaved four-phase step-down dc converter with low switch voltage stress and automatic uniform current-sharing characteristics," *IEEE Trans. Power Electron.*, vol. 31, no.1, pp. 406-417, Jan 2016.

[11] D. S. Oliveira and I. Barbi, " A three-phase ZVS PWM DC/DC converter with asymmetric duty cycle for high power applications," *IEEE Trans. Power Electron.*, vol. 20, no. 2, pp. 370-377, Mar. 2005.

[12] K. R. Sree and A. K. Rathore, "Impulse commutated zero current switching current-fed three-phase DC/DC converter," *IEEE Trans. Ind. Appls.*, 2015, DOI: 10.1109/TIA.2015.2487443.

ZCS Resonant Converter Based Parallel Balancing of Serially Connected Batteries String

Ilya Zeltser, *Member, IEEE*

Power Electronics Department
Rafael Advanced Defense Systems Ltd.
P.O. Box 2250, Haifa 31021, Israel.
ilyaz@rafael.co.il
www.rafael.co.il

Or Kirshenboim, *Student Member, IEEE*, Nadav Dahan, *Student Member, IEEE*, and Mor Mordechai Peretz, *Member, IEEE*

The Center for Power Electronics and Mixed-Signal IC
Department of Electrical and Computer Engineering
Ben-Gurion University of the Negev
P.O. Box 653, Beer-Sheva 84105, Israel.
orkir@post.bgu.ac.il, nadad@post.bgu.ac.il,
morp@ee.bgu.ac.il
www.ee.bgu.ac.il/~pemic

Abstract – **This paper introduces a new topology for parallel balancing of serially connected batteries string. The main advantage of the balancing concept is that energy is transferred only when the cells are unbalanced. As a result, the power losses are significantly reduced since no current circulates through the system when balanced. This has been enabled by a modification of an isolated series-resonant converter operating in DCM and features zero current switching (ZCS). Another attractive feature is that one transformer for two cells is used, as opposed to conventional isolated topologies that require a transformer per cell. The realization is simple and requires simple current polarity detector. Experimental results have been obtained by a prototype of two series connected batteries, which demonstrates the balancing capabilities of the system.**

I. INTRODUCTION

Serially connected batteries strings have been used for many high voltage DC applications. Among them, electric vehicles (EV) [1], hybrid electric vehicles (HEV) [2], plug-in hybrid electric vehicle (PHEV) [3] and other battery powered applications [4], [5]. Due to manufacturing and environmental variances, degradation with aging, internal impedance difference and thermal conditions, the charges transferred to or consumed from each battery are not equal. As a result, the lifetime and efficiency of the batteries string are reduced. Moreover, the overcharge of the batteries in the string can cause explosion or fire in the case of sensitive batteries cells [6]. Therefore, a charge equalizer (i.e., balancing circuit) is essential to reduce the imbalances and consequently to improve the overall performance of the system [7], [8].

Several balancing circuits have been investigated over the recent years [9]-[33]. These can be generally classified into two main categories: passive and active. Passive balancing features

simple design and implementation and relatively low cost [9], but due to inherent energy loss, it is less attractive in terms of energy saving. The active balancing architectures include variety of topologies such as switched-capacitor converters [10]-[12], isolated and non-isolated unidirectional and bidirectional DC-DC converters [13]-[24], and multi-winding transformer-based converters [25]-[27]. The main challenges of the active solutions often relate to implementation issues such as high component count and complex control algorithms [28], [29]. In addition, increasing the balancing speed is traded for quiescent power loss, i.e. losses that exist when no balancing action is required.

Recent studies have reported improved balancing schemes that are based on parallel balancing techniques [30], exhibiting higher equalization speed when compared to series balancing techniques. One particular challenge of the parallel balancing is the complexity in implementing large arrays. This has been addressed by a modularization concept presented in [31]-[34]. There, n batteries string is divided into M modules, each of them is balancing K cells so that $n=M*K$. This concept has established a solid foundation for a parallel balancing architecture, allowing higher efficiency that is pursued in this study.

The objective of the study is to introduce a new active balancing converter topology for series connected batteries string, as detailed in Fig. 1. The concept is realized by a modification of an isolated series-resonant converter such that one transformer is used to balance two cells, reducing the complexity of the system. The converter operates in DCM and therefore ZCS is assured. Balancing occurs only when necessary, i.e. energy does not circulate through the system when the batteries are balanced. As a result, the quiescent power loss is minimal. The converter is controlled by simple control method and can be easily scaled up and modularized.

978-1-4673-9551-9/16 $31.00 © 2016 IEEE

The paper is organized as follows: Section II describes the topology, its principle of operation and the major features of it, Section III delineates the system's implementation and provides design guidelines. Experimental results and conclusion are then provided in Sections IV and V, respectively.

II. PRINCIPLE OF OPERATION

The system in Fig. 1 can be divided into $m = n/2$ double-cell modules, where each module is constructed of a top cell (odd-numbered cells) and a bottom cell (even-numbered cells). Each of the m double-cells is balanced using a balancing module, built of a half-bridge loaded by a resonant tank in series with the primary side of the transformer. The secondary side connects to the bus capacitor via a full-bridge transistor assembly. The bus capacitor is the common voltage for all of the m double-cell modules and acts as a "link" that is used to transfer energy between the cells. As a result of the transformer's isolation, each of the n battery cells can be equalized independently of the action in other cells, that is, no synchronization between the modules is required, and therefore the complexity of the system is significantly reduced. Additionally, due to the transformer's isolation, the voltage stress on the switches is no more than the cell voltage and is independent on the number of cells in the string. The switching frequency f_s is set to be lower than the resonant frequency f_r (i.e. $f_s < f_r$) to allow operation in DCM with ZCS.

The balancing action for each battery cell is divided into two steps: current polarity detection and then directional energy transfer based on the polarity. The operation of the switches is described in Fig. 2 for the upper cell, assisted by the current and voltage waveforms which are depicted in Fig. 3. In the first step, switches S_{m1}, S_{m3}, S_{m5} and S_{m8} are turned on at $t=t_0$, and remain on for $t_0 < t < t_1$ ((Fig. 2(a), Fig. 2(d)), where $t_1 < t_0 + 1/2f_r$, allowing bi-directional current flow. During this interval, the current polarity sensor determines the current direction in the resonant tank. The current direction depends on the difference between the battery cell voltage $V_{cell,n-1}$ and the bus voltage reflected to the primary (V_{bus} for the unity transformer ratio). Cell voltage that is higher than twice the bus capacitor voltage ($V_{cell,n-1} > 2V_{bus}$) results in a positive current i_r, whereas $V_{cell,n-1} < 2V_{bus}$, results in a negative current direction. In case that the cell voltage equals twice the bus capacitor voltage ($V_{cell,n-1}=2V_{bus}$), i.e., no balancing is needed, the current is zero and no energy is transferred in either direction. The current polarity is detected by a sensor at the primary side, as shown in Fig. 2(a) and in Fig. 2(d), that determines the consecutive switching sequence in the second step. It should be noted that the switching configuration at the beginning of the switching cycle ($t_0 < t < t_1$) is similar regardless of the difference between the cell and the bus voltages, that is for both positive and negative current.

In the second step, after the current polarity has been detected, the operation resembles a conventional resonant

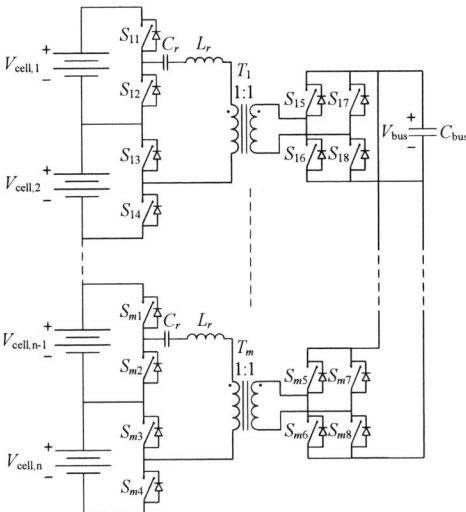

Fig. 1. Battery management system for n serially connected batteries.

converter in DCM. For positive detected current, S_{m5} and S_{m8} turn off at t_1, and their body diodes are conducting until the current becomes zero at $t=t_0+1/2f_r$ (Fig. 2(b)). It should be noted that turning S_{m5} and S_{m8} off at a low, nonzero current (before t_1) further allows ZVS at turn off. The current remains zero until half of the switching period $t_2=t_0+1/2f_s$. At t_2, S_{m1} turns off and, followed by a short dead-time, S_{m2} turns on (Fig. 2(c)). This discharges the resonant capacitor with a negative current during $t_2 < t < t_2 + 1/2f_r$ and then the current is discontinued by the body diodes of S_{m6} and S_{m7} until the next switching cycle at $t_3=t_0+1/f_s$.

For a negative detected current at $t_0 < t < t_1$, the operation is mirrored with respect to the cell and the bus side. S_{m1} is turned off at t_1, as shown in Fig. 2(e). The switches S_{m5}, S_{m8} are turned off and S_{m6}, S_{m7} are turned on (after a short dead-time) at t_2 (Fig. 2(f)). It should be noted that the switching timing can be loosely set. ZCS is naturally obtained by the body diodes as long as the switches have been turned off prior to the zero crossing point [35].

Simulated typical convergence during balancing operation of the system is presented in Fig. 4, where four battery cells (emulated by large capacitances) have been preset to different voltages. As can be observed, the cells voltages converge to the average of the initial voltages, validating the balancing operation of the system. The bus voltage converges to the average of half the voltage levels of the cells in the pack, and is given by

$$V_{bus} = \frac{1}{2N}\sum_{k=1}^{N} V_{cell,k} \qquad (1)$$

To derive the expressions for the system's key waveforms and parameters, in the following analysis, it is assumed that the cell voltage (V_{cell}) and the bus voltage V_{bus} are constant and that the average voltage across the resonant capacitor has reached a steady-state value of half the cell voltage, i.e. $V_{Cr}=0.5V_{cell}$. It is further assumed that the charge capacity of the bus capacitor is significantly lower than the cells' capacity. As a result, the bus voltage rapidly converges to the half of the average cells' voltages as in (1) which translates to a relatively small voltage difference $\Delta V=0.5V_{cell}-V_{bus}$ during steady-state. Due to symmetrical operation, the derivations are applied for the case that $V_{cell}>2V_{bus}$. The resonant current i_r can be expressed as

$$i_r(t) = \begin{cases} \dfrac{0.5V_{cell}-V_{bus}}{\left(1-e^{-\frac{\pi}{4Q}}\right)Z_r}e^{-\frac{2\pi f_r}{2Q}t}\sin(2\pi f_r t) & , t_0 \le t \le t_0 + 1/2f_r \\[2ex] -\dfrac{0.5V_{cell}-V_{bus}}{\left(1-e^{-\frac{\pi}{4Q}}\right)Z_r}e^{-\frac{2\pi f_r}{2Q}t}\sin(2\pi f_r t) & , t_2 \le t \le t_2 + 1/2f_r \quad (2) \\[2ex] 0 & , \text{elsewhere} \end{cases}$$

where, the resonant tank's characteristic impedance Z_r, resonant frequency f_r and the quality factor Q are:

$$Z_r = \sqrt{\frac{L_r}{C_r}};\ f_r = \frac{1}{2\pi\sqrt{L_r C_r}};\ Q = \frac{1}{R}\sqrt{\frac{L_r}{C_r}} \quad (3)$$

and, L_r and C_r are the resonant network components and R is the total parasitic or stray series resistance in the resonant network.

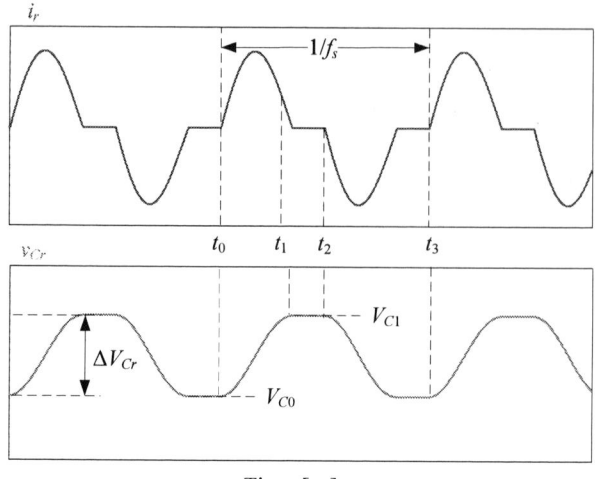

Fig. 3. Typical waveforms of the resonant inductor current and resonant capacitor voltage during balancing for the case when $V_{cell,n-1}>2V_{bus}$.

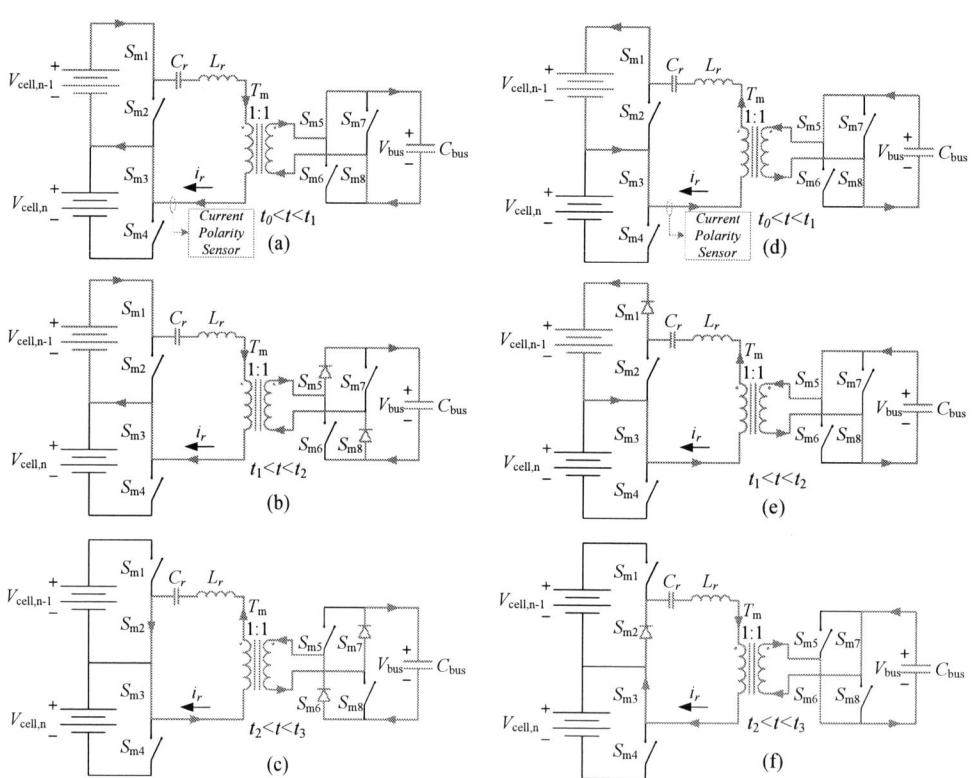

Fig. 2. Modes of operation: (a), (b) and (c) when $V_{cell,n-1}>2V_{bus}$, (d), (e) and (f) when $V_{cell,n-1}<2V_{bus}$. (a) and (d): current polarity detection at $t_0<t<t_1$, (b), (c), (e) and (f): energy transfer.

978-1-4673-9551-9/16 $31.00 © 2016 IEEE

From (2), the peak amplitude of the resonant current and the average current flowing through the cell can be expressed as:

$$I_{\text{peak}} = \frac{0.5V_{\text{cell}} - V_{\text{bus}}}{Z_r} \frac{e^{-\frac{\pi}{4Q}}}{1 - e^{-\frac{\pi}{4Q}}} \ , \tag{4}$$

$$I_{\text{cell}} = \frac{\delta}{\pi} I_{\text{peak}} = \frac{\delta}{\pi} \frac{0.5V_{\text{cell}} - V_{\text{bus}}}{Z_r} \frac{e^{-\frac{\pi}{4Q}}}{1 - e^{-\frac{\pi}{4Q}}} , \tag{5}$$

where $\delta = 0.5 f_s / f_r$. The voltage swing of the resonant capacitor ΔV_{Cr} during one switching cycle $T_s = 1/f_s$ can be expressed as:

$$\Delta V_{C_r} = 2 I_{\text{peak}} Z_r \tag{6}$$

where, V_{C0} and V_{C1} are the initial and final voltages (Fig. 3), respectively, given by:

$$V_{C_0} = 0.5(V_{\text{cell}} - \Delta V_{C_r})$$
$$V_{C_1} = 0.5(V_{\text{cell}} + \Delta V_{C_r}) \tag{7}$$

The power processing efficiency for a given voltage difference of ΔV can be expressed as:

$$\eta = 1 - \frac{P_{\text{loss}}}{P_{\text{cell}}} = 1 - \frac{\pi}{2} \frac{I_{\text{peak}} Z_r}{V_{\text{cell}} Q} = 1 - \frac{\pi}{2} \frac{\Delta V}{V_{\text{cell}}} \frac{e^{-\frac{\pi}{4Q}}}{1 - e^{-\frac{\pi}{4Q}}} \frac{1}{Q}. \tag{8}$$

As mentioned earlier and expresses by (1), the bus voltage converges to half of the cells average voltage. This process can be assumed instantaneous when compared to the rate that cells' voltages vary. As can be seen in (8), the power processing efficiency linearly depends on ΔV, which reflects on the voltage difference between the cells. This implies when balance is achieved the converter exhibits ideal power processing efficiency, a consequence of the fact that no current circulates through the system when balanced. It should be further noted that for a reasonable setting of the resonant parameters of $Q>4$, (8) can be approximated to:

$$\eta = 1 - \frac{\pi}{2} \frac{\Delta V}{V_{\text{cell}}} 1.273 , \tag{9}$$

which highlights that the efficiency depends primarily on the cells' status and is virtually independent on the system parameters. A clear advantage in a balancing system or any system that comprises multiple converters or converters string.

III. IMPLEMENTATION, DESIGN CONSIDERATIONS AND TOPOLOGY EXPANSION

Implementation of each of the m modules in Fig. 1 is straightforward and is depicted in Fig. 5. The battery cell side is built around a conventional resonant half-bridge

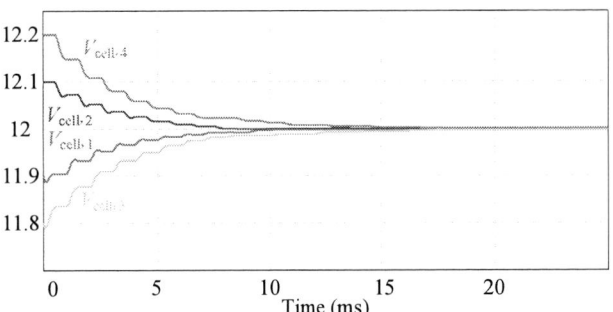

Fig. 4. Simulated convergence of cells with different initial voltage preset.

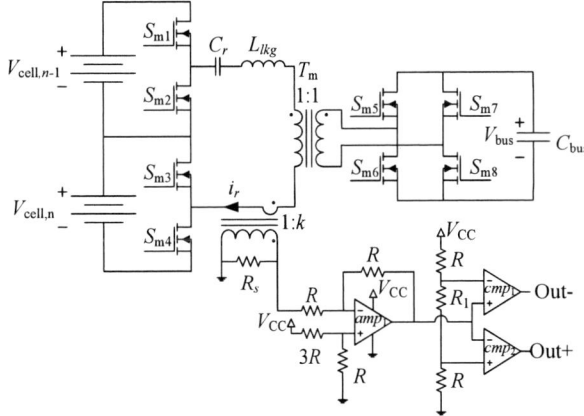

Fig. 5. Implementation of each of the m balancing circuits including the current polarity detection circuitry.

configuration whereas the bus capacitor side is constructed by a standard full-bridge rectifier configuration. As a result, the design of the MOSFET's gates drivers is standard low-side and bootstrapped pairs. The design of the resonant inductance and transformer follows a classical LLC magnetic element design, where L_r can be the leakage inductance L_{lkg} of the transformer which has been estimated as described in [36].

A. Design Considerations

The analysis derived earlier established the main attributes of the converter. Utilizing the above derivations, calculations of the key components of the converter can be established by the following procedure.

Given: Target average current per cell I_{cell} for maximum setting of ΔV; the switching frequency f_s and quality factor $Q>4$.

a. Calculate the expression

$$F(Q) = \frac{e^{-\frac{\pi}{4Q}}}{1 - e^{-\frac{\pi}{4Q}}} \tag{10}$$

b. From (5), (10), and considering $\delta < 1/2$, L_r and C_r can be selected according to

$$L_r < \frac{F(Q)\Delta V}{(2\pi)^2 I_{cell} f_s} \quad (11)$$

$$\left(\frac{2\pi I_{cell}}{F(Q)\Delta V}\right)^2 L_r < C_r < \left(\frac{1}{2\pi f_s}\right)^2 \frac{1}{L_r} \quad (12)$$

c. Verify that the resonant tank's series resistance R satisfies

$$R < \frac{Z_r}{Q}. \quad (13)$$

If (13) does not hold, iterate (11) and (12).

d. The transformer's area of product A'_p (without the leakage inductance consideration) should be selected according to

$$A'_p > \frac{V_{bus} I_{peak}}{\Delta BJK} \sqrt{\frac{f_r}{2f_s}} \quad (14)$$

where ΔB is the magnetic flux density at saturation, J is the current density and K is the winding fill factor.

e. To achieve the desired leakage inductance L_{lkg} the transformer's windings configuration should be designed as illustrated in Fig. 6 (for an E-type magnetic core). As prescribed in [36], the leakage inductance in E core can be estimated by

$$L_{lkg} = \frac{\mu_0 n^2 ATL}{a}\left(\frac{b_1 + b_2}{3} + c\right)[H], \quad (15)$$

where μ_0 is the permeability constant, n is the number of turns in the primary winding, ATL is the average length of turn, a is the winding height, b_1 and b_2 are the thicknesses of the primary and secondary winding, respectively, and c is the distance between the primary and the secondary winding. All lengths are assigned in mm.

In this study, the transformer is with a 1:1 turns ratio and therefore $b_1 = b_2 = b$.

f. Adjust the area product to include the addition of the leakage inductance by

$$A_p = A'_p\left(1 + \frac{c}{2b}\right). \quad (16)$$

B. Practical implementation

The current polarity detector circuitry (Fig. 5) is realized using a current transformer, summing amplifier and two comparators, as depicted in Fig. 5. Three operation modes are detected – positive, negative and zero current. The current is

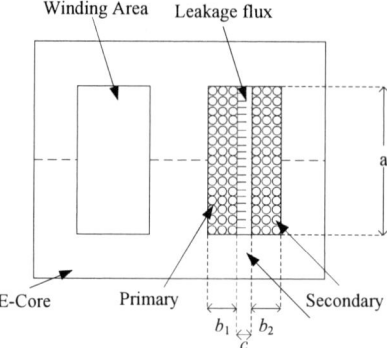

Fig. 6. Conventional transformer winding configuration for an E-core.

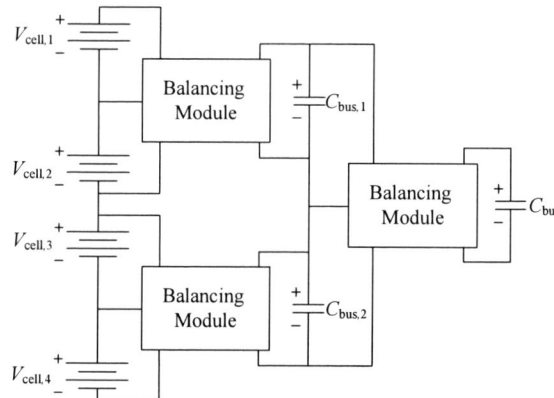

Fig. 7. Modularization of the balancing system for four battery cells.

considered zero if its absolute maximum value, measured within $t_0 < t < t_1$, is less than the threshold determined by the following expression:

$$|I_{th}| = \frac{0.5 V_{CC}}{1 + 2d}\frac{k}{R_s}, \quad (17)$$

where $d = R/R_1$, k is the turns ratio of the current transformer, R_s is the sense resistor at the secondary side of the current transformer, $R \gg R_s$, and $R > R_1$. The outputs of the comparators determine the polarity of i_r as detailed in Table I.

TABLE I – CURRENT POLARITY SENSOR DETECTION

Out+	Out-	Current Polarity
1	0	$i_r > 0$
0	1	$i_r < 0$
0	0	$i_r \approx 0$

The voltage stress of the transistors is the cell voltage for the cell side transistors and the bus capacitor voltage for the bus side. The lower voltage stress allows transistors with lower $R_{DS(on)}$ per silicon area. Current stresses depend on the desired convergence speed.

The isolation between the bus capacitor and the battery cells provided by the transformer enables the topology to be extended, as demonstrated in Fig. 7, for four battery cells. This modularization, as mentioned in [31], provides another stage of balancing and enables the operation at lower voltage stress which can expedite the balancing speed since higher current can be delivered for the same power dissipation.

IV. EXPERIMENTAL VERIFICATION

In order to demonstrate the balancing action and to verify the analysis and simulation results, experiments have been carried out using two large capacitors (emulating the batteries operation, in order to shorten the convergence time) as two cells in series. Table II shows the component types and values used in the experimental prototype. Fig. 8 presents the measured waveforms of the resonant tank during the balancing operation. Based on the principle of operation described earlier, the balancing time is shared equally between the cells. The duration of each cell-balancing has been set to 10 switching cycles. The procedure is demonstrated in Fig. 9, where the resonant current and the capacitor voltage are measured over 20 switching cycle. During the first 10 switching cycles the cell having the higher voltage is balanced (i.e. connects to the bus via the converter), whereas in the next 10 switching cycles the balancing is done for the cell with lower voltage. Convergence of the resonant capacitor voltage to the steady-state value can be observed in Fig. 9, supporting the analysis conjecture that the majority of energy transfer operation is carried under steady-state conditions.

Fig. 10 shows the cells' (realized by large capacitors to allow timely convergence) voltages and the tank's resonant current over a long period of time when cells are pre-charged to different voltages. As can be observed, the voltages of the two cells are equalized one to another and the current decay to zero.

Similar experiment was carried out using two 12 V, 7Ah, Lead-Acid batteries. Fig. 11 shows the batteries voltages over a long period of time when pre-charged to different voltages. As can be observed, the voltages of the two batteries are being balanced up to a small voltage difference.

TABLE II – EXPERIMENTAL PROTOTYPE VALUES

Component	Value
Cells: Batteries	12 V, 7Ah
When realized by capacitors	100 mF
Transformer's leakage inductance L_{lkg}	5 µH
Resonant tank capacitor C_r	5.7 µF
Transformer's magnetizing inductance L_m	3 mH
MOSFETs S_{m1}-S_{m8}	Si4178DY
Bus capacitor C_{bus}	15 mF
Resonant frequency f_r	30 kHz
Switching frequency f_s	20 kHz

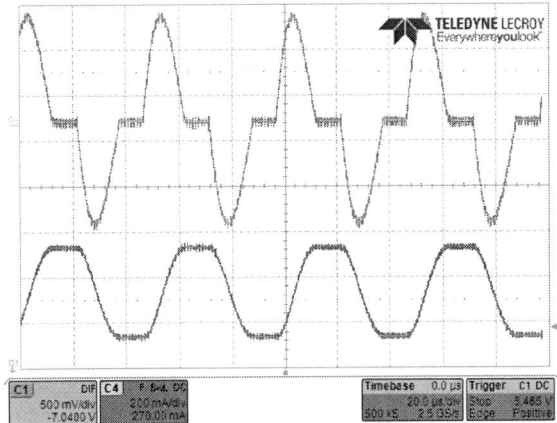

Fig. 8. Experimental results of resonant tank current (top) and resonant capacitor voltage (bottom) during balancing operation. Current 200mA/div, voltage 500mV/div, time scale 20µs/div.

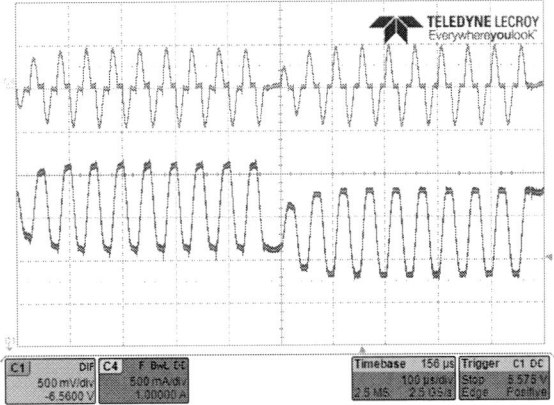

Fig. 9. Resonant tank current (top) and resonant capacitor voltage (bottom) during balancing of upper and lower cells with different voltages, each of them is balanced for 10 switching periods. Current 500mA/div, voltage 500mV/div, time scale 100µs/div.

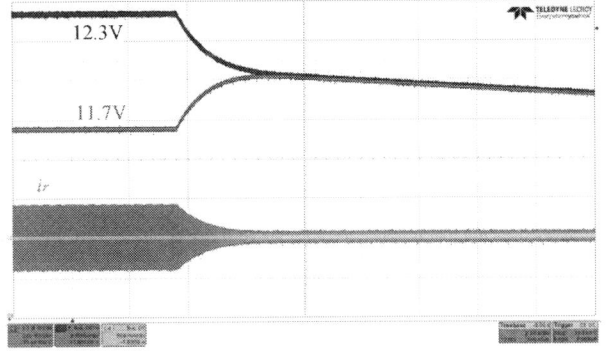

Fig. 10. Experimental results of voltage convergence of two cells with different initial voltages, and the tank's resonant current. Time scale 2s/div, Cells voltages 200mV/div, tank's resonant current 500mA/div.

978-1-4673-9551-9/16 $31.00 © 2016 IEEE

V. CONCLUSION

In this work, a new soft switched isolated balancing topology and operation method have been introduced. The circuit is based on a modified series resonant converter operating in DCM. The new balancing circuit uses one transformer for balancing of two neighboring cells and as a result, less magnetic components are required compared to other isolated topologies. Another significant advantage of the topology is the extremely low quiescent losses due to the native behavior of the converter where no current circulates through the system when the cells are balanced.

The balancing operation is facilitated by a simple current polarity detection and does not require synchronization between modules. Furthermore, the topology can be easily scaled-up and modularized for as many cells as required.

ACKNOWLEDGEMENTS

This research was supported by 'Keren Pazi'.

REFERENCES

[1] A. Emadi, L. Young Joo, and K. Rajashekara, "Power electronics and motor drives in electric, hybrid electric, and plug-in hybrid electric vehicles," *IEEE Trans. Ind. Electron.*, vol. 55, no. 6, pp. 2237–2245, Jun. 2008.

[2] M. Ehsani, G. Yimin, J. M. Miller, "Hybrid electric vehicles: architecture and motor drives," *Proceedings of the IEEE*, vol. 95, no. 4, pp. 719-728, Apr. 2007.

[3] A. Y. Saber and G. K. Venayagamoorthy, "Plug-in vehicles and renewable energy sources for cost and emission reductions," *IEEE Trans. Ind. Electron.*, vol. 58, no. 4, pp. 1229–1238, Apr. 2011.

[4] H. Qian, J. Zhang, J. S. Lai, and W. Yu, "A high-efficiency grid-tie battery energy storage system," *IEEE Trans. Power Electron.*, vol. 26, no. 3, pp. 886–896, Mar. 2011.

[5] B. Gu, J. Dominic, B. Chen, and J. S. Lai, "A high-efficiency single-phase bidirectional AC-DC converter with minimized common mode voltages for battery energy storage systems," in *Proc. IEEE Energy Convers. Congr. Expo. 2013*, pp. 5145-5149, Sep. 2013.

[6] B. T. Kuhn, G. E. Pitel, and P. T. Krein, "Electrical properties and equalization of lithium-ion cells in automotive applications," in *Proc. IEEE Vehicle Power Propuls. Conf.*, pp. 55–59, Sep. 2005.

[7] P. T. Krein and R. S. Balog, "Life extension through charge equalization of lead–acid batteries," in *Proc. Int. Telecommun. Energy Conf. (INTELEC)*, pp. 516–523, 2002.

[8] M. Uno and K. Tanaka, "Influence of high-frequency charge–discharge cycling induced by cell voltage equalizers on the life performance of lithium-ion cells," *IEEE Trans. Veh. Technol.*, vol. 60, no. 4, pp. 1505–1515, May 2011.

[9] J. Cao, M. Schofield and A. Emadi, "Battery balancing methods: A comprehensive review," *IEEE Vehicle Power and Propulsion Conference, VPPC '08.* pp.1,6, Sep. 2008.

[10] A. C. Baughman and M. Ferdowsi, "Double-tiered switched-capacitor battery charge equalization technique," *IEEE Trans. Ind. Electron.*, vol. 55, no. 6, pp. 2277-2285, Jun. 2008.

[11] C. Pascual and P. T. Krein, "Switched capacitor system for automatic series battery equalization" in *Proc. IEEE Appl. Power Electron. Conf. Expo. 1997*, pp. 848-854, Feb. 1997.

[12] M. W. Cheng, Y. S. Lee, R. H. Chen, and W. T. Sie, "Cell voltage equalization using ZCS SC bidirectional converters," in *Proc. Int. Telecommun. Energy Conf.*, pp. 1–6, Oct. 2009.

[13] F. Mestrallet, L. Kerachev, J. C. Crebier and A. Collet, "Multiphase interleaved converter for lithium battery active balancing," *IEEE Trans. Power Electron.*, vol. 29, no. 6, pp. 2874-2881, Jun. 2014.

[14] L. Wang, L. Wang, C. Liao, and J. Liu, "Research on battery balance system applied on HEV," *VPPC '09*, pp. 1788-1791, Sep. 2009.

[15] Z. Nie and C. Mi, "Fast battery equalization with isolated bidirectional DC-DC converter for PHEV applications," *IEEE Vehicle Power and Propulsion Conference, VPPC '08*, pp. 78-81, Sep. 2009.

[16] Y. S. Lee and G. T. Cheng, "Quasi-resonant zero-current-switching bidirectional converter for battery equalization applications," *IEEE Trans. Power Electron.*, vol. 21, no. 5, pp. 1213-1224, Sep. 2006.

[17] M. Uno and K. Tanaka , "Single-switch cell voltage equalizer using multistacked buck-boost converters operating in discontinuous conduction mode for series-connected energy storage cells," *IEEE Trans. Vehicular Technology*, vol. 60, no. 8, pp. 3635-3645, Oct. 2011.

[18] M. Uno and K. Tanaka, "Single-switch multi-output charger using voltage multiplier for series-connected lithium-ion battery/supercapacitor equalization," *IEEE Trans. Ind. Electron.*, vol. 60, no. 8, pp. 3227–3239, Aug. 2013.

[19] G. Oriti, A. L. Julian and P. Norgaard, "Battery management system with cell equalizer for multi-cell battery packs" in *Proc. IEEE Energy Convers. Congr. Expo. 2014*, pp. 900-905, Sep. 2014.

[20] M. Uno and K. Tanaka, "Single-switch cell voltage equalizer using voltage multipliers for series-connected supercapacitors," in *Proc. IEEE Appl. Power Electron. Conf. Expo*, pp. 1266-1272, Feb. 2012.

[21] Y. Yuanmao, K. W. E. Cheng, and Y. P. B. Yeung, "Zero-current switching switched-capacitor zero-voltage-gap automatic equalization system for series battery string", *IEEE Trans. Power Electron.*, vol. 27, no. 7, pp. 3234-3242, Jul. 2012.

[22] C. H. Sung, K. Lee, and B. Kang, "Voltage equalizer for li-ion battery string using LC series resonance," *IECON 2013 - 39th Annual Conference of the IEEE*, pp.1404-1409, Nov. 2013.

[23] A. L. Julian, G. Oriti, M. E. Pfender, "SLR converter design for multi-cell battery charging," in *Proc. IEEE Energy Convers. Congr. Expo.*, pp. 743-748, Sep. 2013.

[24] D. Costinett, K. Hathaway, M. U. Rehman, M. Evzelman, R. Zane., Y. Levron, and D. Maksimovic, "Active balancing system for electric vehicles with incorporated low voltage bus," in *Proc. IEEE Appl. Power Electron. Conf. Expo. 2014*, pp. 3230-3236, Mar. 2014.

[25] S. Li, C. C. Mi and M. Zhang, "A high-efficiency active battery-balancing circuit using multiwinding transformer," *IEEE Trans. Ind. Applications*, vol. 49, no. 1, pp. 198-207, Jan. 2013.

[26] S. H. Park, K. B. Park, H. S. Kim, G. W. Moon, M. J. Youn, "Single-magnetic cell-to-cell charge equalization converter with reduced number of transformer windings," *IEEE Trans. Power Electron.*, vol. 27, no. 6, pp. 2900-2911, Jun. 2012.

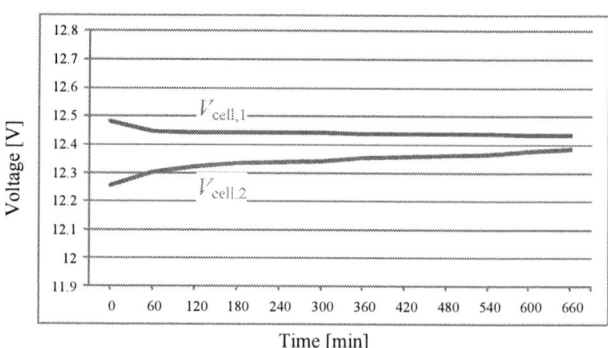

Fig. 11. Experimental results of voltage convergence with two Lead-Acid batteries as cells with different initial voltages.

[27] M. Y Kim, J. H. Kim, G. W. Moon, "Center-cell concentration structure of a cell-to-cell balancing circuit with a reduced number of switches," *IEEE Trans. Power Electron.*, vol. 29, no. 10, pp. 5285–5297, Oct. 2014.

[28] Y. S. Lee and M. W. Cheng, "Intelligent control battery equalization for series connected lithium-ion battery strings," *IEEE Trans. Ind. Electron.*, vol. 52, no. 5, pp. 1297-1307, Oct. 2005.

[29] M. U. Rehman, F. Zhane, M. Evzelman, R. Zane, and D. Maksimovic, "Control of a series-input, parallel-output cell balancing system for electric vehicle battery packs", *IEEE 16th Workshop on Control and Modeling for Power Electronics 2015*, Jul. 2015.

[30] B. Dong, Y. Li and Y. Han, "Parallel architecture for battery charge equalization," *IEEE Trans. Power Electron.*, vol. 30, no. 9, pp. 4906-4913, Sep. 2015.

[31] H. S. Park, C. H. Kim; K. B. Park, G. W. Moon and J. H. Lee, "Design of a charge equalizer based on battery modularization," *IEEE Trans. Vehicular Technology*, vol. 58, no. 7, pp. 3216-3223, Sep. 2009.

[32] H. S. Park, C. E. Kim, C. H. Kim, G. W. Moon and J. H. Lee, "A modularized charge equalizer for an HEV lithium-ion battery string," *IEEE Trans. Ind. Electron.*, vol. 56, no. 5, pp. 1464-1476, May 2009.

[33] C. H. Kim, M. Y. Kim, and G. W. Moon, "A modularized charge equalizer using a battery monitoring IC for series-connected Li-ion battery strings in electric vehicles," *IEEE Trans. Power Electron.*, vol. 28, no. 8, pp. 3779–3787, Aug. 2013.

[34] M. U. Rehman, M. Evzelman, K. Hathaway, R. Zane, G. L. Plett, K. Smith, E. Wood, and D. Maksimovic, "Modular approach for continuous cell-level balancing to improve performance of large battery packs", ," in *Proc. IEEE Energy Convers. Congr. Expo 2014*, pp. 4327-4334, Sep. 2014.

[35] E. Hamo, M. Evzelman, M. M. Peretz, "Modeling and analysis of resonant switched-capacitor converters with dree-wheeling ZCS," *IEEE Trans. Power Electron.*, vol. 30, no. 9, pp. 4952–4959, Sep. 2015.

[36] C. M. T. McLyman, "Winding capacitance and leakage inductance", Chap. 17, *Transformer and Inductor Design Handbook*, 3rd Ed. NY, Marcel Dekker, 2004.

A Novel Topology of High Voltage and High Power Bidirectional ZCS DC-DC Converter based on Serial Capacitors

Lejia Sun, Fang Zhuo, Feng Wang, Tianhua Zhu

School of Electrical Engineering, Xi'an Jiaotong University

Xi'an 710049, China

Email:sunlejia@163.com

Abstract—With the rapid development of renewable energy systems and Microgrid, more and more DC grids of different voltage level need to be connected by the high voltage and high power DC-DC converter. The traditional topology of high voltage and high power DC-DC converter was Phase Shifted Full Bridge cascaded Modules (PSFBM). Due to the inefficiency and difficulty on manufacturing high frequency transformer of PSFBM, a non-isolated topology, Multilevel Modular Capacitor-Clamped DC-DC Converter (MMCCC), was proposed as an alternative. However, the MMCCC also has some inherent disadvantages, especially, a large number of switching devices and no voltage regulation which limit its application in the connection of DC grids requiring high efficiency and accurate control. In this paper, a novel non-isolated high voltage and high power ZCS DC-DC converter based on serial capacitors is proposed. By energy transferring circuit to transfer the energy between adjacent serial capacitors, it can realize the bidirectional DC-DC conversion with less switching devices. Furthermore, by a control system with independent voltage control of each serial capacitor, It can improve the flexibility and accuracy of output voltage and also can realize a higher input or output voltage without limitation in theory. In this paper, the operation principle and design consideration of the novel topology were presented in detail, and a simulation with ten serial capacitors was also built to validated the effectiveness of the novel topology. Finally, a 1kW test equipment with 250V input voltage and 1000V output voltage was developed in laboratory to verify the simulation result.

Keywords—microgrid; transfering; serial; accuracy; limitation

I. INTRODUCTION

Currently, In renewable energy stations and distributed power systems, a lot of DC grids of different voltage level have existed, which require high voltage and high power converter to realize the DC-DC connection and power interaction[1]. The traditional topology, phase shifted full bridge cascaded modules (PSFBM)[2][3], has following disadvantages:

- It involves a large number of isolated high frequency transformers in topology[1][4][5]. The high frequency transformer with high isolation is not only hard to manufacture, but also inefficient due to the obvious skin effect and leakage inductance.

- It involves a large number of switches, which increase

This work was supported by the China National Technology Research and Development Program("863"Program) of Development and Key technologies of Bidirectional Converter Applied in Photovoltaic Micro-gird. No.2015AA050606.

This work was also supported by National Key Basic Research Program of China("973" Program) of Mechanism and Application of Short-Circuit Current Breaking of High Voltage DC Circuit Breaker. No. 2015CB251001.

the cost and decrease the efficiency. For topology with n modules, it requires $8n$ switching devices.

In most cases, considering the transmission efficiency and the cost of DC line, DC-DC converters applied in DC grids connection are non-isolated. Therefore, basic DC-DC converters (Buck, Boost, Buck-Boost) with serial switching devices to withdraw higher voltage was researched[6][7]. But, synchronizing the switching is very difficult because of the inherent differences in the switching characteristics of each device. Recently, a new topology, the Multilevel Modular Capacitor-Clamped DC-DC Converter (MMCCC) was developed[8]-[12], shown in Fig.1, with the advantage of low switching voltage stress and high voltage gain. However, its disadvantages are also obvious:

- The number of switching devices is still excessive. For example, it needs $3n-2$ switching devices for a conversion ratio of n.

- The quality of output voltage waveform is relatively poor, since the MMCCC is a open loop control system with no voltage regulation. The control of each MMCCC switching device is also very simple, just 50% duty-cycle on-off switching.

- The switching loss is serious, since the soft-switching can not be simply realized.

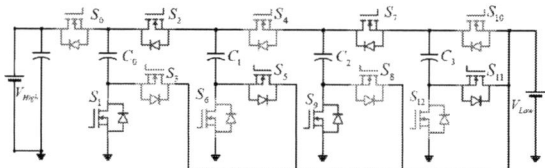

Fig. 1. Topology of the MMCCC ($n = 5$)

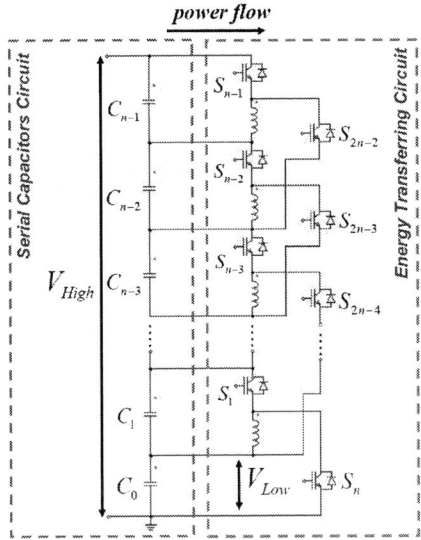

Fig. 2. Proposed novel high voltage DC-DC converter

In order to overcome disadvantages of traditional topology mentioned above, this paper proposed a novel high voltage and high power bidirectional non-isolated DC-DC topology based on serial capacitors, which is shown in Fig.2. It can be divided into tow parts: serial capacitors circuit and energy transferring circuit. The V_{High} and V_{Low} are high DC voltage side and low DC voltage side of the topology. The novel topology abandoned the basic DC-DC circuit such as buck, boost, buck-boost, instead, it using a string of serial capacitors to withdraw the high voltage, and obtaining a low voltage from the lowest potential capacitor. The function of the energy transferring circuit is to transfer the energy between adjacent serial capacitors to support the voltage of each capacitor. Besides benefits of MMCCC, the new topology has following additional advantages:

- The total number of switching device is reduced, which is $2(n-1)$, given a concersion ratio of n.

- The flexibility and accuracy of output voltage are greatly improved by independent voltage control system of each serial capacitor.

- The ZCS can be achieved without extra components, due to the discontinuous inductor current.

II.   OPERATION PRINCIPLE

According to the direction of power flow, the novel topology has tow working modes: buck control mode and boost control mode, which will be explained below separately:

A. Buck Control Mode:

The buck control mode is applied to realizes the buck DC-DC conversion. The energy transferring process is illustrated in Fig.3(a). The red and blue arrows point to the direction of energy transferring. The input voltage is connected to the high voltage side V_{High}, and the output voltage is obtained form the low voltage side V_{Low}. n is the number of serial capacitors. U is the reference voltage of each serial capacitor. When the load is connected to the V_{Low}, the voltage of low potential capacitor C_0 tend to decline and the voltage of other higher potential capacitors(C_{n-1}, C_{n-2}, ...,C_1)tend to increase. By the energy transferring circuit, higher potential capacitors can transfer the energy to capacitor C_0 one by one to support the V_{C_0} to realize the stable output.   The Fig.3(b) is the theoretical waveform when the capacitor C_{n-1} transfers the energy to C_{n-2}. The $V_{GS(S_k)}$ is the driving signal of S_k. f is the switching frequency.   The i_{L_k-MAX} and i_{L_k-MIN} is the maximum and minimum value of inductor current. The energy transferring process can be divided into four stages:

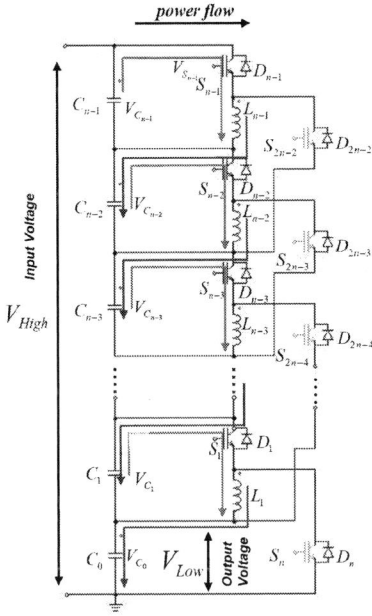

(a) Power flow with buck control

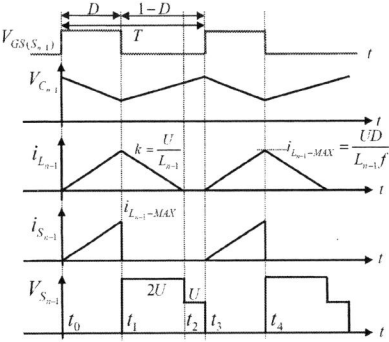

(b) Theoretical waveform with buck control

Fig. 3. Principle of buck control mode

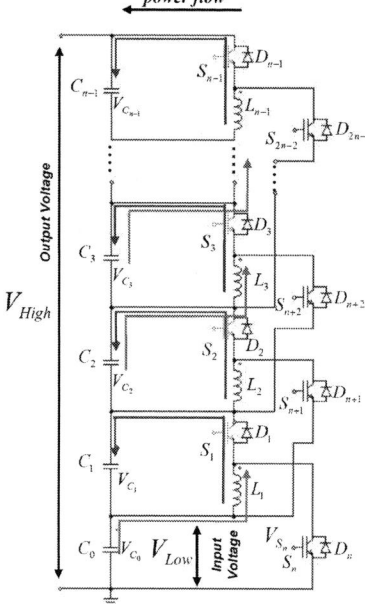

(a) Power flow with boost control

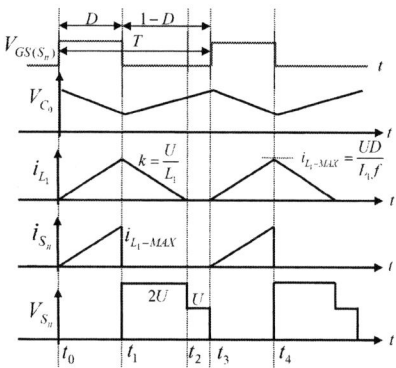

(b) Theoretical waveform with boost control

Fig. 4. Principle of boost control mode

1) $t_0 - t_1$: at time t_0, S_{n-1} is turned on, S_{2n-2} is turned off, the C_{n-1} transfers the energy to L_{n-1}. Then the capacitor voltage $V_{C_{n-1}}$ declines and inductor current $i_{L_{n-1}}$ increases. The increase gradient and the maximum value of $i_{L_{n-1}}$ are calculated as:

$$k = \frac{di_{L_{n-1}}}{dt} = \frac{U}{L_{n-1}} \quad (1)$$

$$i_{L_{n-1}-MAX} = \frac{UD}{L_{n-1}f} \quad (2)$$

2) $t_1 - t_2$: at time t_1, S_{n-1} is turned off. The $i_{L_{n-1}}$ continuously flows through the anti-parallel diode D_{2n-2} and transfers the energy from inductor L_{n-1} to capacitor C_{n-2}. Then

the $i_{L_{n-1}}$ declines and $V_{C_{n-2}}$ increases. The $V_{C_{n-1}}$ is recovered by charging of input voltage V_{High}. The voltage stress of S_{n-1} is the sum of $V_{C_{n-1}}$ and $V_{C_{n-2}}$, which is $2U$. The declined gradient of $i_{L_{n-1}}$ also equals k.

3) $t_2 - t_3$: Due to the anti-parallel diode D_{2n-2}, when the $i_{L_{n-1}}$ reduces to zero, it can not reverse and keep value in zero. The voltage stress $V_{S_{n-1}}$ equals $V_{C_{n-1}}$, which is U.

4) $t_3 - t_4$: At time t_4, the S_{n-1} is turned on again. Since the inductor current $i_{L_{n-1}}$ can not be mutated and increase slowly, it is assumed that the value of $i_{L_{n-1}}$ kept zero during the process of S_{n-1} turning on. The S_{n-1} can be considered to realize ZCS.

The energy transferring process of other adjacent serial capacitors(C_{n-2} and C_{n-3} ,......, C_1 and C_0) is the same as the process between C_{n-1} and C_{n-2}. By these transferring processes, the energy will finally be passed to capacitor C_0 and support V_{Low} to output.

B. Boost Control Mode:

The energy transferring process of boost control mode is illustrated in Fig.4(a). The input voltage is connected to the low voltage side V_{Low}, and the output is obtained form the high voltage side V_{High}. Through the energy transferring circuit, the energy of low potential capacitor C_0 is pumped to the higher potential capacitor $C_1, C_2......C_{n-1}$ to support their voltage. The high voltage V_{High} is outputted from the string of serial capacitors. Fig.4(b) is the waveform of the energy transferring process between C_1 and C_2.

1) $t_0 - t_1$: At time t_0, S_n is turned on, S_1 is turned off, the capacitor C_0 transfers the energy to inductor L_1. Then the inductor current i_{L_1} increases.

2) $t_1 - t_2$: At time t_1, S_n is turned off . The i_{L_1} continuously flow through the anti-parallel diode D_1 and transfers the energy from inductor L_1 to capacitor C_1. Then the i_{L_1} declines and V_{C_1} increases.

3) $t_2 - t_3$: Due to the anti-parallel diode D_1, when the i_{L_1} reduce to zero, it can not reverse and keep value in zero.

4) $t_3 - t_4$: At time t_4, the S_n is turned on again and inductor current i_{L_1} increases.

The energy transferring process of other adjacent serial capacitors is the same as the process between C_1 and C_2. The high voltage output can be finally obtained from the string of serial capacitors

III. DESIGN CONSIDERATION

A. Inductance calculation:

According to the illustration above, in order to realize ZCS, the inductance and capacitance should be carefully designed to make sure the inductor current $(i_{L_1}, i_{L_2}, \ldots\ldots i_{L_{n-1}})$ to be discontinuous.

From the waveform of Fig.4 and Fig.5, it is shown that to make sure the i_{L_k} reach zero, the duty cycle D should below 0.5. P is the rated power of the converter. According to the energy transferring process, the inductance of L_k can be achieved by:

$$\frac{1}{2}L_k(i_{L_k-MAX}^2 - i_{L_k-MIN}^2)f = \frac{n-k}{n}P \gg I_{k-MAX} = (4 - \frac{4k}{n})I_0 \gg L_k = \frac{U^2 n}{8(n-k)fP} \quad (3)$$

B. Capacitance calculation:

In order to share the high voltage V_{High}, the capacitance of each serial capacitor should be the same. The voltage ripple is designed as 4%. T is the switching period. According to the energy transferring process, the capacitance can be achieved by

$$\frac{1}{2}C_k(1.02^2 U^2 - 0.98^2 U^2) = DTP \gg C_k = \frac{25DTP}{U^2} \quad (4)$$

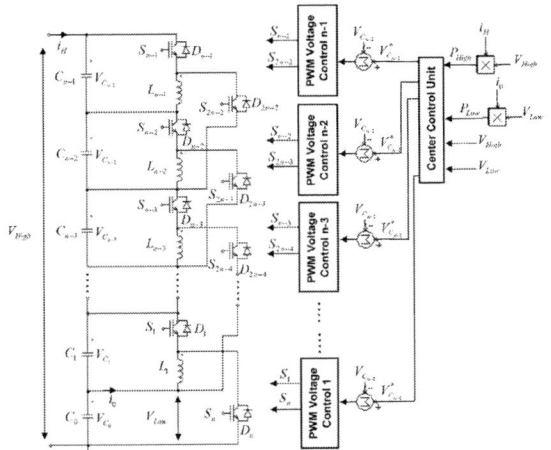

Fig. 6. Diagram of control system

TABLE.I SIMULATION PARAMETERS

Parameters	Value
n	10
P	$100kW$
V_{Low} / V_{High}	$700V / 7000V$
f	$1000Hz$
$C_k (k = 0,1,2\ldots\ldots9)$	$125uF$
$L_k (k = 1,2,3\ldots\ldots9)$	$680uH$, $765uH$, $875uH$, $1020uH$, $1225uH$, $1531uH$, $2041uH$, $3062uH$, $6215uH$
$R(Buck / Boost)$	$4.9\,\Omega / 490\,\Omega$

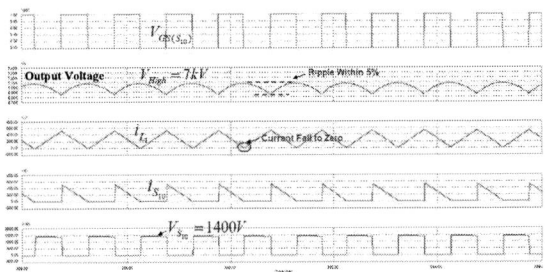

Fig. 7. Simulation waveform with boost control ($P = 100kW$)

IV. CONTROL SYSTEM

The control system of the novel topology is shown in Fig.6. It can be seen that by sampling the current and the voltage of high voltage side and low voltage side ($i_{High}, V_{High}, i_{Low}, V_{Low}$), the value and direction of power flow are calculated to select the buck control mode or boost control mode. Each serial capacitor has independent voltage control system. By separately setting the voltage references $V_{C_{n-1}}^*, V_{C_{n-2}}^*, \ldots\ldots, V_{C_0}^*$, the output voltage can be flexibly and accurately controlled.

V. SIMULATION

In order to confirm that the novel topology can be practical implemented, a stimulation is performed to verify the result by using PSIM software. The simulation parameters are shown in Table I. Fig.7 shows the simulation waveform of boost control mode with the rated power. Fig.8 shows the simulation waveform of buck control mode with half rated power. It can be seen that the bidirectional DC-DC conversion was realized. The simulation waveform is perfectly coincided with the theoretical waveform shown in Fig.3(b) and Fig.4(b). Fig.9 shows the waveform of inductor current ($i_{L_2}, i_{L_4}, i_{L_6}, i_{L_8}$) and capacitor voltage ($V_{C_0}, V_{C_2}, V_{C_4}, V_{C_6}, V_{C_8}$) in buck control mode with the rated power. It can be seen that the voltage of each capacitor can reach its reference value ($700V$), and the current of each inductor is discontinuous to realize the ZCS.

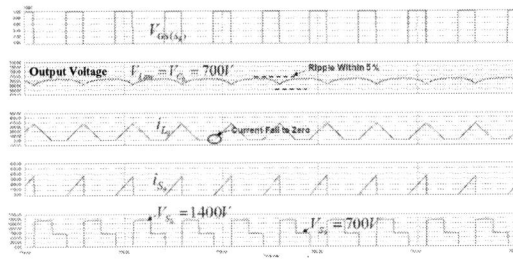

Fig. 8. Simulation waveform with buck control ($P = 50kW$).

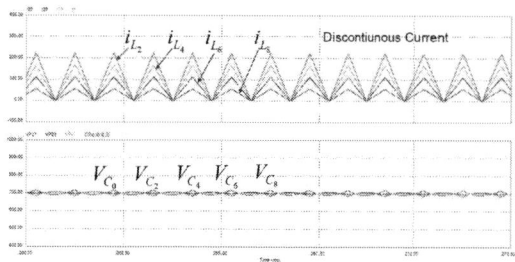

Fig. 9. Simulation waveform of inductor current and capacity voltage in buck control ($P = 100kW$).

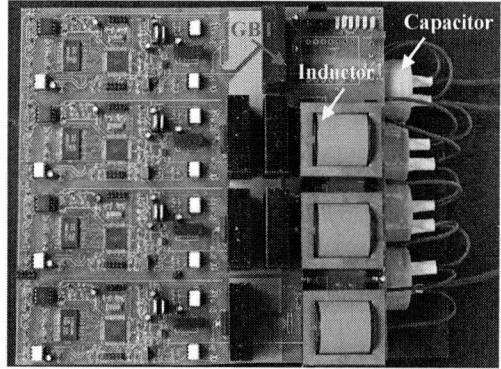

Fig. 10. Photograph of the 1kW test equipment with n=4

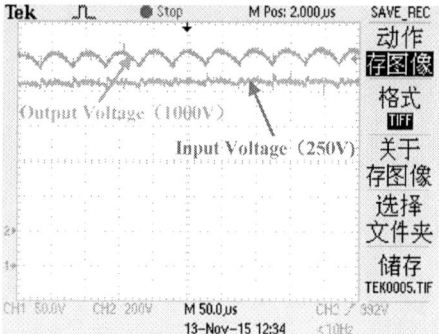

Fig. 11. Waveform of input voltage (CH1: $250V$) and (CH2: $1000V$)

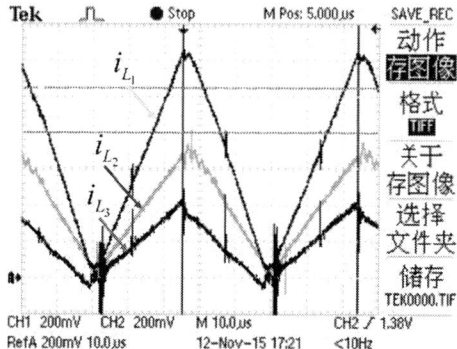

Fig. 12. Waveform of inductor current ($10A/1V$)
$i_{L_1} = 12A$, $i_{L_2} = 7A$, $i_{L_3} = 4A$

VI. EXPERIMENTAL RESULTS

Due to the laboratory condition, a small-scale $1kW$ test equipment of the novel topology in boost control mode is implemented in our laboratory. The parameters is shown in Table □

TABLE. II EQUIPMENT PARAMETERS

Parameters	Value
n	4
P	$1kW$
V_{Low}/V_{High}	$250V/1000V$
f	$20kHz$
R	$1000\,\Omega$

The photograph of test equipment is shown in Fig.10. Fig. 11 shows the waveform of the input voltage($250V$) and output voltage ($1000V$), which indicates the input voltage was boosted four times ($n = 4$). Fig. 12 shows the waveform of each inductor current. It can be seen that the waveform in Fig. 10 and Fig.11 is coincident with the simulation result shown in Fig. 7 and Fig. 9.

VII. CONCLUSION

In this paper, a novel high voltage and high power bidirectional non-isolated ZCS DC-DC converter based on serial capacitors is proposed. Due to the high efficiency and accurate control, it is suitable for connection of DC grids. Compared with the traditional topology, isolated phase shifted full bridge cascaded modules and non-isolated MMCCC, the proposed novel topology can reduce the number of switching device and greatly improve the flexibility and accuracy of output voltage. Furthermore, in the novel topology, the ZCS can be realized by the careful design of inductance to further improve the efficiency.

REFERENCE

[1] Kadri. R, Gaubert. J-P and Champenois. G, " Nondissipative String Current Diverter for Solving the Cascaded DC-DC Converter Connection Problem in Photovoltaic Power Generation System," IEEE Transactions on Power Electronics, vol. 27, no. 3, pp. 1249-1258.

[2] Kenichiro Sano, and Masahiro Takasaki, " A Boost Conversion System Consisting of Multiple DC-DC Converter Modules for Interfacing," in IEEE Energy Conversion Congress and Exposition (ECCE), 2013, pp. 2613-2618.

[3] Huang. B, Shahin. A and Pierfederici. S, "High Voltage Ratio DC–DC Converter for Fuel-Cell Applications," in IEEE Power Electronics Specialists Conference (PESC), 2008, pp. 1277-1283.

[4] Thummala. P, Schneider. H and Zhe Zhang, "Efficiency Optimization Considering the High-Voltage Flyback Transformer Parasitics Using an Automatic Winding Layout Technique," IEEE Transactions on Power Electronics, vol. 30, no. 10, pp. 5755-5768.

[5] Vinnikov. D, Laugis. J and Galkin. I, "Middle-Frequency Isolation Transformer Design Issues for the High-Voltage DC/DC Converter," 2008 Power Electronics Specialists Conference, 2008, pp. 1930-1936.

[6] Barth. T, Semmler. S and Buschendorf. M, "Gate Drive Unit DC-DC Power Supply for Multi-Level Converters or Series Connection of IGBTs with High Voltage Insulation," in IEEE 11th International Multi-Conference on System, Signal and Devices(SSD), 2014, pp. 1-5.

[7] In-Ho Song, Sang-Bong Yoo and Bum-Seok Suh, "A Novel Three-Level ZVS PWM Inverter Topology for High-Voltage DC/DC Conversion Systems with Balanced Voltage Sharing and Wider Load Range," in IEEE Industry Applications Conference, 1996, pp.973-979.

[8] Dong Cao, Fang Zheng Peng, "Multiphase Multilevel Modular DC–DC Converter for High-Current High-Gain TEG Application," IEEE Transactions on Industry Application, vol. 47, no. 3, pp. 1400-1408.

[9] Khan. F. H, Tolbert. L. M, "A Multilevel Modular Capacitor Clamped DC-DC Converter," in IEEE 2006 Industry Applications Conference, 2006, pp. 966-973.

[10] Dong Cao, Shuai Jiang and Fang Zheng Peng, "Optimal Design of a Multilevel Modular Capacitor-Clamped DC–DC Converter," IEEE Transactions on Power Electronics, vol. 28, no. 8, pp. 3816-3826.

[11] Khan. F. H, Tolbert. L. M and Webb. W. E, "Start-Up and Dynamic Modeling of the Multilevel Modular Capacitor-Clamped Converter," IEEE Transactions on Power Electronics, vol. 25, no.2, pp. 519-531.

[12] Peng. F. Z, Wei Qian, Dong Cao, "Recent Advances in Multilevel Converter/Inverter Topologies and Application,' in IEEE Power Electronics Conference (IPEC), 2010, pp.492-501.

Control Scheme for TRIAC Dimming High PF Single-stage LED Driver with Adaptive Bleeder Circuit and Non-linear Current Reference

Weizhong Ma, Xiaogao Xie, Yang Han, Hao Deng
College of Automation
Hangzhou Dianzi University,
Hangzhou, China, 310018
Xiexg@hdu.edu.cn

Abstract- **Control scheme for TRIAC dimming high PF single-stage LED driver with adaptive bleeder circuit and non-linear current reference is studied in this paper. The target of the control scheme is to achieve high efficiency and high compatibility to different kinds of TRIAC dimmers. The control scheme includes adaptive bleeder circuit control, operation modes control, non-linear current reference and output current estimation. The adaptive bleeder circuit regulates the bleeder current according to the phase angle to reduce the loss. The driver operates in boundary conduction model under large phase angle for achieving high efficiency and operates in discontinuous conduction mode with adaptive off-time control for frequency limitation. In order to achieve wide dimming angle, a non-linear current reference circuit is proposed. The output current estimation circuit is applied to estimate the output current so that the switch can be driven directly. Three topologies of TRIAC dimmable driver including floating buck, floating buck-boost and flyback with proposed control schemes are presented in this paper. Detailed theoretical analysis and optimal design considerations are presented. Finally, a 63V/135mA LED driver prototype based on floating buck-boost topology with the proposed control scheme was built up. High dimmer compatibility has been achieved.**

Keywords - **TRIAC dimming, LED driver; adaptive bleeder circuit; non-linear current reference**

I. INTRODUCTION

Compared with the conventional incandescent lights and fluorescent lamps, light-emitting diode (LED) technologies have been widely used in general lighting applications because of the unique advantages [1]. The most convenient way to use LED lighting is to replace the existing lighting fixtures directly, which helps to reduce cost and avoid waste. This alternative trend has developed from local application field especially in our daily in door illumination applications. Many countries have developed the new century lighting plan to save energy, such as the energy-saving industry planning of semiconductor lighting in China and the 21st century light plan in Japan.

While, the LED driver performance has important influence on the quality of light, which includes power factor correct (PFC), driver efficiency, current regulation, and dimming control [2]-[4]. Besides, the constant current output is

necessary for the LED driver, which guarantees the lighting intensity steady. Although there are a lot of dimming methods, the dominant one is still based on TRIAC dimmers [2], [3], [5]-[10].

As we known, the TRIAC dimmer has minimum latching current and holding current requirements [3].When the input voltage is chopped by TRIAC dimmer, the input current oscillates. A severe oscillation will causes the input current to be lower than the holding current of TRIAC. This issue is serious especially when the phase angle is small. To resolve this problem, a simple passive bleeder which consists of a resistor and a capacitor in parallel with the input capacitor has been applied [8], [9]. However, large loss is introduced by this passive bleeder. In order to have more energy saving, an adaptive bleeder circuit is proposed in this paper. The proposed adaptive bleeder circuit regulates the bleeder current according to the phase angle. The bleeder current is large under small phase angle, while it is small under large phase angle. When the phase angle is larger than a certain value, the bleeder circuit is turn off to reduce the loss.

In order to get wide dimming range of TRIAC dimmer LED driver, which consists of two power conversion stages was presented in [10]. While, the single-stage converters have been more and more adopted in general lighting applications, because of the low cost, simple structure and small size of the LED driver [2],[3],[7]. In papers [12]-[13], the variable off time current-mode floating buck converters are proposed as LED driver. The primary side regulated (PSR) flyback converters [2], [3], [14], [15] have been widely used in recent years. The floating buck, floating buck-boost and PSR flyback are preferred topologies for low power LED driver due to their simple structure.

The output current estimation method has been well applied in PSR flyback [2], [3], [14], [15]. For floating buck and floating buck-boost, the output current estimation methods are also required.

The LED driver operates in boundary conduction model to achieve zero voltage switching (ZVS) can reduce the switching losses. However, when the LED driver input voltage is chopped by TRIAC dimmer, the switching frequency will increases. Especially when the phase angle is small, the

978-1-4673-9551-9/16 $31.00 © 2016 IEEE

increased switching frequency will cause large switching losses. In this paper, an adaptive off-time control is applied to limit the frequency.

The phase angle of the input voltage is sent to the controller. Usually, the average value of the phase angle is served as the reference of output current. In this way, the output current is regulated linearly by the phase angle of the input voltage. However, the minimum conduction angle of TRIAC dimmer is varied with the input voltage. The larger the input voltage is, the larger the minimum conduction angle of TRIAC dimmer is. Therefore, the minimum output current increases with the input voltage with linear output current reference, which means the dimming range is reduced. In this paper, the non-linear current reference for output current regulation is proposed and applied to the TRAIC dimming LED driver to increases the dimming range.

Based on the control scheme above, three TRIAC dimmable LED drivers are shown in Fig.1. The detailed description of the proposed control circuit will be illustrated in Section II. The experimental results are shown in Section III. Finally, conclusion will be drawn in Section IV.

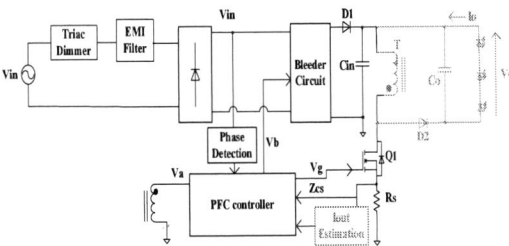

(a) The floating buck-boost driver topology

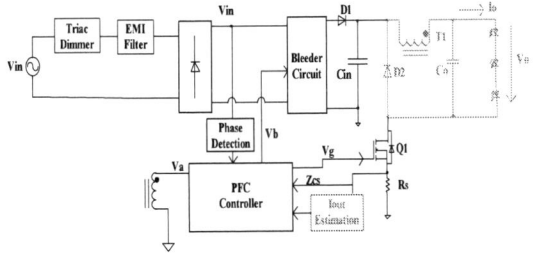

(b) The floating buck driver topology

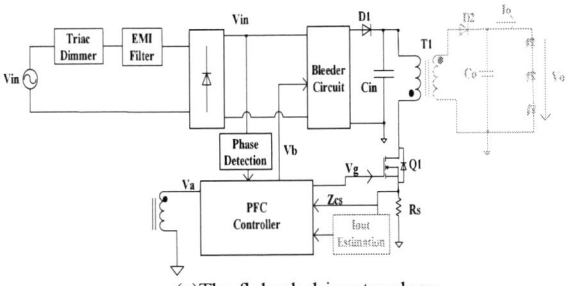

(c)The flyback driver topology

Fig.1 Three TRIAC dimming single-stage LED drivers with the proposed controller

II. PRINCIPLE OF OPERATION

In this section, the single-stage high PF floating buck-boost LED driver with TRIAC dimmer is selected as an example for analysis. Fig.2 shows the detailed PF floating buck-boost LED driver with proposed control scheme. An adaptive bleeder circuit and the non-linear current reference are adopted to improve efficiency and dimming range respectively. Also, an adaptive off-time control method is used to limit the high frequency in light dimming and the output current estimation can help reduce the volume and cost. Besides, the LED driver with the proposed control scheme can achieve high efficiency, wide dimming range and high compatibility to TRIAC dimmers.

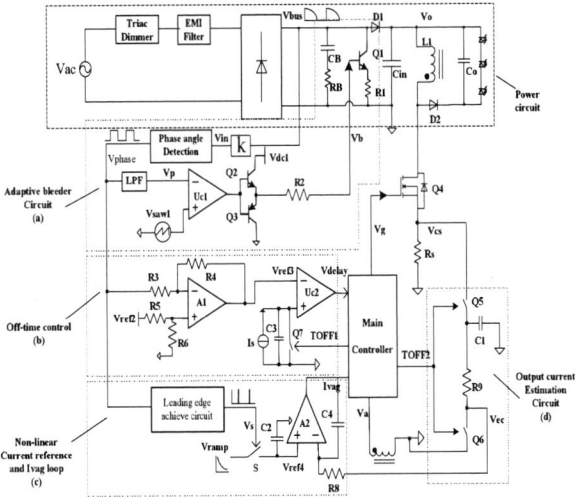

Fig.2 Detailed control schematic of a TRIAC dimmable single-stage buck-boost LED driver

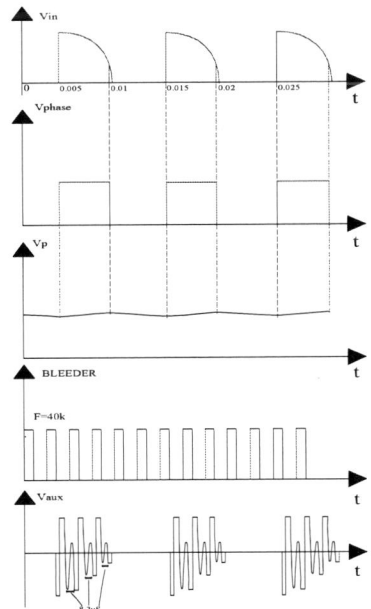

Fig.3 Key waveforms of the proposed controller

2.1 Control of adaptive bleeder circuit

The detailed adaptive bleeder circuit can be seen in Fig.2, which helps to save more energy and eliminate the visible flicker. Fig.3 shows the key waveforms of this block. V_{in} is the input voltage signal of the phase angle detection circuit, which can be derived from the rectified input voltage V_{bus} by a simple resistor divider. The input voltage signal V_{in} compares with a small threshold to generate the pulse signal C_out. V_p represents the average output of the phase angle detection circuit which proportional to the TRIAC dimming angle θ and can be achieved by a low pass filter.

Then, the V_p compares with a sawtooth signal which has a preset frequency and peak-to-peak value to get the bleeder driver signal V_b. The driving signal for Q_1 is a high fixed frequency pulse and its duty cycle is adaptively changed with phase angle. As shown in Fig.4, the sawtooth generator has a fixed frequency of 40 kHz, 0.75 V valley value and 1.65 V peak value. Under small dimming angle, the switching Q_1 has large duty cycle to provide enough holding current for the TRIAC dimmer and avoid flicker. With the phase angle increases, the duty cycle of V_b will decrease and the bleeding power also decreases. When the phase angle is large enough, there will no current flow through the bleeding resistor R_1. Thus, the adaptive bleeder circuit will adjust the holding current according to the phase angle and improve the efficiency under large phase angle.

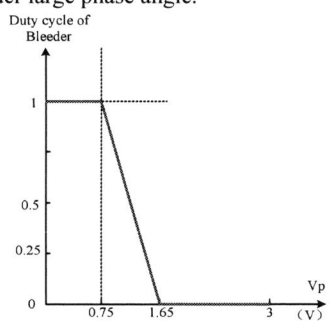

Fig.4 Duty cycle of bleeder transistor

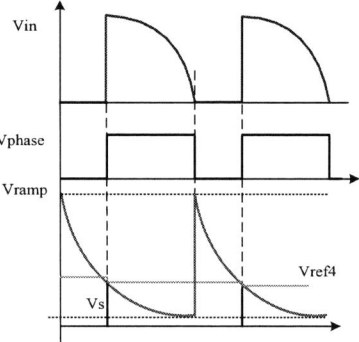

Fig.5 Non-linear current reference

2.2 Non-linear current reference

Exponential non-linear current reference circuit is shown in Fig.2. And the non-linear reference will be achieved in every half line cycle with different phase angle. As shown in Fig.5, the reference decreases rapidly when phase angle is large while it decreases slowly when phase angle is small. Therefore, the output current varies a little with phase angle among small phase angle region. In this way, the dimming range is increased.

2.3 Adaptive off-time control

As shown in Fig.2, the phase angle detection circuit output V_{phase} is filtered with a low-pass filter (LPF) and sent to the inv-input of the differential amplifier. The non-inv-input is achieved by the resistor divider of R5 and R6 with the constant voltage reference Vref2. Then, the output Vref3 of the amplifier A1 which as the off-time control reference to regulate the off-time can be calculated as follows:

$$Vref3 = \frac{R3+R4}{R5+R6} \cdot \frac{R6}{R3} \cdot Vref2 - \frac{R4}{R3} \cdot Vp \qquad (1)$$

The constant current source Is and capacitor C3 form a sawtooth wave generator with fixed slope. And the switch Q7 is used to control the peak value of the sawtooth wave, which decides the value of off-time.

As shown in Fig.6, when Vp is always larger than 1.95V, and the Vref3 is little larger than zero with pre-designed resistances in (1), the LED driver will only operates in boundary conduction model (BCM) to achieve zero voltage switching (ZVS). As the phase angle decrease so that Vp is less than 1.95V, also a positive Vref3 achieved which causes the driver work in DCM. As shown in Fig.6, the off-time has a range of 28us in the DCM and the constant off-time control when the Vp is little than 0.75V. In this way, the frequency of the driver is limited with little switching loss.

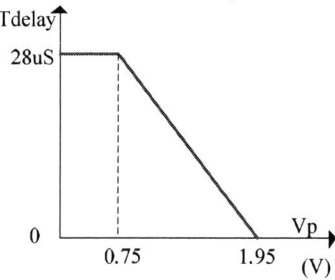

Fig.6 Off-time control

2.4 Output current estimation

In this paper, the output current estimation circuit is shown in Fig.2 and the regulation circuits are necessary which have been well designed in [2], [14]. Then, the secondary diode conduction period is detected from auxiliary winding of the transformer, which generates the control signal TOFF2. According to the voltage of the auxiliary winding, the achieved estimated waveform is a pulse signal with the altitude equal to output current reference Vref4 and the conduction period equal to the secondary diode conduction period. The resistor Rs is used to sample the primary switch current, which is an envelope of the half-sine input waveform.

It combines with the control signal TOFF2 to estimate the output current, and the detailed analysis can be seen in [2]. Obviously, this method is suitable for floating buck-boost converter and flyback converter operates in DCM or BCM.

While, the output current estimation circuit has a little different in the floating buck converter and the detailed schematic is shown in Fig.7. Compared with the circuit shown in Fig.2, there is an additional resistor R9 and the switches Q5, Q6 with reverse control signals. When the switch Q5 is on, the switch Q6 will be off, the signal Vcs which represents the inductor charging state is inserted to the capacitor C1. On the contrary switch control signals, the capacitor C1 and resistors R7, R9 are used to simulate the inductor discharge process. The achieved sample signal can be used to estimate output current through a low pass filter. Then, the other parts of the control schematic stay the same and the proposed control schematic has a high compatibility in LED driver with TRIAC dimmer.

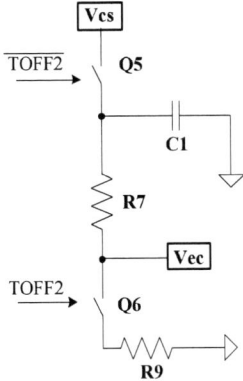

Fig.7 Output current detection circuit in floating buck LED driver

III. EXPERIMENTAL RESULTS

In this paper, a 63V/135mA single-stage LED driver prototype with buck-boost topology is built up to verify the proposed control scheme. Also, about 50 different types of TRIAC dimmers are used to test the compatibility of the driver and the test results shows that the compatibility is over 90%.The key parameters of the prototype are shown in Table I.

Measured input voltage v_{ac}, current i_{ac} and the rectifier output voltage v_{bus} waveforms with a conventional front cut dimmer under different dimming angles are shown in Fig.8 and Fig.9. It is clear that the input current follows the input voltage and the LED driver works very stable with no mis-trigger.

Measured v_{bus} and the bleeder driver signal v_b waveforms under different dimming angles are shown in Fig.10. At the minimum dimming angle, the bleeder driver has a duty cycle about 0.9 which provides enough holding current to avoid the flicker in light dimming. The duty cycle decrease to 0.2 at the dimming angle 105°, and it will be zero when the dimming

TABLE I
KEY PARAMETERS OF THE EXPERIMENTAL PROTOTYPE

Parameter	Designator	Value
AC Input Voltage	Vin	175-265V (RMS)
Output Voltage	Vo	63V
Output Current	Io	135mA
Primary Side Switch	Q4	SD7N60
Transformer Core		EE13
Transformer Turns ratio	n	85:22
Primary Side Inductor	Lp	1.25mH
Output Capacitor	Co	100uF
Current Sense Resistor	Rs	0.5 Ω
Input capacitor	CB	0.15uF/630V
Resistor	RB	510R /1W
Bleeder Resistor	R1	3.6K /2W
Bleeder Switch	Q1	13003F6

angle further increases. Then, the bleeder driver circuit regulates the holding current adaptively, which helps to improve the efficiency under large phase angle.

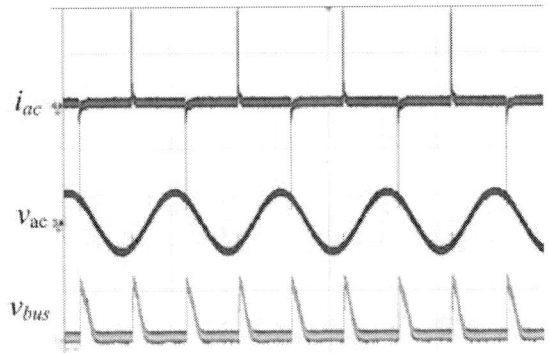

Fig.8 Input waveforms @90Vac & maximum phase angle
(i_{ac}: 500mA/div, v_{ac}:100V/div,v_{bus}:100V/div)

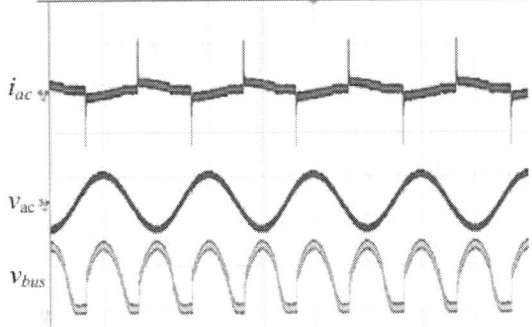

Fig.9 Input waveforms @90Vac & maximum phase angle
(i_{ac}: 500mA/div, v_{ac}:100V/div,v_{bus}:100V/div)

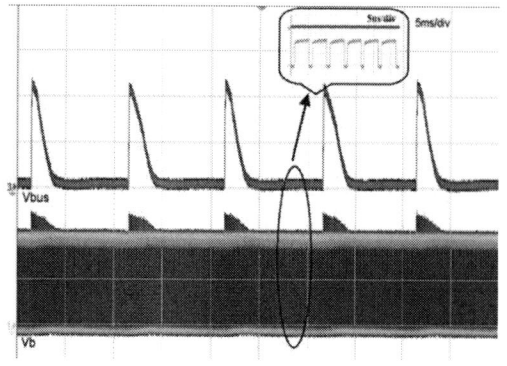

(a) v_b @ 45° phase angle

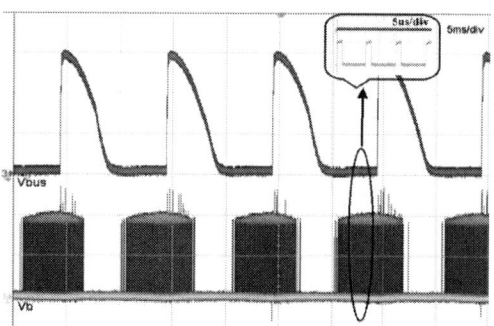

(b) v_b @ 105° phase angle

Fig.10 Bleeder driving signal v_b at different TRIAC dimming angles

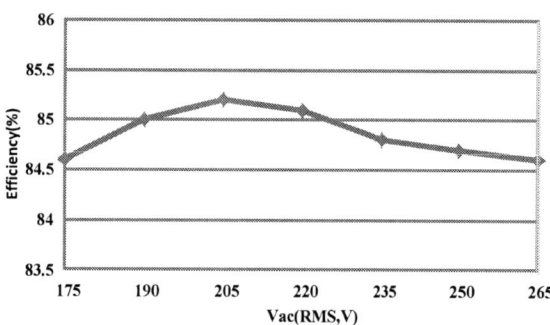

Fig.11 Efficiency versus v_{ac} without TRIAC dimmer

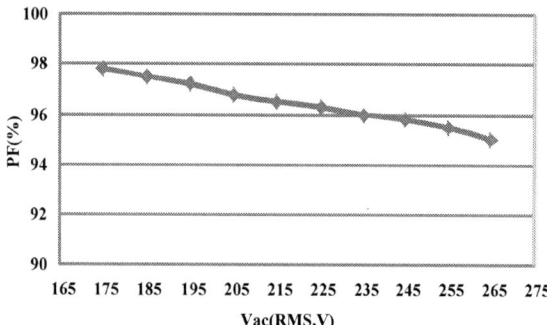

Fig.12 PF versus v_{ac} without TRIAC dimmer

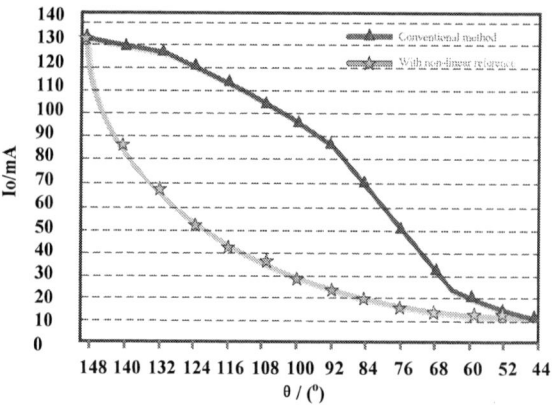

Fig.13 Io versus different phase angle

Measured efficiency curve of the proposed driver under different input voltage without dimmer is shown in Fig.11. And the power factor (PF) versus the input voltage under the no dimming condition is shown in Fig.12. The high PF is achieved within the entire input range. Fig.13 shows the measured output current under different dimming angles. Then, the approximate exponential output current curve has been achieved, which contributes to a comfortable dimming progress and the dimming range is also extended.

IV. CONCLUSION

Control schemes for TRIAC dimming high PF single-stage LED driver with adaptive bleeder circuit and non-linear current reference has been presented in this paper. The proposed control schematic mainly includes four parts, which have been well introduced in section II. Based on the proposed scheme, three common single-stage converters can be used for LED driver with TRIAC dimmer. Some good merits such as high efficiency, wide dimming range and high compatibility to TRIAC dimmers can be achieved. A 63V/135mA laboratory prototype based on floating buck-boost topology and proposed controller was built up. Experimental results verify the theoretical analysis.

Acknowledgment

This work was supported by the National Nature Science Foundation of China under Grant 51377038 and 51407053.

REFERENCES

[1] J. Peck, G. Ashburner and M. Schratz, "Solid state led lighting technology for hazardous environments; lowering total cost of ownership while improving safety, quality of light and reliability,": IEEE, 2011, pp. 1 - 8.

[2] J. Zhang, H. Zeng and T. Jiang, "A Primary-Side Control Scheme for High-Power-Factor LED Driver With TRIAC Dimming Capability," *Power Electronics, IEEE Transactions on,* vol. 27, pp. 4619-4629, 2012.

[3] J. Zhang, T. Jiang, L. Xu, and X. Wu, "Primary Side Constant Power Control Scheme for LED Drivers Compatible with TRIAC Dimmers," *Journal of Power Electronics,* vol. 13, pp. 609-618, 2013.

[4] L. Yan, B. Chen and J. Zheng, "A new TRIAC dimmable LED driver control method achieves high-PF and quality-of-light," in *Applied Power Electronics Conference and Exposition (APEC), 2012 Twenty-Seventh Annual IEEE* Orlando, FL: IEEE, 2012, pp. 969-974.

[5] D. Rand, D. Rand, B. Lehman, B. Lehman, A. Shteynberg, and A. Shteynberg, "Issues, Models and Solutions for Triac Modulated Phase Dimming of LED Lamps," 2007 IEEE Power Electronics Specialists Conference, pp. 1398 - 1404, 2007.

[6] S. Moon, G. Koo and G. Moon, "Dimming-Feedback Control Method for TRIAC Dimmable LED Drivers," *Industrial Electronics, IEEE Transactions on,* vol. 62, pp. 960-965, 2014.

[7] J. T. Hwang, M. S. Jung, D. H. Kim, J. H. Lee, M. H. Jung, and J. H. Shin, "Off-the-Line Primary Side Regulation LED Lamp Driver With Single-Stage PFC and TRIAC Dimming Using LED Forward Voltage and Duty Variation Tracking Control," *Solid-State Circuits, IEEE Journal of,* vol. 47, pp. 3081-3094, 2012.

[8] H. Eom, Y. Kim and Y. Shin, "A TRIAC dimmable driver design for high dimmer compatibility in low power LED lighting," in *Power Electronics and Applications (EPE), 2013 15th European Conference on* Lille: IEEE, 2013, pp. 1-5.

[9] H. Eom, C. Lee, T. Yang, and S. Yang, "Design optimization of TRIAC-dimmable AC-DC converter in LED lighting," in *Applied Power Electronics Conference and Exposition (APEC), 2012 Twenty-Seventh Annual IEEE* Orlando, FL: IEEE, 2012, pp. 831-835.

[10] R. Zhang and H. S. H. Chung, "A TRIAC-Dimmable LED Lamp Driver With Wide Dimming Range," *Power Electronics, IEEE Transactions on,* vol. 29, pp. 1434-1446, 2013.

[11] Y. Wang, Y. Guan, J. Huang, W. Wang, and D. Xu, "A Single-Stage LED Driver Based on Interleaved Buck – Boost Circuit and LLC Resonant Converter," *Emerging and Selected Topics in Power Electronics, IEEE Journal of,* vol. 3, pp. 732-741, 2015.

[12] V. Anghel, C. Bartholomeusz, A. G. Vasilica, G. Pristavu, and G. Brezeanu, "Variable Off-Time Control Loop for Current-Mode Floating Buck Converters in LED Driving Applications," *Solid-State Circuits, IEEE Journal of,* vol. 49, pp. 1571-1579, 2014.

[13] G. Pristavu, A. Vasilica, V. Anghel, and G. Brezeanu, "Less is more — The improved variable OFF time current-mode floating buck controller," in *Semiconductor Conference (CAS), 2013 International* Sinaia: IEEE, 2013, pp. 215-218.

[14] X. Xie, J. Wang, C. Zhao, Q. Lu, and S. Liu, "A Novel Output Current Estimation and Regulation Circuit for Primary Side Controlled High Power Factor Single-Stage Flyback LED Driver," *Power Electronics, IEEE Transactions on,* vol. 27, pp. 4602-4612, 2012.

[15] H. Chou, Y. Hwang and J. Chen, "An Adaptive Output Current Estimation Circuit for a Primary-Side Controlled LED Driver," *Power Electronics, IEEE Transactions on,* vol. 28, pp. 4811-4819, 2012.

978-1-4673-9551-9/16 $31.00 © 2016 IEEE

Three phase converter with galvanic isolation based on loss-free resistors for HB-LED lighting applications

Ignacio Castro, Diego G. Lamar, Manuel Arias, Javier Sebastián and Marta M. Hernando

Departamento de Ingeniería Eléctrica, Electrónica, de Computadores y Sistemas
University of Oviedo
Gijón 33204, Spain
e-mail: castroignacio@uniovi.es

Abstract—This work presents a driver for High-Brightness Light-Emitting Diodes (HB-LED) in three-phase grids, which complies with IEC 1000-3-2 Class C requirements, achieves high Power Factor (PF), low Total Harmonic Distortion (THD), as well as, the capability to achieve full dimming while disposing of the bulk capacitor and having galvanic isolation. The HB-LED driver is based on the use of six four-port cells with their inputs connected to the three-phase network and their outputs connected in parallel. Each one of these cells is a DC/DC converter operating as a Loss-Free Resistor (LFR) based on the concept of a flyback operating in Discontinuous Conduction Mode (DCM). Moreover, it operates in the full range of the European three-phase line voltage, which varies between 380V and 420V, and it supplies an output voltage of 48V with maximum power of 90W.

Keywords—*Three-phase, Power Factor Correction, HB-LED driver, Loss Free Resistor, Capacitorless.*

I. INTRODUCTION

High-Brightness Light-Emitting Diodes (HB-LED) are becoming increasingly ubiquitous across all aspects of illumination products, by offering a lot of advantages over traditional lighting solutions. Furthermore, several commercial and industrial installations around the globe receive primary three-phase power with a wide variety of voltages depending on the country, e.g. line-neutral is 347V in Canada, 480V in the US or 230V in the European Union with the exception of the UK (240V). Hence, one question arises, why is not a specific solution for LEDs used in three-phase grids, in case the three phase grid was available.

HB-LED drivers are normally designed for single-phase universal input voltage supplies (100 to 277V). Therefore, the use of these HB-LED drivers in installations with exclusive access to three-phase normally requires a step-down autotransformer, as well as, access to neutral due to the high voltages in some locations [1]. The use of these step-down autotransformers reduces the efficiency of the whole system greatly due to their electrical efficiency not being higher than ~95% in a best case scenario. Another aspect that needs to be taken into account is the size increase of the power supply [2]. Hence, the necessity of a compact solution especially designed for this specific application.

In prior literature, there are several works dedicated to the study of AC/DC three-phase power supplies, synthesized in [3]. Most of the converters based on a single switch have high power factor (PF) by penalizing the Total Harmonic Distortion (THD)

This work has been supported by the *Spanish Government* under Project MINECO-13-DPI2013-47176-C2-2-R and the *Principality of Asturias* under the grants "*Severo Ochoa*" BP14-140 and BP14-85 and by the Project FC-15-GRUPIN14-143 and by European Regional Development Fund (ERDF) grants.

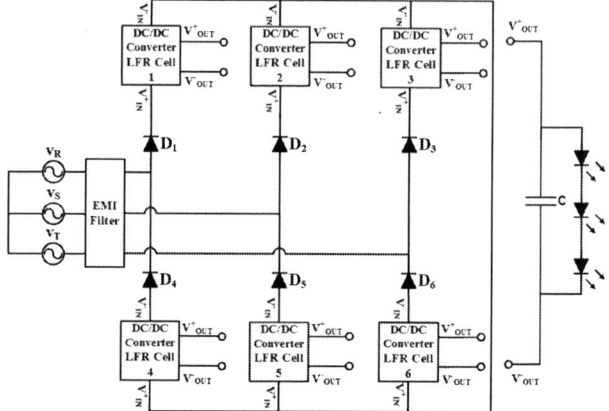

Fig 1. Diagram of the multi-cell three-phase HB-LED driver.

or having the need of high output voltage. In order to have high PF and not to penalize THD, it is possible to use a three-phase driver based on multi-cell loss free resistor (LFR). These drivers are more complex from a control point of view since they add more components and are arguable more expensive, but they have a better trade-off between output voltage and THD. There are only a handful of these converters in literature, based on DC/DC converters such as, DCM flybacks [4] [5], Cùk [6], SEPIC [7], used as LFR cells. This work proposes the use of this type of multi-cell converter as an HB-LED driver. Furthermore, the use of any three-phase converter with unity PF means that neither the input nor the output power is pulsating. Non pulsating power allows not only to remove the electrolytic capacitor in commercial and industrial installations, but also to increase the light quality in these environments, as will be shown in Section IV.

In order to design this driver the IEC 1000-3-2 [8-10] regulation is going to be taken into account. It should be classified as Class A equipment taking into account that it is three-phase equipment, but it should comply with Class C taking into account that it is also lighting equipment. Therefore, the aim of this work will be for the HB-LED driver to comply with the more restrictive of the two, which is Class C.

It should be noted, that some work has been done previously in the field of three-phase dimmable lighting, in this case for fluorescent lamps [11]. Although, it has never been done for HB-LED lighting. In this paper, a compact HB-LED dimmable driver is proposed, based on the idea of [5], by using LFR cells

978-1-4673-9551-9/16 $31.00 © 2016 IEEE

as it is illustrated in Fig. 1, which will be extensively described in Section II. Section III will be dedicated to the control of said converter, and finally Section IV will synthesize the most relevant experimental results. Finally, conclusions and future work will be discussed.

In summary, the use of three-phase AC/DC converters instead of single-phase ones makes possible to remove the most critical device from the point of view of the converter lifespan: the electrolytic capacitor. The price to pay is to connect the LED driver to a 3-wire line instead of a 2-wire one, if available.

II. PROPOSED LED DRIVER

The concept of the HB-LED driver presented in this work is depicted in Fig. 1. It is based on a family of three-phase power supplies which are built with LFR, in order to achieve high quality rectification [4]. The one presented in this work is based on the idea of [5].

Each one of the cells depicted in Fig. 1, is a DC/DC converter working as a LFR. The proposed cells in this case are flyback converters working in Discontinuous Conduction Mode (DCM), as shown in Fig. 2.

As has been stated in [12], a flyback working in DCM supplies a fixed amount of current to the load, which in this case are HB-LED. The LFR value of a DCM flyback converter, which is the basic cell, can be defined by:

$$R_{cell} = \frac{2L}{d^2 T},\tag{1}$$

where d is the duty cycle of the converter, L is the magnetizing inductance of the transformer and T is the switching period. By forcing the flybacks to work in DCM with a fixed duty cycle, it can be assured that each one of the flybacks behaves as a resistor at their input. Hence, each phase will demand a sinusoidal current granting both high PF and low THD.

Henceforth, to simplify the analysis of the HB-LED driver, the LFR cells are going to be considered as ideal resistors. Their value is going to be considered equal not taking into account tolerances that could come from the components of the flyback.

In order for the HB-LED driver to demand a sinusoidal current, each one of the diodes is going to be conducting during half line cycle. Hence, three different diodes are going to be conducting every $\pi/3$ of ωt, depending on the phase voltages (v_R, v_S, v_T), as summarized in Fig. 3. For instance, D_2 conducts during the positive half line cycle of line S, from t_1 to t_4, as stated in Fig. 3. At the time, there are two other diodes conducting.

However, they do not share the same conduction time as D_2 (i.e. [t_1, t_4]) meaning there is a different set of three diodes conducting every $\pi/3$ of ωt.

Therefore, the driver can be divided in three stages for each one of the diodes. These three stages are depicted as an example for diode D_2 in Fig. 4a, b and c, where D_2 is the diode that is conducting from t_1 to t_4 and D_1, D_3, D_4 and D_6 are the diodes that are swapping in between depending on the voltages of phases R and T.

The analysis done for D_2 is equivalent for the rest of the diodes. Hence, if the whole line period where to be considered, the HB-LED driver is equivalent to a star connection (Y), as shown in Fig. 4d, meaning that the input current of each phase (i_N) of the converter is going to be defined by:

$$i_N = \frac{v_N(t)}{R_{cell}} = \frac{v_p}{R_{cell}} \cos(\omega t - \varphi_N),\tag{2}$$

where v_N is the phase-neutral voltage of one of the phases (N defines whether is phase R, S or T), v_p is the peak amplitude of the phase-neutral voltage and φ_N is the phase of the signal.

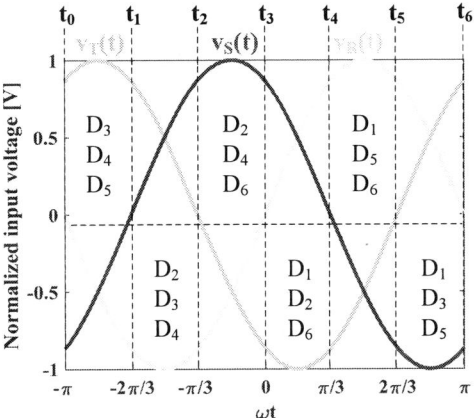

Fig 3. Theoretical conduction of the diodes depending on the phase voltages.

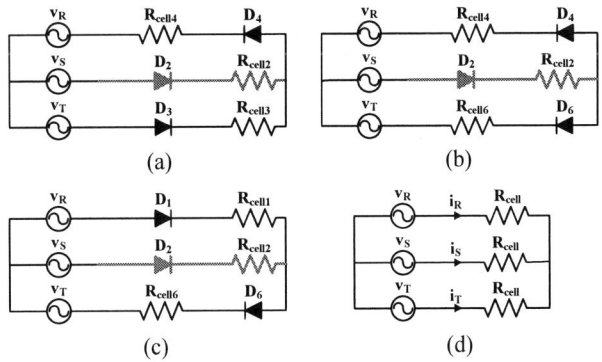

Fig 4. (a) Conduction during [t_1, t_2] (b) Conduction during [t_2, t_3] (c) Conduction during [t_3, t_4] (d) Simplified working behaviour of the driver.

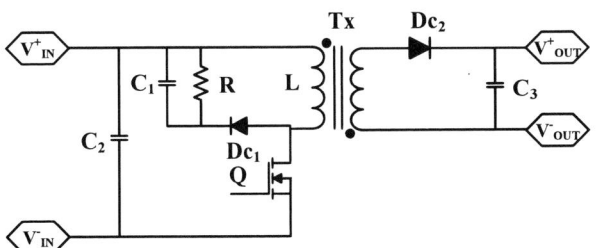

Fig 2. Schematic of the LFR flyback cell.

978-1-4673-9551-9/16 $31.00 © 2016 IEEE   823

Assuring unity factor power correction allows not to have pulsating power at the HB-LED driver input. Therefore, the input power of the converter will be defined by the sum of each phase, as shown in (3), since only three of the six cells are going to be working at the same time.

$$p(t) = \sum_{i=1}^{3} \frac{v_N^2(t)}{R_{cell}} = \frac{3v_p^2}{2R_{cell}} = \frac{3v_p^2 d^2 T}{4L} = P \qquad (3)$$

It should be noted, that the sinusoidal components that come out from the summation in (3) can be removed due to their sum being equal to zero. Hence, the input power of the converter can be defined by a DC component.

From Fig. 4, it can be observed that the line currents undergo the voltage drop corresponding to only three rectifier diodes, which is an important advantage over the topology proposed in [4], where the line currents undergo the voltage drop corresponding to six rectifier diodes.

Regarding the connection of the outputs, every single secondary side of the LFR cells is going to be connected in parallel to the same load. Accordingly, if the simplification from Fig. 4d is taken into account with the parallel output connection, Fig. 5 can be derived. In this figure the basic operation of the converter is explained with the three active cells that are feeding the HB-LED string, keeping in mind the cell behaviour as a power source.

The parallel output connection allows the complete removal of the bulk capacitor due to the non-pulsated power given to the load. Hence, a film capacitor can be used to reduce the switching frequency ripple, which is the theoretical ripple at the HB-LED driver output. Furthermore, being able to eliminate the bulk capacitor from the HB-LED driver, increases dramatically the lifespan of the driver. Particularly important for lighting environments with either difficult access or expensive solutions that need to guarantee a high lifespan of the driver.

Therefore, if non-pulsated power is given to a resistive load that models the HB-LED. A relation can be made between input power (3) and output power, and equation (4) is derived:

$$P = \frac{3v_p^2}{2R_{cell}} = \frac{v_o^2}{R_L} \to M = \frac{v_o}{v_p} = \sqrt{\frac{3R_L}{2R_{cell}}}, \qquad (4)$$

where v_O is the output voltage and R_L is the load resistance. Since v_O is going to be constant, it can be assumed that no ripple should appear at the output current under ideal conditions.

From (4) the duty cycle required to drive the HB-LED driver can be obtained, considering the same PWM signal is going to be driving all the flybacks, as has been stated in (5):

$$d = \frac{2v_o}{v_p} \sqrt{\frac{L}{3R_L T}} \qquad (5)$$

Full dimming on the HB-LED is achieved by reducing the duty cycle, which increases the emulated resistance diminishing the output current while keeping theoretical sinusoidal input current.

From the design point of view, both the theoretical duty cycle (5) and the maximum output power (3) need to be calculated for the required specifications. Afterwards, the LFR flyback cell needs to be designed, as has been explained previously in [12], considering an input voltage in the cells equal to the one of the phase-neutral. Furthermore, the designer needs to keep in mind that each flyback is going to handle one sixth of the input power.

III. CONTROL STRATEGIES

Closed loop operation is mandatory in most applications where a certain voltage or current level needs to be guaranteed at the output. HB-LED drivers are no exception, meaning a certain voltage/current level needs to be assured in order to guarantee not only, good light quality but to avoid harmful effects for human beings in an industrial environment.

From the schematic depicted in Fig. 5, the control of the HB-LED driver can synthesized as a problem of three power supplies connected in parallel to the same load. Many works in previous literature have address the parallelization of powers supplies and are synthesized in [13-15]. From these works a quick conclusion can be extracted: the most optimal way to control the power supplies (LFR cells), would be for each one of them to have their own control current loop. However, in our case of study that

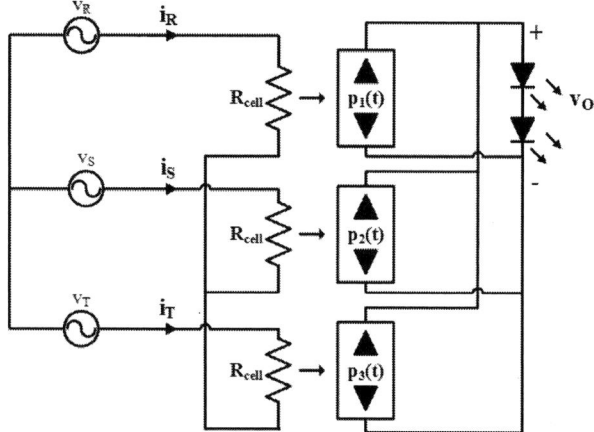

Fig 5. Three-phase HB-LED driver simplified with LFR.

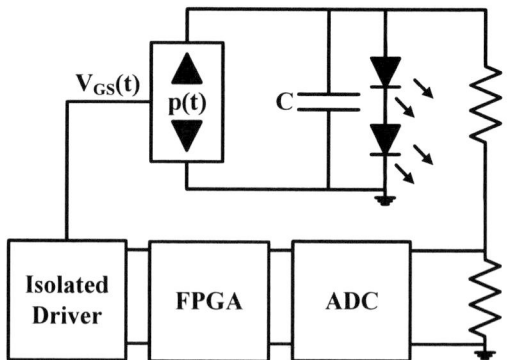

Fig 6. HB-LED driver voltage loop diagram.

means the use of a current sensor for each cell, which leads to six current sensors. This solution would increase not only the price but the complexity of the control. Hence, a voltage loop like the one depicted in Fig. 6 is going to be used. Although a digital control has been implemented, it would be possible to implement an analog control.

It is important to note that the tolerances of the components are going to have an effect over the LFR value (R_{cell}). Especially the tolerance of L which is the most critical component in this sense. This variation of the R_{cell} from one cell to another will have an effect on the output voltage of the converter, as the power processes by each cell will differ causing input power slightly to pulsate. Therefore, an independent current control for each cell would be optimal to reduce the tolerance effects or any unbalance that could come from the three-phase grid. However, since the change between R_{cell} are really small, it does not justify the use of a more complex control.

The voltage loop proposed in Fig. 6, is going to be based on measuring the output voltage by means of a voltage divider. That voltage is going to be digitally converted and processed by an FPGA, in order to generate the digital pulse-width modulation (DPWM) that goes to the isolated driver of each cell. The signal that goes into each isolated driver is $V_{GS}(t)$, and is the same signal for the switch of each cell.

In order to determine the compensator to be used in the HB-LED driver, the plant of the HB-LED driver needs to be calculated using a similar analysis to the one done in [4] by using Ridley's average small signal analysis [16]. Consequently, the starting point for this analysis would be the input power handled by the HB-LED driver and defined by (3) and the equation that can be obtained from the circuit in Fig. 6 defined by (6) considering the HB-LED string as a resistor:

$$p(t) = Cv_o(t)\frac{dv_o(t)}{dt} + \frac{v_o^2(t)}{R_L} \quad (6)$$

By equating both (3) and (6), (7) is derived. It should be noted, that both v_p and d are dependent on time. The first one due to the variations that can occur in a three-phase grid and the second one due to the variation of the duty cycle in order to regulate the output voltage.

$$\frac{3v_p^2(t)d^2(t)T}{4L} - Cv_o(t)\frac{dv_o(t)}{dt} - \frac{v_o^2(t)}{R_L} = 0 \quad (7)$$

After perturbing equation (7), and eliminating the second order and the DC terms, equation (8) is reached. It should be noted that lower case letters have been used for the static analysis and capital letter are going to be used for constant values and to particularize the equation in a determined point.

$$\frac{3V_p^2DT}{2L}\hat{d} + \frac{3V_pD^2T}{2L}\hat{v}_p - CV_o\frac{d\hat{v}_o}{dt} - \frac{2V_o}{R_L}\hat{v}_o = 0 \quad (8)$$

Then, the Laplace transform of (8) needs to be performed in order to yield (9) and (10).

$$G_{v_od}(s) = \left.\frac{\hat{v}_o(s)}{\hat{d}(s)}\right|_{\hat{v}_p=0} = \sqrt{\frac{3R_LT}{4L}}\frac{V_P}{\left(\frac{CR_Ls}{2}+1\right)} \quad (9)$$

$$G_{v_ov_p}(s) = \left.\frac{\hat{v}_o(s)}{\hat{v}_p(s)}\right|_{\hat{d}=0} = \frac{V_o}{V_P\left(\frac{CR_Ls}{2}+1\right)} \quad (10)$$

It should be noted that equations (9) and (10) are valid, if and only if, the compensator has a crossover frequency of less than 300 Hz that guarantees that there is no frequency component having an effect over the control action. These effects appear due to the non-idealities of both the LFR cells and the input voltage not being optimal, as it will be shown in Section IV. It is extremely important, because when closing the loop the driver can vary the value of the R_{cell}, meaning that this will have an impact on the THD. So if rapid changes are allowed the unity factor correction might be compromised meaning that the output current might not be constant.

IV.   EXPERIMENTAL RESULTS

A. Description and basic results.

The HB-LED driver introduced in the previous sections has been designed for a maximum power of 90W. This driver receives a three-phase input of 400V line-to-line and feeds five strings of 12 HB-LED (W42180T2-SW) with their respective equalizing resistor that are equivalent to 1.8A/48V at full load. The switching frequency of each flyback is 100 kHz. Moreover, the LFR cell is based on a flyback converter working in DCM, as the one shown in Fig. 2 and whose components are summarized in Table I. Each one of these flybacks handles one sixth of the total power as it was previously stated, meaning that each one handles 15W. The rest of the components that form the HB-LED driver are summarized in Table II. Finally, Fig. 7 shows a picture of the prototype that was built to validate the concept of this work. It should be noted that the flyback transformers are on the other side of the board.

TABLE I.   COMPONENTS OF THE EXPERIMENTAL LFR CELL PROTOTYPE

Fig. 2 reference	Value
Dc_1	STTH208U
Dc_2	FES8BT-E3/45
C_1	1µF 800V Ceramic Cap.
C_2	100nF 50V Film Cap.
C_3	1µF 50V Film Cap.
R	10.5kΩ
Q	IPP65R225C7
Tx	Coilcraft Z9007-BL

TABLE II.   REST OF COMPONENTS OF THE HB-LED DRIVER

Fig. 1 reference	Value
D_1-D_6	1N4007
C	10µF 100V Film Cap.
FPGA	XC7A100T-1CSG324C

All the tests have been done connected to the real three-phase power grid, which justifies the distortion of V_R in Fig. 8. In the same figure, a snapshot of the oscilloscope is shown where the

978-1-4673-9551-9/16 $31.00 © 2016 IEEE

input currents of the driver, as well as, one of the phase-neutral input voltages of the converter can be observed. As it can be seen, the current (i_R) follows the voltage (v_R) demonstrating that power factor correction is achieved. The phase shift between currents is 6.6 ms (i.e. $2\pi/3$ of ωt), so it can be assumed that power correction in the three phases is achieved. In Fig. 9, it can be observed that both the output current and the output voltage have low ripple with a 10µF film capacitor.

It should be noted that the 300Hz ripple that appears in both the output current and voltage, comes from the fact that the input voltage is not a pure sinusoidal waveform and that the LFR cell is not ideal. Hence, the input power is not constant as it was derived in (3), meaning that a variation should occur at the load as can be observed in Fig. 9.

In order to validate the HB-LED driver, several measurements are going to be taken into account. Firstly, to validate the dimming of the HB-LED driver, three operating

Fig 7. HB-LED driver prototype.

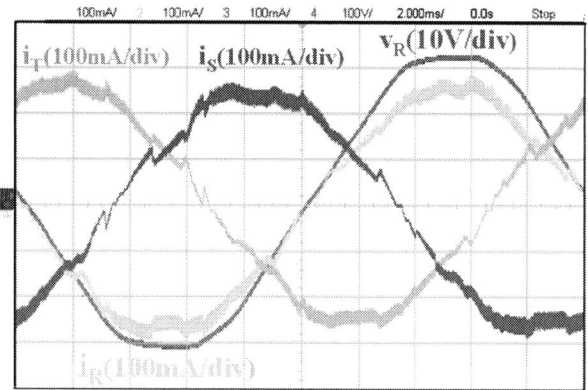

Fig 8. Input currents for all three phases and input voltage of phase R when fully loaded.

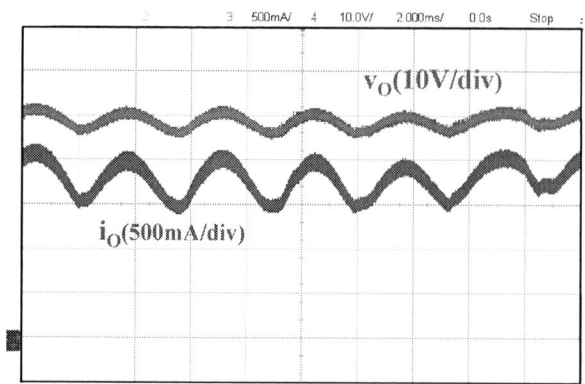

Fig 9. Output current and output voltage when fully loaded.

points are going to be measured by varying the voltage reference of the loop, therefore, varying the output current to 1.8A (fully loaded), 0.9 A and 0.45 A. Secondly, the HB-LED driver needs to work in the full range of the European three-phase voltage line meaning that also three operating points are going to be measured, 380, 400 and 420V.

To correctly analyze the waveforms in both scenarios, the waveforms are going to be extracted from the oscilloscope as data and processed with MATLAB®. The parameters that are going to be extracted for each one of the points are: efficiency, THD, PF, compliance with Class C IEC 1000-3-2 [9-11] for both, and flicker operating recommendation [18,19] for the dimming validation.

B. Dimming validation of the HB-LED driver.

The dimming validation is shown in Table III, which summarizes PF and THD measurements for each phase based on input voltages and input currents extracted from the oscilloscope for the three operating points stated before, which are 1.8A, 0.9A and 0.45A. It can be observed that THD increases and PF decreases, when lowering the output current by diminishing the duty cycle of the driver, due to the larger not conducting periods of the input current. As for the efficiency of the HB-LED driver, in Fig. 10 it can be seen that it decreases by lowering the output current, being roughly 73% at the full dimming point. The efficiency of the HB-LED driver is 88% at full load. It should be noted that the efficiency of the driver was not the aim of this work, since it was proving the concept of feeding HB-LED in three-phase power grids. The efficiency of the HB-LED driver is completely reliant on the efficiency of each cell (flyback). In fact, it will be the efficiency of the cell affected by the voltage drop corresponding to the rectifier diode associated.

TABLE III.   SUMMARY OF THD AND PF FOR THREE DIFFERENT OUTPUT CURRENTS.

Output Current/Phase		1.8A	0.9A	0.45A
R	PF[%]	99.81	99.72	99.56
	THD[%]	5.71	6.97	8.03
S	PF[%]	99.87	99.75	99.59
	THD[%]	4.62	6.96	7.71
T	PF[%]	99.85	99.73	99.58
	THD[%]	5.26	6.86	7.97

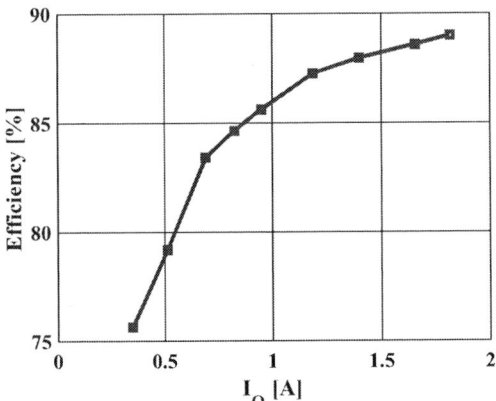

Fig 10. Efficiency of the LED driver versus output current.

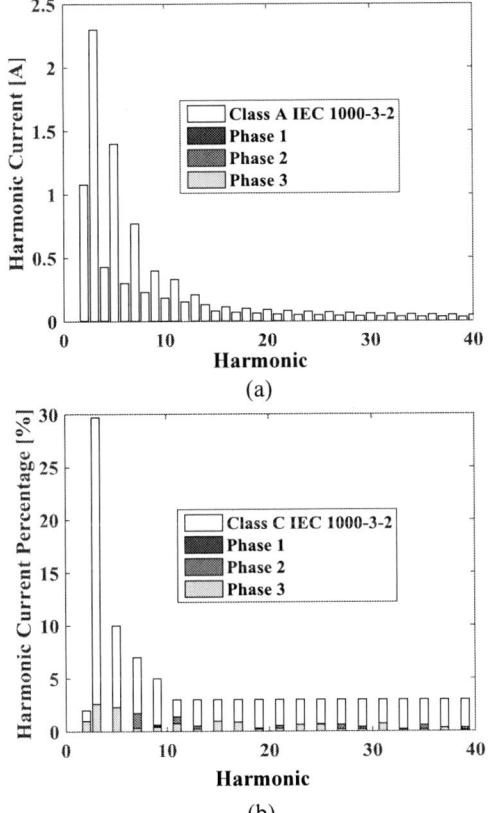

(a)

(b)

Fig 11. Harmonic content of each phase compared with both Class A and Class C limits. (a) Class A comparison. (b) Class C comparison.

The waveforms have also been used to extract the harmonics by using the Fourier series on them. Afterwards, these measurements are compared with Class A IEC 1000-3-2 harmonic limits, since it is the category where a balanced three phase converter, as the one shown in this work, falls in the

regulation. However, the measured harmonics are so low due to the low power processed that they do not appear in Fig. 11a for the 1.8A measurement and they do not appear either for the other two values under study due to their current being even lower. Hence, it seemed interesting to test the compliance of Class C IEC 1000-3-2 harmonic limits for each phase considering the presented power supply as a HB-LED driver. Fig. 11b shows the compliance with said regulation.

To limit the biological effects and detection of flicker in general illumination, the Modulation (%) should be kept within the shaded region defined in [18,19], where Modulation(%) calculation can be define as follows:

$$Modulation\ (\%) = 100 \cdot \frac{(L_{max} - L_{min})}{(L_{max} + L_{min})}, \qquad (11)$$

where L_{max} and L_{min} correspond to the maximum and minimum luminance of each harmonic of the ac component of the output current, respectively.

In this case a proportionality between luminance and the ac component of output current has been assumed. Results of this analysis are shown in Fig. 12 for 1.8A, 0.9A and 0.45A since there is no light at full dimming, considering 50Hz as the fundamental frequency for said analysis. As you can see, all ac harmonic content is within the shaded region, even the ones below 90Hz that are the most crucial. Therefore, good light quality and non-harmful effects can be assumed from the HB-LED driver presented, even in full dimming conditions.

C. Input voltage range validation.

The European three-phase voltage standard is 230V phase-neutral and a variation of 10%, except for the UK, Malta and Cyprus which is 240V [20]. As in this previous subsection three points are going to be considered to validate the operation of the HB-LED driver, in this case those are 380V, 400V and 420V line-to-line RMS. However, only the THD, PF and efficiency are going to be shown since the compliance with class C was already demonstrated in the previous subsection and the variation of the peak value of v_P have next to none effect over

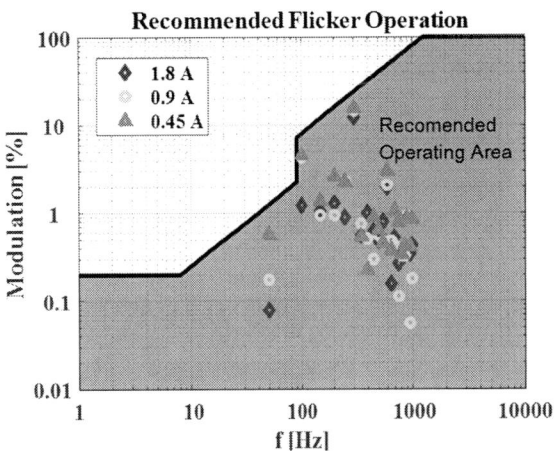

Fig 12. Recommended flicker operation, P1789 [18].

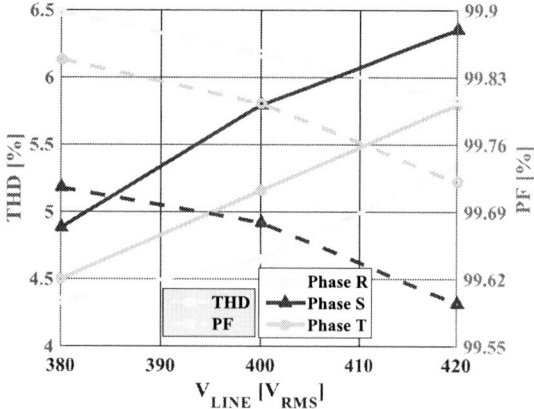

Fig 13. Evolution of the PF, THD versus the line input voltage.

the other parameters. The first two parameters are shown in Fig. 13, where it can be seen that THD increases with the input

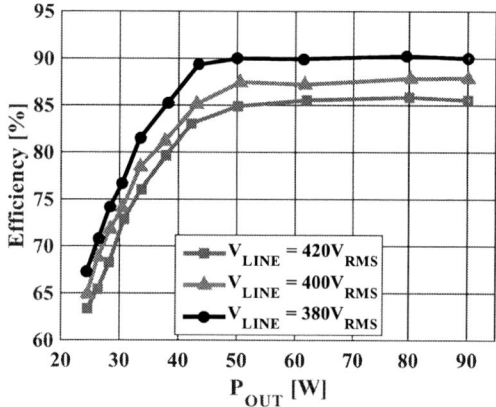

Fig 14. Evolution of the efficiency with input line voltage and the load.

voltage. This is due to the voltage loop lowering the duty cycle in order to regulate the voltage.

As for the efficiency, the results are shown in Fig. 14. In this case the efficiency of the converter have been measured in several operating points using a resistor instead of the HB-LED.

D. Detailed description of the designed voltage loop.

In this subsection the designed compensator for the voltage loop is going to be explained. For that reason, in Fig. 16 there is a detailed diagram of said voltage loop in terms of the Z transform, where H_{AD} represents the gain of the analog/digital converter and H_R represents the gain of the voltage divider.

It is important to note that the compensator is designed for a digital control. Therefore, limit cycling needs to be taken into account in the design of the compensator, in this case by using the definitions that guarantee non limit cycling in PFC [17]. Moreover, the design shown in (12) with a very simple

compensator has a theoretical settling time of roughly 4ms. This compensator satisfies the requirements stated at the end of the Section III while having a phase margin of 80.7° and a crossover frequency of 173Hz (see Fig. 17a and b). Especially important the crossover frequency of less than 300Hz to avoid fast variations of the duty cycle due to the non-ideal ripple of the output.

$$C(z) = \frac{0.2(z + 1)}{(z - 1)} \qquad (12)$$

The tests done to validate the voltage loop are executed with a resistive load going from full load (90W) to half load (45W) and vice versa. These tests are depicted in Fig. 17a and b, where it can be seen that when a load transient is applied, not only is the voltage loop able to correct the output voltage but the settling time matches the 4ms corresponding with the one that was obtained from the theoretical analysis.

It should be noted that these tests are done for the sake of completion of the voltage loop analysis and they are not crucial from the point of view of the stability and dynamic response of a load composed of HB-LED.

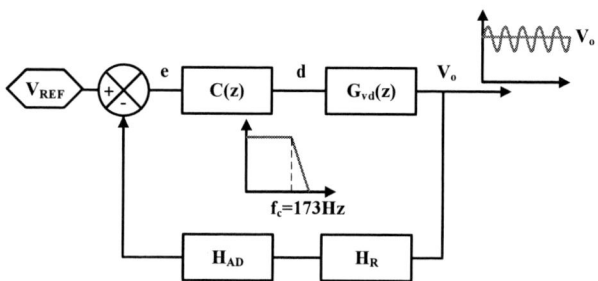

Fig 15. Block diagram of the digital control loop.

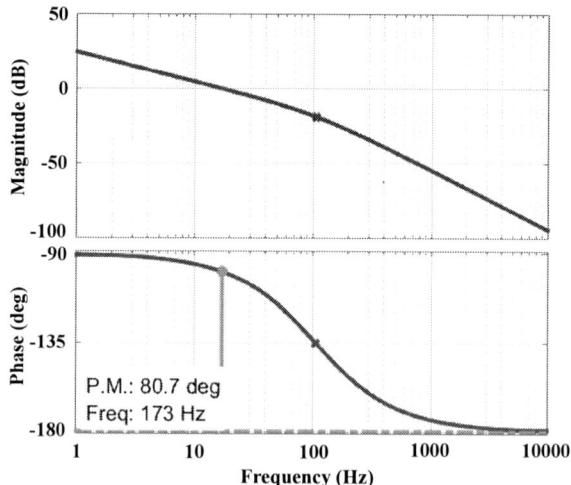

Fig 16. Open loop gain of the HB-LED driver.

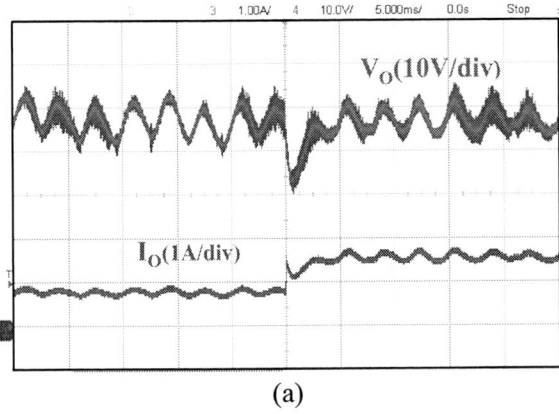

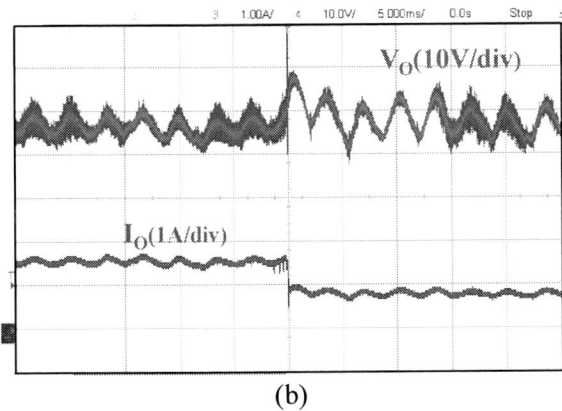

*Fig 17. Load transient response of the HB-LED driver.
(a)Half load to full load. (b) Full load to half load.*

V. CONCLUSIONS AND FUTURE WORK

A three-phase HB-LED driver has been reported and experimentally proven in this work. The HB-LED driver under study provides high PF and low THD and compliance with Class C IEC 1000-3-2. The analysis carried out over different dimming operating shows a non-flicker behaviour from a health point of view, while disposing of the traditional bulk capacitor in power factor correction. The disposal of said capacitor increases greatly the lifespan of the HB-LED driver making it a great solution for lighting in primary three-phase grids.

The driver has been tested in the whole range of European three phase voltage, showing lower efficiencies for the higher values of the range but still being a valid solution, taking into account that THD and PF roughly vary in this analysis. The drawback come in terms of the efficiency being too low when achieving full dimming, so other options for the LFR topology might be studied in the future.

Finally, the voltage loop has been validated with a couple of transient loads with the use of resistive loads. In this regard more work will be done in the future by looking at different controls

or different LFR setups that can help reducing the ripple of both output voltage and current.

REFERENCES

[1] Gray, G.; "Demystifying 347V and 480V Lighting Installations," e-Craftsmen.

[2] "480V to 277V Step-Down Autotransformers For applications up to 375 Watts", GE lighting.

[3] Singh, Bhim; Singh, B.N.; Chandra, A.; Al-Haddad, K.; Pandey, A.; Kothari, D.P., "A review of three-phase improved power quality AC-DC converters," Industrial Electronics, IEEE Transactions on , vol.51, no.3, pp.641,660, June 2004

[4] Singer, S.; Fuchs, A., "Multiphase AC-DC conversion by means of loss-free resistive networks," Circuits, Devices and Systems, IEE Proceedings - , vol.143, no.4, pp.233,240, Aug 1996

[5] "High Power Factor Modular Polyphase AC/DC converters based on Loss-Free Resistors", Paper submitted for APEC '16, Applied Power Electronics Conference and Exposition, 2016.

[6] Kamnarn, U.; Chunkag, V., "Analysis and Design of a Modular Three-Phase AC-to-DC Converter Using CUK Rectifier Module With Nearly Unity Power Factor and Fast Dynamic Response," Power Electronics, IEEE Transactions on , vol.24, no.8, pp.2000,2012, Aug. 2009

[7] Tibola, G.; Barbi, I., "Isolated Three-Phase High Power Factor Rectifier Based on the SEPIC Converter Operating in Discontinuous Conduction Mode," in Power Electronics, IEEE Transactions on , vol.28, no.11, pp.4962-4969, Nov. 2013

[8] Draft of the Proposed CLC Common Modification to IEC 61000-3-2 Document, 2006.

[9] Draft of the Proposed CLC Common Modification to IEC 61000-3-2/A2 Document, 2010.

[10] Electromagnetic Compatibility (EMC)-Part 3: Limits-Section 2: Limits for Harmonic Current Emissions (Equipment Input current < 16 A per Phase), IEC1000-3-2, 1995.

[11] Sabahi, M.; Hosseini, S.H.; Sharifian, M.B.B.; Goharrizi, A.Y.; Gharehpetian, G.B., "A Three-Phase Dimmable Lighting System Using a Bidirectional Power Electronic Transformer," Power Electronics, IEEE Transactions on , vol.24, no.3, pp.830,837, March 2009

[12] Erickson, R.; Madigan, M.; Singer, S., "Design of a simple high-power-factor rectifier based on the flyback converter," Applied Power Electronics Conference and Exposition, 1990. APEC '90, Conference Proceedings 1990, Fifth Annual, vol., no., pp.792,801, 11-16 March 1990

[13] Glaser, J.S.; Witulski, A.F., "Application of a constant-output-power converter in multiple-module converter systems," in Power Electronics Specialists Conference, 1992. PESC '92 Record., 23rd Annual IEEE , vol., no., pp.909-916 vol.2, 29 Jun-3 Jul 1992

[14] Yuehui Huang; Tse, C.K., "Circuit Theoretic Classification of Parallel Connected DC–DC Converters," in Circuits and Systems I: Regular Papers, IEEE Transactions on , vol.54, no.5, pp.1099-1108, May 2007

[15] Yuehui Huang; Tse, C.K., "Classification of parallel DC/DC converters part II: Comparisons and experimental verifications," in Circuit Theory and Design, 2007. ECCTD 2007. 18th European Conference on , vol., no., pp.1014-1017, 27-30 Aug. 2007

[16] Ridley, R.B "Average small-signal analysis of the boost power factor correction circuit", VPEC Seminar Proceedings, 1989, pp. 108-120

[17] Mather, B.A.; Maksimovic, D., "Quantization effects and limit cycling in digitally controlled single-phase PFC rectifiers," in Power Electronics Specialists Conference, 2008. PESC 2008. IEEE , vol., no., pp.1297-1303, 15-19 June 2008

[18] "RP-16-10, Nomenclature and Definitions for Illuminating Engineering", Illuminating Engineering Society.

[19] IEEE Recommended Practices for Modulating Current in High-Brightness LEDs for Mitigating Health Risks to Viewers," in IEEE Std 1789-2015 , vol., no., pp.1-80, June 5 2015

[20] IEC standard voltages, IEC60038, 1983.

978-1-4673-9551-9/16 $31.00 © 2016 IEEE

A ZV-ZCS Electrolytic Capacitor-LessAC/DC Isolated LED Driver with Continous Energy Regulation

John Lam, *IEEE Senior Member*, Nader A. El-Taweel, *IEEE Student Member*
Department of Electrical Engineering and Computer Science
Lassonde School of Engineering
York University
Toronto, ON, M3J 1P3, CANADA
Email: johnlam@cse.yorku.ca, naderelt@yorku.ca

Abstract— Conventional AC/DC LED drivers require a large energy storage capacitor at the output to provide a constant current to the LEDs. In order to minimize the size and cost of the driver circuit, electrolytic capacitors are conventionally used due to its high energy density and low cost. However, electrolytic capacitors are sensitive to operating temperature and have much shorter lifetime than the LED semiconductor devices, which significantly reduces the overall life time of the LED system. Another drawback with the current LED drivers is that the presence of the switching power losses restricts the use of high frequency operation, which results in using bulky passive circuit components in the drivers and significantly reduces the circuit power efficiency. This paper proposes a single-stage high power factor LED driver with almost zero switching losses and without the electrolytic capacitor. In the proposed circuit, discontinuous conduction mode (DCM) boost converter was utilized as a power factor correction (PFC) circuit, where it was integrated with an asymmetrical pulse width modulated (APWM) series resonant converter to form a single stage power conversion unit to drive the LEDs. The proposed circuit is able to achieve zero turn-on and turn-off switching operation and is able to eliminate the conventionally needed electrolytic capacitors by continuously regulating the DC-link voltage. The proposed LED driver was simulated and tested on a 12W design example to confirm that an almost unity power factor and an efficiency of 95% can be achieved.

Keywords— *Light Emitting Diode; Switching Losses; Series Resonant Converter; Power Factor Correction.*

I. INTRODUCTION

Light Emitting Diode (LED) lighting becomes the most environmental friendly and popular light source due to its superior features [1-2]: 1) Long life time up to 100,000 hours of operation. 2) Low wattage light source with high luminous efficacy. 3) It can emit different kind of colors depending on the kind of semiconductor material. According to the U.S department of energy, the LED lighting market share will rise significantly in the near future and it will be accompanied by huge saving in electricity, which can reach 2216 TWh by 2030 [3]. Based on the ENERGY STAR requirements for solid-state lighting, the luminaires power factor should be greater than 0.9 for commercial usage and 0.7 for residential usage [4]. Power supplies that utilize switch-mode operation are essentially non-

linear loads, which cause harmonics and distortion to the input current. A phase shift between the line current and voltage can also be resulted due to the reactive component in the circuit. According to [1], LED drivers are classified into passive and switched mode driver. The passive type uses only passive circuit components, which suffers from heavy weight, bulky size and poor efficiency. Switched mode drivers, on the other hand, are widely used currently due to the fact that the passive components can be significantly reduced by increasing the switching frequency of the driver [5], [6-7]. Switched mode driver can be classified into: (1) single stage, (2) two stage circuit. By employing the two-stage solution, a high power factor LED driver with precisely regulated output current and the possibility of eliminating the electrolytic capacitor can be resulted. However, increasing the number of power processing stages means that more components are required in the overall circuit, which will result in a large size and high cost system [5], [6-8]. Another drawback with the existing LED drivers is the presence of the switching power losses. In order to design a compact and small size LED driver, the switching power losses should be minimized.

It is well-known that LED has a much longer life time compared to that of the electronic driver, due to the use of the unreliable electrolytic capacitor, which has only an average life time of up to 20,000 hours [4-5]. Different methods and circuit topologies to eliminate the electrolytic capacitor in LED drivers have been presented in literature [2], [6] and [8-15]. In this paper, a single-stage LED driver that uses only film capacitors with near unity input power factor and very high power efficiency using energy regulation across the DC-link capacitor is proposed. The rest of the paper is organized as follows: section II discusses the proposed circuit configuration. Section III discusses the circuit controller. The simulation results on a design example are shown in section IV. Finally, the conclusion is included in section V.

II. PROPOSED LED DRIVER CONFIGURATION

AC/DC LED driver requires power factor correction (PFC) circuit to comply with the ENERGY STAR power factor requirements. The driver shall also consist of a DC current

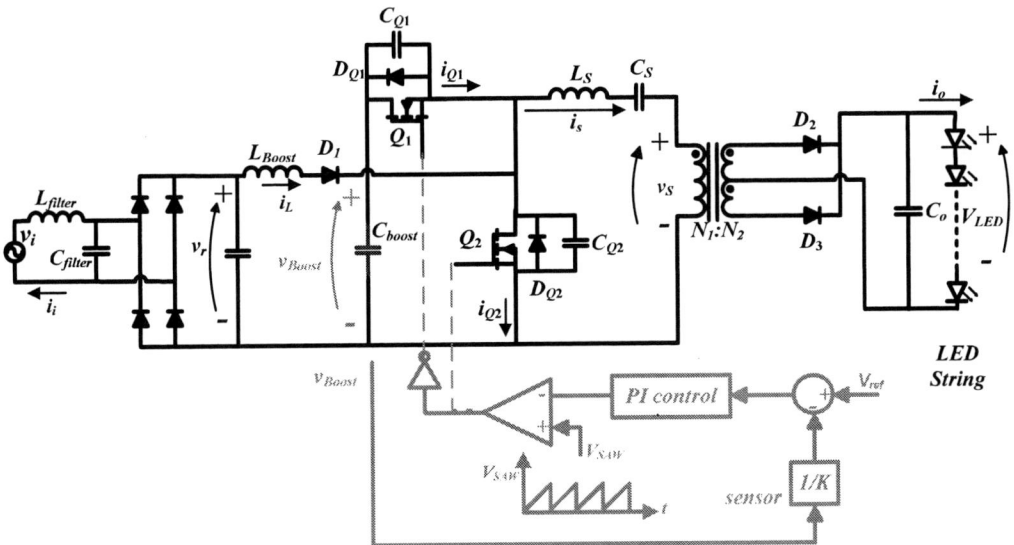

Fig.1 Proposed single-stage LED driver circuit with regulation

regulator to provide constant DC output current [8]. The proposed circuit is a single stage power conversion unit that integrates the discontinuous conduction mode (DCM) boost converter as a power factor correction (PFC) circuit, and the asymmetrical pulse width modulated (APWM) series resonant converter as DC output regulation as shown Fig.1. The APWM series resonant converter was used to achieve soft switching operation so that high frequency switching operation can be realized without affecting the driver efficiency. By operating the proposed driver circuit at high switching frequency, the reactive components size and cost can be substantially reduced.

A. AC/DC Discontinous Conduction Mode (DCM) Boost Converter

The DCM boost converter shown in Fig.2 is known as the voltage follower PFC approach, as this converter ensures that the inductor peak current (I_{Peak}) will always follow the sinusoidal voltage envelope as shown in Fig. 3(b) [16], where V_{in} is the AC input voltage, with V_m represents the peak of the AC input voltage, L is PFC circuit inductor, T_{on} is the converter on switch time and ω_i is the input voltage angular frequency.

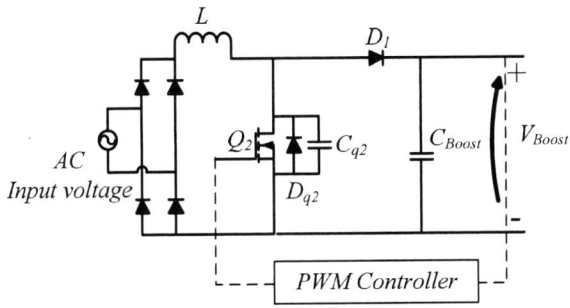

Fig.2 PFC circuit (DCM boost converter)

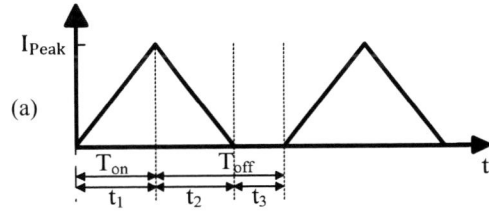

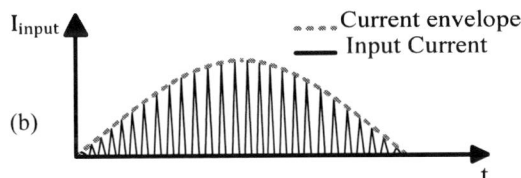

Fig.3 (a) Inductor current within a switching period (b) Half cycle of voltage

$$I_{Peak} = \frac{V_{in}T_{on}}{L} = \frac{V_m T_{on}}{L} Sin(\omega_i t) \qquad (1)$$

The average input current I_{L_Avg} can be calculated as shown in (2)-(5), which is the sum of the average current in the on time (I_{Avg_t1}) and the discharge time period (I_{Avg_t2}) as shown in Fig.3 (a), where, T is the switching cycle time between two successive on switching operation, t_2 is the converter discharge time as shown in (4) and D is the duty cycle and V_{Boost} is the output boosted voltage.

$$I_{L\_Avg} = I_{Avg\_t1} + I_{Avg\_t2} \qquad (2)$$

$$I_{L\_Avg} = \frac{I_{Peak}T_{on}}{2T} + \frac{I_{Peak}t_2}{2T} = K\frac{\alpha Sin(\omega_i t)}{1 - \alpha Sin(\omega_i t)} \qquad (3)$$

$$Since, \ t_2 = \frac{V_{in}T_{on}}{V_{Boost} - V_{in}} \qquad (4)$$

$$where, \ K = \frac{D^2 T V_{Boost}}{2L} \ and \ \alpha = \frac{V_m}{V_{Boost}} \qquad (5)$$

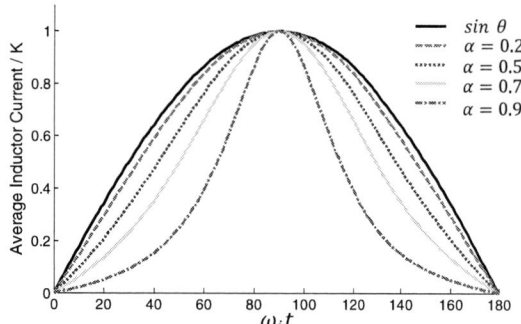

Fig.4 Normalized input (inductor) current for half a line cycle

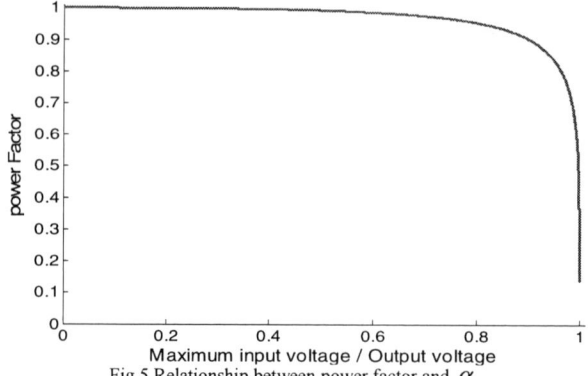

Fig.5 Relationship between power factor and α

According to (3), it can be observed that a close-to-sinusoidal current waveform is obtained when the ratio α decreases. Fig.4 shows the normalized input current for half a line cycle with different values of α. The power factor of the DCM boost circuit is then deduced according to [16] as shown in (6)-(9), where V_{rms} and I_{rms} represents the input RMS voltage and current respectively.

$$P.F = \frac{P_{in}}{V_{rms}I_{rms}} = \frac{X\sqrt{2}}{\sqrt{Y}\pi} \qquad (6)$$

where, $P_{in} = \dfrac{V_m K \alpha Y}{\pi} \qquad (7)$

$$X = \frac{-\pi}{\alpha^2} - \frac{2}{\alpha} + \frac{2}{\alpha^2\sqrt{1-\alpha^2}}\left(\frac{\pi}{2} - \tan^{-1}\left(\frac{-\alpha}{\sqrt{1-\alpha^2}}\right)\right) \qquad (8)$$

$$Y = \frac{\pi}{\alpha^2} + \frac{2}{\alpha(1-\alpha^2)}$$

$$+ \frac{2\alpha^2 - 1}{(1-\alpha^2)\,\alpha^2\sqrt{1-\alpha^2}}\left(\frac{\pi}{2} - \tan^{-1}\left(\frac{-\alpha}{\sqrt{1-\alpha^2}}\right)\right) \qquad (9)$$

Fig.5 shows the relationship between the power factor and α. It can be seen that in order to have almost unity power factor, the coefficient α should be smaller than 0.6, which correlates with the current waveform displayed in Fig.4.

By controlling the DC-link voltage V_{Boost}, the voltage will follow the refrence voltage and capacitor C_{boost} can be potentially reduced, which allows the utlization of film capacitor as the energy storage capacitor. The DCM boost converter output voltage V_{boost} can be determined as shown in (10-12), where D' is the period when Q_2 is off and R is the converter load [17].

$$V_{Boost} = \frac{2V_{in}}{\pi} \times \frac{1 + \sqrt{1 + \dfrac{4D^2}{N}}}{2} \qquad (10)$$

$$N = \frac{2L}{RT} \qquad (11)$$

Such that $N \prec (N_{critical} = DD'^2) \qquad (12)$

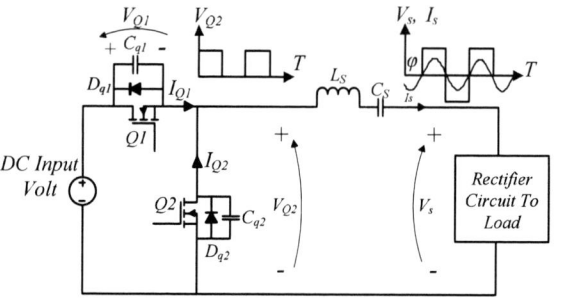

Fig.6 APWM series resonant converter

B. APWM Series Resonant DC/DC Converter

The main advantage of the APWM series resonant converter shown in Fig.6 is that it allows soft switching operation and hence, high switching frequency can be used [18]. Zero switching on is obtained by; passing the current through the diode across the switch prior to the switch turning on so that zero voltage switching (ZVS) turning on is achieved. With the addition of the snubber capacitor across the switch to maintain slow voltage rise across the switch, the capacitor charges when the switch is turning off. This implies that the turn off of the switch is achieved with almost zero switching losses. The small amount of losses during the switching off operation of the switches is charged in the capacitor and then discharged once again in the resonant circuit. The discharge occurs when the switch turns on in the next cycle, just before the current pass through the diode across the same switch. The operation stages of the APWM resonant circuit waveforms are shown in Fig.7 and its different operating stage circuits are shown in Fig.8, which are as following:

$[t_0 < t < t_I]$: At t_0, the gate signal applied to Q_1 is removed and capacitor C_{Q1} starts to charge, at the same time capacitor C_{Q2} starts to discharge.

$[t_I < t < t_{II}]$: At t_I, C_{Q2} is already discharged and its current is carried by switch Q_2 diode and so the switch is turned on under zero voltage switch. Since the gate signal applied to Q_2, i_L and the positive i_s flow through Q_2. In this stage the i_{Q2} is negative as the resonance current is much higher than the inductor i_L current.

978-1-4673-9551-9/16 $31.00 © 2016 IEEE

$[t_{II} < t < t_{III}]$: At t_{II}, switch Q_2 current i_{Q2} changes its polarity, once the different between i_L and positive i_s D_{Q2} becomes positive the current previously flowed through diode now flows through switch Q_2.

$[t_{III} < t < t_{IV}]$: At t_{III}, the resonance current i_s reverse its polarity and the half bridge rectifier diodes are reversed.

$[t_{IV} < t < t_V]$: At t_{IV}, the gate signal applied to Q_2 is removed and capacitor C_{Q2} starts to charge, at the same time capacitor C_{Q1} starts to discharge. i_L also discharges through D_{Q1}.

$[t_V < t < t_{VI}]$: At t_V, C_{Q1} is already discharged and the current is carried by switch Q_1 diode and so the switch is turned on under zero voltage switch. Since the gate signal applied to Q_1, i_L and the negative i_s flow through Q_1.

$[t_{VI} < t < t_{VII}]$: At t_{VI}, L_{Boost} inductor completely discharged and the resonance current i_s reverses its polarity and is supplied from the C_{Boost} capacitor.

The rectifier circuit and the LED load is represented by its equivalent resistance R_{eq} as shown in (13), where R_{LED} is the actual LED string resistance and (N_P/N_S) is the ratio between primary to secondary transformer winding [18-19]. Output DC regulation can be controlled through (13-15), which is a voltage divider across the equivalent LED resistance. The input voltage to the resonant circuit is V_I, the output voltage to the LEDs load is V_{LED}, the angular switching frequency is ω, the angular resonant frequency is ω_s, while L_s and C_s are the series resonant circuit components. The relationship between the voltage gain of the series resonant circuit and the relative operating frequency is shown in Fig.9, with the ZVS region highlighted in the figure.

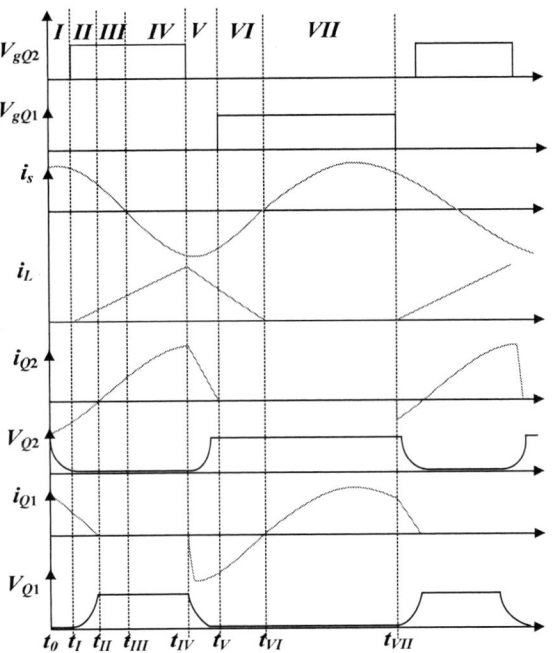

Fig.7 APWM series resonant converter operating waveformes (current waveforms in red)

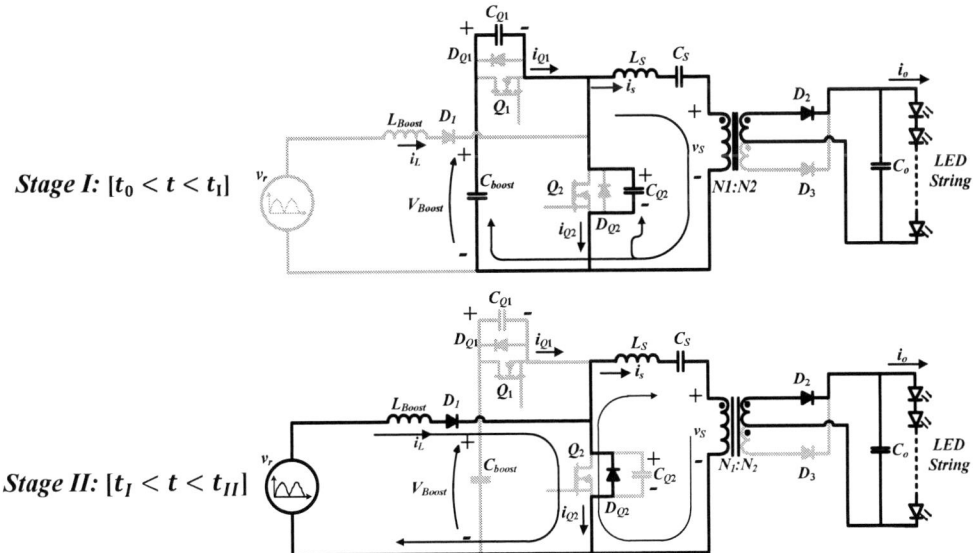

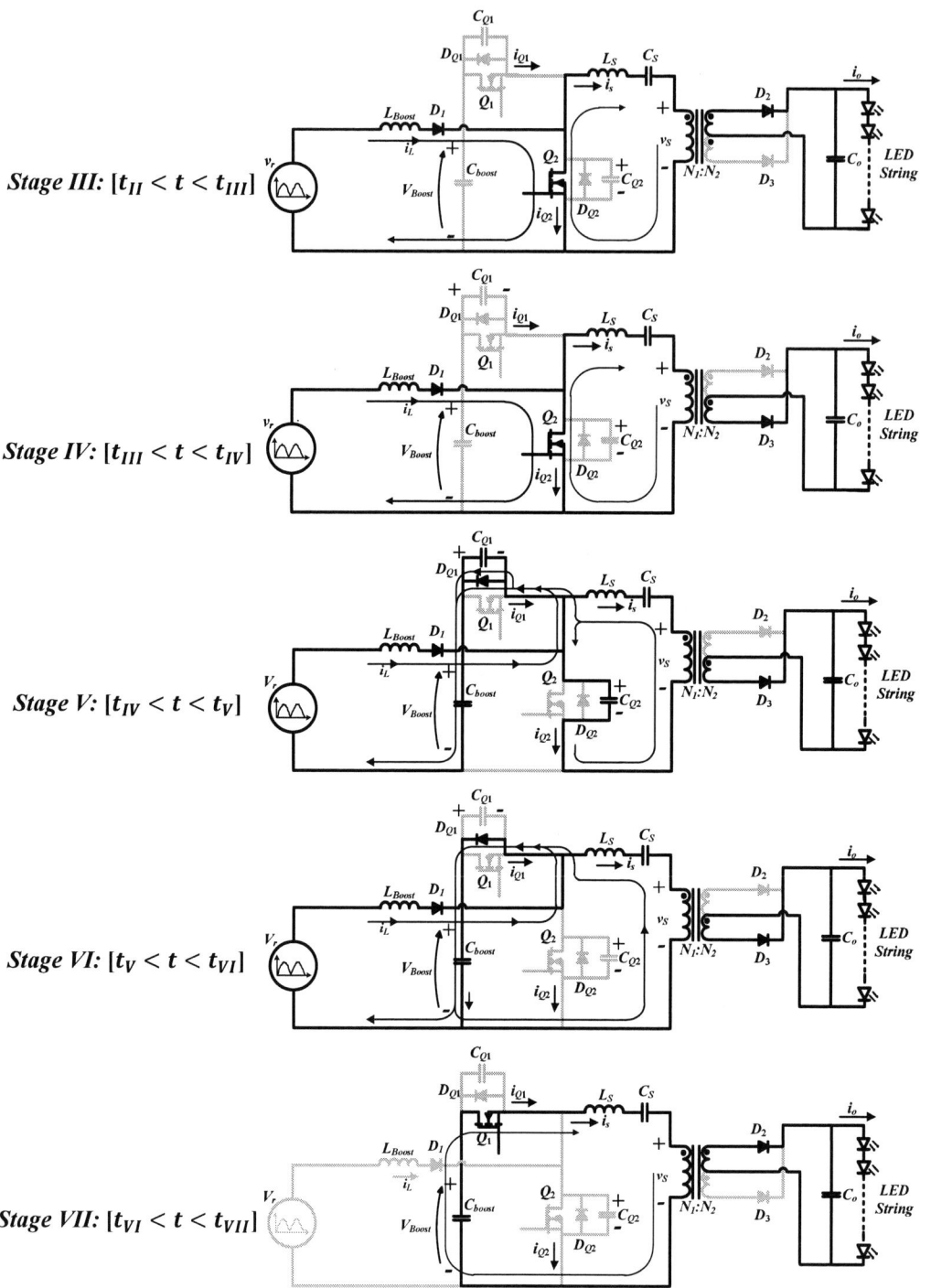

Fig.8 Operating stages of the the proposed LED driver.

$$R_{eq} = \frac{8}{\pi^2} R_{LED} \left(\frac{N_P}{N_S} \right)^2 \qquad (13)$$

By defining, $Q = \dfrac{\omega_s L_s}{R_{eq}}$ (14)

$$\frac{V_{LED}}{V_I} = = \frac{1}{1 + j\frac{\pi^2}{8}\left(\frac{N_P}{N_S}\right)^2 Q\left(\frac{\omega}{\omega_s} - \frac{\omega_s}{\omega}\right)} \qquad (15)$$

By defining ω^* as the relative angular frequency,

$$\omega^* = \frac{\omega}{\omega_s} \qquad (16)$$

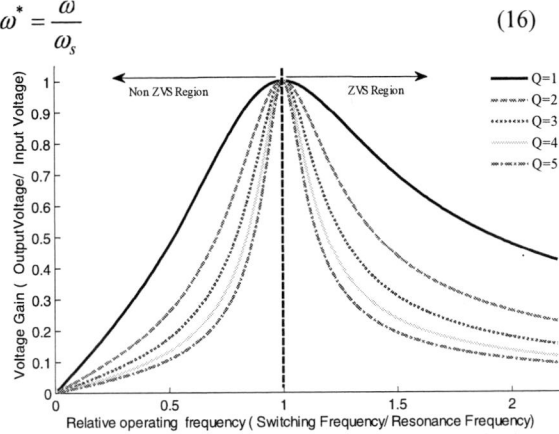

Fig.9 Voltage gain characteristics of series LC resonant circuit

III. FEEDBACK CONTROLLER

A voltage feedback controller is used to control the generated pulse width modulation by regulating the output voltage of the boost PFC circuit (V_{boost}). The feedback is utilized to drive switches Q_1 and Q_2 to control the input voltage to the APWM converter across the C_{boost} capacitor. By regulating V_{boost}, the energy across is indirectly regulated. This concept of regulating the energy across the DC link capacitor to provide constant power in compact fluorescent lighting has been discussed in [20] and [21]. As a result, by changing the duty ratio of Q_1 and Q_2, the low frequency ripple across the DC-link capacitor is reduced, and therefore, the size of C_{boost} can be significantly reduced and film capacitor can be used to replace the unreliable electrolytic capacitor.

V_{boost} is controlled by continuously measuring the voltage and comparing it with a desired reference voltage setting, the error from this comparison will be compensated by a proportional integral (PI) controller in such a way, to ensure the stability of the LED driver. It should be noted that in the dead-time portion highlighted by interval I and V in Fig.7, the snubber capacitor across the switch allows the voltage to rise slowly. Therefore, almost zero turn-off switching losses can be ensured.

IV. DESIGN EXAMPLE AND RESULTS

In order to verify the performance and the stability of the proposed LED driver, the circuit shown in Fig.1 was tested and simulated in PSIM. The system specifications are shown in TABLE I. The Cree® XLamp® XR-E LED has a threshold voltage of 3.5V. The resonant frequency is selected as 180kHz. The Q value is selected as between 3 and 4. The switches (Q_1, Q_2) are IRF840. A snubber capacitor of 90pF

was added to each switch to reduce the turn off losses. The design of the DCM boost converter was performed to provide a voltage gain of 0.56 with the boost output voltage targeted at 300V. The boost inductance (L_{boost}) was calculated based on (7), as the input power should be greater than the LEDs output power as shown below:

$$P_{LED} = V_{LED} \times I_{LED} = 3.5 \times 10 \times 0.34 = 12 \ Watts \qquad (17)$$

Therefore, $P_{in} \succ 12$ and $L_{Boost} \prec 4 \ mH \qquad (18)$

The DCM boost converter output capacitor C_{Boost} was targeted to be less than 10μF, and its ripples will be determined after the designation of the APWM resonance converter parameters. From Fig. 9, the switching frequency should operate above the resonant frequency. In this case, the relative operating frequency is chosen as 1.4. Then, the series resonant inductor is calculated from (14) as follows:

$$R_{LED} = \frac{V_{LED}}{I_{LED}} = \frac{35}{0.34} = 103 \ \Omega \qquad (19)$$

$$L_s = \frac{QR_{eq}}{\omega_s} \approx 0.4mH \qquad (20)$$

And, $C_s = \frac{1}{\omega_s^2 L_s} \approx 2nF \qquad (21)$

Since the APWM converter acts as a load across the boost circuit DC-link capacitor, the impedance Z_{in} (i.e. equivalent impedance of the APWM resonant circuit) must be determined first as shown in (22) to obtain the approximate voltage ripples across C_{boost}, which is 30.6%. To reduce the ripple, a voltage feedback controller with PI control having a gain of 0.8 and a time constant of 0.002 are used in this design. These parameters are determined from PSIM and MATLAB feedback control toolbox.

TABLE I Design Specifications

INPUT VOLT	100-135V$_{rms}$
INPUT FREQUENCY	60 Hz
SWITCHING FREQUENCY	250 kHz
# OF LEDs	10 in series
OUTPUT POWER	12 Watts
LED CURRENT	340 mA
LED PART NUMBER	CREE® XLAMP® XR-E LED

$$|Z_{in}| = \left|R_{eq} + j\left(\omega_s L_s - \frac{1}{\omega_s C_s}\right)\right| = 326\Omega \qquad (22)$$

$$\Delta V_{Boost} = \frac{DV_{Boost}T}{2|Z_{in}|C_{Boost}} = 30.6\% \qquad (23)$$

Fig.10-13 show the simulation results. In Fig.10 and Fig.11, both Q_1 and Q_2 waveforms are presented with labeled ZVS turn-on and zero current switching (ZCS) turn-off. In Fig.12, high power factor can be observed as the line current and line voltage are in phase with almost sinusoidal current waveform at different input voltage values.

Fig.13 shows the LED current with an average value of 340 mA current with low frequency ripples of 11.5% however, the capacitor C_{boost} across the LEDs is 3 μF. Fig. 14 shows the Bode Plot of the closed-loop system, which proves the stability of the proposed LED driver through a closed-loop PI compensation. The gain and the time-constant of the PI controller are 0.8 and 0.002 respectively, and the phase margin is 59.5°.

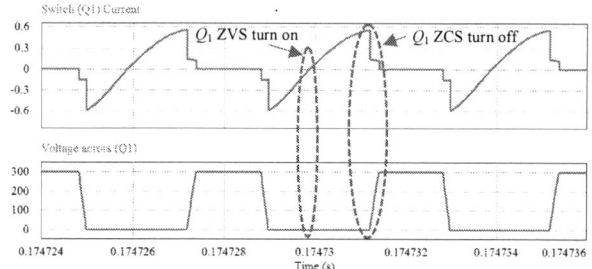

Fig.10 Switch Q_1 current and voltage waveforms at 120 V_{rms}

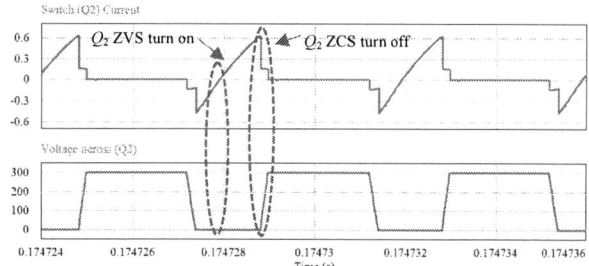

Fig.11 Switch Q_2 current and voltage waveforms at 120 V_{rms}

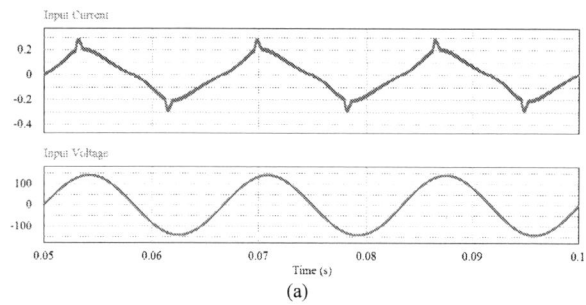

(a)

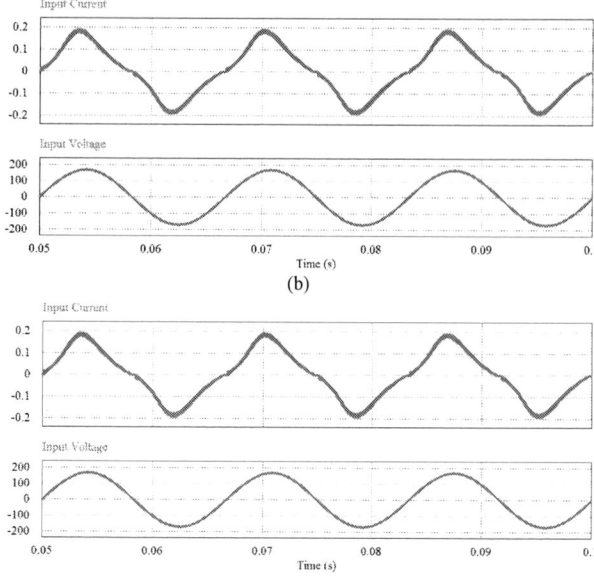

(b)

(c)

Fig.12 Input current and voltage waveform (a) 100 V_{rms} (b) 120 V_{rms} (a) 135V_{rms}

TABLE II shows the performance of the proposed circuit under different voltage values. As shown in TABLE III, the power factor is maintained at close to 0.97 for the specified input voltage range. A 10W experimental prototype has been built in the laboratory based on the circuit parameters obtained earlier and the prototype was tested at 120V_{rms}. Fig. 15 shows the measured input current and voltage. Fig. 16 shows the measured V_{boost} for a few line cycles with energy regulation. C_{boost} used in the prototype is 4 x 1μF (400V) in parallel: ECQ-E4105KF. The input power factor is 0.976 and the efficiency is 91.6%. The voltage ripple across C_{boost} is 9.5%.

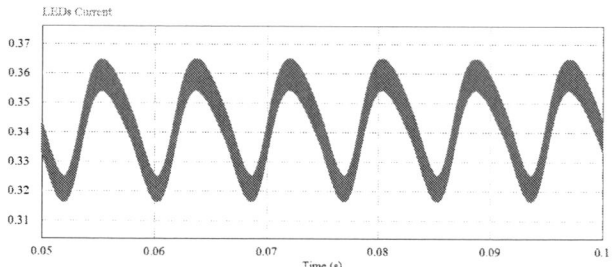

Fig.13 LED current waveform at 120 V_{rms}

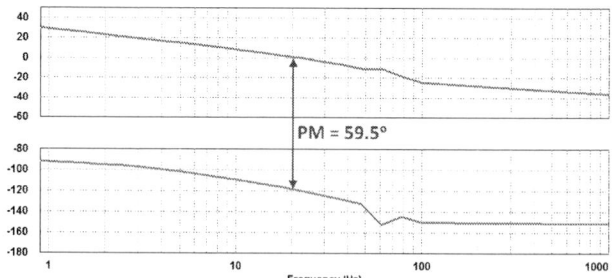

Fig.14: Bode Plot of the closed-loop system

TABLE II Summary of performance

Input Voltage (V_{rms})	Power factor	Efficiency
100	0.9727	94.46%
110	0.9783	94.56%
120	0.9712	94.84%
130	0.9624	95.2%
135	0.971	94.8%

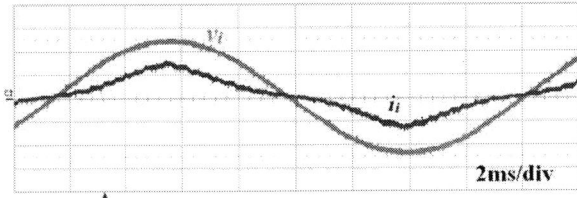

Fig.15: Measured line current and input voltage at 120 V_{rms} (i_i: 0.1A/div, v_i: 70V/div)

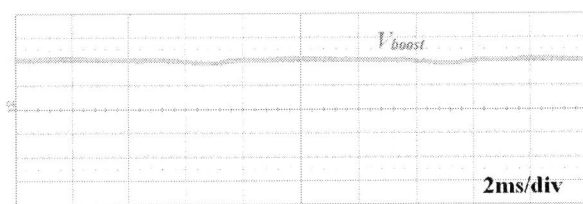

Fig.16: Voltage across C_{boost} at 120 V_{rms} (V_{boost}: 150V/div)

V. CONCLUSION

In conclusion, a long-lifetime, single stage high power factor LED driver with zero switching losses has been presented in this paper. The steady-state analysis and features of the proposed circuit have been described in details in this paper. Results have been provided to highlight the features and performance of the proposed LED circuit. A very high efficiency of 95% with a power factor of 0.97 has been obtained from a 250kHz design example.

VI. REFERENCES

[1] S. Li, S. C. Tan, S. Y. R. Hui, and C. K. Tse, "A review and classification of LED ballasts," *in Proceedings of the 2013 IEEE Energy Conversion Congress and Exposition Energy Convers. Congr. Expo. (ECCE 2013)*, pp. 3102–3109, 2013.

[2] Q. Hu and R. Zane, "Minimizing required energy storage in off-line LED drivers based on series-input converter modules," *IEEE Trans. on Power Electronics*, vol. 26, no.10, Oct. 2011, pp. 2887–2895, 2011.

[3] "Energy Savings Forecast of Solid-State Lighting in General Illumination Applications," Navigant Consulting, Inc. August, 2014.

[4] "ENERGY STAR® Program Requirements for Luminaires," 2011.

[5] Almeida, P.; Camponogara, D.; Dalla Costa, M.; Braga, H.; Alonso, J. "Matching LED and Driver Life Spans: A Review of Different Techniques" *IEEE Industrial Electronics Magazine*, vol. 9, no. 2, June 2015, pp. 36 – 47.

[6] S. Wang and K. Yao, "A Flicker-free Electrolytic Capacitor-less AC-DC LED Driver," *IEEE Trans. on Power Electronics*. vol.27, no. 11, 2012 pp. 4540–4548.

[7] Branas, C.; Azcondo, F.J.; Alonso, J.M. "Solid-State Lighting: A System Review" *IEEE Industrial Electronics Magazine*, vol. 7, no. 4, Dec 2013, pp. 6 - 14

[8] B. Wang, X. Ruan, K. Yao, and M. Xu, "A method of reducing the peak-to-average ratio of LED current for electrolytic capacitor-less AC-DC drivers," *IEEE Trans. on Power Electronics*, vol. 25, no. 3, March 2010, pp. 592–601.

[9] F. Zhang, J. Ni, and Y. Yu, "High power factor AC-DC LED driver with film capacitors," *IEEE Trans. on Power Electronics*, vol. 28, no. 10, Oct. 2013, pp. 4831–4840.

[10] P. S. Almeida, G. M. Soares, D. P. Pinto, and H. a C. Braga, "Integrated SEPIC buck-boost converter as an off-line LED driver without electrolytic capacitors," *in Proceedings of the 2012 IEEE Annual Conference on Industrial. Electronics Society (IECON 2012)*, pp. 4551–4556, 2012.

[11] Y. X. Qin, H. S. H. Chung, D. Y. Lin, and S. Y. R. Hui, "Current source ballast for high power lighting emitting diodes without electrolytic capacitor," *in Proceedings of the 2008 IEEE Annual Conference on Industrial Electronics Society (IECON 2008)*, pp. 1968–1973, 2008.

[12] J. Lam, P. K. Jain, "A Novel Isolated Electrolytic Capacitor-less Single-Switch AC-DC Offline LED Driver with Power Factor Correction," *in Proceedings of the 2014 IEEE Applied Power Electronics Conference &.Exposition (APEC 2014)*, pp. 1356–1361.

[13] P. Fang and Y. F. Liu, "An electrolytic capacitor-free single stage Buck-Boost LED driver and its integrated solution," *in Proceedings of the 2014 IEEE Applied Power Electronics Conference &.Exposition (APEC 2014)*, pp. 1394–1401.

[14] D. T. Nguyen, E. S. Lee, and C. T. Rim, "The LED Driver Compatible with Electronic Ballasts by Variable Switched Capacitor," *in Proceedings of the 2015 IEEE Applied Power Electronics Conference &.Exposition (APEC 2015)*, pp. 877–883.

[15] E. Eloi, E. Mineiro, S. Jr, R. Linhares, P. A. Miranda, and F. L. M. Antunes, "Single Stage Switched Capacitor LED Driver with High Power Factor and Reduced Current Ripple," *in Proceedings of the 2015 IEEE Applied Power Electronics Conference &.Exposition (APEC 2015)*, pp. 906–912.

[16] K.-H. Liu and Y.-L. Lin, "Current waveform distortion in power factor correction circuits employing discontinuous-mode boost converters," *in Proceedings of the 1989 IEEE Power Electronics Specialists Conference (PESEC 1989)*, pp. 825–829.

[17] R. W. Erickson and D. Maksimovic, "Fundamentals of Power Electronics" Second edition, 2001.

[18] Praveen K. Jain, A. St-Martin, Gary Edwards, "Asymmetrical Pulse-Width-Modulated Resonant DC/DC Converter Topologies," *IEEE Trans. on Power Electronics*, vol. 11, no. 3, May 1996 pp.413-422 1996.

[19] R. L. Steigerwald, "Comparison of Half-Bridge Resonant Converter Topologies," *IEEE Trans. on Power Electronics.*, vol. 3, no. 2, 1987, pp. 174–182.

[20] Lam, J.C.W.; Jain, P.K., "A TRIAC Dimmable Single-Switch Electronic Ballast With Power Factor Correction and Lamp Power Regulation," *IEEE Trans. on Power Electronics*, vol.29, no.10, Oct. 2014, pp.5472-5485.

[21] Alonso, J.M.; Calleja, A.J.; Lopez, E.; Ribas, J.; Rico-Secades, M., "A novel single-stage constant-wattage high-power-factor electronic ballast," *IEEE Trans. on Industrial Electronics*, vol.46, no.6, Dec 1999, pp.1148-1158.

High Efficiency and Power Density GaN-Based LED Driver

Eric Faraci, Michael Seeman, Bin Gu, Yogesh Ramadass, Paul Brohlin

GaN Products Group
Texas Instruments
Santa Clara, CA, and Dallas, TX, USA
wefaraci@ti.com

Abstract— LED lighting is becoming more attractive and popular due to its high efficiency, long lifetime, instant brightness and ability to dim. But LEDs are powered by DC, which requires a rectifier to allow it to run off the AC utility grid. These converters must have small form factors and operate in high ambient temperatures due to the close proximity of the LED light. Gallium Nitride (GaN) based switched mode power supplies have the ability to increase switching frequencies to allow for form factors small enough to enable new applications with stricter space requirements. But to achieve this size reduction, the switching frequency must operate in the megahertz range, which requires soft switching techniques to achieve acceptable performance. A small and simple analog hysteresis-controlled valley-switched floating buck converter is proposed to solve these problems. This solution has a minimal component count while maintaining similar performance to existing solutions. A 20W prototype is built to verify the performance of the converter, which is measured to have a size reduction of 1.87x of the power stage when compared to an existing commercial product while achieving an efficiency of 91.2%, an input current THD of 15.9% and PF of 0.976.

Keywords—LED, GaN, HV GaN FET, AC/DC, Valley Switching, High Power Density, Analog Hysteresis control

I. INTRODUCTION

Due to their high efficiency, small size, long lifetime and lack of hazardous materials, LEDs are becoming increasingly popular as a replacement light source for incandescent and fluorescent bulbs. But LEDs are powered by DC, which causes a requirement for a rectifier to allow it to work with the AC utility grid [1, 2]. In order to maximize the benefits of LEDs and make them an attractive alternative to incandescent and fluorescent bulbs, this converter needs to be efficient and small. While silicon-based converters exist that are able to drive the LED while satisfying these requirements [3-5], there is still a desire to further reduce the size to enable new form factors. This work has focused on improving control techniques to further improve performance [6-7] as well as using soft switching methods and newer semiconductor materials [8].

GaN has the ability to solve this problem by reducing the size of the converter by operating at higher switching frequencies, but this increase causes the system's switching

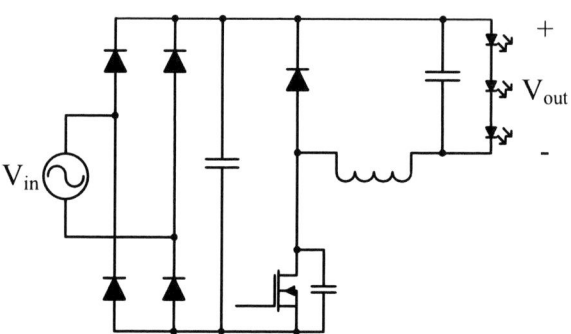

Fig. 1. Quasi-Resonant Floating Buck Converter Power Stage

losses to drastically increase. This problem can be mitigated by implementing soft switching techniques, which use the resonance in the passive elements to reduce the switching loss. However, these topologies typically require larger components to handle the increased circulating currents, additional components and increased complexity of the control to generate the soft switching action, which increase the size of the converter. A quasi-resonant floating buck converter shown in Fig. 1 is an ideal topology to use in this application, since it reduces switching losses by valley switching the FET in a resonant manner [9] without any additional components.

II. VALLEY SWITCHING OPERATION

The quasi-resonant floating buck converter is able to achieve valley switching by operating in discontinuous conduction mode, which causes the inductor to resonate with the output capacitance of the FET. As Fig. 2 shows, when the inductor current I_{LO} fully discharges, the drain-source voltage V_{ds} across the FET rings down approximately twice the output voltage of the LED string due to an LC resonance between the FET output capacitance and output inductor. If the FET is turned on at this point, the turn-on losses are significantly reduced. This operation can be implemented with an SR latch, where the turn on valley switching point is triggered on the set pin S and the current regulation and power factor correction is triggered on the reset pin R. The gate of the FET is controlled by the output of the SR latch.

978-1-4673-9551-9/16 $31.00 © 2016 IEEE

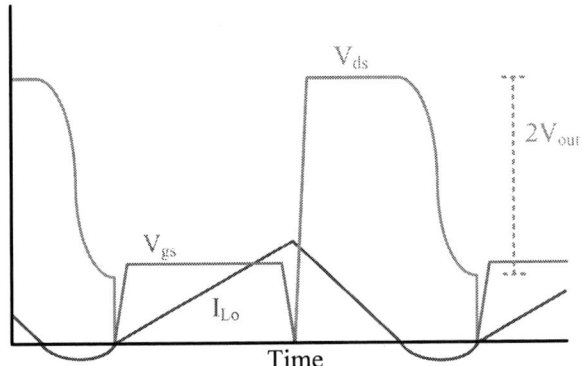

Fig. 2. Key Switching Waveforms of QR Buck

The valley switching point is sensed with a 1 turn auxiliary winding on the output inductor L_O. This is achievable since the turn acts as a transformer, which reflects the voltage seen across the inductor. The voltage seen across the output inductor can be approximated with (1),

$$V_{LO}=V_{IN}\text{-}V_O\text{-}V_Q \qquad (1)$$

where V_{IN} is the input voltage, V_O is the output voltage, and V_Q is the drain to source voltage on the switching FET. Since the input voltage V_{IN} is a low frequency AC term, the output voltage V_O is a DC term, and the FET voltage V_Q is a high frequency term, a high pass filter H_{iL} can eliminate the dependence of V_{IN} and V_O. This creates a control signal V_{iL} that is proportional to only V_Q. By comparing this signal to a fixed DC value V_{Ton}, as shown in Fig. 3, the turn on time of the FET for each switching cycle occurs near the valley point over the input line cycle.

III. Control Design

To regulate the output current to the desired power level and achieve power factor correction, a two-loop control loop is implemented. An inner hysteretic control loop works with the valley control point to regulate the peak output current through a sense resistor $R_{sense,}$ providing a cycle-by-cycle control with the valley switching circuitry. An outer current reference control loop regulates the average output current while

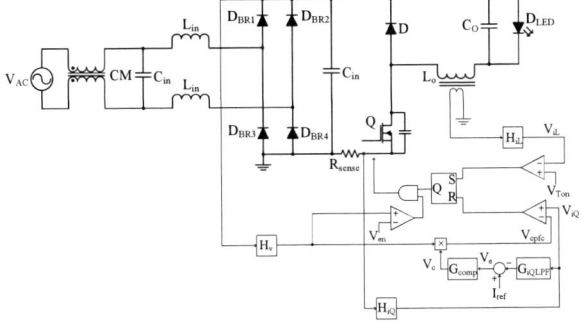

Fig. 3. Control Block Diagram

modulating the load to keep the input current in phase with the input voltage, achieving power factor correction.

A. Inner Hysteresis Control

The cycle-by-cycle regulation of the QR Buck is achieved by implementing a hysteretic SR latch control. The turn on pulse S is set by the valley switching sensing circuitry. To close the inner loop and provide cycle by cycle control, the turn off pulse R is generated by limiting the peak current through the FET. This turns the FET off when the sensed current V_{iQ} exceeds the peak current mode reference V_{cpfc}.

B. Outer Average Control

To properly regulate the output current and achieve PFC, another control loop is required to modulate the current reference V_{cpfc}. The output current is regulated when the average current of the FET is sensed by averaging V_{iQ} with a low pass filter G_{iQLPF}. This sensed average current is subtracted from a reference current and the corresponding error V_e is fed through a compensator to generate the reference current V_c. To prevent interference with the PFC operation, the crossover frequency of this loop is designed to be well below the double line frequency.

C. PFC Control

To achieve PFC, V_c is multiplied by the rectified AC input that is scaled down by a fixed gain H_v. This creates a modified reference signal V_{cpfc} that varies in phase with the input, satisfying input power quality standards while delivering the desired average output current. The delivered current is prevented from having a noticeable flicker on the LED load by using an output capacitor C_O to reduce the output current ripple to an indiscernible amount. Since the turn on point is fixed to achieve valley switching, the adjustment of the turn off point over the line cycle for PFC causes the switching frequency to vary during operation.

IV. Prototype Design

To validate the proposed control technique and to measure the performance gains, a prototype converter was designed and tested. The AC input is selected to be a typical European line voltage of $220V_{rms}$, and the load is a 20W LED string that has a forward voltage drop of about 70V. To satisfy grid interconnect standards, the input power quality is filtered and controlled to meet IEC6100-3-2, which limits the magnitude of the various harmonics of input current for lighting equipment.

A. Power Stage Component Selection

1) Power FET

A 600V 1Ω GaN FET is selected as the power transistor due to its superior switching performance in the megahertz range. By operating at such high speeds the passive filter component size is significantly reduced, dramatically increasing power density.

2) Output Inductor

The first component that is designed is the output inductor L_o, which resonates with the output capacitance of the GaN

FET and determines the switching frequency of the converter. Using SPICE simulations of the GaN FET to determine the resonant times with the non-linear output capacitance, an inductance of 10µH is selected as a tradeoff between efficiency and power density. This led to a switching frequency that varies between 2.5MHz to 4.4MHz. To minimize loss, the inductor was custom built with 20 turns of litz wire, made up of 66 strands of 44 AWG wire in a RM6S/ILP core of 3F45 ferrite.

3) Output Capacitor

The QR Buck can only operate when the input voltage is greater than the output, so a capacitor is required to hold up the output when the AC input is below the 70V LED forward voltage drop. The output capacitor is also required to reduce the output current ripple seen across the LED that comes from the double line frequency variation from the power factor correction. A capacitance of 115µF is found to adequately filter the output to meet these requirements.

4) Power Diode

In order to simplify the topology and keep cost low, a diode is selected for the high side switch. While the reverse recovery of this diode helps the valley switching by adding additional charge to the resonant tank during the resonant action, it delays the switching transition. When operating in the megahertz frequency range this reverse recovery time is significant, causing extensive transition times and delays that hamper operation. For this reason, a Cree C3D1P7060Q SiC Schottky diode was selected due to its almost zero reverse recovery, allowing the fast timing requirements to be satisfied.

5) Input Filter

Input power quality standards are met with a low pass filter on the input. With megahertz switching only an input capacitance of 30nF and inductance of 1.6mH is required to sufficiently filter the input to meet harmonic standards. This filter also handles EMC but was not evaluated in the scope of this work.

B. Controller

1) Turn On

To measure the valley switching point, a one turn auxiliary winding is added to the inductor. At high line input, the 25:1 turn ratio leads to an output voltage that is well above the 5V maximum analog control voltage allowed. Therefore, a resistor divider with capacitor low pass filter is used to step down the voltage to a safe range and eliminate undesired noise. This generates the control voltage V_{iL}, which is then compared to a DC term V_{Ton} with a comparator. When V_{iL} drops below V_{Ton}, the voltage across the FET is at its valley point, and the corresponding S pulse is generated to turn on the FET. As the input voltage varies over the line cycle, the valley switching point varies and the controller tracks the turn on point well.

Experimental results demonstrating this are shown in Fig. 4 and Fig. 5. As these results show, the sensed inductor voltage V_{iL} decreases from above to below the DC V_{Ton} at the valley

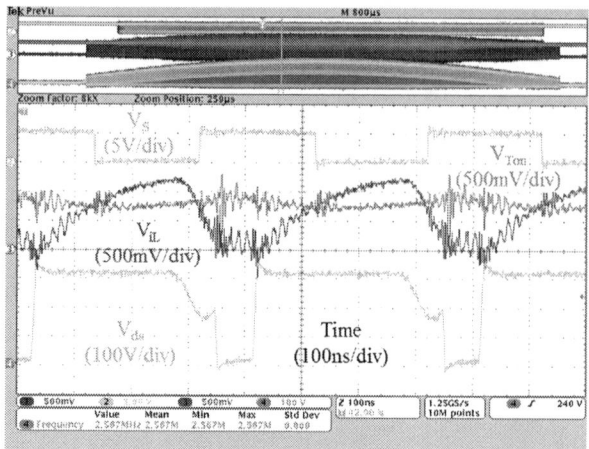

Fig. 4. S Control Waveforms at High Line Input

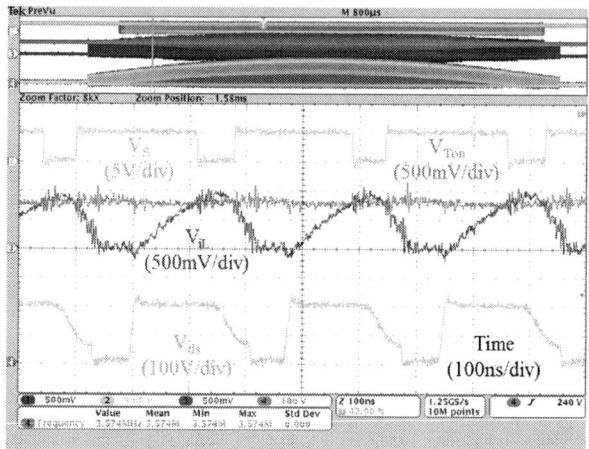

Fig. 5. S Control Waveforms at Low Line Input

point. This transition causes S to go high, which turns the FET on at the valley point as V_{ds} shows.

2) Turn Off

The output current is regulated when the FET current is sensed with a sense resistor. The cycle by cycle current is scaled to a control voltage V_{iQ} with the sense resistor R_{sense} and compared to the control voltage V_{cpfc}. To properly regulate a 20W load, the output current is regulated to about 280mA. This is achieved by filtering V_{iQ} through a 30Hz low pass filter to measure the average current. The current is then regulated with a 1 pole, 1 zero and integrator (type 2) compensator to generate V_c. To prevent interference with the PFC operation, the crossover frequency of this closed loop compensation is about 8Hz, well below the double line frequency.

Experimental results demonstrating this are shown in Fig. 6 and Fig. 7. As these results show, when the sense current through R_{sense}, V_{iQ}, exceeds the reference V_{cpfc}, R goes from low to high. This causes the FET to turn off, as V_{ds} shows.

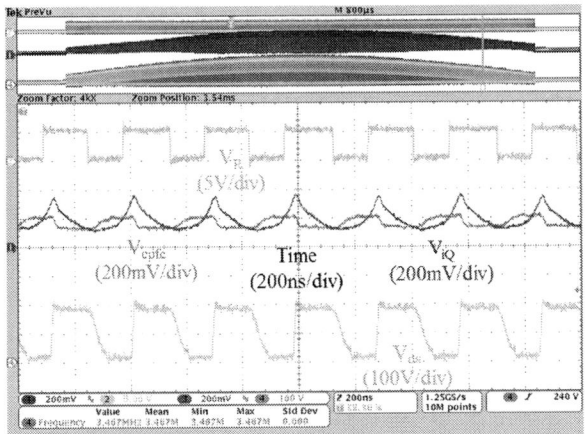

Fig. 6. R Control Waveforms at Low Line Input

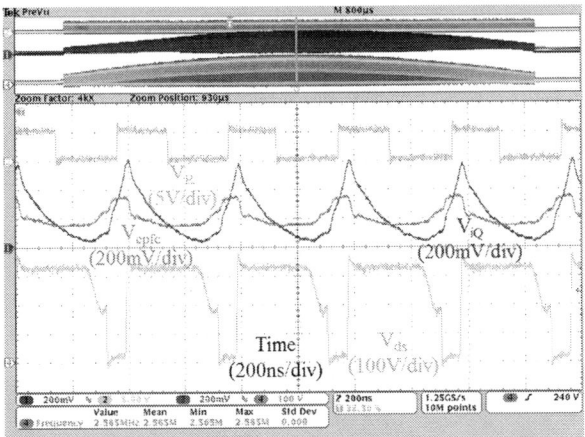

Fig. 7. R Control Waveforms at High Line Input

3) Power Factor Correction

To implement the power factor correction a scaled down rectified input voltage from H_v is multiplied with V_c to create the PFC control voltage V_{cpfc}, which accurately tracks the input voltage. The magnitude of this signal is scaled to give an output current ripple of about 50 percent. This output ripple does not generate any noticeable flicker on the LED load.

As Fig. 6 and Fig. 7 show for switching waveforms and fig. 8 shows for line cycle waveforms, closed loop control with high power factor is achieved. In Fig. 6, when the input is at low line, V_{cpfc} is smaller than in fig. 7, when the input is at high line. This action is most apparent over the switching line cycle I_{in}, which stays in phase with V_{in} while achieving high power factor.

4) SR Implemtation

Over a switching cycle the SR latch inputs go into an invalid S and R high state. This state is typically invalid since the output Q and inverse Q are both low. This is not a problem for the control purposes of the QR Buck converter since only Q is used to drive the GaN. As long as S goes low before R does, as Fig. 9 and Fig. 10 shows, the SR latch will regulate the output and achieve PFC. This adds a constraint on the control

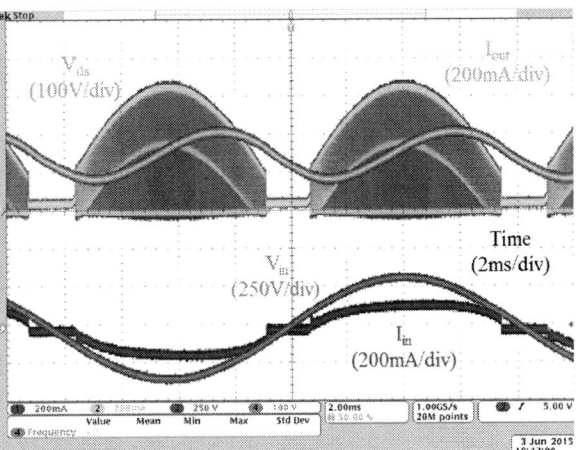

Fig. 8. Line Cycle Waveforms

sense filter H_{iQ}, since if the bandwidth is too small it could delay the R turn off point and violate this rule. This would cause the controller to turn on at points other than the valley and lose regulation. To implement the SR latch, two logic NOR gates are used.

5) Hold Up Control Disable

One limitation of the QR Buck converter is that it cannot operate when the input voltage is smaller than the output voltage. To ensure that there is a smooth transition from operation to hold up mode without significant input current distortion, the FET is held off when the input voltage drops below 100V. This is achieved by comparing the scaled down rectified input voltage to a set threshold V_{en}.

6) Noise Filter

Low pass filters are added to all control signals and ground planes are generously used to maximize shielding and filter out all undesired switching noise that comes from the power stage. These filters are used on low voltage logic signals and the size impact of the converter is negligible. As the experimental waveforms in fig.4, fig. 5, fig. 6, and fig. 7 show, noise is present on all of the control signals. This noise is a combination of actual noise generated by the high slew rate of

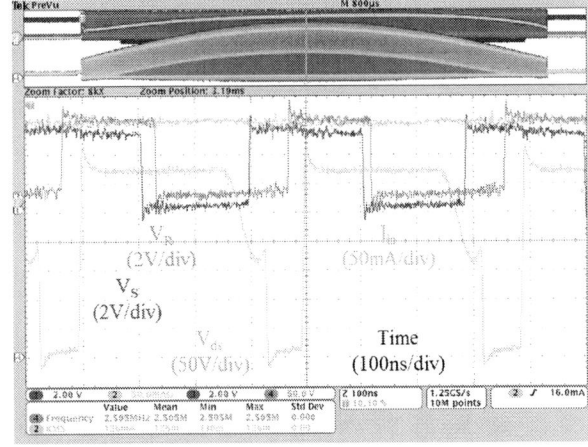

Fig. 9. Key Switching and Control Waveforms at High Line Input

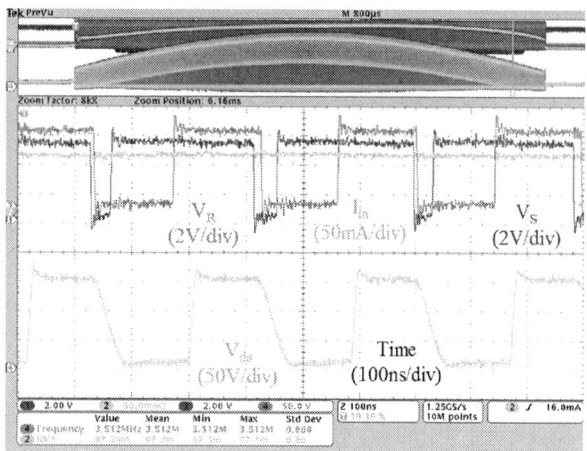

Fig. 10. Key Switching and Control Waveforms at Low Line Input

the GaN FET and measurement noise created by the ground loop inductance of the oscilloscope probe. As the results show the converter operates as expected so this noise is at an acceptable level.

V. Experimental Results

The prototype was built to verify the control technique and validate the performance. The results show the converter was able to successfully drive the LED with an efficiency of 92.9% while meeting the IEC6100-3-2 input power limits for the individual harmonics as shown in Fig. 11 [10]. The measured current total harmonic distortion is 15.9% with a power factor of 0.976.

To evaluate these results, a commercial silicon based product with equal input and output specifications is benchmarked. The same LED string output is used for both converters, along with the same AC input voltage. The QR Buck prototype has a significant power density improvement, with a decreased power stage component size of 0.309 in^3 over the commercial product's 0.579 in^3. The power stage volume includes C_{in}, L_{in}, D_{BR1}, D_{BR2}, D_{BR3}, D_{BR4}, D, Q, L_O, C_O, as well as an overvoltage protection varistor. The controller of the QR Buck is assumed to be an analog integrated circuit that is in a standard SOIC-8 package instead of the several discrete

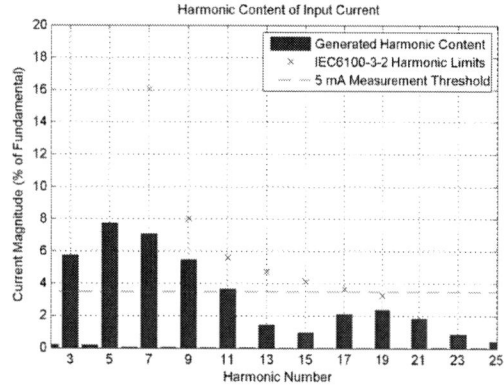

Fig. 11. Measured Input Current Harmonic Contentent Compared to IEC6100-3-2 Harmonic Limits

TABLE I.   Experimental Prototype Results

Design	Switch Freq	Output Power	Power Loss	Efficiency	Power Stage Density
Commerical Product	50kHz	19.2W	1.5W	92.9%	19.80W/in^3
QR Buck Prototype	2.5-4.4 MHz	20.2W	1.8W	91.2%	36.93W/in^3

components used. The efficiency of the QR Buck prototype, while very close, is slightly lower than the commercial product.

VI. Conclusion

As the results show, the size reduction by using GaN and high switching frequency allows for significant size reduction while maintaining similar efficiency and input power quality to existing silicon-based solutions. Commercially available power controllers are not designed to operate in the megahertz switching frequency range that GaN enables. As GaN FETs are further developed and become reliable for applications such as LED drivers, discrete analog controllers, such as the one described and implemented in this paper, will be needed to validate new designs and performance improvements. Once these control techniques are validated, custom optimized analog and digital integrated circuit controllers can be designed to further improve power density and improve performance.

References

[1] Lutron Electronics Co, Inc., "Controlling LEDs," Coopersburg, PA, May 2014. Available: http://www.lutron.com/TechnicalDocumentLibrary/367-2035_LED_white_paper.pdf

[2] US Department of Energy, "General Service LED Lamps," PNNL-SA 87502, April 2012. Available: http://apps1.eere.energy.gov/buildings/publications/pdfs/ssl/led_general-service-lamps.pdf

[3] *LM3444 Datasheet*, Texas Instruments, Dallas, TX, 2013

[4] *TPS92075 Datasheet*, Texas Instruments, Dallas, TX, 2014

[5] *NCL30088-D Datasheet*, On Semiconductor, Phoenix, AZ, 2015

[6] Uprety, S.; Hai Chen; Dongsheng Ma, "Quasi-hysteretic floating buck LED driver with adaptive off-time for accurate average current control in high brightness lighting," *Circuits and Systems (ISCAS), 2011 IEEE International Symposium on* , vol., no., pp.2893,2896, 15-18 May 2011

[7] Anghel, V.; Bartholomeusz, C.; Vasilica, A.G.; Pristavu, G.; Brezeanu, G., "Variable Off-Time Control Loop for Current-Mode Floating Buck Converters in LED Driving Applications," *Solid-State Circuits, IEEE Journal of* , vol.49, no.7, pp.1571,1579, July 2014

[8] Bandyopadhyay, S.; Neidorff, B.; Freeman, D.; Chandrakasan, A.P., "90.6% efficient 11MHz 22W LED driver using GaN FETs and burst-mode controller with 0.96 power factor," *Solid-State Circuits Conference Digest of Technical Papers (ISSCC), 2013 IEEE International* , vol., no., pp.368,369, 17-21 Feb. 2013

[9] B. Sun, and Z. Ye, "PFC THD Reduction and Efficiency Improvement by ZVS or Valley Switching," Dallas, TX, SLUA644 Application Note, April 2012. Avaiable: http://www.ti.com/lit/an/slua644/slua644.pdf

[10] European Power Supply Manufacturers Association, "Harmonic Current Emissions: Guideline to the standard EN 61000-3-2," November 2010. Available: http://www.epsma.org/pdf/PFC%20Guide_November%202010.pdf

A Novel LED Drive System Based on Matrix Rectifier

Baoping Shi[*], Bo Zhou, Jiadan Wei, Xianhui Qin, Yuanyu Yang, Bing Liu
Nanjing University of Aeronautics and Astronautics, Nanjing, China
E-mail: baoping@nuaa.edu.cn

Abstract-This paper presents a novel LED drive system based on matrix rectifier. Matrix rectifier is applied to the LED drive system for the first time as to our knowledge. The novel LED drive system is composed of two stages. The first stage is the matrix rectifier, and the second stage is the multi-port two-transistor forward converters (TTFC) that share a leg. The space vector modulation strategy of matrix rectifier and the modulation strategy of multi-port TTFC matched with rectifier stage are fully studied. In order to maintain the stability of output current, current PI closed-loop control is adopted for LED strings. Both simulation and experiment are carried out to verify the topology and modulation strategy of the proposed LED drive system.

Keywords—matrix rectifier; two-transistor forward converter (TTFC); Space vector modulation

I. INTRODUCTION

With the development of the third generation semiconductor material gallium nitride, high brightness LED brings innovation in the lighting technology, which has received considerable attention due to potentially attractive features such as high efficiency, long lifespan, energy saving and environmental protection[1]. Reliable and stable LED drive circuit is significant for LED strings, which is classified into two kinds: single-stage drive circuit and two-stage drive circuit. The two-stage LED drive circuit is usually connected to a single-phase AC source. Its first stage is the power factor correction circuit to obtain high input power factor [2], and its second stage provides a constant current source to drive the LED strings. Large storage capacitor is necessary to balance the instantaneous input and output power for the existing two-stage LED drive circuit [3], and the electrolytic capacitor is commonly used for its high energy density and low cost. However, the application of electrolytic capacitor will lead to a mismatch between the lifespan of drive circuit and that of LED strings. Because the estimated lifespan of the electrolytic capacitor is only up to approximately 10000 h, whereas the LED strings can last up to at last 50000 h [4]. Recently, several works on electrolytic capacitor-less LED driver circuit has made great progress. A new pulsating current driving to minimize the output ripple is proposed in [5]. In [6-7], a method of injecting the third harmonic current into the input current flow is presented. A novel off-line structure to drive LED strings with low output current ripple is proposed in [8]. A novel power factor correction topology by inserting the valley-fill circuit is presented in [9].The methods proposed above to eliminate the electrolytic capacitor have their own merits, but the large storage capacitor that affects the LED drive system performance still exists.

Besides, the LED drive circuit now available is mainly applied to the single LED string, and the drive power usually is at most dozens of W. Considering some applications such as the landscape lighting, large-area workshop, the underground mall, etc, a large number of LEDs are distributed according to the practical requirements in the light area with unified management and luminance control, so that high power drive is possible. Based on the discussion above, this paper proposes a novel LED drive system eliminating large storage capacitor, which is able to obtain sinusoidal input current and drive multiple independent LED strings, and the concentrated control and power supply is achieved.

This paper is organized as follows: section II presents the structure of proposed LED drive system topology. The modulation strategy and operation principle analysis is presented in section III. Simulation and experiments are implemented and their results are shown in section IV and section V, respectively. Finally, some conclusions are given in section VI.

II. LED DRIVE SYSTEM TOPOLOGY

The topology of proposed LED drive system is shown in Fig. 1. The LED drive system which is powered by three-phase AC power supply eliminates short-life storage capacitor, and all LED strings are controlled independently. The first stage of LED drive system is the matrix rectifier consisting of six bidirectional switches, and the second stage is the multi-port TTFC with multiple independent outputs, which share a common leg to decrease the number of switches and simplify the circuit topology [10-11]. Especially, the unidirectional switch connected to positive pole (P) and the diode connected to negative pole (N) constitute the shared leg, and others are independent legs of each TTFC. Besides, independent transformers, rectifier diodes, free-wheeling diodes and filter inductors are included for each TTFC. Compared with the traditional two-stage LED drive circuits, advantage and innovation are as follows:

1)the matrix rectifier is applied to the LED drive system for the first time , and the proposed LED drive system gets rid of large storage capacitor ;

2)unity input power factor and sinusoidal input current can be achieved after the input filter eliminating high harmonics related to the switching frequency of matrix rectifier;

3)multiple independent LED strings are obtained with three-phase AC power input;

4)the bidirectional switches of matrix rectifier can realize the zero current switch (ZCS) with optimal control.

This work is supported by the National Undergraduate Training Program for Innovation and Entrepreneurship

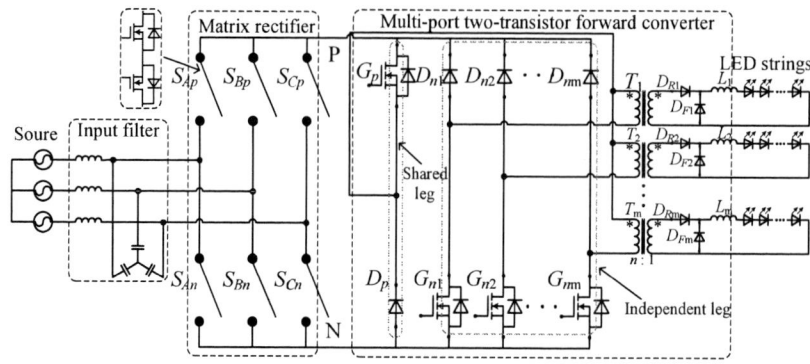

Fig. 1. Topology of proposed LED drive system

III. MODULATION STRATEGY AND OPERATION PRINCIPLE ANALYSIS

The matrix rectifier is operated using the input current space vector modulation, and the distribution diagram of six active current vectors is presented in Fig. 2.

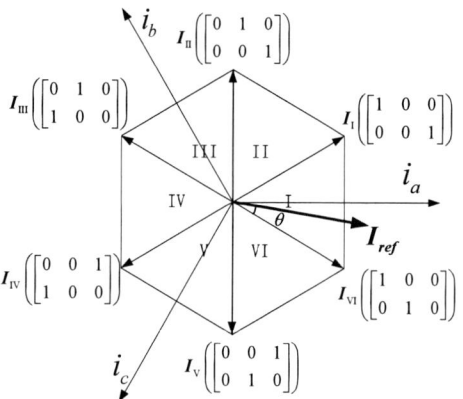

Fig. 2. The space vector diagram

where i_a, i_b, i_c are the input current, and $I_I \sim I_{VI}$ are the six active current vectors. Taking the vector I_I as an example, $I_I = \begin{bmatrix} S_{Ap} & S_{Bp} & S_{Cp} \\ S_{An} & S_{Bn} & S_{Cn} \end{bmatrix} = \begin{bmatrix} 1 & 0 & 0 \\ 0 & 0 & 1 \end{bmatrix}$, which indicates that bidirectional switches S_{Ap}, S_{Cn} are on and other bidirectional switches are off.

The reference input current vector I_{ref} is synthesized by the two nearest active current vectors and zero current vectors in a sampling period. The duty ratio of vectors is determined as follow:

$$\begin{cases} d_1 = m_c \sin\theta \\ d_2 = m_c \sin(\frac{\pi}{3} - \theta) \\ d_0 = 1 - d_1 - d_2 \end{cases} \quad (1)$$

where m_c is the modulation index of the rectifier stage. d_1 and d_2 are the duty ratio of adjacent active current vectors respectively, and d_0 is the duty ratio of zero vectors.

Because the dc-link voltage varies between two values, the average value of the dc-link voltage in one sampling period can be achieved as:

$$U_{PN} = d_1 u_1 + d_2 u_2 \quad (2)$$

where U_{PN} is the average value of the dc-link voltage. u_1 and u_2 are the dc-link voltage respectively corresponded to the duty ratio d_1 and d_2. According to the theory of matrix converter [12-13], in any sampling period, U_{PN} is a constant value that is given as:

$$U_{PN} = d_1 u_1 + d_2 u_2 = \frac{3}{2} m_c V_{im} \cos\varphi_i \quad (3)$$

where V_{im} is the amplitude of input voltage. $\cos\varphi_i$ is set to 1.

To reduce switching losses, zero current vectors are cancelled [14], and the duty ratios of the two active current vectors are given bellow:

$$\begin{aligned} d_m &= d_1 + 0.5 d_0 \\ d_n &= d_2 + 0.5 d_0 \end{aligned} \quad (4)$$

where d_m and d_n represent the amendatory duty ratio of adjacent active vectors, respectively.

The modulation strategy matched with the matrix rectifier is adopted for the multi-port TTFC. Assume that I_{ref} is synthesized by active current vectors I_V and I_{VI} in this sampling period. Taking the shared leg and the first independent leg as an example, the coordinated control between two stages is shown in Fig. 3. The operation principle of other independent legs and modulation periods is similar.

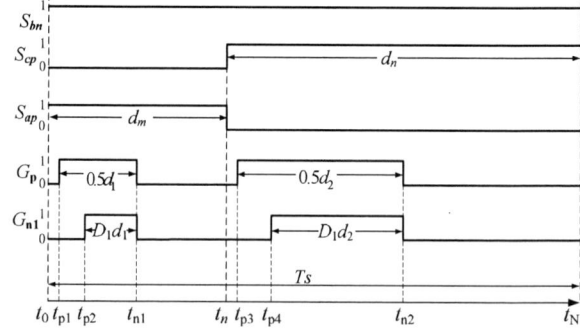

Fig. 3. Diagram of coordinated control

As is shown in the Fig. 3, T_s is a sampling period, and D_1 ($0<D_1<0.5$) represents the duty ratio of switch G_{n1}. Specific value D_1 is decided by current closed-loop adjustment of the first LED string. In the sampling period, bidirectional switch S_{Bn} is turned on and S_{Cn}, S_{An}, S_{Bp} are turned off constantly. From t_0 to t_n, S_{Ap} is on when vector I_{VI} works, and from t_n to t_N, S_{Cp} is on when vector I_V works.

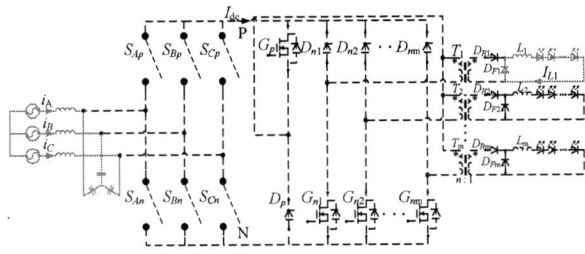

Fig. 4. Circuit state I

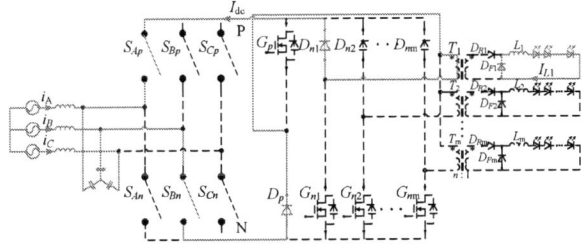

Fig. 5. Circuit state II

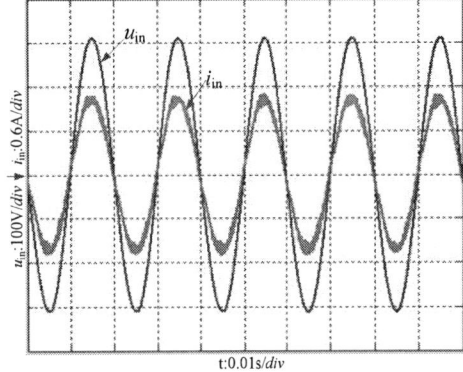

Fig. 6. Circuit state III

Dc-link voltage u_{AB} can be seen as a constant value when vector I_{VI} works because switching frequency is far higher than input voltage frequency. $[t_0\text{-}t_n]$ is divided into three time ranges, namely $[t_0\text{-}t_{p2}]$, $[t_{p2}\text{-}t_{n1}]$, $[t_{n1}\text{-}t_n]$.

1)$[t_0\text{-}t_{p2}]$: Circuit state is shown in Fig. 4. Dc-link current I_{dc} is zero, although G_p is turned on at t_{p1}. The first LED string current I_{L1} freewheels through diode D_{F1}.

2)$[t_{p2}\text{-}t_{n1}]$: G_{n1} is turned on at t_{p2}. G_p and G_{n1} are turned off at t_{n1}, and Fig. 5 shows the circuit state now. Both I_{dc} and magnetic energy of transformer T_1 increase with the power flowing from input source to the first LED string.

3)$[t_{n1}\text{-}t_n]$: Due to $0<D_1<0.5$, supposing that the T_1 magnetic reset finished at t_k ($t_{n1}<t_k<t_n$). During $[t_{n1}\text{-}t_k]$, as is shown in Fig.6, I_{dc} down to zero reversely with magnetic energy feeding back to the input power. During $[t_k\text{-}t_n]$, I_{dc} is zero because T_1 has completed magnetic reset, and the circuit state is shown in Fig. 4. At t_n, current vector I_{VI} switches to I_V, the bidirectional

switch S_{Ap} achieves ZCS. I_{L1} freewheels through diode D_{F1} during $[t_{n1}\text{-}t_n]$.

From t_n to t_N, vector I_V works and dc-link voltage is u_{CB}. Operational principle analysis is similar. At t_N, present sampling period ends with some bidirectional switches achieving ZCS. The period T_{on} when both G_p and G_{n1} are on is given as:

$$T_{on} = D_1 d_1 T_s + D_1 d_2 T_s \tag{5}$$

From (3) and (5), when both G_p and G_{n1} are turned on, average input voltage of the first group TTFC U_1 is recalculated as:

$$U_1 = D_1 d_1 u_{AB} + D_1 d_2 u_{CB} = \frac{3}{2} m_c V_{im} \cos \varphi_i D_1 \tag{6}$$

It can be concluded that U_1 is only related to D_1 in any one modulation period, which is conducive to the stability of the output voltage of the first LED string.

IV. SIMULATION RESULTS

The topology of the proposed LED drive system shown in Fig. 1 has been simulated using the MTALAB software. The simulation parameters are as follows: input phase voltage RMS is 220V; input voltage frequency is 50Hz; the system drives eight independent LED strings connected to 50 LEDs each; current reference value for all LED strings is 350mA; the switching frequency for the first stage is 30kHz and the switching frequency for the second stage is 60kHz; the input filter L_f is 0.5mH, and the C_f is 3μF.

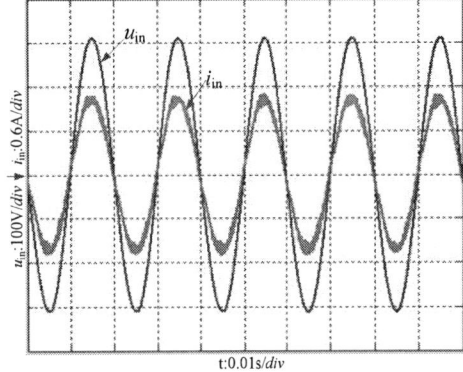

Fig. 7. Simulated waveforms of input voltage and current

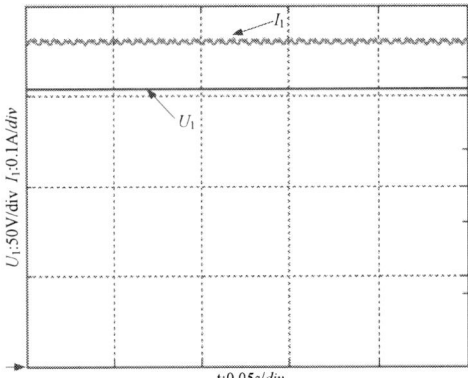

Fig. 8. Simulated waveforms of output voltage and current

978-1-4673-9551-9/16 $31.00 © 2016 IEEE

Fig. 7 shows the simulated waveforms of input voltage and current. Fig. 8 shows the simulated waveforms of output voltage and current. It can be seen that input current i_{in} is sinusoidal in phase with voltage u_{in} and output current I_1 maintains at 350mA with small ripple. The system provides a constant current source with high performance.

V. EXPERIMENTAL RESULTS

To validate the proposed theory and simulated results, the laboratory prototype is built. The control system is developed with the high performance DSP TMS320F28335 produced by Texas Instruments and FPGA XC6SLX4 produced by Xilinx. The power switch MOSFET IPB65R045C7 produced by Infineon has been used to implement the power circuit. Due to limitation of the present experimental conditions, the experimental parameters are as follows: input phase voltage RMS is 50V; input voltage frequency is 50Hz; the system drives four independent LED strings connected to 10 LEDs each; the current reference value for all strings is 300mA; the switching frequency for the first stage is 30kHz and the switching frequency for the second stage is 60kHz; the input filter L_f is 0.5mH, and the C_f is 3μF.

Fig. 9 shows the experimental waveforms of input voltage and current, and the FFT analysis results are shown in Fig. 10. Fig. 11 shows the experimental waveforms of output voltage and current. The results show that input current i_A is sinusoidal with slight harmonics, and the current is in phase with voltage. Output voltage U_o is stable, and output current I_o is fixed to 300mA, realizing constant current source drive for LED strings.

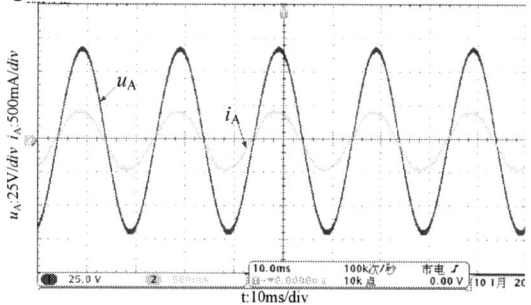

Fig. 9. Experimental waveforms of input voltage and current

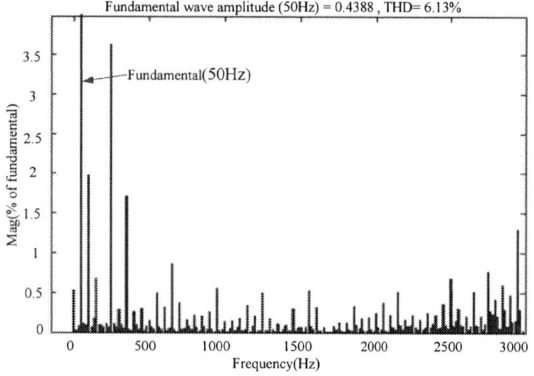

Fig. 10. FFT analysis results of input current

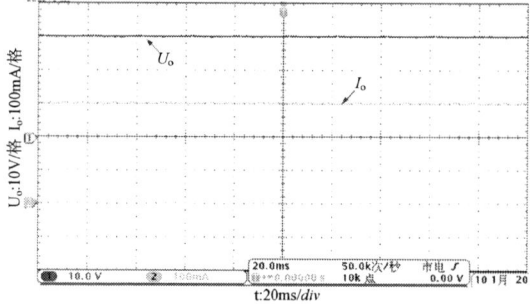

Fig. 11. Experimental waveforms of output voltage and current

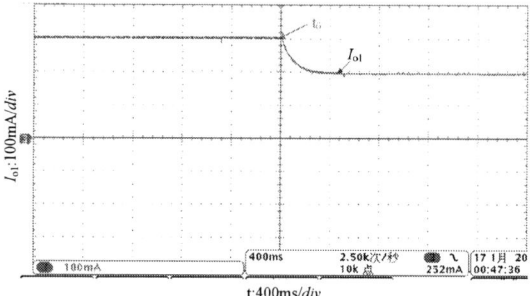

Fig. 12. Experimental waveforms of dynamic response of output current

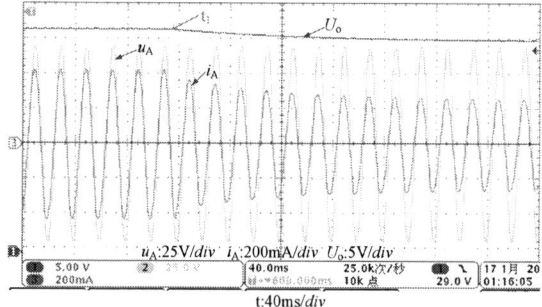

Fig. 13. Experimental waveforms of dynamic response of input current

Fig. 12 shows the experimental waveforms of dynamic response of output current. At t_0, current reference value for the first LED string declines from 300mA to 200mA with other experiment conditions unchanged. From Fig. 12 we can know that output current I_{o1} transits smoothly without serious shake, which shows current closed-loop controller has good dynamic performance and high stability.

Fig. 13 shows the experimental waveforms of dynamic response of input current. At t_1, current reference value for all LED strings declines from 300mA to 200mA with other experiment conditions unchanged. We can see that input current i_A has smooth transition with sinusoidal waveform when load current changes sharply.

Fig. 14 shows the LED drive system prototype. Fig. 15 shows the LED drive system experiment. The current reference values of the four LED strings are respectively 50mA, 100mA, 200mA, 300mA. It can be seen from Fig. 15 that LED luminance increases in sequence from group one to group four, which indicates four LED strings realize independent control of each other.

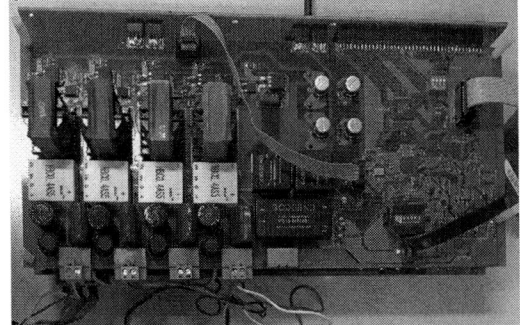

Fig. 14. LED drive system prototype

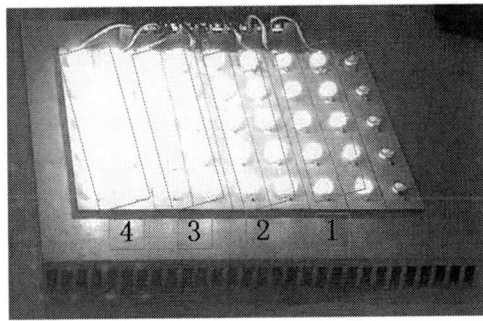

Fig. 15. LED drive system experiment

VI. CONCLUSIONS

This paper proposes a novel LED drive system based on the matrix rectifier, which achieves independent control of multiple LED strings. Large storage capacitor is eliminated compared to the existing two-stage LED drive system. The simulation and experimental results show that sinusoidal input current is obtained and multiple independent LED strings are driven by constant current source, validating the feasibility and rationality of the proposed the drive system. In future work, we will further improve performance of the LED drive system.

ACKNOWLEDGMENT

The authors would like to sincerely thank the National Undergraduate Training Program for Innovation and Entrepreneurship for supporting the research.

REFERENCES

[1] Van der Broeck H, Sauerlander Georg, Wendt M, Power driver topologies and control schemes for LEDs, in *IEEE 2007 Applied Power Electronics Conference*, 2007, pp. 1319-1325.

[2] Myunghyo Ryu, Jonghyun Kim, Juwon Baek, Heung-Geun Kim, New multi-channel LEDs driving methods using current transformer in electrolytic capacitor-less AC-DC Drivers, in *IEEE 2012 Applied Power Electronics Conference*, 2012, pp. 2361-2367.

[3] Fanghua Zhang, IEEE, Jianjun Ni, Yijie Yu, High power factor AC–DC LED driver with film capacitors, *IEEE Trans. Power Electron.*, vol. 28, no. 10, pp. 4831–4840, Oct. 2013.

[4] Report: Lifetime of White LEDs, Energy eficiency and renewable energy, U.S. Dept. Energy, Washington, DC, Sep. 2009.

[5] John C. W. Lam, Praveen K. Jain, A High Power Factor, Electrolytic capacitor-less AC-input LED driver topology with high frequency pulsating output current, *IEEE Transactions on Power Electronics*, vol. 30, no. 2, pp. 943-955, Feb. 2015.

[6] Linlin Gu, Xinbo Ruan, Ming Xu, Means of eliminating electrolytic capacitor in the AC/DC power supplies for LED lighting, *IEEE Transactions on Power Electronics*, vol. 24, no. 5, pp. 1399-1408, May. 2009.

[7] B. Wang, X. Ruan, K. Yao, M. Xu, A method of reducing the peak-to-average ratio of LED current for electrolytic capacitor-less AC–DC drivers, *IEEE Trans. Power Electron.*, vol. 25, no. 3, pp. 592–601, Mar. 2010.

[8] Valipour, H., Rezazadeh, G., Zolghadri, M.R., Noroozi, N, Electrolytic capacitor-less AC-DC LED driver with constant output current and PFC, in *Power Electronics, Drive Systems &Technologies Conference*,2015, pp.107-112

[9] H. Ma, J.-S. Lai, Q. Feng, W. Yu, C. Zheng, Z. Zhao, A novel valley-fill SEPIC-derived power supply without electrolytic capacitor for LED lighting application, *IEEE Trans. Power Electron.*, vol. 27, no. 6,pp. 3057–3074, Jun. 2012.

[10] Chen Yangfei, He Ligao, An interleaved series-parallel combination of two-transistor forward converters, in *IEEE 2006 Industrial Electronics and Applications*, 2006, pp. 1-5.

[11] Shuangjing Yang, Yu Fang; Xun Qiu, Chunying Gong, Voltage sharing control for interleaving series-parallel dual two-transistor forward converter, in *IEEE 2007 Industrial Electronics Society*, 2007, pp. 1896-1900

[12] Zhiping Wang, Yunxiang Xie, Yu Wang, Feedback control strategy for matrix rectifier, in *International Power Electronics and Motion Control Conference*, 2012, pp. 2002-2006

[13] Li Quan-chun, Zhou Bo, Shi Ming-ming, Liu Xiao-yu, Research on VSCF power generation system based on incorporation design of HESM and TSMC, in *International Conference 2010 Electrical and Control Engineering*, 2010, pp. 5558-5561.

[14] Nguyen, T.D. , Hong-Hee Lee, Modulation strategies to reduce common-mode voltage for indirect matrix converters, *IEEE Transactions on Industrial Electronics*, vol. 59, no. 1, pp. 129-140, Jan. 2012.

LLC Synchronous Rectification Using Coordinate Modulation

Mehdi Mohammadi, Navid Shafiei and Martin Ordonez

Electrical and Computer Engineering

The University of British Columbia

Vancouver, BC V6T1Z4 Canada

Email: mehdi.mohammadi.m@ieee.org, navid@ece.ubc.ca, mordonez@ieee.org

Abstract— *LLC* **resonant converters have gained popularity in a variety of applications due to their natural ability to provide soft switching conditions. Although *LLC* features good efficiency, conductive losses in the output rectifier remains a barrier to achieve enhanced efficiency, specially for high output current applications. This paper introduces a new strategy referred to as Coordinate Modulation, which is designed to drive the synchronous rectification MOSFETs. The main challenge to successfully drive synchronous rectifier is to detect/find the rectifier conduction angles with a cost-effective strategy. This paper proposes and analyzes a robust and simple detection method in the $v_{inv} - nv_{sec}$ coordinate plane. The plane allows a remarkable ability to detect conduction angles with two lines, producing proper pulses to drive synchronous rectifier. In order to validate the theoretical analysis, a $1.2KW$ prototype *LLC* converter is implemented and the experimental results using the Coordinate Modulation are presented. At the nominal output power, the converter's efficiency was measured at 97.1%. In comparison with a conventional *LLC* converter, the proposed modulation method increases the efficiency by 3%.**

Keywords-

I. INTRODUCTION

As the use of power converters and the demand for high performance converters grow, it is increasingly necessary to take into account efficiency, electromagnetic interference (EMI), size and weight when designing a converter. Recently, resonant converters have been widely used due to their distinct merits for providing soft switching conditions, reduced EMI and improved efficiency, which allow the switching frequency to increase. Among different types of resonant converters, the *LLC* converter is frequently mentioned for its ability to process the electrical energy in different applications, such as power factor correction (PFC) [1], LED drivers [2], battery chargers [3], and to process the energy produced by fuel cells [4]. Although, the *LLC* converter is capable of providing soft switching conditions, including zero

voltage and zero current switching (ZVS and ZCS) conditions for the semiconductor elements, conductive losses have remained an obstacle to achieving high efficiencies especially in high output current applications. In order to tackle the diminished efficiency due to the conductive losses that are mainly caused by the forward voltage of the rectifier diodes, the use of synchronous rectifiers (SRs) has frequently been addressed in literature [5–18]. The main effort required to use SRs is expended in finding their conduction time and in driving them correctly. A trapezoid synchronized voltage doubler employed in the secondary side of the *LLC* converter's transformer is used to reduce the converter's conductive losses [13]. The advantage of using a voltage doubler SR is that it provides clamped voltage stress for the rectifiers; however, the current rating of SRs becomes twice that of a full bridge or center-tapped rectifier. A current transformer (CT) helps to sense the direction of the current so that the SRs can be driven appropriately. Because, in the applications where SRs are used, the output current is usually high, the current can be sensed through the primary side of the transformer, rather than the secondary side. Since the primary current includes both the current through the series resonant tank and the magnetizing inductor, a current compensating winding can be used to cancel out the effect of the magnetizing inductor's current [14]. Instead of using a CT to drive the SRs, the voltage over the drain to source of the SR can be sensed to reduce the body diode's conduction time [15]. Because of the low resistance of the on state resistor of MOSFETs, the dropped voltage over the drain to source when the SR is in on state is low, which makes the driving circuit of the SRs sensitive to noise [16]. However, by applying a zero-crossing noise filter, the problem of false triggering can be solved [17]. In [18], a hybrid driving method for

978-1-4673-9551-9/16 $31.00 © 2016 IEEE

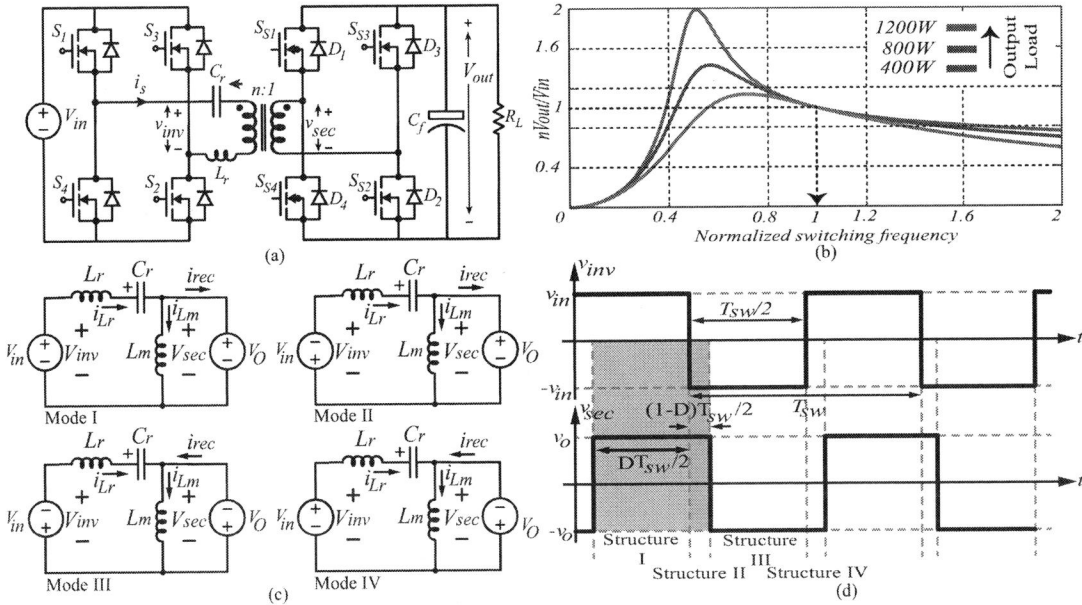

Fig. 1: (a) The *LLC* converter with synchronous rectifier, (b) The converter's gain diagram, (c) The equivalent circuit of each operating mode, (d) the converter's key waveforms.

a full bridge SR of an *LLC* converter is proposed where both the concepts of voltage and current driving methods are engaged. Because, this method uses the converter's transformer, the leakage and magnetizing inductances of the transformer can affect the dynamic response of the SR's driver.

This paper introduces a new modulation method, called coordinate modulation, for synchronizing the rectifier of the *LLC* resonant converter. As mentioned previously, the key to using an SR is to identify the point in time when the SR should conduct. In the proposed method, there is no need to sense the voltage over the drain and source of the SR or to use CTs, since the output voltage of the converter's inverter v_{inv} and the voltage of the transformer's secondary side v_{sec} are sensed. Because the amplitude of these two signals is large enough, in terms of immunity against noise, the proposed method prevents chattering at the instants when the state of the SRs is changed. In order to provide an analytical tool to anticipate the angle in which an SR should be driven, the behavior of the converter in the $v_{inv} - nv_{sec}$ coordinate plane is investigated. However, although the LLC converter is controlled by changing the switching frequency, the analysis of the converter in regards to frequency can be used to analyze the converter while it is using the converter's effective duty cycle D. Therefore,

by using the converter's effective duty cycle, the angle for triggering the SR can be obtained approximately in the $v_{inv} - nv_{sec}$ coordinate plane. Finally, a simple algorithm is introduced to drive the SRs.

This paper is organized in 4 sections. In section II, the *LLC* principle operation and proposed coordinate modulation method are introduced. In order to justify the theoretical analysis, the experimental results are presented in section III. In section IV, the conclusion is presented.

II. THE PROPOSED COORDINATE MODULATION METHOD

Traditionally, LLC converters are analyzed in the gain-frequency plane under different loading conditions, as depicted in Fig. 1 (b). Unfortunately, this traditional analysis does not provide appropriate information required to drive synchronous rectifiers. In this work, the analysis is done by measuring the inverter voltage and the transformer secondary voltage, leading to valuable information about the conduction angles in the rectifier. Therefore, instead of using the gain-frequency plane, the proposed strategy develops a new Coordinate Modulation for SR based on large signal voltages. What follows is the explanation to successfully drive SR after the resonant frequency of the series resonant tank in which the converter has four operating modes as shown in Fig. 1

(c). After the resonant frequency, because of the inductive effect of the resonant tank, a phase delay is produced between v_{inv} and the current through the secondary side of the transformer i_{sec}. Since i_{sec} is in phase with v_{sec}, this feature can be employed to develop the proposed analytical tool and driving algorithm. The key theoretical waveforms of the converter are shown in Fig. 1 (d). Although, in Fig. 1 (a), the LLC converter is shown along with the SRs, in order to discuss the converter's operating modes without the SRs, it is assumed that all the switches of the SR are in off state and only their body diodes conduct. Also, due to the symmetrical performance of the converter, only the first and second modes are discussed here. Before Mode I, it is assumed that the switches S_1 and S_2 are in on state and that the diodes D_1 and D_2 are off, and D_3 and D_4 are on.

Mode I [$t_0 < t < t_1$]: At the beginning of this mode, the diodes D_1 and D_2 turn on under the ZCS condition and cause the voltage of the secondary side of transformer to become $+V_O$. As long as the converter switches S_1 and S_2 are in the on state, the converter remains in Mode I.

Mode II [$t_1 < t < t_2$]: At t_1, the converter switches S_1 and S_2 are turned off and S_3 and S_4 are turned on. Therefore, v_{inv} becomes $-V_{in}$. Because of the phase lag between i_{sec} and v_{inv}, in this mode the diodes D_1 and D_2 remain on until i_{sec} reaches to zero.

At the resonant frequency, because v_{inv} and i_{sec} become in phase, the number of the converter's operating modes reduces to two, since Mode II and Mode IV are eliminated. As the consequence, v_{inv} and v_{sec} become in phase.

According to the discussions above and based on the phase shift between v_{inv} and v_{sec} caused by raising the switching frequency above the resonant frequency, the converter's effective duty cycle D can be defined. As shown in Fig. 1 (d), D is defined when the polarity of both v_{inv} and v_{sec} is the same during half of the switching frequency. The maximum value of D is 1 and it is reached when the switching frequency equals the resonant frequency. By increasing the switching frequency, the phase shift between v_{inv} and v_{sec} increases, causing D to decrease. Therefore, based on this definition, the converter's voltage gain can be explained by using D instead of the switching frequency. In Fig. 2 (a), the

converter's gain over the duty cycle (D) is shown. However, Fig. 2 (a) shows a non-linear performance for different load conditions; compared to the gain-switching frequency plane, it offers a more linear behavior, especially when the switching frequency is close to the resonant frequency. By considering the blue line in Fig. 2 (a), the converter's voltage gain can be calculated as follows:

$$\frac{V_O}{V_{in}} = \frac{(2D-1)}{n} \tag{1}$$

If v_{inv} is drawn over v_{sec}, a $v_{inv} - nv_{sec}$ coordinate plane is obtained where the conduction angles of each rectifier diode can be extracted. The $v_{inv} - nv_{sec}$ coordinate plane is shown in Fig. 2 (b). In the $v_{inv} - nv_{sec}$ coordinate plane, analyzing the converter's performance involves identifying the width of a rectangle the center of which is origin. For two different output voltages, two performances are shown in Fig. 2 (b). In the case that the switching frequency is equal to the resonant frequency, $nV_O = V_{in}$. Therefore, in the $v_{inv} - nv_{sec}$ coordinate plane, the outer rectangle (red rectangle) describes this situation. The points A and B, two edges of the outer rectangle, are connected by a line called the reference line W_r. If the converter meets the reference line W_r, it means that v_{inv} and v_{sec} are in phase and the switching frequency equals the resonant frequency. Because in the buck mode and after the resonant frequency, the maximum gain is 1 (by assuming n to be 1), the angle of W_r is $\pi/4$ radians from the vertical axis. By decreasing the voltage gain, the width of the rectangle decreases. Therefore, for a specific gain below 1, the inner rectangle can be drawn in $v_{inv} - nv_{sec}$ coordinate plane. Similarly to the previous case, if the points A' and B' are connected together through a line called W_O, its angle from the vertical axis becomes less than W_r's angle and can be calculated by use of the equation below:

$$\theta_O = tan^{-1}(\frac{nV_O}{V_{in}}) \tag{2}$$

Because the angle of W_r from the vertical axis is always constant, the difference between the angle of W_r and W_O can determine the phase delay the SRs should have from the inverter's gate pulses. The phase delay between v_{inv} and v_{sec} can be calculated as follows:

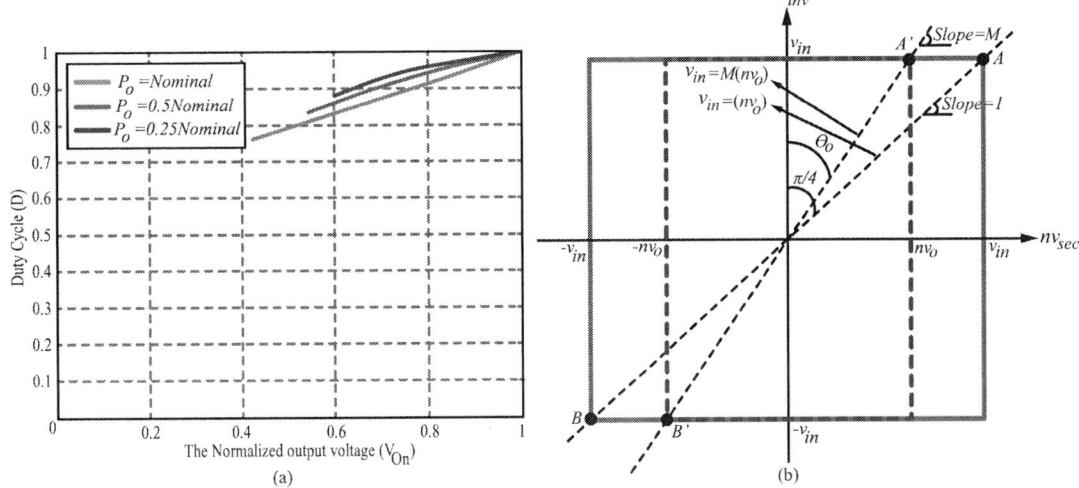

Fig. 2: (a) The converter's effective duty cycle versus gain, (b) the converter's performance in $v_{inv} - nv_{sec}$ coordinate plane.

$$\theta_d = 2[\frac{\pi}{4} - tan^{-1}(\frac{nV_O}{V_{in}})] \quad (3)$$

Based on the equation above, the maximum phase delay between v_{inv} and v_{sec} can be $\pi/2$ radians, at which point the output voltage becomes zero. This process can also be shown in the $v_{inv} - nv_{sec}$ coordinate plane. When the converter's gain is zero, the width of the inner rectangle is zero. Therefore, the angle between the line W_O and the vertical axis becomes zero and causes the angle between W_r and W_O to be $\pi/4$ radians. However, to reach to this point the switching frequency would have to be increased to infinity.

Based on the discussions above, a simple algorithm can now be introduced that is able to drive the SRs appropriately. In simple terms, the SRs only conduct when the polarity of v_{inv} and v_{sec} are the same. Because a full bridge rectifier is used in the LLC converter shown in Fig. 1 (a), when both v_{inv} and v_{sec} are positive, the SRs S_{S1} and S_{S2} conduct, and when both v_{inv} and v_{sec} are negative, the SRs S_{S3} and S_{S4} conduct.

III. EXPERIMENTAL RESULTS

In order to justify the theoretical analysis regarding the proposed coordinate modulation, the experimental results of a $1.2KW$ LLC converter are presented in this section. The input voltage of the converter is $200V$, and the nominal output voltage is $50V$. The resonant frequency is set at $90KHz$. The other converter's parameters are

as follows: $L_r = 39\mu H$, $C_r = 80nF$, $L_m = 110\mu H$, $n = 4$.

In Fig. 3, the performance of the proposed synchronous LLC converter at the nominal output power, where the output voltage is $50V$, is shown in both the time domain (a) and $v_{inv} - nv_{sec}$ coordinate plane (b). In Fig. 3 (b), the lines W_r and W_O are represented by the red and green colors, respectively. Since, in this case, the converter works at the resonant frequency, the angle between the lines W_r and W_O are zero, which means v_{inv} and v_{sec} are in phase. Also, this fact can be seen in the time domain.

In another case, the converter's performance is investigated after the output voltage is reduced to $40V$. In this case, the converter is under a 2Ω resistive load. For this load, to make the output voltage $40V$, the switching frequency should be $100KHz$. Because the switching frequency is greater than the resonant frequency, it is expected that a phase shift will occur between v_{inv} and v_{sec}. From (3), the phase delay should be 0.07π radians. In Fig. 4, the key experimental waveforms of the converter are shown in the time domain and $v_{inv} - nv_{sec}$ coordinate plane. In Fig. 4 (b), the angle between W_O and the vertical axis is 0.215π radians, matches with the theoretical value. Also, in the time domain (Fig. 4 (a)), as it is indicated, the phase delay between v_{inv} and v_{sec} is 0.068π radians, which is very close to the theoretical analysis.

At the nominal output power, the converter's efficiency was measured at 97.1%. The efficiency of an LLC

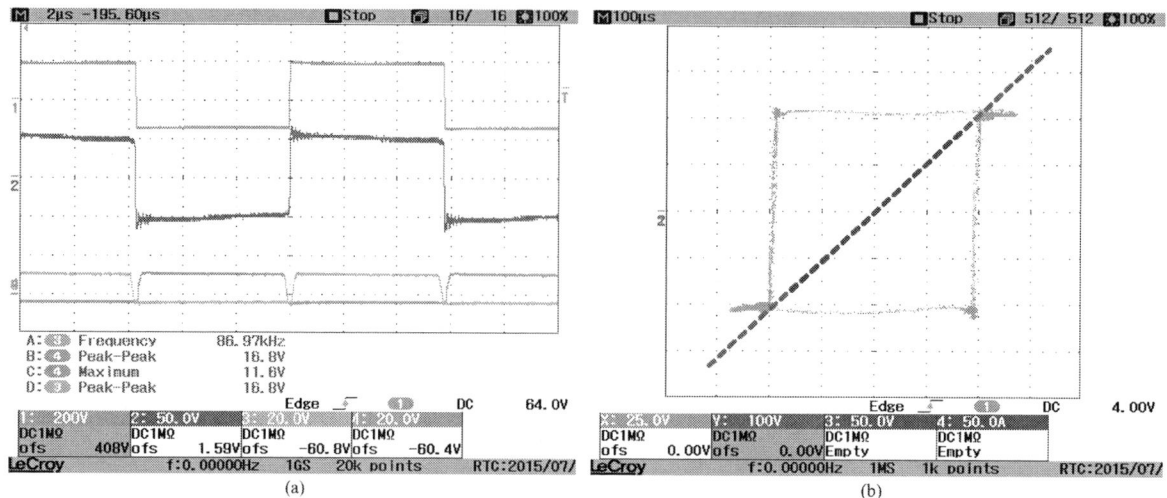

Fig. 3: The converter's performance at the nominal output power, (a) in the time domain, (b) in the v_{inv}-nv_{sec} coordinate plane. CH1: v_{inv}, CH2: v_{sec}, CH3 and CH4: gate pulses of the SRs.

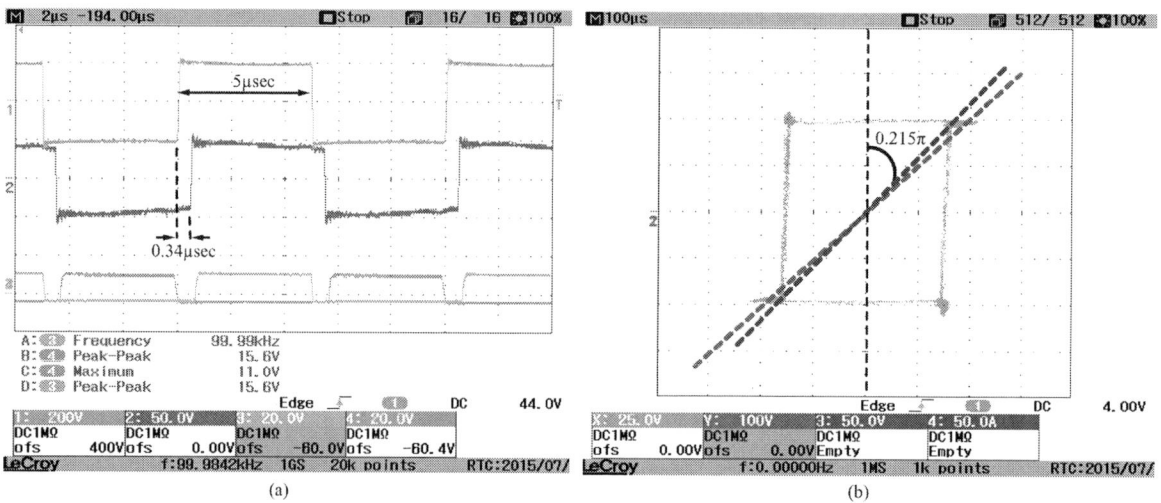

Fig. 4: The converter's performance when $V_O = 35V$, (a) in the time domain, (b) in the v_{inv}-nv_{sec} coordinate plane. CH1: v_{inv}, CH2: v_{sec}, CH3 and CH4: gate pulses of the SRs.

converter using a conventional full bridge rectifier was also measured. The results show that the proposed synchronized LLC converter has increased the efficiency by 3%.

IV. CONCLUSION

The losses in a power electronics converters can be divided into two categories called switching and conductive losses. For the LLC resonant converter, although the issue of the switching losses has been resolved, the conductive losses have prevented the attainment of high efficiencies, especially for high output current applications. Therefore, employing synchronous rectifiers to achieve a higher efficiency is inevitable. In this paper,

a new modulation method, called coordinate modulation, was introduced for synchronizing the output rectifier. The proposed method provides an analytical tool to find the phase delay required for the synchronous rectifier. Also, a simple algorithm was suggested for driving the synchronous rectifiers. To validate the theoretical analysis, a $1.2KW$ prototype LLC converter was implemented. The experimental results show that the proposed modulation method is able to synchronize the rectifier and provide an analytical tool for analyzing the triggering angles. The efficiency of the proposed LLC converter was measured at 97.1%. In comparison with a conventional LLC converter using a full bridge rectifier, the proposed modulation method has increased efficiency by 3%.

REFERENCES

[1] S.-Y. Chen, Z. R. Li and C.-L. Chen, "Analysis and Design of Single-Stage AC/DC LLC Resonant Converter," *IEEE Trans. on Industrial Electron.*, vol. 59, no. 3, pp. 1538-1544, Mar. 2012.

[2] W. Feng, F. C. Lee and P. Mattavelli, "Optimal Trajectory Control of LLC Resonant Converters for LED PWM Dimming," *IEEE Trans. on Power Electronics*, vol. 29, no. 2, pp. 979-987, Feb. 2014.

[3] N. Shafiei, M. Ordonez, M. Craciun, C. Botting and M. Edington, "Burst Mode Elimination in High Power LLC Resonant Battery Charger for Electric Vehicles," *IEEE Trans. on Power Electronics*, IEEE early access, 2015.

[4] J.-Y. Lee, Y.-S. Jeong and B.-M. Han, "An Isolated DC/DC Converter Using High-Frequency Unregulated LLC Resonant Converter for Fuel Cell Applications," *IEEE Trans. on Industrial Electronics*, Vol. 58, No. 7, pp. 2926-2934, Jul. 2011.

[5] J.-W. Kim, and G.-W. Moon, A New LLC Series Resonant Converter with a Narrow Switching Frequency Variation and Reduced Conduction Losses, IEEE Trans. on Power Electron., vol. 29, no. 8, pp. 4278-4287, Aug. 2014.

[6] C. Zhao, M. Chen, G. Zhang, X. Wu, and Z. Qian, A Novel Symmetrical Rectifier Configuration With Low Voltage Stress and Ultralow Output-Current Ripple, IEEE Trans. on Power Electron., vol. 25, no. 7, pp. 1820-1931, Jul. 2010.

[7] H. Miura, S. Arai1, F. Sato, H. Matsuki, and T. Sato, A Synchronous Rectification Using a Digital PLL Technique for Contactless Power Supplies, IEEE Trans. on Magnetics, vol. 41, no. 10, pp. 3997-3999, Oct. 2005.

[8] C. Duan, H. Bai, W. Guo and Z. Nie, Design of a 2.5-kW 400/12-V High-Efficiency DC/DC Converter Using a Novel Synchronous Rectification Control for Electric Vehicles, IEEE Trans. on Transportation Electrification, vol. 1, no. 1, pp. 106-114, Jun. 2015.

[9] M. K. Kazimierczuk, C. Wu, Frequency-Controlled Series-Resonant Converter with Synchronous Rectifier, IEEE Trans. on Aerospace and Electronic Systems, vol. 33, no. 3, pp. 939-948, Jul. 1997.

[10] D. Huang, D. Fu, F. C. Lee and P. Kong, High-Frequency High-Efficiency CLL Resonant Converters With Synchronous Rectifiers, IEEE Trans. on Industrial Electron., vol. 58, no. 8, pp. 3461-3470, Aug. 2011.

[11] D. Huang, S. Ji and F. C. Lee, LLC Resonant Converter With Matrix Transformer, IEEE Trans. on Power Electronics, vol. 29, no. 8, pp. 4339-4347, Aug. 2014.

[12] W. Feng, P. Mattavelli and F. C. Lee, Pulsewidth Locked Loop (PWLL) for Automatic Resonant Frequency Tracking in LLC DCDC Transformer (LLC-DCX), IEEE Trans. on Power Electronics, vol. 28, no. 4, Apr. 2013.

[13] J. Zhang, J. Liao, J. Wang and Z. Qian, A Current-Driving Synchronous Rectifier for an LLC Resonant Converter With Voltage-Doubler Rectifier Structure, IEEE Trans. on Power Electron., vol. 27, no. 4, pp. 1894-1904, Apr. 2012.

[14] X. Wu, G. Hua, J. Zhang, and Z. Qian, A New Current-Driven Synchronous Rectifier for SeriesParallel Resonant (LLC) DCDC Converter, IEEE Trans. on Industrial Electron., vol. 58, no. 1, pp. 289-297, Jan. 2011.

[15] W. Feng, F. C. Lee, P. Mattavelli and D. Huang, A Universal Adaptive Driving Scheme for Synchronous Rectification in LLC Resonant Converters, IEEE Trans. on Power Electron., vol. 27, no. 8, pp. 3775-3781, Aug. 2012.

[16] D. Fu, Y. Liu, F. C. Lee and Ming Xu, A Novel Driving Scheme for Synchronous Rectifiers in LLC Resonant Converters, IEEE Trans. on Power Electron., vol. 24, no. 5, pp. 1321-1329, May 2009.

[17] D. Wang and Y.-F. Liu, A Zero-Crossing Noise Filter for Driving Synchronous Rectifiers of LLC Resonant Converter, IEEE Trans. on Power Electron., vol. 29, no. 4, pp. 1953-1965, Apr. 2014.

[18] J. Zhang, J. Wang, G. Zhang, Z. Qian, A Hybrid Driving Scheme for Full-Bridge Synchronous Rectifier in LLC Resonant Converter, IEEE Trans. on Power Electron., vol. 27, no. 11, pp. 4549-4561, Nov. 2012.

Low Parasitics Planar Transformer for *LLC* Resonant Battery Chargers

Mohammad Ali Saket, Navid Shafiei, and Martin Ordonez
Electrical and Computer Engineering
The University of British Columbia
Vancouver, BC V6T1Z4 Canada
Email: alisaket@ece.ubc.ca, navid@ece.ubc.ca
and mordonez@ieee.org

Marian Craciun, and Chris Botting
Department of Research
Delta-Q Technologies Corp, Burnaby. BC, Canada
Email: cbotting@delta-q.com, mcraciun@delta-q.com

Abstract— Recently, high efficiency *LLC* resonant converters have gained popularity to implement battery chargers and employ traditional wire-wound transformers. Due to the height of traditional cores, the form factor of *LLC* chargers is often plump and bulky. In order to implement slim profile chargers for EVs and portable electronic applications, planar transformers can be used featuring low height, reproducibility, and low thermal resistance. Despite those advantages, planar magnetics suffer from very high parasitic capacitance resulting in severe regulation problems for *LLC* chargers that require wide voltage regulation. This paper analyzes *LLC* regulation issues with planar magnetics and proposes a novel strategy to provide wide regulation by reducing the effects of intra-winding capacitance. The proposed strategy employs a simplified capacitance network and achieves low parasitics to successfully design *LLC* chargers with wide regulation. Experimental results of a $1.2kW$ *LLC* converter confirms that the proposed winding layout not only provides better output voltage regulation in light load conditions, but also increases the overall efficiency of the power converter.

Over the past few years, the trend into high power density and high efficiency switched-mode power supply (SMPS) has spurred interest into planar transformers. In comparison to wire wound transformers, planar transformers provide lots of advantages including lower height, higher efficiency, lower leakage inductance, better heat dissipation, automated manufacturing and repeatability [1], [2], [3]. These benefits make planar transformers very desirable for low profile applications such as electric vehicles and portable electronics, where battery chargers should withstand mechanical stresses [4]. However, planar transformers suffer from high parasitic capacitance. This is a know problem that has been discussed in many previous studies [2], [5], [6], where high values for stray capacitances are reported. Indeed, Overlapping PCB traces and high voltage gradients in traditional winding design form distributed parallel plate capacitors between turns and windings, significantly increasing the parasitic capacitance in planar transformers.

Generally, high inter-winding capacitance between primary and secondary provides a low impedance path for common mode (CM) noise, contributing to EMI issues [4]. On the other hand, intra-winding capacitance gives rise to a high charging current at the transformer input, resulting in a loss

of overall efficiency and in increased peak voltage stress across secondary rectifying devices [7]. In addition, high intra-winding capacitance of planar transformers distorts the voltage conversion ratio of *LLC* resonant converter and impact light-load output regulation [8],[9],[10],[11]. As will be demonstrated in this work, the parasitic capacitance value must be minimized to achieve wide regulation in LLC chargers.

One way to reduce stray capacitance is to remove all overlapping faces between traces. For example, the work in [7] has proposed a new winding layout which removes overlapping faces between primary and secondary traces in order to minimize inter-winding capacitance. The same is applicable for intra-winding capacitance. In this case, overlapping faces between turns of the same winding should be avoided. While removing all overlapping faces will lead to the minimum possible intra-winding capacitance, it will also reduce the track widths by half, thus increasing conduction losses.

In this paper, a novel and high efficiency winding layout is proposed to effectively reduce intra-winding capacitance, while maintaining conduction losses low. The proposed layout is employed in a $1.2KW$ *LLC* resonant converter and results are compared to a typical planar winding layout. Experimental results show that the proposed winding layout not only has better performance in terms of voltage regulation, but also has increased the overall efficiency of converter.

This paper is organized in five Sections. In Section II, the voltage conversion ratio of *LLC* resonant converter considering transformers parasitic elements is presented and analyzed in light load condition. In Section III, a novel winding layout for reducing intra-winding capacitance based on FEA is presented. Experimental results are provided in section IV to verify advantages of the proposed layout. Finally, conclusion is presented in Section V.

I. ANALYSIS OF *LLC* RESONANT CONVERTER CONSIDERING TRANSFORMER PARASITIC ELEMENTS

In order to find the impact of transformer parasitic capacitance on converters performance, the three capacitors model

978-1-4673-9551-9/16 $31.00 © 2016 IEEE

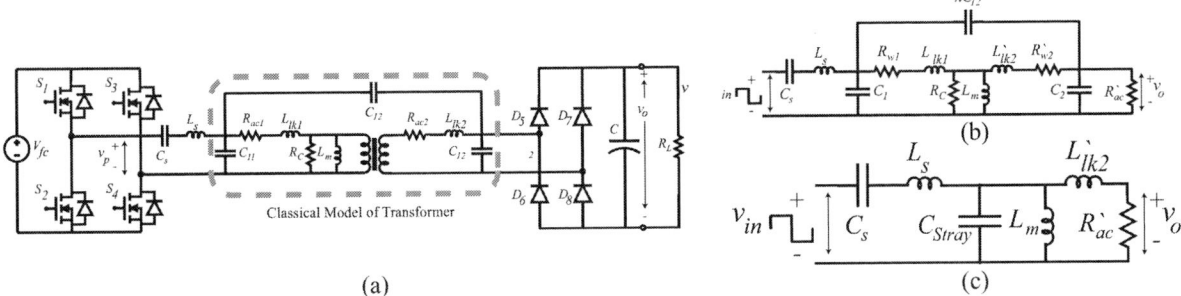

Fig. 1. (a) The *LLC* converter considering parasitics of transformer, (b) AC equivalent circuit refereed to primary, and (c) Simplified AC equivalent circuit.

of transformer is used [9]. Fig. 1 (a) shows the structure of the *LLC* resonant converter with this model.

By applying fundamental harmonic approximation (FHA) to this converter and referring all values to the primary side, the AC equivalent circuit is extracted which is shown in Fig. 1 (b). The values of C_1 and C_2 can be calculated by using (1) and (2). This sophisticated equivalent circuit can be reduced into a much simpler circuit. As long as leakage inductances are small in comparison to the magnetizing inductance- which is a true assumption specially for planar transformers- the three capacitor model of transformer can be reduced into one capacitor model. In this condition, the value of this capacitance is simply the sum of intra-winding capacitances and is presented in (3). In addition, primary leakage inductance can be placed in series with the series inductor which makes it absorbed by the series inductor. Finally, neglecting winding and core resistances, the simplified AC equivalent circuit is shown in Fig. 1 (c). The voltage conversion ratio of the simplified circuit is expressed in equation (4) [8].

$$C_1 = C_{11} + (1-n).C_{12} \tag{1}$$

$$C_2 = n^2.C_{22} - n(n-1).C_{12} \tag{2}$$

$$C_{stray} \approx C_1 + C_2 \tag{3}$$

$$\frac{V_o}{V_{in}} = \frac{1}{2n} \left| \frac{n^2 R_{ac}}{\begin{aligned} j\omega n^2 L_{lk2}&[1 - \frac{\omega_M^2}{\omega^2}(1 - \frac{\omega^2}{\omega_R^2})(1 - \frac{\omega^2}{\omega_{TR}^2})] \\ + n^2 R_{ac}&[1 - \frac{\omega_M^2}{\omega^2}(1 - \frac{\omega^2}{\omega_R^2})(1 - \frac{\omega^2}{\omega_{TR}^2})] \\ &- L_M \frac{\omega_M^2}{\omega^2}(1 - \frac{\omega^2}{\omega_R^2}) \end{aligned}} \right| \tag{4}$$

For experimental verification of the proposed winding layout, a $1.2KW$ *LLC* resonant converter for EV battery charger application is designed. The parameters list of this converter and its transformer are presented in table I. Considering typical value of $1\ \mu H$ for leakage inductances and based on the (4), the voltage conversion ratio of this converter

TABLE I
PROTOTYPE PARAMETERS

Converter Prameters		Transformer Prameters	
Parameters	Value	Parameters	Value
V_{in}	400 V	N_p/N_s	8/4
V_{out}	96 V	Core	ELP 58/11/38
Output Power	1200 W	PCB Thickness	0.4mm
L_s	15 μ H	Copper Thickness	2Oz
C_s	41 nF	Primary Layers	4
L_m	45 μ H	Secondary Layers	5

for various values of stray capacitance in 1% of nominal load is plotted in Fig. 2.

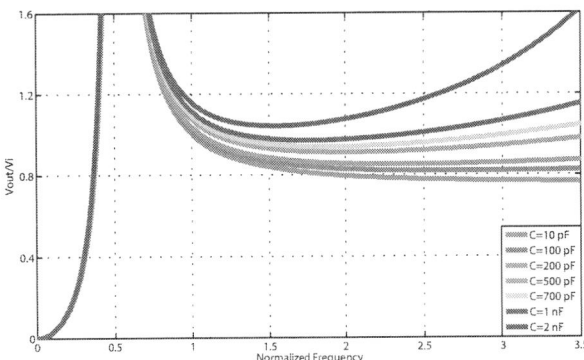

Fig. 2. Voltage conversion ratio based on FHA in presence of transformer's parasitic elements.

Briefly, the LLC resonant converter needs to progressively decrease its voltage conversion ratio in the inductive region to regulate output voltage under operating conditions. From Fig. 2 it is evident that high values of parasitic capacitance distorts the decreasing trend of voltage conversion ratio under light load conditions and lead to an unfortunate increase of output voltage as frequency increases. This problem also has been mentioned in the number of papers [9], [10], [11], [12]. Therefore, the value of intra-winding capacitance should be limited in order to have wide output range.

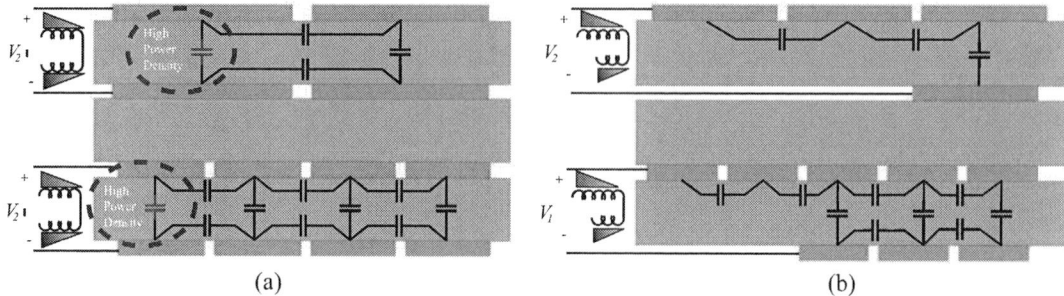

(a)   (b)

Fig. 3. 2D cross section of (a) traditional winding layout and (b) the proposed winding layout. The high power density region is removed in the proposed winding layout by removing overlaps between first and last turn.

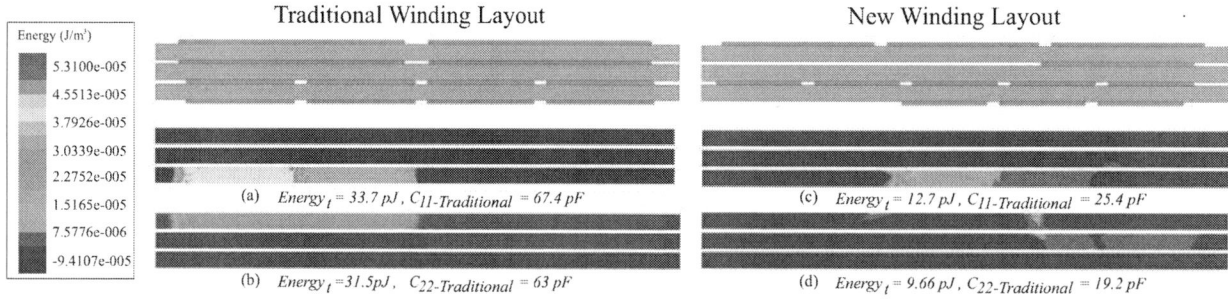

Fig. 4. electrostatic energy from distributing 1V between terminals of (a) primary of the traditional transformer, (b) secondary of the traditional transformer, (c) primary of the proposed transformer, and (d) secondary of the proposed transformer. The intra-winding capacitance is proportional to the value of electrostatic energy and these figures show that the proposed winding layout has led to 63 and 70 percent reduction of intra-winding capacitance in primary and secondary, respectively.

II. THE PROPOSED WINDING LAYOUT FOR REDUCING INTRA-WINDING CAPACITANCE

The main idea in reducing parasitic capacitances in PTs is to avoid overlapping faces between PCB traces. Removing all overlapping faces is a way to minimize stray capacitances. Although doing this will minimize stray capacitances, it will also increase winding resistance and so decrease the efficiency. In order to reduce capacitance without significant increase in the conduction losses, it is necessary to find the most important contributor to the intra-winding capacitance and only removing it. Fig. 3 shows 2D cross section of simplified capacitance network for traditional and the new layout. The red capacitor in the simplified capacitance network of Fig. 3 is in parallel with all other capacitors, so its value is the main contributor to the total intra-winding capacitance. It stores most of electrostatic energy, because it has the biggest voltage gradients between its plates and also it is the largest capacitance due to its big overlapping area.

The above explanations is based on simplified capacitance network. In order to validate this theory, finite element method is used to plot the electrostatic energy distribution and to calculate the parasitic capacitances. Briefly, to find primary intra-winding lumped capacitance, a uniform voltage distribution is applied between terminals of primary winding. The electric energy stored in this condition is associated with the primary intra-winding capacitance. Similar analysis is

needed to find secondary intra-winding capacitance. Finding inter-winding capacitance requires a constant voltages on every turn of one winding and zero voltage on the other winding. The total electric energy in this simulation is represented by mean of the inter-winding capacitance [13]. The condition of these three simulations are presented in Fig. 5. Fig. 6 also shows voltage distribution of secondary winding in the case that $1V$ is applied on its terminals

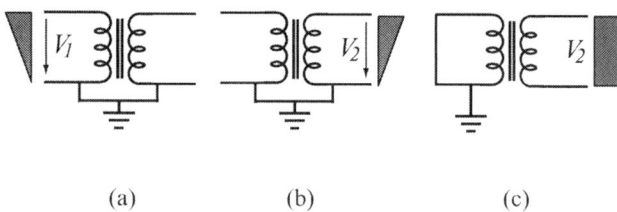

(a)   (b)   (c)

Fig. 5. (a) Analysis required to find the primary capacitance (b) Analysis required to find the secondary capacitance (c) Analysis required to find the inter-winding capacitance. The total electrostatic energy in each analysis is proportional to the value of its lumped capacitance.

Fig. 4 compares electrostatic energy distribution of tradi- tional and the proposed winding in the simple case of one primary-one secondary. These results provide good evidence that the new winding layout effectively reduces intra-winding energy and lead to a huge reduction in the lumped capaci- tance, without essentially removing all face to face traces.

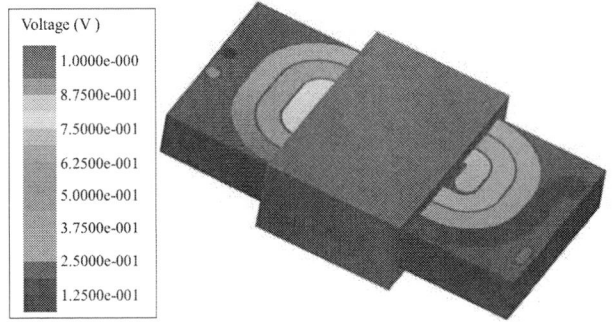

Fig. 6. Linear distribution of 1 Volt between terminals of secondary. This voltage distribution is associated with the analysis of Fig. 5 (b) and the total resulting electrostatic energy is used to find the secondary intra-winding capacitance.

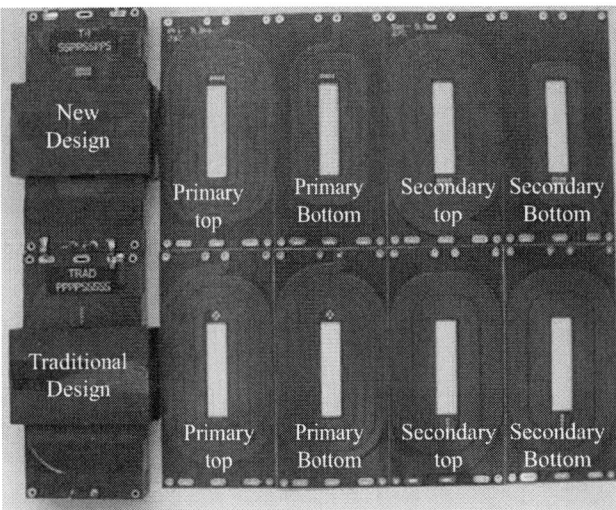

Fig. 7. Prototypes of the new (top row) and traditional transformer (bottom row)

In addition to the winding layout, the structure of transformer also has great effect on the values of parasitic elements. Summarily, while fully-interleaved (FI) structure proposes the minimum ac resistance and leakage inductance for windings, it also increases the inter-winding capacitance. The amount of inter-winding capacitance should be minimized in order to avoid EMI issues. on the other hand, non-interleaved (NI) structure provides the minimum inter-winding capacitance in the expense of higher resistance and leakage inductance. To find the optimum structure that provide good balance between all parasitic parameters, FEA is used to extract parasitic elements in various conditions and finally SSPPSSPPS structure is selected in implementing the proposed transformer.

III. EXPERIMENTAL RESULTS

In order to verify benefits of the proposed winding layout, the new transformer is implemented along with traditional transformer with NI and FI structures. The new and traditional transformer prototypes are shown in Fig. 7. The frequency response of these three transformers are captured using frequency response analyzer to find experimental values of capacitances. Fig. 8 compares frequency response of NI traditional transformer to the new transformer. This figure confirms that the proposed winding layout reduces intra-winding capacitance by 69%. The results are presented in the table II

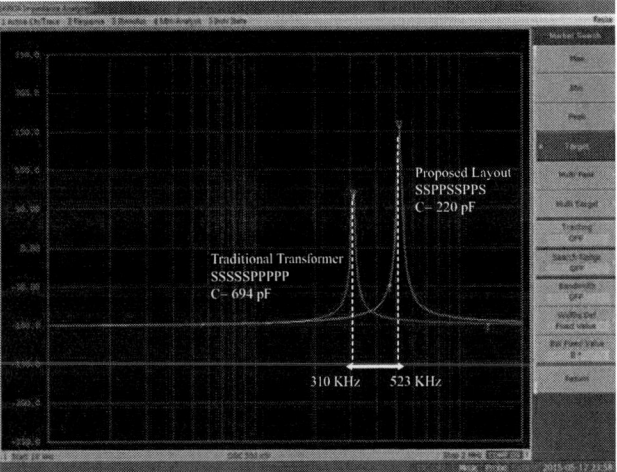

Fig. 8. Frequency response comparison of the new and NI traditional transformer

TABLE II
EXPERIMENTAL RESULTS

Parameters	New	Trad NI	Trad FI
C_{intra}	220 pF	694 pF	620 pF
C_{inter}	360 pF	210 pF	1.2 nF
η_{FL}	96.1%	95.5%	96%
Max θ (C)	78	85	73

Finally, NI and FI traditional transformers along with the proposed transformer are utilized in the LLC resonant converter. Fig. 9 shows experimental voltage conversion ratio

for NI traditional and the proposed transformers in the light load conditions. Fig. 9 confirms that the new transformer can regulate output voltage even in the extremely light load conditions, while traditional transformer cannot regulate output due to its distorted voltage gain characteristics. In addition, efficiency analysis performed on the LLC resonant converter with different transformers. Temperature distribution of three transformers after 30 minutes of working in full load condition is provided in Fig. 10. Results of this analysis show that in spite of a negligible extra conduction loss in the new transformer, the overall efficiency of the converter is increased by 0.1%. In other words, the small increase of conduction loss in the proposed transformer has been completely compensated by the reduced losses on parasitic elements and lower CM noise, leading to higher overall efficiency of the converter.

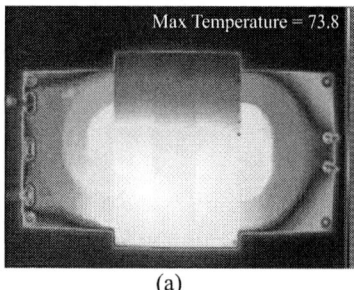 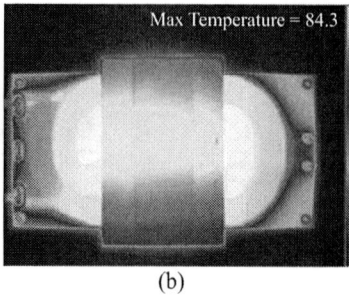 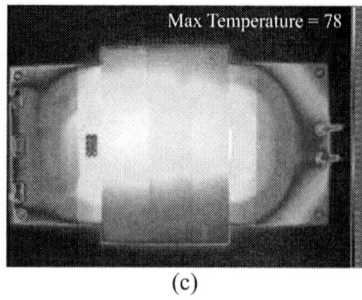

(a)   (b)   (c)

Fig. 10. Temperature distribution after working 30 minutes in the full load condition for (a) FI traditional , (b) NI traditional ,and (c) the proposed transformer

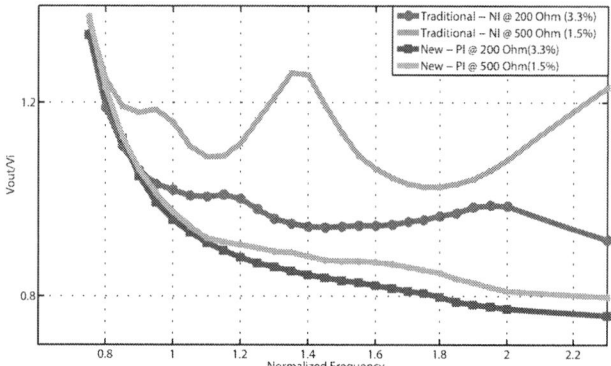

Fig. 9. DC gain characteristics of the *LLC* resonant converter.

IV. CONCLUSIONS

In this paper, a novel and high efficiency winding layout for reducing intra-winding capacitance of PTs is proposed. Aiming to reduce intra-winding capacitance without removing all overlapping faces, finite element approach is used to find electrostatic energy distribution between overlapping layers. Results show that the most of electrostatic energy is stored between the first and last turn of each winding. So instead of removing all overlapping faces to avoid parasitic capacitance, only overlapping area between first and last turns are removed. This method has effectively reduced the amount of intra-winding capacitance from $700pF$ in traditional transformer to $220pF$ in the new design, making the proposed layout suitable for LLC resonant converter with wide output range applications like EV battery chargers. In addition, reducing the intra-winding capacitance of transformer has caused the overall efficiency of converter to increase from 96% to 96.1%.

REFERENCES

[1] R. Prieto, O. Garcia, R. Asensi, J.A. Cobos and J. Uceda, "Optimizing the performance of planar transformers," *in Proc. IEEE Applied Power Electronics Conf. and Expo., 1996,* vol. 1, pp. 415421.

[2] O. Ziwei, O. C. Thomsen, M. A. E. Andersen, "Optimal design and tradeoff analysis of planar transformer in high power DCDC converters," *IEEE Trans. on Industrial Electronics,* vol. 59, no. 7, pp. 2800-2810, July. 2012.

[3] C. Quinn, K. Rinne, T. ODonnell, M. Duffy, and C. O. Mathuna, "A review of planar magnetic techniques and technologies," *in Proc. IEEE APEC, 2001,* pp. 11751183.

[4] M. Pahlevaninezhad, D. Hamza, and P. K. Jain, "An improved layout strategy for common-mode EMI suppression applicable to high-frequency planar transformers in high-power DC/DC converters used for electric vehicles," *IEEE Trans. Power Electronics,* vol. 29, no. 3, pp. 12111228, Mar. 2014.

[5] J. Biela and J. W. Kolar, "Using transformer parasitics for resonant convertersA review of the calculation of the stray capacitance of transformers," *IEEE Trans. on Industry Applications,* vol. 44, no. 1, pp. 223233, Jan./Feb 2008.

[6] L. Dalessandro, F. S. Cavalcante, and J. W. Kolar, "Self-capacitance of high-voltage transformers," *IEEE Trans. on Power Electronics,* vol. 22, no. 5, pp. 20812092, Sep. 2007.

[7] M. Pahlevaninezhad, P. Das, J. Drobnik, P.Jain, A. Bakhshai, G. Moschopoulos, "A novel winding layout strategy for planar transformer applicable to high frequency high power DC-DC converters," *in Proc. Energy Conversion Congress and Exposition (ECCE), 2011,* pp. 3786 - 3791

[8] B.-H. Lee, M.-Y. Kim, C.-E. Kim, K.-B. Park, and G.-W. Moon, "Analysis of LLC resonant converter considering effects of parasitic components,'" *in Proc. 31st Int. Telecommun. Energy Conf., 2009,* pp. 16.

[9] H.-Y. Lu, J.-G. Zhu, and S. Y. R. Hui, "Experimental determination of stray capacitances in high frequency transformers," *IEEE Trans. on Power Electronics* vol. 18, no. 5, pp. 11051112, Sep. 2003.

[10] N. Shafiei, and M. Ordonez, "Improving the regulation range of EV battery chargers with L3C2 resonant converters," *IEEE Transaction on Power Electronics,* vol. 30, no. 6, pp. 3166-3184, May 2012.

[11] J.-H. Kim, C.-E. Kim, J.-K. Kim, and G.-W. Moon, "Analysis for LLC resonant converter considering parasitic components at very light load condition," *in Proc. Power Electronics and ECCE Asia (ICPE & ECCE), 2011,* pp. 1863-1868.

[12] N. Shafiei, M. Ordonez, M. Craciun, C. Botting, and M. Edington, "Burst mode elimination in high power LLC resonant battery charger for electric vehicles," *IEEE Transaction on Power Electronics,* IEEE Early Access.

[13] R. Prieto, J. A. Cobos, O. Garcia, P. Alou, and J. Uceda, "Taking into account all the parasitic effects in the design of magnetic components," *in Proc. IEEE Applied Power Electronics Conf. and Expo., 1998,* vol. 1, pp. 400406.

New Symmetrical Bidirectional L3C Resonant DC-DC Converter with Wide Voltage Range

Minjae Kim, Shinyoung Noh, and Sewan Choi, *IEEE Senior Member*
Department of Electrical and Information Engineering
Seoul National University of Science and **Tech**nology
E-mail: schoi@seoultech.ac.kr

Abstract—This paper proposes a new symmetrical bidirectional L3C resonant converter with wide voltage range. Due to the proposed symmetrical resonant tank with notch resonant characteristic, the proposed converter is capable of not only achieving soft switching of all switches in forward and backward modes but also operating in a narrow switching frequency range even under wide voltage range. Also, due to zero gain at the notch resonant frequency, the proposed converter does not suffer from increased current and voltage stresses of components during start-up. The current rating of the resonant capacitor does not increases as the load increases and can further be reduced by adjusting the tertiary turn ratio of the transformer, making the proposed converter more realizable in high power application. Experimental results from a 3kW prototype are provided to validate the proposed concept.

Keywords—bidirectional resonant converter; symmetrical resonant tank; wide voltage range; soft switching

I. INTRODUCTION

Bidirectional dc-dc converters are used in many application such as electric vehicle chargers, renewable energy systems, uninterruptible power supplies and energy storage systems where high efficiency, high power density and bidirectional power flow capability are required. Several isolated bidirectional dc-dc converter have been proposed and studied in the past decades. The dual active bridge (DAB) converter has attracted lots of research interests due to its simple structure, soft switching capability and high efficiency [1-6]. However, the DAB suffers from high circulating energy, high turn off switching loss and limited soft switching range under wide voltage variation.

Recently, bidirectional resonant converters have been introduced to extend the soft switching range of the DAB [7-13]. A bidirectional series resonant converter (SRC) [7, 8] has simple structure, but the soft switching range is still limited, making it unsuitable for wide voltage range. A bidirectional LLC resonant converter [9, 10] has wider soft switching range than the bidirectional SRC. However, the topology is still a traditional SRC during backward operation, which is not preferred for wide voltage range application. A bidirectional

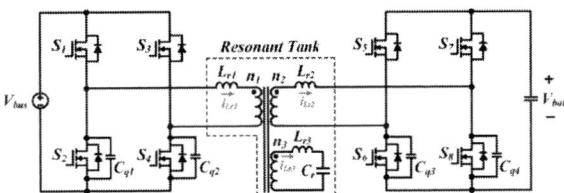

Fig 1. Circuit diagram of the proposed converter.

CLLC resonant converter [11, 12] and a bidirectional LLC resonant converter with extra inductor [13] having a symmetrical resonant tank in both power flows were proposed, but they have high turn off switching loss due to wide switching frequency variation under wide voltage range.

In this paper, a new bidirectional L3C resonant converter is proposed for wide voltage range application. The proposed converter has the following features: 1) symmetrical resonant tank in both power flow 2) soft switching of all switches regardless of voltage and load variation; 3) narrow operating switching frequency range under wide voltage variation; 4) no current and voltage stress during start-up; and 5) easy selection of the resonant capacitor by adjusting a tertiary winding turn ratio of the transformer. Experimental results from a 3kW prototype are provided to validate the proposed concept.

II. PROPOSED CONVERTER

Fig. 1 shows the circuit diagram of the proposed converter. The proposed converter has two full bridge converters which are connected through a L3C resonant tank and a high-frequency transformer. Due to symmetrical structure of the resonant tank, the proposed converter has the same control algorithm and operating characteristics under both forward and backward modes.

A. Operating principles

Fig. 2 shows key waveforms and operation states of the proposed converter for the forward mode. In this mode, switches S_1~S_4 at the dc bus side are operated at variable switching frequency with 50% duty cycle while switches S_5~S_8 at the battery side are kept turned OFF acting as a diode rectifier, and therefore energy on the dc bus side is delivered to the battery side. The proposed converter is able to achieve not

978-1-4673-9551-9/16 $31.00 © 2016 IEEE

only ZVS turn-on of switches $S_1 \sim S_4$ but also soft commutation of switches $S_5 \sim S_8$ due to the resonant operation. Note that the backward mode operation is exactly the same as the forward mode operation due to symmetrical resonant tank structure, and therefore the key waveforms and operation states for backward mode are not shown here.

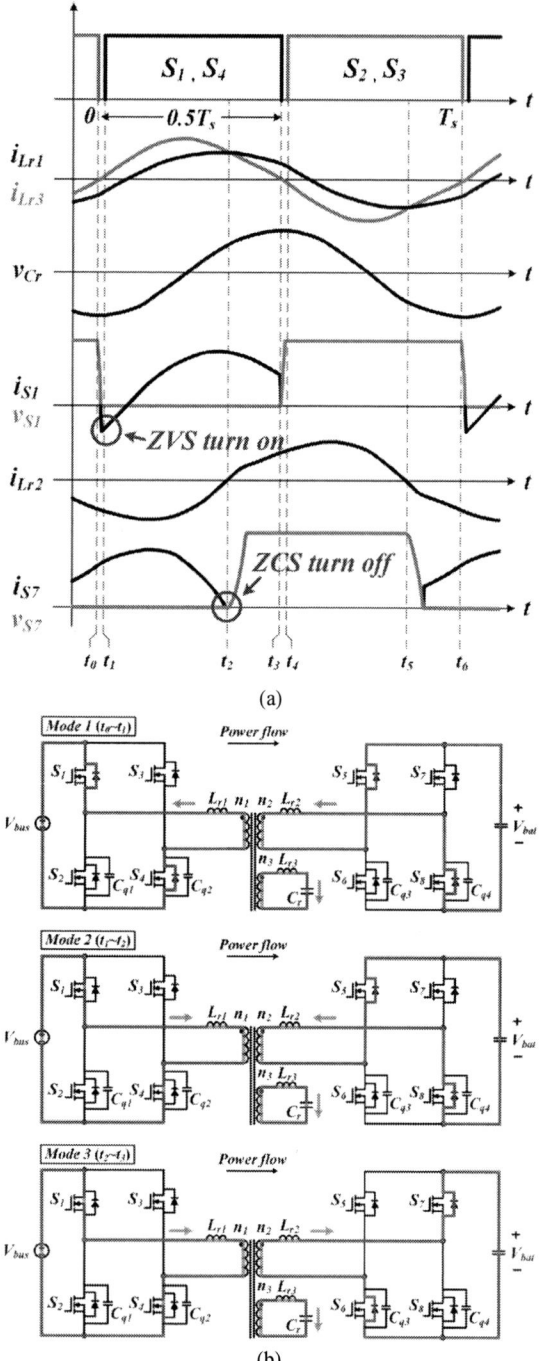

Fig 2. Forward mode operation of the proposed converter (a) Key waveforms (b) Operation states.

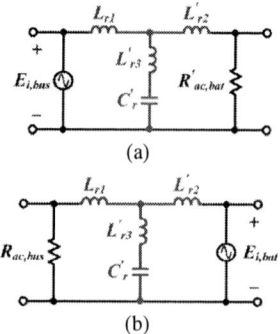

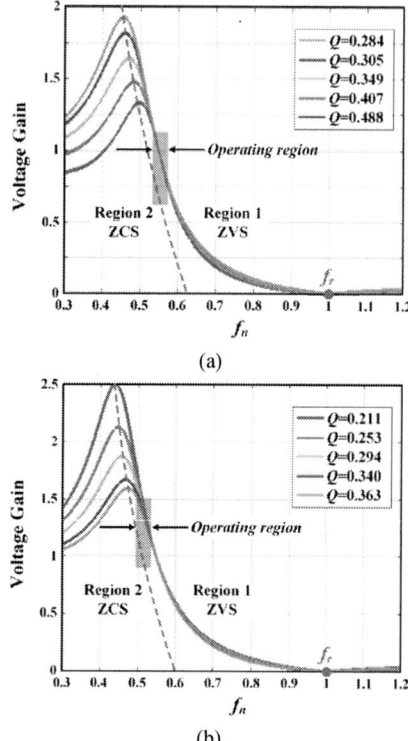

Fig 3. AC equivalent circuit of the proposed converter referring to dc bus side. (a) Forward mode. (b) Backward mode.

Fig 4. Voltage gain of the proposed converter. (a) Forward mode. (b) Backward mode.

B. Voltage gain of the proposed converter

The equivalent circuits for fundamental harmonic approximation (FHA) of forward and backward modes are shown in Fig. 3. The equivalent circuit of the forward mode is identical to that of the backward mode. The FHA method is used to obtain the voltage gain of the proposed converter. In Fig. 3(a), the ac equivalent resistance referring to the dc bus side $R'_{ac,bat}$ can be expressed as follows [11-13]:

$$R'_{ac,bat} = \frac{n_1}{n_2} \frac{8}{\pi^2} R_{L,bat} \qquad (1)$$

978-1-4673-9551-9/16 $31.00 © 2016 IEEE

TABLE I.
COMPARISON OF CHARACTERISTICS OF THE PROPOSED AND CONVERTIONAL SYMMETRICAL CONVERTERS

	# of resonant components	Modulation method	Control variable variation	Soft switching range	Capacitor selection for high power
SRC [7]	1 inductor 1 capacitor	Phase shift modulation	Wide	Narrow	Difficult
CLLC [12]	3 inductors 2 capacitors	Frequency modulation	Wide	Wide	Difficult
LLC [13]	3 inductors 1 capacitor	Frequency modulation	Wide	Wide	Difficult
Proposed	3 inductors 1 capacitor	Frequency modulation	Narrow	Wide	Easy

From Fig. 3(a) the voltage gain of the proposed converter is determined by

$$\frac{V_o}{V_i} = \frac{n_2}{n_1} \frac{1}{\sqrt{\left(\frac{kf_n^2}{f_n^2 - \frac{1}{4\pi^2}} + 1\right)^2 + \left(2\pi kQf_n \left(\frac{n_1}{n_3}\right)^2 \left(2 + \frac{kf_n^2}{f_n^2 - \frac{1}{4\pi^2}}\right)\right)^2}} \quad (2)$$

where $f_r (= \frac{1}{2\pi\sqrt{L_{r3}C_r}})$ is notch resonant frequency, $f_n (= \frac{f_s}{f_r})$ is normalized frequency, $Q(= \frac{1}{R'_{ac,bat}}\sqrt{\frac{L_{r3}}{C_r}})$ is quality factor and

$k(= \frac{n_3^2 L_{r1}}{n_1^2 L_{r3}})$ is the inductance ratio between L_{r1} and L_{r3}.

The voltage gain versus normalized frequency f_n for different values of Q in both modes is shown in Fig. 4. The maximum gain of the proposed converter decreases as Q increases. It should be noted that a load-independent operation with zero gain occurs at the notch resonant frequency f_r in both modes, and therefore not only output voltage regulation under wide voltage and load variation is advantageous in Region 1 but there are no current and voltage stress of components during start-up. Due to symmetric structure of the resonant tank, the gain curve of the backward mode is very similar to that of the forward mode.

C. Comparative analysis

In this section, the proposed converter is compared to the conventional bidirectional resonant converters including SRC [7], CLLC resonant converter [12] and LLC resonant converter [13]. The comparison results are summarized in Table I. Due to the proposed resonant tank with notch characteristic the proposed converter has narrow switching frequency range, which means that the proposed converter is most suitable for wide voltage application. The proposed converter, CLLC converter and LLC converter have wide soft switching range due to frequency modulation. The resonant capacitor of the conventional converters are connected in series with the main power flow path and therefore the current rating of the resonant capacitor increases in proportion to the load power. This makes it difficult for the conventional converter to be applied for

higher power application due to considerable current rating of the resonant capacitor. On the contrary, the resonant capacitor of the proposed converter is connected in parallel with the main power flow path so that the current rating of the resonant capacitor does not change as the load increases. Moreover, the current rating of the resonant capacitor can further be reduced by adjusting the tertiary turn ratio of the transformer. Therefore the proposed converter is most suitable for high power application.

III. EXPERIMENTAL RESULTS

A 3kW prototype of the proposed converter has been built and tested to verify the operating principle, and the experimental results are provided. The system specification used in the experimental is as follows: P_o=3kW, V_{bus}=380V, V_{bat}=250~430V, n_1:n_2:n_3=1:1:1, L_{r1}=159μH, L_{r2}=159μH, L_{r3}=39μH and C_r=65nF.

Figs. 5-8 show the experimental waveforms in both modes. Figs. 5(a)-8(a) show waveforms of i_{Lr1}, i_{Lr2} and i_{Lr3}, which are in close agreement with the ideal waveforms shown in Fig. 2. Figs. 5(b)-8(b) show voltage and current waveforms of switch S_1. Figs. 5(c)-8(c) show voltage and current waveforms of switch S_7. It can be seen that all switches are soft switched in both modes due to the proposed resonant tank. The efficiency of the proposed converter is measured by YOKOGAWA WT3000 and shown in Fig. 9. The maximum efficiencies in forward and backward modes are 96.5% at 3kW(V_{bat}=430V) and 96.6% at 2.1kW(V_{bat}=300V).

IV. CONCLUSION

In this paper, a new bidirectional resonant converter with wide voltage range is proposed for higher power application. Symmetrical and notch characteristics of the resonant tank of the proposed converter lead to wider soft switching range of switches, narrow operating switching frequency range, no current and voltage stresses during start-up. Also, due to parallel connection of the resonant capacitor with the power flow the current rating of the resonant capacitor does not increases as the load increases and can further be reduced by adjusting the turn ratio of the transformer, making the proposed converter more realizable in high power application. Experimental results from a 3kW prototype are provided to

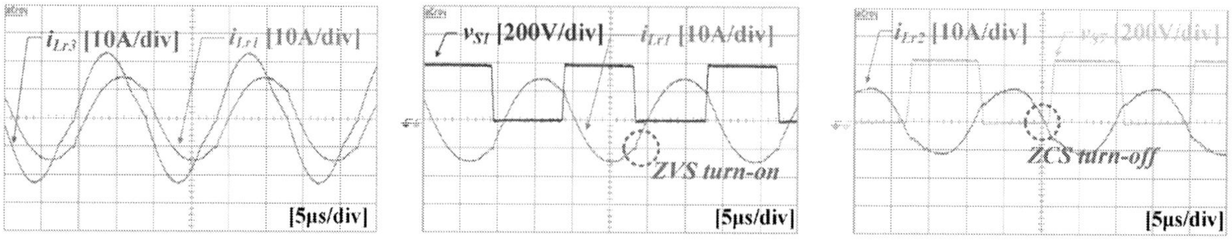

Fig 5. Experimental waveforms at V_{bus} = 380V, V_{bat} = 430V and I_{bat} = 7A in forward mode. (a) Current i_{Lr1} and i_{Lr3}. (b) Switch S_1. (c) Switch S_7.

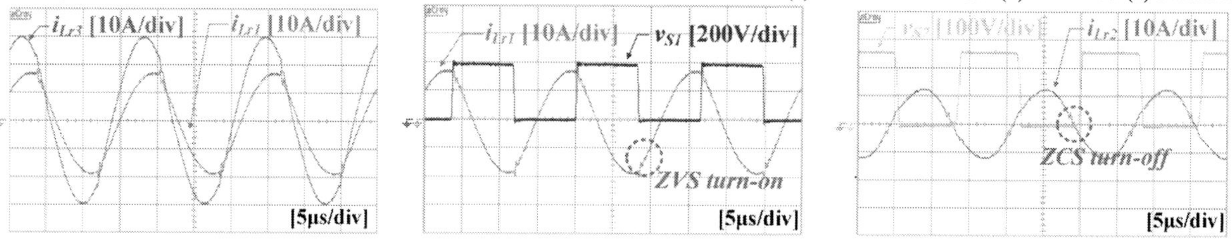

Fig 6. Experimental waveforms at V_{bus} = 380V, V_{bat} = 250V and I_{bat} = 7A in forward mode. (a) Current i_{Lr1} and i_{Lr3}. (b) Switch S_1. (c) Switch S_7.

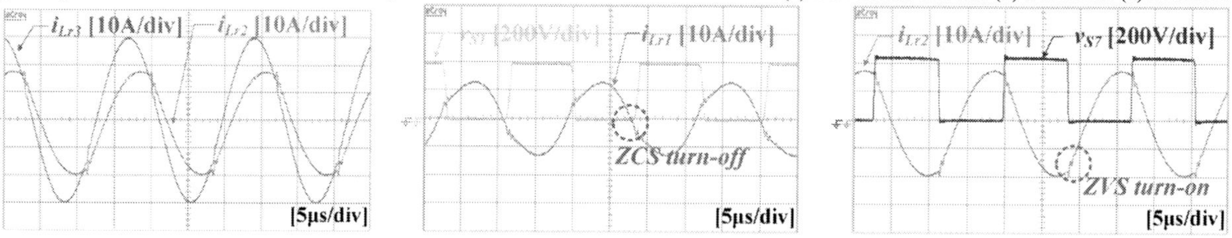

Fig 7. Experimental waveforms at V_{bus} = 380V, V_{bat} = 430V and I_{bat} = 7A in backward mode. (a) Current i_{Lr2} and i_{Lr3}. (b) Switch S_1. (c) Switch S_7.

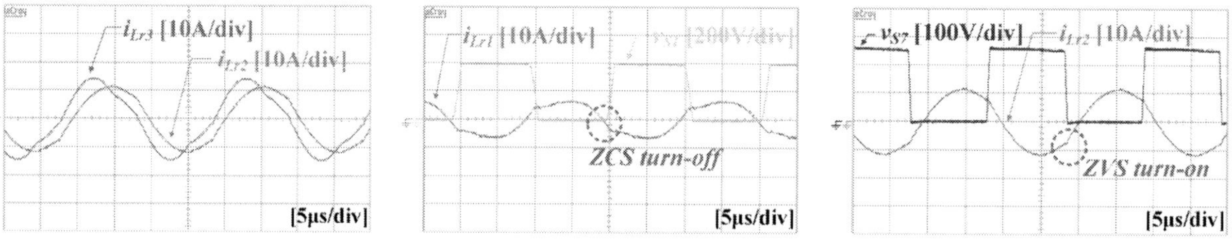

Fig 8. Experimental waveforms at V_{bus} = 380V, V_{bat} = 250V and I_{bat} = 7A in backward mode. (a) Current i_{Lr2} and i_{Lr3}. (b) Switch S_1. (c) Switch S_7.

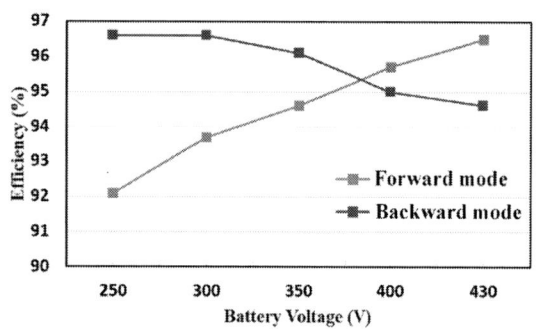

Fig 9. Measured efficiency of the proposed converter at I_o=7A.

validate the proposed concept. The maximum efficiencies in forward and backward modes are 96.5% and 96.6%, respectively.

ACKNOWLEDGMENT

This work was supported by the National Research Foundation of Korea(NRF) grant funded by the Korea government(MSIP) (No. 2014R1A2A2A01003724).

REFERENCES

[1] M. N. Kheraluwala, R. W. Gascoigne, D. M. Divan, and E. D. Baumann, "Performance Characterization of A High-power Dual Active Bridge

DC-to-DC Converter," *IEEE Trans. Ind. Appl.*, vol. 28, no. 6, pp. 1294–1301, Dec. 1992.

[2] J. Zhang, F. Zhang, X. Xie, D. Jiao, and Z. Qian, "A novel ZVS dc/dc converter for high-power applications," *IEEE Trans. Power Electron.*, vol. 19, no. 2, pp. 420–429, Mar. 2004.

[3] S. Inoue and H. Akagi, "A Bidirectional DC–DC Converter for An Energy Storage System with Galvanic Isolation," *IEEE Trans. Power Electron.*, vol. 22, no. 6, pp. 2299–2306, Nov. 2007.

[4] F. Krismer, J. W. Kolar, "Accurate Small-Signal Model for the Digital Control of an Automotive Bidirectional Dual Active Bridge," *IEEE Trans. Power Electron.*, vol. 24, no. 12, pp. 2756-2768, Dec. 2009.

[5] B. Zhao, Q. Song, W. Liu, "Efficiency Characterization and Optimization of Isolated Bidirectional DC–DC Converter Based on Dual-Phase-Shift Control for DC Distribution Application," *IEEE Trans. Power Electron.*, vol. 28, no. 4, pp. 1711-1727, Apr. 2013.

[6] B. Zhao, Q. Song, W. Liu, Y. Sun, "Overview of Dual-Active-Bridge Isolated Bidirectional DC–DC Converter for High-Frequency-Link Power-Conversion System," *IEEE Trans. Power Electron.*, vol. 29, no. 8, pp. 4091-4106, Aug. 2014.

[7] X. Li, A. K. S. Bhat, "Analysis and Design of High-Frequency Isolated Dual-Bridge Series Resonant DC/DC Converter," *IEEE Trans. Power Electron.*, vol. 25, no. 4, pp. 850-862, Apr. 2010.

[8] L. Corradini, D. Seltzer, D. Bloomquist, R. Zane, D. Maksimovic, B. Jacobson, "Minimum Current Operation of Bidirectional Dual-Bridge Series Resonant DC/DC Converters," *IEEE Trans. Power Electron.*, vol. 27, no. 7, pp. 3266-3276, Jul. 2012.

[9] G. Pledl, M. Tauer, D. Buecherl, "Theory of operation, design procedure and simulation of a bidirectional LLC resonant converter for vehicular applications," *IEEE VPPC 2010*, pp.1-5.

[10] S. Abe, J. Yamamoto, T. Zaitsu, T. Ninomiya, "Operating strategy for bi-directional LLC resonant converter with seamless operation," *IEEE ECCE ASIA 2014*, pp.1179-1184.

[11] W. Chen, P. Rong, Z. Lu, "Snubberless Bidirectional DC–DC Converter With New CLLC Resonant Tank Featuring Minimized Switching Loss," *IEEE Trans.Ind. Electron.*, vol.57, no.9, pp.3075-3086, Sept. 2010.

[12] J.-H. Jung, H.S. Kim, M.H. Ryu, J.W. Baek, "Design Methodology of Bidirectional CLLC Resonant Converter for High-Frequency Isolation of DC Distribution Systems," *IEEE Trans. Power Electron.*, vol.28, no.4, pp.1741-1755, April 2013.

[13] T. Jiang, J. Zhang, X. Wu, K. Sheng, Y. Wang, "A Bidirectional LLC Resonant Converter With Automatic Forward and Backward Mode Transition," *IEEE Trans. Power Electron.*, vol.30, no.2, pp.757-770, Feb. 2015.

Influence of the Junction Capacitance of the Secondary Rectifier Diodes on Output Characteristics in Multi-Resonant Converters

Stefan Ditze, Thomas Heckel, Martin März

Fraunhofer Institute for Integrated Systems and Device Technology IISB

Schottkystrasse 10, 91058 Erlangen, Germany

stefan.ditze@iisb.fraunhofer.de

Abstract— Multi-resonant converters like the CLLLC topology are known for their outstanding efficiency and high power density. Little information has however been published about the influences of secondary side diode junction capacitances on the output characteristics of the resonant converter. This paper presents a detailed analysis of these influences in the inductive working range and reviews practical design considerations of the converter. Therefore, experimental results of an inductive power transfer system, using a CLLLC resonant topology, are compared to theoretical time domain solution, showing significant effects of different semiconductor materials and devices on output power. These effects will be discussed and explained in detail by using measured key waveforms.

Keywords—Parasitic components; output rectifier; transistor output capacitance; diode junction capacitance; CLLLC resonant converter; time domain solution

I. INTRODUCTION

Besides the well-established group of galvanically isolated power converters, e.g. forward or push-pull converters, the group of resonant converters are gaining more and more interest in the last years [1]-[3]. The main advantage of the resonant converter is reduced switching losses, due to the soft switching capability of the primary inverter switches. This enables operation at high switching frequencies f_{sw}, resulting in smaller inductive component sizes and hence in low overall converter size and high power density [4], [5]. However, the development of resonant converters is challenging for most designers, due to the resonant operating principle. Especially, the multi resonant behavior of the CLLLC resonant converter (Fig. 1) and also the LLC resonant converter (not shown) presents a considerable obstacle during the design process. Resonant converters can be implemented by employing either unidirectional (Fig. 1(a)) or bidirectional (Fig. 1(b)) switches on the secondary side. Thereby, the unidirectional configuration is often used for applications in galvanically isolated switch-mode power supplies [6], [7]. For systems in which a bidirectional energy transfer is required, e.g. between two DC grids, connection of a HV battery storage to a DC bus, or battery chargers, the bidirectional configuration is applied [8]-[10]. Depending on the present application, both configurations are used in inductive power transfer (IPT) systems [11]-[14].

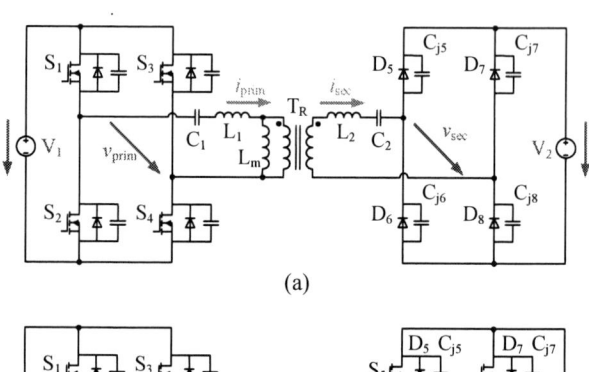

(a)

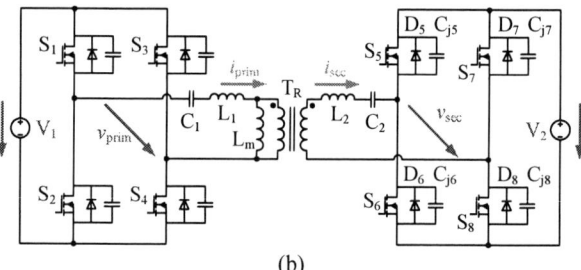

(b)

Fig. 1. Topology of the CLLLC resonant converter with (a) unidirectional switches and (b) bidirectional switches on the secondary side.

Precise knowledge of the characteristics and performance of the resonant converter is therefore necessary in order to achieve the optimum performance of the converter for a given application. As rather straightforward approach resonant converters can be analyzed theoretically by the first harmonic approximation (FHA) method, assuming that only sinusoidal current waveforms are applied on the resonant tank and only ideal components are utilized [15]. A more detailed and advanced analysis is the calculation of the time domain solution (TDS) for steady-state [16], [17]. But it has been shown that there is still a deviation between measurement and TDS calculation [17], which is merely not explainable with resistive losses and/or dead time of the primary inverter. Both methods, FHA and TDS, have the disadvantage that they do not account for the secondary side rectifier diodes junction capacitances and their significant influences on converter output characteristic, which represents the most important point for design considerations. In literature the influence of the

978-1-4673-9551-9/16 $31.00 © 2016 IEEE

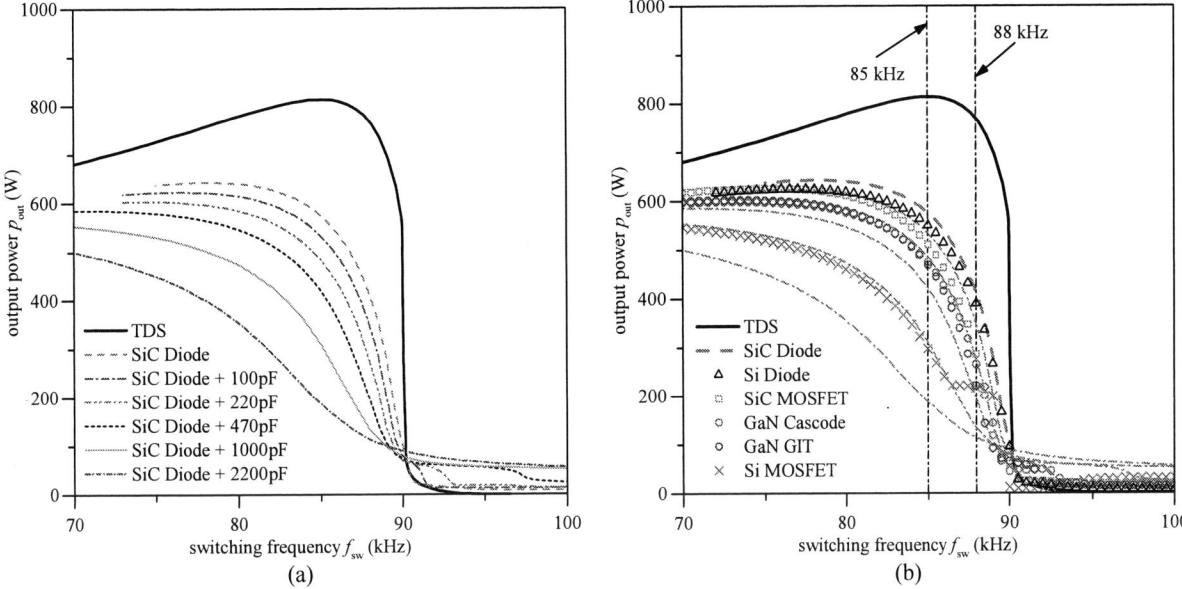

Fig. 2. Measured output power characteristics p_{out} plotted against switching frequency f_{sw} of an IPT system in step-up operation mode with $V_{in} = 200$ V and $V_{out} = 220$ V. Reference measurements using SiC Schottky barrier diode (SiC SBD) as bridge rectifier with additional capacitances parallel to each diode are depicted in (a). Measurements with SiC diode, SiC MOSFET, GaN transistors, Si diode, and Si MOSFET used as secondary side bridge rectifier are shown in (b).

junction capacitance on output power is only rarely discussed. Chen et al. analyze the influence of the secondary side junction capacitances only on zero-voltage switching (ZVS) of the primary inverter switches [18], whereas Kim et al. analyze the influence of secondary side junction capacitance on output voltage, but only at very light load and without differentiation between step-down and step-up operation mode [19]. Other authors discuss the influence of the junction capacitances only rudimentary within the scope of loss analysis in resonant converters [20]-[22].

In this paper the influence of the secondary side rectifier diodes junction capacitance on converter output power is analyzed, by comparing experimental results of an exemplary IPT system, using a CLLLC topology, with TDS calculations. Different devices for unidirectional and bidirectional operating mode and different materials, e.g. SiC, GaN, or Si, are investigated in detail, showing significant effects on converter output power. Based on this knowledge practical design considerations for a series resonant converter are provided for both unidirectional and bidirectional CLLLC topology.

II. MEASUREMENTS OF THE INFLUENCE OF THE RECTIFIER DIODES JUNCTION CAPACITANCE ON P_{OUT}

In this section the output power characteristics in step-up (Fig. 2) and step-down (Fig. 3) operation mode for different Si and wide-bandgap semiconductor devices for an exemplary IPT system are presented and compared to TDS calculations. Depending on the semiconductor device and material the value of the diode junction capacitance C_j and the transistor output capacitance C_{oss} range between several picofarad to a few nanofarad [23]. Furthermore, the capacitance values vary with the applied diode voltage and drain source voltage, respectively. Due to bipolar device effects in Si devices, the

comparison with equivalent capacitance values is just valid for the given operating point as temperature, switching current, and current slope alter the effective capacitance value [24]. Measurements with a SiC Schottky barrier diode (SiC SBD) are carried out as a reference (Fig. 2(a) and Fig. 3(a)). The used SiC SBD performs almost like an ideal diode, because of the low junction capacitance and its unipolar behavior [25], [26]. Adding additional discrete capacitances C_{add} in parallel to each SiC SBD on the secondary side allows estimating the equivalent capacitance value for the different used Si and wide-bandgap semiconductors.

A. Step-up operation mode

First, the output power characteristics in step-up operation mode for a SiC SBD and SiC SBD with additional discrete capacitances (Fig. 2(a)) and different Si and wide-bandgap semiconductor devices (Fig. 2(b)) as secondary side rectifier for an exemplary IPT system with $V_{in} = 200$ V, $V_{out} = 220$ V, and an inductive coupling factor $k = 0.7$ are shown. For comparison, TDS calculations for the exemplary chosen operating point with ideal components are plotted. To begin with, the reference measurements in Fig. 2(a) are described. With increasing values of the additional discrete capacitance, a more substantial deviation from TDS calculation is observed. The measured output power gets significantly lower, which consequently results in a decreased output power capability in step-up mode. Moreover, the switching frequency at which the resonant converter can be regulated down to zero output power shifts towards higher frequencies. Furthermore, for values C_{add} greater than 1000 pF it is not possible to achieve zero output power by increasing the switching frequency f_{sw} for the exemplary IPT system. Because under very light load conditions the stray inductance L_1 (respectively L_2) and the rectifier diodes capacitances form a series resonant LC circuit

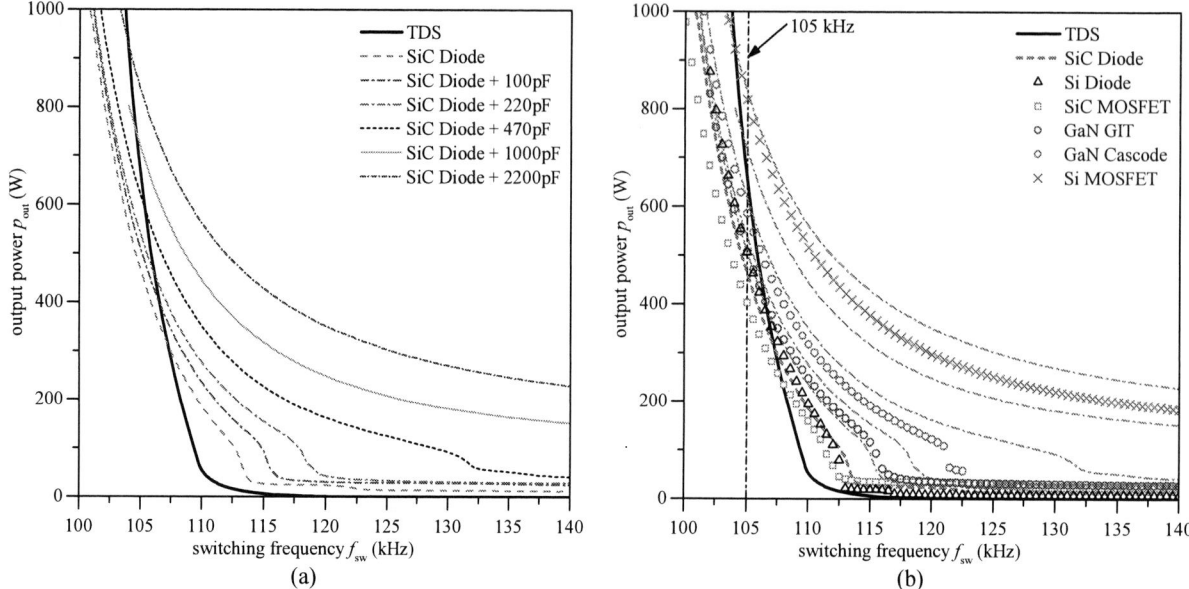

Fig. 3. Measured output power characteristics p_{out} plotted against switching frequency f_{sw} of an IPT system in step-down operating mode with V_{in} = 200 V and V_{out} = 180 V. Reference measurements using SiC SBD bridge rectifier with additional capacitances parallel to each diode are depicted in (a). Measurements with SiC SBD, SiC MOSFET, GaN transistors, Si diode, and Si MOSFET used as secondary side bridge rectifier are shown in (b).

which transfers energy to the secondary side driven by the voltage transition of the primary inverter and prevents to achieve zero output power [19]. In addition, the minimum frequency, where ZVS for the primary inverter is still achievable, decreases also with increasing capacitance values ($C_j + C_{add}$) of the output rectifier diodes. Further, for some reference measurements a bend in the measured curves is observed at switching frequencies greater than 90 kHz, which is explained by a change of the operating mode.

Next, the output power characteristics in step-up operation mode of the different semiconductor devices and materials are described in Fig. 2(b). Here, measurements for a SiC MOSFET (pink squares), GaN transistors (blue and green circles), Si diode (black triangles) and Si super-junction MOSFET (red crosses) as secondary side rectifier are depicted, which represent the state of the art power semiconductors. The investigated semiconductors and their calculated time-related equivalent capacitances $C_{eq,tr}$ using the $C_j(V_R)$, respectively $C_{oss}(V_{ds})$, measurements in the corresponding datasheet are listed in Table I [4]. For comparison the reference measurements with the SiC SBD (red dashes) and SiC SBD with additional discrete capacitances (grey dashed lines) are included. This allows estimating the capacitance values of the different semiconductor devices in the framework of the experimental set-up in step-up operation mode. Further, the comparison to datasheet values is also possible (see Table I).

As expected, the output power characteristics in step-up operation mode deviate substantially for all applied secondary side rectifier semiconductors from TDS calculation (Fig. 2(b)). Obviously, the measured output power gets significantly lower, with increasing values of the capacitance. This indicates that the influence of the secondary side rectifier diodes junction

capacitance on converter output power should not be neglected for the design of a resonant converter. The greatest deviation from TDS calculation is observed for the Si MOSFET (red crosses), due to its large output capacitance (about 1 nF, see Table I), its bipolar body diode and the reverse recovery characteristic of the device. Further, a discontinuity is observed at part load between 86 kHz and 90 kHz, which is not just a change of the operating mode as mentioned before. This special effect of the Si MOSFET device will be explained later

TABLE I. EXTRACTED DATASHEET VALUES FOR STEP-UP AND STEP-DOWN OPERATION MODE OF VARIOUS DEVICES

Device		$C_{eq,tr}$ (pF)	
		Step-Up Mode	Step-Down Mode
C3D20060D[(a)]	SiC diode, 600V	81	89
VS-APU3006-F3[(b)]	Si diode, ultrafast, 600V	45	49
C2M0160120D[(c)]	SiC MOSFET, 1200V, 160mΩ	124	135
-	Normally-off GaN HEMT, 600V, 40mΩ	333	354
TPH3205WS[(d)]	Cascode GaN HEMT, 600V, 52mΩ	428	485
IPW65R110CFD[(e)]	MOSFET, 650V, 110mΩ	1040	1239

a. Cree, Silicon Carbide Schottky Diode Z-Rec™ Rectifier C3D20060D, Datasheet, Rev. C, 2012. *(Derived from C_j, refers to 220V, respectively 180V)*

b. Vishay Semiconductors, Ultrafast Rectifier, 30A FRED Pt® VS-APU3006-F3, Datasheet, Rev. 17-Jul-13. *(Derived from C_j refers to 220V, respectively 180V)*

c. Cree, Silicon Carbide Power MOSFET C2M™ MOSFET Technology C2M0160120D, Datasheet, Rev. June 16, 2015 JH. *(Derived from C_{oss} refers to 220V, respectively 180V)*

d. Transphorm, GaN Power Low-loss Switch TPH3205WS, Datasheet, Rev. C, 2012. *(Derived from C_{oss}, refers to 220V, respectively 180V)*

e. Infineon Technologies AG, 650V CoolMOS™ CFD2 Power Transistor IPW65R110CFD, Datasheet, Rev. 2.6, 2011-09-26. *(Derived from C_{oss} refers to 220V, respectively 180V)*

on the basis of measured current and voltage waveforms of the resonant tank. For the other power semiconductors the deviation from TDS calculation is equally large. As expected from the comparable values of equivalent capacitance $C_{eq,tr}$ from Table I, the measured characteristic of the SiC MOSFET (pink squares) is very similar to the measured reference SiC SBD curve. In contrast to Si counterparts, measurements of the turn-off behavior of the SiC MOSFET body diode have shown a nearly capacitive and unipolar characteristic [27]. Hence, the output power curve of the SiC MOSFET fits well to the reference measurement with an additional capacitance of $C_{add} = 100$ pF. The Si diode is a performance optimized ultrafast diode and shows a great performance with a small junction capacitance (black tringles), similar to the reference SiC SBD. However, at low switching frequencies the reverse recovery effect of the Si diode commences which results in a virtual increase of the diode junction capacitance and the measured curve falls below the reference SiC SBD curve. In contrast to Si semiconductors, GaN high electron mobility transistors (HEMT) offer a much lower gate charge and a smaller junction capacitance due to their lateral structure [28], [29]. Two kinds of GaN HEMT are available which are suitable for power electronic applications: the normally-off (enhancement) GaN HEMT and the cascode GaN HEMT, which utilizes a low voltage Si MOSFET in series with a normally-on (depletion) GaN HEMT. The cascode structure has the advantage of simplified gate driving compared to the single GaN based normally-off device because of the higher driving safety margin (cascode: $v_{GSS} = \pm18$ V, normally-off GaN HEMT: $v_{GSS} = -10$ V...+4 V). The GaN transistors have nearly identical output power curves, only at switching frequencies higher than 90 kHz a discontinuity for the cascode GaN HEMT is observed, similar to the Si MOSFET. This may result from the fact that the body diode of the Si MOSFET conducts if the cascode GaN HEMT is operated in reverse mode. Compared to the datasheet value in Table I the estimated capacitance value from the experimental set-up of the cascode GaN HEMT is lower. Further research has to be conducted to evaluate if this results from e.g. cascode related effects or deviations from the device and the datasheet characteristics.

B. Step-down operation mode

The former described measurements are also performed for the step-down mode for the IPT system with $V_{in} = 200$ V and $V_{out} = 180$ V (Fig. 3). TDS calculations for the exemplary chosen operating point with ideal components are also plotted. In contrast to the step-up mode the measured output power is significantly higher at high switching frequencies and the curve progression of p_{out} rises less steep with decreasing switching frequencies for the reference SiC SBD and the different semiconductors than TDS calculation predicts. Hence, also for step-down operation mode, it is important to consider the secondary side rectifier diodes junction capacitance for the design of a resonant converter. Besides these differences for the step-down mode nearly the same effects as for the step-up mode are observed. With increasing C_{add} the deviation for the reference measurements from TDS gets larger, the output power rises significantly and thus zero output power cannot be achieved at high switching frequencies for values C_{add} greater than 1000 pF (Fig.3 (a)). Further, the above described bends in p_{out} progression are also observed at switching frequencies

greater than 110 kHz and indicate operating mode changes (Fig.3 (a)). Again, for the Si MOSFET the biggest deviation from TDS is observed (Fig. 3(b)). However, at part load a discontinuity is not found for investigated switching frequencies up to 400 kHz. Further, the reverse recovery effect of the Si diode commences also at low switching frequencies, which results in a virtual increase of the diode junction capacitance and thus the measured p_{out} curve rises above the reference SiC SBD curve. For the normally-off GaN transistor a difference in step-down mode is observed. The measurement shows, that the estimated capacitance value from the experimental set-up is much lower than the calculated datasheet value.

III. DISCUSSION AND RESULTS

In order to explain the observed output power characteristics, measured key waveforms of the resonant tank in step-up and step-down operation mode are analyzed.

A. Subintervals and operating modes

In multi-resonant topologies more than one resonant frequency exists which leads to different resonant stages within one switching cycle. These resonant stages, in the following called subintervals, are characterized by the condition of the output rectifier. First, the individual subintervals are briefly discussed (Fig. 4) as this will be important for the following discussion. For a more detailed description please refer to [16] for the LLC and to [17] for the CLLLC topology. Within one

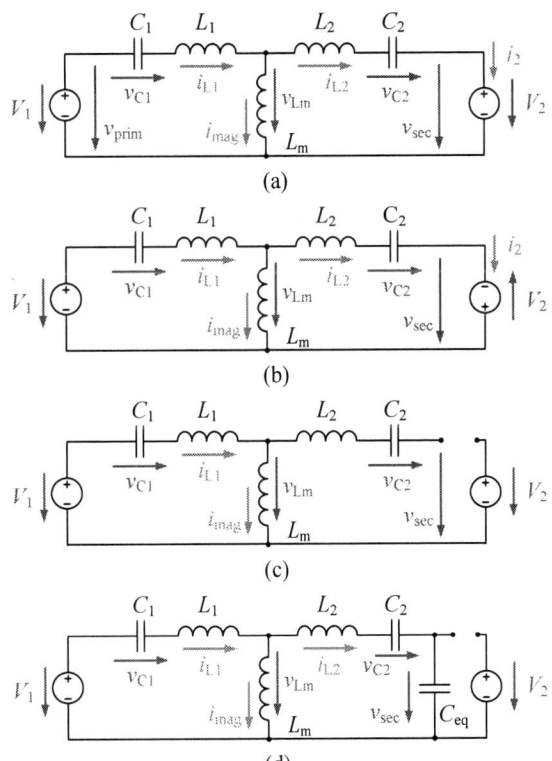

Fig. 4. CLLLC resonant tank equivalent circuits for the first half-cycle switching period. (a) Subinterval P. (b) Subinterval N. (c) Subinterval O. (d) Subinterval D.

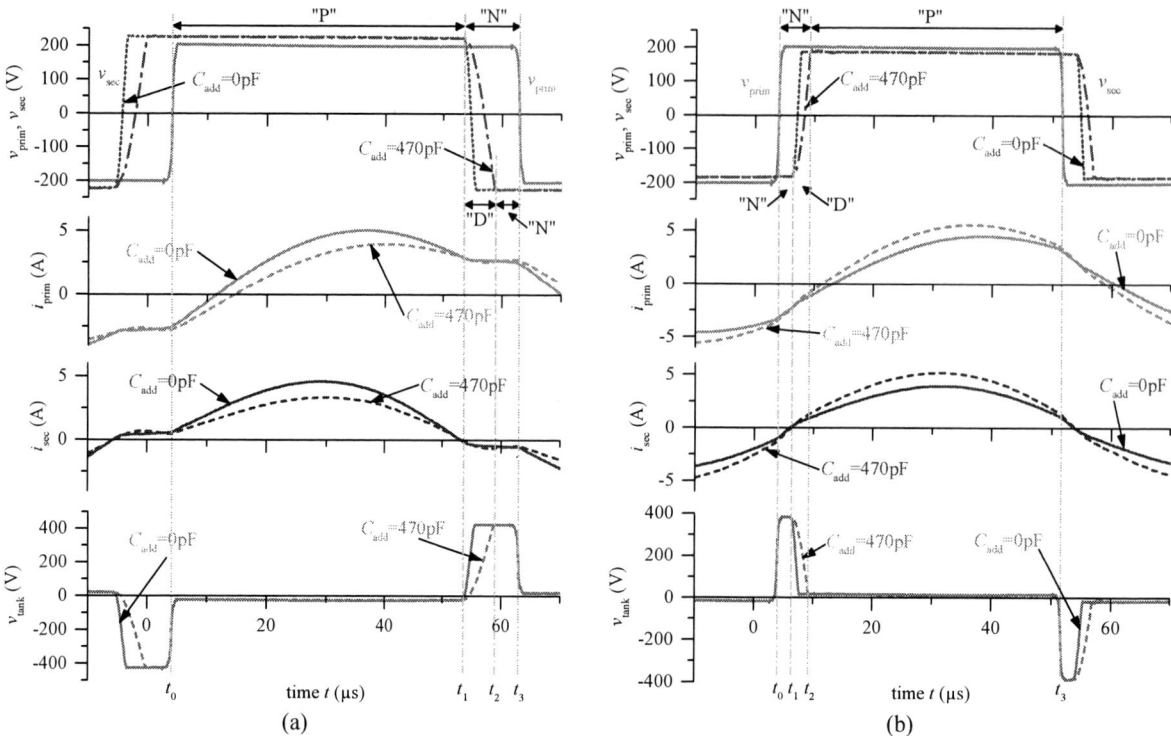

Fig. 5. Measured key waveforms for step-up operation mode with $V_{in} = 200$ V and $V_{out} = 220$ V at $f_{sw} = 85$ kHz (a) and for step-down operation mode with $V_{in} = 200$ V and $V_{out} = 180$ V at $f_{sw} = 105$ kHz (b). For the measurements the reference SiC SBD (solid lines) and the reference SiC SBD with additional capacitance Cadd = 470 pF parallel to each diode are used (dashed lines).

half-cycle of the switching period T_S the polarity of the secondary resonant voltage v_{sec} can be positive (subinterval P), negative (subinterval N), or undefined due to the blocking state of all rectifier diodes (subinterval O). Additional to these subintervals, a new subinterval D has to be introduced, considering the diode junction capacitances. In the subsequent second half-cycle these subintervals occur as well with inverse algebraic signs in resonant tank currents and voltages. In subinterval P, the output rectifier conducts and applies $+V_2$ to the secondary side resonant tank L_m-L_2-C_2 (Fig. 4(a)). The primary current i_{prim} starts to rise in a sinusoidal shape and exceeds the magnetizing current i_{mag}. Therefore, power is delivered to the secondary side as the secondary resonant current $i_{sec} = i_{prim} - i_{mag}$ is positive. Similar to subinterval P, in subinterval N the conducting secondary side rectifier applies $-V_2$ to the secondary resonant tank L_m-L_2-C_2 which reduces i_{prim} and thus i_{sec} in a sinusoidal manner (Fig. 4(b)). Hence, the output power diminishes until i_{sec} crosses zero and subinterval N ends. In contrast to the former discussed subintervals, in subinterval O all rectifier diodes on the secondary side are OFF and thus no current flow into the decoupled output voltage source is established (Fig. 4(c)). The new subinterval D bases on subinterval O, but considers the diode junction capacitance of the applied non-ideal unidirectional or bidirectional switch. In stage D, all rectifier diodes block and decouple V_2 from the resonant tank. However, the secondary resonant tank L_m-L_2-C_2 is connected to the parasitic diode junction capacitance which is charged to

$v_{sec} = \pm V_2$ depending on the preceding subinterval. The diode junction capacitances are discharged (respectively charged) to the output voltage with inverse algebraic sign by the secondary current i_{sec}. Thus, also no output power is transferred to the load and stage D ends as soon as the discharge (respectively charge) process is finished (Fig. 4(d)). Different combinations of the previously discussed subintervals form several operating modes in which the CLLLC converter works in step-up and step-down operation mode. These operating modes are named after the involved subintervals, e.g. in operating mode NP, subinterval P follows after subinterval N. Under the assumption of ideal components and semiconductor devices the subintervals N, P, and O built seven operating modes, namely PN, PON, PO, OPO, and O in step-up operation mode and PN, NP, NOP, OPO, and O in step-down mode. If the subinterval D is also included even more operating modes are possible. The transitions between the different operating modes occur as a function of the switching frequency and load conditions. Hence, the operating mode transition can be observed as the measured or calculated output power p_{out} characteristic alters its gradient.

B. New subinterval D in step-up operation mode

In the following, the new subinterval D is discussed in detail. Figure 5 shows the measured primary v_{prim} and secondary v_{sec} voltage and also the primary i_{prim} and secondary i_{sec} current at the resonant tank for the reference SiC SBD (solid lines) and for the SiC SBD with additional capacitance C_{add} (dashed lines). The step-up operation mode at $f_{sw} = 85$ kHz

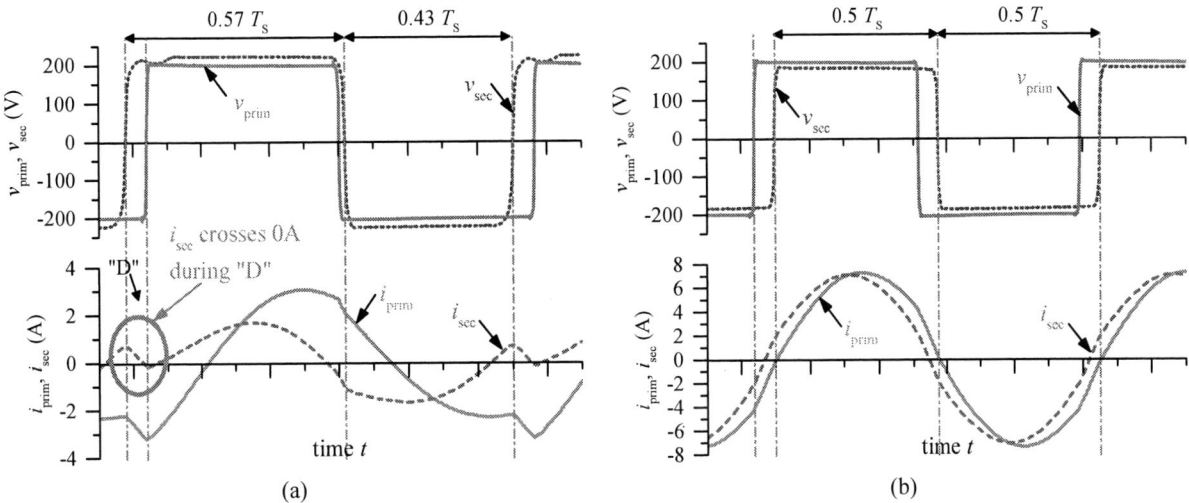

Fig. 6. Measured current and voltage waveforms of the resonant tank for step-up operation mode (a) at $f_{sw} = 88$ kHz and step-down operation mode (b) at $f_{sw} = 105$ kHz for Si super-junction MOSFET (IPW65R110CFD) used as secondary bridge rectifier.

for the SiC SBD in Fig. 5(a) starts with the power transferring subinterval P at $t = t_0$ and is followed by subinterval N at $t = t_1$, in which the power transfer to the secondary side diminishes and energy is stored in the resonant tank [17]. The rising edge of v_{sec} is very steep due to the very small diode junction capacitance and thus the state of the output rectifier abruptly changes. However, increasing the diode junction capacitance leads to the new subinterval D, in which the capacitances are charged and discharged by the secondary resonant current i_{sec}. For the PN operating mode, subinterval D emerges between subinterval P and N as i_{sec} crosses zero at $t = t_1$ and the secondary rectifier begins to alter polarity. Due to the diode junction capacitances an instantaneous transition is not possible compared to ideal devices. At $t = t_1$ all four rectifier diodes begin to block and i_{sec} starts to charge C_{j5} and C_{j8}, resp. discharge C_{j6} and C_{j7} (see Fig. 1). After C_{j6} and C_{j7} are discharged down to the diode forward voltage at $t = t_2$, the diodes D_6 and D_7 begin to conduct and subinterval D ends. No output current and thus no output power is transferred to the secondary side during this subinterval. Moreover, from Fig. 5(a) it can be clearly seen that the insertion of subinterval D shortens in step-up mode the period length of N. Furthermore, in stage D the voltage across the complete resonant network $v_{tank} = v_{prim} - v_{sec}$ slowly rises with the same gradient the diode junction capacitances are discharged. In comparison to the ideal device, the combination of subinterval D and N results in a smaller voltage-time product $v_{tank} \cdot (t_3 - t_1)$ and thus in less stored energy in the resonant tank. In the following subinterval P less energy can be transferred to the load resulting in a decrease of i_{prim}, i_{sec}, and the output power p_{out} for increasing values of C_{add}.

C. New subinterval D in step-down operation mode

In contrast, the subinterval order in step-down operation mode is reversed as shown in Fig. 5(b). Here, at $f_{sw} = 105$ kHz the first half-cycle of the switching period starts with subinterval N at $t = t_0$, followed by subinterval P at $t = t_2$ for the SiC SBD without additional capacitance. Again, for the SiC SBD the secondary resonant voltage v_{sec} rises rapidly from $-V_2$

to $+V_2$. For the SiC SBD with additional capacitance C_{add} the new subinterval D emerges between N and P, starting at $t = t_1$ with the zero-crossing of i_{sec} and ending at $t = t_2$ once the capacitances C_{j5} and C_{j8} are discharged (C_{j6} and C_{j7} are charged, respectively). For this reason, the discharging process in subinterval D extends the length of the preceding subinterval N as can be seen in Fig. 5(b). This results in a larger voltage-time product $v_{tank} \cdot (t_2 - t_0)$ and thus leading to a higher stored energy in the resonant tank and to a higher output power in the subsequent subinterval P. Hence, p_{out} and the peak values of i_{prim} and i_{sec} increase with rising diode junction capacitance values.

D. Special behavior of Si MOSFET

Next, the remarkable behavior of the Si MOSFET is shortly explained. As already described, the super-junction Si MOSFET features a large transistor output capacitance. In addition, the internal capacitances of Si devices exhibit a strong nonlinear charge-voltage (CV) characteristic depending on device technology [23]. In Fig. 6(a) the measured current and voltage waveforms at the observed discontinuity (see Fig. 2(b)) at $f_{sw} = 88$ kHz for step-up operation mode is plotted. Due to part load conditions i_{sec} is low at the beginning of subinterval D, decreases further with a negative slope during D and finally crosses zero (red circle in Fig. 6(a)). This results in an interruption of the charging process of C_{oss} and therefore, the secondary rectifier does not switch polarity at the designated instant. Furthermore, the currently negative secondary resonant current i_{sec} starts to discharge the almost charged transistor output capacitances C_{oss5} and C_{oss8} again, which can be seen in the decrease of v_{sec}. This process is stopped by the rising edge of v_{prim}, which forces i_{sec} to rise. Hence, the following subinterval P is delayed and aborted early by the primary inverter, which switches from $+V_1$ to $-V_1$. The delayed transition of v_{sec} leads to an imbalance of the positive and negative voltage-time product within one period shown in Fig. 6(a). At the examined switching frequency $f_{sw} = 88$ kHz the period length of $v_{sec} = +V_2$ amounts to 57% of the switching period T_S and the length of $v_{sec} = -V_2$ is 43%, respectively.

Thus, the resonant tank waveforms are not symmetric to a half-cycle of the switching period and the output power characteristic is slightly distorted in comparison to TDS calculation. In contrast, in step-down mode (Fig. 6(b)) a similar behavior was not observed. Here the positive and negative voltage-time product within one period is equal.

IV. Design Considerations

Based on the results in this paper, power semiconductors with low diode/transistor output capacitance should be applied in the secondary rectifier. This reduces the effect of the diode junction capacitance (the transistor output capacitance, respectively) on output power and hence TDS calculations can be used for converter design. Therefore, recent ultrafast Si and SiC diodes are suitable for rectifier operation with junction capacitance values in the lower picofarad range. Hence, a higher output power is available in step-up mode utilizing a larger area of the limited output power range. In addition, the regulation in the light load range becomes easier in step-down mode due to the lower output level at high switching frequencies. Moreover, the analyzed wide-bandgap bidirectional switches also show low transistor output capacitances. However, in step-up operation mode the resonant tank has to be designed for a higher nominal output power to compensate for the performance drop introduced by the diode junction capacitances. Thus, the aspired switching frequency range will slightly change compared to TDS calculations. In contrast to step-up operation mode, in step-down mode the influence of the diode junction capacitance is less critical because of the very high output power capability around the series resonance. But, as a result the operating frequency range will slightly change to achieve the desired output power. Moreover, for diodes or transistors with high output capacitance values the regulation of light and very light load conditions can be achieved e.g. by applying pulse skipping control method.

V. Conlusion

This paper demonstrates that the secondary side rectifier diodes junction capacitance has a significant influence on converter output power. The influence is analyzed on the basis of experimental results using a CLLLC topology in an exemplary IPT system. In addition to current Si power semiconductors, wide-bandgap (SiC, GaN) diodes and transistors are analyzed to cover all current state of the art semiconductor devices. The individual operating modes and also voltage and current waveforms of the resonant tank in step-up and step-down operation mode are discussed in detail. Based on these results, practical design considerations for the CLLLC resonant converter are provided.

Acknowledgment

This contribution was supported by the Bavarian Ministry of Economic Affairs and Media, Energy and Technology as a part of the Bavarian project "Leistungszentrum Elektronik-systeme (LZE)". The authors would like to thank Dr. S. Ditze, Cluster of Excellence Engineering of Advanced Materials (EAM) for her valuable scientific input.

References

[1] S. Johnson and R. Erickson, "Steady-state analysis and design of the parallel resonant converter," IEEE Transactions on Power Electronics, vol. 3, no. 1, pp. 93-104, 1988.

[2] R. Liu, C. Lee, and A. Upadhyay, "Experimental study of the LLC-type series resonant converter," in IEEE 1991 Applied Power Electronics Conference and Exposition (APEC), 1991, pp. 31-37.

[3] S. Johnson, A. Witulski, and R. Erickson, "Comparison of resonant topologies in high-voltage DC applications," IEEE Transactions on Aerospace and Electronic Systems, vol. 24, no. 3, pp. 263-274, 1988.

[4] M. Mao, D. Tchobanov, D. Li, and M. Maerz, "Design of a 1 MHz half-bridge CLL resonant converter", IET Power Electronics, vol. 1, no.1, pp. 100-108, March 2008.

[5] A. Hillers, D, Christen, and J. Biela, "Design of a highly efficient bidirectional isolated LLC resonant converter," in IEEE 2012 International Power Electronics and Motion Control Conference (EPE/PEMC), 2012, pp. DS2b.13 1-8.

[6] R. Beiranvand, B. Rashidian, M. R. Zolghadri, S. M. H. Alavi, "Using LLC Resonant Converter for Designing Wide-Range Voltage Source," IEEE Transactions on Industrial Electronics, vol. 58, no. 5, pp. 1746-1756, May 2011.

[7] Z. Zhang, Z. Tang, "Pulse frequency modulation LLC series resonant X-ray power supply," in IEEE 2011 International Conference on Consumer Electronics, Communications and Networks (CECNet)), 2011, pp. 1532-1535.

[8] J. Jung, H. Kim, M. Ryu, J. Baek, "Design Methodology of Bidirectional CLLC Resonant Converter for High Frequency Isolation of DC Distribution Systems," IEEE Transactions on Power Electronics, vol. 28, no. 4, pp. 1751-1755, 2012

[9] W. Chen, P. Rong, and Z. Lu, "Snubberless bidirectional dc-dc converter with new CLLC resonant tank featuring minimized switching loss," IEEE Trans. Ind. Electron., vol. 57, no. 9, pp. 3075–3086, 2010.

[10] Z. U. Zahid, Z. Dalala, L. Jih-Sheng, "Design and control of bidirectional resonant converter for Vehicle-to-Grid (V2G) applications," in IEEE 2014 Annual Conference on IEEE Industrial Electronics Society (IECON), 2014, pp. 1370-1376.

[11] W. Zhang, S.-C. Wong, C. K. Tse, and Q. Chen, "Analysis and comparison of secondary series- and parallel-compensated inductive power transfer systems operating for optimal efficiency and load-independent voltage-transfer ratio," IEEE Transactions on Power Electronics, vol. 29, no. 6, pp. 2979-2990, Jun. 2014.

[12] A. Ecklebe and A. Lindemann, "Analysis and design of a contactless energy transmission system with flexible inductor positioning for automated guided vehicles," in IEEE 2006 Annual Conference on IEEE Industrial Electronics Society (IECON), 2006, pp. 1721-1726.

[13] C. Joffe, S. Ditze, and A. Rosskopf, "A novel positioning tolerant inductive power transfer system," in IEEE 2013 Electric Drives Production Conference (EDPC), 2013, pp. 1-7.

[14] R. M. Miskiewicz, A.J. Moradewicz, and M.P. Kazmierkowski, "Contactless battery charger with bi-directional energy transfer for plug-in vehicles with vehicle-to-grid capability," in IEEE 2011 International Symposium on Industrial Electronics (ISIE), 2011, pp. 1969–1973.

[15] R. Steigerwald, "A comparison of half-bridge resonant converter topologies," IEEE Transactions on Power Electronics, vol. 3, no. 2, pp. 174-182, 1988.

[16] J. Lazar and R. Martinelli, "Steady-state analysis of the LLC series resonant converter," in IEEE 2001 Applied Power Electronics Conference (APEC), 2001, pp. 728-735.

[17] S. Ditze, "Steady-state analysis of the bidirectional CLLLC resonant converter in time domain," in IEEE 2014 Telecommunications Energy Conference (INTELEC), 2014, pp. 1-9.

[18] H. Chen and X. Wu, "Analysis on the influence of the secondary parasitic capacitance to ZVS transient in LLC resonant converter," in IEEE 2014 Energy Conversion Congress and Exposition (ECCE), 2014, pp. 4755-4760.

[19] J.-H. Kim, C.-E. Kim, J.-K. Kim, and G.-W. Moon, "Analysis for LLC resonant converter considering parasitic components at very light load,"

in IEEE 2011 International Conference on Power Electronics (ECCE Asia), 2011, pp. 1863-1868.

[20] B.-H. Lee, M.-Y. Kim, C.-E. Kim, K.-B. Park, and G.-W. Moon, "Analysis of LLC resonant converter considering effects of parasitic components," in IEEE 2009 Telecommunications Energy Conference (INTELEC), 2009, pp. 1-6.

[21] C.-H. Yang, T.-J. Liang, K.-H. Chen, J.-S. Li, and J.-S. Lee, "Loss analysis of half-bridge LLC resonant converter," in IEEE 2013 Future Energy Electronics Conference (IFEEC), 2013, pp. 155-160.

[22] K.-B. Park, B.-H. Lee, G.-W. Moon, and M.-J. Youn, "Analysis on center-tap rectifier voltage oscillation of LLC resonant converter," IEEE Transactions on Power Electronics, vol. 27, no. 6, pp. 2684-2689, Jun. 2012.

[23] R. Elferich, "General ZVS half bridge model regarding nonlinear capacitances and application to LLC design," in IEEE 2012 Energy Conversion Congress and Exposition (ECCE), 2012, pp. 4404-4410.

[24] J. Stahl, D. Kuebrich, T. Duerbaum, A. Leicht, J. Patz, "A fully automated measurement set-up for the determination of the reverse recovery behaviour of ultra-fast diodes," in IEEE 2011 European Conference on Power Electronics and Applications (EPE), 2012, pp. 1-9.

[25] J. Lutz, H. Schlangenotto, H. Scheuermann, R. De Doncker, "Schottky Diodes," in *Semiconductor Power Devices*, 2nd ed. Springer-Verlag Berlin Heidelberg, 2011.

[26] H. Mitlehner, W. Bartsch, M. Bruckmann, K. O. Dohnke, U. Weinert, "The potential of fast high voltage SiC diodes," in IEEE 1997 Power Semiconductor Devices and IC's (ISPSD), 1997, pp. 165-168.

[27] T. Heckel, B. Eckardt, M. März, L. Frey, "SiC MOSFETs in Hard-Switching Bidirectional DC/DC Converters," in Materials Science Forum, Vols. 821-823, pp. 689-692, Jun. 2015.

[28] N. Kaminski, "State of the art and the future of wide band-gap devices," in IEEE 2009 Power Electronics and Applications (EPE), 2009, pp. 1-9.

[29] X. Huang, Z. Liu, Q. Li. F. C. Lee "Evaluation and Application of 600 V GaN HEMT in Cascode Structure," IEEE Transactions on Power Electronics, vol. 29, no. 5, pp. 2453-2461, May 2014.

A Triple Active Bridge DC-DC Converter Capable of Achieving Full-Range ZVS

Ling Jiang, Daniel Costinett

Department of Electrical Engineering and Computer Science
University of Tennessee at Knoxville, USA
Ljiang7@vols.utk.edu

Abstract— **In this paper, a triple active bridge converter is proposed. The topology is capable of achieving ZVS across the full load range with wide input voltage while minimizing heavy load conduction losses to increase overall efficiency. This topology comprises three full bridges coupled by a three-winding transformer. At light load, by adjusting the phase shift between two input bridges, all switching devices can maintain ZVS due to a controlled circulating current. At heavy load, the two input bridges work in parallel to reduce conduction loss. The operation principles of this topology are introduced and the ZVS boundaries are derived. Based on analytical models of power loss, a 200W laboratory prototype has been built to verify theoretical considerations.**

Keywords—soft switching; ZVS; triple active bridge (TAB); DC-DC converter

I. INTRODUCTION

The dual active bridge (DAB) converter [1], has attractive features including bidirectional power flow, low component stresses, isolated power conversion, high power efficiency, and utilization of the transformer leakage inductance as the energy transfer element. Thus, it has played an important role in data center, electric vehicles, and power electronics interfaces for future power systems based on smart-grid technologies [2]-[4].

When using simple phase shift modulation, the DAB is able to achieve high efficiency when operated under soft switching conditions. However, at light load, soft switching is lost, which results in reduced efficiency. Researchers have made great efforts to extend the zero voltage switching (ZVS) range of DAB. In [2], [5]-[8], different switching control strategies are implemented to increase the ZVS range at the expense of increasing control complexity. In [9], a modified unified PWM control method is proposed to extend the ZVS range to zero load. However, it is not applicable for a varying input voltage. To achieve a wider range of ZVS, the magnetizing inductance of the transformer can be reduced, or an external inductor can be added in parallel [10]. However, reducing the shunt inductance results in larger circulating current. Particularly at heavy load, conduction loss will increase significantly. To avoid this problem, implementation of a variable series inductance is introduced in [11]-[13]. The variable inductance is designed to present a small inductance at heavy load to reduce circulating current, while taking a larger value at light load to extend the ZVS range. Nevertheless, this method has little flexibility in controls and requires additional switches. In [11], the method of adjusting switching frequency is proposed, but it increases the design complexity of magnetic components and filters due to operation at various switching frequencies.

This paper proposes a three winding transformer based triple active bridge (TAB) dc-dc converter. Fig. 1 shows the block of this proposed topology. FB_1 (full bridge 1) and FB_2 (full bridge 2) are two input full bridges, while FB_3 (full bridge 3) is the output full bridge. They are coupled by a high frequency transformer, which has three windings (turns ratio is $N_1:N_2:N_3$). Power flow between the three full bridges is mainly controlled by adjusting $\phi_{3,1}$ (phase shift between FB_3 and FB_1) and $\phi_{2,1}$ (phase shift between FB_2 and FB_1).

This topology with an additional bridge provides multiple benefits to converter operation. It is able to have ZVS at entire load range through simple phase-shift control of all bridges. At heavy load, defined as operation mode 1 ($\phi_{2,1}=0$, $0<\phi_{3,1}\leq0.5$) in this paper, both input bridges operate synchronously to reduce conduction loss by sharing the input current equally, as shown in Fig.1(a). At light load as operation mode 2 ($0<\phi_{2,1}\leq\phi_{3,1}\leq0.5$) and mode 3 ($0<\phi_{3,1}<\phi_{2,1}\leq0.5$), one of input full bridges (such as FB_2) reverses power flow and thereby increases circulating current to achieve soft switching , as shown in Fig.1 (b) and (c).

Despite the increase in number of device relative to the DAB, the switching devices in primary bridges only need to process half current of the traditional DAB, allowing the TAB to be implemented with the same total semiconductor area.

The paper is organized as follows. The proposed topology description and analysis is presented in Section II. The ZVS analysis is presented in Section III. The power flow and power loss model are introduced in Section IV. Experimental verification is addressed in V. Conclusions and future work are stated in Section VI.

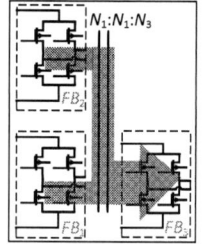 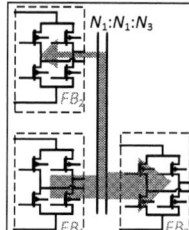 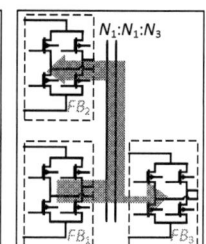

(a) Operation mode1   (b) Operation mode 2   (c) Operation mode 3

Fig. 1. Power flow sketches of proposed TAB at three operation modes

978-1-4673-9551-9/16 $31.00 © 2016 IEEE

II. TOPOLOGY DESCRIPTION AND ANALYSIS

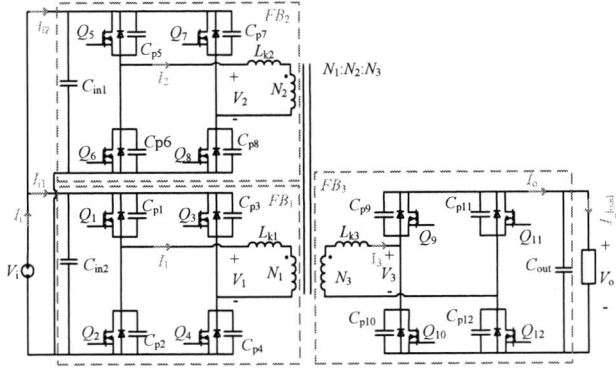

Fig. 2. Proposed triple active bridge dc-dc converter

Fig. 2 shows the schematic of this proposed TAB. Input voltage of this converter is V_i and output voltage is V_o. V_1, V_2 and V_3 are switching node voltage of FB_1, FB_2 and FB_3. And define the current flow through three bridges as I_1, I_2, I_3 respectively. The leakage inductances (L_{k1}, L_{k2} and L_{k3}) of transformer are designed to exchange energy with output capacitors of MOSFETs ($Q_1 \sim Q_{12}$) to achieve ZVS. FB_1, FB_2 and FB_3 complete their ZVS transition in resonant interval dt_1, dt_2 and dt_3 respectively. Similar topologies can be derived where FB_x bridges are replaced with half-bridges or other circuit implementations.

Fig. 3 depicts the equivalent circuit in which both FB_2 and FB_3 are reflected through the transformer to FB_1 for convenience. To facilitate the system analysis, the following assumptions are made.

1. The turns ratios of transformer N_1:N_2:N_3=1:1:1.
2. The leakage inductances of the transformer are identical, $L_{k1}= L_{k2}= L_{k3}=L_k$.
3. Each bridge operates with a duty cycle near 50%.
4. the magnetizing inductance is neglected in theoretical analysis.

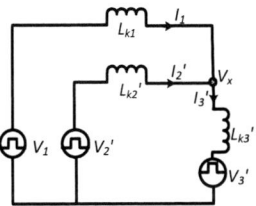

Fig. 3. Primary-reflected equivalent circuit of the TAB converter

Considering phase shift and resonant dead times, the converter exhibits eight different operation modes. Three critical operation modes (mode 1~mode 3) are the focus of this paper, as they exhibit favorable conduction and switching characteristics and jointly span the full load range. The other five modes, whose ZVS commutation intervals exhibit multi-resonant behaviors and only happen in light load are ignored in this paper. Fig.4 presents the main waveforms for V_1, V_2, V_3, I_1, I_2, I_3 and I_o, according to three operation modes.

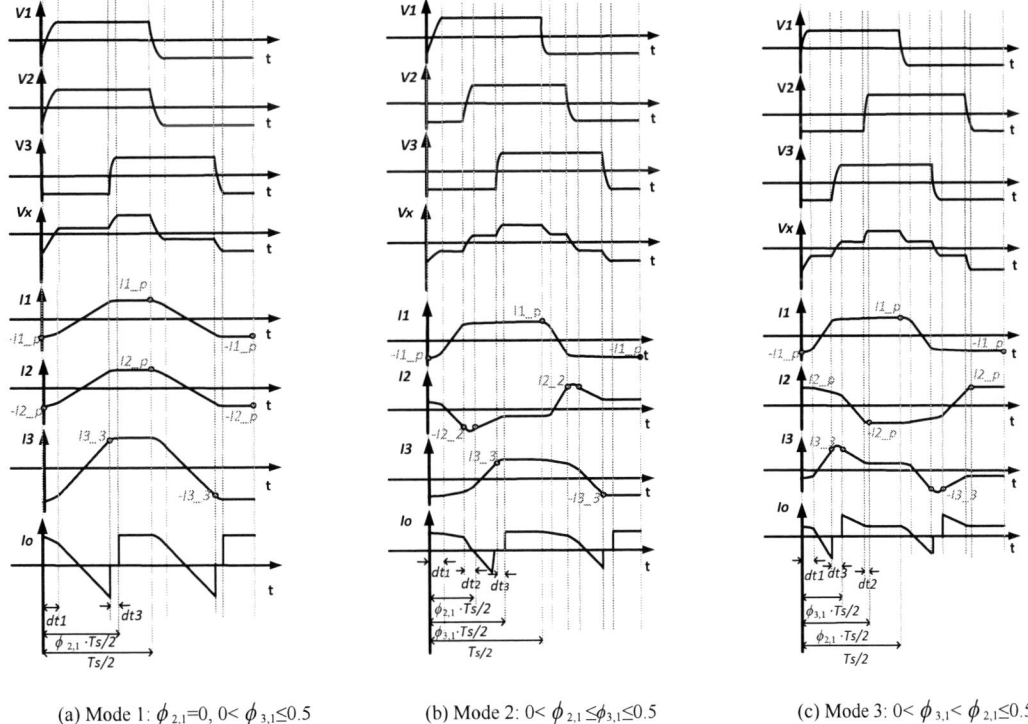

(a) Mode 1: $\phi_{2,1}$=0, 0< $\phi_{3,1} \leq 0.5$ (b) Mode 2: 0< $\phi_{2,1} \leq \phi_{3,1} \leq 0.5$ (c) Mode 3: 0< $\phi_{3,1} < \phi_{2,1} \leq 0.5$

Fig. 4. Ideal operation modes waveforms ($\phi_{3,1}$ and $\phi_{2,1}$ are phase shift; dt_1, dt_2 and dt_3 are dead time; Ts is the switching period)

1) Operation mode 1

Waveforms in operation mode 1 are shown in Fig. 4 (a). This mode is applied when converter operates at heavy load. In this case, FB_1 and FB_2 operate synchronously (i.e. $\phi_{2,1}$=0) to equally share power delivered to FB_3 as stated in in Fig. 1 (a). In this mode, converter operation and waveforms are similar to that of a traditional DAB under phase shift modulation. As shown in Fig.4 (a), I_1 and I_2 are identical and I_3 is sum of I_1 and I_2. As in the DAB, this mode of operation is preferable when output power is sufficient such that I_1 and I_2 are sufficiently large during the switching dead times to fully charge device capacitances C_{px} and achieve ZVS. In [18], besides C_{px}, the parasitic capacitance of transformer is took in to consideration as well. This paper mainly focus on the new topology operation, so parasitic parameters of three winding transformer are ignored to simplify analysis .In addition in this mode, I_{1_rms} is equal to I_{2_rms} and is half value of I_{3_rms}, so total conduction loss can be minimized.

2) Operation mode 2

At light load, the output current I_o reduces. Decreasing I_o leads to an equal reduction in amplitudes of I_1, I_2 and I_3 eventually resulting in loss of soft switching due to insufficient energy available to complete the resonant ZVS transition. To regain ZVS, the converter switches to operation mode 2, shown in Fig. 4 (b). In this mode, the average value of I_2 is near zero, allowing more power to be processed by FB_1 and thereby extending its ZVS range. Bridge 2 processes only a small amount of power as shown in Fig. 1 (b), but has sufficiently high peak current I_{2_2} to maintain soft switching. In addition, bridge 2 has minimum RMS current which is beneficial for high power efficiency.

3) Operation mode 3

Power flow of operation mode 3 is shown in Fig. 1 (c). When load current further reduces, to maintain ZVS of FB_1, FB_2 reverses power flow. As presented the waveform in Fig. 4 (c), the average value of I_2 goes negative. In this mode, FB_2 acts as a parallel load to the input bridge FB_1, increasing circulating power to extend the ZVS range. The RMS value of I_2 is increased, which increases conduction loss, but the increase in circulating current controlled by $\phi_{3,1}$ and $\phi_{2,1}$ to be the minimum value necessary to achieve ZVS of all devices.

III. ZVS ANALYSIS

In this section, converter waveform are analyzed including the resonant switching intervals. Particularly for the converters with high switching frequency, the resonant period can occupy a significant portion of switching period. Thus it must be included for accurate converter analysis [14].

Two conditions need to be met in order to achieve ZVS of all devices in each bridge. First, the stored energy in the bridge's leakage just prior to the switching transition must be sufficient to charge and discharge the output parasitic capacitances of all MOSFETs in it. Second, the dead time must be long enough to allow the resonant transition to complete before devices turn on.

To calculate the required minimum ZVS current and dead time, state plane analysis is used. To save space, only the dt_1 interval in mode 2, shown in Fig. 5, is analyzed in detail in this paper. Assume the nonlinear output capacitance of each device C_{px} is C_{eq} that is calculated according to [15]. Due to the

series/parallel combination of devices, the equivalent resonant capacitance for one full bridge is also equal to C_{eq}.

In Fig. 5 (a), both V_2 and V_3 maintain a constant value of $-V_i$, while V_1 completes resonant commutation from $-V_i$ to $+V_i$ during the dead time of FB_1. Equations defining the circuit behavior are obtained from Fig. 5(a)

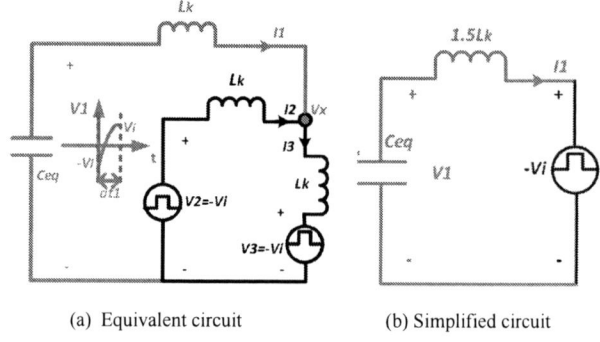

(a) Equivalent circuit   (b) Simplified circuit

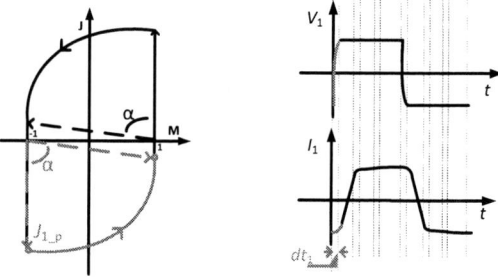

(c) State plane and its corresponding time domain resonant transition

Fig. 5. State plane analysis for resonant transition of dt_1 interval in mode 2

$$L_k \frac{di_1}{dt} = V_1 - V_x \tag{1}$$

$$L_k \frac{di_2}{dt} = -V_i - V_x \tag{2}$$

$$L_k \frac{di_3}{dt} = L_k \frac{d(i_1 + i_2)}{dt} = V_i - V_x \tag{3}$$

Combining (1), (2) and (3),

$$3L_k \frac{di_1}{dt} = 2(V_i + V_1) \tag{4}$$

The circuit in Fig. 5 (a) can be further simplified as Fig. 5 (b) according to (4). Then, the normalized resonant state plane shown in Fig. 5 (c) is used to analyze the resonant behavior during dt_1. The resonant waveforms are normalized according to

$$J_{1\_p} = \frac{I_{1\_p}}{I_{base}} \tag{5}$$

$$M = \frac{V_1}{V_{base}} \tag{6}$$

With $R_0 = (1.5L_k/C_{eq})^{1/2}$, $V_{base}=V_i$, $I_{base}=V_{base}/R_0$.

Analyzing the state plane, the normalized current J_{1_p} is

TABLE I: ZVS CONDITIONS IN DIFFERENT OPERATION MODEL

Items		Mode1	Mode2	Mode3
Inductor current	$I1$	$I_{1\_p} \geq 2 \cdot V_i \cdot \sqrt{\dfrac{C_{eq}}{3Lk}}$	$I_{1\_p} \geq 2 \cdot V_i \cdot \sqrt{\dfrac{C_{eq}}{1.5Lk}}$	$I_{1\_p} > 2 \cdot V_i \cdot \sqrt{\dfrac{C_{eq}}{1.5L_k}}$
	$I2$	$I_{2\_p} \geq 2 \cdot V_i \cdot \sqrt{\dfrac{C_{eq}}{3Lk}}$	$I_{2\_2} > 0$	$I_{2\_p} \geq 2 \cdot V_i \cdot \sqrt{\dfrac{C_{eq}}{1.5L_k}}$
	$I3$	$I_{3\_3} \geq 0$	$I_{3\_3} \geq 0$	$I_{3\_3} \geq 0$
Dead time	dt_1	$dt_1 \geq f(3L_k, I_{1\_p})$	$dt_1 \geq f(1.5L_k, I_{1\_p})$	$dt_1 \geq f(1.5L_k, I_{1\_p})$
	dt_2	$dt_2 \geq f(3L_k, I_{2\_p})$	$dt_2 \geq f(1.5L_k, I_{2\_2})$	$dt_2 \geq f(1.5L_k, I_{2\_p})$
	dt_3	$dt_3 \geq f(1.5L_k, I_{3\_3})$	$dt_3 \geq f(1.5L_k, I_{3\_3})$	$dt_3 \geq f(1.5L_k, I_{3\_3})$

$$J_{1\_p} = \frac{I_{1\_p}}{I_{base}} = \frac{I_{1\_p}}{V_i} \cdot \sqrt{\frac{1.5L_k}{C_{eq}}} \tag{7}$$

To ensure ZVS can be achieved, it is required that

$$J_{1\_p} \geq 2 \tag{8}$$

which is denormalized to obtain the condition on the peak current I_1 just prior to the resonant interval

$$J_{1\_p} \cdot I_{base} \geq 2 \cdot I_{base} \tag{9}$$

$$I_{1\_p} \geq 2 \cdot V_i \cdot \sqrt{\frac{C_{eq}}{1.5L_k}} \tag{10}$$

The required dead time for this transition is given in normalized form by

$$\alpha \geq \frac{\pi}{2} - \tan^{-1}(\frac{\sqrt{J_0 - 4}}{2}) \tag{11}$$

which, when denormalized, gives

$$dt_1 = \frac{\alpha}{w} \geq (\frac{\pi}{2} - \tan^{-1}(\frac{I_{1\_p}}{2 \cdot V_i} \cdot \sqrt{\frac{C_{eq}}{1.5L_k}})) \cdot 1.5L_k \cdot C_{eq} \tag{12}$$

After analyzing all of the resonant transitions of mode 1~mode 3 by state plane analysis, the ZVS boundaries are shown in Fig. 6 is obtained. ZVS conditions in different operation modes are given in Table I. The required dead time is a function of equivalent leakage inductance and current as

$$f(Lk_{eq}, I_0) = (\frac{\pi}{2} - \arctan(\frac{I_0}{2V_i} \cdot \sqrt{\frac{C_{eq}}{Lk_{eq}}})) \cdot \sqrt{Lk_{eq} \cdot C_{eq}} \tag{13}$$

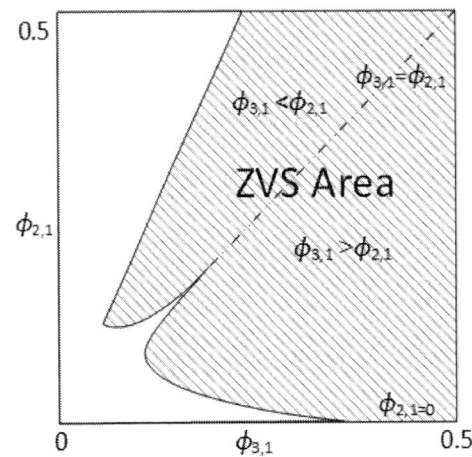

Fig. 6. ZVS Boundary

IV. POWER FLOW AND POWER LOSS MODEL

A. Power Flow Analysis

Power flow in the TAB is controlled by phase shifts, $\phi_{3,1}$ and $\phi_{2,1}$, between the three bridges. Input energy is transferred to the output via leakage inductances L_{k1}, L_{k2} and L_{k3}. According to waveforms of output current I_o in each of the three operating modes, the average output current I_{o_ave} can be derived. By multiplying I_{o_ave} and V_o, output power is calculated for modes 1-3, respectively, as

$$P_o = I_{o\_ave} \cdot V_o = \frac{1}{3} \cdot \frac{V_i \cdot V_o \cdot T_s}{L_k} \cdot \phi_{3,1} \cdot (1 - \phi_{3,1}) \tag{14}$$

$$P_o = \frac{T_s \cdot V_i \cdot V_o}{6L_k^2}[-2\phi_{3,1}^2 + 2\phi_{3,1} + 2\phi_{3,1}\phi_{2,1} - \phi_{2,1}^2 - \phi_{2,1}] \tag{15}$$

$$P_o = \frac{T_s \cdot V_i \cdot V_o}{6L_k} \cdot (2\phi_{3,1} - \phi_{2,1})(1 - \phi_{2,1}) \tag{16}$$

Using the same analysis, power flow through the other two bridges (FB_1 and FB_2) can be obtained.

B. Power Loss Model and Efficiency Calculation

A power loss model is constructed to estimate the efficiency and to provide evidence for selecting an operating trajectory through the three modes. To facilitate analysis, assume all the switching devices use same MOSFET, characterized by on-resistance R_{on}, equivalent linear output capacitance C_{eq} and gate charge Q_g.

1) Conduction Loss

If using I_{x_rms} to present RMS current of FB_x, total device conduction loss is the sum of conduction loss in each full bridge

$$P_{cond} = (I_{1\_rms}^2 + I_{2\_rms}^2 + I_{3\_rms}^2) \cdot 2 \cdot R_{on} \qquad (17)$$

When the converter works at heavy load (operation mode 1), the current flow through each bridge has relationship as

$$I_1 = I_2 = \frac{1}{2} \cdot I_3 \qquad (18)$$

Also

$$I_{1\_rms} = I_{2\_rms} = \frac{1}{2} \cdot I_{3\_rms} \qquad (19)$$

When calculate the RMS current value, resonant transition period is ignored to simplify calculation. Thus, the waveform of I_1 in Fig. 4 (a) is simplified to Fig. 7. The shaded area in the figure represents RMS current, which is calculated according to the method presented in [17].

$$I_{1\_rms} = \frac{V_i \cdot \phi_{3,1} \cdot T_s}{6 L_k} \cdot \sqrt{(1 - \frac{2}{3}\phi_{3,1})} \qquad (20)$$

Combining (19) and (20) into (17), the conduction loss is obtained. The RMS currents and conduction losses in alternate operating modes can be solved using a similar approach, though the result is more complex as (20) no longer holds. Results are omitted here for brevity.

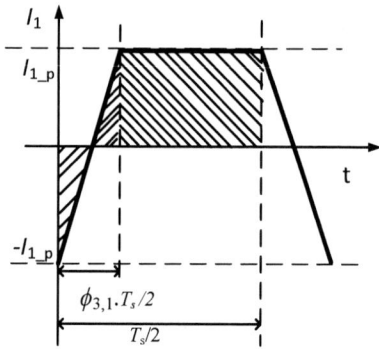

Fig. 7. Current waveform for RMS value calculation

2) Switching Loss

Because of the achievement of ZVS, capacitance loss is negeligible. Overlap losses, though still present are neglected assuming a sufficiently fast gate drive implementation. In this case, only gate charge loss is considered. If Q_g is the gate charge for each device, V_{cc} is the gate driver voltage, and f_s is the switching frequency, the total switching loss is

$$P_{sw} = P_{gate} = 12 \cdot Q_g \cdot V_{cc} \cdot f_s \qquad (21)$$

3) Transformer Loss

Core loss is calculated according to iGSE method [16]. Copper losses that comprise of dc copper loss and high

frequency ac copper loss resulting from skin effect and proximity effect are took into calculation, following the analytical methods disscussed in [17].

C. The Impact of Leackage Inductance On Power Loss

Leakage inductance is a critical parameters in TAB design. Based on equation (14), transferring the same power, larger inductance would require larger phase shift, which results in higher circulating current, motivating the use of a small leakage inductance. As shown the blue curve in Fig. 8 the relation between leakage inductance with conduction loss is analyzed based on a 200W prototype design at heavy power (operation mode1). With inductance increase, the conduction loss increases.

For the traditional DAB, small inductance would result in decreased soft switching range, resulting in a necessary compromise in the selection of L_k. However, in the TAB, the converter can maintain soft switching at light load through controlled circulating current in operating in mode 2 or mode 3. For this converter, the design tradeoffs for L_k present differently; ZVS can be guaranteed at all powers, but the value of L_k dictates the boundaries of the operating modes and the RMS currents within them. Define P_m as the power at this condition: operation mode 2 starts to takes place of mode1. Thus, the smaller P_m means the needed circulating current is smaller, therefore lower loss. Equation (22) present the relation between L_k and P_m, and the red curve in Fig. 8 is derived from (22), showing the impact of leakage inductance on P_m. The values addressing in Fig. 8 is based on 200W prototype design. The value of L_k at cross point of red and blue curves shown in Fig. 8 is optimum option.

$$P_m = \frac{4 V_i \cdot V_o}{3 L_k} \cdot \sqrt{3 L_k \cdot C_{eq}} \cdot (1 - \frac{4}{T_s} \sqrt{3 L_k \cdot C_{eq}}) \qquad (22)$$

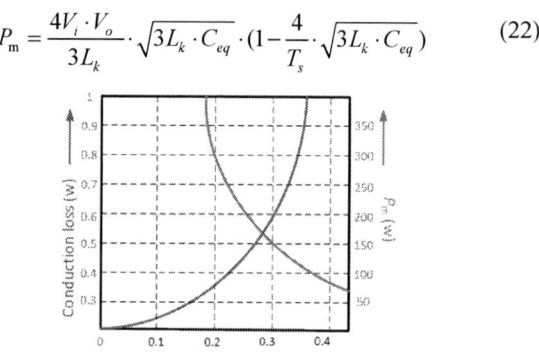

Fig. 8. Optimization of leakage inductance

D. The Optimum Trajectory for Power Flow

The optimum power flow trajectory is defined as the set of phase shifts $\phi_{3,1}$ and $\phi_{2,1}$ which jointly span the full range of output powers while achieving minimum losses at each. For the present design, both input and output voltage are assumed constant. At any given output power, the optimum power flow trajectory is assumed to be the operating point with the smallest circulating current while maintaining ZVS. Though it is possible that a design may achieve higher efficiency if ZVS is sacrificed in favor of reduced circulating currents, this trajectory will result in the best approximation to the optimal prior to selection of specific devices to implement the converter.

Based on the analysis of previous sections, Fig. 9 depicts this theoretical trajectory as the red curves overlaid on the $\phi_{3,1}$-$\phi_{2,1}$-P_{out} surface. The green area shows the output power respect with $\phi_{3,1}$ and $\phi_{2,1}$. The ZVS area, from Fig. 6, is projected onto the surface, enclosed by black dash lines. It can be seen that the optimal trajectory traces the ZVS boundary, indicating that the converter is operating with the minimum permissible circulating current for which ZVS is still obtained.

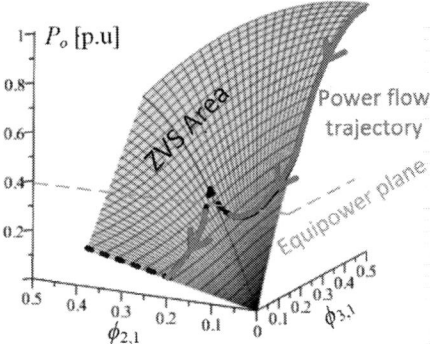

Fig. 9. Power flow trajectory in 3D

V. EXPERIMENTAL VERIFICATION

A. TAB Prototype

A 200W triple active bridge dc-dc converter is built to verify theoretical analysis. The prototype is shown in Fig. 10. This hardware includes three identical full bridges FB_1, FB_2 and FB_3, which are coupled by three winding transformer. A foil winding transformer is designed to minimize leakage inductance. Modulation signals are generated via a Texas Instruments F28069 DSP. Critical parameters of this converter are listed in Table II.

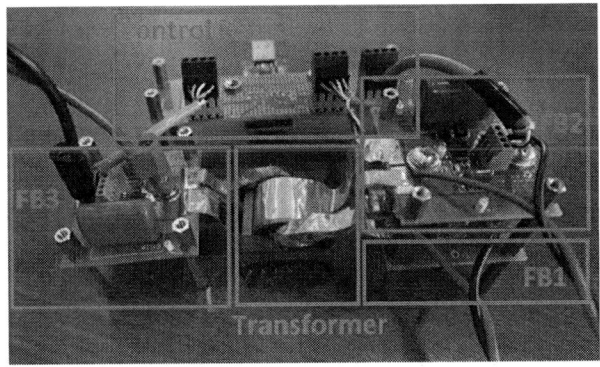

Fig. 10. Hardware of prototype TAB converter

TABLE II: PARAMETERS OF PROTOTYPE

Description	Symbol	Value
Input Voltage	V_i	40V
Output Voltage	V_o	40V
Load Power	P_o	200W
Duty Cycle	$D1, D2, D3$	0.5
Transformer turns	$N_1{:}N_2{:}N_3$	6:6:6
Magnetic core	--	ETD29, 3F35
Magnetizing inductance	L_m	62μF
Leakage Inductance	L_k	0.3μH
Switching frequency	f_s	200kHz
MOSFETs	$Q_1{\sim}Q_{12}$	EPC2001C
Output Capacitance	C_{eq}	560pF

Based on the developed loss model, expected efficiency curve is calculated for this prototype, as shown in Fig. 11. At high power level (mode 1), high efficiency (99%) is expected. And the operating modes vary as described previously to maintain ZVS.

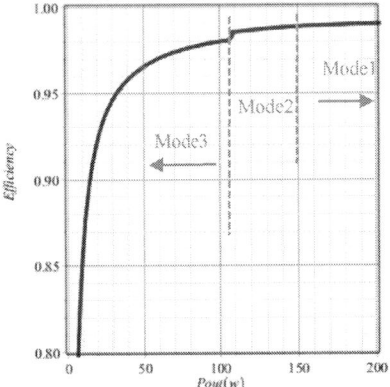

Fig. 11. Calculated efficiency

B. Experimental Results

To illustrate and verify the operating waveforms shown in Fig. 4, experimental waveforms of prototype are shown in Fig. 12.

Fig. 12(a) is the waveform for operation mode 1 at full power. As shown, current flow through FB_1 (I_1) and FB_2 (I_2) are equal and current goes though FB_3 (I_3) is the sum of I_1 and I_2. Fig. 12(b) shows the experimental waveform at operation mode 2. As observed, FB_2 has minimal average current, but maintains a high peak current to maintain ZVS. Fig. 12 (c) shows the tested waveform in operation mode 3 at low output power. To maintain ZVS, FB_2 begins to absorb power from system, increasing the current through FB_1 to maintain ZVS.

978-1-4673-9551-9/16 $31.00 © 2016 IEEE

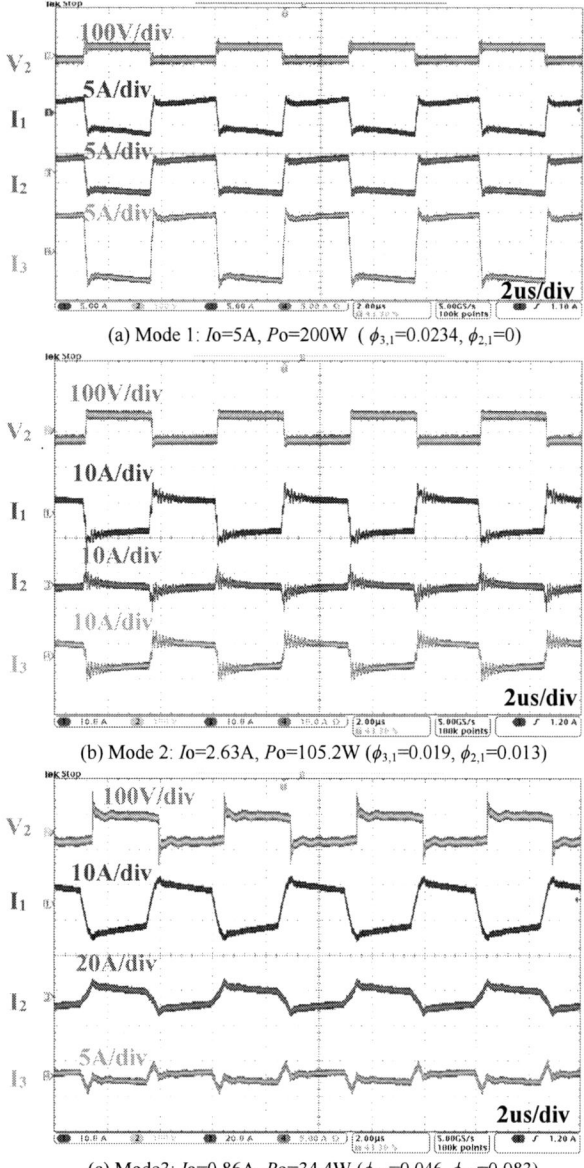

(a) Mode 1: Io=5A, Po=200W ($\phi_{3,1}$=0.0234, $\phi_{2,1}$=0)

(b) Mode 2: Io=2.63A, Po=105.2W ($\phi_{3,1}$=0.019, $\phi_{2,1}$=0.013)

(c) Mode3: Io=0.86A, Po=34.4W ($\phi_{3,1}$=0.046, $\phi_{2,1}$=0.083)

Fig. 12. Experimental waveforms based on three operation modes

Power efficiency is measured and shown in Fig. 13. From 20% to 100% load, the prototype exhibits greater than 90% power stage efficiency, with a maximum approaching 98%. Note that control and other constant losses are not included.

Experimental efficiency is slightly lower than analytical prediction, most notably at light load. There are several reasons for the discrepancy. First, mismatches in input and output voltage, transformer turns ratio, or leakage inductance will cause minor increases in the rms current in each bridge. A similar effect results from the magnetizing current neglected in the analysis. The sum of these effects can be observed when comparing Figs. 4 and 12, noting that currents are not perfectly flat during some intervals, as predicted. Additionally, parasitics such as device inductance are not included in the analysis.

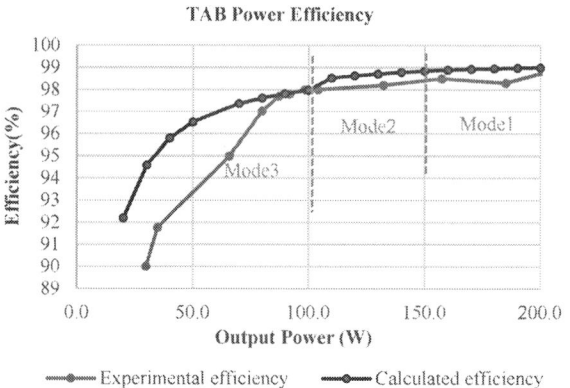

Fig. 13. Experimental efficiency of TAB prototype

VI. CONCLUSION

The operation and analysis of the triple active bridge dc-dc converter is detailed. By adding a third full bridge to the traditional DAB, ZVS operation is maintained at light load without penalty on conduction loss at heavy load, while maintaining the same total semiconductor area. Only simple phase shift modulation is needed for this topology. In this paper, different operation modes, ZVS boundary and power loss model are solved analytically. A prototype is built to verify the theoretical analysis. And experimental results are presented and verified the merits of this proposed topology.

After verifying the feasibility of this topology in 1:1:1 voltage ratio in this paper, more operating conditions will be considered in the future, such as high step down voltage ratio dc-dc converters, or operating points where the conversion ratio is not equal to 1.

ACKNOWLEDGMENT

This work made use of the Engineering Research Center Shared Facilities supported by the Engineering Research Center Program of the National Science Foundation and DOE under NSF Award Number EEC-1041877 and the CURENT Industry Partnership Program.

REFERENCES

[1] R. W. De Doncker, D. M. Divan, and M. H. Kheraluwala, "A three-phase soft-switched high-power-density DC/DC converter for high-power applications," IEEE Transactions on Industry Applications, vol.27, no.1, pp.63, 73, Jan/Feb 1991.

[2] G.G. Oggier, G.O. Garcia, A.R. Oliva, "Modulation strategy to operate the dual active bridge DC-DC converter under soft switching in the whole operating range," IEEE Transactions on Power Electronics, vol.26, no.4, pp.1228, 1236, April 2011.

[3] S. Inoue and H. Akagi, "A bidirectional isolated dc–dc converter as a core circuit of the next-generation medium-voltage power conversion system," IEEE Transactions on Power Electronics, vol. 22, no. 2, pp. 535–542, Mar. 2007.

[4] S. Inoue and H. Akagi, "A bidirectional dc–dc converter for an energy storage system with galvanic isolation," IEEE Transactions on Power Electronics, vol. 22, no. 6, pp. 2299–2306, Nov. 2007.

[5] F. Krismer, S. Round, and J. Kolar, "Performance optimization of a high current dual active bridge with a wide operating voltage range," Power Electronics Specialists Conference, vol., no., pp.1, 7, 18-22 June 2006.

[6] Y. Wang, S. de Haan, and J. Ferreira, "Optimal operating ranges of three modulation methods in dual active bridge converters," Power Electronics and Motion Control Conference, vol., no., pp.1397, 1401, 17-20 May 2009.

[7] M. Ordonez and J. Quaicoe, "Soft-switching techniques for efficiency gains in full-bridge fuel cell power conversion," IEEE Transactions on Power Electronics, vol. 26, no. 2, pp. 482–492, Feb. 2011.

[8] A. Jain and R. Ayyanar, "PWM control of dual active bridge: Comprehensive analysis and experimental verification," IEEE Transactions on Power Electronics, vol. 26, no. 4, pp. 1215–1227, Apr. 2011.

[9] Jun Huang, Yue Wang, Yuan Gao, Wanjun Lei, Ning Li, "Modified unified PWM control to operate the dual active bridge converters under ZVS in the whole load range," ECCE Asia Down Under (ECCE Asia), 2013 IEEE , vol., no., pp.620,625, June 2013.

[10] Everts, J, Krismer, F, Van den Keybus, J, Driesen, J, Kolar, J.W, "Charge-based ZVS soft switching analysis of a single-stage dual active bridge AC-DC converter," Energy Conversion Congress and Exposition (ECCE), 2013 IEEE , vol., no., pp.4820,4829, 15-19 Sept. 2013.

[11] Rodriguez, A.; Vazquez, A.; Lamar, D.G.; Hernando, M.M.; Sebastian, J., "Different Purpose Design Strategies and Techniques to Improve the Performance of a Dual Active Bridge With Phase-Shift Control," IEEE Transactions on Power Electronics, vol.30, no.2, pp.790, 804, Feb. 2015.

[12] G. Guidi, M. Pavlovsky, A. Kawamura, T. Imakubo, and Y. Sasaki, "Improvement of light load efficiency of dual active bridge DC-DC converter by using dual leakage transformer and variable frequency," Energy Conversion Congress and Exposition (ECCE), vol., no., pp.830, 837, 12-16 Sept. 2010

[13] G. Guidi, M. Pavlovsky, A. Kawamura, T. Imakubo, Y. Sasaki, "Efficiency Optimization of High Power Density Dual Active Bridge DCDC Converter", International Power Electronics Conference (IPEC), pp.981-986, June 2010.

[14] Costinett, D., Maksimovic, D., Zane, R., "Design and Control for High Efficiency in High Step-Down Dual Active Bridge Converters Operating at High Switching Frequency", IEEE Transactions on Power Electronics, vol.28, no.8, pp.3931, 3940, Aug. 2013.

[15] Costinett, D.; Maksimovic, D.; Zane, R., "Circuit-Oriented Treatment of Nonlinear Capacitances in Switched-Mode Power Supplies," IEEE Transactions on Power Electronics, vol.30, no.2, pp.985,995, Feb. 2015.

[16] Van den Bossche, A.; Valchev, V.C.; Georgiev, G.B., "Measurement and loss model of ferrites with non-sinusoidal waveforms," Power Electronics Specialists Conference, vol.6, no.2, pp.4814, 4818 June 2004.

[17] Robert W. Erickson, Dragan Maksimovie, Fundamentals of Power Electronics Second Edition. Klumer academic publishers, 2007.

[18] W. Zhang, Z. Xu, Z. Zhang, F. Wang, L. M. Tolbert, and B. J. Blalock, "Evaluation of 600 V cascode GaN HEMT in device characterization and all-GaN-based LLC resonant converter," in Proc. IEEE Energy Conversion Congress and Exposition (ECCE), 2013, pp. 3571 − 3578.

A Novel High Gain Step-Up Resonant DC-DC Converter for Automotive Application

Fei Shang, Mahesh Krishnamurthy
Dept. of Electrical and Computer Engineering
Illinois Institute of Technology
3301 S. Dearborn Street, Chicago, IL 60616
kmahesh@ece.iit.edu

Alexander Isurin
Vanner, Inc.
4282 Reynolds Drive, Hilliard, OH 43026
sashai@vanner.com

Abstract—**Electric power take-off systems in automotives requires the application of low voltage high current step up DC-DC converter with a high voltage gain. Researchers have evaluated enhancements for high efficiency and low cost designs. In this paper, a novel topology is proposed that has the capability of providing a high voltage gain over a wide range of output power. It also has fewer components and lower stress. Fundamental operation of the topology have been introduced and verified by simulation in PSIM. An experimental prototype has been built, which verifies the performance and high efficiency of the circuit topology.**

Keywords—Resonant DC-DC, high gain, step-up, converter, automotive

I. INTRODUCTION

Advancements in battery and power electronics technologies have presented unique opportunities for electrical power take-off (ePTO) system to be realized in vehicles, especially in specialty vehicle such as utility trucks etc, where battery power is used to drive all accessories of the vehicle and the internal combustion engine only provides propulsion to the wheels.[1] The ePTO system leads to a significant increase in the overall system efficiency and reduction in the usage of fossil fuels. Moreover, reduction in the number of belts and brackets can directly result in additional free space near the engine, thus reducing overall weight and cost of vehicle.

For specialty vehicles such as utility trucks, long haul trucks and transit buses, vehicle batteries at 12V, 24V or 48V are used. They are used to provide power to feed loads such as pumps, air conditioner compressor, power tools, pipe fusion, welding and fans, etc. All these loads typically have a 120V or 240V AC voltage interface and their power levels range from hundreds watts to several kW, in some cases may reach 20 kW [2]. Such applications require a high gain step-up converter to raise the low battery voltage to 400VDC for DC-AC conversion and capable of operating at a very high input current (> 200A).

Step-Up DC-DC converters used in automotive applications often involve bridge topology either in the form of full bridge or half bridge, with high frequency transformer providing required step-up ratio as well as meeting isolation requirements. For high power applications, parallel power stages are quite popular [3]-[5]. However, they usually use too many active switches, leading to a high cost but low reliability. Researchers in [6]-[12] have investigated this area for reaching high efficiency with fewer components. However, the disadvantage of using bridge topology is that the transformer primary voltage is almost equal to the input DC voltage. Thus, the RMS current of transformer primary side is even higher than the input DC current. This phenomenon can be even more challenging if it has a low input voltage (<30V) but with a rather high current (>300A) because it requires a very complex and costly design of the transformer with a high turns-ratio (>20) in order to meet this requirements and efficiency cannot be assured. Moreover, the current flowing in the switches of bridge topology is almost the same as the input current. It leads to high conduction loss in the switches so as to lowering the efficiency at full load. In this paper, a novel resonant topology of step-up converter suitable for the low voltage and high current application is proposed. It shows advantages of reaching a high voltage gain at a high output power, but with fewer active switches and lower current stress on the transformer and switches.

II. PROPOSED TOPOLOGY OF STEP-UP DC-DC CONVERTER

Fig. 1 shows the proposed topology of the high gain step-up resonant DC-DC converter. There is only one power stage in this topology, containing three parts: interleaved step-up topology, high frequency transformer and series resonant tank circuit. The interleaved topology, formed by L_1, L_2, C_1, C_2 and $S_1 \sim S_4$, is used to generate the high frequency AC signal to the transformer input. The interleaved step-up topology increases transformer primary voltage by duty ratio D from the input voltage so that the RMS value of primary side current is much lower than the input current and the required turns-ratio is smaller as well. This characteristic significantly benefits the

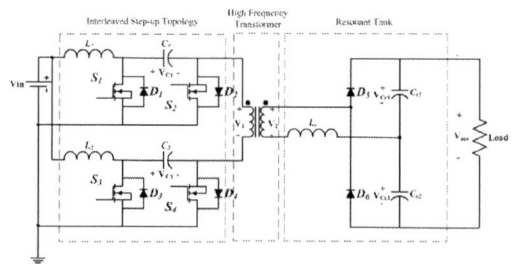

Fig. 1. Propose Topology of step-up resonant DC-DC converter.

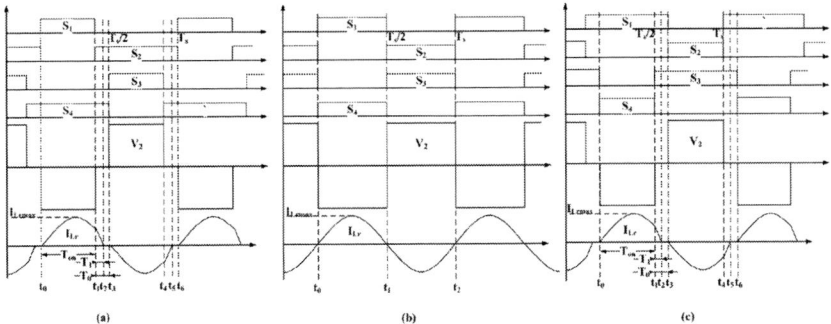

Fig. 2. Theoretical waveforms of gate signals, transformer primary voltage and resonant current at D<0.5(a), D=0.5(b), D>0.5(c).

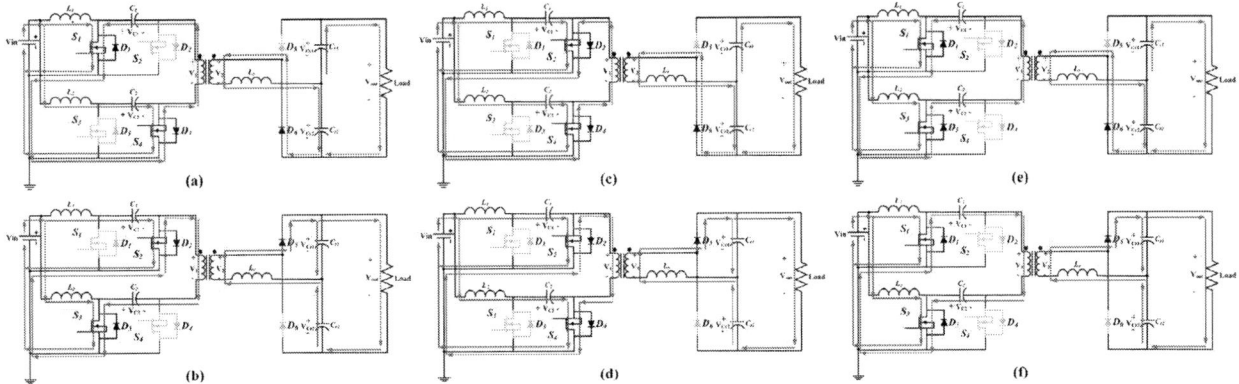

Fig. 3. Equivalent circuits for different intervals.

transformer design for reducing transformer loss and magnetic design complexity. Furthermore, current stress in the switches is reduced to half of the input current, decreasing conduction loss significantly compared to the full bridge topology. Inductor L_r and capacitor C_{r1}/C_{r2} forms a series resonant tank. It provides soft switching capability for the switching devices as well as an additional control of the output voltage.

A. Fundamentals of the Circuit Operation

Regulation of output voltage is realized by controlling the duty ratio D of switches S_1 and S_3. S_2 and S_4 are complementary to S_1 and S_3 respectively. With different values of duty ratio, the steady state operation of the circuit can be described in terms of three modes: $D<0.5$, $D=0.5$ and $D>0.5$. To analyze the circuit operation, it is assumed that input capacitor C_1/C_2 and input inductor L_1/L_2 are large enough so that the voltage across C_1/C_2 and the current in L_1/L_2 are constant. Switching frequency f_s is kept the same as the resonant frequency f_r of the resonant tank.

For $D=0.5$ operation, S_1 and S_3 conduct in each of half cycle as shown in Fig. 2(b). In this mode, the resonant current is in continuous conduction mode (CCM). There are two intervals in this operating mode.

Interval 1 (t_0~t_1): At time t_0, resonant current in the resonant inductor L_r goes across zero starting positive cycle. During the interval, S_1 and S_4 are turned on. The equivalent circuit is shown in Fig. 3(a). Current in inductor L_1 increases through S_1 to store energy while current in L_2 charges C_2 via S_4. C_1 is connected to the transformer through S_1 and S_4. It provides

power through the resonant circuit to the load and the primary side voltage $V_1 = -V_{C1}$. At the end of this interval, the resonant current i_{Lr} reduces to zero at t_1.

Interval 2 (t_1~t_2): At instant t_1, S_1 and S_4 are turned off and S_2 and S_3 are turned on and the equivalent circuit is shown as Fig. 3(b). The circuit works similarly to interval 1. The resonant current i_{Lr} goes to the negative half cycle and finish to zero at the end of the interval. The primary side voltage $V_1 = V_{C2}$.

For $D<0.5$ operation, Fig. 2(a) shows that the resonant current i_{Lr} is not continuous and there is a period when the current is zero. In this mode, the resonant current i_{Lr} is in discontinuous conduction mode (DCM). There are three time period definitions: T_{on} is the time period when the primary side voltage is not zero, while T_0 is the time when the transformer voltage is zero. In this mode, $T_{on}=D*T_s$ and $T_0=(0.5-D)* T_s$. Time T_1 is defined as the time period for resonant current decay to zero during time period T_0. There are six intervals in this operating mode. Using symmetrical operation of the circuit, only positive cycle is analyzed.

Interval 1 (t_0~t_1): During this interval, the circuit operates similar to the interval 1 for $D=0.5$ operation mode. S_1 and S_4 are turned on. Resonant current i_{Lr} rises from zero but ends at instant t_1 instead of back to zero because T_{on} is less than half of the resonant cycle.

Interval 2 (t_1~t_2): At time t_1, S_1 is turned off and S_2 is turned on. During this interval, S_2 and S_4 are turned on together and the equivalent circuit as shown in Fig. 3(c). Current in inductor

L_1 and L_2 charge C_1 and C_2 respectively while the transformer primary side is shorted by S_2 and S_4. Resonant current i_{Lr} decays much faster during this interval and returns to zero after time period T_1 (denoted by instant t_2).

Interval 3 ($t_2 \sim t_3$): S_2 and S_4 are still turned on and current in inductor L_1 and L_2 keeps charging C_1 and C_2. Resonant current stays at zero during this period. Load current is provided by the resonant capacitors C_{r1} and C_{r2}.

For $D>0.5$, as shown in Figure 2(c), the resonant current i_{Lr} is also in DCM mode similar to $D<0.5$. In this mode, $T_{on}=(1-D)*T_s$ and $T_0=(D-0.5)*T_s$. There are also six intervals in this operation mode and interval 1 and 4 are the same as $D<0.5$ operation mode.

Interval 2 ($t_1 \sim t_2$): At time t_1, S_4 is turned off and S_3 is turned on. During this interval, S_1 and S_3 are turned on together and the equivalent circuit as shown in Fig. 3(e). Both L_1 and L_2 store energy from the source and C_1 and C_2 are connected in series to the transformer primary side and the transformer primary side sees a zero net voltage. The resonant current returns to zero after time T_1 (denoted by instant t_2).

Interval 3 ($t_2 \sim t_3$): S_1 and S_3 are still turned on and current in inductor L_1 and L_2 keeps increasing to store energy while resonant current stays at zero. Load current is provided by the resonant capacitors C_{r1} and C_{r2}.

These descriptions explain the fundamental operation of the circuit under different values of duty ratio D. In fact, a small dead-time needs to be added between S_1 and S_2 (also S_3 and S_4) to avoid shoot through of the switches. Circuit operation is not significantly affected by the dead-time, but it gives the capability of soft switching in switches. It is discussed in detail in following sections.

B. Voltage Gain

Due to symmetrical operation of the interleaved topology, it is reasonable to focus on one leg and assume that the other leg generates the same result. The voltage across input inductor L_1 during $D*T_s$ is V_{in} while it is V_c-V_{in} during $(1-D)*T_s$. Applying the volt-second balance rule to the input inductor L_1, V_c can be estimated.

$$V_C = \frac{V_{in}}{1-D} \tag{1}$$

The amplitude of transformer primary voltage V_1 is given as V_C and secondary voltage V_2 can be calculated as $V_2 = NV_1 = NV_{in}/(1-D)$, where N is the transformer turns ratio $N = N_2/N_1$.

For $D=0.5$ operation mode, since the positive cycle and negative cycle are symmetrical, we only analyze the positive cycle. During interval 1, the following differential equations can be induced from the equivalent circuit.

$$-V_2 + L_r \frac{di_{L_r}}{dt} + v_{C_{r2}} = 0$$
$$i_{L_r} = 2C_{r2} \frac{dv_{C_{r2}}}{dt} \tag{2}$$

Solving the equations gives expressions of $v_{Cr2}(t)$ and $i_{Lr}(t)$ as

$$v_{C_{r2}}(t) = (V_{C_{r2}\min} - V_2)\cos\omega_r t + V_2$$
$$i_{L_r}(t) = \frac{V_2 - V_{C_{r2}\min}}{Z}\sin\omega_r t \tag{3}$$

where $\omega_r = 2\pi f_r = 1/\sqrt{L_r C_r}$ and $Z = \sqrt{L_r/C_r}$. Resonant capacitance $C_r = 2C_{r2} = 2C_{r1}$. V_{Cr2min} and V_{Cr2max} is the minimum and maximum voltage across the resonant capacitor. $V_{C_{r2}\max} = v_{C_{r2}}(T_s/2) = 2V_2 - V_{C_{r2}\min}$. From the equivalent circuit, it can be deduced that

$$V_{out} = V_{C_{r2}\max} + V_{C_{r2}\min} \tag{4}$$

Thus the voltage gain under this mode is given by

$$\frac{V_{out}}{V_{in}} = \frac{2N}{1-D} \tag{5}$$

For $D<0.5$ and $D>0.5$ operation modes, the resonant circuit works in DCM and the resonant current waveforms are identical for both modes. Thus, we only analyze $D<0.5$ mode and the other mode generates identical results. During interval 1, the circuit works similarly to $D=0.5$ mode, thus the expressions of $v_{Cr2}(t)$ and $i_{Lr}(t)$ are the same as equation (3). But it should be noted that the period ends at T_{on} instead of $T_s/2$. During interval 2, since $V_2=0$, the differential equations are given by

$$L_r \frac{di_{L_r}}{dt} + v_{C_{r2}} = 0 \tag{6}$$

Solving this equation with the initial condition of $v_{Cr2}(T_{on})$ and $i_{Lr}(T_{on})$ the expressions can be achieved for this interval as

$$v_{C_{r2}}(t) = (V_{C_{r2}\min} - V_2)\cos\omega_r t + V_2\cos\omega_r(t - T_{on})$$
$$i_{L_r}(t) = \frac{V_2 - V_{C_{r2}\min}}{Z}\sin\omega_r t - \frac{V_2}{Z}\sin\omega_r(t - T_{on}) \tag{7}$$

where t ranges from T_{on} to $T_{on}+T_1$. And V_{Cr2max} is given by

$$V_{C_{r2}\max} = v_{C_{r2}}(T_{on} + T_1)$$
$$= (V_{C_{r2}\min} - V_2)\cos\omega_r(T_{on} + T_1) + V_2\cos\omega_r T_1 \tag{8}$$

The average current I_{Lr} can be calculated as

$$I_{L_r} = \frac{2}{T_s}\{\int_0^{T_{on}} \frac{V_2 - V_{C_{r2}\min}}{Z}\sin\omega_r t\, dt$$
$$+ \int_{T_{on}}^{T_{on}+T_1} [\frac{V_2 - V_{C_{r2}\min}}{Z}\sin\omega_r t - \frac{V_2}{Z}\sin\omega_r(t - T_{on})]dt\}$$
$$= \frac{1}{\pi Z}\{V_2\left[\cos\omega_r T_1 - \cos\omega_r(T_{on} + T_1)\right]$$
$$- V_{C_{r2}\min}\left[1 - \cos\omega_r(T_{on} + T_1)\right]\} \tag{9}$$

From the equivalent circuit it can be induced

$$I_{L_r} = 2I_{out} = \frac{2V_{out}}{R} = \frac{2(V_{C_{r2}\max} + V_{C_{r2}\min})}{R} \tag{10}$$

where R is the effective load resistance. Combining (8)-(10) the voltage gain relationship under DCM mode is given by

$$\frac{V_{out}}{V_{in}} = \frac{[\cos\omega_r T_1 - \cos\omega_r(T_{on} + T_1)]}{1 + 2\pi Z/R - (1 - 2\pi Z/R)\cos\omega_r(T_{on} + T_1)}\frac{2N}{1-D} \tag{11}$$

And T_1 is acquired by solving $i_{Lr}(T_{on}+T_1)=0$.

$$T_1 = \frac{\cos^{-1}\left(\frac{1-2\pi Z/R}{1+2\pi Z/R} \cos \omega_r \frac{T_{on}}{2} \right)}{\omega_r} - \frac{T_{on}}{2} \qquad (12)$$

From equation (5) and equation (11), it can be concluded that the voltage gain of the circuit varies with the duty ratio D and effective load resistance R as shown in Fig. 4. Compared to the conventional boost converter, the proposed converter shows a higher voltage gain even without considering the transformer turns-ratio. This is mainly because the rectifier diode and the resonant capacitor form a voltage doubler circuit [13] to allow an extra boost of the voltage. It also benefits the transformer design because it leads to a lower turns-ratio requirement. The resonant circuit gives the more freedom in controlling the output voltage. This attribute is especially important when the input source is a low voltage battery, because its terminal voltage varies widely from fully charged status to the cut-off voltage of the battery.

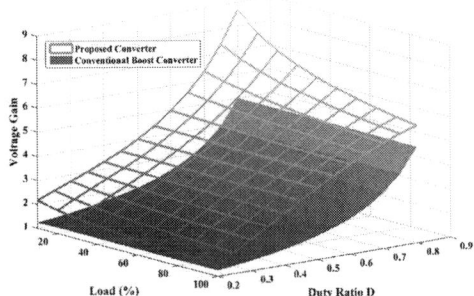

Fig. 4. Voltage gain comparison of the proposed converter without transformer turns-ratio and conventional boost converter.

III. STRESS AND POWER LOSS EVALUATION

A. Stress calculation for switches

For the proper design of a DC-DC converter, it is important to evaluate the maximum voltage and current stress over the devices, since it determines the selection of the switching devices and their associated costs. Extreme duty ratios should be avoided since it increases the voltage stresses on the switching devices. Moreover, high duty ratio also increases the current stress of the switching devices because it leads to higher RMS current and higher power loss during one period, but with reduced time to cool and for the heat to dissipate. Low duty ratios may not be optimal choice either, because it tends to a lower boost ratio thus increase the transformer turns-ratio. Thus it is reasonable to design the converter where the duty ratio is around 0.5.

This paper focuses on the design of a step-up converter for a 24V battery that has a constant 400V output voltage and 2.5kW output power. The typical input voltage range of a 24V battery is 20~30V. Rated voltage of 24V is selected to work under $D=0.5$ operation to give the 400V output voltage,

leading to a 25/6 turns-ratio of the transformer according to (5). Duty ratio D is calculated for other input voltages according to (11) and the results lies in the range of 0.38 to 0.6 over the entire load range of 10% to 100%.

According to the equivalent circuit shown in Fig. 3, the voltage across the power MOSFET $S_1 \sim S_4$ is the capacitor voltage V_C when the switch is turned off. The voltage can be evaluated according to equation (1) and the maximum voltage stress of the power MOSFET is approximate 50V when input voltage is 20V. The voltage across the rectifier diodes in the secondary side is the same as the output voltage thus its voltage stress is 400V. Applying a safety margin, 75V power MOSFET and 600V rectifier diode are considered for the device selection.

Current stress across switching devices varies with duty ratio D in different operating modes. During $D=0.5$ operation mode, S_1 conducts during interval 1 and S_2 conducts during interval 2 and according to equivalent circuit the following equations can be acquired.

$$\begin{aligned} i_{S_1}(t) &= i_{L_1}(t) + Ni_{L_r}(t) & 0 < t \le T_s/2 \\ i_{S_2}(t) &= -i_{L_1}(t) - Ni_{L_r}(t) & T_s/2 < t \le T_s \end{aligned} \qquad (13)$$

Similarly, for $D<0.5$ operation mode,

$$i_{S_1}(t) = i_{L_1}(t) + Ni_{L_r}(t) \qquad\qquad 0 < t \le T_{on}$$

$$i_{S_2}(t) = \begin{cases} -i_{L_1}(t) - Ni_{L_r}(t) & T_{on} < t \le T_{on}+T_1 \\ -i_{L_1}(t) & T_{on}+T_1 < t \le T_s/2 \\ -i_{L_1}(t) - Ni_{L_r}(t) & T_s/2 < t \le T_s/2+T_{on}+T_1 \\ -i_{L_1}(t) & T_s/2+T_{on}+T_1 < t \le T_s \end{cases} \quad (14)$$

and for $D>0.5$ operation mode

$$i_{S_1}(t) = \begin{cases} i_{L_1}(t) + Ni_{L_r}(t) & 0 < t \le T_{on}+T_1 \\ i_{L_1}(t) & T_{on}+T_1 < t \le T_s/2 \\ i_{L_1}(t) + Ni_{L_r}(t) & T_s/2+T_{on} < t \le T_s/2+T_{on}+T_1 \\ i_{L_1}(t) & T_s/2+T_{on}+T_1 < t \le T_s \end{cases} \quad (15)$$

$$i_{S_2}(t) = -i_{L_1}(t) - Ni_{L_r}(t) \qquad T_s/2 < t \le T_s/2+T_{on}$$

Using equation (13)-(15), the expression of the power MOSFET current can be acquired. The RMS current is calculated and the current stress is shown in Fig. 5.

According to Fig. 5, S_1 has a higher current stress than S_2. This is mainly because for most of time, $i_{L1}(t)$ and $i_{Lr}(t)$ are in reverse direction when flowing through S_2 as shown in Fig. 3. But for current in S_1, both currents flow in the same direction, which increases the current stress of S_1. Moreover, there is a clear increase in current stress of S_2 with increasing input voltage. The reason for this is that as input voltage increases, duty ratio D decreases and S_2 conducts for a longer time in one period. This unbalanced current stress in S_1 and S_2 grants freedom in selection of the switching devices. Lower current rating devices can be used for S_2 and no device parallel is needed, thereby achieving lower cost.

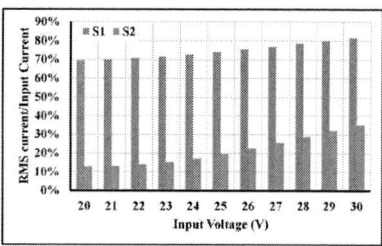

Fig. 5. Current stress in switching devices.

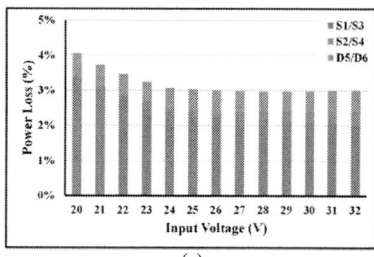

(a)

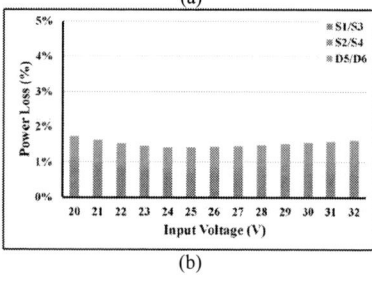

(b)

Fig. 6. Power loss for the swtiching devices. (a) no device parallel; (b) two devices in parallel for S_1/S_3.

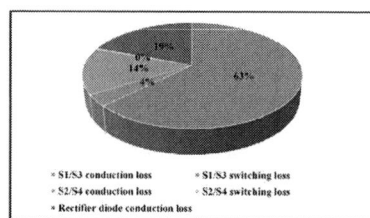

Fig. 7. Power loss distribution over switches.

B. Power loss

Power loss of the switching devices, including the MOSFETs and rectifier diodes, accounts for majority of the power loss in the converter and it consists of conduction loss and switching loss [14]. The conduction loss of the power MOSFET is mainly caused by the on-resistance R_{dson} and the RMS current flowing through. Switching loss is caused by the turn-on and turn-off transitions of the power MOSFET and reverse recovery effect of the body diode [15].

In the proposed converter, S_1/S_3 shows a higher RMS current than S_2/S_4, thus it generates higher conduction loss.

Paralleling devices for S_1/S_3 may be a good choice for reducing the conduction loss. Switching losses are related to the voltage and current at turn-on and turn-off of the switches [14]. In the proposed converter, due to the dead-time between S_1/S_3 and S_2/S_4, the body diode conducts first before S_2/S_4 are turned on which allows ZVS of S_2/S_4. Unlike the full-bridge topology, ZVS can also be achieved even at a very low load (10%). On the secondary side, the rectifier diode operates in ZCS because of the resonant circuit. Therefore, there is no switching loss and reverse recovery effect, and only conduction loss is considered.

For the experimental prototype, Infineon power MOSFET BSC042NE7NS3 G is selected for $S_1\sim S_4$ and STM diode STTH30L06 is selected as the secondary side rectifier diode. Power loss are calculated and shown in Fig. 6. Two devices are connected in parallel for S_1/S_3, while there are no parallel devices for the other switches. Therefore, the peak power loss significantly drops from 4.09% to 1.85%. This is mainly because more than 60% of the total power loss comes from the conduction loss of S_1/S_3 as shown in Fig. 7.

IV. SIMULATION AND EXPERIMENTAL RESULTS

Simulation of the converter was carried out using PSIM for verification of the circuit operation. Fig. 8 shows the waveform when circuit operating at $D=0.5$. An experimental prototype was built with $V_{in}=24V$, $V_{out}=400V$ and $P_{out}=3kW$. Fig. 9 shows the waveform of gate signal of S_1, voltage across S_1 and S_2 and the resonant current i_{Lr} at 24V input. The duty ratio was kept at 0.5 and the real output power is 2.4kW.

Efficiency measurements were carried out at different output power levels for an input voltage range from 20V~30V. As can be seen in Fig. 10, efficiency of the converter shows a consistency of over 94% with input voltage at different power level.

V. CONCLUSIONS

In this paper, a novel topology for step-up resonant DC-DC converter with a high voltage gain has been proposed. It has the advantages of lower stress on transformer and switches, low complexity in transformer design, wide operating range and fewer number of components compared with the bridge topology DC-DC converters. Soft switching of power MOSFETs and diodes helps maintain the high efficiency of the circuit. Simulations in PSIM and experimental results from the prototype have been used to validate the effectiveness of the circuit topology and confirm the high efficiency.

REFERENCES

[1] Transportation Electrification: Utility Fleets Leading the Charge. Edison Electric Institute. [Online]. Available: http://www.eei.org/issuesandpolicy/electrictransportation/FleetVehicles/Documents/EEI_UtilityFleetsLeadingTheCharge.pdf

[2] A. Cook, "The road to electrification for specialty vehicles," in Vehicular Electronics and Safety, 2008. ICVES 2008. IEEE International Conference on, pp. 103-107, Sept 2008.

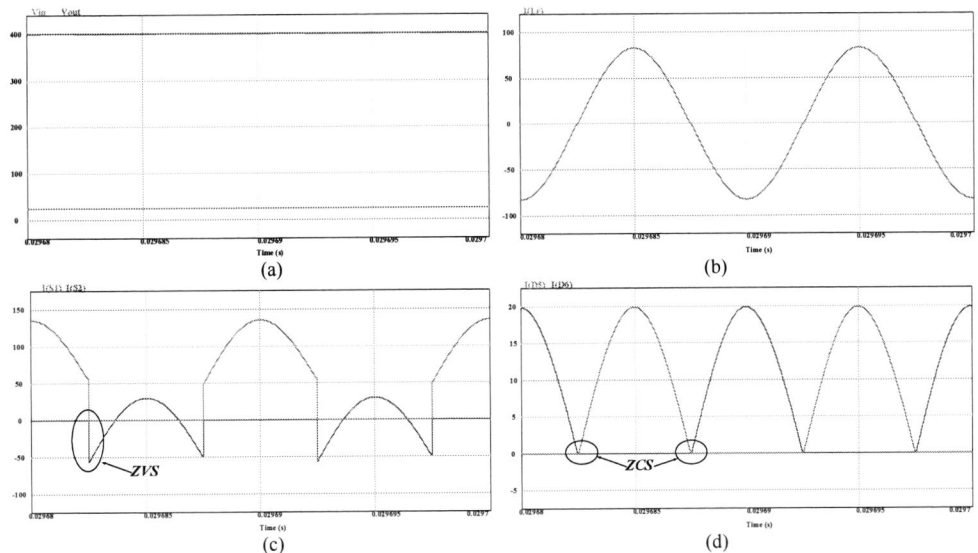

Fig. 8. Simulation waveforms for D=0.5. (a) input voltage and output voltage; (b) resonant current; (c) current in S_1 and S_2; (d)Current in rectifier diode.

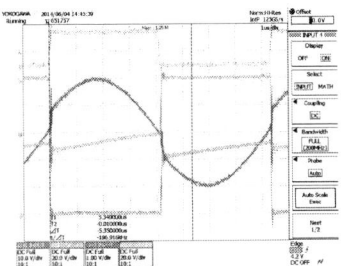

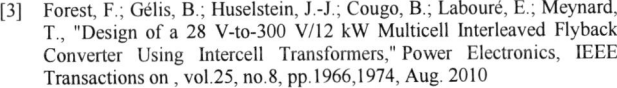

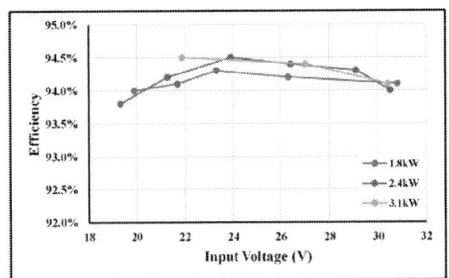

Fig. 9. Experimental waveform at 2.4kW. CH1-V_{GS} for S_1, CH2- V_{DS} for S_1, CH3-resonant current, CH4- V_{DS} for S_2.

Fig. 10. Efficiency at various output power levels.

[3] Forest, F.; Gélis, B.; Huselstein, J.-J.; Cougo, B.; Labouré, E.; Meynard, T., "Design of a 28 V-to-300 V/12 kW Multicell Interleaved Flyback Converter Using Intercell Transformers," Power Electronics, IEEE Transactions on , vol.25, no.8, pp.1966,1974, Aug. 2010

[4] Bal, Satarupa; Rathore, Akshay K; Srinivasan, Dipti, "Modular snubberless bidirectional soft-switching current-fed dual 6-pack (CFD6P) DC/DC converter," Energy Conversion Congress and Exposition (ECCE), 2014 IEEE , vol., no., pp.2043,2050, 14-18 Sept. 2014

[5] Zhan Wang; Hui Li, "A Soft Switching Three-phase Current-fed Bidirectional DC-DC Converter With High Efficiency Over a Wide Input Voltage Range," Power Electronics, IEEE Transactions on , vol.27, no.2, pp.669,684, Feb. 2012

[6] Garcia, O.; Flores, L.A.; Oliver, J.A.; Cobos, J.A.; de la Pena, J., "Bidirectional DC-DC Converter For Hybrid Vehicles", Power Electronics Specialists Conference, 2005. PESC '05. IEEE 36th, 16- 16 June 2005 Page(s):1881 – 1886

[7] Lizhi Zhu, "A Novel Soft-Commutating Isolated Boost Full-bridge ZVS-PWM DC-DC Converter for Bi-directional High Power Applications", Power Electronics Specialists Conference, 2004. PESC 04. 2004 IEEE 35th Annual, Volume 3, 20-25 June 2004 Page(s):2141 - 2146 Vol.3

[8] Kunrong Wang; Zhu, L.; Dayu Qu; Odendaal, H.; Lai, J.; Lee, F.C., "Design, implementation, and experimental results of bi-directional full-bridge DC/DC converter with unified soft-switching scheme and soft-starting capability," Power Electronics Specialists Conference, 2000. PESC 00. 2000 IEEE 31st Annual, vol.2, no., pp.1058, 1063 vol.2, 2000

[9] Su-Jin Jang, Tae-Won Lee, Won-Cbul Lee, Chung-Yuen Won, "Bidirectional DC to DC Converters for Fuel Cell Generation System",

Power Electronics Specialists Conference, 2004. PESC 04. 2004 IEEE 35th Annual Volume 6, 20-25 June 2004 Page(s):4722 - 4728 Vol.6

[10] Xinyu Xu, Ashwin M Khambadkone, Ramesh Oruganti, "A Soft Switched Back-to-Back Bi-directional DC/DC Converter with a FPGA based Digital Control for Automotive applications", Industrial Electronics Society, 2007. IECON 2007. 33rd Annual Conference of the IEEE, Page(s): 262-267

[11] L.A. Flores, O. García, J.A. Oliver, J.A. Cobos, "High-Frequency Bi-Directional DC/DC Converter Using Two Inductor Rectifier", IEEE Industrial Electronics, IECON 2006 - 32nd Annual Conference on 6-10 Nov. 2006 Page(s):2793 – 2798

[12] Isurin, A.; Cook, A., "A novel resonant converter topology and its application," Power Electronics Specialists Conference, 2001. PESC. 2001 IEEE 32nd Annual, vol.2, no., pp.1039, 1044 vol.2, 2001

[13] Salehi, S.; Gharehpetian, G.B.; Monfared, J.M.; Taheri, M.; Moradi, H., "Analysis and design of current-fed high step up quasi-resonant DC-DC converter for fuel cell applications," in Power Electronics, Drive Systems and Technologies Conference (PEDSTC), 2013 4th , vol., no., pp.442-447, 13-14 Feb. 2013

[14] Fei Shang; Arribas, A.P.; Krishnamurthy, M., "A comprehensive evaluation of SiC devices in traction applications," in Transportation Electrification Conference and Expo (ITEC), 2014 IEEE , vol., no., pp.1-5, 15-18 June 2014

[15] Arribas, A.P.; Fei Shang; Krishnamurthy, M.; Shenai, K., "Simple and Accurate Circuit Simulation Model for SiC Power MOSFETs," in Electron Devices, IEEE Transactions on , vol.62, no.2, pp.449-457, Feb. 2015

Series Injection Enabled Full ZVS Light Load Operation of a 15kV SiC IGBT Based Dual Active Half Bridge Converter

Awneesh Tripathi, Sachin Madhusoodhanan, Krishna Mainali
Kasunaidu Vechalapu, Subhashish Bhattacharya
Department of Electrical and Computer Engineering
North Carolina State University, Raleigh, NC 27606
{aktripat}, {sbhatta4}@ncsu.edu

Abstract—The 15 kV SiC IGBT has second higher dv/dt turn-off slope above the punch-through level resulting in EMI. Increasing gate-resistance also slows the first dv/dt causing increased switching loss. A snubber capacitor assisted turn-off solves these issues for a high power dual active bridge (DAB) converter based on this device, but the light load turn-on ZVS becomes hard to achieve. This paper proposes a series injection enabled triangular current shaping at the light load turn-off instant in the DAB to create enough current for smooth free-wheeling transition of device voltage during the dead-time period for ZVS turn-on. The proposed technique is validated through simulations followed by experiments on a medium voltage DAB hardware implementation of this technique.

I. INTRODUCTION

The Dual Active Bridge (DAB) is a high power density isolated dc-dc converter suited as an important component for medium-voltage (MV) renewable energy integration system [1], [2]. It is turn-on zero voltage switching (ZVS) soft switched converter providing dc voltage transformation and isolation. The control simplicity and higher MTBF reliability requirements make a single phase Dual Active Half Bridge (DAHB) popular over other DAB topologies as it requires fewer switches [3]. However it has the disadvantages of relatively higher ripple voltage and lower power density compared to a three-phase DAB [4]. The DAB may lose the ZVS at light loads due to small freewheeling current during complementary switch turn-off. The 15 kV SiC IGBT has considerable parasitic capacitance especially below punch-through and also its dv/dt is made relatively lower (30 kV/μs) with higher gate resistance to avoid EMI problems in this region [5]. Therefore a dead-time in the range of 3.8 μs is kept. To maintain turn-on ZVS at light loads there should be sufficient current to discharge the parasitic before the dead-time ends. Three-phase DAB with smoother current near zero-crossings is more difficult to maintain ZVS compared to a single phase DAB with trapezoidal current [6]. There are good number of solutions proposed to solve light load DAB ZVS problem [7]–[9]. The general light load solutions may not be suitable for this MV DAB. A 15 kV SiC IGBT being a research

device has still challenges posed that need to be specifically solved due to the voltage magnitude and its characteristics. Above punch-through the parasitic cap reduces in 15 kV SiC IGBT, this results in EMI and increased turn-off losses even if the turn-on ZVS is maintained [10]. If a snubber cap is used in parallel with IGBT for solution, the turn-on ZVS is more difficult to achieve as the light load current needed during turn-off of the complementary switch to discharge the added cap within dead-time, is high [6]. This problem discourages snubber solution for 15 kV SiC IGBT based DAB since non-ZVS turn-on will abruptly short circuit the snubber cap and damage the devices.

A series compensated DAB for optimum ZVS control in wide load and voltage range is proposed in [11]. The control approach in that paper is to keep DAB current fundamental vector in middle of voltage vectors. The objective is to maintain ZVS at light load even in presence of significant voltage deviation. This paper uses the same topology to solve above mentioned problem for a snubbered 15 kV SiC IGBT based MV topology but presents a different objective and control approach. Also the series transformer turn-ratio is different here as higher depth of compensation is required to be able to induce a high current with small series duty-ratio. Since a high turn-off current is required in DAB MV pole such as to freewheel and discharge the turning on device & snubber cap, the series circuit only adds a suitable polarity triangular current independent of loading. It is not used to balance the voltage vectors here as done in the previous work.

This paper is organized in following sections. Section II presents the series enabled full ZVS DAHB converter topology and operation. Its simulation is also presented at rated condition. A simple parameter selection for series circuit based on simulations are presented in section III. The series compensation makes the turn-on soft in all loading conditions and the snubber cap makes the turn-off soft. The energy stored in the cap during turn-off is regenerated into DAB during next turn-on [12]. Thus this topology is fully soft-switched especially for MV side under all conditions. The 15 kV SiC IGBT based MV half bridge experimental results and thermal

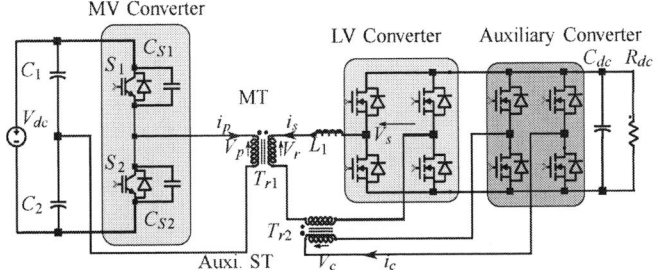

Fig. 1: Schematic of Series Enabled Full ZVS DAB

TABLE I: Rated Parameters of MV simulation and MV DABC Setup

Parameters	MV Hardware & Simulation Value[Units]
Input dc Voltage[V_{dc}]	6 kV
Output dc Voltage[V_o]	300 V
External Inductance[L_1]	80 μH, 0.2 Ω
Snubbers C_{s1}, C_{s2}	0.55 nF
T_{r1} Turns-ratio	10 : 1
T_{r1} Leakage Inductance (LV side)	8 μH, 13 mΩ
T_{r1} Mutual Inductance (MV side)	293 mH
T_{r1} Shunt Cap (MV side)	250 pF
T_{r2} Turns-ratio	1 : 1
T_{r2} Leakage Inductance	23 μH, 8 mΩ
T_{r2} Mutual Inductance	22 mH
C_{dc}	10000 μF
Rated R_{dc}	20 Ω
C_1, C_2	120 μF
Real Power Transfer	5 kW
T_{r1} Core Loss	50 W
T_{r2} Core Loss	20 W
Switching-frequency	10 kHz

measurements highlighting the effect of snubber, are presented in section IV. The MV results for a series compensated 15 kV SiC IGBT based DAHB to enable the snubbered operation are also presented. The paper is finally concluded in section V.

II. Series Enabled Full ZVS DAHB Topology

As shown in Fig. 1, a two-level half-bridge based on 15 kV/40A SiC-IGBT, is connected to the MV side of a medium frequency (MF) DAB transformer operating at 10 kHz [11]. A 1200 V SiC-MOSFET based two-level H-

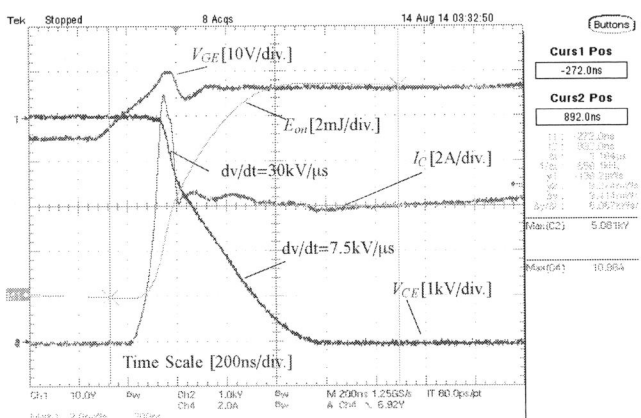

Fig. 2: 15 kV/40A SiC IGBT DP Test Turn-on

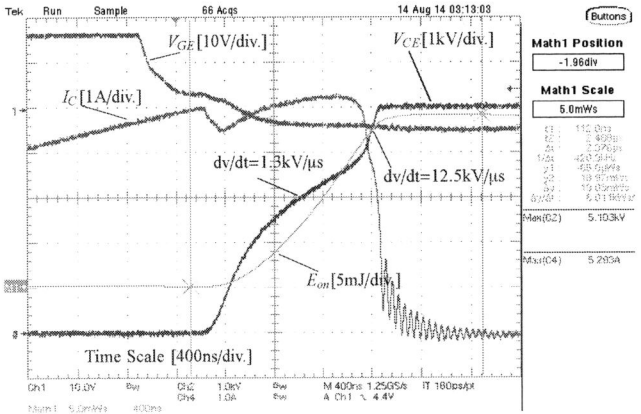

Fig. 3: 15 kV/40A SiC IGBT DP Test Turn-off

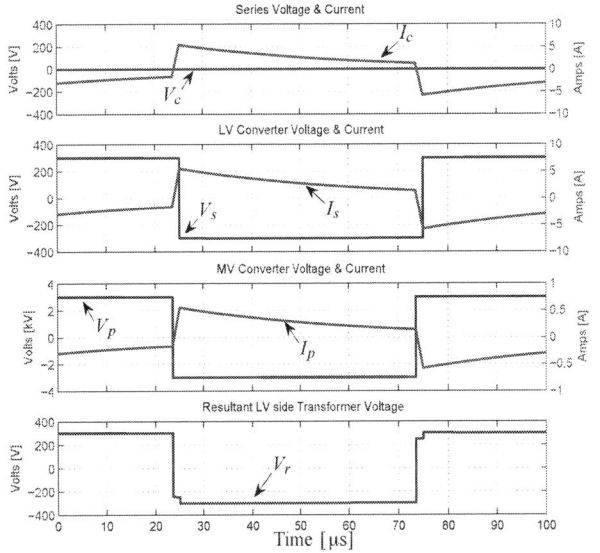

Fig. 4: Simulation at Light Load without Series Injection or Snubber

bridge is connected to the low voltage (LV) side of the transformer. A series transformer rated for lower secondary voltage and power, injects a controlled voltage pulse in series of the secondary DAB circuit as shown in Fig. 1. This series transformer is excited by an additional LV H-bridge composed of low current rating devices.

Fig. 2 and Fig. 3 show the Double Pulse (DP) test results on the 15 kV/40A SiC IGBT co-pack module during turn-on and turn-off respectively. The gate resistances are selected such that the below punch-through dv/dt is limited to a practical value for EMI consideration. The snubber slows dv/dt throughout without active gating and without resulting in increased losses [12]. With DAB switching frequency at 10 kHz and given dv/dt, the dead-time is selected as 3.8 μs. Fig. 4 shows the ideal DAHB simulation for light load. The phase lead of primary MV converter voltage w.r.t secondary voltage causes power flow to maintain LV dc bus. Fig. 5 shows the DAHB simulation considering converter dead-times for light load. This is approximate MATLAB/PLECS simulation without considering device parasitics. But a realistic notch in

978-1-4673-9551-9/16 $31.00 © 2016 IEEE

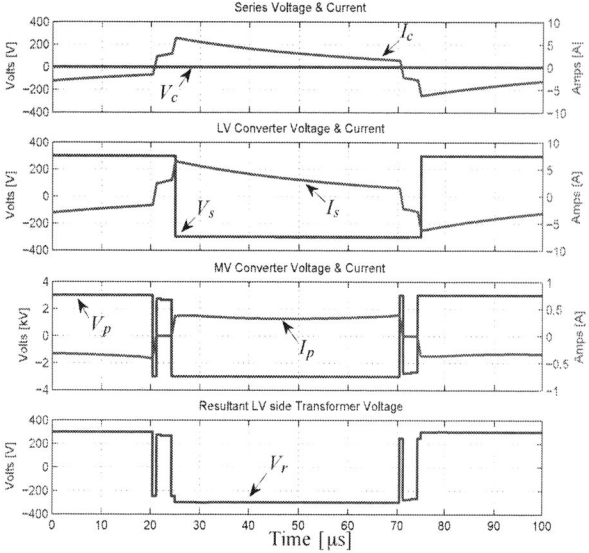

Fig. 5: Simulation at Light Load without Series Injection or Snubber but with Dead-time

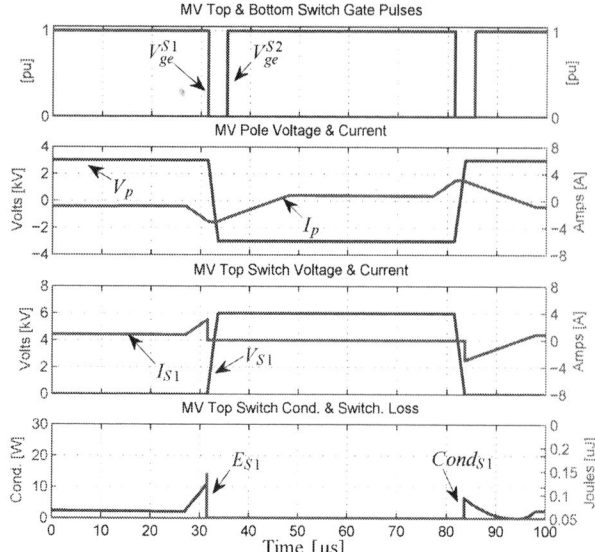

Fig. 7: MV side Simulation at Light Load with Full ZVS Compensation

TABLE II: Symbols Used

Parameters	Symbol
Ser. Tra. Turns	N_s
Peak MV Current	I_p
MV dc bus	V_{dc}
Series duty	D_s
DAB Inductance	L_d
Dead-time	t_d
Tot. Snubber Cap.	C_s^*

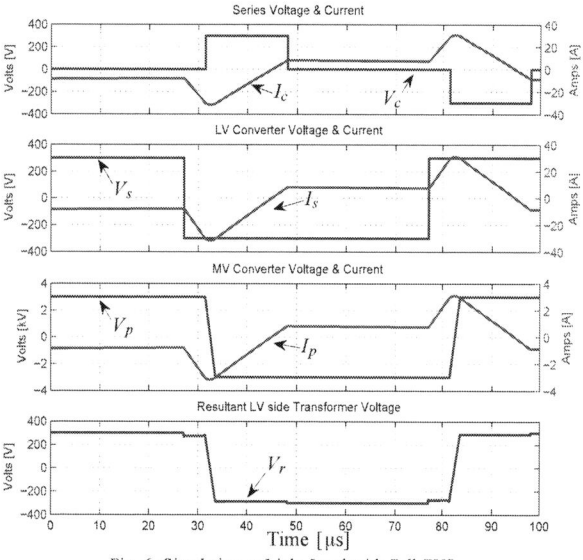

Fig. 6: Simulation at Light Load with Full ZVS

MV voltage can be seen that occurs as there is not enough lagging current for freewheeling and maintaining zero voltage across the turning on device. The series voltage pulse with certain duty-ratio when added near zero phase of the primary, generates a triangular freewheeling current as shown in the simulation result in Fig. 6. Since the turn-on ZVS is now enforced, a snubber cap is used in this simulation resulting in slower dv/dt loss-less transitions. There is no considerable loss associated with the series circuit as it only conducts for small duration. It is excited from LV side with either a half bridge or full fridge depending on output dc cap configurations and series size-weight constraints. The DAB LV converter might lose ZVS due to the injection but it is acceptable as it does not result in considerable loss addition or EMI. Since the DAB

inductors are rated for full load current, there is no problem posed by the triangular current peak in the original design. The parameters of this presented MV topology and its hardware are shown in Table I [6].

III. DESIGN PARAMETERS

To design the series transformer turn-ratio and pulse duty-ratio, the snubber cap, maximum discharge time, maximum current and DAB inductance parameters are taken into account. As shown in (1), there is a minimum peak current requirement for a given snubber cap value across the MV switches for ZVS and loss-less operation. Also in (2), dependence of this current achieved through this method on the series duty-ratio is shown. The symbols are described in Table II. The peak current during turn-off instant is dependent on the DAB loading and a series induced value. This induced value is effective more at light load when DAB contributed current is small.

$$t_d > C_s^* V_{dc}/I_p \qquad (1)$$

$$I_p = I_d + D_s V_{lv}/[L_d N_s f_s] \qquad (2)$$

Fig. 7 shows the simulation of MV switch waveforms considering the snubber for the 15 kV/40A SiC IGBT co-pack at light load with series injection. Fig. 8 shows the waveforms without series injection and without any snubber. In practice there is always some parasitic cap that may not discharge

978-1-4673-9551-9/16 $31.00 © 2016 IEEE

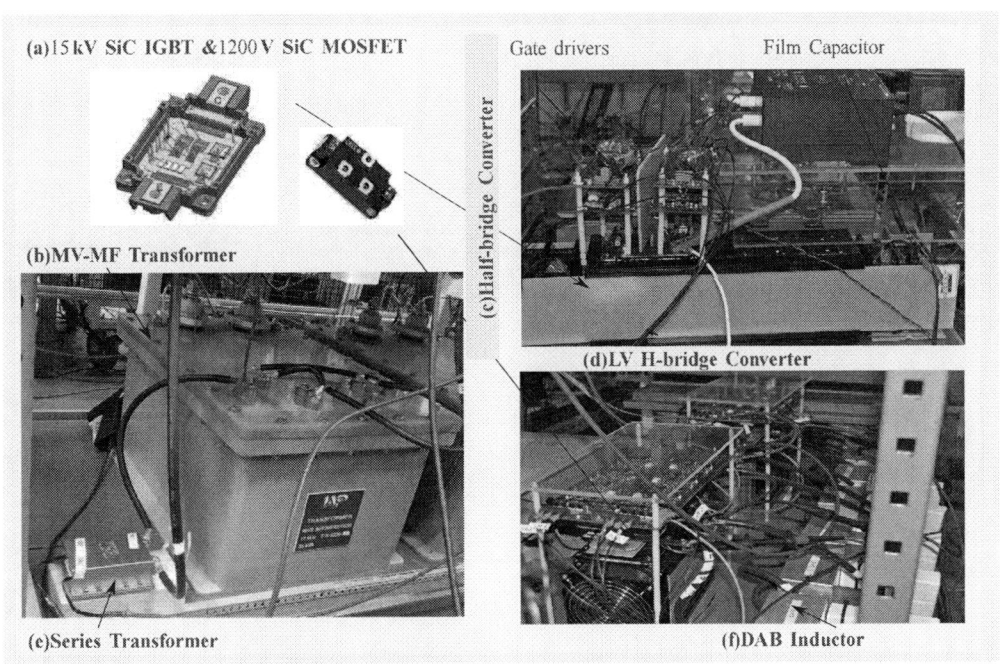

(a)15 kV SiC IGBT &1200 V SiC MOSFET

(b)MV-MF Transformer

(c)Half-bridge Converter

Gate drivers   Film Capacitor

(d)LV H-bridge Converter

(e)Series Transformer

(f)DAB Inductor

Fig. 10: Photograph of 6kV MV DAHB Setup for the ZVS Test

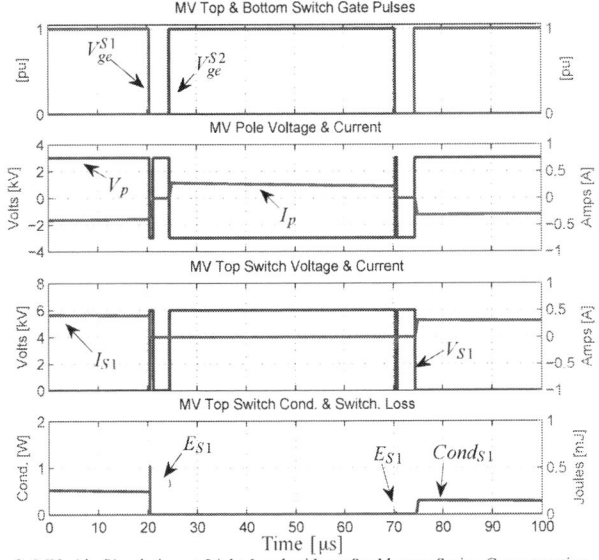

Fig. 8: MV side Simulation at Light Load without Snubber or Series Compensation

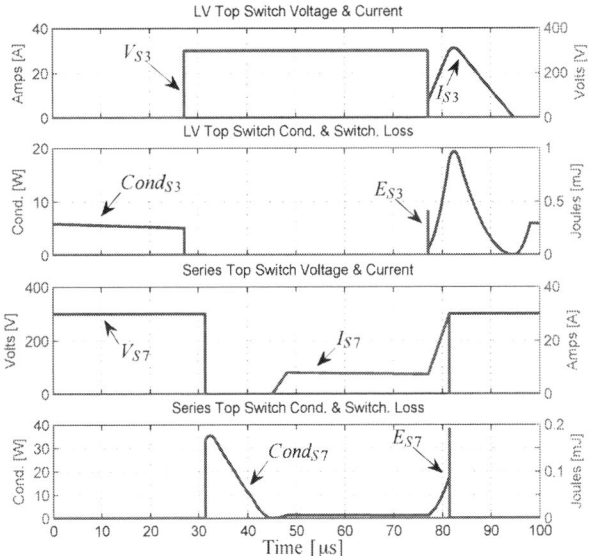

Fig. 9: LV side Simulation at Light Load with Full ZVS Compensation Circuit

in regenerative manner, resulting in loss during turn on at light load. Fig. 9 shows the DAB LV switch waveforms for light load and with full ZVS topology. It can be seen that ZVS is lost but as mentioned earlier it has no considerable disadvantage. For just MV side ZVS the LV dc bus could have been decreased at light load, but this results in reactive current in the DAB generating conduction losses. The overall efficiency of this full ZVS topology is expected to be highest.

IV. EXPERIMENTAL RESULTS

Fig. 10 shows the MV experimental setup for the MV topology with parameters shown in Table I. It shows the topology

components such as series transformer, main transformer, MV half-bridge & LV converters and the DAB external inductor. For showing the benefit of snubber in terms of turn-off loss and dv/dt reduction for these IGBTs, the half-bridge is run with stand-alone inductive load to create high current for turn-on ZVS [12]. Fig. 11 shows the experimental result for an inductively loaded 15 kV SiC IGBT based half-bridge at 7kV dc operation without snubber cap. Fig. 12 shows the expanded view of the same result in same condition. It can be seen that dv/dt is 14.6 kV/μs after the punch-through. Fig. 13 shows the result in same condition but with snubber of 1.1 nF. The

978-1-4673-9551-9/16 $31.00 © 2016 IEEE

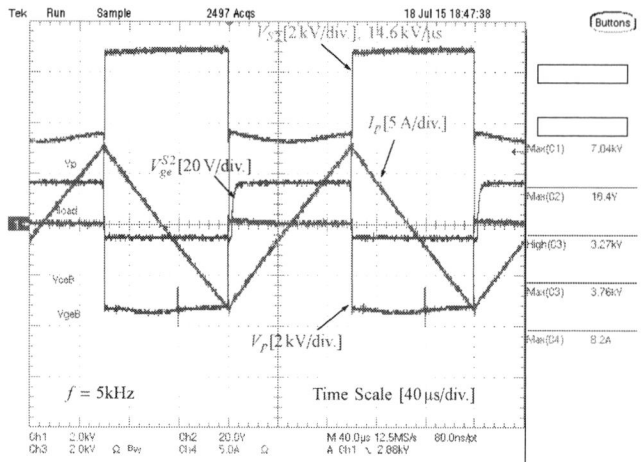

Fig. 11: 15 kV SiC IGBT 7 kV Half-bridge without Snubber

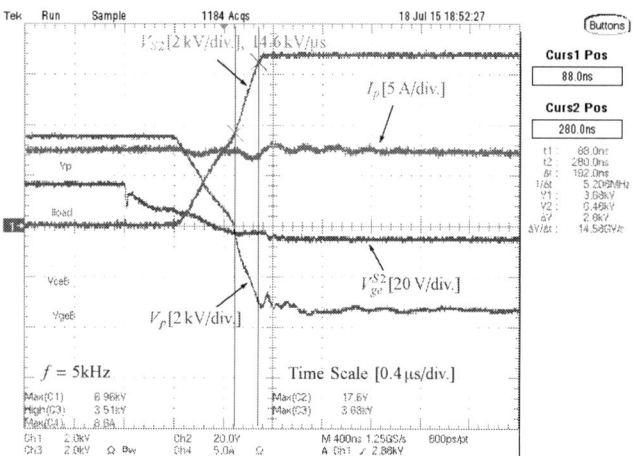

Fig. 12: Expanded 7 kV Half-bridge result without Snubber

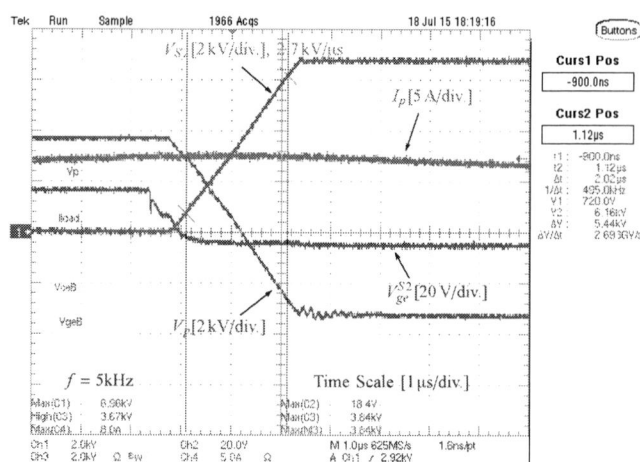

Fig. 13: Expanded 7 kV Half-bridge result with 1.1 nF Snubber

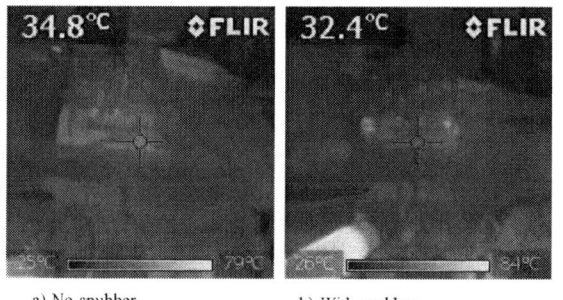

a) No snubber   b) With snubber

Fig. 14: 20 min Thermal Runs without & with Full ZVS

dv/dt is 2.7 kV/µs as desired. Also the temperature recorded after 20 min heat run test is found to be lower on the same heat-sink point as shown in thermal photograph in Fig. 14. This indicates that losses due to turn-off switching have reduced. Here the turn-on ZVS was created for both tests by inductive load to achieve the required peak-current during turn-off instant. Thus the loss in snubber case is contributed mainly by conduction and the final turn-off loss is very small. For a non-compensated MV DAB, there might be turn-on loss as well due to loss of ZVS at light load since there is always significant device capacitance to discharge. For implementing snubber in the DAHB, the inductive load is replaced by the DAHB transformer connected with secondary and series converter. Fig. 15 shows the 3 kV DAHB experimental result without snubber or series compensation. Fig. 16 shows the the same result without snubber but with series compensation using $5 : 1$ N_s series transformer and 16.7% duty. Fig. 17 shows the the same result but with series compensation using $1 : 1$ N_s series transformer. Now the series duty is kept 8.3% compared to the simulations as this provides sufficient peak triangular current. Fig. 18 shows the the same result

with 150 pF snubber and series compensation using $1 : 1$ N_s series transformer. It can be seen there is dv/dt reduction in the pole voltage but the turning-on device voltage reduces to zero before the dead-time ends. This mode of operation also corresponds to maximum efficiency. But the overall loss is not measured in this paper to verify it. Fig. 19 shows the comparison of dv/dt for half-bridge voltage and current for Fig. 17 and Fig. 18 cases. It can be seen that there are slower dv/dt near zero pole voltage in middle of higher dv/dt

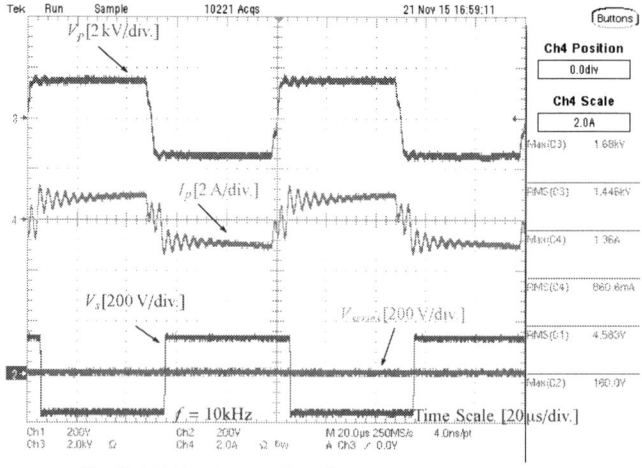

Fig. 15: 3 kV DAHB operation without series addition or Snubber

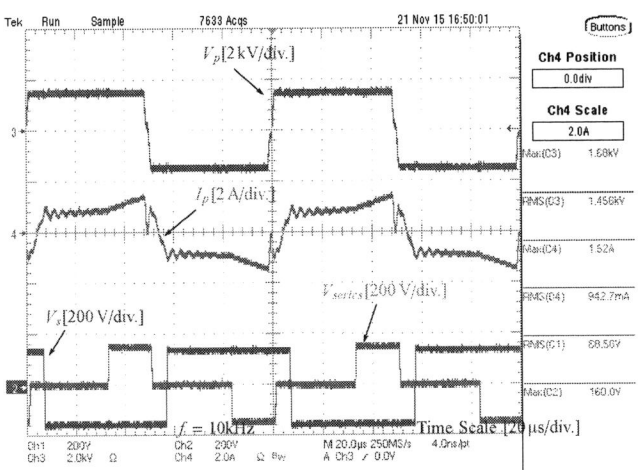

Fig. 16: 3 kV DAHB operation without Snubber with 5 : 1 N_s

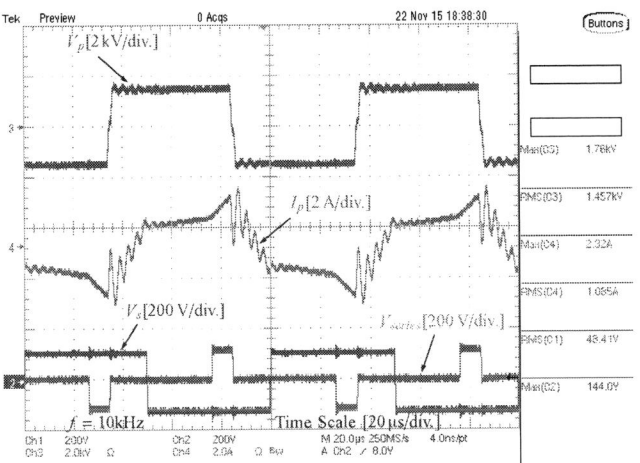

Fig. 17: 3 kV DAHB operation without Snubber with 1 : 1 N_s

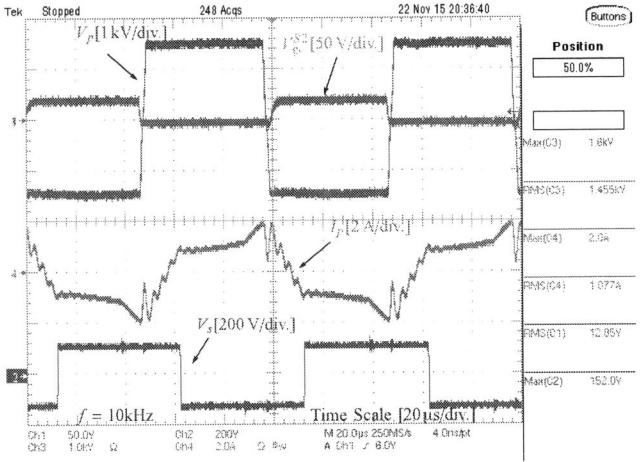

Fig. 18: 3 kV DAHB operation with 150 pF Snubber and 1 : 1 N_s

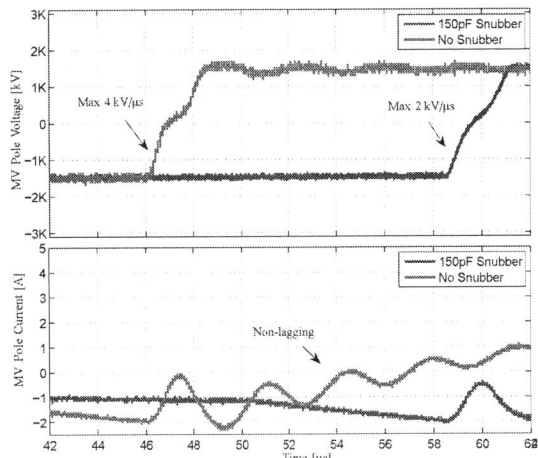

Fig. 19: Experimental dv/dt comparison between with and without snubber

transitions. After using the snubber, the difference between two dv/dt's is brought down but the total discharge time increases. This implies that it may be happening due to non-linear device capacitance. The energy stored in snubber is regenerated in the next discharge cycle. Thus the snubber decreases EMI together with the turn-off losses.

V. CONCLUSIONS

15kV SiC IGBT based simple DAB topology is very promising for high power and medium-voltage application. The snubber cap addition can improve the EMI and turn-off performance of the MV device above punch-through level. But it introduces the challenges for turn-on ZVS operation especially at light load. A simple and inexpensive series circuit consisting a fractional rating LV transformer and low current LV switches, can be utilized to shape the current near zero-crossing to meet the turn-on ZVS criteria. This additions makes this DAB topology low EMI prone and fully soft-switched.

VI. ACKNOWLEDGMENT

The information, data, or work presented herein was funded in part by the Office of Energy Efficiency and Renewable Energy (EERE), U.S. Department of Energy, under Award Number DE-EE0006521 with North Carolina State University, PowerAmerica Institute. This work made use of FREEDM ERC shared facilities supported by NSF under award no. EEC-0812121.

VII. DISCLAIMER

The information, data, or work presented herein was funded in part by an agency of the United States Government. Neither the United States Government nor any agency thereof, nor any of their employees, makes any warranty, express or implied, or assumes any legal liability or responsibility for the accuracy, completeness, or usefulness of any information, apparatus, product, or process disclosed, or represents that its use would not infringe privately owned rights. Reference herein to any specific commercial product, process, or service by trade name, trademark, manufacturer, or otherwise does not necessarily constitute or imply its endorsement, recommendation, or favoring by the United States Government or any agency thereof. The views and opinions of authors expressed herein do not necessarily state or reflect those of the United States Government or any agency thereof.

REFERENCES

[1] R. W. A. A. DeDoneker, D. M. Divan, and M. H. Kheraluwala, "A three-phase soft-switched high-power-density dc/dc converter for high-power applications", *IEEE Trans. Ind. Appl.*, vol.27, no.1, pp. 797-806, Jan./Feb. 1991.

[2] M. H. Kheraluwala, R. W. Gascoigne, D. M. Divan, and E. D. Baumann, "Performance characterization of a high-power dual active bridge dc-to-dc converter," *IEEE Transactions on Industry Applications*, vol. 28, no. 6, pp. 1294–1301, 1992.

[3] A. Tripathi, K. Mainali, D. Patel, S. Bhattacharya, and K. Hatua, "Control and performance of a single-phase dual active half bridge converter based on 15 kV SiC IGBT and 1200v SiC MOSFET," in *IEEE Applied Power Electronics Conference and Exposition (APEC)*, Mar. 2014, pp. 2120–2125.

[4] D. Segaran, D. G. Holmes, and B. P. McGrath, "Comparative analysis of single and three-phase dual active bridge bidirectional DC-DC converters," in *Australasian Universities Power Engineering Conference (AUPEC)*, pp. 1-6, 2008.

[5] A. Kadavelugu, S. Bhattacharya, S. Leslie, Sei-Hyung Ryu, D. Grider, and K. Hatua. "Understanding dv/dt of 15 kV SiC N-IGBT and Its Control Using Active Gate Driver." *IEEE Energy Conversion Congress and Exposition (ECCE)*, 221320, 2014.

[6] A. Tripathi, K. Mainali, D. Patel, A. Kadavelugu, S. Hazra, S. Bhattacharya, K. Hatua, "Design Considerations of a 15-kV SiC IGBT-Based Medium-Voltage High-Frequency Isolated DC-DC Converter," *IEEE Transactions on Industry Applications*, vol. 51, no. 4, pp. 3284-3294, July 2015.

[7] G. Guidi, M. Pavlovsky, A. Kawamura, T. Imakubo, Y. Sasaki, "Improvement of Light Load Efficiency of Dual Active Bridge DC-DC Converter by Using Dual Leakage Transformer and Variable Frequency",*IEEE Energy Conversion and Congress Exposition (ECCE)*, pp. 830-837, 2010.

[8] G. Ortiz, H. Uemura, D. Bortis, J. W. Kolar, and O. Apeldoorn, "Modeling of Soft-Switching Losses of IGBTs in High-Power High-Efficiency Dual-Active-Bridge DC/DC Converters." IEEE Trans. on Electron Devices, vol. 60, no. 2, pp. 587-597, February 2013.

[9] H. Bai and C. Mi, "Eliminate reactive power and increase system efficiency of isolated bidirectional dual-active-bridge DC-DC converters using novel dual-phase-shift control," *IEEE Transactions on Power Electronics*, vol. 23, no. 6, pp. 2905–2914, 2008.

[10] A. Kadavelugu, S. Bhattacharya, S.-H. Ryu, E. Van Brunt, D. Grider, S. Leslie, "Experimental switching frequency limits of 15 kV SiC N-IGBT module," in *Power Electronics Conference (IPEC-Hiroshima 2014-ECCE-ASIA), 2014 International.* IEEE, 2014, pp. 3726–3733.

[11] A. Tripathi, K. Mainali, and S. Bhattacharya, "A series compensation enabled ZVS range enhancement of a dual active bridge converter for wide range load conditions," in *2014 IEEE Energy Conversion Congress and Exposition (ECCE)*, Sept. 2014, pp. 5384–5391.

[12] K. Vechalapu, A. Tripathi, K. Mainali, B. J. Baliga, and S. Bhattacharya, "Soft switching characterization of 15 kV SiC n-IGBT and performance evaluation for high power converter applications," in *Energy Conversion Congress and Exposition (ECCE), 2015 IEEE.* IEEE, 2015, pp. 4151–4158.

Soft Switching for Half Bridge Current Doubler for High Voltage Point of Load Converter in Data Center Power Supplies

Yutian Cui, Weimin Zhang, Leon M. Tolbert, Daniel J. Costinett, Fred Wang, Benjamin J. Blalock
Center for Ultra-Wide Area Resilient Electric Energy Transmission Networks (CURENT)
Department of Electrical Engineering and Computer Science
The University of Tennessee
Knoxville, TN 37996-2250
ycui7@utk.edu

Abstract— In this paper, a single stage system which converts 400 V to 1 V within one stage and performs as the high voltage point of load (HV POL) converter for data centers is proposed. A load dependent soft switching method has been proposed for half bridge current doubler with simple auxiliary circuit. The operation principles of the soft switching converter have been analyzed in detail. A lossless RCD current sensing method is used to sense the output current value to reduce the auxiliary circuit loss and turn off loss of secondary side devices as load reduces to achieve higher efficiency. Experimental efficiency has been tested to prove the proposed method can increase the converter's efficiency in both heavy and light load condition. A prototype of the half bridge current doubler circuit has been built to verify the theory.

I. Introduction

Nowadays, high voltage (400 V) DC distribution architecture for power supply of data centers has shown superior efficiency compared to the traditional AC power supply architecture [1]. After rectifying the three phase 480 V AC into 400 V DC, there are several popular architectures between the 400 V bus voltage and 1 V: (1) intermediate bus architecture (IBA) shown in Fig. 1 [1], (2) factorized power architecture (FPA) proposed by Vicor [2], (3) high frequency AC link (HFAC) architecture [3], etc. These architectures have more than one power stage connected in series. The efficiency of the overall power supply chain becomes the product of each power stage's efficiency [4] [5]. Therefore, a power supply with one single power stage is expected to be one way to improve power conversion efficiency [4] [5].

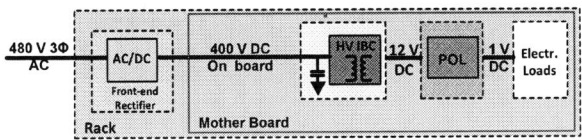

Fig. 1. IBA architecture.

A six phase input series and output parallel connected interleaved controlled converter is selected as the single power stage acting like the high voltage (HV) point of load (POL) converter as shown in Fig. 2 [4, 5]. Several advantages of ISOP connected converter can be obtained: 1) it is easy to implement interleaved control signal which helps to reduce the output inductor value and volume, 2) transient response can be

improved with smaller output inductor, and 3) ISOP control ensures stable operation of the system even with existence of mismatch in the converters [6, 7]. Because of the advantages, ISOP structure is used in this study.

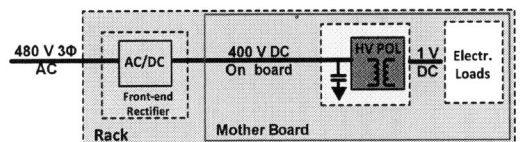

Fig. 2. HV POL architecture.

In this paper, the converter used in ISOP is selected to be half bridge current doubler as discussed in Section II. Next, a soft switching methods with dependency on load condition is proposed and analyzed. In the end, experimental results are given to verify the proposed methods.

II. Topology Selection

The half bridge current doubler circuit shown in Fig. 3 is a widely used switch mode power supply topology because of its limited number of devices and no circulating current compared with the full bridge converter [8]. Two control methods are commonly used for half bridge converter: 1) symmetrical control where Q_1 and Q_2 are controlled with identical signals with 180° and 2) asymmetrical control where Q_1 and Q_2 are controlled with complementary gate signals [8-13].

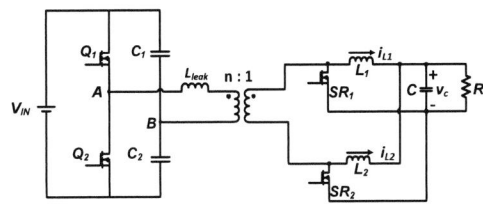

Fig 3. Half bridge current doubler.

Asymmetrical control leads to uneven current and voltage stress on all the devices, and the transformer is biased as well. Also, the leakage inductance causes duty cycle loss and increased RMS current [8-11]. Soft switching is lost under lighter load when the energy stored in leakage inductance is

978-1-4673-9551-9/16 $31.00 © 2016 IEEE

not enough to discharge the capacitors. Symmetrical control offers the same current and voltage stress and unbiased operation of the transformer; therefore, it simplifies circuit design. The major drawback of symmetrical switching converter is that the transformer leakage inductance resonates with the output junction capacitances of Q_1 and Q_2 when both of them are *OFF*, and Q_1 and Q_2 are hard switched. The resonance can be mitigated by a snubber circuit [10, 11], however, with a planar transformer, the losses induced by resonance are smaller than snubber circuit losses; therefore, there is no need to implement a snubber circuit.

The second issue is hard-switching primary side devices and reverse recovery of secondary side MOSFETs' body diodes. Therefore, a soft switching half bridge current doubler with load dependent control method is proposed in this paper as shown in Fig. 2 to reduce switching losses. Soft switching has been studied in various publications [13-16]. In [13, 14], a similar method has been proposed for a buck converter with constant charging time of auxiliary inductor as in Fig. 5, and an extra snubber capacitor has been placed in parallel with S_1. The turn off current of bottom switch Q_2 ($I_{Q2,OFF}$ in Fig. 6) increases as load decreases. Therefore, the turn off of Q_2 would increase, which is solved by the snubber capacitor. However, the snubber capacitor is not feasible for a half bridge converter as it induces snubber loss. Therefore, a load dependent charging time of auxiliary inductor has been proposed in this paper. Also, the losses in the snubber circuit reduce as load a decrease, which helps the light load efficiency.

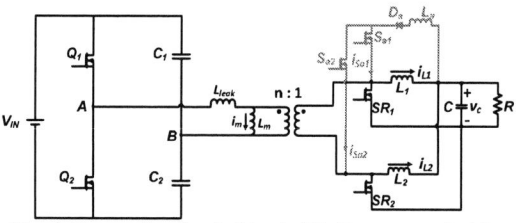

Fig. 4. Proposed soft switching half bridge current doubler.

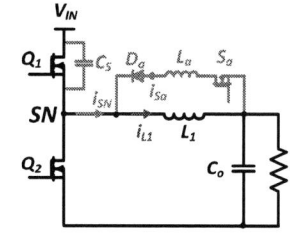

Fig. 5. Soft switching buck converter [13,14].

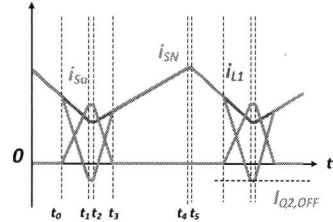

Fig. 6. Waveforms of soft switching buck converter [13,14].

III. OPERATING PRINCIPLE

Fig. 7 is the operational waveform of the half bridge current doubler with the proposed control scheme. The operation can be divided into 12 intervals. Because of symmetry of the half bridge converter, t_0 to t_5 and t_6 to t_{11} are essentially the same. Only the first three intervals (t_0 to t_3) of the proposed method differ from conventional methods, and will be discussed here.

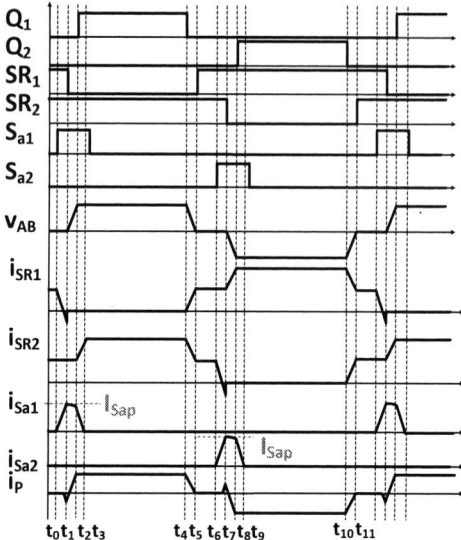

Fig. 7. Operational waveforms of proposed method.

1. Interval 1: t_0 to t_1

Before t_0, two inductor currents i_{L1} and i_{L2} are freewheeling through SR_1 and SR_2. At time t_0, S_a is turned on, as a positive voltage is applied across L_a, current in the direction of the arrow is built up. The equivalent circuit is shown in Fig. 6. In order to achieve zero voltage switching of primary side devices and avoid reverse recovery of SRs' body diode, the current direction of i_{SR1} is negative with respect to the current direction in Fig. 8. The current i_{sa} can be calculated as

$$i_{sa} = \frac{V_o - V_{Da}}{L_a}(t - t_0) \qquad (1)$$

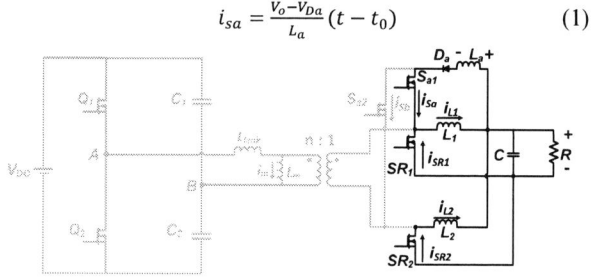

Fig. 8. Equivalent circuit during t_0 to t_1.

2. Interval 2: t_1 to t_2

At time t_1, SR_1 is turned off. As the current is flowing from drain to source when SR_1 is turned off, the body diode conduction of SR_1 is prohibited. The difference between i_{Sa} and i_{L1} is forced to flow through the capacitances including

SR_1, Q_1 and Q_2. The requirement to achieve soft switching is expressed in (2).

$$\tfrac{1}{2}L_{Sa}(I_{sap}-\tfrac{1}{2}I_{L1})^2 \geq C_{OSS_p}V_{in}^2 + \tfrac{1}{2}C_{OSS_{SR}}V_{SR}^2 \qquad (2)$$

SR device is turned off with current direction from drain to source; therefore, the body diode reverse recovery loss will be eliminated. However, active turn off loss is added to the converter compared with the conventional method. The turn off loss of the SR device can be calculated as (3). If constant charging time is used without adding turn off snubber capacitor, turn off loss will increase almost linearly as load becomes lighter (I_o reduces). Adding the load dependent charging time control, lower value of I_{sap} is charged so that almost constant ($I_{sap}-I_o$) and turn off loss is achieved in (3) as load reduces. The current value at the end of this interval is I_{sa2}.

$$P_{SRoff} = \tfrac{1}{2}V_{SR}(I_{sap}-I_o)t_{off,sw}f_s - \tfrac{1}{2}V_{SR}^2 C_{SR}f_s \qquad (3)$$

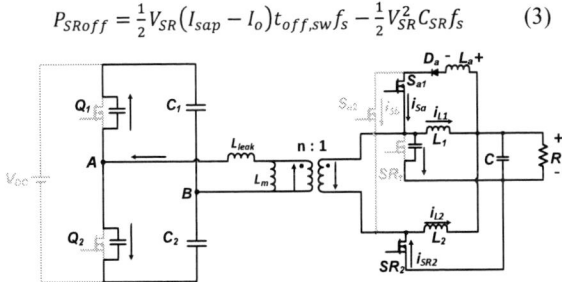

Fig. 9. Equivalent circuit during t_1 to t_2.

3. Interval 3: t_2 to t_3

At time t_2, Q_1 is turned on with zero voltage across it. The equivalent circuit is shown in Fig. 10. When Q_1 is turned on, a positive voltage is applied across the transformer while a negative voltage is applied across L_a. Therefore, the current I_{sa} starts to reduce linearly as (4). When the current reduces to zero, the diode D_a starts to block the voltage and stops conducting current. This interval ends when the current reduces to zero in the auxiliary branch. The time for the current to reduce to zero can be calculated as well. After that, S_a can be turned off with zero current.

$$i_{sa} = I_{sa2} - \frac{\frac{V_{in}}{2n}-V_o-V_{D_a}}{L_a}(t-t_2) \qquad (4)$$

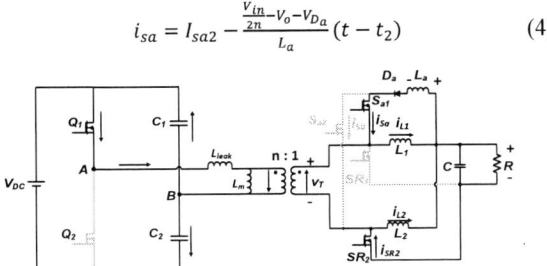

Fig. 10. Equivalent circuit during t_2 to t_3.

IV. LOAD DEPENDENT CONTROL SCHEME

Fig. 11 is the control circuit for soft switching of the half bridge current doubler. A commercial half bridge controller IC is used to generate the control signals. A large dead time

between SR devices' turn off (SR_{1C} and SR_{2C} in Fig. 12) and primary devices' turn on is programmed on the controller IC so that the falling edge of SR_{1C} and SR_{2C} can be used to generate gate signals for S_{a1} and S_{a2}. In order to consider the load condition, a comparator between averaged inductor current and the RC delayed falling edge of SR_{1C} and SR_{2C} is used to generate the turn on signal for S_{a1} and S_{a2}. A larger load will trigger the comparator earlier to create a longer charging time for L_a. Also, a falling edge delay circuit is implemented so that SR_1 and SR_2 will have a suitable dead-time.

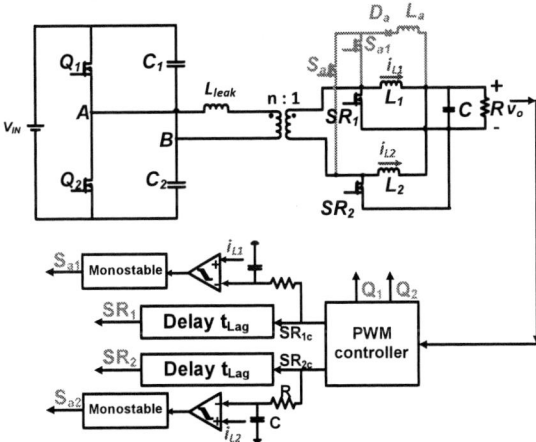

Fig. 11. Load dependent control scheme of soft switching.

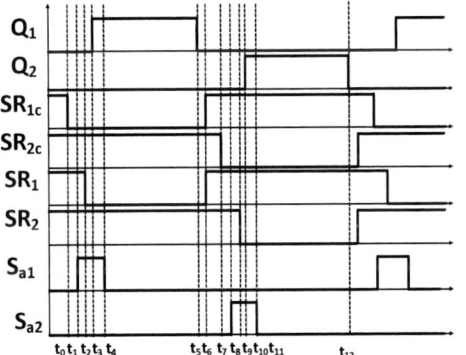

Fig. 12. Gate signal sequence for half bridge current doubler.

In order to eliminate current shunt losses, a lossless current sensing circuit shown in Fig. 13 is implemented. When $R_{CS}C_{CS} = L/DCR$ is satisfied, the voltage across the capacitor, C_{CS}, is proportional to the inductor current [17,18].

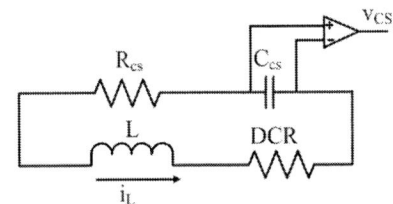

Fig. 13. RCD current sensor.

978-1-4673-9551-9/16 $31.00 © 2016 IEEE

V. CONVERTER DESIGN

A half bridge current doubler with specification shown in Table I has been designed and built in hardware prototype, which can be used as power supply circuit [19].

Table I. Specification of half bridge current doubler.

Power rating	30 W
Input voltage	66 V nominal
Input voltage ripple	± 1%
Output voltage	0.98 V – 1.02 V (1 V nominal)
V_{out} ripple	10 mv peak to peak

1. Transformer design

The transformer turns ratio is selected to be 6:1 so that the converter can operate normally when the input voltage is above 200 V. As the leakage inductance is detrimental to the converter with symmetrical control, therefore, interleaved transformer winding structure shown in Fig.13 is selected [20, 21]. The leakage inductance L_{leak} is 85 nH and the AC winding resistance of the transformer primary side's winding is 75.8 mΩ based on measurement.

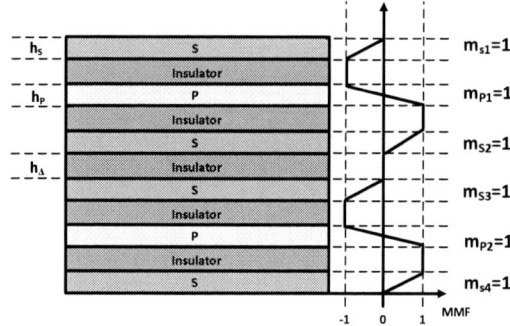

Fig. 13. Interleaved transformer winding structure in half bridge current doubler.

2. Device selection

Even though the primary side device is being turned on with zero voltage across it, it takes more energy to discharge it. With larger output junction capacitances, the equivalent impedance of LC network when both Q_1 and Q_2 are not conducting is becoming smaller, which will lead to larger oscillation current. Also, the RMS losses in devices and transformer will increase. Based on this, MOSFETs with following parameters are selected in Table II as primary side MOSFETs as Q_1 and Q_2 in Fig. 3.

Table II. Primary side device's specification.

$R_{ds(on)}$(mΩ)	C_{oss}(pF)	Q_g (nC)	C_{rss}(pF)
66	160 pF	6.5	3

The major loss of SR device is conduction loss, therefore, low $R_{ds(on)}$ is the priority when selecting SR devices. Table III summarizes the key parameters of SR devices.

Table 1SR device's specification.

$R_{ds(on)}$ (mΩ)	C_{oss}(pF)	Q_g (nC)	C_{rss}(pF)
1.0	3600	67	180

3. Auxiliary inductor design

Based on the selected devices, an auxiliary inductor at 10 nH is selected and designed with PCB winding and air core. Based on Q3D simulation, the inductance is 9.26 nH.

VI. TESTING RESULTS

Figs. 14 through 19 show the ZVS behavior under different load condition with variable charging time which is indicated by the overlap between secondary side devices' and auxiliary devices' gate signal. From the experimental results, it can be clearly seen that ZVS of primary side devices can be achieved over entire load condition and the charging time is dependent on load condition, prevention charging the auxiliary inductor to a unnecessary large value and increasing conduction loss of the auxiliary MOSFET and diode.

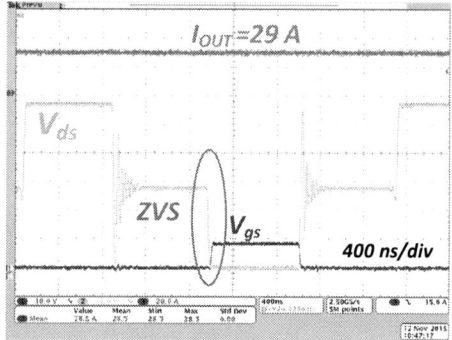

Fig. 14. ZVS for primary device with 29 A load current.

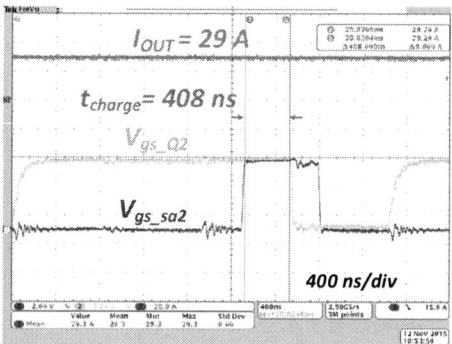

Fig. 15. Charging time for auxiliary inductor with 29 A load current.

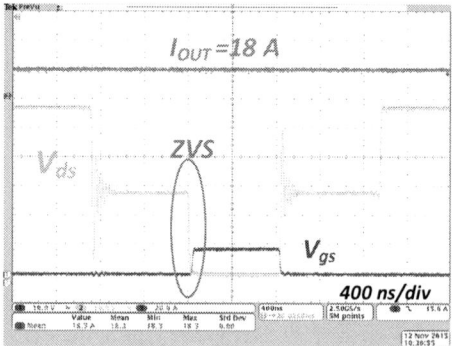

Fig. 16. ZVS for primary device with 18 A load current.

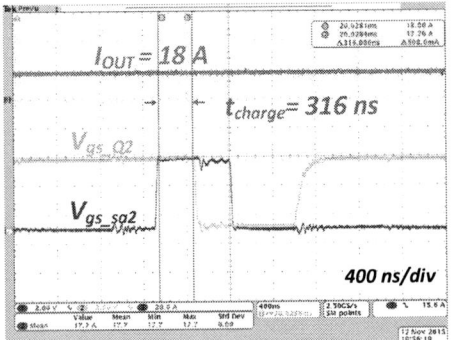

Fig. 17. Charging time for auxiliary inductor with 18 A load current.

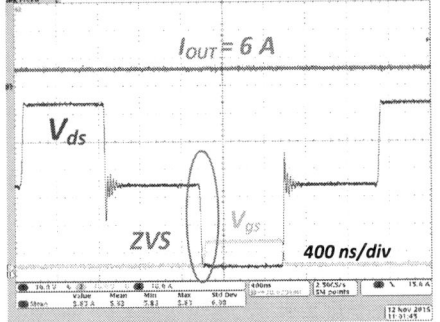

Fig. 18. ZVS for primary device with 6 A load current.

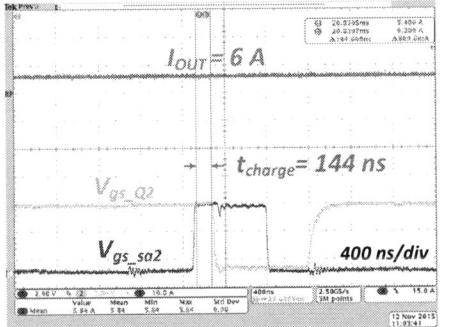

Fig. 19. Charging time for auxiliary inductor with 6 A load current.

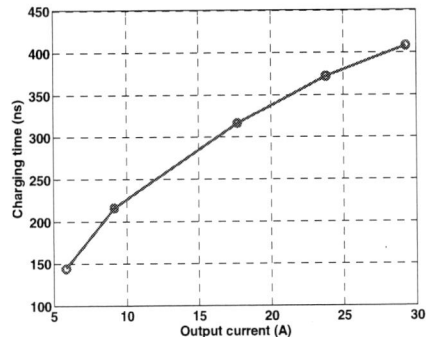

Fig. 20. Experimental charging time under different load condition.

Fig. 20 shows the charging time of auxiliary inductor under different load conditions based on testing results. Charging time reduces with load becoming lighter as designed. One thing to notice is that in experiments, it actually requires longer charging time to achieve ZVS in all the load condition, therefore, longer charging time is implemented in the experiment compared with analysis. However, this difference does not impact the load dependent operation, it simply means a larger I_{sap} is needed to obtain ZVS in Fig. 11.

Fig. 21 illustrates the efficiency curve of different control methods, including conventional hard switching, constant charging of auxiliary inductor and proposed load dependent charging method. It clearly shows that the proposed method has higher efficiency compared with the other two methods in the entire load range.

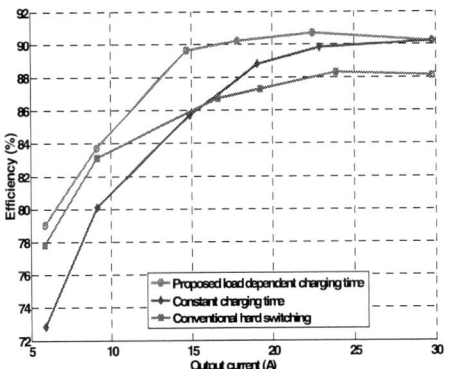

Fig. 21. Experimental efficiency comparison among proposed load dependent charging time, constant charging time and conventional hard switching.

Fig. 22 shows the transient performance of the proposed method when load changes from 100 % to 10 %. It can be seen from the figure that the charging time reduces with reduced load current. It can be seen that it takes several switching cycles for the control to respond to load change, mainly because inductor current is used for the load condition in the controller, which is not as fast as the output current to the load.

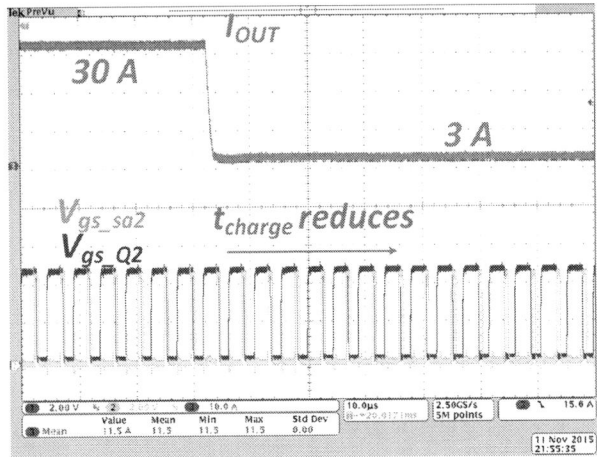

Fig. 22. Transient of proposed load dependent charging method.

978-1-4673-9551-9/16 $31.00 © 2016 IEEE

VII. CONCLUSION

Symmetrical control of half bridge current doubler provides smaller duty cycle loss and even current and voltage stress compared with asymmetrical control. The major drawback of symmetrical control is hard switch over entire range. Therefore, a load dependent soft switching method has been proposed in this paper to achieve ZVS over entire load range with increased light load efficiency compared with existing method.

Detailed operation under different load conditions has been studied to show that the proposed control method is feasible regardless of load condition. With a simple RCD current sensing circuit, a load dependent pulse width controller can be achieved. The converter design has been performed including device selection and transformer design.

Experimental results indicate that the proposed method can achieve soft switching of primary side devices and helps to improve efficiency in the entire load condition. It also can respond to load changing automatically with several switching cycles.

ACKNOWLEDGMENT

This work made use of the Engineering Research Center Shared Facilities supported by the Engineering Research Center Program of the National Science Foundation and DOE under NSF Award Number EEC-1041877 and the Center for Ultra-wide-area Resilient Electric Energy Transmission Networks (CURENT) Industry Partnership Program.

REFERENCES

[1] A. Pratt, P. Kumar, T. V. Aldridge, "Evaluation of 400V DC distribution in telco and data centers to improve energy efficiency," in *29th International* Telecommunications *Energy Conference,* Sept. 30 2007-Oct. 4, 2007, pp. 32-39.

[2] http://www.vicorpower.com/

[3] http://www.idt.com/landing/coolRAC-Technology

[4] Y. Cui, L. M. Tolbert, "High step down ratio (400 V to 1 V) phase shift full bridge DC/DC converter for data center power supplies with GaN FETs," *in IEEE Workshop on Wide Bandgap Devices and Applications,* Oct. 2013, Columbus, Ohio, pp. 23-27.

[5] Y. Cui, W. Zhang, L. M. Tolbert, F. Wang, B. J. Blalock, "Direct 400 V to 1 V converter for data center power supplies using GaN FETs," *IEEE Applied Power Electronics Conference and Exposition (APEC),* pp.1698-1704, Mar. 2014.

[6] R. Giri, V. Choudhary, R. Ayyanar, N. Mohan, "Common-duty-ratio control of input-series connected modular DC-DC converters with active input voltage and load-current sharing," *IEEE Transactions on Industry Applications,* vol.42, no.4, pp.1101-1111, July-Aug. 2006.

[7] V. Choudhary, E. Ledezma, R. Ayyanar, R. M. Button, "Fault tolerant circuit topology and control method for input-series and output-parallel modular DC-DC converters," *IEEE Transactions on Power Electronics,* vol.23, no.1, pp.402-411, Jan. 2008.

[8] Bo Yang, "Topology investigation of front end DC/DC converter for distributed power system," Ph. D. dissertation, Virginia Tech, 2003.

[9] J. H. Liang, P. C. Wang, K. C. Huang, C. L. Chen, Y. H. Leu, T. M. Chen, "Design optimization for asymmetrical half-bridge converters," *2001 IEEE Applied Power Electronics Conference and Exposition,* 2001, pp.697-702.

[10] I. O. Lee, G. W. Moon, "A new asymmetrical half-bridge converter with zero dc-offset current in transformer," *IEEE Transactions on Power Electronics,* vol.28, no.5, pp.2297-2306, May 2013.

[11] Y. C. Hung, F. S. Shyu, C. J. Lin; Y. S. Lai, "Design and implementation of symmetrical half-bridge DC-DC converter," *2003 Power Electronics and Drive Systems,* 2003, pp.338-342.

[12] Y. Panov, M. M. Jovanovic, "Design and performance evaluation of low-voltage/high-current DC/DC on-board modules," *IEEE Transactions on Power Electronics,* vol.16, no.1, pp.26-33, Jan 2001.

[13] E. Adib, H. Farzanehfard, "Zero-voltage-transition PWM converters with synchronous rectifier," *IEEE Transactions on Power Electronics,* vol.25, no.1, pp.105-110, Jan. 2010.

[14] M. R. Mohammadi, H. Farzanehfard, "Analysis of diode reverse recovery effect on the improvement of soft-switching range in zero-voltage-transition bidirectional converters," *IEEE Transactions on Industrial Electronics,* vol. 62, no.3, pp.1471-1479, March 2015.

[15] L Fu, X Zhang, F Guo, J Wang, "A phase shift controlled current-fed quasi-switched-capacitor isolated dc/dc converter with GaN HEMTs for photovoltaic applications," *IEEE Applied Power Electronics Conference and Exposition (APEC),* 2015, pp. 191-198.

[16] F Guo, L Fu, X Zhang, C Yao, H Li, J Wang, "A family of dual-input DC/DC converters based on quasi-switched-capacitor circuit," *IEEE Energy Conversion Congress and Exposition (ECCE),* 2014, pp. 5281-5287.

[17] E. Dallago, M. Passoni, G. Sassone, "Lossless current sensing in low-voltage high-current DC/DC modular supplies," *IEEE Transactions on Industrial Electronics,* vol.47, no.6, pp.1249-1252, Dec 2000.

[18] Y. Zhang, F. Zheng, P. Wang, Y. Qiao, X. Yang; J. Wang, "A current-sensing circuit for DC–DC converters with bootstrap characteristics," *IEEE Transactions on Industrial Electronics,,* vol.61, no.8, pp.4183-4192, Aug. 2014.

[19] Y. Cui, W. Zhang, L. M. Tolbert, D. J. Costinett, F. Wang, B. J. Blalock, "Two phase interleaved ISOP connected high step down ratio phase shift full bridge DC/DC converter with GaN FETs," *2015 IEEE Applied Power Electronics Conference and Exposition (APEC),* 2015, pp.1414-1419.

[20] O. Ziwei, O. C. Thomsen, M. A. E. Andersen, "Optimal design and tradeoff analysis of planar transformer in high-power DC–DC converters," *IEEE Transactions on Industrial Electronics,* vol. 59, no. 7, pp. 2800-2810, July 2012.

[21] W. Zhang, Y. Long, Y. Cui, F. Wang, L. M. Tolbert, B. J. Blalock, S. Henning, J. Moses, and R. Dean, "Impact of planar transformer winding capacitance on Si-based and GaN-based LLC resonant converter," *IEEE Applied Power Electronics Conference and Exposition (APEC),* pp. 1668-1674, March 2013.

An Algorithm to Analyze Circulating Current for Multi-Phase Resonant Converter

Hongliang Wang, *Senior Member, IEEE*, Yang Chen, Zhiyuan Hu, Laili Wang, Tianshu Liu, Wenbo Liu, Yan-Fei Liu, *Fellow, IEEE*
Department of Electrical and Computer Engineering
Queen's University, Kingston, Canada
hongliang.wang@queensu.ca, yang.chen@queensu.ca, zhiyuan.hu@queensu.ca,
l.l.wang@queensu.ca, tianshu.liu@queensu.ca, Wenbo.liu@queensu.ca,
yanfei.liu@queensu.ca

Jahangir Afsharian and Zhihua (Alex) Yang
Murata Power Solutions
Toronto, Canada
jafsharian@murata.com, ZYang@murata.com

Abstract— an algorithm based on new circulated current model is proposed to analyze the circulating current in the multi-phase LLC converter. First of all, a virtual-open concept is introduced and LLC phases are decoupled from each other, Thus, A circulating current model can be achieved from fundamental harmonic approximation (FHA). Secondly, the ac voltages on the equivalent resistor of different phases are assumed same magnitude but different angle viewing from the primary-side. From mentioned above, current sharing error can be calculated, two types calculated results are shown under two-phase independent LLC converter and two-phase common inductor LLC converter. Simulation results from different input voltage levels, different resonant component tolerances, and different total load are compared. A two-phase 600W prototype is built to verify the effectiveness of the proposed analysis method

I. INTRODUCTION

LLC resonant converter has been widely used due to high performance, such as the zero voltage switching (ZVS) for the primary-side MOSFET and zero current switching (ZCS) for the secondary-side diodes [1, 2]. For high power applications, multiphase parallel technique is a good choice to reduce the high current stress [3-5].

However, due to the resonant component tolerances (such as less than 5%), LLC converter may have unequal output currents among phases [6-8]. Small component tolerances can cause drastic current imbalance. Thus, an accurate method is needed to analyze the inter-phase circulating current.

There are several literatures analyzing circulating current [9-12]. In [9-10], these methods rely on simulation to get the circulating current information. In [11], the secondary rectifier circuits are equivalent to two ac resistors for the two-phase LLC converter. However, the ac voltages on the equivalent resistors are assumed to have same magnitude and angle. The rectifier circuit is equivalent three ac resistor voltages for three-phase LLC converter, it also exist coupling parameter

for each phase, Thus, it is hard to get decoupled circulating current for each phase [12].

In this paper, an algorithm to analyze the inter-phase circulating current for multi-phase LLC resonant converter is proposed. The primary ac voltages of phases are of same voltage magnitude but different angle. The circulating current model can be derived by viewing the primary ac voltage of each phase decoupled. Circulating current error can be calculated with any combination of component tolerances. Section II describes circulating current analysis method. Section III shows experimental results of a 600W prototype; and Section IV gives the conclusion.

II. CIRCULATING CURRENT ANALYSIS METHOD

Mathematical model of LLC converter is needed for analyzing the current sharing characteristics of multi-phase LLC converter. For simplicity, two-phase LLC converter is analyzed in this paper. Conventional two-phase LLC resonant converter is shown in Fig.1. Lr, Cr, Lm are respectively the resonant inductor, resonant capacitor, magnetizing inductor of Phase one. Parameters aLr, bCr and cLm are the resonant inductor, resonant capacitor, magnetizing inductor of phase two. Of phase two, in which a, b and c indicate the resonant component tolerances as compared with Phase one. n is transformer turn ratio. iLr1, iLr2, Irect1, Irect2, Io1, and Io2 are the resonant current, rectifier current and load current of two units. Fig. 1(b) shows the equivalent circuit of the two-phase LLC resonant converter. The output resistor R is divided into R_1 and R_2. The values of R_1 and R_2 are decided by the steady-state load current, considering the output DC voltage V_O is well regulated and same for the two phase. The impedance error k is defined in Eq. (1). Fig. 1(c) shows the FHA equivalent circuit of Fig.1 (b).

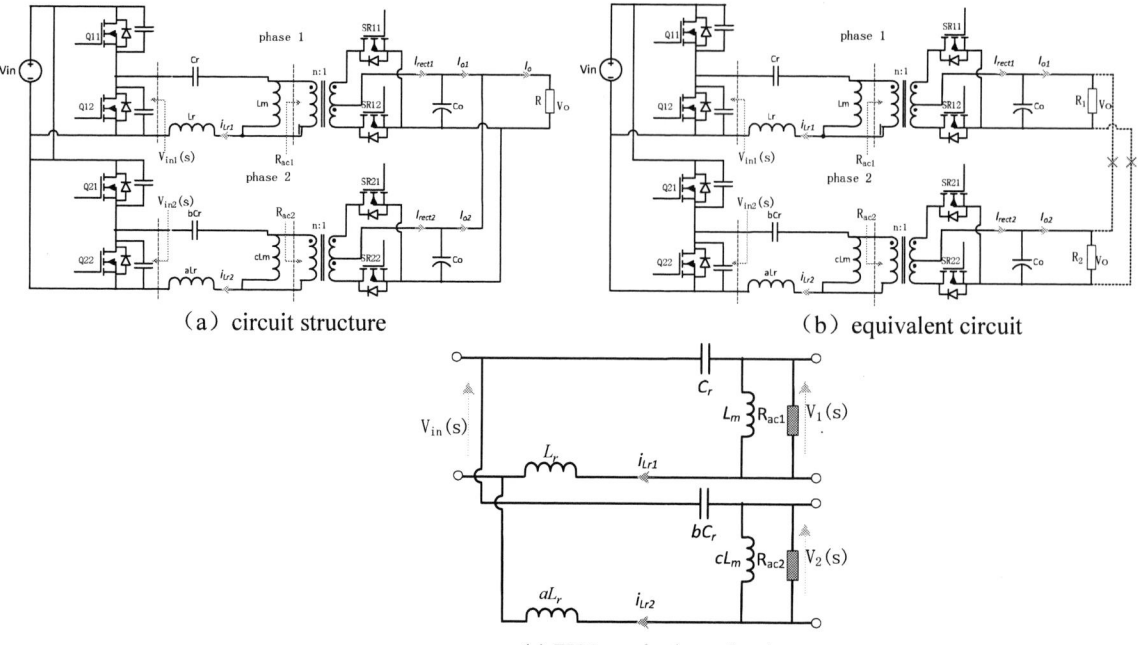

(a) circuit structure   (b) equivalent circuit

(c) FHA equivalent circuit

Fig.1 two-phase independent LLC resonant converter

The impedance error k is defined in Eq. (1). Fig. 1(c) shows the FHA equivalent circuit of fig.1 (b).

$$R_1 = \frac{1}{k}R, R_2 = \frac{1}{(1-k)}R, k \in [0,1] \qquad (1)$$

Two-phase common inductor LLC converter is shown in Fig.2 [13]. Similar analysis, the equivalent circuit and FHA circuit are shown in Fig.2 (a), Fig.2 (b).

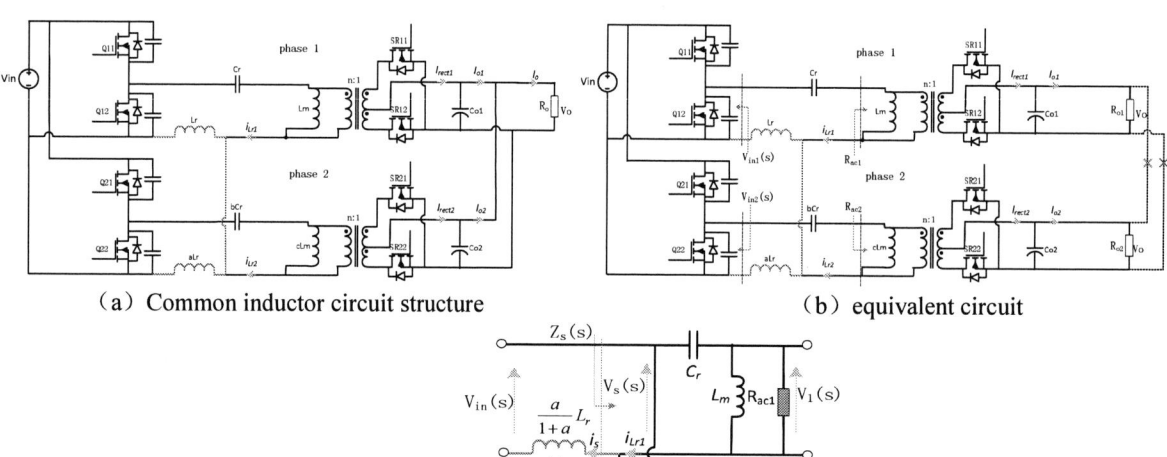

(a) Common inductor circuit structure   (b) equivalent circuit

(c) FHA equivalent circuit

Fig.2 Two-phase common inductor LLC resonant converter

V1 and V2 are of same magnitude but different angle. The ac voltage angles are always different at parameter tolerances. The relationship is shown Eq. (2).

$$|V_1(s)| = |V_2(s)| \qquad (2)$$

The ac loads Rac1 and Rac2 are defined in Eq. (3).

$$R_{ac1} = \frac{8n^2}{\pi^2}R_1, R_{ac2} = \frac{8n^2}{\pi^2}R_2 \qquad (3)$$

For two-phase independent LLC converter, the transfer function $V_1(s)$, $V_2(s)$ are in Eq. (4) from Fig.1(c).

$$\begin{cases} V_1(\text{s}) = G_1(\text{s})V_{in}(\text{s}) \\ V_2(\text{s}) = G_2(\text{s})V_{in}(\text{s}) \end{cases} \quad (4)$$

For two-phase common inductor converter, the transfer function $V_3(s)$, $V_4(s)$ are in Eq. (5) from Fig.2(c).

$$\begin{cases} V_3(\text{s}) = \dfrac{R_{ac1}//sL_m}{R_{ac1}//sL_m + 1/sC_r}(V_{in}(s) + V_{Lr}(s)) \\ V_4(\text{s}) = \dfrac{R_{ac2}//scL_m}{R_{ac2}//scL_m + 1/sbC_r}(V_{in}(s) + V_{Lr}(s)) \end{cases} \quad (5)$$

According to Eq. (1) to (5), the follow relationship can be found:

$$Ak^2 + Bk + C = 0 \quad (6)$$

The coefficient under two-phase independent LLC converter is shown in Eq. (7),

$$\begin{aligned} A &= \omega^2(1-b^2)c^2L_m^2 - \omega^4(2ab-2b^2)c^2L_rL_m^2C_r \\ &+ \omega^6(a^2-1)b^2c^2L_r^2L_m^2C_r^2 \\ B &= -2\omega^2c^2L_m^2 + 4\omega^4abc^2L_rL_m^2C_r \\ &- 2\omega^6a^2b^2c^2L_r^2L_m^2C_r^2 \\ C &= \omega^2c^2L_m^2 - 2\omega^4abc^2L_rL_m^2C_r + \omega^6a^2b^2c^2L_r^2L_m^2C_r^2 \\ &+ (1-b^2c^2)R^2 - \omega^2[(2ab-2b^2c^2)L_r + (2bc-2b^2c^2)L_m]C_rR^2 \\ &+ \omega^4(ab-bc)[(ab+bc)L_r^2 + 2bcL_rL_m]C_r^2R^2 \end{aligned} \quad (7)$$

The coefficient under two-phase common inductor is shown in Eq. (8),

$$\begin{cases} A = \omega^2(1-b^2)c^2L_m^2 \\ B = -2\omega^2c^2L_m^2 \\ C = \omega^2c^2L_m^2 + (1-b^2c^2)R^2 - 2\omega^2(bc-b^2c^2)L_mC_rR_{ac}^2 \end{cases} \quad (8)$$

Then, impedance error k can be expressed as in Eq. (9),

$$k = \begin{cases} -\dfrac{C}{B} & A=0, B\neq 0 \\ \dfrac{-B\pm\sqrt{B^2-4AC}}{2A} & A\neq 0, \sqrt{B^2-4AC}\geq 0 \end{cases} \quad and\ k\in[0,1] \quad (9)$$

The impedance error k is effective when k is between 0 and 1. Otherwise only one unit will provide the load power.

According to (9),

The load current sharing error σ is defined in (10),

$$\sigma = abs(\frac{I_{01}-I_{02}}{I_{01}+I_{02}}) = abs(1-2k), \text{k}\in[0,1] \quad (10)$$

The resonant current sharing error is defined in (11)

$$\sigma_{resonant} = \frac{abs(rms(i_{Lr1}) - rms(i_{Lr2}))}{abs(rms(i_{Lr1}) + rms(i_{Lr2}))} \quad (11)$$

Table.1 shows the resonant parameters of the Phase #1, serving as the reference, to which the component tolerances of Phase #2 will be compared.

Tab.1 Parameters

Lr	Cr	Lm	n	fr	Vo	Po(total)
29μH	12nF	95μH	20	270KHZ	12V	600W

As the resonant inductance, resonant capacitance is increased of phase two, the resonant frequency is reduced, thus, the gain value of phase two will be increased if the load resistor doesn't changed based on virtual open concept. Actually, phase two will provide more power than phase one. And then keep the same between phase one and phase two. Similarly, phase two provide more power than phase on with magnetic inductance of phase two increasing. Thus, the worst situation is that parameters a, b, c is increased or deduced at same time.

Fig.3 shows load current sharing error with different parameter tolerance under two-phase independent LLC converter. Fig.3 (a), (b), (c) shows the current sharing error at +5% Lr, +5% Cr, +5% Lm, respectively. The current sharing error is almost reduced with total load current increasing and input voltage increasing. The worst situation of current sharing error is shown in Fig.3 (d), which three resonant parameters have +5% increasing. The current sharing error is 60% at 50A load current, 400V input voltage. Fig.4 shows the load current sharing error with different parameter tolerance under two-phase common inductor LLC converter. Fig.3 (a), (b), (c) shows the current sharing error at +5% Lr, +5% Cr, +5% Lm, respectively.

The current sharing error is reduced with total load current increasing and input voltage increasing. The worst situation of current sharing error is shown in Fig.3 (d), which three resonant parameters have +5% increasing. The current sharing error is about 5% at 50A load current, 400V input voltage.

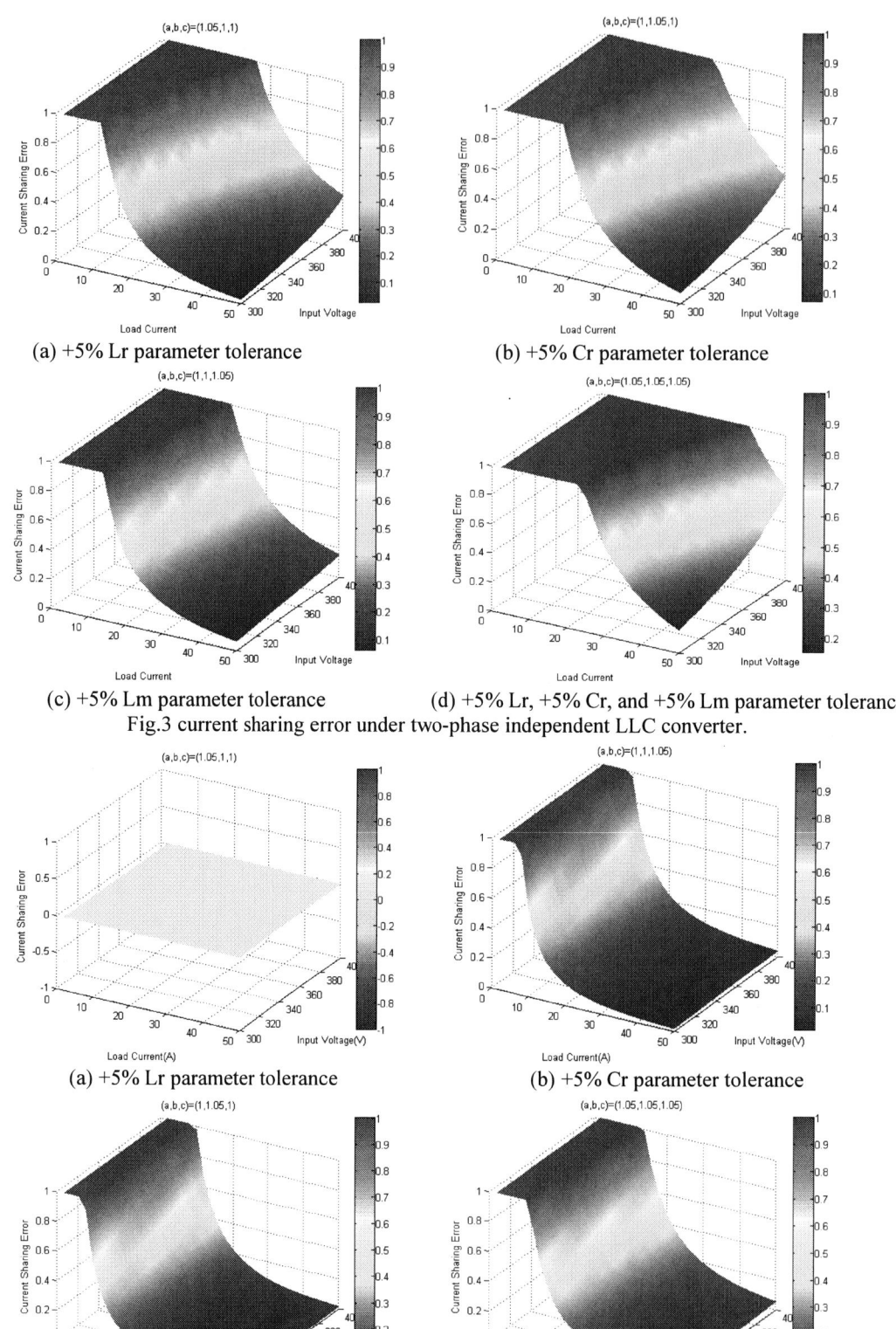

(a) +5% Lr parameter tolerance   (b) +5% Cr parameter tolerance

(c) +5% Lm parameter tolerance   (d) +5% Lr, +5% Cr, and +5% Lm parameter tolerance

Fig.3 current sharing error under two-phase independent LLC converter.

(a) +5% Lr parameter tolerance   (b) +5% Cr parameter tolerance

(c) +5% Lm parameter tolerance   (d) +5% Lr, +5% Cr, and +5% Lm parameter tolerance

Fig.4 current sharing error under two-phase common inductor LLC converter.

III. SIMULATION AND EXPERIMENT RESULTS

From Section II, the simulation results under 400V input voltage, 60A load current are shown in Fig.5, which parameters (a, b, c) equals to (1.05, 1.05, 1.05).

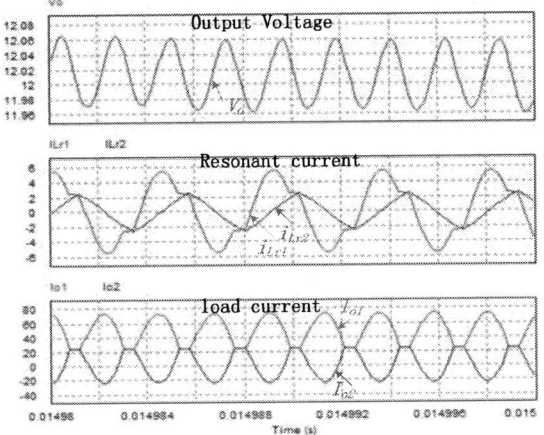

(a) Waveforms under two-phase independent LLC converter

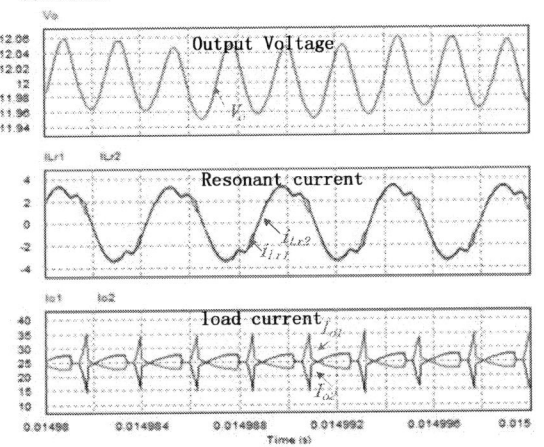

(b) Waveforms under two-phase common inductor LLC converter

Fig.5 simulation waveforms at the worst case.

Fig.5 shows simulation waveform under the worst case (a, b, c) = (1.05, 1.05, 1.05). Fig. 5(a) shows the simulation waveform under two-phase independent LLC converter. Output voltage is 12V. There is large difference between resonant current of each phase. Load current of each phase is 48.5A, 1.5A. Thus, almost only phase one provides total load power. The simulation waveform of two-phase common inductor LLC converter is shown in Fig.5 (b).The resonant current and load current are been shared.

As output capacitor are connected together in industry product, it is hard to get the load current of each phase. However, the resonant current of each phase can also influence the current sharing performance. Thus, the resonant current sharing error can be used to estimate circulating current performance based on Fig.5.

2*300W LLC converter is built to verify of proposed method. To compare the different component tolerances. The circuit diagram is shown in Fig.5 (a). The prototype parameters are shown in Table 2.

Tab.2 Prototype parameters

Switching frequency	180kHz-270kHZ
Input Voltage	340V-400V
Output Voltage	12V
Output Power	300W × 2
Transformer Ratio n	20:1
Output Capacitance	1790μF
Series Capacitance(Cr)	12nF +5%
Resonant Inductance(Lr)	22.5μH(Phase1) 24.5μH(Phase2)
Leakage Inductance(Le)	6μH(Phase1) 6.5μH(Phase2)
Magnetizing Inductance(Lm)	95μH(Phase1) 92μH(Phase2)

Fig.6 show simulation waveforms of two-phase independent LLC converter without current sharing at 15A, 25A total load. Actually, the designed rated current value is

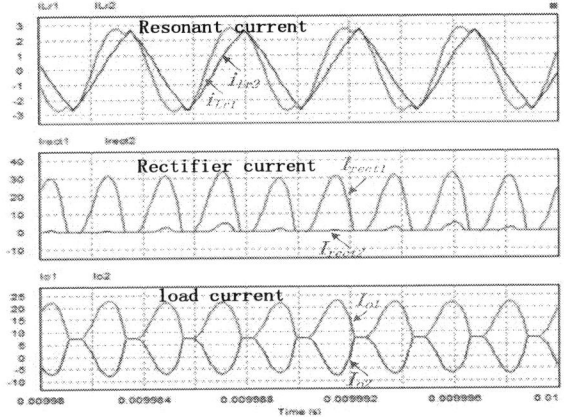

(a) Steady state at 15A load

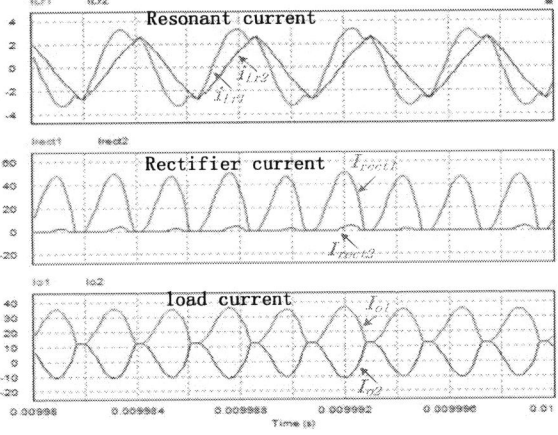

(b) Steady state at 25A load

Fig. 6 simulation waveform of two-phase independent LLC converter

25A for each phase, which means that it doesn't provide 50A for two phase parallel converter. In other words, when the total load current is larger than 25A, the second phase loads current will exceed the rated current according to Fig.3.To escape the overcurrent of each phase, the total maximum 25A current experiment is done without current sharing technology method. As the output voltage has switching frequency ripple, The load current I_{o2} has a high frequency ripple to charge or discharge the output capacitor C_{o2}. Thus, it has negative high frequency current or positive high frequency current. The average load current average value is zero. Thus, only phase one provides the load power.

Fig.7 shows simulation waveforms of two-phase common inductor LLC converter at 15A, 25A, 50A total load.

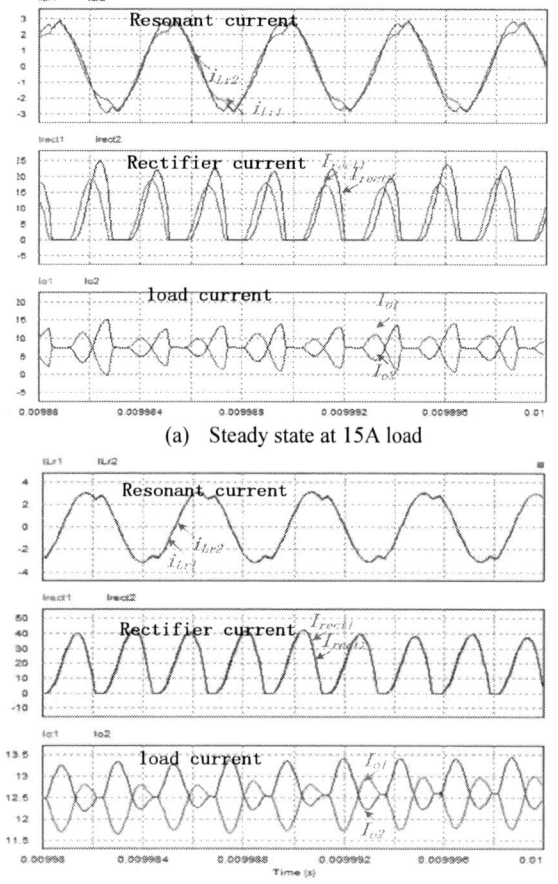

(a) Steady state at 15A load

(b) Steady state at 25A load.

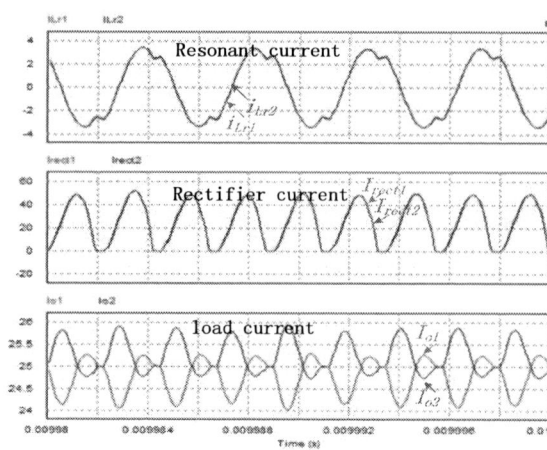

(c) Steady stat at 50A load

Fig.7 simulation waveform of two-phase common inductor LLC converter

The load current difference is reduced from 15A to 3A between Fig.6 (a) and Fig.7 (a). The load current difference is reduced from 25A to 0.5A between Fig.6 (b) and Fig. 7(b). Fig.7 (c) shows the good load sharing for total 50A load. As the output voltage has switching frequency ripple, The load current I_{o2} has a high frequency ripple to charge or discharge the output capacitor C_{o2}. Thus, it has negative high frequency current or positive high frequency current. The average load current average value is almost same in Fig.7 (b), (c). Thus, almost two phase units provide same load power.

The resonant current, rectifier current are almost same for two phases. Thus, the load current is shared by two phases. It is believed that good resonant inductor current sharing guarantees good load current sharing as indicated according to Fig.6, Fig.7. It is hard to sense the load current of two phase because output capacitor are connected together closely. The resonant current sharing error is used to estimate the circulating current performance in experiment results.

Fig.8 shows the experiment waveform of two-phase conventional LLC converter. Channel 1 is the output voltage. Channel 3, channel 4 are the resonant current of two phases. The resonant current i_{Lr1} is almost triangulate waveform, which means phase one almost doesn't provide the power for output load. Fig.9 shows the experiment waveform of two-phase proposed LLC converter. The resonant current i_{Lr1} and i_{L2} is almost same. A very small angle difference between them is shown at different load.

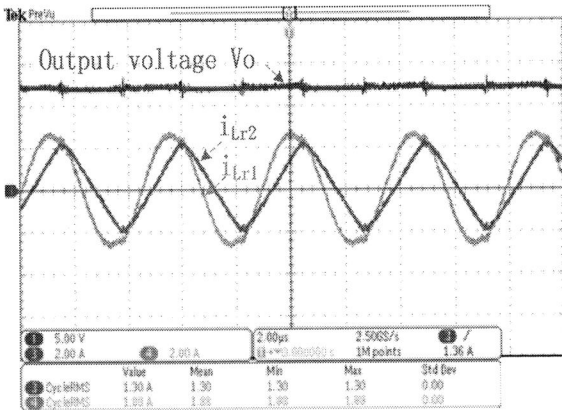

(a) Steady state at 180W load

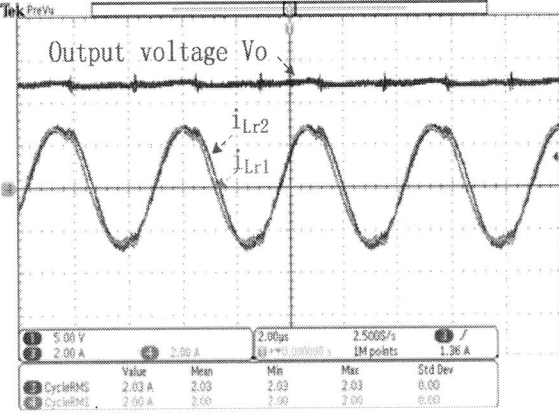

(b) Steady state at 300W load

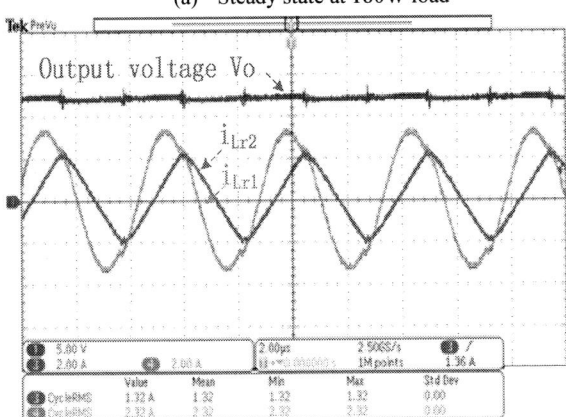

(b) Steady state at 300W load

Ch1: output voltage; Ch3: resonant current of phase two;
Ch4: resonant current of phase one.

Fig.8 experiment waveform of two-phase conventional LLC converter

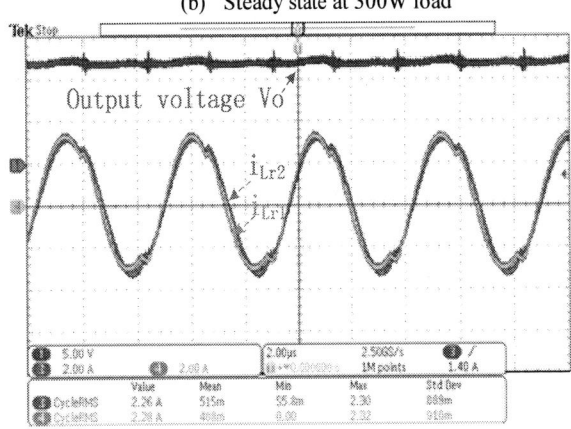

(c) Steady state at 600W load

Ch1: output voltage; Ch3: resonant current of phase two;
Ch4: resonant current of phase one.

Fig.9 experiment waveform of two-phase proposed LLC converter

To express circulating current according to Eq. (11), the resonant current and resonant current sharing errors are shown in Fig.10, Fig.11.

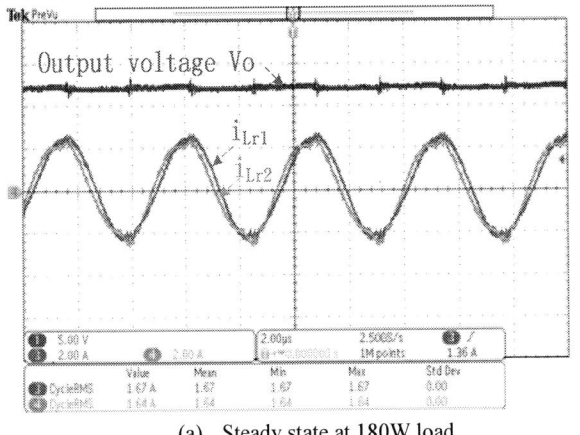

(a) Steady state at 180W load

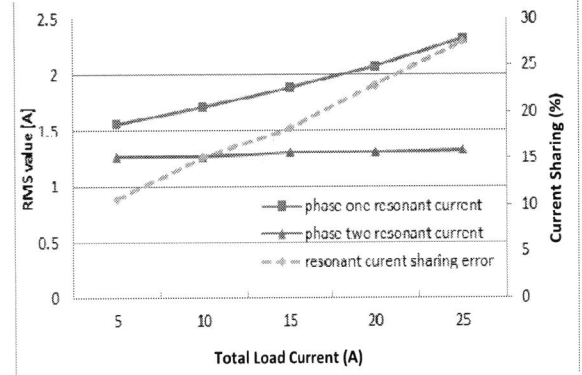

Fig.10 resonant current of two-phase conventional LLC converter

978-1-4673-9551-9/16 $31.00 © 2016 IEEE

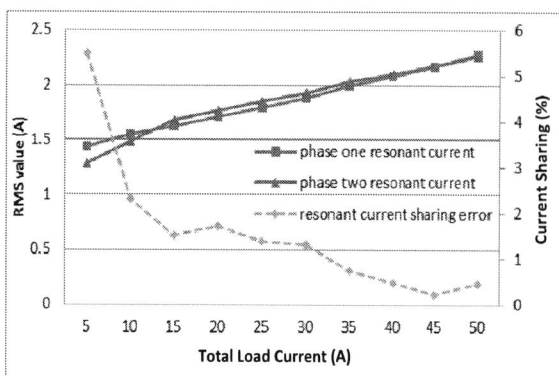

Fig.11 resonant current of two-phase proposed LLC converter

The resonant current sharing error increases from 10% to 28% with load power from 5A to 25A according to Fig.10. The resonant current sharing error is reduced from 2.3% to 0.44% when load power changes from 5A to 50A based on Fig.11.

IV. CONCLUSION

An analysis algorithm to analyze the circulating current is proposed for multi-phase LLC converter. The equivalent ac resistor voltage is of the same magnitude but different angle based on FHA. The detailed analysis according to the proposed method shows that the worse situation is same direction of deviation between resonant capacitor and magnetizing inductance (one is increased, the other is reduced). There is significant reduction from two-phase independent LLC converter to two-phase common inductor LLC converter according to analysis algorithm. Two-phase LLC converter prototype with 300W each phase is built to verify the effectiveness of proposed analysis method. Simulation and experiment results shows that resonant current sharing error is reduced from 28% to 0.44% when load power changes is 50A.This analysis method can be extended any phases that is more than two, and any resonant converters, such as series resonant converter, parallel resonant converter, LCC and so on.

References

[1] Y. Bo, "Topology Investigation for Front End DC/DC Power Conversion for Distributed Power System," Virginia Polytechnic Institute and Stage University, 2003.

[2] Y. Z. Y. Zhang, D. X. D. Xu, M. C. M. Chen, Y. H. Y. Han, and Z. D. Z. Du, "LLC resonant converter for 48 V to 0.9 V VRM," 2004 IEEE 35th Annu. Power Electron. Spec. Conf. (IEEE Cat. No.04CH37551), vol. 3, 2004.

[3] M. T. Zhang, M. M. Jovanović, and F. C. Y. Lee, "Analysis and evaluation of interleaving techniques in forward converters," IEEE Transactions on Power Electronics, vol. 13, no. 4. pp. 690–698, 1998.

[4] R. Hermann, S. Bernet, Y. Suh, and P. K. Steimer, "Parallel connection of integrated gate commutated thyristors (IGCTs) and diodes," IEEE Trans. Power Electron., vol. 24, no. 9, pp. 2159–2170, 2009.

[5] J. Rabkowski, D. Peftitsis, and H. P. Nee, "Parallel-operation of discrete SiC BJTs in a 6-kW/250-kHz DC/DC boost converter,"

IEEE Trans. Power Electron., vol. 29, no. 5, pp. 2482–2491, 2014.

[6] Z. Hu, Y. Qiu, Y. F. Liu, and P. C. Sen, "An interleaving and load sharing method for multiphase LLC converters," Conf. Proc. - IEEE Appl. Power Electron. Conf. Expo. - APEC, no. 1, pp. 1421–1428, 2013.

[7] H. Figge, T. Grote, N. Froehleke, J. Boecker, and P. Ide, "Paralleling of LLC resonant converters using frequency controlled current balancing," PESC Rec. - IEEE Annu. Power Electron. Spec. Conf., pp. 1080–1085, 2008.

[8] B. C. Kim, K. B. Park, and G. W. Moon, "Analysis and design of two-phase interleaved LLC resonant converter considering load sharing," in 2009 IEEE Energy Conversion Congress and Exposition, ECCE 2009, 2009, pp. 1141–1144.

[9] E. Orietti, P. Mattavelli, G. Spiazzi, C. Adragna, and G. Gattavari, "Current sharing in three-phase LLC interleaved resonant converter," 2009 IEEE Energy Convers. Congr. Expo. ECCE 2009, pp. 1145–1152, 2009.

[10] E. Orietti, P. Mattavelli, G. Spiazzi, C. Adragna, and G. Gattavari, "Analysis of multi-phase LLC resonant converters," 2009 Brazilian Power Electron. Conf. COBEP2009, pp. 464–471, 2009.

[11] B. C. Kim, K. B. Park, C. E. Kim, and G. W. Moon, "Load sharing characteristic of two-phase interleaved LLC resonant converter with parallel and series input structure," 2009 IEEE Energy Convers. Congr. Expo. ECCE 2009, pp. 750–753, 2009.

[12] E.Kim, K.Lee, B. Chung, "A novel topology of LLC resonant converter with two resonant tanks for power conditioning system," IEEE Annu. Power Electron. Spec. Conf., pp. 1698–1703, 2010.

[13] H. Wang, Y. Chen, Y. Liu, J. Afsharian and Z. Yang."A common inductor multi-phase LLC resonant converter," 2015 IEEE Energy Convers. Congr. Expo. ECCE 2015, pp. 548-555, 2015.

978-1-4673-9551-9/16 $31.00 © 2016 IEEE

A Simple Active Damping Method for Active Power Filters

Huawei Yuan, Xinjian Jiang
State Key Laboratory of Power System
Tsinghua University
Beijing, China
yhw13@mails.tsinghua.edu.cn

Abstract—Active Power Filters (APFs) are widely applied devices in the power quality field for control flexibility and effective attenuation of the harmonic, reactive and negative components of the grid current and voltage. As the interface connecting an APF with the grid, an LCL-filter is preferred over an L-filter for its smaller size and lower price. However, the LCL-filter is a third-order system and hence causes stability problems by its resonance phenomenon. To suppress resonances, either active or passive damping methods should be conducted. Passive methods are simple but result in more energy loss and reduce system's efficiency. Meanwhile, active methods result in no energy loss but often require more sensors and complicate the controller. In this paper, a simple active damping method for APFs without additional sensors is proposed. The method is analyzed theoretically and verified with simulations and experiments.

Keywords—Active Power Filters; LCL-filters; active damping; without additional sensors; current feedback

I. INTRODUCTION

As the information technologies develop at a rapid rate especially in recent decades, more and more electrical loads sensitive to power quality, like different kinds of electronic equipment, instruments and computer systems, are widely used in the power system. Those electrical loads demand a power supply of high quality. The appliance of Active Power Filters (APFs) is an effective way of improving power quality for its high compensation precision and fast response. In an APF system, the technique of pulse width modulation (PWM) are applied and hence brings the switching frequency component into the output voltage and current of the APF. Therefore, an interface filter is required to interconnect the APF and the power grid. Generally, an LCL-filter is used as the interface filter to mitigate the switching component caused by the PWM technique owing to its better attenuation ability in high frequency over an L-filter. As a result, using an LCL-filter can decrease the inductance of the interface filter, in contrast to an L-filter, and thereby save the cost and the size of the system, and improve the dynamic response.

Despite the advantages of the LCL-filter, a severe drawback should be considered, that is, as a third-order system, the LCL-filter causes the stability problem by its resonant nature. The stability problem in the system must be resolved

for the operation of the APF. Many methods are carried out to suppress the resonances, categorized as active damping methods (ADMs) and passive damping methods (PDMs). PDMs simply add a series or parallel resistor to the inductor or the capacitor of the LCL-filter and are relatively straightforward [1]-[3]. The simplicity of the method still widely used in industrial applications, but it increase energy loss of the system and sacrifices the efficiency and some attenuation abilities in high frequency. Conversely, ADMs increase system's damping in resonant frequency through some additional control algorithms instead of using any resistors and thus do not have the weaknesses of PDMs. Although additional algorithms, to a certain extent, result in more complexity in control, the advantages of ADMs attract scholars to conduct various studies on ADMs.

There are two commonly mentioned ADMs known as the virtual resistor method and the capacitor current feedback method. The former analyzes the LCL-filter with real resistors and obtains similar damping effect by transforming the resistors into the controller. Since no real resistors are installed, the damping made through control is called virtual resistor [4]-[6]. The latter detects the filter capacitor current and uses it as a feedback inside the main current control loop to mitigate the resonances [7]-[8]. Both methods require additional sensors and hence increase the cost of the system. In addition to those two methods, other ADMs have been proposed to keep the system with an LCL-filter in stable operation. [9] proposes an ADM based on the capacitor voltage feedback with a Lead-Lag correction module in feedback path. The method takes processing delay into consideration, but the system is analyzed in a continuous model and the Lead-Lag module is not easy to design. [10] comes up with an ADM using a series correction module and do not need any additional sensors. The parameter of the series correction module is optimized online by genetic algorithm, which is too complex to be practically used in industrial application at the present stage. [11] proposes a line current derivative feedback based ADM. This method requires the information of the grid-side output current and thus at least two more current sensors should be installed. As it is claimed in [11], time delay of the system can undermine the effectiveness of the method. [12] uses the high frequency components of the filter capacitor voltage as the feedback to damp the resonances caused by the LCL-filter. This method adds no sensors, but the damping effect relies on the

978-1-4673-9551-9/16 $31.00 © 2016 IEEE

parameters of the LCL-filter, and the time delay is not considered. [13]-[15] introduce some other ADMs for PWM converters without adding any current or voltage sensors. However, more should be taken into consideration when those methods are applied in APF cases due to wider bandwidth of the current loop in APF system. In this paper, a simple ADM is proposed for APFs and requires no additional sensors. Simulations and experiments are conducted to verify the validity of the proposal.

II. PRINCIPLE OF ACTIVE POWER FILTERS WITH AN LCL-FILTER

A. Topology of the Active Power Filter System

The topology of an APF system constituted by the power grid, the APF and the electrical load is shown in Fig.1, where L_1, L_2 and C stands for the grid-side inductor, the APF-side inductor and the filter capacitor of the LCL-filter respectively. L_g is the equivalent inductor of the power grid and C_d is the DC-link capacitor of the APF. The APF consists of a three-phase bridge and a DC-link capacitor. Both the APF and the electrical load are connected to the grid at the common point (CP). When the APF is working, it detects the harmonic, reactive and negative components of the load current and injects a neutralizing current into the grid so that only the active current enters the grid.

The single phase equivalent circuit of the system can be acquired as shown in Fig.2, where the load is represented by a current source. Variables used in this paper and their directions are marked in Fig.2. u_g, u_{CP}, u_c and u_{APF} stand for the grid voltage, the CP voltage, the filter capacitor voltage and the output voltage of the APF respectively. i_1, i_2 and i_c stand for the grid-side current, the APF-side current of the APF and the capacitor current. When analyzing the system model, L_g is sometimes merged into L_1, but in a common power grid, L_g is rather small so that taking u_g as u_{CP} results in a tiny error which is tolerable. Therefore, this paper does not specifically differentiate the values of u_g and u_{CP}.

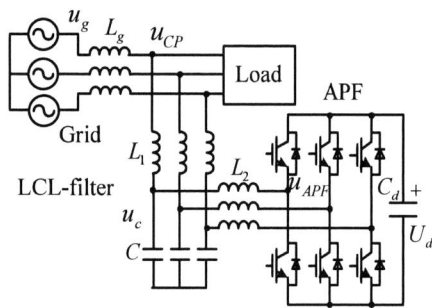

Fig. 1. Topology of the APF system constituted by the power grid, the APF and the electrical load

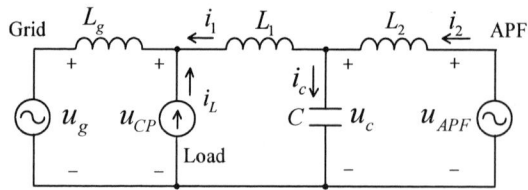

Fig. 2. Single phase equivalent circuit of the APF system

B. Modeling of the LCL-filter

The state variable equations of the LCL-filter can be obtained from the single phase equivalent circuit of the APF system as

$$\dot{x}(t) = Ax(t) + Bu(t) \tag{1}$$

where the vector x(t), u(t) and the matrix A, B are specified as

$$x(t) = \begin{bmatrix} i_1(t) & i_2(t) & u_c(t) \end{bmatrix}^T \tag{2}$$

$$u(t) = \begin{bmatrix} u_{APF}(t) & u_{CP}(t) \end{bmatrix}^T \tag{3}$$

$$A = \begin{bmatrix} 0 & 0 & 1/L_1 \\ 0 & 0 & -1/L_2 \\ -1/C & 1/C & 0 \end{bmatrix} \tag{4}$$

$$B = \begin{bmatrix} 0 & -1/L_1 \\ 1/L_2 & 0 \\ 0 & 0 \end{bmatrix} \tag{5}$$

The block diagram of the LCL-filter can be derived from (1)-(5) and is shown in Fig.3, from which it is easy to obtain the transfer function of the LCL-filter as the following equation

$$G_{LCL}(s) = \frac{I_1}{U_{APF}} = \frac{1}{L_1 L_2 C s^3 + (L_1 + L_2)s} \tag{6}$$

It is obvious that the LCL-filter is a three-order system with a pole at the origin and a pair of conjugate complex poles at the imaginary axis. The imaginary axis coordinates of the conjugate poles correspond to the resonant frequency of the LCL-filter, which is shown in (7).

$$f_{res} = \frac{1}{2\pi} \sqrt{\frac{L_1 + L_2}{L_1 L_2 C}} \tag{7}$$

The bode diagram of the LCL-filter is shown in Fig.4, where an L-filter with equal total inductance is also displayed for comparison.

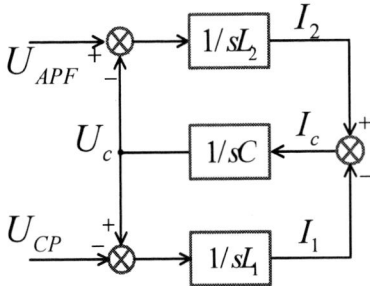

Fig. 3. Block diagram of the LCL-filter

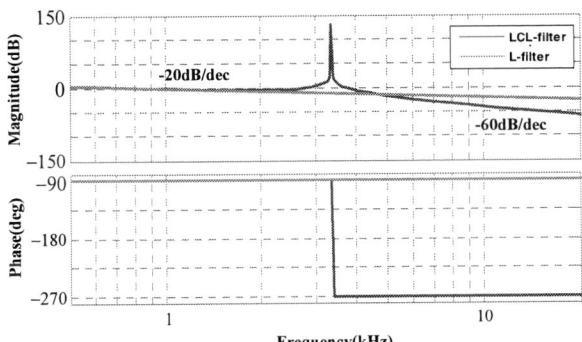

Fig. 4. Bode diagram of the LCL-filter

The bode diagram of the LCL-filter shows that the LCL-filter and the L-filter share similar characteristics in low frequency due to the high impedance of the capacitor to a low frequency signal, whereas the LCL-filter has better attenuation ability (-60dB/dec) over the L-filter (-20dB/dec) in high frequency owing to the higher order of the LCL-filter. Nonetheless, Fig.4 obviously indicates a resonant peak of the LCL-filter at the frequency of f_{res}, which severely undermines the system's stability. To acquire a stable system, damping methods, either passive ones or active ones, are required.

C. Control Scheme of the Active Power Filter

For an APF system, there are two control goals, that is, a stable DC-link voltage and an output current that well tracks the reference. In order to achieve the control goals, a two-loop control scheme is always adopted with an inner loop to control the APF current and an outer loop to stabilize the DC-link voltage. In such a control scheme, either an APF-side current or a grid-side current can be chosen for APF control. However, it can be learned from Fig.4 that the resonant peak remains a positive magnitude near the phase of -180 degree, so the system is instable if the grid-side current is chosen as the controlled variable without any damping methods. By contrast, taking the APF-side current as a feedback sets the resonant part out of the control loop and contributes to the system's stability. Therefore, in this paper, the controlled current in the inner loop is specified as the APF-side current, rather than as the grid-side current. The block diagram of the two-loop APF control scheme is shown in Fig.5.

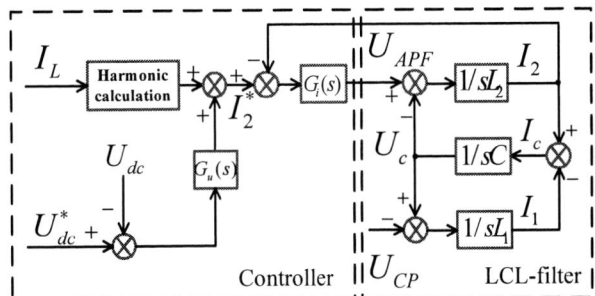

Fig. 5. Block diagram of the two-loop APF control scheme

III. ACTIVE DAMPING METHOD

Since PDMs weaken the performance of the LCL-filter and undermine the system's efficiency, ADMs are adopted and discussed in this section. Considering the time delay of the digital controller, a discrete model of the system should be analyzed. The continuous model displayed in (1)-(5) is discretized as (8)-(12), where T stands for the sampling period of discretization. The third power of T and above are ignored because the sampling interval is quite a short period of time.

$$x(k) = e^{AT}x(k-1) + \int_0^T e^{At}dt \cdot Bu(k) = Cx(k-1) + Du(k) \quad (8)$$

$$x(k) = \begin{bmatrix} i_1(k) & i_2(k) & u_c(k) \end{bmatrix}^T \quad (9)$$

$$u(k) = \begin{bmatrix} u_{APF}(k) & u_{CP}(k) \end{bmatrix}^T \quad (10)$$

$$C = \begin{bmatrix} 1 & 0 & T/L_1 \\ 0 & 1 & -T/L_2 \\ -T/C & T/C & 1 \end{bmatrix} \quad (11)$$

$$D = \begin{bmatrix} 0 & -T/L_1 \\ T/L_2 & 0 \\ T^2/L_2C & T^2/L_1C \end{bmatrix} \quad (12)$$

The block diagram of the discretized LCL-filter model is displayed in Fig.6, where the time delay of the digital controller is considered. Tc represents the time of the control cycle and k is the number of control cycles of the delay.

From the discrete model of the LCL-filter, the following equations can be obtained

$$I_2(z) = H_1(z)U_{APF}(z)z^{-kT_c/T} + H_2(z)U_{CP}(z) \quad (13)$$

$$H_1(z) = \frac{TL_1C + T^3 - (2TL_1C + L_1T^3/L_2)z^{-1} + (T_1L_1C + L_1T^3/L_2)z^{-2}}{L_1L_2C + T^2L - (3L_1L_2C + T^2L)z^{-1} + 3L_1L_2Cz^{-2} - L_1L_2Cz^{-3}} \quad (14)$$

$$H_2(z) = \frac{T^3(1 + z^{-1} - z^{-2})}{L_1L_2C + T^2L - (3L_1L_2C + T^2L)z^{-1} + 3L_1L_2Cz^{-2} - L_1L_2Cz^{-3}} \quad (15)$$

$$L = L_1 + L_2 \qquad (16)$$

The bode diagram of the open-loop system from U_{APF} to I_2 is shown in Fig.7. If there is no time delay in the system, the phase corresponding to the positive resonant peak is between -90 degree and 90 degree. It can be inferred that the feedback of I_2 in the inner control loop can suppress the peak and be conducive to the system's stability. However, the time delay caused by the digital controller cannot be avoided, and the phase characteristic drastically changes as the time delay gets longer. Generally, if the controller is well designed, it is typical that the time delay of the digital controller equals to one control cycle, so the delay of one control cycle is discussed in this paper.

Fig.7 indicates that the phase of the positive peak is near -180 degree in this case, which means that the system's stability is threatened by the resonance due to the time delay. To keep the system stable during its operation and, meanwhile, to avert the undesirable use of additional sensors, an ADM based on the APF current feedback is proposed in this paper. In order to limit the complication of the controller, a simple proportion coefficient K_D is used in the feedback path of the APF current as shown in Fig.8. The open-loop discrete system function from U_{in} to I_2 is given with (17).

$$H_3(z) = \frac{I_2(z)}{U_{in}(z)} = \frac{H_1(z)z^{-T_c/T}}{1 - H_1(z)K_D z^{-T_c/T}} \qquad (17)$$

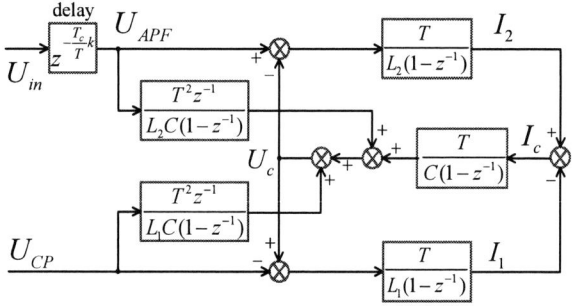

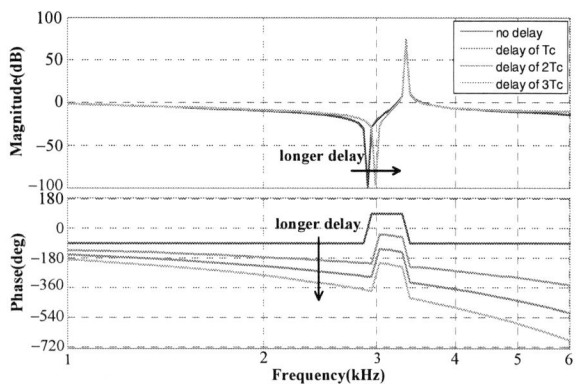

Fig. 6. Block diagram of the discretized LCL-filter

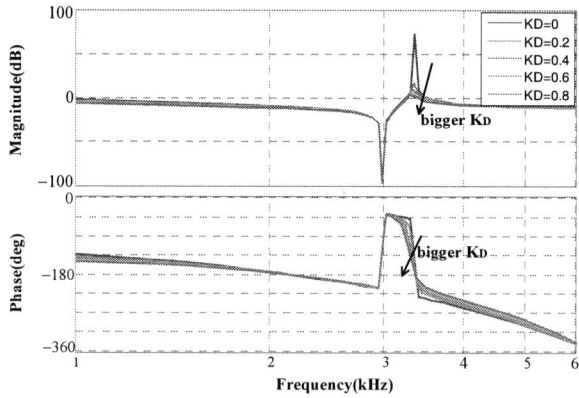

Fig. 7. Bode diagram of the discretized open-loop system

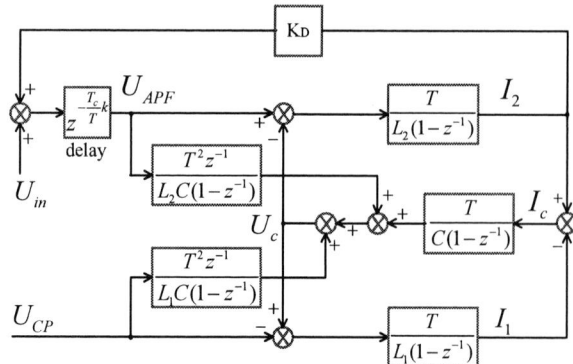

Fig. 8. Block diagram of the APF current feedback based ADM

According to the discrete system function, the bode diagram from U_{in} to I_2 can be derived as it is shown in Fig.9, where the effect of different coefficient K_D in the feedback path is displayed. It indicates that the APF current feedback is beneficial to the suppression of the positive resonant peak and hence makes the system more stable, and that a bigger K_D results in a more stable system. In a real system, the system's parameters can possibly be imprecise and keep changing during its operation, and thus the exact resonant frequency of the system may not be acquired. However, it is not a problem in this method because it is unnecessary to detect the resonant frequency of system.

Nevertheless, it should be noticed that the current feedback path simultaneously has influence on the whole spectrum of the current loop, both in high frequency and in low frequency, so some compensation measures need to be taken in low frequency band and a relatively small K_D is preferred. The control scheme of the APF system with the proposed ADM is shown in Fig.10.

IV. SIMULATION AND EXPERIMENT VERIFICATION

A. Simulation Results

Fig. 9. Effect of the APF current feedback with the time delay set as one control cycle

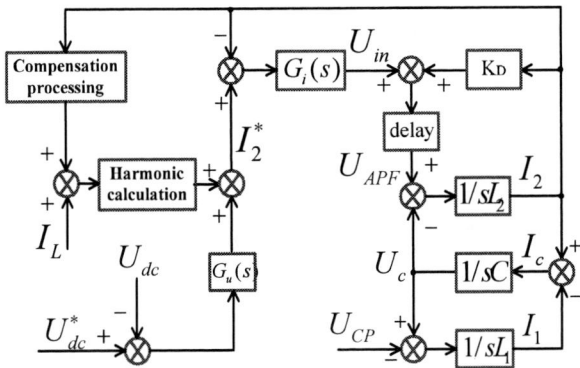

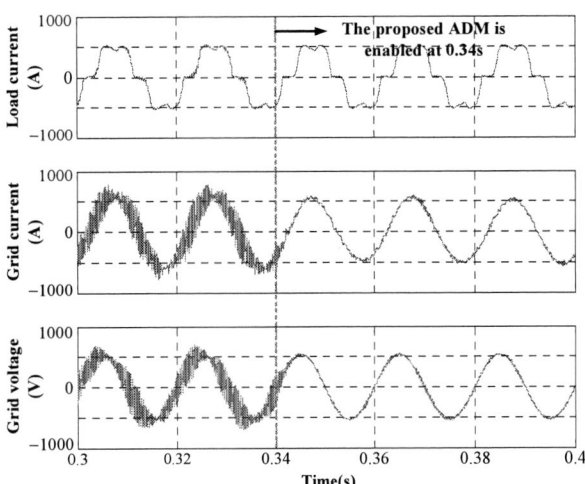

Fig. 10. Block diagram of the APF system control scheme with the proposed ADM

Fig. 11. Simulation Results of the proposal

To verify the proposed ADM, an APF system model is constructed with Matlab/Simulink. A three-phase diode rectifier, with a resistor and a small inductor on its DC side, is chosen as the nonlinear load. The control scheme with the proposed ADM shown in Fig.10 is adopted in the APF control. In the simulation model, the time delay of the digital controller is considered and set as one control cycle. The simulation parameters are listed in TABLE 1. The feedback coefficient K_D of the ADM is chosen as 0.5.

The result of the simulation is displayed in Fig.11 where the three waveforms stand for the load current, the grid current and the grid voltage respectively. Since the system is three-phase symmetrical, only the waveform of one single phase is displayed. It can be seen from Fig.11 that the load current contains a great amount of harmonics, but the grid current is almost sinusoidal owing to the compensation function of the APF. However, the resonance phenomenon caused by the LCL-filter is very obvious before 0.34s because no ADMs are adopted. When the proposed method is enabled at 0.34s, the resonance is quickly mitigated within 5ms. The simulation result indicates the validity of the proposal.

TABLE I.   PARAMETERS IN SIMULATIONS AND EXPERIMENTS

Parameter	Value
Grid line-to-line voltage	380 V
Rated APF current	100 A
Switching frequency	12.8kHz
DC-link voltage	780V
DC-link capacitor	10000uF
APF-side inductor	0.15mH
Grid-side inductor	0.05mH
Filter capacitor	60uF

B. Experiment Results

Besides the simulation, experiments were conducted to verify the proposed ADM. The parameters were the same with those shown in TABLE 1. The ADM feedback coefficient was chosen as 0.3. For the safety reason and to avert the operation of the over-current protection, the comparative experiments were conducted on low current condition. In the comparative experiments, the APF was controlled to produce a reactive current without and with the proposed method. Fig.12 shows the results of the experiments. The result illustrates that the resonance of the system is damped by the proposed ADM obviously. After the comparative experiments, the APF control scheme along with the proposed ADM was tested in the APF function experiment. In the APF function experiment, the APF was controlled to compensate the harmonics of a nonlinear load, which was a three-phase diode rectifier with a resistor on its DC side. The APF was operating under the rated condition and the result of the function experiment is shown in Fig.13, where the three waveforms represent the load current, the APF current and the grid current respectively. Fig.13 indicates that the APF with the proposed control scheme well compensated the harmonics of the nonlinear load and nearly no resonance phenomenon was found in the grid current. The total harmonic distortion (THD) of the load current is 23.5% whereas that of the grid current is 3.5%, due to the compensation of the APF (with only the harmonics of 25th and lower orders being calculated). The results of the experiment verify the effectiveness of the proposed method for APFs.

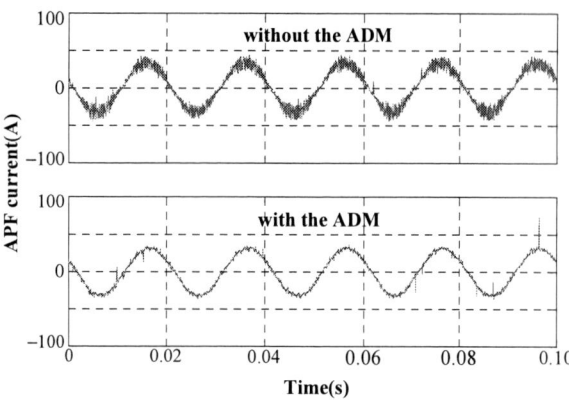

Fig. 12. Results of the comparative experiments

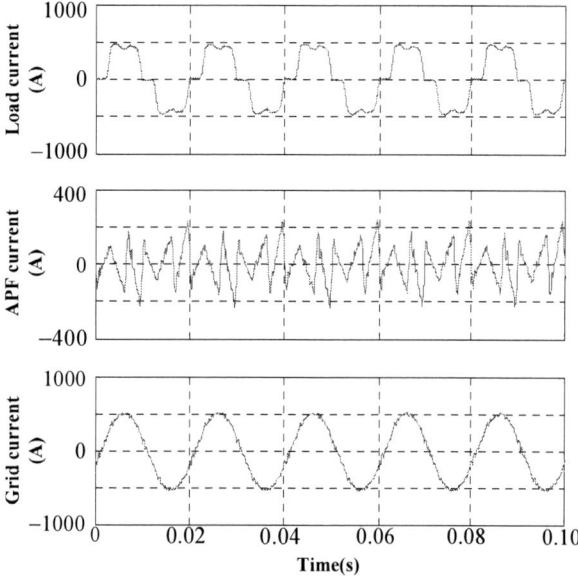

Fig. 13. Compensation result of the APF with the proposed ADM

V. CONCLUSION

This paper analyzes the model of an APF system with an LCL-filter. The frequency domain based analysis indicates the resonance nature of the LCL-filter which can cause stability problems. By discretizing the model of the system, the time delay of the digital controller is taken into consideration and a simple method is proposed to damp the resonance. The method is based on the APF current feedback and a simple proportion coefficient is used in the feedback path. The proposed method requires no additional sensors and avoids

complicating the APF control. Simulations and experiment results verify the validity and effectiveness of the proposal.

REFERENCES

[1] M. Liserre, F. Blaabjerg, S. Hansen, "Design and control of an LCL-filter-based three-phase active rectifier," IEEE Transactions on Industry Applications, vol. 41, no. 5, pp. 1281-1291, Sept. 2005.

[2] W. Wu, Y. He, T. Tang, F. Blaabjerg, "A new design method for the passive damped LCL and LLCL filter-based single-phase grid-tied inverter," IEEE Transactions on Industrial Electronics, vol. 60, no. 10, pp. 4339-4350, May 2013.

[3] C. Chen, Z. Wang, Y. Zhang, G. Li, Y. Wu, "A novel passive damping LCL-filter for active power filter," in IEEE 2014 International Transportation Electrification Conference Asia-Pacific, 2014, pp. 1-5.

[4] P. A. Dahono, "A control method to damp oscillation in the input LC filter," in IEEE 2002 Power Electronics Specialists Conference, 2002, pp. 1630-1635.

[5] C. Wessels, J. Dannehl, F. W. Fuchs, "Active damping of LCL-filter resonance based on virtual resistor for PWM rectifiers—Stability analysis with different filter parameters," in IEEE 2008 Power Electronics Specialists Conference, 2008, pp. 3532-3538.

[6] S. Nuilers, B. Neammanee, "Control performance of active damp LCL filter of three phase PWM boost rectifier," in IEEE 2010 Electrical Engineering/Electronics Computer Telecommunications and Information Technology, 2010, pp. 259-263.

[7] E. Twining, D. G. Holmes, "Grid current regulation of a three-phase voltage source inverter with an LCL input filter," IEEE Transactions on Power Electronics, vol. 18, no. 3, pp. 888-895, May 2003.

[8] G. Zeng, T. W. Rasmussen, L. Ma, R. Teodorescu, "Design and control of LCL-filter with active damping for Active Power Filter," in IEEE 2010 International Symposium on Industrial Electronics, 2010, pp. 2557-2562.

[9] V. Blasko, V. Kaura. "A novel control to actively damp resonance in input LC filter of a three-phase voltage source converter," in IEEE Transactions on Industry Applications, vol. 33, no. 2, pp. 542-550, March 1997.

[10] M. Liserre, A. D. Aquila, F. Blaabjerg, "Genetic algorithm-based design of the active damping for an LCL-filter three-phase active rectifier," in IEEE Transactions on Power Electronics, vol. 19, no. 1, pp. 76-86, Jan. 2004.

[11] A. M. Hava, T. Lipo, W. L. Erdman, "Utility interface issues for line connected PWM voltage source converters: a comparative study," in IEEE 1995 Applied Power Electronics Conference and Exposition, 1995, pp. 125-132.

[12] M. Malinowski, M. P. Kazmierkowski, W. Szczygiel, S. Bernet, "Simple sensorless active damping solution for three-phase PWM rectifier with LCL filter," in IEEE 2005 Industrial Electronics Society Conference, 2005, pp.

[13] W. Yao, Y. Yang, X. Zhang, F. Blaabjerg, "Digital notch filter based active damping for LCL filters," in IEEE 2015 Applied Power Electronics Conference and Exposition, 2015, pp. 2399-2406.

[14] R. Pena-Alzola, M. Liserre, F. Blaabjerg, T. Kerekes, "Self-commissioning notch filter for active damping in three phase LCL-filter based grid converters," in IEEE 2013 European Conference on Power Electronics and Applications, 2013, pp. 1-9.

[15] C. P. Dick, S. Richter, M. Rosekeit, J. Rolink, R.W. De Doncker, "Active damping of LCL resonance with minimum sensor effort by means of a digital infinite impulse response filter," in IEEE 2007 European Conference on Power Electronics and Applications, 2007, pp. 1-8.

Simultaneous Voltage and Current Compensation of the 3-Phase Electric Spring with Decomposed Voltage Control

Shuo Yan[1], Tianbo Yang[1]

[1]Department of Electrical & Electronic Engineering
The University of Hong Kong
Hong Kong
Email: yanshuo@connect.hku.hk

C. K. Lee[1], Siew-Chong Tan[1], S.Y. Ron Hui[1,2]

[2]Department of Electrical & Electronic Engineering
Imperial College London
London, United Kingdom
Email: ronhui@eee.hku.hk

Abstract—A decomposed voltage control is proposed to integrate two favorable functions of the 3-ph electric spring (ES) to stabilize the mains voltage and to reduce power imbalance. As compared with its precedent counterpart, the proposed control can expand the usage of the 3-ph ES, simplify the control implementation, and enable the ES to conduct multiple tasks at one time. The decomposition of the 3-ph ES voltage into d and q components allows the 3-ph ES to conduct decoupled real and reactive current compensations. The further separation of the d component enables the 3-ph ES to simultaneously compensate the mains voltage and the real current of the critical load. The proposed method consists of three loops designed respectively for voltage regulation, real power balance, and reactive power compensation. Experimental results demonstrate that the multi-tasking of the 3-ph ES is made attainable by the proposed control, with current balancing and voltage regulation as the primary objectives and power factor correction as a favorable byproduct.

Keywords—Electric spring; power balance; voltage stabilization; decomposed voltage control

I. INTRODUCTION

The integration of Renewable Energy Sources (RESs) to power grids helps to relieve the reliance on fossil fuel for the production of electricity. However, the increasing penetration of RESs could destabilize the power grids and bring new challenges to power quality improvement. Recent report and survey show that when intermittent RESs reach a certain percentage of total power generation, the frequency and voltage regulations in power grids would become serious problems [1], [2]. So far no consent has been reached on the maximum penetration level of RESs, but the majority of conclusions have highlighted that the installed capacity of RESs must be carefully weighted.

To secure the operation of electrical loads in power grid, modified controls have been proposed to address the conceivable issues brought by the high penetration of RESs. Some controllers [3], [4] are designed to limit the maximum

This project is supported by the Hong Kong Research Grant Council under a Theme-based Project: T23-701/14-N.

feed-in power of RESs in order to avoid overloading. A considerable number of research findings [5]–[7] indicate that the RES system with reactive power control can help to regulate the grid voltage, which would enable inverter-based RESs to be more beneficial to the grid. Control strategies such as droop control are proposed to enforce the inverter-based RESs in sharing the task of voltage and frequency regulations [8]–[11]. Alternative publications speak highly of the installation of energy storage in the future grid systems, in spite of its immediate problems such as cost and lack of proper control [12]–[15]. An extensive list of applications of energy storage can be found. Examples include voltage support, frequency regulation, peak shaving, and power quality enhancement.

A close examination on these strategies can reveal their mutual attribute that the intermittent power of RES is targeted to be either shaved or stored. Thus, they can be categorized as the "supply-side management" with the incentive to control power generations so as to stabilize the power grid. However, the shift from traditional power generation to less predictable RESs leads to the need for a change of the control philosophy. Following the new control paradigm of the future smart grids that "load demand following power generation" [16], [17], various "demand-side management" techniques have been proposed. Examples include energy storage [18], [19], on and off load control [20], real time pricing [21], and electric vehicles (EVs) [22].

The Electric Spring (ES) has been introduced to solve various power quality issues including voltage and frequency instability [23]–[25], power quality improvement [26], [27], and power imbalance [28]. Reference [23] introduces the first version of ES with two operating modes to conduct voltage stabilization. Since the original control decomposes the mains voltage into two perpendicular vectors, the noncritical load voltage can only be reduced. The original control enables the ES to be more effective in supporting the mains voltage than in suppressing it. The second version of ES is associated with a dc voltage source (that can be a battery) and can provide both active and reactive power compensations [29]. The 3-ph ES proposed in [28] can reduce line current imbalance by compensating both real and reactive power and stabilize the mains voltage by compensating reactive power. However, its capability is not fully exerted due to the separate implementations of voltage stabilization and current balancing control. Additionally, the Genetic Algorithm (GA) based control system may not provide fast response to load changes.

Apart from using reactive power, the control scheme proposed in [30] utilizes the real power in regulating the mains voltage. The supply voltage of the noncritical load can be increased and decreased so that its power can be manipulated to perturb the voltage drop across the distribution line. Comparing with the first ES equipped with reactive power compensation [23], the ES with real power compensation [30] possesses an equal ability to support and suppress the mains voltage. Building on such advancement, this paper introduces a decomposed voltage control to facilitate the 3-ph ES to simultaneously address two crucial power system issues, namely the stabilization of mains voltage and the balance of line currents.

The adverse effect caused by voltage instability has been extensively investigated in previous studies [1], [2]. Voltage reductions can increase the loss in transmission line and transformers, while the increased voltage reduces the lifetime of equipment and brings excessive heating in loads. The induced light flicker causes general discomfort to individuals. Additionally, the power grid is endangered by power imbalance. Unbalanced line current causes uneven mains voltages among the three phases due to the asymmetric voltage drops. The consequential circulating currents further deteriorate the power quality. Increased power loss, excessive neutral current, increased investment in infrastructure, and decreased system efficiency are other possible impacts caused by power imbalance [28].

By integrating three control loops in a single decomposed voltage controller, the 3-ph ES is equipped with the new ability to simultaneously stabilize the mains voltage and balance the line current, with power quality improvement as a byproduct. The potential of the 3-ph ES is thus much exerted. Experimental results are provided to verify the intended functions of the proposed decomposed voltage control.

II. PRINCIPLE OF THE 3-PHASE ELECTRIC SPRING

A. Hardware Implementation

The architecture and implementation of the 3-ph ES discussed in [28] is adopted here and shown in Figure 1. The topology of the 3-ph ES takes the form of a 3-ph inverter in star connection. *LC* filters are used so that the 3-ph ES can operate as a voltage source. Three 1-ph isolation transformers are used to couple compensation voltages into grid. Their primary sides are connected to the output of the 3-ph ES, whilst the secondary sides are in series with the 3-ph noncritical load. The batteries on the DC link allow the 3-ph ES to exchange real power with the grid. Under this

implementation, the ES voltage can dynamically change the supply voltage of the noncritical loads to adapt their power for control purpose. In previous references, the ES has been successfully implemented to reduce the line current imbalance, stabilize the mains voltage, dampen the frequency fluctuation, and improve the power quality.

The configuration of a 3-ph inverter as an ES is unique and different from existing power electronic facilities such as active filter, STATCOM, and series Reactive Power Compensator. The series combination of the ES and the noncritical load makes it possible to change the power consumption of the noncritical load, while no such functionality can be found in other facilities. In this way, the noncritical load (when associated with an ES) becomes an adaptive/smart load that can participate in the regulation of power grid.

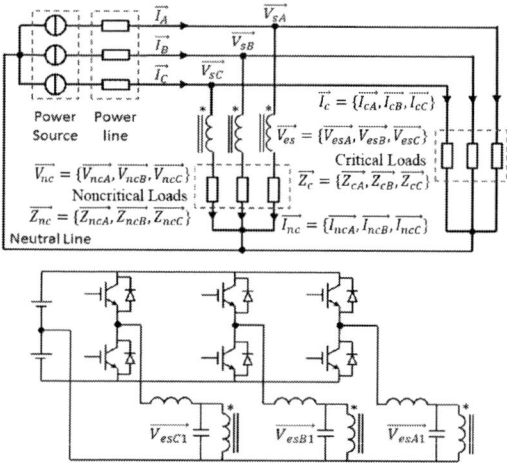

Fig. 1. Topology of a 3-ph ES [28].

B. Operating Principle

The operation of a 3-ph ES relies on its power dampening and boosting effect on the noncritical load. From the perspective of power compensation, the 3-ph ES possesses eight operating modes including four primary modes (resistive mode, negative-resistive mode, inductive mode, and capacitive mode) and four secondary modes (resistive plus inductive mode, negative-resistive plus inductive mode, resistive plus capacitive mode, negative-resistive plus capacitive mode). The specific operating mode of the 3-ph ES is determined by the phasor relationship between the ES voltage and the noncritical load current.

The 3-ph ES can work as a 3-ph symmetric voltage source or three 1-ph inverters. In particular, when operating to balance the line current, the 3-ph ES generates unbalanced compensation voltage to fill up the asymmetric power of the critical load. Each phase of the 3-ph ES possesses a particular operating mode. In the case of voltage stabilization, the ES voltage is fixed to be either 90° leading or lagging the noncritical load current. All three phases follow an identical compensation profile when only reactive power is provided by the ES. It should be noted that the 3-ph ES with this control method can only reduce the power of the noncritical load and thus shows a good ability to support the voltage while a limited one to suppress voltage.

The usefulness of a 3-ph ES is dependent on a few factors, including the power and voltage ratings of the 3-ph ES, the power rating of noncritical load and the turns ratio of isolation transformer. A few important points are listed here for better understanding: i) For the same power and voltage rating of 3-ph ES, a higher power rating of the noncritical load allows a larger compensation capacity; ii) with the same power rating of noncritical load, higher voltage and power ratings of the 3-ph ES give a larger compensation capacity; iii) Using a large step-up turns ratio of the isolation transformer would allow a reduction in the voltage rating of the ES.

III. UTILIZATION OF THE ES VOLTAGE

Examining the noncritical load current when the ES is in place can help to understand the principle of the 3-ph ES for current and voltage regulation. Based on the circuit diagram given in Fig. 1, the vector equation of the noncritical load current is given in (1). For simplicity of analysis, the noncritical load is assumed to be resistive. In a Cartesian coordinate system with a fundamental frequency of 50 Hz, (1) can be decomposed into $d-q$ form as in (2) and (3). For simultaneous voltage and current compensation, the d components of the ES voltages are further decomposed into (4).

$$\begin{cases} \vec{I}_{ncA} = \dfrac{\vec{V}_{sA} - \vec{V}_{esA}}{R_{ncA}} \\[2mm] \vec{I}_{ncB} = \dfrac{\vec{V}_{sB} - \vec{V}_{esB}}{R_{ncB}} \\[2mm] \vec{I}_{ncC} = \dfrac{\vec{V}_{sC} - \vec{V}_{esC}}{R_{ncC}} \end{cases} \tag{1}$$

Equations (3) and (4) indicate that the 3-ph ES can conduct multiple tasks at one time. The first part of the d component of the ES voltage ($V_{esA/B/C\_d\_v}$) is used to stabilize the mains voltage. This part of the voltage in each phase varies in the same fashion to change the noncritical load current so that the mains voltage can be supported and suppressed. The second part of the d component of the ES voltage in each phase ($V_{esA/B/C\_d\_i}$) changes independently to compensate for the unbalanced real current of the critical load. The q component of each ES voltage ($V_{esA/B/C\_q}$) generates reactive current to compensate the reactive current of the critical load. The function of $V_{esA/B/C\_q}$ is two-fold: (i) the reactive current of critical loads can be balanced, and (ii) the power factor of the

critical load can be improved when the 3-ph ES partially or fully provides reactive current for critical loads.

The idea of controlling the 3-ph ES to conduct both voltage and current compensation can be collectively understood as:

- $V_{esA/B/C\_d\_i}$ and $V_{esA/B/C\_q}$ in each phase give independent compensations to balance the real and reactive current of the critical load.

- $V_{esA/B/C\_q}$ improves the power factor of the critical load by providing reactive current.

- The balanced system is further compensated by $V_{esA/B/C\_d\_v}$ in a symmetric fashion. This part of the ES voltages boosts or reduces the power consumption of the noncritical load to match the fluctuating power of RESs.

$$\begin{cases} I_{ncA\_d} = \dfrac{V_{sA\_d} - V_{esA\_d}}{R_{ncA}} \\[2mm] I_{ncB\_d} = \dfrac{V_{sB\_d} - V_{esB\_d}}{R_{ncB}} \\[2mm] I_{ncC\_d} = \dfrac{V_{sC\_d} - V_{esC\_d}}{R_{ncC}} \end{cases} \tag{2}$$

$$\begin{cases} I_{ncA\_q} = \dfrac{V_{sA\_q} - V_{esA\_q}}{R_{ncA}} \\[2mm] I_{ncB\_q} = \dfrac{V_{sB\_q} - V_{esB\_q}}{R_{ncB}} \\[2mm] I_{ncC\_q} = \dfrac{V_{sC\_q} - V_{esC\_q}}{R_{ncC}} \end{cases} \tag{3}$$

$$\begin{cases} I_{ncA\_d} = \dfrac{V_{sA\_d} - (V_{esA\_d\_v} + V_{esA\_d\_i})}{R_{ncA}} \\[2mm] I_{ncB\_d} = \dfrac{V_{sB\_d} - (V_{esB\_d\_v} + V_{esB\_d\_i})}{R_{ncB}} \\[2mm] I_{ncC\_d} = \dfrac{V_{sC\_d} - (V_{esC\_d\_v} + V_{esC\_d\_i})}{R_{ncC}} \end{cases} \tag{4}$$

IV. DECOMPOSED VOLAGE CONTROL OF THE 3-PHASE ES

The theoretical analysis indicates the possibility of operating the 3-ph ES to conduct both voltage and current compensations. In this section, a detailed discussion on the controller is presented. PI control is used in the 3-ph ES with fast and accurate response. Fig. 2 shows the block diagram of the proposed control loops. The 3-ph power grid is handled as three 1-ph power grids with the consideration of 120° phase shift. In each phase, three PI control loops are set up for different control objectives. The orange loop conducts the task of stabilizing the mains voltage. It accepts the error of the d

978-1-4673-9551-9/16 $31.00 © 2016 IEEE

component of the mains voltage and uses the PI controller to generate the first part of ES voltage reference, $V_{esA/B/C\_d\_v\_ref}$.

The green loop carries out the task of balancing the real current of the critical loads. The real currents of the critical load in three phases are averaged to generate the real current reference of the noncritical load. The PI controller receives the error of the real current of the noncritical load and computes the second part of the ES voltage reference, $V_{esA/B/C\_d\_i\_ref}$.

The above two parts are added up to form the d component of the ES voltage reference, $V_{esA/B/C\_d\_ref}$. The blue loop shows the compensation of the reactive current of the critical load. Another PI controller is implemented here to derive the third part of the ES voltage reference, $V_{esA/B/C\_q\_ref}$. After the synchronization, the complete sinusoidal reference of ES voltage can be derived.

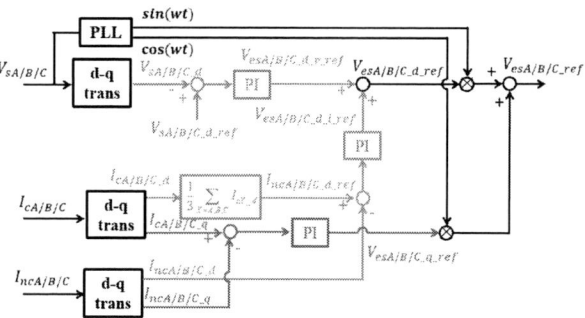

Fig. 2. The proposed decomposed voltage control of the 3-ph ES.

V. EXPERIMENTAL RESULTS

In this section, the usefulness of the decomposed voltage control of the 3-ph ES is practically illustrated. Two sets of experiments are conducted to test the operation of the 3-ph ES to provide both voltage and current compensation. In the first experiment, the 3-ph ES is programmed to balance the line current and support the mains voltage. Two control objectives are tested individually and then collectively. The second experiment repeats the function of the 3-ph ES with decomposed voltage control to balance the line current and tests its ability to suppress the main voltage. The individual and collective operations of both functions are considered.

A. Voltage Support and Current Balance

In this experiment, the functions of the 3-ph ES with decomposed voltage control to balance the line current and to support the mains voltage are verified. A small-capacity experimental bench is set up, where a 3-ph ES, a 3-ph asymmetric critical load, and a 3-ph symmetric noncritical load are included. The specifications can be found in TABLE I. The power supply is programmed to give a mains voltage at the point of common coupling (PCC) to be slightly lower than the nominal value ($|V_n|$ = 120 V). In the uncompensated system, the unbalanced loading of the critical load gives a neutral current of 0.52 A in RMS. The 3-ph line current in all three stages of operation can be found in TABLE II.

TABLE I. SPECIFICATIONS OF HARDWARE SETUP

Specifications of Loads			
	Phase A	**Phase B**	**Phase C**
Critical Load	68 // j 116 Ω (resistive plus inductive)	113 // j 116 Ω (resistive plus inductive)	81 // j 116 Ω (resistive plus inductive)
Noncritical Load	42 Ω (resistive)	42 Ω (resistive)	42 Ω (resistive)
Specifications of the Grid			
Power Source	Vg = 120 V (RMS)		
Transmission Line	0.3 Ω, 2.4 mH		

TABLE II. 3-PH LINE CURRENTS IN THREE STAGES

	Phase A	**Phase B**	**Phase C**
I Balance	4.3∠-5° A	4.2∠117° A	4.2∠236° A
V Support	2.9∠-18° A	2.1∠96° A	2.5∠219° A
I Balance & V Support	3.4∠-3° A	3.4∠118° A	3.3∠237° A

The 3-ph ES is firstly activated to balance the line current. For $t=0$ to 60 s, the mains voltage remains to be 2 V lower than the nominal value, as shown in Fig. 3. During the same time interval, the neutral current is reduced to 0.14 A, a 70% reduction from the uncompensated value, as given in Fig. 4. The effectiveness of 3-ph ES to balance the line current is further demonstrated by the 3-ph line current given in TABLE II. In the first stage of current balance ($t=0$ to 60 s), the RMS values of the 3-ph line currents match each other, and the phase angles are in a symmetric state with a phase difference between Phase A and B as 122° and the one between Phase B and C as 119°. The phase angles of the 3-ph line currents further indicate that the power factor of the load bank in each phase is 0.99. It can be further observed in the steady-state waveforms given in Fig. 5(a) that the supply voltage of noncritical load is deliberately tuned to be asymmetric. The change of the supply voltage of the noncritical load can actively increase or reduce the power of the noncritical load to make up for the unbalanced power of the critical load. Specifically in this example, the supply voltage of Phase A is reduced from 118.0 V to 111.7 V in RMS, the supply voltage of Phase B is increased from 118.0 V to 145.7 V in RMS, and the supply voltage of Phase C remains close to the mains voltage. In the meantime, the measurements in Fig. 6(a) reveal that the 3-ph ES changes the 3-ph noncritical load current to be leading (i.e. capacitive) so that the reactive power of the critical load comes from the 3-ph ES rather than the power supply.

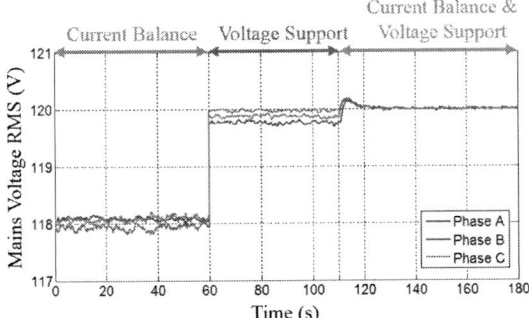

Fig. 3. The compensated mains voltage.

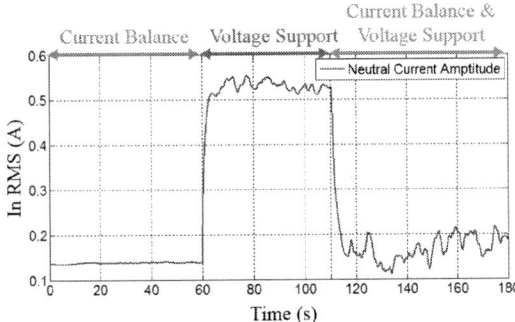

Fig. 4. The compensated neutral current.

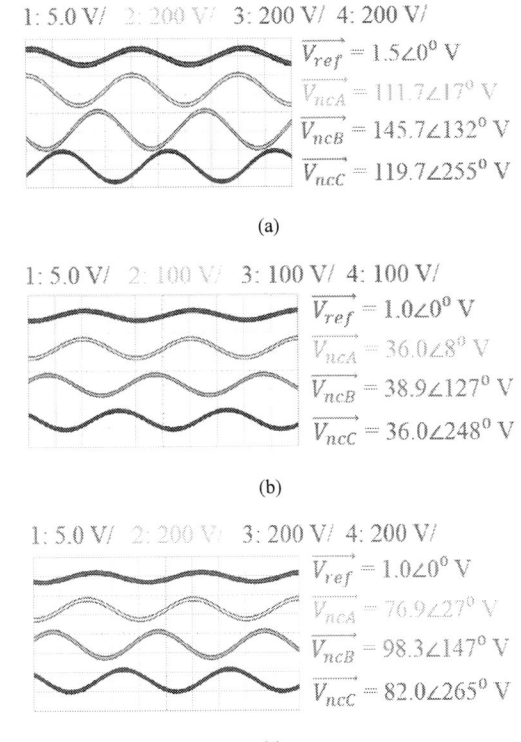

Fig. 5. The steady-state waveform of noncritical load voltages in (a) current balance, (b) voltage support, and (c) current balance and voltage support.

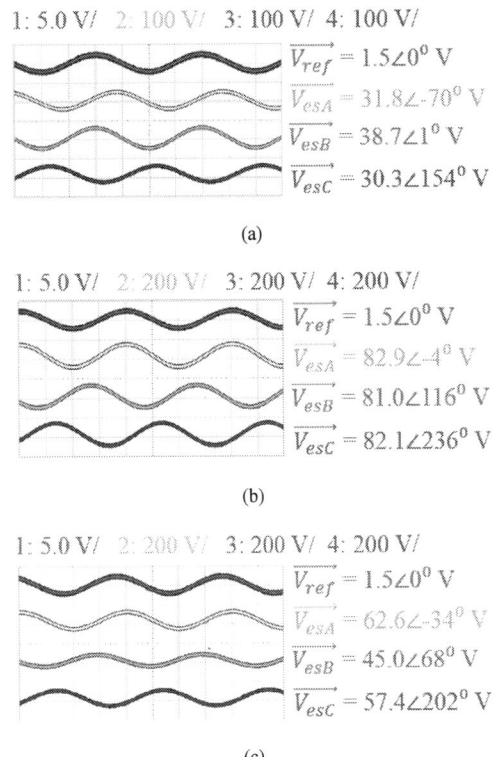

Fig. 6. The steady-state waveform of 3-ph ES volages in (a) current balance, (b) voltage support, and (c) current balance and voltage support.

In the second interval (t = 60 to 110 s), the function of current balance is turned off while the function of voltage support is activated. During the inverval, as shown in Fig. 3, the mains voltage in the three phases are all supported to the nominal value. The neutral current goes back to its undampened value ($|I_n|$ = 0.52 A), as shown in Fig. 4. In TABLE II, the 3-ph line current returns to the asymmetric state. The RMS values of the 3-ph line currents are no longer identical and the phase angles do not possess 120° phase shift. To support the mains voltage, the power of the noncritical load must be shedded so that the voltage drop across the distribution line can be reduced. Fig. 5(b) shows that the supply voltage of the 3-ph noncritical load in RMS is reduced down to 36.0 V, 38.9 V, and 36.0 V, giving a 70% power reduction. It should be noted that since the mains voltage is symmetric, the 3-ph ES exerts symmetric compensation to support the mains voltage. The steady-state measurements given in Fig. 6(b) show that the 3-ph ES no longer delivers reactive power to the critical load. The reduced power factor of the load bank indicates that power source provides the reactive power to critical load for t = 60 to 110 s.

In the final stage (t = 110 to 180 s), both the functions of current balance and voltage support are activated. The results in Fig. 3 show that the mains voltage remains to be compensated to its norminal value. Meanwhile, as shown in Fig. 4, for t = 110 to 180 s, the neutral current is reduced to

0.18 A. From the results given in TABLE II, it can be found that the 3-ph line currents have the same RMS value, and the phase angles show a 120° phase shift. The phase angles of the 3-ph line currents further indicate that power factor of the load band is compensated to be unity. The asymmetric loading of the noncritical load shown in Fig. 5(c) indicates that the 3-ph ES carries out compensation to support the mains voltage and simultaneously conducts extra compensation to trim the unbalanced critical load current. The 3-ph ES voltage given in Fig. 6(c) shows that the ES injects reactive power to the critical load to improve the power factor of the load bank and concurrently reduces the power of noncritical load to support the mains voltage.

B. Voltage Suppression and Current Balance

TABLE III.   SPECIFICATIONS OF HARDWARE SETUP

Specifications of Loads			
	Phase A	Phase B	Phase C
Critical Load	83 // j 116 Ω (resistive plus inductive)	98 // j 116 Ω (resistive plus inductive)	68 // j 116 Ω (resistive plus inductive)
Noncritical Load	48 Ω (resistive)	48 Ω (resistive)	48 Ω (resistive)
Specifications of the Grid			
Power Source	Vg = 124 V (RMS)		
Transmission Line	0.4 Ω, 3.6 mH		

TABLE IV.   3-PH LINE CURRENTS IN THREE STAGES

	Phase A	Phase B	Phase C
I Balance	4.1∠-2° A	4.2∠117° A	4.1∠238° A
V Suppress	5.6∠-11° A	5.4∠109° A	5.9∠230° A
I Balance & V Suppress	5.7∠-2° A	5.6∠119° A	5.8∠238° A

The second experiment demonstrates the ability of the 3-ph ES to suppress the mains voltage and balance the line currents. Specifications of hardware are given in TABLE III. The experiment consists of three stages. In the first stage, only the function of the current balance is activated. The 3-ph is switched on to compensate the unbalanced loading of the critical load. The second stage tests the use of 3-ph ES to suppress the mains voltage. The power source is programmed to lift the mains voltage to be higher than its nominal values ($|V_n|$ = 120 V) when the 3-ph ES is operated to bring down the voltage. In the third stage, the 3-ph ES is activated to simultaneously balance the line currents and suppress the mains voltage. The power factor of the load bank is still compensated to be unity as a favorable byproduct of the control. The 3-ph line currents in all three stages can be found in TABLE IV.

In the first stage, the mains voltage remains to be uncompensated as 122.5 V in RMS, given in Fig. 7. In the meantime, the results for t = 0 to 60 s shown in Fig. 8

demonstrate the effectiveness of the decomposed voltage control in mitigating the unbalanced condition. As shown in Fig. 8, when the 3-ph ES is activated, the neutral current is brought down from 0.43 A to 0.11 A in RMS, a 74% reduction. The values of the line currents in TABLE IV further validate the function of the 3-ph ES to balance 3-ph line currents. The 3-ph line currents show identical values in RMS and 120° phase shift. Besides, the phase angles of 3-ph line currents indicate the power factor of load bank is unity. The reallocation of the loading in all three phases further proves the usefulness of 3-ph ES in current balancing. As shown in Fig. 9(a), the supply voltage of the 3-ph noncritical load is changed to be asymmetric to compensate the unbalanced power. The measurements given in Fig. 10(a) show that the 3-ph ES operates to reallocate the power of the noncritical load and simultaneously to deliver reactive power to the critical load.

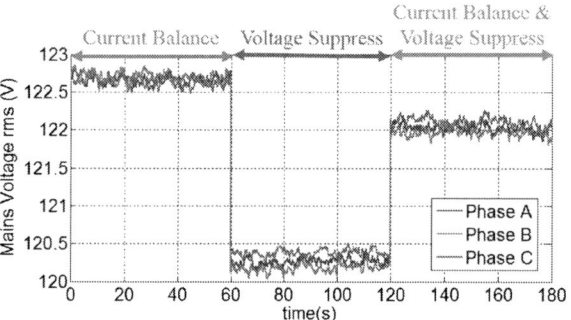

Fig. 7.   The compensated mains voltage.

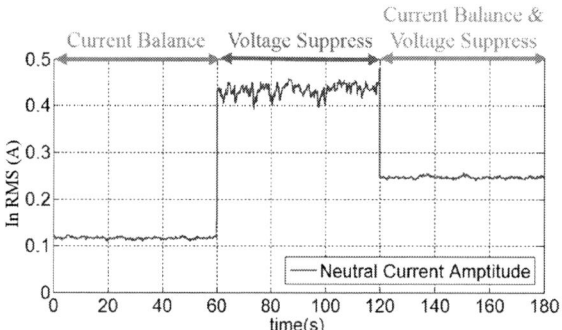

Fig. 8.   The compensated neutral current.

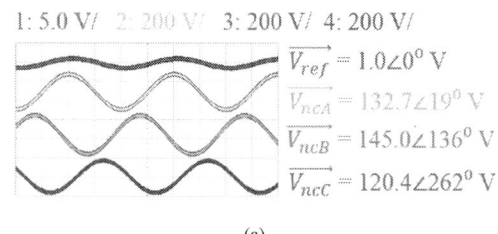

(a)

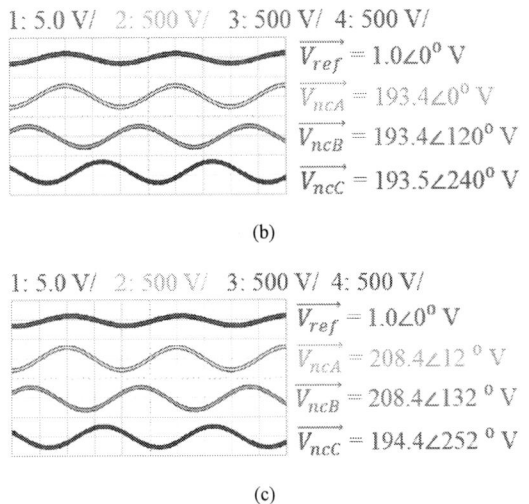

(b)

(c)

Fig. 9. The steady-state waveform of noncritical load voltages in (a) current balance, (b) voltage suppress, and (c) current balance and voltage suppress.

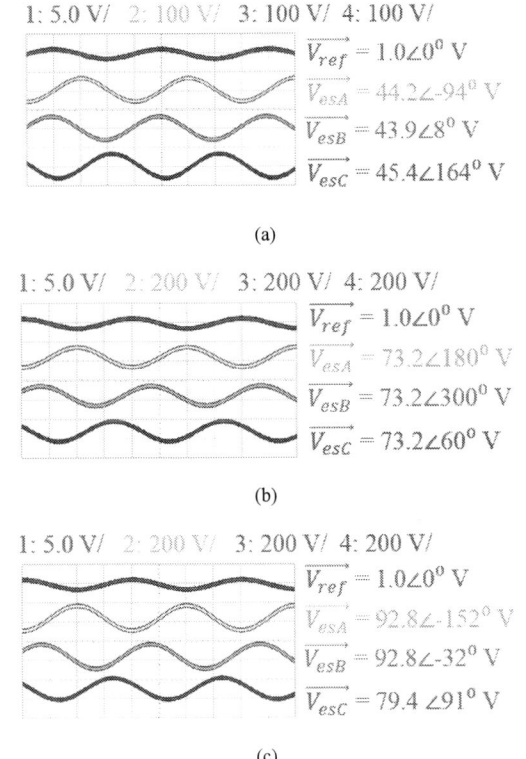

(a)

(b)

(c)

Fig. 10. The steady-state waveform of 3-ph ES voltages in (a) current balance, (b) voltage suppress, and (c) current balance and voltage suppress.

The result in Fig. 7 for t = 60 to 120 s proves that the 3-ph ES is competent in suppressing the mains voltage. The mains voltage is brought down to the nominal value of 120 V in RMS. In Fig. 8, the neutral current of the system goes back to the uncompensated state when the 3-ph ES is triggered solely

to suppress the mains voltage. As shown in TABLE IV, the 3-ph line currents are no longer symmetric, and the power factor of the load bank is reduced to 0.94, 0.91, and 0.93 for Phase A, B, and C respectively. The measurements given in Fig. 9(b) indicate that the power of the noncritical load is increased so that the voltage drop across the distribution line can be increased to reduce the mains voltage. The measurements in Fig. 10(b) show that the 3-ph ES no longer provides the reactive power to the critical load.

In the final stage, the 3-ph ES is activated to balance the line currents and to suppress the mains voltage for t = 120 to 180 s. The results in Fig. 7 show the effectiveness of the 3-ph ES to suppress the mains voltage. Correspondingly, Fig. 8 gives the suppressed neutral current. The 3-ph line currents in TABLE IV show the alleviation of line current imbalance and the improvement of power factor of the load bank. Fig. 9(c) shows that the real power of the noncritical load is again adapted into an asymmetric state. An equal amount of reallocated power in the three phases is used to bring down the mains voltage while the rest unsymmetrical part is left to balance the line currents. The reactive capacity of the 3-ph ES is fully utilized to supply the critical load with reactive power. Fig. 10(c) shows that the 3-ph ES boosts the power of the noncritical load and delivers the reactive power to the critical load. In this final stage, the 3-ph ES conducts multiple functions. The performance of balancing the line currents and suppressing the voltage are less effective than that of the ES when being activated to fulfill only a single task.

VI. CONCLUSIONS

The increasing penetration of renewable energy source requires the adjustment of the control philosophy of power grids. Besides the management of feed-in power and the installation of energy storage, the use of adaptive/smart loads can facilitate the regulation of the power grids and offer extra functions for system operation. The ES has been extensively developed for various applications, among which voltage regulation and power balance are of vital importance. The proposed decomposed voltage control intends to incorporate these two functions into a single controller. Building on the framework of decomposing the ES voltage into $V_{esA/B/C\_d\_v}$, $V_{esA/B/C\_d\_i}$, $V_{esA/B/C\_q}$, the capacity of the 3-ph ES can be extended so that multiple tasks can be conducted simultaneously. Experimental results demonstrate that the 3-ph ES with the decomposed voltage control is a promising technology to simultaneously address issues of voltage fluctuation and power imbalance.

ACKNOWLEDGMENT

The authors are grateful to the Hong Kong Research Grant Council for its support of the Theme-based Research Project (T23-701/14-N).

References

[1] F. Katiraei and J. R. Agüero, "Solar PV integration challenges," IEEE Power Energy Mag., vol. 9, no. 3, pp. 62–71, May-Jun. 2011.

[2] Mohamed A. Eltawil and Zhengming Zhao, "Grid-connected photovoltaic power systems: Technical and potential problems–A review," Renewable and Sustainable Energy Reviews, vol. 14, no. 1, pp. 112–129, Jan. 2010.

[3] M. Datta, T. Senjyu, A. Yona, T. Funabashi, and C. Kim, "A coordinated control method for leveling PV output power fluctuations of PV-diesel hybrid systems connected to isolated power utility," IEEE Trans. Energy Convers., vol. 24, no. 1, pp. 153–162, Mar. 2009.

[4] Y. H. Yang, F. Blaabjerg, and H. Wang, "Constant power generation of photovoltaic systems considering the distributed grid capacity," in Proc. 29th Annu. IEEE APEC, Mar. 2014, pp. 379–385.

[5] F. Blaabjerg , R. Teodorescu , M. Liserre and A. Timbus "Overview of control and grid synchronization for distributed power generation systems", IEEE Trans. Ind. Electron., vol. 53, no. 5, pp. 1398 -1409, Oct. 2006.

[6] A. Camacho, M. Castilla, J. Miret, R. Guzman, and A. Borrell, "Reactive power control for distributed generation power plants to comply with voltage limits during grid faults," IEEE Trans. Power Electron., vol. 29, no. 11, pp. 6224–6234, Nov. 2014.

[7] A. Camacho, M. Castilla, J. Miret, A. Borrell, and L. de Vicuna, "Active and reactive power strategies with peak current limitation for distributed generation inverters during unbalanced grid faults," IEEE Trans. Ind. Electron., vol. 62, no. 3, pp. 1515–1525, Mar. 2015.

[8] K. De Brabandere, B. Bolsens, J. Van den Keybus, A. Woyte, J. Driesen, and R. Belmans, "A voltage and frequency droop control method for parallel inverters," IEEE Trans. Power Electron., vol. 22, no. 4, pp. 1107–1115, July 2007.

[9] J. M. Guerrero, J. C. Vasquez, J. Matas, L. G. Vicuna, and M. Castilla, "Hierarchical control of droop-controlled AC and DC microgrids—A general approach toward standardization," IEEE Trans. Ind. Electron., vol. 58, no. 1, pp. 158–172, Jan. 2011.

[10] J. Rocabert, A. Luna, F. Blaabjerg, and P. Rodriguez, "Control of power converters in AC microgrids," IEEE Trans. Power Electron., vol. 27, no. 11, pp. 4734–4749, Nov. 2012.

[11] F. Guo, C. Wen, J. Mao, and Y. D. Song, "Distributed secondary voltage and frequency restoration control of droop-controlled inverter-based microgrids," IEEE Trans. Ind. Electron., vol. 62, no. 7, pp. 4355–4364, Jul. 2015.

[12] S. Vazquez, S. M. Lukic, E. Galvan, L. G. Franquelo, and J. M. Carrasco, "Energy storage systems for transport and grid applications," IEEE Trans. Ind. Electron., vol. 57, no. 12, pp. 3881–3895, Dec. 2010.

[13] A. Nourai and D. Kearns, "Batteries included: Realizing smart grid goals with intelligent energy storage," IEEE Power Energy Mag., vol. 8, no. 2, pp. 49–54, Mar./Apr. 2010.

[14] C. A. Hill, M. C. Such, D. Chen, J. Gonzalez, and W. M. Grady, "Battery energy storage for enabling integration of distributed solar power generation," IEEE Trans. Smart Grid, vol. 3, no. 2, pp. 850–857, Jun. 2012.

[15] J. M. Guerrero, P. C. Loh, T. L. Lee, and M. Chandorkar, "Advanced control architectures for intelligent microgrids –Part II: Power quality, energy storage, and AC/DC microgrids," IEEE Trans. Ind. Electron., vol. 60, no. 4, pp. 1263–1270, Apr. 2013.

[16] P. Varaiya, F. Wu and J. Bialek, "Smart operation of smart grid: Risk-limiting dispatch", Proc. IEEE, vol. 99, no. 1, pp. 40–57, Jan. 2011.

[17] I. Koutsopoulos and L. Tassiulas "Challenges in demand load control for the smart grid", IEEE Netw., vol. 25, no. 5, pp. 16–21, Sep.–Oct. 2011.

[18] R. Zamora and A. K. Srivastava, "Controls for microgrids with storage: Review, challenges, and research needs," Renewable and Sustainable Energy Reviews, vol. 14, no. 7, pp. 2009–2018, Sep. 2010.

[19] S. C. Smith, P. K. Sen, and B. Kroposki, "Advancement of energy storage devices and applications in electrical power system," in Proc. IEEE Power and Energy Soc. General Meet., Jul. 2008, pp. 1–8.

[20] S. C. Lee , S. J. Kim and S. H. Kim "Demand side management with air conditioner loads based on the queuing system model", IEEE Trans. Power Syst., vol. 26, no. 2, pp. 661–668, May 2011.

[21] P. Palensky and D. Dietrich "Demand side management: Demand response, intelligent energy systems, and smart loads", IEEE Trans. Ind. Informat., vol. 7, no. 2, pp. 1551–3203, Aug. 2011.

[22] E. Sortomme, M. M. Hindi, S. D. J. MacPherson and S. S. Venkata "Coordinated charging of plug-in hybrid electric vehicles to minimize distribution system losses", IEEE Trans. Smart Grid, vol. 2, no. 1, pp. 198–205, Mar. 2011.

[23] S. Y. R. Hui, C. K. Lee, and F. F. Wu, "Electric springs–a new smart grid technology," IEEE Trans. on Smart Grid, vol. 3, no. 3, pp. 1552–1561, Sep. 2012.

[24] X. Luo, Z. Akhtar, C. K. Lee, B. Chaudhuri, S. C. Tan, and S. Y. R. Hui, "Distributed voltage control with electric springs: Comparison with STATCOM," IEEE Trans. Smart Grid, vol. 6, no. 1, pp. 209–219, Aug. 2014.

[25] X. Chen, Y. H. Hou, S. C. Tan, C. K. Lee, and S. Y. R. Hui, "Mitigating voltage and frequency fluctuation in microgrid using electric springs," IEEE Trans. Smart Grid, vol. 6, no. 2, pp. 508–515, Aug. 2014.

[26] P. Kanjiya, and V. Khadkikar, "Enhancing power quality and stability of future smart grid with intermittent renewable energy sources using electric springs," International Conference on Renewable Energy Research and Applications., pp. 918–922, 2013.

[27] S. Yan, S. C. Tan, C. K. Lee, and S. Y. R. Hui, "Electric spring for power quality improvement," in Proc. IEEE Appl. Power Electron. Conf. Expo., 2014, pp. 2140–2147.

[28] S. Yan, S. C. Tan, C. K. Lee, B. Chaudhuri, and S. Y. Ron Hui, "Electric springs for reducing power imbalance in three-phase power systems," IEEE Trans. on Power Electron., vol. 30, no. 7, pp. 3601–3609, Aug. 2015.

[29] S.C. Tan, C.K. Lee and S.Y.R. Hui, "General steady-state analysis and control principle of electric springs with active and reactive power compensations," IEEE Trans. on Power Electron., vol. 28, no. 8, pp. 3958–3969, Aug. 2013.

[30] K. T. Mok, S. C. Tan, and S. Y. R. Hui, "Decoupled power angle and voltage control of electric springs," IEEE Trans. Power Electron., vol. 31, no. 2, pp. 1216–1229, Feb. 2016.

Self-Synchronization Operation of Global Synchronous Pulsewidth Modulation with Communication Fault Tolerant and Simplified Calculation Capabilities

Tao Xu, Feng Gao
School of Electrical Engineering
Shandong University
Jinan, China

Abstract—The distributed inverters are generally integrated into power grid without switching sequence coordinated control capability since they are equipped with their own micro controllers to command the output quantities. The switching ripples of all distributed inverters will then be randomly accumulated at the point of common coupling. The recently proposed global synchronous pulsewidth modulation (GSPWM) method however can significantly attenuate the accumulated switching ripples by periodically coordinating the distributed inverters at low synchronization frequency. GSPWM can be applied to reduce filter or switching frequency of coordinated inverters under high reliable communication system. While to increase the system robustness, this paper deeply analyzes the performance of GSPWM when the communication channels fail, especially lose the synchronous signals. And then the self-synchronization method is proposed to avoid the breakdown of GSPWM. Moreover, the self-synchronization operation could benefit to explore a simplified method to calculate the sending frequency of synchronous signals, which can greatly release the calculation burden. Finally, experimental results are presented to verify the performance of the proposed self-synchronization operation of GSWPM.

Keywords—self-synchronization; global synchronous pulsewidth modulation; communication fault tolerant operation; distributed inverters

I. INTRODUCTION

The distributed inverters have been broadly implemented in many applications, e.g. PV plants, wind plants, microgrids and etc., to connect the distributed power sources to power grid [1-2]. The accumulated switching current ripples at the point of common coupling (PCC) are randomly distributed from the maximum allowed value to the minimum value in theory because the switching sequences of distributed inverters cannot be controlled coordinately. As long as the switching current ripple of individual inverter is qualified, the accumulated current ripple at PCC will be qualified either. Reference [3] proposed the global synchronous pulsewidth modulation (GSPWM) method to intentionally control the current quality at PCC even when the switching ripples of individual inverter severely exceed the maximum allowed limitation. Global

synchronous unit (GSU) and communication channels are necessary in implementation and the general structure is shown in Fig. 1. Reference [4] deeply analyzes the performance of GSPWM including the current ripple analysis under steady state and its variation under transient state. The GSPWM could therefore be assumed in the applications with massive distributed inverters to reduce output filter or switching frequency of individual inverter meanwhile satisfy the output quality requirement at PCC. Reference [4] also proposed a method to dynamically adjust the switching sequence of inverters to attenuate the negative effect when one or more inverters cease or start over their operation. All of these analyses are based on the reliable communication channels.

When the GSPWM method is assumed in practice, the possibility of communication failure must be carefully

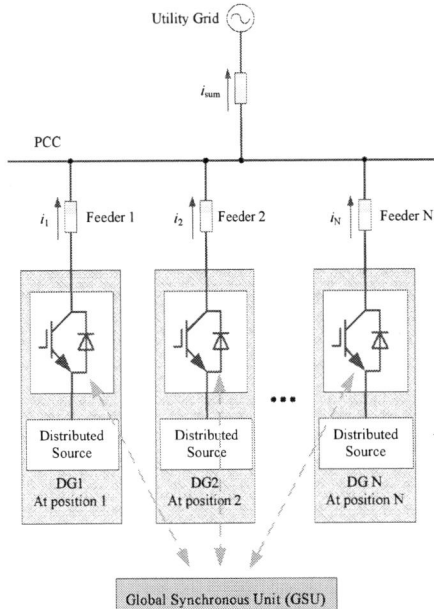

Fig. 1. Illustration of distributed inverters connected to a PCC with communication channel and GSU.

considered. Aiming to handle the communication failure, this paper deeply analyzes the operational characteristics of GSPWM when communication fails and proposes a self-synchronization method to ensure GSPWM operate for a relative long time without communication. Besides, the self-synchronization method can help simplify the calculation process of sending frequency of synchronous signals [3]. The detailed simplified calculation method is fully elaborated in this paper. Finally, experimental results verified the performance of self-synchronization operation of GSPWM.

II. BASIC PRINCIPLE OF GSPWM

Most of inverters are controlled by digital signal processer (DSP) or other digital controllers. Crystal inside the digital controller is an important element, which resonates at a fixed frequency. All of the operational process in the digital controller and the PWM generation function are synchronous with the crystal resonant frequency. But the crystal resonant frequency could be significantly influenced by its working condition [5]. Hence the PWM frequency is not exactly equal to what it has been set. Such phenomenon does not influence the operational performance of individual inverter. But when synchronizing PWM sequences of multiple inverters, the unequal crystal frequency among multiple digital controllers will unavoidably induce the PWM deviation. Therefore, the distributed inverters do not have a synchronous pulsewidth modulation before [3]. Such phenomenon can be effectively illustrated in Fig. 2, where, the PWM sequences of two identical PWM controlled inverters will deviate along with the time progress and the accumulated output current ripple band will change along with the time progress either due to the different crystal resonant frequencies. Taking two inverters with bipolar modulation mode, f_c=10kHz, f_0=50Hz, V_d=200V, L=2mH as an example, the THD of accumulated current is drawn in Fig. 3. It is noted that the minimum THD value occurs when the phase shift between two identical inverters is 180°, being similar as the interleaved PWM in a multi-leg inverter [6-8].

It is superior to fix the optimal phase shifted angles among distributed inverters in practical operation. The global synchronous PWM method assumes the commonly used communication channels and a global synchronous unit (GSU) to achieve this objective. Fig. 1 shows the configuration of distributed inverters with communication and GSU. The communication channel can be RS485, RS232, power line carrier (PLC) and optical fiber. The choice of communication channel is determined by working condition, cost and etc. The GSU could work in one digital controller among distributed inverters or independently. The basic realization of GSPWM is shown in Fig. 4. And the corresponding processes are:

1) Controller 2 will send its parameters to Controller 1 through the preset communication channel (RS485 in Fig. 4).

2) Controller 1 receives the parameters sent from Controller 2. Controller 1 will calculate optimal phase shift angle (e.g. $\varphi_{\text{PWM2best}}$=180°) and the sending frequency of synchronous signal (e.g. f_{syn}=3Hz) according to the inverter parameters [3].

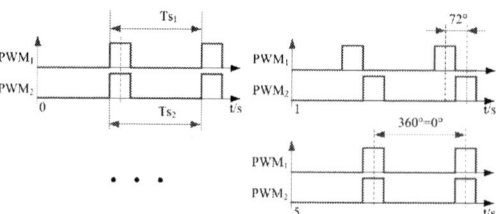

Fig. 2. Illustration of phase shift between two PWM sequences.

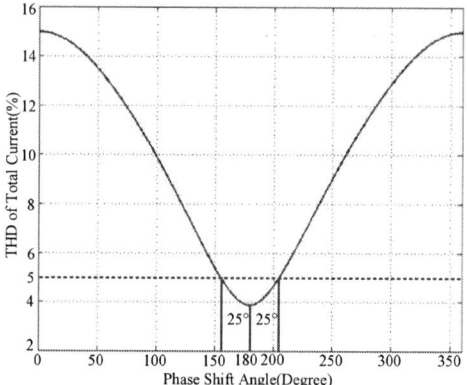

Fig. 3. The trajectory of THD when phase shift angle changes from 0° to 360° (2 inverters using bipolar modulation mode).

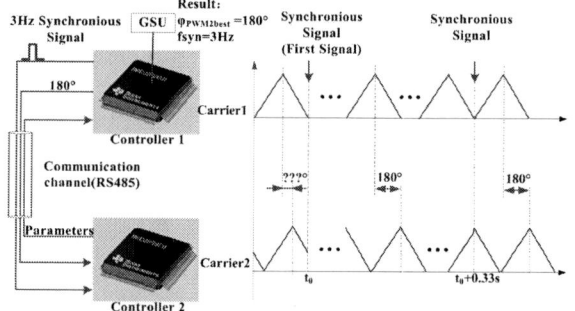

Fig. 4. Illustration of working principle with GSU and communication channel.

3) Controller 1 sends the calculation result of $\varphi_{\text{PWM2best}}$=180° to Controller 2.

4) Controller 1 sends the synchronous signals to Controller 2 with frequency f_{syn}. The synchronous signal is shown in Fig. 4, and it is sent when the carrier counter of Controller 1 is zero.

5) Controller 2 saves its carrier phase shift 180° in its memory and change its carrier phase shift gradually until its phase shift become 180°.

III. PRINCIPLE AND REALIZATION OF SELF-SYNCHRONIZATION OPERATION

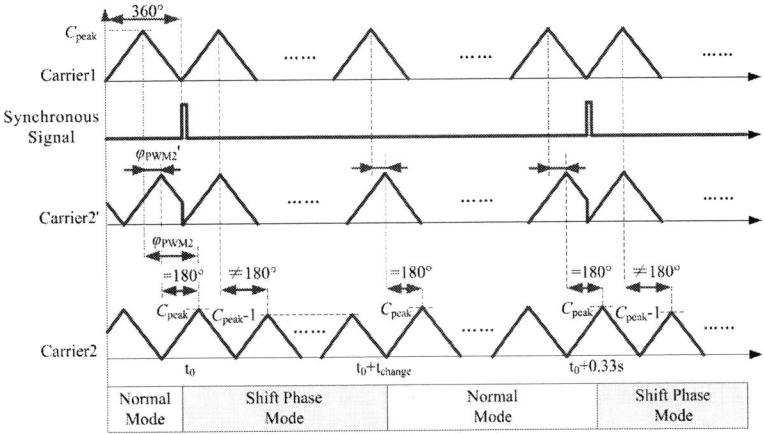

Fig. 5. Operation process of carriers.

According to the analysis in Section 2, it is known that GSPWM can fix the optimal phase shifted angles among distributed inverters with the synchronous signals. When losing the synchronous signals, the phase shifted angles among multiple PWM sequences will deviate from the optimal positions. The self-synchronization method in this paper can effectively fix the optimal phase shifted angles in theory when the communication fails.

This section also uses two identical inverters as the analysis example and assumes $f_{DSP1} > f_{DSP2}$. f_{DSP1}, f_{DSP2} are operating frequency of DSP1, DSP2, respectively. When the communication channels operate normally, the details of operation process proposed in [3] is shown in Fig. 5, where Carrier 1 is the PWM carrier of inverter 1, Carrier 2 is the PWM carrier of inverter 2, Carrier 2' is a fast responded reference for Carrier 2 and C_{peak} is the peak value of counter for generating carriers. Once inverter 2 receives the synchronous signals, Carrier 2' will change suddenly and Carrier 2 will change gradually to reach the optimal phase shift after t_{change} as shown in Fig. 5 [4]. When communication channels operate normally, the change of φ_{PWM2}' and φ_{PWM2} is shown in Fig. 6, where φ_{PWM2}' and φ_{PWM2} represent the phase shifted angles of Carrier 2' and Carrier 2, respectively. φ_{PWM2}' can change suddenly to follow Carrier 1 because Carrier 2' is only the reference of Carrier 2. Carrier 2 changes its phase shift gradually according to Carrier 2' and the red line shows the change of φ_{PWM2}. $\Delta\varphi_{PWM2}$ is the maximum phase shift deviation of φ_{PWM2}. It is seen that φ_{PWM2} will slowly reach its optimal position to avoid the output distortion. However, doing so, $\Delta\varphi_{PWM2}$ can only be reduced by increasing the sending frequency f_{syn} of synchronous signals. The THD of i_{sum} will still meet the requirement because this deviation is taken into consideration when calculating f_{syn} [3], but need high f_{syn}, which may challenge the communication channels. Another disadvantage is shown in Fig. 7, where when the communication fails, φ_{PWM2} will deviate from $\varphi_{PWM2best}$ rapidly and then GSPWM fails. In specific, when losing several synchronous signals, $\Delta\varphi_{PWM2}$ will exceed the deviation range. The proposed self-synchronization operation in this paper can help fix the optimal phase shifted angles when communication fails, which contains two procedures referring to the calculation

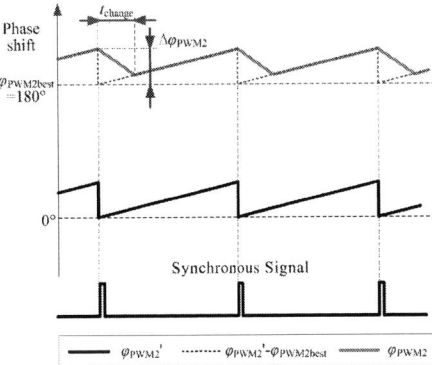

Fig. 6. Change of φ_{PWM2}' and φ_{PWM2} in normal operation.

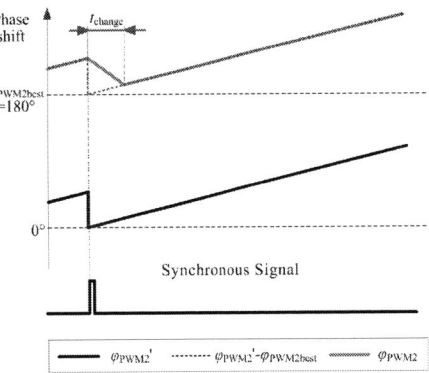

Fig. 7. Change of φ_{PWM2}' and φ_{PWM2} in fault condition.

of frequency difference between f_{DSP1} and f_{DSP2}, and self-synchronization operation of Carrier 2.

A. Calculation of Resonant Frequency Difference

In initial, the synchronous signals are sent with the preset frequency f_{syn}' before the inverters output the desired currents. f_{syn}' should make sure that φ_{PWM2}' will not change for more than 360° in Fig. 8 in order to precisely calculate the frequency difference. f_{syn}' can be calculated as follows. f_{syn} is the calculated sending frequency during normal operation after the start process and its calculation method has been proposed in [3].

$$\Delta\varphi_{1s} = \frac{2Err_{\text{cry}}}{T_s} \times 360° \tag{1}$$

$\Delta\varphi_{1s}$ is the phase shift per second. When $Err_{\text{cry}} = 10\text{PPM} = 10^{-5}$ and $T_s = 10^{-5}$, $\Delta\varphi_{1s}$ will be 72°. $\Delta\varphi_{\max}$ is 360°. So the minimum of f_{syn}' is

$$f_{\text{synmin}}' = \frac{1}{\Delta\varphi_{\max}/\Delta\varphi_{1s}} = 0.2\text{Hz} \tag{2}$$

f_{syn}' is chosen as 2Hz in this example. Fig. 8 shows the illustration of frequency difference calculation. In detail, f_{DSP1} can be calculated as:

$$f_{DSP1} = \frac{2C_{peak}W_1}{T_{syn}'} = 2C_{peak}W_1 f_{syn}' \tag{3}$$

Where, T_{syn}' is the period of sending synchronous signals, W_1 is the quantity of the peaks of Carrier 1 in T_{syn}'.

And the resonant frequency difference between two controllers can be calculated according to different cases. W_2 is the quantity of the peaks of Carrier 2' in T_{syn}'. Dir represents the counting direction of Carrier 2' when controller 2 receives the synchronous signal. $Dir=1$ or $Dir=0$ represents the case where counting number of Carrier 2' is increased or decreased, respectively.

Case 1: $W_2 = W_1$; $Dir=0$
$$f_{DSP2} = \frac{2C_{peak}W_2 - \Delta C}{T_{syn}'} = (2C_{peak}W_2 - \Delta C) f_{syn}' \tag{4}$$

Case 2: $W_2 = W_1 - 1$; $Dir=1$
$$f_{DSP2} = \frac{2C_{peak}W_2 + \Delta C}{T_{syn}'} = [2C_{peak}W_2 + \Delta C] f_{syn}' \tag{5}$$

Case 3: $W_2 = W_1 + 1$; $Dir=0$
$$f_{DSP2} = \frac{2C_{peak}W_2 - \Delta C}{T_{syn}'} = [2C_{peak}W_2 - \Delta C] f_{syn}' \tag{6}$$

Case 4: $W_2 = W_1$; $Dir=1$
$$f_{DSP2} = \frac{2C_{peak}W_2 + \Delta C}{T_{syn}'} = [2C_{peak}W_2 + \Delta C] f_{syn}' \tag{7}$$

Where, ΔC is the sudden change of counter in carrier 2' when controller 2 receives the synchronous signal, and it can be calculated easily by adding simple program in DSP.

The frequency difference Δf is:

$$\Delta f = f_{DSP2} - f_{DSP1} \tag{8}$$

B. Self-Synchronization Operation

In theory, φ_{PWM2}' can be zero by changing the peak value of carrier counter from C_{peak} to $f_{\text{DSP2}}C_{\text{peak}}/f_{\text{DSP1}}$, and $f_{\text{DSP2}}C_{\text{peak}}/f_{\text{DSP1}}$ is independent from f_{syn}. In practice, $f_{\text{DSP2}}C_{\text{peak}}/f_{\text{DSP1}}$ may not be integer so it cannot be applied to DSP directly. Moreover, $|C_{\text{peak}}-f_{\text{DSP2}}C_{\text{peak}}/f_{\text{DSP1}}|$ is smaller than 1 because f_{DSP2} approximates f_{DSP1}. For example, $f_s=10\text{kHz}$, $f_{\text{syn}}'=2\text{Hz}$. The

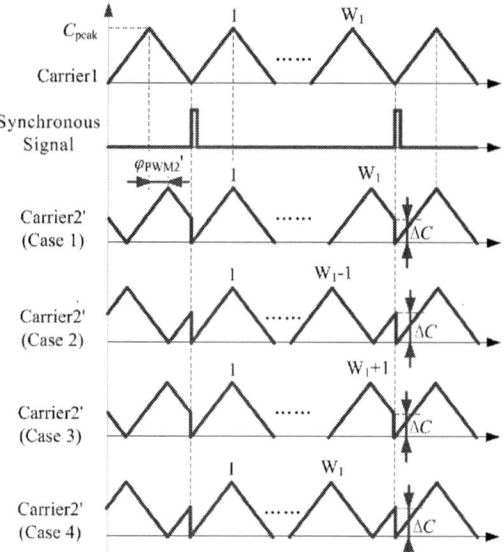

Fig. 8. Illustration of calculating frequency difference.

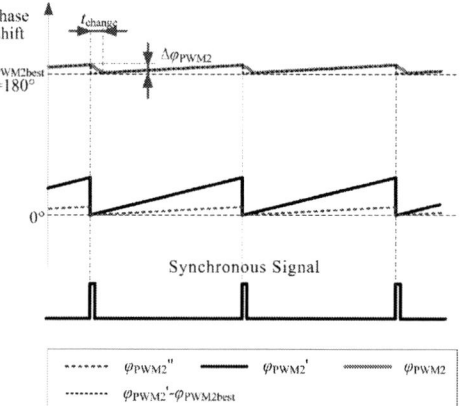

Fig. 9 Change of φ_{PWM2}', φ_{PWM2}'', φ_{PWM2} in normal operation when use Self-Synchronization.

calculation result in DSP 1 is $W_1 = f_s/f_{\text{syn}}' = 5000$. In DSP 2, it detects that $W_2 = W_1 = 5000$, $Dir=0$, $\Delta C = 1900$. According to formula (3) to (7), the calculated results are $f_{\text{DSP1}} = 150\text{MHz}$ and $f_{\text{DSP2}} = 149.9962\text{MHz}$. $|C_{\text{peak}}-f_{\text{DSP2}}C_{\text{peak}}/f_{\text{DSP1}}| = 0.19$. φ_{PWM2}' can be zero when C_{peak} change from 7500 to 7499.81. It is impractical because C_{peak} can only be integer. So we use the following method to approach the expected value.

1) Carrier 2' will change suddenly during normal operation as shown in Fig. 9. When the communication fails, Carrier 2' will still change suddenly by self-synchronization as shown in Fig. 10. Taking case 1 in Fig. 8 as an example, where upon DSP 2 detects $W_2 = W_1$, $Dir=0$, it will reset counter to zero when counter is ΔC. Similar operation can be applied to other cases.

2) To reduce $\Delta\varphi_{\text{PWM2}}$ and t_{change}, Carrier 2'' as the new reference for Carrier 2' is assumed. And φ_{PWMM} is the phase shifted angle between Carrier 2'' and Carrier 1. D represents the

adjustment interval in Carrier 2" for precisely approaching the wanted C_{peak}.

$$D = \left[\left|\frac{1}{C_{peak} - f_{DSP2}C_{peak}/f_{DSP1}}\right|\right] + 1 \qquad (9)$$

Where, [x] is the integer part of number x. For example, [3.8]=3. In case 1 and case 2, the peak value of Carrier 2" remains C_{peak} for D-1 carrier period and then changes as C_{peak}-1 for one carrier period. In case 3 and case 4, the peak value of Carrier 2" remains C_{peak} for D-1 carrier period and then changes as C_{peak}+1 for one carrier period. And they are shown in Fig. 11. In the preceding example, D=[5.2632]+1=6.

3) Carrier 2 will change its phase angle according to Carrier 2" instead of Carrier 2'.

When the parameters of inverters are different, Self-Synchronization can still be employed.

IV. SIMPLIFIED METHOD TO CALCULATE SENDING FREQUENCY OF SYNCHRONOUS SIGNALS

In [3], the basic method of calculating f_{syn} was proposed by finding the maximum value of an objective function: $I_{hsum}=f(\varphi_{PWM1......}\varphi_{PWMN})$

Subject to: $\varphi_{PWMMbest}-\Delta\varphi_{Mmax} \leq \varphi_{PWMM} \leq \varphi_{PWMMbest}+\Delta\varphi_{Mmax}$, for M=1,...,N

$$I_{hsum} = \sqrt{\begin{aligned}&\sum_{f=0}^{\infty}\left(\sum_{M=1}^{N}I_{hMf}\cos\left(\varphi_{hMf}+\theta_{hMf}\right)\right)^2 \\ &+\sum_{f=0}^{\infty}\left(\sum_{M=1}^{N}I_{hMf}\sin\left(\varphi_{hMf}+\theta_{hMf}\right)\right)^2\end{aligned}} \qquad (10)$$

$$= f\left(\varphi_{PWM1},...,\varphi_{PWMN}\right)$$

Where, Double-integral Fourier analysis [9-10] is applied when calculating $I_{hsum}=f(\varphi_{PWM1......}\varphi_{PWMN})$. The calculation processes are to:

1) Choose communication channel. Let $f_{syn}= f_{synmax}$, and f_{synmax} is the maximum frequency of sending synchronous signals with the chosen communication channel;

2) Calculate the $\Delta\varphi_{Mmax}$ with f_{syn} and $\Delta\varphi_{1s}$;

3) Use particle swarm optimization (PSO) [11-15] method to find the maximum THD when the range of φ_{PWMM} is from $\varphi_{PWMMbest}$ - $\Delta\varphi_{Mmax}$ to $\varphi_{PWMMbest}+\Delta\varphi_{Mmax}$.

4) determine whether the maximum THD is lower than the maximum allowable THD. If not, increase the communicate rate and repeat 1).

This process is complex because the exact range of φ_{PWMM} is unknown and PSO must be employed to solve N dimension problem. And this is the most time consuming part in GSPWM. The process of calculating f_{syn} can be simplified with known f_{DSPM}. φ_{PWMM} can be expressed as:

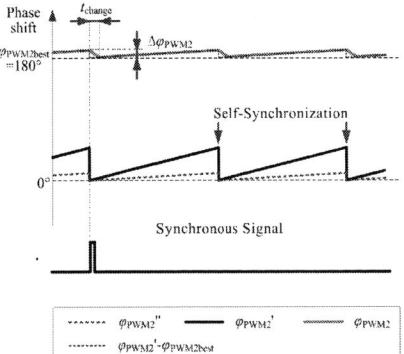

Fig. 10 Change of φ_{PWM2}', φ_{PWM2}", φ_{PWM2} in fault condition when use Self-Synchronization.

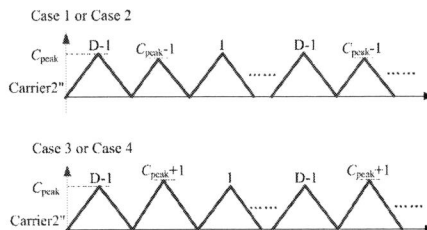

Fig. 11. Illustration of changing carrier 2"

$$\varphi_{PWMM}(t)=\begin{cases}\varphi_{PWMMbest}+\dfrac{\left(f_{DSP1}-f_{DSP2}-2f_s/D\right)t}{2C_{peak}}\cdot360° \\ \left(t_{change}<t<\dfrac{1}{f_{syn}}\right) \\ \\ \left[\varphi_{PWMMbest}+\varphi_{PWMM}"\left(\dfrac{1}{f_{syn}}\right)\right. \\ \left.+\dfrac{\left(f_{DSP1}-f_{DSP2}-2f_s\right)t}{2C_{peak}}\cdot360°\right] \\ \left(0<t<t_{change}\right)\end{cases}$$

$$(11)$$

Where,

$$\varphi_{PWMM}"(t)=\frac{\left(f_{DSP1}-f_{DSP2}-2f_s/D\right)t}{2C_{peak}}\cdot360° \quad 0<t<\frac{1}{f_{syn}} \qquad (12)$$

$$t_{change}=\frac{\left(f_{DSP1}-f_{DSP2}-2f_s/D\right)}{2\left(1-1/D\right)f_sf_{syn}} \qquad (13)$$

The processes include:

Step 1) and 4) are same as that before.

2) To generate formula $\varphi_{PWMM}(t)$.

3) To solve the problem:

Find the maximum of an objective function:
$I_{\text{hsum}} = f(\varphi_{\text{PWM1}}(t) \ldots \varphi_{\text{PWMN}}(t))$

Subject to: $0 \leq t \leq 1/f_{\text{syn}}$

This is a one-dimension problem. It consumes much less time than previous methods. When the parameters of inverters are different, this method can still be employed.

V. EXPERIMENTAL VERIFICATION

The constructed experimental prototype has four distributed inverters with their own independent DC sources, H-bridge circuits, output filters and digital controllers as the picture shown in Fig. 12. All these inverters are connected to an emulated grid using a programmable AC source AMETEK-CI-4500LS, whose RMS value of output voltage is 110V and output frequency is 50Hz. The parameters of these four inverters are shown in Table 1.

Firstly, four inverters operate with GSPWM but without self-synchronization method. And all the communication channels fail at 6 second. Oscilloscope is employed to record the waveform of i_{sum} per 0.1 second. Matlab is employed to calculate the THD of each current. Fig. 13 shows the change of THD in 30 seconds by employing GSPWM but without using self-synchronization method. When all of the communication channels fail at 6s and then all inverters lose the synchronous signals, and THD of total current i_{sum} increases significantly and changes as time processes.

Then, self-synchronization operation method is employed and all communication channels fail at 7 second. The above method to calculate the THD is also employed. The THD of i_{sum} is always very small even under communication fault condition, and the THD trajectory is shown in Fig. 14. Fig. 15 shows the experimental current waveforms when employing self-synchronization method. Fig. 15(b) shows that the phase shift will not deviate from the optimal phase shift angle.

VI. CONCLUSION

This paper deeply analyzes the performance of GSPWM when the communication fails especially loses the synchronous signals. The self-synchronization method is proposed to avoid the breakdown of GSPWM when losing the synchronous signals. Moreover, a simplified method to calculate the sending frequency of synchronous signals is proposed. Finally, the experimental results verified the performance of GSPWM when using the self-synchronization operation method.

VII. REFERENCES

[1] F. Blaabjerg, Z. Chen and S. B. Kjaer, "Power electronics as efficient interface in dispersed power generation systems," *IEEE Trans. Power Electron.*, vol.19, no. 5, pp. 1184- 1194, Sep.2004.

[2] F. Blaabjerg, R. Teodorescu, M. Liserre, and A. V. Timbus, "Overview of control and grid synchronization for distributed power generationsystems," *IEEE Trans. Ind. Electron.*, vol. 53, no. 5, pp. 1398–1409, Oct. 2006.

[3] T. Xu and F. Gao, "Global Synchronous Pulse Width Modulation of Distributed Inverters," in *Proc. ECCE Asia'2015*, 2015.

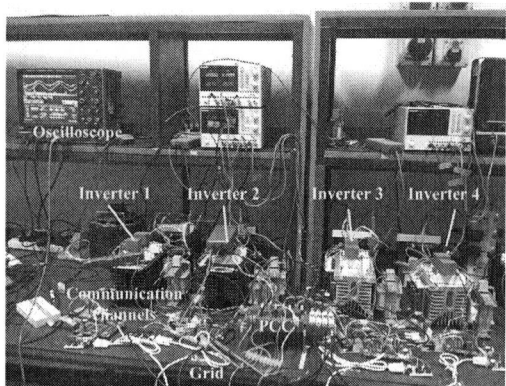

Fig. 12. Photograph of experimental prototype.

TABLE I.   PARAMETERS OF TWO EXPERIMENTAL INVERTERS

Parameter	Inverter 1 - Inverter 4
Switchting Frequency/kHz	≈10
DC Voltage/V	200
Filter/mH	3.5
Active Power/W	150
Modulation Mode	Bipolar
IGBT	Infineon FF100R12RT4
Utility Grid	50Hz 110V

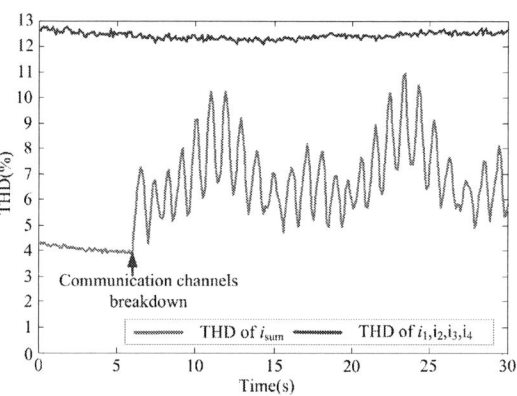

Fig. 13. Experimental THD in 30 seconds without using self-synchronization method.

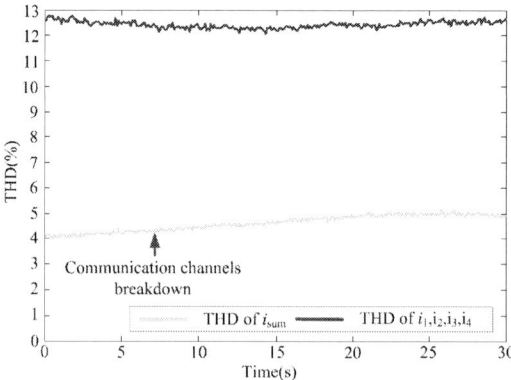

Fig. 14. Experimental THD in 30 seconds by using self-synchronization method.

[4] T. Xu, F. Gao, W. Duan and R. Wei, "Performance Analysis of Global Synchronous Pulsewidth Modulation for Distributed Inverters," in *Proc. ECCE'2015*, 2015.

[5] Randall W. Rhea, "Discrete Oscillator Design: Linear, Nonlinear, Transient, and Noise Domains," *Artech House*, 2014, pp. 332-334.

[6] B. Cougo, T. Meynard, and G. Gateau, "Parallel three-phase inverters: Optimal PWM method for flux reduction in intercell transformers," *IEEE Trans. Power Electron.*, vol. 26, no. 8, pp. 2184–2191, Aug. 2011.

[7] M. A. Abusara, S. M. Sharkh, "Design and control of a grid-connected interleaved inverter," *IEEE Trans. Power Electron.*, vol.28, no. 2, pp. 748 - 764, Feb.2013.

[8] P. J. Grbovic, "Closed form analysis of N-cell interleaved two-level DC-DC converters: The DC bus capacitor current stress," in Proc. ECCE Asia'2013, 2013, pp. 122-129.

[9] S. Jayawant and J. Sun, "Double-integral Fourier analysis of interleaved pulse width modulation," in Proc. Comput. Power Electron. Workshop, 2006, pp. 34–39.

[10] D. Holmes and T. Lipo, "Pulse width modulation for power converters : principles and practice," Wiley-IEEE Press, 2003.

[11] J. Kennedy and R. C. Eberhart, "Particle swarm optimization," in *Proc. IEEE Int. Conf. Neural Netw.*, 1995, pp. 1942-1948.

[12] R. Eberhart and J. Kennedy, "A new optimizer using particle swarm theory," in *Proc. MHS'1995*, 1995, pp. 39-43.

[13] Y. del Valley, G. K. Venayagamoorthy, S. Mohagheghi, J.-C. Hernandez, and R. G. Harley, "Particle swarm optimization: Basic concepts, variants and applications in power systems," *IEEE Trans. Power Del.*, vol. 12, no. 2, pp. 171-195, Apr. 2008.

[14] J. Kennedy, "The particle swarm: Social adaptation of knowledge," in *Proc. IEEE Int. Conf. Evol. Comput.*, 1997, pp. 303-308.

[15] E. Ozcan and C. Mohan, "Particle swarm optimization: Surfing the waves," in *Proc. IEEE Congress Evol. Comput.*, 1999, vol. 3, pp.1939-1944.

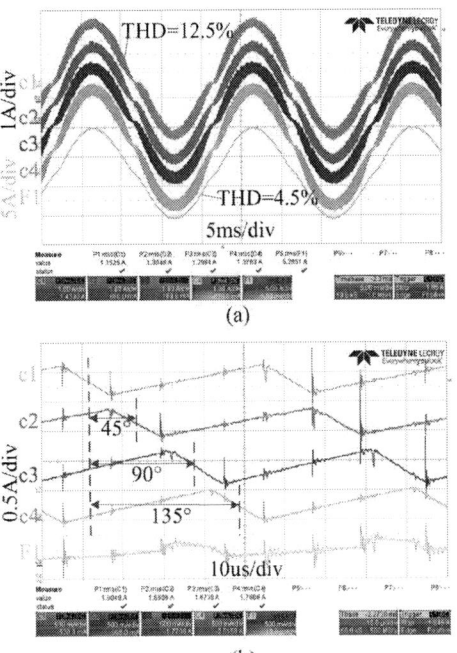

Fig. 15. (a) Experimental current waveform of (from TOP to BOTTOM) i_1(C1), i_2(C2), i_3(C3), i_4(C4) and i_{sum}(F1) with GSPWM. (b) Zoomed view of (a)

Design Considerations and Predictive Direct Current Control of Active Regenerative Rectifiers for Harmonic and Current Ripple Reduction

Alberto Berzoy and A. A. S. Mohamed IEEE Students, and Osama Mohammed IEEE Fellow
Florida International University
Energy Systems Research Laboratory
Miami, Florida
mohammed@fiu.edu

Abstract—This paper proposes a general and an improved methodology for the design of the passive elements of Active Regenerative Rectifiers (ARR). The proposed strategy for the L-filter is based on the current area error in the time domain and on the high order current harmonics in the frequency domain. This strategy is independent of the type of control technique and it can be implemented for different modulation techniques. Moreover, it can be extended to different converters and, under certain small considerations, to Active Power Filters (APF). For purposes of the technique explanation and practical demonstration, a two-level three-phase voltage source converter (VSC) with PWM is used in this work. Also in this paper, a predictive direct current control technique (PDCC) is presented. The control technique provides fast dynamic, high performance under diverse operation conditions and parameter independency. The combination between the proposed design and control technique lends to significant reduction on the current harmonics content and ripple. Furthermore, simulation and experimental results corroborate the ability of the proposed design procedure and the accuracy the PDCC.

Keywords—*Design, predictive control, active reganerative rectifier, harmonic content, PWM.*

I. INTRODUCTION

ARR have been used in several applications [1] and it is recently widely used in micro-grids where bidirectional power flow is necessary. It has the capacity to control both DC voltage and power factor (PF) while working in rectifying and regenerative mode. ARR has the ability to obtain sinusoidal input current with total harmonic distortion (THD) less than 5% (IEEE Std. 519-2014), unity PF and bidirectional active power transfer. For these reasons many researches have been conducted in the ARR's control [2]–[12], however, few papers dedicate to the design issues [13]–[19] and even less both [20]. Basically the ARR design includes the selection of the AC L-filter and DC-link capacitor. To our knowledge, the published work in this area does not agree for the best ARR design technique [13]–[18], [20]. The reason for this disagreement is the high complexity of the modulation techniques, the vast diversity of voltage source converters and their high nonlinearity, the abundant types of load applications, the different control objectives and finally the different types of coupling filters (e.g. L and LCL).

Some literatures propose to use LCL–filter [14], [16], [20] because, first it can reduce the high order harmonics at the grid side as its high frequency response increases the attenuation to 60dB/decade compared with the 20dB/decade of the L filter and second to avoid very large values of inductances in the L-filter. Also, the high gain of VSC controllers (bang-bang, direct power control or un-tuned PIs) cause this requirement (large L) to attenuate the current ripple. Nevertheless a poor design of the LCL-filter can cause lower attenuation compare to that expected or even increase the distortion due to oscillation effects. Among other drawbacks of the LCL-filter:

- It introduces phase-shift between line current and line-to-neutral voltage, lowering the PF. Thus, proper control considerations and claims for additional investigation of the correct orientation of the reference frame are required. Stability problems, also need to be correctly addressed. Moreover, as consequence LCL-filter increase the number of transducers for the control purposes [20].

- It may cause resonance problems, which needs to be avoided or minimized by means of damping resistance.

- It shows high current ripple in the inner inductor which results in high amount of iron losses (hysteresis and eddy current losses) which implies the thin laminated iron cores or even powder core [21].

In conclusion the LCL-filter is more complex, difficult to design, it can cause stability problems, it is less cost effective than a single L-filter and it needs more elements and transducers, then is more prone to failure.

The diverse set of ARR design methodologies can be classified into two categories: time domain [13], [17], [18] and frequency domain [20], [17]. Both of them have advantages and missing considerations in the other domain, however under certain assumptions it is possible to conciliate the two design domains. Thus, this paper presents a generic ARR design strategy that can be extended for different converters topologies, modulations, load applications and, under certain conditions, can be implemented in APF. The proposed design strategy defines the optimum range for the choke inductance (L-filter) and the DC link capacitor. However, and optimum design alone is not sufficient for the minimization of current

978-1-4673-9551-9/16 $31.00 © 2016 IEEE

harmonic content and ripple in the ARR. Thus, a non-linear current control technique (predictive direct control) is proposed. This technique is an extension of the one presented in [22] to provide fast dynamic, high performance and parameter independency. For purposes of the design technique development and control, a two-level three-phase VSC with PWM is used. The proposed design and control methodologies are verified by simulations and experimental tests for the ARR.

II. ACTIVE REGENERATIVE RECTIFIER

The entire block diagram of the ARR is shown in Fig 1, and its state-space representation is given in (1):

$$\frac{di_L}{dt} = \frac{1}{L}(\mathbf{v_s} - R\,\mathbf{i_L} - \mathbf{v_c})$$
$$\frac{dV_{dc}}{dt} = \frac{1}{C}\left(\mathbf{S_w i_L}^t \mp i_o\right) \tag{1}$$

where $\mathbf{v_s} = [v_{sa}\ v_{sb}\ v_{sc}]$ is the voltage source space vector, $\mathbf{i_L} = [i_{La}\ i_{Lb}\ i_{Lc}]$ is the line current space vector that is bi-directional (from AC to DC and vice versa), $\mathbf{v_c} = [v_{ca}\ v_{cb}\ v_{cc}]$ is the converter voltage space vector and $\mathbf{S_w} = [S_a\ S_b\ S_c]$ is the switching function where $S_x = \begin{cases} 0 & \text{off} \\ 1 & \text{on} \end{cases}$ models the VSC states, the super index t means transpose vector. By applying one of the six non-zero vectors [22] of the VSC, and neglecting losses in (1), the inductor voltage per phase can be obtained as:

$$v_{Lx} = \pm \begin{cases} v_{sx} - \frac{m}{3}V_{dc} & \text{on state} \\ v_{sx} + \frac{m}{3}V_{dc} & \text{off state} \end{cases} \tag{2}$$

where $x = \{a, b, c\}$, $m = \{-1, 2\}$ depending on which of the eight possible switching states of the converter is applied and the $\pm$ sign indicates rectifier or regenerative mode. Based on $v_L = L\frac{di_L}{dt}$, assuming high switching frequency (linear variation of the currents), selecting $m = 2$, as the worst case, using the regenerative mode and using center PWM, then the current variation can be stated as in (3).

$$\Delta i_{Lx} = \begin{cases} -\frac{1}{L}\left(v_{sx} - \frac{2}{3}V_{dc}\right)d_x T_s & \text{on state} \\ -\frac{1}{L}\left(v_{sx} + \frac{2}{3}V_{dc}\right)(1 - d_x)T_s & \text{off state} \end{cases} \tag{3}$$

where $T_s = 1/10k$ Hz is the switching time period and d_x is the duty cycle of phase x.

III. ACTIVE REGENRATIVE RECTIFIER DESIGN

The procedure for the ARR design requires the selection of the: DC-link voltage (V_{dc}), desired DC link voltage ripple (V_{ripple}), desired line current ripple (I_{ripple}), desired line current THD, AC grid peak voltage supply (V_{sx}) and line peak current (I_{Lx}) or active power transfer (P_{ac}). Also, some assumptions are required such as: balanced sinusoidal voltage supply, unity PF operation (achieved by the controller), much higher switching frequency (f_s) than the grid frequency (f_e) and negligible associated internal coil resistance (R).

A. L-filter Design (Choke Inductance)

The design of the choke inductance is not a simple task as it serves two purposes: short term energy storage for the

boost-type converter (VSC) and keep the switching noise away from the power-grid. Basically the switching noise in the line current is determined by the size of the inductors together with the VSC control strategy. Thus, a design procedure of the line filter is presented. The procedure is a time domain based strategy as an extension of the one presented in [18], where the current variation in T_s is ΔI (it considers the "on" and "off" state, where "on" increases and "off" decreases the i_{Lx}, which is already considered by (3)). The current area error (S_{error}) and the current ripple (I_{ripple}) are defined in Fig. 2 and (4).

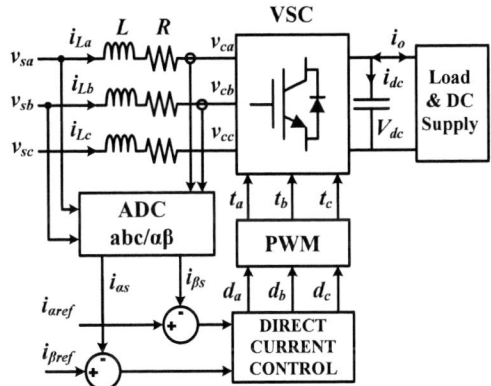

Fig. 1: Three-phase controlled rectifier

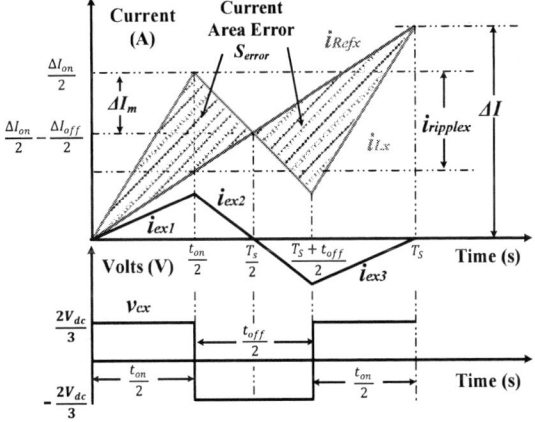

Fig. 2: Error current calculation method.

$$\Delta I = \Delta i_{Lx(on)} + \Delta i_{Lx(off)} = \Delta i_{Lx(on)} + 2\Delta I_m$$
$$S_{error} = \frac{T_s}{2}\left[\left(\frac{1-D}{2}\right)|\Delta I| + |\Delta I_m|\right] \tag{4}$$
$$I_{ripple} = \left(\frac{1-D}{2}\right)|\Delta I| + |\Delta I_m|$$

Introducing (3) in (4), it can be deduced that the maximum area error and current ripple occurs at $d_x = 50\%$, and the minimum limit value of the L-filter can be defined as in (5).

$$I_{ripple} > \frac{T_s . V_{dc}}{6\,L} \quad \rightarrow \quad L > \frac{V_{dc}}{6\,.f_s\,.I_{ripple}} \tag{5}$$

By considering the fundamental components of all the ARR state space vectors, the relationship between the AC side and DC side vectors of VSC is shown in Fig.3. The ARR can

978-1-4673-9551-9/16 $31.00 © 2016 IEEE

operate from the origin **O** to the circle's origin. In Fig. 3(a) the ARR is not achieving unity power factor, it is consuming active (p) and reactive power (q). In Fig. 3(b) the ARR is working at point B, achieving unity power factor, therefore it is consuming only active power (p) in rectifier mode. When the ARR is operating at point D, it is in regenerative mode (PF = -1).

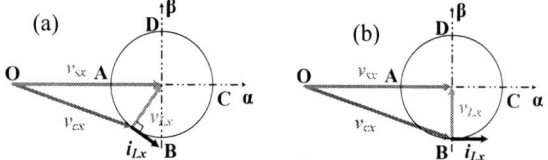

Fig. 3: Operation modes of the ARR and voltage relatioship between AC and DC side of the VSC. (a) $PF \neq 1$ (b) $PF = 1$

In Fig. 3(b), the voltage vectors are in permanent regime where $v_{sx} = V_{sx} \angle 0$, $v_{cx} = \pm \frac{2V_{dc}}{3} \angle \cos^{-1}\left(\frac{V_{sx}}{\frac{2V_{dc}}{3}}\right)$, $v_{Lx} = V_{Lx} \angle \frac{\pi}{2}$ and $V_{Lx} = |X_L| I_{Lx} = \omega_e L I_{Lx}$. Thus:

$$V_{Lx} = \sqrt{V_{cx}^2 - V_{sx}^2} = \sqrt{\left(\frac{2V_{dc}}{3}\right)^2 - V_{sx}^2} \quad (6)$$

From (6) it can be deduced that the maximum inductance drop voltage in both rectification and regeneration is as in (7).

$$\omega_e L I_{Lx} < \sqrt{\frac{4}{9}V_{dc}^2 - V_{sx}^2} \quad \rightarrow \quad L < \frac{\sqrt{4V_{dc}^2 - 9V_{sx}^2}}{3\omega_e I_{Lx}} \quad (7)$$

Then finally, the optimal range of inductor design is:

$$\frac{V_{dc}}{6 \cdot f_s \cdot I_{ripple}} < L < \frac{\sqrt{4V_{dc}^2 - 9V_{sx}^2}}{3\omega_e I_{Lx}} \quad (8)$$

Moreover, a power flow stability analysis is considered following [23], thus, the maximum critic reactance can be defined for the voltage stability as:

$$X_L < \frac{V_{sx}}{2P} \quad \rightarrow \quad L < \frac{V_{sx}}{2P\omega_e} \quad (9)$$

where $P = \frac{V_{sx}V_{cx}}{X_L} \sin(\theta)$ is the averaged active power, θ is the phase-shift between the supply voltage vector and the converter voltage vector. Noticing that $\frac{V_{sx}}{2P\omega_e} < \frac{\sqrt{4V_{dc}^2 - 9V_{sx}^2}}{3\omega_e I_{Lx}}$, the final optimal design range for the L-filter is:

$$L_{min} = \frac{V_{dc}}{6 \cdot f_s \cdot I_{ripple}} < L < \frac{V_{sx}}{2P\omega_e} = L_{max} \quad (10)$$

To comply with the current distortion limits for systems rated 120 V_{rms} through 69 kV_{rms} on the IEEE std. 519-2014 a complementary frequency domain procedure is developed. Harmonic modeling of the line current is performed. And the Fourier Transform (FT) for the resultant line current waveform for the phase x is presented. Considering the above assumptions the line current can be divided into two signals:

$$i_{Lx} = i_{Refx} + i_{ex} \quad (11)$$

Where i_{Refx} is the current reference at $f_e = 60$ Hz and i_{ex} is the current error (current ripple) that comes from the switching activity at $f_{sw} = 10$ kHz (Fig 2). Performing Fourier Series (FS) analysis for i_{ex} and applying the superposition, the line

current i_{Lx} can be obtained. Performing FS analysis a general formula for FS coefficients can be obtained as in (12). The DC components and a_n coefficients disappear as i_{ex} has odd symmetry.

$$b_n = \frac{2\left[\int_0^{\frac{t_{on}}{2}} i_{ex1} s1\, dt + \int_{\frac{t_{on}}{2}}^{\frac{T_s + t_{off}}{2}} i_{ex2} s1\, dt + \int_{\frac{T_s + t_{off}}{2}}^{T_s} i_{ex3} s1\, dt\right]}{T_s} \quad (12)$$

where $s1 = \sin(n\omega_s t)dt$, $\omega_s = 2\pi f_s$ and i_{ex} is:

$$i_{ex} = \begin{cases} \frac{K_G}{d_x T_s} t & 0 < t < \frac{t_{on}}{2} \\ \frac{-K_G}{(1-d_x)T_s} t + \frac{K_G}{2(1-d_x)} & \frac{t_{on}}{2} < t < \frac{T_s + t_{off}}{2} \\ \frac{K_G}{d_x T_s}(t - T_s) & \frac{T_s + t_{off}}{2} < t < T_s \end{cases} \quad (13)$$

$$K_G = \Delta i_{Lx(on)}(1 - d_x) + \Delta i_{Lx(off)} d_x$$

Substituting the current variations (3) in (13) and evaluating the integrals in (12), i_{Lx} can be expressed as:

$$i_{Lx} = I_{Lx} \sin(\omega_e t) - \sum_{n=1}^{\infty} \frac{16V_{dc}}{3Lf_s (2\pi n)^2} \sin(\pi n D) s1 \quad (14)$$

where n stands for the harmonic number and $\omega_e = 2\pi f_e$.

Thus, the inductance design can be computed considering that at high frequencies the converter acts as a harmonic generator and the grid as a short-circuit [20], so the inductance can be calculated as in (15) [16], [20], [21], [24]:

$$L > \max\left(\frac{v_{cx}(n)}{n\omega_{sw}i_{ex}(n)}\right) \quad \text{for } n>2 \quad (15)$$

where $v_{cx}(n)$ is the converter voltage at the terminal of the phase x and it is a square waveform function as shown in Fig. 2. Its representation in FS is given in (16).

$$v_{cx} = \frac{2}{3}V_{dc}(2d_x - 1) + \sum_{n=1}^{\infty} \frac{8V_{dc}}{3n\pi} \sin(\pi n D) \cos(n\omega_s t) \quad (16)$$

B. DC-link Capacitor Design

The DC link capacitor provides an intermediate energy storage, which decouples the AC side of the VSC from the DC side. In the rectifier mode, the DC side is a load consuming power and the variation on the DC link depend in the control strategy of the VSC, but in the regenerative mode the instantaneous difference in active power (from the AC and DC side) is stored in C which causes the DC link voltage to vary. Hence, the value of C is determined by the constraint on the maximum allowed DC-link voltage ripple (V_{ripple}).

The design of the DC link capacitor is an extension for the procedure in [18]. In [18] the design considered a passive load in the DC side. But this paper proposes a methodology that considers a load or/and power supply (DC generator, battery, Photovoltaic Panel, etc) in the DC side.

The voltage-current relationship of a capacitor, it can be written as:

$$V_{dc}(t_f) - V_{dc}(t_i) = \frac{1}{C}\int_{t_i}^{t_f} i_{dc}(t)dt \quad (17)$$

where $t_f = t_i + T_s$, t_i is the initial time and $i_{dc}(t)$ is the capacitor current (Fig. 1) given in (18).

$$i_{dc}(t) = \begin{cases} S_w i_L t - i_o & \text{rectifier mode} \\ -S_w i_L t + i_o & \text{regenerative mode} \end{cases} \quad (18)$$

In the rectifier mode i_o is the current towards the load, in the regenerative mode i_o is the current from the DC source.

978-1-4673-9551-9/16 $31.00 © 2016 IEEE

By considering one specific instant, the switch function of one phase is one and the other two phases is zero (e.g. $[S_a\,S_b\,S_c]=[1\,0\,0\,]$ for the on state, then $[S_a\,S_b\,S_c]=[0\,1\,1\,]$ for the off state). Analyzing (17), the maximum DC-link ripple is at the minimum slope in the sinusoidal wave which is at the peak ($\omega_e t=\pi/2$). Assuming $d_x=1$ as the worst case which means that there is no PWM. Assuming constant i_o and solving (17), the capacitor minimum value can be defined. At no load C_{min} is defined. Thus, the design range is as in (19).

$$C_{min}=\frac{I_{Lx}}{f_s V_{ripple}}<C<\frac{I_{Lx}+i_o}{f_s V_{ripple}}=C_{max} \tag{19}$$

IV. Predictive Direct Current Control

The line current switching noise depends on the combination of an optimal L-filter and control design. The consolidation of this two techniques results in the cancellation of the current ripple to comply with the IEEE std. 519 and the L size reduction. The L-filter design procedure assumed that the current control was achieving its goal. Plethora of research can be found about the ARR control [2]–[6], [25], [7]–[10], [26], [11], [12], nevertheless a niche area is the predictive technique [2], [22], [27]–[29]. Direct power control (DPC) [3]–[6] can be explained by means of non-linear theory or predictive control [30] as it is done in [2] using the output regulation subspaces. However, DPC controller require the computation of the instantaneous active and reactive power. The control technique presented in [22], also known as predictive direct current control (PDCC) [26], [11], [31], [32], is a reduction of DPC where no power computation is required and keeping the control performance.

A. Predicitve Direct Current Control Description

The state space representation in (1) can be transformed to the stationary reference frame by the Clarke transformation. The representation in vector form is:

$$\frac{di_{L\alpha\beta}}{dt}=\frac{1}{L}\left(v_{s\alpha\beta}-R\,i_{L\alpha\beta}-v_{c\alpha\beta}\right)$$
$$\frac{dV_{dc}}{dt}=\frac{1}{C}\left(S_{w\alpha}i_{L\alpha}-S_{w\beta}i_{L\beta}\mp i_o\right) \tag{20}$$

where $f_{\alpha\beta}=f_\alpha+jf_\beta$, $f=\{v_s,i_L,v_c\}$. A discrete version of (20) is obtained by the following first order approximation:

$$\Delta i_{L\alpha\beta(k)}=\frac{T_s}{L}\left(v_{s\alpha\beta(k)}-R\,i_{L\alpha\beta(k)}-v_{c\alpha\beta(k)}\right) \tag{21}$$

where $f_{\alpha\beta}(k)$ represents the present sample value of the state vector $f_{\alpha\beta}$. It can be defined for every state space vector:

$$f_{\alpha\beta(k+1)}=f_{\alpha\beta(k)}+\Delta f_{\alpha\beta(k)} \tag{22}$$

Thus using (21) and (22), in a lossless inductor, it can be estimated the next line current value depending on the seven VSC switching states:

$$i_{L\alpha\beta(k+1)}=i_{L\alpha\beta(k)}+\frac{T_s}{L}\left(v_{s\alpha\beta(k)}-v_{c\alpha\beta(k)}\right) \tag{23}$$

Defining the line current error as:

$$e_{\alpha\beta(k)}=i_{L\alpha\beta\_ref(k)}-i_{L\alpha\beta(k)} \tag{24}$$

where $i_{L\alpha\beta\_ref(k)}=G v_{s\alpha\beta(k)}$, G is a scale factor used to build the current reference from the measure line voltage. For $G>0$ the operation mode of the VSC is rectifier and for $G<0$ is a regenerator.

Assuming sinusoidal and balance voltage supply, the voltage space vector at the next cycle of control can also be estimated as:

$$v_{s\alpha\beta(k+1)}=v_{s\alpha\beta(k)}e^{j\omega_e T_s} \tag{25}$$

Then, using the current reference definition, (23) and (25) in next cycle of (24), the error in the next cycle of control can be obtained:

$$e_{\alpha\beta(k+1)}=G v_{s\alpha\beta(k)}e^{j\omega_e T_s}-i_{L\alpha\beta(k)}+\frac{v_{s\alpha\beta(k)}-v_{c\alpha\beta(k)}}{L/T_s} \tag{26}$$

1) PDCC (No modulation): VSC Vector Selection

From (26), seven errors can be computed at every cycle of control depending on the seven switching states of the VSC. Thus, the control law is the minimum of a cost quadratic function (27) that indicates which switching state is the best:

$$J_{(k+1)\_n}=k_\alpha e_{\alpha(k+1)\_n}^2+k_\beta e_{\beta(k+1)\_n}^2 \tag{27}$$

where the subscript $n=\{1,2\ldots,7\}$ indicating the switching state of the VSC. The switching state n that achieves the minimum error is the best selection and k_α and k_β are the cost function weights.

2) PDCC (PWM): Optimum VSC Vector's Computation

Observing (26), it can be seen that the current error in the next cycle of control can be equal to zero:

$$G v_{s\alpha\beta(k)}e^{j\omega_e T_s}-i_{L\alpha\beta(k)}+\frac{v_{s\alpha\beta(k)}-v_{c\alpha\beta(k)}}{\frac{L}{T_s}}=0 \tag{28}$$

From where, it can be deduced the action of control, *input space* or $v_{c\alpha\beta(k)}$ as:

$$v_{c\alpha\beta(k)}=v_{s\alpha\beta(k)}-\frac{L}{T_s}\left(G v_{s\alpha\beta(k)}e^{j\omega_e T_s}-i_{L\alpha\beta(k)}\right) \tag{29}$$

It is worth to mention that the universe search of VSC switching vectors in this case is infinite as it uses modulation. The selection of the modulation can be arbitrary, however in this paper we selected a classical centered PWM.

V. PDCC Simulation and Experimental Results

For verification purposes of the L-filter design and control accuracy, simulations and experimental results are conducted and compared. The simulations were performed in Simulink MatLab while the experiment implementation was tested in a test bench using on Dspace 1104 as it is shown in Fig. 4. The ARR parameters used in the simulation and the experimental tests are indicated in table I. The configuration of the ARR is shown in Fig. 1, where the DC side is composed by a load resistance in parallel with a DC source. The DC source is composed by a three-phase bridge diode rectifier supplied by an AC supply through a step-up transformer to reach line-line voltage equal to 346 V_{rms}. This voltage will set the DC link in 490 V for purposes of bidirectional active power flow.

A. PDCC Simulation and Experimental Results: Optimum VSC's vector computation

The PDCC used in this section is the optimum VSC's vector computation described in section IV-B-2. The complete simulation and experiment time is 6 sec. The simulation step time of 10^{-5} sec. (100 kHz) while the acquisition frequency in the experiments was 20 kHz. In both simulation and experiments a centered PWM of 10kHz is used. The control coding, in the simulation, was programmed to run every 10 steps time of the simulation time (that means at 10 kHz) synchronized with the PWM. Thus, fair comparison can be achieve between simulation and experimental results. The control is programmed in the most basic MatLab language inside of an embedded MatLab function to emulate a micro-processor with low level programming language.

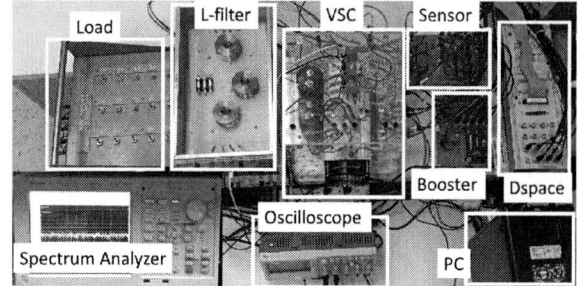

Figure 4: The experimental test setup

TABLE I.   THE SETUP PARAMETERS OF ARR

Parameters	Value
AC line-line voltage (V_{LNx})	208 V_{rms}
DC voltage supply (V_{dc})	490 V
Line current (I_{Lx})	G=0.04 of V_{LN}
Switching frequency (f_s)	10 kHz
DC Load resistance (R)	125 Ω
L-Filter (L)	12 mH
DC link capacitor (C)	1 mF
IGBT rating	1200 V, 50 A

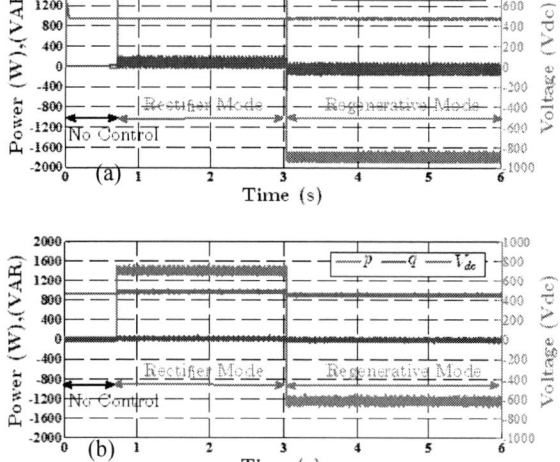

Fig. 5: Active power, reactive power and DC link voltage of the ARR. (a) Simulation and (b) Experimental.

The control is started at 0.715 sec. in rectification mode and then switched to regeneration mode at 3.03 sec. In Fig. 5(a) and 5(b) it can be noticed the three modes of operation. In Fig. 5(a), during the no control mode, there is no active or reactive power flow. During the rectification mode, 1600 watts of active power flows from the AC to the DC side, keeping the reactive power in zero, which means unity power factor operation. In the regenerative mode, 1600 watts flows from

the DC to the AC side also maintaining the instantaneous reactive power in zero (PF = - 1). The same performance can be observed in In Fig. 5(b), however, the power flow is 1400 watts due to the losses in the IGBTs and associated resistances of the capacitor, inductances and wires.

From, Fig. 5(a) and (b) it can be observed the fast transient dynamic of the optimum VSC's vector computation control. In both cases there is no overshooting and almost immediate response time at the mode transitions. Moreover, in Fig. 6 and 7 it can be shown the line current for the two transition moments: control starting and control mode change. The figures present the actual line current, the reference line current and the line-neutral voltage of the phase A for the simulation and the experiments respectively.

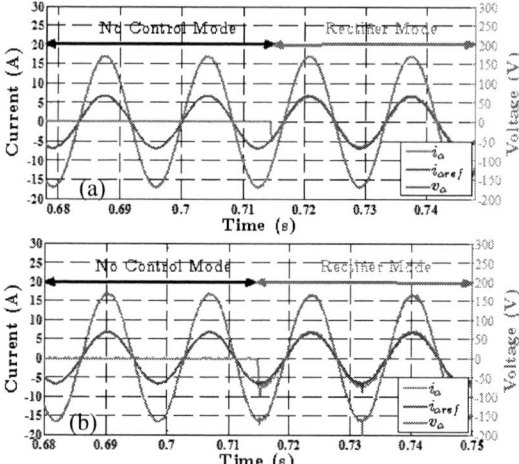

Fig. 6: Control starting: line current, reference line current and line-neutral voltage of phase A. (a) Simulation and (b) Experimental.

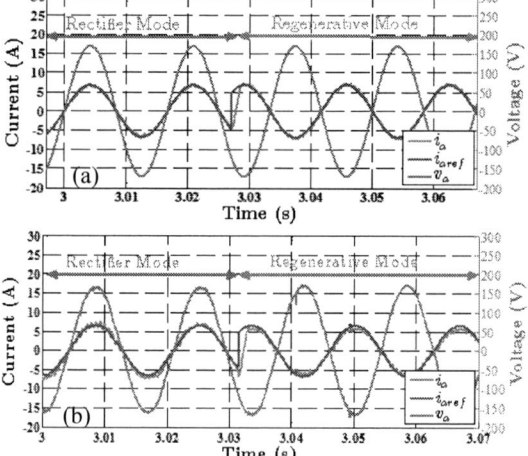

Fig. 7: Control mode transient from rectification to regeneration: line current, reference current and line voltage of phase A. (a) Simulation and (b) Experimental.

In Fig. 6(a) and (b), previous to the control starting, the line current is zero as the DC power supply is higher that the AC line-to-line voltage, thus the reverse diodes in the VSC are off. Notice, once the control start in rectification mode the actual line current tracks perfectly the current reference

($i_{La\_ref} = Gv_{sa}, G = 0.04$) achieving unity power factor operation (in both simulation and experiment). In Fig. 7(a) and (b) the control mode change from rectification to regeneration is presented. It can be observed that before the change the current is in phase with the line-neutral voltage (PF=1) and after the control change the actual line current and current reference ($i_{La\_ref} = Gv_{sa}, G = -0.04$) is 180 degrees phase shift (PF=-1). It can be notice the high correlation between simulation and experimental results, however, slightly higher ripple can notice in the experimental results. The quantification of the ripple will be studied in the next section.

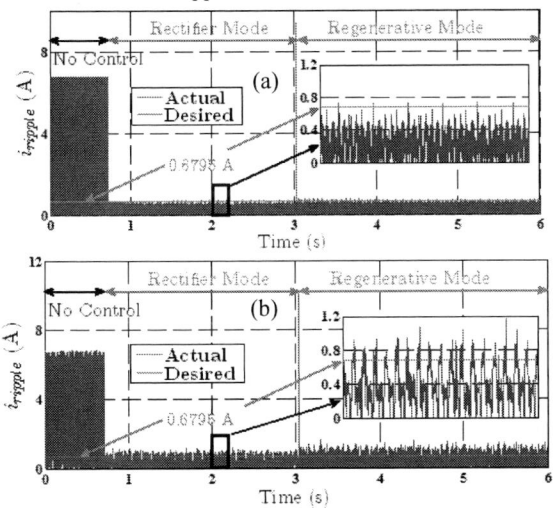

Fig. 8: Current error or current line ripple of phase A. (a) Simulation and (b) Experimental.

In Fig. 8, the actual line current ripple of the phase A ($i_{ripplea}$) and the desired ripple (I_{ripple}) is presented for the whole time of the simulation and experiment. The line current ripple is defined in Fig. 2 and is computed as $i_{ripplea} = i_a - i_{aref}$ at every cycle of control (10 kHz). In the simulation the data acquisition is at 100 kHz, a 10 points decimation process is programmed and synchronized with the PWM pulses. In the experimental results a 2 points decimation process were performed. The desired ripple is calculated from the design methodology. The maximum ripple I_{ripple} is obtained from the minimum inductance in (10). Thus, using the values in table I, it can be obtained $I_{ripple} = \frac{V_{dc}}{6 f_s L} = 0.6795\ A$. Notice that in the simulation (Fig. 8(a)) the current ripple is almost always below the desired value of design, however in the experimental results (Fig. 8(b)) the current ripple is sometimes over the design. Several considerations need to be taken into account:

- The worst case inductance design is chosen (L_{min}), thus $i_{ripplea}$ will be the maximum for the designed I_{ripple}.
- The design process gives a range of inductance. If we increase L (closer to L_{max}), then the actual current ripple decreases.
- The actual current ripple also depends on the control algorithm, which was assumed in perfectly operation for the design procedure.

- Errors of calibration in the acquisition and transducers nonlinearity affects the control performance.
- Parasitic components in the complete setup elements are not fully considered.
- PWM dead time used in the experiments is not consider neither in the simulation nor in the analytical design.

VI. ARR DESIGN CASES RESULTS

In order to corroborate the L-filter design procedure, 4 cases of inductance design has been chosen. The 4 cases are shown in Table II, Fig 9 and 10, for which four current ripples are calculated using the worst case (L_{min}) scenario in (10), $V_{dc} = 400V$, $I_{Lx} = 0.03V_{LN}$, $V_{LN} = 170\ V_{rms}$ and $f_s = 10\ kHz$. Notice that L_{max} is always the same as it does not depend on the current ripple for the stability issues.

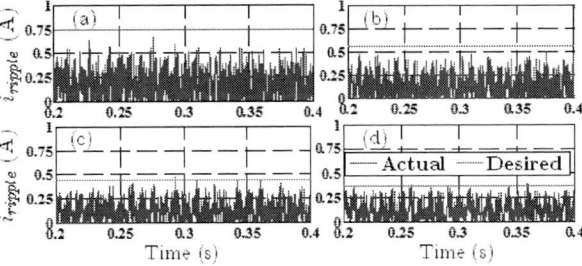

Fig. 9: Simulated line current ripple and desired current ripple. (a) L=9mH (b) L=12mH, (c) L=15mH and (d) L=18mH

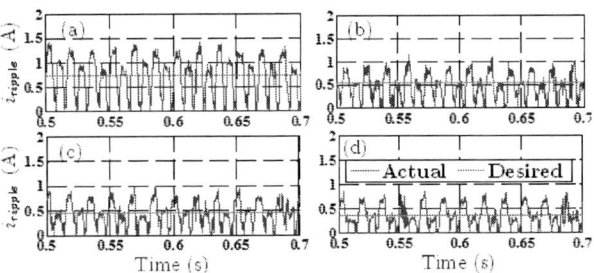

Fig. 10: Experimental results of line current ripple and desired current ripple. (a) L=9mH (b) L=12mH, (c) L=15mH and (d) L=18mH

In Fig. 9 it can be observed a zoom of the actual line current ripple from the simulation for the 4 inductance design cases. In the 4 cases, the actual current ripple ($i_{ripplea}$) is almost always below the desire current ripple (I_{ripple}). In Fig. 10 the same analysis is presented for the experimental results. In this figure the actual current ripple goes over the desire value due to reason discussed before. Table II summarizes the information shown in Figs. 9 and 10, where the average current ripple is computed. From these results it can be concluded that the experimental current ripples are in average below the desired value of design.

TABLE II.   COMPARITIVE RESULTS DESIRED CURRENT RIPPLE.

L (mH)		I_{ripple}	i_{ripple} Simulation		i_{ripple} Experimental	
Min.	Max.	Desired	Max.	Mean	Max.	Mean
9	22	0.74	0.7308	0.0224	1.431	0.7237
12	22	0.555	0.5702	0.0181	1.159	0.4649
15	22	0.444	0.5020	0.0153	0.9758	0.4284

| 18 | 22 | 0.369 | 0.4648 | 0.0138 | 0.8649 | 0.3582 |

Finally to verify the complementary frequency domain design, the second design case from table II (with a desired current ripple of 0.555 A) is chosen. A MatLab script is written to compute the harmonic model of (14) for the current ripple and (16) for the VSC voltage. Fig. 11(a) demonstrates the FS modelling of the current ripple and the VSC converter waveform for $d_x = 45\%$. It is chosen 45% to show the apparition of the even harmonics. In a symmetric 50% duty-cycle waveform, only odd harmonics appears.

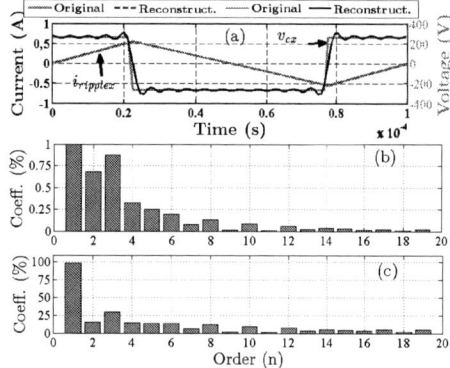

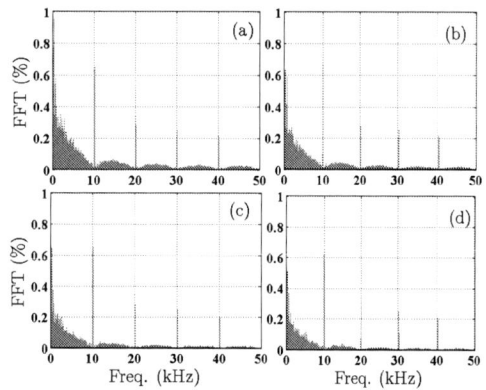

Fig. 11: (a) Current ripple, VSC voltage and their reconstruction in FS (b) Current ripple FS Coefficients (c) VSC voltage FS Coefficients.

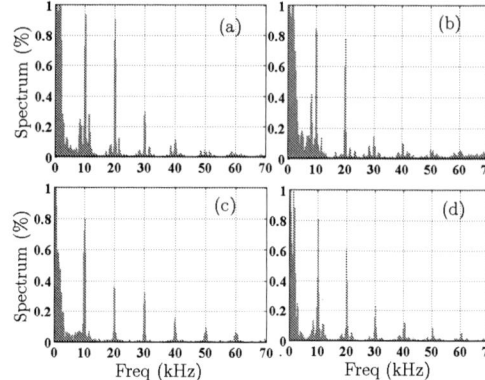

Fig. 12: Simulated results of line current spectrum. (a) L=9mH (b) L=12mH, (c) L=15mH and (d) L=18mH

In Fig. 11(b) and 11(c) the FS coefficients of the current ripple and VSC converter are presented. In this FS it is used 20 coefficients. From this analysis and using (15) it can be deduced that a minimum inductance of 12mH need to be designed for achieving the current ripple of 0.555 A (Fig. 11(a)). Exactly the same value was found with the time domain analysis shown in table II, demonstrating thus the two methods are complementary.

The FFT (Matlab function) and spectrum analysis (from a Tektronix RSA 5103A Real Time Signal Analyzer) of the actual line current of the phase A is presented in Fig. 12 and 13 for the simulation and the experimental tests respectively. In these figures the 4 inductance cases design are presented. It can be noticed that the second harmonic of f_s appears as expected. Due to the acquisition resolution in the simulation

(100 kHz) the maximum frequency shown is until the fifth harmonic. Moreover, higher frequencies are more attenuated and they have very small effect.

Fig. 13: Experimental results of line current spectrum. (a) L=9mH (b) L=12mH, (c) L=15mH and (d) L=18mH

A summary of the results in Fig. 11(b), 12 and 13 for the case of L=12mH can be observed in table III. From this table a very good correlation between simulation and experimental results can be observed (as well as from Figs. 12 and 13). The theoretical results has a huge discrepancy in the first harmonic, however the remaining harmonics are more correlated. This result difference comes from the assumptions considered in the frequency domain analysis, however, for practical issues it was demonstrated that the method is completely equivalent to the time domain for the inferior limit of the L-filter.

TABLE III.   COMPARITIVE LINE CURRENT HARMONICS FOR L=12MH.

Harm. order f_s	Theoretical (%)	Simulation (%)	Experimental (%)
1	8.72102	0.658	0.846
2	0,682134	0.348	0.781
3	0,874149	0.252	0.148
4	0,324374	0.215	0.104
5	0,249742	0.187	0.044

A summary for the DC-link capacitor design, using (19) and the values from the case 2 of table II, can found in table IV.

TABLE IV.   COMPARITIVE RESULTS DESIRED DC-LINK VOLTAGE RIPPLE

C (mF)		V_{ripple}	ΔV_{dc} Simulation		ΔV_{dc} Experimental	
Min.	Max.	Desired	Max.	Mean	Max.	Mean
1	1.667	0.5	0.66473	0.0259	5.01	1.2
0.75	1.228	0.679	0.98276	0.0480	-	-
0.5	0.8257	1.01	0.88648	0.0277	-	-
0.25	0.4108	2.03	1.30760	0.0356	-	-

VII.   CONCLUSIONS

A general design methodology for the ARR is developed. The strategy is based on the current area error and the current harmonic analysis in the time and frequency domain respectively. A comparative analysis of the design techniques was performed based on simulation and experimental test. It was demonstrated that the design analysis achieved the criteria of the current ripple in the time domain and the harmonic content in the frequency domain as well. The dynamic performance and accuracy of the PDCC was tested and verified by means of simulated and experimental results. It was corroborated its fast response, no time delay, no overshooting and moreover control of the active and reactive power without any power computation.

REFERENCES

[1] M. Liserre, A. Dell'Aquila, and F. Blaabjerg, "An overview of three-phase voltage source active rectifiers interfacing the utility," in *Power Tech Conference Proceedings, 2003 IEEE Bologna*, 2003, vol. 3, p. 8 pp. Vol.3–.

[2] G. Escobar, A. M. Stankovic, J. M. Carrasco, E. Galvan, and R. Ortega, "Analysis and design of direct power control (DPC) for a three phase synchronous rectifier via output regulation subspaces," *IEEE Trans. Power Electron.*, vol. 18, no. 3, pp. 823–830, May 2003.

[3] T. Noguchi, H. Tomiki, S. Kondo, and I. Takahashi, "Direct power control of PWM converter without power-source voltage sensors," *IEEE Trans. Ind. Appl.*, vol. 34, no. 3, pp. 473–479, May 1998.

[4] J. Ge, Z. Zhao, L. Yuan, T. Lu, and F. He, "Direct Power Control Based on Natural Switching Surface for Three-Phase PWM Rectifiers," *IEEE Trans. Power Electron.*, vol. 30, no. 6, pp. 2918–2922, Jun. 2015.

[5] Y. Zhang and C. Qu, "Direct Power Control of a Pulse Width Modulation Rectifier Using Space Vector Modulation Under Unbalanced Grid Voltages," *IEEE Trans. Power Electron.*, vol. 30, no. 10, pp. 5892–5901, Oct. 2015.

[6] J. A. Restrepo, J. M. Aller, A. Bueno, J. C. Viola, A. Berzoy, R. G. Harley, and T. G. Habetler, "Direct Power Control of a Dual Converter Operating as a Synchronous Rectifier," *IEEE Trans. Power Electron.*, vol. 26, no. 5, pp. 1410–1417, May 2011.

[7] M. Curkovic, K. Jezernik, and R. Horvat, "FPGA-Based Predictive Sliding Mode Controller of a Three-Phase Inverter," *IEEE Trans. Ind. Electron.*, vol. 60, no. 2, pp. 637–644, Feb. 2013.

[8] Y. Zhang and C. Qu, "Model Predictive Direct Power Control of PWM Rectifiers Under Unbalanced Network Conditions," *IEEE Trans. Ind. Electron.*, vol. 62, no. 7, pp. 4011–4022, Jul. 2015.

[9] S. Kwak and J.-C. Park, "Model-Predictive Direct Power Control With Vector Preselection Technique for Highly Efficient Active Rectifiers," *IEEE Trans. Ind. Inform.*, vol. 11, no. 1, pp. 44–52, Feb. 2015.

[10] J. A. Restrepo, J. M. Aller, J. C. Viola, A. Bueno, and T. G. Habetler, "Optimum Space Vector Computation Technique for Direct Power Control," *IEEE Trans. Power Electron.*, vol. 24, no. 6, pp. 1637–1645, Jun. 2009.

[11] J. Rodriguez, J. Pontt, C. A. Silva, P. Correa, P. Lezana, P. Cortes, and U. Ammann, "Predictive Current Control of a Voltage Source Inverter," *IEEE Trans. Ind. Electron.*, vol. 54, no. 1, pp. 495–503, Feb. 2007.

[12] M. Malinowski, M. Jasinski, and M. P. Kazmierkowski, "Simple direct power control of three-phase PWM rectifier using space-vector modulation (DPC-SVM)," *IEEE Trans. Ind. Electron.*, vol. 51, no. 2, pp. 447–454, Apr. 2004.

[13] Z. Chen, Y. Luo, Y. Zhu, and G. Shen, "Analysis and design of three-phase rectifier with near-sinusoidal input currents," in *Power Electronics and Motion Control Conference, 2009. IPEMC '09. IEEE 6th International*, 2009, pp. 1703–1–1703–7.

[14] Y. Lang, D. Xu, S. R. Hadianamrei, and H. Ma, "A Novel Design Method of LCL Type Utility Interface for Three-Phase Voltage Source Rectifier," in *Power Electronics Specialists Conference, 2005. PESC '05. IEEE 36th*, 2005, pp. 313–317.

[15] L. A. Moran, J. W. Dixon, and R. R. Wallace, "A three-phase active power filter operating with fixed switching frequency for reactive power and current harmonic compensation," *IEEE Trans. Ind. Electron.*, vol. 42, no. 4, pp. 402–408, Aug. 1995.

[16] Y. Tong, F. Tang, Y. Chen, F. Zhou, and X. Jin, "Design algorithm of grid-side LCL-filter for three-phase voltage source PWM rectifier," in *2008 IEEE Power and Energy Society General Meeting - Conversion and Delivery of Electrical Energy in the 21st Century*, 2008, pp. 1–6.

[17] N.-Y. Dai and M.-C. Wong, "Design considerations of coupling inductance for active power filters," in *2011 6th IEEE Conference on Industrial Electronics and Applications (ICIEA)*, 2011, pp. 1370–1375.

[18] A. Berzoy, A. Elsayed, T. Youssef, and O. A. Mohammed, "Improved design of controlled rectifier for reduced ripple resulting from integration of DC loads to AC systems," in *2014 IEEE PES General Meeting | Conference Exposition*, 2014, pp. 1–5.

[19] Y. Konishi, M. Ishibashi, and M. Nakaoka, "Three-phase current-source soft-switching PWM rectifier for high-power applications and its design considerations," in *Power Electronics and Variable Speed Drives, 1998. Seventh International Conference on (Conf. Publ. No. 456)*, 1998, pp. 133–138.

[20] M. Liserre, F. Blaabjerg, and S. Hansen, "Design and control of an LCL-filter-based three-phase active rectifier," *IEEE Trans. Ind. Appl.*, vol. 41, no. 5, pp. 1281–1291, Sep. 2005.

[21] A. Carlsson, *The back to back converter: control and design*. Sweden: Department of Industrial Electrical Engineering and Automation Lund Institute of Technology, 1998.

[22] A. Berzoy, J. Viola, and J. Restrepo, "Voltage space vector's computation for current control in three phase converters," *Rev. Fac. Ing.-Univ. Antioquia*, vol. 64, pp. 45–56, 2012.

[23] *Power System Analysis: Short-Circuit Load Flow and Harmonics, Second Edition*, 2 edition. Boca Raton: CRC Press, 2011.

[24] M. Bojrup, *Advanced Control of Active Filters in a Battery Charger Application*. SWEDEN: Department of Industrial Electrical Engineering and Automation (IEA) Lund Institute of Technology (LTH), 1999.

[25] A. Bouafia, F. Krim, and J.-P. Gaubert, "Direct power control of three-phase PWM rectifier based on fuzzy logic controller," in *IEEE International Symposium on Industrial Electronics, 2008. ISIE 2008*, 2008, pp. 323–328.

[26] D. Nedeljkovic, M. Nemec, K. Drobnic, and V. Ambrozic, "Direct current control of active power filter without filter current measurement," in *International Symposium on Power Electronics, Electrical Drives, Automation and Motion, 2008. SPEEDAM 2008*, 2008, pp. 72–76.

[27] J. A. Restrepo, J. M. Aller, A. Bueno, J. C. Viola, A. Berzoy, R. G. Harley, and T. G. Habetler, "Direct Power Control of a Dual Converter Operating as a Synchronous Rectifier," *IEEE Trans. Power Electron.*, vol. 26, no. 5, pp. 1410–1417, May 2011.

[28] S. Vazquez, J. I. Leon, J. A. Sanchez, E. Galvan, J. M. Carrasco, L. G. Franquelo, E. Dominguez, and G. Escobar, "Optimized Direct Power Control Strategy using Output Regulation Subspaces and Pulse Width Modulation," in *IECON 2006 - 32nd Annual Conference on IEEE Industrial Electronics*, 2006, pp. 1896–1901.

[29] J. M. Carrasco, E. Galvan, G. Escobar, A. M. Stankovic, and R. Ortega, "Direct active and reactive power control (DPQ) for a three phase synchronous rectifier," in *Power Electronics Specialists Conference, 2000. PESC 00. 2000 IEEE 31st Annual*, 2000, vol. 2, pp. 737–742 vol.2.

[30] Q. Lu, Y. Sun, and S. Mei, *Nonlinear Control Systems and Power System Dynamics*. Springer Science & Business Media, 2001.

[31] C. Townsend, C. Rowe, T. Summers, and T. Wylie, "Predictive current control of an Active Harmonic Filter," in *Power Engineering Conference, 2008. AUPEC '08. Australasian Universities*, 2008, pp. 1–6.

[32] J. Scoltock, T. Geyer, and U. K. Madawala, "A Model Predictive Direct Current Control Strategy With Predictive References for MV Grid-Connected Converters With -Filters," *IEEE Trans. Power Electron.*, vol. 30, no. 10, pp. 5926–5937, Oct. 2015.

A Robust Controller for Medium Voltage AC Collection Grid for Large Scale Photovoltaic Plants Based on Medium Frequency Transformers

Bahaa Hafez
Member, IEEE

Prasad Enjeti
Fellow, IEEE

Electrical and Computer Engineering Department
Texas A&M University
College Station, Texas, USA
enjeti@tamu.edu

Shehab Ahmed
Senior Member

Electrical and Computer Engineering Department
Texas A&M University at Qatar
Doha, Qatar
shehab.ahmed@qatar.tamu.edu

Abstract—A controller is proposed for a medium voltage medium frequency (MVMF) AC collection grid for utility scale Photovoltaic (PV) farms. A new MVMF AC collection grid was presented in previous work. It enhanced the system modularity, increased the system efficiency and reduced the passive component sizes. The MVMF AC collection grid comprises modules of inverter and medium frequency transformer that are series connected to construct the MF-AC collection grid. An AC/AC cycloconverter is used to interface the MF-AC collection grid to the low frequency utility grid. This paper presented an easy yet robust and fast controller for the proposed MVMF AC collection grid. An 80MW PV power plant located in Eggebek/Germany is used as a design example. The proposed controller is presented in details followed by simulation and hardware results to proof the fast/robust response of the proposed controller.

Keywords—Medium voltage medium frequency collection grid, Utility scale PV farms, Medium frequency transformer

I. INTRODUCTION

The U.S. PV penetration in 2013 is about 1.1% of the total renewable energy generation that is 15% of US total energy generation, the US department of energy is expecting that PV penetration to increase by 5% to reach a total of 20% in 2050 [1]. The utility scale PV farms are major contributor in the PV renewable energy penetration, their installations in 2014 first quarter is 4 times in their same quarter installations in 2013. In 2014/Q1 the pre-contracted installation was 2 times the installed [2]. A new structure for utility scale PV farms based on medium frequency transformer concept was introduced in [3]. It was shown that the proposed structure improves the system efficiency, enhance system modularity, reduce system size and installation cost. The proposed structure was built based on data from a state of the art PV farm in Eggebek Germany [4].

Different structures can be implemented to harness the sun power. Central and multi-string structures are the most economical for megawatt scale PV farms and results in the lowest watts per dollar. Economic studies were carried out to determine the optimum range for the inverter unit that should be used to construct a megawatts scale PV farm and the investigation concludes that based on existing inverter technology at present, the economical size of inverters range from 8 kW to 100 kW[5].

The electrical (galvanic) isolation between the PV panels and the utility grid is one requirement for the PV farm. The transformer provides the required isolation in two options, either at low frequency or at high frequency within the intermediate DC stage that is part of the voltage conversion stage. The transformer at the low frequency will have bigger size compared to the transformer on the high ; however, adding the extra conversion stage will decrease the overall system efficiency. Some countries such as UK, Italy and US require galvanic isolation but other countries such as Germany and Spain galvanic isolation is not mandated[6].

The DC connection, based on high frequency(HF) transformer of the semi string PV farm structure is investigated and compared considering different metrics such as system modularity, efficiency, reliability, cost and DC connectivity. It was shown that these systems are promising solutions however they suffer from high cost due to the usage of pricy core materials [7].

Smart transformers (ST) have intermitted DC link. A tradeoff between the transformer operating frequency and the price of the core materials should always be considered. In the current scenario, the optimum answer was to build the inner farm collection grid using medium frequency voltage as in [8]. [9] investigated similar solution but in wind farms. Increasing the collection grid frequency improves the overall efficiency plus it decreases the system passive components size. However, the main limiting factor for this is the cable losses and reactive power consumption.

Figure 1 shows the proposed structure. The 80 MW PV farm is divided into six Rings, each process 13.6 MW. Each Ring comprises six sectors where each sector proceses 1.5MW. Seven 3-phase voltage source inverter/medium frequency transformers sets harvest the solar energy and generate 600V/1000Hz voltage. The transformer output windings are cascaded to construct the MVMF voltage ac collection grid. A 3- phase cycloconverter converts the MV-MF voltage to the grid voltage. This paper presents a simple yet robust controller topology for the proposed structure, the details of the controller topology is presented in the following section followed by comprehensive simulation and hardware results.

978-1-4673-9551-9/16 $31.00 © 2016 IEEE

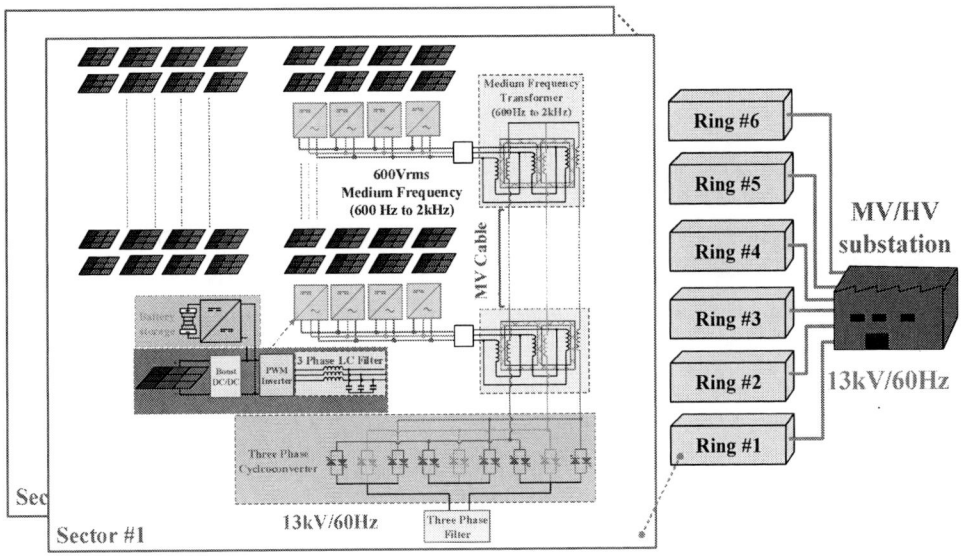

Figure 1. Layout for the proposed structure for utility scale PV farm based on medium frequency

II. PROPOSED CONTROLLER STRUCTURE

One sector equivalent block diagram is shown in Figure 2. Equations (1)-(4) show the system constraints. The current in all transformers outputs must be equal because the transformers primary windings' are series connected (1). The MF voltage is the sum of all the transformers output (2) and the total power injected to the grid is the power available by each inverter set minus the system losses (3). The fourth constrain is imposed by the cycloconverter buck nature, the MF voltage must be greater than the grid voltage by $\sqrt{3}$ (4). This will result in the inverter minimum voltage in (5) after considering the transformer turns ratio. The controller strategy for each sector has two parts. The first is the 3-phase inverter controller and the second is the cycloconverter controller. The inverter controller is chosen to be voltage controlled and it is responsible for maintaining the inverter voltage while the cycloconverter controller is closed loop power and is responsible for the MF grid current. The current will be determined based on the available power. In case there is enough power, the MF grid current increases and if the available power decreases the current will decrease. It is assumed that a communication link shares the power availability information between the different inverters and the cycloconverter.

$$I_1 = \cdots I_n = I = I_{MF} \tag{1}$$

$$\sum_{i=1}^{n} V_i = V_{MF} \tag{2}$$

$$\sum_{i=1}^{n} P_i = P_g - P_{loss} \tag{3}$$

$$V_{MF} = \sqrt{3} V_g \tag{4}$$

$$V_{InvOutMin} = \frac{V_{CylcoMin}}{n_{Inv} * N_{trans}} = \frac{\sqrt{3} * V_g}{n_{Inv} * N_{trans}} \tag{5}$$

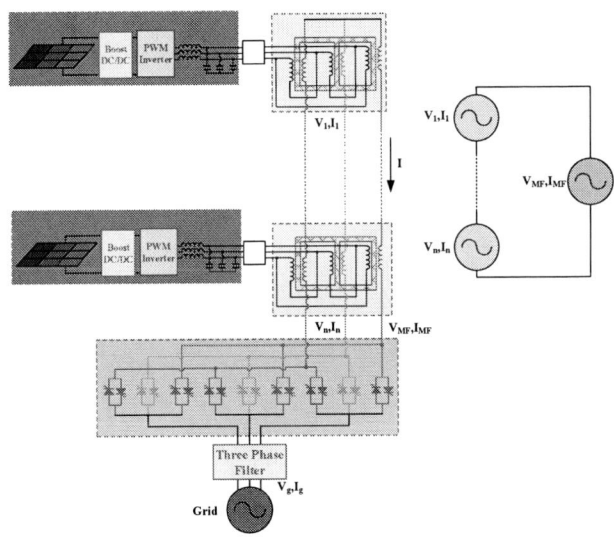

Figure 2. Simplified equivalent block diagram for the proposed controller

The PV inverter is chosen to be closed loop voltage controller implemented in dq frame(Synchronous frame) and it should be heavily relying on its integral part, this controller should not have any fast sudden changes to ensure the stability of the system [10]. The controller commanded voltage is calculated using the inverter PV algorithm as in Figure 3. The first step for the algorithm is to determine the maximum available power at the inverter and send it to the cycloconverter controller. In case of faulty condition and the fault in the inverter feeder, the inverter will command zero voltage. The inverter will command voltage either in normal operation or if

978-1-4673-9551-9/16 $31.00 © 2016 IEEE

it is not the faulty one. A Droop like controller will determine the output voltage of the inverter. If the available power at the inverter is higher than 85% of the nominal power then the applied voltage will be directly proportional with available power. If the available power is lower than the 85% of the nominal power, then the converter voltage is set to $V_{InvOutMin}$. The maximum voltage output is limited by the converter maximum overvoltage limit which can be 10%. The 85% limit of power is calculated from the minimum voltage limit that is imposed by the cycloconverter operation (5), number of inverters and transformer turns ratio. This control strategy allows for maximum power tracking and maintains the voltage in the required level for valid operation and is simple.

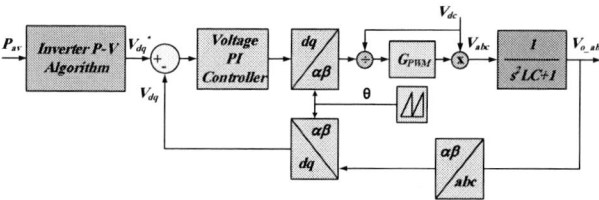

Figure 3. Block diagram for the PV inverter voltage controller

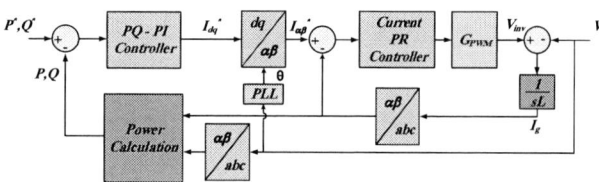

Figure 4. Block diagram for the cycloconverter controller

The cycloconverter controller will act as a power load (i.e. closed loop power) and its controller block diagram is shown in Figure 4.The commanded power to the controller is the total sum of all the available inverter powers from all the sector inverters, then after subtracting the system losses, the resultant power is injected to the grid at unity power factor.

This will ensure that the commanded current is always within the capability of each inverter and the inverter voltage controller will ensure that each inverter is operating at maximum efficiency and maintaining its voltage limit. The inner controller current loop is implemented in the stationary reference frame using proportional resonant controller[10]. The command for this controller is the active and reactive power errors that are calculated using the commanded instantaneous counter [11].

III. SIMULATION RESULTS

A sector with three inverter/transformer section and one cycloconverter is simulated using PLECS. The total power processed through this setup is 450kW where each inverter/transformer section process 150kW.

To test the controller robustness, two main cases were simulated. The first is for normal condition operation where the inverters are imposed to different penetration levels where its results are summarized in Figure 5 and Figure 6. The second is for a fault condition is imposed on Inverter#3 and its results are summarized Figure 7and Figure 8.

For normal operation, the power is changed by 0.25PU which is indicated by the voltage command change in Figure 5(a), (d) and (g). It can be noted that the inverters voltages are following the power change and the inverter currents are the same Figure 5(c) - (f) and (i). The bigger the available power the bigger the commanded voltage.

The MF collection grid voltages/currents are in Figure 6(a)-(b) and the cycloconverter active/reactive power command tracking are shown in Figure 6(c). It can be seen that the normal operation of MF grid is maintained during different power levels even when the change is very fast which proves the robustness of the proposed control topology. The injected grid current is changing according to the available power as in Figure 6(h).

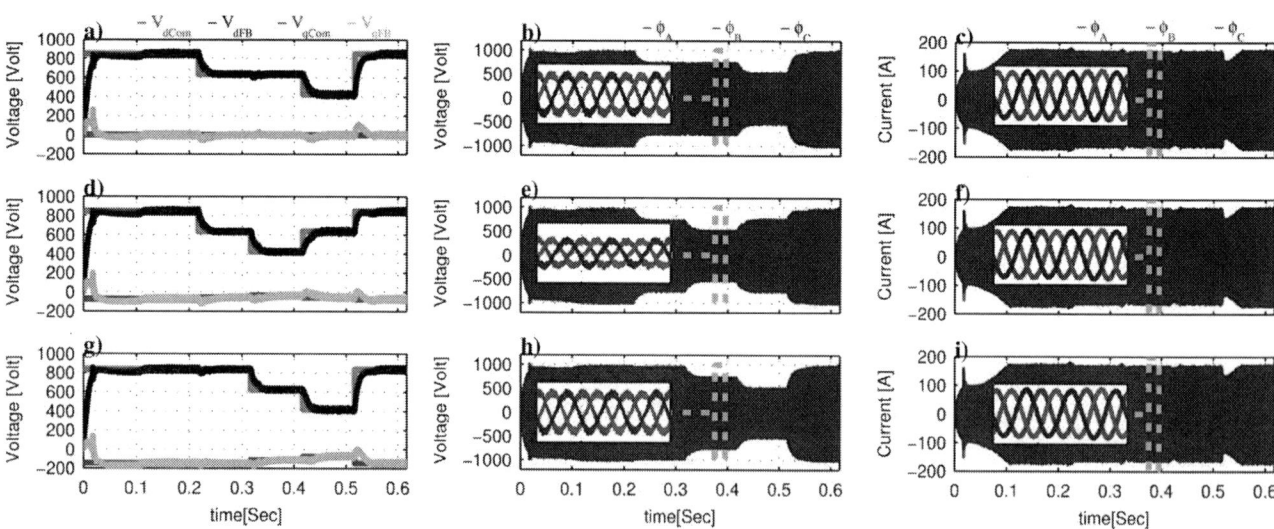

Figure 5. Time domain waveforms for (a) Inverter 1 synrhronous dq voltage command tracking (b) Inverter 1 line voltage (c) Inverter 1 line currrent (d) Inverter 2 synrhronous dq voltage command tracking (e) Inverter 2 line voltage (f) Inverter 2 line currrent (g) Inverter 3 synrhronous dq voltage command tracking (h)

978-1-4673-9551-9/16 $31.00 © 2016 IEEE

Inverter 3 line voltage (i) Inverter 3 line currrent during different PV peneration levels at normal operation

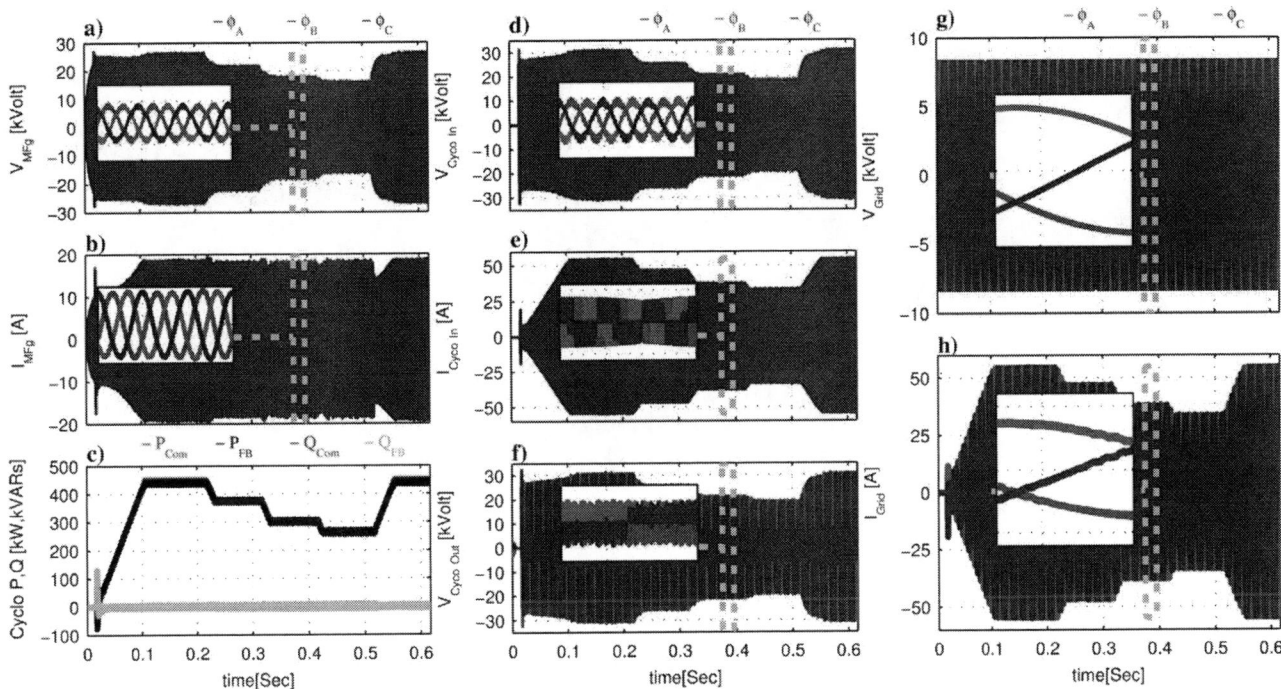

Figure 6. Time domain waveforms for (a) ac collection grid voltage (b) ac collection grid current (c) cycloconverter active and reactive power command tracking (d) cyclconverter input voltage (e) cyclcoconverter input current (f) cyclconverter output voltage (g) utility grid voltages (h) utility grid injected current during different PV peneration levels at normal operation

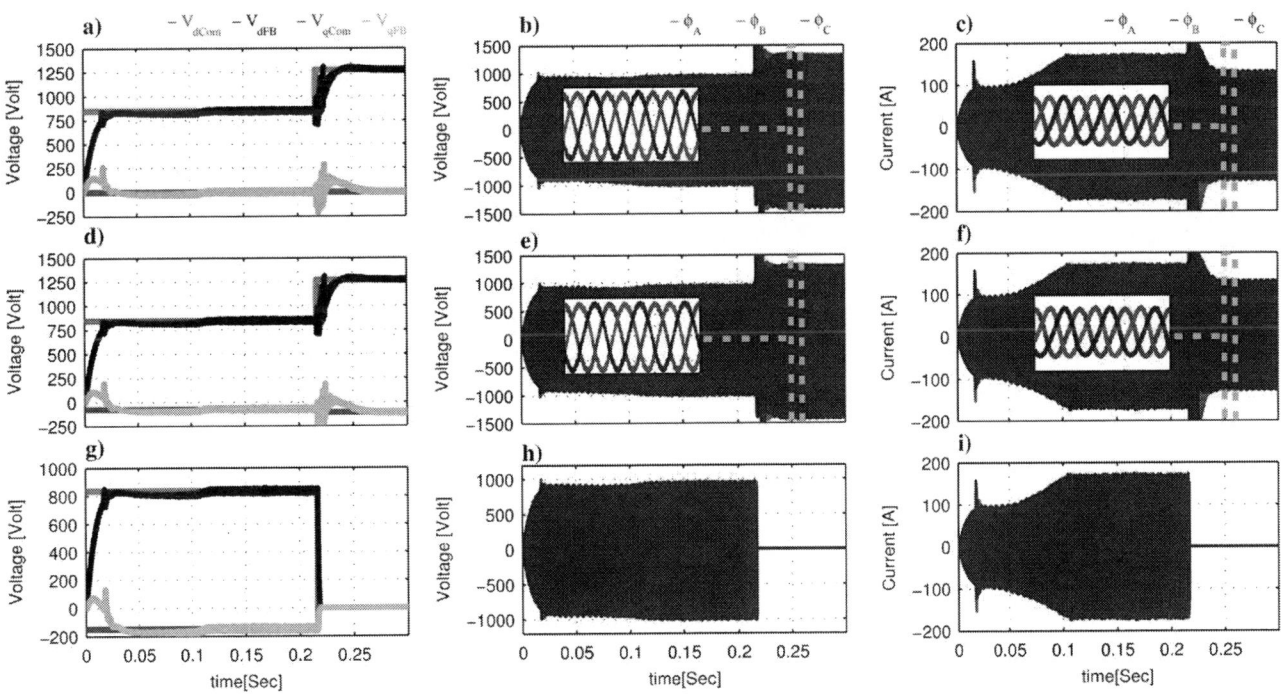

Figure 7. Time domain waveforms for (a) Inverter 1 synrhronous dq voltage command tracking (b) Inverter 1 line voltage (c) Inverter 1 line currrent (d) Inverter 2 synrhronous dq voltage command tracking (e) Inverter 2 line voltage (f) Inverter 2 line currrent (g) Inverter 3 synrhronous dq voltage command tracking (h) Inverter 3 line voltage (i) Inverter 3 line currrent during a fault on inverter 3

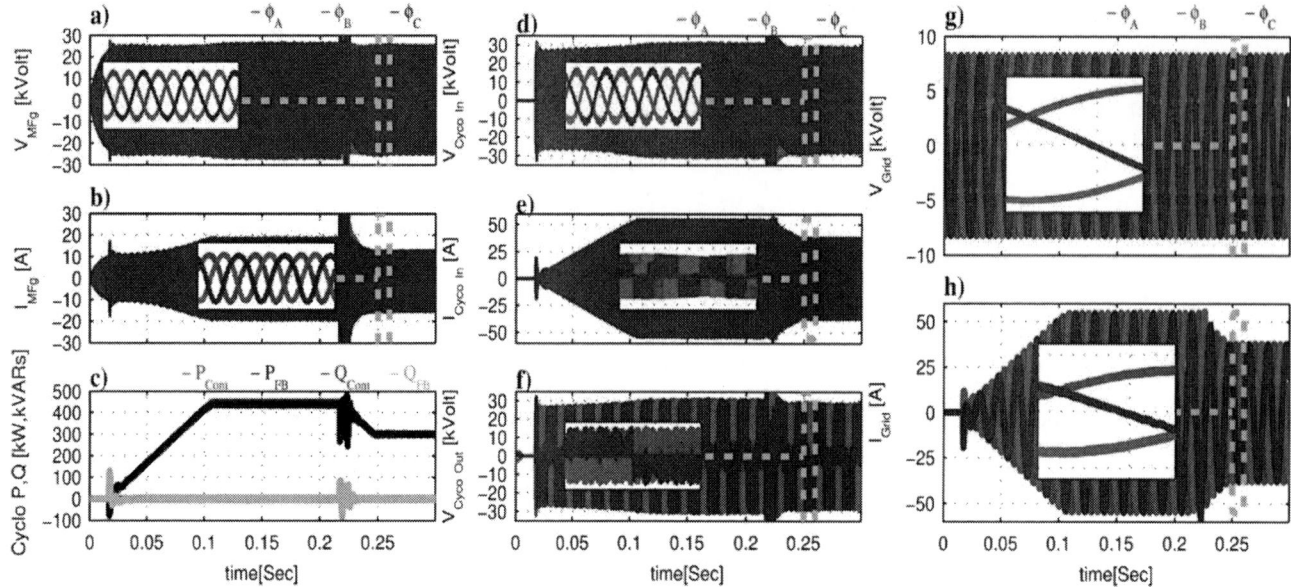

Figure 8. Time domain waveforms for (a) ac collection grid voltage (b) ac collection grid current (c) cycloconverter active and reactive power command tracking (d) cyclconverter input voltage (e) cyclcoconverter input current (f) cyclconverter output voltage (g) utility grid voltages (h) utility grid injected current during a fault on inverter 3

Figure 7 and Figure 8 summarized the results for the same system but in faulty condition where the third inverter will be switched out. It is assumed that the other two inverters could generate an extra voltage of 0.5PU to support the grid voltage after the outage of the faulty inverter. This assumption is valid only in this simulation case because the target is to test the controller robustness, however, in the real condition; there will be seven inverters, so if one goes out, the rest should provide an extra 1/6 PU voltage.

The voltage closed loop command tracking for the three inverters is shown in Figure 7(a),(d) and(g). As can be seen the system controller will withstand the fault status and increase the two inverter voltages to maintain the MF collection grid voltage and sustain the collection grid operation. The inverters voltage and currents are in Figure 7(b),(c),(e),(f),(h)and(i). It should be noted that the total voltage stays the same however; due to the power decrease the collection grid (i.e. medium frequency) current decreased and the two inverter currents followed this command while they are having the same current value.

The cycloconverter, at the same time, decreased the commanded power to the sum of the available to maintain the normal operation as indicated by its command tracking performance in Figure 8(c). The input voltage, input current and output switching voltage of the cycloconverter are shown in Figure 8(d),(e) and(f). Finally, the grid voltage and injected currents are shown in Figure 8(g) and(h) and they proof the seamless robust operation of the collection grid system during the fault.

IV. HARDWARE RESULTS

This section presents the hardware results for the proposed two structures. It first starts with the utility scale PV structure and explore different operating scenarios that include steady

state vs. transient, normal vs. faulty. These different scenarios are used to fully test the proposed controller and verify its robustness.

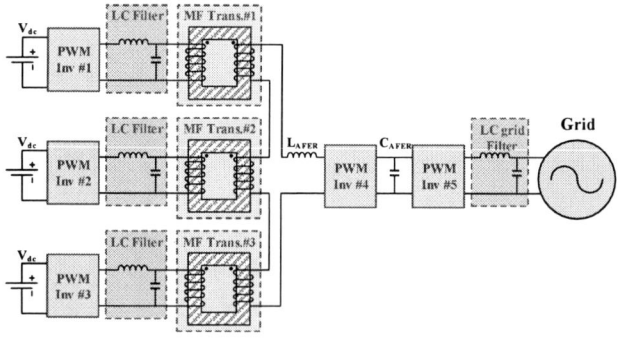

Figure 9. Block digarm for the experimental setup

A block diagram for the lab setup for utility scale PV is shown in Figure 9, it comprises five PWM based inverter modules. Inv#1, Inv#2 and Inv3 represent the PV connected inverters and they have the same voltage closed loop controller. This controller has only one control loop that is closed on the voltage and it is not responsible for the current. The output of each of these inverters is connected to the input of a medium frequency transformer. The transformers' outputs are series connected to construct the MF collection grid. Inv#4 and Inv#5 form the AC to AC converter that is responsible to interface the MF voltage to the utility grid. Inv#4 acts as an Active Front End Rectifier [12]. It forces the MF grid current to be in phase with the MF grid voltage. It has two control loops, the inner is current control loop and the outer is voltage control loop. The outer voltage control loop is controlling the voltage over the C_{AFER} to a fixed value and in the same time the inner current loop controls the input current from the MF grid to achieve this target. Inv#5 is the final stage that is controlling the injected grid current. It forces the injected current to be in phase with the grid voltage using a PR controller.

A TI TMS320F28335 DSP is used as the system controller[13]. It is programmed using Matlab/Simulink embedded coder toolbox[14]. A modular three phase inverter setup is used as a building block to construct the two proposed structures; it comprises SK13GD063 Semikron power module and SKHI61R Semikron gate driver circuit[15, 16]. ValueCAN3 is used for real time communication with the DSP that provides a way to debug/tune the controller and change the command to implement the different operating scenarios [17] . Figure 10 shows the lab setup used to test the proposed controller structure and Table 1 summarized the circuit parameters.

Figure 10. Lab setup for testing the proposed controller

TABLE 1. PV SETUP CIRCUIT PARAMETERS

Parameter	Value
DSP switching frequency	10kHz
PV Inverters DC link voltage	50V
Medium frequency	200Hz
AFER DC link voltage	160V
Grid voltage (peak)	100V
Grid frequency	60Hz
Grid current (peak)	2.5A
L (LC filter)	220uH
C (LC filter)	5uF
L_{AFER}	6mH
C_{AFER}	580uF
L(LC grid filter)	1mH
C(LC grid filter)	60uF

Table 2 summarized the different scenarios that the collection grid is operated at. It shows each inverter commanded voltage during each case when it will be applied in the cases to come below. For example; during normal inverter voltage case Inv#1, Inv#2 and Inv#3 will be commanded 40V and this will result in MF grid voltage of 150V. In all cases the DC link voltage across C_{AFER} has a constant command of 160V and the grid voltage is kept constant at 100V and the injected current is kept constant at 2.5A.

TABLE 2. PV DIFFERENT CASES VOLTAGES

Case	Inv #1	Inv #2	Inv #3	MF Voltage
Normal Inverter Voltage	50	50	50	160
One Zero Inverter Voltage	75	0	75	160

Figure 11 shows the case when the normal voltage cases are imposed on the structure. It can be noted that the whole grid architecture performs as expected. All the PV inverter grid voltages produce synchronized voltages that are summed together to build the MF grid and the collected power is injected to the utility grid.

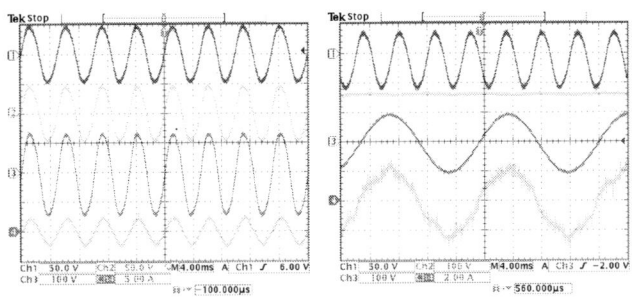

—Inv1 V—Inv 2 V—MF grid V—MF grid I   —AFER Vdc—Inv 3 V— Grid V— Grid I

Figure 11. Steady state time domain waveforms for normal inverter voltage case

The time domain waveforms for the transient periods between the two cases are shown. Figure 12 shows the waveforms for normal to zero and Figure 13 shows the waveforms for zero to normal. The shown transient hardware results prove the robustness of the proposed controller topology as it can handle the loss of power from one inverter and get it back again while maintaining the power continuity.

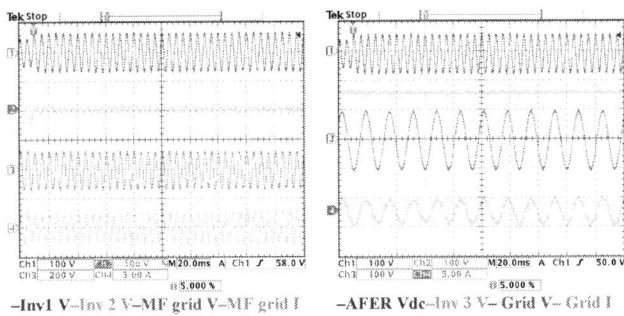

—Inv1 V—Inv 2 V—MF grid V—MF grid I   —AFER Vdc—Inv 3 V— Grid V— Grid I

Figure 12. Normal to Zero transient time domain waveforms for the five inverters

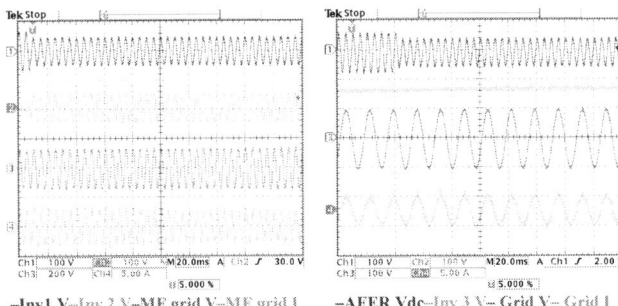

—Inv1 V—Inv 2 V—MF grid V—MF grid I   —AFER Vdc—Inv 3 V— Grid V— Grid I

Figure 13. Zero to Normal transient time domain waveforms for the five inverters

V. CONCLUSION

A robust controller topology for Medium Frequency Medium voltage PV farms is presented in this paper. The proposed controller is adopted to control a new utility scale PV structures based on medium frequency concept. The controller topology is built around the system physical constraints. It leaves the voltage responsibility to the inverter controllers and current/power responsibility to the cycloconverter controller. The presented simulation/hardware results prove the robustness of the proposed controller topology during different scenarios.

VI. REFERENCES

[1] S. Esterly and R. Gelman, *2013 Renewable Energy Data Book*: US Department of Energy, Energy Efficiency & Renewable Energy Laboratory, 2013.

[2] Solar Energy Industries Assocation, "Solar Market Insight Report 2013 Year in Review," 2013.

[3] B. Hafez, H. S. Krishnamoorthy, P. Enjeti, U. Borup, and S. Ahmed, "Medium voltage AC collection grid for large scale photovoltaic plants based on medium frequency transformers," in *Energy Conversion Congress and Exposition (ECCE), 2014 IEEE*, 2014, pp. 5304-5311.

[4] Danfoss Solar. (June 2012), Optimized system layout , Online Available: http://www.danfoss.com/NR/rdonlyres/6507E2ED-D210-422C-976B-6E8DC943B65D/0/DKSIPM204A102_Eggebek_TechCaseStudy_WEB.pdf

[5] Z. Moradi-Shahrbabak, A. Tabesh, and G. R. Yousefi, "Economical design of utility-scale photovoltaic power plants with optimum availability," *Industrial Electronics, IEEE Transactions on*, vol. 61, pp. 3399-3406, 2014.

[6] T. Kerekes, *Analysis and modeling of transformerless photovoltaic inverter systems*: Videnbasen for Aalborg UniversitetVBN, Aalborg UniversitetAalborg University, Det Teknisk-Naturvidenskabelige FakultetThe Faculty of Engineering and Science, Institut for EnergiteknikDepartment of Energy Technology, 2009.

[7] R. Pena-Alzola, G. Gohil, L. Mathe, M. Liserre, and F. Blaabjerg, "Review of modular power converters solutions for smart transformer in distribution system," in *Energy Conversion Congress and Exposition (ECCE), 2013 IEEE*, 2013, pp. 380-387.

[8] X. Li, X. Ai, and Y. Wang, "Study of single-phase HFAC microgrid based on Matlab/Simulink," in *Electric Utility Deregulation and Restructuring and Power Technologies (DRPT), 2011 4th International Conference on*, 2011, pp. 1104-1108.

[9] S. Gierschner, H.-G. Eckel, and M.-M. Bakran, "A competitive medium frequency AC distribution grid for offshore wind farms using HVDC," in *Power Electronics and Applications (EPE), 2013 15th European Conference on*, 2013, pp. 1-10.

[10] J. Irwin, M. P. Kazmierkowski, R. Krishnan, and F. Blaabjerg, *Control in power electronics: selected problems*: Academic press, 2002.

[11] H. Akagi, E. H. Watanabe, and M. Aredes, *Instantaneous power theory and applications to power conditioning* vol. 31: John Wiley & Sons, 2007.

[12] R. Labaki, B. Kedjar, and K. Al-Haddad, "Single-Phase Active Front End Converter with series compensation," in *Industrial Electronics, 2006 IEEE International Symposium on*, 2006, pp. 769-774.

[13] T. Instruments. (2012), TMS320F28335, TMS320F28334, TMS320F28332, TMS320F28235, TMS320F28234, TMS320F28232 Digital Signal Controllers (DSCs) Data Manual [Online]. Available: http://www.ti.com/lit/ds/symlink/tms320f28335.pdf

[14] MathWorks(accessed on May 31,2015). Embedded Coder Generate C and C++ code optimized for embedded systems [Online]. Available: http://www.mathworks.com/products/embedded-coder/

[15] Semikron. (2006), SK13GD063 [Online]. Available: http://shop.semikron.com/out/media/ds/SEMIKRON_DataSheet_SK_13_GD_063_24508201.pdf

[16] Semikron. (2007), SKHI 61 (R) [Online]. Available: http://shop.semikron.com/out/media/ds/SEMIKRON_DataSheet_SKHI_61_R_L6100061.pdf

[17] Intrepidcs(accessed on May 31,2015). ValueCAN3 [Online]. Available: http://store.intrepidcs.com/ValueCAN3-DW-2-Channel-p/vcan-dw3.htm

Optimal Low Switching Frequency Pulse Width Modulation of Current-Fed Five-Level Inverter for Solar Integration

Gnana Sambandam K*, *Student Member, IEEE*, Akshay K. Rathore*, *Senior Member, IEEE*,
Amarendra Edpuganti, *Student Member, IEEE* and Dipti Srinivasan*, *Senior Member, IEEE*
Department of Electrical and Computer Engineering
*Solar Energy Research Institute of Singapore
National University of Singapore, 117583, Singapore
gnana@u.nus.edu; eleakr@nus.edu.sg; amarendra@u.nus.edu; dipti@nus.edu.sg;

Abstract—Current fed multilevel converters have several advantages in terms of high power capability, transformer less operation, short circuit protection and excellent quality of waveforms. For high power applications, these converters are required to be operated under low device switching frequency to necessitate thermal constraints of the semiconductor devices. However, the converter output currents under low device switching frequency results in higher total harmonic (THD) distortion. Synchronous optimal pulse-width modulation (SOP) technique is an emerging low device frequency modulation technique which has been successfully implemented for voltage source multilevel inverters without compromising on total harmonic distortion. This paper proposes and implements a modified SOP technique for five level current source inverter topology. The experimental results obtained from a low power prototype demonstrated its effectiveness.

Index Terms—Multilevel inverters, current source inverter, solar power integration, synchronous optimal pulsewidth modulation

I. INTRODUCTION

In recent years, the solar energy generation has seen exponential growth due to continual decline in the cost of photovoltaic (PV) cells. At present, several LSPV power plants have been installed around the world by growing group of companies [1]. The total installed capacity has reached up to 177 GW by the end of 2014 [2]. The major component of LSPV system is the power conditioning system for solar to grid integration. The power conditioning stage has two basic components such as dc-dc converter that extracts maximum power from photo-voltaic modules and dc-ac inverter which integrates the solar power to grid. The dc-ac inversion can be achieved by voltage-source or current-source inverters. As the photo-voltaic cells are current source in nature, current source inverter is a suitable inverter for LSPV system. For multi-megawatt LSPV plant, simple two-level inverters are not sufficient to meet the total power. Paralleling such two-level inverters can able to share the power among inverters, but they do not bring any additional benefit in terms of power quality. This has been overcome by the introduction of multilevel converters for PV applications [3].

Current-source multilevel converter have been proposed in literature based on the principle of duality between voltage and current sources [4]. These inverters achieve high operating

current capability with low or medium current semiconductor devices. In addition, current-source multilevel inverters offer advantages in terms of high voltage gain, reduced transformer turns ratio, lower duty cycle loss, inherent short-circuit protection and so on [5], [6].

For high power applications, the converter has to be operated at low device frequency (< 1 kHz) to limit the switching loses in semiconductor devices. However, low device switching frequency operation leads to higher harmonic distortion of output currents. Therefore, the challenge is to minimize the harmonic distortion of output current at low device switching frequency operation. Classical modulation techniques such as sinusoidal pulse-width modulation (SPWM) and space vector modulation (SVM) techniques require higher device switching frequency to achieve better quality of output current waveforms [7]. Some of the notable low device switching frequency modulation techniques include selective harmonic elimination (SHE) technique, staircase modulation technique, synchronous optimal pulse-width modulation technique (SOP) and model predictive control (MPC) technique. Among them, SOP and MPC techniques have better steady-state as well as dynamic performance [8]. However, key issue of MPC technique is higher computational burden.

SOP is an emerging low device switching frequency modulation technique that has been successfully implemented in commercial medium voltage (MV) drives [9]. It is based on an off-line optimization technique to compute switching patterns that minimize harmonic distortion of inverter output currents. Till now, implementation of SOP technique has been limited to voltage-fed (buck) multilevel converter (MLC) topologies [10]–[12]. The state-of-the-art SOP technique requires some modifications when applied to CSI topology due to voltage boost and current source operation. This leads to additional constraints on switching commutations. Therefore, a modified SOP technique should be proposed to handle the operational constraints of CSI topology.

The contribution of the paper are as follows: 1) analysis and implementation of modified SOP technique to handle constraints of CSI topology by utilizing a special conversion method, 2) low total harmonic distortion (THD) of output current within grid integration standards, while limiting the

978-1-4673-9551-9/16 $31.00 © 2016 IEEE

device switching frequency to 350 Hz.

The contents of the paper are organized as follows: Circuit topology and operation of five-level (5L) CSI is discussed in Section II; Modulation and control of 5L CSI with modified SOP technique for current source operation is proposed in Section III; Experimental results are presented in Section IV to validate the proposed technique.

II. TOPOLOGY AND OPERATION OF 5L CSI

The topology of 5L CSI is shown in Fig. 1. It consists of two three-level (3L) CSI supplied with separate equal constant current sources of magnitude I_{dc}. Each 3L inverter has three phase legs and each leg has one top, S_{qxp} and one bottom S_{qxn} semiconductor device with series power diodes D_{qxp}, and D_{qxn} ($x \in a, b, c$ and q represents inverter 1 or 2). The series diode provides a reverse voltage blocking capability and unidirectional input-current flow for the inverter. The outputs of 3L inverters are connected in parallel to produce output current i_x with five levels $2I_{dc}$, I_{dc}, 0, $-I_{dc}$, and $-2I_{dc}$. At the output side, CSI requires three phase capacitors C_{af}, C_{bf}, and C_{cf}, to assist in the commutation of power semiconductor devices. In addition, capacitor also acts as a harmonic filter, improving the grid current waveforms.

The operation of CSI imposes following constraints [13] :

1) continuous path for input DC current i_{dc}
2) inverter output current i_x should be defined

By applying these constraints for 5L CSI, the switching states and output currents of 3L and 5L CSI can be obtained as shown in Table I and Table II respectively. There are 19 possible 5L output combination and each combination is achieved by one or more switching combination of two 3L CSIs. These redundancies can be utilized for achieving equal commutation of semiconductor devices. The method of utilizing redundant states is explained in detail in Section III-B5. It should also be observed from the Table II that the summation of three phase current levels is zero for each switching state.

III. MODULATION AND CONTROL

The modulation and control of 5L CSI involve identifying optimal reference signal using off-line computation and assign the optimal pulses for each semiconductor device. SOP technique is utilized to obtain optimal quasi-sine reference waveform and a conversion method is needed to incorporate constraints on the reference waveform as mentioned in Section II. Conversion method and SOP optimization technique are given next.

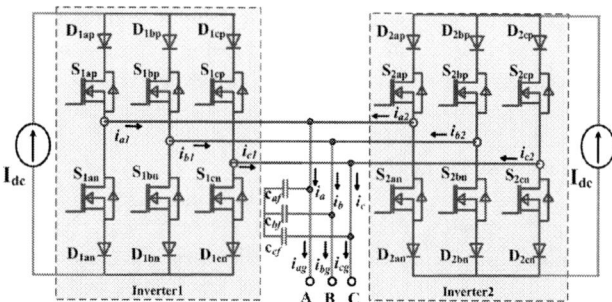

Fig. 1: 5L CSI topology.

TABLE I
SWITCHING STATES AND OUTPUT CURRENT OF 3L CSI

State	I_a	I_b	I_c	S_{ap}	S_{an}	S_{bp}	S_{bn}	S_{cp}	S_{cn}
1	0	I_{dc}	$-I_{dc}$	0	0	1	0	0	1
2	I_{dc}	$-I_{dc}$	0	1	0	0	1	0	0
3	I_{dc}	0	$-I_{dc}$	1	0	0	0	0	1
4	$-I_{dc}$	0	I_{dc}	0	1	0	0	1	0
5	$-I_{dc}$	I_{dc}	0	0	1	1	0	0	0
6	0	$-I_{dc}$	I_{dc}	0	0	0	1	1	0
7	0	0	0	1	1	0	0	0	0
8	0	0	0	0	0	1	1	0	0
9	0	0	0	0	0	0	0	1	1

TABLE II
SWITCHING STATES AND OUTPUT CURRENT OF 5L CSI

State	I_a	I_b	I_c	$(3L_1 state, 3L_2 state)_{option}$
1	0	0	0	$(7,7)_1$ or $(8,8)_2$ or $(9,9)_3$
2	0	I_{dc}	$-I_{dc}$	$(7,1)_1$ or $(1,7)_2$
3	0	$2I_{dc}$	$-2I_{dc}$	$(1,1)_1$
4	0	$-I_{dc}$	I_{dc}	$(7,6)_1$ or $(6,7)_2$
5	0	$-2I_{dc}$	$2I_{dc}$	$(6,6)_1$
6	I_{dc}	0	$-I_{dc}$	$(8,3)_1$ or $(3,8)_2$
7	I_{dc}	I_{dc}	$-2I_{dc}$	$(3,1)_1$ or $(1,3)_2$
8	I_{dc}	$-I_{dc}$	0	$(9,2)_1$ or $(2,9)_2$
9	I_{dc}	$-2I_{dc}$	I_{dc}	$(2,6)_1$ or $(6,2)_2$
10	$2I_{dc}$	0	$-2I_{dc}$	$(3,3)_1$
11	$2I_{dc}$	$-I_{dc}$	$-I_{dc}$	$(2,3)_1$ or $(3,2)_2$
12	$2I_{dc}$	$-2I_{dc}$	0	$(2,2)_1$
13	$-I_{dc}$	0	I_{dc}	$(8,4)_1$ or $(4,8)_2$
14	$-I_{dc}$	I_{dc}	0	$(9,5)_1$ or $(5,9)_2$
15	$-I_{dc}$	$2I_{dc}$	$-I_{dc}$	$(1,5)_1$ or $(5,1)_2$
16	$-I_{dc}$	$-I_{dc}$	$2I_{dc}$	$(4,6)_1$ or $(6,4)_2$
17	$-2I_{dc}$	0	$2I_{dc}$	$(4,4)_1$
18	$-2I_{dc}$	I_{dc}	I_{dc}	$(4,5)_1$ or $(5,4)_2$
19	$-2I_{dc}$	$2I_{dc}$	0	$(5,5)_1$

$3L_1 \rightarrow$ inverter-1 $3L_2 \rightarrow$ inverter-2

A. Conversion method

The objective of the conversion method is to derive realizable waveform from any quasi-sine reference wave in order to include the operation constraints mentioned in Section II, while maintaining the objective of the reference waveform. It should be noticed from Table II, that at any instant of time, the sum of three phase output current levels should be zero. For example, the inverter needs to be modulated to produce 5L three-phase output current waveforms, $i_{a,ref}$, $i_{b,ref}$, and $i_{c,ref}$, as shown in Fig. 2. The equation that governs the constraint is given by,

$$i_{a,ref} + i_{b,ref} + i_{c,ref} = 0 \qquad (1)$$

where, $i_{a,ref}$, $i_{b,ref}$, and $i_{c,ref}$ are reference currents for three phases with 120° rotation. Consider a small time interval Δt, the sum of these three waveforms during this interval gives a non-zero value -1. Therefore, these references cannot be utilized for operating 5L CSI.

Fourier-series analysis demonstrates that the sum of three waveforms with 120° phase shift will have only third order harmonic components $(3, 6, 9, 12, ..)$. Hence, in order to sat-

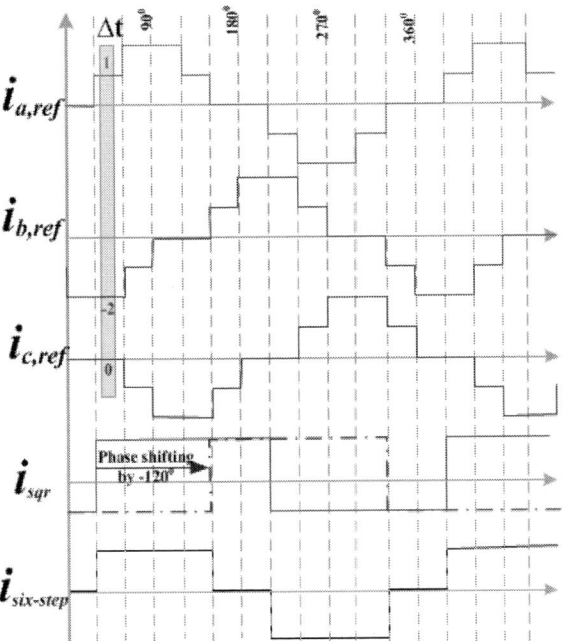

Fig. 2: Reference waveforms considered for analysing conversion method.

isfy (1), the current reference should not contain third order harmonics. Elimination of third order harmonic component on any quasi reference waveform can be achieved by delaying it to 120^0 and subtracting the delayed signal from actual waveform. For example, consider square wave i_{sqr} as reference, it is then phase shifted by 120^0 as shown in Fig. 2, and subtracted to obtain reference signal, $i_{six-step}$. This six-step waveform can be directly utilized as a reference for 3L CSI to produce maximum fundamental ($m=1$). It could be observed that $i_{six-step}$ is a 3L reference signal obtained from 2L square waveform. Similarly for 3L reference waveform this method will produce 5L reference signal and so on. Hence for 5L CSI, this conversion method requires a 3L optimal switching angles.

B. Modified SOP for current source operation

SOP technique is the combination of synchronous PWM and optimization. The technique is proposed for 3L VSI topology [14], and later a generalized methodology has been developed for voltage-fed multilevel converters with any number of voltage levels [11]. Synchronous here refers to the ratio between the device switching frequency f_s and operating frequency f should be an integer. This is utilized to eliminate subharmonic frequencies, which are undesirable in many applications. The first step in SOP technique is to determine the number of switching angles for each steady state operating point and then optimization is performed to obtain switching angles that minimize the harmonic distortion of inverter output currents. In the last step, optimal switching angles are assigned to each semiconductor device to realize optimal current waveforms based on a systematic procedure.

The classical SOP technique requires some modifications in order to modulate and control 5L CSI. One possible option is to modify the optimization algorithm to directly obtain 5L optimal switching angles that satisfies operational constraints

of 5L CSI topology. Another option is to modify the last step of SOP technique to convert 3L optimal switching angles into 5L optimal switching angles and assign them to each power semiconductor device [4]. In this paper, the second approach has been implemented for SOP modulation of 5L current-fed multi-level converter topology.

1) Mathematical analysis: Consider 3L reference current waveform i_{3L} shown in Fig. 3 with switching angles at α_1 to α_5 in a quarter period. Half-wave and quarter-wave symmetries is introduced to eliminate all even order harmonics in the switching pattern. Using Fourier series analysis, harmonic components of i_{3L} can be obtained as,

$$i_{k,3L} = \frac{4I_{peak}}{k\pi}\left(\sum_{i=1}^{N_\alpha} s(i)cos(k\alpha_i)\right) \quad (2)$$

where, I_{peak} represents the peak value of reference current waveform, k is the harmonic order ($k=3,5,7...$), $i_{k,3L}$ is the amplitude of kth harmonic current component of i_{3L}, N_α is the number of switching angles for 3L waveform in a quarter period, $s(i)$ represents the slopes of switching transients at switching angles α_i, $s(i)=(-1)^{i+1}$ for three level waveform. The total harmonic rms current of i_{3L} is given by,

$$i_{h,3L} = \sum_k \sqrt{i_{k,3L}^2} \quad (3)$$

$$i_{h,3L} = \frac{4I_{peak}}{\pi}\sqrt{\sum_k \frac{1}{k^2}\left(\sum_{i=1}^{N_\alpha} s(i)cos(k\alpha_i)\right)^2} \quad (4)$$

The switching angles are optimized to reduce the harmonic distortion. In order to eliminate the system parameters in optimization function, distortion factor d is obtained as follows [14],

$$d = \frac{i_{h,3L}}{i_{h,sqr}} \quad (5)$$

where $i_{h,3L}$ represents the total harmonic content of 3L reference current at a given operating point and $i_{h,sq}$ represents the total harmonic content of square waveform I_{sqr} in Fig. 3. The total harmonic contents of I_{sqr} is given by,

$$i_{h,sqr} = \frac{4I_{peak}}{\pi}\sqrt{\sum_k \frac{1}{k^2}} \quad (6)$$

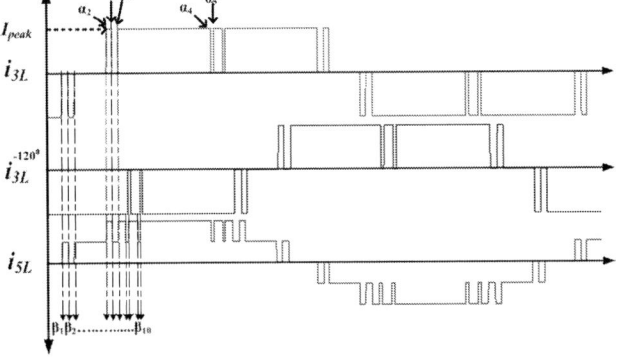

Fig. 3: 3L and 5L waveforms with optimal switching angles.

After simplifying (5) and (6), the final expression for d is obtained as,

$$d = \frac{\sqrt{\sum_k \frac{1}{k^2}\left(\sum_{i=1}^{N_\alpha} s(i)cos(k\alpha_i)\right)^2}}{\sqrt{\sum_k \frac{1}{k^2}}} \qquad (7)$$

From (7), it should be noted that, d is dependent only on k, $s(i)$ and α_i. SOP technique utilizes optimization algorithm for determining 3L switching angles α_i to minimize the value of d in (7) by including only harmonics of k=5,7,11,.... It should be noted that in the distortion factor calculation, the third order components are not included in the conventional SOP objective function calculation. However these 3L switching angles can not be utilized for operating CSI. Hence these 3L angles need to be converted into 5L angles by utilizing the conversion method explained in Section III-A. For example, consider the 3L angles α_1, α_2, ...α_5, of waveform i_{3L} in Fig. 3 are the optimal switching angles obtained from SOP technique. On applying the conversion method, these five angles are converted into ten 5L angles β_1, β_2,....β_{10}. The 5L waveform i_{5L} with the resultant ten angles is shown in Fig. 3.

In general, the conversion of N_α number of 3L angles returns/obtains N_β number of 5L angles as N_β =$(2 * N_\alpha)$.

The relation between device switching frequency f_s and selection of pulse number N_α is given by

$$N_\alpha = \left[\text{floor}(\frac{f_s}{f}) - 1\right] \qquad (8)$$

The harmonic component of i_{5L} is obtained as, $i_{k,5L} = \sqrt{3}i_{k,3L}$ if k=5,7,11,... and $i_{k,5L} = 0$ if k=3,6,9,.... The maximum value of fundamental component on inverter output current can be obtained if 5L CSI is modulated with reference signal as six-step waveform $i_{six-step}$ in Fig. 2, i.e. $m = 1$. The harmonic content of $i_{six-step}$ is given by $i_{k,six-step} = \sqrt{3}i_{k,sqr}$ for k=5,7,11,.... The distortion factor d with respect to 5L waveform is given by,

$$d = \frac{i_{h,5L}}{i_{h,six-step}} \qquad (9)$$

$$d = \frac{\sqrt{\sum_k \frac{1}{k^2}\left(\sum_{i=1}^{N_\alpha} s(i)cos(k\alpha_i)\right)^2}}{\sqrt{\sum_k \frac{1}{k^2}}}, k = 5,7,11,\ldots \qquad (10)$$

Hence, the distortion factor is unchanged with the conversion operation.

SOP technique requires optimal switching patterns to be calculated off-line for all steady-state operating points. The modulation index m is defined as,

$$m = \frac{i_{1,3L}}{i_{1,sq}} \qquad (11)$$

where $i_{1,sqr}$ is the amplitude of fundamental component of I_{sqr} and $i_{1,3L}$ is the fundamental amplitude of i_{3L} at the given operating point.

To obtain the desired fundamental amplitude of inverter output current, the switching angles should satisfy the following equality constraint.

$$m = \frac{i_{1,3L}}{i_{1,sqr}} = \left(\sum_{i=1}^{N_\alpha} s(i)cos(k\alpha_i)\right) \qquad (12)$$

2) Inverter Control: A detailed signal flow graph that explains control algorithm is shown in Fig. 4. Angle-selection in the control flow utilizes the magnitude of the reference current vector i_{ref} to select optimal 5L pattern $P(m, N_\alpha)$ that consists of optimized switching angles along with switching transitions $s(i)$. The optimal 5L pattern $P(m, N_\beta)$, phase angle of reference current vector and fundamental frequency f, are given as input to modulator which generates 5L switching state vector $i_k^{(5L)}$. The state-selector utilizes the redundant states in Table I to minimize the overall switching transition and produces the gating signals for each semiconductor device of CSI. The mechanism of redundant state selection can be referred to Section III-B5.

3) Optimization Algorithm: The optimized angles are generated for $0 < m < 1$ by modified SOP optimization algorithm in order to minimize d. The flowchart of modified SOP optimization algorithm is shown in Fig. 5. The constraints of optimization for CSI are as follows:

a) To allow minimum turn ON times and OFF times of power semiconductor devices a sufficient gap (10 μs) between consecutive switching angles should be provided;

b) To maintain current modulation index value that satisfies the relation (12).

The step-by-step procedure of proposed modified SOP algorithm are as follows:

i) Calculate number of 3L pulses required for desired device switching frequency f_s using (8),

ii) For each modulation index m, MATLAB function "randn" is used to generate the initial values of switching angles for further optimization while satisfying the relation (12).

iii) The gradient method 'FMINCON'of MATLAB built-in function is used for obtaining optimized switching

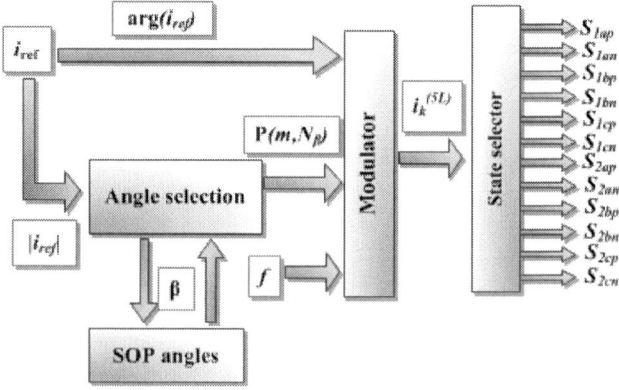

Fig. 4: Modified SOP control Flow for 5L CSI.

978-1-4673-9551-9/16 $31.00 © 2016 IEEE

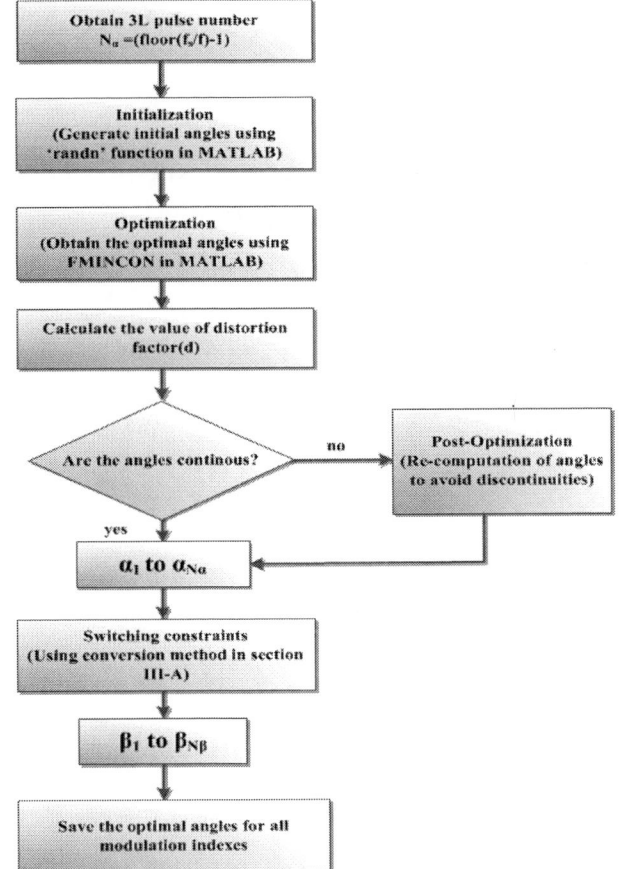

Fig. 5: Modified SOP optimization method.

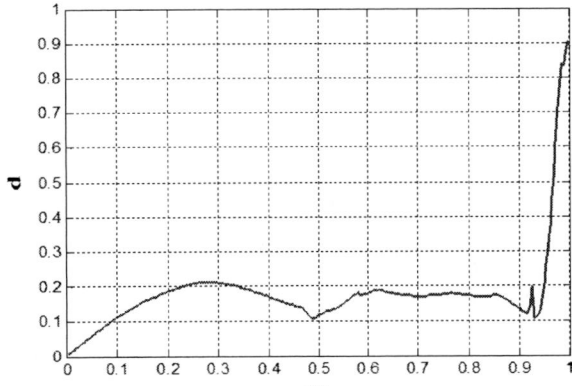

Fig. 6: Distortion factor d versus modulation index m for 5L CSI.

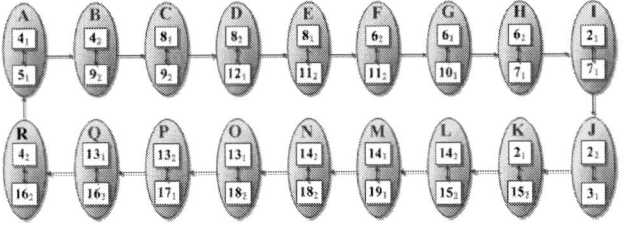

Fig. 7: 5L CSI redundant switching states sequencing

TABLE III
DEVICES AND PARAMETERS

Device/ Parameter	Part number/ Value
Converter power	1200 W
Input DC Current	3.1 A
Switching frequency	350 Hz
AC output Frequency	50 Hz
Output capacitor	30μF
Load Resistance	15.6 Ω
Load Inductance	10 mH
Power Diode	STTH30R04-Y
	I_f 30 A, V_{RRM} 400 V
Half-bridge module	SK 25 GB 12T4
	V_{CES} 1200 V, I_C 25 A
Six pack driver	SKHI 61R

patterns for each modulation indexes. For distortion-factor calculations, only harmonic components up to 100 are considered.

iv) The optimization loop runs for modulation index range ($0 < m < 1$).

v) If switching angles for consecutive modulation index values differ by more than 5 degrees, post-optimization is performed starting with optimized switching angles as initial values. Due to post-optimization, transients in output currents are reduced but the distortion factor d is slightly compromised.

vi) The final optimal switching angles are stored as complete patterns $P(m, N_\beta)$ in a DSP and they are retrieved during real time operation depending on the output current to be delivered to the grid.

4) Optimization Results: The distortion factor for 5L CSI for $0 < m < 1$ is shown in Fig. 6. It should be observed that distortion factor approached unity at $m = 1$. This is because inverter output will be similar to six-step waveform $i_{six-step}$ shown in Fig. 2. In addition, due to reduced harmonic distortion and harmonic orders are shifted to higher frequencies, the proposed modified SOP technique reduces output filter size needed for CSI in Fig. 1.

5) Sequencing: The redundant switching states can be referred from Table II. A sequencing technique shown in Fig. 7 has been developed. There are 18 sequence blocks (A, B, C, ... R) have been identified for a complete fundamental cycle

of switching pattern for the range of modulation index 0.5 to 1. Each block has two 5L CSI switching states with preassigned redundant options. It can also be observed, the redundant options are equally distributed among blocks such that the inverters shares the total power equally. When state selector receives $i_k^{(5L)}$, state selector checks the current sequence block for choosing the redundant option. Transition from one sequence block to another block occurs if the next state is not in current sequence block. For example, if the state selection is assigned with block A and the next state identified from the modulator is 9 then block B is selected as current block.

IV. EXPERIMENTAL RESULTS

A low power prototype of 1.2 kW has been setup as shown in Fig. 8, for controlling 5L CSI using proposed SOP technique. Separate PCB boards has been designed for implementing

Fig. 8: Experimental setup of 5L CSI.

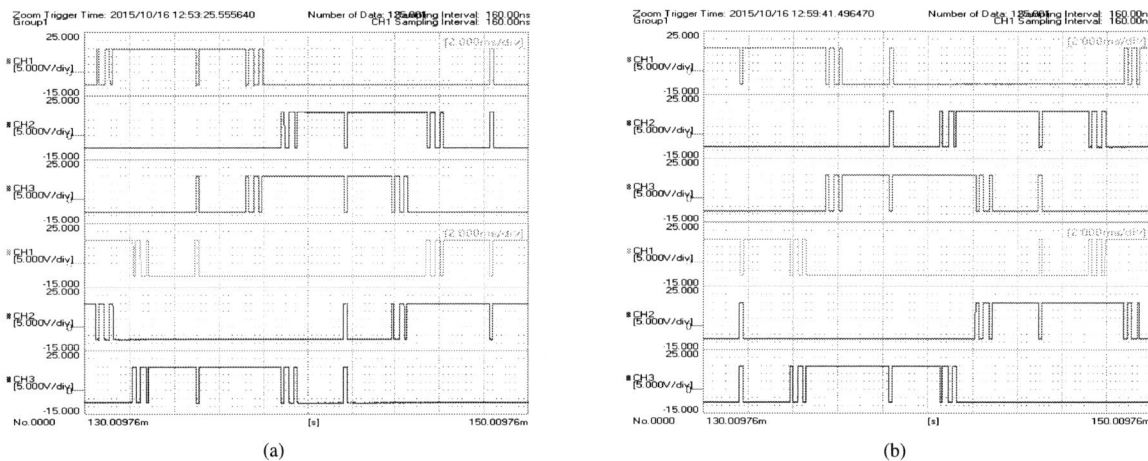

(a)   (b)

Fig. 9: Inverter gating pulses (Y-axis: 5 V/div; X-axis: 2 ms/div) for $m = 0.9294$. (a) Inverter-1 gating pulses for switches S_{1ap} to S_{1cn} (top to bottom). (b) Inverter-2 gating pulses for switches S_{2ap} to S_{2cn} (top to bottom).

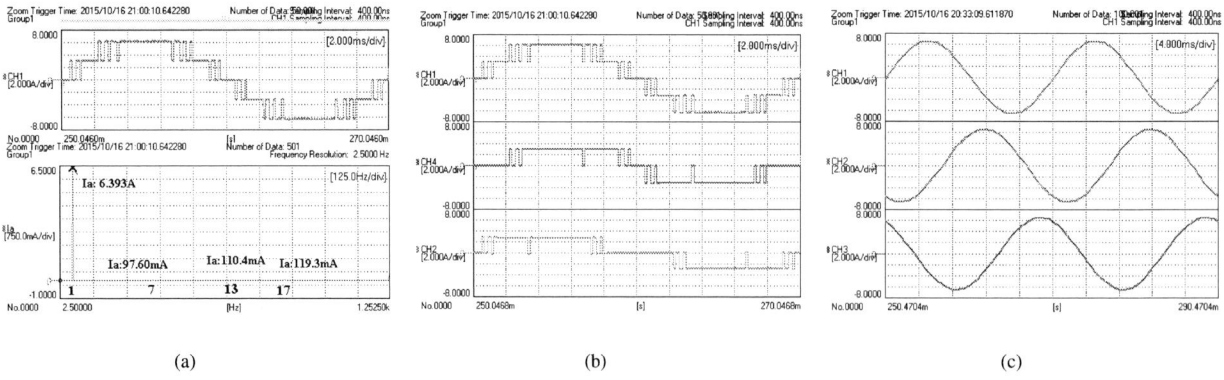

(a)   (b)   (c)

Fig. 10: Experimental results for $m = 0.9294$ (a) 5L output current, i_a (Y-axis: 2 A/div, X-axis: 2 ms/div) and its FFT spectrum (Y-axis: 0.75 A/div, X-axis: 125 Hz/div). (b) Phase-A 5L current, i_a, inverter-1 output current, i_{a1} and inverter-2 output current, i_{a2} (Y-axis :2A/div, X-axis: 2 ms/div). (c) Filtered currents, i_{ag}, i_{bg}, i_{cg} (Y-axis: 1.25 A/div, X-axis: 8 ms/div).

two 3L CSI. Each leg of the 3L CSI was implemented by using IGBT half-bridge power module SK25GB12T4 and a six-pack driver SKHI 61R from Semikron was used as a driver. The switching signals were programmed on a Texas

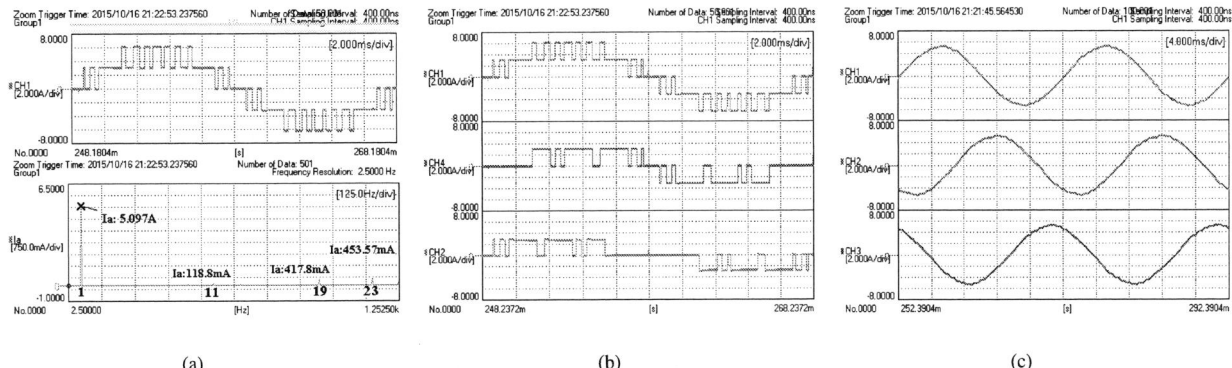

Fig. 11: Experimental results for $m = 0.749$ (a) 5L output current, i_a (Y-axis: 2 A/div, X-axis: 2 ms/div) and its FFT spectrum (Y-axis: 0.75 A/div, X-axis: 125 Hz/div). (b) Phase-A 5L current, i_a, inverter-1 output current, i_{a1} and inverter-2 output current, i_{a2} (Y-axis :2A/div, X-axis: 2 ms/div). (c) Filtered currents, i_{ag}, i_{bg}, i_{cg} (Y-axis: 1.25 A/div, X-axis: 8 ms/div).

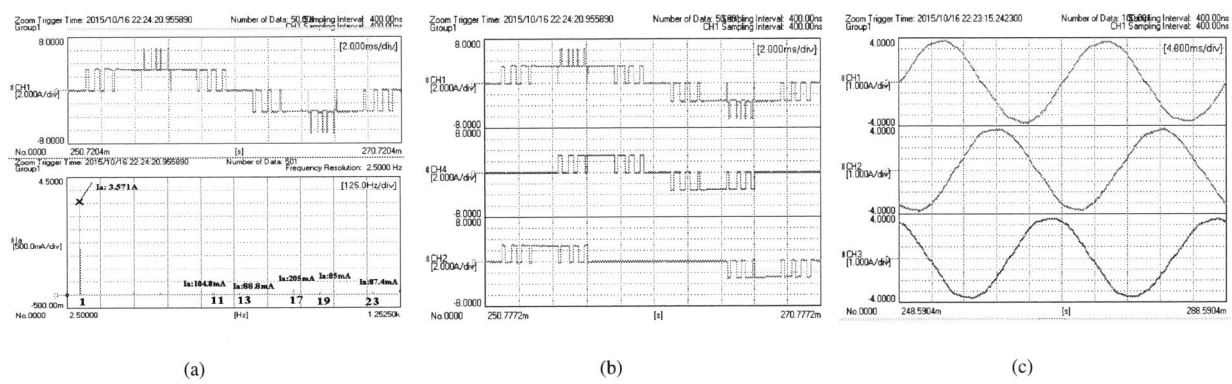

Fig. 12: Experimental results for $m = 0.5216$ (a) 5L output current, i_a (Y-axis: 2 A/div, X-axis: 2 ms/div) and its FFT spectrum (Y-axis: 0.5 A/div, X-axis: 125 Hz/div). (b) Phase-A 5L current, i_a, inverter-1 output current, i_{a1} and inverter-2 output current, i_{a2} (Y-axis : 1 A/div, X-axis: 2 ms/div). (c) Filtered currents, i_{ag}, i_{bg}, i_{cg} (Y-axis: 1.25 A/div, X-axis: 8 ms/div).

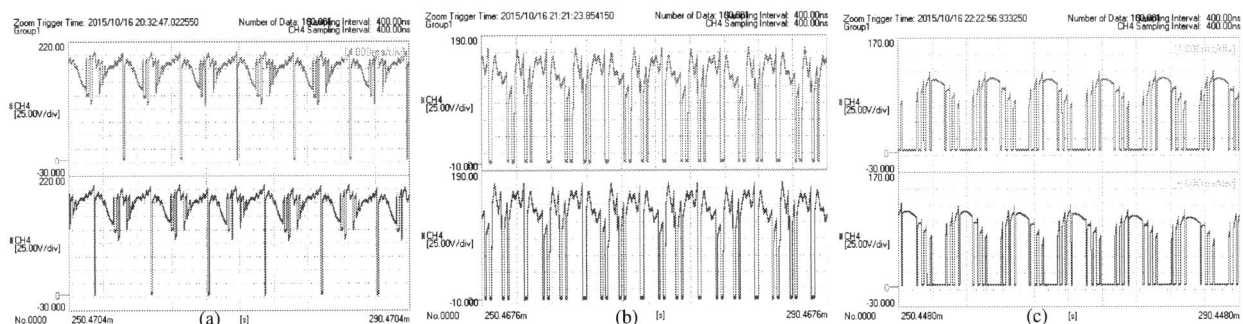

Fig. 13: Inverter-1 and Inverter-2 input voltages of 5L CSI (Y-axis : 25 V/div, X-axis: 4 ms/div) for (a). $m = 0.9249$. (b) $m = 0.749$. (c) $m = 0.5216$.

TABLE IV
3L SOP SWITCHING ANGLES

m	α_1	α_2	α_3	α_4	α_5	α_6
0.9294	9.04	11.48	15.71	19.57	22.29	88.63
0.749	21.35	40.80	46.00	51.96	55.73	86.16
0.5216	15.58	48.24	54.55	60.34	64.56	73.13

Instrument TMS320F28335 and a sampling frequency of 20 kHz is utilized. Table III shows the list of major components along with their parameters.

Proposed modified SOP technique is utilized for generating optimal switching angles at three different operating points: ($m = 0.9294$, $N_\alpha = 6$), ($m = 0.749$, $N_\alpha = 6$) and ($m = 0.5216$, $N_\alpha = 6$). The 3L and 5L angles for these three operating points are shown in Table IV and V, respectively.

The pulse number for the operating points are maintained at $N_\beta=10$, so the device switching frequency of all semiconductor devices should be equal to 350 Hz (8). The gating signals for all the twelve semiconductor devices of CSI for operating point $m = 0.9294$ are shown in Fig. 9 (a)-(b), and it is clear from the waveforms that each semiconductor devices is turned ON and OFF for 7 times within one fundamental cycle, i.e.

978-1-4673-9551-9/16 $31.00 © 2016 IEEE

TABLE V
5L SOP SWITCHING ANGLES

m	β_1	β_2	β_3	β_4	β_5	β_6	β_7	β_8	β_9	β_{10}	β_{11}	β_{12}
0.9294	7.72	10.44	14.31	18.54	20.98	39.05	41.49	45.72	49.59	52.31	58.65	61.38
0.749	8.67	10.81	16.01	21.98	25.74	51.36	56.17	63.86	70.81	76.01	81.98	85.74
0.5216	14.44	18.25	24.57	30.35	34.57	43.14	45.59	76.89	78.25	84.57	85.46	89.68

switching frequency is 350 Hz.

The waveforms of (1) inverter output current and its FFT spectrum, (2) output current for each 3L CSI, (3) filtered three phase output currents pertaining to operating point (m = 0.9294, N_α = 6, f = 50Hz) are shown in Fig.10 (a)-(c), respectively. İt could be noticed from the FFT spectrum that the lower order harmonic components such as 5th, 7th, 13th, 19th and 23rd of the inverter output current are infinitesimal compared to the amplitude of fundamental. The magnitude of these harmonic components are also shown in Fig. 10 (a). It should also be noticed that the even order harmonic and third order harmonic components are eliminated. The filtered line current i_{ag} of 5L CSI is nearly sinusoidal although the device switching frequency is reduced to 350 Hz. The THD of the filtered current is obtained as 1.36%.

Similar observations about inverter output current, FFT spectrum and filtered currents for operating points m = 0.749 and m = 0.5216 are made from Fig. 11 to Fig. 12. It should be noticed that lower order harmonic components ($<$1 kHz) of inverter output current are infinitesimal for these operating points. The THD of output filtered currents for these two operating points are obtained as, 1.93%, and 2.79%, respectively. These values are well below ($<$ 5%) the PV grid integration standard [15].

The waveforms of input voltage for each 3L CSI are shown in Fig.13 for all three operating points. It should be noticed that there are no voltage spikes during switching transitions. This has been achieved by providing sufficient overlap (1μs) between switching transition. It should also be noticed that, input voltage is zero for short intervals of time, which is due to turning on both top (S_{xp}) and bottom (S_{xn}) semiconductor devices of one phase leg.

V. CONCLUSION

High power applications like large scale solar power plants, low device switching frequency operation is needed in order to satisfy the thermal constraint of semiconductor devices and efficient operation of the inverter. A modified synchronous optimal pulse-width modulation technique has been proposed, analysed and implemented for controlling five-level voltage boost current-fed multilevel inverter topology at low device switching frequency. A simple conversion method has been introduced to include operational constraints of CSI on the optimal switching angles. A laboratory prototype has been designed, developed and tested at 1200 W to validate the proposed technique. Experimental results demonstrated the effectiveness of the proposed method and from the experi-mental results, it should be noticed that the inverter output

current is nearly sinusoidal. The THD of line current has been maintained below 5% at all operating points of power flow without compromising on device switching frequency.

REFERENCES

[1] (2014, August) Top 400 Solar Contractors. [Online]. Available: http://www.solarpowerworldonline.com/2014-top-400-solar-contractors/
[2] Internation Energy Agency, "2014 Snapshot of Global PV Markets," Rep. IEA PVPS T1-26:2015.
[3] V. Vekhande and B. Fernandes, "Central multilevel current-fed inverter with module integrated DC-DC converters for grid-connected PV plant," in *IEEE Energy Convers. Cong. and Expo. (ECCE '13)*, Sept 2013, pp. 1933–1940.
[4] Z. Bai and Z. Zhang, "Conformation of Multilevel Current Source Converter Topologies Using the Duality Principle," *IEEE Trans. Power Electron.*, vol. 23, no. 5, pp. 2260–2267, Sept 2008.
[5] N. Vazquez, H. Lopez, C. Hernandez, E. Vazquez, R. Osorio, and J. Arau, "A Different Multilevel Current-Source Inverter," *IEEE Trans. Ind. Electron.*, vol. 57, no. 8, pp. 2623–2632, 2010.
[6] P. G. Barbosa, H. A. C. Braga, M. C. B. Rodrigues, and E. C. Teixeira, "Boost current multilevel inverter and its application on single-phase grid-connected photovoltaic systems," *IEEE Trans. Power Electron.*, vol. 21, no. 4, pp. 1116–1124, 2006.
[7] J. Holtz and X. Qi, "Optimal Control of Medium-Voltage Drives-An Overview," *IEEE Trans. Ind. Electron.*, vol. 60, no. 12, pp. 5472–5481, Dec 2013.
[8] A. Edpuganti and A. Rathore, "A survey of low-switching frequency modulation techniques for medium-voltage multilevel converters," in *IEEE Ind. Appl. Soc. Annu. Meeting (IAS '14)*, Oct 2014, pp. 1–8.
[9] P. Torri, G. da Cunha, T. Boller, A. Rathore, J. Holtz, and N. Oikonomou, "Optimal pulse width modulation for multi-level inverter systems," Apr. 20 2011, eP Patent App. EP20,090,171,698. [Online]. Available: http://www.google.com/patents/EP2312739A1?cl=en
[10] T. Boller, J. Holtz, and A. Rathore, "Neutral-Point Potential Balancing Using Synchronous Optimal Pulsewidth Modulation of Multilevel In-verters in Medium-Voltage High-Power AC Drives," *IEEE Trans. Ind. Appl.*, vol. 50, no. 1, pp. 549–557, Jan 2014.
[11] A. K. Rathore, J. Holtz, and T. Boller, "Generalized Optimal Pulsewidth Modulation of Multilevel Inverters for Low-Switching-Frequency Con-trol of Medium-Voltage High-Power Industrial AC Drives," *IEEE Trans. Ind. Electron.*, vol. 60, no. 10, pp. 4215–4224, Oct 2013.
[12] T. Boller, J. Holtz, and A. Rathore, "Optimal Pulsewidth Modulation of a Dual Three-Level Inverter System Operated From a Single DC Link," *IEEE Trans. Ind. Appl.*, vol. 48, no. 5, pp. 1610–1615, Sept 2012.
[13] B. Wu, J. Pontt, J. Rodriguez, S. Bernet, and S. Kouro, "Current-Source Converter and Cycloconverter Topologies for Industrial Medium-Voltage Drives," *IEEE Trans. Ind. Electron.*, vol. 55, no. 7, pp. 2786–2797, July 2008.
[14] J. Holtz, "Pulsewidth modulation for electronic power conversion," *Proc. IEEE*, vol. 82, no. 8, pp. 1194–1214, Aug 1994.
[15] "IEEE Application Guide for IEEE Std 1547(TM), IEEE Standard for Interconnecting Distributed Resources with Electric Power Systems," *IEEE Std 1547.2-2008*, pp. 1–217, April 2009.

Design and Implementation of D-Σ Digital Controlled Multi-function Inverter to Achieve APF, Active Power Injection and Rectification

T.-F. Wu[1], H.-C. Hsieh[2], L.-C. Lin[1] and C.-H. Chang[1]

[1,3]Elegant Power Electronics Applied Research

Laboratory (EPEARL)

Department of Electrical Engineering

National Tsing Hua University

Hsinchu, Taiwan, ROC

E-mail: tfwu@ee.nthu.edu.tw

[2]Elegant Power Application Research Center,

Department of Electrical Engineering,

National Chung Cheng University,

Chia-Yi, Taiwan,ROC

E-mail: wellian@ms22.url.com.tw

Abstract—**There has been a growing demand of using multi-function inverters for grid-connected systems applied to nonconventional energy sources, such as solar, wind and so on. In addition to power quality conditioning, the inverter can also be used for bidirectional active power exchange with a three-phase four-wire grid. Therefore, the inverter acts as a multi-function compensator. The functions of the proposed inverter system include active power injection, rectification and active power filtering (APF) (including phase power balancing). This paper presents design and implementation of a three-leg split-capacitor shunt multi-function inverter with division-summation (D-Σ) digital control. The adopted D-Σ digital control can accommodate filter inductance variation, reducing core size significantly, and its control laws can be derived directly to cancel the variation effects of dc-bus voltage, switching period and filter inductance. An average power method is adopted in this paper for determining fundamental currents at the source side. In the design and implementation, the inductances corresponding to various inductor currents were estimated at the startup and stored in the microcontroller for scheduling loop gain cycle by cycle, which can insure system stability. Measured results from a three-phase four-wire inverter have confirmed the analysis and discussion.**

Index Terms—**average power, division-summation (D-Σ) digital control, and three-phase Active power filter.**

I. INTRODUCTION

In recent years, the development of electrical systems for renewable energy sources, such as solar, wind, and etc., was experiencing dramatic growth. These systems either work in stand-alone or grid-connected mode[1]. If they operate in grid-connected mode, the power generated typically needs to be exchanged with the grid through a dedicated power electronic interface. The generated power generally feeds local single-phase or three-phase loads over a three-phase four-wire distribution feeder. Conventionally, an active power filter (APF) [2] and [3] is used to perform power conditioning tasks, such as harmonic elimination, reactive power compensation, phase power balancing, and so on. In addition, the APF can also be used to perform bidirectional active power exchange with the grid.

The harmonic current generated by power electronic components and equipment is regarded as one of the major problems that deteriorates in power quality. Harmonic standards have been therefore set up worldwide, such as IEEE-519 [4] and IEC-61000 [5]. Electronic products are required to comply with the standards to ensure high power quality. In particular, harmonic components in power distribution systems can cause excessive voltage and current distortion, resulting in instability, abnormal operation or damage to electric components. Typical inverter topology is a three-phase four-wire system with split capacitors which can be also used for eliminating leakage ground current. With the split capacitors, the middle point is connected to the neutral point of the ac grid, and its common-mode voltage can be clamped by the large dc-bus capacitance, which can eliminate leakage current effectively.

The current control based on space-vector pulse width modulation (SVPWM) has been widely applied to the three-phase inverter[6]. Typical conventional control schemes include: 1) proportional-integral (PI) control [7], 2) deadbeat control [8]-[10], and 3) hysteretic control [11]-[13], which were designed for current-error compensation. However, the three-phase inductances are assumed constant for simplifying state equations and abc to dq frame transformation, while the inductance varies with current level, resulting in inductor current fluctuation at high power applications. Moreover, grid voltage harmonics and three-phase voltage imbalance will complicate the dq transformation, limiting its wide applications [14]-[16]. In literature, a D-Σ digital control has been designed and implemented [17]-[21], which can overcome the limitations of the conventional abc to dq frame transformation and cover the effect of wide filter-inductance variation.

In this paper, the D-Σ digital control applied for a shunt-type multi-function inverter is adopted. Based on the control, this paper further discusses how to determine current commands for compensating harmonic current due to loads. Experimental results measured from a three-phase four-wire inverter are presented to confirm the analysis and discussion of the proposed control.

978-1-4673-9551-9/16 $31.00 © 2016 IEEE

II. D-Σ Digital Control

A power circuit diagram of a three-phase four-wire multi-function inverter with split capacitors is shown in Fig. 1. The adopted D-Σ digital control can take into account the wide inductance variation to tune loop gain cycle by cycle. In the following, the control laws for the multi-function inverter are derived and presented.

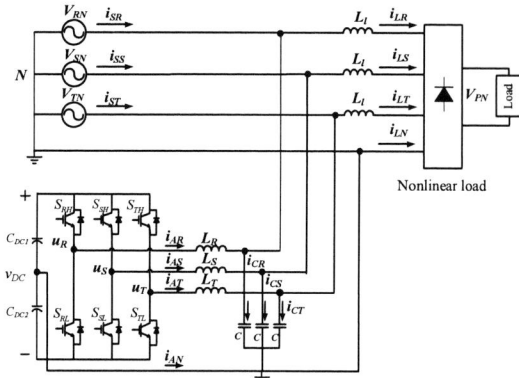

Fig. 1. Circuit diagram of a three-phase bi-directional inverter with split capacitors.

A. D-Σ Digital Control Based on SPWM

The three-phase inverter with split capacitors can be equivalent to three single-phase half-bridge inverters, as shown in Fig. 2. The state equation of the equivalent inverter can be described as

$$\frac{di_{AX}}{dt} = \frac{u_X - v_{XN}}{L_X}, \tag{1}$$

where u_X is the switching state voltage which changes with the state of the switch, v_{DC} is the dc-bus voltage, and i_{AX} and L_X stand for the inductor current and the inductor of each phase, respectively.

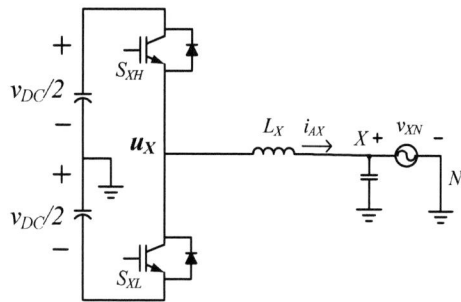

Fig. 2. An equivalent circuit of the three-phase inverter with split capacitors for each phase.

During one switching period T_s, either the upper switch or the lower switch is turned on, which will lead to two switching states. Thus, equation (1) can be expressed as follows:

$$i_{AXH} = \frac{\frac{v_{DC}}{2} - v_{XN}}{L_X} \cdot d_X T_s \tag{2}$$

and

$$i_{AX} = \frac{-\frac{v_{DC}}{2} - v_{XN}}{L_X} \cdot (1 - d_X) T_s, \tag{3}$$

where d_X and $1-d_X$ are the duty ratios of the upper switch and the lower switch, respectively, and i_{AXH} and i_{AXL} are the inductor current variations during the upper and the lower switch turn-on time intervals, respectively. By summarizing the above two state equations, we can have

$$\Delta i_{AX} = \frac{v_{DC} d_X - \frac{v_{DC}}{2} - v_{XN}}{L_X} \cdot T_s. \tag{4}$$

It can be seen that the output controlled variable Δi_{AX} consists of one control variable d_X and two voltage variables v_{XN} and v_{DC}. By taking partial derivative of Δi_{AX} with respect to the three variables, the plant can be expressed as follows:

$$G_{P, \frac{\partial \Delta i_{AX}}{\partial d_X}} = \frac{T_S V_{DC}}{L_X}, \tag{5}$$

$$G_{P, \frac{\partial \Delta i_{AX}}{\partial v_{XN}}} = -\frac{T_S}{L_X} \tag{6}$$

and

$$G_{P, \frac{\partial \Delta i_{AX}}{\partial v_{DC}}} = -\frac{T_S}{2L_X}. \tag{7}$$

Note that voltage V_{DC} in (5) is the dc component of v_{DC} ($=V_{DC} + \hat{v}_{DC}$) and it is also the average value of dc-bus voltage over one switching cycle. To cancel the variation effects of the system parameters, such as L_X and T_S, control gain G_C of the duty ratio to current error i_e ($=I_{ref} - i_{fb}$) can be derived to satisfy the following equation:

$$G_{C, \frac{\partial \Delta d_X}{\partial i_{AX}}} \cdot G_{P, \frac{\partial \Delta i_{AX}}{\partial d_X}} = 1, \tag{8}$$

By substituting (5) into (8), the control gain can be obtained below:

$$G_{C, \frac{\partial \Delta i_{AX}}{\partial d_X}} = \frac{L_X}{T_S V_{DC}}, \tag{9}$$

The overall duty-ratio control signal includes operating point and the compensated value. The duty-ratio control signal is the product of control gain and current error, and the operating point of the control signal can be derived from the condition $\Delta i_{AX} = 0$, and then the state equation can be rewritten as follows:

$$0 = d_{X0} v_{DC} - v_{XN} - \frac{1}{2} v_{DC}, \tag{10}$$

, and the operation point of the control signal can be expressed as

$$d_{X0} = \frac{1}{v_{DC}}\left(v_{XN} + \frac{1}{2}v_{DC}\right), \tag{11}$$

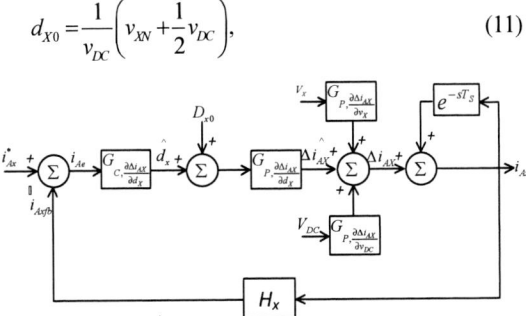

Fig. 3. Equivalent control block diagram of a phase in a three-phase inverter.

An overall control block diagram of a phase in a three-phase inverter is shown in Fig. 3. It can be observed that the control block includes one feedback variable i_{fb} and two feedforwards V_X and V_{DC}. With the proposed D-Σ digital control, the designed controllers can eliminate the impacts of system parameters (i.e. L_X and T_S) on the plant $G_{P,\frac{\partial \Delta i_{AX}}{\partial d_X}}$, insuring current tracking accuracy. According to Fig. 3, the duty-ratio control law can be then expressed as

$$d_X = \frac{L_X i_e}{T_S V_{DC}} + d_{X0}. \tag{12}$$

Over one switching cycle, the ac component of the dc-bus voltage, $\hat{v}_{DC}$, is very close to zero, so as v_{DC}/V_{DC} can be approximated to unity. Thus, (12) can be rewritten as

$$d_X = \frac{L_X i_e}{T_S V_{DC}} + \frac{v_{XN}}{V_{DC}} + \frac{1}{2}. \tag{13}$$

The duty-ratio control law of each phase is independent of the others, so as the discussed inverter can track sinusoidal reference current of each phase individually.

B. Stability Analysis

The control block diagram of the system is shown in Fig. 4 and the stability analysis is conducted based on the bode plot of the loop gain.

In circuit implementation, the feedback gain H_X is unity. Since the delay time of the described procedure is limited to about 2 us, which is far less than the sampling period (50.5 us), and therefore it can be regarded as an ideal condition. In addition, the delay time of feedback integrator of the output current i_{AX} is a switching period. The control block diagram of the single-phase system can be simplified to a single-loop diagram shown in Fig. 4.

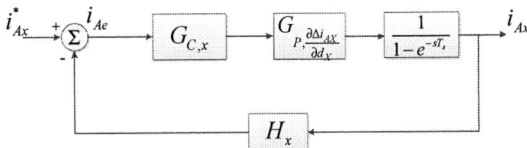

Fig. 4. Control block diagram of a single-phase system

Then, the loop gain $\mathcal{L}$ can be calculated as follows:

$$\mathcal{L} = G_C \times G_{P,\frac{\partial \Delta i_{AX}}{\partial d_X}} \times \frac{1}{1-e^{-sT_s}} \times H_X \tag{14}$$

Substituting G_C in (9), G_{Px} in (5) and H_X into (14), the loop gain $\mathcal{L}$ can be obtained as shown below:

$$\mathcal{L} = \frac{1}{1-e^{-sT_s}}. \tag{15}$$

Taking the first two terms of the Taylor series of the exponential function in (15) can have

$$\mathcal{L} = \frac{1}{sT_s}. \tag{16}$$

Due to the fact that (16) is equivalent to a first-order low-pass filter, it is always stable. Then the transfer function of input to output current of the system can be derived as follows:

$$G = \frac{1}{sT_s+1}. \tag{17}$$

A bode plot of the system transfer function G is shown in Fig. 5. It can be seen that the bandwidth is 20 k rad / sec and the gain before this bandwidth is always 0 dB, showing a tight tracking capability. That is the output inductor current can track the input current command tightly in a switch cycle.

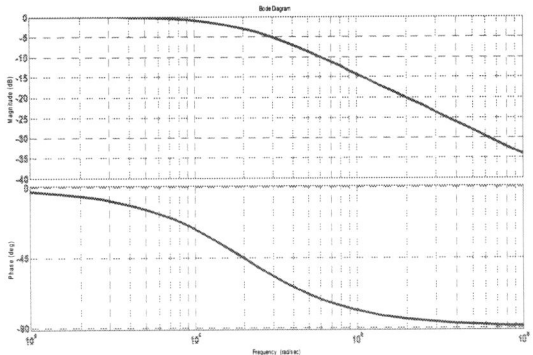

Fig. 5. Bode plot of the system transfer function G.

III. COMMAND DETERMINATION AND DC-BUS VOLTAGE REGULATION FOR APF

A shunt APF is connected at the point of common coupling to compensate current harmonics and reactive power generated by loads. It is composed of electrolytic capacitors, a three-leg inverter and an LC filter, as shown in Fig. 1. A block diagram for determining reference current and with D-Σ digital control to determine gate signals is shown in Fig. 6. The total current command (i_{ref}^*) consists of two components which are the reference currents (i_{abc}^*) and the dc-bus voltage regulation current (i_{dc}^*) needed to compensate the power loss during inverter operation. The reference currents can be derived from the average power method, and the dc-bus voltage regulation current is obtained by the PI controller.

In the following, determination of the reference current for the APF, control law derivation for the plant and generation of the gate signals for the inverter are presented.

A. Determination of Compensation Currents

Determination of the source fundamental currents is critical in an APF system. Assuming the three-phase voltages and the three-phase loads are balanced, the grid voltage can be expressed as follows:

$$v_{ac}(t) = V_1 \sin(\omega t) + V_2 \sin(2\omega t) + \cdots + V_h \sin(h\omega t), \quad (18)$$

where $v_{ac}(t)$ is instantaneous voltage. With a nonlinear load, the load current containing harmonic components will be:

$$i_{Lac}(t) = I_1 \sin(\omega t + \theta_1) + I_2 \sin(2\omega t + \theta_2) + \cdots + I_h \sin(h\omega t + \theta_h), \ (19)$$

where $\theta_1 \sim \theta_n$ are the power factor angles for each harmonic order. The grid voltage consists of fundamental component and harmonic components. Thus, the active power of the load includes fundamental power and harmonic power. According to the power definition, the active power (P_{av}) of the three phase loads can be determined from the following equation:

$$P_{av} = \frac{1}{T_l}\int_0^{T_l} v_{ac,R}(t)\cdot i_{Lac,R}(t)dt + \frac{1}{T_l}\int_0^{T_l} v_{ac,S}(t)\cdot i_{Lac,S}(t)dt + \frac{1}{T_l}\int_0^{T_l} v_{ac,T}(t)\cdot i_{Lac,T}(t)dt$$

$$= P_{1,R} + \sum_{n=2}^{h} P_{n,R} + P_{1,S} + \sum_{n=2}^{h} P_{n,S} + P_{1,T} + \sum_{n=2}^{h} P_{n,T} \quad , \quad (20)$$

where T_l is the line period, $P_{1,RST}$ are the three phase fundamental powers and the $\sum_{n=2}^{h} P_{n,RST}$ denote the three phase total harmonic powers. The reference current can be determined as follows:

$$i_{abc}^* = \frac{P_{av}}{3 \times V_{RN}^1}, \quad (21)$$

where V_{RN}^1 is the amplitude of the fundamental voltage which can be determined by DFT algorithms [22], [23]. With this reference-current determination, phase power balancing can be also achieved.

B. DC-bus Voltage Regulation

Dc-bus voltage is controlled by a PI controller. It can regulate the dc-bus voltage and compensate the total losses of the converter, such as switching loss, wiring loss and inductor loss, etc. A dc-bus voltage control block diagram is shown in Fig. 7. The compensated dc current to dc-bus voltage transfer function G_p can be derived as follows: [24], [25]

$$G_p(s) = \frac{v_{dc}}{i_{dc}^*} = \frac{3V_{ac}}{sC \cdot V_{dc}}, \quad (22)$$

where V_{ac} is the grid voltage and C is the dc bus capacitance, and the controller is given as

$$G_c(s) = k_p + \frac{k_i}{s}. \quad (23)$$

The open-loop G_o gain can be then obtained by the product of G_p and G_c,

$$G_o = k_p \cdot \frac{s + \frac{k_i}{k_p}}{s} \cdot \frac{3V_{ac}}{sC \cdot V_{dc}}, \quad (24)$$

where k_p and k_i are the proportional and integral gains, respectively. By Mason's gain formula and pole-zero cancelation method, the input to output transfer function can be derived as follows:

$$G_{cl} = \frac{v_{dc}}{v_{dc}^*} = \frac{3V_{ac} \cdot k_p / C \cdot V_{dc}}{s + 3V_{ac} \cdot k_p / C \cdot V_{dc}}. \quad (25)$$

The closed-loop bandwidth $\omega_n = 2\pi \times 360$ rad/s is

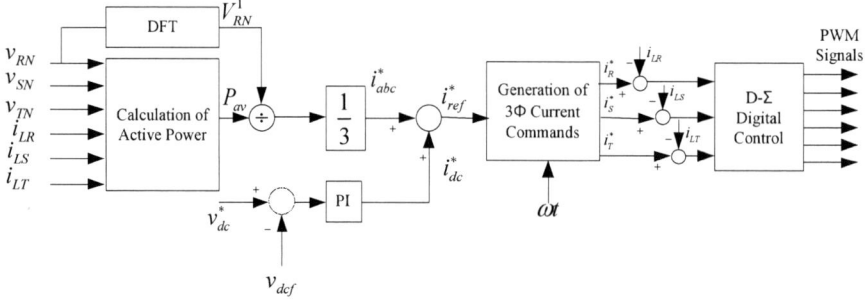

Fig. 6. Block diagram for determining reference current and with D-Σ digital control to determine gate signals.

selected and parameters k_p and k_i can be obtained as follows:

$$k_p = \omega_n \cdot C \cdot v_{dc} / 3V_{ac},$$

and

$$k_i = 0.$$

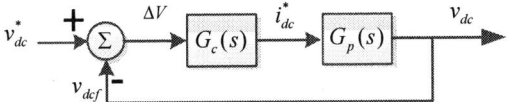

Fig. 7. DC-bus voltage control block diagram.

IV. EXPERIMENTAL RESULTS

The proposed digital current control for multi-function system has been verified by a three-phase four-wire inverter. Based on the aforementioned specifications and analysis, parameter design of the power stage is summarized in Table I. The phase voltage is 127 V, the line frequency is 60 Hz and the dc-link voltage is 380 V. The inverter inductance varies from 2 mH to 650 μH per phase, the filter capacitance is 10 μF and the switching frequency is 20 kHz. An APF inverter with a Renesas-based prototype shown in Fig. 8 was implemented. The APF consists of six IGBT and diode discrete components from Fairchild and Cree, respectively.

Fig. 8. Photograph of the designed three-phase APF prototype.

TABLE I. SYSTEM PARAMETERS OF THE EXPERIMENT SET-UP.

Elements	Parameters	Values
AC Source	Phase voltage	127 V_rms
	DC-bus voltage	380 V
	line frequency	60 Hz
APF Inverter	Filter inductor	2 mH to 650 μH
	Filter capacitor	10 μF
	Switching frequency	20 kHz
Nonlinear Load	Load inductor (L_L)	2 mH
	Load capacitor (C_L)	141 μF
	Resistive Load (R_L)	100 ~ 25 Ω

Besides, the inverter is connected to an LC filter with the filter capacitance of 10 μF and the inductance of

2 mH at zero current. With D-Σ digital control, wide inductance variation is allowed; therefore, the inductance varies from 2 mH to 650 μH when the inverter current is operated from 0 A to 20A. A plot of inductance versus its inductor current is shown in Fig. 9. If a system control, such as a conventional abc to dq frame transformation approach, cannot cover wide inductance variation, the allowed minimum inductance is typically selected to be 80 % (the range is 100 % ~ 80 %). Its core size is much larger than that with wide range, 100 % ~ 32 %, as shown in Fig. 10. The inductor size of CK1625060 is about 3.5 times CK778060.

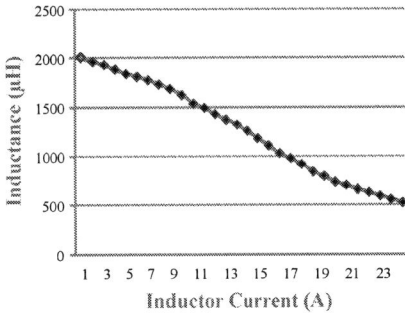

Fig. 9. Measured inductance of an inductor constructed with a Mega-Flux core (60μ) corresponding to inductor currents 0-20 A.

Fig. 10. Photograph of the two types of inductors: CK778060 and CK1625060.

Measured current waveforms without considering inductance variation are shown in Fig. 11, where i_{LR} is load current, i_{SR} is source current, i_{AR} is the compensation current, V_{RN} is the phase voltage and the power of load is 4 kW. Before compensation, the source current is distorted. After providing compensation by the APF, the source current should be nearly sinusoidal. However, it can be seen that the compensated source current exists distortion, since the inductance variation has not been considered yet. Spectrum of the source current is shown in Fig. 12, and its total harmonic distortion (THD) is 7.15%.

978-1-4673-9551-9/16 $31.00 © 2016 IEEE

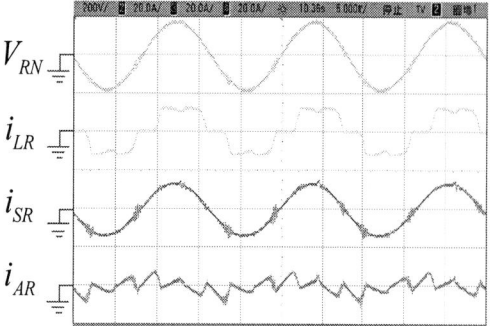

Fig. 11. Measured waveforms of APF currents and phase voltage without considering inductance variation under 4 kW. (i_{SR} THD: 7.15%)

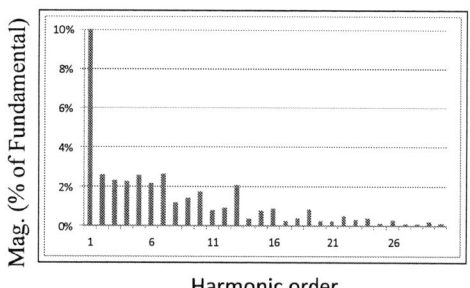

Fig. 12. Measured spectrum of source current without considering inductance variation.

The measured results of compensation current, load current, source current and line to line voltage under APF mode with unbalanced load are shown in Fig. 13. Before compensation, the source current is distorted. After compensation, the source current is nearly sinusoidal but still unbalanced.

Fig. 14 shows the source current waveforms under APF mode with unbalanced load current. It can be observed that after compensation by the APF, the source currents are balanced. Thus the implemented APF can also achieve load power balancing function.

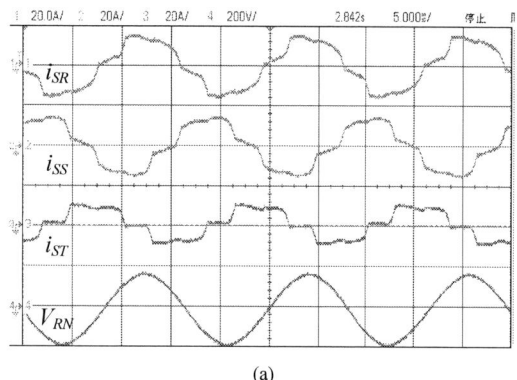

(a)

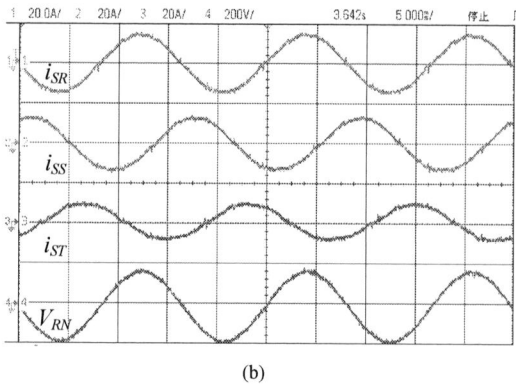

(b)

Fig. 13. Measured waveforms of the three currents: (a) load currents and (b) source current under unbalanced load condition

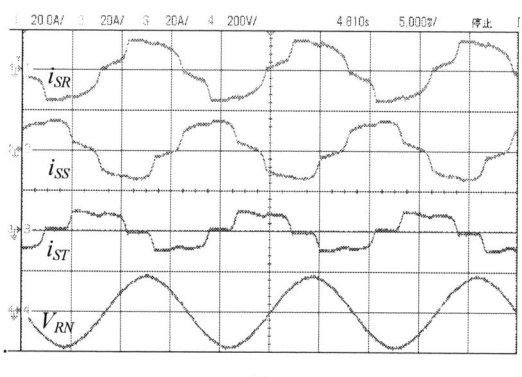

(a)

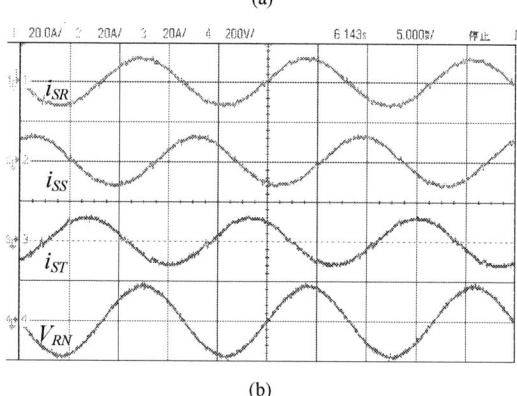

(b)

Fig. 14. Measured waveforms of the three currents: (a)load currents and (b)balanced source current under unbalanced load condition.

The measured inductor current waveforms in both rectification and active power injection modes are shown in Fig. 15(a) and Fig. 15(b), respectively. It can be seen that the inductor current waveforms are almost sinusoidal.

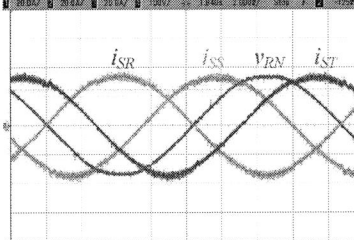

(a) rectification mode

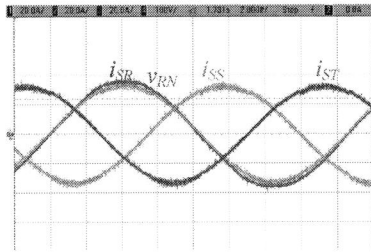

(b) active power injection mode

Fig. 15. Measured waveforms of the inductor currents in (a) rectification mode and (b) active power injection mode.

V. CONCLUSIONS

In this paper, the D-Σ digital control for multi-function inverter application has been presented and verified with a three-phase four-wire inverter system. The reference current determination algorithms and D-Σ digital control have been discussed. By considering wide induction variation, the controller can tune loop gains cycle by cycle. In addition, the operation of the multi-function inverter in three modes: APF mode (including phase power balancing), rectification mode and active power injection mode has been demonstrated. Experimental results have also verified the feasibility of the D-Σ digital controlled multi-functional inverter system.

REFERENCES

[1] Sawant, Rajendra R., and Mukul C. Chandorkar. "A Multifunctional Four-Leg Grid-Connected Compensator", *IEEE Transactions on Industry Applications*, 2009.

[2] M. Singh, V. Khadkikar, A. Chandra, and R. K. Varma, "Grid Interconnection of Renewable Energy Sources at the Distribution Level With Power-Quality Improvement Features," *IEEE Trans. on Power Delivery*, vol. 26, no. 1, pp. 307-315, January 2011.

[3] M. Ucar, S. Ozdemir and E. Ozdemir, "A Combined Series-Parallel Active Filter System Implementation Using Generalized Non-Active Power Theory," *Applied Power Electronics Conference and Exposition (APEC)*, pp. 367-373, Feb. 2010.

[4] "IEEE Recommended Practices and Requirements for Harmonic Control in Electrical Power Systems," *IEEE Std. 519-1992*, pp. 1-112, 1993.

[5] Limits for Harmonic Current Emissions (Equipment Input Current up to and including 16A Per Phase), IEC 61000-3-2, International Standard, 2002.

[6] J. Wang, F. Peng, Q. Wu, Y. Ji and D.Yaping, "A Novel Control Method for Shunt Active Power Filters Using SVPWM", *Proceedings of the 39th Annual Industry Application conference*, Oct. 2004.

[7] S.Rahmani, N.Mendalek and K. Al-Haddad, "Experimental Design of a Nonlinear Control Technique for Three-Phase Shunt Active Power Filter", *IEEE Trans. on Industrial Electronics*, Vol. 57, No. 10, pp3364-3375, Sep. 2010.

[8] K. Nishida, T. Ahmed and N. Nakaoka, "Robust Deadbeat Current Control with Adptive Predictor for Three-Phase Voltage-Source Active Power Filter", *Proceedings of the 39th Annual Industry Application conference*, Oct. 2004.

[9] K. Nishida and N. Nakaoka, "Deadbeat Current Control with Adaptive Predictor for Three-phase Voltage-Source Active Power Filter" *Proceeding of 2004 35th Annual IEEE Power Electronics Specialists Conference (PSEC)*, vol.2, pp.1010-1016, 2004.

[10] Y. He, J. Liu, J. Tang, Z. Wang and Y.Zou, "An Novel Deadbeat Control Method for Active Power Filters with Three-Level NPC Inverter", *Proceeding of the 2008 IEEE PESC*, 2008.

[11] Q. Sun, J.Ji, H.Tian and G. Chen, "Design of Active Power Filter for Low Voltage and High Current Switching Power Supply", *Proceeding of the 2011 APPEEC*, pp.1-4, 2011.

[12] Syamnaresh Garlapati and Rajesh Gupta, "Shunt Active Power Filter as Front End Converter for DC Loads", *Proceeding of the 2012 IICPE*, pp.1-6, 2012.

[13] Harsha Vanjani , "Performance Analysis of Three-phase Four-wire Shunt Active Power Filter" *Proceeding of the 2014 ICROIT*, pp. 496-500, 2014.

[14] D. N. Zmood and D.G. Holmes, "Stationary Frame Harmonic Reference Generation for Active Filter Systems," *IEEE Trans. on Industrial Applications*, vol. 38, no. 6, pp. 1591-1599, Nov./Dec. 2002.

[15] P. Mattavelli, "A Close-Loop Selective Harmonic Compensation for Active Filters," *IEEE Trans. on Industrial Applications*, vol. 37, no. 1, pp. 81-89, Jan./Feb. 2001.

[16] M. Liserre, R. Teodorescu, and F. Blaabjerg, "Multiple Harmonics Control for Three-Phase Grid Converter Systems with the Use of PI-RES Current Controller in A Rotating Frame," *IEEE Trans. on Power Electronics*, vol. 21, no. 3, pp. 836-841, May. 2006.

[17] T.-F. Wu, C.-H. Chang, L.-C. Lin, Y.-C. Chang, and Y. R. Chang, "Two-Phase Modulated Digital Control for Three-Phase Bidirectional Inverter With Wide Inductance Variation", *IEEE Trans. Power Electron*, Vol. 28, No. 4, pp.1598-1607, April 2013.

[18] T.-F. Wu, L.-C. Lin, C.-H. Chang, Y.-L. Lin, and Y.-R. Chang, "Current Improvement for a 3Φ Bi-directional Inverter with Wide Inductance Variation", *Proceeding on the ICPE & ECCE Asia*, pp. 1777-1784, May 2011.

[19] T.-F. Wu, C.-H. Chang, L.-C. Lin and H.-C. Hsieh, "D-Σ Digital Control for A Three-Phase Transformerless Bi-directional Inverter with Wide Inductance Variation", *Proceeding of the 2013 ECCE Asia*, pp. 73-79, June 2013.

[20] Wu, T.-F., H.-C. Hsieh, and C.-H. Chang. "D-ormerless Bi-directional Inverter with WiPower Filter", *2015 9th International Conference on Power Electronics and ECCE Asia (ICPE-ECCE Asia)*, 2015.

[21] Wu, T.-F., C.-H. Chang, and L.-C. Lin. "SVPWM-based D-Σ digital control for 3☐ grid-connected inverter with wide inductance variation", *2014 IEEE Energy Conversion Congress and Exposition (ECCE)*, 2014.

[22] J. Bruce, "Discrete Fourier Transforms, Linear Filters, and Spectrum Weighting," *IEEE Trans. on Audio and Electroacoustics*, vol. 16, pp. 495-499, Dec 1968.

[23] T. Komrska, J. Žák, and Z. Peroutka, "Control Strategy of Active Power Filter with Adaptive FIR Filter-based and DFT-based Reference Estimation," *Power Electronics Electrical Drives Automation and Motion (SPEEDAM), 2010 International Symposium on*, pp. 1524-1529, June 2010.

[24] P. Acuna, L. Moran, M. Rivera, J. Dixon, and J. Rodriguez, "Improved Active Power Filter Performance for Renewable Power Generation Systems," *IEEE Trans. on Power Electronics*, vol. 29, no. 2, pp. 687-694, 2014.

[25] S. Srianthumrong and H. Akagi, "A Medium-Voltage Transformerless AC/DC Power Conversion System Consisting of a Diode Rectifier and a Shunt Hybrid Filter," *IEEE Trans. on Industry Applications*, vol. 39, no. 3, May/June 2003.

Page Intentionally Left Blank

Operation and Analysis of an Improved Transformer-less Unified Power Flow Controller

Yang Liu[1], Shuitao Yang[1,2] and Fang Zheng Peng[1]

[1]Dept. of Electrical and Computer Engineering, Michigan State University, East Lansing, MI 48824, USA
[2]Ford Motor Company, Dearborn, MI 48124, USA
Email: liuyan19@msu.edu

Abstract—In this paper, operation and analysis for an improved transformer-less unified power flow controller (UPFC) is presented. As is well known, the conventional UPFC that consists of two back-to-back inverters requires bulky and often complicated zigzag transformers for isolation and reaching high power rating with desired voltage waveforms. To overcome this problem, a completely transformer-less UPFC based on an innovative configuration of two cascade multilevel inverters (CMIs) has been proposed in [1]. Although the new transformer-less UPFC offers several advantages over the traditional technology, such as transformer-less, light weight, high efficiency, low cost and fast dynamic response, its performance has been limited for some operation points and in some scenarios, the rating of the transformer-less UPFC suffers a lot. In light of this, an improved transformer-less UPFC is proposed in this paper. The benefits of the improved UPFC include, besides the advantages of transformer-less UPFC already possessed, same functionality of original structure, no more hardware needed, more flexible operation and less converter rating. This paper presents rating analysis and operation principle for this new transformer-less T-shape UPFC (TUPFC).

Keywords—Flexible AC Transmission Systems (FACTS), Unified Power Flow Controller (UPFC), Transformer-less, Cascade Multilevel Inverter.

I. INTRODUCTION

The unified power flow controller (UPFC) is able to control, simultaneously or selectively, all the parameters affecting power flow in the transmission line (i.e., voltage magnitude, impedance, and phase angle) [2-4]. The conventional UPFC consists of two back-to-back connected voltage source inverters (VSIs) that share a common dc link, as shown in Fig. 1. UPFC is the most versatile and powerful flexible ac transmission systems (FACTS) device. It can effectively reduce congestions and increase the capacity of existing transmission lines. This allows the overall system to operate at its theoretical maximum capacity. The basic control methods, transient analysis, and practical operation considerations for UPFC have been investigated in [5-7].

Yet, UPFC is seldom used for power flow control owing to the high cost and size of the bulky zigzag transformers. This high-voltage, high-power inverters have to use bulky and complicated zigzag transformers to reach their required VA ratings and desired voltage waveforms. The zigzag transformers are: 1) very expensive (30-40% of total system cost); 2) lossy (50% of the total power losses); 3) bulky (40% of system real estate area and 90% of the system weight); and 4) prone to failure [8].

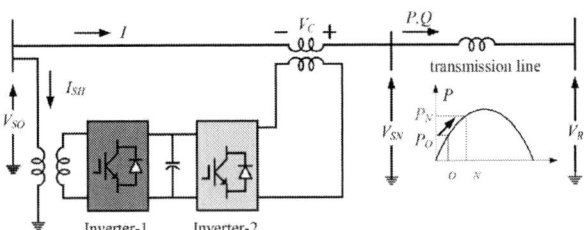

Fig. 1. The conventional unified power flow controller.

To eliminate the transformer completely, a new transformer-less UPFC based on an innovative configuration of two CMIs has been proposed in [1]. The system configuration is shown in Fig. 2(a) and main system parameters for a 13.8-kV/ 2-MVA prototype is shown in Table I. As shown in Fig. 2(a), the transformer-less UPFC consists of two CMIs, one is series CMI, which is directly connected in series with the transmission line; while the other is shunt CMI, which is connected in parallel to the sending end after series CMI. Each CMI is composed of a series of cascaded H-bridge modules as shown in Fig. 2(b). Fig. 2(c) shows the phasor diagram of the new transformer-less UPFC. The transformer-less UPFC has significant advantages over the traditional UPFC such as highly modular structure, light weight, high efficiency, high reliability, low cost, and a fast dynamic response. The basic operation principle, operation range, and required VA rating for series and shunt CMIs have been studied in [1].

However, as has been discussed in [1], the transformer-less UPFC exists some operation limits and in some scenarios, the shunt CMI will have huge injection current. One of the suggested solution, from control perspective, in [1] is releasing the reactive power restriction (allow $Q_c \neq 0$), or paraphrase, the intermediate bus voltage restriction along the transmission line. The introduced controllable reactive power (Q_c) will minimize the required shunt CMI current rating and series CMI voltage rating. Furthermore, the basic modulation and control implementation for the transformer-less UPFC has been discussed in [9, 10]. Fundamental frequency modulation (FFM) has been designed for the transformer-less UPFC and extremely low total harmonic distortion (THD) can be achieved without any extra filters.

This paper presents another solution from topology point of view and even less converter rating is required. This is known in this paper as transformer-less T-shape UPFC (TUPFC). The circuit configuration and phasor diagram is shown in Fig. 3. The improved transformer-less TUPFC consists of three CMIs. One is series CMI-1 (part A), which is directly connected in

978-1-4673-9551-9/16 $31.00 © 2016 IEEE

series with transmission line near the sending-end bus; the shunt CMI is connected in parallel to the sending end after the series CMI-1. After that, the series CMI-2 (part B) is series-connected with the transmission line near the receiving-end bus. Since the elimination of the common dc bus, the connection of the system is more flexible. If series CMI's module number in original transformer-less UPFC is equal to the sum of series CMI-1 plus CMI-2 in TUPFC, the total converter rating of the new transformer-less TUPFC is greatly reduced. Consequently, no more hardware is needed for the new configuration just the wire reconfiguration. The new improved transformer-less TUPFC offers more control freedom and less converter rating.

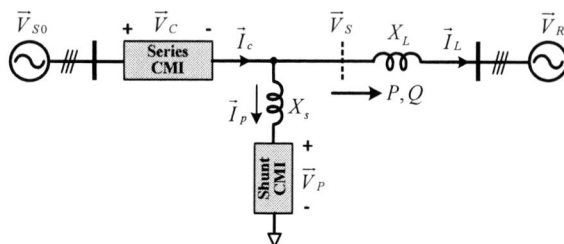

(a) System Configuration of Transformer-less UPFC.

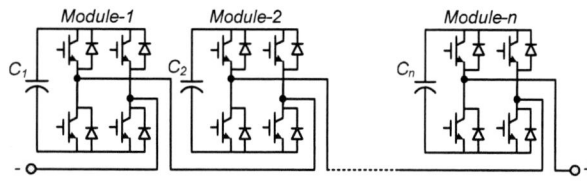

(b) One phase of the cascaded multilevel inverter (CMI).

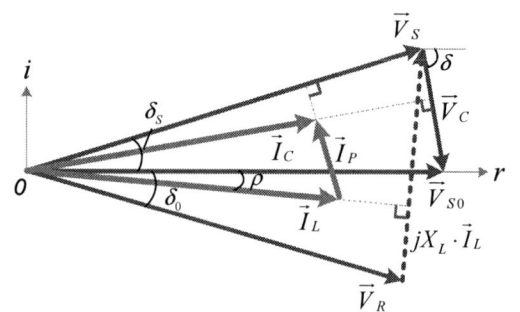

(c) Phasor diagram of the transformer-less UPFC.

Fig. 2. New transformer-less UFPC, (a) System configuration of transformer-less UPFC , (b) One phase of the cascaded multilevel inverter, (c) Phasor diagram of the transformer-less UPFC.

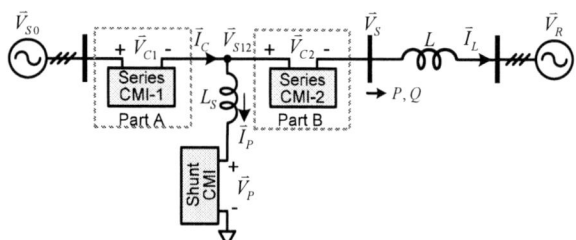

(a). System configuration of improved transformer-less T-shape UPFC.

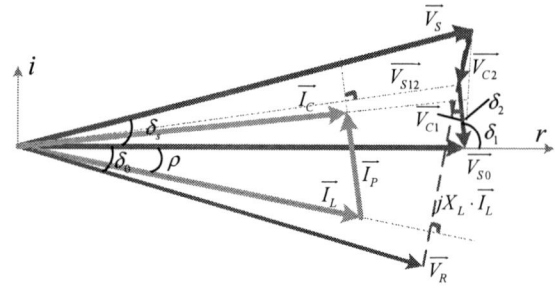

(b). Phasor diagram of the improved transformer-less T-shape UPFC.

Fig. 3. Improved transformer-less T-shape UFPC, (a) System configuration and (b) Phasor diagram.

TABLE I. MAIN SYSTEM PARAMETERS FOR 13.8-KV/ 2-MVA PROTOTYPE

Parameters	Value
System power rating	2 MVA
V_{s0} and V_R *rms*	13.8 kV
Max series CMI current, I_C *rms*	84 A
Max shunt CMI current, I_P *rms*	42 A
V_{dc} (Shunt)	600 V
V_{dc} (Series)	600 V
H-bridge capacitance	2350 μF
No. of H-bridges per phase (Shunt)	20
No. of H-bridges per phase (Series)	8
No. of H-bridges per phase (Series CMI-1)	4
No. of H-bridges per phase (Series CMI-2)	4

II. OPERATION PRINCIPLE OF THE IMPROVED TRANSFORMER-LESS T-SHAPE UPFC

With the unique configuration of the series and shunt CMIs, the transformer-less TUPFC has some new features:

1) Unlike the conventional back-to-back dc link coupling, the transformer-less TUPFC requires no transformer, thus it can achieve low cost, light weight, small size, high efficiency, high reliability, and fast dynamic response;

2) The shunt inverter is connected parallel in between the two series inverters CMI-1 and CMI-2, forms a configuration like a T-shape, which is distinctively different from the traditional UPFC. Each CMI has its own dc capacitor to support dc voltage;

3) There is no active power exchange between the three CMIs and all dc capacitors are floating;

4) The new TUPFC uses modular CMIs and their inherent redundancy provides greater flexibility to system design and higher reliability.

A. Steady-state Models of TUPFC system

Due to the unique system configuration, the basic operation principle of the transformer-less TUPFC is quite different from conventional UPFC. Fig. 3(b) shows a phasor-diagram explanation of the transformer-less UPFC, where $\overrightarrow{V_{s0}}$ and $\overrightarrow{V_R}$ are the original sending- and receiving-end voltage, respectively. Here, $\overrightarrow{V_{s0}}$ is aligned with real axis, which means phase angle of $\overrightarrow{V_{s0}}$ is zero. The series CMI-1 and CMI-2 is controlled to corporately generate a desired voltage

$\overrightarrow{V_C} = \overrightarrow{V_{C1}} + \overrightarrow{V_{C2}}$ for obtaining the new sending-end voltage $\overrightarrow{V_S}$, which in turn, controls active and reactive power flows over the transmission line. Meanwhile, the shunt CMI injects a current $\overrightarrow{I_P}$ to the new sending-end bus to make zero active power into three CMIs, i.e., to make the series CMI-1 current $\overrightarrow{I_C}$, series CMI-2 current $\overrightarrow{I_L}$ and the shunt CMI current $\overrightarrow{I_P}$ be perpendicular to their voltages $\overrightarrow{V_{C1}}$, $\overrightarrow{V_{C2}}$ and $\overrightarrow{V_{S12}}$, respectively. As a result, both series and shunt CMIs only need to provide the reactive power. In such a way, it is possible to apply the CMIs to the transformer-less TUPFC with floating dc capacitors for H-bridge modules.

With the above mentioned restrictions, there is

$$\vec{V}_{S12} \cdot \vec{I}_P \equiv 0, \ \vec{V}_{C1} \cdot \vec{I}_C \equiv 0, \ \vec{V}_{C2} \cdot \vec{I}_L \equiv 0 \qquad (1).$$

The transmitted active power P and reactive power Q over the line with the transformer-less TUPFC can be expressed as

$$
\begin{aligned}
P + jQ &= \vec{V}_R \cdot \left(\frac{\vec{V}_{S0} - \vec{V}_{C1} - \vec{V}_{C2} - \vec{V}_R}{jX_L} \right)^* \\
&= \left(-\frac{V_{S0}V_R}{X_L}\sin\delta_0 + \frac{V_{Ceq}V_R}{X_L}\sin(\delta_0 - \delta') \right) + \\
&\quad j\left(\frac{V_{S0}V_R\cos\delta_0 - V_R^2}{X_L} - \frac{V_{Ceq}V_R}{X_L}\cos(\delta_0 - \delta') \right)
\end{aligned}
\qquad (2),
$$

where, $V_{Ceq} = \sqrt{V_{C1}^2 + V_{C2}^2 - 2V_{C1}V_{C2}\cos\left(\frac{\pi}{2} - \rho - \delta_1\right)}$,

$$\delta' = \delta_1 - \delta_2, \ \delta_2 = \arcsin\left(\frac{\sin\left(\frac{\pi}{2} - \rho - \delta_1\right)\cdot V_{C2}}{\sqrt{V_{C1}^2 + V_{C2}^2 - 2V_{C1}V_{C2}\cos\left(\frac{\pi}{2} - \rho - \delta_1\right)}} \right),$$

and symbol * represents the conjugate of a complex number; δ_0 is the phase angle of the receiving-end voltage $\overrightarrow{V_R}$; $\overrightarrow{V_{Ceq}}$ is the equivalent injected voltage of series CMI-1 plus series CMI-2; δ' is the phase angle of the equivalent series CMI injected voltage $\overrightarrow{V_{Ceq}}$; X_L is the equivalent transmission line impedance. The original active and reactive powers, P_0 and Q_0 with the uncompensated system (without the TUPFC, or $V_{Ceq}=0$) are

$$
\begin{cases}
P_0 = -\dfrac{V_{S0}V_R}{X_L}\sin\delta_0 \\[2mm]
Q_0 = \dfrac{V_{S0}V_R\cos\delta_0 - V_R^2}{X_L}
\end{cases}
\qquad (3).
$$

The net differences between the original (without the TUPFC) powers expressed in equation (3) and the new (with the TUPFC) powers in equation (2) are the controllable active and reactive powers, P_C and Q_C by the TUPFC and can be expressed as,

$$
\begin{cases}
P_C = \dfrac{V_{Ceq}V_R}{X_L}\sin(\delta_0 - \delta') \\[2mm]
Q_C = -\dfrac{V_{Ceq}V_R}{X_L}\cos(\delta_0 - \delta')
\end{cases}
\qquad (4).
$$

Because both amplitude V_{Ceq} and phase angle δ' of the TUPFC injected voltage $\overrightarrow{V_{C1}}$ and $\overrightarrow{V_{C2}}$ can be any values as commanded, the new TUPFC provides a full controllable range of $(-V_{Ceq}V_R/X_L)$ to $(+V_{Ceq}V_R/X_L)$ for both active and reactive powers, P_C and Q_C, which are advantageously independent of the original sending-end voltage and phase angle δ_0. In summary, equations (2) to (4) indicate that the new transformer-less TUPFC has the same functionality as the conventional UPFC. Furthermore, as will be discussed later, it needs less converter rating for the same operation range as original transformer-less UPFC owing to one more control freedom.

B. Dynamic Models of TUPFC system

The dynamic models for the improved transformer-less TUPFC are based on synchronous (dq) reference frame. The phase angle of original sending-end voltage V_{s0} is obtained from a digital phase-locked loop (PLL), which is used for abc to dq transformation.

The dynamic models for the whole system shown in Fig. 3(a) will be divided into several parts. Firstly, we can get the dynamic model from the new sending-end bus $\overrightarrow{V_S}$ to receiving-end bus $\overrightarrow{V_R}$

$$
\begin{cases}
V_{Sd} = V_{Rd} - X_L \cdot I_{Lq} + L\dfrac{d}{dt}I_{Ld} \\[2mm]
V_{Sq} = V_{Rq} + X_L \cdot I_{Ld} + L\dfrac{d}{dt}I_{Lq}
\end{cases}
\qquad (5).
$$

Since the new sending-end voltage V_S is equal to original sending-end voltage V_{S0} minus series CMI-1 and CMI-2 injected voltage V_{c1} and V_{c2}, thus we have

$$
\begin{cases}
V_{C2d} = V_{Sd} - V_{S12d} \\
V_{C2q} = V_{Sq} - V_{S12q}
\end{cases}
\qquad (6)
$$

and

$$
\begin{cases}
V_{C1d} = V_{S12d} - V_{S0d} \\
V_{C1q} = V_{S12q} - V_{S0q}
\end{cases}
\qquad (7).
$$

Furthermore, the model from the new sending-end $\overrightarrow{V_S}$ to shunt CMI is

$$
\begin{cases}
V_{Pd} = V_{S12d} - X_S \cdot I_{Pq} + L_S\dfrac{d}{dt}I_{Pd} \\[2mm]
V_{Pq} = V_{S12q} + X_S \cdot I_{Pd} + L_S\dfrac{d}{dt}I_{Pq}
\end{cases}
\qquad (8).
$$

The current relationship for the three CMIs is

$$\begin{cases} I_{Ld} = I_{Cd} + I_{Pd} \\ I_{Lq} = I_{Cq} + I_{Pq} \end{cases} \quad (9).$$

The restriction for the reactive power for the three CMIs, as shown in (1), can be expressed in dq reference frame as

$$(V_{Sd} - V_{C2d}) \cdot I_{Pd} + (V_{Sq} - V_{C2q}) \cdot I_{Pq} = 0 \quad (10),$$

$$k_1 V_{C2d}(I_{Ld} - I_{Pd}) + k_2 V_{C2q}(I_{Lq} - I_{Pq}) = 0 \quad (11),$$

$$V_{C2d} I_{Ld} + V_{C2q} I_{Lq} = 0 \quad (12),$$

where, $V_{c1d} = k_1 * V_{c2d}$ and $V_{c1q} = k_2 * V_{c2q}$,

$$k_1 V_{C2d} I_{Ld} - k_2 V_{C2d} I_{Lq} = \left[k_1 V_{C2d} - \frac{k_2 V_{C2q}(V_{Sd} - V_{C2d})}{V_{Sq} - V_{C2q}} \right] \cdot I_{pd} \quad (13).$$

$$I_{Pq} = -\frac{V_{Sd} - V_{C2d}}{V_{Sq} - V_{C2q}} \cdot I_{Pd} \quad (14).$$

From the above equations, all of the parameters in improved transformer-less TUPFC can be obtained once the desired P^* and Q^* is given.

In summary, there are two critical steps for the operation of TUPFC: a) calculation of injected voltage $\overrightarrow{V_{C1}}$ and $\overrightarrow{V_{C2}}$ for series CMI-1 and CMI-2 according to active/reactive power command over the transmission line, and b) calculation of injected current $\overrightarrow{I_P}$ for shunt CMI to guarantee zero active power into all series and shunt CMIs.

III. CONVERTER RATING COMPARISON BETWEEN TWO TRANSFORMER-LESS UPFCS

The required converter rating for the operation of original transformer-less UPFC and transformer-less TUPFC is presented in this section. The main purpose of a UPFC is to control active power flow, while maintaining reactive power flow minimal over long distance lines. Thus, reactive power flow is usually compensated locally near the receiving end. If we restrict the voltage variation along the transmission line ($V_s \equiv 1$ pu) for all the bus, the current rating could be huge according to analytical analysis in [1]. An example of this operation range is shown in Fig. 4.

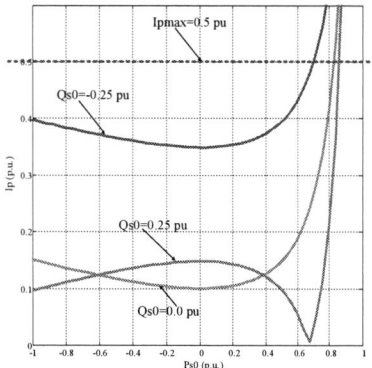

Fig. 4. The relationship between shunt CMI current I_p and the original power P_{s0} with equipment constraint (I_{pmax}=0.5 pu) and voltage constraint (Q_c=0 pu) for original transformer-less UPFC.

As can be seen from Fig. 4, the original transformer-less UPFC has limited operation especially when the original active is already large (>0.6 pu). An effective way to get rid of this operation limitation is by changing the controllable reactive power operation range, in other words, allow the intermediate voltage V_s, V_{s12} varied within a certain limitation. Fig. 5 shows the operation of the original transformer-less UPFC if the intermediate bus has an allowable changing magnitude with $\pm10\%$.

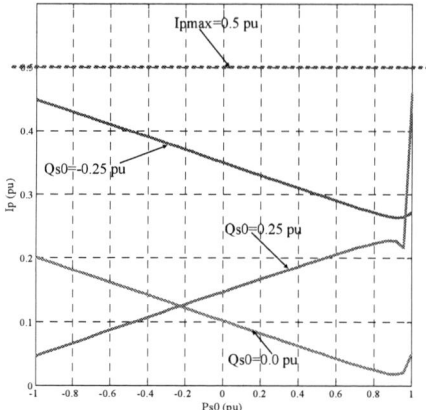

Fig. 5. The relationship between shunt CMI current I_p and the original power P_{s0} with equipment constraint (I_{pmax}=0.5 pu) and (Q_c=0.1 pu) for original transformer-less UPFC.

Whereas, if the new improved transformer-less TUPFC is employed, an even lower shunt current can be obtained. Owing to the one more control freedom brings about by series CMI-1 and CMI-2. An objective function is utilized to achieve minimum shunt current thus can get minimum total converter ratings.

As can be seen from Fig. 6, the current rating for the shunt CMI is significantly reduced while not compromising the transmission line voltage variation. Furthermore, Fig. 8-Fig. 11 show the comparison between original transformer-less UPFC and improved T-shape UPFC. As can be seen from these results, due to the one more control freedom introduced by T-shape transformer-less UPFC, the shunt converter rating has been greatly reduced while the series converter rating is almost the same. Consequently, the total converter rating is reduced. There is no more hardware requirement for the T-shape transformer-less UPFC, just by changing wire configuration, new operation range can be obtained.

It should be noted that the new configuration will not influence the series CMI rating and it is advantageously small because it is mainly determined by the line impedance and original real power flow. For example, 0.1-pu impedance line requires only 0.1-pu series voltage even for a 1-pu net power change. So, even for the worse case, P_0 change from -1pu to 1pu, around 0.2-pu series voltage compensation is enough. However, in order to support the operation of the transformer-less UPFC, for the reactive power limitation, the shunt CMI in some scenario is huge value but this can be well solved with the help of T-shape transformer-less UPFC.

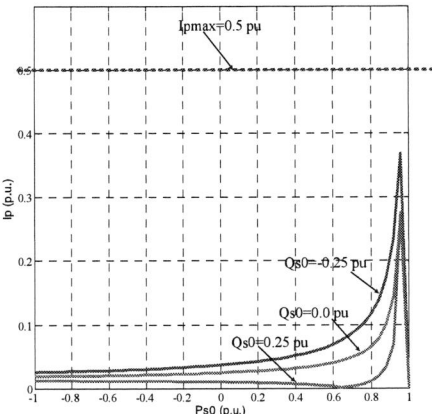

Fig. 6. The relationship between shunt CMI current I_p and the original power P_{s0} with equipment constraint (I_{pmax}=0.5 pu) and (Q_c=0 pu) for improved transformer-less TUPFC.

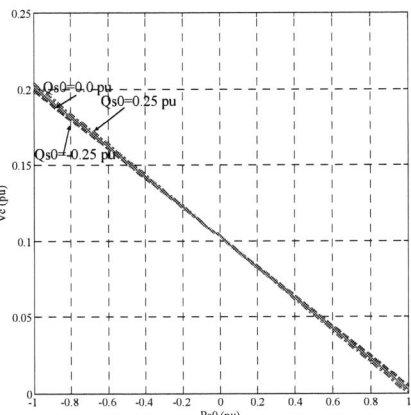

Fig. 7. The relationship between series CMI voltage V_C and the original power P_{s0} with equipment constraint (V_{Cmax}=0.5 pu) and (Q_c=0.1 pu) for original transformer-less UPFC.

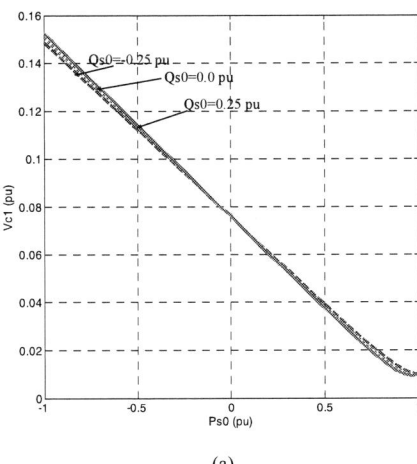

(a)

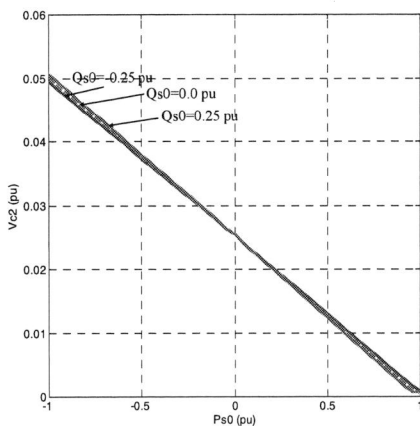

(b)

Fig. 8. The relationship between series CMIs voltage (a). V_{C1} and (b). V_{C2} with the original power P_{s0} with equipment constraint (V_{Cmax}=0.5 pu) for improved transformer-less TUPFC.

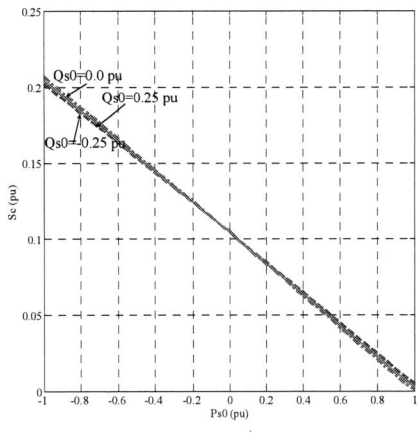

(a)

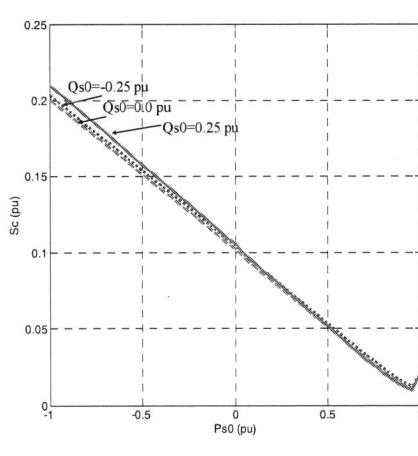

(b)

Fig. 9. The comparison between series converter rating S_c and the original power P_{s0} with equipment constraint (I_{Cmax}=2 pu & V_{Cmax}=0.5 pu) for (a). original transformer-less UPFC and (b). improved transformer-less TUPFC.

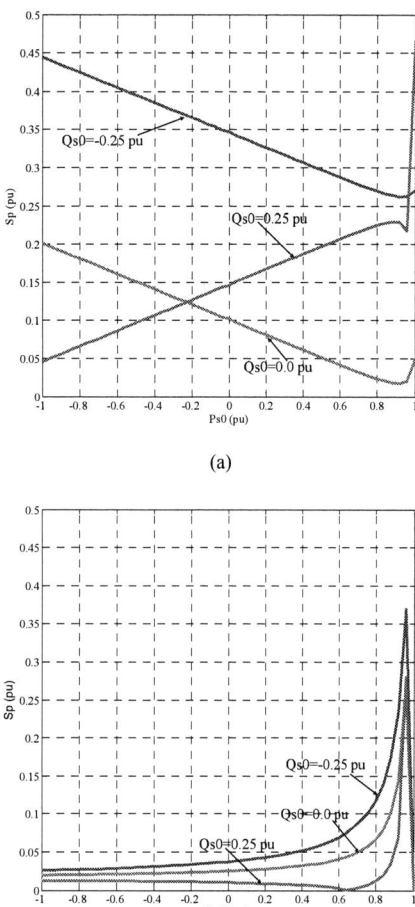

(a)

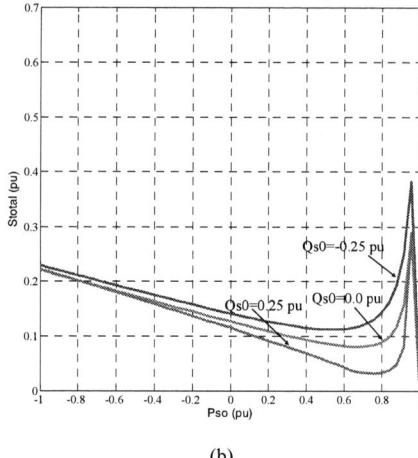

(b)

Fig. 11. The comparison between total converter rating S_{total} and the original power P_{s0} over the transmission line with equipment constraint (I_{Cmax}=2 pu & V_{Cmax}=0.5 pu & I_{pmax}=0.5 pu & V_{pmax}=1.1 pu) for (a). original transformer-less UPFC and (b). improved transformer-less TUPFC.

IV. CONCLUSION

This paper presents the operation and analysis for an improved T-shape transformer-less UPFC. The steady-state and dynamic model have been developed and basic operation principle has been analyzed. Furthermore, its converter rating has been compared with original transformer-less UPFC. The improved transformer-less TUPFC has the following features: 1) same hardware requirement and more control freedom; 2) reduced shunt current rating and reduced total converter rating compared to original transformer-less UPFC 3)able to achieve independent active and reactive power flow control over the transmission line. Due to the flexible implementation, the improved transformer-less TUPFC can be installed anywhere in the grid to maximize/optimize energy transmission over the existing grids and reduce transmission congestion. The experimental verification at 13.8 kV/ 2 MVA prototype is left for future work.

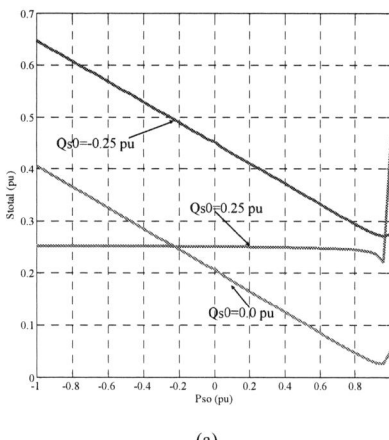

(b)

Fig. 10. The comparison between shunt converter rating S_p and the original power P_{s0} with equipment constraint (I_{pmax}=0.5 pu & V_{pmax}=1.2 pu) for (a). original transformer-less UPFC and (b). improved transformer-less TUPFC.

ACKNOWLEDGMENT

The authors gratefully acknowledge the funding support of Advanced Research Project Agency-Energy (ARPA-E), Department of Energy (DoE), United States to design, prototype and test the transformer-less UPFC prototype.

REFERENCES

[1] F. Z. Peng, S. Zhang, S. T. Yang, D. Gunasekaran, and U. Karki, "Transformer-less unified power flow controller using the cascade multilevel inverter," in *2014 International Power Electronics Conference (IPEC-Hiroshima 2014 - ECCE-ASIA)*, 2014, pp. 1342-1349.

[2] N. G. Hingorani and L. Gyugyi, "Understanding FACTS: concept and technology of flexible AC transmission systems.," 2000.

[3] L. Gyugyi, C. D. Schauder, S. L. Williams, T. R. Rietman, D. R. Torgerson, and A. Edris, "The unified power flow controller: a new approach to power transmission control," *IEEE Transactions on Power Delivery*, vol. 10, pp. 1085-1097, 1995.

[4] A. Rajabi-Ghahnavieh, M. Fotuhi-Firuzabad, M. Shahidehpour, and R. Feuillet, "UPFC for Enhancing Power System Reliability," *IEEE Transactions on Power Delivery*, vol. 25, pp. 2881-2890, 2010.

(a)

[5] H. Fujita, Y. Watanabe, and H. Akagi, "Control and analysis of a unified power flow controller," *IEEE Transactions on Power Electronics,* vol. 14, pp. 1021-1027, 1999.

[6] L. M. Liu, P. C. Zhu, Y. Kang, and J. Chen, "Power-Flow Control Performance Analysis of a Unified Power-Flow Controller in a Novel Control Scheme," *IEEE Transactions on Power Delivery,* vol. 22, pp. 1613-1619, 2007.

[7] S. Kannan, S. Jayaram, and M. M. A. Salama, "Real and reactive power coordination for a unified power flow controller," *IEEE Transactions on Power Systems,* vol. 19, pp. 1454-1461, 2004.

[8] F. Z. Peng, J. S. Lai, J. W. McKeever, and J. VanCoevering, "A multilevel voltage-source inverter with separate DC sources for static VAr generation," *IEEE Transactions on Industry Applications,* vol. 32, pp. 1130-1138, 1996.

[9] S. T. Yang, Y. Liu, X. R. Wang, D. Gunasekaran, U. Karki, and F. Z. Peng, "Modulation and Control of Transformer-less UPFC," *IEEE Transactions on Power Electronics,* vol. PP, pp. 1-1, 2015.

[10] S. T. Yang, S. Zhang, X. R. Wang, D. Gunasekaran, and F. Z. Peng, "Optimization of fundamental frequency modulation for cascaded multilevel inverter based transformer-less UPFC," in *2014 IEEE Energy Conversion Congress and Exposition (ECCE)*, 2014, pp. 4647-4652.

Design Consideration of Converter Based Transmission Line Emulation

Bo Liu, Shuoting Zhang, Sheng Zheng, Yiwei Ma, Fred Wang, Leon M. Tolbert

Center for Ultra-wide-area Resilient Electric Energy Transmission Networks (CURENT)
Department of Electrical Engineering and Computer Science
The University of Tennessee
Knoxville, TN 37996-2250, USA
bliu16@vols.utk.edu

Abstract—Ac transmission line emulator is the bridge to interconnect ac systems to fulfill the grid emulation function, where all the emulator elements such as generators, loads and lines are implemented by universal three-phase voltage source converters. In this paper, three design issues are addressed. First, the impact of ac voltage switching noise on the performance of a transmission line emulator in terms of steady state and dynamic accuracy is described, and an improved sampling algorithm is presented. Then, a new dc offset controller is proposed to mitigate the induced dc current flow by sampled dc offset noise, to guarantee the normal operation of ac line emulator. Furthermore, the stability issues regarding different emulation schemes are analyzed, providing a metric to predict the feasible impedance range that a line emulator can reach and to choose the proper emulation strategy for a specific system. Finally, experimental results obtained from a multi-converter based hardware testbed verify the design schemes.

Keywords—transmission line; emulation; dc offset control; sampling; stability

I. INTRODUCTION

Multi-converter based hardware emulation system for ac grid application has been discussed in several papers [1-4]. The hardware emulation implements all the models and function in universal inverters, running the real power with suitable scaling to mimic the real grid system operation. Thus it provides more field information, showing the system behaviors in different time scales, and can naturally mimic the real communication and control hierarchy among different sub-area systems in real-time. Generally, each power converter emulates a certain type of grid infrastructure such as generator, load, wind farm and etc., and joints with each other modularly.

In this interconnected emulation system, the transmission line is the main device to interface with different emulators. The simplest way is to use passive L, R and C components to mimic the transmission lines [5]. However, it is quite inconvenient to change the line parameters. One has to replace the LCR each time for different studies, which is time-consuming and not cost-effective. A more efficient way of implementing such a device is to adopt power converters as a line emulator. Thus, one smart programmable line emulator is competent to perform tasks with variable line parameters.

Its accuracy and dynamic response are crucial for precisely emulating the power transfer behaviors. Two types of emulation schemes are discussed in [4]. The power flow based emulation is easy to be implemented, but more suitable for steady state and slow dynamic applications, such as the automatic runback application reported in [6] to push a new stable set point for HVDC. The time domain model based emulation, on the other hand, as illustrated in Fig. 1, samples the instantaneous input voltages at both terminals and calculates the instantaneous current to be regulated based on (1) and (2), and therefore is capable of emulating fast transients and is desirable for system emulation.

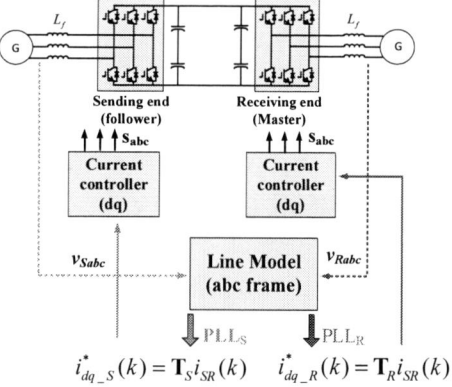

Fig. 1. Control scheme of time domain line emulation.

$$\frac{di_{SR}(t)}{dt} = \frac{v_S(t) - v_R(t)}{L} \tag{1}$$

$$i_{SR}(k) = \frac{\tau}{2L}\left[v_S(k) + v_S(k-1) - v_R(k) - v_R(k-1)\right] + i_{SR}(k-1) \tag{2}$$

where τ is the discrete step size, and "S" "R" denote the sending end and receiving end.

However, this approach is very sensitive to voltage sampling noise. Any undesirable voltage noise will induce faulty current commands and drive the emulator into the wrong track. Two types of noises are observed: 1) switching harmonic voltage, which is unique in the multi-interconnected converter system with only L filters, and 2) dc offset from sensor circuit or analog/digital conversion (ADC). Unfortunately, a strong filter may eliminate most noise but it introduces phase delay, which is not acceptable to meet emulation accuracy and high bandwidth requirement. This is similar to three-phase power factor correction (PFC) control, where the current switching noise cannot be simply filtered much due to its negative impact on phase margin for a high bandwidth control design.

A good sampling algorithm with correct sampling instant is therefore important for sampling quality. Typically, center-aligned (symmetrical) PWM generation is preferred for three-phase converters [7-9]. The sampling instant is the median of the rising edge or falling edge, which naturally ensures the average value of the sampling signal. Reference [9] also proposes a hybrid sampling scheme for the high modulation index case. However, unlike the PFC with a grid with minimum harmonics, in the emulation system, the sampling switching noises of both voltage and current are also dependent on its interfaced converters. These converters are controlled independently following the modular concept, resulting in asynchronized PWM patterns. This "more noisy" system with asynchronized time domains results in the sampling and control to be more difficult. Thus, the analysis and a simple solution for this noisy emulation environment are provided in this paper.

The dc offset issue for this time domain line model is similar to the scenario in motor control, where the motor flux feedback is estimated from the pure integration of the instantaneous ac voltage. A trivial dc drift in the voltage sampling will rapidly and eventually lead to a significant high dc offset flux. Many efforts have been made in motor applications, and a literature review is found in [10]. In [11-14], different low pass filter (LPF) structures are studied to replace the integrator, which damps the dc bias to a certain level but introduces steady-state error in both phase and magnitude, especially at low frequency operation. And in order to compensate this error, complicated compensation algorithms are added [15]. Closed loop based offset compensator is proposed in [16] to aid the flux estimator, which however requires the flux reference information.

Reference [10] further combines the dc compensation and integrator into a simplified LPF form without this information, which theoretically can achieve zero error in the specific frequency but suffers inevitable amplification of high frequency harmonics. From the authors' prospective, the past works mainly focus on the offset elimination, flux estimation accuracy and response time at the stationary specific frequency point but still considers the motors' variable frequency operation. There exists a lack of the whole picture of the steady and dynamic performance for these controls in the full frequency spectrum, and

consequently it is not clear of the true impacts imposed to the original system. In this paper, the function of decreasing the sensitivity of the pure integrator to its input bias is reconsidered as a high pass filter (HPF) from the frequency domain perspective. That means the dc bias noise will be attenuated and blocked in the integration path, and only the favored ac frequencies can pass through. With that, a closed loop dc offset control based HPF structure is proposed to adaptively cancel the dc offset with only input information.

The last issue this paper will address is if there is an impedance limit that the transmission line emulator cannot exceed, to keep it in the stable operation, and where the stability boundary is. The analysis and stability criteria for different emulation schemes are proposed.

II. SAMPLING NOISE: IMPACT AND SOLUTION

A. Swiching Harmonics in Ac Voltage Sampling

In a multi-converter emulation system with only L filter, switching harmonics cannot find a bypass path, and thus are fully present at the ac terminals as illustrated in Fig. 2. Meanwhile, voltage sampling is a key process in the line emulation system, as illustrated in (2). Strong filters that introduce several degrees of phase delay at the fundamental frequency are thus not acceptable. With limited filtering, if still sampling the noisy voltage once per switching cycle, the resulting signal is found distorted due to the noticeable switching noise, as shown in Fig. 3. This can be explained by the well-known aliasing effect [17-19].

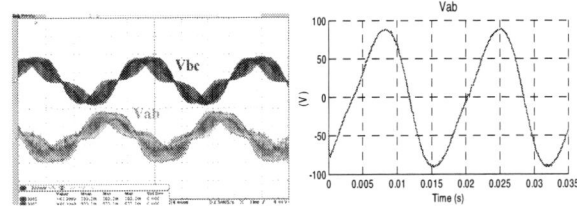

Fig. 2. (left) Input ac voltages with high switching harmonic noises.
Fig. 3. (right) One sampling per switching.

Thus, the minimum sampling frequency should be double of the switching frequency, followed with 2-point moving average filter. However, a very low frequency deviation occurs in Fig. 4 (a), which can be more clearly seen from phase-locked-loop (PLL) test result under dq frame in Fig. 4 (b).

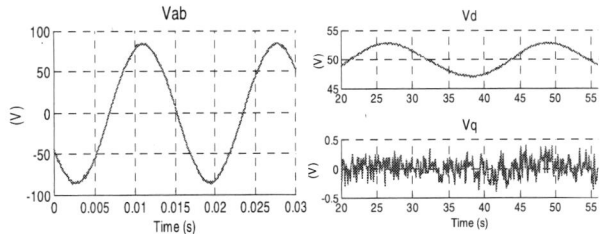

Fig. 4. (a) Double sampling per switching cycle.

Fig. 4. (b) Low frequency variation under dq frame.

$$v = V_m \sin(\omega_0 t) + \Delta V_m \sin((\omega_0 + \Delta\omega)t) \xrightarrow{abc2dq} \quad (3)$$
$$v_d = V_m + \Delta V_m \sin(\Delta\omega t)$$

The reasons can be listed as follows: First, PWM harmonic voltage has certain sidebands and noticeable 2nd switching frequency component, thus aliasing effects may not be fully avoided with only double sampling. From moving filter point, these noises are also not effectively filtered due to the limited attenuation resulting from the lowest depth $N=2$ where N is the filter order. Moreover, due to the non-uniform magnitude attenuation of front-end analog filter over the whole frequency range, the asynchronized PWM voltage sampling (converters are individually controlled and filters have significant phase shift at high frequency), and also the variable duty cycles over each line cycle, voltage ripples in two adjacent switching cycles as illustrated in Fig. 6 (a) are not fully cancelled in the moving average process and instead have a slight amplitude difference. Eventually, this leads to a low amplitude periodic noise along with the line cycle in Fig. 5, and introduces a low frequency oscillation in grid frequency and line power flow.

Therefore, a sampling scheme with four-point per switching cycle plus $N=4$ moving average filter is proposed. With that, all the 2nd f_s ripple and part of side harmonics are filtered from the moving filter prospective. From time domain view, the average of four samplings per switching cycle is also more immune to the PWM voltage asynchronization and non-uniform noise attenuation.

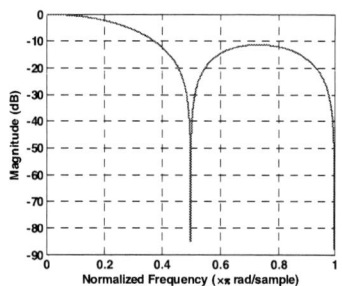

Fig. 5. Frequency response of $N=4$ moving average filter.

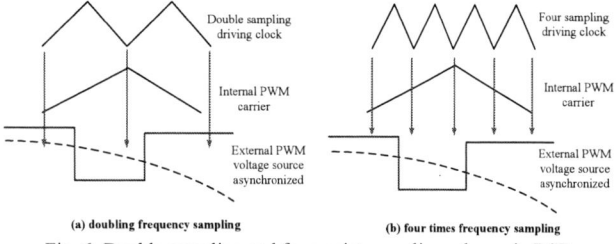

Fig. 6. Double sampling and four-point sampling scheme in DSP.

B. Dc Offset Issue in Ac Current Sampling

1) Basic concept of dc offset controller

As shown in (2), for the simple lossless line model, i.e. an inductor L, current reference is generated from the voltage based pure integrator, which however is prone to the dc drift of input voltage sampling resulting from the

asymmetry in sensor circuit or imbalance in A/D channel. According to the test, this dc drift may also vary with time, therefore it cannot be easily compensated in an open loop manner. A closed loop compensator is preferred. The proposed controller is illustrated in Fig. 8, which can also be regarded as a dc offset estimator. The dc component of the output current is adopted as the feedback signal, obtained via an adaptive notch filter with its notching point designed at line frequency. In fact, this is the key difference from the approach in [10], where the derivative of motor flux is used instead of its dc component as the feedback. The combination of L-LR model is because the lossless L model suffers the severest impact by dc bias noise, while LR model is more practical in emulating real transmission lines and also less sensitive to dc bias. Therefore, by adopting L model as a dc bias removal loop, it provides a sensitive bias indicator to LR model emulation, enabling the better dc elimination.

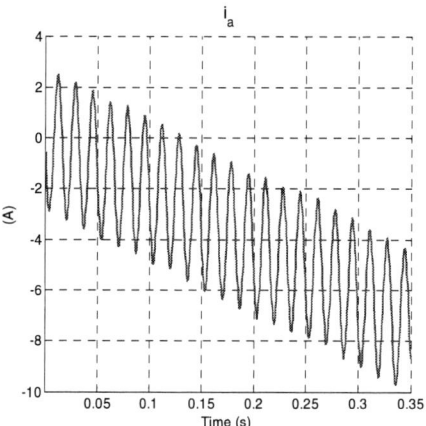

Fig.7. DC current drift due to small dc bias noise

A PI regulator then generates the corresponding counterpart of dc offset noise, to force the input of the integrator to be zero bias. The transfer function of dc offset control loop is given in (4).

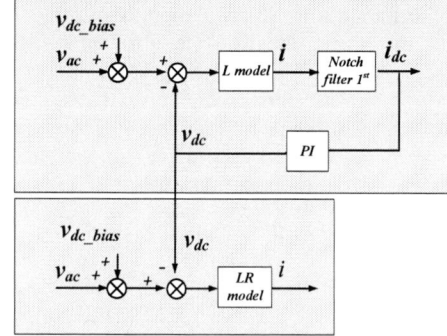

Fig. 8. The block diagram of proposed dc offset controller

$$G_{dc\_offset\_loop}(s) = \frac{\hat{v}_{dc}}{\hat{v}_{dc\_bais}} = \frac{\frac{1}{SL} \cdot NF(s) \cdot PI(s)}{1 + \frac{1}{SL} \cdot NF(s) \cdot PI(s)} \approx \frac{\frac{1}{SL} \cdot PI(s)}{1 + \frac{1}{SL} \cdot PI(s)} \quad (4)$$

$$= \frac{k_p s + k_i}{s^2 L + k_p s + k_i} = \frac{k_p s + k_i}{s^2 + 2\xi\omega_n s + \omega_n^2}$$

where PI parameters are derived as shown in (5) following the normalized second-order system design,

$$k_p = 2\xi\omega_n L$$
$$k_i = \omega_n^2 L \tag{5}$$

It is preferred to select

$$\omega_n < \omega_0 / d \tag{6}$$

$$\xi = 0.5 \sim 1 \tag{7}$$

where d is the ratio of line frequency ω_0 to corner frequency ω_n. d should be selected to achieve the desirable bandwidth.

As discussed in section I, the dc offset removal should function as a HPF, with the least impact on the signals in the high frequency spectrum, to ensure the emulation system operated with zero error during the fast transients. By reorganizing the whole control block diagram from input voltage to output current, its equivalent HPF feature can be obtained,

$$\frac{\hat{i}}{\hat{v}_{ac}} = \frac{1}{1 + \frac{1}{SL} \cdot NF(s) \cdot PI(s)} \frac{1}{SL + R} \tag{7}$$

$$G_{HPF\_LLRNF}(s) = \frac{1}{1 + \frac{1}{SL} \cdot NF(s) \cdot PI(s)} \tag{8}$$

If only LR model is adopted in the dc bias controller, the following transfer function will be derived,

$$G_{HPF\_LRNF}(s) = \frac{1}{1 + \frac{1}{SL + R} \cdot NF(s) \cdot PI(s)} \tag{9}$$

From the HPF characteristic curves of the two models in Fig. 10, it is clear that both can achieve unit gain and zero phase lag above the target frequency with this proposed dc removal scheme, while the L-LR companion model provides extra 20 dB/dec attenuation at dc and low frequencies, which verifies the previous discussion.

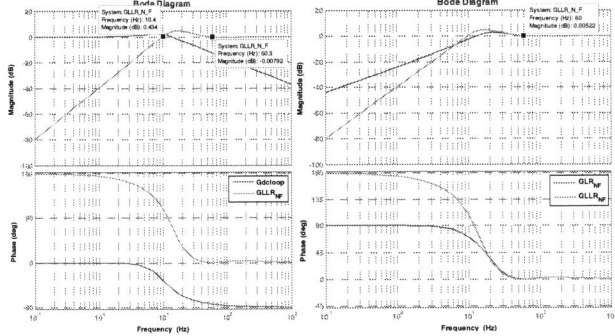

Fig. 9. DC offset controller.   Fig. 10. Equivalent HPF characteristics.

2)   Adaptive offset control for variable frequency operation

In the previous analysis, to extract dc bias noise, a notch filter with resonance frequency at 60 Hz is used. However,

this may introduce static errors in a variable frequency system or during system transients. To ensure the unit filter gain and zero phase lag in high frequency range from HPF prospective, an adaptive notch filter is therefore implemented in the offset controller. One simple approach is to adopt the PLL output frequency as the varying resonance frequency of the notch filter.

$$NF(s) = \frac{s^2 + \omega_r^2}{s^2 + 2\xi_r\omega_r s + \omega_r^2} \tag{10}$$

where,

$$\omega_r = \omega_{PLL} \tag{11}$$

Another benefit from the adaptive design would be consistent offset control dynamics under variable operation frequencies, as determined by (6).

3)   Controller design considering external line impedance

The above control only assumes the line emulator interfacing with the ideal ac system, where no external line impedance is considered. Now, if it connects two typical ac grids with serial impedances, as illustrated in Fig. 14, the dc offset control loop has to be modified as depicted in Fig. 11. L_g and R_g represent the equivalent external line impedance, and $Tcc(s)$ denotes the emulator current control, which is approximately unity gain within the current loop bandwidth.

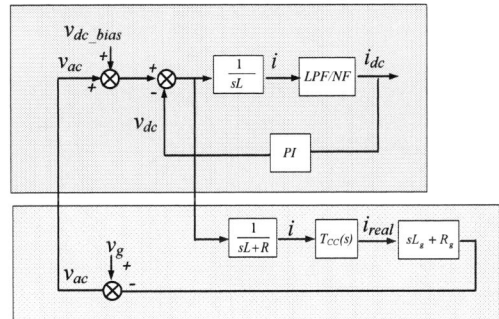

Fig. 11. DC offset control block diagram considering external line impedances.

It is easy to find that HPF transfer function still holds the same form as (8), but the dc offset loop introduces an extra term, i.e. an impedance ratio L_g/L. Consequently, the PI parameters should be altered as shown in (13).

$$G_{dc\_offset\_loop} \approx \frac{k_p s + k_i}{(1 + \frac{L_g}{L})s^2 + 2\xi\omega_n s + \omega_n^2} \tag{12}$$

$$k_p = 2\xi\omega_n L(1 + \frac{L_g}{L})$$
$$k_i = \omega_n^2 L(1 + \frac{L_g}{L}) \tag{13}$$

III. STABILITY CRITERIA OF LINE EMULATOR

A. General Emulation Schemes

Before addressing the stability issues, different line emulation schemes will be briefly discussed.

Typically, we can categorize ac systems configurations into three groups as listed on the left side of Fig. 12. In system (a), both the sending end and receiving end contains a generator, thus it can treated as a voltage source. In system (b), the receiving end has only power load, thus it will be a current source. In system (c), the sending end is a current source. Meanwhile, to emulate the transmission line, although scheme (a) on the right side of Fig. 12 is adopted in [4], there are actually three types, i.e. (a) current/current source pair, (b) current/voltage source pair, and (c) voltage/current source pair.

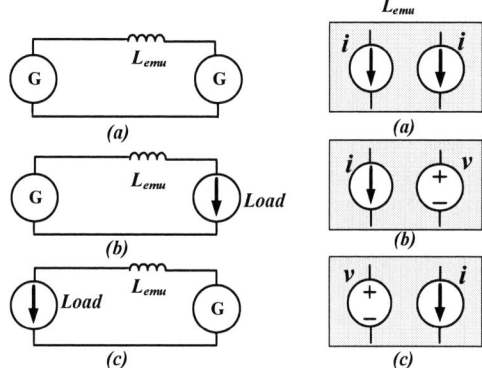

Fig. 12. General system configurations and emulation schemes.

To link the two-area system via a line emulator, there will be nine combinations in total. However, intuitively and also for the sake of stability concern, it is better to couple the voltage source and current source, i.e. using a voltage source to interface with a current source, vice versa. Following this principle, three emulation schemes are provided in Fig. 13. Which scheme should be adopted is of importance to the system design, implementation and stability. In a typical two area system, where both ends usually have generators and loads, system (a) is more common. However, in some special operation conditions, where one end has lost all generators and has only loads, scheme (b) or (c) should be better. Regardless of the system differences, one advantage of scheme (a) is that since the emulator always interfaces the two ends as voltage controlled current sources, only current control mode is needed, while the other two schemes may have to combine together and transition from one control mode to the other control mode to fulfill operation as conditions change. For simplicity and brevity, only scheme (a) is considered in this paper.

As for the scheme (a), there are also two types of emulation schemes. The first is a phasor domain based model and the other is a time domain based model. In the following sections, the stability issue in these two schemes will be addressed.

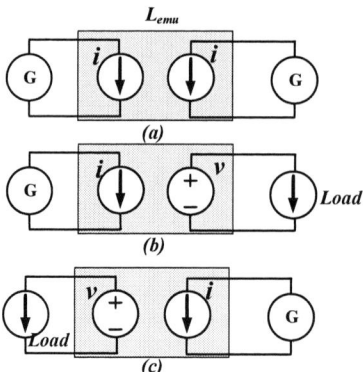

Fig. 13. Preferred emulation schemes.

B. Stability Issue in Phasor Domain Based Emulation

Ideally, if the current controlled line emulator only interfaces with the ideal grid with zero impedance, then this system is pretty robust and no stability issue arises. However, as illustrated in the simplified equivalent two-area transmission line system in Fig. 14, a converter based emulator is not an isolated autonomic system. Instead, its current flow is determined by the configurations of two interface voltages. Hence, its stability has to consider the interaction with the external impedance network, specifically if the grid has its own inductive impedances. In addition, for a digital implementation, different discrete methods may also lead to significant differences in terms of the stability region.

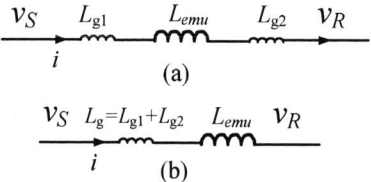

Fig. 14 Simplified equivalent transmission line.

To better illustrate the stability issue, first, the phasor domain based model will be analyzed in brief. Here, power flow at the two ends is calculated based on the steady state phasor vectors [4]. Therefore, only slow dynamics can be observed. Even so, if the external impedance is considered, the system may still lose stability.

By controlling the power, line current can be regulated according to (14). From phasor vector relationship shown in Fig. 15, we can characterize the emulation model as an nth-order discrete equation (15), where n is the ratio of time delay resulting from the slow steady model over the discrete step.

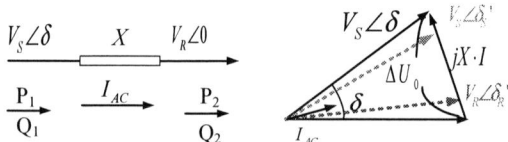

Fig. 15. Simplified phasor domain ac line emulation.

$$I = \left(\frac{S_1}{V_S}\right)^* = \frac{P_1 - jQ_1}{V_S} \qquad (14)$$

$$I_k = \frac{\Delta U_0 - jX_g I_{k-n}}{jX_{emu}} = (1 + \frac{L_g}{L_{emu}})I_0 - \frac{L_g}{L_{ref}}I_{k-n} \qquad (15)$$

To ensure the convergence of (15), its eigenvalue should obey

$$\lambda = \left| \sqrt[n]{\frac{L_g}{L_{emu}}} \right| < 1 \qquad (16)$$

which means if the external line impedance is larger than that of the emulated line, the emulation system will lose its convergence, leading to current oscillations. Moreover, due to the slow control law in the phasor domain, the stability issue is even more severe. As an example, if the impedance ratio is 0.5, but n=20, then λ will become 0.966, indicating a much smaller stability boundary.

C. Stability Issue in Time Domain Based Emulation

The time domain based model is more useful for dynamic study. This paper will address its stability issue with three common digital implementations i.e. forward-Euler, backward-Euler, and trapezoidal discrete. Different from EMTP based software simulation, which adopts implicit discrete methods and solves full-element equations in large dimension matrices to attain better convergence and accuracy, hardware based emulation with modular configurations focusing on real-time emulation does not know any external element information, and has to rely on its real-time measurements from the interface to build up the real-time emulation models and consequently react. Thus, it has to use the current and past samplings as inputs to generate the next-step outputs due to the mechanism of PWM converter control, i.e. only explicit implementation is practical. Therefore, backward-Euler discrete will recede to forward-Euler, and trapezoidal approach also has to be modified as shown from (17) to (18) in order to correctly represent the real modeling processing.

$$i(k) - i(k-1) = \frac{\tau}{2L_{emu}}[v_{emu}(k) + v_{emu}(k-1)] \qquad (17)$$

$$i(k) - i(k-1) = \frac{\tau}{2L_{emu}}[v_{emu}(k-1) + v_{emu}(k-2)] \qquad (18)$$

where,

$$v_{emu}(k) = v_S(k) - v_g(k) - v_R(k) \qquad (19)$$

and different from (18) for the emulated main inductor, the external inductance voltage drop is the physical line voltage drop varying instantaneously. Thus, an analog signal should be modeled as

$$v_g(k) = L_g \frac{i(k) - i(k-1)}{\tau} \qquad (20)$$

From (18) ~ (20), the discrete current differential equation is obtained showing a second-order system,

$$[i(k) - i(k-1)] + \frac{L_g}{2L_{emu}}[i(k-1) - i(k-2)] \qquad (21)$$

$$+ \frac{L_g}{2L_{emu}}[i(k-2) - i(k-3)] = 0$$

And its eigenvalues are,

$$\lambda_{1,2} = \frac{-\alpha \pm j\sqrt{4\alpha - \alpha^2}}{2} \qquad (22)$$

where,

$$\alpha = \frac{L_g}{2L_{emu}} \qquad (23)$$

To maintain stability, the criterion is provided,

$$|\lambda| = \begin{cases} \sqrt{\alpha} & (0 < \alpha < 4) \\ \sqrt{\frac{\alpha^2}{2} - \alpha} & (\alpha > 4) \end{cases} < 1 \qquad (24)$$

And the final solution is

$$\alpha = \frac{L_g}{2L_{emu}} < 1 \qquad (25)$$

D. Stability Issue in Time Domain Based Emulation with Pure Active Load

In the above analysis, two ac grids are only considered as two voltage sources; however, in a typical system, there will be definitely certain active loads at the receiving end, since generators are usually not assumed to take in power. To match with the experimental condition and also to illustrate the generic stability trend when connecting a load, a simple two generators, one active load emulation system is configured, as shown in Fig. 16.

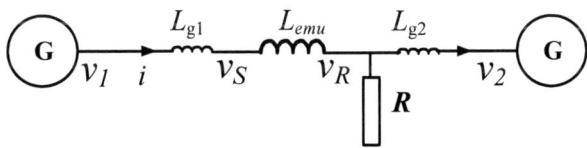

Fig. 16. A simple two-area system with active load.

The discrete-time approximation procedure is similar to the above case. The only difference is the current of the right side inductor L_{g2} now becomes

$$i_2 = i - v_R / R \qquad (26)$$

And

$$v_R = v_1 - (L_{g1} + L_{emu})di / dt \qquad (27)$$

$$v_{Lg2} = L_{g2}\frac{di}{dt} - (L_{g1} + L_{emu})\frac{d^2i}{dt^2} \qquad (28)$$

To ensure the true stability status, it is recommended to use symmetric secondary derivative. The final homogeneous second-order differential equation can be derived as,

978-1-4673-9551-9/16 $31.00 © 2016 IEEE

$$\left(1 + \frac{L_{g2}(L_{g1} + L_{emu})}{2L_{emu}R}\right)[i(k) - i(k-1)] + \frac{L_g}{2L_{emu}}[i(k-1) - i(k-2)]$$

$$+ \left(\frac{L_g}{2L_{emu}} - \frac{L_{g2}(L_{g1} + L_{emu})}{2L_{emu}R}\right)[i(k-2) - i(k-3)] = 0 \qquad (29)$$

Though the coefficient terms are a bit complicated in this case, the stability zone is actually extended, due to the resistive damping from circuit perspective.

IV. EXPERIMENTAL VERIFICATION

A system study is carried out in CURENT's Hardware Testbed (HTB) emulation platform shown in Fig. 17, which is a scaled-down two-area Kundur system.

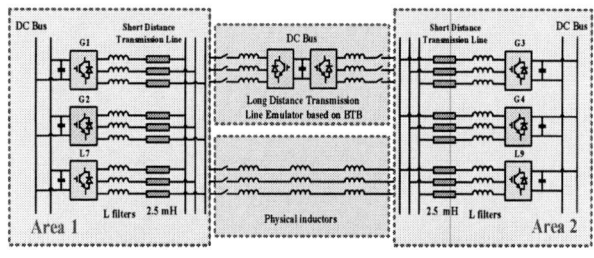

Fig. 17. Schematic and test bench of HTB two-area systems.

First, after adopting four-point sampling, as we can see from the line power flow in Fig. 18, low frequency power ripple and frequency ripple are eliminated. The line emulator result matches the physical inductor result.

Fig. 19 shows the line current result in a lossless line. Without dc offset control, even very small dc bias in the sampling signal can lead to significant dc current flow. After enabling the proposed control, as shown in Fig. 20, the dc current bias is fully eliminated.

Another emulation test is also carried out, aiming at evaluating the emulation accuracy and dynamics when line impedance has abrupt changes. The test is done by tripping one set of two paralleled transmission lines, i.e. doubling the line impedance. A comparison is made among Simulink simulation with line inductor model, HTB test with line emulator with and without dc offset control. In this case, the line is modeled with a fairly high ESR, thus the steady state dc current is low. From the results in Fig. 21, it is clear that a good match exists between simulation and HTB line emulation. To be clarified, the small ripples observed in

HTB test results are measurement noises due to power converter switching operations.

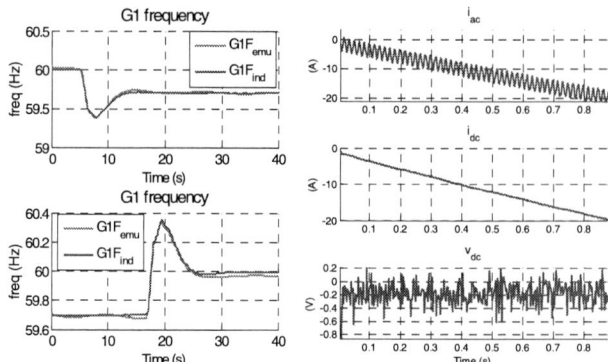

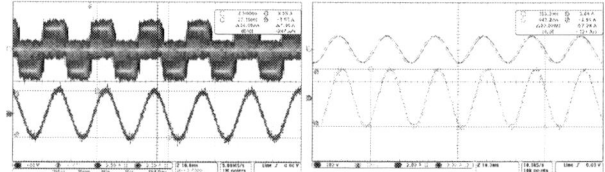

Fig. 18. (left) Line frequency during power step up and down transients.
Fig. 19. (right) Small dc bias noise leads to high dc current drift.

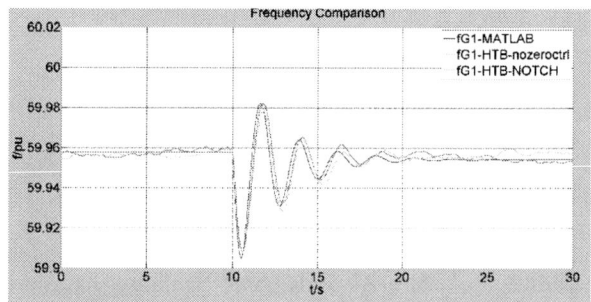

Fig. 20. Zero-dc current flow of line emulation.
Ch1&2: line-line voltages at both ends. Ch3&4: currents at both ends.
Left: without filter. Right: with filter in oscilloscope.

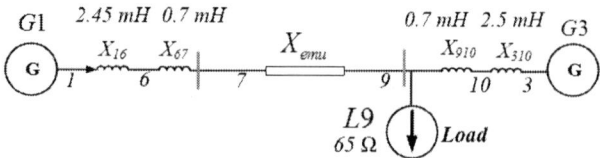

Fig. 21. Generator 1 frequency under line tripping condition.

To verify the stability analysis, four tests with emulation line impedance step change from 10.7 mH, 5 mH, 2.5 mH and 1mH are compared, based on the scenario analyzed in Section III part D. Also, with this information, the root locus of (29) can be obtained in Fig. 23. It marks that below 1.1 mH, line emulator will enter its instability zone. This matches with the following test result. As we can see from Fig. 24, when inductance reduces to 1 mH, current oscillation occurs, indicating the transmission line emulation system loses its stability.

Fig. 22. System schematic and configuration for stability test.

978-1-4673-9551-9/16 $31.00 © 2016 IEEE

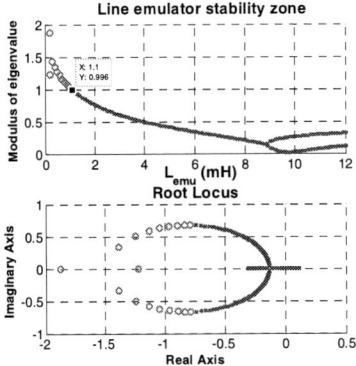

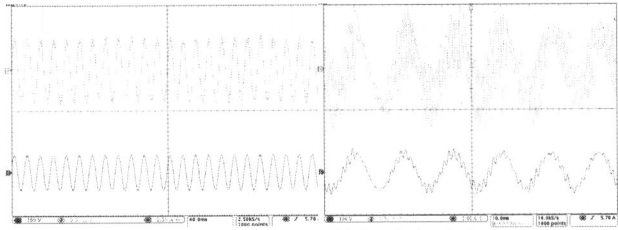

Fig. 23. Derived system stability zone with test parameters.

(a) Emulation inductance = 2.5 mH. (b) Emulation inductance = 1 mH.

Fig. 24. Emulation stability test under different line inductances.

Ch1: ac line current. Ch4: ac line-line voltage.

V. CONCLUSION

In this paper three design issues, sampling issue of input voltage with high switching harmonics, sampling issue with dc offset and emulation stability issue, are discussed to address the design and implementation challenges of an accurate, rapid responding and stabilized transmission line emulation system. The proposed mitigation approaches for two sampling issues are effective. The stability analysis for different emulation schemes, different system networks and their corresponding stability criteria are presented and are promising as a design guideline. Experimental results in a full converter based two-area transmission system validate the analysis and approaches. Future work will focus on extending its stability margin.

ACKNOWLEDGMENT

This work was supported primarily by the Engineering Research Center Program of the National Science Foundation and Department of Energy under NSF Award Number EEC-1041877 and the CURENT Industry Partnership Program.

REFERENCES

[1] J. Wang, L. Yang, Y. Ma, X. Shi, X. Zhang, L. Hang, K. Lin, L. M. Tolbert, F. Wang, and K. Tomsovic, "Regenerative power converters representation of grid control and actuation emulator," *IEEE Applied Power Electronics Conference and Exposition (APEC)*, 2012, pp. 2460-2465.

[2] L. Yang, X. Zhang, Y. Ma, J. Wang, L. Hang, K. Lin, L. Tolbert, F. Wang, and K. Tomsovic, "Hardware implementation and control design of generator emulator in multi-converter system," *IEEE*

Applied Power Electronics Conference and Exposition (APEC), Mar. 2013, pp. 2316-2323.

[3] L. Yang, X. Zhang, Y. Ma, J. Wang, L. Hang, K. Lin, L. Tolbert, F. Wang, and K. Tomsovic, "Stability analysis of inverter based generator emulator in test-bed for power systems," *IEEE Energy Conversion Congress and Exposition (ECCE)*, Sep. 2013, pp. 5410-5417.

[4] B. Liu, S. Sheng, Y. Ma, F. Wang, L. M. Tolbert, " Control and implementation of converter based ac transmission line emulation", *IEEE Applied Power Electronics Conference and Exposition (APEC)*, Mar. 2015, pp. 1807-1814.

[5] H. W. Dommel, "Digital computer solution of electromagnetic transients in single and multiphase networks," *IEEE Trans. Power App. and Sys.*, vol. PAS-88, no. 4, pp. 388-399, Apr. 1969.

[6] M. Marz, K. Copp, D. Dickmander, J. Danielsson, M. Bahrman, F. Johansson, et.al, "Mackinac HVDC converter automatic runback utilizing locally measured quantities," *Cigré 2014*, no. 425.

[7] D. M. VandeSype, K. D. Gusseme, F. M. L. L. DeBelie, A. P. VandenBossche, and J. A. A. Melkebeek, "Small-signal z-domain analysis of digitally controlled converters," *IEEE Trans. Power Electron.*, vol. 21, no. 2, pp. 470-478, Mar. 2006.

[8] U. Drofenik and J. W. Kolar, "Comparison of not synchronized sawtooth carrier and synchronized triangular carrier phase current control for the VIENNA rectifier I," in *Proc. IEEE Int. Symp. Ind. Electron. (ISIE 1999)*, Bled, Slovenia, vol. 1, pp. 13-19.

[9] D. M. Van de Sype, K. D. Gusseme, A. P. Van den Bossche, and J. A. A. Melkebeek, "A sampling algorithm for digitally controlled boost PFC converters," *IEEE Trans. Power Electron.*, vol. 19, no. 3, pp. 649-657, May 2004.

[10] D. Stojic, M. Milinkovic, S. Veinovic and I. Klasnic, "Improved stator flux estimator for speed sensorless induction motor drives", *IEEE Trans. Power Electron.*, vol. 30, no. 4, pp. 2363-2371, April. 2015.

[11] J. Hu and B. Wu, "New integration algorithms for estimating motor flux over a wide speed range," *IEEE Trans. Power Electron.* vol. 13, no. 5, pp. 969–977, Sep. 1998.

[12] N. R. N. Idris and A. H. M. Yatim, "An improved stator flux estimation in steady-state operation for direct torque control of induction machines," *IEEE Trans. Ind. Appl.*, vol. 38, no. 1, pp. 110–116, Jan./Feb. 2002.

[13] M. Hinkkanen and J. Luomi, "Modified integrator for voltage model flux estimation of induction motors," *IEEE Trans. Ind. Electron.*, vol. 50, no. 4, pp. 818-820, Aug. 2003.

[14] M. Comanescu and L. Xu, "An improved flux observer based on PLL frequency estimator for sensorless vector control of induction motors," *IEEE Trans. Ind. Electron.*, vol. 53, no. 1, pp. 50-56, Feb. 2006.

[15] Y. Wang and Z. Deng, "Improved stator flux estimation method for direct torque linear control of parallel hybrid excitation switched-flux generator," *IEEE Trans. Energy Convers.*, vol. 27, no. 3, pp. 747-756, Sep. 2012.

[16] C. Lascu and G. D. Andreescu, "Sliding-mode observer and improved integrator with DC-offset compensation for flux estimation in sensorless controlled induction motors," *IEEE Trans. Ind. Electron.*, vol. 53, no. 3, pp. 785-794, Jun. 2006.

[17] G. Verghese and V. Thottuvelil, "Aliasing effects in PWM power converters," *IEEE Power Electron. Spec. Conf.*, 1999, pp. 1043-1049.

[18] A. M. A. Amin, M. I. El-Korfolly, and S. Mohammed,"Aliasing effects on uniformly sampled PWM," *Power System Conference*, 2008, pp. 348-353.

[19] T. Roinila, M. Vilkko, and T. Suntio, "Fast loop gain measurement of a switched-mode converter using a binary signal with a specified fourier amplitude spectrum," *IEEE Trans. Power Electron.*, vol. 24, no. 12, pp. 2746-2755, Dec. 2009.

Short-Circuit Characterization of 10 kV 10A 4H-SiC MOSFET

Emanuel-Petre Eni, Szymon Bęczkowski, Stig Munk-Nielsen, Tamas Kerekes, Remus Teodorescu
Department of Energy Technology
Aalborg University
Aalborg, Denmark
{epe,sbe,smn,tak,ret}@et.aau.dk

Abstract—**The short-circuit capability of a power device is highly relevant for converter design and fault protection. In this paper a 10kV 10A 4H-SiC MOSFET is characterized and its short circuit withstand capability is studied and analyzed at 6 kV DC-link voltage. The test setup for this study is also introduced as its design, especially the inductance in the switching loop, can affect the experimental results. The study aims to present insights specific to the device which are different from that of silicon (Si) based devices. During the short-circuit operation, MOSFET saturation current, $I_{D,sat}$, increases for a few microseconds before decreasing gently. Degradation of the device can be observed at pulses longer than 5.9μs. The SiC MOSFET failed after-turn off, after a pulse of 8.6μs, due to an increase in the leakage current.**

Keywords— 4H-SiC MOSFET, Short-Circuit, Reliability, Device Characterization

I. INTRODUCTION

In order to guarantee long operation times and availability of the converter, power switching devices are expected to have a guaranteed minimum short-circuit (SC) time. This means they are required to withstand short-circuit operation for a sufficient amount of time in order for the control circuit to detect and interrupt the abnormal operation and avoid destruction or even degradation of the device.

In the case of 1.2kV 4H-SiC devices different short-circuit failure mechanism and testing circuits have been reported in literature, including theoretical analysis of the temperature dependency [1]-[4]. Due to their wide bandgap properties, the devices have shown superior performance at high stress conditions and have proven to have higher short-circuit capabilities when compared to their Si counterparts [5], [6]. In the case of 10kV SiC devices no such investigations have been found in literature.

With the advancement in SiC fabrication, 10kV SiC-based devices have been demonstrated [7], [8] and their possible application in high power converters and solid state transformers have been investigated [9]-[12], proving to be a feasible alternative to classical Si IGBT devices for high power, high voltage applications.

The aim of this work is to investigate the short-circuit capability of 10kV SiC MOSFET when such a transient occurs at turn-on. In order to evaluate such an event a test circuit for 10kV 4H-SiC MOSFETs short-circuit characterization is explained. The device behavior during such stressful events is investigated in order to obtain an idea of the device reliability when it comes to short-circuit events and its design limits.

II. DC CHARACTERIZATION

The device under test (DUT) is a 10kV 10A SiC MOSFET with a die size of 8.11mm×8.11mm and a specific on-resistance, $R_{DS,on\text{-}sp}$, of 263mΩ•cm^2 for a gate-source voltage, V_{GS}=18V at room temperature. The DC characteristics were measured with a curve tracer for junction temperatures ranging from 25°C to 150°C in steps of 25°C. A full characterization of the devices has been presented in [13].

As it can be seen in Fig. 1, for a V_{GS} above 17V the difference in on-state resistance of the device gets smaller and it is difficult to observe the transition of the device into saturation. Fig. 2 and Fig. 3 show, respectively, the output and the transfer characteristics of a 10kV 10A SiC MOSFET

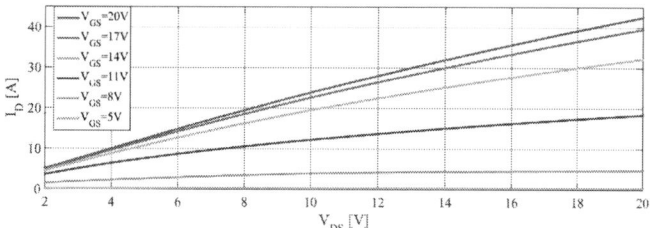

Figure 1: Typical output characteristics at room temperature [13]

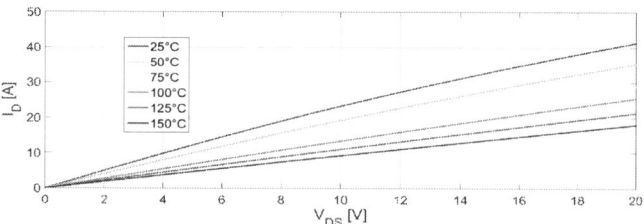

Figure 2: Output characteristics for V_{GS}=20V at different temperatures [13]

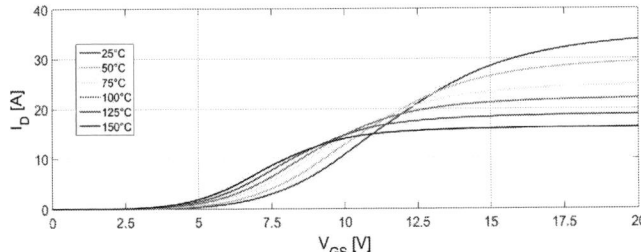

Figure 3: Transfer characteristics at different temperatures [13]

measured at different temperature values. The output characteristics refer to a V_{GS}= 20V and highlight that the on-state resistance, $R_{DS,on}$, increases with temperature, showing a positive temperature coefficient. The transfer characteristics are represented for a drain-source bias voltage,V_{DS}= 30V.

From Fig. 3 it can be observed that at elevated temperatures the drain current, I_D, increases for lower gate-source voltages (e.g. V_{GS}<10V) but it decreases with temperatures for higher gate voltages (e.g. V_{GS}= 20V). This is a result of the two components of the on-state resistance, $R_{DS,on}$: channel resistance, R_{ch}, which has a negative temperature coefficient and the resistance of the drift region, R_{Drift}, which presents positive temperature coefficient. For lower gate-source voltages (e.g. V_{GS}<10V), R_{ch} is the dominant part of $R_{DS,on}$, and at low temperatures the reduced inversion charge in the channels results in a high R_{ch}. As the temperature increases, the channel mobility is improved, thus R_{ch} is reduced. At higher gate-source voltages (e.g. V_{GS}= 20V) the induced inversion charge in the channel is high enough to diminish the impact of R_{ch} on the total resistance and R_{Drift} and becomes the dominant one, which increases with the temperature increase.

Though at high temperature during normal operation R_{Drift} could reduce I_D, it will not be able to contain the current during short-circuit transients when the SiC device is biased at high-voltage.

Overall, the device shows a positive temperature coefficient, $\alpha_T=\partial I_D/\partial T$, for a wide range of the gate-source voltage and temperature. A positive α_T is specific to thermally unstable behavior, where the device saturation current might increase with temperature at higher V_{GS} [4]. For Si-based MOSFETs the unstable behavior has been associated with a reduction of the Safe Operating Area (SOA) at high V_{DS} values, values, due to current crowding phenomena and forming of hot spots [14], [15]. Si device failure is associated in other studies, to the activation and thermal runaway of the parasitic BJT structure or the device becoming intrinsic [14]-[16].

Thermal stability (or in this case instability) in SiC is defined based on two concurrent effects dependent on temperature increase in the chip: on one hand $R_{DS,on}$ increase while on the other hand the threshold voltage, V_{th}, decreases. In SiC power MOSFETs, due to manufacturing limitations and

specific charge carriers mobility, the decrease of V_{th} is the dominant effect and transfer characteristics with positive α_T is common for all SiC power MOSFETs, independent on voltage ratings or manufacturers [4].

In the case of SiC MOSFETs the thermal unstable behavior during short-circuit has already been investigated and can be associated with an initial improvement in electron mobility (due to Coulomb scattering being dominant at lower temperatures), and an increase in the saturation current as the internal temperature rises. As the junction temperature surpasses a certain temperature, the mobility in the SiC MOSFET will decrease due to increased effect of other scattering mechanisms (phonon scattering, e.g.) [2], [17].

III. TEST SETUP

In order to evaluate the performance of the DUT, a custom build test setup capable of supporting DC-link voltages of up to 15kV and can deliver currents of hundreds of amps [18]. Due to the fast switching times of the device and the high current slope which can be achieved at turn-on and turn-off during short circuit, undesirable voltage spikes can appear across the device, thus it is desirable to reduce the stray inductance to a very low level (nH range).

The test setup is a versatile solution based on the classic double pulse tester which allows for both characterization of the DUT switching behavior and, by removing the diode and inductor, short circuit investigation under clamped and unclamped situation while allowing for control of device junction temperature.

The busbar design is presented in Fig. 4 and Fig. 5. Fig. 4 (a) provides a 3D view of the setup, composed of 0.25mm Mylar insulation, rated for 20kV, sandwiched in between two 2mm aluminum plates which form the positive and negative terminals of the DC-link. Under the busbar 4 capacitors, with a total capacitance of 24µF, are placed symmetrical around the DUT in order to equally distribute the current on the entire

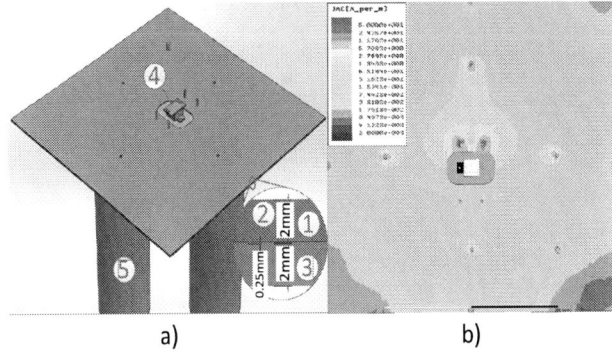

Figure 4: Busbar design of the power circuit: a) CAD representation :1) 2mm aluminum DC+ plate; 2) 0.25mm 20kV Mylar insulation; 3) 2mm Aluminum DC-plate; 4) DUT; 5) DC-link capacitors; b) FEM analysis of current distribution in busbar [18]

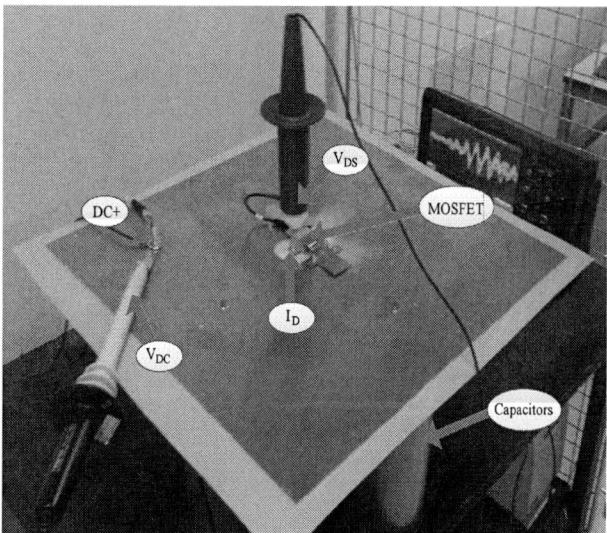

Figure 5: Short-Circuit test setup [18]

surface of the plates, as it can be seen in Fig. 4 (b). In order to avoid excessive holes in the busbar plates, the negative terminals of the capacitors were welded to the bottom plate, thus keeping the structure of the busbar as integral as possible. The finished setup, configured for short-circuits characterization, with the low capacitance and high bandwidth measuring devices attached is shown in Fig. 5.

The design of the busbar has been performed with 3D FEM modeling tool in order to obtain the very low stray inductance and laboratory measurements have confirmed a inductance within the entire setup of only 9nH [18].

IV. EXPERIMENTAL RESULTS

In the test setup the DUT is connected across a large input capacitance with a very low parasitic inductance in the commutation path. The pulse width of the V_{GS} was varied while the V_{DS} was kept constant at 6kV in order to identify the maximum short-circuit energy and short-circuit time the device can handle.

Because no previous investigations for such a device have been presented until now in literature, and in order to observe the behavior of the device during different length of the SC, the initial width of the pulse was chosen arbitrarily at $t_{SC,init}=$ 500ns. This ensures the device has sufficient time to turn on. Afterwards the pulse width was increased in steps of 100ns until catastrophic failure was observed. The V_{GS} was chosen at −5V/+18V in order to avoid stresses of the gate oxide. A pause of at least three minutes between pulses was introduced in order to allow the device to cool down.

The first short-circuit results are shown in Fig. 6. The saturation current kept rising up to 190A at which point the MOSFET was turned off. The total energy dissipated in the

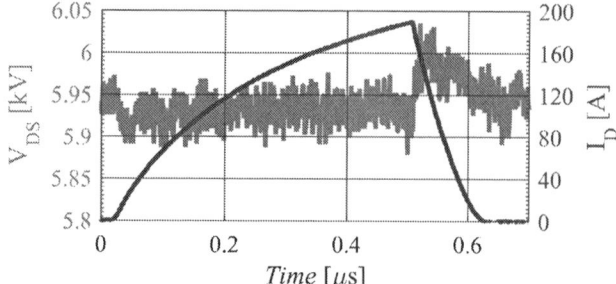

Figure 6: SC voltage and current for $t_{sc} = 0.5 \mu s$

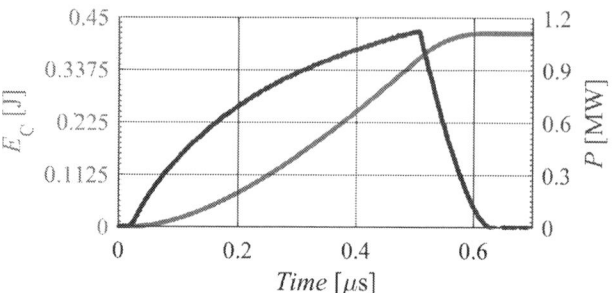

Figure 7: SC energy and power for $t_{sc} = 0.5 \mu s$

MOSFET was 0.42J as shown in Fig. 7. As expected, the device is showing a positive α_T, which is a result of the temperature increase in the device during the pulse due to the dissipated energy. This temperature increase will reduce V_{th}, and allow I_D to increase even further, showing characteristics of a thermally unstable behavior.

The pulse width was increased in steps of 100ns up to 2.4μs, which is shown in Fig. 8. After this point the device starts transitioning from thermally unstable to thermally stable behavior due to the transient temperature increase which will bring the device into thermally stable operation, and the drain current will show a negative temperature coefficient. The total dissipated energy for this pulse was 3.24 J as it can be seen in Fig. 9.

After 2.4μs into the short-circuit, the device started showing thermally stable behavior which is associated to a negative α_T. This is a result of a decrease in electron mobility, μ, with temperature, and subsequently a decrease in I_D as the

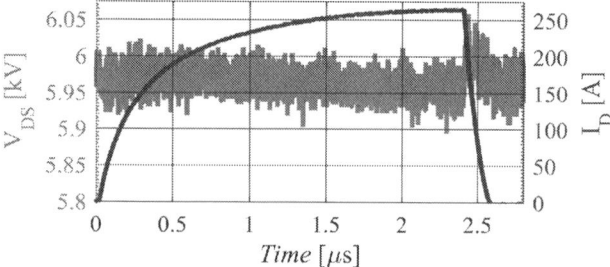

Figure 8: SC voltage and current for $t_{sc} = 2.4 \mu s$

978-1-4673-9551-9/16 $31.00 © 2016 IEEE

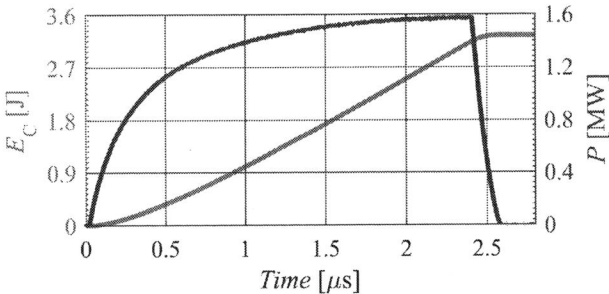

Figure 9: SC energy and power for $t_{sc} = 2.4\,\mu s$

internal temperature rises as it can be observed from Fig. 10. The total energy dissipated by the chip for a pulse of 5.9μs was 8.34J (Fig. 11), with a peak I_D current of 266A. An interesting observation was made after the pulse length was increased even further, where the peak I_D current started to decrease due to aging and degradation of the device and increase of the on-state resistance. The metallization layer reconstruction is the most observed degradation phenomena during repetitive high temperature stresses. The aluminum metallization layer degrades due to thermo-mechanical stresses generated by high temperature swings and operation, and has been associated with increase in on-state resistance and decrease in saturation currents [19]-[21].

The test was continued until failure was observed for a pulse of 8.6μs. The failure occurred after turn off and was associated with an increase in the leakage current which stressed the device above the thermal limit. The I_D related to the last test is shown in Fig. 12. The short-circuit energy

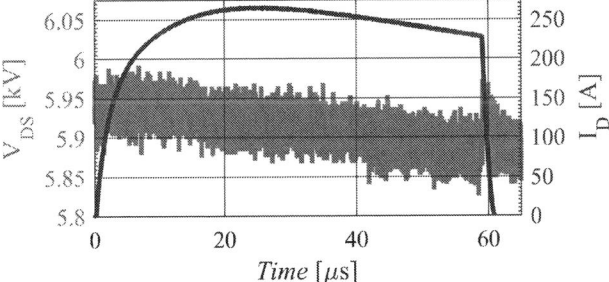

Figure 11: SC voltage and current for $t_{sc} = 5.9\,\mu s$

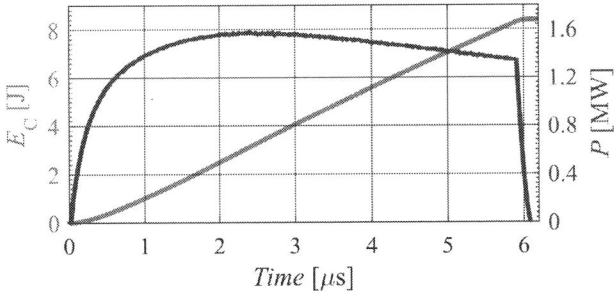

Figure 13: SC energy and power for $t_{sc} = 5.9\,\mu s$

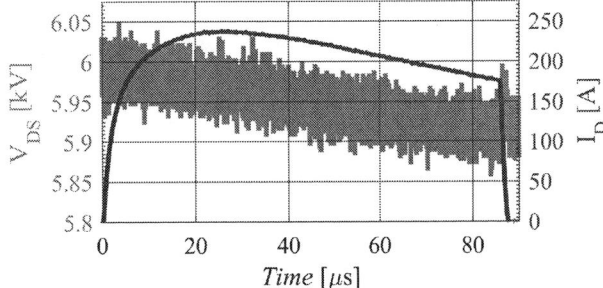

Figure 12: SC voltage and current for $t_{sc} = 8.6\,\mu s$

Figure 10: SC energy and power for $t_{sc} = 8.6\,\mu s$

dissipated in the last test was 10.7J (Fig. 13) for a maximum current of 238A. This concluded that the device safely survived a pulse length of 8.5μs with a short-circuit energy of 10.65J.

V. DISCUSSION AND CONCLUSION

In this paper a 10kV 10A 4H-SIC MOSFET short-circuit capability is investigated for the first time in order to observe the short-circuit behavior and limitations. A custom built test circuit designed to evaluate such a device under a controlled and repetitive situation was presented.

A theoretical analysis of the device behavior was proposed, based on literature review. A more precise understanding and validation would require higher knowledge about the device characteristics and is beyond the scope of this work.

The device is subjected to consecutively increasing pulses in order to closely observe the limitations in terms of pulse length, degradation and maximum energy. The device safely survived pulses of up to 8.5μs and energies up to 10.65J. The devices failed after a pulse of 8.6μs and an energy of 10.7J. This was a failure after turn-off, related to an increase in the leakage current which made the device exceed its thermal limits.

It was observed that the maximum short-circuit current is approximately 26 times the rated one. A behavior similar to that of the 1.2kV SiC devices has been observed, where the device shows a positive temperature coefficient of the drain current and a thermally unstable behavior. This has been attributed to the electron mobility. At lower temperatures the main scattering mechanism is Coulomb scattering which allow the mobility to increase with an increase in temperature, as a certain temperature is surpassed other scattering mechanism (acoustic scattering, e.g.) became dominant and $I_{D,sat}$ will be reduced. This reduction in the drain current is associated with a thermally stable behavior.

This results allow for a detailed study of a short-circuit protection circuit and reaction times and even an in-depth thermal analysis of the device behavior during short circuit.

ACKNOWLEDGMENT

The authors would like to thank Wolfspeed for their help and support during this study.

REFERENCES

[1] M. Bouarroudj-Berkani, D. Othman, S. Lefebvre, S. Moumen, Z. Khatir and T.B. Sallah, "Ageing of SiC JFET transistors under repetitive current limitation conditions," *Microelectronics reliability*, vol. 50, no. 9, pp. 1532-1537 2010.

[2] X. Huang, G. Wang, Y. Li, A.Q. Huang and B.J. Baliga, "Short-circuit capability of 1200V SiC MOSFET and JFET for fault protection," *in Proc. Twenty-Eighth Annual IEEE Applied Power Electronics Conference and Exposition (APEC)*, 2013, pp. 197-200.

[3] K. Wada, S. Nishizawa and H. Ohashi, "Design and implementation of a non-destructive test circuit for SiC-MOSFETs," *in Proc Power Electronics and Motion Control Conference (IPEMC)*, 2012, pp. 10-15.

[4] A. Castellazzi, T. Funaki, T. Kimoto and T. Hikihara, "Thermal instability effects in SiC power MOSFETs," *Microelectronics Reliability*, vol. 52, no. 9, pp. 2414-2419 2012.

[5] Y. Nakao, S. Watanabe, N. Miura, M. Imaizumi and T. Oomori, "Investigation into short-circuit ruggedness of 1.2 kV 4H-SiC MOSFETs," in *Materials Science Forum*, 2009, pp. 1123-1126.

[6] N. Boughrara, S. Moumen, S. Lefebvre, Z. Khatir, P. Friedrichs and J. Faugières, "Robustness of SiC JFET in short-circuit modes," *IEEE Electron Device Lett.*, vol. 30, no. 1, pp. 51-53 2009.

[7] S. Ryu, S. Krishnaswami, B. Hull, J. Richmond, A. Agarwal and A. Hefner, "10 kV, 5A 4H-SiC Power DMOSFET," *in Proc. IEEE International Symposium on Power Semiconductor Devices and IC's, (ISPSD 2006)*, 2006, pp. 1-4.

[8] S. Ryu, S. Krishnaswami, M. O'Loughlin, J. Richmond, A. Agarwal, J. Palmour and A.R. Hefner, "10-kV, 123-mΩ· cm 2 4H-SiC power DMOSFETs," *Electron Device Letters, IEEE*, vol. 25, no. 8, pp. 556-558 2004.

[9] Jun Wang, Yu Du, S. Bhattacharya and A.Q. Huang, "Characterization, modeling of 10-kV SiC JBS diodes and their application prospect in X-ray generators," *in Proc. IEEE Energy Conversion Congress and Exposition (ECCE 2009)*, 2009, pp. 1488-1493.

[10] H. Mirzaee, S. Bhattacharya, A. De and A. Tripathi, "Design comparison of high power medium-voltage converters based on 6.5 kV

si- IGBT/si-PiN diode, 6.5 kV si-IGBT/SiC-JBS diode, and 10 kV SiC-MOSFET/SiC-JBS diode," *IEEE Transactions on Industry Applications*, vol. PP, no. 99, pp. 1-1 2014.

[11] J. Wang, X. Zhou, J. Li, T. Zhao, A.Q. Huang, R. Callanan, F. Husna and A. Agarwal, "10-kV SiC MOSFET-based boost converter," *IEEE Transactions on Industry Applications*, vol. 45, no. 6, pp. 2056-2063 2009.

[12] L. Yang, T. Zhao, J. Wang and A.Q. Huang, "Design and analysis of a 270kW five-level DC/DC converter for solid state transformer using 10kV SiC power devices," *in Proc. IEEE Power Electronics Specialists Conference(PESC 2007)*, 2007, pp. 245-251.

[13] E. . Eni, B.I. Incau, T. Kerekes, R. Teodorescu and S. Munk-Nielsen, "Characterisation of 10 kV 10 A SiC MOSFET," *in Proc. IEEE ACEMP - OPTIM - ELECTROMOTION JOINT CONFERENCE*, Side, Turkey; 2015.

[14] A. Castellazzi, V. Kartal, R. Kraus, N. Seliger, M. Honsberg-Riedl and D. Schmitt-Landsiedel, "Hot-spot meaurements and analysis of electro-thermal effects in low-voltage power-MOSFET's," *Microelectronics Reliability*, vol. 43, no. 9, pp. 1877-1882 2003.

[15] A. Castellazzi, V. Kartal, R. Kraus, N. Seliger, M. Honsberg-Riedl and D. Schmitt-Landsiedel, "Hot-spot meaurements and analysis of electro-thermal effects in low-voltage power-MOSFET's," *Microelectronics Reliability*, vol. 43, no. 9, pp. 1877-1882 2003.

[16] A. Castellazzi and G. Wachutka, "Low-voltage PowerMOSFETs used as dissipative elements: electrothermal analysis and characterization," *in Proc. IEEE Power Electronics Specialists Conference (PESC 2006)*, 2006, pp. 1-7.

[17] M. Denison, M. Pfost, K. Pieper, S. Märkl, D. Metzner and M. Stecher, "Influence of inhomogeneous current distribution on the thermal SOA of integrated DMOS transistors," *in Proc. Power Semiconductor Devices and ICs (ISPSD'04)*, 2004, pp. 409-412.

[18] V. Tilak, K. Matocha and G. Dunne, "Electron-scattering mechanisms in heavily doped silicon carbide MOSFET inversion layers," *IEEE Transactions on Electron Devices*, vol. 54, no. 11, pp. 2823-2829 2007.

[19] E. Eni, T. Kerekes, C. Uhrenfeldt, R. Teodorescu and S. Munk-Nielsen, "Design of low impedance busbar for 10 kV, 100A 4H-SiC MOSFET short-circuit tester using axial capacitors," *in Proc. IEEE POWER ELECTRONICS FOR DISTRIBUTED GENERATION SYSTEMS, (PEDG 2015)*, AACHEN, 2015.

[20] M. Berkani, S. Lefebvre and Z. Khatir, "Saturation current and on-resistance correlation during during repetitive short-circuit conditions on SiC JFET transistors," *IEEE Transactions on Power Electronics*, vol. 28, no. 2, pp. 621-624 2013.

[21] S. Yang, D. Xiang, A. Bryant, P. Mawby, L. Ran and P. Tavner, "Condition monitoring for device reliability in power electronic converters: A review," *Power Electronics, IEEE Transactions on*, vol. 25, no. 11, pp. 2734-2752 2010.

[22] D. Martineau, T. Mazeaud, M. Legros, P. Dupuy and C. Levade, "Characterization of alterations on power MOSFET devices under extreme electro-thermal fatigue," *Microelectronics Reliability*, vol. 50, no. 9, pp. 1768-1772 2010.

Record-low 10mΩ SiC MOSFETs in TO-247, Rated at 900V

Vipindas Pala, Gangyao Wang, Brett Hull, Scott Allen, Jeff Casady and John Palmour

Wolfspeed, A Cree Company
Research Triangle Park, North Carolina
Vipindas.pala@wolfspeed.com

Abstract—We demonstrate a 900V SiC MOSFET with a record-low ON resistance of 10 mΩ in a TO-247 package. Due to their record low specific ON resistance, and a low $R_{DS,ON}$ temperature coefficient, 900V SiC MOSFETs can far exceed the current densities of IGBTs in discrete packages. SiC MOSFETs also have a robust, low reverse recovery body diode which makes them suited to hard-switched and soft-switched topologies with bi-directional conduction. They can also achieve superior light load efficiencies due to knee-less conduction in both 1st and 3rd quadrant, low switching losses and low body diode reverse recovery losses.

Keywords—Silicon Carbide, SiC, MOSFET

I. INTRODUCTION

The advent of renewable energy, such as solar and wind energy, has led to more use of localized DC power generation and storage. In concert, more electrified vehicles of all types are emerging to meet the need for CO_2 emission reduction and higher efficiency demanded by leading markets in Europe, North America [1], and Asia. These electrified vehicles are also expected to greatly increase the need for increased DC power distribution and storage to fit seamlessly with more localized renewable energy which is generated and stored locally. These emerging markets increase the need for dramatically improved and more reliable DC-DC converters, and DC-AC inverters, to efficiently and cheaply transform the energy from distributed source to load. In addition to these emerging markets, traditional markets remain that share the same needs, including telecom power supplies, server power supplies, and PC power supplies. All of these markets have a need for higher-power converters and inverters, which offer high power density, high efficiency, and low cost.

From a semiconductor component viewpoint, in many of these markets the power transistors must be able to operate with low-losses at higher frequencies than are conventionally used, for example from 40kHz to >100kHz in PV string inverters, whereas many inverters today are <25kHz. The need for higher-frequency switching dictates the need for either MOSFETs with very low conduction losses or IGBTs with very low switching losses; and especially to maintain low conduction and switching losses over temperature in these applications. For resonant topologies found in on-board chargers and telecom power supplies, having MOSFETs with the lowest $R_{DS,ON}$ per TO-247 package is often a requirement, again, with emphasis on the $R_{DS,ON}$ over temperature. Super-junction MOSFETs are rapidly approaching theoretical limits

for specific ON resistance and improvements on existing technology is likely limited to a factor of two at most [2]. Lateral Gallium Nitride transistors have much improved switching characteristics, but are comparable to silicon superjunction MOSFETs with respect to conduction losses per unit semiconductor area, and hence are not expected to bring significant gains in current densities. Silicon IGBTs offer a very attractive solution in terms of current densities because they have low conduction losses at high power, high temperature operating conditions. However in light load conditions, IGBT losses increase due to their higher switching losses and the knee voltage drop.

Silicon Carbide MOSFETs offer an attractive solution in these respects– they can achieve switching speeds that are comparable to lateral GaN power transistors, but can also achieve much higher current densities due to their lower specific ON resistances so that the forward drop at high temperatures is lower than that of state-of-the-art silicon IGBTs. In addition, SiC MOSFETs also have a rugged, low reverse recovery body diode that eliminates the need for an external diode. In this abstract, we present for the first time, the lowest $R_{DS,ON}$ power MOSFET of >600V, in a TO-247 package, which is a 900V, 10 mΩ SiC MOSFET. The room-temperature 10 mΩ measured value is ½ that of state-of-the-art offerings in Si [3], and at 150˚C the $R_{DS,ON}$ is only 30% that of the Si offerings.

II. STATIC PERFORMANCE

The 900V SiC MOSFET presented in this paper has a specific ON resistance (normalized to active area) of 2.4 mΩ•cm², which is lower than state-of-the-art 650V silicon superjunction MOSFETs by a factor of 4 at 25 °C, and a factor of 6 at 150 °C junction temperature. A low specific ON resistance leads to lower semiconductor costs by minimizing the die area. The small chip sizes can reduce the ON resistance and power densities that are achievable for a certain package type. In this instance, we have achieved a record low ON resistance of 10 mΩ in a TO-247 package, with a chip size of 4.4 x 7.3 mm². The ON resistance of the 900V SiC MOSFET compared to the lowest $R_{DS,ON}$ 650V silicon MOSFET is shown in Fig.1. The SiC MOSFET has a typical value of 17.5mΩ at 25˚C, increasing to 40mΩ at 150˚C. For the 900V SiC MOSFET, also in a TO-247, the comparative values are 10 mΩ at 25˚C, increasing to only 15 mΩ at 150˚C.

978-1-4673-9551-9/16 $31.00 © 2016 IEEE

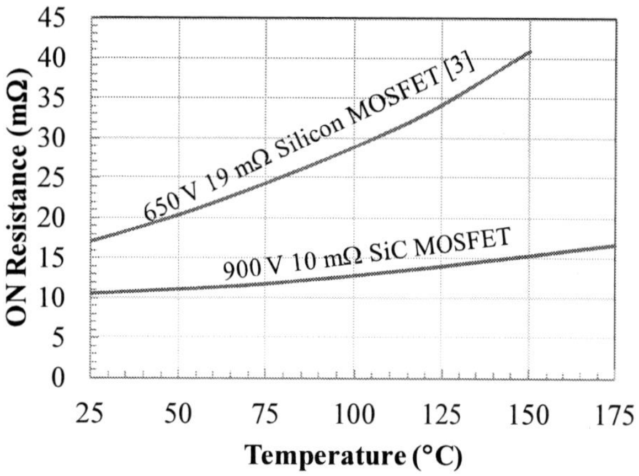

Fig. 1. Comparison of $R_{DS,ON}$ over temperature of two power MOSFETs in TO-247 style package

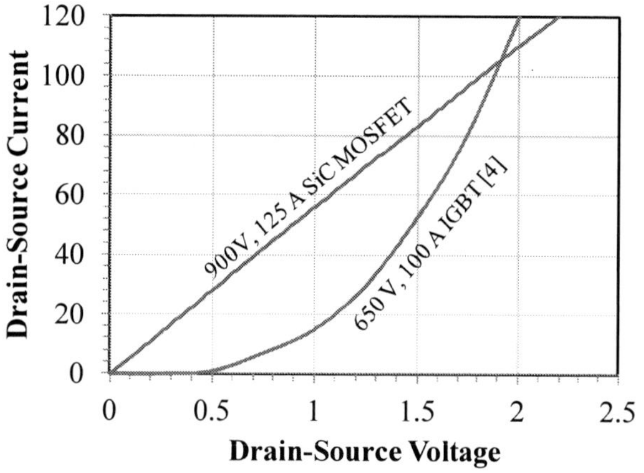

Fig. 2. Comparison of forward characteristics of the 900V, 10 mΩ SiC MOSFET in a TO-247 package and a 650V 100 A IGBT [3] at 175 °C.

Due to their low $R_{DS,ON}$ temperature coefficient, the 900V SiC MOSFET has a low forward voltage drop even at a junction temperature of 175 °C,, thus making it possible to achieve extremely high power densities, previously only possible with IGBTs. Fig.2 compares the forward characteristics of the 900V 10 mΩ SiC MOSFET and a 5th generation 650 V, 100 A IGBT at 175 °C [4]. Even in applications which cannot take advantage of extremely high frequencies, such as electric drive train in electric or hybrid-electric vehicles, the conduction losses (and switching losses), must be kept very low over temperature and over wide load range. In this application, traditionally served by a Si IGBT, switching to a SiC MOSFET allows the knee-voltage barrier of the IGBT to be eliminated, with improved light-load efficiencies as a result. Fig. 2 shows that at high current densities, SiC MOSFET and Si IGBT have similar voltage drop, and therefore their conduction losses are comparable at full load. However, at light load conditions SiC MOSFETs have a large advantage in terms of reducing losses.

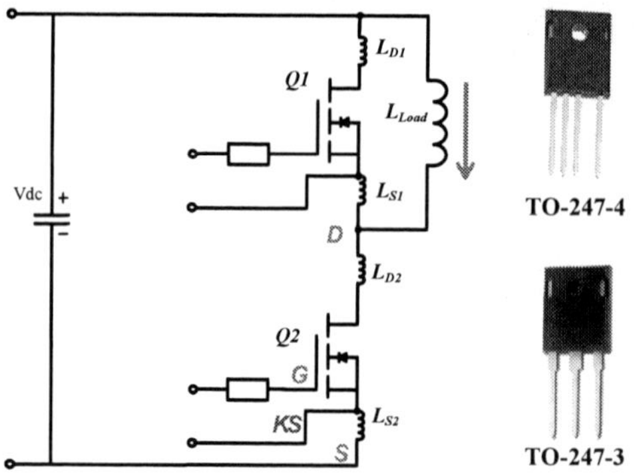

Fig. 3. Schematic of the TO-247-3 and TO-247-4 packages and the switching test set-up

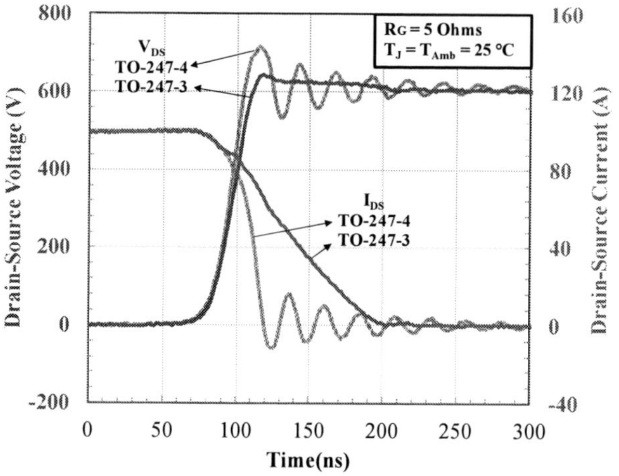

Fig. 4. Comparison of Turn-OFF for 900V, 10 mΩ SiC MOSFET in TO-247-3 and TO-247-4 packages (R_G=5Ohm, V_{GS}=-4V/+15V)

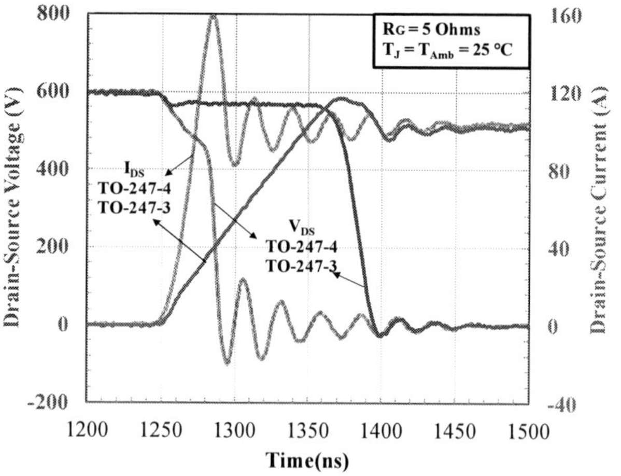

Fig. 5. Comparison of Turn-ON for 900V, 10 mΩ SiC MOSFET in TO-247-3 and TO-247-4 packages (R_G=5Ohm, V_{GS}=-4V/+15V)

III. DYNAMIC PERFORMANCE

The 900V 10 mΩ SiC MOSFET chip is capable of extremely fast switching transitions; however in a three-lead TO-247-3 package, the source inductance in the gate driver loop will limit the switching speed. As shown in Fig. 3, we have evaluated the SiC MOSFET dynamic performance in a TO-247-3 package and TO-247-4 package. The four lead TO-247-4 package has a separate source return pin for the gate driver (equivalent circuit is shown as the bottom device in Fig. 3). Therefore during the switching transient, the gate bias $V_{G,KS}$ is not affected by the voltage drop in the source inductance L_{S2} introduced by the *di/dt* of the drain-source current.

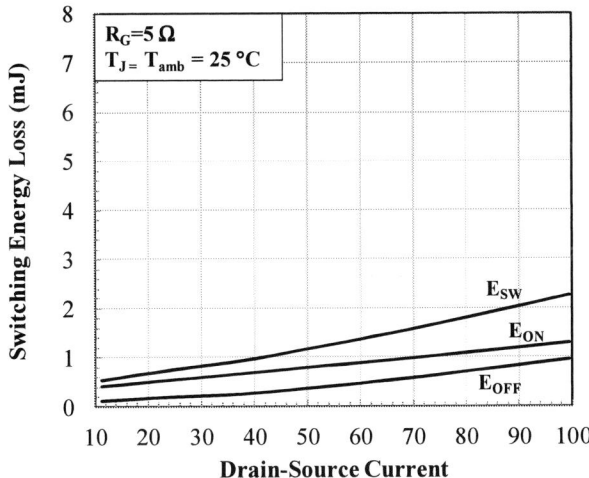

Fig. 6. Switching Energy losses at 25 °C for the 900V for 900V, 10 mΩ SiC MOSFET in TO-247-4 package (R$_G$=5Ohm, V$_{GS}$=-4V/+15V, V$_{DD}$=600V)

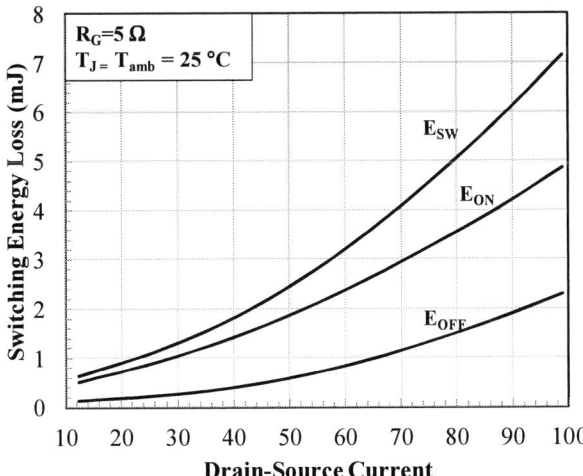

Fig. 7. Switching Energy losses at 25 °C for the 900V for 900V, 10 mΩ SiC MOSFET in TO-247-3 package (R$_G$=5Ohm, V$_{GS}$=-4V/+15V, V$_{DD}$=600V)

Fig. 4 and Fig. 5 show the switching transient comparisons for both packages under same test conditions. The 900V 10

mΩ SiC MOSFET in TO-247-4 package exhibits much faster switching transitions. Fig. 6 and Fig. 7 show the switching energy loss comparison for the two package types as a function of current. At 100 A the TO-247-4 package has 3x lower switching loss compared to the TO-247-3. It can be observed that packaging has a bigger impact for higher current, due to the larger magnitude of source inductance voltage drop.

IV. BODY DIODE PERFORMANCE

900V 10 mΩ MOSFET also exhibits extremely low body diode recovery losses, which makes them ideally suited to topologies where bi-directional conduction is required. Silicon super-junction MOSFETs suffer from excessive reverse recovery losses in these topologies and cannot be used. IGBTs on the other hand, do not have reverse conduction and have to be paired with an external or integrated reverse diode. The body diode characteristics of the SiC MOSFET at 175 °C are shown in Fig. 8. In body diode mode (V$_{GS}$=-4V), the knee voltage is -2.5V, and the conduction losses are quite high. However by turning on the gate to +15V using synchronous rectification, the knee can be overcome and symmetric conduction in both forward and reverse directions can be obtained. This can improve the overall system efficiency considerably when power is transferred in both directions - in battery chargers or in drive and regenerative cycles of an electric vehicle or hybrid drive-train for example.

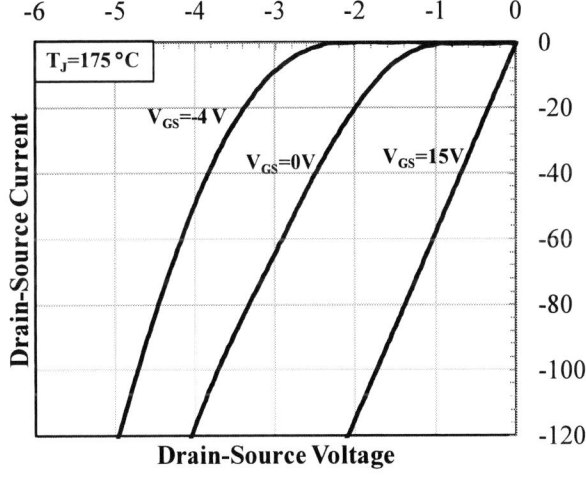

Fig. 8. 3rd Quadrant characteristics of the 900V, 10 mΩ MOSFET at 175 °C

The SiC MOSFET body diode also exhibits excellent dynamic performance. Fig. 9 shows the body diode recovery waveforms for the device at 25 °C and at 150 °C. When tested at V$_R$=600V, I$_F$=90A and di/dt=1200A/us, the reverse recovery charge (Q$_{RR}$) is 510 nC at 25 °C and the reverse recovery time is only 56 ns. The device exhibits only a moderate increase in reverse recovery at 150 °C. This indicates that most of the losses are due to the stored energy in the output capacitance (E$_{OSS}$) and not due to bipolar recombination (E$_{REC}$), as shown in Fig. 10. The stored energy component can be recovered in resonant topologies, further reducing the overall diode loss in the system. Furthermore, there is no observed body diode *di/dt* limit for the SiC MOSFET. The 900V 10 mΩ SiC MOSFET body diode has been tested up to a *di/dt* of 5900A/us and no failures were observed.

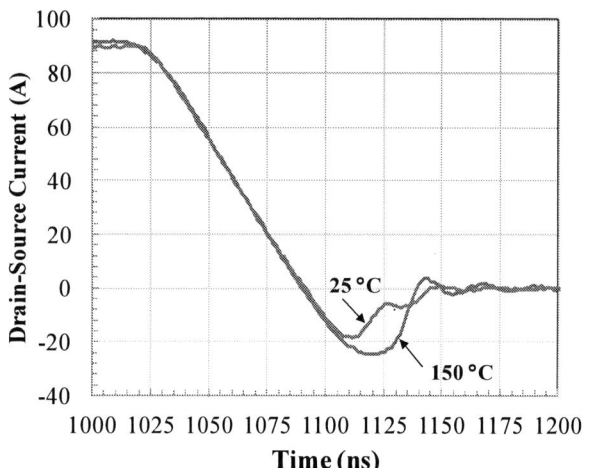

Fig. 9. Reverse recovery waveforms of the 900V, 10 mΩ MOSFET at 25 °C and 150 °C

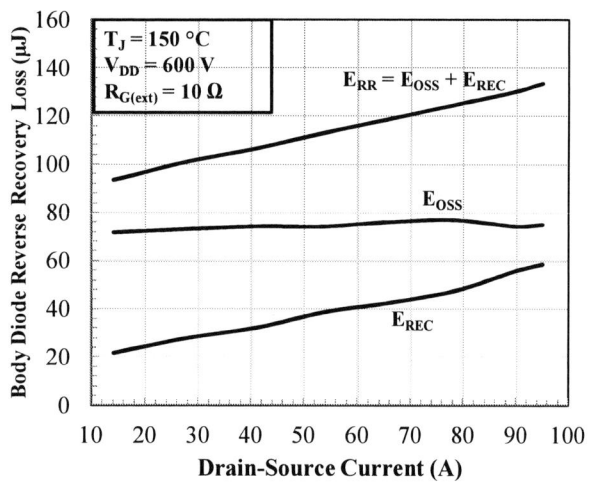

Fig. 10. Components of body diode recovery loss at 150 °C extracted from switching measurements

V. SUMMARY

The 10 mΩ 900V SiC MOSFETs achieve a record low $R_{DS,ON}$ in a TO-247 package with a chip size of only 32 mm^2. This is enabled by the low specific ON resistance of SiC MOSFET technology, which can enable large gains in current density, power density and efficiency compared to silicon MOSFETs and IGBTs. SiC MOSFETs have superior high temperature performance and a robust, low recovery body diode which makes them suited to hard switched topologies with bi-directional conduction. They also have much reduced low switching losses and knee-less conduction in first and third quadrant compared to IGBTs, thus simultaneously enabling high frequency, high power density, lower component count and higher efficiencies.

ACKNOWLEDGMENT

The information, data, or work presented herein was funded in part by the Office of Energy Efficiency and Renewable Energy (EERE), U.S. Department of Energy, under Award Number DE-EE0006920.

DISCLAIMER

The information, data, or work presented herein was funded in part by an agency of the United States Government. Neither the United States Government nor any agency thereof, nor any of their employees, makes any warranty, express or implied, or assumes any legal liability or responsibility for the accuracy, completeness, or usefulness of any information, apparatus, product, or process disclosed, or represents that its use would not infringe privately owned rights. Reference herein to any specific commercial product, process, or service by trade name, trademark, manufacturer, or otherwise does not necessarily constitute or imply its endorsement, recommendation, or favoring by the United States Government or any agency thereof. The views and opinions of authors expressed herein do not necessarily state or reflect those of the United States Government or any agency thereof.

REFERENCES

[1] http://www.epa.gov/oms/climate/regulations.htm

[2] M. Treu, E. Vecino, M. Pippan, O. Haberlen, G. Curatola, G. Deboy, M. Kutschak and U. Kirchner, "The role of silicon, silicon carbide and gallium nitride in power electronics," IEDM, 2012, pp. 147-150.

[3] Infineon 650V, 19mΩ MOSFET, part number IPZ65R019C7, http://www.infineon.com/dgdl/Infineon-IPZ65R019C7-DS-v02_00-en.pdf.

[4] Infineon 650V 100 A IGBT, Part No. IGZ100N65H5, http://www.infineon.com/dgdl/Infineon-IGZ100N65H5-DS-v02_01-EN.pdf.

Performance evaluation of multiple Si and SiC solid state devices for circuit breaker application in 380VDC delivery system

Kai Tan, Pengkun Liu, Xijun Ni, Chang Peng, Xiaoqing Song and Alex Q. Huang
FREEDM Systems Center, Department of Electrical and Computer Engineering
North Carolina State University
Raleigh, NC

Abstract—The DC power delivery system is becoming an appealing research topic and real world solution due to its higher energy efficiency compare with AC delivery system. It has already been applied in data centers, commercial buildings, electrical vehicle charge stations and micro grid systems, etc. Among many new issues that need to be addressed for the DC power delivery system, ultra-fast and accurate protection is one of them. The 1200V SiC devices has been developed by many manufacturers in recent years which makes them good candidates in DC circuit breaker application. In this paper, the 380V DC circuit breaker employing solid state devices for DC power delivery system has been proposed and introduced. The criteria of device characteristics particularly for DC solid state circuit breaker application is discussed and defined. Characteristics of 4 different Si and SiC solid state devices in similar power rating have been compared based on defined criteria. The pros and cons of different devices candidates is introduced with test results for DC circuit breaker application.

Keywords—380V DC system; solid state circuit breaker; SiC devices; device characteristic; performance evaluation

I. INTRODUCTION

Today, 120V (or 220V in some countries) AC power systems are widely used worldwide. However, low voltage DC (LVDC) distribution is becoming more and more utilized in electrical systems thanks to the development of power electronics technology. For power distribution, more and more DC loads appears in our daily life, such as LED lights, computers, communications and numerous types of batteries used in consumer electronics. Currently, these loads need a power rectifier stage to transfer AC power to DC power which increases the cost and power loss. For massive electronics DC load system, such as a DC data centers, DC buildings, electrical vehicle charge stations, etc.[1]–[3], the power loss will become significant which makes a direct DC distribution system an attractive high efficiency option[4].

At present, the off the shelf circuit breakers are based on traditional mechanical DC circuit breaker technology such as Miniature Circuit Breaker (MCB) and Molded Case Circuit Breaker (MCCB). Many electrical equipment manufacturers offer them commercially with different current ratings for

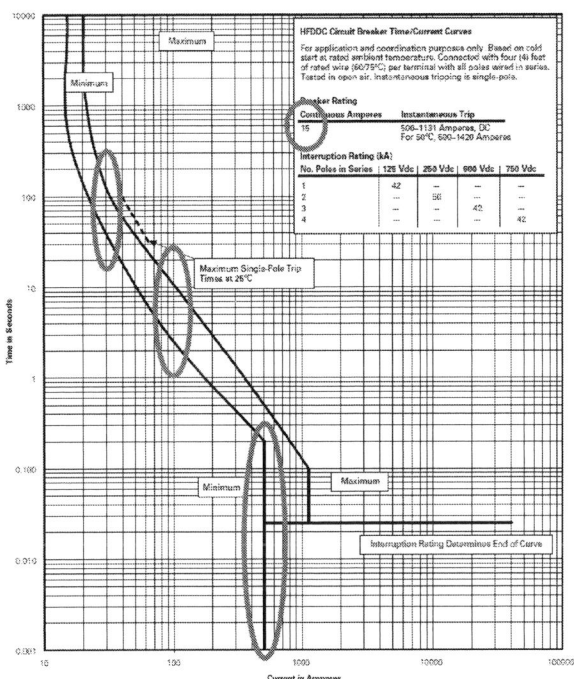

Figure 1. Trip time curve of a 15A MCCB DC circuit breaker [7]

600V or 750V DC applications [5]. However, one issue for these conventional circuit breaker is the trip speed. Fig. 1 [6] shows a trip time curve of a MCCB with nominal 15A current rating. In this figure, the instantaneous trip current for this 15A MCCB is 500-1400A and the trip time for 100A is 2.5-10s.

In DC distribution system, there are a lot of capacitive components in each end users to keep the bus voltage stiff with small ripple and filter the high frequency harmonics caused by switching converters. The inrush current is one issue for capacitance which may cause unexpected circuit breaker trip when the user is connected into the DC distribution system. Computer servers and other electronic loads behave like a capacitive load when they are plugged into the 380V DC bus.

978-1-4673-9551-9/16 $31.00 © 2016 IEEE

A large inrush current will flow through the contacts to charge capacitor voltage to 380V. Therefore it performs like a short circuit current due to the low impedance of the initially un-energized capacitors. The inrush current is higher for larger capacitive loads. The peak value of this inrush current can be 150A to 450A for 0.68uF to 6.8uF capacitors [7] which induces a large current stress on conductors and will damage the weak joint point on the current loop. On the other hand, a large inrush current may also cause a voltage sag on the 380V DC bus due to the transient low impedance. This voltage sag may cause under voltage lock out on other equipment connected on the bus.

The second issue caused by these capacitive components is the discharge current for low impedance or short circuit. Once a fault occurs, the fault current will also be large and di/dt rate is very high due to the discharge of the energized capacitive components on the DC bus. If a low impedance fault occurs on one load, the input capacitor will be discharged by the fault and further generates a low impedance for other loads in neighborhood. This current can be several hundred amperes with milliseconds rise speed [8]. The large and fast discharge current will affect other users until the low impedance fault is cleared. It also may damage the circuit breakers with repeat arc [9].

Therefore it is urgently needed to improve the circuit breaker technology with enhanced reliability, speed and performance for DC system protection. In this paper, low voltage DC solid state circuit breaker are proposed for the protection in DC distributions systems, and candidates of solid state devices are evaluated for this specific applications. With the advent of high performance SiC power devices, solid state circuit breakers potentially have much lower steady state losses and higher transient conduction and interruption capabilities. Therefore it is imperative to have an updated review of device selections for low voltage DC circuit breaker applications, and compared newly emerged commercial SiC devices with already sophisticated Si devices in terms of both steady state and transient characteristics.

II. SOLID STATE CIRCUIT BREAKER

Considering all aspects mentioned above in DC power delivery system, a solid state circuit breaker for 380V/15A nominal power rating is proposed and shown in Fig. 2. The circuit breaker use two devices parallel to share the 15A during steady state due to the $R_{DS(ON)}$ and thermal dissipation limitation. More devices can be paralleled for larger power ratings with similar strategy.

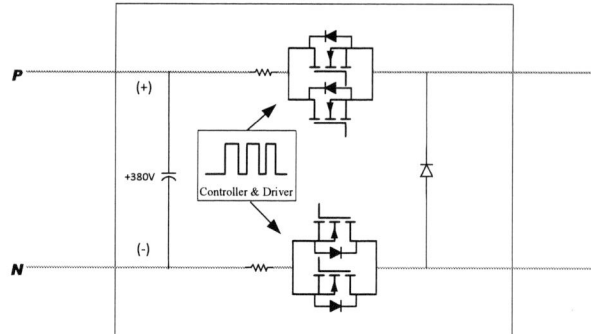

Figure 2. Proposed circuit breaker for 380V/15A unidirectional application

As stated above, the capacitance in each end user could cause inrush current and discharge fault current. The proposed operation modes and flowchart is shown in Fig.3. It can be identified as latch off mode, inrush hiccup startup mode, normally on mode, dual threshold over current trip and over temperature trip.

A. Latching OFF mode

Before t_2 and after t_7, the whole circuit breaker is latched off. The devices are controlled in open circuit to trip the main circuit due to over current or over temperature fault. The fault indicator is high to indicate fault condition.

B. Hiccup startup mode

At the moment of t_2, the circuit breaker is setup to start manually or remotely with fault been cleared. The sum of drain to source voltage on devices of two bus is equals 380V because the output bus capacitor voltage is zero. The devices are still in off state with only leakage current. From t_2, the control circuit initializes the designed hiccup startup and generates a low V_{GS} close to threshold V_T to ensure the devices works in the saturation region to limit the current with a short pulse (t_2-t_3). The output capacitor is charged a train of pulses like this and its voltage is increased. The drain to source voltage on devices is still large which makes power stress large during the

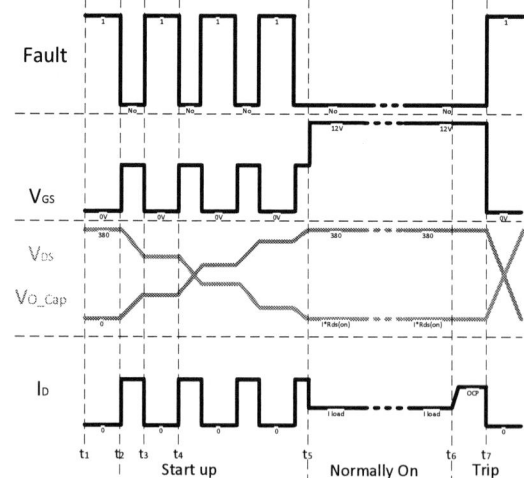

Figure 3. Proposed operation strategy for 380V DC circuit breaker

pulse because it operates in saturation . Therefore the pulse width and the value of limited current need to be designed within the safe operation area (SOA) of the NMOS device to avoid the device failure due to the excessive power dissipation. After each short pulse, the device is off again from t3-t4 to control the average power stress and cool the device down. The voltage on the device and capacitor remain the same in this period.

The current on the drain is sensed with close loop control to a given value with adaptive VGS. The on-off hiccup process continues until the voltage of the output capacitor is close to the bus voltage. Then the control circuit fully turns on the NMOS device with 12V V_{GS} at the moment of t5. Because the drain to source voltage is already close to zero, there is no inrush current, the normal load current flows continuously.

C. Normally ON mode

The solid state circuit breaker works in normal conduction mode. The devices keeps fully turn on to keep lowest $R_{DS(ON)}$ so the power loss on the solid state device is minimized.

D. Dual threshold over current trip

The strategy for tripping the over current fault is implemented by controller to monitor over current or short fault. These two thresholds I_{Limit} and I_{SC} are adopted to trip the real over current fault and filter nuisance transient noises.

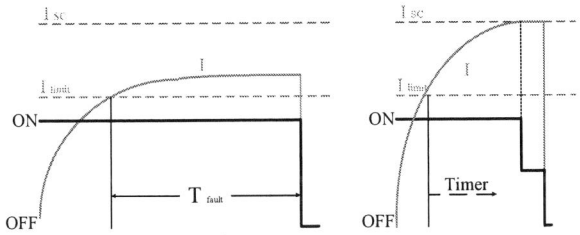

Figure 4. Dual thresholds off strategy for over current protection

- I_{Limit} – Fault timer runs if $I_{Load} > I_{Limit}$ and the circuit breaker trip when timer runs out of set up $T_{(fault)}$ which is usually been used to filter nuisance transient noises.

- I_{SC} – Short circuit threshold (typically 1.3 to 2.5 times of I_{Limit}). I_{SC} is programmed as a current limiting value to control the MOSFET back to saturation before final preset trip time to limit the capacitor discharge current and isolate the low impedance fault with affecting others.

III. DEVICE CHARACTERISTICS ANALYSIS

One of most crucial component in DC solid state circuit breaker is the embedded solid state device. Traditionally, Si devices including MOSFET, IGBT and etc. are considered the best options among all solid state devices in low voltage range. [10]-[12] And a lot of researchers proposed and developed many new design based on SiC devices due to the development of power devices technology in recent years. [13][14] For developing the proposed 380V/15A solid state circuit breaker using paralleled devices, the criteria for choosing the solid state

device in proposed design is discussed. From analysis in previous section, the most important characteristics of solid state device for 380V DC circuit breaker application are listed:

A. Static conduction resistance:

Low static conduction resistance is one of most important parameter for circuit breaker to reduce the steady state thermal stress and power loss in normally on mode. It directly related with the efficiency of circuit breaker. For Si IGBT, the initial $V_{CE(SAT)}$ (about 0.7~1V for 600V IGBT) makes the power loss larger than paralleled MOSFET in low current range (<20A) for 380V DC circuit breaker application. It needs larger heatsink or forced air cooling for IGBT than natural convection. The SiC devices have better static conduction resistance due to the physics characteristic of wide-band gap material compare with Si. Many manufacturers has released 1200V SiC devices which is good candidate for circuit breaker application. This value is related with die size which need to be specified for evaluation.

B. Thermal impedance in steady and transient state:

The steady state thermal resistance can be calculated with:

$$R_{th} = \frac{L}{kA} \qquad (1)$$

L is the thickness of the material (cm), A is the area of the material (cm^{-2}), and k is the thermal conductivity (W·cm^{-1}·K^{-1}).

For the SiC, this value is about 4.9 and Si is about 1.5. [15][16] This makes the steady state thermal resistance of SiC device much better than Si device with same L and A. It also means that the die size of SiC device can be saved to achieve equivalent thermal resistance with larger die size Si device.

For the same heatsink design and power loss, the device junction temperature will be lower if the steady state thermal resistance is smaller. It means that the smaller steady state thermal resistance can work with more compact heatsink design to save the cost.

For the transient thermal impedance, the thermal capacitance plays an important role. The larger thermal capacitance makes the transient thermal impedance smaller to have lower junction temperature rise in short pulse during hiccup startup. It will reduce the risk of device failure caused by thermal issue during pulses hiccup startup.

The thermal resistance also contribute to transient state. The cooling down period can be shorter with smaller thermal resistance to shorten the startup duration.

C. Temperature coefficient in steady and transient state:

The steady state temperature coefficient is important to achieve evenly current sharing for paralleled operation. The MOSFET device is better than IGBT on this characteristic makes it can be easily used for multiple MOSFET parallel operation to share the large steady state current and thermal stress.

For the hiccup startup, the saturation current is a function of drain to source voltage, gate voltage and junction temperature:

$$I_D = f(T_J, V_{GS}, V_{DS}) \qquad (2)$$

The paralleled devices have same V_{DS} and V_{GS}. However, the threshold voltage VT of devices varies in $\pm 1V$ makes the initial I_D on each devices are not equal. The larger saturation I_D brings larger thermal stress and higher junction temperature with self-heating.

$$P_1 = I_{D1} \cdot V_{DS} > I_{D2} \cdot V_{DS} = P_2 \qquad (3)$$

This higher junction temperature will affect the value of I_D due to (2). The temperature coefficient of T_J and V_T need to be negative feedback to share the I_D to have equally T_J.

D. Temperature range:

The SiC devices can work with much higher junction temperature than Si device, which makes SiC devices have more margin for hiccup startup, over current and over temperature trip and lower risk in failure due to the high junction temperature.

E. Drive circuit power consumption:

The power consumption of driver circuit is a part of total power loss besides the device conduction loss. The normally ON feature of SiC JFET from USCi makes it can consumes least power on driver circuit with 0V VGS. The current driven BJT will consume more power in driver circuit compare with voltage driven MOSFET.

IV. CANDIDATE ANALYSIS AND EXPERIMENTAL VERIFICATION

The single device experimental results for proposed DC solid state circuit breaker in section II is shown below to demonstrate the designed operation strategy.

Figure 5 shows the experimental waveforms of Mode A-C as in section II. The specifications is 380V with 47uF output capacitor. The device in the test waveform is Si CoolMOS. Five hiccup pulses with 10A close loop I_D is adopted to charge the capacitor from 0V to 380V in about 70ms. The close loop controlled adaptive gate voltage is lower than 6V to ensure the I_D is limited as 10A.

Figure 6 shows the experimental waveforms of I_{limit} threshold over current trip with timer. The programmed timer is 680us. The I_{limit} for single device is 5A. The controlled MOSFET is turned off after the $I_D = 5.4A$ with about 684us.

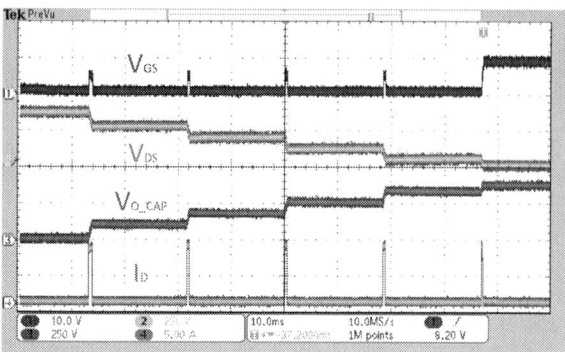

Figure 5. Operation modes from hiccup startup to normally on

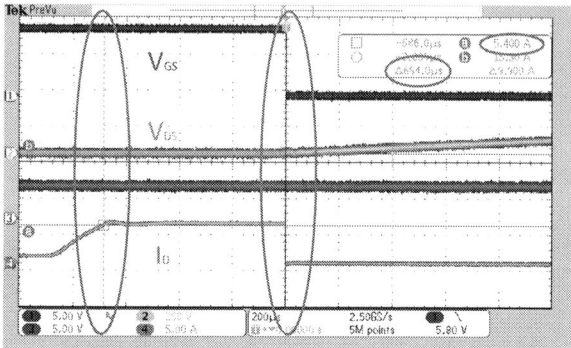

Figure 6. Ilimit threshold over current protection with timer

The drain to source voltage start to increase because the output capacitor voltage start to decrease with 5.4A load connected.

Figure 7 shows the test waveform of short circuit over current protection. The I_D is controlled with adaptive V_{GS} once the short circuit current is larger than 9A. This current saturation protection for low impedance or short circuit fault can isolate the downstream fault without affecting the upstream bus voltage and other loads neighborhood on the bus.

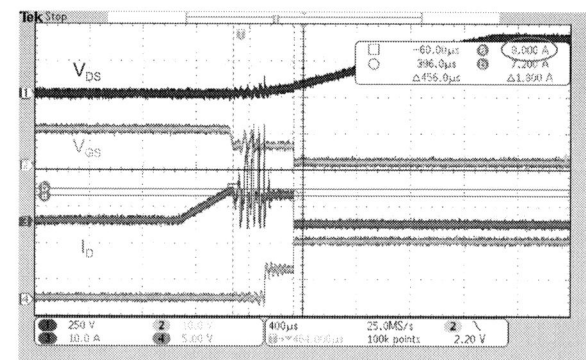

Figure 7. Short circuit current saturation protection

For device paralleled operation similar with single device operation close loop control strategy, 4 devices are considered as candidates in 380V DC solid state circuit breaker application. The characteristics listed in section III is evaluated among candidates with considering the die size. [17]-[20]

Table 1 lists the 4 devices and their basic characteristics for further evaluation.

TABLE I. CHARACTERISTICS OF FOUR SOLID STATE DEVICES

TYPE	MANUF ACTUR ER	PART NO.	$V_{(BR)DSS}$ (V)	$R_{DS}(ON)$ @25 °C($M\Omega$)	DIE SIZE (MM^2)
Si MOSFET	Infineon	IPW65R019C7	650	19	68.29
Si MOSFET	ST	STY145N65M5	650	14	121.88
SiC MOSFET	WOLFS PEED	C2M0025120D	1200	25	26.02
SiC JFET	USCi	UJN1205K	1200	45	9.42

A. Static conduction resistance

As stated above in section III, the $R_{DS(ON)}$ can be evaluated with considering its die size, the specified value is listed in table 2.

TABLE II. $R_{DS,SP(ON)}$ COMPARISON OF 4 CANDIDATES

TYPE	MANUF ACTUR ER	PART NO.	$R_{DS(ON)}$ @25 °C($M\Omega$)	DIE SIZE (CM^2)	$R_{DS,SP(ON)}$@ 25°C($M\Omega*$ CM^2)
Si MOSFET	Infineon	IPW65R019C7	19	0.6829	12.9751
Si MOSFET	ST	STY145N65M5	14	1.2188	17.0632
SiC MOSFET	WOLFS PEED	C2M0025120D	25	0.2602	6.505
SiC JFET	USCi	UJN1205K	45	0.0942	4.239

From Table. 2, the SiC JFET is best candidate considering the die size. Value of SiC devices are 1/3 to 1/4 of Si devices. This make SiC devices very competitive in circuit breaker application with smaller die size and lower power loss compare with Si devices.

B. Thermal impedance in steady and transient state

The typical steady state thermal resistance is listed in table 3. And specified value is calculated with die size.

TABLE III. THERMAL IMPEDANCE OF 4 CANDIDATES

TYPE	MANUF ACTUR ER	PART NO.	$R_{TH,J-C}$ (°C/W)	DIE SIZE (CM^2)	$R_{TH,J-C}$ (SP)(°C/W *CM^2)
Si MOSFET	Infineon	IPW65R019C7	0.28	0.6829	0.1912
Si MOSFET	ST	STY145N65M5	0.2	1.2188	0.2438
SiC MOSFET	WOLFS PEED	C2M0025120D	0.24	0.2602	0.0624
SiC JFET	USCi	UJN1205K	0.65	0.0942	0.0613

The calculated results is close to previous analysis in section III. The SiC devices has larger thermal conductivity makes its thermal resistance is about 1/3 to 1/4 of Si devices with same die size or SiC devices can have similar thermal resistance with 1/3 to 1/4 die size compare with Si devices.

C. Temperature coefficient in steady and transient state

The all four candidates have positive $R_{DS(ON)}$-T_J coefficient which is good for steady state current sharing with paralleled operation. It is shown in figure 8.

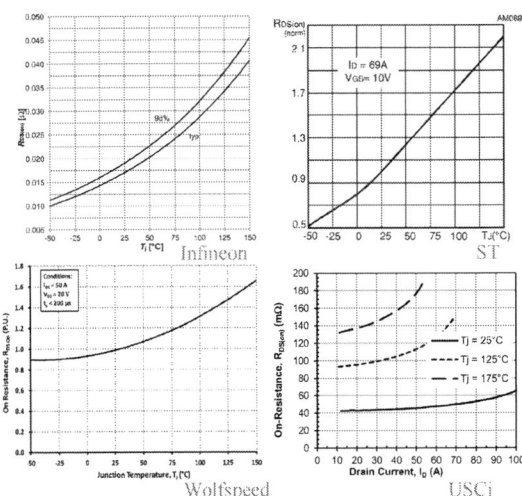

Figure 8. positve $R_{DS(ON)}$-T_J coefficient of 4 candidates for steady state

As analyzed above, the transient I_D-T_J coefficient is very important during hiccup startup with multiple devices parallel. Figure 9 shows the test setup with different junction temperature and a typical saturation current waveform with short pulse test with minimum self-heating. The ceramic heat plate is used to heat the junction and a very short pulse (10us) is tested to avoid the temperature rise due to internal power dissipation.

In figure 10, all four devices saturation current is listed under different junction temperature for same V_{GS} in each device itself. The SiC JFET from USCi is the only device has the negative coefficient, which makes it the best option for dynamic current sharing during hiccup startup mode.

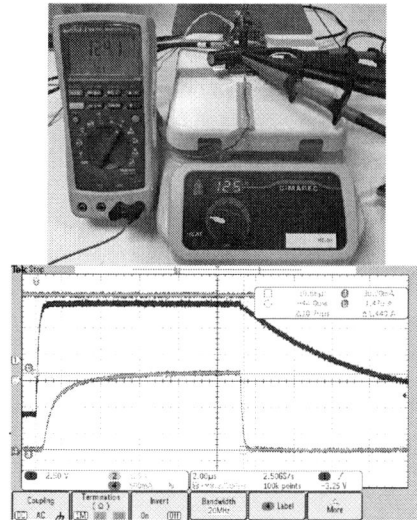

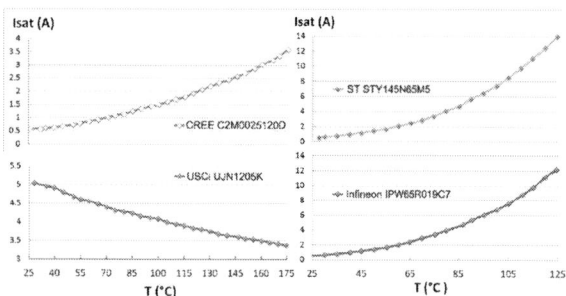

Figure 9. (a) Short pulse saturation test setup (b) Typical short pulse test with saturation current

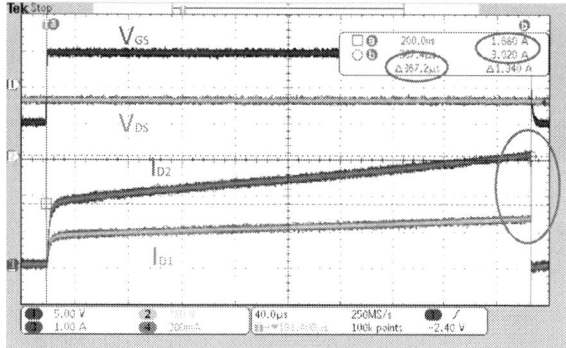

Figure 10. Saturation current vs. Junction temperature

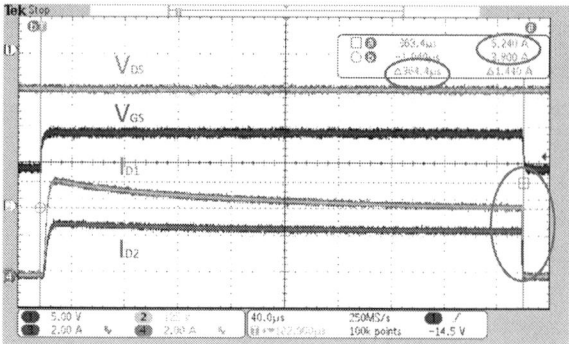

Figure 12. Long pulse saturation current sharing of paralleled SiC JFET

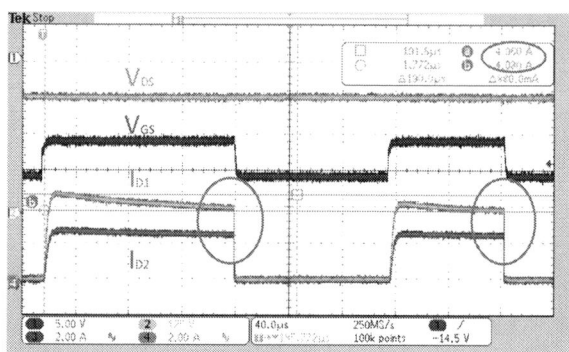

Figure 13. Multiple pulse saturation current sharing of paralleled SiC JFET

Figure 11 to 13 shows the differences. The Fig. 11 shows the current sharing of paralleled Si CoolMOS^{TM} under long pulse self-heating. The saturation current will increase and the current differences between two paralleled MOSFET will become larger, which is dangerous for hiccup startup with multiple devices parallel. The other two devices is similar as Fig. 11.

It is observed that both Si and SiC MOSFETs show increases current with a constant gate voltage when tested under saturation condition. This is primarily because the threshold voltage dependency on the temperature: V_{GS} decreases with increased temperature. Therefore, Neither Si nor SiC MOSFETs are good to limit inrush current during transient when they are required to operate in the saturation mode. However, it is noticeable that Si MOSFETs have a much stronger temperature dependency compared to the SiC MOSFET. At 175 °C, saturation currents in the Si MOSFETs have increased to 12-14 A compared to 0.5 A at 25 °C.

The figure 12 shows the current sharing of paralleled SiC JFET under long pulse self-heating. The saturation current will decrease and the current differences between two paralleled SiC JFET will become smaller. It is a very good feature for hiccup startup with multiple devices parallel. Fig. 13 shows the continuous pulses current sharing of SiC JFET.

Figure 11. Long pulse saturation current sharing of paralleled Si CoolMOS^{TM}

978-1-4673-9551-9/16 $31.00 © 2016 IEEE

On the contrary, the SiC JFET has shown negative temperature dependency of the saturation current, which is desirable for current sharing during hiccup startup operation in solid state circuit breakers.

The SiC devices in candidates obviously have wider temperature range for safe operation compare with Si devices. And the SiC JFET normally on characteristic consumes lowest driver power compare with other three voltage driven MOSFET.

The concern for SiC JFET is also its normally on characteristic. The circuit breaker should be better in OFF state if the grid is from powered off to recover. The standard related with 380V DC solid state circuit breaker application need to be further defined with the whole system.

V. CONCLUSIONS AND FUTURE WORK

In sum, the paralleled solid state device circuit breaker is proposed and analyzed in this paper. Four different devices with small $R_{DS(ON)}$ are listed with their parameters. The criteria of characteristic for 380V DC circuit breaker is investigated and discussed. Four devices are selected as candidates for 380V DC circuit breaker application. Comparison and evaluation of four devices has been analyzed and verified with experiment. The SiC devices are better than Si MOSFET considering the influence of Die size. Negative I_{SAT}-T_J coefficient during startup and 0 VGS normally on of SiC JFET makes it the best candidate among 4 devices. The SiC devices performs much better and should be the better solution for DC circuit breaker application in future.

REFERENCES

[1] D. J. Becker and B. J. Sonnenberg, "DC microgrids in buildings and data centers," *2011 IEEE 33rd Int. Telecommun. Energy Conf.*, pp. 1–7, 2011.

[2] Yuzhi Zhang; Umuhoza, J.; Yusi Liu; Farnell, C.; Mantooth, H.A.; Dougal, R., "Optimized control of isolated residential power router for photovoltaic applications," in Energy Conversion Congress and Exposition (ECCE), 2014 IEEE , vol., no., pp.53-59, 14-18 Sept. 2014.

[3] M. Ton and B. Fortenbery, "DC Power for Improved Data Center Efficiency," no. March, pp. 1–74, 2008.

[4] Cuzner, R.M.; Venkataramanan, G., "The Status of DC Micro-Grid Protection," Industry Applications Society Annual Meeting, 2008. IAS '08. IEEE, vol., no., pp.1,8, 5-9 Oct. 2008

[5] Schneider-Electric, "Dual Rated AC/DC Circuit Breakers - Dual Rated AC/DC Circuit Breakers," datasheet, May 2014

[6] EATON, "DC (direct current) Series molded-case circuit breakers," Trip Curves TC01215003E, May 2014

[7] B. Davies, "Analysis of inrush currents for DC powered IT equipment," *2011 IEEE 33rd Int. Telecommun. Energy Conf.*, pp. 1–4, 2011.

[8] R. M. Cuzner and G. Venkataramanan, "The status of DC micro-grid protection," *Conf. Rec. - IAS Annu. Meet. (IEEE Ind. Appl. Soc.*, pp. 1–8, 2008.

[9] Siu, A., "Discrimination of miniature circuit breakers in a telecommunication DC power system," Telecommunications Energy Conference, 1997. INTELEC 97., 19th International , vol., no., pp.448,453, 19-23 Oct 1997

[10] Wei Liu; Huang, A.Q., "A novel high current solid state power controller," Industrial Electronics Society, 2005. IECON 2005. 31st Annual Conference of IEEE, vol., no., pp.5 pp.,, 6-10 Nov. 2005

[11] Boudreaux, R.R.; Nelms, R.M., "A comparison of MOSFETs, IGBTs, and MCTs for solid state circuit breakers," in Applied Power Electronics Conference and Exposition, 1996. APEC '96. Conference Proceedings 1996., Eleventh Annual , vol.1, no., pp.227-233 vol.1, 3-7 Mar 1996

[12] Jin-young Kim; Seung-soo Choi; In-dong Kim, "A novel reclosing and rebreaking DC solid state circuit breaker," in Power Electronics and ECCE Asia (ICPE-ECCE Asia), 2015 9th International Conference on , vol., no., pp.1282-1288, 1-5 June 2015

[13] Shen, Z.J.; Sabui, G.; Zhenyu Miao; Zhikang Shuai, "Wide-Bandgap Solid-State Circuit Breakers for DC Power Systems: Device and Circuit Considerations," in Electron Devices, IEEE Transactions on , vol.62, no.2, pp.294-300, Feb. 2015

[14] Shen, Z.J.; Zhenyu Miao; Roshandeh, A.M., "Solid state circuit breakers for DC micrgrids: Current status and future trends," in DC Microgrids (ICDCM), 2015 IEEE First International Conference on , vol., no., pp.228-233, 7-10 June 2015

[15] Bhat, K.P.; Yuan-Bo Guo; Yang Xu; Baltis, T.; Hazelmyer, D.R.; Hopkins, D.C., "Results for an Al/AlN composite 350°C SiC solid-state circuit breaker module," in Applied Power Electronics Conference and Exposition (APEC), 2012 Twenty-Seventh Annual IEEE , vol., no., pp.2492-2498, 5-9 Feb. 2012

[16] Yoder, M.N., "Wide bandgap semiconductor materials and devices," in Electron Devices, IEEE Transactions on , vol.43, no.10, pp.1633-1636, Oct 1996

[17] Infineon Technologies AG, CoolMOS™C7 IPW65R019C7 Datasheet [EB/OL]. Rev 2.1, (2013-04-18).

[18] ST Microelectronics Corporation, STY145N65M5 Datasheet [EB/OL]. Rev 2, (2013-01).

[19] WOLFSPEED, Inc. C2M0025120D Datasheet [EB/OL]. Rev A.

[20] United Silicon Carbide, Inc. UJN1205K Datasheet [EB/OL]. Rev 1.0.

Evaluation of High Voltage Cascode GaN HEMTs in Parallel Operation

He Li, Xuan Zhang, Lucheng Wen, John Alex
Brothers, Chengcheng Yao, Ke Zhu, Jin Wang
Center for High Performance Power Electronics
The Ohio State University
Columbus, OH, USA

Liming Liu, Jing Xu, Joonas Puukko
ABB Inc.
Raleigh, USA

Abstract — Paralleling devices is an effective way to achieve higher power applications while still having the convenience brought by discrete devices. However, very few papers investigate the challenges of paralleling Gallium Nitride high electron mobility transistors (GaN HEMTs) in cascode configuration, especially the potential failure modes and its related mechanisms. In this paper, a comprehensive study on paralleled high voltage cascode GaN HEMTs is presented. The influence of paralleling cascode GaN HEMTs on the circuit's stray inductance is studied. Potential operation failure modes and the mechanisms of the cascode GaN HEMTs parallel operation were analyzed in detail. The Ansoft Q3D FEA tool and SPICE-based simulation model were used together to quantify the impacts of the circuit and device mismatch on the paralleled GaN HEMTs operation. The SPICE model is validated by the experimental results

Keywords — GaN HEMT; cascode structure; parallel operation; reliability

I. INTRODUCTION

Compared to the enhancement mode (E-mode) GaN device, the depletion mode (D-mode) GaN device has a simpler device structure. However, the normally-on property of the D-mode GaN device is not preferred in applications because of the complicated driver designs and potential shoot-through risks. In order to fully utilize the advantages offered by the GaN HEMT, manufactures like Transphorm, RFMD and Infineon started to provide high voltage normally-off GaN HEMTs in the cascode structure to the research institutes and market. The schematic of the D-mode GaN HEMT in cascode structure is shown in Fig. 1. The one drawback is that they are mostly available in lower current ratings. For this reason, there is a great desire to parallel them for higher power applications [1]-[2].

The paralleling of IGBTs and MOSFETs is well understood and has been implemented in several different applications [3]-[4]. The study of device and circuit mismatch on the paralleling of SiC MOSFETs has been reported [5]-[6]. The problem with the previous studies on the device parallel operation is that they cannot be directly applied to cascode high voltage high switching speed Wide Bandgap (WBG) devices, especially the cascode GaN HEMT. In fact, very

limited information on parallel operation of cascode GaN HEMTs has been reported so far. In 2013, a four cascode GaN HEMTs based boost converter was demonstrated [7]. The symmetry of the main loop layout was emphasized. In 2015, a four cascode GaN HEMTs based half-bridge circuit was demonstrated [8]. An additional coupled inductor was used to compensate the current imbalance between devices.

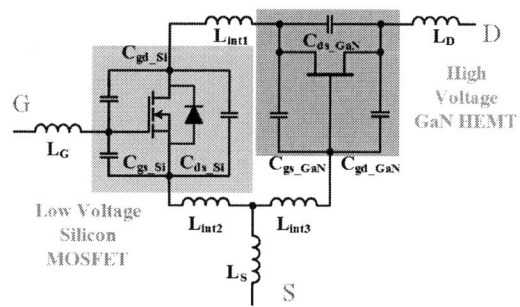

Fig. 1. GaN HEMT in cascode configuration.

In this paper, a comprehensive study on paralleling high voltage cascode GaN HEMTs is presented. Section II shows the influence of the parallel devices on circuit parasitic inductance. Section III discusses the potential failure modes of the parallel cascode GaN HEMTs operation and the mechanisms. Section IV investigates the sensitivity of the circuit and device mismatch on the cascode GaN HEMTs parallel operation. Section V concludes the paper.

II. THE INFLUNCE OF PARALLEL DISCRETE GAN HEMTs ON CIRCUIT PARASITIC INDUCTANCE

The parasitic inductance of the PCB traces have been recognized to be very critical to the GaN HEMTs safe and efficient operation [9]. Several compact layouts for paralleling bare die E-mode GaN devices have been demonstrated with minimized parasitic inductances [2]. Due to the unique structure of the cascode device, the To-220/247 and the PQFN are still the popular packages for them. With the high switching speed of the GaN HEMT and relatively larger package, the PCB layout of paralleled discrete cascode GaN HEMTs was more challenging. In order to demonstrate the influence, two versions of the Double Pulse Tester (DPT)

978-1-4673-9551-9/16 $31.00 © 2016 IEEE

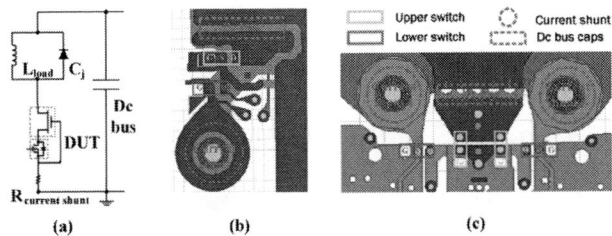

Fig.2. DPT circuit (a) schematic, (b) single GaN HEMT DPT layout (c) parallel GaN HEMTs layout

TABLE I.   DPT CIRCUIT PARASITIC INDUCTANCE

Parasitic inductance	Single GaN HEMT DPT circuit	Parallel GaN HEMTs DPT circuit
+Dc bus to upper device	0.05 nH	0.45 nH
Switching node	1.98 nH	2.84 nH
Lower device to –dc bus	1.32 nH	3.22 nH
Total PCB trace	3.35 nH	6.51 nH

were designed with the To-220 package device.

In the single GaN HEMT version, the circuit main power loop had a vertical structure [10]. The main power loop traces in the top and adjacent middle layer were able to be overlapped with each other to minimize the loop area. In the parallel GaN HEMTs version, the DPT circuit main loop traces were in the same layer. Both of the layouts, as shown in Fig. 2b and 2c, were analyzed by the FEA tool, and the extracted circuit parasitic inductances are listed in Table 1. Even though the position of the coaxial current shunts does influence the PCB layout, the general trend still remains the same. Increasing the number of devices adds more constrains to the main loop layout and leads to a larger and even unsymmetrical main power loop.

III. POTENTIAL FALUIUER MODES AND MEHCNISIMS OF CASCODE GAN HEMTs PARALLEL OPERATION

In this section, selected static parameters for the cascode GaN HEMT were first evaluated. Next, the gate signal mismatch was taken as the example to explain the potential failure modes and mechanisms which could happen during the parallel operation of cascode GaN HEMTs. Lastly, both the SPICE based simulation and experiment were conducted to verify the analysis.

A. Critical Static Parameters

The device parameter variation was examined at the beginning. In total 20 cascode GaN HEMT samples from the same production batch were selected. Since the gate-to-source threshold voltage (Vth) and the static drain-to-source on-resistance (Rds_on) are the most critical static parameters for the paralleled devices dynamic and static current sharing, both of the parameters were measured by using curve tracer. The

measurement results showed that very small variations in the Vth ($< 3\%$) and Rds_on ($< 4\%$) existed among all the samples. This also means that device parameter variation is not a safety concern for parallel operation of cascode GaN HEMT with latest device fabrication technology.

Then the temperature influence on cascode GaN HEMTs Vth and Rds_on were evaluated across the entire operation temperature range. Similar to the Si MOSFET and the SiC MOSFET, the Rds_on of the cascode GaN HEMT has a positive temperature coefficient (PTC) (Fig. 3a). It is beneficial for the paralleled devices with different conduction losses to reach a thermal equilibrium point and maintain good current sharing during conduction. For the cascode GaN HEMT in this study, the amplitude of the Rds_on positive temperature coefficient is similar to the SiC MOSFET [11], but much larger than the E-mode GaN HEMTs [12]. The comparison between the SiC MOSFET, the cascode GaN HEMT and the E-mode GaN HEMT is shown in Table II. In practical case, a common heat spreader or heatsink for paralleled devices will be helpful to leverage the temperature imbalance among parallel devices. However, the gate threshold voltage (Vth) of the cascode GaN HEMT has a negative temperature coefficient (NTC), which means the device has higher junction temperature will turn-on first under the parallel operation condition. If this occurs, this device would take the higher share of the dynamic current and have a higher turn-on loss. Since the turn-on loss dominates the cascode GaN HEMT switching loss, early turn-on will enlarge the current differences among all the paralleled. The measured Vth under different device junction temperatures is shown in Fig. 3b. The comparison among different types of WBG devices is shown in Table II.

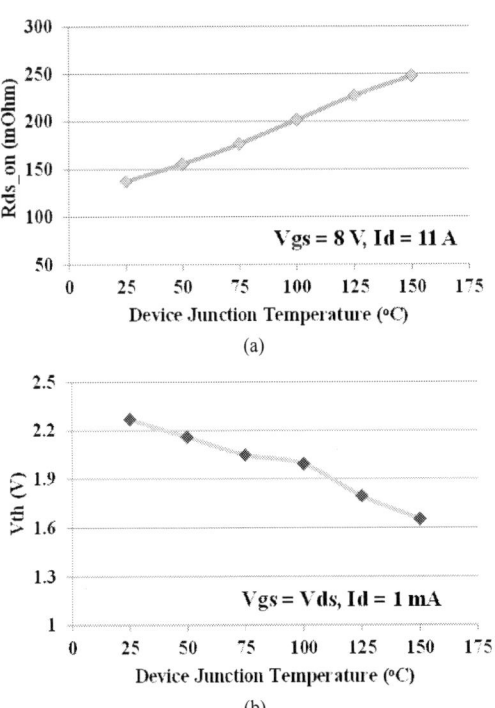

Fig.3. Static parameters at different junction temperatures (a) Rds_on (b) Vth.

TABLE II.   COMPARISION OF SELECTED WBG DEVICES

Device	Vth temperature coefficient (mV/°C)	Rds_on temperature coefficient (mOhm/°C)
1200 V SiC MOSFET	-5.49	+0.75
600 V GaN based Gate Injection Transistor	-0.34 to -0.65	+0.37 to +0.42
650 V E-mode GaN HEMT	≈ 0	+0.11
600 V Cascode GaN HEMT	-5.50	+0.9

B. Gate Signal Mismatch

During the turn-on transient of the GaN HEMTs parallel operation, a couple nanoseconds gate signal mismatch could cause severe dynamic current imbalance between devices due to the high switching speed. The entire turn-on transient can be divided into two stages. In the first stage, the unbalanced current is built up due to the different turn-on time of the paralleled GaN HEMTs. The drain current of device x is formulated with Eq. (1). Circulating current ΔI between the parallel devices is formulated with Eq. (2).

$$I_{dX}=g_{m\_GaNX}(V_{gs\_GaNX}-V_{th\_GaNX})+C_{ds\_GaNX}\frac{dv_{ds\_GaNX}}{dt}$$
$$-C_{gd\_GaNX}\frac{d(V_{gs\_GaNX}-V_{ds\_GaNX})}{dt} \qquad (1)$$

$$\Delta I \text{ (circulating)} = I_{d1}-I_{d2} \qquad (2)$$

In the second stage, an internal voltage oscillation inside

the cascode structure occurs during each D-mode GaN HEMT's turn-on transient. Due to the low impedance of the GaN HEMT's turn-on current charging loop, the parasitic inductance of the package's internal wire bonds and the GaN HEMT's input capacitance will create an oscillation without effective damping. High unbalanced dynamic current could make the voltage oscillation in the gate-to-source voltage of the GaN HEMT (Vgs_GaN) in this stage even worse. The circuit schematic of the two stages are shown in Fig. 4a and 4b respectively.

The phenomenon of dynamic current sharing imbalance during turn-on and turn-off transients are quite different due to the unique cascode structure current source turn-off mechanism [13]. This will cause the oscillation of the turn-off transient to be much shorter in time but larger in amplitude. In the cascode structure, a low voltage Si MOSFET is often used to control the high speed depletion mode GaN HEMT. By properly selecting the Si MOSFET, it will not only control the Vgs_GaN, but also provide overvoltage protection to the Vgs_GaN. In an ideal case, the Vgs_GaN would follow the drain-to-source voltage of the Si MOSFET, and never exceed the breakdown voltage of Si MOSFET (Vds_bk) [14]. However, due to the limitation of the packaging method [15], the internal parasitic inductance of the wire bonds between the Si MOSFET and GaN HEMTs may cause voltage ringing on to the Vgs_GaN, as shown in Eq. 3.

$$-V_{gs\_GaN1}=V_{ds\_Si1}+V_{Lint1}+V_{Lint2}+V_{lint3} \qquad (3)$$

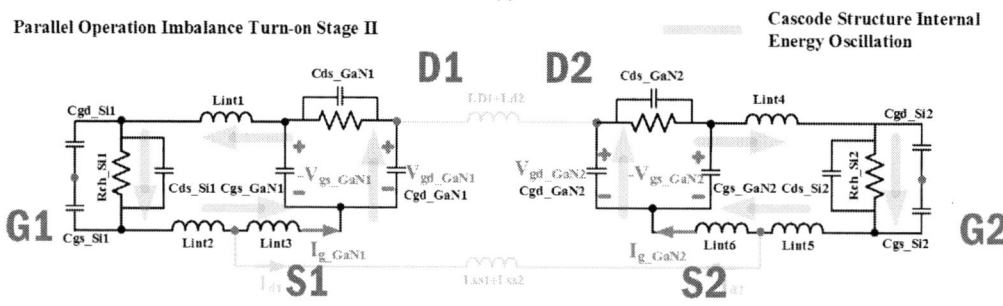

(a)

(b)

Fig. 4. The unbalanced paralleled cascode GaN HEMTs turn-on transient (a) dynamic unbalance current build up stage (b) cascode structure internal voltage oscillation stage.

Parallel Operation Imbalance Turn-off

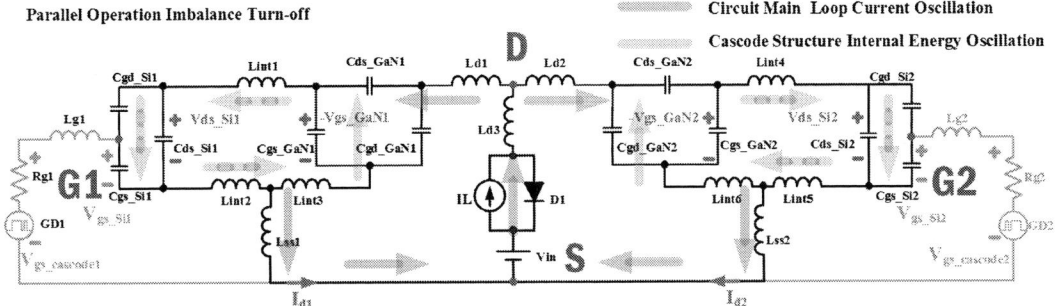

Fig. 5: Paralleled cascode GaN HEMTs turn-off transient current oscillation.

The amplitude of the ringing is mainly decided by the di/dt of the main loop current. Unbalanced dynamic current will cause higher voltage oscillations on the internal parasitic inductance of the device package, which would not only extend the Si MOSFET avalanche time, but also exceed the minimum gate-to-source voltage of the D-mode GaN HEMT. The schematic of the oscillation loop is shown in Fig. 5.

C. Analysis Verification

A SPICE based lumped device model was provided by the device manufacture and further developed for cascode GaN HEMT internal waveform observation. A comparison between the simulated device's hard-switching waveforms and experimental waveforms is shown in Fig. 6. All the PCB traces tray inductances were extracted by FEA based simulation and all the gate drive circuit components impedances were simulated with the impedance analyzer extraction results.

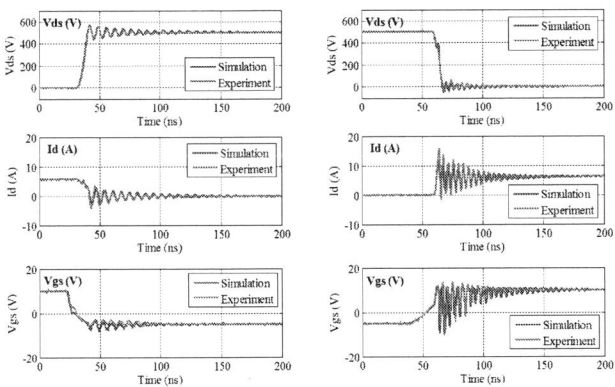

Fig. 6. Cascode GaN devices parallel operation and simulation model verification.

Then, the same modeling method was applied to develop a bridge circuit model with paralleled device pair. The circuit schematic is shown in Fig. 7a. A phase-leg bridge circuit prototype was designed and tested with the same schematic. The lower two devices had separate gate drive circuits and chips. For each gate drive circuit, the gate signal could be controlled independently. The circuit schematic and the prototype are shown in Fig. 7b.

A 4 ns delay was finely tuned and added into one of the paralleled device's gate signal during the experiments. The turn-on transient waveform showed a 130 ns dynamic unbalanced current. The turn-off transient showed a 50 ns

severe current oscillation with a larger amplitude, which validates parts of the oscillation analysis in previous section. The test conditions were 500 V dc bus voltage (Vdc) with 16 A total drain-to-source current (Id). The upper device was the SiC Shockey diode C3D04060A. The switching transient waveforms is shown in Fig. 8a and 8b.

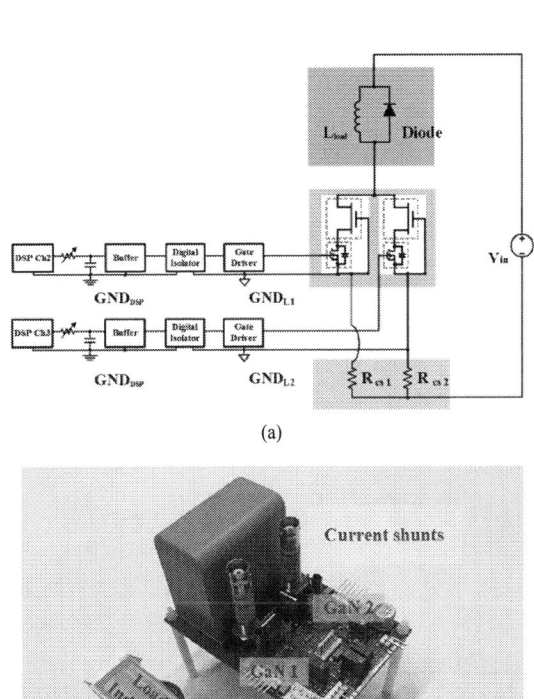

(a)

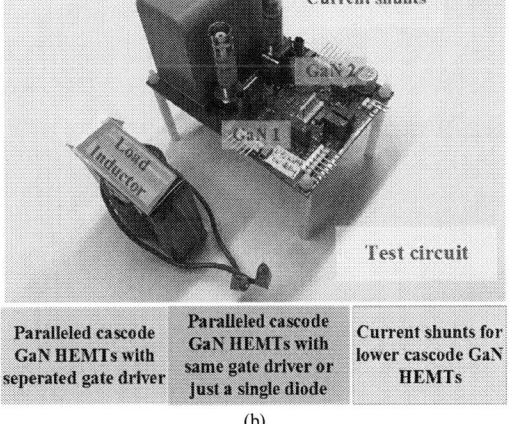

(b)

Fig. 7. Parallel device bridge circuit (a) schematic and the (b) prototype.

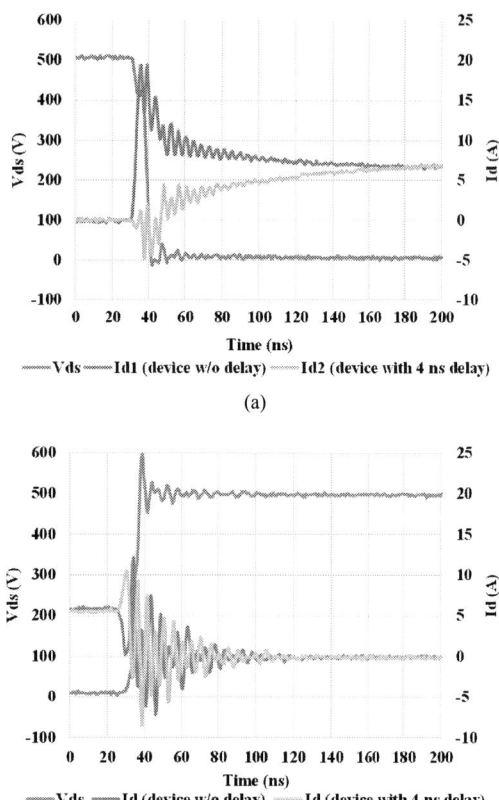

(a)

(b)

Fig. 8 Experiential results of 4 ns gate signal delay between paralleled cascode GaN HEMTs, (a) hard-switching turn-on, and (b) hard-switching turn-off.

The SPICE based simulation model was modified to be able to observe the cascode structure's internal waveforms. The simulation conditions were same with the experimental conditions. The simulation result of the turn-off transient is shown in Fig. 9. It verified that the larger dynamic current imbalance leads to around 10 ns of Si MOSFET avalanche and larger GaN HEMT gate-to-source voltage under shoot.

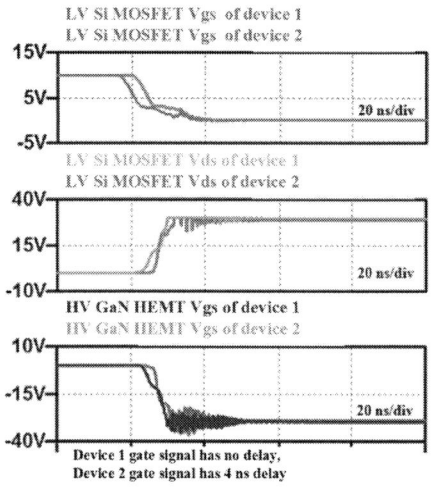

Fig 9. Simulated parallel cascode GaN HEMTs turn-off transient with different gate signal.

IV. CIRCUIT AND DEVICE PARAMETER SENSITIVITY OF CASCODE GaN HEMTs PARALLEL OPERATION

The unbalanced dynamic current could also be caused by the circuit level parasitic inductance mismatch and the component level variation. In this study, the impact from the main power circuit layout, the gate drive circuit and the device parameter mismatch were evaluated separately by running a series circuit worst-case simulation. In each round of the simulation, the selected parameter was given a 10% tolerance. The rest of the parameters remained the same. The simulations were based on the SPICE model of the aforementioned parallel cascode GaN HEMTs bridge circuit.

The device model provided by the device manufacturer was used as the base line model. A few parameters were modified based on the original model to achieve better performance. The base line circuit's parasitic inductance was extracted from the PCB layout with the FEA tool. The critical main loop inductance was already listed in Table 1. In general, no mismatch between the devices, PCB layout and gate signal was introduced in the base line model. The entire parallel structure was perfectly symmetric.

The gate drive delay range was confined by 10% of the propagation delay value of gate driver ISL89163. The gate drive chip output voltage was 10 V/-5 V with a 16 Ohm turn-on resistance and a 2 Ohm turn-off resistance. Eight parameters of the cascode GaN HEMTs were studied separately including LV Si MOSFET and HV GaN HEMT parasitic capacitances and gate threshold voltages. Four circuit level parameter mismatches were evaluated including gate signal, external gate resistance, stray inductance from switching node to separate drain pins (Ld), gate loops inductance (Lg) and common source inductances (Lss). The simulation conditions were 500 V Vdc and 16 A Id for the device pair. Four critical signals inside the cascode structure were monitored as shown in Table III. It shows the level of sensitivity of the cascode GaN HEMT to different circuit and device mismatches.

TABLE III.   CIRCUIT AND DEVICE PARAMETER SENSITIVITY

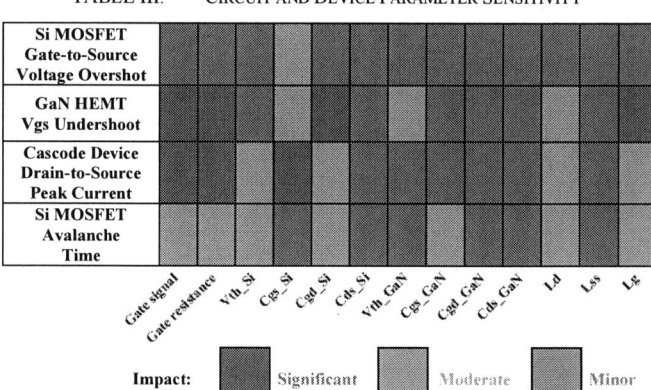

Based on the simulation results, it was concluded that the mismatch of the gate drive circuit has the most significant impact on cascode GaN HEMTs parallel operation. In the practical design, a single gate driver with symmetric gate loop layouts and separate same value gate resistor for each switch will provide the best current and power. The unshared switching node to drain pins circuit lay out needs to be as symmetric as possible. The common source inductance is very critical as well, however, this inductance usually can be easily minimized by separating the gate loop and main current loop from the device's source pin. The value of the Lss is only decided by the device package itself instead of the PCB layout. Even though the simulation results shows the mismatch of certain device parameters has a significant impact, these variations can be avoided by using same production batch devices from a high quality GaN power transistor manufacture.

V. CONCLUSION

Based on the above analysis of the paralleled cascode GaN HEMTs operation, it can be concluded that 1) parallel discrete devices leads to higher loop inductance and stringent layout requirements, 2) off-the-shelf cascode GaN HEMTS have sufficiently small device variation which helps to avoid the parallel operation failure caused by the device mismatch, 3) with a few nanoseconds mismatched gate drive signal, significant dynamic current imbalance will exist during both the turn-on and the turn-off stage with different behavior and mechanisms, 4) a SPICE model based simulation was performed to evaluate the circuit and device parameters sensitivity to the cascode GaN HEMTs parallel operation. It turns out that the symmetry of the gate drive circuit has the highest priority. Our experimental results show that paralleling cascode GaN devices can be achieved with a symmetric gate drive circuit layout and same gate drive signal, and 5) the analysis of the cascode GaN HEMTs parallel operation could be extended to other types of WGB devices in the cascode configuration.

ACKNOWLEDGMENT

The authors would like to thank Transphorm Inc. and RFMD for providing the GaN power transistor samples and technical support.

REFERENCES

[1] X. Lingxiao, *et al*, "The optimal design of GaN-based Dual Active Bridge for bi-directional Plug-in Hybrid Electric Vehicle (PHEV)

charger," *in Proc. Appl. Power Electron. Conf. and Expo. (APEC)*, vol., no., pp.602-608, 15-19 March 2015.

[2] D. Reusch, J. Strydom, "Improving Performance of High Speed GaN Transistors Operating in Parallel for High Current Applications," in *Proc. of Int. Exhib. and Conf. for Power Electron., Intell. Motion, Renew. Energy and Energy Manage. (PCIM)*, vol., no., pp.1-8, 20-22 May 2014.

[3] J. B. Forsythe, "Paralleling of Power MOSFETs For Higher Power Output," *IR Applicat. Note*, [Online]. Available: http://www.irf.com/technical-info/appnotes/para.pdf.

[4] C. Nan; *et al*, "Dynamic Characterization of Parallel-Connected High-Power IGBT Modules," *IEEE Trans. Ind. Appl.*, vol.51, no.1, pp.539-546, Jan.-Feb. 2015.

[5] W. Gangyao, *et al*, "Dynamic and static behavior of packaged silicon carbide MOSFETs in paralleled applications," *in Proc. Appl. Power Electron. Conf. and Expo. (APEC)*, vol., no., pp.1478-1483, 16-20 March 2014.

[6] Y. Xue, *et al*, "Active current balancing for parallel-connected silicon carbide MOSFETs," *in Proc. IEEE Energy Convers. Congr. Expo. (ECCE)*, vol., no., pp.1563-1569.

[7] Y. F. Wu, "Paralleling High-speed GaN Power HEMTs for Quadrupled Power Output," *in Proc. Appl. Power Electron. Conf. and Expo. (APEC), 2013 IEEE*, vol., no., pp.211-214, 17-21 March 2013.

[8] Z. Wang; *et al*, "Paralleling GaN HEMTs for diode-free bridge power converters," *in Proc. IEEE Appl. Power Electron. Conf. and Expo. (APEC)*, vol., no., pp.752-758, 15-19 March 2015.

[9] Z. Chen, "Characterization and modeling of high-switching-speed behavior of SiC active devices," *M.S. thesis*, Dept. Electrical Eng., Virginia Tech., Blacksburg, VA, USA, 2009.

[10] D. Reusch, J. Strydom, "Understanding the Effect of PCB Layout on Circuit Performance in a High Frequency Gallium Nitride Based Point of Load Converter," *IEEE Trans. Power Electron.*, vol.29, no.4, pp. 2008-2015, April 2014.

[11] C. DiMarino, *et al*, "High-temperature characterization and comparison of 1.2 kV SiC power MOSFETs," *Proc. IEEE Energy Convers. Congr. Expo. (ECCE)*, vol., no., pp.3235-3242, 15-19 Sept. 2013.

[12] H. Li, *et al*, "Evaluation of 600 V GaN Based Gate Injection Transistors for High Temperature and High Efficiency Applications," *accepted in Wide Bandgap Power Devices and Applications (WiPDA)*, 2015.

[13] X. Huang, , *et al*, "Analytical loss model of high voltage GaN HEMT in cascode configuration," *IEEE Trans. Power Electron.*, vol. 29, no. 5, pp. 2208–2219, May 2014.

[14] Z. Liu, "Characterization and Failure Mode Analysis of Cascode GaN HEMT," M.S. thesis, Dept. Electrical Eng., Virginia Tech., Blacksburg, VA, USA, 2015.

[15] Z. Liu, *et al*, "Package parasitic inductance extraction and simulation model development for the high-voltage cascode GaN HEMT," *IEEE Trans. Power Electron.*, vol. 29, no. 4, pp. 1977–1985, Apr. 2014.

A New Driving Concept for Normally-on GaN Switches in Cascode Configuration

Bernhard Zojer

Infineon Technologies Austria AG
Villach, Austria
bernhard.zojer@infineon.com

Abstract — Normally-on high-voltage (HV) power transistors are usually operated in a series connection with low-voltage (LV) MOSFETs to ensure safe operation. In the widely used cascode configuration (CC) the status of the combined switch is controlled via the MOSFET gate, whereas the alternative direct drive (DD) method controls the gate of the HV switch directly and utilizes the LV transistor as a safety switch only. Both concepts have their respective benefits: CC allows simple standard driving schemes, while DD excels in low switching losses. This paper investigates a third approach: by controlling both HV and LV gates an optimization of switching performance particularly in hard-switched half-bridge applications can be achieved. However, a straight-forward implementation of such a "dual drive" (2D) concept would obviously require a sophisticated driving scheme to allow independent control of both gates. The key idea of this paper is thus the substitution of a separate HV gate driver by a charge pump circuit connected to the LV gate driver. The new concept finally modifies CC by simply adding 3 passive components (resistor, capacitor, diode – "RCD" concept), yet it significantly lowers both losses and voltage stress. In a nutshell, the proposed cascode is able to combine direct drive switching performance with standard driving schemes.

Keywords— Gallium Nitride transistor, HEMT, normally-on, reverse (diode) operation, cascode configuration, direct drive, dual drive, RCD drive

I. INTRODUCTION

Gallium Nitride (GaN) high electron mobility transistors (HEMTs) are the most promising candidates for next-generation power switches. Their small terminal capacitances together with the lack of physical p/n junctions makes them ideally suited for new attractive power topologies ([1] – [3]). Unfortunately, the basic, simplest and thus best-performing GaN HEMT is "normally-on". That means, for switching it "off" a gate drive voltage below the negative threshold voltage V_{th} is required. If such a voltage is not available, e.g. during start-up or under fault conditions, short circuits might result that could cause malfunction or even destruction of the power system.

To overcome this problem, the normally-on HV GaN transistor is usually operated in a series connection with a MOSFET of lower voltage rating. Fig. 1 depicts a somewhat generalized circuit of this kind utilizing an n-channel LV MOSFET. The GaN gate G_1 is connected to a voltage V_{G1} referenced to the MOSFET source. It is important to

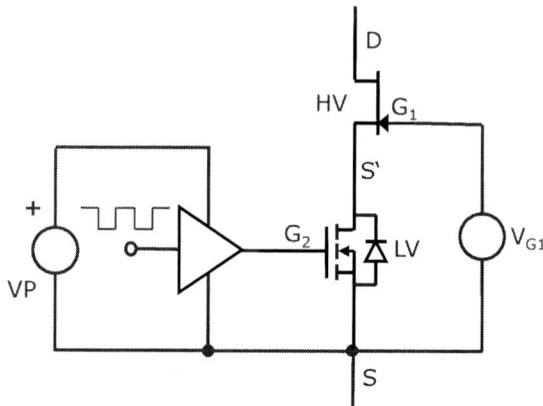

Fig. 1   Generalized classic GaN cascode configuration

understand that this circuit can be controlled from gate G_2 for any value of V_{G1} higher than V_{th} (regardless of V_{th} polarity).

As state of the art the transistor combination of Fig. 1 is usually controlled utilizing one of the following two basic methods.

A. The Classic Cascode Configuration (CC)

This is today the most common concept with G_2 acting as the control terminal and G_1 connected directly to the source terminal ($V_{G1}=0$). The structure is compatible with conventional driving schemes utilizing a single positive driving voltage VP to drive G_2 and thus it is very popular. As a serious drawback, however, CC suffers from increased switching losses and/or voltage stress.

Many recent publications deal in detail with the complex switching dynamics of the cascode circuit ([4] – [8]). Only the most important effects shall be summarized here without complicate calculations. It is evident, that in the conducting state both transistors are "on" when the MOSFET is "on" due to the connection of the GaN gate that causes a gate-to-source voltage close to zero. This is sufficiently above V_{th} and results in full conduction of the HEMT.

In a steady blocking state both transistors have to be "off", because only the HV transistor is able to block the full operating voltage. But as only the MOSFET is actively switched off, the final state has to be reached by charging node S' until $V_{G1S'} = -V_{S'S}$ falls below the GaN threshold V_{th},

978-1-4673-9551-9/16 $31.00 © 2016 IEEE

thereby switching off the HV transistor, too. It is mainly this phase of charging the capacitances connected to node S', i. e. the gate/source capacitance of the GaN HEMT in parallel with the drain/source capacitance of the MOSFET, that is responsible for additional switching losses. In a half-bridge topology of two cascodes that are hard-switched between 0 and a high voltage V_H these losses per switching cycle can be estimated to be

$$\Delta E \sim - (C_{GS1,GaN} + C_{DS1,MOS}) \cdot V_{th} \cdot V_H \qquad (1)$$

As typically the transistor terminal capacitances are strongly nonlinear, $C_{GS1,GaN}$ and $C_{DS1,MOS}$ denote here the average terminal capacitances in the first switching phase with the voltage across the capacitances rising from 0 to V_{th}.

In the following second switching phase the GaN HEMT is "off" and V_{DS} increases to its final value thereby further charging S' via capacitive voltage division until the steady state is reached. A simplified calculation yields

$$V_{S'S} \sim -V_{th} + Q_{oss,GaN} / (C_{GS2,GaN} + C_{DS2,MOS}) \qquad (2)$$

with $C_{GS2,GaN}$, $C_{DS2,MOS}$ and $Q_{oss,GaN}$ denoting the respective terminal capacitances and output charge, now averaged over the voltage range from $-V_{th}$ to V_H. From eq. (1) and (2) losses can be traded for voltage stress essentially by varying $C_{DS,MOS}$ via the MOSFET size. In some cascode solutions a small transistor is chosen to limit capacitive losses and $V_{S'S}$ is allowed to exceed the breakdown voltage; this, however, leads to repetitive MOSFET avalanche. Although the energies involved are small, a careful investigation of potential reliability risks associated with such a mode of operation is required.

B. The Direct-Driven GaN HEMT (DD)

The alternative to the cascode, i.e. controlling the GaN HEMT and keeping the MOSFET always "on" during normal switching operation, is known as direct drive (DD) or safety FET configuration ([9] – [10]); it is characterized by the lowest possible switching losses. With respect to Fig. 1 now V_{G1} is switched between 0 and a negative voltage $-VN < V_{th}$, while V_{G2} is kept constant at VP. If VN is not available, the structure can be switched off via G_2. A dedicated gate driver comprising two separate output stages with different supply voltages is required to be able to control both gates. The negative gate drive voltage has to be sufficiently below V_{th} to guarantee a safe turn-off under all conditions and to avoid the well-known spurious turn-on effects [11].

II. THE DUAL DRIVE (2D) CONCEPT

A. Basic idea

From the above, a new switching scheme (dual drive, 2D) can be considered that allows a direct transition from "on" to the final "off" state by actively switching off both gates, thereby avoiding the first switching phase described in the previous chapter, i.e. the charging of node S'. In the 2D concept thus the GaN gate/source capacitance $C_{GS1,GaN}$ is charged by driving G_1 negative via a gate drive supply VN (Fig. 2) rather than driving S' positive via the power loop in CC. This avoids the cascode-specific additional charging losses

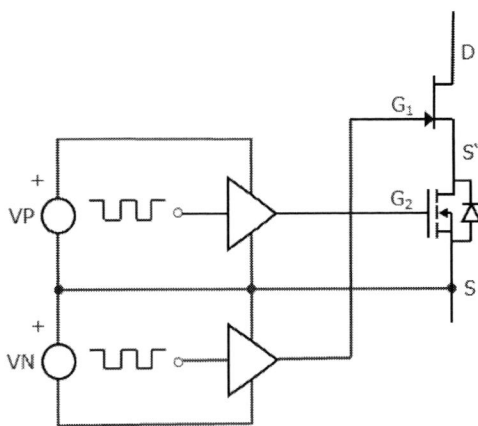

Fig. 2   Multimode GaN cascode configuration

as given in eq. (1) and besides lowers the voltage stress for node S' (eq. (2)). Compared with the DD concept, the negative gate drive voltage can be closer to or even above V_{th}, as a safe "off" state is always guaranteed by the MOSFET. This results in a reduced voltage drop in reverse conduction (diode) mode, as for a GaN HEMT in "off" state this voltage drop is given by the voltage difference between V_{th} and -VN [12].

Fig. 2 represents a simple circuit that is able to support each of the discussed operating modes and in particular enables dual drive. Both drivers are related to source S as the common reference node. This is possible without sacrificing switching performance or running into stability problems, if critical parasitic inductances are sufficiently low. As an example, the inductance between HEMT source and MOSFET drain should be kept below a few tenths of nH, a value that can be practically reached with reasonable package concepts.

B. Target application

It has to be pointed out again, that one of the main advantages of GaN switches is the lack of any intrinsic p/n junction. That means, unlike Superjunction MOSFETs, GaN HEMTs can be operated at both current polarities without suffering from excessive diode recovery charge and thus they are well suited to take over both switch and diode functionality as required in many power applications. The advantages of the 2D concept become evident if two of such composed switches are operated in a half-bridge topology with switch and diode functionality, resp.

In Fig. 3 a boost configuration as it is typically used in Power Factor Correction (PFC) stages with a low-side switch and a high-side diode driven from a constant current source I_{Load} is assumed. During the "on" transient at time t_{on}, I_{Load} is commutated from the diode to the switch and vice versa at t_{off}. The respective gate drive signals of Fig. 4 exhibit the difference between CC and 2D drive concept. In the first cycle (CC) only the MOSFET gates are switched; signal G_{2Sin} defines the switching transients at times t_{on1} and t_{off1}, resp. As usual, the diode-operated cascode is controlled by an inverse non-overlapping signal G_{2Din} that is switched off at time t_1 shortly before t_{on1}. G_{1Din} stays high. In the second switching cycle,

978-1-4673-9551-9/16 $31.00 © 2016 IEEE

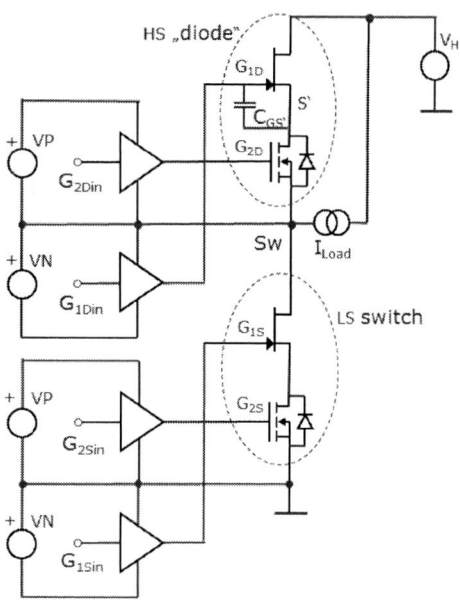

Fig. 3   GaN cascodes in half-bridge boost configuration

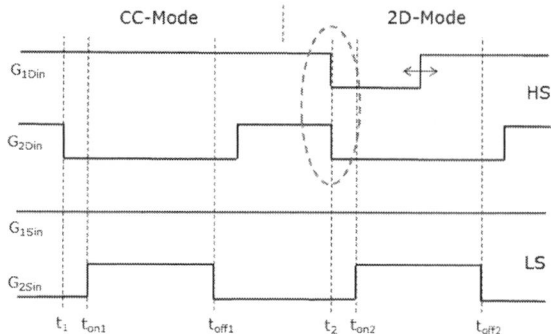

Fig. 4   Half-bridge gate control signals in CC and 2D mode

however, the GaN control signal G_{1Din} is switched to "off" state together with G_{2Din}. As a result $C_{GS'}$ is charged from the gate side prior to the second "on" event at t_{on2} by applying the negative gate driver supply voltage VN to G_{1D} at time t_2. This completely eliminates the first phase of the "off" transient and the associated losses and voltage increase at S'. Switching G_{1D} back to zero during the diode blocking state is uncritical in

timing. In Fig. 4 the switch is operated in cascode mode for both cycles, it could however also be operated in DD or 2D mode, whichever mode is most advantageous in a given application.

C. Simulation results

To prove the validity of the 2D concept, a half-bridge configuration according to Fig. 3 has been simulated. The transistor models and parameters have been derived from an experimental cascode structure consisting of a 650V / 100mΩ GaN HEMT with a threshold voltage of -7V and a 30V / 2mΩ MOSFET out of Infineon's OptiMOS™ family. Both transistors are mounted in an 8 by 8 mm² leadless SMD package with very low inductances of the internal connections.

The simulated waveforms for a load current of 10A are given in Fig. 5. V_H is 400V, gate drive supply voltages are +12

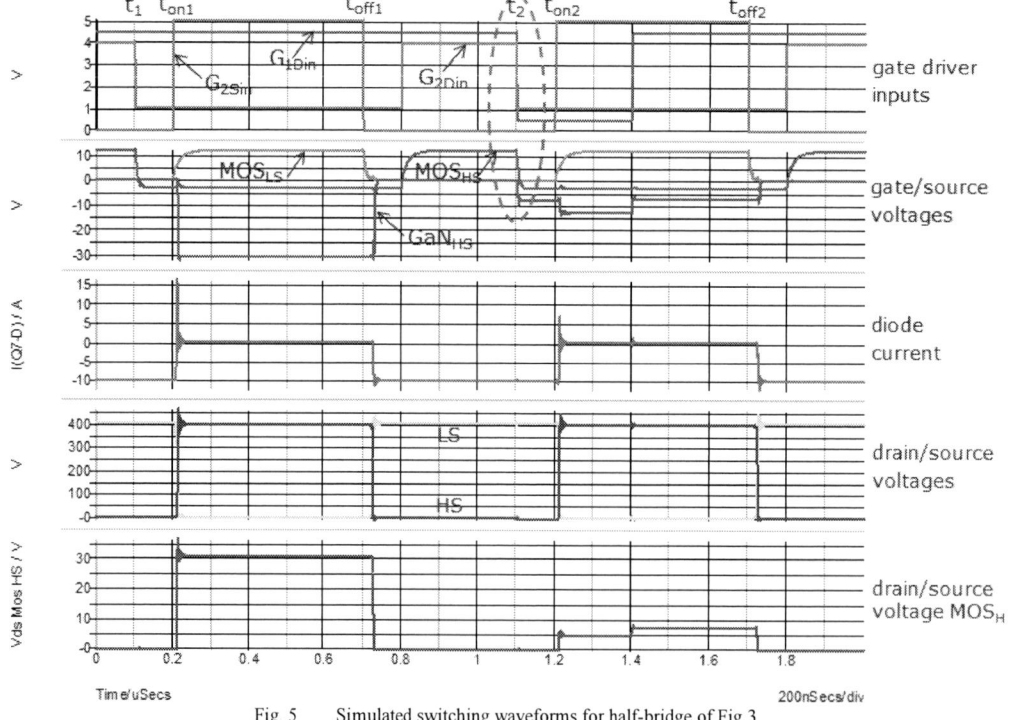

Fig. 5   Simulated switching waveforms for half-bridge of Fig.3

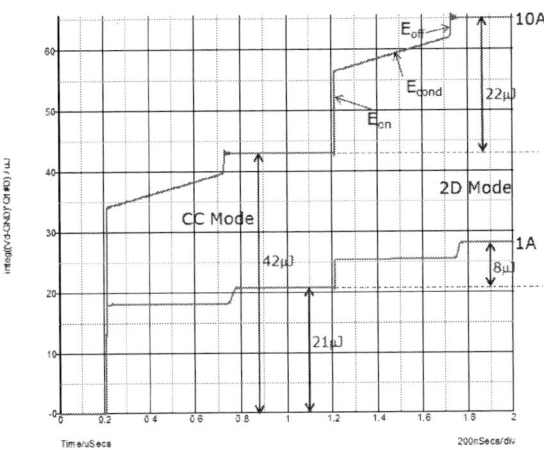

Fig. 6   Switching losses calculated from waveforms of Fig.5

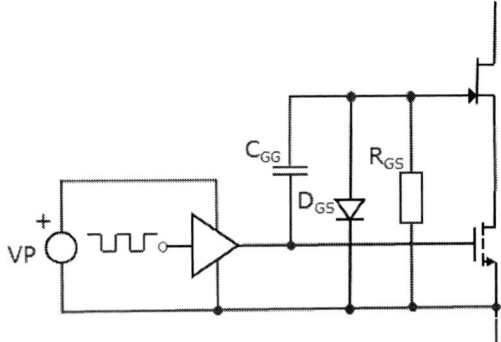

Fig. 7   Proposed RCD circuit

and -8V, resp. The gate drive signals in the first graph are chosen according to Fig. 4, the only difference between the two cycles being the characteristic active switching of the GaN gate G_{1D} at time t_2 in 2D mode. The second graph of Fig. 5 shows the resulting gate/source voltages. As explained, node S' is charged to a high voltage in CC mode, leading to a high negative V_{GS} of -30V for the diode operated switch compared with -10 to -15V in 2D mode. The reduced voltage stress is also clearly seen in the waveform for the drain/source voltage of the MOS transistor (last graph in Fig. 5).

The lower losses in 2D operation are reflected by the smaller transient current peak in the diode current graph. The expected reduction in losses is clearly visible in Fig. 6. Here switching energies are calculated from the waveforms of Fig. 5 by integration of the product of switch current and voltage. Losses are significantly reduced in 2D mode, from 21 to 8µJ for 1A, and from 42 to 22µJ for 10A of load current (including 5µJ of conduction losses E_{cond}). The simulated differences fit quite well to the estimation of eq. (1); the main contribution results from the MOSFET's drain/source capacitance with its low-voltage value of several nF. It should also be pointed out that the absolute loss level is very low regarding the given application; the switching transients take place within 3 to 4 ns.

III.   THE RCD CONCEPT

A. Basic idea

In the 2D concept the switching times of MOS and GaN gate can be chosen independently. However, even for simultaneous switching of both gates the essential benefits are maintained. Unfortunately, the different threshold voltages and thus gate drive levels required by MOS and GaN switches prohibit utilizing a single gate driver, as long as only direct coupling of driver to gate is considered. This is why we propose a modification of the cascode circuitry substituting the original HV gate connection of the cascode by the three passive components depicted in Fig. 7 ("RCD" concept).

Here the GaN gate is capacitively coupled to the driver. Due to the clamping action of diode D_{GS} the coupling capacitance C_{GG} is charged to approximately the positive drive voltage VP during the "on" state. When the driver output is switched back to zero, the charge stored in C_{GG} gets redistributed and shifts the GaN gate/source voltage to a negative value. In other words, the gate/source capacitance is again precharged from the gate driver supply. The circuit behaviour is similar to the direct dual gate drive described in chapter II., however the proposed method is compatible with standard driving schemes. The resistor R_{GS} replaces the HV gate connection of the classic cascode, thus all safety considerations remain unchanged. Besides, R_{GS} allows discharging of C_{GG} during the off-phase to lower the voltage drop of the diode operated device in reverse "off" operation. As a summary the RCD concept is able to combine all the respective benefits of currently existing solutions.

B. Simulation results

For a verification of the concept, also the RCD circuit has been simulated in a boost half-bridge configuration similar to that of Fig.3. Both half-bridge devices use the circuit of Fig. 7 and the simulation models are again based on the experimental cascode described above. Beside the package parasitics also inductances of the passive components have been included in the simulation model with realistic values. For the diode a BAV70 model is used. The coupling capacitance C_{GG} is 3nF, whereas R_{GS} has been chosen to be 500Ω. Gate driving voltage is 13V.

The simulated switching waveforms for a load current of 10A are shown in Fig. 8. As above the simulation includes two switching cycles to directly demonstrate the improvements enabled by the RCD circuit. During the first 1µs-cycle the diode is shorted and the circuit thus behaves like a classic cascode. After 1µs the short is opened and the second cycle thus reflects the performance of the new concept in a direct comparison.

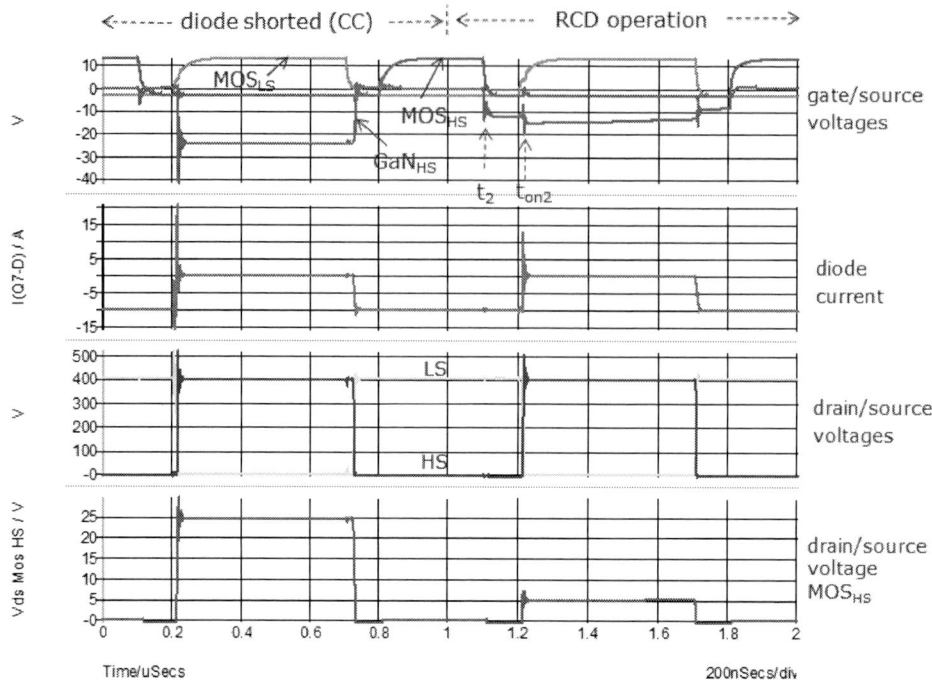

Fig. 8   Simulated switching waveforms for half-bridge with RCD circuits

The first graph in Fig. 8 shows the gate/source voltages of the switched MOSFETs. Besides also the gate/source voltage of the high-side GaN transistor is depicted. Here the difference between the two concepts is evident with the high negative voltage in "off" state in the first cycle. With the RCD components activated, V_{GS} of the HS GaN switch goes negative with the falling gate drive edge at time t_2. The reduced voltage stress is obvious in the MOS V_{DS} waveform (last graph). The loss reduction enabled by the RCD circuit is again due to the smaller switching charge as reflected in the transient current peaks in the diode current graph (13 vs. 20A). And in Fig. 9 again the resulting switching energies have been calculated directly from the waveforms of Fig. 8. As expected, the results are similar to those for the dual drive. Switching losses are significantly reduced by the RCD circuit, from 23 to 8 μJ of total switching energy for 1A, and from 38 to 25μJ for 10A of load current.

IV.   CONCLUSIONS

Normally-on GaN HEMTs are often operated in cascode configuration, i.e. with a series connected low-voltage MOSFET. In this paper a circuit modification has been proposed based on an analysis of the involved switching mechanisms. The new RCD circuit achieves a switching performance similar to directly driven solutions and is compatible with standard driving schemes. Performance results have not yet been proven experimentally, but are so far based on simulations. However, as the available measurement results for both cascoded and directly driven transistors are in very good agreement with simulation, the confidence in the validity of the presented results is high.

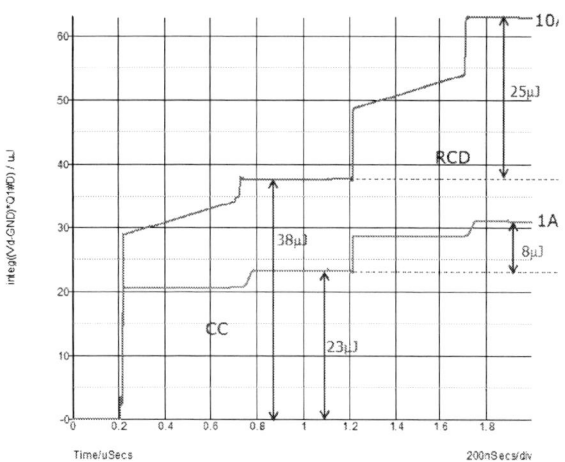

Fig. 9   Switching losses calculated from waveforms of Fig.8

REFERENCES

[1] L. Huber, Y. Jang, M. Jovanovic, "Performance Evaluation of Bridgeless PFC Rectifiers", IEEE Trans. Power Electr., May 2008, pp. 1381-1390

[2] J. Shin, J. Baek, B. Cho, "Bridgeless Isolated PFC Rectifier Using Bidirectional Switch and Dual Output Windings", APEC 2011, Dig. Tech. Papers, pp. 2879 - 2884

[3] X. Huang, Z. Liu, Q. Li, F. Lee, "Evaluation and Application of 600V GaN HEMT in Cascode Structure", APEC 2013, Dig. Tech. Papers, pp. 1279 - 1286

[4] X. Huang, Q. Li, Z. Liu, F. Lee, "Analytical loss model of high-voltage GaN HEMT in Cascode Configuration", ECCE 2013, pp. 3587 – 3594

[5] X. Huang, W. Du, F. Lee, Q. Li, Z. Liu, "Avoiding Si MOSFET Avalanche and Achieving Zero-Voltage-Switching for Cascode GaN Devices", to be published in IEEE Trans. Power Electr.

[6] L. Carrasco, A. Forsyth, "Energy Analysis and Performance Evaluation of GaN Cascode Switches in an Inverter Leg Configuration", APEC 2015, Dig. Tech. Papers, pp. 2424 - 2431

[7] J. Roig, F. Bauwens, A. Banerjee et al., "Unified Theory of Reverse Blocking Dynamics in High-Voltage Cascode Devices", APEC 2015, Dig. Tech. Papers, pp. 1256 - 1261

[8] W. Zhang, F. Wang, L. Tolbert, B. Blalock, D. Costinett, "Investigation of Soft-switching Behavior of 600V Cascode GaN HEMT", APEC 2014, Dig. Tech. Papers, pp. 2865 - 2872

[9] K. Norling, C. Lindholm, D. Draxelmayr, "An Optimized Driver for SiC JFET-Based Switches Enabling Converter Operation with more than 99% Efficiency", IEEE JSSC 2012, pp. 3095 - 3104

[10] M. Seeman, S. Bahl, D. Anderson, G. Shah, "Advantages of GaN in a High-Voltage Resonant LLC Converter", APEC 2014, Dig. Tech. Papers, pp. 476 – 483

[11] Z. Zhang, W. Zhang, F. Wang, L.Tolbert, B. Blalock, "Analysis of the Switching Speed Limitation of Wide Bandgap Devices in a Phase-log Configuration", APEC 2014, Dig. Tech, Papers, pp. 3950 – 3955

[12] Z. Dong, Z. Zhang, X. Ren, X. Ruan, Y. Liu, "A Gate Drive Circuit with Mid-level Voltage for GaN Transistors in a 7MHz Isolated Resonant Converter", APEC 2015 Dig. Tech. Papers, pp. 731 - 736

Avoiding Divergent Oscillation of Cascode GaN Device under High Current Turn-off Condition

Weijing Du, Xiucheng Huang, Fred C. Lee, Qiang Li, and Wenli Zhang

Center for Power Electronics Systems, the Bradley Department of Electrical and Computer Engineering
Virginia Polytechnic Institute and State University, Blacksburg, VA, 24061, USA

Abstract— Cascode structure is widely used for high voltage normally-on GaN devices. However, the capacitance mismatch between the high voltage GaN device and the low voltage normally-off Si MOSFET may induce several undesired features, such as Si MOSFET reaches avalanche during turn-off, and high voltage GaN device loses ZVS turn-on condition internally during soft-switching turn-on process in every switching cycle. This paper presents another issue associated with the capacitance mismatch in the cascode GaN devices. Divergent oscillation could occur at high current turn-off condition, and eventually destroy the device. The intrinsic reason of this phenomenon is analyzed in detail in this paper. A simple solution is proposed by adding an additional capacitor whose position is critical and should be optimized. Experimental results validate the theoretical analysis, and show that the proposed method improves device performance significantly under high current turn-off condition.

Keywords—Gallium nitride device, cascode, capacitance mismatch, avalanche, divergent oscillation

I. INTRODUCTION

High efficiency and high power density are always being pursued rigorously in advanced power conversions for all forms of consumer electronics. High switching frequency is the major catalyst for size reduction, but it is seemly at the detriment of efficiency when using silicon-based devices. The emerging gallium nitride (GaN) power devices have a much lower gate charge and lower output junction capacitance than silicon MOSFETs, and therefore are capable of operating the switching frequency 10 times higher. GaN devices seem to be the game changing devices in the applications such as point of load converters, off-line switching power supplies, battery charger and motor drives [1]-[7]. To realize the benefits of GaN devices resulting from significantly higher operating frequency, a number of issues have to be addressed, such as converter topology, magnetics, control, packaging, and thermal management.

The high voltage GaN devices can be categorized into two types: enhancement mode and depletion mode. One of the desirable features for enhancement mode GaN devices is gate overdrive protection. The recommended gate drive voltage is close to the breakdown voltage which requires dedicated driving circuit design to avoid device failure. For depletion mode GaN devices, a low-voltage Si MOSFET is typically connected in series to control the on-off state of high voltage GaN device, which is well known as cascode structure and is shown in Fig. 1 with all junction capacitors been marked. The external driving circuit is simple with sufficient margin. The evaluations and applications of cascode GaN have been studied in literatures [8]-[17]. Compared to the enhancement mode GaN devices, the cascode GaN devices have smaller turn-off

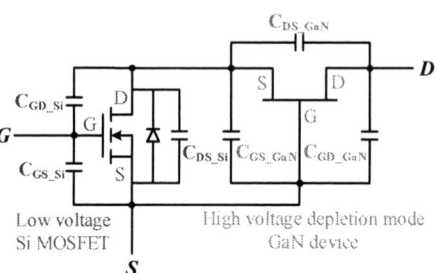

Fig. 1 Cascode structure for a high voltage depletion-mode GaN device with all junction capacitances

loss at higher current condition due to intrinsic current source turn-off mechanism [12, 14]. This feature makes cascode GaN devices very attractive in PWM converter with high turn-off current applications, such as totem-pole bridgeless PFC [6, 15].

In the cascode configuration, interaction between the two devices may result in instability due to the large package parasitic inductance [18]. Advanced package method, such as chip-on-chip technique, can dramatically reduce the package inductance and avoid instability [8]. On the other hand, junction capacitances of the two devices also play an important role in the dynamic performance of the cascode device. The capacitance charge of high voltage GaN device is typically larger than the low-voltage silicon MOSFET. Capacitance mismatch between the high voltage GaN and the low voltage Si MOSFET may induce several undesired features, such as Si MOSFET reaches avalanche during turn-off transition in every switching cycle, and the high voltage GaN switch loses zero-voltage switching (ZVS) turn-on internally even when a ZVS technique is applied [16]. These issues can be solved by adding an additional capacitor in parallel with the drain-source terminals of Si MOSFET to compensate the capacitance mismatch [16]. This solution may slightly increase the turn-off switching loss, but in general, the switching loss is negligible due to intrinsic current source turn-off mechanism [12,13].

This paper presents another issue related to the capacitance mismatch in the cascode GaN devices which would cause device and circuit failure. The parasitic ringing during turn-off transition may trigger the GaN device to internally turn on at high current condition. The parasitic resonant impedance network changes when GaN is turned on and it absorbs more energy from source. The resonant amplitude gradually increases and becomes divergent oscillation. This phenomenon may result in cascode GaN device and other circuit components failure. This failure mode can be theoretically avoided by the same method proposed in literature [16].

978-1-4673-9551-9/16 $31.00 © 2016 IEEE

However, the packaging of the additional capacitor is critical to the failure. The detailed illustration of the divergent oscillation is presented in section II. The analysis of different packaging impact on the failure mode is presented in section III. Finally, the theoretical analysis is validated by experimental results shown in section IV. It is worthwhile to point out that the theoretical analysis and proposed solution are applicable to all kind of cascode devices.

II. ANALYSIS OF DIVERGENT OSCILLATION ISSUE

To better understand the divergent oscillation issue, Si MOSFET avalanche and GaN internally turn on mechanism in the cascode device with capacitance mismatch should be briefly reviewed first. A boost converter is used as an example which is shown in Fig. 2. The bottom switch is the cascode GaN device.

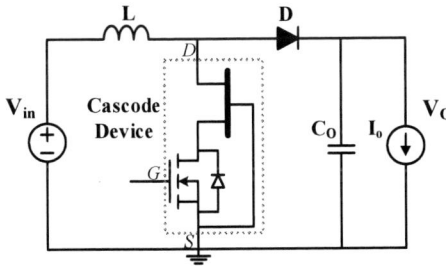

Fig. 2 Boost converter with cascode GaN device

A. Si MOSFET avalanche and GaN internally turn on mechanism in cascode structure [16]

The key waveforms of Si MOSFET reaching avalanche during turn off transition is shown in Fig. 3 and the simplified equivalent circuits are shown in Fig. 4. The turn-off signal is applied at t_0, and C_{OSS_Si} and C_{GS_GaN} are charged up in parallel by turn off current flowing through the channel of GaN until GaN is pinched off at t_2 when V_{DS_Si} which is also V_{GS_GaN} with inverse polarity reaches $-V_{TH\_GaN}$. During stage II ($t_2 \sim t_3$), C_{DS_GaN} is charged in series with C_{OSS_Si} and C_{GS_GaN}. When V_{DS_Si} is driven to avalanche voltage V_A at t_3, V_{DS_GaN} only rises up to V_1, which is lower than its steady state voltage V_O-V_A. The amount of charge stored in C_{DS_GaN} during $t_2 \sim t_3$ is defined as Q_{II} which is the same with the charge stored in C_{OSS_Si} and C_{GS_GaN}. After t_3, Si MOSFET stays in the avalanche status. C_{DS_GaN} is charged through avalanche path independently and the amount of charge during stage III is defined as mismatched charge Q_{III}. This amount of charge flows through the avalanche path, and it causes additional loss which is proportional to the mismatch charge, switching frequency, and avalanche voltage.

The ZVS turn-on transition is the reverse process of turn-off transition. The key waveforms are shown in Fig. 5 and the equivalent circuits are shown in Fig. 6. During $t_0 \sim t_1$, C_{DS_GaN} is discharged in series with C_{OSS_Si} and C_{GS_GaN}. When V_{DS_Si} decreases to V_{TH_GaN} at t_1, V_{DS_GaN} only decreases to V_2. Total charge being discharged during this period is Q_{II}, which is exactly the same as the charge stored during stage II of turn-off process. After t_1, GaN switch is triggered to turn on internally, and remaining charge Q_{III} is dissipated through the channel directly. After all the charge been dissipated, the remaining

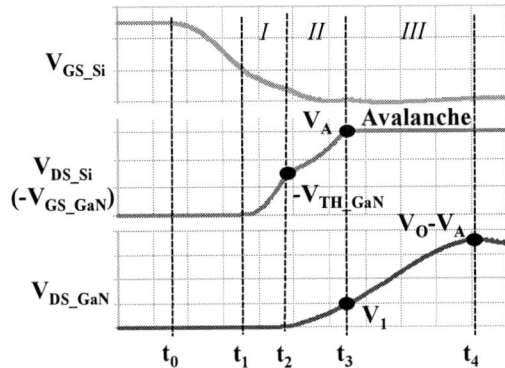

Fig.3 Voltage waveforms of cascode GaN device during turn-off transition period

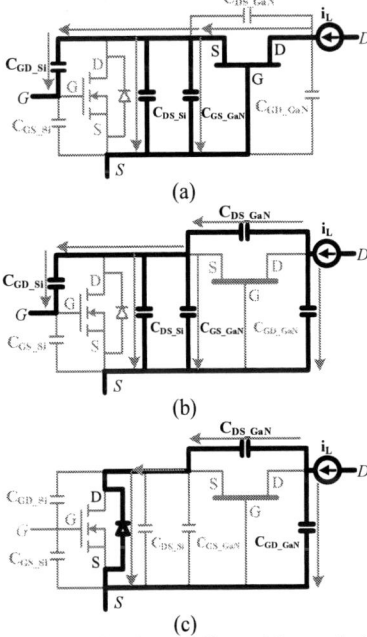

Fig. 4 Equivalent circuit of turn-off transition period (a) Stage I: t_1-t_2 (b) Stage II: t_2-t_3 (c) Stage III: t_3-t_4

external current continues to discharge C_{OSS_Si} and C_{GS_GaN} to almost 0V. The waveform of V_{DS_GaN} makes the terminal waveform of the cascode GaN device appear to have ZVS turn-on. However, the majority of the energy stored in C_{DS_GaN} is actually dissipated internally due to a mismatch in charge. This phenomenon always occurs, regardless of what kind of ZVS techniques are applied. The internal switching loss is related to the mismatch charge and switching frequency.

B. Divergent oscillation issue of cascode GaN device under high current turn-off condition

The Si MOSFET avalanche and GaN internal switching loss may increase the additional loss significantly. The divergent oscillation caused by the capacitance mismatch may be fatal to the device and circuit. The key waveforms of divergent oscillation are shown in Fig. 7 and the simplified equivalent circuits are shown in Fig. 8.

978-1-4673-9551-9/16 $31.00 © 2016 IEEE

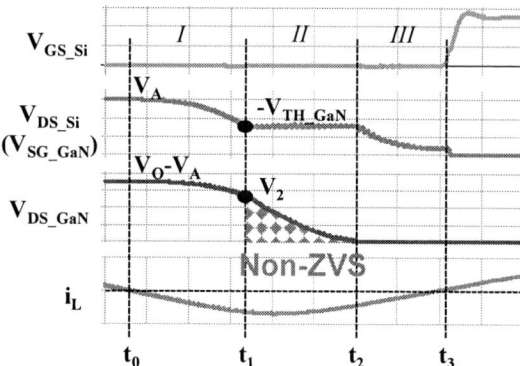

Fig. 5 Voltage waveforms of cascode device during ZVS turn on transition period

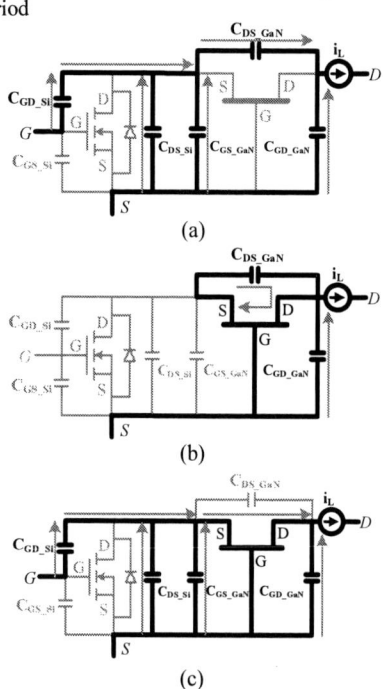

Fig. 6 Equivalent circuit of ZVS turn-on transition period (a) Stage I: t_0-t_1 (b) Stage II: t_1-t_2 (c) Stage III: t_2-t_3

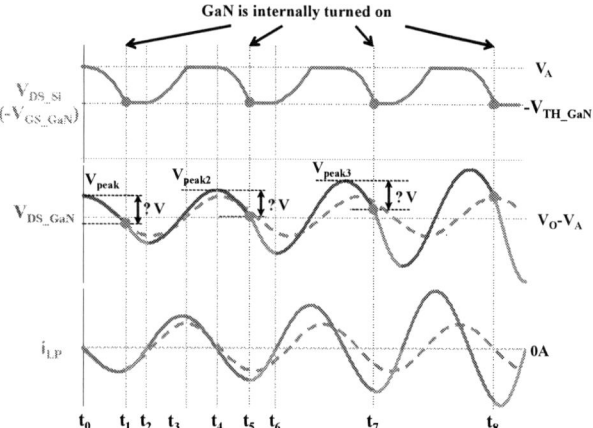

Fig. 7 Key waveforms of divergent oscillation during ringing period right after the cascode GaN device been turned off

Turn off process is the same as mentioned above in section II-A. After Si MOSFET reaches avalanche, V_{DS_GaN} is charged up to V_{peak}, and turn off transition is over. Junction capacitance resonates with L_P and the initial voltage of V_{DS_GaN} and V_{DS_Si} is V_{peak} and V_A respectively. The initial oscillation current of L_P is 0A. Ideal oscillation waveforms are the dash lines shown in Fig. 7. Detailed operation are described as follows:

[t_0-t_1]: L_P resonates with two capacitance branches. Branch I is C_{DS_GaN} in series with C_{OSS_Si} and C_{GS_GaN}. Equivalent capacitance value of this branch C_{equ} is shown in equation (1). Branch II is C_{GD_GaN}. The equivalent circuit of this period is shown in Fig. 8(a). C_{DS_GaN} is discharged in series with C_{OSS_Si} and C_{GS_GaN} by part of the loop inductance current i_{LP} during this stage, and very similar to the stage I of ZVS turn on process shown in Fig. 6a. State equations of this period is shown in equation (2), in which V_{DS} is the drain-source terminal voltage of bottom switch, and R_L is the total loop resistance including on-resistance of top switch and current shunt. With large enough ringing amplitude, at t_1 instant, V_{DS_Si} reaches $-V_{TH\_GaN}$, while V_{DS_GaN} only drops ΔV, which is smaller than the ideal resonant peak-peak amplitude. The charge removed in capacitance branch I during this stage is Q_{II} which is the same charge stored during stage II of turn off process mentioned above.

$$C_{equ} = \frac{(C_{OSS\_Si} + C_{GS\_GaN}) \cdot C_{DS\_GaN}}{C_{OSS\_Si} + C_{GS\_GaN} + C_{DS\_GaN}} \tag{1}$$

$$\begin{cases} (C_{equ} + C_{GD\_GaN}) \cdot \dfrac{dV_{DS}}{dt} = i_{LP} \\ V_{in} - L_P \cdot \dfrac{di_{LP}}{dt} - i_{LP} \cdot R_L = V_{DS} \\ V_{DS} = V_{GD\_GaN} \end{cases} \tag{2}$$

[t_1-t_2]: After t_1, GaN is internally turned on and C_{DS_GaN} is by-passed by GaN channel directly. Therefore, the loop inductance L_P only resonates with branch II C_{GD_GaN} and the resonant structure changes to Fig. 8(b). State equations of this period is shown in equation (3). The resonant period reduces due to smaller capacitance, and characteristic impedance becomes larger. With the same current and voltage initial conditions at t_1, larger characteristic impedance means larger resonant amplitude of capacitance voltage. So, the voltage drop of V_{DS} increases compared with ideal case. V_{GS_GaN} remains almost constant during this stage with the same reason described in [16]. Although C_{DS_GaN} does not participate in oscillation during [t_1-t_2], V_{DS_GaN} and V_{GD_GaN} still satisfy the KVL law which is shown in equation (4). Therefore, at t_2, i_{LP} reaches 0A, and V_{DS_GaN} reaches the valley point which is lower than the ideal case.

$$\begin{cases} C_{GD\_GaN} \cdot \dfrac{dV_{DS}}{dt} = i_{LP} \\ V_{in} - L_P \cdot \dfrac{di_{LP}}{dt} - i_{LP} \cdot R_L = V_{DS} \end{cases} \tag{3}$$

$$V_{DS\_GaN} = V_{GD\_GaN} - |V_{TH\_GaN}| = V_{DS} - |V_{TH\_GaN}| \tag{4}$$

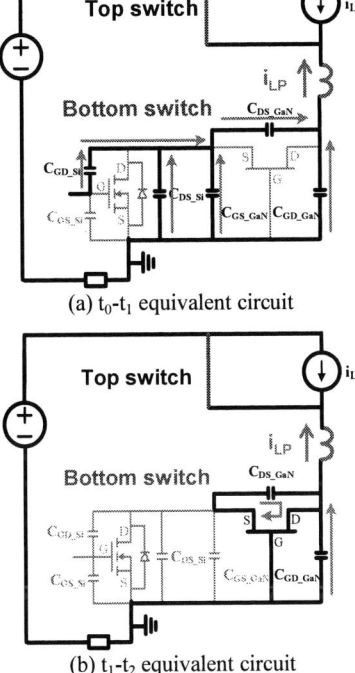

(a) t_0-t_1 equivalent circuit

(b) t_1-t_2 equivalent circuit

Fig. 8 Equivalent oscillation circuit during ringing period

[t_2-t_3]: After t_2, the next oscillation period starts. The value of V_{DS_Si}, V_{DS_GaN} and i_{LP} at t_2 instant become the initial conditions of the next oscillation cycle. V_{DS_Si} reaches avalanche at t_3, and the charge stored in C_{DS_GaN} during this period is Q_{II}, which is the same as the charge removed during [t_0-t_1].

[t_3-t_4]: V_{DS_Si} stays in avalanche region and V_{DS_GaN} continues increasing to the second peak value V_{peak2}. V_{peak2} should be larger than V_{peak}, and the resonant current also increases due to lower voltage initial condition.

[t_4-t_5]: This stage is similar to stage [t_0-t_1]. V_{DS_Si} and V_{DS_GaN} decrease simultaneously, and V_{DS_Si} reaches $-V_{TH\_GaN}$ when Q_{II} is removed by resonant current at t_5. GaN is internally turned on again and relatively earlier than stage [t_0-t_1] due to higher resonant current.

[t_5-t_6]: Resonant structure changes to Fig. 8(b) again. Higher resonant current amplitude results in even lower valley voltage of V_{DS_GaN} at t_6.

In the following oscillation periods, GaN is internally turned on in every oscillation cycle and each turn on instant moves earlier compared to previous cycle. The ringing amplitude of V_{DS_GaN} and i_{LP} increase cycle by cycle due to resonant structure change, and the oscillation eventually becomes divergent.

III. EFFECTIVE SOLUTION TO AVOID DIVERGENT OSCILLATION FOR CASCODE GAN DEVICES

Based on the analysis above, the fundamental reason of the divergent oscillation issue is the capacitance mismatch between Si MOSFET and GaN in cascode configuration. It is triggered

by parasitic ringing at certain amplitude which is determined by turn-off current, loop inductance and junction capacitance.

One way to avoid the divergent oscillation is to parallel RC snubber circuit to suppress the voltage spike and damp the parasitic ringing which is a common practice for devices switching at high current. The simulation results comparison is shown in Fig. 9. It is obvious that the divergent oscillation occurs at 31A without RC snubber circuit, while the devices can safely switch 70A and above with RC snubber circuit. The power dissipation on the RC snubber circuit is around 10uJ which is acceptable for the switching frequency lower than few hundreds of kHz.

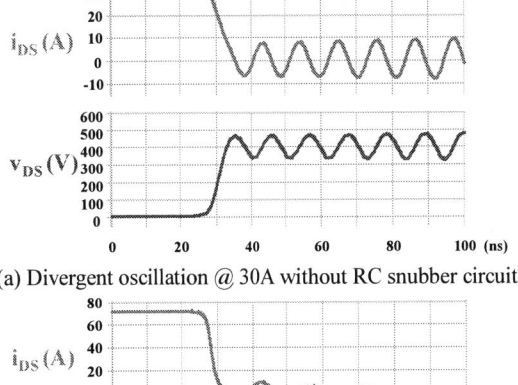

(a) Divergent oscillation @ 30A without RC snubber circuit

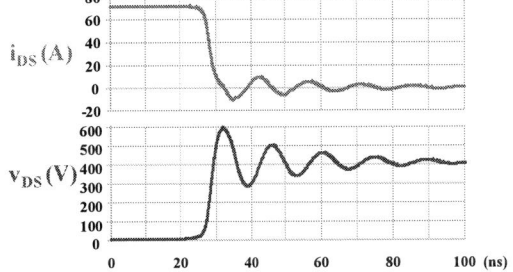

(a) Turn-off current =70A with RC snubber circuit

Fig. 9 Simulation results comparison

The power dissipation becomes significant at MHz frequency range which requires bulky resistance. Moreover, the Si MOSFET avalanche and GaN internal switching loss cannot be solve by adding snubber circuit, which also induce significant loss at high frequency.

A fundamental solution to compensate the capacitance mismatch by adding an additional capacitor C_X is proposed in literature [16]. The Si MOSFET avalanche and GaN internal switching loss can be effectively eliminated. This method will not increase the driving loss and only has very little impact on the turn off loss. As mentioned in section II-A, the total mismatched charge in the cascode device is Q_{III}. The required minimum value of C_X should satisfies the expression as shown below:

$$C_X \geq \frac{Q_{III}}{V_A - V_{TH\_GaN}}$$

(5)

Considering all the parasitic inductance induced by adding C_X, the position of C_X is critical to achieve optimal performance. Fig. 10 shows two possible packaging diagram which integrating the external capacitor C_X.

978-1-4673-9551-9/16 $31.00 © 2016 IEEE

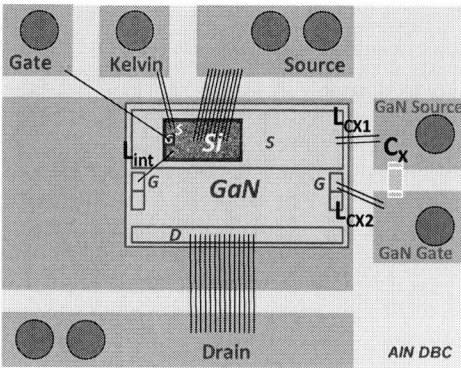

(a) Packaging A

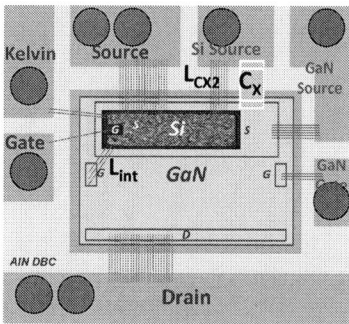

(b) Packaging B

Fig. 10 Two possible packaging approach with C_X

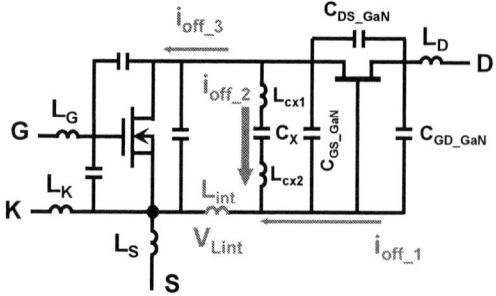

(a) Packaging A

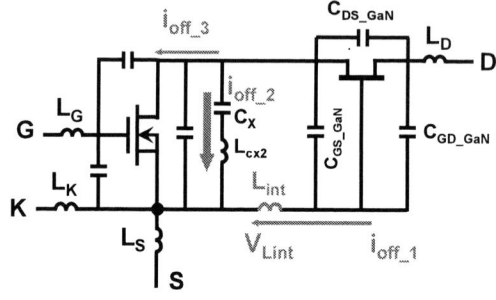

(b) Packaging B

Fig. 12 Current distribution in different packaging

The Si MOSFET is on top of GaN switch source pad in order to reduce inter-connect parasitic inductance. The C_X is on side of GaN switch in packaging A and on top of GaN switch source pad in packaging B, respectively. As shown in the diagram of packaging A, one terminal of the external capacitor C_X is connected to GaN source pad through L_{CX1}. The other terminal is connected to one of the GaN switch gate pad through L_{CX2}, and then connected to the source pad of Si MOSFET through L_{int}. It is noticed that L_{int} is connected through the other gate pad of GaN switch for easy wire bonding consideration. Actually the packaging A is used to demonstrate the benefits of the proposed solution in literature [16]. In packaging B, one terminal of C_X is connected directly to the source pad of GaN switch, and the other terminal is connected to the source pad of Si MOSFET through L_{CX2}. It is obvious that packaging B has much smaller parasitic inductance. The equivalent circuit considering the parasitic inductance is shown in Fig. 11.

Different positions of C_X may result in different current flowing through the parasitic inductance, and therefore impact the voltage across V_{GS_GaN}. In packaging A, C_X is on the right side of L_{int}. During turn-off transition, most of the turn off current flows through L_{int} which is shown in Fig. 12(a). i_{off_1} is the turn-off current flowing through C_{GD_GaN} and C_{GS_GaN}. i_{off_2} is the current charging C_X. i_{off_3} is the current flowing through Si MOSFET branch. C_X value is typically much larger than any other junction capacitance, therefore, i_{off_2} is much higher than i_{off_3}. It is worthwhile to point out that C_{DS_GaN} is typically much larger than C_{GD_GaN} due to the nonlinear characteristic of GaN junction capacitances. As a result, i_{off_2} is also much larger than i_{off_1}. In packaging B, C_X is on the left side of L_{int}. Only i_{off_1} flows through Lint during turn off period, while i_{off_2} directly flow to the source of Si MOSFET. Equivalent circuit is shown in Fig. 12(b).

It is obvious that packaging B reduces the current flowing through L_{int} which induces less voltage drop V_{Lint}. V_{GS_GaN} is the sum of V_{DS_Si} and V_{Lint}. Smaller V_{Lint} results in smaller resonant amplitude in the GaN gate loop, and as a result, V_{GS_GaN} is far from touching turn-on threshold value which is the trigger of divergent oscillation. Therefore, packaging B is preferred for cascode GaN operating at high current turn-off condition.

IV. EXPERIMENT VALIDATION AND DISCUSSION

To validate the analysis of divergent oscillation and the proposed solution for cascode GaN device, two cascode GaN devices are fabricated with the two packaging diagram mentioned in section III, as shown in Fig. 13. The 600V normally-

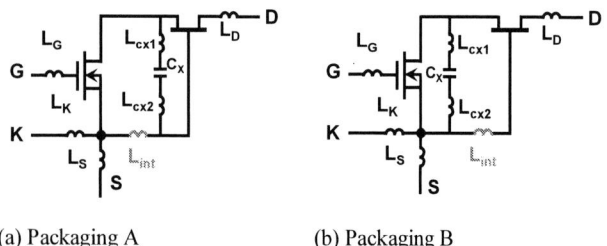

(a) Packaging A   (b) Packaging B

Fig. 11 Equivalent circuit of two different packaging approaches

978-1-4673-9551-9/16 $31.00 © 2016 IEEE   1006

on GaN switch with 35A continuous current capability is provided by Transphorm Inc. The threshold voltage is around -20V and the maximum source-gate voltage is around -40V. Therefore a 30V Si MOSFET is selected to control the on/off state of the GaN as well as to protect the GaN gate by clamping the source-gate voltage of the GaN to the avalanche value of the Si MOSFET, which is 30V. Another criterion for choosing the Si MOSFET is to minimize the driving loss at high frequency. Therefore the junction capacitance of this 30V Si MOSFET is usually much smaller than the GaN switch. The estimated mismatch charge is about 33nC, and therefore, a 3.6nF capacitor is added according to (5).

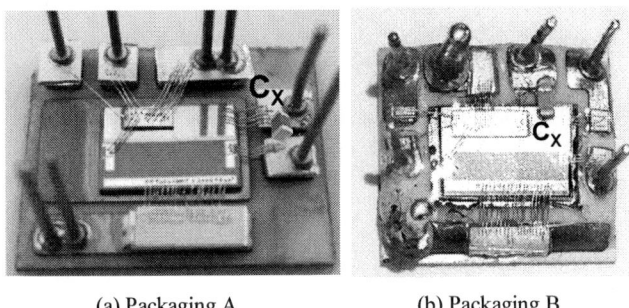

(a) Packaging A   (b) Packaging B

Fig. 13 Prototypes of two different packaging approaches

The Si MOSFET avoids avalanche and the GaN switch achieves true ZVS with the additional capacitor .The experimental verification will not be repeat here. The evaluation of high current turn off condition is carried on double-pulse-test circuit. Input voltage is 400V. Fig.14 shows the experimental waveforms of different packaging approach. Fig. 14(a) shows that the packaging A without C_X occurs divergent oscillation at 30A turn off condition. The packaging A with C_X can switch to slightly higher turn off current, but it still occurs divergent oscillation at 33A turn-off condition as shown in Fig. 14(b). The major reason is the position of C_X as mentioned in section III. The prototype using packaging B approach can successfully switch under 40A turn off condition without oscillation as shown in Fig. 14(c).

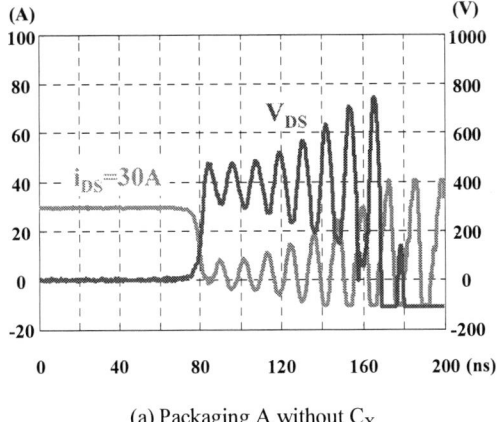

(a) Packaging A without C_X

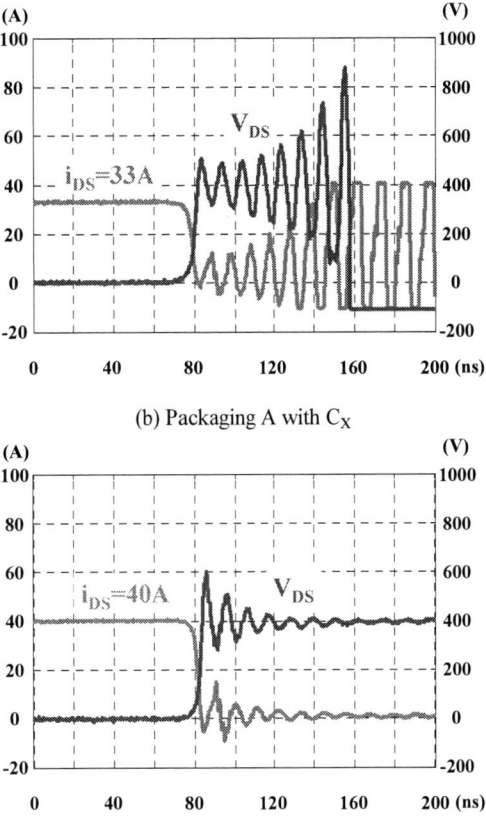

(b) Packaging A with C_X

(c) Packaging B with C_X

Fig. 14 Experimental results with different packaging approach

It is noticed that the voltage spike is close to 600V due to relatively large power loop inductance in order to accommodate the shunt resistance (SSDN-10, T&M Research Products Inc.) on the PCB board. The turn-off current can go up to 48A by removing the shunt resistance as shown in Fig. 15. The red curve is the inductor current. The parasitic ringing period reduces from 12ns to 8.5ns based on Fig. 14(c) and Fig. 15. It indicates about 50% reduction in the power loop inductance by removing the shunt resistance. The voltage spike is about 570V.

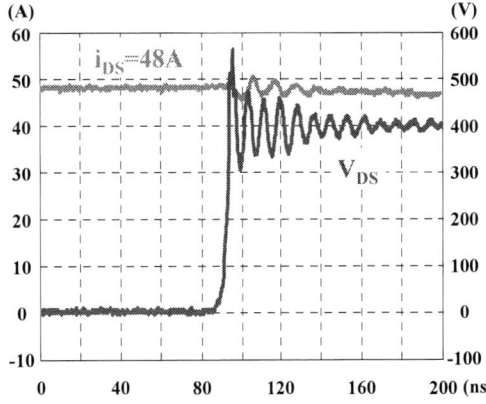

Fig. 15 Experiment with packaging B, without shunt resistance

In order to further reduce the power loop inductance, a half-bridge module is fabricated, as shown in Fig. 16. The loop inductance reduces about 50% compared to the one with two discrete devices. The devices are capable to switch at 54A turn-off current, as shown in Fig. 17. The proposed solution of adding C_X at right position can significantly extend the high turn-off current capability.

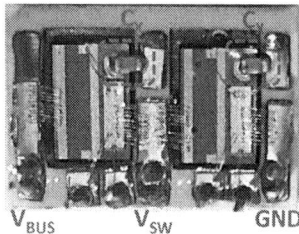

Fig. 16 Half bridge module of cascode GaN devices with packaging B approach

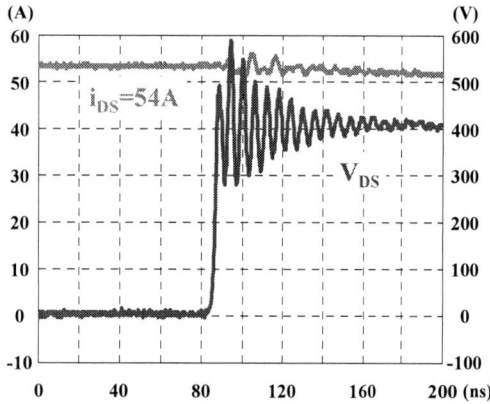

Fig. 17 Experiment with half bridge module of cascode GaN devices with packaging B approach

V. CONCLUSION

Cascode structure is widely used for high voltage normally-on GaN devices. However, the capacitance mismatch between GaN switch and Si MOSFET may result in divergent oscillation issue at high current turn-off condition. The amplitude of voltage ringing formed by loop inductance and junction capacitance may exceed the threshold and trigger GaN internally turn on mechanism during ringing period after cascode device is turned off, and therefore leads to a divergent oscillation. Adding an additional capacitor C_X in parallel with Si MOSFET can compensate the mismatched charge and solve the problem. Considering all the parasitic inductances induced by adding C_X, the position of C_X is critical to achieve optimal performance. Experiments results validate the theoretical analysis and effectiveness of proposed method. The proposed method extends the high current turn-off capability and improves device performance significantly.

It is worthwhile to point out that the theoretical analysis and proposed solution are applicable to all kind of cascode devices.

ACKNOWLEDGMENT

This material is based upon work supported by the U.S. Department of Energy, Office of Energy Efficiency and Renewable Energy, under the project of High Efficiency High Density GaN Based 6.6kW Bi-Directional On-board Charger for PEVs, through Delta Products Corp.

This work was supported in part by the Power Management Consortium in CPES, Virginia Tech.

This work was conducted with the use of GaN device samples by Transphorm of the CPES Industry Consortium Program.

REFERENCES

[1] M. A. Khan, G. Simin, S. G. Pytel, A. Monti, E. Santi, and J. L. Hudgins, "New developments in gallium nitride and the impact on power electron-ics," in *proc. IEEE PESC*, 2005, pp. 15-26.

[2] N. Kaminski, "State of the art and the future of wide band-gap devices," in Proc. IEEE Power Electronics and Applications, 2009, pp. 1–9.

[3] Y. Wu, M. J. Mitos, M. Moore, and S. Heikman, "A 97.8% Efficient GaN HEMT boost converter with 300W output power at 1 MHz," *IEEE Elec-tron Device Letters*, vol. 29, no. 8, pp. 824–826, Aug. 2008.

[4] B. Hughes, J. Lazar, S. Hulsey, D. Zehnder, D. Matic, and K. Boutros, "GaN HFET switching characteristics at 350V-20A and synchronous boost converter performance at 1MHz," in *proc. IEEE APEC*, 2012, pp 2506-2508.

[5] J. Delaine, P. Olivier, D. Frey, and K. Guepratte, "High frequency DC-DC converter using GaN device," in *proc. IEEE APEC*, 2012, pp 1754-1761.

[6] L. Zhou, Y-F. Wu, and U. Mishra, "True Bridgeless Totem-pole PFC based on GaN HEMTs", *PCIM Europe* 2013, pp.1017-1022.

[7] X. Zhang, C.Yao, X. Lu, E. Davidson, M. Sievers, M. J. Scott, P. Xu, and J. Wang, "A GaN transistor based 90W AC/DC adapter with a buck-PFC stage and an isolated Quasi-switched-capacitor DC/DC stage," in *proc. IEEE APEC*, 2014, pp 109-116.

[8] Z. Liu, X. Huang, W. Zhang, F.C. Lee, Q. Li, "Evaluation of high-voltage cascode GaN HEMT in different packages," *in proc. IEEE APEC*, 2014, pp.168-173.

[9] S. She, W. Zhang, X. Huang, W. Du, Z. Liu, F.C. Lee, and Q. Li, "Thermal analysis and improvement of cascode GaN HEMT in stack-die structure," *in proc. IEEE ECCE 2014*, pp.5709-5715.

[10] Z. Liu, X. Huang, F.C. Lee, and Q. Li, "Package parasitic inductance extraction and simulation model development for the high-voltage cascode GaN HEMT,"*IEEE Trans. on Power Electron.*, vol.29, no.4, pp.1977-1985, Apr. 2014.

[11] X. Huang, Z. Liu, Q. Li, and F C. Lee, "Evaluation and application of 600 V GaN HEMT in cascode structure," IEEE Trans. on Power Electron., vol. 29, no. 5, pp.2453-2461, May. 2014.

[12] X. Huang; Q. Li; Z. Liu, and F.C. Lee, "Analytical loss model of high voltage GaN HEMT in cascode configuration," *IEEE Trans. on Power Electron.* vol.29, no.5, pp.2208-2219, May 2014.

[13] X. Huang, Z, Liu, F. C. Lee, and Q. Li, "Characterization and enhancement of high-votlage cascode GaN devices," IEEE Trans. on Electron Devices, vol. 62, no. 2, pp.270-277, Feb. 2015.

[14] X. Huang, T. Liu, B. Li, F. C. Lee, and Q. Li, "Evaluation and applications of 600V/650V enhancement-mode GaN devices," in *Proc. IEEE WiPDA*, 2015, to be published.

[15] Z. Liu, F. C. Lee, Q. Li, and Y. Yang, "Design of GaN-based MHz totem-pole PFC rectifier," in *proc. IEEE ECCE*, 2015, pp 682-688.

[16] X. Huang, W. Du, F. C. Lee, Q. Li, and Z. Liu, "Avoiding Si MOSFET Avalanche and Achieving Zero-Voltage-Switching for Cascode GaN Devices", *IEEE Trans. on Power Electron.*, vol. 31, no. 1, pp.593-600, Jan. 2016.

[17] W. Zhang, X. Huang, Z. Liu, F. C. Lee, S. She, W. Du, and Q. Li, "A New Package of High-Voltage Cascode Gallium Nitride Device for Megahertz Operation," *IEEE Trans. on Power Electron.*, vol. 31, no. 2, pp.1344-1353, Feb. 2016.

[18] B. J. Baliga, and M. S. Adler, "Composite circuit for power semiconductor switching," European patent EP 0063749 B1, 1982.

[19] R. Siemieniec, G. Nobauer, and D. Domes, "Stability and performance analysis of a SiC-based cascode switch and an alternative solution," Microelectronics Reliability, nol. 52, no. 3, pp 509-518, Mar, 2012.

Temperature-Dependent Turn-On Loss Analysis for GaN HFETs

Edward A. Jones, Fred Wang, Daniel Costinett, Zheyu Zhang
Center for Ultra-wide-area Resilient Electric Energy Transmission Networks (CURENT)
Dept. of Electrical Engineering and Computer Science
University of Tennessee
Knoxville, TN
edward.jones@utk.edu

Ben Guo
United Technologies Research Center
East Hartford, CT
guob@utrc.utc.com

Abstract— **Enhancement-mode GaN HFETS enable efficient high-frequency converter design, but this technology is relatively new and exhibits different characteristics from Si or SiC MOSFETs. GaN performance at elevated temperature is especially unique. Turn-on time increases significantly with temperature, and turn-on losses increase as a result. This phenomenon can be explained based on the relationships between junction temperature and GaN device transconductance, and between transconductance and turn-on time. An analytical relationship between temperature and turn-on loss has been derived for the 650-V GS66508 from GaN Systems, and verified with experimental results. Based on this relationship, a detailed model is developed, and a simplified scaling factor is proposed for estimating turn-on loss in e-mode GaN HFETs, using room-temperature switching characterization and typically published datasheet parameters.**

Keywords— *GaN; Gallium Nitride; wide bandgap; double pulse test; device characterization; turn-on loss; transconductance*

I. INTRODUCTION

GaN heterojunction field-effect transistors (HFETs), also known as high electron mobility transistors (HEMTs), take advantage of the dense layer of electrons (2DEG) formed at the heterojunction between GaN and AlGaN to provide fast switching with lower switching losses than conventional Si MOSFETs. But in order to take advantage of these new power devices for high-frequency converter design, it is important to understand their unique characteristics and limitations [1].

One distinction between Si, SiC, and GaN semiconductors is their behavior at elevated junction temperature. Both Si and SiC exhibit little change in switching loss at higher temperature. But enhancement-mode GaN devices have been shown to experience higher turn-on loss at elevated junction temperature, requiring a robust cooling design to prevent thermal runaway [2],[3]. This points to a fundamental difference between GaN and other power semiconductor materials.

The effect of temperature on GaN transconductance has been well-studied in prior literature, and is generally attributed to the decrease in 2DEG density and quantum well depth [4]. This work will show that the increased turn-on losses can be explained by the negative temperature coefficient of transconductance for these GaN devices. A relationship will

be derived between junction temperature and turn-on loss, and verified experimentally with the GaN Systems GS66508 [5]. The results published previously in [2] demonstrate that turn-on and turn-off overlap switching losses increase with temperature, but the increase in turn-on loss is much more significant and should not be ignored in GaN-based converter design.

Performing DPT at elevated temperature is very challeging with GaN devices, because of the chip-scale or PQFN packaging commonly used rather than bulkier TO-220 or TO-247 packages that are easier to heat. Elevating the temperature of the GaN device with a hot plate requires a connection on the top-side of the device or through the PCB, neither of which is convenient in a DPT. Alternatively, placing the entire board in an oven to heat the device compromises the other components, as well as affecting the integrity of measurement probes. Therefore, it is more desirable to perform the DPT at room temperature and scale the measured losses based on available static parameters.

The switching transients and losses of Si power MOSFETs have been well studied, but the interactions between device and board parasitics makes it difficult to analytically predict switching losses without experimental data [6]. For this work, DPT experiments were performed at elevated temperatures, and a temperature-dependent model is proposed to predict the turn-on losses as junction temperature increases.

II. THEORETICAL ANALYSIS OF THE TURN-ON TRANSIENT

The typical turn-on transient for a phase leg with GaN HFETs or other fast-switching devices is shown in Fig. 1.

Initially, the load current flows through the synchronous device, either in a body diode, antiparallel diode, or the GaN diode-like reverse channel. This stage continues until the active device gate voltage reaches its threshold, and the active device channel begins to conduct in the saturation region.

Next, the load current commutates from the synchronous device to the active device during the current rise time, t_{cr}. Here, the voltage on the synchronous device is clamped to zero, so it cannot begin to block any voltage yet. The parasitic power loop inductance blocks some of the dc voltage during this subinterval, determined by $L_{loop}di/dt$. With a dense board

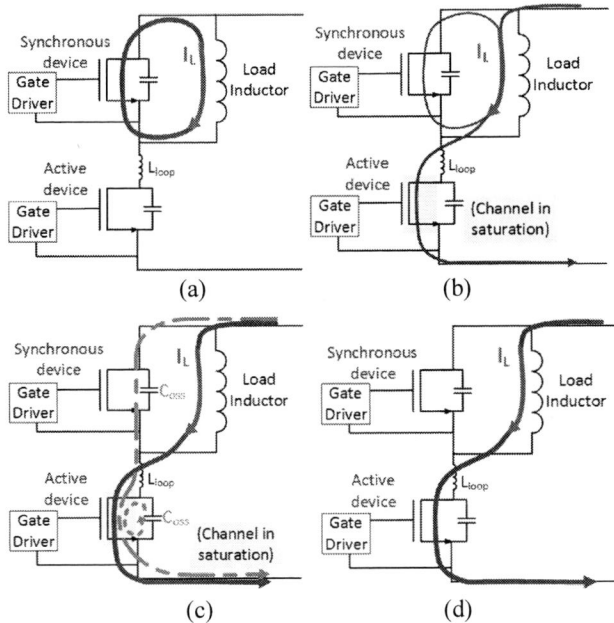

Fig. 1. Current in a phase leg during a hard turn-on transient with fast-switching devices. (a) Before turn-on transient and during turn-on delay time; (b) During current rise time; (c) During voltage fall time; (d) After turn-on transient is complete.

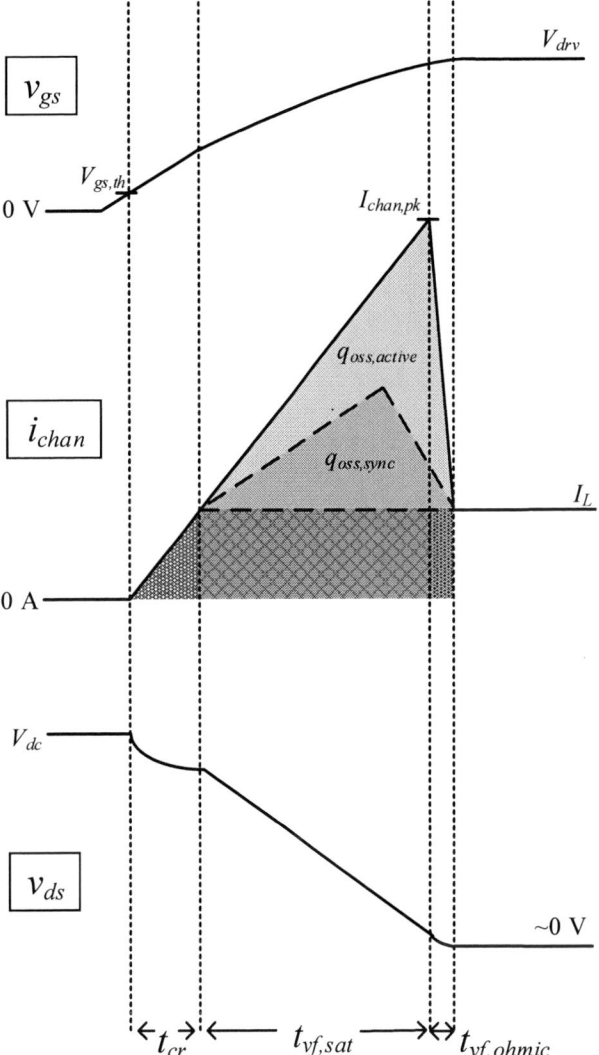

Fig. 2. Turn-on transient of the active device.

design, this drop should be relatively small compared to V_{dc}. Once the current is fully commutated to the active device channel, the synchronous device can block dc voltage, and the voltage fall time t_{vf} begins. This point is also defined as the moment when v_{gs} reaches the Miller voltage, V_M.

During the voltage fall time, the output capacitance in the active device is discharged into its saturated channel, and this energy is lost. Simultaneously, the output capacitance of the synchronous device is charged with current from the dc bus and decoupling capacitors, and this displacement current flows through the saturated active device channel and incurs additional loss. Once the switch-node voltage reaches 0 V, the turn-on transient is complete. At this time, there may be some ringing between C_{oss} and L_{loop} as the switch-node voltage settles to $I_L R_{ds\text{-}on}$. In this analysis, the consequences of this ringing are neglected, because experimental results for e-mode GaN have shown that the net energy losses during this time are very small [2].

Fig. 2 shows the theoretical turn-on waveforms divided into these subintervals. The lightly shaded area shows the charge associated with displacement current in the output capacitors of the two devices, which contributes a fixed amount of turn-on energy loss for a given dc voltage, regardless of load current or temperature. The magnitude of this displacement current component determines the rate of change of voltage in both devices, as given by

$$\frac{dv_{ds}}{dt} = \frac{i_{coss}}{C_{oss}} . \tag{1}$$

The area with hatched shading shows the charge associated with conduction of the load current through the

saturated channel during the turn-on transient, which contributes energy loss that is dependent on the time duration of the transient. Because it is operating in the saturation region, the total current in the active device channel is determined by the gate voltage, threshold voltage, and device transconductance as in

$$i_{chan,sat} = g_{fs}(v_{gs} - V_{gs,th}) . \tag{2}$$

The channel current can also be described as the sum of the load current and the two C_{oss} displacement currents,

$$i_{chan,act} = I_L + i_{coss,sync} + i_{coss,act} . \tag{3}$$

When v_{ds} has dropped to nearly zero at the end of $t_{vf,sat}$, the device exits saturation and enters the ohmic region. Following this, there is a short subinterval $t_{vf,ohmic}$ during which the voltage and current both drop. Nearly all of the C_{oss} charge displacement occurs during $t_{vf,sat}$, so this ohmic subinterval is very short and is ignored for this analysis. The derived equations for current rise time and voltage fall time are given in (4) and (5), respectively.

$$t_{cr} = \frac{I_L}{g_{fs}\left(\frac{dv_{gs}}{dt}\right)} \qquad (4)$$

$$t_{vf,sat} = \sqrt{\frac{4\,q_{oss}}{g_{fs}\left(\frac{dv_{gs}}{dt}\right)}} \qquad (5)$$

The losses due to the displacement of C_{oss} stored energy in the active and synchronous devices are independent of turn-on time. The time-independent loss in the active device due to the synchronous device C_{oss} can be calculated from (6). The loss component due to the discharge of the active device C_{oss} into its own channel is given in (7) [7].

$$E_{on,Coss,sync} = \int_0^{V_{dc}} C_{oss}(v) \times (V_{dc} - v)\,dv . \qquad (6)$$

$$E_{on,Coss,active} = \int_0^{V_{dc}} C_{oss}(v) \times v\,dv . \qquad (7)$$

The remaining losses during t_{cr} and $t_{vf,sat}$ are strongly dependent on the duration of the turn-on transient. During t_{cr}, the voltage is nearly constant at V_{dc} while the channel current rises linearly up to the load current. The time-dependent energy loss during this time can be calculated as

$$E_{on,cr} = \frac{1}{2} t_{cr}(V_{dc}I_L) . \qquad (8)$$

During $t_{vf,sat}$, the channel conducts the full load current while the voltage drops from V_{dc} to nearly 0 V. The voltage during $t_{vf,sat}$ is determined by the nonlinear C_{oss} characteristic. If the voltage drop is initially approximated as linear, the time-dependent losses during this interval can be estimated with the simplified relationship

$$E_{on,td,vf} \approx \frac{1}{2} t_{vf,sat}(V_{dc}I_L) . \qquad (9)$$

The total turn-on loss can be described as the sum of these four components, but the component given in (7) is generally considered as turn-off loss. This energy is actually lost during the turn-on transient, but it can only be experimentally measured during the turn-off transient, when the energy is stored in the active device C_{oss}. Therefore, the conventional definition of turn-on loss can be described as the sum of (6), (8), and (9), as

$$E_{on} = E_{on,Coss,sync} + E_{on,cr} + E_{on,td,vf} . \qquad (10)$$

III. TEMPERATURE-DEPENDENCE OF THE TURN-ON TRANSIENT

For a given device, the turn-on speed is mainly determined by operating voltage and current, the C_{oss} characteristic, gate driver speed dv_{gs}/dt, and device transconductance g_{fs}. If dv_{gs}/dt and g_{fs} are both relatively high, the device channel will conduct more current, and the C_{oss} charge will be displaced quickly. However, GaN is distinct from Si and SiC in that its transconductance has a significant negative temperature coefficient. For the device under test

(DUT) considered in this analysis, GaN Systems GS66508, g_{fs} was experimentally measured with a curve tracer over the rated temperature range. The calculation for average g_{fs} is given in (11), and the results over the device temperature range are shown in Fig. 3 along with a second-order curve fit.

$$g_{fs} = \frac{I_{dss(vgs=5V)} - I_{dss(vgs=2V)}}{(3\,V)} \qquad (11)$$

The average transconductance drops by about one half over the rated temperature range of the DUT. According to (2), this also reduces the channel current while the device is operating in saturation, and consequently lengthens t_{cr} and $t_{vf,sat}$. As given in (4)-(10), this increases the current rise time and $E_{on,cr}$ by a factor of 2, and the voltage fall time and $E_{on,td,vf}$ by a factor of $\sqrt{2}$.

With elevated temperature, there is a special case in which the analysis in Section II no longer holds. If the $t_{vf,sat}$ becomes long enough that the gate voltage v_{gs} reaches V_{drv} while the device is still displacing C_{oss} charge, then the gate voltage can no longer increase. The channel current will be limited by the driving voltage after that time, as determined by (2). Therefore, the voltage fall time can no longer be determined from (5). Instead, it must be modeled as two distinct subintervals, $t_{vf,sat1}$ and $t_{vf,sat2}$.

This will not generally occur at room temperature, because the recommended driving voltage and room temperature g_{fs} typically allow for a saturation channel current much higher than the rated maximum load current. This special case will occur when

$$t_{cr} + t_{vfsat} > \frac{(V_{drv} - V_{gs,th})}{dv_{gs}/dt} . \qquad (12)$$

The turn-on transient for this special case is shown in Fig. 4. During the first subinterval, (2) still applies, but during the second subinterval, the channel current will be held constant at

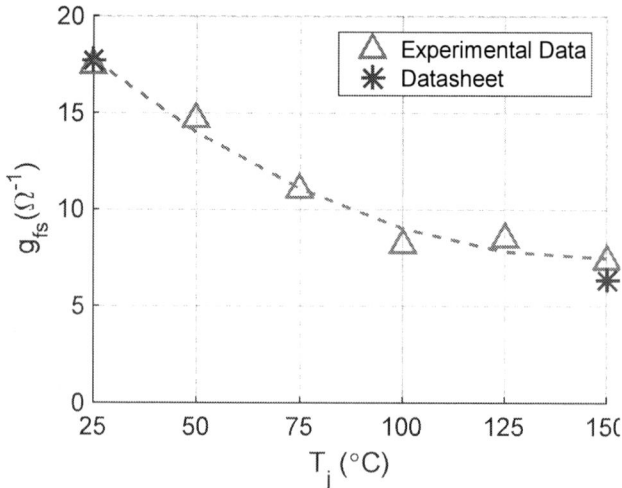

Fig. 3 Average transconductance of the DUT, calculated using (11) based on static testing performed in [2] and the transfer characteristics published in [5].

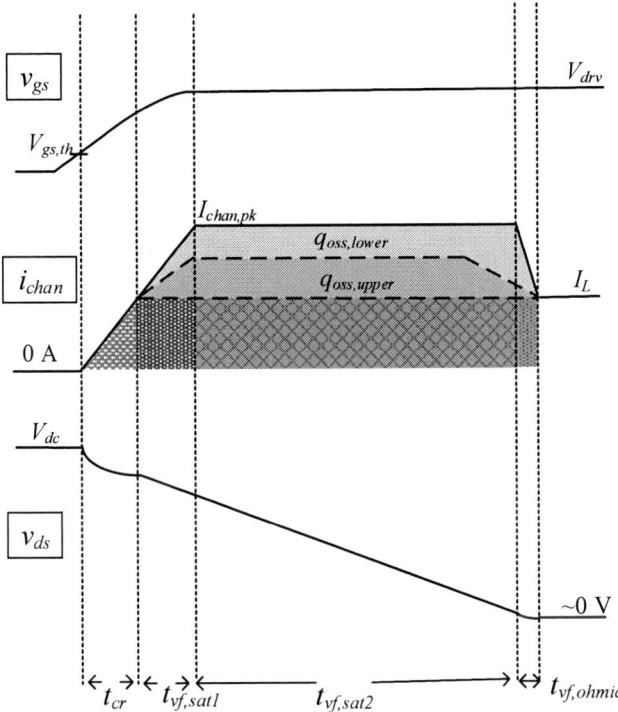

Fig. 4. Turn-on transient analysis of the active device for the special case when (12) is true.

$$i_{chan,tvfsat2} = g_{fs}(V_{drv} - V_{gs,th}) . \qquad (13)$$

Therefore, the displacement current will be fixed during this subinterval, rather than increasing as in $t_{vf,sat1}$. The C_{oss} charge will be displaced at a fixed, non-increasing rate, and the new subinterval $t_{vf,sat2}$ can become quite long. The durations of $t_{vf,sat1}$ and $t_{vf,sat2}$ can be calculated as derived in (14)-(17).

$$t_{vf,sat1} = \frac{V_{drv} - V_{gs,th}}{dv_{gs}/dt} - t_{cr} \qquad (14)$$

$$q_{sat1} = \frac{1}{2} g_{fs} \frac{dv_{gs}}{dt} t_{vf,sat1}^2 \qquad (15)$$

$$q_{sat2} = 2q_{oss} - q_{sat1} \qquad (16)$$

$$t_{vf,sat2} = \frac{q_{sat2}}{g_{fs}(V_{drv} - V_{gs,th}) - I_L} \qquad (17)$$

Since the voltage is still dropping from V_{dc} to 0 V during the combined voltage fall time, the linear simplification can still approximate the t_{vf} energy loss as

$$E_{on,vf} \approx \frac{1}{2}(V_{dc}I_L)(t_{vf,sat1} + t_{vf,sat2}) . \qquad (18)$$

IV. CONSIDERATION OF NONLINEAR C_{OSS}

For the analysis in Sections II and III, the voltage drop during the voltage fall time was treated as linear. This was necessary for the theoretical understanding of temperature

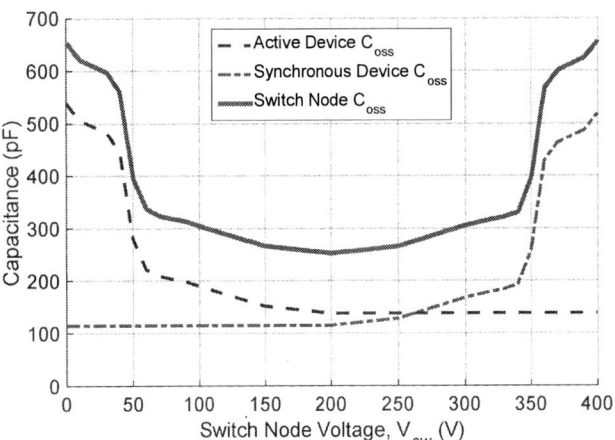

Fig. 5. Output capacitance at the switch-node, with the DUT switching at 400 V.

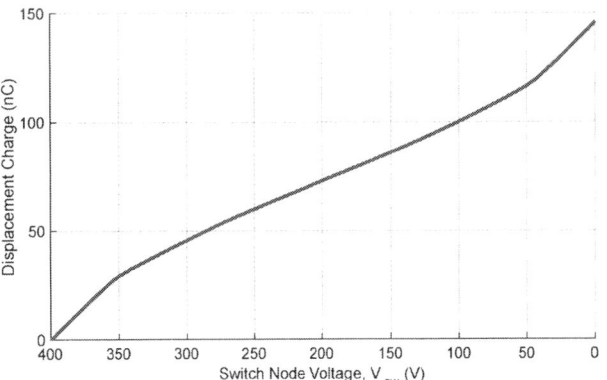

Fig. 6. Charge displaced in the active and synchronous devices as the switching node voltage falls from 400 V.

dependence, but the nonlinear C_{oss} characteristic can be used to improve the accuracy of the model. The linear approximation is inherently an underestimate, because the voltage is exponentially decreasing, and the average voltage during t_{vf} should be higher than $V_{dc}/2$.

The C_{oss} characteristic is given in the datasheet for most devices. For the DUT in this analysis, C_{oss} was measured with a precision impedance analyzer and also published in [5]. Fig. 5 shows the total switch-node capacitance, which is the sum of the active device and synchronous device C_{oss} as the switch-node voltage falls from 400 V to 0 V and is calculated by

$$C_{sw}(v_{sw}) = C_{oss}(v_{sw}) + C_{oss}(V_{dc} - v_{sw}) . \qquad (19)$$

The C_{oss}-related switching losses can be calculated from this curve using (7) and (8), and the total charge displaced q_{sw} can be calculated as a function of switch-node voltage using

$$q_{sw}(v_{sw}) = \int_{v_{sw}}^{V_{dc}} C_{sw}(v)dv . \qquad (20)$$

The total displacement charge in both devices is equal to $2q_{oss}$ and can be found from the q_{sw} curve when $V_{sw}=0$ V.

Accounting for this nonlinear capacitance complicates the turn-on loss analysis, because the calculation requires integration of switch-node voltage over time as in

$$E_{on,\text{td},vf} = I_L \int_0^{t_{vf,sat}} v_{sw}(t)dt \qquad (21)$$

where $v_{sw}(t)$ can be determined from a combination of the v_{sw} vs. q_{sw} characteristic given by (20) and the equation

$$q_{sw}(t) = \frac{1}{2} g_{fs} \frac{dv_{gs}}{dt} t^2 . \qquad (22)$$

This calculation is more accurate, but requires a software tool such as MATLAB to solve. The C_{oss} characteristic can be imported and interpolated, then a discrete-time interval can be used to calculate q_{sw} and $E_{on,\text{td},vf}$.

Similarly, (18) can be revised to

$$E_{on,\text{td},vfsat1} = I_L \int_0^{t_{vf,sat1}} v_{sw}(t)dt \qquad (23)$$

where $v_{sw}(t)$ can be determined from (20) and (22), and

$$E_{on,\text{td},vfsat2} = I_L \int_0^{t_{vf,sat2}} v_{sw}(t)dt \qquad (24)$$

where $v_{sw}(t)$ can be determined from (20) and

$$q_{sw}(t) = q_{sat1} + \left[g_{fs}(V_{drv} - V_{gs,th}) - I_L \right] t . \qquad (25)$$

V. ASSUMPTIONS FOR THE MODEL

A. Average transconductance

In the preceding analysis, the average transconductance was used, however g_{fs} is actually variable during the transient and increases as a function of the gate voltage v_{gs}. To investigate this relationship, the transconductance was measured dynamically for the DUT, using a double pulse test circuit as described in [2]. In a typical DPT, the externally measured v_{gs} is not an accurate representation of the true internal v_{gs} due to the voltage drop across the internal R_g and L_{gs} of the device package. For this dynamic g_{fs} characterization, an external gate resistance of 1 kΩ was used so that the voltage drop across the internal device impedance was negligible in comparison. Using these v_{gs} and I_{dss} measurements, a dynamic g_{fs} vs. v_{gs} characteristic was determined during the current rise time. Fig. 7 shows the g_{fs} calculated from two turn-on transients and two turn-off transients, demonstrating the positive correlation between g_{fs} and v_{gs}. However, the average g_{fs} between 2 V and 5 V matches the static measurement shown in Fig. 3 at 25 °C.

B. Constant gate current

The slope dv_{gs}/dt was approximated as a constant value for this analysis, implying a constant gate current during the transient. However, this slope will drop exponentially as the gate voltage approaches the driving voltage. Because g_{fs} increases with gate voltage and dv_{gs}/dt decreases, the product $g_{fs}dv_{gs}/dt$ remains fairly constant during the transient. It would improve the accuracy of the model to consider both dv_{gs}/dt and g_{fs} as v_{gs}-dependent variables, but this would add considerable complexity. Instead, it is effective to use the average values for both.

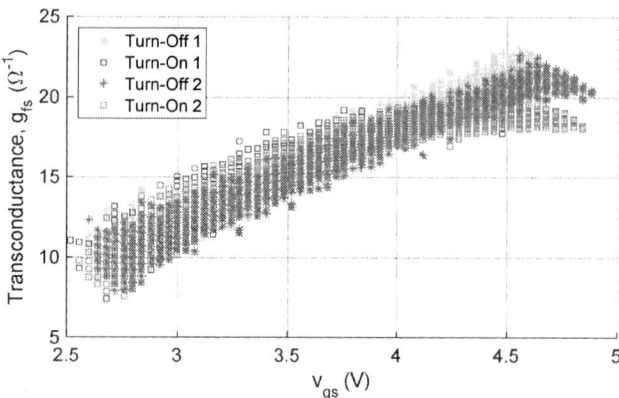

Fig. 7. Transconductance of the device under test measured during dynamic testing at 25 °C.

C. Negligible Miller effects

During the voltage fall time, the Miller capacitance C_{gd} will also discharge as C_{ds} discharges. The C_{oss} characteristic represents the sum of these two capacitances in parallel. While the C_{ds} component can discharge directly into the channel, the displacement current that discharges C_{gd} into the channel may also begin to discharge C_{gs}. In conventional devices, this is referred to as the Miller plateau, since the gate voltage is relatively constant during this time. In a GaN device, a fast gate drive circuit can minimize the Miller effect, so that dv_{gs}/dt is only slightly reduced during the voltage fall time $t_{vf,sat}$. This requires a gate driver IC that can provide sufficient current, and the total impedance of the gate loop must be minimized with low resistance and a dense layout. Common-source inductance is also critical, so a Kelvin gate connection should be used when possible.

Even if these conditions are met, there may still be significant ringing in the v_{gs}, v_{ds}, and i_{chan} waveforms at the end of $t_{vf,sat}$, because C_{gd} gets much larger at low voltage. The ringing on C_{gd} will cause the device to alternate between the ohmic and saturation regions several times before C_{gd} has fully discharged. However, as long as it only occurs near the end of t_{vfsat}, the losses due to this ringing are generally not very significant.

D. Constant threshold voltage

For some power devices, the threshold voltage $V_{gs,th}$ may be affected by temperature. However, the DUT and most GaN devices have a very stable threshold voltage over temperature, so it was not considered as a significant factor in this analysis.

E. Low power loop inductance

The parasitic power loop inductance L_{loop} is usually minimized for GaN-based converters with a dense board design and optimal device packaging. However, L_{loop} still causes an initial drop in V_{sw} during t_{cr}, which may not be negligible. However, this drop occurs on the active device when its voltage is high and its capacitance is at its lowest, and the magnitude of L_{loop} is not affected by temperature, so the effects of L_{loop} were neglected in this analysis.

F. Cross-conduction loss

Lastly, the additional loss in the saturated channel due to cross-conduction in the synchronous device was not considered in this analysis. The DPT experiments for this work used a gate-source shorted synchronous device to minimize cross conduction effects. Additionally, [8] showed that the cross conduction component of turn-on loss is not noticeably impacted by temperature. Cross conduction should not be ignored for turn-on loss estimation, but the magnitude of this loss can be considered separately from the temperature-dependent losses.

VI. LINEAR APPROXIMATION

In GaN-based converter design, it is common to perform a DPT with the chosen device, driver, and layout to characterize the switching losses at the intended operating voltage and current condition. It is therefore desirable to scale these results for the intended operating temperature, without the need to elevate device temperature during testing. Although the detailed model could be used for this purpose, it is useful to consider the linearized model described in (8)-(10). Because the majority of overlap loss occurs during the voltage fall time, the turn-on loss can be estimated for a given temperature using a simple scaling factor derived from (5) and (9), with the approximation

$$E_{on,Tj} \approx E_{on,Coss,sync}$$
$$+ k_{Eon,Tj}\left(E_{on,25C} - E_{on,Coss,sync}\right) \quad (26)$$

where $E_{on,25C}$ is the experimentally measured turn-on loss at the operating current and voltage, $E_{on,Coss,sync}$ is the loss calculated with (6) based on the nonlinear C_{oss} characteristic, and $k_{Eon,Tj}$ is the scaling factor

$$k_{Eon,Tj} = \sqrt{\frac{g_{fs,25C}}{g_{fs,Tj}}} \quad . \quad (27)$$

This scaling factor will typically be between 1 and 2 depending on the GaN device under test, and it can be easily determined from the device datasheet. Typical datasheets show the transfer characteristic at various gate voltages, with 25 °C junction temperature and a higher T_j such as 125 °C or 150 °C. Average transconductance can be calculated from these characteristics using (11) at each temperature, and g_{fs} can be determined at the intended operating temperature with linear interpolation.

This approximation requires no tuning or simulation, but its accuracy heavily depends on the experimental turn-on loss data at room temperature and the C_{oss} characteristic, as well as consideration for any additional capacitance from the PCB and test setup. It is also more limited in application than the detailed model, because it assumes that the voltage fall time is longer than the current rise time. With high current and very low voltage, the scaling factor may begin to approach $k_{Eon,Tj}^2$.

VII. EXPERIMENTAL VERIFICATION

Double pulse testing (DPT) was performed on the DUT at junction temperatures ranging from 25 °C to 150 °C, with an 8 V gate drive and 0 Ω external gate resistance. Experimental

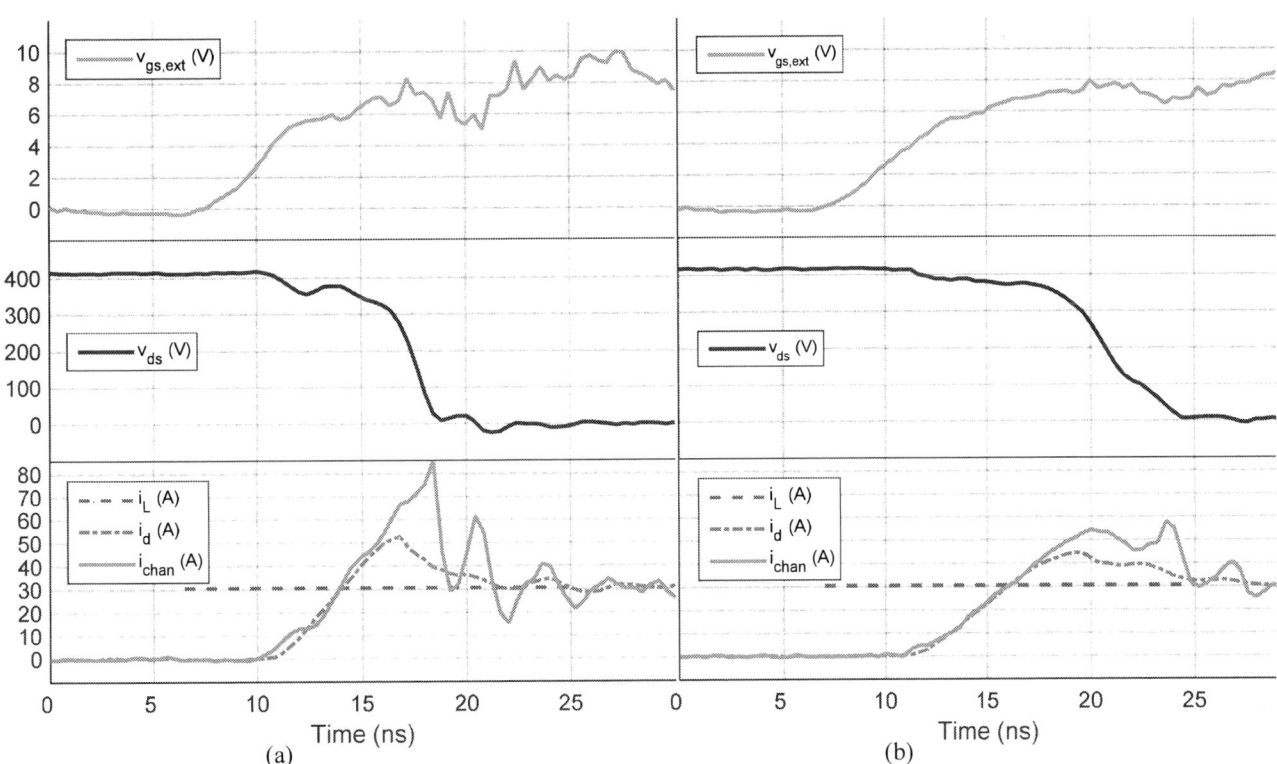

Fig. 8. Experimental turn-on transient waveforms of the active device at 400 V and 30 A, (a) 25 °C and (b) 150 °C.

test setup, results, and device characterization are detailed in [2]. The internal C_{oss} current of the active device was calculated using (1) and added to the measured drain current to estimate the channel current waveform. It is important to note that the measured external gate voltage is only an approximation of the true internal gate voltage, so it is not equivalent to the v_{gs} used in this theoretical analysis. Rather

than rely on this waveform, the value for average dv_{gs}/dt was tuned based on experimental data for turn-on time and loss, and determined to be approximately 1 V/ns in the 400 V switching case.

Fig. 8 shows the turn-on transients at 400 V and 30 A, with junction temperatures of 25 °C and 150 °C respectively. At 25 °C, the device exits the saturation region as the gate

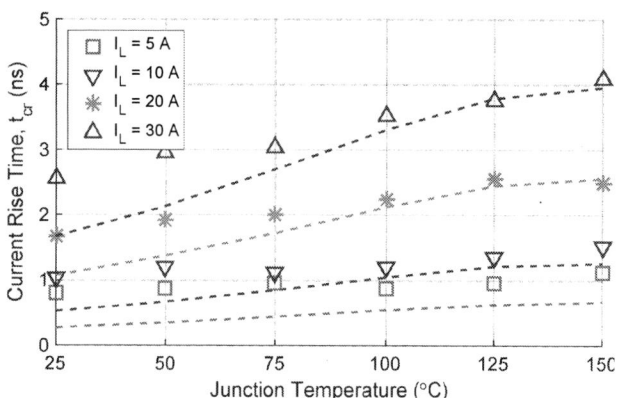

Fig. 9. Experimental results for current rise time of the DUT at 400 V, with the dashed lines showing calculated t_{cr} based on the detailed analytical model.

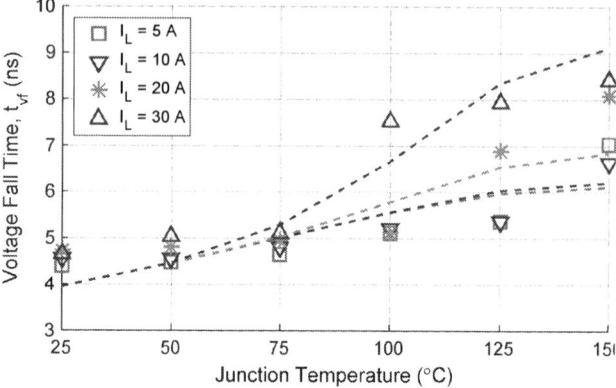

Fig. 10. Experimental results for voltage fall time of the DUT at 400 V, with the dashed lines showing calculated $t_{vf,sat}$ based on the detailed analytical model.

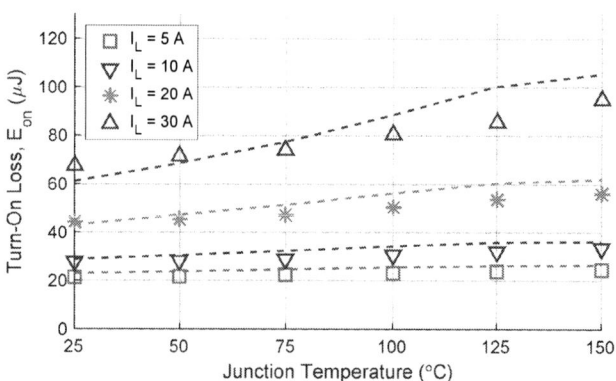

Fig. 11. Experimental results for E_{on} of the DUT at 400 V, with the dashed lines showing predictions by the detailed analytical model.

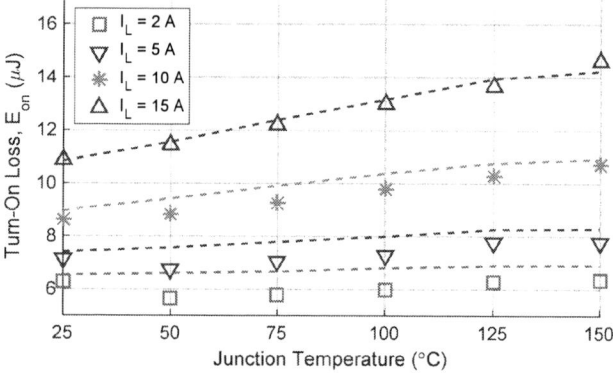

Fig. 12. Experimental results for E_{on} of the DUT at 200 V, with the dashed lines showing predictions by the detailed analytical model.

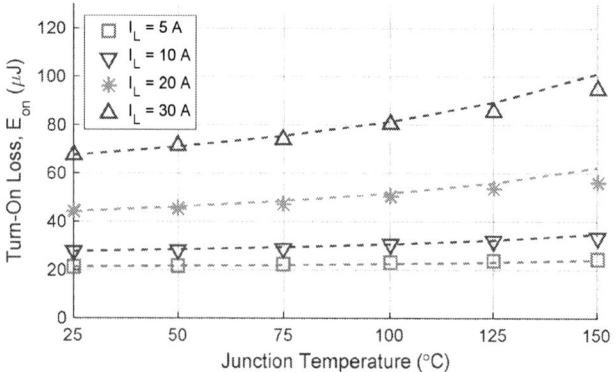

Fig. 13. Experimental results for E_{on} of the DUT at 400 V, with the dashed lines showing the linear approximations calculated using published datasheet parameters and DPT data at 25 °C.

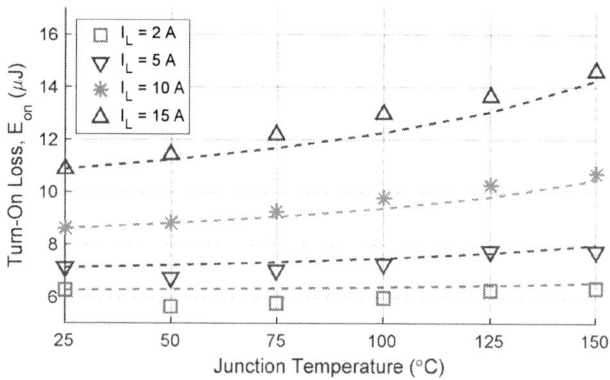

Fig. 14. Experimental results for E_{on} of the DUT at 200 V, with the dashed lines showing the linear approximations calculated using published datasheet parameters and DPT data at 25 °C.

978-1-4673-9551-9/16 $31.00 © 2016 IEEE

voltage is still increasing, reaching a peak channel current of 85 A before the device enters the Ohmic region and the channel current quickly drops to I_L. At higher temperature, v_{gs} reaches the driving voltage before all of the C_{oss} charge has been displaced, in the special case described in Section III. The channel current is then held at ~50 A until the C_{oss} charge has been displaced and the device drops into the ohmic region. In both waveforms, the ringing due to C_{gd} is very pronounced as the device enters the ohmic region, but the net energy loss during this ringing period is low, since v_{ds} has already dropped to nearly 0 V.

Using the derived model, switching times and energy loss were calculated across the current and temperature range at 200 V and 400 V. Experimental results were not used in these calculations, except for tuning dv_{gs}/dt as previously mentioned. Figs. 9-12 compare the experimental results with those predicted by the model. The calculated energy loss matches well with experimental results, although it is shown to be slightly less accurate in some cases. With low voltage or high load current, the assumptions made in Section V begin to break down, especially regarding the Miller effect and the significance of the initial voltage drop across the loop inductance. However, the model matches experimental results very well if the average dv_{gs}/dt is tuned appropriately.

Figs. 13 and 14 show the linear approximations based on (26) and (27), at 400 V and 200 V respectively. This approximation matches experimental results even more closely than the detailed theoretical model, because it is directly based on the DPT results at 25 °C. However, this approximation relies on even more assumptions than the detailed analytical model, so it may not be as accurate for all e-mode GaN devices and converter designs as it is for the DUT considered here.

VIII. CONCLUSION

The root cause of temperature-dependent turn-on loss in GaN devices has been attributed to decreased transconductance. Using theoretical analysis and experimental results, a turn-on loss model was developed to predict the effect of elevated junction temperature at any given current and voltage condition for a GaN device, based on the average transconductance over temperature. A detailed analytical model was developed, as well as a simplified linear approximation. The scaling factor $k_{Eon,Tj}$ allows for typical room temperature DPT switching loss results to be adjusted for any operating temperature using only the transfer characteristic published on the device datasheet. This approximation was shown to be both simple and accurate.

Accurate and comprehensive loss calculation is critical for effective thermal design, to prevent thermal runaway. Because turn-off loss does not change much with temperature, and T_j vs. R_{ds-on} trends are readily available, turn-on loss is the final component needed for calculating total losses. The proposed model closes the loop on GaN-based converter loss characterization, enabling a more robust design.

ACKNOWLEDGMENT

This work made use of facilities supported in part by the Engineering Research Center Program of the National Science Foundation and Department of Energy under NSF Award Number EEC-1041877 and the CURENT Industry Partnership Program. Financial support was also provided by the Bredesen Center for Interdisciplinary Research and Graduate Education.

The authors would like to acknowledge the support and collaboration of Dr. Eugene Solodovnik and Dr. Kamiar Karimi from The Boeing Company; Julian Styles, Di Chen, and Greg Klowak from GaN Systems; Eric Persson, Dr. Wenduo Liu, and Dr. Vickie Zhou from Infineon; and Xiucheng Huang, Bin Lu, and Zhengyang Liu from CPES at Virginia Tech.

REFERENCES

[1] E.A. Jones, F. Wang, B. Ozpineci, "Application-based review of GaN HFETs," in *Proc. IEEE Workshop on Wide Bandgap Power Devices and Applications* (WiPDA), Knoxville, TN, 13-15 Oct. 2014, pp. 24-29.

[2] E.A. Jones, F. Wang, D. Costinett, Z. Zhang, B. Guo, B. Liu, R. Ren, "Characterization of an enhancement-mode 650-V GaN HFET," in *Proc. IEEE Energy Conversion Congress and Exposition* (ECCE), Montreal, CA, Sept. 2015.

[3] M. Danilovic, Z. Chen, R. Wang, F. Luo, D. Boroyevich, P. Mattavelli, "Evaluation of the switching characteristics of a Gallium-Nitride transistor," in *Proc. IEEE Energy Conversion Congress and Exposition* (ECCE), Phoenix, AZ, 17-22 Sept. 2011, pp. 2681-2688.

[4] R. Yahyazadeh, Z. Hashempour, G. Abdollahi, B. Beghdarghi, "Effect of high temperature on the transconductance of AlGaN/AlN/GaN high electron mobility transistors (HEMT)," in *International Journal of Academic Research*, vol. 5, iss. 4, Jul. 2013, pp. 78-86.

[5] GaN Systems, "GS66508P-E05, 650V enhancement mode GaN transistor, Rev. 151016" preliminary datasheet, 2015.

[6] M. Rodriquez, A. Rodriguez, P.F. Miaja, D.G. Lamar, J.S. Zuniga, "An insight into the switching process of power MOSFETs: an improved analytical losses model," in *IEEE Transactions on Power Electronics*, vol. 25, no. 6, June 2010, pp. 1626-1640.

[7] D. Costinett, D. Maksimovic, R. Zane, "Circuit-Oriented Treatment of Nonlinear Capacitances in Switched-Mode Power Supplies," in *IEEE Transactions on Power Electronics*, vol. 30, no. 2, Feb. 2015, pp. 985-995.

[8] E.A. Jones, F. Wang, D. Costinett, Z. Zhang, B. Guo, "Cross conduction analysis for enhancement-mode 650 V GaN HFETs in a phase-leg topology," in *Proc. IEEE Workshop on Wide Bandgap Power Devices and Applications* (WiPDA), Blacksburg, VA, 3 Nov. 2015.

Analysis of Parasitic Elements of SiC Power Modules with Special Emphasis on Reliability Issues

Diane-Perle Sadik, Juan
Colmenares, Hans-Peter Nee
Department of Electrical Energy
Conversion
KTH Royal Institute of Technology
Stockholm, Sweden

Konstantin Kostov
Acreo, Swedish ICT
Kista, Sweden

Florian Giezendanner, Per Ranstad
Alstom Power Sweden AB,
Växjö, Sweden

Abstract — **Commercially available Silicon Carbide (SiC) MOSFET power modules often have a design based on existing packages previously used for silicon insulated-gate bipolar transistors. However, these packages are not optimized to take advantage of the SiC benefits, such as, high switching speeds and high-temperature operation. The package of a half-bridge SiC MOSFET module has been modeled and the parasitic elements have been extracted. The model is validated through experiments. An analysis of the impact of these parasitic elements on the gate-source voltage on the chip has been performed for both low switching speeds and high switching speeds. These results reveal potential reliability issues for the gate-oxide if higher switching speeds are targeted.**

Keywords—Silicon Carbide (SiC), MOSFET, Power Modules, Reliability, Multichip Modules.

I. INTRODUCTION

Silicon carbide (SiC) metal-oxide semiconductor field-effect transistors (MOSFETs) have undergone a great development over the past few years [1]. With the emerging 3rd Generation of SiC MOSFETs, the oxide-layer stability seems to have improved. The main advantages of SiC MOSFETs are the higher maximum switching frequency, lower on-state resistance, and higher maximum operating temperature capability compared to silicon (Si) insulated-gate bipolar transistors (IGBTs).

However, high-speed switching may be compromised due to the strong impact of parasitic inductances and capacitances on switching characteristics of power semiconductor devices [2]–[4]. The design of commercially available SiC modules is often based on existing packages previously designed for Si-IGBTs devices. The design and packaging issues for such a module are presented in [5]. It was found that the difference in placement of the switching devices results in a difference in parasitic inductances for each of the switching devices in the power module. Higher switching speeds and lower ringing can reduce the switching losses, that are a major component of the overall losses of power converters operating at high frequencies [6]. As recently presented in [2], it is possible to build "ultra-low" inductance power modules by mean of an anti-parallel placement of the switching cells inside the module. More recently, it was shown that switching losses can

be reduced by improving the peak-current sourcing capability of the gate-drive circuit [7]. Similarly, an active gate driver suppressing parasitic-induced cross-talk of SiC devices in a phase-leg configuration has been presented in [8].

This paper is assessing the impact of parasitic inductances in the gate loop of the SiC MOSFETs inside the power module housing and the impact of these parasitic inductances on reliability. In fact, it is possible to measure the *external values* of the gate-source voltage and current as well as the drain-source voltage and current from the terminals of the power modules. However, very little has yet been investigated on what occurs *inside*. The objective of this paper is to determine whether there are higher transients inside the housing of the power module compared to what is observable from the outside. Based on this information, it may then be possible to predict whether the internal gate-loop inductance has an impact on the reliability of SiC Power MOSFETs. Higher voltages on the gate can strongly influence the lifetime of the gate oxide of SiC Power MOSFETs [9].

First, in Section II, an extraction of the internal parasitic elements of a commercially available SiC MOSFETs power module has been realized using the software ANSYS Q3D. The extracted parasitic elements are then used to develop an electrical simulation model. Experimental measurements are performed in order to verify the accuracy of the developed model (Section III). The measurements are first performed at the module terminals. Measurements are also performed directly on a chip inside the module in order to obtain a *detailed verification* of the accuracy of the model. Moreover, in Section IV, the impact of the gate-loop inductance on the reliability of SiC Power MOSFETs is discussed based on the simulation model. Finally, Section V concludes this paper.

II. MODELING AND EQUIVALENT CIRCUIT

The investigated module in this paper, is the commercially available CAS300M12BM2 1.2 kV/ 300 A from Cree. It is a half-bridge SiC Power MOSFET module encapsulated in a standard 62 mm package. This package is an existing packaged used for Si-IGBT based modules. Fig. 1 shows a side view of the module reconstruction in ANSYS Q3D. This

978-1-4673-9551-9/16 $31.00 © 2016 IEEE

software has previously been used for estimating the stray inductance of silicon IGBTs [10]–[12] and [2], [5] for the design of an ultra-low inductance SiC MOSFET module. The parasitic elements of the module are extracted from the geometry and material properties of the actual module and the results are capacitance, inductance and resistance matrices at a given frequency. The module is then simulated with LTSPICE using the extracted parasitic elements. Fig. 2 shows a top view of the module where V_{ds} and V_{gs} indicate the location where the on-chip measurements where performed. The extracted parameters reveal that the main inductance in the gate loop is 3 – 4 times higher for the chips situated in the lower switch position compared with those situated in upper switch position. This issue has been previously mentioned in [5] and can be understood by looking at Fig. 2, where the lower switch position is situated on the right half of the module. The gate signal has thus to cross the whole module to reach the chip gate pad situated at the far right of the module. The simulation model is verified with two tests. The following section presents these tests in detail.

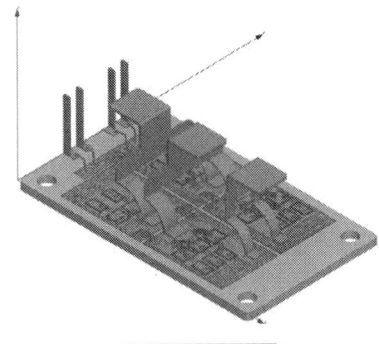

Fig. 1. Q3D model of the CAS300M12BM2

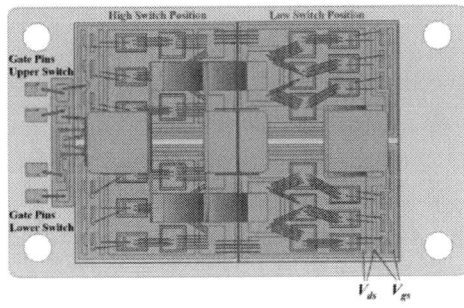

Fig. 2. Top view of CAS300M12BM2

III. EXPERIMENTAL VERIFICATIONS OF THE MODEL

In order to verify the model, two tests have been performed. The first one is a standard double-pulse test (DPT) where the drain-source voltage and gate-source voltage of the lower SiC MOSFET position are measured at the corresponding module terminals. The second one is a single-pulse test where the same voltages are measured directly on the chip as shown in Fig. 2.

In both simulations, the stray inductance L_{g_stray} corresponds to the parasitic inductance of the leads between the gate-driver output and the gate and source terminals of the module. The stray inductance of the dc bus bars has been estimated such that the frequency of the oscillations in the simulation matches the frequency of the oscillations in the experiments. Doing so, it was found that the amplitudes (current and voltage) of the oscillations in the simulation match the ones obtained from the experiments.

A. Double-Pulse Test – measurements on the terminals

The first test is a double-pulse test with the measurement done at the module terminals. In the simulation model, the measurement points are identical to those in the experiment. The parameters of the double-pulse test circuits are given in TABLE I. The double-pulse test measurements are shown in Fig. 3(a) and Fig. 3(b) for the turn-on and turn-off, respectively. The simulation results for the same conditions are shown in Fig. 4(a) and Fig. 4(b). In this case, the experimental results confirm the simulation. It can be noted that the frequency and the amplitude of the oscillations in the simulation are similar to those in the experimentals.

TABLE I.   DOUBLE-PULSE TEST PARAMETERS

Vds	Vgs	Cdc	Rg	L	Lg_stray
500 V	-5/+20 V	400 µF	5 Ω	5 mH	17 nH

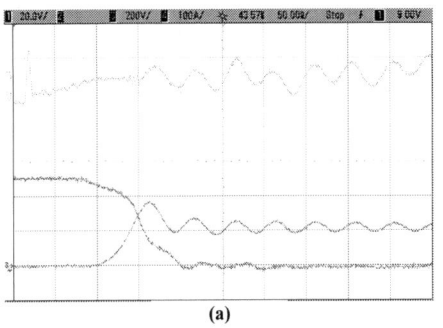

(a)

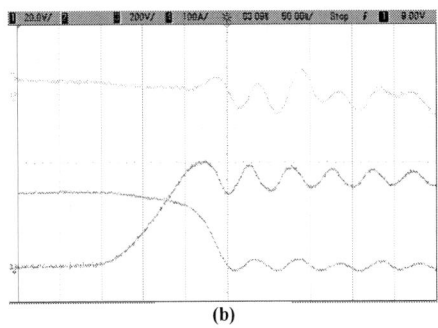

(b)

Fig. 3. Double-pulse test of the SiC MOSFET module, lower switch position. Measurements at the module terminals. Gate-source voltage (yellow, 20 V/div), drain-source (purple 200 V/div), , drain current (pink, 100 A/div). Time base (50 ns/div). (a) turn-on transients, (b) turn-off transients.

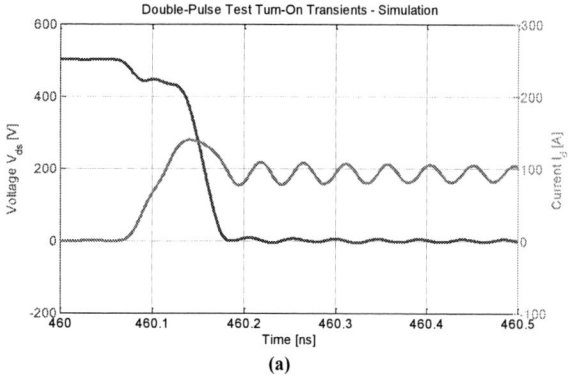

(a)

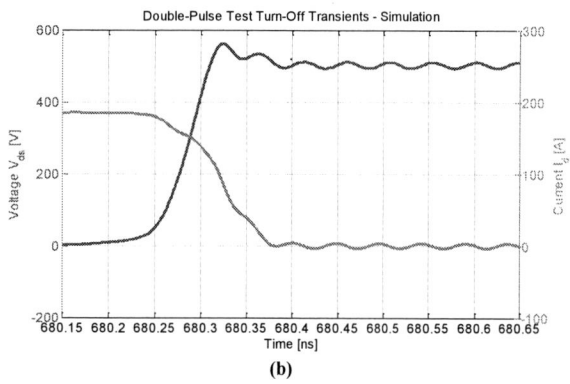

(b)

Fig. 4. Simulation results of the SiC MOSFET Module. Drain-source voltage (blue, 200 /div), drain-source current (red, 100A/div). Time base 50 ns. (a) Turn-on (b) Turn-off transients measured at the module terminals.

B. Single-Pulse Test – measurements on the chip

To obtain a *detailed verification* of the model, a single-pulse test was performed "on-chip". To measure transients directly "on the chip", a special setup has been realized. The module has been opened and the voltage probes are inserted through the silicone gel at the locations indicated in Fig. 2. The measured chip is the 6th chip in the model which is the furthest away from the gate-source terminals of the power module. The drain current is measured at the phase terminal. The setup parameters are stated in TABLE II. , where the parasitic inductance of the bus bars has been estimated. For safety reasons, the transients are measured at 200 V_{dc} / 100 A for V_{ds} and Id. Similarly, the on-chip V_{gs} measurement are done at 100 V_{dc} / 50 A. The simulation is done according to the parameters stated in Table II.

The results of the single-pulse tests are shown in Fig. 5 and Fig. 6, where the upper figure is the simulation and the lower one is the measurement. Fig. 5 shows the drain-source voltage V_{ds} on the chip and the drain current I_d. The measurement in Fig. 6b is the gate-to-source voltage V_{gs6} measured as shown in Fig. 2. The measurement of V_{gs6} has not been realized exactly on the gate pad which probably explains the additional oscillations observed in Fig. 6b.

TABLE II. ON-CHIP SINGLE-PULSE TEST PARAMETERS

Test	Vds	Vgs	Cdc	Rg	L	Lg_stray
Vds Id	100 V	-5/+20 V	200 μF	5 Ω	21 μH	18 nH
Vgs	200 V	-5/+20 V	200 μF	5 Ω	21 μH	18 nH

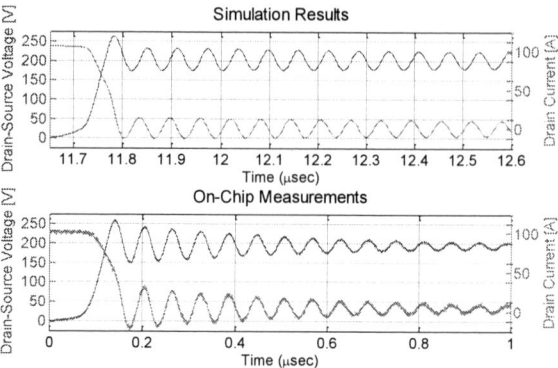

Fig. 5. Single-Pulse test – Turn-off transients of the drain-to-source voltage V_{ds} and the drain current I_d (a) simulation (b) "on-chip" measurements. Drain-source voltage (blue, 100 V/div), drain-current (green, 100 A/div). Time base 100 ns.

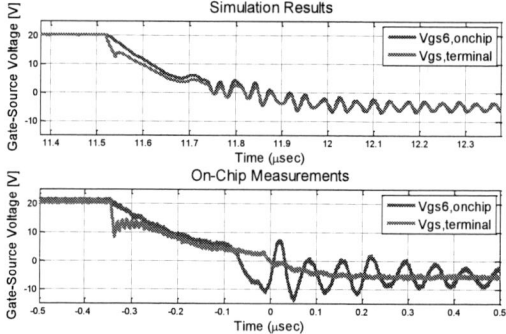

Fig. 6. Single-Pulse test – Turn-off transients of the gate-to-source voltage V_{gs6} (a) simulation, measured as in Fig. 2 and at the gate-source terminals (b) experimental "on-chip" measurements (see Fig. 2) and at the gate-source terminals.

IV. DISCUSSION AND RELIABILITY CONCERN

Comparing the results from the simulations and experiments, it is found that the agreement is good. The model will now be used to investigate reliability issues of the gate-oxied which may arrise if the gate resistor is reduced or if the gate-loop inductance is increased. In real applications where high efficiency is targeted, a fast switching performance is needed in order to reduce the switching losses. However, if this is done, oscillations are observed at the gate-source terminals as shown in Fig. 7. The simulation results show that the overshoot and the oscillations observed at the gate-source terminal are not present directly on the chip (see Fig. 8 and Fig. 9). These results are explained in detail in the following paragraphs.

978-1-4673-9551-9/16 $31.00 © 2016 IEEE

To observe the effect of a gate-resistor reduction or an increase of the inductance in the gate-loop, three different simulations are performed. First, a nominal case matching the characteristics of the driver used to validate the simulation model in the previous section is investigated. Secondly, a simulation with a reduced gate resistance (recommended in the device data-sheet) and unchanged gate-driver parasitic inductance is studied. Finally, the gate-driver parasitic inductance was increased to investigate the case where the driver is not situated close to the device. The two latter cases are regarded as "worst-case" scenarios. In fact, a higher voltage-overshoot and possibly higher ringing are expected. The simulation parameters for these three cases are given in TABLE III.

TABLE III.   SIMULATION PARAMETERS

Simulation parameters	Vds	Vgs	Rg	Lg_stray
1. Nominal case	540 V	-5/+20 V	5 Ω	20 nH
Worst cases 2. low gate resistance 3. high gate driver inductance	 540 V 540 V	 -5/+20 V -5/+20 V	 2.5 Ω 5 Ω	 20 nH 100 nH

A. First case – nominal conditions

The gate-loop switching transients of the module in Case 1 are shown in Fig. 8; where the blue and light blue traces show the gate-source voltage of a chip far and close from the gate terminal of the module, respectively. Vgs_3 is situated far from the gate terminal of the module and Vgs_6 is closer. The red line shows the gate-voltage seen from gate terminal, outside the module. The two gate-voltages on the chip are overlapping in the figure. The gate-loop voltage measured at the pin oscillates and has an overshoot of 21 V peak. No under-shoot is observed at turn-off. Taking into consideration the results of the simulation, the gate voltage on the chip does not suffer from this kind of oscillation. Experimental results from the gate-source voltage (yellow) and the gate-source current (green) at turn-on and turn-off are shown in Fig. 7. It corresponds to what is observed in the simulations. The cross talk in the upper switch position resulting from the switching transients of the lower switch position is shown in Fig. 9. During the turn-on transient of the lower switch position, the gate voltages of the upper switches Vgs_9 and Vgs_{12} are reaching -1.3 and -1.6 V respectively. As for the lower switch, the gate-loop voltage measured at the pin oscillates witch an envelope reaching values exceeding the device ratings. During the turn-off transient of the lower switch position, Vgs_9 and Vgs_{12} have an under-shoot of -9.81 V and -9.05 V, respectively. This is below the recommended values (-5 /+20 V) but still within the maximum ratings (-10 V/+25 V). At this point, it can be observed that there is an unbalance between the gate voltages of the individual chips in the module. This unbalance may result in a reduced lifetime of chips exposed to higher voltages in the same module.

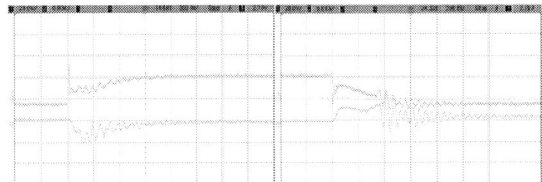

Fig. 7.   Double-Pulse Test of the SiC MOSFET module. Gate-source voltage (yellow) and gate-source current of the lower switch position (green), (a) turn-on transients, (b) turn-off transients

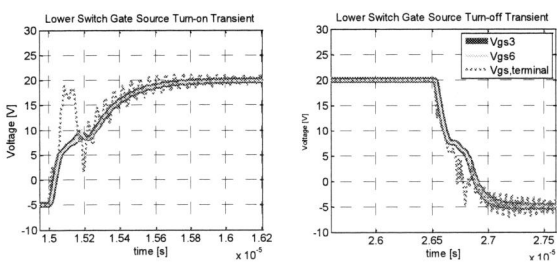

Fig. 8.   Switching transients on the gate of the lower switch position under nominal conditions – lower switch position

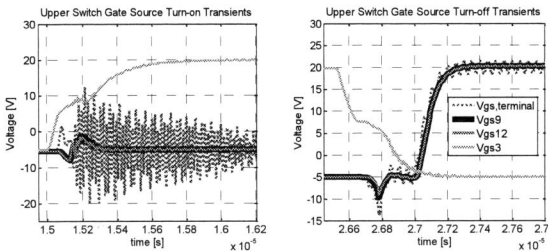

Fig. 9.   Cross-talk of the upper-switch position at turn-on and turn-off of the lower switch position – nominal conditions

B. Second case – low external gate resistance

The gate-loop switching transients of the module when the switching speed is increased are shown in Fig. 10. The voltage seen at the gate-source terminal outside the module reveals an overshoot of 26.6 V exceeding the voltage rating of the gate-oxide. However, the voltages on chips of the position switch of the module (Vgs_3 and Vgs_6 in the figure) remain within the recommended values (-5 V /+ 20 V). Fig. 11 shows the effect of reduced external gate resistance on the cross-talk from the lower switch position on the upper-switch position. The gate voltages of the upper switches Vgs_9 and Vgs_{12} are reaching -2.49 and -2.51 V respectively during the turn-on transient. During the turn-off transient of the lower switch position, Vgs_9 and Vgs_{12} have an under-shoot of -9.37 V and -8.65 V, respectively. These values are lower than for the nominal case. The cross talk is slightly reduced. Fig. 11 also reveals that an increase of the switching speed result in an overshoot at the turn-on of the upper switch. The peak values of Vgs_9 and Vgs_{12} are 22.32 V and 22.24 V, respectively. These values exceed the recommended values without exceeding the maximum tolerated values of the gate-source voltage. Exceeding the gate-source voltage recommended values may

978-1-4673-9551-9/16 $31.00 © 2016 IEEE

lead to a lifetime reduction. According to [13], for this case, the lifetime may be reduced by a factor 10 approximately.

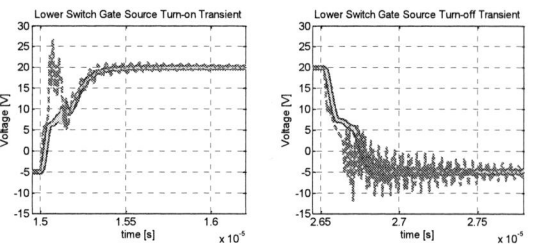

Fig. 10. Switching transients on the gate of the lower switch position – lower switch position, case 2.

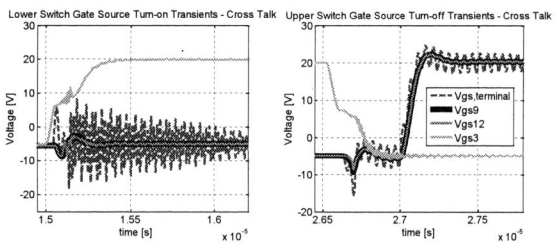

Fig. 11. Cross-talk of the upper-switch position at turn-on and turn-off of the lower switch position, case 2.

C. Third case – high inductance between gate drive and module

The gate-loop switching transients of the module when the inductance between the gate-driver and the gate terminal is increased are shown in Fig. 12 and Fig. 13. In this case, a lower overshoot at the gate-source terminal of the module is observed. The peak value is 25 V. Similarly, no overshoot is observed on the "on-chip" gate-source voltages. The gate voltages of the upper switches Vgs_9 and Vgs_{12} are reaching -1.8 V and -2.4 V respectively during the turn-on transient. During the turn-off transient of the lower switch position, Vgs_9 and Vgs_{12} have an under-shoot of -10.28 V and -9.4 V, respectively. Thus, Vgs_9 exceeds the maximum ratings allowed in the datasheet. Although no publication refers to lifetime reduction in the case the absolute maximum negative voltage is exceeded, the authors believe that this case also presents a reliability issue.

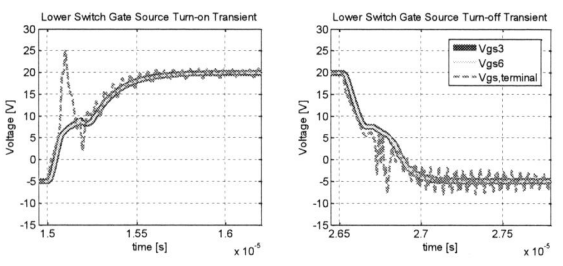

Fig. 12. Switching transients on the gate of the lower switch position – case 3

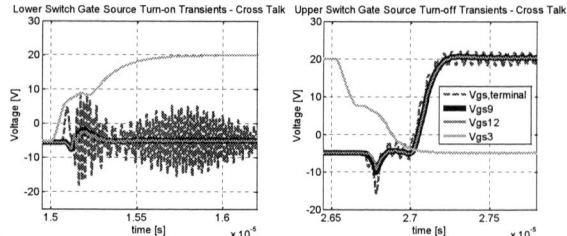

Fig. 13. Cross-talk of the upper-switch position at turn-on and turn-off of the lower switch position, case 3.

D. Discussion

From the observation done in the present section, it can be concluded that the design of SiC MOSFET modules based on existing packages previously designed for Si-IGBTs devices may not be optimal to fully utilize the high-switching speeds offered by the SiC technology. In fact, in order to remain in a safe-operation area and limit the cross-talk between the lower and upper switch positions, one has to limit the switching speed and connect the driver as close as possible to the module to avoid further parasitic inductance. The authors believe that there is a need to develop low-inductive packages adapted to the high-switching speeds of the new SiC technology. A reduction of the gate-loop inductance is desired to increase the reliability, decrease the cross-talk, but also to increase the switching speeds and thus decrease the switching losses.

V. CONCLUSIONS

A SiC MOSFET power module available on the market has been modeled with ANSYS Q3D and validated experimentally. The validity of the model has been confirmed by on-chip measurements. An analysis of the impact of parasitic elements on the gate-source voltages inside the module has been conducted, revealing a difference between the measured voltage at the gate terminal and the gate voltage on the chip. The oscillations observed at the gate-source terminal are not present at a chip level. The high over-shoot on the lower switch position gates observed at the terminal during turn-on is not present on the chip. However, when increasing the switching speeds, it was found that the gate-oxide was exposed to values exceeding the recommended operation values. This could lead to reliability issues, particularly when high-switching speeds are targeted. Moreover, a poorly designed gate-drive connection leading to high parasitic inductance in the gate loop can also be harmful for the device immunity. For reliability reasons, the gate drive design may have to compromise on the switching speeds but also minimize the distance between the gate driver output and the gate-source terminal of the 62 mm module. There is thus a need to develop low-inductive packages adapted to the fast switching SiC technology.

For future work, the on-chip measurements will be performed on the upper-switch position of the module where an overshoot and undershoot on the chip are observed in the simulations. Further analysis on gate-source voltage unbalances and its impact on the current sharing in the module has to be investigated as well. Finally, the authors intend to realize similar work on commercially available SiC MOSFET modules with the same 62 mm package from different suppliers.

ACKNOWLEDGMENTS

The authors would like to thank the Swedish Energy Agency and Vinnova for the financial support of the work.

REFERENCES

[1] H.-P. Nee, J. W. Kolar, P. Friedrichs, and J. Rabkowski, "Editorial: Special Issue on Wide Bandgap Power Devices and Their Applications, 2014," *IEEE Trans. Power Electron.*, vol. 29, no. 5, pp. 2153–2154, May 2014.

[2] K. Takao and S. Kyogoku, "Ultra low inductance power module for fast switching SiC power devices," in *2015 IEEE 27th International Symposium on Power Semiconductor Devices IC's (ISPSD)*, 2015, pp. 313–316.

[3] J. Colmenares, D. Peftitsis, J. Rabkowski, D.-P. Sadik, and H.-P. Nee, "Dual-Function Gate Driver for a Power Module With SiC Junction Field-Effect Transistors," *IEEE Trans. Power Electron.*, vol. 29, no. 5, pp. 2367–2379, May 2014.

[4] A. Lemmon, M. Mazzola, J. Gafford, and C. Parker, "Instability in Half-Bridge Circuits Switched With Wide Band-Gap Transistors," *IEEE Trans. Power Electron.*, vol. 29, no. 5, pp. 2380–2392, May 2014.

[5] A. Dutta, S. Wang, J. Zhou, S. S. Ang, J.-C. Chang, and C.-S. Chen, "The design and fabrication of a 50KVA 450A silicon carbide power electronic module," in *2013 4th IEEE International Symposium on Power Electronics for Distributed Generation Systems (PEDG)*, 2013, pp. 1–5.

[6] Z. Zhang, B. Guo, F. Wang, L. M. Tolbert, B. J. Blalock, Z. Liang, and P. Ning, "Impact of ringing on switching losses of wide band-gap devices in a phase-leg configuration," in *2014 Twenty-Ninth Annual IEEE Applied Power Electronics Conference and Exposition (APEC)*, 2014, pp. 2542–2549.

[7] A. Lemmon, R. Graves, and J. Gafford, "Evaluation of 1.2 kV, 100A SiC modules for high-frequency, high-temperature applications," in *2015 IEEE Applied Power Electronics Conference and Exposition (APEC)*, 2015, pp. 789–793.

[8] Z. Zhang, F. Wang, L. M. Tolbert, B. J. Blalock, and D. J. Costinett, "Active gate driver for fast switching and cross-talk suppression of SiC devices in a phase-leg configuration," in *2015 IEEE Applied Power Electronics Conference and Exposition (APEC)*, 2015, pp. 774–781.

[9] B. Hull, S. Allen, Q. Zhang, D. Gajewski, V. Pala, J. Richmond, S. Ryu, M. O'Loughlin, E. Van Brunt, L. Cheng, A. Burk, J. Casady, D. Grider, and J. Palmour, "Reliability and stability of SiC power mosfets and next-generation SiC MOSFETs," in *2014 IEEE Workshop on Wide Bandgap Power Devices and Applications (WiPDA)*, 2014, pp. 139–142.

[10] J. Z. Chen, L. Yang, D. Boroyevich, and W. G. Odendaal, "Modeling and measurements of parasitic parameters for integrated power electronics modules," in *Nineteenth Annual IEEE Applied Power Electronics Conference and Exposition, 2004. APEC '04*, 2004, vol. 1, pp. 522–525 Vol.1.

[11] S. Li, L. M. Tolbert, F. Wang, and F. Z. Peng, "P-cell and N-cell based IGBT module: Layout design, parasitic extraction, and experimental verification," in *2011 Twenty-Sixth Annual IEEE Applied Power Electronics Conference and Exposition (APEC)*, 2011, pp. 372–378.

[12] L. Popova, R. Juntunen, T. Musikka, M. Lohtander, P. Silventoinen, O. Pyrhonen, and J. Pyrhonen, "Stray inductance estimation with detailed model of the IGBT module," in *2013 15th European Conference on Power Electronics and Applications (EPE)*, 2013, pp. 1–8.

[13] S. Allen, "Silicon Carbide MOSFETs for High Powered Modules", *CREE InC.*, March 2013.

978-1-4673-9551-9/16 $31.00 © 2016 IEEE

Static and Dynamic Characterization of GaN HEMT with Low Inductance Vertical Phase Leg Design for High Frequency High Power Applications

Nidhi Haryani, Xuning Zhang, Rolando Burgos, Dushan Boroyevich

Center for Power Electronics Systems (CPES)
Virginia Tech., Blacksburg, VA 24061 USA

Abstract— In this paper, detailed static characterization of the new 650 V/ 30 A GaN device from GaN Systems is presented. The on-resistance and output capacitance of this device are considerably low making GaN a viable option for high switching frequency (in the range of MHz) medium power applications (> 3 kW). A comprehensive examination of device characteristics and variation of device parasitics depending on operating conditions is presented. Due to the high slew rate of these devices, it is important to incorporate galvanic isolation between gate driver input and output stage as recommended in *GaN Systems'* application note[1]. Also, the gate loop and power loop stray inductances and capacitances need to be minimized to curtail the switching noise. To achieve this, vertical power loop design with very small parasitic inductance is proposed. The dynamic characterization carried with this design shows reduction of voltage overshoot to less than half compared to the lateral power loop design from [1].

Keywords— *Gallium Nitirde (GaN), High-electron-mobility transistor (HEMT), Gate driver, Isolation, Miller capacitance, Vertical power loop, Lateral power loop*

I. Introduction

Wide band-gap GaN-based devices are emerging as an attractive candidate for high-efficiency power driving systems [3], [4], owing to their high breakdown voltage, low on-resistance, fast switching, and high temperature operation. As shown in Fig. 1(a), the theoretical limit of GaN breakdown voltage is very high with very low specific on-resistance. The 600 V cascode GaN-on-Si structure has been dominant in medium power applications for a long time but the cascode structure has issues of charge imbalance between the low voltage Si and high voltage normally-on GaN, lower reliability and high common source inductance [5].

The enhancement mode GaN switches available until recently had lower breakdown voltages suitable only for low voltage applications. However, 650 V/ 30 A e-mode GaN switches are now available commercially. These devices are very small (as shown in Fig. 1(b)) and have low package inductance but to extract the full benefits of these devices, it is important to understand the static and dynamic characteristics of the device. The preliminary datasheet of the device provides

the vital electrical characteristics' information only at a few data points which is not sufficient. The static characteristics of the device are presented in detail in section II and III. Low temperature characterization is discussed in section II and it is showed that the device has very low on-resistance as compared to state-of-the art GaN devices. In section III, high temperature static characterization procedure is discussed and the results are examined in detail.

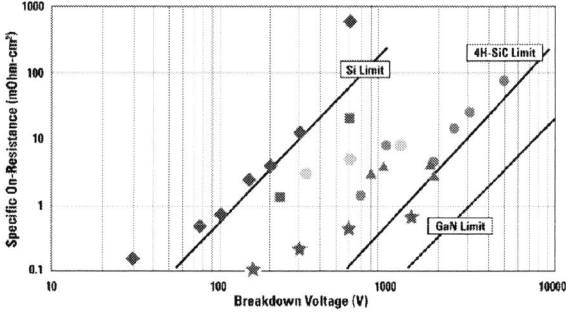

(a)

(b)

Fig. 1. (a) Specific on-resistance vs breakdown voltage for different semiconductor materials showing that the theoretical limit for GaN is the highest (b) 650V/ 30 A GaN device from GaN Systems with very low package inductance and small package size.

In section 1V, a low parasitic inductance vertical power loop design and its switching characterization is presented. As compared to the reference lateral power loop design from [1], it is shown that this design has considerably lower switching noise which opens up the possibility of reduction of external gate resistance leading to faster switching transient and lower turn-on (E_{on}) and turn-off (E_{off}) energy losses.

II. STATIC CHARACTERIZATION

A. Characterization Procedure

The equivalent gate circuit of GaN HEMT is shown in Fig. 2(a). As mentioned above, the datasheet [2] only provides the on-resistance and the parasitic capacitors' values at a few data points which is not sufficient to fully understand the device behavior. In order to gain a deep insight of the device switching and thermal performance, it is important to measure each parasitic element accurately. The parasitic on-resistance (R_{dson}) varies with drain current (I_d) and device junction temperature (T_j). The on resistance also changes with the gate source voltage (V_{gs}) applied on the device as the enhancement mode operation requires a minimum positive gate-source voltage on the device. On-resistance is measured during enhancement mode of the device as in most power electronics application the MOSFET is operated in linear mode to minimize the parasitic resistance.

The parasitic capacitances namely gate-source capacitance (C_{gs}), drain source capacitance (C_{ds}) and gate-drain capacitance (C_{gd}) as shown in Fig. 2(a) depend on the drain-source voltage V_{ds}. The static characterization was performed using a curve tracer (Agilent B1505A) as shown in Fig. 2(b). The tests which need a high current capability (>1 A) like the output characteristics, transfer characteristics etc. were done by connecting the curve tracer to an ultrahigh current expander (Agilent N1256A). The tests which needed a higher voltage (> 60 V) such as capacitance measurement were done by connecting the curve tracer to a high voltage bias tee (Agilent N1260A). The bias tee can be used for testing upto 3 kV. The device is mounted on the socket with gate, source and drain connections as shown in Fig. 2(c).

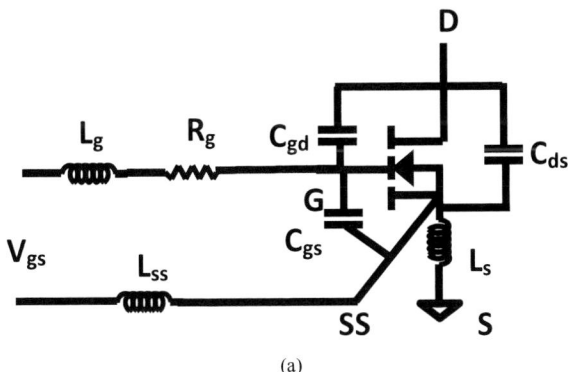

(a)

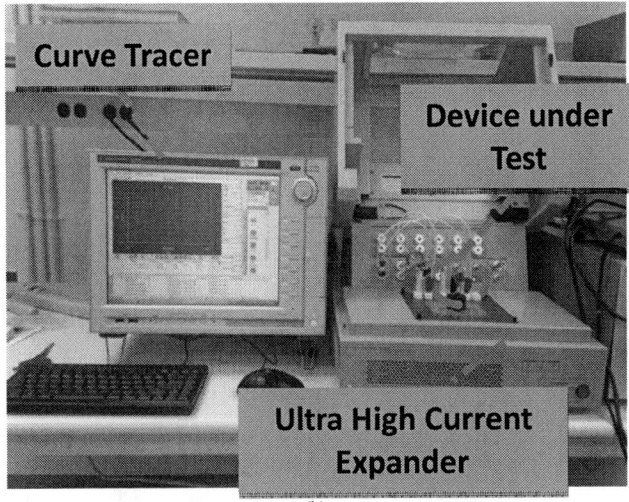

(b)

(c)

Fig. 2. (a) Switching circuit of GaN HEMT (b) Experimental setup of static characterization (c) Ultra high current expander fixture showing device under test and drain, source and gate connections.

B. I-V Charateristics

Output Characteristics

The output forward characteristics were measured by sweeping V_{ds} from 0-5 V at V_{gs} varying from 1 V- 10 V as the breakdown V_{gs} limit from [2] is +/- 10 V. The gate-source voltage is kept constant for one V_{ds} sweep. The curve tracer has separate force and sense connections to apply voltage and current and to sense them separately. This Kelvin sense connection increases the accuracy of measurement. The pulse width of V_{gs} for measurement is kept very low (0.001 %) to avoid self-heating of the device. Also, it is assumed that the junction temperature will be similar to case temperature because of the small duty ratio. A gate resistor of 100 Ω was selected on the curve tracer fixture to avoid any voltage and current overshoot on the device during turn-on and turn-off.

The forward output characteristics of the 650 V/ 30 A GaN HEMT at case temperature (T_c) 25 $^\circ$C are shown in Fig. 3(a). The device starts to conduct very small current at V_{gs} 2 V and is fully enhanced at 5 V. The device saturates at I_d 44 A for V_{gs} 7 V-10 V. The reverse output characteristics as shown in Fig. 3(b) are measured by varying V_{ds} from -5-0 V and V_{gs} varying from -5-10 V. An important thing to note during reverse

978-1-4673-9551-9/16 $31.00 © 2016 IEEE

conduction is that the GaN HEMT doesn't have a body diode anti-parallel to the device but during reverse conduction, the channel starts to conduct when

$$V_{sd} > V_{sg} + V_{gsth} \qquad (1)$$

where

V_{sd}- negative of applied drain-source voltage
V_{sg}- negative of applied gate-source voltage
V_{gsth}- gate-source threshold voltage

This phenomenon ensures that GaN HEMT will have very low on resistance during reverse conduction and hence lower reverse conduction losses as compared to silicon carbide (SiC) and silicon (Si) MOSFETs. Also, since enhancement mode GaN has no body diode, it has no reverse recovery losses which is a very big advantage over Silicon carbide (SiC) and cascode GaN devices. The reverse characteristics are similar to forward characteristics when applied gate source voltage is greater than 4 V. The device saturates at I_d -44 A for V_{gs} 7 V-10 V. The results shown in this paper are for one sample of the device although to verify the consistency of the tests the experiments were conducted on various samples.

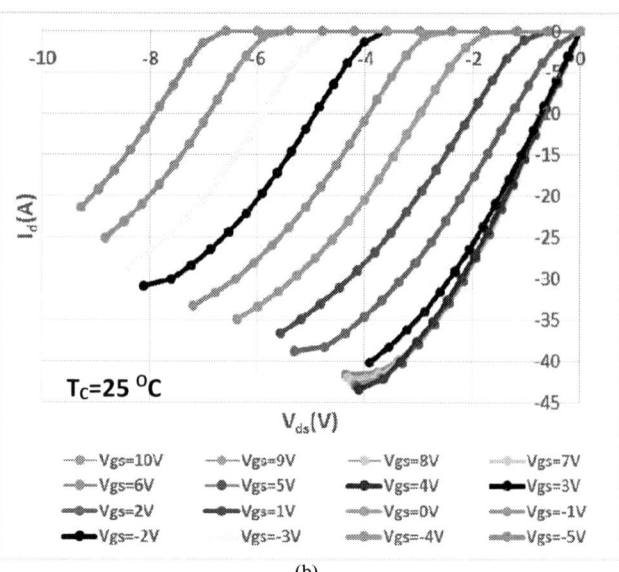

(b)

Fig. 3. (a) Forward characteristics of GaN HEMT and (b) Reverse characteristics of GaN HEMT at case temperature 25 °C

The device on-resistance is measured by the curve tracer by measuring the slope of the I-V curve. The resistance of the wires and connectors is compensated by subtracting the value from the measured on resistance. The measured forward on-resistance (R_{dson}) at I_d 9 A and V_{gs} 7 V is 55 mΩ as shown in Fig. 4(a) while in [2] it is mentioned to be 52 mΩ. The measured reverse on-resistance (R_{dson}) at I_d -9 A and V_{gs} 7 V is 54.5 mΩ. This shows that this device has a very low on resistance which makes it ideal for high power applications. The cascode GaN device with similar voltage and current ratings has three times on resistance at similar current levels. Also, it can be seen from Fig. 4 that on resistance (both forward and reverse) is almost constant for I_d 0 - 30 A.

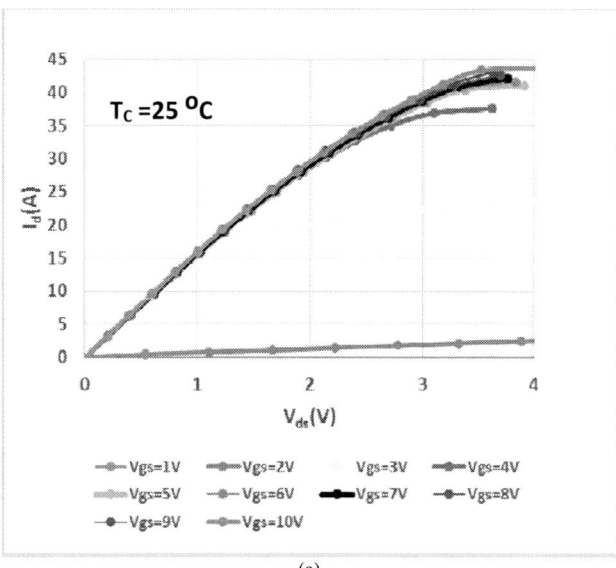

(a)

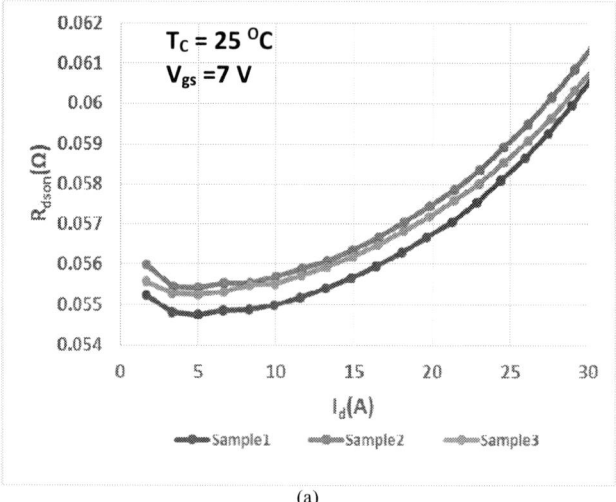

(a)

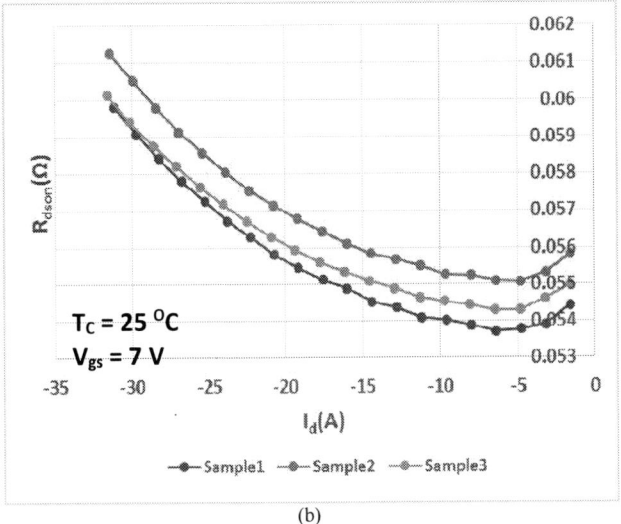

(b)

Fig. 4. (a) Forward on resistance vs drain current and (b) Reverse conduction resistance vs drain current of GaN HEMT at V_{gs} 7 V and T_c 25 °C

Transfer Characteristics

The transfer characteristics are measured by varying V_{gs} from 0-5 V at a constant V_{ds} bias of 7 V. The device starts to conduct when $V_{gs} > V_{gsth}$ as shown in Fig. 5(a). The gate threshold voltage is 1.6 V from [2] and from the measurement, it is found to be 1.8 V. The threshold voltage is the applied gate-source voltage at which the drain current I_d is 7 mA under the operating conditions of $V_{ds} = V_{gs}$. The drain current at which the threshold voltage is defined is recommended by the device manufacturers. The measured trans conductance is 15 S while in [2], it's mentioned to be 13 S.

Blocking Capability

The device blocking capability can usually be evaluated by two parameters: BV_{DSS} –the drain-source breakdown voltage, and I_{DSS} – the drain leakage current at the rated voltage. For MOSFET, BV_{DSS} is usually defined as the voltage which produces 250 µA drain leakage current with the gate-source terminals shorted. Under this criterion, BV_{DSS} is usually measured to be higher than the rated voltage. However, under elevated temperatures, devices may break down even before the leakage current reaches 250 µA. For mature and rugged devices, all the single cells within the device are uniform in their blocking capability. These devices also have built-in avalanche capability that helps them survive from short-time unexpected over-voltage stress. However, for certain prototype DUTs, some single cells are weaker than the others, which will break down first under high voltage stress, and the resultant damage is usually not recoverable. In this case, the DUT is destroyed (i.e. can no longer block voltage) even if the output characteristics still look fine. To avoid possible damages to the GaN DUTs, I_{DSS} is adopted in this work to evaluate the device blocking capability instead of BV_{DSS}, and is measured under the High Voltage Mode of the curve tracer. Measuring I_{DSS} is relatively safer because the leakage current is obtained under rated V_{DS} voltage with the channel pinched-

off. The drain leakage current is measured to be 2.4 µA at 650 V V_{ds} and T_C 25 °C as shown in Fig. 5(b).

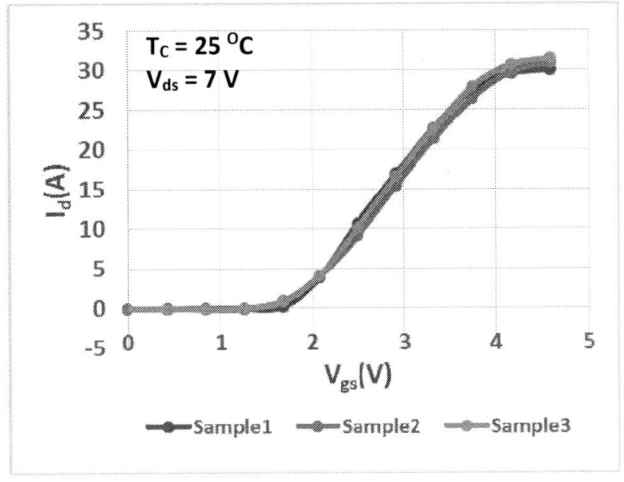

(a)

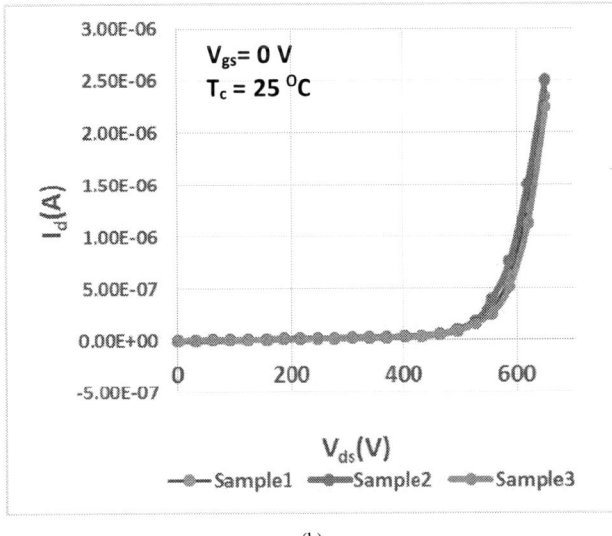

(b)

Fig. 5. (a) Transfer characteristics of GaN HEMT at V_{ds} 7 V and T_c 25 °C. (b) Drain leakage current vs drain source voltage with device pinched off and T_c 25 °C

CV Curves

As discussed above the device parasitic capacitances are termed as C_{ds} (drain-source capacitance), C_{gs} (gate-source capacitance) and C_{gd} (gate-drain capacitance). The input capacitance (C_{iss}), output capacitance (C_{oss}) and reverse transfer capacitance (C_{rss}) are given by:

$$C_{iss} = C_{gs} + C_{gd} \qquad (2)$$

$$C_{oss} = C_{gd} + C_{ds} \qquad (3)$$

$$C_{rss} = C_{gd} \qquad (4)$$

The capacitances are measured by using the high voltage bias tee (Agilent N1260A) with the curve tracer. The C_{gd}

measurement block diagram is shown in Fig. 6(a). Only the AC measurement signal passing through C_{dg} component is measured by the Im current meter, and converted to C_{dg}. The AC current passing through C_{ds} flows in AC Guard, and it does not affect the C_{dg} component. Also the potential of L (low) input is almost GND level, and the C_{gd} current branching off to C_{gs} component can be ignored since the potential between C_{gs} is close to zero volts. However, at high frequency the impedance of AC guard increases which may cause some signal leaks to C_{gs} affecting measurement accuracy.

For C_{oss} measurement, the gate and source of the device are both connected to L. The measured C_{oss}, C_{rss} and C_{iss} from experiment are 62 pF, 7 pF and 182 pF respectively at V_{ds} 400 V. These measurements match well with the capacitance values in the preliminary datasheet. These capacitances are quite lower as compared to state of the art SiC devices. The results are shown in Fig. 6(b).

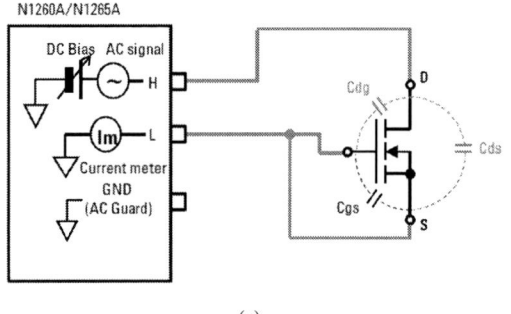

(a)

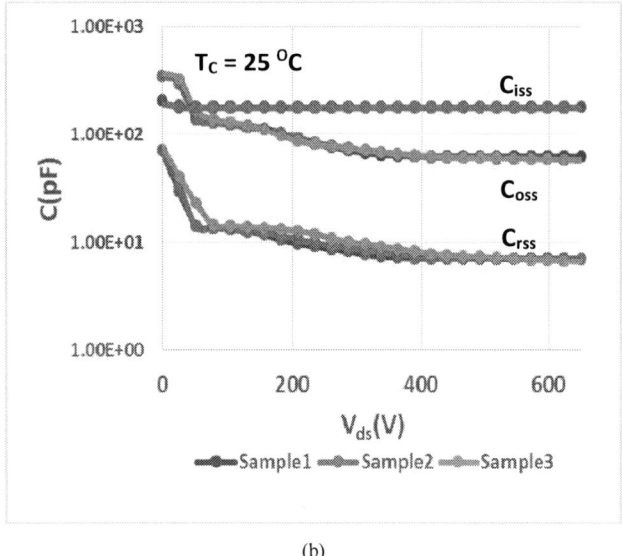

(b)

Fig. 6. (a) Block diagram for C_{dg} measurement (b) Parasitic capacitances vs drain source voltage. These capacitances are measured at a frequency of 1 MHz

III. HIGH TEMPERATURE STATIC CHARACTERIZATION

A lot of device parameters such as on-resistance, threshold voltage, blocking capability etc. change a lot with junction temperature. The device on resistance increases with temperature which leads to increase in power loss on the device and hence temperature increases further. The leakage currents also increase with increase in junction temperature. This is due to the intrinsic carrier concentration increasing exponentially with temperature. The high temperature static characterization setup is shown in Fig. 7. The device is heated on a hot plate, a thermocouple is used to measure the device case temperature which is connected to the curve tracer. The device begins to saturate at lower drain currents for the same gate-source voltage as shown in Fig. 8(a), the GaN HEMT saturates at 30 A for V_{gs} 7-10 V. The device is fully enhanced at V_{gs} 7 V at T_C 100 ^{O}C as compared to 5 V at 25 ^{O}C as shown in Fig. 8(a) and (b). Hence, the recommended gate driver voltage for this device is 7 V which is 3 V lower than the gate breakdown voltage. The trans conductance of the device at 100 ^{O}C is half of the trans conductance at 25 ^{O}C as shown in Fig. 8(c).

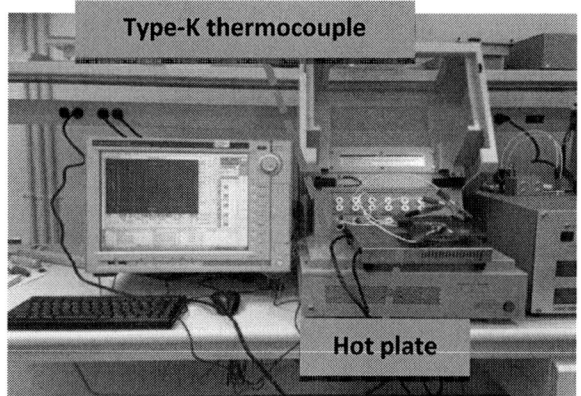

Fig. 7. High temperature static characterization setup

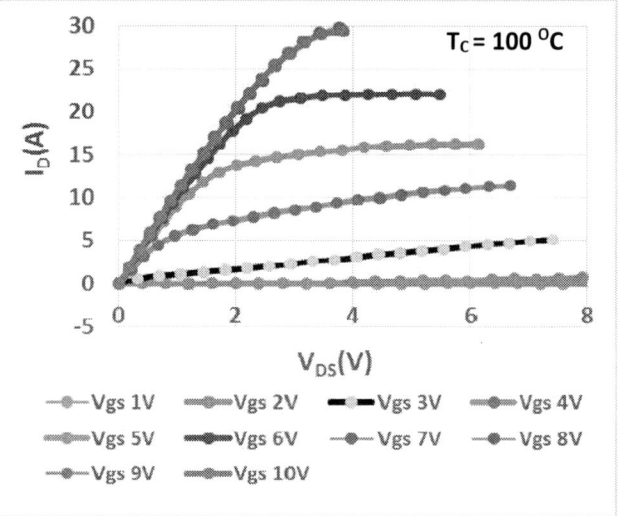

(a)

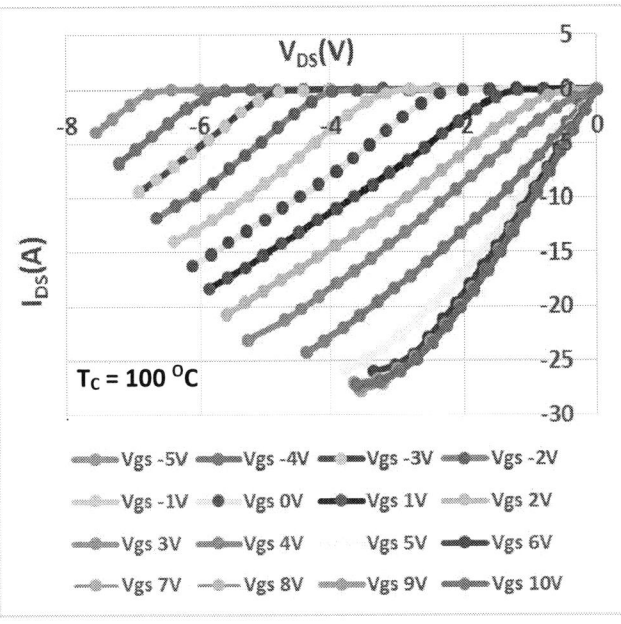

(b)

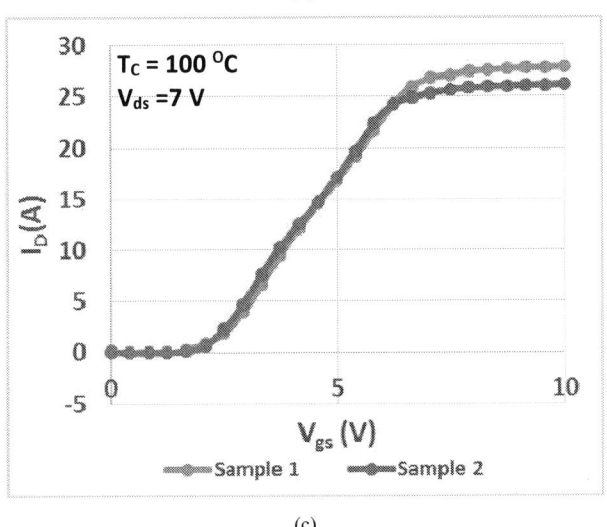

(c)

Fig. 8. (a) Output forward characteristics (b) Reverse characteristics (c) Transfer characteristics at case temperature 100 °C

As can be seen from Fig. 8(a) and (b), the slope of I-V curve during enhancement mode has greatly reduced which signifies a large increase in R_{dson}. Forward on resistance increases to 136 mΩ at I_d 9 A and T_C 144 °C as compared to 54.5 mΩ at T_C 25 °C. Similarly, the reverse on resistance also increases to 129 mΩ at 144 °C case temperature and -9 A I_d. The normalized on-resistance vs temperature is shown in Fig. 9.

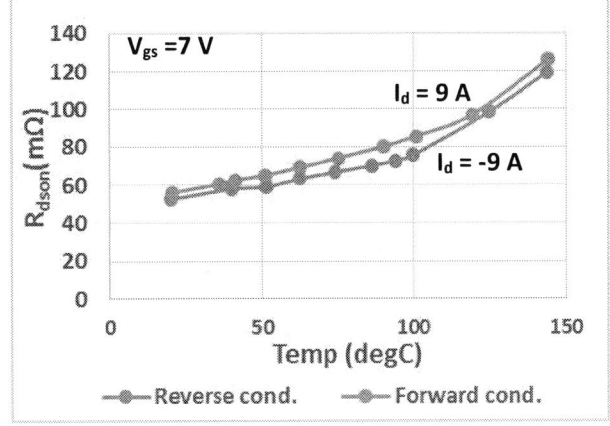

Fig. 9. Normalized forward conduction and reverse conduction on resistance vs temperature

IV. DYNAMIC CHARACTERIZATION

As discussed in introduction section, for high frequency applications, it is very important to design the phase leg PCB layout with low loop inductance. GaN Systems' GS66508 has very low package inductance (drain and source inductances are each 0.2 nH from [2]), hence to fully utilize this advantage a low inductance vertical power loop is proposed in this paper. In this phase leg design, the top/ high side switch is mounted on the top layer of the PCB whereas the bottom/ low side switch is mounted on the bottom layer of the PCB as shown in Fig. 11. The power loop inductance is thus limited only by device package inductance, dimensions of the device and the PCB. The phase leg schematic is shown in Fig. 12(a) with inductor across top switch. As compared to the reference lateral power loop layout (the two devices of the phase leg are mounted side-by-side on the same layer of PCB as shown in Fig. 10(a)) in [1], this design has lower power loop inductance, minimized coupling between gate loop and power loop as the gate loop is perpendicular to the power loop. Whereas in the reference design the gate loop is parallel to the power loop and the AC and DC terminals are coupled with each other as shown in Fig. 10(b). The proposed design has minimal coupling between AC and DC terminals of the phase leg.

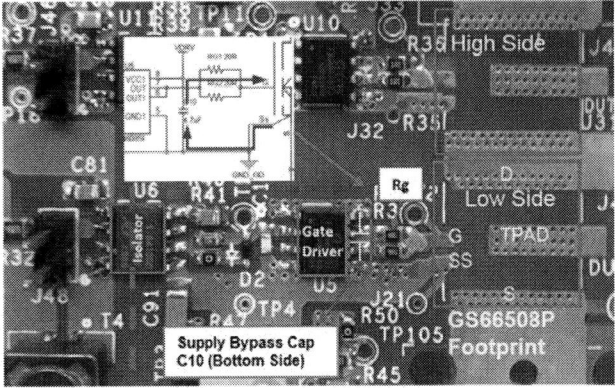

(a)

978-1-4673-9551-9/16 $31.00 © 2016 IEEE

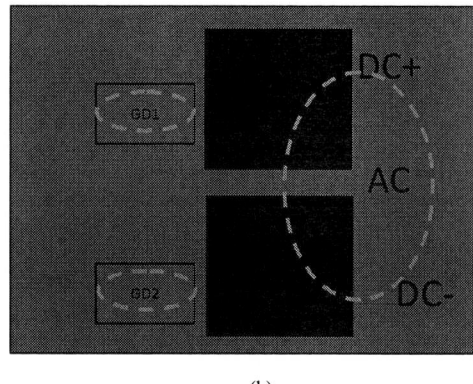

(b)

Fig. 10 (a) Refernce lateral phase leg design from [1] (b) Power loop and gate loop coupling in reference design

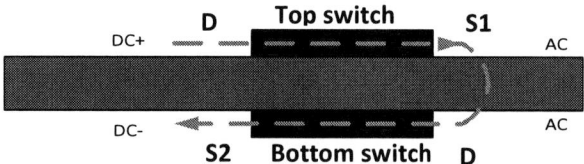

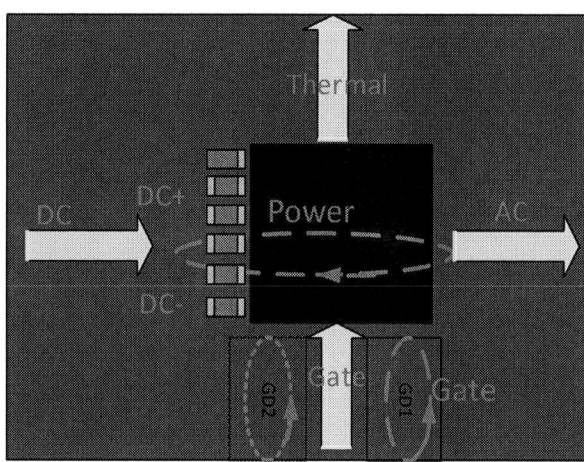

Fig. 11. Vertical power loop design

The double pulse test of the proposed design was conducted with the load inductor connected across the top switch as shown in Fig. 12(a). The external gate resistance is 10 Ω as recommended in [1]. The bottom switch was turned on and off (with 5 μs dead time) while the top switch was kept off. The gate driver is designed for on state V_{gs} 7 V and off state V_{gs} 0 V. The results at 400V DC voltage, 30 A inductor current are shown in Fig. 12(b). The V_{ds} overshoot measured was 68.8V (17.2%) whereas from the double pulse test results in [1] at similar conditions, the V_{ds} overshoot is 188 V (47 %). Thus the vertical design can reduce the voltage overshoot to less than half as compared to the lateral power loop design. This low overshoot paves the way for reduction of gate resistance, faster switching and hence lower energy losses. V_{ds} was measured with high voltage differential probe (bandwidth 200 MHz) as well as passive probe (bandwidth 200 MHz) and the results were found to be similar. The V_{gs} rise time is less

than 20 ns and peak V_{gs} measured is 9 V. The high noise in V_{gs} might also be due to absence of accurate measuring points on the board. Also, the device current was not measured as the board was not designed to accommodate a current shunt. For the future, accurate measuring points will be added on board to measure V_{gs} and device current I_d.

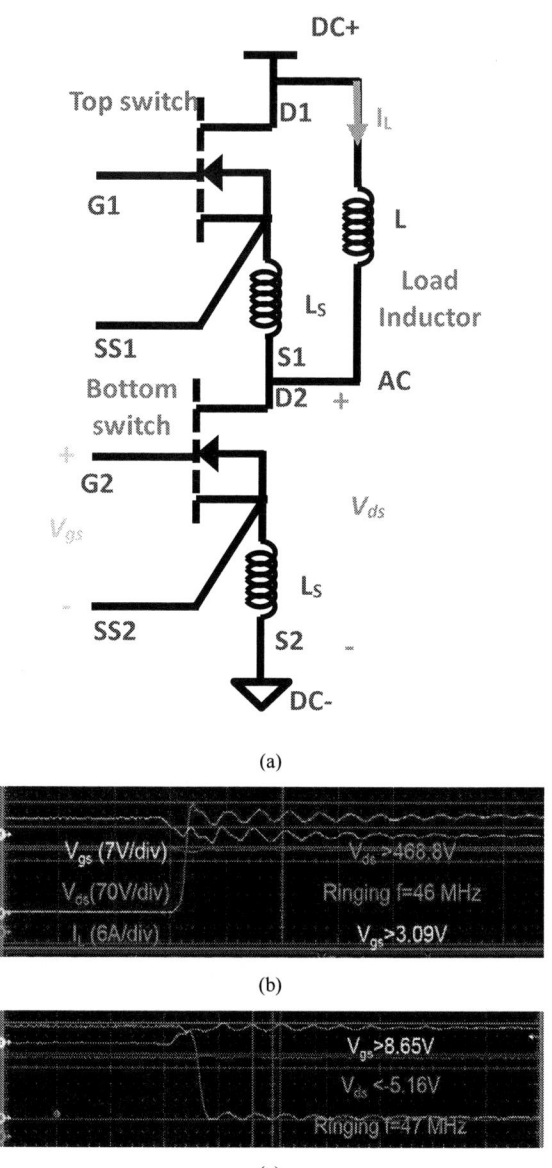

(a)

(b)

(c)

Fig. 12 (a) Phase leg schematic for DPT with inductor across top switch (b) Turn-off transient and (c) Turn on transient at 400 V, 30 A

V. CONCLUSION AND FUTURE WORK

The detailed static characterization of the low package inductance 650 V/ 30 A GaN HEMT at room temperature and high temperature is presented supplementing the datasheet, which only has a few data points mentioned. The parasitic capacitances and the on resistance of the device are very small but the resistance increases to 144 mΩ at case temperature 140

^{O}C. The dynamic characterization of the proposed low inductance vertical layout is presented, the voltage overshoot in this case is less than half of the voltage overshoot in the lateral phase leg design from [1]. Hence, reduction of external gate resistance will be considered in the future. Also, a double pulse test board has been designed with measuring points to accurately capture the high frequency switching transient of V_{ds}, V_{gs} and I_d in the vertical phase leg design. The switching loses will be measured in the future with this board and will be compared to the reference phase leg design.

REFERENCES

[1] Application note- "How to Drive GaN Enhancement Mode Power Switching Transistors", GaN Systems inc., 2014.

[2] Preliminary datasheet- "GS66508P-E03- 650V enhancement mode GaN transistor", GaN Systems inc., 2014.

[3] T. Uesugi and T. Kachi, "GaN power switching devices for automotive applications," in Proc. Int. Conf. CS Matech, Tampa, FL, May 18–21, 2009.

[4] M. A. Briere, "Commercial 600 V GaN-based power devices coming of age," Power Electron. Technol., pp. 26–28, Jul. 2011.

[5] Z. Liu, X. Huang, FC Lee, Q. Li, "Investigation of Package Influence on High Voltage Cascode GaN HEMT with Simulation Model," CPES review 2-13-2013, Milpitas, CA

[6] Application note- "SPICE Model for GaN HEMT- usage guidelines and examples", GaN Systems inc., 2014.

[7] Delaine J., Jeannin P., Frey D., Guepratte K., "High frequency DC-DC Converter using GaN device," in Proc. IEEE-APEC' 2012 Ann. Meet., pp.1754-1761

A Family Of Single-Phase Current Source Converters With Double Outputs

Louelson A. Costa[*][†], Maurício B. R. Corrêa[†], Montiê A. Vitorino[†], Gutemberg G. dos Santos[†] and
Darlan A. Fernandes[§]
[*]Post-Graduate Program in Electrical Engineering – PPgEE – COPELE
[†]Department of Electrical Engineering
Federal University of Campina Grande (UFCG), Campina Grande – PB – Brazil
Emails: louelson@gmail.com; [mbrcorrea; vitorino; gutemberg]@dee.ufcg.edu.br
[§]Department of Electrical Engineering
Federal University of Paraiba (UFPB), Joao Pessoa – PB – Brazil
Email: darlan@cear.ufpb.br

Abstract—**This work introduces a family of Current Source Converters (CSCs) that allows to control single-phase AC converters with two outputs independently sharing the same DC-bus. It is presented four different topologies which have their own advantages and disadvantages compared to each other. The comparison criterias are the total number of switches, the number of conducting switches per switching cycle, quadrant operations and DC-bus current. One of the topologies needs to always have four conducting switches with full-quadrant operation (named Fq4c), while another topology (Lq2c) needs only two, however with limited-quadrant operation. The last two presented topologies have three conducting switches in a switching period, one with full-quadrant operation (Fq3c) and another with limited-quadrant operation (Lq3c). Also it is presented the Scalar Pulse Width Modulations (SPWMs) to control each converter with switching optimization. Details of the topologies, operation and modulation logic are presented. Simulations and experimental results are provided to validate the theoretical approach.**

I. INTRODUCTION

The main drawback of the Current Source Converters (CSCs), when compared to the Voltage Source Converters (VSCs), are the conducting losses. Since the CSCs have an inductor on its DC-bus and its switches need a reverse blocked capability, which is achieved by connecting a series diode with an IGBT, the CSCs would be less attractive than the VSCs. However, the evolution of the magnetics materials [1], allowing to build more dense and efficient inductors [2], allied with the development of the RBIGBT [3] (which replaces the series diode with the IGBT by a single component, reducing the conducting losses) are bringing the CSCs and VSCs to the same level in terms of conducting losses. Despite that, in [4] can be seen that while the conducting losses are dominant in the CSCs, the switching losses are dominant in the VSCs, and above a threshold frequency (10kHz in the referred study) the CSCs are more efficient than the VSCs, concluding that the CSC are preferable in high power and high frequency applications.

Despite the conducting and switching losses, the CSCs have an inherent characteristic of voltage boosting (thus, current bucking), which is a greatly attractive in some applications

like photovoltaic or fuel cell power conditioning [5], [6]. Specifically for the dual output family, some applications might be imagined, such as AC-AC converters, uninterruptible power supply (UPS), universal power filter (UPF), single-phase power decoupling, motor drive, multi-level converter, etc. That said, this paper gathers a family of Dual Outputs CSC (DOCSC) sharing the same DC-bus (see Fig. 1), and makes a comparative study of the four presented topologies: Lq2c, Fq4c, Lq3c and Fq3c. The nomenclature is defined as follows: the fixed letters 'q' and 'c' mean 'quadrant' and 'conducting switches', respectively; since the changeable letters L/F means 'limited'/'full'. For example, Fq4c represents a full-quadrant converter with four conducting switches.

The first two converters presented aiming to achieve a dual output from a single DC-Bus are formed by a three-phase CSC (Lq2c, Fig. 1(c)), and two H-bridge single-phase CSCs connected in series (Fq4c, Fig. 1(a)). However, while the Lq2c has only two conducting switches per switching cycle, it has a limited-quadrant operation. On the other hand, the Fq4c has a full-quadrant operation but four conducting switches per cycle.

Aiming to reduce the number of conducting switches per cycle, the Lq3c is proposed in [7], see Fig. 1(d). The Lq3c replaces the four central switches from the Fq4c by only two switches with a combined logic of modulation. However, it has a limited-quadrant operation, even though it has only three conducting switches per cycle, down from four (Fq4c).

In [8], a topology of CSC with three conducting switches per cycle and full-quadrant operation is proposed, here it is called Fq3c (Fig. 1(b)), it consists of an improvement for the Lq3c topology, the central switches are reorganized in a way that the full-quadrant operation can be achieved. Since the modulation guarantees a single conducting switch per row, a total of three conducting switches per cycle is achieved. Table I shows a brief comparison between the CSCs in terms of switches, conducting switches and conducting losses.

In terms of hardware, the Lq2c CSC topology proposed in [9] is also a foundation for this work. Topology Fq4c does not appear in literature, however it is an intuitive and conventional topology. The topologies Lq2c and Lq3c present limitations of

TABLE I
NUMBER OF SWITCHES AND NUMBER OF CONDUCTING SWITCHES
COMPARISON.

Converter	Fq4c	Fq3c	Lq2c	Lq3c
Number of switches	8	8	6	6
Number of conducting switches	4	3	2	3
Quadrants	Full	Full	Limited	Limited
Relative Conducting losses (%)	100	75	50	75

operation and they may not be able to work independently in all quadrants, as can be seen in Table III and in Fig. 3(b). The Fq4c and Fq3c converters allows independent control for two CSCs sharing the same DC-bus, which is explained by Table II and in Fig. 3(a). It is worth noting that power switches h_j and l_j ($j = 0, 1, 2, 3$) have their complementary switches defined by $\bar{h}_j$ and $\bar{l}_j$, since the switches m_k ($k = 0, 1, 2, 3, 4$) are complementary among them, i.e., they are driven one per time, always having one (only one) switch conducting. This shows that the complementarity of the switches on the CSCs occurs horizontality regarding to the switches of the legs.

In this paper, an evaluation of the four shown single-phase CSC with double outputs is presented, with simulation and experimental results. Also, a modulation approach (SPWM) is presented to improve the usage of the DC-bus current.

II. OPERATION

The operation of each converter is discussed in this section. Since they are CSCs, the DC-bus current is modulated, thus the amplitude of the output current is lower than the input current (DC-bus), on the other hand the output voltage is higher then the input voltage, that is the characteristic of voltage boosting of the CSCs. Let us define D the duty cycle of its respective indexed switch, given as $D = t/T_{sw}$, where t is the conducting time over a switching period T_{sw}.

A. Operation of the Lq2c

It is defined that the currents in the AC side of the converter, in a switching period, is

$$i_1 = (D_{h_1} - D_{l_1})I_{dc} \tag{1}$$
$$i_2 = (D_{h_2} - D_{l_2})I_{dc} \tag{2}$$

The voltage through the DC-bus of the Lq2c is defined as:

$$V_{bus} = (D_{h_1} - D_{l_1})v_1 + (D_{h_2} - D_{l_2})v_2. \tag{3}$$

To satisfy the complementarity of the converter, the switch conduction time starts right after the end of the conduction time of the previous switch, where the total conduction time can not overtime the switching period T_{sw}, which means that:

$$t_{h_1} + t_{h_2} + t_{h_3} = T_{sw} \tag{4}$$
$$t_{l_1} + t_{l_2} + t_{l_3} = T_{sw} \tag{5}$$

In Fig. 2 is shown how the sequence of the switch conduction time occurs for a switching period T_{sw} with centered pulse.

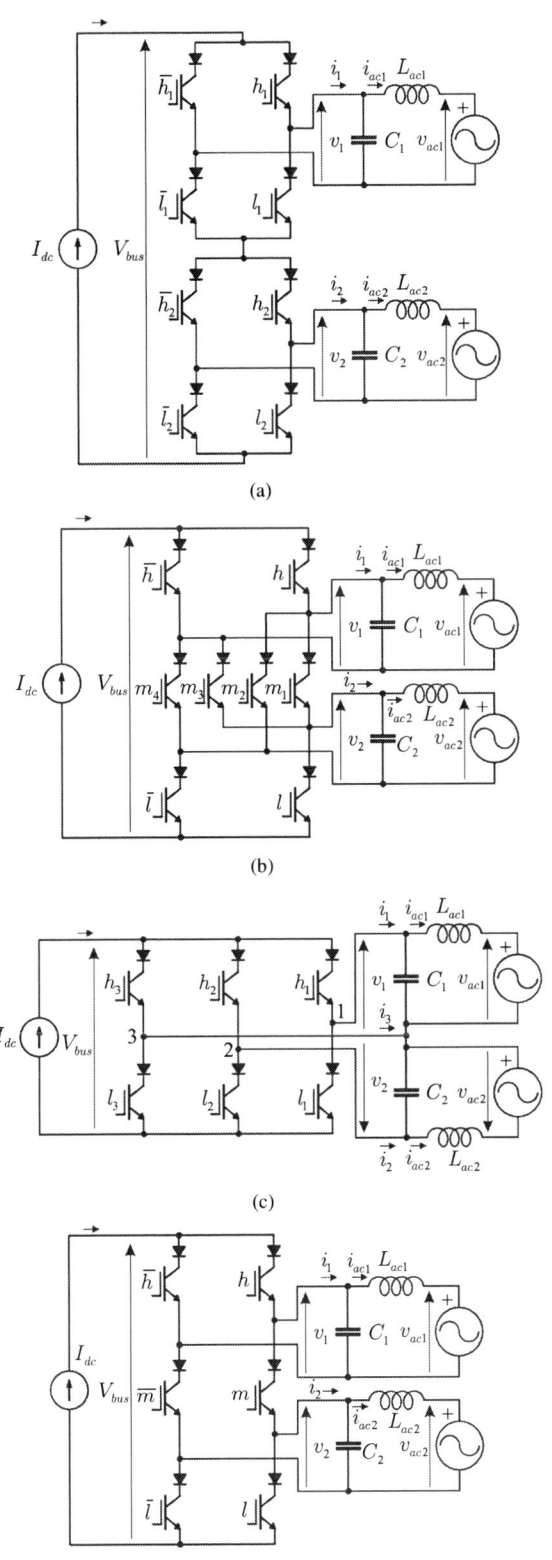

(a)

(b)

(c)

(d)

Fig. 1. Single-phase CSCs with double outputs: (a) Fq4c, (b) Fq3c, (c) Lq2c and (d) Lq3c.

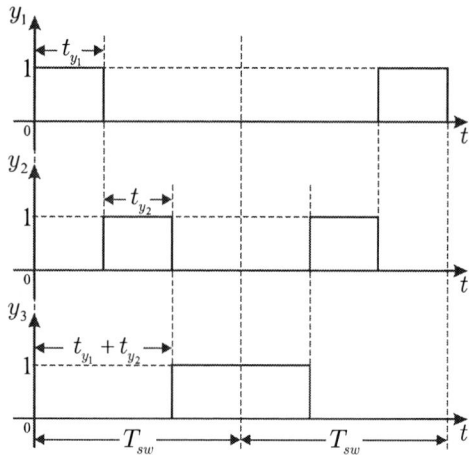

Fig. 2. Sequence and switch conduction time for the PWM of the Lq2c, for $y = h, l$.

B. Operation of the Fq4c

For the Fq4c, the average output currents (i_1 and i_2) and DC input voltage (V_{bus}) over one switching period are given by:

$$i_1 = (D_{h_1} - D_{l_1})I_{dc}, \tag{6}$$

$$i_2 = (D_{h_2} - D_{l_2})I_{dc}, \tag{7}$$

$$V_{bus} = (D_{h_1} - D_{l_1})v_1 + (D_{h_2} - D_{l_2})v_2, \tag{8}$$

where,

$$0 \le D_h \le 1, \tag{9}$$

$$0 \le D_l \le 1. \tag{10}$$

A simple SPWM block diagram can be used to controle the Fq4c, with the horizontal complementarity. Each converter (the top one and the bottom one) can be controlled independently, achieving a full-quadrant operation. Thus, the Fq4c is the simplest configuration both in hardware and SPWM logic, since it can be seen as two separated H-bridge single-phase CSCs.

C. Operation of the Lq3c

The operation of the Lq3c (described in [7]) derives from the modulation of the Fq4c, but with some changes in the SPWM block diagram in order to correctly control the middle switches. Taking the average current on the left (or right) switches of the converter, can be written:

$$i_h = D_h I_{dc}, \tag{11}$$

$$i_m = D_m I_{dc}, \tag{12}$$

$$i_l = D_l I_{dc}. \tag{13}$$

Which the output currents can be obtained as:

$$i_1 = i_h - i_m = (D_h - D_m)I_{dc}, \tag{14}$$

$$i_2 = i_m - i_l = (D_m - D_l)I_{dc}. \tag{15}$$

Similarly, the DC-bus voltage is written as:

$$V_{bus_1} = (D_h - D_m)v_1,$$
$$V_{bus_2} = (D_m - D_l)v_2,$$
$$V_{bus} = V_{bus_1} + V_{bus_2}. \tag{16}$$

Furthermore, two auxiliary variables are used to connect the D_h, D_m and D_l to the references duty cycles D_1 and D_2.

$$D_1 = (D_h - D_m) = i_1/I_{dc}, \tag{17}$$

$$D_2 = (D_m - D_l) = i_2/I_{dc}, \tag{18}$$

where,

$$0 \le D_h \le 1, \tag{19}$$

$$0 \le D_m \le 1, \tag{20}$$

$$0 \le D_l \le 1. \tag{21}$$

Thus, D_1 and D_2 are bounded to $[-1, 1]$.

D. Operation of the Fq3c

Since de Fq3c have a new set-up on the middle switches, its operation needs some different understatement. As will be seen in, a new logic is implemented for the switching control of the central switches. As follows from [8], the equations of the output currents and DC-bus voltages are:

$$i_1 = (D_h - (D_{m_1} + D_{m_2}))I_{dc}, \tag{22}$$

$$i_2 = ((D_{m_1} + D_{m_3}) - D_l)I_{dc}, \tag{23}$$

and

$$v_{bus_1} = (D_h - (D_{m_1} + D_{m_2}))v_1, \tag{24}$$

$$v_{bus_2} = ((D_{m_1} + D_{m_3}) - D_l)v_2. \tag{25}$$

And the total voltage across the DC-bus is given by:

$$V_{bus} = v_{bus_1} + v_{bus_2}. \tag{26}$$

Since the horizontal complementarity is mandatory, the SPWM of the switches from the middle of the converter have to follow the rule

$$\sum_{j=1}^{4} m_j = 1. \tag{27}$$

Considering,

$$D_1 = (D_h - (D_{m_1} + D_{m_2})), \tag{28}$$

$$D_2 = ((D_{m_1} + D_{m_3}) - D_l), \tag{29}$$

where,

$$0 \le D_h \le 1, \tag{30}$$

$$0 \le D_{m_j} \le 1, \tag{31}$$

$$0 \le D_l \le 1, \tag{32}$$

for $j = 1, 2, 3, 4$. Thus, D_1 and $D_2 \in [-1, 1]$.

TABLE II

VECTOR POSSIBILITIES FOR THE OUTPUT CURRENTS OF FULL-QUADRANT
DOUBLE CSCs.

Fq4c				Fq3c			Currents		State/Vector
h_1	l_1	h_2	l_2	h	m_k	l	i_1	i_2	
0	1	1	0	0	1	0	$-I_{dc}$	I_{dc}	0
1	1	1	0	1	1	0	0	I_{dc}	I
0	1	1	0	0	1	1	$-I_{dc}$	0	II
1	1	1	1	1	1	1	0	0	III
0	1	0	0	0	2	0	$-I_{dc}$	0	IV
1	1	0	0	1	2	0	0	0	V
0	1	0	1	0	2	1	$-I_{dc}$	$-I_{dc}$	VI
1	1	0	1	1	2	1	0	$-I_{dc}$	VII
0	0	1	0	0	3	0	0	I_{dc}	VIII
1	0	1	0	1	3	0	I_{dc}	I_{dc}	IX
0	0	1	1	0	3	1	0	0	X
1	0	1	1	1	3	1	I_{dc}	0	XI
0	0	0	0	0	4	0	0	0	XII
1	0	0	0	1	4	0	I_{dc}	0	XIII
0	0	0	1	0	4	1	0	$-I_{dc}$	XIV
1	0	0	1	1	4	1	I_{dc}	$-I_{dc}$	XV

TABLE III

VECTOR POSSIBILITIES FOR THE OUTPUT CURRENTS OF
LIMITED-QUADRANT DOUBLE CSCs.

Lq2c		Lq3c			Currents		State/Vector
h_j	l_j	h	m	l	i_1	i_2	
1	1	0	0	0	0	0	0
3	2	0	0	1	0	$-I_{dc}$	I
2	1	0	1	0	$-I_{dc}$	I_{dc}	II
3	1	0	1	1	$-I_{dc}$	0	III
1	3	1	0	0	I_{dc}	0	IV
1	2	1	0	1	I_{dc}	$-I_{dc}$	V
2	3	1	1	0	0	I_{dc}	VI
2	2	1	1	1	0	0	VII
3	3	-	-	-	0	0	VIII

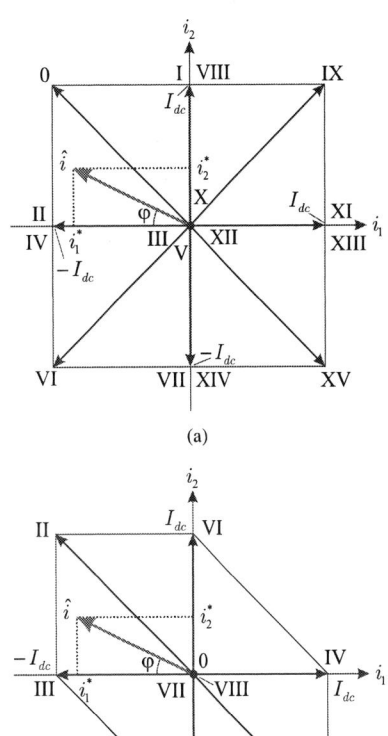

Fig. 3.   Vector map for the CSCs: (a) Fq4c and Fq3c, (b) Lq2c and Lq3c.

III. VECTORIAL ANALYSIS AND SCALAR PULSE WIDTH MODULATION FOR THE CSCs

Tables II and III show all the vector possibilities for the AC output currents of the full-quadrant and limited-quadrant, respectively, where m_k, h_j and l_j represent the ON state when $k = 0, 1, 2, 3, 4$ or $j = 0, 1, 2, 3$, and OFF state otherwise, indicating that only one of the m_k switches is driven per switching period.

The vector possibilities shown in Tables II and III can be drawn in a Cartesian coordinate plane $i_1 \times i_2$, called vector map. The vector map with all the possible vectors for the proposed CSC can be seen in Fig. 3. Any output converter's current written as a point (i_1, i_2) must be inside the boundaries of vector map to avoid overmodulation. Thus, topologies Lq2c and Lq3c have limited-quadrant operation. Since the topologies Fq4c and Fq3c can generate all the vectors, they allow full-quadrant operation. Fig. 3 shows an instantaneous rotating vector $\hat{i}$, composed by i_1 and i_2, that can be written in a Euller's representation, where I and φ are its magnitude and angle, respectively. Thus:

$$\hat{i} = i_1 + ji_2, \tag{33}$$

$$I = \sqrt{i_1^2 + i_2^2}, \tag{34}$$

$$\cos(\varphi) = \frac{i_1}{I}, \tag{35}$$

$$\sin(\varphi) = \frac{i_2}{I}. \tag{36}$$

The modulation presented in this paper allows the two single-phase converters to work independently/simultaneously and improves the usage of the DC-bus current, decreasing its value.

In Fig 4 is shown the logic implementation of the SPWM for all CSC topologies.

IV. SIMULATION AND EXPERIMENTAL RESULTS

For simulation (using Simulink®) and experimental validations, the parameters from Table IV were used. Since the system is in open-loop, the DC voltage source is adjusted at a voltage that provides the desired DC current. Due to hardware limitation from the laboratory, the DC-bus current is limited up to 3A. For the experimental results, the converters were controlled by a DSP TMS320F28335 from Texas Instruments.

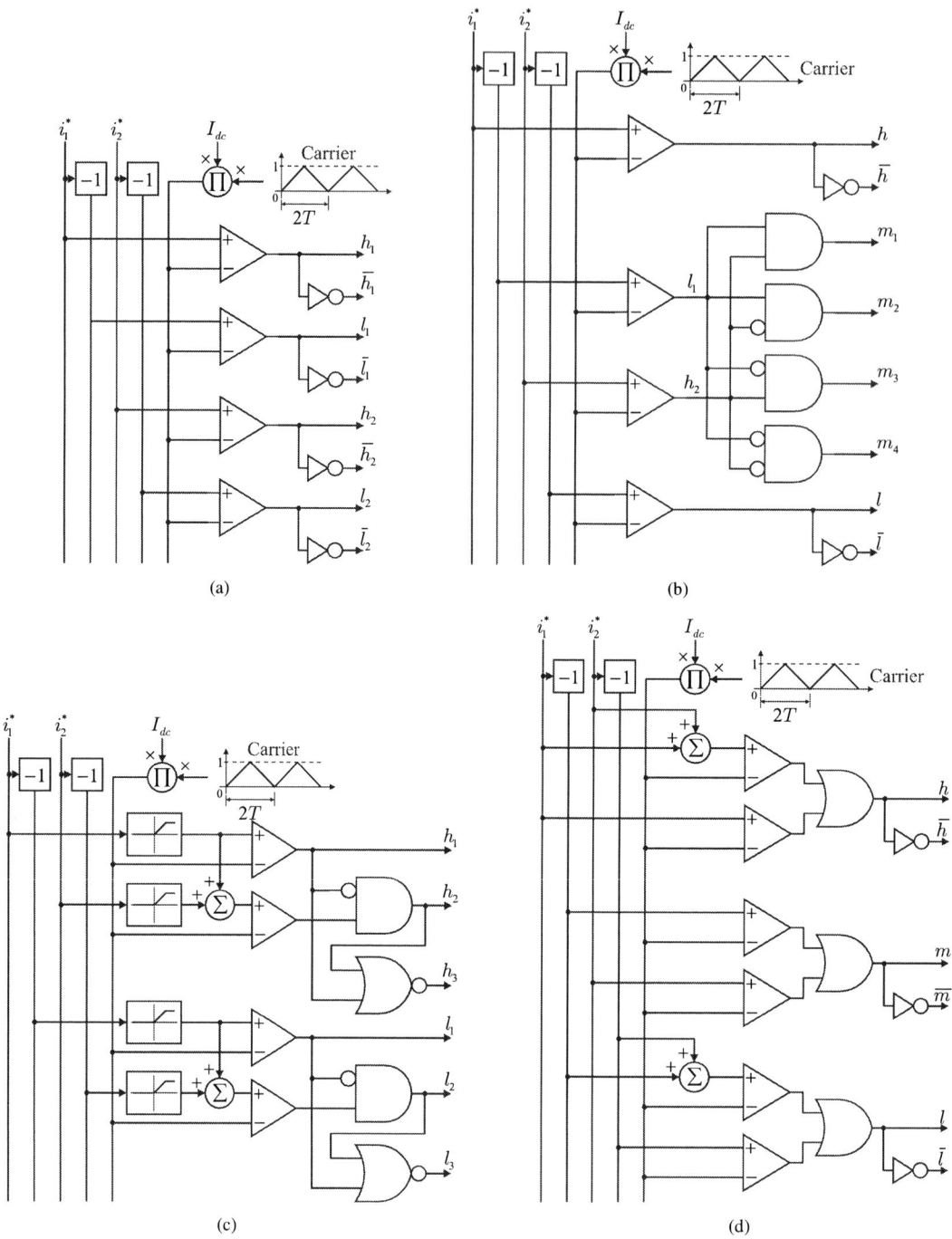

Fig. 4. SPWM implementation for topologies: (a) Fq4c, (b) Fq3c, (c) Lq2c and (d) Lq3c.

Simulation validations for Lq2c and Fq4c are presented in Figs. 5 and 6, respectively. The SPWM references were chosen so that the resultant vector of the output currents will not present overmodulation.

The signal characteristics utilized during the simulation of the Lq2c are: $m_{ac_1} = m_{ac_2} = 1$, $f_{ac_1} = f_{ac_2} = 60$Hz and $\phi_{ac_1} - \phi_{ac_2} = -120°$, where m_{ac_j} $(j = 1, 2)$ is the modulation index defined as I_{ac_j}/I_{dc}, f_{ac_j} is the reference frequency

and ϕ_{ac_j} is the phase displacement between the signals. The waveforms for this topology can be seen in Fig. 5(a). It can be seen that the resultant vector reaches the boundary of the vector map (Fig. 5(b)), but there is no overmodulation. However, that is the result for a specific situation, a change on the frequency, phase or amplitude of the currents could cause an overmodulation problem. In a graphical representation: the resultant vector would be outside the hexagon from the vector

978-1-4673-9551-9/16 $31.00 © 2016 IEEE

TABLE IV

PARAMETERS FOR SIMULATION AND EXPERIMENTAL VALIDATIONS.

Parameter	Value
DC Current (I_{dc})	3A
DC-bus inductor (L_{dc})	67mH
Filter inductor (L_{ac_j})	1.5mH
Filter capacitor (C_j)	15μF
Load (R_j)	3.6Ω
Switching frequency (f_{sw})	10kHz
Simulation step (T_s)	2μs

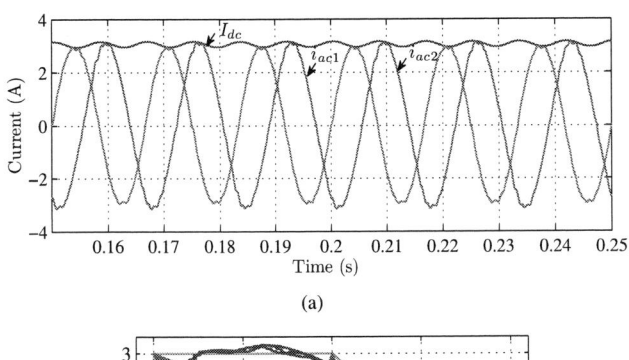

(a)

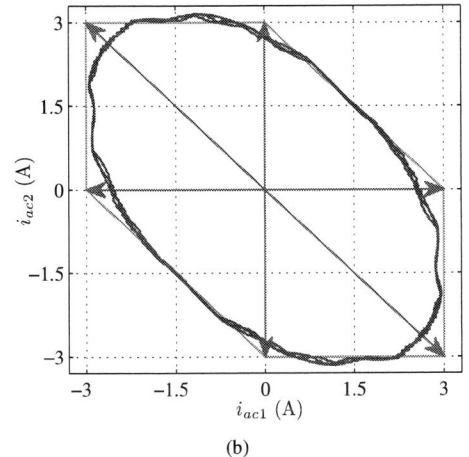

(b)

Fig. 5. Simulation results for Lq2c CSC: (a) Input and output currents, (b) Simulated Lissajous' figure result drawn in a $i_{ac1} \times i_{ac2}$ plane.

map from Fig. 3(b).

To validate the theoretical approach of the Fq4c, the simulation results are shown in Fig. 6. The input and output currents are shown in Fig. 6(a), for $f_{ac_1} = 60$Hz and $f_{ac_2} = 50$Hz. It is possible to see that the converter is able to generate the currents independently. Figure 6(b) shows that the converter is able to generate any two AC currents without interference from each other.

For experimental validation, a Lq2c CSC was built using RBIGBTs (IXRP15N120, from IXYS), whose results are shown in Fig. 8 and implementation in Fig. 7(a). Figure 8(a) shows the generated DC-bus current and AC voltages, which represent the behavior of their respective currents due the restive loads. Figure 8(b) shows that the converter is able to generate any AC currents without overmodulation.

Moreover, a Fq4c/Lq3c was also built (see Fig. 7(b)), but with diodes (Hyperfast diode RHRP1560, from Fairchild

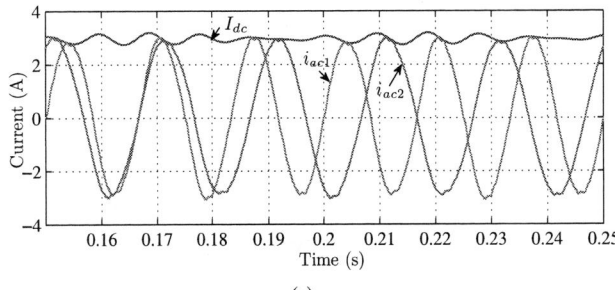

(a)

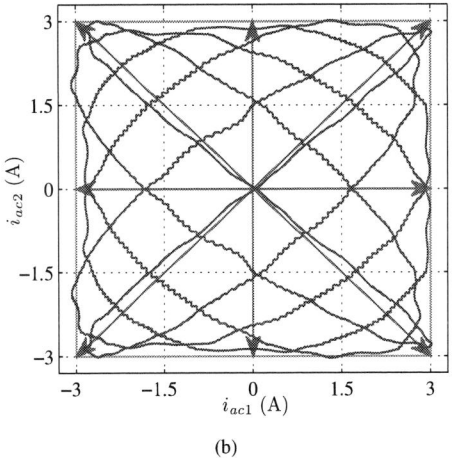

(b)

Fig. 6. Simulation results for Fq4c CSC: (a) Input and output currents, (b) Simulated Lissajous' figure result drawn in a $i_{ac1} \times i_{ac2}$ plane.

Semiconductor) connected in series with IGBTs (IRG4PC40U, from International Rectifier). Due to the fact that the Fq4c is the Lq3c with two switches crossing the middle points of the converter, the Lq3c can be obtained by simply forcing the switches m_2 and m_3 on a permanent OFF-state.

Since the Lq3c is limited, the reference signals used on the experimental result are: $m_{ac_1} = m_{ac_2} = 1$, $f_{ac_1} = f_{ac_2} = 60$Hz and $\phi_{ac_1} - \phi_{ac_2} = -120^\circ$. Its experimental results are shown in Fig. 9. As expected, the resulting vector from the output currents will not pass the boundary of the hexagon that delimits the limited-quadrant operation area.

The experimental result of the Fq3c can be seen in Fig. 10, both time domain and Lissajous' figure. In this experiment, the references are: $m_{ac_1} = m_{ac_2} = 1$, $f_{ac_1} = f_{ac_2} = 60$Hz and $\phi_{ac_1} - \phi_{ac_2} = 90^\circ$. It can be seen that the Lissajous' figure occupies the whole vector map, validating the full-quadrant operation. Since there is no quadrant limitation, the DC-Bus has a lower current than that would be necessary for a limited-quadrant converter.

In a limited-quadrant converter, to achieve a 3A output current (peak value), the DC-Bus current should be at least 6A, due to the quadrant-limitation. In other words, the limited-quadrant converter (to avoid overmodulation) should have a modulation index (m_{ac_j}) limited by 0.5, while the full-quadrant converters have no restrictions.

Despite the experimental results have the same modulation

978-1-4673-9551-9/16 $31.00 © 2016 IEEE 1037

(a)

(b)

Fig. 7. Experimental setup. (a) Implemented Lq2c. (b) Implemented Fq4c/Lq3c.

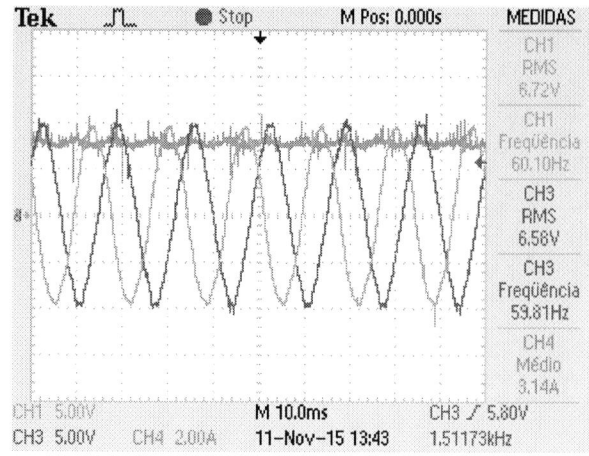

(a)

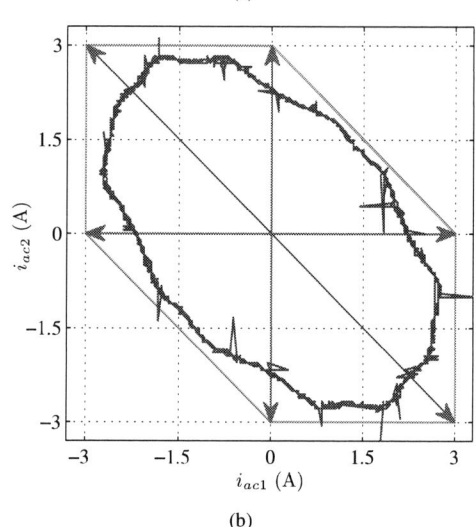

(b)

Fig. 8. Experimental results for Lq2c CSC. (a) Waveforms: CH1: v_{ac1}, CH3: v_{ac2} and CH4: I_{dc}. (b) Experimental Lissajous' figure result drawn in a $i_{ac1} \times i_{ac2}$ plane.

index and frequency, all the converters can generate output currents with different amplitudes and frequencies, as can be seen in the simulation result of the Fq4c. For a more intuitive Lissajous' plot, same amplitude and frequency were chosen.

V. CONCLUSION

Single-phase AC converters are widely applied. However, VSCs converters are more used then CSCs due the losses issues and lack of researches in CSCs. Nevertheless, this paper presents four topologies of single-phase CSC with double outputs that share the same DC-bus, which have their own advantages and disadvantages compared to each other, such as optimization of DC-bus current, total number of switches and number of conducting switches (see Table I). Results show that the converter can control two AC output currents independently without interference between converters. By the stipulated criteria, the Fq3c is the best converter, since its full-quadrant operation overcomes the disadvantage of three conducting switches per switching cycle. Several applications can be imagined for these CSCs topology, such as AC-AC converters, uninterruptible power supply (UPS), universal power filter (UPF), single-phase power decoupling, motor drive, multi-level converter and etc. Details involving SPWM approach and control strategy were presented. Then, the theoretical approach was validated by simulation and experimental results.

ACKNOWLEDGMENT

The authors gratefully acknowledge the financial support in part by the PPgEE/UFCG, by the National Council for Scientific and Technological Development (CNPq), by the Coordination of Improvement of Higher Education Personnel (CAPES), and by the Brazilian Electric Power Plans S.A. (ELETROBRAS).

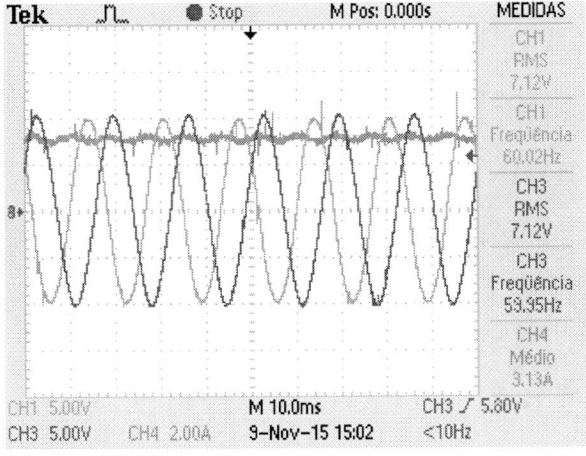

(a)

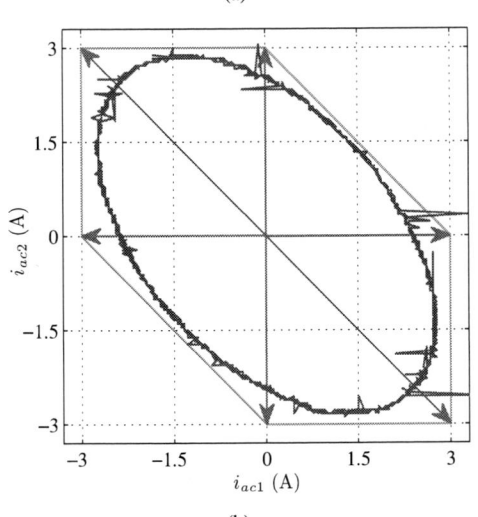

(b)

Fig. 9. Experimental results for Lq3c CSC: (a) Waveforms: CH1: v_{ac1}, CH3: v_{ac2} and CH4: I_{dc}. (b) Experimental Lissajous' figure result drawn in a $i_{ac1} \times i_{ac2}$ plane.

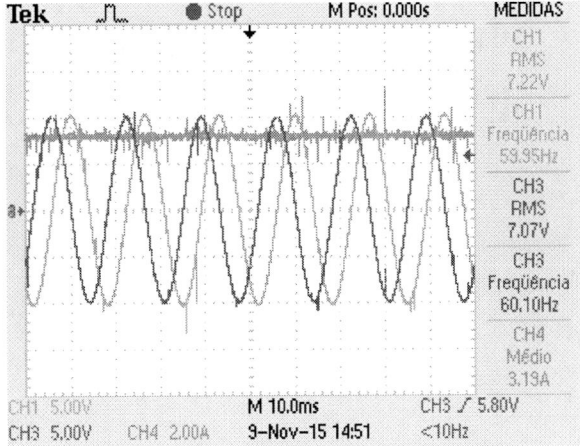

(a)

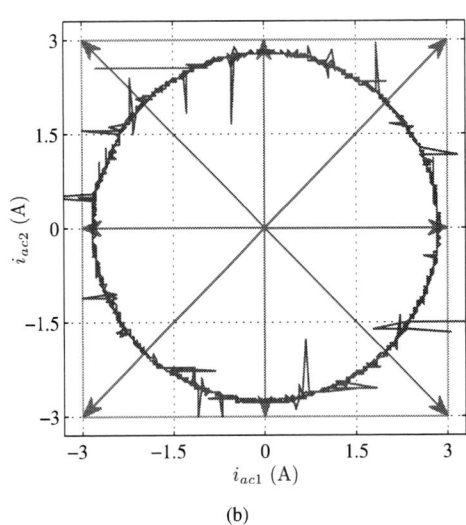

(b)

Fig. 10. Experimental results for Fq3c CSC: (a) Waveforms: CH1: v_{ac1}, CH3: v_{ac2} and CH4: I_{dc}. (b) Experimental Lissajous' figure result drawn in a $i_{ac1} \times i_{ac2}$ plane.

REFERENCES

[1] Q. Huang, Y. Song, X. Sun, L. Jiang, and P. Pong, "Magnetics in smart grid," *Magnetics, IEEE Transactions on*, vol. 50, no. 7, pp. 1–7, July 2014.

[2] H. Chan, K. Cheng, T. Cheung, and C. Cheung, "Study on magnetic materials used in power transformer and inductor," in *Power Electronics Systems and Applications, 2006. ICPESA '06. 2nd International Conference on*, Nov 2006, pp. 165–169.

[3] T. Naito, M. Takei, M. Nemoto, T. Hayashi, and K. Ueno, "1200v reverse blocking igbt with low loss for matrix converter," in *Power Semiconductor Devices and ICs, 2004. Proceedings. ISPSD '04. The 16th International Symposium on*, may 2004, pp. 125 – 128.

[4] S. Azmi, K. Ahmed, S. Finney, and B. Williams, "Comparative analysis between voltage and current source inverters in grid-connected application," in *Renewable Power Generation (RPG 2011), IET Conference on*, Sept 2011, pp. 1–6.

[5] B. Sahan, A. Vergara, N. Henze, A. Engler, and P. Zacharias, "A single-stage pv module integrated converter based on a low-power current-source inverter," *Industrial Electronics, IEEE Transactions on*, vol. 55, no. 7, pp. 2602–2609, July 2008.

[6] C. Klumpner, "A new single-stage current source inverter for photovoltaic and fuel cell applications using reverse blocking igbts," in *Power Elec-*

tronics Specialists Conference, 2007. PESC 2007. IEEE, june 2007, pp. 1683 –1689.

[7] M. Vitorino, L. Hartmann, D. Fernandes, E. Silva, and M. Correa, "Single-phase current source converter with new modulation approach and power decoupling," in *Applied Power Electronics Conference and Exposition (APEC), 2014 Twenty-Ninth Annual IEEE*, March 2014, pp. 2200–2207.

[8] M. Vitorino, M. Correa, L. Costa, L. Hartmann, and D. Fernandes, "Double four-quadrants single-phase current source converter sharing the same dc-bus," in *Energy Conversion Congress and Exposition (ECCE), 2014 IEEE*, Sept 2014, pp. 2801–2808.

[9] F. Zapata, L. Salazar, and J. Espinoza, "Analysis and design of a single-phase pwm current source rectifier with neutral leg," in *Industrial Electronics Society, 1998. IECON '98. Proceedings of the 24th Annual Conference of the IEEE*, vol. 1, aug-4 sep 1998, pp. 519 –524 vol.1.

Multiple-Output Boost Resonant Inverter for High Efficiency and Cost-Effective Induction Heating Applications

Hector Sarnago, Oscar Lucia, and Jose M. Burdio
Department of Electronic Engineering and Communications
Universidad de Zaragoza
Zaragoza, Spain
hsarnago@unizar.es

Abstract— Induction heating is an efficient and high performance heating method which already has a significant market penetration and economic impact due to its benefits inherent to contactless energy transfer systems such as quickness, cleanness, and safety.

In order to improve the efficiency and performance of modern induction heating systems, this paper presents a multiple-output boost resonant inverter. The proposed topologies significantly improves the converter efficiency due to the reduced current level while being able to supply several induction loads. Consequently, a high-performance and cost-effective implementation is achieved. The proposed converter has been tested through a 3.6-kW dual-output prototype, proving the feasibility of this proposal.

Keywords—Resonant power conversion, efficiency, home appliances, induction heating.

I. INTRODUCTION

High efficiency [1-4] and high performance [5-8] are the two main pillars that have made induction heating (IH) technology a successful product with a market share higher than 50% in many countries. Considering the mass-market orientation, those features must be achieved in a cost-effective way, creating a challenging power electronic design.

State of the arte technology feature single-output resonant converters which are combined to build 3-4 burner induction hobs [9]. However, current high performance IH design trends pursues the design of flexible multi-coil systems [10-14] (Fig. 1 (a)). Such systems, require the design of a specific power converter able to supply multiple IH loads. Usually, the half-bridge series resonant inverter [15, 16] (Fig. 1 (b)) is used to its good trade-off between cost and performance. Moreover, the use of a boost half-bridge resonant inverter has also been proposed to increase the efficiency [17-19]. However, such solutions cannot be directly applied to multi-coil systems due to its high cost. The aim of this paper is therefore to propose a multiple-output boost resonant inverter able to supply multi-coil induction heating systems with high performance and efficiency in a cost-effective way.

(a)

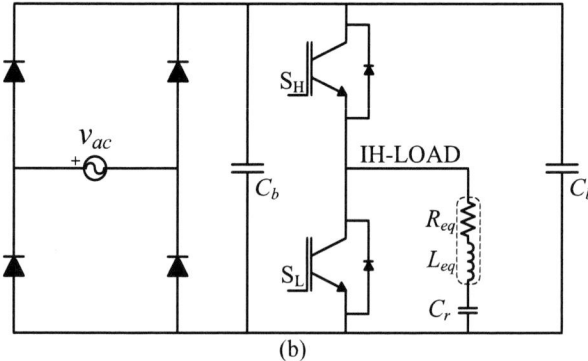

(b)

Fig. 1.   Multi-coil induction heating system (a) and half-bridge series resonant inverter (b).

The remainder of this paper is organized as follows. Section II introduces the proposed multiple-output boost resonant inverter topology. Section III presents the main analytical and simulation results detailing the converter operation. Section IV summarizes the main experimental results and, finally, Section V draws the main conclusions of this paper.

978-1-4673-9551-9/16 $31.00 © 2016 IEEE

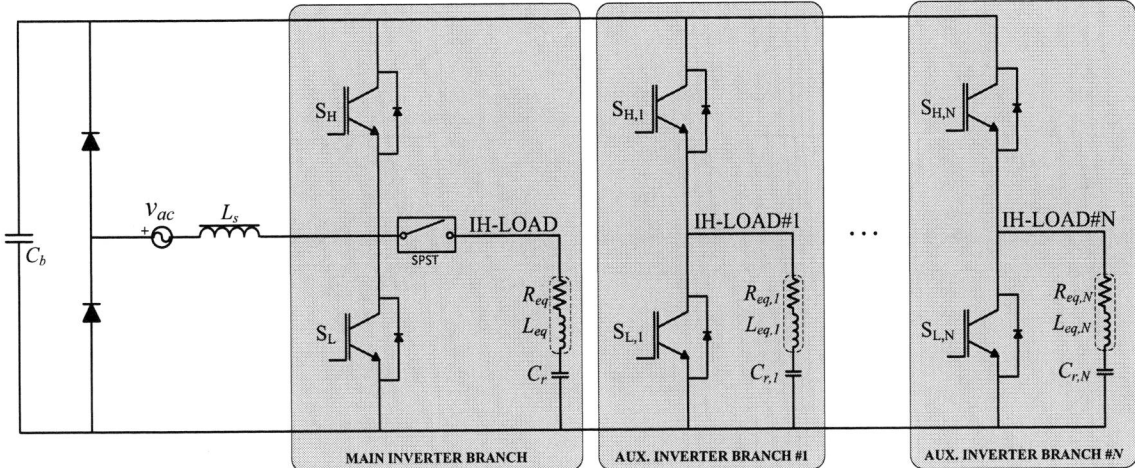

Fig. 2.   Multiple-output boost resonant inverter for multi-coil induction heating systems.

II.   MULTIPE-OUTPUT BOOST RESONANT INVERTER

The proposed multiple-output boost resonant inverter is shown in Fig. 2. It is composed of a main inverter branch following the half-bridge structure which perform both the boost and inverter function, and several auxiliary inverter branches, which operate as.standard series-resonant half-bridge inverters. Each induction heating load is modeled as the series connection of a resistor and an inductor, R_{eq}, L_{eq}, and a resonant capacitor is added, C_r, to complete the resonant tank. Besides, the boost inductor L_s is connected to the main inverter branch medium point, operating as a boost inverter. An electromechanical relay is used to disconnect the main load in case that it do not need to be activated, enabling the boost converter to continue normal operation.

The main benefit of such approach is that a variable bus voltage is obtained, enabling higher efficiency and improved output power control. Moreover, since the components of the main inverter branch, including the boost inductor L_s are common, a cost-effective implementation is achieved. When the power converter needs to operate without the main inverter load, a relay is used to turn off the load while enabling controlling the bus voltage.

III.   ANALYTICAL AND SIMULATION RESULTS

The proposed converter can be easily analyzed as the combination of a boost converter plus several parallel connected series resonant inverters, where the bus voltage is controlled by the first stage. Consequently, the bus voltage common for the auxiliary inverter branches can be calculated as:

$$V_b = \frac{1}{1-D}V_{ac} \qquad (1)$$

where D is the main inverter duty cycle and V_{ac} is the mains voltage.

In this converter, the maximum output power in the main inverter branch is obtained at the resonant frequency that is calculated as follows:

$$f_o = \frac{1}{2\pi\sqrt{L_{eq}C_{eq}}}. \qquad (2)$$

Thus, the maximum output power as a function of the duty cycle can be therefore calculated as:

$$P_{1,f_s=f_o} = \left[2\left(\frac{V_{ac}}{1-D}\right)^2 \middle/ \pi^2 R_{eq}\right]sen^2\left(\pi D\right). \qquad (3)$$

It is important to note that, since the main inverter branch operates as a boost converter, the higher the duty cycle the higher the output power. On the other, hand, the output power at the resonant frequency for the remaining auxiliary branches can be calculated as

$$P_{1,f_s=f_o} = \left[2\left(\frac{V_{ac}}{1-D}\right)^2 \middle/ \pi^2 R_{eq}\right]sen^2\left(\pi D_i\right), \qquad (4)$$

where the maximum output power for a given bus voltage, i.e. a given D, is obtained for $D_i=0.5$ and can be reduced from this point. Fig. 3 shows the output power as a function of the switching frequency and duty cycle for the common (a) and auxiliary branches (b). In this figure, it can be seen that the proposed converter enables a full control of the output power in the complete reange, which is a mandatory feature for domestic induction heating. In this figures, the zero voltage switching region is included, which ensures the efficient and safe converter operation. Next section detail the experimental prototype and main experimental results used to prove the proper converter operation.

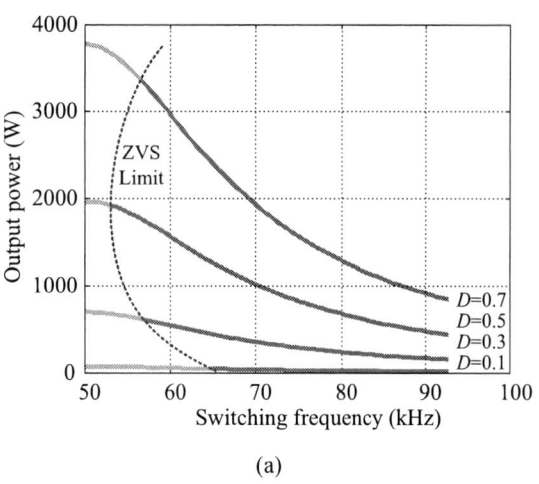

(a)

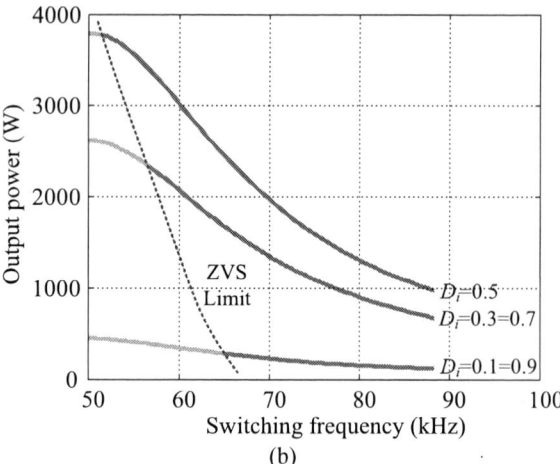

(b)

Fig. 3. Simulation results: output power control for the main branch (a) and for the auxiliary branches (b).

IV. EXPERIMENTAL RESULTS

In order to prove the feasibility of the proposed power converter, a 3.6 kW experimental prototype for induction heating applications has been designed and implemented (Fig. 4). The proposed converter is a dual-output inverter controlled by a FPGA which enables USB communication with a PC [20-22]. The inductor current as well as the mains voltage are measured to compute the output power and the main induction heating operation. This enables proper output power control as well as ensuring the converter reliability. Besides, two electromechanical relays enables the induction load reconfiguration for extended flexibility and improved efficiency and output power control.

Fig. 5 shows the experimental measurements for different duty cycles in the main inverter branch. As it shown, the dc-ink voltage is controlled by means of the duty cycle, controlling the overall output power, whereas the applied power to the IH-load can be varied according to the switching frequency. Compare to conventional resonant inverters, this approach enables minimizing the current through the switching devices and inductors, improving the converter efficiency.

Finally, Fig. 6 shows the experimental measurements of the proposed converter operating with two loads simultaneously. These results prove the proper converter operation and established the proposed multiple-output boost resonant inverter as a new state-of-the-art topology in terms of efficiency and performance for domestic induction heating applications.

Fig. 4. Multiple-output boost resonant inverter for multi-coil induction heating systems.

V. CONCLUSIONS

In this paper, a multiple-output boost resonant inverter has been presented. The proposed topology achieves high efficiency due to the reduced current through the inductor and devices, while allowing a high performance due to the accurate output power control in each load. Moreover, the proposed topology allows reducing the number of inductors required, enabling a significant cost reduction essential for this application. A 3.6-kW dual-output converter has been designed and implemented obtaining satisfactory results and proving the feasibility of this proposal. As a conclusion, the proposed topology enables the implementation of high-efficiency and cost-effective resonant inverters further improving domestic induction heating applications.

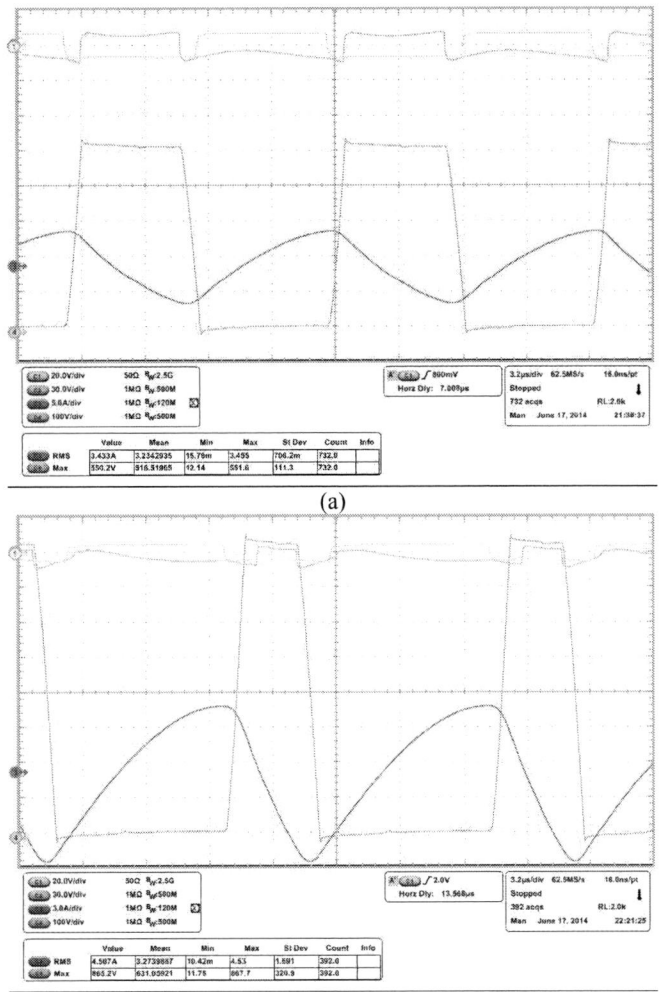

(a)

Fig. 5. Experimental measurements for different duty cycles, showing the effect of the variable dc-link voltage. Results for D=0.5 (a) and D=0.7 (b). From top to bottom: gating signals, inverter voltage and IH load current.

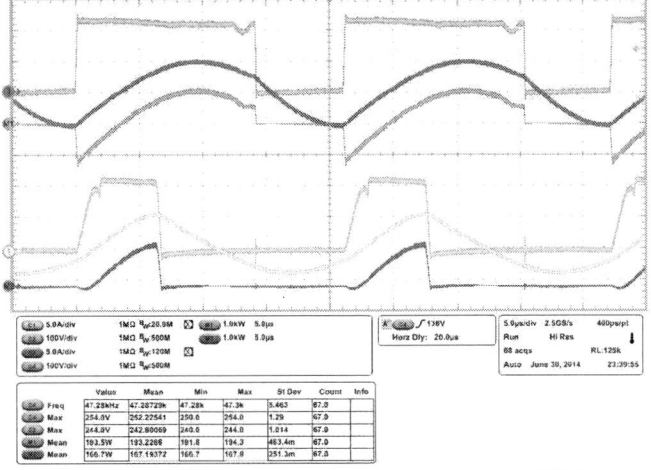

Fig. 6. Experimental measurements for the case of N=2 active loads at the same frequency. The output power is controlled via the duty cycle. From top to bottom: inverter output voltage and load current for the main branch and the auxiliary branch. The instantaneous output power trace has been included in both loads.

ACKNOWLEDGMENT

This work was partly supported by the Spanish MINECO under Project TEC2013-42937-R, Project CSD2009-00046, and Project RTC-2014-1847-6, by the DGA-FSE, by the University of Zaragoza under Project JIUZ-2014-TEC-08, and by the BSH Home Appliances Group.

REFERENCES

[1] H. Sarnago, O. Lucía, A. Mediano, and J. M. Burdío, "High efficiency parallel quasi-resonant current source inverter featuring SiC MOSFETs for induction heating systems with coupled inductors," *IET Power Electronics,* vol. 6, pp. 183-191, 2013.

[2] H. Sarnago, A. Mediano, and O. Lucía, "High efficiency ac-ac power electronic converter applied to domestic induction heating," *IEEE Transactions on Power Electronics,* vol. 27, pp. 3676-3684, August 2012.

[3] O. Lucía, J. M. Burdío, I. Millán, J. Acero, and L. A. Barragán, "Efficiency oriented design of ZVS half-bridge series resonant inverter with variable frequency duty cycle control," *IEEE Transactions on Power Electronics,* vol. 25, pp. 1671-1674, July 2010.

[4] H. Sarnago, O. Lucía, A. Mediano, and J. M. Burdío, "Analysis and design of high-efficiency resonant inverters for domestic induction heating applications," *International Journal of Applied Electromagnetics and Mechanics,* vol. 44, pp. 201-208, 2014.

[5] O. Lucía, J. Acero, C. Carretero, and J. M. Burdío, "Induction heating appliances: Towards more flexible cooking surfaces," *IEEE Industrial Electronics Magazine,* vol. 7, pp. 35-47, September 2013.

[6] V. Esteve, J. Jordán, E. Sanchis-Kilders, E. J. Dede, E. Maset, J. B. Ejea, *et al.,* "Comparative Study of a Single Inverter Bridge for Dual-Frequency Induction Heating Using Si and SiC MOSFETs," *IEEE Transactions on Industrial Electronics,* vol. 62, pp. 1440-1450, 2015.

[7] V. Esteve, J. Jordan, E. Sanchis Kilders, E. Dede, E. Maset, J. Ejea, *et al.,* "Comparative study of a single inverter bridge for dual-frequency induction heating using Si and SiC MOSFETs," *IEEE Transactions on Industrial Electronics,* vol. PP, pp. 1-1, 2014.

[8] V. Esteve, J. Jordan, E. Sanchis Kilders, E. Dede, E. Maset, J. Ejea, *et al.,* "Improving the reliability of series resonant inverters for induction heating applications," *IEEE Transactions on Industrial Electronics,* vol. 61, pp. 2564-2572, May 2014.

[9] O. Lucía, J. M. Burdío, I. Millán, J. Acero, and D. Puyal, "Load-adaptive control algorithm of half-bridge series resonant inverter for domestic induction heating," *IEEE Transactions on Industrial Electronics,* vol. 56, pp. 3106-3116, August 2009.

[10] O. Lucía, J. M. Burdío, L. A. Barragán, J. Acero, and C. Carretero, "Series resonant multi-inverter with discontinuous-mode control for improved light-load operation," *IEEE Transactions on Industrial Electronics,* vol. 58, pp. 5163-5171, November 2011.

[11] O. Lucía, J. M. Burdío, L. A. Barragán, J. Acero, and I. Millán, "Series-resonant multiinverter for multiple induction heaters," *IEEE Transactions on Power Electronics,* vol. 24, pp. 2860-2868, November 2010.

[12] O. Lucía, C. Carretero, J. M. Burdío, J. Acero, and F. Almazán, "Multiple-output resonant matrix converter for multiple induction heaters," *IEEE Transactions on Industry Applications,* vol. 48, pp. 1387-1396, July/August 2012.

[13] F. Forest, S. Faucher, J.-Y. Gaspard, D. Montloup, J.-J. Huselstein, and C. Joubert, "Frequency-synchronized resonant converters for the supply of multiwindings coils in induction cooking appliances," *IEEE Transactions on Industrial Electronics,* vol. 54, pp. 441-452, February 2007.

[14] F. Forest, E. Labouré, F. Costa, and J.-Y. Gaspard, "Principle of a multi-load/single converter system for low power induction heating," *IEEE Transactions on Industrial Electronics,* vol. 15, pp. 223-230, March 2000.

[15] H. Sarnago, O. Lucía, A. Mediano, and J. M. Burdío, "Analytical model of the half-bridge series resonant inverter for improved power conversion efficiency and performance," *IEEE Transactions on Power Electronics,* vol. 30, pp. 4128-4143, August 2015.

[16] T. Mishima, C. Takami, and M. Nakaoka, "A new current phasor-controlled ZVS twin half-bridge high-frequency resonant inverter for induction heating," *IEEE Transactions on Industrial Electronics,* vol. 61, pp. 2531-2545, May 2014.

[17] H. Sarnago, O. Lucía, A. Mediano, and J. M. Burdío, "Direct ac-ac resonant boost converter for efficient domestic induction heating applications," *IEEE Transactions on Power Electronics,* vol. 29, pp. 1128-1139, March 2014.

[18] N. A. Ahmed and M. Nakaoka, "Boost-half-bridge edge resonant soft switching PWM high-frequency inverter for consumer induction heating appliances," *IEE Proceedings on Electric Power Applications,* vol. 153, pp. 932-938, 2006.

[19] B. Saha, K. Soon Kurl, N. A. Ahmed, H. Omori, and M. Nakaoka, "Commercial frequency AC to high frequency AC converter with boost-active clamp bridge single stage ZVS-PWM inverter," *IEEE Transactions on Power Electronics,* vol. 23, pp. 412-419, January 2008.

[20] O. Jiménez, O. Lucía, I. Urriza, L. A. Barragán, and D. Navarro, "Power measurement for resonant power converters applied to induction heating applications," *IEEE Transactions on Power Electronics,* vol. 29, pp. 6779-6788, December 2014.

[21] O. Jiménez, O. Lucía, I. Urriza, L. A. Barragán, and D. Navarro, "Analysis and implementation of FPGA-based online parametric identification algorithms for resonant power converters," *IEEE Transactions on Industrial Informatics,* vol. 10, pp. 1144-1153, May 2014.

[22] O. Jiménez, O. Lucia, I. Urriza, L. A. Barragán, P. Mattavelli, and D. Boroyevich, "FPGA-based gain-scheduled controller for resonant converters applied to induction cooktops," *IEEE Transactions on Power Electronics,* vol. 29, pp. 2143-2152, April 2014.

Development of 2-kW interleaved DC-capacitor-less single-phase inverter system

Runruo Chen, Hulong Zeng, Deepak Gunasekaran, Yunting Liu, and Fang Z. Peng
Department of Electrical and Computer Engineering
Michigan State University
East Lansing, MI 48824 USA
chenrunr@msu.edu

Abstract—This paper presents the development and experimental performance of a 2-kW interleaved DC-capacitor-less single-phase inverter. By utilizing an AC capacitor to absorb the ripple power pulsating at twice the line frequency, the proposed interleaved DC-capacitor-less single-phase inverter significantly reduced the total capacitor size. The adoption of SiC MOSFET with 216-kHz switching frequency and the interleaved structure further reduce the filtering components size. A comparison of existing topologies for inverter application and the design considerations of interleaved PWM scheme and integrated coupled inductors are described in details. Experimental results are provided to verify the performance and feasibility of the proposed inverter system.

Keywords—inverter; AC capacitor; SiC MOSFET; interleave; coupled inductor

I. INTRODUCTION

H-bridge inverter has been widely used for single phase inverter system. However, conventional H-bridge inverter requires a bulky DC capacitor decoupling the unbalance AC and DC power to maintain a smooth DC link voltage. The bulky DC capacitor become a big limitation to achieve high power density system.

In order to reduce the DC capacitor bank, a great deal of research has been done[1-10]. The basic idea is using extra energy-storage components such as capacitor or inductor, which permits much larger fluctuation of the voltage or current, to balance the 2ω ripple power. Because the energy density of capacitor is much higher than that of inductors, the topologies based on capacitive energy storage are investigated [2-7]. Fig. 1 shows the four typical voltage source inverter topologies with extra energy storage capacitor for ripple power.

The key point is how to minimize the energy-storage capacitor, and devices' voltage/current stress. One effective way to evaluate the topology and control method is to compare the energy requirement of energy-storage capacitor, E_C, which is the indication of the capacitor size [1] and the total device power rating, *TDPR*, which is the indication of how much total silicon area is needed for the semiconductor devices. For single-phase inverter application, the topology in Fig. 1 (d) using SVPWM control shows advantages in terms of E_C and *TDPR* and the detailed analysis and comparison is shown in section II.

In order to further reduce the conduction loss and the input/output current ripple, the two channel interleaved structure is adopted. Since the system in Fig. 1 (d) is controlled as an unbalance three phase system[7], the THD of line-to-line voltage is not optimized by using interleaved phase-shifted PWM method for single phase inverter. In order to achieve the better THD performance, an enhanced interleaved phase-shifted PWM method for three phase system is adopted[8].

For interleaved operation of inverters, one shortcoming is circulating currents. The inversely coupled inductor can substantially reduce the circulating currents because it can build up high impendence to the circulating path[9-10]. In order to achieve higher power density, the inversely coupled inductor and the filtering inductor is integrated by sharing the core and windings. The design of the integrated magnetic component is shown in Section III.

Finally, an interleaved DC-capacitor-less inverter system shown in Fig. 3 is proposed. The proposed system utilizes an additional phase leg and an AC capacitor, C_{ac} to absorb the ripple power pulsating at twice the line frequency (2ω ripple power). By adopting space vector PWM (SVPWM) control, the voltage and current stress of the switches can be minimized. By using the SiC MOSFET, the switching frequency is boosted to 216 kHz, and the filter size is further minimized by integrating the magnetic components.

II. COMPARSION OF TOPLOGIES FOR INVERTER APPLICATION

Fig. 1 shows the four typical voltage source inverter topologies with extra energy storage capacitor for 2ω ripple power. Based on the power of the system, the minimal capacitor energy storage requirement, Ec_{min} can be calculated by [1]:

$$Ec_{min} = \frac{S}{2\pi f_o} , \qquad (1)$$

where S is the apparent power of the system and f_o is the line frequency.

For conventional H-bridge, energy-storage capacitor is the DC capacitor; and $Ec_{H-bridge}$ can be calculated by:

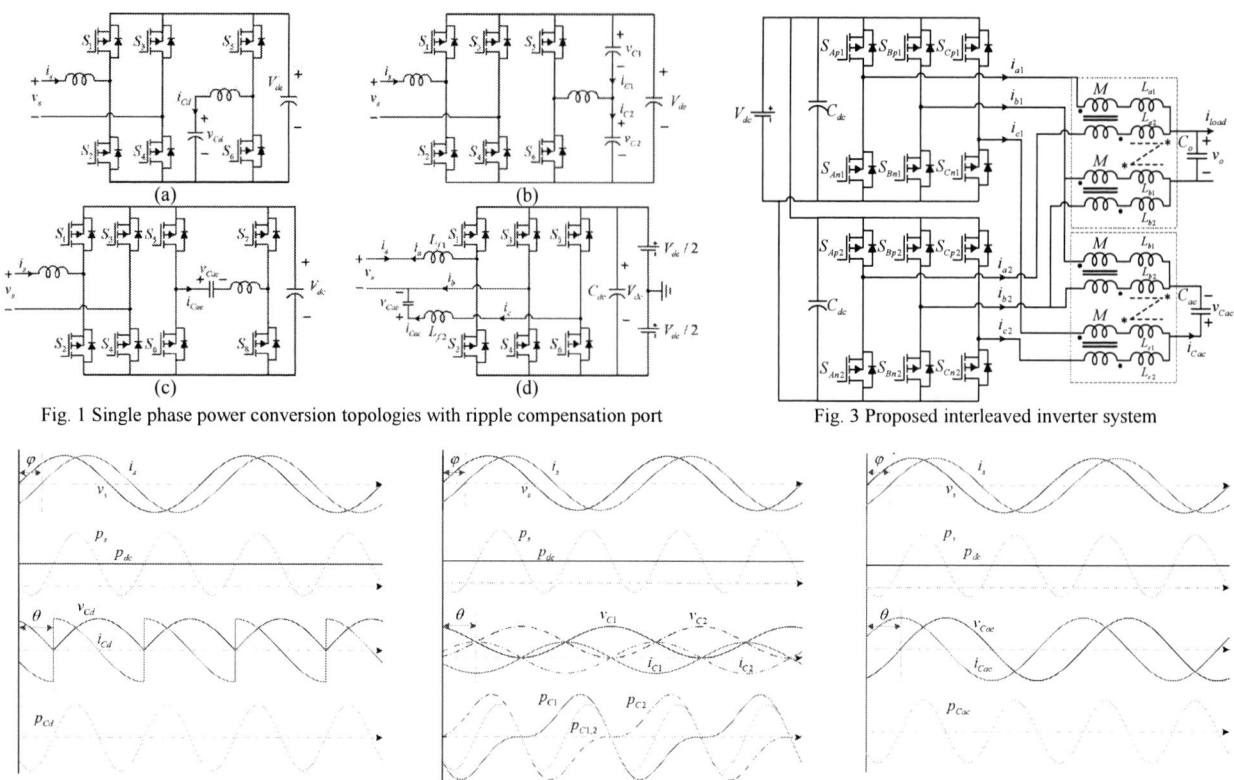

Fig. 1 Single phase power conversion topologies with ripple compensation port

Fig. 3 Proposed interleaved inverter system

Fig. 2 Key waveforms of single phase power conversion topologies with ripple compensation port

$$Ec_{H-bridge} = \frac{1}{2} C_{dc\_H} V_{dc}^2 = \frac{S \cdot V_{dc}}{4\pi f_o \Delta V_{dc}} = \frac{V_{dc}}{2\Delta V_{dc}} Ec_{min} , \quad (2)$$

where V_{dc} is the average voltage across the DC capacitor, ΔV_{dc} is the allowed peak-to-peak DC voltage ripple. And the E_C will be 20 times of Ec_{min} if the DC voltage ripple is designed within 2.5% of V_{dc}.

The *TDPR* of H-bridge can be calculated by:

$$TDPR_{H-bridge} = 4V_{dc} \cdot \sqrt{2}I_s = 8S . \quad (3)$$

In Fig. 1 (a), the energy-storage capacitor, C_d is connected to DC link by a bi-directional buck converter, and the capacitor voltage is unipolar [2,3]. If the capacitor voltage reference is a full-wave rectified sine wave, the 2ω ripple power can be transferred to the capacitor as it shown in Fig. 2 (a). The energy storage requirement, Ec is Ec_{min}, and *TDPR* is 1.5 times of $TDPR_{H-bridge}$.

However, the full-wave rectified sine reference, which contains rich harmonics, is difficult for the control system to track. Although it is possible to lower the harmonic content in the reference by increasing the energy-storage margin (therefore the voltage does not go down to zero), this will compromise full utilization of the energy-storage capacitor. Therefore, in real application, Ec will be larger than the ideal case, Ec_{min}.

In Fig. 1 (b), the PWM rectifier system consists of an H-bridge and a half-bridge circuit for ripple power compensation

[4]. The voltage/current waveform and power are shown in Fig. 2 (b). The power of *C1* and *C2* is controlled equal to the grid 2ω ripple power. The advantage is the *C1* and *C2* are also used for the DC filter capacitor. However, Ec is 4 times of Ec_{min} and *TDPR* is 2 times of $TDPR_{H-bridge}$.

A single-phase PWM rectifier with the power decoupling ripple-port shown in Fig. 1 (c) is proposed in [5]. By adding any extra H-bridge to interface the energy-storage capacitor, the capacitor works in AC mode and the voltage/current waveforms are sinusoidal as it shown in Fig. 2 (c). Ec is Ec_{min} and *TDPR* is 2 times of $TDPR_{H-bridge}$.

In [6,7], the PWM rectifier system shown in Fig. 1 (d) consists of an H-bridge and an additional phase leg connected to an AC capacitor, C_{ac}. The AC capacitor's voltage/current waveforms are same as circuit in Fig. 1 (c). However by adopting SVPWM to control the system as an unbalance three phase system, the voltage and current stress will be the same as the H-bridge system, Ec is Ec_{min} and *TDPR* is 1.5 times of $TDPR_{H-bridge}$.

The energy storage requirement, Ec and total device power rating *TDPR* for circuit in Fig. 1 (a) ~ (d) is summarized in Table I. Among the four topologies, circuit in Fig. 1 (d) has the minimal Ec and *TDPR*, in other words, the capacitor size and cost of semiconductor devices is minimized.

The proposed interleaved DC-capacitor-less inverter system shown in Fig.3 has a paralleled interleaved inverter shown in Fig.1 (d). The phase-shift PWM scheme and

978-1-4673-9551-9/16 $31.00 © 2016 IEEE   1046

integrated coupled inductor design for the interleaved structure are presented in the following sections.

Table I COMPARSION OF EC AND TDPR FOR DIFFERENT TOPOLOGIES

Topologies	Ec	$TDPR$
Fig.1 (a)	$>Ec_{min}$	$1.5\ TDPR_{H\text{-}bridge}$
Fig.1 (b)	$4Ec_{min}$	$2\ TDPR_{H\text{-}bridge}$
Fig.1 (c)	Ec_{min}	$2\ TDPR_{H\text{-}bridge}$
Fig.1 (d)	Ec_{min}	$1.5\ TDPR_{H\text{-}bridge}$

III. PHASE-SHIFTED PWM SCHEME

The interleaving technique is applied to achieve higher equivalent switching frequency and reduce the filter size. The Thevenin-equivalent output voltage between phase A and phase B becomes:

$$v_{ab\_eq} = \frac{v_{a1} + v_{a2} - v_{b1} - v_{b2}}{2}. \qquad (4)$$

For conventional interleaved single phase inverter system, phase-shifted PWM scheme adopts the two carriers (v_{carr1}, and v_{carr2}) with phase angles of 0°, 90° respectively; and the output voltage can achieve best THD. As it shown in Fig.4 (a), the equivalent voltage switching only happens between adjacent levels.

However for the proposed the system, where SVPWM is adopted, although the interleaving technique yields the best attainable phase output voltage in terms of THD with 180°

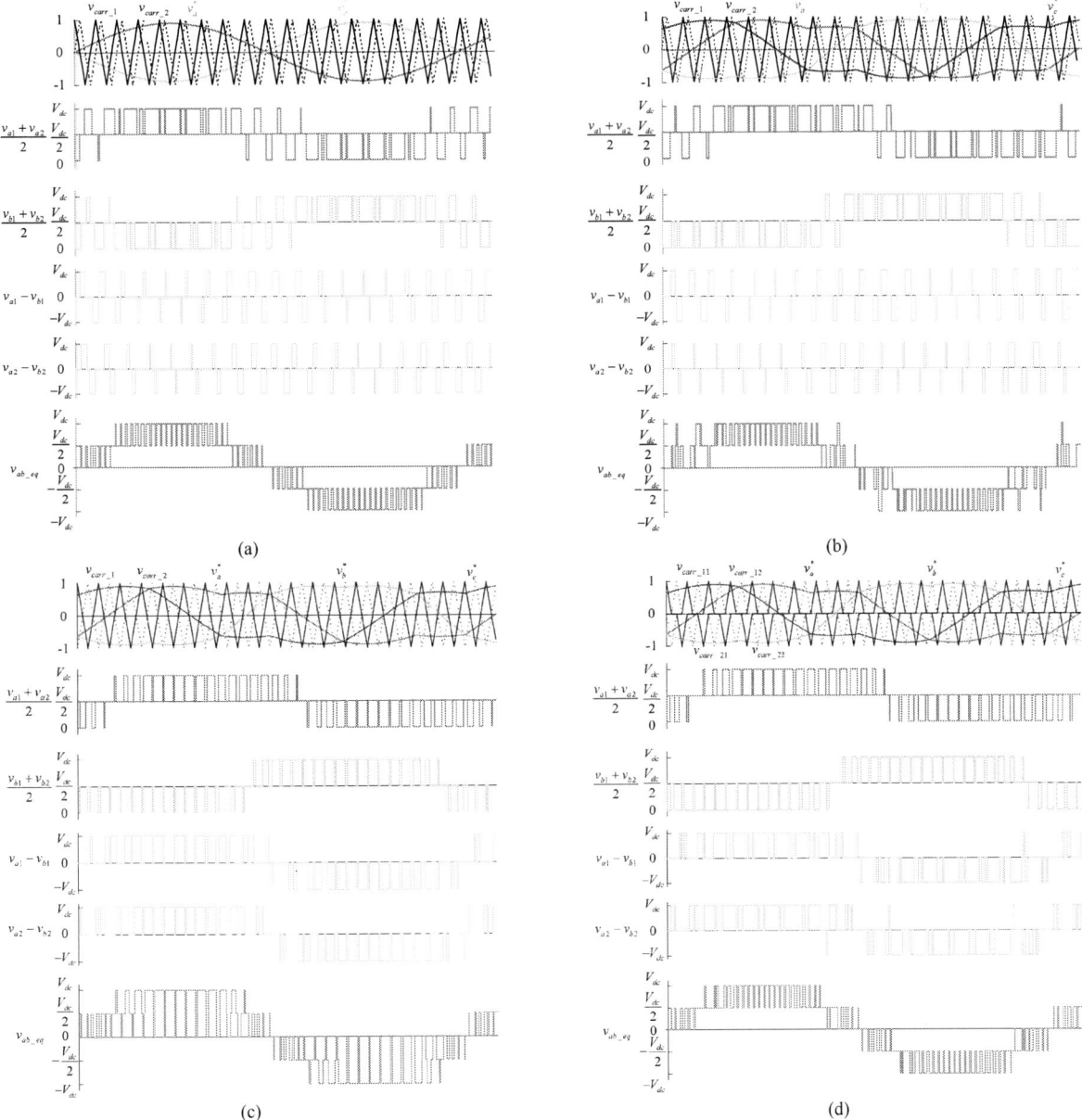

Fig.4 Key waveforms for different PWM carriers' scheme

phase shift of the PWM carriers, it is not the case for line-to-line output voltages [8].

Fig.4 (b) and (c) illustrates the cases with 90° and 180° phase shift of PWM carriers for the proposed system respectively. The reference signals (v_a^*, v_b^*, and v_c^*) are compared to their respective carrier signal to set the ON–OFF state of the switches. In Fig.4 (b), the phase angles for the two carriers (v_{carr1}, and v_{carr2}) are 0°, 90°; and In Fig.4 (c), the phase angles for the two carriers (v_{carr1}, and v_{carr2}) are 0°, 180°. However, for both cases, during certain intervals, the equivalent lint-to-line voltage, v_{ab_eq} is switching among three adjacent levels.

In order to achieve better line-to-line voltage THD performance, phase-shifted PWM carrier disposition method [2] is adopted, which provides the equivalent line-to-line voltage, v_{ab_eq}. As it shown in Fig.4 (b), two sets of carriers are used. The phase angles for the two carriers in set 1 (v_{carr11}, and v_{carr12}) are 0°, 180° respectively; the phase angles for the two carriers in set 2 (v_{carr21}, and v_{carr22}) are 90°, 270° respectively. When the reference signals large than 0, set 1 carries are used; When the reference signals smaller than 0, set 2 carries are used. As a result, V_{ab_eq} is only switching between adjacent levels.

IV. INTERGATED COUPLED INDUCTOR DESIGN

For two-level paralleled inverter system, interleaving operation can reduce the overall input or output current ripple. However, interleaving operation will create significant circulating currents between the paralleled two phases [9-10]. The inversely coupled inductor can substantially reduce the circulating current between the interleaved two phases.

An integrated coupled inductor can be built to achieve the purpose of suppressing circulating current and filtering output current at the same time, as indicated by the dashed box in Fig. 3. The currents through the inductors are split into common mode (CM) component and differential mode (DM) component. For phase A, the current of each leg will be:

$$
\begin{aligned}
i_{a1} &= 0.5(i_{a\_CM} + i_{a\_DM}), \\
i_{a2} &= 0.5(i_{a\_CM} - i_{a\_DM}).
\end{aligned}
\tag{5}
$$

The DM current sees high impedance produced by the mutual inductor, M. The DM voltage, v_{x1}-v_{x2} ($x=a,b,c$) in each phase is shown Fig. 4 (d). When DM voltage is given, the higher DM inductance, M means the smaller DM current. The maximum DM peak current, i_{DM_max} can be calculated by:

$$
i_{DM\_max} = \frac{V_{dc}}{4 f_{sw}(2M + L_a + L_b)}
\tag{6}
$$

The CM current is determined by the load current and output ripple current. Considering the dynamic response, the allowed maximum CM current, i_{CM_max} is designed to be 125% of the maximum load peak current.

The 3D structure of the integrated coupled inductor and the DM/CM flux path are shown in Fig. 5 and Fig. 6 respectively. The turn number of the winding is designed to be 23; the DM inductance and CM inductance are designed to be:

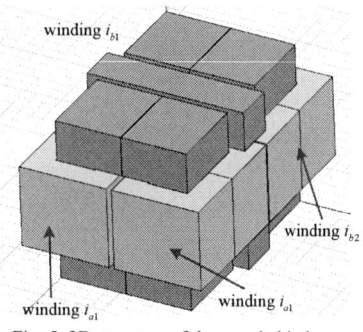

Fig. 5 3D structure of the coupled inductor

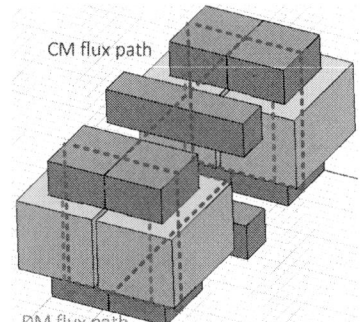

Fig.6 DM/CM flux path

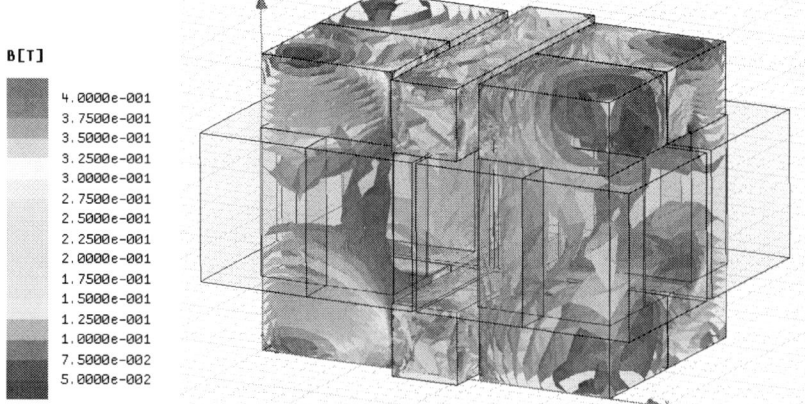

Fig.7 Simulation results

$$M = 500 \ \mu H,$$
$$L_{a1} = L_{a2} = L_{b1} = L_{b2} = 50 \ \mu H. \tag{7}$$

The maximum flux density happens when the DM current and the CM current both reach the maximum value. Fig. 7 shows the simulation results. When $i_{DM\_max} = 0.4A$ and $i_{CM\_max} = 10A$, the maximum flux density reaches 0.4T.

V. EXPERIEMNTAL RESULTS

The key parameters for the 2-kW inverter system are listed below in Table II.

For conventional H-bridge system, if the allowed DC voltage ripple is 2%, the DC capacitance is 2 mF (20 pu). By using the proposed inverter system, the DC capacitor is 20 μF (0.2 pu), and the AC capacitor is 100 μF (1 pu). Compared to conventional H-bridge system, the total size of the capacitor (including DC capacitor and AC capacitor) is reduced by 16 times.

Fig. 8 shows the 2-kW prototype which consists of main power board, DSP controller board, AC capacitor and integrated coupled inductors.

Fig. 9~11 shows the experimental waveforms. The equivalent line to line voltage shown in Fig. 9 jumps only between adjacent levels, which is consistent with the theoretical analysis.

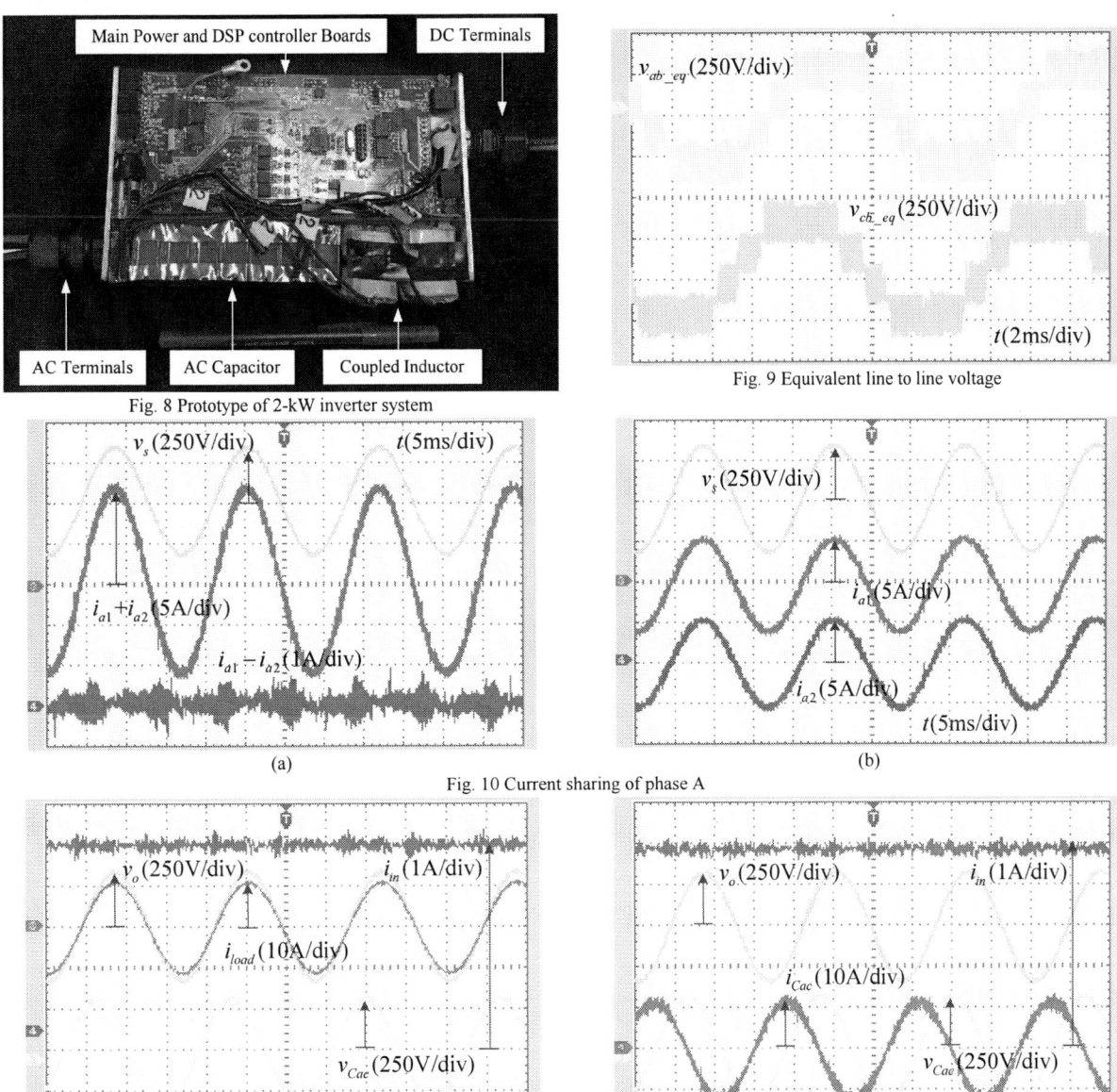

Fig. 8 Prototype of 2-kW inverter system

Fig. 9 Equivalent line to line voltage

Fig. 10 Current sharing of phase A

Fig. 11 Full load experimental waveforms

TABLE II KEY PARAMETERS

DC voltage, V_{dc}	400 V
DC capacitance, C_{dc}	20 µF (0.2 pu)
switching frequency, f_s	216 kHz
AC output voltage, V_s	240 VAC (60 Hz)
AC capacitance, C_{ac}	100 µF (1 pu)
sampling frequency, f_{sample}	54 kHz

Fig. 10 shows the current sharing experimental waveforms of phase A. The current difference between different channel is within 1A, indicating that the integrated coupled inductor has effectively suppressed the circulation current.

Fig. 11 shows the full load experimental waveforms (pf = 1). The DC input current ripple is within 0.5A, indicating that the AC capacitor has effectively absorb the 2ω ripple power.

Fig. 12 shows the efficiency test of the system. For rated power, the system can achieve the efficiency of 96.2%.

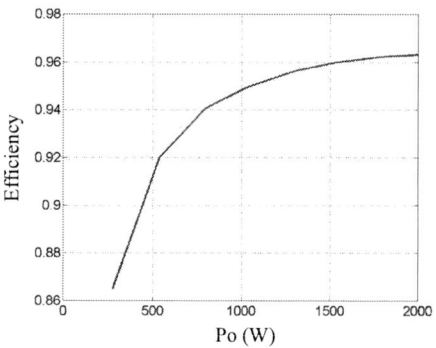

Fig. 12 Efficiency test

REFERENCES

[1] P. T. Krein, R. S. Balog, and M. Mirjafari, "Minimum Energy and Capacitance Requirements for Single-Phase Inverters and Rectifiers Using a Ripple Port," Power Electronics, IEEE Transactions on, vol. 27, pp. 4690-4698, 2012.

[2] W. Ruxi, F. Wang, D. Boroyevich, R. Burgos, L. Rixin, N. Puqi, and K. Rajashekara, "A High Power Density Single-Phase PWM Rectifier With Active Ripple Energy Storage," Power Electronics, IEEE Transactions on, vol. 26, pp. 1430-1443, 2011.

[3] W. Ruxi, F. Wang, L. Rixin, N. Puqi, R. Burgos, and D. Boroyevich, "Study of Energy Storage Capacitor Reduction for Single Phase PWM Rectifier," in Applied Power Electronics Conference and Exposition, 2009. APEC 2009. Twenty-Fourth Annual IEEE, 2009, pp. 1177-1183.

[4] Y. Tang and F. Blaabjerg, "A Component-Minimized Single-Phase Active Power Decoupling Circuit With Reduced Current Stress to Semiconductor Switches," Power Electronics, IEEE Transactions on, vol. 30, pp. 2905-2910, 2015.

[5] S. Harb and R. S. Balog, "Single-phase PWM rectifier with power decoupling ripple-port for double-line-frequency ripple cancellation," in Applied Power Electronics Conference and Exposition (APEC), 2013 Twenty-Eighth Annual IEEE, 2013, pp. 1025-1029.

[6] L. Hongbo, Z. Kai, Z. Hui, F. Shengfang, and X. Jian, "Active Power Decoupling for High-Power Single-Phase PWM Rectifiers," Power Electronics, IEEE Transactions on, vol. 28, pp. 1308-1319, 2013.

[7] Runruo Chen; Yunting Liu; Fang Zheng Peng, "DC Capacitor-Less Inverter for Single-Phase Power Conversion With Minimum Voltage and Current Stress," Power Electronics, IEEE Transactions on , vol.30, no.10, pp.5499,5507, Oct. 2015

[8] G. J. Capella, J. Pou, S. Ceballos, G. Konstantinou, J. Zaragoza, and V. G. Agelidis, "Enhanced Phase-Shifted PWM Carrier Disposition for Interleaved Voltage-Source Inverters," Power Electronics, IEEE Transactions on, vol. 30, pp. 1121-1125, 2015.

[9] Mingkai Mu; Lee, F.C.; Yang Jiao; Sizhao Lu, "Analysis and design of coupled inductor for interleaved multiphase three-level DC-DC converters," Applied Power Electronics Conference and Exposition (APEC), 2015 IEEE , vol., no., pp.2999,3006, 15-19 March 2015

[10] Pit-Leong Wong; Peng Xu; Yang, B.; Lee, F.C., "Performance improvements of interleaving VRMs with coupling inductors," Power Electronics, IEEE Transactions on , vol.16, no.4, pp.499,507, Jul 2001

Single Stage Transformer Isolated High Frequency AC Link Based Open End Drive

Srikant Gandikota and Ned Mohan
Department of Electrical and Computer Engineering
University of Minnesota, Minneapolis,Minnesota 55455
Email: gand0088@umn.edu, mohan@umn.edu

Abstract—An open end drive configuration utilizing a transformer isolated single phase high frequency ac link is presented. A high frequency ac link enables the use of a compact and light weight transformer to provide voltage transformation and galvanic isolation while the open end configuration leads to an improved utilization of the ac link. The desired output is synthesized using dual cycloconverters modulated using a space vector technique that has been modified to work with a high frequency ac link and leads to single stage conversion. Any change in the switch states is restricted to the zero crossings of the ac link to achieve zero voltage switching in the cycloconverters. The topology described also has the advantage of achieving natural commutation of leakage energy in the transformer windings and avoids the need for additional circuits to achieve the same. The proposed drive is simulated using MATLAB/Simulink and the results are discussed. A hardware prototype has been constructed to demonstrate the implementation of the proposed drive.

I. INTRODUCTION

The majority of the variable speed drives used in industry utilize voltage source inverters (VSI) which have been thoroughly investigated over the years. The output voltages and currents in such drives can be generated by modulating the fixed DC link using space vector PWM techniques (SVPWM) [1]. While this method is widely used, it causes unwanted bearing currents which lead to the eventual failure of the drive. Open end winding configuration of electric motors utilizing dual VSIs has been proposed which offers several advantages over the conventional single ended drive such as improved utilization of the DC bus and common mode voltage suppression [2], [3]. However, a major drawback of the DC link based VSI configuration is that the electrolytic capacitors used to construct the DC link tend to be bulky and degrade with time. To avoid the need for such capacitors and improve the reliability of the drive, open end drive configurations based on matrix converters have been investigated in [4], [5]where the matrix converters are modulated using synchronous rotating vectors to generate the desired outputs. Alternately, inverters utilizing a high frequency AC link offer an attractive solution for use in open end drives by eliminating the need for bulky electrolytic capacitors along with the added advantages of allowing the use of compact and light weight transformers to achieve galvanic isolation and voltage transformation.

High frequency ac link converters have been investigated with renewed interest in recent years and several circuit topologies have been published in this regard. Reference [6] proposed a high frequency ac link power distribution scheme utilizing a parallel LC network. The design and implementation of a high frequency ac link motor drive utilizing this idea

has been described in [7]. An inverter for PV applications using a parallel partial resonant ac link has been reported in [8]. However, the use of transformers in a high frequency ac link presents the problem of overvoltages appearing across the transformer windings caused due to the leakage inductance as explained in [9]–[12]. A source based commutation technique as been explained in [10] to overcome this problem. The use of a resonant tank for the natural commutation of leakage energy is described in [13], [14] while [15] describes a partial resonant commutation technique to overcome this problem. A multi-level asymmetrical transformer with interleaved windings was used to minimize the leakage inductance and reduce the problem of leakage energy commutation in [16].

This paper presents an open end drive configuration that uses dual cycloconverters to generate the necessary drive voltages from a transformer isolated high frequency ac link. It combines the advantages of an open end winding configuration and a high frequency ac link. The open end configuration leads to an improved utilization of the ac link while achieving constant common mode voltage across the motor windings to suppress circulating currents. The use of bulky electrolytic capacitors is avoided and it does not have the problem of leakage energy commutation which eliminates the need for lossy voltage clamping circuits. By restricting any change in the switch states of the cycloconverter to the zero crossing regions of the link, zero voltage switching is ensured in the cycloconverter.

In this paper, Section II describes the circuit topology. The principle of operation and the modulation of the converter is explained in Section III. The commutation method used for bidirectional switches is discussed in Section IV. A 20kW open end drive configuration using an ac link at 15 kHz has been simulated in MATLAB/Simulink to show the operation. A preliminary experimental prototype has also been constructed to demonstrate the operation of such a drive and relevant results are presented in Section V.

II. DESCRIPTION OF THE CONVERTER

The proposed high frequency ac link open end drive is shown in Fig 1. An H-bridge comprising of switches S_A through S_D is used to generate a high frequency AC waveform at a frequency f_l from a DC voltage source. A high frequency transformer with a turns ratio of n_t is connected in the link and L_{lkg_p} and L_{lkg_s} are the leakage inductances of the transformer windings. A parallel tank comprising of inductor L and capacitor C is connected across the secondary winding

978-1-4673-9551-9/16 $31.00 © 2016 IEEE

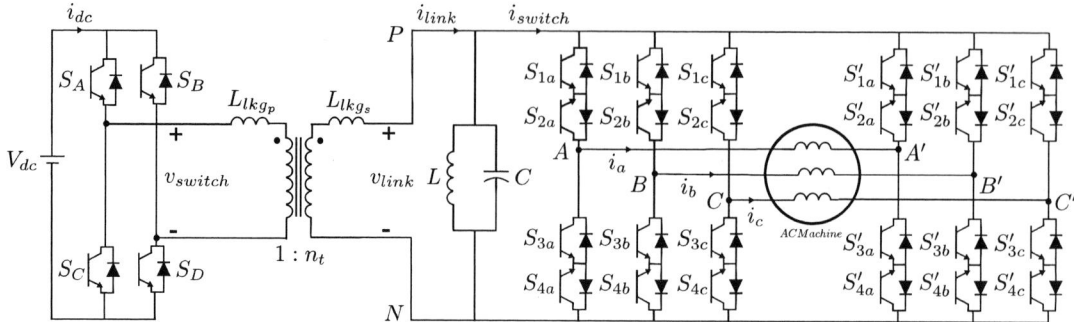

Fig. 1: High frequency transformer Isolated AC link open end drive

of the transformer. The resonant frequency of the tank, given by $f_r = \dfrac{1}{2\pi\sqrt{LC}}$ is designed to be equal to f_l. The terminals of the open end machine are fed from two cycloconverters, C1 and C2, comprising of switches S_{1a} through S_{4c} and S'_{1a} through S'_{4c} respectively. These switches need to be capable of blocking the alternating link voltage and hence should have bidirectional voltage blocking capability. This can achieved by connecting two IGBT's in a common emitter scheme as shown. An open end winding configuration in an induction machine is obtained by opening the neutral point of the stator windings in a regular induction machine. In the following sections, it is assumed that the switches are ideal and the magnetizing inductance of the transformer is very large and can be neglected.

III. PRINCIPLE OF OPERATION

The switching scheme of the proposed drive comprises of modulating the H-bridge to generate a high frequency ac link and the subsequent single stage conversion of the ac link to obtain the desired voltages at lower frequenices by using the cycloconverters.

A. Modulation of the H Bridge

Selective Harmonic Elimination(SHE) technique [17] is used to generate a high frequency voltage v_{switch}, with a fundamental frequency that is the same as f_r. The presence of a tank comprising of a parallel combination of inductance and capacitance affects the switching strategy of the H-bridge.

The impedance of the parallel combination of an inductor and capacitor is given according to (1). From the tank impedance characteristics in Fig. 1, it can be seen that the combination offers the maximum impedance at f_r and very low impedance at other frequencies. This characteristic means that such a tank offers a low impedance path to any harmonics in i_{switch} while offering a high impedance path to the component at the frequency f_r.

$$Z(\omega) = \frac{\omega L}{1 - (\dfrac{\omega}{\omega_r})^2} \qquad (1)$$

If the H-bridge is switched to generate a simple high frequency square wave, the resulting waveform would contain

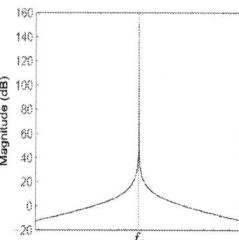

Fig. 2: Plot of tank impedance versus frequency

the third,fifth and other odd harmonics which can lead to circulating currents that flow between the DC source and the tank. Consequently, the transformer and the switches in the H-bridge have to designed to handle this extra current which also results in additional losses.

This problem is overcome by the use of Selective Harmonic Elimination(SHE) switching strategy. The switching instants are pre-calculated to ensure that the harmonics upto the 17th multiple of f_l in v_{switch} are eliminated while the peak value of the fundamental component equals V_{dc}. The switched voltage,v_{switch} applied to the primary winding of the transformer is shown in Fig. 6a. The resulting voltage at the secondary terminals of the transformer with a turns ratio of n_t has the amplitude given by $\hat{v}_{link} = n_t V_{dc}$ and is shown in Fig. 6a.

B. Open End Drive Operation

As previously described, an open-end winding induction motor is realized by opening the neutral point of the stator windings of a regular induction motor. This provides access to three additional motor terminals which are fed from a separate three phase converter. When utilizing a DC-link based open end drive, the two pairs of motor terminals are fed by voltage source inverters(VSI). The space vector modulation technique for an open drive using dual VSI's has been described in [2]. A major advantage of the open end configuration is that it requires the use of only half the bus voltage to generate a given output voltage. The possible voltage vectors of the VSI's can be represented using the space vector diagram as shown in Fig. 3a and 3b. The switch states to obtain the voltage

978-1-4673-9551-9/16 $31.00 © 2016 IEEE

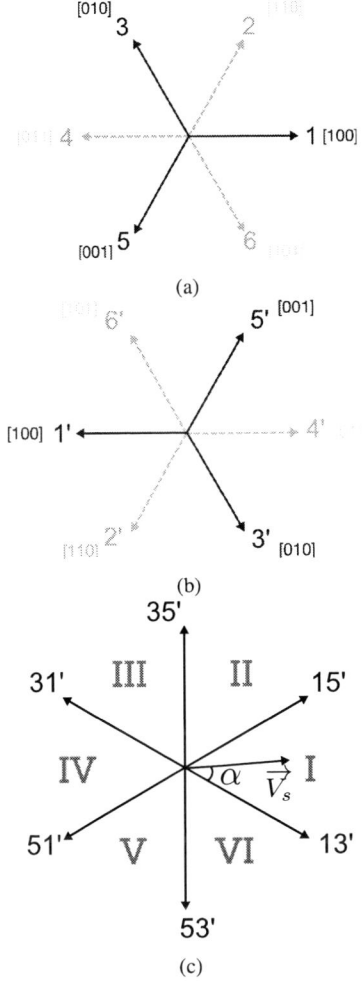

(a)

(b)

(c)

Fig. 3: (a) & (b) Space vector diagrams of the cycloconverters showing the vectors with similar common mode voltage (c) Resultant voltage vectors of the open end configuration

vector are also shown. The common mode voltage of a switch combination, V_{CM}, is defined as (2).

$$V_{CM} = \frac{V_{AN} + V_{BN} + V_{CN}}{3} \qquad (2)$$

It can be seen that the common mode voltage generated by the active vectors of a VSI with a DC bus value of V_{DC} is either $\frac{2V_{DC}}{3}$ or $\frac{V_{DC}}{3}$. If two active vectors with different common mode voltage are applied on either end of the open end drive, it causes circulating currents to flow through the machine windings. This is undesirable and leads to a reduction in efficiency. Rather than using isolated voltage sources, the modulation strategy is restricted to utilizing active vectors with the same common mode voltage to avoid circulating currents.

C. Modulation of the Cycloconverters

A constant common mode voltage across the phase windings can be achieved by utilizing vectors which have the same common mode voltage as described in [3]. It can be seen that

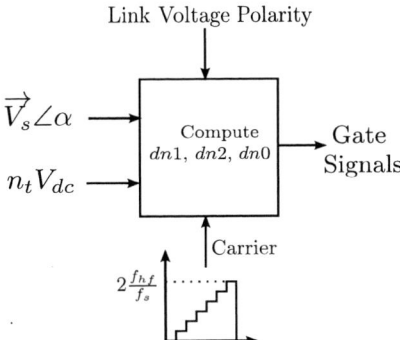

Fig. 4: Generation of gate signals

vectors **1,3,5** possessing the switch states $S_a S_b S_c$ as 100, 010 and 001 respectively have the same V_{CM}. The switch state '100' means the top bidirectional switch in Phase A is ON and the bottom bidirectional switches in Phase B and C are ON. The zero vector is constructed using one of the combinations of **11',33',55'** to ensure that V_{CM} remains constant. It should be noted that using the vectors **2,4,6** will also yield similar results. The resulting voltage vectors possible by using the **1,3,5** vectors and **1',3',5'** vectors that generate the same common mode voltage on either end of the motor windings are shown in Fig. 3c along with the desired output voltage space vector, $\vec{V_s}(t)$. The vector 13' means that cycloconverter 1 has a switch state of 100 while cycloconverter 2 has a switch state of 010.

Depending on the sector in which $\vec{V_s}(t)$ lies, the voltage is synthesized by applying two adjacent active vectors and a zero vector for specific time periods within a given sampling period. For the ac link case, this is implemented by considering the volt-seconds available in one sinusoidal half-cycle of the link as the fundamental unit to synthesize the output during a sampling period T_s. The volt-seconds in one half cycle of the link voltage are given as $\varphi = \frac{n_t V_{dc}}{\pi f_l}$ as shown in [13]. For a given vector V_{mn} where $m \in 1, 3, 5$ and $n \in 1', 3', 5'$, let S_{mx} and S_{nx} be switch states of the cycloconverter legs respectively where $x \in a, b, c$. The volt-seconds of the vector V_{mn} are calculated as shown in (3).

$$\vec{V_{mn}} = \varphi((S_{ma} - S_{na})e^{j0} + (S_{mb} - S_{nb})e^{j\frac{2\pi}{3}} + \\ (S_{mc} - S_{nc})e^{j\frac{4\pi}{3}}) \qquad (3)$$

$$|\vec{V_s}|\angle\alpha T_s = V_1 dn_1 + V_2 dn_2 + V_0 dn_0 \qquad (4)$$

The output voltage is synthesized during the time T_s by applying an integral number of half-cycles of two adjacent active vectors, say dn_1 and dn_2, and using a zero vector for the remaining time given by dn_0 as given by (5). By applying an integral number of half cycles, it is ensured that the output switches always change state when the voltage across them is very low and resulting in negligible switching loss. From (5), the resulting length of the vectors in Fig. 3c is calculated as

$\frac{\sqrt{3}n_t V_{dc}}{\pi f_l}$ and it can be shown that the maximum value of m_v, where $m_v = \frac{\hat{V}_o}{n_t V_{dc}}$, of this configuration is 0.6366 which is much higher than the maximum value of m_v of a similar single end configuration which is 0.3675. Thus, using an open end configuration allows the generation of larger output voltages from a given ac link and reduces the component stresses.

$$dn_1 = \pi \frac{\hat{V}_o f_l}{n_t V_{dc} f_s} \sin(60^0 - \alpha);$$
$$dn_2 = \pi \frac{\hat{V}_o f_l}{n_t V_{dc} f_s} \sin(\alpha); \qquad (5)$$
$$dn_0 = n - dn_1 - dn_2;$$
$$\text{where} \quad f_s = \frac{1}{T_s}; \quad \hat{V}_o = \frac{2}{3}|\overrightarrow{V_s}|; \quad \text{and} \quad n = 2\frac{f_l}{f_s}$$

The obtained values of dn_1, dn_2 and dn_0 are then compared with a sawtooth carrier with discrete levels ranging from 0 to n as shown in Fig. 4 to get the appropriate gate signals. The switch states of the cycloconverters are reversed every time the link voltage polarity changes. The current i_{switch} depends on the vector combination applied and the values of the three phase currents at that instant. Assuming the load current to be inductive and at a much lower frequency than f_l, the effect of reversing the switch states is the polarity reversal of i_{switch} resulting in the switched waveforms as shown in Fig. 6a. From Fig. 2, it can be seen that the tank offers a low impedance path for the harmonics in i_{switch} while offering a very high impedance to the link frequency component. The resulting transformer current i_{link} is shown in Fig. 6a and does not contain any abrupt changes in currents.

IV. COMMUTATION

The cycloconverters are constructed using IGBT's connected in the common emitter configuration to obtain switches capable of blocking voltage of either polarity. Consequently, when switching between the top and bottom bidirectional switch in a leg, four step commutation technique needs to be used to ensure that there is always a safe path for the phase currents to flow [10]. Based on the polarity of the phase current, the entire process of transferring the current from the top switch to the bottom switch is accomplished in four discrete steps. This is illustrated in Fig 5 where the states of the four individual devices S_{1x}, S_{2x}, S_{3x} and S_{4x} that constitute one bidirectional leg are shown during the transition from ON state to OFF and vice versa. The states to the left of the dotted line refer to the case when $i > 0$ i.e current is flowing out of the leg while states to the right of the dotted line occur when the current is flowing into the leg.

V. RESULTS

The converter described above was simulated in the $SIMULINK$ environment by considering a transformer with leakage inductance of 20 μH and a link frequency of 15kHz and supplying 20kW of power at a load power factor of 0.8. The value of V_{dc} was assumed to be 600V and n_t=1. The modulation index was set to 0.56 to generate a line voltage of

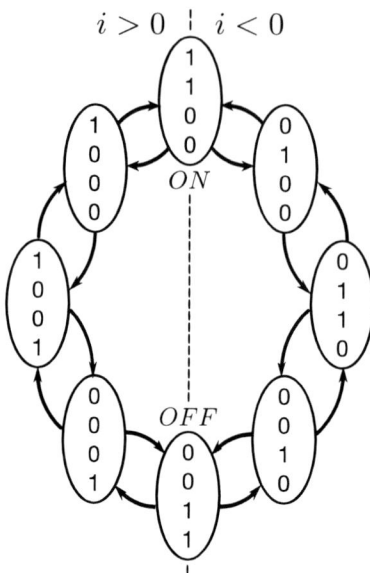

Fig. 5: Four step commutation

415V(RMS). The parallel tank is constructed using an inductor of 28 μH and a capacitance of 4μF. The plot of i_{switch} in Fig. 6a shows different current levels corresponding to the application of different active vectors and the corresponding i_{link}. The output currents and the voltage $V_{AA'}$ are shown in Fig. 6b. It can be observed that the phase voltage is composed of sinusoidal half cycles of the ac link.

A hardware prototype was constructed using STGB7NC60HDT4 IGBTs configured in a common emitter configuration for switches S_{1a} through S'_{4c} and SCH2080KE power MOSFETs for switches S_A-S_D. The DC bus was set at 50V to achieve a no-load peak link voltage of 100V. A modulation index of 0.56 was assumed and a 30Hz output was generated across a static RL load with R=10 Ω and L=45.8 mH. Current sensors are used to sense the polarity of load currents to implement four step commutation which is necessary to switch the bidirectional switches. The results are shown in Fig. 7

A. Transformer design

TABLE II: Electrical Parameters

Parameter	Value
Power	20kW
V_{dc}	600V
V_{out}	415V(RMS)
L	28μH
C	4μF
L_{Lkg}	20μH
n_t	1

The high frequency transformer with turns ratio 2:1 was

TABLE I: Specifications

Power	Voltage, V_p	Frequency, f	Current Density, J	B_{max}	k_w
2.5 kW	500 V	15 kHz	4.5 A/mm^2	0.85 T	0.35

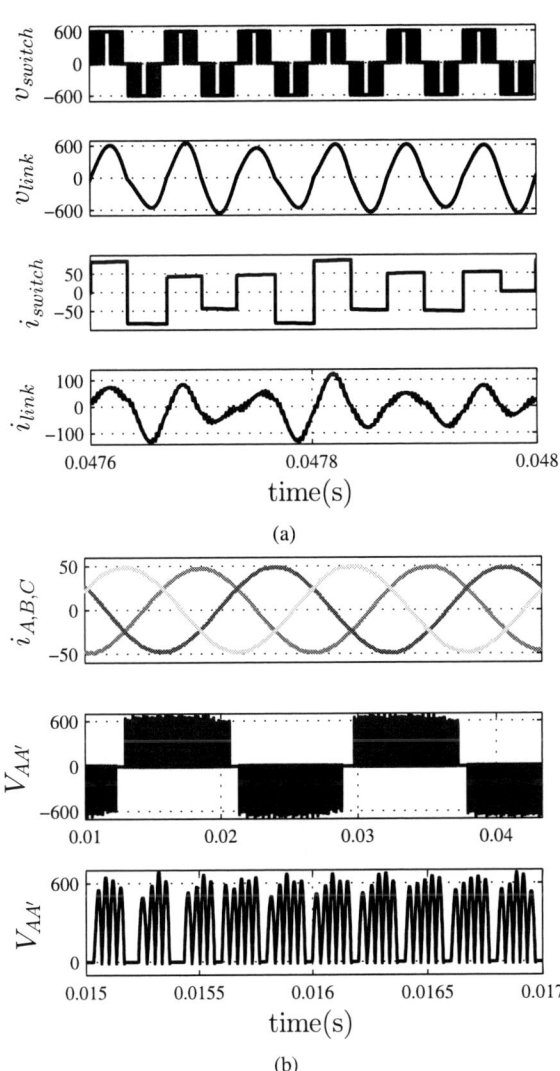

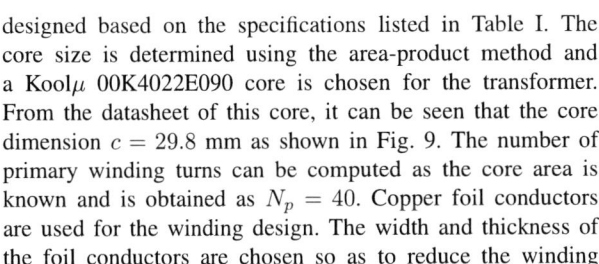

(b)

Fig. 6: (a) Link voltages and currents (b) Plot of output phase currents and phase voltage

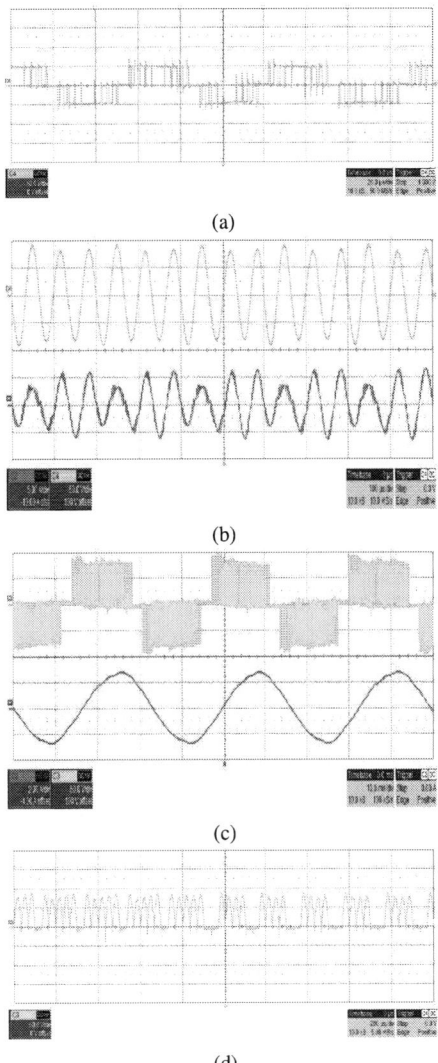

Fig. 7: (a) Voltage across the primary winding (b) AC Link voltage and current (c) Output phase a current and voltage (d) Phase a voltage on an expanded time scale

designed based on the specifications listed in Table I. The core size is determined using the area-product method and a Koolµ 00K4022E090 core is chosen for the transformer. From the datasheet of this core, it can be seen that the core dimension $c = 29.8$ mm as shown in Fig. 9. The number of primary winding turns can be computed as the core area is known and is obtained as $N_p = 40$. Copper foil conductors are used for the winding design. The width and thickness of the foil conductors are chosen so as to reduce the winding losses.

$$p\Delta = \frac{N_p I_p}{J \eta_1 \delta c} \qquad (6)$$

Substituting the values in (6), the $p\Delta$ product is equal to 3.07. $\eta_1 = 0.9$ is assumed for all computations. Here, I_p is the current on the primary winding and δ is the skin depth. According to [18], $\Delta=0.1535$ gives the conductor thickness to be $d=0.083$ mm when the frequency is 15 kHz. Based

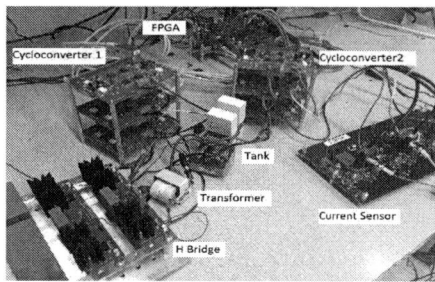

Fig. 8: Hardware prototype

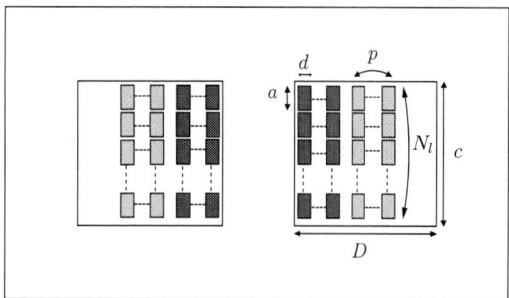

Fig. 9: Multi-turn multi-layer foil winding of a transformer [18]

on the current density J, the width of the foil conductor can be computed and the primary foil conductor dimensions are chosen as 0.003" X 0.5" inch. A similar computation for secondary windings yields the secondary foil conductor dimensions to be 0.003" X 1" inch. The windings are arranged as one section of 20 layers of primary winding wherein two primary turns are arranged in each layer followed by one section of secondary windings having 20 layers, for the ease of winding.

VI. CONCLUSION

The paper described a novel transformer isolated drive configuration which incorporates the benefits of high frequency ac link inverters and open end drives. Single stage conversion is achieved without the problem of leakage energy commutation by the use of the parallel resonant tank. Further, zero voltage switching is achieved in the cycloconverters which results in improved efficiency. Circulating currents in the open end windings are avoided by utilizing voltage vectors possessing the same common mode voltage to synthesize the output voltages. Moreover, the open end configuration leads to a higher overall voltage gain and the high frequency nature of the link yields a compact configuration with increased power density. The sinusoidal current through the transformer is advantageous as it avoids additional conduction losses that would have resulted from the harmonics in a square waveform. Results from the computer simulation and the hardware prototype are in agreement.

REFERENCES

[1] M. Kazmierkowski, L. Franquelo, J. Rodriguez, M. Perez, and J. Leon, "High-performance motor drives," *Industrial Electronics Magazine, IEEE*, vol. 5, no. 3, pp. 6–26, Sept 2011.

[2] E. Shivakumar, K. Gopakumar, S. Sinha, A. Pittet, and V. Ranganathan, "Space vector pwm control of dual inverter fed open-end winding induction motor drive," in *Applied Power Electronics Conference and Exposition, 2001. APEC 2001. Sixteenth Annual IEEE*, vol. 1, 2001, pp. 399–405 vol.1.

[3] M. Baiju, K. Mohapatra, R. Kanchan, and K. Gopakumar, "A dual two-level inverter scheme with common mode voltage elimination for an induction motor drive," *Power Electronics, IEEE Transactions on*, vol. 19, no. 3, pp. 794–805, May 2004.

[4] R. Baranwal, K. Basu, and N. Mohan, "Carrier-based implementation of svpwm for dual two-level vsi and dual matrix converter with zero common-mode voltage," *Power Electronics, IEEE Transactions on*, vol. 30, no. 3, pp. 1471–1487, March 2015.

[5] R. Gupta, K. Mohapatra, A. Somani, and N. Mohan, "Direct-matrix-converter-based drive for a three-phase open-end-winding ac machine with advanced features," *Industrial Electronics, IEEE Transactions on*, vol. 57, no. 12, pp. 4032–4042, Dec 2010.

[6] P. Sood and T. Lipo, "Power conversion distribution system using a high-frequency ac link," *Industry Applications, IEEE Transactions on*, vol. 24, no. 2, pp. 288–300, 1988.

[7] S. Sul and T. Lipo, "Design and performance of a high-frequency link induction motor drive operating at unity power factor," *Industry Applications, IEEE Transactions on*, vol. 26, no. 3, pp. 434–440, May 1990.

[8] M. Amirabadi, A. Balakrishnan, H. Toliyat, and W. Alexander, "High-frequency ac-link pv inverter," *Industrial Electronics, IEEE Transactions on*, vol. 61, no. 1, pp. 281–291, Jan 2014.

[9] L. Hui, B. Ozpineci, and B. Bose, "A soft-switched high frequency nonresonant link integral pulse modulated dc-ac converter for ac motor drive," in *Industrial Electronics Society, 1998. IECON '98. Proceedings of the 24th Annual Conference of the IEEE*, vol. 2, 1998, pp. 726–732 vol.2.

[10] K. Basu and M. Ned, "A high frequency link single stage pwm inverter with common-mode voltage suppression and source based commutation of leakage energy." *Power Electronics, IEEE Transactions on*, vol. PP, no. 99, pp. 1–1, 2013.

[11] M. Matsui, M. Nagai, M. Mochizuki, and A. Nabae, "High-frequency link dc/ac converter with suppressed voltage clamp circuits-naturally commutated phase angle control with self turn-off devices," *Industry Applications, IEEE Transactions on*, vol. 32, no. 2, pp. 293–300, 1996.

[12] I. Yamato and N. Tokunaga, "Power loss reduction techniques for three phase high frequency link dc-ac converter," in *Power Electronics Specialists Conference, 1993. PESC '93 Record., 24th Annual IEEE*, 1993, pp. 663–668.

[13] S. Gandikota and N. Mohan, "High frequency ac-link transformer isolated three phase inverter with natural commutation of leakage energy," in *Power Electronics, Machines and Drives (PEMD 2014), 7th IET International Conference on*, April 2014, pp. 1–5.

[14] S. Nath and N. Mohan, "A solid sate power converter with sinusoidal currents in high frequency transformer for power system applications," in *Industrial Technology (ICIT), 2011 IEEE International Conference on*, March 2011, pp. 110–114.

[15] S. Gandikota and N. Mohan, "A new leakage energy commutation technique for single stage high frequency link inverters," in *Power Electronics, Drives and Energy Systems (PEDES), 2014 IEEE International Conference on*, Dec 2014, pp. 1–6.

[16] K. Iyer and N. Mohan, "Multi-level converter to interface low voltage dc to 3-phase high voltage grid with medium frequency transformer isolation," in *Industrial Electronics Society, IECON 2014 - 40th Annual Conference of the IEEE*, Oct 2014, pp. 4683–4689.

[17] P. Enjeti, P. Ziogas, and J. Lindsay, "Programmed pwm techniques to eliminate harmonics: a critical evaluation," *Industry Applications, IEEE Transactions on*, vol. 26, no. 2, pp. 302–316, 1990.

[18] K. Iyer, W. Robbins, and N. Mohan, "Design and comparison of high frequency transformers using foil and round windings," in *Power Electronics Conference (IPEC-Hiroshima 2014 - ECCE-ASIA), 2014 International*, May 2014, pp. 3037–3043.

978-1-4673-9551-9/16 $31.00 © 2016 IEEE

A Quasi-Z-source Integrated Multi-port Power Converter with Reduced Capacitance for Switched Reluctance Motor Drives

Fan Yi, *Student Member, IEEE* and Wen Cai, *Student Member, IEEE*
Department of Electrical Engineering
The University of Texas at Dallas
Richardson, US
Email: fan.yi@utdallas.edu, wen.cai@utdallas.edu

Abstract—**This paper presents a quasi Z-source integrated multiport converter (ZIMPC) for switched reluctance motor (SRM) drives to reduce the dc link capacitance. In conventional SRM drives, employing multi-phase asymmetrical H-bridge (ASHB) topology, large capacitors are necessary to absorb the transient energy during phase current commutation. However, electrolytic capacitors would affect the lifetime, cost and power density of the drive system. With switch multiplexing technique, a Z-source integrated multiport power converter is derived to achieve power ripple reduction using relatively small capacitance. Corresponding control method is designed and developed for the proposed SRM drive. Also, the ZIMPC can boost the equivalent phase exciting voltage and widen the constant power speed range (CPSR). At last, simulation and experimental results verify the feasibility of the proposed ZIMPC and its superior performance with smaller capacitance as compared with ASHB.**

Keywords—capacitance minimization; motor drive; multiport converter; quasi-Z-source; switch multiplexing; switched reluctance motor (SRM)

I. INTRODUCTION

Switched reluctance motor (SRM) [1, 2] has several advantages as compared to other competing machines including but not limited to low cost, fault-tolerance, wide range of speed and ability to operate under harsh environment [3], making it a serious contender for electric propulsion and industrial drive systems. However, because of current

commutation in various phases, the energy required by SRM changes greatly, which would lead to large current ripple from the dc source. In electric propulsion systems with battery packs as dc source, such large current ripple will cause excessive temperature rise and adversely affect the performance and lifetime of the battery packs [4, 5]. Conventionally, large electrolytic capacitors are used to restrain this power ripple and to keep the input current smooth. However, large electrolytic capacitors are not desired since the lifetime of them is much shorter than that of semiconductors and other passive components [6-8], and they occupy large volume and need cooling arrangements.

In order to cancel the undesirable power ripple without using large electrolytic capacitors, many researches have been launched to adopt either passive or active compensation approaches. The current ripple generation and propagation in fuel cell power system are analyzed and a dual-loop control method to reduce the current ripple through the fuel cell is proposed in [9]. Authors of [10] add a LC branch at the midpoint of the transformer primary winding to achieve ripple control. For renewable energy applications, an extra circuit is utilized to replace the electrolytic capacitors in [11, 12], thereby achieving power decoupling with small capacitance requirement. Despite of the feasibility of all these methods, they mainly focus on fixed-frequency (double-utility frequency, 100/120-Hz) power ripple reduction, while

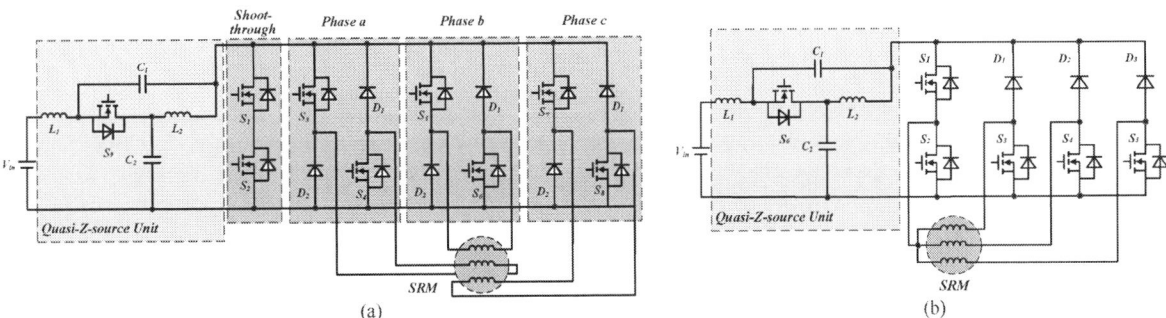

Fig. 1. The topologies of SRM drive with quasi-Z-source unit: (a) ASHB + Quasi-Z-source unit, (b) the proposed ZIMPC.

978-1-4673-9551-9/16 $31.00 © 2016 IEEE

the current ripple in SRM drives varies with the speed of the motor. This limits the usage of the presented solutions in SRM drives. Further, power mismatch during current commutation period is nonlinear and uncertain which would generate many harmonics instead of only double-utility frequency power ripple. Wide-region harmonic requires a better control method to compensate. Third, capacitor design is difficult because of wide motor speed range which is an advantage of SRM. Hence, it is desirable to propose a new solution with ripple control for SRM drives. The method based on commutation elimination is discussed in [13] to overcome the disadvantages of high dc-link capacitance requirement. Based on the idea of modifying the turn-off and turn-on procedure, a direct dc-link voltage control scheme was developed in [14, 15]. Similarly, a modified switching technique is proposed to replace hysteresis control for dc-link capacitor minimization and its feasibility is verified in [16]. However, the current rating for switches and diodes would increase by following their method. By employing various torque ripple reduction method, including both machine design [17, 18] and control strategies [19-22], the power ripple can also be reduced. However, the extent to which this can be achieved is always limited by the efficiency of the motor [23]. A specific current profile is proposed for SRM to reduce the input current pulsation in [24], but this method is only applicable to pseudo-sinusoidal SRMs.

This paper focuses on SRM drive and proposes a novel power converter called quasi-Z-source integrated multiport power converter (ZIMPC), which is shown in Fig. 1(b), to replace the conventional asymmetrical H-bridge (ASHB) topology. The first Z-source inverter was proposed in 2002 [25]. A series of semi-Z-source/quasi-Z-source/trans-Z-source inverters have been presented and studied in [26-29]. The bidirectional operation of quasi-Z-source inverter for electric vehicle is studied since in [30]. With the Z-source unit, Z-source inverters can achieve voltage boost and enlarge the input voltage range, which makes them suitable for performance improvement. Moreover, the capacitors in Z-source unit can be used to compensate power ripple caused by current commutation, thereby smoothing the current from the dc source. Compared with existing motor drives or capacitance reduction solutions, the presented ZIMPC can achieve three advantages: First, the capacitance required in this topology can be reduced and their voltage rating will be decreased as well, reducing size and cost of the drive; Second, the input current is controlled at virtually a constant value and it would not be affected by current ripple which is more amicable for dc source especially battery pack in electric vehicle applications; Third, this topology is applicable for various SRMs directly such as multi-phase SRM or double-stator SRM because it's not necessary to change motor structure. Advanced power flow control for ZIMPC is introduced which can reduce the impact of multi-frequency input current harmonic. In addition, with ZIMPC, it is possible to maintain low capacitor voltages for low operating speeds to reduce power losses, while boosting the capacitor voltages and the equivalent phase exciting voltages at high

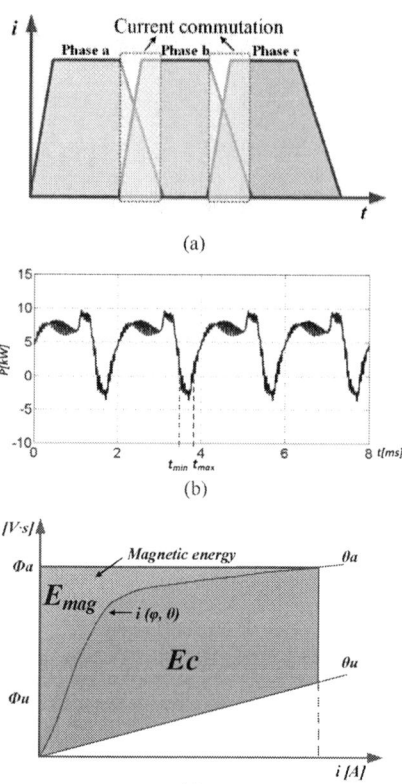

Fig. 2. Power ripple generated by SRM because of current commutation: (a) current vs. time; (b) power vs. time; (c) magnetic energy stored in SRM.

speeds to ensure that enough torque is at hand to widen the constant power speed range (CPSR). At last, experimental results are presented to verify the feasibility of such topology and the proposed control solution.

II. POWER RIPPLE ANALYSIS IN SRM DRIVE SYSTEM

Using ASHB, typical waveform of phase current injected to SRM is shown in Fig. 2(a). The waveform of the input power in SRM in low speed region is shown in Fig. 2(b). The relationship between magnetic energy stored in one phase of SRM and its current is illustrated in Fig. 2(c). The motor absorbs large amount of energy when phase is being energized, and it requires little or even returns some energy when phase is not energized or being de-energized. However, it is desired that the battery pack provide constant energy. As a result, the ripple power should be provided or absorbed by another buffer such as the capacitor on the dc bus.

In order to calculate the required capacitance, a theoretical study on the value of the dc bus capacitor based on the voltage ripples can be performed. The maximum voltage ripple appears when a phase of the stator is being de-energized. This is due to the high inductance at the aligned position and the large amount of magnetic energy stored in the phase winding due to the high current.

Based on the electrical dynamics of SRM phases, the voltage equation for each phase can be written as

978-1-4673-9551-9/16 $31.00 © 2016 IEEE

$$v_i = i_i R_i + \frac{\partial \varphi_i}{\partial i_i}\frac{di_i}{dt} + \frac{\partial \varphi_i}{\partial \theta_i}\frac{d\theta_i}{dt} \qquad (1)$$

where, v_i and i_i represent the voltage of the ith phase of the motor and the current through it, respectively. φ_i is the flux linkage for the ith phase.

Generally, in constant torque region, during current commutation period only two phases are energized and the other phases are in idle mode. So the power of SRM can be calculated as (2).

$$p_{total} = p_a + p_b = i_a^2 R_a + i_a \frac{\partial \varphi_a}{\partial i_a}\frac{di_a}{dt} + i_a \frac{\partial \varphi_a}{\partial \theta_a}\frac{d\theta_a}{dt}$$
$$+ i_b^2 R_b + i_b \frac{\partial \varphi_b}{\partial i_b}\frac{di_b}{dt} + i_b \frac{\partial \varphi_b}{\partial \theta_b}\frac{d\theta_b}{dt} \qquad (2)$$

Equation (2) shows that the power fluctuates greatly as the rotor position and phase current of SRM change. In order to achieve low ripple in the current of the dc source, capacitors in the converter have to be employed to smooth the input power. The requirement for capacitance can be estimated by analyzing energy storage characteristics. During the current commutation period, at most two phases are active, however the energy stored in the trailing winding is limited since the inductance is small at unaligned position. It would not cause significant difference in transferred energy and capacitance requirement. Therefore, only one phase is considered for energy storage characteristics analysis, without the loss of generality. Here, theoretical analysis from electrical point of view is done for energy ripple assessment. The stored magnetic energy for one phase reaches maximum value when the rotor reaches the aligned position and the phase angle is equal to θ_a. The energy stored in the magnetic field with the assumption of un-saturated operation can be derived using

$$E_{mag} = \int_{\varphi(\theta_a)}^{0} i(\varphi, \theta_a) d\varphi = \frac{1}{2}\frac{\partial \varphi_a}{\partial i_a}i^2 \qquad (3)$$

where: $\partial \varphi_a / \partial i_a$ is the inductance at the aligned position which can be defined as L_a. When the current commutation period starts, the phase is demagnetized. Most of the stored magnetic energy would be converted to electric energy and fed back to the dc bus capacitor or the source. The remaining energy is converted to mechanical energy. Meanwhile, the upcoming stator phase is being magnetized as well. If the period of the overlap is short, one is able to treat the maximum magnetic energy as the energy ripple ($E_{mag} \approx E_r$). Therefore, the dc bus voltage would increase accordingly. Assuming all the energy is absorbed by the capacitors and the voltage increase is Δv_{dc}, the relationship can be expressed as:

$$\frac{1}{2}C(v_{dc} + \Delta v_{dc})^2 - \frac{1}{2}Cv_{dc}^2 = \frac{1}{2}\frac{\partial \varphi_a}{\partial i_a}i^2 \qquad (4)$$

where, v_{dc} is the nominal voltage of the dc bus and C is the capacitance connected to the dc bus. According to this equation, the minimum capacitance can be estimated as

$$C_{min} = \frac{\partial \varphi_a}{\partial i_a}\frac{i^2}{(\Delta v_{dc} + 2v_{dc})\Delta v_{dc}} \qquad (5)$$

In general, the voltage ripple is desirable to be smaller than 5% of the nominal voltage in ASHB converter because the capacitors are in parallel with the dc source, thus, the capacitor can be calculated using

$$C_{min} = \frac{L_a i^2}{0.1 v_{dc}^2} \qquad (6)$$

Equation (6) also reveals the two methods to reduce the required capacitance even though the power ripple caused by SRM is tangible, namely increasing the average voltage of capacitor (v_{dc}) and increasing the voltage ripple (Δv_{dc}).

III. PROPOSED QUASI-Z-SOURCE BASED SRM DRIVE TOPOLOGY

In order to pursue the two methods summarized above, it is proposed to add a quasi-Z-source unit in front of ASHB. The topology based on quasi-Z-source and ASHB is shown in Fig. 1(a). Because the asymmetrical leg cannot achieve shoot-through mode, an extra leg is necessary. Therefore, it needs nine switches and six diodes for a three-phase SRM. Subsequently, switch multiplexing technique is employed here to reduce switch number so as to simplify the topology.

Switch multiplexing is an effective technique to simplify topologies for multiport systems and to achieve better performance. Since SRM drive is also a multiport converter, its topology can be simplified by adopting switch multiplexing. Switch multiplexing technique is used to share some of the semiconductors from two or more topologies so as to decrease the number of switches and to simplify the overall system. It is worth pointing out that switch multiplexing is generally accompanied with a modified modulation strategy. Some specific applications of switch multiplexing technique have been demonstrated in [31-34]. By multiplexing switches, the shoot-through leg and one leg from each phase can be combined into one leg. Based on this, the quasi-Z-source integrated multiport converter is derived and shown in Fig. 1(b).

Shoot-through time is necessary with quasi-Z-source unit, which means that the two switches in the first leg are both turned on for this time period. For the first leg, the boost ratio of the converter is determined by the shoot through time while the average voltage of the midpoint is determined by the non-shoot-through duty cycle along with the boost ratio. For other legs, the duty cycle controls the phase current. The switch in each leg is turned on to energize the phase winding connected to this leg, or turned off to de-energize it.

The proposed topology has several major advantages:

1)   The capacitor is separated from the dc source. Hence the dc source voltage would not be affected by the capacitor voltage.

2)   Both the average voltage and the ripple voltage of the capacitor can be large which will decrease the capacitance value as much as possible.

3) The current through the dc source is filtered by an inductor, thus no electrolytic capacitor is necessary any more in parallel with the input dc source. Moreover, the required dc source voltage is relaxed and can be lower than what the SRM needs since it doesn't act as a dc link. This is of high importance for electric propulsion drive applications.

4) The equivalent dc link voltage is flexible. By increasing the dc link voltage, the time for energizing and de-energizing of each phase decreases so that conduction angle could be increased and larger average torque could be obtained. The constant power speed range is widened which is especially appealing for automotive and high speed application. The voltage range of dc source is expanded to lower level as well.

5) There are only six switches, three diodes, two inductors and two small capacitors in ZIMPC for three-phase SRM which can achieve cost reduction and enhance the power density of the motor drive. Furthermore, if the SRM has n phases, n+1 legs would be necessary so that n+3 switches and n diodes are required totally.

6) The wire harness connecting to the SRM is less bulky since only n+1 cables are required for an n-phase motor.

IV. CONTROL STRATEGY

Even though it is possible to achieve power ripple compensation by replacing ASHB with ZIMPC, a proper control strategy should be employed. Firstly, the ripple of the input current must be restrained. Secondly, the voltage of the capacitors (v_{c1} & v_{c2}) need to be regulated within a certain range considering capacitor voltage rating and SRM performance. Thus, multi-objective power flow control is necessary when using ZIMPC. In this section, the operational modes of ZIMPC are analyzed. Then, the equivalent model of the overall system with ZIMPC and SRM is developed. After that, an advanced multi-objective control method is presented with current ripple reduction for dc source and speed/torque control for SRM.

A. Mode analysis

The modulation sequence and important waveforms have been shown in Fig. 3. Shoot-through is employed in the common leg connected to all three phases of the SRM. For the Z-source unit, S_6 is turned off in shoot-through mode and is turned on in nonshoot-through mode. Hysteresis current control is employed for the remaining part, which is the same as discussed in [35].

Based on the modulation sequence shown above, there are five modes for each phase. The waveforms for other phases are similar and can be analyzed using the same principle. The modes are listed below:

Mode 1: S_1 and S_6 are turned on, S_2 and S_3 are turned off. The voltage drops across the two inductors are negative ($v_{in} - v_{c2}$ and $-v_{c1}$). Two inductors are being de-energized. The capacitors are being charged and their voltages increase. Meanwhile, the phase current, in the motoring mode, decreases because of the back electromotive force (back EMF, v_{emf}) as shown in Fig. 4(a).

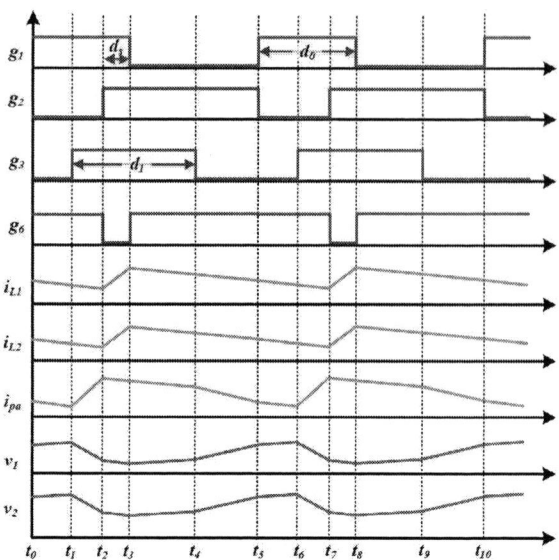

Fig. 3. The modulation sequence and some important waveforms of the modified topology.

Mode 2: S_1, S_3 and S_6 are turned on, S_2 is turned off. The current of two inductors continue to decrease. The voltage of two capacitors decreases by the virtue of a negative current. Meanwhile, the phase current increases because of the positive voltage drop, $v_{c1} + v_{c2} - v_{emf}$. Fig. 4(b) shows the corresponding equivalent circuit.

Mode 3: As shown in Fig. 4(c), S_1, S_2 and S_3 are turned on, and S_6 is turned off. The voltage drops between two inductors are $v_{in} + v_{c1}$ and v_{c2}, respectively. Thus, the inductor current increases and the capacitors are being discharged in this mode. This is also the shoot-through mode for the Z-source unit.

Mode 4: S_2 and S_6 are turned on, S_1 and S_3 are turned off as Fig. 4(d) shows. The two inductors are being discharged. The voltages of the two capacitors are decreasing with current $-i_{pa}$. The voltage drop across the phase winding is $-v_{emf} - v_{c1} - v_{c2}$, hence the phase current decreases.

Mode 5: In this mode (Fig. 4(e)), S_2, S_3 and S_6 are turned on, S_1 is turned off. This mode is similar to Mode 1. Both of the two inductors are being discharged and the voltage of the two capacitors increases. In addition, the stator phase of SRM is being de-energized with $-v_{emf}$.

Based on the mode analysis above, it can be seen that the phase can be energized with the average voltage drop as shown in (7). In (7), v_{c2} is the voltage of the capacitor C_2, d_s is the of shoot-through duty cycle, d_0 is the duty cycle of S_1 and $(1-d_k)$ is the duty cycle of S_{k-2}. From (7), the maximum voltage drop is $v_{in}(1-d_s)/(1-2d_s)$. By regulating the duty cycle d_s, the exciting voltage can be higher than the input voltage v_{in}.

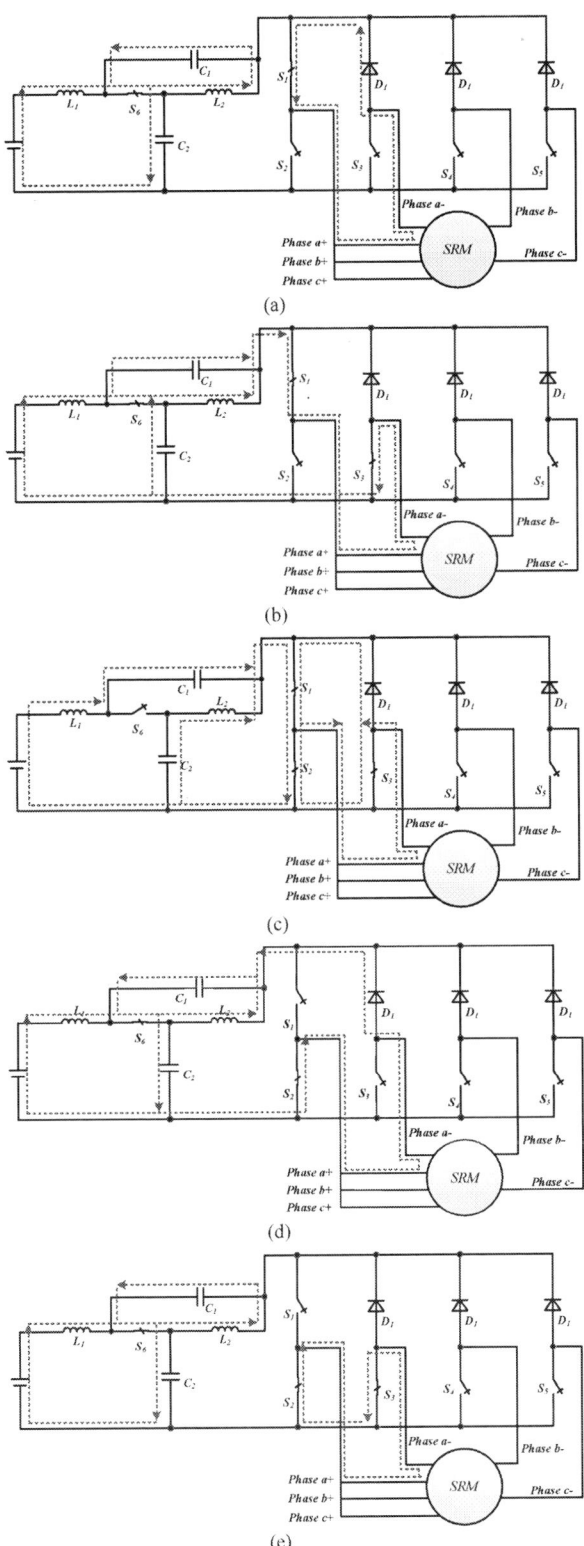

(a)

(b)

(c)

(d)

(e)

Fig. 4. Five modes of ZIMPC when energizing phase a: (a) mode 1; (b) mode 2; (c) mode 3; (d) mode 4; (e) mode 5.

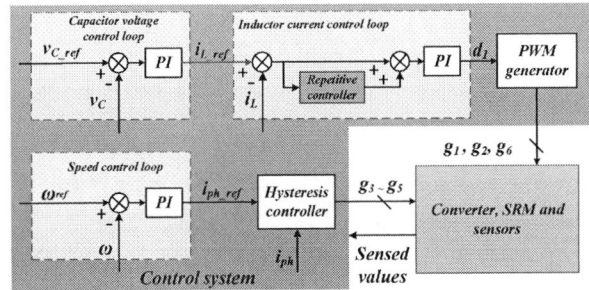

Fig. 5. Control strategy of QZSIMPC as SRM drives.

$$\overline{v}_{pk} = v_{c2}\left(d_0 - d_k\right) = v_{in}\frac{1-d_s}{1-2d_s}\left(d_0 - d_k\right) \qquad (7)$$

B. Multi-objective control method

The proposed multi-objective power flow control method has been illustrated in Fig. 5. There are two control branches, one is ZIMPC control including capacitor voltage and input current, and the second one is SRM control with speed/torque and phase current. In ZIMPC control branch, the average voltage of capacitor C_2 is detected and regulated at a constant value. The voltage command is decided based on the requirement of speed (higher capacitor voltage for C_2 is needed for high speed operation of SRM). The output of the capacitor voltage control loop is the input current reference. With the detected input current i_{L1}, the duty cycle of the shoot-through time (d_s) is regulated to make the input current track the reference. In this inner loop, a repetitive controller [36] is inserted in series with inductor current controller to further reduce the low-frequency ripple which cannot be restrained using conventional controller alone. On the other hand, in SRM control branch, the speed or torque is expected to be regulated. The output of this control loop is the reference for phase currents. Subsequently the phase current is regulated using a hysteresis controller with a satisfactory dynamic performance. In Fig. 5, g_1 and g_2 are the PWM signals for S_1 and S_2, respectively, g_6 is the PWM signal for S_6. Meanwhile, g_3, g_4 and g_5 are the signals for S_3, S_4 and S_5.

C. Disscusion

The proposed ZIMPC and its power flow control method as SRM drives have been described. While, it is worth pointing out that ZIMPC also allows the SRM to work as a generator. In one electric cycle, the electrical angel of the rotor increases from 0° to 360°. For motoring operation, the phase current command equals i_{ref} when the rotor goes from unaligned position to aligned position. When the rotor passes the aligned position, the phase current reference changes to 0. For generating operation, the phase current is kept at 0 before the rotor reaches aligned position. Then, the phase current command becomes i_{ref}. The rest of the control system remains the same, and the capacitor voltage as well as the source current can still be regulated.

V. EXPERIMENTAL RESULTS

Two prototypes based on the conventional ASHB topology and the proposed ZIMPC have been built to compare their performances. Here, a 2-stack 2-phase SRM is driven by

978-1-4673-9551-9/16 $31.00 © 2016 IEEE

TABLE I. PARAMETERS OF THE MOTOR DRIVES

Asymmetrical H-bridge converter (ASHB)	
Parameter	**Value**
DC bus voltage	125 V
Switches	IPL65R130C7
Diodes	RFUS20NS6STL
DC bus capacitor	1100 μF
Quasi-Z-source Integrated multiport converter (ZIMPC)	
Parameter	**Value**
Input voltage	40-100 V
Switches	IPL65R130C7
Diodes	RFUS20NS6STL
Switching frequency	20-kHz
Inductor (L_1)	1 mH, 14 mΩ
Inductor (L_2)	1 mH, 14 mΩ
Capacitor (C_1)	220 μF
Capacitor (C_2)	220 μF
Capacitor voltage (v_{c1})	40-90 V
Capacitor voltage (v_{c2})	70-150 V

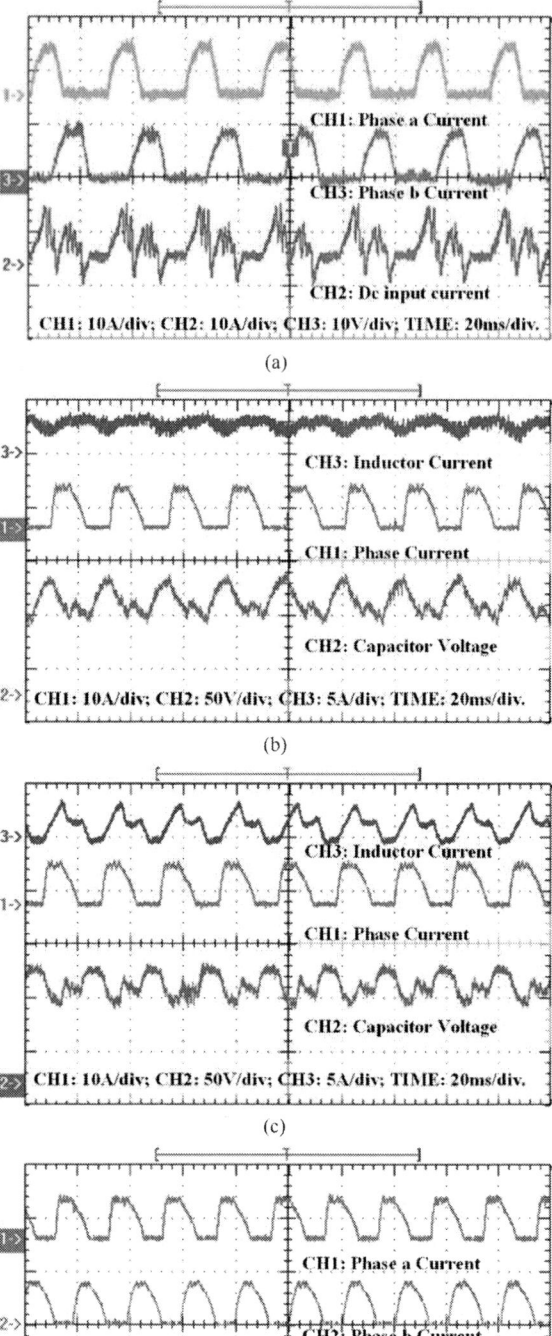

Fig. 6. Experimental results.

the converters so as to test the feasibility of ZIMPC and its advantages. Both the number of stator poles and rotor poles are equal to 8 in each stack. The dc alternator is coupled with the SRM and a power resistor is connected to the output of dc alternator, which acts as load. Considering the voltage and current rating, MOSFET IPL65R130C7 from Infineon is used as power switches and RFUS20NS6STL from Rohm semiconductor is used as power diodes. For the control part, the designed power flow control scheme is implemented using ARM-based microprocessor XMC4400 from Infineon. Similar to the simulation, the switching frequency is selected at 20-kHz. The detailed hardware parameters of the prototypes are listed in Table I.

For the ASHB SRM drive, the waveform of the two phase currents and input current are shown in Fig. 6(a). The two phases are driven alternatively according to the rotor position. In this drive, five 220μF electrolytic capacitors are connected in parallel to the dc bus and the dc source. Even though a large capacitance is used, the input current is still pulsatory and the peak-to-peak current is 13.8A. Furthermore, the current becomes negative when the dc bus capacitor doesn't absorb enough energy during the period of phase winding de-energizing.

The ZSIMPC is tested with the same SRM and the waveforms are shown in Fig. 6(b)-(d). Two 100μF capacitors are used as the two capacitor in the quasi-Z-source unit. If the proposed control method is not utilized, meaning that the shoot-through duty cycle is fixed, the input current is similar to that of ASHB, as shown in Fig. 6(b). When using the proposed control method, the input current ripple is greatly reduced as shown in Fig. 6(c). Additional phase current waveforms when the proposed topology is used with the proposed control method are provided in Fig. 6(d). The waveforms reveal that the peak-to-peak current is less than

2.6A which is only about 19% of the peak-to-peak input current with ASHB converter. The frequency distributions of the input current in Fig. 6(a)-(c) are shown in Fig. 7.

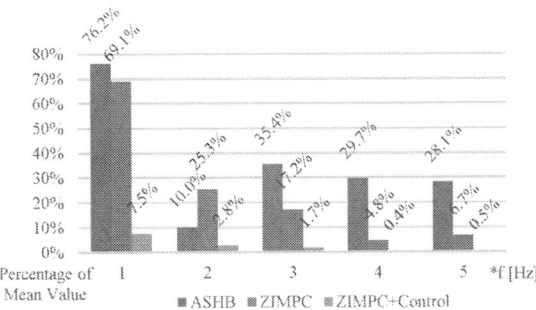

Fig. 7. Frequency distributions of the input current using different topologies and control method.

VI. CONCLUSIONS

A quasi-Z-source integrated multiport converter (ZIMPC) is proposed as SRM drives in this paper for capacitance reduction and enabling a wide-speed-range operation. By using this topology in place of asymmetrical H-bridge converter (ASHB), the requirement for capacitance to provide the power ripple during current commutation can be reduced significantly. This, in turn, can expand the converter's lifetime and enhance its power density. The current ripple through the input source is also reduced by controlling the shoot-through time of the quasi-Z-source unit. In addition, performance during high-speed operation can be improved with higher equivalent exciting voltage given by the quasi-Z-source unit, enabling a wider CPSR. With the operation modes and the equivalent model of the overall system, an advanced power flow control method is developed. The validity and performance of the proposed ZIMPC is verified by the simulation and experimental results. Comparative results reveal that the capacitor can compensate for 90% of periodical transient power ripple with only 40% of the original capacitance as compared to ASHB for SRM drives.

ACKNOWLEDGMENT

This work is done in the Renewable Energy and Vehicular Technology (REVT) Laboratory, the University of Texas at Dallas. The authors would like to thank the founding director Dr. Babak Fahimi for his support.

REFERENCES

[1] M. Abbasian, M. Moallem, and B. Fahimi, "Double-Stator Switched Reluctance Machines (DSSRM): Fundamentals and Magnetic Force Analysis," *Energy Conversion, IEEE Transactions on*, vol. 25, pp. 589-597, 2010.

[2] P. N. Materu and R. Krishnan, "Steady-state analysis of the variable-speed switched-reluctance motor drive," *Industrial Electronics, IEEE Transactions on*, vol. 36, pp. 523-529, 1989.

[3] B. Fahimi, A. Emadi, and R. B. Sepe, "A switched reluctance machine-based starter/alternator for more electric cars," *Energy Conversion, IEEE Transactions on*, vol. 19, pp. 116-124, 2004.

[4] F. Savoye, P. Venet, M. Millet, and J. Groot, "Impact of Periodic Current Pulses on Li-Ion Battery Performance," *Industrial Electronics, IEEE Transactions on*, vol. 59, pp. 3481-3488, 2012.

[5] L. Long and P. Bauer, "Practical Capacity Fading Model for Li-Ion Battery Cells in Electric Vehicles," *Power Electronics, IEEE Transactions on*, vol. 28, pp. 5910-5918, 2013.

[6] L. Gu, X. Ruan, M. Xu, and K. Yao, "Means of Eliminating Electrolytic Capacitor in AC/DC Power Supplies for LED Lightings," *IEEE Trans. Power Electrons.*, vol. 24, pp. 1399-1408, 2009.

[7] J. L. Stevens, J. S. Shaffer, and J. T. Vandenham, "The service life of large aluminum electrolytic capacitors: effects of construction and application," *IEEE Trans. Ind. Appl.*, vol. 38, pp. 1441-1446, 2002.

[8] B. Gu, J. Dominic, J. Zhang, L. Zhang, B. Chen, and J.-S. Lai, "Control of electrolyte-free microinverter with improved MPPT performance and grid current quality," in *Applied Power Electronics Conference and Exposition (APEC), 2014 Twenty-Ninth Annual IEEE*, 2014, pp. 1788-1792.

[9] L. Changrong and L. Jih-Sheng, "Low Frequency Current Ripple Reduction Technique With Active Control in a Fuel Cell Power System With Inverter Load," *IEEE Trans. Power Electrons.*, vol. 22, pp. 1429-1436, 2007.

[10] J. I. Itoh and F. Hayashi, "Ripple Current Reduction of a Fuel Cell for a Single-Phase Isolated Converter Using a DC Active Filter With a Center Tap," *IEEE Trans. Power Electrons.*, vol. 25, pp. 550-556, 2010.

[11] W. Rong-Jong and L. Chun-Yu, "Active Low-Frequency Ripple Control for Clean-Energy Power-Conditioning Mechanism," *IEEE Trans. Ind. Electrons.*, vol. 57, pp. 3780-3792, 2010.

[12] C. Wen, L. Bangyin, D. Shanxu, and J. Ling, "An Active Low-Frequency Ripple Control Method Based on the Virtual Capacitor Concept for BIPV Systems," *IEEE Trans. Power Electrons.*, vol. 29, pp. 1733-1745, 2014.

[13] X. Liu, Z. Q. Zhu, M. Hasegawa, A. Pride, R. Deohar, T. Maruyama, et al., "Dc-link capacitance requirement and noise and vibration reduction in 6/4 switched reluctance machine with sinusoidal bipolar excitation," in *IEEE Energy Conversion Congress and Exposition (ECCE)*, 2011, pp. 1596-1603.

[14] C. R. Neuhaus and R. W. De Doncker, "DC-link voltage control for Switched Reluctance Drives with reduced DC-link capacitance," in *IEEE Energy Conversion Congress and Exposition (ECCE)*, 2010, pp. 4192-4198.

[15] C. R. Neuhaus, N. H. Fuengwarodsakul, and R. W. De Doncker, "Control Scheme for Switched Reluctance Drives With Minimized DC-Link Capacitance," *IEEE Trans. Power Electrons.*, vol. 23, pp. 2557-2564, 2008.

[16] W. Suppharangsan and J. Wang, "Experimental validation of a new switching technique for DC-link capacitor minimization in switched reluctance machine drives," in *IEEE International Electric Machines & Drives Conference (IEMDC)*, 2013, pp. 1031-1036.

[17] L. Guangjin, S. Ojeda, S. Hlioui, E. Hoang, M. Lecrivain, and M. Gabsi, "Modification in Rotor Pole Geometry of Mutually Coupled Switched Reluctance Machine for Torque Ripple Mitigating," *Magnetics, IEEE Transactions on*, vol. 48, pp. 2025-2034, 2012.

[18] P. C. Desai, M. Krishnamurthy, N. Schofield, and A. Emadi, "Novel Switched Reluctance Machine Configuration With Higher Number of Rotor Poles Than Stator Poles: Concept to Implementation," *Industrial Electronics, IEEE Transactions on*, vol. 57, pp. 649-659, 2010.

[19] R. Mikail, I. Husain, Y. Sozer, M. S. Islam, and T. Sebastian, "Torque-Ripple Minimization of Switched Reluctance Machines Through Current Profiling," *Industry Applications, IEEE Transactions on*, vol. 49, pp. 1258-1267, 2013.

[20] S. K. Sahoo, S. Dasgupta, S. K. Panda, and X. Jian-Xin, "A Lyapunov Function-Based Robust Direct Torque Controller for a Switched Reluctance Motor Drive System," *Power Electronics, IEEE Transactions on*, vol. 27, pp. 555-564, 2012.

[21] H. J. Brauer, M. D. Hennen, and R. W. De Doncker, "Control for Polyphase Switched Reluctance Machines to Minimize Torque Ripple and Decrease Ohmic Machine Losses," *Power Electronics, IEEE Transactions on*, vol. 27, pp. 370-378, 2012.

[22] X. D. Xue, K. W. E. Cheng, and S. L. Ho, "Optimization and Evaluation of Torque-Sharing Functions for Torque Ripple Minimization in Switched Reluctance Motor Drives," *Power Electronics, IEEE Transactions on*, vol. 24, pp. 2076-2090, 2009.

[23] Y. Jin, B. Bilgin, and A. Emadi, "An Offline Torque Sharing Function for Torque Ripple Reduction in Switched Reluctance Motor Drives," *Energy Conversion, IEEE Transactions on*, vol. 30, pp. 726-735, 2015.

[24] D. Le, G. Bin, J. S. J. Lai, and E. Swint, "Control of pseudo-sinusoidal switched reluctance motor with zero torque ripple and reduced input

current ripple," in *Energy Conversion Congress and Exposition (ECCE), 2013 IEEE*, 2013, pp. 3770-3775.

[25] F. Z. Peng, "Z-source inverter," in *Industry Applications Conference, 2002. 37th IAS Annual Meeting. Conference Record of the*, 2002, pp. 775-781 vol.2.

[26] J. Anderson and F. Peng, "Four quasi-Z-Source inverters," in *Power Electronics Specialists Conference, 2008. PESC 2008. IEEE*, 2008, pp. 2743-2749.

[27] L. Yuan, J. Shuai, J. G. Cintron-Rivera, and P. Fang Zheng, "Modeling and Control of Quasi-Z-Source Inverter for Distributed Generation Applications," *IEEE Trans. Ind. Electrons.*, vol. 60, pp. 1532-1541, 2013.

[28] C. Dong, J. Shuai, Y. Xianhao, and P. Fang Zheng, "Low-Cost Semi-Z-source Inverter for Single-Phase Photovoltaic Systems," *IEEE Trans. Power Electrons.*, vol. 26, pp. 3514-3523, 2011.

[29] W. Qian, F. Z. Peng, and H. Cha, "Trans-Z-source inverters," in *Power Electronics Conference (IPEC), 2010 International*, 2010, pp. 1874-1881.

[30] G. Feng, F. Lixing, L. Chien-Hui, L. Cong, C. Woongchul, and W. Jin, "Development of an 85-kW Bidirectional Quasi-Z-Source Inverter With DC-Link Feed-Forward Compensation for Electric Vehicle Applications," *Power Electronics, IEEE Transactions on*, vol. 28, pp. 5477-5488, 2013.

[31] W. Cai, L. Jiang, B. Liu, S. Duan, and C. Zou, "A power decoupling method based on four-switch three-port DC/DC/AC converter in DC microgrid," in *IEEE Energy Conversion Congress and Exposition (ECCE)*, 2013, pp. 4678-4682.

[32] A. Kwasinski, "Identification of Feasible Topologies for Multiple-Input DC-DC Converters," *IEEE Transactions on Power Electronics*, vol. 24, pp. 856-861, 2009.

[33] L. Yan, R. Xinbo, Y. Dongsheng, L. Fuxin, and C. K. Tse, "Synthesis of Multiple-Input DC/DC Converters," *IEEE Transactions on Power Electronics*, vol. 25, pp. 2372-2385, 2010.

[34] W. Cai and F. Yi, "Topology simplification method based on switch multiplexing technique to deliver DC-DC-AC converters for microgrid applications," in *Energy Conversion Congress and Exposition (ECCE), 2015 IEEE*, 2015, pp. 6667-6674.

[35] W. Cai and F. Yi, "An Integrated Multi-Port Power Converter with Small Capacitance Requirement for Switched Reluctance Motor Drive," *Power Electronics, IEEE Transactions on*, vol. PP, pp. 1-1, 2015.

[36] F. Yi and W. Cai, "Repetitive control-based current ripple reduction method with a multi-port power converter for SRM drive," in *Transportation Electrification Conference and Expo (ITEC), 2015 IEEE*, 2015, pp. 1-6.

A Fault-Tolerant Topology of T-Type NPC Inverter with Increased Thermal Overload Capability

Jiangbiao He[*], Nathan Weise[*], Lixiang Wei[**], Nabeel A.O. Demerdash[*]

[*]Department of Electrical and Computer Engineering
Marquette University
Milwaukee, Wisconsin, 53233
jiangbiao@ieee.org

[**]Standard Drives Business
Rockwell Automation
Mequon, Wisconsin, 53092
lwei@ra.rockwell.com

Abstract—Reliability of multilevel power converters has received increased attention over the past few years, due to the relatively high component failure probability caused by the large number of switching devices utilized in the converter topologies. Thus, the fault-tolerant operation of multilevel converters plays an essential role in safety-critical industrial applications. This paper introduces an improved fault-tolerant inverter topology based on the conventional three-level T-Type neutral-point-clamped (NPC) inverter. Unlike other existing three-phase four-leg fault-tolerant inverter topologies, the fault-tolerant T-Type inverter topology proposed here can significantly enhance the overall thermal overload capability of the inverter, in addition to providing desired fault-tolerant capability. The operating principle of this proposed fault-tolerant T-Type inverter is detailed in this paper, and simulation results are presented to verify the advantages of this improved inverter topology.

Keywords—Neutral-point clamped inverters; fault-tolerant operation; overload capability; three-phase four-leg inverter.

I. INTRODUCTION

T-Type neutral-point-clamped (NPC) inverters have been regarded as a very promising breed of high-performance multilevel inverters in low-voltage industrial applications. Compared to the conventional I-Type NPC inverters [1], advantages of T-Type NPC inverters include the relatively lower number of switching devices involved in the circuit topology and higher converter efficiency under certain operating conditions [2][3]. However, like other types of multilevel converters, T-Type NPC inverters are not immune to switching device faults. For instance, solid-state switch open-circuit or short-circuit faults, which would cause catastrophic system failures if no fault-tolerant solution is provided, especially when such inverters are applied in safety-critical applications, such as electric/hybrid vehicles (EVs/HEVs), uninterruptable power supplies (UPSs), solar inverters, and the like.

Although the T-Type NPC inverter has certain inherent fault-tolerant capability due to its unique topology, as reported in [4], the output voltage will be derated significantly under the proposed fault-tolerant operation strategy. Such voltage derating is not accepted in certain applications (e.g., UPSs, EVs/HEVs, etc.) where rated voltage is a stringent requirement. One more existing solution for the fault-tolerant operation of T-Type NPC inverters is mainly achieved by paralleling one or three redundant inverter legs, as reported in [5], which does ensure a

rated voltage output under inverter faulty condition, but at a much higher system cost with decreased efficiency due to much more additional semiconductor devices employed. As a matter of fact, most of the redundant semiconductor devices in the existing fault-tolerant topology idle, under healthy conditions, in the circuit without contribution to system performance, while conversely degrading system efficiency due to additional switching and conduction losses [5]. Therefore, it would be of great significance to improve the inverter topology to obtain certain fault-tolerant characteristics, while guaranteeing rated output voltages under faulty conditions and with acceptable efficiency degradation under healthy condition.

In this paper, a novel fault-tolerant solution is introduced based on the conventional T-Type NPC inverters. The fault-tolerant topology is shown in Fig. 1. In this topology, a redundant inverter leg is added to provide a hardware backup in case of any switch faults occurring in one of the original three phase legs of the T-Type inverter. Moreover, this redundant leg can share overload current with all the original phase legs of the T-Type topology. Thus, the overall overload capability of the T-Type inverter can be improved without oversizing all the power devices of the inverter. During the utilization of the fourth leg for current sharing, a quasi zero voltage switching (ZVS) PWM strategy will be adopted to mitigate the overall device losses.

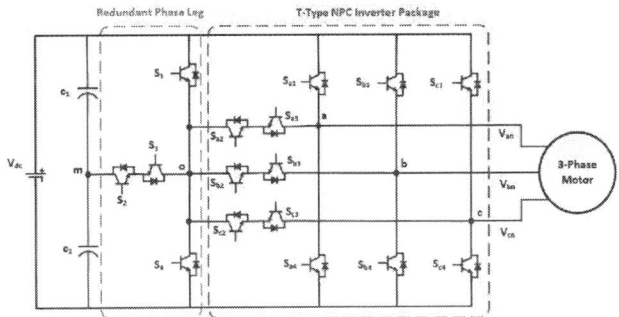

Fig. 1 The proposed fault-tolerant topology of the three-phase three-level T-Type NPC inverter.

The remainder content of this paper is organized as follows. In section II, the fault tolerant strategy for each fault scenario of the switching devices will be introduced. In Section III, the operating principle of utilizing the fourth inverter leg for sharing overload current under healthy condition will be explained. In

978-1-4673-9551-9/16 $31.00 © 2016 IEEE

Section IV, a quasi zero-voltage switching (ZVS) strategy which can reduce the device losses contributed by the fourth leg will be presented. The simulation results focusing on the thermal stress mitigation and overload current sharing will be given in Section V. Finally, conclusions will be summarized in Section VI.

II. FAULT-TOLERANT OPERATION OF THE PROPOSED T-TYPE NPC INVERTER TOPOLOGY

Considering the circuit symmetry of the T-Type NPC inverter, only four cases of switch faults are analyzed in the following discussion to represent all the possible switch fault scenarios that could happen in such inverter. Here, it is assumed that only single device fault takes place in the inverter. Although the fault scenario analysis and fault-tolerant solutions discussed below only focus on IGBT devices in the T-Type inverter, they are also applicable for the related free-wheeling diode faults, which will be elaborated on in future work.

A. Case I: Open-Circuit fault in IGBT S_{a1}

Once an open-circuit fault in IGBT S_{a1} is identified, Phase-A leg of the T-Type inverter will not be able to produce a positive voltage due to the disconnection to the positive dc bus. Under such scenario, IGBT S_{a1} will be replaced by S_1 from the redundant phase leg through switching on IGBT S_{a2}, while all other IGBTs on the redundant leg are kept off, as illustrated in Fig. 2 (a). As can be seen, during such fault-tolerant operation, the three-phase inverter can still output rated voltage, but will be operated as a two-level one. The similar fault-tolerant solutions can be applied for open-circuit faults in other IGBTs S_{x1} (where x=b or c) and S_{x4} (where x=a, b, and c).

B. Case II: Short-circuit fault in IGBT S_{a1}

Generally, a short-circuit failure mode in IGBT modules concludes with an open-circuit mode due to the large short-circuit current and rapid accumulated heat dissipation in IGBT bond wires or soldering joints if no fast protection actions are available (typically short-circuit withstanding time is within 10 μs). When a short-circuit fault occurs in IGBT S_{a1}, assuming that the large short-circuit current will change the short-circuit fault mode into open-circuit fault by melting the bond wires or cracking the soldering joint in the IGBT package, then the fault-tolerant solution to such fault scenario will be the same as that for open-circuit fault which is explained in Case I.

C. Case III: Open-Circuit fault in IGBT S_{a2}

If an open-circuit fault happens in IGBT S_{a2}, the output terminal of the Phase-A leg will not have access to the dc-bus middle point. This indicates that the three-level T-Type NPC inverter will have to be operated as a two-level inverter by only using S_{x1} and S_{x4} for fault-tolerant operation (the same as the conventional two-level inverter topology). Under such situation, all the switches in the redundant leg and all the bi-directional switches (S_{x2} and S_{x3}, where, x=b, or c) will be switched off, as shown in Fig. 2 (b). There is no derating for the rated and maximum voltage output during post-fault operation, but the harmonic distortion will be higher under two-level modulation compared to that under three-level modulation. The similar fault-tolerant solutions can be applied for open-circuit faults in other IGBTs S_{x2}/S_{x3} (where x=a, b, and c).

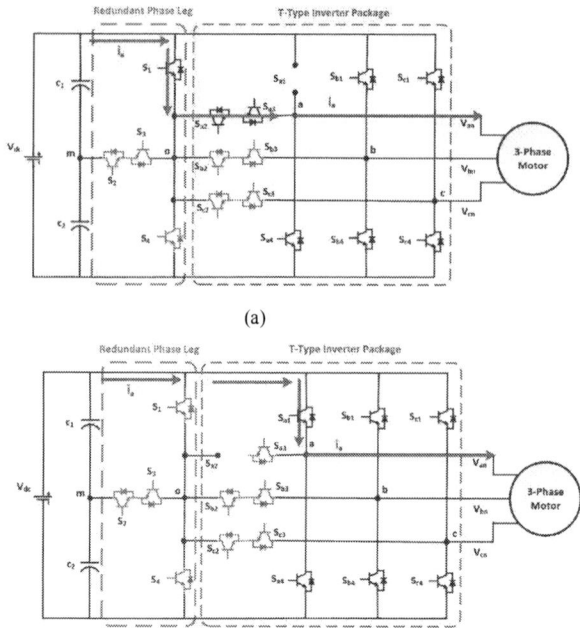

(a)

(b)

Fig. 2 Current flow illustration during fault-tolerant operation when (a) Switch S_{a1} has an open-circuit fault (b) Switch S_{a3} has an open-circuit fault.

D. Case IV: Short-circuit fault in IGBT S_{a2}

If a short-circuit fault in IGBT S_{a2} is determined, its complimentary switch Sa3 has to be switched off due to the loss of reverse blocking capability, and the three-level inverter will be operated as a conventional two-level inverter by only using S_{x1} and S_{x4} for post-fault operation, which is similar to Case III discussed above. A similar fault-tolerant strategy can be applied for short-circuit faults in IGBTs S_{x2} (where, x=b, or c) and S_{x3} (where, x=a, b, or c).

In summary, as can be seen, when any of the IGBTs in the T-Type inverter has a fault, there is no derating required during the fault-tolerant operating region of the inverter. However, such inverter has to be modulated as a two-level one under faulty condition, which implies a slightly higher harmonic distortion in the output currents and voltages compared to these under three-level healthy operation. However, reliability of the inverters has a much higher priority than the increase of harmonic distortion, particularly for these inverters used in safety-critical applications.

III. IMPROVED T-TYPE NPC INVERTER THERMAL OVERLOAD CAPABILITY

It is known that three-level NPC inverters exhibit output voltage waveforms similar to two-level inverters at low amplitude modulation indices (i.e., Ma≤0.5) [6]. Then, the bi-directional switches (S_{x2} and S_{x3} shown in Fig. 3, x=a, b, c) under low modulation indices, instead of being used to output any zero voltage states, can be used to conduct the redundant phase leg to the inherent three legs to share overload current. For instance, when the upper switch S_{a1} in Phase-A leg of the T-Type inverter commutates a large positive load current (current flows from source to load), the switch S_1 in the redundant leg can be used to share the current with S_{a1}. Such current sharing is achieved by

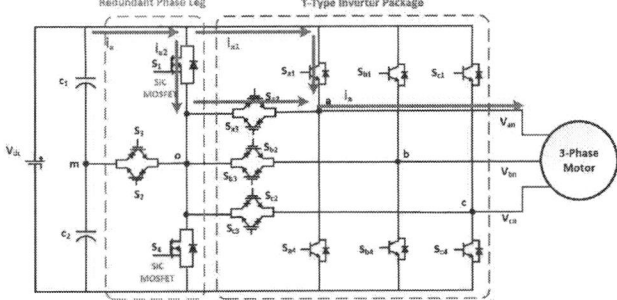

Fig. 3 The redundant phase leg shares load current in Phase-A.

the conduction of S_{a1}. The current sharing and flow direction are illustrated in Fig. 3. Similarly, when the switch S_{a4} in Phase-A leg commutates a large negative current (current flows from load to source), the switch S_4 in the redundant leg will be used to share the overload current with S_{a4} through the conduction of the switch S_{a3}. The similar principle applies for using the redundant leg for sharing overload current with Phase-B and Phase-C.

At higher modulation index (i.e., $M_a>0.5$), the redundant leg can also be used to share the overload current with other original phase legs. However, such current sharing capability will make the three-level operate as a two-level, which will cause more harmonic distortions in the output voltages, compared to the normal operation of a three-level T-Type inverter. The reason is that the directional switches (S_{x2} and S_{x3} shown in Fig. 3, x=a, b, c) have to be kept in turn-on state to parallel the redundant switch (S_1 or S_2) with the related main switches (S_{x1} and S_{x4}, where x=a, b, or c) for sharing the overload current. On such occasions, the switches S_2 and S_3 have to be turned off, which makes the access to the neutral-point voltage unavailable. In other words, the utilization of the redundant leg for overload current sharing under high modulation indices is at the cost of relatively larger harmonic distortions in the output voltages and currents.

However, it should be noted that the current sharing may not be applicable for all the three phase legs at the same time. For the sake of explanation, one can define the positive output voltage ($V_{dc}/2$) of a phase leg as "1", negative output voltage (-$V_{dc}/2$) as "-1", and zero output voltage as "O". The load current sharing cannot be achieved simultaneously for states "1", "0", and "-1" for different phase legs. This is because of the fact that the load current sharing for both "1" and "-1" switching states for two different legs will involve the simultaneous turn-on of switches S_1 and S_2 in the redundant leg, which will short circuit the dc-bus. Similarly, if the load current sharing is achieved during the "1" or "-1" switching state, the switches S_2 and S_3 have to be kept in OFF state, which indicates the loss of access to the dc-bus mid-point for achieving the switching state "0" in other phase legs at the same time. Nevertheless, the load current sharing can be applied to the phase leg which undertakes the larger current during the commutation. For instance, during the switching state (1, 1, -1), the switch S_{c4} in Phase-C leg will experience larger current. Thus, the switches S_4 and S_{c2} can be turned on to share the share load current with S_{c4}. Moreover, for the zero vector states, namely, (P, P, P) and (N, N, N), the load current sharing can be applied in order to share the load current with all the three phase legs. The hexagonal space vector diagram for the T-Type inverter is given in Fig. 4, and the

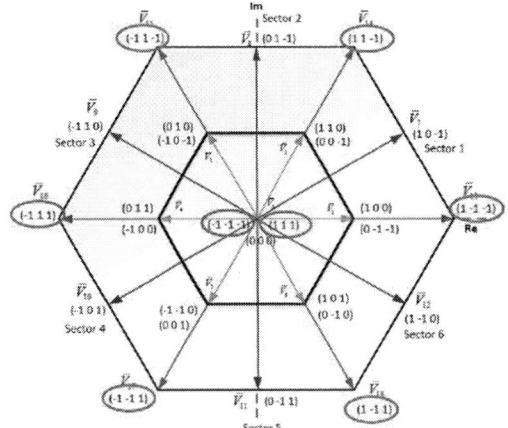

Fig. 4 space vector diagram for the T-Type inverter.

TABLE-1 SWITCHING PATTERN FOR LOAD CURRENT SHARING OF THE THREE-PHASE T-TYPE NPC INVERTER.

Voltage space vector	Devices to share the load current	Switching action for load current sharing
(1, -1, -1)	S_{a1} and S_1	Turn-on S_1 and S_{a2}
(1, 1, -1)	S_{c4} and S_4	Turn-on S_4 and S_{c3}
(-1, 1, -1)	S_{b1} and S_1	Turn-on S_1 and S_{b2}
(-1, 1, 1)	S_{a4} and S_4	Turn-on S_4 and S_{a3}
(-1, -1, 1)	S_{c1} and S_1	Turn-on S_1 and S_{c2}
(1, -1, 1)	S_{b4} and S_4	Turn-on S_4 and S_{b3}
(1, 1, 1)	S_{a1}, S_{b1}, S_{c1} and S_1	Turn-on S_1, S_{a2}, S_{b2}, S_{c2}
(-1, -1, -1)	S_{a4}, S_{b4}, S_{c4} and S_4	Turn-on S_4, S_{a3}, S_{b3}, S_{c3}

switching states where the load current sharing is applied in this paper is circled in red in Fig. 4. It can be seen that, among the total 27 switching states for the normal operation of a three-level T-Type inverter, all the six large voltage space vectors and two of the zero voltage vectors are selected for load current sharing in this paper. The specific switching pattern for these selected voltage vectors are given in Table-1.

IV. QUASI ZVS SWITCHING PATTERN

One inherent concern with the T-Type NPC inverter is that, the losses distribution among devices in the inverter is unbalanced. Especially, when the inverter is operating at high modulation indices, most of the device losses are dissipated from the main switches S_{x1} and S_{x4} [2]. To alleviate the high junction temperatures in these main IGBT modules, a quasi ZVS strategy is used to transfer the switching losses from S_{x1}/S_{x4} to the switches on the redundant phase leg and bi-direction switches in the T-Type inverter. Specifically, when the redundant phase leg is determined to share the overload current with other legs, the switching devices S_1/S_4 and the bi-directional switching devices (S_{x2}/S_{x3}) are switched on prior to the turn-on of the main switches (S_{x1}/S_{x4}), to provide a very low on-state voltage for the subsequent switching-on of the main switches (S_{x1}/S_{x4}). Regarding the switching off, the main switches (S_{x1}/S_{x4}) are turned off prior to the redundant switches S_1/S_4 and bi-directional switches (S_{x2}/S_{x3}) for similarly achieving quasi-ZVS soft switching. Such switching sequence is illustrated in Fig. 5. More details about such ZVS switching pattern are introduced in [7-9].

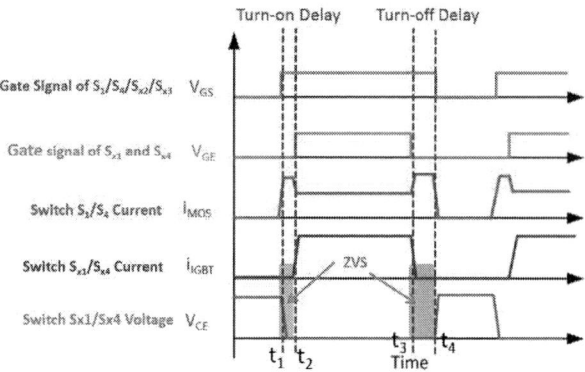

Fig. 5 Quasi-ZVS switching pattern for the fault-tolerant T-Type inverter.

V. SIMULATION ANALYSIS

Since the fault-tolerant operation of the proposed topology presented in Section II is quite straightforward, thus only the simulation on the load current sharing and the proposed quasi-ZVS switching pattern will be carried out in this section.

A. Load Current Sharing

To increase the load current sharing ratio and restrict additional device losses from the redundant phase leg, SiC MOSFETs are used in this work for switches S_1 and S_4 on the redundant leg, and a commercial modular three-level T-Type inverter package is employed where the RB-IGBTs are integrated as the bi-directional switches in the package due to their low on-state voltage drop [10]. To illustrate the load current sharing between the redundant leg and one of the inherent inverter legs, the I-V curve comparison between S_{a1} and another resultant device formed by switches S_1 and S_{a2} is shown in Fig. 6. As can be seen, if the T-Type NPC inverter under normal operation is sized to drive continuous load current of 40A, the redundant leg can share an additional load current of 10A considering the same resultant on-state voltage between the two parallel paths depicted in Fig. 3. This indicates that the total load current becomes 50A, which is an increase of 25% compared to the inherent current rating of 40A.

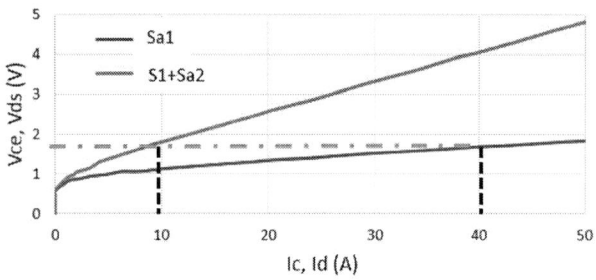

Fig. 6 Device output characteristics comparison.

B. Thermal Modeling of the Fault-Tolerant T-Type Inverter

To examine the mitigation on device junction temperatures due to the using of the redundant leg during the load current sharing, device thermal modeling and the related thermal RC network from junction to ambient have been established for the proposed fault-tolerant T-Type inverter in PLECS. The device conduction loss is calculated based on the follows equations:

$$u_{ce}(t) = u_{ce0} + r_{on} \cdot i_c(t) \tag{1}$$

$$p_{con}(t) = u_{ce}(t) \cdot i_c(t) = u_{ce0} \cdot i_c(t) + r_{on} \cdot i_c^2(t) \tag{2}$$

where, u_{ce0} is the voltage drop of the IGBT chip at near-zero forward current; r_{on} is the on-state resistance; $i_c(t)$ is the instantaneous IGBT collector current; $p_{con}(t)$ is the instantaneous conduction losses of the IGBT.

If the average IGBT current value is designated as I_{av}, and the RMS value of the IGBT current is I_{rms}, then the average IGBT conduction losses can be expresses as:

$$P_{avcon} = \frac{1}{T_{sw}} \int_0^{T_{sw}} p_{con}(t) dt \tag{3}$$

$$P_{avcon} = \frac{1}{T_{sw}} \int_0^{T_{sw}} (u_{ce0} \cdot i_c(t) + r_{on} \cdot i_c^2(t)) dt \tag{4}$$

$$P_{av\_con} = u_{ce0} \cdot I_{av} + r_{on} \cdot I_{rms}^2 \tag{5}$$

where, P_{av_con} is the average IGBT conduction losses, and T_{sw} represents the switching period.

Similarly, the conduction losses of the antiparallel diodes are calculated based on the following equations:

$$u_d(t) = u_{d0} + r_d \cdot i_d(t) \tag{6}$$

$$p_{d\_con}(t) = u_d(t) \cdot i_d(t) = u_{d0} \cdot i_d(t) + r_d \cdot i_d^2(t) \tag{7}$$

$$P_{davcon} = \frac{1}{T_{sw}} \int_0^{T_{sw}} P_{dcon}(t) dt \tag{8}$$

$$P_{davcon} = \frac{1}{T_{sw}} \int_0^{T_{sw}} (u_{d0} \cdot i_d(t) + r_d \cdot i_d^2(t)) dt \tag{9}$$

$$P_{d\_av\_con} = u_{d0} \cdot I_{d\_av}(t) + r_d \cdot I_{d\_rms}^2 \tag{10}$$

where, u_d is the voltage drop across the diode at certain dc current, and u_{d0} refers to the approximated voltage drop of the diode at near-zero forwarding current. r_d is the slop resistance of the diode chip, and i_d is the current flowing through the diode. Also, p_{d_con} and $P_{d\_av\_con}$ represents the instantaneous and average conduction losses in the diode, respectively. I_{d_av} and I_{d_rms} refer to the average diode current and RMS value of the diode current, respectively.

Switching loss of semiconductor devices depends on the turn-on and turn-off energy at specific operating points (current, blocking voltage, and junction temperature). The mathematical model for calculating IGBT instantaneous switching loss, $P_{sw}(t)$, is given as follows:

$$P_{sw}(t) = (E_{on}(i_c, v_{ce}, T_j) + E_{off}(i_c, v_{ce}, T_j)) \cdot f_{sw} \tag{11}$$

where, $E_{on}(i_c, v_{ce}, T_j)$ and $E_{off}(i_c, v_{ce}, T_j)$ refer to the switching-on energy and switching-off energy at the specific levels of the currents, voltages and junction temperatures, respectively, while f_{sw} is the switching frequency.

978-1-4673-9551-9/16 $31.00 © 2016 IEEE

In order to analyze the thermal performance of the fault-tolerant T-Type inverter, thermal equivalent RC network from device junction to module case, from case to heatsink, and from heatsink to the ambient needs to be modeled. A fourth-order Foster model describing the junction to case of IGBT modules is adopted in this work, which is shown in Fig. 7 (a). The thermal impedance from device case to heatsink as well as from heatsink to ambient is also considered here, as illustrated in Fig. 7 (b). It

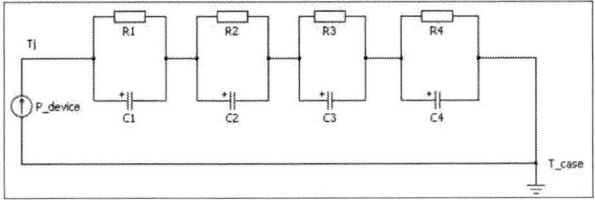

(a)

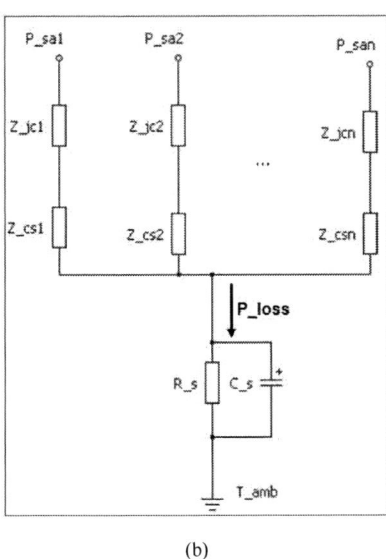

(b)

Fig. 7 Thermal RC network modeling (a) junction-to-case fourth-order Foster thermal network for each power device (IGBTs/diodes) (b) thermal RC network of the whole three-level NPC inverter.

should be mentioned that the parameters of the fourth-order Foster model describing the junction-to case thermal conduction of IGBTs and diodes are extracted from their transient thermal impedance curves given by the manufacture datasheet [14][15].

Through the modeling of device conduction losses, switching losses, as well as the thermal RC network, the variation of IGBT and diode junction temperatures and the related lifetime of the NPC inverter can be predicted.

C. Simulation Results

In this subsection, simulation results will be presented to verify the performance of the proposed fault-tolerant T-Type inverter. Main parameters used in this simulation are given in Table-2. The junction temperature profiles of the IGBT S_{a1} with/without the current sharing from the redundant phase leg under the same thermal overload capability are shown in Fig. 7. Also, it can be observed in Figs. 7 (a)-(b) that the swing magnitude of the junction temperature in S_{a1} is reduced from 7°C

to 5°C by taking advantage of this redundant leg, leading to improved power cycling lifetime of the related power device [11-13].

The load current sharing between the redundant IGBT S_1 and the main IGBT S_{a1} is shown in Fig. 8, from which it shows the current sharing at the turn-on instant, turn-off instant, as well as the current sharing under parallel conduction mode. It should be noted that not all switching states can utilize the redundant leg for current sharing, which is the reason for the discontinuous current sharing between switches S_1 and S_{a1} shown in Fig. 9. The three-phase load currents during load current sharing and the employed ZVS switching pattern are shown in Fig. 10.

TABLE-2 MAIN PARAMETERS USED IN THE SIMULATION

Simulation Parameter	Value	Unit
Input line-to-line voltage (RMS)	480	Volt
Input line frequency	60	Hz
DC-bus capacitance	2000	μF
Switching frequency	4000	Hz
Modulation index	0.95	/
Inverter output filter inductance	2.5	mH
Inverter output filter capacitance	10	μF
Ambient temperature used in the inverter thermal modeling	60	°C

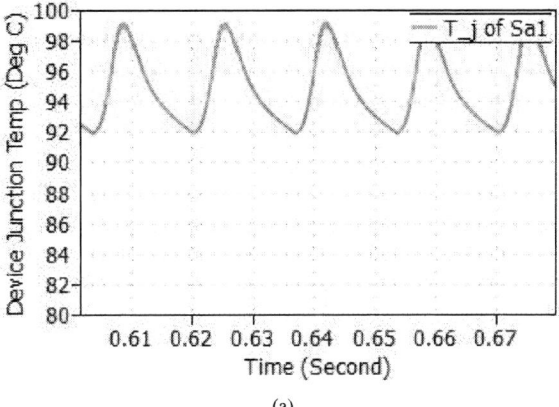

(a)

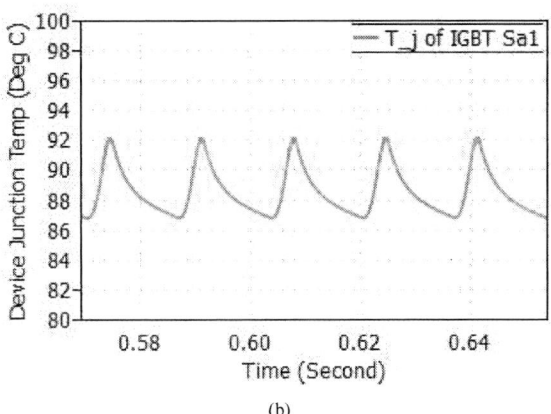

(b)

Fig. 8 Comparison of the junction temperature profiles of the IGBT S_{a1} under heavy load conditions with/without using the redundant phase leg for overload current sharing (a) without using the redundant leg (b) with using the redundant leg.

978-1-4673-9551-9/16 $31.00 © 2016 IEEE   1069

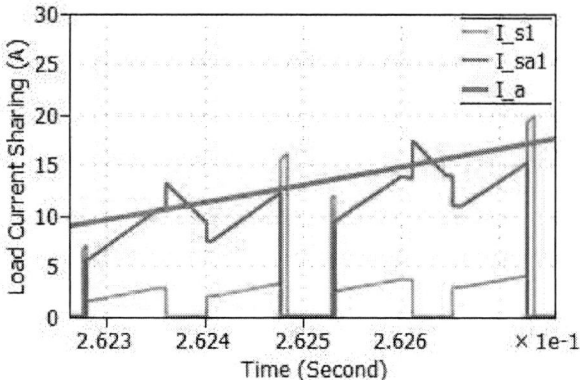

Fig. 9 Load current sharing between the switches S_1 and S_{a1}.

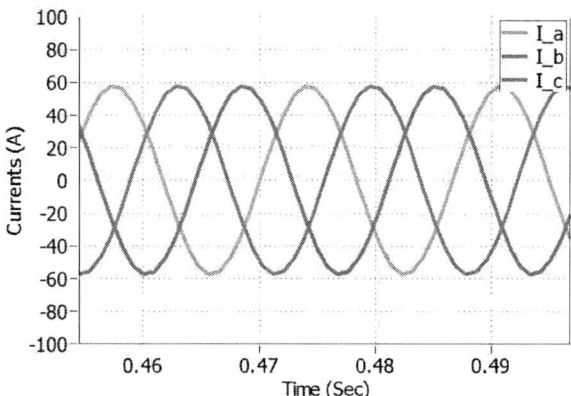

Fig. 9 Three-phase load currents of the four-leg T-Type inverter during load current sharing operation.

VI. CONCLUSIONS

This paper introduced an improved fault-tolerant inverter topology based on the conventional T-Type NPC inverter. According to the analysis and simulation results presented above, the following conclusions can be drawn as follows:

1) The proposed fault-tolerant inverter topology provides desired fault-tolerant solutions to device open-circuit and short-circuit faults in the T-Type inverters. During post-fault operation of any of the device faults, the inverter is still able to output the same maximum and rated voltage/power as that under normal operation. In other words, no derating is required during fault-tolerant operation, although the inverter is working as a two-level one.

2) Under normal healthy condition, the redundant inverter leg helps to share the overload current with the inherent three inverter legs, and therefore can significantly improve the inverter thermal overload capability. Since a normal T-Type inverter exhibits output voltage waveforms similar to a conventional two-level inverter under low modulation indices ($Ma \leq 0.5$), there is no penalty in harmonic distortion in the output voltages under such scenario. Moreover, the redundant leg can also be utilized for load current sharing at high modulation indices ($Ma > 0.5$) to relieve large thermal stress on main switches (S_{x1}/S_{x4}), but the harmonic distortion in the output

voltages will be slightly higher compared to these output from a three-level modulation.

Presently, a 30-kVA adjustable speed drive prototype is being developed in the laboratory based on this proposed fault-tolerant inverter topology. Experimental results as well as more complete analysis and simulation results will be presented in future work to verify the fault-tolerant operations and other enhanced performance of this novel fault-tolerant topology.

ACKNOWLEDGMENT

The authors would like to express thanks to the U.S. National Science Foundation (NSF) for the financial support of the work in this paper through NSF-GOALI Grant No. 1028348. Thanks should also be given to Plexim, Inc. for the access to the PLECS simulation software used in this paper.

REFERENCES

[1] A. Nabae, I. Takahashi, and H. Akagi, "A new neutral-point-clamped PWM inverter," IEEE Tran. on Industry Applications, vol. IA-17, no. 5, pp. 518-523, 1981.

[2] M. Schweizer and J. W. Kolar, "Design and implementation of a highly efficient three-level T-Type converter for low-voltage applications," *IEEE Trans. on Power Electronics*, vol. 28, no. 2, pp.899-907, Feb. 2013.

[3] S. Chen, S. Ogawa, and A. Iso, "High-power IGBT modules for 3-level power converters," FUJI Electric Review, vol. 59, no. 4, 2013.

[4] U. Choi, F. Blaabjerg, and K. Lee, "Reliability improvement of a T-Type three-level inverter with fault-tolerant control strategy," IEEE Trans. on Power Electronics, vol. 30, no. 5, pp.2660-2673, May 2015.

[5] W. Zhang, et al, "A fault-tolerant T-Type three-level inverter system," in *Proc. of IEEE 2014 Applied Power Electronics Conference (APEC)*, 2014, pp. 274-280.

[6] L. M. Tolbert, F. Z. Peng, and T. G. Habetler, "Multilevel PWM Methods at low modulation indices," *IEEE Trans. on Power Electronics*, vol. 15, no. 4, pp.719-725, Jul. 2000.

[7] J. He, T. Zhao, X. Jing, and N.A.O. Demerdash, "Application of wide bandgap devices in renewable energy systems-benefits and challenges," in *Proc. of IEEE 2014 International Conference on Renewable Energy Research and Application (ICRERA)*, 2014, pp. 749-754.

[8] T. Zhao and J. He, "An optimal switching pattern for "SiC+Si" hybrid device based voltage source converters," in *Proc. of IEEE 2015 Applied Power Electronics Conference (APEC)*, 2015, pp. 1276-1281.

[9] J. He, "Health condition monitoring and fault-tolerant operation of adjustable speed drives," Ph.D. dissertation, Marquette University, Milwaukee, Wisconsin, 2015.

[10] Fuji Electric, "Application note: three-level modules with authentic IGBT modules," [Online] https://www.fujielectric.com/, 2015.

[11] M. Held, et al, "Fast power cycling test for IGBT modules in traction applications," in proceedings of 1997 International conference on power electronics and drive systems, May 26-29, 1997, pp. 425-430.

[12] L. Wei, J. McGuire and R. A. Lukaszewski, "Analysis of PWM frequency control to improve the lifetime of PWM inverter," *IEEE Trans. on Industry Applications,* vol. 47, no. 2, pp. 922-929, 2011.

[13] J. He, L. Wei, and N. A.O. Demerdash, "Power cycling lifetime improvement of three-level NPC inverters with an imrpoved DPWM method," Conference proceedings of the 2016 IEEE Applied Power Electronics Conference and Exposition (APEC), Long Beach, CA, 2015.

[14] [Online] Fuji Electric, Datasheet of IGBT Module 12MBI50VX, http://www.fujielectric.com/, 2015.

[15] [Online] Cree Inc, Datasheet of Cree MOSFET C2M0040120D http://www.cree.com/, 2015.

A novel analysis and design method of phase lead filters in repetitive controllers for pulse-width modulated inverters

Shunfeng Yang[1,2], Peng Wang[1], Yi Tang[1], Michael Zagrodnik[3], Xiaolei Hu[2] and King Jet Tseng[1]

Email: syang012@e.ntu.edu.sg epwang@ntu.edu.sg yitang@ntu.edu.sg

[1] School of Electrical and Electronic Engineering, Nanyang Technological University, Singapore
[2] Rolls-Royce@NTU Corporate Lab, Nanyang Technological University, Singapore
[3] Rolls-Royce Singapore Pte Ltd, Singapore

Abstract—Repetitive control strategies have been commonly applied in pulse-width modulated (PWM) inverters for many industrial applications. The phase delay in the overall control system, e.g. the delay caused by the digital duty cycle calculation, PWM generation and the repetitive controller, has to be compensated by elaborately designed phase lead filters in order to prevent control performance deterioration and system instability. Nevertheless, quantificational analysis and practical design of the time advance unit in such filters is hardly to be found in existing literatures. To provide explicit analysis of the phase lead filters, this paper presents a novel design method of the time advance unit in repetitive controllers, which can predict the overall system stability margin and improve the control performance. In order to verify its correctness and effectiveness, the proposed method is simulated with PLECS and implemented experimentally to show the accurate stability margin calculation, as well as excellent steady state performance and dynamic response of the control scheme.

Keywords—*Pulse-width Modulated (PWM) Inverter; Repetitive Control; Phase Lead Filter; Time Advance Unit*

I. INTRODUCTION

Pulse-width modulated (PWM) voltage source inverters (VSIs) are widely adopted as an important component in nowadays power conversion applications such as uninterruptible power supplies (UPSs) [1], grid-tied inverters [2] and active power filters (APFs) [3, 4]. In these applications, a well-designed PWM inverter should be able to regulate the output voltage or current accurately according to the reference sinusoidal signal with low total harmonics distortion (THD) and fast dynamic response, and the control target should be achieved regardless of the presence of nonlinearities both in the loads and the inverters themselves. In order to address this, various control principles have been reported including deadbeat control [5], multi-loop feedback control [6] and repetitive control [7-12].

Considerable attention has been devoted onto the repetitive control scheme due to its outstanding effectiveness on periodic error elimination and periodic disturbance rejection, according to the internal model principle [13]. In practice, the phase delay caused inevitably by the digital control system as well as PWM generation might affect the system performance by changing characteristics such as the resonant frequency and system damping ratio [14, 15] and even lead to system instability. A phase lead filter is usually incorporated in repetitive control schemes to compensate the phase delay in the overall system and more importantly ameliorate the system stability. The inverse transfer function of the inverter with precisely known system parameters can be synthesized as the phase lead filter in repetitive controllers [16], but it is difficult to be implemented in real applications. An alternative is to use a filter to shape the magnitude characteristics of the system for better harmonic rejection and a time advance unit to compensate the phase delay caused by the filter and the plant [3, 10, 17-21]. However, the design process of the time advance unit in these papers is implicit, either based on comparing curves in Matlab [10, 17, 18] or trial-and-error through experimental check [3, 19]. The design of the phase lead compensator is discussed in [21] and the main focus is the design of the repetitive gain rather than the time advance unit. The analytical design process of the time advance unit is rarely discussed in literatures.

In this paper, a repetitive controller scheme with simple multi-loop control for a constant-voltage-constant-frequency (CVCF) PWM inverter is presented to eliminate the harmonics in the output voltage caused by nonlinear loads. Detailed mathematical analysis and an explicit design method regarding the time advance unit in the phase lead filter will be proposed. In addition to ensuring the overall system stability, the proposed method also suggests the stability range and estimates the best time advance unit that can provide the largest phase margin and excellent closed-loop performance. Simulations with PLECS and experimental results based on a single phase PWM inverter with uncontrolled rectifier loads are presented to prove the effectiveness of the repetitive control scheme and time advance unit design method, as well as the correctness of the theoretic analysis.

This work was conducted with support from the National Research Foundation (NRF) Singapore under the Corp Lab@University Scheme.

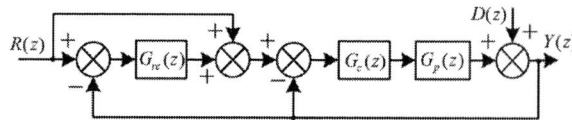

Fig. 1 Block diagram of repetitive control scheme.

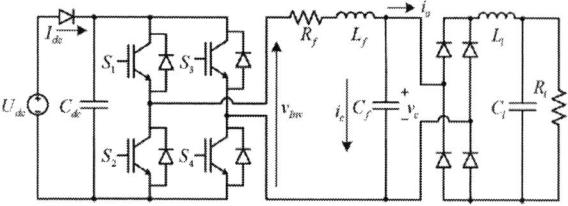

Fig. 2 Setup of a single-phase PWM inverter.

II. REPETITIVE CONTROL SCHEME FOR PWM INVERTERS

A. Reptitive Control Scheme

The block diagram of the repetitive control scheme used in PWM inverters is shown in Fig. 1, where $R(z)$ is the reference, $Y(z)$ is the output, $D(z)$ is the disturbance, $G_{rc}(z)$, $G_c(z)$ and $G_p(z)$ are the z-domain transfer functions of the repetitive controller, the conventional inverter controller and the plant respectively.

$G_c(z)$ might be a single- or multi-loop controller, depending on different applications, i.e. in APF or UPS applications. $G_{rc}(z)$ can also be designed to cope with odd, even or both harmonics according to compensation requirements. Due to the space limit, only the repetitive control scheme for a CVCF PWM inverter is discussed and analyzed in this paper.

B. Multi-loop Controller for the CVCF PWM Inverter

The setup of a single-phase CVCF PWM inverter is shown in Fig. 2. The H-bridge based single phase PWM inverter is connected to the dc bus capacitor C_{dc} with average voltage U_{dc}. The output terminal of the PWM inverter is equipped with an LC filter, L_f and C_f. A diode rectifier is employed as a nonlinear load. R_l, L_l and C_l are resistance, inductance and capacitance of the nonlinear load respectively. An equivalent resistor R_f is employed to represent the damping effects from the inverter losses, dead time, and equivalent series resistance of the filter inductor L_f, etc. A conventional multi-loop controller for such a CVCF inverter is shown in Fig. 3. The outer loop is the capacitor voltage control loop, and the inner loop is chosen as the capacitor current feedback control for better disturbance rejection [6]. $G_{vc}(s)$ and and $G_{ic}(s)$ represent the voltage and current controllers respectively. The output voltage of the

PWM inverter is denoted as v_{inv}. v_c and i_c represent the capacitor voltage and current respectively. The load current i_o is considered as a disturbance in this system.

In this system, the computation delay is one sampling period T_s [22] in the control loop and can be expressed as

$$D_e(s) = e^{-sT_s} \tag{1}$$

where T_s is the sampling period of the control system.

The PWM inverter is modelled as a gain of 1 for low-frequency control analysis. The PWM modulation is treated as a zero-order hold (ZOH) block $G_h(s)/T_s$. The 1.5 sampling periods delay caused by the control loop and the ZOH block might deteriorate the performance of the control system [15] and consequently lead to system instability. The effects of digital delay has to be taken into account while designing a repetitive controller for such inverter systems.

The closed-loop transfer function of the system can be obtained as

$$v_c = \frac{G_{v_c}(s)G_{i_c}(s)D_e(s)G_h(s)*v_c^* - (L_f s + R_f)T_s * i_o}{C_f s(L_f s + R_f)T_s + G_{i_c}(s)D_e(s)G_h(s)[G_{v_c}(s) + C_f s] + T_s} \tag{2}$$
$$= H(s)v_c^* - D(s)i_o$$

where $H(s)$ is the reference tracking stability of the conventional control system, and $D(s)$ represents the system ability to reject the load current influence at fundamental and harmonic frequencies.

C. Repetitive Control Scheme for the CVCF PWM Inverter

In such a CVCF PWM inverter in Fig. 2, a repetitive controller is normally implemented to track the reference and eliminate periodic harmonics effectively. The overall control system of the inverter is shown in Fig. 4. The repetitive controller is indicated in the orange dashed block, where K_r is the repetitive control gain, N_s is the number of samples in one repetitive period, $Q(z)$ is employed to improve the repetitive control system robustness, and $G_f(z)$ is the phase lead filter contributing to the stability of the overall closed-loop system. The transfer function of the repetitive controller can be derived from Fig. 4 and expressed as

$$G_{rc}(z) = \frac{K_r G_f(z)}{z^{N_s} - Q(z)} \tag{3}$$

The closed-loop transfer function of the overall control system can be obtained by feeding the new voltage reference $v_c^* = v_c^* + (v_c^* - v_c) \times G_{rc}(z)$ into (2), as

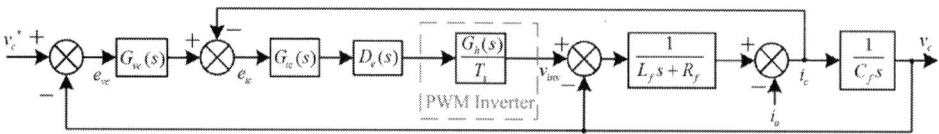

Fig. 3 Block diagram of conventional multi-loop controller.

978-1-4673-9551-9/16 $31.00 © 2016 IEEE

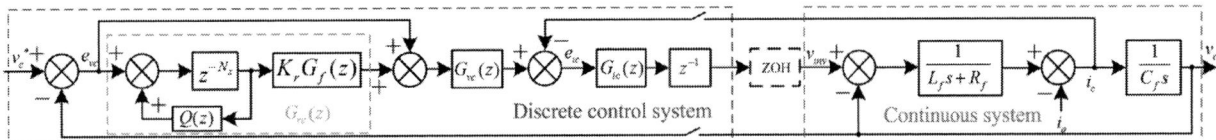

Fig. 4 Block diagram of the overall control system.

$$
\begin{aligned}
v_c &= \frac{H(z)[G_{rc}(z)+1]}{1+G_{rc}(z)H(z)}v_c^* - \frac{D(z)}{1+G_{rc}(z)H(z)}i_o \\
&= \frac{H(z)[z^{N_s}-Q(z)+K_rG_f(z)]}{z^{N_s}-Q(z)+K_rG_f(z)H(z)}v_c^* \\
&\quad - \frac{D(z)(z^{N_s}-Q(z))}{z^{N_s}-Q(z)+K_rG_f(z)H(z)}i_o
\end{aligned}
\tag{4}
$$

where $H(z)$ and $D(z)$ are the z-domain versions of $H(s)$ and $D(s)$ respectively. The tracking error of this closed-loop control system can be expressed

$$
\begin{aligned}
E(z) &= v_c^* - v_c = \frac{[1-H(z)][z^{N_s}-Q(z)]}{z^{N_s}-Q(z)+K_rG_f(z)H(z)}v_c^* \\
&\quad + \frac{D(z)(z^{N_s}-Q(z))}{z^{N_s}-Q(z)+K_rG_f(z)H(z)}i_o
\end{aligned}
\tag{5}
$$

From both (4) and (5), one can find that the closed-loop system is stable if all the roots of its characteristic function $z^{N_s}-Q(z)+K_rG_f(z)H(z)$ are placed inside of the unit circle centered at the origin of the z-plane. For simplification, one sufficient condition [23] expressed as

$$
|Q(z)-K_rG_f(z)H(z)| < 1, \quad \forall z = e^{j\omega T_s}, 0 < \omega T_s < \pi
\tag{6}
$$

The inequality in (6) shows that $Q(z)$, K_r and $G_f(z)$ all affect the system stability, which makes them important in repetitive controller design. $Q(z)$ is usually designed to be a low pass filter with the features of unit gain at low frequencies and no phase delay. K_r is normally selected as large as possible for smaller damping ratio and faster transient response [19], as long as the stability condition can be satisfied.

Furthermore, the phase delay in the feedback control system has to be carefully compensated by a phase lead filter to prevent the overall system instability. If the phase lead filter is designed to be $G_f(z)=1/H(z)$, perfect cancellation is achieved. However, the perfect cancellation is not always possible in practical applications due to the system uncertainties, such as inaccurate parameters of the inverter and plant and load disturbances in the system. Alternatively, $G_f(z)$ is normally designed to be a low pass filter $S(z)$ embedded with a pure time advance unit z^k, as

$$
G_f(z) = z^k S(z)
\tag{7}
$$

The low pass filter $S(z)$ can reject high frequency harmonics and disturbance, while z^k can compensate the phase delay caused by $S(z)$ and $H(z)$.

In general, in the design process of a repetitive controller, $Q(z)$, $S(z)$ and K_r are firstly designed according to the system requirements, e.g. tracking error, fast dynamics, harmonics elimination, and resonance prevention. Then, the time advance unit is employed to ensure the stability of the overall system.

III. MATHEMATICAL ANALYSIS OF THE TIME ADVANCE UNIT

In previous studies, the time advance unit in z^k is obtained through empirical design, e.g. by comparing the phase response of $S(z)H(z)$ and z^k in Matlab or a trial-and-error approach through experiments. In this paper, the role of z^k in the overall system stability will be clearly explained, and an explicit analytical description for selecting the value of k are detailed.

Substituting (7) into (6), the sufficient condition of the system stability at a particular frequency ω_p can be described geometrically as in Fig. 5 (a). $Q(z)$ is represented as the phasor $Q(e^{j\omega_p T_s})$ whose magnitude is $|Q(e^{j\omega_p T_s})|$ and phase angle is $\theta_Q(e^{j\omega_p T_s})$. The overall system is sufficiently stable if the end of phasor $P(e^{j\omega_p T_s}) = K_rS(e^{j\omega_p T_s})H(e^{j\omega_p T_s}) = |P(e^{j\omega_p T_s})| \angle \theta_P(e^{j\omega_p T_s})$ is placed inside of the red dashed unit circle, whose center is at the end of the phasor $Q(e^{j\omega_p T_s})$. Without the phase compensation, the end of $P(e^{j\omega_p T_s})$ might be placed outside of the red dashed circle, as indicated by the solid blue arrow in Fig. 5 (a). This will result in an unstable closed-loop system.

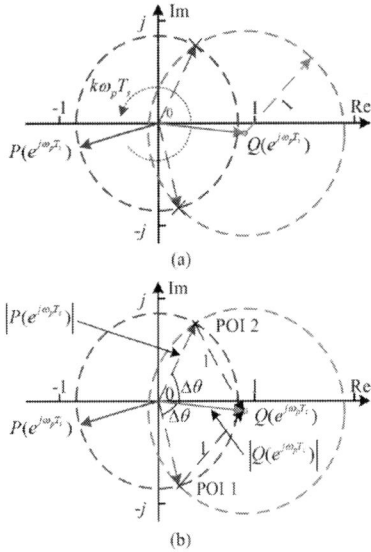

Fig. 5 Geometric illustration of the stability condition.

By introducing the time advance unit into the repetitive controller, $P(e^{j\omega_p T_s})$ is rotated around the origin anticlockwise by $k\omega_p T_s$ radian while keeping the same magnitude. It is predictable that the system will become stable at frequency ω_p if $P(e^{j\omega_p T_s})$ can be rotated properly, and its end is eventually located inside of the red dashed circle. The locus of the end of the rotating $P(e^{j\omega_p T_s})$ is indicated by the blue dashed circle centered at the origin.

As shown in Fig. 5 (b), the intersection points of the two dashed circles reveal the value of k that can make the system critically stable, and the available range of k can be obtained by solving these points.

The angles that $P(e^{j\omega_p T_s})$ should be rotated to intersect with the red dashed circle can be written as

$$\begin{cases} \Delta\theta_1(e^{j\omega_p T_s}) = \theta_Q(e^{j\omega_p T_s}) - \Delta\theta(e^{j\omega_p T_s}) - \theta_P(e^{j\omega_p T_s}) \\ \Delta\theta_2(e^{j\omega_p T_s}) = \theta_Q(e^{j\omega_p T_s}) + \Delta\theta(e^{j\omega_p T_s}) - \theta_P(e^{j\omega_p T_s}) \end{cases} \quad (8)$$

where $\Delta\theta(e^{j\omega_p T_s})$ indicated in Fig. 5 (b) is obtained from

$$\Delta\theta(e^{j\omega_p T_s}) = \arccos\left(\frac{\left|P(e^{j\omega_p T_s})\right|^2 + \left|Q(e^{j\omega_p T_s})\right|^2 - 1}{2\left|P(e^{j\omega_p T_s})\right|\left|Q(e^{j\omega_p T_s})\right|} \right) \quad (9)$$

$$1 - \left|Q(e^{j\omega_p T_s})\right| \le \left|P(e^{j\omega_p T_s})\right| \le 1 + \left|Q(e^{j\omega_p T_s})\right| \quad (10)$$

Note that (9) is solvable only if (10) is satisfied, which means the intersection exists. The inequality $|P(e^{j\omega_p T_s})| < 1 - |Q(e^{j\omega_p T_s})|$ indicates that the end of phasor $P(e^{j\omega_p T_s})$ is always inside the unity circle and the system is always stable at ω_p. $|P(e^{j\omega_p T_s})| > 1 + |Q(e^{j\omega_p T_s})|$ might lead to system instability despite the phase compensation.

K_r can be selected according to the equation as

$$0 < K_r \le \frac{1 + \left|Q(e^{j\omega_p T_s})\right|}{\left|S(e^{j\omega_p T_s})H(e^{j\omega_p T_s})\right|} \quad (11)$$

The range of k in the time advance unit can be derived as

$$\frac{\Delta\theta_1(e^{j\omega_p T_s})}{\omega_p T_s} \le k_1 \le k \le k_2 \le \frac{\Delta\theta_2(e^{j\omega_p T_s})}{\omega_p T_s} \quad (12)$$

where k, k_1 and k_2 are nonnegative integers. Furthermore, the overall system has the maximum phase margin as well as relatively high tracking accuracy and disturbance rejection at ω_p if $Q(e^{j\omega_p T_s})$ and the rotated $P(e^{j\omega_p T_s})$ are in phase, according to Fig. 5 and (5).

$$k = \frac{\theta_Q(e^{j\omega_p T_s}) - \theta_P(e^{j\omega_p T_s})}{\omega_p T_s} \quad (13)$$

Equation (13) suggests the value of k that can provide the maximum system stability margin. Equation (12) shows that how much phase margin is left when a proper k is selected. Moreover, the time advance unit has to be designed to guarantee the system stability at all frequencies up to Nyquist

TABLE I. INVERTER SYSTEM PARAMETERS

Parameters	Values
DC link voltage: U_{dc}	200 V
Output voltage (RMS):	110 V
Rated output frequency: f_o	50 Hz
Filter inductance: L_f	0.8 mH
Filter capacitance: C_f	30 uF
Equivalent Resistor: R_f	1.8 Ω
Load inductance: L_l	0.5 mH
Load capacitance: C_l	470 uF
Load Resistor: R_l	80 Ω
Switching frequency: f_c	20k Hz
Sampling frequency: f_s	20k Hz

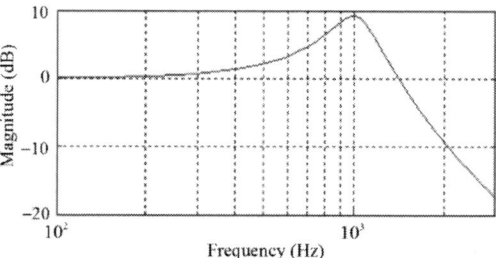

Fig. 6 Amplitude-frequency response of the inverter.

frequency, and the detailed design process will be discussed in the case study.

IV. CASE STUDY AND CONTROL SYSTEM DESIGN

The detailed parameters of the inverter system are listed in Table I. The amplitude-frequency response of the inverter with the LC filter can be plotted as in Fig. 6, where the resonant frequency of this system is obviously around 1.027 kHz. Since v_{inv} in Fig. 2 is actually in PWM waveform pattern, it contains the information of the reference signal as well as high frequency harmonics. The harmonics and the disturbances from nonlinear loads, along with the delays in the system, might cause oscillations around the resonant frequency, which has to be carefully considered while designing a controller for it.

The control system for this CVCF PWM inverter is synchronized with the sampling frequency f_s. Voltage and current controllers are designed to be simple proportional compensators, and the gains are chosen to be $K_{pv}=0.2$ and $K_{pi}=4$ respectively, with the consideration of the system stability of both feedback control loops and high controller gains.

In the repetitive controller, N_s is assigned as $N_s=f_s/f_o$ and $Q(z)$ is designed as $Q(z)=0.25z+0.5+0.25z^{-1}$ with the cutoff frequency of 3.63 kHz. The phase lead filter expressed in (7) is adopted and $S(z)$ is designed as a second-order low pass filter with the nature frequency of 800 Hz. In this case, the amplitude-frequency response of $S(z)H(z)$ as shown in Fig. 7 is

978-1-4673-9551-9/16 $31.00 © 2016 IEEE

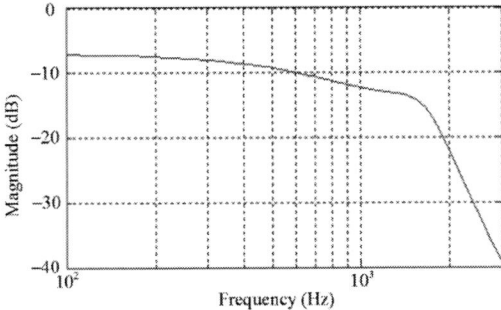

Fig. 7 Amplitude-frequency response of $S(z)H(z)$.

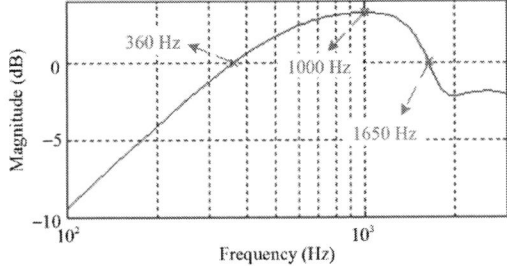

Fig. 8 Amplitude-frequency response of $Q(z) - K_rS(z)H(z)$

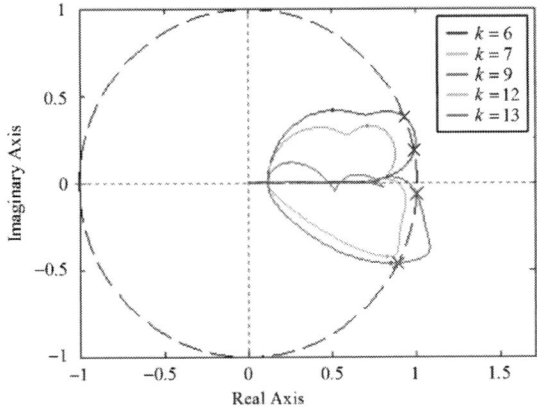

Fig. 9 Nyquist locus of $Q(z) - K_rz^kS(z)H(z)$

achieved, and it is preferable for the repetitive controller design. The magnitude characteristics in Fig. 7 suggests the relatively slow dynamics of the repetitive control scheme. Fortunately, the dynamic response is able to be easily compensated by choosing a larger repetitive control gain such as $K_r = 2$.

The condition in (6) cannot be satisfied at all frequencies up to Nyquist frequency without the time advance unit, as illustrated in Fig. 8. This system might be unstable in the range of 360 Hz to 1650 Hz, since the magnitude of $Q(z) - k_rS(z)H(z)$ is higher than 0 dB at these frequencies, and the worst case occurs around the resonant frequency (1 kHz) of the plant.

At low frequencies, the system is normally not so sensitive to the time advance unit and the stability condition (6) is nearly always met. And at frequencies higher than 2.15 kHz, the system is always stable because of $|P(e^{j\omega_pT_s})| < 1 - |Q(e^{j\omega_pT_s})|$. Therefore, the time advance unit that satisfies the inequality in (12) at both 1 and 1.65 kHz is sufficient to yield the stability range of the overall system. The range of k can be calculated as $6.1 < k < 12.2$, and the approximately maximum stability margin is achieved when $k=9$ is adopted. The Nyquist locus of $Q(z) - k_rz^kS(z)H(z)$ as in Fig. 9 indicates that the stability condition is satisfied while $7 \le k \le 12$, which is in accordance with the mathematical analysis.

V. SIMULATION AND EXPERIMENTAL RESULTS

The repetitive control scheme as well as the proposed design method of the time advance unit for the phase lead filter have been demonstrated and verified in simulations with PLECS and experiments based on the single-phase CVCF PWM inverter configuration shown in Fig. 2 and the system parameters listed in Table I.

A. Simulation Results

Fig. 10 shows the simulated steady-state capacitor voltage and load current waveforms, when the output voltage is regulated by the open-loop control and the repetitive control scheme respectively. The time advance unit is set to be z^9, as suggested by the previous analysis for maximum stability margin. It can be seen in Fig. 10 that the voltage distortion caused by the nonlinear load current are almost completely removed and the capacitor voltage is nearly pure sinusoidal. The THD of output voltages under these two control schemes are 4.6% and 0.3% respectively.

B. Experimental Results

The experimental results obtained from an inverter prototype built in the laboratory are shown in Fig. 11 to Fig. 13. Fig. 11 shows the effectiveness of the repetitive control scheme to eliminate the harmonics caused by the nonlinear load. The

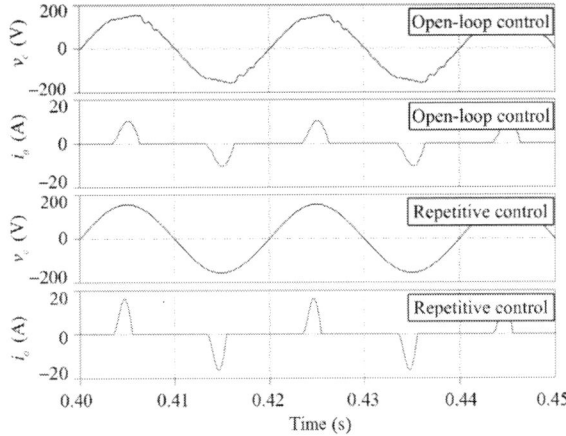

Fig. 10 Capacitor voltage and output current waveforms under open-loop and repetitive control scheme in PLECS.

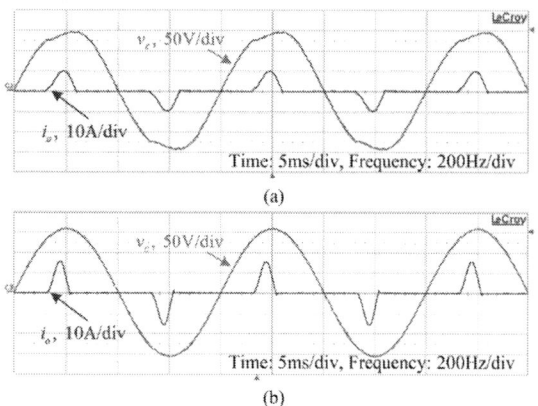

(a)

(b)

Fig. 11 Experimental capacitor voltage and output current waveforms under:
(a) open-loop control scheme; (b) repetitive control scheme.

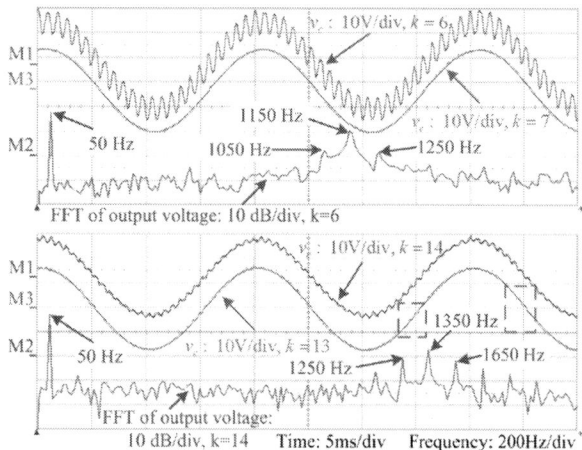

Fig. 12 Inverter output voltage with different values of k.

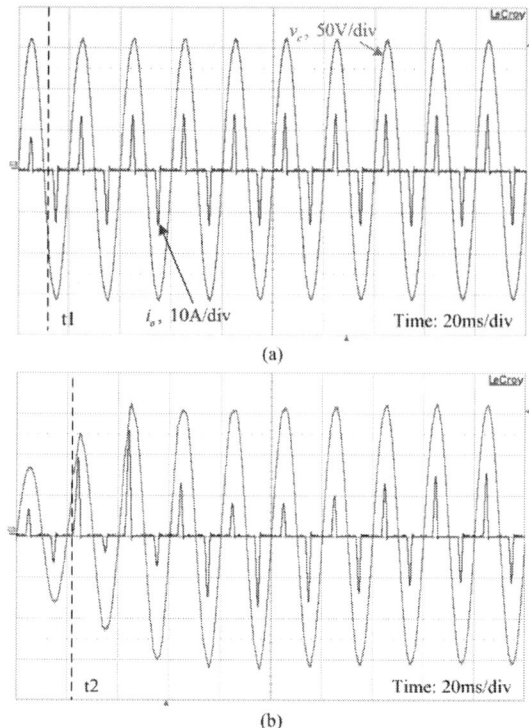

(a)

(b)

Fig. 13 Experimental capacitor voltage and output current waveforms under:
(a) Load step change; (b) Voltage reference step change.

time advance step k is set to be 9 for largest stability margin. It is evident that the output voltage is almost pure sinusoidal when the repetitive controller is enabled, and the FFT analysis indicates that low frequency harmonics are effectively suppressed due to the repetitive controller. The THD of the output voltage is 1.1%.

The correctness of the mathematical design method for the time advance unit is proved as in Fig. 12, where the no-load stability is checked and the dc bus voltage is reduced to 50 V for safety purpose. The overall system starts to oscillate when k is selected as 6 or 14, while it is stable if $k=7$ or $k=13$ is adopted. The phase delay compensation is insufficient for relatively low frequencies close to the resonant frequency (1050 Hz to 1250 Hz) when $k=6$. On the other hand, the phases of relatively high frequencies (1350 Hz to 1650 Hz) are overcompensated if k is selected as 14. Note that the experimental prototype is almost critically stable at $k=13$, and very slight oscillations can be observed as indicated in the red dashed blocks in Fig. 12. This result is slightly different from the calculated stability range, but it is still reasonable because the system damping effects at high frequencies are generally

much higher than that of lower frequency ones in practical applications. The system passive damping can significantly contribute to the system stability, particularly at high frequencies.

In Fig. 13, the dynamics of the inverter are demonstrated with $k=9$. The load resistance is suddenly changed from 200 Ω to 100 Ω at time t1, and the output voltage of the inverter is scarcely affected by the load variation as shown in Fig. 13 (a). The magnitude of the output voltage reference steps from 80 V to 155 V at time t2 in Fig. 13 (b), and the output voltage of the inverter tracks the reference within 3 fundamental periods. The output voltage regulated by the repetitive control scheme has excellent dynamics without serve transients.

VI. CONCLUSION

This paper has presented a repetitive control scheme to suppress the harmonics in the output voltage of CVCF PWM inverters caused by nonlinear loads. A novel method for analyzing and designing the time advance unit in a repetitive controller is proposed, which makes the explicit analytical calculation of the time advance unit possible rather than existing implicit methods. Based on the proposed method, not only is the system stability guaranteed, but the maximum phase margin, good tracking accuracy and disturbance rejection can be obtained as well. The effectiveness of the repetitive control scheme and the proposed phase lead filter design method are verified in both simulation and experiments. The theoretical analysis agrees very well with the experimental results obtained from the CVCF PWM inverter prototype. The results show that the harmonics in the output voltage can be successfully suppressed, and the system stability range is consistent with the prediction of the mathematical calculation. The system dynamic response under step loads and voltage changes can be very fast and smooth.

REFERENCES

[1] B. Tamyurek, "A High-Performance SPWM Controller for Three-Phase UPS Systems Operating Under Highly Nonlinear Loads," *IEEE Trans. Power Electron.*, vol. 28, pp. 3689-3701, 2013.

[2] Y. Yang, K. Zhou, H. Wang, F. Blaabjerg, D. Wang, and B. Zhang, "Frequency Adaptive Selective Harmonic Control for Grid-Connected Inverters," *IEEE Trans. Power Electron.*, vol. 30, pp. 3912-3924, 2015.

[3] Z. Zou, K. Zhou, Z. Wang, and M. Cheng, "Frequency-Adaptive Fractional-Order Repetitive Control of Shunt Active Power Filters," *IEEE Trans. Ind. Electron.*, vol. 62, pp. 1659-1668, 2015.

[4] T. Yi, L. Poh Chiang, W. Peng, C. Fook Hoong, G. Feng, and F. Blaabjerg, "Generalized Design of High Performance Shunt Active Power Filter With Output LCL Filter," *IEEE Trans. Ind. Electron.*, vol. 59, pp. 1443-1452, 2012.

[5] T. Yokoyama and A. Kawamura, "Disturbance Observer-Based Fully Digital Controlled PWM Inverter for CVCF Operation," *IEEE Trans. Power Electron.*, vol. 9, pp. 473-480, Sep 1994.

[6] P. C. Loh and D. G. Holmes, "Analysis of multiloop control strategies for LC/CL/LCL-filtered voltage-source and current-source inverters," *IEEE Trans. Ind. Appl.*, vol. 41, pp. 644-654, 2005.

[7] D. Chen, J. Zhang, and Z. Qian, "An Improved Repetitive Control Scheme for Grid-Connected Inverter With Frequency-Adaptive Capability," *IEEE Trans. Ind. Electron.*, vol. 60, pp. 814-823, 2013.

[8] K. Zhou, D. Wang, B. Zhang, and Y. Wang, "Plug-In Dual-Mode-Structure Repetitive Controller for CVCF PWM Inverters," *IEEE Trans. Ind. Electron.*, vol. 56, pp. 784-791, 2009.

[9] K. Zhou and D. Wang, "Digital repetitive controlled three-phase PWM rectifier," *IEEE Trans. Power Electron.*, vol. 18, pp. 309-316, 2003.

[10] K. Zhang, Y. Kang, J. Xiong, and J. Chen, "Direct repetitive control of SPWM inverter for UPS purpose," *IEEE Trans. Power Electron.*, vol. 18, pp. 784-792, 2003.

[11] K. Zhou and D. Wang, "Digital repetitive learning controller for three-phase CVCF PWM inverter," *IEEE Trans. Ind. Electron.*, vol. 48, pp. 820-830, 2001.

[12] P. C. Loh, Y. Tang, F. Blaabjerg, and P. Wang, "Mixed-frame and stationary-frame repetitive control schemes for compensating typical load and grid harmonics," *Power Electronics, IET*, vol. 4, pp. 218-226, 2011.

[13] B. A. Francis and W. M. Wonham, "The internal model principle of control theory," *Automatica*, vol. 12, pp. 457-465, 1976.

[14] S. G. Parker, B. P. McGrath, and D. G. Holmes, "Regions of Active Damping Control for LCL Filters," *IEEE Trans. Ind. Appl.*, vol. 50, pp. 424-432, 2014.

[15] D. Pan, X. Ruan, C. Bao, W. Li, and X. Wang, "Capacitor-Current-Feedback Active Damping With Reduced Computation Delay for Improving Robustness of LCL-Type Grid-Connected Inverter," *IEEE Trans. Power Electron.*, vol. 29, pp. 3414-3427, 2014.

[16] M. Tomizuka, "Zero Phase Error Tracking Algorithm for Digital-Control," *Journal of Dynamic Systems Measurement and Control-Transactions of the Asme*, vol. 109, pp. 65-68, Mar 1987.

[17] Z. Wang, C. Xie, C. He, and G. Chen, "A waveform control technique for high power Shunt Active Power Filter based on repetitive control algorithm," in *Applied Power Electronics Conference and Exposition (APEC), 2010 Twenty-Fifth Annual IEEE*, 2010, pp. 361-366.

[18] S. Yang, B. Cui, F. Zhang, and Z. Qian, "A Robust Repetitive Control Strategy for CVCF Inverters with Very Low Harmonic Distortion," in *Applied Power Electronics Conference, APEC 2007 - Twenty Second Annual IEEE*, 2007, pp. 1673-1677.

[19] K. Zhou, K. S. Low, D. Wang, F. Luo, B. Zhang, and Y. Wang, "Zero-phase odd-harmonic repetitive controller for a single-phase PWM inverter," *IEEE Trans. Power Electron.*, vol. 21, pp. 193-201, 2006.

[20] H. L. Broberg and R. G. Molyet, "A New Approach to Phase Cancellation in Repetitive Control," *Ias '94 - Conference Record of the 1994 Industry Applications Conference/Twenty-Ninth Ias Annual Meeting, Vols 1-3*, pp. 1766-1770, 1994.

[21] B. Zhang, D. Wang, K. Zhou, and Y. Wang, "Linear Phase Lead Compensation Repetitive Control of a CVCF PWM Inverter," *IEEE Trans. Ind. Electron.*, vol. 55, pp. 1595-1602, 2008.

[22] B. P. McGrath, S. G. Parker, and D. G. Holmes, "High-Performance Current Regulation for Low-Pulse-Ratio Inverters," *IEEE Trans. Ind. Appl.*, vol. 49, pp. 149-158, 2013.

[23] G. F. Franklin, J. D. Powell, and A. Emami-Naeini, *Feedback Control of Dynamic Systems*. MA: Addison-Wesley, 1991.

Research on the Filter of Load Side Converter in BDFG Based Ship Shaft Power Generation System

Meilin Wang, Hua Lin, Hongbin Yang, Xingwei Wang
State Key Laboratory of Advanced Electromagnetic Engineering and Technology
Huazhong University of Science and Technology
Wuhan, 430074, China
wangmeilinmeilin@163.com

Abstract—In stand-alone brushless doubly fed generator (BDFG) based shaft power generation system, the output voltage of BDFG is strong influenced by the filter of load side converter (LSC) because of the coupling caused by the stator power winding leakage inductance. In order to eliminate the harmonic of the voltage, the LC-L filter based on the leakage inductance is put forward in this paper. The relationship between the filter parameters and performance, and the limitation of filter parameters for positive damping characteristic without extra active or passive damping has been analyzed. Then the design criterion and steps are presented and a filter based on a 64-kw BDFG shaft power generation system is designed. Finally, the validity of the filter design is verified by experiment.

Keywords—stand-alone; coupling; possitive damping; stator power leakage inductance; PCC voltage

I. INTRODUCTION

With the international oil price rising, shaft power generation system which takes full use of reserve ship host power becomes an effective way of energy saving [1]. Compared with the traditional synchronous generator with full power phase compensator used in this system, the converter only deals with the slip power which makes the converter small in size and cheap for BDFG based system. Besides, it reduces the maintenance cost by canceling brush and slip ring compared with DFIG, which is more suitable in ship shaft power generation system. To adapt to the situation where the speed of ship can be higher or lower than the synchronous speed, a back-to-back PWM converter is used in this system. The converter connected to the stator power winding is named as LSC [2] and the other one connected to the stator control winding is named as machine side converter (MSC).

Traditionally, L filter can be used as the output filter of LSC for its simplicity and reliability [3]. But in this stand-alone BDFG shaft power generation system, there is strong coupling between the LSC and BDFG because of the existence of the stator power winding leakage inductance. Furthermore, the output voltage will be strong influenced by the PWM voltage of LSC because the stator power winding leakage inductance is too large to be ignored. To achieve the better harmonic attenuation, a larger filter inductance is required in this

situation. But it will lead to large volume and weight of filter, large voltage drop of the inductance and bad current tracking property. So a higher order filter is needed in this system. A capacitor is usually placed between the filter inductance and BDFG [4-7] for the better harmonic attenuation. Therefor the stability problem of the system is caused by the resonance of the inductance and capacitance. However, so far in most researches, there is still lack of the design for filter capacitance and the inherent damping characteristic analysis considering the stator power leakage inductance.

This paper addresses the filter design problem of LSC in stand-alone BDFG shaft power generation system. Considering the stator power winding leakage inductance, to reduce the volume and weight of the filter LC-L topology is adopted. Different from the conventional LCL filter design, not only the harmonic of the current that injected to the BDFG is limited because it will influence the operation of BDFG. But the PCC voltage which is used as load voltage needs to meet the THD requirement. And the stability problem caused by the resonance of the filter is analyzed too. Then, the design criterion and steps of the filter is put forward which considers the harmonic attenuation of both the current and output voltage, and the damping characteristic that takes the digital control effect into account at the absence of the extra active or passive damping method. Finally, the validity of the filter design is verified by the experiment based on a 64-kW BDFG shaft power generation system.

II. TOPOLOGY ANALYSIS AND SELECTION OF THE FILTER FOR LSC

Fig.1 shows the circuit of LSC. Without loss of generality, BDFG is considered as the combination of electromotive force e and stator power winding leakage inductance L_σ. LSC supplies constant DC voltage for MSC while the MSC supplies excitation voltage for BDFG. At the same time, the LSC and MSC are decoupled through the capacitance C_{dc}. So the circuit for LSC analysis does not include MSC. Meanwhile, the parasitic resistance is small and both the parasitic resistance and the resistance of the load will attenuate the instability of system, the worst situation that leaves out the parasitic resistance and unloaded is considered here.

This work was supported by the Key Delta Science and Technology Educational Development Program under Grant DREM2015001.

978-1-4673-9551-9/16 $31.00 © 2016 IEEE

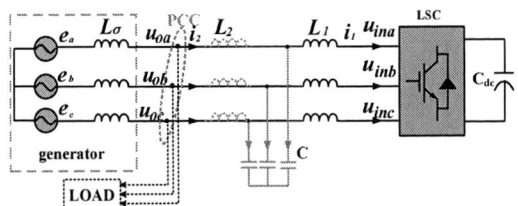

Fig.1 circuit of LSC in BDFG generation system

u_{in} is the AC input voltage of LSC, i_2 is the load side current, i_1 is the converter side current and u_o is the voltage of PCC. With the simplest L filter, the PCC voltage can be expressed as:

$$u_o = \frac{L_1}{L_1 + L_\sigma} e + \frac{L_\sigma}{L_1 + L_\sigma} u_{in} \qquad (1)$$

From formula (1), the harmonic of PCC voltage comes from u_{in}. So the ratio of $L_\sigma/(L_1+L_\sigma)$ determines the percent of the harmonic in PCC voltage. To achieve a better harmonic attenuation, the small coefficient is preferred.

Since L_σ can reach the same magnitude as filter inductance L_1, the output voltage u_o is strong influenced by the high frequency harmonic caused by the load side converter. Because of the disadvantages of increasing the value of filter inductance, to eliminate the harmonic to achieve the requirement of power supply quality, a higher order filter LCL filter is considered. The current i_2 and voltage u_o with respect to u_{in} can be deduced as:

$$\begin{cases} \dfrac{u_o}{u_{in}} = \dfrac{L_\sigma}{L_\sigma + L_2} \dfrac{(L_\sigma + L_2)}{L_1(L_\sigma + L_2)Cs^2 + L_1 + (L_\sigma + L_2)} \\[3mm] \dfrac{i_2}{u_{in}} = \dfrac{1}{L_1(L_\sigma + L_2)Cs^3 + (L_1 + L_\sigma + L_2)s} \end{cases} \qquad (2)$$

Because L_2 is in series with L_σ, the filter inductance L_2 can be replaced by the leakage inductance. In this way, the LC-L structure is formed by LC filter and stator power winding leakage inductance L_σ which means $L_2=0$ in formula (2). The LC-L filter can achieve the same harmonic attenuate rate as ordinary LCL filter for both u_o and i_2. Because the stator power leakage inductance is determined by the BDFG, so only the filter parameters L_1 and C need to be designed.

III. THE ANALYSIS OF THE FILTER DESIGN CRITERION

In order to figure out the design criterion of LC-L filter parameters. Not only the relationship between the filter parameters and its performance but the relationship between the filter parameters and its damping characteristics has been analyzed here. It is aimed to achieve a stable system with good filter performance at the absence of the extra damping method.

A. The Influence of Filter Parameters on Filter Performance

Since the filter is designed to attenuate the harmonic of switching frequency and its sideband harmonics. As the main harmonic, the harmonic attenuation at switching frequency is analyzed here.

According to formula (2), the harmonic attenuation of output voltage u_o and converter output current i_2 at switching frequency can be deduced as formula (3).

$$\begin{cases} \left| \dfrac{u_{os}}{u_{ins}} \right| = \dfrac{L_\sigma}{L_1 L_\sigma C \omega_s^2 - (L_1 + L_\sigma)} \\[3mm] \left| \dfrac{i_{2s}}{u_{ins}} \right| = \dfrac{1}{L_1 L_\sigma C \omega_s^3 - (L_1 + L_\sigma)\omega_s} \end{cases} \qquad (3)$$

So the harmonic attenuation relations with filter parameters L_1 and C when the stator power winding leakage inductance is determined can be shown in Fig.2.

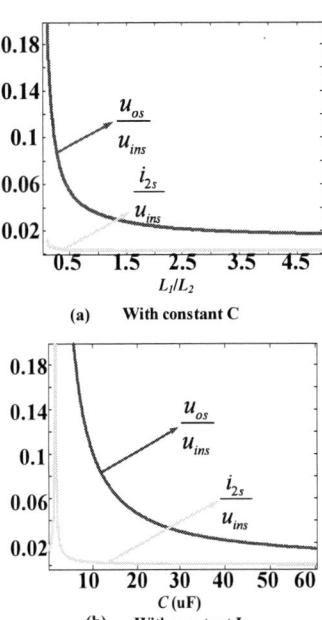

(a)   With constant C

(b)   With constant L_1

Fig.2 Harmonic attenuate ratio with respect to the filter parameters

In Fig.2, the better harmonic attenuation will be achieved when either the L_1 or C increases. But the harmonic will increase when the capacitance increases from a small value. Since the resonance frequency becomes more close to the switching frequency when the capacitance increases from a small value.

From Fig.2 (a), the harmonic attenuation of i_2 keeps nearly constant as L_1 increasing after a certain value. Compared with Fig.2 (b), increasing C brings larger harmonic attenuation of u_o. To reduce filter volume and weight, this paper suggests achieving the expected harmonic attenuation of u_o by increasing C instead of L_1.

B. The Influence of Filter Parameters on the Damping Characteristic of the Controlled Plant

The conventional voltage-oriented double loop control is adopted by LSC where the current cross decoupling and PCC voltage feed-forward decoupling is used for better dynamic process. Considering that the voltage loop is decoupling with the current loop and it is same as the ordinary control. Only the current loop is analyzed in this paper. The control block diagram of the current loop control at d axis can be shown in Fig.3 (a).

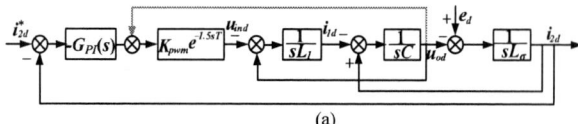

(a)

Fig.3 Control block diagram of the inner current loop for LC-L structure based LSC

Different from conventional LCL filter, the voltage used for feed-forward is PCC voltage. So it becomes a feed-back path which changes the damping characteristic of the controlled plant [8-9].

In order to get the transfer function of the controlled plant, the simplified control block diagram is shown in Fig.3 (b).

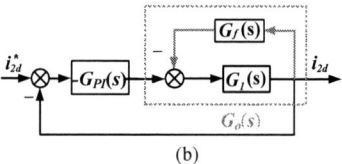

(b)

Fig.3 Simplified control block diagram of the inner current loop for LC-L structure based LSC

The transfer function of the controlled plant in this system can be expressed as:

$$G_o(s) = -\frac{K_{pwm}e^{-1.5sT}}{L_1 L_\sigma C s^3 + (L_1 + L_\sigma)s - L_\sigma s e^{-1.5sT}} \tag{3}$$

From formula (3), there is additional delay part in the characteristic function of the controlled plant compared with the ordinary LCL filter. Since it is not a rational function, the analytical expression of poles is not easy to get. Although the approximate solution may help to some extent, it will lead to wrong result some times.

This paper tries to figure it out by physical significance. To reveal the physical significance of the feed-back path, an equivalent block diagram is shown in Fig. 3 (c).

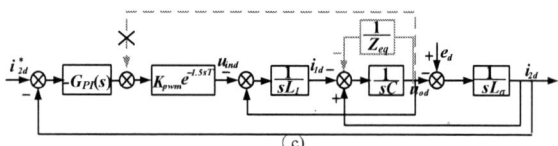

(c)

Fig.3 Control block diagram of the inner current loop for LC-L structure based LSC for physical significance

It shows that the voltage feedback can be considered as equivalent impedance z_{eq} in parallel with the filter capacitor.

$$z_{eq} = -L_1 s e^{1.5sT_s} = R_{eq} + \frac{1}{j\omega C_{eq}} \tag{4}$$

$$= L_1\omega\sin(1.5\omega T_s) + \frac{1}{j\omega \dfrac{1}{L_1\omega^2\cos(1.5\omega T_s)}}$$

Both the equivalent resistance R_{eq} and capacitance C_{eq} are frequency dependent. And R_{eq} is positive when the resonance happens at the frequency smaller than $(1/3)\omega_s$, otherwise it will be negative.

The positive R_{eq} can be used to damp the resonance peak just as the passive damping method do when the resonance frequency design properly. At the same time, the equivalent capacitance C_{eq} will shift the resonance peak. The actual resonant frequency ω_r' can be expressed as formula (5).

$$\omega_r' = \sqrt{\frac{L_1 + L_\sigma}{L_1 L_\sigma\left(C + \dfrac{1}{L_1\omega^2\cos(1.5\omega T_s)}\right)}} \tag{5}$$

To achieve positive damping at the resonance frequency, the actual resonance frequency ω_r' should be smaller than $(1/3)\omega_s$. So the limitation for LC-L structure to achieve positive damping can be deduced as (6).

$$\omega_r = \sqrt{\frac{L_1 + L_\sigma}{L_1 L_\sigma C}} < \sqrt{\left(\frac{\omega_s}{3}\right)^2 - \frac{1}{L_1 C}} \tag{6}$$

The stability of the controlled plant is determined by the damping characteristic. To verify it, the nyquist diagram of the open loop transfer function of the controlled plant $G_{o\text{-}o}(s)$ in two different damping characteristics situations is shown in Fig.4.

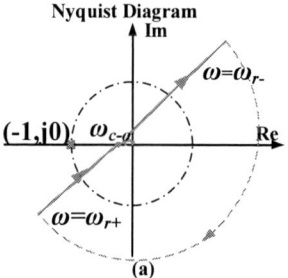

(a)

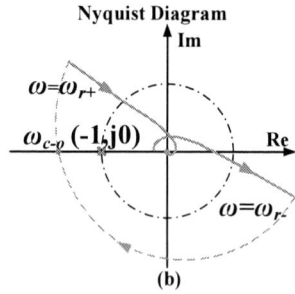

(b)

Fig.4 The nyquist diagram of the open loop controlled plant transfer function $G_{o\text{-}o}(s)$

978-1-4673-9551-9/16 $31.00 © 2016 IEEE   1080

The open loop transfer function of the controlled plant can be expressed as:

$$G_{o-o}(s) = -\frac{K_{pwm}L_\sigma e^{-1.5sT}}{L_1 L_\sigma C s^2 + (L_1 + L_\sigma)} \tag{7}$$

According to nyquist criterion, the controlled plant at positive damping situation as shown in Fig.4 (a) is stable because the formula (7) has no right half poles. And it is unstable at negative damping characteristic as shown in Fig.4 (b). When the controlled plant shows negative damping characteristic, the system will show non-minimum-phase phenomenon. In this way, the range of filter parameters to achieve system stability is small and it is more complicated to design controller parameters. The robustness is worse than the situation with positive damping [11]. So the LC-L filter in this paper is designed to achieve positive damping.

IV. STEPS FOR FILTER DESIGN AND A DESIGN EXAMPLE

According to the former analysis, the criterion for filter design in this paper not only considers the harmonic attenuation but the damping characteristic. It is aimed to put forward the filter design criterion and steps that achieve both the positive damping characteristic and good harmonic attenuation of u_o and i_2 based on the conventional filter design

1) Determine the region of the filter inductance L_1.

According to former analysis, the L_1 has small effect on the harmonic attenuation of i_2 when L_1 is not too small. So only the ordinary design criterions need to be considered in L_1 design.

For the safety of the switching devices, the ripple of the current through the converter needs to be limited. So the lower limitation can be expressed as:

$$L_{\min} = \frac{(2U_{dc} - 3u_{gd})u_{gd}T}{2U_{dc}\Delta i_{\max}} \tag{8}$$

In order to achieve the four quadrant operation for the LSC, the voltage drop in filter inductance needs to be limited. So a up limit of L_1 is get. At the same time, larger filter inductance will lead to bad tracking performance so another up limit is necessary for the insuring of control performance. The up limit of L_1 can be expressed as:

$$L_{\max} = \min\left[\frac{3u_{gd}(U_{dc}/\sqrt{3} - u_{gd})}{2\omega q}, \frac{2U_{dc}}{3\omega I_m}\right] \tag{9}$$

2) Choosing a smaller value for L_1

According to Fig.2, increasing the filter inductance L_1 will do little affect for the harmonic attenuation of i2 after $L_1/L_\sigma > 0.5$. So for enough margins, L_1 is usually designed little larger than the lower limit in formula (8).

3) Determine the region of the filter capacitance C

According to the former analysis, a lower limit of filter capacitance to achieve the positive damping characteristic can be deduced from formula (6) when the filter inductance L_1 and the stator power winding leakage inductance L_σ have been

determined. The lower limit of the filter capacitance can be expressed as:

$$C > \frac{9L_1 + 18L_\sigma}{\omega_s^2 L_1 L_\sigma} \tag{10}$$

Besides the damping characteristic limitation, the resonance frequency of LC-L needs to be designed at the range of ($10\omega_o$, $0.5\omega_s$). When the filter inductance L_1 decided, a lower and an up limit can be calculated as:

$$\frac{4(L_1 + L_\sigma)}{\omega_s^2 L_1 L_\sigma} < C < \frac{L_1 + L_\sigma}{(10\omega_0)^2 L_1 L_\sigma} \tag{11}$$

For BDFG shaft power generation system, one advantage is the smaller capacity of the converter. The larger filter capacitance can achieve better harmonic attenuation but will lead to larger capacity of the converter which is not allowed in this system. So the filter capacitance in this system is limited by the power limitation 5%.

$$C < \frac{5\%P}{3U^2\omega} \tag{12}$$

To conclude the limitations of the filter capacitance above, the simplified boundary of the filter capacitance can be expressed as:

$$\frac{9L_1 + 18L_\sigma}{\omega_s^2 L_1 L_\sigma} < C < \min\left[\frac{5\%P}{3U^2\omega}, \frac{L_1 + L_\sigma}{(10\omega_0)^2 L_1 L_\sigma}\right] \tag{13}$$

4) Choosing the capacitance value according to the requirement of the THD for u_o.

The THD of u_o can be expressed as:

$$THD_{uo} = \frac{u_{oh}}{u_{o1}} = \frac{u_{oh}}{u_{inh}}\frac{u_{inh}}{u_{in1}}\frac{u_{in1}}{u_{o1}} \approx \frac{u_{oh}}{u_{inh}}THD_{uin} \cdot 1 \approx k\frac{u_{os}}{u_{ins}}THD_{uin} \tag{14}$$

The subscript h represents the total harmonic voltage and 1 represents the fundamental voltage. Because the voltage drop of the filter inductance L_1 is small compared with the rated voltage of the system so $u_{in1} \approx u_{o1}$. And THD_{uin} is fixed when the modulation mode and ratio is designed for a determined system.

Because the main harmonic component is the harmonic of switching frequency, the harmonic voltage ratio u_{oh}/u_{inh} can be represented by u_{os}/u_{ins} times a regulated coefficient k. the coefficient k is larger than 1 but usually smaller than 2 here.

So when the modulation mode and the system is determined, the THD_{uo} is proportional to u_{os}/u_{ins}. According to former analysis, the harmonic voltage ratio can be expressed by the function of the filter parameters as formula (2). So the desired C can be calculated as formula (15).

$$C = \frac{1}{L_1\omega_s^2}\left(\frac{L_1}{L_\sigma} + 1 + k\frac{THD_{uin}}{THD_{uo}}\right) \tag{15}$$

If the filter capacitance calculated from (15) is larger than the up limit in step (3), L_1 should be re-chosen from step (2). Increase L_1 to a value when the filter capacitance calculated from formula (15) is smaller than the up limit in step (3).

Otherwise, choose the larger one as the value of the filter capacitance between formula (15) and the lower limit in step (3).

Then the LC-L filter design based on a 64-kW BDFG based shaft power generation system is described below. Table 1 shows the system parameters.

Table 1 system parameters

Parameters	Values	Parameters	Values
f_0/Hz	50	u_o(line)/V	380
f_s/Hz	5000	L_σ/mH	1.3
V_{dc}/V	740	C_{dc}/F	0.1

According to step (1), the up and lower limit can be calculated as 0.9mH $<L_1<$3.6mH. Because the lower limit of L_1 is 0.9mH which is larger than $0.5L_\sigma$. The good harmonic attenuation of i_2 can be achieved in that range. For enough margins, L_1 is designed to be 1.2mH.

When the L_1 is designed as 1.2mH, the region of capacitance can be deduced. According to formula (13) in the step (3), the range of the filter capacitance is calculated as 22.2μf $<C<$64μf.

According to step (4), to determine the value of the filter capacitance, the THD of the output voltage of LSC THD_{uin} needs to be calculated first. In order to achieve enough harmonic attenuation, the correct coefficient k is designed to be 2 here. With the requirement of the THD of the load voltage witch is smaller than 3%, the capacitance is calculated as:

$$C \approx \frac{1}{0.0012 \cdot (2 \cdot \pi \cdot 5000)^2}\left(\frac{0.0012}{0.0013}+1+2\frac{0.79}{0.03}\right) \approx 45\mu F \quad (16)$$

The filter capacitance calculated from formula (16) is 45μf which is in the region of 22.2μf $<C<$64μf. So it is not necessary to re-choose L_1 from step (2). The result can be used as the final value of the filter capacitance.

V. EXPERIMENT

The experiment is carried out at a 64-kw BDFG shaft power generation system whose parameters are shown in table 1. In order to verify the filter performance, the experiment at the full load situation is carried out. Both the THD of the current i_2 and PCC voltage u_o are analyzed. For the positive damping characteristic, the dynamic process from full load to unloaded situation is carried out too. And the experiment is carried out at the situation where the filter capacitance is out of the range for positive damping characteristic.

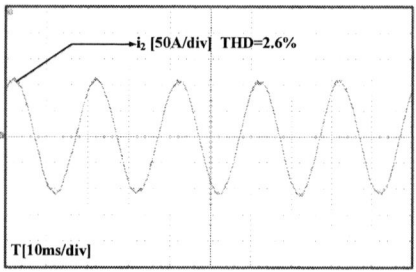

Fig.5 the waveform of i_2 at full load situation

Fig.5 shows the waveform of i_2 when the system is full load, the THD of it is 2.6%.

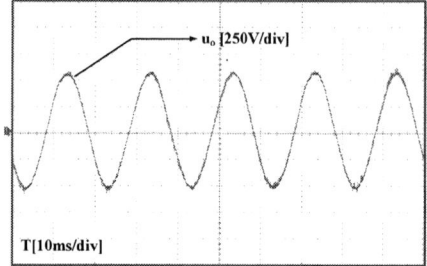

Fig.6 the waveform of u_o at unloaded situation

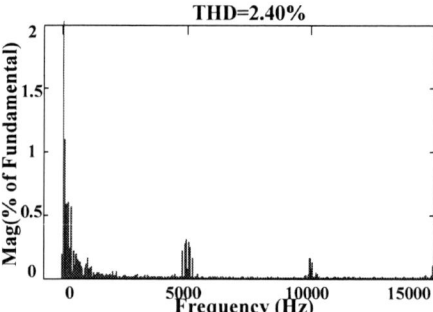

Fig.7 the spectrum of u_o at unloaded situation

The waveform of u_o when it is unloaded is shown in Fig.6. For the power supply quality verification, the spectrum analysis is shown in Fig.7. It shows that THD is smaller than 3%.

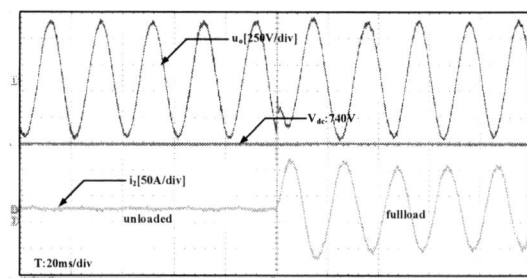

Fig.8 the dynamic process from unloaded to full load situation

To verify the inherent positive damping characteristic, the dynamic process is given out as Fig.8. The result shows that system is table at the process from unloaded to full load situation at the absence of extra active or passive damping method.

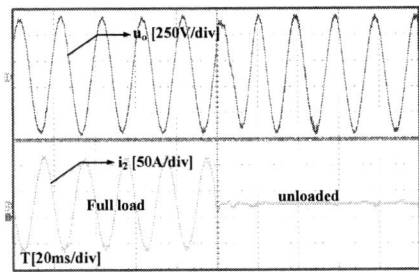

Fig.9 the dynamic process from full load to unloaded situation

And the dynamic process from full load to unloaded situation shown in Fig.9 further verifies the stability without extra damping.

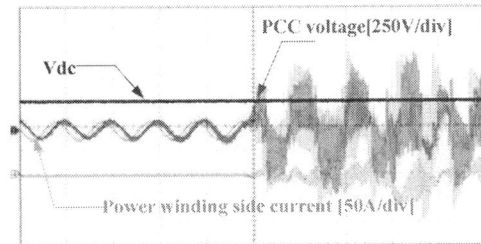

Fig.10 the situation when the filter capacitance is out of the range of the positive damping

Fig.10 gives the experiment result when the filter parameters do not meet the positive damping limitation. The system cannot maintain stable in this situation because of the resonance between the inductance and the capacitance.

VI. CONCLUSIONS

Instead of using LCL filter to achieve higher harmonic attenuation, this paper puts forward to adopt LC filter combined with the stator power winding leakage inductance to form LC-L structure. Compared with the grid-connected situation, the additional criterion for filter design in this stand-alone situation is analyzed.

It puts forward that the harmonic attenuation of both the current that injected to BDFG and PCC voltage is important in this system. Based on the analysis of the relationship between the filter performance and its parameters, this paper puts forward the design principle for filter parameters. In order to reduce the volume and weight of the filter, this paper puts forward to achieve the THD requirement of PCC voltage by increasing the capacitance instead of inductance. The relationship between THD of PCC voltage and the capacitance has been analyzed.

For the instability problem caused by the resonance of LC-L, the inherent damping characteristic that takes the digital effect into account has been analyzed. Finally, the limitation for filter parameters to achieve both positive damping characteristic and required performance is deduced. And the experiment verified that the system can be stable by the proper design of the filter without extra active or passive damping method.

ACKNOWLEDGMENT

Meilin Wang would like to thank Engineer Xinmin Liu for his great help in this study and the help from the classmates at the same laboratory.

REFERENCES

[1] H. Liu, W. Fan, F. Long and N. Shen. "Shaft Power Generation System Based on BDFG," *IEEE Conference and Expo Transportation Electrification Asia-Pacific* (ITEC Asia-Pacific), 2014, pp.1-4.

[2] N.T Nguyen and H.H Lee. "An effective harmonic elimination for DFIG feeding non-linear loads in stand-alone operation," *IECON 2012-38th Annual Conference on IEEE Industrial Electronics Society*, 2012, pp. 3527-3532.

[3] A. A. Rockhill, M. Liserre, R. Teodorescu, and P. Rodriguez, "Grid-filterdesign for a multimegawatt medium-voltage voltage-source inverter,"*IEEE Trans. Ind. Electron*, Vol. 58, no. 4, pp. 1205–1217, 2011.

[4] Phan, V. and Lee, H. "Control Strategy for Harmonic Elimination in Stand-Alone DFIG Applications With Nonlinear Loads," *IEEE Transactions on Power Electronics.* 26(9):2662-2675, 2011.

[5] Pattnaik, M. and Kastha, D. "Harmonic Compensation With Zero-Sequence Load Voltage Control in a Speed-Sensorless DFIG-Based Stand-Alone VSCF Generating System," *IEEE Transactions on Industrial Electronics.* 60(12):5506-5514, 2013.

[6] Phan, V. and Lee, H. "Performance Enhancement of Stand-Alone DFIG Systems With Control of Rotor and Load Side Converters Using Resonant Controllers," *IEEE Transactions on Industry Applications.* 48(1):199-210, 2012.

[7] Rishabh Dev Shukla and Prof. R. K. Tripathi. "Topologies for Stand-alone DFIG base d Wind Energy Conversion System," *2012 2nd International Conference on Power*, 2012.

[8] J. He, W.L. Yun, D. Bosnjak and B. Harris, "Investigation and resonances damping of multiple PV inverters," *Applied Power Electronics Conference and Exposition (APEC)*, 2012 Twenty-Seventh Annual IEEE, 2012, pp. 246-253.

[9] Changzhou Yu, Xing Zhang, Fang Liu, et al, "A general active damping method based on capacitor voltage detection for grid-connected inverter," *ECCE Asia Downunder (ECCE Asia)*, 2013, pp.829-835

[10] Tang, Y., Yao, W., Loh, P. and Blaabjerg, F. "Design of LCL-Filters with LCL Resonance Frequencies beyond the Nyquist Frequency for Grid-Connected Converters," *IEEE Journal of Emerging and Selected Topics in Power Electronics*, 2015.

[11] Meilin Wang, Hua Lin, Fei Gong, Hongbin Yang, Xingwei Wang, "Stability Control for LC-L Filter Based Load Side Converter of BDFG Ship Shaft Power Generation System without Extra Damping" *IEEE Energy Conversion Congress and Exposition (ECCE)*, Seventh Annual IEEE, 2015, pp.5525-5530.

Investigation of Common Mode Current Related DC-Bus Overvoltage in Multiple Converter Systems

Jiangbiao He, Zoran Vrankovic, Patrick E. Ozimek, Craig Winterhalter

Rockwell Automation
Mequon, Wisconsin, 53092
jiangbiao@ieee.org

Abstract—Active front end (AFE) converters have been extensively used in industrial motor-drive systems due to their energy regenerative capability and low harmonic distortions. However, the Pulse Width Modulation (PWM) switching operation in AFE converters generates higher common-mode voltage than their non-regenerative diode front-end (DFE) counterparts. Such significant common-mode voltage can be converted to common-mode current through common-mode capacitance in the system. This large common-mode current can cause overvoltage of the dc-bus of a DFE converter system where both AFE and DFE converter systems share the same ac source. This paper investigates such overvoltage phenomena occurring in a DFE dc-bus in a multiple converter system. Mechanism of such dc-bus overvoltage will be explained, and both simulation as well as experimental results are presented to confirm the theoretical analysis. Moreover, a dc-bus clamping circuit is recommended to clamp the DFE dc-bus voltage, thus any overvoltage that could trip the DFE drive can be avoided.

Keywords—Overvoltage; common-mode current; active front end converter; multiple converter systems.

I. INTRODUCTION

Voltage source inverters commonly used in motor drive systems generate significant common mode voltages on their motor outputs. When long shielded motor cables are used, this common mode voltage is converted into common mode current by the parasitic capacitance from motor phase wires to ground. This common mode current can cause the drive system to fail electromagnetic compatibility (EMC) standards (e.g., IEC/EN 61800-3 EMC Compliance) if this common mode current returns to the inverter through the AC input. One common way to mitigate this effect is to install high frequency common mode capacitors (commonly called Y-capacitors) from each of the DC Bus voltage rails to ground, as shown in Fig. 1. This provides a low impedance path for the common mode current to return to the source of the noise (which is the IGBTs) bypassing the AC line input. The higher the value of Y-capacitance, the larger the attenuation in the common mode seen on the AC line input.

However, issues can occur when such a motor drive is fed from an active front end converter (AFE). AFE converters have been increasingly applied in industrial motor drives for obtaining high performance such as regenerative operation, low harmonic distortion of the input current, as well as wide-range boosted dc-bus voltage [1-3]. The AFE converter will generate

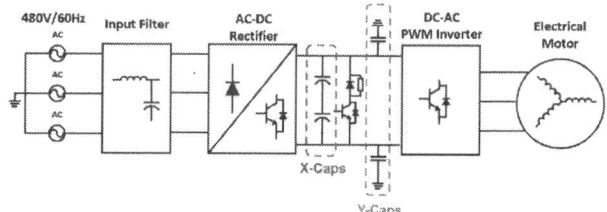

Fig. 1 Common-mode capacitors (Y-Caps) and differential-mode capacitors (X-Caps) in a typical adjustable-speed drive system.

common mode voltage on the DC bus output due to the PWM switching of the converter. This common mode voltage applied to the DC bus will be converted into common mode current by the Y-capacitors in the inverter drive. To reduce the size of the input filter inductance for current ripple attenuation, AFE rectifiers are generally switched at a high frequency (e.g., above 10 kHz). However, such high switching frequency operation of AFE rectifiers can result in higher common-mode current in the drive system.

If the Y-capacitors in the inverter drive were designed with a high value to maximize the amount of EMC filtering, the additional common-mode current flowing through the Y-capacitors caused by the AFE switching may exceed the ratings of the capacitor. This can result in large temperature rise and decreased capacitor lifetime.

In addition to AFE converters, PWM inverter drives can be supplied from passive diode front end (DFE) converters. DFE converters are commonly used when the application does not result in significant regenerative energy. It is possible to have both AFE and DFE fed inverter drive systems operating from the same AC source as illustrated in Fig. 2. Each system would have a single bus power supply (AFE or DFE) and one or more inverter drives sharing a common DC bus. Under such a condition, the common mode current generated by the AFE system can return to the source through the inverter Y-capacitors in the DFE system. As the current passes through the DFE system, it can charge the DC bus capacitors and cause the DC bus voltage to rise (or called pump-up) to a dangerous level unless detected and prevented by some other means in the system. Therefore, it is of great significance to investigate the mechanism of such common-mode current related dc-bus overvoltage phenomenon, in order to better design/protect the multiple converter systems from damaging during applications.

978-1-4673-9551-9/16 $31.00 © 2016 IEEE

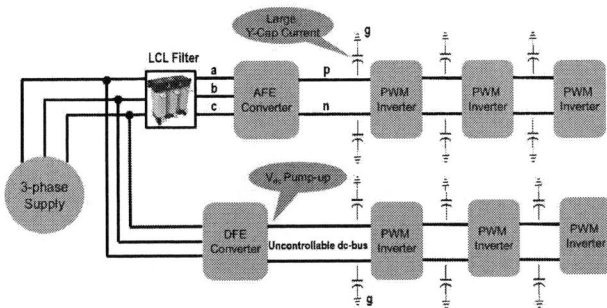

Fig. 2 Common-mode current related dc-bus overvoltage and Y-capacitor overcurrent in multiple converter systems.

The remainder content of this paper is organized as follows. In Section II, the mechanism of the dc-bus overvoltage caused by the common-mode current will be explained. In Section III, simulation results obtained from the modeling in ANSYS Simplorer will be presented to demonstrate the influence of common-mode current on the dc-bus overvoltage and Y-capacitor overheating. In Section IV, experimental results acquired from an industrial multiple converter system will be given to confirm the analysis and simulation results. In Section V, a dc-bus voltage clamping circuit is recommended to mitigate the voltage pump-up issue. Finally, conclusions will be drawn in Section VI.

II. MECHANISM OF THE COMMON MODE CURRENT RELATED DC-BUS OVERVOLTAGE PHENOMENON

In the system defined in Fig. 2, the common-mode voltage generated on the ac side of the AFE rectifier, $v_{com(ac)}$, can be expressed as:

$$v_{com(ac)} = \frac{v_{ag}+v_{bg}+v_{cg}}{3} \qquad (1)$$

where, v_{ag}, v_{bg}, and v_{cg} are the voltage potentials between the input ac terminals (a, b, and c) of the AFE converter and the neutral point of the three-phase source.

Correspondingly, common-mode current, $i_{com(ac)}$ can be defined as the sum of the line currents:

$$i_{com} = i_a + i_b + i_c \qquad (2)$$

where, i_a, i_b, and i_c are the input ac currents of the AFE converter.

Common-mode voltage generated on the dc side of the AFE rectifier, $v_{com(dc)}$, can be expressed as:

$$v_{com(dc)} = \frac{v_{pg}+v_{ng}}{2} \qquad (3)$$

where, v_{pg} is the voltage potential between the positive dc-bus and the system ground, while v_{ng} is the voltage potential between the negative dc-bus and the system ground.

Equation (3) can also be written as follows:

$$v_{com(dc)} = \frac{1}{2}\left(\frac{v_{dc}}{2} + v_{mg} + v_{mg} - \frac{v_{dc}}{2}\right) = v_{mg} \qquad (4)$$

where, v_{dc} is the dc-bus voltage, and v_{mg} is the voltage between the dc-bus mid-point and the system ground.

The common-mode current flowing through the dc-bus Y-capacitors can be expressed as:

$$i_{com(dc)} = i_{i_{C_{y1}}} + i_{C_{y2}} = C_{y1}\frac{dv_{C_{y1}}}{dt} + C_{y2}\frac{dv_{C_{y2}}}{dt} \qquad (5)$$

where, $i_{C_{y1}}$ and $i_{C_{y1}}$ refer to the common-mode current in the common-mode capacitors, C_{y1} and C_{y2}, on the dc-bus. $v_{C_{y1}}$ and $v_{C_{y2}}$ are the voltages across the two common-mode capacitors, C_{y1} and C_{y2}.

As can be seen in Equation (5), the peak value of the DC common mode current flowing through the y-caps is proportional to dv/dt, which is determined by the rise time of the IGBTs in the AFE converter. Accordingly, the RMS value of the common mode current through the Y-Capacitors is proportional to both the switching speed (i.e., IGBT rise time) as well as the switching frequency of the AFE converter.

The mechanism for the voltage pump-up in the DFE system is shown in Fig 3. Some of the common mode current will enter the DC bus of the DFE system through its inverters Y-Caps. This current will charge the DC Bus capacitors as it passes through the system and returns to the AC line through the DFE rectifier diodes. This path is low impedance because the AC inputs of both systems are tied together downstream of the source transformer.

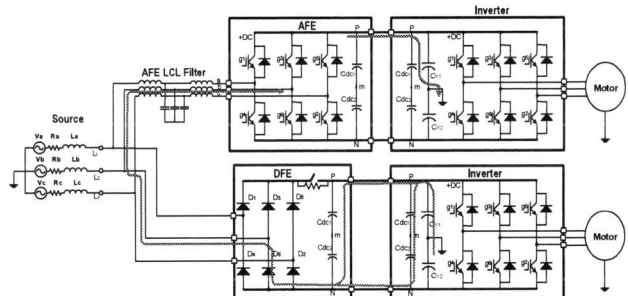

Fig. 3 Basic circuit schematic of the multiple converter system.

III. SIMULATION RESULTS

To investigate the influence of common-mode current produced from the AFE rectifier on the performance of the drive system, simulation has been carried out in ANSYS Simplorer. This simulation was created to model a system similar to the one depicted in Fig 2. Both an AFE and a DFE converter system are connected to the same AC source downstream from a shared transformer. The ac supply is three-phase 480V, 60Hz, with neutral point solidly grounded. In the simulation, the topology of the AFE converter is a three-phase two-level voltage source converter, and the switching frequency is set at 10 kHz. Voltage-oriented closed-loop control strategy combined with the space vector pulse width modulation (SVPWM) [4] is used for obtaining unity power factor and regulating the AFE dc-bus voltage at 690V. The DFE converter topology is a three-phase full bridge diode rectifier. The AFE converter has up to six PWM inverters connected in parallel on the shared DC Bus generated by the AFE converter. The DFE converter system was simulated with a single PWM inverter connected to its shared DC bus. Each PWM inverter in both systems has two Y-capacitors with the same ratings.

In this simulation, none of the PWM inverters in either the AFE or DFE system have their inverters enabled. This is the worst case because the PWM switching of the inverters in the DFE system will consume a small amount of DC bus current which will mitigate the voltage pump-up caused by the AFE system common mode current.

A. Variation of Y-Capacitor Current with AFE System CM Capacitance

According to the simulation data shown in Fig. 4 (a), the common mode current per Y-capacitor increases when the quantity of the inverters is reduced from six to one. The RMS value of the common-mode current per Y-capacitor is 1.3A for six inverters and 3.4A for a single PWM inverter. As the total common mode capacitance of the system decreases, the dv/dt of the IGBT switching increases. This higher dv/dt appears across the remaining Y-capacitors and causes an increase in current through each capacitor.

The highest current per Y-capacitor will be seen in the condition of a single PWM inverter in the AFE converter system. In such a condition, it is possible to exceed the current rating of the capacitor and cause excessive heating which will reduce the lifetime of the part and potentially lead to premature failure.

However, even though the current per Y-capacitor will decrease as the total Y-capacitance increases, the total common mode current generated by the AFE system will increase with increasing total common mode capacitance, which is shown in Fig. 4 (b).

B. Variation of DFE System Voltage Pump-Up with AFE System CM Capacitance

To investigate the voltage pump-up phenomenon, the DC bus voltage of the DFE system was simulated. Fig. 5 shows that the DC bus Voltage of the DFE system will increase as the amount of inverters in the AFE system increases. Similarly, the DFE DC bus voltage is also pumped up when the quantity of the PWM inverters in the DFE system increases, as shown in Fig. 6. Again, the reason contributing to the dc-bus voltage pump-up is due to the increased common-mode capacitance from the Y-capacitors in the dc-bus of each PWM inverter module.

However, such dc bus voltage pump-up phenomenon can be eliminated by disconnecting the AFE system Y-capacitors from ground as illustrated in Fig. 7. When the Y-capacitors are disconnected from ground, the DFE System dc bus voltage pump-up phenomenon is eliminated as shown in Fig 8. This is because the remaining parasitic common-mode capacitance after removing the Y-capacitors is not significant enough to cause the pump-up issues. It can be seen in Fig. 8 that the DFE dc-bus voltage, depicted in the blue line, almost stays unchanged

when the common-mode capacitors are disconnected from the dc-bus rails to the ground.

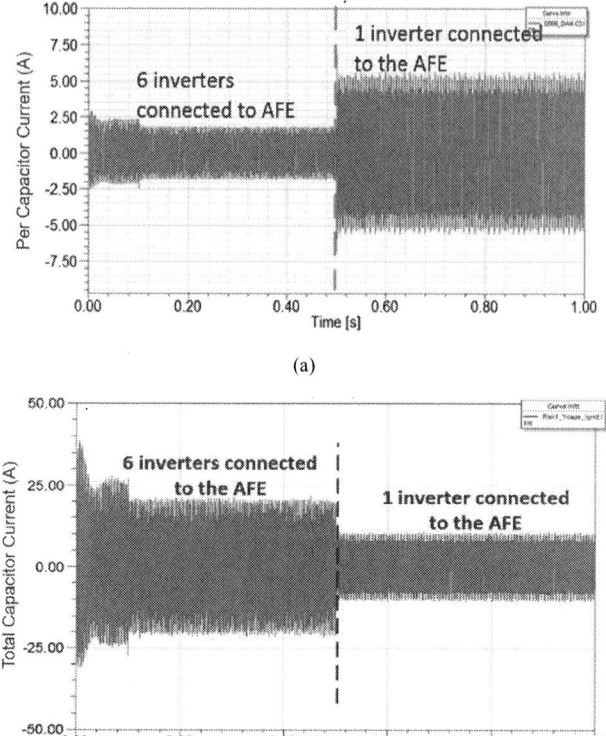

(a)

(b)

Fig. 4 Variation of the common-mode current versus the change of the common-mode capacitance (a) common-mode current in each individual Y-capacitor (b) total common-mode current flowing to the ground (charging the DFE dc bus).

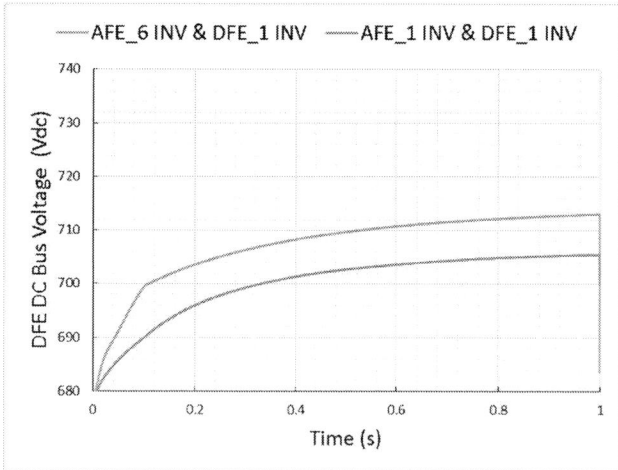

Fig. 5 DC bus voltage pump-up on DFE System inverters as a function of number of inverters connected in AFE system (the blue curve shows the DFE bus voltage when there are six PWM inverters connected with the AFE converter and one PWM inverter connected with the DFE converter, while the grey curve shows the DFE bus voltage when there is only one PWM inverter connected with the AFE and DFE converters).

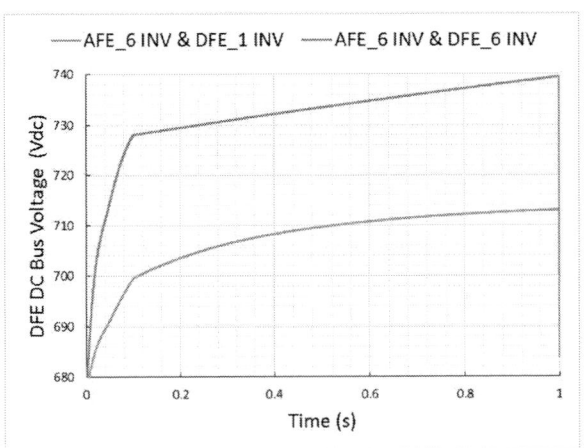

Fig. 6 DC bus voltage pump-up on DFE System inverters as a function of number of inverters connected in DFE system (the blue curve shows the DFE bus voltage when there are six PWM inverters connected with the AFE converter and another six PWM inverters connected with the DFE converter, while the grey curve shows the DFE bus voltage when there are six PWM invevrters connected with the AFE converter but only one PWM inverter connected with the DFE converters).

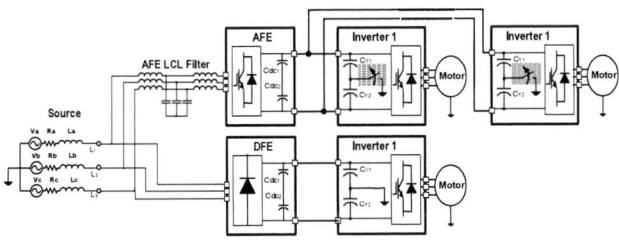

Fig. 7 Illustration of removal of the ground jumpers to disconnect the Y-capacitors from ground.

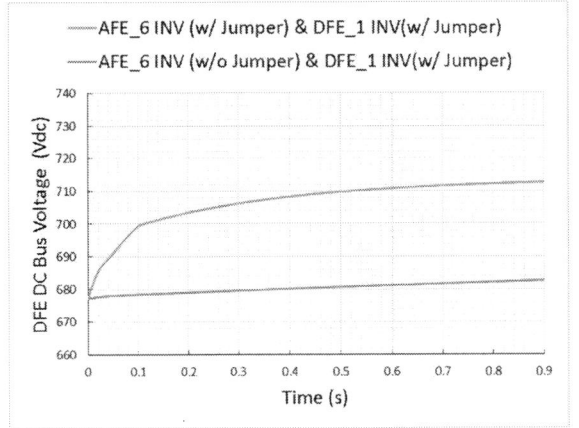

Fig. 8 DC bus voltage pump-up on DFE System inverters with and without the AFE System Inverter Y-caps connected to ground.

IV. EXPERIMENTAL RESULTS

In this section, experimental results are presented to show the detrimental effect of common-mode current on the dc-bus overvoltage of the DFE converter and the temperature rise in the common-mode capacitors on the AFE dc-bus. In this testing, the multiple converter system consists of two sets of drives sharing the same three-phase power supply (480V/60Hz). One set of drive is powered by an AFE-based regenerative dc supply, the other is DFE-based non-regenerative bus supply, as shown in Fig. 9. Two 150 μF metallized polypropylene film (MPPF) capacitors from WIMA are utilized on the dc-bus of every PWM inverter connected with the AFE and DFE converters. Also, in the experiments, only one PWM inverter is connected in each of the AFE and DFE systems. The switching frequency used in the AFE rectifier is 10 kHz, and the related dc-bus voltage is controlled at 690V. The measured DFE dc-bus voltage is close to the theoretical value of 678V (varying between 672~674V due to certain voltage drop in the circuit) when the AFE rectifier is powered off, which will be suddenly boosted to a higher voltage value when the AFE rectifier is powered on. The measured common-mode current flowing through the Y-capacitors is 5.1A (RMS), as shown in Fig. 10, which is much larger than the permissible ripple current of 2.74A [5]. Moreover, six thermal sensors are placed around the surface of the Y-capacitors to measure the external temperature of the Y-capacitors and the ambient temperature, as shown in Fig. 11. The measured external temperature rise of the common-mode capacitor is 13.6 °C, as shown in Fig. 12, which is also higher than the targeted temperature rise of 10 °C to meet the rated lifetime [5].

In the experiments, the DC bus voltage of the DFE converter system was measured under different conditions of the AFE converter and the availability of the Y-capacitors on the AFE dc-bus rails. The measured DC-bus voltage values (RMS) are given in Table I. It can be seen that the DFE dc-bus voltage experienced a higher voltage of 691V under the operation of the AFE converter and the presence of the common mode capacitors, compared to other measured data. This supports the assertion that the common mode current generated from the switching operation of the AFE converter boosts the DFE dc-bus voltage through the propagation paths of Y-capacitors.

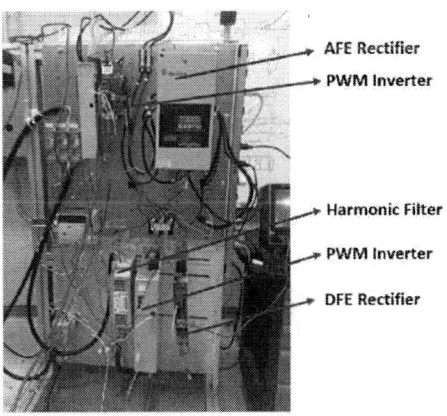

Fig. 9 An industrial multiple converter drive system.

TABLE I. MEASURED DFE DC-BUS VOLTAGE (RMS) UNDER VARIOUS CONDITIONS OF THE AFE CONVERTER AND Y-CAPACITORS

With/Without the Y-Capacitors Connected	Operation Status of the AFE Converter	
	AFE Turn-on	*AFE Turn-off*
With Y-Capacitors	691V	674V
Without Y-capacitors	672V	672V

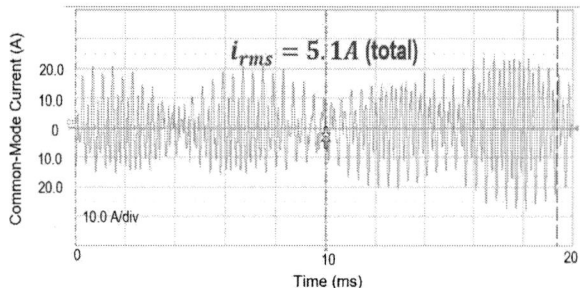

Fig. 10 Measured total common-mode current in the two Y-capacitors on the AFE dc-bus.

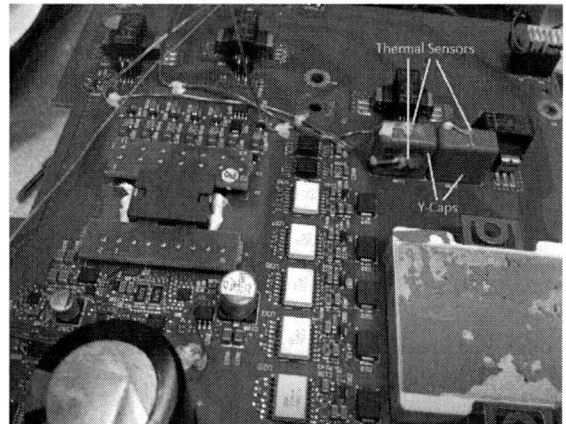

Fig. 11 Measurement of the temperature rise in the common-mode Y-capacitors on the dc-bus of the PWM inverters connected with the AFE converter.

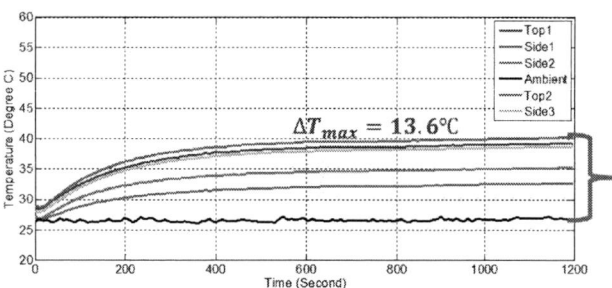

Fig. 12 Measured temperature rise in one single MPPF Y-capacitor in the PWM inverter connected with the AFE converter.

V. A RECOMMENDED PASSIVE DC-BUS OVERVOLTAGE CLAMPING CIRCUIT

The issue of excessive temperature rise in dc-bus common-mode capacitors can be solved by either disconnecting these capacitors through the ground jumpers, or oversizing the Y-capacitors to withstand the high common-mode current. If the temperature rise issue is solved by oversizing the Y-capacitors, this issue of dc-bus voltage pump-up in DFE systems remains and may be more severe. One way to mitigate this dc-bus voltage pump-up is using a voltage clamping circuit on the DFE dc-bus.

In this paper, a passive clamping circuit is recommended to clamp the DFE dc-bus voltage to the nominal value. The details of such dc-bus voltage clamping circuit was introduced in reference [6]. The clamping circuit topology is shown in Fig. 13, in which it can be seen that the clamping circuit consists of two diodes clamped between the positive dc bus and negative dc-bus. A clamping capacitor (Csn) is connected between the neutral of the two diodes and the ground, in which the snubber capacitance is generally selected to be much higher than the common-mode capacitance of the dc-bus. A discharging resistor (Rsn) is connected in parallel with the clamping capacitor. During normal operation, the voltage potential of the ground is always lower than the positive dc-bus voltage but higher than the negative dc-bus voltage, which will make both the clamping diodes reverse biased and consequently the voltage across the snubber circuit is zero. However, once the voltage potential of the ground is either higher than the positive dc bus or lower than the negative dc bus, one of the clamping diodes starts to conduct. As a result, the potential of the ground will be clamped to one of the two dc-bus voltages. Based on this voltage clamping circuit, the simulated DFE bus voltage and AFE bus voltage obtained in ANSYS Simplorer under the same conditions of the multiple converter system are shown in Fig. 14, in which it can be seen that the DFE dc-bus voltage is clamped at the nominal value of 678V. Therefore, any overvoltage tripping in the multiple converter system can be avoided in operation. For more details of the operating principle and experimental results of using this clamping circuit for damping the bus overvoltage, see reference [6].

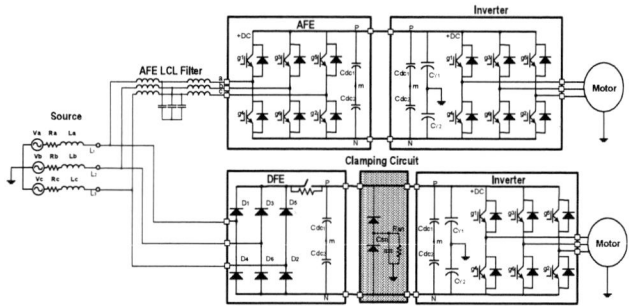

Fig. 13 The multiple converter system with the recommended dc-bus voltage clamping circuit on the DFE converter.

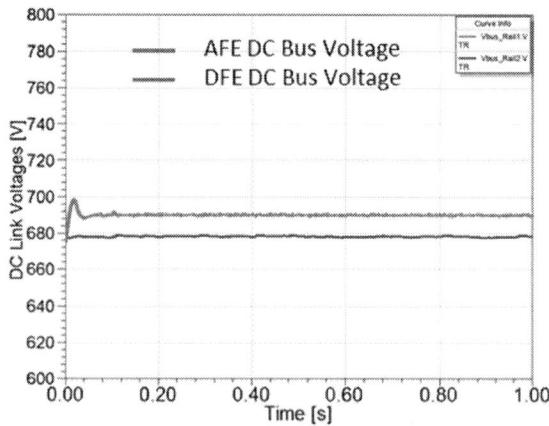

Fig. 14 Simulation results of the AFE and DFE dc-bus voltages in the multiple converter system based on the recommended dc-bus voltage clamping circuit.

978-1-4673-9551-9/16 $31.00 © 2016 IEEE   1088

VI. CONCLUSIONS

This paper investigated the dc-bus overvoltage issue caused by common-mode current in a multiple converter system. According to the analysis, simulation, and experimental results, a few conclusions can be drawn as follows:

(1) The quantity of the PWM inverters connected to the AFE converter determines the total common-mode capacitance which will determine the switching speed of the IGBTs in the AFE converter. As the switching speed increases, the dv/dt seen by the Y-Capacitors in the AFE system will increase, resulting in higher common-mode current per capacitor. When very few PWM inverters are connected to the AFE, the common-mode current flowing through each dc-link Y-capacitor become quite large, which leads to accumulative heating and temperature rise in such capacitors. Under such scenario, the Y-capacitors on the AFE dc-bus have to be disconnected via the ground jumpers or sized to handle the significant common-mode current.

(2) As the quantity of PWM inverters connected to the AFE system increases, the total common mode current generated by the AFE system will increase in spite of the current per Y-capacitor decreasing.

(3) The common-mode current generated by the Y-capacitors in the AFE system will circulate through the DC of any passive DFE converter system connected to the same AC source. This common-mode current will result in DC bus voltage pump-up of the DFE system and could lead to nuisance tripping or system damage in extreme cases.

(4) A passive dc-bus clamping circuit is recommended to clamp the DFE dc-bus voltage, therefore dc-bus overvoltage tripping caused by the common-mode current from the AFE converter can be avoided.

Particularly, with the application of the emerging wide bandgap switching devices and operation of high-frequency switching operation in next-generation AFE converters [7-9], the aforementioned dc-bus overvoltage in the associated DFE converters and the overcurrent in the Y-capacitors will be more severe. Therefore, special attention has to be paid to mitigate the harmful effects of the generated common-mode current. Solutions based on active manipulation of the PWM schemes for the AFE converters to attenuate the related common-mode current will be explored in future work.

REFERENCES

[1] J. R. Rodriguez, J. W. Dixon, J. R. Espinoza, J. Pontt, and P. Lezana, "PWM regenerative rectifiers: state of the art," *IEEE Trans. on Industrial Electronics*, vol. 52, no. 1, pp. 5-22, Feb. 2005.

[2] M. Liserre, A. Dell' Aquila, and F. Blaabjerg, "An Overview of three-phase voltage source active rectifiers interfacing the utility," in Proceedings of *2003 Power Tech Conference*, Bologna, Italy, 2003.

[3] A. Sayed-Ahmed, L. Wei, and B. Seibel, "Industrial regenerative motor-drive systems," in Proceedings of *2012* Twenty-Seventh Annual IEEE Applied Power Electronics Conference and Exposition (APEC), Orlando, FL, 2012.

[4] M. Malinowski, M. P. Kazmierkowski, and A. M. Trzynalowski, "A comparative study of control techniques for PWM rectifiers in AC adjustable speed drives," *IEEE Trans. on Power Electronics*, vol. 18, no. 6, November 2003.

[5] Application note of WIMA MKP-X2 R Metallized Polypropylene film capacitors, [Online] *www.wima.com*, January 2014.

[6] L. Wei, Z. Liu, and G.L. Skibinski, "DC-bus voltage clamp method to prevent overvoltage failures in adjustable speed drives," *IEEE Trans. on Industry Applications*, vol. 47, no. 4, pp. 1778-1785, May. 2011.

[7] T. Zhao, J. Wang, A.Q. Huang, A. Agarwal, "Comparisons of SiC MOSFET and Si IGBT based motor drive systems," Conference Record of 2007 IEEE IAS Annual Meeting, pp. 331-335, 2007

[8] H. A. Mantooth, M. D. Glover, P. Shepherd, "Wide Bandgap Technologies and Their Implications on Miniaturizing Power Electronic Systems", IEEE Journal of Emerging and Selected Topics in Power Electronics, vol.2, no. 3, pp. 374 - 385, 2014

[9] J. He, T. Zhao, X. Jing and and N. Demerdash, "Application of Wide Bandgap Devices in Renewable Energy Systems – Benefits and Challenges", in proceedings of International Conference on Renewable Energy Research and Applications (ICRERA), Milwaukee, WI, pp. 749-754, Oct 2014.

High Frequency AC Inductor Analysis and Design for Dual Active Bridge (DAB) Converters

Zhe Zhang and Michael A. E. Andersen
Department of Electrical Engineering
Technical University of Denmark
Kgs. Lyngby, DENMARK
zz@elektro.dtu.dk and ma@elektro.dtu.dk

Abstract—The dual active bridge (DAB) converter is an isolated bidirectional dc-dc topology which is the most critical part for the power conversion systems such as solid-state transformers (SST). This paper focuses on analysis and design of high frequency ac inductors which are the power interfacing component in DAB converters or DAB's derivative topologies for transferring energy between the primary and secondary sides. The DAB converter's operation principles, and the corresponding voltage and current stresses over its ac inductor are analyzed. Hereby, six diverse winding arrangements are studied in order to find a design having the lowest ac resistance and core loss. Core loss is calculated by both GSE and iGSE methods, and then the results are compared under two operating conditions. Based upon the finite element method (FEM) simulation, winding losses are investigated. Finally, the case in which the core loss and the winding loss are almost equal is selected as the optimal one. The experimental results are presented to verify the validity of the analysis and design.

Keywords—ac inductor; converter; dual active bridge; FEM; winding loss

I. INTRODUCTION

Bidirectional dc-dc converters have been attracting attention and becoming a research focus in the recent years, along with the ceaseless development and application of renewable energy systems in which a proper energy storage element and its corresponding power electronic converters are needed [1]-[6]. The dual active bridge converter (DAB), as an isolated bidirectional converter, has been proposed for over two decades. Due to its simple circuitry, symmetrical structure, low number of magnetic components and absence of bulky dc inductors, the DAB converter is considered to be a suitable candidate to carry out voltage conversion and galvanic isolation with high power density and high power efficiency, in particular, for high power applications such as solid-state transformers, electric tractions and battery interfacing circuits [7]-[9].

With regard to magnetic components of DAB converters, as a DAB topology illustrated in Fig. 1, only one high frequency transformer is adopted and its leakage inductance can be used as L_{ac} to interface the power transferring between the primary and secondary sides of the high frequency transformer. However, for many relatively low power applications in which

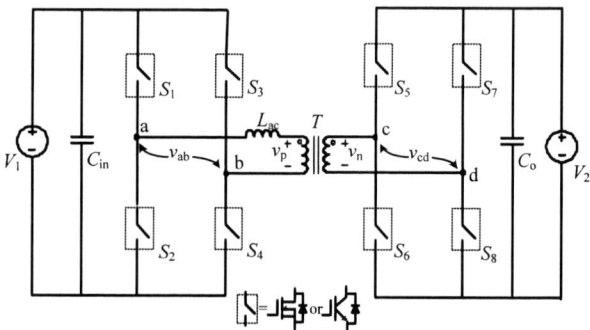

Fig. 1: Topology of the dual active bridge (DAB) converter.

if the input and output voltages are high, an external inductor to provide larger interfacing inductance is needed in order to adjust the output power as well as to extend the zero voltage switching (ZVS) operation. Of course, rather than using an extra inductive component to get larger inductance, the leakage inductance can be increased by weaken the coupling effect between the transformer windings, but the price is that the transformer winding ac resistance cannot be minimized effectively and more severe radiated EMI issues may occur. Therefore, it is more efficient to utilize external inductors instead. Diverse from filtering inductors in Buck-derived or Boost-derived dc-dc converters, here, a high frequency ac inductor which operates at the switching frequency is needed in DAB converters. For ac inductors that only carry high frequency ac current without any dc offset, the core selection and winding design have the issues very similar to that of transformer design but with one more difficulty, i.e. there are no secondary windings which can be used to reduce the proximity effect by interleaving. For the core selection, ac flux component causes core losses and potentially contributes to eddy current losses if air gaps and windings are in a close proximity. Iron powered cores cannot be used due to their relatively high core loss compared to their ferrite counterparts; on the other hand, a ferrite core with a lumped gap will have unacceptably high winding losses due to the fringing effect. These issues actually challenge designing highly efficient ac inductor, on which this paper will focus.

This paper is organized as follows. After this introduction, the special design requirements for ac inductors employed in

978-1-4673-9551-9/16 $31.00 © 2016 IEEE

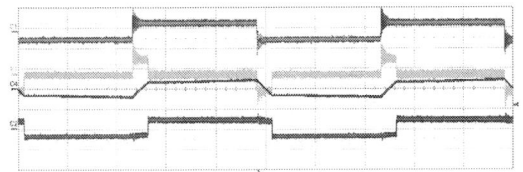

Fig. 2: Experimental waveforms of DAB converter. Ch1: v_{ab} (100V/div), Ch2: v_p (100 V/div), F1: v_L (100 V/div) and Ch4: i_L (50 A/div).

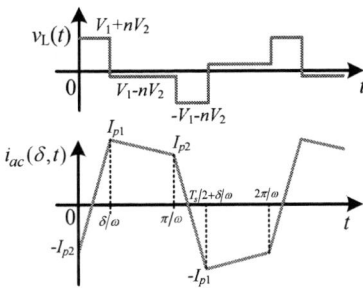

Fig. 3: Typical waveforms of ac inductor of DAB converters.

DAB converters are introduced in Section II. By using finite element method (FEM) simulation, ac inductors with six variant winding arrangements are analyzed and compared in Section III. The measurement results are presented in Section IV. Finally, the conclusion is given in Section V.

II. AC INDUCTOR IN DAB CONVERTERS

In the cases, in particular for the low power applications (below 10 kW), the leakage inductance of the high frequency transformer is not sufficient to keep DAB converters operating under ZVS conditions and the deadband effect will emerge that leads to even worse EMI noise. Therefore, an external ac inductor is needed.

A. Typical operation principle of DAB converters

As the DAB converter shown in Fig. 1, the bidirectional switch cells, i.e. S_1~S_8, are used to control reversible power between the terminals V_1 and V_2, with a phase-shift modulation. The primary and secondary full bridges with 50% duty cycle to achieve a rectangular voltage v_{ab} and v_{cd}, which have a phase-shift angle δ used as a power regulation parameter. By controlling the phase shift with a range of $-\pi \leq \delta \leq \pi$, the transferred power is then regulated as,

$$P = P_1 = P_2 = \frac{n \cdot V_1 \cdot V_2 \cdot \delta \cdot (\pi - |\delta|)}{2\pi^2 f_s L_s} \quad (1)$$

where L_s represents inductance of the ac inductor L_{ac}; f_s is the switching frequency, and n is the transformer turn ratio defined as $N_p{:}N_s$.

Therefore from the typical experimental voltage and current waveforms of the ac inductor as illustrated in Fig. 2, the more specifically theoretical waveforms, i.e. the voltage across the ac inductor and the ac inductor current, can be plotted in Fig. 3.

As derived in [2], if $V_1 = nV_2$, the ac inductor current i_{Lac} can be expressed by Fourier series,

$$i_{Lac} = \sum_{n=1,3,5...}^{\infty} \frac{4}{m^2 \pi \omega L_s n} \left(\sin \frac{m\delta}{2} \right) \cdot \sin \left(m\omega t - \frac{m\delta}{2} \right). \quad (2)$$

Whereas, if $V_1 \neq nV_2$, a more general expression of i_{Lac} can be obtained.

$$i(\delta, t) = \sum_{m=1,3,5...}^{\infty} \frac{4}{m^2 \pi \omega L_s} \cdot [nV_2 \cdot \cos m\delta - V_1] \cdot \cos m\omega t$$

$$+ \sum_{m=1,3,5...}^{\infty} \frac{4}{m^2 \pi \omega L_s} \cdot nV_2 \cdot \sin m\delta \cdot \sin m\omega t. \quad (3)$$

From (3), it can be seen that there is no even-order harmonics existing in the ac current waveform and the harmonic magnitudes vary with the phase-shift angle and the diversity between V_1 and nV_2.

Moreover, the average power that the DAB converter can deliver actually only depends on the products of the fundamental components of the ac voltage and current at the switching frequency and their harmonics with the same orders. The average power then can be calculated.

$$P_{avg} = \sum_{n=1,3,5...}^{\infty} \frac{4V_1 \cdot V_2}{m^3 \pi^2 \omega L_s n} \sin(m\delta). \quad (4)$$

Based on (2)-(4), it is clear that for DAB converters the smaller phase-shift angle and the smaller current slope during $\delta/\omega \sim \pi/\omega$ can give less reactive power which circulates in the converter and causes extra loss. Otherwise, more complex modulation schemes such as a phase-shift plus phase-shift method has to be used in order to attenuate the *rms* ac current at a certain average or active power required [10]-[12].

B. Analysis of ac inductor

Based on the typical operating waveforms illustrated in Fig. 3, the flux linkage at the beginning of core saturation is given in (5)

$$\lambda_s = \frac{\Delta\lambda_{\max}}{2} = N_L A_c B_s = \begin{cases} \dfrac{2V_1\delta + \pi(nV_2 - V_1)}{4\pi f_{s\min}} & if\ V_1 \leq nV_2 \\[3mm] \dfrac{2nV_2\delta + \pi(V_1 - nV_2)}{4\pi f_{s\min}} & if\ V_1 > nV_2 \end{cases}. \quad (5)$$

where N_L, A_c and B_s represent the number of turns of the ac inductor, the cross-sectional area of the inductor core and the flux density of the core material, respectively. Accordingly, based upon the pre-defined input and output voltages, output power and switching frequency, the needed minimum cross-sectional area and the number of turns can be calculated and selected to avoid saturating the inductor.

The copper loss or winding loss of ac inductors can be calculated in (6).

$$P_{L,cond} = R_{L,ac} \cdot I_{Lac}^2 \quad (6)$$

where $R_{L,ac}$ represents ac resistance of the winding, and I_{Lac} is the ac *rms* current calculated from (3).

The inductor core losses are estimated using the improved Generalized Steinmetz Equation (iGSE) [13], [14].

$$P_v = \frac{1}{T} \int_0^T k_i \left| \frac{dB}{dt} \right|^\alpha (\Delta B)^{\beta - \alpha} dt . \quad (7)$$

where P_v is the time-average power loss per unit volume, ΔB is the peak-to-peak flux density, k, α and β are the material parameters, and

$$k_i = \frac{k}{(2\pi)^{\alpha-1} \int_0^{2\pi} |\cos\theta|^\alpha 2^{\beta-\alpha} d\theta} . \quad (8)$$

From (5)-(8), the dimension of the ac inductor and its associated losses can be calculated and evaluated.

III. CASE STUDY

Nowadays, the DAB converter and the isolated full-bridge bidirectional boost converter (IFB³C) are the two main topologies which can provide galvanic isolation and voltage conversion. In [15]-[17], an IFB³C is investigated to overcome the challenges such as achieving high efficiency over a wide input voltage range for reversible solid oxide electrolyser cells (SOECs). Hereby, in this paper, the same specifications for the IFB³C, as listed in Table I, are used to design the DAB converter and also its ac inductor.

A. Core loss of the ac inductor

Based on the parameters listed in Table I and using (7) and (8), the time-average core loss is:

$$P_v = \frac{2 \cdot k_i}{T} \left[\left| \frac{V_1 + nV_2}{N_L \cdot A_c} \right|^\alpha \left(\frac{V_1 + nV_2}{N_L \cdot A_c} \cdot \frac{\delta}{\omega} \right)^{\beta-\alpha} \cdot \frac{\delta}{\omega} + \left| \frac{V_1 - nV_2}{N_L \cdot A_c} \right|^\alpha \left(\frac{V_1 - nV_2}{N_L \cdot A_c} \cdot (\frac{T}{2} - \frac{\delta}{\omega}) \right)^{\beta-\alpha} \left(\frac{T}{2} - \frac{\delta}{\omega} \right) \right] \quad (9)$$

Adopting N87 E64/10/50 EI ferrite core in this study, the core loss can be evaluated by both GSE and iGSE methods. When $V_1 = nV_2 = 55$ V, the core loss as a function of the phase shift angle δ, with varying number of turns, is plotted in Fig. 4. Apparently, the larger number of turns, the higher core loss is.

TABLE I. SPECIFICATIONS FOR DC-DC CONVERTER DESIGN

Low voltage (LV) side	30-80 VDC
High voltage side	700-800 VDC
Rated power	3 kW
Transformer turn ratio, n	1:14
Ac inductance (LV side)	3 uH
Switching frequency, f_s	100 kHz
Core material	N87 Ferrite
K_c	16.9
α	1.25
β	2.35
B_{peak}	0.49 T

Additionally, the core loss is underestimated by using GSE method, in particular when the phase shift angle is close to $\pi/2$.

Moreover, the core loss is calculated with V_1 of 60 V and V_2 of 700 V by iGSE, which means $V_1 \neq nV_2$ but $V_1 + nV_2$ is kept the same, and hereby the result is plotted in Fig. 5. It can be seen that: due to the second term in the right hand side of (9) the core loss will be even worse; and a large diversity locates in the range with a smaller δ. The core loss as a function of δ and voltage of V_2 can be plotted in Fig. 6. When δ is smaller than $\pi/4$, which is the recommended operating range, the core loss will be minimized under the condition of $V_1 = nV_2$. With lower input voltage, for instance $V_1 = 30$V, the core loss can be even smaller than that of $V_1 = nV_2$ when the phase-shift angle is larger. When V_1 is the maximum input voltage of 80V, due to the voltage mismatch compared to nV_2, the core loss increases significantly.

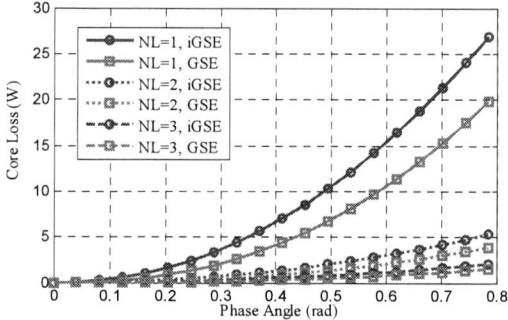

Fig. 4: Core loss as a function of δ with $V_1 = nV_2$.

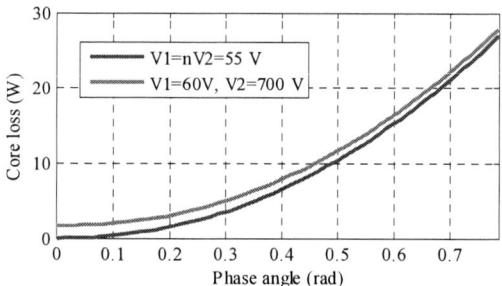

Fig. 5: Core loss with $V_1 \neq nV_2$.

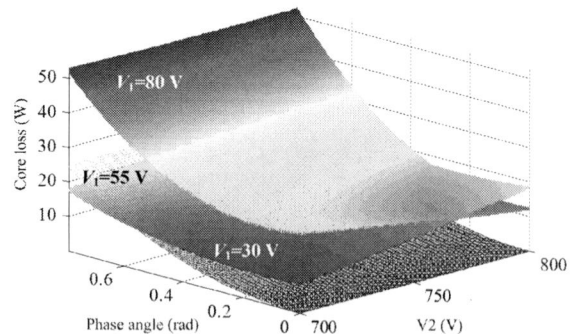

Fig. 6: Core loss as a function of δ and V_2.

B. Windling loss of the ac inductor

The ac inductor here has no dc current offset so that is diverse from inductors used as a dc output filter in Buck or Boost converters. Therefore, reduction of ac resistance is the most important issue for obtaining low winding loss. Based on the selected planar core, six winding arrangements defined in Table II respectively, , are compared as illustrated in Fig. 7, in which the yellow bars represent windings, and the ferrite cores are gaped only on the center legs. With Maxwell 2D, these six winding configurations are evaluated and compared in terms of ac resistance. Case 1 to Case 4, and Case 5 and Case 6, can be categorized into two groups, depending on the number of turns.

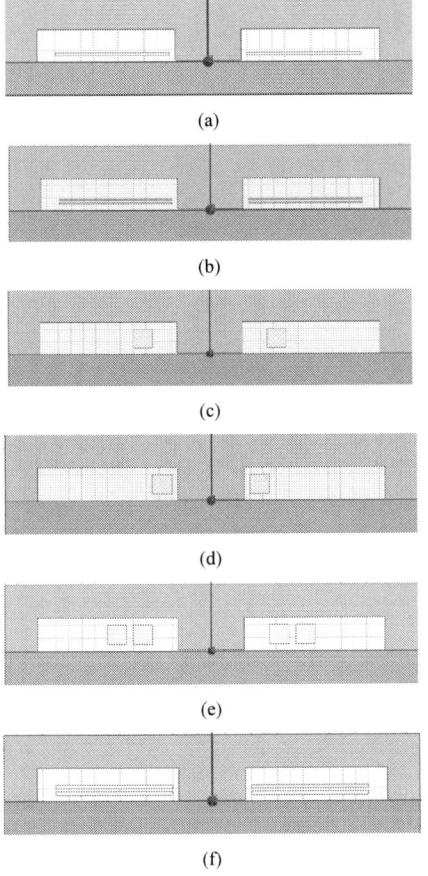

Fig. 7: Winding arrangements. (a) Case 1, (b) Case 2, (c) Case 3, (d) Case 4, (e) Case 5, and (f) Case 6.

TABLE II. WINDING ARRANGEMENT

Case nr.	Winding Configures
Case 1	Single rectangular flat wire
Case 2	Paralleled rectangular flat wires with same cross-section as that in case 1
Case 3	Single square wire with same cross-section as that in case 1 and away from the air-gap
Case 4	Single square wire with same cross-section as that in case 1 and close to the air-gap
Case 5	Two-turn square wire
Case 6	Two-turn rectangular flat wire

In Fig. 8, where only the left parts of the ac inductors are shown due to the symmetrical structure, the FEM analysis results are presented. It can be seen that with a single winding, there are no proximity effect, and accordingly the ac resistance is much smaller than that in the cases of two-turn winding configuration. But the price that must pay for using this single turn inductor with low ac resistance is a relatively large core cross-section area which is needed to avoid core saturation.

Moreover, the fringing flux, due to the air gap in the center leg, can affect the winding loss badly, so the winding must be put at least twice of gap length away from the gap. Finally, the ratios of R_{ac}/R_{dc} are compared in Fig. 9.

C. Ac inductor design

With the winding cross-section area of 10 mm², the winding losses in the six cases are given in Fig. 10. Comparing to the core loss shown in Fig.5, the cases with a single turn structure can fulfill the optimization criteria, i.e. the core loss is approximately equal to the winding loss, and hereby an ac inductor with only one turn can be designed. If a larger inductance is needed, more ferrite EI cores can be connected in cascade to obtain a larger effective core area, which results in a larger inductance and can avoid saturation effectively.

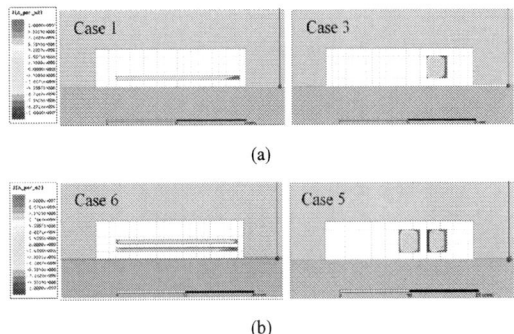

(a)

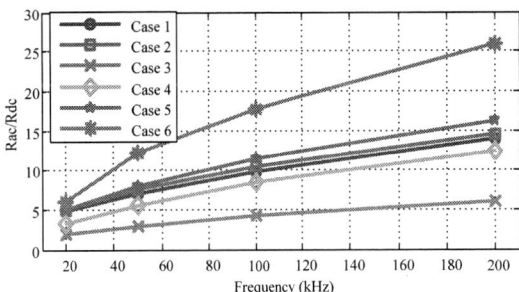

(b)

Fig. 8: FEM simulation. (a) Single-turn winding structure, and (b) two-turn winding structure.

Fig. 9: Comparison of R_{ac}/R_{dc}.

IV. EXPERIMENTAL RESULTS

To verify the proposed ac inductor design, a 3 kW DAB converter has been prototyped. The typical waveforms, i.e. the primary and secondary ac voltages and the ac inductor current of the converter operating at 100 kHz, are given in Fig. 11.

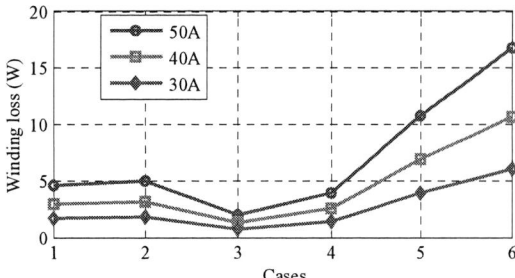

Fig. 10: Winding losses.

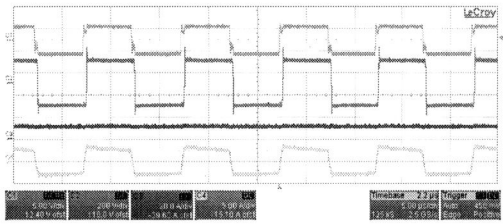

Fig. 11: Experimental waveforms of the DAB converter. CH1: the primary side ac voltage, CH2: the secondary ac voltage, CH3: the input dc current, and CH4: the ac current at the secondary or high voltage side.

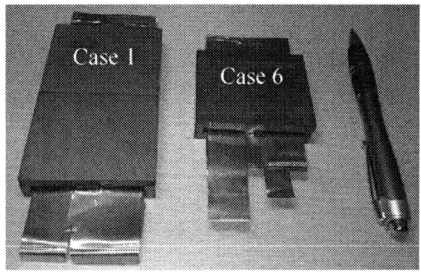

Fig. 12: Ac Inductors of Case 1 and Case 6.

For the purpose of performance comparison, two ac inductors based on the Case1 and Case 6 respectively are built and their photographs are given in Fig. 12, following the parameters listed in Table III. By using Agilent 4294A Precision Impedance Analyzer, the constructed ac inductors are measured in terms of inductance and ac resistance. For the Case 1, the measured inductance and ac resistance at 100 kHz are shown in Fig. 13 (a), in which the values are highlighted with a square in red line. Moreover, the ac resistances of these two inductors at various switching frequencies, i.e. 50 kHz, 100 kHz and 200 kHz are plotted in Fig. 13(b). It can be seen that the Case 1 has smaller ac resistance than Case 6 over the whole frequency band.

TABLE III. AC INDUCTOR DESIGN

	Case 1	Case 6
Inductance	3.2 µH	3.2 µH
Core	Planar EI E64/50/10-3C90	
Air gap	0.19 mm	0.46 mm
Number of turns	1	2

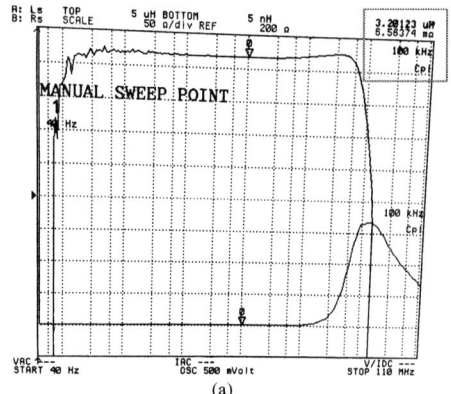

(a)

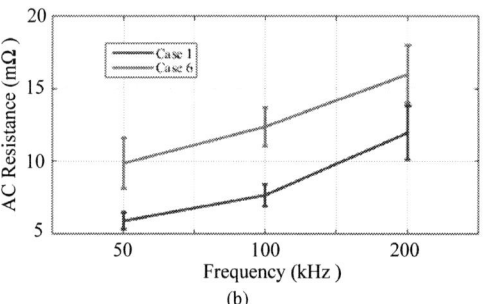

(b)

Fig. 13: Inductance and ac resistance measurement. (a) Measured inductance and ac resistance of Case 1 ac inductor at 100 kHz, and (b) ac resistance comparison of Case 1 and Case 6 at various operating frequencies.

V. CONCLUSION

This paper studied the design of ac inductors employed in the widely used DAB converters. The design difficulty for ac inductors in DAB converters, which operate with a wide input voltage range, has been pointed out and analyzed. Core loss is calculated by the iGSE approach, and the core loss due to input and output voltage mismatch is evaluated. Six diverse winding arrangements and their design are studied and compared with a FEM analysis by Maxwell 2D, and it is found that the ac inductors with a single-turn winding configuration can achieve lowest total losses.

REFERENCES

[1] L. Zhu "A novel soft-commutating isolated boost full-bridge ZVS-PWM DC-DC converter for bidirectional high power applications", *IEEE Trans. Power Electron.*, vol. 21, no. 2, pp.422 -429, 2006

[2] Z. Zhang, O. C. Thomsen, M. A. E. Andersen, "Optimal design of a push-pull-forward half-bridge (PPFHB) bidirectional DC–DC converter with variable input voltage," *IEEE Trans. on Ind. Electron.*, vol.59, no. 7, pp. 2761-2771, Jul. 2012.

[3] S. Jalbrzykowski , A. Bogdan and T. Citko "A dula full-bridge resonant class-E bidirectional dc-dc converter", *IEEE Trans. Ind. Electron.*, vol. 58, no. 9, pp.3879-3883, 2011.

[4] R. Pittini, Z. Zhang, and M. A. E. Andersen, "Isolated full bridge boost dc-dc converter designed for bidirectional operation of fuel

cells/electrolyzer cells in grid-tie applications," *EPE'13-ECCE Europe*, 2013.

[5] Pan Xuewei and Rathore, A.K. "Novel interleaved bidirectional snubberless soft-switching current-fed full-Bridge voltage doubler for fuel-cell vehicles", *IEEE Trans. Power Electron.*, vol. 28, no. 12, pp. 5535 - 5546, Dec. 2013.

[6] Z. Zhang, Z. Ouyang, O. C. Thomsen, M. A. E. Andersen, "Analysis and design of a bidirectional isolated dc–dc converter for fuel cells and supercapacitors hybrid system," *IEEE Trans. Power Electron.*, vol. 27, no. 2, pp. 848 - 859, Feb. 2012.

[7] R. W. De Doncker, D. M. Divan, and M. H. Kheraluwala, "A three-phase soft-switched high-power desity dc/dc converter for high power applications," *IEEE Transactions on Industry Application*, vol. 27, no. 1, pp.63-67, 1991.

[8] S. Inoue and H. Akagi, "A bidirectional dc–dc converter for an energy storage system with galvanic isolation," *IEEE Transactions on Power Electronics*, vol. 22, no. 6, pp. 2299-2306, 2007.

[9] J. Ge, Z. Zhao, L. Yuan and T. Lu, "Energy feed-forward and direct feed-forward control for solid-state transformer," *IEEE Trans. Power Electron.*, vol. 30, no. 8, pp.4042-4047, Aug. 2015.

[10] G. G. Oggier , G. O. Garcia and A. R. Oliva "Modulation strategy to operate the dual active bridge dc-dc converter under soft switching in the whole operating range", *IEEE Trans. Power Electron.*, vol. 26, no. 4, pp.1228-1236, Apr. 2011.

[11] F. Krismer and J. W. Kolar "Efficiency-optimized high current dual active bridge converter for automotive applications", *IEEE Trans. Ind. Electron.*, vol. 59, no. 7, pp.2745 -2760 2012

[12] B. Zhao, Q. Song and W. Liu "Efficiency characterization and optimization of isolated bidirectional dc-dc converter based on dual-phase-shift control for dc distribution application," *IEEE Trans. Power Electron.*, vol. 28, no. 4, pp.1711-1727, Apr. 2013.

[13] J. Muhlethaler, J. Biela, J. W. Kolar and A. Ecklebe, "Improved core-loss calculation for magnetic components employed in power electronics systems," *IEEE Trans. Power Electron.* vol.27, no.2, pp.964-973, Feb. 2012.

[14] W. G. Hurley and W. H. Wolfle, "Transformers and inductors for power electronics," Wiley, 2013.

[15] R. Pittini, "High efficiency reversible fuel cell power converter," *PhD dissertation*, Techninical University of Denmark, 2015.

[16] R. Pittini, Z. Zhang, and M. A. E. Andersen, "High current planar magnetics for high efficiency bidirectional dc-dc converters for fuel cell applications," in *the 29th Annual IEEE Applied Power Electronics Conference and Exposition (APEC)*, pp. 2641-2648, Fort Worth, TX, USA, 2014.

[17] R. Pittini, M. C. Mira, Z Zhang, A Knott, and M. A. E. Andersen, "Analysis and comparison based on component stress factor of dual active bridge and isolated full bridge boost converters for bidirectional fuel cells systems," in *Power Electronics and Application Conference and Exposition (PEAC)*, pp. 1026-1031, 2014.

A Comprehensive Assessment of PM Motor Topology Impact on Magnet Defect Fault Signatures

Mohsen Zafarani, Taner Goktas, Bilal Akin

Department of Electrical Engineering, University of Texas at Dallas, Dallas, TX 75080 USA
mohsen.zafarani@utdallas.edu, taner.goktas@utdallas.edu, bilal.akin@utdallas.edu

Abstract— This paper presents a detailed study on topology dependent magnet defect fault signatures in permanent magnet (PM) motors. A new mathematical approach is introduced to analyze the fault signature behaviors in PM motors with different design specs such as combinations including winding types, number and location of stator coils / magnet defects etc. Thanks to the flexible features of the proposed method, it is possible to analyze various winding configurations (including full, short and fractional pitch, single and double layer, delta and star connection) and combination of multiple magnet defect effects on the motor variables' frequency spectrum. It is shown that depending on the motor topology, the corresponding spectrum can be different from others and the generalized fault models may not be valid. In addition, current or back-emf spectrums do not include any fault related harmonics in some cases or the fault components showing up in the stator back-emf do not show up in the stator current. Comparative 2-D Finite Element (FE) simulation and experimental results justify the theoretical magnet defect fault analysis and show the efficacy of proposed approach.

Keywords—Magnetic equivalent circuit, Permanent magnet motor, Magnet defect fault, Broken magnet, PM motor topology

I. INTRODUCTION

Permanent magnet (PM) motors are one of the most attractive ones among others as they are efficient, have higher power density and accurate dynamic performance [1]. Due to this growth in the applications, condition monitoring of such motors in industry has been becoming an emerging technology to improve the overall drive reliability [1-7]. Several fault types have been analyzed in the literatures and various techniques are proposed to detect the faults such as: vibration monitoring [2], search coils [3], stator current sequence components [4], [5], motor current signature analysis [6], [7], residual stator current which is the difference between the actual and estimated current [8], [9], instantaneous power [10], [11], time delay embedding [12], stator inductances [13], [14] are some proper ones. Though these kinds of studies significantly contribute to fault diagnosis, the reasons behind the differences between the current/back-emf spectrums of various motor have not been investigated. This is because most of the studies focus on a specific motor, that's why the results are not generic and do not cover all types of surface mount PM motor topologies. In this study, a mathematical approach is proposed to analyze the effect of magnet defect fault on motor variables. Through the theoretical and experimental

tests, it is justified that the proposed approach can provide details regarding topology dependencies and yield consistent results with finite element tools and experiments. Motor design specs including different number of pole and slot, numbers of defected magnet and different winding connections are discussed. 2-D finite element method is used to compare the findings and analyze the effect of fault. An experimental setup has been prepared to validate the findings and simulation results.

II. ANALYSIS OF MAGNET FAULTS SIGNATURES AND TOPOLOGY DEPENDENCIES

Unless evenly distributed throughout the rotor, the magnet defects and demagnetization issues create the well-known fault signature pattern given in (1).

$$f_{f,k} = \left(1 \pm \frac{k}{p/2}\right)f_s = kf_m \quad k = 1, 2, 3, \ldots \quad (1)$$

These magnetic rotor faults introduce asymmetry on the rotor flux distribution cycling at every mechanical period. This asymmetry is reflected on the back-emf and current waveforms and creates signatures in the associated frequency spectrums at mechanical frequency. However, the frequency spectrum of

TABLE I. DIFFERENT NUMBER OF POLE, SLOT AND WINDING CONFIGURATIONS STUDIED IN THIS PAPER

N_s	p	Layer	N_s/p	CP	Pitch
9	6	Double	1.5	2	Fractional
9	8	Double	1.125	1	Fractional
12	8	Double	1.5	2	Fractional
12	8	Single	1.5	2	Fractional
12	10	Double	1.2	1	Fractional
12	10	Single	1.2	1	Fractional
18	6	Double	3	3	Full
18	6	Single	3	3	Full
18	12	Double	1.5	2	Fractional
36	6	Double	6	6	Full
36	6	Single	6	6	Full

back-emf and current waveforms can be different depending in machine design schemes, number and location of the faulty magnets. Even the identical faults exhibit different spectrums with some missing or additional components depending on the motor structure. In this section, these cases are studied elaborately so that one can predict the machine specific spectrum content of the faulty motor and develop diagnosis algorithms based on that to minimize false alarms. In order to analyze the topology dependency of magnet faults, a number of 3-phase permanent magnet synchronous motors (PMSMs) with different number of pole and slot are selected for various theoretical case studies. Different types of winding configurations are taken into account including full, short and fractional coil pitch, single and double layer to investigate the effect of all these on the fault signatures. The motors analyzed with different pole, slot and winding configuration combinations are given in Table I. The method presented in this paper is based on the partitioned magnetic equivalent circuit (MEC).

III. PARTITIONED MEC IMPLEMENTATION

The magnetic equivalent circuit calculations are meant to obtain electromagnetic system variables theoretically. In order to understand the effects of magnet faults and dynamics of the corresponding signatures, basic MEC calculations are reorganized and resorted. This enables comprehensive understanding of each stator coils and rotor magnets interaction and its reflection on the induced back-emf voltage. This approach is very practical and useful to characterize any magnet defect in various PM motors topologies. In addition, it is used to investigate the electromagnetic reflections of multiple faulty magnets on the stator flux linkage and back emf, which is important to discern variable spectrums of motors with multiple magnet faults that are located at various rotor positions. First, the stator flux linkages due to each individual magnet are calculated separately using MEC tools and then the overall flux linkage is obtained by superposing these components. In order to elaborate and justify the implementation, two PM motors including a permanent magnet surface mounted (PMSM) motor with 9-slot/8-pole, double layer winding and an interior permanent magnet (IPM) with 36-slot/6-pole, single layer winding are used. The specifications of these two motors are summarized in Table II and the magnetic equivalent circuit of the PMSM is shown in Fig. 1. In the following discussions, just one magnet and one coil are modeled before superposition. The more detail of the MEC method is presented in [16]-[18] and the results are given in (2).

$$
B_g = \begin{cases} B_{g1} & \theta_r - \dfrac{\theta_{PM}}{2} \leq \theta \leq \theta_r + \dfrac{\theta_{PM}}{2} \\ B_{g2} & other \end{cases} \tag{2}
$$

$$
B_{g1} = \frac{\left(1 - \left(\theta_{PM} / 2\pi\right)\right) B_r}{1 + \mu_{rPM}\left(g / l_{PM}\right) + \left(\mu_{rPM} - 1\right)\left(\theta_{PM} / 2\pi\right)} \tag{3}
$$

TABLE II. SPECIFICATIONS OF THE TEST MOTORS

	PMSM		IPM	
Rated voltage	133	V	360	V
Rated current	2.7	A	1.72	A
Rated torque	1.27	Nm	3.95	Nm
Rated speed	3000	rpm	1800	rpm
Number of poles	8		6	
Number of stator slots	9		36	
Number of stator turn	71		68	
R_s	2.25	Ω	4.28	Ω
Slot/Pole	9slots/8poles		36slots/6poles	

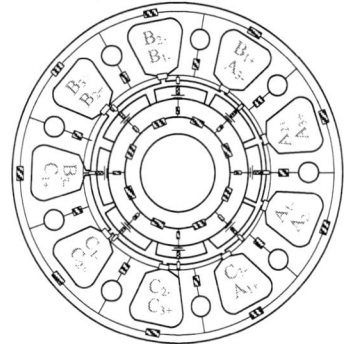

Fig. 1 Magnetic equivalent circuit of the test motor

$$
B_{g2} = \frac{\left(\theta_{PM} / 2\pi\right) B_r}{1 + \mu_{rPM}\left(g / l_{PM}\right) + \left(\mu_{rPM} - 1\right)\left(\theta_{PM} / 2\pi\right)} \tag{4}
$$

where B_{g1} and B_{g2} are the flux densities in the two air-gap sections and B_g is the function of the air-gap flux density with respect to the stator space angle θ, θ_r is rotor position, θ_{PM}, l_{PM}, μ_{rPM} and B_r are PM angle dimension, length, relative permeability and residual flux density, respectively. The flux linkage of each coil given in (5) represents the flux which passes the air-gap and links that particular coil.

$$
\lambda_{coil}(\theta_r) = NL_{st}R_s \int_{\theta=-\frac{\theta_t}{2}}^{\theta=\frac{\theta_t}{2}} B_g(\theta,\theta_r)d\theta \tag{5}
$$

where λ_{coil} and N are the flux linkage and the number of turns of each coil.

A. Back-Emf Calculation and Broken Magnet Fault Modeling

Back-emf in PMSMs is the induced voltage across the stator phases due to relative change in rotor PM flux linking the stator coils. Mathematical definition of back-emf which is given in (6) as the rate of the flux linkage change in stator coils when the terminals are disconnected and the rotor is run at constant speed.

$$e_a = \frac{d\lambda_a}{dt} = \frac{d\theta_r}{dt}\frac{d\lambda_a}{d\theta_r} = \omega_r\frac{d\lambda_a}{d\theta_r} \tag{6}$$

where e_a is the induced voltage across the stator coil or back-emf, λ_a is stator flux linkage, ω_r and θ_r are the electrical rotor speed and position respectively. As any fault on the rotor affects the flux density in the air-gap and is reflected to the back-emf, it is an ideal variable to analyze the rotor or stator faults. According to (6), when the motor operates at constant speed, any signature introduced into flux linkage will be reflected to the back-emf at the same frequency with $90°$ phase shift. In order to investigate the effect of magnet defect fault, the equations derived in [16] is used to analyze the effect of motor topology on fault signatures. In this method, the induced voltage in each stator phase is calculated through (7).

$$e_a = \sum_{n=-\infty}^{+\infty} k_{mn}k_{san}E_n e^{jn\theta_m} \tag{7}$$

where n is the number of mechanical component, k_{san} and k_{mn} are coil coefficient and magnet coefficient for the nth mechanical harmonic which are calculated through (8) and (9), θ_m is the mechanical rotor position. In this equation E_n is the amplitude of nth mechanical component of the induced voltage in one coil of phase-a due to the first only one rotor magnet neglecting the effect of other coils and magnets, D_z is the coil direction factor which is defined based on the magnetic field vectors. If the magnetic field vector of the zth coil is in the same direction with the first coil, then it is 1, otherwise it is -1. .

$$k_{san} = \sum_{z=1}^{N_{coil}} D_z e^{jn(z-1)\theta_z} \tag{8}$$

$$k_{mn} = \sum_{i=1}^{p}(-1)^{i-1}e^{jn(i-1)\frac{360}{p}} \tag{9}$$

The induced voltage in other phases is also calculated through (7). The only difference between different phases is coil coefficient. In this case, all three coils in the phase-b is located at $120°$ beyond the corresponding coil in the phase-a. This difference is inserted by changing just coil coefficient in other phases which is shown in (10).

$$\begin{cases} k_{sbn} = \sum_{z=1}^{N_{coil}} D_z e^{jn(z-1)\theta_z} e^{jn120} \\ k_{scn} = \sum_{z=1}^{N_{coil}} D_z e^{jn(z-1)\theta_z} e^{jn240} \end{cases} \tag{10}$$

This coefficient represents the phase difference between two stator phases for different harmonics. In the following sections, this coefficient will be used to explain the difference between stator winding connections.

B. PM Fault Modelling

Magnet faults including demagnetization or broken magnet cause reduction in PM flux and consequently affect the stator phase back emf. Since this phenomenon corresponds to the rotor magnets, it affects the magnet coefficient k_{mn} in (9). In this study, fault severity factor is inserted into the magnet coefficient to model magnet fault. This parameter shows the condition of each magnet which becomes zero when the magnet is totally defected, one for healthy ones and fractional number in between. The magnet coefficient given in (9) is re-written as (11) to point out the fault severity.

$$k_{mn} = \sum_{i=1}^{p}(-1)^{i-1}k_{fi}e^{jn(i-1)\frac{360}{p}} \tag{11}$$

where k_{fi} $\square$ [0-1] is the fault severity index of ith rotor magnet. In the next sections, this coefficient is used to investigate the effect of location and number of faulty magnets on the motor variables. The multiple of magnetic and coil coefficients determine the frequency content of the back-emf waveform. It also calculates the phase different between the stator phases for various harmonics. Based on these two characteristics, this multiplication is called construction filter and is shown with R_{an} for phase-a (12).

$$R_{an} = k_{mn}k_{san} \tag{12}$$

IV. DIFFERENT NUMBER OF POLE AND SLOT

Table III shows the construction filter magnitude for different fault harmonics in the case of double layer winding and different number of pole and slot combinations.

TABLE III. NORMALIZED CONSTRUCTION FILTER MAGNITUDE FOR DIFFERENT NUMBER OF POLES AND SLOTS (DOUBLE LAYER WINDING)

		$|R_{an}|$											
N_s	P	1/(p/2)	2/(p/2)	3/(p/2)	4/(p/2)	5/(p/2)	6/(p/2)	7/(p/2)	8/(p/2)	9/(p/2)	10/(p/2)	11/(p/2)	12/(p/2)
9	6	0	0	1	0	0	0.2	0	0	1	0	0	0.2
9	8	0.026	0.032	0.099	1	0.142	0.099	0.032	0.026	0.049	0.026	0.032	0.69
12	8	0	0	0	1	0	0	0	0.142	0	0	0	1
12	10	0.029	0	0.081	0	1	0	0.111	0	0.081	0	0.03	0
18	6	0	0	1	0	0	0	0	0	1	0	0	0
18	12	0	0	0	0	0	1	0	0	0	0	0	0.091
24	8	0	0	0	1	0	0	0	1	0	0	0	1
36	6	0	0	1	0	0	0	0	0	0.732	0	0	0

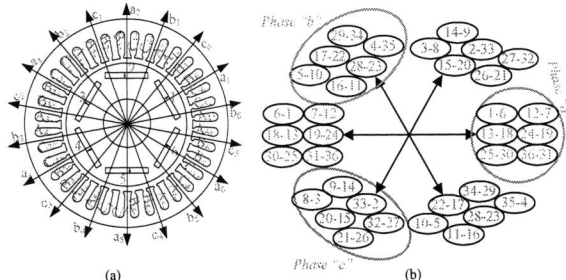

Fig. 2. A three phase, 36-slot/6-pole double layer winding motor a) Magnetic field vector b) winding layout

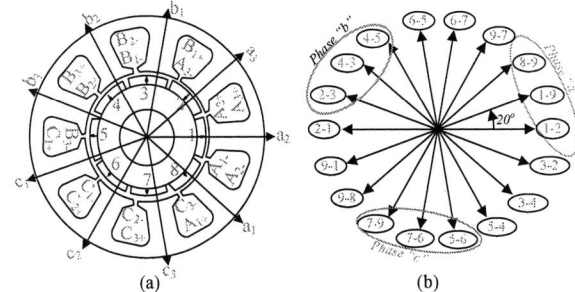

Fig. 3. A three phase, 9-slot/8-pole double layer winding motor a) Magnetic field vector b) winding layout

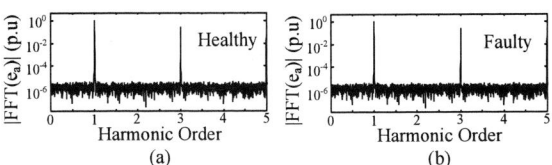

Fig. 4. Stator back-emf spectrum at rated speed for the 36-slot/6-pole IPM motor MEC results a) Healthy b) One magnet defect

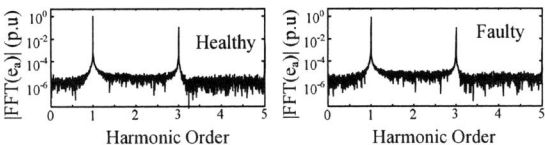

Fig. 5. Stator back-emf spectrum at rated speed for a three phase, 36-slot/6-pole single layer winding, FE results a) Healthy b) One magnet cracked

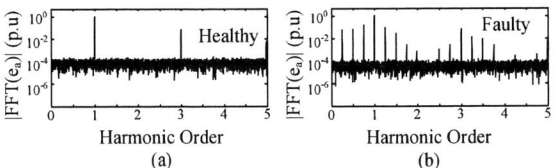

Fig. 6. Stator back-emf spectrum at rated speed for the 9-slot/8-pole PMSM motor, MEC results a) Healthy b) One magnet defect

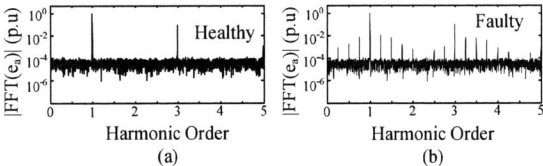

Fig. 7. Stator back-emf spectrum at rated speed for the 9-slot/8-pole PMSM motor, FE results a) Healthy b) One magnet defect

As shown here, in the cases where the number of slot and pole ratio is multiple of 1.5, the fault related signatures cannot be observed. When a motor has a three phase double layer winding and the number of slot/pole ratio is multiple of 1.5, all coils in the same phase are mechanically located in the space such that mechanical phase shift with respect to each other leads to cancellation of all fault harmonics. In order to verify these findings, the two motors with specifications given in Table II are investigated through the proposed MEC method. The magnetic field vectors and winding layout of these two motors are shown in Fig. 2 and Fig. 3. Since for the IPM motor the ratio of number of pole and slot is multiple of 3, it is expected all coils in the same phase do not have any phase difference, electrically. Physically, all coils in one phase are located in the space in a way that they have 120° mechanical degree respect to each other (Fig. 2b). Therefore, the coil coefficient for all mechanical components is zero except than those which are multiple of p/2 which are multiple of electrical frequency. Similarly, for single layer winding, when the number of slot/pole ratio is multiple of 3 all fault harmonics are canceled due to the coils space locations. The phase back-emf voltage of the IPM obtained

through MEC and FE tools for healthy and faulty conditions (when 25% of one magnet is removed) are shown in Fig. 4 and Fig. 5 respectively. As the ratio of pole and slot is multiple of 3, it is expected not to observe any fault related signatures in the spectrum even if there is a broken magnet. These two figures also show that the theoretical findings summarized in Table III match with the simulation results. In order to verify the theoretical findings, partitioned MEC and finite element tools are used to model the test motors given in Table II. Fig. 4 and Fig. 5 depicted to compare the PMSM phase back emf for healthy and faulty conditions (when 7% of one magnet is removed) using the proposed MEC approach and FE tool, respectively. As shown in these figures, in the healthy case, back-emf spectrum contains only the fundamental component and odd multiples of fundamental electrical frequency generated by the flattened flux linkage waveform, whereas, in the faulty case, fault harmonics at the multiples of mechanical frequency show up in the spectrum as pointed out in Table III. These figures show that the theoretical findings summarized in Table III match with the simulation results and the proposed method can estimate the behavior of a specific faulty motor accurately.

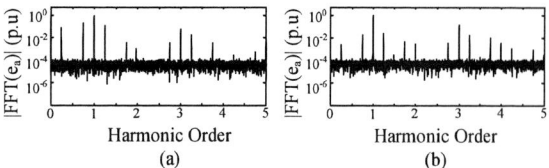

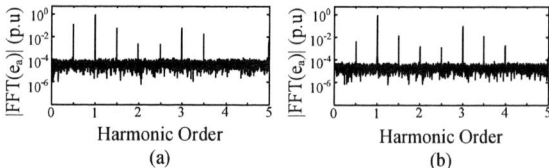

Fig. 8. Phase "a" back emf spectrum when magnet number 1 and 3 of the 9-slot/8-pole PMSM are defected a) MEC result b) FE result

Fig. 9. Phase "a" back emf spectrum when magnet number 1 and 5 of the 9-slot/8-pole PMSM are defected a) MEC result b) FE result

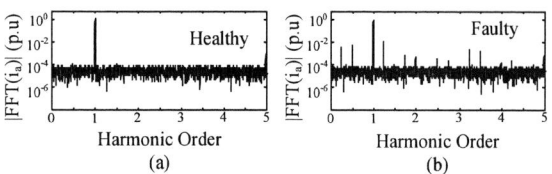

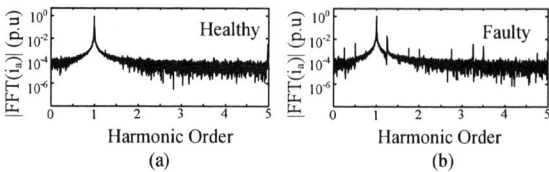

Fig. 10. Stator current spectrum at rated speed and rated torque for a 9/8 PM motor, FE results a)Healthy b) One magnet cracked

Fig. 11. Stator current spectrum at rated speed and rated torque for a 9/8 PM motor, experimental results a) Healthy b) One magnet cracked

V. NUMBER AND LOCATION OF MAGNET DEFECTS

The number and location of the magnet defects affect the signature content in the motor variables which can easily be analyzed through partitioned MEC. During the analysis multiple defects, it is assumed that the defects are identical to each other. When there is only one defect or partially demagnetized magnet, all components defined in (1) appear in the back-emf spectrum. In the case of two or more defected magnets, depending on the location of the magnets, the spectrum can be different. This situation is observed when the severities of fault in all faulty magnets are exactly the same. This can be explained through magnet coefficient given in (26). Fig. 8 compares the simulation and MEC results of a defected magnet fault when the 1st and 3rd magnets are broken. As it is shown in Fig. 8, in this scenario the odd multiples of 0.5th harmonic do not show up in the spectrum. Similarly, Fig. 9 shows the same comparison when the 1st and 5th magnets are defected. In this case, the odd multiple of 0.25th harmonic do not show up in the back emf spectrum.

VI. WINDING CONFIGURATION

The fault related harmonics created by magnet faults in the back-emf are the source of the harmonics in the stator phase current and both variables' spectrums should be very similar to each other. However, there are some components which show up in the back emf but not in the stator current. This is partially due to the winding connections where it is essential to understand the phase shift between the fault signatures in any motor phases. Fig. 10 shows the stator current spectrum for healthy and faulty motor with 9 slots and 8 poles. The stator winding configuration used in the simulations is star connection with floating neutral point. In this configuration, the sum of three phase current is zero as there is no path for other harmonics to flow. As shown in the simulation results, in the healthy condition the stator current does not have any harmonic except than the fundamental

component. In the case of faulty motor, all harmonics except than 0.75th harmonic and multiples of it show up as these components are in the same phase and can't flow in star winding configuration. Fig. 11 shows the experimental results for stator current at the same conditions. In this case, even in the healthy case there are some harmonics due to inherent asymmetries and unbalances caused by manufacturing imperfections. But the behavior of faulty components is the same as the simulation results.

VII. EXPERIMENTAL SETUP

In order to justify the analytical and simulation results, one rotor magnet is cut vertically using a fine cutter and is tested under different speed and torque conditions. For this purpose, an experimental setup is prepared which is shown in Fig. 12. The setup has a surface mounted PM motor which is coupled with a hysteresis dynamometer to apply adjustable load and measuring shaft torque and speed. To adjust the speed and torque, the

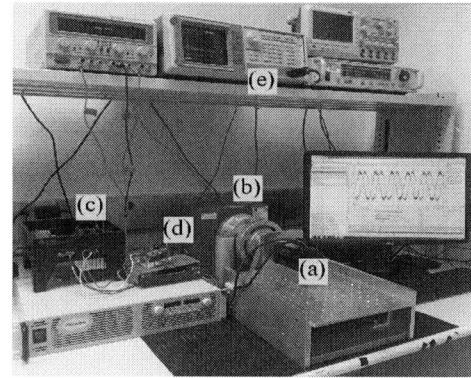

Fig. 12 Experimental setup a) PMSM b) Dynamometer c) Motor drive d)Data acquisition card e) Spectrum analyzer

dynamometer has controller which is able to set these two variables using PID controller. A miniature bellow coupling is used to suppress vibrations on the motor shaft. To track the fault signatures behavior online, an FFT spectrum analyzer is used. An NI USB-6351 data acquisition with 1MS/s sampling rate and 16 bit ADC resolutions is used to collect the current data for post processing.

VIII. CONCLUSIONS

In this paper, the effect of PM motor topology on the magnet defect fault signatures is investigated through modified MEC which can individually calculate the effect of coil and defected magnet on the back emf. Several motors with different number of pole and slot combinations are analyzed through this method. Design parameters such as full, fractional and short pitch winding, the number and location of magnet defect and stator winding connection are also included for a complete analysis. It is shown that in the case of single layer winding, when the number of slot and pole ratio is multiple of 3, magnet defect fault does not create any additional signature in the stator back emf and current. The same case is observed in the motors with double layer when the ratio is multiple of 1.5. It is also shown that the fault signatures have dependency on the stator windings connection. Depending on the motor topology; some fault related harmonics do not show up in the current while they do appear in the back emf. Two different motor with different number of slot and pole are simulated using FE analysis and the results are compared with the MEC results. Also, a surface mounted PM motor is experimentally tested to prove the accuracy of the proposed method.

References

[1] Romeral, L.; Urresty, J.C.; Riba Ruiz, J.-R.; Garcia Espinosa, A., "Modeling of Surface-Mounted Permanent Magnet Synchronous Motors With Stator Winding Interturn Faults," Industrial Electronics, IEEE Transactions on , vol.58, no.5, pp.1576,1585, May 2011

[2] Tao Sun; Ji-Min Kim; Geun-Ho Lee; Jung-Pyo Hong; Myung-Ryul Choi, "Effect of Pole and Slot Combination on Noise and Vibration in Permanent Magnet Synchronous Motor," Magnetics, IEEE Transactions on , vol.47, no.5, pp.1038,1041, May 2011

[3] Torregrossa, D.; Khoobroo, A.; Fahimi, B., "Prediction of Acoustic Noise and Torque Pulsation in PM Synchronous Machines With Static Eccentricity and Partial Demagnetization Using Field Reconstruction Method," *Industrial Electronics, IEEE Transactions on*, vol.59, no.2, pp.934,944, Feb. 2012

[4] Capolino, G.-A.; Antonino-Daviu, J.A.; Riera-Guasp, M., "Modern Diagnostics Techniques for Electrical Machines, Power Electronics, and Drives,"*Industrial Electronics, IEEE Transactions on*, vol.62, no.3, pp.1738,1745, March 2015

[5] Cruz, S.M.A.; Mendes, A.M.S.; Cardoso, A.J.M., "A New Fault Diagnosis Method and a Fault-Tolerant Switching Strategy for Matrix Converters Operating With Optimum Alesina-Venturini Modulation," *Industrial Electronics, IEEE Transactions on*, vol.59, no.1, pp.269,280, Jan. 2012

[6] Estima, J.O.; Marques Cardoso, A.J., "A New Algorithm for Real-Time Multiple Open-Circuit Fault Diagnosis in Voltage-Fed PWM Motor Drives by the Reference Current Errors," *Industrial Electronics, IEEE Transactions on* , vol.60, no.8, pp.3496,3505, Aug. 2013

[7] Fournier, E.; Picot, A.; Regnier, J.; Yamdeu, M.T.; Andrejak, J.; Maussion, P., "Current-Based Detection of Mechanical Unbalance in an Induction Machine Using Spectral Kurtosis With Reference,"*Industrial Electronics, IEEE Transactions on*, vol.62, no.3, pp.1879,1887, March 2015

[8] Raminosoa, T.; Gerada, C.; Galea, M., "Design Considerations for a Fault-Tolerant Flux-Switching Permanent-Magnet Machine,"*Industrial Electronics, IEEE Transactions* on, vol.58, no.7, pp.2818,2825, July 2011

[9] Choi, S.; Pazouki, E.; Baek, J.; Bahrami, H.R., "Iterative Condition Monitoring and Fault Diagnosis Scheme of Electric Motor for Harsh Industrial Application,"*Industrial Electronics, IEEE Transactions* on, vol.62, no.3, pp.1760,1769, March 2015

[10] Yao Duan; Toliyat, H., "A review of condition monitoring and fault diagnosis for permanent magnet machines," Power and Energy Society General Meeting, 2012 IEEE , vol., no., pp.1,4, 22-26 July 2012

[11] Arthur, N.; Penman, J., "Induction machine condition monitoring with higher order spectra,"*Industrial Electronics, IEEE Transactions on*, vol.47, no.5, pp.1031,1041, Oct 2000

[12] Ojeda, X.; Li, G.J.; Gabsi, M., "Fault diagnosis using vibration measurements of a Flux-Switching permanent magnet motor," *Industrial Electronics* (ISIE), 2010 IEEE International Symposium on, vol., no., pp.2091,2096, 4-7 July 2010

[13] Penman, J.; Sedding, H.G.; Lloyd, B.A.; Fink, W. T., "Detection and location of interturn short circuits in the stator windings of operating motors," *Energy Conversion, IEEE Transactions on* , vol.9, no.4, pp.652,658, Dec 1994

[14] Urresty, J.; Riba, J.; Romeral, L., "A back-emf based method to detect magnet failures in PMSMs," *Magnetics, IEEE Transactions on*, vol.49, no.1, pp.591,598, Jan. 2013

[15] Khan, M.; Azizur Rahman, M., "*Development and implementation of a novel fault diagnostic and protection technique for ipm motor drives,*" *Industrial Electronics, IEEE Transactions on*, vol.56, no.1, pp.85,92, Jan. 2009

[16] Zafarani, M.; Goktas, T.; Akin, B., "*A Comprehensive Analysis of Magnet Defect Faults in Permanent Magnet Synchronous Motors,*" in Industry Applications, IEEE Transactions on, vol.PP, no.99, pp.1-1

[17] M. Zafarani, T. Goktas, B. Akin, "A Simplified Numerical Approach to Analyze Magnet Defects in Permanent Magnet Synchronous Motors", in Energy Conversion Congress and Exposition (ECCE), 2015 IEEE , vol., no., pp.6117-6121, 20-24 Sept. 2015

[18] D.C. Hanselman, *Brushless permanent-magnet motor design.* New York: McGraw-Hill, 1994.

High Frequency Modeling for Transformer Common Mode Noise Coupling Path Based on Multiconductor Transmission Line Theory

Peipei Meng
School of Automation, Wuhan University of Technology
Wuhan, Hubei 430070, China
Email: cathleen@zju.edu.cn

Xiangming Zhang (Corresponding author)
National Key Laboratory of Science and Technology on Vessel Integrated Power System Technology, Naval University of Engineering, Wuhan, Hubei 430033, China
Email: 15827460499@139.com

Abstract—**Transformer is an important common mode (CM) noise coupling path in isolated power converters. But the EMI behavior of transformer is difficult to predict because of the complicated high frequency parasitics. This paper proposes a transformer Common-mode (CM) noise coupling path modeling approach based on multiconductor transmission line (MTL) theory. The MTL model can include the complicated magnetic and electrical couplings among the windings, the distribution of parasitic parameters, and the high frequency losses caused by skin and proximity effects. Physical structure oriented model is established in Maxwell 2D to extract the high frequency parasitic parameters of the transformer based on finite element method (FEM). With the MTL model the CM noise caused by the converter can be well predicted in the whole conducted EMI frequency range from 150 kHz to 30 MHz, which is validated by measured results with EMI receiver.**

Keywords—EMI/EMC, common mode noise, multiconductor transmission line theory, finite element method, transformer.

I. INTRODUCTION

Power converters always generate noises due to their fast switching transitions, which usually cause electromagnetic interference (EMI) problems. The complex EMI prediction through accurate modeling and characterizing the EMI noise coupling mechanism becomes a hot topic for effectively managing the electromagnetic compatibility (EMC) performance of power converters. The fast switching of the semiconductor power devices is the major cause of the EMI noises in a power converter, but some parasitic elements in the circuit are important propagation paths for EMI noises. Both noise sources and propagation paths need to be characterized in detail to diagnose the EMI generated by a converter before EMI design.

Several effective methods have been developed for noise source modeling, including the device physical parameters based modeling [1][2], the device terminal characteristics based modeling [3][4], and etc. Compare with the physical parameter based approach, which requires intimate knowledge of the device structure and is somewhat time consuming, the

Supported by "the Fundamental Research Funds for the Central Universities" (WUT: 2013-IV-088).

terminal characteristics based approach is simpler and more flexible.

The noise propagation paths can be extracted by experimental measurements [5-8] and/or by numerical analysis (finite-element method, partial-element equivalent circuit, etc.) [9-11]. Mathematical simulation or network computation approaches works well for parasitic modeling, but more complicated circuit structure is needed to improve its accuracy. Measurement method is much simpler, but it merely tries to emulate measurements on the converter without necessarily relating to the physical structure.

Transformer is one of the most important common mode (CM) noise coupling paths in isolated switching converters. Lumped circuit models are commonly used to describe the CM noise coupling path of the transformer [12-14]. However, the lumped circuit model would be invalid in high frequencies due to the influences of the high frequency distributed parasitic parameters and the complicate couplings among the windings. The high frequency characteristics of transformer have been studied [15-17] for years. But the CM noise coupling mechanism through the transformer still remains mysterious for the designers.

This paper applies the multiconductor transmission line (MTL) theory to model the CM noise coupling through the transformer, and extracts the high frequency parasitic parameter matrices of the transformer with finite element method (FEM) based tool – Ansoft Maxwell 2D. With the physical structure based MTL model, all the high frequency couplings and distribution characteristics of the transformer can be modeled. CM noise predicting model of a Flyback converter is also established. Evaluation is made between the measured CM noise and the predicted result, which verifies the theoretical analysis in this paper.

II. MODELING PROCEDURE AND TEST SETUP

A 32 W, 16 V / 2A Flyback converter is studied as a typical example. The transformer in this converter uses a PQ 26/20 core (material: PC40) from TDK, with 24 turns of primary windings and 4 turns of secondary windings. The primary windings are winded as the inner layer, and the secondary windings are winded as the outer layer, as illustrated in Fig.1.

978-1-4673-9551-9/16 $31.00 © 2016 IEEE

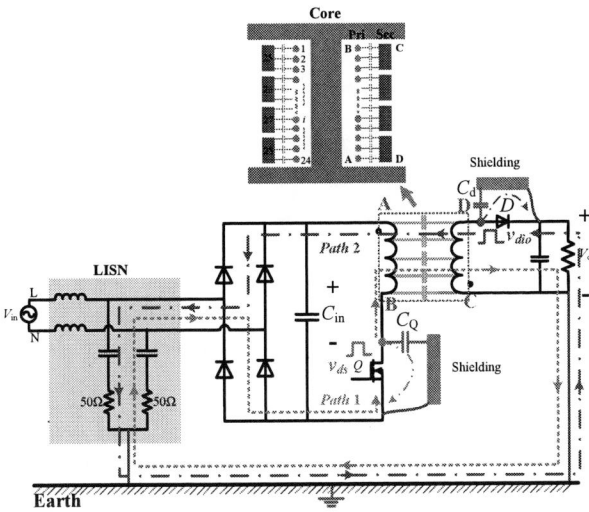

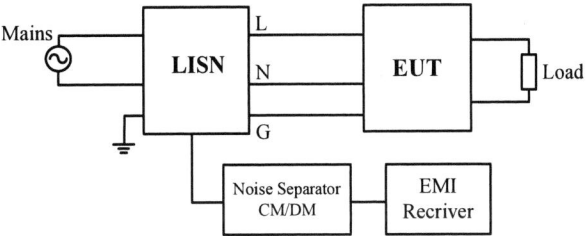

Fig. 1 The main CM noise coupling paths in Flyback converter and Sectional view of the transformer used in this converter

The semiconductor devices working on fast-switching model are the main sources of EMI noise. In this converter, the MOSFET Q and diode D working at switching model are the main noise sources. All components in the circuit would serve as paths for noise propagation. It mainly has two CM noise coupling paths in Fig.1, *Path 1* and *Path 2*. The MOSFET Q and diode D are the noise sources of *Path 1* and *Path 2* respectively. Both of the noises sources and the propagation paths should be characterized and modeled in detail to predict the CM noise generated by the converter.

To extract the spectrum of noise sources, the time domain waveforms of the semiconductor devices are measured by oscilloscope; then the noise source spectrum is calculated by Fast Fourier transform (FFT) followed by IF filtering with the consideration of the 9 kHz bandwidth (in Band B: 150 kHz − 30 MHz) of EMI receiver specified by the EMC standards (e.g. CISPR 16).

The analysis of noise propagation paths are divided into two parts: the magnetic components and the non-magnetic components. The noise coupling through the non-magnetic components, including the power lines, PCB routing, capacitors, grounding lines, etc, is relatively simple. In this paper the non-magnetic components are modeled with lumped model, and the parameters are extracted by impedance analyzer (Agilent 4395A).

The noise propagation mechanism through the magnetic component is complex, and the lumped circuit model of the transformer would be invalid in high frequencies due to the distribution of parasitic parameters, magnetic and electrical coupling among the windings, and the high frequency parameters, etc. In this paper the multiconductor transmission line (MTL) theory is applied to model the magnetic component (the transformer T in Fig. 1). The high frequency parasitic parameter matrices of the magnetic component are extracted with finite element method (FEM) based tool – Ansoft Maxwell 2D. With the physical structure based MTL

model, all the high frequency couplings and distributions characteristics of the magnetic component can be modeled.

Finally, the CM noise propagation transfer function of each CM noise coupling path can be identified, combine with the spectrum of noise sources, the CM noise caused by the converter can be predicted. Evaluation is also made between the measured CM noise and the predict result. The test setup for conducted CM noise measurements consists of an EMI receiver, a CM/DM noise separator, a LISN, and the EUT(Equipment Under Test), as shown in Fig. 2.

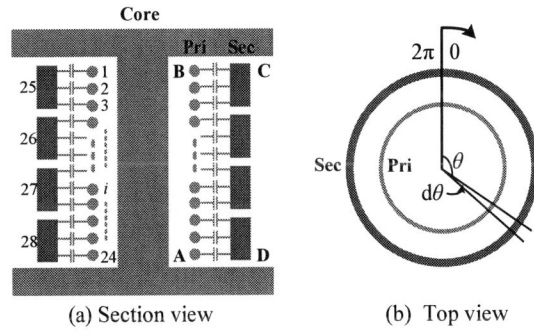

Fig. 2 CM/DM noise emission test setup

III. TRANSFORMER MODELING BASED ON MTL MODEL

In this section, the multiconductor transmission line (MTL) theory is applied to model the transformer in order to reveal the CM noise coupling mechanism throng the transformer, and predict the CM noise caused by the converter. The CM noise coupling path associated with each noise source should be identified, and the noise propagation transfer function should be evaluated for noise predicting.

Core

| 25 | 26 | 27 | 28 | Pri Sec | B C | A D | 2π 0 | Sec Pri θ dθ |

(a) Section view   (b) Top view

Fig. 3 Section view and top view of the transformer

A. Radian MTL equations for transformer modeling

The transformer in the Flyback converter in Fig. 1 provides a CM noise coupling path from primary-side to secondary-side, or on the contrary likewise. Both of the section view and the top view of the transformer are shown in Fig. 3. It can be found that the diameters of the windings in the outer layer and in the inner layer are different. So, the traditional MTL theory can't be applied directly to model the transformer with spiral windings. To solve this problem, this paper proposes the radian based MTL equations, in which the voltages and currents are defined as the function of radian position θ ($0 \leq \theta < 2\pi$), and per-unit-radian parameters are applied. In the radian based MTL model, each turn of the

978-1-4673-9551-9/16 $31.00 © 2016 IEEE

transformer winding is treated as a transmission line, at the start of each turn, $\theta = 0$; at the end of each turn, $\theta = 2\pi$. The voltage and current are defined as a function of θ.

The radian based MTL equation is expressed in frequency domain (phasor) by matrix form as

$$\begin{cases} \dfrac{d}{d\theta}\dot{\mathbf{V}}(\theta) = -\mathbf{Z}_\theta\,\dot{\mathbf{I}}(\theta) \\[2mm] \dfrac{d}{d\theta}\dot{\mathbf{I}}(\theta) = -\mathbf{Y}_\theta\,\dot{\mathbf{V}}(\theta) \end{cases} \tag{1}$$

where, the $n \times 1$ (n is the total number of windings, $n = 28$) column voltage vector $\dot{\mathbf{V}}(\theta) = [\dot{V}_1(\theta)\cdots\dot{V}_i(\theta)\cdots\dot{V}_n(\theta)]^{\mathrm{T}}$ and current vector $\dot{\mathbf{I}}(\theta) = [\dot{I}_1(\theta)\cdots\dot{I}_i(\theta)\cdots\dot{I}_n(\theta)]^{\mathrm{T}}$ are the voltages and currents of the windings at position θ ($0 \le \theta \le 2\pi$) respectively. $\dot{V}_i(\theta)$ and $\dot{I}_i(\theta)$ are the phasor equivalent of the signals as $\dot{X}(j\omega) \Leftrightarrow x(t) = \mathrm{Re}\{\dot{X}e^{j\omega t}\}$. The $n \times n$ per-unit-radian impedance matrix $\mathbf{Z}_\theta$, and per-unit-radian admittance matrix $\mathbf{Y}_\theta$ are given by $\mathbf{Z}_\theta = \mathbf{R}_\theta + j\omega\mathbf{L}_\theta$, $\mathbf{Y}_\theta = \mathbf{G}_\theta + j\omega\mathbf{C}_\theta$. In which, the per-unit-radian resistance matrix $\mathbf{R}_\theta$, per-unit-radian inductance matrix $\mathbf{L}_\theta$, per-unit-radian conductance matrix $\mathbf{G}_\theta$ and the per-unit-radian capacitance matrix $\mathbf{C}_\theta$ are the parasitic parameters of the transformer which can model the losses and couplings of the windings.

B. The general solution for MTL equations

The radian based MTL equation in (1) has the general solution as

$$\begin{cases} \dot{\mathbf{V}}(\theta) = \mathbf{T}_\mathrm{V}(e^{-\gamma\theta}\mathbf{V}_\mathrm{m}^+ + e^{\gamma\theta}\mathbf{V}_\mathrm{m}^-) \\[2mm] \dot{\mathbf{I}}(\theta) = \mathbf{Y}_\mathrm{C}\mathbf{T}_\mathrm{V}(e^{-\gamma\theta}\mathbf{V}_\mathrm{m}^+ - e^{\gamma\theta}\mathbf{V}_\mathrm{m}^-) \end{cases} \tag{2}$$

where the characteristic admittance matrix $\mathbf{Y}_\mathrm{C} = \mathbf{Z}_\theta^{-1}\mathbf{T}_\mathrm{V}\gamma\mathbf{T}_\mathrm{V}^{-1} = \mathbf{Y}_\theta\mathbf{T}_\mathrm{V}\gamma^{-1}\mathbf{T}_\mathrm{V}^{-1}$. $\mathbf{T}_\mathrm{V}$ is composed of $2n$ linearly independent eigenvectors of $\mathbf{Z}_\theta\mathbf{Y}_\theta$. $\gamma^2 = diag\{\gamma_1^2,\ \gamma_2^2\ \cdots\ \gamma_n^2\} = \mathbf{T}_\mathrm{V}^{-1}\mathbf{Z}_\theta\mathbf{Y}_\theta\mathbf{T}_\mathrm{V}$. $\mathbf{V}_\mathrm{m}^+$ and $\mathbf{V}_\mathrm{m}^-$ are the $2n$ undetermined constants which need to be evaluated by the boundary conditions of the MTL.

C. Boundary condition and special solution for MTL equations

The boundary conditions of the transformer need to be considered to solve the special solution of the MTL equations. Consider *Path 1* in the Flyback converter in Fig. 1, the terminal connections and external circuit condition of the transformer can be illustrated as shown in Fig. 4. The transformer has four ports (A, B, C and D), 24 primary windings (1, 2...24) and 4 secondary windings (25, 26, 27, 28). Each turn of the windings in the transformer is treated as one transmission line.

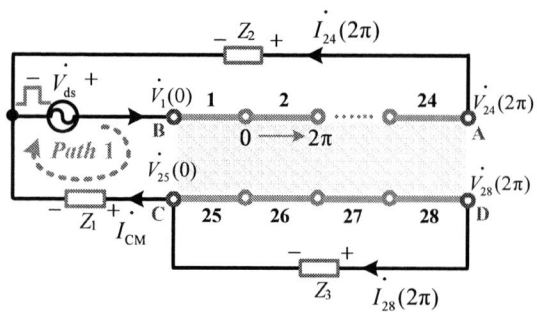

Fig. 4 Terminal connections and external circuit condition of the transformer associated with Path 1

V_{ds} is the drain-to-source voltage of MOSFET Q, which is the noise source in *Path 1*. The non-magnetic components in the converter are modeled by lumped model. Z_1 is the lumped impedance between point C of the transformer and the source terminal of the MOSFET, including 25 Ω resistance of the LISN and the route impedance; Z_2 is the lumped impedance for primary side current, including the input filtering capacitor C_{in} and the route impedance; Z_3 is the lumped impedance for secondary side current, including the output filtering capacitor and the route impedance.

$\dot{V}_i(\theta)$ and $\dot{I}_i(\theta)$ are the voltage and current of the ith winding ($i = 1, 2...28$) at position θ ($0 \le \theta < 2\pi$). The boundary conditions of the MTL in Fig. 4 can be determined by

$$\begin{cases} \dot{V}_i(2\pi) = \dot{V}_{i+1}(0), & i = 1,\ 2,\ 3...23,\ 25,\ 26,\ 27 \\[1mm] \dot{V}_{24}(2\pi) - \dot{V}_{28}(2\pi) = \dot{I}_{24}(2\pi)\cdot Z_2 - I_{\mathrm{CM}}\cdot Z_1 - \dot{I}_{28}(2\pi)\cdot Z_3 \\[1mm] \dot{V}_{25}(0) - \dot{V}_{28}(2\pi) = -\dot{I}_{28}(2\pi)\cdot Z_3 \\[1mm] \dot{I}_i(2\pi) = \dot{I}_{i+1}(0), & i = 1,\ 2,\ 3...23,\ 25,\ 26,\ 27 \\[1mm] \dot{I}_1(0) = \dot{I}_{24}(2\pi) + I_{\mathrm{CM}} \\[1mm] \dot{I}_{25}(0) = \dot{I}_{28}(2\pi) - I_{\mathrm{CM}} \end{cases} \tag{3}$$

Equ. (3) has $2n$ ($n = 28$) independent equations, so the $2n$ constants $\mathbf{V}_\mathrm{m}^+$ and $\mathbf{V}_\mathrm{m}^-$ in (4) can be solved. Hereby the special solution of the MTL equations can be solved and the impedances of the transformer can be obtained.

Based on Fig. 4 and combine the special solution of MTL equations, the effective CM noise coupling path impedance of the transformer in *Path 1* can be evaluated by

$$Z_{\mathrm{trans\_CM\_1}} = \dfrac{\dot{V}_{\mathrm{BC}}}{\dot{I}_{\mathrm{CM}}} = \dfrac{\dot{V}_1(0) - \dot{V}_{25}(0)}{\dot{I}_{\mathrm{CM}}} \tag{4}$$

Notice that, the result in (4) won't contain $\dot{I}_{\mathrm{CM}}$, because it would be reduced in the result.

978-1-4673-9551-9/16 $31.00 © 2016 IEEE

The CM noise coupling path impedance of *Path 1* can be evaluated by

$$Z_{CM\_Path1} = \frac{\dot{V}_{ds}}{\dot{I}_{CM}} = \frac{\dot{V}_{BC}}{\dot{I}_{CM}} + Z_1 = Z_{trans\_CM\_1} + Z_1 \quad (5)$$

IV. PARAMETER EXTRACTION AND NOISE SOURCE MODELING

A. Parameter extraction with Finite Element Analysis

To establish the MTL model of the transformer, the high frequency parasitic parameter matrixes of the transformer need to be extracted, including the per-unit-radian resistance matrix $\mathbf{R}_\theta$, per-unit-radian inductance matrix $\mathbf{L}_\theta$, per-unit-radian conductance matrix $\mathbf{G}_\theta$ and the per-unit-radian capacitance matrix $\mathbf{C}_\theta$. In this paper, the parasitic parameter matrixes of the transformer are extracted by finite element method (FEM) tool – Ansoft Maxwell 2D. The simulation model of the transformer established is shown in Fig. 5, which is based on the cylindrical model in Maxwell 2D. The capacitance and conductance matrixes are solved in the solution type of AC Conduction, the resistance and inductance matrixes are solved in the solution type of Eddy Current. Due to the limited space, the values of the matrixes are not listed in this paper.

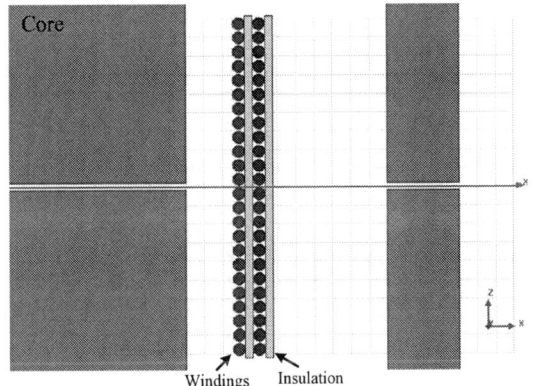

Fig. 5 Maxwell 2D simulation model for parameter matrixes extraction

B. Noise Source Spectrum Analysis

The simulation approach for EMI noise predicting need to emulate the complete behavior of an EMI receiver, which has to fulfill the requirements that are given by the international standard CISPR 16-1-1[18]. The spectrum analysis method of "time domain EMI measurements" specified in [19] is adopted in this paper for noise source modeling. The measurement and analysis sketch is shown in Fig. 6.

Firstly, the time domain waveforms of the noise sources (v_{ds}, v_{dio}) are measured by digital storage oscilloscope Tektronix TDS5054 (500 MHz bandwidth and 5 GHz real time sampling rate, four-input channels). The time domain waveforms of v_{ds} and v_{dio} are shown in Fig. 7. The time domain voltage waveform should be calculated for half-period of the mains, so the total recording time equals to 10 ms (mains frequency f_m=50 Hz).

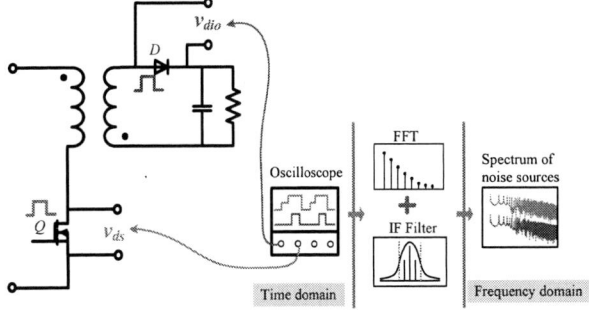

Fig.6 Noise source measurement and analysis sketch

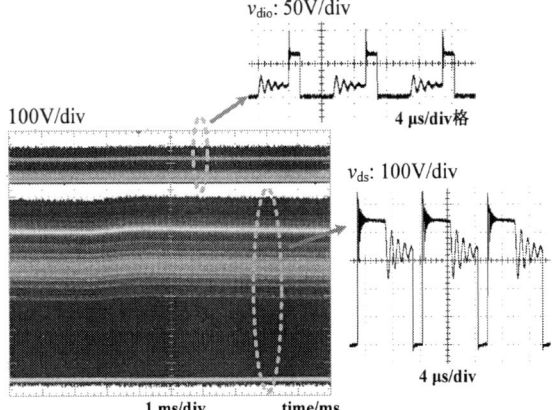

Fig.7 Time domain waveforms of noise sources

Then, the noise source spectrum is calculated by Fast Fourier transform (FFT) followed by Gaussian window function filtering to model the intermediate frequency (IF) bandwidth of EMI receiver. In CISPR 16-1-1, different IF filters are defined depending on the frequency range of interest. In Band B (150 kHz – 30 MHz), the 6 dB bandwidth is 9 kHz. More detail analysis can be referred in [19-21]. Both of the magnitude and phase information of the noise sources are calculated, the magnitude spectrums of v_{ds} and v_{dio} are shown in Fig. 8.

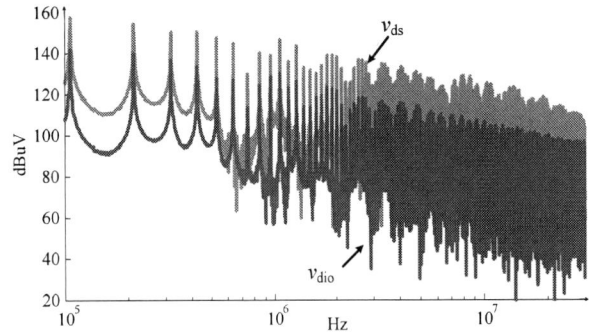

Fig. 8 Spectrums of noise sources

V. Implementation and Comparison with Measurements

Combine the parasitic parameters extracted by Ansoft Maxwell 2D and the lumped route impedance Z_1, Z_2 and Z_3 extracted by impedance analyzer (Agilent 4395A), as shown in Table. I, the impedance of $Z_{\text{trans\_CM\_1}}$ and $Z_{\text{CM\_Path1}}$ in (4) and (5) can be evaluated which are drawn in Fig. 9.

Table I Impedance of Z_1, Z_2 and Z_3

		R_1	L_1
$Z_1 = 25\,\Omega + R_1 + j\omega L_1$		0.5 Ω	1.4 µH
	C_{in}	R_2	L_2
$Z_2 = 1/j\omega C_{\text{in}} + R_2 + j\omega L_2$	100 µF	0.3 Ω	20 nH
	C_{out}	R_3	L_3
$Z_3 = 1/j\omega C_{\text{out}} + R_3 + j\omega L_3$	136 µF	0.06 Ω	57 nH

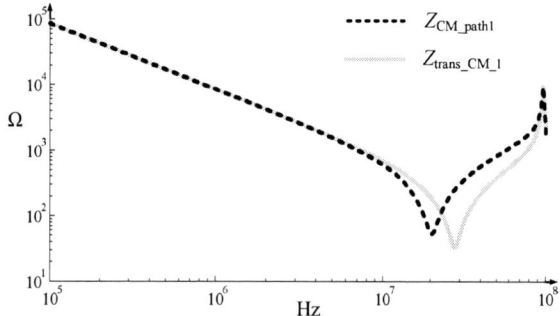

Fig. 9 Calculated impedance of $Z_{\text{trans\_CM\_1}}$ and $Z_{\text{CM\_Path1}}$

Similarly, the CM noise coupling path impedance of *Path* 2 ($Z_{\text{CM\_Path2}}$) can also be evaluated. Then the total CM noise of the Flyback converter can be calculated by

$$\dot{V}_{\text{com}} = \frac{25 \cdot \dot{V}_{\text{ds}}}{Z_{\text{CM\_Path1}}} - \frac{25 \cdot \dot{V}_{\text{dio}}}{Z_{\text{CM\_Path2}}} \qquad (6)$$

where $\dot{V}_{\text{ds}}$ and $\dot{V}_{\text{dio}}$ are the spectrum of noise sources which are obtained in Section IV.

Both the calculated CM noise spectrum with the MTL model and the tested CM noise spectrum with EMI receiver are drawn in Fig. 10, which show good coincidence with each other. The envelope of the calculated spectrum with lumped model of the transformer [14] is also drawn in Fig. 10, but it is only valid below 7 MHz.

The comparison between the calculated spectrum and tested spectrum in Fig. 10 validated the transformer MTL model proposed in this paper. With the proposed model, both of the EMI performance of the transformer and the converter can be predicted. The MTL model can also be used to guide the design of a transformer before it is physically constructed. Thus, the transformer can be designed and optimized for a given converter to achieve the desired EMI performance.

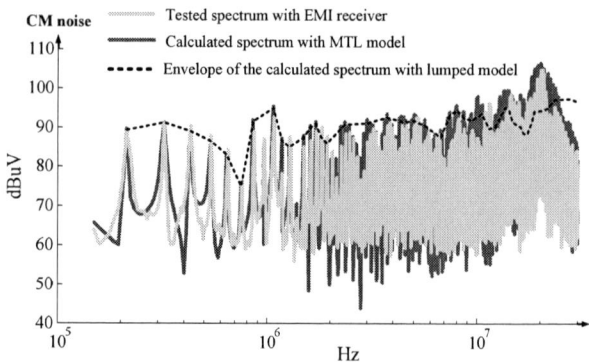

Fig. 10 CM noise: Comparison between the tested spectrum with EMI receiver and the calculated spectrum with MTL model

VI. Conclution

A novel transformer CM noise coupling path modeling approach based on Multiconductor transmission line theory is proposed in this paper. This new model provides a practical and accurate transformer modeling approach for engineers. Compare with the transformer modeling approach based on lumped circuit model without any attempt to ascertain the structure of the transformer, the proposed model can provide a full feature of the CM noise coupling mechanism through the transformer and it is more accurate. Based on the MTL model proposed in this paper, the CM noise spectrum of a Flyback converter is calculated and validated by measurements.

References

[1] A R Hefner. An improved understanding for the transient operation of the power insulated gate bipolar transistor (IGBT)[J] . IEEE Trans. Power Electron., 1990, 5(5): 459–468.

[2] Tsung-Po Chen, "Common-Mode Ripple Current Estimator for Parallel Three-Phase Inverters," *IEEE Trans. Power Electron.*, vol. 24, no. 5, pp. 1330-1339, May 2009.

[3] Qian Liu, Fred Wang, and Dushan Boroyevich, "Conducted-EMI Prediction for AC Converter Systems Using an Equivalent Modular–Terminal–Behavioral (MTB) Source Model," *IEEE Trans. Ind. Appl.*, vol. 43, no. 5, pp. 1360-1370, Sep./Oct. 2007.

[4] D. Gonzalez, J. Gago, and J. Balcells, "New simplified method for the simulation of conducted EMI generated by switched power converters,"IEEE Trans. Ind. Electron., vol. 50, no. 6, pp. 1078–1084, Dec. 2003.

[5] H. Zhu, A. R. Hefner, Jr., and J.-S. Lai, "Characterization of power electronics system interconnect parasitics using time domain reflectometry," *IEEE Trans. Power Electron.*, vol. 14, no. 4, pp. 622–628, Jul. 1999.

[6] Meng Jin, Ma Weiming, "Identification of Essential Coupling Path Models for Conducted EMI Prediction in Switching Power Converters," *IEEE Trans. Power Electron.*, vol. 21, no. 6, pp. 1795–1803, Nov. 2006.

[7] Denis Labrousse, Bertrand Revol, and Francois Costa, "Common-Mode Modeling of the Association of N-Switching Cells: Application to an Electric-Vehicle-Drive System," *IEEE Trans. Power Electron.*, vol. 24, no. 11, pp. 2852–2859, Nov. 2010.

[8] Jettanasen C., Costa F., Vollaire C., "Common-Mode Emissions Measurements and Simulation in Variable-Speed Drive Systems," *IEEE Trans. Power Electron.*, vol. 24, no. 11, pp. 2456-2464, Nov. 2009.

[9] Jithesh V, Pande D C. A review on computational EMI modeling techniques[C]. In: Proc. of Int. Conf. on Electromagnetic Interference and Compatibility, 2003: 159-166.

[10] Yong Li, Longfu Luo, Christian Rehtanz, Can Wang, and Sven Rüberg, "Simulation of the Electromagnetic Response Characteristic of an Inductively Filtered HVDC Converter Transformer Using Field-Circuit Coupling," IEEE Transactions on Industrial Electronics, vol. 59, no. 11, pp. 4020-4031, November 2012.

[11] Florian Giezendanner, Juergen Biela, Johann Walter Kolar, and Stefan Zudrell-Koch, "EMI Noise Prediction for Electronic Ballasts," *IEEE Trans. Power Electron.*, vol. 25, no. 8, pp. 2133–2141, Aug. 2010.

[12] Pengju Kong, Shuo Wang and Fred C. Lee, "Reducing Common Mode EMI Noise in Two-switch Forward Converter," in *Proc. IEEE ECCE*, 2009, pp. 3622 – 3629.

[13] Zengyi Lu and Wei Chen, "Common Mode EMI Noise Reduction Technique by Noise Path Configuration of High Frequency Power Transformer," in *Proc. IEEE IPEMC*, 2009, pp. 954 – 956.

[14] Peipei Meng, Yuwen Shen, Junming Zhang, Henglin Chen, Zhaoming Qian. Characterizing Noise Source and Coupling Path in Flyback Converter for Common-mode Noise Prediction[C]. in *Proc. IEEE APEC*, 2011：1704 - 1709.

[15] Ziwei Ouyang, Ole C. Thomsen, and Michael A. E. Andersen, "Optimal Design and Tradeoff Analysis of PlanarTransformer in High-Power DC–DC Converters," IEEE Transactions on Industrial Electronics, vol. 59, no. 7, pp. 2800-2810, July 2012

[16] Davoudi A., Chapman P. L., Jatskevich J., Khaligh A., "Reduced-Order Modeling of High-Fidelity Magnetic Equivalent Circuits," *IEEE Trans. Power Electron.*, vol. 24, no. 12, pp. 2847-2855, Dec. 2009.

[17] Sabahi M., Goharrizi A. Y., Hosseini S. H., Sharifian M. B. B., Gharehpetian G. B., "Flexible Power Electronic Transformer," *IEEE Trans. Power Electron.*, vol. 25, no. 8, pp. 2159-2169, Aug. 2010.

[18] CISPR16-1-1, Specification for Radio Disturbance and Immunity Measuring Apparatus and Methods Part 1-1: Radio Disturbance and Immunity Measuring Apparatus — Measuring Apparatus. Switzerland: Int. Electrotech. Commiss., 2003.

[19] Stephan Braun, Thomas Donauer, and Peter Russer, "A Real-Time Time-Domain EMI Measurement System for Full-Compliance Measurements According to CISPR 16-1-1," *IEEE Trans. Electromagnetic Compatibility*, vol. 50, no. 2, pp. 259–267, May. 2008.

[20] S. Braun, F. Krug, and P. Russer, "A novel automatic digital quasi-peak detector for a time domain measurement system," in *Proc. IEEE Int. Symp. Electromagn. Compat. Digest*, 2004, vol. 3, pp. 919–924.

[21] Zhang Xiangming, Zhao Zhihua, Meng Jin, Zhang Lei, Chen Junquan, Pan Qijun, "EMI spectrum analysis based on FFT with consideration of measurement bandwidth effect," in *Proc. of the Chinese Society of Electrical Engineering*, vol. 30, no. 36, pp. 117 – 122, Dec. 2010.

Leakage Flux Modelling of Multi-Winding Transformer using Permeance Magnetic Circuit

Min Luo and Drazen Dujic
Power Electronics Laboratory
École Polytechnique Fédérale de Lausanne (EPFL)
Lausanne CH-1015, Switzerland
min.luo@epfl.ch, drazen.dujic@epfl.ch

Jost Allmeling
Plexim GmbH
Zürich CH-8005, Switzerland
allmeling@plexim.com

Abstract—**The size and position difference of the windings determine the leakage flux path and give rise to unbalance of the short-circuit impedances, which strongly affects the transient behaviour of the transformer. This work proposes a new approach of modelling with magnetic equivalent circuit, making use of the transformer geometry and permeance magnetic equivalent circuit, which is suitable for system-level simulation in terms of complexity. Equivalent model requires limited number of parameters and for verification purposes, FEM simulations as well as measurement on the experimental prototype have been performed.**

I. Introduction

Multi-winding transformers are adopted in power electronics system for transferring power between multiple terminals. Different size and position make the leakage coupling between some windings stronger than the others, which give rise to unbalanced short-circuit impedance values or non-symmetrical conditions during operation.

Finite element method (FEM) is able to accurately model the leakage field, however it is not suitable for transient simulation of large power electronics system due to the high computational effort. For this purpose electrical equivalent circuit with coupled inductors is commonly used. This type of model requires hardware test to identify the magnetisation inductance L_m and the leakage inductance L_σ. If more than two windings are present, the short-circuit test needs to be enumerated among all the winding pairs, which can be time consuming or in some cases impractical.

Magnetic circuit is closely related to geometry and material parameters, which is intuitive and usually easier to obtain compared to the hardware test results. Existing publications based on this methodology have modelled the iron core in quite deep details level like [1], [2] and [3], in which however the leakage path was rather roughly represented. Authors of [4] included the leakage flux path into the magnetic circuit, and proposed a way to quantitatively calculate the parameters. But this approach was only validated by a FEM-simulated single-winding inductor with simple topology. In the work of [5] authors aimed at more complicated cases, where the leakage flux path was also integrated into the magnetic circuit and the result was compared with FEM simulation of multi-winding transformers. The work of [6] further on verified [5]'s approach via experimental test. Typical winding arrangements like multilayer- (Fig. 1a) and multidisk- (Fig. 1b) structures

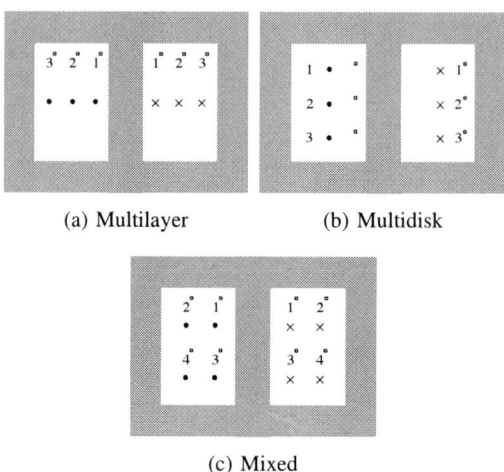

(a) Multilayer (b) Multidisk

(c) Mixed

Fig. 1: Typical winding structures considered in the existing literature

have been investigated, as well as a mixed case (Fig. 1c). But the topology shown in Fig. 2 has not been discussed, where the height of the inner layer winding is much larger than the others. This topology is typical for the phase-shift transformers used in multi-pulse rectifier systems, which interface power grid to medium voltage power electronics system like the cascaded H-bridge converter [7].

Moreover, in both publications the magnetic circuit was finally transformed back to electrical equivalent with coupled inductors before parametrisation. The model in [5] required to obtain the short-circuit impedances of all the winding pairs via experimental measurement, in order to derive the coupled inductance matrix of the electrical equivalent circuit. Making use of the approach from [8], authors of [6] managed to calculate single short-circuit impedance based on geometry information. However the characteristic of the core was not modelled in a straight forward way. Instead of obtained from the geometry and material characteristic or open circuit test, the information about main flux path in the core was indirectly derived from the short-circuit impedance, where the core characteristic only has slight affect. This may result in significant error when simulating light load conditions when the main flux path inside the core dominates the transient behaviour, especially when

978-1-4673-9551-9/16 $31.00 © 2016 IEEE   1108

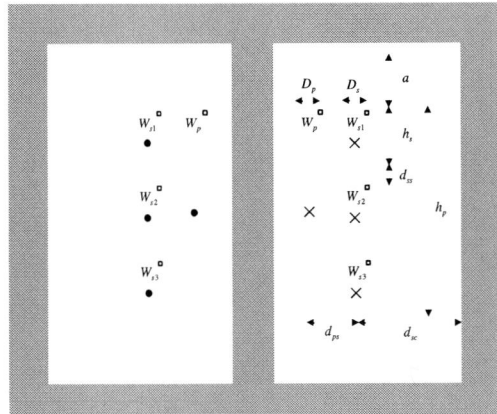

Fig. 2: Winding structure considered in this work, with the height of primary winding significantly larger than the secondary ones

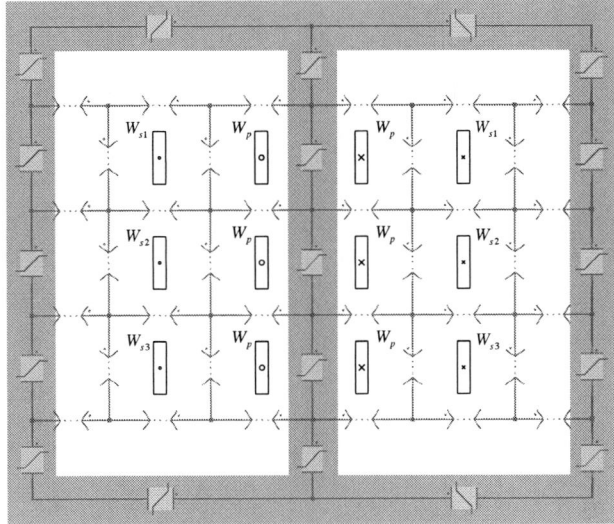

Fig. 3: Preliminary permeance circuit of the transformer, where the flux path of the core and air are illustrated

taking into account the nonlinearity of the core material.

This paper deals with direct magnetic circuit modelling of the special transformer topology in Fig. 2 with a large number of winding, which is directly integrated into electric circuit simulation without being transformed back to electrical equivalent. The permeance-capacitor representation proposed by [9] and implemented by [10] is adopted, considering its high efficiency for circuit simulation. The core- and the leakage-flux path are modelled separately in the magnetic circuit, which provides the possibility to accurately represent the core's characteristic. In particular, the emphasis is placed on the modelling of various leakage flux paths, that may be encountered in these type of transformers. Especially when the windings' geometry has repetitive property, the parametrisation of the model can be simplified. For parameter identification, core geometry and material characteristic together with only a limited number of short circuit test are required. FEM simulations and a small scale prototype transformer are used to confirm validity of proposed modelling. Single-phase case is considered in this work.

II. ANALYSIS

Assuming that all secondary windings W_{sn} ($n = 1...3$) of the transformer in Fig. 2 are h_s high and D_s thick, they are allocated along the middle core limb with the same vertical interval d_{ss}. d_{ss} is usually significantly smaller than h_s and D_s in these types of transformers. The height of the primary winding h_p at the inner layer is equal to the distance from the top of W_{s1} to the bottom of W_{s3}. This can be interpreted into a magnetic circuit demonstrated in Fig. 3 assuming that both the air- and core-flux are flowing through the defined permeance channels. The permeances filled with yellow color represent flux path inside the core, while the others represent the leakage flux path through the air. Please note that the long primary winding W_p has been separated into 3 sections aligned with each of the secondary windings. If all the windings are wounded close enough to the middle limb of the transformer core (as is usually the case in practice) the horizontal leakage permeances above- or below the windings can be assumed to

be short circuit. Moreover the physically depicted windings can be interpreted into lumped-components by means of inserting several fictitious windings W_s' and W_s'', as required by the magnetic equivalent circuit. The fictitious windings have the same number of turns as the physical ones and wired in opposite directions, so that they cancel each other and do not have netto influence on the system. The permeance circuit with fictitious windings is demonstrated in Fig. 4.

Combining the physical W_s and the fictitious W_s' into single winding components on the left- and right-side respectively, then placing W_s'' on the middle limb, the permeance network can be transformed into the form as shown in Fig.

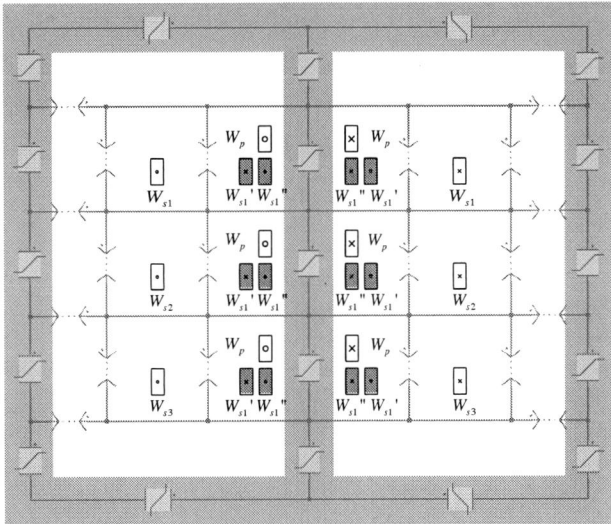

Fig. 4: Permeance circuit with horizontal permeance neglected and insertion of fictitious windings

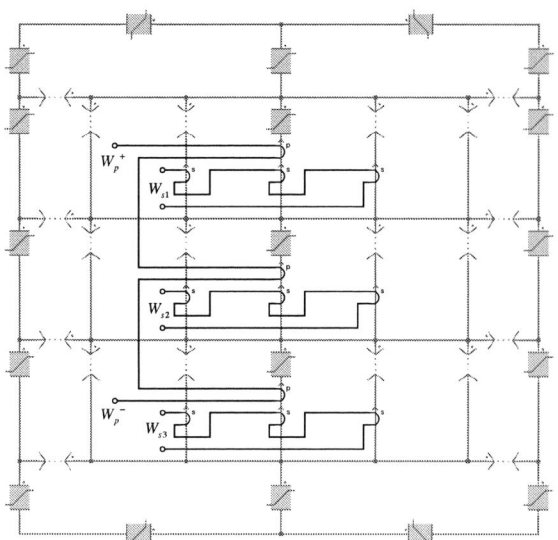

Fig. 5: Permeance circuit with lumped-winding components to interface electrical circuit

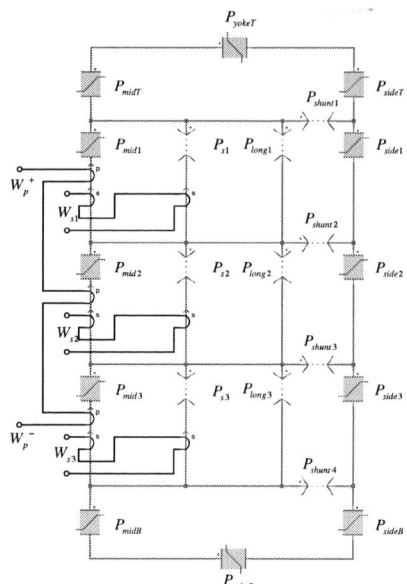

Fig. 7: Mirrored permeance network used to model multi-winding single phase transformer

5, which is able to be directly connected to electrical circuit model. The lumped winding components are implemented using the gyrator structure proposed by [9], which serves as interface between the electrical and magnetic circuit shown in Fig. 6. On one hand, the electrical voltage measured on the winding terminal is divided by the number of turns N and fed into the magnetic circuit as a flux rate source ($d\Phi/dt$). The flux rate source will charge or discharge the magnetic permeances which behave like a capacitor. On the other hand, the "voltage" measured on the magnetic terminal (MMF) is also divided by N and fed into the electrical circuit as current source.

Please note that all the primary winding sections as well as the secondary windings together with their fictitious counterparts are connected in series respectively in the electrical circuit. For this single phase transformer, the left side can be mirror to the right side due to the symmetry about the middle limb of the core, so that the model can be further simplified as shown in Fig. 7. Here the core permeances are calculated as

$$P = \frac{\mu_r \mu_0 A}{l} \qquad (1)$$

The equivalent cross-section area A and magnetic path length l can be calculated from the core geometry according to [11]. For this case, the length of P_{midn} and P_{siden} is equal to $h_s + d_{ss}$, while that of $P_{midT(B)}$ and $P_{sideT(B)}$ equal to a.

The permeance $P_{yokeT(B)}$ includes both the yoke and corner. Should nonlinearity of the core material be considered, the relative permeance μ_r will be a function of the magnetomotive force F, which is the magnetic "voltage" measure on the individual permeance. The approach proposed by [12] can also be applied to model the corner area of the core more accurately.

Up to this stage only the leakage permeances P_s, P_{long} and P_{shunt} are still unknown. Considering the fact that all the secondary windings have the same size and are located homogeneously along the middle limb, all P_{sn}, P_{longn} and P_{shuntn} ($n = 1, 2, 3$) are assumed to have the same value respectively. Since it is difficult to accurately define the geometry of the air flux path, short circuit test will be carried out to identify the three unknown parameters. In the next section, it will demonstrated that the same structure can be easily extended to the case with more than 3 secondary windings.

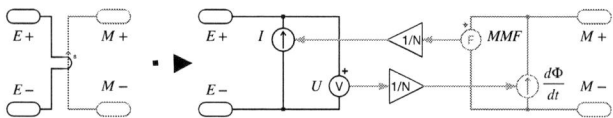

Fig. 6: Gyrator structure of the lumped-winding component

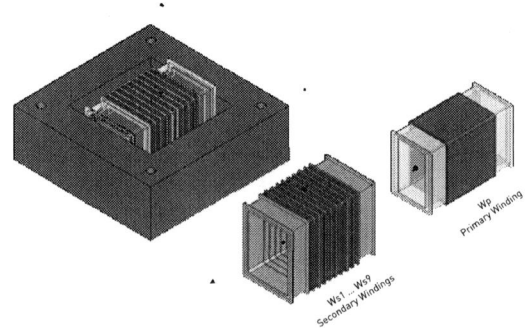

Fig. 8: Multiwinding transformer prototype

978-1-4673-9551-9/16 $31.00 © 2016 IEEE

Fig. 9: Flux distribution with W_p excited and W_{s5} shorted

Fig. 11: Flux distribution with W_{s1} excited and W_{s9} shorted

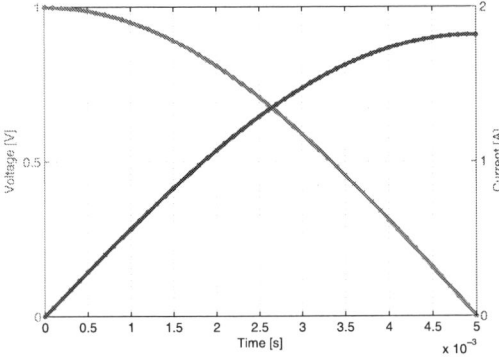

Fig. 10: W_p current and voltage with W_p excited, W_{s5} shorted

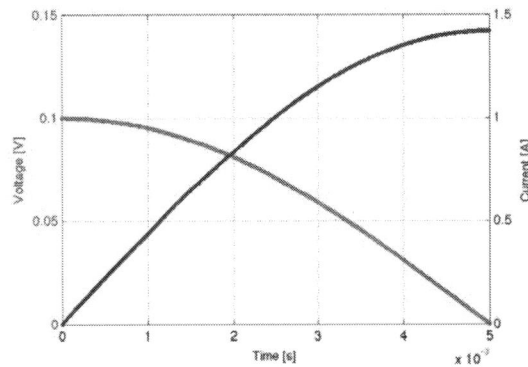

Fig. 12: W_p current and voltage with W_{s1} excited, W_{s9} shorted

III. VALIDATION

In order to validate the fidelity of the proposed modelling's approach, a low-power, single-phase transformer prototype has been built up as depicted in Fig. 8. The prototype three-limb transformer core (50mm thick) is made of standard silicon steel laminations type "EI 150/150" of material "M330-35A". The primary winding (W_p) has $N_p = 193$ turns which spreads along the middle limb with $h_p = 53mm$ height and $1mm$ thick. The nine secondary windings (W_{s1} - W_{s9}) have Ns = 30 turns each and are installed outside the primary winding, $h_s = 5mm$ high and $2mm$ thick, stacking vertically over eachother with $d_{ss} = 1mm$ interval in between.

A 2-dimensional FEM model has been made in COMSOL according to the real geometry, considering its accuracy in simulating leakage flux distribution. Two kinds of schemes are simulated in FEM. Firstly, sinusoidal voltage excitation with $1V$ amplitude is applied on primary winding W_p with single secondary winding shorted. Fig. 9 shows the flux density distribution at the current peak from FEM simulation, and Fig. 10

shows time-domain winding current and voltage waveform of 1/4 AC period, in the scheme when the fifth secondary winding W_{s5} is shorted. The short-circuit impedance is calculated via dividing the peak value of the voltage by the current

$$Z_k = \hat{V}_{Wp}/\hat{I}_{Wp} \qquad (2)$$

Resistivity of the windings and laminations is neglected in this model, so that Z_k is purely inductive, therefore the short-circuit inductance can be calculated directly as

$$L_k = Z_k/(2\pi f) \qquad (3)$$

Secondly, sinusoidal voltage with $0.1V$ amplitude is applied on one secondary winding, while another secondary winding is shorted. The short impedance is calculated as well by $\hat{V}_{Ws}/\hat{I}_{Ws}$. Fig. 11 and Fig. 12 display the simulation result when W_{s1} is excited and W_{s9} shorted.

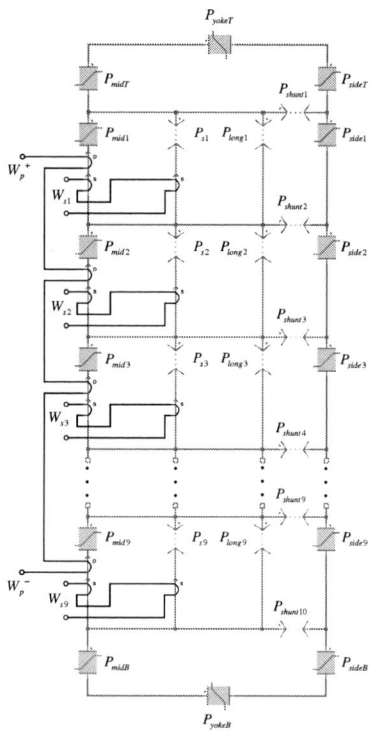

Fig. 13: Permeance circuit model of the prototype transformer

Afterwards the model based on permeance magnetic circuit is built up in the system-level simulation platform PLECS for power electronics, making use of its magnetic component library [10]. This mode is essentially an extension of the transformer example from the last section as shown in Fig. 13. As discussed in section II, the permeances representing the core (P_{midn}, P_{siden} ($n = 1...9$), $P_{midT(B)}$, $P_{sideT(B)}$, $P_{yokeT(B)}$) can be directly parametrised using the geometrical and material information, whose values are listed in TABLE. I.

TABLE I: Core permeance values $[Wb/A]$

P_{midn}	$P_{midT(B)}$	P_{siden}	$P_{sideT(B)}$	$P_{yokeT(B)}$
$3.13 \cdot 10^{-4}$	$1.04 \cdot 10^{-4}$	$6.26 \cdot 10^{-4}$	$2.09 \cdot 10^{-4}$	$3.02 \cdot 10^{-5}$

The magnetisation characteristic (B-H curve) of the material "M330-35A" can be specified to the core permeances using atan fit to representing the saturation effect. Moreover, if the hysteresis loop is available from experimental test, Preisach model with Lorentzian distribution function proposed by [13] can be directly applied to the permeance implementation as well. Since focus of this work is on leakage flux, only the characteristic in the linear area of the B-H curve has been assigned to the permeances, while the effect of the nonlinearity like saturation and hysteresis will be discussed in the future work.

Apart from the core permeances, the 3 values of the leakage permeances (P_s, P_{long}, P_{shunt}) is obtained via fitting of test results (FEM or experimental). The short-circuit impedance

values of 4 short-circuit test schemes are used for the parameter fitting, which essentially compose optimisation problem to minimise a quadratic objective function, described as

$$f(P_s, P_{long}, P_{shunt}) = \sum_{i=1}^{4} (\frac{L_{ki,Test} - L_{ki,Sim}}{L_{ki,Test}})^2 \quad (4)$$

- $L_{k1,Test}$ measured with W_p excited and W_{s1} shorted
- $L_{k2,Test}$ measured with W_p excited and W_{s5} shorted
- $L_{k3,Test}$ measured with W_{s1} excited and W_{s2} shorted
- $L_{k4,Test}$ measured with W_{s1} excited and W_{s9} shorted

Commonly used gradient based- or evolutionary algorithms can be applied to solve this problem. Initial values are configured in order to make the algorithm converge fast, which in this case can be calculated as

$$P_{s0} = 2 \cdot \frac{\mu_r \mu_0 D_c d_{ps}}{h_s + d_{ss}} \quad (5)$$

$$P_{long0} = 0 \quad (6)$$

$$P_{shunt0} = 2 \cdot \frac{\mu_r \mu_0 D_c (h_s + d_{ss})}{d_{sc}} \quad (7)$$

$D_c = 50mm$ is the thickness of the core, while other geometrical parameters present in the equation above have been defined in Fig. 2. The "fminsearch" function provided by MATLAB (referred as unconstrained nonlinear optimisation) is adopted for this case, the resulted permeance values are listed in TABLE II.

TABLE II: Leakage permeance values fitted to FEM $[Wb/A]$

P_s	P_{long}	P_{shunt}
$100 \cdot 10^{-9}$	$13.9 \cdot 10^{-9}$	$30 \cdot 10^{-9}$

For verification, other schemes are simulated with the leakage permeance values obtained from parameter fitting, and compared to the FEM result. The short-circuit inductances from FEM and PLECS circuit simulation are illustrated in Fig. 14 and Fig. 15. On one hand, Fig. 14 accounts for the scheme with the primary winding W_p excited and single secondary winding W_{sn} ($n = 1...9$) shorted. Except for the schemes that are used for parameter fitting of the leakage permeances (fit point 1 and 2), the short-circuit inductance values of the other test point matches quite well between FEM and PLECS. This U-shape curve reveals the fact that the more centralised the secondary winding is located with respect to the primary winding, the stronger is the coupling and thus the lower the short-circuit impedance will be. On the other hand, Fig. 15 accounts for the case when the secondary winding W_{s1} at the very top is excited, while the other 8 secondary windings are shorted one by one. Also in this case FEM and PLECS shows good match, and the rising curve demonstrates that the closer the secondary winding are located to each other, the better the coupling and the lower the short-circuit inductance value will

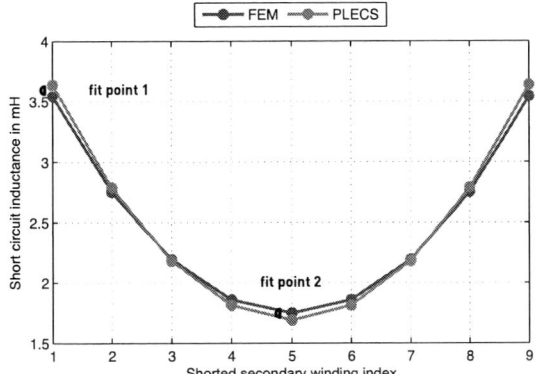

Fig. 14: Comparison to FEM with W_p excited, W_s shorted

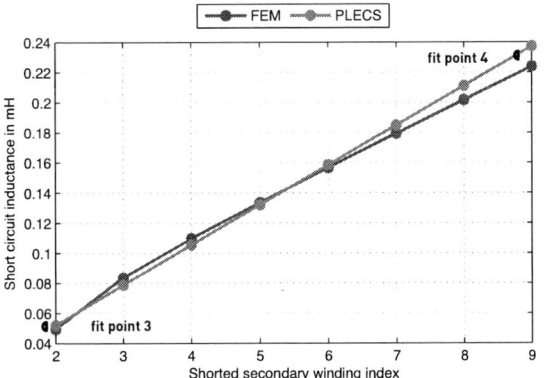

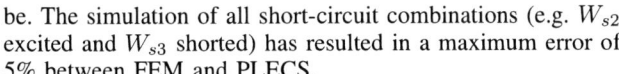

Fig. 15: Comparison to FEM, W_{s1} excited, other W_s shorted

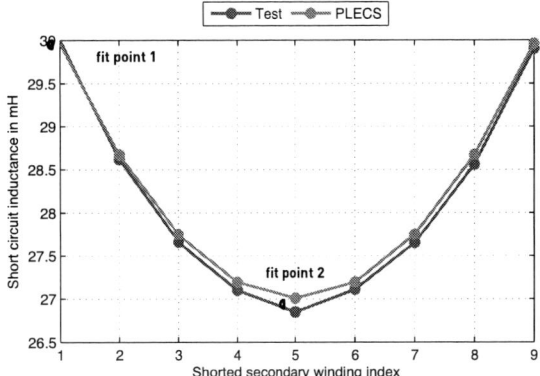

Fig. 16: Comparison to test with W_p excited, W_s shorted

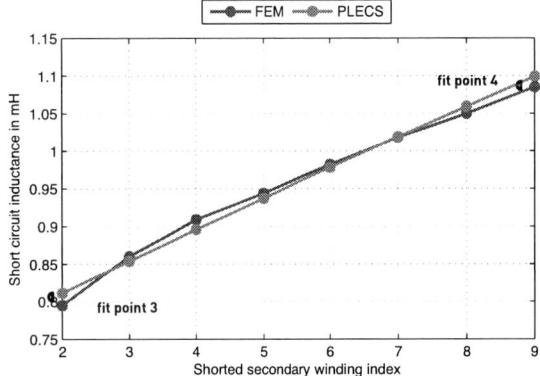

Fig. 17: Comparison to test, W_{s1} excited, other W_s shorted

be. The simulation of all short-circuit combinations (e.g. W_{s2} excited and W_{s3} shorted) has resulted in a maximum error of 5% between FEM and PLECS.

As a further step, the transformer prototype was tested experimentally as shown in Fig. 18. The bode analyser BODE 100 from Omicron has been used to measure the short circuit impedance at 50Hz, the inductive part of the impedance values was extracted for the parameter fitting and verification. The transformer prototype has undertaken the same short-circuit schemes as by FEM simulation. According to the four test points, the permeance magnetic circuit model has been fitted again as shown in Fig. 16 and Fig. 17. The leakage permeance values are listed in TABLE III

TABLE III: Leakage permeance values fitted to test $[Wb/A]$

P_s	P_{long}	P_{shunt}
$3000 \cdot 10^{-9}$	$440 \cdot 10^{-9}$	$50 \cdot 10^{-9}$

Attention needs to be paid that the short circuit inductance values from hardware test are significantly larger than that from FEM simulation. This is due to the fact that part of the windings exposes outside the window area of the core, which

was not taken into account by the 2-D FEM model. This part of flux path can be modelled as an additional permeance network as depicted in Fig. 19, which is characterised by two groups of additional permeances P_{sAn} and P_{longAn}. Since P_{sAn} and

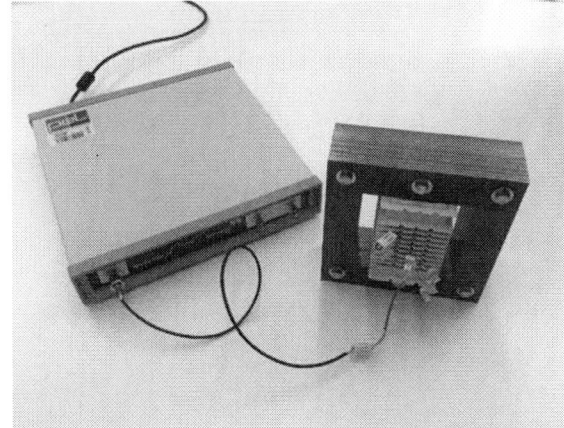

Fig. 18: Test environment of the transformer prototype

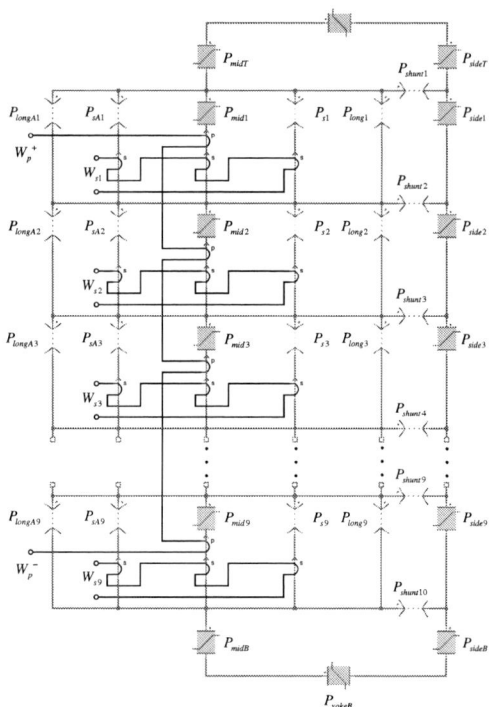

Fig. 19: Permeance circuit model of the prototype transformer including the air flux path outside the window area

P_{longAn} are linear and can be merged in to P_{sn} and P_{longn}, the final magnetic circuit should still have the same structure as in Fig. 13. In this way, the proposed model is still able to match the experimental test scheme quite well with an maximum error of 3%.

IV. CONCLUSION

This paper has demonstrated the modelling of multi-winding transformer using permeance-based magnetic circuit, which can be seamlessly integrated into system-level simulation of power electronic, with direct interface to the electrical circuit. Via making use of information about the repetitive geometry, the parameter identification process of the leakage flux path can be much simplified. The result from the proposed model shows good match to both FEM simulation and hardware measurements, under two kinds of

short-circuit test schemes. Future work will be invested in extending this approach to 3-phase case with different winding configurations (e.g. star, delta, mixed delta), and connection of power electronics converters on the secondary winding will be also a next step to further verify the fidelity.

REFERENCES

[1] C. Marxgut, J. Muehlethaler, F. Krismer, and J. W. Kolar, "Multiobjective optimization of ultraflat magnetic components with pcb-integrated core," in *IEEE Transactions on Power Electronics*, vol. 28, no. 7, 2013, pp. 3591–3602.

[2] I. Hernandez, F. de Leon, J. Canedo, and J. C. Olivares-Galvan, "Modelling transformer cores joints using gaussian models for the magnetic flux density and permeability," in *IET Electric Power Applications*, vol. 4, 2010, pp. 761–771.

[3] J. van Vlerken and P. G. Blanken, "Lumped modelling of rotary transformers, heads and electronics for helical-scan recording," in *IEEE Transactions on Magnetics*, vol. 31, no. 2, 1995, pp. 1050–1055.

[4] J. Cale, S. D. Sudhoff, and L.-Q. Tan, "Accurately modeling ei core inductor using a high-fidelity magnetic equivalent circuit approach," in *IEEE Transactions on Magnetics*, vol. 42, no. 1, 2006, pp. 40–46.

[5] C. Alverez-Marino, F. de Leon, and X. M. Lopez-Fernandez, "Equivalent circuit for the leakage inductance of multiwinding transformers: unification of terminal and duality models," in *IEEE Transactions on Power Delivery*, vol. 27, no. 1, 2012, pp. 353–361.

[6] M. Lambert, M. Matinez-Duro, J. Mahseredjian, F. de Leon, and F. Sirois, "Transformer leakage flux models for electromagnetic transients: Critical review and validation of a new model," in *IEEE Transactions on Power Delivery*, vol. 29, no. 5, 2014, pp. 2180–2188.

[7] M. D. Manjrekar, P. K. Steimer, and T. A. Lipo, "Hybrid multilevel power conversion system: A competitive solution for high-power applications," in *IEEE Transactions on Industry Applications*, vol. 36, no. 3, 2011, pp. 834–841.

[8] M. Lambert, F. Sirois, M. Matinez-Duro, and J. Mahseredjian, "Analytical calculation of leakage inductance for low-frequency transformer modelling," in *IEEE Transactions on Power Delivery*, vol. 28, no. 1, 2013, pp. 507–515.

[9] D. Hamill, "Lumped equivalent circuits of magnetic components: the gyrator-capacitor approach," in *IEEE Transactions on Power Electronics*, vol. 8, 1994, pp. 97–103.

[10] J. Allmeling, W. Hammer, and J. Schönberger, "Transient simulation of magnetic circuits using the permeance-capacitance analogy," in *Control and Modeling for Power Electronics (COMPEL), IEEE 13th Workshop on*, 2012.

[11] E. C. Snelling, *Soft Ferrite: Properties and Applications*. London Iliffe Books Ltd., 1969.

[12] M. Luo and D. Dujic, "Permeance based modelling of the core corners considering magnetic material nonlinearity," in *Annual Conference of the IEEE Industrial Electronics Society (IECON)*, 2015, pp. 950–955.

[13] B. Azzerboni, E. Cardelli, G. Finocchio, and F. L. Foresta, "Remarks about preisach function approximation using lorentzian function and its identification for nonoriented steels," in *IEEE Transactions on Magnetics*, vol. 39, no. 5, 2003, pp. 3028–3030.

Modeling Magnetic Devices using SPICE: Application to Variable Inductors

J. Marcos Alonso[1], Gilberto Martínez[1,2], Marina Perdigão[3,4], Marcelo Cosetin[5], Ricardo N. do Prado[5]

(1) University of Oviedo, Electrical Eng. Dept., Campus de Viesques, Gijón, Asturias, Spain.
(2) Continental Automotive R&D; ID HMI Hardware Development Dept., Jalisco, México.
(3) Instituto de Telecomunicações, University of Coimbra, DEEC, Coimbra.
(4) IPC, Instituto Superior de Engenharia de Coimbra, ISEC, DEE, Portugal.
(5) Federal University of Santa Maria, Group of Intelligence in Lighting (GEDRE), Brazil.

marcos@uniovi.es; gilbertomar9@hotmail.com; perdigao@isec.pt; mcosetin@gedre.ufsm.br; ricardo@gedre.ufsm.br

Abstract—In this paper a methodology to develop SPICE-based models of complex magnetic devices is presented. The proposed methodology is based on a reluctance equivalent circuit (REC), which allows the user to study both the magnetic and electric behavior of the structure under any operating conditions. The different elements required to implement the reluctance model, namely, constant reluctances, variable reluctances and windings, are implemented using SPICE behavioral modeling. These elements can thus be used to build a complete model for any magnetic device. The modeling process is illustrated with a particular example for a variable inductor. Simulations and experimental results are presented and compared to evaluate the accuracy and usefulness of the proposed modeling procedure.

I. INTRODUCTION

Modeling magnetic devices, such as inductors and transformers, has been a particularly important topic in power electronics. The great advances achieved in the last decades in the power electronics area in terms of miniaturization, efficiency improvement and reliability, would have been impossible without new magnetic materials, devices and modeling techniques. Understanding the real behavior of magnetic devices is a key issue to improve its performance, and that of the whole power electronics converter. Additionally, in today's power electronics applications, more complex magnetic structures are being employed, such as variable inductors (VI) and transformers (VT), saturable inductors (SI) and transformers (ST), etc. [1]-[8].

One of the more accurate and painless ways to study the behavior of magnetic structures is by using computer simulations. Basically, there are two methods to address this task: (i) Finite Elements Analysis (FEA) and (ii) SPICE-based behavioral models.

Surely, FEA provides the best accuracy, especially when using 3-D models, but it takes considerable time to develop a 3-D model of a magnetic device. Moreover, the simulation time using FEA is very long; it can take many hours or even days to obtain the final results. Convergence problems may arise, breaking the simulation and wasting many hours of time. Any change or improvement in the model means a new simulation, making the analysis process quite painful and strenuous. On the other hand, computer simulation based on SPICE-like models results much more friendly; simulations can be done in seconds, or minutes at the most. Of course, accuracy is lower than using FEA, nonetheless good enough for many applications.

In this paper, a SPICE-based model for computer simulation of any magnetic structure is presented. The proposed model is inspired in several models previously presented in the literature [9]-[14], which employed a reluctance or permeance equivalent circuit to simulate both the electrical and magnetic behavior of the magnetic structure. Particularly interesting is the methodology used in PSIM simulator [14]. However, the authors have found it too closed to allow the users to attain full benefit of it. This paper explores a similar methodology based on a reluctance model in which the reluctances depend on the magnetic flux level. A variable magnetic permeability, which is derived from the $B - H$ curve of the magnetic material, is used to model the change of the reluctance. In this way, the different elements present in any magnetic structure, such as variable reluctances, air gaps, constant reluctances and windings, will be modelled separately. Thus, they can be used to build an equivalent circuit of any magnetic structure. The model can be employed under any operating conditions, including dc and ac operation in any winding. In addition, the model will be able to provide electric and magnetic outcomes as voltage, current, magnetic flux, magnetic flux density, inductance, and so forth.

The original motivation to develop this work has been the investigation of the behavior of both saturable and controllable inductors and transformers. These devices have been studied and employed in power electronics applications for many years now with encouraging results [1]-[23]. Particularly, controllable (variable) inductors are complex magnetic devices

---

This work has been supported by Spain national government and Asturias regional government under research grants ENE2013-41491-R and GRUPIN14-076, respectively.

that deal with several windings under ac and dc superposed operation. In these devices, the reluctance of some sections of the magnetic material experiences a great variation owing to the dc component of the magnetic flux, even though the ac component remains under the saturation level. This makes a conventional analysis of these devices very laborious, while by using computer simulations it can be carried out very easily. Also, computer simulations allow the designer to change core geometry, magnetic properties, number of turns, air gap length, current and voltage levels, etc., making design and redesign processes extremely straightforward.

In Section II a review of the variable inductor structure that will be used as a modeling example is carried out. Section III presents the modeling of the different elements of the reluctance equivalent model of a magnetic device. Section IV shows the implementation of the variable inductor SPICE model using the proposed methodology. Section V presents simulation and experimental results. Finally, the conclusions of the paper are provided in Section VI.

II. REVIEW OF THE SELECTED VARIABLE INDUCTOR STRUCTURE

In this paper, the variable inductor (VI) structure based on a double E core, as shown in Fig. 1, will be employed as an example to illustrate the use of the proposed modeling technique.

The detailed operation of this device has been presented in previous literature [5][7][8]. A summary of its structure and operation is as follows. A double-E core with a gapped center arm is used, while there is no gap in the left and right arms. The main coil is wound on the center arm with a given number of turns (N_p), which will implement the main inductor. Two auxiliary windings are placed on the left and right arms of the core, with an equal number of turns (N_{dc}) and with reversed polarity, so that the ac voltages generated across them tend to cancel each other when connected in series. Actually, as demonstrated in [10], a full cancellation is not possible due to the non-linear behavior of the magnetic material $B - H$ curve around the saturation knee. This is one of the aspects that would be very difficult to tackle by an analytical study. However, it is possible to deal with this issue quite straightforwardly by computer simulation.

In the VI structure, a dc current is injected through the auxiliary windings. This generates a dc flux bias that circulates mainly through the outer part of the core, because the center arm is gapped and exhibits a much higher reluctance. On the other hand, the ac flux circulating through the main winding splits into the outer arms, as shown in Fig. 1a. The dc flux is used to bias the operating point of the magnetic material within the $B - H$ curve, thus modifying the reluctances and changing the value of the inductance seen from the main winding terminals.

This device can be modelled by using the reluctance circuit shown in Fig. 2. In this circuit, $\mathfrak{R}_c$, $\mathfrak{R}_l$ and $\mathfrak{R}_r$ represent the reluctances of the center, left and right arms respectively and, in a general case, their value depend on the dc operating point

of the magnetic material, and can therefore be expressed as a function of the magnetic permeability of the material. $\mathfrak{R}_g$ represents the air gap reluctance and can be assumed as a constant.

The other components in the circuit are the voltage sources F_{dc} and F_p, which represent the magnetomotive forces (MMF) created by the auxiliary and main windings respectively, which are given by the turns by current product. Next section will present how the different elements of the reluctance model can be implemented following the methodology presented in this work.

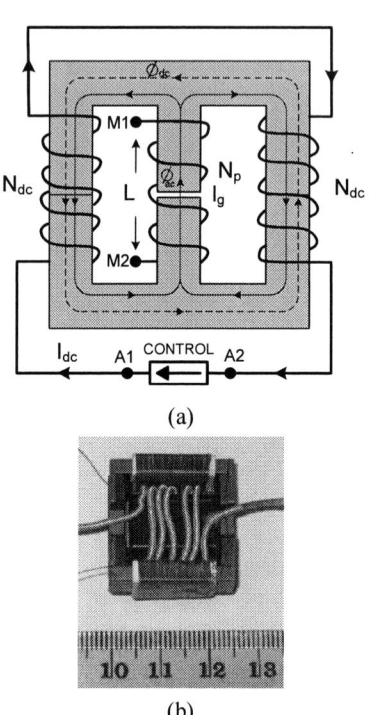

(a)

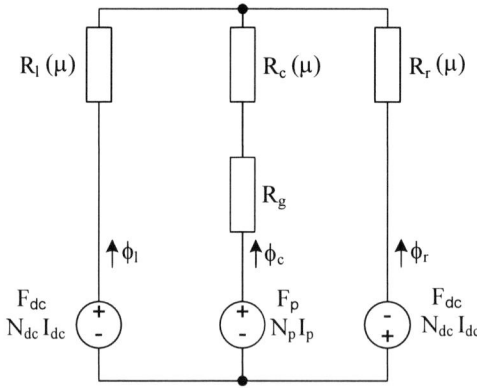

(b)

Fig. 1.   Physical structure of a VI implemented in an EE core.

Fig. 2.   Reluctance equivalent circuit of the VI shown in Fig. 1.

III. BASIC ELEMENTS OF THE RELUCTANCE EQUIVALENT CIRCUIT

The basic elements of a reluctance equivalent circuit are three: (i) constant reluctances, which model an air gap or any other non-ferromagnetic material used in the core structure; (ii) variable reluctances, used to model the non-linear behavior of the magnetic material of the core, and (iii) windings, which model the interaction between the electric and magnetic quantities involved in the behavior of any magnetic device. Once these elements are independently modelled for computer simulation, they can be used to implement any magnetic structure model, disregarding its complexity.

A. Constant Reluctance Model

A constant reluctance is used to model the behavior of a non-ferromagnetic section of the magnetic device. In the reluctance equivalent circuit, a constant reluctance is modelled by a resistor, whose value is defined as shown in (1).

$$\mathfrak{R}_0 = \frac{l_0}{\mu_0 A_0 \nu}, \tag{1}$$

where l_0, A_0 are the length and section of the constant reluctance element, $\mu_0 = 4\pi 10^{-7} H/m$ is the permeability of free space, and ν is the fringing coefficient, which is equal to 1 when the fringing effect is disregarded. However, a value slightly higher than 1 will usually render a better accuracy. Note that reluctance units are H^{-1}.

B. Variable Reluctance Model

A variable reluctance models the non-linear behavior of the magnetic material employed in the device. It can be expressed as shown in (2):

$$\mathfrak{R}_m(\mu) = \frac{l_m}{\mu(B) A_m}, \tag{2}$$

where l_m, A_m are the length and section of the variable reluctance element, and $\mu(B)$ is the absolute magnetic permeability of the material, expressed as a function of the magnetic flux density B.

As can be seen in (2), in this type of element the reluctance is a function of the magnetic permeability of the material, which in turn depends on the dc operating point within the material $B - H$ curve. Therefore, at this point it is necessary to find a model for the $B - H$ characteristic of the magnetic material, from which the permeability can be obtained.

In this work, after testing other possibilities, the Brauer's model of the $B - H$ curve has been selected [24][25]. This model defines the relationship between the magnetic field intensity H and the magnetic flux density B, by expressing H as a function of B as shown in (3).

$$H(B) = \left(k_1 e^{k_2 B^2} + k_3\right) B, \tag{3}$$

where k_1, k_2 and k_3 are the Brauer's model constants for each magnetic material.

For the sake of simplicity, in this first work the hysteresis effect is being neglected. Also, it must be noted that (3) gives the $B - H$ relationship only in the first quadrant. It must be extended to the third quadrant taking into account that the curve is symmetrical with respect to the origin of coordinates. This is very easily done in SPICE using the built-in function "*if*".

From (3) it is very simple to obtain by differentiation the differential permeability of the magnetic material. The result is given in a closed form by (4). Also, note that (4) is valid for positive and negative values of B, because the $B - \mu$ curve is symmetrical with respect to the vertical axis.

$$\mu(B) = \left[k_1(1 + 2k_2 B^2)e^{k_2 B^2} + k_3\right]^{-1}. \tag{4}$$

As an example, which will also be used to test the proposed model against experimental measurements, the N87 material from TDK-EPCOS has been modelled [26]. The Brauer's model coefficients were derived from the graphical information given by the manurfacturer's datasheet [26], obtaining the $B - H$ curve data points and using a mathematical software for curve-fitting. The resulting values are shown in Table I. Fig. 3 shows a comparison between manufacturer's datasheet information and Brauer's model, both for N87 material at 25 ºC. With this approximation, the maximum error was obtained at $B = 0.3 T$ and was calculated as 26%.

TABLE I. BRAUER'S COEFFICIENTS FOR N87 MATERIAL AT 25ºC

Parameter	Value
k_1	$0.062\ Am^{-1}T^{-1}$
k_2	$42.995\ T^{-2}$
k_3	$302.904\ Am^{-1}T^{-1}$

Fig. 4 shows the implementation of the variable reluctance component in SPICE, particularly in LTspice, which is the software that has been selected in this work [27]. As can be seen in Fig. 4, a behavioral voltage source E_m is used to implement a resistive behavior that models the variable reluctance. The behavioral current source G_{mb} generates a current equal to the flux density in the magnetic element, which is transformed into a voltage at node B_m. The behavioral current source G_{mu} generates a current equal to the actual permeability of the magnetic material by implementing (4). Finally, the voltage at node u_m is used in the expression of E_m to define the component reluctance by using (2).

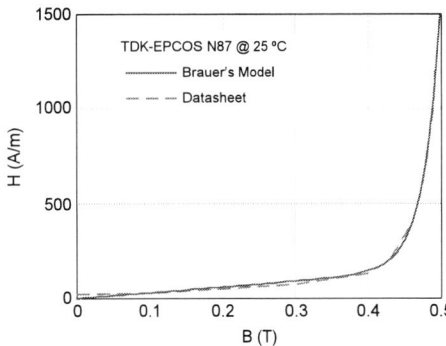

Fig. 3. B-H curve of N87 material at 25 ºC. Comparison between datasheet information and Brauer's model.

```
.func Hu(B) { ( k1*exp(k2*B*B) + k3 )*B } ; H-B curve, 1st quadrant
.func H(B) { if( B>=0, Hu(B), -Hu(-B) ) } ; H-B curve, 1st & 3rd quadrants
.func u(B) { 1/( (1+2*k2*B*B)*exp(k2*B*B) + k3 ) } ; differential permeability
```

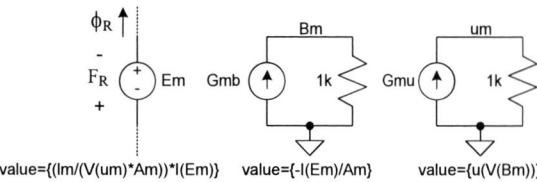

value={(Im/(V(um)*Am))*I(Em)}   value={-I(Em)/Am}   value={u(V(Bm))}

Fig. 4.   Implementation of a variable reluctance in LTspice.

C. Winding Model

A winding model must represent the electrical and magnetic interaction within the magnetic device structure. Neglecting losses, the winding model can be expressed by (5) and (6):

$$\mathcal{F}_w(t) = N_w \cdot i_w(t), \tag{5}$$

$$v_w(t) = N_w \cdot \frac{d\emptyset_w(t)}{dt} = N_w A_w \frac{dB_w(t)}{dt}, \tag{6}$$

where $\mathcal{F}_w$ is the magnetomotive force (MMF) created by the winding inside the magnetic core, N_w is the winding number of turns, v_w and i_w are the winding voltage and current, $\emptyset_w$ and B_w are the magnetic flux and magnetic flux density in the core respectively, and A_w is the area of the core.

Fig. 5 shows the SPICE implementation of the winding model. A voltage dependent voltage source EV_w is used to implement the relationship between voltage and flux as given by (6). The behavioral voltage source EF_w is employed in the magnetic part of the model to generate the corresponding MMF according to (5). The behavioral current source GV_w generates a current equal to the magnetic flux, which is time derived using the inductance L_w with a value equal to the number of turns of the winding (N_w) so that (6) can be implemented. The 1 mΩ resistance placed in series with GV_w is used to avoid convergence issues during simulation.

IV. IMPLEMENTATION OF THE VARIABLE INDUCTOR MODEL

By using the three basic elements presented in previous section, the reluctance equivalent circuit of any magnetic structure can be implemented for SPICE simulation. In this section, a model developed for the VI shown in Fig. 1 will be presented. This particular VI has been used in a previous work for the controlling of the output voltage in dc-dc resonant converters [17].

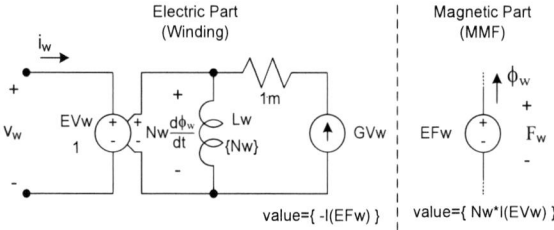

value={ -I(EFw) }   value={ Nw*I(EVw) }

Fig. 5.   A winding model implemented in LTspice.

Fig. 6 illustrates the dimensions of the EFD25 core used to calculate average lengths and sections of the magnetic paths. Table I gathers the values of all the parameters related to the VI under study.

Fig. 7 illustrates the complete electrical diagram of the VI implemented in LTspice. As can be seen, it is divided in two main parts: magnetic and electric.

The magnetic part includes the reluctance model, where the variable reluctances R_l, R_c and R_r of Fig. 2 are implemented through the voltage sources E_5, E_3 and E_4 respectively. The magnetic flux density and permeability on each arm are calculated by current sources G_5, G_3 (central arm), G_8, G_6 (right arm) and G_{11}, G_9 (left arm). Reluctances R_{sl}, R_{sc} and R_{sr} are used to measure the flux on each arm and avoid voltage-loop issues during simulation. Their value of 1 H^{-1} is very small compared to the other series reluctances and therefore have no effect on the structure. Voltage source E_1 implements the magnetic part of the main winding of the VI, while voltage sources E_7 and E_6 implement the magnetic part of the left and right auxiliary windings respectively.

The electric part of the model includes the implementation of the electric part of the three windings of the VI. Thus, sources E_2, G_1 and L_1 implement the behavior of the main winding, E_8, G_{12} and L_2 correspond to the right auxiliary winding and E_9, G_{13} and L_3 to the left auxiliary winding. Note that some additional small value resistors $(R_3, R_6, R_7, R_8, R_{13})$ are required to avoid voltage-loop and convergence issues.

The circuit shown in Fig. 7 can be employed to perform a dc transient simulation. Thus, the main winding is supplied with a 1 V dc voltage source with a 10 Ω series resistance. Auxiliary windings are supplied by using a dc current source with an output impedance of 100 kΩ. An important point that must be highlighted is that the auxiliary windings must be supplied with a dc source with a high output impedance, just as it is done in a real application. In this way, any interaction of the auxiliary source on the VI will be avoided.

The behavioral current source G_2 is used to calculate the inductance of the main winding by multiplying the flux of the main winding by its number of turns and dividing by its current.

The dc operating point simulation (.dc directive in SPICE) of the circuit shown in Fig. 7 will provide the steady state values of all magnetic and electric variables of the VI, including the inductance value, given as a voltage at node L_SPICE.

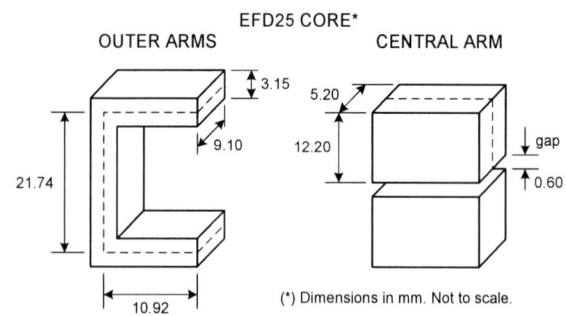

Fig. 6.   EFD25 core dimensions.

TABLE II. VARIABLE INDUCTOR DATA

Core material	N87 TDK-EPCOS	Air gap length	0.6 mm
Core type	EFD25	Estimated fringing factor (ν)	1.06
Expected inductance range	1.2 µH – 4.5 µH	Central arm average length (lc)	24.4 mm
Main winding turns (N_p)	6	Central arm section (A_c)	59.3 mm²
Intended operation regime	Sinusoidal, 500 kHz	Outer arms average length (l_1)	43.6 mm
Main winding peak current	6.0 A	Outer arms section (A_1)	28.7 mm²
Bias windings turns (N_{dc})	65		

V. SIMULATION AND EXPERIMENTAL RESULTS

The first simulation carried out was a dc operating point using the circuit shown in Fig. 7, in which a small current of 0.1 A is injected to the main winding in steady state, while the auxiliary windings carry a dc current of 0.3 A. Fig. 7 itself shows the simulation results by means of the voltage labels on each relevant node. As can be seen, the resulting inductance

value is 2.04 µH (voltage at node L_SPICE). The magnetic flux density at central, right and left arms is obtained as voltage at nodes B_c, B_r and B_l, giving 572.7 µT, 327.8 mT and -328.9 mT, respectively. Therefore, the central arm is operating with almost zero flux density (origin of the $B - H$ curve), while the right and left arms are operated at a given dc flux density level owing to their biasing by the auxiliary windings. The small difference of flux density between them is due to the small flux level (0.57 mT) generated by the main winding, which adds to the left arm, while subtracts to the right arm.

The magnetic permeability of central, right and left arms is obtained as voltages at nodes u_c, u_r and u_l, respectively. The resulting values are 3.29 mH/m, 745.8 µH/m and 723.0 µH/m, for central, right and left arms respectively. As expected, the permeability of the central arm is higher because of the lower flux density; the material of the central arm is operated around the origin of the $B - H$ curve. On the other hand, the permeability of right and left arms is much lower, because they operate at a much higher flux density level, around the knee of the $B - H$ curve.

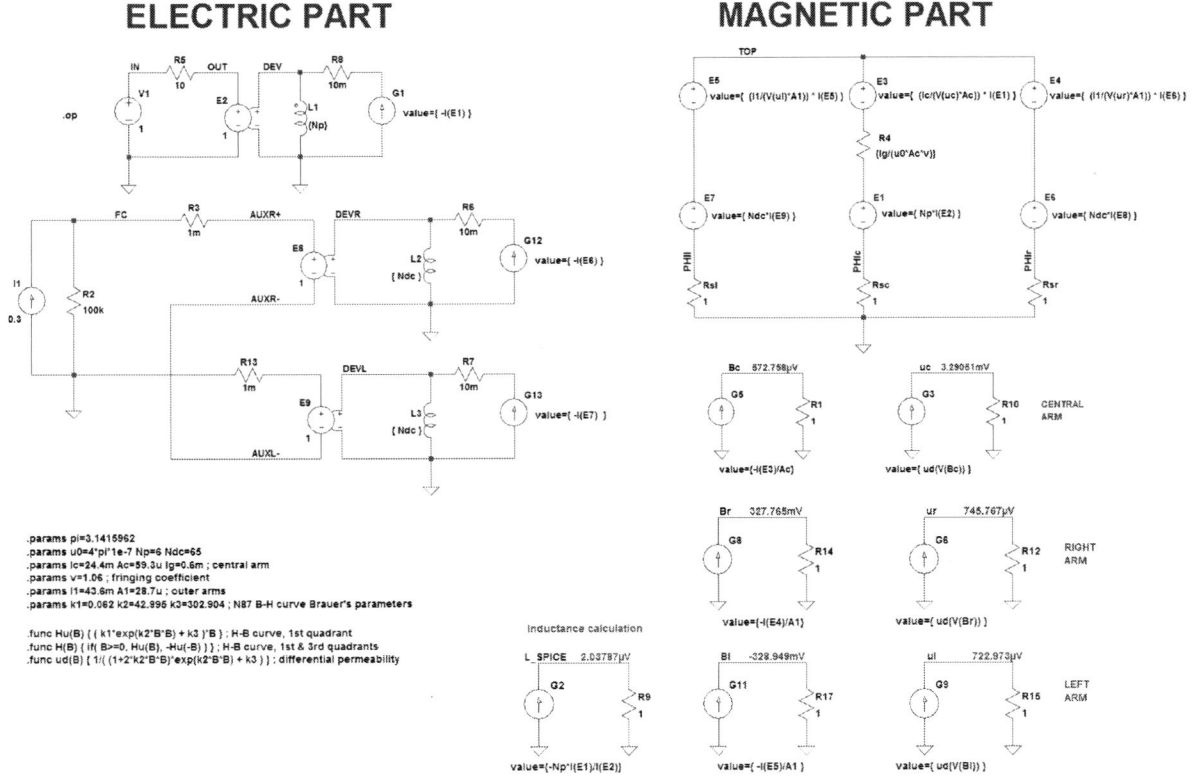

Fig. 7. Complete electrical model of the VI implemented in LTspice.

The above-presented simulation under dc operation is also intended to obtain the small-signal value of the VI inductance with very fast simulations. This can be done by performing a dc sweep analysis over the dc bias current source, shown as I_1 in Fig. 7. Thus, Fig. 8 shows the simulation results in comparison with the experimental results obtained at the laboratory by using an impedance analyzer under small signal measurement of the inductance at a frequency of 1 kHz. Experimental results were added into the simulation for the sake of comparison by using a voltage controlled source with a look-up table (not shown in Fig. 7). Relative errors of -11% at 0.125 A bias current and +25% at 0.45 A bias current can be measured.

The difference between simulation and experimental results can be justified by two causes: (i) tolerances during the magnetic material fabrication, producing that the actual material $B - H$ curve is not exactly the same as that given in the manufacturer's datasheet, and (ii) the error in the approximation of the $B - H$ curve by the selected equation, in this case the Brauer's model.

As an example, Fig. 9 shows the waveforms obtained from the same circuit shown in Fig. 7 when a transient simulation is performed. The inductor voltage and current, and the induction inside the 3 arms of the structure are shown in this figure. As can be seen, it takes some time to the induction in right and left arms to reach their steady-state values, 50 ns approximately. This time is related to the dynamic response of the VI model. In the proposed model this time is affected by the output resistance of the current source that supplies the auxiliary winding. In the example under study, an output resistance of 100 $k\Omega$ has been selected in order to have a dynamic response fast enough so that it can be used up to frequencies of several hundred of kHz. Nevertheless, this is a point that requires further investigation. The VI should be dynamically modelled, theoretically or experimentally, and the time constant should be adjusted according to the obtained results.

Fig. 10 illustrates the electrical diagram used for the simulation of a dc-ac converter with VI control. The converter has an input voltage of 20 V, applying a 10 V peak voltage square waveform to the $L - R$ circuit. The switches are operated at 500 kHz with 200 ns dead time. The load resistance is 11 Ω. In the following, simulation and experimental results corresponding to this circuit are presented.

Fig. 11 shows the simulation and experimental results corresponding to the square wave voltage and the output voltage across the load resistance for a VI bias current of 0.25 A. Fig. 12 illustrates the voltage across the VI and the output voltage for the same value of the VI bias current. As can be seen, simulation and experimental results match well.

Fig. 13 shows the simulation and experimental results corresponding to the VI bias windings. As can be seen, bias windings reflect voltages with peak values around 100 V. Also, a small difference can be seen comparing the voltages of left and right bias windings. As commented before, this difference is due to the non-linear behavior of the magnetic material around the knee of the $B - H$ curve. However, the difference is higher in the simulation than in the experiments. This

disagreement could be due to the response of the oscilloscope probe, which could not be able to follow so high a dv/dt. Also, the high voltage ripple superposed to the experimental waveform does not appear in the simulation. This ripple could also appear due to the effect of the oscilloscope voltage probe or other parasitic elements not considered by the model. The voltage probe used in these measurements was a differential voltage probe model 700924 from Yokowaga with 1:100 attenuation selected.

Fig. 14 shows the simulation results from the circuit in Fig. 10 corresponding to the instantaneous flux density in left, right and center arms of the structure, and also to the instantaneous inductance of the VI. All of them obtained from the corresponding nodes in the model. VI bias current is 0.25 A. Experimental results are not available for this variables due to the difficulty of measuring dc levels of flux density inside a magnetic material. Nevertheless, the simulation results are in accordance with the expected ones.

Finally, in order to test the possibility of controlling the output voltage by means of the bias current on the VI, Fig. 15 presents the simulation and experimental results of the rms voltage in the load resistance of the circuit in Fig. 10 when varying the dc bias current of the VI. As can be seen, simulation and experimental results exhibit a similar curve. Again, the difference between simulation and experiments can be justified by the error in the modeling of the magnetic material $B - H$ curve.

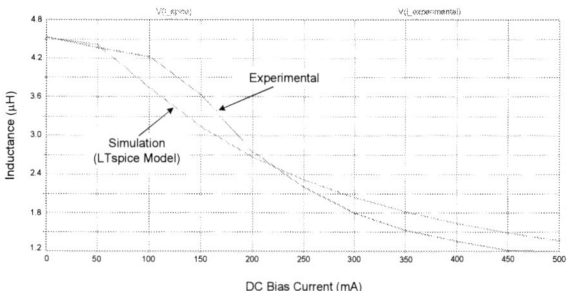

Fig. 8. Small-signal inductance of the VI. Comparison between simulation and experimental results.

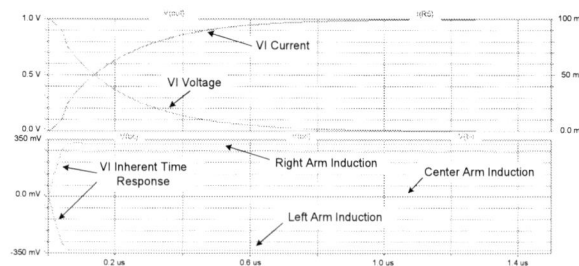

Fig. 9. Simulations results of the VI model under a dc start-up transient.

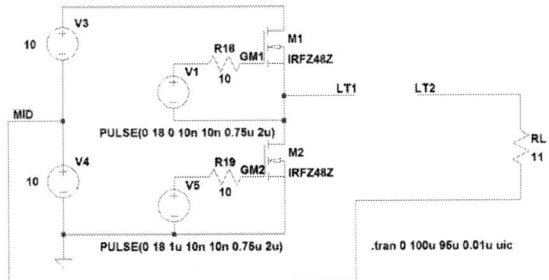

Fig. 10. Electric diagram for the simulation of an L-R inverter with variable inductor. The variable inductor is placed across terminals LT1 and LT2 according to the schematic shown in Fig. 7.

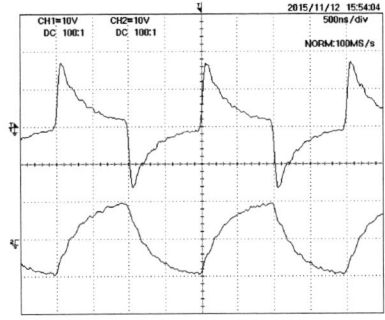

(b) 10 V/div, 500 ns/div

Fig. 12. VI voltage (top) and output voltage (bottom) of the circuit shown in Fig. 10: (a) simulation, (b) experimental. VI bias current 0.25 A.

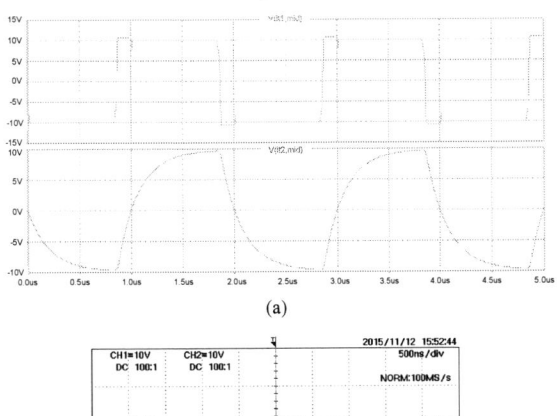

(a)

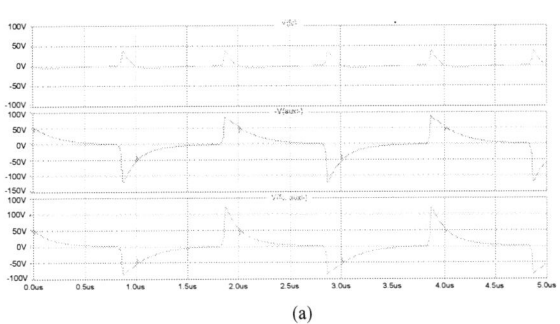

(a)

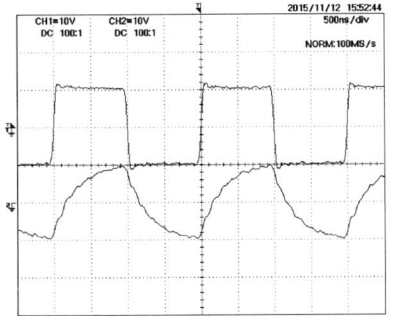

(b) 10 V/div, 500 ns/div

Fig. 11. Square voltage (top) and output voltage (bottom) of the circuit shown in Fig. 10: (a) simulation, (b) experimental. VI bias current 0.25 A.

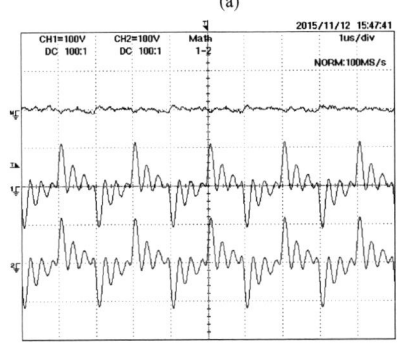

(b) 100 V/div, 500 ns/div

Fig. 13. Top: total voltage across VI bias windings. Middle: voltage across left bias winding. Bottom: voltage across right bias winding. All of them corresponding to the circuit shown in Fig. 10. (a) simulation, (b) experimental. VI bias current 0.25 A.

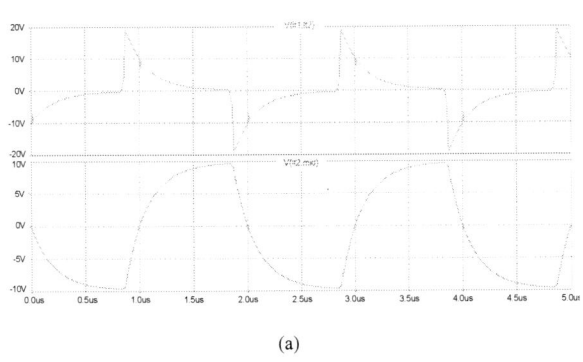

(a)

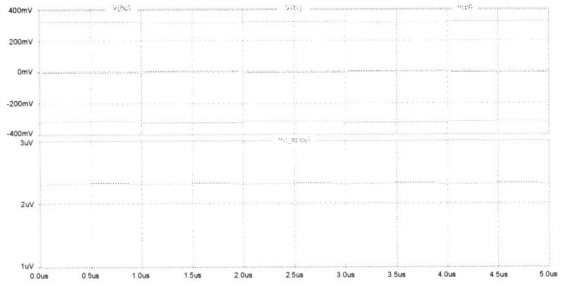

Fig. 14. Simulation results of the circuit in Fig. 10 with VI bias current of 0.25 A. Top: flux density in each VI arm. Right arm (top), center arm (middle) and left arm (bot.). Scales: 200 mT/div. 500 ns/div. Bottom: inductance of the VI as obtained from the model. Scales: 1 μH/div, 500 ns/div.

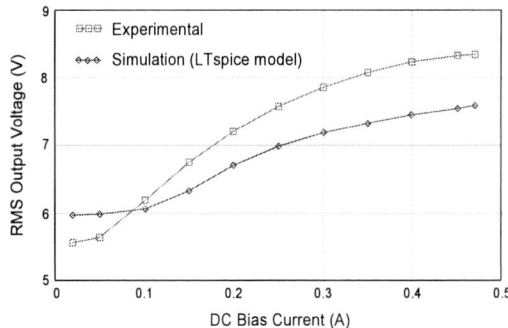

Fig. 15. Comparison between experimental and simulation results. RMS output voltage as a function of VI dc bias current for the converter shown in Fig. 10.

VI. CONCLUSIONS

This paper has presented a modeling technique for magnetic devices based on SPICE behavioral modeling. The three basic elements required for developing a reluctance-based equivalent circuit have been implemented in SPICE. These elements can easily be integrated in a SPICE library in order to simplify its use in circuit simulation. It has been explained how by using these three elements any magnetic structure can be modeled and simulated providing magnetic, electric and inductance results. Besides, the proposed modeling technique can be used under any operating conditions with good accuracy, because the non-linear behavior of the magnetic material is taken into consideration. A modeling example of a VI with an EE structure has been presented and simulated, showing good approximation with experimental results. Simulation and experimental results for a VI-controlled dc-ac inverter have also been presented. The comparison between simulated and experimental results rendered good expectancies for the proposed modeling methodology.

APPENDIX I

LTspice models presented in this work are available to interested readers at the web site indicated below. See resources section. www.unioviedo.net/ate/marcosaa/

REFERENCES

[1] Kislovski, A. S.; "Quasi-linear controllable inductor," Proc. of the IEEE, Vol. 75, No. 2, Feb. 1987, pp. 267-269.

[2] Kislovski, A. S.; "Linear variable inductor in dc current sensors utilized in telecom solar battery chargers," Proc. on International Telecom. Energy Conf., 1989, pp. 1-3.

[3] Kislovski, A. S.; "Linear variable inductor (LVI) in single-phase off-line telecom rectifiers," Proc. of IEEE Int. Telecom. Energ. Conf. (INTELEC), 1995, pp. 93-98.

[4] Birx, Daniel L., and Louis L. Reginato; "Saturable inductor and transformer structures for magnetic pulse compression." U.S. Patent No. 4,928,020. 22 May 1990.

[5] Medini, D., Ben-Yaakov S.; "A current-controlled variable-inductor for high frequency resonant power circuits," Applied Power Electronics Conference and Exposition, pp. 219-225, vol. 1, 13-17 Feb. 1994.

[6] Wölfle, Werner Hugo, and William Gerard Hurley; "Quasi-active power factor correction with a variable inductive filter: theory, design and practice." Power Electronics, IEEE Transactions on 18.1 (2003): 248-255.

[7] Alonso, J. M.; Dalla Costa, M. A.; Cardesín, J.; Garcia, J.; "Magnetic dimming of electronic ballasts", Electronic Letters, vol. 41, n° 12, June 2005.

[8] Perdigão, M. S.; Alonso, J. M.; Vaquero, D.G.; Saraiva, E. S.; "Magnetically Controlled Electronic Ballasts With Isolated Output: The Variable Transformer Solution," Industrial Electronics, IEEE Transactions on , vol. 58, no. 9, pp. 4117-4129, Sept. 2011.

[9] Hamill, D.C.; "Gyrator-capacitor modeling: a better way of understanding magnetic components," Applied Power Electronics Conference and Exposition, vol. 1, pp.326-332, 1994.

[10] Rozanov, E.; Ben-Yaakov, S.; "Analysis of current-controlled inductors by new SPICE behavioral model," HAIT Journal of Sc. and Eng. B, Vol. 2, Iss. 3-4, pp. 558-570, 2005.

[11] Ngo, K.D.T.; "Subcircuit modeling of magnetic cores with hysteresis in PSpice," Aerospace and Electronic Systems, IEEE Transactions on , vol. 38, no. 4, pp. 1425-1434, Oct. 2002.

[12] Perdigão, M.S.; Alonso, J.M.; Dalla Costa, M.A.; Saraiva, E.S.; "A variable inductor MATLAB/Simulink behavioral model for application in magnetically-controlled electronic ballasts," Power Electronics, Electrical Drives, Automation and Motion, International Symposium on, pp. 349-354, 11-13 June 2008.

[13] Perdigão, M. S.; "Research and Development on New Control Techniques for Electronic Ballasts based on Magnetic Regulators," Ph. D. dissertation, University of Coimbra, 2011.

[14] "Tutorial on How to Define the Saturable Core Element." PSIM Software. Powersim Inc. July 2006.

[15] Alonso, J. M.; Dalla Costa, M. A.; Rico-Secades, M.; Cardesín, J.; Garcia, J.; "Investigation of a New Control Strategy for Electronic Ballasts Based on Variable Inductor," IEEE Trans. on Industrial Electronics, Vol. 55, N° 1, pp. 3-10, January 2008.

[16] Alonso, J. M.; Perdigao, M. S.; Gacio, D.; Campa, L.; Saraiva, E.S.; "Magnetic control of DC-DC resonant converters provides constant frequency operation," Electronics Letters , vol. 46, no. 6, pp. 440-442, March 18, 2010.

[17] Alonso, J.M.; Perdigão, M.S.; Vaquero, D.G.; Calleja, A.J.; Saraiva, E.S.; "Analysis, Design, and Experimentation on Constant-Frequency DC-DC Resonant Converters With Magnetic Control," Power Electronics, IEEE Transactions on , vol. 27, no. 3, pp. 1369-1382, March 2012.

[18] Perdigao, M.S.; Alonso, J.M.; Dalla Costa, M.A.; Saraiva, E.S.; "Comparative Analysis and Experiments of Resonant Tanks for Magnetically Controlled Electronic Ballasts," Industrial Electronics, IEEE Transactions on, vol.55, no.9, pp. 3201-3211, Sept. 2008.

[19] Perdigao, M.S.; Trovao, J.P.F.; Alonso, J.M.; Saraiva, E.S.; "Large-Signal Characterization of Power Inductors in EV Bidirectional DC–DC Converters Focused on Core Size Optimization," Industrial Electronics, IEEE Transactions on , vol.62, no.5, pp.3042-3051, May 2015.

[20] Perdigao, M.S.; Alonso, J.M.; Saraiva, E.S.; "Magnetically-controlled dimming technique with isolated output," Electronics Letters, vol. 45, no. 14, pp. 756-758, July 2009.

[21] Perdigao, M.S.; Alonso, J.M.; Dalla Costa, M.A.; Saraiva, E.S.; "Using Magnetic Regulators for the Optimization of Universal Ballasts," Power Electronics, IEEE Transactions on , vol.23, no.6, pp. 3126-3134, Nov. 2008.

[22] Perdigão, M.S.; Menke, M.; Seidel, A.R.; Pinto, R.A.; Alonso, J.M.; "A review on variable inductors and variable transformers: Applications to lighting drivers," IEEE Trans. on Industry Applications, Jan. 2016.

[23] Pinto, R.A.; Alonso, J.M.; Perdigao, M.S.; da Silva, M.F.; do Prado, R.N., "A new technique to equalize branch currents in multiarray LED lamps based on variable inductor," Industry Applications Society Annual Meeting, IEEE , pp. 1-9, 5-9 Oct. 2014.

[24] Brauer, John R.; "Simple equations for the magnetization and reluctivity curves of steel," Magnetics, IEEE Transactions on, vol.11, no.1, pp.81,81, Jan 1975.

[25] Hülsmann, T.; "Nonlinear Material Curve Modeling and Sensitivity Analysis for MQS-Problems," Master Thesis, Faculty of Electrical, Information and Media Engineering, Bergische Universitat Wuppertal, 2012.

[26] "Ferrite Cores and Accessories", Datasheet, EPCOS, September 2006.

[27] "LTspice IV. Getting Started Guide," Linear Technology, 2011.

Investigation of a Thermal Model for a Permanent Magnet Assisted Synchronous Reluctance Motor

Joseph Herbert, AKM Arafat, Dr. Guo-Xiang Wang, Seungdeog Choi

Department of Electrical & Computer Engineering, Department of Mechanical Engineering,
University of Akron
Akron, Ohio 44325-3904

Abstract—The use of High Power Density (HPD) electric machines in applications such as Electric Vehicles amplifies the need for their optimal thermal design in conjunction with their electromagnetic design. Permanent Magnet Synchronous Reluctance Machines (PMaSynRMs) are specialized Interior Permanent Magnet Synchronous Machines where the magnet content in the flux barrier paths is reduced resulting in a relatively economical design. This paper provides the overview of a study of a thermal model for a Five Phase Permanent Magnet Synchronous Reluctance Machine with two flux barrier paths. The proposed thermal model is then verified with experimental results and FEA simulations. Once developed, its use for on-line temperature estimation or integrated with an optimization algorithm will be considered as future work.

Index Terms—Thermal Model, Permanent Magnet Synchronous Reluctance Motor

I. INTRODUCTION

With the growing trend of the use of electric motors in applications such as electric vehicles and aircrafts, there is a corresponding growing desire for the need of high power density machines [1]. These machines are designed to be able to provide high power but are of considerably smaller size. The obstacle in the implementation and the design of such motors is that with the reduction in the size of the motor there is additionally a reduction in the surface area of heat dissipation. This results in machines where there are significant temperature rises in critical parts of the machine.

A lot of research has already been conducted in the study of the temperature vulnerability of electric machines with regards to the impact of high temperatures on machine components such as insulators and magnets and machine performance parameters such as losses, efficiency etc. The reduction of the lifetime of insulators as a function of temperature is well known leading to the thumb rule of 10 which states that with every $10°C$ temperature rise in the insulation there is a reduction in the lifetime of the insulator and the machine by half [2].The increase in the resistance of the conductors as a function of temperature is also an acknowledged phenomena which results in an increase in the I^2R copper losses of the machine.At the same time the reduction of the magnetic losses of permanent magnet machines with an increase in the temperature necessitates the need for the determination of an optimal temperature of operation for the maximum possible efficiency of the motor [3] [4].

The most critical impact of temperature on permanent magnet machines however lies in the demagnetization of the magnets which adversely effects on the operation of the machine [5]. For machines at constant load, it has been determined that as a result of demagnetization, the consequent increase in the current further adds to the I^2R copper losses and an additional reduction of the efficiency of the machine as well.

This concern over the impact of temperature rise in critical components of electric machines has lead the establishment of lumped thermal models for the evaluation of temperature at the corresponding sections and to determine the critical operation and design parameters. The common procedures adopted to conduct thermal analysis are Finite Element Analysis (FEA), Computation Fluid Dynamics (CFD) and lastly the lumped parameter thermal model [6]. Well established FEA and CFD softwares exist today which provide a very high level of resolution in determining the temperature gradient of the various portions of the electric machines with relatively high accuracy. The disadvantage to both these methods lies in the simulation time. These softwares, though extremely rigorous are unfortunately time consuming further warranting the need for a quick and relatively accurate method of temperature estimation.

Lumped parameter modeling is a reduced finite element method where various geometries of the machine are lumped together and are represented by equivalent thermal resistances. These resistances are then evaluated as an equivalent electrical network, to determine the temperature in the form of voltage at the various nodes of the circuit - Each corresponding to a particular point on the machine. Over the past two decades, thermal lumped parameter models of electric machines have been constantly evolving. The earliest literature for the thermal modeling of electric machines can be traced back to the thermal modeling of an induction motor [7]. This model was then closely followed up by a model G. Kylander [8]. The study conducted by Kylander also involved the development of a thermal model of an induction motor advocating the use of a simplified thermal model with fewer nodes to determine the temperatures. This simplified model has since been established a frame of reference for most lumped parameter thermal models of electric machines conducted so far. The model proposed in [9] for a surface mounted permanent magnet motor was among the first thermal modeling conducted for

978-1-4673-9551-9/16 $31.00 © 2016 IEEE

permanent magnet machines. Using this model as a frame of reference, a thermal model for an synchronous motor with multi barrier interior permanent magnets was developed [10]. The challenge in the modeling of the Interior Permanent Magnet Synchronous Motors (IPMSM) was the modeling of the rotor as the non-homogeneous configuration between the core and the permanent magnet portions required the estimation of an equivalent model to withstand the best possible accuracy while determining an easy method for temperature estimation. In order to accomplish this the model proposed estimating the equivalent thermal model of the rotor in the form of cylindrical segments with varying radii from the center of the rotor and angular spans based on the width of the magnetic portions. The thermal model of a synchronous reluctance machine proposed by Mahmoudi in [11] the multi barriers of the rotor did not contain any material for the thermal conduction of heat and so the modeling was confined to the rotor portions only. Superficially, the modeling of the stator and the remaining portions of internal rotor motors has almost been stabilized based on the approaches adopted by Kylander in [8] but the major modifications and challenges with the modification of rotor designs are in their thermal modeling, especially in the case of interior permanent magnet design motors.

The PMaSynRM is a special type of IPMSM where the amount of magnet in the flux barrier paths are reduced. As a matter of fact, the current model is more of a hybrid of a synchronous reluctance motor and an IPMSM where one of the flux barrier paths are empty and the other is filled with reduced magnets to provide a relative magnetic torque, which are normally much lesser than the reluctance torque. The thermal modeling of these types of motors has been conducted by [12], [13]. The rotor configuration of the motor considered in these studies is however different from the configuration of the model being evaluated in this paper. The relative complexity of the rotor configuration considering the air in one flux barrier and the air and magnet composition in the next, required a careful evaluation of the rotor heat flow characteristics. This necessitates the need for a different approach to the thermal modeling as certain areas under the rotor can be considered as inconsequential to the thermal flow because of the high thermal resistance of air.

The proposed model is focused on determining acceptable levels of rotor and magnet temperatures which poses the challenge in this study. This is needed as the NdFeB magnets being employed in this motor are highly susceptible to demagnetization which higher temperatures and would limit the temperature of operation of the machine severely. The study conducted in this paper involves the first in a series of steps to determine a thermal model for a PMaSynRM with the confuguration as shown in the Fig. 1.

The following sections will describe the PMaSynRMs and the motor specifications followed by an understanding of the thermal model, beginning with the loss modeling considered. The following subsections will discuss the thermal modeling of the motor resistances finally followed up with the experimental and simulation results for the verification of the

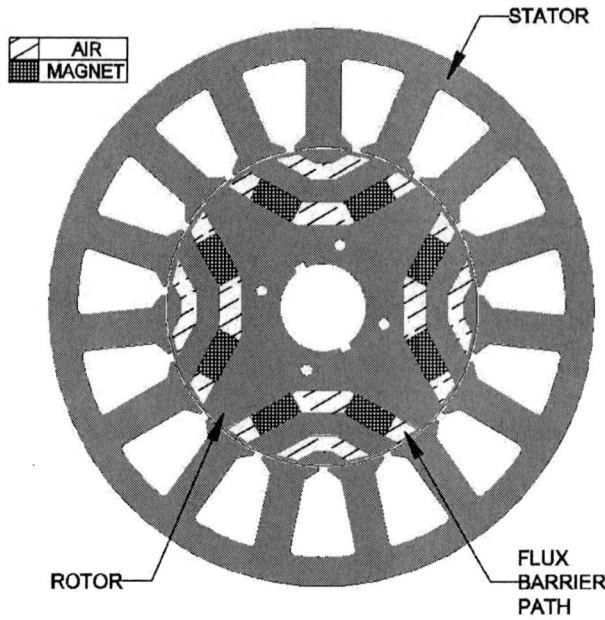

Fig. 1. Motor Layout.

proposed model.

II. PERMANENT MAGNET ASSISTED SYNCHRONOUS RELUCTANCE MOTORS

The PMaSynRM, as discussed, is a hybrid machine that consists of an IPMSM with reduced magnet content. These machines are more economical as a result of the reduced magnet content. The machine under consideration was designed using an Lumped Magnetic Model and optimized using a DES algorithm for a five phase operation and low torque ripple [14]. The winding and the control of this motor is done for five phases to provide a better fault tolerant control. From a thermal modeling perspective, this is taken care of by considering the apparent extra conductor content in the motor and the additional copper losses considering all the five phases and currents. The use of a five phase system is very useful in the fault tolerant control of motors, an area that is already being covered for this motor [15].

The layout and the specifications of the PMaSynRM under consideration is as shown in the Fig. 1 and table I respectively.

III. LUMPED PARAMETER THERMAL MODEL

An overview of the Lumped Thermal Model considered for this motor is as shown in the Fig. 2. The overall approach to the model is similar to the approaches considered in [8–11], with regards to the thermal modeling of the stator slots, windings and core. The modifications required for the five phase system has been incorporated and will be discussed under the loss analysis in the following subsections.

The machine, being a prototype design, was initially of open drip proof (ODP) design. However, suitable modifications

TABLE I
MOTOR SPECIFICATIONS

MOTOR PARAMETER	VALUE
Number of slots/poles	15
Rated current (rms)(A)	15.17
Rated voltage (rms) (V)	67
Power (kW)	3
Rated speed (rpm)	1800
Rated Torque	15
Phases	5

were made during the experimental phase to incorporate more of a totally enclosed design to use the equations employed in [8–11] in the determination of the thermal resistance between the internal air and end winding. Other areas of contention in the thermal modeling such as the thermal resistance between the frame and core was considered as twice the stator frame resistance instead as a worst case situation as proposed by [9]. A cross section of the motor superimposed on the considered thermal network is as shown in the Fig. 3.

The convection thermal resistances illustrating the interactions between the internal air and the end windings, frame and rotor surface and the air gap are all solved considering the empirical formulae considered in [8–11]. The air gap thermal resistance is a temperature and speed dependent quantity. Based on the speed of the rotor, the fluid flow may be laminar or turbulent which would correspondingly effect the calculation of this thermal resistance. Additionally, the empirical formula considered for the air gap the depends on the kinematic viscosity of the air in the path. The viscosity is dependent on the air density and the dynamic viscosity of the air, both of which are temperature dependent. This therefore requires the circuit would have to be iteratively solved so that the impact of the calculated temperature rise on the air gap thermal resistance is considered in the determination of the steady state temperature for the given operating conditions.

The current thermal network considered is only considering steady state temperature estimations. The transient and dynamic thermal model will be considered in future works.

A. Loss Analysis

The thermal circuit is generated by using thermal resistances which are dependent on design parameters and motor operation and the losses which are modeled as current sources. The main losses in electric machines which are being considered in the thermal model are as mentioned below.

1. Copper Losses or Winding Losses
2. Eddy current Losses
3. Core Losses
4. Mechanical or Friction Losses

Copper Losses: As discussed, the copper losses for the stator windings were modified for a five phase system with the losses calculated as shown in (1) where I_{RMS}^2 is the per

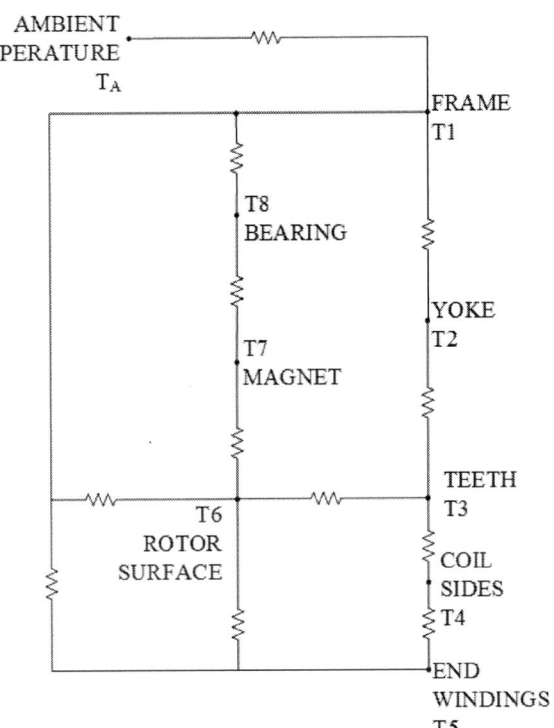

Fig. 2. Lumped Thermal Network.

phase RMS current of the windings and R is the per phase resistance of the windings.

$$P_{Copper} = 5I_{RMS}^2 R \qquad (1)$$

As the losses generated in the end windings show a different dynamic as opposed to that in the coil sides, the modeling of the equivalent resistances of these sections is done differently. This is because the dissipation of the heat to the internal air by the end windings is different when compared to the heat transfer along the tooth to the yoke by the coil sides. As these losses are dependent on the resistances of the conductors which are inherently a temperature dependent quantity, the impact of the temperature rise on the conductors and the corresponding impact on the copper losses will have to be considered iteratively as well. The modified equation for the copper losses as a function of the temperature would be as per (2).

$$P_{Copper} = 5I_{RMS}^2 R(1 + \alpha \Delta T) \qquad (2)$$

Where R would correspond to the resistance of the conductors at a temperature T, α corresponds to the temperature coefficient of resistance of the conductor material and ΔT corresponds to the temperature rise above the temperature T.

Eddy Current and Core Losses: Eddy current losses are the losses introduced into a magnet due to the induction of currents in the transverse direction of the machine along the magnets. The core losses are the losses setup due to the

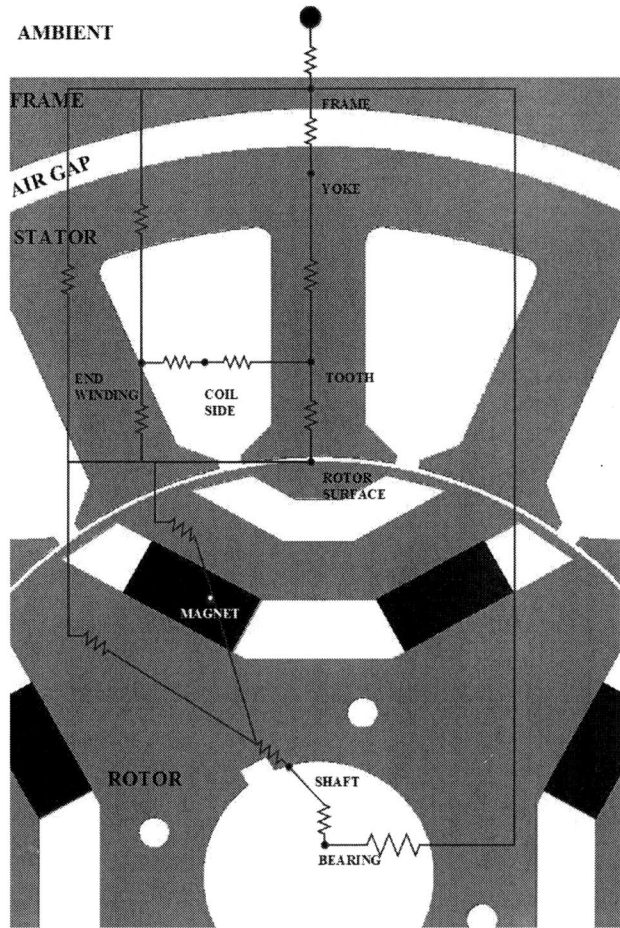

Fig. 3. Lumped Thermal Model against the motor cross section.

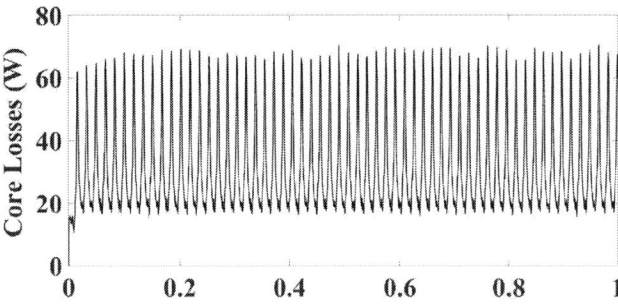

Fig. 4. Motor Core Losses at rated current Irms = 15.5A using FEA Electromagnetic software.

magnetizing and the demagnetizing of the cores with the varying magnetic field. The temperature dependency for these losses was considered negligible. The Eddy current losses and the core losses for the model were estimated using an electromagnetic FEA simulation software for this study. The results of the eddy current losses and core losses obtained for rated the I_{rms} of 15.5A are shown in the Fig. 4 and Fig. 5.

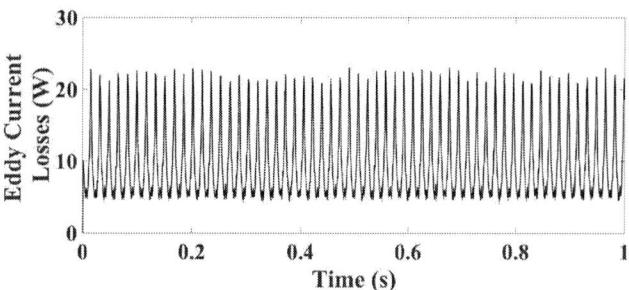

Fig. 5. Motor Eddy Current Losses at Irms = 15.5A using FEA Electromagnetic software.

Mechanical Losses: The mechanical losses are a result of friction and windage losses in the machine and are normally very negligible. These losses have not been considered as an independent current source and instead any stray additional losses obtained in the experimental phase of this study have been combined with the stator losses.

B. Determination of Thermal Resistances

The thermal resistances of the motor are calculated based on the equivalent thermal conductive resistances based on the fundamental Fourier's law of heat transfer, where the thermal resistances are determined by the formula (3)

$$R_{thermal} = \frac{l}{\lambda A} \qquad (3)$$

Where l is the direction along which the heat flows, λ is the thermal conductivity of the material and A is the area along which the heat transfer is taking place. Due to the lamination structure of the stator, the heat flow in the axial direction is neglected and the heat flow is only considered in the radial directions for the stator and the rotor cores.

The general formula for equivalent thermal convective resistances is as per the equation 4

$$R_{thermal} = \frac{1}{hA} \qquad (4)$$

Where h is known as the heat transfer coefficient. In this model all the heat transfer coefficients of the equivalent convective thermal resistances is based on the empirical formula considered in [8], [9].

C. Determination of Thermal Resistances for Rotor

The determination of the resistance of the rotor was the key challenge being approached in this research. The current structure of the rotor as shown in the Fig. 3 with two flux barrier paths and with two magnets in one of the paths, provides a relatively complex thermal structure to be studied. In studying the structure of the rotor, there were three portions to be considered in the modeling - the core, the magnets and lastly the air portions. The thermal conductivity of these three materials are as shown in the Table II.

From the thermal conductivity data in the table II and from (3) it is clear that the equivalent thermal resistances of the

TABLE II
THERMAL CONDUCIVITIES OF ROTOR MATERIALS

ROTOR MATERIAL	THERMAL CONDUCTIVITY λ (W/m-K)
Rotor Core	16.2
Rotor Magnet (NdFeB)	9
Air	0.0243

portions being modeled as air will be significantly higher than the resistances of the remaining portions. This modeling is concurrent with the thermal understanding that air is inherently a thermal insulation and would not provide a path of easy conductance of heat in the thermal circuit. As the design of the rotor was made with two flux barrier paths one only consisting of air while the other consisted of air interleaved with small magnet portions to direct the flux, it is clear that the upper barrier consisting of only air would represent a very high resistance and can therefore be approximated as an open circuit in the thermal modeling of the rotor. This hypothesis is further acknowledged in the Steady State Thermal Simulation conducted by an FEA Simulation. An FEA simulation result showing the distribution of temperature clearly shows this detail as can be seen in the Fig. 6 where the insulating influence of the temperature above the rotor flux barrier path with air demonstrates an open circuit behavior to the direction of heat flow. The heat flow from the surface of the rotor to the shaft is clearly prohibited by the flux path with air, implying the rotor circuit when implemented should only consider a path along the region between the flux paths and through the magnet and another below the lower flux path.

Based on this, the modeling of the rotor was done as shown in the Fig. 7. The flow of heat in the insulated air flux barrier path portion is neglected and therefore the thermal modeling of that portion of the rotor is completely neglected for simplification of the rotor thermal resistance determination. The remaining resistances are calculated considering an eighth of the rotor portion as the rotor is symmetric. The resistance for each section was calculated using the equation (3) with l corresponding to the flow of heat in each segmented section as shown in the Fig. 7 and A corresponding to the area under consideration for each segment considering the dimensions and the thickness of each lamination.

The resistances R_1 and R_11 are considered as cylindrical segments and the equivalent resistances are calculated using the formula for two cylinders at two different temperatures given by the equation (5).

$$R_{thermal} = \frac{ln\frac{r_{out}}{r_{in}}}{2\pi\lambda L} \frac{2\pi}{2p\theta} \qquad (5)$$

where r_{out} is the outer radius and r_{in} is the inner radius of the cylindrical segment under consideration. L is the active length of the core, λ is the the thermal conductivity of the

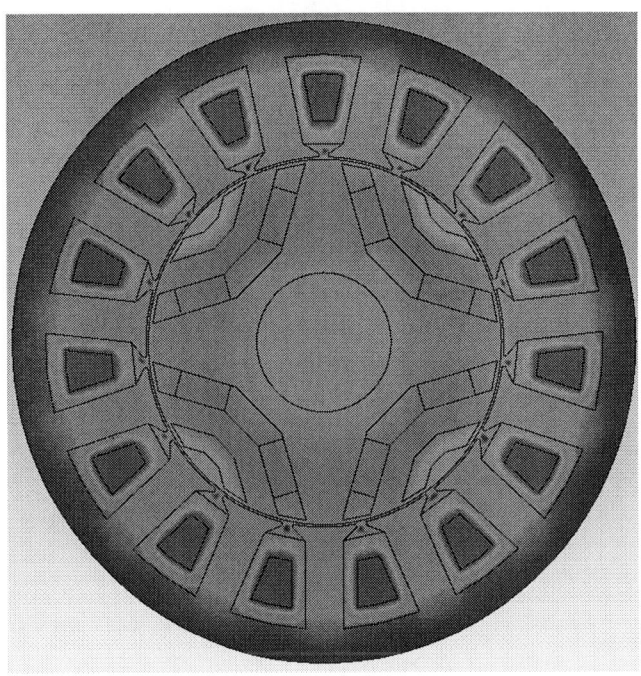

Fig. 6. A temperature distribution in the rotor illustrating the influence of the air in the flux barrier path on the heat flow.

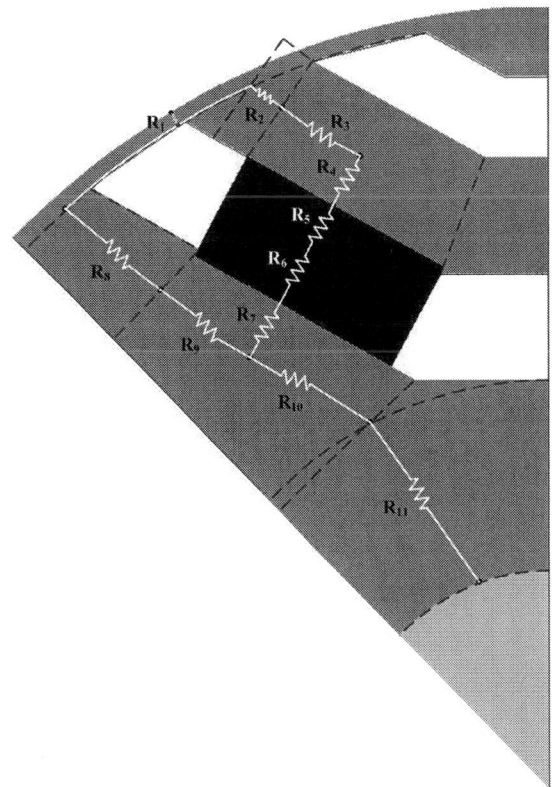

Fig. 7. Thermal model of rotor segment.

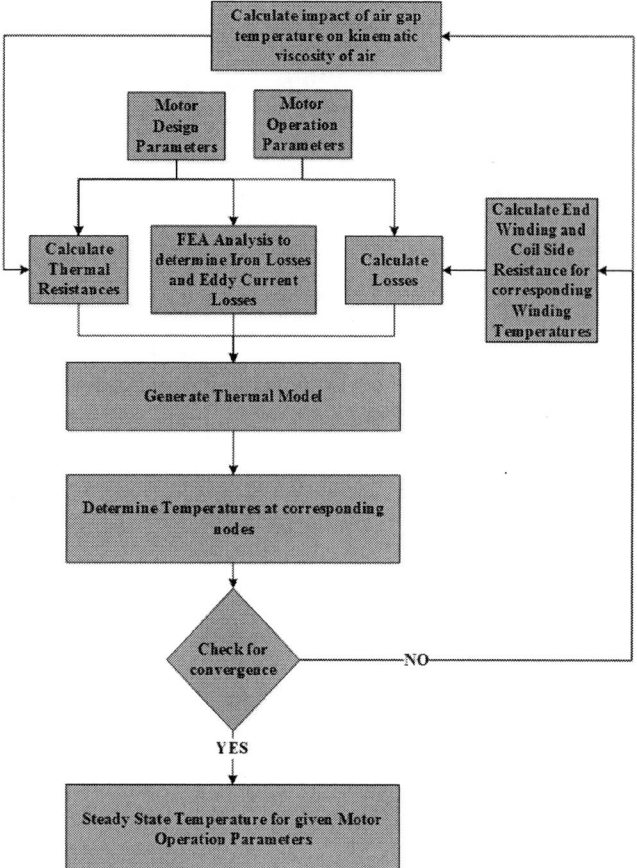

Fig. 8. Flow chart for the solving of the thermal model.

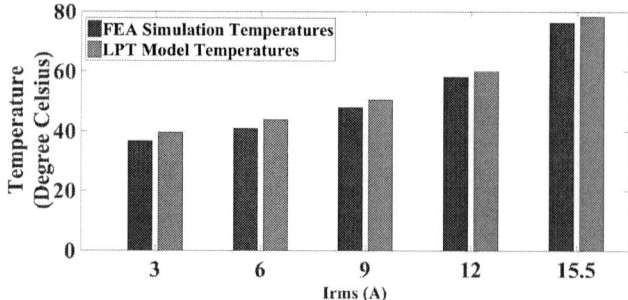

Fig. 9. Comparison of rotor magnet estimated temperatures obtained from the FEA simulation and the LPT model for different I_{rms} currents at a constant speed

material, p is the number of pole pairs and θ is the angular span of the corresponding cylindrical segment. In the thermal modeling of interior permanent magnet motors considered in [10], all the flux barrier paths were made into equivalent cylindrical segments and the equivalent thermal resistances were calculated using (5). The reason this approach was avoided in this paper was because the aim of the study in progress is to determine a model with the highest possible resolution of the magnets and the estimation was avoided to counter any possible reduction in accuracy.

D. Evaluation of the Thermal Network

Once the equivalent thermal resistances were determined, a network for the steady state temperature estimation of the system was setup. In the study conducted so far, the thermal resistances were calculated and the corresponding thermal network and temperatures were estimated using a circuit simulation software. The normal approach for resolving this network in most thermal models [8–11] involves the realization of the network in the form of a matrix which is solved iteratively till the point of convergence at the steady state.

In the evaluation of the circuit the air gap resistance between the stator tooth and the rotor surface is calculated

till convergence as it is a convective resistance and its heat transfer coefficient h which depends on the temperature of the air gap as well as the speed of the rotor [9]. At the same time, the copper losses of the machine are dependent on the temperature dependent resistances which will increase with increasing estimated temperatures. The impact of the variation of these two temperature dependent quantities till convergence will require the iterative running of the thermal resistance estimating software and the network evaluation for the estimation of the temperatures. A flowchart of the running of this iterative method is as shown in the Fig. 8.

IV. Verification of the Thermal Network

In order to verify the thermal network setup two approaches were considered. As the rotor is not accessible, the thermal network for the rotor would have to be verified using an FEA Simulation for Steady State Thermal Analysis. The stator portions are accessible and have been mounted with Type K thermocouples and real time temperature data was recorded for a particular motor operation of speed.

A. Verification with FEA Simulation

The FEA simulations conducted for the analysis of the temperature of the rotor, were conducted considering a constant speed output and by varying the current. The Fig. 9 shows the temperatures estimated for the magnet of the rotor for different currents considering a constant speed output of 1800 rpm. The values obtained through the LPT model show close correlation with those obtained using the FEA simulation softwares. The Fig. 10 shows the temperature distribution for a motor RMS current of 6A.

B. Experimental Setup

The experimental setup for the verification of the stator is as shown in the Fig. 11. The temperature data was collected with the help of Type K thermocouples mounted on the stator tooth, the stator coil sides and embedded in the stator end windings. The data collected was stored with the help of a Data Acquistion device. The plots of the transient temperature rise for these three nodes obtained via the DAQ are as shown in the Fig. 12 for $I_{rms} = 3.1$A.

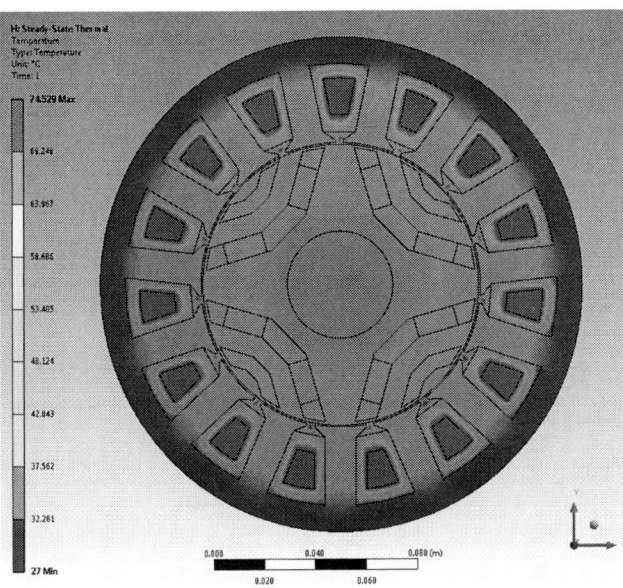

Fig. 10. Temperature distribution for I_{rms}=6A.

Fig. 11. Experimental setup using Type K Thermocouples.

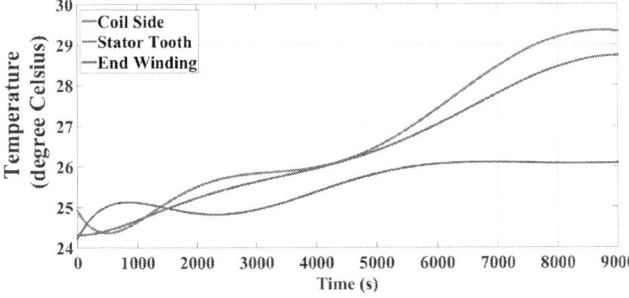

Fig. 12. Experimental temperatures obtained using Type K Thermocouples for I_{rms}=3.1 A.

A comparison of the steady state temperatures obtained using the LPT and the experimental data for these nodes at the operating condition of $I_{rms} = 3.1A$ is shown in the Table III.

TABLE III
COMPARISON OF STATOR TEMPERATURES

PART	EXPERIMENT	LPT MODEL
STATOR TOOTH	26	25.4
COIL SIDE	29.2	28.6
END WINDING	28.7	28.5

V. CONCLUSION & FUTURE WORK

The proposed thermal modeling of the rotor of the given configuration of the PMaSynRM motor is found to be a good approximate for the estimation of temperatures. The steady state temperatures determined from the LPT model have been verified experimentally for the stator and the estimated rotor magnet temperatures determined from the LPT model are in close approximation to that of the FEA simulations conducted. The suggested theory neglecting the thermally isolated segment of the rotor enclosed by the flux barrier path in the estimation of the rotor magnet temperature is a good consideration in the simplification of the complexity of the model. The approach can further be used in future studies with similar rotor configurations where flux barrier paths of similar configuration thermally isolate portions of the rotor for simplification of thermal model estimation. Further modifications in the future will attempt at improving the accuracy of the model as well as attempt at further simplifying the process of model parameter determination. The determination of a transient model and will also have to considered along with an overall sensitivity analysis.

REFERENCES

[1] P. Masson and C. Luongo, "High power density superconducting motor for all-electric aircraft propulsion," *Applied Superconductivity, IEEE Transactions on*, vol. 15, no. 2, pp. 2226–2229, June 2005.

[2] E. L. Brancato, "Estimation of lifetime expectancies of motors," *Electrical Insulation Magazine, IEEE*, vol. 8, no. 3, pp. 5–13, May 1992.

[3] J. Schutzhold and W. Hofmann, "Analysis of the temperature dependence of losses in electrical machines," in *Energy Conversion Congress and Exposition (ECCE), 2013 IEEE*, Sept 2013, pp. 3159–3165.

[4] C. Debruyne, S. Derammelaere, J. Desmet, and L. Vandevelde, "Temperature dependency of the efficiency of line start permanent magnet machines," in *Electrical Machines (ICEM), 2012 XXth International Conference on*, Sept 2012, pp. 1846–1853.

[5] T. Sebastian, "Temperature effects on torque production and efficiency of pm motors using ndfeb magnets," *Industry Applications, IEEE Transactions on*, vol. 31, no. 2, pp. 353–357, Mar 1995.

[6] A. Boglietti, A. Cavagnino, D. Staton, M. Shanel, M. Mueller, and C. Mejuto, "Evolution and modern approaches for thermal analysis of electrical machines," *Industrial Electronics, IEEE Transactions on*, vol. 56, no. 3, pp. 871–882, March 2009.

[7] P. Mellor, D. Roberts, and D. Turner, "Lumped parameter thermal model for electrical machines of tefc design," *Electric Power Applications, IEE Proceedings B*, vol. 138, no. 5, pp. 205–218, Sep 1991.

[8] G. Kylander, *Thermal modelling of small cage induction motors*. Chalmers University of Technology, 1995.

[9] J. Lindström, "Thermal model of a permanent-magnet motor for a hybrid electric vehicle," *Licentiate thesis, Chalmers University of Technology, Göteborg, Sweden*, 1999.

[10] A. EL-Refaie, N. Harris, T. Jahns, and K. Rahman, "Thermal analysis of multibarrier interior pm synchronous machine using lumped parameter model," *Energy Conversion, IEEE Transactions on*, vol. 19, no. 2, pp. 303–309, June 2004.

[11] M. Mahmoudi, "Thermal modelling of the synchronous reluctance machine," 2012.

[12] S. Nategh, O. Wallmark, and M. Leksell, "Thermal analysis of permanent-magnet synchronous reluctance machines," in *Power Electronics and Applications (EPE 2011), Proceedings of the 2011-14th European Conference on*, Aug 2011, pp. 1–10.

[13] S. Nategh, O. Wallmark, M. Leksell, and S. Zhao, "Thermal analysis of a pmasrm using partial fea and lumped parameter modeling," *Energy Conversion, IEEE Transactions on*, vol. 27, no. 2, pp. 477–488, June 2012.

[14] J. Baek, S. Bonthu, S. Kwak, and S. Choi, "Optimal design of five-phase permanent magnet assisted synchronous reluctance motor for low output torque ripple," in *Energy Conversion Congress and Exposition (ECCE), 2014 IEEE*, Sept 2014, pp. 2418–2424.

[15] A. K. M. Arafat and S. Choi, "Fault tolerant control of five-phase permanent magnet assisted synchronous reluctance motor based on dynamic current phase advance," in *Energy Conversion Congress and Exposition (ECCE), 2015 IEEE*, Sept 2015, pp. 1208–1214.

978-1-4673-9551-9/16 $31.00 © 2016 IEEE

Design Procedure for Multi-Phase External Rotor Permanent Magnet Assisted Synchronous Reluctance Machines

Sai Sudheer Reddy Bonthu
IEEE, Student Member

Seungdeog Choi
IEEE, Member

Department of Electrical and Computer Engineering
University of Akron
United States
E-mail: sb158@zips.uakron.edu, schoi@uakron.edu

Abstract— This paper presents the optimal design procedure to develop multi-phase external rotor permanent magnet assisted synchronous reluctance machines (EPMa-SynRMs). With its higher torque density and higher power density compared to the internal rotor PMa-SynRM, external rotor PMa-SynRM is best suitable in applications to electric bikes and aircrafts. Enormous amount of research has been done in optimizing internal rotor PM machines. However, an efficient optimization technique to develop a five-phase EPMa-SynRM is not presented in the literature. The optimization of the EPMa-SynRMs is important to provide better performance characteristics and controllability in terms of lower back-EMF harmonics and cogging torque for critical applications. In this study, a detailed analysis on developing magnetic equivalent circuit for the multi-phase EPMa-SynRM is presented. Differential evolution (DE) optimization algorithm is utilized to develop the optimal models for five-phase EPMa-SynRM. The effects of rotational forces on the rotor in both internal and external rotor PMa-SynRMs are analytically studied. A thermal model for the proposed EPMa-SynRM structure is presented. Initial simulation results for stress and thermal heat flow for the proposed designs are presented. Furthermore, electromagnetic finite element simulation results such as back-EMF, flux linkage, cogging torque, and their harmonics are presented for the developed five-phase EPMa-SynRM model. The best design which has lower back-EMF harmonics and cogging torque is chosen to fabricate and conduct experimental tests.

Keywords— Permanent magnet (PM) motors, reluctance machines, multi-phase machines, finite element method, electromagnetic forces, design optimization, cogging torque

Nomenclature

L_d	d-axis inductance
L_q	q-axis inductance
λ_d	d-axis flux linkage
λ_q	q-axis flux linkage
λ_{PM}	permanent magnet flux linkage
ω_r	rotor angular speed
r_{out}	outer radius of rotor
r_{in}	inner radius of rotor
l	Stack length of the machine
B_t	flux density at teeth
l_m	length of the permanent magnet
h_{m1}	width of the inner flux barrier
h_{m2}	width of the outer flux barrier
s_h	slot height
s_w	slot width
s_o	slot opening width
t_w	tooth width
g_k	air gap sections with respect to flux barriers where $k \in \{1,2,3.....n\}$
r_{gk}	air gap reluctances
θ	angular span
θ_{fk}	flux barrier edge angles from center
θ_{mk}	flux barrier angles from center
Φ_{gk}	air gap flux sources
$\Phi_{m,}\ \Phi_b$	permanent magnet flux sources
r_{mk}	reluctances in flux barriers
f_{qrk}	magnetic voltage source
N	number of flux barriers
f_{qsk}	stator magnetic voltage source
$r_{stk},\ r_{syk}$	stator core reluctances
$r_{ryk},\ r_{gk}$	rotor core and air gap reluctances
m	rotating mass
p	number of pole pairs

978-1-4673-9551-9/16 $31.00 © 2016 IEEE

I. INTRODUCTION

EXTERNAL rotor permanent magnet assisted synchronous reluctance motors (EPMa-SynRMs) possess high torque density compared to the internal rotor permanent magnet (PM) motors. In specific, for applications such as electric bikes and air crafts, EPMa-SynRM is a better substitute to the internal rotor motor due to its advantages such as in-wheel drive capability, larger air gap radius, and high power density. Moreover, EPMa-SynRM has more stator slot area which leads to decreased flux leakage and losses [1, 2].

Traditionally, brushless PM motors have been utilized for the inwheel applications in electric bikes [3, 4]. However, most of these machine designs have surface mounted PMs to the rotor and they have sleeves to keep them intact. An EPMa-SynRM would be a better solution for such applications as the PMs are imbedded into the rotor providing better mechanical stability.

Recently, multi-phase internal rotor PMa-SynRMs have been intensively researched [5, 6]. Five-phase internal rotor PMa-SynRM has been promising due to its advantages such as lower torque ripple, lower back-EMF harmonics, and higher reliability. Adding additional phases to a three-phase EPMa-SynRM improves its fault tolerantly cability and efficiency. Particularly, in applications such as electric bikes and air crafts, a five-phase EPMa-SynRM can be a better alternative to the conventional three-phase motors.

An optimized five-phase EPMa-SynRM can provide improved performance characteristics in terms of fault tolerant capability, cogging torque, back-EMF harmonics, losses, and reliability. Moreover, lower back-EMF harmonics will reduce the losses and lower cogging torque provides smooth and safer operation of the EPMa-SynRM at higher speeds. Most of the discussions on the optimization have been based on the machines with three-phase and five-phase internal rotor machines [5, 6]. In some of the recent studies, optimization techniques based on the optimal split ratios to achieve maximum torque per volume for external rotor PM machines has been presented [8]. External rotor ferrite based synchronous reluctance motor (Fa-SynRM) has been developed in [9] for electric two wheeler application. However, these studies have been limited to three-phase external rotor PM machines. There has been little research on the design of five-phase EPMa-SynRM to the best of author's knowledge.

In this study, multi-phase external rotor PMa-SynRM structures have been presented. Based on optimal design process of 3kW three-phase and five-phase internal rotor PMa-SynRM models, three-phase [10] and five-phase external rotor PMa-SynRM designs have been proposed. Detailed multi-physics analyses with respect to electromagnetic, structural and thermal analyses have been conducted on the designs to distinguish of the EPMa-SynRM design in terms of stress on rotor and heat flow. With same design specifications, a five-phase EPMa-SynRM design is developed which has lower back-EMF harmonics and cogging torque compared to the three-phase internal and external rotor PMa-SynRMs. The five-phase EPMa-SynRM is modeled and optimized using lumped parameter modeling (LPM) and differential evolution (DE) algorithm respectively. Extensive simulations are conducted on the optimal models to analyze the performance characteristics such as flux linkage, cogging torque, back-EMF harmonics and losses.

II. FIVE-PHASE EPMA-SYNRM DESIGN AND OPTIMIZATION

The proposed five-phase EPMa-SynRM is developed based on the design procedure that has been adopted to develop three-phase and five-phase internal rotor PMa-SynRMs [6]. Fig. 1 shows the design procedure followed in this study. Initially, a LPM model has been developed for the machine based on its magnetic equivalent circuits. The DE algorithm then optimizes the developed LPM models and gives the optimal solutions for further analysis. These optimal machine models are exported to 2-dimensional finite element analysis for analyzing the machine performance characteristics.

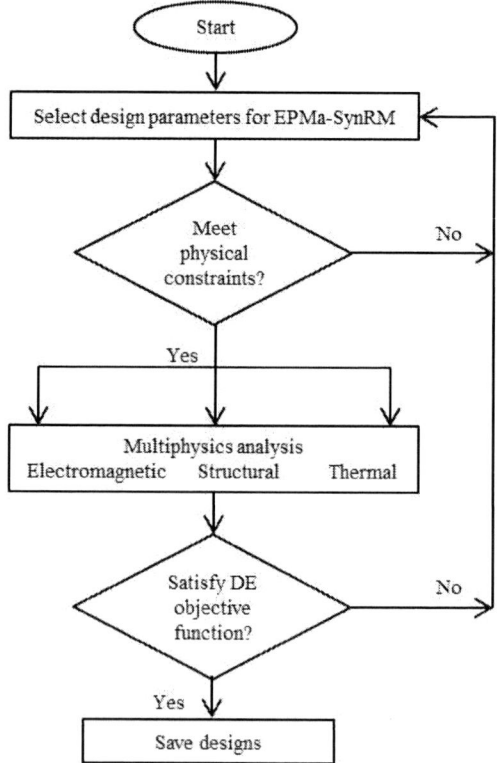

Fig. 1. Flow chart of LPM based design of EPMa-SynRM.

A. LPM model of multi-phase EPMa-SynRM

The q- and d-axis magnetic equivalent circuits of multi-phase EPMa-SynRMs are shown in Fig. 2 (a) and (b) respectively. The permanent magnet flux in EPMa-SynRMs is oriented along the q-axis and the d-axis is aligned 90^0 electrical to the q-axis with magnets in the first layer of the flux barriers. They represent the equivalent q-axis circuit with constant saturation and d-axis nonlinear magnetic equivalent circuit model replicating the nonlinear magnetic core characteristics. The flux paths in q- and d- axis for EPMa-SynRMs have been investigated based on the MECs of internal rotor PMa-SynRM structure [6, 11]. The parameters in the flux path of d- and q-axis magnetic equivalent circuits are explained in the nomenclature. Fig. 3. depicts the design parameters utilized to develop the stator and rotor structure for the multi-phase EPMa-SynRM design. It can be clearly observed in Fig. 3 that the d_{mk} values for $k = 1,2,3$ are determined based on the flux barrier span in the rotor. This affects the values of reluctances in flux barriers and air gap sections as shown in equations (1) to (6).

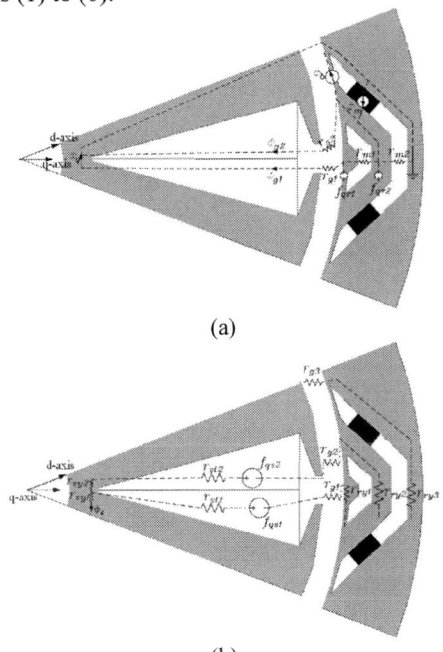

(a)

(b)

Fig. 2. Equivalent circuits required for five-phase EPMa-SynRM (a) q-axis and (b) d-axis.

The air gap sections with respect to the inner and outer flux barriers are divided into three parts as we have two flux barriers and are mentioned as g_1, g_2, and g_3. The reluctances in flux barriers and the airgap sections can be explained as follows:

$$r_{mk} = \frac{2d_{mk}A_s}{gA_{mk}} \tag{1}$$

$$r_{gk} = \frac{\Delta\alpha_s}{\Delta\alpha_k} \tag{2}$$

where $k \in \{1,2,3.....n\}$.

The magnetic voltage source in the rotor can be given as follows:

$$f_{qrk} = [(1/r)_{kj}]^{-1}[f_{qsj}/r_{gj}]_{j=1\ to\ k} \tag{3}$$

The q-axis inductance can be calculated based on the stator and rotor sources. The angle α_k represents the angular distance of the rotor surface between the flux paths.

$$L_{qm} = (1 - \frac{4}{\pi}\sum_k f_{qsk}.f_{qrk}.\Delta\alpha_k).L_{dm} \tag{4}$$

Similarly, the saturable core reluctances for each cross section of stator and rotor core can be calculated as follows:

$$r_{mk} = \frac{l_k}{\mu B_k A_k} \tag{5}$$

The reluctances in the airgaps are calculated with respect to the angular distance α_k as follows:

$$r_{gk} = \frac{g}{\mu_o(\alpha_k - \alpha_{k-1})rl} \tag{6}$$

The stator MMF source for the kth segment can be defined as follows:

$$f_{qsk} = \frac{1}{\alpha_k - \alpha_{k-1}}\int_{\alpha_{k-1}}^{\alpha_k}(f_{qs}\sin\alpha)\,d\alpha \tag{7}$$

For this given excitation, saturable d-axis flux linkage and inductance are

$$\lambda_{ds} = N_a k_{a1}\phi_d \tag{8}$$

$$L_{dm} = \frac{\lambda_{ds}}{I_d} \tag{9}$$

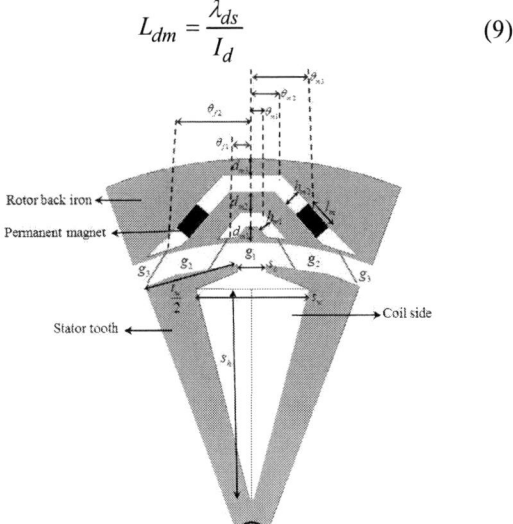

Fig. 3. EPMa-SynRM per pole cross section with design parameters.

B. Multi-physics analysis

In this paper, the multi-phase EPMa-SynRM models are exported to multi-physics simulation environment to conduct electromagnetic, structural and thermal analysis. In Fig. 4, an attempt has been made to show all the three effects on the stator and rotor of the EPMa-SynRM design for one pole. Materials used for PMs, stator and rotor cores are listed in Table I.

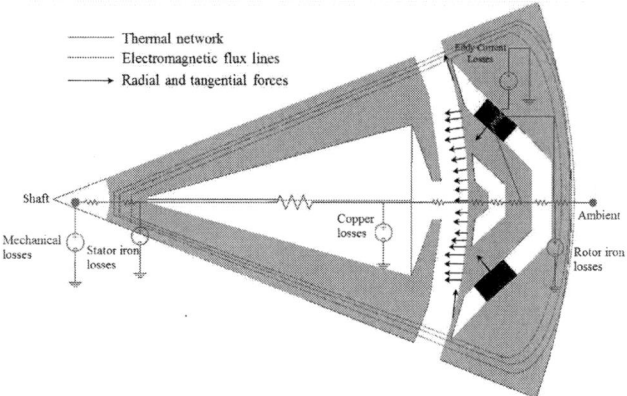

Fig. 4. Multi-physics analysis on the multi-phase EPMa-SynRM design.

In electromagnetic analysis, the flux density B(T) can be calculated by flux lines (φ) per unit area (A) as shown in equation (10). Magnetic flux density plot for one pole cross section has been presented in the Fig. 5 (a).

$$B = \frac{\varphi}{A} \qquad (10)$$

In structural analysis of the electric motor [12-14], when the rotor is rotating, the centrifugal force will result in the deflection of rotor stacks. The static deflection, σ can be calculated as follows:

$$[-\omega_r^2 + k] = 0$$

$$\omega_r = \sqrt{\frac{k}{m}} \qquad (11)$$

$$\sigma = \frac{mg}{k} \qquad (12)$$

where $k = 48\frac{EI}{L^3}$ is the stiffness, $I = \frac{\pi D^4}{64}$ is the moment of inertia, L is the span considered, and EI is the bending stiffness. Equivalent stress on the EPMa-SynRM rotor has been shown in Fig. 5 (b) at 1800rpm rotational velocity. As shown in equation (11), at a fixed rotational velocity the stiffness is directly proportional to the mass. The rotor mass in the EPMa-SynRM has been considerably increased by around 30-50% compared to that of the three-phase and five-phase internal rotor PMa-SynRM designs. This results in reduced deflection due to the centrifugal forces.

Based on the thermal analysis of interior permanent magnet motors [15], the thermal model network of the proposed multi-phase EPMa-SynRM design can be represented as shown in Fig. 6 (a). The heat flow from the frame flows through the flux barriers followed by stator

tooth, coil side and end windings. From the stator tooth, heat flows through the shaft followed by bearings and back to the frame. Internal air heat flow has been presented between rotor surface, end windings and the frame. Initial heat flow from the coil side to the rotor has been shown in Fig. 6 (b). The temperature of the coil sides is set at 24^0C.

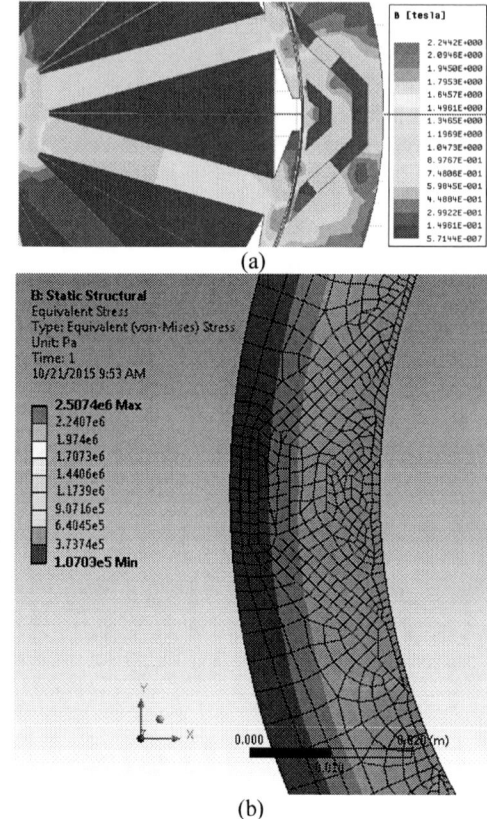

(a)

(b)

Fig. 5. FEM results for (a) flux plot and (b) equivalent stress.

The thermal losses include mechanical losses in the shaft, stator iron losses in the stator core, rotor iron losses in the rotor core, eddy current losses, and copper losses in the coil as shown in Fig. 4. Each thermal resistance, R_{th} can be calculated using the general equation as follows:

$$R_{th} = \frac{\ln(\frac{r_{out}}{r_{in}})}{2\pi\lambda l} * \frac{2\pi}{2p\theta} \qquad (12)$$

TABLE I
PROPERTIES OF MATERIALS

Part	Material	Mass density (kg/m3)
PMs	NdFeB	7501
Coil	Copper	8933
Stator/rotor core	S40 steel	7770

978-1-4673-9551-9/16 $31.00 © 2016 IEEE

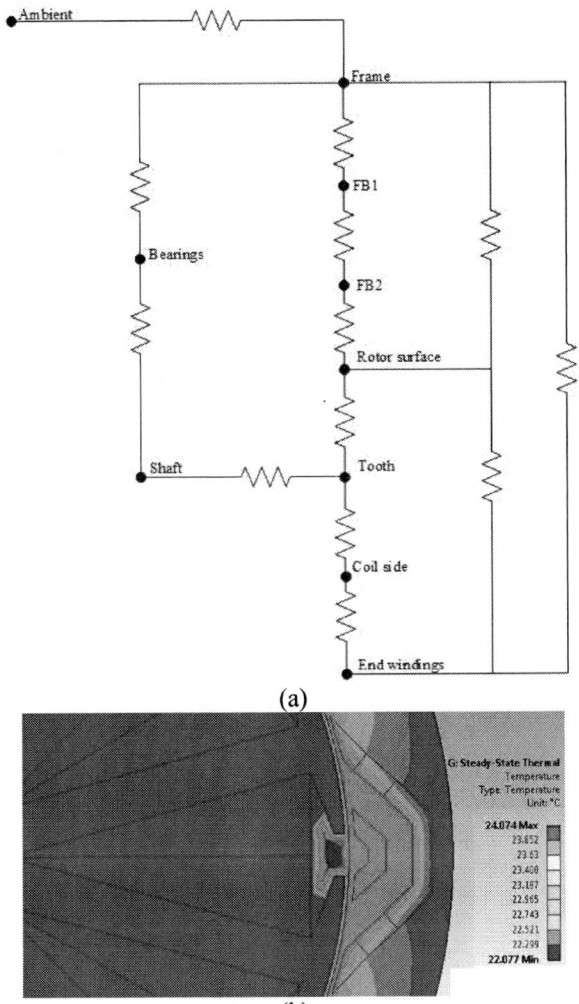

(a)

(b)

Fig. 6. (a) analytical thermal model of the EPMa-SynRM and (b) simulation of heat flow.

C. Optimization using DE algorithm

Multi-phase EPMa-SynRMs are modeled using LPM and optimized using DE algorithm with the given design parameter values. Optimized three-phase and five-phase internal rotor PMa-SynRMs LPM models developed previously in [5] are shown in Fig. 7 (a) and (b) respectively. Fig. 7. (c) and (d) shows the proposed three-phase and five-phase EPMa-SynRMs FEA models respectively. Equation (13) depicts the objective function for optimizing multi-phase EPMa-SynRM models.

$$Objective\ function = \sqrt{(k_1 \times Machine\ loss^2 + k_2 \times Torque\ ripple^2)} \quad (13)$$

where k_1 and k_2 are the weighting coefficients which are added to modify the multiple criteria in the objective function. For the multi-phase EPMa-SynRMs k_1 and k_2 are set to 0.6 and 0.4 respectively to have higher efficiency and lower torque ripple. Machine loss in the EPMa-SynRM designs has been given top priority since copper losses in

EPMa-SynRM are high compared to the internal rotor PMa-SynRM due to the increased rotor core and slot area. Thorough iterative process has been implemented to run the optimization with this objective function to identify the best candidate solution for the motor. Some of the design parameters chosen for DES are shown in Table II. The optimal solutions for each design parameter value are selected to fabricate the motors.

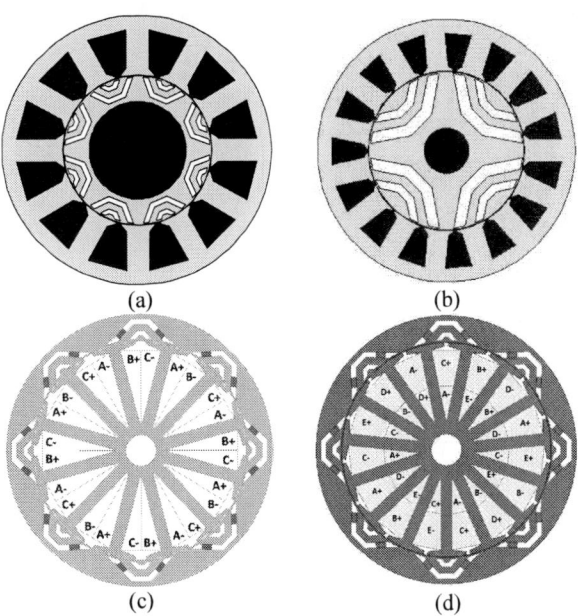

(a) (b)

(c) (d)

Fig. 7. Optimized LPM models of (a) three-phase internal rotor PMa-SynRM (b) five-phase internal rotor PMa-SynRM and proposed FEA models for (c) three-phase external rotor PMa-SynRM (d) five-phase EPMa-SynRM model.

TABLE II
DESIGN PARAMETERS IN DES

Design parameter in LPM	Five-phase EPMa-SynRM	
	Minimum for DES	Maximum for DES
Number of poles	5	50
Number of slots	12	60
Stator back iron depth [mm]	5	20
Slot area (mm²)	200	500
Outer diameter of stator (mm)	160	190
Thickness ratio of rotor bridge	0.1	5

III. RESULTS

In this section, initial finite element electromagnetic simulation results at 1800rpm for proposed 3kW three-phase and five-phase EPMa-SynRM designs have been presented in terms of their flux linkages, back-EMFs, and cogging torque. Fig. 8 (a) shows the flux linkage waveforms for three-phase and five-phase EPMa-SynRMs. It has been observed that the five-phase EPMa-SynRM has produced 50% higher peak flux linkage compared to that of the three-phase EPMa-SynRM.

978-1-4673-9551-9/16 $31.00 © 2016 IEEE

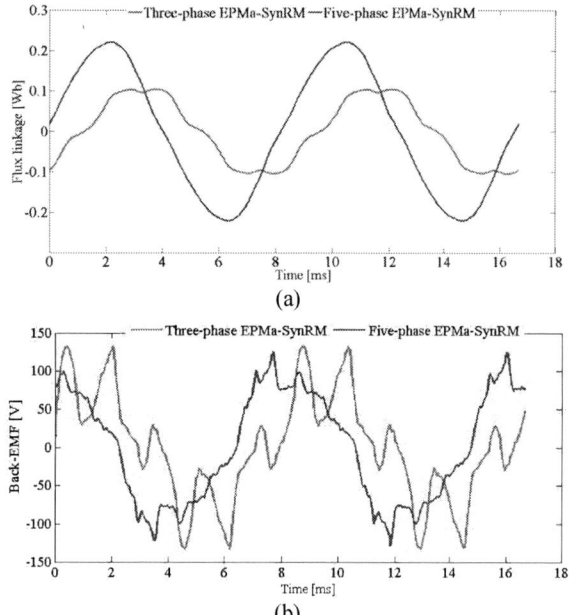

(a)

(b)

Fig. 8. FEA results for five-phase and three-phase EPMa-SynRMs (a) flux linkage and (b) back-EMF.

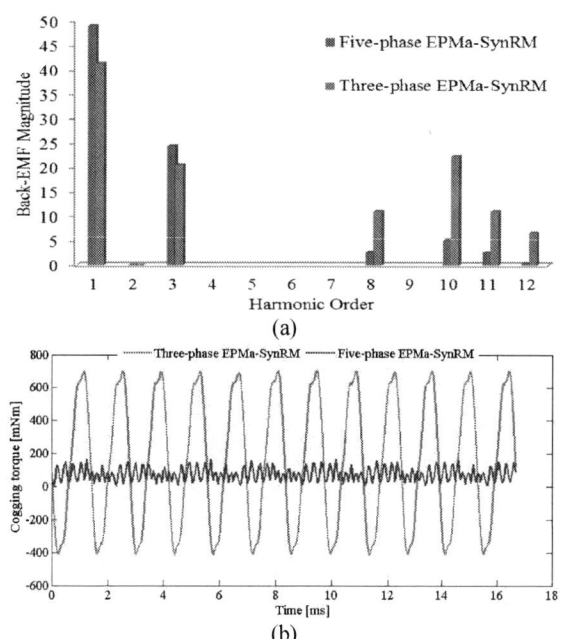

(a)

(b)

Fig. 9. FEA results for five-phase and three-phase EPMa-SynRMs (a) back-EMF harmonics and (b) cogging torque.

Fig. 8 (b) shows the back-EMF waveforms for three-phase and five-phase EPMa-SynRM models. As shown in Fig. 9 (a), the higher order back-EMF harmonics in five-phase EPMa-SynRM are lower compared to that of three-phase EPMa-SynRM design. It can be deduced that the harmonic losses will be lower. In Fig. 9 (b), it can be observed that the cogging torque has reduced by around

85% in five-phase EPMa-SynRM when compared to that of three-phase EPMa-SynRM.

Proposed five-phase EPMa-SynRM will be further optimized in LPM using DE algorithm. Best design configuration will be selected to fabricate the multi-phase EPMa-SynRMs. Fig. 10 shows the fabricated internal rotor PMa-SynRMs which have been initially developed in LPM and further optimized using DE strategy. Multi-phase EPMa-SynRMs proposed in this study will be fabricated and tested to compare their performance characteristics with these machines. The best optimal solutions obtained from the DE optimizer will be utilized to fabricate the above proposed three-phase and five-phase EPMa-SynRM designs.

Fig. 10. Fabricated five-phase and three-phase internal rotor PMa-SynRM with optimized 4 pole and 8 pole rotors respectively.

IV. CONCLUSION

In this paper, the design and optimization procedure for a multi-phase external rotor PMa-SynRM has been discussed. Initially, LPM model for the multi-phase EP Ma-SynRM has been presented in detail. Then the multi-phase analysis on the multi-phase EPMa-SynRM has been presented. The analytical models of the stress and thermal networks have been discussed in detail. Electromagnetic, structural and thermal analyses have been conducted on the designed multi-phase EPMa-SynRM and initial simulation results are presented to validate the theory. It has been observed that the mass of the rotor in the EPMa-SynRM has been increased by around 30-50% which results in lower deflection of the rotor due to centrifugal forces.

The process of optimizing the developed LPM models based on DE algorithm has been discussed. The optimal models proposed in the study are exported to 2D FEA for detailed analysis of their performance characteristics. Initial simulation results and reference test setup used for internal rotor PMa-SynRM has been presented. The optimal designs of multi-phase EPMa-SynRMs, based on their LPMs, have to be obtained to validate the effectiveness of the optimizer. As supported by the simulation results, the fabricated optimal design of five-phase EPMa-SynRM promises better performance characteristics in terms of lower back-EMF harmonics and lower cogging torque. Fabricated five-phase EPMa-SynRM design and experimental setup for testing will be discussed in future along with their intended application in detail. Experimental results for three-phase and five-phase internal rotor PMa-SynRMs will be compared with the three-phase and five-phase EPMa-SynRMs to validate the lower stresses due to increased rotor mass and their effective application in the in-wheel applications.

V. REFERENCES

[1] K. Boughrara, R. Ibtiouen, D. Zarko, O. Touhami, and A. Rezzoug, "Magnetic field analysis of external rotor permanent-magnet synchronous motors using conformal mapping," *IEEE Transactions on Magnetics*, vol. 46, no. 9, pp. 3684-3693, 2010.

[2] M. Lukaniszyn and R. Wrobel, "A study on the influence of permanent magnet dimensions and stator core structures on the torque of the disc-type brushless DC motor," Elect. Eng., vol. 82, pp. 163-171, 2000.

[3] R. Wrobel and P. H. Mellor "Design considerations of a direct drive brushless machine with concentrated windings," *IEEE Transactions on Energy Conversion*, vol. 23, no. 1, pp. 1-8, 2008.

[4] S. S. Nair, "Design aspects of high torque density-low speed permanent magnet motor for electric two wheeler applications," *Proc. IEEE IEMDC*, pp. 768-774, 2013.

[5] I. Boldea, L. Tutelea, and C. I. Pitic, "PM-Assisted Reluctance Synchronous Motor/Generator (PM-RSM) for Mild Hybrid Vehicle: Electromagnetic Design," *IEEE Transactions on Industrial Applications*, vol. 40, no. 2, pp. 492-498, 2004.

[6] S. Bonthu, J. Baek and S. Choi, "Comparison of optimized permanent magnet assisted synchronous reluctance motors with three-phase and five-phase systems," in *Energy Conversion Congress and Exposition (ECCE)*, Pittsburgh, PA, 2014.

[7] Y.B. Deshpande, H.A. Toliyat, S.S. Nair, S. Jabez Dhinagar, S. Immadisetty, and S. Nalakath, "High-torque-density single tooth-wound bar conductor permanent-magnet motor for electric two wheeler application," *IEEE Transactions on Industry Applications*, vol. 51, no. 3, pp. 2123-2135, 2015.

[8] W. Chu, Z. Zhu and Y. Shen, "Analytical optimisation of external rotor permanent magnet machines," *IET Electrical Systems in Transportation*, vol. 3, no. 2, pp. 41-49, 2013.

[9] D. Yateendra and T. Hamid A, "Design of an outer rotor ferrite assisted synchronous reluctance machine (Fa-SynRM) for electric two wheeler application," in *Energy Conversion Congress and Exposition (ECCE)*, Pittsburgh, PA, 2014.

[10] S. Bonthu, S. Choi, A. Gorgani and K. Jang, "Design of permanent magnet assisted synchronous reluctance motor with external rotor architecture," in *IEMDC*, Coeur d'Alene, 2015.

[11] E. Lovelace, T. M. Jahns and J. H. Lang, "A saturating lumped-parameter model for an interior PM synchronous machine," *IEEE Transactions on Industrial Applications*, vol. 38, no. 3, pp. 645 - 650, 2002.

[12] J.F. Gieras, "Design of permanent magnet brushless motors for high speed applications," in *Electrical Machines and Systems (ICEMS), 2014 17th International Conference on*, pp. 1-16, 2014.

[13] J. S. Rao, Rotor dynamics, 3rd edition, New Delhi, New Age Int Publishers, 1983.

[14] S. Zeze, Y. Kai, T. Todaka and M. Enokizono, "Vector magnetic characteristic analysis of a PM motor considering residual stress distribution with complex-approximated material modeling," *IEEE Transactions on Magnetics*, vol. 48, no. 11, pp. 3352-3355, 2012.

[15] A.M. EL-Refaie, N.C. Harris, T.M. Jahns, K.M. Rahman, "Thermal analysis of multibarrier interior PM synchronous machine using lumped parameter model," *IEEE Transactions on Energy Conversion*, vol. 19, no. 2, pp. 303-309, 2004.

Applicability and Limitations of an *M2Spice*-assisted "Planar-Magnetics-in-the-Circuit" Simulation Approach

Samantha J. Gunter[1], Minjie Chen[1], Stephanie A. Pavlick[1], Rose A. Abramson[1],
Khurram K. Afridi[2], and David J. Perreault[1]

[1]Massachusetts Institute of Technology, Cambridge, MA, 02139, USA
[2]University of Colorado Boulder, Boulder, CO, 80309, USA

Abstract—Planar magnetics design for power electronics naturally involves many tradeoffs, especially in the selection of the core size, winding structure and printed circuit board stackup. "Magnetics-in-the-circuit" SPICE simulations can facilitate quick magnetics design evaluation and iteration. This paper introduces and evaluates a "planar-magnetics-in-the-circuit" simulation approach with an *M2Spice* software tool, which has been developed based on an earlier presented Modular Layer Model (MLM) analysis approach [1]. *M2Spice* converts the magnetics geometry into a SPICE netlist, which can be simulated with other circuit elements in a power converter under a unified setup. This paper presents an analysis of the applicability and limitations of this approach across wide frequency bands, followed by an evaluation of the accuracy of the SPICE simulation results (by comparing the simulation results to finite-element-modeling (FEM) results and experimental measurements). Multiple planar magnetics prototypes are designed, modeled, simulated, built, and measured, with results reported and discussed.

I. INTRODUCTION

PLANAR magnetics offer low profile, good thermal characteristics, high power density, high repeatability and the ease of realizing complex winding structures [2]–[4]. As frequency increases, accurate modeling of planar magnetics becomes both important and increasingly challenging, mostly due to the impact of skin- and proximity-effects. In previous work, efforts have been made to estimate the loss [5]–[9], extract parasitics [10], and investigate the current distribution in windings [11]–[20]. Numerical methods (e.g., finite-element-modeling) and discretization-based experimental measurements are widely applicable, but are often difficult to use for design optimizations that involve many tradeoffs.

Following earlier modeling work that utilizes modularized sub-circuit cells to represent planar layers [12]–[20], an analytical model, named the Modular Layer Model (MLM), that is widely applicable to modeling planar magnetics, has been recently presented in [1]. The MLM is developed based on a minimum set of assumptions - the 1-D and the magneto-quasistatic (MQS) assumptions. It utilizes analytical solutions for field and current relationships and captures the relationship between variables in the electromagnetic domain and variables in the circuit domain using KVL and KCL equations. The current distribution and field strength can be rapidly found using SPICE simulations, and be visualized in the time-domain when the magnetic device is connected to an external circuit.

S. Gunter and M. Chen have equal contribution to this paper.

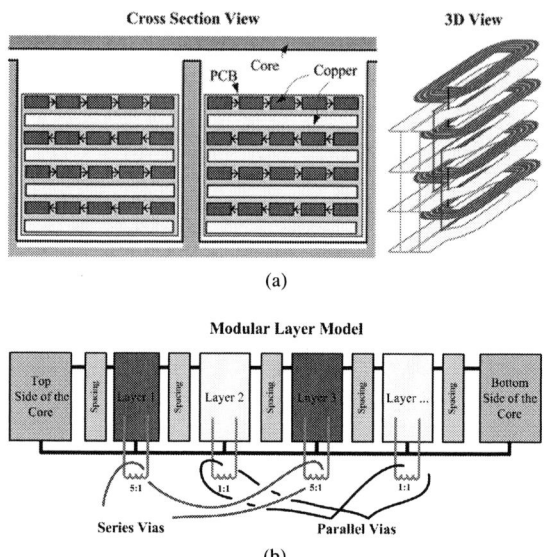

Fig. 1. Modular Layer Model-based modeling approach presented in [1]. (a) Physical structure consisting of many layers. (b) The full structure broken into many sub-sections.

With additional assumptions and approximations, "1-D-single-frequency" magnetic models can be extended to cover 2-D or wide-frequency-band cases by curve-fitting methods (linearization), or discretization methods such as those used in [13], [15]; at the price of additional approximation and increased computational requirements. In many cases, it may be preferable to maintain the simplicity of the MLM, while having a clear estimation about inaccuracies owing to violating modeling limits (e.g., non 1-D and non-sinusoidal effects), and thus avoid the computationally-demanding and non-intuitive discretization process, and keep the complexity manageable for analytical analysis. It was shown in [1] that the MLM can provide reasonable accuracy in predicting the port impedances without extensive modeling of non 1-D effects in single-frequency planar magnetics designs with sinusoidal waveforms. In this paper, we focus on evaluating the applicability and limitations of the proposed method in circuit simulations in the presence of substantial non 1-D and non-sinusoidal conditions, without using linearization or discretization.

The remainder of this paper is organized as follows: section II provides a brief overview of the MLM modeling

978-1-4673-9551-9/16 $31.00 © 2016 IEEE

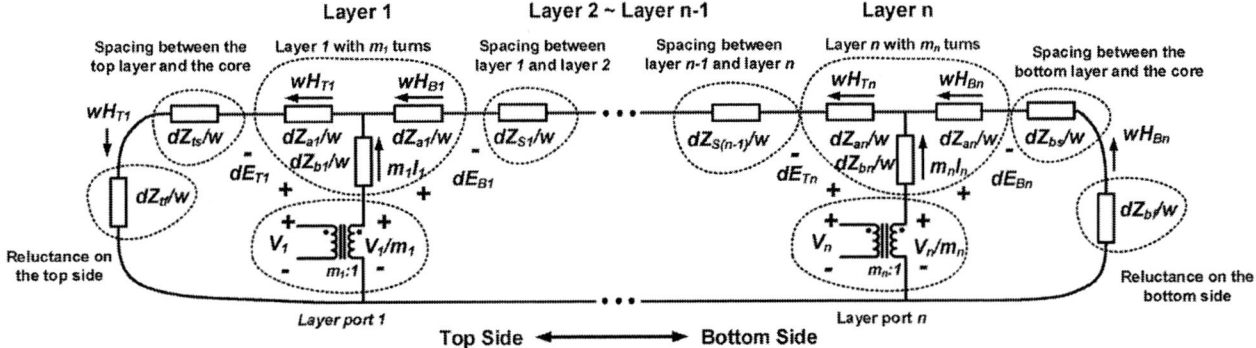

Fig. 2. Lumped circuit model consisting of many modular sub-circuits, each representing a portion of the physical planar magnetic structure.

approach and the *M2Spice* software tool. The frequency dependent behavior of the MLM approach is analyzed in section III. Section IV-A investigates the applicability and limitations of the MLM time-domain simulation approach with a single-frequency experimental setup that has substantial non-1D effects (e.g. fringing effects, edge effects). In section IV-B, the investigation is further extended to a practical magnetics design that is simulated and tested together with a dc-dc converter. Section V summarizes the applicability and limitations of the proposed "planar-magnetics-in-the-circuit" simulation approach. Finally, section VI concludes the paper.

II. MLM AND *M2Spice* OVERVIEW

Here we briefly introduce the modular layer model (MLM) presented in [1]. A multilayer planar magnetic structure as shown in Fig. 1a can be modeled as a combination of multiple circuit blocks as shown in Fig. 1b, and represented by a lumped circuit model comprising many iterative sub-circuit blocks as shown in Fig. 2. Each sub-circuit block represents a portion of the magnetic structure, including the magnetic reluctance on top of the layer stack, the conductor layers, the spacings, and the magnetic reluctance below the layer stack. This model can analytically capture skin- and proximity-effects in the windings under 1-D and MQS assumptions [1]. The cross and through variables in the lumped circuit model, i.e., values of E and H in Fig. 2, represent the current density and the field strength in the magnetic structure. By simulating this lumped circuit model with the external driving circuits and probing the current flowing through the corresponding sub-circuit in the SPICE simulation, the current flowing through each layer can be visualized and evaluated in SPICE.

A software tool – *M2Spice* – that can rapidly compute the element values, and automatically generate a SPICE netlist has been created to minimize the additional effort of using the model. Figure 3 shows its user interface. This tool is open-sourced by the authors[1] and is utilized for the studies here. Fig. 4 shows the information flow in the *M2Spice*-assisted design approach. The magnetic geometry information is first processed by *M2Spice*, which produces a netlist for a subcircuit that captures the electrical behavior of the magnetic component. This netlist is combined with a netlist

[1]*M2Spice* [Online]. Available: http://www.rle.mit.edu/per/M2Spice/.

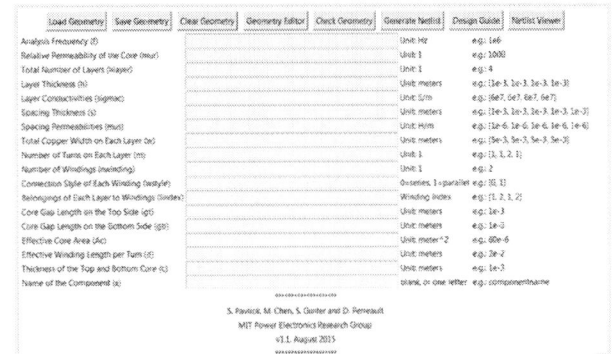

Fig. 3. Screenshot of the user-interface of the *M2Spice* tool.

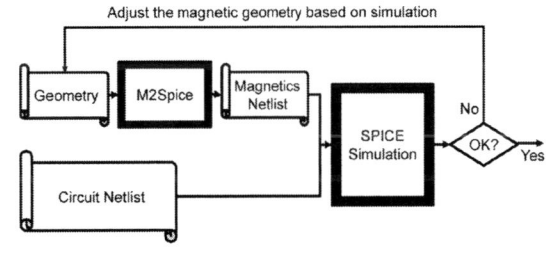

Fig. 4. Information flow of the *M2Spice*-based magnetics design approach.

that represents other elements in the circuit (e.g., capacitors, resistors, switching devices, etc.) and fed into a SPICE simulation platform (e.g., *LTSpice*). Based on the simulated circuit performance, designers can very quickly adjust the geometry of the magnetic component (e.g., layer thickness, interleaving patterns, core shapes, etc.), and iterate the design.

III. FREQUENCY DEPENDENT IMPEDANCE VALUES

This section analyzes the applicability and limitations of using the MLM in circuit simulations. Assuming the permeability and permitivity of all materials stay constant across the full operating range, many elements in the MLM are frequency independent, e.g. those representing the spacing and the magnetic core (Z_{tf}, Z_{ts}, Z_{S1}–$Z_{S(n-1)}$, Z_{bs}, and Z_{bf} in Fig. 2), as well as the wire connections describing how each layer is connected to other layers. There is generally no limitation in using these element values in SPICE simulations.

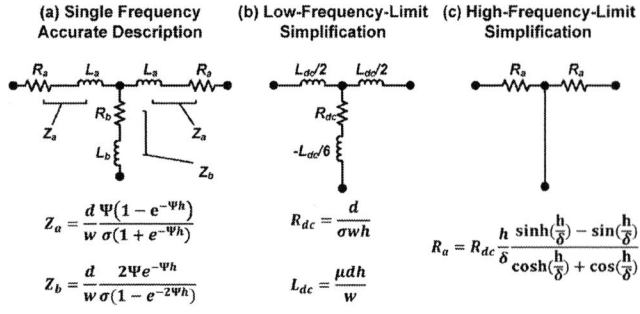

(a) Single Frequency Accurate Description   **(b) Low-Frequency-Limit Simplification**   **(c) High-Frequency-Limit Simplification**

$$Z_a = \frac{d}{w}\frac{\Psi(1-e^{-\Psi h})}{\sigma(1+e^{-\Psi h})}$$

$$Z_b = \frac{d}{w}\frac{2\Psi e^{-\Psi h}}{\sigma(1-e^{-2\Psi h})}$$

$$R_{dc} = \frac{d}{\sigma w h}$$

$$L_{dc} = \frac{\mu d h}{w}$$

$$R_a = R_{dc}\frac{h}{\delta}\frac{\sinh(\frac{h}{\delta})-\sin(\frac{h}{\delta})}{\cosh(\frac{h}{\delta})+\cos(\frac{h}{\delta})}$$

Fig. 5. (a) Single frequency T network of the MLM and (b) its Low-Frequency-Limit (LFL) simplification (i.e., when $h \ll \delta$), and (c) its High-Frequency-Limit (HFL) simplification (i.e., when $h \gg \delta$).

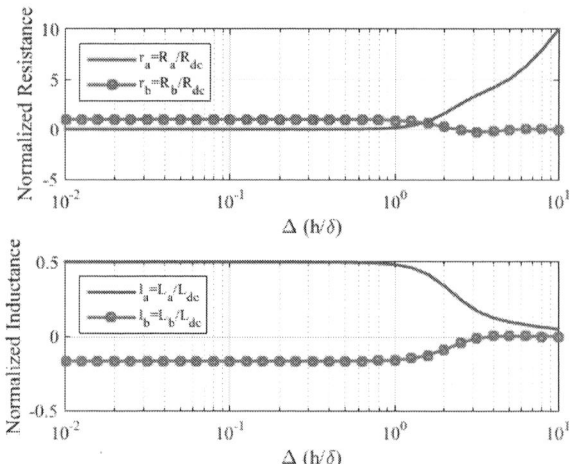

Fig. 6. Normalized impedance values of the elements in the T network of Fig. 5a as functions of Δ.

Elements related to the conductor layers are frequency dependent. As derived in [1] and shown in Fig. 5, the accurate netlist representing each conductor layer consists of two impedances, Z_a, and Z_b. They can be calculated using:

$$\begin{cases} Z_a = \frac{d\Psi(1-e^{-\Psi h})}{w\sigma(1+e^{-\Psi h})} \\ Z_b = \frac{2d\Psi e^{-\Psi h}}{w\sigma(1-e^{-2\Psi h})} \end{cases} \tag{1}$$

where d is the conductor length per turn, w is the total width of the copper on this layer, h is the thickness, $\Psi = \frac{1+j}{\delta}$, where $\delta = \sqrt{\frac{2}{\omega\mu\sigma}}$ is the skin depth of the conductor, μ is its permeability, σ is the conductivity, and ω is the angular frequency. In the netlist generated by *M2Spice*, each complex impedance is represented by a resistor and an inductor, which can have positive or negative values, i.e., $Z_a = R_a + \omega L_a j$, and $Z_b = R_b + \omega L_b j$, as shown in Fig. 5a.

Since Z_a and Z_b are both frequency dependent, R_a, R_b, L_a, and L_b are all frequency dependent. As derived in Appendix I, when $\omega \to 0$, R_a approaches zero; L_a has a limit of $\frac{\mu d h}{w}$; R_b has a limit of $\frac{d}{\sigma w h}$; and L_b has a limit of $-\frac{\mu d h}{6w}$. We name these element values as Low-Frequency-Limit (LFL) element values. And when $\omega \to +\infty$, R_b, L_a, and L_b all approach zero, and $R_a = \frac{d}{\sigma w \delta}\frac{\sinh(h/\delta)-\sin(h/\delta)}{\cosh(h/\delta)+\cos(h/\delta)}$. We name these element values as High-Frequency-Limit (HFL) element values.[2]

R_a, L_a, R_b, and L_b are normalized to their LFL values to investigate their frequency dependent characteristics. We define the dc resistance of a copper layer with thickness h, width w, and length d, as R_{dc} equals $\frac{d}{w\sigma h}$; and the inductance of a spacing with thickness h, width w, and length d, as L_{dc} equals $\frac{\mu d h}{w}$. Defining $\Delta = h/\delta$, then

$$\begin{cases} r_a = \frac{R_a}{R_{dc}} = \mathbb{R}e\left(\frac{(1+j)\Delta(1-e^{-(1+j)\Delta})}{1+e^{-(1+j)\Delta}}\right) \\ l_a = \frac{L_a}{L_{dc}} = \mathbb{I}m\left(\frac{j(1-e^{-(1+j)\Delta})}{(1+j)\Delta(1+e^{-(1+j)\Delta})}\right) \\ r_b = \frac{R_b}{R_{dc}} = \mathbb{R}e\left(\frac{2(1+j)\Delta e^{-(1+j)\Delta}}{1-e^{-2(1+j)\Delta}}\right) \\ l_b = \frac{L_b}{L_{dc}} = \mathbb{I}m\left(\frac{2je^{-(1+j)\Delta}}{(1+j)\Delta(1-e^{-2(1+j)\Delta})}\right) \end{cases} \tag{2}$$

$\mathbb{R}e$ represents the real part of a complex value, and $\mathbb{I}m$ represents the imaginary part of a complex value. r_a, l_a, r_b, and l_b are plotted as functions of Δ in Fig. 6. Note $\Delta = 1$ indicates the frequency at which the layer thickness equals the skin depth ($h=\delta$).

Figure 6 indicates that for non-sinusoidal waveforms with harmonics distributed in multiple frequencies, if a majority of its harmonic components are in the low frequency ($\Delta<1$) range (including dc), using the netlist generated at the fundamental frequency of this circuit is accurate, because the element values stay relatively constant across the whole frequency range. If a majority of its harmonic components are distributed at multiple frequencies in the high frequency ($\Delta>3$) range, then a single frequency netlist cannot capture the wide-band behaviors of the magnetic components. As discussed in Appendix II, under this situation, concepts of effective frequency [21] can be utilized, which offers conservation of loss according to Fourier Analysis.

IV. EXPERIMENTAL VERIFICATION

It is clear that the MLM approach – with layer stacking and frequency dependent impedances included – can always provide more information than many conventional magnetic models. An interesting question is how accurate an MLM-based time-domain simulation is in predicting the current distribution and loss in the magnetics (when the magnetics are simulated together with the circuit). To answer this question, we investigate the applicability and limitations of the MLM-based time-domain simulation method by comparing the SPICE simulation results with experimental measurements

---

[2]The LFL and HFL values can be utilized in further simplified MLM modeling (but still capture a major amount of information, especially the loss) as illustrated in Fig. 6: if $h \gg \delta$ for the frequency range of interest, HFL element values should be used; if $h \ll \delta$, LFL element values should be used. Intuitively, at low frequencies, the inductive behavior and the dc resistance dominates the conductor behavior; and at high frequencies, the reactive energy stored in the conductor diminishes, leaving only the ac resistance dominating the conductor behavior.

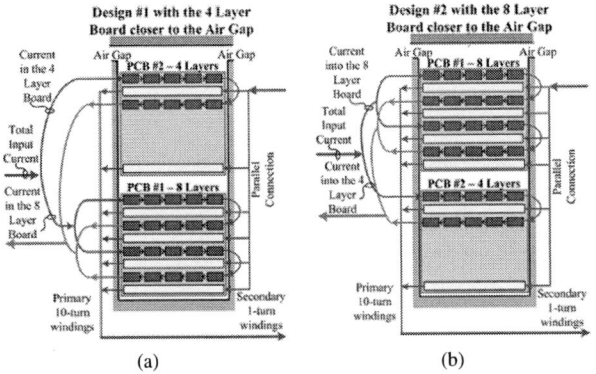

Fig. 7. Cross-section view of the layer stack of the two coupled inductors (a) Design #1 and (b) Design #2. Note that the dimension of this stackup is not to scale.

Physical Prototype/Lumped Circuit Model

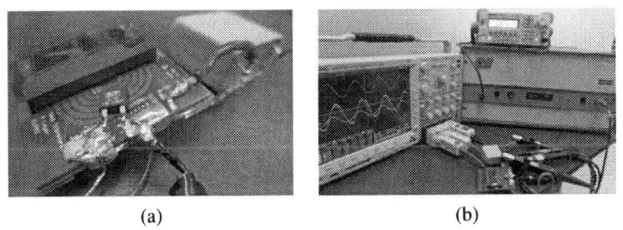

Fig. 8. Circuit structure of the test setup. The leakage and magnetizing inductances are included in the dashed circuit block.

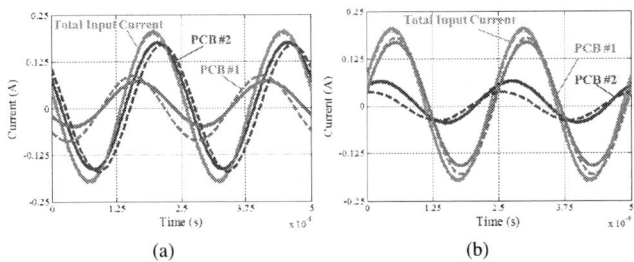

Fig. 9. (a) Photo of the prototype. (b) Experimental setup.

in two setups. The first setup investigates a coupled inductor design with sophisticated winding patterns, but only with a single frequency component. The second setup is the design of multiple magnetic components in a newly presented circuit architecture that has switching harmonics [22].

A. Design Evaluation in a Purely Sinusoidal Setup

In this setup, the simulation and experimental waveforms of two 10:1 coupled inductors with identical cores but different layer stack-ups are compared. Figure 7 shows the cross-section view of the two coupled inductors. The winding structure of the two coupled inductors are manufactured with two 72 mil printed-circuit boards (PCB #1 and #2), which are linked by a single EPCOS ELP43 core (N49 material) with a distributed air gap on top of both the center and side legs. PCB #1 has eight copper layers and PCB #2 has four copper layers. All layers are 4 oz copper layers. In Design #1, the 4-layer board is placed on top of the 8-layer board, and is closer to the air

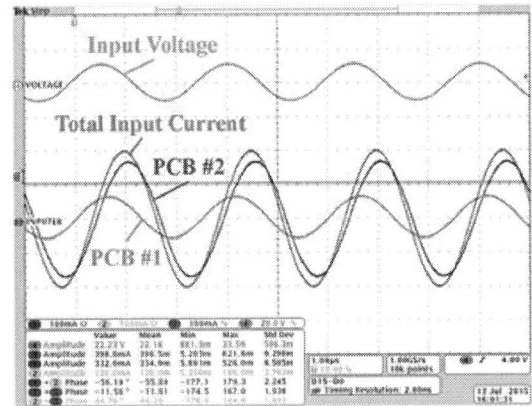

Fig. 10. Measured waveforms for Design #1 at 400 kHz.

Fig. 11. Simulation and experimental waveforms showing the current sharing between the two primary windings at 400 kHz. (a) Design #1: PCB #2 (4 layer) on top. (b) Design #2: PCB #1 (8 layer) on top. Solid line: experimental waveform. Dashed line: simulated waveform.

gap (Fig. 7a); in Design #2, the 8-layer board is placed on top of the 4-layer board, and is closer to the air gap (Fig. 7b). The odd layers (from the top) of both PCBs each have five spiral series-connected turns. Layers 1&3 and 5&7 of PCB #1, and layers 1&3 of PCB #2 are each connected into a series-tied pair to formulate three 10-turn windings (red layers of Fig. 7). These three 10-turn windings are then connected in parallel as primary windings. The even layers of both PCBs each have a single turn and are connected in parallel to form a six layer 1-turn secondary winding (yellow layers of Fig. 7).

Figure 8 shows the schematic of the test circuit. A film capacitor (not shown) is connected in series with the input source to block the dc component. The current in the 4-layer PCB, in the 8-layer PCB and the total input current (I_{4L}, I_{8L} and I_{in}) are observed using TCP202 and TCP0030 current probes. Since the two PCBs are stacked and are linked with a single core, they ideally have identical flux linkage and identical voltage-drop per turn. The different ac impedance of the two boards will create unbalanced current sharing between their copper layers. Intuitively, at direct current (dc) or low frequency alternate current (ac), the 8-layer board has a lower resistance and will always carry more current. However, high frequency ac current will tend to flow in the board that is closer to the gap, redistributing the current and creating interesting phenomena that we seek to visualize in time-domain simulations.

Figure 9a shows the prototype planar coupled inductor with

978-1-4673-9551-9/16 $31.00 © 2016 IEEE

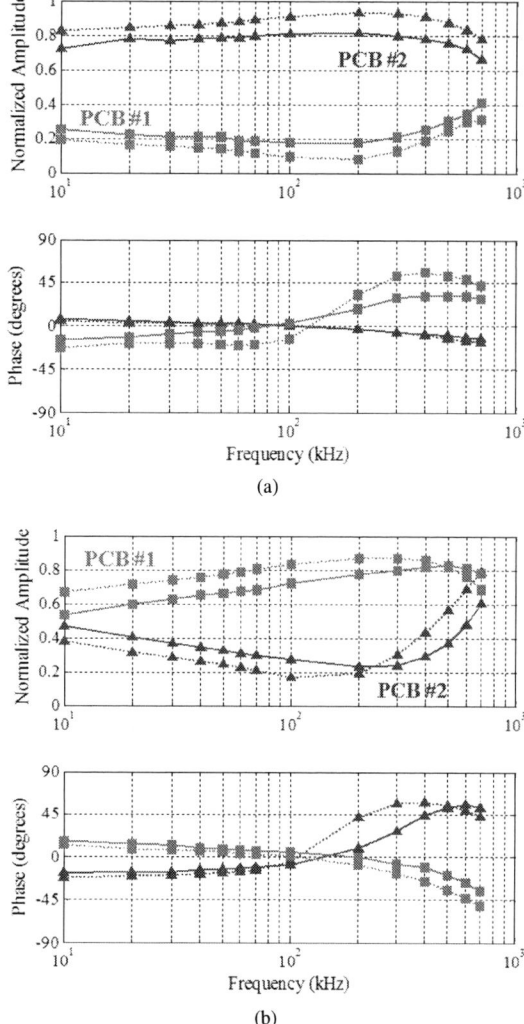

(a)

(b)

Fig. 12. Comparison of the amplitude and phase of the currents for the experimental (**solid lines**) and simulated results (**dashed lines**). (a) Design #1: PCB #2 (4 layer) on top. (b) Design #2: PCB #1 (8 layer) on top. The amplitude and phase of each PCB is normalized to the input current (200 mA). The phase is positive if the PCB current leads the input current.

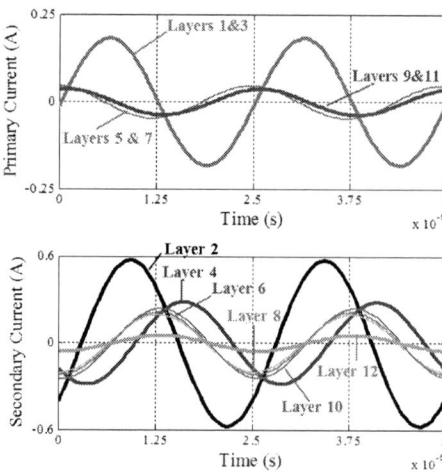

Fig. 13. Simulated current distribution in all 12 layers of Design #1 at 400 kHz using time-domain *LTSpice* simulations based on the *M2Spice* subcircuit. The layers are numbered from top to bottom with layer 1 closest to the gap. Each primary winding consists of two layers.

design information (the known system geometry only), without information gathered after the prototype was built.

Figure 10 shows the measured waveforms of Design #1 at 400 kHz. Figure 11 compares the simulated and measured current waveforms of the two layer stacks operating at 400 kHz. The simulated waveforms match well with the experimental waveforms. The time-domain SPICE simulation based on the MLM as generated by *M2Spice* can accurately predict the current distribution between two parallel windings at different frequencies. The time-domain simulation is capable of accurately quantifying the phenomenon that the PCB closer to the air gap carries more current regardless of its dc resistance.

Figure 12 compares the simulated and measured amplitudes and phases of the current flowing through the two layer stacks across a frequency range of 10 kHz–700 kHz (with netlists separately generated for each frequency). As frequency increases, it can be seen that the phase shift between the two PCB currents also increases, especially when the frequency is above 100 kHz. As a result of the increased phase shift, the amplitude increases in the bottom PCB, indicating circulating currents and higher overall rms current.

The experimental measurement can only reveal the current sharing between the two PCBs (grouping multiple layers). An experimental effort to determine the current distribution in all 12 layers is extremely challenging, because any additional current measurement infrastructure needs to be accurately modeled and calibrated, and will involve new approximations. While hard to do experimentally, the current distribution in the 12 layers can be easily visualized using SPICE time-domain simulations. Fig. 13 shows the current distribution in the 12 layers in Design #1 of Fig. 7a. As expected, the series-connected pairs have identical current distribution (layers 1&3, layers 5&7 and layers 9&11), and the parallel layers unevenly share the current with dramatic phase shift.

To validate the current distribution shown in simulation, we predict the current distribution using Ansoft Maxwell 2D simulations. To emulate the condition that the magnetic

the blocking capacitor and terminating resistor. Fig. 9b shows the experimental bench setup. The gain of the power amplifier is tuned to make the total input current 200 mA (peak). The 1 ohm load resistor has a constant resistance across the full frequency range.

Using the MLM-based approach, the geometry of the magnetic circuit including the two planar winding structures is processed by *M2Spice* to generate the SPICE netlist. The cross winding capacitance is modeled separately with EQS methods [6]. The layer-to-layer capacitance of the prototype lumped to the secondary is calculated to be 150 nF. The netlist is then imported into *LTSpice* and simulated. The three currents are then modeled using time-domain simulations. In *LTSpice*, the device is constantly driven by a 200 mA ac current source and the frequency is swept from 10 kHz to 700 kHz (with a netlist generated for each frequency). We intentionally isolated the modeling and the measurement process, i.e., all modeling efforts are rigorously developed based on a priori

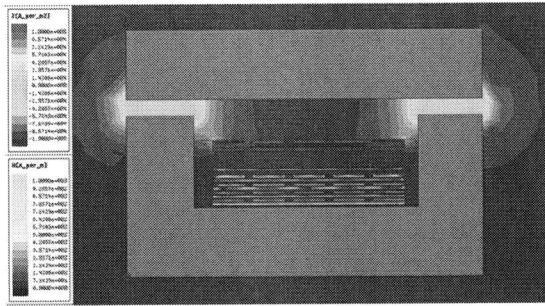

Fig. 14. FEM simulation results for an instantaneous current and field distribution of Design #1 with the 4-layer board on top when the total primary side current is 200 mA at 0 degrees.

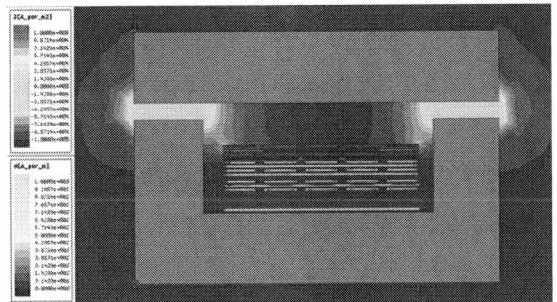

Fig. 15. FEM simulation results for an instantaneous current and field distribution of Design #2 with the 8-layer board on top when the total primary side current is 200 mA at 0 degrees.

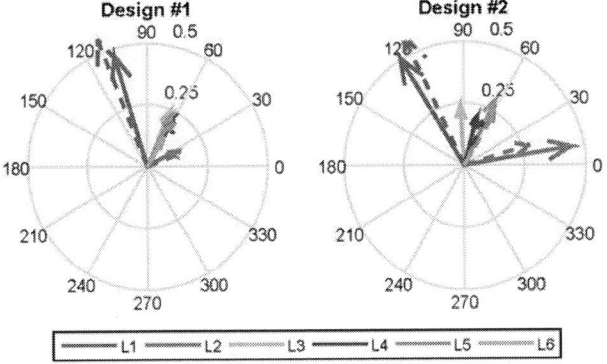

Fig. 16. Polar plot of the SPICE and FEM predicted current sharing in all 6 paralleled secondary layers in Design #1 and Design #2 when the total primary side current is 200 mA at 0 degrees. Solid arrows: FEM results; Dashed arrows: SPICE results. Different colors indicate different layers: L1–top secondary layer; L6–bottom secondary layer.

component is connected to the circuit, the external drive current for the magnetic component is pre-determined by SPICE simulation, and the current distribution inside the layers is solved by FEM. The FEM-simulated current and field distribution in Design #1 and Design #2 are shown in Fig. 14–15, and the SPICE predicted and FEM predicted amplitude and phase of the current in all 6 secondary layers are compared in Fig. 16. The current distribution predicted by *M2Spice* matches well with that predicted by the FEM tool, even with the presence of substantial 2-D effects (including both fringing

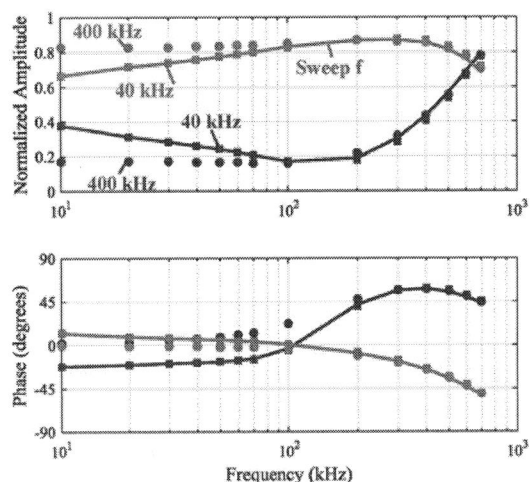

Fig. 17. Comparing the simulation results of Design #2 using single-frequency netlists (square: 40 kHz, or circle: 400 kHz) to visualize the current distribution across a wide frequency range (10 kHz–700 kHz), and the results of using different netlists for different frequencies (solid curve: Sweep f).

TABLE I
SPICE SIMULATION AND FEM PREDICTED TOTAL LOSS AT 400 KHz.

Total Loss in All Layers # (mW)	SPICE	FEM
Design #1, 12 layers in total, 4 layer top, 8 layer bottom	10.511	7.997
Design #2, 12 layers in total, 8 layer top, 4 layer bottom	10.511	8.007
Design #3, 8 layers in total, with the 8 layer PCB	9.759	4.287
Design #4, 4 layers in total, with the 4 layer PCB	10.122	4.667

fields and vertical spacings among primary layers).

To investigate the limitation of the model in predicting wide-band behaviors, two single-frequency netlists (40 kHz and 400 kHz) are used to simulate the current distribution across a wide frequency range (10 kHz–700 kHz). Fig. 17 compares the simulated results with simulations that use different netlists for different frequencies (solid line, labeled as "Sweep f"). The netlist generated for 40 kHz can well model the current distribution across a wide frequency range up to the frequency when $\Delta = 1$ (for 4 oz copper, this frequency is 226 kHz), while the netlist generated for 400 kHz operation only shows a close match only if the frequency is in the hundreds of kHz. This limitation matches the analysis presented in Section III.

Unfortunately, a well matched turn-level current distribution does not guarantee good prediction of loss. 2-D factors change the current density along the width of the conductor – an effect which is not captured by the 1-D method. Moreover, estimating the loss in the time domain naturally involves integration, which brings in quantization noise in SPICE simulations. We compared the loss of Design #1 and Design #2 when they are simulated in SPICE[3] and FEM[4] under the setup of Fig. 8. We also took the 4 layer PCB away from Design #1 and the 8 layer PCB away from Design #2 to investigate

[3]LTSpice v4.23h – Solver=alternate, Trtol=1, Max thread=2.
[4]Ansoft Maxwell 2D v16.0 – EddyCurrent Simulation

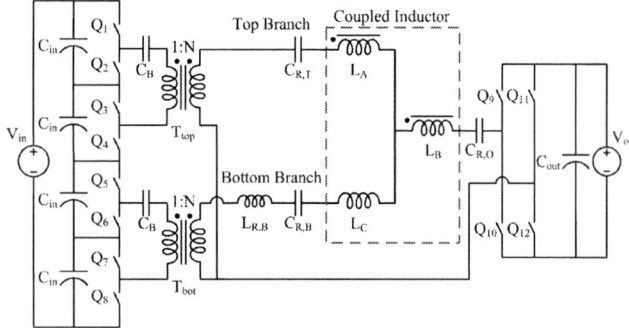

Fig. 18. Proposed implementation of the ICN converter [22]. The ICN incorporates harmonic filtering through the use of resonant tanks as well as two equal but opposite impedances. This structure contains four magnetic structures - two transformers, a coupled inductor (indicated by the red-dashed box), and a resonant inductor.

the impacts of fringing fields to the total loss. Table I lists the predicted losses. Although the loss values are slightly mismatched in magnitude, SPICE simulation does indicate an interesting trend that designs with less copper layers may actually perform better in terms of loss in certain setups (even without considering the fringing effects). This is correct though it is contradictory to the common misconception that adding more copper layers usually helps to reduce the loss. The trend can be likely be explained by proximity effect and the resultant phase shifting of the layer currents.

B. Design Evaluation in a Switched-Mode-Power-Converter

This section investigates the applicability and limitations of an MLM-based simulation approach in a recently published power converter containing sophisticated planar magnetic structures - the Impedance Control Network (ICN) resonant converter [22] as shown in Fig. 18. It has two transformers, a resonant inductor, and a coupled inductor. This converter is designed to operate with an input voltage of 260 V–410 V, an output voltage of 12 V, and a rated output power of 300 W. The operating frequency is set to 500 kHz. Detailed operation of the ICN converter is presented in [22]. Using this converter as a test platform, we tried to visualize the current distribution in the magnetic components using *M2Spice*, and investigate the applicability of the MLM approach for predicting the loss of the magnetics when it is simulated with the full converter.

We first investigate the capability of SPICE simulation in predicting current distribution. The ICN converter has two transformers. The winding of each transformer was built using two copies of PCB #1 used in section IV, which are coupled with a single ELP43 core with N49 material (the two PCBs are paralleled in the same manner as described in Fig. 8 but now both PCBs are the 8 layer version). The PCBs both have 4 oz copper layer thickness (the frequency when h equals Δ is 226 kHz). Here we focus on the current sharing in the two PCBs under two operating modes - 260 V and 410 V as shown in Figs. 19–20. A close match was found between measured and simulated results. When the input voltage is 260 V, the total input current was quite sinusoidal and well shared in the two boards (similar amplitude and phase). When the input

TABLE II
SPICE SIMULATION PREDICTED LOSS FOR VARIOUS TRANSFORMER INTERLEAVING PATTERNS. P: PRIMARY LAYERS WITH 5 SERIES CONNECTED TURNS; S; SECONDARY LAYERS WITH 1 SINGLE TURN. "PSPS" REPRESENTS A FOUR LAYER INTERLEAVING PATTERN: THE 1st AND 3rd LAYERS ARE TWO PRIMARY LAYERS WITH 5 TURNS, THE 2nd AND 4th LAYERS ARE TWO SECONDARY LAYERS WITH 1 TURN.

Layer Stack	SPICE Predicted Loss
SSPP	3.8068 W
PPSS	3.7261 W
SPPS	2.9148 W
PSSP	2.8963 W
SPSP	2.8678 W
PSPS	2.8465 W
SSPPSSPP	2.2267 W
PPSSPPSS	2.1986 W
PSSPPSSP	2.0331 W
SPPSSPPS	2.0291 W
SPSPSPSP	1.9593 W
PSPSPSPS	1.9548 W
PSPSPSPSPSPSPSPS	1.9733 W
SPSPSPSPSPSPSPSP	1.9565 W

voltage is 410 V, the current is more trapezoidal and there is a phase shift between the current injected into the top PCB and that in the bottom PCB, indicating circulating current, and additional loss. Other time-domain details, such as the soft-switching transition, and the high-order harmonics and their patterns, all matched well. As mentioned before, unlike in the experimental setup, it is easy to see the current distribution in all of the layers using *M2Spice* as shown in Fig. 21. By looking at the current in all of the primary layers, it can be seen that the phase shift between the two PCBs is a result of uneven current sharing in the topmost primary winding as opposed to all remaining windings. This winding is closest to the gap and therefore is the most different. It is verified in this example that in a practical design with sophisticated winding patterns and non-sinusoidal circuit driving patterns, the netlist generated by *M2Spice* can accurately predict the current distribution in planar magnetics in time domain simulations.

A further investigation is to use SPICE simulation in predicting the loss in magnetics, while the magnetic component is driven by the actual circuit. By measuring the current and voltage waveforms in SPICE, one is able to predict the loss in the magnetic components, and thus make appropriate design choices based on the loss information. For example, we used SPICE simulations to determine the most cost-effective number of fully interleaved primary-secondary pair sets in the transformers of Fig. 18. The losses predicted by SPICE for different design variations are listed in Table II. Increasing the layer count from 4 to 8 gives us substantial reduction in predicted loss (>40%). However, negligible improvements were found when the layer count increased beyond 8, which was verified in the experimental setup that was used to generate Figs. 19–20. This is the reason why the transformer layer count was chosen as 8 in [22].

In the ICN converter, the two windings in the coupled

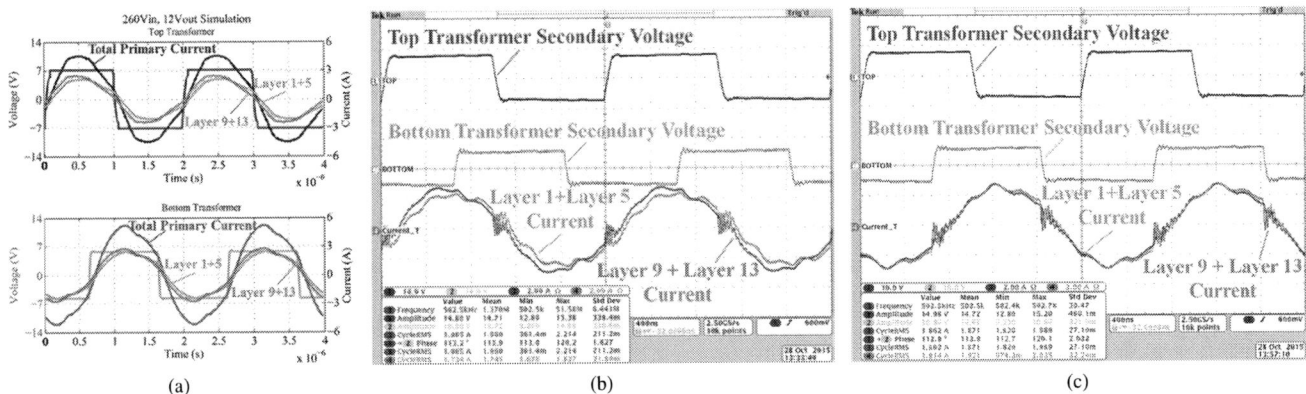

Fig. 19. (a) Simulated and (b)-(c) measured current sharing of the two PCBs in the ICN converter when the input voltage of the converter is 260 V. The current for the top transformer is shown in (b) and the current for the bottom transformer is shown in (c).

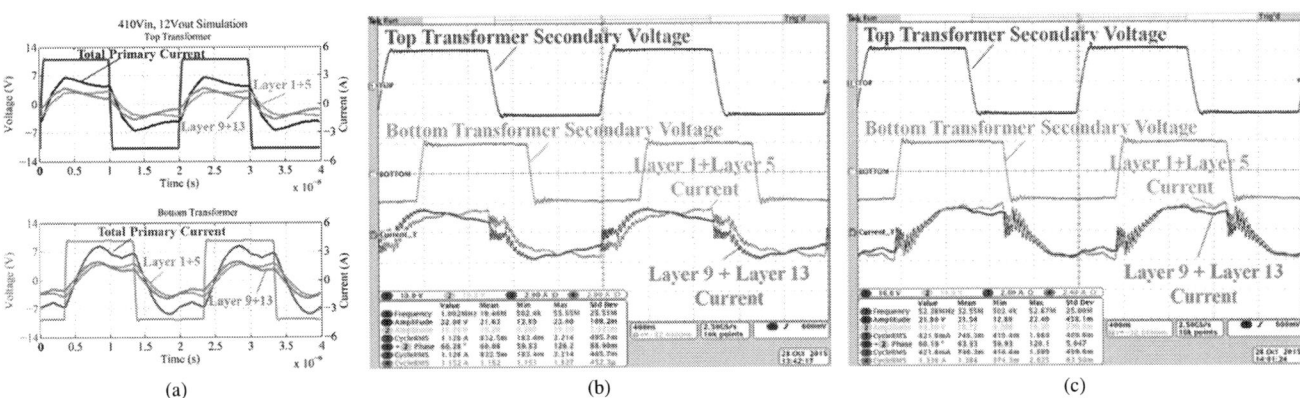

Fig. 20. (a) Simulated and (b)-(c) measured current sharing of the two PCBs in the ICN converter when the input voltage of the converter is 410 V. The current for the top transformer is shown in (b) and the current for the bottom transformer is shown in (c.)

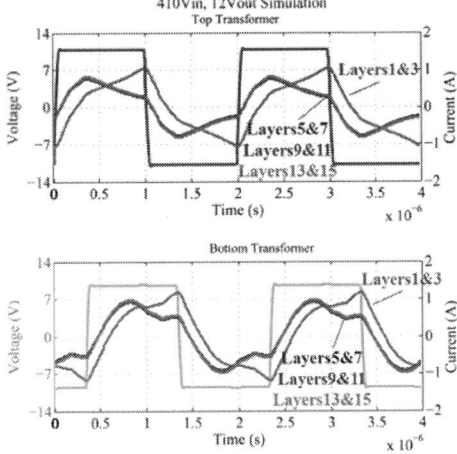

Fig. 21. Simulated current distribution in all 8 primary layers of the transformers at an input voltage of 410 V using the *M2Spice*-derived magnetics model. Significant harmonic waveform components are present. The layers are numbered from top to bottom with layer 1 closest to the gap. Each primary winding consists of two layers.

inductor are effectively driven by two current sources. The amplitude and phase of each current source are highly dependent on the circuit behavior. It is difficult to model the circuit and

the coupled inductor separately as in conventional magnetics modeling approaches, or FEM. *M2Spice* allows the magnetic component to be evaluated together with the circuit in SPICE simulations. In the test setup, this inductor is implemented with the Coilcraft Planar Transformer Prototyping Kit [23]. The core is a PL140 core, and the primary and secondary windings[5] are implemented with single-turn stamps. Each winding is created by multiple parallel stamps. To avoid the impact of temperature rise in experiments, the converter is operated in a low-power mode with 100 V input voltage, 5 V output voltage, and 10 A output current, and the total loss of the converter with different coupled inductor implementations is measured. The predicted and measured losses in multiple experimental setups are shown and compared in Table III. It can be seen that SPICE simulations do predict the trend that using parallel layers reduces the loss, having the secondary layer closer to the gap reduces the loss, and a more efficient coupled inductor design can help to reduce the total converter loss. At the same time, the quantitative predictions of total converter loss are not very accurate, in part because the SPICE model does not include many practical details (e.g., PCB trace losses, etc.).

---

[5]The primary (P) winding is denoted as the single turn forming L_A in Fig. 18 and the secondary (S) winding is denoted as the single turn forming L_B in Fig. 18.

TABLE III
SPICE SIMULATION PREDICTED LOSS IN THE COUPLED INDUCTOR AND TOTAL SYSTEM, EXPERIMENTAL MEASURED LOSS IN THE TOTAL SYSTEM.

	SPICE		Experimental
Interleaving	Inductor Loss	Total Converter Loss	Total Converter Loss
PS	0.91973 W	4.749 W	14.008 W
SP	0.89358 W	4.638 W	12.721 W
PSPS	0.72098 W	4.455 W	11.047 W
SPSP	0.61108 W	4.382 W	10.627 W

V. SUMMARY OF THE APPLICABILITY AND LIMITATIONS

Applicability:

1) *M2Spice*-assisted circuit simulations can usually provide more information than conventional magnetic models.

2) *M2Spice*-assisted circuit simulations allow magnetic components to be analyzed together with external drive circuits – a function that conventional magnetics modeling approaches and commercial FEM tools[6] do not offer.

3) *M2Spice*-assisted circuit simulations are highly applicable in visualizing the current distribution in planar magnetics even with substantial non-1D or non-sinusoidal factors in switched-mode-power-converters. It can also usually provide useful qualitative loss comparison information in a rapid design iteration process.

Limitations:

1) *M2Spice*-assisted circuit simulations may not be able to accurately predict the current distribution if the circuit has wide-frequency-range harmonic components.

2) *M2Spice*-assisted circuit simulations cannot always accurately predict the absolute loss in magnetic components having very broadband excitation, for a number of reasons, but including violations of the 1-D constraints under which the model is developed.

VI. CONCLUSIONS

"Magnetics-in-the-circuit" time-domain SPICE simulations can facilitate quick circuit design iterations. This paper introduced a software tool –*M2Spice* – and investigates its applicability and limitations in evaluating various planar magnetics designs with "planar-magnetics-in-the-circuit" time domain simulations. It is verified that time-domain SPICE simulations with *M2Spice*-generated netlists can accurately predict current sharing in multiple layers and parallel layers, even if significant non-1D effects are present. If there is one fundamental frequency component dominating, simulating the circuit with a netlist generated at a single frequency can offer substantial efficiency in visualizing the current distribution. SPICE simulation cannot accurately predict loss if there are significant non-1D effects and/or broadband excitation waveforms, but can illustrate the loss trend for different interleaving patterns, and thus can assist evaluation of different planar magnetics implementations.

---

[6]Some FEM tools offer "magnetic-in-circuit" joint-design functions, but the design options are usually limited to a few commonly-used topologies.

ACKNOWLEDGEMENTS

The authors gratefully acknowledge C. R. Sullivan of Dartmouth College for the helpful discussion, and the support provided for this work by Texas Instruments, by the National Science Foundation under NSF award number 1307699, and by a SkolTech-MIT research program.

APPENDIX I: DC LIMITS OF THE ELEMENTS IN THE T NETWORK

When $\omega \to 0$, $\delta \to \infty$, and $\Psi \to 0$. The first-order Taylor series expansion of $e^{-\Psi h}$ around $\Psi = 0$ is $1 - \Psi h$. Using this approximation, the limit of Z_a as $\Psi \to 0$ is

$$\lim_{\Psi \to 0} Z_a = \frac{\mu dh}{2\text{w}}\omega j \qquad (3)$$

Therefore, Z_a can be approximated as an inductor with its inductance equal to $L_{a,dc} = \frac{\mu dh}{2\text{w}}$. The dc resistance of this branch is zero ($R_{a,dc} = 0$). As an intuitive verification, assuming there is no current in the conductor, the inductance of a spacing that has a thickness of h is $L_s = \frac{\mu dh}{\text{w}}$. $L_{a,dc}$ is effectively one half of L_s.

Taking the third-order Taylor series expansion of $e^{-\Psi h}$ and $e^{\Psi h}$ around $\Psi = 0$ will yield:

$$Z_b \approx \frac{d(1 - j\frac{h^2\omega\mu\sigma}{6})}{\text{w}\sigma h(1 + \frac{(h^2\omega\mu\sigma)^2}{36})} \qquad (4)$$

$$\lim_{\omega \to 0} Z_b = \frac{d}{\text{w}\sigma h} - \frac{\mu dh}{6\text{w}}\omega j \qquad (5)$$

Thus, when $\omega \to 0$, Z_b can be approximated as a resistance of $R_{b,dc} = \frac{d}{\text{w}\sigma h}$ in series with an inductance of $L_{b,dc} = -\frac{\mu dh}{6\text{w}}$.

APPENDIX II: NON-SINUSOIDAL WAVEFORMS AND EFFECTIVE FREQUENCY

For systems with a single frequency (e.g. resonant converters with sinusoidal waveforms), a single frequency netlist is well-suited to time-domain simulations. For non-sinusoidal waveforms with harmonics, we still propose to use a single frequency netlist for time-domain simulations to retain the simplicity and convenience of the MLM method. As shown in Fig. 6, when $\Delta < 1$, r_a, r_b, l_a and l_b stay relatively constant across wide frequency range, and can be directly used for wide spectrum simulation. When $\Delta > 1$, r_b, l_a and l_b all rapidly approach zero. As a result, the key frequency dependent element is r_a, which can be rewritten as:

$$r_a = \Delta \frac{\sinh \Delta - \sin \Delta}{\cosh \Delta + \cos \Delta}. \qquad (6)$$

The value of r_a is plot in Fig. 22 in log scale. When $\Delta < 2$, r_a is proportional to Δ^4, indicating that the resistance is proportional to f^2 (proximity effects are more significant than skin effects). When $\Delta > 3$, r_a is proportional to Δ, indicating that the resistance is proportional to $\sqrt{f}$ (skin effects are more significant than proximity effects).

Non-sinusoidal current waveforms (waveforms having one or more significant harmonics) can be treated by Fourier analysis utilizing the effective frequency concept described by

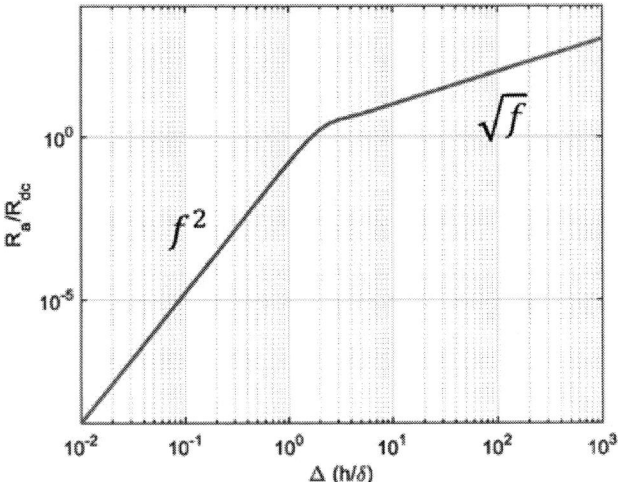

Fig. 22. Normalized R_a value in the T network as a function of Δ. R_a is the most frequency sensitive element in the network, and can be used to select the effective frequency. When $\Delta < 1$, R_a is proportional to f^2 (Δ^4). When $\Delta > 1$, R_a is proportional to $\sqrt{f}$ (Δ).

Eq. 26 in [21]. This way of choosing the effective frequency is derived based on a conservation of the loss, which is the major optimization goal for many designs. In the MLM model, r_a and r_b are loss contributing components. For a waveform shape with a majority of its harmonic components distributed in the low frequency ($\Delta < 1$) range (this case includes waveforms comprising significant dc components), since r_a is proportional to f^2 and r_b, l_a, and l_b stay constant, the effective frequency, F_{eff}, that can be used to generate the single-frequency netlist can be selected as:

$$F_{eff|\Delta<1} = \sqrt{\frac{\Sigma_{j=0}^{\infty} I_j^2 f_i^2}{\Sigma_{j=0}^{\infty} I_j^2}} \qquad (7)$$

I_j is the rms value of the respected harmonic component, and f_j is the respected harmonic frequency. Simplified methods of calculating F_{eff}, and the Fourier coefficients of many typical waveforms are provided in [7].

For a waveform shape with a majority of its harmonic components distributed in the high frequency range (i.e., $\Delta > 3$) range, r_a is proportional to $\sqrt{f}$, and r_b, l_a, l_b rapidly approach zero. As a result, the effective frequency used for generating the netlist can be selected as:

$$F_{eff|\Delta>3} = \left(\frac{\Sigma_{j=0}^{\infty} I_j^2 \sqrt{f_i}}{\Sigma_{j=0}^{\infty} I_j^2}\right)^2 . \qquad (8)$$

For cases when harmonic components are not all located in either the $\Delta < 1$ or $\Delta > 3$ region, since r_b also changes across frequencies (drops from 1 to zero as Δ increases from 1 to 3) and may have significant impact, we do not suggest one single effective frequency for the time-domain simulation. Nevertheless, one can always simulate the magnetics at different frequencies (with netlists generated for different frequencies), and apply superposition to these single frequency results.

REFERENCES

[1] M. Chen, M. Araghchini, K. K. Afridi, J. H. Lang, C. R. Sullivan, and D. J. Perreault, "A systematic approach to modeling impedances and current distribution in planar magnetics," *IEEE Trans. Power Electron.*, vol.31, no.1, pp. 560-580, January, 2016.

[2] M. T. Quirke, J. J. Barrett, and M. Hayes, "Planar magnetic component technology-a review," *IEEE Trans. Compon. Hybrids, Manuf. Technol.*, vol.15, no.5, pp.884-892, Oct. 1992.

[3] C. R. Sullivan, D. V. Harburg, J. Qiu, C.G. Levey, and D. Yao, "Integrating magnetics for on-chip power: a perspective," *IEEE Trans. Power Electron.*,vol.28, no.9, pp.4342-4353, September 2013.

[4] Z. Ouyang and M. Andersen, "Overview of planar magnetic technology fundamental properties", *IEEE Trans. Power Electron.*, vol.29, no.9, September 2014.

[5] P. L. Dowell, "Effects of eddy currents in transformer windings," *Proc. of the Institution of Electrical Engineers*, vol.113, no.8, pp.1387-1394, August 1966.

[6] A. F. Goldberg, J. G. Kassakian, and M. F. Schlecht, "Issues related to 1-10-MHz transformer design," *IEEE Trans. Power Electron.*, vol.4, no.1, pp.113-123, January 1989.

[7] W. G. Hurley, E. Gath, and J. G. Breslin, "Optimizing the AC resistance of multilayer transformer windings with arbitrary current waveforms," *IEEE Trans. Power Electron.*, vol.15, no.2, pp.369-376, Mar. 2000.

[8] Y. Han, G. Cheung, A. Li, C. R. Sullivan, and D. J. Perreault, "Evaluation of magnetic materials for very high frequency power applications," *IEEE Trans. Power Electron.*, vol.27, no.1, pp.425-435, Jan. 2012.

[9] J. A. Ferreira, "Improved analytical modeling of conductive losses in magnetic components," *IEEE Trans. Power Electron.*, vol.9, no.1, pp.127-131, Jan 1994.

[10] R. W. Erickson and D. Maksimovic, "A multiple-winding magnetics model having directly measurable parameters," *Proc. of the IEEE Power Electron. Special. Conf. (PESC)*, vol.2, pp.1472-1478, 17-22, May 1998.

[11] J.H. Spreen, "Electrical terminal representation of conductor loss in transformers," *IEEE Trans. Power Electron.*, vol.5, no.4, pp.424-429, Oct. 1990.

[12] D. C., Hamill, "Lumped equivalent circuits of magnetic components: the gyrator-capacitor approach," *IEEE Trans. Power Electron.*, vol.8, no.2, pp.97-103, Apr 1993.

[13] J.-P. Keradec, B. Cogitore, and F. Blache, "Power transfer in a two winding transformer: from 1-D propagation to an equivalent circuit," *IEEE Trans. Magn.*, vol.32, no.1, pp.274-280, Jan. 1996.

[14] J. M. Lopera, M. Pernia, J. Diaz, J. M. Alonso, and F. Nuno, "A complete transformer electric model, including frequency and geometry effects," *Proc. of the IEEE Power Electron. Special. Conf. (PESC)*, vol.2, pp.1247-1252, June 1992.

[15] J. M. Lopera, M.J. Prieto, A. M. Perna, and F. N. Nuno, "A multiwinding modeling method for high frequency transformers and inductors," *IEEE Trans. Power Electron.*, vol.18, no.3, May 2003.

[16] A., Besri, X. Margueron, J.-P. Keradec, and B. Delinchant, "Wide frequency range lumped element equivalent circuit for HF planar transformer," *Proc. of the IEEE Power Electron. Special. Conf. (PESC)*, no., pp.766-772, 15-19 June 2008.

[17] A.M. Pernia, F. Nuno, and J.M. Lopera, "1D/2D transformer electric model for simulation in power converters," *Proc. of the IEEE Power Electron. Special. Conf. (PESC)*, pp.1043-1049 vol.2, 18-22, Jun 1995.

[18] R. Asensi, J. A. Cobos, O. Garcia, R. Prieto, and J. Uceda, "A full procedure to model high frequency transformer windings," *Proc. of the IEEE Power Electron. Special. Conf. (PESC)*, pp.856-863 vol.2, 20-25 Jun 1994.

[19] P. G., Blanken, "A lumped winding model for use in transformer models for circuit simulation," *IEEE Trans. Power Electron.*, vol.16, no.3, pp. 445-460, May 2001.

[20] X. Margueron, A. Besri, Y. Lembeye and J.-P. Keradec, "Current sharing between parallel turns of a planar transformer: prediction and improvement using a circuit simulation software," *IEEE Trans. on Ind. Applica.*, vol.46, no.3, May-June 2010.

[21] C. R. Sullivan, "Optimal choice for number of strands in a Litz-wire transformer winding," *IEEE Trans. Power Electron.*, vol.14, no.2, pp.283-291, Mar 1999.

[22] S. J. Gunter, K. K. Afridi, D. M. Otten, R. A. Abramson, and D. J. Perreault, "Impedance Control Network Resonant Step-Down DC-DC Converter Architecture," *Proc. of the IEEE Energy Conv. Congr. and Expo. (ECCE)*, Montreal, Canada, September 2015.

[23] PL140 Series SMT Planar Transformers Datasheets [Online], Coilcraft, Inc., Available: http://www.coilcraft.com/pdfs/pl140.pdf.

Solution of Input Double-Line Frequency Ripple Rejection for High-Efficiency High-Power Density String Inverter in Photovoltaic Application

Xiaonan Zhao, Lanhua Zhang, Rachael Born, Jih-Sheng Lai
Future Energy Electronics Center
Virginia Tech
Blacksburg, VA, USA
xiaonanzhao@vt.edu

Abstract— **One important type of power conditioning systems in photovoltaic (PV) application is the string inverter which requires small input voltage and current ripple. In addition, high-efficiency and high-power density are also the critical requirements for string inverter system. So the input voltage and current ripple cannot be easily rejected by paralleling large conventional electrolytic capacitors in the inverter input side only, since the huge electrolytic capacitors will limit the efficiency and power density. This paper presents a control strategy to limit the input double-line frequency ripple with a high-efficiency and high-power density front-end buck-type dc/dc converter. The double-line frequency energy is stored at the DC bus capacitor with the control strategy employed and the input current and voltage ripple are highly reduced compared with traditional electrolytic capacitor filter. Experimental results based on a 2-kW 6.6 inch³ hardware prototype with 99.6% peak efficiency are provided to validate the proposed double-line frequency ripple rejection control method.**

Keywords—string inverter; two-stage inverter; high power density; double line frequency rejection; proportional-resonant controller.

I. INTRODUCTION

In the past years, the demand of renewable energy is steadily increasing due to the limited availability of fossil fuels and their negative impact on the environment [1]. Solar energy as one type of the renewable energy has been fast growing because of its great potential of installations and price decreasing [2]. One type of PV conditioning systems is string inverter as shown in Fig.1. Several modules connected in series to form a PV string and each string is connected with a single phase grid-tie inverter as referred to "string inverter". The string inverter is widely utilized because of higher maximum power point tracking (MPPT) efficiency compared with centralized inverter and lower cost and easier to maintain compared with microinverter [3]. However, the string inverter also needs to face the common issue existing in PV conditioning systems which is the double-line frequency energy in the input side. The problem is caused by the unbalance instantaneous power between the source and load [4]. This double-line frequency energy causes a double-line frequency current and voltage ripple on the PV panel output side[5], [6], which could cause some other issues, for example, the MPPT capability is affected or it may also be hurt to the AC

side regulation [7]-[9]. The traditional solution of the double-line frequency energy problem is to parallel a series of huge electrolytic capacitors in the inverter input side, which heavily hurts the efficiency, increases the volume and limits the inverter lifetime [10]-[11].

Considering the efficiency and power density are the two of most significant characteristics of the string inverter, the best solution of rejecting double-line frequency energy should affect the efficiency and power density as least as possible. There have been many methods, as referred to active power decoupling, proposed to solve the double-line frequency energy in the literatures [12]-[18]. Different circuits and control technologies were used in the different topologies. However, each of them faces the problem of low efficiency, control complexity or low system dynamic performance [19].

This paper presents one power decoupling solution with two-stage inverter design. A novel control strategy is employed in front-end buck-type dc/dc to provide high rejection capability at double-line frequency. As a result, the double-line frequency energy is stored in the dc bus instead of the input side. Compared with the traditional capacitor-only solution, the input voltage and current ripples are only 4.4 % and 0.8 % respectively. Moreover, the volume of this front-end dc/dc converter is only 6.6 inch³ and its peak efficiency can reach to 99.6 %.

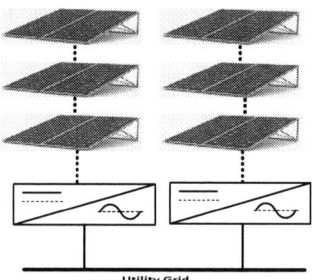

Fig. 1 PV power conditioning system: string inverter

II. INPUT RIPPLE ANALYSIS

The double-line frequency ripple in the DC side reflected from the AC side for the inverter system is shown in the Fig.2 (a). The instantaneous current as expressed in (2) can be analyzed by the balance of input and output instantaneous power as shown in the equation (1). From the instantaneous input current expression in equation (2), it is clear that the input current carries double-line frequency harmonics with relatively high amplitude.

$$p(t) = v_{ac}(t) \cdot i_{ac}(t) = V_{ac-rms} I_{ac-rms} \left(1 - \cos(2\omega t)\right) \quad (1)$$

$$i_{inv}(t) = \frac{p(t)}{V_{dc}} = \frac{V_{ac-rms} I_{ac-rms}}{V_{dc}} - \frac{V_{ac-rms} I_{ac-rms}}{V_{dc}} \cos(2\omega t) \quad (2)$$

The ac circuit model for input side current ripple is shown in the Fig.2 (b). By employing Kirchhoff's current law (KCL) and Kirchhoff's voltage law (KVL), the ripple relationship between voltage and current ripples and input side capacitor value can be derived in equation (3) and (4). It is clear that large capacitors will be needed if meet the ripple limitations. For example, if the inverter requires 3% voltage ripple and 20% current ripple, the input electrolytic capacitor should be at least 1.5 mF and the voltage ratings of these capacitors are 450-V or higher considering the light load conditions.

$$I_{in-p} = \frac{P_{ac}}{V_{dc}} \frac{1/(j\omega C_{dc})}{R_s + 1/(j\omega C_{dc})} \quad (3)$$

$$V_{dc-p} = \frac{P_{ac}}{2\omega C_{dc} V_{dc}} \quad (4)$$

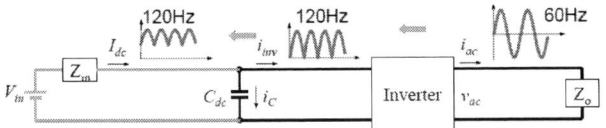

(a) Single-Phase Current Ripple Propagations

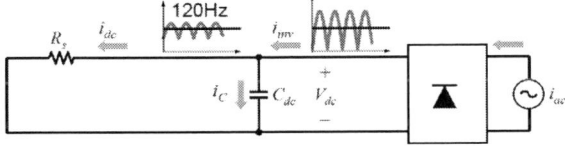

(b) Equivalent AC Circuit Model for Ripple Current

Fig. 2 Double-line frequency ripple issue in the inverter system

III. HIGN EFFICIENCY AND HIGH POWER DENSITY DC/DC CONVERTER DESIGN AND MODELING

As shown in the Fig.3, a four phase interleaved continuous-conduction-mode (CCM) inverse buck converter is employed to serve as a front-end dc/dc stage in two-stage inverter system [20], [21]. This part mainly illustrates the hardware design for high efficiency and high power density and the converter modeling.

A. Hardware Deisgn Considerations.

The target of the converter hardware design is high efficiency, high power density and accurate sensing circuit design for the input ripples rejection control. The considerations are listed as following.

(1) With the availability and price decreasing of wide-band gap devices, the converter can achieve higher efficiency using wide-band gap devices compared with traditional silicon devices

for their faster switching and smaller parasitic capacitance. In this paper's converter, Transphorm's TPH3002LD and Cree's C3D1P7060Q are employed with high switching frequency of 400 kHz to achieve the highest efficiency. The other details should also be considered, for example, minimizing gate-driving loop, reducing the parasitic train-to-source capacitance and slowing down the turn-on speed.

(2) The 3D view of dc/dc stage is shown in the Fig.4 with the main components marked. To fully utilize the space, every component arrangement should corporate with dc/ac stage and control board design. In addition, the inverter copper case also serves as the heat sink to further reduce the volume. The dc/dc stage is mounted to the top lid.

(3) To limit the input side ripple, the inductor current of the dc/dc converter should be sensed and controlled which requires accurate sensing of the inductor current. Here, LEM' current transducer (CT) HO-NP-0000 series are employed, since this series CT are not be easily affected by the external power supply fluctuation and temperature.

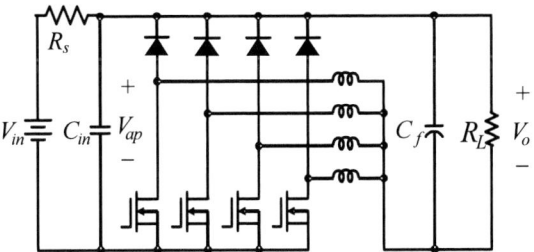

Fig. 3 Topology of four phase interleaved inverse buck converter

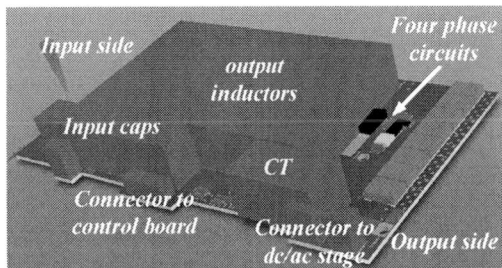

Fig. 4 3D view of front-end dc/dc converter

B. Converter Modeling

To design the controller, the first step is to model the converter and derive the open-loop transfer function. In this paper, traditional state space model method for CCM buck converter is utilized [22] considering the parasitic resistor of inductor, r_L, and parasitic resistor of dc-bus capacitor, R_c. And R_s is the resistor in series in the input side as shown in the Fig.3.

The control-to-output voltage and the control-to-inductor current transfer functions are expressed in (5) and (6) respectively. And control-to-inductor current transfer function is plotted in Fig.5, where the red lines are under the full load condition and the blue lines are under the light load condition.

978-1-4673-9551-9/16 $31.00 © 2016 IEEE

$$G_{vd} = \frac{\hat{v}_o}{\hat{d}} = \left(V_{ap} - DI_o \frac{R_s / C_{in}}{s \cdot R_s + 1 / C_{in}}\right) \frac{\left(\frac{\left(s \cdot R_c + 1 / C_f\right) R_L}{s \cdot \left(R_L + R_c\right) + 1 / C_f}\right)}{\left(D^2 \frac{R_s / C_{in}}{s \cdot R_s + 1 / C_{in}} + s \cdot L_f + r_L + \frac{\left(s \cdot R_c + 1 / C_f\right) R_L}{s \cdot \left(R_L + R_c\right) + 1 / C_f}\right)} \tag{5}$$

$$G_{id} = \frac{\hat{i}_L}{\hat{d}} = \frac{s \cdot \left(R_L + R_c\right) + 1 / C_f}{\left(s \cdot R_c + 1 / C_f\right) R_L} \cdot G_{vd} \tag{6}$$

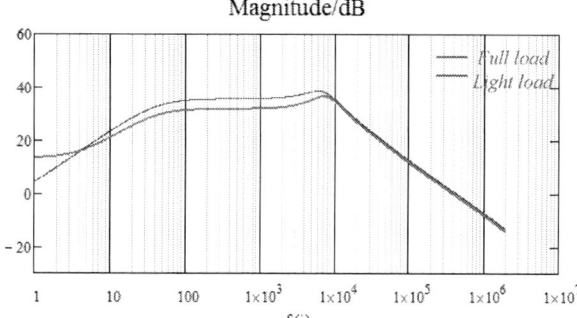

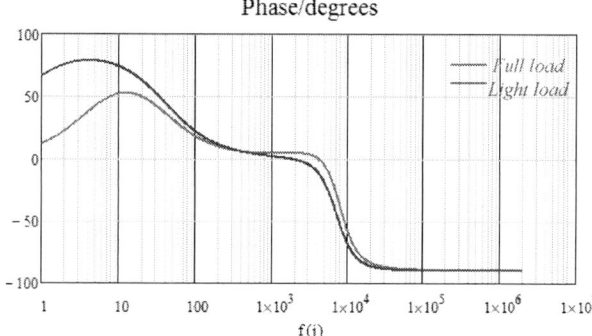

Fig. 5 Bode plot of control-to-inductor current transfer function

IV. CONTROLLER DESIGN

A. Control Diagram Introduction

There are two functions for the controller of front-end dc/dc converter: (1) regulating the dc-bus voltage for the dc/ac inverter, (2) limiting the input double-line frequency voltage and current ripples by controlling the inductor current. The control diagram is shown in the Fig.6.

An internal current loop and external voltage loop work together to minimize the double-line frquency input ripple. The external voltage loop senses the voltage of the dc-bus capacitor, C_f, and passes it through a notch filter to eliminate the double-line frequency ripple. The voltage compensator, a traditional PI controller, then generates a current reference, i_{ref}, for the internal current loop. Ideally, this current reference, i_{ref}, is a constant value. The internal current loop controls the current of the inductor based on the current reference, i_{ref}. The total inductor current of four phase, i_L, is sensed. A proportional-resonant (PR) controller is used for the internal current loop to boost the gain at double-line frequency and to increase the output impedance of the double-line frequency ripple. So with this control strategy,

the controller of dc/dc converter can not only eliminate the input ripples but also regulate the dc bus voltage.

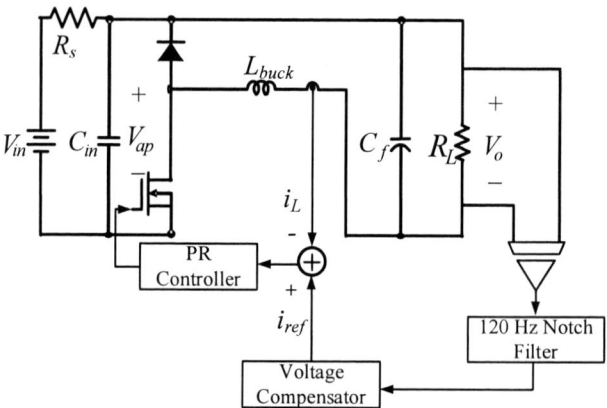

Fig. 6 Control diagram of dc/dc converter

B. PR Controller Design in Current Loop

Quite a few nonlinear current control methods were proposed recently [23]-[26] and some of them can fulfill fast transition and high control accuracy [25]-[26]. But the implementation of these nonlinear control method are too complex to be employed in this system. To provide a high gain injection at double-line frequency and inspired by [27], a PR controller is employed. The structure of PR controller is shown in the equation (7), consisting of a proportional part and a part with one zero and two complex-conjugate poles as referred to resonant controller. The bode plot of this PR controller is shown in the Fig.7, where we can see that it provides a high gain at double-line frequency.

$$G_{PR}(s) = K_p + K_I \frac{2\omega_c s}{s^2 + 2\omega_c s + \omega_0^2} \tag{7}$$

Fig.8 shows the control block diagram of current loop, where H_{ii} is the inductor current sensing gain which is a constant gain during low frequency range. The current loop gain is expressed in (8) and the transfer function from the reference current to the inductor current is expresses in (9).

$$T_i = G_{id} G_{PR} H_{ii} \tag{8}$$

$$G_{ii} = \frac{\hat{i}_L}{\hat{i}_{ref}} = \frac{T_i}{T_i + 1} \tag{9}$$

Fig.9 shows the bode plots of current loop gain T_i compared with the bode plot of open loop control-to-inductor current transfer function, G_{id}. Due to the gain characteristic of the PR controller, the current loop gain has a gain boost at double-line frequency. Taking a detailed look at the gain at double-line frequency, the higher gain boost, the closer to one of G_{ii} which means the smaller error between reference current and real inductor current. So, to ensure elimination of the double-line frequency current ripple through the inductor, the reference current has to carry no double-line frequency ripple information.

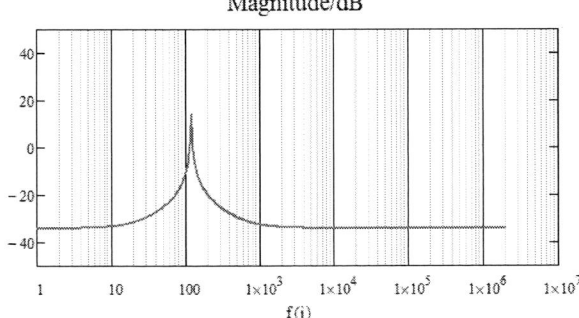

Magnitude/dB

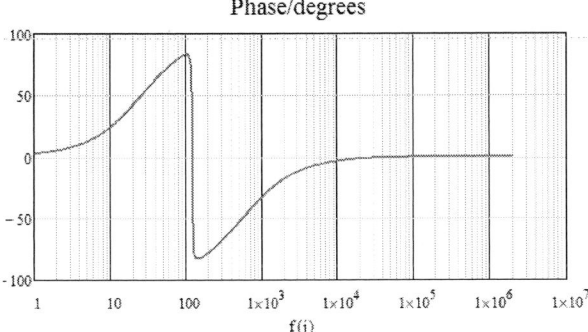

Phase/degrees

Fig. 7 Bode plot of PR controller

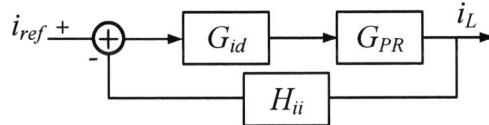

Fig. 8 Control block diagram of current loop

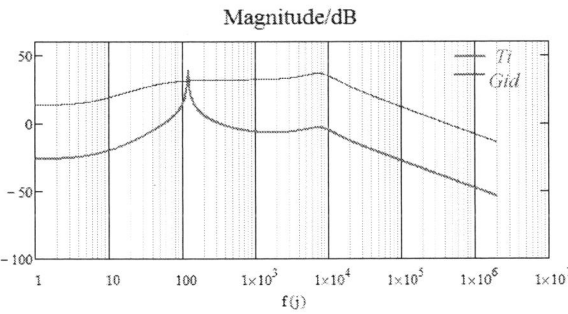

Magnitude/dB

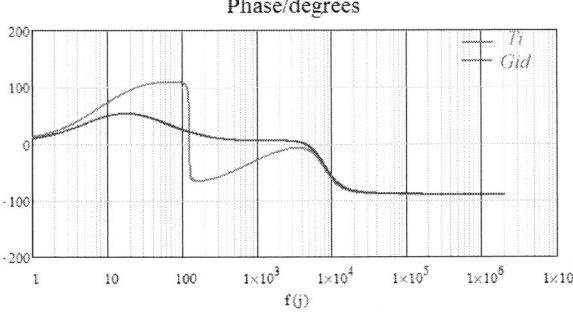

Phase/degrees

Fig. 9 Bode plot of current loop gain T_i

C. Controller Design in Voltage Loop

From the conclusion of the current loop controller design, no double-line frequency ripple in the reference current needs to be guarantee. Notch filter is an ideal choice to eliminate the double-line frequency ripple from the sensed dc bus voltage. The structure of notch filter is shown in (10).

$$G_{notch}(s) = \frac{s^2 + s/60 + \omega_0^2}{s^2 + \omega_o s + \omega_0^2} \tag{10}$$

Where ω_0 is the angular double-line frequency. The bode plots comparison between notch filter and low pass filter is shown in the Fig. 10. The blue lines are the bode plot of notch filter where the red lines are the bode plot of low pass filter. The gain of the notch at double-line frequency is much lower than that of low pass filter. So the notch filter is employed here to eliminate the double-line frequency ripple of sensed dc-bus voltage.

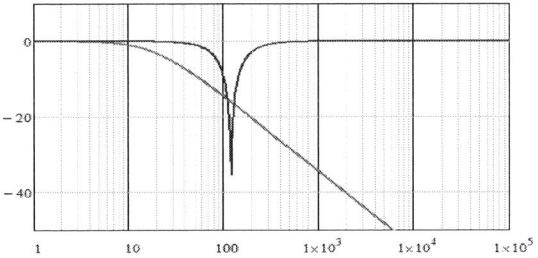

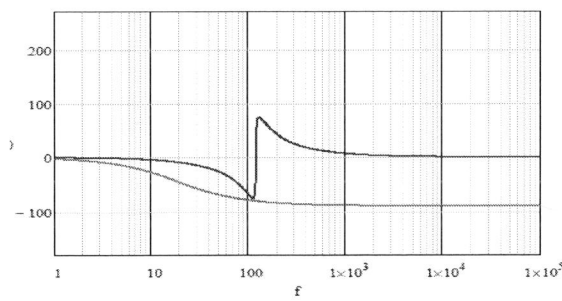

Fig. 10 Bode plots comparison between notch filter and low pass filter

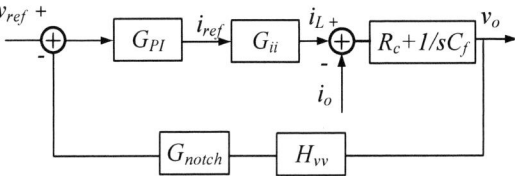

Fig. 11 Control block diagram of voltage loop

Fig.11 shows the control block diagram of voltage loop, where H_{ii} is the dc-bus voltage sensing gain, G_{PI} is the PI controller as expressed in (11). PI controller is to provide a high gain at dc and to regulate the dc-bus voltage. The voltage loop is very slow compared with inner current loop.

$$G_{PI} = K_p + K_I/s \tag{11}$$

To derive the voltage loop gain, the perturbation of output current, i_o, should be ignored. The voltage loop gain is

978-1-4673-9551-9/16 $31.00 © 2016 IEEE

expressed in (12) and the transfer function from the reference voltage to dc-bus voltage is expressed in (13).

$$T_v = H_{vv} G_{notch} G_{PI} G_{ii} \cdot (R_c + 1/sC_f) \qquad (12)$$

$$G_{vv} = \frac{\hat{v}_o}{\hat{v}_{ref}} = \frac{T_v}{T_v + 1} \qquad (13)$$

Ignoring the perturbation of reference voltage, v_{ref}, the output impedance in closed loop can be derived in (14) according to the Fig.11.

$$Z_{o\_c} = \cfrac{1}{\cfrac{1}{R_c + 1/sC_f} - H_{vv} G_{notch} G_{PI} G_{ii}} \qquad (14)$$

Due to the high gain rejection of notch filter, the gain of $H_{vv} G_{notch} G_{PI} G_{ii}$ at double line frequency is very low, ideally, even close to zero. So the gain of $H_{vv} G_{notch} G_{PI} G_{ii}$ can be ignored and the impedance in closed loop at double-line frequency can be approximated as (15). The equivalent circuit can be treated as the converter branch and dc-bus capacitor branch in parallel as shown in the Fig.12. The output impedance in closed loop is shown in (16) based on the Fig.12 equivalent circuit. Combined with (15) and (16), it is clear that the approximated closed loop impedance is equal to the impedance Z_2 which means impedance Z_1 is infinite large at double-line frequency. So all of the double-line frequency ripple will go through the dc-bus capacitor branch rather than the converter branch. The input ripples will be highly reduced and the double-line frequency energy will be stored in the dc-bus capacitors.

$$Z_{o\_c} = R_c + 1/sC_f \qquad (15)$$

$$Z_{o\_c} = Z_1 // Z_2 \qquad (16)$$

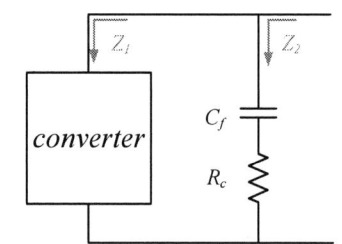

Fig. 12 Equivalent circuit for impedance analysis

V. EXPERIMENTAL RESULTS

A 2-kW four phase interleaved buck type dc/dc converter prototype is built to validate the proposed controller. The normal input voltage of the converter is designed at 400-V and the normal dc-bus voltage is designed at 365-V. Fig. 13 shows the photograph of the hardware prototype. The dimensions of the prototype are 4.7'' length, 2.4'' width and 0.59'' height. The efficiency of this dc/dc converter can reach to 99.6% under the full load condition. The two-stage string inverter prototype is shown in the Fig. 14. DSP TMS320F28069 is employed as the digital controller. The inverter testing set up is pictured in the Fig.15, where the input side is in series with a 10 Ohms resistor and the load resistor is 28.8 Ohms. The output of the voltage source is 450-V and the inverter output is 240-V RMS with 60-Hz.

Fig.16 and Fig.17 show the input voltage and current ripples comparison with and without the prosed control scheme under the full load condition. Without the proposed controller employed as shown in Fig.16, the input current ripple is 2.3-A in red waveform and voltage ripple is 28.9-V in yellow waveform. The percent of the current ripple is 7.2 % and the percent of the voltage ripple is 47.6 %. In the Fig.17, the input current ripple is 0.22-A and the percentage of the current ripple is 4.4 %, while the voltage ripple is 3.4-V and the percentage of the voltage ripple is 0.8 %. Under the full load condition, the input voltage and current ripple are highly decreased with the proposed controller.

Fig. 13 Photograph of dc/dc converter hardware prototype

Fig. 14 Hardware prototype of 2-KW high-efficiency and high-power density inverter

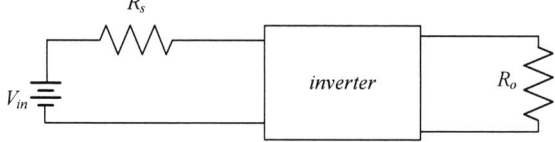

Fig. 15 Inverter testing set up

A comparison between traditional electrolytic capacitor-only solution and proposed control strategy is shown in the Table.1.

The benefits of the proposed controller are listed as following:

(1) The electrolytic capacitor that requires for the storage energy is dropped from 1.5 mF to 470 uF to meet the 20 % input current ripple and 3 % voltage ripple requirement. The electrolytic capacitor is in the dc-bus side with the proposed control strategy.

(2) Because of the buck-type converter, the dc-bus voltage is regulated at 365-V, which decreases the electrolytic capacitor voltage rating requirements and further shrinks the volume of this capacitor. The capacitor voltage ratings of the traditional capacitor-only solution should be at least 450-V, since under the light load condition, for example, 10 % load condition, the input voltage is as high as 430-V.

(3) The results with proposed controller bring the percent of the input ripple well below the 20% and 3% requirements to 4.4% and 0.8% for current and voltage ripples, respectively.

(4) The total converter volume with proposed controller is less than one third of capacitor-only solution when suppose that the capacitor-only solution uses three 680 uF 500-V rated electrolytic capacitors with 1.575" D and 2.756" H dimensions.

VI. CONCLUSION

For string inverter in PV application, the traditional electrolytic-capacitor only solution to drop the inverter input double-line frequency ripples greatly limits efficiency and power density. In this paper, a control strategy is proposed to limit the input voltage and current double-line frequency ripples applying in a high-efficiency and high-power density front-end dc/dc converter. The hardware design considerations are introduced in the first part. As a result, the volume of this front-end dc/dc converter is only 6.6 inch^3 and its peak efficiency can reach to 99.6%. The proposed controller illustrated in the second part consists of a current loop with PR controller to push the input double-line frequency energy to the dc-bus capacitor and a voltage loop with notch filter and PI controller to assist the current loop limiting the input double-line frequency ripples and regulate the dc bus voltage. Compared with the traditional capacitor-only solution, the input voltage and current ripples are only 4.4% and 0.8% respectively while the original current and voltage ripple without the proposed controller are 47.6 % and 7.2 % .

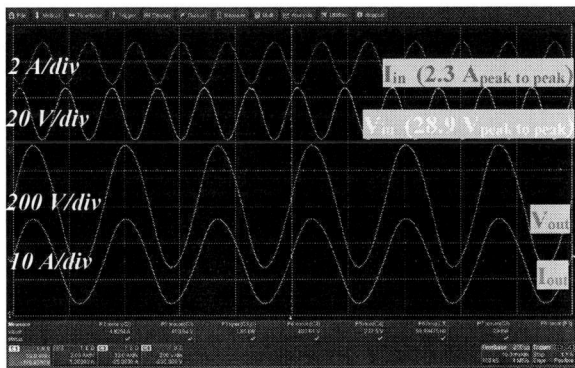

Fig. 16 Waveforms without proposed controller

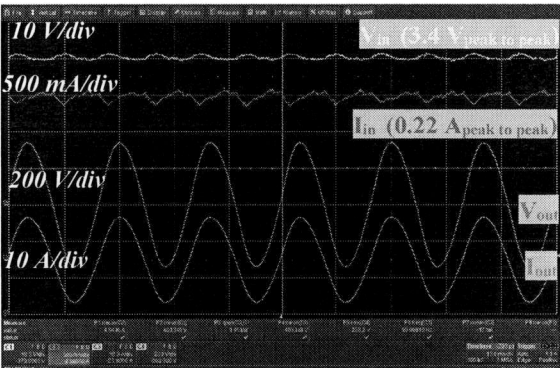

Fig. 17 Waveforms with proposed controller

Table 1. Comparison between capacitor-only and proposed control strategy

	Capacitor-only	FEEC DC/DC stage
Energy Storage Required	>1.5mF	0.47mF
Capacitor Voltage Rating	>450V	400V
Effectiveness	Reduces input voltage and current ripples to the required 3% and 20%	Reduces input voltage and current ripples to less than 0.8% and 4.4%
Energy Buffering Volume*	20.5 in^3	6.6 in^3

*Capacitor-only assumption uses (3) 680uF 500V rated electrolytic capacitors with 1.575" D and 2.756" H

REFERENCES

[1] B. K. Bose, "Global Warming: Energy, Environmental Pollution, and the Impact of Power Electronics," *IEEE Ind. Electron. Mag.*, vol. 4, no. 1, pp. 6-17, Mar. 2010.J. Clerk Maxwell.

[2] "Global market outlook for photovoltaics, 2013-2017," www.epia.org.K.

[3] X. Zhao, L. Zhang, X. Cui, C. Zheng, C.-Y. Lin, Y.-C. Liu, J.-S. Lai, "A High-efficiency Hybrid Series Resonant DC-DC Converter with Boost Converter as Secondary for Photovoltaic Applications", in *IEEE ECCE 2015*, pp5462-5467, Sep. 2015.

[4] J.-S. Lai, "Power conditioning circuit topologies," *IEEE Ind. Electron. Mag.*, vol. 3, pp. 24-34. Jun. 2009.

[5] C. Liu and J.-S. Lai, "Low frequency current ripple reduction technique with active control in a fuel cell power system with inverter load," in *IEEE Transactions on Power Electronics*, vol. 22, no. 4, pp. 1429-1436, July 2007.

[6] S. B. Kjaer, J. K. Pedersen, and F. Blaabjerg, "A review of single phase grid-connected inverters for photovoltaic modules," *IEEE Trans. Ind. Appl.*, vol. 41, pp. 1292-1306, Sep./Oct. 2005.

[7] C. Hutchens, W.S. Yu and J. S. Lai, "Modeling and control of charge-pumped reboost converter for PV Applications," In *COMPEL'10, IEEE 12th Workshop on*, 2010.

[8] B. Gu, J. Dominic, and J.-S. Lai, "Modeling and control of a high boost ratio PV module dc-dc converter with double grid line ripple rejection," in Proc. of *IEEE COMPEL Workshop, Salt Lake City, Utah*, June 2013, pp. 1–4.

[9] T. LaBalla, "A High-Efficiency Hybrid Resonant Microconverter for Photovoltaic Generation Systems", Ph.D disseratoin, Chapter 5, 2014.

[10] R. G. Rojo and D. Olalla, "DC -Link Capacitors for Industrial Applications," 2009.

[11] S. Harb, M. Kedia, H. Zhang, and R. S. Balog, "Microinverter and string inverter grid -connected photovoltaic system - A comprehensive study,"

in *2013 IEEE 39th Photovoltaic Specialists Conference* (PVSC), 2013, pp. 2885 -2890.

[12] H. Hu, S. Harb, X. Fang, D. Zhang, Q. Zhang, Z. J. Shen, and I. Batarseh,"A three-port flyback for PV micro-inverter applications with power pulsation decoupling capability," *IEEE Trans. Power Electron.*, vol. 27, no. 9, pp. 3953–3964, Sep. 2012.

[13] S. Harb, H. Haibing, N. Kutkut, I. Batarseh, and Z. J. Shen, "A three-port photovoltaic (PV) micro-inverter with power decoupling capability," *in Proc. 26th Annu. IEEE Appl. Power Electron. Conf. Expo.*, Mar. 2011, pp. 203–208.

[14] F. Shinjo, K. Wada, and T. Shimizu, "A single-phase grid-connected inverter with a power decoupling function," *in Proc. IEEE Power Electron. Spec. Conf.*, Jun. 2007, pp. 1245–1249.

[15] T. Hirao, T. Shimizu, M. Ishikawa, and K. Yasui, "A modified modulation control of a single-phase inverter with enhanced power decoupling for a photovoltaic ac module," *in Proc. Eur. Conf. Power Electron. Appl., 2005*, pp. 1–10.

[16] Y.-M. Chen and C.-Y. Liao, "Three-port flyback-type single-phase microinverter with active power decoupling circuit," *in Proc. IEEE Energy Convers. Congr. Expo.*, Sep. 2011, pp. 501–506.

[17] Wen Cai, Bangyin Liu, Shanxu Duan and Ling Jiang, "An Active Low-Frequency Ripple Control Method Based on the Virtual Capacitor Concept for BIPV Systems," *IEEE Transactions on Power Electronics*, vol. 29, no. 4, pp. 1733-1745, 2014.

[18] Wen Cai, Ling Jiang, Bangyin Liu, Shanxu Duan and Changyue Zou, "A power decoupling method based on four-switch three-port DC/DC/AC converter in DC microgrid," *IEEE Transactions on Industry Applications*, vol. 51, no. 1, pp. 336-343, 2015

[19] H. Hu, S. Harb, N. Kutkut, I. Batarseh and Z. J. Shen, "A Review of Power Decoupling Techniques for Microinverters With Three Different Decoupling Capacitor Locations in PV Systems", *IEEE Transaction on Power Electroncis,* Vol. 28, No. 6, June 2013.

[20] J.-S. Lai and C. Liu, "Multiphase soft switched DC/DC converter and active control   technique for fuel cell ripple current elimination," U.S. Patent #7,518,886.

[21] J.-S. Lai, C. Lin, Y. Liu, L. Zhang and X. Zhao," Design Optimization for Ultrahigh Efficiency Buck Regulator using Wide Bandgap Devices", *in ECCE 2015*, pp. 4797 – 4803, sep. 2015.

[22] R. W. Erickson and D. Maksimovic, *Fundamentals of Power Electronics*, 2nd ed. New York: Springer, 2001.

[23] K. M. Smedley, L. Zhou, and C. Qiao, "Unified constant-frequency integration control of active power filters-steady-state and dynamics," *IEEE Trans. Power Electron.*, vol. 16, no. 3, pp. 428-436, May 2001.

[24] D. Maksimovic, Y. Jang, and R. W. Erickson, "Nonlinear-carrier control for high-power-factor boost rectifiers," *IEEE Trans. Power Electron.*, vol. 11, pp. 578-584, July 1996

[25] L. Zhang, B. Gu, J. Dominic, B. Chen, C. Zheng, and J. -S. Lai, "A dead-time compensation method for parabolic current control with improved current tracking and enhanced stability range," *IEEE Trans. Power Electron.*, vol. 30, no. 7, pp. 3892-3902, Jul. 2015

[26] L. Zhang, B. Gu, and J. -S. Lai, "The implementation of parabolic current control for dual-carrier PWM", in *Proc. of IEEE APEC 2015.*, Mar. 2015

[27] B. Gu, J. Dominic, J. Zhang, L. Zhang, B. Chen, and J. -S. Lai, "Control of electrolyte free microinverter with improved MPPT performance and grid current quality," in *Proc. of IEEE APEC 2014*, pp.1788-1792, Mar. 2014.

978-1-4673-9551-9/16 $31.00 © 2016 IEEE

Fractional-order Phase Lead Compensation for Multi-rate Repetitive Control on Three-phase PWM DC/AC Inverter

Zhichao Liu[*1], Bin Zhang[*2], IEEE senior member, and Keliang Zhou[†], IEEE senior member
*College of Engineering and Computing
University of South Carolina, Columbia, SC, USA 29208
Email: [1]zhichao@email.sc.edu, [2]zhangbin@cec.sc.edu
[†]School of Engineering
University of Glasgow, Glasgow G12 8QQ, Scotland, United Kingdom
Email: keliang.zhou@glasgow.ac.uk

Abstract—For constant voltage constant frequency pulse-width modulation (PWM) inverter system, repetitive control (RC) can achieve zero steady-state tracking error for any periodic signal. Multi-rate repetitive control (MRC), which is featured by a fast system sampling rate and a reduced RC rate, is able to lower CPU computation load while achieving low tracking error and fast convergence speed. To accurately compensate the phase lag of MRC, this paper proposes a fractional-order phase lead compensation solution to further improve the tracking performance. Implemented with a Lagrange polynomial, the fractional-order phase lead compensator has more accurate and flexible phase lead compensation than traditional phase lead compensator. Experimental results are provided to show the effectiveness of the proposed fractional-order phase lead compensation.

Index Terms: repetitive control, DC/AC inverter, multi-rate repetitive control, phase lead compensation, Lagrange interpolation polynomial

I. INTRODUCTION

The constant voltage constant frequency (CVCF) PWM inverters are wildly used in programmable AC power source, uninterrupted power supply, and other industry facilities [1]–[6]. High quality power sources require low total harmonic distortion (THD), fast transient response, and low steady state error [7]. Repetitive control [8], based on internal model principle, is an effective way to improve inverter performance [9].

To reduce CPU processing load, multi-rate repetitive control (MRC) was introduced [10]. In MRC scheme, the converter samples with a high sampling rate, while RC processes the data with a reduced low rate. Previous work shows that MRC is able to maintain the convergence speed and THD as good as conventional RC.

Linear phase lead compensation [11] for RC is an effective way to improve RC convergence speed, tracking accuracy, and stability. The phase lead filter z^n, where z is z-transform operator and n is the phase lead, can compensate the effect of the phase delay from plant and feedback controller. For a signal with frequency ω, phase lead compensator z^n provides a phase lead $\theta = n \times (\omega/\omega_N) \times 180°$, where ω_N represents Nyquist frequency, to compensate the system phase lag.

Although the phase lead z^n with an integer step n in conventional RC is able to compensate the phase lag well enough for many applications [12]–[15], it is insufficient for phase compensation on MRC, especially those applications with a high frequency reference signal such as on-board AC power or programmable AC power source. With a downsampling ratio m, the integer phase lead compensator z_m^n ($z_m = z^m$ indicates the RC rate is m times slower than the system sampling rate) provides a phase lead $\theta = n \times m(\omega/\omega_N) \times 180°$. For the signal with frequency ω, its phase lead compensation resolution will reduce m times. Imprecise phase lead may lower control performance or even make the system unstable.

To address this issue, an MRC scheme with fractional-order phase lead compensation is proposed for CVCF PWM inverter system. The fractional-order phase lead is achieved with a 2nd-order finite impulse response (FIR) filter implemented by Lagrange interpolation polynomial.

The paper is organized as follows: Section II introduces the multi-rate RC and formulates fractional-order phase lead. Section III develops the fractional-order phase lead compensation for three-phase PWM inverter system. Section IV presents experimental results, analysis, and comparison, which is followed by concluding remarks in Section V.

II. FRACTIONAL-ORDER PHASE LEAD COMPENSATION FOR MRC

A. Conventional RC

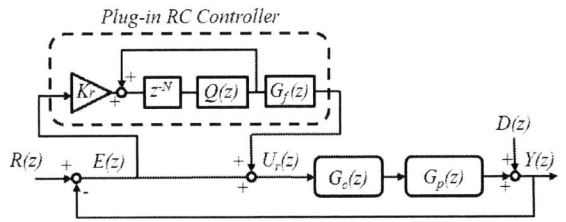

Fig. 1. Control system with plug-in RC controller

A typical digital control system with plug-in RC controller [16] is shown in Fig. 1, where $R(z)$ is the reference input, $G_c(z)$ is the conventional controller, $G_p(z)$ is the plant model, $D(z)$ is the disturbance, $Y(z)$ is the system output. Plug-in RC controller is a feed-forward controller consisted of RC gain K_r, period delay z^{-N}, and robustness filter $Q(z)$, stability filter $G_f(z)$.

RC transfer function G_r in Fig. 1 is:

$$G_r(z) = \frac{U_r(z)}{E(z)} = K_r \frac{z^{-N}Q(z)}{1 - z^{-N}Q(z)}G_f(z) \qquad (1)$$

where $E(z) = R(z) - Y(z)$ is the system tracking error, $N = f_s/f \in \mathbb{N}$ with f_s being sample frequency, f being the fundamental frequency of reference signal $R(z)$.

From frequency characteristics of (1), RC has infinity magnitude gain at reference signal fundamental and harmonic frequencies when $Q(z) = 1$. Therefore, RC can realize zero steady-state error tracking for periodic signal in closed-loop control.

B. Multi-rate RC

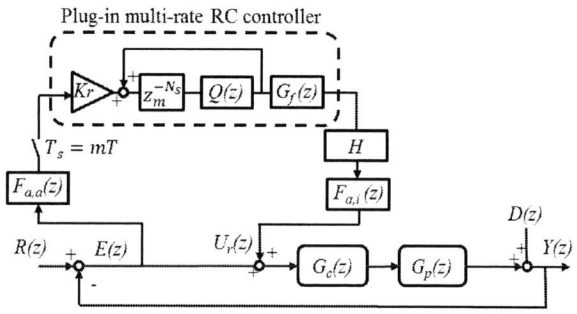

Fig. 2. Multi-rate RC system

Multi-rate repetitive control is able to reduce the CPU processing load and keep the high sampling frequency. In conventional RC controller, processor stores N error data samples from the current cycle for next cycle calculation. For multi-rate RC shown in Fig. 2, the N error samples are downsampled m times to reduce the computation time by increasing RC sampling time T_s m times:

$$T_s = mT; z_m = e^{sT_s} = e^{msT} = z^m \qquad (2)$$

The MRC transfer function G_{MRC} is:

$$G_{MRC}(z_m) = K_r \frac{z_m^{-N_s}Q(z_m)}{1 - z_m^{-N_s}Q(z_m)}G_f(z_m) \qquad (3)$$

To prevent aliasing and imaging caused by downsampling and upsampling in MRC, two low-pass filters $F_{a,a}(z)$ and $F_{a,i}(z)$ are implemented with the zero-phase window filter [10]. A zero-order holder H is used to hold the values between every two samples. Previous work shows that MRC is able to maintain the convergence speed and THD as good as conventional RC.

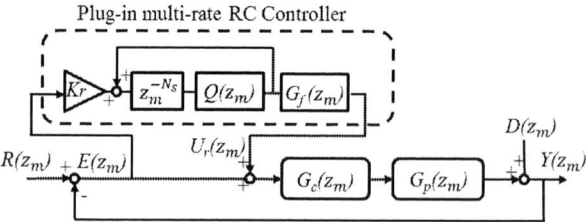

Fig. 3. Equivalent single-rate RC

C. Multi-rate RC Stability Analysis

To analyze the system with two different sampling rates, MRC is presented by an equivalent plug-in RC structure [16] shown in Fig. 3, through transforming the entire system with the same sampling rate f_c/m [17].

The overall transfer function of the equivalent single-rate RC shown in Fig. 3 is:

$$\frac{Y(z_m)}{R(z_m)} = \frac{[1 - Q(z_m)z_m^{-N_s}(1 - K_rG_f(z_m))]G(z_m)}{1 - Q(z_m)z_m^{-N_s}[1 - K_rG_f(z_m)G(z_m)]} \qquad (4)$$

$$\frac{Y(z_m)}{D(z_m)} = \frac{\bar{S}(z_m)[(1 - Q(z_m)z_m^{-N_s})]}{1 - Q(z_m)z_m^{-N_s}[1 - K_rG_f(z_m)G(z_m)]} \qquad (5)$$

where $G(z_m) = G_c(z_m)G_s(z_m)/[1 + G_c(z_m)G_s(z_m)]$, $\bar{S}(z_m)$ is RC rate counterpart of $S(z) = 1/(1 + G_c(z)G_s(z))$.

The error transfer function is derived as follows:

$$
\begin{aligned}
E(z_m) = &\frac{(1 - Q(z_m)z_m^{-N_s})(1 - G(z_m))}{1 - Q(z_m)z_m^{-N_s}(1 - K_rG_f(z_m)G(z_m))} \\
&\times [R(z_m) - D(z_m)]
\end{aligned} \qquad (6)
$$

The overall system holds the stability conditions: the closed-loop feedback system $G(z_m)$ is stable, and

$$| Q(z_m)(1 - K_rG_f(z_m)G(z_m)) | < 1, z_m = e^{j\omega_m};$$
$$\forall 0 < \omega_m < \frac{\pi}{T_s} \qquad (7)$$

If the frequency of reference $r(t)$ and disturbance $d(t)$ approaches $\omega_l = 2\pi l f$ with $l = 0, 1, 2, ..., L_s$ ($L_s = N_s/2$ for even N_s and $L_s = (N_s - 1)/2$ for odd N_s), then $z_m^{-N_s} = 1$. From (6) with the assumption that $Q(z_m) = 1$, we have:

$$\lim_{\omega \to \omega_l} \|E(z_m)\| = 0 \qquad (8)$$

Equation (8) indicates that this scheme can suppress all harmonics of the periodic reference signal up to L_s, even in the presence of unmodeled dynamics.

D. Design of Fractional-order Phase Lead Compensator

The phase lead compensator $G_f(z_m)$ improves the system stability by providing phase lead to cancel out the phase lag of $G(z_m)$ [18]. Ideally, it can be implemented as the inverse of the system model $G(z_m)$ [19]. In practice, it is difficult to obtain the inverse of the closed-loop system accurately due to various plant model uncertainties.

In the proposed design, a fractional-order phase lead compensation is design as:

$$G_f(z_m) = z_m^\gamma = z_m^{N_i + F} \tag{9}$$

where $\gamma \in \mathbb{R}^+$ is the lead step, N_i is the integer part of γ, F is its fractional part.

The fractional-order phase lead F can be approximated by a 2nd-order Lagrange interpolation polynomial FIR filter as:

$$z_m^\gamma \approx (1 - F)z_m^{N_i} + F z_m^{N_i+1} \tag{10}$$

Such a lead step γ will produce a linear phase lead:

$$\theta = \gamma \frac{\omega}{\omega_N^m} 180° \tag{11}$$

where ω_N^m is the equivalent MRC system Nyquist frequency.

With this $G_f(z_m)$ filter, (7) can be rewritten as:

$$\mid 1 - K_r z_m^\gamma G(z_m) \mid < \frac{1}{\|Q(z_m)\|} \tag{12}$$

Substitute $z_m = e^{j\omega_m}$, where ω_m is equivalent system phase velocity, where ω_m, the equivalent closed-loop transfer function can be expressed as $G(e^{j\omega_m}) = N_g(e^{j\omega_m})exp(j\theta_g(e^{j\omega_m}))$, with $N_g(e^{j\omega_m}$ and $\theta_g(e^{j\omega_m})$ being its magnitude and phase characteristics. The robustness filter $Q(z_m)$, with the form as $Q(z_m) = az_m + (1 - 2a) + az_m^{-1}$, can be expressed as $Q(e^{j\omega_m}) = N_q(e^{j\omega_m})$, with $N_q(e^{j\omega_m})$ being its magnitude characteristics. Note that $Q(z_m)$ is a zero phase low pass filter, so its phase characteristics is always 0. Then (12) becomes:

$$\mid 1 - K_r N_g(e^{j\omega_m})e^{j(\theta_g(e^{j\omega_m}) + \gamma\omega_m)} \mid < \frac{1}{N_q(e^{j\omega_m})} \tag{13}$$

Since K_r and $N_g(e^{j\omega_m})$ are both positive, we can get:

$$0 < K_r < \frac{1 - N_q^2(e^{j\omega_m})}{N_q^2(e^{j\omega_m})(K_r N_g^2(e^{j\omega_m}))} + \frac{2cos(\theta_g(e^{j\omega_m}) + \gamma\omega_m)}{N_g(e^{j\omega_m})} \tag{14}$$

The first item on the right-hand side of (14) is a non-negative value because $0 \le N_q(e^{j\omega_m}) \le 1$. Then, (14) will be satisfied if the following condition holds:

$$0 < K_r \le \frac{2cos(\theta_g(e^{j\omega_m}) + \gamma\omega_m)}{N_g(e^{j\omega_m})} \tag{15}$$

From (15), the frequency bandwidth condition can be obtained as:

$$\mid \theta_g(e^{j\omega_m}) + \gamma\omega_m \mid < 90° - \varepsilon \tag{16}$$

where ε is a small positive constant to enhance the system robustness.

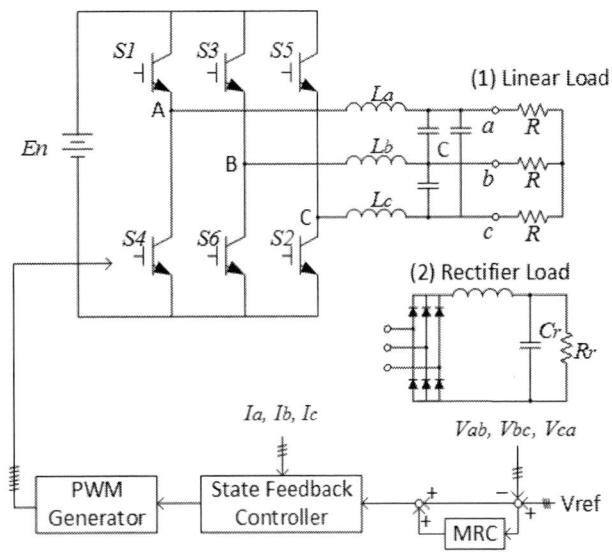

Fig. 4. Three phase MRC-controlled PWM inverter system

III. MRC CONTROLLED DC/AC INVERTER

A. Modelling of Three Phase PWM Inverter

The state space representation of three-phase CVCF PWM inverter shown in Fig. 4 can be described in (17).

$$\begin{pmatrix} 1 & 0 & -1 & 0 & 0 & 0 \\ -1 & 1 & 0 & 0 & 0 & 0 \\ 0 & -1 & 1 & 0 & 0 & 0 \\ 0 & 0 & 0 & 1 & -1 & 0 \\ 0 & 0 & 0 & 0 & 1 & -1 \\ 0 & 0 & 0 & -1 & 0 & 1 \end{pmatrix} \begin{pmatrix} \dot{v}_{ab} \\ \dot{v}_{bc} \\ \dot{v}_{ca} \\ \dot{i}_{La} \\ \dot{i}_{Lb} \\ \dot{i}_{Lc} \end{pmatrix}$$

$$= \begin{pmatrix} \frac{-1}{2RC_f} & 0 & \frac{1}{2RC_f} & \frac{1}{C_f} & 0 & 0 \\ \frac{1}{2RC_f} & \frac{-1}{2RC_f} & 0 & 0 & \frac{1}{C_f} & 0 \\ 0 & \frac{1}{2RC_f} & \frac{-1}{2RC_f} & 0 & 0 & \frac{1}{C_f} \\ \frac{-1}{L_f} & 0 & 0 & 0 & 0 & 0 \\ 0 & \frac{-1}{L_f} & 0 & 0 & 0 & 0 \\ 0 & 0 & \frac{-1}{L_f} & 0 & 0 & 0 \end{pmatrix} \begin{pmatrix} v_{ab} \\ v_{bc} \\ v_{ca} \\ i_{La} \\ i_{Lb} \\ i_{Lc} \end{pmatrix} \tag{17}$$

$$+ \begin{pmatrix} 0 & 0 & 0 \\ 0 & 0 & 0 \\ 0 & 0 & 0 \\ \frac{1}{L_f} & 0 & 0 \\ 0 & \frac{1}{L_f} & 0 \\ 0 & 0 & \frac{1}{L_f} \end{pmatrix} \begin{pmatrix} V_{AB} \\ V_{BC} \\ V_{CA} \end{pmatrix}$$

where the state variables v_{ab}, v_{bc} and v_{ca} are output line-to-line voltage, i_{La}, i_{Lb} and i_{Lc} are each phase inductor current; the state input are V_{AB}, V_{BC} and V_{CA}, which are controlled by phase duty cycle are between E_n and $-E_n$, are PWM-modulated voltage; R, C_f, L_f are load resistor, filter capacitor, and filter inductor of each phase.

When three phase loads are balanced, equation (17) can be transform into $\alpha\beta$ frame [20] as:

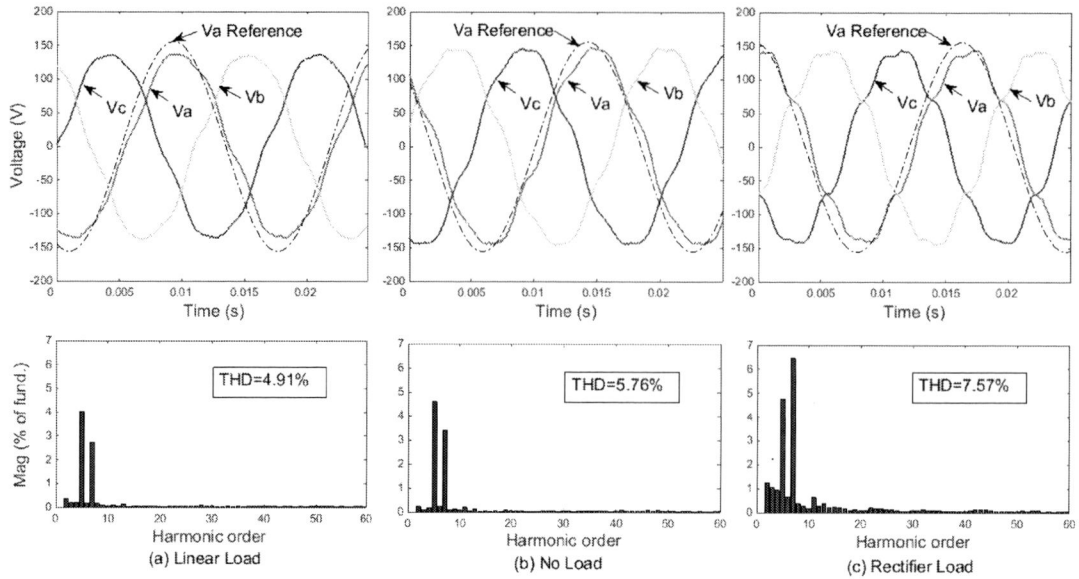

Fig. 5. Feedback control response and harmonic distribution

$$
\begin{pmatrix} \dot{v}_\alpha \\ \dot{i}_\alpha \\ \dot{v}_\beta \\ \dot{i}_\beta \end{pmatrix} = \begin{pmatrix} \frac{-1}{3RC_f} & \frac{1}{3C_f} & 0 & 0 \\ \frac{-1}{L_f} & 0 & 0 & 0 \\ 0 & 0 & \frac{-1}{3RC_f} & \frac{1}{3C_f} \\ 0 & 0 & \frac{-1}{L_f} & 0 \end{pmatrix} \begin{pmatrix} v_\alpha \\ i_\alpha \\ v_\beta \\ i_\beta \end{pmatrix}
$$
$$
+ \begin{pmatrix} 0 & 0 \\ \frac{E_n}{L_f} & 0 \\ 0 & 0 \\ 0 & \frac{E_n}{L_f} \end{pmatrix} \begin{pmatrix} u_\alpha \\ u_\beta \end{pmatrix} \tag{18}
$$

where state varibles v_α, v_β, i_α and i_β are output voltage and inductor current under $\alpha\beta$ frame. Vector $[u_\alpha \quad u_\beta]^T$ is the corresponding control vector. Equation (18) can be obviously treated as two independent sampled-data subsystems as:

$$
\begin{bmatrix} v(k+1) \\ i(k+1) \end{bmatrix} = \begin{pmatrix} \varphi_{11} & \varphi_{12} \\ \varphi_{21} & \varphi_{22} \end{pmatrix} \begin{bmatrix} v(k) \\ i(k) \end{bmatrix} + \begin{pmatrix} g_1 \\ g_2 \end{pmatrix} u(k) \tag{19}
$$

where $v = v_\alpha$ or v_β, $i = i_\alpha$ or i_β, $u = u_\alpha$ or u_β; and the coefficients $\varphi_{11} = 1 - T_s/(3RC_f) + T_s^2/(18R^2C_f^2) - T_s^2/(6L_fC_f)$, $\varphi_{12} = T_s/(3C_f) - T_s^2/(18RC_f^2)$, $\varphi_{21} = -T_s/L_f + T_s^2/(6RL_fC_f)$, $\varphi_{22} = 1 - T_s^2/(6L_fC_f)$, $g_1 = E_nT_s^2/(6L_fC_f)$, $g_2 = E_nT_s/L_f$.

B. State Feedback Controller

The state feedback controller based on system (19) has the form as:

$$
u = -KX(k) + gv_{ref}(k) = -k_1v(k) - k_2i(k) + gv_{ref}(k) \tag{20}
$$

where k_1, k_2 and g are controller parameters, v_{ref} is the reference sinusoidal voltage.

TABLE I
SYSTEM PARAMETERS

Parameter	Value	Parameter	Value
DC bus voltage E_n	300 V	Linear load R	100 Ω
Inductor filter L_f	3 mH	Capacitor filter C_f	20 μF
Rectifier capacitor C_r	60 μF	Rectifier inductor L_r	6 mH
Rectifier resistance R_r	200Ω	Multi-rate ratio m	8
PWM frequency f_c	12 kHz	Sampling frequency f_c	12 kHz
Feedback controller k_1	0.0033	Feedback controller k_2	0.0067
Feedback controller g	0.0092	Reference voltage RMS	110V
Reference voltage f	60 Hz		

With the state feedback controller (20), closed-loop system becomes:

$$
\begin{bmatrix} v(k+1) \\ i(k+1) \end{bmatrix} = \begin{pmatrix} \varphi_{11} - g_1k_1 & \varphi_{12} - g_1k_2 \\ \varphi_{21} - g_2k_1 & \varphi_{22} - g_2k_2 \end{pmatrix} \begin{bmatrix} v(k) \\ i(k) \end{bmatrix}
$$
$$
+ \begin{pmatrix} g_1g \\ g_2g \end{pmatrix} u(k) \tag{21}
$$

The pole of feedback control system can be assigned by adjusting coefficients k_1 and k_2. The transfer function can be rewritten as:

$$
H(z) = \frac{m_1z + m_2}{z^2 + p_1z + p_2} \tag{22}
$$

where $p_1 = -(\varphi_{22} - g_2k_2) - (\varphi_{11} - g_1k_1)$, $p_2 = (\varphi_{11} - g_1k_1)(\varphi_{22} - g_2k_2) - (\varphi_{12} - g_1k_2)(\varphi_{21} - g_2k_1)$, $m_1 = g_1k$, and $m_2 = g_2k - g_1k(\varphi_{22} - g_2k_2)$.

With the parameters in Table I, the closed loop system model can be derived as:

$$
G(z) = \frac{0.0501z + 0.0476}{z^2 - 1.785z + 0.858} \tag{23}
$$

978-1-4673-9551-9/16 $31.00 © 2016 IEEE   1158

The feedback control system gets stable with the poles at $0.8927 \pm 0.2477i$ in the unit cycle. With this feedback controller, Fig. 5 shows the tracking performance and THD for three different types of loads. It shows that the major error component appears on the $5th$ and $7th$ harmonic frequencies.

C. MRC with Fractional-order Phase Lead Compensator

To improve the tracking performance, MRC controller with multi-rate ratio of 8 is employed, which reduces RC computation load to 1/8 of conventional RC.

The phase lead compensation is obtained by a 2nd-order FIR filter that is able to realize any real number phase lead. For phase lead step $\gamma = 0.5$, the FIR filter is given by $G_f(z_m) = 0.5 + 0.5z_m$.

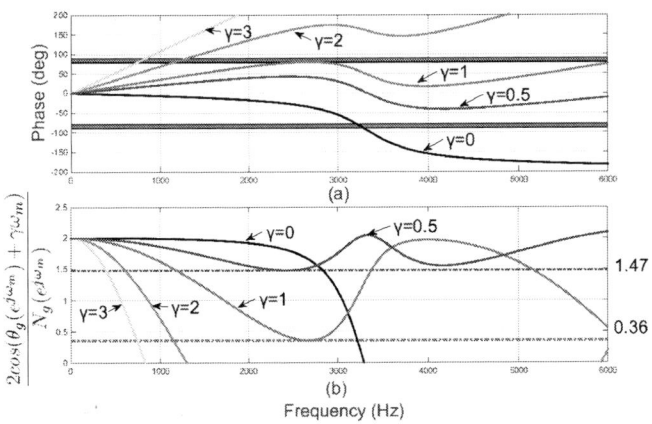

Fig. 6. phase lead compensations comparison: (a) frequency bandwidth (b) K_r upper limit

Fig. 6 (a) and (b) show the phase lead compensation bandwidth and k_r range in the entire Nyquist frequency. Based on conditions (16), the phase after compensation should stay in $\pm 75°$ with $\varepsilon = 15°$. While condition (15) indicates that the upper limit of the gain K_r must be positive but smaller than the minimum value of the curve $2cos(\theta_g(e^{j\omega_m}) + \gamma\omega_m)/N_g(e^{j\omega_m})$. Fig. 6 (a) shows that for both no phase lead ($\gamma = 0$) and excessive phase lead ($\gamma = 2, 3$), the phase lead compensation results in limited controllable bandwidth of about 3250Hz, 1300Hz and 800Hz, respectively. One the other hand, the phase lead compensation with $\gamma = 0.5$ and $\gamma = 1$ stay in $\pm 75°$ for the entire Nyquist frequency, which means it can eliminate periodic signals and disturbances to achieve zero error tracking. From Fig. 6 (b), the gain K_r of the fractional-order compensator $\gamma = 0.5$ has an upper limit of 1.47 for the entire Nyquist frequency, compared with 0.36 for $\gamma = 1$. For $\gamma = 0, 2$ and 3, it is not possible to hold K_r positive for high frequency error components. This result shows that fractional-order phase compensation leads to a wider stable frequency bandwidth with a much larger gain K_r.

IV. EXPERIMENT

In the experiments, the multi-rate RC controller and state feedback controller are generated in MATLAB and implemented by dSPACE controller board. The output voltage and

inductance current waveform are recorded for system control and performance analysis.

A. Experiment Setup

Fig. 7 shows the experimental setup, which consists of a dSPACE-controlled three-phase PWM IGBT full-bridge device, LC filter, DC power source, load resistances, and sampling circuit. The feedback and MRC controllers are designed in Matlab Simulink and implemented by a dSPACE DS1103 control board to control three-phase full-bridge IGBT converter.

Fig. 7. Experiment setups

The low-pass filter in RC is $Q(z) = 0.25z + 0.5 + 0.25z^{-1}$, anti-aliasing and anti-imaging filters $F_{a,a}(z)$ and $F_{a,i}(z)$ [21] are chosen as $0.1z + 0.8 + 0.1z^{-1}$.

B. Experiment Results

The MRC experiments with integral and fractional order phase lead compensation are conducted under three different loads: linear load, no load, and rectifier load. Under each load condition, experiments are carried out with MRC gain K_r

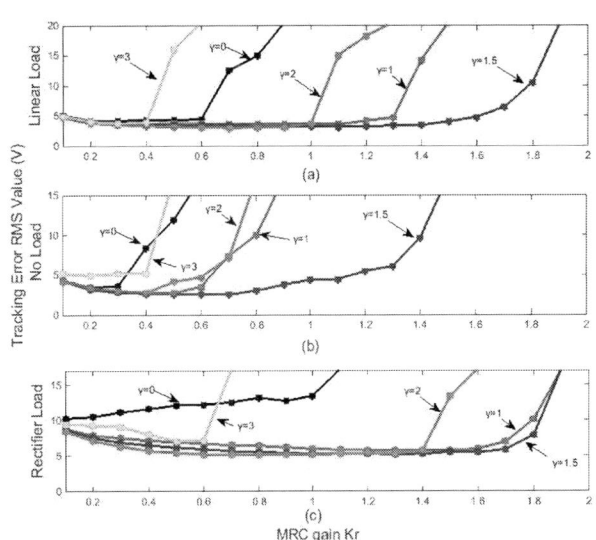

Fig. 8. tracking error RMS value of integral and fractional order phase lead under linear load, no load, and rectifier load

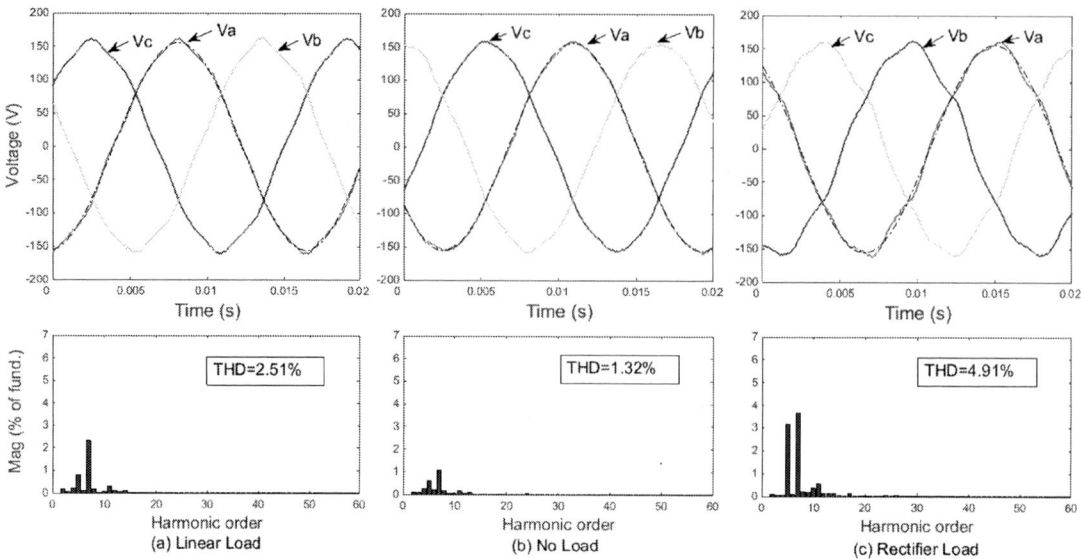

Fig. 9. Voltage steady-state response and V_a harmonic distribution under fractional-order phase lead compensation $\gamma = 1.5$ with MRC gain K_r at 0.7

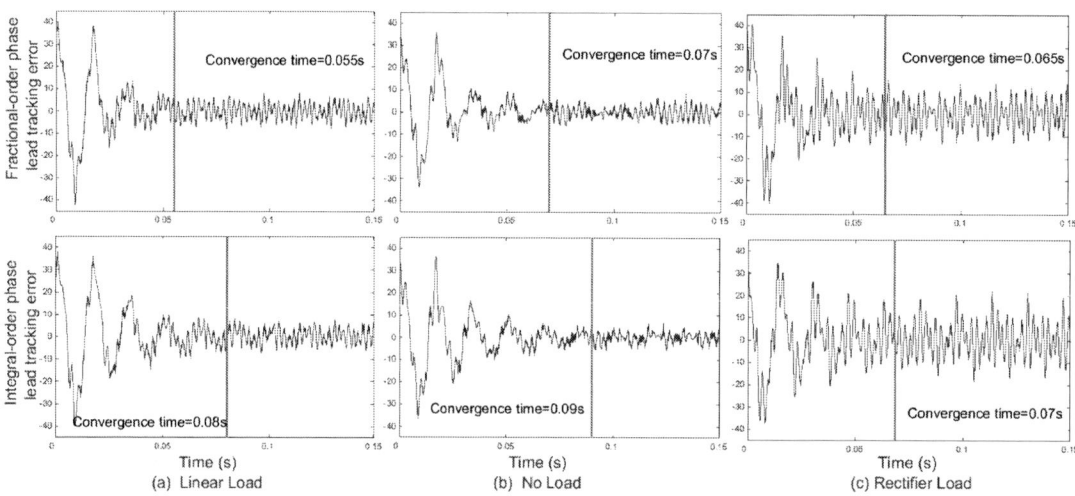

Fig. 10. tracking error RMS value under linear load: (a) fractional-order phase lead $\gamma = 1.5$ with $K_r = 0.7$, (b) fractional-order phase lead $\gamma = 1$ with $K_r = 0.5$

increased from 0.1 by 0.1 until it diverges for the integral-order phase lead compensation of $\gamma = 0, 1, 2, 3$ and fractional-order phase lead compensation $\gamma = 1.5$. The MRC gain K_r is positively related with system transient response speed when the system is stable. When the MRC gain K_r is too large, the system response will diverge based on the analysis on (15) and (16). Thus, our goal is to find MRC gain K_r stable range for all different situations.

Fig. 8 shows tracking error root-mean-square (RMS) value under linear load, no load and rectifier load:

1) Linear load: MRC controller with phase lead compensation is applied when system reaches a steady state under state feedback control. The steady-state responses, in terms of tracking error RMS value, are shown in Fig. 8. From

Fig. 8 (a), tracking error RMS value of phase lead $\gamma = 1.5$ decreases when MRC gain K_r grows from 0.1 to 0.6 and keeps a low level within 3.5V until K_r increases to 1.5, which means MRC can keep stable with the K_r range from 0 to 1.4. When K_r is larger than 1.4, system cannot restrain some frequency disturbance at steady state and the disturbance will increase when K_r grows. For the same situations, MRC with the integral-order phase lead $\gamma = 1$ can keep a stable K_r range within 1.0, and $\gamma = 2, 3$ at 1.0 and 0.4, respectively. MRC without any phase lead $\gamma = 0$ has the K_r stable range from 0 to 0.5.

2) no load: Under no load situation, system output terminals keep in open circuit. From Fig. 8 (b), MRC with phase lead $\gamma = 1.5$ also has the largest stable range from 0 to 0.7

TABLE II
STEADY-STATE PERFORMANCE COMPARISON

	Fractional-order phase lead ($\gamma = 1.5$, $K_r = 0.7$)			Integral-order phase lead ($\gamma = 1$, $K_r = 0.5$)		
	THD	RMS Error	Peak Error	THD	RMS Error	Peak Error
Linear Load	2.51%	3.41 V	9.72 V	3.03%	3.75 V	11.01 V
No load	1.32%	2.54 V	9.05 V	1.37%	2.75 V	11.02 V
Rectifier Load	4.91%	5.67 V	18.48 V	5.97%	6.89 V	20.92 V

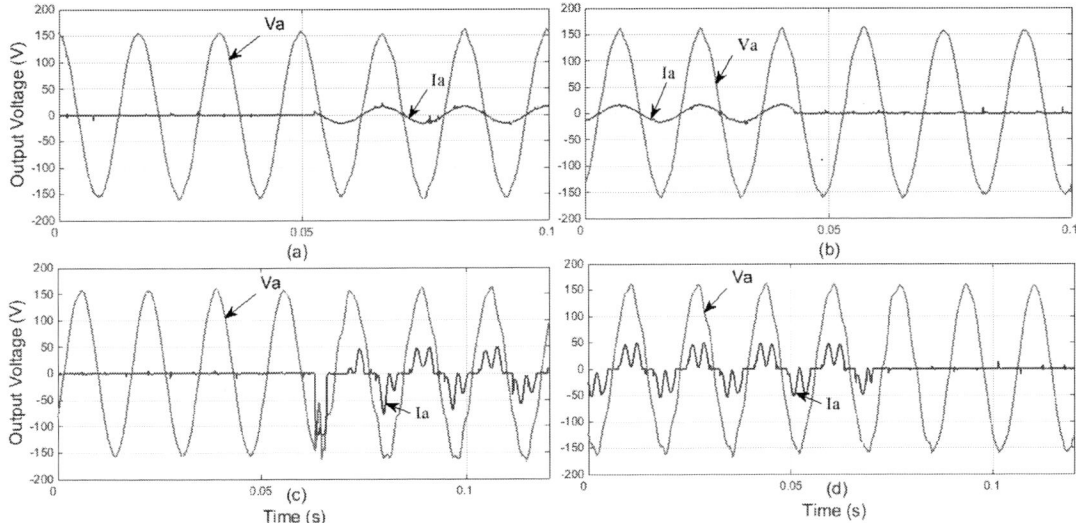

Fig. 11. MRC with fractional-order phase lead compensation under sudden load change:(a) load change from no load to linear load (b) load change from linear load to no load (c) load change from no load to rectifier load (d) load change from rectifier load to no load

with tracking error RMS 2.54V at 0.7. The other MRCs with integral-order phase lead $\gamma = 1, 2, 3$ and without phase lead $\gamma = 0$ get stable range respectively at 0.5, 0.4, 0.4, and 0.3.

3) Rectifier load: Rectifier load consists of a three-phase rectifier bridge, LC filter and a 200Ω resistance. Fig. 8 (c) shows tracking error RMS value variation. MRC with phase lead $\gamma = 1.5$ and 1 has the largest stable range at 1.6. MRC with $\gamma = 2$ and 3 are at 1.4 and 0.6. MRC without phase lead can only keep stable at $K_r = 0.1$.

4) Summary of experiments: Based on the results in Fig. 8 and analysis, the MRC with fractional-order phase lead has better stable range for gain K_r in three different loads. For keeping system stable under all three kinds of loads, MRC with fractional-order phase lead $\gamma = 1.5$ can have the largest gain at 0.7. On the contrary, integral-order phase lead has the largest gain at 0.5 with $\gamma = 1$.

C. Results Comparison

From the experiment results in section B, the largest stable MRC gain K_r is obtained as 0.7 for fractional-order phase lead compensation $\gamma = 1.5$ and 0.5 for integral-order $\gamma = 1$. Fig. 10 shows MRC transient response comparison under three different loads between fractional and integral order phase lead. The fractional-order phase lead compensation has convergence time 0.055s, 0.07s and 0.065s for linear load, no load and rectifier load. On the contrary, the convergence

time of integral-order phase lead is 0.08s, 0.09s, and 0.07s. Because fractional-order phase lead compensation increases MRC gain K_r stable range, transient state performance gets correspondingly improvement.

For steady-state performance, Table II shows stable-state performance comparison between fractional and integral order phase lead compensation. From the table, fractional-order phase lead $\gamma = 1.5$ with $K_r = 0.7$ has better performance on THD, RMS error, and peak error for all three kinds of loads than integral-order phase lead compensation, which means response under fractional-order phase lead compensation has better wave integrity and voltage accuracy. The steady-state performance of fractional-order phase lead $\gamma = 1.5$ with $K_r = 0.7$ is shown in Fig. 9.

D. MRC with fractional-order phase lead compensation under sudden load change

Fig. 11 shows that the MRC with fractional-order phase lead compensation operates under sudden step load switch. It is clear from the diagrams that output voltages do not vary too much and recover from the sudden step load changes within two cycles. The experiments proves that MRC with fractional-order phase lead compensation has enough robustness to sudden load changes. Moreover, from Fig. 11, we notice that the plug-in MRC controller and state feedback controller are complementary: state feedback control offers instantaneous dynamic response to the sudden load change, but its tracking accuracy is relatively low; the plug-in MRC controller can significantly reduce the tracking errors with a fast convergence speed.

V. CONCLUSION

Fractional-order phase lead compensation can offset system lag influence accurately when applying multi-rate repetitive control. Comparing with traditional integral-order compensation, fractional-order compensation provides more accurate phase compensation. Experiments under different load situations show that, with more accurate phase compensation, multi-rate repetitive control not only increases system response speed, but also achieves better tracking performance and a low THD.

REFERENCES

[1] Z. Zou, K. Zhou, Z. Wang, and M. Cheng, "Fractional-order repetitive control of programmable ac power sources," *IET Power Electronics*, vol. 7, pp. 431–438(7), February 2014.

[2] Y.-Y. Tzou, R.-S. Ou, S.-L. Jung, and M.-Y. Chang, "High-performance programmable ac power source with low harmonic distortion using dsp-based repetitive control technique," *Power Electronics, IEEE Transactions on*, vol. 12, no. 4, pp. 715–725, Jul 1997.

[3] K. Chang, I. Shim, and G. Park, "Adaptive repetitive control for an eccentricity compensation of optical disk drivers," *Consumer Electronics, IEEE Transactions on*, vol. 52, no. 2, pp. 445–450, May 2006.

[4] A. Garcia-Cerrada, O. Pinzon-Ardila, V. Feliu-Batlle, P. Roncero-Sanchez, and P. Garcia-Gonzalez, "Application of a repetitive controller for a three-phase active power filter," *Power Electronics, IEEE Transactions on*, vol. 22, no. 1, pp. 237–246, Jan 2007.

[5] G. Ramos and R. Costa-Castello, "Power factor correction and harmonic compensation using second-order odd-harmonic repetitive control," *Control Theory Applications, IET*, vol. 6, no. 11, pp. 1633–1644, July 2012.

[6] K. Kaneko and R. Horowitz, "Repetitive and adaptive control of robot manipulators with velocity estimation," *Robotics and Automation, IEEE Transactions on*, vol. 13, no. 2, pp. 204–217, Apr 1997.

[7] V. Khadkikar, A. Chandra, A. Barry, and T. Nguyen, "Power quality enhancement utilising single-phase unified power quality conditioner: digital signal processor-based experimental validation," vol. 4, no. 3, pp. 323–331, March 2011.

[8] Haneyoshi, T., A. Kawamura, and R. Hoft, "Waveform compensation of pwm inverter with cyclic fluctuating loads," *Industry Applications, IEEE Transactions on*, vol. 24, no. 4, pp. 582–589, Jul 1988.

[9] R. W. Longman, "Iterative learning control and repetitive control for engineering practice," *Int. J. Control*, vol. 73, no. 10, pp. 930–954, 2000.

[10] B. Zhang, K. Zhou, and D. Wang, "Multirate repetitive control for pwm dc/ac converters," *Industrial Electronics, IEEE Transactions on*, vol. 61, no. 6, pp. 2883–2890, June 2014.

[11] Y. Ye, K. Zhou, B. Zhang, D. Wang, and J. Wang, "Performance improvement of repetitive controlled pwm inverters: A phase-lead compensation solution," *International Journal of Circuit theory and applications*, vol. 38, pp. 453–469, Nov 2010.

[12] B. Zhang, D. Wang, K. Zhou, and Y. Wang, "Linear phase lead compensation repetitive control of a cvcf pwm inverter," *Industrial Electronics, IEEE Transactions on*, vol. 55, no. 4, pp. 1595–1602, April 2008.

[13] Y. Ye, K. Zhou, B. Zhang, D. Wang, and J. Wang, "High-performance repetitive control of pwm dc-ac converters with real-time phase-lead fir filter," *Circuits and Systems II: Express Briefs, IEEE Transactions on*, vol. 53, no. 8, pp. 768–772, Aug 2006.

[14] Y. Yang, K. Zhou, M. Cheng, and B. Zhang, "Phase compensation multiresonant control of cvcf pwm converters," *Power Electronics, IEEE Transactions on*, vol. 28, no. 8, pp. 3923–3930, Aug 2013.

[15] Y. Yang, K. Zhou, H. Wang, F. Blaabjerg, D. Wang, and B. Zhang, "Frequency adaptive selection harmonic control for grid-connected inverters," *IEEE Transactions on Power Electronics*, vol. 30, no. 7, pp. 3912–3924, 2015.

[16] K. Zhou and D. Wang, "Digital repetitive controlled three-phase pwm rectifier," *Power Electronics, IEEE Transactions on*, vol. 18, no. 1, pp. 309–316, Jan 2003.

[17] C. James, N. Sadegh, and A.-P. Hu, "Synthesis, stability analysis, and experimental implementation of a multirate repetitive learning controller," in *Advanced Intelligent Mechatronics, 1999. Proceedings. 1999 IEEE/ASME International Conference on*, 1999, pp. 653–658.

[18] T. Inoue, "Practical repetitive control system design," in *Decision and Control, 1990., Proceedings of the 29th IEEE Conference on*, vol. 3, Dec 1990, pp. 1673–1678.

[19] D. Chen, J. Zhang, and Z. Qian, "An improved repetitive control scheme for grid-connected inverter with frequency-adaptive capability," *Industrial Electronics, IEEE Transactions on*, vol. 60, no. 2, pp. 814–823, Feb 2013.

[20] K. Zhou and D. Wang, "Digital repetitive learning controller for three-phase cvcf pwm inverter," *Industrial Electronics, IEEE Transactions on*, vol. 48, no. 4, pp. 820–830, Aug 2001.

[21] B. Zhang, D. Wang, Y. Wang, Y. Ye, and K. Zhou, "Practical repetitive control system design," in *2008 IFAC World Congress*, July 2008, pp. 12 468–12 473.

A Robust Modified Model Predictive Control (MMPC) Based on Lyapunov Function for Three-Phase Active-Front-End (AFE) Rectifier

M. Parvez* and S. Mekhilef
Power Electronics and Renewable
Energy Research Laboratory (PEARL)
Dept. of Electrical Engineering,
University of Malaya
50603 Kuala Lumpur, Malaysia
E-mail: parvez0602025@gmail.com*
saad@um.edu.my

Nadia M. L. Tan
Dept. of Electrical Power Engineering
Universiti Tenaga Nasional
Kajang 43000, Malaysia
E-mail: nadia@uniten.edu.my

Hirofumi Akagi
Dept. of Electrical and Electronic
Engineering
Tokyo Institute of Technology
Tokyo 152-8552, Japan
E-mail: akagi@ee.titech.ac.jp

Abstract-This paper proposes a robust modified model predictive control (MMPC) algorithm based on Lyapunov function for a three-phase active-front-end (AFE) rectifier in order to improve the system performance as well as ensure the stability, robustness, and fast dynamic response. The proposed control technique utilizes the discrete behavior of the converter considering the unavoidable quantization errors between the controller and the control actions selected from the finite control set of the AFE rectifier. The proposed MMPC method increases the dynamic response by avoiding the calcultation of future current for every switching states. Moreover, the nonlinear system stability of the proposed MMPC technique is ensured by a 2.5 kW experimental system and the direct Lyapunov method. Detailed robustness analysis with experimental parameter variations are provided to show the efficacy of the proposed Lyapunov function based MMPC technique.

Keywords-Modified model predictive control (MMPC); active front end (AFE) rectifier; stability analysis; robustness analysis; ac-dc power conversion.

I. INTRODUCTION

Rectifiers are the most widely used converters in the field of power electronics for its transformation capability of alternating current to direct current [1]. Among several types of rectifier topologies, the active-front-end (AFE) rectifier has become an important part of various industry applications as well as household appliances due to its bidirectional-power-flow and grid synchronization capabilities [2]. The AFE rectifier needs to be highly efficient, robust and stable for preventing high THD in input current, low power factor, ac voltage distortion, ripple in DC current and DC voltage pulsations [3]. In the recent works, several control techniques such as voltage-oriented control (VOC) scheme [4], virtual-flux-oriented PWM control [5], direct power control (DPC) [6], direct torque control (DTC) [7], fuzzy-logic control [8] and sliding mode nonlinear control [9] have been proposed to improve the efficiency and performance of the AFE rectifier.

Nowadays, a fully digital control strategy based on model predictive control (MPC) has become an attractive mode of control technique for AFE rectifier comparing with all the classical solutions discussed above due to its simple and intuitive concept with no PWM blocks [10-12]. Moreover, MPC algorithm is very easy to configure with constraints and non-linearity and also for practical implementation [13, 14]. Due to these advantages, the MPC algorithm has been extensively applied in active front-end rectifier [15, 16], indirect matrix converter [17], three-level converter [18] and voltage source inverter [19] etc. In spite of these boundless advantages, MPC faces computational burden due to solving the underlying optimization problem of the discrete manipulated variables [20, 21]. Hence, computational issues have become very important for long prediction horizons. As a result, the modification in the existing MPC techniques is required for minimizing the amount of calculations as well as the computational time. Furthermore, a complete research considering practical nonlinearity issues with explicit addressing of stability and robustness issues of MPC is missing in the present state-of-art.

This paper proposes a modified model predictive control (MMPC) method based on Lyapunov function for AFE rectifier to reduce the computational time, ensure stability & robustness, and improve system performances. The rest of the paper is organized in the following manner. The proposed Lyapunov function based MMPC control strategy for the active-front-end (AFE) rectifier is elaborately described in section II. The stability and robustness of the proposed MMPC algorithm are analyzed with a 2.5 kW experimental system and described in section III. Section IV presents the comparative evaluation of the proposed Lyapunov function based MMPC with conventional MPC. Finally the conclusions are drawn in section V.

II. PROPOSED MMPC TECHNIQUE BASED ON LYAPUNOV FUNCTION FOR AFE RECTIFIER

Fig. 1 shows the three-phase active front end (AFE) rectifier topology and its proposed MMPC control technique, which consists of six IGBT-Diode (S_1-S_6) switches. The AFE rectifier is connected with the three-phase voltage supply (v_s) using the line filter inductances (L_s) and resistances (R_s). A dc capacitor (C_{dc}) is connected across the resistive load to reduce dc voltage ripple.

978-1-4673-9551-9/16 $31.00 © 2016 IEEE

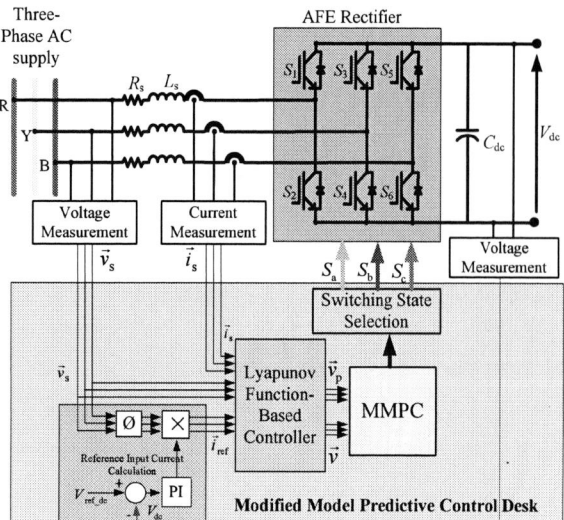

Fig. 1: AFE rectifier topology and its MMPC block diagram.

A. AFE Rectifier Model

In order to avoid short circuit, the two switches in each leg of the AFE rectifier should be operated in a complementary mode. Hence, the gating signals S_a, S_b, and S_c determine the switching states of the three-phase AFE rectifier are,

$$S_a = \begin{cases} 1, & S_1 \text{ is on and } S_2 \text{ is off} \\ 0, & S_1 \text{ is off and } S_2 \text{ is on} \end{cases} \quad (1)$$

$$S_b = \begin{cases} 1, & S_3 \text{ is on and } S_4 \text{ is off} \\ 0, & S_3 \text{ is off and } S_4 \text{ is on} \end{cases} \quad (2)$$

$$S_c = \begin{cases} 1, & S_5 \text{ is on and } S_6 \text{ is off} \\ 0, & S_5 \text{ is off and } S_6 \text{ is on} \end{cases} \quad (3)$$

Therefore, the switching function vector ($\vec{S}$) of the AFE rectifier can be expressed as:

$$\vec{S} = \frac{2}{3}(S_a + \vec{\omega} S_b + \vec{\omega}^2 S_c) \quad (4)$$

where, $\vec{\omega} = e^{j2\pi/3}$, is the unity vector.

The output voltage space vector ($\vec{v}_{AFE}$) of the AFE rectifier can be presented with phase to neutral voltages (v_{ao}, v_{bo}, and v_{co}) as:

$$\vec{v}_{AFE} = \frac{2}{3}(v_{ao} + \vec{\omega} v_{bo} + \vec{\omega}^2 v_{co}). \quad (5)$$

This output voltage space vector ($\vec{v}_{AFE}$) can also be related to the dc bus voltage (V_{dc}) and the switching vector ($\vec{S}$) as:

$$\vec{v}_{AFE} = \vec{S} \times V_{dc}. \quad (6)$$

There are eight possible voltage vectors that can be obtained from the eight consequence switching states of the

TABLE I. VOLTAGE SPACE VECTORS OF THE AFE RECTIFIER

Switching States			Voltage space vector
S_a	S_b	S_c	$\vec{v}_{AFE}$
0	0	0	$\vec{v}_1 = 0$
0	0	1	$\vec{v}_2 = -(1/3)V_{dc} - j(\sqrt{3}/3)V_{dc}$
0	1	0	$\vec{v}_3 = -(1/3)V_{dc} + j(\sqrt{3}/3)V_{dc}$
0	1	1	$\vec{v}_4 = -(2/3)V_{dc}$
1	0	0	$\vec{v}_5 = (2/3)V_{dc}$
1	0	1	$\vec{v}_6 = (1/3)V_{dc} - j(\sqrt{3}/3)V_{dc}$
1	1	0	$\vec{v}_7 = (1/3)V_{dc} + j(\sqrt{3}/3)V_{dc}$
1	1	1	$\vec{v}_8 = 0$

switching signals, which are listed in Table I. The AFE rectifier is connected to the three-phase ac voltage source through the input filter as shown in Fig. 1. By applying Kirchhoff's voltage law at the ac side of the rectifier, the relationship between the three-phase ac supply voltage and rectifier input voltage vectors are,

$$\vec{v}_s = L_s \frac{d\vec{i}_s}{dt} + R_s \vec{i}_s + \frac{2}{3}\left(v_{ao} + \vec{\omega} v_{bo} + \vec{\omega}^2 v_{co}\right)$$
$$- \frac{2}{3}\left(v_{no} + \vec{\omega} v_{no} + \vec{\omega}^2 v_{no}\right). \quad (7)$$

The space-vector model of three-phase ac voltage ($\vec{v}_s$) and current ($\vec{i}_s$) can be derived from phase voltage and current as:

$$\vec{v}_s = \frac{2}{3}(v_{sa} + \vec{\omega} v_{sb} + \vec{\omega}^2 v_{sc}) \quad (8)$$

and

$$\vec{i}_s = \frac{2}{3}(i_{sa} + \vec{\omega} i_{sb} + \vec{\omega}^2 i_{sc}), \quad (9)$$

where v_{sa}, v_{sb}, and v_{sc} are phase voltages; i_{sa}, i_{sb} and i_{sc} are phase currents of three-phase ac voltage source.

Therefore, the input current dynamics of the AFE rectifier can be evaluated from (5) and (7) as:

$$\frac{d\vec{i}_s}{dt} = -\frac{R_s}{L_s}\vec{i}_s + \frac{1}{L_s}\vec{v}_s - \frac{1}{L_s}\vec{v}_{AFE}. \quad (10)$$

B. Discrete Time Model for Prediction Horizon

Predictive control technique utilizes the discrete nature of the switching devices and finite number of valid switching states of the power converter. It is important to derive a discrete time model for the power converter system because the predictive controller is formulated in discrete time domain. For $(k-1)T_s \le t \le (k)T_s$, with T_s being the sampling time, the system model derivative dx/dt can be expressed from the Euler approximation to increase fast dynamic response as,

978-1-4673-9551-9/16 $31.00 © 2016 IEEE

$$di_{\mathrm{s}}/dt \approx \left(i_{\mathrm{s}}(k) - i_{\mathrm{s}}(k-1)\right)/T_{\mathrm{s}}. \tag{11}$$

Using one step advance of the above approximation, the discrete time model of predictive input currents for the next $(k+1)$ sampling instant of the AFE rectifier can be derived as follows,

$$\vec{i}_{\mathrm{s}}(k+1) = \frac{1}{L_{\mathrm{s}} + R_{\mathrm{s}}T_{\mathrm{s}}} \left\{ \begin{array}{l} L_{\mathrm{s}}\vec{i}_{\mathrm{s}}(k) + \\ T_{\mathrm{s}}\left[\vec{v}_{\mathrm{s}}(k+1) - \vec{v}_{\mathrm{AFE}}(k+1)\right] \end{array} \right\} \tag{12}$$

C. Determination of the proposed Lyapunov-Function-Based MMPC Method

The proposed MMPC directly applies the voltage vector constrained in the finite set. Hence, the future voltage vector ($\vec{v}(k+1)$) is expressed with the voltage vector generated by the AFE rectifier ($\vec{v}_{\mathrm{AFE}}(k+1)$) and the unavoidable quantization error vectors ($\delta(k+1)$) as:

$$\vec{v}(k+1) = \vec{v}_{\mathrm{AFE}}(k+1) + \delta(k+1) \tag{13}$$

To determine the Lyapunov-function-based control algorithm for the modification of conventional MPC method, it is important to analyse the AFE rectifier with control point of view. Therefore, the future-current error vector of the input current dynamics (12) can be rewritten as:

$$\begin{aligned} \vec{i}_{\mathrm{error}}(k+1) &= \vec{i}_{\mathrm{s}}(k+1) - \vec{i}_{\mathrm{ref}}(k+1) \\ &= \frac{1}{L_{\mathrm{s}} + R_{\mathrm{s}}T_{\mathrm{s}}} \left\{ \begin{array}{l} L_{\mathrm{s}}\vec{i}_{\mathrm{s}}(k) + \\ T_{\mathrm{s}}\left[\vec{v}_{\mathrm{s}}(k+1) - \vec{v}_{\mathrm{AFE}}(k+1)\right] \end{array} \right\}. \\ &- \vec{i}_{\mathrm{ref}}(k+1) \end{aligned} \tag{14}$$

It is necessary to find a control function such that the current tracking error asymptotically converges to zero. Lyapunov function is used for the specific application and is taken as:

$$L(k) = \left(\frac{1}{2}\right)[\vec{i}_{\mathrm{error}}(k)]^{\mathrm{T}}[\vec{i}_{\mathrm{error}}(k)]. \tag{15}$$

From (14) and (15) the rate of change of the Lyapunov function can be expressed as:

$$\begin{aligned} \Delta L(k) &= L\left(\vec{i}_{\mathrm{error}}(k+1)\right) - L\left(\vec{i}_{\mathrm{error}}(k)\right) \\ &= \frac{1}{2}\left[\left(\frac{1}{R_{\mathrm{s}}T_{\mathrm{s}} + L_{\mathrm{s}}}\right) \times \left(L_{\mathrm{s}}\vec{i}_{\mathrm{s}}(k) + T_{\mathrm{s}}\begin{pmatrix} \vec{v}_{\mathrm{s}}(k+1) \\ -\vec{v}_{\mathrm{AFE}}(k+1) \\ -\delta(k+1) \end{pmatrix}\right) \\ & \qquad -\vec{i}_{\mathrm{ref}}(k+1) \end{array} \right]^{\mathrm{T}} \\ & \times \left[\left(\frac{1}{R_{\mathrm{s}}T_{\mathrm{s}} + L_{\mathrm{s}}}\right) \times \left(L_{\mathrm{s}}\vec{i}_{\mathrm{s}}(k) + T_{\mathrm{s}}\begin{pmatrix} \vec{v}_{\mathrm{s}}(k+1) \\ -\vec{v}_{\mathrm{AFE}}(k+1) \\ -\delta(k+1) \end{pmatrix}\right) \\ & \qquad -\vec{i}_{\mathrm{ref}}(k+1) \end{array} \right] \\ & \quad -\frac{1}{2}[\vec{i}_{\mathrm{error}}(k)]^{\mathrm{T}}[\vec{i}_{\mathrm{error}}(k)] \end{aligned} \tag{16}$$

In order to make an effective control algorithm for converging the tracking error to zero; the rate of change of the Lyapunov function (ΔL) always needs to be negative. Therefore, the discrete voltage vector at next sampling instant ($\vec{v}(k+1)$), which assures the rate of change of the Lyapunov function (16) is negative, which is written as:

$$\vec{v}(k+1) = \frac{L_{\mathrm{s}}}{T_{\mathrm{s}}}\vec{i}_{\mathrm{s}}(k) + \vec{v}_{\mathrm{s}}(k+1) - \frac{R_{\mathrm{s}}T_{\mathrm{s}} + L_{\mathrm{s}}}{T_{\mathrm{s}}}\vec{i}_{\mathrm{ref}}(k+1) \tag{17}$$

D. Cost Function

The conventional MPC is generally based on current oriented control techniques, in which the current tracks the reference value by utilizing the discrete behavior of the converter [22]. In this conventional MPC method, the future current of the converter is calculated for each of eight possible switching states and the state that minimizes the cost function is selected for firing the power switches. One of the major limitations of this conventional MPC algorithm is high execution time delay caused by the large number of calculation for the current prediction. In order to reduce the execution time, the conventional MPC algorithm is modified based on the Lyapunov function. The proposed MMPC is based on voltage oriented control, in which the optimum possible future voltage vector ($\vec{v}_{\mathrm{p}}(k+1)$) of the converter is directly selected to track the calculated reference voltage vector of (17) by utilizing the modified cost function as:

$$g(k+1) = \left\| \vec{v}(k+1) - \vec{v}_{\mathrm{p}}(k+1) \right\|. \tag{18}$$

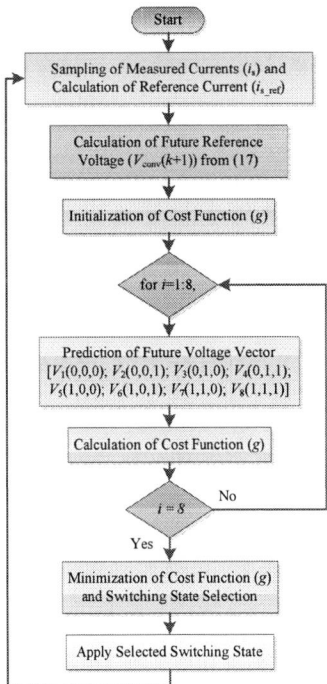

Fig. 2: Control algorithm of proposed Lyapunov function based MMPC technique.

E. Proposed Control Strategy

Fig. 1 shows the proposed control strategy of the MMPC algorithm which is a one-step modification of the conventional FCS-MPC control scheme [12, 23]. The reference voltage vector ($\vec{v}(k+1)$) of the rectifier is calculated from the measured current ac current ($\vec{i}_s(k)$) and the reference current ($\vec{i}_{ref}(k)$) by utilizing (17). The overall control algorithm of the proposed Lyapunov function based MMPC is presented in Fig. 2. The proposed Lyapunov function based MMPC method satisfies the following steps:

Step 1: Measurement of the AC currents ($\vec{i}_s(k)$) and calculation of the reference current ($\vec{i}_{ref}(k)$).

Step 2: Calculation of the future reference voltage vector ($\vec{v}_{ref}(k+1)$) from the measured current AC current ($\vec{i}_s(k)$) and the reference current ($\vec{i}_{ref}$) by utilizing (17).

Step 3: Estimation of the initial cost function value as ($g_{init}(k+1)=1e^{10}$) from the calculation in the kth sampling period.

Step 4: Prediction of the optimum possible future voltage vector ($\vec{v}_p(k+1)$) from Table I.

Step 5: Evaluation of the cost function ($g(k+1)$) by using (18).

Step 6: Optimization of the cost function and selection of the appropriate switching state that minimizes the cost function.

Step 7: Application of the selected switching state of the AFE rectifier for firing the semiconductor switches.

III. Stability and Robustness Analysis with Experimental System

The performance of the MPC controlled AFE rectifier is investigated with 2.5 kW experimental configurations. The experimental verification is carried out by using the rapid prototyping and real-time interface system dSPACE with DS1104 control card which consist of Texas Instruments TMS320F240 sub-processor and the Power PC 603e/250 MHz main processor. The experimental system of the proposed MMPC controlled AFE rectifier is presented in Fig. 3. The parameters listed in Table II are employed for experimentation.

A. Stability Analysis with Nonlinear Experimental Model

The stability analysis of proposed MMPC controlled AFE rectifier is performed with nonlinear experimental system by considering the nonlinear criteria such as: the unsymmetrical three-phase supply, existence of higher harmonics in converter input voltage, cross-coupling effect and time delay variation.

The voltage is measured with differential probe [PINTEK DP-25] and the current with current transducer [LEM LA 25-NP]. The current transducer output voltage gain is set at 1.0 V output to measure 1.0 A current. The dc bus reference voltage (V_{dc_ref}) is varied from 270 V to 320 V to check the nonlinear

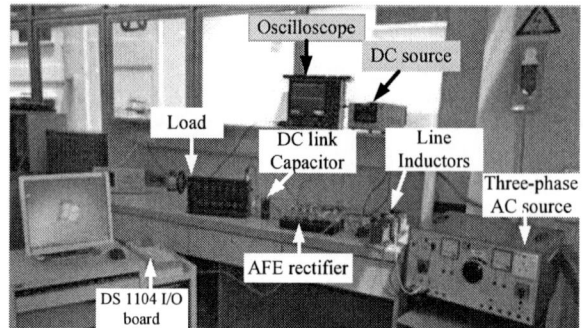

Fig. 3. Experimental system of AFE rectifier with proposed MMPC.

TABLE II: Experimental Parameters

Parameters values and unit			
Variables and Parameters	*Symbols*	*Values*	*Unit*
Power rating	P	2.5	kW
Three-phase supply voltage	v_s(rms)	110	V
Grid frequency	f_s	50	Hz
DC bus voltage	V_{dc}	320	V
Input filter inductance	L_s	5	mH
Input filter resistance	R_s	0.1	Ω
Sampling time	T_s	50	μs
Load resistance	R_{Load}	50	Ω
Capacitor value	C	1000	μF

stability of the proposed Lyapunov function based MMPC algorithm for AFE rectifier. The three-phase ac current variation drawn by the rectifier due to the V_{dc_ref} voltage variation is presented in Fig. 4(a). The result in Fig. 4(a) shows that, the output phase current is accurately tracking the reference value and reached its steady state condition within 0.02s, which verifies the stability and effectiveness of the MMPC algorithm. Moreover, the stability of the proposed Lyapunov function based MMPC method for AFE rectifier can be further confirmed with the variation of dc-link voltage and current as shown in Fig. 4(b).

B. Stability Analysis with Direct Lyapunov Method

The stability of the proposed MMPC technique for the AFE rectifier is investigated with direct Lyapunov method. The direct Lyapunov method gives the following stability criteria for a continuous function $L\left(\vec{i}_{error}(k)\right)$, in which the solutions of current dynamics in the rectifier mode (12) is uniformly and ultimately bounded [24]:

$$L\left(\vec{i}_{error}(k)\right) \geq c_1 \left|\vec{i}_{error}(k)\right|^l, \qquad \forall \vec{i}_{error}(k) \in G$$

$$L\left(\vec{i}_{error}(k)\right) \leq c_2 \left|\vec{i}_{error}(k)\right|^l, \qquad \forall \vec{i}_{error}(k) \in \Gamma \quad (19)$$

$$L\left(\vec{i}_{error}(k+1)\right) - L\left(\vec{i}_{error}(k)\right) < -c_3 \left|\vec{i}_{error}(k)\right|^l + c_4$$

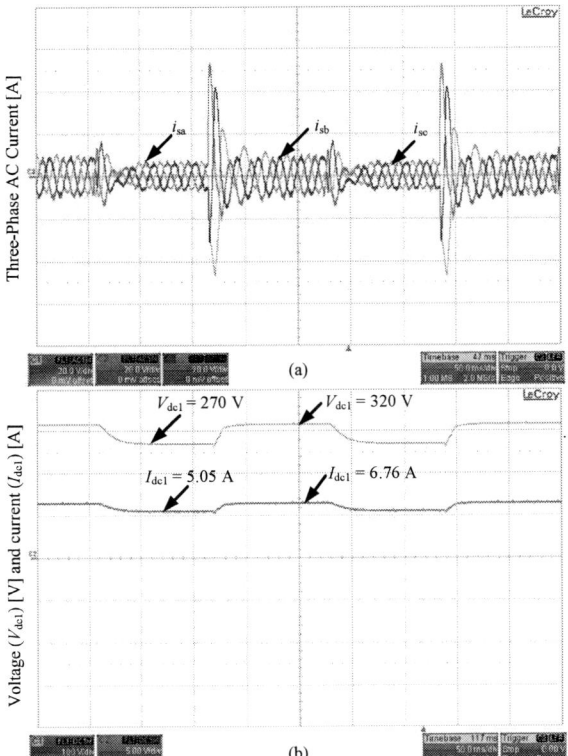

Fig. 4. The stability investigation via experimental system: (a) Three-Phase ac Current (i_{sa}) and (b) dc Bus voltage (V_{dc}) vs current (I_{dc}), with V_{dc_ref} voltage change of the AFE rectifier.

where, c_1, c_2, c_3 and c_4 are positive constants, $l \geq 1$, $G \subseteq R^n$ is positive control invariant set and $\Gamma \subset G$ is a compact set.

By applying the value of future voltage vector ($\vec{v}(k+1)$) of the rectifier (17), the rate of change of the Lyapunov function can be written as follows:

$$\Delta L(k) \leq -\frac{1}{2}[\vec{i}_{error}(k)]^T[\vec{i}_{error}(k)] + \frac{1}{2}\left(\frac{T_s}{R_s T_s + L_s}\right)^2 \varphi^2 \quad (20)$$

Therefore, the stability condition (19) is satisfied by the constant values as:

$$c_1 = c_2 = 1$$

$$c_3 = \frac{1}{2} \quad (21)$$

$$c_4 = \frac{1}{2}\left(\frac{T_s}{R_s T_s + L_s}\right)^2 \varphi^2$$

As a result, all signals in the Lyapunov function based close loop modified model predictive controlled AFE rectifier system are ultimately uniformly bounded. Then, the rate of change of the Lyapunov function in (20) is:

$$\Delta L(k) \leq -2c_3 L\left(\vec{i}_{error\_inv}(k)\right) + c_4. \quad (22)$$

This inequality implies that, as time increases, the current control error vectors converge to the compact set as:

$$A = \left[\left\|\vec{i}_{error}\right\| \middle| \left\|\vec{i}_{error}\right\| \leq \sqrt{c_4/c_3}\right]. \quad (23)$$

Thus, all signals of the proposed MMPC control method in (17) are uniformly and ultimately bounded.

C. Robustness Analysis

The control variables of nonlinear practical power converters are always affected by state measurement errors. Generally, these state measurement errors are caused by the filter parameter variations. Therefore, it is important that the closed-loop control techniques of nonlinear power converters should be robust with parameter variation [25, 26]. The robustness of the proposed MMPC algorithm is analyzed with the percentage deviation of current error ($\vec{i}_{error} = \vec{i}_s - \vec{i}_{ref}$) with different values of filter parameters, and presented in Fig. 5, which shows that the maximum deviation of current tracking error of about 8% is negligible. Therefore, it is possible to demonstrate that the proposed control strategy is robust and can operate efficiently even under parameter variations.

IV. COMPARATIVE EVALUATION OF PROPOSED MMPC WITH CONVENTIONAL MPC

One of the major advantages of the proposed Lyapunov function based MMPC algorithm is its very low execution time. The proposed MMPC algorithm reduces the amount of calculations required to predict a future variable by half compared with the conventional MPC method. A comparative evaluation between conventional MPC and proposed Lyapunov function based MMPC has been performed and summarized in Table III. The proposed MMPC algorithm operates as a voltage mode control technique which reduces the execution time by minimizing the amount of calculation compared to conventional current mode MPC algorithm. Moreover, the average switching frequency of the MMPC algorithm is also less than the conventional MPC method.

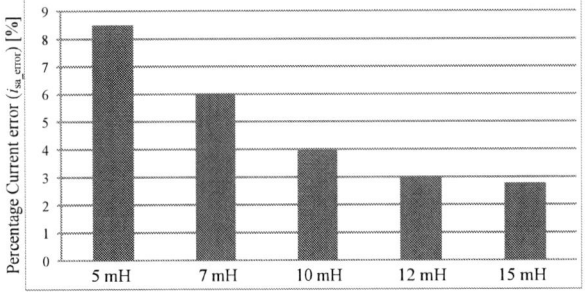

Fig. 5. Robustness analysis of the proposed MMPC algorithm against parameter variation.

TABLE III: COMPARISON BETWEEN CONVENTIONAL MPC AND PROPOSED LYAPUNOV-FUNCTION-BASED MMPC

Item description	Conventional MPC	Proposed MMPC
Control mode	Current mode control	Voltage mode control
Switching frequency	Variable switching frequency	Average switching frequency is less than MPC
Execution time	3.95 μs	3.24 μs
System stability	Impossible to analyze with direct Lyapunov method	Ensure the system stability with direct Lyapunov method

Finally, the proposed MMPC ensures the system stability with direct Lyapunov stability analysis method which is impossible for conventional MPC.

V. CONCLUSION

In this paper, a Lyapunov function based MMPC algorithm is proposed to control the AFE rectifier. Lyapunov function based MMPC is a powerful control algorithm in the field of power converters which provides fast dynamic response, nonlinear system stability and robustness. The stability of this control algorithm is analysed with a nonlinear experimental system and direct Lyapunov method. The result confirms that, the MMPC is stable for the operation of the AFE rectifier. Moreover, the proposed control technique is robust in case of parameter uncertainities. The results associated in this investigation are very encouraging and will continue to play a strategic role in the improvement of modern high performance AFE rectifier.

ACKNOWLEDGMENT

The authors wish to thank the financial support from the University of Malaya through HIR-MOHE project UM.C/HIR/MOHE/ENG/24 and UMRG project No. RP006E-13ICT.

REFERENCES

[1] J. Rodriguez and P. Cortes, "Control of an Active Front-End Rectifier," *Predictive Control of Power Converters and Electrical Drives*, pp. 81-98.

[2] J. R. Rodríguez, L. Dixon, J. R. Espinoza, J. Pontt, and P. Lezana, "PWM regenerative rectifiers: state of the art," *IEEE Trans. Ind. Electron.*, vol. 52, pp. 5-22, 2005.

[3] B. Singh, S. Gairola, B. N. Singh, A. Chandra, and K. Al-Haddad, "Multipulse AC–DC converters for improving power quality: a review," *IEEE Trans. Power Electron.*, vol. 23, pp. 260-281, 2008.

[4] J. Dannehl, C. Wessels, and F. W. Fuchs, "Limitations of Voltage-Oriented PI Current Control of Grid-Connected PWM Rectifiers With LCL Filters," *IEEE Trans. Ind. Electron.*, vol. 56, 2009.

[5] M. Malinowski, M. P. Kazmierkowski, and A. M. Trzynadlowski, "A comparative study of control techniques for PWM rectifiers in AC adjustable speed drives," *IEEE Trans. Power Electron.*, vol. 18, pp. 1390-1396, 2003.

[6] T. Noguchi, H. Tomiki, S. Kondo, and I. Takahashi, "Direct power control of PWM converter without power-source voltage sensors," *IEEE Trans. Ind. Application*, vol. 34, pp. 473-479, 1998.

[7] I. Takahashi and Y. Ohmori, "High-performance direct torque control of an induction motor," *IEEE Trans. Ind. Application*, vol. 25, pp. 257-264, 1989.

[8] A. Bouafia, F. Krim, and J.-P. Gaubert, "Fuzzy-logic-based switching state selection for direct power control of three-phase PWM rectifier," *IEEE Trans. Ind. Electron.*, vol. 56, pp. 1984-1992, 2009.

[9] J. Hu, L. Shang, Y. He, and Z. Zhu, "Direct active and reactive power regulation of grid-connected DC/AC converters using sliding mode control approach," *IEEE Trans. Power Electron.*, vol. 26, pp. 210-222, 2011.

[10] P. Cortés, M. P. Kazmierkowski, R. M. Kennel, D. E. Quevedo, and J. Rodríguez, "Predictive control in power electronics and drives," *IEEE Trans. Ind. Electron.*, vol. 55, pp. 4312-4324, 2008.

[11] M. Parvez Akter, S. Mekhilef, N. ML Tan, and H. Akagi, "Modified Model Predictive Control of a Bidirectional AC-DC Converter Based on Lyapunov Function for Energy Storage Systems," *IEEE Trans. Ind. Electron.*, 2015, DOI: 10.1109/TIE.2015.2478752.

[12] J. Rodriguez, M. Kazmierkowski, J. Espinoza, P. Zanchetta, H. Abu-Rub, H. Young, *et al.*, "State of the Art of Finite Control Set Model Predictive Control in Power Electronics," *IEEE Trans. Power Electron.*, 2013.

[13] M. P. Akter, S. Mekhilef, N. M. L. Tan, and H. Akagi, "Model Predictive Control of Bidirectional AC-DC Converter for Energy Storage System," *Journal of Electrical Eng. & Tech.*, vol. 10, pp. 165-175, 2015.

[14] M. P. Akter, S. Mekhilef, N. M. L. Tan, and H. Akagi, "Stability and Performance Investigations of Model Predictive Controlled Active-Front-End (AFE) Rectifiers for Energy Storage Systems," *Journal of Power Electron.*, vol. 15, pp. 202-215, 2015.

[15] M. Parvez, S. Mekhilef, N. M. L. Tan, and H. Akagi, "Model predictive control of a bidirectional AC-DC converter for V2G and G2V applications in electric vehicle battery charger," in *Proc. IEEE Transportation Electrification Conf. and Expo (ITEC)*, 2014, pp. 1-6.

[16] S. Muslem Uddin, P. Akter, S. Mekhilef, M. Mubin, M. Rivera, and J. Rodriguez, "Model predictive control of an active front end rectifier with unity displacement factor," in *Proc. IEEE Int. Conf. Circuits and Systems (ICCAS)*, 2013, pp. 81-85.

[17] M. Uddin, S. Mekhilef, M. Mubin, M. Rivera, and J. Rodriguez, "Model Predictive Torque Ripple Reduction with Weighting Factor Optimization Fed by an Indirect Matrix Converter," *Electric Power Components and Systems*, vol. 42, pp. 1059-1069, 2014.

[18] R. Vargas, P. Cortés, U. Ammann, J. Rodríguez, and J. Pontt, "Predictive control of a three-phase neutral-point-clamped inverter," *IEEE Trans. Ind. Electron.*, vol. 54, pp. 2697-2705, 2007.

[19] V. Yaramasu, M. Rivera, M. Narimani, B. Wu, and J. Rodriguez, "Model Predictive Approach for a Simple and Effective Load Voltage Control of Four-Leg Inverter with an Output LC Filter," *IEEE Trans. Ind. Electron.*, 2014.

[20] P. Cortes, J. Rodriguez, C. Silva, and A. Flores, "Delay compensation in model predictive current control of a three-phase inverter," *IEEE Trans. Ind. Electron.*, vol. 59, pp. 1323-1325, 2012.

[21] D. E. Quevedo, R. P. Aguilera, and T. Geyer, "Predictive control in power electronics and drives: Basic concepts, theory, and methods," in *Advanced and Intelligent Control in Power Electronics and Drives*, ed: Springer, 2014, pp. 181-226.

[22] M. Parvez, S. Mekhilef, N. M. L. Tan, and H. Akagi, "An improved active-front-end rectifier using model predictive control," in *Applied Power Electronics Conference and Exposition (APEC), 2015 IEEE*, 2015, pp. 122-127.

[23] S. Vazquez, J. Leon, L. Franquelo, J. Rodriguez, H. Young, A. Marquez, *et al.*, "Model predictive control: A review of its applications in power electronics," *IEEE Ind. Electron. Magazine*, vol. 8, pp. 16-31, 2014.

[24] J.-J. E. Slotine and W. Li, *Applied nonlinear control* vol. 199: Prentice-Hall Englewood Cliffs, NJ, 1991.

[25] T. Geyer, R. P. Aguilera, and D. E. Quevedo, "On the stability and robustness of model predictive direct current control," in *Proc. IEEE Int. Conf. Industrial Technology (ICIT), 2013*, 2013, pp. 374-379.

[26] M. Rivera, V. Yaramasu, J. Rodriguez, and B. Wu, "Model predictive current control of two-level four-leg inverters—Part II: experimental implementation and validation," *IEEE Trans. Power Electron.*, vol. 28, pp. 3469-3478, 2013.

Adaptive Reference Model Predictive Control for Power Electronics

Yun Yang[*], Siew-Chong Tan[*]

[*]Department of Electrical and Electronic Engineering
The University of Hong Kong, Hong Kong
Email: sctan@eee.hku.hk

Shu-Yuen (Ron) Hui[*][†]

[†]Department of Electrical and Electronic Engineering
Imperial College London, U.K.

Abstract—An adaptive reference model predictive control (ARMPC) approach is proposed as an alternative means of controlling power converters in response to the issue of steady-state residual errors presented in power converters under the conventional model predictive control (MPC). Differing from other methods of eliminating steady-state errors of MPC based control, such as MPC with integrator, the proposed ARMPC is designed to track the so-called virtual references instead of the actual references. Subsequently, additional tuning is not required for different operating conditions. In this paper, ARMPC is applied to a single-phase full-bridge voltage source inverter (VSI). It is experimentally validated that ARMPC exhibits strength in substantially eliminating the residual errors in environment of model mismatch, load change, and input voltage change, which would otherwise be present under MPC control. Moreover, it is experimentally demonstrated that the proposed ARMPC shows a consistent erasion of steady-state errors, while the MPC with integrator performs inconsistently for different cases of model mismatch after a fixed tuning of the weighting factor.

Keywords—Adaptive reference model predictive control (ARMPC); model predictive contro (MPC); steady-state errors; virtual reference; voltage source inverter (VSI); MPC with integrator.

I. INTRODUCTION

Model predictive control (MPC) is a process control in which the future control inputs and system response are predicted at regular intervals with respect to a performance index using a system model [1]. MPC has been found to be well suited for the control of power converters [2]–[13]. In practice, most predictive models are designed based on the so-called nominal model of the power converter, which is a simplified approximated model of the actual power converter since the precise description of the actual model in state-space form is impossible. Besides, power converters may have varying parameters of temperature, operating condition, and lifetime, which are susceptible to external disturbance. This leads to a parametric mismatch between the actual system and the predictive system, which results in the actual system to operate with non-negligible steady-state errors [14] and may

violate the required regulation standards. Therefore, it is necessary to compensate the control set such that this error can be mitigated. Several schemes have been proposed to overcome such steady-state errors [10]–[12]. In [10], a Kalman filter is added with MPC to account for unmeasured load variations and to achieve zero steady-state errors. In [11], the proposed adaptive robust predictive current control (ARPCC) removes the current error works in parallel with the deadbeat algorithm. In [12], MPC with integrator is proposed to reduce the steady-state errors both at the sampling instant and during the intersampling. To integrate the merits of the strategies proposed in [10]–[12] and also extensively simplifying the algorithm, a method that adaptively compensate the control set to mitigate the steady-state errors, which is achieved through the application of trajectory-based control [15]–[18] to the existing framework of MPC, is proposed in this paper. This method is known as adaptive reference model predictive control (ARMPC).

ARMPC has two important components. The first is the virtual references. The predictive strategy in ARMPC exists only to track the virtual references. The other component is the virtual multiple-input-multiple-output (MIMO) system, which derives the virtual references as outputs that are based on the state trajectories in the state plane. The measured change of trajectories reflects the instantaneous status of the system including the variation of the external disturbance or the presence of a model parameter mismatch. The model of the virtual MIMO is flexible. The selection of the algorithm is dependent on the requirement of the output performance and natural properties of the controlled plant. Compared with MPC with Kalman filter, ARMPC can be easily realized by a simple algorithm, using techniques of linear fitting, polynomial fitting, and Gaussian fitting, etc., to achieve the necessary steady-state performance of the system. This enables the implementation of ARMPC using simple micro-controllers. Besides, ARMPC is free of tuning. All the adaptive process is automatically done via measurement and online calculation. This is not the case for the MPC with integrator, which requires tuning of the weighting factor for the integral term and ARPCC needs tuning for the observe gain L_0 and the adaption integrator gain k_i. For large-scale nonlinear model mismatch, the system under the control of MPC with integrator has inferior steady-state performance than that under the control of ARMPC.

In this paper, ARMPC is applied to a single-phase full-bridge voltage source inverter (VSI) for illustration. The advantages of ARMPC over the conventional MPC and MPC with integrator in steady-state performance are experimentally provided. While the illustration of the use of ARMPC in this paper is focused only on the VSI, the proposed ARMPC is widely applicable to all power electronics.

II. STEADY-STATE TRACKING ISSUES OF MPC

Most previous works on VSI controlled by MPC are limited to current control and RL load [4]–[7]. However, for most applications, MPC based on voltage control is more applicable for VSI with LC filter (see Fig. 1).

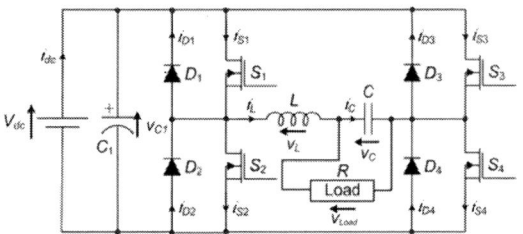

Fig. 1. Single-phase full-bridge inverter with output LC filter.

With this, the discrete model of VSI using the Euler's forward method [19] will be

$$\begin{cases} i_L(k+1) = i_L(k) - \dfrac{T_s}{L} v_C(k) + \dfrac{V_{dc} T_s}{L} u(k) \\ v_C(k+1) = \left(1 - \dfrac{T_s}{RC}\right) v_C(k) + \dfrac{T_s}{C} i_L(k) \end{cases}, \quad (1)$$

where T_s is the sampling time, i_L is the inductor current, v_C is the capacitor/load voltage, V_{dc} is the DC input voltage, and u is the control signal. Using the two-step-ahead prediction [11], the predicted voltage at the sampling time $k+1$ for the MPC at the sampling time k is

$$\hat{v}_C(k+1) = \left(1 - \frac{2T_s}{RC} + \frac{T_s^2}{R^2C^2} - \frac{T_s^2}{LC}\right) v_C(k-1)$$

$$+ \left(\frac{2T_s}{C} - \frac{T_s^2}{RC^2}\right) i_L(k-1) + \frac{T_s^2 V_{dc}}{LC} u(k-1). \quad (2)$$

Then, the expression of the cost function can be formulated as

$$J = \underbrace{\Gamma_y \left(v_{\text{ref}}(k+1) - \hat{v}_C(k+1)\right)^2}_{\text{Primary term}} + \underbrace{\Gamma_u \left(u(k) - u(k-1)\right)^2}_{\text{Secondary term}}, \quad (3)$$

where $v_{\text{ref}}(k+1)$ is the actual reference at sampling time $k+1$, and Γ_y and Γ_y are the weighting factors.

Then, the trajectory for tracking the reference can be recovered by solving

$$u^*(k) = \underset{u \in \{-1,0,1\}}{\arg \min} J, \quad (4)$$

and the corresponding control diagram can be depicted as shown in Fig. 2.

In trajectory-based control, the natural trajectories of a system will ideally correspond to its actual state-space model [15]. If model parameter mismatch exists, the natural trajectories will be changed (see Fig. 3). If the operating

trajectory is made to follow a nominal predictive law, the performance of the controlled system will be deteriorated. As shown in Fig. 4(a), when a model mismatch occurs, the state trajectory moves from circle A to circle B. Fig. 4(b) shows that after the change, the output v_C is incapable of tracking the reference accurately. The results show that under MPC, steady-state errors exist when the model mismatch of the system occurs.

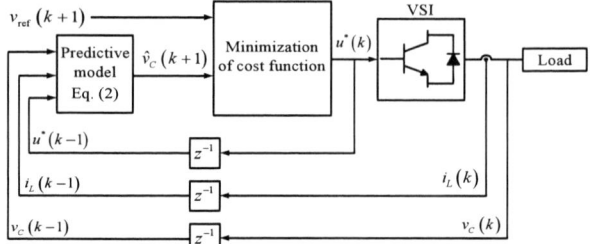

Fig. 2. Model predictive voltage control block diagram of VSI.

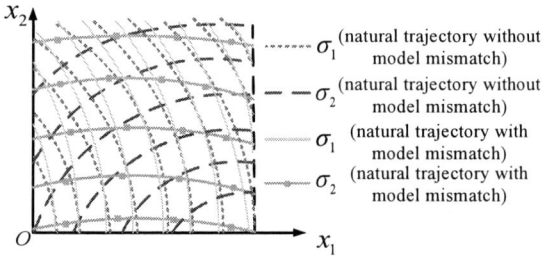

Fig. 3. Model predictive voltage control block diagram of VSI.

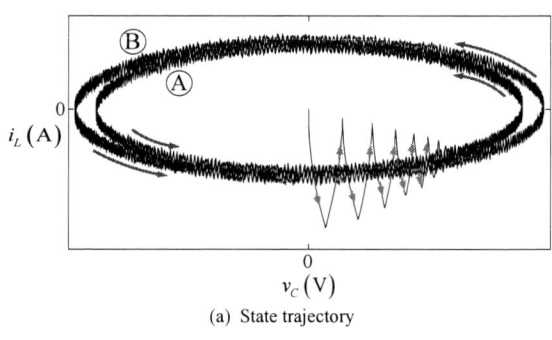

(a) State trajectory

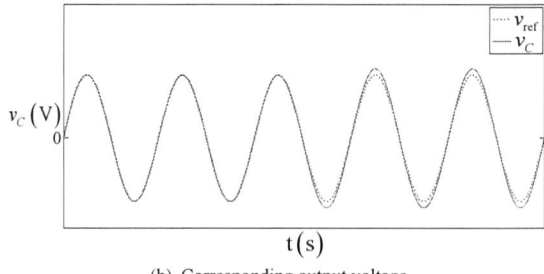

(b) Corresponding output voltage

Fig. 4. State trajectory and the corresponding output voltage of VSI before and after the appearance of model mismatch.

978-1-4673-9551-9/16 $31.00 © 2016 IEEE

III. PROPOSED ARMPC FOR VSI

Consider the illustration given in Fig. 5 which shows an extreme case where model mismatch exists throughout the entire operation of the VSI. After the appearance of a model mismatch, the root mean square (RMS) value of the output of MPC is shifted from point A to B to C to D to E and then to F. However, if this RMS value can be made to track a set of so-called virtual references y'_{ref1} to y'_{ref4} instead of the actual reference while still based on the nominal state-space model of the system, the steady-state errors in the system controlled by the MPC can be eliminated. Therefore, acquisition of the virtual references is critical in the actualization of ARMPC.

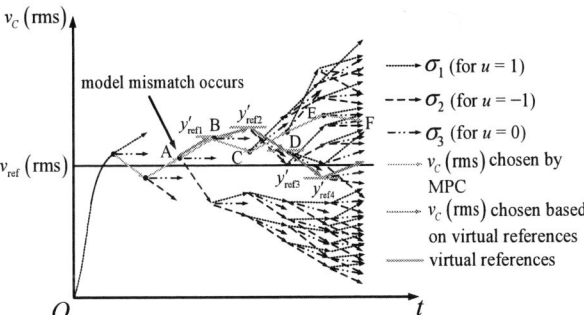

Fig. 5. v_C(rms) and virtual references of the system.

The approach of deriving a virtual MIMO system for generating the virtual references is proposed. The function of the virtual MIMO system can be briefly explained using Fig. 6. Steady-state errors exist between the operating trajectory and the actual reference trajectory. If the virtual reference trajectory C inside the actual reference trajectory A is tracked, the operating trajectory can be relocated to operate on the actual reference trajectory A instead of the trajectory B. Conversely (not shown in the figure), if the operating trajectory is located inside the actual reference trajectory, the virtual reference trajectory which is outside the actual reference trajectory will be tracked. Hence, the virtual MIMO system can robustly make the operating trajectory to move in accordance with the actual reference trajectory. In terms of implementation, the virtual MIMO system is built up of a series of past data which can be the positions of the state variables and virtual references.

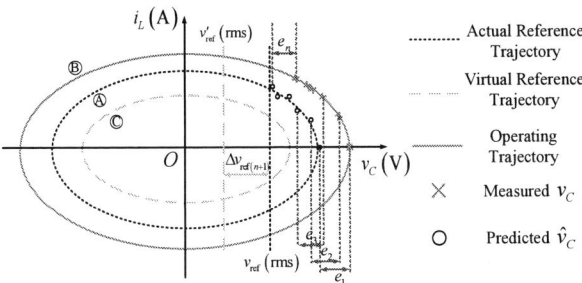

Fig. 6. Virtual MIMO system in the state plane.

For instance, at sampling time k, the error between the predicted output voltage $\hat{v}_C(k)$ derived by (2) and the measured output voltage $v_C(k)$ is defined as

$$e(k) = \hat{v}_C(k) - v_C(k), \qquad (5)$$

and $2n$ past data of these errors

$$\mathbf{e} = \begin{bmatrix} e(k-2n+1) & e(k-2n+2) & \dots & e(k) \end{bmatrix} \quad (6)$$

are stored and adopted for the virtual MIMO system. Meanwhile, the corresponding variation of virtual references for the last n sampling moment are denoted as

$$\Delta\mathbf{v}_{ref} = \begin{bmatrix} \Delta v_{ref}(k-2n+1) & \Delta v_{ref}(k-2n+2) & \dots & \Delta v_{ref}(k) \end{bmatrix}. (7)$$

In this paper, considering the scale of the model mismatch in the experiments, the virtual MIMO system is achieve via linear fitting. Then, the virtual MIMO system can be mathematically described as

$$\Delta\mathbf{v}_{ref}^{\mathbf{T}} = \begin{bmatrix} e(k-2n+1) & e(k-2n+2) & \cdots & e(k-n) \\ e(k-2n+2) & e(k-2n+3) & \cdots & e(k-n+1) \\ \vdots & \vdots & \ddots & \vdots \\ e(k-n) & e(k-n+1) & \cdots & e(k-1) \end{bmatrix} \mathbf{K}^{\mathbf{T}}, (8)$$

where $\mathbf{K} = \begin{bmatrix} K_1 & K_2 & \dots & K_n \end{bmatrix}$ are the weighting factors of the errors on the variation of the virtual references. Due to the number of elements in matrix $\Delta\mathbf{v}_{ref}$ and $\mathbf{K}$ being the same, $\mathbf{K}$ can be derived from (8) as

$$\mathbf{K} = \Delta\mathbf{v}_{ref} \left(\begin{bmatrix} e(k-2n+1) & e(k-2n+2) & \cdots & e(k-n) \\ e(k-2n+2) & e(k-2n+3) & \cdots & e(k-n+1) \\ \vdots & \vdots & \ddots & \vdots \\ e(k-n) & e(k-n+1) & \cdots & e(k-1) \end{bmatrix}^{-1} \right)^{\mathbf{T}}. (9)$$

Then, (9) is applied with $\begin{bmatrix} e(k-n+1) & e(k-n+2) & \dots & e(k) \end{bmatrix}$ to obtain the variation of the virtual reference $\Delta v_{ref}(k+1)$ at sampling time $k+1$ as

$$\Delta v_{ref}(k+1) = \mathbf{K}\begin{bmatrix} e(k-n+1) & e(k-n+2) & \dots & e(k) \end{bmatrix}^{\mathbf{T}}. (10)$$

Then, $\Delta v_{ref}(k-2n+1)$ in matrix $\Delta\mathbf{v}_{ref}$ in (8) and the first row of the error matrix $\begin{bmatrix} e(k-2n+1) & e(k-2n+2) & \dots & e(k-n) \end{bmatrix}$ are replaced by $\Delta v_{ref}(k+1)$ and $\begin{bmatrix} e(k-n+1) & e(k-n+2) & \dots & e(k) \end{bmatrix}$ respectively to derive a new value of $\mathbf{K}$ for the next iteration. This rolling process will instantaneously gain the information of model mismatch of the system and produce variation of the virtual references in real time. The flowchart of the algorithm for the VSI with ARMPC is shown in Fig. 7.

IV. EXPERIMENTAL VERIFICATION AND DISCUSSION

The experiment is conducted on a Texas Instruments' C2000™ Solar DC/AC VSI. The control used is Texas Instruments' digital signal processor (DSP) F28069 Piccolo controlCARD. The DC power supply is California Instruments' programmable source CSW5550. The specifications of the VSI are given in Table I. All the values of dominant components are provided by the TI Company, which can be found in the datasheets [20]. The loads used are three incandescent bulbs, at nominal power of 20 W and nominal voltage of 110 V. The number of the stored data $\Delta \mathbf{v}_{\mathrm{ref}}$ for the virtual MIMO system is 3. The weighting factors $\Gamma_y = 0.9$ and $\Gamma_u = 0.1$ are used. The switching frequency is 20 kHz for all the scenarios examined in the experiment. The values of v_C for all the cases in this paper are measured by the Keysight's 34401A Multimeter.

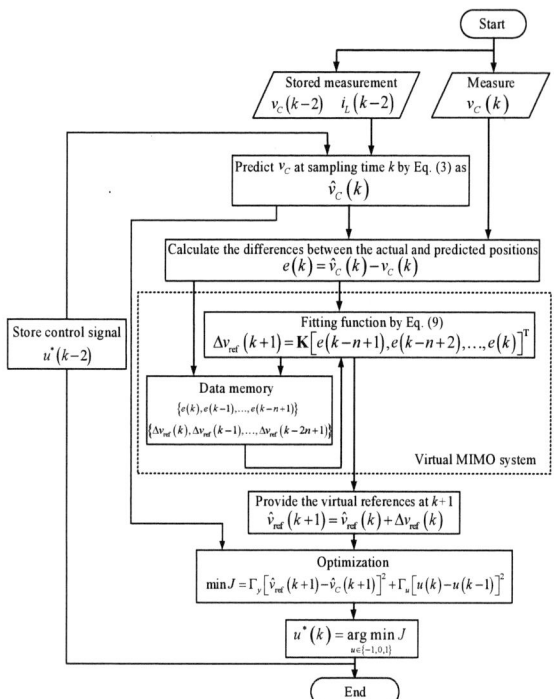

Fig. 7. Flowchart of ARMPC for the VSI.

TABLE I. SPECIFICATIONS OF THE NOMINAL MODEL OF VSI

Parameters	Values	Parameters	Values
V_{dc}	165 V	C_1	470 μF
C	1 μF	L	7 mH
v_{ref}(RMS)	110 V		

The VSI is initially controlled by the MPC with the nominal model. The RMS value of v_C being measured by the multimeter is around 106.001 V at steady state, which is a 3.64% deviation from the reference when VSI is controlled by MPC. However, when controlled by ARMPC, the steady-state RMS value of v_C is about 110.042 V, which is 0.04% deviation from the reference. In reality, model mismatch exists between the nominal model based on the specifications of VSI provided by datasheet and the actual VSI. If a controller employing the MPC is still designed based on the nominal model, steady-state errors will exist. If ARMPC is adopted, the steady-state errors caused by the model mismatch are eliminated. Meanwhile, the MPC with integrator being proposed in [12] is adopted with a proper tuning of K_i=0.25. The steady-state residuals are nearly removed as shown in Fig. 8(c).

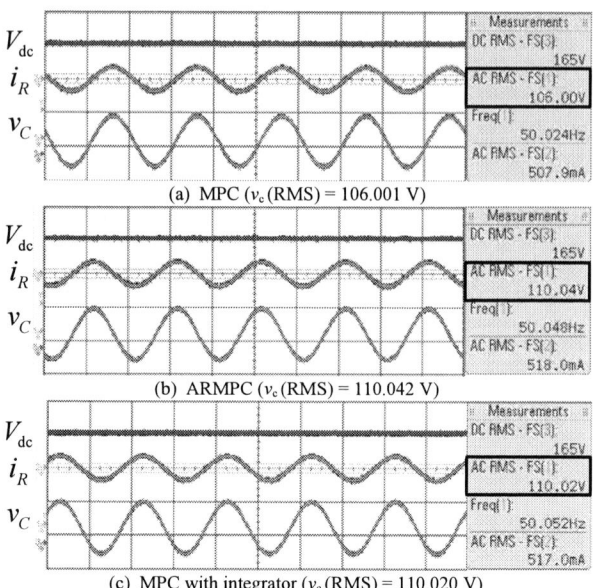

Fig. 8. Steady-state performance of VSI controlled by (a) MPC, (b) ARMPC, and (c) MPC with integrator based on the nominal model.

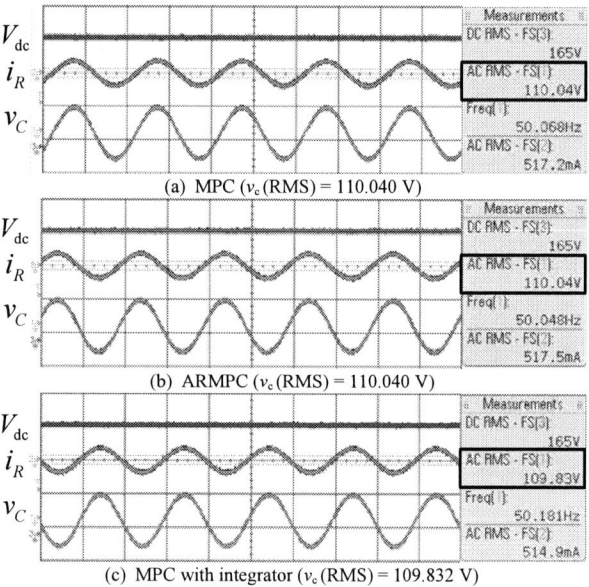

Fig. 9. Steady-state performance of VSI controlled by (a) MPC, (b) ARMPC, and (c) MPC with integrator based on the optimal model.

On the other hand, it is possible to tune by trail-and-error (since precise model is unavailable), an optimal predictive model for MPC and the results of the VSI controlled by MPC based on the optimal model can be obtained as shown in Fig. 9(a). Apparently, MPC with the optimal model has an excellent regulation performance and the RMS value of v_c is about 110.040 V at steady state, which is about 0.04% deviation from the reference. Meanwhile, the VSI controlled by ARMPC and MPC with integrator using the same optimal model can also have an equivalently good performance with the RMS value of v_c being 110.040 V and 109.832 V at steady state, which is about 0.04 % and 0.15 % deviation from the reference, as can be seen in Fig. 9(b) and Fig. 9(c), respectively.

Then, assume that one bulb is burnt out. So, one 20 W bulb is removed, leaving only two 20 W bulbs as loads. Even with the optimal model, the RMS value of v_c is about 112.511 V at steady state, which is 2.28 % deviation from the reference when VSI is controlled by MPC, as can be seen in Fig. 10(a). However, the RMS value of v_c is about 110.059 V at steady state, which is about 0.05 % deviation from the reference when VSI is controlled by ARMPC with the optimal model, as shown in Fig. 10(b). Fig. 10(c) presents the steady-state RMS value of v_c is 108.843 V under the control of MPC with integrator, which is about 1.05 % deviation from the reference, being slightly larger than the deviation of ARMPC.

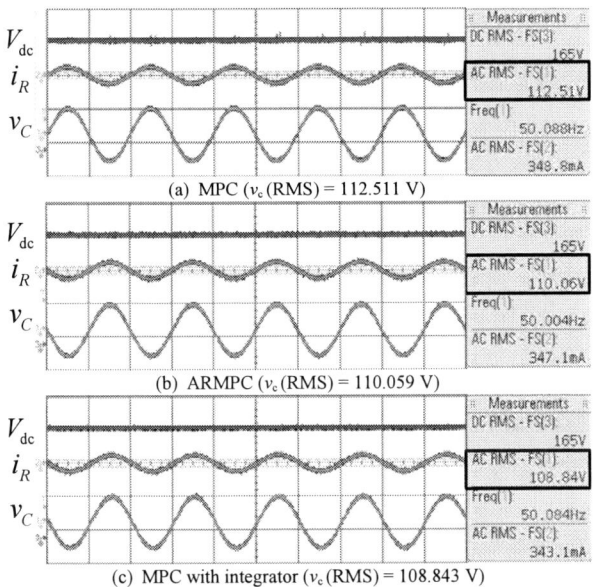

Fig. 10. Steady-state performance of VSI with a lighter load (two bulbs) controlled by (a) MPC, (b) ARMPC, and (c) MPC with integrator based on the optimal model.

The comparisons of steady-state performance of VSI when the load changes from three 20 W bulbs to one 20 W bulb between MPC, ARMPC, and MPC with integrator are also conducted (waveforms are not shown in the paper). With the optimal model, the RMS value of v_c is about 113.091 V at steady state, which is 2.81 % deviation from the reference when VSI is controlled by MPC. However, the RMS value of v_c is about 110.060 V at steady state, which is about 0.05 %

deviation from the reference when VSI is controlled by ARMPC with the optimal model. As for the MPC with integrator, the RMS value of v_c is 1.67 % deviation from reference. Obviously, VSI being controlled by ARMPC has better steady-state performance than VSI being controlled by MPC and MPC with integrator. MPC with integrator with a selected tuning $K_i = 0.25$ can guarantee one optimal operating condition to have almost zero steady-state error, but gradually lost its advantage over the conventional MPC when the operation is shifted far away from the optimal operating point. Hence, ARMPC can be perceived as a type of auto-tuning MPC with integrator, which achieves almost zero steady-state error of VSI over a wide operating range.

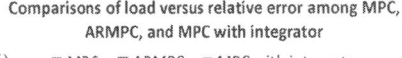

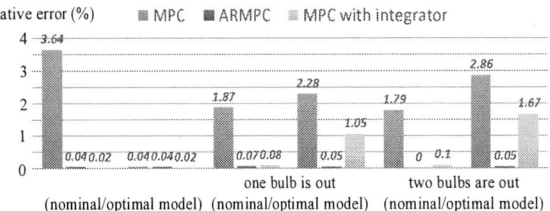

Fig. 11. Comparative bar-charts of the load versus relative error of v_c between the VSI controlled by MPC, ARMPC, and MPC with integrator based on both optimal and nominal models.

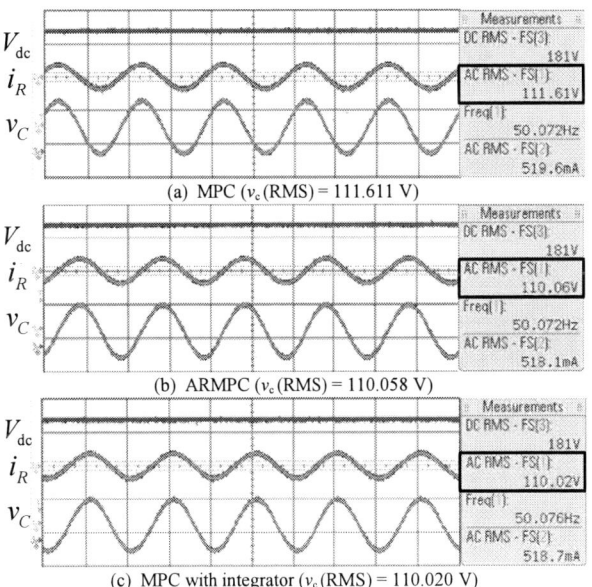

Fig. 12. Steady-state performance of VSI with V_{dc} change (165 V to 180 V) controlled by (a) MPC, (b) ARMPC, and (c) MPC with integrator based on the optimal model.

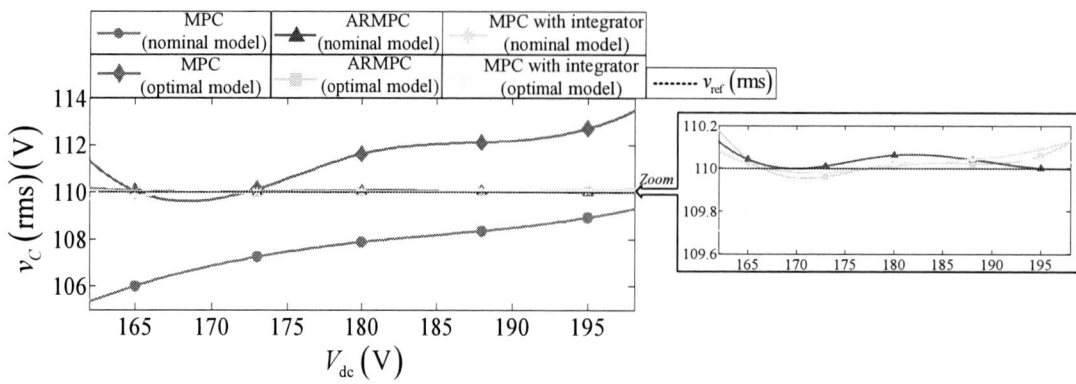

Fig. 13. Comparative curves of V_{dc} versus v_c(RMS) between the VSI controlled by MPC, ARMPC, and MPC with integrator with both optimal and nominal model.

TABLE II.   COMPREHENSIVE COMPARISONS OF Vc(RMS) BETWEEN MPC, ARMPC, AND MPC WITH INTEGRATOR

Case	MPC	Error	ARMPC	Error	MPC with integrator	Error
Nominal model	106. 001 V	**3.64 %**	110.042 V	**0.04 %**	110.020 V	**0.02 %**
Optimal model	110. 040 V	**0.04 %**	110. 040 V	**0.04 %**	109. 832 V	**0.15 %**
One bulb is out (optimal model)	112. 511 V	**2.28 %**	110. 059 V	**0.05 %**	108. 843 V	**1.05 %**
Two bulbs are out (optimal model)	113. 091 V	**2.81 %**	110. 060 V	**0.05 %**	108. 159 V	**1.67 %**
One bulb is out (nominal model)	107. 940 V	**1.87 %**	110. 081 V	**0.07 %**	109.170 V	**0.08 %**
Two bulbs are out (nominal model)	108. 031 V	**1.79 %**	109. 995 V	**0.00 %**	108. 933 V	**0.10 %**
V_{dc} is changed to 173 V (optimal model)	110. 122 V	**0.11 %**	110. 011 V	**0.01 %**	109. 994 V	**0.00 %**
V_{dc} is changed to 180 V (optimal model)	111. 611 V	**1.46 %**	110. 058 V	**0.05 %**	110. 020 V	**0.02 %**
V_{dc} is changed to 188 V (optimal model)	112. 114 V	**1.92 %**	110. 040 V	**0.04 %**	110. 041 V	**0.04 %**
V_{dc} is changed to 195 V (optimal model)	112. 707 V	**2.46 %**	110. 002 V	**0.00 %**	110. 043 V	**0.04 %**
V_{dc} is changed to 173 V (nominal model)	107. 251 V	**2.50 %**	109. 963 V	**0.04 %**	109. 980 V	**0.02 %**
V_{dc} is changed to 180 V (nominal model)	107. 882 V	**1.93 %**	110. 023 V	**0.02 %**	110. 019 V	**0.02 %**
V_{dc} is changed to 188 V (nominal model)	108. 361 V	**1.49 %**	110. 020 V	**0.02 %**	110. 041 V	**0.04 %**
V_{dc} is changed to 195 V (nominal model)	108. 922 V	**0.98 %**	110. 057 V	**0.05 %**	110. 091 V	**0.08 %**

Additional verification for the conclusions made can also be found in the comparative results of the load change (one bulb out and two bulbs out) by MPC, ARMPC, and MPC with integrator based on the nominal model. To clearly identify the differences between three controllers, a bar-chart based comparison is given in Fig. 11. Apparently, the VSI controlled by ARMPC will always have a better steady-state performance than the VSI controlled by MPC and MPC with integrator when the load deviates from the nominal value. More importantly, it is shown that ARMPC is highly tolerant to parameter shifts and the model it adopts in this situation does not affect its steady-state control performance.

Then, experiments are performed on the VSI based on the change of V_{dc} to study the effect of input voltage change on the output control performance. First, V_{dc} is changed from 165 V to 180 V. Fig. 12(a) shows that with the optimal model, the RMS value of v_c with MPC with 180 V input is about 111.611

V at steady state, which is a 1.46 % deviation from the reference when VSI is controlled by MPC at 165 V. However, in Fig. 12(b), the RMS value of v_c is about 110.058 V at steady state, which is about 0.05 % deviation from the reference when VSI is controlled by ARMPC. In Fig. 12(c), when VSI is regulated by MPC with integrator, the steady-state error is also nearly 0.02 % deviation from the reference. Apparently, VSI controlled by ARMPC and MPC with integrator having better steady-state performance than VSI being controlled by MPC is once again verified. Besides, note that MPC with integrator possess comparable off-set compensation ability to ARMPC for the change of V_{dc}. Then, several values of V_{dc} changing from 165 V to 195 V at 5 V intervals are used for the same experiment.

The curves are shown in Fig. 13. As shown, the VSI controlled by ARMPC will always have a better stead-state performance than the VSI controlled by MPC when the input voltage changes. Besides, the performance of ARMPC is unaffected by the model used. Meanwhile, MPC with integrator also performs well at steady state when the input voltage is varied.

Table II gives a comparison of the RMS value of the output voltage v_c(RMS) of the VSI with MPC, ARMPC, and MPC with integrator for all different cases. Except for the optimal model case where both MPC and ARMPC have a relative error of 0.04 %, almost all the other cases for MPC have a relative error of greater than 1 % with the largest error being 3.64 %. For ARMPC, all errors are kept within 0.1 %. MPC with integrator is also capable of regulating the relative error within 0.1 % when the input voltage fluctuates, but a load change will deteriorate its steady-state performance to a maximum of 1 %.

V. CONCLUSIONS

It has been explained that the method of model predictive control (MPC) applied for controlling power converters will induce non-negligible steady-state errors of the controlled outputs when there is model mismatch. Therefore, various methods, such as MPC with Kalman filter and MPC with integrator have been applied to eliminate the steady-state residues. In this paper, an alternative approach, known as adaptive reference model predictive control (ARMPC) that is based on the trajectory-based control theory and built upon the framework of MPC, is proposed as an alternative means of controlling power electronics of such nature. Instead of tracking the actual references, ARMPC is designed to track the so-called virtual references. As compared with conventional methods, particularly MPC with Kalman filter, the proposed ARMPC can be easily implemented by a low-performance inexpensive digital controller. As an illustration, the proposed ARMPC has been applied to a single-phase full-bridge voltage source inverter (VSI). The experimental results show that the ARMPC will always give a better steady-state performance than the MPC when there exists a model mismatch, load change, and input voltage change. Besides, as compared with MPC with integrator, ARMPC is free of tuning and operates efficiently over a wider operating range.

REFERENCES

[1] E. F. Camacho and C. Bordons, *Model Predictive Control*, Springer Verlag, 1999.

[2] S. Kouro, P. Cortés, R. Vargas, U. Ammann, and J. Rodríguez, "Model predictive control – a simple and powerful method to control power converters," *IEEE Trans. Ind. Electron.*, vol. 56, no. 6, pp. 1826-1838, June 2009.

[3] J. B. Rawlings and D. Q. Mayne, *Model Predictive Control: Theory and Design*, Nob Hill Publishing, 2009.

[4] J. Rodríguez, J. Pontt, C. A. Silva et al., "Predictive current control of a voltage source inverter," *IEEE Trans. Ind. Electron.*, vol. 54, no. 1, pp. 495-503, Feb. 2007.

[5] P. Cortés, J. Rodríguez, D. E. Quevedo, and C. Silva, "Predictive current control strategy with imposed load current spectrum," *IEEE Trans. Power Electron.*, vol. 23, no. 2, pp. 612-618, Mar. 2008.

[6] R. Vargas, J. Rodríguez, U. Ammann, and P. Wheeler, "Predictive current control of an induction machine fed by a matrix converter with reactive power control," *IEEE Trans. Ind. Electron.*, vol. 55, no. 12, pp. 4362-4371, Dec. 2008.

[7] R. Vargas, P. Cortés, U. Ammann, J. Rodríguez, and J. Pontt, "Predictive control of a three-phase neutral-point-clamped inverter," *IEEE Trans. Ind. Electron.*, vol. 54, no. 5, pp. 2697-2705, Oct. 2007.

[8] P. Cortés, G. Ortiz, J. I. Yuz et al., "Model predictive control of an inverter with output LC filter for UPS applications," *IEEE Trans. Ind. Electron.*, vol. 56, no. 6, pp. 1857-1883, June 2009.

[9] C. E. Garcia, D. M. Prett, and M. Morari, "Model predictive control: theory and practice – a survey," *Automatica*, vol. 25, no. 3, pp. 335-348, May 1989.

[10] T. Geyer, G. Papafotiou, R. Frasca, and M. Morari, "Constrained optimal control of the step-down dc-dc converter," *IEEE Trans. Ind. Electron.*, vol. 23, no. 5, pp. 2454-2464, Sept. 2008.

[11] J. M. Espi, J. Castello, and R. Garcia-Gil, "An adaptive robust predictive current control for three-phase grid-connected inverters," *IEEE Trans. Ind. Electron.*, vol. 58, no. 8, pp. 3537-3546, Aug. 2011.

[12] R. P. Aquilera, P. Lezana, and D. E. Quevedo, "Finite-control-set model predictive control with improved steady-state performance," *IEEE Trans. Ind. Inform.*, vol. 9, no. 2, pp. 658-667, May 2013.

[13] T. Geyer, D. E. Quevedo, "Performance of multistep finite control set model predictive control for power electronics", *IEEE Tran. Power Electron.*, vol. 30, no. 3, pp. 1633-1644, Mar. 2015.

[14] J. Rodríguez and P. Cortés, Predictive Control of Power Converters and Electrical Drives, NJ: Wiley, 2012.

[15] P. T. Krein, *Elements of Power Electronics*, Oxford: Oxford University Press, 1998.

[16] K. W. Chan, H. S. Chung, and S. Y. (Ron) Hui, "A genearalized theory of boundary control for single-phase multilevel inverter using second-order switching surface," *IEEE Tran. Power Electron.*, vol. 24, no. 10, pp. 2298-2313, Oct. 2009.

[17] S. C. Tan, Y. M. Lai, and C. K. Tse, *Sliding Mode Control of Switching Power Converters – Techniques and Implementation*, Boca Raton: CRC, 2012.

[18] M. Ordonez, M. T. Iqbal, and J. E. Quaicoe, "Selection of a curved switching surface for buck converters," *IEEE Tran. Power Electron.*, vol. 21, no. 4, pp. 1148-1153, Jul. 2006.

[19] H. Abu-Rub, J. Guziniski, Z. Krzeminski, and H. Toliyat, "Predictive current control of voltage-source inverter," *IEEE Trans. Ind. Electron*, vol. 51, no. 3, pp. 585-593, June 2004.

[20] "C2000 solar DC/AC single phase inverter schematic," Texas Instruments, USA, 17 Jun. 2014.

Power Switch Lifetime Extension Strategies for Three-Phase Converters

Serkan Dusmez, *Student Member, IEEE*, Enes Ugur, *Student Member, IEEE*, and Bilal Akin, *Senior Member, IEEE*

Electrical and Computer Science Department, Power Electronics and Drives Laboratory
University of Texas at Dallas, Richardson, TX, USA
serkan.dusmez@utdallas.edu, enes.ugur@utdallas.com, bilal.akin@utdallas.edu

Abstract— The recent advances and reports on failure precursors of power switches have led to estimation of lifetime as well as developing secondary control schemes to increase the lifetime of the converters. In this study, a secondary control scheme to extend the lifetime of the converter based on the identified failure precursors such as the on-state resistance variation for power MOSFETs and collector emitter voltage variation for IGBTs is proposed for three-phase converters. The controller switches the modulation scheme from SVPWM to discontinuous PWM, and adjusts the switching frequency, once the tracked failure precursor reaches to the defined threshold value. The tradeoff between the total harmonic distortion and lifetime extension at different modulation techniques and switching frequencies are addressed.

Keywords—failure diagnosis, power MOSFET, PWM modulation, health monitoring, power converter, lifetime extension, IGBT, inverter prognosis.

I. INTRODUCTION

The power switches of power electronics converters undergo both mechanical and environmental stresses which causes wear out over time. Among these, thermal swings experienced by the device due to power cycling are the major contributors to increased rate of package related failures [1]-[5]. For instance, the temperature swing amplitude of the traction drive can be up to 80°C degree in electric trains [6], which significantly reduces the reliability of the packaging of the power switches.

The failure mechanisms for the thermal stress have been studied on the custom built test-beds [4], [7]-[11]. Most commonly, bond-wire failure is observed in power modules with direct-bond-copper (DBC) layer, and die attach solder degradation/delamination is observed in discrete packages. The die attach degradation typically causes changes in thermal impedance as well as electrical resistance of the device, which can be observed from the on-state resistance variation of power MOSFETs and collector-emitter voltage of IGBTs. The on-state resistance of a discrete power MOSFET increases with a specific pattern when the failure mechanism is die attach solder [12]. Similarly, when a bond-wire lifts off in a

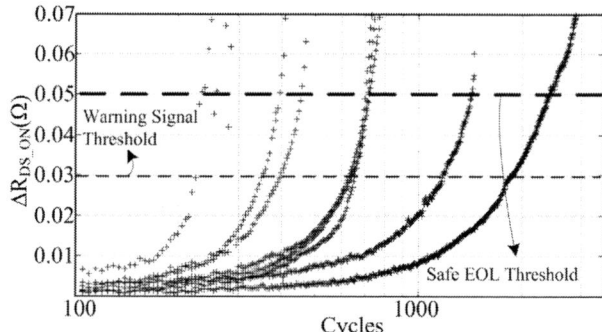

Fig. 1. ΔR_{ds_on} variation of thermally aged power MOSFETs in the exponential region.

power module, there is a sudden jump in the collector-emitter voltage of an IGBT module [4],[13]-[15]. These failure precursors can be monitored to trigger an early warning signal once they reach a predefined threshold value.

In order to demonstrate the on-state resistance variation with respect to thermal aging, the experimental data presented in [11] has been recalled in this paper. In Fig. 1, the on-state resistance variations of 400V/11A discrete package MOSFETs which are aged under three different thermal swings are given. It has been found that the on-state resistances of the switches increase exponentially to a value after which the resistance jumps to a much higher value and exhibits unstable resistance [9],[12]. After conducting a number of experimental tests on the chosen power MOSFETs, 50mΩ difference in on-state resistance of a healthy switch has been defined as a safe threshold value for the device under test (DUT) where reliability is concerned.

By tracking the variations in the failure precursors, it is possible to further extend the converters lifetime in mission critical applications by changing the modulation/switching scheme. By doing this, the thermal stress on the degraded switches can be reduced at the expense of degraded efficiency or current waveform quality. However, not much research was found on the topic of switch life extension schemes. Many efforts have been devoted to more reliable systems, which can

978-1-4673-9551-9/16 $31.00 © 2016 IEEE

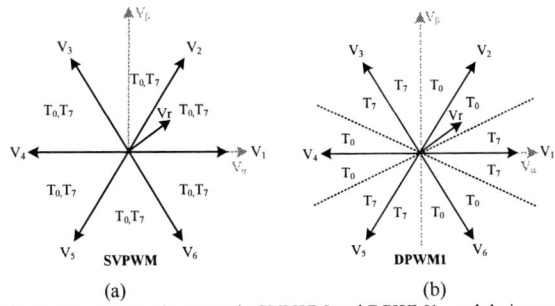

(a)   (b)

Fig. 2. Zero-vector placement in SVPWM and DPWM1 modulation schemes.

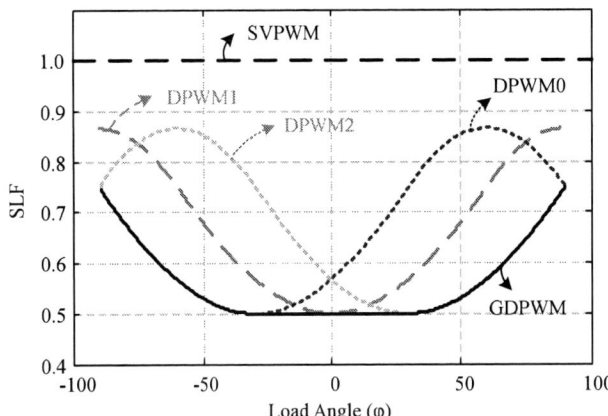

Fig. 3. SLF plot for SVPWM and DPWM at different load angles.

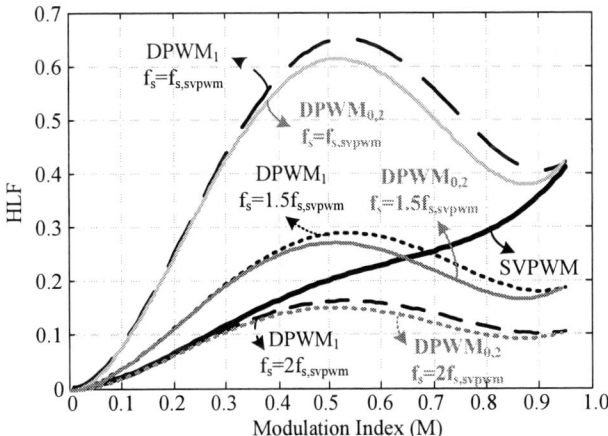

Fig. 4. HLF plot for SVPWM and DPWM modulation schemes at different modulation indexes.

continue to operate in case of failure [16]. The simplest way to ensure maximum reliability for the power converter is to include redundant switches but this approach is expensive [17]-[18]. Majority of the work done on life extension strategies are on controlling the junction temperature that prevents the maximum junction temperature to exceed maximum continuous temperature limit defined by the manufacturer, which can cause a sudden failure [19]-[26]. In [24]-[26], the switching frequency reduction technique has been employed to keep the junction temperature at safe level during the start-up process of an ac machine till synchronization is achieved. However, the purpose of these studies is to keep the junction temperature under control during only low frequency operation without monitoring the health of the devices. Also, the effects of switching frequency reduction on the current waveform and converter's lifetime have not been explored.

In this study, an active lifetime extension strategy based on the switching frequency and modulation scheme adjustment is proposed for three-phase converters through evaluating the variations in the failure precursors. When the measured failure precursor reaches the predetermined threshold value, the thermal stress or, in other words, power dissipation on the switches are reduced by either 1) decreasing the switching frequency, 2) changing the zero-vector placement. The effects of modulation and switching frequency adjustment on the harmonic distortion factor as well as lifetime of converter are investigated.

II. INVESTIGATION ON THE LIFETIME EXTENSION STRATEGIES

In typical adjustable speed drive inverters, the PWM modulation controller involves more than one type of modulation scheme to achieve better current quality according to the given modulation index. On the other hand, the switching frequency is usually optimized for the system considering the harmonic distortion, switching losses, device switching characteristics, EMI filters, etc. at the design stage and remains constant throughout the operation. With the recent findings on the failure precursors indicating package related wear-out, it is possible to monitor the health of the switches, which in turn necessitates new tasks concerning the reliability of the inverter to be integrated to the PWM

modulation controller.

In order to increase the lifetime of the converter, both changing the modulation from space-vector-PWM (SVPWM) to discontinuous PWM (DPWM), and adopting variable switching frequency are considered in this paper. Different than the conventional control algorithm, the modulation controller determines the zero-vector placement and switching frequency according to the condition of the switches with respect to the sector in which the reference vector resides.

A. Continuous and Discontinous Modulation Techniques

In SVPWM, two active vectors and two zero vectors are used to synthesize the reference voltage vector in each sector. The time durations of zero vectors V_0 (000) and V_7 (111) are equally distributed in each switching period as shown in Fig. 2(a), where $t_0 = t_7 = T_0/2$. This distribution provides a better current waveform, basically lower harmonics at switching frequency and its sidebands under low modulation indexes; however, exhibits higher switching losses. To reduce the switching losses and improve the current waveform quality, the modulation scheme is typically switched to DPWM at higher modulation indexes together with switching frequency

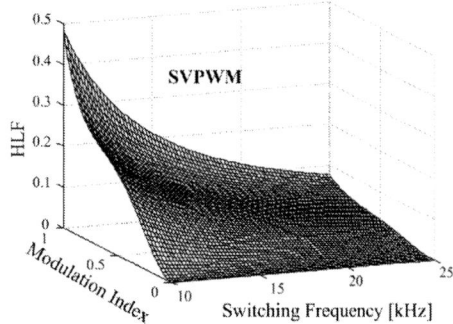

Fig. 5. HLF surface of SVPWM at different modulation indexes and switching frequencies.

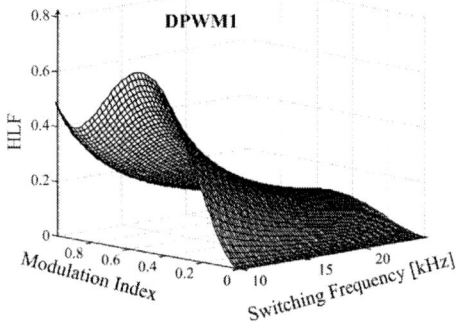

Fig. 6. HLF surface of DPWM1 at different modulation indexes and switching frequencies.

adjustment whenever possible. In DPWM techniques, only one zero vector is used in every 60° sectors. The placement of zero vectors leads to various different DPWMs. For instance, -30° shift in the defined sectors results in DPWM1 strategy as shown in Fig. 2(b), where the upper leg switches are clamped to "1" when leg voltages are highest, and clamped to "0" when they are lowest.

B. Switching Loss Comparison

Switching loss is one of the major losses that rise the junction temperature of the power devices. It increases proportionally to the dc bus voltage and instantaneous load current, and it can be calculated by taking the average over a line cycle. In SVPWM, the switching loss of a switch is expressed as

$$P_{SVPWM}\left(V_{dc}, I_M\right) = \frac{2V_{dc}I_M}{\pi} \tag{1}$$

For comparison with switching losses that in DPWM, a switching loss factor (SLF) is defined.

$$SLF = \frac{P_{DPWM}\left(V_{dc}, I_M, \varphi\right)}{P_{SVPWM}\left(V_{dc}, I_M\right)} \tag{2}$$

where φ denotes the load angle. For DPWM techniques, the switching loss is a function of load angle as the switches are clamped to "1" during a 60° interval in which switching losses become zero. This suggests that DPWM1 provides minimum switching loss when load angle is 0°. Likewise, DPWM0 and

DPWM2 provide optimum loss reduction when load angle is -30° and +30°, respectively. Eliminating switching losses for 60° interval introduces a boundary to the integral of loss calculation. The generalized formula for calculating the SLF is expressed as [27].

$$SLF_{DPWM} = \begin{cases} \frac{\sqrt{3}}{2}\cos\left(\frac{4\pi}{3}/3 + \psi - \varphi\right), & -\frac{\pi}{2} \le \varphi \le -\frac{\pi}{2} + \psi \\ 1 - \frac{1}{2}\sin\left(\frac{\pi}{3} + \psi - \varphi\right), & -\frac{\pi}{2} + \psi \le \varphi \le \frac{\pi}{6} \\ \frac{\sqrt{3}}{2}\cos\left(\frac{\pi}{3} + \psi - \varphi\right), & \frac{\pi}{6} + \psi \le \varphi \le \frac{\pi}{2} \end{cases} \tag{3}$$

The SLF function is plotted in Fig. 3. As proposed in [28], the 60° clamping interval can be shifted with respect to load angle within -30° and +30° to achieve minimum switching losses. This method is called generalized DPWM and denoted as GDPWM in Fig. 4. Thus, it is assumed that switching losses of DPWM are half of that in SVPWM for the same switching frequency operation for a wide load angle operation range. This figure of merit suggests that theoretically the switching frequency in DPWM can be increased to twice of that in SVPWM for the same overall switching losses. However, it should be noted that the instantaneous power loss increases the device junction temperature which in turn impacts the turn-on and turn-off characteristics and changes the switching energy losses. The junction temperature effect on the switching losses is neglected in Eq. (1) and Eq. (3).

C. Harmonic Distortion Comparison

The major harmonics in a balanced three-phase inverter are typically observed around the switching frequency. The harmonic distortion can be found by analyzing the ripple current within a switching cycle as in [29]. The switching harmonics are a function of applied dc bus voltage, load impedance, frequency and modulation index. Considering that the load is inductive, and inverter operates under the same dc bus voltage, a harmonic loss factor (HLF) that can be used to compare the harmonic losses of SVPWM and DPWM can be defined. The ratio of total harmonic distortion (THD) can be found by taking the root square of HLF. In SVPWM technique, the integration over the positive cycle results in

$$HLF_{SVPWM}\left(f_s, M\right) = \frac{1}{f_s^2}\left(\begin{array}{c} \frac{3}{2}\left(\frac{\pi M}{4}\right)^2 - \frac{4\sqrt{3}}{\pi}\left(\frac{\pi M}{4}\right)^3 \\ +\frac{9}{8}\left(\frac{3}{2} - \frac{9\sqrt{3}}{8\pi}\right)\left(\frac{\pi M}{4}\right)^4 \end{array}\right) \tag{4}$$

Likewise, the HLF for DPWM techniques are found as

$$HLF_{DPWM1}\left(f_s, M\right) = \frac{1}{f_s^2}\left(\begin{array}{c} 6\left(\frac{\pi M}{4}\right)^2 - \left(\frac{45}{2\pi} + \frac{4\sqrt{3}}{\pi}\right)\left(\frac{\pi M}{4}\right)^3 \\ +\left(\frac{27}{8} + \frac{27\sqrt{3}}{32\pi}\right)\left(\frac{\pi M}{4}\right)^4 \end{array}\right) \tag{5}$$

978-1-4673-9551-9/16 $31.00 © 2016 IEEE

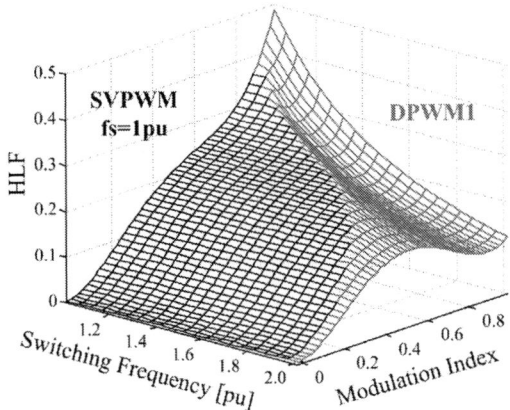

Fig. 7. Optimal HLF surface when $f_{s,SVPWM}$=1pu and $1pu \leq f_{s,DPWM1} \leq 2pu$.

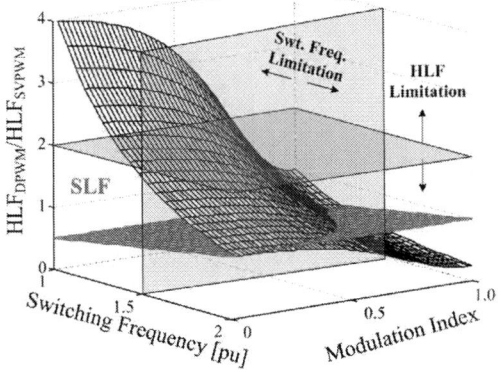

Fig. 8. HLF and SLF surfaces of DWPM1 with respect to SVPWM under switching frequency and HLF limitations i.e. $HLF_{DWPM1} \leq 2\ HLF_{SVPWM}$ and $1pu \leq f_{s,DPWM1} \leq 1.5pu$.

$$HLF_{DPWM0,2}\left(f_s,M\right) = \frac{1}{f_s^2}\left(\begin{array}{l} 6\left(\dfrac{\pi M}{4}\right)^2 - \left(\dfrac{35\sqrt{3}}{2\pi}\right)\left(\dfrac{\pi M}{4}\right)^3 \\ + \left(\dfrac{27}{8} + \dfrac{81\sqrt{3}}{64\pi}\right)\left(\dfrac{\pi M}{4}\right)^4 \end{array}\right) \quad (6)$$

HLF is plotted in Fig. 4 according to the modulation index for SVPWM and DPWM. As it can be observed from the plots, there is no significant difference in terms of harmonic distortion in between DPWM methods, particularly when $0 \leq M \leq 0.4$. In case generalized DPWM method is used, HLF plot would stand in between the curves of DPWM1 and DPWM0,2; thus, the comparison will be made according to DPWM1 herein after.

With respect to above discussion on the switching losses, the HLF of DPWM when switching frequency is increased to twice of that in SVPWM is also plotted for a fair comparison. As it can be seen from the figure, SVPWM performs superior in terms of harmonic distortion for the same switching frequency. When the switching frequency in DPWM is increased to twice in the SVPWM, the HLF of DPWM becomes better in whole modulation range, while the switching losses remain the same as of SVPWM as depicted in

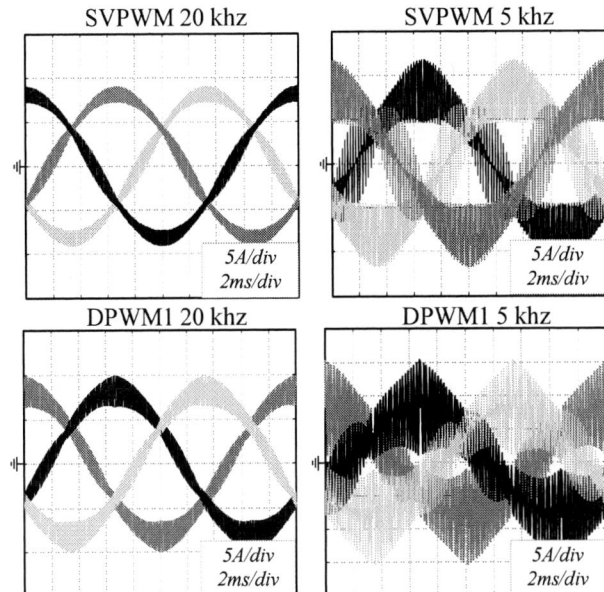

Fig. 9. Current waveform at different switching frequencies under SVPWM and DPWM1 techniques.

Fig. 3. The HLF surfaces with respect to modulation index and switching frequencies have been plotted in Fig. 5 and Fig. 6 for SVPWM and DPWM1, respectively.

D. Lifetime Extension Strategies

The optimal HLF surface that can be achieved when switching frequency of SVPWM is kept at 1pu and DPWM1 is swept from 1pu to 2pu, is shown in Fig. 7. It should be noted that there is no switching frequency limitation in this surface within the defined limits. However, in typical power converters both the lower and upper limits of switching frequency are set considering the power losses, EMI filters, audible noise level and switching capability of the device and so on. Thus, it is not always viable to increase the switching frequency to 2pu. As seen from Fig. 4 and Fig. 7, SVPWM provides better current waveform quality and is preferable at modulation indexes less than 0.65 when the upper switching frequency is limited to 1.5pu. However, as the tracked failure precursors reach to defined threshold values; 1) the switching frequency can be reduced, 2) the modulation technique can be switched to DPWM. Both of these methods aim to reduce the power loss of the devices. Nevertheless, they are not preferable under normal operating conditions when the devices are classified as "healthy", as these adjustment methods deteriorate the current waveform and increases the harmonic distortion.

The selection between SVPWM and DPWM considering a feasible switching frequency range is essentially dependent on the harmonic distortion or harmonic losses limitation and the desired power loss reduction which has an exponential relation with the end-of-life (EOL) of the switches. In Fig. 8, the SLF of DPWM and HLF_{DPWM}/HLF_{SVPWM} are plotted with given

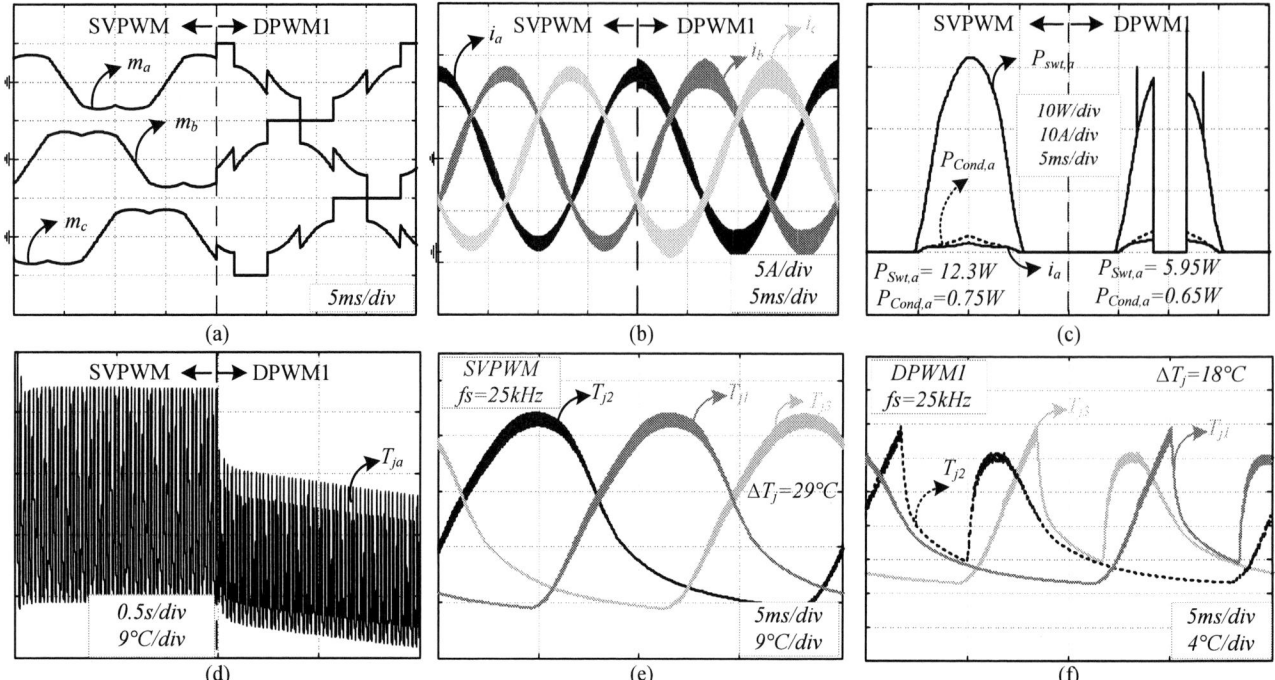

Fig. 10. Simulation results of three-phase converter when modulation technique is switched from SVPWM to DPWM1; (a) modulation, (b) phase currents, (c) power losses, (d) junction temperatures, (e) zoom-in profile of junction temperatures in SVPWM at steady-state, (f) zoom-in profile of junction temperatures in DPWM1 at steady-state.

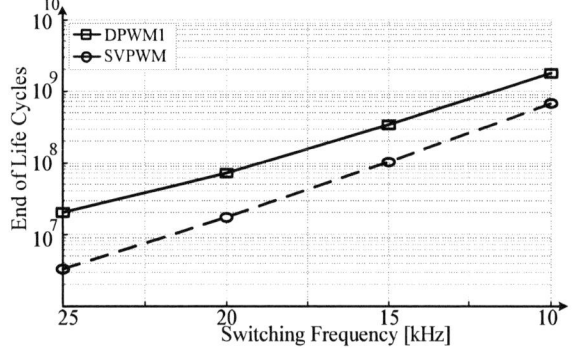

Fig. 11. EOL results for SVPWM and DPWM1 under variable switching frequency.

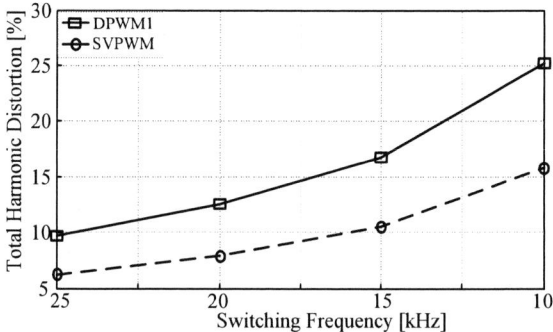

Fig. 12. THD results for SVPWM and DPWM1 under variable switching frequency.

frequency and HLF limitation planes. The figure illustrates an application with following limitations; $HLF_{DWPM1} \leq 2$ HLF_{SVPWM} and $1pu \leq fs_{,DPWM1} \leq 1.5pu$. The SLF plot resembles the power loss that can be achieved and HLF_{DPWM}/HLF_{SVPWM} shows the increased HLF ratio compared to that of SVPWM. As it is apparent, higher switching frequency leads to lower HLF, while switching power losses decrease with reduced switching frequency. Once the limitations are defined, the selection criterion is mostly dependent on the desired level of reliability.

III. SIMULATION AND EXPERIMENTAL RESULTS

The simulations are conducted in PLECS software which can combine electrical and thermal quantities. The parameters of the used three-phase converter is as follows; $L=1mH$,

$P_o=2250W$, $V_{dc}=450V$, IGBT rating: 1200V/13A, R_{th} of the heat sink=3.4W/K, $T_{amb}=25°C$. The Foster thermal network coefficients of the switch are; $R_{th1}=0.187$ (K/W), $\tau_1=0.173$ (s) $R_{th2}=0.575$(K/W), $\tau_2=0.0275$ (s), $R_{th3}=0.589$(K/W), $\tau_3=0.00257$ (s) $R_{th4}=0.35$(K/W), $\tau_4=0.000271$ (s). The output power is intentionally selected high for the chosen switch type in order to highlight the junction temperature reduction with the proposed approaches.

In the first simulation, the switching frequency is reduced in each modulation technique to illustrate the current waveform distortion. As seen from Fig. 9, the worst current waveform is achieved when the converter is switched at 5kHz under DWPM1 technique. In Fig. 10, the simulation results of modulation, phase currents, power losses, and junction temperatures are shown for SVPWM and DPWM1 at the same

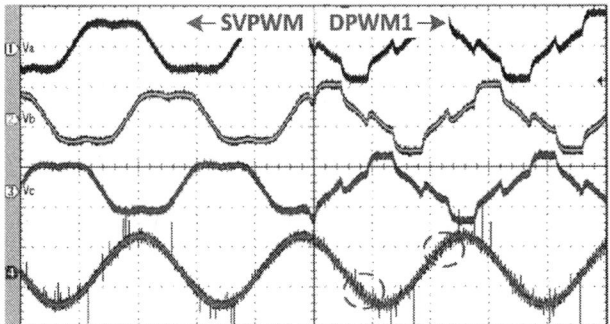

Fig. 13. Experimental results when modulation scheme is switched from SVPWM and DPWM1.

switching frequency. Basically, the power losses are continuous for the half period and proportional to load current in continuous PWM strategies, whereas it is chopped when the phase current is maximum in DPWM. The switching losses decrease from 12.3W to 5.95W. The junction temperatures corresponding to each modulation strategies are shown at steady state in Fig. 10(e)-(f). The thermal swing amplitudes of the switches are equal to 29°C and 18°C, when SVPWM and DPWM are used, respectively. As it can be seen, the losses per switch and junction temperature significantly decrease at the expense of distorted current waveform as soon as DPWM1 is engaged. Fig. 11 shows the end-of-life (EOL) of the switches, which is calculated offline from the junction temperature of the switches using Coffin-Manson model, and Fig. 12 presents the THD results. These two curves are for the output power and modulation index used in the simulation. In practice, the selection between the switching frequency and modulation technique should be made according to these figures of merits for the given application.

In Fig. 13, the experimental results of given modulation references and phase current are given when the modulation technique is switched from SVPWM to DPWM1. This figure illustrates the increased harmonic distortion in case DPWM1 is used as highlighted with circles.

IV. CONCLUSION

In this study, two approaches are presented to increase the lifetime of the three-phase converters. The methods are proposed as secondary control strategies, and engaged after the failure precursor reaches the defined threshold value. The proposed modulation controller adjusts the zero-vector placement as well as the switching frequency in order to reduce the power losses on the switches. The proposed approach results in degraded current waveform quality but increases the remaining lifetime of the switches as switching losses can be significantly reduced. This tradeoff needs to be well analyzed for given application. In this paper, a framework has been shown to address this tradeoff. The selection of the switching frequency and modulation technique is strictly dependent on the desired level of reliability and application limitations.

V. ACKNOWLEDGEMENT

This work was supported in part by the TXACE/SRC under TASK 1836.154 and NSF Grant 1454311. Authors would like to thank Priority Labs for providing the failure analysis results which form the basis of this work.

REFERENCES

[1] K. Li, G.Y. Tian, L. Cheng, A. Yin, W. Cao, and S. Crichton, "State Detection of Bond Wires in IGBT Modules Using Eddy Current Pulsed Thermography," *IEEE Trans on Power Electron.*, vol. 29, no. 9, pp. 5000-5009, Sept. 2014.

[2] S. Yang, D. Xiang, A. Bryant, P. Mawby, L. Ran, and P. Tavner, "Condition Monitoring for Device Reliability in Power Electronic Converters: A Review," *IEEE Trans on Power Electron.*, vol. 25, no. 11, pp. 2734-2752, Nov. 2010.

[3] W. Kexin, D. Mingxing, X. Linlin, and L. Jian, "Study of Bonding Wire Failure Effects on External Measurable Signals of IGBT Module," *IEEE Trans on Device and Materials Reliability*, vol. 14, no. 1, pp. 83-89, March 2014.

[4] V. Smet, F. Forest, J,-J. Huselstein, F. Richardeau, Z. Khatir, S. Lefebvre, and M. Berkani, "Ageing and Failure Modes of IGBT Modules in High-Temperature Power Cycling," *IEEE Trans on Ind. Electron.*, vol. 58, no. 10, pp. 4931-4941, Oct. 2011.

[5] H. Oh, B. Han, P. McCluskey, C. Han, and B.D. Youn, "Physics-of-Failure, Condition Monitoring, and Prognostics of Insulated Gate Bipolar Transistor Modules: A Review," *IEEE Trans on Power Electron.*, vol. 30, no. 5, pp. 2413-2426, May 2015.

[6] M. Held, P. Jacob, G. Nicoletti, P. Scacco, and M. H. Poech, "Fast power cycling test of IGBT modules in traction application," in *Proc. Int. Conf. Power Electron. Drive Syst.*, 1997, pp. 425–430.

[7] N. Patil, J. Celaya, D. Das, K. Goebel, and M. Pecht, "Precursor Parameter Identification for Insulated Gate Bipolar Transistor (IGBT) Prognostics," *IEEE Trans on Reliability*, vol. 58, no. 2, pp. 271-276, June 2009.

[8] M. Held, P. Jacob, G. Nicoletti, P. Scacco, M.-H. Poech, "Fast power cycling test of IGBT modules in traction application," in *Proc. Power Electronics and Drive Systems*, pp. 425-430, 26-29 May 1997.

[9] J. Celaya, A. Saxena, P. Wysocki, S. Saha, and K. Goebel, "Towards Prognostics of Power MOSFETs: Accelerated Aging and Precursors of Failure", in *Proc. Annual Conference of the Prognostics and Health Management Society*, pp. 1-10, 2010.

[10] J. R. Celaya, N. Patil, S. Saha, P. Wysocki, and K. Goebel, "Towards Accelerated Aging Methodologies and Health Management of Power MOSFETs (Technical Brief)", in *Proc. Annual Conference of the Prognostics and Health Management Society*, pp. 1-8, 2009.

[11] S. Dusmez and B. Akin, "An accelerated thermal aging platform to monitor fault precursor on-state resistance", in Proc. *IEEE International Electric Machines and Drives Conference (IEMDC)*, pp. 1-6, 10-13 May 2015.

[12] S. Dusmez, and B. Akin, "Remaining useful lifetime estimation for degraded power MOSFETs under cyclic thermal stress", *in Proc. IEEE Energy Conversion Conference & Expo.*, pp. 3846-3851, 2015.

[13] Y. Xiong, X. Cheng, Z. J. Shen, C. Mi, H. Wu, and V. K. Garg, "Prognostic and Warning System for Power-Electronic Modules in Electric, Hybrid Electric, and Fuel-Cell Vehicles," *IEEE Trans on Ind. Electron.*, vol. 55, no. 6, pp. 2268-2276, June 2008.

[14] G. Coquery and R. Lallemand, "Failure criteria for long term accelerated power cycling test linked to electrical turn off SOA on IGBT module. A 4000 h test on 1200 A-3300 V module with AlSiC base plate," Microelectron. Rel., vol. 40, pp. 1665–1670, Aug./Oct. 2000.

[15] M. Tounsi, A. Oukaour, B. Tala-Ighil, H. Gualous, B. Boudart, and D. Aissani, "Characterization of high-voltage IGBT module degradations under PWM power cycling test at high ambient temperature," *Microelectron. Rel.*, vol. 50, pp. 1810–1814, Sep./Nov. 2010.

[16] W. Zhang, D. Xu, P. N. Enjeti, H. Li, J. T. Hawke, H. S. Krishnamoorthy, "Survey on Fault-Tolerant Techniques for Power Electronic Converters," *IEEE Trans on Power Electron.*, vol. 29, no. 12, pp. 6319-6331, 2014.

[17] S. Kwak, T. Kim and G. Park, "Phase-Redundant-Based Reliable Direct AC/AC Converter Drive for Series Hybrid Off-Highway Heavy Electric

Vehicles," *IEEE Trans on Veh. Technol.*, vol. 59, no. 6, pp. 2674 – 2688, Jul. 2010.

[18] A. L. Julian, and G. Oriti, "A Comparison of Redundant Inverter Topologies to Improve Voltage Source Inverter Reliability", *IEEE Trans on Ind. Appl.*, vol. 43, no. 5, pp. 1371 – 1378, Sept/Oct. 2007.

[19] G. Chen, J. Zhang, M. Zhu, N. Dai, X. Cai, "Adaptive thermal control for power fluctuation to improve lifetime of IGBTs in multi-MW medium voltage wind power converter" in *Proc. International Power Electronics Conference, (IPEC2014)*, Hiroshima, Japan, May 2014, pp. 1496–1500.

[20] T. Bruckner, S. Bernet and H. Guldner, "The Active NPC Converter and Its Loss-Balancing Control", *IEEE Trans on Ind. Electron.*, vol. 52, no. 3, pp. 855-868, June 2005.

[21] G. Riedel, N. Oikonomou, R. Schmidt und D. Cottet, "Active lifetime extension - Demonstrated for voltage source converters", in *Proc. IEEE 26th Convention of Electrical and Electronics Engineers in Israel (IEEEI)*, pp. 530-534, Nov. 2010.

[22] T. Phan, N. Oikonomou, G. Riedel, M. Pacas, "PWM for active thermal protection in three level neutral point clamped inverters," in *Proc. IEEE ECCE Asia*, pp. 906-911, June 2013.

[23] C. Nesgaard and M.A.E. Andersen, "Optimized load sharing control by means of thermal reliability management," in *Proc. Power Electronics Specialists Conference (PESC) 2004*, vol. 6, pp. 4901-4906, Aachen, Germany, June 2004.

[24] D. A. Murdock, J. E. R. Torres, J. J. Connors, and R. D. Lorenz, "Active Thermal Control of Power Electronic Modules," *IEEE Trans on Ind. Appl.*, vol. 42, no. 2, pp. 552 – 558, Mar/Apr. 2006.

[25] V. Blasko, R. Lukaszewski, and R. Sladky, "On line thermal model and thermal management strategy of a three phase voltage source inverter," in Proc. *Industry Applications Conference, 1999. Thirty-Fourth IAS Annual Meeting. Conference Record of the 1999 IEEE*, pp. 1423-1431, 1999.

[26] L. Wei, J. McGuire, R. A. Lukaszewski, "Analysis of PWM Frequency Control to Improve the Lifetime of PWM Inverter," *IEEE Trans on Ind. Appl.*, vol. 47, no. 2, pp. 922-929, Mar/Apr. 2011.

[27] A. M. Hava, R. J. Kerkman, and T. A. Lipo, "Simple Analytical And Graphical Methods For Carrier-Based PWM-VSI Drives," *IEEE Trans on Power Electron.*, vol. 14, no. 1, pp. 49-61, Jan 1999.

[28] A. M. Hava, R. J. Kerkman, and T. A. Lipo, "A High-Performance Generalized Discontinuous PWM Algorithm," *IEEE Trans on Ind. Appl.*, vol. 34, no. 5, pp. 1059-1071, Sep/Oct 1998.

[29] D. G. Holmes, A. T. Lipo, *Pulse Width Modulation for Power Converters: Principles and Practice*. Wiley-IEEE Press, Oct. 2003.

Current Controller Modeling for an Interleaved Boost with Voltage Multiplier Cells for PV Applications

Alessandro Pevere, Urmimala Chatterjee, Johan Driesen
Department of Electrical Engineering
KU Leuven – EnergyVille
Leuven, Belgium
email: alessandro.pevere@kuleuven.be

Abstract— **Nowadays, due to the increasing interest on low voltage renewable energy sources, like those based on photovoltaic modules and fuel-cells, an intensive academic and industrial research has been focusing on high step-up ratio DC-DC converters. One simple way to implement a high gain converter is to combine a boost topology at the input stage with a number of voltage multiplier cells, in order to step-up the output voltage beyond the conventional boost capability. Since in low-voltage (renewable energy) applications the input current is typically high, interleaved topologies are employed to reduce the device stress and the current ripple by separating power into multiple phases. Current sharing in paralleled converters can be ensured by adopting current-mode control techniques.**

The aim of this paper is to design a current controller deriving the duty-to-input current small signal model for the interleaved boost with one voltage multiplier cell, comparing it with the conventional interleaved boost. This result is further used to estimate the current function when the number of cells increases. The derived model has been validated by PLECS simulations and by experimental measurements based on a 500 W-interleaved boost prototype with two voltage multiplier cells, digitally controlled by DSP TMS320F28377.

I.   INTRODUCTION

High step-up converters are able to efficiently interface low-voltage high-current energy sources with the utility grid gaining interest for applications in almost all power electronic systems that include batteries as energy storage elements. In addition, they have been more and more frequently utilized for processing renewable energy sources, such as photovoltaic (PV) panels and fuel-cells. One typical application is the two-stage solution for photovoltaic micro-inverters, where the first stage has to boost the panel voltage (20−40V) to a higher voltage dc-link (360-400V). Among the non-isolated topologies investigated in the literature [1]-[10], the ones that employ single-phase [8]-[9] or multi-phase boost [6]-[7] converters cascaded with voltage multiplier cells have been considered. These converters are interesting for several reasons, i.e. the input current extracted from the power source is continuous, the switch voltage stress is reduced below the output voltage, and the topology is very simple. In particular, the interleaved boost with voltage multiplier cells, providing current sharing between two parallel phases is well suited for medium power applications. The design of a proper control loop for these converters requires the knowledge of their dynamic behavior. The small-signal model developed in [7] for the interleaved boost converter with voltage multiplier was used to derive the duty-to-output voltage transfer function considering only one multiplier cell.

Output of the PV module fluctuates due to the constantly changing solar irradiation. Therefore, a variable input range converter with an input voltage controller is required to utilize PV energy efficiently. Following design requirements are taken into account to make a module-level distributed converter such as wide input range, high step-up, compact and cost-effective converter. In our previous paper [13], design requirements of an intra-module DC-DC converter for PV application have been discussed and a converter has been built accordingly.

In this work, a non-isolated high gain DC-DC converter with voltage multiplier cells is developed as per the design requirements. The converter is with wide input voltage and current range and capable to boost high till 360V DC-link voltage that it can fed to AC. The aim is to use a single stage power conversion between low voltage outputs of PV module to DC-link voltage. The controller is designed for this converter and validated by experiments. In this paper, the duty-to-input current small signal model is derived for the interleaved boost with one voltage multiplier cell, comparing it with the conventional interleaved boost. The result is further used to estimate the input current transfer function when the number of cells is doubled. The model has been validated by PLECS-based simulations and a prototype based on a 500 W-interleaved boost with two voltage multiplier cells has been developed to prove experimental verification.

The paper is organized as follows: a review of the interleaved boost converter is reported in Section II as well as its control and current function small signal. In Section III, a suitable boost topology for a PV-module converter, i.e. interleaved boost with voltage multiplier cells, is presented and the main operation of the converter is described in Section IV. The small signal model for this boost topology is discussed in detail in Section V, including most of the analytical steps. Some simulation results are finally presented in Section VI in order to validate the model. Finally, in Section VII a prototype and experimental results are shown. Conclusions and future work are reserved for section VIII.

978-1-4673-9551-9/16 $31.00 © 2016 IEEE

II. INTERLEAVEED BOOST CONVERTER

Interleaved multiphase converters are usually employed because of a certain number of beneficial features, such as the reduction of the power device stress by splitting the overall power into multiple phases, reduction of the input filter size by increasing the equivalent frequency and reduction/cancellation of the current ripple effect. This last aspect could also lead to other beneficial technology changes, such as the replacement of aluminum electrolytic capacitors with film or ceramic capacitors, therefore providing improvement of the equivalent series resistance, power density and better dynamic response. In Fig. 1, a standard interleaved boost topology has been reported. Basically, the converter is composed by two boost converters in parallel, each composed by an inductor, a switch and a diode. The power switches S_1 and S_2 are driven with 0° and 180° phase-shift gate signals respectively, in order to take advantage of the input current ripple cancellation.

A. Control loop

Current sharing in paralleled converters can be ensured by adopting current-mode control techniques [1]. Peak current mode control mode control for interleaved boost converter has been reported in [2]. Average current-mode control is well established and it has advantages of achieving higher bandwidths when compared to voltage mode control [3]. Average current control is free from instability problems unlike peak current mode control which requires slope compensation to make it stable at duty ratio greater than 50%. Steady state gain and noise immunity are also high in average current control.

Normally, in an interleaved boost converter the average current on each phase is controlled by using independent current loops, whereas output or input voltage (depending on type of application) is regulated with an external voltage loop as shown in Fig. 2. Current reference I^* to the inner current loops is provided by the outer voltage controller [10].

The steps involved in mathematical modeling of average current-mode controlled interleaved boost converter are explained in [3]. The state space averaging technique has been used in order to obtain the small signal model of the interleaved

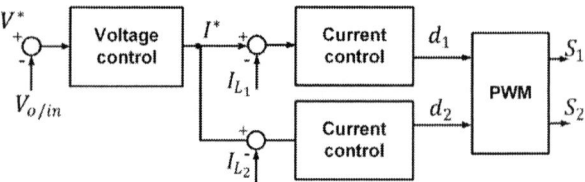

Fig. 2. Typical control scheme for an interleaved boost converter.

boost converter. In this case, the model is derived under the assumption that all the converter elements are ideal. However for the proposed topology with voltage multiplier cells, also leakage elements will be taken into account in Section IV.

B. Small Signal Model

By introducing sinusoidal disturbance in duty $\hat{d}$, each phase duty-to-inductor current transfer function $G_{id}(s)$ is reported as [3]:

$$G_{id}(s) = \frac{\hat{\imath}_{L_1}(s)}{\hat{d}(s)} = \frac{V_o}{R \cdot D'^2} \frac{\left(1 + \frac{R \cdot C_o}{2} s\right)}{\left(1 + \frac{L_{eq}}{D'^2 \cdot R} s + \frac{L_{eq} \cdot C}{D'^2} s^2\right)} \quad (1)$$

where $L_{eq} = L_1 \| L_2 = \frac{L}{2}$ is the equivalent inductance and $D' = 1 - D$.

The obtained transfer function (1) is similar to duty-to-inductor current transfer function of the conventional boost converter [3]. It can be observed that the equivalent inductance L_{eq} is reduced to half, compared with the case of the conventional boost where it is simply L. The reduced effective inductance results in occurrence of right half plane zero at higher frequencies compared to conventional boost converter. This ensures that higher bandwidths can be achieved in an interleaved boost converter compared to its counterpart, resulting in a faster controller.

III. INTERLEAVED BOOST WITH VOLTAGE MULTIPLIER CELLS

Non-isolated classical boost can provide high step-up voltage gain, but with the penalty of high voltage and current stress, high duty-cycle operation and limited dynamic response. The diode reverse recovery current can reduce the efficiency when operating with high current and voltage levels. There are some non-isolated dc–dc converters operating with high static gain, as the quadratic boost converter, but additional inductors and filter capacitors must be used and the switch voltage is high [5]. An alternative for the implementation of high step-up structures is with the use of the voltage multiplier cells (based on capacitors and diodes) integrated with classical non-isolated dc–dc converters [6]-[9].

The voltage multiplier cells (VMC) also can be integrated with interleaved converters, resulting in a modular structure well suited for high output voltage and high input current

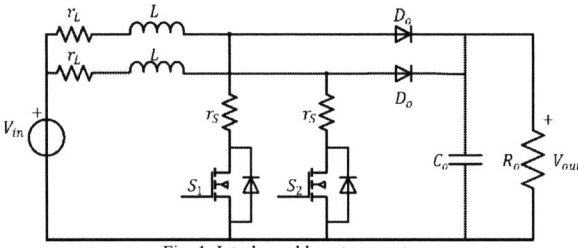

Fig. 1. Interleaved boost converter.

978-1-4673-9551-9/16 $31.00 © 2016 IEEE   1184

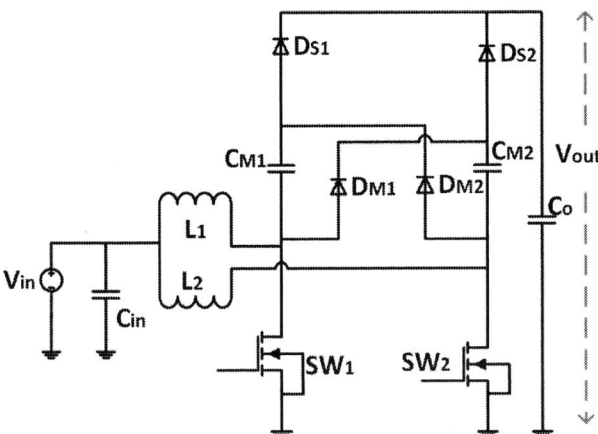

Fig. 3. Interleaved boost converter with one voltage multiplier stage.

applications. Depending on the number of VMCs, the voltage ratio M between input and output voltages can be written by the following equation:

$$M = \frac{V_o}{V_{in}} = \frac{(N+1)}{D'} \qquad (2)$$

where N indicates the number of voltage multiplier cells.

In Fig. 3 the circuit scheme of an interleaved boost converter with only one voltage multiplier cell has been reported. The main parasitic elements, i.e. the DCR resistance r_L of the inductors, the series resistance r_C of the capacitors and of the ON resistance r_S of the power switches are not highlighted in figure but they will be considered in the model. When the converter operates in continuous conduction mode (CCM) with

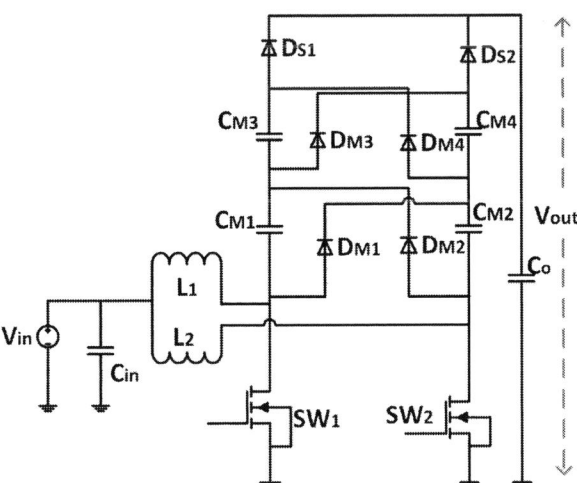

Fig. 4. Interleaved boost converter with two voltage multiplier stages.

duty-cycle above 0,5, the ideal (lossless) converter can be shown to have high voltage gain with reduced switch voltage stress [7]. In fact, the voltage ratio becomes $M = \frac{2}{D'}$ and the voltage stress seen by the transistors reduces to $V_{S_1} = V_{S_2} = \frac{V_o}{2}$.

When the number of VMCs is further increased to two, the converter topology can be modified as the circuit scheme proposed in Fig. 4. This is not the only way to realize an interleaved boost with double cells [6], but this configuration is the one with less components. In this case the voltage gain is three-times the one for a conventional boost ($M = \frac{3}{D'}$) and the voltage stress in the switches is one third ($V_{S_1} = V_{S_2} = \frac{V_o}{3}$).

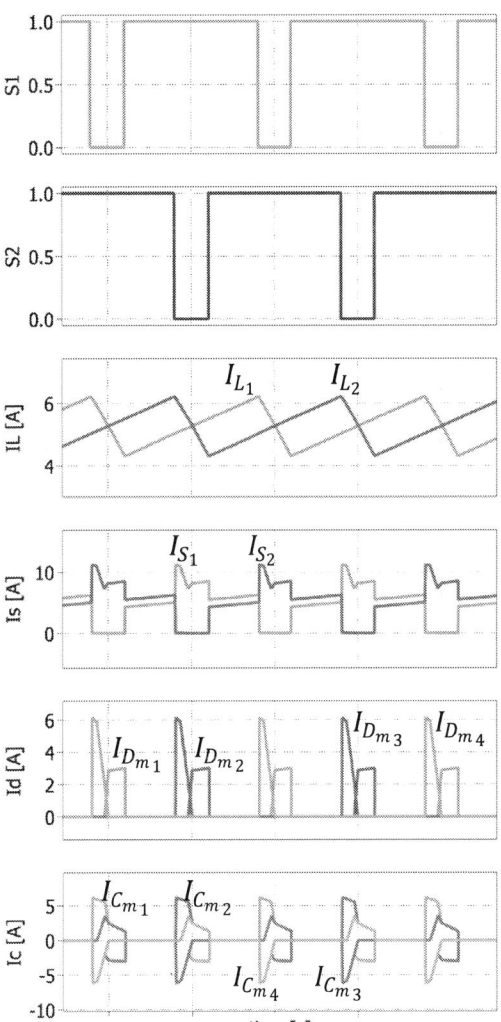

Fig. 5. Converter main waveforms.

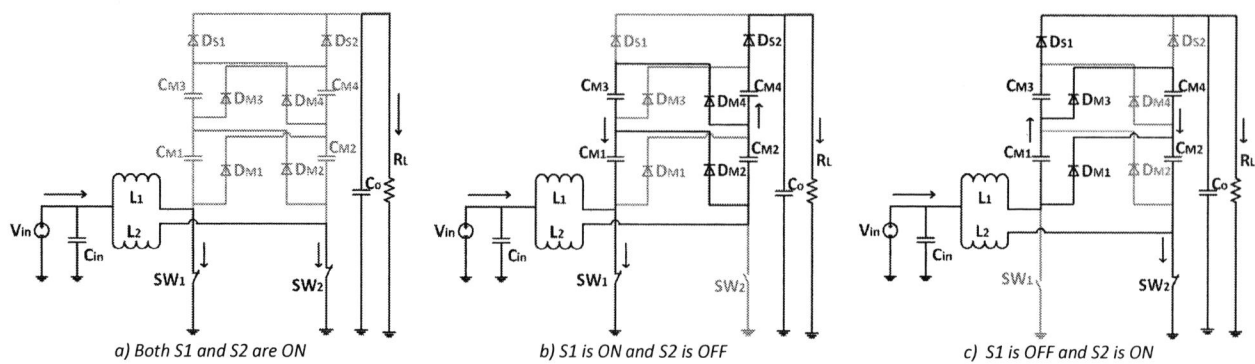

Fig. 6. Converter sub-circuits during a switching period.

IV. CONVERTER OPERATION

In this section, a brief description of the converter operation during a switching period is given. The main converter waveforms for a duty-cycle $D > 0,5$ are shown in Fig. 5, i.e. control signals of S_{W_1} and S_{W_2}, interleaved inductors currents, currents flowing through the switches, multiplier cell diodes and capacitors currents.

When both the switches are closed (as shown in Fig. 6a), the inductors are charged with the input voltage. D_{M_1}, D_{M_2}, D_{M_3}, D_{M_4}, D_{S_1} and D_{S_2} are all OFF and consequently no current is flowing through the voltage multiplier stage. In this interval, the load is supplied by the output capacitor.

When S_{W_1} is closed and S_{W_2} is open (as presented in Fig. 6b), D_{M_2}, D_{M_4} and D_{S_2} are conducting. The energy stored in the inductor L_2 is transferred to the load thorough D_{S_2} and also to C_{M_2}, C_{M_4}, D_{M_2} and D_{M_4}. It can be observed in Fig. 6b that the multiplier capacitors C_{M_2} and C_{M_3} are connected in parallel and in series with C_{M_1} and C_{M_4}. Thus, the output voltage will be three times the voltage across the single multiplier capacitor, as already anticipated in the previous section.

When S_{W_1} is open and S_{W_2} is closed (as shown in Fig. 6c), the same behavior but in an opposite way of the previous sub-period happens. The energy stored in the inductor L_1 is transferred to the load thorough D_{S_1} and also to C_{M_1}, C_{M_3}, D_{M_1} and D_{M_3}. In this case, C_{M_1} and C_{M_4} are connected in parallel and in series with C_{M_2} and C_{M_3}.

V. SMALL SIGNAL MODEL FOR AN INTERLEAVED BOOST WITH VMCS

The design of a proper control loop for these converters requires the knowledge of their dynamic behavior. In [7] a small signal model and the duty-to-output voltage transfer function have been derived for the case of $N = 1$. Hereafter, the duty-to-input current transfer function will be calculated, discussing the relationship with the conventional boost one and proposing an equivalent reduced model. Moreover, a simplified approach to predict the transfer function in the case of $N = 2$ will be discussed too.

Referring to the symmetric system derived in [7] for the interleaved boost with $N = 1$, the following system equation has to be considered for small signal model:

$$\frac{d\hat{x}}{dt} = \bar{\bar{A}}_s\hat{x} + \bar{\bar{B}}_s\hat{u} + \bar{\bar{F}}_s\hat{d} \qquad (3)$$

where $\hat{x} = [i_L \; u_C \; u_o]^T$ is the perturbed system vector, $\hat{u} = [u_{in} \; i_o]^T$ is the perturbed input and $\hat{d}$ is the perturbed duty cycle.
The matrices $\bar{\bar{A}}_s$, $\bar{\bar{B}}_s$ and $\bar{\bar{F}}_s$ can be written as:

$$\bar{\bar{A}}_s = \begin{bmatrix} -\dfrac{r_e}{L} & 0 & -\dfrac{D'}{2L} \\ 0 & -\dfrac{2}{C \cdot r_{Ce}} & \dfrac{1}{C \cdot r_{Ce}} \\ \dfrac{D'}{C_o} & \dfrac{2}{C_o \cdot r_{Ce}} & \dfrac{1}{R_{oe} \cdot C_o} \end{bmatrix} \qquad (4)$$

$$\bar{\bar{B}}_s = \begin{bmatrix} \dfrac{1}{L} & 0 \\ 0 & 0 \\ 0 & -\dfrac{1}{C_o} \end{bmatrix} \qquad \bar{\bar{F}}_s = \begin{bmatrix} \left(\dfrac{r_{Ce}}{R_o}+1\right)\dfrac{U_o}{2L} \\ 0 \\ -\dfrac{U_o}{D' \cdot C_o \cdot R_o} \end{bmatrix}$$

where $r_e = r_L + r_S + \dfrac{r_C}{2} \cdot D$, $r_{Ce} = \dfrac{r_C}{D'}$ and $R_{oe} = \dfrac{R_o \cdot r_{Ce}}{R_o + r_{Ce}}$.
A closed form expression of the control-to-output voltage transfer function $G_{id}(s)$ is now derived from system in (3).

$$G_{id}(s) = \frac{\hat{\imath}_{L_1}(s)}{\hat{d}(s)}$$
$$= \frac{[(s - a_{22})(s - a_{33}) - a_{23}a_{32}]f_1 + a_{13}(s - a_{22})f_3}{(s - a_{11})(s - a_{22})(s - a_{33}) - a_{31}a_{13}(s - a_{22}) - a_{32}a_{23}(s - a_{11})}$$

$$(5)$$

where a_{ij} and f_i are the elements of matrices $\bar{\bar{A}}_s$ and $\bar{\bar{F}}_s$. Now, substituting the real elements of (4) in (5) and re-arranging, $G_{id}(s)$ can be rewritten with the following expression;

$$G_{id}(s) = G \frac{(1 + z_1 \cdot s + z_2 \cdot s^2)}{(1 + p_1 \cdot s + p_2 \cdot s^2 + p_3 \cdot s^3)} \quad (6)$$

where:

$$G = \frac{U_o}{R_o \cdot D'^2} \frac{\left(2 + \frac{r_{Ce}}{R_o}\right)}{\left(1 + \frac{2r_e}{R_o \cdot D'^2}\right)}$$

$$z_1 = \frac{R_o \cdot C_o}{2} \frac{\left[\left(1 + \frac{1}{R_o}\right)\left(1 + \frac{C \cdot r_{Ce}}{2C_o \cdot R_{oe}}\right) + \frac{C \cdot r_{Ce}}{2C_o \cdot R_o}\right]}{\left(1 + \frac{r_{Ce}}{2R_o}\right)}$$

(7)

$$p_1 = \frac{L}{2R_o \cdot D'^2} \frac{\left[4 + 2\frac{R_o \cdot r_e}{L}\left(\frac{C}{R_{oe}} + 2C_o\right) + \frac{C \cdot R_o \cdot r_{Ce} \cdot D'^2}{L}\right]}{\left(1 + \frac{r_{Ce}}{2R_o \cdot D'^2}\right)}$$

$$p_2 = \frac{L \cdot C_o}{2D'^2} \frac{\left[4 + 2C \cdot r_{Ce}\left(\frac{r_e}{L} + \frac{1}{C_o \cdot R_{oe}}\right)\right]}{\left(1 + \frac{r_{Ce}}{2R_o \cdot D'^2}\right)}$$

Due to space limitations, the expressions of z_2 and p_3 have not been reported here. Since their values are close to zero, the order of the numerator and denominator polynomials in (6) can be thus reduced of one order each, leading to the approximated expression:

$$G_{id}(s) \approx G \cdot \frac{(1 + z_1 \cdot s)}{(1 + p_1 \cdot s + p_2 \cdot s^2)} \quad (8)$$

Moreover, if the parasitic elements are also neglected in (6), expression (7) becomes:

$$G_{id}(s) \approx 2 \frac{U_o}{R_o \cdot D'^2} \frac{\left(1 + \frac{R_o \cdot C_o}{2} \cdot s\right)}{\left(1 + 4\frac{L}{2R_o \cdot D'^2} \cdot s + 4\frac{L}{2R_o \cdot D'^2} \cdot s^2\right)} \quad (9)$$

Then, looking at the approximated duty-to-input current transfer function (9) for an interleaved boost with one VMC ($N = 1$) and comparing it with the case of the conventional interleaved boost of (1), some important conclusions can be done, i.e. the gain has a multiplying factor 2, the coefficient z_1 is not affected by the VMC and p_1 and p_2 are 4 times higher. As a consequence, the duty-to-input current transfer function for an interleaved boost with N VMCs can be thought to be approximated by the following general form:

$$G_{id}(s)$$
$$\approx k_G \frac{U_o}{R_o \cdot D'^2} \frac{\left(1 + \frac{R_o \cdot C_o}{2} \cdot s\right)}{\left(1 + 2k_D \frac{L}{2R_o \cdot D'^2} \cdot s + 2k_D \frac{L}{2R_o \cdot D'^2} \cdot s^2\right)} \quad (10)$$

where k is a coefficient depending of the number of VMCs. For $N = 1$ it has been analytically proofed that $k = 1$, whereas for $N = 2$ the coefficient k will be derived through simulations in the next section.

VI. SIMULATION ANALYSIS

The model derived in Section V has been validated by extensive simulations using MATLAB™ and the standalone version of PLECS®. The main parameters of the interleaved boost converter are reported in Appendix.

In Fig. 7, the Bode plot (magnitude and phase) of the duty-to-input current transfer function G_{id} for $N = 1$ is shown, obtained from full-order of equation (6), from reduced order of equation (8), neglecting parasitic elements with equation (9) and from simulations. The simulation points of Fig. 4 has been calculated using the small signal Analysis Tool of PLECS. One can notice that all analytical models well predict the simulated frequency response. As predictable, when the parasitic elements are taken into account in the model (equations (6)(9) and simulations), the peak in correspondence of the double pole is dumped.

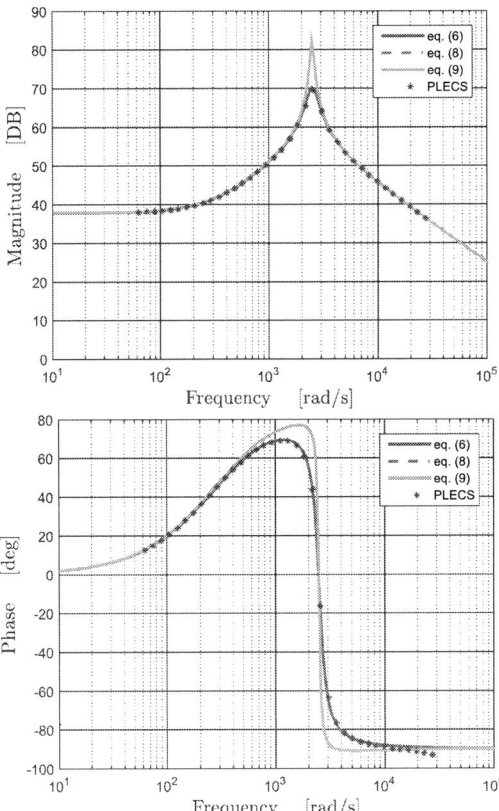

Fig. 7. Interleaved boost converter with one voltage multiplier stage.

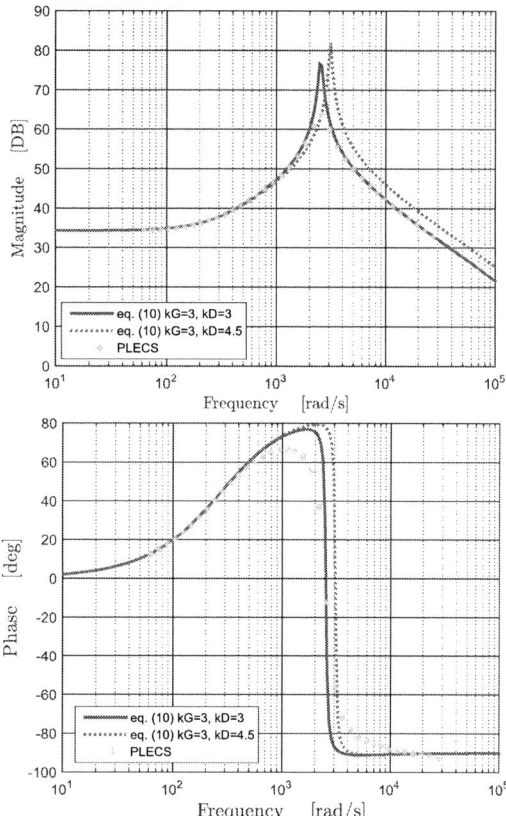

Fig. 8. Interleaved boost converter with 2 voltage multiplier stages.

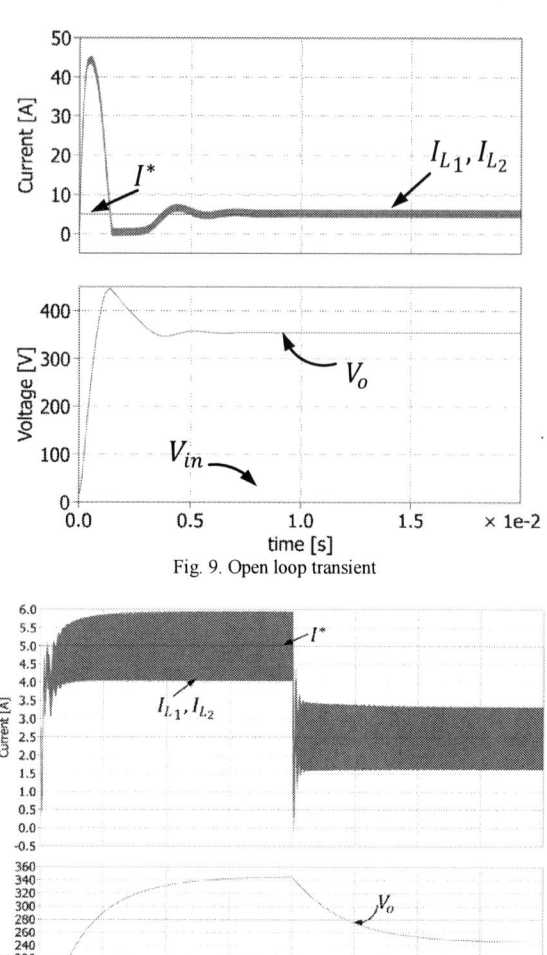

Fig. 9. Open loop transient

Fig. 10. Input current step response for $N = 2$ with current control.

In Fig. 8, the Bode plot (magnitude and phase) of the duty-to-input current transfer function G_{id} in the case of a $N = 2$ is shown, using prediction of equation (10) and PLECS simulations. One can notice that $k_G = 3$ is the right choice for the gain coefficient. If the same value is then used for k_D, the actual transfer function is not well approximated. In this case, the best choice for the pole coefficient is $k_D = 4.5$, as clearly highlighted in Fig. 8. How to analytically estimate these values is not trivial and further investigations are needed, being therefore the topic of a future paper.

The model derived in Section V has been used to calculate the PI controller parameters for the converter with 2 VMCs given a certain bandwidth. Then, a complete model of the converter and its digital control have been implemented in PLECS simulation tool.

A duty-cycle step from 0 to 0,8 is presented in Fig. 9. As expected, during an open loop start-up of the converter, a relatively high current peak occurs because the current is not controlled.

Finally, in Fig. 10 an input current reference step response is shown while the designed average current controller is active for both the converter phases. One can notice that the inductor currents are fast and well controlled.

VII. PROTOTYPE AND EXPERIMENTAL RESULTS

A 500W-intereaved boost prototype with two voltage multiplier cells has been built (as shown in Fig. 9). The power converter board is controlled by an embedded sensing and control PCB, based on real-time microcontroller TMS320F28337, used for the experimental verification of the model.

Fig. 11. Interleaved boost converter.

In the experiment of Fig. 12, input current (blue line) and output voltage (purple line) during an open loop transient from duty-cycle 0 to 0,75 are reported. In this operating transient the input voltage has been fixed to 40 V and the peak of the current can go up to 28 A.

In Fig. 13, the inductor currents and their gate signals are reported during steady-state conditions, showing the interleaved operation of the two converter phases at $100kHz$.

The experimental results of Fig. 14 and Fig. 15 present a current reference step response for two different control parameter designs. The input voltage is $40V$, the output resistance is 280Ω and the current reference for each phase is $5,5A$.

In Fig. 14, the PI parameters used for the current controller are $K_P = 20 \cdot 10^{-3}$ and $K_I = 100$. As shown by the figure the current peak has been reduced and the reference is reached after around $4ms$.

In Fig. 15, the integral part is increased to $K_I = 250$ while the proportional coefficient is kept constant. One can notice that the input current takes less time to reach the reference in the cost of more transient oscillations and higher initial peak.

VIII.   CONCLUSIONS AND FUTURE WORK

In this paper, the current controller modelling for an interleaved boost converter with voltage multiplier cells has been investigated. The small signal model of the duty-to-input current for the converter with one multiplier cell has been analytically derived. Based on this model, a prediction for a boost converter with two voltage multiplier cells has been proposed and verified with Matlab and PLECS-based simulations. A prototype digitally controlled by DSP TMS320F28377 has been built in order to compare the

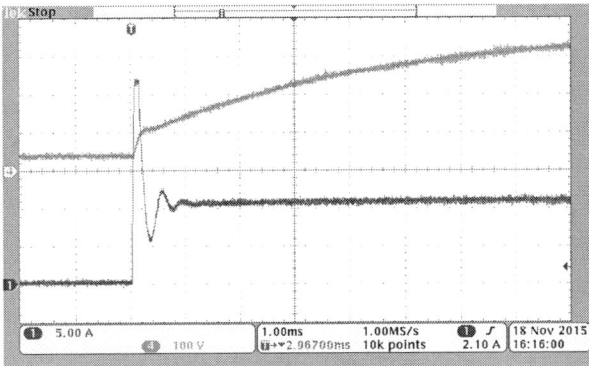

Fig. 12. Output voltage and input current during open loop transient.

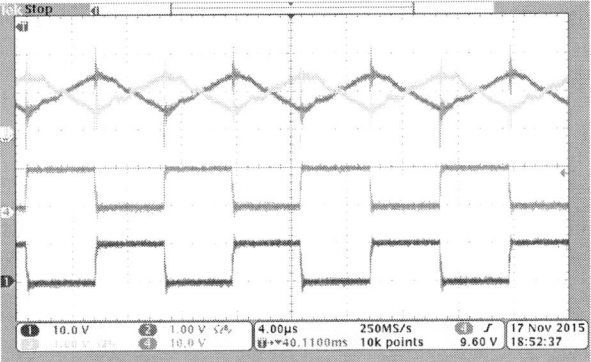

Fig. 13. Inductor currents and gate signals at steady-state condition.

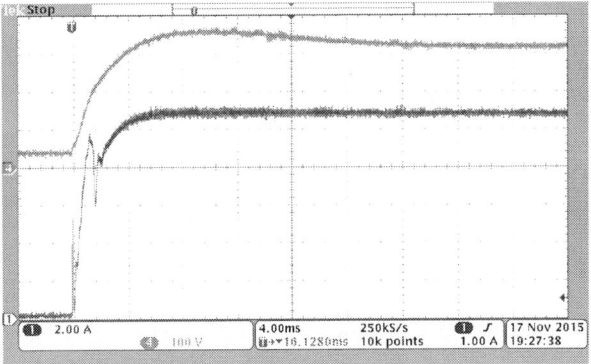

Fig. 14. Output voltage and input current during a closed-loop transient.

experimental results against the analytical and simulation models.

As a future work, a detailed procedure to estimate the control parameters will be investigated as well as a full analytical model for the converter with two voltage multiplier cells will be derived.

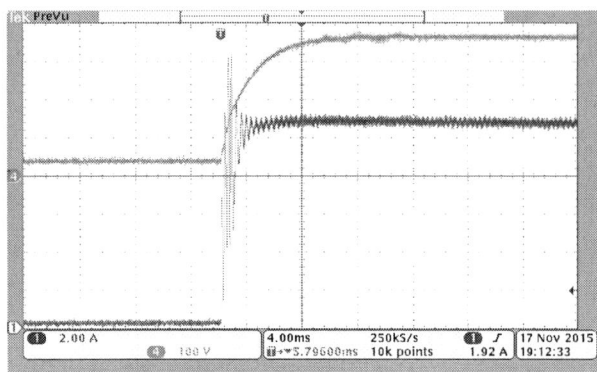

Fig. 15. Output voltage and input current during a closed-loop transient.

ACKNOWLEDGMENT

This work is supported by the 'SMART PV' project under IWT-SBO grant. This work also acknowledges the help of team in ESAT-ELECTA in KU Leuven.

APPENDIX

SIMULATION PARAMETERS OF THE INTERLEAVED BOOST CONVERTER.

Parameter	Symbol	Value
Input voltage	V_{in}	24 [V]
Output voltage	V_{out}	360 [V]
Output capacitance	C_o	14 [μF]
ESR of C_o	r_C	100 $m\Omega$
Phase inductance	$L_1 = L_2$	100 [μH]
DCR of L_1 @100kHz	r_L	25 [$m\Omega$]
On resistance switch	r_S	11 [$m\Omega$]
Switching frequency	f_{sw}	100 [kHz]

REFERENCES

[1] Garcera, G.; Pascual, M.; Figueres, E., "Robust average current-mode control of multimodule parallel DC-DC PWM converter systems with improved dynamic response," Industrial Electronics, IEEE Transactions on , vol.48, no.5, pp.995,1005, Oct 2001.

[2] Sanchis, E.; Maset, E.; Ferreres, A.; Ejea, J.B.; Esteve, V.; Jordan, J.; Garrigos, A.; Blanes, J.M., "High-Power Battery Discharge Regulator for Space Applications," Industrial Electronics, IEEE Transactions on , vol.57, no.12, pp.3935,3943, Dec. 2010.

[3] Fernandez, A.; Tonicello, F.; Aroca, J.; Mourra, O., "Battery discharge regulator for space applications based on the boost converter," Applied Power Electronics Conference and Exposition (APEC), 2010 Twenty-Fifth Annual IEEE , vol., no., pp.1792,1799, 21-25 Feb. 2010.

[4] Kolluri, S.; Narasamma, N.L., "Analysis, modeling, design and implementation of average current mode control for interleaved boost converter," Power Electronics and Drive Systems (PEDS), 2013 IEEE 10th International Conference on , vol., no., pp.280,285, 22-25 April 2013.

[5] Cabral, J.B.R.F.; Lemes da Silva, T.; Vidal Garcia Oliveira, S.; De Novaes, Y.R., "A new high gain non-isolated DC-DC boost converter for photovoltaic application," Power Electronics Conference (COBEP), 2013 Brazilian , vol., no., pp.569,574, 27-31 Oct. 2013.

[6] Prudente, M.; Pfitscher, L.L.; Emmendoerfer, G.; Romaneli, E.F.; Gules, R., "Voltage Multiplier Cells Applied to Non-Isolated DC–DC Converters," Power Electronics, IEEE Transactions on , vol.23, no.2, pp.871,887, March 2008.

[7] Spiazzi, G.; Buso, S.; Sichirollo, F.; Corradini, L., "Small-signal modeling of the interleaved boost with voltage multiplier," Energy Conversion Congress and Exposition (ECCE), 2012 IEEE , vol., no., pp.431,437, 15-20 Sept. 2012.

[8] Dupont, F.H.; Rech, C.; Gules, R.; Pinheiro, J.R., "Analysis and design of a control approach for a boost converter with voltage multiplier cell," Power Electronics Conference (COBEP), 2011 Brazilian , vol., no., pp.444,450, 11-15 Sept. 2011.

[9] Dupont, F.H.; Rech, C.; Gules, R.; Pinheiro, J.R., "Reduced-Order Model and Control Approach for the Boost Converter With a Voltage Multiplier Cell," Power Electronics, IEEE Transactions on , vol.28, no.7, pp.3395,3404, July 2013.

[10] Mira, M.C.; Knott, A.; Thomsen, O.C.; Andersen, M.A.E., "Boost converter with combined control loop for a stand-alone photovoltaic battery charge system," Control and Modeling for Power Electronics (COMPEL), 2013 IEEE 14th Workshop on , vol., no., pp.1,8, 23-26 June 2013.

[11] Prasanna, U.R.; Rathore, A.K., "Small signal analysis and control design of current-fed full-bridge isolated dc/dc converter with active-clamp," in Industrial Electronics (ISIE), 2012 IEEE International Symposium on , vol., no., pp.509-514, 28-31 May 2012.

[12] Ye ZhiHao; Wu Xiaobo, "Compensation Loop Design of a Photovoltaic System Based on Constant Voltage MPPT," in Power and Energy Engineering Conference, 2009. APPEEC 2009. Asia-Pacific , vol., no., pp.1-4, 27-31 March 2009.

[13] Chatterjee, U.; Driesen, J., "Intra-Module DC-DC converter: topology selection and analysis," in 28th European PV Solar Energy Conference and Exhibition (EU PVSEC), pp. 3419-3423, October 2013.

New Active Capacitor Voltage Balancing Method for Five-Level Stacked Multicell Converter

Arash Khoshkbar Sadigh

Extron Electronics

Anaheim, USA

Vahid Dargahi and Keith Corzine

Microgrid and Power Electronics Laboratory

Holcombe Department of Electrical and Computer Engineering

Clemson University, Clemson, USA

Abstract— This paper proposes a new active capacitor voltage balancing method for five-level stacked multicell converter which is fully implemented using the logic-form equations. The proposed active voltage balancing method measures load current and flying capacitor voltages in order to generate the appropriate switching states to produce the required output voltage level as well as to balance the flying capacitor voltages at their reference values. Output voltage of the stacked multicell converter controlled with proposed active voltage balancing method can be modulated with any pulse-width-modulation techniques such as phase-shifted-carrier or level-shifted-carrier pulse-width-modulation. The advantage of the proposed active voltage balancing method in comparison with other control techniques is its simplicity without requiring complex computations. Simulation results of the five-level SM converters are presented to verify the performance of proposed active voltage balancing method.

Keywords— Active voltage balancing method; multilevel converter; level-shifted PWM; phase-shifted PWM; stacked multicell converter.

I. INTRODUCTION

The well-established, enabling, and promising breeds of power electronics converters enabling power conversion for industrial and electrical applications are: two-level voltage source converter (2L-VSC), multilevel VSC (ML-VSC), and the pulse width modulated current source converter (PWM-CSC) [1]–[3]. Nowadays, ML-VSCs are promising, viable, and high-efficient converter topologies for medium-voltage high-power applications [4]–[11].

Among the multilevel topologies, the flying capacitor multicell (FCM) converter and its sub-topology referred as stacked multicell (SM) converter have been the focus of the considerable industrial and academic research attention due to their substantial advantages for medium-voltage market such as the natural voltage balancing and ease of switching-power-cell extension for achieving high number of voltage levels [12]–[16]. These topologies are particularly attractive because of the inherent natural flying capacitor (FC) voltage balancing process. In order to generate the regularly stepped levels of chopped input-voltage at the output side of FCM or SM converters and also to guarantee safe operation of the converters, it is mandatory for all of the FCs to get stabilized at their target voltage values [17].

The voltage balancing of FCs which ensures the safe operation of the FC-based multicell converters is a concern of paramount importance in FC-based topologies. Generally there are two major switching PWM techniques to maintain

the voltage balancing for such converters [18]. The first approach relies on the natural balancing property and open-loop control of the FCM and SM converters by applying an appropriate and feasible switching strategy. Last analysis of natural balancing in frequency domain for FCM [19], [20], and SM [21]–[23] converters proved that natural capacitor voltage balancing mechanism depends on harmonic content of the sinusoidal reference waveform, switching frequency, overlap of harmonic groups of switching functions, and also load impedance. Determination of the converter poles and zeros with an intention of probing the FC-based converter's performance using the powerful strategies such as root locus approach is accomplished in the frequency domain in [24], [25].

Since the natural voltage balancing of flying-capacitor-based converters depends on the current ripple of the output current, balancing mechanism might be weak for practical use in some applications where load impedance at switching frequency and above is high, and hence, total current ripple is consequently low [26]. For such applications, notably ac motor drive fields, manipulating converter's load impedance through adding multiple balance booster circuits resonating at the multiples of the switching frequency speeds up natural balancing mechanism at the cost of the converter's increased cost and power loss [27].

To resolve this issue and to make FC-based multilevel converters much more applicable, active voltage balancing strategy is utilized to regulate FC voltages at their desired values. This strategy utilizes voltage and/or current sensors, and uses these measurements to balance voltage across FCs [28]–[31]. In literature, several active FC voltage balancing methods are presented to control FCM and SM converters [32]–[34]. In [28], the proposed active voltage balancing control measures the voltage of each FC and compares with the reference voltage value. The error signal passes through proportional-integrator (PI) controller for each FC to modify the modulation index of each associated FC to balance it. In [29], a method for FCM converter based active rectifier is presented which provides a unity power factor (PF) for input current and regulates main dc output voltage and balances the FC voltages. In [33], an active voltage balancing method based on space vector modulation (SVM) is proposed for three-cell four-level FCM converter which is applicable only for three-phase applications. The methods presented in [35], [36] is based on the optimization of the cost function to determine the switching states which requires a complex and time-consuming computations. Besides, methods presented in [33], [34] are joint-phase techniques which are applicable only in three-phase applications. In [37], a method utilizing the redundant switching states and optimal level-transitions is proposed to balance FCs voltages and control an SMC fed induction machine. Digital sliding-mode observer for active control of SM converter is studied in [38]. In [39], an active voltage balancing method based on the minimizing a cost function is proposed to regulate the FC voltages in an FCM converter. Another optimization-based approach for active voltage balancing control of SM converter is presented in [40] which modifies the FC voltage deviation cost functions in order to have a reduced switching losses.

So far, most of the active voltage balancing techniques for FCM and SM converters are based on the optimization of FCs' voltage deviation functions as cost functions which requires complex computations for generating gate signals. Other active voltage balancing techniques are based on the SVM method which is only applicable to three-phase cases. This paper proposes an active capacitor voltage balancing method for five-level SM converter which is based on the logic equations. The proposed method does not require any complex computations since it is not based on optimization of any cost functions. The proposed active capacitor voltage balancing method is implemented using only logic-form equation which is faster than the cost function optimization method. It is worth mentioning that proposed active voltage balancing method in this paper can be implemented using any kind of pulse width modulation techniques like phase-shifted carrier (PSC) or level-shifted-carrier (LSC). Thus, the proposed method is adaptable for each of the individual phases, and consequently, is applicable in both of the single-phase and three-phase applications.

II. PROPOSED LOGIC-FORM-EQUATION BASED ACTIVE VOLTAGE BALANCING METHOD

In this paper, a novel active voltage balancing method which is based on the logic-form equations is proposed to control the FCs voltages of five-level SM converter at their reference value as well as to generate the required voltage levels at converter output. The advantages of the proposed method are that it does not need computation to minimize any complex cost functions and it is easy to implement since it is based on the logic-form equations.

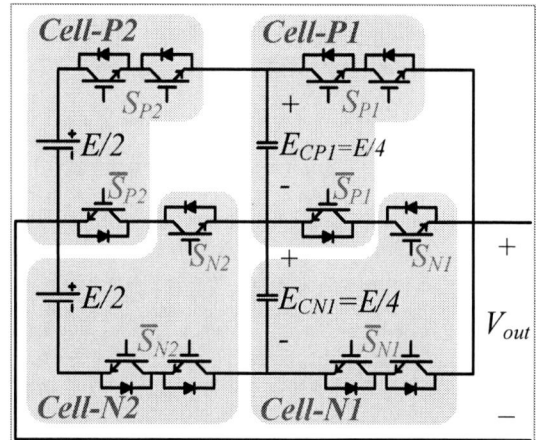

Figure 1. General topology of a four-cell five-level SM converter.

To elaborate proposed active voltage balancing method which is fully implemented utilizing logic-form equations, required parameters are defined as eq. (1) to (4) accordingly to facilitate the determination of switching states for power switches.

$$I = \begin{cases} 1 & i_{out}(t) \geq 0 \\ 0 & i_{out}(t) < 0 \end{cases} \tag{1}$$

$$\Delta V_{Cj} = \frac{E_{Cj}}{E_{Cj}\big|_{Reference}} - 1 \tag{2}$$

$$\Delta_{Cj} = \begin{cases} 1 & \Delta V_{Cj} \geq 0 \\ 0 & \Delta V_{Cj} < 0 \end{cases} \tag{3}$$

$$Y_{i,j} = \begin{cases} 1 & |\Delta V_{Ci}| \geq |\Delta V_{Cj}| \\ 0 & |\Delta V_{Ci}| < |\Delta V_{Cj}| \end{cases} \tag{4}$$

where variable I determines direction of load current, and E_{Cj} and $E_{Cj}\big|_{Reference}$ are measured and reference voltages of capacitor C_j (j is $P1$ and $N1$) at each sampling moment, respectively.

A four-cell five-level SM converter is shown in Figure 1 and its state of power switches is illustrated in Table 1. As listed in Table 1, the voltage variation of each FC, i.e., $\Delta E_{Cx} = E_{Cx}(t + \Delta t) - E_{Cx}(t)$, is determined for each switching state by taking into account the direction of the output current. As it is illustrated, all levels except $+E/2$ and $-E/2$ have redundancies which can be used to charge or discharge FCs to balance their voltages at their reference values. Therefore, the states of switches should be selected properly according to the direction of output current and voltage of FCs in order to generate the desired output voltage level as well as to balance FCs voltages. As shown in Figure 2, the sinusoidal reference waveform (V_Ref) is intersected with the triangle carriers to generate the staircase reference waveform of output. It is worth mentioning that the triangle carriers can be arranged based on any type of PWM methods like as PSC or LSC ones. The LSC-PWM switching strategy includes the phase-disposition (PD), the phase-opposition-disposition (POD), and alternate-phase-opposition-disposition (APOD). In Figure 2, the PD LSC-PWM method is utilized.

According to Table 1, generating switching states for voltage levels of $+E/2$ and $-E/2$ is straightforward since there is only one possible state for switches. Therefore, $S_{P1}, S_{P2}, \bar{S}_{N1}, \bar{S}_{N1}$ are 1 For level $+E/2$, no matter how much is the voltage of C_{P1} and C_{N1}. This is why the first term of eq. (5) and (7) is L_{+2}. However, generating states of switches for levels $+E/4$ and $+E/4$ needs more considerations. In order to generate switching states for level $+E/4$, the FC voltage of cell #P1 (E_{CP1}) is measured and compared with its reference voltage which is a quarter of dc link voltage, i.e., $E_{CP1,Ref} = 0.25E$. If FC voltage of cell #P1 (E_{CP1}) is higher than its reference voltage, it needs to be discharged. Therefore, switches states of S_{P1}, S_{P2} is selected as $(0,1)$ and $(1,0)$ for the positive and negative output current, respectively, to generate level $+E/4$ and discharge FC of cell #P1 (C_{P1}). On the other hand, if FC voltage of cell #P1 (E_{CP1}) is lower than its reference voltage, it needs to be charged. So, switches states of S_{P1}, S_{P2} is selected as $(1,0)$ and $(0,1)$ for positive and negative output current, respectively, to generate level $+E/4$ and charge the FC of cell #P1 (C_{P1}). It should be mentioned that switching states of $\bar{S}_{N1}$, S_{N1} are always $(1,1)$ for positive output voltages. Thus, S_{P1} is 1 for level $+E/4$ (i.e., L_{+1}) if one of the following conditions happens otherwise it is 0:

- current is positive ($I=1$) and C_{P1} needs to be discharged ($\Delta_{CP1}=1$) which equals $I \cdot \Delta_{CP1}$.

- current is negative ($\bar{I}=1$) and C_{P1} needs to be charged ($\bar{\Delta}_{CP1}=1$) which equals $\bar{I} \cdot \bar{\Delta}_{CP1}$.

Finally, it is needed to consider all above two cases for S_{P1} at level $+E/4$. This is why the second term of eq. (5) is $L_{+1} \cdot (I.\Delta_{CP1}+\bar{I}.\bar{\Delta}_{CP1})$ which is the logic-form equation and interpretation of the above explained approach. Moreover, according to (7), S_2 is always opposite of S_1 for level $+E/4$ (i.e., L_{+1}). This is why the second term of eq. (7) is opposite of the second term of eq. (5). This approach is used to obtain all logic-form equations in this paper for each switch at each level.

TABLE 1. STATES OF POWER SWITCHES AND CHARGING/DISCHARGING PROCESS OF FLYING CAPACITORS IN FIVE-LEVEL SM CONVERTER.

Output Voltage	State of switches $(S_{N2}, S_{N1}),(S_{P2}, S_{P1})$	ΔE_{CN1}	ΔE_{CP1}	i_{out}	Number of States
$+E/2$	$(1,1),(1,1)$	0	0	*	1
$+E/4$	$(1,1),(0,1)$	0	-	+	2
		0	+	-	
	$(1,1),(1,0)$	0	+	+	
		0	-	-	
0	$(1,1),(0,0)$	0	0	*	1
$-E/4$	$(0,1),(0,0)$	-	0	+	2
		+	0	-	
	$(1,0),(0,0)$	+	0	+	
		-	0	-	
$-E/2$	$(0,0),(0,0)$	0	0	*	1

* means any direction of i_{out}

+ means $i_{out} > 0$

- means $i_{out} < 0$

Now let's consider that the output current is positive and it is required to generate level $+E/4$. The first option of $[(1,1),(0,1)]$ decreases FC voltage of cell #P1 (E_{CP1}) by amount of, say K pu, while the second option of the $[(1,1),(1,0)]$ increases FC voltage of cell #P1 (E_{CP1}) by

amount of K pu. Therefore, it is required to determine which option should be selected to control the FC voltage of cell #P1 (E_{CP1}). In order to select the proper option, Δ_{CP1} should be checked. If Δ_{CP1} is one, it means that voltage deviation of FC voltage of cell #P1 ($\Delta V_{CP1,pu}$) in per-unit (unit is the associated reference voltage of each FC) is positive indicating that FC voltage of cell #P1 (E_{CP1}) is higher than its reference value. Thus, FC of cell #P1 (C_{P1}) needs to be discharged. Since it is assumed that the current is positive, the first option of the $[(1,1),(0,1)]$ should be selected otherwise, the second option of the $[(1,1),(1,0)]$ should be selected. But if Δ_{CP1} is zero, it means that voltage deviation of FC voltage of cell #P1 ($\Delta V_{CP1,pu}$) in per-unit is negative indicating that the FC voltage of cell #P1 (E_{CP1}) is lower than its reference value. Thus, FC of cell #P1 (C_{P1}) needs to be charged. Since it is assumed that the current is positive, the second option of the $[(1,1),(1,0)]$ should be selected otherwise, the first option of $[(1,1),(0,1)]$ should be selected. The same approach can be applied to control the FC voltage of cell #N1 (E_{CN1}) whenever it is required to generate level $-E/4$.

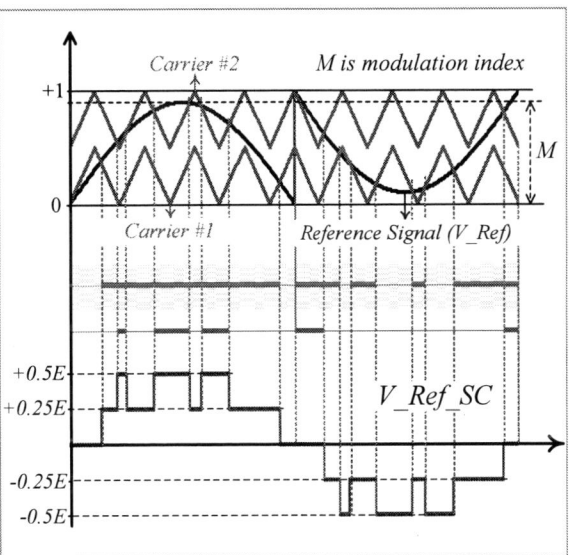

Figure 2. Phase-disposition LSC-PWM strategy to produce the staircase output voltage reference (V_Ref_SC) for four-cell five-level SM converter.

TABLE 2. DEFINITION OF THE REQUIRED VARIABLES FOR PROPOSED ACTIVE VOLTAGE BALANCING METHOD IN FIVE-LEVEL SM CONVERTER.

Level (per unit)	L_{-2}	L_{-1}	L_0	L_{+1}	L_{+2}
$+E/2$	0	0	0	0	1
$+E/4$	0	0	0	1	0
0	0	0	1	0	0
$-E/4$	0	1	0	0	0
$-E/2$	1	0	0	0	0

978-1-4673-9551-9/16 $31.00 © 2016 IEEE

According to the explained procedure to balance the FC voltages and defined variables in Table 2, following logic-form-equations must be applied to control the five-level SM converter. The switching state of the power switch S_{P1} is as follows:

$$S_{P1} = L_{+2} + L_{+1} X_{SP1,+1} \tag{5}$$

where:

$$X_{SP1,+1} = \Delta_{CP1} I + \overline{\Delta}_{CP1} \overline{I} \tag{6}$$

The switching state of power switch S_{P2} is as follows:

$$S_{P2} = L_{+2} + L_{+1} X_{SP2,+1} \tag{7}$$

where:

$$X_{SP2,+1} = \Delta_{CP1} \overline{I} + \overline{\Delta}_{CP1} I \tag{8}$$

The switching states of the power switches S_{N2}, S_{N1} are as follows:

$$S_{N1} = L_{+2} + L_{+1} + L_0 + L_{-1} X_{SN1,-1} \tag{9}$$

$$S_{N2} = L_{+2} + L_{+1} + L_0 + L_{-1} X_{SN2,-1} \tag{10}$$

where:

$$X_{SN1,-1} = \Delta_{CN1} I + \overline{\Delta}_{CN1} \overline{I} \tag{11}$$

$$X_{SN2,-1} = \Delta_{CN1} \overline{I} + \overline{\Delta}_{CN1} I \tag{12}$$

Using the proposed active capacitor voltage balancing method which is based on the above mentioned logic-form equations, five-level SM converter can be modulated based on the any kind of the PWM methods while its FC voltages are actively balanced at their reference values.

III. SIMULATION RESULTS

In this section, some simulation results are illustrated to verify the performance of the proposed logic-form-equation based active voltage balancing method. In following, five-level SM converters are simulated in the PSCAD/EMTDC and the obtained results are provided. The main parameters used in the simulations are given in Table 3.

TABLE 3. MAIN CIRCUIT PARAMETERS OF THE SIMULATED FIVE-LEVEL SM CONVERTER.

System Parameters	Values
System frequency	50 Hz
dc link voltage (E)	5.4 kV
Resistive inductive load (R-L)	10Ω - 20mH
Capacitance of FCs in five-level SM converter (C_{P1}, C_{N1})	1 mF
PD LSC-PWM carrier frequency	1.95 kHz
Modulation index (M)	0.98

The converter output voltage, load current (increased 10 times), and FC voltages in four-cell five-level SM converter are shown in Figure 3. As it is shown in Figure 3, voltages

of FCs are stabilized at their reference value (i.e., 1350V) verifying performance of the proposed logic-form-equation based active voltage balancing method. To investigate the dynamic behavior of the proposed switching method and its ability to balance the voltages across the FCs with undesired initial voltages, SM converter is tested with undesired initial voltages across the FCs. As it is depicted in Figure 4 which reflects transient and steady states of the FC voltages, the proposed active voltage balancing method being triggered at $t=0.02s$ regulates the FC voltages correctly at their reference values.

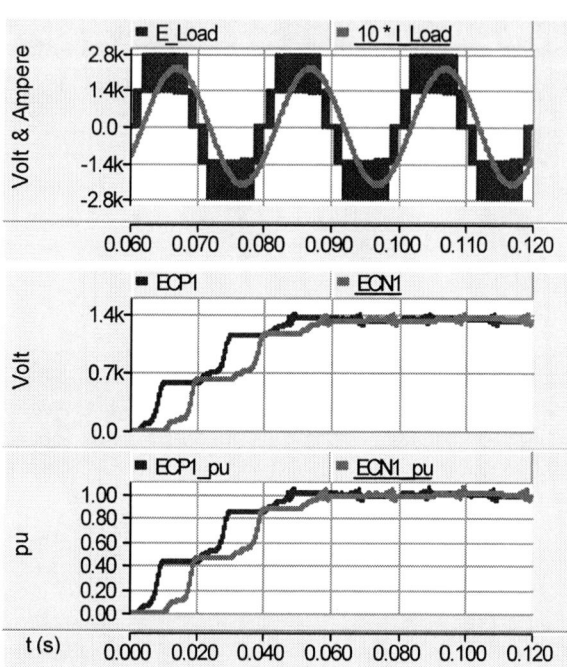

Figure 3. Simulation results: converter output voltage, converter load current (increased 10 times), and FCs voltages in four-cell five-level SM converter.

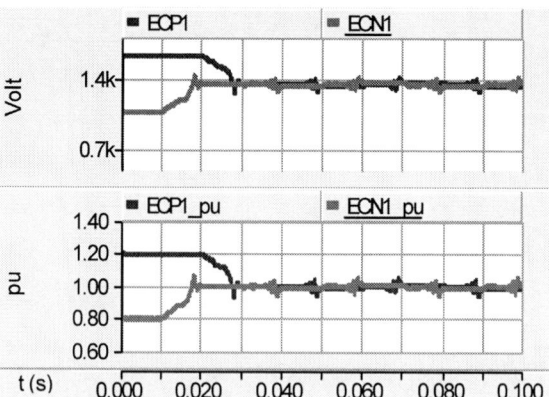

Figure 4. Simulation results: Voltage balancing of FCs with undesired initial voltage in four-cell five-level SM converter using the proposed active voltage balancing method.

As depicted in the Figure 5(a) which demonstrates the frequency spectrum of line-to-neutral output voltage of five-level SM converter with THD of 27%, it contains harmonic clusters around $\left(k \times 1.95^{kHz}\right)^{th}$ harmonics where k is an integer number [41]. As it is seen, the highest-amplitude harmonics of the line-to-neutral voltage is related to 39th harmonic, *i.e.*, 1.95 kHz, which is a triplen harmonics. Due to cancellation of 39th as well as all the other triplen harmonics in the line-to-line output voltage, the PD LSC-PWM switching strategy possesses a superior line-to-line THD [42] characteristics. Figure 5(b) illustrates the frequency spectrum of the line-to-line output voltage with THD of 15%. This confirms that the proper selection of the triangle carrier waveform frequency leads to an optimum THD of the line-to-line output voltage. For instance, if the triangle carrier frequency was selected as 2.05 kHz, the harmonic component of 41th harmonics would have the highest amplitude. So it would not be eliminated in line-to-line output voltage since it is not a triplen harmonics. In this case, the THD of the line-to-neutral and line-to-line output voltage would be 27% and 25%, respectively.

The dynamic performance of the proposed logic-based active capacitor voltage balancing method for five-level SM converter under different modulation indices ranging from 0.1 to 1 with intervals of 0.1 is illustrated in Figure 6. As it is shown, the proposed method does not have any restriction regarding the value of modulation index.

balancing method is implemented based on the logic-form-equations, and modulates the converter output voltage based on the PD-LSC PWM in order to optimize the THD of the line-to-line output voltage and also to balance the voltage of FCs. Besides, the effect of the selecting a proper frequency for triangle-carriers on THD of line-to-line output voltage was discussed. Additionally, the dynamic performance and stability of the proposed method under different modulation indices were illustrated. It was concluded that the proposed method does not have any restriction regarding the value of the converter modulation index. Moreover, the performance of the proposed active control method for balancing the FC voltages correctly during converter transients with undesired initial voltages across the FCs is verified.

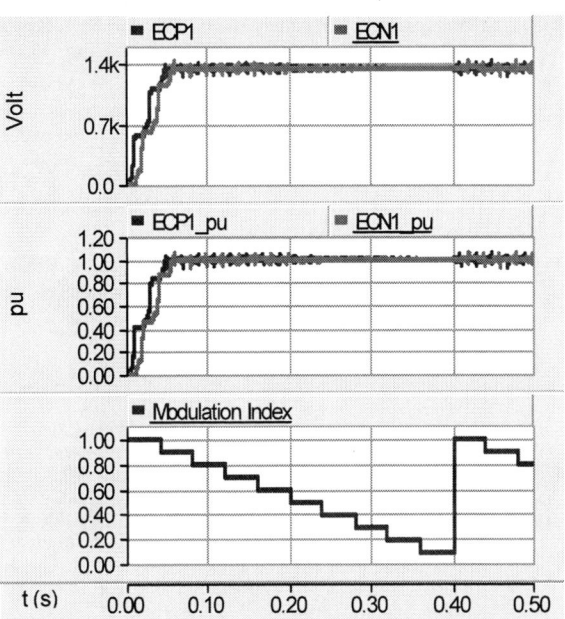

Figure 6. Dynamic performance of the proposed active voltage balancing method to regulate FCs voltages under different modulation indexes in five-level SM converter.

REFERENCES

[1] S. Kouro, J. Rodriguez, B. Wu, S. Bernet, and M. Perez, "Powering the Future of Industry: High-Power Adjustable Speed Drive Topologies," *IEEE Ind. Appl. Mag.*, vol. 18, no. 4, pp. 26–39, Jul. 2012.

[2] B. Wu, J. Pontt, J. Rodríguez, S. Bernet, and S. Kouro, "Current-source converter and cycloconverter topologies for industrial medium-voltage drives," *IEEE Trans. Ind. Electron.*, vol. 55, no. 7, pp. 2786–2797, Jul. 2008.

[3] H. Abu-Rub, J. Holtz, J. Rodriguez, and Ge Baoming, "Medium-Voltage Multilevel Converters—State of the Art, Challenges, and Requirements in Industrial Applications," *IEEE Trans. Ind. Electron.*, vol. 57, no. 8, pp. 2581–2596, Aug. 2010.

[4] S. Rivera, S. Kouro, B. Wu, S. Alepuz, M. Malinowski, P. Cortes, and J. Rodriguez, "Multilevel Direct Power Control—A Generalized Approach for Grid-Tied Multilevel Converter Applications," *IEEE Trans. Power Electron.*, vol. 29, no. 10, pp. 5592–5604, Oct. 2014.

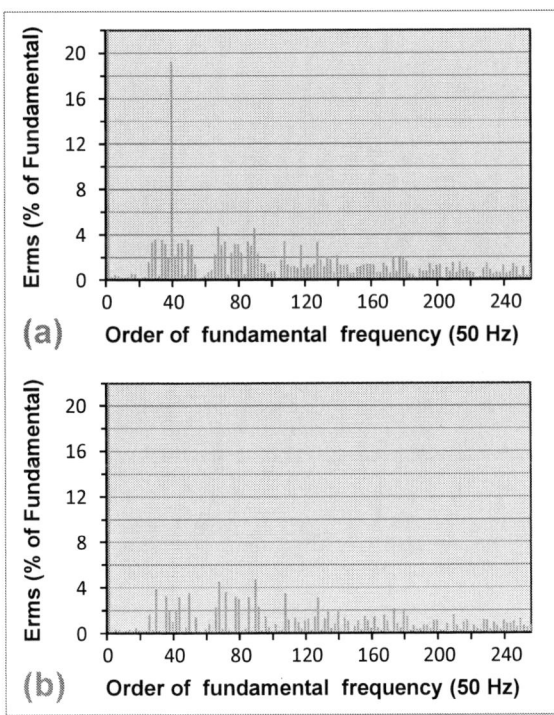

Figure 5. Frequency spectrum of five-level (line-to-neutral) SM converter controlled with proposed method: (a) FFT of output line-to-neutral voltage; (b) FFT of output line-to-line voltage.

IV. CONCLUSION

Stacked multicell inverter is a well-known breed of FC-based multilevel converters. This paper presented a novel active capacitor voltage balancing method for five-level SM converter to stabilize and balance flying capacitor voltages at their reference voltages. Proposed active capacitor voltage

[5] S. Kouro, M. Malinowski, K. Gopakumar, J. Pou, L. G. Franquelo, J. Rodriguez, M. A. Pérez, and J. I. Leon, "Recent Advances and Industrial Applications of Multilevel Converters," *IEEE Trans. Ind. Electron.*, vol. 57, no. 8, pp. 2553–2580, Aug. 2010.

[6] M. Aleenejad, H. Iman-Eini, and S. Farhangi, "Modified space vector modulation for fault-tolerant operation of multilevel cascaded H-bridge inverters," *IET Power Electron.*, vol. 6, no. 4, pp. 742–751, Apr. 2013.

[7] M. Aleenejad, R. Ahmadi, and P. Moamaei, "A modified selective harmonic elimination method for fault-tolerant operation of multilevel cascaded H-bridge inverters," in *2014 Power and Energy Conference at Illinois (PECI)*, 2014, pp. 1–5.

[8] J. Rodriguez, L. G. Franquelo, S. Kouro, J. I. Leon, R. C. Portillo, M. A. M. Prats, and M. A. Perez, "Multilevel Converters: An Enabling Technology for High-Power Applications," *Proc. IEEE*, vol. 97, no. 11, pp. 1786–1817, Nov. 2009.

[9] M. Aleenejad, P. Moamaei, H. Mahmoudi, and R. Ahmadi, "Unbalanced Selective Harmonic Elimination for fault-tolerant operation of three phase multilevel Cascaded H-bridge inverters," in *2015 IEEE Applied Power Electronics Conference and Exposition (APEC)*, 2015, pp. 1589–1594.

[10] J. Rodriguez, S. Bernet, B. Wu, J. O. Pontt, and S. Kouro, "Multilevel Voltage-Source-Converter Topologies for Industrial Medium-Voltage Drives," *IEEE Trans. Ind. Electron.*, vol. 54, no. 6, pp. 2930–2945, Dec. 2007.

[11] V. Dargahi, A. Khoshkbar Sadigh, M. Abarzadeh, S. Eskandari, and K. Corzine, "A New Family of Modular Multilevel Converter Based on Modified Flying-Capacitor Multicell Converters," *IEEE Trans. Power Electron.*, vol. 8993, pp. 1–1, 2014.

[12] V. Dargahi, A. K. Sadigh, G. K. Venayagamoorthy, and K. Corzine, "Hybrid double flying capacitor multicell converter and its application in grid-tied renewable energy resources," *IET Gener. Transm. Distrib.*, vol. 9, no. 10, pp. 947–956, Jul. 2015.

[13] A. K. Sadigh, V. Dargahi, and K. A. Corzine, "New Multilevel Converter Based on Cascade Connection of Double Flying Capacitor Multicell Converters and Its Improved Modulation Technique," *IEEE Trans. Power Electron.*, vol. 30, no. 12, pp. 6568–6580, Dec. 2015.

[14] V. Dargahi, A. Khoshkbar Sadigh, M. Abarzadeh, M. R. A. Pahlavani, and A. Shoulaie, "Flying Capacitors Reduction in an Improved Double Flying Capacitor Multicell Converter Controlled by a Modified Modulation Method," *IEEE Trans. Power Electron.*, vol. 27, no. 9, pp. 3875–3887, Sep. 2012.

[15] V. Dargahi and S. Dargahi, "Analytical modelling of single-phase stacked multicell multilevel converters exploiting Kapteyn (Fourier–Bessel) series," *IET Power Electron.*, vol. 6, no. 6, pp. 1220–1238, Jul. 2013.

[16] A. K. Sadigh, S. H. Hosseini, M. Sabahi, and G. B. Gharehpetian, "Double flying capacitor multicell converter based on modified phase-shifted pulsewidth modulation," *IEEE Trans. Power Electron.*, vol. 25, pp. 1517–1526, 2010.

[17] V. Dargahi and A. Shoulaie, "Capacitors voltage balancing modeling in three phase flying capacitor converters with booster," in *2012 3rd Power Electronics and Drive Systems Technology (PEDSTC)*, 2012, pp. 103–108.

[18] G. Gateau, M. Fadel, P. Maussion, R. Bensaid, and T. A. Meynard, "Multicell converters: active control and observation of flying-capacitor voltages," *IEEE Trans. Ind. Electron.*, vol. 49, no. 5, pp. 998–1008, Oct. 2002.

[19] R. H. Wilkinson, T. A. Meynard, and H. du Toit Mouton, "Natural Balance of Multicell Converters: The Two-Cell Case," *IEEE Trans. Power Electron.*, vol. 21, no. 6, pp. 1649–1657, Nov. 2006.

[20] R. H. Wilkinson, T. A. Meynard, and H. du Toit Mouton, "Natural Balance of Multicell Converters: The General Case," *IEEE Trans. Power Electron.*, vol. 21, no. 6, pp. 1658–1666, Nov. 2006.

[21] M. Ben Smida and F. Ben Ammar, "Modeling and DBC-PSC-PWM Control of a Three-Phase Flying-Capacitor Stacked Multilevel Voltage Source Inverter," *Ind. Electron. IEEE Trans.*, vol. 57, no. 7, pp. 2231–2239, Jul. 2010.

[22] B. P. McGrath, T. Meynard, G. Gateau, and D. G. Holmes, "Optimal Modulation of Flying Capacitor and Stacked Multicell Converters Using a State Machine Decoder," *IEEE Trans. Power Electron.*, vol. 22, no. 2, pp. 508–516, Mar. 2007.

[23] S. Dargahi, M. Sabahi, S. Eskandari, E. Babaei, and V. Dargahi, "Flying-capacitor stacked multicell multilevel voltage source inverters: analysis and modelling," *IET Power Electron.*, vol. 7, no. 12, pp. 2969–2987, Dec. 2014.

[24] B. P. McGrath and D. G. Holmes, "Analytical Modelling of Voltage Balance Dynamics for a Flying Capacitor Multilevel Converter," *IEEE Trans. Power Electron.*, vol. 23, no. 2, pp. 543–550, Mar. 2008.

[25] B. P. McGrath and D. G. Holmes, "Analytical Determination of the Capacitor Voltage Balancing Dynamics for Three-Phase Flying Capacitor Converters," *IEEE Trans. Ind. Appl.*, vol. 45, no. 4, pp. 1425–1433, Jul. 2009.

[26] B. P. McGrath and D. G. Holmes, "Enhanced Voltage Balancing of a Flying Capacitor Multilevel Converter Using Phase Disposition (PD) Modulation," *IEEE Trans. Power Electron.*, vol. 26, no. 7, pp. 1933–1942, Jul. 2011.

[27] B. P. McGrath and D. G. Holmes, "Natural Capacitor Voltage Balancing for a Flying Capacitor Converter Induction Motor Drive," *IEEE Trans. Power Electron.*, vol. 24, no. 6, pp. 1554–1561, 2009.

[28] M. Khazraei, H. Sepahvand, K. A. Corzine, and M. Ferdowsi, "Active Capacitor Voltage Balancing in Single-Phase Flying-Capacitor Multilevel Power Converters," *IEEE Trans. Ind. Electron.*, vol. 59, no. 2, pp. 769–778, Feb. 2012.

[29] M. Khazraei, H. Sepahvand, M. Ferdowsi, and K. A. Corzine, "Hysteresis-Based Control of a Single-Phase Multilevel Flying Capacitor Active Rectifier," *IEEE Trans. Power Electron.*, vol. 28, no. 1, pp. 154–164, Jan. 2013.

[30] A. Shukla, A. Ghosh, and A. Joshi, "Hysteresis Current Control Operation of Flying Capacitor Multilevel Inverter and Its Application in Shunt Compensation of Distribution Systems," *IEEE Trans. Power Deliv.*, vol. 22, no. 1, pp. 396–405, Jan. 2007.

[31] A. Shukla, A. Ghosh, and A. Joshi, "Improved Multilevel Hysteresis Current Regulation and Capacitor Voltage Balancing Schemes for Flying Capacitor Multilevel Inverter," *IEEE Trans. Power Electron.*, vol. 23, no. 2, pp. 518–529, Mar. 2008.

[32] P. Lezana, R. Aguilera, and D. E. Quevedo, "Model Predictive Control of an Asymmetric Flying Capacitor Converter," *IEEE Trans. Ind. Electron.*, vol. 56, no. 6, pp. 1839–1846, Jun. 2009.

[33] S. Choi and M. Saeedifard, "Capacitor voltage balancing of flying capacitor multilevel converters by space vector PWM," *IEEE Trans. Power Deliv.*, vol. 27, no. 3, pp. 1154–1161, 2012.

[34] Jing Huang and K. A. Corzine, "Extended operation of flying capacitor multilevel inverters," *IEEE Trans. Power Electron.*, vol. 21, no. 1, pp. 140–147, Jan. 2006.

[35] A. M. Y. M. Ghias, J. Pou, and V. G. Agelidis, "Voltage-Balancing Method for Stacked Multicell Converters Using Phase-Disposition PWM," *IEEE Trans. Ind. Electron.*, vol. 62, no. 7, pp. 4001–4010, Jul. 2015.

[36] A. M. Y. M. Ghias, J. Pou, M. Ciobotaru, and V. G. Agelidis, "Voltage-balancing method using phase-shifted PWM for the flying capacitor multilevel converter," *IEEE Trans. Power Electron.*, vol. 29, no. 9, pp. 4521–4531, Sep. 2014.

[37] A. Bennani, T. Meynard, and G. Gateau, "Direct torque control for stacked multicell (SMC) VSI fed induction machine," in *Power Electronics and ...*, 2005, pp. 1–10.

[38] A.-M. Lienhardt, G. Gateau, and T. A. Meynard, "Digital Sliding-Mode Observer Implementation Using FPGA," *IEEE Trans. Ind. Electron.*, vol. 54, no. 4, pp. 1865–1875, Aug. 2007.

[39] A. M. Y. Mohammad Ghias, J. Pou, V. Agelidis, and M. Ciobotaru, "Optimal Switching Transitions Based Voltage Balancing Method for Flying Capacitor Multilevel Converters," *IEEE Trans. Power Electron.*, vol. 8993, no. c, pp. 1–1, Apr. 2014.

[40] A. M. Y. Mohammad Ghias, J. Pou, and V. Agelidis, "On Reducing Power Losses in Stack Multicell Converters withOptimal Voltage Balancing Method," *IEEE Trans. Power Electron.*, pp. 1–1, 2014.

[41] B. P. McGrath and D. G. Holmes, "An analytical technique for the determination of spectral components of multilevel carrier-based PWM methods," *IEEE Trans. Ind. Electron.*, vol. 49, no. 4, pp. 847–857, 2002.

[42] D. G. Holmes and B. P. McGrath, "Opportunities for harmonic cancellation with carrier-based PWM for two-level and multilevel cascaded inverters," *IEEE Trans. Ind. Appl.*, vol. 37, no. 2, pp. 574–582, 2001.

Gate Signal Jitter Elimination and Noise Shaping Modulation for High-SNR Class-D Power Amplifiers

M. Mauerer, A. Tüysüz and J. W. Kolar

Power Electronic Systems Laboratory, ETH Zürich, Switzerland
Email: mauerer@lem.ee.ethz.ch

Abstract—**Two important aspects of switched-mode (Class-D) amplifiers providing a high signal to noise ratio (SNR) for mechatronic applications are investigated. Signal jitter is common in digital systems and introduces noise, leading to a deterioration of the SNR. Hence, a jitter elimination technique for the transistor gate signals in power electronic converters is presented and verified. Jitter is reduced tenfold as compared to traditional approaches to values of 25 ps at the output of the power stage. Additionally, digital modulators used for the generation of the switch control signals can only achieve a limited resolution (and hence, limited SNR) due to timing constraints in digital circuits. Consequently, a specialized modulator structure based on noise shaping is presented and optimized which enables the creation of high-resolution switch control signals. This, together with the jitter reduction circuit, enables half-bridge output voltage SNR values of more than 100 dB in an open-loop system.**

I. INTRODUCTION

Low-noise switched-mode amplifiers are required in different industry applications, amongst them integrated circuit manufacturing, where such amplifiers are used to drive various kinds of actuators of high-precision motion systems [1]. These systems have to provide positioning accuracies in the nanometer range, which requires the amplifiers to provide output currents of extremely low noise and high precision in order to avoid disturbing forces or torques in the actuators. It is estimated that the output current signal to noise ratio (SNR) of the power amplifiers needs to increase by approximately 20 dB every five years in order to keep track with the industry's growing requirements [2]. Therefore, a current SNR in excess of 110 dB is desired. State-of-the-art systems usually achieve SNR values in the range of 80 − 100 dB [2], [3].

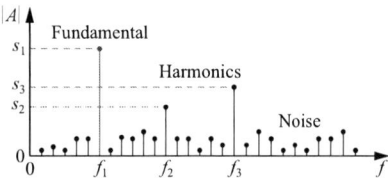

Fig. 1: Discrete amplitude spectrum of a noisy, periodic signal. The signal to noise ratio (SNR) relates the fundamental power $p(s_1)$ to the noise power $p(n)$ in a certain frequency range.

As the amplifiers target mechatronic applications, their output must be of low noise and high precision only in a limited frequency range from DC to ≈ 10 kHz, as the mechanical systems provide sufficient damping at higher frequencies and are thus less sensitive to high-frequency signal distortion or noise in the actuator's currents.

Digital control of such amplifiers is desired for the reasons of development and maintainability. Accordingly, this paper deals with two main aspects that are required to enable digitally controlled low-noise power amplifiers: jitter, which deteriorates the achievable SNR, and digital modulators, which, although operating with a coarse duty cycle resolution in order to achieve a given PWM switching frequency, are still capable of creating gate signals with a high SNR.

The SNR relates the power of a desired signal $p(s_1)$ to the power of unwanted signals $p(n)$, such as noise, in a given frequency range. The desired signal can be, e.g., the fundamental frequency component of an amplifier output voltage- or current signal, which can also contain low-order harmonics and noise. **Fig. 1** exemplarily illustrates the spectrum of a periodic signal that contains the desired fundamental with amplitude s_1, some integer harmonics s_x, and wideband noise. The signal's SNR is defined as:

$$\text{SNR} = 10 \log_{10} \left(\frac{p(s_1)}{p(n)|_{s_2, s_3, s_x = 0}} \right), \qquad (1)$$

where the noise power $p(n)$ is calculated in a given frequency range $[f_a \ldots f_b]$ and the harmonics are excluded from the noise power calculation. If they were included, the figure would be called signal to noise and distortion ratio (SINAD) [4]. This work uses the SNR, as the investigated processes (jitter and PWM) cause few harmonics (if at all) of negligible power. The Fast Fourier Transform can be used to obtain the signal and noise powers from simulations or measurements [5], [6].

Although mechatronic amplifiers need to provide output currents of high SNR to create low-noise torques in the actuators, this work analyzes the SNR of the switching stage's output voltage in order to present a concise analysis of distortion and noise originating from the fundamental building blocks of the power converter, as a distorted voltage finally translates to current distortion in filter inductors or actuators.

Section II presents sources of jitter in power electronic converters and proposes a simple new circuit for generating isolated, low-jitter gate signals. The circuit is also immune to fast common-mode voltage transients across the isolation barrier and its functionality is experimentally verified. **Section III** details advanced modulation structures which outperform conventional pulse-width modulators. Their performance is also demonstrated with a prototype system. Finally, **Section IV** summarizes the achieved results and presents an outlook.

978-1-4673-9551-9/16 $31.00 © 2016 IEEE

II. SIGNAL JITTER AND ITS ELIMINATION IN A GATE DRIVER CIRCUIT

Electrical signal jitter has various classifications and, due to its often stochastic nature, can be modeled like noise [7]. **Fig. 2** shows a periodic digital signal with jitter, which expresses itself as the time-dependent deviations Δt_i of the signal's edge positions from their ideal positions [8].

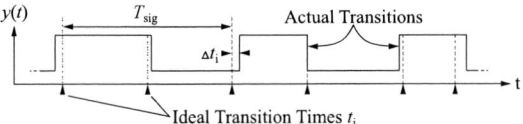

Fig. 2: A digital, periodic signal $y(t)$ shows jitter if its edges occur at time instants different from the ideal transition times. The time difference Δt_i between actual and ideal edge is often of stochastic nature.

In digital signals, where information is encoded in the time instants of its edges, jitter alters the information and introduces wideband noise [4]. For high-precision PWM applications with targeted SNRs in excess of $100\,\text{dB}$ and pulse frequencies $f_{\text{PWM}} = 1/T_{\text{PWM}}$ in the range of $50 \ldots 200\,\text{kHz}$, even jitter figures ($\Delta t_i$) in the picosecond range can introduce significant noise and limit the SNR as the equation from [9] shows:

$$\text{SNR}_{\text{max}} = 20 \log_{10} \left(\frac{m}{4\sqrt{2} \cdot T_{\text{Jit,RMS}}} \sqrt{\frac{T_{\text{PWM}}}{f_{\text{BW}}}} \right) \quad (2)$$

Fig. 3 illustrates eq. (2) for different pulse frequencies f_{PWM}. The bandwidth f_{BW} in which the noise power is assessed is $10\,\text{kHz}$ (cf. eq. (1)) and the modulation index m of the PWM modulator approaches 1 ($0 \leq m \leq 1$). $T_{\text{Jit,RMS}}$ is the RMS jitter and denotes the stationary RMS value of the time deviations Δt_i. As these deviations follow a stochastic distribution, a sufficient amount of samples, depending on the underlying stochastic process, must be considered in order to correctly calculate their RMS value.

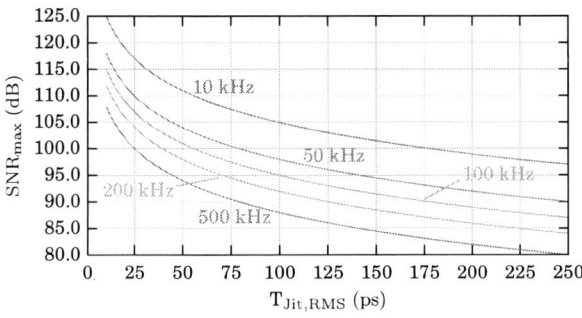

Fig. 3: Plots of eq. (2): Maximum achievable SNR for a given PWM frequency in dependence of the RMS jitter. The modulation index is $m \to 1$ and the bandwidth for which the SNR is calculated is $10\,\text{kHz}$.

A. Sources of Jitter in Power Electronics

Fig. 4 depicts a half-bridge (HB) leg which is a typical building block of Class-D amplifiers and consists of two power transistors, T_1 and T_2 and their respective gate driver circuits. Both drivers require an isolated supply and an isolated gate control signal path in order to allow any control system reference potential between the potentials U_{POS}, U_{NEG} of the DC link rails.

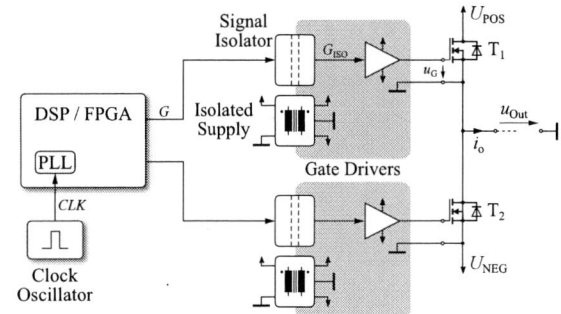

Fig. 4: Conventional half-bridge topology with two power transistors, their isolated gate drivers and a digital gate signal generator.

Every functional element in the gate signal path is a potential source of jitter. In this regard, the clock oscillators of digital signal processors (DSPs) or field programmable gate arrays (FPGAs), both being widely used to generate the gate control signals, as well as the signal routing in an FPGA, add jitter. However, the most prominent jitter source is the signal isolator, which is commonly realized using optocouplers or non-optical digital signal isolators[1].

B. Gate Driver with Reduced Jitter

Existing jitter minimization techniques reduce the jitter of a (periodic) clock signal with phase locked loops (PLL) or filters, and are usually applied in low-power digital-to-analog converters [11]. This work proposes the circuit depicted in **Fig. 5** to reduce the jitter of the (potentially non-periodic) gate control signals which includes a basic concept proposed in [12], but extends it by a system that renders the gate drivers immune to fast output voltage transients, i.e., it prevents faulty switching actions of the power transistors in case the signal isolator emits erroneous signals during short output voltage transients of high $\mathrm{d}u/\mathrm{d}t$.

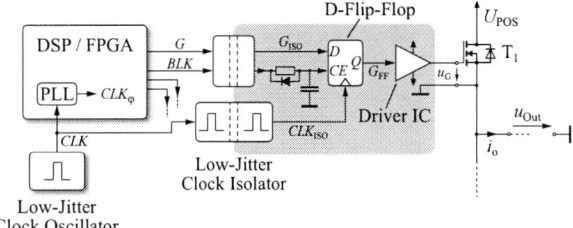

Fig. 5: Improved gate driver which significantly reduces jitter and is immune to output voltage transitions with a high $\mathrm{d}u/\mathrm{d}t$. (The isolated supply and the second identical driver are not drawn).

---

[1]Digital signal isolators feature shorter propagation delay times, stricter propagation delay tolerances and a higher $\mathrm{d}u/\mathrm{d}t$ withstand capability than optocouplers and are thus expected to perform better in low-distortion applications where the signal integrity is of importance [10].

The key component of this circuit is a low-jitter clock isolator, which, in this work, is made of a small signal transformer made for clock frequencies greater than $100\,\text{MHz}$, driven by a high-speed differential driver. The circuit transmits the clock signal to the gate driver where it clocks a flip-flop which resynchronizes the gate control signal to this low-jitter clock CLK_{ISO}. **Fig. 6** depicts the timing waveforms of this method. In order to meet the setup and hold times of the flip-flop, the clock used for the DSP/FPGA, CLK_φ, must potentially be shifted in phase relative to the isolated clock CLK_{ISO}, which can be accomplished by, e.g., using the FPGA's on-chip clock manager.

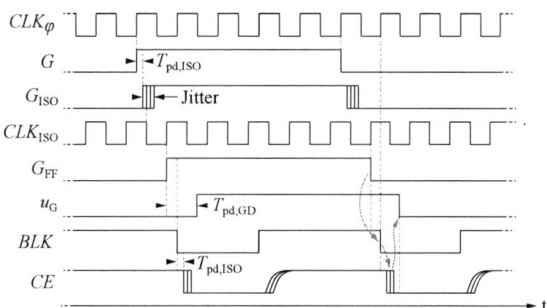

Fig. 6: Timing waveforms of the proposed circuit. CLK_φ is used by the digital control system. *BLK* and *CE* are used to render the system immune to bridge leg output transitions with a high du/dt. $T_{\text{pd,ISO}}$ and $T_{\text{pd,GD}}$ are the propagation delays of the signal isolator and the gate driver IC. The required sequencing of G_{FF}, *BLK* and *CE* is indicated with arrows.

C. Output Transient Withstand Capability

In order to reduce distortion, amplifier output signals ($u_{\text{Out}}(t)$ in **Fig. 4**) require a short output transition time to approximate the ideal square waveform [13]. However, the signal isolators can only withstand a certain du/dt, caused by bridge leg output voltage transitions, across their isolation barrier without producing erroneous output signals (usually $< 50\,\text{kV}\,\mu\text{s}^{-1}$ [14]). The circuit presented in **Fig. 5** is designed to prevent faulty control signals to be transmitted to the driver IC during fast output voltage transients. The method utilizes a blanking signal, *BLK*, which is forwarded to the clock-enable (*CE*) input of the flip-flop via an RC low-pass filter. As long as the *CE* input is low, the flip-flop conserves its state regardless of its potentially faulty input (G_{ISO}).

The low-pass filter prevents temporary (ns range) faulty signals that are potentially emitted by the signal isolator during output transients from affecting the *CE* input. In order to be able to quickly pull the *CE* input low, a diode is placed across the low-pass filter. The waveforms of this method are also illustrated in **Fig. 6**. The required sequencing of the control signals G_{FF}, *BLK* and *CE* is indicated with arrows. It is essential that the *CE* input is pulled low before the gate driver changes the gate voltage u_{G}, as afterwards, the function of the signal isolator might not be guaranteed during the output transient. This can be achieved as the gate driver has a given propagation delay of usually $5 - 30\,\text{ns}$ during which the *CE* input is pulled low. The blanking sequence is always executed simultaneously by both gate driver circuits in a half-bridge

whenever any of the two bridge leg gate drivers changes its transistor's gate voltage, as the output voltage transient may occur only when the complementary transistor switches, depending on the bridge leg's output current direction.

This system limits the minimum dead time (i.e., the time required, after a transistor has switched, before the complementary transistor may be switched, to prevent a short-circuit of the DC link rails) and minimum duty cycle, as time must be allotted for the blanking cycle and the low-pass filter, during which the bridge leg's output state cannot be changed. Experiments show that with a clock frequency of $100\,\text{MHz}$, the minimum dead time is $\approx 50\,\text{ns}$.

D. Jitter Measurements

The circuit from **Fig. 5** has been implemented in a hardware demonstrator (without the power transistors) in order to verify its basic functionality. **Fig. 7** shows the system which features two isolated gate drivers, an FPGA and a low-jitter clock oscillator which serves as the FPGA's clock source. The isolated supplies of the gate drivers create adjustable positive and negative supply voltages.

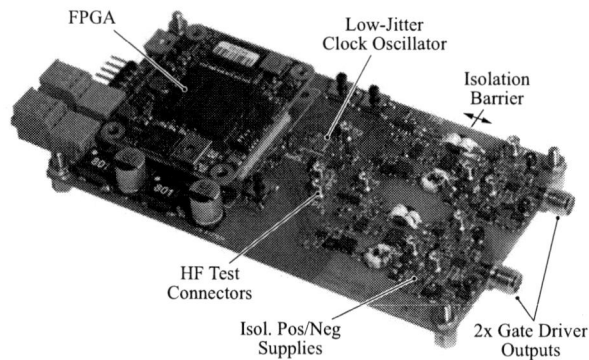

Fig. 7: Hardware demonstrator featuring two isolated gate drivers with the jitter reduction circuitry. The clock isolator comprises a $5.95\,\text{mm}$ *Fair-Rite 5943000101* toroid core with $N_1 = N_2 = 4$ turns, being driven by an LVDS driver. The digital signal isolator is an *SI8620*, the gate driver IC is an *LM5114* and the clock oscillator an *FXO-HC536R-100*. The flip-flop is a *74LVC1G79GV* and its *CE* functionality is implemented using an AND-gate at its clock input.

Using this prototype, the signal jitter (Δt_i) along the gate signal path is measured. **Fig. 8** illustrates such a measurement in the time domain for the rising edges of the signals G_{ISO} and G_{FF} (cf. **Fig. 5**), with each edge transition being made visible using the oscilloscope's infinite screen persistence. A high-bandwidth oscilloscope must be used to minimize the influence of its own time base jitter, which is usually in the low ps range [15].

For each signal transition, the time deviation from the ideal edge position (Δt_i, cf. **Fig. 2**) is recorded and the RMS value of the deviations is calculated. **Tab. I** lists the RMS jitter values of different signals along the gate signal path. For these measurements, a square wave is created by the FPGA and transmitted to the gate drivers. The clock frequency of the low-jitter oscillator is $100\,\text{MHz}$. Using eq. (2), the maximum achievable SNR, which is calculated using the measured jitter

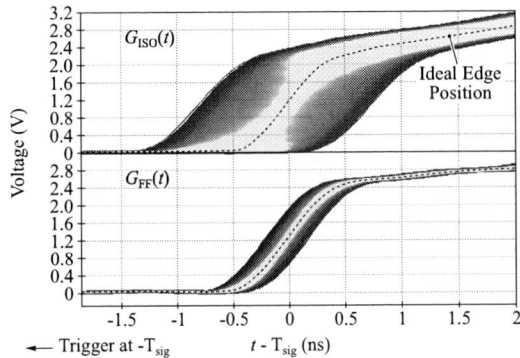

Fig. 8: Persistent oscilloscope waveforms of two different signals (G_{ISO} and G_{FF}, cf. **Fig. 5**), visualizing their jitter. Brighter colors represent more signal transitions. The oscilloscope has a 1 GHz analog bandwidth and samples with 10 GHz (*R&S RTO1014*).

values, is also listed ($f_{PWM} = 100\,kHz$, $f_{BW} = 10\,kHz$, cf. **Fig. 3**). It is evident that the signal isolator adds the most jitter (215 ps) to the gate signal. The clock isolator only contributes 5.7 ps to the clock jitter. The remaining jitter contributions arise from the re-synchronization flip-flop and the gate driver IC.

Signal	$T_{Jit,RMS}$ (ps)	SNR_{max} (dB)
CLK	3.3	124.6
CLK_{ISO}	9.0	115.9
G	33.5	104.4
G_{ISO}	248	87.1
G_{FF}	24.7	107.1
u_G	25.1	107.0
$^2u_{Out}$ (400 V)	25.6	106.8

TABLE I: Measured RMS values of the time differences Δt_i of signals along the gate signal's path. The max. achievable SNR is derived from eq. (2) and calculated for a PWM frequency of 100 kHz, a bandwidth of 10 kHz and a modulation index of $m \to 1$.

Due to the stochastic nature of jitter, a visualization with a histogram is typically done. **Fig. 9** depicts the jitter (Δt_i) of G_{ISO} and G_{FF}. It reveals that the jitter of G_{FF} can be approximated by a normal distribution with a standard deviation of 24.7 ps. This does not apply to the distribution of G_{ISO}, which shows a more triangular shape.

These measurements prove the necessity and functionality of the proposed circuit. As expected, the signal isolator adds a high amount of jitter to the signal. After the resynchronization flip-flop, the RMS jitter is reduced tenfold and the achievable SNR can be increased from 87 dB to 107 dB.

E. Transient Withstand Capability Verification

As section II-C explains, the proposed gate driver circuit is capable of withstanding an output voltage transient with a high du/dt across its isolation barrier without producing erroneous switch signals. In order to verify this functionality, a half-bridge comprising two *GaN Systems GS66508T* gallium nitride (GaN) enhancement-mode HEMTs (high electron mobility transistor,

---

^{2}Power transistors are connected to the drivers and the jitter of u_{Out} is measured, cf. sec. II-E

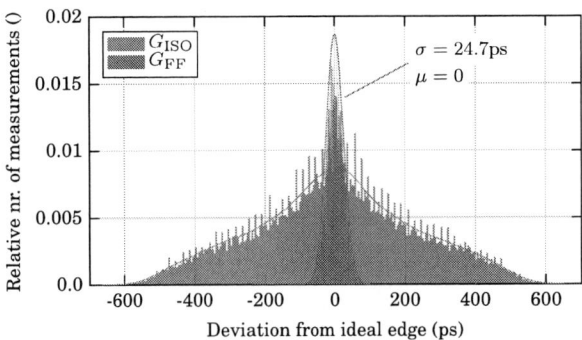

Fig. 9: Histogram of two signals with jitter as measured on the hardware demonstrator. G_{ISO} is the signal after the digital isolator and G_{FF} is the same signal after the re-synchronization flip-flop.

650 V, 30 A, 55 mΩ) is connected to the hardware demonstrator (cf. **Fig. 7**). This setup allows the generation of half-bridge output voltage transitions with a high du/dt in both double-pulse experiments as well as under a constant operation in a buck converter. **Fig. 10** (a) and (b) illustrate the hard- and soft-switched output voltage transitions respectively as they were obtained during double-pulse experiments with a half-bridge output current of 20 A. The du/dt values observed are in the

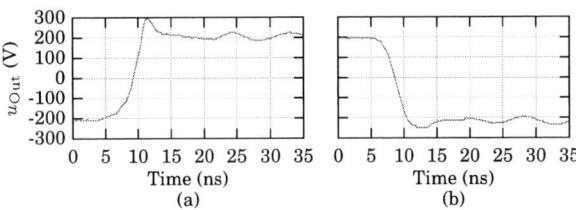

Fig. 10: GaN Half-bridge output voltage transitions with a load current of 20 A. (a): Hard-switched transition. $du/dt_{(10\%-90\%)}$: $83.2\,kV\,\mu s^{-1}$. Max. du/dt: $221\,kV\,\mu s^{-1}$. (b): Soft-switched transition. $du/dt_{(10\%-90\%)}$: $-93.2\,kV\,\mu s^{-1}$. Max. du/dt: $-115\,kV\,\mu s^{-1}$.

range of $80 - 220\,kV\,\mu s^{-1}$ and easily exceed the signal isolator's rating of $50\,kV\,\mu s^{-1}$ [14]. Consequently, an erroneous behavior of this device is likely. This demonstrates the need for an isolated signal transmission system which is capable of dealing with fast voltage transitions across its isolation barrier.

In order to verify the reliability of the gate driver circuit, the same half-bridge has also been continuously operated in a buck converter ($P = 1.9\,kW$, $U_1 = 400\,V$, $U_2 = 200\,V$, $f_{PWM} = 100\,kHz$, $\eta_{tot} = 98.5\%$). During this operation, with a load current of $\approx 10\,A$, the RMS jitter of the switched, 400 V output voltage remained at $\approx 26\,ps$, and no faulty switching actions occurred. However, fast switching transients across the signal isolator's dielectric can lead to its deterioration during longer operation due to capacitive displacement currents. A low-capacitance common-mode filter in series to the signal isolator input can potentially reduce these dielectric currents [16].

This chapter demonstrated the functionality and performance of the proposed gate driver circuit. In the following, a digital modulation scheme is presented which creates high-resolution (low-noise) switch control signals that utilize the full potential of the low-jitter gate drivers.

III. NOISE SHAPING PWM

This chapter addresses specialized digital signal processing techniques that allow the creation of high-resolution (and hence, low-noise) pulse width modulated (PWM) switch control signals.

A. Digital PWM Resolution Limits

In digital systems, PWM signals are commonly created with, as **Fig. 11** (a) shows, a counter that counts (with the clock frequency f_{Clk}) during one switching period ($T_{\text{PWM}} = 1/f_{\text{PWM}}$) from zero to a predefined value (*TOP*), and, in case of double-sided modulation, back down to zero. Whenever the counter value is lower than a compare value *CMP*, the PWM output is high and otherwise, it is low (assuming two switching levels), leading to a modulation of the output signal by *CMP* [17].

Each counter step represents a possible PWM duty cycle and hence, the duty cycle is amplitude-quantized with the number of available counter steps. As the duty cycle is only changed once every switching period, the PWM signal is also time-quantized (i.e., sampled, with f_{PWM}). The achievable SNR of the PWM signal is limited due to the amplitude-quantization error present in signals of finite amplitude resolution [18]. The SNR of a sinusoidal, amplitude-quantized and sampled signal with a resolution of n bits can be approximated by:

$$\text{SNR} = 6.02n + 1.76 \text{ dB}, \quad (3)$$

where the frequency range is from DC to $f_{\text{PWM}}/2$ [4]. Eq. (4) emphasizes the relationships between the PWM resolution, the PWM frequency f_{PWM} and the counter frequency f_{Clk} (cf. **Fig. 11** (a)):

$$\text{PWM}_{\text{Res}}(\text{Bits}) = n = \frac{\ln(TOP)}{\ln(2)} = \frac{\ln\left(\frac{f_{\text{Clk}}}{2f_{\text{PWM}}}\right)}{\ln(2)}. \quad (4)$$

This shows that for high-SNR PWM signals and a given PWM frequency, a high digital clock frequency is required, e.g., for a PWM signal with $f_{\text{PWM}} = 20\,\text{kHz}$ and a resolution of $n = 16$ bits, a clock frequency $f_{\text{Clk}} > 2.6\,\text{GHz}$ is necessary. This is unfeasible for practical implementations as high-resolution counters cannot easily be implemented in digital hardware at such high clock rates. The limit for configurable logic is usually below 500 MHz [19]. Therefore, the following section presents a DSP technique which can create low-noise digital signals despite using low-resolution signals.

B. Noise Shaping

In audio engineering, high-resolution amplitude-quantized signals with up to 24 bits need to be fed to a PWM modulator in Class-D amplifiers with switching frequencies in excess of 300 kHz [20], [21]. This is accomplished by using a nonlinear DSP technique named oversampled noise shaping (NS), which can convert a high-resolution, amplitude-quantized signal to a signal of lower amplitude resolution such that a counter-based PWM modulator can be used while still achieving little quantization noise in a given frequency band (i.e., the baseband) [22], [23]. These techniques shift the quantization noise created by the reduction of signal resolution to higher frequencies, where they can be attenuated by a low-pass filter or where they are of no concern for the target application.

Fig. 11 (b) illustrates a simplified structure of a noise shaping system. The reference signal has a high amplitude resolution of m bits. The digital quantizer reduces the resolution to n bits such that $n < m$. The quantization error is fed back into the system and its spectrum is shaped by a transfer function H such that the quantization noise is significantly reduced in the baseband, resulting in a low-noise output. This signal can now be fed to a PWM modulator with a resolution of also n bits.

Assuming that the quantizer adds the quantization noise additively, its spectral behavior in a noise shaping system can be described by the so-called noise transfer function (NTF), whereas the influence of the noise shaper on the desired signal can be described by the signal transfer function (STF) [22]. The block diagram in **Fig. 12** visualizes this approach. These transfer functions are implemented by the noise shaper's internal structure. **Fig. 12** also illustrates an exemplary NTF. The baseband ranges from DC to 10 kHz in which the amplitude-quantization noise gets attenuated by $\approx$75 dB. At higher frequencies, the quantization noise increases (high-pass characteristic).

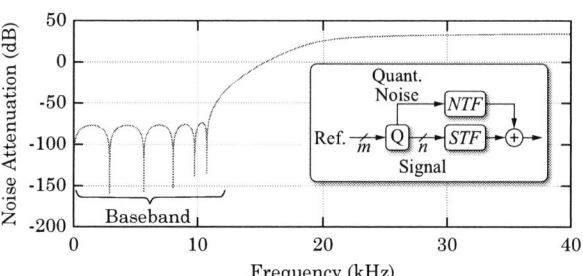

Fig. 12: Amplitude plot of an exemplary noise transfer function (NTF) with 11 poles and 11 zeros. The high-pass characteristic shifts the quantization noise to higher frequencies.

For a typical mechatronic amplifier, noise above $\approx$10 kHz is not critical as the inertia of the mechanical systems introduces sufficient attenuation. Hence, such NS techniques can be applied with a baseband ranging from DC to $\approx$10 kHz.

Fig. 13 depicts the simulated output signal response of a noise shaper to an arbitrary, varying input signal in the time domain. In this case, the signal transfer function is $STF = 1$, which is visualized by the perfect tracking performance of the noise shaper's output. The difference of the input and output

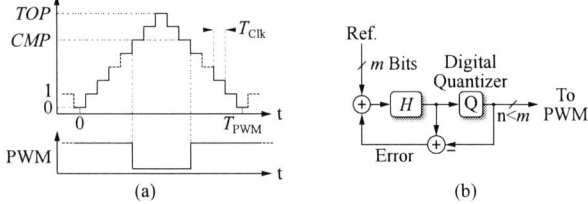

Fig. 11: (a): Counter-based PWM modulator. Each counter step represents a possible duty cycle. (b): Simplified noise shaping principle. The quantization error is fed back into the system whereas its frequency components are shaped by the transfer function H.

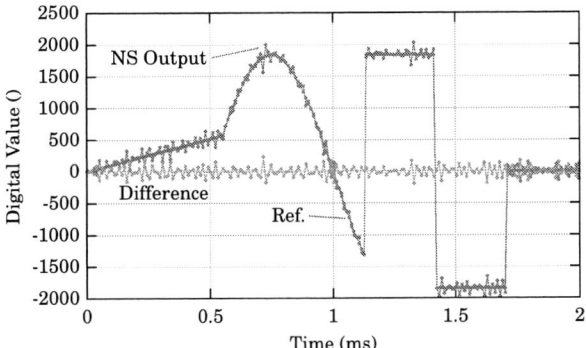

Fig. 13: Exemplary time-domain behavior of a noise shaper. The input signal has an amplitude resolution of 12 bits and the output of the noise shaper is 9 bits wide. The output is rescaled (multiplied by 2^{12-9}) such that the input and output are directly comparable.

is the quantization error whose spectrum is shaped by the NTF (cf. **Fig. 12**). As a noise shaper employs feedback, the system can potentially become unstable. Due to their nonlinear nature (due to the quantizer and digital saturation or overflow effects), noise shapers can only be well investigated using computer simulations. This is done to assess its performance or stability with different input signals.

Noise shaping can be employed in different ways to obtain the switch control signals. **Fig. 14** illustrates two possible signal flow paths in a noise shaping DSP system. The goal of the structure is to replicate the reference signal with a physical signal (i.e., voltage or current) in the amplifier's power stage. The reference signal is created digitally (e.g., by an overlaying amplifier control system) with a high resolution of m bits and a sampling rate f_S. This sampling rate is usually lower than the PWM modulator's sampling rate f_{PWM} and consequently, there must be a digital upsampling block which performs interpolation and increases the reference signal's sampling rate to f_{PWM}, which is also the noise shaper's sampling rate. The ratio between f_{PWM} and f_S is the oversampling ratio (OSR). After the upsampling block, there are two options of how to create the power transistor's switching signal, which can have k levels ($k \geq 2$), depending on the power stage topology.

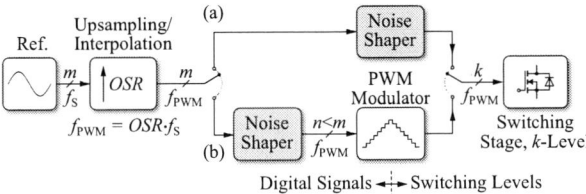

Fig. 14: Two possible signal flow paths in a digital, noise shaping modulation system for power electronic converters. (a): The noise shaper directly creates the switching signals. (b): The noise shaper feeds a PWM modulator which creates the switching levels.

In the first option, the noise shaper outputs a very low-resolution signal which is directly used as the switch control signal, e.g., a 1-bit signal in the case of a $k = 2$-level switching stage (cf. **Fig. 14** (a)). In the other option (cf. **Fig. 14** (b)), the noise shaper forms an output signal with a resolution of several

bits (e.g., $n = 8$) that is fed to a counter-based PWM modulator which creates the switch signals as illustrated in **Fig. 11** (a). This lower-resolution modulator can easily be implemented in digital hardware with clock frequencies not exceeding several hundred MHz (cf. eq. (4)).

Computer simulations with common noise shaper implementations [24] have been performed in order to determine which of the two options results in the lowest baseband (DC-10 kHz) quantization noise. The results show that option (b) generally performs better. This is due to the fact that the noise shapers with the low output resolution as used in **Fig. 14** (a) become unstable more easily than noise shapers with a higher output resolution. This limits their ability to attenuate the quantization noise sufficiently.

In the following, the structure of the evaluated noise shaper and its performance are presented in more detail.

C. Noise Shaper Implementation

Different common noise shaping structures [24] have been simulated with a signal flow path as illustrated in **Fig. 14** (b) and compared with respect to their achievable performance and stability. The structure presented in [25], named noise-coupled noise shaper (NCS), performed best and simulations verify its robust performance. **Fig. 15** (a) illustrates this noise shaper structure.

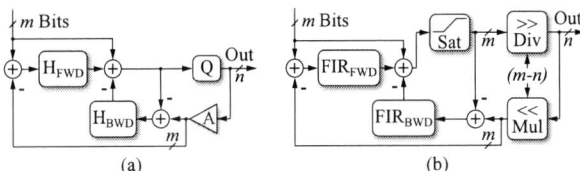

Fig. 15: (a): The noise-coupled noise shaping structure (NCS) used in this work. The transfer functions H_{FWD} and H_{BWD} define the noise transfer function (NTF) and can be implemented as FIR filters. (b): Digital implementation of the noise-coupled shaper.

The signal transfer function of this structure is $STF = 1$, which means that the desired signal passes the noise shaper without alteration. The NTF is given by

$$NTF = \frac{1 - H_{BWD}}{1 + H_{FWD}}. \tag{5}$$

If H_{FWD} and H_{BWD} have just zeros but no poles, they can be implemented as FIR filters. Furthermore, H_{BWD} then determines the zeros of the NTF and H_{FWD} its poles. Consequently, if the NTF is given, H_{BWD} and H_{FWD} are found by comparing the coefficients. The order of the noise shaper is determined by the number of coefficients in H_{FWD} and H_{BWD}. A higher-order noise shaper shows a better noise suppression, but is more prone to unstable behavior [22]. The gain block A (cf. **Fig. 15** (a)) simply rescales the quantizer output to the corresponding value in the quantizer's input range (multiplication by 2^{m-n}).

Fig. 15 (b) illustrates the digital implementation of the NCS. Only simple structures like FIR filters or bit shifts (division/multiplication by powers of 2) are required. The quantization noise is introduced when the quantizer (the divider) shifts the lower bits out of the signal (division without remainder).

D. Noise Shaper Optimization

The poles and zeros of the NTF can be optimally arranged such that the noise level is minimized in the baseband [22], [24]. Furthermore, the order of the noise shaper is an additional degree of freedom which influences the performance. However, the stability of different NS configurations with different input signals must be assessed using computer simulations. Such optimizations and stability assessments have been performed for the noise-coupled shaper (cf. **Fig. 15**) with the aim of finding stable configurations (i.e., zero/pole placement and order) with the best possible quantization noise suppression from DC to $10\,\mathrm{kHz}$.

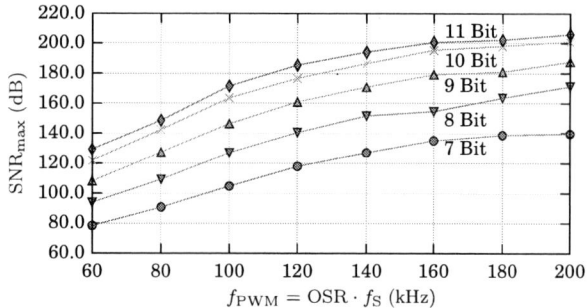

Fig. 16: Best achievable SNR of stable noise-coupled shapers with different output resolutions from DC to $10\,\mathrm{kHz}$. The reference input signal is a $80\,\mathrm{Hz}$ sinus with a resolution of 35 bits and a modulation index of $m = 0.9$.

Fig. 16 illustrates the optimization results. The best achievable SNR of the noise-coupled shapers is plotted for different OSR and noise shaper output bit widths n. The different noise shaper configurations feature orders ranging from 7 to 14.

E. Noise Shaper Verification

The NCS structure presented above has been implemented in an FPGA in order to verify its functionality. The testbench presented in **Fig. 7** is used together with the half-bridge setup described in section II-E to create a low-distortion and low-noise digital output signal. The supply of the half-bridge is fed by batteries ($9\,\mathrm{V}$) to minimize external noise sources and it is buffered by large DC link capacitors in order to minimize supply harmonics [13]. The noise shaper has an order of 11, an output resolution of $n = 9$ bit and the PWM switching frequency is $f_{\mathrm{PWM}} = 97.7\,\mathrm{kHz}$. The digital reference signal is a $170\,\mathrm{Hz}$ sinus with a resolution of 26 bits and a modulation index of $m = 0.85$. The noise shaper is optimized for a baseband from DC to $\approx 10\,\mathrm{kHz}$. The reference signal is sampled with f_{PWM} in order to eliminate the need for an upsampling/interpolation stage.

Fig. 17 shows the computer simulated amplitude spectra of the signals after this noise shaper and after the PWM modulator ($k = 2$ switching levels). The spectra are rescaled such that the fundamentals match in amplitude. At frequencies higher than $10\,\mathrm{kHz}$, the quantization noise increases significantly, as designed. The SNR of the noise shaper output signal from DC to $10\,\mathrm{kHz}$ is $138\,\mathrm{dB}$. The spectrum of the PWM output contains two harmonics. These originate from the non-naturally sam-

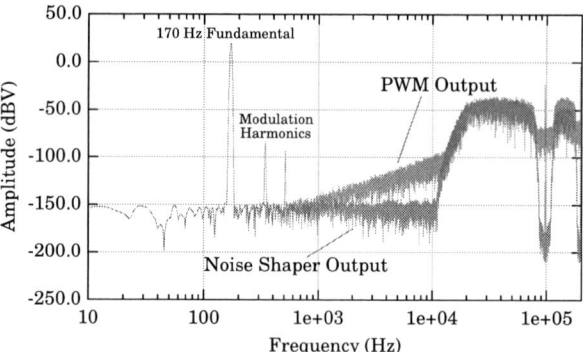

Fig. 17: Simulated amplitude spectra of the digital signal at the noise shaper output and the 1-bit switch signal from the PWM modulator's output. Rescaled in amplitude to match the fundamentals.

pling PWM modulator, as the duty cycle is updated once every switching period [17]. Pre-distortion techniques can potentially be used to reduce these harmonics [20], [26]. Additionally, the noise floor of the PWM output starts to rise at frequencies lower than $10\,\mathrm{kHz}$. This is an effect caused by intermodulation distortion at the PWM modulator's nonlinearities which leads to noise being folded back into the baseband [20]. The SNR of the PWM output is $97.8\,\mathrm{dB}$ and its total harmonic distortion (THD, [13]) is $-103\,\mathrm{dB}$.

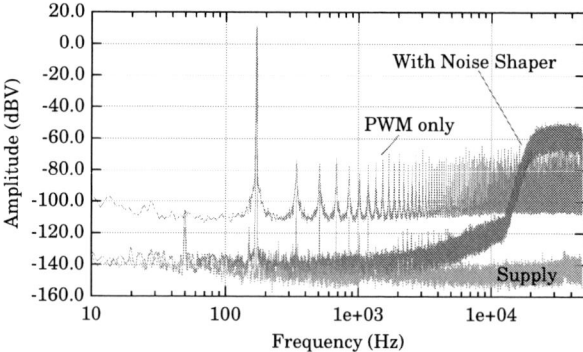

Fig. 18: Spectra obtained from a switched half-bridge output operated with and without the noise-coupled noise shaper (cf. **Fig. 15** and sec. III-E). The supply spectrum is measured at the input of the half-bridge.

In order to measure such low-noise signals, specialized equipment is required (*R&S UPV* Audio Analyzer). **Fig. 18** depicts the measurement of this noise shaper performed on the hardware demonstrator at the half-bridge output. If only a 9-bit PWM modulator would be used without noise shaping, the quantization noise would be significant and the SNR in the baseband from DC to $10\,\mathrm{kHz}$ would only be $65.8\,\mathrm{dB}$. As expected from the simulation results, the noise shaper can increase this figure to $97\,\mathrm{dB}$. There are, however, more harmonics in the noise-shaped spectrum than predicted by the simulations. This is due to the nonzero switching times of the power transistors, dead time and the non-ideal supply [13]. The THD of the noise-shaped signal is $-102\,\mathrm{dB}$ (considering only the first two harmonics), which is a close match with the simulation. Note that an RMS jitter of ≈ 26 ps is still present in the switched output waveform. This limits the best achievable SNR in the 10 kHz baseband to 106 dB (cf. eq. (2)), which is close to what is shown by the measurement.

In order to compare the measurement with the simulation, as the spectra are computed with different FFT windows and lengths, **Fig. 19** shows the power spectrum densities (PSD) of the measured signal and the simulation, which allows a direct comparison of the noise levels [6]. At frequencies from DC to ≈ 2 kHz, the noise floor of the measurement is higher than the simulation's due to the remaining RMS jitter of ≈ 26 ps at the half-bridge output. Otherwise, the measurement matches the simulation well which verifies the effectiveness of the noise shaper.

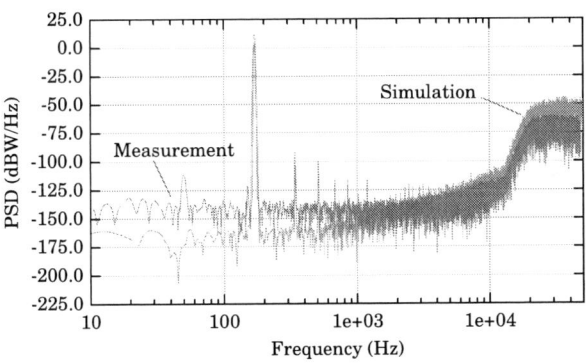

Fig. 19: Power spectrum densities of the measured half-bridge output and its computer simulation. Reference impedance: 1 Ω.

IV. Conclusion & Outlook

Two systems which allow the creation of high-SNR, low-noise, PWM modulated output signals for a switched-mode amplifier are presented. A special gate driver can significantly reduce jitter at the output of a 400 V GaN half-bridge from values of ≈ 250 ps to values under 30 ps while being immune to repeated common-mode transients in excess of $200 \, \text{kV} \, \mu\text{s}^{-1}$ across its isolation barrier.

A digital modulator based on noise shaping, which reduces quantization noise in a given frequency band through feedback action, can create high-SNR PWM signals which exceed the noise performance of traditional modulators. The quantization noise is shifted to higher frequencies, where it can be low-pass filtered and where it is of no concern to the mechatronic actuators. Using these techniques, open-loop half-bridge output voltage SNR figures of nearly 100 dB (DC-10 kHz) are achieved.

These results are obtained in an open-loop voltage output system. The overall distortion at the amplifier output (in both voltage and current) can be improved by employing feedback, which attenuates unwanted harmonics or noise. As an output current with an SNR in excess of 110 dB is desired by the industry, specialized low-noise and highly linear current sensors are required. Furthermore, in order to obtain a complete high-power, and low-noise mechatronic amplifier system, the supply of the power stage must provide an output of low noise and

low parasitic series impedance in order to minimize distortion at the amplifier output [13].

References

[1] R. Munnig Schmidt, G. Schitter and J. van Eijk, *The Design of High Performance Mechatronics*. Delft University Press, 2011, ISBN: 978-1-60750-825-0.

[2] J.M. Schellekens, "A Class of Robust Switched-Mode Power Amplifiers with Highly Linear Transfer Characteristics," *PhD Thesis, University of Eindhoven*, 2014, ISBN: 978-94-6259-400-5.

[3] L. Risbo and T.Mørch, "Performance of an All-Digital Power Amplification System," in *Proc. of the 104th Audio Engineering Society Convention*, May 1998.

[4] W. Kester, *The Data Conversion Handbook*. Elsevier, 2005, ISBN: 0-7506-7841-0.

[5] W.B. Davenport, Jr., and W.L. Root, *An Introduction to the Theory of Random Signals and Noise*. McGraw-Hill, Lincoln Laboratory, Massachusetts Institute of Technology, 1958.

[6] G. Heinzel, A. Rüdiger and R. Schilling, "Spectrum and Spectral Density Estimation by the Discrete Fourier Transform (DFT), including a Comprehensive List of Window Functions and some New Flat-Top Windows," *Max-Planck-Institut für Gravitationsphysik*, 2002.

[7] M. Shimanouchi, "An Approach to Consistent Jitter Modeling for Various Jitter Aspects and Measurement Methods," in *Proc. of the International Test Conference*, 2001, pp. 848–857.

[8] M. Peng Li, *Jitter, Noise, and Signal Integrity at High-Speed*. Prentice Hall, 2007, ISBN: 978-0-13-242961-0.

[9] B. Kelleci, E. Sanchez-Sinencio and A.I. Karsilayan, "THD+Noise Estimation in Class-D Amplifiers," in *IEEE International Symposium on Circuits and Systems (ISCAS)*, May 2007, pp. 465–468.

[10] K. Gingerich and C. Sterzik, "The ISO72x Family of High-Speed Digital Isolators," January 2006, TI Application Report SLLA198.

[11] M.J. Hawksford, "Dynamic Jitter Filtering in High-Resolution DSM and PWM Digital-to-Analog Conversion," in *Proc. of the Audio Engineering Society Convention*, Feb 1994.

[12] M.A. Alexander, "PWM Re-Clocking Scheme to Reject Accumulated Asynchronous Jitter," June 2014, US Patent US8742841 B2.

[13] M. Mauerer, A. Tüysüz and J.W. Kolar, "Distortion Analysis of Low-THD/High-Bandwidth GaN/SiC Class-D Amplifier Power Stages," in *Proc. of the Energy Conversion Congress and Exposition (ECCE)*, Sept 2015, pp. 2563–2571.

[14] Silicon Labs, "Si861x/2x Data Sheet," Rev. 1.6.

[15] Rohde&Schwarz, "Jitter Analysis with the R&S RTO Digital Oscilloscope." August 2013, Application Note 1TD03_2e.

[16] J.W. Kolar, J. Miniböck and M. Baumann, "Three-Phase PWM Power Conversion - The Route to Ultra High Power Density and Efficiency," in *Proc. of the CPES Annual Seminar/Industry Review*, April 2003.

[17] D.G. Holmes and T.A. Lipo, *Pulse Width Modulation for Power Converters - Principles and Practice*. Wiley-IEEE, 2003, ISBN: 978-0-471-20814-3.

[18] W.R. Bennett, "Spectra of Quantized Signals," in *Bell System Technical Journal*, 1948.

[19] *Xilinx LogiCORE IP Binary Counter v11.0 - DS215*, Xilinx, 2011.

[20] P. Craven, "Toward the 24-bit DAC: Novel Noise-Shaping Topologies Incorporating Correction for the Nonlinearity in a PWM Output Stage," *Journal of the Audio Engineering Society*, vol. 41, no. 5, pp. 291–313, 1993.

[21] J.M. Goldberg and M.B. Sandler, "Noise Shaping and Pulse-Width Modulation for an All-Digital Audio Power Amplifier," *Journal of the Audio Engineering Society*, vol. 39, no. 6, pp. 449–460, 1991.

[22] S. R. Norsworthy, R. Schreier and G.C. Temes, *Delta-Sigma Data Converters*. Wiley-IEEE, 1996, ISBN: 0-7803-1045-4.

[23] M. Norris, L.M. Platon, E. Alarcon and D. Maksimovic, "Quantization Noise Shaping in Digital PWM Converters," in *Proc. of the IEEE Power Electronics Specialists Conference (PESC)*, June 2008, pp. 127–133.

[24] R. Schreier and G.C. Temes, *Understanding Delta-Sigma Data Converters*. Wiley-IEEE, 2005, ISBN: 0-471-46582-2.

[25] K. Lee, M. Bonu and G.C. Temes, "Noise-coupled $\Delta\Sigma$ ADCs," *Electronics Letters*, vol. 42, no. 24, pp. 1381–1382, November 2006.

[26] M.O.J. Hawksford, "Dynamic Model-Based Linearization of Quantized Pulse-Width Modulation for Applications in Digital-to-Analog Conversion and Digital Power Amplifier Systems," *Journal of the Audio Engineering Society*, vol. 40, no. 4, pp. 235–252, 1992.

Analysis and Compensation of Inverter Nonlinearity for Three-Level T-Type Inverters

Hyeon-Sik Kim, Yong-Cheol Kwon, Seung-Jun Chee, and Seung-Ki Sul

Department of Electrical and Computer Engineering
Seoul National University
Seoul, Korea
hyeonsik@eepel.snu.ac.kr, dydcjfe@eepel.snu.ac.kr, cheesj80@eepel.snu.ac.kr, sulsk@plaza.snu.ac.kr

Abstract—This paper analyzes the inverter nonlinearity effect resulting in issues such as narrow pulses and the even order harmonics in three-level T-type inverters. These issues make the compensation of the inverter nonlinearity be difficult. Based on the analysis of the output voltage distortion, carrier-based PWM methods to avoid these issues and to balance dc-link voltages simultaneously are proposed using the concept of the offset voltage. The proposed PWM methods can be easily implemented by adding appropriate offset voltages to output voltage references. Also, a compensation method to alleviate inverter nonlinearity effects is proposed based on the modeling of the inverter nonlinearity. The effectiveness of proposed methods is verified by experimental results. Through the proposed algorithms, not only even harmonics but also 5th and 7th harmonic components of current are conspicuously reduced. At the same time, the neutral voltage of the inverter can be balanced effectively by the proposed PWM methods.

Keywords—*Inverter nonlinearity; three-level topology; T-type inverter; offset voltage; neutral point (NP) voltage control*

I. INTRODUCTION

Multi-level Pulse-Width Modulation (PWM) Voltage Source Inverters (VSIs) are getting attentions thanks to their advantages over two-level VSIs such as better harmonic characteristics, smaller *dv/dt*, and higher efficiency. Among many multi-level inverters, three-level inverters such as Neutral Point Clamped (NPC) topology and T-type topology have been widely used because of their relatively simple control and technical maturity. Compared to NPC topology, T-type topology is more preferred particularly in low voltage applications where the conduction losses could be minimized due to the reduced number of switching semiconductors [1].

However, due to increased number of switching states, the effect of the inverter nonlinearity in three-level topology is more complicated compared to that in two-level topology. Inverter nonlinearities coming from the dead time, parasitic capacitors, and voltage drop across switching devices provoke the distortion of the output voltage of the inverter and degrades overall performances of the variable speed electric machine drive system.

There have been many researches on the inverter nonlinearity compensation for two-level inverter [2]-[9]. In the early literatures [2]-[5], inverter nonlinearity effects had been compensated by adding a compensation voltage to voltage references [2]-[3] or adjusting the length of gating pulses [4]-[5]. The effect of parasitic capacitances [5]-[7] and voltage

drop across switching devices [2], [7] had been included to improve the compensation accuracy. Also, the inverter nonlinearity in multi-level inverters and its compensation have been addressed in several papers [10]-[12]. However, the analysis of inverter nonlinearity effects was done similarly to that of two-level inverter. And, the most of previous works have not considered the special characteristics of three-level T-type inverters.

The majority of them didn't cover output voltage distortion induced by, so called, narrow pulse problem [8]-[9] that occurs when pole voltage references are near the edges of PWM carriers. Since this problem becomes more severe in the case of multi-level inverters, it should be definitely be considered and compensated. In [12], the narrow pulse problem in over-modulation region was considered on the basis of Space Vector PWM (SVPWM). Nevertheless, SVPWM method was complicated to accommodate the compensation algorithm against the narrow pulse problem because the on-time of each switch of the inverter should be geometrically computed. Compared to SVPWM method, the carrier-based PWM method has been known to be equivalent to SVPWM but simple to implement and could avoid narrow pulse problem instinctively. Also, even order harmonics from the nonlinearity of T-type inverters have not been covered properly, although it causes considerable distortion of the output voltage.

In this paper, inverter nonlinearity effects of three-level T-type inverter resulting in the narrow pulse and even order harmonic issues are addressed in detail. Based on the analyses, a carrier-based PWM technique and a compensation method to alleviate inverter nonlinearity effects are proposed. In proposed PWM methods, not only output voltage linearity but also controllability of Neutral Point (NP) voltage are considered. By adjusting essential parameters in the proposed PWM methods,

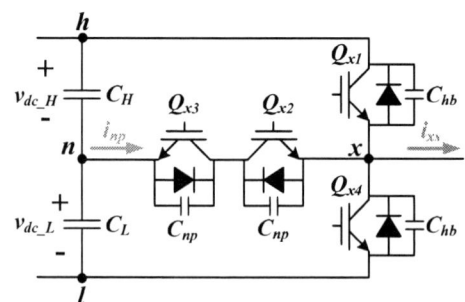

Fig. 1. One leg of a T-type inverter.

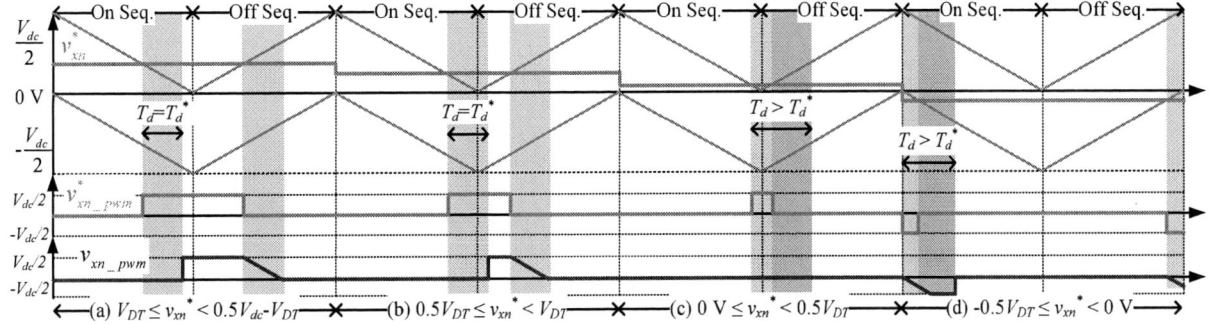

Fig 2. Dead time effect due to narrow pulses.

NP balancing could be realized at the same time. Several experimental results are provided to verify the validity of the proposed method.

II. ANALYSIS OF INVERTER NONLINEARITY EFFECTS IN THREE-LEVEL T-TYPE INVERTERS

Fig. 1 shows one leg of a T-type inverter, where x denotes an arbitrary phase among a, b, and c. Because of the junction capacitance of the semiconductor switches, there are naturally parasitic capacitors connected in parallel with the half-bridge switches (C_{hb}) and bidirectional switches (C_{np}). For the output pole voltage, v_{xn}, three switching states can be defined as follows: "H" state when phase current, i_{xs}, flows through the upper switch, "M" state, through the bidirectional switch, and "L" state, through the lower switch.

There are two main effects that provoke output voltage error, δv_{xn}, defined as (1), where v_{xn}^* indicates a pole voltage reference, and v_{xn}, actual pole voltage. Note that all the variables in (1) are per-switching-cycle average quantities.

$$\delta v_{xn} = v_{xn}^* - v_{xn}. \tag{1}$$

A. Dead time effect due to narrow pulses

At first, the voltage distortion from the narrow pulse can be depicted as Fig. 2, where the dead time is set as T_d. Fig. 2 shows generation of x-phase pole voltage in a three-level inverter based on level-shifted carrier waves, considering the dead time effect. In this figure, the values with subscript 'pwm' indicate instantaneous quantities and V_{DT} is defined as $V_{DT} = T_d / T_{sw} \cdot V_{dc}$. If v_{xn}^* is between V_{DT} and $(0.5V_{dc}-V_{DT})$ as shown in Fig. 2(a), the dead time effect in three-level inverters is very similar with that in two-level inverters. At on-sequence, where the switches are turning on, the output voltage is simply delayed by T_d, which leads to volt-sec loss in the inverter output. At off-sequence, where the switches are turning off, the output voltage doesn't fall instantly after the gating signal is turned off because of the parasitic capacitors, which leads to volt-sec gain in the inverter output. In this condition, the falling rate is determined by the parasitic capacitances and the current during the dead time. When v_{xn}^* is set between $0.5V_{DT}$ and V_{DT} as shown in Fig. 2(b), the dead time at on-sequence affects the output voltage at off-sequence. However, the dead time effect in this case is the same as that in the previous case for a switching-period in the average manner. However, if v_{xn}^* is close to 0V as shown in Fig. 2(c), the actual pole voltage v_{xn_pwm} is clamped to 0V because the two dead time durations at

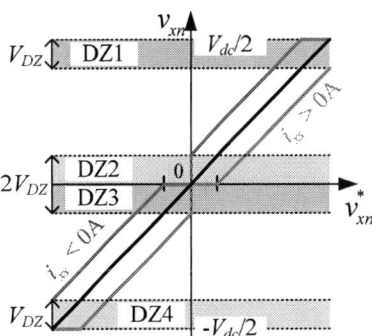

Fig. 3. Relation between v_{xn}^* and i_{xs} disregarding parasitic capacitors.

on and off sequences are overlapped and then turn-on signal for the switch is cut off. This phenomenon is called narrow pulse problem [8]. If v_{xn}^* is negative and is getting closer to the peak of the carrier, larger distortion occurs in the output voltage as shown in Fig. 2(d).

Fig. 3 shows the distortion of the output voltage assuming that the parasitic capacitor doesn't exist. In this figure, $V_{DZ}=0.5V_{DT}$ and four regions, denoted as DZ1-DZ4 whose widths are determined by V_{DZ}, indicate dead zones where the output voltage can't be synthesized properly. DZ1 and DZ3 arise with the positive phase current, whereas DZ2 and DZ4 arise with the negative phase current. This narrow pulse problem has already been investigated for two-level inverters [8]-[9]. However, this problem occurs in two-level inverters only under high Modulation Index (MI) operation where v_{xn}^* goes near the peak or valley of the carrier wave, e.g. in the case near or at over-modulation range. However, in three-level inverters, such problem could occur even under low MI operation since DZ2 and DZ3 are located at near zero voltage, which hasn't been sufficiently covered in the previous literatures.

Considering all the above-mentioned behaviors, the voltage error induced by the dead time, δv_{xn_DT}, can be derived from the difference between $v_{xn\_pwm}^*$ and v_{xn_pwm} within a switching period. δv_{xn_DT} at outside of the dead zones can be expressed as (2), where C_{eff} is effective parasitic capacitance which is defined as (3). Note that δv_{xn_DT} in (2) is an averaged value in a switching period. δv_{xn_DT} according to i_{xs} and v_{xn}^* with $f_{sw}=10$ kHz, $T_d=3$ μs, and $V_{dc}=310$ V is illustrated in Fig. 4. It is worth noting that δv_{xn_DT} is not only a function of the current but also affected by the voltage reference itself.

978-1-4673-9551-9/16 $31.00 © 2016 IEEE

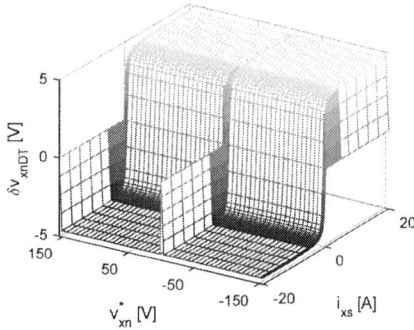

Fig. 4. δv_{xn_DT} according to i_{xs} and $v_{xn}{}^*$.

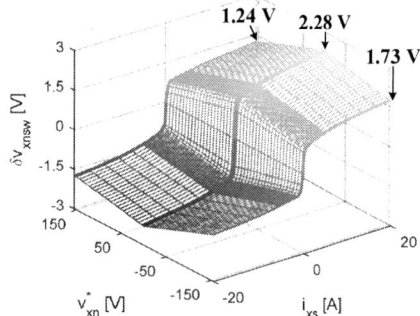

Fig. 5. δv_{xn_SW} according to i_{xs} and $v_{xn}{}^*$.

$$\delta v_{xn\_DT} = \begin{cases} -\dfrac{1}{8}\dfrac{C_{eff}V_{dc}^2}{T_{sw}}\dfrac{1}{i_{xs}} - \dfrac{T_d}{T_{sw}}\dfrac{V_{dc}}{2} & (i_{xs} \le -I_{crit}) \\[2mm] \dfrac{T_d}{T_{sw}}\dfrac{T_d}{2C_{eff}}i_{xs}. & (-I_{crit} \le i_{xs} < I_{crit}) \\[2mm] -\dfrac{1}{8}\dfrac{C_{eff}V_{dc}^2}{T_{sw}}\dfrac{1}{i_{xs}} + \dfrac{T_d}{T_{sw}}\dfrac{V_{dc}}{2} & (I_{crit} \le i_{xs}) \end{cases} \quad (2)$$

$$C_{eff} = 2C_{hb} + C_{np} . \quad (3)$$

B. Voltage drop of switching devices

In T-type inverters, pole voltage error induced by voltage drop of switching devices, δv_{xn_SW}, depends on the switching state. As shown in Fig. 1, the current should flow through the two switches, Q_{x2} and Q_{x3}, for "M" state, while, only one switch is in the current conduction path at "H" and "L" states. For this reason, δv_{xn_SW} at "M" state is distinctively large compared with that at other states. Considering the duty ratio, per-switching-cycle average value of δv_{xn_SW} can be modeled as a function of i_{xs} and $v_{xn}{}^*$. Fig. 5 shows δv_{xn_SW} according to i_{xs} and $v_{xn}{}^*$ in the case of 1200V, 80A T-type IGBT module (VINCOTECH 10-FZ12NMA080SH01-M260F), which is used in the experiments. In this figure, it can be seen that the voltage error becomes larger as the magnitude of $v_{xn}{}^*$ is getting smaller. The reason is that the duty ratio of the bidirectional switch increases as the magnitude of $v_{xn}{}^*$ is getting reduced to zero.

Total voltage error is simply a sum of δv_{xn_DT} and δv_{xn_SW}, i.e., $\delta v_{xn} = \delta v_{xn\_DT} + \delta v_{xn\_SW}$. Because δv_{xn_SW} takes large portion in δv_{xn} compared to the case of two-level inverters, the dependency of δv_{xn_SW} on $v_{xn}{}^*$ should be considered carefully.

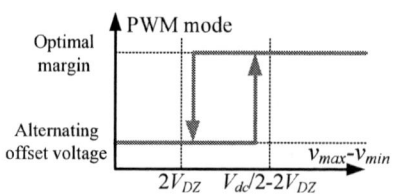

Fig. 6. Transition of PWM mode.

III. Proposed Scheme

As described in Fig. 3-4, it is physically impossible to synthesize v_{xn} in the dead zones. For proper synthesis of the output voltage, pole voltage references, $\mathbf{v}_{abcn}^* \equiv \begin{bmatrix} v_{an}^* & v_{bn}^* & v_{cn}^* \end{bmatrix}^T$, should be located outside of region where the dead time effect is dominant, so called as the dead zones. In the carrier-based PWM method, this can be implemented by adding an offset voltage which is also called zero sequence voltage, v_{sn}^*, to phase voltage references, $\mathbf{v}_{abcs}^* \equiv \begin{bmatrix} v_{as}^* & v_{bs}^* & v_{cs}^* \end{bmatrix}^T$. This operation is expressed as (4).

$$v_{xn}^* = v_{xs}^* + v_{sn}^* . \quad (4)$$

In the proposed scheme, PWM method to avoid the dead zones is changed according to $v_{max}-v_{min}$ as shown in Fig. 6, where v_{max} and v_{min} are defined as (5). For stable transition, hysteresis is applied in the transition function. Under the assumption that $\mathbf{v}_{abcn}^*$ are properly located off the dead zones, the inverter nonlinearity can be compensated using a compensation function according to i_{xs} and $v_{xn}{}^*$.

$$\begin{aligned} v_{max} &= \max(\mathbf{v}_{abcs}^*). \\ v_{min} &= \min(\mathbf{v}_{abcs}^*). \end{aligned} \quad (5)$$

A. Technique to avoid dead zone using alternating offset voltage PWM

When $\mathbf{v}_{abcs}^*$ are small in low MI operation, shifting all of $\mathbf{v}_{abcs}^*$ to the middle of a carrier is the only way to avoid the dead zones. By doing so, the output voltage can be generated by only two switching states, "H" and "M" states or "M" and "L" states. By excluding one switching state, three-level inverters can operate like two-level inverter in low MI operation. One of the most popular settings of the offset voltage for two-level VSIs can be expressed as (6), which is known as equivalent to three-phase symmetry SVPWM [13]. This method will be called as Symmetrical Continuous PWM (SCPWM) hereafter. Since $\mathbf{v}_{abcn}^*$ generated by (6) cross the dead zones when using three-level inverters, it can be modified to positive offset voltage, (7), or negative offset voltage, (8), which also minimizes sideband harmonics at the carrier frequency. Fig. 7(a) shows the waveform of $\mathbf{v}_{abcn}^*$ using (7).

$$v_{sn}^* = -\frac{v_{max} + v_{min}}{2} . \quad (6)$$

$$v_{sn}^* = -\frac{v_{max} + v_{min}}{2} + \frac{v_{dc\_H}}{2} . \quad (7)$$

$$v_{sn}^* = -\frac{v_{max} + v_{min}}{2} - \frac{v_{dc\_L}}{2} . \quad (8)$$

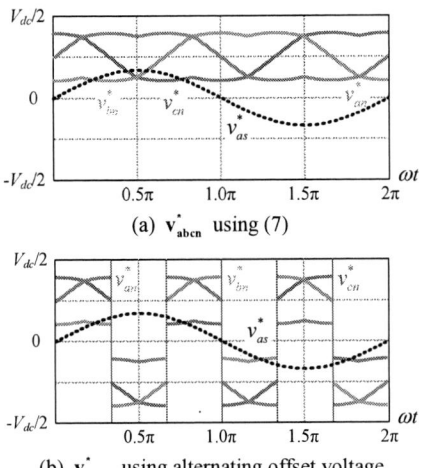

(a) $\mathbf{v}_{abcn}^*$ using (7)

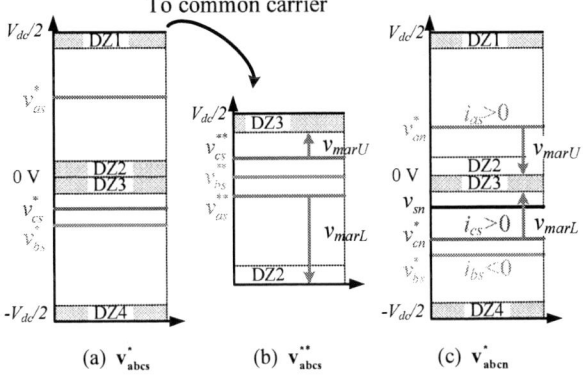

(b) $\mathbf{v}_{abcn}^*$ using alternating offset voltage

Fig. 7. Waveform of $\mathbf{v}_{abcn}^*$ at low MI.

(a) $\mathbf{v}_{abcs}^*$ (b) $\mathbf{v}_{abcs}^{**}$ (c) $\mathbf{v}_{abcn}^*$

Fig. 8. Principle of optimal margin PWM.

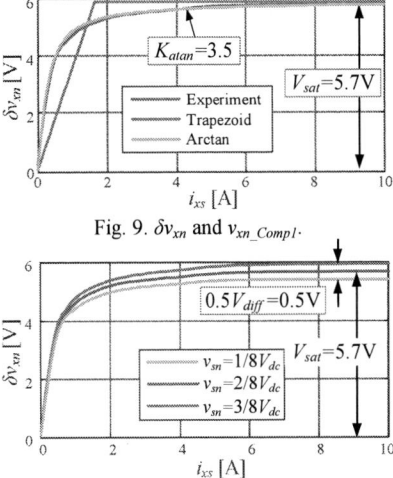

Fig. 9. δv_{xn} and v_{xn_Comp1}.

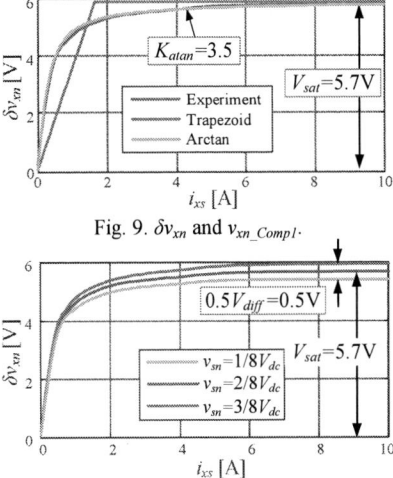

Fig. 10. Experimentally measured δv_{xn} according to v_{sn}.

However, applying either (7) or (8) provokes even order harmonics in the output voltage. Even order harmonics come up due to the asymmetry of δv_{xn_SW} shown in Fig. 5. Since δv_{xn_SW} becomes larger as v_{xn}^* is closer to zero, v_{xn}^* in Fig. 7(a) suffers more voltage loss at valley than at peak, which breaks half-wave symmetry of the output voltage, i.e., $\delta v_{xn}(\theta){\neq}{-}\delta v_{xn}(\theta{-}180°)$. To satisfy $\delta v_{xn}(\theta){=}{-}\delta v_{xn}(\theta{-}180°)$, v_{sn}^* can be changed between (7) and (8) alternately with 120° period as depicted in Fig. 7(b). Using this method, namely Alternating Offset Voltage PWM (AOVPWM), even order harmonics in the output voltage can be successfully diminished.

B. Technique to aviod dead zone using optimal margin PWM

Over a certain MI, $\mathbf{v}_{abcn}^*$ can reach the dead zones even though AOVPWM is applied. In this case, one or two of $\mathbf{v}_{abcn}^*$ have to be located in upper carrier region and the other has to be located in lower carrier region. In the region with middle and high MI, the optimal margin PWM (OMPWM) method can be used to avoid the dead zones.

As a basic principle, OMPWM optimizes the voltage margin between $\mathbf{v}_{abcn}^*$ and the dead zones which can be utilized for balancing the upper and lower capacitor voltages of dc-link

of the three-level inverters. Fig. 8 shows the setting $\mathbf{v}_{abcn}^*$ in OMPWM method. If $\mathbf{v}_{abcs}^*$ is given as in Fig. 8(a), negative phase voltages are shifted to the upper carrier region by applying (9) as seen in Fig. 8(b). In the figure, v_{marU} and v_{marL} indicate upper and lower voltage margin, respectively. The dead zones are affected by the polarity of the current. For example, if $i_{as}>0$ and $i_{cs}>0$ in Fig. 8(b), DZ3 is effective but DZ2 isn't. In this case, v_{marU} and v_{marL} can be computed by (10). In order to avoid the dead zones and to secure maximum voltage margin, it is appropriate to set $|v_{marU}|{=}|v_{marL}|$. For maximum voltage margin, the offset voltage can be set as (11). Fig. 8(c) shows $\mathbf{v}_{abcn}^*$ after applying (11). It can be found that the magnitudes of v_{marU} and v_{marL} become the same in the figure.

$$v_{xs}^{**} = v_{xs}^* + V_{dc}/2 . \tag{9}$$

$$v_{marU} = V_{dc}/2 - V_{DZ} - v_{cs}^{**}. \tag{10}$$
$$v_{marL} = 0\,\mathrm{V} - v_{as}^{**}.$$

$$v_{sn}^* = \frac{v_{marU} + v_{marL}}{2} . \tag{11}$$

C. Compensation of inverter nonlinearity

Fig. 9 shows experimentally measured δv_{xn} according to i_{xs}, with v_{sn} set as (7) to locate the $\mathbf{v}_{abcn}^*$ off the dead zones. The performance of the inverter nonlinearity compensation is determined by the shape of compensation function, v_{xn_Comp}. Among many compensation functions such as a step function, a trapezoidal voltage, and an arctangent function, the arctangent function defined as (12) is selected for the compensation function since it exhibits great agreement with the actual δv_{xn} as shown in Fig. 9. v_{xn_Comp1} is suitable for conventional NPC topology because the difference of δv_{xn_SW} between switching states would be negligible.

$$v_{xn\_Comp1} = V_{sat} \cdot \frac{2}{\pi} \arctan(K_{atan} \cdot i_{xs}) . \tag{12}$$

In T-type inverter, δv_{xn_SW} is not only a function of the current but also affected by pole voltage reference. To consider asymmetric δv_{xn_SW}, the compensation function should be modified like v_{xn_Comp2} as shown in (13), where V_{diff} means the difference of δv_{xn} between "M" state and "H" or "L" state. To get V_{diff} experimentally, δv_{xn} should be extracted in various v_{sn} at outside of the dead zones. Fig. 10 shows experimentally measured δv_{xn} according to v_{sn}. The resistance of switch modules is excluded when δv_{xn} is extracted because it cannot be distinguished from that of load connected to the inverter.

$$v_{xn\_Comp2} = (V_{sat} + V_{diff}(\frac{|v_{xn}^*|}{0.5V_{dc}} - 0.5)) \cdot \frac{2}{\pi} \arctan(K_{a\tan} \cdot i_{xs}). \quad (13)$$

IV. VOLTAGE BALANCING CONTROL OF PROPOSED SCHEME

For three-level inverters, NP current causes the voltage difference between capacitors in the dc-link. NP balancing algorithms are widely discussed in the past literatures [14]-[16]. But none of them have dealt with the compensation of inverter nonlinearity effects because the offset voltage selection was only based on NP balancing objective. v_{xn}^* determined by conventional NP balancing algorithms could be located in the dead zones. Hence, NP balancing control should consider inverter nonlinearity effects to minimize the distortion of the output voltage.

Conventional NP balancing methods aren't compatible with the proposed PWM methods because both algorithms use the offset voltage simultaneously for different purposes. This problem is not severe for three-level converters with active front-ends such as boost PWM converter where either the offset voltage of PWM converter or that of PWM inverter could be used to balance NP potential. However, considering passive front ends such as diode rectifier, voltage balancing controls incorporated with the proposed PWM methods should be devised for maintaining voltage balance at NP. Proposed NP controllers in this paper are optimized for each PWM method. It has a low impact on original PWM algorithms at slight dc-link unbalance conditions and is easy to implement. And, it reveals better THD performance than other NP balancing algorithms.

A. Analysis of NP current versus offset voltage

NP current averaged over a switching period, i_{np} shown in Fig. 1, can be analytically calculated [14]-[15]. Measured dc-link voltages, v_{dc_H} and v_{dc_L} defined in Fig. 1, are used to synthesize proper v_{xn}^* even when the difference between high- and low-side dc-link voltages, $\delta v_{dc} \equiv v_{dc\_H} - v_{dc\_L}$, exists. i_{np} can be calculated in three-phase system as (14), where D_{xn} is duty ratio to neutral point which is defined as (15).

$$i_{np} = D_{an}i_a + D_{bn}i_b + D_{cn}i_c. \quad (14)$$

$$D_{xn} = \begin{cases} 1 + v_{xn}/v_{dc\_L} & (v_{xn} < 0) \\ 1 - v_{xn}/v_{dc\_H} & (v_{xn} \geq 0) \end{cases}. \quad (15)$$

i_{np} varies according to v_{sn}, so adding a proper offset voltage is a key factor to minimize δv_{dc}. The relationship between i_{np} and v_{sn} can be derived as (16), where v_{max}, v_{mid}, and v_{min} are defined as the maximum, medium, and minimum pole voltages, respectively. i_{max}, i_{mid}, and i_{min} are the currents of corresponding

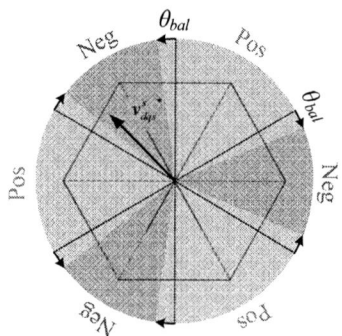

Fig. 11. Principle of NP balancing for AOVPWM.

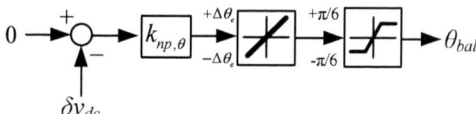

Fig. 12. Block diagram of NP controller for AOVPWM.

phases. To simplify the analysis, the output power, P_{out} defined as (17), is assumed to be a constant over an electrical rotating period.

$$i_{np} = \begin{cases} -\dfrac{P_{out}}{v_{dc\_H}} & (v_{sn} \geq -v_{min}) \\[2mm] -\dfrac{P_{out}}{v_{dc\_H}} + (\dfrac{1}{v_{dc\_H}} + \dfrac{1}{v_{dc\_L}})(v_{min} + v_{sn})i_{min} & (-v_{mid} \leq v_{sn} < -v_{min}) \\[2mm] \dfrac{P_{out}}{v_{dc\_L}} - (\dfrac{1}{v_{dc\_H}} + \dfrac{1}{v_{dc\_L}})(v_{max} + v_{sn})i_{max} & (-v_{max} \leq v_{sn} < -v_{mid}) \\[2mm] \dfrac{P_{out}}{v_{dc\_L}} & (v_{sn} < -v_{max}) \end{cases} \quad (16)$$

$$P_{out} = v_{max}i_{max} + v_{mid}i_{mid} + v_{min}i_{min}. \quad (17)$$

B. NP controller for alternating offset voltage PWM

All pole voltages are shifted to one of the carrier region for AOVPWM. And, when the positive offset voltage, (7), is applied, all pole voltages are positive, i.e., $v_{sn} \geq -v_{min}$. But when the negative offset voltage, (8), is applied, all pole voltages are negative, i.e., $v_{sn} < -v_{max}$. In either condition, i_{np} can't be changed by shifting v_{sn}. But, the polarity of i_{np} is different in the case of (7) and (8). And, NP balancing can be achieved by controlling the ratio of each offset voltage.

Fig. 11 shows the principle of NP balancing for AOVPWM. NP current averaged over an electrical rotating period, $\bar{i}_{np}$, can be adjusted by balancing angle, θ_{bal}. $\bar{i}_{np}$ can be expressed as a linear function of θ_{bal} in (18).

$$\bar{i}_{np} = \frac{1}{2\pi} \int_0^{2\pi} i_{np}(\theta) d\theta$$
$$= \frac{P_{out}}{2}(\frac{1}{v_{dc\_L}} - \frac{1}{v_{dc\_H}}) - \frac{3P_{out}}{\pi}(\frac{1}{v_{dc\_L}} + \frac{1}{v_{dc\_H}})\theta_{bal}. \quad (18)$$

NP controller for AOVPWM, namely θ_{bal} controller, consists of proportional gain, rate limiter, and limiter as shown

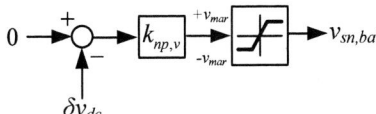

Fig. 13. Block diagram of NP controller for OMPWM.

in Fig. 12. The rate limiter is chosen to suppress rapid variation of offset voltage in low rotating speed. When NP controller is operated continuously, δv_{dc} would converge to zero voltage, i.e. $v_{dc\_H} = v_{dc\_L}$. Then, the relation between $\bar{i}_{np}$ and θ_{bal} can be simplified and the gain of controller, $k_{np,\theta}$, can be set as (19), where ω_{vc} is the bandwidth of NP control loop.

$$k_{np,\theta} = -\frac{\pi}{12} C_{dc} V_{dc} \frac{\omega_{vc}}{P_{out}} . \qquad (19)$$

C. NP controller for optimal margin PWM

Under OMPWM, there can be only on or two positive pole voltages since v_{sn} is limited as $-v_{max} \le v_{sn} < -v_{min}$. In this v_{sn} range, i_{np} can be changed by shifting v_{sn}. Fig. 13 shows the block diagram of NP controller for OMPWM. Balancing offset voltage, $v_{sn,bal}$, is determined by proportional controller with limiter. It has similar structure with the dc common-mode voltage injection algorithm [16]. But the limiter is added to constraint $v_{sn,bal}$ within v_{mar} to keep pole voltage references out of the dead zones. The proportional gain, $k_{np,v}$, can be set in consideration of the performance of NP balancing and current

Fig. 14. Experimental Setup.

TABLE I. PARAMETERS OF INDUCTION MOTOR

Parameter	Value
Rated Power	3.7 kW
Pole number	4
R_s / R_r	0.22 Ω / 0.3 Ω
L_m	63.62 mH
L_{ls} / L_{lr}	2.44 mH / 2.44 mH
Rated torque	11.5 N·m

harmonics. It can be set as -0.5 because it locates pole voltage references center of dc-link voltage. Proposed NP controllers can be utilized not only for motoring but also regenerative operation by changing sign of output variables, θ_{bal} or $v_{sn,bal}$.

V. EXPERIMENTAL RESULTS

The experimental setup for the verification of the proposed scheme has been built as shown in Fig. 14. The induction machine under test is controlled by a three-level T-type inverter. The parameters of the induction machine are shown in Table I. The target machine is simply driven by V/f control.

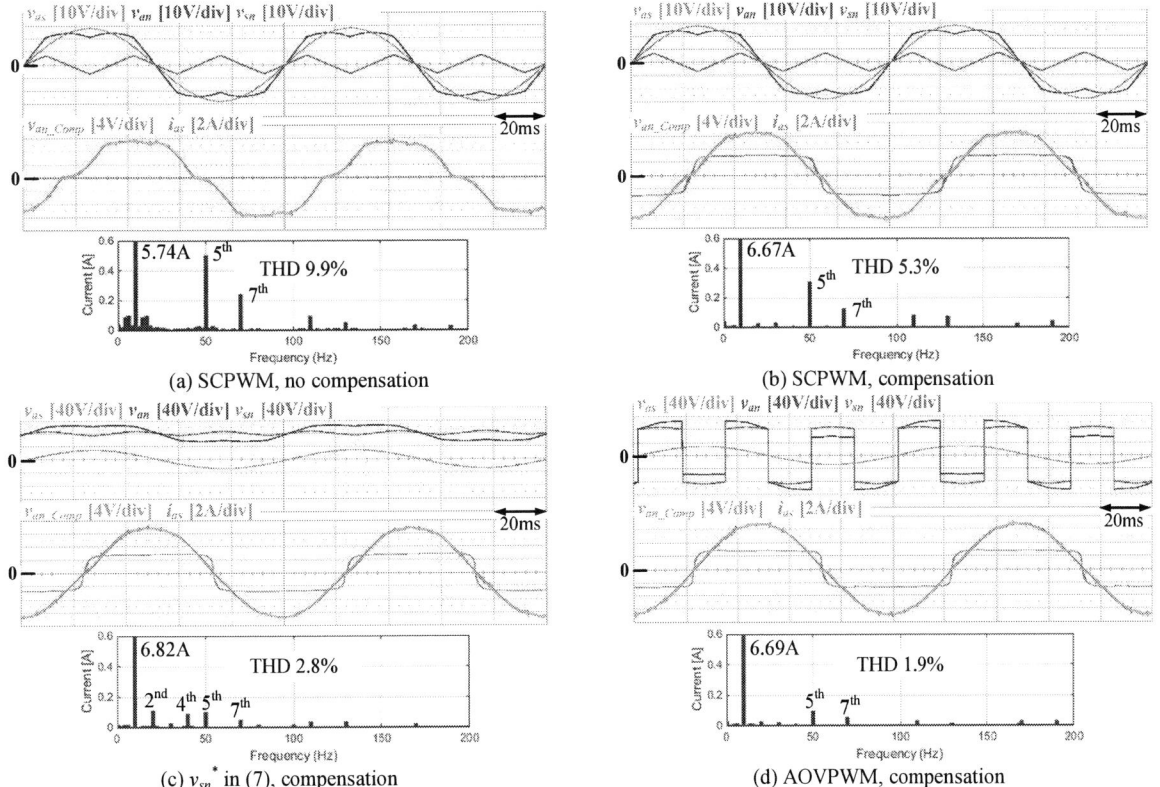

(a) SCPWM, no compensation

(b) SCPWM, compensation

(c) v_{sn}^* in (7), compensation

(d) AOVPWM, compensation

Fig. 15. Experiment 1: Waveforms and harmonic spectrum of i_{as} under no load, 10Hz frequency.

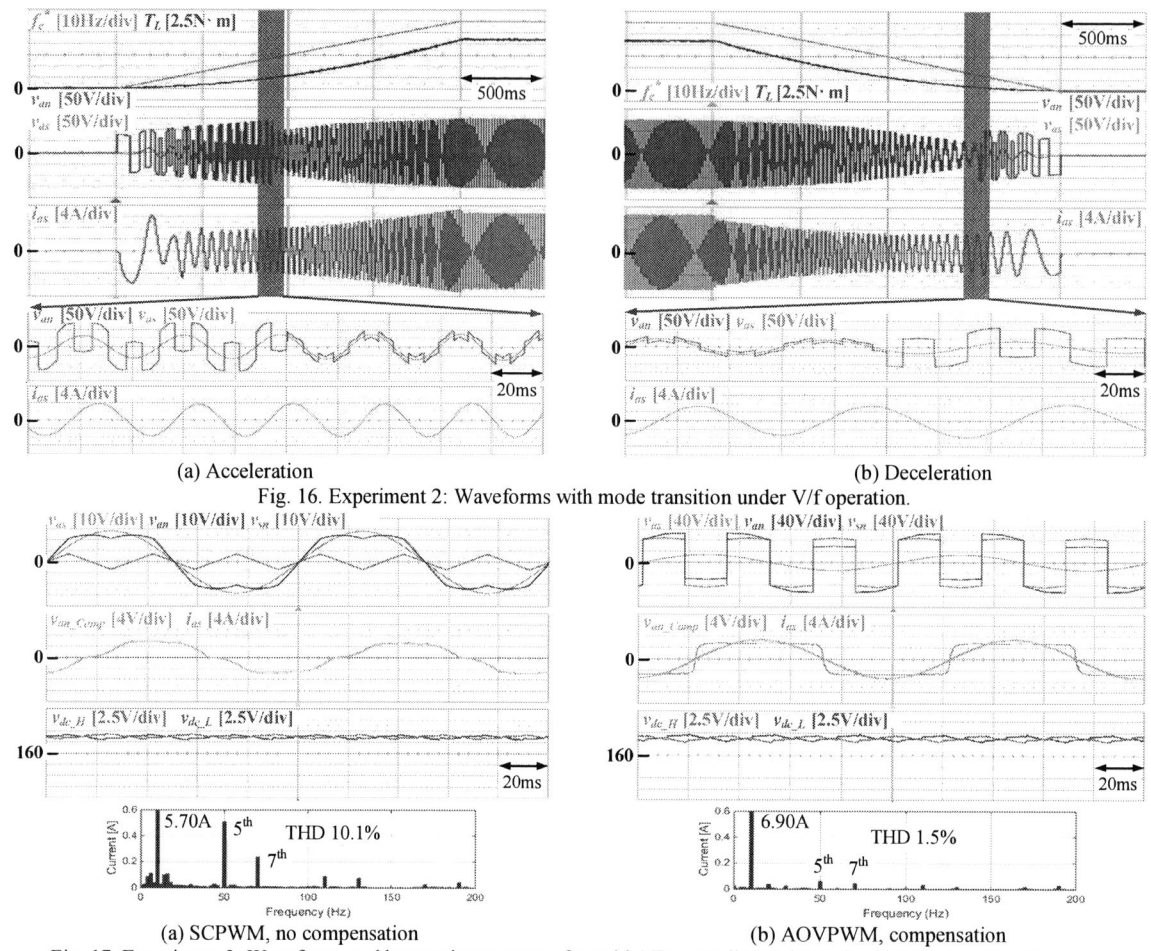

(a) Acceleration   (b) Deceleration

Fig. 16. Experiment 2: Waveforms with mode transition under V/f operation.

(a) SCPWM, no compensation   (b) AOVPWM, compensation

Fig. 17. Experiment 3: Waveforms and harmonic spectrum of i_{as} with NP controller under no load, 10Hz operating frequency.

The SMPMSM is controlled by a two-level inverter for applying load torque to the motor under the test. All control algorithms are implemented digitally in a DSP board. Each dc-link capacitance, switching frequency, and preset dead time are 4000μF, 10 kHz, and 3μs, respectively. The sampling frequency is set to 20 kHz.

A. Performance of inverter nonliearity comepnsation

Fig. 15 shows waveform and harmonic spectrum of i_{as} according to PWM methods under no load and 10Hz operating frequency. Each dc-link voltage is fixed to 155V by a DC supply. In Fig. 15(a) where no compensation is applied, i_{as} is severely distorted due to the inverter nonlinearity. Even after applying the compensation method as shown in Fig. 15(b), 5th and 7th harmonics in i_{as} are not eliminated because the voltage disturbance induced by the dead zones can't be compensated. Applying v_{sn}^* in (7), 5th and 7th harmonics of the current are conspicuously reduced as shown in Fig. 15(c). However, additional even order harmonics arise in i_{as} due to the voltage disturbance induced by aforementioned asymmetric δv_{xn_SW}. Applying AOVPWM, thanks to alternatively shifted pole voltages, even order harmonics of the current are eliminated as shown in Fig. 15(d).

Fig. 16 shows the transition between the two proposed PWMs. The load torque is set to be proportional to the square of the rotational speed to emulate fan and pump load. During acceleration, PWM mode is changed from AOVPWM mode to OMPWM mode when v_{max}-v_{min} = 0.5V_{dc}-4V_{DZ} as shown in Fig. 16(a). Fig. 16(b) shows that OMPWM mode is changed to AOVPWM mode when v_{max}-v_{min} = 6V_{DZ}. The transition points during acceleration and deceleration are set differently by applying hysteresis for stable transition. The distortion of the current is well compensated even at the transition point.

B. Effectiveness of NP balancing control

The dc-link capacitor is connected to line-to-line 220V_{rms} 60Hz grid through three-phase diode rectifier. NP current is controlled by the proposed NP controllers. When SCPWM is applied, NP controller for OMPWM excluding v_{mar} limiter is used to compare the performance of proposed algorithms. NP controller gain, ω_{vc} and $k_{np,v}$, are set to 1Hz and -0.5.

Fig. 17 shows the effectiveness of AOVPWM with θ_{bal} controller under no load and 10Hz operating frequency. In Fig. 17(a) where SCPWM is applied without the compensation, dc-link voltages are well balanced. However, i_{as} is severely distorted due to the inverter nonlinearity. Applying AOVPWM

978-1-4673-9551-9/16 $31.00 © 2016 IEEE

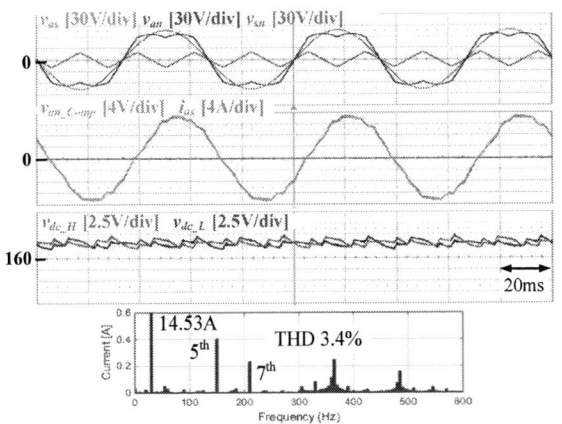 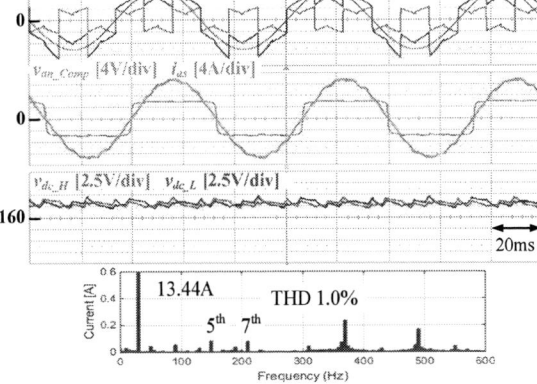

(a) SCPWM, no compensation   (b) OMPWM, compensation

Fig. 18. Experiment 4: Waveforms and harmonic spectrum of i_{as} with NP controller under full load, 30Hz operating frequency.

and the compensation method, 5th and 7th harmonics of the current are conspicuously reduced without losing the voltage balancing control performance as shown in Fig. 17(b).

Fig. 18 shows the effectiveness of OMPWM under full load torque and 30Hz operating frequency with $v_{sn,bal}$ controller. Using the proposed scheme, 5th and 7th harmonic components of i_{as} are remarkably reduced compared to those in Fig. 18(a), maintaining the voltage balancing control performance.

VI. CONCLUSION

In this paper, inverter nonlinearity effects in three-level T-type inverters have been analyzed. Also, a scheme consisting of PWM techniques and compensation method to alleviate inverter nonlinearity effects have been proposed. The nonlinear effects are classified two parts, namely, one due to dead time and the other one due to voltage drop of switching devices. The effects are addressed and the remedy against each part has been proposed and discussed. Also, NP controllers to keep the voltage balancing of dc-link are introduced to incorporate proposed PWM methods. Finally, the validity and effectiveness of the proposed scheme are verified by experimental results. Using the proposed AOVPWM and OMPWM, not only even harmonics but also 5th and 7th harmonic currents are remarkably reduced in three-level T-type inverter. In addition, NP voltage can be balanced with proposed NP controllers. The proposed scheme could be extended to multi-level and multi-leg topology.

REFERENCES

[1] M. Schweizer and J. W. Kolar, "Design and Implementation of a Highly Efficient Three-Level T-Type Converter for Low-Voltage Applications," *IEEE Trans. Power Electron.*, vol. 28, no. 2, pp. 899–907, Feb. 2013.

[2] J. W. Choi and S. K. Sul, "Inverter output voltage synthesis using novel dead time compensation," *IEEE Trans. Power Electron.*, vol. 11, no. 2, pp. 221–227, Mar. 1996.

[3] A. R. Munoz and T. A. Lipo, "On-line dead-time compensation technique for open-loop PWM-VSI drives," *IEEE Trans. Power Electron.*, vol. 14, no. 4, pp. 683–689, Jul. 1999.

[4] D. Leggate and R. J. Kerkman, "Pulse-based dead-time compensator for PWM voltage inverters," *IEEE Trans. Ind. Electron.*, vol. 44, no. 2, pp. 191–197, Apr. 1997.

[5] J. S. Kim, J. W. Choi, and S. K. Sul, "Analysis and compensation of voltage distortion by zero current clamping in voltage-fed PWM

inverter," in *Proc. IPEC'95 Conf.*, Yokohama, Japan, Apr. 3–7, 1995, pp. 265–270.

[6] Z. Zhang and L. Xu, "Dead-Time Compensation of Inverters Considering Snubber and Parasitic Capacitance," *IEEE Trans. Power Electron.*, vol. 29, no. 6, pp. 3179–3187, Jun. 2014.

[7] N. Bedetti, S. Calligaro, and R. Petrella, "Self-Commissioning of Inverter Dead-Time Compensation by Multiple Linear Regression Based on a Physical Model," *IEEE Trans. Ind. Appl.*, vol. 51, no. 5, pp. 3954–3964, Sep. 2015.

[8] W. S. Oh, Y. T. Kim, and H. J. Kim, "Dead time compensation of current controlled inverter using space vector modulation method," in Proceedings of 1995 International Conference on Power Electronics and Drive Systems. PEDS 95, 1995, pp. 374–378.

[9] A. C. Oliveira, C. B. Jacobina, and A. M. N. Lima, "Improved dead-time compensation for sinusoidal PWM inverters operating at high switching frequencies," *IEEE Trans. Ind. Electron.*, vol. 54, no. 4, pp. 2295–2304, Aug. 2007.

[10] D. Zhou and D. G. Rouaud, "Dead-time effect and compensations of three-level neutral point clamp inverters for high-performance drive applications," *IEEE Trans. Power Electron.*, vol. 14, no. 4, pp. 782–788, 1999.

[11] S. R. Minshull, C. M. Bingham, D. A. Stone, and M. P. Foster, "Compensation of Nonlinearities in Diode-Clamped Multilevel Converters," *IEEE Trans. Ind. Electron.*, vol. 57, no. 8, pp. 2651 –2658, Aug. 2010.

[12] X. Li, S. Dusmez, B. Akin, and K. Rajashekara, "Vector based dead-time compensation for a three-level T-type converter," in *2015 IEEE Applied Power Electronics Conference and Exposition (APEC)*, 2015, pp. 1534–1540.

[13] D. W. Chung, J. S. Kim, and S. K. Sul, "Unified voltage modulation technique for real-time three-phase power conversion," *IEEE Trans. Ind. Appl.*, vol. 34, no. 2, pp. 374–380, 1998.

[14] J. Shen, S. Schroder, B. Duro, and R. Roesner, "A Neutral-Point Balancing Controller for a Three-Level Inverter With Full Power-Factor Range and Low Distortion," *IEEE Trans. Ind. Appl.*, vol. 49, no. 1, pp. 138–148, Jan. 2013.

[15] C. Wang and Y. Li, "Analysis and Calculation of Zero-Sequence Voltage Considering Neutral-Point Potential Balancing in Three-Level NPC Converters," *IEEE Trans. Ind. Electron.*, vol. 57, no. 7, pp. 2262–2271, Jul. 2010.

[16] C. Newton and M. Sumner, "Neutral point control for multi-level inverters: theory, design and operational limitations," in *IAS '97. Conference Record of the 1997 IEEE Industry Applications Conference Thirty-Second IAS Annual Meeting*, 1997, vol. 2, pp. 1336–1343.

Front-End Isolated Quasi-Z-Source DC-DC Converter Modules in Series for Photovoltaic High-Voltage DC Applications

Yushan Liu[1], Member, *IEEE*, Haitham Abu-Rub[1,2], *Senior Member, IEEE*, Baoming Ge[3], *Member, IEEE*
[1]Electrical and Computer Engineering Program, Texas A&M University at Qatar, Doha 23874, Qatar
[2]Qatar Environment and Energy Research Institute, Hamad Bin Khalifa University (HBKU), Doha, Qatar
[3]Department of Electrical Engineering, Texas A&M University, College Station, TX 77843, USA
Email: yushan.liu@qatar.tamu.edu, haitham.abu-rub@qatar.tamu.edu, baomge@gmail.com

Abstract— A quasi-Z-source modular cascaded converter (qZS-MCC) is proposed for high-voltage (HV) dc integration of photovoltaic (PV) power. The qZS-MCC comprises front-end isolated qZS half-bridge (HB) dc-dc converter submodules (SMs) in series. By the qZS-HB handling PV voltage and power variations, a unified duty cycle is applicable for the front-end isolation converter of all SMs. Resultantly, the proposed system improves the quasi-Z-source cascaded multilevel inverter and the modular multilevel converter based PV counterparts in terms of no double-line-frequency pulsating power so as to low qZS impedance, HV dc collection of PV power thus to reduce conversion stages for dc transmission, and overcoming the limit of series-output voltage with simple galvanic isolation. Operating principle and power loss evaluation of the qZS-MCC are presented. Parameter design guidelines and simulation are addressed based on a 60-kW SM; experimental results are carried out on a downscaled prototype as a proof-of-concept, demonstrating the validity of the proposed system.

Keywords— *Photovoltaic power system, dc-dc power conversion, quasi-Z-source converter, galvanic isolation.*

I. INTRODUCTION

Nowadays, the ever-increasing installed capacity of the large-scale photovoltaic (PV) power plant, ranging from few hundreds of kilowatts up to several hundreds of megawatts, is in consequence of the rapid growing demand for sustainable energy, the market competition, and the ability of reducing the cost per watt ratio, etc. [1]. The quasi-Z-source cascaded multilevel inverter (qZS-CMI) has spurred growing attention in PV applications recently due to the features of [2]-[5]: 1) the single-stage qZS H-bridge inverter module independently handling dc-link voltage balance and dc-ac power conversion; 2) the improved reliability benefiting from the short-through feasibility; 3) the direct connection of medium-voltage (MV) grid without the bulky, costly, and lossy line-frequency (LF) transformer that the central inverter uses; 4) the distributed MPPT and low harmonic spectra inherited from the traditional CMI; 5) one-third reduction of the module number and efficiency enhancement with respect to the traditional CMI, etc.

However, due to the single-phase inverter topology of H-bridge modules, the double-line-frequency (DLF) power oscillates at the dc link of each module, no matter the traditional CMI [6] or qZS-CMI [2], [3]. Contributions have been dedicated to handle the undesirable DLF ripple in the following categories: 1) a passive way utilized the qZS inductance and capacitance to buffer the DLF ripple within the engineering tolerant range [2], [3], but the resultant large impedance values brought in obstacles to the high-power implementation of qZS-CMI PV system; 2) a hybrid pulsewidth modulation made the dc-link voltage envelope cope with the ac voltage, thus to not limit the dc-link DLF ripple and reduce the requirement of qZS impedance [7], however, a large PV terminal capacitance was applied to prevent the ripple power from propagating to the PV panel; 3) elaborate controls were explored to minimize the qZS network [8], [9], leading to complicated control design; 4) an active power filter integrated single-phase qZSI stored the dc-side DLF ripple in APF's *LC* branch by means of ac [10], but will greatly increase complexity and component counts in building the qZS-CMI.

In view of PV's natural dc characteristic, the modular multilevel converter (MMC) was deployed for PV's high-voltage (HV) applications, expecting to minimize the power conversion stages and reduce the power loss on transmission lines [11], [12]. However, the non-isolated one by PV panels in series accessing the dc link of dc-ac MMC has restricted voltage level, which is within one kilovolt (kV) due to the PV-panel insulation demand [11]; high-frequency (HF) isolation converters were coupled into each half-bridge (HB) dc-dc submodule (SM), so that series connection of numbers of such SMs is able to achieve HV (over 1 kV) [12], but the rising of dc-bus voltage level will significantly burden the computation and gating signal because output-side HBs are controlled to balance dc-link voltages and isolation converters also require controls to track MPPT.

This paper contributes to propose a qZS modular cascaded converter (qZS-MCC) for HV dc applications of PV systems. The qZS-MCC is a series connection of front-end isolated qZS-HB dc-dc converter SMs. By inserting shoot-through events into the post-stage HB dc-dc converter to balance dc-link voltages and distribute powers with the

---

This paper was made possible by NPRP-EP grant # [X-033-2-007] from the Qatar National Research Fund (a member of Qatar Foundation). The statements made herein are solely the responsibility of the authors.

modulation index, the qZS-HB SM no longer desires the extra shoot-through comparison as that in the qZSI or back-end isolated qZS dc-dc converter [3], [13], and a unified duty cycle without control is executable for all front-end isolation converters. The qZS-MCC demonstrates free of DLF ripple, HV dc collection/transmission of PV power, and simple galvanic isolation with respect to qZS-CMI and MMC based PV counterparts. The latter two have been investigated in detail, but the dc performed qZS-MCC has not yet been disclosed.

II. PROPOSED QZS-MCC FOR HV DC COLLECTION OF PV POWER

Fig. 1 shows the proposed qZS-MCC for the dc collection of PV power. Fig. 1 (a) shows the system configuration. A HV output of qZS-MCC is fulfilled by front-end isolated qZS-HB dc-dc PV converter SMs in series, as Fig. 1 (b) shows. Several qZS-MCC based PV systems can simply combine at the dc terminal (V_{dc}), and further interface with the MV-ac grid or serve an MV-ac motor through a dc-ac MMC.

Fig. 2 shows gating signals and typical waveforms of the two qZS-HB SMs in cascade. A $2\pi/n$ phase shift between carriers aims at interleaving of adjacent SMs. The sawtooth carrier is performed to avoid additional switching introduced by shoot-through events [2].

The unified duty cycle of D_0=0.5 for upper switches S_{fk1} and 1-D_0 for lower ones S_{fk2}, $k \in \{1, ..., n\}$, is performed to the front-end HB of all SMs. Then the input voltage and current of qZS network are [14]

$$v_{ink} = v_{PVk}/D_0 = 2v_{PVk} , \ i_{ink} = D_0 i_{PVk} = i_{PVk}/2 \quad (1)$$

where v_{PVk} and i_{PVk} denote PV voltage and current of kth SM.

The qZS-MCC's output voltage is the summation of

$$v_o = \sum_{k=1}^{n} v_{ok} = \sum_{k=1}^{n} S_k \hat{v}_{DCk} \quad (2)$$

with its average value of

$$V_{dc} = \sum_{k=1}^{n} V_{ok} = \sum_{k=1}^{n} M_k \hat{v}_{DCk} \quad (3)$$

where S_k denotes the switching function of the kth SM, V_{ok} denotes the average dc output voltage of the kth SM, and M_k denotes the modulation index of the kth SM, $k \in \{1, ..., n\}$.

The qZS inductors are charging in the ST and discharging in the non-ST independently in each qZS-HB, as typical waveforms of i_{in1} and i_{in2} show in Fig. 2. The dc-link peak voltage, and voltages of capacitor C_1 and C_2 of the kth SM are, respectively

$$\hat{v}_{DCk} = \frac{1}{1-2D_k} v_{ink} , \ V_{C1k} = \frac{1-D_k}{1-2D_k} v_{ink} , \ V_{C2k} = \frac{D_k}{1-2D_k} v_{ink} \quad (4)$$

where D_k denotes the ST duty cycle of the kth SM.

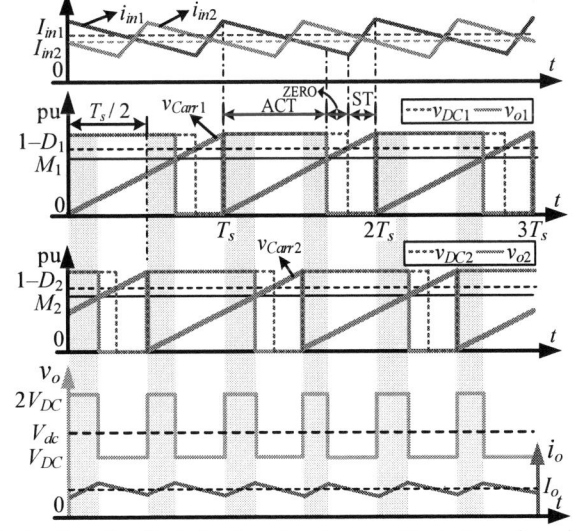

Fig. 2. Operating illustration for gating signals and typical waveforms by using two SMs formed qZS-MCC.

III. PARAMETER DESIGN GUIDELINES

A. Power Loss Analysis

The power loss of the qZS-HB SM includes the HB switching loss, HB conduction loss, qZS diode loss, and qZS inductors and capacitors loss. Each HB switch is an IGBT or MOSFET with its anti-parallel diode.

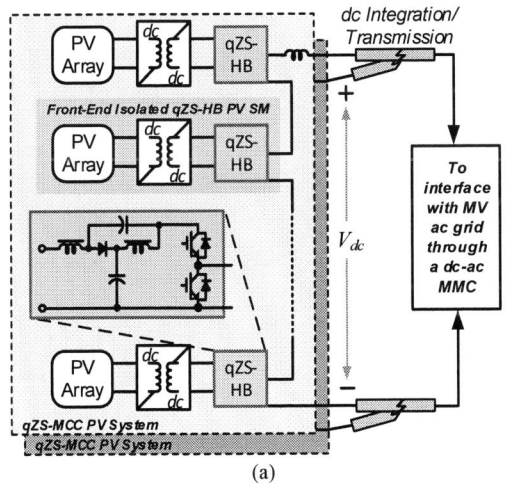

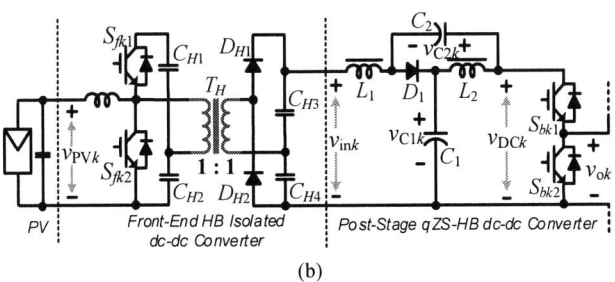

Fig. 1. Proposed qZS-MCC PV system. (a) System configuration; (b) front-end isolated qZS-HB PV SM.

Using SM_1 for illustration, in the ACT, as Fig. 2 shows, the S_{b11} is on and S_{b12} is off; the current through S_{b11} equals to the qZS-MCC's output current i_o; the qZS diode is on with the current of $2i_{in}-i_o$. In the ZERO, the anti-parallel diode of S_{b12} is free-wheeling for the qZS-MCC's output current i_o, and the qZS-diode current i_{D1} equals to $2i_{in}$. In the ST, the qZS diode is blocked due to negative voltage; S_{b11} and S_{b12} are on, with currents $2i_{in}$ and $2i_{in}-i_o$, respectively.

The switching event of the qZS-HB power devices is summarized as: 1) S_{b11} turns on and off one time in each control cycle, where the switching current are i_o and $2i_{in}$, respectively, and the switching voltage is the dc-link peak voltage V_{DC}; 2) the on and off events of S_{b12} are at the same voltage of V_{DC}, but with zero-current switching (ZCS) at the turn-on event and the current of $2i_{in}-i_o$ at the turn-off event; 3) the anti-parallel diode of S_{b12} holds zero-voltage switching (ZVS) at the reverse recovery event; 4) the reverse recovery loss of qZS diode D_1 occurs at the current of $2i_{in}$ and voltage of V_{DC}. It is noticeable that there is no extra switching when performing ST; moreover, the ZCS for transistor and ZVS for its anti-parallel diode in the lower switch are obtained.

1) HB Switching Loss

From the above summary, the qZS-HB switching loss is computed by

$$P_{SW} = \frac{1}{T_s} \int_0^{T_s} \frac{f_s V_{DC}}{V_{ref} I_{ref}} \left[i_o E_{OFF} + 2i_{in} E_{ON} + (2i_{in}-i_o)E_{OFF} \right] dt$$
$$= \frac{2I_{in} f_s V_{DC}}{V_{ref} I_{ref}} (E_{ON} + E_{OFF}) \tag{5}$$

where f_s denotes the switching frequency; E_{ON} and E_{OFF} denote the turn-on and turn-off energy, respectively; V_{ref} and I_{ref} denote the reference of switching voltage and current.

2) HB Conduction Loss

The conduction loss of a switch is generally expressed by

$$P_{CON} = I_{AVE}V_{ON} + I_{RMS}^2 R_{ON},$$

$$I_{AVE} = \frac{1}{T_s} \int_0^{T_s} i_{CON}(t)d(t)dt, \quad I_{RMS}^2 = \frac{1}{T_s} \int_0^{T_s} i_{CON}^2(t)d(t)dt \tag{6}$$

where V_{ON} denotes the threshold voltage of an IGBT or anti-parallel diode; R_{ON} denotes the conduction resistance of an IGBT or diode; I_{AVE} and I_{RMS} denote the average and RMS current through the IGBT or diode; $i_{CON}(t)$ denotes the conduction current; $d(t)$ denotes the duty cycle function.

From the aforementioned analysis, the conducted current and duty cycle for the IGBT and its anti-parallel diode of the upper and lower switches are, respectively

$$i_{T1}(t) = \begin{cases} i_o & d(t) = M \\ 2i_{in} & d(t) = D \end{cases}, \quad i_{T2}(t) = [2i_{in}-i_o \quad d(t) = D],$$

$$i_{SD2}(t) = [i_o \quad d(t) = 1-M-D] \tag{7}$$

Therefore, the HB conduction loss can be calculated through (6) and (7).

3) qZS Diode Power Loss

According to the analysis, the reverse recovery power and conducted current of qZS diode D_1 are, respectively

$$P_{REC,D1} = \frac{1}{T_s} \int_0^{T_s} \frac{f_s V_{DC}}{V_{refz} I_{refz}} [2i_{in} E_{DREC}] dt = \frac{2I_{in} E_{DREC} f_s V_{DC}}{V_{refz} I_{refz}} \tag{8}$$

$$i_{D1}(t) = \begin{cases} 2i_{in}-i_o & d(t) = M \\ 2i_{in} & d(t) = 1-M-D \end{cases} \tag{9}$$

where E_{DREC} denotes the reverse recovery energy of D_1; V_{refz} and I_{refz} denote the reference of switching voltage and current. Substituting (9) into (6), the conduction loss of qZS diode D_1 can be computed.

4) Power Loss of qZS Inductors and Capacitors

The calculation of power loss on qZS inductors and capacitors is similar to that of the qZSI [10], [11], wherein the difference is that the qZS-HB capacitor current will be $i_{in}-i_o$ in the ACT, $-i_{in}$ in the ST, and i_{in} in the ZERO. Thus, the RMS capacitor current is estimated by

$$I_{CZ} = \sqrt{\frac{1}{T_s} \int_0^{T_s} \left[(i_{in}-i_o)^2 \cdot M + (-i_{in})^2 \cdot D + (i_{in})^2 \cdot (1-M-D) \right] dt}$$
$$= \sqrt{(I_{in})^2 - 2I_{in}I_o M + (I_o)^2 M} \tag{10}$$

B. Parameters Definition of a High-Power SM

Due to the DLF ripple free, the switching frequency is of critical importance for the value and size of passive components in this high-power SM. The 1.2-kV/193-A all SiC HB power module CAS120M12BM2 is therefore adopted, which is appropriate for higher switching frequency without significant prejudice on the efficiency and robustness to temperature. Accordingly, the 600-V/60-kW qZS-HB SM is defined.

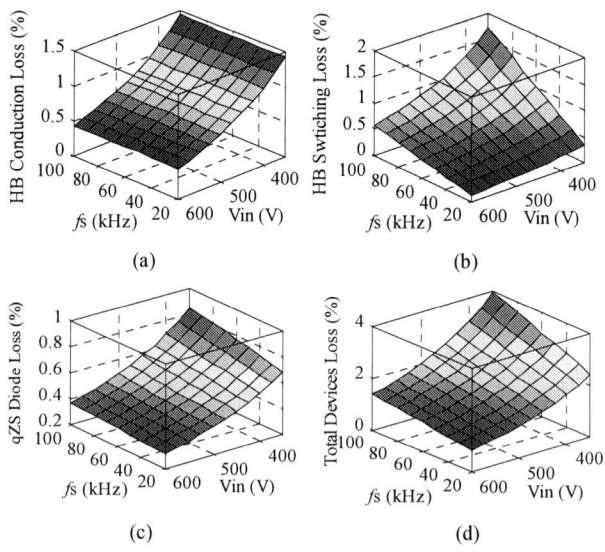

Fig. 3 Power losses of qZS-HB devices versus input voltage v_{in} and switching frequency f_s. (a) HB conduction loss; (b) HB switching loss; (c) qZS diode loss; and (d) total loss of power devices.

978-1-4673-9551-9/16 $31.00 © 2016 IEEE

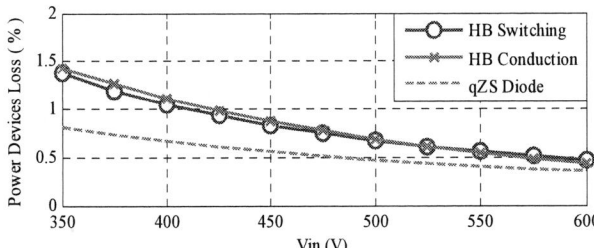

Fig. 4 The HB conduction loss, HB switching loss, and qZS diode loss versus the input voltage of the qZS-HB at the defined specifications.

Fig. 3 shows the HB conduction loss, HB switching loss, qZS diode loss, and total loss of power devices versus input voltage v_{in} and switching frequency f_s. Taking trade-offs of power devices' loss and passive components' values, an 80-kHz switching frequency is adopted. The resultant qZS capacitance and inductance are 137 μF and 83 μH, respectively.

Fig. 4 shows the estimated power loss of the 60-kW qZS-HB SM at the designed specifications, which presents a 98.5% efficiency at the full power. Through soft switching to alleviate the power loss of the front-end isolation converter, a highly efficient qZS-MCC system is achievable.

IV. SIMULATION AND DOWNSCALED EXPERIMENTAL RESULTS

A qZS-MCC PV system in series of four 60-kW front-end isolated qZS SMs is simulated in PSIM. Table I lists system specifications according to the design in Section III, which presents a 2.4-kV/240-kW system. A resistive load R_L, as Fig. 2 (a) shows, is adjusted by $R_L = V_{dc}^2/P_o$ to keep a constant series-output dc-bus voltage V_{dc}=2.4 kV when the system total power P_o changes.

TABLE I. SYSTEM SPECIFICATIONS IN SIMULATION.

Parameters	Values
Rated power of one SM, P_m	60 kW
MPPT voltage range of the PV array, v_{PVk}	175~300 V
qZS-HB input voltage range, v_{ink}	350~600 V
Average dc output voltage of one SM, V_{ok}	600 V
Cascaded module numbers, n	4
Series-output dc-bus voltage, V_{dc}	2.4 kV
Switching frequency of front-end HB	20 kHz
Switching frequency of post-stage HB	80 kHz
qZS inductance, L_1 and L_2	84 μH (42 μH coupling)
qZS capacitance, C_1 and C_2	150 μF
Filter inductance of qZS-MCC, L_F	300 μH

A. Simulation Results

All four qZS-HB SMs work at modulation index M_k=0.8, with D_0=0.5 for the front-end HB. Then from (4)-(7), the kth qZS-HB input voltage v_{ink}=480 V, input current I_{ink}=100 A, PV-array voltage v_{PVk}=240 V, PV-array current i_{PVk}=200 V, dc-link peak voltages V_{DC}=750 V, ST duty cycle D_k=0.18, and the qZS-MCC's output current I_o=80 A in calculation.

Fig. 5 shows simulation results of the SM$_1$'s PV-array current i_{PV1} and qZS-HB input current i_{in1}; SM$_1$'s PV-array voltage v_{PV1}, qZS-HB input voltage v_{in1}, and dc-link voltage v_{DC1}; qZS-MCC's output voltage v_o and dc-bus voltage V_{dc}; as well as load current i_o. The variables of other SMs are the same as SM$_1$. It can be seen that a 2.4-kV series-output voltage is achieved without DLF ripple in SM's dc side, regardless of low qZS inductance and capacitance values.

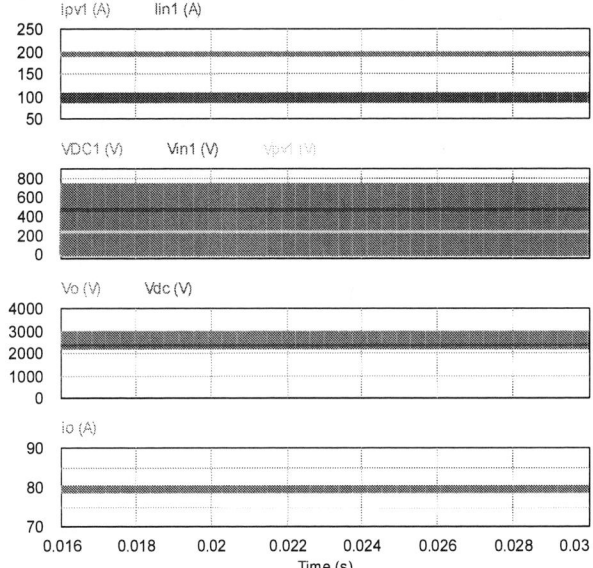

Fig. 5. Simulation results of the 240-kW qZS-MCC-based PV power system: SM$_1$'s PV-array current i_{PV1} and qZS-HB input current i_{in1}; SM$_1$'s PV-array voltage v_{PV1}, qZS-HB input voltage v_{in1}, and dc-link voltage v_{DC1}; qZS-MCC's output voltage v_o and dc-bus voltage V_{dc}; load current i_o.

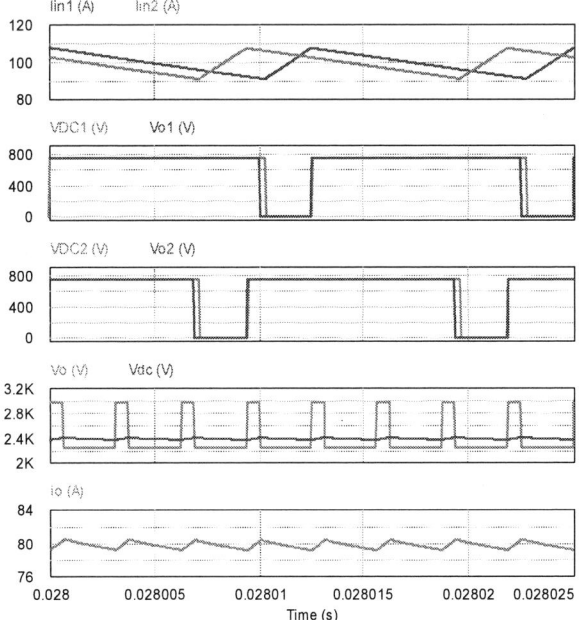

Fig. 6. Results in two control cycles when all SMs at same condition: SM$_1$ and SM$_2$'s qZS-HB input currents i_{in1} and i_{in2}; SM$_1$'s dc-link voltage v_{DC1} and output voltage v_{o1}; SM$_2$'s dc-link voltage v_{DC2} and output voltage v_{o2}; qZS-MCC's output voltage v_o and dc-bus voltage V_{dc}; and load current i_o.

Fig. 6 shows the SM_1 and SM_2's qZS-HB input currents i_{in1} and i_{in2}, SM_1's dc-link voltage v_{DC1} and output voltage v_{o1}, SM_2's dc-link voltage v_{DC2} and output voltage v_{o2}, qZS-MCC's output voltage v_o and dc-bus voltage V_{dc}, as well as qZS-MCC's output current i_o in two control cycles.

From Fig. 6, it can be seen that each qZS-HB operates at its own states independently. The qZS inductors are charging in the ST and discharging in the non-ST, with the peak-to-peak current ripple within 20%. In addition, the qZS-MCC's output voltage v_o presents oscillations in the amplitude of dc-link peak voltage V_{DC} and frequency of four-times switching frequency f_s, as a result of the $4f_s$-times charging-discharging event to the output current i_o. The average values of qZS-HB input currents, qZS-MCC's output current, and dc-bus voltage in simulation results are in accord with the calculation.

B. Experimental Results

A 2-kW downscaled qZS-MCC prototype, consisted of two isolated qZS-HB SMs, has been built as the proof of concept. Different conditions of v_{PV1}=36 V and v_{PV2}=40 V are applied to the two SMs. From (1) and (3), the SM_1's qZS-HB input voltage v_{in1}=72 V and modulation index M_1=0.72; those of SM_2 are v_{in1}=80 V and M_2=0.8; thereby, the two SMs' balanced dc-link peak voltage will be V_{DC}=131.6 V. And from (4), SM_1 and SM_2's D_1=0.226 and D_2=0.196, respectively. Fig. 7 shows experimental results. It can be obtained that the average i_{in1} and i_{in2}, i_o, and V_{dc} agree with calculations.

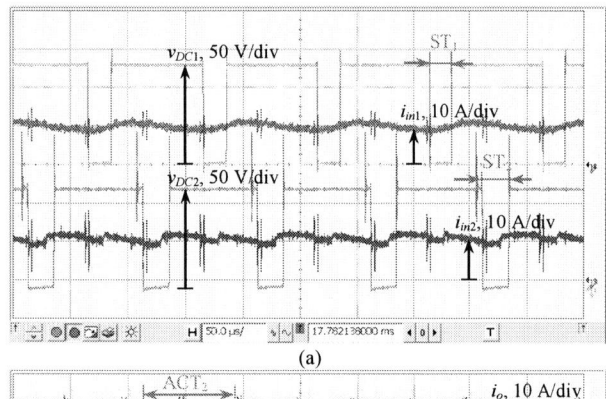

(a)

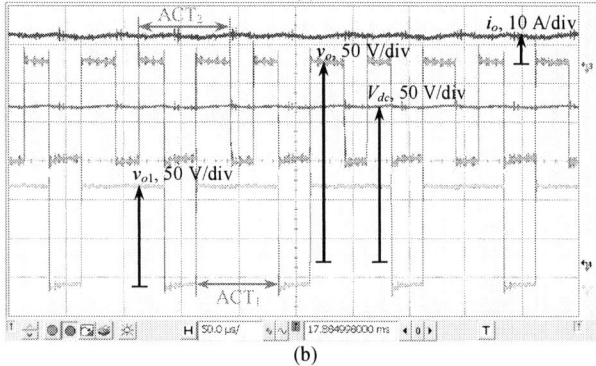

(b)

Fig. 7. Experimental results of the two SMs at different conditions. (a) SM_1's qZS-HB input current i_{in1} and dc-link voltage v_{DC1}, and SM_2's qZS-HB input current i_{in2} and dc-link voltage v_{DC2}; (b) qZS-MCC's output current i_o and voltage v_o, dc-bus voltage V_{dc}, and SM_1's output voltage v_{o1}.

Moreover, from the two SMs' ST time intervals, ST_1 and ST_2 in Fig. 7 (a), as well as ACT intervals, ACT_1 and ACT_2 in Fig. 7 (b), it can be seen that the SM_1 performs smaller duty cycle of the ACT and larger duty cycle of the ST than those of the SM_2. As a result, the same dc-link peak voltages of the two SMs, seen from v_{DC1} and v_{DC2} in Fig. 7 (a), and the balanced oscillation of output voltage v_o are fulfilled even though the two SMs' source voltage and power are different. It demonstrates the feasibility that the modulation index and ST duty cycle of the post-stage qZS-HB handle the voltage and power difference among SMs, while the front-end HB isolation converters share a constant duty cycle.

V. CONCLUSION

This paper proposed front-end isolated qZS-HB SMs in series formed the qZS-MCC for HV dc applications of PV power systems. The operating principle and power loss evaluation of qZS-MCC were analyzed; the specification was designed based on 60-kW SM. Simulation results of a four SMs formed 240-kW configuration and experimental results carried out on a downscaled prototype presented the validity of the proposed topology. Low control complexity of this modular cascaded system was fulfilled through the post-stage qZS-HB dealing with voltage and power difference among SMs, while the unified and constant duty cycle for all front-end HB isolation converters. In addition to the properties of HV dc integration, galvanic isolation, free of DLF ripple, and modular structure, the qZS-MCC provided a competitive candidate for large-scale PV power systems.

REFERENCES

[1] H. Choi, M. Ciobotaru, M. Jang, V.G. Agelidis, "Performance of medium-voltage dc-bus PV system architecture utilizing high-gain dc-dc converter," *IEEE Trans. on Sustain. Energy*, vol.6, no.2, pp.464-473, April 2015.

[2] Y. Zhou, L. Liu, H. Li, "A high-performance photovoltaic module-integrated converter (MIC) based on cascaded quasi-Z-source inverters (qZSI) using eGaN FETs," *IEEE Trans. Power Electron.*, vol.28, no.6, pp.2727-2738, June 2013.

[3] D. Sun, B. Ge, X. Yan, D. Bi, H. Zhang, Y. Liu, H. Abu-Rub, L. Ben-Brahim, F. Peng, "Modeling, impedance design, and efficiency analysis of quasi-Z source module in cascade multilevel photovoltaic power system," *IEEE Trans. Ind. Electron.*, vol.61, no.11, pp.6108-6117, Nov. 2014.

[4] Y. Xue, B. Ge, F.Z. Peng, "Reliability, efficiency, and cost comparisons of MW-scale photovoltaic inverters," *2012 IEEE Energy Conversion Congress and Exposition (ECCE)*, 15-20 Sept. 2012, pp.1627-1634.

[5] D. Sun, B. Ge, H. Zhang, et al. "An energy stored quasi-Z source cascaded multilevel inverter based photovoltaic power generation system," *The Twenty-Ninth Annual IEEE Applied Power Electronics Conference and Exposition (APEC)*, pp.1747-1751, 2014.

[6] Y. Shi, R. Li, H. Li, Y. Xue, "High-frequency-link based grid-tied pv system with small DC-link capacitor and low-frequency ripple-free maximum power point tracking," *IEEE Trans. Power Electron.*, doi: 10.1109/TPEL.2015.2411858.

[7] H. Zhang, B. Ge, Y. Liu, D. Sun, H. Abu-Rub, F. Peng, "A hybrid modulation method for single-phase quasi-Z source inverter," *2014 IEEE Energy Conversion Congress and Exposition (ECCE)*, pp.4444-4449, 14-18 Sept. 2014.

[8] B. Ge, H. Abu-Rub, Y. Liu, R. Balog, "Minimized quasi-Z source network for single-phase inverter", *2015 Thirtieth Annual IEEE Applied Power Electronics Conference and Exposition (APEC)*, pp.806-811, 15-19 March 2015.

[9] Y. Zhou, H. Li, H. Li, X. Lin, "A capacitance minimization control strategy for single-phase PV quasi-Z-source inverter", *2015 Thirtieth Annual IEEE Applied Power Electronics Conference and Exposition (APEC)*, pp.1730-1735, 15-19 March 2015.

[10] B. Ge, H. Abu-Rub, Y. Liu, R. Balog, "An active filter method to eliminate dc-side's low-frequency power ripples for single-phase quasi-Z-source inverter", *2015 Thirtieth Annual IEEE Applied Power Electronics Conference and Exposition (APEC)*, pp.827-832, 15-19 March 2015.

[11] J. Mei, B. Xiao, K. Shen, L.M. Tolbert, J. Zheng, "Modular multilevel inverter with new modulation method and its application to photovoltaic grid-connected generator," *IEEE Trans. Power Electron.*, vol.28, no.11, pp.5063-5073, Nov. 2013.

[12] J. Echeverria, S. Kouro, M. Perez, H. Abu-Rub, "Multi-modular cascaded DC-DC converter for HVDC grid connection of large-scale photovoltaic power systems," *2013-39th Annual Conference of the IEEE Industrial Electronics Society (IECON)*, pp.6999-7005, 2013.

[13] I. Roasto, D. Vinnikov, J. Zakis, O. Husev, "New shoot-through control methods for qZSI-based DC/DC converters," *IEEE Trans. Ind. Informat.*, vol.9, no.2, pp.640-647, May 2013.

[14] S. Jiang, D. Cao, F.Z. Peng, Y. Li, "Grid-connected boost-half-bridge photovoltaic micro inverter system using repetitive current control and maximum power point tracking," *2012 Twenty-Seventh Annual IEEE Applied Power Electronics Conference and Exposition (APEC)*, pp.590-597, Feb. 2012.

Analysis of Non Detection Zone for Multiple Distributed PCS Based on Equivalent Single PCS Using Reactive Power Approach

Byeong-Heon Kim and Seung-Ki Sul
Dept. of Electrical and Computer Engineering
Seoul National University
Seoul, Republic of Korea
heon13@snu.ac.kr

Abstract— In this paper, the methodology to analyze Non-Detection Zone (NDZ) of islanding detection algorithm for multiple Power Conditioning Systems (PCSs) connected to grid is proposed. The PCSs with renewable energy source or battery energy storage system are equipped with anti-islanding functionality and the method using frequency positive-feedback is one of the most popular methods to detect islanding condition. Because it is figured out that the positive feedback of grid frequency is only valid on reactive current, not the phase of current or active current, the reactive power according to the frequency drift algorithm is only considered for NDZ analysis in the proposed analysis. Additionally, the reactive power component by frequency drift anti-islanding can be linearized in normal operation range of grid frequency. By incorporated with this concept, Equivalent Single PCS (ESPCS), which equivalently supplies whole reactive current of multiple PCSs, is proposed and this could emulate the operation of multiple parallel PCSs without loss of the detection capability. The NDZ is analyzed by ESPCS and is compared to the NDZ of parallel PCSs. Through the computer simulation and hardware experiments, it is confirmed that NDZ based on ESPCS and operation dynamics is well matched to those of multiple PCSs.

Keywords—Anti-islanding; frequency drift; positive feedback; reactive power; multiple inverters; non-detection zone

I. INTRODUCTION

The islanding is defined as a condition where a portion of the utility power system, which is energized by one or more local electric power systems through the associated point of common coupling (PCC), is electrically separated from the remainder of utility power system [1], [2]. The anti-islanding function is the strategy to detect islanding condition by each distributed resource by hardware setup and/or software implementation. Because the islanding would cause a severe threat to the grid operation, anti-islanding functionality is mandatory feature for distributed power conditioning systems (PCSs) connected to the grid [1]-[3].

The anti-islanding scheme implemented in software of PCS can be classified as passive and active methods. Among active methods, the frequency drift method (FDM) based on positive-feedback by grid frequency is one of the most popular methods

and equipped in many commercially available grid connected inverters [4]. The Active Frequency Drift (AFD), Sandia Frequency Shift (SFS) and Slip-Mode frequency Shift (SMS) are the representative algorithm for FDM, which are developed originally for single-phase photovoltaic (PV) applications. These methods control the phase of inverter output current, which is modulated by function of grid frequency. Because the frequency is drift away after islanding due to the positive feedback loop, the FDMs can detect islanding by triggering Over/Under Frequency (OUF) relay [5], which is the basic requirement of protection for grid connected PCSs [1]-[3]. The Non-Detection Zone (NDZ) for FDMs, which is the performance index for anti-islanding, is commonly depicted on load plane of quality factor and resonant frequency of islanding load with assuming that the load has passive components (parallel R, L, C) [5, 6].

The conventional phase criterion is a powerful tool to analyze the islanding operation of the inverter equipped frequency drift anti-islanding scheme. By the results on [7], the non-detection zone (NDZ) of multiple inverters can be analyzed. However, because of the complexity of phase relationship due to trigonometric functions, the equation for analyzing NDZ also becomes complex. This complexity increases as the number of PCSs connected to PCC is getting larger.

In this paper, an analysis method, which overcomes the complexity of phase criteria, has been proposed. In the previous literatures, it is figured out that the reactive power of PCS, not the phase of current of PCS, causes the frequency drift after islanding [4, 8-10]. And, for analyzing NDZ it is sufficient that reactive power component of current is considered. The reactive power of the PCS, which is caused by the frequency drift algorithm, can be represented by a linear polynomial under normal operation region of grid frequency. To analyze NDZ of multiple PCSs, the Equivalent Single PCS (ESPCS), which equivalently injects the whole reactive power made by whole PCS system, is proposed. By merging the linearized reactive components of each PCS to single reactive power, the whole PCSs can be emulated by an ESPCS. The NDZ of multiple PCSs also could be simply obtained through the proposed ESPCS concept and detail explanation is

978-1-4673-9551-9/16 $31.00 © 2016 IEEE

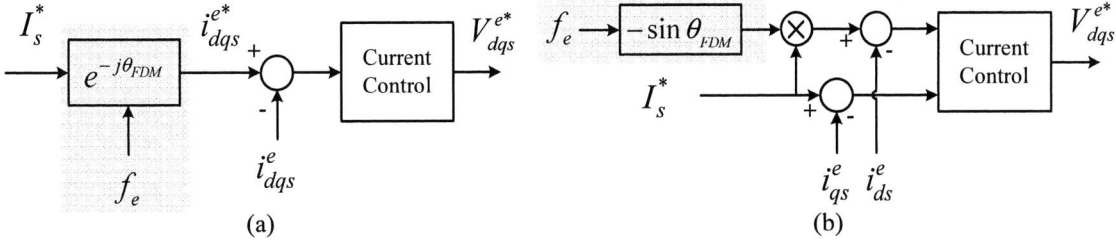

Fig. 1. Implementation of frequency drift anti-islanding in three phase system
(a) Simple rotation of current reference (b) Injection of reactive component only

demonstrated in followed sections. The computer simulation and experimental results show the effectiveness of proposed concept of ESPCS.

II. LINEARIZED REACTIVE POWER COMPONENT OF PCS BY FREUQENCY DRIFT ANTI-ISALDNING

In single-phase applications with constant current regulation, the magnitude and the phase of current are possible candidates to construct positive feedback loop. The conventional anti-islanding algorithm based on the frequency drift, such as AFD, SFS or SMS, uses the phase of current reference to build the frequency positive feedback loop. In three-phase system, the variation of current reference due to the phase made from anti-islanding can be represented as in (1) and (2). This variation leads to the rotation of the current reference vector as shown in Fig. 1 (a) [6, 10], with assumption that the power factor is controlled as unity.

$$i_{ds}^{e*} = -\sin\theta_{FDM} \times I_s^* \tag{1}$$

$$i_{qs}^{e*} = \cos\theta_{FDM} \times I_s^* \tag{2}$$

where, θ_{FDM} represents the phase function of current reference made by the frequency drift algorithms. And I_s^* represents the magnitude of original current reference produced by outer loop or reference generator. The current is represented in the synchronously rotating reference frame. In this paper, the d-axis component means the reactive power component and the q-axis component means the active power component. If the q-axis current has positive value, PCS supplies active power to the grid. For the d-axis current, capacitive current has positive value and inductive current has negative value.

In the conventional implementation, the frequency is fed-back to both active and reactive components. However, the relationship between the reactive power of PCS and the grid frequency after islanding is only reason of frequency drift as in [10]. That is, the only reactive component of current reference in (1) would drift frequency to outward of normal frequency range. In result, not active but reactive component should be injected for a positive feedback as shown in Fig. 1 (b). With using the injection of the reactive power component, the reactive power reference can be represented as (3). The equivalent reactive current reference is derived as (4).

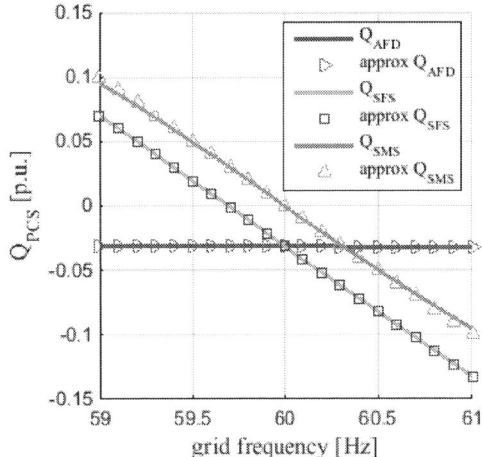

Fig. 2. Reactive power of PCS vs. Grid frequency

$$Q_{DG}^* = g\left(f_e - f_g\right)P_{DG}^* = -\sin\theta_{FDM} \times P_{DG}^* \tag{3}$$

$$i_{ds}^{e*} = g\left(f_e - f_g\right)I_s^* = -\sin\theta_{FDM} \times I_s^* \tag{4}$$

where, function noted by $g\left(f_e - f_g\right)$ represents the generalized form of the positive feedback anti-islanding. f_e represents the estimated grid frequency by phase-locked-loop (PLL) and f_g represents the nominal grid frequency.

With Taylor series expansion of (3) and (4), (5) can be derived approximately representing the reactive power reference of PCSs. Because of linearity, high-order terms could be neglected.

$$Q_{DG}^* \approx \left[a_0 + a_1\left(f_e - f_g\right)\right]P_{DG}^* \tag{5}$$

This approximation could be understood through the Fig. 2. In Fig. 2, several reactive power references for conventional method are plotted. In the Table 1, for several frequency drift algorithms, the phase function of current reference and equivalent coefficients of (5) are shown. The method using

978-1-4673-9551-9/16 $31.00 © 2016 IEEE   1221

Table 1. Coefficient of linearized reactive power of PCS

	θ_{FDM} [5]	a_0	a_1
AFD	$\pi f t_z$	$-\pi f_g t_z$	$-\pi t_z$
SFS	$\dfrac{\pi}{2}\left[cf_0 + k\left(f_e - f_g\right)\right]$	$-\dfrac{\pi}{2}cf_0$	$-\dfrac{\pi}{2}k$
SMS	$\theta_m \sin\left(\dfrac{\pi}{2}\dfrac{f_e - f_g}{f_m - f_g}\right)$	0	$-\theta_m \dfrac{\pi}{2\left(f_m - f_g\right)}$
PRI	-	$\begin{cases} K_1 & Period\,1 \\ K_2 & Period\,2 \end{cases}$	0

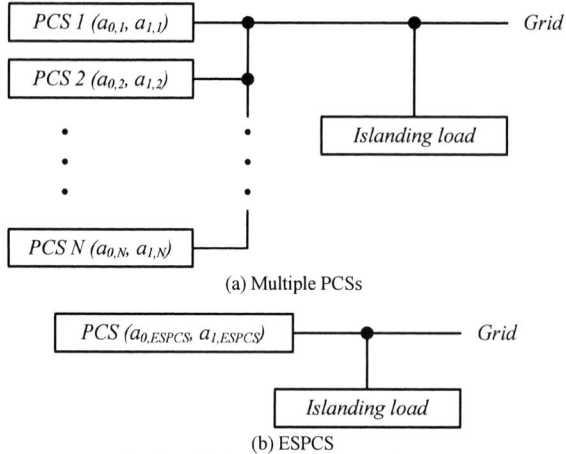

(a) Multiple PCSs

(b) ESPCS

Fig. 4. Equivalent Single PCS to analyze NDZ

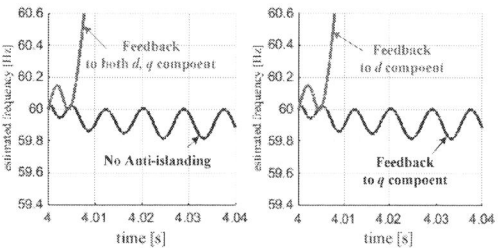

(a) Single phase PCS case

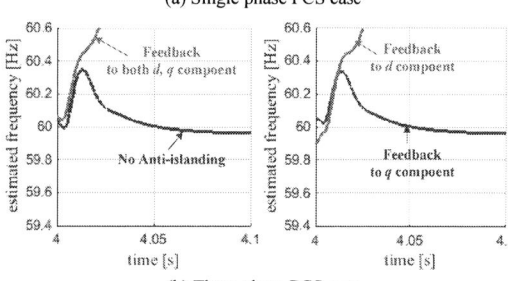

(b) Three phase PCS case

Fig. 3. Test of islanding detection capability with various feedback components

III. ANALYSIS OF NON-DETECTION ZONE FOR MULTIPLE DISTRIBUTED PCSs BY EQUIVALENT SINGLE PCS

In [7], for analyzing NDZ of Multi-PCSs, the phase of 'the equivalent active frequency drifting islanding detection method' is calculated. However, because of complexity of the relation between the each current of inverter, the equivalent phase becomes complex as number of inverters connected to the grid is getting larger. The main reason of this complexity is the trigonometric function. This complexity makes to extend the analysis to arbitrary number of inverters be difficult.

As discussions in the section II, in this paper, by considering only reactive current of each inverter, an Equivalent Single PCS (ESPCS) has been suggested and this alleviates the burden of adding other PCSs to the grid. Similar concept is already proposed in [4], and that has shown the performance of islanding detection for parallel inverters with same anti-islanding algorithm.

From the view of load or grid side, the multiple PCSs supply reactive power which is the sum of reactive power of each PCS in whole power system. It can be assumed that the ESPCS, which has same power capacity with total sum of the multiple PCSs, supply this whole reactive power equivalently. In Fig. 4, ESPCS representing multiple PCSs is depicted. Equation (6) represents the reactive power reference of the ESPCS. The subscript k represents index of each PCS.

$$Q^*_{DG,ESPCS} = \Sigma_k \left(\left[a_{0,k} + a_{1,k}\left(f_e - f_g\right)\right]P^*_{DG,k}\right) \qquad (6)$$

If the power ratio of each PCS to whole PCSs is defined as the weighting factor, w_k, (6) can be rewritten as (7).

$$\begin{aligned} Q^*_{DG,ESPCS} &= \Sigma_k \left(w_k\left[a_{0,k} + a_{1,k}\left(f_e - f_g\right)\right]\right)P^*_{DG,ESPCS} \\ &= \left[\left(\Sigma_k w_k a_{0,k}\right) + \left(\Sigma_k w_k a_{1,k}\right)\left(f_e - f_g\right)\right]P^*_{DG,ESPCS} \\ &= \left[a_{0,ESPCS} + a_{1,ESPCS}\left(f_e - f_g\right)\right]P^*_{DG,ESPCS} \end{aligned} \qquad (7)$$

Periodic Reactive power Injection (PRI) is also represented [11]. These coefficients can be applied for the linearization of (4).

The Fig. 3 shows the simulation results of test of islanding detection capability with SFS algorithm. In the case of single-phase PCS, the active and reactive powers are decomposed by using d-q synchronous reference frame. Islanding has happened at 4 s. Both single phase and three phase, the feedback of the active power component cannot drift grid frequency under islanding condition and shows same dynamics to the result with no anti-islanding. The only positive feedback-loop to reactive component results frequency drift and can detect islanding condition.

That is, not the phase of current reference, but the reactive current can only drift the grid frequency with positive feedback.

Because of the linearity, the calculation of equivalent reactive power of ESPCS is much simpler than equivalent phase in [7]. In the case of not using frequency drift algorithm, for example, passive anti-islanding or other anti-islanding methods injecting current or voltage of arbitrary frequency, the weighting factor can be set to zero because the reactive power from that method is zero. This assumption is more intuitive than [7], which modifies the quality factor of islanding load considering the active power of inverter not using FDM.

To obtain NDZ on load plane, the convergent frequency after islanding should be identified. The frequency is the crossover point of reactive power curve of PCS and of islanding load. If the parallel RLC load is assumed, under islanding condition, the load has reactive power curve as (8). This could be linearized around operation frequency [8].

$$Q_{load} = P_{DG}Q_f\left(\frac{f_{res}}{f_{is}} - \frac{f_{is}}{f_{res}}\right) \approx -2P_{DG}\frac{Q_f}{f_{res}}\left(f_{is} - f_{res}\right) \quad (8)$$

where, Q_f is the quality factor of islanding load and f_{res} is the resonant frequency of the load [1]. f_{is} is the grid frequency at the PCC after islanding.

From (7) and (8), with assumption of the reactive power matching between PCS and load under islanding, the crossover point can be calculated as (9).

$$f_{cross} = \left(a_{1,ESPCS} + 2\frac{Q_f}{f_{res}}\right)^{-1}\left(-a_{0,ESPCS} + a_{1,ESPCS}f_g + 2Q_f\right). (9)$$

With the crossover frequency, the instability condition is also considered [5, 8]. The instability criterion is given by (10). If this criterion is met, the grid frequency will be diverged after islanding and the islanding can be detected by OUF relay.

$$a_{1,ESPCS} < -2P_{DG}\frac{Q_f}{f_{res}} \quad (10)$$

The NDZ for ESPCS can be defined as load condition that couldn't satisfy (10) and has the crossover frequency which resides in normal frequency range after islanding. This NDZ is based on steady-state concept. Based on (10), the gain of anti-islanding can be designed to detect islanding under given local load condition. When the additional PCS is installed, if the weighted feedback gain is enough to detect the islanding with the local load, then the additional PCS doesn't need to equip FDM. And, the existing PCS equipped the injection of reactive power could detect the islanding operation with the additional PCS.

IV. VERIFICATION

In Fig. 5, the conceptual diagrams of circuit used in simulation and experiment are depicted. For parallel PCSs, three independent PCSs are operated with local islanding load,

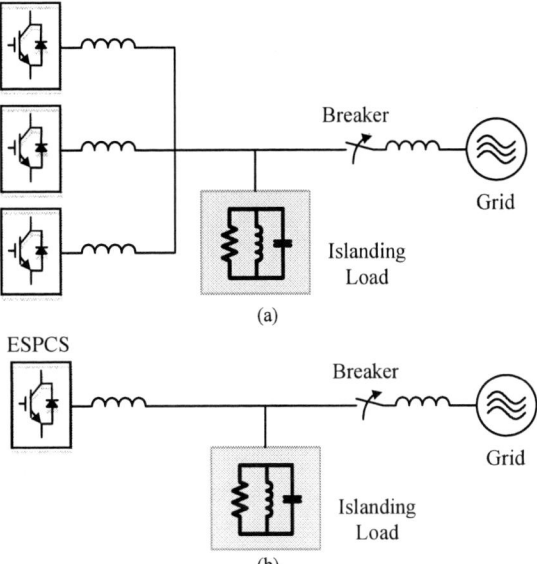

Fig. 5. Circuit diagram for verification

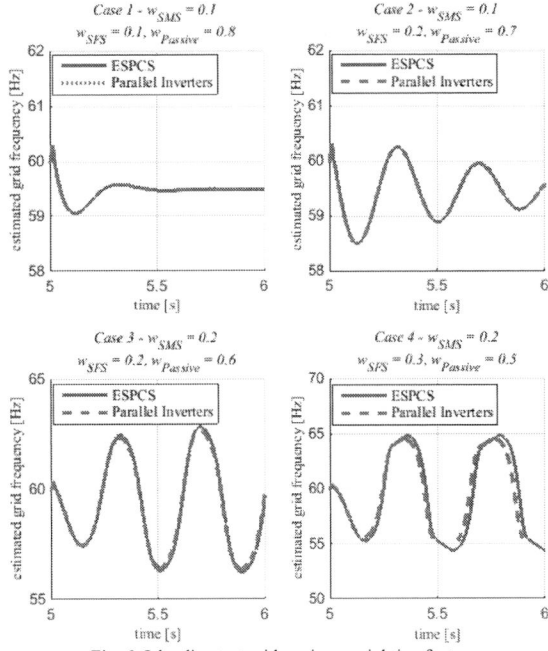

Fig. 6. Islanding test with various weighting factor

while ESPCS is emulated by single PCS. Islanding condition is simulated by opening of the breaker connected to the grid.

A. Simulation Results

By the computer simulation, three parallel-connected inverters and the single inverter with the proposed concept of ESPCS are under test. The control algorithms are implemented by Matlab/Simulink in discrete model and the electric circuit

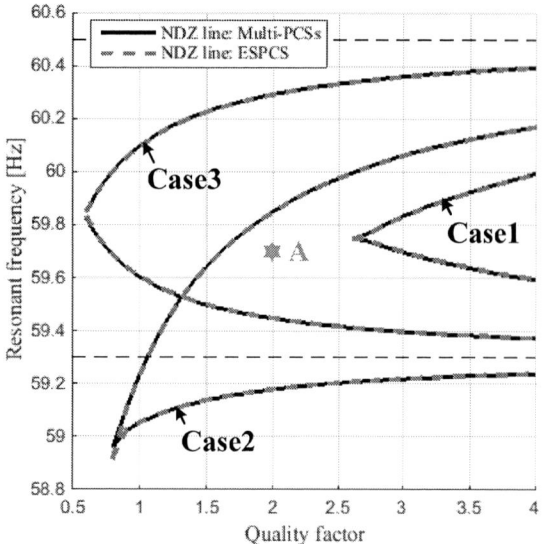

Fig. 7. NDZ for experimental setup

Table 2. Weighting factor for case study

Case	AFD	SFS	SMS	Passive
1	0.1	0.6	0.3	0
2	0.7	0.2	0.1	0
3	0	0.1	0.1	0.8

As seen in all results of Fig. 6, the frequency drift property of ESPCS is well matched to that of three parallel inverters. In case 1 and 2, the estimated grid frequency is converged and islanding couldn't be detected. However, in another case, the estimated frequency has large oscillation and goes outward of the abnormal operation region would trip OUF relay.

That is, the PCSs operated in parallel can be replaced by an ESPCS whose capacity is same to the whole inverters and the NDZ of parallel PCSs can be easily identified through this ESPCS.

B. Experimental Results

To verify the discussions in section III, islanding test with the hardware experimental setup is carried on. The test circuit, which is commonly specified in international standards such as [1]-[3], is constructed. The total power of whole system is set to 5 kW. The islanding load contains three phase passive components (parallel R, L, C). The quality factor of islanding load in the experiment was set as 2; The 'A' point in Fig. 7 represents this load condition. Due to the component tolerance, the resonant frequency has been identified as 59.7 Hz.

For the condition used in experiment, the NDZ analyzed by ESPCS could be represented as Fig. 7. Each PCS is equipped with different anti-islanding methods. In Table 2, the weighting factor of each PCS for three cases is represented. For AFD, the

simulation is done by simulation language PLECS [12]. The trip function caused by software frequency relay is deactivated in simulation for observing the behavior of PCS frequency after islanding.

For three parallel-connected inverters, the SFS, SMS and passive method (voltage/frequency relay) are equipped in each PCS. Fig. 6 shows the result of islanding test. Islanding occurs at 5 s. The weighting factor for each case is represented in the title of each figure. The dynamics of current regulator is neglect and the same PLL algorithm in [13] is equipped.

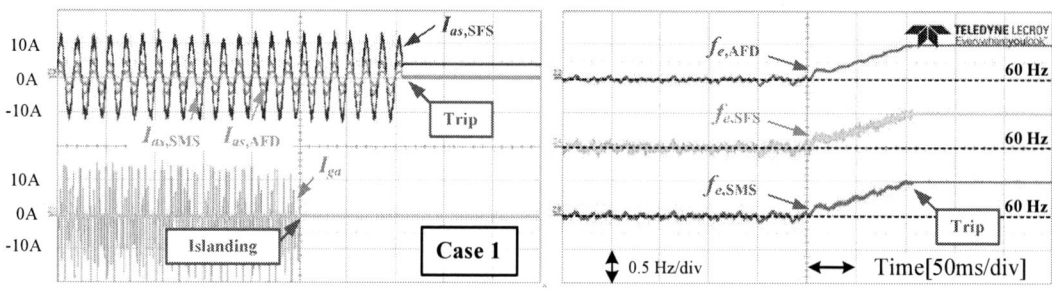

Figure 8. Experimental results – three PCSs (Case 1)

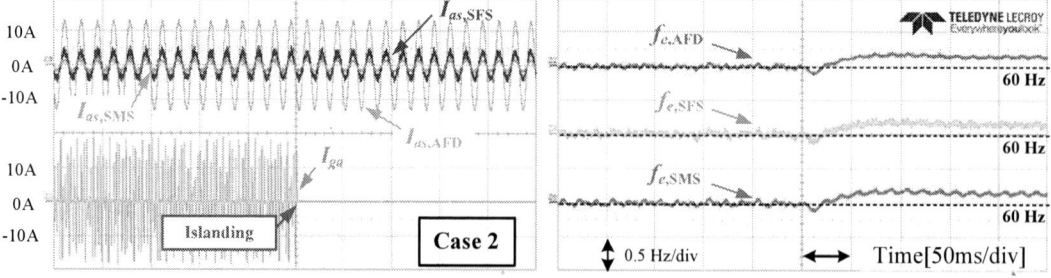

Figure 9. Experimental results – three PCSs (Case 2)

978-1-4673-9551-9/16 $31.00 © 2016 IEEE

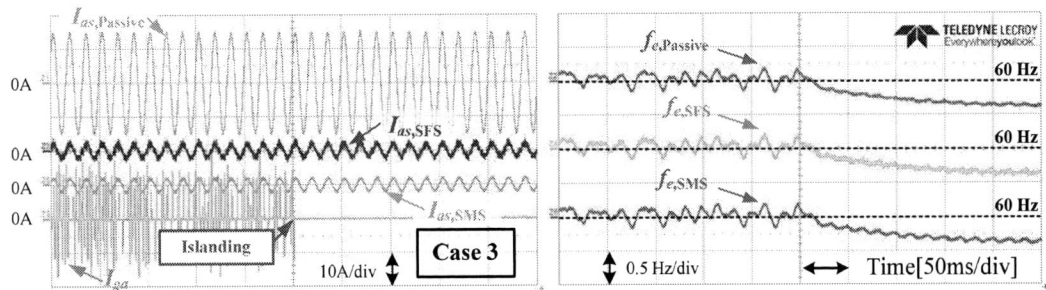

Figure 10. Experimental results – three PCSs (Case 3)

chopping factor is set to 0.02. The feedback gain of SFS and SMS is adjusted to detect islanding load with quality factor 3 when the PCS is operated separately. The weighting factors and positive feedback gain for case 2 and 3 are set not to be able to detect islanding condition under quality factor 2. And case 1 is set to detect islanding by FDM. For comparison, the NDZs simulated by multiple PCSs are also plotted. In Fig. 7, the NDZ of multiple PCSs obtained by the proposed ESPCS is overlaid and it is almost coincide with NDZ lines with Multi-PCSs.

The experimental results are shown in Fig. 8 ~ 13. Fig. 8 ~ 10 show the test results for the case of three parallel PCSs and Fig. 11 ~ 13 show the results for single PCS with reactive power made by proposed ESPCS concept. For showing the amount of active power mismatch, voltage magnitude at PCC is also plotted in result of ESPCS. Because the active power is adjusted to match, the PCC voltage after islanding varies under 10% with respect to the nominal value and this value is inside the normal voltage range. The frequency relay is triggered under 59.3 Hz or above 60.5 Hz that satisfies the international regulations.

According to the Fig. 7, the position of load condition on load plane doesn't reside in the NDZ for case 1 and it is expected to detect islanding for this case. And, Fig. 8 and Fig 11 clearly demonstrate that the frequency of both three PCS and ESPCS cases drift to trigger OUF relay due to the over frequency. Three parallel PCSs supply the weighted active current and total current is equal to that of ESPCS. Because the same PLL structure is adopted, the estimated frequencies in three parallel PCSs show same drift dynamics after islanding.

On the other hand, for the case 2 and 3, the load condition used in the experiments resides in the NDZs. As expected, the islanding couldn't be detected by anti-islanding and system continuously supplies power to load, because the reactive power slope is not large enough to drift frequency and the crossover frequency is in the normal operation region.

As experimentally verified, by proposed concept of ESPCS, the operation of multiple PCSs after islanding can be forecasted accurately.

V. CONCLUSION

In this paper, an analysis tool to investigate NDZ for multiple PCSs has been proposed. Based on the fundamental

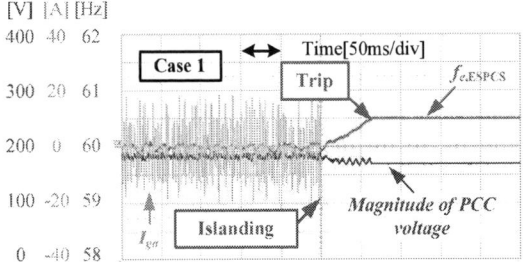

Figure 11. Experimental results – ESPCS (Case 1)

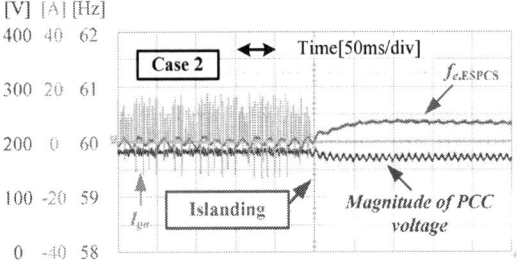

Figure 12. Experimental results – ESPCS (Case 2)

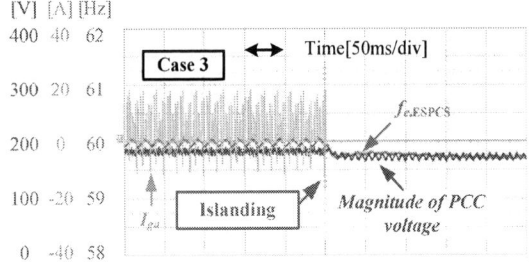

Figure 13. Experimental results – ESPCS (Case 3)

characteristics of the frequency drift, the only reactive power components of each PCS are considered to analyze NDZ. These reactive power components of each PCS are summed and represented by an ESPCS which supply this total sum of the reactive power. To sum the reactive power effectively, the

978-1-4673-9551-9/16 $31.00 © 2016 IEEE

reactive power produced by the frequency drift is linearized and it is shown that this linearization is valid on the normal operation region of grid frequency. The coefficient of linear function for conventional FDM also calculated. This approximation has made summing up of the reactive power of each PCS without trigonometric function possible and reduces the burden of computation compared to phase criteria.

With linearized reactive power of the islanding load, the NDZ of ESPCS was also obtained by the instability criterion and the crossover frequency after islanding. This NDZ can emulate the NDZ of parallel PCSs. Through the instability criterion, the setting of the feedback gain to detect islanding under given local load would be possible.

The results of the simulation and experiment have confirmed that the proposed ESPCS can emulate the operation and the NDZ of the multiple PCSs accurately.

REFERENCES

[1] *IEEE Recommended Practice for Utility Interface of Photovoltaic (PV) Systems*, IEEE Standard 929-2000.

[2] *IEEE Standard for Interconnecting Distributed Resources with Electric Power Systems*, IEEE Standard 1547-2003 (R2008).

[3] *Inverters, Converters, Controllers and Interconnection System Equipment for Use with Distributed Energy Resources*, UL 1741-2005.

[4] B.-H. Kim and S.-K. Sul, "Anti-islanding accelerated by grid unbalance component," in *proceeding of 9th International Conf. on Power Electron. and ECCE Asia (ICPE-ECCE Asia 2015)*, 2015, pp. 1268-1275.

[5] L. A. Lopes and H. Sun, "Performance assessment of active frequency drifting islanding detection methods," *IEEE Trans. Energy Conversion*, vol. 21, pp. 171-180, 2006.

[6] X. Wang, W. Freitas, and W. Xu, "Dynamic non-detection zones of positive feedback anti-islanding methods for inverter-based distributed generators," *IEEE Trans. Power Delivery*, vol. 26, pp. 1145-1155, 2011.

[7] L. A. Lopes and Y. Zhang, "Islanding detection assessment of multi-inverter systems with active frequency drifting methods," *IEEE Trans. Power Delivery*, vol. 23, pp. 480-486, 2008.

[8] H. H. Zeineldin, "A–Droop Curve for Facilitating Islanding Detection of Inverter-Based Distributed Generation," *IEEE Trans. Power Electronics*, vol. 24, pp. 665-673, 2009.

[9] Y. Jin, Q. Song, and W. Liu, "Anti-islanding protection for distributed generation systems based on reactive power drift," in *35th Annual Conference of IEEE Industrial Electronics 2009 (IECON 2009)*, 2009, pp. 3970-3975.

[10] B.-H. Kim and S.-K. Sul, " Comparison of Non-Detection Zone of Frequency Drift Anti-islanding with Closed-loop Power Controlled Distributed Generators," in *proceeding of IEEE International Future Energy Electronics Conf. (IEEE IFEEC 2015)*, 2015, pp. 46-50.

[11] B.-G. Cho, Y. Lee, S.-J. Chee, and S.-K. Sul, "A bilateral reactive power injection method for islanding detection of grid-connected converters in distributed generation units," in *proceeding of IEEE 6th International Symposium on Power Electronics for Distributed Generation Systems (PEDG 2015)*, 2015, pp. 625-632.

[12] http://www.plexim.com/

[13] Y. Park, S.-K. Sul, W.-C. Kim, and H.-Y. Lee, "Phase-locked loop based on an observer for grid synchronization," *IEEE Trans. Ind. Appl.*, vol. 50, pp. 1256-1265, 2014.

Optimal Power Scheduling for a Grid-Connected Hybrid PV-Wind-Battery Microgrid System

Adriana Luna
Nelson Diaz
Mehdi Savaghebi
Juan C. Vasquez
Josep M. Guerrero
Department of Energy Technology
Aalborg University, Aalborg, Denmark
Email: acl@et.aau.dk, nda@et.aau.dk,
mes@et.aau.dk, juq@et.aau.dk, joz@et.aau.dk
http://www.microgrids.et.aau.dk

Kai Sun
Tsinghua University
Tsinghua, China
Email: sun-kai@mail.tsinghua.edu.cn

Guoliang Chen
and Libing Sun
Shanghai Solar Energy
& Technology Co., Ltd.

Abstract—In this paper, a lineal mathematical model is proposed to schedule optimally the power references of the distributed energy resources in a grid-connected hybrid PV-wind-battery microgrid. The optimization of the short term scheduling problem is addressed through a mixed-integer linear programming mathematical model, wherein the cost of energy purchased from the main grid is minimized and profits for selling energy generated by photovoltaic arrays are maximized by considering both physical constraints and requirements for a feasible deployment in the real system. The optimization model is tested by using a real-time simulation of the model and uploaded it in a digital control platform. The results show the economic benefit of the proposed optimal scheduling approach in two different scenarios.

I. INTRODUCTION

A microgrid is an aggregation of distributed energy resources (DER) such as renewable energy sources (RES), loads and energy storage systems (ESS) as controllable entities which may operate in grid-connected or islanded mode. Hierarchical operation has been defined in order to standardize the control of the microgrid. In this sense, primary, secondary and tertiary controllers have been defined in order to provide adequate voltage and current regulation (primary control), restoring any steady-state error introduced by primary control (secondary control), and regulate the power flow between DER or between several clusters of microgrids (tertiary control) [1]. On top of that an Energy Management System (EMS) defines set points for the previous control layers in order to achieve optimal dispatch regarding specific objectives and limits generation capacity when is in islanded mode or power exchange with the utility grid while operates in grid-connected mode, buying and selling the shortage or surplus of power to or from the main grid [2].

A full-scale demonstrative, research-oriented microgrid has been installed in Shanghai, China for evaluating the performance of several EMS strategies, oriented to optimize the operation of the microgrid by scheduling the power of the distributed energy resources (Fig. 1) [1] (www.meter.et.aau.dk). The proposed EMS aims to minimize the cost of buying energy

from the main grid and maximize the revenue due to the photovoltaic (PV) energy generation, while preserving the lifetime of the ESS based on batteries by avoiding overcharge, excessive discharge and discharge cycles. In this proposal, a mixed integer linear programming (MILP) model is used in order to obtain an optimal power dispatch regarding economic issues. The proposed optimization model can be deployed easily in real microgrids sites since it is linear, simple and can be synthesized in commercial optimization software such as GAMS. This fact is a remarkable advantage compared to others optimization strategies [2] [4, 5].

II. MICROGRID DESCRIPTION

The system under test is a 200 kW PV-wind-battery microgrid (Fig. 2). For the case study presented in this paper, the microgrid will operate connected to the main grid. Because of that, the main grid impose the voltage and frequency conditions at the point of common coupling. Meanwhile, all the inverters will operate in current control mode as grid-following units [3].

The microgrid is composed by 6 PV generators all of them will operate by following local maximum power point tracking (MPPT) strategies. On top of that, the EMS schedules the startup or shutdown signals for PV generators. On the contrary, the two wind generators can operate in interactive and non-interactive way since their power generation may be curtailed from the EMS [4].

Moreover, in order to enhance the performance of the ESS based on batteries a two stage procedure which avoids battery overcharge is recommended by the battery manufacturers. Therefore, the battery control should be complemented with a voltage-limiting strategy. In this case, the ESS has two control operation modes: limited current charge and constant voltage charge. At the first stage, the ESS is charged based in the power reference defined by the EMS. It is important to say that the EMS considers maximum power ratings of the ESS in order to limit the battery current. On the other hand, the second stage (constant voltage charge) is activated once the battery array reach a threshold value. At this stage, the ESS

978-1-4673-9551-9/16 $31.00 © 2016 IEEE

Fig. 1. Real site microgrid in Shanghai-China.

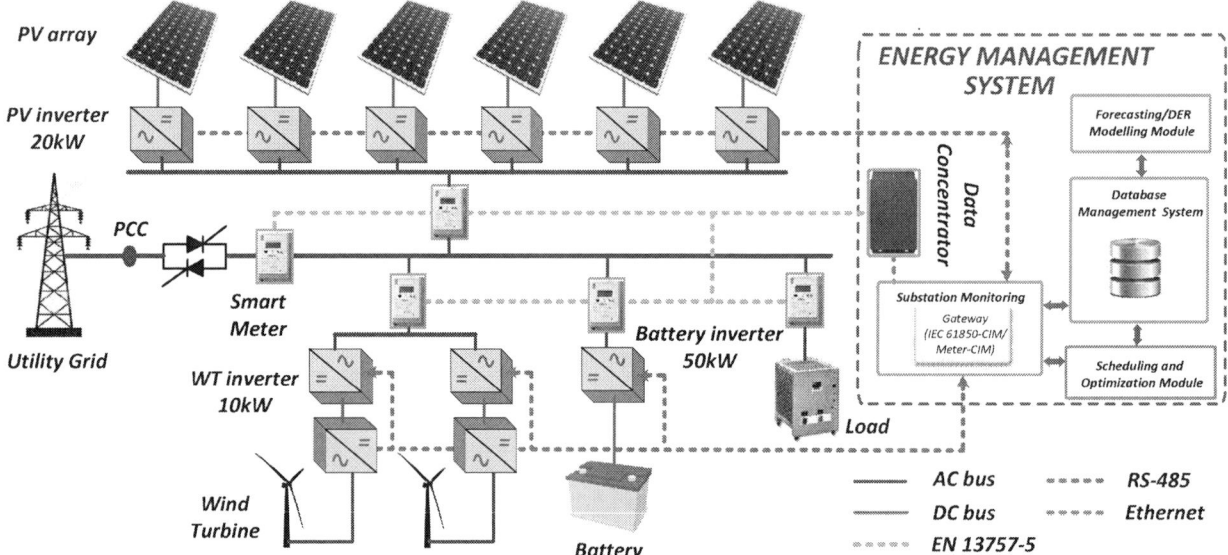

Fig. 2. Scheme of the microgrid.

only takes from the microgrid as much power as it needs in order to keep the battery voltage in a constant value [5], [4].

Another important point is to avoid excessive discharge of the battery. At this sense, the state of charge (SoC) of the battery is restricted to 50% for ensuring larger life-time of the battery array [5], [6]. For achieving this objective the EMS should ensure proper scheduling of the other distributed energy resources including the main grid in order to get an optimal commitment between battery charge and the energy bought from the grid. Additionally, the EMS optimizes the operation of the system in order to reduce the number of discharging cycles. Also, the microgrid supplies a local resistive load of 5 kW.

Apart from that, the microgrid is complemented with a EMS which schedules the operation of the distributed energy resources by considering optimal operational objectives [2]. The proposed EMS has been designed as a modular architecture in which each module performs different functions independently and exchanges information through the database system, as can be seen in Fig. 2. In this paper we will presented specifically the model implemented in the *scheduling and optimization module*.

TABLE I.   SETS OF THE OPTIMIZATION MODEL

Name	Description	Definition
t	discrete time slot	$\{1, 2, ...T\}$
i	PV units	$\{PV1, PV2, ..., PV6\}$
j	WT units	$\{WT1, WT2\}$

III. OPTIMIZATION PROBLEM

In this proposed approach, a Mixed Integer Linear Programming (MILP) problem is defined in order to minimize the cost of buying energy from the utility grid and maximize the revenue obtained by selling energy generating by the PVs.

A. Statements

The indexes, parameters and variables used in this model are summarized in tables I, II and III, respectively and will be presented during the description of the model.

The optimization problem is formulated assuming discrete time representation [7]. Thereby, the time horizon corresponds to $T * \Delta t$. Additionally, the values of the power are considered equal to the average value for each time interval and have been scaled in per units (p.u.) with S_{base} as the base.

TABLE II.   PARAMETERS OF THE MODEL

Name	Description	Value
T	Number of time slots	24 (h)
Δt	Duration of interval	1 (h)
n_i	Number of PV arrays	6
n_j	Number of WT	2
$C_{sell}(t)$	Cost of selling energy	1.147 (Yuan/kWh)
$C_{buy}(t)$	Cost of buying energy (night)	0.307 (Yuan/kWh)
	Cost of buying energy (day)	0.617 (Yuan/kWh)
$P_{PV_{max}}(i,t)$	Max power for PV arrays	
	for $i = \{PV1$ to $PV4\}$	0.733 (p.u.)
	for $i = \{PV5, PV6\}$	0.585 (p.u.)
$P_{WT_{max}}(j,t)$	Max WT power $\forall j$	0.3923 (p.u.)
$P_L(t)$	Power required by the load	0.1 (p.u.)
$P_{bat_{max}}$	Maximum power of battery	1 (p.u.)
$P_{bat_{min}}$	Minimum power of battery	-1 (p.u.)
$P_{grid_{max}}$	Max. power bought from utility	1 (p.u.)
S_{base}	Scaled base	5 kW
SoC_{max}	Maximum SoC	100 (%)
SoC_{min}	Minimum SoC	50 (%)
$SoC(0)$	Initial Condition of SoC	70 (%)
Cap_{bat}	Capacity of the battery	1 p.u.
χ	Penalty cost to the battery	$C_{buy}(t) * 0.1$(Yuan/kWh)

TABLE III.   VARIABLES OF THE MODEL

Name	Description
Decision variable	
$Totalcost$	Objective function
Scheduled variables	
$X_{PV}(i,t)$	ON/OFF commands for PV arrays
$P_{WT}(j,t)$	Power of the WTs
$P_{bat}(t)$	Power of the battery
$P_{buy}(t)$	Power bought from the utility
$P_{sell}(t)$	Power sold to the utility
Auxiliar variables	
$X_{grid}(t)$	Status bought/sold
$SoC(k,t)$	State of charge

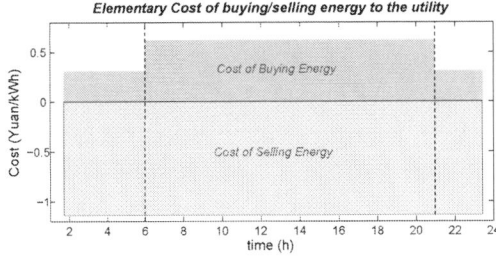

Fig. 3.   Elementary costs of buying and selling energy to the utility in Shanghai [8]

On the other hand, the cost of buying for microgrids in Shanghai [8] is established in a time of use (ToU) scheme where the consumers is persuaded to change its consumption behavior by means of fixed differentiate tariff during the day and the night (Fig. 3). Additionally, the grid company only buys energy generated by PV arrays, so the generation provided by WT can be used just for local demand.

B. Definition of the Model

This proposal aims to minimize the cost of buying energy from the utility grid and maximizing the revenue obtained by generating energy from the PVs. In this way, the following objective function has been defined,

$$
\begin{aligned}
Totalcost = & \sum_{t=1}^{T} \{P_{buy}(t) * \Delta t\} * C_{buy}(t) \\
& - \sum_{t=1}^{T} \{P_{sell}(t) * \Delta t\} * C_{sell}(t) \\
& + \sum_{t=1}^{T} \left[\frac{SoC_{max} - SoC(t)}{100} \right] * \chi
\end{aligned}
\tag{1}
$$

The first two terms in 2 fulfills the requirements of the optimization problem for the time horizon. Additionally, the third term is included as a penalty for not charging the battery in order to take advantage of the surplus energy that should be curtailed.

The constraints of the optimization problem start with the energy balance.

$$
\begin{aligned}
& \{P_{buy}(t) * \Delta t - P_{sell}(t) * \Delta t\} + \\
& \sum_{i=1}^{n_i} X_{PV}(i,t) * P_{PV_{max}}(i,t) * \Delta t + \\
& \sum_{j=1}^{n_w} P_{WT}(j,t) * \Delta t + P_{bat}(t) * \Delta t = P_L(t) * \Delta t
\end{aligned}
\tag{2}
$$

The first two terms are related to the energy absorbed from the grid and injected to the grid respectively. In turns, the third expression is the energy provided by the PV arrays. In this case, the variable $X_{PV}(i,t)$ allows to manage the energy by sending ON/OFF commands. The next term corresponds to the energy supplied by the WT. After that, the energy of the battery is included. To complete the balance, the energy required by the load has to equal to the addition of the previous expressions.

The boundaries related to the power of the WTs are,

$$
0 < P_{WT}(j,t) < P_{WT_{max}}(j,t)
\tag{3}
$$

Regarding the utility, the constraints are,

$$
0 \leq \quad P_{buy}(t) \quad \leq P_{grid_{max}}(t) * X_{grid}(t)
\tag{4}
$$

$$
0 \leq \quad P_{sell}(t) \quad \leq \sum_{i=1}^{n_i} P_{PV_{max}}(i,t) * (1 - X_{grid}(t))
\tag{5}
$$

Besides, the constraints related to the battery are

$$
SoC(t) = SoC(t-1) - \varphi * P_{bat}(t) * \Delta T
\tag{6}
$$

$$
\sum_{t=1}^{T} SoC(t) - SoC(t-1) > 0
\tag{7}
$$

$$
SoC_{min} < SoC(t) < SoC_{max}
\tag{8}
$$

$$
P_{bat_{min}} < P_{bat}(t) < P_{bat_{max}}
\tag{9}
$$

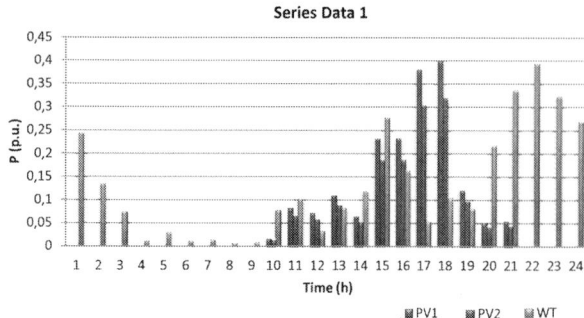

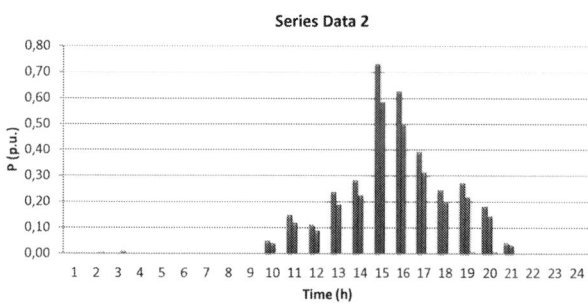

Fig. 4. RESs series data. Top to down: scenario 1 (average RESs generation) and scenario 2 (low WT generation and high PV generation).

IV. RESULTS

A. Scheduling Results

The software GAMS is used as algebraic model language which configures the use of the solver CPLEX in order to compute the power references obtained for the proposed MILP. The optimization model is run assuming an initial condition of SoC of 70%. Additionally, two scenarios of RES generation are considered as shown in Fig. 4, with an average and a low WT generation. In the figure, the profile *PV1* corresponds to the generation of the arrays PV1 to PV4, while the profile *PV1* is the one of the arrays PV5 and PV6, and the profile *WT* is the same for WT1 and WT2.

1) Scenario 1. Average generation: The scheduled variable related to the PV arrays is shown in Figs. 5. As can be seen, the arrays are not turned off during the day at any time since this energy is sold to the grid at good price, and they are disconnected during the night because there are no generation at that times.

Figure 6 presents the WT power profile. As can be seen, there are surplus of energy during the day and just part of the energy is used while some curtailment is deployed.

On the other hand, the profiles of selling and buying power to the utility are shown in Fig. 7 and 8. It is possible to see in Fig. 7 that all the generated PV energy is sold to the utility whereas Fig. 8 shows that the local generation is enough to supply the load and thus, it is not needed to buy energy from the grid.

Regarding the battery, the scheduled power profile and the consequently SoC are presented in Figs. 9 and 10. As can be seen, the battery is discharged when the power provided for

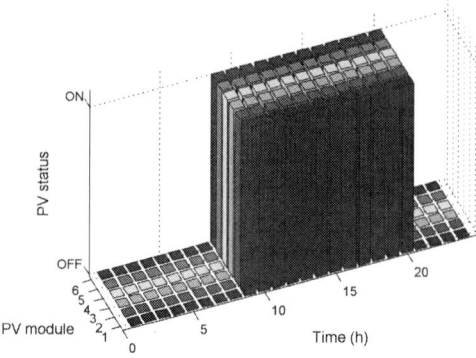

Fig. 5. Scheduled ON/OFF commands for PV modules.

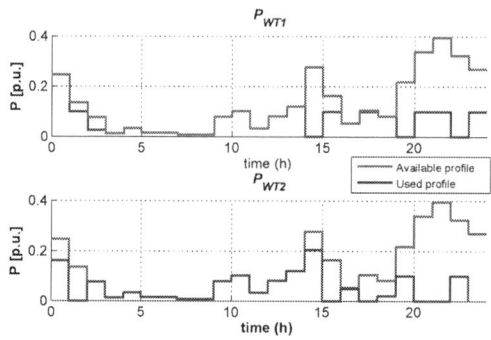

Fig. 6. Power references for WTs in scenario 1. Red line: available power. Blue line: Used power

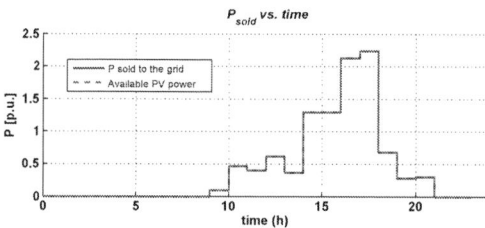

Fig. 7. Scheduled power sold to the utility in scenario 1. Red dashed line: available PV power. Blue solid line: scheduled power to be sold to the utility

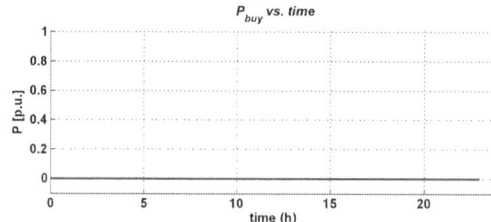

Fig. 8. Scheduled power bought from the utility in scenario 1

the WT is low while it is charged when there is high power in the WT.

2) Scenario 2. Low WT generation and High PV generation: In this case the PV profile for the PV array is identical

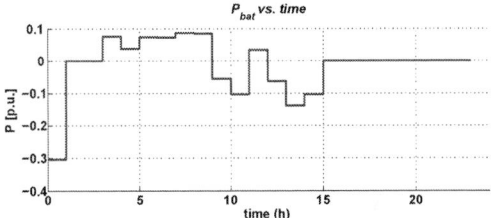

Fig. 9. Scheduled power of the battery in scenario 1

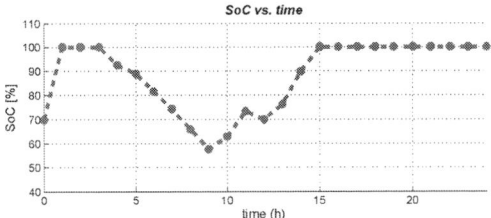

Fig. 10. Expected State of Charge of the battery in scenario 1.

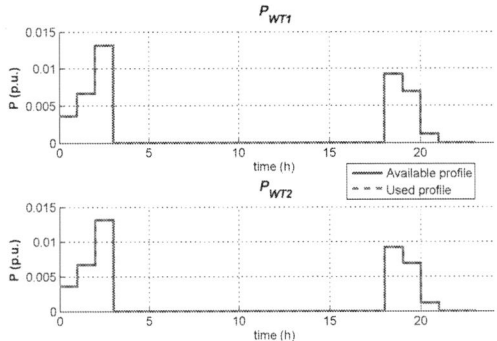

Fig. 11. Power references for WTs in scenario 2. Red dashed line: Used power. Blue solid line: available power

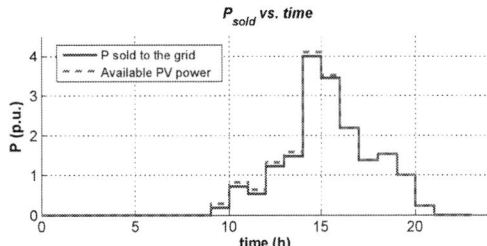

Fig. 12. Scheduled power sold to the utility in scenario 2. Red dashed line: available PV power, Blue solid line: scheduled power to be sold to the utility

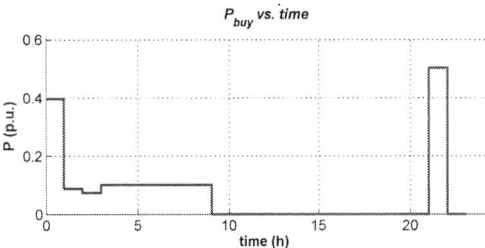

Fig. 13. Scheduled power bought from the utility in scenario 2

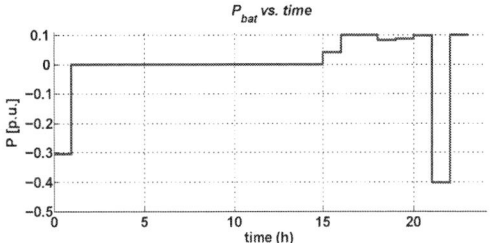

Fig. 14. Pbat

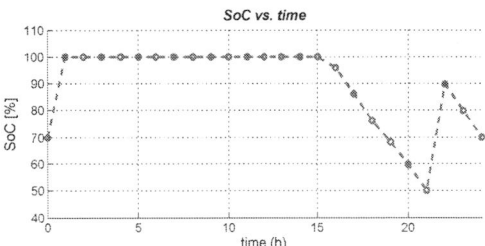

Fig. 15. Expected State of Charge of the battery in scenario 2.

to the previous case (see Fig. 5) since the energy produced by the PV can be sold to the utility without restrictions.

On the other hand, the scheduled profile of the WT is presented in Fig. 11. It is possible to see that in this scenario there is no surplus of energy, so it is not required to curtail energy at any time during the day.

Regarding the utility, the profiles of selling and buying energy are shown in Fig. 12 and 13. As can be seen, part of the energy generated by the PV arrays is used in the local consumption and, at the same time, it is required to buy energy from the utility because the energy provided by the WT is not enough to supply the demand.

Additionally, the scheduled power profile and the expected SoC of the battery are presented in Fig. 14 and 15, respectively.

Note that, the optimization model schedules to buy energy during the time when the elementary cost of buying energy from the main grid is lower, rather than using the energy available in the battery. After that, the scheduling hold charged the battery to use the energy at the end of the day when the

PV arrays do not generate energy.

Finally, Table IV summarizes the revenue for the user in each case in which negative values indicate earning money and positive values mean must pay money. It is possible to see that for average generation it is expected profits of 582 Yuan in one day. Also, when there is high generation of PV, the user gets more profits even in this case when it is required to buy energy during some times during the day due to the low power generation by WTs.

TABLE IV. REVENUE FOR THE USER (COST FOR BUYING MINUS COST FOR SELLING)

Scenario	Cost (Yuan)
S 1 (Average generation)	-582.81
S 2 (High PV and low WT generation)	-995.97

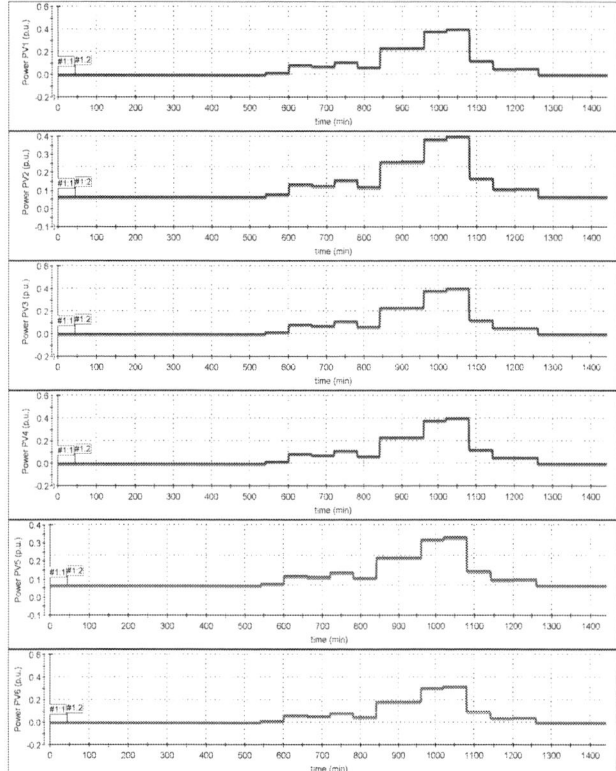

Fig. 16. Power profile of each PV array in real-time simulation considering scenario 1. Top to bottom: PV1 to PV6 power profiles. From PV1 to PV4: blue line: available power, green line: used power. For PV5 and PV6: red line: available power, green line: used power

B. Real time simulation

In order to test the proposed optimization model, a real time simulation of the microgrid is deployed considering the case study. To be more precise, the obtained results are included in a detailed Simulink model of the microgrid and subsequently uploaded in a digital control platform in one of the setups at the Research Microgrid Laboratory of Aalborg University[9].

To deploy the simulation of one day, the data were time-scaled i.e. one hour corresponds to one minute simulation. In the same way, the capacity of the battery and the elementary cost of the grid were scaled. The real time simulation is performed by using average values of available power in deterministic scenarios in order to test the performance of the optimization model.

1) Scenario 1. Average generation: The real time simulation results related to the first scenario are presented in Fig. 16, 17, 18, 19 and 20.

The available and used power profile of the PV arrays are

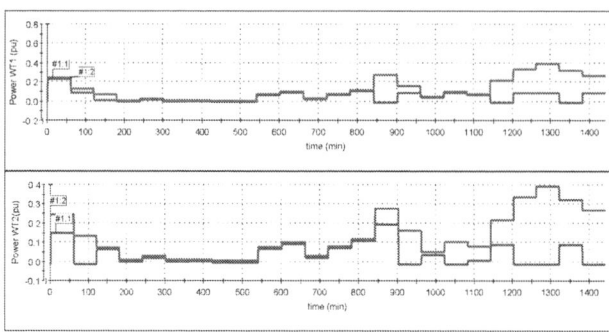

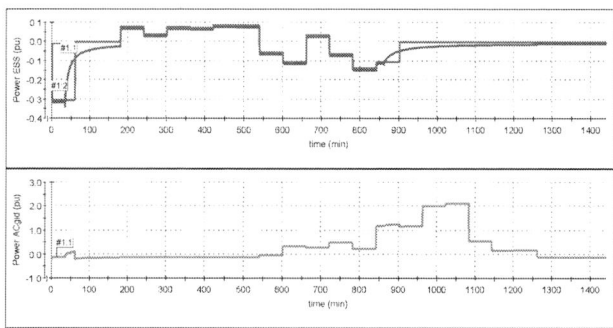

Fig. 17. Power profile of each WTs in real-time simulation considering scenario 1. Red line: available power, green line: used power

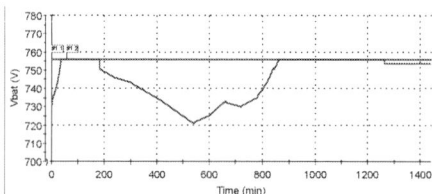

Fig. 18. Power profile of the battery and the utility in real-time simulation considering scenario 1. For battery power profile (top): scheduled profile (red line) and obtained profile (green line).

Fig. 19. Battery voltage obtained in real-time simulation considering scenario 1. Threshold voltage (green line) and battery voltage (blue line).

shown in Fig. 16. As can be seen, the first four PV arrays produce more energy (blue lines) than the next two PV arrays (red lines) since the first ones have a slightly higher power rating. Besides, the available and used profiles of each PV array are equal due to they have been scheduled to be activated during the generation hours (Fig. 5).

Meanwhile, the available and used power profile of the WT are presented in Fig. 17. It is possible to see that the powers are set to follow the scheduled profiles presented in Fig. 6, where part of the energy has been curtailed.

Regarding the battery, in the first frame of Fig. 18, in Fig. 19, and in Fig. 20 are presented the power profile, voltage and SoC, respectively.

Particularly, the first frame of Fig. 18 shows the scheduled power profile (red line) and the power profile obtained in the real-time simulation (green line) of the battery. As can be seen,

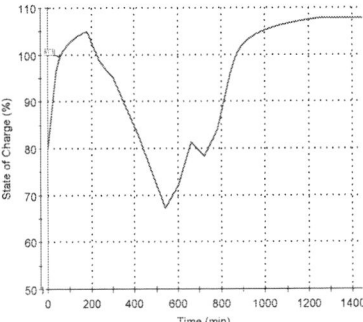

Fig. 20. SoC of the battery in real-time simulation considering scenario 1.

the battery follows the scheduled references most of the time during the day. In fact, these profiles differ when the battery voltage reaches the voltage threshold (Fig. 19) and thus, it is considered charged. In this case, the battery should control the voltage in this value to avoid damage due to over-voltage. For this reason, the battery stop following the power reference until the available energy is not enough to hold the voltage threshold. Nevertheless, the power profile keeps the tendency scheduled by the optimization model. Accordingly, the SoC of the battery (Fig. 20) are similar to the expected SoC (Fig. 10).

Moreover, the power profile exchanged with the utility is presented in the second frame of Fig. 18. In this case, the profile is positive when microgrid injects power to the grid (sells) and negative when absorbs power (buys). The behavior of this profile follows accurately the behavior expected in the optimization stage (see Fig. 7 and 8).

2) Scenario 2. Low WT generation and High PV generation: The real time simulation results related to the second scenario are presented in Fig. 21, 22, 23, 24 and 25.

As in the previous scenario, all the energy generated by the PV arrays are used as shown in Fig. 21. In turns, the power profile of the WT follows the scheduled reference without curtail available energy.

Regarding the battery, in the first frame of Fig. 23, in Fig. 24, and in Fig. 25 are presented the power profile, voltage and SoC, respectively.

The scheduled power profile (red line) and the power profile obtained in the real-time simulation (green line) of the battery are shown in the first frame of Fig. 23. It is possible to see that these profiles differ only when the battery is charged (when voltage reaches the voltage threshold in Fig. 24). Nevertheless, the power profile tends to be as the scheduled profile by the optimization model. Besides, the SoC of the battery (Fig. 25) has the same tendency as the expected SoC (Fig. 15).

Furthermore, the second frame of Fig. 18 presents the power profile exchanged with the utility. Once again, the profile is positive when microgrid sells power to the utility and negative when buys power. The behavior of this profile follows accurately the behavior expected in the optimization stage (see Fig. 7 and 8).

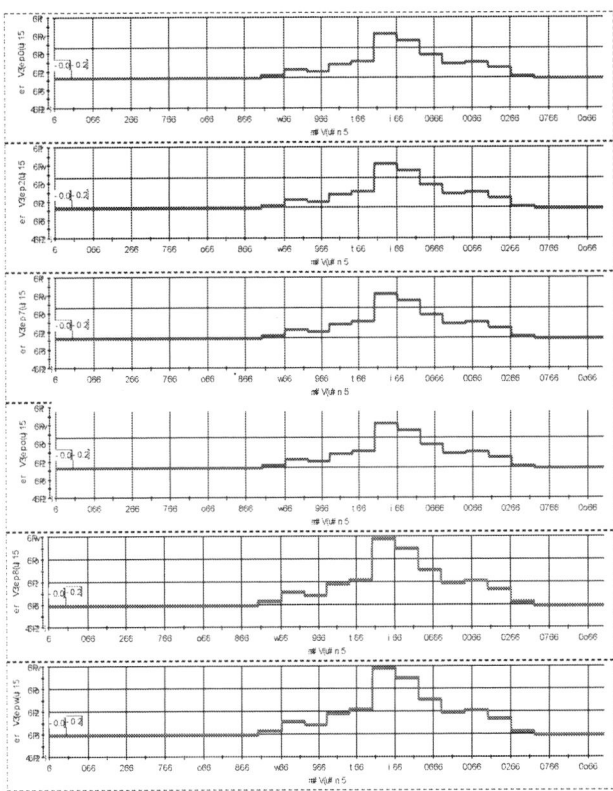

Fig. 21. Power profile of each PV array in real-time simulation considering scenario 2. Top to bottom: PV1 to PV6 power profiles. From PV1 to PV4: blue line: available power, green line: used power. For PV5 and PV6: red line: available power, green line: used power

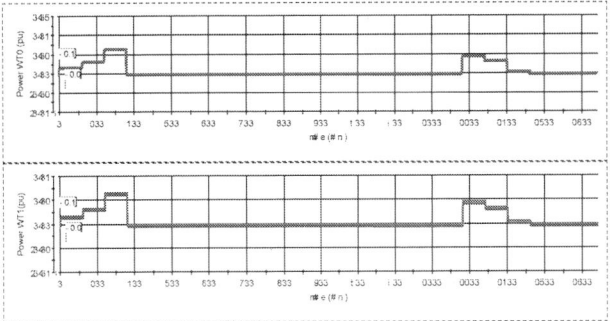

Fig. 22. Power profile of each WTs in real-time simulation considering scenario 2. Red line: available power, green line: used power

V. CONCLUSION AND FUTURE WORKS

A linear model for a grid-connected hybrid PV-Wind-Battery microgrid system has been proposed in order to minimize cost of buying energy from the grid and maximize profits for selling energy generated by PVs. The optimization strategy has been defined as a mixed integer lineal model to set optimal power references for the distributed resources of the microgrid. The strategy has been tested by considering two scenarios with different generation profiles. The behavior

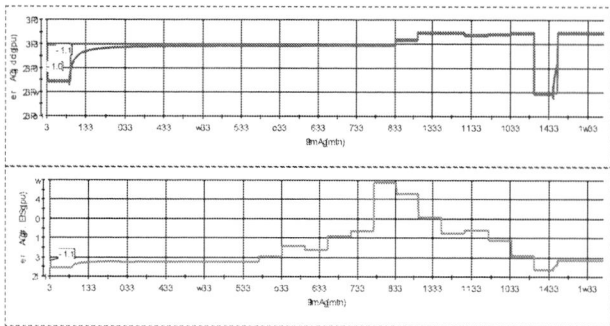

Fig. 23. Power profile of the battery and the utility in real-time simulation considering scenario 2. For battery power profile (top): scheduled profile (red line) and obtained profile (green line).

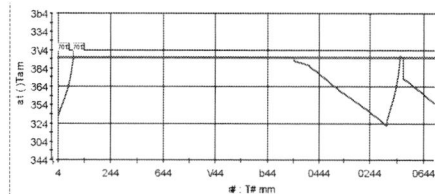

Fig. 24. Battery voltage obtained in real-time simulation considering scenario 2. Threshold voltage (green line) and battery voltage (blue line).

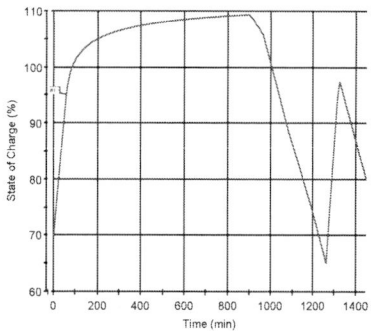

Fig. 25. SoC of the battery in real-time simulation considering scenario 2.

of the variables obtained in real-time simulation follows the expected features given by the optimization stage. As future work, demand side management should be considered to take advantage of the surplus renewable energy.

ACKNOWLEDGMENT

This work was supported by the Energy Technology Development and Demonstration Program through the Sino-Danish Project Microgrid Technology Research and Demonstration (www.meter.et.aau.dk).

REFERENCES

[1] Microgrid technology research and demonstration. Aalborg University. [Online]. Available: www.meter.et.aau.dk

[2] M. Iqbal, M. Azam, M. Naeem, A. Khwaja, and A. Anpalagan, "Optimization classification, algorithms and tools for renewable energy: A review," *Renewable and Sustainable Energy Reviews*, vol. 39, no. 0, pp. 640 – 654, 2014.

[3] D. Wu, F. Tang, T. Dragicevic, J. Vasquez, and J. Guerrero, "Autonomous active power control for islanded ac microgrids with photovoltaic generation and energy storage system," *IEEE Transactions on Energy Conversion*, vol. 29, no. 4, pp. 882–892, Dec 2014.

[4] F. Katiraei, R. Iravani, N. Hatziargyriou, and A. Dimeas, "Microgrids management," *Power and Energy Magazine, IEEE*, vol. 6, no. 3, pp. 54–65, May 2008.

[5] I. S. C. C. 21, "Guide for optimizing the performance and life of lead-acid batteries in remote hybrid power systems," *IEEE Std 1561-2007*, pp. C1–25, 2008.

[6] F. Marra and G. Yang, "Decentralized energy storage in residential feeders with photovoltaics," in *Energy Storage for Smart Grids*, P. D. Lu, Ed. Boston: Academic Press, 2015, pp. 277 – 294.

[7] W. Shi, X. Xie, C.-C. Chu, and R. Gadh, "Distributed optimal energy management in microgrids," *IEEE Transactions on Smart Grid*, vol. 6, no. 3, pp. 1137–1146, May 2015.

[8] S. C. C. by letter Date: 2012-12-21. Shanghai region tariff table. [Online]. Available: http://www.sheitc.gov.cn/dfjf/637315.htm

[9] Intelligent microgrid laboratory. Aalborg University. [Online]. Available: http://www.et.aau.dk/department/laboratory-facilities/intelligent-microgrid-lab/

High Efficiency Power Converter for a Doubly-fed SOEC/SOFC System

Kevin Tomas-Manez
Technical University of Denmark
Department of Electrical Engineering
Email: ktma@elektro.dtu.dk

Alexander Anthon
Technical University of Denmark
Department of Electrical Engineering
Email: jant@elektro.dtu.dk

Zhe Zhang
Technical University of Denmark
Department of Electrical Engineering
Email: zz@elektro.dtu.dk

Abstract—Regenerative fuel cells (RFC) have become an attractive technology for energy storage systems due to their high energy density and lower end-of-life disposal concerns. However, high efficiency design of power conditioning unit (PCU) for RFC becomes challenging due to their asymmetrical current-power characteristics that are dependent on the operation mode (energy storage / energy supply). This paper proposes a new PCU architecture for grid-tie RFC with which the RFC's asymmetrical characteristic becomes less critical and thus a much more symmetrical power rating of the dc-dc converter for both operating modes is possible. This paper discusses the design considerations for this novel PCU, and verifies its operation principle with Matlab/Simulink simulations. Experimental results on a tailored dc-dc converter confirm the design simplifications for high efficiency operation along the entire power operating range of the RFC as well as the utilization of the same control strategy design for the two RFC operating modes.

Keywords—*Bidirectional fuel cells, power conditioning, Interleaved boost converter, renewable energies, grid tie*

I. INTRODUCTION

Over the last years renewable energies have experienced a strong development to become alternatives for conventional energy resources, among others due to the global awareness on limited fossil fuel resources and a widespread sensibility towards the environmental impacts. However, large scale integration of renewable energy sources present an important drawback because of their highly irregular and mostly unpredictable production [1], which causes high dynamics on the grid infrastructure and thus, electrical grid reliability is lowered. Large scale energy storage systems are therefore a potential solution to improve grid system reliability and stability when supplied by renewable energy sources [2]. With the utilization of information technologies to the grid system, consumers' behavior can become much more predictable, and therefore contribute to a better load regulation and an increased grid reliability [1]. The combination of renewable energy sources including energy storage systems and information technology systems can establish an equilibrium between energy production and demand, because surplus energy from renewable energy sources can be stored for later use to support the grid demands when necessary [3].

Regenerative or bidirectional fuel cells (RFC) could be an attractive technology for such large scale energy storage systems by means of hydrogen storage. Their particular benefits over traditional batteries are their high energy density of the fuel and their lower environmental disposal concerns [4]. In

[5] an energy storage system based on RFC which couples the electricity grid with the natural gas grid is presented showing the potential of RFC for large scale energy storage.

Grid-tie energy storage systems based on RFC require power converter units to couple the grid with the RFC, to accommodate voltage levels and to regulate the system power flow. For traditional grid-tie fuel cell systems (not RFC), high efficiency PCU ranging from 96.8 % to 98 % have been demonstrated [6]. However, RFCs present a very asymmetrical current-power characteristic depending on the operating mode (i.e. energy storage or energy supply), leading to wide power rating spans with which the power conditioning unit (PCU) has to operate at high efficiency.

Different approaches to improve system performance and efficiency have been addressed. For instance, it has been verified in [7] that the efficiency of bidirectional dc-dc converters for grid-tie RFC systems can be greatly improved with the utilization of wide bandgap semiconductors such as SiC MOSFETs. Another way towards increasing efficiency is to reduce the switching energies of power devices in the RFC bidirectional dc-dc converters by means of soft switching as shown in [8]. Research on magnetics optimization based on planar magnetics can further increase both efficiency and power density [9]. Other investigations aim to compensate the slow dynamics of RFC by the inclusion of auxiliary sources. For instance, in [10] a dc-dc converter for FCs with an additional battery-based energy storage and a bidirectional dc-dc converter has proven to successfully support rapid load demand, and in [16] a grid-tie multiple port converter for fuel cells with super-capacitors as an auxiliary source has been proven to clearly increase the dynamic response compared to systems without auxiliary energy storage element.

Research performed to date is based on traditional PCU architectures for energy storage systems, as the one shown in Fig. 3. However, they are limited to the power rating symmetry characteristic of the PCU itself. Power symmetry implies that the power rating of the system is equal regardless the power flow direction. Considering the asymmetrical and wide power range of RFC, the design of high efficiency PCU for the entire operating area becomes very challenging. Furthermore, due to the power asymmetry of RFC, cooling effort for the PCU can be oversized depending on the operation mode, which greatly challenges high power density design. Due to the aforementioned drawbacks with the RFC systems, this paper presents a novel PCU architecture aiming for a much more symmetrical power rating of the dc-dc converter in both

978-1-4673-9551-9/16 $31.00 © 2016 IEEE

operation modes in order to simplify the high efficiency converter design. Design considerations are discussed, principle operation modes verified by simulations and experimentally confirmed on a 5 kW interleaved boost converter

II. BIDIRECTIONAL SOEC/SOFC TECHNOLOGY

Solid oxide electrolyzer cells/fuel cells (SOEC/-SOFC) are a kind of fuel cell technology which uses hydrogen as a fuel and takes advantage of the released energy during the reaction of hydrogen and oxygen to produce electricity. It has been proven that solid oxide cells (SOCs) have the capability to operate in bidirectional mode [11], also referred as regenerative mode. When hydrogen is used a fuel, reaction of hydrogen and oxygen is produced and an amount of energy is released generating electricity when connected to a load (SOFC operation). On the other hand, when current is forced to flow through the SOCs the electrolysis of water is produced and hydrogen is generated (SOEC operation).

Design and system specifications of a grid-tie power conditioning units for SOEC/SOFC systems are defined according to the current-voltage (I-V) characteristics of the SOEC/SOFC system, which is set by the number of stacked cells. However, I-V characteristics of SOEC/SOFC are highly dependent on operating temperature, fuel, pressure and degradation as explained in [12], [13]. This dependency must be considered in mature prototyping stages for long-term operation, specially the degradation ratio since it can strongly degrade the operating voltage ratings [12], [13]. The I-V characteristic for a single SOEC/SOFC provided by SOCs manufacturers is shown in Fig. 1, which will be the starting point of the PCU design.

As shown in the I-V curve from Fig. 1, SOCs operate at very low voltage and high currents, and for that reason it is necessary to stack many cells in series. Due to the immaturity of SOEC/SOFC technology, currently there are still mechanical limitations to obtain high power SOEC/SOFC systems, among others due to the maximum number of stackable cells, current density limitations, operating temperature, etc. Currently, the maximum current capability in SOEC mode is up to 60 A and up to 30 A in SOFC mode.

For the proposed system design in this work, fours stacks in series with 75 cells per stack has been chosen in agreement with SOCs manufacturers. Fig. 2a shows the I-V characteristics and Fig. 2b shows the current-power (I-P) characteristics for the SOEC/SOFC stacks system, which has been extrapolated from the I-V curve of a single cell Fig. 1. From the I-P curve in Fig. 2b, it can be seen that the power rating in SOFC mode is lower than in SOEC mode, which is due to the variation of the internal resistance and current direction [11]. This results in a very asymmetrical I-P characteristic, which leads to a wide operating power range for the PCU and thus, high efficiency design for the whole power range becomes challenging.

III. GRID-TIE POWER CONDITIONING UNIT FOR SOEC/SOFC SYSTEMS

A. Traditional PCU architecture for SOEC/SOFC systems

Several power conditioning topologies can be used to realize SOEC/SOFC systems, and may differ depending on the particular application, system requirements and stacks structure

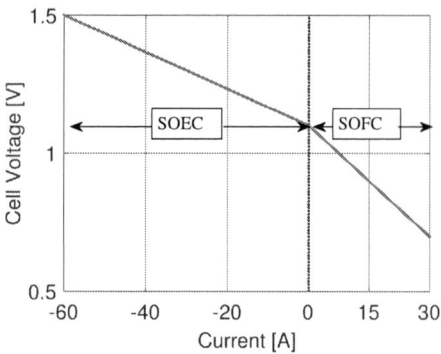

Fig. 1. Current-voltage characteristics of a single Solid Oxide Electrolyzer/Fuel Cell

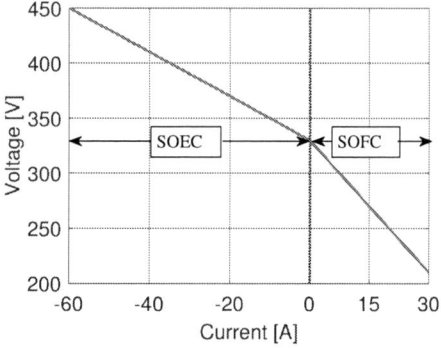

(a) Current-voltage characteristics.

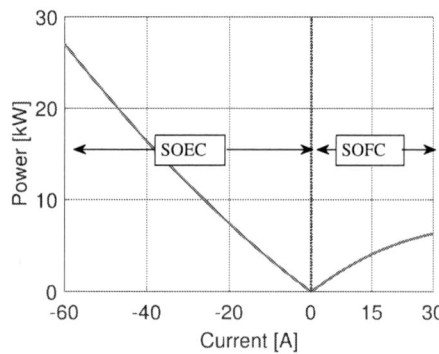

(b) Current-power characteristics.

Fig. 2. SOEC/SOFC stacks composed by 4 stacks in series with 75 cells/stack

[14]. However, the simplified architecture from Fig. 3 can be defined as a basis. System configuration is based on the parallel connection of an ac-dc converter, a dc-dc converter and the SOEC/SOFC stacks. Bidirectional power flow of the power converter units is required to allow the RFCs to operate in both modes. In particular, power flows from the grid to the SOCs when operating in SOEC mode (energy storage) and from the SOCs to the grid when operating in SOFC mode (energy generation). The dc-dc converter regulates the SOCs power flow and sets the required voltage level for the dc-link of the inverter.

TABLE I.   TRADITIONAL PCU ARCHITECTURE: SOEC/SOFC AND
DC-DC CONVERTER SPECIFICATIONS

Specification	SOEC	SOFC	dc-dc converter
Voltage [V]	330 - 450	210 - 330	210 - 450
Current [A]	0 - 60	0 - 30	0 - 60
Power rating [W]	27000	6300	27000
Power flow	from the grid	to the grid	bidirectional

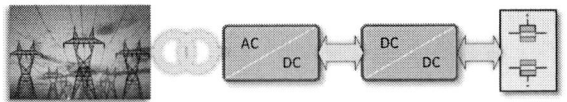

Fig. 3.   Traditional PCU architecture for SOEC/SOFC systems.

According to the SOEC/SOFC stacks characteristics from Fig. 2, the system specifications for the dc-dc converter and the SOEC/SOFC stacks using the traditional PCU architecture are specified in Table I. With the traditional architecture from Fig. 3, a 27 kW power rated dc-dc converter would be required to cover the whole power range in both operation modes. This clearly leads to an oversized dc-dc power converter when operating in SOFC mode, which only requires a 6.3 kW rated system. Therefore, a different PCU architecture that counteracts the SOEC/SOFC power asymmetry would greatly simplify the converter design.

B. Novel doubly-fed SOEC/SOFC system

The novel PCU architecture presented in this work aims to achieve a much more symmetrical power operating range of dc-dc converter. The architecture is based on a dynamic PCU which connection varies according to the operation mode by the utilization of two single pole double through (SPDT) relays as subsequently explained. The proposed PCU architecture for both operation modes is shown in Fig. 4.

Under **SOFC mode**, shown in Fig. 4a, SOC stacks are connected in parallel to the dc-dc converter, which is the same scenario as with the traditional PCU architecture. Power is transferred from the SOCs to the output of the dc-dc converter P_{out} through the dc-dc converter. Therefore the power rating of the dc-dc converter P_{conv} is equal to the power generated by the SOFCs P_{sofc}, and the dc-dc converter electrical characteristics are defined as in equations 1, 2 and 3.

$$V_{conv(sofc)} = V_{sofc} \tag{1}$$

$$I_{conv(sofc)} = I_{sofc} \tag{2}$$

$$P_{conv(sofc)} = P_{soc} = V_{sofc} \cdot I_{sofc} \tag{3}$$

Under **SOEC mode**, shown in Fig. 4b, SOC stacks are connected in series with the dc-dc converter, and these two are then connected in parallel with the ac-dc converter. Energy required for the electrolysis of water and hydrogen generation is supplied from the grid through the ac-dc converter. For this operating mode the power rating of the dc-dc converter P_{conv} is the power difference between the output power of the dc-ac converter P_{inv} and the power consumed by the SOECs P_{soec}. The electrical characteristics of the dc-dc converter are expressed with equations 4, 5 and 6. From these equations, it can be inferred that in this scenario the power rating of the

dc-dc converter will be reduced compared to the traditional PCU architecture. This is illustrated in the following design example.

$$V_{conv(soec)} = V_{inv} - V_{soec} \tag{4}$$

$$I_{conv,(soec)} = -I_{soec} \tag{5}$$

$$P_{conv(soec)} = P_{inv} - P_{soec} = V_{conv(soec)} \cdot I_{soec} \tag{6}$$

A 3-phase grid with $V_{line} = 230\,\text{V}$ is considered. Then the voltage at the output of the ac-dc converter is calculated with equation 7 as shown in [15].

$$V_{inv} = \sqrt{2} \cdot \sqrt{3} \cdot V_{line} \approx 560\,\text{V} \tag{7}$$

According to the SOEC/SOFC stacks characteristics from Fig. 2, system specifications for the dc-dc converter and the SOEC/SOFC stacks using the proposed PCU architecture can be redefined using equations 1-6 and with the specifications from Table II. Calculations show that using a dc-dc converter rated at 6.6 kW, a 27 kW-SOEC/6.3 kW-SOFC system can be realized using the proposed PCU architecture. In other words, in SOFC mode the dc-dc converter maximum power is $P_{conv(sofc)} = 6.3\,\text{kW}$ while in SOEC mode maximum power is $P_{conv(soec)} = 6.3\,\text{kW}$, resulting not only in a much more symmetrical dc-dc converter I-P characteristic, but also clearly a four times lower rated power system and thus a reduced cooling effort.

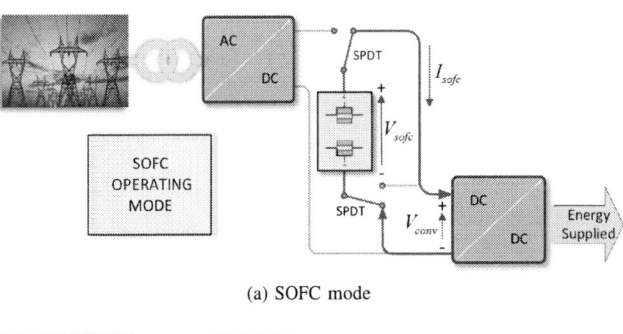

(a) SOFC mode

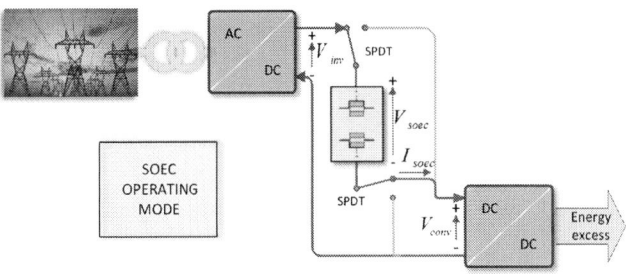

(b) SOEC mode

Fig. 4.   Novel doubly-fed power conditioning architecture for SOEC/SOFC.

TABLE II.   NOVEL PCU ARCHITECTURE: SOEC/SOFC AND DC-DC
CONVERTER SPECIFICATIONS

Specification	SOEC	SOFC	dc-dc converter
Voltage [V]	330 - 450	210 - 330	110 - 330
Current [A]	0 - 60	0 - 30	0 - 60
Power rating [W]	27000	6300	6600
Power flow	from the grid	to the grid	unidirectional

978-1-4673-9551-9/16 $31.00 © 2016 IEEE

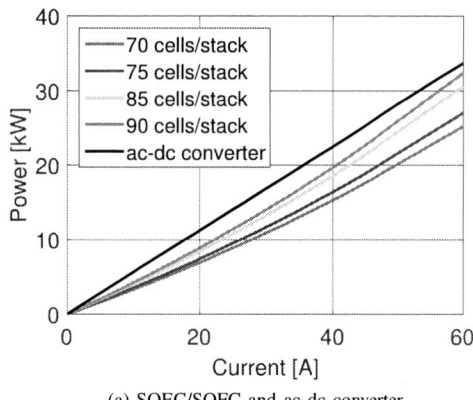

(a) SOEC/SOFC and ac-dc converter

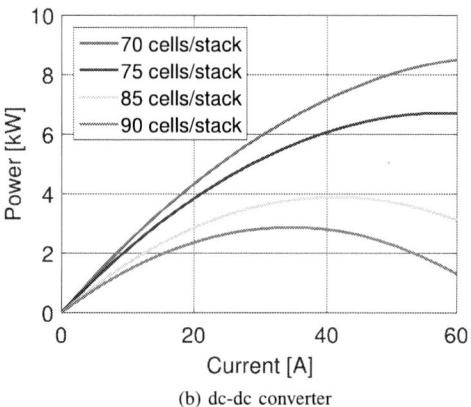

(b) dc-dc converter

Fig. 5. Current-power characteristics for different number of cells per stack with four stacks in series.

The proposed system is subsequently analyzed when operating under SOEC mode with different number of SOEC/SOFC cells by calculating the dc-dc converter specifications as previously shown. Fig. 5a shows the I-P characteristics of the SOEC/SOFC stacks and Fig. 5b shows the I-P characteristics for the dc-dc converter. Fig. 5 shows that by increasing the number of cells the power consumed by the SOEC stacks increases. Since the voltage across the SOEC stacks moves towards the inverter voltage with an increasing number of cells, the voltage at the input of the dc-dc converter decreases. This causes a significant reduction of the power rating of the dc-dc converter.

In order to describe the circuit operation in more detail, simulations of the entire PCU including a closed-loop SOEC/SOFC current control have been performed with a full bridge rectifier connected to the 3-phase grid and a boost dc-dc converter. Two ideal SPDT relays are used to switch from one operating mode to another, which occurs at very low frequencies because the operating mode is related to the grid power excess and power demand rather than conventional PWM mode of the dc-dc converter itself. Simulations are performed according to the specifications from Table II. Fig. 6a shows the current through the SOEC/SOFC system I_{conv}, whereas Fig. 6b shows the inverter voltage V_{inv}, the voltage across the SOEC/SOFC stacks V_{soec} / V_{sofc} and the input voltage of the dc-dc converter V_{conv}. Fig. 6c shows the inverter

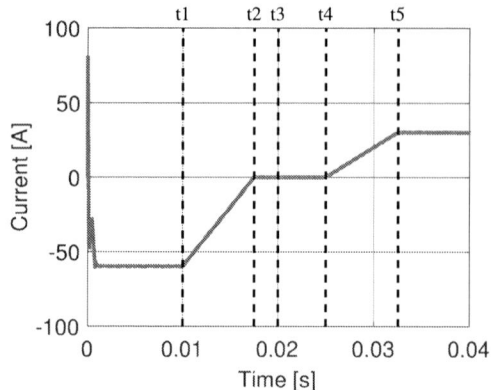

(a) Current through the SOEC/SOFC stacks ((+) for SOEC and (-) for SOFC)

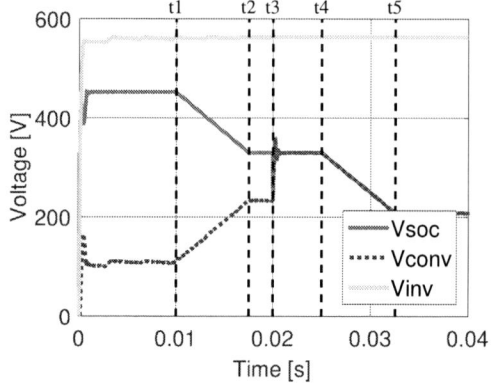

(b) Voltage at the rectifier output, SOCs and input of dc-dc converter

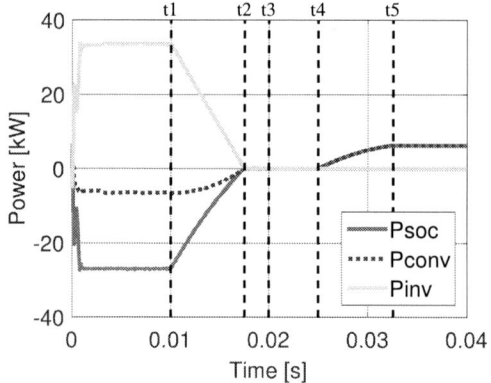

(c) Power at the rectifier output, SOCs and dc-dc converter input

Fig. 6. Simulation results

power P_{inv}, the power of the SOEC/SOFC stacks P_{soec} / P_{sofc} and the power of the dc-dc converter P_{conv}.

Step 1 (before t_1): System is in steady state, operating in SOEC mode, as depicted in Fig. 4b. Current reference is regulated to the maximum SOEC current capability $I_{soec,max} = -60$ A. The input voltage of the dc-dc converter V_{conv} is the difference between inverter voltage and SOEC stacks voltage as expressed by equation 4, and thus P_{conv} corresponds to

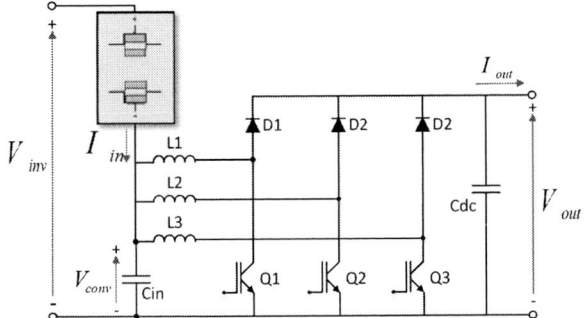

Fig. 7. Three stages interleaved boost converter schematic operating in SOEC mode.

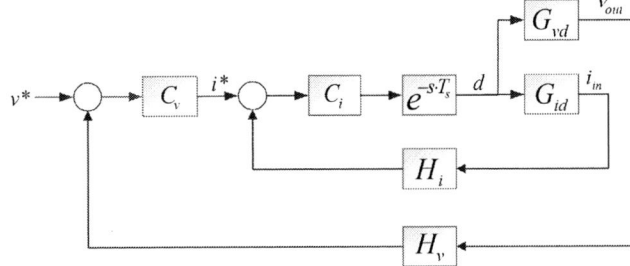

Fig. 8. Block diagram of the double-loop control strategy.

$P_{inv} - P_{soec}$ (Eq. 6). Note that P_{soec} is negative in Fig. 6c, because SOCs are consuming power.

Step 2 ($t_1 - t_2$): Before switching the operating mode, current reference decreases to zero with a certain slope to prevent any over-voltages on the SOEC/SOFC stacks. At the end of this period, voltage across the SOEC/SOFC stacks reaches the open-circuit voltage.

Step 3 (t_3): The SPDT relays are switched, and the system connection is changed to SOFC mode as shown in Fig. 4a. A step variation of V_{conv} occurs since the input voltage condition changes from Eq. 4 to Eq. 1.

Step 4 ($t_3 - t_4$): Due to the V_{conv} step variation, voltage oscillations occur across SOEC/SOFC stacks and V_{conv}. A short period of time with zero current is held to stabilize the system.

Step 5 ($t_4 - t_5$): Current reference is driven with a certain slope to the maximum SOFC current capability $I_{sofc,max} = 30$ A.

Step 6 ($t_5 - end$): Systems is in steady state, operating in SOFC mode, where $V_{conv} = V_{sofc}$ and $P_{conv} = P_{sofc}$, as expressed with Eq. 1 and Eq. 3, respectively.

IV. DC-DC CONVERTER

The main focus of this paper is the verification of the feasibility of the proposed PCU, for that reason a tailored dc-dc converter with a closed-loop control has been implemented for the defined SOEC/SOFC system in table II.

A. Power stage

A three-stages Interleaved Boost Converter (IBC) is used in this work as shown in Fig. 7 to test the proposed PCU architecture. The IBC topology is a widely accepted topology for fuel cell power conditioning systems due to its benefits [17]-[20]. SOCs lifetime can be dramatically reduced with large ripple current [21], thus by means of interleaving the inductor currents, the input current ripple amplitude can be reduced [20], thus greatly improving the lifespan of SOCs. Moreover, the output voltage frequency is also increased, therefore reducing the voltage ripple and allowing a reduction of dc-link capacitors [20].

Converter's component details are given in Table III. Output voltage is set to 600 V in order to reach a proper dc-link voltage level to feed energy to the grid and the switching frequency is 25 kHz.

TABLE III. CONVERTER'S COMPONENT DETAILS

Inductors $L_1 - L_3$	1 mH KoolMμ core
DC-Link cap. (I) C_{DC}	20 μF/800 V Film Cap.: 2 in parallel
DC-Link cap. (II) C_{DC}	12 nF/800 V Film Cap.: 3 in parallel
Input cap. C_{in}	50 μF/800 V Film Cap.
IGBTs $Q_1 - Q_3$	IKW25N120H3 1200 V/25 A
Diodes $D_1 - D_3$	IDH08S120 1200 V/7.5 A

B. Closed-loop control strategy

A DSP-based closed-loop control system has been implemented. It is intended to design and apply the established control loop strategy for both operating modes (SOEC and SOFC) by considering the input-output I-V characteristics of the dc-dc converter in the design process.

The closed-loop control strategy is designed in order to reach and keep the 600 V output voltage and to keep the input power of the dc-dc converter inside the allowed limits of the SOC stacks.

The classical double-loop control strategy represented in Fig. 8 is applied [22], where G_{id} refers to the duty cycle-to-input current transfer function and G_{vd} refers to the duty cycle-to-output voltage transfer function. This controller requires the measurement of two variables, i.e. the dc output voltage and current at the input of the converter, which are filtered by second-order filters from the measurement circuits, H_i and H_v, such that the high frequency components are properly attenuated. Classical proportional-integral (PI) controllers C_v and C_i are used for voltage and current compensation. In addition, a soft start-up procedure has been integrated to the control system, so during the converter turn-on voltage overshoots are avoided. The soft start-up consists of a reference signal v_{ref} with a certain slope from 0 to the desired reference.

Referring to the double-loop control, the fast inner loop is used to regulate the average input current using the controller C_i. The output voltage is regulated with the slower outer loop with the controller C_v. To obtain the closed-loop system stability, the inner loop bandwidth has to be larger than the outer loop [22]. Due to the slow dynamics of the SOCs the

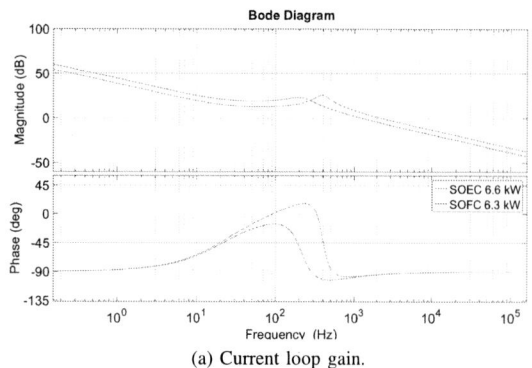

(a) Current loop gain.

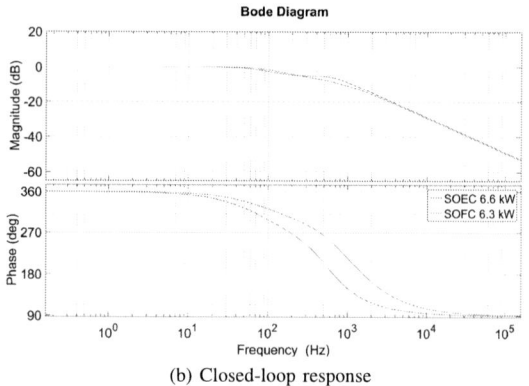

(b) Closed-loop response

Fig. 9. Control system Bode Plots

outer loop bandwidth has to be kept small and the inner current loop needs to assure reaching the faster transient response of the dc-dc converter.

State-space average model for the IBC is developed as performed in [23], [24], to thereafter calculate the system transfer functions shown in Eq. 8 and Eq. 9.

$$G_{id} = \left.\frac{\tilde{i}_{in}(s)}{\tilde{d}(s)}\right|_{\tilde{v}_{conv}(s)=0} = \frac{V_{conv}}{R \cdot D'^2} \cdot \frac{1 + s\frac{1}{R \cdot C_{dc}}}{1 + s\frac{L}{R \cdot D'^2} + s^2\frac{L \cdot C_{dc}}{D'}} \tag{8}$$

$$G_{vd} = \left.\frac{\tilde{v}_c(s)}{\tilde{d}(s)}\right|_{\tilde{v}_{conv}(s)=0} = \frac{V_{conv}}{D'} \cdot \frac{1 - s\frac{L}{R \cdot D'^2}}{1 + s\frac{L}{R \cdot D'^2} + s^2\frac{L \cdot C_{dc}}{D'}} \tag{9}$$

Where $D' = 1 - D$, $R = V_{out}/I_{out}$ and $L = L_1 = L_2 = L_3$.

PI controller for inner current loop is designed for full load condition obtaining the bode diagram shown in Fig. 9a. The loop is closed at the cross-over frequency $f_c = 1.25\,\text{kHz}$ with a phase margin of $83°$ for the SOEC mode at full load and at $f_c = 2.26\,\text{kHz}$ with a phase margin of $86°$ for the SOFC mode on full load condition. Once the stability of the inner current loop is obtained, the outer voltage loop controller is calculated for full load condition and the closed-loop response shown in Fig. 9b is obtained.

V. EXPERIMENTAL RESULTS

Experimental tests have been carried out independently for each operating mode. Tests under SOFC mode have been executed simply using a laboratory power supply at the dc-dc converter input and an output resistive load emulating the power demand from the dc-link side. Tests under SOEC mode have been carried out using a dc source to supply the ac-dc converter output voltage and an electronic load in fixed voltage mode with a dynamic resistance emulating the SOEC stacks power consumption. Fig. 10 shows the diagrams representing the experimental set-up.

Interleaved inductor currents and input current waveforms are shown in Fig. 11. From these results it is verified that the input current ripple is greatly reduced with the interleaving technique, therefore demonstrating an attractive topology for RFC systems.

A. Efficiency measurements

Efficiencies of the dc-dc converter are measured by using a power analyser PPA5530 from N4L. The efficiencies have been measured under SOEC and SOFC operating modes independently according to dc-dc converter I-P and I-V characteristics derived from the SOEC/SOFC stacks characteristics from Fig. 2. The results are shown in Fig. 12, clearly demonstrating that since the dc-dc converter can be rated for a similar power level in both operation modes, high efficiencies in both modes are possible. Note that efficiency curves are close to each other resulting in similar thermal stress for both modes which can simplify the heat sink design.

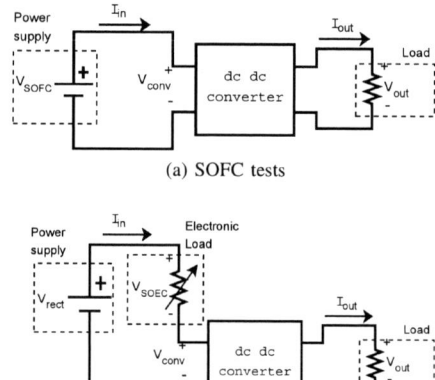

(a) SOFC tests

(b) SOEC tests

Fig. 10. Experimental set-up

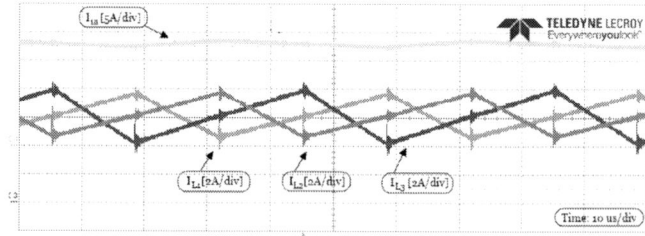

Fig. 11. Experimental results: 3 stages IBC inductors current (I_{L1}, I_{L2} and I_{L3} and input current I_{in}.

978-1-4673-9551-9/16 $31.00 © 2016 IEEE

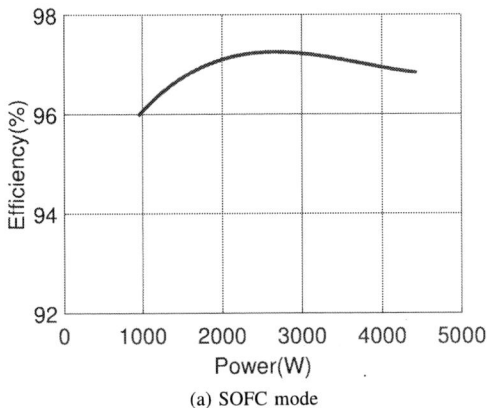

(a) SOFC mode

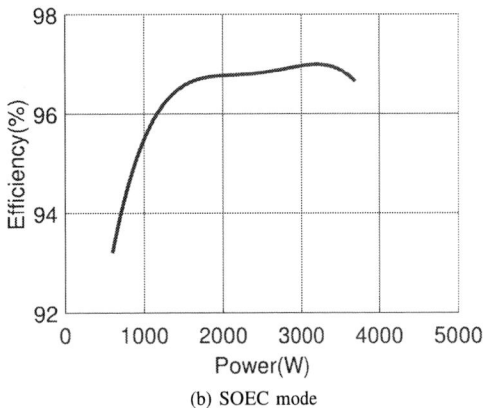

(b) SOEC mode

Fig. 12. Experimental and theoretical efficiency curves

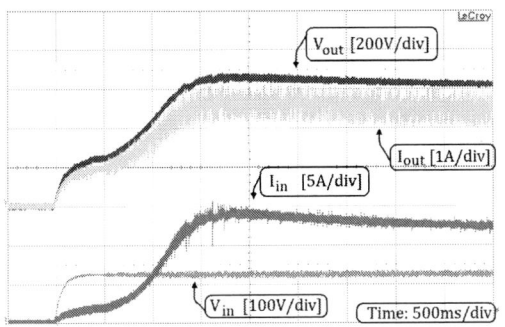

Fig. 13. Soft start-up test

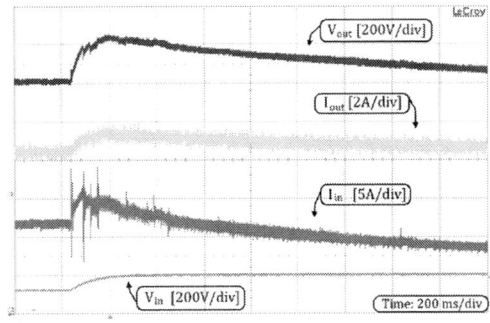

Fig. 15. SOFC voltage 80 V step-up

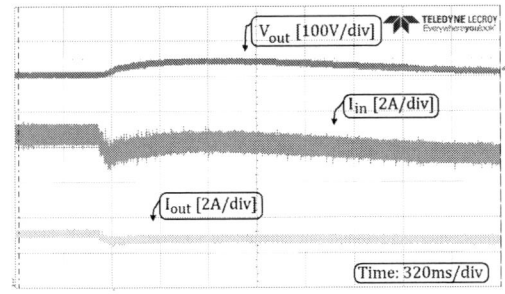

Fig. 14. 10% load step-up

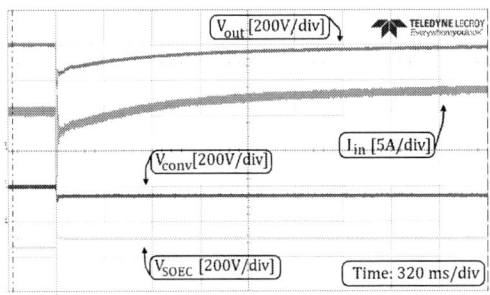

Fig. 16. SOEC voltage 50 V step-up

B. Closed-loop tests

To verify the proposed PCU including its control design is suitable for both operating modes, closed-loop measurements are performed.

Fig. 13 shows the main converter waveforms during the system turn-on when a step from 0-120 V occurs at the input voltage. Steady-state condition is reached in approximately 4 s, with a very smooth output voltage response having a small overshoot of 20 V. Fig. 14 shows the system response for an output load step of 10 % of the rated load, where the output voltage and input current have no oscillations and a small undershoot is appreciated. Fig. 15 shows the system response for SOFC voltage (V_{in}) 80 V step-up, where the inner

current loop suffers oscillations but the outer voltage loop has negligible oscillations and in 2 s steady-state is reached. Fig. 16 shows the system response for SOEC voltage 50 V step-up. Notice that the inverter voltage has been limited to 450 V to protect the voltage supply from over rated current. Similarly as with the previous test, few oscillations in the current are present at the step time, but the output voltage has a smooth response without noticeable oscillations, and reaching a steady-state condition within 2.2 s. As explained throughout the system operation principle and simulations, the inverter voltage is shared across the SOEC and the dc-dc converter, leading to a reduced power rating of the dc-dc converter.

VI. CONCLUSION

This paper has presented the design considerations for a novel PCU for grid-tie SOEC/SOFC system. A 6.7 kW dc-dc converter has been implemented which aims to be able to regulate 27 kW-SOEC/6.3 kW-SOFC stacks, with efficiencies of up to 97 % in SOEC mode and 97.3 % in SOFC mode. Furthermore, it has been verified that the power rating reduction of the dc-dc converter in SOEC mode leads to a more symmetrical I-P characteristic of dc-dc converter which eases the design for a high efficiency converter, leading to similar efficiency curves for both operation modes of the SOEC/SOFC system. This not only results in similar losses in both modes, but can also be beneficial in terms of simplified heat sink design of the dc-dc converter power stage. Closed-loop experimental tests show that with a dual-loop control strategy, system robustness in terms of steady-state and transient performance is reliable under both operating modes SOEC and SOFC at the same time. Thus, the proposed PCU architecture can be a very attractive alternative for high efficiency RFC systems.

REFERENCES

[1] Z.H. Rather, Z. Chen and P. Thgersen *Challenges of Danish Power Systems and Their Possible Solutions*, IEEE International Conference on Power System Technology (POWERCON), 2012

[2] Y. Xu and C. Singh *Power System Reliability Impact of Energy Storage Integration With Intelligent Operation Strategy*, IEEE Transactions on smart grid, VOL.5, NO.2, March 2014.

[3] B. Dollinger and K. Dietrich *Storage Systems for Integrating Wind and Solar Energy in Spain*, IEEE International Conference on Renewable Energy Research and Applications (ICRERA), October 2013.

[4] G.L. Soloveichik *Regenerative Fuel Cells for Energy Storage*, Proceeding of the IEEE, VOL.102, NO.6, June 2014.

[5] Y. Redissi, H. Er.rbib and C. Bouallou *Storage and restoring the electricity of renewable energies by coupling with natural gas grid*, IEEE Renewable and Sustainable Energy Conference (IRSEC), 2013 International.

[6] M. Nymand and M.A.E. Andersen *High-Efficiency Isolated Boost DCDC Converter for High-Power Low-Voltage Fuel-Cell Applications*, IEEE Transactions on industrial electronics, VOL.57, NO.2, February 2010.

[7] R. Pittini, A. Anthon, Z. Zhang and M.A.E. Andersen, *Analysis and Comparison on a Grid-Tie Fuel Cell Energy Storage System Based on Si and SiC Power Devices.*, IEEE International Power Electronics and Applications Conference and Exposition (PEAC), 2014.

[8] F.Z. Peng, H. Li, G.J. Su and J.S. Lawler *A New ZVS Bidirectional DCDC Converter for Fuel Cell and Battery Application.*, IEEE Transactions on power electronics, VOL. 19, NO. 1, January 2004.

[9] R. Pittini, Z. Zhang and M.A.E. Andersen, *High Current Planar Magnetics for High Efficiency Bidirectional DC-DC Converters for Fuel Cell Applications.*, Twenty-Ninth Annual IEEE, Applied Power Electronics Conference and Exposition (APEC), 2014

[10] M. Jang and V.G. Agelidis, *Grid-Interfaced Fuel Cell Energy System Based on A Boost-Inverter with A Bi-Directional Back-Up Battery Storage.*, IEEE Energy Conversion Congress and Exposition (ECCE), 2010

[11] A. Brisse,J. Schefold, C. Stoots and J.O'Brien *Electrolysis Using Fuel Cell Technology*, , In: W. Lehnert and R. Steinberger-Wilckens, Innovation in Fuel Cell Technologies, RSC Publishing, Cambridge, 263-286, 2010

[12] G-B. Jung1, L.H. Fang, C.Y. Lin, X.V. Nguyen, C.C. Yeh, C.Y. Lee, J.W. Yu, S.H. Chan, W.T. Lee, S.W. Chang and I.C. Kao *Electrochemical Performance and Long-Term Durability of a Reversible Solid Oxide Fuel Cell*, International Journal of Electrochemical Science, 2015.

[13] A. Brisse,J. Schefold and G. Corre *Performance and lifetime of Solid Oxide Electrolyzer Cells and Stacks*, July 2015.

[14] Z. Zhang, R. Pittini, M.A.E. Andersen and O.C.Thomsen, *A review and design of power electronics converters for fuel cell hybrid system applications.* Renewable Energy Research Conference (RERC 2012), Technoport 2012.

[15] H. Qi, Y. Wu and Y. Bi, *The Main Parameters Design Based On Three-phase Voltage Source PWM Rectifier of Voltage Oriented Control.*, International Conference on Information Science, Electronics and Electrical Engineering (ISEEE), April 2014.

[16] Z. Zhang, R. Pittini, M.A.E. Andersen and O.C. Thomsen, *A Review and Design of Power Electronics Converters for Fuel Cell Hybrid System Applications.*, Sharing Possibilities and 2nd Renewable Energy Research Conference (RERC2012), Technoport 2012

[17] J.E. Valdez-Resendiz, A. Claudio-Sanchez, G.V. Guerrero-Ramirez, C. Aguilar-Castillo, A. Tapia-Hernandez, J. Gordillo-Estrada, *Interleaved High-Gain Boost Converter with Low Input-Current Ripple for Fuel Cell Electric Vehicle Applications.*, IEEE, International Conference on Connected Vehicles and Expo (ICCVE), 2013.

[18] O. Hegazy, J.V. Mierlo and P. Lataire, *Analysis, Modeling, and Implementation of a Multidevice Interleaved DC/DC Converter for Fuel Cell Hybrid Electric Vehicles.*, IEEE, Transaction on power electronics, Vol. 27, No. 11, November 2012.

[19] K. Senthilkumar, M.S.K. Reddy, D. Elangovan and R. Saravanakumar *Interleaved isolated boost converter as a front-end converter for fuel cell applications.*, IEEE, International Conference on Electrical Energy Systems (ICEES), January 2014.

[20] B.A. Miwa, D.M. Otten and M.F. Schlecht, *High efficiency power factor correction using interleaving techniques.*, IEEE, Applied Power Electronics Conference and Exposition (APEC), 1992.

[21] G. Fontes, C. Turpin, R. Saisset, T. Meynard and S. Astier, *Interactions between fuel cells and power converters: Influence of current harmonics on a fuel cell stack.*, 1 IEEE Power Electronics Specialists Conference, June 2004.

[22] O. Ellabban, O. Hegazy, J. Van Mierlo, P. Lataire, *Dual loop digital control design and implementation of a DSP based high power boost converter in fuel cell electric vehicle.*, 12th International Conference on Optimization of Electrical and Electronic Equipment (OPTIM), 2010.

[23] H. Xu, E. Qiao, X. Guo, X. Wen and L. Kong *Analysis and Design of High Power Interleaved Boost Converters for Fuel Cell Distributed Generation System.*, IEEE, Power Electronics Specialists Conference, June 2005.

[24] B. Bryant and M.K. Kazimierczuk *Small-signal duty cycle to inductor current transfer function for boost PMM DC-DC converter in continuous conduction mode.*, Proceedings of the International Symposium on Circuits and Systems, May 2004.

978-1-4673-9551-9/16 $31.00 © 2016 IEEE

A Hierarchical Active Balancing Architecture for Li-ion Batteries[*]

Han-Dong Gui[1], Zhiliang Zhang[1], Dong-Jie Gu[1], Yang Yang[1], Zhouyu Lu[1] and Yan-Fei Liu[2], *Fellow, IEEE*

[1]Jiangsu Key Laboratory of New Energy Generation and Power Conversion
Nanjing University of Aeronautics and Astronautics, Nanjing, Jiangsu, P. R. China
[2]Department of Electrical and Computer Engineering
Queen's University, Kingston, Ontario, Canada, K7L 3N6
{ghdcat, zlzhang, dongjiegu, yyang, luzhouyu} @nuaa.edu.cn and yanfei.liu@queensu.ca

Abstract— **This paper proposes a hierarchical battery balancing architecture for the series connected lithium-ion batteries. The battery cells are grouped into different packs and the bottom layer is the Adjacent Cell-to-Cell structure consisting of the packs. The top layer is connected to different packs and can deliver the energy from one pack to any other pack bi-directionally, leading to high flexibility. A multi-directional multi-port converter is proposed to serve as the top layer. With the hierarchical architecture, the balanced energy transfer of the cells in different packs can be decoupled, which avoid the repeated charging and discharging during the balancing process. This is beneficial for lengthening the battery lifetime and increasing the State-of-Health (SOH). Moreover, the proposed architecture can lower the current rating of the balancing circuits, which helps decrease the required cost and improve the system efficiency. The experimental results verified the benefits of the proposed architecture.**

Keywords----Battery balancing; Cell-to-Cell; multi-port converter; hierarchical layer.

I. INTRODUCTION

With the wide use and development of electrical vehicles (EV), hybrid electric vehicles (HEV) and energy storage units for renewable energy systems, lithium-ion battery plays a more and more important role because of its advantages such as high energy density, low self-discharge rate, and no memory effect. However, the multiple single cells should be connected in series to meet the voltage requirement of the above applications. Unfortunately, due to the manufacturing inconsistence, environment difference, degradation with the aging and performance characteristics variance, there can be imbalance among the series-connected battery cells, which significantly reduces the energy storage capacity, battery lifetime and safety. Hence, balancing is necessary for the series-connected batteries [1], [2].

Many different balancing methods have been developed and summarized. Among them, the active balancing gains more attention because of the short balancing speed, low energy waste and less heat generation [2]-[15]. From the view of system architecture, it can be categorized into three types: Adjacent Cell-to-Cell (A-C2C) architecture [3]-[5], Direct Cell-to-Cell (D-C2C) architecture [6]-[8] and Cell-to-Pack (C2P) architecture [9]-[11]. For the A-C2C architecture, the balancing circuit has simple structure and low voltage stress of the components. But the adjacent circuits are influenced with each other, leading to complex control and unnecessary energy loss. For the D-C2C architecture, each cell can be balanced independently but only two selected cells can be balanced at the same time, which is not suitable for large number of batteries. For the C2P architecture, different balancing circuits can operate independently but when the target cell is balanced by the mean of discharging, it is also simultaneously charged through the battery pack, which is equivalent to a repeated charging and discharging process for the cell. This phenomenon not only results in energy loss but also harms the state of health (SOH) of the battery. Also, this architecture usually requires the same number of transformers as the number of the cells and the high turns ratio leads to high voltage stress of the semiconductor devices, which results in complex circuit structure, low efficiency and high cost.

In order to solve the above problems, this paper proposes a hierarchical architecture to optimize the balancing performance.

II. PROPOSED HIERARCHICAL ACTIVE BALANCING ARCHITECTURE

Fig. 1 shows the proposed hierarchical architecture, the bottom layer is the A-C2C structure and n battery cells are grouped into m packs with p cells in each pack. In the top layer, each pack is connected to an additional balancing circuit and all the circuits are coupled by a multi-winding transformer. The balanced energy can be exchanged between different packs independently in the top layer with high efficiency.

A. Top Layer

Fig. 2 shows the structure of the top layer. A master battery management system (BMS) calculates the required balancing current of all the balancing circuits in the top and bottom layer. The master BMS directly controls the power flow of the m balancing circuits in the top layer and sends the current reference through CAN interface to the slave BMS in each pack.

The proposed balancing circuits applied in the top layer form a multi-directional multi-port converter. Fig. 3 demonstrates two examples of the possible power flow in the

---

[*]This work is supported by Natural Science Foundation of China (51577089) and Lite-On Research Funding.

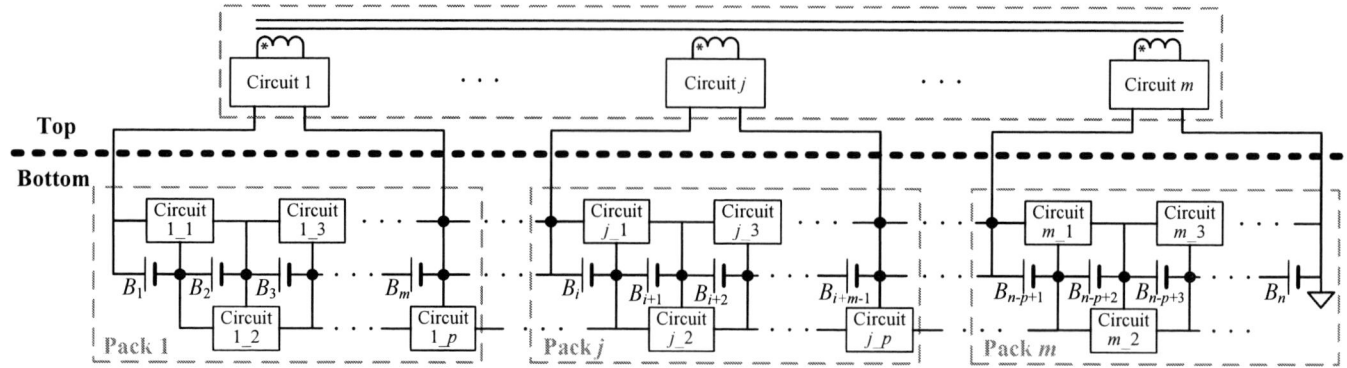

Fig. 1 Proposed hierarchical architecture

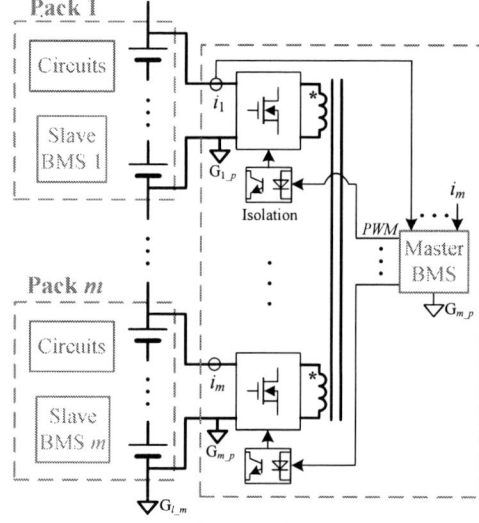

Fig. 2 Structure of top layer

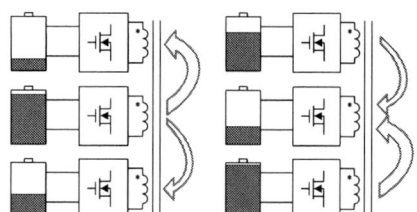

Fig. 3 Two examples of power flow in top layer

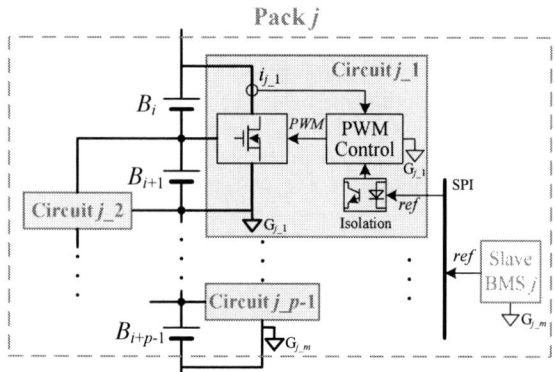

Fig. 4 Structure of bottom layer

converter. It is noted that energy can be delivered by the multi-winding transformer from one pack to any other pack bi-directionally and each pack can serve as both source and load. So this converter owns the advantage of high flexibility of the power flow, which is desired by the balancing system. A multi-port multi-directional converter based on half bridge structure is proposed and applied here and is analyzed in detail in Section III.

B. Bottom Layer

Fig. 4 shows the structure of the bottom layer. The slave BMS detects the operation of the batteries and receives the balancing current reference of each balancing circuit from the master BMS. Each balancing circuit is regulated independently by a local PWM control IC following the balancing current reference sent by the slave BMS through SPI interface.

C. Current Control of the Balancing System

The SOC of ith battery cell can be expressed as

$$SOC_i(t) = SOC_i(t_0) + \frac{\int_{t_0}^{t_0+t} \alpha_i I_i(t)dt}{C_i} \tag{1}$$

where $SOC_i(t_0)$ is the initial SOC, α_i is the current acceptance coefficient, C_i is the battery capacity, and $I_i(t)$ is the charging current.

Define K_i equals to C_i/α_i and it represents the variance of battery characteristic and leads to the imbalance. The aim of the control is to introduce a balancing current $I_{b\_i}(t)$ to eliminate the effect of the difference in K_i and keep all the $SOC(t)$ at the same level.

For the top layer, the required balancing current of each circuit can be expressed as

$$\begin{bmatrix} I_{cT\_1}(t) \\ I_{cT\_2}(t) \\ \vdots \\ I_{cT\_m}(t) \end{bmatrix} = \begin{bmatrix} -1 & 0 & \cdots & 0 \\ 0 & -1 & \cdots & 0 \\ \vdots & \vdots & \ddots & \vdots \\ 0 & 0 & \cdots & -1 \end{bmatrix} \begin{bmatrix} \dfrac{\overline{K_1 \cdot \Delta SOC_1(t)}}{T_b} \\ \dfrac{\overline{K_2 \cdot \Delta SOC_2(t)}}{T_b} \\ \vdots \\ \dfrac{\overline{K_m \cdot \Delta SOC_m(t)}}{T_b} \end{bmatrix} \tag{2}$$

where $I_{cT\_j}(t)$ is the required current of jth balancing circuit in the top layer, T_b is the required balancing time, $\overline{\Delta SOC_j(t)}$ is the difference between the average SOC of the cells in jth pack and that of all the cells in the system. $\overline{K_j}$ is defined as

$$\overline{K_j} = \frac{\sum_{i=1}^{p} \dfrac{C_{j\_i}}{\alpha_{j\_i}}}{p} \tag{3}$$

where C_{j_i} and α_{j_i} is the battery capacity and current acceptance coefficient of ith cell in jth pack respectively, p is the number of cells in one pack.

For the bottom layer, Fig. 5 shows the power flow of the bottom layer and each dot represents a battery cell. The required current of the ith balancing circuit in the bottom layer can be expressed as

$$I_{cB\_i} = \begin{cases} I_{b\_i} & i=1 \\ I_{b\_i} + I_{cB\_i-1} & i>1 \end{cases} \tag{4}$$

where I_{cB_i} is the required current of the ith balancing circuit, I_{b_i} is the required balancing current of the ith cell.

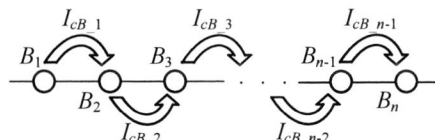

Fig. 5 Power flow in bottom layer

Based on (4), the required balancing current of each circuit can be expressed as

$$\begin{bmatrix} I_{cB\_1}(t) \\ I_{cB\_2}(t) \\ \vdots \\ I_{cB\_n-1}(t) \end{bmatrix} = \begin{bmatrix} -1 & 0 & \cdots & 0 \\ -1 & -1 & \cdots & 0 \\ \vdots & \cdots & \cdots & \vdots \\ -1 & -1 & \cdots & -1 \end{bmatrix} \begin{bmatrix} \dfrac{K_1 \cdot \Delta SOC_1(t)}{T_b} \\ \dfrac{K_2 \cdot \Delta SOC_2(t)}{T_b} \\ \vdots \\ \dfrac{K_{n-1} \cdot \Delta SOC_{n-1}(t)}{T_b} \end{bmatrix}$$

$$+ \begin{bmatrix} a_{11} & a_{12} & \cdots & a_{1m} \\ a_{21} & a_{22} & \cdots & a_{2m} \\ \vdots & \vdots & \ddots & \vdots \\ a_{(n-1)1} & a_{(n-1)2} & \cdots & a_{(n-1)m} \end{bmatrix} \begin{bmatrix} I_{cT\_1}(t) \\ I_{cT\_2}(t) \\ \vdots \\ I_{cT\_m}(t) \end{bmatrix} \tag{5}$$

where $\Delta SOC_i(t)$ is the difference between the SOC of ith cell and the average SOC of all the cell in the system, the coefficient a_{ij} is

$$a_{ij} = \begin{cases} -1 & (j-1)p < i \le jp \\ 0 & else \end{cases} \tag{6}$$

Thus, from (2) and (5), the required current of all the circuits in both top and bottom layer can be calculated and the BMS in the system is able to regulate the balancing current to complete the balancing process.

D. Advantage of the Proposed Architecture

In the C2P architecture, when a cell is balanced by the means of discharging, part of the balancing current would in turn charge the cell through the battery pack at the same time. This repeated charging and discharging is harmful for the battery SOH. The top layer in the proposed architecture can

realize free energy transfer between any two packs through the multi-winding transformer. The battery cells in different packs are totally decoupled and the balancing current would not return to the pack, so the repeated charging and discharging can be avoided and the battery lifetime can be extended. Besides, compared to the C2P architecture with multiple high turns ratio transformers, the number of transformer in the top layer is significantly reduced by selecting proper m, so the cost can be decreased and the efficiency can be optimized due to the similar voltage level of each winding.

In the conventional A-C2C balancing system, there is only one way to transfer energy from one cell to another in this system and every two adjacent balancing circuits are influenced by each other. For example, B_1 is the only overcharged cell and the energy stored in B_1 is $E+n\Delta E$, while the energy of all other cells is E. In order to balance the energy of each cell to the same level that equals to $E+\Delta E$, the balancing circuit between B_i and B_i+1 has to transfer $(n-i)\Delta E$ from B_i to B_i+1. So in order to balance the only overcharged cell, all balancing circuits have to operate at the same time, resulting in considerable energy loss.

In the proposed structure, the top layer supplies an additional energy path among the packs and the cells can be balanced inside each pack independently. So this architecture eliminates the coupled influence between the cells in different packs. Take the aforementioned situation as the example, if the n cells are divided into m packs, the first pack transfers $(n/l)\Delta E$ to other packs through the top layer. The cells in 1st pack can be balanced inside the pack and the circuits in other packs do not need to operate, which reduces the energy loss of the system.

Furthermore, the current rating of the balancing circuit in the bottom layer can be significantly decreased. Assume that I_{b_i} conforms to uniform distribution

$$I_{b\_i} \sim U(-I_{b\_max}, I_{b\_max}) \tag{7}$$

where I_{b_max} is the maximal possible balancing current of the batteries.

It can be proved that for a series connected battery string with n cells, I_{cB_i} conforms to normal distribution

$$I_{cB\_i} \sim N(0, \theta n^2 I_{b\_max}^2)$$

$$F(I_{cB\_i}) = \frac{1}{\sqrt{2\pi \cdot \theta n^2 I_{b\_max}^2}} e^{-\frac{I_{cB\_i}^2}{2\theta n^2 I_{b\_max}^2}} \tag{8}$$

where θ is a constant value and $F(I_{cB\_i})$ is the probability density of I_{cB_i}.

From (8), it can be observed that the standard deviation is proportional to n, which means that with the same I_{b_max}, larger cell number in a string would introduce wider range of I_{cB_i}.

Fig. 6 shows the graph of $F(I_{cB\_i})$. The blue dotted line is the conventional A-C2C architecture with n cells while the red line is the proposed architecture with n cells and they are grouped into three packs. In order to meet the balancing requirement in 90% random cases, the current rating of the balancing circuit should be designed at point A for the blue line and B for the red line, and A is about three times larger than B. So the current rating of the balancing circuit in the bottom layer can be reduced to one-third by applying the proposed architecture, which is essential for decreasing cost and improving efficiency.

978-1-4673-9551-9/16 $31.00 © 2016 IEEE

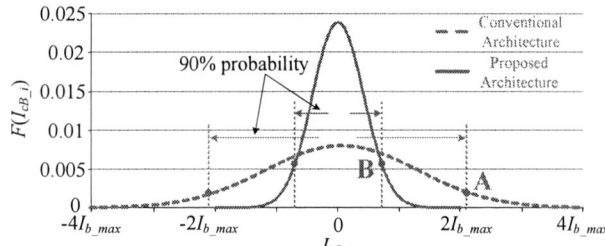

Fig. 6 Probability density of required balancing current

III. BALANCING CIRCUITS IN THE PROPOSED ARCHITECTURE

A. Proposed Multi-directional Multi-port Converter in Top Layer

In order to realize the aforementioned function of the balancing circuit in the top layer, a multi-directional multi-port converter based on half bridge structure is proposed.

Fig. 7 illustrates the topology of the converter. Each battery pack is connected to a half bridge as the balancing circuit shown in Fig. 1. T is a multi-winding transformer and L_{rj} is the leakage inductance that serves as the energy transfer element. Phase Shift Modulation is applied to control the value and direction of the power flow in different circuits.

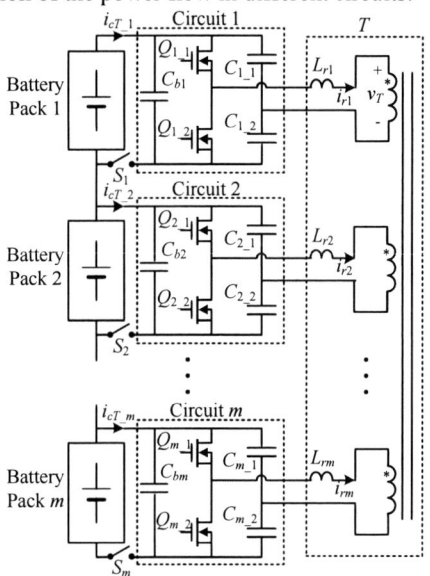

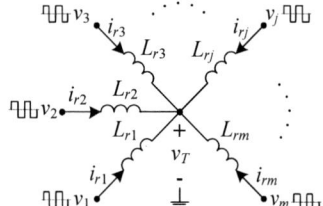

Fig. 7 Topology of the proposed converter

In the proposed converter, each half bridge generates a square wave voltage and its amplitude equals to the half of the voltage of the battery pack. Fig. 8 presents the equivalent model of the proposed converter.

Fig. 8 Equivalent model of the proposed converter

Assume that each battery pack owns the same voltage V_b and each leakage inductance is the same as L_r. Fig. 9 shows the key waveforms of the proposed converter when $m=3$. The current of different circuits I_{cT_j} equals to the average value of the shaded area.

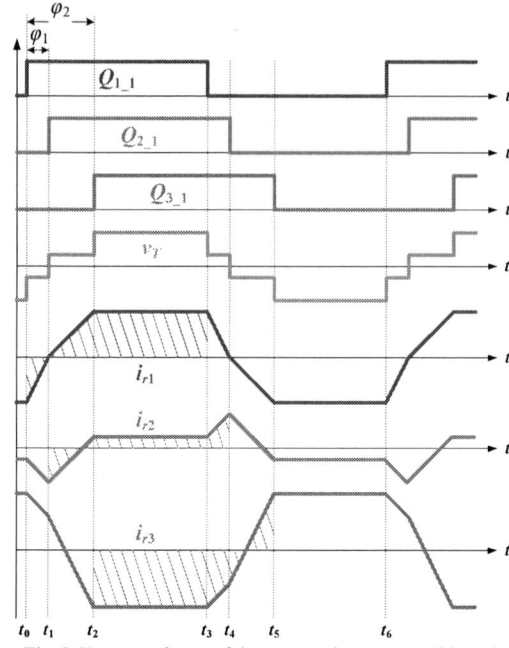

Fig. 9 Key waveforms of the proposed converter with $m=3$

The peak to peak value of the current on leakage inductance ΔI_{rj} is given by

$$
\begin{bmatrix} \Delta I_{r1} \\ \Delta I_{r2} \\ \Delta I_{r3} \\ \vdots \\ \Delta I_{r(m-1)} \\ \Delta I_{rm} \end{bmatrix} = \frac{V_b}{4\pi L_r \cdot f_{sT}} M \begin{bmatrix} \varphi_1 \\ \varphi_2 - \varphi_1 \\ \varphi_3 - \varphi_2 \\ \vdots \\ \varphi_{m-2} - \varphi_{m-3} \\ \varphi_{m-1} - \varphi_{m-2} \end{bmatrix} \quad (9)
$$

$$
M = \begin{bmatrix} \dfrac{2(m-1)}{m} & \dfrac{2(m-2)}{m} & \cdots & \dfrac{4}{m} & \dfrac{2}{m} \\ \dfrac{2}{m} & \dfrac{2(m-2)}{m} & \cdots & \dfrac{4}{m} & \dfrac{2}{m} \\ \vdots & \vdots & \ddots & \vdots & \vdots \\ \dfrac{2}{m} & \dfrac{4}{m} & \cdots & \dfrac{2(m-2)}{m} & \dfrac{2}{m} \\ \dfrac{2}{m} & \dfrac{4}{m} & \cdots & \dfrac{2(m-2)}{m} & \dfrac{2(m-1)}{m} \end{bmatrix} \quad (10)
$$

where φ_j is the phase shift angle between Circuit $j+1$ and Circuit 1, f_{sT} is the switching frequency.

Based on (9), the relationship between the phase shift angles and the current of each circuit can be derived as (11). Therefore, the balancing current in the top layer can be regulated by controlling the phase shift angles.

$$
\begin{bmatrix} I_{cT\_1} \\ I_{cT\_2} \\ I_{cT\_3} \\ \vdots \\ I_{cT\_m-1} \\ I_{cT\_m} \end{bmatrix} = \begin{bmatrix} -\dfrac{\Delta I_{r1}}{4} \\ -\dfrac{\Delta I_{r2}}{4} \\ -\dfrac{\Delta I_{r3}}{4} \\ \vdots \\ -\dfrac{\Delta I_{r(m-1)}}{4} \\ -\dfrac{\Delta I_{rm}}{4} \end{bmatrix} + \frac{V_b}{32\pi^2 L_r \cdot f_{sT}} M \left(\begin{bmatrix} \varphi_1^2 \\ \varphi_2^2 \\ \varphi_3^2 \\ \vdots \\ \varphi_{m-2}^2 \\ \varphi_{m-1}^2 \end{bmatrix} + diag \left(M \begin{bmatrix} \varphi_1 & & & \\ & \varphi_2 - \varphi_1 & & \\ & & \ddots & \\ & & & \varphi_{m-1} - \varphi_{m-2} \end{bmatrix} \right) \times \right.
$$

$$
\left. \begin{bmatrix} \pi - \varphi_1 & 0 & \varphi_2 - \varphi_1 & \cdots & \varphi_{m-2} - \varphi_{m-3} & \varphi_{m-1} - \varphi_{m-2} \\ \pi - \varphi_2 & \pi + \varphi_1 - \varphi_2 & 0 & \cdots & \varphi_{m-3} - \varphi_{m-4} & \varphi_{m-2} - \varphi_{m-3} \\ \pi - \varphi_3 & \pi + \varphi_1 - \varphi_3 & \pi + \varphi_2 - \varphi_3 & \cdots & \varphi_{m-4} - \varphi_{m-5} & \varphi_{m-3} - \varphi_{m-4} \\ \vdots & \vdots & \vdots & \ddots & \vdots & \vdots \\ \pi - \varphi_{m-2} & \pi + \varphi_1 - \varphi_{m-2} & \pi + \varphi_2 - \varphi_{m-2} & \cdots & 0 & \varphi_2 - \varphi_1 \\ \pi - \varphi_{m-1} & \pi + \varphi_1 - \varphi_{m-1} & \pi + \varphi_2 - \varphi_{m-1} & \cdots & \pi + \varphi_{m-2} - \varphi_{m-1} & 0 \end{bmatrix} \right) \quad (11)
$$

B. Buck Boost Converter in Bottom Layer

The conventional buck boost converter is applied to serve as the balancing circuit in the bottom layer because of its simple structure and the ability to control the operating current bi-directionally. Fig. 10 illustrates its topology.

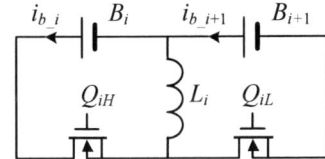

Fig. 10 Topology of buck boost converter in bottom layer

IV. EXPERIMENTAL RESULTS AND DISCUSSION

To verify the analysis of the proposed architecture, a hardware platform is built. In the system, 24 battery cells are connected in series and the rated capacity of each cell is 200 Ah. They are grouped into 3 packs with 8 cells in each pack. The maximal charging/discharging current of the batteries is 200 A. The maximal variance of different battery characteristic is 5%. The current rating of the balancing circuit in the bottom layer is set as 10 A. According to (8), 98% imbalance cases can be included and well solved. Similarly, the current rating of the balancing circuit in the top layer is 3.6 A. Table I lists the parameters of the balancing circuits in the top and bottom layer. Fig. 11 shows the photos of the prototype.

Table I Circuit parameters of converters in top and bottom layer

	Top layer	Bottom layer
Input/output voltage	24~36 V	3.2~4.2 V
Output power	130 W	42 W
Balancing current	-3.6~3.6 A	-10~10 A
Switching frequency	100 kHz	120 kHz
Inductance	1 µH (L_{rj})	10 µH (L_i)
MOSFETs	BSC039N06NS	SiR416DP
Controller	STM32F407VG	UCC35705

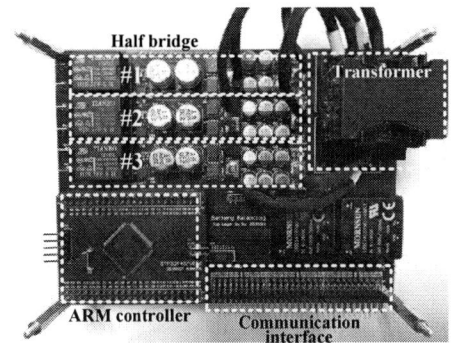

(a) Top layer

(b) Bottom layer

Fig. 11 Photo of prototype

Fig. 12 and Fig. 13 show the experimental waveform of the balancing current in the top and bottom layer. In Fig. 12, the balancing current of the three packs is 3.6 A, -1 A and -2.6 A respectively. In Fig. 13, the balancing current of the two cells is 10 A and -10 A. Thus, both the balancing circuits in the top and bottom layer can operate properly, which verifies the effectiveness of the proposed architecture.

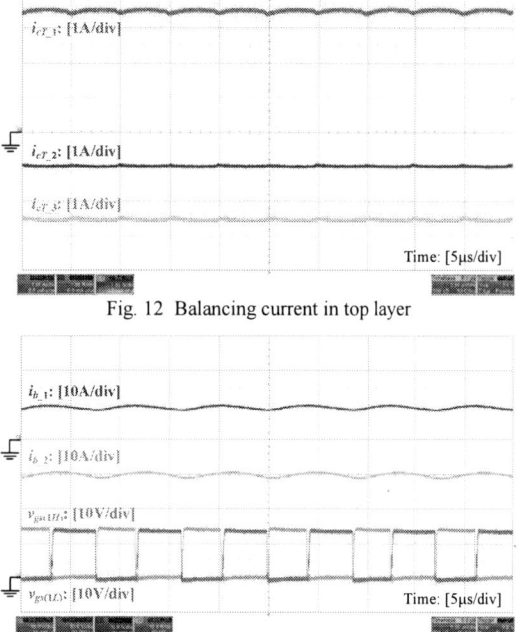

Fig. 12 Balancing current in top layer

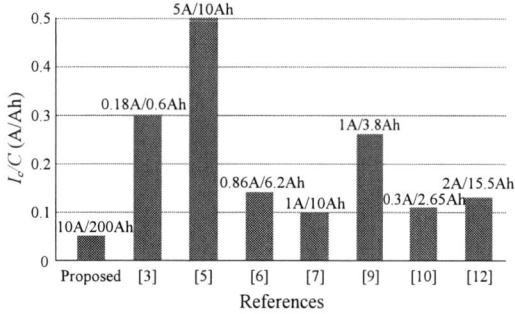

Fig. 13 Balancing current in bottom layer

Fig. 14 illustrates the current rating density of the balancing circuit, which is the ratio of the balancing current rating and the battery capacity. Compared to other literature, the minimal and maximal reduction of the current rating is 50% and 90% respectively by the proposed architecture. So the system applied with the proposed architecture needs far less current than other architecture when the same battery is balanced.

Fig. 14 Comparison of current rating density in different architecture

V. CONCLUSION

In this paper, a hierarchical architecture is proposed to improve the performance of the balancing system for series connected batteries. In the proposed architecture, the batteries in the bottom layer are applied with the Adjacent Cell-to-Cell structure and they are grouped into different packs. A multi-directional multi-port converter is proposed to form the top layer and it can deliver the energy from one pack to any other pack bi-directionally. The top layer decouples the battery cells in different packs and avoids the repeated charging and discharging problem in the Cell-to-Pack architecture. Thus the battery lifetime can be extended. Moreover, the proposed architecture

can reduce the required current rating of the balancing circuits. According to the experimental results, the effectiveness of the balancing circuits in the proposed architecture is validated and the current rating of the balancing circuits with the proposed architecture is much lower than that with other architecture.

REFERENCES

[1] Z. Zhang, Y. Y. Cai, Y. Zhang, D. J. Gui and Y. F. Liu, "A distributed architecture based on micro-bank modules with self-reconfiguration control to improve the energy efficiency in the battery energy storage system," *IEEE Trans. Power Electron.*, vol. 31, no. 1, pp. 304-317, Jan. 2016.

[2] J. G.-Lozano and E. R. Cadaval, "Battery equalization active methods," *J. Power Sources*, vol. 246, pp. 934–949, 2014.

[3] Y. Yuanmao, K. W. E. Cheng and Y. P. B. Yeung, "Zero-current switching switched-capacitor zero-voltage-gap automatic equalization system for series battery string," *IEEE Trans. Power Electron.*, vol. 27, no. 7, pp. 3234–3242, Jul. 2012.

[4] M. Uno and K. Tanaka, "Single-switch cell voltage equalizer using multistacked buck–boost converters operating in discontinuous conduction mode for series-connected energy storage cells," *IEEE Trans. Veh. Technol.*, vol. 60, no. 8, pp. 3635–3645, Oct. 2011.

[5] T. Phung, A. Collet and J.-C. Crebier, "An optimized topology for next-to-next balancing of series-connected lithium-ion cells," *IEEE Trans. Power Electron.*, vol. 29, no. 9, pp. 4603–4613, Sep. 2014.

[6] Y. Shang, C. Zhang, N. Cui and J. M. Guerrero, "A cell-to-cell battery equalizer with zero-current switching and zero-voltage gap based on quasi-resonant LC converter and boost converter," *IEEE Trans. Power Electron.*, vol. 30, no. 7, pp. 3731-3747, Jul. 2015.

[7] F. Baronti, G. Fantechi, R. Roncella and R. Saletti, "High-efficiency digitally controlled charge equalizer for series-connected cells based on switching converter and super-capacitor," *IEEE Trans. Ind. Informat.*, vol. 9, no. 2, pp. 1139–1147, May. 2013.

[8] F. Mestrallet, L. Kerachev, J. C. Crebier and A. Collet, "Multiphase interleaved converter for lithium battery active balancing," *IEEE Trans. Power Electron.*, vol. 29, no. 6, pp. 2874–2881, Jun. 2014.

[9] C.-S. Lim, K.-J. Lee, N.-J. Ku, D.-S. Hyun and R.-Y. Kim, "A modularized equalization method based on magnetizing energy for a series-connected lithium-ion battery string," *IEEE Trans. Power Electron.*, vol. 29, no. 4, pp. 1791–1799, Apr. 2014.

[10] A. M. Imtiaz and F. H. Khan, "Time shared flyback converter' based regenerative cell balancing technique for series connected Li-ion battery strings," *IEEE Trans. Power Electron.*, vol. 28, no. 12, pp. 5960–5975, Dec. 2013.

[11] M. Einhorn, W. Guertlschmid, T. Blochberger, R. Kumpusch, R. Permann, F. V. Conte, C. Kral and J. Fleig, "A current equalization method for serially connected battery cells using a single power converter for each cell," *IEEE Trans. Veh. Technol.*, vol. 60, no. 9, pp. 4227–4237, Nov. 2011.

[12] M. Y. Kim, J. H. Kim and G. W. Moon, "Center-cell concentration structure of a cell-to-cell balancing circuit with a reduced number of switches", *IEEE Trans. Power Electron.*, vol. 29, no. 10, pp. 5285–5297, Oct. 2014.

[13] F. Deng and Z. Chen, "A control method for voltage balancing in modular multilevel converters," *IEEE Trans. Power Electron.*, vol. 29, no. 1, pp. 66–76, Jan. 2014.

[14] H. Chen, L. Zhang and Y. Han, "System-theoretic analysis of a class of battery equalization systems: mathematical modeling and performance evaluation," *IEEE Trans. Veh. Technol.*, vol. 64, no. 4, pp. 1445–1457, Apr. 2015.

[15] S. Li, C. Mi, and M. Zhang, "A high-efficiency active battery-balancing circuit using multiwinding transformer," *IEEE Trans. Ind. Appl.*, vol. 49, no. 1, pp. 198–207, Jan. 2013.

A Series-DG Based Autonomous Islanding Microgrid

Beihua Liang[1], Yun Wei Li[2], *Senior Member, IEEE*, Jinwei He[1], *Member, IEEE* and Chengshan Wang[1]

1. School of Electrical Engineering and Automation, Tianjin University, Tianjin, China
2. Dept. Electrical and Computer Engineering, University of Alberta, Edmonton, AB, T6G 2V4, Canada
Email: hjinwei@ualberta.ca

Abstract—A series distributed generation (DG) units based islanding microgrid configuration and the corresponding power sharing method are proposed in this paper. Unlike a conventional islanding microgrid where parallel DG units are connected to the point of common coupling (PCC) to supply electricity to loads in a cooperative manner, this paper discusses the possibility of using series connected low voltage (LV) converters as an alternative solution. It has been demonstrated that the DC/DC boost conversion in the back stage of a conventional DG unit can be removed from this proposed microgrid. In addition, via a very simple power factor-frequency inverse droop control, an accurate real and reactive power sharing can be simultaneously achieved in multiple series DG units. Note that the effectiveness of the proposed power sharing control does not rely on the communications between DG units or the knowledge of detailed microgird circuitry parameters. Therefore, the challenging power sharing problems in conventional islanding microgrids is perfectly solved with this proposed configuration and the control method.

Keywords—*inverse droop control, islanding microgrid, power sharing, power factor control.*

I. INTRODUCTION

The increasing interests in renewable energy power generation and energy storage system has popularized the application of LV microgrid [1], aiming to provide more reliable and flexible electricity to loads in a power distribution network [2-8]. The microgrid is usually composed of a few parallel DG units with power electronics interfacing converters. For many renewable energy systems, such as a PV system, a back stage boost converter is usually used to stabilize the DC link voltage to a range that is suitable for grid integration. However, this DC/DC conversion essentially increases the cost and decreases the efficiency of the DG system. In addition, some DG units may be far away from PCC [2]. Thus, the power loss of the long DG feeder is a concern.

When multiple parallel DG units are working together, the power sharing between them is another issue [2-4]. To reduce the costs on the communication and control system, the real power-frequency droop control and the reactive power-voltage magnitude droop for the operation of synchronous generators are also applied to control the power sharing between DG

units. However, this method is not always effective, particularly for weak microgrids with nontrivial voltage drop on DG feeders. Specifically, the convectional droop control may cause a few problems, such as power sharing errors and instable operation [2]. To address these problems, a few enhanced methods have been proposed in recent literature including the virtual impedance control method [3], the low bandwidth communication aided droop control [7], and the control method using virtual real and reactive power or virtual frequency and voltage magnitude [10]. Although these methods can improve the power control performance of an islanding microgrid, it is necessary to note that they are usually complicated and the effectiveness of these methods often relies on the detailed knowledge of the microgrid parameters, such as the feeder impedance, PCC voltage, or communication delays.

Motivated by the fact that the conventional droop control still has some limitations, an alternative series-connected DG units based microgrid configuration is proposed. To share load demand between series DG units, a very simple power factor-frequency inverse droop control is also proposed. With this power control method, the PCC load demand is accurately shared by multiple interfacing converters. Verification results are provided to validate the performance of the proposed system.

II. THE PROPOSED MICROGRID

Fig. 1 shows a conventional microgrid topology with parallel DG units. In each DG unit, a DC power source, either from the renewable energy source or the energy storage source, is connected to the LV DC link. The high voltage (HV) DC link and the LV DC link are interconnected by a DC/DC boost converter. The LC filter capacitor voltage in the output of the DG unit is controlled by the DC/AC converter. In this system, the PCC load demand shall be properly shared by parallel DG units. For the sake of simplicity, this paper assumes that all DG units are at the same power rating and the load demand shall be equally shared by these DG units.

To share the PCC load demand between DG units, the frequency and the voltage magnitude are determined by the well-understood droop control.

978-1-4673-9551-9/16 $31.00 © 2016 IEEE

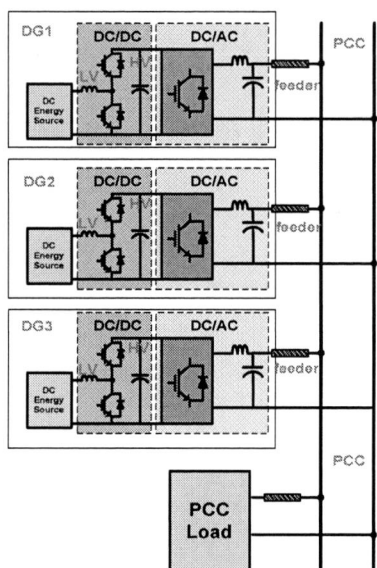

Fig. 1. Diagram of parallel-converter based microgrid.

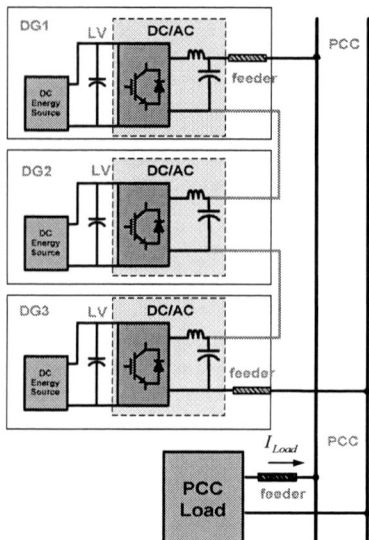

Fig. 2. Diagram of series-LV-converter based microgrid.

However, this topology is not always the most suitable choice due to the following reasons:

1)A DC/DC boost stage is placed between the DC energy source and the DC/AC converter. This DC/DC power conversion stage essentially increases the cost of the DG unit and reduces the DG unit efficiency.

2)The accuracy of power sharing using droop control is affected by the feeder impedance. According to [2], the reactive power sharing may have significant errors in a week microgrid.

3)Each DG unit must be connected to the PCC to deliver power to the islanding microgrid. When remotely installed DG units are integrated into the system, the power loss of the feeder is high.

In order to overcome those drawbacks, the proposed microgrid and the corresponding power sharing method are presented in this section.

A. Series-Converter based Microgrid

Fig. 2 demonstrates the configuration of the proposed system. In contrast to the conventional DG unit as shown in Fig. 1, the LV DC energy source in this case is directly connected to the DC/AC interfacing converter. Accordingly, the inverted AC voltage is significantly lower than the suitable range for microgrid integration. To address this problem, the output AC voltages of these LV converters are connected in series and the sum of the output voltage is used to supply electricity to the load at PCC.

B. Power Converter Control Scheme

The simplified equivalent circuit of the microgrid is shown in Fig. 3, where it can be seen that all DG units have the same line current I_{Load}. However, the output DG voltage can be different, due to a fact that the multiple DG units are not synchronized when only fixed magnitude and fixed frequency voltage reference is used in each DG unit.

The corresponding phasor diagram of the series-DG based microgrid is shown in the left half of Fig. 4. When multiple DG units have different voltage phase angles, the magnitude of PCC voltage is lower than the rated value. In addition, the load demand is not equally shared by these DG units. The corresponding output power diagram of DG units is shown in the right half of Fig. 4. As each DG unit has the same voltage magnitude and line current magnitude, multiple DG units have the same apparent power. However, the voltage angle differences between these DG units cause the unequal sharing of both real and reactive power. This phenomenon is obviously not the same as the droop control of parallel inverter based microgrid, where the real power sharing is always accurate at the steady-state but reactive power sharing may

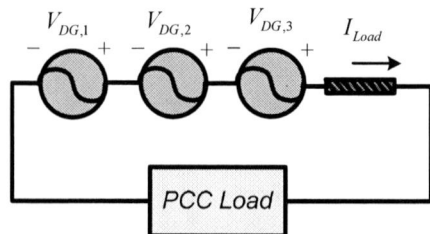

Fig. 3. Equivalent circuit of series-LV-converter based microgrid.

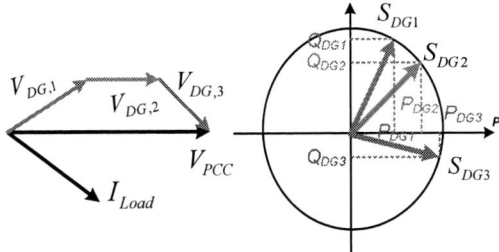

Fig. 4. Phasor diagram of series-LV-converter based microgrid.

have some errors.

To avoid the use of communications between DG units, it is preferred to use only the locally measured signals to compensate the power sharing errors. Further considering that multiple series DG units have same apparent power and the power factors of these DG units can be not the same, a power factor-frequency inverse droop control is proposed here to ensure an accurate power sharing as:

$$\omega_{DG} = \omega^* + D_{PF} \cdot \cos(\theta) \qquad (1)$$

where D_{PF} is the inverse droop coefficient and $\cos(\theta)$ is the power factor of the DG unit. As the increase of reference frequency can cause a decrease of power factor and the apparent power of the DG unit is always the same as others, only an inverse droop control can be used to ensure proper power sharing of the system.

When the frequency reference is obtained in (1), the instantaneous voltage reference of a DG unit is determined as:

$$V_{ref} = \sqrt{2} E \cos(\omega_{DG} t) \qquad (2)$$

where E is the nominal RMS value of the voltage reference.

Afterwards, the double-loop voltage controller as shown in [5] is selected for the LC filter capacitor voltage tracking as:

$$I_{ref} = G_{Outer}(s) \cdot (V_{ref} - V_C) \qquad (3)$$
$$= (k_p + \sum_{h=1,3,\dots 11} \frac{2 k_{i,h} \omega_c s}{s^2 + 2\omega_c s + (h \cdot \omega^*)^2}) \cdot (V_{ref} - V_C)$$

$$V_{out} = G_{Inner}(s) \cdot (I_{ref} - I_o) = k_{inner} \cdot (I_{ref} - I_o) \qquad (4)$$

where V_C is the measured filter capacitor voltage, k_p is the proportional gain, $k_{i,h}$ is the resonant controller gain at the order h, ω_c is the cut-off frequency of a quasi-resonant controller, k_{inner} is the proportional gain of the inner loop controller, I_o is the measured interfacing output current, and V_{out} is the reference voltage for PWM processing.

The control diagram of a DG unit is shown in Fig. 5. First, the real and the reactive power of the DG unit are measured and the power factor of the system is calculated. The power factor is then used as an input of the power factor-frequency

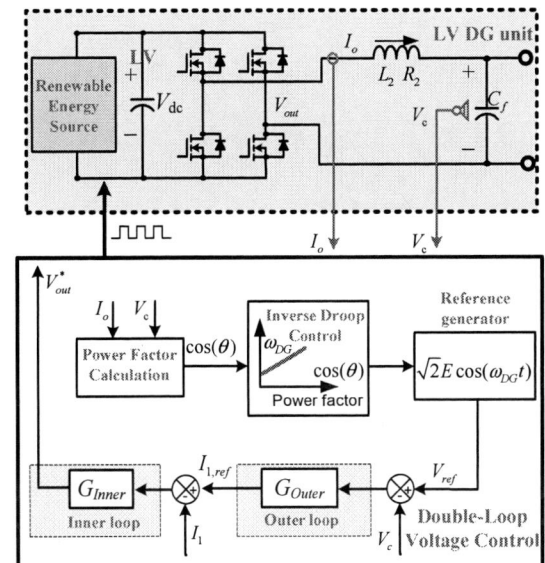

Fig. 5. Control Scheme of a DG unit.

TABLE I. Parameters of the Simulated & Experimental System

Circuit Parameter	Value
Rated voltage	PCC voltage 120V/60Hz DG voltage 40V/60Hz
DC link voltage	95V
LC filter	L_1=2mH; C_f=30uF;
Switching Frequency	10kHz
Dead-time	2.25 μsec
Control parameter	Value
Outer control loop	$K_{i,h}$=(25, h=1; 20, h=3; 20, h=5)
Inner control loop	K_{inner}=20

inverse droop control. When the voltage reference is obtained, a double-loop controller is employed for an accurate LC filter capacitor voltage tracking.

III. VERIFICATION RESULTS

Experimental results are obtained from a single-phase microgrid model. Three series DG units at the same power rating are used to form the microgrid. In the test, the DC link voltage of each DG unit is only 95V. The detailed control and circuit parameters can be seen from Table I and results are shown from Fig. 6 to Fig. 8.

First, the power sharing performance before and after the adoption of the proposed method can be seen in Fig. 6. When fixed frequency and magnitude voltage reference is applied to all DG units, the real and reactive power sharing has significant errors as shown in the beginning of this test. After the application of the proposed inverse droop method, it can be seen that power sharing errors can effectively compensated at the steady-state.

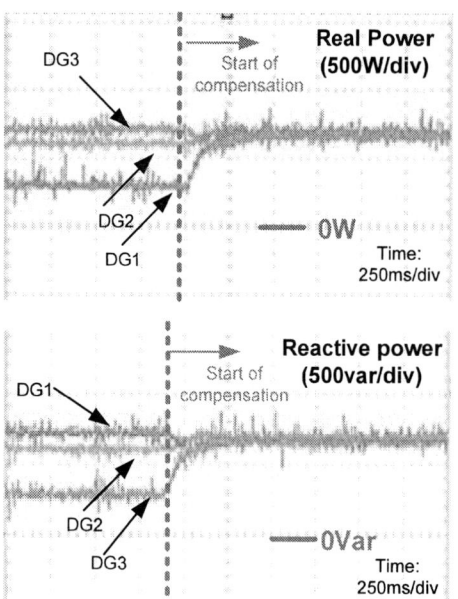

Fig. 6. Real and reactive power sharing performance of series-DG based microgrid.

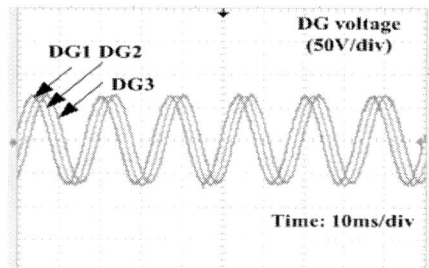

Fig. 7. DG voltage waveforms before using the proposed method.

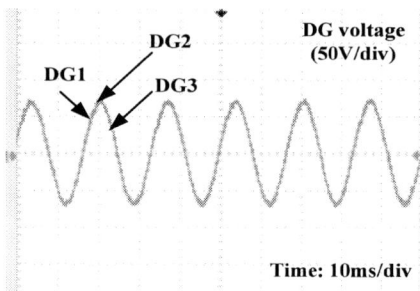

Fig. 8. DG voltage waveforms after using the proposed method.

The waveform of DG voltage before the use of inverse droop control is shown in Fig. 7. Due to the initial voltage phase angle differences between these DG units, the waveforms of DG units are obviously not the same.

When the proposed method is applied to this islanding microgrid, the corresponding waveforms of DG units are shown in Fig. 8. In this case, the DG voltages are almost identical at the steady-state.

IV. CONCLUSION

The aim of this paper is to reconsider the topology and the control schemes of islanding microgrids. Unlike a conventional mcirogrid where the load demand sharing is achieved by parallel interfacing converters, the proposed method uses a few series-connected DG units to form an islanding micogrid. It is found that with a simple power factor-frequency inverse droop control, the proposed microgrid can easily achieve an accurate real and reactive power sharing between series-connected interfacing converters. Note that only the principle of the proposed microgrid is discussed in this paper. The following aspects will be studied in the future of our research:

1)Ride-through operation of the proposed microgrid system when one or a few series interfacing converters are by-passed.

2)The possibility of using both parallel and series converters to further enhance the power control performance of an islanding microgrid.

References

[1] Y. W. Li, D. M. Vilathgamuwa, and P. C. Loh, "Design, analysis and real-time testing of a controller for multibus microgrid system," IEEE Trans. Power Electron, vol. 19, pp. 1195-1204, Sep. 2004.

[2] Y. W. Li and C. N. Kao, "An accurate power control strategy for power-electronics-interfaced distributed generation units operating in a low-voltage multibus microgrid," IEEE Trans. Power Electron., vol. 24, pp. 2977-2988, Dec. 2009.

[3] J. M. Guerrero, L. G. Vicuna, J. Matas, M. Castilla, and J. Miret, "Output impedance design of parallel-connected UPS inverters with wireless load sharing control," IEEE Trans. Ind. Electron., vol. 52, no. 4, pp. 1126-1135, Aug. 2005.

[4] J. M. Guerrero, L. G. Vicuna, J. Matas, M. Castilla, and J. Miret, "A wireless controller to enhance dynamic performance of parallel inverters in distributed generation systems," IEEE Trans. Power Electron., vol. 19, no. 4, pp. 1205-1213, Sep, 2004.

[5] J. He and Y. W. Li, "An enhanced microgrid load demand sharing strategy," IEEE Trans. Power Electron., vol. 19, no. 5, pp. 1184-1194, May. 2004.

[6] Q. Zhang, "Robust droop controller for accurate proportional load sharing among inverters operated in parallel," IEEE Trans. Ind. Electron., vol. 60, no. 4, pp. 1281–1290, Apr. 2013.

[7] M. Savaghebi, A. Jalilian, J. C. Vasquez, J. M. Guerrero, "Autonomous voltage unbalance compensation in an islanded droop controlled microgrid," IEEE Trans. Ind. Electron., vol. 60, no. 4, pp. 1390–1402, Apr. 2013.

[8] Q-C.Zhong, "Harmonic droop controller to reduce the voltage harmonics of inverters," IEEE Trans. Ind. Electron., vol. 60, no. 3, pp. 936–945, Mar. 2013.

[9] T. L. Vandoorn, B. Renders, L. Degroote, B. Meersman, L. Vandevelde, "Controllable harmonic current sharing in islanded microgrids: DG units with programmable resistive behavior toward harmonics," IEEE Trans. Power Del., vol. 27, no. 3, pp.1405-1414, Jul. 2012.

[10] Y. Li and Y. W. Li "Power Management of Inverter Interfaced Autonomous Microgrid Based on Virtual Frequency-Voltage Frame", IEEE *Trans.* on Smart Grid, vol. 2, pp. 30-40, Mar. 2011.

An Enhanced Droop Control Scheme for Resilient Active Power Sharing in Paralleled Two-Stage PV Inverter Systems

Hongpeng Liu[1], Yongheng Yang[2], Xiongfei Wang[2], Poh Chiang Loh[2], Frede Blaabjerg[2], Wei Wang[1], and Dianguo Xu[1]

1. Department of Electrical Engineering, Harbin Institute of Technology, Harbin, China
2. Department of energy Technology, Aalborg University, Aalborg, Denmark

Abstract—**Traditional droop-controlled systems assume that the generators are able to provide sufficient power as required. This is however not always true, especially in renewable systems, where the energy sources (e.g., photovoltaic source) may not be able to provide enough power (or even loss of power generation) due to the intermittency. In that case, unbalance in active power generation may occur among the paralleled systems. Additionally, most droop-controlled systems have been assumed to be a single dc-ac inverter with a fixed dc input source. The dc-dc converter as the front-end of a two-stage photovoltaic (PV) system has not been considered. In this paper, an enhanced droop scheme is thus proposed to address those issues, and the proposed scheme can enable resilient active power sharing in parallel two-stage PV inverter systems. Furthermore, a small-signal analysis for the proposed droop control strategy is carried out. Experiments have verified the effectiveness of the proposed droop control scheme.**

Keywords—**microgrid; two-stage PV systems; single-phase systems;droop control;active power sharing; small-signal model**

I. INTRODUCTION

With the increased concerns on environment and cost of energy, Renewable Energy Sources (RES) are becoming major Distributed Generation (DG) sources in Microgrids (MGs) [1]-[3]. One of the most promising RESs is Photovoltaic (PV) energy source. It has some advantages such as abundance and cleanness with a still declining price [4]-[5]. Because of solar power intermittency, MG-connected PV systems however require special treatment in order to maintain the power balance between generation and load in an islanded mode.

In the case of islanded mode operation, the interfacing inverters are usually governed by a droop control method so that they mimic the behavior of synchronous generators operating in parallel [6]. Furthermore, the prior-art work has mainly focused on single-stage converters. This makes the droop control schemes not directly applicable to two-stage converters including PV systems, which are expected to play an even active role in the future MGs. Compared to single-stage converter, two-stage configuration allows flexible installation number of PV panel and convenient controller design [7]-[8]. Nevertheless, a dc-link performed by a capacitor is required in those cases for power decoupling. However, the dc-link voltage is an important issue during transients.

On the other hand, several research studies have been witnessed in literature, where droop control schemes are employed for MG-connected PV systems [9]-[13]. For example, in [9] and [10], the transition from grid-connected mode to islanded mode (or reverse) is accomplished by reconfiguring droop controllers. Compared to the schemes using a single-common control structure during the transition, reconfiguring the droop controller is not an optimized solution. However, such common control schemes are still not well discussed. In [11], a PV inverter with Consortium for Electric Reliability Technology Solutions (CERTS) droop control has been proposed to maintain the dc-link voltage stability during load transients and seamless transfer between different operation modes. However, this method is only studied on single-stage systems. In [13], a universal controller is proposed to achieve Maximum Power Point Tracking (MPPT), droop control, and dc-link voltage regulation without reconfigurations of the controller. Although the droop controller can regulate the voltage and frequency in the MGs, the PV inverters always operate at maximum power point, which limits the loading range. In addition, the above approaches may result in a slow response to track the maximum power point of PV panels due to the droop control scheme.

Moreover, in most of the droop-controlled systems in the literature, the active power provided by the generator (e.g., PV panels) has always been assumed as sufficient as demanded by the local loads [14]. However, in practice, this is not true, especially in renewable energy sources like PV sources are employed. Due to the high weather-dependency, the energy resource may not be able to provide adequate power to the loads. In that case, the power shortage of one of the energy sources leads to power imbalance among the parallel inverter systems and further frequency excursions in the droop-controlled systems, thus potentially causing instability. Thus, it calls for advanced droop control schemes in order to ensure a stable operation of the MGs.

In light of the above issues, an enhanced droop control scheme for two-stage PV inverter systems has been proposed in this paper. The proposed droop control scheme can adaptively adjust the droop curves according to the load demands, and thus it enables resilient power sharing among the parallel inverter systems. Experiments performed on a MG

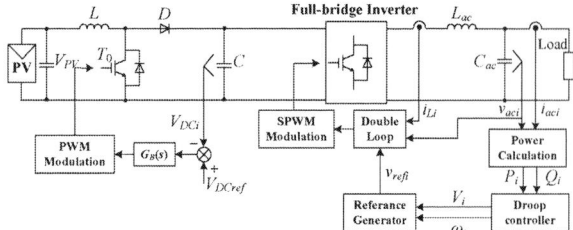

Fig. 1. Schematic and traditional control diagram of a two-stage single-phase droop-controlled PV inverter system.

system consisting of two two-stage single-phase PV systems have validated the effectiveness of the proposed scheme in terms of resilient power sharing.

II. CONTROL STRUCTURE OF PARALLELED PV SYSTEM

The parallel MG system is consisted of two two-stage PV inverters. The PV inverter is a two-stage configuration, where the front-end stage is a dc-dc boost converter and the rear-end stage is a full-bridge inverter. Fig. 1 shows the schematic and the overall traditional control scheme of the two-stage single-phase PV inverter system. The boost converter is controlled by a single-loop voltage controller denoted as $G_B(s)$. Input to $G_B(s)$ is the difference between the measured dc-link voltage V_{DCi} and its reference V_{DCref}. In order to nullify this difference, $G_B(s)$ is usually a Proportional-Integral (PI) controller, which may not be necessary for the proposed droop control scheme, which will be elaborated in the following section. On the other hand, the output voltage v_{aci} and current i_{aci} of the rear-end inverter have been measured for computing P_i and Q_i. When the power transmission line is mainly inductive, the conventional P-ω droop and Q-V droop controls can be accomplished with (1) and (2). Accordingly, it is possible to map out the desired ω_i and V_i, from which the demanded voltage reference v_{refi} is computed for the double-loop controller.

$$\omega_i = \omega_{0i} - k_{pi}(P_i - P_{0i}) \tag{1}$$

$$V_i = V_{0i} - k_{qi}(Q_i - Q_{0i}) \tag{2}$$

where $i = 1, 2$ represents the number of the two PV inverters in the droop controlled system. ω_{0i} and V_{0i} are the rated angular frequency and rated output voltage amplitudes of the PV inverter, P_{0i} and Q_{0i} are the rated active and reactive powers, and k_{pi} and k_{qi} are the active power and reactive power droop coefficients, respectively.

III. ENHANCED DROOP CONTROL

For the paralleled PV inverter system, when the ambient condition varies especially during clouds-passing periods, the output power of the PV panels will fluctuate. Fig. 2 shows the P-ω interactions due to the variations of the PV panel output power. When ignoring the power losses and assuming the active power droop coefficients for both inverter systems are the same, inverter 1 and 2 should operate at point a with $P_1 = P_a$. At a time interval, the PV maximum power P_{PVmax} of inverter 1 is reduced due to the passing clouds (i.e., $P_{PVmax} < P_a$ corresponding to point b_1 in Fig. 2). In this case, the power delivered to the dc-link by the boost converter 1 is less than the power drawn by inverter 1. If there is not sufficient storage to buffer the power changes, the entire PV system may go into instability due to the power imbalance. Therefore, by lowering

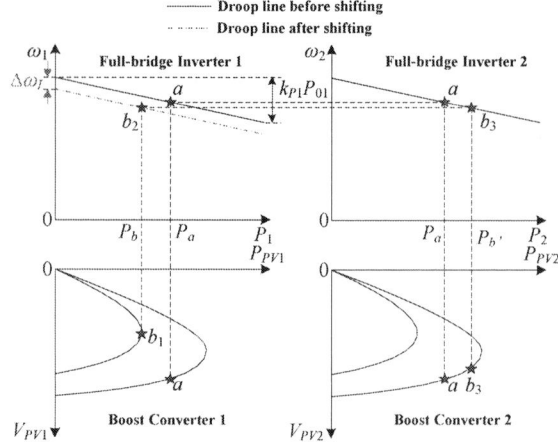

Fig. 2. P-ω interactions during changes of PV power.

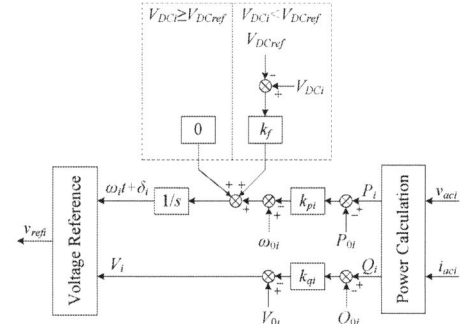

Fig. 3. Improved droop control scheme.

the droop line by $\Delta\omega_I$ as it is shown in Fig. 2 (see point b_2), the dc-link voltage will drop until the inverter power is reduced to $P_1 = P_b = P_{PVmax}$. Meanwhile, the droop controller pushes the frequency of inverter 2 downward, increasing the output power. Inverter 2 operates at a new point b_3 with $P_2 = P_{b'}$, which supplies the rest of active power.

The principle of this enhanced droop control scheme is simple. However, the challenging issue is how to determine the value of $\Delta\omega_I$, and the simplest way is to automatically generate $\Delta\omega_I$ by feeding the power imbalance to a PI controller. In the steady state, the power imbalance will be nullified to avoid charging or discharging the dc-link capacitor endlessly. However, the integral action of the controller will still output a finite $\Delta\omega_I$ for lowering the droop line in Fig. 2. Unfortunately, the integral action also creates an indefinite solution with the zero input error mapped to any output that can stabilize the system. Such single-to-multiple mapping is not desired in the parallel PV systems, which should instead be controlled by a proportional droop scheme without any integral term. The PI controller with power imbalance as input is thus not pursued further even though it is technically feasible.

Instead, the dc-link voltage V_{DCi} is allowed to vary within a very small range denoted as $V_{DCmin} \leq V_{DCi} \leq V_{DCref}$. The lower limit of the variation range must at least be higher than the maximum inverter output voltage to avoid over-modulation. The necessary expression governing it can be derived from (2) as $V_{DCmin} \geq V_{0i} + k_{qi}Q_{0i}$ by substituting $Q_i = 0$. Then, the required $\Delta\omega_I$ can be calculated from (3) based on the proportional droop principle as

$$\Delta\omega_I = \begin{cases} k_f\left(V_{DCi} - V_{DCref}\right), & V_{DCi} < V_{DCref} \\ 0, & V_{DCi} \geq V_{DCref} \end{cases}$$

$$-k_{pi}P_{0i} \leq \Delta\omega_I \leq 0 \tag{3}$$

$$k_f = k_{pi}P_{0i}\Big/\left(V_{DCref} - V_{DCi}\right)$$

where k_f is the droop coefficient to calculate $\Delta\omega_I$. Some insights drawn from (3) are summarized as follows:

- The offset $\Delta\omega_I$ varies in the range of $-k_{pi}P_{0i} \leq \Delta\omega_I \leq 0$. The lower limit $-k_{pi}P_{0i}$ is obtained by subtracting the maximum of (1) from its minimum $(\omega_{0i} - (\omega_{0i} + k_{pi}P_{0i}))$.

- The actual V_{DCi}, when fallen below V_{DCref}, represents a lack of PV capacity (e.g. insufficient irradiation) for restoring the dc-link voltage. Hence, the inverter power must be lowered by introducing a negative $\Delta\omega_I$.

- The actual V_{DCi}, when above V_{DCref}, represents an excess PV capacity (e.g. strong irradiation), which can definitely be lowered to meet the lower load demand. No offset ($\Delta\omega_I = 0$) is thus added to adjust the inverter power.

- As V_{DCi} is now required to vary, controller $G_B(s)$ for the boost converter in Fig. 1 cannot be a PI controller. It should be a proportional droop controller with gain given by $(V_{DCref} - V_{DCmin})\,/\,P_{0i}$.

- The coefficient k_f will usually be high if only a narrow dc-link voltage range is permitted because of the large dc-link capacitor C included for decoupling purposes.

The improved P-ω droop expression is then rewritten as (4), where the first term is the conventional droop expression for power sharing, and the second term is for regulating the internal power limitation within the generator.

$$\omega_i = \underbrace{\omega_{0i} - k_{pi}(P_i - P_{0i})}_{} + \underbrace{k_f\left(V_{DCi} - V_{DCref}\right)}_{} \tag{4}$$

The Q-V expression, on the other hand, remains unchanged since internal active power variation will not limit the amount of reactive power that the rear-end inverter can deliver. Fig. 3 shows the diagram of the improved droop control scheme.

IV. SMALL-SIGNAL ANALYSIS OF PROPOSED ENHANCED DROOP CONTROL

In this section, a small-signal model based on (2) and (4) is carried out to analyze the behavior and the stability of the paralleled inverter systems with the proposed droop scheme.

A. Model of Each Inverter

In order to ensure the proposed droop control laws, it is necessary to measure the output power of each inverter. In general, the instantaneous power components are passed through low-pass filter [15]. Hence, the average power P_{avgi} and Q_{avgi} in the s-domain, are given by

$$P_{avgi} = \frac{\omega_f}{s + \omega_f}P_i(s) \tag{5}$$

$$Q_{avgi} = \frac{\omega_f}{s + \omega_f}Q_i(s) \tag{6}$$

where $P_i(s)$ and $Q_i(s)$ are the instantaneous active power and reactive power, respectively. ω_f is the cut-off frequency of the low-pass filter. Due to the low-pass filter, the inner voltage and current control bandwidth are much higher than the outer power loop. Thus, it can be assumed that the dynamics of the inner loops can be neglected [16].

Applying the perturbing terms to (2), (4), (5) and (6) gives

$$\Delta\omega_i = -\frac{k_p\omega_f}{s + \omega_f}\Delta P_i + k_f\Delta V_{DCi} \tag{7}$$

$$\Delta V_i = -\frac{k_q\omega_f}{s + \omega_f}\Delta Q_i \tag{8}$$

where Δ means a small perturbation around the equilibrium points.

Considering (7) and (8) in the time-domain,

$$\Delta\dot{\omega}_i = -\omega_f\Delta\omega_i - k_p\omega_f\Delta P_i + k_f\Delta\dot{V}_{DCi} + k_f\omega_f\Delta V_{DCi} \tag{9}$$

$$\Delta\dot{V}_i = -\omega_f\Delta V_i - k_q\omega_f\Delta Q_i \tag{10}$$

where a dotted variable represents the derivative with respect to time.

The variation of the energy E_i in the dc-link capacitor is related to the dc-link voltage V_{DCi} by

$$E_i = \int P_i(t)dt = \frac{1}{2}CV_{DCref}^2 - \frac{1}{2}CV_{DCi}^2 \tag{11}$$

where C is the dc-link capacitance.

Therefore, the linear relationship between the dc-link voltage and the power is given by

$$\Delta V_{DCi} = -\frac{1}{mCs}\cdot\Delta P_i = -\frac{k_{DC}}{s}\cdot\Delta P_i \tag{12}$$

where m is the equilibrium point of V_{DCi}.

Therefore, (9) can be rewritten as

$$\Delta\dot{\omega}_i = -\omega_f\Delta\omega_i - \left(k_p\omega_f + k_fk_{DC}\right)\Delta P_i + k_f\omega_f\Delta V_{DCi} \tag{13}$$

In the d-q reference frame, the vector form of the inverter output voltage can be obtained as

$$\overline{V}_i = V_{di} + jV_{qi} \tag{14}$$

where $V_{di} = V_i\cos(\delta)$, $V_{qi} = V_i\sin(\delta)$ and $\delta_i = \arctan(V_{qi}/V_{di})$. As a result, the following equation can be obtained as

$$\Delta\delta_i = m_{di}\Delta V_{di} + m_{qi}\Delta V_{qi} \tag{15}$$

where $m_{di} = -V_{qi}\big/\left(V_{di}^2 + V_{qi}^2\right)$ and $m_{qi} = V_{di}\big/\left(V_{di}^2 + V_{qi}^2\right)$.

Since $\Delta\omega_i = s\Delta\delta_i$, the time-domain form can be obtained as

$$\Delta\omega_i = m_{di}\Delta\dot{V}_{di} + m_{qi}\Delta\dot{V}_{qi} \tag{16}$$

The output voltage amplitude can be obtained as

$$V_i = \left| V_{di} + j V_{qi} \right| = \sqrt{V_{di}^2 + V_{qi}^2} . \qquad (17)$$

which can be linearized as

$$\Delta V_i = n_{di} \Delta V_{di} + n_{qi} \Delta V_{qi} \qquad (18)$$

where $n_{di} = V_{di} \big/ \sqrt{V_{di}^2 + V_{qi}^2}$ and $n_{qi} = V_{qi} \big/ \sqrt{V_{di}^2 + V_{qi}^2}$.

The state variable "$\Delta \dot{V}$" becomes

$$\Delta \dot{V}_i = n_{di} \Delta \dot{V}_{di} + n_{qi} \Delta \dot{V}_{qi} \qquad (19)$$

According to (10), (12), (13), (16) and (19), Eq. (20) can be obtained, which describes the behavior of each inverter

$$\begin{bmatrix} \Delta \dot{V}_{DCi} \\ \Delta \dot{\omega}_i \\ \Delta \dot{V}_{di} \\ \Delta \dot{V}_{qi} \end{bmatrix} = A_i \begin{bmatrix} \Delta V_{DCi} \\ \Delta \omega_i \\ \Delta U_{di} \\ \Delta U_{qi} \end{bmatrix} + B_i \begin{bmatrix} \Delta P_i \\ \Delta Q_i \end{bmatrix} \qquad (20)$$

where

$$A_i = \begin{bmatrix} 0 & 0 & 0 & 0 \\ k_f \omega_f & -\omega_f & 0 & 0 \\ 0 & \dfrac{n_{qi}}{m_{di} n_{qi} - m_{qi} n_{di}} & \dfrac{m_{qi} n_{di} \omega_f}{m_{di} n_{qi} - m_{qi} n_{di}} & \dfrac{m_{qi} n_{qi} \omega_f}{m_{di} n_{qi} - m_{qi} n_{di}} \\ 0 & \dfrac{n_{di}}{m_{qi} n_{di} - m_{di} n_{qi}} & \dfrac{m_{di} n_{di} \omega_f}{m_{qi} n_{di} - m_{di} n_{qi}} & \dfrac{m_{di} n_{qi} \omega_f}{m_{qi} n_{di} - m_{di} n_{qi}} \end{bmatrix}$$

and

$$B_i = \begin{bmatrix} k_{DC} & 0 \\ -k_p \omega_f - k_f k_{DC} & 0 \\ 0 & \dfrac{m_{qi} k_q \omega_f}{m_{di} n_{qi} - m_{qi} n_{di}} \\ 0 & \dfrac{m_{di} k_q \omega_f}{m_{qi} n_{di} - m_{di} n_{qi}} \end{bmatrix} .$$

B. Model of the Parallel Inverter System

Next, a parallel system composed of two inverters is analyzed, as shown in Fig. 4. The output voltage of two inverters are $\dot{V}_1 = V_{d1} + j V_{q1}$ and $\dot{V}_2 = V_{d2} + j V_{q2}$, respectively. The equivalent line impedance and the load are $r+jX$ and R_L+jX_L, respectively.

According to Kirchhoff's Current Law, the output current of each inverter can be obtained as

$$\begin{cases} \dot{I}_1 = \dfrac{\dot{V}_1 \left[(r + R_L) + j(X + X_L) \right] - \dot{V}_2 (R_L + jX_L)}{(r + jX)^2 + 2(R_L + jX_L)(r + jX)} \\[4mm] \dot{I}_2 = \dfrac{\dot{V}_2 \left[(r + R_L) + j(X + X_L) \right] - \dot{V}_1 (R_L + jX_L)}{(r + jX)^2 + 2(R_L + jX_L)(r + jX)} \end{cases} \qquad (21)$$

Eq. (21) can be rewritten as

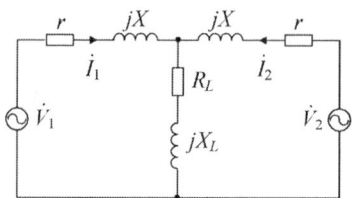

Fig. 4. Equivalent circuit of the parallel system.

$$\begin{cases} \dot{I}_1 = I_{d1} + j I_{q1} = Y_{11} V_{d1} + Y_{12} V_{q1} + Y_{13} V_{d2} + Y_{14} V_{q2} \\ \qquad\quad + j \left(Y_{21} V_{d1} + Y_{22} V_{q2} + Y_{23} V_{d2} + Y_{24} V_{q2} \right) \\ \dot{I}_2 = I_{d2} + j I_{q2} = Y_{13} V_{d1} + Y_{14} V_{q1} + Y_{11} V_{d2} + Y_{12} V_{q2} \\ \qquad\quad + j \left(Y_{23} V_{d1} + Y_{24} V_{q2} + Y_{21} V_{d2} + Y_{22} V_{q2} \right) \end{cases} \quad (22)$$

where

$$Y_{11} = \left(r^3 + 3R_L r^2 + X^2 r + 2X_L^2 r + 2R_L^2 r \right.$$
$$\left. + 2X_L Xr + X^2 R_L \right) \big/ Y_D ,$$

$$Y_{12} = \left(Xr^2 + X_L r^2 + 2R_L Xr + 2R_L^2 X + 3X_L X^2 \right.$$
$$\left. + 2X_L^2 X + X^3 \right) \big/ Y_D ,$$

$$Y_{13} = \left(-R_L r^2 - 2X_L^2 r - 2X_L Xr - 2R_L^2 r + R_L X^2 \right) \big/ Y_D ,$$

$$Y_{14} = \left(X_L r^2 - 2R_L Xr - 2R_L^2 X - 2X_L^2 X - X_L X^2 \right) \big/ Y_D ,$$

$$Y_{21} = \left(-X_L r^2 - Xr^2 - 2R_L Xr - X^3 - 3X_L X^2 \right.$$
$$\left. - 2R_L^2 X - 2X_L^2 X \right) \big/ Y_D ,$$

$$Y_{22} = \left(r^3 + 3R_L r^2 + X^2 r + 2X_L^2 r + 2R_L^2 r \right.$$
$$\left. + 2X_L Xr + R_L X^2 \right) \big/ Y_D ,$$

$$Y_{23} = \left(-X_L r^2 + 2R_L Xr + X_L X^2 + 2R_L^2 X + 2X_L^2 X \right) \big/ Y_D ,$$

$$Y_{24} = \left(-R_L r^2 - 2X_L Xr - 2X_L^2 r - 2R_L^2 r + R_L X^2 \right) \big/ Y_D ,$$

$$Y_D = X^4 + 4X_L X^3 + \left(4R_L r + 4R_L^2 + 4X_L^2 + 2r^2 \right) X^2$$
$$+ 4r^2 X_L X + 4R_L^2 r^2 + 4X_L^2 r^2 + r^4 + 4r^3 R_L .$$

Eq. (22) can be equivalent to the matrix as

$$\begin{bmatrix} I_{d1} \\ I_{q1} \\ I_{d2} \\ I_{q2} \end{bmatrix} = \begin{bmatrix} Y_{11} & Y_{12} & Y_{13} & Y_{14} \\ Y_{21} & Y_{22} & Y_{23} & Y_{24} \\ Y_{13} & Y_{14} & Y_{11} & Y_{12} \\ Y_{23} & Y_{24} & Y_{21} & Y_{22} \end{bmatrix} \begin{bmatrix} V_{d1} \\ V_{q1} \\ V_{d2} \\ V_{q2} \end{bmatrix} \qquad (23)$$

which can be simply written as $[I] = [Y][V]$. By linearizing (23), the corresponding equation can be obtained as

$$[\Delta I] = [Y][\Delta V] . \qquad (24)$$

The active power and reactive power of each inverter can be defined as

$$\begin{cases} P_i = V_{di} I_{di} + V_{qi} I_{qi} \\ Qi = V_{qi} I_{di} - V_{di} I_{qi} \end{cases} . \qquad (25)$$

It implies that

$$\begin{bmatrix} \Delta P_1 \\ \Delta Q_1 \\ \Delta P_2 \\ \Delta Q_2 \end{bmatrix} = \begin{bmatrix} I_{d1} & I_{q1} & 0 & 0 \\ -I_{q1} & I_{d1} & 0 & 0 \\ 0 & 0 & I_{d2} & I_{q2} \\ 0 & 0 & -I_{q2} & I_{d2} \end{bmatrix} \begin{bmatrix} \Delta V_{d1} \\ \Delta V_{q1} \\ \Delta V_{d2} \\ \Delta V_{q2} \end{bmatrix} \qquad (26)$$

$$+ \begin{bmatrix} V_{d1} & V_{q1} & 0 & 0 \\ V_{q1} & -V_{d1} & 0 & 0 \\ 0 & 0 & V_{d2} & V_{q2} \\ 0 & 0 & V_{q2} & -V_{d2} \end{bmatrix} \begin{bmatrix} \Delta I_{d1} \\ \Delta I_{q1} \\ \Delta I_{d2} \\ \Delta I_{q2} \end{bmatrix}$$

which in its symbolic form is

$$[\Delta S] = [I][\Delta V] + [V][\Delta I]. \qquad (27)$$

Substituting (24) in (26) results in

$$[\Delta S] = ([I] + [V][Y])[\Delta V]. \qquad (28)$$

According to (20) and (28), the following equation can be obtained

$$\left[\Delta \dot{X} \right] = \begin{bmatrix} A_1 & 0 \\ 0 & A_2 \end{bmatrix} [\Delta X] + \begin{bmatrix} B_1 & 0 \\ 0 & B_2 \end{bmatrix} ([I] + [V][Y]) \begin{bmatrix} \Delta V_{d1} \\ \Delta V_{q1} \\ \Delta V_{d2} \\ \Delta V_{q2} \end{bmatrix} \quad (29)$$

where

$$\left[\Delta X \right] = \begin{bmatrix} \Delta V_{DC1} & \Delta \omega_1 & \Delta V_{d1} & \Delta V_{q1} & \Delta V_{DC2} & \Delta \omega_2 & \Delta V_{d2} & \Delta V_{q2} \end{bmatrix}^{\mathrm{T}}.$$

$$\begin{bmatrix} \Delta V_{d1} \\ \Delta V_{q1} \\ \Delta V_{d2} \\ \Delta V_{q2} \end{bmatrix} = \begin{bmatrix} 0 & 0 & 1 & 0 & 0 & 0 & 0 & 0 \\ 0 & 0 & 0 & 1 & 0 & 0 & 0 & 0 \\ 0 & 0 & 0 & 0 & 0 & 0 & 1 & 0 \\ 0 & 0 & 0 & 0 & 0 & 0 & 0 & 1 \end{bmatrix} \begin{bmatrix} \Delta V_{DC1} \\ \Delta \omega_1 \\ \Delta V_{d1} \\ \Delta V_{q1} \\ \Delta V_{DC2} \\ \Delta \omega_2 \\ \Delta V_{d2} \\ \Delta V_{q2} \end{bmatrix} = [k] \begin{bmatrix} \Delta V_{DC1} \\ \Delta \omega_1 \\ \Delta V_{d1} \\ \Delta V_{q1} \\ \Delta V_{DC2} \\ \Delta \omega_2 \\ \Delta V_{d2} \\ \Delta V_{q2} \end{bmatrix}$$

$$(30)$$

Substituting (30) in (29) gives

$$\left[\Delta \dot{X} \right] = \left\{ \begin{bmatrix} A_1 & 0 \\ 0 & A_2 \end{bmatrix} + \begin{bmatrix} B_1 & 0 \\ 0 & B_2 \end{bmatrix} ([I] + [V][Y])[k] \right\} [\Delta X] \quad (31)$$

Thus, the state space matrix M is

$$M = \begin{bmatrix} A_1 & 0 \\ 0 & A_2 \end{bmatrix} + \begin{bmatrix} B_1 & 0 \\ 0 & B_2 \end{bmatrix} ([I] + [V][Y])[k] \qquad (32)$$

C. Eigenvalue Analysis

The system response can be investigated through root locus plots defined by (31). The parameters of inverters are as follows: output voltage $\dot{V}_1 = \dot{V}_2 = 218 + j30\,\mathrm{V}$, load impedance $Z_L = 50 + j0.2\,\Omega$, transmission-line impedance $Z = 0.2 + j1.8\,\Omega$, and cut-off frequency of low-pass filter $\omega_f = 3.141\,\mathrm{rad/s}$.

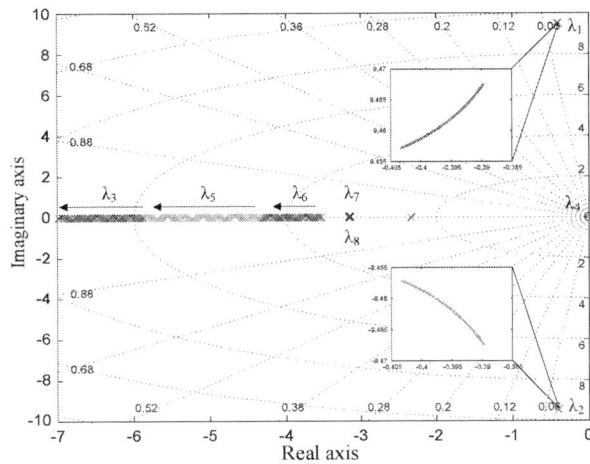

Fig. 5. Root locus diagram for $0.0001 \leq k_q \leq 0.01$.

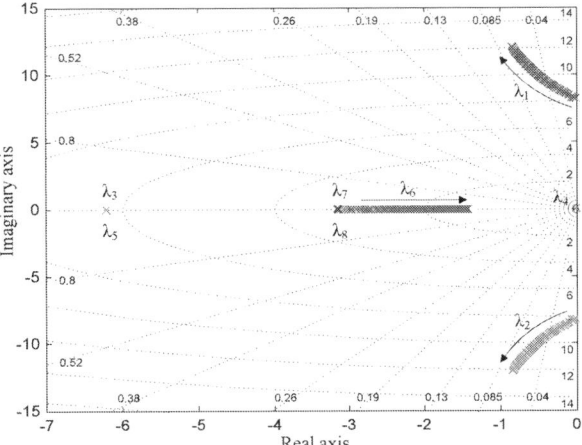

Fig. 6. Root locus diagram for $0.00001 \leq k_p \leq 0.001$.

Fig. 5 shows the root locus diagram with $k_p = 0.0003$, $k_f = 0.001$ and k_q varying from 0.0001 to 0.01. The eigenvalue appears in the origin because the system matrix is singular, and hence it does not impact the dynamics. The only two poles (λ_1 and λ_2) mainly determine the dynamic response and three poles (λ_3, λ_5 and λ_6) will change along with the reactive power droop coefficient k_q. The system is stable in the range of concern since the poles remain in the left half of the s-plane.

Fig. 6 shows the root locus diagram with $k_q = 0.008$, $k_f = 0.001$ and k_p varying from 0.00001 to 0.001. With k_p increasing, λ_6 is moved towards the origin and it becomes more dominant. It can be seen that the system dynamic performance can be improved for power regulation by increasing k_p. However, a large value of k_p can result in the ac bus frequency deviation in steady-state. Thus, k_p is determined considering a trade-off of the dynamic and the steady-state responses.

Fig. 7 shows the root locus diagram with $k_p = 0.0003$, $k_q = 0.008$ and k_f varying from 0.0001 to 0.05. The system becomes faster by increasing k_f since the poles become the dominant poles, while the effect of the real poles decreases. This could result in significant oscillations in the dc-link voltage. Even when k_f is bigger than 0.0316, the system is still not stable.

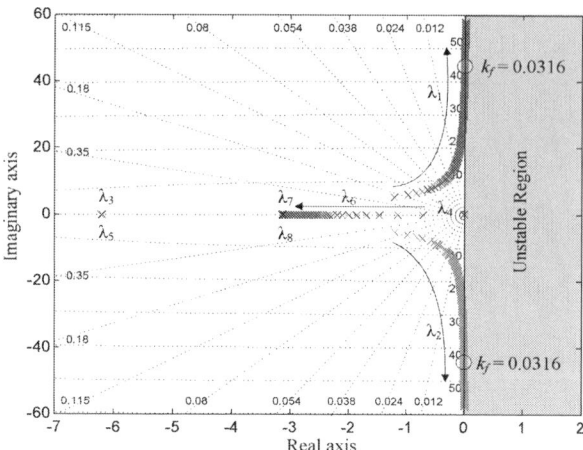

Fig. 7. Root locus diagram for $0.0001 \leq k_f \leq 0.05$.

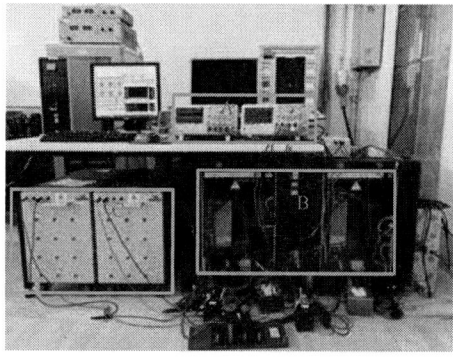

Fig. 8. Experimental setup of the MG system consisting of two parallel two-stage inverters with the proposed droop control scheme: A – oscilloscopes, B – inverters, and C – loads, where the control schemes are implemented in a dSPACE DS 1103 system.

V. EXPERIMENTAL RESULTS

In order to verify the proposed droop control scheme for resilient power sharing among parallel inverters, experiments have been performed on a MG system, which consists of two single-phase two-stage PV inverter systems, referring to Fig. 1. The hardware setup is shown in Fig. 8, with its parameters being given in Table I. A dSPACE DS 1103 system was used to implement the control algorithms.

TABLE I
PARAMETERS OF TESTED SYSTEM.

Parameter	Values
PV panel voltage variation (V_{PV})	200 V–300 V
Rated output voltage amplitude (V_{0i})	220 V(RMS)
Rated frequencies of the PV inverter (f_{0i})	50 Hz
DC-link reference voltage(V_{DCref})	400 V
DC-link voltage capacitance (C)	940 μF
Boost inductance (L)	4 mH
Output filter inductance (L_{ac})	6 mH
Output filter capacitance (C_{ac})	10 μF
Boost and inverter switching frequencies (f_s)	10 kHz
Active droop coefficient (k_{pi})	0.0003 rad/(s·W)
Reactive droop coefficient (k_{qi})	0.008 V/Var

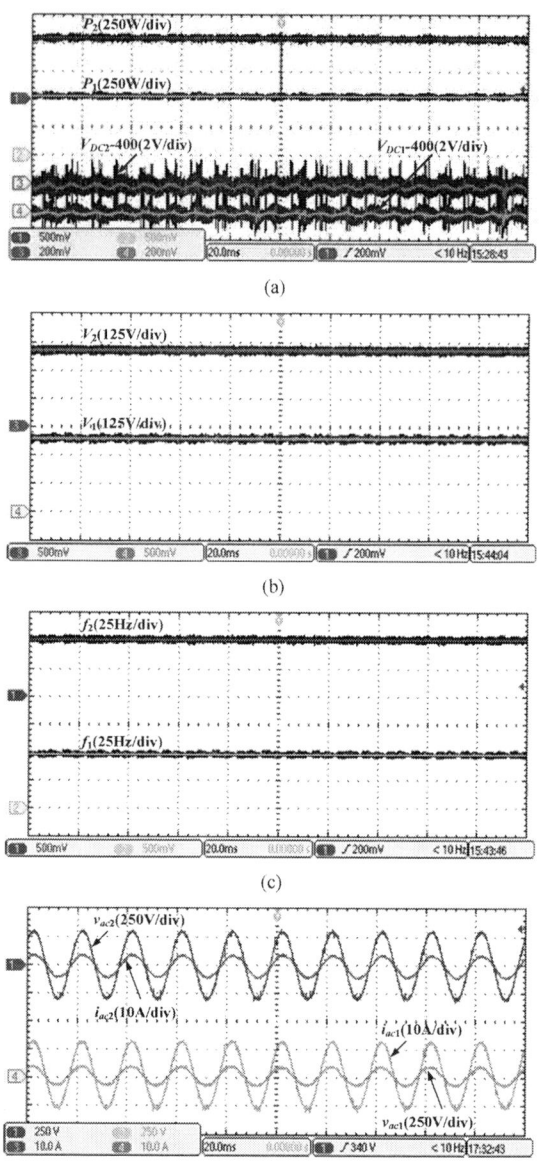

Fig. 9. Experimental results of the parallel inverter system with the enhanced droop control: (a) output active power P_1, P_2 and dc-link voltage V_{DC1}-400, V_{DC2}-400, (b) ac output voltage amplitude V_1, V_2, (c) output voltage frequency f_1, f_2, and (d) ac output voltage v_{ac1}, v_{ac2} and ac output current i_{ac1}, i_{ac2}.

Fig. 9 shows the experimental results with the enhanced droop control scheme. When the output active power of two inverters is enough to supply the load, the proposed droop scheme ensures that the active power of the load is shared evenly between the inverters.

Fig. 10 presents the experimental results with the traditional droop control. When the energy resource of inverter 1 is not able to provide adequate power to the load, the power shortage will lead to power imbalance among the parallel inverter systems and frequency excursions in the droop-controlled based on (1).

978-1-4673-9551-9/16 $31.00 © 2016 IEEE   1258

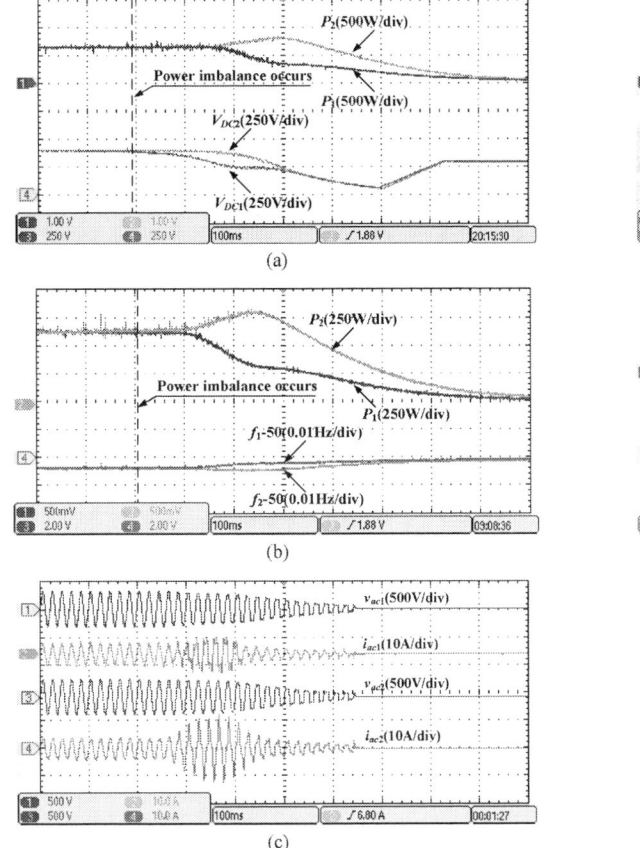

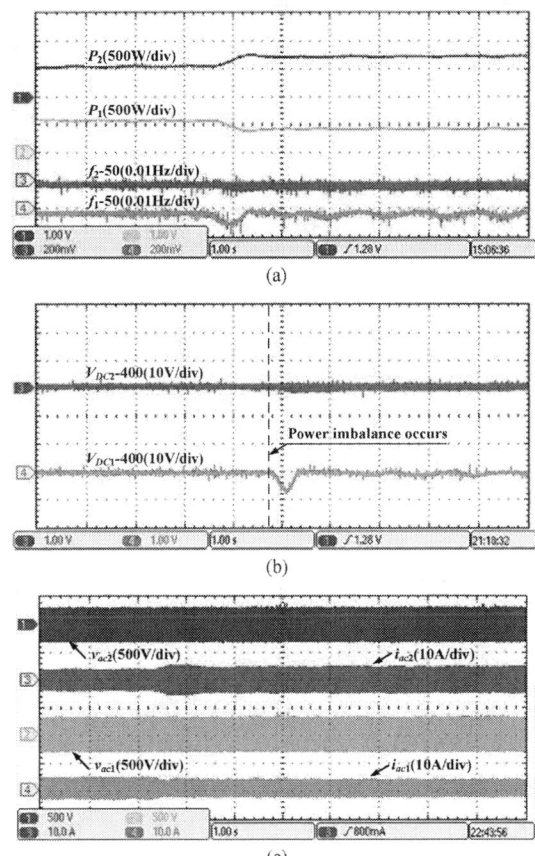

Fig. 10. Experimental results of the parallel system when the traditional droop control is employed: (a) output active power P_1, P_2 and dc-link voltage V_{DC1}, V_{DC2}, (b) output active power P_1, P_2 and output voltage frequency f_1-50, f_2-50, and (c) ac output voltage v_{ac1}, v_{ac2} and ac output current i_{ac1}, i_{ac2}.

Fig. 11. Experimental results of the active power fluctuations with k_f= 0.01: (a) output active power P_1, P_2 and output voltage frequency f_1-50, f_2-50, (b) dc-link voltage V_{DC1}-400, V_{DC2}-400 and (c) ac output voltage v_{ac1}, v_{ac2} and ac output current i_{ac1}, i_{ac2}.

Fig. 11 shows the power fluctuations of inverter 1 with k_f = 0.01, where the PV source power has been changed from 800 W to 400 W. In this case, inverter 1 cannot supply sufficient active power to the load and thus the dc-link voltage drops, as it is shown in Fig. 11(b). By adjusting the droop curve according to Fig. 2, the output active power of inverter 1 decreases and the inverter 2 supplies the rest of the active power. Finally, the parallel system can achieve the power balancing between supply and demand. Fig. 12 shows the power fluctuations of inverter 1 with k_f = 0.001. It can be seen that the settling time is longer and the dc-link voltage fluctuates more significantly. Fig. 13 shows the experimental results of load variations. In all, the experiments have demonstrated that the proposed droop control scheme can realize a resilient power sharing among inverters.

VI. CONCLUSION

An improved droop control scheme has been proposed in this paper for paralleled two-stage PV inverter systems in order to balance the active power between supply and demand. By varying the dc-link voltage within a reasonable range, the proposed droop control scheme allows lowering the droop line

in the inverter, which thus achieves the active power balance. In contrast to the conventional PI-based controller for the boost converter of a two-stage system, only proportional controller has been employed in the proposed droop control strategy. Moreover, based on the small signal model, several root locus plots as a function of the system parameters can be obtained in order to design the optimal droop coefficients. The experimental results have validated the effectiveness of the proposed scheme in terms of resilient power sharing among droop-controlled paralleled PV inverters. More important, the proposed droop control scheme can also be extended to other parallel two-stage inverter systems, where the entire system is able to supply the demanded power, but one of the inverters has limited power during operation.

ACKNOWLEDGEMENT

This work was supported by National Natural Science Foundation of China (51477033), Harbin Science and Technology Innovation Talent Research Special Funds (RC2014QN007004), and Lite-On Power Electronics Technology Research Fund (PRC20151382). The authors would like to acknowledge for the support of the project.

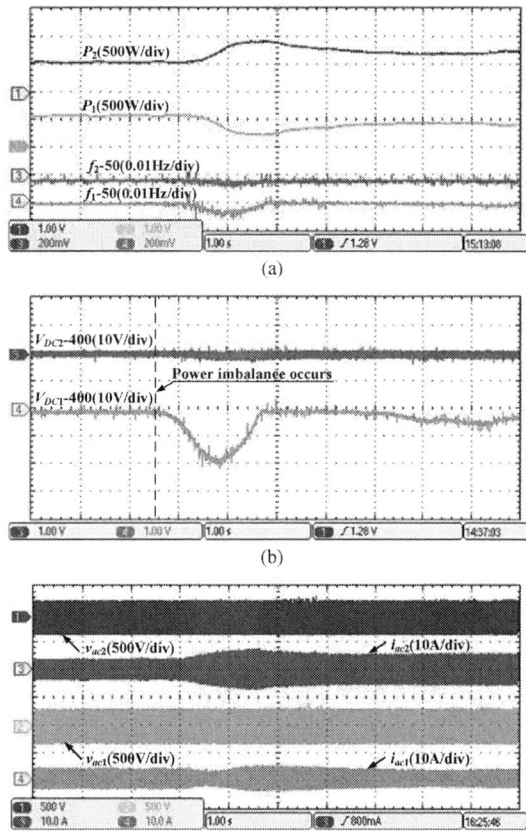

Fig. 12. Experimental results in the case of the active power fluctuation with $k_f = 0.001$, when the enhanced droop control scheme is adopted: (a) output active power P_1, P_2 and output voltage frequency f_1-50, f_2-50, (b) dc-link voltage V_{DC1}-400, V_{DC2}-400 and (c) v_{ac1}, v_{ac2} and ac output current i_{ac1}, i_{ac2}.

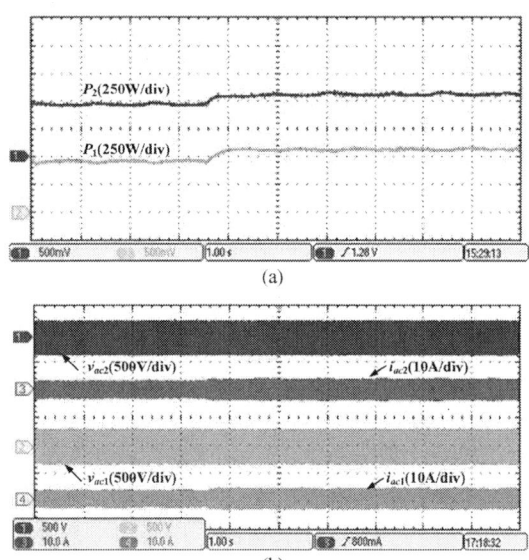

Fig. 13. Experimental results in the case of load variations, when the enhanced droop control scheme is adopted: (a) output active power P_1, P_2 and (b) ac output voltage v_{ac1}, v_{ac2} and ac output current i_{ac1}, i_{ac2}.

REFERENCES

[1] M. Amirabadi, A. Balakrishnan, H. A. Toliyat, and W. C. Alexander, "High-frequency ac-link PV inverter," *IEEE Trans. Ind. Electron.*, vol. 61, no. 1, pp. 281–291, Jan. 2014.

[2] H. Mahmood, D. Michaelson, and J. Jiang, "Accurate reactive power sharing in an islanded microgrid using adaptive virtual impedances," *IEEE Trans. Power Electron.*, vol. 30, no. 3, pp. 1605–1617, Mar. 2015.

[3] Y. Tao, Q. W. Liu, Y. Deng, X. H. Liu, and X. N. He, "Analysis and mitigation of inverter output impedance impacts for distributed energy resource interface," *IEEE Trans. Power Electron.*, vol. 30, no. 7, pp. 3563–3576, Jul. 2015.

[4] T. K. S. Freddy, N. A. Rahim, W. P. Hew, and H. S. Che, "Comparison and analysis of single-phase transformerless grid-connected PV inverters," *IEEE Trans. Power Electron.*, vol. 29, no. 10, pp. 5358–5369, Oct. 2014.

[5] R. G. Wandhare, and V. Agarwal, "Novel integration of a PV-wind energy system with enhanced efficiency," *IEEE Trans. Power Electron.*, vol. 30, no. 7, pp. 3638–3649, Jul. 2015.

[6] J. W. He, Y. W. Li, and F. Blaabjerg, , "An enhanced islanding microgeid reactive power, imbalance power ,and harmonic power sharing scheme," *IEEE Trans. Power Electron.*, vol. 30, no. 6, pp. 3389–3401, Jun. 2015.

[7] N. E. Zakzouk, A. K. Abdelsalam, A. A. Helal, and B. W. Williams, "DC-link voltage sensorless control technique for single-phase two-stage photovoltaic grid-connected system," in *Proc. ENERGYCON*, 2014, pp. 58–64.

[8] D. Debnath, and K. Chatterjee, "Two-stage solar Photovoltaic-based stand-alone scheme having battery as energy storage element for rural deloyment," *IEEE Trans. Ind. Electron.*, vol. 62, no. 7, pp. 4148–4157, Jul. 2015.

[9] D. Velaso DE La Fuente, C. L. T. Rodriguez, G. Garcero, E. Figueres, and R. O. Gonzalez, "Photovoltaic power system with battery backup with grid-connected and islanded operation capabilities," *IEEE Trans. Ind. Electron.*, vol. 60, no. 4, pp. 1571–1582, Apr. 2013.

[10] C. Trujillo Rodriguez, D. Velasco De La Fuente, G. Garcera, E. Figueres, and J. A. Guacaneme Moreno, "Reconfigurable control scheme for a PV microinverter working in both grid-connected and island modes," *IEEE Trans. Ind. Electron.*, vol. 60, no. 4, pp. 1582–1595, Apr. 2013.

[11] W. Du, Q. R. Jiang, M. J. Erickson, and R. H. Lasseter, "Voltage-source control of PV inverter in a CERTS microgrid," *IEEE Trans. Power Del.*, vol. 29, no. 4, pp. 1726–1734, Aug. 2014.

[12] A. Elrayyah, Y. Sozer, and M. E. Elbuluk, "Modeling and control design of microgrid-connected PV-based sources," *IEEE Trans. Power Electron.*, vol. 2, no. 4, pp. 907–919, Dec. 2014.

[13] A. Elrayyah, Y. Sozer, and M. Elbuluk, "Microgrid-connected PV-based sources: a novel autonomous control method for maintaining maximum power," *IEEE Ind. Appl. Mag.*, vol. 21, no. 2, pp. 19–29, Mar./Apr. 2015.

[14] S. V. Iyer, M. N. Belur, and M. C. Chandorka, "A generalized computational method to determine stability of a multi-inverter microgrid," *IEEE Trans. Power Electron.*, vol. 25, no. 9, pp. 2420–2432, Sep. 2010.

[15] W. R. Issa, M. A. Abusara, and S. M. Sharkh, "Control of transient power during unintentional islanding of microgrids," *IEEE Trans. Power Electron.*, vol. 30, no. 8, pp. 4573–4584, Aug. 2015.

[16] X. Q. Guo, Z. G. Lu, B. C. Wang, X. F. Sun, L. Wang, and J. M. Guerrero, "Dynamic phasors-based modeling and stability analysis of droop-controlled inverters for microgrid applications," *IEEE Trans. Smart Grid*, vol. 5, no. 6, pp. 2980–2987, Nov. 2014.

Voltage Closed-Loop Virtual Synchronous Generator Control of Full Converter Wind Turbine for Grid-Connected and Stand-Alone Operation

Yiwei Ma, Liu Yang, Fred Wang, Leon M. Tolbert
Center for Ultra-Wide-Area Resilient Electric Energy Transmission Networks (CURENT)
Department of Electrical Engineering and Computer Science
The University of Tennessee
Knoxville, TN 37996-2250, USA
yma13@utk.edu

Abstract—One way to incorporate the increasing amount of wind penetration is to control wind turbines to emulate the behavior of conventional synchronous generators. This paper presents a comprehensive virtual generator control for the full converter wind turbine considering the power balance. The voltage closed-loop virtual synchronous generator control of the wind turbine allows it to work under both grid-connected and stand-alone condition. Power system control and power dispatch can also be realized through the control. The power balance of the wind turbine system is achieved by controlling the rotor speed of the turbine according to the loading condition. The optional integration of the short term turbine level energy storage is also considered. Experimental results on emulation testbed are presented to demonstrate the feasibility and effectiveness of the proposed control method.

Keywords—wind turbine; virtual synchronous generator; MPPT; grid-connected; stand-alone

I. INTRODUCTION

With more than 370 GW total capacity installed worldwide by 2014, wind power is one of the largest and most promising renewable energy sources [1]. The U.S. Department of Energy has envisioned that wind power will supply 20% of all U.S. electricity by 2030 [2]. Most wind turbines (WTs) nowadays are controlled to output the maximum power available and not react to the grid disturbances, and the power fluctuation is balanced by conventional synchronous generators (SGs). However the case will be different when the penetration becomes larger with more traditional SGs replaced by the WTs. The loss of system controllability and decrease of the system inertia will cause fluctuations in grid voltage and frequency. In addition, the uncertainty and variation of the wind will also pose challenges to system operators [3].

Thus, grid support control functions from the wind turbine will be very important to be integrated in future grid. For this purpose, various kinds of grid support controls have been developed for the variable speed wind turbines (VSWTs). With appropriate control, the VSWTs will be able to change the active and reactive power injections, in the case of grid disturbances [4-6]. However these VSWTs still cannot be regulated and dispatched like the traditional SGs by the system operators.

Authors from [7-11] seek to develop controls for the grid interfacing power electronics converters to mimic the behaviors of SGs. With the same output behavior as a conventional SG, the replacement units can guarantee stable operation of the grid. However, these virtual synchronous generator (VSG) controls are mostly focused on the grid interfacing controls. Energy storage is assumed to be able to support all the power balance of the renewable generation system, which is not always the case. The control coordination between the renewable generation and the grid integration for the VSG controls still remains challenge.

This paper develops a comprehensive VSG control strategy for a full converter wind turbine (FCWT) system considering the power balance among all the components in the generation system, considering with and without turbine level energy storages. The structure of the paper is organized as follows: Section II introduces the studied FCWT system; Section III presents the controls for each component in the system; and experiment results are given in Section IV.

II. STRUCTURE AND BASIC CONTROL PRINCIPLES

Also known as Type-4 WT, FCWT adopts an ac/dc/ac converter configuration to transfer the captured energy from the wind generator to the grid, as shown in Fig. 1(a). Since short term turbine level energy storage has been proposed and tested in the field to help with power smoothing and grid support [12, 13], it is reasonable to assume a minute-level storage is connected to the dc-link of the system to help with maintaining the dispatch command, as shown in Fig. 1(b). The VSG control can be applied for both WT systems.

As the output characteristics of the WT is determined by the emulated generator, grid side converter (GSC) decides the output power by the grid condition and system dispatch. The other converters will then maintain the power balance of the WT generation system.

In the Fig. 1(a) system without energy storage, the dc voltage is regulated by the machine side converter (MSC). It controls the wind generation to match the load condition of the GSC by adjusting the turbine rotating speed; and in Fig. 1(b), the storage side converter (SSC) controls the dc voltage, and MSC can operate based on the state of charge (SoC) in a similar manner. The detailed control strategies will be presented next in Section III.

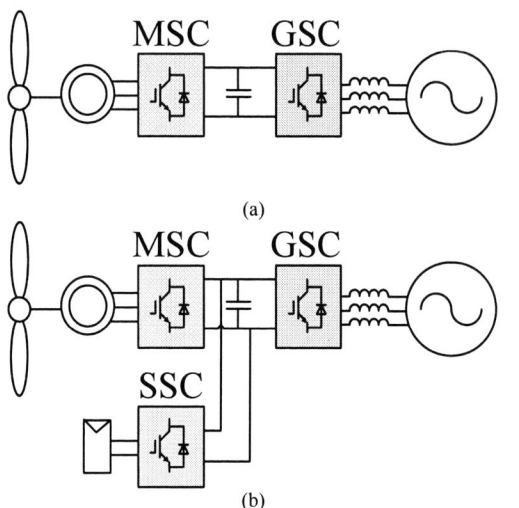

(a)

(b)

Fig. 1. FCWT system: (a) without energy storage; (b) with energy storage.

III. DETAILED CONTROL STRATEGY

A. Grid Side Converter Control

In order to mimic the behavior of a synchronous generator, the GSC measures the output currents i_{dq} of the WT, and feeds them into the emulated generator's model. The voltage and frequency outputs from the model V_{dq}^* and f are closed-loop controlled by the GSC, thus it can be considered to supply steady voltage source to the power system or directly to the loads in the stand-alone operation. The detailed voltage closed-loop control strategy can be implemented by single loop or double loop with inner current control loop, and will not be considered in this paper. The overall control diagram is shown in Fig. 2(a).

The output voltage reference is calculated through the electrical model of the generator, shown in Fig. 2(b), where e_{fd} is the field winding voltage, x_d and x_q are the synchronous reactances, x_d' and x_q' are the transient reactances, T_{do}' and T_{qo}' are the transient open-circuit time constants, E_d' and E_q' are transient back EMFs, u_t is the terminal voltage amplitude, U_{tref} is the voltage reference given by the operator, K_A and T_e are the gain and time constant of the excitation system, and E_{fmax} and E_{fmin} are the maximum allowable field voltage.

The output frequency is determined by the mechanical model of the virtual generator, demonstrated in Fig. 2(c). P_{cmd} is the power reference given by the system dispatch, M and D are the inertia constant and the mechanical friction caused damping factor, ω_r is the rotor speed, ω_s is the synchronous speed and the integrated angle θ is given to the park transformation of the converter control. In addition to the generator model, a power limiter is proposed to limit the power output during low-wind scenarios and prevent power back-flow to the WT. This will be discussed in Section III. D. Power balance part.

The GSC is synchronized with the grid frequency by using the swing equation provided by the mechanical model of the virtual generator, without the help of the phase locked loop (PLL). The inertia-frequency response can be intrinsically emulated through the VSG control. The parameters also allow tuning to achieve equal or even better performance than the traditional SGs.

In addition, power system controls developed for SGs can also be implemented, including: automatic generation control (AGC), automatic voltage regulator (AVR), power system stabilizer (PSS), etc. [14].

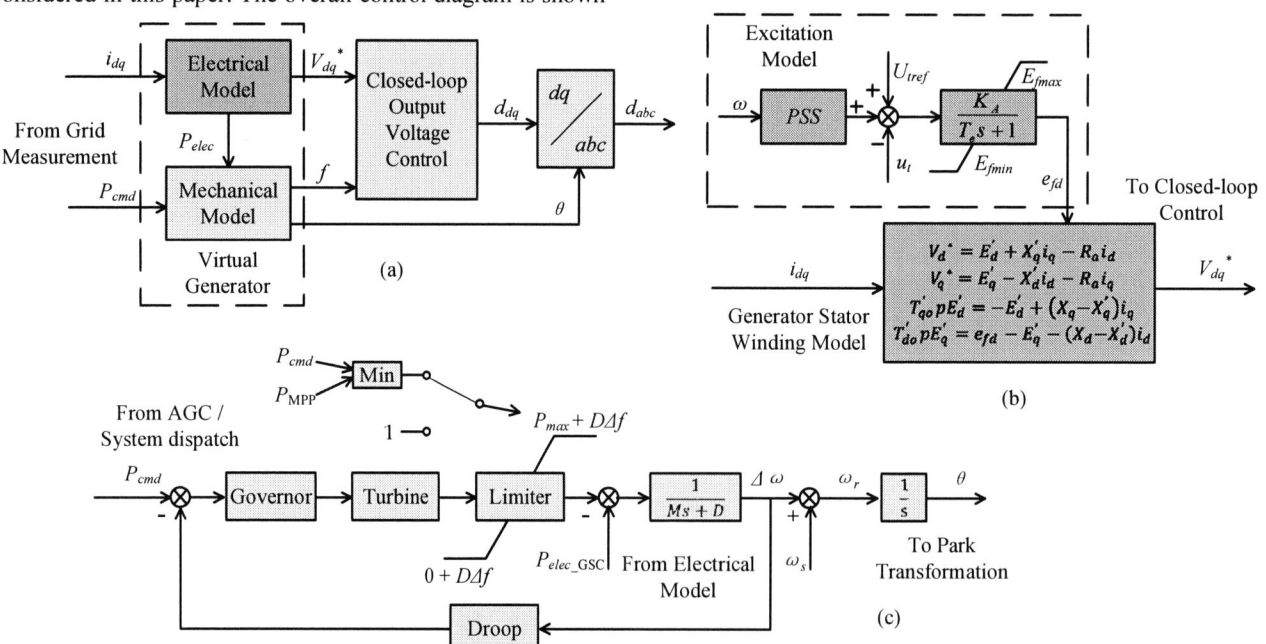

Fig. 2. GSC control: (a) overall control diagram; (b) electrical model; (c) mechanical model.

B. Storage Side Converter Control

If the energy storage system is included as shown in Fig. 1(b), the SSC will take over the dc voltage regulation. The control diagram is shown in Fig. 3, where V_{dc} is the dc-link voltage of the FCWT, $V_{dc}*$ is the voltage reference, and $I_{dc\_stor}*$ is the current reference of the energy storage converter. Thus the dc voltage can be maintained with allowable SoC. The storage is considered ideal, and the detailed model and current controls will not be further discussed in this paper.

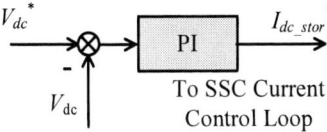

Fig. 3. SSC control.

The energy storage assumed in this paper acts as an energy buffer between the grid interfacing converter GSC and the wind generation converter MSC. It will absorb or release the energy difference between the wind generation and the output power. The energy balance strategy to maintain the SoC will be discussed later in Section III. D part.

C. Machine Side Converter Control

MSC controls the rotor speed of the turbine, and thus the power balance can be achieved by moving the operating point along the wind characteristic curve. For system without storage in Fig. 1(a), the control block (a) in Fig. 4 is enabled. The dc voltage can be maintained in the desirable range, allowing some variations [15, 16]. For the system with storage in Fig. 1(b), dc voltage is regulated by SSC. SoC of the storage can be used as the input of the speed loop, as shown in control block (b) in Fig. 4.

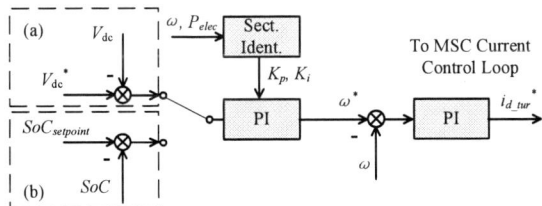

Fig. 4. MSC control: (a) for system without storage; (b) for system with storage.

The speed reference $\omega*$ is generated by a PI controller. The PI parameters are chosen based on the operating condition of the wind turbine [16]. As shown in Fig. 5, there are two possible operating points A and B, given a certain desired output wind power: A is to the left of the maximum power point (MPP), with positive curve slope, and requires positive PI parameters; while B is to the right with negative curve slope, and requires negative PI parameters. Note that operating point A is in Sector 1 and B is in Sector 2.

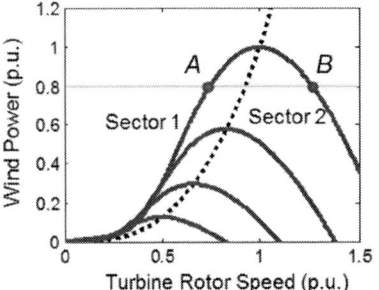

Fig. 5. Wind characteristic curve.

Thus the identification of the operating point sector is important for the control and operation of the MSC, otherwise wrong PI parameters will be applied. Intuitively, the operating sector can be recognized by the relationship of the changing direction of the wind power P_w and rotor speed ω, as shown in (1). However, since wind turbines usually have relatively large inertia constant, the actual wind power is hard to be estimated. Misidentification will be caused by the noises and measurement error of the rotor speed.

$$\begin{cases} \Delta P_w \cdot \Delta \omega > 0 & \text{Sector 1} \\ \Delta P_w \cdot \Delta \omega < 0 & \text{Sector 2} \end{cases} \quad (1)$$

Therefore identification criterion (2) is proposed in this paper, by using the rotor speed and the electric power output of the turbine P_e, and the pre-determined MPP curve $f(\omega)$, as shown by the dashed curve in Fig. 5. This curve is usually provided by the manufacturer and used for the traditional maximum power point tracking (MPPT) control, and it is usually close to the cubic function of the rotor speed ($f(\omega) \approx k\omega^3$). Thus when the electric power output of the wind turbine is above the supposed MPP curve, it is working in Sector 1 and vice versa.

$$\begin{cases} P_e > f(\omega) & \text{Sector 1} \\ P_e < f(\omega) & \text{Sector 2} \end{cases} \quad (2)$$

The electric power output P_e is the difference of the wind power and the kinetic energy absorbed by the turbine rotor. Misidentification happens when the turbine is accelerating in Sector 1, or decelerating in Sector 2, in other words, when the turbine is moving toward the MPP. Since the $\partial P_w / \partial \omega$ is close to 0 around the MPP, the misidentification will not cause significant variation of the wind input power. When the operating point reaches closer to the MPP, the criterions will identify it to be the opposite sector more often. This helps to slow down the speed control loop. When the identification is constantly switching between the sectors, it can be recognize that the WT is already working at the MPP, and no additional wind power can be made available for the load.

D. Power Balance Considerations

By implementing the electrical and mechanical model of the SG, the output power of the WT is determined by the grid and load condition. It is possible that the output power exceeds the maximum available power from the wind, and it is also possible that the GSC will absorb power when the grid frequency is higher than the nominal value.

A power limiting function is then required for these two possible scenarios, as shown in Fig. 2(c). P_{max} is the maximum power allowed in steady state, while the minimum power output is 0, keeping it from going negative. $D\Delta f$ is the steady-state error of the inertia block, and it is added to the limits in order to compensate the error. Thus in grid-connected condition, the implemented limiter will keep the GSC electric power output P_{elec_GSC} within the thresholds in steady-state.

P_{max} is set to be the smaller value between the power dispatch value P_{cmd} and the maximum available wind power P_{MPP}, as it is preferable to output its maximum power when it cannot fulfill the demand during low wind period. P_{MPP} is determined as the WT reaches MPP through the sector identification explained in Section II. C. If the command is set to be the rated power, the MPP limiter would always be in effect. In this case, the WT can achieve the MPPT control, while maintaining generator dynamics during transients. In this paper, it is denoted as VSG MPPT mode.

During standalone operation, the WT output power is solely determined by the load. In this situation, if the output electrical power of the virtual generator becomes larger than the P_{max} provide by the limiter, the output frequency will be greatly decreased. Under Frequency Load Shedding (UFLS) mechanism will then be triggered and reduce the system load until the generation-load system reaches equilibrium.

With the help of the energy storage in the Fig. 1 (b) system, the WT can maintain the dispatched power for a certain period, even if the available power is less than the output power set point. As it is only providing the difference between the demand and the generation, a short-term storage is able to carry out the dispatched power for a longer period. A SoC threshold is set to keep the storage from being overdrawn. Once below the threshold, the WT will transition to the VSG MPP mode, stop following the system dispatch and output the available power from the wind. The system operators will have to reset the power output set point according to the forecast. When the wind power input become larger than the demanded power, the storage will begin to charge, and allow to fulfill the system dispatch command again.

In addition, pitch control is used to spill the excessive power when the wind provides more power than needed, the control adopted in the VSG control is shown in Fig. 6.

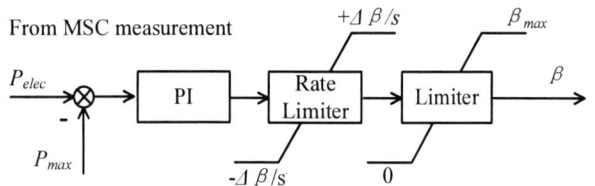

Fig. 6. Wind turbine blade pitch control.

IV. EXPERIMENTAL VERIFICATION

The experiments were conducted in a real-time scaled-down power electronics based power system emulation platform, as shown in Fig. 7, where all the components in the system are emulated by reconfigurable converters in a paralleled connection [17-19]. These converters can be configured and programmed to behave like traditional generators, static and dynamic loads, wind turbines, and energy storage in real-time. The experimental results are recorded by LabVIEW, which commands and communicates with all the components in the experiment system.

Fig. 7. Experimental platform of the emulation system

Fig. 8 shows a FCWT without storage in grid-connected operation. The power output of the wind turbine can follow the system dispatch. During high wind speed period, the pitch control is enabled to limit the wind power; when the wind speed is low and cannot provide enough power to the power command, the power limiter is enabled and transition to the VSG MPPT mode.

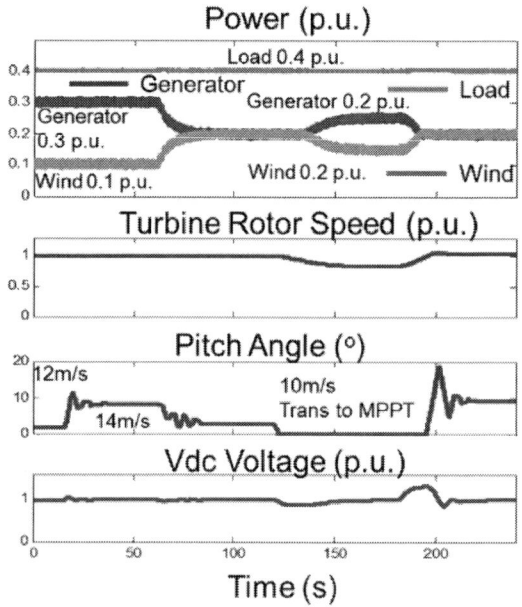

Fig. 8. Grid-connected operation experimental results.

Fig. 9 shows the experimental results with the FCWT transition between the VSG MPPT mode and the traditional MPPT mode. Both control modes can track the wind variation and output maximum power. But the behavior is different when there is large disturbance in the grid, as the load step change in Fig. 9. Traditional MPPT will not change its power injection, while VSG MPPT will provide the inertia response in the first few seconds, and help with the frequency control of the system.

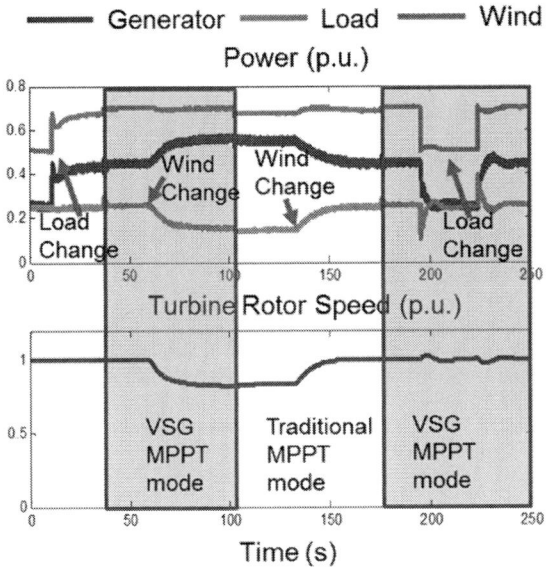

Fig. 9. Grid-connected operation experimental results with mode trasitions

Fig. 10 shows the experimental result with WT in stand-alone operation. The WT can adjust the rotor speed and pitch angle according to the load condition. If there is no UFLS protection when too much load is needed, the WT will stall as expected.

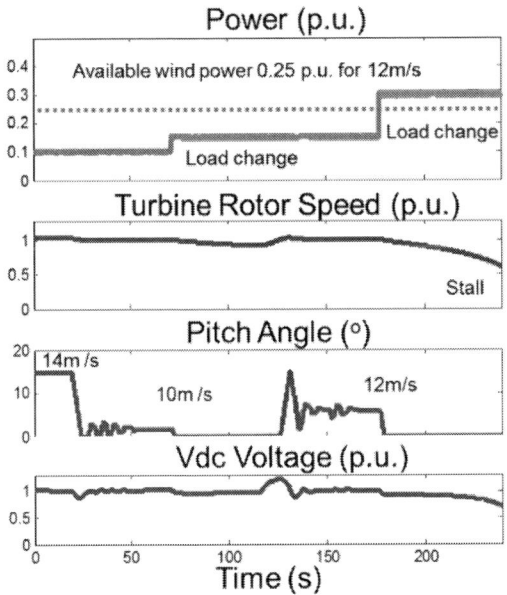

Fig. 10. Stand-alone operation experimental results.

Fig. 11 shows the performance of the system with storage under variable wind condition, the time sequence of the wind speed is generated by NREL TurbSim [20]. In the beginning,

the output power is limited to the maximum available power, and the storage charges when the wind power is higher than the system dispatch; once the *SoC* is high enough and reaches the predetermined threshold, the WT transitions to normal operation and carries out the dispatch commands.

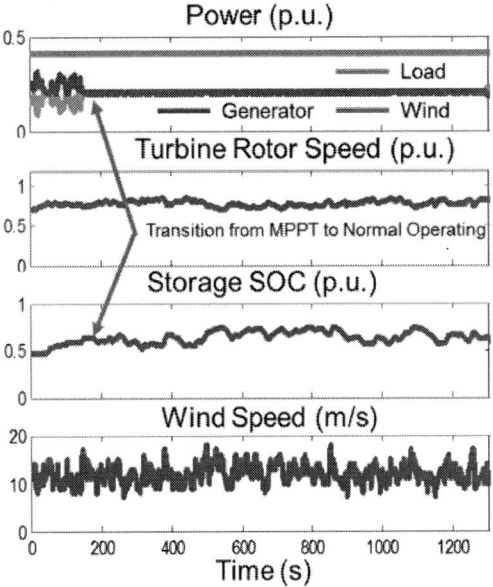

Fig. 11. Experimental results with energy storage

V. CONCLUSIONS

This paper develops the VSG control for FCWT, considering all parts in the generation system, including the grid interfacing converter GSC, wind generation converter MSC, and the optional short term dc-link energy storage. The proposed control allows the WTs to behave like the traditional SG in the grid-connected and stand-alone conditions. MPPT control function can be achieved when the available wind power is low, while maintaining the dynamic responses of the generator. System level dispatch and controls can also enabled. The power balance is achieved by turbine speed control though MSC.

The experiments are conducted in a power electronics based power system emulation platform, results demonstrate the feasibility and effectiveness of the control.

For future work, low voltage ride through (LVRT) capability and black start strategy of the proposed VSG control method will be considered.

ACKNOWLEDGMENT

This work was supported primarily by the Engineering Research Center Program of the National Science Foundation and the Department of Energy under NSF Award Number EEC-1041877 and the CURENT Industry Partnership Program.

References

[1] *Renewables 2014 Global Status Report*, REN21, [Online]. Available: http://www.ren21.net

[2] *Wind Vision: A New Era for Wind Power in the United* States, US DOE, [Online]. Available: http://www.energy.gov/windvision

[3] F. Blaabjerg and K. Ma, "Future on Power Electronics for Wind Turbine Systems," *IEEE J. Emerg. Select. Topics Power Electron.*, vol. 1, no. 3, pp.139 -152, 2013.

[4] G. Ramtharan, J. B. Ekanayake, and N. Jenkins, "Frequency Support from Doubly Fed Induction Generator Wind Turbines," *IET Renew. Power Gen.*, vol. 1, no. 1, pp. 3–9, Mar. 2007.

[5] N. R. Ullah, T. Thiringer, and D. Karlsson, "Temporary Primary Frequency Control Support by Variable Speed Wind Turbines—Potential and Applications," *IEEE Trans. Power Syst.*, vol. 23, no. 2, pp. 601–612, May 2008.

[6] F. M. Hughes, O. Anaya-Lara, N. Jenkins, and G. Strbac, "Control of DFIG-Based Wind Generation for Power Network Support," *IEEE Trans. Power Syst.*, vol. 20, no. 4, pp. 1958–1966, Nov. 2005

[7] Q. C. Zhong and G. Weiss, "Synchronverters: Inverters that Mimic Synchronous Generators," *IEEE Trans. Ind. Electron.*, vol. 58, no. 4, pp. 1259–1267, Apr. 2011

[8] J. Driesen and K. Visscher, "Virtual Synchronous Generators," in *Proc. IEEE Power and Energy Society General Meeting*, 2008, pp. 1-3.

[9] Y. Chen, R. Hesse, D. Turschner, and H. P. Beck, "Improving the Grid Power Quality Using Virtual Synchronous Machines," in *Proc. International Conference on Power Engineering, Energy and Electrical Drives (POWERENG)*, 2011, pp. 1-6.

[10] M. Torres and L. A. C. Lopes, "Virtual Synchronous Generator Control in Autonomous Wind-Diesel Power Systems," in *Proc. IEEE Elect. Power Energy Conf.*, 2009, pp.1-6.

[11] C. Li, R. Burgos, I. Cvetkovic, D. Boroyevich, L. Mili, and P. Rodriguez, "Evaluation and Control Design of Virtual-Synchronous-Machine-Based STATCOM for Grids with High Penetration of Renewable Energy," in *Proc. IEEE Energy Conversion Congress and Exposition (ECCE)*, 2014, pp. 5652-5658.

[12] M. Liserre , R. Cardenas , M. Molinas and J. Rodriguez "Overview of Multi-MW Wind Turbines and Wind Parks," *IEEE Trans. Ind. Electron.*, vol. 58, no. 4, pp.1081-1095, 2011.

[13] N. W. Miller (2014, July) *GE Experience with Turbine Integrated Battery Energy Storage* [Online]. Available: http://www.ieee-pes.org/presentations/gm2014/PESGM2014P-000717.pdf

[14] P. Kundur, *Power System Stability and Control*. New York: McGraw-Hill, 1994.

[15] F. D. Kanellos and N. D. Hatziargyriou, "Control of Variable Speed Wind Turbines Equipped with Synchronous or Doubly Fed Induction Generators Supplying Islanded Power Systems," *Renewable Power Generation*, vol. 3, no. 1, pp. 96–108, 2009.

[16] X. Yuan, F. Wang, R. Burgos, Y. Li, and D. Boroyevich, "DC-link Voltage Control of Full Power Converter for Wind Generator Operating in Weak Grid Systems," *IEEE Transactions on Power Electronics*, vol.24, no. 9, pp. 2178 – 2192, 2009.

[17] J. Wang, L. Yang, Y. Ma, X. Shi, X. Zhang, L. Hang, K. Lin, L. M. Tolbert, F. Wang and K. Tomsovic, "Regenerative Power Converters Representation of grid Control and Actuation Emulator," in *Proc. IEEE Energy Conversion Congress and Exposition (ECCE)*, Sep. 2012, pp. 2460–2465.

[18] L. Yang, X. Zhang, Y. Ma, J. Wang, L. Hang, K. Lin, L. M. Tolbert, F. Wang, and K. Tomsovic, "Hardware Implementation and Control Design of Generator Emulator in Multi-Converter System," in *Proc. IEEE Applied Power Electronics Conference and Exposition (APEC)*, Mar. 2013, pp. 2316–2323.

[19] Y. Ma, L. Yang, J. Wang, F. Wang, and L. M. Tolbert, "Emulating Full-Converter Wind Turbine by a Single Converter in a Multiple Converter Based Emulation System," *in Proc. IEEE Applied Power Electronics Conference and Exposition (APEC)*, 2014, pp. 3042-3047.

[20] N. Kelley and B. Jonkman, (2015, Apr.) TurbSim: A stochastic, full-field, turbulence simulator primarialy for use with InflowWind/AeroDyn-based simulation tools [Online]. Available: https://nwtc.nrel.gov/TurbSim

[21] P. Rodriguez, I. Candela, and A. Luna, "Control of PV Generation Systems Using the Synchronous Power Controller," *in Proc. IEEE Energy Conversion Congress and Exposition (ECCE)*, Sep. 2013, pp. 993-998.

DC Voltage Ripple Quantification for a Flywheel-Battery Based Hybrid Energy Storage System

Christopher R. Lashway, Ahmed T. Elsayed, *Student Members, IEEE* and Osama A. Mohammed, *Fellow, IEEE*

Abstract— **Flywheel energy storage has started attracting more attention as an energy storage means, but certain impediments face their deployment such as a high self-discharging rate and power quality issues. A potential solution is to combine flywheels with another energy storage types to form a Hybrid Energy Storage System (HESS). In this paper, a new method is established to perform power quality analysis and DC voltage ripple quantification in an HESS connected solely to a DC bus. Previous efforts have analyzed voltage and current ripple using an AC frequency reference, but these techniques are ineffective when the system does not contain an AC connection. Extensive laboratory testing and verification is conducted to characterize a flywheel-battery based HESS with different battery contribution levels. A correlation is made between the required battery support and resulting DC voltage ripple. Due to the nature of a flywheel operating at various speeds, a new Machine Speed Multiple (MSM) frequency reference is used as a profiling tool corresponding to the harmonic number in AC systems. Using the MSM in conjunction with the Discrete Fourier Transform, a voltage ripple frequency table is produced to highlight the target frequencies which must be reduced. A quantitative analysis identifies an overall reduction of voltage ripple magnitudes as a result of current injection from the battery.**

Keywords— **Flywheel; FESS; HESS; Power Quality.**

I. INTRODUCTION

FLYWHEEL energy storage systems (FESS) have attracted new research attention in several applications, one of which is in future Naval platforms where loads require significantly high power under short time spans [1]. Although the concept of the flywheel can be dated back to the early 20th century, their assembly and particularly the applications to which they are used have changed dramatically. The FESS operates by mechanically storing kinetic energy in a rotating mass. Flywheels can be designed for low or high speed operations though the low-speed application has advantages in a lower cost due to the integration of proven technologies [2].

Although low-speed flywheels have seen a wide range of usage, high speed operations have gained recent attention due to developments in relevant technology. Magnetic levitation, the introduction of composite materials, low-loss machinery, and power electronic switches have driven this progression. Through the replacement with new types of magnetic or superconducting materials, the theoretical energy storage

This work was partially supported by grants from the Office of Naval Research and the Department of Energy. The authors are with the Energy Systems Research Laboratory, ECE Department, Florida International University, Miami, FL 33174 USA (e-mail: mohammed@fiu.edu).

capacities can be increased as well [3],[4]. In [5], a new model for a flywheel is proposed where a superconducting magnetic bearing together with a permanent magnet is introduced. The system was able to increase the rotational speed of the flywheel and suppress vibrational aspects of the rotor while also reducing the cost of cooling the motor.

While the materials to construct a flywheel are being improved, so have their power electronic and control integration algorithms in an aim to provide solutions to the intermittency associated with most renewable energy sources. In [6], a FESS is connected to a wind farm through a solid state transformer to store excess wind power when generation levels are high and provide a restoration measure during the time when the wind is at a deficit. Similar solutions have been proposed for usage in solar power applications [7].

Initial studies identified the flywheel to be a tool in improving land-grid power quality. This has since been extended to include its application on shipboard power systems as well. [8] focuses on voltage and power stability improvements by identifying a reduction in peak-to-peak transients, however the introduction of harmonics, or ripple frequencies as a result of operation was not considered. Previously in [6], a Distribution Static Synchronous Compensator (DSSC) is coupled with a flywheel to provide active and reactive power assistance as well as power factor correction and voltage control. Although the harmonics were addressed as a consequence of the control, the final system did not provide a full harmonic analysis.

A majority of the networks identified are hybrid power systems containing both an AC and a DC connection. In addressing harmonics, the same fundamental frequency from the AC connection can be applied to that of the analysis when referring to DC harmonic components. However, this technique becomes ineffective when the system does not contain an AC connection. The emerging electric vehicle, in most cases, presents a purely DC bus [9],[10]. The interconnection of FESS and inducing of ripple voltages and currents present a much different problem in these scenarios. In [10], an active power filter (APF) is designed to combat some of these issues, however the frequency spectra is only viewed in terms of overall Total Harmonic Distortion (THD) reduction, as opposed to identifying specific common frequency multiples which should be targeted when applying filtering methods. The onboard battery typically identifies a simple solution to this issue without the need for an APF. However, the required energy from the battery versus that of the flywheel must be quantified in order to adequately solve this problem while preventing an increased system cost.

In this work, a new metric is identified to specifically evaluate ripple voltage frequencies associated with a DC-only system. As this metric is identified, the power quality issues associated with a hybrid energy storage system (HESS) composed of a lead acid battery bank and FESS will be evaluated. This paper is organized as follows: a discussion over the mechanics involved in FESS is discussed in Section II. Section III provides an investigation into power quality issues associated with flywheel operation. Section IV provides extensive experimental testing to quantify the reduction of ripple frequencies in the system and Section V concludes this study.

II. FLYWHEEL ENERGY STORAGE SYSTEMS

The common FESS consists of a high inertia rotating mass and an electric machine which can operate simultaneously as a motor during charge and generator during discharge. One of the disadvantages of the flywheel is its high self-discharging rate. Thus, the rotating parts are enclosed in a vacuum system to reduce friction losses. In the flywheel electromechanical system, the rotor is accelerated to high speed in order to store kinetic energy, a process which resembles charging. The amount of the stored energy E is then based on the common physics equation:

$$E = \frac{1}{2} J \omega_m^2 \qquad (1)$$

where the moment of inertia J and square of the speed ω_m^2 designate the energy density. As the speed of the flywheel is adjusted, the shaft power is adjusted proportionally. The shaft power P is then:

$$P = -J \omega_m \frac{d\omega_m}{dt} \qquad (2)$$

It can be concluded from (1) and (2) that the energy storage capacity is a function of the operation of the flywheel over a wide range of speeds. One can also recognize that a steady-state operating point is not typical under the normal operation of the flywheel. In this study, the ripple frequency levels in a hybrid flywheel-battery combination is investigated. The ripple levels are determined at different levels of contribution from each energy storage entity.

III. INVESTIGATING POWER QUALITY ISSUES

Power quality metrics for the flywheel have begun to surface designating voltage ripple limitations to be no more than 2% of the DC value [11]. Three significant factors are involved in the production of induced ripple frequencies on the DC machine: non-homogenous flux distribution across the air gap, evenly-spaced commutator segments, and uniform slots on the stator [12]. Non-homogenous fields in the air gap are of typically low order. Conversely, slot ripple frequencies are typically of high order and obtained from the number of S slots and P poles.

A. Machine Speed Multiple

In most HESS, the fundamental frequency is designated to conduct harmonic analysis, but in a DC-only system these components can appear ambiguous. Furthermore, the flywheel

TABLE I
HAMPDEN DYN-300X DC MOTOR SPECIFICATIONS

	Nominal Voltage	125 V
	Nominal Power	3 hp
	Slots	36
	Commutators	72
	Poles	4
	Armature Current	19 A
	Rated Speed	1400 RPM

must operate at different speeds to share a load with the battery bank. To establish a metric to place meaning on specific ripple frequencies, a machine speed multiple (MSM) value has been defined where the fundamental is equal to the mechanical rotation speed of the flywheel f_m. Using the MSM, a correlation is made between the spectral responses of the machine at each speed multiple. The MSM is calculated by these three factors based on poles P, slots S, and commutator segments C. The *MSM* general formula is:

$$MSM = \begin{cases} 2P(1+k) \\ C(1+k) \\ \left(\frac{2kS}{P} \pm 1 \right) \end{cases} \text{ for } k = 0,1,2 \dots \qquad (3)$$

where k an integer. Target flywheel ripple frequencies f_{fw} can then be determined by:

$$f_{fw} = MSM f_m \qquad (4)$$

In this experiment, the DC machine taking the place of the flywheel is a Hampden DYN-300X. Its specifications are outlined above in Table I.

B. Mathematical Modeling

The flywheel HESS is tested under four different levels of current contribution. Traditionally, the use of the flywheel under normal operating conditions would be expected to initially expend a great deal of energy following a relatively short discharge period. This is investigated in the following experimental study, where a series-connected lead acid battery bank consisting of 10-12 V lead acid batteries at 110 Ah of capacity are connected in parallel to the flywheel [13]. The system under study is as shown in Fig. 1 where the battery and flywheel are combined together to form a HESS. A buck-boost DC-DC converter is placed between the flywheel and the DC bus to increase the output voltage from the flywheel during discharge while reducing the DC bus voltage to the flywheel during charging.

The value of the DC-DC converter inductor is chosen based on the desired ripple current. It is usually recommended to operate the system at a ripple current equal to less than 20% of the average inductor current. A higher input V_{in} or output V_{out} voltage also increases the ripple current as depicted in Eq.(5) where V_{in} and V_{out} are the input and output voltages of the flywheel driving converter, respectively, f_s is the switching frequency, and L is coil inductance.

$$\Delta I_L = \frac{1}{f_s L} V_{out} \left(1 - \frac{V_{out}}{V_{in}} \right) \qquad (5)$$

Smaller inductance values will result in a higher output current slew rate and improve the load transient response of

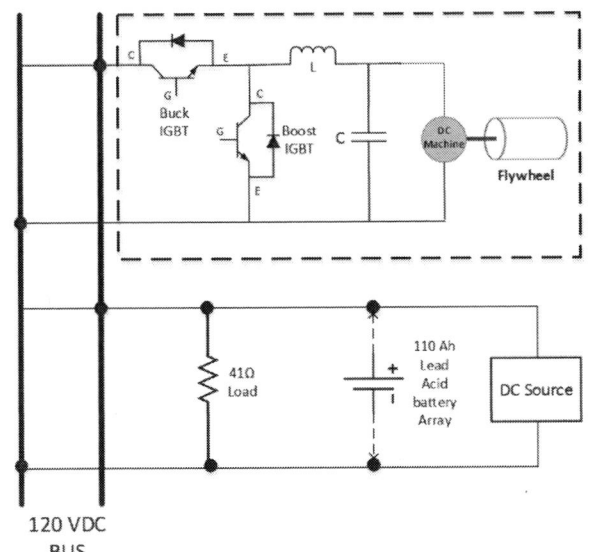

Fig. 1. Hybrid Energy Storage System (HESS)

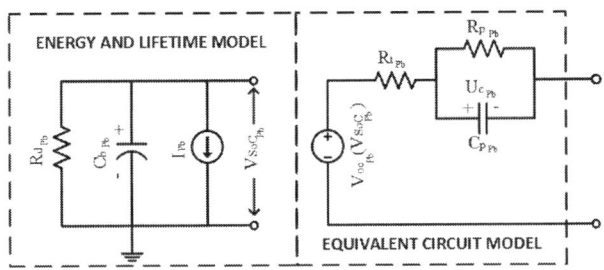

Fig. 2. Electrical Lead Acid Battery Model

the converter, but will result higher output voltage ripple. Conversely, larger inductance values reduce the ripple current as well as core magnetic hysteresis losses, but will increase the size and weight of the converter. Moreover, it significantly limits the amount of transferred power.

As an alternative solution, capacitors are standard for storage and smoothening of the DC output in power converters. A comprehensive electrical lead acid battery model is shown in Fig. 2 [14]. This model partitions the battery into two parts: an energy and lifetime model as well as an equivalent circuit model which models the voltage-current characteristics. The energy and lifetime model reveals that the energy stored inside a lead acid battery (denoted by the "Pb" subscript) can be represented as a large capacitance $C_{b_{Pb}}$ in parallel with a large self-discharge resistor $R_{d_{Pb}}$. The self-discharge resistor has been neglected in this work since all tests are of short duration. The virtual capacitance $C_{b_{Pb}}$ can therefore be estimated using the battery capacity $E_{b_{Pb}}$ (in Ah):

$$C_{b_{Pb}} = 3600 V_{Pb_{FC}} E_{b_{Pb}} \qquad (6)$$

where $V_{Pb_{FC}}$ represents the full charge voltage of the battery module. This lead acid battery array has a 110 Ah capacity with an expected full charge voltage of 128 V yielding a extremely large capacitance.

The equivalent circuit model portion, represented by the Randles circuit, describes the behavior of the lead acid battery

under load where $R_{t_{Pb}}$ represents the electrolytic resistance and $R_{P_{Pb}}$ and $C_{P_{Pb}}$ represent the polarization resistance and capacitance governing the impulse response which can be determined through electrochemical impedance spectroscopy. Although $C_{b_{Pb}} \gg C_{P_{Pb}}$, both play a role in smoothening the ripple current and voltage. This capacitance will behave as an additional filter in parallel with the flywheel converter.

C. Experimental Test Setup

To conduct the experimental testing, a National Instruments NI-9206 DAQ with 32 analog input channels in conjunction with the LabVIEW Development Platform was used [15]. As shown in Fig. 1, both the flywheel and lead acid battery are supplying a resistive load drawing approximately 3 A at 120 VDC. The ripple levels are determined at 5%, 25% and 50% contribution amounts from the battery. The remaining load current is supplied from the flywheel. During this test, the inertia coupled to the machine was large enough to supply the load over the entire test period (which is short). i.e. the flywheel does not require charging during the test.

Voltage and current were measured through the use of 2 LEM Hall Effect sensors [16],[17]. In order to ensure high precision, a noise bias test was conducted first. To ensure the highest fidelity in frequency measurements, the NI-9206 frequency and sampling rate are maximized at 10 kHz. Under the Nyquist criterion, the configuration provides an accurate Discrete Fourier Transform (DFT) of the signal to 5 kHz [18]. Assuming $f(v)$ to be the continuous voltage signal under 5000 samples, the DFT or $F[n]$ is:

$$F[n] = \sum_{k=0}^{N-1} V[k] e^{-j\frac{2\pi n}{NT}} \text{ where } 0 \le n < N - 1 \qquad (7)$$

where $v[k]$ is the discrete voltage sequence, N is the sample window of 5000, and n is the sampling frequency.

IV. EXPERIMENTAL TESTING

The HESS is then tested under 4 different levels of contribution. Traditionally, the use of the flywheel in an electric vehicle application would be expected to initially expend a great deal of energy following a relatively short discharge period. This is comprehensively investigated in the following experimental study, where a series lead acid battery bank of 10 batteries at 120 VDC is connected in parallel. The hardware investigation of this study is one of its important contributions, since most of the machine models in commercially available software (such as MATLAB) neglect the effect of the internal construction of the machine. The DC output of a machine is represented as a clean, pure DC source which is impractical. The following demonstrates battery support at 5%, 25%, and 50% of the total energy to the load.

A. Flywheel System Only

An initial test is conducted where the DC motor is connected directly to the load. The bus voltage is regulated near the terminal voltage of the battery bank, or approximately 116 VDC. The machine speed in this case was 773 RPM (12.88 Hz). Shown in Fig. 3, both the voltage and current waveforms reveal a high level of noise, however a closer

TABLE II
POWER QUALITY ANALYSIS OF VOLTAGE AT THE LOAD

MSM	ID	Flywheel Only 773 RPM (12.88Hz)		95% Flywheel 5% Battery 772 RPM (12.86 Hz)			75% Flywheel 25% Battery 751 RPM (12.52 Hz)			50% Flywheel 50% Battery 729 RPM (12.15 Hz)		
		f	$\%V_{DC}$	f	$\%V_{DC}$	$\Delta\%V_{DC}$	f	$\%V_{DC}$	$\Delta\%V_{DC}$	f	$\%V_{DC}$	$\Delta\%V_{DC}$
8	A	103	87.9910	103	7.1228	-80.868	100	1.0808	-86.910	97	0.2579	-87.733
20	S	257	4.3640	258	0.2716	-4.092	251	0.0820	-4.282	243	0.0075	-4.356
38	S	489	6.1330	490	0.5282	-5.604	477	0.0421	-6.090	462	0.0122	-6.120
52	S	669	5.6630	671	0.2309	-5.432	652	0.0579	-5.605	632	0.0054	-5.657
64	A	824	53.4310	825	1.0518	-52.379	803	0.4834	-52.947	778	0.1049	-53.326
72	C	927	240.5100	929	3.4926	-237.017	903	1.2199	-239.290	875	0.5039	-240.006
80	A	1030	142.2700	1032	2.1355	-140.134	1004	0.5905	-141.679	973	0.1183	-142.151
92	S	1184	17.3290	1186	0.0816	-17.247	1154	0.0375	-17.291	1119	0.0093	-17.319
106	S	1364	14.1000	1367	0.3774	-13.722	1330	0.0353	-14.064	1289	0.0080	-14.092
126	S	1622	2.1500	1625	0.0435	-2.106	1581	0.0033	-2.146	1532	0.0052	-2.144
144	C	1854	29.9800	1857	0.1632	-29.816	1807	0.0249	-29.955	1751	0.0046	-29.975
162	S	2085	6.4480	2089	0.0049	-6.443	2032	0.0038	-6.444	1970	0.0032	-6.444
178	S	2291	7.3700	2296	0.0691	-7.300	2233	0.0039	-7.366	2164	0.0039	-7.366
196	S	2523	3.6320	2528	0.0197	-3.612	2459	0.0038	-3.628	2383	0.0014	-3.630
216	C	2780	9.2500	2786	0.1811	-9.068	2710	0.0579	-9.192	2626	0.0113	-9.238
268	S	3450	2.5890	3456	0.0067	-2.582	3363	0.0021	-2.586	3259	0.0006	-2.588
288	C	3707	8.4010	3714	0.0329	-8.368	3613	0.0030	-8.398	3502	0.0026	-8.398

CODES FOR DESCRIBING RIPPLE FREQUENCY CAUSAL ID: A (Non-homogeneous Flux across air gap), S (Slot), C (Commutator)

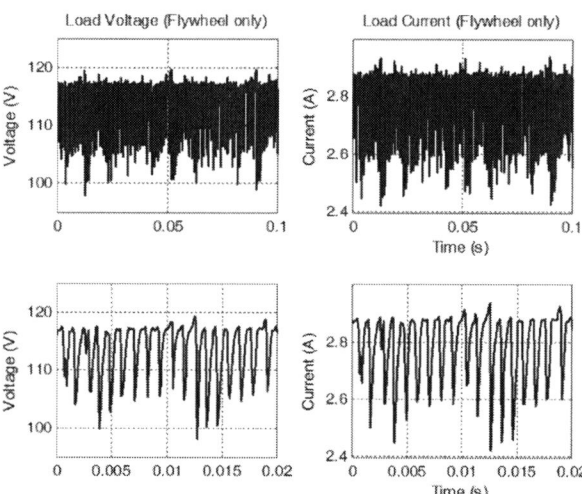

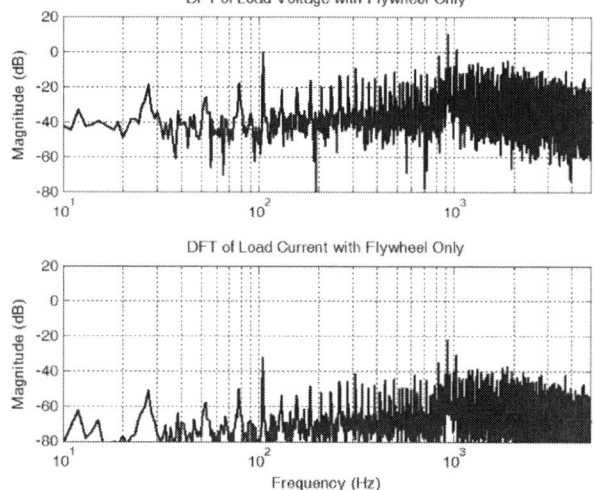

Fig. 3. Voltage & Current of the Flywheel

Fig. 4. DFT of the Flywheel Voltage and Current

inspection reveals detectable periodic contents. A close-up of the voltage and current below in Fig. 3 reveal a quasiperiodic square wave. This major component has a direct relationship to the number of commutator segments in the machine. Fig. 4 depicts the DFT of both the voltage and current waveforms shown under a dB-scale. A close correlation is identified between the two waveforms, which is to be expected under a linear load. The major difference is a shift in their biases. The current ripple frequencies are -30 dB lower than that of the voltage which places them below the 2%, or -17 dB noise threshold. For this reason, as the HESS system is connected, only the voltage will be analyzed.

A frequency spectra of the voltage in linear scale is shown in Fig. 4 where 17 ripple frequencies exceed the 2% threshold. To profile ripple frequencies, the MSM is used and is shown in linear scale below in Fig. 5. Shown above in Table II, these frequencies are identified with respect to its classification "ID". The ID column classifies the MSM into one of 3 categories discussed in Section III which are a result of DC machine power quality factors. IDs are correlated to their

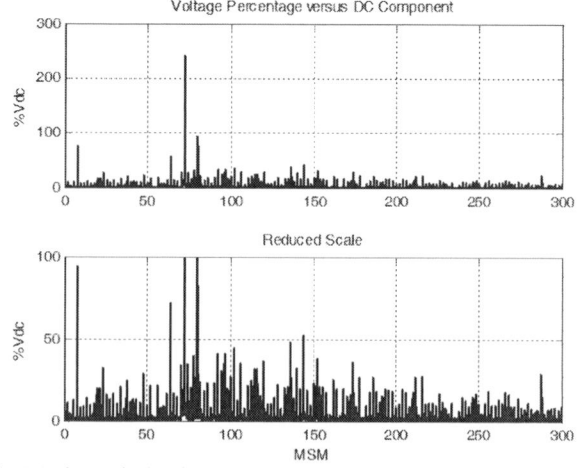

Fig. 5. Voltage Ripple Distortion versus DC (%)

MSM and related to: a slot ripple frequency (S), non-homogenous flux across the air gap (A), or a result of the commutator (C).

The highest ripple frequency is present at the commutator multiple, or 72, producing 240% of the DC component. This is shown on the top plot in Fig. 5 where the remaining ripple frequency magnitudes are significantly lower than that of the commutator. This scale is reduced below in Fig. 5 to highlight the remaining components. In Table II, each MSM investigated is shown in terms of each flywheel-battery combination. As the flywheel speed varies, the frequency shifts in a linear fashion across the chart confirming the geometric correlation associated with each of the causal IDs.

B. 95% Flywheel / 5% Battery Contribution

The first test quantifies ripple frequency reductions under a minimum battery current. In this case, the flywheel is sourcing 95% (2.65 A) while the battery bank contributes 5% (139 mA). The machine speed is held close to that of the flywheel only case at 772 RPM (12.86 Hz). This already reveals a huge impact in reducing the magnitude of the ripple frequencies under a similar speed. Fig. 6 depicts the original spectra to 5 kHz in black and the new HESS spectra in red while Fig. 7 depicts the ripple voltage frequencies on the linear MSM scale where the 2% threshold is identified. One can observe a 10 dB decrease in the overall ripple frequency noise bias. In Table II, a column depicts the associated MSM frequencies as well as the new voltage percentages at each multiple with respect to the fundamental. A drastic reduction is observed in the commutator MSM where its percentage is reduced from 87% to 7%. Fig. 8 highlights the ripple voltage percentage versus the DC component as compared to the 2% threshold. A close-up depicts all frequencies which have fallen below this reference.

C. 25% Battery / 75% Flywheel Contribution

In this case, the flywheel speed is reduced to allow for the battery bank current to begin injecting 25% of the load current. The flywheel current is reduced to 2.10 A (75%) while the battery bank increases its loading to 660 mA (25%). This results in a reduction of the machine speed to around 751 RPM (12.52 Hz). Figs. 9 and 10 once again depict the original spectra to 5 kHz in black and the new HESS spectra under 25% battery contribution in red. Fig. 9 features the magnitude reduction of each ripple voltage frequency in log scale where Fig. 10 displays the ripple voltage percentage versus the DC component over the new linear MSM scale. The 2% threshold marker is once again shown in blue.

Under a relatively small level of current, the overall spectral comparison reveals a 20 dB magnitude decrease in the overall ripple bias. Table II confirms that all ripple voltage percentages have been further reduced to meet the 2% threshold as depicted in the figure. Fig. 11 highlights the ripple voltage percentage versus the DC component as compared to the 2% threshold. A close-up depicts all frequencies have been reduced by >0.5% from the threshold.

D. 50% Flywheel / 50% Battery Contribution

The final case provides a shared HESS energy supply case where both the flywheel and battery are supporting 1.35 A

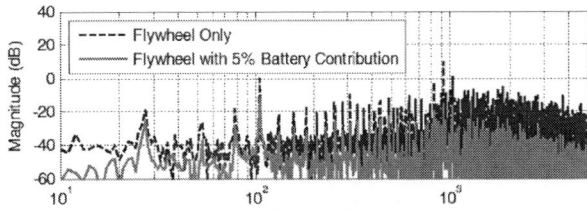
Fig. 6. DFT of Load Voltage with 5% Battery Contribution

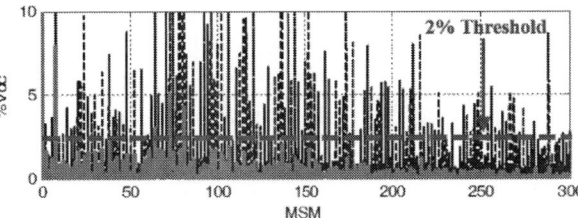
Fig. 7. Voltage (%) versus DC Component under 5% Battery Contribution

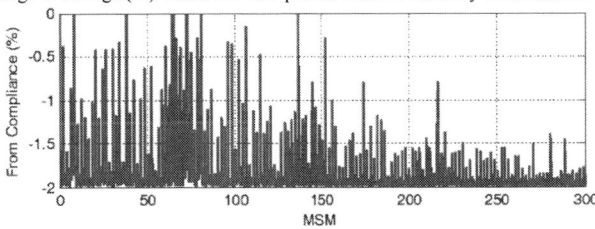
Fig. 8. Departure from 2% Compliance under 5% Battery Contribution

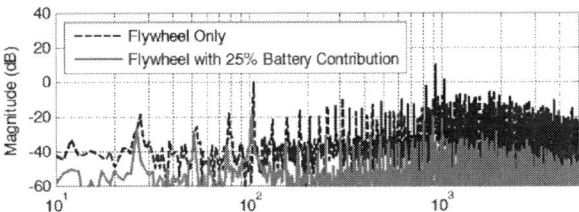
Fig. 9. DFT of Load Voltage with 25% Battery Contribution

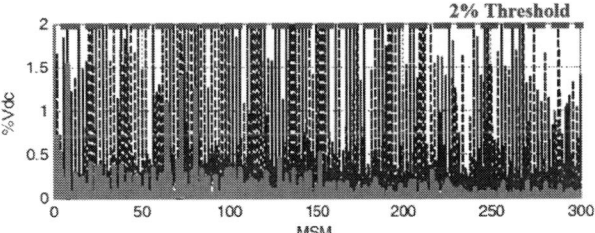
Fig. 10. Voltage (%) versus DC Component under 25% Battery Contribution

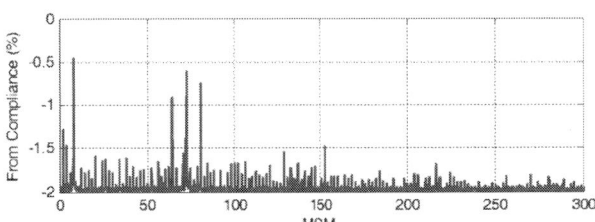
Fig. 11. Departure from 2% Compliance under 25% Battery Contribution

(50%), respectively. The machine speed is dropped to 729 RPM (12.15 Hz) to maintain the energy output required. Figs. 12 and 13 depict the original spectra to 5 kHz in black and the new HESS spectra under 50% battery contribution in red. The overall spectral bias is shown in Fig. 12 revealing a 30 dB reduction from the case with the flywheel only. Fig. 13

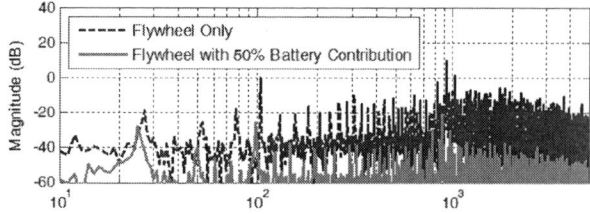

Fig. 12. DFT of Load Voltage with 50% Battery Contribution

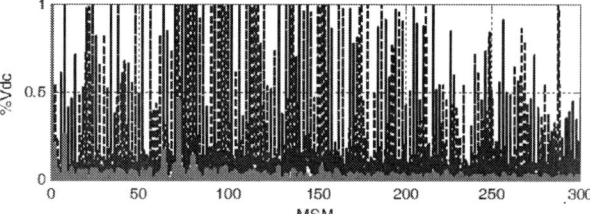

Fig. 13. Voltage (%) versus DC Component under 50% Battery Contribution

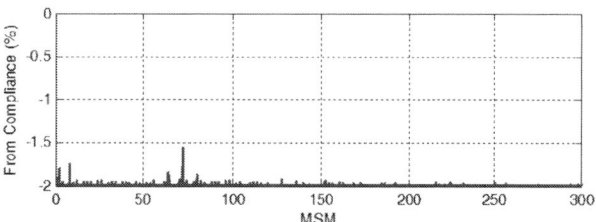

Fig. 14. Departure from 2% Compliance under 50% Battery Contribution

displays the ripple voltage percentage versus the DC component over a linear MSM scale. Since no frequencies are near 2%, a threshold marker is not shown in this figure.

Although a 30 dB reduction is observed from the base flywheel case, when viewing this in the linear MSM scale in Fig. 13, one can observe that this reduction does not provide a notable advantage except for highly sensitive applications. From Table II, one can see that all ripple frequencies have once again been reduced below 1%, but only the 8th and 72nd MSM magnitudes have a noticeable decrease. Shown in Fig. 13, only 4 spectral frequencies (other than the DC component) are easily identified with the remaining components below 0.1%. Fig. 14 highlights the ripple voltage percentage versus the DC component as compared to the 2% threshold where the peak ripple voltage magnitude falls 1.5% below the threshold. This case proves that extracting more than 25% energy from the battery bank is unnecessary to significantly improve the power quality of the HESS.

V. CONCLUSION

This paper presents an intensive investigation into the power quality issues associated with the flywheel energy storage system. Voltage and current ripple frequencies induced while connected to a 3 A load is investigated. Common geometric and electromagnetic causes generating these frequencies are discussed. Multiples of the flywheel rotation speed are interpreted to explain the presence and reduction of voltage ripple under 3 HESS cases involving a battery bank connected in parallel with the flywheel. This study identifies the ripple frequency reduction at different machine speed multiples (MSM) and highlights that with only a small contribution from a battery bank, a significant improvement can be made in the power quality delivered to

the load voltage. Using the common standard threshold of a <2% voltage ripple frequency magnitude referred to the DC component, a 5% battery contribution provides enough support where only 3 frequencies remain above the threshold. As the battery contribution is increased to 25%, all voltage ripple frequencies reduce to below 2%. A final case designates half the energy sourced from the battery but demonstrates that this significant energy demand is not necessary to maintain good power quality. Using the MSM as a frequency profiler, an advanced controller can correlate the target voltage ripple frequencies to the required design specification. In light of this study, a best combination and control scheme can be determined to reduce the overall voltage ripple frequencies for the HESS.

REFERENCES

[1] J. McGroarty, J. Schmeller, R. Hockney and M. Polimeno, "Flywheel energy storage systems for electric start and an all-electric ship," *IEEE Electric Ship Technologies Symposium*, PP. 400-406, 2005.

[2] S. Samineni, B. K. Johnson, H. L. Hess, and J. D. Law, "Modeling and analysis of a flywheel energy storage system for Voltage sag correction," IEEE Trans. Ind. Appl., vol. 42, no. 1, pp. 42–52, Jan. 2006.

[3] A. V. Filatov and E. H. Maslen, "Passive magnetic bearing for flywheel energy storage systems," *IEEE Trans. on Magn.*, vol. 37, no. 6, pp. 3913–3924, Nov. 2001.

[4] F. N. Werfel, U. Floegel-Delor, T. Riedel, R. Rothfeld, D. Wippich, B. Goebel, G. Reiner, and N. Wehlau, "A Compact HTS 5 kWh/250 kW Flywheel Energy Storage System," IEEE Trans. Appl. Supercond., vol. 17, no. 2, pp. 2138–2141, Jun. 2007.

[5] M. Subkhan and M. Komori, "New Concept for Flywheel Energy Storage System Using SMB and PMB," IEEE Trans. Appl. Supercond., vol. 21, no. 3, pp. 1485–1488, Jun. 2011.

[6] G. O. Suvire and P. E. Mercado, "Improvement of power quality in wind energy applications using a DSTATCOM coupled with a Flywheel Energy Storage System," in Power Electronics Conference, 2009. COBEP '09. Brazilian, 2009, pp. 58–64.

[7] S. Ye and B. Sun, "Application of Flywheel Battery in Solar Power System," in International Conference on Energy and Environment Technology, 2009. ICEET '09, 2009, vol. 1, pp. 533–536.

[8] C. Xie and C. Zhang, "Research on the Ship Electric Propulsion System Network Power Quality with Flywheel Energy Storage," in Power and Energy Engineering Conference, 2010 Asia-Pacific, 2010, pp. 1–3.

[9] M. Jiang, J. Salmon, and A. M. Knight, "Design of a permanent magnet synchronous machine for a flywheel energy storage system within a hybrid electric vehicle," in Electric Machines and Drives Conference, 2009. IEMDC '09. IEEE International, 2009, pp. 1736–1742.

[10] L. Zhang, L. Li, W. Cui, and S. Li, "Study on improvement of micro-grid's power quality based on APF and FESS," in 2012 IEEE Innovative Smart Grid Technologies - Asia (ISGT Asia), 2012, pp. 1–6.

[11] Emerson, "Leibert FS Flywheel Energy Storage System Guide Specifications," SL-29310_REV05_03-10, Columbus, OH.

[12] R. Haiduck, Y. Baghzouz, and R. F. Boehm, "Slot-harmonic effects on DC voltage of generator-rectifier set," in 9th International Conference on Harmonics and Quality of Power, 2000., vol. 2, pp. 565–569 vol.2.

[13] UPG, "UB121100 12V 110 Ah SLA Battery Spec Sheet," No. 4582, Carrollton, TX, Available online at: http://upgi.com.

[14] M. Chen and G. A. Rincon-Mora, "Accurate electrical battery model capable of predicting runtime and I-V performance," *IEEE Transactions on Energy Conversion*, vol. 21, no. 2, pp. 504–511, Jun. 2006.

[15] *NI 9206: 16-Channel Analog Input Module for Fuel Cells*, National Instruments, Austin, TX, Feb. 2008.

[16] *Current Transducer LA 25-NP*, LEM USA, Milwaukee, WI, Nov. 2011.

[17] *Voltage Transducer LV 25-P*, LEM USA, Milwaukee, WI, Nov. 2011.

[18] S. Sumathi, "Inputs and Outputs," in *LabVIEW based Advanced Instrumentation Systems*, London, England: Springer Berlin Heidelberg, 2007, ch. 4, sec. 4.3.3, pp. 180.

Adaptive Loss Reduction Charging Strategy Considering Variation of Internal Impedance of Lithium-Ion Polymer Batteries in Electric Vehicle Charging Systems

Nari Kim, Jung-Hoon Ahn, Dong-Hee Kim, and Byoung-Kuk Lee [†]

Department of Electrical and Computer Engineering
Sungkyunkwan University
Suwon, South Korea
E-mail: bkleeskku@skku.edu

Abstract—This paper presents a charging strategy to minimize losses in a lithium-ion polymer battery for an electric vehicle. The proposed charging method which considers the variation of battery internal impedance according to state of charge and C-rate significantly reduces charging losses. Experiment results for a cell prove that the proposed charging method reduces loss by 41.9 % compared to a conventional constant-current charging. Moreover, a 3.3 kW prototype on-board charger is built to investigate the entire reduction of losses in an electric vehicle charging system which includes a 12 kWh battery pack considering capacity deviation. Additionally, the feasibility of the proposed charging strategy is examined according to the charger capacity, and it is shown that the effect is enhanced with increasing charger capacity. It is also revealed that the loss reduction due to the proposed strategy is equivalent to an improved efficiency of 3.35% on a 52.8 kW charger with an average efficiency of 94.49%.

Keywords—charging strategy; loss reduction; lithium-ion batteries; electric vehicle charging systems; on-board chargers

I. INTRODUCTION

As increasing attention has been put on eco-friendly cars in recent times, a significant number of studies to reduce losses in the charging system of electric vehicles (EVs) have been conducted. The charging system of an EV is comprised mainly of a charger and a battery pack, and the efficiency of the system is determined largely by the ratio of the stored energy in the battery to the input energy of the charger. However, while most studies focus on improving the efficiency of a charger itself [1-10], there has been little research on the analysis of total losses that occur in an entire charging system which includes not only a charger but also its corresponding batteries.

Lithium-ion batteries have internal impedance which varies with the state of charge (SOC) and C-rate [11, 12]. However,

constant-current (CC) charging, a conventional method of charging a battery, releases a constant charging current, therefore it does not consider the variation of internal resistance. On the contrary, charging with a variable current corresponding to the variation of internal impedance can minimize losses in a battery. Recently, based on this idea, a study has been attempting to increase charging efficiency [13]. However, this previous work does not consider the variation of battery internal resistance according to C-rate and capacity deviation between cells in a pack. Moreover, loss analysis of an entire charging system with a practical charger topology is not discussed.

In this paper, an effective charging strategy considering the variation of internal impedance in a battery according to its SOC and C-rate is proposed in order to increase the efficiency of a charging system of an EV. With the help of the proposed method, the total loss analysis of a charging system which includes an actual 3.3 kW on-board charger and a battery pack considering capacity deviation is also examined. Furthermore, informative analyses on loss variation in a charging system according to charger capacities from 3.3 kW to 52.8 kW are presented.

II. ADAPTIVE CHARGING STRATEGY FOR LITHIUM-ION POLYMER BATTERIES

A. Adapitve Charging Strategy

Internal impedance of a lithium-ion battery varies with SOC. Charging with a small current for high resistance and with a large current for low resistance can reduce ohmic loss in a battery which is expressed as

$$ohmic\ loss = I^2 R\,t \qquad (1)$$

978-1-4673-9551-9/16 $31.00 © 2016 IEEE

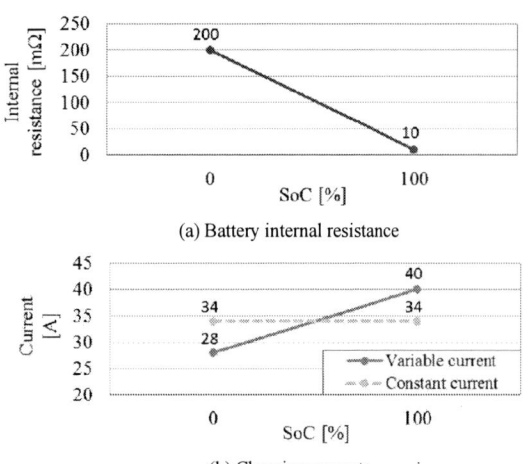

(a) Battery internal resistance

(b) Charging currents

Fig. 1. Example of charging with a variable current

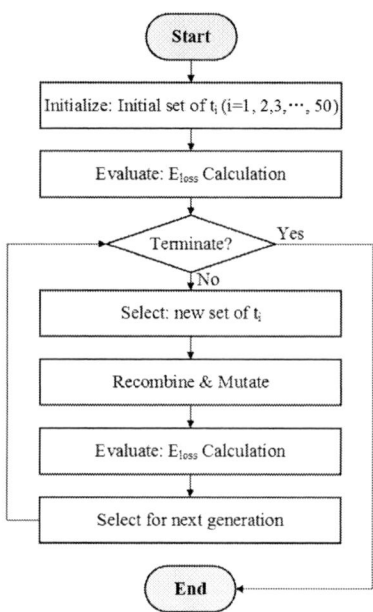

Fig. 2. The flow chart of the EA.

where *I*, *R* and *t* indicate charging current, internal resistance, and charging time, respectively. Fig. 1 presents a brief example of the idea that a variable charging current according to the amplitude of internal resistance can decrease charging loss. Assuming that the battery capacity is 34 Ah, it is shown that charging with a variable current results in loss reduction of 5.85 % compared to CC charging.

For the actual implementation of the idea, the current profile should be divided into multiple interval currents. As the number of interval currents rises, the amount of computational work as well as the resolution of the charging profile increases. Therefore, in order to avoid a burden on computations while maintaining a certain level of the resolution, 50 interval currents compose a charging profile with 2% steps of SOC from SOC 0% to 100%. The proposed adaptive charging profile consists of a set of 50 interval currents which minimize the sum of interval losses due to the internal resistance which is expressed as

$$E_{loss} = \Sigma\, I_i^2 R_i t_i \quad (i = 1, 2, 3, \dots, 50) \qquad (2)$$

where I_i, R_i and t_i represent intervals of charging current, internal resistance, and charging time, respectively. Each interval charging current is determined by its interval charging time as given by

$$I_i = capacity \cdot \Delta SOC / t_i \qquad (3)$$

where ΔSOC for each interval is constant during charging. Since a charging profile contains a large number of currents which should fulfill the multiple requirements such as total charging time and minimum total loss, a stochastic computing algorithm, rather than a deterministic algorithm, is appropriate to calculate a charging profile. Thus, an evolutionary algorithm (EA), which is a stochastic algorithm suitable for global optimization of a complex problem [14], is then implemented to compute optimal currents. Fig. 2 demonstrates the flow chart of the EA implemented in this paper. As charging loss is

repeatedly calculated whenever a new set is created through recombination and mutation processes, a charging profile with minimum loss is achieved.

B. Experiment Results

1) Experiment setup: The specification of the cell tested in this paper is presented in Table I. The test battery shown in Fig. 3 is a KOKAM lithium-ion polymer battery cell with a rated capacity of 34 Ah and the range of the available voltage is between 3.0 V and 4.2 V. The test bench is set up as shown

Fig. 3. The test battery cell.

TABLE I.   SPECIFICATION OF THE BATTERY CELL

Manufacturer	Type	Capacity	Nominal voltage
KOKAM	Lithium-ion polymer	34 Ah	3.7 V

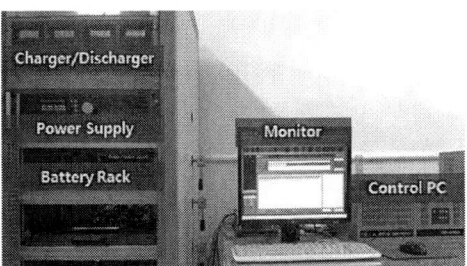

Fig. 4. The test bench.

in Fig. 4. It includes a 4-channel charger/discharger and a control computer. Experiments are conducted at room temperature (22 °C – 24 °C).

2) Modeling of lithium-ion battery cell: An electrical circuit model is considered for a battery cell since it is suitable for characterizing battery current and voltage. In this paper, the second-order RC ladder model shown in Fig. 5 is selected for the equivalent circuit of a battery cell, which is an optimal model for both accuracy and computational complexity [15]. Based on the model, parameter identification is performed to extract internal impedance of the battery. The test prodecure is as follows: The battery cell is charged by $\Delta SOC = 10\%$ from SOC 0% to 100% with three different C-rates (1 C, 0.5 C and 0.25 C). After being charged by $\Delta SOC = 10\%$, it is left in relaxation for an hour which is assumed to be enough time to reach the equilibrium of the battery voltage. Parameters of the battery is then obtained from modeling the relaxation effect as shown in Fig. 6. Fig. 7 reveals the extracted internal internal impedance which varies according to SOC and C-rate.

3) Loss comparison: The proposed charging profiles are calculated using the internal impedance through the EA. Fig. 8

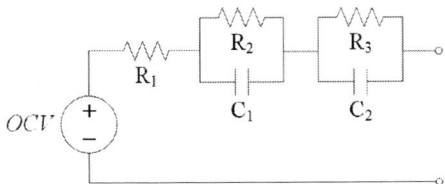

Fig. 5. The second-order RC model.

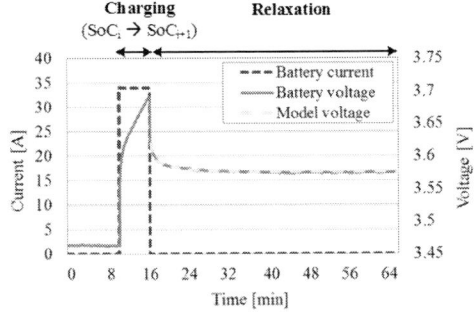

Fig. 6. Parameter test.

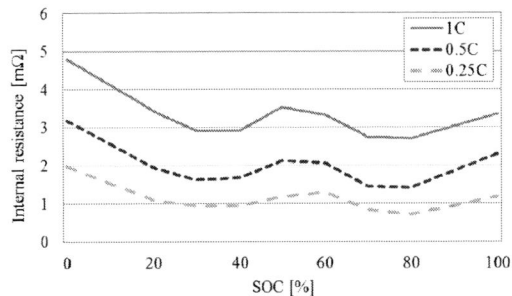

Fig. 7. Battery internal resistance according to SOC and C-rate.

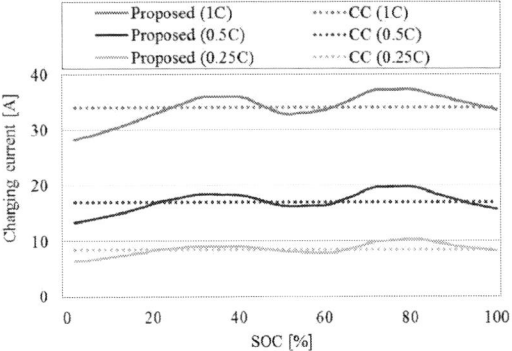

Fig. 8. Charging profiles according to SOC and C-rate.

illustrates the proposed charging profiles for different charging times as a result of the EA. Each proposed profile is compared with its corresponding CC profile which has the same average charging time. In order to validate the proposed charging strategy, capacity tests are conducted to compare losses between the proposed and CC charging. Charging losses are calculated by the difference between charging and discharging energy. Once the battery voltage reaches the maximum available value while being charged, the battery begins charging in constant-voltage (CV) charging for both the proposed and the CC charging. In this paper, only the energy spent prior to CV charging mode for both charging methods is compared to calculate loss since loss which occurs during CV mode is approximately identical in both the proposed and the CC charging. The results presented in Table II demonstrate that the proposed adaptive method decreases battery losses by 41.9%, 31.7% and 14.8% compared to CC charging at

TABLE II.   CELL LOSS ACCORDING TO THE CHARGING METHODS

Charging time	Average C-rate	Charging method	Losses [Wh]	Energy saving [%]
1 hour	1 C	Proposed	3.696	41.9
		CC	6.362	
2 hours	0.5 C	Proposed	3.746	31.7
		CC	5.488	
4 hours	0.25 C	Proposed	3.889	14.8
		CC	4.564	

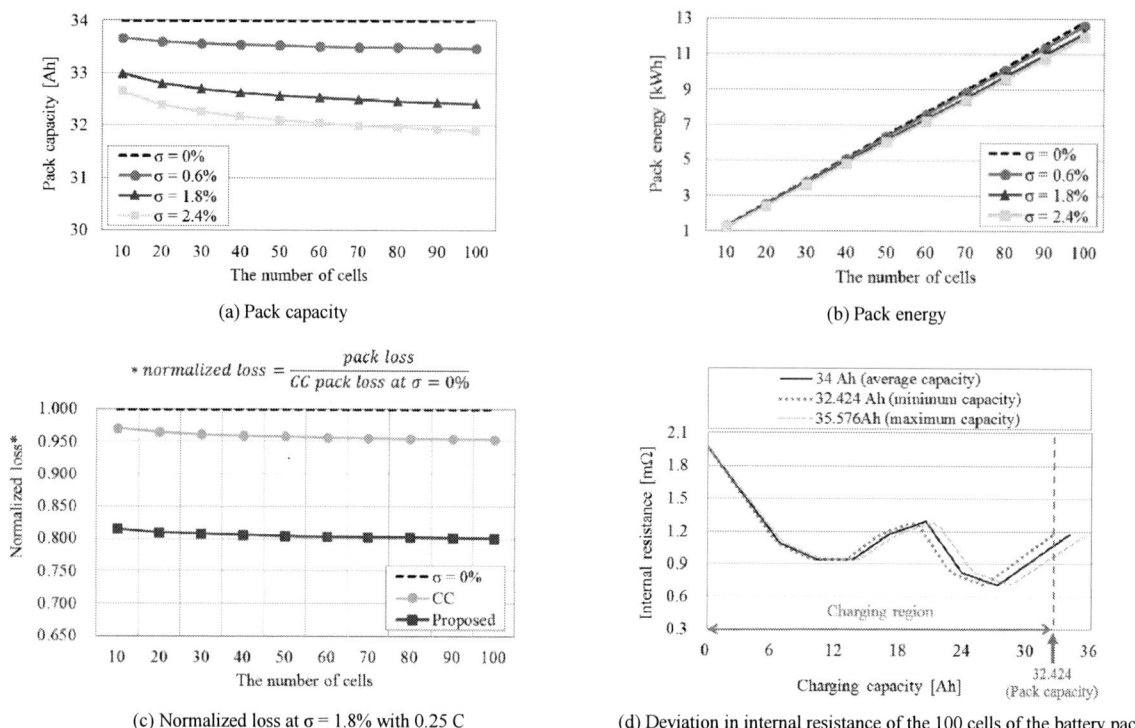

(a) Pack capacity

(b) Pack energy

(c) Normalized loss at σ = 1.8% with 0.25 C

(d) Deviation in internal resistance of the 100 cells of the battery pack

Fig. 9. Characteristics of the battery pack with capacity deviation according to the number of cells.

charging times of 1 hour, 2 hours and 4 hours, respectively. It is revealed that a shorter charging time with larger currents has a greater energy saving compared to a longer charging time with smaller currents.

III. ESTIMATION OF BATTERY PACK LOSSES CONSIDERING CAPACITY DEVIATION

In order to investigate losses in an EV charging system, cells of the same type can be connected in series to create a battery pack with an energy of 12 kWh, which is equal to the battery energy utilized in a Toyota Scion iQ EV. Here, it is assumed that only capacity deviation which occurs during manufacturing processes between cells is considered and the capacity of batteries follows a normal distribution $N (m, \sigma^2)$. The average of the batteries capacity, m, is 34 Ah and the standard deviation for the battery capacity, σ, is assumed to be 1.8%, a value of σ that is generally accepted for capacity deviation [16, 17].

Figs. 9 (a) and (b) illustrate that pack capacity and pack energy fall further from the results of σ = 0% as the number of cells or capacity deviation increases. It has been calculated from the figures that 100 cells should be connected to achieve the target energy of the pack. Fig. 9 (c) shows pack loss at σ = 1.8% normalized by the loss caused by the CC charging at σ = 0% under the assumption that every cell is charged from SOC 0%. As shown in Fig. 9 (d), the difference in the internal resistances between the 100 cells in the pack occurs due to the capacity deviation. The minimum and maximum capacities of the cells in the pack are 32.424 Ah and 35.576 Ah, respectively, and the capacity of the battery pack is determined to the minimum capacity. Thus, losses in the pack which consists of 100 cells according to charging time are calculated as shown in Table III.

TABLE III. PACK LOSSES ACCORDING TO THE CHARGING METHODS

Charging time	Average C-rate	Charging method	Normalized losses	Pack losses [Wh]	Reduced energy [Wh]	Energy saving [%]
1 hour	1 C	Proposed	0.5469	347.9	253.4	41.9
		CC	0.9452	601.4		
2 hours	0.5 C	Proposed	0.6425	352.6	167.7	32.2
		CC	0.9481	520.3		
4 hours	0.25 C	Proposed	0.8006	365.4	36.7	16.0
		CC	0.9532	435.0		

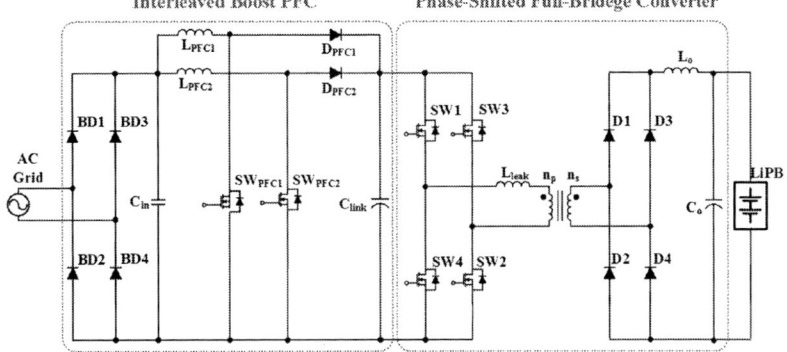

(a) Schematic of the on-board charger

(b) Hardware prototype of the on-board charger

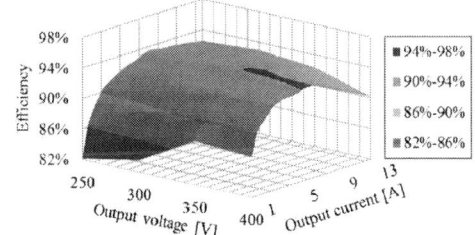

(c) Measured efficiency of the on-board charger

Fig. 10. The 3.3 kW phase-shifted full bridge-type on-board charger.

IV. TOTAL LOSSES IN AN ELECTRIC VEHICLE CHARGING SYSTEM

A 3.3 kW on-board charger optimized for the conventional CC charging is designed and tested to examine losses in an EV charging system. The charger consists of an interleaved boost power factor correction and a phase-shifted full-bridge converter as shown in Fig. 10 (a). The prototype of the charger

and its measured efficiency are displayed in Figs. 10 (b) and (c), respectively.

Losses in the charging system including the 3.3 kW charger and the battery pack are shown in Fig. 11. The proposed charging method reduces total losses, including charger and battery losses, by 60.8 Wh. Furthermore, in order to inspect losses according to charger capacity using the

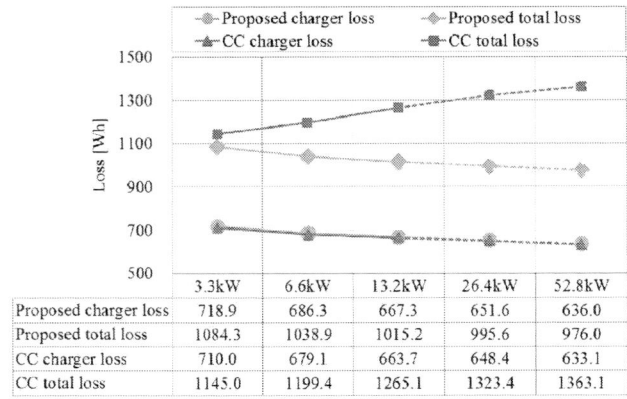

	3.3kW	6.6kW	13.2kW	26.4kW	52.8kW
Proposed charger loss	718.9	686.3	667.3	651.6	636.0
Proposed total loss	1084.3	1038.9	1015.2	995.6	976.0
CC charger loss	710.0	679.1	663.7	648.4	633.1
CC total loss	1145.0	1199.4	1265.1	1323.4	1363.1

(a) Charging loss comparison between the charging methods

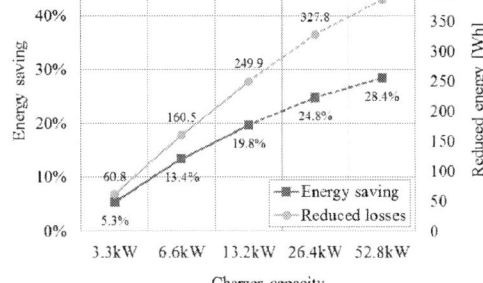

(b) Loss reduction and energy saving by the proposed method

Fig. 11. Loss comparison and energy saving in the electric vehicle charging system according to charger capacity.

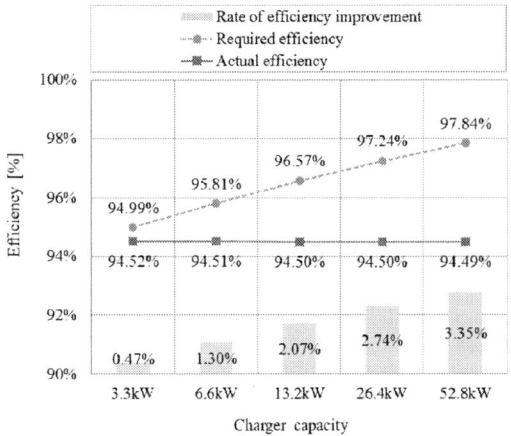

Fig. 12. Required efficiency of the on-board charger according to the charger capacity for the equivalent loss reduction by the proposed method.

experiment data, systems with a 6.6 kW and a 13.2 kW charger, formed by connecting multiple 3.3 kW chargers in parallel, are analyzed. Fig. 11 shows that charger losses decrease by a low amount as charger capacity increases. However, while battery losses grow with an increase in charger capacity for a CC charging system, causing total losses to rise, the battery losses fall with an increase in charger capacity for the proposed strategy, resulting in a widening of the gap between the total losses of the two charging methods.

According to this tendency, losses in a fast charger which has a capacity of 52.8 kW can be projected as shown in Fig. 11 with the dashed lines, in which the proposed method reduces total losses by 387.1 Wh. It is also shown that energy saving is improved from 5.3% using the 3.3 kW charger to 28.4% using the 52.8 kW charger. Fig. 12 shows the efficiency that the on-board charger should reach in order to obtain the same levels of loss reduction that the proposed strategy does. In the figure, the solid line indicates the actual efficiency of the on-board chargers and the dashed line represents the required efficiency that the charger should reach for the equivalent loss reduction the proposed charging method does. It is speculated that efficiency should be enhanced by 0.47% for a 3.3 kW charger with an average efficiency of 94.52% and 3.35% for a 52.8 kW charger with an average efficiency of 94.49%.

V. CONCLUSION

An adaptive charging strategy to minimize battery losses by considering the variation of the internal impedance in a lithium-ion polymer battery has been proposed. Experiment results have demonstrated that the proposed method decreases losses by 41.9%, 31.7% and 14.8%, for C-rates of 1 C, 0.5 C and 0.25 C, respectively. Moreover, the entire charging system of an electric vehicle with a 3.3 kW on-board charger and a 12 kWh battery pack has been investigated to analyze the loss reduction in the system using the proposed method. The 3.3 kW charger reduces total losses by 5.3% and this loss tends to decrease with increasing charger capacity, forecasting that a

52.8 kW charger will reduce total losses by 28.4%. Analysis shows that the proposed strategy is equivalent to an efficiency improvement by 0.47% of a 3.3 kW charger and 3.35% of a 52.8 kW charger. It is expected that a fast charger will have a greater benefit with the proposed charging strategy over slow chargers.

ACKNOWLEDGMENT

This work was supported by the Industrial Strategic Technology Development Program (No.10053711) funded by the Ministry of Trade, Industry & Energy (MI, Korea).

REFERENCES

[1] B. Hughes, J. Lazar, S. Hulsey, M. Musni, D. Zehnder, A. Garrido, R. Khanna, R. chu, S. Khalil, and K. Boutros, "Normally-off GaN-on-Si multi-chip module boost converter with 96% efficiency and low gate and drain overshoot", in IEEE 2014 Applied Power Electronics Conference and Exposition, 2014, pp. 484-487.

[2] B. Whitaker, A. Barkley, Z. Cole, B. Passmore, D. Martin, T.R. McNutt, A.B. Lostetter, J. Lee, and K. Shiozaki, "A high-density, high-efficiency, isolated on-board vehicle battery charger utilizing silicon carbide power devices", IEEE Transactions on Power Electronics, vol. 29, no. 5, pp. 2606-2617, May 2014.

[3] I. Lee, and G. Moon, "Half-bridge integrated ZVS full-bridge converter with reduced conduction loss for electric vehicle battery chargers", IEEE Transactions on Industrial Electronics, vol. 61, no. 8, pp. 3978-3988, August 2014.

[4] Z. Chen, Y. Wang, X. Ma, and T. Yuan, "A novel ZVZCS phase-shifted full-bridge converter with secondary-side energy storage inductor used for electric vehicles", in IEEE 2014 Industrial Electronics Society, 2014, pp. 5019-5025.

[5] J. Lai, H. Miwa, W. Lai, N. Tseng, C. Lee, C. Lin, and Y. Shih "A high-efficiency on-board charger utilitzing a hybrid LLC and phase-shift DC-DC converter", in IEEE 2014 Intelligent Green Building and Smart Grid, 2014, pp. 1-8.

[6] H. Wang, S. Dusmez, and A. Khaligh, "Design considerations for a level-2 on-board PEV charger based on interleaved boost PFC and LLC resonant converters", in IEEE 2013 Transportation Electrification Conference and Expo, 2013, pp. 1-8.

[7] W. Yu, H. Wan, J. Lai, H. Miwa, W. Lai, N. Tseng, C. Lee, C. Lin, and Y. Shih, "High efficiency isolated DC-DC converter combining resonant and phase-shifted topologies for electrical vehicle chargers", in IEEE 2013 Future Energy Electronics Conference, 2013, pp. 175-180.

[8] B. Gu, J. Lai, N. Kees, and C. Zheng, "Hybrid-switching full-bridge DC-DC converter with minimal voltage stress of bridge rectifier, reduced circulating losses, and filter requirement for electric vehicle battery chargers", IEEE Transactions on Power Electronics, vol. 28, no. 3, pp. 1132-1144, March 2013.

[9] D. Zhu, Y. Tang, C. Jin, P. Wang, and F. Blaabjerg, "An efficiency improved single-phase PFC converter for electric vehicle charger applications", in IEEE 2013 Industrial Electronics Society, 2013, pp. 4558-4563.

[10] D. Gautam, F. Musavi, M. Edington, W. Eberle, and W. Dunford, "A zero voltage switching full-bridge DC-DC converter for an on-board PHEV battery charger", in IEEE 2012 Transportation Electrification Conference and Expo, 2012, pp. 1-6.

[11] J. Kim, S. Lee, J. Lee, and B. Cho, "A new direct internal resistance and state of charge relationship for the Li-ion battery pulse power estimation", in IEEE 2007 International Conference on Performance Engineering, 2007, pp. 1173-1178.

[12] H. Zhang, R. Lu, C. Zhu, and Y. Zhao, "On-line measurement of internal resistance of lithium ion battery for EV and its application research", International Journal of u-and e-Services, Science and Technology, vol. 7, no. 4, pp. 301-310, August 2014.

[13] Z. Chen, B. Xia, C.C. Mi, and R. Xiong, "Loss minimization-based charging strategy for lithium-ion battery", in IEEE 2014 Energy Conversion Congress and Exposition, 2014, pp. 4306-4312.

[14] A.E. Eiben and J.E. Smith, Introduction to Evolutionary Computing, Springer, Berlin, 2003.

[15] H. Zhang and M.Y. Chow, "Comprehensive dynamic battery modeling for PHEV applications", in IEEE 2010 Power and Energy Society General Meeting, 2010, pp.1-6.

[16] M. Dubarry, N. Vuillaume, and B.Y. Liaw, "From Li-ion single cell model to battery pack simulation", in IEEE 2008 Control Applications, 2008, pp. 708-713.

[17] W. Zhen-Po, L. Peng, and W. Li-Fang, "Analysis on the capacity degradation mechanism of a series lithium-ion power battery pack based on inconsistency of capacity", Chinese Physics B, vol. 22, no. 8 (088801), August 2013.

A Pulse Width Modulated LLC Type Resonant Topology Adpated to Wide Output Voltage Range

Haoyu Wang, *Member, IEEE*
School of Information Science and Technology
ShanghaiTech University
Shanghai, China
wanghy@shanghaitech.edu.cn

Abstract—In a two stage LLC based plug-in electric vehicle (PEV) onboard charger, the dc link voltage is usually kept constant while the battery pack voltage varies in a wide range. This makes it extremely difficult to optimally design the second stage pulse frequency modulated (PFM) LLC resonant converter. In this paper, a modified pulse width modulated (PWM) LLC resonant topology is proposed and investigated in PEV charging applications. The proposed converter is able to achieve wide output voltage range, while maintaining the switching frequency of the full bridge constant and equal to the resonant frequency. In comparison with the conventional LLC topology, only one auxiliary MOSFET is added. Zero-voltage-switching (ZVS) is realized among all power MOSFETs. A 1.2 kW converter prototype, generating 250V-420V output voltages from the 390 V dc link, is analyzed, designed, and simulated to verify the proof of concept.

Keywords— fixed frequency; LLC; PWM; ZVS; onboard charging; PEV

I. INTRODUCTION

A typical PEV onboard charger consists of two stages: a) the first stage ac/dc converter for rectification and power factor correction, and b) the second stage dc/dc converter for voltage/current regulation and galvanic isolation[1]–[3]. LLC resonant topology has attractive features, such as: a) wide zero voltage switching (ZVS) range, b) wide output voltage range, and c) reduced electromagnetic interferences [4]. Therefore, it is deemed as a good candidate to implement the second stage dc/dc converter of the PEV onboard chargers [5]–[7].

In [8], a 10 kW LLC resonate converter was designed to charge a 250 V- 450 V battery pack and achieved 96% peak load efficiency at 420 V. However, when the output voltage deviates from 420 V, the conversion efficiency degrades fast. Similar phenomenon is also observed in the other literatures [9]–[13]. This is because LLC topology only demonstrates optimum efficiency performance around the resonant frequency (f_r) between its resonant inductor (L_r) and capacitor (C_r). However, in order to be adapted to the wide voltage gain variation range, the switching frequency (f_s) must swing in a wide range and deviate from f_r. This phenomenon can be observed clearly from Fig. 1, which shows the dc voltage gain of LLC converter adapted to wide output voltage range. As shown, the desired output voltage window [V_{min}, V_{max}] on the gain axis is mapped to an ultra-wide switching frequency range

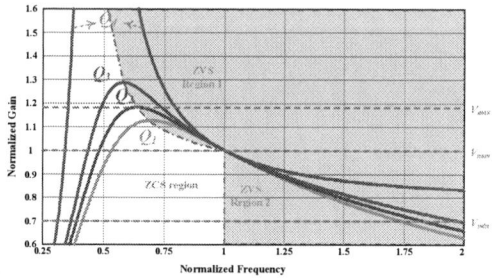

Fig. 1. DC voltage characteristics of conventional LLC converter adapted to wide output voltage range.

in the frequency axis. Therefore, it is extremely difficult to optimally design the LLC converter.

To alleviate this problem and to narrow down the switching frequency range of LLC topology, different alternative techniques were investigated in literature [14]–[16]. In [14], phase shift controlled MOSFETs are introduced to the secondary side of LLC topology, mainly for compensating the switching frequency deviation during the hold-up time. In [15], the first stage ac/dc PFC converter of the LLC based PEV charger is designed to have variable output dc link voltage. This dc link voltage linearly follows the battery pack voltage. Therefore, the LLC switching frequency variation is significantly narrowed down. In [15], the LLC topology is integrated with a two-phase interleaved boost circuit. The latter provides the boost feature to the modified circuit and successfully squeezes the switching frequency of the LLC converter.

In this paper, a novel pulse width controlled LLC topology is proposed. The proposed converter partially utilizes the voltage doubler architecture on the secondary side rectification stage. By inserting an actively controlled MOSFET and a passive controlled diode to the voltage doubler rectifier, the output voltage of the converter can be modulated by the duty cycle of the auxiliary MOSFET. Therefore, the main LLC topology can operate at its resonant frequency. This proposed converter demonstrates benefits including: a) optimum efficiency of the main LLC power circuit; b) ZVS turning-on of auxiliary MOSFET; c) ZCS turning-off the secondary side diodes; d) reduced circuit control complexity; and e) reduced circulating current and conduction losses.

978-1-4673-9551-9/16 $31.00 © 2016 IEEE

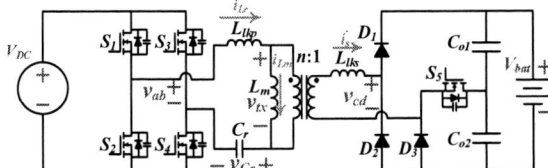

Fig. 2. Proposed PWM LLC topology.

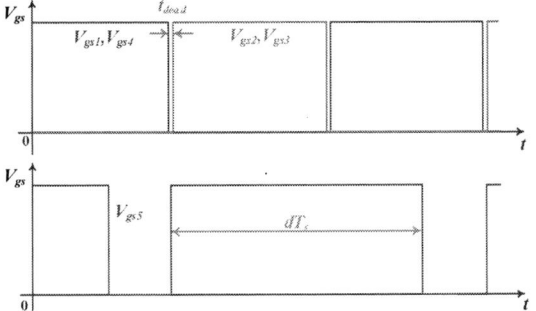

Fig. 3. Switch pattern of the actively controlled MOSFETs.

II. PROPOSED PWM LLC TYPE DC/DC CONVERTER

A. Topology Description

The schematic of the proposed PWM LLC resonant converter is demonstrated in Fig. 2. This topology is derived from the full bridge LLC topology with voltage doubler rectification stage. However, compared with the voltage doubler rectification, an auxiliary diode-MOSFET bridge (D_3 and S_5) is added on the secondary side (see Fig. 2). By actively controlling the duty ratio of S_5, the output voltage changes accordingly. Therefore, this converter can be modulated by the pulse width of M_5.

Moreover, the transformer is designed to have a secondary side leakage inductance, L_{lks}. This modification adds those benefits to the circuit: a) the sharp current variation slope on the secondary side semiconductors is it mitigated; b) the ZCS turning off of diode D_1 is facilitated; and c) the pulse width modulation capability of the voltage gain is enhanced.

B. Operation Principle

In the proposed converter, the primary side full bridge functions as a constant frequency square wave generator. The upper and lower power MOSFETs are turned on and off complementarily with certain deadband (t_{dead}), to prevent the circuit shoot through. This switching frequency is fixed and equal to the resonant frequency between the transformer primary leakage inductance and the resonant capacitance:

$$f_s = \frac{1}{2\sqrt{L_{lkp}C_r}} \quad (1)$$

The switching pattern of MOSFETs S_1-S_5, is plotted in Fig. 3. It should be noted that the switching frequency of MOSFET S_5 is equal to f_s. Therefore, the primary side MOSFETs can be guaranteed with both ZVS and optimized circulating currents. On the secondary side, the turning on moment must be

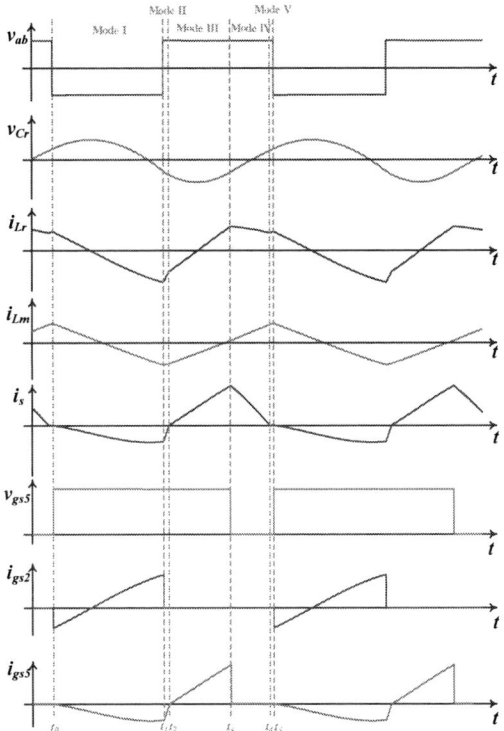

Fig. 4. Steady state operation waveforms of the proposed converter.

synchronized within the deadband, when the MOSFETs in the primary side full bridge inverter are all off. Typically, the duty cycle of M_5, d, should be constrained within the range of [0.5, 1]. This is mainly due to two facts associated with $d < 0.5$: a) the auxiliary MOSFET loses its ZVS feature; and b) the output will be a constant voltage; the circuit loses its PWM feature.

The steady state waveforms of the proposed converter is plotted in Fig. 4. In each switching period, the steady state operation can be divided into 5 operation modes. Those operation modes are given in Fig. 5.

Mode I: [t_0, t_1). Mode I begins in the deadband at t_0 when S_5 is tuned on. In the beginning of Mode I, i_{Lr} charges and discharges the output capacitors of S_1-S_4. Thus, v_{ab} is inverted from V_{DC} to $-V_{DC}$. After t_0, S_2 and S_3 are both turned on. During Mode I, L_r resonates with C_r. The secondary side current, i_s, is negative and flows through D_2 and the body diode of the auxiliary MOSFET, D_{S5}. It should be noted that: $t_1 - t_0 = T_s/2$.

Mode II: [t_1, t_2). When the switch pattern of S_1-S_4 is inverted at t_1, the circuit operation enters into Mode II. Similarly, at the beginning of Mode II, i_{Lr} discharges and charges the output capacitors of S_1-S_4. Thus, v_{ab} is inverted from $-V_{DC}$ to V_{DC}. D_{S5} and D_2 continue to conduct in mode II. During Mode II, V_{DC} on the primary side and $-V_{Co2}$ on the secondary side jointly enforce a steep di/dt on the secondary side leakage inductor L_{lks}. Thus, i_s increases from negative to zero in a short time period, which marks the end of Mode II.

978-1-4673-9551-9/16 $31.00 © 2016 IEEE   1281

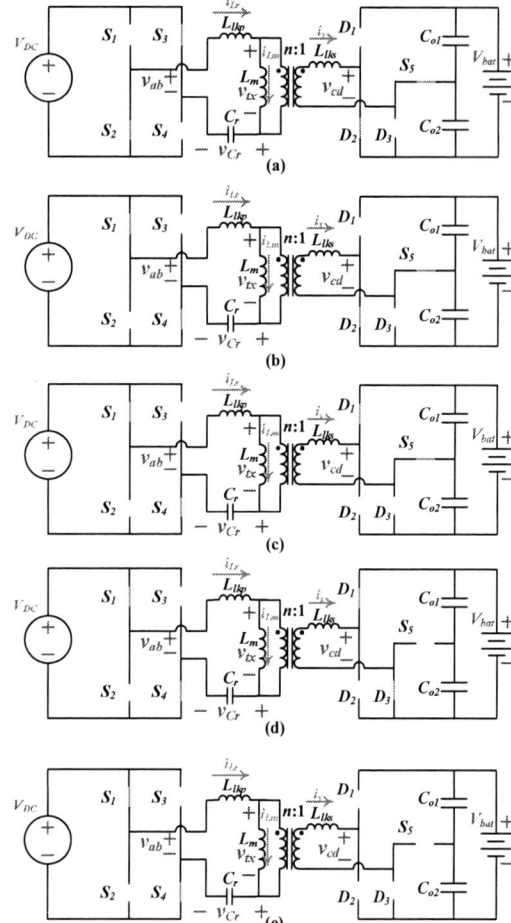

Fig. 5. Operation modes of the converter; a) Mode I: $t_0 \leq t < t_1$; b) Mode II, $t_1 \leq t < t_2$; c) Mode III, $t_2 \leq t < t_3$; d) Mode IV, $t_3 \leq t < t_4$; e) Mode V, $t_4 \leq t < t_5$.

Mode III [t_2, t_3]. Mode III starts when i_s reaches zero at t_2. After t_2, the current conduction on the secondary side is switched from D_{S5} and D_2 to D_1 and S_5. Therefore, S_5 is turned on at ZVS. V_{Co1} is connected to the secondary side of the transformer. In Mode III, the effect of V_{DC} on i_s is partially cancelled by V_{Co1}. Thus, although i_s continues to increase, the growth rate is much slower; this can been clearly seen in Fig. 4. $S_{2,3,5}$, and D_1 are the conducting semiconductors. Mode III ends at t_3 when S_5 is turned off. It should be noted that: $t_3 - t_1 = (d - 0.5)T_s$.

Mode IV [t_3, t_4]. Mode IV starts as S_5 is turned off. Since the inductor current is continuous, its path is switched from S_5 to D_3. Hence, V_{bat} is connected to the secondary side of the transformer directly. Since V_{bat} is much larger than V_{Co1}, it induces i_s to decrease in this operation mode, as can be seen in Fig. 4. Mode IV ends when i_s reaches zero.

Mode V [t_4, t_5]. At t_4, i_s reaches zero and the circuit enters into Mode V. Since M_5 is still off, i_s stays at zero. Thus, all the semiconductors on the secondary side are off. Mode V ends

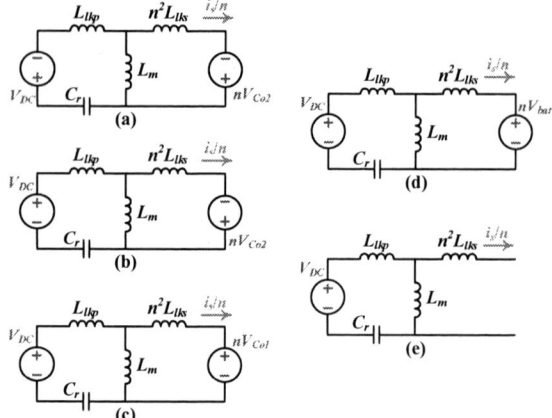

Fig. 6. Equivalent circuit model of the converter; a) Mode I: $t_0 \leq t < t_1$; b) Mode II, $t_1 \leq t < t_2$; c) Mode III, $t_2 \leq t < t_3$; d) Mode IV, $t_3 \leq t < t_4$; e) Mode V, $t_4 \leq t < t_5$.

when the switch pattern of S_1-S_4 is inverted again, which marks the start of next switching period.

III. CIRCUIT MODELING AND ANALYSIS

A. Equivalent Circuit Model

Based on the steady state analysis in Section II-B, the equivalent circuit model of the proposed topology can be obtained. Assuming that both output filtering capacitances (C_{o1} and C_{o2}) are sufficiently large, the voltage on each capacitor could be considered as a constant. Therefore, C_{o1} and C_{o2} can be equivalent to two constant voltage sources, whose voltages are denoted as V_{Co1} and V_{Co2}, respectively. Moreover, the voltage source and the impedance on the secondary side of the transformer can be pushed to the primary side with ratios n and n^2, respectively. The resultant equivalent circuit model of the converter is plotted in Fig. 6. The circuit model is composed by an input voltage source, a two port impedance network, as well as an output voltage source. The output voltage source varies step-by-step with the transitions of operation modes.

B. Piecewise Linear Current Approximation

As mentioned earlier, the full bridge inverter generates a square wave, v_{ab}. Its Fourier series is expressed as,

$$v_{ab}(t) = \sum_{n=1,3,5,7..} \frac{4V_{DC}}{n\pi} \sin\left(2\pi n f_s\right) \qquad (2)$$

f_s is equal to the resonant frequency between C_r and L_{lkp}. This means that the impedance of C_r and L_{lkp} is equal to zero at f_s. Thus, the first harmonic component of the square wave is transmitted to the magnetizing inductor without degradation. Conventionally, first harmonic approximation technique is widely used to model the LLC topology operating at its resonant frequency [17].

In this case, to facilitate the circuit analysis, it is assumed that the voltage applied to L_m is approximately equal to the input voltage source without degradation. Hence, the magnetizing inductor current i_{Lm} decreases at the rate of $-V_{dc}/L_m$ in Mode I; while i_{Lm} increases at the rate of V_{DC}/L_m in

978-1-4673-9551-9/16 $31.00 © 2016 IEEE 1282

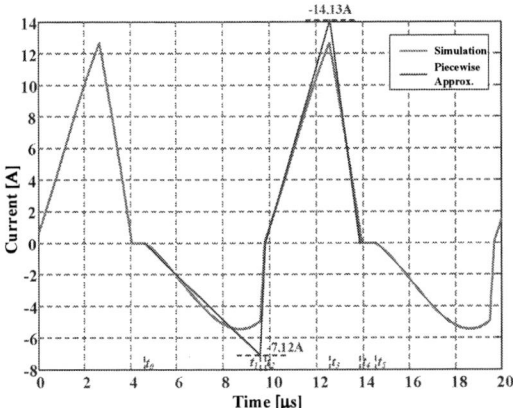

Fig. 7. Piecewise linear approximation of i_s and its comparison with the simulated waveform.

Modes II-V. The accuracy of this approximation can be roughly evaluated from the linearity of i_{Lm}, as shown in Fig. 4.

Based on this approximation, i_s changes piecewise linearly. The rate di_s/dt changes step-by-step with the transition of operation modes. In Mode I,

$$i_s(t) = \frac{-V_{DC}/n + V_{Co2}}{L_{lks}}(t - t_0) \tag{3}$$

Since the duration of mode I is $T_s/2$, $i_s(t_1)$ can be easily calculated based on Eq. (3). It should be noted that $i_s(t_1)$ denotes the peak current of D_2.

In Mode II, the input voltage source revert its polarity, V_{DC} and V_{Co2} jointly push $i_s(t)$ to zero very fast.

$$i_s(t) = \frac{V_{DC}/n + V_{Co2}}{L_{lks}}(t - t_1) + i_s(t_1) \tag{4}$$

The duration of Mode II is the initial current $i_s(t_1)$ divided by the slope of $i_s(t)$ in this mode, as,

$$t_2 - t_1 = \frac{V_{DC}/n - V_{Co2}}{V_{DC}/n + V_{Co2}} \frac{T_s}{2} \tag{5}$$

In Mode III, the output voltage source is transited to nV_{Co1}, $i_s(t)$ begins to increase from zero.

$$i_s(t) = \frac{V_{DC}/n - V_{Co1}}{L_{lks}}(t - t_2) \tag{6}$$

Since the turning off of S_5 denotes the end of Mode III, the duration of Mode III is $dT_s - t_2$. It should be noted that $i_s(t_3)$ denotes the peak current of D_1 and S_5.

In Mode IV, the secondary side of transformer sees the sum of V_{Co1} and V_{Co2}.

$$i_s(t) = \frac{V_{DC}/n - V_{bat}}{L_{lks}}(t - t_3) + i_s(t_3) \tag{7}$$

The duration of Mode IV is the initial current $i_s(t_3)$ divided by the slope of $i_s(t)$ in this mode, as,

$$t_4 - t_3 = \frac{V_{DC}/n - V_{Co2}}{V_{bat} - V_{DC}/n}(dT_s - t_2) \tag{8}$$

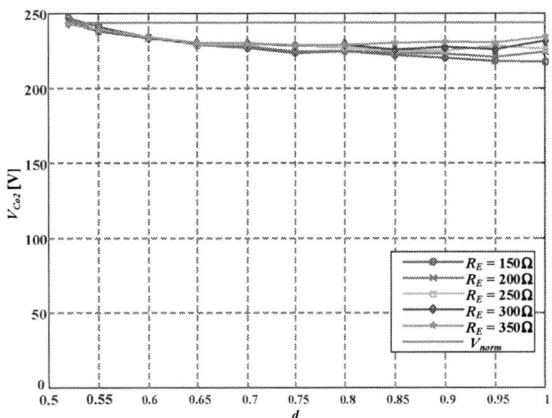

Fig. 8. Simulated V_{Co2} versus duty cycle under different effective load conditions and their comparison with V_{norm}.

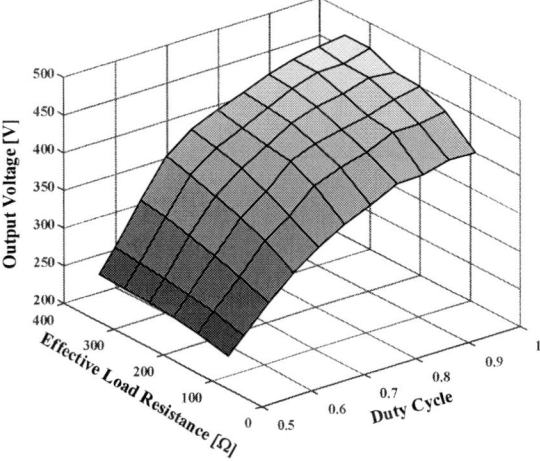

Fig. 9. Voltage gain versus equivalent R_L and d.

In Mode V, the voltage applied to L_{lks} is equal to zero. The duration of Mode V is $T_s - t_4$.

$$i_s(t) = 0 \tag{9}$$

The data obtained based on equations (3-9) and the simulated data are plotted and compared in Fig. 7 to verify the accuracy of the proposed piecewise linear current approximation method. As shown in Fig. 7, t_2 and t_4 agrees very well with the simulation. While there is a small error on $i_s(t_1)$ and $i_s(t_3)$; this error is considered as acceptable.

C. Voltage Gain

According to the law of energy conservation and assuming the converter is ideal without intermediate energy losses, the input energy is equal to the energy delivered to the battery pack. Therefore,

$$\int_0^{t_0+T_s} v_{ab}(t)i_s(t)\,dt = V_{bat}I_{bat}T_s \tag{10}$$

By applying the piecewise linear current approximation model discussed in Section III-B, one can obtain,

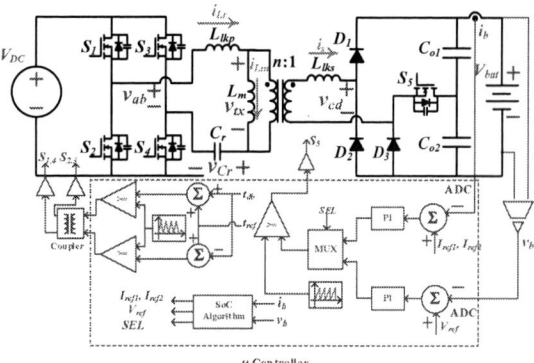

Fig. 10. Digital control diagram of proposed PWM LLC type converter in PEV battery charging applications.

$$\frac{i_s(t_1)V_{Co2}(t_2-t_1)}{2} + \frac{i_s(t_3)V_{Co1}(t_3-t_2)}{2} + \frac{i_s(t_3)V_{bat}(t_4-t_3)}{2} = V_{bat}I_{bat}T_s$$

(11)

Where $V_{bat} = V_{Co1} + V_{Co2}$, and I_{bat} is determined by the charging power. It should be noted that the proposed analysis methodology is also valid to the resistive load.

Based on Eq. (11), by substituting the previous derived expressions, there are still 3 unknowns left, which are V_{Co1}, V_{Co2}, and V_{DC}, respectively. Therefore, it is difficult to theoretically predict the voltage gain of the circuit based on the proposed model. On the other hand, it is interesting to note that V_{Co2} is a weak function of both the effective load and the duty cycle. Indeed, V_{Co2} is always close to the normalized voltage V_{norm}, as defined,

$$V_{norm} = \frac{V_{DC}}{n}$$

(12)

This phenomenon can be clearly observed from the Fig. 8, which provides the simulated data of V_{Co2} versus the duty cycle under different load conditions. According to Fig. 8, V_{Co2} slightly decreases with the increase of duty cycle; it is always slightly lower than but close to V_{norm}. Also, V_{Co2} demonstrate very good load independence performance. Thus, one can consider V_{Co2} is a constant which is approximately equal to V_{norm}. By adopting this approximation, the theoretical voltage gain can be easily calculated.

Fig. 9 plots the output voltage versus the effective load resistance, and the duty cycle. The data is obtained from software simulation, which is based on the design parameters provided in Table I. As shown in Fig. 9, the output voltage is a strong function of the duty cycle and a weak function of the effective load resistance. This means that the voltage gain can be easily regulated by the pulse width modulation.

IV. CONTROLLER DESIGN

The digital control scheme of the proposed converter is plotted in Fig. 10. As demonstrated in the diagram, control of the circuit in the application of PEV battery charging can be implemented using a microcontroller. On the primary side of

TABLE I

DESIGN PARAMETERS OF THE PROPOSED CONVERTER

Quantity	Symbol	Parameter
Input voltage	V_{DC}	390V
Output voltage	V_{out}	250V-420V
Rated power	P_{max}	1.2 kW
Primary side inductor	L_{lkp}	13.8 μH
Secondary side inductor	L_{lks}	13.8 μH
Magnetizing inductor	L_m	158 μH
Resonant capacitor	C_r	180 nF
Output capacitor 1	C_{o1}	100 μF
Output capacitor 2	C_{o2}	100 μF
Switching frequency	f_s	100 kHz
Transformer turns ratio	n	1.6

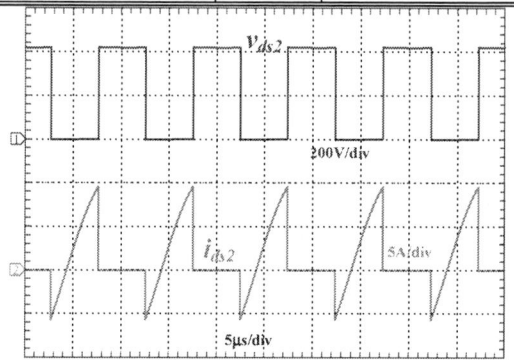

Fig. 11. ZVS turning-on of primary side MOSFET.

the transformer, those four MOSFETs are driven by four channel PWM signals issued by the microcontroller. No feedback is required to control the full bridge inverter. On the secondary side, the transducers sample the battery charging current, charging voltage information and send it to the microcontroller. While the CPU estimate the state of charge and determine the charging mode. In constant current charging mode, the current PI control loop is activated to maintain constant charging current. While in constant voltage charging mode, the voltage PI control loop is activated to maintain constant charging voltage. In comparison with the conventional LLC topology which utilizes the pulse frequency modulation, the control scheme in this proposed converter is much simpler. Moreover, this control algorithm can also be implemented using analog controller. This controller is easy to be integrated into an application specific integrated circuit.

V. RESULTS

Based on the proposed topology and the circuit analysis, a converter rated at 1.2 kW is designed and simulated to charge a 250 V - 420 V battery pack. The input is 390 VDC, which is the typical output voltage of the ac/dc power factor correction stage. The design parameters are provided in Table I. The simulation is conducted in Simulink.

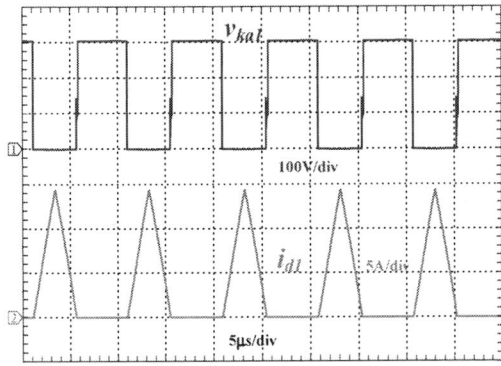

Fig. 12. ZCS turning-off of secondary side diode D_1.

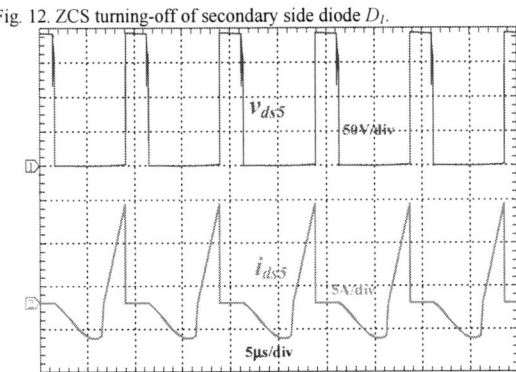

Fig. 13. ZVS turning-on of auxiliary MOSFET.

Fig. 11 demonstrates the voltage and current waveforms of the primary side MOSFET, S_2. As shown, the body diode conducts before the conduction of S_2, which provides zero voltage condition for the turning on of the MOSFET channel. Therefore, ZVS is achieved during the MOSFET turn-on process. Similar waveforms apply to $S_{1,3,4}$, too.

Fig. 12 shows the voltage and current waveforms of the secondary side diode D_1. As shown, D_1 is turned off with small di/dt, which significantly mitigates the diode reverse recovery problems. With regards to D_3, it has the identical turning off process as D_1 with small di/dt. However, it should be noted that D_2 suffers from the large di/dt of i_s during operation mode II. The slope is provided in Eq. 4. Typically, D_2 should be designed using fast recovery power diode.

Fig. 13 illustrates the voltage and current waveforms of auxiliary MOSFET, S_5. As shown, the body diode of the S_5 conducts before the conduction of MOSFET channel, which guarantees the MOSFET ZVS turning-on.

VI. CONCLUSIONS

In this paper, a PWM LLC type resonant converter is proposed for use in PEV onboard chargers. The proposed converter demonstrates the benefits of a) optimized LLC switching frequency; b) ZVS of all MOSFETs; c) ZCS turning-off the rectification diodes; d) reduced circuit control complexity; and e) reduced circulating current. A 1.2 kW

converter prototype is designed to verify the proof of concept. The proposed converter topology is not only useful for PEV battery charging application, but also valid for applications where wide voltage gain range is desired.

References

[1] A. Khaligh and S. Dusmez, "Comprehensive Topological Analysis of Conductive and Inductive Charging Solutions for Plug-In Electric Vehicles," *IEEE Trans. Veh. Technol.*, vol. 61, no. 8, pp. 3475–3489, Oct. 2012.

[2] D. S. Gautam, F. Musavi, W. Eberle, and W. G. Dunford, "A Zero-Voltage Switching Full-Bridge DC--DC Converter With Capacitive Output Filter for Plug-In Hybrid Electric Vehicle Battery Charging," *IEEE Trans. Power Electron.*, vol. 28, no. 12, pp. 5728–5735, Dec. 2013.

[3] H. Wang and A. Khaligh, "Interleaved SEPIC PFC converter using coupled inductors in PEV battery charging applications," in *2015 IEEE Applied Power Electronics Conference and Exposition (APEC)*, 2015, pp. 586–591.

[4] W. Feng, F. C. Lee, and P. Mattavelli, "Optimal Trajectory Control of Burst Mode for LLC Resonant Converter," *IEEE Trans. Power Electron.*, vol. 28, no. 1, pp. 457–466, Jan. 2013.

[5] X. Fang, H. Hu, F. Chen, U. Somani, E. Auadisian, J. Shen, and I. Batarseh, "Efficiency-Oriented Optimal Design of the LLC Resonant Converter Based on Peak Gain Placement," *IEEE Trans. Power Electron.*, vol. 28, no. 5, pp. 2285–2296, May 2013.

[6] X. Fang, H. Hu, Z. J. Shen, and I. Batarseh, "Operation Mode Analysis and Peak Gain Approximation of the LLC Resonant Converter," *IEEE Trans. Power Electron.*, vol. 27, no. 4, pp. 1985–1995, Apr. 2012.

[7] H. Wang, A. Hasanzadeh, and A. Khaligh, "Transportation Electrification: Conductive Charging of Electrified Vehicles," *IEEE Electrif. Mag.*, vol. 1, no. 2, pp. 46–58, Dec. 2013.

[8] W. Guo, H. K. Bai, G. Szatmari-voicu, A. Taylor, J. Patterson, and J. Kane, "A 10kW 97%-efficiency LLC Resonant DC/DC Converter with Wide Range of Output Voltage for the Battery Chargers in Plug-in Hybrid Electric Vehicles," in *IEEE Transportation Electrification Conference and Expo*, 2012, pp. 8–11.

[9] F. Musavi, M. Craciun, D. S. D. S. Gautam, W. Eberle, and W. G. W. G. Dunford, "An LLC Resonant DC–DC Converter for Wide Output Voltage Range Battery Charging Applications," *IEEE Trans. Power Electron.*, vol. 28, no. 12, pp. 5437–5445, Dec. 2013.

[10] J. Deng, S. Li, S. Hu, C. C. Mi, and R. Ma, "Design Methodology of LLC Resonant Converters for Electric Vehicle Battery Chargers," *IEEE Trans. Veh. Technol.*, vol. 63, no. 4, pp. 1581 – 1592, 2014.

[11] H. Hu, X. Fang, F. Chen, Z. J. Shen, and I. Batarseh, "A Modified High-Efficiency LLC Converter With Two Transformers for Wide Input-Voltage Range Applications," *IEEE Trans. Power Electron.*, vol. 28, no. 4, pp. 1946–1960, Apr. 2013.

[12] R. Beiranvand, B. Rashidian, M. R. Zolghadri, and S. M. H. Alavi, "Using LLC Resonant Converter for Designing Wide-Range Voltage Source," *IEEE Trans. Ind. Electron.*, vol. 58, no. 5, pp. 1746–1756, 2011.

[13] R. Beiranvand, B. Rashidian, M. R. Zolghadri, and S. M. Hossein Alavi, "A Design Procedure for Optimizing the LLC Resonant Converter as a Wide Output Range Voltage Source," *IEEE Trans. Power Electron.*, vol. 27, no. 8, pp. 3749–3763, Aug. 2012.

[14] J.-W. Kim and G.-W. Moon, "A New LLC Series Resonant Converter with a Narrow Switching Frequency Variation and Reduced Conduction Losses," *IEEE Trans. Power Electron.*, vol. 29, no. 8, pp. 4278–4287, Aug. 2014.

[15] H. Wang, S. Dusmez, and A. Khaligh, "Maximum Efficiency Point Tracking Technique for LLC-Based PEV Chargers Through Variable DC Link Control," *IEEE Trans. Ind. Electron.*, vol. 61, no. 11, pp. 6041–6049, Nov. 2014.

[16] X. Sun, Y. Shen, Y. Zhu, and X. Guo, "Interleaved Boost-Integrated LLC Resonant Converter With Fixed-Frequency PWM Control for Renewable Energy Generation Applications," *IEEE Trans. Power Electron.*, vol. 30, no. 8, pp. 4312–4326, Aug. 2015.

[17] H. Wang, S. Dusmez, and A. Khaligh, "A Novel Approach to Design EV Battery Chargers Using SEPIC PFC Stage and Optimal Operating Point Tracking Technique for LLC Converter," in *2014 IEEE Applied Power Electronics Conference and Exposition - APEC 2014*, 2014, pp. 1683–1689.

A Series Resonant Circuit for Voltage Equalization of Series Connected Energy Storage Devices

Yanqi Yu, Raed Saasaa and Wilson Eberle
School of Engineering
The University of British Columbia
Kelowna, Canada
Email: yanqi.yu@alumni.ubc.ca

Abstract - **In this paper, a novel cell voltage equalizer using a series LC resonant converter topology is proposed for a series connection of energy storage devices, namely battery, or supercapacitor cells. The objective of the target project is to design an active voltage equalization circuit that is low cost, small in size and achieves a short voltage equalization time. The target application is electric vehicles. Compared to existing solutions, the series LC resonant converter eliminates the complexity of multi-winding transformers and it can balance series connected energy storage devices quickly by transporting energy successively between the lowest and highest charge cell pairs. The circuit operation and a theoretical analysis is provided. Simulation and experimental results are presented for a circuit balancing three supercapacitors demonstrating from an initial voltage difference of 527 mV to a final difference of 10 mV in 900 seconds.**

Keywords—Supercapacitor; resonant converter; voltage equalization; voltage deviation.

I. INTRODUCTION

With the world's continued economic development, electrochemical energy storage devices, including batteries and capacitors are increasingly being adopted for industrial and consumer applications, particularly in electric vehicles (EVs). These energy storage devices are appealing because of their small scales and flexibility to manufacture.

Recently, supercapacitors have garnered increasing popularity because of their unique features such as extremely high capacitance, wide operating temperature range and virtually infinite number of cycles [1]. Perhaps most significantly, they have much higher specific power ratings than most existing batteries, so that they can easily satisfy highly dynamic loads. Therefore, they are the optimal device for absorbing and delivering sudden power surges [2]. For a typical electric vehicle application, supercapacitors can be used as a temporary energy storage device in conjunction with a fuel cell, or battery pack [3]. Theoretically speaking, supercapacitors can be used as the sole energy source in EV applications. However, their low specific energy rating prohibits them from long distance applications. Finally, supercapacitors have the potential to reduce battery thermal stress and prolong battery life by working in parallel with the batteries to absorb braking energy and supply the energy during vehicle acceleration [4].

Similar to lithium battery cells, due to manufacturing limitations, individual supercapacitor voltages are inherently low (around 2.1 to 2.7 volts). Therefore, supercapacitors are usually configured in series in order to reach practical working voltage level [5]-[9]. The voltage characteristics of supercapacitors depends on many factors, including capacitance tolerance, self-discharge rates and leakage currents [10],[11]. Among these causes, capacitance tolerance play a major role in the voltage deviation between series connected cells. Even for supercapacitors manufactured under the same procedure, their capacitances are not identical because of their different chemical characteristics depending on the temperatures they are exposed to [12].

Since supercapacitors are connected in series, the current flowing through a string is the same for all cells. Due to cell tolerance, the series current, will yield an imbalance in the individual supercapacitor voltages in a string [13]. As a result, during charging, supercapacitors with higher capacitance will be at lower voltages and not reach their full capacity and those with lower capacitance will be at higher voltages, and potentially can be damaged due to overvoltage. Therefore, control and protection circuits may be required to regulate the individual voltage of each supercapacitor in a string.

Voltage equalization techniques have been proposed to alleviate the voltage imbalance issue for series connected energy storage cells. Generally, these voltage equalization techniques can be divided into two groups, passive and active voltage equalization [14]-[16]. Passive voltage equalization methods employ energy-consuming devices such as resistors to remove excess energy from energy storage cells with higher voltages. The excess energy is dissipated in the resistors as heat [17]. These circuits are relatively simple, but lossy. On the other hand, active voltage equalization methods remove excess charge from supercapacitors with higher voltages and then transport the extra charge to the ones with lower voltages through adjacent supercapacitors [18].

Many active cell voltage balancing circuits have been proposed. Among them, switched capacitor topologies and transformer based topologies are the most common. Switched capacitor topologies are appealing in terms of their small package sizes and relatively low costs [19]-[21]. However, their voltage balancing times are long since the extra energy is transported among connected cells back and forth. On the other hand, multi-winding transformer based topologies tend to balance supercapacitor cell voltages by preferentially concentrating all the dedicated energy into the cells with

978-1-4673-9551-9/16 $31.00 © 2016 IEEE

comparatively low voltages [22]-[24]. This feature enables short equalization times. Nonetheless, the bulky size of transformers as well as high costs yield significant challenges for their potential application in EVs.

In this paper, a single series LC resonant tank equalizer is presented. The converter is tested with supercapacitors, but can be used for the voltage equalization of other energy storage devices, including lithium-ion batteries. The series LC resonant converter is switched near the LC resonant frequency, where the impedance in the circuit is minimized, allowing the converter to exchange energy quickly. The single LC resonant tank performs a similar function as in the active multi-winding transformer circuits, providing out of series energy transfer paths among supercapacitors. But, in contrast to transformer based schemes, the proposed circuit reduces the converter cost sharply and can be manufactured in a much smaller package. In the case where hundreds of supercapacitors are series connected, the proposed circuit can be easily modified by stacking additional sets of switches in parallel with existing levels of switches, instead of adding extra transformer windings.

This paper is organized as follows. In Section II, the single series LC resonant tank equalizer converter circuit configuration is introduced and analyzed. In Section III, the equivalent circuit and mathematical derivation are presented. Simulation and experimental waveforms for the converter circuit and voltage equalization process are presented in Section IV to verify the converter operation. The conclusions are presented in Section V.

II. PROPOSED VOLTAGE EQUALIZATION TOPOLOGY

The proposed series LC resonant tank equalizer circuit topology is presented Fig. 1. The circuit includes the energy storage device string (ES$_1$, ES$_2$... ES$_n$) with each energy storage cell in the string having four associated switches connected to the resonant tank. The anti-series switches are used to prevent body diode conduction since low voltage MOSFETs cannot block a negative voltage due to the MOSFET intrinsic body diode. The four switches with each energy storage cell are controlled by the same signal. Three diodes D$_1$, D$_{2a}$, D$_{2b}$, are configured in series and parallel with the LC resonant tank in order to provide a continuous path for the resonant inductor current. The resonant tank includes inductor, L$_r$ and capacitor C$_r$. In order to minimize the voltage equalization time, the switching frequency is set to be close to the resonant frequency, where the impedance of the converter is minimized allowing maximum current in the energy flow path in each switching cycle. This results in the maximum possible energy being transported in each switching cycle, enabling a short voltage balancing time to be achieved.

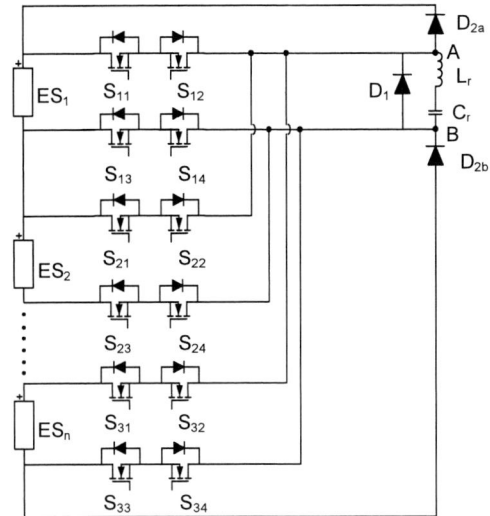

Fig. 1. Propsed energy storage cell voltage equalization converter circuit.

Operation of the circuit requires additional control and voltage detection circuitry (not shown). When the highest voltage energy storage cell and the lowest voltage cell are selected in the string by a voltage detection circuit, the switches associated with the higher voltage cell are triggered on resulting in charge flowing from this cell to the series LC resonant tank. The resonant tank temporarily stores the energy and then releases the energy to the lower voltage cell, therefore increasing its voltage in the second phase of the switching cycle.

III. MODES OF OPERATION

Since in the general case, the proposed voltage equalizer only balances the two cells with the greatest voltage difference, the proposed voltage equalization circuit operation is presented for two cells. The following analysis is conducted assuming that energy storage cells ES$_1$ and ES$_n$ have voltage values with V$_{ES1}$ is greater than V$_{ESn}$.

The circuit operation for the proposed converter can be divided into four modes, described in the following paragraphs. Fig. 2 illustrates current paths for the operation of equalization converter. The waveforms for all four modes are provided in Fig. 3.

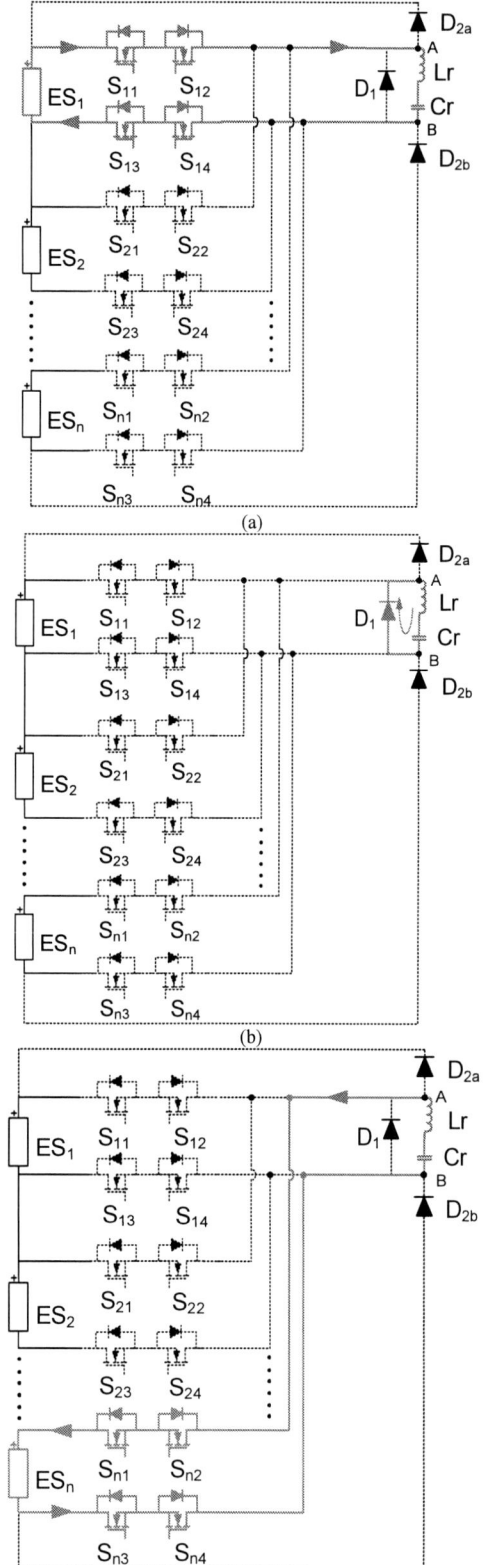

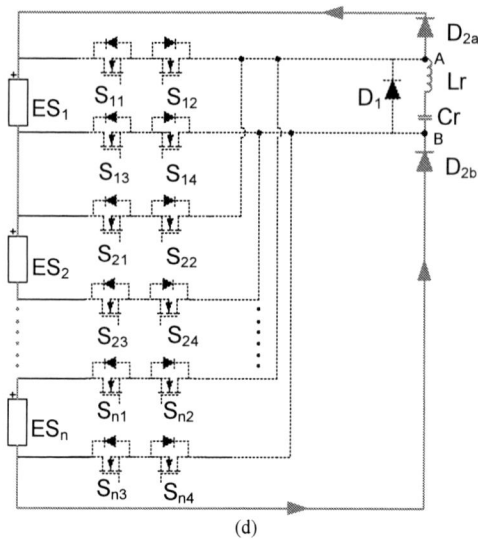

Fig. 2. Current paths of the proposed converter. (a) Mode 1: Energy transportation from the high voltage cell to the LC resonant tank. (b) Mode 2: Energy path when resonant inductor releases its stored energy. (c) Mode 3: Energy transportation from the resonant tank to the low voltage cell. (d) Mode 4: Energy transfer path when resonant inductor releases its stored energy in the opposite direction.

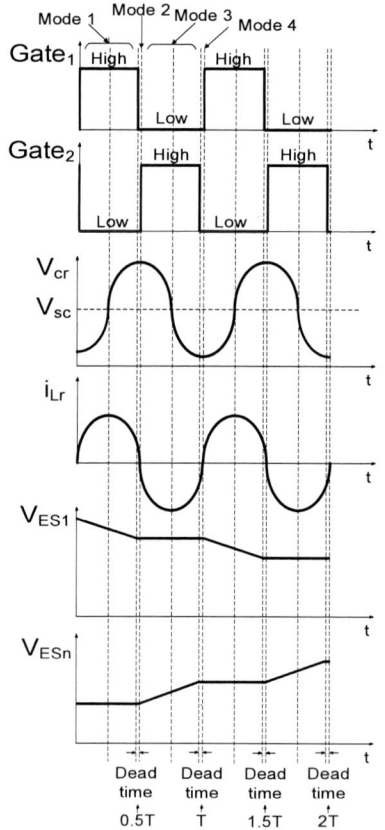

Fig. 3. Analytical waveforms of resonant capacitor voltage, resonant inductor current and supercapacitor voltages in the four modes of operation

During the first half switching cycle, since the voltage of ES_1 is higher than that of ES_n, all four switches associated with ES_1 are turned on. Therefore, ES_1 is connected to the LC resonant tank directly. The charging current flows through the resonant inductor and capacitor, resulting in an increase of the resonant inductor current and resonant capacitor voltage. When the resonant capacitor voltage reaches the same voltage level as ES_1, the voltage across the resonant inductor becomes zero, and the current flowing through the inductor reaches its positive peak value. After this critical point, ES_1 continues to charge the resonant capacitor and the ES_1 voltage is less than the resonant capacitor voltage. Therefore, the voltage across the resonant inductor is now negative. This negative voltage contributes to a gradual decrease of the inductor current. At the moment when all the switches connected to ES_1 are turned off, mode 2 begins.

In mode 2, since the inductor current cannot change its direction immediately, diode D_a is forced on in order to provide a path for the resonant inductor current. When the inductor current returns to zero, the resonant capacitor voltage reaches its positive peak value. The resonant capacitor holds this voltage until the beginning of the next half of the switching cycle.

Mode 3 begins when the switches associated with ES_n conduct. During the second half of switching cycle, the energy transport direction is reversed. The resonant capacitor releases all its reserve energy to ES_n. The resonant inductor current flows in the opposite direction, gradually bringing its voltage back to zero. When the resonant capacitor voltage reaches the same level as ES_n, the resonant inductor current reaches its negative peak. Then the resonant capacitor voltage further decreases, resulting in a low current flowing in the circuit. When the switches associated with ES_n are turned off, mode 4 begins.

In mode 4, the resonant inductor current still cannot change its direction and therefore it forces the conduction of diodes D_{2a} and D_{2b}. The resonant inductor current flows through diode D_{2a}, the energy storage device string and diode D_{2b} back to resonant capacitor. The analysis of the circuit operation for modes 3 and 4 is similar to the analysis in the first half of the switching cycle, i.e. modes 1 and 2. This energy transport process continues in successive switching cycles until the voltage difference of the two energy storage cells reaches a pre-selected target value.

During a switching cycle, the energy storage device cell voltages are changing. However, the voltage variations are small and can be assumed to be constant during one switching cycle. The following analysis assumes that the resistance in the circuit can be neglected.

IV. SIMULATION AND EXPERIMENTAL RESULTS

In order to verify the validity of the theoretical analysis, a simulation analysis was conducted using the power electronics simulation software PSIM from Powersim Technologies. A three-supercapacitor voltage equalization converter system model was built for the topology. Each of the three supercapacitors had a capacitance of 300 F and the voltages of SC_1, SC_2 and SC_3 were initially set to 2.50V, 2.30V and

2.00V, respectively. The sets of switches were controlled by pulse width modulation signals with a duty cycle of 49% each. In each switching period, only two sets of switches were operated. These two sets of switches were manipulated by complementary signals. A dead time of 1% of the switching period was selected to provide a path for the resonant inductor current between the complementary switching actions. The resonant inductor was selected to be 2.2 μH and the resonant capacitor value was 10 μF. The initial voltage of the resonant capacitor was set to zero. Therefore, the resonant frequency was calculated to be 33.9 kHz. In order to operate near resonance, the switching frequency was set to 34 kHz.

The general trend of voltage equalization process is illustrated in Fig. 4. The initial voltage difference of 500 mV between supercapacitors SC_1 and SC_3 is almost completely eliminated, with deviation of less than 10 mV after 900 seconds.

In order to verify the circuit operation, an experimental setup using three supercapacitors was built. Fig. 5 illustrates printed circuit board prototype built for the proposed topology. The components were selected using the voltage and current ratings from the simulation results. Table I. shows the key circuit components selected for the prototype.

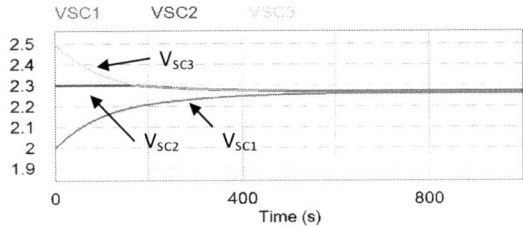

Fig. 4. Voltage equalization characteristics of series connected three-supercapacitor string.

Fig. 5. Experimental prototype of the proposed voltage equalization converter for three 300F supercapacitor system.

TABLE I. Component List

Component	Value and part number
Supercapacitor ES_1-ES_n	300 F (XB3550)
Resonant inductor L_r	2.2 µH (XAL5030MEB)
Resonant capacitor C_r	10 µF (12063D106KAT2A)
MOSFET switches S_{11}-S_{n4}	IRF8721

Three Cooper Bussman PowerStor supercapacitors, XB3550, were selected because of their tolerance of +/- 10%, low equivalent series resistance of 7 mΩ and wide operating temperature range of -25°C to 70°C. The capacitance for each supercapacitor is 300F and the rated voltage is 2.5V. The IRF8721 MOSFET switches were used. The resonant inductor was selected to be 2.2 µH and the resonant capacitor was 10 µF. The initial unbalanced voltages for the three supercapacitors were 2.5V, 2.3V, 2.0V, respectively.

The experimental measured resonant inductor current and resonant capacitor voltage waveforms are illustrated in Fig. 6. The waveform for the resonant inductor current is sinusoidal, with zero average value, as expected. The waveform of the resonant capacitor voltage is also sinusoidal with its average voltage being the average voltage of the supercapacitors being balanced, e.g. 2.25V.

Fig. 7 shows the voltage equalization profile from zero to 990s for the three supercapacitor system. The initial voltage difference of 527 mV was gradually alleviated. The largest deviation between Vsc1 and Vsc3 was reduced to 10 mV when the voltage equalization process terminated, as illustrated in Fig. 8. The total time duration for the tested voltage equalization process was 15 minutes (900 seconds), demonstrating the relatively fast speed of the voltage equalization process.

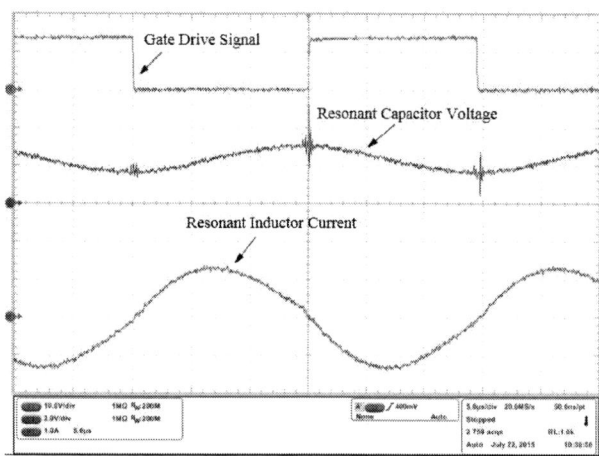

Fig. 6. Measured resonant inductor current and resonant capacitor waveforms during the initial operation of the converter. Gate drive signal scale = 10V/div; Resonant voltage scale =2.0V/div; Resonant current scale = 1.0A/div; Time scale = 10µs/div.

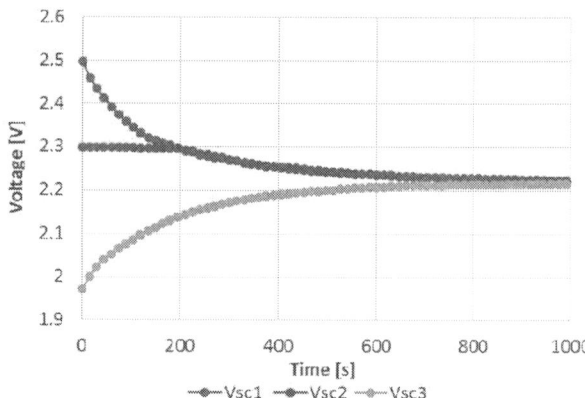

Fig. 7. Measured voltage equalization process supercapacitor cell voltages from zero to 990 seconds for the three 300F supercapacitor system.

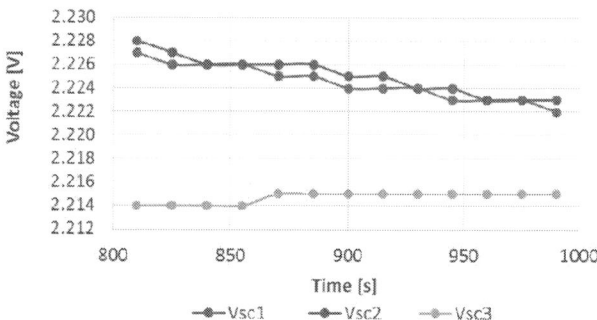

Fig. 8. Measured cell voltages from 810 seconds to 990 seconds during the completion of the equalization process; process terminates at a maximum 10 mV deviation between Vsc2 and Vsc3 at 900 seconds

V. CONCLUSIONS

Energy storage cells, including supercapacitors and lithium-ion battery cells are usually connected in series to meet the high bus voltage requirements for EV applications. Due to cell tolerances, this leads to the voltage imbalance in the string. This paper proposes a novel series resonant converter used as a voltage equalizer for energy storage device cells. The circuit operation was presented along with simulation and experimental results to verify the operation of the proposed equalizer. Experimental results demonstrated that the converter could equalize the voltages for three cells in 900 seconds, starting from a difference of 527 mV to a final voltage difference of 10 mV.

REFERENCES

[1] R. Lu, L. Tian, C. Zhu and H. Yu, "A new topology of switched capacitor circuit for the balance system of ultra-capacitor stacks," in *Proceeding of IEEE Vehicle Power and Propulsion Conference*, 2008.

[2] J. M. Timmermans, P. Zadora, Y. Cheng, J. Van Mierlo and P. Lataire, "Modeling and design of supercapacitors as peak power unit for hybrid electric vehicles," in *Proceeding of IEEE Vehicle Power and Propulsion Conference*, 2005.

[3] W. Gao, "Performance comparison of a fuel cell-battery hybrid power train and a fuel cell-ultracapacitor hybrid power train," *IEEE Trans. Veh. Technol.*, vol. 54, no. 3, pp. 846-855, 2005.

[4] A. M. Imtiaz and F. H. Khan, "Time shared fly-back converter based regenerative cell balancing technique for series connected Li-Ion battery strings," *IEEE Trans. Power Electron.*, vol. 28, no. 12, pp. 5960-5975, 2013.

[5] M. Uno and K. Tanaka, "Single-switch cell voltage equalizer using multi-stacked buck-boost converters operating in discontinuous conduction mode for series-connected energy storage cells," *IEEE Trans. Veh. Technol.*, vol. 60, no. 8, pp. 3635-3645, 2011.

[6] A. Rufer and P. Barrade, "A supercapacitor-based energy-storage system elevators with soft commutated interface," *IEEE Trans. Ind. Appl.*, vol. 38, no. 5, pp. 1151-1159, 2002.

[7] N. Tan, S. Inoue, A. Kobayashi and H. Akagi, "Voltage balancing of a 320-V, 12-F Electric Double-Layer Capacitor bank combined with a 10-kW bidirectional isolated DC--DC converter," *IEEE Trans. Power Electron.,*, vol. 23, no. 6, pp. 2355-2365, 2008.

[8] H. Khant, K. Matsui and M. Hasegawa, "Proposal and improvements of voltage equalizers for EDLCs," in *Proceeding of IEEE International Conference on Power Electronics and Drive Systems*, 2013.

[9] X. Gai, S. Yang, W. Yang and J. Huang, "Analysis on equalization circuit topology and system architecture for series-connected Ultra-capacitor," in *Proceeding of IEEE Vehicle Power and Propulsion Conference*, 2008.

[10] D. Linzen, S. Buller and E. Karden, "Analysis and evaluation of charge-balancing circuits on performance, reliability, and lifetime of super-capacitor systems," *IEEE Trans. Ind. Appl.*, vol. 41, no. 5, pp. 1135-1141, 2005.

[11] A. Xu, S. Xie and X. Liu, "Dynamic voltage equalization for series-connected ultra-capacitors in EV/HEV applications," *IEEE Trans. Veh. Technol.*, vol. 58, no. 8, pp. 3981-3987, 2009.

[12] S. Lambert, V. Pickert, J. Holden, W. Li and X. He, "Overview of super-capacitor voltage equalization circuits for an electric vehicle charging application," in *Proceeding of IEEE Vehicle Power and Propulsion Conference*, 2010.

[13] M. Uno and A. Kukita, "Double-switch equalizer using parallel- or series-parallel-resonant inverter and voltage multiplier for series-connected super-capacitors," *IEEE Trans. Power Electron.*, vol. 29, no. 2, pp. 812-828, 2014.

[14] M. Daowd, N. Omar and P. Van Den Bossche, "Passive and active battery balancing comparison based on MATLAB simulation," in *Proceeding of IEEE Vehicle Power and Propulsion Conference*, 2011.

[15] J. Cao, N. Schofield and A. Emadi, "Battery balancing methods: A comprehensive review," in *Proceeding of IEEE Vehicle Power and Propulsion Conference*, 2008.

[16] Z. Kong, C. Zhu, R. Lu and S. Cheng, "Comparison and evaluation of charge equalization technique for series connected batteries," in *Proceeding of IEEE Power Electronics Specialists Conference*, 2006.

[17] M. Einhorn, W. Roessler and J. Fleig, "Improved performance of serially connected Li-ion batteries with active cell balancing in electric vehicles," *IEEE Trans. Veh. Technol.*, vol. 60, no. 6, pp. 2448-2457, 2011.

[18] J. Shin, G. Seo, C. Chun and B. Cho, "Selective fly-back balancing circuit with improved balancing speed for series connected Lithium-ion batteries," in *Proceeding of IEEE International Power Electronics Conference(IPEC)*, 2010.

[19] H. Park, C. Kim, K. Park, G. Moon and J. O. Lee, "Design of a Charge Equalizer Based on Battery Modularization," *IEEE Trans. Veh. Technol.*, vol. 58, no. 7, p. 3216– 3223, 2009.

[20] A. Baughman and M. Ferdowsi, "Double-Tiered Switched-Capacitor Battery Charge Equalization Technique," *IEEE Trans. Ind. Electron.*, vol. 55, no. 6, p. 2277–2285, 2008.

[21] P. A. Cassani and S. S. Williamson, "Feasibility Analysis of a Novel Cell Equalizer Topology for Plug-In Hybrid Electric Vehicle Energy-Storage Systems," *IEEE Trans. Veh. Technol.*, vol. 58, no. 8, p. 3938–3946, 2009.

[22] M. Tang and T. Stuart, "Selective buck-boost equalizer for series battery packs," *IEEE Trans. Aerosp. Electron. Syst.*, vol. 36, no. 1, p. 201–211, 2000.

[23] H. Park, C. Kim, G. Moon and J. Lee, "Two-Stage Cell Balancing Scheme for Hybrid Electric Vehicle Lithium-Ion Battery Strings," in *Proceeding of IEEE Power Electronics Specialists Conference (PESC)*, 2007.

[24] N. Kutkut, H. Wiegman, D. Divan and D. N. , "Design considerations for charge equalization of an electric vehicle battery system," *IEEE Trans. Ind. Appl.*, vol. 35, no. 1, pp. 28-35, 1999.

978-1-4673-9551-9/16 $31.00 © 2016 IEEE

Implementation of 3.3-kW GaN-Based DC-DC Converter for EV On-Board Charger with Series-Resonant Converter that Employs Combination of Variable-Frequency and Delay-Time Control

Yungtaek Jang, Milan M. Jovanović, Juan M. Ruiz, Misha Kumar, and Gang Liu[1, 2]

Power Electronics Laboratory, Delta Products Corporation, 5101 Davis Drive, Research Triangle Park, NC, USA
[1] Electrical Engineering, Fudan University, Shanghai 200433, People's Republic of China
[2] Delta Power Electronics (Shanghai) Co. Ltd, 201209, People's Republic of China

Abstract—An isolated dc-dc series-resonant converter that is controlled by a combination of variable-frequency and secondary-side-switch delay-time control is employed as the output stage of the on-board charger module (OBCM) that operates with a wide battery-voltage range. The delay-time control which is implemented by the modulation of secondary-side switches is used to assist the conventional variable-switching-frequency control of primary switches to reduce the switching-frequency range. By substantially reducing the switching-frequency range and utilizing gallium nitride (GaN) switches, the overall operating frequency is increased to reduce the sizes of the passive components, and hence, increase power density.

The performance evaluation of the proposed series-resonant converter with delay-time control was done on a 3.3-kW prototype delivering energy from 400-V bus, which is the output of the PFC front end, to a battery operating with voltage range between 180 V and 430 V. The prototype circuit exhibits the maximum full-load efficiency of 97.3% with a switching frequency variation from 144 kHz to 175 kHz over the entire output-voltage range.

I. INTRODUCTION

To maximize the range of electric vehicles (EVs) and plug-in hybrid electric vehicles (PHEVs), it is necessary to utilize the maximum available energy from the battery pack. As a result, the battery-pack voltage varies in a wide range which makes the design of high-efficiency high-power-density on-board charger modules (OBCMs) extremely challenging. Generally, resonant converters with variable switching-frequency control are extensively used in state-of-the-art power supplies that offer the highest power densities and efficiencies [1]-[11]. However, variable switching-frequency control is seen as a drawback of a resonant converter especially in applications with a wide input-voltage and/or output-voltage range [12]-[15]. Specifically, as the input or output voltage range increases, the control frequency range also increases so that driving and magnetic component losses also increase, thereby reducing conversion efficiency. Recently, a new control technique that significantly improves the performance of a series resonant converter that operates with a wide input-voltage range and/or a wide output-voltage

range by substantially reducing their switching-frequency range has been introduced [16]. Reduction in the switching frequency range is achieved by controlling the output voltage with a combination of variable-frequency feedback control and the open-loop delay-time control. Variable-frequency control is used to control the primary switches of the series resonant converter, while delay-time control is used to control secondary-side rectifier switches provided in place of diode rectifiers.

In this paper, the introduced control method is applied to a dc-dc series-resonant converter employed as the output stage of an OBCM operating with a wide output-voltage range. By substantially reducing the switching-frequency range and utilizing gallium nitride (GaN) switches, the overall operating frequency is increased to reduce the sizes of the passive components, and hence, increase power density without sacrificing its performance. The performance of the proposed dc-dc converter was verified on a 3.3-kW prototype operating with a 400-V input and an output that varies between 180 V and 430 V.

II. SERIES-RESONANT CONVERTER WITH COMBINATION OF VARIABLE-FREQUENCY CONTROL AND SECONDARY-SIDE-SWITCH DELAY-TIME CONTROL

Figure 1 illustrates the proposed control method in a series-resonant converter with secondary synchronous rectifiers. As illustrated in Fig. 1(a), output voltage regulation is achieved using a combination of variable-frequency feedback control and open-loop delay-time control. Specifically, variable-frequency control is applied to primary switches S_{P1}-S_{P4}, and delay-time control is applied to secondary-side switches S_{S2} and S_{S3}. Figure 1(b) shows ideal gate waveforms of primary switches S_{P1}-S_{P4}, drain and gate waveforms of secondary switches S_{S2} and S_{S3}, and resonant inductor current i_{LR} in the resonant converter of Fig. 1(a). As shown in Fig. 1(b), switches in the same leg of the primary full-bridge operate in a complementary fashion with a small dead time between their commutations to achieve zero-voltage switching (ZVS). The delay-time control is

978-1-4673-9551-9/16 $31.00 © 2016 IEEE

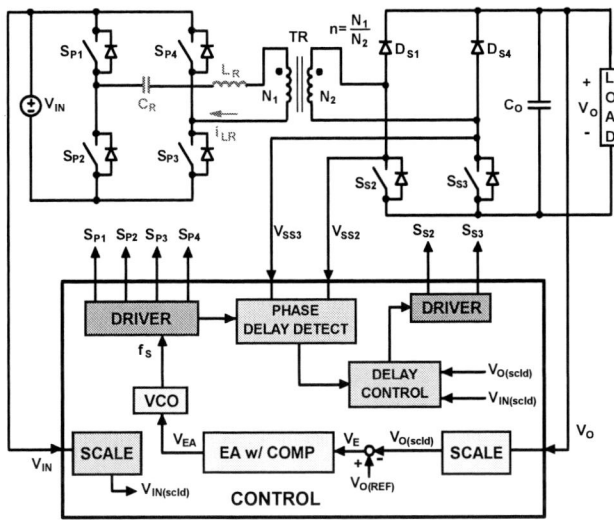

Fig. 1(a)

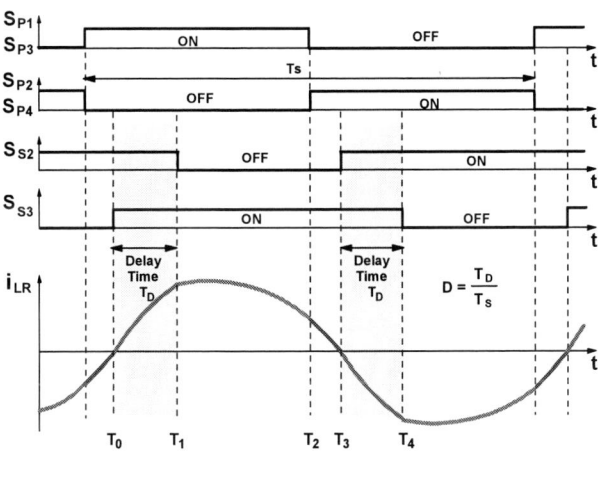

Fig. 1(b)

Fig. 1. Proposed series resonant converter with additional secondary switch control: (a) circuit diagram, (b) control waveforms.

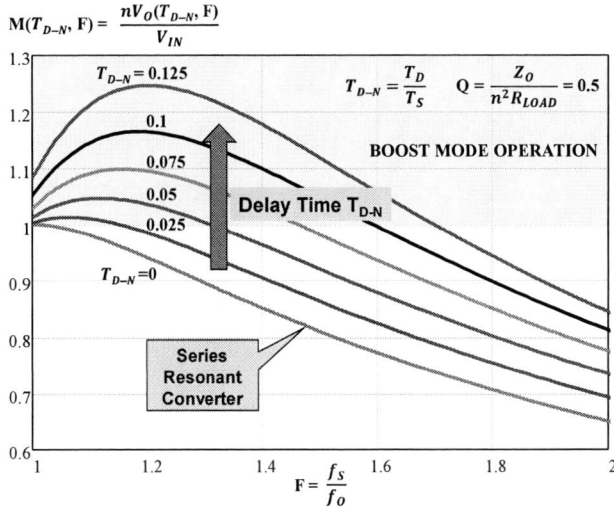

Fig. 2. Input-to-output voltage gain of proposed converter for Q=0.5 [16]. It should be noted that converter operates as series resonant converter when delay time T_{D-N} is set to be zero.

higher turns ratio in the transformer to reduce primary conduction losses and (ii) a higher magnetizing inductance to reduce circulating (i.e., magnetizing) current loss. Because of its boost property, in typical applications, the delay-time control is used in the middle and high output-voltage and/or low input-voltage range.

Dc-conversion ratio $M=nV_O/V_{IN}$ of the series-resonant converter with proposed delay-time control is derived in [16] by using state-plane analysis. For a given Q and normalized delay time $T_{D-N}=T_D/T_S$, dc conversion ratio $M = nV_O/V_{IN}$ can be numerically calculated and plotted as function of normalized frequency $F = f_S/f_O$, as shown in Fig. 2. When delay time T_D is zero, the converter characteristic is the same as that of a conventional series resonant converter. As delay time T_D increases, dc gain M increases and exhibits a boost characteristic. Because delay time T_D depends on input or output voltage, a microcontroller-based control implementation is preferred since the delay time can be easily programmed.

Figure 3 shows detailed control waveforms of the proposed series resonant converter with delay-time control shown in Fig. 1(a). In this implementation, as shown in Fig. 3, primary switches S_{P2} and S_{P4} and secondary switch S_{S3} turn on together at $t=T_0$ (waveforms ① and ⑦), whereas primary switches S_{P1} and S_{P3} and secondary switch S_{S2} turn on together at $t=T_3$ (waveforms ② and ⑪). To implement the delay-time control, the zero crossing of resonant inductor current i_{LR} at $t=T_1$ should be detected for gating of switch S_{S3}, whereas the gating of switch S_{S2} requires zero-crossing detection of the inductor current at $t=T_4$. Generally, the sensing of the zero crossings of resonant current i_{LR} can be done by using a current transformer. However, at light loads, the magnitude of resonant current i_{LR} is too small to be used for reliable detection of the zero crossings. As a result, in this paper, the drain-to-source voltage waveforms of secondary-

implemented by delaying the turn-off of switches S_{S2} and S_{S3} with respect to corresponding zero crossings of resonant current i_{LR} so that both switches S_{S2} and S_{S3} conduct during delay-time intervals [T_0-T_1] and [T_3-T_4] and short the secondary of transformer TR. Because of the shorted secondary of the transformer, the voltage across resonant tank C_R-L_R during delay-time interval [T_0-T_1] is V_{IN} instead of V_{IN}−nV_O which is the case with no time-delay control. Therefore, with the delay-time control, a higher voltage is applied across the resonant inductor and, consequently, a higher amount of energy is stored in resonant inductor L_R. Therefore, at the same input voltage and switching frequency, secondary-side delay-time control provides a higher output voltage compared to the conventional frequency control. This boost characteristic makes optimizing circuit performance possible by selecting: (i) a

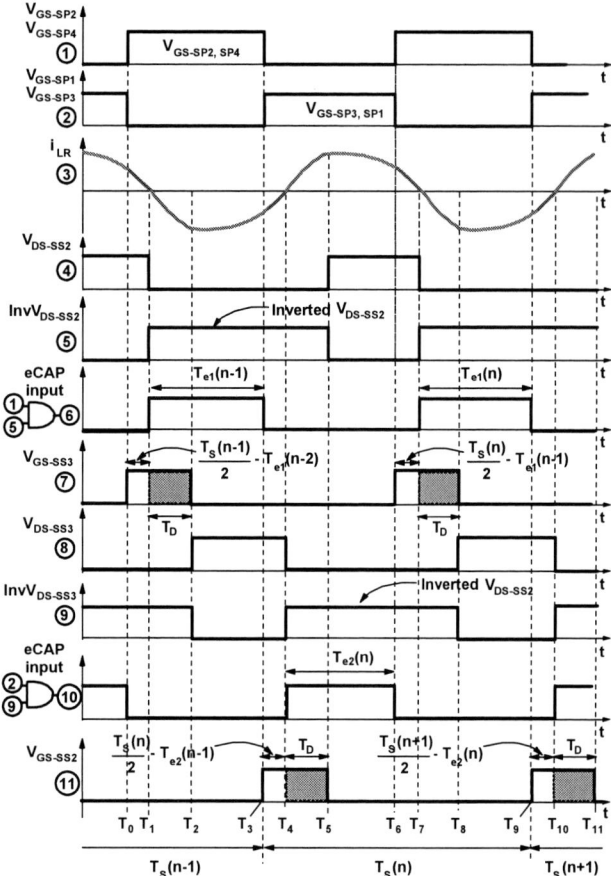

Fig. 3. Detailed control waveforms of proposed series resonant converter shown in Fig. 1.

side switches S_{S2} and S_{S3} are used to indirectly determine the zero crossings.

This zero-current-detection method is based on the fact that at the zero crossings of the secondary current, the drain-to-source voltage of the secondary–side switch experiences an abrupt change without commutation delay. Specifically, for the zero crossings that occur when the secondary current changes from positive to negative, such as that at $t=T_1$, the drain-to-source voltage of switch S_{S2}, V_{DS-SS2} (waveform ④), changes from V_O to zero because of the commutation of the secondary current from rectifier D_{S1} to antiparallel diode of switch S_{S2}. Similarly, for the zero crossings that occur at negative-to-positive secondary-current transitions, e.g., at $t=T_4$, the drain-to-source voltage of switch S_{S3}, V_{DS-SS3} (waveform ⑧), changes from V_O to zero because of the commutation of the secondary current from rectifier D_{S4} to antiparallel diode of switch S_{S3}.

In the digital implementation of the controller, the zero crossings can be determined by the time differences between the primary switch commutation instants and the instants the drain-to-source voltage of the corresponding secondary-side switch transitions from V_O to zero, i.e., by calculating the

durations of time intervals $[T_1\text{-}T_0]$, $[T_4\text{-}T_3]$, $[T_7\text{-}T_6]$, etc. For the calculation of these time intervals at positive-to-negative current transitions, the inverted signal of drain-to-source voltage V_{DS-SS2} (waveform ⑤) and gate-to-source voltage V_{GS-SP2} (waveform ①) are processed by the AND gate as shown in the waveform ⑥ of Fig. 3. The output signal of the AND gate is read by Enhanced Capture (eCAP) Module of the microcontroller (TMS320F28069) that captures the pulse width and stores its value as T_{e1}. During the next switching cycle, the duration of the time intervals $[T_1\text{-}T_0]$, $[T_7\text{-}T_6]$, etc. is calculated by subtracting T_{e1} from one half of the current switching period T_S, i.e., as $T_S[n]/2\text{-}T_{e1}[n\text{-}1]$. Finally, the gate pulse width of switch S_{S3} is determined by the sum of the calculated time interval ($T_S/2\text{-}T_{e1}$) and required delay time T_D, as shown in the waveform ⑦ of Fig. 3. It should be noted that although the duration of these time intervals changes over the input voltage, output voltage, and load range, they can be considered to be near constant during a single switching cycle since the voltage and load changes are much slower than the switching period. For gating of switch S_{S2}, the time interval between primary switch gate transition at $t=T_3$ and zero crossing of resonant inductor current i_{LR} at $t=T_4$ should be calculated. For the calculation of these time interval the inverted signal of drain-to-source voltage V_{DS-SS3} (waveform ⑨) and gate voltage V_{GS-SP1} (waveform ②) are processed by the AND gate as shown in the waveform ⑩ of Fig. 3. The gate pulse width of switch S_{S2} is determined by the sum of the calculated time interval ($T_S[n]/2\text{-}T_{e2}[n\text{-}1]$) and required delay time T_D, as shown in the waveform ⑪.

It should be noted that in the control implementation in Fig. 3, secondary–side switches S_{S2} and S_{S3} are conducting only during time intervals that implements the delay-time control, i.e., they are not used as synchronous-rectifier switches. As a result, the body diodes of the switches are utilized as output rectifiers. However, the control method in Fig. 3 can also be extended to synchronous implementation operation by extending the conduction of the secondary-side switches beyond that required by the delay-time control.

III. DESIGN CONSIDERATIONS

For performance evaluation, the proposed dc-dc converter for EV on-board charger has been designed and built according to the following key specifications:

Input voltage V_{IN}:	400 V_{DC}
Output voltage V_O:	180-430 V_{DC}
Maximum output current $I_{O\text{-}MAX}$:	11 A
Maximum output power $P_{O\text{-}MAX}$:	3.3 kW
Efficiency η:	>96% above 50% load
Dimension:	250 mm × 180 mm × 75 mm

A. Switching Frequency Selection

It is well understood that switching frequency selection is based on the trade-off between efficiency and size, i.e., power density. In this design, the minimum frequency is set at 140 kHz, whereas the maximum frequency is limited to 350 kHz to meet power density for the specified dimension.

B. Transformer Design

The turns ratio n=N_1/N_2=20/16 of transformer TR is chosen to make input-to-output control characteristic M be 1 when the output voltage is approximately 320 V with 400-V input. As a result, the delay time of the secondary switching is set to zero when the converter operates from 180 V to approximately 320 V output so that the proposed converter operates as a series resonant converter. However when the output voltage increases above 320 V, the controller starts monotonically increasing the delay time to provide the boost characteristic to maintain the output voltage regulation with the selected turns ratio of transformer TR.

The transformer has following specifications:
Core: A pair of PQ4040-PC40 ferrite cores.
Primary winding: N_1 = 20 turns, Litz wire (435 strands /AWG #40).
Secondary winding: N_2 = 16 turns, Litz wire (435 strands /AWG #40).
Air gap: 0.02 mm.

The measured magnetizing and leakage inductances are 1.05 mH and 1.6 µH, respectively. The maximum flux density at steady state operation is approximately 0.23 T, which gives plenty of margin from the saturation flux of the ferrite core.

C. Resonant Capacitor Design

A high-frequency film capacitor is a suitable candidate for resonant capacitor C_R because of its cost and long-term reliability. However, its maximum permissible ac voltage is inversely proportional to frequency. For example, a 33-nF, 2000-VDC, FKP 1 type film capacitor from WIMA has the ac-voltage rating of 700-$V_{AC\text{-}RMS}$ at line-frequency of 50/60 Hz and only 180 $V_{AC\text{-}RMS}$ at approximately 140-kHz.

The peak voltage of the resonant capacitor can be determined by recognizing that during a half switching period, the positive resonant-inductor current continuously charges the resonant capacitor so that the voltage across resonant capacitor C_R increases from its negative to its positive peak, i.e., changes for $2V_{CR\_PK}$. The relationship between the resonant-capacitor voltage change and stored charge q_{CR}^{tot} during the half switching cycle is given by

$$q_{CR}^{tot} = \int_{T_0}^{T_3} i_{LR}dt = \int_{T_0}^{T_1} i_{LR}dt + \int_{T_1}^{T_3} i_{LR}dt = C_R \cdot 2V_{CR\_PK} \ (1)$$

Since the average of the resonant inductor current i_{LR} over a half switching period is equal to the output (load) current reflected to the primary, stored charge q_{CR}^{tot} during the half switching cycle can be calculated as

$$q_{CR}^{tot} = \int_{T_1}^{T_3} i_{LR}dt = \frac{I_O}{n} \cdot \frac{T_S}{2} \tag{2}$$

From equations (1)-(2), it follows that

$$V_{CR\_PK} = \frac{I_O \cdot T_S}{4nC_R} = \frac{I_O}{4nC_R f_S}, \tag{3}$$

where n=N_1/N_2 is the turns ratio of transformer TR.

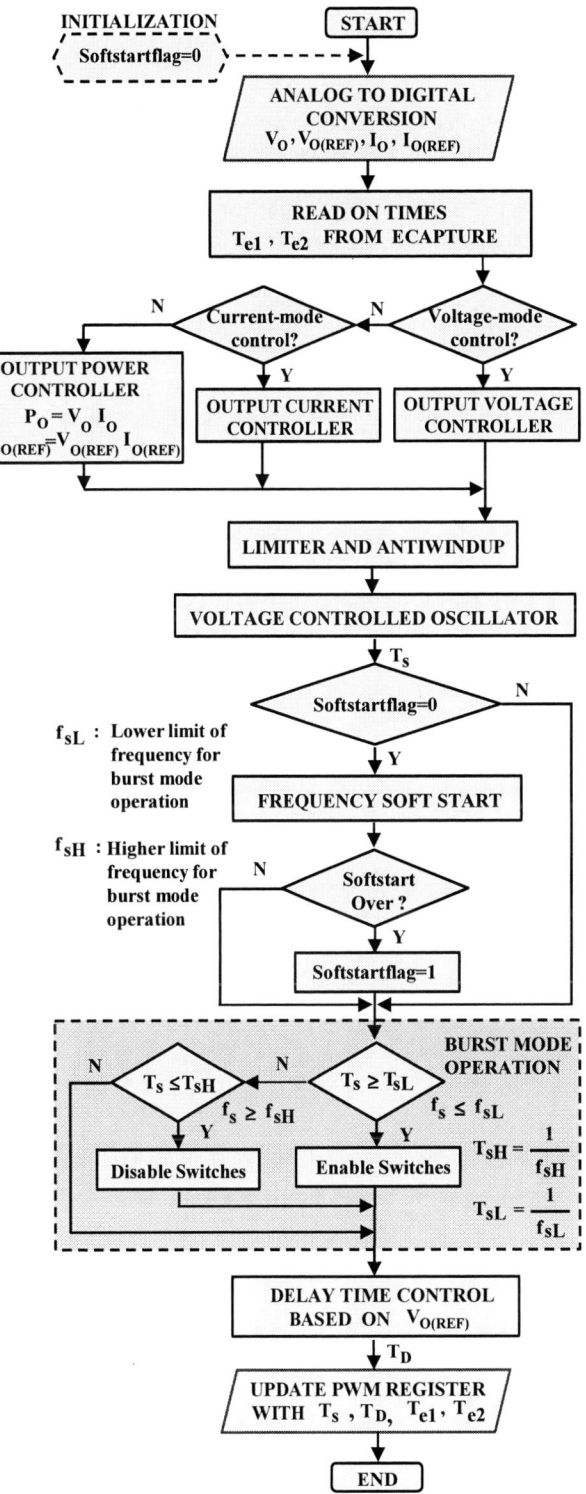

Fig. 4.   Flow chart of proposed control scheme.

The maximum of capacitor peak voltage V_{CR_PK} occurs at minimum frequency (minimum input voltage) and full load. The peak voltage across a 33-nF capacitors is calculated as

$$V_{CR_{PK}} = \frac{I_{O(max)}}{4nC_R f_{S(min)}} = \frac{11}{4 \cdot \frac{20}{16} \cdot 33 \cdot 10^{-9} \cdot 140 \cdot 10^3}$$
$$= 476 \ V$$

Since the voltage waveform across the resonant capacitors is a sine wave shape, rms voltage across the serially connected two capacitors is approximately 336 V. To keep the maximum voltage stress of capacitor within the 180-$V_{AC\text{-}RMS}$ limit, two sets of series-connected two 33-nF, 2000-VDC, 700-$V_{AC\text{-}RMS}$, FKP 1 type film capacitors are connected in parallel. As a result, total capacitance of resonant capacitor C_R is 33 nF.

D. Resonant Inductor Design

To guarantee the above-resonant-frequency operation with ±10% inductance tolerance of L_R and ±5% capacitance tolerance of C_R, the resonant-tank inductor value is calculated by assuming resonant frequency of 130 kHz which is 10 kHz lower than the minimum specified switching frequency. For the 130-KHz resonant frequency and 33-nF resonant capacitor value, the required inductance of the resonant-tank inductor is approximately 46 µH. To obtain this inductance, the resonant inductor was built using a pair of ferrite cores (PQ-40/40, PC40) with a 3.8-mm gap in all three legs. The winding was implemented with 28 turns of

Litz wire (435 strands of AWG#40) to reduce the fringing-effect-induced winding loss near the gap of the inductor core. For this inductor design, the maximum flux density which occurs at full load and the minimum switching frequency is approximately 0.28 T.

E. Semiconductor Device Selection

Because the voltage stresses of primary switches S_{P1} - S_{P4} and secondary switches S_{S1} - S_{S4} are approximately equal to input voltage V_{IN} and output voltage V_O, respectively, i.e., they are below 450 V, it is necessary to use switches that are rated at least 600-V to maintain the desirable design margin of 20%. In the prototype circuit a TPH3205WT GaN MOSFET (V_{DS} = 600 V, R_{DS} = 0.051 Ω, C_{OSS}=108 pF, Q_{rr}=138 nC) from Transphorm was used for each switch. It should be noted that the body diode of the selected switch has relatively small reverse recovery charge.

F. Control Implementation

The output control of the converter was implemented by a TMS320F28069 digital controller from TI. To implement the battery-charging profile, a constant-current, constant-power, and constant-voltage output control is employed. The flow chart of the employed control is shown in Fig. 4. It should be noted that converter starts with frequency soft start, i.e., the switching frequency starts from the maximum and gradually reduces until the output voltage reaches the desired level. Moreover, the converter enters burst-mode operation at very light load.

IV. EXPERIMENTAL RESULTS

The performance of the proposed converter with the delay-time control shown in Fig. 1 was evaluated on a 3.3-kW prototype circuit that is designed to operate from a 400-V input and deliver power over 180-430-V output voltage range as described in Section III. Figure 5 shows a view of the on-board charger module with the cover removed. The on-board charger in Fig. 5 consists of the proposed dc-dc stage and the front-end PFC stage which is not discussed in this paper.

Figure 6 shows the circuit diagram along with component specifications. It should be noted that all primary and secondary switches are GaN devices. Since in this implementation secondary switches S_{S1} and S_{S2} are not operated as synchronous rectifiers, their body diodes are utilized as output rectifiers.

Figure 7 shows the measured waveforms of gate and drain voltages of primary switch S_{P2}, resonant current i_{LR}, and resonant capacitor voltage V_{CR} of the experimental circuit when it delivers full power at 430-V, 320 -V, and 180-V output. The waveforms show ZVS of primary switch S_{P2}. Although Fig. 7 only shows waveforms of switch S_{P2}, the waveforms of all other primary switches are similar to that of switch S_{P2} and achieve ZVS as well. As shown in Fig. 7, during a half switching period, the positive resonant-inductor current continuously charges the resonant capacitor so that the voltage across resonant capacitor C_R increases

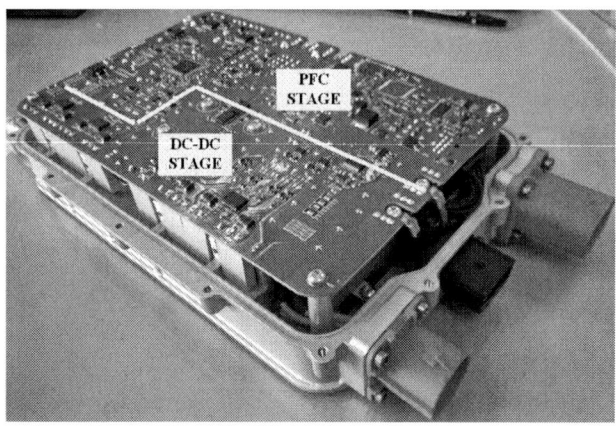

Fig. 5.   Experimental prototype circuit.

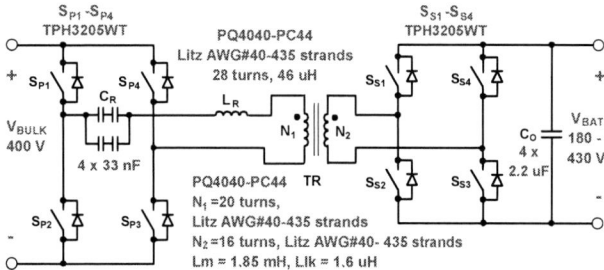

Fig. 6.   Prototype circuit diagram with details of employed power components. It should be noted that all primary and secondary switches are GaN devices

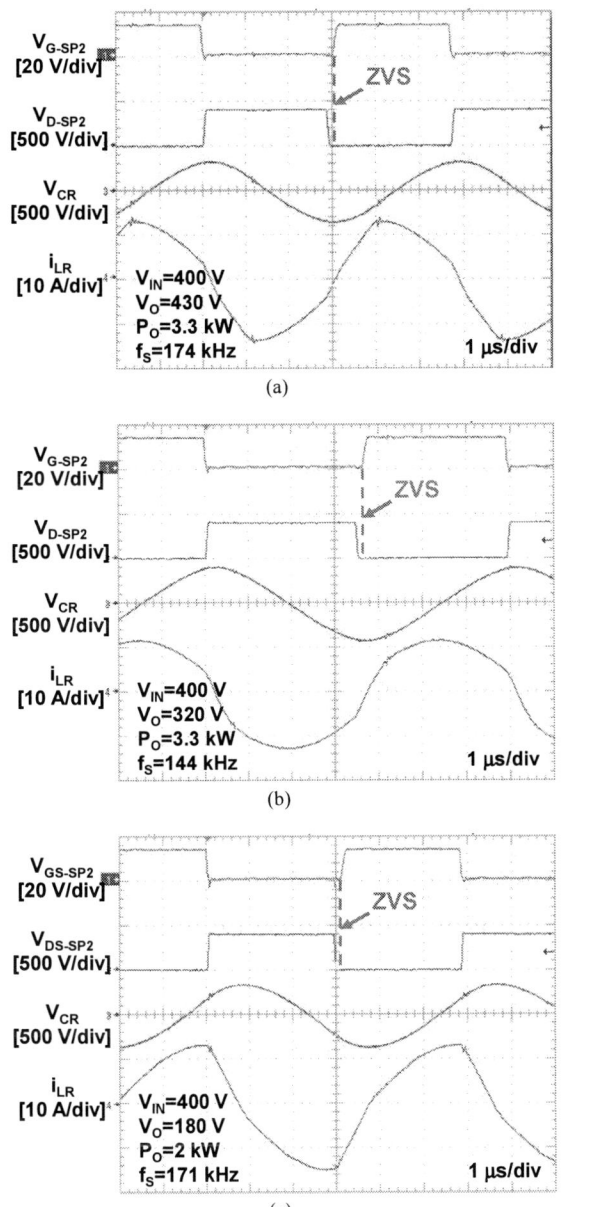

Fig. 7. Measured drain and gate voltage waveforms of primary switch S$_{P2}$, voltage waveforms of resonant capacitor C$_R$, and current waveform of resonant inductor L$_R$ for output voltages: (a) 430 V; (b) 320 V; and (c) 180 V. Time scale is 1 μs/div.

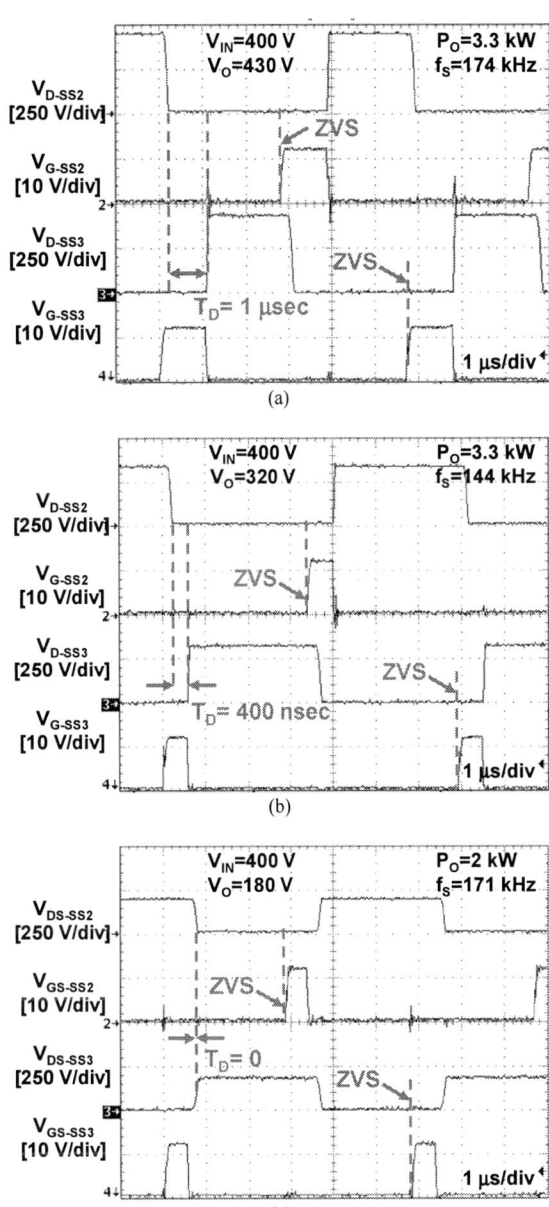

Fig. 8. Measured drain and gate voltage waveforms of secondary switches S$_{S2}$ and S$_{S3}$ for output voltages: (a) 430 V; (b) 320 V; and (c) 180 V. Time scale is 1 μs/div.

from its negative to its positive peak, i.e., changes for 2V$_{CR\_PK}$. The maximum peak capacitor voltage occurs at 320 V output as shown in Fig. 7(b), which is approximately 400 V. As a result, the rms voltage across each 33-nF resonant capacitor is approximately 140 V at 144 kHz.

Figure 8 shows the measured waveforms of gate and drain voltages of secondary switches S$_{S2}$ and S$_{S3}$ of the experimental circuit when it delivers full power at 430-V, 320-V, and 180- V output. As shown in Fig. 8, all the secondary switches operate with ZVS. Figure 8 also shows

delay time T$_D$. Delay time T$_D$ is the period when both drain voltages of switches S$_{S2}$ and S$_{S3}$ are zero, i.e., both switches conduct and the secondary winding of TR is shorted. It should be noted that the turns ratio of transformer TR is chosen to make input-to-output control characteristic M equal to 1 when the output voltage is approximately 320 V. However to properly regulate the output voltage with component tolerances as well as under a transient condition, delay-time T$_D$ is introduced from 260 V output. As a result, the delay time of the secondary switching is set to zero when the converter operates from 180 V to approximately 260 V output. When the output voltage increases above 260 V, the controller increases the delay time to provide the boost characteristic to maintain the output voltage regulation with the selected turns ratio of transformer TR.

Figure 9 shows measured efficiency of the prototype converter as a function of the output voltage. It should be noted that the converter exhibits the best full-load efficiency when the output voltage is between 240-340 V, which is the operating range it most frequently operates. Specifically, the converter exhibits the maximum full-load efficiency of 97.3% at 280-V output. Figure 10 shows the measured full-load switching frequency of the experimental prototype as a function of the output voltage. The measured full-load switching frequencies are in a 144-kHz to 175-kHz range over the entire output-voltage range, which enables more effective optimization of all the magnetic components and filter capacitors. As seen from Fig. 11, the delay-time control is activated above 260-V output.

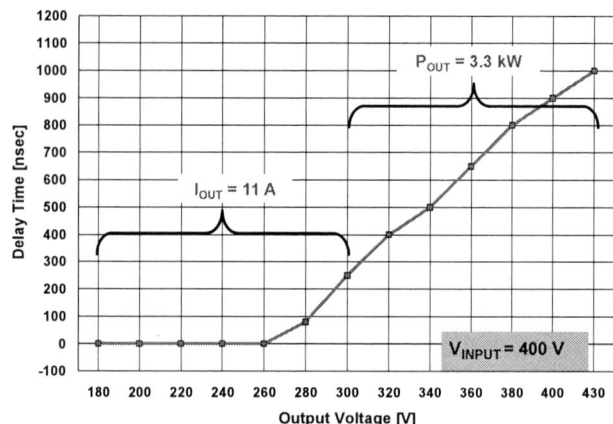

Fig. 11. Measured delay-time T_D of experimental prototype as function of output voltage.

V. SUMMARY

In this paper, an isolated dc-dc series-resonant converter that is controlled by a combination of variable-frequency and secondary-side-switch delay-time control has been introduced. The proposed converter is designed as the output stage of an OBCM that operates with a wide battery-voltage range. The delay-time control which is implemented by the modulation of secondary-side switches is used to assist the conventional variable-switching-frequency control of primary switches. The performance evaluation of the proposed series-resonant converter with delay-time control was done on a 3.3-kW prototype delivering energy from 400-V bus, which is the output of the PFC front end, to a battery operating with voltage range between 180 V and 430 V. The prototype circuit exhibits the maximum full-load efficiency of 97.3% with a full-load switching-frequency variation from 144 kHz to 175 kHz over the entire output-voltage range.

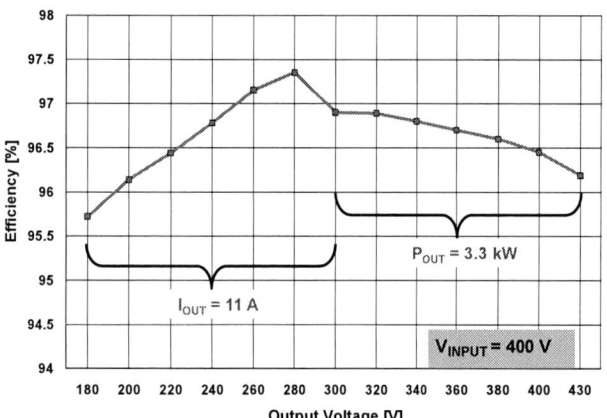

Fig. 9. Measured efficiencies of the experimental prototype as functions of output voltage.

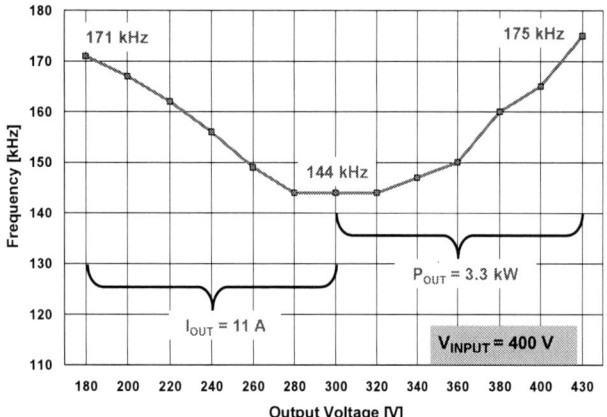

Fig. 10. Measured switching frequency of experimental prototype as function of output voltage. Frequency range is significantly narrowed by delay-time control.

REFERENCES

[1] M. Carrasco, E. Galvan, G. Escobar, R. Ortega, and A. M. Stankovic, "Analysis and experimentation of nonlinear adaptive controllers for the series resonant converter," IEEE Trans. Power Electron., vol. 15, no. 3, pp. 536–544, May 2000.

[2] B. Yang, R. Chen, and F.C. Lee, "Integrated magnetics for LLC Resonant Converter," in IEEE Applied Power Electronics Conf. Rec. 2002, pp. 346-351.

[3] B. Lu, W. Liu, Y. Liang, F.C. Lee, and J.D. Van Wyk, "Optimal design methodology for LLC resonant converter," in IEEE Applied Power Electronics Conf. Rec. 2006, pp. 533-538.

[4] G. Ivensky, S. Bronshtein, and A. Abramovitz, "Approximate analysis of resonant LLC DC-DC converter," IEEE Trans. Power Electron., vol. 26, no. 11, pp. 3274–3284, Nov. 2011.

[5] H. Molla-Ahmadian, A. Karimpour, N. Pariz, and F. Tahami, "Hybrid modeling of a DC–DC series resonant converter: Direct piecewise affine approach," IEEE Trans. Circuits Syst. I, Fundam. Theory Appl., vol. 18, no. 5, pp. 3112–3120, Jul. 2012.

[6] X. Fang, H. Hu, Z. J. Shen, and I. Batarseh, "Operation mode analysis and peak gain approximation of the LLC resonant converter," IEEE Trans. Power Electron., vol. 27, no. 4, pp. 1985–1995, Apr. 2012.

[7] R. Beiranvand, B. Rashidian, M.R. Zolghadri, S.M.H. Alavi, "A design procedure for optimizing the LLC resonant converter as a wide output range voltage source," IEEE Transactions on Power Electronics, vol. 27, No. 8, pp. 3749-3763, August 2012.

[8] F. Musavi, M. Cracium, D.S. Guatam, W. Eberle, and W.G. Dunford, "An LLC resonant DC-DC converter for wide output voltage range battery charging applications," IEEE Transactions on Power Electronics, vol. 28, No. 12, pp. 5437-5445, December 2013.

[9] J. Deng, S. Li, S. Hu, C.C. Mi, and R. Ma, "Design Methodology of LLC Resonant Converters for Electric Vehicle Battery Chargers," IEEE Transactions on Vehicular Technology, vol. 63, No. 4, pp. 1581-1592, May 2014.

[10] M. Momeni, H.M. Kelk, and H. Talebi,, "Rotating switching surface control of series-resonant converter based on a piecewise affine model," IEEE Transactions on Power Electronics, vol. 30, No. 3, pp. 1762-1772, March 2015.

[11] Z. Fang, T. Cai, S Duan, and C. Chen, "Optimal Design Methodology for LLC Resonant Converter in Battery Charging Applications Based on Time-Weighted Average Efficiency," IEEE Transactions on Power Electronics, vol. 30, No. 10, pp. 5469-5483, October 2015.

[12] F.S. Tsai, P. Materu, and F.C. Lee, "Constant-frequency clamped-mode resonant converters," IEEE Transactions on Power Electronics, vol. 3, No. 4, pp. 460-473, October 1988.

[13] B. Yang, P. Xu, and F.C. Lee, "Range winding for wide input range front end DC/DC converter," in IEEE Applied Power Electronics Conf. Rec. 2001, pp. 476-479.

[14] B.C. Kim, K.B. Park, and G.W. Moon, "Asymmetric PWM control scheme during hold-up time for LLC resonant converter," IEEE Transactions on Industrial Electronics, vol. 59, no. 7, pp. 2992-2997, July 2012.

[15] I.H. Cho, Y.D. Kim, and G.W. Moon, "A half-bridge LLC resonant converter adopting boost PWM control scheme for hold-up state operation," IEEE Transactions on Power Electronics, vol. 29, No. 2, pp. 841-850, February 2014.

[16] Y. Jang, M.M. Jovanovic, J.M. Ruiz, and G. Liu, "Series-Resonant Converter with Reduced-Frequency-Range Control," in IEEE Applied Power Electronics Conf. Rec. 2015, pp. 1453-1460.

978-1-4673-9551-9/16 $31.00 © 2016 IEEE

Dual Active Bridge-Based Full-Integrated Active Filter Auxiliary Power Module for Electrified Vehicle Applications with Single-Phase Onboard Chargers

Ruoyu Hou, *Student Member, IEEE*, and Ali Emadi, *Fellow, IEEE*
Electrical and Computer Engineering Department
McMaster University, Hamilton, ON, Canada
E-mail: hour@mcmaster.ca, emadi@mcmaster.ca

Abstract—In single-phase AC onboard chargers for electrified vehicles, second-order harmonic current and corresponding ripple voltage exist on the DC bus. The low-frequency harmonic current is normally filtered using a bulk capacitor or an additional active filter (AF) circuit. This presents an obstacle for improving the power density as well as reducing the cost. In this paper, a dual active bridge (DAB)-based full-integrated active filter auxiliary power module (AFAPM) is presented so that when the high-voltage (HV) battery is charging, the DAB converter acts as an active filter; and when the low-voltage (LV) battery needs to be charged, the DAB converter works as a LV battery charger. Therefore, the required DC-link capacitance in the HV battery can be reduced significantly and, hence, the cost and volume of the dual-voltage system in the electrified vehicle applications can be reduced. The effectiveness of the proposed method is verified by simulation results.

Index Terms—Active filters, auxiliary power modules, battery chargers, dual active bridge, electrified vehicles, harmonic current, system integration.

I. INTRODUCTION

Transportation electrification is a paradigm shift from less-efficient internal combustion engine (ICE) based vehicles towards more-efficient and cleaner electrified vehicles to enable a more sustainable transportation system. Electrification can occur in both vehicular propulsion and non-propulsion loads [1]. The dual-voltage charging system is now widely applied in the electrified vehicle applications [2]. A low-voltage (LV) battery charger auxiliary power module (APM) is placed between the high-voltage (HV) system and LV system to convert the HV power into LV power and, thus, feed the 12V non-propulsion loads. The APM also provides galvanic isolation between the HV and LV systems, which protects the LV system from HV [3]. Different isolated DC/DC converter topologies for the LV battery charger APM in the electrified vehicle applications are compared in [4].

In the meanwhile, in electrified vehicles, the AC HV charging system is commonly a single-phase onboard charger mounted inside the vehicle and is connected to the grid when the vehicle is plugged in [5][6]. In the AC HV battery charger,

significant second-order harmonic current exist on the DC-link after the power factor correction (PFC) boost converter [7]. This harmonic current is introduced by the fluctuation of the instantaneous input power. As a battery charger, an upper bound on the AC ripple current is typically defined as 10% of the rated battery current to avoid degradation of the battery life [8]. Traditionally, this pulsating harmonic energy is mainly stored in a HV DC-link capacitor bank, which has a low lifetime and potential safety issues if electrolytic capacitors are used; if film capacitors are applied, this high voltage, high capacitance capacitor bank occupies significant volume and weight, which becomes the major power density barrier of the battery charger [9]. So far, researchers have made large efforts to reduce the DC-link capacitance for the AC single-phase chargers [7][10][11]. Another method to eliminate the second-order harmonic on the DC-link is to add an active power filter in parallel [11]. However, extra harmonic energy storage components, power switches as well as their corresponding driver circuits and heat sink have to be applied. In [12] and [13], we evaluated the integrated active filter auxiliary power module (AFAPM) concept and proposed a primary-integrated AFAPM topology in the electrified vehicles. With this method, the AFAPM converter is not only a LV battery charger, but also an active filter for the HV battery charger when the HV battery is charging.

This paper presents a proposed dual active bridge (DAB)-based full-integrated AFAPM for the electrified vehicle applications. With the proposed method, the active filtering is achieved and the bulk DC-link capacitor in the single-phase AC HV battery chargers can be reduced significantly without extra switches and harmonic energy storage components. Therefore, the cost and volume of the dual-voltage system in the electrified vehicle applications can be reduced.

The rest of the paper is organized as follows: Section II discusses the DAB converter design for the LV battery charger APM. The selections of transformer's turn ratio and the leakage inductance are presented. In Section III, the proposed DAB feed-forward active filtering control is presented so that the DAB can be used as an active filter to assimilate the second-order harmonic from the HV battery

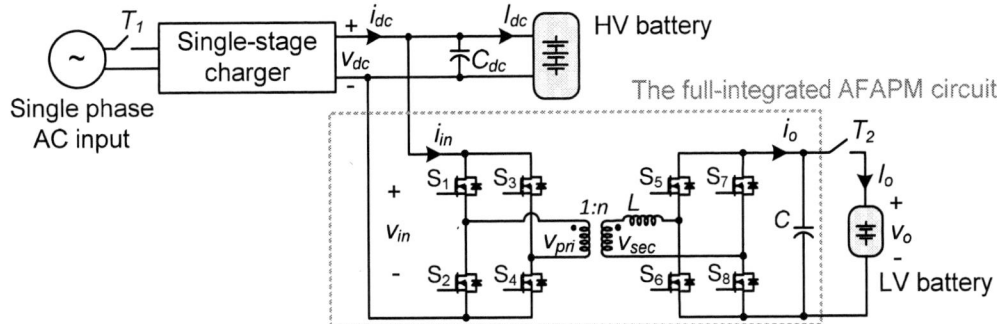

Fig. 1: The dual-voltage charging system with proposed DAB-based full-integrated AFAPM

charger's DC link. Section IV presents the integrated output capacitor requirements for both the LV battery charging and active filtering modes. Section V shows the simulation results as well as integration results from switches VA ratings aspect. Conclusions are given in Section VI.

II. DAB-BASED LV BATTERY CHARGER APM

DAB is a promising topology for isolated and bi-directional power conversion due to zero voltage switching (ZVS) for both primary and secondary bridges, utilization of parasitic, and fixed switching frequency [14]-[16]. In this paper, the single-phase-shift (SPS) modulation is applied to the DAB for simplicity. For electrified vehicle applications, a general dual-voltage charging system specification can be found in Table I. Generally, the onboard AC level 2 HV battery charger is sized around 3.3kW to 6.6kW [5][17]. The LV battery charger is sized around 2.4kW to provide sufficient power for the 12V non-proplusion loads [3]. The dual-voltage charging system with proposed DAB-based full-integrated AFAPM is shown in Fig. 1. The DAB works as a LV battery charger when the vehicle is running on the road; and works as an active filter when the vehicle at charging station and the HV battery is charging. The different AFAPM-based dual-voltage system structures and their corresponding system controls can be found in [12][13] for more details.

Table I: A general dual-voltage system specification

Parameters	Value	Parameters	Value
HV battery charger output power	4kW	LV battery charger output power	2.4kW
HV battery DC bus voltage V_{dc}	400V	HV battery voltage V_{in}	200V-400V
AC Input voltage	240V	LV battery voltage V_o	12V
AC supply frequency f	60Hz	LV battery charging current I_o	200A

The small signal analysis of the DAB can be found in [18]-[20]. Typically, when designing the DAB converter, we need to consider the transformer turn ratio and duty cycle range to achieve ZVS; and consider the leakage inductance to achieve maximum power.

The DAB soft switching constrains can be found in [15][20]. The limits for the duty cycle d to maintain ZVS can be obtained as,

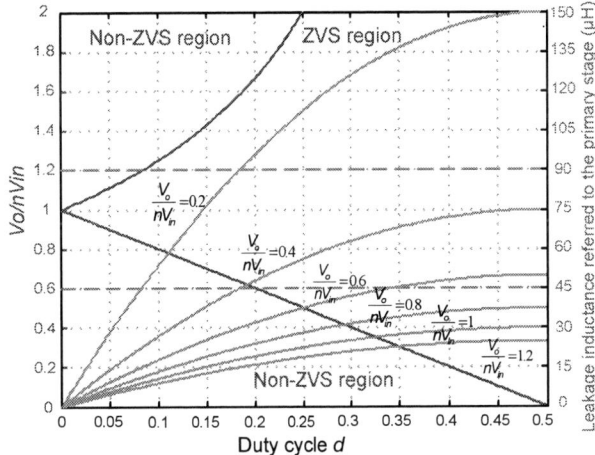

Fig. 2: DAB-based LV battery charger design map

$$\begin{cases} d > 0.5 - \dfrac{nV_{in}}{2V_o}, if\ V_o \geq nV_{in} \\ \\ d > 0.5 - \dfrac{V_o}{2nV_{in}}, if\ V_o \leq nV_{in} \end{cases} \quad (1)$$

where n is the transformer turn ratio. A more comprehensive analysis for the necessary and sufficient conditions to achieve ZVS is conducted in [18].

Another important design issue is the leakage inductance L. The supplied power equation of DAB can be drawn as follows [16],

$$P = \frac{nV_oV_{in}}{2Lf_s}d(1-d) \quad (2)$$

where f_s is the switching frequency, d is the duty cycle, and L is the leakage inductance referred to the secondary stage. From (2), it is clear that P is inversely proportional to the L. Hence, there is a trade-off between peak current and maximum output power for the leakage inductance selection.

Considering both the soft switching and leakage inductance constrains, the DAB-based LV battery charger design map can be drawn in Fig. 2. Clearly, with the V_o/nV_{in} closer to 1, the range to achieve ZVS is wider. Hence, in this paper, 1/20 is chosen as the transformer turn ratio n. As the HV battery V_{in} varies between 200V to 400V, it gives us a V_o/nV_{in} between around 0.6 and 1.2. Hence, in order to be able to provide 2.4kW maximum power to the LV battery, the leakage inductance L referred to the secondary stage is chosen as 62.5nH.

In addition, another design consideration with SPS modulation is the well-known circulating current [21]. Typically, with smaller duty cycle applying to the converter, the less circulating current will be produced. In fact, this reactive power can be eliminated by some other control methods, like dual-phase-shift (DPS) or extended-phase-shift (EPS), etc. Readers can refer to paper [21][22] for more details.

III. PROPOSED DAB ACTIVE FILTERING CONTROL

The single-phase input AC voltage and current can be written as,

$$\begin{cases} u_s(t) = U_s sin\omega t \\ i_s(t) = I_s sin\omega t \end{cases} \tag{3}$$

where U_s and I_s are the voltage and current peak values, respectively; and ω is the supply angular frequency. Accordingly, the supply power P_{in} from the AC source is,

$$P_{in} = u_s(t)i_s(t) = P_o - P_r = \frac{U_s I_s}{2} - \frac{U_s I_s}{2} cos2\omega t \tag{4}$$

It can be observed that the first part P_o is the power feed to the DC load, and the second part P_r is the harmonic power caused by the second-order harmonic. Clearly, the amplitude of the ripple power is equal to the value of DC output power. In this paper, since it's a 4kW HV battery charger, the peak ripple power is also equal to 4kW. As the HV DC charging voltage and current equal to 400V and 10A respectively, we can consider the second-order harmonic current i_r as,

$$i_r = I_{dc} cos 2\omega t \tag{5}$$

where I_{dc} is 10A. In fact, in order to assimilate the harmonic current into the AFAPM, this harmonic current i_r is exactly the input current to the AFAPM. Hence, the key to control the DAB as an active filter to assimilate this harmonic is to find out the relationship between the input current and the duty cycle, and then let the average input current equal to the harmonic current i_r.

In this paper, feed-forward control is applied. The duty cycle is calculated directly based on the input current and other corresponding parameters. Fig. 3 shows the typical waveforms of a DAB converter in the buck phase and boost phase, respectively. For both buck and boost phases, the inductor current I_1 and I_2 can be expressed as,

$$I_1 = \frac{1}{4Lf_s}[nV_{in}(2d-1) + V_o] \tag{6}$$

$$I_2 = \frac{1}{4Lf_s}[nV_{in} + V_o(2d-1)] \tag{7}$$

During the buck phase, in order to calculate the average input current, the shaded areas A_1 and A_2 need to be obtained as shown in Fig. 3a. This requires knowing the expression of interval Δt_x. The $tan\theta$ needs to be calculated, and thus, Δt_x can be obtained easily. The $tan\theta$ and Δt_x can be obtained as,

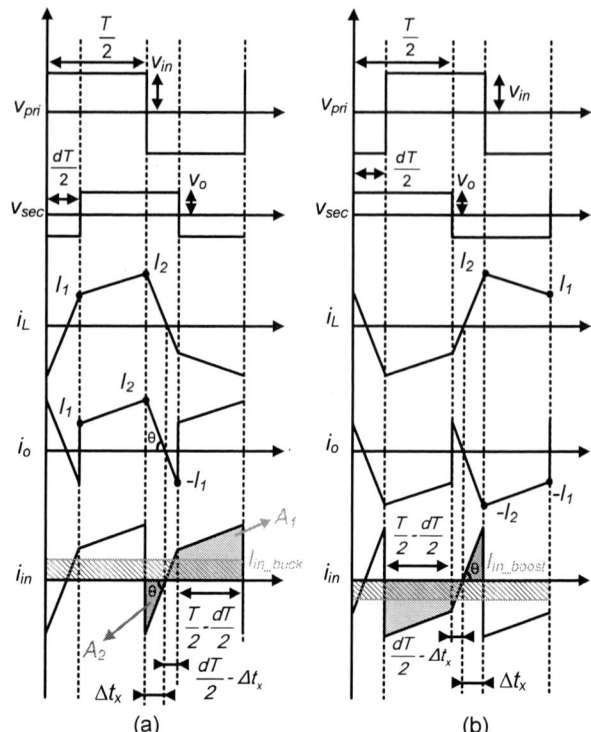

Fig. 3: Typical waveforms of DAB, (a) buck charging phase, (b) boost discharging phase

$$\tan\theta = \frac{I_1 + I_2}{\frac{dT}{2}} = \frac{I_2}{\Delta t_x} \tag{8}$$

$$\Delta t_x = \frac{[nV_{in} + V_o(2d-1)]}{4f_s(nV_{in} + V_o)} \tag{9}$$

Hence, the average input current can be expressed as,

$$I_{in\_buck} = \frac{A_1 - A_2}{\frac{T}{2}} = \frac{nV_o}{2Lf_s}(d - d^2) \tag{10}$$

With the same procedure, the average input current during the boost phase can be derived as,

$$I_{in\_boost} = \frac{nV_o}{2Lf_s}(d^2 - d) \tag{11}$$

After the calculation, the duty cycle for the buck and boost phase can be obtained as,

$$d_{buck} = \frac{1 - \sqrt{1 - \frac{8Lf_s I_{in\_buck}}{nV_o}}}{2} \tag{12}$$

$$d_{boost} = \frac{1 - \sqrt{1 + \frac{8Lf_s I_{in\_boost}}{nV_o}}}{2} \tag{13}$$

Then, the relationship between the harmonic current i_r and duty cycle is found. During the buck charging phase, I_{in_buck} is equal to i_r and d_{buck} is the duty cycle; during the boost discharging phase, I_{in_boost} is equal to i_r and d_{boost} is the duty cycle applied to the converter.

IV. INTEGRATED OUTPUT CAPACITOR REQUIREMENTS

As the DAB converter is used to achieve both LV battery charging and active filtering functions, its output capacitor bank needs to meet both requirements as well. They are the ripple percentage requirement during the LV battery charging mode and harmonic energy storage requirement during the active filtering mode. By calculating and combining these two requirements together, we can obtain the integrated output capacitor requirements.

A. Ripple percentage and capacitor requirements for LV battery charging

Part of the DAB output capacitance analysis has been done in [16]. However, the authors only discuss the condition when $V_o > nV_{in}$. As we choose the V_o/nV_{in} smaller than 1, the output voltage analysis for the case $V_o < nV_{in}$ is presented in this section. Furthermore, it is interesting to observe when the duty cycle increases, and thus, the reactive power increases, the value of I_1 will be greater than I_o, as shown in Fig. 4. This will also affect the voltage ripple calculation. Therefore, another equation needs to be derived to calculate the voltage ripple for the scenario when $I_1 > I_o$.

When $I_1 < I_o$, from Fig. 4a, the capacitor current can be obtained as,

$$\begin{cases} I_o + \dfrac{nV_{in} - V_o}{L} \cdot \Delta t_1 = I_2 \\ \\ I_2 + \dfrac{(-nV_{in} - V_o)}{L} \cdot \Delta t_2 = I_o \end{cases} \quad (14)$$

where the output current I_o and output voltage V_o can be found as,

$$\begin{cases} I_o = \dfrac{nV_{in}}{2Lf_s}(1 - d^2) \\ \\ V_o = \dfrac{nV_{in}R}{2Lf_s}(1 - d^2) \end{cases} \quad (15)$$

where R is the load resistance. In the meanwhile, the voltage ripple can be derived as,

$$\Delta V_o = \frac{\Delta t_1 + \Delta t_2}{C} \cdot \frac{I_2 - I_o}{2} \quad (16)$$

Substitute (7), (14) and (15) to (16), the peak-to-peak voltage ripple percentage $\Delta V_o\%$ can be obtained as,

$$\Delta V_o\% = \frac{[2f_s L(2d^2 - 2d + 1) + Rd(1-d)(2d-1)]^2}{8f_s C[4f_s{}^2 L^2 Rd(1-d) - R^3 d^3(1-d)^3]} \times 100\% \quad (17)$$

When $I_1 > I_o$, from Fig. 4b, the capacitor current can be obtained as,

$$I_o - \frac{nV_{in} + V_o}{L} \cdot \Delta t_3 = -I_1 \quad (18)$$

In the meanwhile, the voltage ripple can be expressed as,

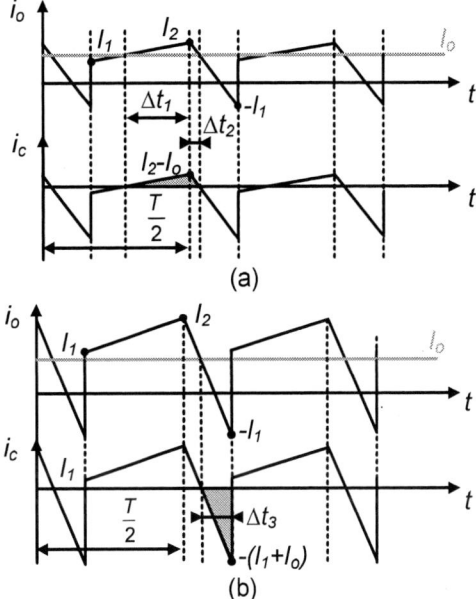

(a)

(b)

Fig. 4: Current ripple waveform for $V_o < nV_{in}$, (a) $I_1 < I_o$, (b) $I_1 > I_o$

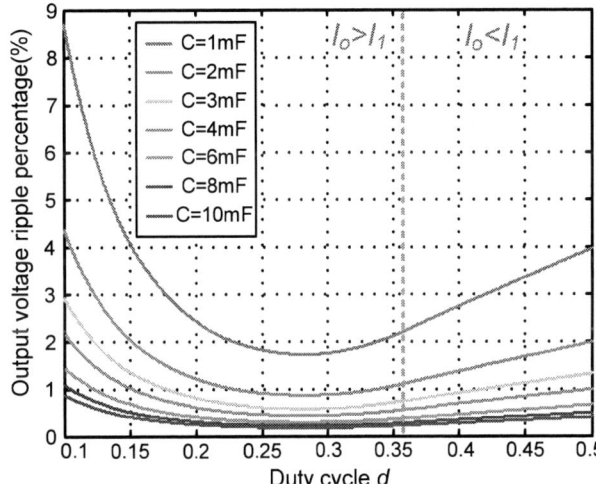

Fig. 5: Ripple percentage requirements for LV battery charging

$$\Delta V_o = \frac{\Delta t_3}{C} \cdot \frac{I_1 + I_o}{2} \quad (19)$$

Substitute (6), (15) and (18) to (19), the peak-to-peak voltage ripple percentage $\Delta V_o\%$ can be obtained as,

$$\Delta V_o\% = \frac{[2f_s L(4d - 2d^2 - 1) + Rd(1-d)]^2}{32f_s{}^2 LC[R^2 d^2(1-d)^2 + 2f_s LRd(1-d)]} \times 100\% \quad (20)$$

From (17) and (20), the voltage ripple percentage $\Delta V_o\%$ under different capacitances can be drawn in Fig. 5. If the voltage ripple percentage requirement for the LV battery is 1%, 10mF capacitance needs be applied for the worst case. Actually, the output current ripple is the biggest drawback for the single-phase DAB converter. By applying three-phase DAB converter [14] or multiple DAB converters to interleave the output current, the output current ripple can be reduced significantly and, thus, smaller output capacitor for the LV battery charger is required.

978-1-4673-9551-9/16 $31.00 © 2016 IEEE   1303

B. Harmonic energy storage and capacitor requirements for active filtering

During active filtering mode, as the transformer is used to transfer all the energy only, the output capacitor needs to be able to store all the second-order harmonic energy from the HV battery charger. The energy stored in the capacitor is,

$$E_r = \frac{1}{2} C V_c^2 \tag{21}$$

If we suppose the harmonic power goes to the capacitor can be expressed as a function of $sin2\omega t$, the harmonic power P_r can be rewritten as,

$$P_r = P_{r\_peak} sin2\omega t = \frac{1}{2} C \frac{dV_c^2}{dt} \tag{22}$$

where, P_{r_peak} is equal to 4kW. Solving the differential equation, we can obtain the capacitor voltage V_c as,

$$V_c = \sqrt{V_{c\_max}^2 - \frac{P_{r\_peak}}{\omega C}(1 + \cos 2\omega t)} \tag{23}$$

It is clear that the capacitor voltage is dependent on the peak ripple power P_{r_peak}, the capacitance C, the frequency ω and the maximum peak voltage V_{c_max}. If we want to store all the harmonic energy in the capacitor, V_{c_max} and the capacitance C need to satisfy following equation,

$$C \geq \frac{2P_{r\_peak}}{\omega V_{c\_max}^2} \tag{24}$$

Hence, with different peak voltage requirements, diverse capacitance requirements can be made shown in Fig. 6.

For the conventional DAB-based LV battery charger, the voltage stress on the secondary stage switches is equal to the LV battery's DC link, which may varies between 11V to 16V. However, from Fig. 6, we can see that with around 16V peak voltage, the required capacitance is relatively large. Therefore, during the active filtering mode, lifting the 16V peak voltage is more reasonable and easy to achieve. However, the peak voltage level cannot be lifted too high. Considering the case if 60V peak voltage ripple is selected, then secondary-stage switches with around 120V voltage rating may need to be applied. Switch with higher voltage rating yields larger on-state resistance. As in the LV battery charging mode, the AFAPM converter needs to produce around 200A current to the LV battery, the increase in the on-state resistance of switch will lead to higher conduction loss on the secondary stage, and thus, impact the APM converter's efficiency. In this paper, we select 60V voltage rating switches on the secondary. Hence, around 35V peak voltage ripple can be chosen. Then the corresponding required capacitance will be 17.3mF.

As the peak voltage is set, the larger capacitance is, the smaller voltage ripple will be produced. The relationship between capacitance and voltage ripple can be drawn in Fig. 7. However, when the AFAPM works as an active filter, the output capacitor is disconnected from the LV battery as shown in Fig. 1. Hence, there are no voltage ripple requirements on that capacitor during that time period. From volume and cost aspects, the 17.3mF capacitor is sufficient to store all the harmonic energy regardless of the voltage ripple on it.

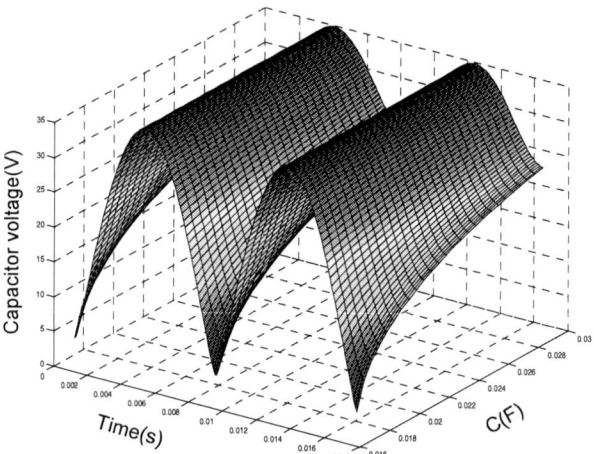

Fig. 7: Capacitor voltage ripple variations under fixed 35V peak voltage

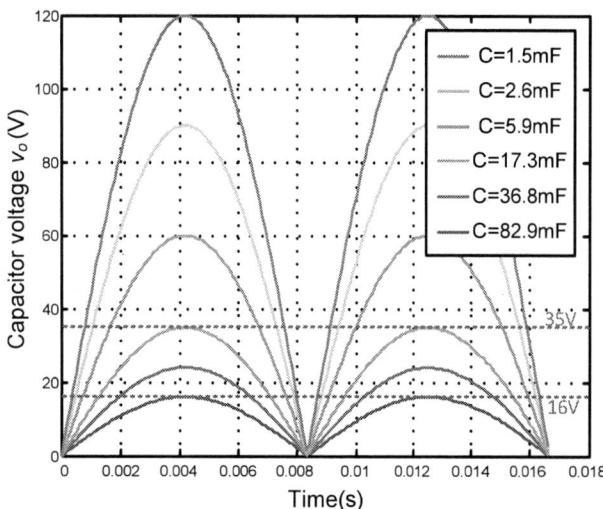

Fig. 6: Harmonic energy storage requirements for active filtering

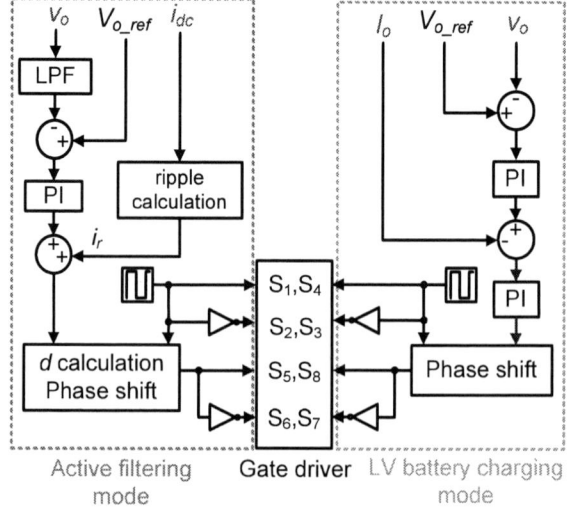

Fig. 8: Dual-mode controller block diagram

V. SIMULATION RESULTS

A simulation study is conducted in MATLAB/Simulink. The dual-mode controller block diagram for the proposed full-integrated AFAPM is shown in Fig. 8. The SPS modulation is applied to both modes. In the DAB converter, both full bridge voltages are 50% duty cycle square wave with a phase shift d. In the active filtering mode, to prevent the overcharging of the output capacitor, another voltage loop is used to control the average value of the capacitor voltage, where a low pass filter is applied. The simulation parameters are shown in the Table II. A 100μF capacitor C_{dc} is still needed on the HV DC-link to assimilate the high frequency harmonics.

Table II: Simulation parameters

Parameters	Value	Parameters	Value
Capacitance C	17.3mF	Turn ratio n	1/20
Inductance L	62.5nH	Switching Frequency f_s	100kHz
Capacitance C_{dc}	100μF	AC supply Frequency f	60Hz

Fig. 9 shows the LV output voltage and current, when the DAB is operating at the LV battery charging mode. The output voltage V_o is 12V and output current I_o is 200A. As the 17.3mF capacitor bank is applied at the output of DAB converter, this yields a relatively small charging current ripple for the LV battery.

Fig. 10 shows the voltage and current of the HV battery charger and DAB converter, when the DAB is operating at the active filtering mode. Before 0.5s, the DAB is working. The voltage and current ripple on the HV battery charger's DC-link are relatively small. The second-order harmonic energy in the HV battery charger is transferred and stored into the DAB's output capacitor. The voltage on the DAB's output capacitor fluctuates between 0V to 35V. After 0.5s, the DAB is turned off, and thus, all the harmonic energy is back to the HV DC-link.

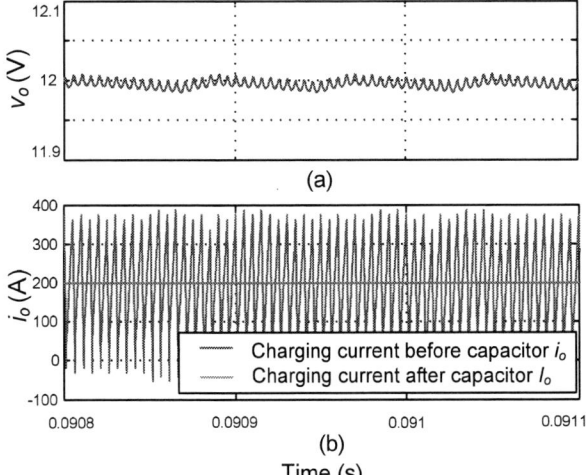

Fig. 9: Simulation results in LV battery charging mode, (a) output voltage v_o, (b) output current i_o

Fig. 11 is the switch requirements of the DAB converter at LV battery charging APM mode and active filtering mode, respectively. Clearly, for the input stage switches of AFAPM, the maximum voltage stress is 400V for both modes; and the maximum RMS current for both would be around 10A. Hence, the input stage switches VA ratings for AF and APM are exactly the same. However, during the active filtering mode, as we lifting the peak output capacitor's voltage to 35V, the output stage switches voltage stress has to be lifted to 35V rather than 16V. Hence, 30V MOSFETs need to be replaced by 60V MOSFETs on the secondary stage.

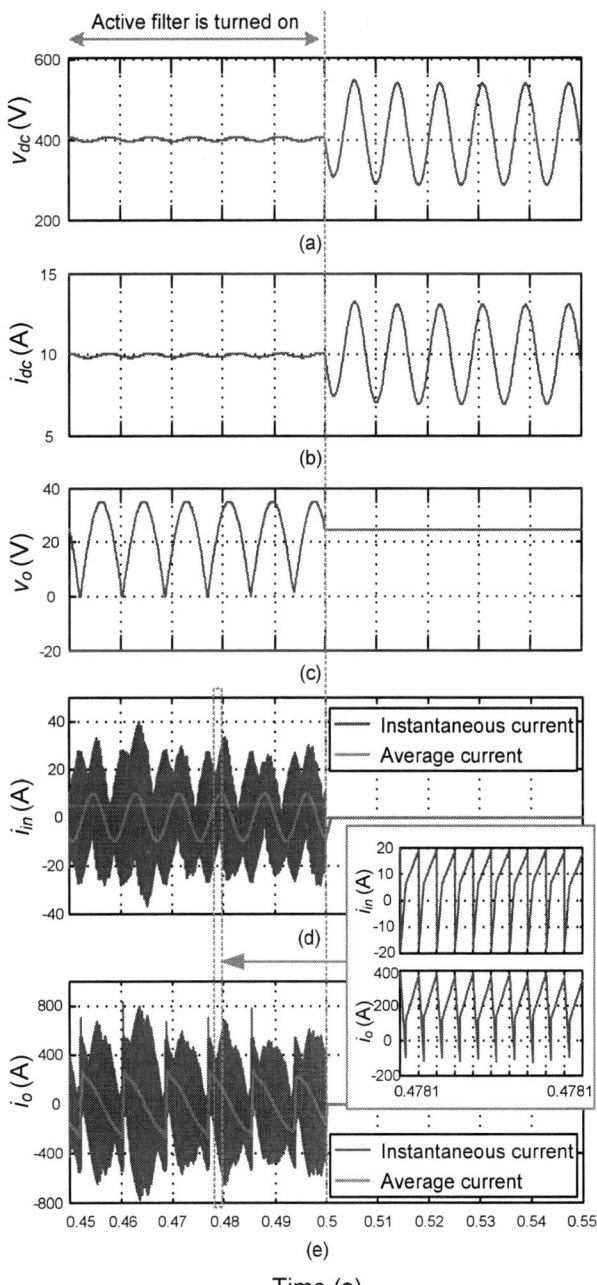

Fig. 10: Simulation results of active filtering mode, (a) HV output voltage v_{dc}, (b) HV output current i_{dc}, (c) LV output voltage v_o, (d) LV input current i_{in}, (e) LV output current i_o

Switch requirements of DAB-based full-integrated AFAPM

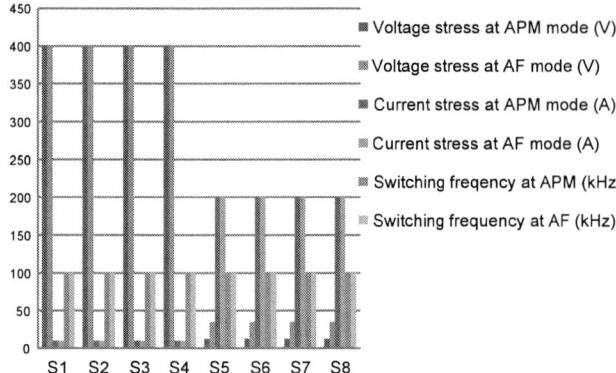

Fig. 11: Switch requirements of DAB-based full-integrated AFAPM

VI. CONLUSIONS

In this paper, a new DAB-based full-integrated AFAPM is proposed. The DAB converter acts as an active filter to assimilate the second-order harmonic current when the HV battery is charging; and works as a LV battery charger when the LV battery is charging. The active filtering function is achieved by using the DAB's power switches, leakage inductor, and output capacitors. Therefore, the required DC-link capacitance in the HV battery charger can be reduced significantly without extra active filtering components. As a result, the cost and volume of the dual-voltage charging system in the electrified vehicle applications can be reduced.

ACKNOWLEDGMENT

This research was undertaken, in part, thanks to funding from the Canada Excellence Research Chairs (CERC) Program.

REFERENCES

[1] B. Bilgin, P. Magne, P. Malysz, Y. Yang, V. Pantelic, M. Preindl, A. Korobkine, W. Jiang, M. Lawford, and A. Emadi, "Making the case for electrified transportation," *IEEE Transactions on Transportation Electrification*, vol. 1, no. 1, pp. 4-17, June 2015.

[2] A. Emadi, S. S. Williamson, and A. Khaligh, "Power electronics intensive solutions for advanced electric, hybrid electric, and fuel cell vehicular power systems," *IEEE Transactions on Power Electronics*, vol. 21, no. 3, pp. 567-577, May 2006.

[3] R. Hou, P. Magne, and B. Bilgin, *Chapter 9: Low-Voltage Electrical Systems for Nonpropulsion Loads*, in *Advanced Electric Drive Vehicles*, Edited by A. Emadi, pp. 317-328. Boca Raton, FL: CRC Press, ISBN: 978-1-4665-9769-3, 2014.

[4] R. Hou, P. Magne, B. Bilgin, and A. Emadi, "A topological evaluation of isolated DC/DC converters for auxiliary power modules in electrified vehicle applications," in *Proc. 2015 IEEE Applied Power Electronics Conference and Exposition*, Charlotte, NC, Mar. 2015, pp. 1360-1366.

[5] F. Musavi, *Chapter 13: Fundamentals of Chargers*, in *Advanced Electric Drive Vehicles*, Edited by A. Emadi, pp. 439-463. Boca Raton, FL: CRC Press, ISBN: 978-1-4665-9769-3, 2014.

[6] A. Khaligh and S. Dusmez, "Comprehensive topological analysis of conductive and inductive charging solutions for plug-in electric vehicles," *IEEE Transactions on Vehicular Technology*, vol. 61, no. 8, pp. 3475-3489, Oct. 2012.

[7] K. Yoo, K. Kim, and J. Lee, "Single- and three-phase PHEV onboard battery charger using small link capacitor," *IEEE Transactions on Industrial Electronics*, vol. 60, no. 8, pp. 3136-3144, Aug. 2013.

[8] H. Wen, W. Xiao, X. Wen, and P. Armstrong, "Analysis and evaluation of DC-link capacitors for high-power-density electric vehicle drive systems," *IEEE Transactions on Vehicular Technology*, vol. 61, no. 7, pp. 2950-2964, Sept. 2012.

[9] H. Wang and F. Blaabjerg, "Reliability of capacitors for DC-link applications in power electronic converters–An overview," *IEEE Transactions on Industry Applications*, vol. 50, no. 5, pp. 3569-3578, Sept. 2014.

[10] L. Xue, Z. Shen, D. Boroyevich, P. Matavelli, and D. Diaz, "Dual active bridge-based battery charger for plug-in hybrid electric vehicle with charging current containing low frequency ripple," *IEEE Transactions on Power Electronics*, vol. 30, no. 12, pp. 7299-7307, Dec. 2015.

[11] R. Wang, F. Wang, D. Boroyevich, R. Burgos, R. Lai, P. Ning, and K. Rajashekara, "A high power density single-phase PWM rectifier with active ripple energy storage," *IEEE Transactions on Power Electronics*, vol. 26, no. 5, pp. 1430-1443, May 2011.

[12] R. Hou and A. Emadi, "Evaluation of integrated active filter auxiliary power modules in electrified vehicle applications," in *Proc. 2015 IEEE Energy Conversion Congress and Exposition*, Montreal, Canada, Sept. 2015, pp. 6314-6319.

[13] R. Hou and A. Emadi, "Integrated active power filter auxiliary power modules for electrified vehicle applications with single-phase on-board chargers," in *Proc. 2015 IEEE Transportation Electrification Conference and Expo*, Dearborn, MI, June 2015, pp. 1-6.

[14] R. W. A. A. DeDoncker, D. M. Divan, and M. H. Kheraluwala, "A three-phase soft-switched high-power-density DC/DC converter for high-power applications," *IEEE Transactions on Industry Applications*, vol. 27, no. 1, pp. 63-73, Jan. 1991.

[15] R. T. Naayagi, A. J. Forsyth, and R. Shuttleworth, "High-power bidirectional DC-DC converter for aerospace applications," *IEEE Transactions on Power Electronics*, vol. 27, no. 11, pp. 4366-4379, Nov. 2012.

[16] C. Mi, H. Bai, C. Wang, and S. Gargies, "Operation, design and control of dual H-bridge-based isolated bidirectional DC-DC converter," *IET Power Electronics*, vol. 1, no. 4, pp. 507–517, Dec. 2008.

[17] D. S. Gautam, F. Musavi, M. Edington, W. Eberle, and W. G. Dunford, "An automotive onboard 3.3-kW battery charger for PHEV application," *IEEE Transactions on Vehicular Technology*, vol. 61, no. 8, pp. 3466-3474, Oct. 2012.

[18] A. Rodriguez, A. Vazquez, D. G. Lamar, M. M. Hernando, and J. Sebastian, "Different purpose design strategies and techniques to improve the performance of a dual active bridge with phase-shift control," *IEEE Transactions on Power Electronics*, vol. 30, no. 2, pp. 790-804, Feb. 2015.

[19] F. Krismer and J. W. Kolar, "Accurate small-signal model for the digital control of an automotive bidirectional dual active bridge," *IEEE Transactions on Industrial Electronics*, vol. 24, no. 12, pp. 2756-2768, Dec. 2009.

[20] A. R. Alonso, J. Sebastian, D. G. Lamar, M. M. Hernando, and A. Vazquez, "An overall study of a dual active bridge for bidirectional DC/DC conversion," in *Proc. 2010 IEEE Energy Conversion Congress and Exposition*, Atlanta, GA, Sept. 2010, pp. 1129-1135.

[21] H. Bai and C. Mi, "Eliminate reactive power and increase system efficiency of isolated bidirectional dual-active-bridge DC-DC converters using novel dual-phase-shift control," *IEEE Transactions on Power Electronics*, vol. 23, no. 6, pp. 2905-2914, Nov. 2008.

[22] B. Zhao, Q. Song, W. Liu, and Y. Sun, "Overview of dual-active-birdge isolated bidirectional DC-DC covnerter for high-frequency-link power-conversion system," *IEEE Transactions on Power Electronics*, vol. 29, no. 8, pp. 4091-4106, Aug. 2014.

All-SiC Inductively Coupled Charger with Integrated Plug-in and Boost Functionalities for PEV Applications

M. Chinthavali, O. C. Onar, S. L. Campbell, and L. M. Tolbert

Power Electronics and Electric Machinery Group, Oak Ridge National Laboratory, Oak Ridge, TN 37831
chinthavalim@ornl.gov, campbellsl@ornl.gov, onaroc@ornl.gov, tolbertlm@ornl.gov

Abstract— So far, the charging functionality for vehicles has been integrated either into the traction drive system or to the dc-dc converters in plug-in electric vehicles (PEV). This study features a unique way of combining the wired and wireless charging functionalities with vehicle side boost converter and maintaining the isolation to provide a hybrid plug-in and wireless charging solution to the plug-in electric vehicle users. The proposed integrated charger combined with SiC technology shows the end-to-end and dc-to-dc system efficiencies of 85.9% and 88.9% for wireless charging mode, and 88.8% and 92.4% for the wired charging mode of operation.

Keywords—wireless power transfer, on-board charger, integrated charger, wireless charging, boost converter, plug-in electric vehicle charger.

I. INTRODUCTION

With the increased growth in electric vehicles, there is a need for reduction in weight, size, and cost while increasing the efficiency and power rating of the on-board chargers (OBCs). Integrating the function of charging into the traction drive systems of PEVs has been proven to significantly reduce the volume and weight and increase the power density of the of OBCs [1]-[6]. Since all the components are on-board, the power levels are typically limited because of the size and weight constraints on the vehicle; therefore, charging times are usually long. Higher power levels can be achieved with off-board chargers; however, both the on-board and off-board chargers featuring plug-in options inherently pose a danger of electric shock. Furthermore, depending on the plug and cable ratings, they are heavy and inconvenient. Additionally, there is always a possibility of forgetting to plug-in and running out of charge.

Some of the challenges associated with plug-in chargers can be eliminated by implementing wireless power transfer (WPT) technology. WPT is a safe, convenient and flexible charging method for PEVs. However, having the wired (plug-in) and wireless charging systems available at the same time is critical, especially before the wireless charging infrastructure is readily available. Along with convenience, the other aspects of the charging technology that need to be addressed are the cost and the efficiency of the charger.

Integrated chargers generally will provide huge cost benefits by utilizing same components for different functionalities. By integrating vehicle side components with wireless charging functionalities, substantial cost benefits can be achieved [7], [8]. In this integration, multi-functional operations can be achieved since charging and boost modes of operation are independent. However, because of the high number of power conversion states, the other challenge for wireless integrated topologies is the total system efficiency. According to the Society of Automotive Engineers (SAE) J2954 standard development activity [9], greater than 85% efficiency is expected from utility outlet to the vehicle high voltage (HV) bus in wireless charging applications. Currently, silicon (Si) based devices, comprised mostly of integrated gate bipolar transistors (IGBTs) and PiN diodes, are being used in charging and traction drive systems. With the limitations of Si based technology, it becomes extremely difficult to achieve the desired efficiency, especially at higher frequencies and power levels. WBG device technology with lower losses and higher operating temperature can be utilized to address this issue and prove to be a vital solution for the wireless charging technology.

It is evident, from the literature survey, that the topologies with integrated dc-dc converter and charging functionality have no isolation in the system [1]-[6], [10]-[19]. Isolation is an important feature that is required for user interface systems that have grid connections to prevent potential shock hazards and also ground faults and therefore isolation is a major limitation that needs to be addressed along with the integrated functionality. There are two solutions that can be used to essentially address this issue. One is a traditional high frequency (HF) isolation transformer that can be added to the integrated system on board. However, this adds an additional power stage, complexity, component cost, and weight to the car, and minimizes the impact of the integrated functionality. The other solution that can be utilized is an inherently isolated wireless charging system that can provide the required isolation from the grid. The WPT system has a significant benefit compared to other systems in terms of the component count that can be added to the car. The secondary system in the car can be as simple as a LC resonant coil with a rectifier and a filter capacitor. The rest of the system is located on the grid side stationary unit that has several stages of power conversion. Both the above-mentioned solutions are very attractive and have their pros and cons. However, there is a

---

This manuscript has been authored by the Oak Ridge National Laboratory, operated by the UT-Battelle, LLC under Contract No. DE-AC05-00OR22725 with the U.S. Department of Energy. The United States Government retains and the publisher, by accepting the article for publication, acknowledges that the United States Government retains a non-exclusive, paid-up, irrevocable, world-wide license to publish or reproduce the published form of this manuscript, or allow others to do so, for United States Government purposes. The Department of Energy will provide public access to these results of federally sponsored research in accordance with the DOE Public Access Plan (http://energy.gov/downloads/doe-public-access-plan).

U.S. Government work not protected by U.S. copyright

unique way of combining these functionalities and providing the best solution to EV and PHEV users. The topology proposed in this study integrates the wireless charging system and the boost converter of the traction drive system while moving the high frequency isolation stage off-board in addition to the inherent airgap. This architecture has two levels of isolation with multifunctional operation flexibility, such as charging and boosting function. The proposed system also utlizes Silicon carbide devices to improve the overall efficieny and makes the system more comparable with on-board charger systems efficiencies in a commercial EVs. The overall system architecture and the details of the design trade-offs compared to an on-board charging system are discussed in the following sections.

II. SYSTEM DESCRIPTION

Figure 1 illustrates the integrated stationary wireless charging circuit topology that is proposed. The topology has five stages of power conversion from the wall to the vehicle battery in the wireless charging mode and four stages in the wired charging mode. The boost mode of operation essentially remains the same as it is in the on-road vehicles.

A. Wireless Charging Mode of Operation

Figure 2 demonstrates the circuit topology for the wireless mode of operation. Utility ac power is converted to controllable dc voltage by the active front-end rectifier with power factor correction (PFC). Adjustable dc voltage is applied to the input of the high frequency (HF) full-bridge inverter with a controlled duty ratio. The HF stage delivers excitation current to a series tuned primary coil for magnetic field generation that is linked to the secondary coil on the vehicle across the air gap. Voltage induced at the secondary is rectified, filtered, and delivered to the vehicle HV battery. The integrated wireless charger has an advantage compared to the conventional topology. The boost inductor in series with the battery pack will minimize the battery current ripples, and with the capacitor, it acts as a low pass filter. This feature will increase the reliability and the lifetime of the battery. WPT applications may require inclusion of a HF transformer to provide electrical isolation of the WPT primary pad and cabling from the utility. Isolation requirements and grid power quality are currently being discussed by SAE J2954 standards development committee [9]. In addition to the isolation, the HF transformer can be utilized to step down the HF inverter output voltage that enables the inverter to be operated at high voltage and low currents and results in higher efficiencies for the system.

B. Wired Charging Mode of Operation

The HF transformer solution also provides the flexibility of using the system as a wired charger. This can be achieved by simply using a relay system that can be operated to disconnect the resonant coil system and connect the output of the HF transformer to the on-board section of the integrated charger system. The wired charging mode of operation has four power conversion stages. This mode of operation will enable the EV users to use the plug-in charger wherever there is no wireless charging option. The efficiency of the system is compared to the wireless mode of operation in Section VI. Figure 2 shows the circuit topology for the wireless mode of operation.

C. Boost Mode of Operation

The boost mode operation simply utilizes a single stage switch and a diode combined with an inductor designed for boost operation. The wireless and wired charging power stages are completely isolated by using the relays in the system. This topology is completely based on the two independent functions of the integrated systems, (1) charging and (2) boost, which do not require simultaneous operation as shown in Fig. 4.

III. SIMULATIONS OF THE PROPOSED SYSTEM

A simulation model was built using PSIM which is an implementation of the overall system depicted in Fig. 1. The resonant coil system is modeled as a loosely coupled transformer model. The HF transformer is modeled as an ideal transformer with series leakage inductance. The magnetizing branch inductance, L_m, is assumed to be too high for this highly coupled core based HF transformer and therefore has not been included in the model. The control system that was implemented for this model is discussed in detail in [20], [21]. The only difference is the addition of the transformer model which affects the controller gains. The parameters used for the simulation are shown in Table I. The inductances of the coils were obtained from the measured values of the prototypes built. The turns-ratio of the transformer is designed to be 10:2. This was enough to step the voltage of the inverter output and accommodate the designed resonant system gain of 3.1 at 22 kHz. This turns ratio also enabled control over a wide range of power and a boost voltage of up to 600 V max.

A. Wireless Charging Mode of Operation

Figures 5 and 6 show the complete system response to a 6.6 kW charge power request at the battery terminals. The current that the controlled rectifier draws from the grid is in phase with the grid voltage with close to unity power factor. The reference and actual measured grid current waveforms are plotted in Fig. 5 (a) in order to demonstrate the performance of the current control loop.

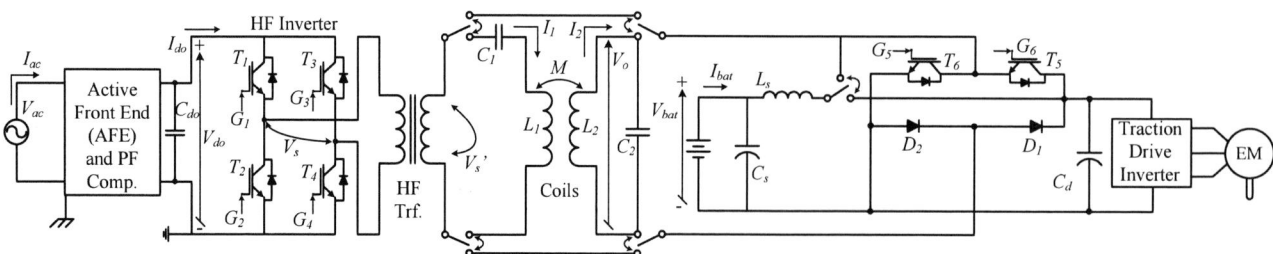

Fig. 1. Circuit architecture of the proposed integrated charger.

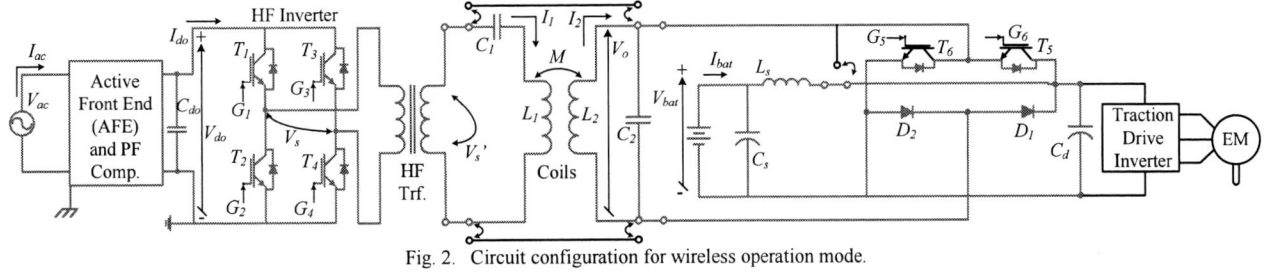

Fig. 2.   Circuit configuration for wireless operation mode.

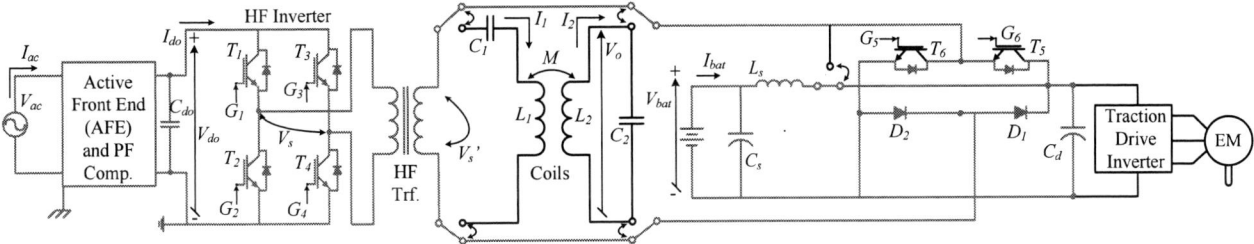

Fig. 3.   Circuit configuration for wired operation mode.

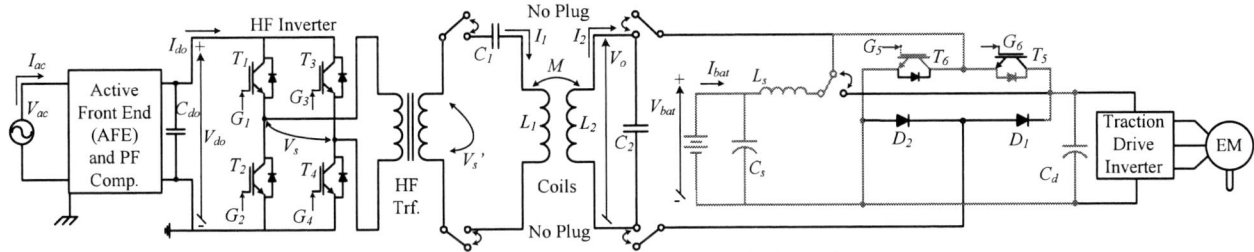

Fig. 4.   Circuit configuration for boost mode of operation.

In order to supply this amount of power, the control system automatically boosts and regulates the dc-link voltage to ~450V as shown in Fig. 5 (b). The inverter output voltage, primary coil voltage, and the transformer output voltage in Fig. 6 (a) and (b) show the impact of the total transformer leakage inductance. The voltage has a transient curve because of the drop across the leakage inductance and the RMS value of the voltage across the coil is reduced. The battery voltage and battery current with detailed zoom show the resulting ripples in Fig. 6 (c). Figure 6 (c) also shows the amount of power transfer to the load side (~6.6kW).

TABLE I. SIMULATION MODEL PARAMETERS

Parameter	Symbol	Value	Unit
AC input voltage	V_{ac}	220	[V_{ac}]
DC link voltage	V_{do}	350-600 (controlled)	[V_{dc}]
Primary tuning capacitor	C_1	0.45	[μF]
Primary coil inductance	L_1	115.79	[μH]
L_1 internal resistance	R_{L1}	122.7	[mΩ]
Secondary coil inductance	L_2	132.6	[μH]
L_2 internal resistance	R_{L2}	150.8	[mΩ]
Secondary tuning capacitor	C_2	0.40	[μF]
Coupling factor	k	0.23	#
Distance between the coils	z	160	[mm]
Filtering capacitor	C_o	680	[μF]
Battery nominal voltage	V_b	310	[V_{dc}]
Battery target power	P_b	6600 (target)	[W]

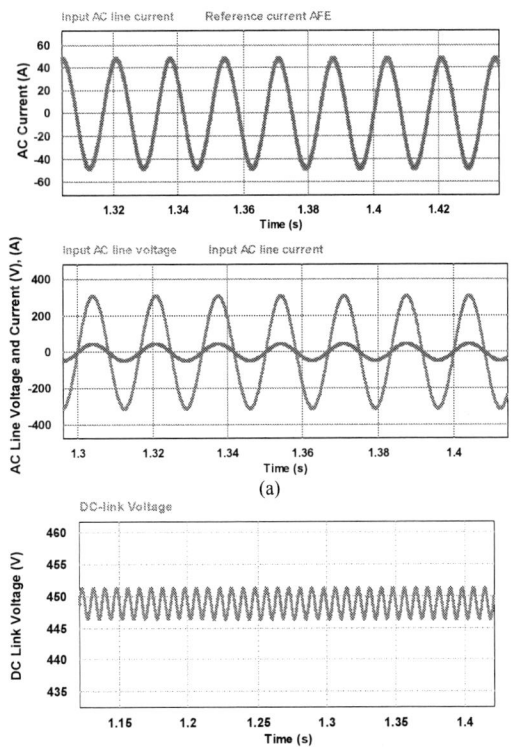

(a)

(b)

Fig. 5.   Simulation result: AC line voltage and current (a) and the boosted DC link voltage (b).

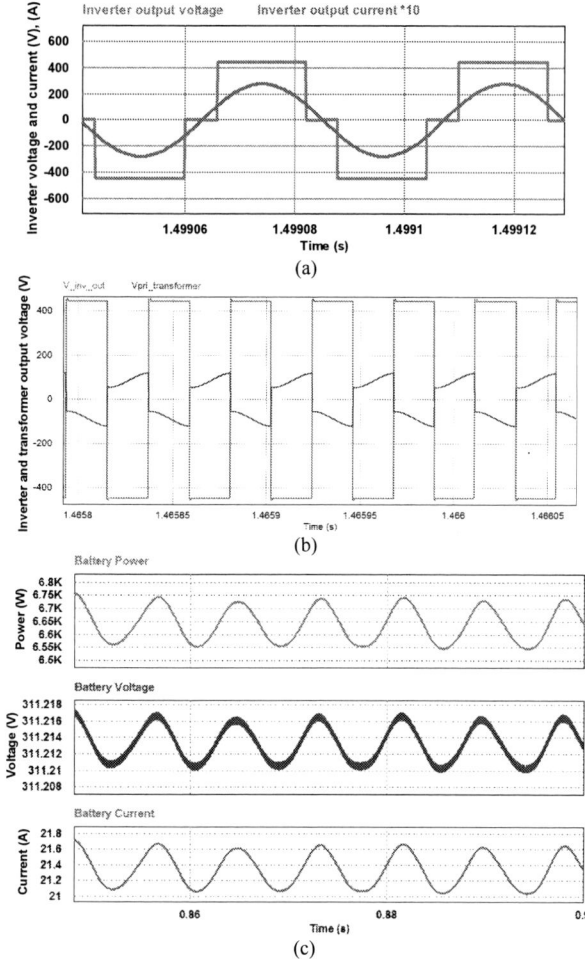

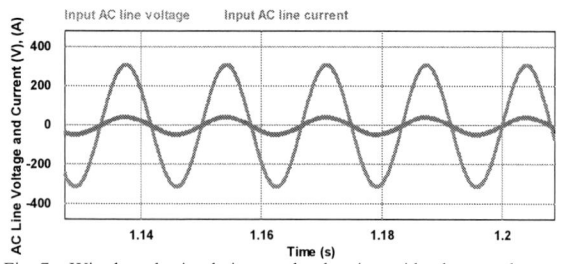

Fig. 6. Simulation results: Inverter output voltage and current (a), inverter output voltage and primary coil voltage (transformer output) (b), vehicle battery power, voltage and current (c) waveforms.

B. Wired Charging Mode of Operation

The wired mode was simulated without the resonant tank and the parameters of the components except the transformer were kept the same. The turns ratio of the transformer was changed to 8:5 to account for the abesnce in the resonant gain and the corresponding design parameters were used.

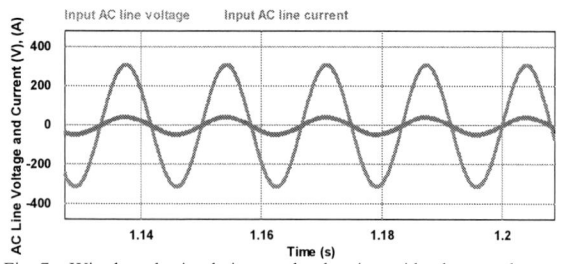

Fig. 7. Wired mode simulation results showing grid voltage and current.

The control system strategy for the wired mode was kept the same. Unity power factor is achieved with the wired mode as well. AC line grid voltage and current are plotted in Fig. 7 whereas Fig. 8 shows the DC link voltage without coils. It is seen from Fig. 8 that the rectfied boosted voltage is higher

than that of the wireless mode. This is due to the fact that system requires higher voltage since there is no resonating circuit after the inverter and there is no resonant voltage gain and therefore a higher voltage level is needed for the same level of power transfer to the load. The inverter and the transformer output voltages and currents are shown in Fig. 9. Finally, Fig. 10 shows the amount of power transfer to the load side (~6.6 kW). Again this verifies that the closed loop control system is working and the output power is regulated to the designed value.

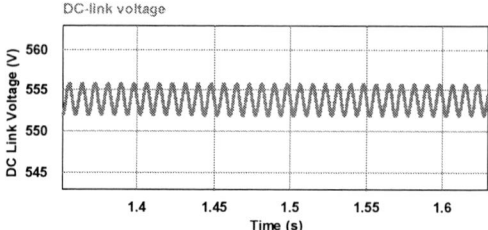

Fig. 8. DC link voltage in wired mode of operation.

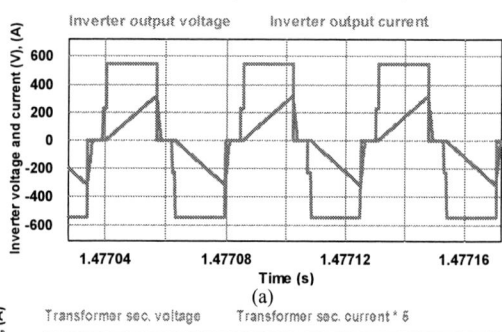

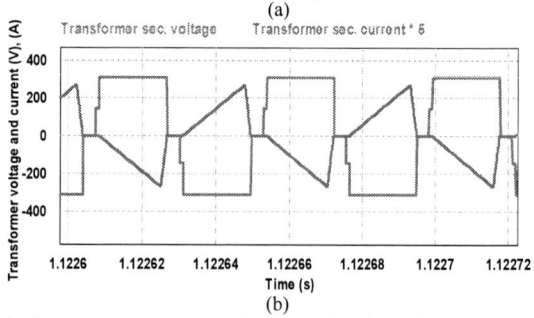

Fig. 9. Inverter output voltage and current (a) and transformer output voltage and current (b).

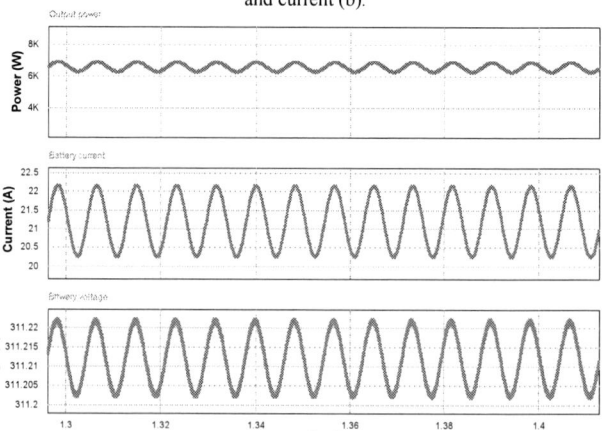

Fig. 10. Battery power, current, and voltage under wired mode.

IV. EXPERIMENTAL TEST SETUP AND RESULTS

The active front-end rectifier was built using the commercially available 1200V, 100A SiC MOSFET-based phase-leg modules. The SiC modules were used because of the high voltage in the dc-link. The details of the design are presented in [20]. The HF inverter was built using the same SiC devices. Also, vehicle side rectifier consist of the internal SiC diodes of these devices The inverter and the PFC were cooled using a water ethylene glycol mix and the receiver side rectifier was cooled using fan based heat sink. The control system of the inverter is implemented within a TMS320F28335PGFA DSP module from Texas Instruments. The primary side (grid-side) power stage is given in Fig. 11, including the front-end rectifier (a) and the HF inverter (b). The experimental setup of the system with SiC-based active-rectifier including primary (bottom) and secondary coil (top) and in the background the HF transformer, resonant tuning capacitors and vehicle side rectifier with a small filter capacitor are shown in Fig. 12 (a) and (b).

 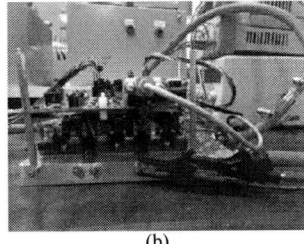

(a)   (b)

Fig. 11. SiC-based active front-end rectifier with PFC (a) and the SiC. H-bridge inverter (b).

A. Wireless Mode of Operation Experimental Results

The experimental set-up was used to evaluate the wireless mode first. The transformer was wound for a turns-ratio of 10:2 as per the design and the load voltage was set to 311 V_{dc}. The ac input voltage was ramped up to 220 V_{rms} and the corresponding power to the load was recorded. The required power output was changed in steps and the efficiencies were calculated. The desired 6.56 kW power transfer to the vehicle battery for an input power of 7.63 kW to the system was achieved. This power level corresponds to ~85.9% overall efficiency from AC grid to the vehicle battery terminals. The voltage and current waveforms for the AC input and DC link to the inverter are given in Fig. 13 whereas the inverter, transformer, coil, and rectifier output waveforms are given in Fig. 14. The detailed stage-by-stage power flow of the system is illustrated in Fig. 15 (highlighted with frame).

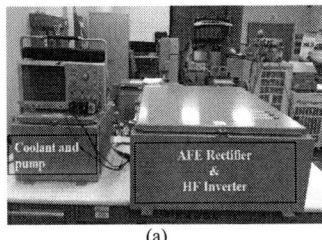 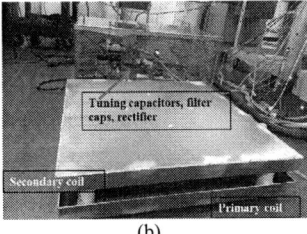

(a)   (b)

Fig. 12. Experimental test setup: PFC converter integrated with the high frequency power inverter and enclosed in the stationary box with the cooler (a) and primary and secondary coils, tuning capacitors, and vehicle side rectifier and filter (b).

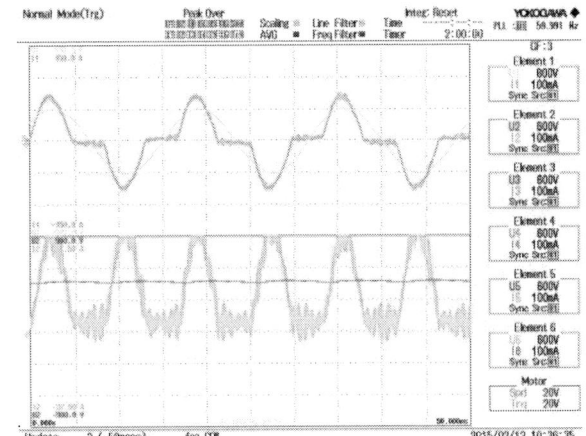

Fig. 13. Experimental voltage and current waveforms for the AC input and the DC link to the inverter.

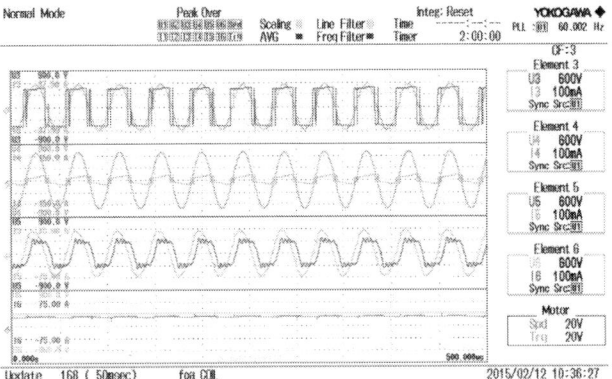

Fig. 14. Experimental voltage and current waveforms for the inverter, transformer, coil, and rectifier outputs.

% change items	Element 1	Element 2	Element 3	Element 4	Element 5	Element 6
Urms [V]	219.70	469.57	459.38	91.20	276.12	311.18
Irms [A]	39.09	20.98	18.15	89.75	35.74	21.13
Udc [V]	-0.00	469.54	-1.09	0.19	0.03	311.18
Idc [A]	-0.96	15.71	-0.09	-0.08	0.08	21.09
P [W]	7.639k	7.379k	7.087k	7.018k	6.657k	6.563k
λ [ ]	0.8894	0.7491	0.8502	0.8575	0.6746	0.9981
φ [°]	D27.21	D41.49	D31.77	G30.96	D47.58	D3.50
Itpk [A]	76.16	48.29	29.28	135.53	54.02	31.18
Gap_V [V]	357.09					

Fig. 15. The stage-by-stage power of the system after each power conversion.

There are several important parameters that can be analyzed to understand the impact of each component and also to validate the simulation results. The dc-link boost voltage is 459 V_{dc} which closely matches with the simulated value of 452 V_{dc} for the same system parameters. One more important parameter is the leakage inductance of the transformer and its effect on the turns-ratio. The turns-ratio from the experiments is 10:2, validating the designed value. This turns-ratio although is an expected value for the design, the resonant tank load makes the parameters in the transformer more critical. The most important parameter is the leakage inductance and the turns-ratio associated with that design which will make the voltage drop across the leakage inductance. This will affect the RMS value of the voltage on the secondary side. So for this design the turns-ratio being close shows that voltage-drop across the leakage inductance for the transformer is negligible. This can also be observed in the waveforms of the transformer secondary voltage as shown in Fig. 14. The value of leakage inductance is 45 μH and the current is around 18 A_{rms}. The inductive reactance is dependent on the resonant frequency. So for higher frequencies the voltage drop could be even higher which makes the impact of the transformer parameters even more critical. As shown in Fig. 15, stage by stage cascaded efficiencies are found as follows: Active front-end rectifier: 96.6%, inverter: 96%, transformer: 99%, coils: 94.9%, and rectifier: 98.6%. This results in 85.9% end-to-end efficiency and 88.9% dc-to-dc efficiency in wireless charging mode. The proposed system is also tested at different power levels; i.e., at 0.93, 2.35, 3.21, 4.01, 4.79, 5.54, 6.28, 6.56 kW, and the efficiency of the each power conversion stage is obtained with respect to the different levels of the charging power delivered to the load. The efficiency vs. power curves of the PFC, inverter, coils, and the rectifier are presented in Fig. 16 (a) and (b).

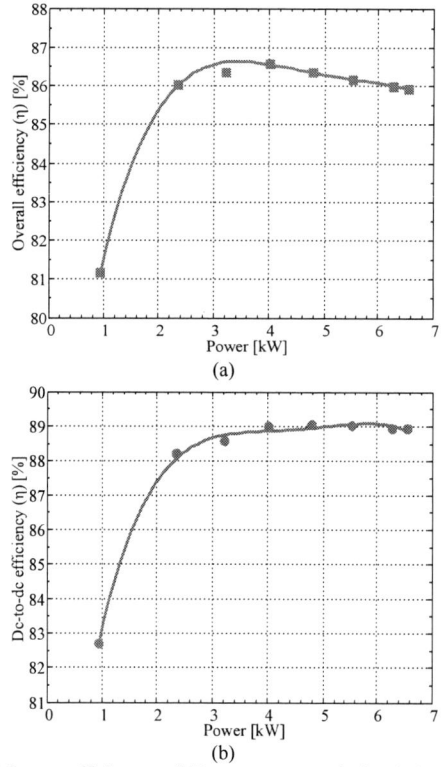

Fig. 16. System efficiency at different power transfer levels for wireless operation: (a) Overal efficiency and (b) dc-to-dc efficiency.

B. Wired Mode of Operation Experimental Results

The experimental set-up that was used to evaluate the wireless mode was changed to the wired mode test by changing the transformer and removing the resonant network. The transformer was wound for a turns-ratio of 8:5 as per the design and the load voltage was set to 311 V_{dc}. The ac input voltage was ramped up to 220 V_{rms} and the required power output was changed in steps. The dc-link boost voltage is 555.76 V_{dc} which closely matches with the simulated value of 554 V_{dc} for the same system parameters. Unlike the wireless mode of operation the effect of leakage inductance is negligible. The desired 6.7 kW power transfer to the vehicle battery for an input power of 7.58 kW to the system was achieved. This power level corresponds to ~88.76% overall efficiency from AC grid to the vehicle battery terminals. The detailed stage-by-stage power output of the system is illustrated in Fig. 17. The voltage and current waveforms for the AC input and DC link to the inverter are given in Fig. 18, whereas the inverter, transformer, coil, and rectifier output waveforms are given in Fig. 19.

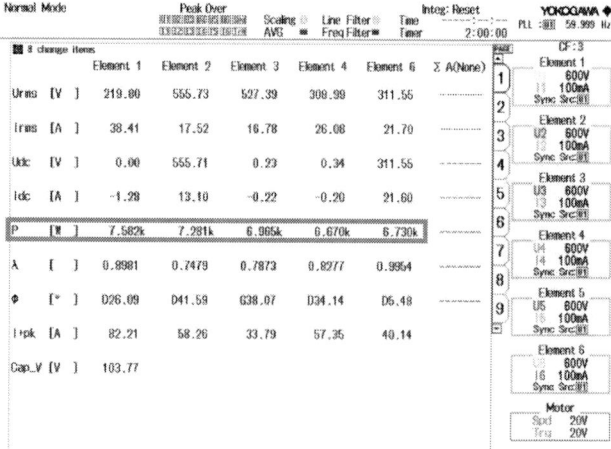

Fig. 17. Detailed stage-by-stage powers of the wired charging operation mode.

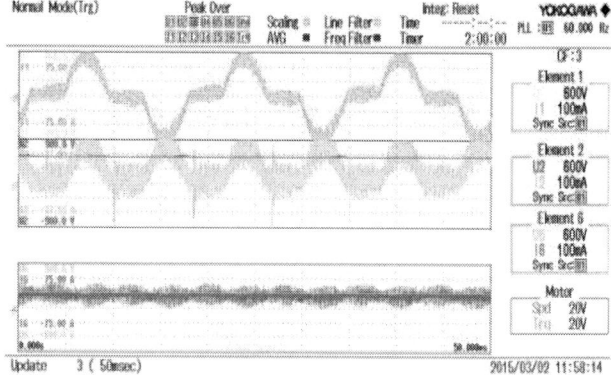

Fig. 18. Experimental voltage and current waveforms for the AC input and DC link to the inverter.

According to Fig. 17, dc-to-dc (from inverter input to the vehicle battery) and the overall end-to-end efficiency values of each power conversion stages are as follows: active front-end rectifier: 96%, inverter: 95.7%, transformer: 97.6%, rectifier: 99%. The proposed system is also tested at different power levels; i.e., 1.42, 2.29, 3.24, 4.03, 5.6, and 6.73 kW and the efficiency of the each power conversion stage is obtained with respect to the different levels of the charging power delivered to the load, as shown in Fig. 20 (a) and (b).

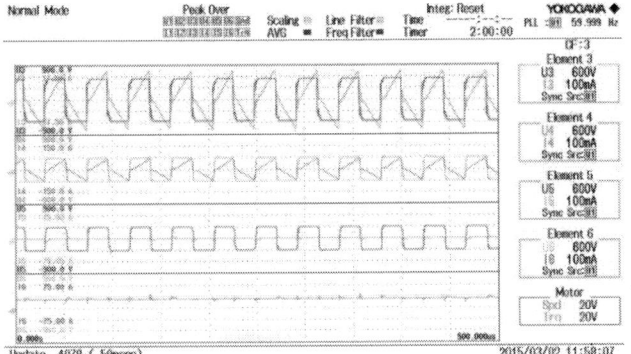

Fig. 19. Experimental waveforms for the inverter, transformer, and rectifier outputs.

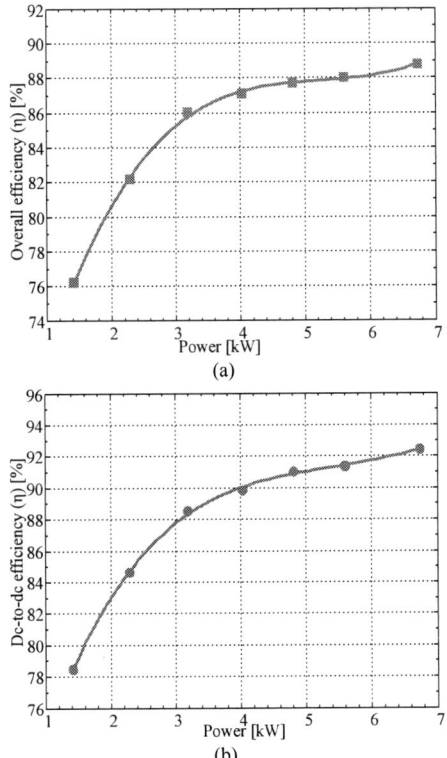

Fig. 20. System efficiency at different power transfer levels for wired operation: Overal efficiency (a) and dc-to-dc efficiency (b).

The integrated charger is tested in wired and wireless modes with different transformer turns-ratios because of the difference in the voltage gain of the resonant tank, the input and output voltage requirements. In order to optimize the design with one single transformer turns-ratio the system had to be tested with 10:2 transformer for the wired mode and 8:5 transformer for the wireless mode. For the wired mode in order to obtain the desired 311 V at the load with the 10:2 transformer, the input dc-link voltage has to be around 1500 V_{rms}. This voltage is above the ratings of the devices being used so this turns-ratio will not be suitable for the wired mode.

However, for the wireless mode the 8:5 turns-ratio will work in both operating modes, wired and wireless, with some reduction in the overall efficiency. This is because the lower turns-ratio will cause an increase in the current from 35 A_{rms} to 58 A_{rms} on the primary side of the transformer for the same output power.

For the repeated tests, the efficiencies are shown in Fig. 21 with 8:5 transformer turns-ratio. The overall efficiency of the system dropped from 85.9% to 84.4% whereas the dc-to-dc efficiency dropped from 88.9% to 87.4%. There is another effect on the overall system performance because of the increase in current. The voltage drop across the leakage inductance of the transformer will be much higher than the previous case because of the increase in current. The leakage inductance for this transformer was 23.4 μH. At the same resonant frequency as in the 10:2 turns-ratio case the drop was less. With the resonant voltage gain considered, the effective turns-ratio is 2.4 instead of 1.6 (8:5).

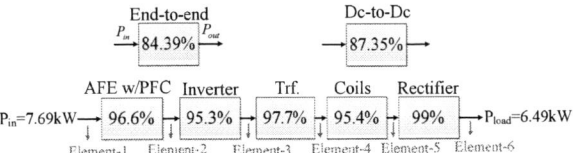

Fig. 21. Stage by stage efficiency for wireless operation with 8:5 transformer turns ration (repeated test).

V. CONCLUSIONS

A new integrated charger topology with wired and wireless charging and boost function is presented. According to the experimental test results, the end-to-end and dc-to-dc efficiencies of the system were 85.9% and 88.9% for wireless charging mode whereas these efficiency values were 88.8% and 92.4% for the wired charging mode of operation. Based on the references performance of commercially available OBCs from manufacturers, Chevy Volt OBC is 89.6% efficient at level-2 and 86.8% efficient at level-1 [22] and Nissan Leaf OBC efficiency is 90.9% at level-2 [23]. The proposed integrated charger, with both the wired and wireless charging functionalities, is on par with the performance of these commercially available units. In order to have the system interoperable with both wired and wireless charging modes, the transformer turns ratio had to be changed to 8:5 instead of 10:2 since the wired option does not have the natural voltage gain effect of the resonant wireless charging. In this case, the same transformer turns ratio provided both wired and wireless charging functionalities without any additional changes in the system but with slight efficiency reduction due to the higher current levels that the primary circuit had to deliver.

REFERENCES

[1] D. Thimmesch, "An SCR inverter with an integral battery charger for electric vehicles," *IEEE Transactions on Industry Applications*, vol. IA-21, no. 4, pp.1023–1029, July/August 1985.

[2] S. –K. Sul and S. J. Lee, "An integral battery charger for electric vehicles," *IEEE Transactions on Industry Applications*, vol. 31, no. 5, pp. 1096-1099, Sept./Oct. 1995.

[3] L. Tang and G. J. Su, "A Low-Cost, Digital-Controlled Charger for Plug-In Hybrid Electric Vehicles," in *Proc., IEEE Energy Conversion Conference & Exposition (ECCE)*, pp. 3923–3929, September 2009, San Jose, CA.

[4] G. J. Su and L. Tang, "An Integrated Onboard Charger and Accessory Power Converter for Plug-in Electric Vehicles," in *Proc., IEEE Energy Conversion Congress and Exposition (ECCE)*, pp. 1592–1597, September 15–19, 2013, Denver, CO.

[5] G. J. Su and L. Tang, "A New Integrated Onboard Charger and Accessory Power Converter for Plug-in Electric Vehicles," in *Proc., IEEE Energy Conversion Congress and Exposition (ECCE)*, pp. 4790–4796, September 2014, Pittsburgh, PA.

[6] G. J. Su and L. Tang, "An integrated onboard charger and accessory power converter using WBG devices," in *Proc., IEEE Energy Conversion Congress and Exposition (ECCE)*, pp. 6306–6313, September 2015, Montreal, Canada.

[7] M. Chinthavali, O. C. Onar, S. L. Campbell, and L. M. Tolbert, "Isolated wired and wireless battery charger with integrated boost converter for PEV applications," in *Proc., IEEE Energy Conversion Congress & Exposition (ECCE)*, pp. 607-614, September 2015, Montreal, Canada.

[8] M. Chinthavali, O. C. Onar, S. L. Campbell, and L. M. Tolbert, "Integrated charger with wireless charging and boost functions for PHEV and EV applications,". in *Proc., IEEE Transportation Electtification Conference and Expo, (ITEC)*, pp. 1-8, June 2015, Dearborn, MI.

[9] SAE J2954 Standards Development for Wireless Charging of Electric and Plug-in Hybrid Vehicles, Standard development currently in progress, Available online: http://standards.sae.org/wip/j2954/

[10] W. E. Rippel and A. G. Cocconi, "Integrated motor drive and recharge system," *US Patent 5099186*, March 1992.

[11] D. G. Woo, G. Y. Choe, J. Kim, B. Lee, J. Hur, G. Kang, "Comparison of integrated battery chargers for plug-in electric vehicles: Topology and control," in *Proc. IEEE International Electric Machines & Drives Conference (IEMDC)*, pp. 1294-1299, May 2011, Niagara Falls, CA.

[12] L. Tang and G. J. Su, "A low-cost, digitally-controlled charger for plug-in hybrid electric Vehicles," in *Proc., IEEE Energy Conversion Congress and Exposition (ECCE)*, pp. 3923-3929, September 2009, San Jose, CA.

[13] L. De-Sousa and B. Bouchez, "Method and electric combined device for powering and charging with compensation means," *International Patent WO 2010/057893 A1*, 2010.

[14] S. Haghbin, "An isolated integrated charger for electric or plug-in hybrid vehicles," *Thesis for The Degree of Licentiate of Engineering*, Chalmers University of Technology Göteborg, Sweden, 2011.

[15] A. Bruyère, L. De Sousa, B. Bouchez, P. Sandulescu, X. Kestelyn, and E. Semail, "A Multiphase Traction/Fast-Battery-Charger Drive for Electric or Plug-in Hybrid Vehicles," in *Proc. IEEE Vehicle Power and Propulsion Conference (VPPC)*, pp. 1-7, September 2010, Lille, France.

[16] Y. J. Lee, A. Khaligh, and A. Emadi, "Advanced integrated bidirectional ac/dc and dc/dc converter for plug-in hybrid electric vehicles," *IEEE Transactions on Vehicular Technology*, vol. 58, no. 8, pp. 3970–3980, October 2009.

[17] D. C. Erb, O. C. Onar, and A. Khaligh, "An integrated bidirectional power electronic converter with multi-level AC–DC/DC–AC converter and non-inverted buck-boost converter for PHEVs with minimal grid level disruptions," in *Proc., IEEE Vehicle Power and Propulsion Conference (VPPC)*, pp. 1–6, September 2010, Lille, France.

[18] H. Chen, X. Wang, and A. Khaligh, "A single stage integrated bidirectional ac/dc and dc/dc converter for plug-in hybrid electric vehicles," in *Proc. IEEE Vehicle Power and Propulsion Conference (VPPC)*, pp. 1–6, September 2011, Chicago, IL.

[19] S. Dusmez and A. Khaligh, "A novel low cost integrated on-board charger topology for electric vehicles and plug-in hybrid electric vehicles," in *Proc., Applied Power Electronics Conference and Exposition (APEC)*, pp. 2611–2616, Feburary 2012, Orlando, FL.

[20] M. S. Chinthavali, O. C. Onar, J. M. Miller, and L. Tang, "Single-phase active boost rectifier with power factor correction for wireless power transfer applications," in *Proc., IEEE Energy Conversion Congress and Exposition (ECCE)*, pp. 3258-3265, September 2013, Denver, CO.

[21] J. M. Miller and O. C. Onar, Short Course on Wireless Power Transfer (WPT) Systems, *IEEE Transportation Electrification Conference and Expo (ITEC)*, June 2013, Dearborn, MI.

[22] B. M. Conlon, T. Blohm, M. Harpster, A. Homes, and M. Palardy, "The next generation "Voltec" extended range EV propulsion system," *SAE International Journal of Alternative Powertrains*, vol. 4, no. 2, pp. 1-12, 2015-01-1152, April 2015.

[23] Idaho National Laboratory, "Nissan Leaf – VIN 0356, Advanced Vehicle Testing – Baseline Testing Reults," Available online: http://avt.inel.gov/pdf/fsev/fact2011nissanleaf.pdf

978-1-4673-9551-9/16 $31.00 © 2016 IEEE

Switching Condition and Loss Modeling of GaN-Based Dual Active Bridge Converter for PHEV Charger

Lingxiao Xue, Dushan Boroyevich
Center for Power Electronics System
Virginia Tech
Blacksburg, VA, USA
lxxue@vt.edu

Paolo Mattavelli
DTG-University of Padova, Italy

Abstract— The dual active bridge converter is examined as the DC/DC stage of a battery charger for the plug-in hybrid electric vehicle where both load voltage and load current have very wide operating range. The load current for sinusoidal charging scheme could swing down to zero. To optimize the battery charger under this charging scheme, a DAB loss model over wide load range is needed. Traditional DAB analysis focus on the loss model within the full zero-voltage-switching (ZVS) region where all DAB switches have ZVS. Beyond this region, the switching conditions of the DAB switches are more complicated with mixed states of ZVS, hard-switching (HS), or partial zero-voltage-switching (PZVS). This paper investigates the DAB operating and switching conditions with the consideration of the resonant transition and deatime, and identified the boundaries among hard-switching, partial zero-voltage-switching, and zero-voltage-switching conditions. Based on this, accurate DAB loss model can be derived. The loss model was verified by experimental measurement of a 500 kHz GaN DAB converter.

Keywords—Dual Active Bridge; PHEV Charger; Gallium Nitride; High Frequency

I. INTRODUCTION

In a PHEV battery charger, the DC link capacitor occupies a large portion of the total volume, even if other passive components can be significantly shrunk by using wide-band gap semiconductor devices at high frequency, mainly because the required capacitance is largely determined by ripple power at two times the line frequency. Sinusoidal charging scheme, in which the doule-line frequency power is diverted into the battery, is proposed and implemented to significantly reduce the DC link size [1, 2]. In both works, the charging current of the isolated DC-DC converter were controlled as a sinusoidal shape that swings from around zero to a high amplitude. In [3], different charging waveforms are further examined to tradeoff the DC link size and DC-DC converter efficiency. The controlled charging currents of those schemes become more arbitrary but their amplitudes all vary over wide range in every double-line frequency cycle. This fluctuating output current, together with the wide output ranges that a battery charger requires, brings the DC-DC converter working across different switching conditions, in which zero-voltage-

switching(ZVS), hard-switching(HS), and partial zero-voltage-switching(PZVS) will generally happen.

Dual active bridge (DAB) converter is selected as the DC-DC stage of the battery charger. Most DAB literatures focus on the operation and loss modeling only under ZVS conditions[4]. Traditional soft-switching region derivation can only distinguish the full ZVS region while other regions are unclear. This paper will describe the operation of this topology at wide load current and voltage conditions, with special emphasis placed on the case of low load current, where HS or PZVS occurs. The region identification of ZVS, HS, and PZVS is important to the converter design and optimization. This process is challenging because the phase shift value becomes comparable to the deadtime of the two switches in a phase leg so that the classical governing equations of DAB in [5] become inaccurate. Also, with such low phase shift, the impact of the resonant transition between the DAB commutation inductor and the switch output capacitance is not negligible.

In this paper, a derivation of DAB operating principle will be conducted to account for the wide load voltage range and the extremely wide current range, with the considerations of deadtime, resonant transition, and soft-switching and hard-switching boundaries.

II. DUAL ACTIVE BRIDGE OPERATING PRINCIPLE

The DAB converter of a battery charger is shown in Fig. 1 with the device parasitic capacitors. The AC/DC stage is ignored and the input and output voltages of the DAB converter are V_{dc} and V_b, respectively.

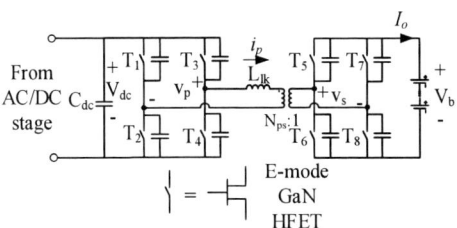

Fig. 1. DAB topology with parasitic capacitors

The dual active bridge converter has eight active switches, which provide considerable freedom for modulating the input voltage into a form that can facilitate power transfer to the output by exciting the commutation inductor. While each switching node of the phase legs can be either connected to the positive or the negative DC rail at different instants, both the primary and secondary AC terminal voltages V_p and V_s can generate three voltage levels, namely, V_{dc} (or V_b), -V_{dc} (or -V_b), or zero. The duration of each level can also be programmed to generate different voltage patterns. An additional degree of modulation freedom is the relative time difference between the two AC voltages V_p and V_s. Considering all the possible combinations of the time of the falling and rising edges of the voltages V_p and V_s, 12 different switching modes can be generated [6]. Those transition edges can be carefully designed to improve the converter's performance, in ways such as reducing RMS current and extending ZVS range [6-8]. However, the online calculation requires high occupation of the DSP resource, so it is hard to implement in a high switching frequency system. In this work, we will use the simple phase shift modulation method because of the targeted high switching frequency.

With phase shift modulation, each device is driven with a 50% duty cycle and the two-phase legs at the same side have a 180-degree phase shift, so that both of the full bridge output voltages (V_p and V_s) are 50% of the duty cycle square wave, as shown in Fig. 2. When there is a phase shift φ_1 between V_p and V_s, the inductor current i_p will rise and fall according to the sum of the input and output DC voltages during this phase shift interval. The power delivery equations can be found in various papers [5, 9], and will be only briefly repeated here. The typical DAB waveforms without considering the effect of deadtime and parasitic capacitances are shown in Fig. 2.

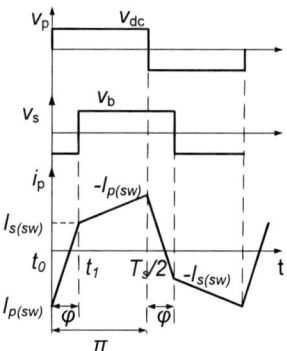

Fig. 2. DAB waveforms ignoring deadtime and parasitic capacitances

By implementing the derivation in [10], we can obtain the three critical inductor current values as

$$I_{p(sw)}(\varphi) = \frac{V_{dc}}{2\pi f_s L_{lk}}(-d \cdot \varphi - \frac{\pi}{2} + d \cdot \frac{\pi}{2}), \ d = \frac{N_{ps}V_b}{V_{dc}} \quad (1)$$

$$I_{s(sw)}(\varphi) = \frac{V_{dc}}{2\pi f_s L_{lk}}(\varphi - \frac{\pi}{2} + d \cdot \frac{\pi}{2}), \ d = \frac{N_{ps}V_b}{V_{dc}} \quad (2)$$

$$I_o(\varphi) = \frac{N_{ps}V_{dc} \cdot \varphi \cdot (\pi - \varphi)}{2\pi^2 f_s L_{lk}} \quad (3)$$

in which N_{ps} is the transformer turns ratio from the primary to secondary sides, V_{dc} is the DC link voltage, φ is the phase shift angle, f_s is the switching frequency, V_b is the battery voltage, and L_{lk} is the leakage inductance of the DAB transformer. I_o is the DAB output current.

In fact, current equations (1) to (3) still show high accuracy even when the effects of deadtime and resonant transition are considered. As long as the phase shift φ between the AC voltages of the two full bridges is fixed, the AC voltages (V_p and V_s) that excite the inductor are determined, which directly decides the output current according to (3). Note that the phase shift φ is between the AC voltages of the two full bridges, not the corresponding gate signals. The effect of deadtime is demonstrated as the timing difference between the gate signal edge and its corresponding phaseleg mid-point voltage. To be specific, in a hard-switching case, the mid-point voltage of a phaseleg aligns with the turn-on edge of the gate signal, while, in a soft-switching case, it aligns with the turn-off edge. In other words, it is the mid-point voltage of the phaseleg that directly decides the power transfer, not the gate signals. As a result, the gate timing needs to be adjusted, depending on the switching transition (HS, ZVS, or pZVS), in order to generate the right AC voltage across the inductor(V_p and V_s) that can outputs a designated DAB current I_o. With a closed-loop control, this is accomplished by a negative feedback, while in the open-loop case, fine-tuning is needed.

Resonant transition will slow down the slew rate of the AC voltages (V_p and V_s), but the waveform currents $I_{s(sw)}$ and $I_{p(sw)}$ in (1) and (2) are still good to use. This is because the resonant transition is much shorter in time than the main power transfer interval (π-φ), so it has very little influence on the power transfer. However, the resonant time is comparable to the low phase shift time at low load condition, which this paper focuses on. The resonant time, in this case, makes the gate timing more sensitive and needs to be considered and modeled.

In addition, to model the converter loss, the switching condition boundaries between HS, ZVS, and pZVS need to be identified, which strongly relies on the gate signal timing with deadtime and resonant transition considered. The analysis will be conducted as follows.

III. DAB OPERATION CONSIDERING DEADTIME AND RESONANT TRANSITION

A. Switching Condition Derivation for Primary-Side Bridge

Fig. 3 provides an expanded view of the second phase shift interval φ (starting from Ts/2) in Fig. 2, with careful consideration of the deadtime and resonant transition, since their impact on DAB operation is the most significant around the phase shift interval. We take the current-falling stage as an example, but the results will apply to the current-rising stage as well, due to the half-cycle symmetry of the waveforms. The analysis is based on positive power flow where the secondary-side bridge lags the primary-side bridge, but this idea can be easily extended to the reversed power flow. In Fig. 3, the gate signals $V_{g,p}$ and $V_{g,s}$ and the drain-to-source voltages of the representing switches $V_{ds(T2)}$ and $V_{ds(T5)}$. Inductor current i_p is also plotted to better explain the operation. The switches are numbered according to Fig. 1.

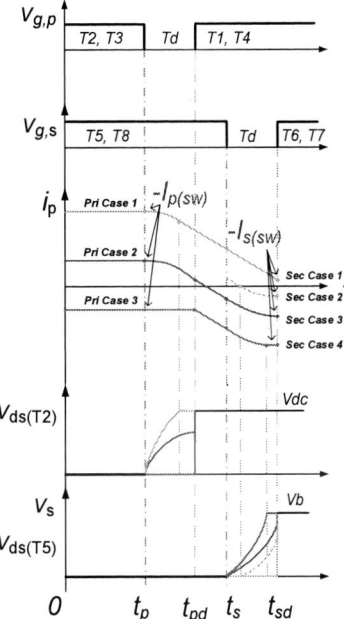

Fig. 3. DAB operating waveforms considering the deadtime and resonant transition

The analysis starts with the state where T2 and T3 are on at the primary bridge, and T5 and T8 are on at the secondary side. The initial inductor current equals to $-I_{p(sw)}$ in (1) before T2 and T3 are turned off at t_p. The captured interval between 0 and t_p is merely for demonstration purposes and its duration is extremely short, so the current level is depicted as constant. From the instant that the primary conducting switches T2 and T3 are turned off at t_p, to the instant that the secondary conducting switches T5 and T8 are turned off, there are three operational cases, depending on the initial current level and the circuit component parameters at the primary side. During this interval, no switching action happens at the secondary side, so the secondary side only exhibits as a constant voltage source to the primary side.

Primary-Side Case 1: ZVS

In this case, the initial inductor current is positive, so when T2 and T3 are turned off, the inductor will resonate with the output capacitances of all four switches at the primary side. The equivalent circuit is shown in Fig. 4.

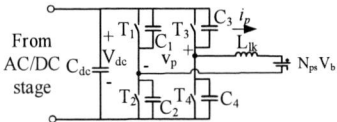

Fig. 4. Equivalent circuit of the resonant transition for the primary side bridge

It is assumed that all the switches are the same with also the same output capacitances. To deal with the nonlinearity of the output capacitance, the equivalent capacitance obtained from the output charge characterization in [11] is used. The initial condition of the inductor current and the capacitor voltage are $-I_{p(sw)}$ and 0, respectively. Then the solutions of the resonant circuits are

$$v_{C2,pri\_res}(t) = (V_{dc} - N_{ps}V_b) \cdot \frac{1 - \cos[\omega_{pri}(t - t_p)]}{2} \qquad (4)$$
$$+ Z_{pri} \cdot (-I_{p(sw)}) \cdot \frac{\sin[\omega_{pri}(t - t_p)]}{2}$$

$$i_{p,pri\_res}(t) = \frac{V_{dc} - N_{ps}V_b}{Z_0} \cdot \sin[\omega_{pri}(t - t_p)] \qquad (5)$$
$$+ (-I_{p(sw)}) \cdot \cos[\omega_{pri}(t - t_p)]$$

where

$$\omega_{pri} = \frac{1}{\sqrt{L_{lk}C_{oss}}}, Z_{pri} = \sqrt{\frac{L_{lk}}{C_{oss}}} \qquad (6)$$

The subscripts "pri_res" designate that the quantities are correspondent to the resonant transition at the primary side. Also, note that both ω_{pri} and Z_{pri} are the functions of the V_{dc} voltage, due to the nonlinearity of the output capacitance of C_{oss}. In case 1, the capacitor voltage v_{c2} can resonate from 0 volts to the DC bus voltage V_{dc} by the end of the dead time T_d, and then the switches T1 and T4 will conduct current in the reverse direction with a low voltage drop. Therefore, zero-voltage-switching can be achieved when T1 and T4 are turned on later. Once the capacitor voltage reaches V_{dc}, the inductor current will decrease linearly by the excitation of $V_{dc}+N_{ps}V_{dc}$, which yields

$$i_{p,pri\_lin}(t) = i_{p,pri\_lin\_ini} - \frac{V_{dc} + N_{ps}V_b}{L_{lk}} \cdot (t - t_{pri\_lin\_ini}) \qquad (7)$$

where i_{p,pri_lin_ini} is the initial current of the linear function and $t_{pri\_lin\_ini}$ is the instant when the capacitor voltage reaches the bus voltage. Both values can be solved from (4) and (5) by equating (4) to V_{dc}. The $t_{pri\_lin\_ini}$ solution for Case 1 is designated at t_{pri_res1} where

Table 1. Summary of primary side cases

	Boundary Conditions	**Inductor current at ts:** $i_p(t_s) = i_{p,pri\_lin}(t_s)$
Pri. Case 1	$-I_{p(sw)} > 0$, $v_{C2,pri\_res}(t_p + T_d) \geq V_{dc}$	$i_{p,pri\_res}(t_{pri\_res1}) - (V_{dc} + N_{ps}V_b)/L_{lk} \cdot (t_s - t_{pri\_res1})$
Pri. Case 2	$-I_{p(sw)} > 0$, $v_{C2,pri\_res}(t_p + T_d) < V_{dc}$	$i_{p,pri\_res}(t_{pd}) - (V_{dc} + N_{ps}V_b)/L_{lk} \cdot (t_s - t_{pd})$
Pri. Case 3	$-I_{p(sw)} \leq 0$	$-I_{p(sw)} - (V_{dc} + N_{ps}V_b)/L_{lk} \cdot (t_s - t_{pd})$

$$t_{pri\_res1} = t\big|_{v_{C2,pri\_res}(t)=V_{dc}} \qquad (8)$$

Primary-Side Case 2: PZVS

In this case, similar resonance happens between the DAB inductor and the output capacitances of the switches. The governing equations will still be (4) and (5). However, the inductor energy is not enough to bring the capacitor voltage of T2 and T3 to the DC bus voltage by the end of the deadtime Td. Accordingly, the residual charge of T1 and T4 will be dissipated during the turn-on at t_{pd}, and partial-ZVS occurs. After t_{pd}, the inductor current changes linearly following (7). Note that we have ignored the case in which the inductor current has resonated below zero before the deadtime ends. The deadtime is selected to be slightly lower than one-quarter of the resonant period $T_d < \pi\sqrt{L_{lk}C_{oss}}/2$ so that the capacitor voltage in (4) always be rising and no voltage bumping as described in [12] occurs to cause higher loss.

Primary-Side Case 3: HS

In this case, the initial inductor current at t_p is negative. This means at turn-off of the switches T2 and T3, the current is flowing from to source to drain in synchronous rectifier mode. Therefore, the turn-off of the gate will not directly cause the drain-to-source voltage swing, and thus the inductor will be excited by the same voltage. The current change during the deadtime is negligible due to low voltage slew rate ($V_{dc} - N_{ps}V_b$) / L_{lk}. so it is assumed to be constant. At t_{pd}, when T1 and T4 are turned on in hard-switching conditions, the Vds(T2) rises with a much higher slew rate than that of the turn-off transition. As a result, the inductor current remains almost unchanged until the end of the deadtime t_{pd}. After t_{pd}, the inductor current will decrease in the form of (7).

Table 1 summarizes the boundary conditions for all the three cases. We can also find the inductor current at the instant t_s, $i_p(t_s)$, when the switches T5 and T8 from the secondary-side full-bridge are turned off. Those values will serve as the initial conditions for the secondary-side analysis in the next section.

B. Switching Condition Derivation for Secondary-Side Bridge

At t_s, the switches T5 and T8 from the secondary full-bridge are turned off. There are four operational cases, depending on the initial value of the inductor current at t_s (in Table 1) and the circuit parameters. It is worth mentioning that the primary-side cases do not directly correspond to the secondary-side cases, but their effects on the inductor current at t_s do determine which secondary-side case the converter will be operated in. For example, a DAB converter whose primary switches operate in primary-side case 1 can either have its secondary switches fall into the secondary-side case 3 or secondary-side case 4, depending on the inductor current at t_s.

Secondary-Side Case 1: HS

In this case, the initial inductor at the turn-off of T5 and T8 will remain positive even by the end of the deadtime of the secondary side switches (instant t_{sd}). The inductor current will decrease following

$$i_{p,sec\_lin}(t) = i_p(t_s) - \frac{V_{dc} + N_{ps}V_b}{L_{lk}} \cdot (t - t_s) \qquad (9)$$

The subscript "sec" designates that the quantities are related to secondary side transition. The switch voltage does not change from t_s to t_{sd} because the current will continue flowing through T5 and T8 in the SR mode.

Secondary-Side Case 2: PZVS with Positive Initial Current

In this case, the initial current at the turn-off of T5 and T8 is positive, but the current is reduced to negative thanks to the high voltage excitation that equals to ($V_{dc} + N_{ps}V_b$) / L_{lk}. This is different from the Primary-side case 3 where the voltage excitation is much lower, and the current change is negligible. The time required for the inductor current to fall to zero is calculated as

$$t_{sec\_lin2} = i_p(t_s) \cdot \frac{L_{lk}}{V_{dc} + N_{ps}V_b} \qquad (10)$$

When the current falls to zero, the resonance process starts. The equivalent circuit for the secondary side resonance is shown in Fig. 5. Note that the primary quantities have been referred to the secondary side.

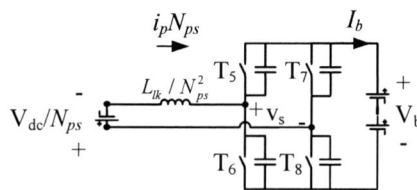

Fig. 5. Equivalent circuit of the resonant transition for secondary side bridge

The initial capacitor voltage is zero, and if assuming the initial inductor current is i_{p,sec_ini}, the solutions of the resonant transition can be obtained, they are

$$
\begin{aligned}
v_{C5,sec\_res}&(t, t_{sec\_ini}, i_{p,sec\_ini}) \\
&= (\frac{V_{dc}}{N_{ps}} + V_b) \cdot \frac{1 - \cos[\omega_{sec}(t - t_{sec\_ini})]}{2} \\
&+ Z_{sec} \cdot (-i_{p,sec\_ini}) \cdot \frac{\sin[\omega_{sec}(t - t_{sec\_ini})]}{2}
\end{aligned} \qquad (11)
$$

$$
\begin{aligned}
i_{p,sec\_res}&(t, t_{sec\_ini}, i_{p,sec\_ini}) \\
&= \frac{1}{Z_{sec}} \cdot (\frac{V_{dc}}{N_{ps}} + V_b) \cdot \sin[\omega_{sec}(t - t_{sec\_ini})] \\
&+ (-i_{p,sec\_ini}) \cdot \cos[\omega_{sec}(t - t_{sec\_ini})]
\end{aligned} \qquad (12)
$$

Where

$$\omega_{sec} = \frac{N_{ps}}{\sqrt{L_{lk}C_{oss}}}, Z_{sec} = \frac{1}{N_{ps}}\sqrt{\frac{L_{lk}}{C_{oss}}} \qquad (13)$$

and t_{sec_ini} is the start time of the resonant process. For secondary side case 2, we know $t_{sec\_ini} = t_s + t_{sec\_lin2}$, and the initial inductor current $I_{p,sec\_ini} = 0$. Therefore, at the end of the deadtime or the instant of t_{sd}, the inductor current will be $i_{p,sec\_res}(t_{sd}, t_s + t_{sec\_lin2}, 0)$, and the capacitor voltage will be $v_{c5,sec\_res}(t_{sd}, t_s + t_{sec\_lin2}, 0)$.

In this case 2, if the deadtime is long enough, the capacitor voltage should finally reach the battery voltage, even if the initial inductor energy is zero. However, this fixed deadtime will be too long for other conditions where the inductor does have enough energy. This will lead to high reverse conduction losses. Adaptive deadtime could be implemented, but will not be used in this work due to increased complexity. Instead, we selected the deadtime whose value equals to the primary side switches. The length of the deadtime is not enough to swing the switch capacitance voltage up to the full battery voltage given zero initial current of the inductor. What is more, even when the inductor has some low initial energy at the beginning of the resonant transition, it might not be sufficient to fully charge and discharge the switch capacitances, as will be quantified in the following section: Secondary Side Case 3.

Secondary-Side Case 3: PZVS with Negative Initial Current

In this case, the initial current at t_s is negative. This will directly initiate the resonant transition governed by the same equations of (11) to (13) only with the initial current $i_{p,sec\_ini} = i_p(t_s)$. Within the deadtime, the initial inductor energy is not enough to fully charge the capacitor so that the final capacitor voltage will be $v_{c5,sec\_res}(t_{sd}, t_s, i_p(t_s))$. Accordingly, the inductor current at the end of the deadtime will be $i_{p,sec\_res}(t_{sd}, t_s, i_p(t_s))$.

Secondary-Side Case 4: ZVS

The last case happens when the inductor has enough initial energy at t_s, which can bring the capacitor voltage to the battery voltage V_b by the end of the deadtime. The key equations are still (11) to (13). The time required to fully charge the capacitor can be obtained by equating (11) to V_b, and the solution is

$$t_{sec\_res4} = t \mid_{v_{C5,sec\_res}(t)=V_b} \tag{14}$$

At the rest of the deadtime, T6 and T7 will conduct in the reverse direction and generate conduction loss. The inductor current will change linearly as a result of the excitation of V_{dc}-$N_{ps}V_{dc}$, but this change is very small, and so it is neglected.

Table 2 summarizes the boundary conditions for all the four secondary-side cases.

C. Switching Condition Boundaries

In a battery charger application, the DAB converter operates in the wide range of battery voltage and charging current. This wide load range introduces the DAB converter into different switching conditions as derived in the previous section. It will be beneficial to show the different conditions on the load profile map of the battery current and battery voltage, so it is clear under which condition the charger is operated. In this section, the boundary lines between different operating conditions are plotted, and the load profile map is divided into several regions accordingly.

For the primary side, the two boundaries between HS and PZVS, and between PZVS and ZVS can be obtained by making the items in the first column of Table 1 to zero.

For the secondary side, the calculations are more difficult because, as will be described in the next section, the phase shift of gate signals (or the instant t_s in Fig. 3) requires correction to account for the switching delays. The delays will impact the derivation of the secondary side currents. A simplification is made by ignoring the Secondary side case 2, and directly using Is(sw) = 0 as the boundary between HS and PZVS. The boundary between partial ZVS and ZVS is obtained by

$$v_{C5,sec\_res}(t_{sd}, t_s, i_p(t_s)) = V_b \tag{15}$$

in which the initial current

$$i_p(t_s) = -I_{p(sw)} - \frac{V_{dc} + N_{ps}V_b}{L_{lk}} \cdot (t_s - t_{pd} - \frac{T_d}{2}) \tag{16}$$

One-half of the deadtime is assumed as the switching delay because at the boundary of ZVS and PZVS the output capacitance is charged to the full DC voltage using the entire deadtime. Also note that t_s is the real phase shift between the two excitation voltages V_p and V_s across the DAB inductor, not the gate signals. The impact of the timing difference between the gate signals and the excitation voltages due to switching delay will be detailed in the following section. Solving the battery current i_b with respect to the battery voltage V_b from the above boundary equations yields the boundary solutions. An example plot of the boundary solutions is shown in Fig. 6. The DAB parameters used in this chart are summarized in Table 3. Those parameters match with the measurement conditions in the section V. The output capacitance of the GaN switch is obtained from the curve fitting of the characterization results from [11, 13].

The blue regions are related to the primary side, while the red regions are related to the secondary side. We can see that the primary switches experience hard-switching (HS), partial-zero-voltage-switching (PZVS), and finally zero-voltage-switching (ZVS) from the left-lower corner to the right-upper

Table 2. Summary of secondary side cases

	Boundary Conditions	Inductor current at t_{sd}: $i_p(t_{sd})$
Sec. Case 1	$i_p(t_s) > 0, i_{p,sec\_lin1}(t_{sd}) > 0$	$i_{p,sec\_lin}(t_{sd})$
Sec. Case 2	$i_p(t_s) > 0, i_{p,sec\_lin1}(t_{sd}) \leq 0$	$i_{p,sec\_res}(t_{sd}, t_s + t_{sec\_lin2}, 0)$
Sec. Case 3	$i_p(t_s) \leq 0, v_{C5,sec\_res}(t_{sd}) < V_b$	$i_{p,sec\_res}(t_{sd}, t_s, i_p(t_s))$
Sec. Case 4	$i_p(t_s) \leq 0, v_{C5,sec\_res}(t_{sd}) \geq V_b$	$i_{p,sec\_res}(t_{sd} - t_{sec\_res4}, t_s, i_p(t_s))$

corner. Similarly, the secondary switches experience HS, PZVS and ZVS from the right-lower corner to the left-upper corner. The different operating conditions of primary and secondary switches combine with each other and form eight distinct regions. It is clear that at higher battery currents, it is easier to achieve soft-switching.

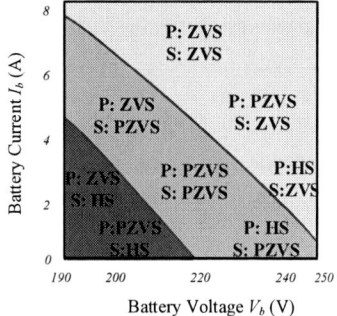

Fig. 6. DAB operation regions in term of switching conditions.

Table 3. DAB parameters used to plot the switching condition chart

Input voltage V_{dc} (V)	220
Battery voltage V_b (V)	190 - 250
DAB inductance L_{lk} (uH)	3.25
DAB transformer turns ratio	1:1
Switching frequency (kHz)	500

D. Phase Shift Correction on the Gate Signals

From Table 2, we can also find the inductor current at the instant t_{sd}, $i_p(t_{sd})$, when the switches T6 and T7 from the secondary-side full-bridge are turned on. Those values should equal to the current value of $i_p(t_{sd}) = -I_{s(sw)}$ in Fig. 3. However, we will see later in this section that this is not exactly true without the timing correction of the gate signals in the modeling process.

As has been briefly mentioned in Section II, the DAB topology is equivalent to that of two AC voltages from the primary and secondary full bridges applying across the DAB inductor, and it is the phase shift φ between those two AC voltages that directly determines the delivered power. This phase shift value, depending on switching conditions (HS, ZVS, or PZVS), may not equal to the phase shift between the corresponding gate signals (designated as t_s-t_p in Fig. 3), which we relied on in deriving the models of the 3 primary cases and 4 secondary cases. In the following paragraphs, we will examine the cause of the discrepancy between those two phase shift values and the solution to correct the phase shift of gate signals to increase the model accuracy.

The reasons for the discrepancy can be explained by Fig. 7. Here we take the primary side switch T2 as an example. When the turn-off pulse is applied to the gate of a power switch T2, its drain-to-source voltage $V_{ds(T2)}$ cannot swing immediately, and a delay is inevitable. The delay depends on the switching conditions as zero-voltage-switching (ZVS), hard-switching (HS) and partial-zero-voltage-switching (Partial ZVS).

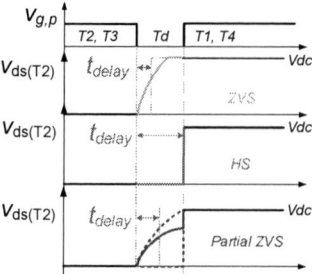

Fig. 7. Correcting the delay between the gate signals to the drain-to-source voltages

If ZVS is achieved, the $V_{ds(T2)}$ swings to the bus voltage within the deadtime. When this voltage is delayed, the applied excitation to the DAB inductor is delayed. Since the excitation voltage changes within the ZVS transition, we use half of the resonant transition time in (8) as the delay time $t_{pri\_res1}/2$. In this way, equivalently a step voltage is used to represent the original resonant waveform by providing a similar amount of charge. In the case of HS, the full deadtime T_d is used as the delay. Special attention is paid to the partial ZVS case, where two extreme points are identified: the capacitor fully charged or not charged at all within the deadtime. For the first point, the delay is approximated by half of the deadtime. For the second point, the delay time equals to the entire deadtime, the same as that of the hard-switching case. The battery current of the two points can be solved by the boundary conditions in Table 1, respectively. A linear interpolation anchored by the two extreme points is used to fit the partial ZVS cases in between

$$t_{delay,p,PZVS} = \frac{(T_d - T_d/2) \cdot (i_b - I_{b,pri\_bdr\_PZVS}(V_b))}{I_{b,pri\_bdr\_HS}(V_b) - I_{b,pri\_bdr\_PZVS}(V_b)} + \frac{T_d}{2} \quad (17)$$

in which $I_{b,pri\_bdr\_HS}(V_b)$ is the battery current at the boundary of HS and partial-ZVS regions, and $I_{b,pri\_bdr\_PZVS}(V_b)$ is the battery current at the boundary of partial-ZVS and ZVS regions.

This mechanism of time delay also applies to the secondary side. For the ZVS case, we cannot directly use half the resonant transition time in (14) as the delay time, because $i_p(t_s)$ needs to be corrected first using the right phase shift. In order to solve the correct transition time, the following two equations are used:

$$v_{C5,sec\_res}(t_x, t_s, i_p(t_s)) = V_b \quad (18)$$

$$i_p(t_s) = -I_{p(sw)} - \frac{V_{dc} + N_{ps}V_b}{L_{lk}} \cdot (t - t_{pd} - \frac{t_x}{2}) \quad (19)$$

where t_x is the transition time of the secondary side to be solved. Note that here we assumed the primary side is doing hard switching. This assumption is valid at low battery current where hard-switching occurs at the primary side, and also at high battery current where resonant delay is much lower than the phase shift time.

Similarly to that of the primary side, the HS case has a delay time of one deadtime T_d. For the partial ZVS case, two extreme points are identified from the two boundary lines among HS, PZVS and ZVS cases. The resultant delay time for PZVS is

$$t_{delay,s} = \frac{(T_d - T_d/2) \cdot (i_b - I_{b,sec\_bdr\_PZVS}(V_b))}{I_{b,sec\_bdr\_HS}(V_b) - I_{b,sec\_bdr\_PZVS}(V_b)} + \frac{T_d}{2} \quad (20)$$

in which $I_{b,sec\_bdr\_HS}(V_b)$ is the battery current at the boundary of HS and PZVS regions for secondary side switches, and $I_{b,sec\_bdr\_PZVS}(V_b)$ is the battery current at the boundary of partial-ZVS and ZVS regions for secondary side switches.

With the delay time considered, the currents at turn-on of the secondary switches at t_s in Table 1 will be updated. Accordingly, the switching currents of the secondary side $i_p(t_{sd})$ in Table 2 are also changed. The modeled secondary side currents $i_p(t_{sd})$, which locates at the end of the derivation chain in Fig. 3, can be used to verifiy the validity of the derivation process against $-I_{s(sw)}$ value in (2). The $-I_{s(sw)}$ value is treated as the "real" current because it is not impacted by the deadtime and only weakly impacted by the resonant transition. The results are shown in Fig. 8.

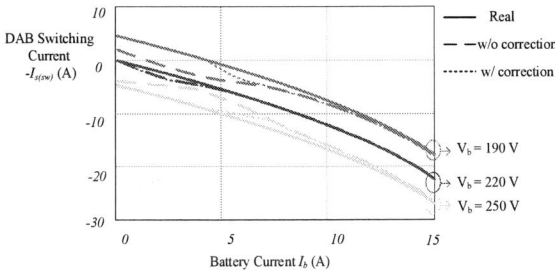

Fig. 8. Model comparison on DAB secondary-side switching current

Compared to $-I_{s(sw)}$ (solid line), the uncorrected $i_p(t_{sd})$ (dashed line) shows high discrepancy while the corrected $i_p(t_{sd})$ (dot line) has significantly reduced discrepancy. The corrected model shows good match both at low and high battery currents. Relatively high error is observed in medium current range due to the assumption of primary-side hard-switching when deriving the secondary side boundary between ZVS and PZVS.

IV. GAN-BASED DAB LOSS MODELING

With the switching condition boundaries identified, it becomes possible to model the GaN-based DAB converter loss. The conduction loss is

$$p_{con\_DAB} = i_{lk(rms)}^2 \cdot 2R_{ds(on)} \cdot (1 + N_{ps}^2) \quad (21)$$

where

$$R_{ds(on)} = \begin{cases} R_{ds(on),ZVS}(V_{sw}) \; if \; ZVS \\ R_{ds(on),HS}(V_{sw}) \; if \; HS \; and \; PZVS \end{cases} \quad (22)$$

This means that the dynamic on-resistance of the GaN module changes needs to be considered according to different switching conditions. The modeled GaN device conduction loss is shown in Fig. 9.

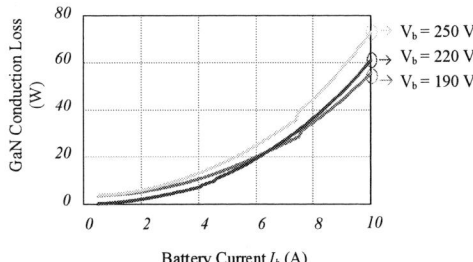

Fig. 9. Modeled GaN device conduction loss of the DAB converter

For the GaN module, the turn-off loss is negligible due to the very short overlap between channel current and drain-source voltage, so only turn-on energy is taken into account. The energy loss models for HS, ZVS, and partial-ZVS cases are different. The turn-on energy is summarized as the following equation.

$$E_{sw}(V_{sw}, I_{sw}) = \begin{cases} C_{oss,eq}(V_{sw}) \cdot V_{sw} \cdot V_{sw} + 0.5 \cdot t_{sw} \cdot V_{sw} \cdot I_{sw} \; if \; HS \\ C_{oss,eq}(V_{sw}) \cdot V_{sw} \cdot V_{sw} - 0.5 \cdot t_{sw} \cdot V_{sw} \cdot I_{sw} \; if \; PZVS \\ V_f(I_{sw}) \cdot I_{sw} \cdot (T_d - t_{res}) \; if \; ZVS \end{cases} \quad (23)$$

in which V_{sw} and I_{sw} are the switching voltage and current that can be derived from the equations (4) to (6) and (11) to (13); t_{sw} is the switching time; and $C_{oss,eq}(V_{sw})$ is the equivalent output capacitance at V_{sw}. Both of them can be obtained from device characterization mentioned in [11].

In the case of ZVS, there is no turn-on loss. However, as the output capacitance is fully discharged before the deadtime ends, this switch will conduct in "diode" mode with a high voltage drop for enhancement-mode GaN. We put this deadtime-conduction loss in the switching energy equation for more concise and clear formulation. In the case of HS and PZVS, the consumed energy consists of two parts, the capacitance-related energy loss and the voltage-current overlapping loss. But the overlapping energy has different signs in the two cases due to opposite load current directions. The switching loss is then derived by multiplying the switching energy by device numbers and switching frequency as

$$p_{sw\_DAB} = 4E_{sw,pri}f_s + 4E_{sw,sec}f_s \quad (24)$$

The modeled GaN switching loss is shown in Fig. 10.

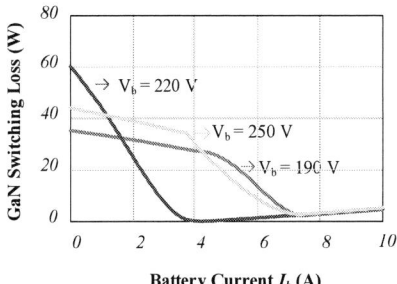

Fig. 10. Modeled GaN device switching loss of the DAB converter

The transformer winding AC resistance is measured by an impedance analyzer. An FFT transform on the DAB current

978-1-4673-9551-9/16 $31.00 © 2016 IEEE 1321

waveform provides the harmonic components, and then the winding loss is modeled by adding up the Ohm loss of each harmonic frequency. Harmonics up to the eleventh order are considered due to the steep edges of the DAB current waveform.

The transformer core loss is characterized by measurement [14]. The measurement excited the cores with the same voltage sources for the primary and secondary sides. A linear interpolation between the measurement points then derives the core loss model.

The total modeled DAB losses are compared with the measurement results of a GaN DAB converter (parameters in Table 3), as shown in Fig. 11. We can see a reasonable match between the model and the measurement.

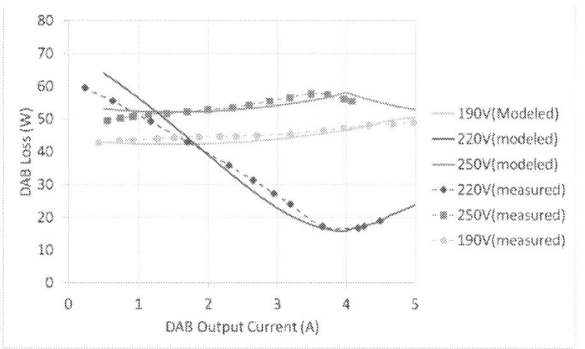

Fig. 11. Comparison between the measured loss and modeled loss for a DAB converter at the switching frequency of 500 kHz

V. CONCLUSION

The dual active bridge converter is examined as the DC/DC stage of a battery charger for the plug-in hybrid electric vehicle. The DAB design is challenging considering the wide battery voltage range. If sinusoidal charging is implemented, the output current will fluctuate at double line frequency, which further makes the design difficult. This paper investigated the DAB operating and switching conditions with the consideration of the resonant transition and deatime, and identified the boundaries among hard-switching, partial zero-voltage-switching, and zero-voltage-switching conditions. The derived boundaries divided the battery voltage and current map into eight regions where the switching condition of both primary side and secondary side switches are clearly identified. Based on this, accurate DAB loss model can be derived. The loss model was verified by experimental measurement of a 500 kHz GaN DAB converter.

REFERENCES

[1] X. Lingxiao, D. Diaz, S. Zhiyu, L. Fang, P. Mattavelli, and D. Boroyevich, "Dual active bridge based battery charger for plug-in hybrid electric vehicle with charging current containing low frequency ripple," in *Applied Power Electronics Conference and Exposition (APEC), 2013 Twenty-Eighth Annual IEEE*, 2013, pp. 1920-1925.

[2] X. Lingxiao, S. Zhiyu, D. Boroyevich, P. Mattavelli, and D. Diaz, "Dual Active Bridge-Based Battery Charger for Plug-in Hybrid Electric Vehicle With Charging Current Containing Low Frequency Ripple," *Power Electronics, IEEE Transactions on,* vol. 30, pp. 7299-7307, 2015.

[3] X. Lingxiao, P. Mattavelli, D. Boroyevich, B. Hughes, and R. Burgos, "Efficiency optimized AC charging waveform for GaN bidirectional PHEV Battery Charger," in *Power Electronics and Applications (EPE'14-ECCE Europe), 2014 16th European Conference on*, 2014, pp. 1-10.

[4] D. Costinett, N. Hien, R. Zane, and D. Maksimovic, "GaN-FET based dual active bridge DC-DC converter," in *Applied Power Electronics Conference and Exposition (APEC), 2011 Twenty-Sixth Annual IEEE*, 2011, pp. 1425-1432.

[5] R. W. A. A. De Doncker, D. M. Divan, and M. H. Kheraluwala, "A three-phase soft-switched high-power-density DC/DC converter for high-power applications," *Industry Applications, IEEE Transactions on,* vol. 27, pp. 63-73, 1991.

[6] F. Krismer and J. W. Kolar, "Closed Form Solution for Minimum Conduction Loss Modulation of DAB Converters," *Power Electronics, IEEE Transactions on,* vol. 27, pp. 174-188, 2012.

[7] G. G. Oggier, R. Ledhold, G. O. Garcia, A. R. Oliva, J. C. Balda, and F. Barlow, "Extending the ZVS Operating Range of Dual Active Bridge High-Power DC-DC Converters," in *Power Electronics Specialists Conference, 2006. PESC '06. 37th IEEE*, 2006, pp. 1-7.

[8] J. Everts, F. Krismer, J. Van den Keybus, J. Driesen, and J. W. Kolar, "Optimal ZVS Modulation of Single-Phase Single-Stage Bidirectional DAB AC–DC Converters," *Power Electronics, IEEE Transactions on,* vol. 29, pp. 3954-3970, 2014.

[9] M. N. Kheraluwala, R. W. Gascoigne, D. M. Divan, and E. D. Baumann, "Performance characterization of a high-power dual active bridge," *Industry Applications, IEEE Transactions on,* vol. 28, pp. 1294-1301, 1992.

[10] F. Krismer, "Modeling and optimization of bidirectional dual active bridge DC-DC converter topologies," Diss., Eidgenössische Technische Hochschule ETH Zürich, Nr. 19177, 2010, 2010.

[11] X. Lingxiao, M. Mingkai, D. Boroyevich, and P. Mattavelli, "The optimal design of GaN-based Dual Active Bridge for bi-directional Plug-IN Hybrid Electric Vehicle (PHEV) charger," in *Applied Power Electronics Conference and Exposition (APEC), 2015 IEEE*, 2015, pp. 602-608.

[12] L. Jin, C. Zheng, S. Zhiyu, P. Mattavelli, L. Jinjun, and D. Boroyevich, "An adaptive dead-time control scheme for high-switching-frequency dual-active-bridge converter," in *Applied Power Electronics Conference and Exposition (APEC), 2012 Twenty-Seventh Annual IEEE*, 2012, pp. 1355-1361.

[13] X. Lingxiao, S. Zhiyu, M. Mingkai, D. Boroyevich, R. Burgos, B. Hughes, *et al.*, "Bi-directional PHEV battery charger based on normally-off GaN-on-Si multi-chip module," in *Applied Power Electronics Conference and Exposition (APEC), 2014 Twenty-Ninth Annual IEEE*, 2014, pp. 1662-1668.

[14] M. Mingkai, L. Qiang, D. J. Gilham, F. C. Lee, and K. D. T. Ngo, "New Core Loss Measurement Method for High-Frequency Magnetic Materials," *Power Electronics, IEEE Transactions on,* vol. 29, pp. 4374-4381, 2014.

Analysis of Cascaded Multi-Output-Port Converter for Wireless Plug-in Hybrid/On-Board EV Chargers

Erdem Asa[1,3], Kerim Colak[2], Dariusz Czarkowski[3]

[1]Hevo Power Inc., New York, USA
[2]Istanbul Ulasim A.S., Istanbul, Turkey
[3]New York University, Polytechnic School of Engineering, New York, USA

Abstract— This paper presents a new cascaded connection of multi-output-port converter for Plug-in Hybrid Electric and On-Board Electric Vehicles, which is suitable for dynamic wireless power transfer energy storage systems such as battery, ultra-capacitor, and load. In order to acquire the system power transfer integration between output ports, the central control stability is analyzed by phase shift function between converter ports. The theoretical calculation of the converter is explored describing the system model with a constant frequency under the different load conditions. The performance of the proposed converter is confirmed with experimental results using 8 inch air gap coreless transformer. The converter is tested at 120 V input and creates an output of 0-100 V and 0-50 V in each output port at a full power of 700 W.

Keywords— *cascaded, multi-output, phase shift, control, wireless, EV charger*

I. INTRODUCTION

For charging electrical vehicles (EVs), the contactless charging techniques is a viable choice as mentioned in the literature [1]-[2] such as low risk hazardous, weather proof, anti-vandalism, etc.. The wireless charging process of EVs can be carried out when vehicles are on roads, called dynamic charging [3]; or parked, refer to stationary charging [4]. A dynamic EV inductive power transfer system is basically consisted of two stages; a transmitter platform on the ground and receiver side in the vehicle as shown in Fig. 1. The first stage, power line transmitter is basically main part to deliver energy into the second stage. The dc output voltage connected the different loads is provided by the second stage with high frequency rectifier, and various non-isolated dc / dc converters.

Due to the mobility of the charging target, the constant continuous energy transfer is the most important issue for dynamic charging systems. The energy transfer performance is deteriorated because of the misalignment between the transmitter and receiver coils. Hence, the converter topology development should be taking into account with emphasis on controllability, high power ratio, and high efficiency, having

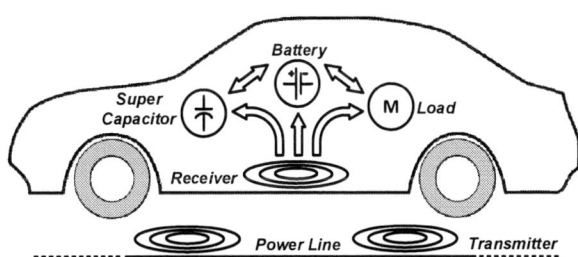

Fig. 1. In-motion dynamic EV wireless power transfer system.

become the main focus of research by both academia and industry recently [5]-[6].

Many researchers have proposed numerous control strategies and circuit topologies for inductive power transfer (IPT) systems. The most common power control strategies are carried from the primary [7], secondary [8]-[10], or both primary and secondary sides [11]-[12]. The study of the IPT charging system while driving has been recommended by a few papers [13]-[17]. The University of Auckland developed the first dynamic charging IPT system [13]. Korea Rail Research Institute has shown a great progress achieving up to 1 MW output power for the dynamic charging of trains [14]. KAIST established a dynamic charging system for their online electric vehicles (OLEV) [15]. Oak Ridge National Laboratory investigated power transfer characteristics for dynamic charging applications [16]. Also, hybrid energy storage systems (HESS) have been proposed by many authors. Using a multi-port power electronics interface, integration of inductively coupled power transfer and hybrid energy storage system is proposed for battery powered EVs in [17].

In this work, a cascaded connection of multi-output-port converter is studied for in-motion dynamic IPT systems. A phase shifted modulation technic is used in order to provide the power flowing between output ports. The topology is explored with different load conditions at the constant frequency and variable output voltage in each output port. The converter model controllability is analyzed and the system mathematical

978-1-4673-9551-9/16 $31.00 © 2016 IEEE

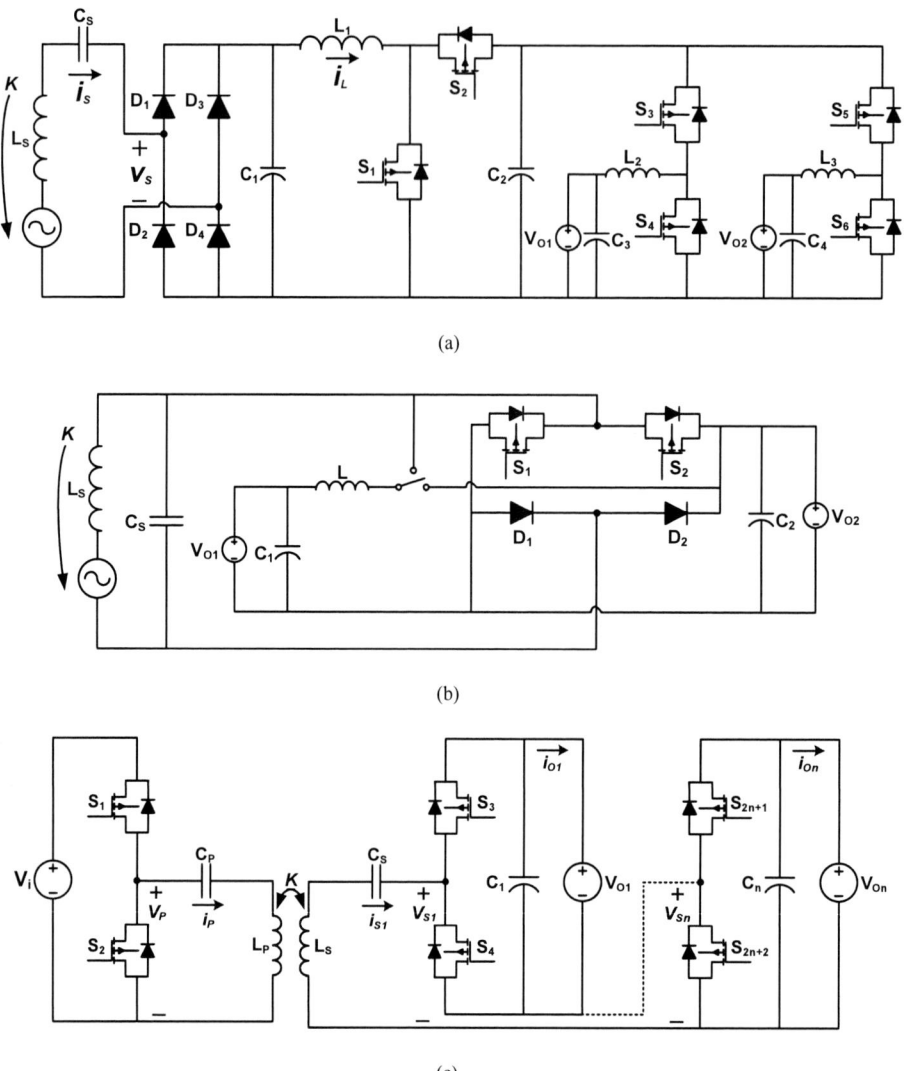

(a)

(b)

(c)

Fig. 2. The multi-output converter technolgies, a) the proposed integrated charger with wireless charging for PHEVand EV applications [16], b) integration of inductively coupled power transfer and hybrid energy storage system [17], c) the proposed cascaded connected multi-output-port converter topology.

calculation is derived. The system performance is confirmed with experimental results at 8 inch air gaps, 150 kHz operating frequency, and a 700 W load in laboratory conditions. The experimental results show that the proposed converter is an appropriate topology for different type of energy storage of dynamic IPT systems such as battery, ultra-capacitor, etc. at various power capacities. A related circuit system and the theoretical analysis with verification of the converter are discussed more detailed in the following chapters.

II. ANALYSIS OF THE CASCADED MULTI-OUTPUT-PORT CONVERTER

A central control of multiport power electronics interface is recommended [17] with IPT system where the proposed system is consisted of a series of bidirectional single phase switching dc-dc converters as seen in Fig. 2(a). The integrated charging topology [18] is presented in Fig. 2(b). The receiver side rectifier stage is integrated with the dc-dc boost converter which utilizes a single stage switch and a diode combined with an inductor. The investigated phase shift principle of a multi-output-port IPT circuit topology for the dynamic wireless power transfer is shown in Fig. 2(c). It comprises a half bridge resonant inverter in the primary side, a cascaded multi-port converter in the secondary side, an air gap coreless transformer, and a resonant capacitors in each impedance matching network. The output voltage and power of the system are managed by a phase-shift angle of the transistors between output ports in terms of different load conditions.

978-1-4673-9551-9/16 $31.00 © 2016 IEEE

Since each output port is phase-shifted by an angle ϕ, the power switches of the each converter can be replaced by multi voltage sources $V_{S1},...,V_{Sn}$ respectively. Also, the input of the output converters is square-wave voltages. The wireless power link can be represented as two coupled inductors and two resonant capacitor connected in primary and one capacitor in the secondary as shown in Fig. 3.

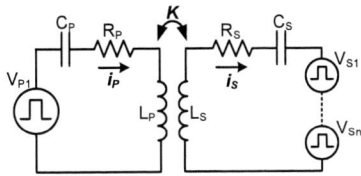

Fig. 3.   The equivalent circuit model of the proposed converter.

In this model, output voltage sources are $V_{S1},...,V_{Sn}$, and C_P, C_S are resonant capacitors. Here, the resonant capacitors are equal $\{C_P=C_S\}$. The two coupled inductors L_P and L_S can be equivalently modeled as a transformer with proper leakage L_L and magnetizing inductance L_M. To simplify the wireless transformer, both coils are assumed to be identical and equal to L. Then, the model can be equivalently represented by following equations.

$$L_M = K\sqrt{L_P L_S} = KL \tag{1}$$

$$L_L = L - L_M = (1-K)L \tag{2}$$

The each output of converter circuits generate a square wave voltages at the input points during the three different cycle. Using the Fourier analysis, the each output port of the converter can be stated as

$$v_{S1}(t) = \begin{cases} V_{O1}, & -\phi_1 < \omega t \le \pi - \alpha \\ 0, & \pi - \alpha < \omega t \le 2\pi - \phi_1 \end{cases} \tag{3}$$

$$v_{Sn}(t) = \begin{cases} V_{On}, & -\phi_n < \omega t \le \pi - \alpha \\ 0, & \pi - \alpha < \omega t \le 2\pi - \phi_n \end{cases} \tag{4}$$

where α is the secondary side current phase angle. The fundamental components of the voltages $v_{P1,1}(t)$ and $v_{S1,1}(t),..,v_{Sn,1}(t)$ can be stated in the phasor domain as

$$V_{P1,1} = \frac{2}{\pi} V_{i1} \tag{5}$$

and

$$V_{S1,1} = V_{m1} e^{j\left(\frac{\alpha+\phi_1}{2}\right)} \tag{6}$$

$$V_{Sn,1} = V_{mn} e^{j\left(\frac{\alpha+\phi_n}{2}\right)} \tag{7}$$

where $\phi_1,..., \phi_n$ describes the phase difference in each leg between upper and lower switches. The variables $V_{m1},..., V_{mn}$ are,

$$V_{m1} = \frac{2}{\pi} V_{O1} cos\left(\frac{\alpha+\phi_1}{2}\right) \tag{8}$$

$$V_{mn} = \frac{2}{\pi} V_{On} cos\left(\frac{\alpha+\phi_n}{2}\right) \tag{9}$$

The output voltage for the n port of the converter can be simplified as

$$V_S = \sum_{u=1}^{n} V_{Su} \tag{10}$$

The voltage transfer function of the system $|M_V|=|V_O/V_I|$ can be derived as

$$|M_V| = \left| \frac{\sum_{u=1}^{n} V_{Su}}{V_I} \right| = \left| \frac{\sum_{u=1}^{n} V_{Ou} e^{j\left(\frac{\alpha+\phi_n}{2}\right)}}{V_I} \right| \tag{11}$$

III.   THE POWER FLOW CONTROL ANALYSIS

Power flow controllability to the output and between ports is analyzed considering the simplified equivalent circuit of Fig. 3. The corresponding fundamental currents of the contactless system in the primary and secondary winding can be expressed as

$$I_P = \frac{V_{P1,1} R_S - \sum_{u=1}^{n} V_{mu} e^{j\left(\frac{\alpha+\phi_u}{2}\right)} (j\omega L_M)}{R_P R_S + (\omega L_M)^2} \tag{12}$$

$$I_P^* = \frac{V_{P1,1} R_S + \sum_{u=1}^{n} V_{mu} e^{-j\left(\frac{\alpha+\phi_u}{2}\right)} (j\omega L_M)}{R_P R_S + (\omega L_M)^2} \tag{13}$$

and

$$I_S = \frac{\sum_{u=1}^{n} V_{mu} e^{j\left(\frac{\alpha+\phi_u}{2}\right)} R_P - V_{P1,1} (j\omega L_M)}{R_P R_S + (\omega L_M)^2} \tag{14}$$

$$I_S^* = \frac{\sum_{u=1}^{n} V_{mu} e^{-j\left(\frac{\alpha+\phi_u}{2}\right)} R_P + V_{P1,1} (j\omega L_M)}{R_P R_S + (\omega L_M)^2} \tag{15}$$

The system power flow is investigated for n different sources that have the same converter parameters. The power flow for the n port of the converter can be investigated by writing the primary side power flow as

$$P_P = Re\left\{ \frac{V_{P1,1} I_P^*}{2} \right\} \tag{16}$$

$$P_P = \frac{V_{P1,1}^2 R_S}{2[R_P R_S + (\omega L_M)^2]} + \frac{\sum_{u=1}^{n} V_{P1,1} V_{mu} (\omega L_M) sin\left(\frac{\alpha+\phi_u}{2}\right)}{2[R_P R_S + (\omega L_M)^2]} \tag{17}$$

The power flow can be described between output ports as

$$P_{S1} = Re\left\{\frac{V_{S1,1}I_S^*}{2}\right\} \quad (18)$$

$$P_{S1} = \frac{\sum_{u=1}^{n} V_{m1}V_{mu}R_P \cos\left(\frac{\phi_1-\phi_u}{2}\right)}{2[R_PR_S+(\omega L_M)^2]} - \frac{V_{P1,1}V_{m1}\omega L_M \sin\left(\frac{\alpha+\phi_1}{2}\right)}{2[R_PR_S+(\omega L_M)^2]} \quad (19)$$

$$P_{Sn} = Re\left\{\frac{V_{Sn,1}I_S^*}{2}\right\} \quad (20)$$

$$P_{Sn} = \frac{\sum_{u=1}^{n} V_{mn}V_{mu}R_P \cos\left(\frac{\phi_n-\phi_u}{2}\right)}{2[R_PR_S+(\omega L_M)^2]} - \frac{V_{P1,1}V_{mn}\omega L_M \sin\left(\frac{\alpha+\phi_n}{2}\right)}{2[R_PR_S+(\omega L_M)^2]} \quad (21)$$

The total output power can be calculated with the following equation

$$P_S = \sum_{u=1}^{n} P_{Su} \quad (22)$$

For the two source configurations, the each source power and total output power can be evaluated as following

$$P_{S1} = \frac{V_{m1}^2 R_P}{2[R_PR_S+(\omega L_M)^2]} - \frac{V_{P1,1}V_{m1}\omega L_M \sin\left(\frac{\alpha+\phi_1}{2}\right)}{2[R_PR_S+(\omega L_M)^2]} \quad (23)$$

$$P_{S2} = \frac{V_{m2}^2 R_P}{2[R_PR_S+(\omega L_M)^2]} - \frac{V_{P1,1}V_{m2}\omega L_M \sin\left(\frac{\alpha+\phi_2}{2}\right)}{2[R_PR_S+(\omega L_M)^2]} \quad (24)$$

The total output power is

$$P_S = \frac{(V_{m1}^2+V_{m2}^2)R_P}{2[R_PR_S+(\omega L_M)^2]} - \frac{(V_{m1}+V_{m2})V_{P1,1}\omega L_M \sin\left(\frac{\alpha+\phi_2}{2}\right)}{2[R_PR_S+(\omega L_M)^2]} \quad (25)$$

IV. EXPERIMENTAL RESULTS

The proposed wireless converter technology is consisted of a two port converter in the receiver side and it is designed for 700 W, 120 V input voltage considering 10 Ω and 20 Ω output load conditions. The coreless transformer is tested with 8 inch distances between coils which results in 0.3 coupling factor. The component values of the prototype and parameters are presented in Table I.

TABLE I

Symbol	PARAMETER	Value
V_i	dc input voltage	120 V
P_O	maximum output power	700 W
C_P	primary side capacitor	40 nF
C_S	secondary side capacitor	40 nF
L_P	primary coil self-inductances	25 μH
L_P	secondary coil self-inductances	25 μH
d	square coil dimension	2.5 x 2.5 feet
n	coil turn number	4
f_{sw}	operating frequency	150 kHz
C	filter capacitor	10 μF

The system proposed controller is described in Fig. 4. Phase-shift function forms the output power at the each output load terminal based on the reference value as shown in the scheme. The power management can also be adjusted regulating the phase shift between output ports. The control modulation is implemented using field programmable gate array (FPGA) Spartan 6 XC6SLX9 and microprocessor STM32F407VG. The FPGA mainly ensures PWM phase-shift function with sensing the phase in each output port. Then, the phase-shift modulation is carried out to the MCU.

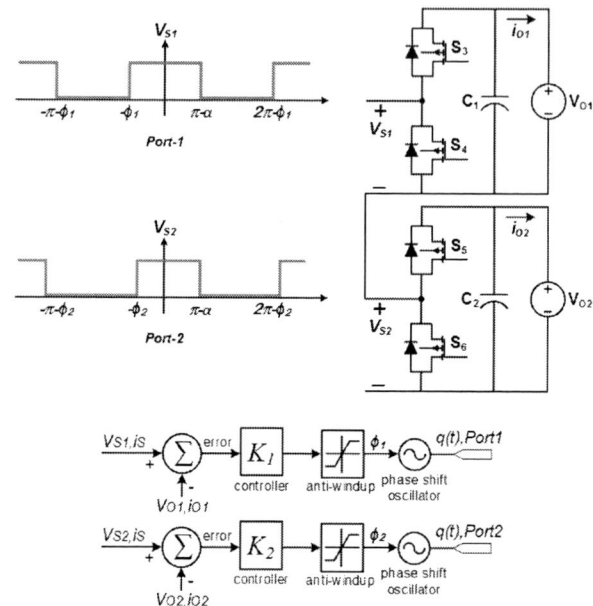

Fig. 4. The system controller model with phase-shift modulation.

In order to see the performance of the system, a two-output-port converter topology is considered to experimentally verify the proposed converter analysis. The output power characteristics with phase shift angle and different load conditions are realized in Fig. 5. The phase shift angle is performed from 0° through 360° degrees between Port-1 and Port-2 converter to obtain the output power waveform at the constant operating frequency 150 kHz. The output power flow between converter output ports is achieved by different phase shift values considering the different load conditions 10Ω and 20Ω as seen in figures.

Fig. 6 shows the voltage and current waveforms for the Port-1 and Port-2 at different phase shifts. As seen selected waveforms in the figure, the phase-shift angle between converter ports controls the output power in each converter output by changing the converter input voltage magnitude. The output power flow also can be arranged from the first output port to the second port or from the second port to the first port by a proper phase shift modulation.

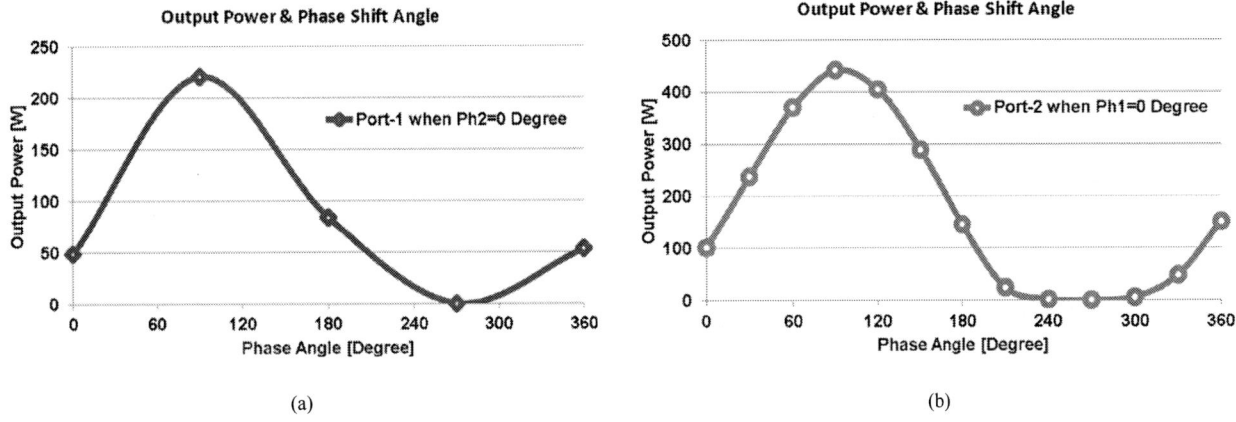

(a)   (b)

Fig. 5. The output power characteristics with phase shift angle for a) 10 Ω and b) 20 Ω.

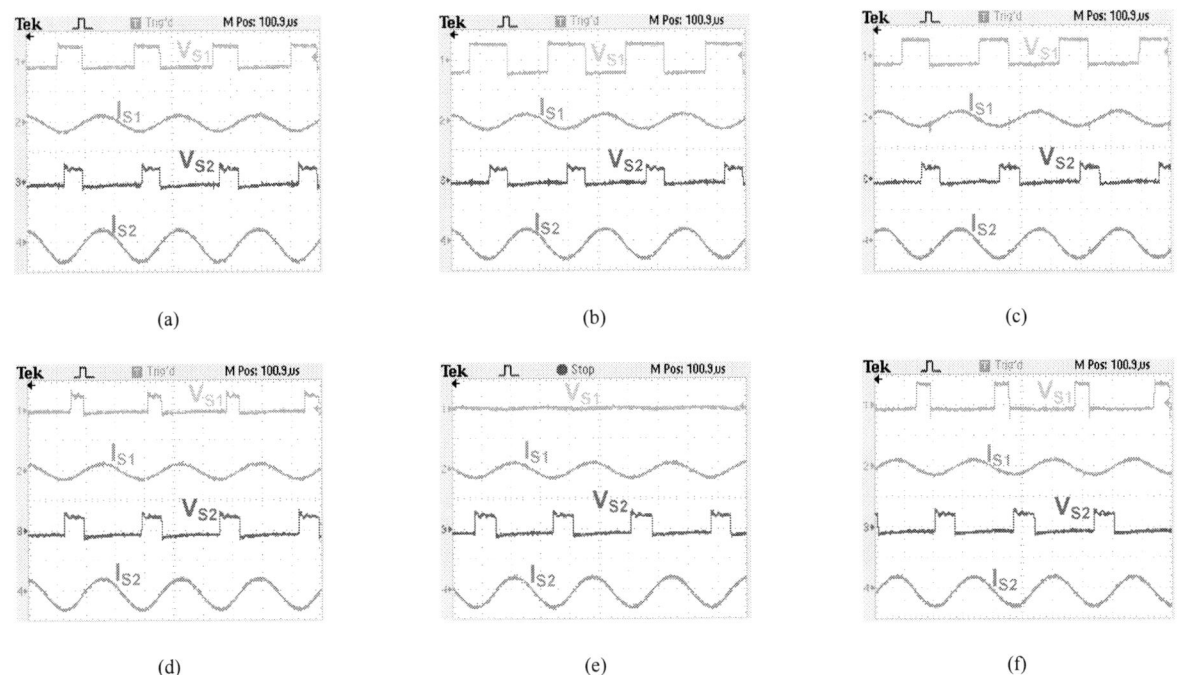

Fig. 6. The output Port-1 voltage (V_{S1}), current (I_{S1}) and the output Port-2 voltage (V_{S2}), current (I_{S1}): the Port-1 is at 10 Ω and the Port-2 is at 20 Ω. a, b, c for V_{S1} *(100V/div)*, I_{S1} *(5A/div)*, V_{S2} *(40V/div)*, I_{S2} *(5A/div)*; d, e, f for V_{S1} *(40V/div)*, I_{S1} *(5A/div)*, V_{S2} *(40V/div)*, I_{S2} *(5A/div)*.

V. CONCLUSIONS

In this study, a new cascaded connected multi-output-port converter is presented for dynamic wireless EV charging applications. The power flow analysis and transfer function equation of the converter are explained by describing equivalent circuit of the system. The experimental results are performed with different unbalanced output load conditions and 0 to 100 V in Port-1 and 0 to 50 V in Port-2 output range at total output power of 700 W. The proposed multi-port converter topology provides a power flow between output converter ports with the phase shift angle. The analysis results show that the different voltage and power level of combinations of energy storage systems (battery, ultra-capacitor, etc.) can be integrated with the load by the proposed converter.

REFERENCES

[1] E. Asa, K. Colak, M. Bojarski, D. Czarkowski, "A Novel Multi-Level Phase-Controlled Resonant Inverter with Common Mode Capacitor for Wireless EV Chargers," IEEE Transportation Electrification Conference and Expo (ITEC), pp.1-6, Jun. 2015.

[2] K. Colak, M. Bojarski, E. Asa, D. Czarkowski, "A Constant Resistance Analysis and Control of Cascaded Buck and Boost Converter for Wireless EV Chargers," IEEE Applied Power Electronics Conference and Exposition (APEC), pp.3157-3161, Mar. 2015.

[3] J. M. Miller, P. T. Jones, J. M. Li, O. C. Onar, "ORNL Experience and Challenges Facing Dynamic Wireless Power Charging of EV's," IEEE Circuits and Systems Magazine, vol.15, no.2, pp.40-53, Secondquarter 2015.

[4] M. Bojarski, E. Asa, M. T. Outeiro, D. Czarkowski, "Control and Analysis of Multi-level Type Multi-phase Resonant Converter for Wireless EV Charging," 41th Annual Conference of IEEE Industrial Electronics Society (IECON), pp.5008-5013, Nov. 2015.

[5] K. Colak, E. Asa, M. Bojarski, D. Czarkowski, O. C. Onar, "A Novel Phase Shift Control of Semi-Bridgeless Active Rectifier for Wireless Power Transfer" IEEE Transaction on Power Electronics, vol.30, no.11, pp.6288-6297, Nov. 2015.

[6] E. Asa, K. Colak, M. Bojarski, D. Czarkowski, "A Novel Phase Control of Semi Bridgeless Active Rectifier for Wireless Power Transfer Applications," IEEE Applied Power Electronics Conference and Exposition (APEC), pp.3225-3231, Mar. 2015.

[7] M. Bojarski, E. Asa, D. Czarkowski, "A 25 kW Industrial Prototype Wireless Electric Vehicle Charger," IEEE Applied Power Electronics Conference and Exposition (APEC), pp.1-6, Mar. 2016.

[8] K. Colak, E. Asa, M. Bojarski, D. Czarkowski, "A Novel Common Mode Multi-phase Half-wave Semi-synchronous Rectifier for Inductive Power Transfer Applications," IEEE Transportation Electrification Conference and Expo (ITEC), pp.1-6, Jun. 2015.

[9] E. Asa, K. Colak, D. Czarkowski, "Analysis of Multi-Output Half-Wave Semi-Synchronous Rectifier with a Uniform Magnetic Field Transmitter," IEEE Applied Power Electronics Conference and Exposition (APEC), pp. 1-5, March 2016.

[10] K. Colak, E. Asa, D. Czarkowski, "A Novel Phase Control of Single Switch Active Rectifier for Inductive Power Transfer Applications,"

IEEE Applied Power Electronics Conference and Exposition (APEC), pp.1-6, Mar. 2016.

[11] E. Asa, K. Colak, D. Czarkowski, B. Tamyurek, "An Efficiency Analysis of Bi-directional DC/DC Converter for Wireless Energy Transfer Applications," IEEE Energy Conversion Congress and Exposition (ECCE), pp.594-598, Sep. 2015.

[12] K. Colak, E. Asa, D. Czarkowski, H. Komurcugil, "A Novel Multi-level Bi-directional DC/DC Converter for Inductive Power Transfer Applications," 41th Annual Conference of IEEE Industrial Electronics Society (IECON), pp.3827-3831, Nov. 2015.

[13] G. A. J. Elliott, J. T. Boys, A. W. Green, "Magnetically Coupled Systems for Power Transfer to Electric Vehicles," IEEE Proceedings of Power Electronics and Drive Systems, pp.797-801, Feb. 1995.

[14] J. H. Kim, B. S. Lee, J. H. Lee, S. H. Lee, C. B. Park, S. M. Jung, S. G. Lee, K. P. Yi, J. Baek, "Development of 1-MW Inductive Power Transfer System for a High-Speed Train," IEEE Transactions on Industrial Electronics, vol.62, no.10, pp.6242-6250, Oct. 2015.

[15] J. Huh, S. W. Lee, W. Y. Lee, G. H. Cho, C. T. Rim, "Narrow-Width Inductive Power Transfer System for Online Electrical Vehicles," IEEE Transactions on Power Electronics, vol.26, no.12, pp.3666-3679, Dec. 2011.

[16] J. M. Miller, P. T. Jones, J. M. Li, O. C. Onar, "ORNL Experience and Challenges Facing Dynamic Wireless Power Charging of EV's," IEEE Circuits and Systems Magazine, vol.15, no.2, pp.40-53, Secondquarter 2015.

[17] M. McDonough, "Integration of Inductively Coupled Power Transfer and Hybrid Energy Storage System: A Multiport Power Electronics Interface for Battery-Powered Electric Vehicles," IEEE Transactions on Power Electronics, vol.30, no.11, pp.6423-6433, Nov. 2015.

[18] M. Chinthavali, O. C. Onar, S. L. Campbell, L. M. Tolbert, "Integrated Charger with Wireless Charging and Boost Functions for PHEV and EV Applications," IEEE Transportation Electrification Conference and Expo (ITEC), pp.1-8, Jun. 2015.

978-1-4673-9551-9/16 $31.00 © 2016 IEEE

Comparative Analysis of High Step-down Ratio Isolated DC/DC Topologies in PEV Applications

Zhiqing Li, *Student Member, IEEE* and Haoyu Wang, *Member, IEEE*

School of Information Science and Technology

ShanghaiTech University

Shanghai, China

wanghy@shanghaitech.edu.cn

Abstract—In plug-in electric vehicles (PEVs), an isolated dc/dc converter with high step-down ratio is required to supply power to the onboard appliances. The wide battery pack voltage range on the input side brings significant challenges to the design and optimization of this dc/dc converter. This study comprehensively compares LLC resonant topology, series-resonant topology with delay-time control (DTC-SRC), and LLC cascaded with buck topology (LLC-Buck) in terms of the power density, system efficiency and control complexity. It is found that the LLC-Buck topology maintains the best overall conversion efficiency performance. A 400 W LLC-Buck converter prototype is designed and tested to verify the proof of concept.

Keywords—Delay-time control; high step-down; LLC; LLC-Buck; PEV; SRC; zero voltage switching (ZVS)

I. INTRODUCTION

With the benefits of low carbon emission and reduced urban air pollution, PEVs have gained their worldwide popularity [1]–[3]. PEV is typically equipped with a high voltage battery pack and a low voltage battery providing power to the traction motor and the onboard appliances, respectively [4]. A high step down ratio isolated dc/dc converter is required to convert the high voltage from the propulsion battery pack to a low voltage defined by the auxiliary battery (12V/14V). For safety reasons, galvanic isolation is required for this application. The output voltage of the high voltage battery pack varies in a wide range (e.g. 250V-420V) with the change of battery State-of-Charge (SOC) [5], [6]. Thus, this dc/dc converter must be adapted to the wide output voltage range of the propulsion battery pack.

Pulse frequency modulated (PFM) resonant topologies are good candidates to achieve this dc/dc conversion. This is mainly due to their attractive features, such as ZVS, high conversion efficiency, high power density and low electromagnetic interference (EMI) [7], [8]. However, the wide input voltage range is mapped to a wide voltage conversion ratio range. This conversion ratio range is mirrored to a wide frequency variation range in the gain-frequency map of resonant topologies. When the switching frequency of resonant circuits increases beyond the resonant frequency, driving losses, switching losses and core losses of magnetic components will increase[9], [10]. When the switching frequency of resonant circuits decreases below the resonant frequency, the size of magnetic cores, current stresses, conduction losses will

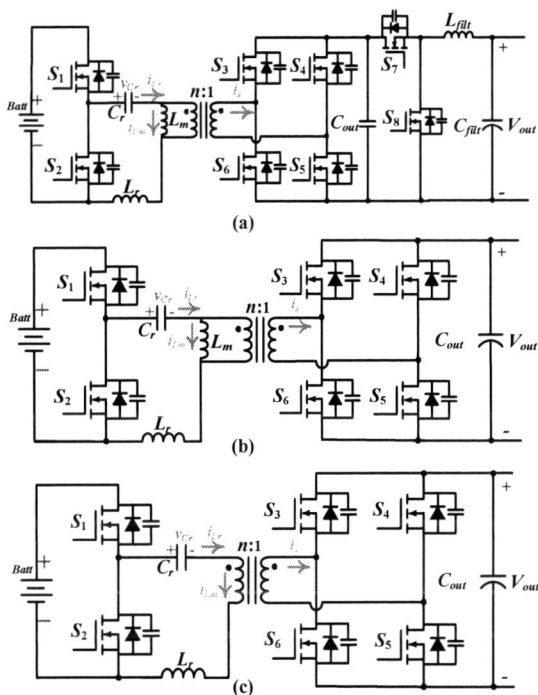

Fig. 1. High step-down ratio isolated dc/dc converters for PEV applications: (a) LLC-Buck (b) LLC(c) DTC-SRC.

increase. Therefore, reducing the switching frequency range of resonant converters with a wide input-voltage range brings significant benefits. Recently, many studies have been done to address this issue [11]–[14]. In [11], Reduction in the switching frequency range is achieved by controlling the output voltage with a combination of variable-frequency feedback control and open-loop delay-time control. In [12], switching frequency reduction is implemented with a two-phase interleaved boost converter and a full-bridge LLC resonant converter. In [13], a switch-controlled capacitor modulated LLC converter is presented to solve the load-sharing problem with constant switching frequency. in [14], one leg of the full-bridge diode rectifier is replaced with synchronous rectifier (SR) switches. When the input voltage of the converter decreases, the SR switches are controlled using phase shifted switching signals to obtain a higher gain without decreasing the switching

978-1-4673-9551-9/16 $31.00 © 2016 IEEE

frequency. However, these proposed studies achieve reduced switching frequency range at the cost of large components count or complicated control method. Hence, a study of applicable topologies and a comprehensive comparison regarding to power density, system efficiency and control complexity is necessary.

This paper lists three applicable promising topologies for this specific application. They are LLC-Buck, LLC [15], [16], and DTC-SRC [11] respectively. The schematics of those three converters are plotted in Fig. 1. It should be noted that in order to facilitate the comparison and to realize the open-loop delay-time control of DTC-SRC topology, half bridge inverter and full bridge synchronous rectifier are uniformly adopted in the secondary side of those three circuits.

In this paper, a mathematic model is proposed to analyze and to compare those three converters. The effectiveness of the switching frequency reduction of resonant converters is evaluated. The applicable occasions for those three converters are discussed. The power density, system conversion efficiency performances and control complexity are compared using theoretical analysis and experimental results.

II. CONVERTER OPERATING PRINCIPLES

The normalized voltage gain of LLC converter is either higher or lower than unity, depending on the relationship between its switching frequency and the resonant frequency of L_r and C_r. If properly designed, this topology can achieve ZVS of the primary switches and zero-current-switching (ZCS) of secondary rectifier.

DTC-SRC topology has the same circuit configuration as the conventional SRC converter. However, the control technique is different. The voltage regulation is achieved by the combination of variable-frequency feedback control and open-loop delay-time control. When the input voltage decreases to be lower than the threshold, open-loop delay-time control is activated. During the delay-time interval, the secondary side of the transformer is shorted and sees a zero voltage. Therefore, this control technique introduces boost feature to the circuit, which contributes to a boosted output voltage. Meanwhile, combined with the frequency variation feedback control, this converter can also achieve a wide voltage gain range with reduced frequency window.

LLC-Buck topology is a two stage converter. In the first stage, constant frequency half bridge LLC topology is adopted to provide unity normalized gain and galvanic isolation. In the second stage, a buck topology is introduced to provide the voltage regulation performance. Moreover, this bidirectional buck converter is able to achieve power flow from the 12 V onboard low voltage battery to pre-charge the intermediate dc bus capacitor. Accordingly, the surge current problem of LLC converter in the startup stage can be mitigated.

III. COMPARATIVE ANALYSIS OF THREE CONVERTERS

A. Critical Circuit Parameters Analysis

The input voltage of this step-down isolated dc/dc converter varies with the SOC of high voltage battery pack. In this study,

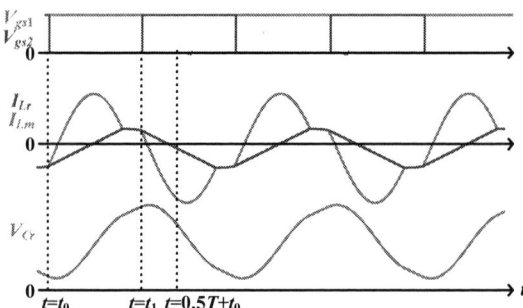

Fig. 2. Theoretical waveforms of the LLC converter working below resonant frequency.

the input voltage range is set as 250 V to 420 V and the output voltage is set as 14 V. Comparative analyses of three converters at full load (400 W) are based on the losses breakdown, conversion efficiency, power density and control complexity.

To estimate the power losses and efficiency of the converters, an accurate steady state analysis is necessary. In LLC-Buck topology, since the switching frequency is matched with the resonant frequency (f_{r1}) between L_r and C_r, the steady state current expression can be easily derived [16]. The resonant tank current, i_{Lr}, is a sinusoidal wave as defined,

$$i_{Lr}(t) = \sqrt{2} I_{rms,p} \sin(2\pi f_{r1} + \phi) \qquad (1)$$

where, $I_{rms,p}$ is the RMS current, and ϕ is the initial phase angle. The equivalent resistance of buck stage is R_L/D^2 (D is the duty cycle of S_7). The RMS current value can be represented by the equation,

$$I_{rms,p} = \frac{1}{8} \frac{DV_O}{nR_L} \sqrt{\frac{2n^4 R_L^2 T^2}{L_m^2 D^4} + 8\pi^2} \qquad (2)$$

where, n is the transformer turns ratio, V_o is the output voltage, R_L is the load resistance, T is the switching cycle and L_m is magnetizing inductance. The magnetizing inductor current changes linearly and its peak value can be represented as,

$$I_{pk} = \frac{nV_o}{DL_m} \frac{T}{4} \qquad (3)$$

ϕ can be derived from,

$$\sqrt{2} I_{rms,p} \sin(\phi) = \frac{nV_o}{DL_m} \frac{T}{4} \qquad (4)$$

The formulas for both the resonant tank current and magnetizing inductor current are obtained. Thus, the secondary side current can be easily derived. Besides, in LLC-Buck topology, the dc gain is,

$$Gain = \frac{D}{2n} \qquad (5)$$

In LLC converter, when the switching frequency deviates from the resonant frequency, the circuit characteristics are complicated to analyze. Usually, first harmonics approximation (FHA) is applied to give an approximate description of the voltage gain. However, a more accurate dc gain and current stresses model is required to calculate the conversion efficiency.

Taking advantage of numerical calculation stool, a more precise analysis method is proposed in [17]. According to the relationship between circuit parameters, equation set can be constructed and solved by mathematic software. Taking the LLC topology [as seen in Fig. 1 (b)] as an example. Its theoretical waveforms with switching frequency (f_s) below f_{r1} are plotted in Fig. 2. There are two operation modes in half switching cycle. The first mode is denoted as x. x begins at $t = 0$ and circuit works like a SRC with L_r resonating with C_r. The second mode is denoted as y. y begins at $t = t_1$ and secondary side of transformer is equivalent to an open circuit. The resonant tank contains $L_m + L_r$ and C_r. The resonant inductor current, magnetizing inductor current and resonant capacitor voltage in both modes are shown as,

$$i_{Lm,x}(t) = I_{Lm0} + \frac{nV_o}{L_m}t \tag{6}$$

$$i_{Lr,x}(t) = I_{Lr0}\sin\left(2\pi f_{r1}t + \alpha\right) \tag{7}$$

$$i_{Lm,y}(t) = i_{Lr,y}(t) = I_{Lr1}\sin\left(2\pi f_{r2}t + \beta\right) \tag{8}$$

$$v_{Cr,x}(t) = V_{in} - i_{Lr,x}(t)^{'}L_r - nV_o \tag{9}$$

$$v_{Cr,y}(t) = V_{in} - i_{Lr,y}(t)^{'}\left(L_r + L_m\right) \tag{10}$$

where, I_{Lr0} is the initial value of resonant inductor current, I_{Lm0} is the initial value of magnetizing inductance current, f_{r2} is the resonant frequency between $L_r + L_m$ and C_r, α is the initial phase angel of $i_{Lr,x}(t)$, β is the initial phase angel of $i_{Lr,y}(t)$.

The difference between resonant tank current and magnetizing inductor current is the current transferred to the load. Moreover, according to mutual relationships among equations (6-10), a new group of equations are derived as follows,

$$i_{Lm,x}(0) = i_{Lr,x}(0) \tag{11}$$

$$i_{Lm,x}(t_1) = i_{Lr,x}(t_1) \tag{12}$$

$$i_{Lm,x}(t_1) = i_{Lr,y}(t_1) \tag{13}$$

$$i_{Lr,y}\left(\frac{T}{2}\right) = -i_{Lm,x}\left(\frac{T}{2}\right) \tag{14}$$

$$v_{Cr,x}(t_1) = v_{Cr,y}(t_1) \tag{15}$$

$$v_{Cr,y}\left(\frac{T}{2}\right) + v_{Cr,x}(0) = \frac{V_{in}}{2} \tag{16}$$

$$n\int_0^{t_1}\left(i_{Lr,x}(t) - i_{Lm,x}(t)\right) = \frac{V_o}{R_L}\frac{T}{2} \tag{17}$$

This equation group can be solved and the accurate current stresses and dc gain can be calculated accordingly.

The theoretical waveforms of DTC-SRC topology under delay time control mode are plotted in Fig. 3. The expressions for dc gain and current stresses are also unavailable. Similar

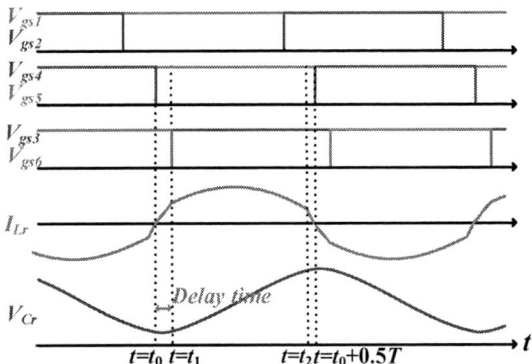

Fig. 3. Theoretical waveforms of the DTC-SRC converter.

method can be used to establish the equation group to calculate the current stresses and dc gain.

B. Efficiency Analysis

Power loss models representing switching losses, driving losses, conduction losses, and magnetic core losses, are presented and evaluated for each converter.

1) Switching losses

The turn-on and turn-off transitions of semiconductor devices take tens of nanoseconds to microseconds. The overlap of MOSFET current and voltage during the switching transient results in switching losses which are proportional to switching frequency. While in an appropriately designed LLC converter, ZVS of the primary switches and ZCS of secondary rectifier can be achieved. For the turn-off transitions of primary side MOSFETs, the channel is turned off sufficiently fast while the output capacitance (C_{oss}) is slowly charged from zero to V_{in}. Thus, the overlap between channel current and drain source voltage is negligible. The energy loss during turning off transition can be neglected.

In LLC-Buck topology, the switching losses of S_7 and S_8 in buck stage should be considered. One can approximate that the MOSFET voltage $V(t)$ and current $I(t)$ are both piecewise-linear. Hence, the extra switching losses can be expressed as,

$$P_{sw} = \frac{1}{2}V_{ds}I_{ds}\left(t_{on} + t_{off}\right)f_s \tag{18}$$

where, I_{ds} is the peak value of $I(t)$, V_{ds} is the peak value of $V(t)$. While practically, the switching frequency of buck stage is typically much lower than that of the LLC stage. Meanwhile, because the voltage stress of S_7 and S_8 is low, the turn on time and turn off time can be several nanoseconds, the switching losses of buck stage can be ignored.

In DTC-SRC topology, L_r and C_r are in series to form a series resonant tank. While this resonant tank is in series with the load. Switching frequency is higher than f_{r1} to facilitate ZVS. Therefore, the switching losses of DTC-SRC can also be neglected.

2) Driving losses

A gate charge Q_g is required to turn on the MOSFET. When the MOSFET is turned off, Q_g is released to the ground

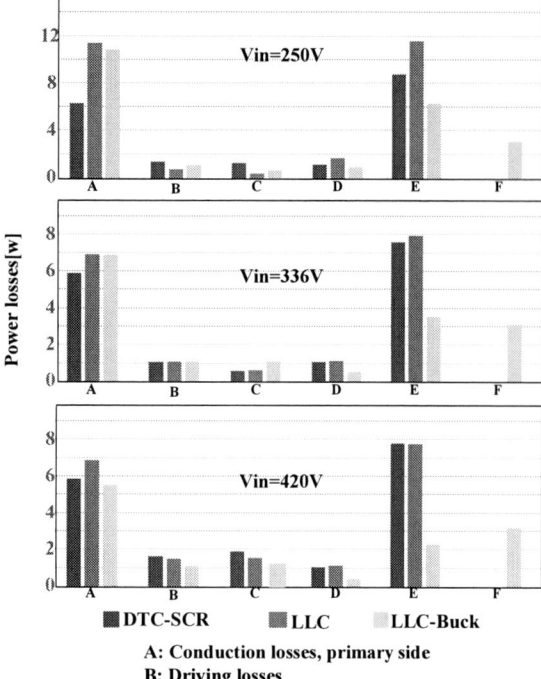

A: Conduction losses, primary side
B: Driving losses
C: Transformer core losses
D: Transformer copper losses
E: Conduction losses, secondary side
F: Buck conduction losses

Fig. 4. The calculated power loss distribution of three converters (P_o = 400 W).

and the stored energy is lost. Thus, the driving losses is approximated as,

$$P_{dri} = f_s V_{gs} Q_g \qquad (19)$$

According to Eq. (19), P_{dri} is associated with the semiconductor selection and switching frequency.

In LLC topology, a wide f_s range is required. When f_s is increased to achieve a lower voltage gain, the driving losses would increase and vice versa. In LLC-Buck topology, f_s is fixed, so the driving losses is also fixed. In DTC-SRC topology, by changing the frequency, the impedance of resonant tank changes accordingly. According to the principle of voltage divider, voltage divided by the load changes. If V_{in} increases, f_s increases and the driving losses increase.

3) Conduction losses

The conduction losses are the dominant loss in low voltage, high current applications. The main contributors to the conduction losses are the ON-state resistances of switches, and the magnetic component copper losses. The conduction losses are proportional to the square of component root-mean-square

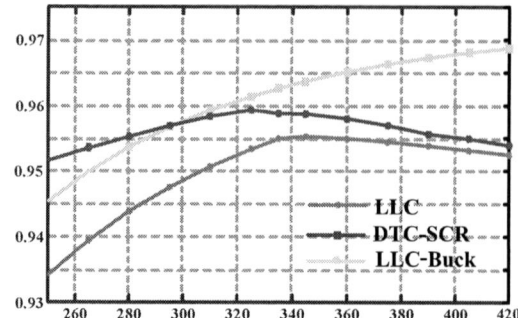

Fig.5. Calculated efficiency of three converters for varying input voltage (P_o = 400 W).

(RMS) current ($I_{RMS}{}^2$). The conduction losses can be described as,

$$P_{con} = I_{rms,p}{}^2 \sum_{pri} R_{on} + I_{rms,s}{}^2 \sum_{sec} R_{on} \qquad (20)$$

It should be noted that the RMS currents on both sides of those three converters can be calculated based on the circuit analysis described in Section III-A.

4) Core Losses

Core losses, or iron losses, are caused by the alternating magnetic field in the magnetic core. This magnetic field variation generates hysteresis losses, eddy current losses and residual losses. The core loss for a given material is a function of operating frequency (f) and total flux swing (ΔB) as,

$$P_{fe} = k f^x \Delta B^y \qquad (21)$$

where, k, x, and y are all the coefficients. The calculation of core losses is usually empirical. Curve fitting is typically used to determine the coefficients. Practically, this curve can be obtained from the datasheet. In those three topologies, the voltage on the primary side of transformer is a square wave. ΔB can be expressed as,

$$\Delta B = \frac{E_i T_n}{2 n A_e} \qquad (22)$$

where E_i is the amplitude of applied voltage, T_n is the half of the switching period, n is the coil number on the primary side, and A_e is the effective cross-sectional area. T_n is determined by the operating points. In LLC and DTC-SRC topologies, E_i equals nV_o. In LLC-Buck topology, E_i equals $V_{in}/2$ approximately.

5) System efficiency

Switching losses, driving losses, conduction losses, core losses are all taken into consideration to calculate the system efficiency. The system efficiency can be depressed as,

$$\eta = \frac{P_o}{P_{in}} = \frac{P_{in} - P_{sw} - P_{dri} - P_{con} - P_{fe}}{P_{in}} \qquad (23)$$

TABLE I

DESIGN PARAMETERS OF THE THREE CONVERTERS

Quantity	Symbol	LLC-Buck	LLC	DTC-SRC
Input voltage	V_{in}	250V-420V	250V-420V	250V-420V
Output voltage	V_{out}	14V	14V	14V
Rated power	P_{max}	400W	400W	400W
Resonant inductor	L_r	30μH	30μH	30μH
Magnetizing inductor	L_m	120μH	120μH	
Resonant capacitor	C_r	20nF	20nF	20nF
Output capacitor	C_{out}	470μF	470μF	470μF
Primary switching frequency	f_s	200 kHz	200kHz	200kHz
Transformer turns ratio	n	8	12	12
Filter inductor	L_{filt}	50μH		
Filter capacitor	C_{filt}	470μF		
Switching frequency of buck	f_{buck}	20kHz		

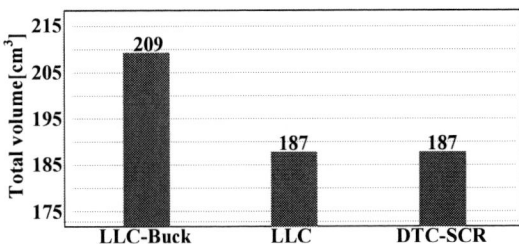

Fig. 6. The total volume of three converters (P_o = 400W).

Fig. 7. Photo of the LLC-Buck converter.

Three operation points: a) V_{in} = 250 V, V_o = 14 V, P_o = 400 W; b) V_{in} = 336 V, V_o = 14 V, P_o = 400 W; and c) V_{in} = 420 V, V_o = 14 V, P_o = 400 W are defined as the benchmark points to make the comparison. The calculated various power loss distributions in those three points of those three converters are demonstrated in Fig. 4. It is obvious that conduction losses in both sides are the dominant power losses. When the input voltage increases, the RMS current decreases. Thus, the conduction losses decrease. The extra conduction losses of buck stage of LLC-Buck topology is a fixed value and is determined by the output voltage and current.

Fig. 5 summarizes the calculated efficiency of three converters versus input voltage. In LLC topology, the maximum efficiency point is located around f_{r1}. In LLC-Buck topology, the system efficiency increase with the increase of input voltage. Because of the step-down feature of the buck stage, a transformer with reduced turns ratio can be used. This brings the benefits of reduced RMS current and increased efficiency. In comparison with LLC topology, there is no magnetic inductor circulating current in the DTC-SRC topology. Thus, the circulating losses associated with L_m is zero. DTC-SRC topology demonstrates higher efficiency than LLC topology.

According to the analysis, LLC-Buck topology is the most efficient topology in comparison to LLC topology and DTC-SRC topology in most conditions. The design parameters are provided in Table I.

C. Power density

The system volume is evaluated based on the actual hardware design at the rated power. The sizes of power MOSFETs, transformer, capacitors, inductors, auxiliary power supply, and the heat sink are considered as the main contributors to the system volume. It should be noted that the two extra MOSFETs, one filter capacitor and one filter inductor in the buck stage of LLC-BUCK topology degrade its power density.

The voltage and current stresses and conduction losses should be considered to evaluate the sizes of the power MOSFETs and the heat sink. The size of magnetic component is constrained by the flux saturation. Since the transformer primary side sees a square wave, this design equation should be satisfied,

$$B_s n A_e > E_i T_n \qquad (24)$$

To avoid the magnetic saturation, the inductor should be designed based on,

$$B_s n A_e > L I_p \qquad (25)$$

where, I_p is the peak current value through the inductor. As shown in Fig. 6, LLC-Buck topology has a larger volume than the others.

D. Control Complexity

The LLC-Buck topology comes with two additional MOSFETs and their driving pulses. However, its operating frequency is fixed as a constant no matter how the input voltage value changes. It is easier to generate fixed-frequency pluses than variable-frequency pluses. Another difference between the control methods of those three converters is the synchronous rectification realization on the secondary side. In LLC-Buck topology, the secondary side current is synchronized with square wave on the primary side. Thus, the secondary side current zero-crossing detection is unnecessary to realize the synchronous rectification. In LLC topology, when fs is lower than f_{r1}, phase commutation of the secondary current

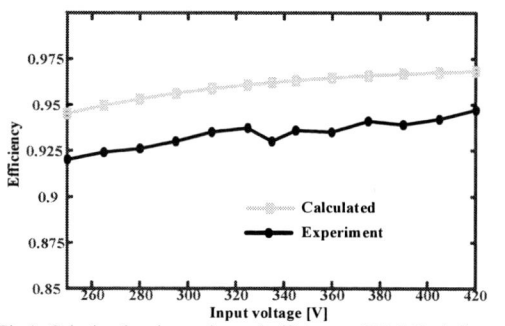

Fig.8. Calculated and experimental efficiency of LLC-Buck for varying input voltage (P_o = 400 W).

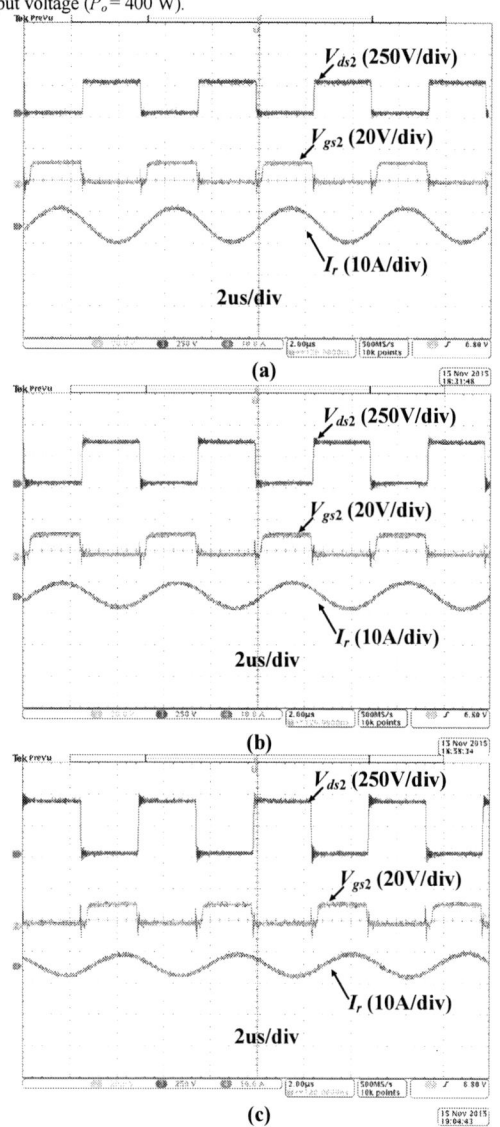

Fig. 9. Experimental steady-state waveforms of LLC-Buck: (a) V_{in} = 250V, P_o = 400 W (b) V_{in} = 336 V, P_o = 400 W (c) V_{in} = 420V, P_o = 400W.

occurs in the dead band. The secondary side current zero-

crossing detection is also unnecessary.

However, when f_s is higher than f_{r1}, since there exists a phase shift between the square wave and the secondary side current, the secondary side current zero-crossing detection becomes necessary. In DTC-SRC topology, to realize the delay time control, zero-crossing detection is also required. From the perspective of zero-crossing detection, the LLC-Buck topology is easier to be implemented than the others.

IV. EXPERIMENTAL RESULTS

According to the efficiency analysis, LLC-Buck topology has the best overall efficiency performance in comparison to LLC and DTC-SRC topologies. A 400 W, 250 V-420 V to 14 V LLC-Buck converter prototype has been designed as the proof of concept, as shown in Fig. 7.

Calculated and experimental efficiency curves of LLC-Buck topology for varying input voltage are plotted in Fig. 8. Experimental steady-state waveforms of LLC-Buck topology in three benchmark points are demonstrated in Fig. 9. As can be seen from efficiency curves, the experimental efficiency data is worse than the calculated data. This is because of the difference between the actual environment and theoretical data. With the increase of the input voltage, the conversion efficiency increases. This trend agrees well with the theatrical predication. This efficiency increase is due to the RMS current decrease. Therefore, the dominate power losses are the conduction losses.

V. CONCLUSION

In this paper, LLC, LLC-Buck, and DTC-SRC topologies are compared in terms of power density, conversion efficiency, and control complexity in the PEV auxiliary power supply applications. LLC topology have good power density. However, it suffers from wide switching frequency range. When f_s deviates from f_r, the conversion efficiency degrades fast. Also PFM is more complicated than PWM. DTC-SRC topology utilizes advanced delay time control to reduce the switching frequency range and to boost the conversion efficiency. However, this control brings drawbacks of higher control complexity. LLC-Buck demonstrates the best overall conversion efficiency and lowest control complexity with a little sacrifice on the converter power density.

REFERENCES

[1] A. Y. Saber and G. K. Venayagamoorthy, "One Million Plug-in Electric Vehicles on the Road by 2015," in *2009 12th International IEEE Conference on Intelligent Transportation Systems*, 2009, pp. 1–7.
[2] A. R. Salisa, N. Zhang, and J. G. Zhu, "A Comparative Analysis of Fuel Economy and Emissions Between a Conventional HEV and the UTS PHEV," *IEEE Trans. Veh. Technol.*, vol. 60, no. 1, pp. 44–54, Jan. 2011.
[3] H. Wang, A. Hasanzadeh, and A. Khaligh, "Transportation Electrification: Conductive Charging of Electrified Vehicles," *IEEE Electrif. Mag.*, vol. 1, no. 2, pp. 46–58, Dec. 2013.
[4] A. Emadi and K. Rajashekara, "Power Electronics and Motor Drives in Electric, Hybrid Electric, and Plug-In Hybrid Electric Vehicles," *IEEE Trans. Ind. Electron.*, vol. 55, no. 6, pp. 2237–2245, Jun. 2008.
[5] K. W. E. Cheng, B. P. Divakar, H. Wu, K. Ding, and H. F. Ho, "Battery-management system (BMS) and SOC development for electrical vehicles," *IEEE Trans. Veh. Technol.*, vol. 60, no. 1, pp. 76–88, 2011.
[6] C. P. Zhang, J. C. Jiang, W. G. Zhang, and F. Wen, "Determination of SOC

of a battery pack used in pure electric vehicles," *Asia-Pacific Power Energy Eng. Conf. APPEEC*, pp. 5–8, 2012.

[7] B. Yang, F. C. Lee, a. J. Zhang, and G. H. G. Huang, "LLC resonant converter for front end DC/DC conversion," *APEC. Seventeenth Annu. IEEE Appl. Power Electron. Conf. Expo.*, pp. 1108–1112, 2002.

[8] P. Caldeira, R. Liu, D. Dalal, and W. J. Gu, "Comparison of EMI performance of PWM and resonant power converters," *Proc. IEEE Power Electron. Spec. Conf. - PESC '93*, pp. 134–140, 1993.

[9] H. Wang, S. Dusmez and A. Khaligh, "Design and Analysis of a Full-Bridge LLC-Based PEV Charger Optimized for Wide Battery Voltage Range," *IEEE Trans. Veh. Technol.*, vol. 63, no. 4, pp. 1603–1613, May 2014.

[10] H. Wang, S. Dusmez, and A. Khaligh, "Maximum Efficiency Point Tracking Technique for LLC Based PEV Chargers Through Variable DC Link Control," *IEEE Trans. Ind. Electron.*, vol. 61, no. 11, pp. 6041–6049, Nov. 2014.

[11] Y. Jang, M. M. Jovanovi, J. M. Ruiz, and G. Liu, "Series-Resonant Converter with Reduced- Frequency-Range Control," vol. 1, pp. 1453–1460, 2015

[12] X. Sun, Y. Shen, W. Li, and H. Wu,"A PWM and PFM Hybrid Modulated Three-Port Converter for a Stand-Alone PV/Battery Power System," *IEEE J. Emerg. Sel. Top. Power Electron.*, vol. 6777, no. c, pp. 1–1, 2015.

[13] Z. Hu, Y. Qiu, L. Wang, and Y.-F. Liu, "An Interleaved LLC Resonant Converter Operating at Constant Switching Frequency," *IEEE Trans. Power Electron.*, vol. 29, no. 6, pp. 2931–2943, 2014.

[14] J.-W. W. Kim and G.-W. W. Moon, "A New LLC Series Resonant Converter with a Narrow Switching Frequency Variation and Reduced Conduction Losses," *IEEE Trans. Power Electron.*, vol. 29, no. 8, pp. 4278–4287, Aug. 2014.

[15] Jee-Hoon Jung and Joong-Gi Kwon, "Theoretical Analysis and Optimal Design of LLC Resonant Converter," in *2007 European Conference on Power Electronics and Applications*, 2007, pp. 1–10.

[16] B. Lu, W. Liu, Y. Liang, F. C. Lee, and J. D. Van Wyk, "Optimal Design Methodology for LLC Resonant Converter," *TwentyFirst Annu. IEEE Appl. Power Electron. Conf. Expo. 2006 APEC 06*, pp. 533–538, 2006.

[17] X. Fang, H. Hu, Z. J. Shen, and I. Batarseh, "Operation Mode Analysis and Peak Gain Approximation of the LLC Resonant Converter," *IEEE Trans. Power Electron.*, vol. 27, no. 4, pp. 1985–1995, Apr. 2012.

978-1-4673-9551-9/16 $31.00 © 2016 IEEE

DC to Single-phase AC Voltage Source Inverter with Power Decoupling Circuit based on Flying Capacitor Topology for PV System

Hiroki Watanabe, and Keisuke Kusaka
Department of Electrical engineering
Nagaoka University of Technology, NUT
Niigata, Japan
hwatanabe@stn.nagaokaut.ac.jp

Keita Furukawa, Koji Orikawa and Jun-ichi Itoh
Department of Electrical engineering
Nagaoka University of Technology, NUT
Niigata, Japan
itoh@vos.nagaokaut.ac.jp

Abstract— A novel power decoupling circuit using a flying capacitor topology is proposed in this paper. The inverters, which are connected to a single-phase grid, have single-phase power fluctuation at the twice the grid frequency. Thus, bulky electrolytic capacitor is generally used as DC link capacitor. In the proposed circuit, the power fluctuation is compensated by the flying capacitor in the flying capacitor DC-DC converter. The proposed circuit does not need an additional magnetic component in comparison with the conventional system, which has a boost chopper and an inverter. The proposed converter is experimentally tested with a 1-kW prototype. A voltage ripple at twice the frequency on the inverter DC voltage is suppressed from 35.1% to 4.6%. Moreover, the maximum efficiency of 94.5% with an output power of 1.0 kW is achieved. Finally, the design method of the boost-up inductor for proposed circuit is estimated by experiment. As a result, the error of the ripple current between design and measurement value is 4.7%.

Keywords—Photovoltaic; grid connected inverter; active power decoupling; flying capacitor DC-DC converter; single-phase power ripple.

I. INTRODUCTION

In recent years, the Photovoltaic (PV) system is active researched as a sustainable power solution. In order to supply the solar power to single-phase grid, a Power Conditioning system (PCS) is used. In general, the PCS is constructed by a boost chopper, and a Voltage Source Inverter (VSI) [1]. The boost chopper is required the function as the boost-up and maximum power point tracking (MPPT). The VSI is needed for the grid connection. However, a conventional system needs the large electrolytic capacitor on DC link due to the single-phase power fluctuation. The electrolytic capacitor has a limited lifetime, approximately less than 7000 hours at 105 degree Celsius operating temperature although the lifetime of a PV panel is namely 25 years. As a result, periodical maintenance is required on the conventional system.

In order to remove the bulky electrolytic capacitor, the active power decoupling method has been studied [2]-[6]. These topologies can reduce the capacitance of the DC link

capacitor. However, the extra magnetic components and switching devices are required. These additional components decrease the efficiency, and increase the circuit volume.

On the other hand, the general boost chopper has the large magnetic component as the boost-up inductor for energy buffer. This is because the general boost chopper has to store the energy for boost-up of the input voltage by large inductor only. As a result, the inductance should be designed widely.

In order to reduce the volume of the boost-up inductor, the flying capacitor type DC-DC converter is proposed. The flying topology has the boost-up inductor and flying capacitor, and these components is used for boost-up of the input voltage. In generally, the energy density of the capacitor is large in comparison with the inductor. Thus, the inductance on the flying capacitor DC-DC converter can be reduced more than the general boost-up chopper.

In this paper, a novel power decoupling circuit that uses the flying capacitor topology is proposed in order to solve above mentioned problems. The proposed circuit has the function of the boost-up and single-phase power fluctuation compensation. In the decoupling control strategy, the flying capacitor is charged and discharged at the twice grid frequency. As a results, the single-phase power fluctuation is compensated without the large DC link capacitor. The advantages of the proposed circuit are follows: 1) the additional magnetic component for active power decoupling is not required; 2) the inductance of the boost-up inductor can be reduced in comparison with conventional boost chopper. 3) Simple configuration and easy control.

This paper is organized as follows: first, the configuration of the proposed circuit is shown. Next, the principle of the power decoupling control is described. In addition, the design method of the boost-up inductor is explained, and the inductance between the boost chopper and proposed circuit is estimated. Moreover, the operation of the proposed circuit is demonstrated by the experiment. From the experimental results, the voltage ripple at twice the grid frequency on the inverter

DC voltage is suppressed from 35.1% to 4.6%. Finally, the maximum efficiency of 94.5% with the output power of 1.0 kW is achieved.

II. CIRCUIT TOPOLOGY

A. General configuration of the AC grid connection system

Fig. 1 shows a general configuration of the DC to single-phase AC grid connection system. The input voltage V_{in} is boosted up to more than the peak grid voltage V_{ac} by the boost chopper. When the PV panel is connected to the single-phase grid by VSI, the single-phase power fluctuation which has twice of the power grid frequency in the output side. It is one of the cause to decay the control performance of MPPT. Thus, the large electrolytic capacitor C_{dc} is connected to the DC line, and instantaneous difference power between input and output side is compensated by charge and discharge of C_{dc}. However, the life time of the conventional system is limited by the large electrolytic capacitor.

B. Grid connection system with conventional active power decoupling

Fig. 2 shows the circuit configuration with the conventional active buffer decoupling circuit. In this system, the power decoupling circuit is connected to the DC line instead of the large electrolytic capacitor. This topology is possible to compensate the power fluctuation with small capacitor C_{buf}. Thus, this system is smaller size and longer life-time compared with the conventional circuit in Fig.1. This is because the small film or ceramic capacitor can be used instead of the electrolytic capacitor. However, the additional inductor and switching devices are needed. These components lead to the low efficiency, and increase the circuit volume.

C. Grid connection system with the proposed circuit

Fig.3 shows the grid connection system with the proposed circuit based on flying capacitor DC-DC converter topology [7]. The proposed circuit has capability to compensate the single-phase power fluctuation using the flying capacitor C_{fc} which is controlled for active power decoupling. The advantages of the proposed circuit are follows: 1) Active power decoupling can be achieved without additional magnetic components; 2) the inductance of the boost-up inductor can be reduced in comparison with the general boost-up chopper. Especially, the boost-up operation and power decoupling capability can be achieved by the flying capacitor DC-DC converter. Thus, the total efficiency is improved in comparison with the conventional circuit in Fig.2. This is because the additional circuit for power decoupling is not required in the proposed system.

III. PRICIPLE OF THE ACTIVE POWER DECOUPLING

Fig. 4 shows the principle of the power decoupling operation between the DC and AC sides [8-19]. When both the output voltage and current waveforms are sinusoidal, the instantaneous output power p_{out} is expressed as

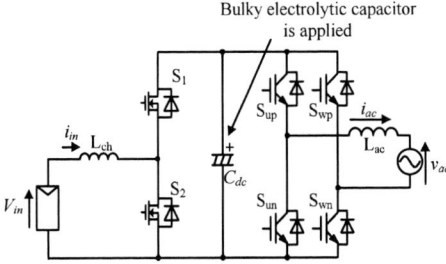

Fig.1. General configuration for DC to single-phase AC grid connection system. This circuit has a large electrolytic capacitor C_{dc} on the DC link part. C_{dc} is usually used the bulky electirolytic capacitor.

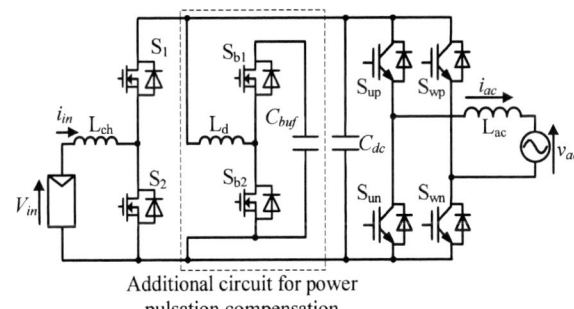

Fig.2 Power conditioner system using the conventional active power decoupling method. This circuit needs an additional inductor and switching devices. It cause low efficiency and large circuit volume.

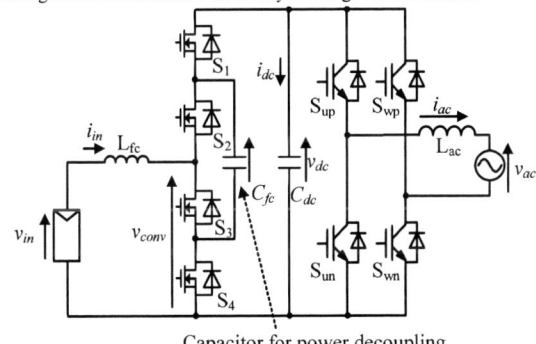

Fig.3. Proposed circuit using the flying capacitor DC-DC converter topology. This circuit use the flying capacitor C_{fc} in order to compensate the single-phase power fluctuation. Thus, the additional component is not needed.

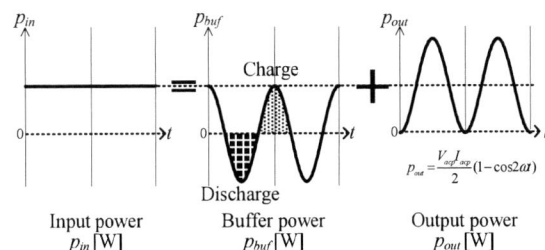

Fig.4. Compensation principle of the single-phase power fluctuation. The power ripple that contains frequency twice of the grid frequency appears at DC link voltage. In order to compensate the power fluctuation at input DC side, buffer power p_{buf} is fluctuated by flying capacitor C_{fc}.

978-1-4673-9551-9/16 $31.00 © 2016 IEEE   1337

$$p_{out} = \frac{V_{acp}I_{acp}}{2}(1 - \cos 2\omega t) \qquad (1)$$

Where, V_{acp} is the peak voltage, I_{acp} is the peak current, and ω is the angular frequency of the output voltage. From (1), the power ripple, that contains twice frequency of the power grid, appears at DC link.

In order to absorb the power ripple, the instantaneous power p_{buf}, should be controlled by

$$p_{buf} = \frac{1}{2}V_{acp}I_{acp}\cos 2\omega t \qquad (2)$$

where, the polarity of the p_{buf} is defined as positive when the flying capacitor C_{fc} discharges. Note that the active power of the flying capacitor C_{fc} is zero. Due to the power decoupling, the input power is matched to the output power. Thus, the relationship between the input and output power is expressed as

$$p_{in} = \frac{1}{2}V_{acp}I_{acp} = V_{IN}I_{IN} \qquad (3)$$

IV. CONTROL STRAGEGY FOR THE PROPOSED CIRCUIT

Fig.5 shows the operation mode of the flying capacitor DC-DC converter. The flying capacitor C_{fc} is charged or discharged for boost-up and compensation of the power fluctuation. In the proposed control, each switching duty is decided by relationship between input and output power in Fig.1. When the input power is large in comparison with the instantaneous output power, the on duty of charge mode set widely. On the other hand, when the input power is small, the on duty of discharge mode should be set widely.

Fig.6 shows the control block diagram for the flying capacitor DC-DC converter. In the proposed control, the Automatic Current Regulator (ACR) and Automatic Voltage Regulator (AVR) are implemented to control the generative power P_{in} of PV and charge or discharge of C_{fc}. Note that, P_{in} is controlled by MPPT, and the input current i_{in}^{*} is calculated at the maximum power point. Thus, the input current i_{in} is regulated on the maximum power point by ACR.

In the flying capacitor control, the flying capacitor voltage v_{fc} is controlled to vary at the twice of the grid frequency in order to compensate the single-phase power fluctuation by AVR. The voltage reference v_{fc}^{*} is expressed as

$$v_{fc}^{*} = \sqrt{\left(\frac{v_{dc}}{2} - \frac{p_{out}}{\omega C_{fc}V_{dc}}\right)^{2} - \frac{p_{out}}{\omega C}\{\sin(2\omega t) - 1\}} \qquad (4)$$

where, v_{dc} is the DC link voltage, p_{out} is the output power, C_{fc} is the capacitance of the flying capacitor, ω is the angular frequency. From the flying capacitor voltage control, the power

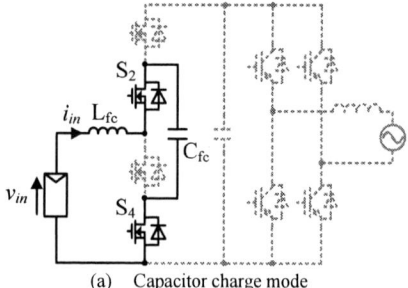

(a) Capacitor charge mode

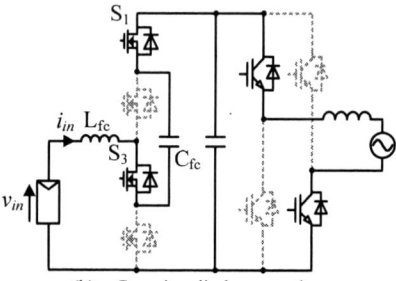

(b) Capacitor discharge mode

Fig.5. Operation mode of the flying capacitor DC-DC converter. In order to achieve the power decoupling, each duty period is adjusted by the single-phase power fluctuation.

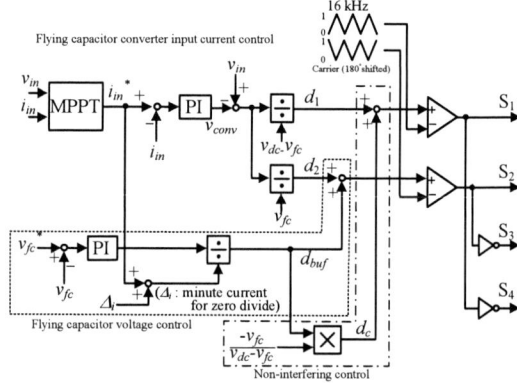

Fig.6. Control block diagram for the flying capacitor DC-DC converter. This is circuit is controlled based on the input curren and flying capacitor voltage control.

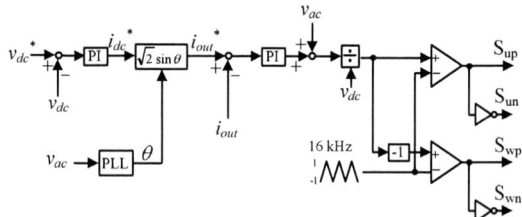

Fig.7. Control block diagram for VSI. The inverter output current control and DC link voltage control is applied for grid connection.

fluctuation compensation duty reference d_{buf} is calculated. Note that, the switching signal S_1 to S_4 is decided by the some duty reference and triangle waveform on the modulation part. However, d_{buf} interfere with the control performance of ACR. As the results, the input current is fluctuated at the twice of the single-phase grid frequency. In order to solve this problem, the control decoupling reference d_c is added. The d_c is expressed as

$$d_c = -\frac{v_{fc}}{v_{dc} - v_{fc}} d_{buf} \qquad (5)$$

Fig.7 shows the control block diagram of the VSI. In order to control the DC link voltage V_{dc}, reference value of AVR is set more than the grid voltage. Finally, the Phase Looked Loop (PLL) is applied to ensure that the phase angle of the inverter output current is identical to the grid.

V. COMPONENT DESIGN FOR PROPOSED CIRCUIT

A. Flying capacitor C_{fc}

The flying capacitor C_{fc} is designed from p_{out} which is the electric storage energy to compensate the power fluctuation. The C_{fc} is expressed as

$$C_{fc} = \frac{2P_{out}}{\omega(V_{c\max}^2 - V_{c\min}^2)} \qquad (6)$$

where, P_{out} is output power, V_{cmax} is maximum flying capacitor voltage, V_{cmin} is minimum flying capacitor voltage, ω is the angular frequency. In order to reduce the C_{fc}, the difference value between V_{cmax} and V_{cmin} should be set widely. In other words, the peak voltage increase. Thus, C_{fc} is needed to design less than the blocking voltage.

B. Boost-up inductor L_{fc}

The boost-up inductor L_{fc} is generally designed from ripple current of the inductor current i_L. In the conventional circuit, it is easy to design the boost-up inductor because the duty reference and peak to peak value of the ripple current is constant. However, the duty reference is not constant in the proposed circuit. This is because the flying capacitor voltage v_{fc} is fluctuated in order to achieve the active power decoupling. It is mean the duty reference and peak to peak value of the ripple current swing. Thus, in the proposed circuit, the boost-up inductor L_{fc} is designed when the current ripple is maximum value.

When the active power decoupling is not applied, the inductance of L_{fc} is expressed as

$$L_{fc} = \alpha \frac{V_L}{4 f_{sw} \Delta I} \qquad (7)$$

where, α is the duty ratio, V_L is inductor voltage, f_{sw} is the switching frequency, ΔI is the ripple ratio of the inductor current. In comparison with the boost-up chopper, the inductance of the flying capacitor DC-DC converter can be designed by one quarter. This is because the ripple current frequency become the twofold of the switching frequency, and the flying capacitor voltage V_{fc} is controlled at the half of DC link voltage V_{dc}. However in the proposed circuit, the duty ratio and the V_L is swung due to the power decoupling control. Thus, the boost-up inductor of the proposed circuit is designed by the some duty reference d_1 and d_2 [20].

The duty reference of the d_1 and d_2 are expressed as

$$d_1 = \frac{V_{in}}{V_{dc}} \{1 - \cos(4\pi ft)\} \qquad (8)$$

$$d_2 = \frac{V_{in}}{V_{dc}} \left\{1 + \left(\frac{V_{dc}}{v_{fc}} - 1\right)\cos(4\pi ft)\right\} \qquad (9)$$

where, the duty reference d_1 and d_2 is included the frequency component. This is given by the power decoupling duty d_{buf} and the control decoupling reference d_c. From the equation as (8) and (9), the boost-up inductor of the proposed circuit is

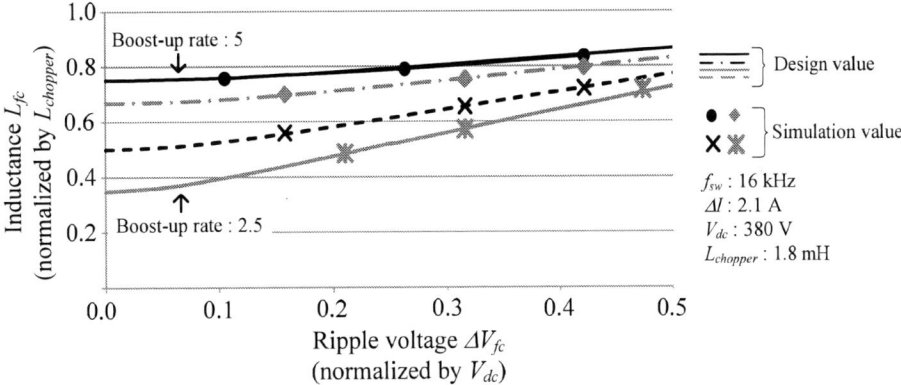

Fig.8 Relationship between the ripple voltage of the flying capacitor and inductance of the boost-up inductor.

expressed as

$$L_{fc} = \frac{\max\left[\max\left\{\left(V_{dc} - v_{fc} - V_{in}\right)d_1\right\}, \max\left\{\left(v_{fc} - V_{in}\right)d_2\right\}\right]}{f_{sw}\Delta I} \quad (10)$$

Fig.8 shows the relationship between the boost-up inductor and ripple voltage. Where, the switching frequency f_{sw} is 16 kHz, DC link voltage V_{dc} is 380 V, the ripple current ΔI is 2.1 A, and the inductance of the general boost-up chopper is 1.8 mH. In order to consider the validity of the designing equation by (10), the design value and simulation results are estimated. In addition, the boost-up rate is changed from 2.5 to 5. Moreover, the boost-up inductor L_{fc} is normalized by the inductance of the boost-chopper $L_{chopper}$.

According to Fig.8, the design value and simulation result are also matched. Thus, the validity of the design for boost-up inductor is checked. In addition, the inductance rate is under than 1p.u. at the all conditions. It is mean the boost-up inductor can be design under than the inductance of the general boost-up chopper. Moreover, the boost-up inductor can be most reduced when the boost-up ratio is 2.5. This is because a lot of energy for boost-up is needed when the boost-up ratio is high. Finally, when the peak to peak voltage of the flying capacitor ΔV_{fc} is increased, the boost-up inductor L_{fc} become large. This is because the buffer power for the single-phase power fluctuation is increased.

VI. SIMURATION AND EXPERIMENT

A. Simularion result

In this chapter, the validity of the single-phase power fluctuation compensation is revealed by the simulation.

Table 1 shows the simulation parameter. Fig.9 shows the simulation results of the proposed circuit. In the simulation, the operation of MPPT is not estimated. The rated power of the proposed circuit is 1 kW, and the large capacitor is not set to the simulation parameter.

Fig.9 (a) shows the operation waveform without the power decoupling operation. According to Fig.9 (a), the DC link voltage v_{dc} and inverter output current i_{ac} are distorted by the single-phase power fluctuation. In the proposed circuit, the capacitance of the DC link capacitor C_{dc} is small. Thus, when the single-phase power fluctuation is not compensated, the DC link voltage v_{dc} is included the twice grid frequency component. As a result, the controlled performance for the inverter output current control is decayed. From these reason, the bulky electrolytic capacitor is needed on the DC link part in the conventional circuit. Due to use the electrolytic capacitor, the life-time is limited. In addition, the circuit volume is increased.

Fig.9 (b) shows the operation waveform with the power decoupling operation. According Fig.9 (b), the DC link ripple voltage is reduced by 96.2%. In addition, the flying capacitor voltage v_{fc} is fluctuated by the twice grid frequency. Moreover, the flying capacitor average voltage is balanced at the half value of the DC link voltage. Finally, the inverter output

Table.1 Simulation parameter

Input voltage	V_{in}	100 V
Input current	I_{in}	10 A
Switching frequency	f_{sw}	16 kHz
Flying capacitor	C_{fc}	180 µF
DC link capacitor	C_{buf}	20 µF
Boost-up inductor	C_{filter}	2 mH
Grid voltage	$v_{\_grid}$	200 V$_{rms}$
Grid frequency	$f_{\_grid}$	50 Hz
Output power	P_{out}	1 kW
Angular frequency for ACR	ω_{nacr}	4000 rad/s
Angular frequency for AVR	ω_{navr}	400 rad/s

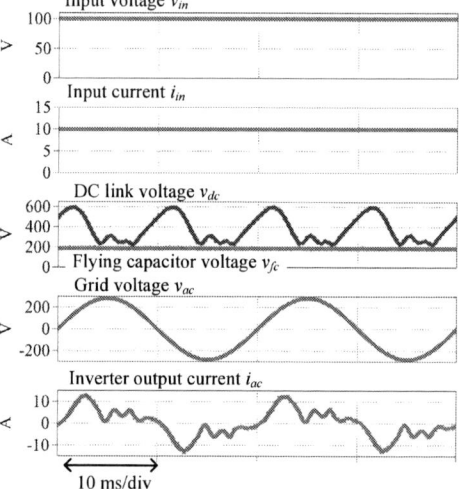

(a)   Without power decoupling

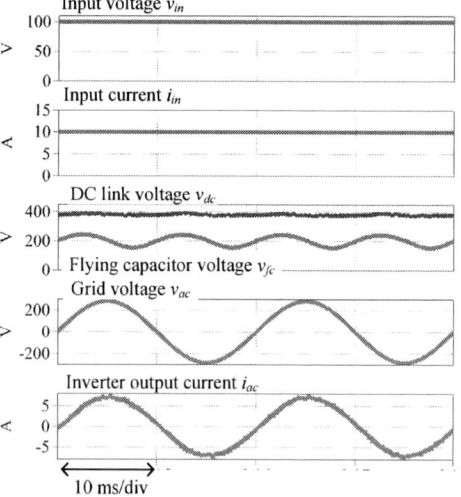

(b)   With power decoupling

Fig.9 Simulation result of the proposed circuit.

current becomes sinusoidal waveform. From these results, the single-phase power ripple is compensated with the small capacitor. Especially, the proposed circuit is not required the additional circuit and electrolytic capacitor in order to compensate the power decoupling.

Fig.10 shows the harmonic analysis result of the DC link voltage V_{dc}. According to Fig.10, the second order harmonic component is reduced by 82.1% compared to the without power decoupling. Thus, the single-phase power fluctuation is compensated. The reason that the second order harmonic component exist is because; (i) the compensation value is not enough at 1 kW (ii) the phase of the buffer power for single-phase power fluctuation compensation is not matching to the single-phase power grid. This will be improved in the future work.

B. Experimental result

In order to demonstrate the validity of the proposed circuit, a 1 kW class prototype circuit is tested. Fig. 11 show the experimental results in order to confirm the availability of the proposed decoupling control. In this experiment, MPPT control is not introduced for simplicity.

According to Fig. 11 (a), the input current i_{in} is constant. However, the inverter DC voltage v_{dc} is fluctuated at twice of the single-phase grid frequency. This is because the instantaneous difference value between input and output power is compensated by the DC link capacitor C_{dc}. As a result, the Total Harmonic Distortion (THD) of the output current i_{ac} is increased.

According to Fig. 11 (b), the flying capacitor voltage v_{fc} is fluctuated by the proposed control, and the v_{dc} is constant. From the results, the single-phase power fluctuation is compensated by the flying capacitor C_{fc}. In addition the flying capacitor average voltage is balanced at the half value of the DC link voltage. Finally, the output current i_{ac} became a sinusoidal waveform by the proposed control. Especially, the proposed circuit can achieve the active power decoupling without the additional circuit. Thus, the efficiency and power density can be improved.

Fig.12 shows the harmonic analysis of the DC link voltage. The second order harmonic component is reduced by 89.9% compared to without the power decoupling control. However, the second order harmonic component is still included on the DC link voltage. This is because the power fluctuation compensation value has the error by the actual single-phase power fluctuation.

Fig.13 shows the efficiency characteristics of the proposed circuit. From this result, maximum efficiency is 94.5% when the output power is 1 kW. However, the efficiency at the light load decrease. In order to improve the efficiency, the loss analysis for the proposed circuit is conducted.

Finally, the ripple current of the boost-up inductor is estimated. Fig.14 shows the ripple current ΔI. In this experiment, peak to peak value of the ripple current ΔI is designed at 2.1 A. In addition, the inductance of the boost-up inductor L_{fc} is designed from equation as (10). As a result, the

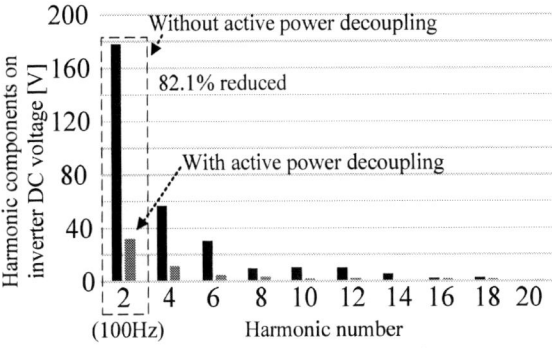

Fig.10 Harmonic analysis result in simulation.

Table.2 Experimental parameter

Switching frequency	f_{sw}	16	kHz	
Capacitances	C_{fc}	180	μH	
	C_{dc}	60	μH	
Inductances	L_{fc}	2	mH	
	L_{ac}	2.8	mH	(%Z=2.3%)
Input voltage	V_{in}	150	V	
Inverter DC voltage	V_{dc}	300	V	
MOSFETs	S_1-S_4	IXYS, IXFN132N50P3		
IGBTs	S_{up}, S_{un} S_{vp}, S_{vn}	Fuji electric, 2MBI100VA		

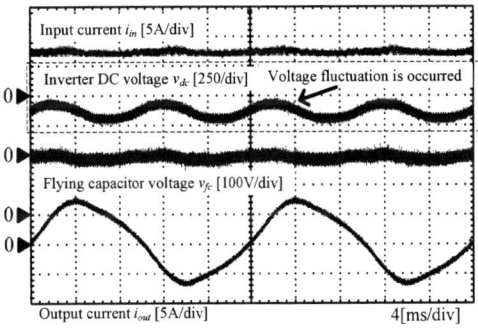

(a)　Without power decoupling

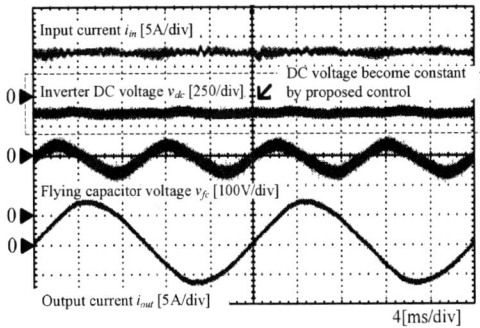

(b)　With power decoupling
Fig.11 Experimental result

error between the actual and design value is 4.7%. Therefore, the validity of the design method is demonstrated by experiment.

VII. CONCLUSION

In this paper, the novel power decoupling circuit that uses the flying capacitor topology is proposed in order to solve above mentioned problems. The proposed circuit has the function of the boost-up and single-phase power fluctuation compensation. In the decoupling control strategy, the flying capacitor is charged and discharged at the twice grid frequency. As a results, the single-phase power fluctuation is compensated without the large DC link capacitor. The advantages of the proposed circuit are follows: 1) the additional magnetic component for active power decoupling is not required; 2) the inductance of the boost-up inductor can be reduced in comparison with conventional boost chopper. 3) Simple constitution and easy control. In addition, the validity of the single-phase power fluctuation compensation is revealed by the simulation and experiment.

From the experimental results, the flying capacitor v_{fc} voltage is fluctuated by the proposed control, and v_{dc} is constant. From the results, the single-phase power fluctuation is compensated by the flying capacitor C_{fc}. In addition the flying capacitor average voltage is balanced at the half value of the DC link voltage. Finally, the output current i_{out} became a sinusoidal waveform by the proposed control. Especially, the proposed circuit can achieve the active power decoupling without the additional circuit. Thus, the efficiency and power density can be improved.

In future works, the converter loss will be estimated in order to improve the efficiency. In addition, optimization of the switching frequency will be estimated. Moreover, the power density both the conventional circuit and proposed circuit will be compared.

REFERENCES

[1] T. Boutot and L. Chang, "Development of a single-phase inverter for small wind turbins," in Proc. IEEE Electrical and Computer Engineering Canadien Conf. (CCECE'98), pp. 305-308 (1998)

[2] Y. Xue, L. Chang, S. B. Kjaer, Bau, J. Bordonau, T. Shimizu, "Topologies of Single-phase Inverters for Small Distributed Power Generators: An Overview," IEEE Transactions on Power Electronics, Vol.19, No.5, pp. 1305-1314 (2004)

[3] H. Hu, S. Harb, X. Fang, D. Zang,Q, Zhang, Z. J. Shen: "A Three-port Flyback for PV Microinverter Applications With Power Pulsation Decoupling Capability", IEEE Trans, Vol. 29, No. 9, pp. 3953-3964 (2012).

[4] H. Hu, S. Harb, N. Kutkut, I. Batarseh, Z. J. Shen, "Power Decoupling Techniques for Micro-inverters in PV Systems — a Review," Energy Conversion Congress and Exposition 2010, pp.3235-3240, pp. 12-16 (2010)

[5] T. Shimizu, K. Wada, N. Nakamura: "Flyback-Type Single-Phase Utility Interactive Inverter With Power Pulsation Decoupling on the DC Input for an AC Photovoltaic Module System", IEEE Trans., Vol. 21, No. 5, pp. 1264-1272 (2006)

[6] F.Shinjo, K.Wada, T.Shimizu:"A Single-Phase Grid-Connected Inverter with a Power Decoupling Function " PESC 2007, pp.1245-1249, (2007)

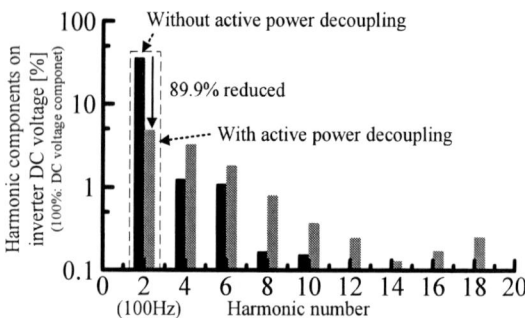

Fig.12 Harmonic analysis result in experiment.

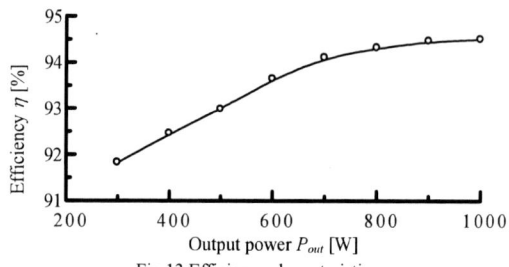

Fig.13 Efficiency characteristics.

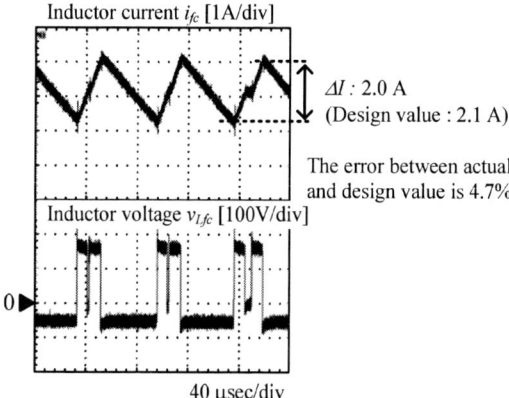

Fig.14 Estimation of the ripple current of the boost-up inductor.

[7] K. Kusaka, H. Watanabe, K. Furukawa, J. Itoh: "Power Decoupling Circuit with Flying Capacitor DC-DC Converter", JIASC2015, Vol. , No. 1-78, pp. (2015)

[8] F. Schimpf, L. Norum: "Effective Use of Film Capacitors in Single-Phase PV-inverters by Active Power Decoupling", IECON 2010, Vol. , No. , pp. 2784-2789 (2010)

[9] Kuo-Hen Chao, Po-Tai Cheng : "Power decoupling methods for single-phase three-poles AC/DC converters " ECCE 2009, pp3742-3747,(2009)

[10] Ruxi Wang, F.Wang, Rixin Lai, Puqi Ning, R.Burgos, D.Boroyevich : " Study of Energy Storage Capacitor Reduction for Single Phase PWM Rectifier " APEC 2009, pp1177-1183,(2009)

[11] Chaia-Tse Lee, Yen-Ming Chen, Li-Chung Chen, Po-Tai Cheng: "Efficiency Improvement of a DC/AC Converter with the Power Decoupling Capability", IEEE APEC 2012 Vol. , No. , pp. 1462-1468 (2012)

[12] Y. Ohnuma, J. Itoh: "A Single-Phase Current-Source PV Inverter With Power Decoupling Capability Using an Active Buffer", , Vol. 51, No. 1, pp. 531-538 (2015)

[13] T. Kitano, M. Matsui: "Reduction of Filter Capacitance for a Single Phase PWM Converter with DC Active Filter Function", Annual Conference of IEEJ, No.4-10-4-11(1996) (in Japanese)

[14] M. Calais, J. Myrzik, T. Spoone, and V. G. Agelidis, "Inverters for single-phase grid connected photovoltaic systems-an overview," in Proc. 2002 IEEE 33rd Annual Power Electronics Specialists Conference, 2002. pesc 02, vol. 2, pp. 1995–2000, (2002)

[15] S. Kjaer and J. Pedersen, "A review of single-phase grid-connected inverters for photo-voltaic modules," IEEE Trans. Ind. Ap., vol. 41, no. 5, pp. 1292-1306, (2005)

[16] A. C. Kyritsis, N. P. Papanikolaou, E. C. Tatakis, "A novel Parallel Active Filter for Current Pulsation Smoothing on Single Stage Grid-connected AC-PV Modules", European Conference on Power Electronics and Applications 2007 (2007)

[17] B. Singh, B. N. Singh, A. Chandra, K. Al-Haddad, A. Pandey and D. P. Kothari, "A review of single-phase improved power quality ACDC converters", IEEE Trans. Ind. Electron., vol. 50, No. 5, pp. 962- 981, (2003)

[18] C. Rodriguez and G. Amaratunga, "Long-lifetime power inverter for photovoltaic AC modules," IEEE Trans. Ind. Elec., vol. 55, no. 7, pp. 2593-2601, (2008)

[19] J. Itoh, H.Watanabe, K.Koiwa, Y. Ohnuma: "Experimental verification of single-phase inverter with power decoupling function using boost chopper", EPE '13-ECCE Europe, the 15th European Conference on Power Electronics and Applications, Vol. , No. , pp. (2013)

[20] K.Furukawa, H.Watanabe, K.Kusaka, J.Itoh: "", JHES2015, Vol. , No. A3-13, pp. (2015)

GaN FET and hybrid modulation based differential-mode inverter

Sudip K. Mazumder[*,**], Ankit Gupta[*], Shirish Raizada[*], Harshit Soni[*], Nikhil Kumar[*],
Paromita Mazumder[**], and Parijat Bhattachaarjee[**]

[*]University of Illinois, Chicago, IL: 60607
mazumder@uic.edu

[**]NextWatt LLC, Hoffman Estates, IL: 60192
parijatb@gmail.com

Abstract— **Recently differential-mode inverter has been introduced, which has shown significant promise for low-power applications due to its modularity, symmetry, and reduced component count. Initially a continuous modulation scheme was adopted which has now been superseded using hybrid modulation that introduces discontinuity in modulation or yields topological switching that results in enhanced efficiency, which is captured in this digest. To further improve the energy-conversion efficiency, GaN HFETs (or enhancement-mode HEMTs) is incorporated that yields further improvements in loss reduction.**

I.   INTRODUCTION

A differential-mode single-phase inverter [1] as illustrated in Fig. 1 has a wide application base [3]-[7], as they yields plurality of advantages including but not limited to the following:

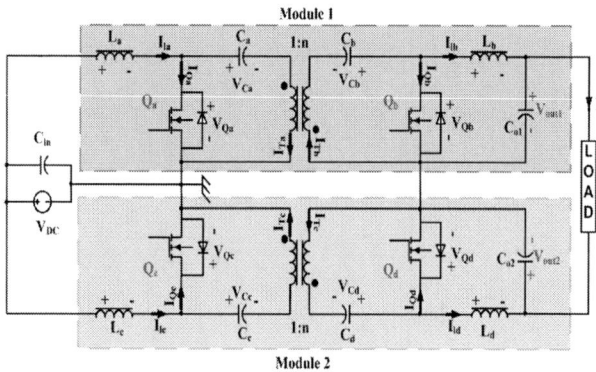

Fig. 1. Illustration of the single-phase differential-mode-inverter topology, which comprises two modules (Module 1 and Module 2).

- No high-side devices;
- Yields gate-driver simplicity;
- Reduced device component count;
- Simple control implementation
- Transformer magnetizing inductance unaffected by low-frequency current ripple due to symmetrical blocking capacitors;
- High-density system realizable due to above and easy extension to magnetics coupling;
- Both dc/ac as well as ac/dc power flow possible with control changes only due to converter's topological symmetry.

The operational detail of the differential-mode inverter, comprising essentially two dc/dc modules, is captured in Section II. It comprises two possible modes of operation.

One (see Fig. 3) that is based on continuous modulation scheme that necessitates continuous high-frequency activation of both the converter modules leading to higher circulating power, higher device breakdown-voltage requirement, and enhanced device losses. The other scheme (shown in Fig. 4) is based on hybrid modulation that enables high-frequency operation of only one converter module at a given time, thereby reducing losses and eliminating circulating power. The experimental results outlining these improvements are captured in Section VI. The improvement in the energy-conversion efficiency using the hybrid-modulation scheme is tangible. Notwithstanding, the primary-side and the secondary-side active devices incur conduction and switching losses under hard-switching condition. To reduce these device conduction and switching losses, gallium nitride (GaN) based heterostructure field-effect transistor (HFET) is being pursued. These GaN HFETs incorporate ballistic transport due to quantum well, as exhibited in Fig. 2 [2], which yields lower inverter operational losses. This is captured in Section IV and the relative improvements compared to the Si-based differential-mode inverter are shown.

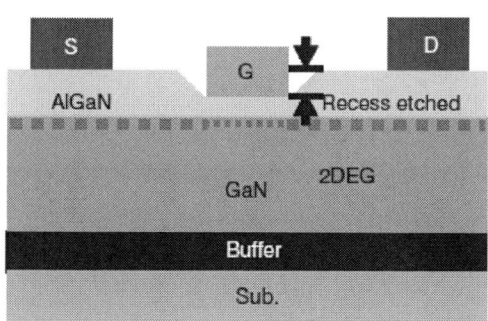

Fig. 2. Illustration of the ballistic transport of electrons (marked in orange) due to the heterostructure design.

II.   HYBRID MODULATION BASED DIFFERENTIAL-MODE INVERTER

Fig. 3 demonstrates the continuous-modulation-scheme-based operating modes of the differential-mode inverter. It is apparent from Fig. 3 that at any given time, both the dc/dc converter modules of the inverter are operating at high frequency, which as outlined earlier, leads to circulating power and enhanced loss. The latter issues are mitigated using the hybrid modulation scheme, as illustrated in Fig. 4.

978-1-4673-9551-9/16 $31.00 © 2016 IEEE

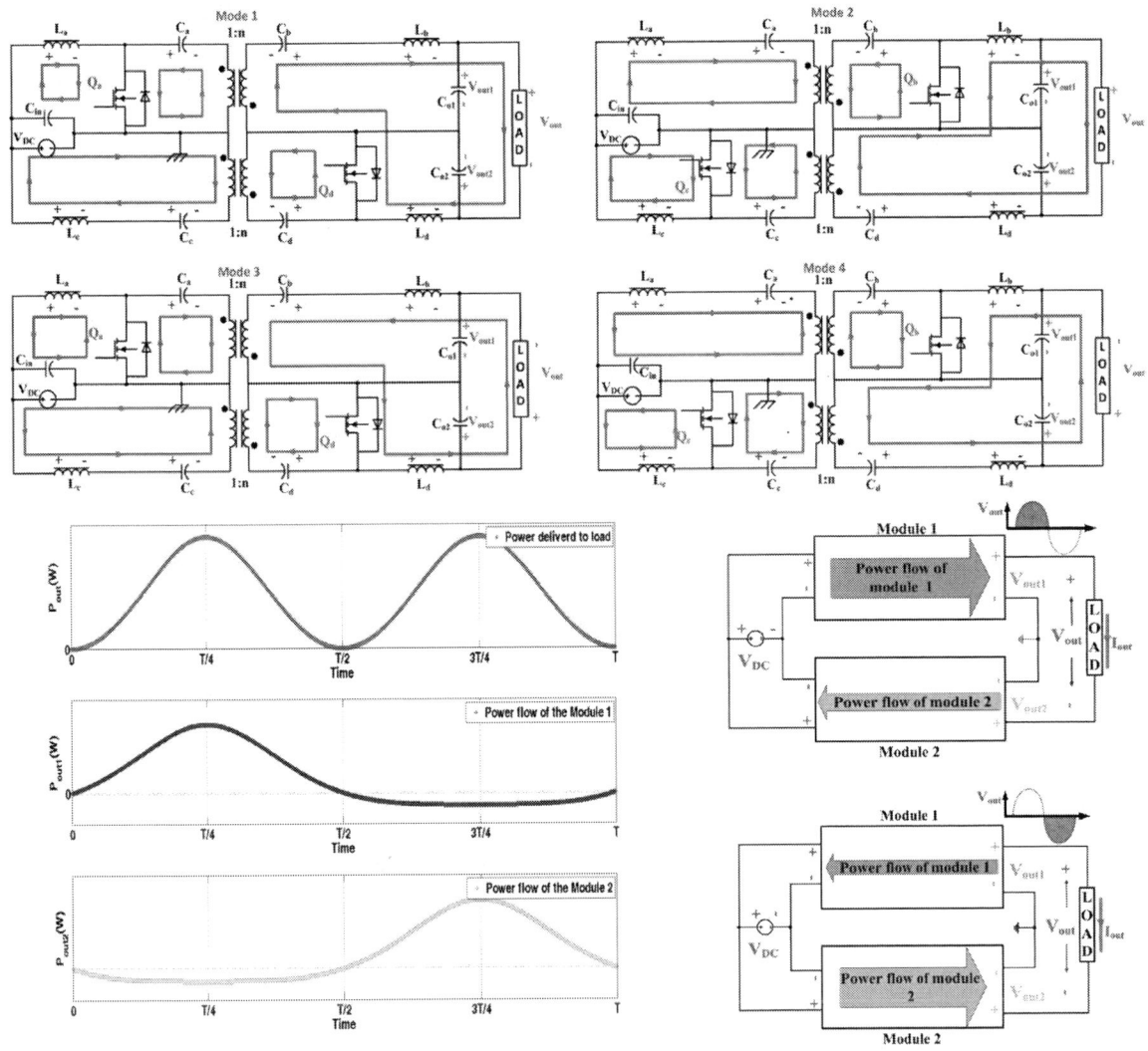

Fig. 3. (Top) Modes of operation of the differential-mode single-phase inverter operating using continuous-modulation scheme. (Bottom) Illustrations of instantaneous power flow of the inverter and its two modules during the positive and negative halves of a line cycle.

It is obvious from the figure that hybrid modulation – based on topological switching – yields high-frequency switching at any given time only for one of the inverter modules [7]. Thus, because only one of the inverter modules is operating at high frequency, circulating power is eliminated, overall loss is reduced, and topological order is reduced as well. However, as evident from Fig. 4, the hybrid-modulation-based inverter still incurs conduction and some switching losses. The initial approach for the inverter is based on Si MOSFET. The corresponding results are shown in Section III and in Section IV the improvement using GaN HFETs are demonstrated.

III. SI-MOSFET BASED EXPERIMENTAL DIFFERENTIAL-MODE-INVERTER

Fig. 5 shows the experimental Si-MOSFET based differential-mode 500-W inverter. The critical comparative results for the inverter using the hybrid-modulation scheme

(referred to as HMS) as against the continuous-modulation scheme (referred to as CMS) are captured in Figs. 6-8.

IV. GAN-HFET BASED DIFFERENTIAL-MODE-INVERTER PERFORMANCE AND COMPARISON WITH SI-MOSFET BASED INVERTER

Fig. 9 presents the calculated value of major losses incurring in the hybrid-modulation based hard-switched differential-mode experimental inverter (shown in Fig. 5) using Si MOSFETs for various discrete input voltage and output power levels. As can be observed from the sectioned bar graph in Fig. 9, snubber and the device switching losses contribute about 50% to 60% of the total power loss at all output power and input voltage levels.

While the former losses can be mitigated using currently-available lossless snubber or active snubber solutions, the latter requires the adoption of either soft-switching schemes

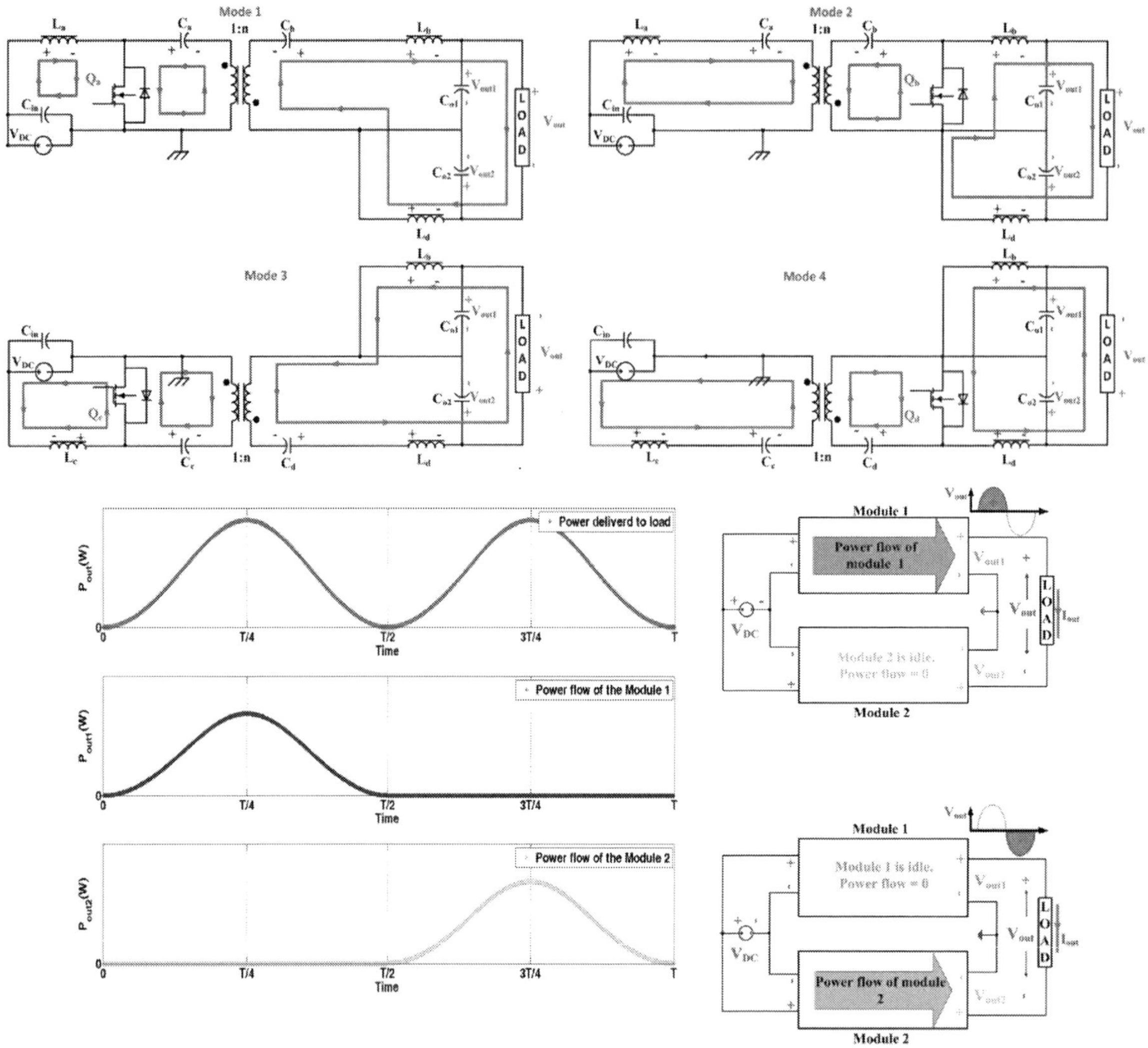

Fig. 4. (Top) Modes of operation of the differential-mode single-phase inverter operating using hybrid-modulation scheme. (Bottom) Illustrations of instantaneous power flow of the inverter and its two modules during the positive and negative halves of a line cycle.

Fig. 5. Experimental prototype of the single-phase differential-mode inverter. It shows the TMS320F28335 DSP based digital controller on the top feeding the inverter power stage at the bottom.

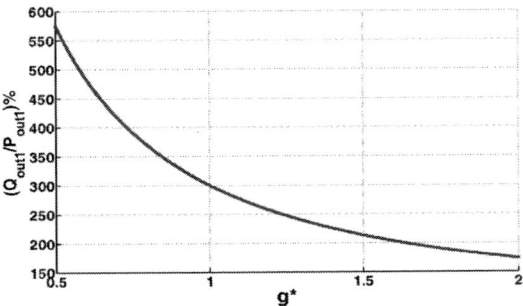

Fig. 6. For Module 1, the ratio of the reactive (circulating) power to the active power for varying peak normalized dc-voltage gain for the experimental differential-mode inverter when it is operated using CMS. The inverter has no circulating-power when it is operated using HMS.

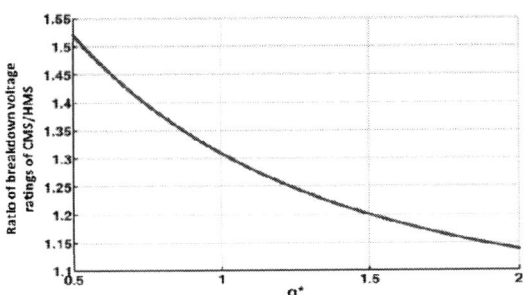

Fig. 7. Ratio of the breakdown voltages required for CMS vs HMS schemes as a function of the dc-voltage gain of the experimental inverter. It clearly shows the improvement with HMS.

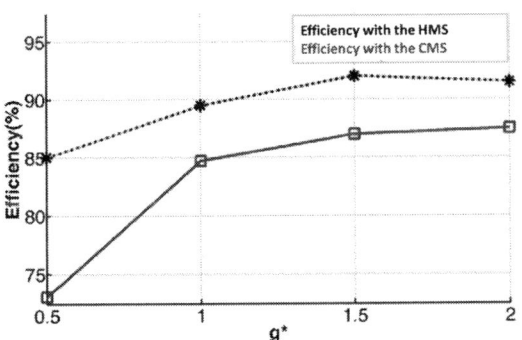

Fig. 8. Improvement in experimental efficiency of the Si-MOSFET based differential-mode inverter shown in Fig. 5.

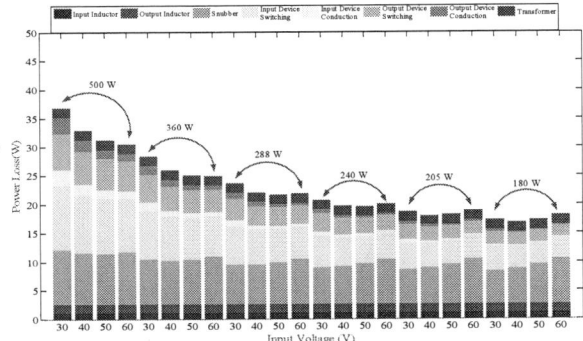

Fig. 9. Sectioned bar graph representing numerically calculated losses incurring in a hybrid modulation based differential-mode inverter for various discrete input voltage and output power levels is presented. Loss calculations are done considering a Si device on primary side (IXFX160N30T [8] and SiC device on the secondary side (CMF20120D).

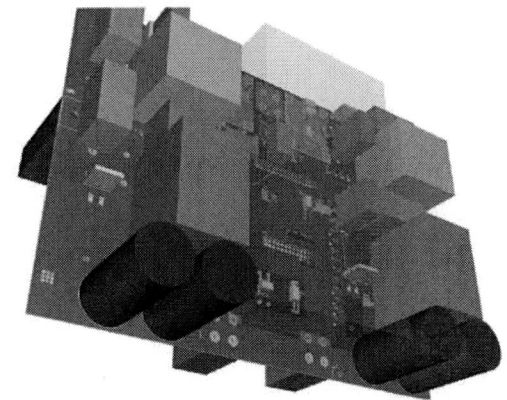

Fig. 10. The Altium 3D based prototype GaN-HFET-based differential-mode inverter design is presented. The experimental inverter unit will be presented in the final paper. The experimental Si-based inverter is already presented in Fig. 5. The block in the middle is spaced for the DSP controller unit following the same design shown in Fig. 5.

or replacement of the Si MOSFETs with devices such as GaN HFET's compared to similar rated state-of-the-art Si devices, GaN devices offer significant improvement in gate charge, on-state resistance, output capacitance, and package size. This in turn help in realizing enhanced efficiency and achieving higher power density at the same output power.

Next, an efficiency comparison is presented between hybrid-modulation based inverter having GaN HFETs and Si MOSFETs devices. Calculations were done for the following devices listed in Table I. The Altium 3D prototype GaN inverter design is presented in Fig. 10. Fig. 11 and Fig. 12 provide projected efficiency curves for the inverter by using devices listed in Table I. It is observed that the inverter using GaN devices can achieve 3% more efficiency then the inverter using state-of-art Si devices.

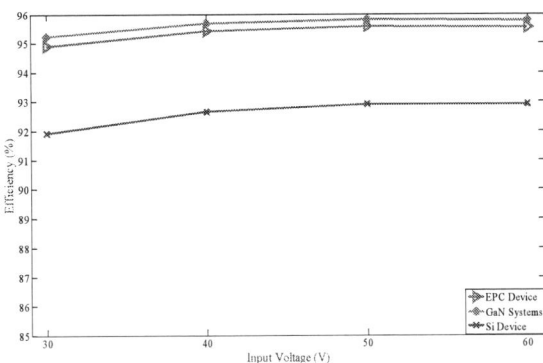

Fig. 11. Projected efficiency for a 500 W hybrid-modulation based differential-mode inverter with variable input voltage, using EPC [9] Device (primary-EPC2025; Secondary- EPC2027), GaN Systems [10] Device (Primary- GS66506T, Secondary- GS66516T) and Si device on primary (IXFX160N30T) and SIC device on Secondary (CMF20120D).

TABLE I. SI AND GAN DEVICE DETAILS.

Primary side device	Secondary side device
EPC2025:300V,4A	EPC2027:450V,4A
GS66506T:650V,22A	GS66516T:650V,60A
IXFX160N30T:300V,160A	CMF20120D:1.2kV,42A

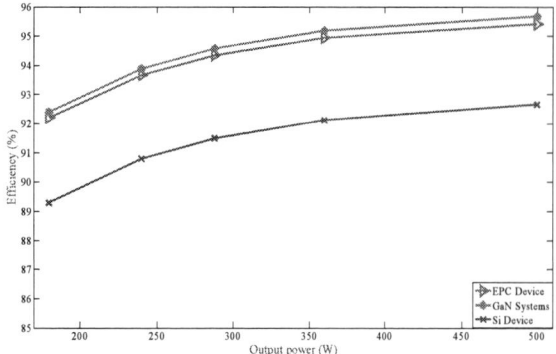

Fig. 12. Projected efficiency for a hybrid-modulation based differential-mode inverter having an input voltage of 40 V, using EPC Device (primary-EPC2025; Secondary- EPC2027), GaN Systems Device (Primary- GS66506T, Secondary- GS66516T) and Si device on primary (IXFX160N30T) and SIC device on Secondary (CMF20120D).

V. DESIGN CONSIDERATION USING GaN HFET DEVICES

GaN transistors generally work similar to a power MOSFET, but owing to its low device capacitance, turn on resistance and small device package one can attain much higher switching speed and high power density. On one hand, these features strongly put forward the candidature of GaN devices to be the future of power switches. On the other, careful circuit layout is required to fully harness their benefits. This section outlines some design techniques and provide guidelines to avoid common pitfalls that can arise due to GaN performance capabilities.

For most GaN devices available in the market today absolute maximum voltage limit that can be applied to the gate with respect to source is specified. For the enhanced mode GaN system devices this limit is specified as +/- 10 V. Exceeding these limits can destroy the gate terminal of the device permanently, hence one need to avoid any event which can lead to a voltage greater then +/- 10 V to be applied to the gate terminal of the device. One of the common solution to this problem is to add a Zener diode at the input of gate terminal of the device. This solution works well for small switching frequency, but as the switching frequency of the device id increased, Zener diode adds a nonlinear, voltage and frequency dependent parasitic capacitance to the RLC gate loop and can sometimes introduce gate oscillations along with increasing the gate drive and switching losses of the device.

As the switching frequency is increased, effect of parasitic inductance on circuit performance escalates and careful circuit layout is required to minimize its effects. Because of the extremely low device capacitance, gate driver loop is most suitable high frequency ringing between the parasitic inductance and gate capacitance of the device. Hence, limiting the size of gate driver loop is of critical importance. Benefits of providing a separate source-sense (Kelvin) connection for gate loop are well known. Availability of these separate source-sense pins provide a separation between the gate driver loop and the power loop inside the device and minimize the adverse effects of parasitic inductance.

Fig. 13 shows the primary and secondary side gate driver design layout for an isolated 500 W differential-mode single-phase inverter. GaN 1, 2 and 3 are the GaN systems GS66508P 30 A, 650 V bottom side cooled GaN transistors and Gate_driver 1 and 2 are IXYS, IXDn609si 9 A, high speed gate drivers. To limit the oscillations ferrite bit FB1, 2 and 3 and 8.2 V cutoff Zener diodes DZ1, 2 and 3 are used at the input of the gate terminal connection of the GaN devices. It is to be noted that, the layout of the two parallel GaN devices on the primary side of the circuit is kept symmetrical to have identical rising and falling times for both the devices.

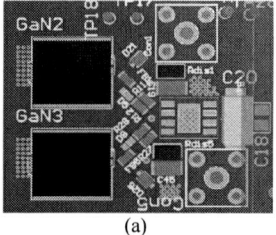
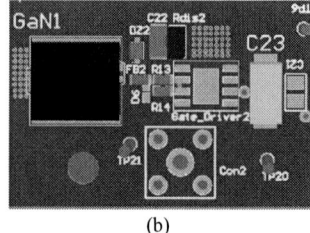

(a)   (b)

Fig. 13. PCB layout for (a) primary side and (b) secondary side gate driver circuit for an isolated 500 W differential-mode single-phase inverter.

VI. EXPERIMENTAL RESULTS FOR THE GaN INVERTER

This section presents the experimental results of GaN based differential-mode single-phase inverter. The experimental prototype as shown in Fig. 14 is being operated at 500 W with an output voltage of 120 V RMS. GaN system's GS66508P 30 A, 650 V GaN transistors were selected as the switching device. Fig. 15 shows the transient response of the inverter at an input voltage of 40 V dc. As can be inferred from the result, the inverter attains its steady state in the 2nd line cycle and has a fast transient response. Fig. 16 outlines the inverter performance under load transient conditions. Parametric plot of THD vs output power for the GaN based DMCI is provided in Fig. 17. It can be observed that the THD of the inverter varies between 1.4% at 300 W and 2.7% at 570 W, which is well within satisfactory limits. The peak output voltage in the THD result is set at 120V RMS and the input voltage is set to 40V dc.

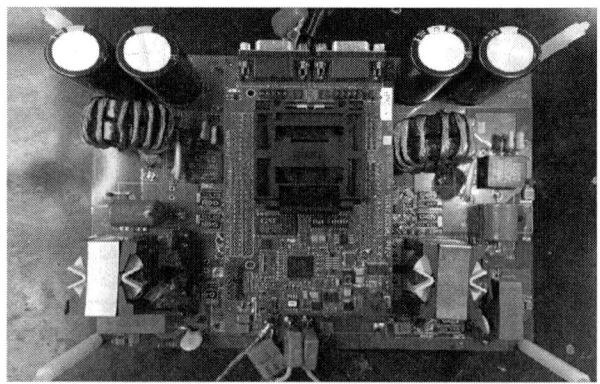

Fig. 14. The experimental prototype of the GaN inverter.

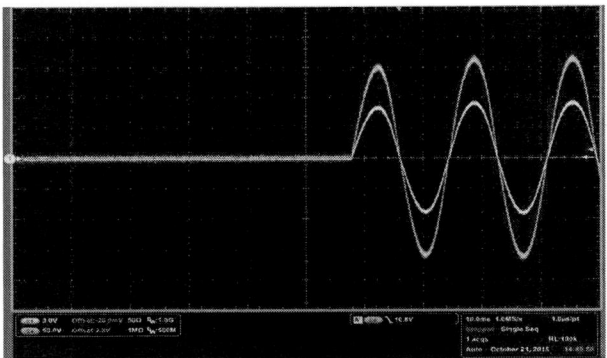

Fig. 15. Start-up response of the GaN inverter at 500 W operating with an input voltage of 40 V. Channel 1 (C1) represents the load current and Channel 4 (C4) represents the output voltage.

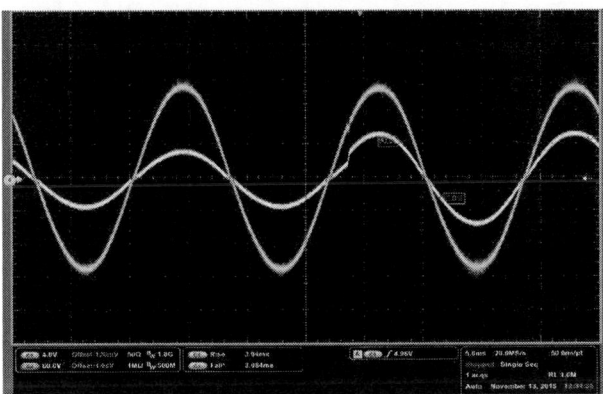

Fig. 16. Transient response of the GaN inverter under load switching condition. Load is switched from 300 W to 500 W with an input voltage of 40 V. Channel 1 (C1) represents the load current and Channel 4 (C4) represents the output voltage.

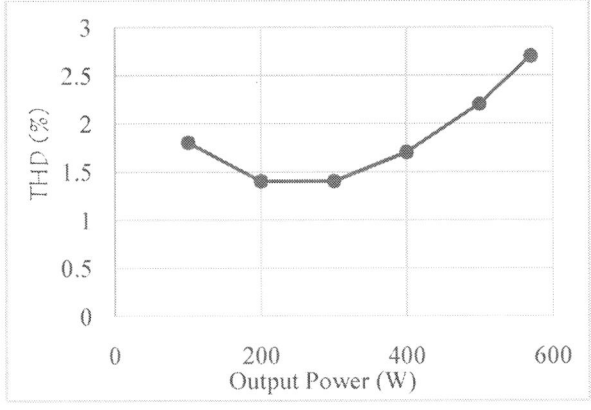

Fig. 17. Plot of THD (%) vs output power of the GaN inverter at 40 V dc input voltage.

VII. CONCLUSIONS

The hybrid-modulation scheme-based differential-mode single-phase inverter, owing to the topological switching that enables order reduction yields tangible improvement in efficiency, reduced breakdown voltage rating, and eliminates single-stage direct-power-conversion solar circulating power.

The incorporation of GaN HFETs in the same design leads to an increase in power density and further improvement in the performance of the inverter owing to the improved on-state and switching characteristics of the GaN devices. Aspect of design such as gate driver design and circuit layout are discussed and performance comparison is provided between the Si and GaN based differential-mode single-phase inverter. The experimental results show satisfactory steady-state, startup, and load transient performances of the inverter under closed-loop control. Future goals for this project is to facilitate grid connection and integrating all magnetic devices onto one core.

ACKNOWLEDGEMENT

This work was supported in part by NSF Award No. 1448181. However, any opinions, findings, conclusions, or recommendations expressed herein are those of the authors and do not necessarily reflect the views of the NSF.

REFERENCES

[1] S.K. Mazumder, "Differential mode inverters for renewable and alternative energy systems", filed by the University of Illinois at Chicago with the USPTO on 06/23/11 with patent application number 61/500,344.

[2] N. Ikeda, J. Li, S. Kato, M. Masuda, and S. Yoshida, "A novel GaN device with thin AlGaN/GaN heterostructure for high-power applications", Furukawa Review, No. 29 2006.

[3] S.K. Mazumder and S. Mehrnami, "A low-device-count single-stage direct-power-conversion solar microinverter for microgrid," in *IEEE 3rd Int. Symp on Power Electron. For Distributed Generation Systems (PEDG), 25-28 Jun. 2012, pp.725-730.*

[4] R.D. Middlebrook and S. Ćuk, "Advances in switched-mode power conversion", vols. I, II, and III. TESLACO, 1983

[5] S.K. Mazumder, R.K. Burra, R. Huang, and V. Arguelles, "A low-cost single-stage isolated differential Cuk inverter for fuel cell application", *IEEE Power Electronics Specialists Conference, Rhodes, Greece, 2008, Pg 4426-4431.*

[6] A. Gupta, N.Kumar and S.K Mazumder, "Frequency-dependent criterion for mitigation of transmission-line effects in a high-frequency distributed power systems" *Energy Conversion congress and exposition (ECCE), Sep. 2015, PP 4624- 4631.*

[7] S. Mehrnami and S.K. Mazumder "Discontinuous modulation scheme for a Differential-Mode Ćuk Inverter" *IEEE Trans. Power Electron., vol. 30, Issue: 3, pp. 1242-1254, March 2015.*

[8] http://www.ixys.com/

[9] http://epc-co.com/epc/

[10] http://www.gansystems.com/

Thermal and Electrical Co-Design of a Modular High-Density Single-Phase Inverter Using Wide-Bandgap Devices

Steven Chung[1], Miad Nasr[1], David Guirguis[1], Masafumi Otsuka[1], Shahab Poshtkouhi[1], David K. W. Li[1],
Vishal Palaniappan[1], David Romero[1], Cristina Amon[1], Ray Orr[2], Olivier Trescases[1]

[1]University of Toronto, 10 King's College Road, Toronto, ON, M5S 3G4, Canada

[2]Solantro Semiconductor, Ottawa, ON, K2E 7Y1, Canada

Abstract—This paper explores the multi-disciplinary design challenges in building a 240 VAC, 2 kVA modular single-phase inverter with high power-density using wide-bandgap transistors. The compromise between the electrical and mechanical design is extremely important in any high-density power converter. In this work the electrical and mechanical systems were iteratively co-designed using detailed 3D thermal and air-flow simulations. Custom copper heat-sinks and heat-pipes were developed for optimal thermal management. The inverter uses three soft-switching sub-inverters in parallel, which are controlled using a novel digital Hysteretic Current Mode Control (HCMC) scheme. To achieve a flat high efficiency curve with low THD over a wide load range, two operating modes are used: 1) Boundary Conduction Mode (BCM) with a slight negative inductor valley current for soft-switching, and 2) Continuous Conduction Mode (CCM) to limit the required saturation current in the inductors. The design of an active power decoupling scheme to minimize input capacitance is also discussed. The designed single-phase inverter has a volume of 33.1 in³ and resulting theoretical power-density of 60.3 W/in³ at 2 kW load. A measured efficiency of 97.7% is achieved for a single sub-inverter with 4.5% THD at 632.7 W.

I. INTRODUCTION

Emerging wide-bandgap, high-voltage Gallium Nitride (GaN) and Silicon Carbide (SiC) devices offer the opportunity to shrink inverters by scaling up the switching frequency, f_s, and reducing the size of passive components [1]–[5]. Wide-bandgap devices have been demonstrated for a wide-range of low and medium-power Power Factor Correction (PFC) ac-dc converters below 10 kW [6]–[8]. Four major challenges are addressed in this work to achieve a high density:

1) precise current-mode control to balance efficiency, size and Total Harmonic Distortion (THD),
2) active thermal management,
3) 120 Hz power decoupling, and
4) EMI mitigation.

The target outcome of this work is a compact single-phase inverter design. This paper covers the electrical and mechanical co-design of the inverter, as well as the initial testing and integration. The electrical architecture and control schemes of the inverter are discussed in Section II. The mechanical design, methodology and thermal simulations are presented in Section III. Initial experimental results of efficiency and Total Harmonic Distortion (THD) optimization of a single sub-inverter are presented in Section IV.

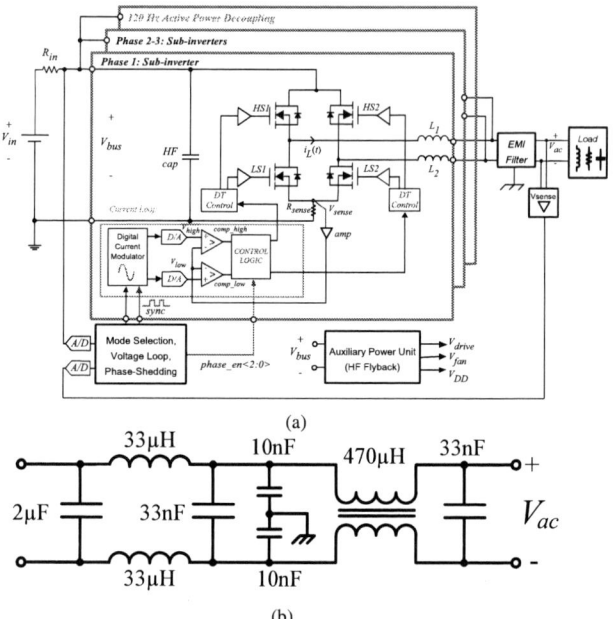

Fig. 1. The modular inverter's (a) system architecture composed of three sub-inverters and one active power decoupling block and (b) the EMI filter block.

II. ELECTRICAL DESIGN

The proposed system architecture is shown in Fig. 1(a). A modular approach is used to optimize the mechanical design by distributing the thermal losses throughout the inverter's volume. Three independently controlled sub-inverters, each designed for a rated power of 0.67 kVA, are connected in parallel and share a common EMI filter to generate a low-THD ac output voltage, V_{ac}. The master controller provides a 60Hz synchronization pulse to the slave controllers to allow for phase locking. The digital current reference is communicated by the master to the slave sub-inverters as part of the outer voltage loop. An interface board is used to route multiple control signals and to distribute various low-voltage supplies needed for the digital controllers, sensors and data-converters. The sub-inverters can be turned off at light loads for improved efficiency, which is commonly referred to as phase-shedding in multi-phase dc - dc converters [9], [10]. The EMI filter is shown in Fig. 1(b).

Both GaN and SiC technologies are considered for the power transistors. The chosen 650 V, 55 mΩ GaN MOSFET has an extremely low package volume of 41.6 mm^3 and a superior $R_{on}Q_{gate}$ figure-of-merit (FOM) of 357 mΩ·nC [11]. The chosen 900 V, 65 mΩ SiC MOSFET has a volume of 722.5 mm^3 and $R_{on}Q_{gate}$ FOM of 1950 mΩ·nC [12]. The GaN device was used in the initial design and thermal modeling and was subsequently replaced with the SiC power devices during the testing phase for improved reliability.

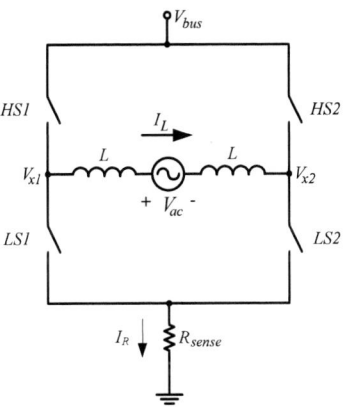

Fig. 2. Simplified inverter topology with common low-side sense resistor, R_{sense}, for current mode control (the EMI filter effect has been ignored).

A. Sub-Inverter Design and Control

Each sub-inverter power-stage is a full-bridge topology that can be used in four-quadrant operation, as shown in Fig. 2. Unlike conventional sinusoidal Pulse-Width-Modulation (SPWM), each sub-inverter operates in Hysteretic Current Mode Control (HCMC) with a variable switching frequency, which is more common in PFC applications [13]. The ideal switching operation is illustrated in Fig. 3. The inductor current, I_L, is sensed using a single low-side shunt resistor, R_{sense}. Both the rising and falling inductor currents can be detected through resistor current I_R. When the inductor current reaches the desired peak or valley current, the switches toggle. The inductor slopes are given by

$$m_1 = \frac{V_{bus} - V_{ac}}{2L}, \tag{1}$$

$$m_2 = \frac{V_{bus} - (-V_{ac})}{2L}. \tag{2}$$

Each sub-inverter operates in one of two possible modes during the ac line cycle: 1) Boundary Conduction Mode (BCM) or 2) Continuous Conduction Mode (CCM). When operating in BCM, a slight negative valley current, I_{valley}, is imposed to achieve zero-voltage switching turn-on on both switching transitions through the resonance of the switching node capacitance and main inductor. The dead times, t_{d1} and t_{d2}, are used to allow such resonant soft-switching, as shown in Fig. 3(a). During the transition period, the sensed voltage across R_{sense} instantaneously changes polarity as one set of

switches commutates. In CCM, the inductor current does not change polarity throughout the switching period. During a positive line cycle, the sensed resistor current is positive when inductor current rises, but is inverted when inductor current falls, as shown in Fig. 3(b). In CCM, the soft-switching can only be maintained for the peak inductor switching transition. A small dead time, t_{min}, is used at the inductor valley switching transition to avoid free-wheeling diode conduction.

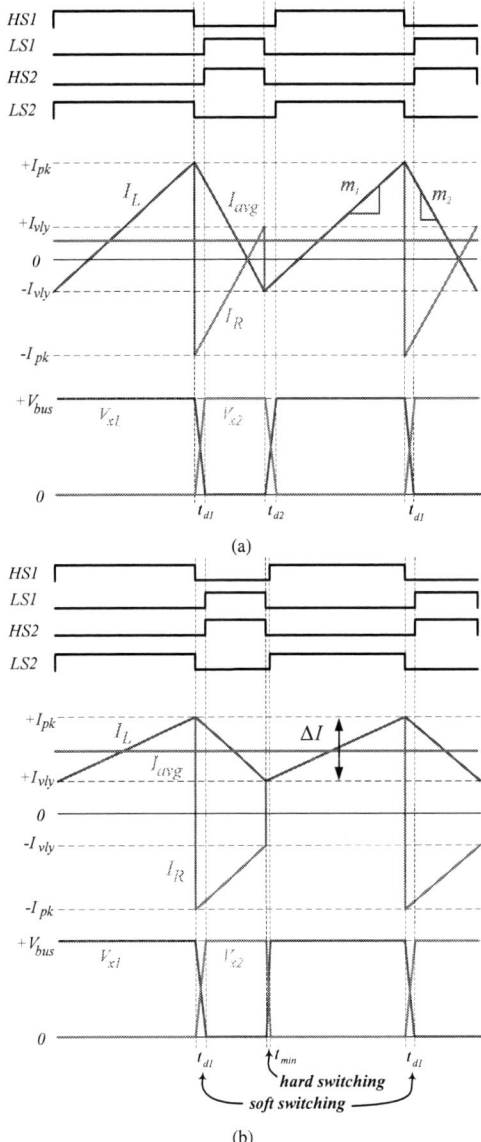

Fig. 3. Ideal switching waveforms for (a) BCM and (b) CCM operation.

The off-grid inverter controller is designed with the assumption of linear loads, hence the average inductor current, I_{avg}, is sinusoidal. To maintain a sinusoidal average current, I_{pk} is modulated according to

$$I_{pk} = 2(I_{avg} + I_{vly}), \tag{3}$$

where I_{vly} is a constant negative current in BCM. In CCM, sinusoidal current is maintained by setting a fixed ripple current, ΔI, around I_{avg} such that

$$I_{pk} = I_{avg} + \frac{\Delta I}{2}, \tag{4}$$

and

$$I_{vly} = I_{avg} - \frac{\Delta I}{2}, \tag{5}$$

as shown in Fig. 3(b).

The inductors are designed to optimize 1) the volume, 2) the core losses, 3) the saturation current, 4) the inter-winding capacitance and 5) the inductance based on the variable-frequency HCMC scheme. Type P61 high-frequency Ferrite cores are used for this design. Each sub-inverter employs two 105 μH custom inductors. Each inductor is designed to have a saturation current of 6 A to limit the core losses at higher ripple currents, which limits the average inductor current to slightly less than 3 A in BCM. To reach average currents above 3 A, the sub-inverter must operate in CCM at the high current peaks and troughs and in BCM through the zero crossings, which is referred to as dual-mode operation. The ideal inductor current for this dual-mode, BCM/CCM, operation is shown in Fig. 4. The resonant transitions during the dead-times can cause significant distortion with this open-loop approach, however the use of low-capacitance wide-bandgap devices significantly suppresses this effect.

At lower output power levels, operating purely in BCM results in the lowest switching losses, at the expense of higher peak current, higher RMS conduction losses and higher core losses in the inductors. As load power increases, inductor core losses increase with inductor ripple currents. In dual-mode operation, the transition to CCM limits the saturation current and minimizes the inductor core losses at the highest current levels, but causes hard-switching in the MOSFETs for half of the switching transitions. The sub-inverter can maintain higher efficiency over the load range by operating in dual-mode operation.

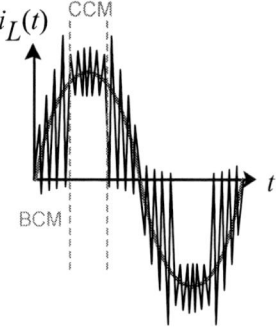

Fig. 4. Ideal dual-mode BCM/CCM operation during the ac line cycle.

The major advantage of the dual-mode scheme is that the trade-off between efficiency and THD can be optimized at

every power level. Additionally, the inverter has a variable switching frequency based on HCMC operation, which helps to reduce the EMI filter size due to the spread-spectrum effect [14]. The frequency is limited to 500 kHz for control stability and to limit switching losses. The boundary between BCM and CCM is optimized according to the operating conditions based on simulations and testing results.

The analog current sensing circuit is realized with a high frequency amplifier whose output is fed to two comparators with a 4.5 ns propagation delay. The comparator trigger levels are digitally controlled on a cycle-by-cycle basis using a pair of 12-bit digital-to-analog converters (DACs), and the comparator outputs are fed to a digital controller for precise timing control. A pre-calibrated digital feed-forward scheme, which accounts for the input and output voltage variations, can be used to precisely set the dead-times with 7 ns resolution in all operating modes to minimize the MOSFET switching losses under all conditions.

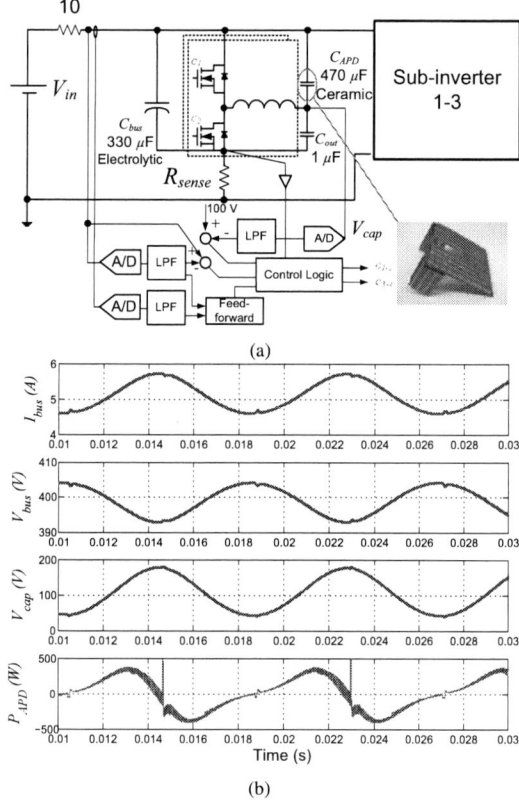

Fig. 5. (a) APD architecture and high-level control scheme. (b) Simulated performance at full load showing an input voltage ripple below 3%.

B. Active Power Decoupling for 120 Hz Component

Assuming an equivalent source resistance, R_{in}, of 10 Ω, a dc input voltage of 450 Va and an ac load power of 2 kW, an input capacitance of 1.32 mF is required to keep the input voltage and current ripple below 3% and 20%, respectively. It is well known that this capacitance can be drastically reduced by using Active Power Decoupling (APD) methods, where the 120 Hz ripple is steered into a secondary capacitor having a much larger ripple voltage [15]–[17]. The losses associated with this scheme can easily outweigh the capacitance volume reduction. A Partial Power Processing (PPP) approach is proposed for the APD in this work, as shown in Fig. 5(a).

The APD capacitor-bank is realized using $222 \times 1\,\mu$F, 450 V ceramic capacitors connected in parallel in a 3D structure. The main advantage of this series-parallel approach is the reduced processed power requirement compared to the conventional parallel or series-connected approach [18]. The double control loop strategy uses a feedback-loop as well as feedforward path in order to 1) keep the bus voltage ripple to a minimum, and 2) maintain a target dc bias voltage (100 V) on the node between C_{out} and C_{APD}. An electrolytic bus capacitor of 330 μF is used together with the APD capacitor bank.

Based on this design, the APD processes only $\approx$300 W RMS or roughly 3$\times$ lower than a conventional design, at the full 2 kW load power, as shown in Fig. 5(b). The value of C_{bus} can be further reduced by processing more power in the APD stage. In this design, the chosen capacitor values and thermal overhead imposed by the losses in the APD stage gives the best trade-off for optimized volume. Experimental validation of the APD stage is outside the scope of this paper.

III. THERMAL DESIGN

The main design objectives for the thermal design are to
1) achieve a uniform heat distribution inside the enclosure,
2) take advantage of the two main heat transfer modes, convection and radiation.

The design approach maximizes heat removal through forced convection, and distribute it evenly across the outer surfaces. This leads to maximized free convection and radiation in order to limit the temperature of outer enclosure and exhaust air to 60°C.

A vertical tower shape was chosen, as shown in Fig. 6(a), with an aspect ratio that optimizes both free convection from the outer surfaces as well as the time-of-residence of the air flow inside the device. Active cooling through forced convection, custom heat spreaders and copper heat-pipes were used to achieve a volume of a volume of 33.1 in³. This results in a power-density of 60.34 W/in³ with the 650 V GaN devices, assuming 100 W of total losses at the full load of 2 kW, corresponding to an efficiency of 95%.

A. Enclosure-level Cooling

The enclosure is made from pure copper with optimized air inlets and outlets. The aspect ratios were chosen to increase available surface area from the side faces for convection and radiation, while leaving adequate height for proper forced convection. For better thermal management, the device is divided vertically into two stages: the EMI filter and bus capacitance are placed at the bottom, and the power boards and inductors are placed in the upper stage. The PCB of the auxiliary power supply separates the two stages; it

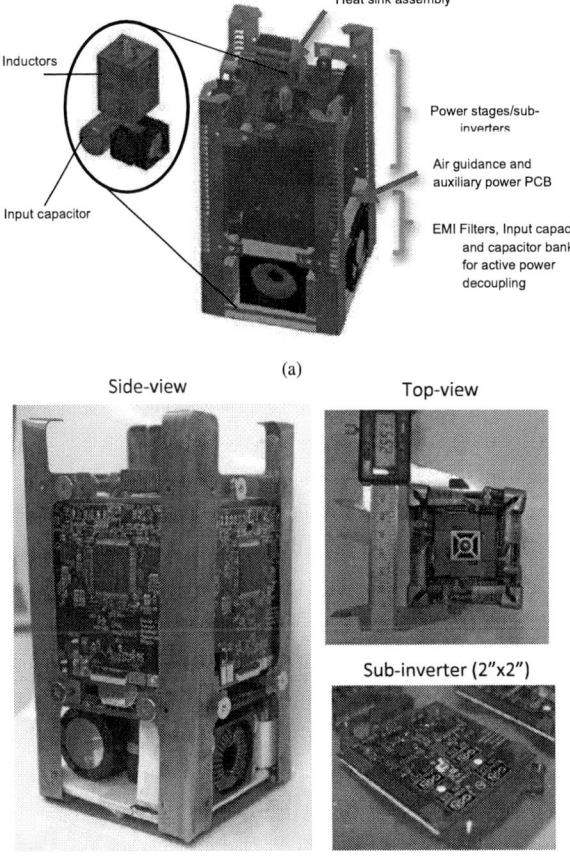

Fig. 6. (a) 3D CAD model of the outer and inner assemblies (top fan removed to show the internal design). (b) Partially assembled inverter and power-stage.

has a circular cut-out in the center to allow air flow. A high-performance heat-pipe is placed at the center of the inverter, as shown in Fig. 6, to balance heat in the two stages, improving heat distribution uniformity inside the box.

B. Power-board and Passive Component Cooling

Each sub-inverter is implemented with two back-to-back 2×2 inch PCBs; a control board and a power board. A special PCB fabrication process was used to reduce the power board thickness to 0.5 mm while maintaining 4 oz./ft² copper weight to improve back-side cooling and reduce the conduction losses. A modular concept was used for efficient component space utilization, uniform heat distribution, and to ease the coordination of the multi-disciplinary electrical-thermal design effort. All power boards are mounted vertically to the sides of the enclosure and attached to low profile, pure copper heat-sinks with embedded heat-pipes. The eight power inductors attach to a central heat sink with spacers. The EMI filter coils are attached to a small heat-sink that connects to a central heat pipe. Power transistors are cooled from the bottom side via the PCB to the heat-sink and from the upper side via a thermally conductive heat spreader. An exhaust air fan on

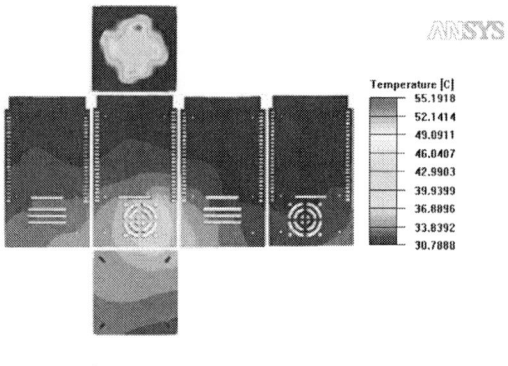

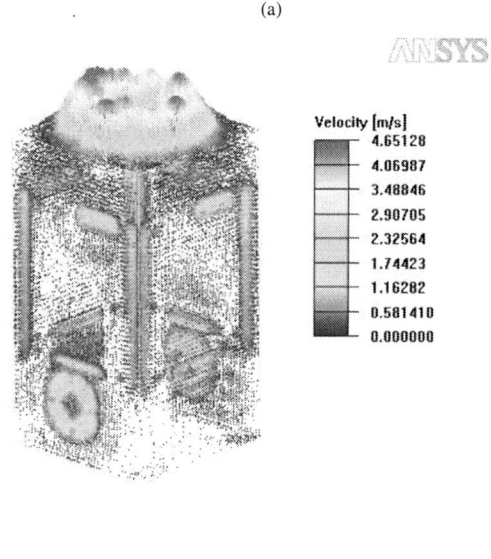

Fig. 7. (a) Simulated temperature contours for the outer enclosure at the rated power. (b) Air velocity 2 mm outside the box.

the top of the box, and two smaller intake fans at the bottom are used for active cooling. The fans consume 3.5 W in total. Extensive thermal simulations of the full system were done in ANSYS, including all major power-dissipating electronic components, a faithful representation of internal components and air flow paths, and the power flow curves of the fans.

Thermal simulations show that with a maximum distributed power loss of 100 W, the surface temperature of the enclosure would be less than 40°C, with the maximum surface temperature of 55.2°C at the air outlet as shown in Fig. 7(a). The air velocity vectors reach 4.65 m/s at the air outlet as shown in Fig. 7(b).

IV. EXPERIMENTAL RESULTS

The experimental results are based on the 900V, 65 mΩ SiC devices. A single sub-inverter with SiC transistors is shown

in Fig. 8. A bus voltage of 400 V is used throughout the measurements. The trade-off between high efficiency and low THD is explored in detail.

An FPGA controller is used for maximum flexibility in this initial project phase, however the long-term objective is to deploy the controller on a highly integrated ASIC that include all the on-chip mixed-signal circuits for high-performance current-mode control [19].

Fig. 8. Single power-stage with SiC devices mounted to heat-sink.

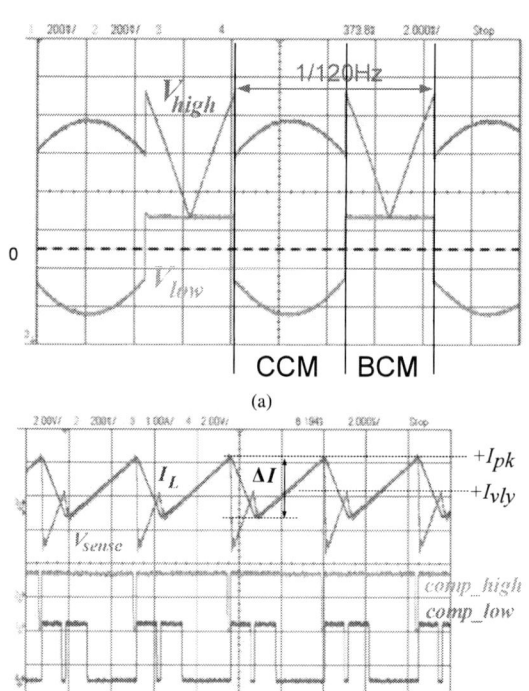

Fig. 9. Current mode waveforms (a) peak and valley envelopes for inductor current for a line cycle in dual-mode (BCM/CCM) operation and (b) I_L and I_R with corresponding comparator output signals in BCM operation.

A. Inverter implementation

The ideal inductor waveforms shown in Fig. 4 and Fig. 3 are realized with a pair of 12-bit DACs that shape the inductor

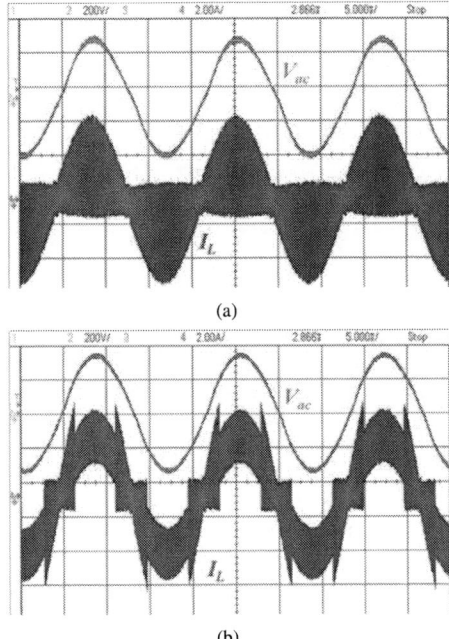

(a)

(b)

Fig. 10. Sub-inverter operation (a) in BCM at 330 W (47% rated power) and (b) in dual-mode BCM/CCM operation at 632.7 W (95% rated power).

peak and valley current envelopes. The V_{high} and V_{low} DAC outputs that shape I_{pk} and I_{vly} over the ac line cycle are shown in Fig. 9(a) for dual-mode operation. The inductor current I_L and the sensed I_R signal V_{sense}, along with corresponding comparator outputs, are shown in Fig. 9(b).

To avoid switching frequencies above 500 kHz at the zero crossings, a zero-current region is imposed, and ΔI is maintained to 2 A. The presence of the zero-current region introduces nonlinearity to (3) that adds undesirable THD to I_{avg}.

A measured voltage THD of 5.3% is achieved for a sub-inverter processing 330W, 47% of rated power, in BCM operation, as shown in Fig. 10(a). At high load power, the sub-inverter spends a smaller portion of the line cycle within the zero-current region, resulting in a lower THD. A voltage THD of 4.5% is achieved for a sub-inverter processing 632.7 W, 95% of load power, in dual-mode operation as shown in Fig. 10(b).

The measured output voltage THD versus load power is shown in Fig. 11(a), where I_{pk} directly follows the modulation defined by (3). This modulation is referred to as Scheme 1. The THD is $\geq$ 6% at 177.7 W and falls below 5% above 200 W. The average THD is 4.2%, while a THD of 3% is achieved at the highest load level of 600 W. The sub-inverter efficiency is shown in Fig. 11(a). The weighted CEC efficiency is 95.4%, and the efficiency falls below 95% above 550 W.

B. Efficiency optimization

Two avenues for inverter efficiency optimization are explored through:

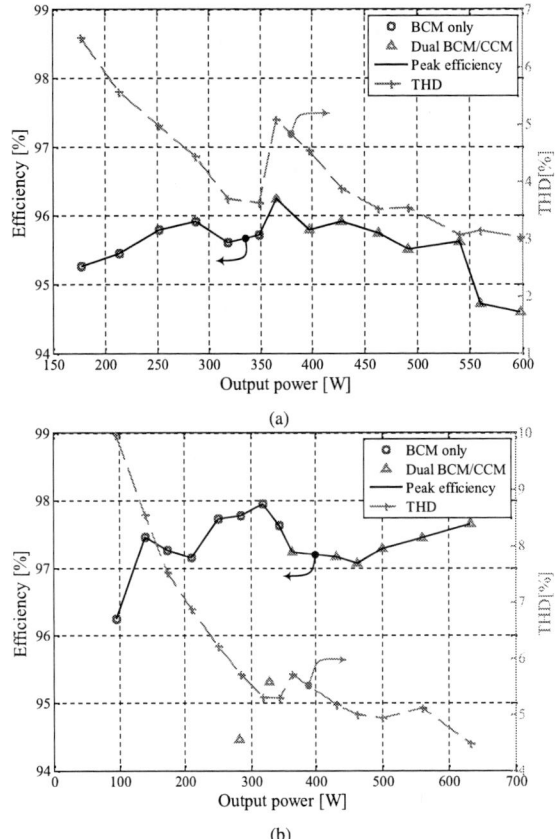

(a)

(b)

Fig. 11. Efficiency and THD versus load in pure BCM and dual BCM/CCM operation for I_{avg} modulation for (a) Scheme 1, optimized for THD and (b) Scheme 2, optimized for efficiency.

1) modifying the I_{avg} modulation to trade off THD for efficiency, and
2) optimizing the transition between BCM and CCM modes.

The trade-off between THD and efficiency can be explored by modifying (3) to

$$I_{pk} = 2(I_{avg}) + C, \qquad (6)$$

where C, an adjustable constant, replaces $2I_{vly}$ in the original equation. For the same I_{avg}, a C value of less than $2I_{vly}$ lowers I_{pk} and inductor ripple current, ΔI, which results in less inductor core losses and higher THD. This is referred to as Scheme 2. With the adjustment, a 3% increase in efficiency is achieved at 600 W load power while maintaining sub-5% THD.

The efficiency and voltage THD versus load power are shown in Fig. 11(b) for Scheme 2. The CEC efficiency for Scheme 2 is 1.9% higher than for Scheme 1 CEC efficiency, however the average THD in Scheme 2 is 6.31%.

As expected, BCM is more efficient at low output power due to the soft-switching, while CCM operation is more efficient for high loads with smaller inductor ripple currents, as

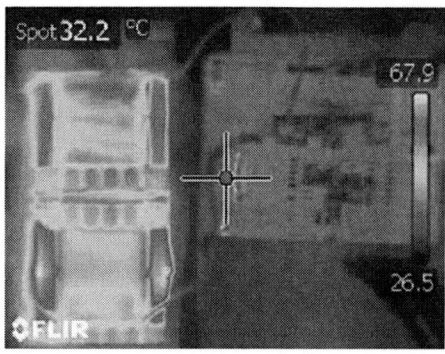

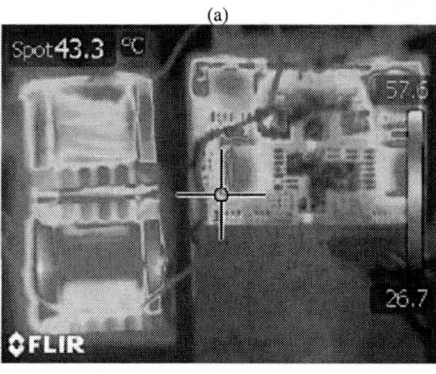

Fig. 12. Thermal capture of inductors (left) and power-stage (right) (a) processing 170-W, 25% of rated power in BCM operation only and (b) processing 600-W, 90% of rated power in dual-mode BCM/CCM operation.

shown in Fig. 11(b). The optimum transition from pure BCM operation to dual BCM/CCM operation occurs around 350 W, which corresponds to a ΔI of approximately 6 A, the inductor saturation current. For load power below 350 W, efficiency in dual-mode operation drops below 96%.

The measured thermal profile of the sub-inverter shows that the inductor core losses make up a majority of losses in BCM operation, as depicted in Fig. 12(a). The SiC MOSFETs remain below 40°C, while the inductor cores heat to 67.9°C. When the sub-inverter processes 632.7 W (95.9% rated power) at 97.7% efficiency in dual-mode operation, the MOSFETs heat up to 57.6°C due to partial hard-switching, and the inductors cool to near 50°C, as shown in Fig. 12(b).

The sub-inverter controller can optimize for efficiency and THD by modifying the I_{avg} modulation scheme depending on the load power, while maintaining the BCM/CCM transition level at 350 W. By utilizing scheme 1 and 2 for under and above 320 W load power, respectively, the sub-inverter achieves CEC efficiency of 96.5% with an average THD of 4.8%, as shown in Table. I. The input voltage is measured for a single sub-inverter with a 330 μF electrolytic capacitor on V_{bus} and a 10 Ω resistor in series with V_{in} and V_{bus}, without APD.

The parallel operation of two sub-inverters is demonstrated in both pure BCM and dual-mode operation as shown in Fig. 13. Future work will include efficiency and THD

TABLE I
EXPERIMENTAL SiC SUB-INVERTER SPECIFICATIONS

Parameter	Value	Unit
Input voltage, V_{in}	400	V_{DC}
Output voltage, V_{ac}	240	V_{RMS}
Peak output power	667	W
Input voltage ripple	15	V_{pp}
CEC efficiency	96.5	%
Peak efficiency (318.8 W)	97.9	%
Average voltage THD	4.8	%

optimization with three sub-inverters in parallel, with the option to use phase-shedding to optimize efficiency over the load range.

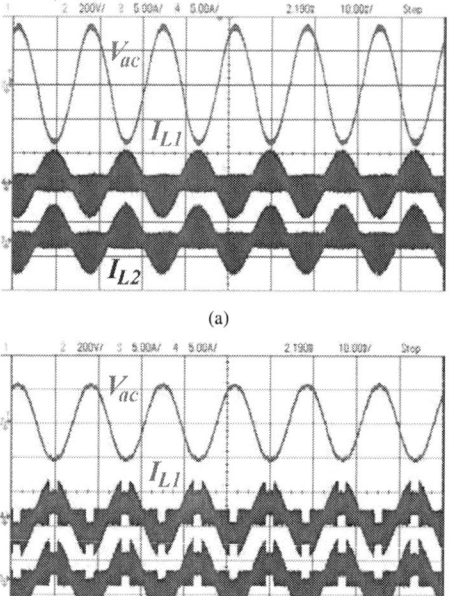

Fig. 13. Parallel sub-inverter operation in (a) pure BCM operation and (b) dual-mode operation.

V. CONCLUSIONS AND FUTURE WORK

A HCMC dual-mode scheme was presented and demonstrated. An efficiency of 97.7% is achieved for 632.7 W (95% of rated load) with 4.5% THD for a single SiC based sub-inverter. A weighted CEC efficiency of 96.5% and an average THD of 4.8%, are achieved for the load power between 100 W and 632 W. The original GaN based inverter has a volume of 33.1 in^3 and can achieve a theoretical power density of 60.3 W/in^3 while maintaining an enclosure temperature below 60°C, based on thermal simulations. Adapting the original GaN-based mechanical design to accommodate the larger SiC transistors results in an increased volume to 39.6 in^3, and a decreased power-density of 50.5 W/in^3. An optimized mechanical platform for the modular inverter was designed and assembled. Future work includes experimental realization of the full three parallel sub-inverter solution, as well as implementation of the active power decoupling circuit using the same SiC power-stage and inductors.

ACKNOWLEDGEMENTS

This work was supported by Solantro Semiconductor and the Ontario Centres of Excellence.

REFERENCES

[1] M. Acanski, J. Popovic-Gerber, and J. Ferreira, "Comparison of si and gan power devices used in pv module integrated converters," in *Energy Conversion Congress and Exposition (ECCE), 2011 IEEE*, Sept 2011, pp. 1217–1223.

[2] X. Huang, Z. Liu, Q. Li, and F. Lee, "Evaluation and application of 600 v gan hemt in cascode structure," *Power Electronics, IEEE Transactions on*, vol. 29, no. 5, pp. 2453–2461, May 2014.

[3] K. Shirabe, M. Swamy, J.-K. Kang, M. Hisatsune, Y. Wu, D. Kebort, and J. Honea, "Efficiency comparison between si-igbt-based drive and gan-based drive," *Industry Applications, IEEE Transactions on*, vol. 50, no. 1, pp. 566–572, Jan 2014.

[4] B. Whitaker, A. Barkley, Z. Cole, B. Passmore, T. McNutt, and A. Lostetter, "High-frequency ac-dc conversion with a silicon carbide power module to achieve high-efficiency and greatly improved power density," in *Power Electronics for Distributed Generation Systems (PEDG), 2013 4th IEEE International Symposium on*, July 2013, pp. 1–5.

[5] J. Biela, M. Schweizer, S. Waffler, and J. Kolar, "Sic versus si evaluation of potentials for performance improvement of inverter and dc dc converter systems by sic power semiconductors," *Industrial Electronics, IEEE Transactions on*, vol. 58, no. 7, pp. 2872–2882, July 2011.

[6] D. Han, A. Ogale, S. Li, Y. Li, and B. Sarlioglu, "Efficiency characterization and thermal study of gan based 1 kw inverter," in *Applied Power Electronics Conference and Exposition (APEC), 2014 Twenty-Ninth Annual IEEE*, March 2014, pp. 2344–2350.

[7] J. Choi, D. Tsukiyama, Y. Tsuruda, and J. Rivas, "13.56 mhz 1.3 kw resonant converter with gan fet for wireless power transfer," in *Wireless Power Transfer Conference (WPTC), 2015 IEEE*, May 2015, pp. 1–4.

[8] L. Liu, H. Li, Y. Zhao, X. He, and Z. Shen, "1 mhz cascaded z-source inverters for scalable grid-interactive photovoltaic (pv) applications using gan device," in *Energy Conversion Congress and Exposition (ECCE), 2011 IEEE*, Sept 2011, pp. 2738–2745.

[9] J.-T. Su and C.-W. Liu, "A novel phase-shedding control scheme for improved light load efficiency of multiphase interleaved dc-dc converters," *Power Electronics, IEEE Transactions on*, vol. 28, no. 10, pp. 4742–4752, Oct 2013.

[10] A. Costabeber, P. Mattavelli, and S. Saggini, "Digital time-optimal phase shedding in multiphase buck converters," *Power Electronics, IEEE Transactions on*, vol. 25, no. 9, pp. 2242–2247, Sept 2010.

[11] "GaN Systems Product Datasheet," GaN Systems, available at http://www.gansystems.com/gs66508p.php.

[12] "CREE Product Datasheet," CREE, available at http://www.wolfspeed.com/ /media/Files/Cree/Power/Data

[13] R. Fernandes and O. Trescases, "A multi-mode 1 mhz pfc front end with digital peak current modulation," *Power Electronics, IEEE Transactions on*, vol. PP, no. 99, pp. 1–1, 2015.

[14] D. Stone and B. Chambers, "Effect of spread-spectrum modulation of switched mode power converter pwm carrier frequencies on conducted emi," *Electronics Letters*, vol. 31, no. 10, pp. 769–770, May 1995.

[15] H. Hu, S. Harb, N. Kutkut, I. Batarseh, and Z. Shen, "Power decoupling techniques for micro-inverters in pv systems-a review," in *Energy Conversion Congress and Exposition (ECCE), 2010 IEEE*, Sept 2010, pp. 3235–3240.

[16] S. Kjaer, J. Pedersen, and F. Blaabjerg, "A review of single-phase grid-connected inverters for photovoltaic modules," *Industry Applications, IEEE Transactions on*, vol. 41, no. 5, pp. 1292–1306, Sept 2005.

[17] H. Hu, S. Harb, N. Kutkut, I. Batarseh, and Z. Shen, "A review of power decoupling techniques for microinverters with three different decoupling capacitor locations in pv systems," *Power Electronics, IEEE Transactions on*, vol. 28, no. 6, pp. 2711–2726, June 2013.

[18] B. Pierquet and D. Perreault, "A single-phase photovoltaic inverter topology with a series-connected energy buffer," *Power Electronics, IEEE Transactions on*, vol. 28, no. 10, pp. 4603–4611, Oct 2013.

[19] T. K. Gachovska, G. Scarlatescu, T. Lipan, A. Chung, C. Gerolami, R. Orr, N. Radimov, T. Reinberger, M. Varlan, and S. Dell'Aera, "High density control circuit integration for low cost grid tied inverter," in *Energy Conversion Congress and Exposition (ECCE), 2015 IEEE*, Sept 2015, pp. 466–469.

Reactive Power Compensation with Improvement of Current Waveform Quality for Single-Phase Buck-Type Dynamic Capacitor

Xinwen Chen, Ke Dai, Chen Xu, Ziwei Dai, Li Peng

State Key Laboratory of Advanced Electromagnetic Engineering and Technology

Huazhong University of Science and Technology

1037 Luoyu Road, Wuhan 430074, China

chenxw_hust@sina.cn, daike@hust.edu.cn, xuchen_hust@sina.com, ziweidai@foxmail.com, pe105@hust.edu.cn

Abstract—Due to relatively low cost and flexible performance, Dynamic Capacitor (D-CAP) is often designed to implement dynamic reactive power compensation, whose output current might distort under the effects of background harmonic voltage from the grid or non-ideal PWM mode caused by non-linear switch characteristics. This paper presents a reactive power compensation control strategy with waveform quality amelioration of output capacitive current for single-phase Buck-type D-CAP. Through establishing basic control equations, reactive power compensation principle and equivalent model of single-phase Buck-type D-CAP are analyzed. In order to suppress output harmonic current, duty ratio of Buck-type D-CAP is generated by means of introducing Even Harmonic Modulation (EHM). Then a selective current control strategy in multiple synchronous reference frames is adopted to track reactive power compensation current reference with nearly zero steady-state error and alleviate output current distortion with demanded harmonic components. Finally, a series of experimental results from single-phase Buck-type D-CAP laboratory prototype are provided to verify the validity of dynamic reactive power compensation and almost sinusoidal output current waveform optimizing method.

Keywords—Buck-type dynamic capacitor; reactive power compensation; current waveform improvement; even harmonic modulation; selective current tracking control.

I. INTRODUCTION

Due to increasing residential and industrial electricity demand, various single- or three-phase reactive power loads such as air-conditioning, electric-motor and transformer are connected to power system. These loads consume large amounts of reactive power, leading to poor power factor and incremental system losses. At the same time, reactive current flow through inductive transmission line would cause a drop of the voltage from the Point of Common Coupling (PCC) to power ground and even make it below threshold value [1][2]. In order to solve these problems, reactive power compensation is vitally important to steady and reliable operation of whole power system, which could be mainly implemented by static and dynamic compensation technologies [3][4].

Fixed or mechanically switched power capacitors are usually installed in parallel through grid system for power factor correction, which could not meet rapidly changing reactive power demand derived from the frequent variation of loads [5]. Some advanced compensators based on DC/AC inverter such as STATCOM [6]-[8] could realize dynamic reactive power compensation, but their reliabilities are mainly influenced on DC-link electrolytic capacitors, which should be replaced periodically with large maintenance costs. Compared with above compensators, dynamic capacitor, which is designed through renovating power capacitor with direct AC/AC converter, is a simple and economical solution for dynamic reactive compensation [9]-[11]. D-CAP could effectively compensate reactive current based on the grid current detection [12]-[14]. However, unwanted harmonic components from D-CAP output current might be generated under the action of background harmonic voltage of grid voltage or voltage drop of switches, which even causes system stability problem due to series or parallel harmonic resonance. This paper presents a single-phase Buck-type dynamic capacitor to generate desired reactive power compensation current by means of inductive load current (or given reactive reference current) tracking control and suppress itself harmonic current components simultaneously. The basic circuit and principle of Buck-type D-CAP is analyzed in section II. And corresponding control system is also designed to implement reactive power compensation. In section III, a reactive power compensation control with the selective mitigation of output current distortion is proposed, which is realized in multiple synchronous reference frames for zero steady-state current tracking error. A series of laboratory results from single-phase Buck-type D-CAP are presented to validate the effectiveness of proposed reactive power compensation control mothed in Section IV. Finally, concise conclusions are given in Section V.

II. CIRCUIT PRINCIPLE AND BASIC CONTROL METHOD OF BUCK-TYPE D-CAP

A. Principle of Buck-type D-CAP

Basic circuit structure of Buck type D-CAP is shown in Fig. 1(a), which includes mainly power capacitor C and Buck type Thin AC converter. Power capacitor C is in series with a little buffer inductor L_B so that inrush current of power capacitor could be limited when switches are turned on or off. Front-end filter is used to suppress high-order harmonic components of D-CAP output current, which is composed of filtering inductor L_F and filtering capacitor C_F. When D-CAP works normally, the switches S_1 and S_2 are turned on complementarily. Two basic operating states of switch circuit are shown in Fig. 1(b), where duty ratios of switches S_1 and S_2 are D and $1-D$ respectively. Due to forward and reverse flows of capacitor current, S1 and S2 should be designed as bidirectional switches. As shown in Fig. 1(c), bidirectional switch could be composed of two common-collector or common-emitter IGBTs in reversed series. In this paper, bidirectional switch S_1 is composed of two common-collector IGBTs in series while S_2 is made up of two common-emitter IGBTs in series.

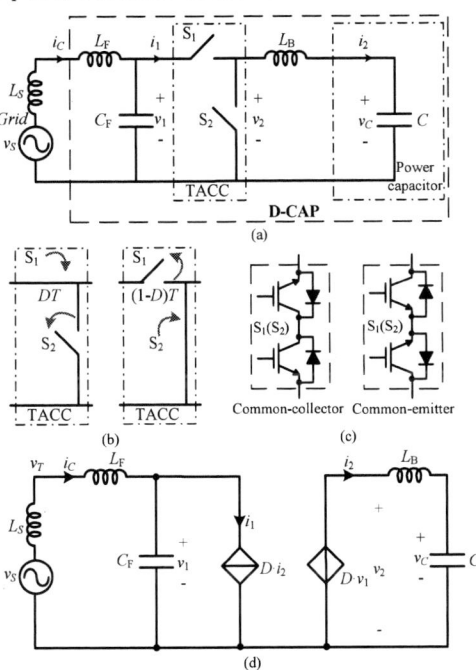

Fig. 1. Single-phase Buck-type D-CAP (a) Main circuit (b) Two operating states of switch circuit (c) Bidirectional switches (d) Equivalent circuit model

In Fig. 1(a), it is defined that $i_C(t)$ is D-CAP output current, $v_1(t)$ is the voltage of filtering capacitor C_F, $i_1(t)$ is converter-side current flowing through the switch S_1, $D(t)$ is duty ratio of S_1, $v_2(t)$ is port voltage of the branch composed of buffer inductor L_B and power capacitor C, and $i_2(t)$ is the current of power capacitor C. In order to analyze the principle of Buck type D-CAP, some assumptions are made as follow.

1) Switching frequency is high enough, so the values of all voltage and current quantities could be represented with their averages in a switching period.

2) $L_B C$ and $L_F C_F$ filters is regarded as ideal low-pass filter, so it is reasonable that capacitor voltage $v_C(t) \approx v_2(t)$ and D-CAP output current $i_C(t) \approx i_1(t)$ within their passband.

According to the circuit principle of Buck-type converter, it is reasonable to deduce corresponding relation equations.

$$v_2(t) = D(t)v_1(t) \tag{1}$$

$$i_2(t) = C\frac{dv_2(t)}{dt} = Cv_1(t)\frac{dD(t)}{dt} + D(t)C\frac{dv_1(t)}{dt} \tag{2}$$

$$i_1(t) = D(t)i_2(t) = Cv_1(t)D(t)\frac{dD(t)}{dt} + D^2(t)C\frac{dv_1(t)}{dt} \tag{3}$$

According to (1)-(3), equivalent circuit model of Buck type D-CAP could be described in Fig. 1(d). Port voltage $v_2(t)$ is regarded as controlled voltage source while converter-side current $i_1(t)$ represents controlled voltage source. If $D(t)$ is equivalent to a constant D_0 in every switching period, thus $D(t)$ could be decomposed into a sum of constant term and sinusoidal terms by means of Fourier series expansion.

$$D(t) = D_0 + \sum_{n=1,3,5}^{\infty} \frac{2}{n\pi}\sin(D_0 n\pi)\cos(n\omega_s t) \tag{4}$$

Where ω_s is switching angular-frequency. According to (1)-(4), Port voltage $v_2(t)$, capacitor current $i_2(t)$ and converter-side current $i_1(t)$ in Fig. 1(d) could be further expressed respectively.

$$v_2(t) = D_0 v_1(t) + \sum_{n}^{\infty} \frac{2\sin(D_0 n\pi)}{n\pi}\cos(n\omega_s t)v_1(t) \tag{5}$$

$$i_2(t) = D_0 C\frac{dv_1(t)}{dt} - \sum_{n}^{\infty} \frac{2\omega_s C\sin(D_0 n\pi)}{\pi}v_1(t)\sin(n\omega_s t)$$
$$+ \sum_{n}^{\infty} \frac{2C\sin(D_0 n\pi)}{n\pi}\cdot\cos(n\omega_s t)\cdot\frac{dv_1(t)}{dt} \tag{6}$$

$$i_1(t) = D_0^2 C\frac{dv_1(t)}{dt} - \sum_{n}^{\infty} \frac{2D_0\omega_s C\sin(D_0 n\pi)}{\pi}v_1(t)\sin(n\omega_s t)$$
$$+ \sum_{n}^{\infty} \frac{4D_0 C\sin(D_0 n\pi)}{n\pi}\cdot\cos(n\omega_s t)\cdot\frac{dv_1(t)}{dt}$$
$$-[\sum_{n}^{\infty} \frac{2\sin(D_0 n\pi)}{n\pi}\cos(n\omega_s t)] \tag{7}$$
$$\times[\sum_{n}^{\infty} \frac{2\omega_s C\sin(D_0 n\pi)}{\pi}v_1(t)\sin(n\omega_s t)$$
$$+ \sum_{n}^{\infty} \frac{2C\sin(D_0 n\pi)}{n\pi}\cdot\cos(n\omega_s t)\cdot\frac{dv_1(t)}{dt}]$$

It could be indicated from (7) that converter-side current $i_1(t)$ includes harmonic components whose frequencies are around switching frequency or odd times of switching frequency. Because cut-off frequency of $L_F C_F$ low-pass filter is far lower than switching frequency, D-CAP output current $i_C(t)$ could be obtained through eliminating high-order harmonic components of converter-side current $i_1(t)$.

$$i_C(t) = D_0^2 C\frac{dv_1(t)}{dt} \tag{8}$$

According to (8), Buck-type D-CAP could be equivalent to controlled capacitor $D_0^2 C$ when duty ratio $D(t)$ is adjusted as a constant in every switching period. So Buck-type D-CAP could realize dynamic reactive power compensation. Generally, assuming that converter-side current i_1 would entirely flow into grid system within passband of $L_F C_F$ low-pass filter, so D-CAP output current $i_C(t) = i_1(t)$ at any duty ratio $D(t)$.

$$i_C(t) = Cv_1(t)D(t)\frac{dD(t)}{dt} + D^2(t)C\frac{dv_1(t)}{dt} \tag{9}$$

B. Basic Control System of Buck-type D-CAP

Based on above principle analysis, Buck-type D-CAP could compensate reactive power dynamically for power factor correction by means of adjusting duty ratio $D(t)$. Control

978-1-4673-9551-9/16 $31.00 © 2016 IEEE

system of D-CAP for reactive power compensation is shown in Fig. 2. A PI controller in *dq* synchronous rotation frame of fundamental frequency is used to track reactive current reference with zero steady-state error. In order to obtain *α*-axis and *β*-axis components of reference current i_C^* and feedback current i_C in *αβ* stationary reference frame, i_C^* and i_C are directly used to generate *α*-axis components $i_{C\alpha}^*$ and $i_{C\alpha}$ respectively, and meanwhile *β*-axis components $i_{C\beta}^*$ and $i_{C\beta}$ are extracted respectively from i_C^* and i_C through time-delay of a quarter of fundamental period *T*/4. Then corresponding current components in *αβ* stationary reference frame would be transformed to obtain q-axis reactive components i_{Cq}^* and i_{Cq} in *dq* synchronous rotation frame, which are regulated by using PI controller to generate duty ratio reference *D*.

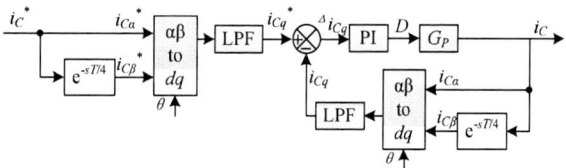

Fig. 2. Control system of Buck-type D-CAP for reactive power compensation

III. REACTIVE POWER COMPENSATION CONTROL WITH IMPROVEMENT OF OUTPUT CURRENT WAVEFORM QUALITY

A. Harmonic Current Compensation Based on Even Harmonic Modulation of Duty Ratio

Due to background harmonic voltage from the grid and voltage drop of switches, D-CAP output current $i_C(t)$ would include low-order odd harmonic current components, which should be suppressed to improve its waveform quality. Based on Even Harmonic Modulation (EHM), when duty ratio *D*(t) includes even harmonic component, D-CAP could generate controlled harmonic current in order to offset itself corresponding harmonic current component. For example, if input voltage $v_1(t)=V_m\sin(\omega t+\varphi_1)$ and duty ratio $D(t)=D_0+D_2\sin(2\omega t+\square_2)$, output current $i_C(t)$ could be deduced according to (9) as follow.

$$
\begin{aligned}
i_C(t) =\ & \omega C V_m \left(D_0^2 + D_2^2/2 \right) \sin\left(\omega t + \varphi_1 + \pi/2 \right) \\
& + \omega C V_m\, 2 D_0 D_2 \sin\left(3\omega t + \phi_2 + \varphi_1 \right) \\
& + \omega C V_m \frac{1}{4} D_2^2 \sin\left(3\omega t + 2\phi_2 - \varphi_1 + \pi/2 \right) \\
& + \omega C V_m \frac{3}{4} D_2^2 \sin\left(5\omega t + 2\phi_2 + \varphi_1 - \pi/2 \right)
\end{aligned} \quad (10)
$$

When 2nd harmonic component is added to duty ratio *D*(t), D-CAP output current $i_C(t)$ would include 3th and 5th harmonic components. Moreover, 2nd harmonic component of duty ratio *D*(t) mainly influences on 3th harmonic component of D-CAP output current $i_C(t)$. Generally, when duty ratio $D(t)=D_0+D_n\sin(n\omega t+\square_n)$, (*n*=2,3,4,...), D-CAP output current $i_C(t)$ could be deduced in (11). According to (11), when duty ratio *D*(t) has *n* order harmonic component, D-CAP output current $i_C(t)$ would include *n*-1, *n*+1, 2*n*-1 and 2*n*+1 order harmonic currents at the same time. Moreover, *n*+1 order harmonic component of D-CAP output current $i_C(t)$ could obtain maximum amplitude through duty ratio modulation including *n* order harmonic component.

$$
\begin{aligned}
i_C(t) =\ & \omega C V_m \left(D_0^2 + D_n^2/2 \right) \sin\left(\omega t + \varphi_1 + \pi/2 \right) \\
& + \omega C V_m \frac{n-2}{2} D_0 D_n \sin[(n-1)\omega t + \phi_n - \varphi_1 + \pi] \\
& + \omega C V_m \frac{n+2}{2} D_0 D_n \sin[(n+1)\omega t + \phi_n + \varphi_1] \\
& + \omega C V_m \frac{n-1}{4} D_n^2 \sin[(2n-1)\omega t + 2\phi_n - \varphi_1 + \pi/2] \\
& + \omega C V_m \frac{n+1}{4} D_n^2 \sin[(2n+1)\omega t + 2\phi_n + \varphi_1 - \pi/2]
\end{aligned} \quad (11)
$$

Therefore, in order to suppress *n*+1 order harmonic component of D-CAP output current $i_C(t)$, it need to regulate *n* order harmonic component of duty ratio *D*(t), thus D-CAP could generate controlled *n*+1 order harmonic current with the same amplitude and reversed phase of compensated harmonic current.

B. Proposed Reactive Power Compensation Control Strategy

System configuration diagram with single-phase Buck-type D-CAP is shown in Fig. 3(a). D-CAP is designed to generate reactive power compensation current and suppress its own harmonic current simultaneously. Control system of D-CAP is given in Fig. 3(b), in which a selective current tracking control is implemented in multiple synchronous reference frames. In order to obtain reference current components in *αβ* stationary reference frame, inductive load current i_L (or given reactive reference current i_C^*) is directly used to generate *α*-axis reference current components and *β*-axis reference current component is extracted from i_L (or i_C^*) through time-delay of a quarter of fundamental period *T*/4. Meanwhile, D-CAP output current i_C is directly used to generate *α*-axis feedback current component and *β*-axis feedback current component is also extracted from i_C through time-delay of a quarter of fundamental period *T*/4. Then *α*-axis and *β*-axis reference current components from i_L (or i_C^*) are transformed to fundamental reference frame through $C_{2s/2r}$ transform in order to extract reactive reference current component i_{Cq1}^*. In addition, *α*-axis and *β*-axis feedback current components from i_C^* are also transformed to fundamental reference frame through $C_{2s/2r}$ transform in order to extract reactive feedback current component i_{Cq1}. Thus fundamental duty ratio D_0 would be generated through the regulation of PI$_1$ controller. Similarly, *α*-axis and *β*-axis feedback current components from i_C is also transformed to hth harmonic reference frame to generate hth harmonic current components i_{Cdh} and i_{Cqh}, which would be adjusted by PI$_h$ controller in order to obtain (*h*-1)th harmonic duty ratio D_{h-1} through *h*-1 order reverse $C_{2s/2r}$ transform. Finally, all duty ratio references are added to obtain total duty ratio reference *D*. Thus D-CAP output current i_C could be regulated to track reactive reference current and alleviate itself distortion simultaneously. Transformation matrix from *αβ* stationary reference frame to *dq* synchronous rotation reference is as bellow.

$$
C_{2s/2r} = \begin{bmatrix} (-1)^n \sin(h\omega t) & -\cos(h\omega t) \\ \cos(h\omega t) & (-1)^n \sin(h\omega t) \end{bmatrix} \quad (12)
$$

Where *n*=0 at *h*=4*m*-3 while *n*=1 at *h*=4*m*-1 (*m* is positive integer).

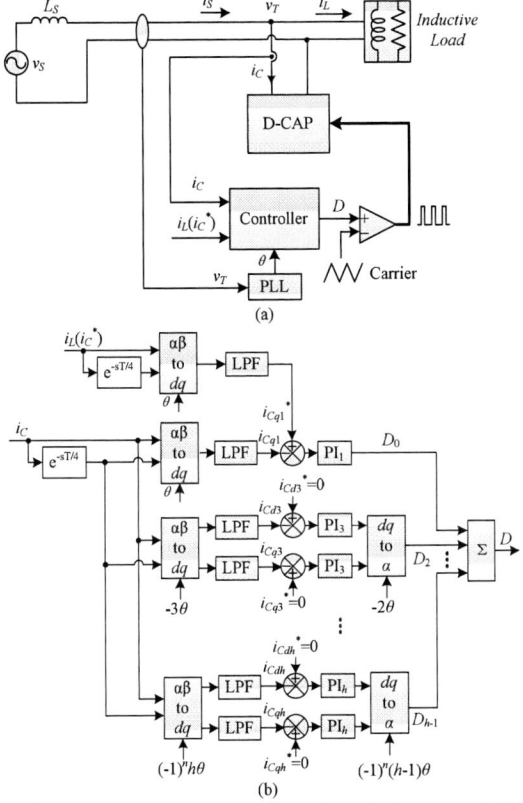

Fig. 3. Single-phase power system configuration with Buck-type D-CAP (a) Total system structure (b) Control system of Buck-type D-CAP

IV. EXPERIMENT RESULTS

A laboratory test setup with single-phase Buck-type 11kVar D-CAP has been carried out to verify the effectiveness of proposed reactive power compensation control strategy. Total system configuration is shown in Fig. 3 and system parameters are given in Table I. $L_F C_F$ low-pass filter is designed to suppress harmonic components of D-CAP output current $i_C(t)$ around switching frequency, and its cut-off frequency is about 0.1-0.5 times of switching frequency. Buffer inductor L_B is designed to decrease the ripple of power capacitor current i_2, which should be less than 20% of rated power capacitor current. D-CAP is operated with open-loop or closed-loop state of output current.

Table I System parameters

AC Grid	Grid voltage v_S	220V
	Grid frequency f_S	50Hz
	Grid inductance L_S	300μH
D-CAP	Buffer inductance L_B	160μH
	Filtering capacitor C_F	80μF
	Filtering inductance L_F	180μH
	Power capacitor C	660μF
	Switching frequency f_w	9.6kHz

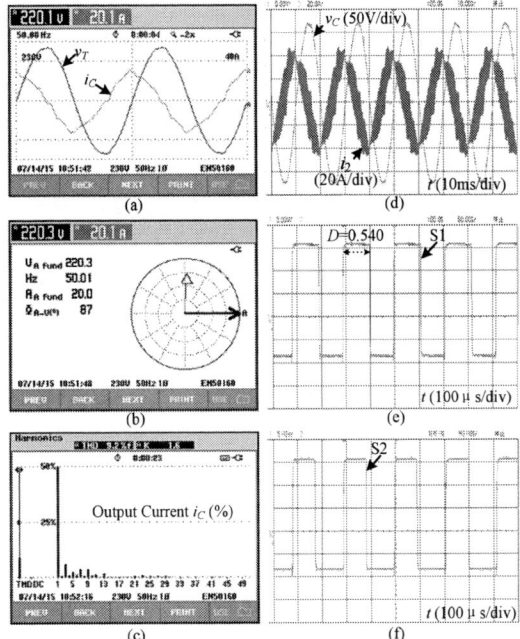

Fig. 4. Steady-state laboratory results when D-CAP is in open-loop operation with D=0.54

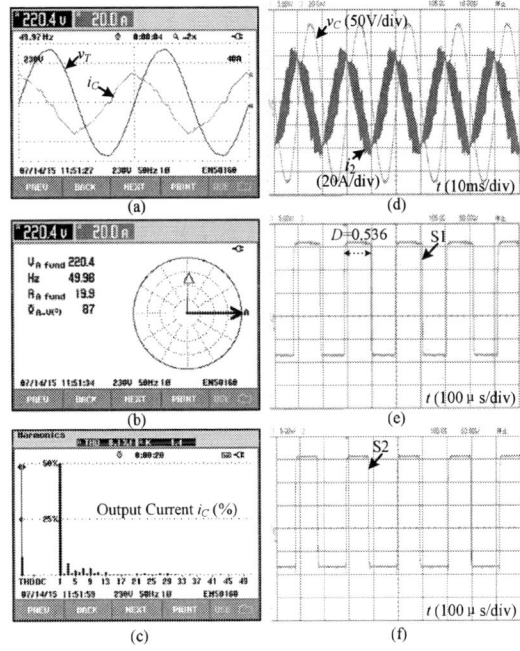

Fig. 5. Steady-state laboratory results at i_C^*=20A when D-CAP is in closed-loop operation to generate only reactive compensation current

When duty ratio D is a constant 0.54 in every switching period, Fig.4 shows that D-CAP would generate about 20.1A capacitive output current i_C, whose phase-angle is leading 87° of PCC voltage v_T. However, its output current has been distorted, whose THD is about 9.2%. When D-CAP is in closed-loop operation to only generate 20A capacitive reactive current, Fig.5 indicates that output current still includes 3th, 5th, 7th and 9th low-order harmonic components. Its THD is 8.1%, which could not meet the demand of power quality standards.

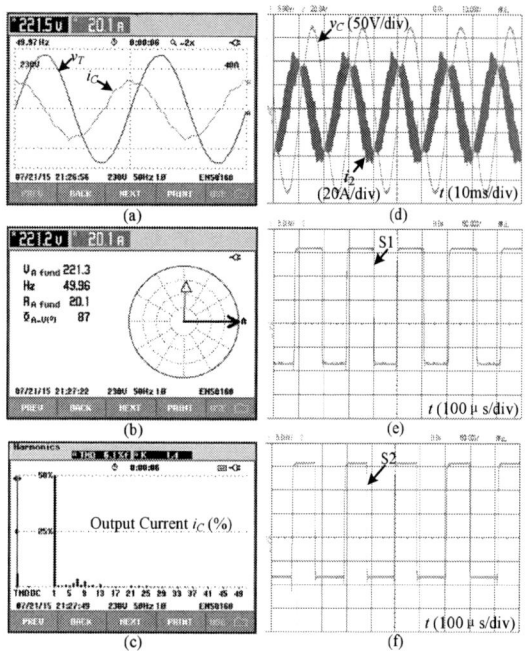

Fig. 6. Steady-state laboratory results at i_C^*=20A when D-CAP generates compensation current and suppresses 3th and 5th harmonic components to improve current waveform quality simultaneously

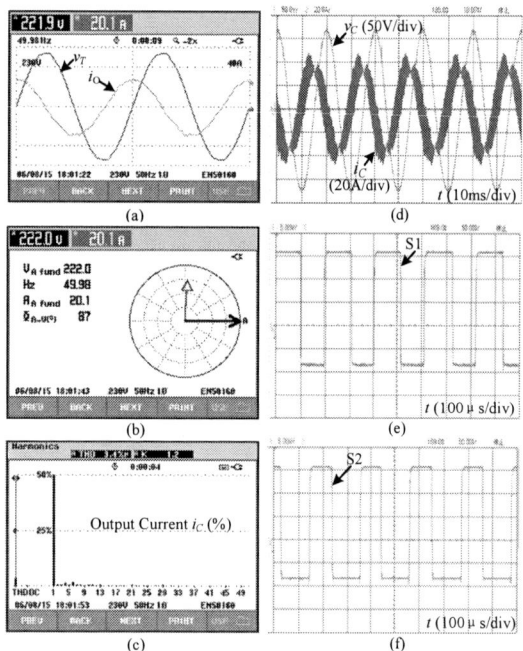

Fig. 7. Steady-state laboratory results at i_C^*=20A when D-CAP generates compensation current and suppresses all 3th-13th characteristic harmonics to improve current waveform quality simultaneously

Then D-CAP is operated to generate given 20A capacitive reactive current and suppress itself characteristic harmonic current components simultaneously. Fig. 6 reveals that THD of output current i_C is reduced to 6.1% when 3th and 5th harmonic currents are suppressed, while its THD is down to 3.4% in Fig.7 when all 3th-13th characteristic harmonic currents are suppressed further. Meanwhile, Fig.7 shows that steady-state error of D-CAP output current is about 1%. Therefore, proposed control strategy has excellent performance for both reactive power compensation and improvement of output current waveform quality. When a step change of reference current i_C^* occurs, dynamic experiment results are presented in Fig. 8. Reference current i_C^* changes from 0A to 20A in Fig. 8(a) while reference current i_C^* changes from 20A to 0A in Fig. 8(b). When D-CAP does not work, only front-end filter $L_F C_F$ is connected in power grid and about 5A output current is generated due to filtering capacitor C_F. No matter reference current i_C^* changes from 0A to 20A or from 20A to 0A, Fig. 8 reveals that output current i_C would reaches steady state in almost two fundamental periods (40ms).

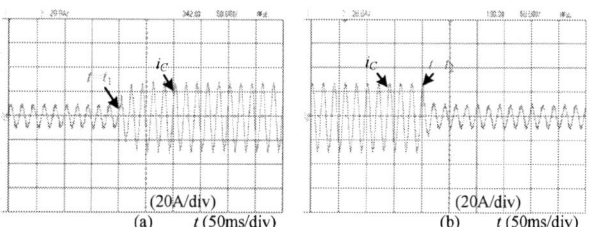

Fig. 8. Dynamic laboratory results under step change of i_C^* (a) from 0A to 20A (b) from 20A to 0A

V. CONCLUSION

This paper focuses on reactive power compensation control strategy with improvement of output current waveform quality for Buck-type D-CAP. Basic principle analysis infers that D-CAP could be equivalent to continuously adjustable capacitor when duty ratio ranges from 0 to 1. Thus dynamic reactive power compensation for D-CAP could be implemented by means of adjusting duty ratio. Due to background harmonic voltage from the grid and non-ideal PWM mode caused by voltage drop of switches, D-CAP output current would include some low-order harmonic components, which could not meet the demand of power quality standards. Through reactive power compensation control with selective improvement of output current waveform quality, D-CAP output current distortion could be greatly mitigated. A lot of experimental results reveal that proposed Buck-type D-CAP could generate desired capacitive compensation current, whose low-order harmonic current components have been suppressed. Meanwhile, steady-state error of output current could also be limited to about 1%.

In future studies, Buck type D-CAP could also be used for three-phase power system to compensate reactive current dynamically. Furthermore, if a series of corresponding even harmonic components are added to duty ratio $D(t)$, D-CAP would be used to compensate harmonic currents from nonlinear load and damp harmonic resonance in power system simultaneously.

ACKNOWLEDGMENT

This work is a part of the project "Multimode Resonance Mechanism and Corresponding Multifunction Active Damping Control Technique for Power Electronic Hybrid Systems" (51277086), sponsored by National Natural Science Foundation of China. The authors would like to thank the support of National Natural Science Foundation of China.

REFERENCES

[1] H. Fujita, and H. Akagi, "Voltage-regulation performance of a shunt active filter intended for installation on a power distribution system," *IEEE Trans. Power Electron.*, vol. 22, no. 3, pp. 1046-1053, May 2007.

[2] R. Majumder, "Reactive power compensation in single-phase operation of microgrid," *IEEE Trans. Ind. Electron.*, vol. 60, no. 4, pp. 1403-1416, Apr. 2013.

[3] *IEEE Guide for Application of Shunt Power Capacitors*, IEEE Std. 1036, 2010.

[4] A. Prasai, J. Sastry, and D. Divan, "Dynamic Var/harmonic compensation with inverter-less active filters," in *Proc. IEEE Ind. Appl. Socie. Annu. Meeting*, Edmomton, Alberta, Canada, Oct. 2008, pp. 1–6.

[5] P. Ladoux, A. Lowinsky, P. Marino, and G. Raimondo, "Reactive power compensation in railways using active impedance concept," in *Proc. SPEEDAM Conf.*, Jun. 2010, pp. 1362-1367.

[6] Q. Song, W. Liu, and Z. Yuan, "Multilevel optimal modulation and dynamic control strategies for STATCOMs using cascaded multilevel inverters," *IEEE Trans. Power Del.*, vol. 22, no. 3, pp. 1937–1946, Jul. 2007.

[7] B. Singh and J. Solanki, "A comparison of control algorithms for DSTATCOM," *IEEE Trans. Ind. Electron.*, vol. 56, no. 7, pp. 2738-2745, Jul. 2009.

[8] H. Mohammadi P. and M. Tavakoli B., "A Transformer-less medium-voltage STATCOM topology based on extended modular multilevel converters," *IEEE Transactions on Power Electronics*, vol. 26, no. 5, pp. 1534-1545. May. 2011.

[9] H. Akagi, "Active harmonic filters," *Proc. IEEE*, vol. 93, no. 12, pp. 2128–2141, Sep./Oct. 2005.

[10] D. Divan and J. Sastry, "Inverter-less active filters—A new concept in harmonic and VAR compensation," in *Proc. IEEE PESC*, Jun. 2007, pp. 2926–2932.

[11] F. Dijkhuizen, and M. Godde, "Dynamic capacitor for HV applications," in *Proc. IEEE ECCE*, Sep. 2010, pp. 1511-1518.

[12] Q. Liu, Y. Deng, X. He, "A novel AC-AC shunt active power filter without large energy storage elements," in *Proc. EPE Conf.*, Aug. 2011, pp. 2016-2024.

[13] A. Prasai, J. Sastry, and D. M. Divan, "Dynamic capacitor (D-CAP): an integrated approach to reactive and harmonic compensation," *IEEE Trans. Ind. Appl.*, vol. 46, no. 6, pp. 2518-2525, Nov./Dec. 2010.

[14] A. Prasai, and D. M. Divan, "Control of dynamic capacitor," *IEEE Trans. Ind. Appl.*, vol. 47, no. 1, pp. 161-168, Jan. 2011.

Circulating Current Reduction for a D-Σ Digital Controlled Transformerless UPS

T.-F. Wu, T.-H. Shiu, P.-H. Lin, L.-C. Lin, and [#]J.-W. Huang

Elegant Power Electronic Applied Research
Laboratory (EPEARL)
Department of Electrical Engineering
National Tsing Hua University
Hsinchu, Taiwan.
Email: tfwu@ee.nthu.edu.tw

[#]Green Energy and Environment Research
Laboratories
System Application Department,
Photovoltaic Technology Division
Industrial Technology Research Institute
Hsinchu, Taiwan.

Abstract — This paper presents a three-phase transformerless uninterruptible power supply (UPS) with sinusoidal pulse width modulation (SPWM) based division-summation (D-Σ) digital control. A transformerless UPS controls the power flow between dc link and utility grid, as well as tracks the ac reference voltage. The proposed control law derived with D-Σ digital approach takes into account the effects of dc-link voltage fluctuation, grid-voltage distortion and inductance variation due to different current levels. Thus, distortion of input current and filter inductor core size can be reduced significantly. However, circulating current may flow through the common ground between the input power factor corrector (PFC) and the output three-phase four-wire inverter. The derived control law based on SPWM can suppress this circulating current and regulate output voltages tightly. Experimental results measured from a three-phase transformerless UPS have confirmed the analysis and discussion of the proposed control approach.

Index Term — *D-Σ digital control, circulating current, transformerless UPS*

I. INTRODUCTION

With highly developed technology and utilization of sophisticated equipment, power quality becomes more and more important. A UPS is used as a backup power and supplies reliable and high quality power to critical loads. Three-phase inverter topologies are typically employed for a UPS to generate low-distortion sinusoidal voltage when there is a power failure [1]. Thus, the control of a UPS inverter plays a dominant role in reaching high power reliability and quality.

Three-phase three-wire inverters are suitable for supplying balanced loads and their control algorithms are usually based on two-dimensional space vector pulse width modulation (2-D SVPWM). However, there is no path for neutral current in this three-wire topology; therefore it cannot supply unbalanced loads. Besides, 2-D SVPWM cannot be applied to a three-phase four-wire inverter. Thus, three-dimensional space vector pulse width modulation (3-D SVPWM) was presented to deal with this limitation [2]-[5]. Moreover, conventional *abc* to *dq* frame transformation will yield inaccurate re-

sults due to grid-voltage harmonics and variable inductance. When inductance varies with current, it might result in instability at high power applications. The D-Σ digital control based three-phase four-wire inverter with load impedance estimation and repetitive control [6], [7] therefore was introduced to overcome the aforementioned limitations. Besides, it can achieve fast response and low harmonic distortion, and eliminate steady-state error [8], [9]. In this paper, the D-Σ digital control [9], [10], is extended to a transformerless UPS application. The proposed SPWM based D-Σ digital control can track sinusoidal reference current in each phase independently and eliminate circulating current significantly, which has not been discussed in literature.

II. D-Σ DIGITAL CONTROL

Fig. 1 shows a configuration of the studied tansformerless UPS, including a power stage and a control stage. Two-phase modulation (TPM) is generally considered to be a feasible control strategy because it can reduce switching loss and have better voltage utilization. However, it requires more complicated algorithms to obtain control laws and results in high circulating current. Therefore, the SPWM is adopted for its simple algorithm and to reduce circulating current. For an SPWM converter, one switching cycle is divided into several time intervals and each time interval is corresponding to a switching-state voltage. The average switching-state voltage over one switching cycle equals to the sum of all products of duty ratio and switching-state voltage, which simplifies control-law derivation. The D-Σ digital controller is then developed to cancel all the variation effects of the system parameters in the plant. The three-phase inverter shown in Fig. 1 can be equivalent to three individual single-phase inverters, in which the average circuit model of each inverter is shown in Fig. 2(a). By KVL, the control laws of the three-phase three-wire inverter can be expressed as follows:

$$D_{P,R}(n+1) = \frac{L_{P,R}\Delta i_{P,LR}(n+1)}{v_{DC}T_s} + \frac{v_{P,RN}(n+1)}{v_{DC}} + \frac{v_{P,NO}}{v_{DC}}, \quad (1)$$

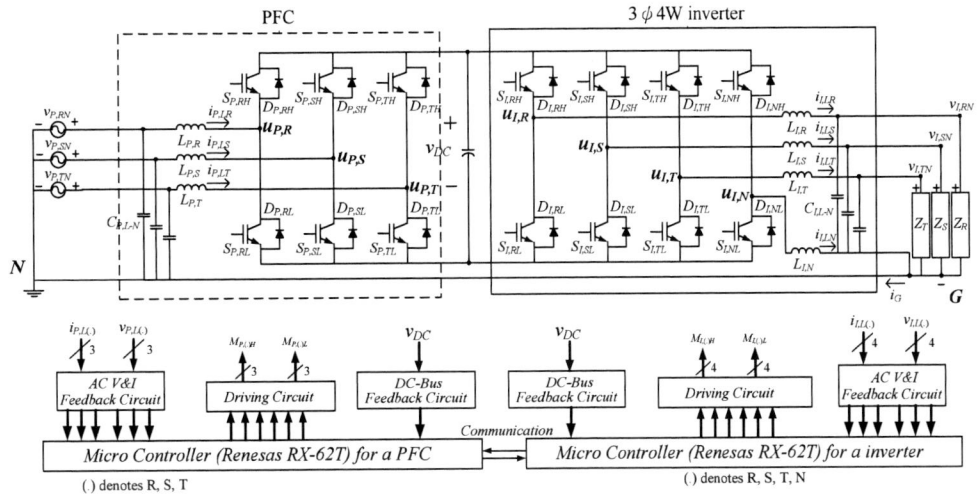

Fig. 1. Configuration of the studied tansformerless UPS.

$$D_{P,S}(n+1) = \frac{L_{P,S}\Delta i_{P,LS}(n+1)}{v_{DC}T_s} + \frac{V_{P,SN}(n+1)}{v_{DC}} + \frac{v_{P,NO}}{v_{DC}}, \quad (2)$$

and

$$D_{P,T}(n+1) = \frac{L_{P,T}\Delta i_{P,LT}(n+1)}{v_{DC}T_s} + \frac{V_{P,TN}(n+1)}{v_{DC}} + \frac{v_{P,NO}}{v_{DC}}. \quad (3)$$

Then, the three-phase voltages are divided into six sectors to derive the average voltage at the neutral point in each sector. Since the average voltages at the neutral point in six sectors are identical, this paper shows only the derivation of that in the first sector.

There are four switching states $V_0(0,0,0)$, $V_1(1,0,0)$, $V_2(1,1,0)$, and $V_7(1,1,1)$ in the first sector. The voltages at the neutral point in each switching state are shown below:

$$v_{nN0} = 0, \quad (4)$$

$$v_{nN1} = \frac{1}{3}v_{DC}, \quad (5)$$

$$v_{nN2} = \frac{2}{3}v_{DC}, \quad (6)$$

and

$$v_{nN7} = v_{DC}. \quad (7)$$

Thus, the three-phase quantities can be transformed into the d-q quantities by Park's transformation. According to volt-second balance, the time intervals T_0, T_1, T_2, and T_7 can be determined from the reference voltage $Ve^{j\theta}$ and they are shown as follows:

$$T_0 = \frac{T_S}{2v_{DC}}(v_{DC} - 2V\cos\theta), \quad (8)$$

$$T_1 = \frac{3VT_S}{2v_{DC}}(\cos\theta + \frac{\sqrt{3}}{3}\sin\theta), \quad (9)$$

$$T_2 = \frac{VT_S}{v_{DC}}(-\sqrt{3}\sin\theta), \quad (10)$$

and

$$T_7 = T_S - T_0 - T_1 - T_2. \quad (11)$$

As a result, the average voltage at the neutral point in the first sector can be obtained as

$$v_{NO} = \frac{1}{T_S}[T_1 \times v_{nN1} + T_2 \times v_{nN2} + T_7 \times v_{nN7}] = \frac{v_{DC}}{2}. \quad (12)$$

Therefore, the control laws of the three-phase three-wire inverter can be shown as follows:

$$D_{P,R}(n+1) = \frac{L_{P,R}\Delta i_{P,LR}(n+1)}{v_{DC}T_s} + \frac{v_{P,RN}(n+1)}{v_{DC}} + \frac{1}{2}, \quad (13)$$

$$D_{P,S}(n+1) = \frac{L_{P,S}\Delta i_{P,LS}(n+1)}{v_{DC}T_s} + \frac{V_{P,SN}(n+1)}{v_{DC}} + \frac{1}{2}, \quad (14)$$

and

$$D_{P,T}(n+1) = \frac{L_{P,T}\Delta i_{P,LT}(n+1)}{v_{DC}T_s} + \frac{V_{P,TN}(n+1)}{v_{DC}} + \frac{1}{2}, \quad (15)$$

where $\Delta i_{P,LX}(n+1)$ denotes the current variation over one switching period.

In order to derive the control laws for the three-phase four-wire inverter, the basic concept of the above discussion is adopted and it is shown in Fig. 2(b). The derivation based on KVL divides one switching period into four time intervals and each of which corresponds to a switching-state voltage. The average switching-state voltage over one switching period can be obtained consequently. In addition, the average voltage at the neutral point can be also obtained with the procedures mentioned above. Consequently, the control laws can be expressed as follows :

$$D_{I,R}(n+1) = \frac{L_{I,R}\Delta i_{I,LR}(n+1)}{v_{DC}T_s} + \frac{v_{I,RN,ref}(n+1)}{v_{DC}} + \frac{1}{2}, \quad (16)$$

$$D_{I,S}(n+1) = \frac{L_{I,S}\Delta i_{I,LS}(n+1)}{v_{DC}T_s} + \frac{v_{I,SN,ref}(n+1)}{v_{DC}} + \frac{1}{2}, \quad (17)$$

$$D_{I,T}(n+1) = \frac{L_{I,T}\Delta i_{I,LT}(n+1)}{v_{DC}T_s} + \frac{V_{I,TN,ref}(n+1)}{v_{DC}} + \frac{1}{2}, \quad (18)$$

and

$$D_{I,N}(n+1) = \frac{L_{I,N}\Delta i_{I,LN}(n+1)}{v_{DC}T_s} + \frac{1}{2}, \quad (19)$$

where $\Delta i_{I,LX}(n+1)$ denotes the current variation over one switching period.

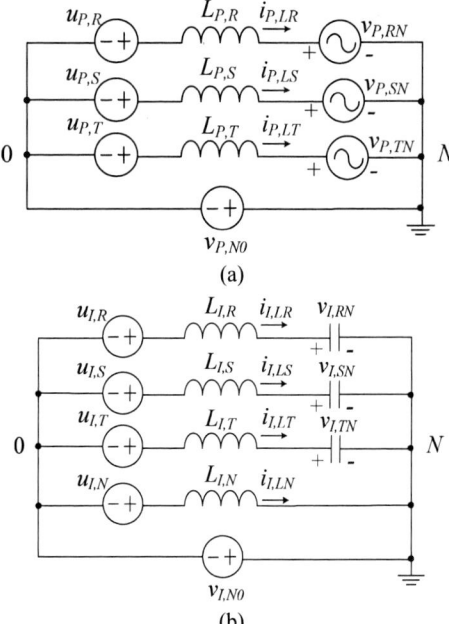

(a)

(b)

Fig. 2. Average circuit models of (a) a three-phase three-wire inverter, and (b) a three-phase four-wire inverter.

III. REDUCTION of CIRCULATING CURRENT

The configuration of a transformerless UPS with a common ground can comply with safety regulation, but it brings up a circulating-current issue, which is related to the following expression:

$$i_{I,LR} + i_{I,LS} + i_{I,LT} + i_{I,LN} = i_G. \tag{20}$$

If current i_G can be controlled to flow through the fourth leg of the $3\phi 4W$ inverter, circulating current will be zero. The current command $i_{I,LN,ref}$ of the fourth leg and the current error $\Delta i_{I,LN}$, therefore, are determined and expressed as follows:

$$i_{I,LN,ref} = -(i_{I,LR} + i_{I,LS} + i_{I,LT}), \tag{21}$$

and

$$\Delta i_{I,LN} = i_{I,LN,ref} - i_{I,LN,fb}. \tag{22}$$

Accordingly, a compensation for the current error can eliminate the circulating current except switching current ripple.

IV. SIMULATED AND EXPERIMENTAL RESULTS

In the simulation and experiment, the specifications and designed parameters used are collected in Table 1. Derivation of the average voltage at the neutral point was based on the fundamental grid-voltage only. However, in practice, there always exists harmonic voltage in the grid. To verify how the proposed control with feedback can take care of the grid harmonic voltages, several simulations with certain harmonic combinations were conducted. Fig. 3 and Fig. 4 show the simulated results of the three-phase three-wire inverter under the grid with different harmonic components. The simulated results of the three-phase four-wire inverter under the resistive load with different harmonic components are shown in Fig. 5

and Fig. 6. Table 2 and Table 3 summarize the simulated results with more test conditions. It can be observed that even with high grid-voltage THD, the input inductor current and output voltage still can be controlled with low THD components ($<3\%$).

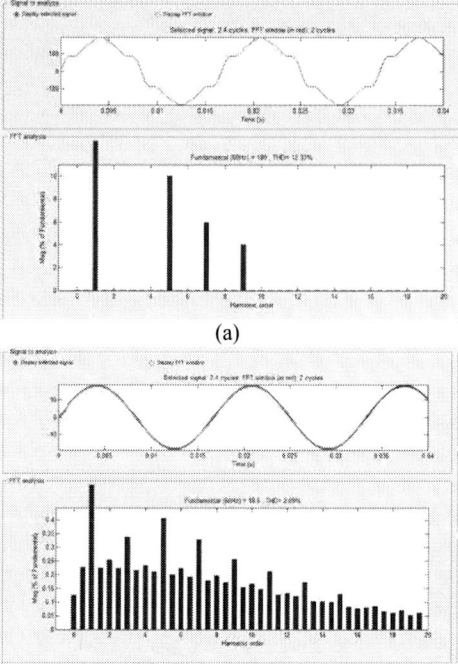

(a)

(b)

Fig. 3. (a) Simulated waveform and FFT analysis of grid voltage containing: $V_{5th} = 10\%$, $V_{7th} = 6\%$ and $V_{9th} = 4\%$, and (b) those of the input inductor current.

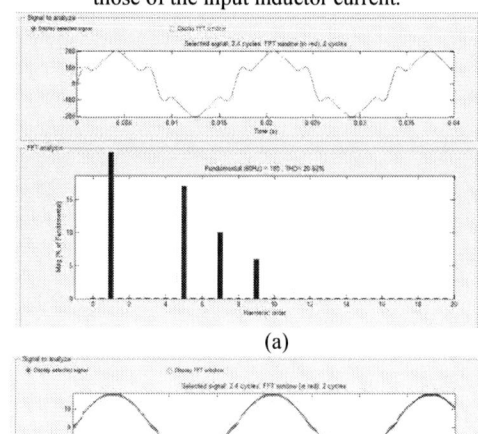

(a)

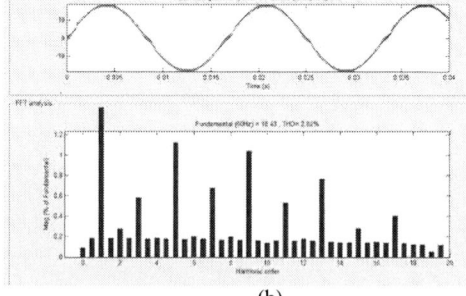

(b)

Fig. 4. (a) Simulated waveform and FFT analysis of grid voltage containing: $V_{5th} = 17\%$, $V_{7th} = 10\%$ and $V_{9th} = 6\%$, and (b) those of the input inductor current.

Table 1
Specifications and designed parameters of the UPS

AC grid voltage (3 ϕ 3W)	$v_{P,RS}$, $v_{P,ST}$ and $v_{P,TR}$	220 V$_{rms}$
DC bus voltage	v_{DC}	380 ± 20 V
Output voltage (3 ϕ 4W)	$v_{I,RN}$, $v_{I,SN}$ and $v_{I,TN}$	127 V$_{rms}$
Maximum rated power	P_{max}	10 kVA
Line frequency	f_l	60 Hz
Switching frequency	f_S	20 kHz
Power factor	PF	0.7 leading/lagging
Inductor	$L_{P,R}$, $L_{P,S}$, and $L_{P,T}$ for 3 ϕ 3W $L_{I,R}$, $L_{I,S}$, $L_{I,T}$, and $L_{I,N}$ for 3 ϕ 4W	2000μH ~ 250μH
Output filter capacitor	$C_{P,L-N}$ and $C_{I,L-N}$	10μF

Table 2
Tests of input current distortion under various grid voltage distortions for the three-phase three-wire inverter.

	V_{5th}(%)	V_{7th}(%)	V_{9th}(%)	$V_{P,THD}$(%)	$I_{P,THD}$(%)
Case 1	2.1	1.5	0.7	2.67	1.96
Case 2	2.5	1.7	0.8	3.13	1.97
Case 3	5	3.5	1.5	6.29	1.99
Case 4	10	6	4	12.33	2.09
Case 5	15	8	5	17.72	2.44
Case 6	17	10	6	20.62	2.82

Table 3
Tests of output voltage distortion under various grid voltage distortion and resistive load.

	V_{5th}(%)	V_{7th}(%)	V_{9th}(%)	$V_{P,THD}$(%)	$V_{I,THD}$(%)
Case 1	2.1	1.5	0.7	2.67	2.05
Case 2	2.5	1.7	0.8	3.13	2.05
Case 3	5	3.5	1.5	6.28	2.06
Case 4	10	6	4	12.33	2.15
Case5	15	8	5	17.72	2.7
Case 6	17	10	6	20.62	3.1

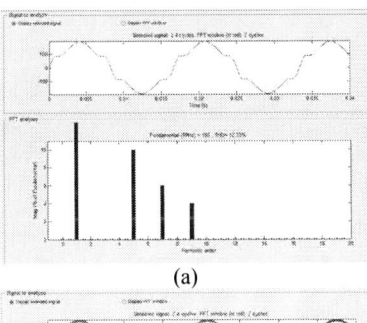

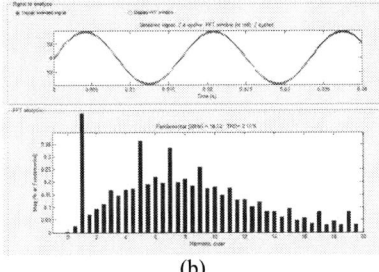

(a)

(b)

Fig. 5. (a) Simulated waveform and FFT analysis of grid voltage containing: V$_{5th}$ = 10%, V$_{7th}$ = 6% and V$_{9th}$ = 4%, and (b) those of the output inductor current.

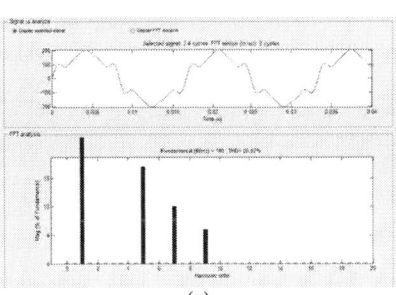

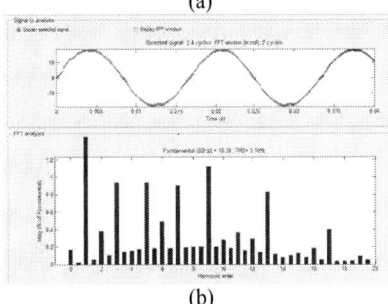

(a)

(b)

Fig. 6. (a) Simulated waveform and FFT analysis of grid voltage containing: V$_{5th}$ = 17%, V$_{7th}$ = 10% and V$_{9th}$ = 6%, and (b) those of the output inductor current.

A three-phase transformerless UPS was built in the lab. The nominal three-phase voltages are 127 V_{rms} and the frequency is 60 Hz. The inverter inductance varies from 2000 μH to 250 μH per phase and the switching frequency is 20 kHz. Fig. 7 shows the measured waveforms of input inductor currents and input voltage. The proposed transformerless UPS tracks the ac current reference precisely not only in the steady state but also during the transition period shown in Fig. 8. Besides, the proposed control laws have been confirmed under different types of loads and the output voltages are with low total harmonic distortion (THD). Fig. 9 shows the output voltages and output inductor current under resistive load. The measured THD of the output voltage is about 2.73%. The waveform of output voltages and output inductor current under inductive load is shown in Fig. 10 in which the THD of the output voltage is around 2.76%. Fig. 11 shows the output voltages and output inductor current under capacitive load. The measured THD of the output voltage is about 2.79%. The waveform of output voltages and output inductor current under unbalanced load is shown in Fig. 12 in which the THD of the output voltage is 1.84%. Fig. 13 shows the output voltages and output inductor current under rectified load. The measured THD of the output voltage is about 2.85%.

Fig. 14 shows measured circulating current, phase currents and output voltages. The circulating current is composed of two different frequencies, namely beat and switching frequencies. This phenomenon affects phase inductor currents and output phase voltages simultaneously. Thus, a synchronizing PWM circuit is added to synchronize two microprocessors and eliminate the fluctuation of output inductor currents and output voltages. The improved waveforms are shown in Fig. 15(a) and (b), and the expanded waveforms of Fig. 15(a) are shown in Fig. 16, which shows only switching frequency component.

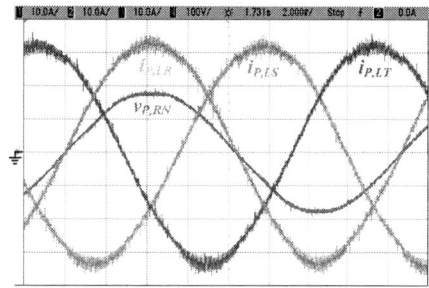

($i_{P,LR}$, $i_{P,LS}$ and $i_{P,LT}$:10 A/div ; $v_{P,RN}$: 100 V/div ; time:2 ms/div)
Fig. 7. Measured waveform of input inductor currents and input voltage.

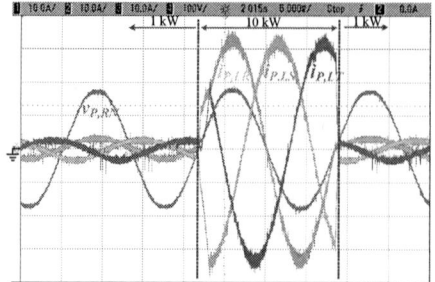

($i_{P,LR}$, $i_{P,LS}$ and $i_{P,LT}$:10 A/div ; $v_{P,RN}$: 100 V/div ; time:2 ms/div)
Fig. 8. Measured waveform of the transition from 1 kW to 10 kW and from 10 kW to 1 kW.

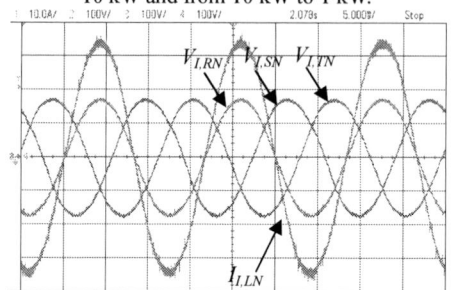

($v_{I,RN}$, $v_{I,SN}$ and $v_{I,TN}$:100 V/div ; $i_{I,LR}$: 10 A/div ; time:5 ms/div)
Fig. 9. Measured waveform of output voltages and output inductor current under resistive load ($V_{I,THD}$ = 2.73%).

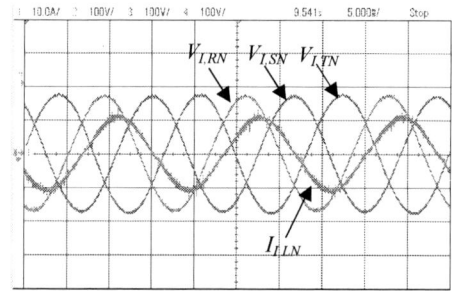

($v_{I,RN}$, $v_{I,SN}$ and $v_{I,TN}$:100 V/div ; $i_{I,LR}$: 10 A/div ; time:5 ms/div)
Fig. 10. Measured waveform of output voltages and output inductor current under inductive load.
(PF = 0.82 lagging and $V_{I,THD}$ = 2.76%)

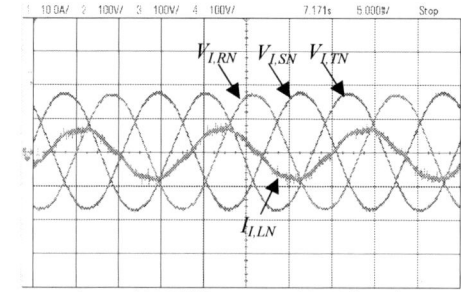

($v_{I,RN}$, $v_{I,SN}$ and $v_{I,TN}$:100 V/div ; $i_{I,LR}$: 10 A/div ; time:5 ms/div)
Fig. 11. Measured waveform of output voltages and output inductor current under capacitive load.
(PF = 0.77 leading and $V_{I,THD}$ = 2.79%)

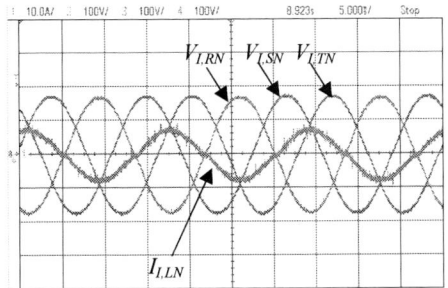

($v_{I,RN}$, $v_{I,SN}$ and $v_{I,TN}$:100 V/div ; $i_{I,LR}$: 10 A/div ; time:5 ms/div)
Fig. 12. Measured waveform of output voltages and output inductor current under unbalanced load. ($V_{I,THD}$ = 1.84%)

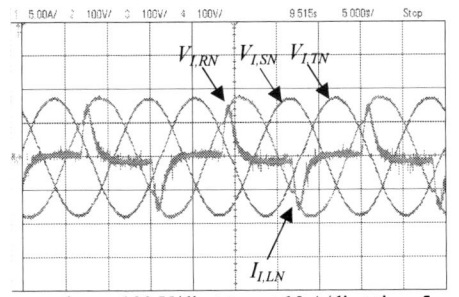

($v_{I,RN}$, $v_{I,SN}$ and $v_{I,TN}$:100 V/div ; $i_{I,LR}$: 10 A/div ; time:5 ms/div)
Fig. 13. Measured waveform of output voltages and output inductor current under rectified load. ($V_{I,THD}$ = 2.85%)

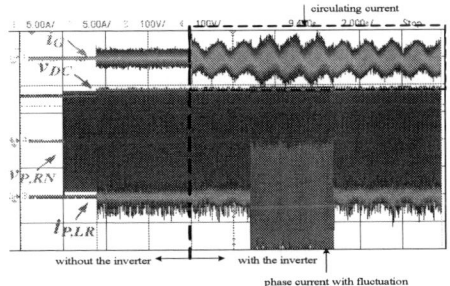

(a)
(i_G:5 A/div; $i_{P,LR}$:5 A/div; v_{DC}:100 V/div; $v_{P,RN}$:100 V/div; time:2 s/div)

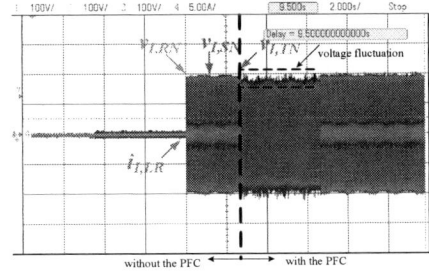

(b)
($v_{I,RN}$, $v_{I,SN}$ and $v_{I,TN}$:100 V/div;$i_{I,LR}$:5 A/div; time:2 s/div)
Fig. 14. Measured waveforms of (a) circulating current (including beat and switching frequency components), and (b) inverter output voltage showing voltage fluctuation.

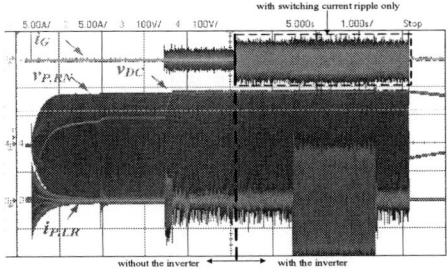

(a)
(i_G:5 A/div; $i_{P,LR}$A/div; v_{DC}:100 V/div; $v_{P,RN}$:100 V/div; time:1 s/div)

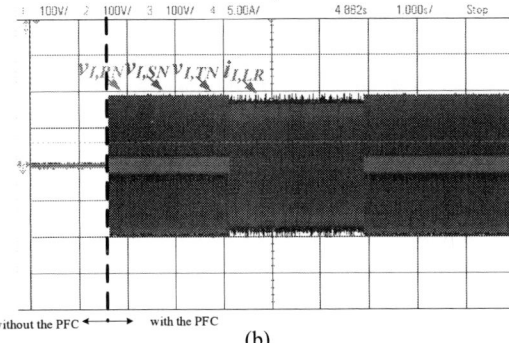

(b)
($v_{I,RN}$, $v_{I,SN}$ and $v_{I,TN}$:100 V/div; $i_{I,LR}$:5 A/div; time:1 s/div)
Fig. 15. Measured waveforms of (a) the improved PFC circulating current and (b) the improved inverter output voltage.

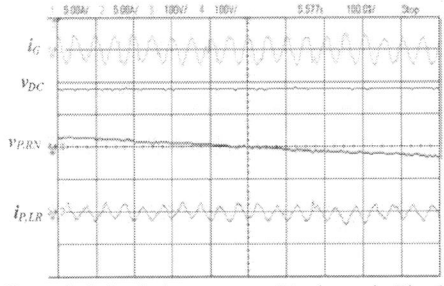

Fig. 16. Expanded circulating current of i_G shown in Fig. 13(a).

V. CONCLUSIONS

This paper has proposed a D-Σ digital controlled transformerless UPS which regulates the power flow between dc link and utility grid, as well as stabilizes the output voltage. The control laws based on SPWM are simple and can track sinusoidal reference current in each phase precisely and independently. The circulating current between PFC and inverter stage can be reduced significantly, containing only switching current ripples. At the same time, a synchronizing PWM circuit is added to synchronize the microprocessors on the both stages and thus, eliminate the fluctuation of output inductor current and output voltage. A 10 kVA prototype of the tranformerless UPS system has verified the feasibility of proposed control methods under different types of load.

ACKONWLEDGEMENT

The authors would like to thank Industrial Technology Research Institute for sponsoring this research.

REFERENCES

[1] J.-M. Guerrero, L.-G. de Vicuna, and J. Uceda, "Uninterruptible power supply systems provide protection," *IEEE Ind. Electron. Mag.*, vol. 1, no. 1, pp. 28–38, Spring 2007.

[2] M. Zhang, D. Atkinson, B. Ji, M. Armstrong, M. Ma, "A Zero-Sequence Component Injected PWM Method with Reduced Switching Losses and Suppressed Common-Mode Voltage for A Three-Phase Four-Leg Voltage Source Inverter," *Proceedings of IECON 2012*, pp.5068–5073, 25-28 Oct. 2012.

[3] M. Zhang, D. Atkinson, B. Ji, M. Armstrong, M. Ma, "A Near-state Three-Dimensional Space Vector Modulation for a Three-Phase Four- Leg Voltage Source Inverter," *IEEE Trans. Power Electron.*, vol. 29, pp. 5715–5726, Nov. 2014.

[4] R. Zhang, V. H. Prasad, D. Boroyevich, and F. C. Lee, "Three-Dimensional Space Vector Modulation for Four-Leg Voltage-Source Converters," *IEEE Trans. Power Electron.*, vol. 17, no. 3, pp. 314–326, May 2002.

[5] Tashackori, S. Hosseini, M. Sabahi, and T. Nouri, "A Three-Phase Four-Leg DVR Using Three Dimensional Space Vector Modulation," *Proceedings of ICEE*, pp.1-5, 14-16 May 2013.

[6] Y. Cho and J. S. Lai, "Digital plug-in repetitive controller for single-phase bridgeless PFC converters," *IEEE Trans. Power Electron.*, vol. 28, no. 1, pp. 165–175, Jan. 2013.

[7] S. Jiang, D. Cao, Y. Li, J. Liu, and F. Z. Peng, "Low-THD, fast-transient, and cost-effective synchronous-frame repetitive controller for three-phase UPS inverters," *IEEE Trans. Power Electron*, vol. 27, no. 6, pp. 2994–3005, Jun. 2012.

[8] T.-F. Wu, C.-H. Chang, L.-C. Lin, Y.-C. Chang and Y.-R. Chang, "Two-phase modulated digital control for three-phase bi-directional inverter with wide inductance variation," *IEEE Trans. on Power Electron.*, vol. 28, pp. 1598–1607, Apr. 2013.

[9] T.-F. Wu, L.-C. Lin, C.-H. Chang, Y.-R. Chang, Y.-D. Li "Load Impedance Estimation and Iterative-Learning Control for a Three-Phase Four-Wire Inverter," *IEEE ECCE Asia 2013*, pp. 1275-1281, June 2013.

[10] T.-F. Wu, C.-H. Chang, L.-C. Lin, G.-R. Yu and Y.-R. Chang, "A D-Σ Digital Control for Three-Phase Inverter to Achieve Active and Reactive Power Injection," *IEEE Trans. on Power Electron.*, vol. 61, no. 8, pp. 3879–3890, Aug. 2014.

A Multi-Function Three-Level Dynamic Voltage Corrector with Wide Correction Range and Short Circuit Fault Isolation

Jiankun Cao, Pengling Ding, Haichun Liu and Shaojun Xie

College of Automation Engineering, Nanjing University of Aeronautics and Astronautics
Nanjing 210016, China
E-mail: jiankun1818@163.com, dingpengling@126.com, nuaalhc@nuaa.edu.cn, eeac@nuaa.edu.cn

Abstract—The dynamic sag corrector is used at the sensitive loads terminal in order to avoid the grid voltage sag. This paper focuses on both the correction of grid voltage swell/sag in large range and the quality of compensated voltage. A wide correction range dynamic voltage corrector is proposed for solving above problems with small cost. It possesses dual functions of both the wide range voltage correction and the short circuit fault isolation. The operation principle and the voltage correction range are analyzed. The increase of the DC bus voltage broadens the voltage sag correction range, while the realization of the voltage swell correction benefits from the bidirectional energy flow. The output inductor current control completes both the current limitation and the fault isolation at the moment of the load short circuit. The experimental results verify the validity of the theoretical analysis. The proposed dynamic voltage corrector is suitable for the voltage stabilization of ground power unit, production lines or aerospace equipment under the centralized power supply. It has the characteristics of small size, light weight, good dynamic performance and high quality of output waveform.

Keywords—*wide correction range; short circuit fault isolation; voltage sag/swell corrector; double closed loop control; energy regeneration*

I. INTRODUCTION

The point of common coupling (PCC) would suffer disturbances of the voltage sag or swell inevitably under the centralized power supply system. Therefore, the voltage corrector should be equipped at the power supply terminal of sensitive loads for guaranteeing the power supply quality [1-2]. The existing scheme like dynamic voltage restorer (DVR) compensates the voltage at the power supply terminal through transformer serial-in [3-5]. In order to reduce the volume and weight some literatures propose the DVR without transformer, but the isolated power supply still needs to be considered for the transformerless DVR [6-7]. In recent years, dynamic sag corrector (DySC) was researched in [8-20] for meeting the light-weight requirement. DySC realizes both the self-powered without transformer and the voltage compensation of 50% sag.

At present, most researches of DySC are focused on the mitigating performance of the voltage sag condition. The main reason is because during the voltage sag condition, some of the sensitive loads are often automatically disconnected from the

supply system for devices protection. However, a prolonged and excessive voltage swell is also damaging to the equipment such as the aircraft avionics or the precision grade production line. Hence, it is necessary in finding ways to mitigate the effects of the voltage swell.

At present, the key point of the voltage swell compensation by using DySC is that the regenerated energy will lift the voltage of the DC bus inside the DySC continuously. The phase-shift control of the compensated voltage is researched in [11-16]. The zero active power compensation technology is used for avoiding the DC bus voltage rising. Although the amplitude and phase of the compensated voltage can be adjusted timely, the phase of output voltage is changed with the voltage sag depth. This method will also make both the compensated voltage and the equipment capacity increase.

Another technical index for evaluating DySC is the quality of the compensated voltage waveform. When the depth of voltage sag is shallow, the inverter stage of DySC operates with a very low modulation index (MI). The quality of compensated voltage is poor, which increases the cost of output filter. In [17]-[20] the transformers are used at the input or output side for increasing the MI of the inverter stage and broadening the depth of the voltage sag. However, the adding transformers weaken the advantages of small size and simple structure for DySC. In [8] the controllable power switches are connected on the DC bus in series for decreasing the DC bus voltage. But the controllable power switches will be broken down by overvoltage if there is any inductive reactance in the circuit loop. In [9-10] the diodes in rectifier stage are replaced by thyristors whose firing angles are controlled for decreasing the voltage of DC bus.

The load short circuit, which influences the security and stability of power grid operation, is one of the serious faults under the centralized power supply system. If the fault isolation action of the load short circuit is not timely, the voltage at the PCC will sag sharply until the fault is recovered. During this time, the other equipment connected with the PCC will be influenced. The voltage sag at the PCC is not permissible for the power supply of some sensitive loads.

Because the action time of the circuit breaker is show, the fault isolation is not timely and the limitation of the short

978-1-4673-9551-9/16 $31.00 © 2016 IEEE

circuit current cannot be realized. In [21-24] the solid state fault current limiter is proposed. A large inductor is series in the power supply terminal by the actions of power switches at the moment of the load short circuit. The short circuit current will be limited and the voltage at the PCC is not influenced. But the volume and weight of the power supply terminal is increased by using the solid state fault current limiter. In [25] a multi-function solid state fault current limiter is presented for both the voltage compensation and the short circuit current limiting. The filter inductor is used as the current limiting inductor. The limited current value depends on the equivalent impedance of the filter inductor, which is short of the flexibility.

This paper proposes a multi-function dynamic voltage corrector with both the wide correction range and the short circuit fault isolation, which can correct both the swell and sag of grid voltage by bidirectional energy control. According to the comparison of the existing DySC scheme, the quality of the compensated voltage is improved by the multi-level technology. The device can also isolate the fault of the load short circuit rapidly by the output inductor current limitation. The verification of the proposed device was carried out through the experimental works with a prototype of 3kVA rating.

II. OPERATION PRINCIPLE

DySC is a two-stage converter essentially including the grid side converter and the load side converter with the DC bus in common, shown in Fig. 1. The grid side converter transmits power grid energy to DC bus. The load side converter controls the waveform of compensated voltage. In order to connect between the compensated voltage and power grid voltage in series directly, there is a point of common connection between the grid side converter and the load side converter. So, the topology of two converters in single phase DySC should be half-bridge type for connection. The capacitance mid-point of the DC bus is the point of common connection.

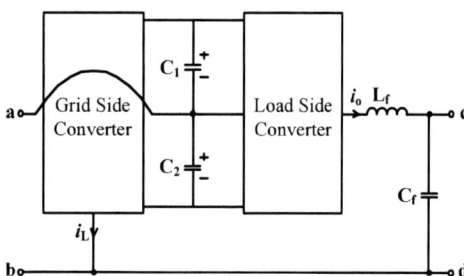

Fig. 1. Equivalent circuit diagram of DySC.

Based on the above characteristic a novel dynamic voltage corrector with wide correction range is proposed. The topology is shown in Fig. 2. The uncontrolled rectifier is replaced by a PWM rectifier and the two-level topology of the inverter stage is changed into three-level topology. The proposed scheme not only realizes the correction of voltage swell and sag with high waveform quality, but also isolates the fault of the load short circuit rapidly. Combining with the control scheme the operations of the load side converter and the grid side converter are analyzed as follows:

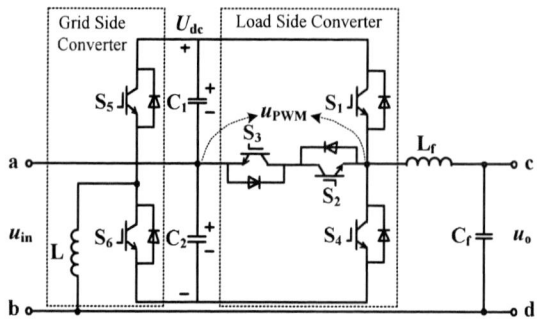

Fig. 2. Topology of the proposed dynamic voltage corrector.

A. Voltage correction mode

When the swell or sag of the power grid voltage happens, the load side converter takes the stable and accurate output voltage as the control target. The output voltage and the output inductive current are sampled for voltage and current double closed loop control shown in Fig. 3. The phase of the voltage reference is followed with the power grid by phase-locked loop. The output voltage is controlled by the outer loop and the output inductive current is controlled by the inner loop. When the sag of power grid voltage happens, the phase of the compensated voltage is same with the gird voltage. The energy of voltage sag compensation is supplied by the DC bus. When the swell of power grid voltage happens, the phase of the compensated voltage is opposite with the gird voltage. The energy of the voltage swell compensation flows back to the DC bus.

When the swell or sag of power grid voltage occurs, the grid side converter takes the stable of DC bus voltage and the waveform quality of grid side current as the control targets. Meanwhile, the voltage balance of capacitance mid-point should be considered. The DC bus voltage and the current of the grid side inductor are sampled for voltage and current double closed loop control shown in Fig. 4. The reference of the grid side inductive current is phase-locked for rectifier or inverter with unity power factor. The voltage of DC bus is controlled by the outer loop and the inductive current of grid side is controlled by the inner loop. The voltage difference between C_1 and C_2 is added to the reference of the inner loop. The balance of the capacitance mid-point is designed in the grid side converter for avoiding the DC bias in compensated voltage. The grid side converter can realize bi-directional energy transmission. When the sag of the power grid voltage happens, the energy is transmitted from the power grid to the DC bus. When the swell of the power grid voltage happens, the energy is regenerated from the DC bus to the power grid.

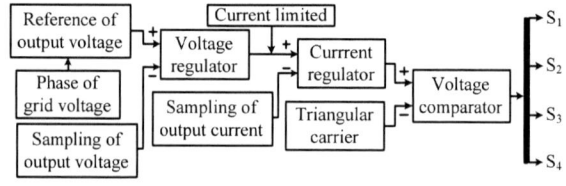

Fig. 3. Control diagram of the load side converter.

978-1-4673-9551-9/16 $31.00 © 2016 IEEE

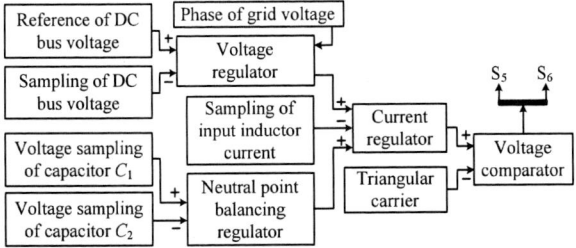

Fig. 4. Control diagram of the grid side converter.

B. Isolation of short circuit fault

When the load short circuit happens, the outer loop regulator of the load side converter is saturated. The load side converter is controlled by the single current loop. The current reference of the grid side inductor is restricted by the current limited value. So the current of the grid side inductor is limited. According to the output impedance characteristic the load side converter releases both active and reactive power. The reactive power is absorbed by the output inductor of the load side converter. The active power is regenerated to the DC bus and then absorbed by the power grid with unity power factor. The power grid voltage is not influenced by the load short circuit because of the load side converter. According to the analysis above, the isolation of the load short circuit fault can be realized naturally without changing the existing control strategy.

In addition, the three-level structure is used in the load side converter. Comparing with the THD of two-level SPWM, the THD of three-level SPWM waveform has 50% reduction at least. With the decreasing of the MI, the superiority of the three-level will be more obvious on the waveform quality. When the depth of the compensation is shallow for the load side converter, the waveform quality of the compensated voltage is improved significantly.

III. ANALYSIS OF VOLTAGE CORRECTION RANGE

A. Correction range of the voltage sag

The correction range of the voltage sag is limited by the maximum input current of the grid side converter named I_{in_max}. With the increasing of I_{in_max}, the size and loss of the inductance L and the current stresses of the power switches S_5 and S_6 in the grid side converter would be enlarged.

When the voltage sag happens, the compensation energy is the input power of the grid side converter. So the effective value of the minimum power grid voltage U_{in_min} is:

$$U_{in\_min} = \frac{P_o U_o}{P_o + U_o I_{in\_max}} \qquad (1)$$

Where P_o is the power of the sensitive load and U_o is the effective value of the total output voltage after compensation.

According to (1) when P_o and U_o are confirmed, the depth of the voltage sag compensation is associated with I_{in_max}. When P_o=3000W and U_o=115V, the relation between the compensation depth and I_{in_max} is shown in Fig. 5. With the

increase of the compensation depth, the maximum input current of the grid side converter raises rapidly. For example, if I_{in_max} is allowed to 52A, the sustainable compensation depth is 67%. While I_{in_max} is allowed up to 148A, the attainable compensation depth is as high as 85%. The maximum compensation depth of the traditional DySC is only 50%.

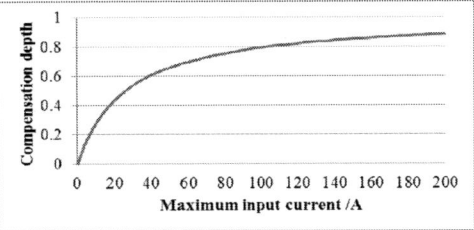

Fig. 5. Relation between the compensation depth and the maximum input current.

B. Correction range of the voltage swell

The correction range of the voltage swell is limited by the DC bus voltage named U_{dc}. When the DC bus voltage is ensured and the grid side converter is working at the maximum MI, the voltage waveform after the grid side inductor can approximate to square wave. According to the Fourier polynomial of square wave the effective value of the maximum power grid voltage is shown as following:

$$U_{in\_max} = \frac{\sqrt{2}}{\pi} U_{dc} \qquad (2)$$

The following four factors should be considered for the selection of U_{dc}:

1) The voltage stress of the power switches is determined by the U_{dc}.

2) The MI of the load side converter should be raised for increasing the THD of the compensated voltage.

3) U_{dc} is not suggested to be higher than peak-to-peak value of the power grid voltage.

4) The appropriate voltage transmission ratio should be considered for the grid side converter according to U_{in_min}.

In conclusion, the compensation depth of the voltage sag for the proposed device is higher than the performance of the traditional DySC. What's more, the voltage swell can also be corrected.

IV. EXPERIMENTAL RESULTS

Taking the centralized 400Hz electric power system for example, the proposed dynamic voltage corrector is applied to the 400Hz power supply terminal. A prototype of 3kVA single phase is developed for verifying the functions of both the grid voltage sag/swell correction and the fault isolation of the load short circuit. The Key parameters of the circuit are shown in Table I.

The DSP, which is TMS320F28035 by Texas Instruments, is used for the digital control. The topology and control scheme are according to the above analyze. The power switches is

IGBT with anti-parallel diode. The type of the power switches is IKW75N60T by Infineon.

TABLE I.   KEY PARAMETERS OF THE CIRCUIT

Items	Values
Input inductance of the grid side converter L	1.6mH
Capacitance of the DC bus C_1 and C_2	4.4mF
Output filtering inductance L_f	420μH
Output filtering capacitance C_f	21μF
Switching frequency f_s	12kHz
Rated output voltage U_o	115V/400Hz
Input voltage U_{in}	38V-180V/400Hz ($0.33U_o$-$1.23U_o$)
DC bus voltage U_{dc}	400V
Apparent power of Load S	3kVA
Rated output voltage U_o	115V/400Hz

The experimental results of the voltage corrector are shown in Fig. 6. The voltage sag of the power grid is 67% and then the grid voltage recovers to normal, shown in the Fig. 6(a). The times of the voltage jumps are selected at the peak for the worst condition. The experimental waveforms show that the output voltage after compensation is not influenced by the voltage sag of the power grid. The response time is limited within 1 cycle. The DC bus voltage is stabilized near 400V. The output pulse voltage of the load side converter u_{PWM} keeps the same phase as the grid side voltage for the voltage sag compensation. The voltage swell of the power grid is 23% and then the grid voltage recovers to normal, shown in the Fig. 6(b). The u_{PWM} is opposite with the grid side voltage's for the voltage swell compensation. The energy is regenerated back to the power grid by the grid side converter, which is consistent to the theoretical analysis.

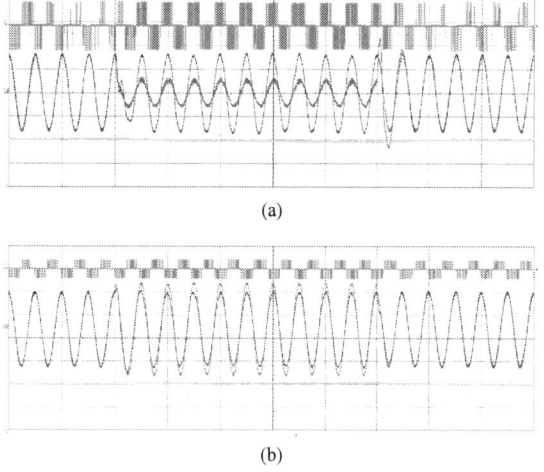

(a)

(b)

Fig. 6. Experimental waveforms of the voltage correction. (a) Voltage sag correction waveforms. (b) Voltage swell correction waveforms. Time: 5ms/div. From top to bottom: output pulse voltage of load side converter u_{PWM}(200V/div for (a) and 500V/div for (b)), grid side voltage u_{in}(red, 100V/div), load side voltage u_o(blue, 100V/div) and DC bus voltage U_{dc}(200V/div).

The relevant waveforms at the moment of the load short circuit are shown in Fig. 7. When the load short circuit happens, the output voltage is close to 0. The peak value of the output current is restricted by the limited value. The DC bus voltage lifts during the short circuit fault, but the DC bus voltage still keeps stable under 500V after a few cycles. According to Fig. 7(b) the fault of the load short circuit does not influence the power grid voltage. The grid side inductive current is also limited. The fault isolation of the load short circuit is realized in the proposed dynamic voltage corrector.

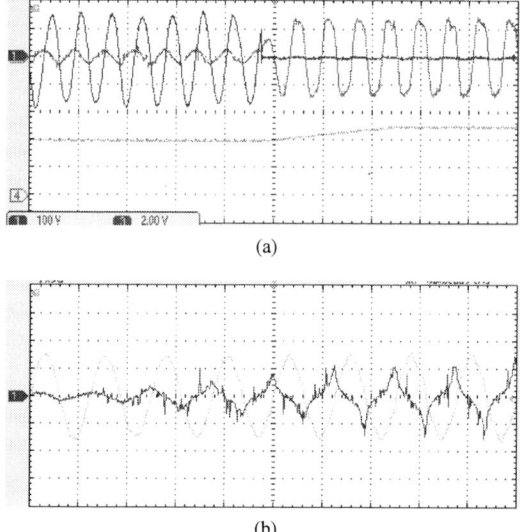

(a)

(b)

Fig. 7. Experimental waveforms of the load short circuit. (a) Time: 4ms/div. From top to bottom: load side voltage u_o(blue, 100V/div), load side inductor current i_o(50A/div) and DC bus voltage U_{dc}. (b) Time: 2ms/div. From top to bottom: grid side voltage u_{in}(light blue, 100V/div), grid side inductor current i_L(blue, 50A/div).

V. CONCLUSION

This paper presents a wide correction range dynamic voltage corrector derived from the traditional DySC. The compensation range is greatly widened by the PWM scheme of the grid side converter. The overvoltage of the DC bus for the grid voltage swell correction is solved by the energy regeneration scheme. The waveform quality of the compensated voltage is improved by using multi-level technology. The proposed dynamic voltage corrector is also with the function of the load short circuit isolation. The solid-state fault current limiter or other similar devices can be left out for reducing the size of the power supply terminal. It is suitable for the voltage stabilization of aircrafts, production lines or aerospace equipment under the centralized power supply. It possesses small size, light weight, high compensation efficiency, good dynamic performance and quick fault isolation of the load short circuit.

ACKNOWLEDGMENT

The work of this paper is supported by the funding of the Fundamental Research Funds for the Central Universities

(NS2015033), the National Natural Science Foundation of China (51477077) and the Jiangsu Innovation Program for Graduate Education (CXZZ12_0150, the Fundamental Research Funds for the Central Universities).

REFERENCE

[1] M.E. Elbuluk, "Potential starter/generator technologies for future aerospace application," IEEE Aerospace and Electronic Systems Magazine, vol. 11, no. 10, pp. 17-24, 1996.

[2] Uffe Borup, Bo Vork Nielsen and Frede Blaabjerg, "Compensation of cable voltage drops and automatic identification of cable parameters in 400Hz ground power units," IEEE Transactions on Industry Applications, vol. 40, no.5, pp. 1281-1286, 2004.

[3] Jayaprakash P., Singh B., Kothari D.P., Chandra A. and Al-Haddad K., "Control of reduced-rating dynamic voltage restorer with a battery energy storage system," IEEE Transactions on Industry Applications, vol. 50, no.2, pp. 1295-1303, 2014.

[4] Rauf A.M. and Khadkikar V., "An enhanced voltage sag compensation scheme for dynamic voltage restorer," IEEE Transactions on Industrial Electronics, vol. 62, no.5, pp. 2683-2692, 2015.

[5] Kaniewski J., Fedyczak Z. and Benysek G., "AC voltage sag/swell compensator based on three-phase hybrid transformer with buck-boost matrix-reactance chopper," IEEE Transactions on Industrial Electronics, vol. 61, no. 8, pp. 3835-3846, 2014.

[6] Songcen Wang, Guangfu Tang, Kunshan Yu and Jianchao Zheng, "Modeling and control of a novel transformer-less dynamic voltage restorer based on H-Bridge cascaded multilevel Inverter," International Conference on Power System Technology, pp. 358-366, 2006.

[7] Eng Kian Kenneth Sng, S. S. Choi, and D. Mahinda Vilathgamuwa, "Analysis of series compensation and DC-link voltage controls of a transformerless self-charging dynamic voltage restorer," IEEE Transactions on Power Delivery, vol.19, no. 3, pp. 1511-1518, 2004.

[8] Li Yanfang, Chen Zenglu, Zhan Pei, Chen Pei and Chen Zhanbin, "DC link voltage control of a new type of series voltage sag compensator," Power System Technology, vol. 33, no. 6, pp. 50-54, 2009.

[9] Yong Lu, Guochun Xiao, Bo Lei, Xuanlv Wu and Sihan Zhu, "A transformerless active voltage quality regulator with the parasitic boost circuit," IEEE Transactions on Power Electronics, vol. 29, no. 4, pp. 1746-1756, 2014.

[10] Guochun Xiao, Zhiliang Hu, Lei Zhang and Zhaoan Wang, "Variable DC-bus control strategy for an active voltage quality regulator," 24th Annual IEEE Applied Power Electronics Conference and Exposition, pp. 1558-1563, 2009.

[11] Chen Zenglu and Tian Miaomiao, "A novel simplified voltage sag compensator topology and its control strategies," Power System Technology, vol. 36, no. 4, pp. 144-148, 2012.

[12] Choi S.S., Li J.D., Vilathgamuwa D.M., "A generalized voltage compensation strategy for mitigating the impacts of voltage sags/swells," IEEE Transactions on Power Delivery, vol. 20, no. 3, pp. 2289-2297, 2005.

[13] Tsai M.T., "Analysis and design of a cost-effective series connected AC voltage regulator," IEE Proceedings Electric Power Applications, vol. 151, no. 1, pp. 107-115, 2004.

[14] Guochun Xiao, Zhiliang Hu, Changhe Nan and Zhaoan Wang, "DC-link voltage pumping-up analysis and phase shift control for a series active voltage regulator," 37th IEEE Power Electronics Specialists Conference, pp. 1-5, 2006.

[15] Ming Tsung Tsai, "Design of a compact series-connected AC voltage regulator with an improved control algorithm," IEEE Transactions on Industrial Electronics, vol. 51, no. 4, pp. 933-936, 2004.

[16] Vilathgamuwa D.M. and Wijekoon H.M., "Control and analysis of a new dynamic voltage restorer circuit topology for mitigating long duration voltage sags," 37th IAS Annual Meeting. Conference Record of the Industry Applications Conference, vol.2, pp. 1105-1112, 2002.

[17] Silva S.M. and Filho B.J.C., "Component-minimized voltage sag compensators," 37th IAS Annual Meeting, Conference Record of the Industry Applications Conference, vol. 2, pp. 883-889, 2002.

[18] Divan D., Bendre A., Kranz W. and Schneider R., "Dual source dynamic sag correctors - a cost effective topology for enhancing the reliability of dual source systems," IEEE 38th Annual Industry Applications Conference, vol. 2, pp. 940-947, 2003.

[19] Brumsickle W.E., Schneider R.S., Luckjiff G.A., Divan D.M. and Mcgranaghan M.F., "Dynamic sag correctors: cost-effective industrial power line conditioning," IEEE Transactions on Industry Applications, vol. 37, no. 1, pp. 212-217, 2001.

[20] Bhadkamkar A., Bendre A., Schneider R., Kranz W. and Divan D., "Application of zig-zag transformers in a three-wire three-phase dynamic sag corrector system," IEEE 34th Annual Power Electronics Specialist Conference, vol. 3, pp. 1260-1265, 2003.

[21] Abramovitz A. and Smedley K.M., "Survey of solid-state fault current limiters," IEEE Transactions on Power Electronics, vol. 27, no.6, pp. 2770-2782, 2012.

[22] Ghanbari T. and Farjah E., "Development of an efficient solid-state fault current limiter for microgrid," IEEE Transactions on Power Delivery, vol. 27, no. 4, pp. 1829-1834, 2012.

[23] Hongliang Wang, Xiumei Yue, Xuejun Pei and Yong Kang, "Improved software current-limiting protection strategy for starting the high-power motor," International Conference on Electrical Machines and Systems, pp. 1-4, 2009.

[24] FAN Yu, JIANG Daozhuo and LÜ Wentao, "A novel bridge-type solid-state fault current limiter based on limiting reactance current control," Power System Technology, vol. 38, no. 3, pp. 776-781, 2014.

[25] TU Chunming, DENG Shu, GUO Cheng, WEI Chengzhi, YAO Peng and JIANG Fei, "A novel multi-function solid-state fault current limiter and its analysis of current-limiting filter selection," Power System Technology, vol. 38, no. 6, pp. 1634-1638, 2014.

Effects and Analysis of Minimum Pulse Width Limitation on Adaptive DC Voltage Control of Grid Converters

Bo Sun, Ionut Trintis, Stig Munk-Nielsen, Josep M. Guerrero

Department of Energy Technology, Aalborg University,

Aalborg, Denmark

{sbo, itr, smn, joz}@et.aau.dk

Abstract— This paper presents an adaptive dc-link voltage controller for the purpose of decreasing the operating dc-link voltage in a back-to-back converter system. An outer loop to control the maximum modulation (Mmax) index is added to the conventional dc-link voltage controller, and hence the system can online adapt its dc-link voltage reference based on the system operating state. Furthermore, the relationship between current THD, minimum pulse width limitation and Mmax index reference is analyzed in order to obtain an optimized performance of the system. Finally, the dynamic performance and analysis of controller is investigated by experimental results.

Keywords—dc/ac converter; dc voltage adaptive control; Minimum Pulse Width Limitation

I. INTRODUCTION

Wind energy is one of the fastest-emerging renewable alternatives for the electrical supply, and wind turbines are installed with exponential capacity over the past years around the world [1]. The general state-of-the-art wind turbine system is shown in Figure 1. Back to back (B2B) converters are widely applied to realize the controllability and power processing. Two vector control structures are implemented on generator-side and grid-side converters, respectively. The objective of the control scheme for the generator-side converter is to obtain the optimal power tracking for maximum energy capture from the wind by adjusting the speed of the wind turbine. The control of grid-side converter is to coordinate the energy balance between grid and generator, and it can be also responsible for controlling reactive power flow between grid and converter [4].

The grid converter synchronizes with the grid voltage, and the control normally includes an inner current loop and an outer dc-link voltage controller. Normally the dc-link voltage reference is fixed, and set to a level that allows the converter operation over the entire power and grid voltage range. However, decreasing the energy stored in the dc-link can benefit the efficiency of both grid and generator converters due to reduced switching losses [4]. Also for some applications that converter are installed at high altitude, the effect of reduced energy stored in dc-link improves the reliability by decreasing the failure rate based on cosmic rays [5]. The energy storage in dc-link can be reduced by adaptively adjusting the dc-link voltage reference according to different system states. In [6], a method based on calculation of an offline table for the dc-link voltage reference was introduced. However, the performance of this offline calculation method depends on the accuracy of the system model, which is practically affected by various factors, such as operating temperature of the generator, grid voltage, grid impedance and grid filter impedance.

Adaptively adjusting the dc-link voltage reference online can eliminate the calculation errors based on the involved systems modelling. In [4], an outer loop is added to the dc-link voltage controller, which adapts the dc-link voltage reference by controlling the maximum modulation (Mmax) index. The controller shows an improved efficiency implemented in a wind turbine converter. However the Mmax index is simply set as 0.999 in that work. In a practical system, Minimum pulse-width (MPW) limitation is normally imposed on power converter for the purpose of protection but introduce the current distortion. It influences the modulation and needs to be taken into account.

This paper improves the adaptive dc voltage controller proposed in [4] for grid converter, and further studies the effects of MPW limitation in order to optimize the performance of controller. The rest of paper is organized as follows: Section II demonstrates the Mmax adaptive dc voltage control strategy

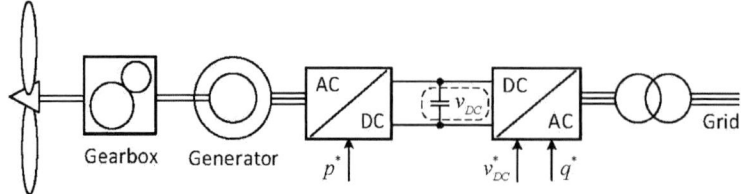

Fig. 1 Wind turbine system with full scale converter

in detail, and gives an explanation on the Mmax index detection and effects of MPW limitation. Section III shows the experimental results and the parameters guidance of Mmax controller on a 2 kW converter prototype. Finally, Section IV gives the conclusion and future research directions.

II. ADAPTIVE DC VOLTAGE CONTROL

To achieve an online adaptation of the dc-link voltage reference, a control loop can be implemented as shown in Fig. 2. Instead of setting a fixed dc-link voltage reference, the Mmax index reference is controlled taking the feedback from the input modulation index to the modulator.

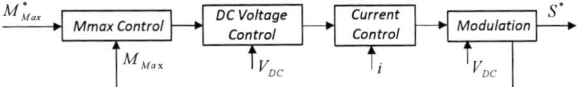

Fig. 2 Proposed adaptive dc-link voltage controller

A. Control structure

The detailed control structure for a typical grid converter with LCL filter is shown in Fig. 3. The inner loop is current control loop which is decoupled in synchronous reference frame (d-q frame) [8]:

$$\begin{bmatrix} i_d \\ i_q \end{bmatrix} = \frac{v_{DC}}{L} \begin{bmatrix} d_d \\ d_q \end{bmatrix} + \begin{bmatrix} -\dfrac{R}{L} & \omega \\ -\omega & -\dfrac{R}{L} \end{bmatrix} \begin{bmatrix} i_d \\ i_q \end{bmatrix} - \frac{1}{L_{line}} \begin{bmatrix} v_{gd} \\ v_{gq} \end{bmatrix} \quad (1)$$

where i_q and i_d, d_q and d_d are DC-like currents and duty ratios aligned with q and d rotating axes, respectively, while R and L are per phase resistance and inductance of the ac filter including line impedance.

The typical out loop contains only a dc voltage regulator with the smaller bandwidth than current loop. Here, a *Mmax*

index regulator is added to voltage control loop by detecting the Mmax index to modulator. Hence, the output dc-link voltage reference will adapt to the converter system state, and at all power levels, the operating dc-link voltage is reduced significantly.

B. Mmax Index detection and reference

According to the results in [4], the Mmax index reference is direct proportional with the grid voltage, current, impedance and additional harmonic voltage respectively and inverse proportional with the dc-link voltage. Because Mmax index is in relationship with so many parameters, it is optimal to online detect it. In this paper, the Mmax value of modulation duty ratio is updated in each fundamental period. In this way, Mmax index is the maximum value of duty ratios of three phases.

In [4], the reference of Mmax index is directly set as 1. However, the Mmax is also affected by system states, such as grid condition, transferred power and MPW limitation. Among those factors, MPW limitation directly influences the maximum modulation index. Therefore, the Mmax index cannot be set as the ideal value of "1" in a practical system.

C. Minimum Pulse Width limitation

MPW limitation is usually imposed by the manufacturers of the active power electronic devices used in converters.

Such limitation has to be enforced to prevent the introduction of power stress on the devices, especially in high power applications [7]. Commonly there are three MPW methods, as is shown in Fig. 4.

- Pulses dropped if they are lower than a minimum width

- Pulses are hold as minimum width value if they are lower than minimum width.

- Pulses are dropped if they are lower than half MPW, but held to MPW value if they are slightly higher than the MPW.

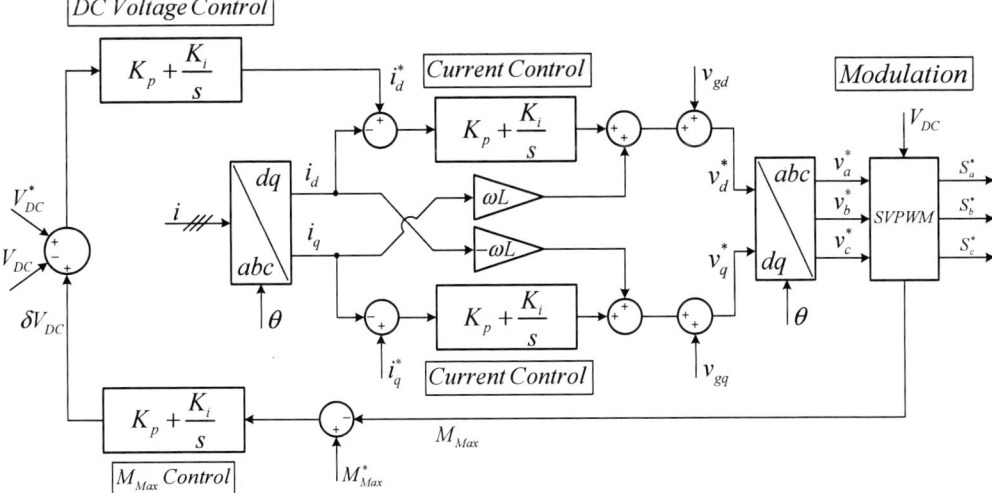

Fig. 3 Grid converter d-q adaptive control structure

This paper the third condition is applied. Setting a maximum modulation index reference above the voltage reserve for the MPW filter can introduce additional distortion on the output current. Therefore the maximum Mmax reference should be experimentally determined in order to comply with the harmonic standards.

Finally, the modulation scheme and MPW limitation is presented in fig. 5.

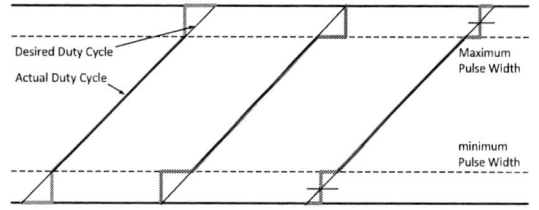

Fig. 4 minimum pulse width limitation methods

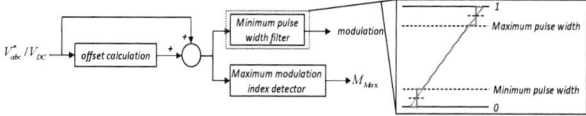

Fig. 5 Modulation and MPW limitation

III. EXPERIMENTAL RESULTS

The simulated control from Fig.3 was implemented in a dspace1006 control platform. The converter prototype is a two-level 2.2 kW Danfoss converter connected to the 400 V grid. The test setup is shown in Fig.6. The grid LCL filter had a total inductance of 3.6 mH and a delta connected capacitor bank of 9 μF. The converter is modulated with continuous space vector pulse width modulation, switching and sampling at a frequency of 10 kHz.

A. Dynamic test

The dynamic operation of proposed Mmax controller is showed in Fig.7. The converter starts operating at a fixed dc-link voltage reference of 600 V, operating at light load in rectifying mode. The Mmax controller was enabled at 1.8 s with reference as 0.985. The dc-link voltage is regulated at an adaptive reference of around 565 V with a response of 1 s and the dc-link voltage is reduced by 35V.

B. MPW limitation test

To study and analyze the effect of MPW limitation on the Mmax controller, the experimental tests are implemented respectively when MPW value is between 1% and 4% of the switching period and Mmax ref is between 98% and 100%. The dc power source input is kept constant at 2 kW to simulate the wind generator. Fig. 8 shows the experimental results in MPW value is 0.02. Fig.8a shows the Mmax ref increase from 0.98 to 1 with a step of 0.005. The MPW limitation can be observed in Fig.8b when Mmax ref is 0.98. Converter-side current and zoomed spectrums are presented in Fig.8c.

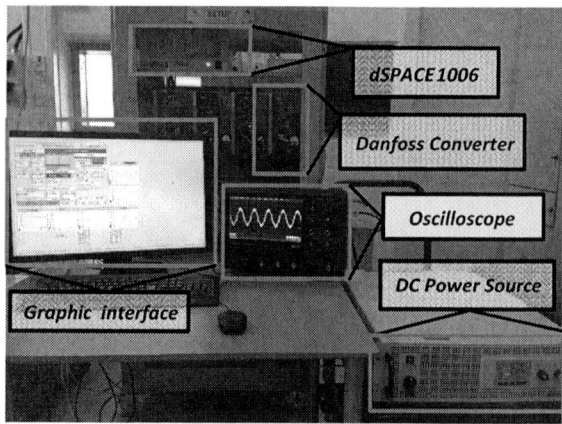

Fig.6 Experimental test setup.

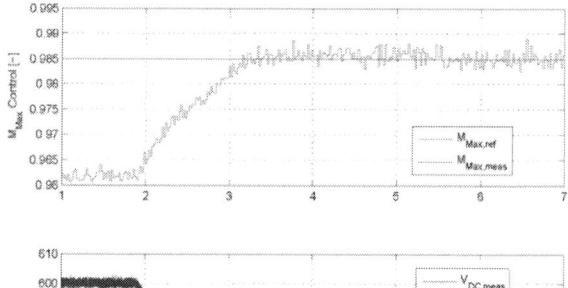

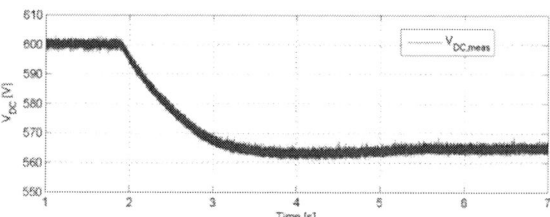

Fig. 7 dynamic operation of *Mmax* controller

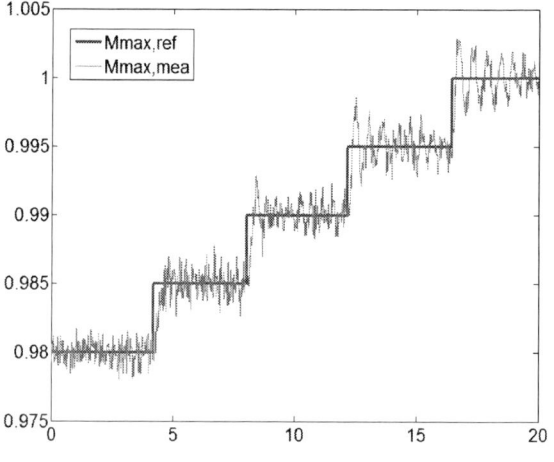

a. *Mmax* reference in MPW=0.02

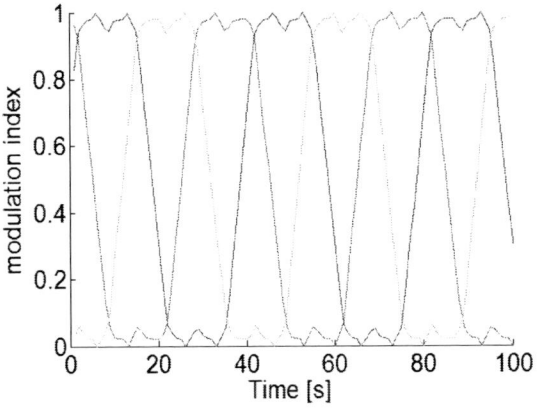

b. modulation index in *Mmax* ref=0.995, MPW=0.02

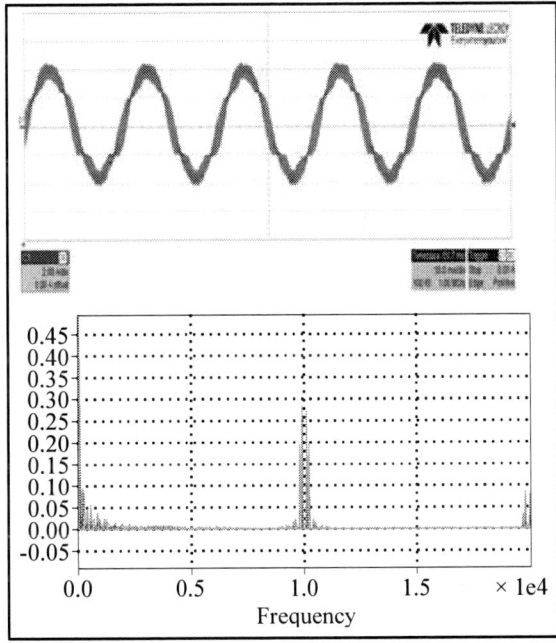

c. Converter-side current and zoomed spectrum in *Mmax* ref=0.995,

MPW=0.02

Fig.8. Experimental results of MPW=0.02

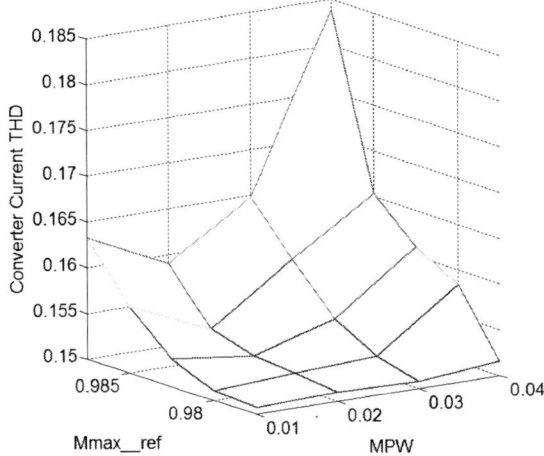

Fig.9 Converter-side Current THD in Different System States

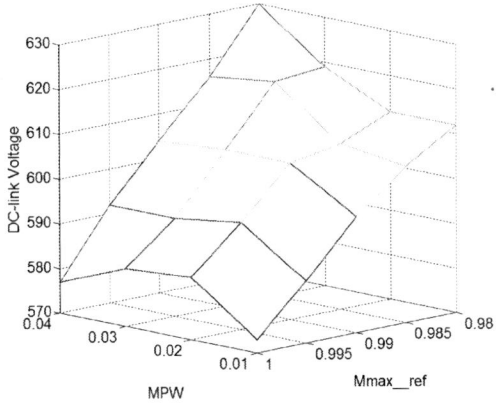

Fig.10 dc-link voltage in different system state

TABLE I. Converter-side Current THD in Different System States

Mmax_ref / MPW	0.980	0.985	0.990	0.995	1.000
0.01	15.1%	15.1%	15.3%	15.7%	16.3%
0.02	15.1%	15.2%	15.2%	15.4%	15.9%
0.03	15.1%	15.0%	15.5%	16.0%	16.5%
0.04	15.2%	15.8%	16.1%	16.5%	18.4%

TABLE II. dc-link voltage in different system state (V)

Mmax_ref / MPW	0.980	0.985	0.990	0.995	1.000
0.01	612	605	596	584	573
0.02	612	607	605	594	584
0.03	619	618	605	592	583
0.04	630	616	604	592	577

Finally 20 groups of data are obtained and imported to MATLAB to analyze the relationship between current THD, dc voltage, MPW limitation and Mmax ref. The current THD and voltage in different system states are listed in Table I and II, and plotted in Fig. 9 and 10.

It is observed from Fig. 9 and 10 that the THD of converter-side current increase slightly with the Mmax ref increasing when MPW value is below 0.03, while THD boosts critically more than 3% when MPW value is over 0.03. It can be also observed that, dc link voltage decreases with the Mmax ref increasing and reaches its minimum value to when Mmax is 1.

From the experimental results, it can be concluded that the MPW increases the harmonics and reduces the effective applied voltage, thus requires higher dc link voltage for the given modulation.

As a trade-off, it may be recommended that the Mmax ref is set as "1" to obtain the minimum dc link voltage when MPW is from 0.01 to 0.03; when the MPW value is larger than 0.03, the Mmax ref can be reduced to decrease the current distortion.

IV. CONCLUSION AND FUTURE WORK

An improved and simple controller that adaptively regulates the dc-link voltage is used for B2B converter in wind turbine system in this paper.. The dynamic of operation was investigated and the relationship between current THD, dc voltage, MPW limitation and Mmax ref are analyzed in detail by experimental results on a 2kW prototype. The Mmax reference is chosen in trade-off between minimum dc voltage and current THD.

As future works, the Mmax controller operation with discontinuous PWM schemes will be investigated, and stability of the proposed control strategy will be analyzed.

REFERENCES

[1] Blaabjerg, F.; Ke Ma, "Future on Power Electronics for Wind Turbine Systems," *IEEE J. Emerg. Sel. Topics Power Electron.*, vol.1, no.3, pp.139-152, Sept. 2013.

[2] Friedli, T.; Kolar, J.W.; Rodriguez, J.; Wheeler, P.W.; , "Comparative Evaluation of Three-Phase ACAC Matrix Converter and Voltage DC-Link Back-to-Back Converter Systems,", *IEEE Trans. Ind. Electron.*, vol.59, no.12, pp.4487-4510, Dec. 2012.

[3] Massing, J.R.; Stefanello, M.; Grundling, H.A.; Pinheiro, H., "Adaptive Current Control for Grid Connected Converters With LCL Filter," *IEEE Trans. Ind. Electron.*, vol.59, no.12, pp.4681-4693, Dec. 2012.

[4] Trintis, I.; Munk-Nielsen, S.; Abrahamsen, F.; Thoegersen, P.B., "Efficiency and reliability improvement in wind turbine converters by grid converter adaptive control," *Power Electronics and Applications (EPE)*, Lille, France 2013.

[5] Consentino, G.; Laudani, M.; Privitera, G.; Pace, C.; Giordano, C.; Hernandez, J.; Mazzeo, M., "Effects on power transistors of Terrestrial Cosmic Rays: Study, experimental results and analysis," *Applied Power Electronics Conference and Exposition (APEC)*, Fort worth, US, 2014.

[6] Dayaratne, U.I.; Tennakoon, S.B.; Shammas, N.Y.A.; Knight, J.S., "Investigation of variable DC link voltage operation of a PMSG based wind turbine with fully rated converters at steady state," *Power Electronics and Applications (EPE)*, Birmingham, UK, 2011.

[7] B. Welchko, S. Schulz, and S. Hiti, "Effects and compensation of dead-time and minimum pulse-width limitations in two-level PWM voltage source inverters," in *Proc. 2006 IEEE Ind. Appl. Conf.*, 8–12 Oct. 2006, vol. 2, pp. 889–896.

[8] V. Blasko and V. Kaura, "A new mathematical model and control of three-phase ac-dc voltage source converter," Power Electron., IEEE Trans. on, vol. 12, no. 1, pp. 116–123, 1997.

Improved Three-Phase Micro-Inverter Using Dynamic Dead Time Optimization and Phase-Skipping Control Techniques

S. Milad Tayebi, Xianmin Mu and Issa Batarseh
Department of Electrical Engineering and Computer Science
University of Central Florida
Orlando, USA
tayebi@knights.ucf.edu

Abstract— This paper introduces two efficiency improvement techniques for a grid-tied micro-inverter with current mode control zero voltage switching (ZVS) output stages. The first technique is dynamic dead time optimization wherein PWM dead times are dynamically adjusted as a function of load current. The second method is advanced phase-skipping control which distributes power on individual phases depending on the available input power from PV source. Neither of the techniques require any additional components and both can be easily implemented in the digital controller firmware. The two techniques were designed and implemented in a 400W three-phase micro-inverter prototype and the experimental results confirm practical implementation of these techniques and demonstrate that significant efficiency improvement can be achieved.

Index Terms—Three-phase inverters, DC-AC micro-inverters, CEC weighted efficiency, Zero Voltage Switching (ZVS), phase skipping, dead time optimization.

I. INTRODUCTION

Micro-inverters with typical power levels from 200 to 400W are widely used in photovoltaic (PV) system architectures due to improved energy harvesting, high reliability and simple installation. Employing high switching frequency and soft switching techniques reduces cost by shrinking the size of passive components and improves efficiency by reducing switching losses. Soft switching can be improved by minimizing MOSFET body diode conduction time.

Fig. 1 depicts a single-phase half-bridge inverter with a controller providing duty cycles and dead times for the MOSFETs. The inverter MOSFET gates are driven by alternating PWM signals so that when one MOSFET is on, the other is off. Dead time is the state where both MOSFETs are off and one MOSFET's body diode is conducting. Since body diode conduction losses are higher than MOSFET conduction losses, efficiency is maximized when body diode conduction time is small. Body diode conduction time is directly proportional to dead time, therefore it is advantageous to select a dead time value that minimizes body diode conduction time without resulting in shoot-through current.

Excessive dead time can also have a negative impact on inverter efficiency especially at higher operating frequencies where the time is large percentage of the conduction time.

This work was supported in part by the U.S. Department of Energy under Grant DE-EE0003176, FESC and in part by the National Science Foundation (NSF) under Grant ECCS-1156633.

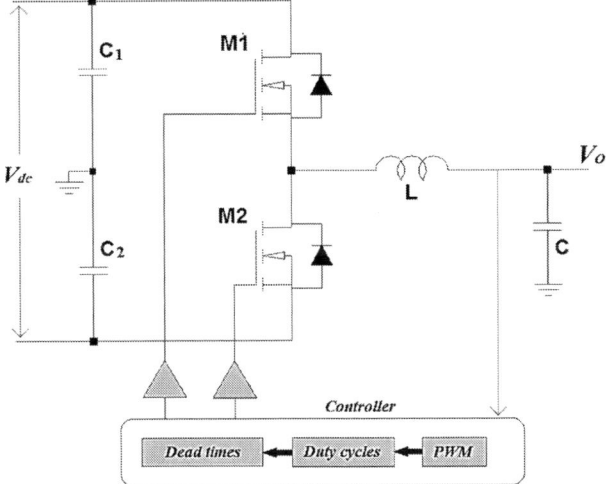

Fig. 1. Single-phase half-bridge inverter with the corresponding controller.

Typically, a fixed dead time is determined based on the worst-case operating conditions such as load, voltage and temperature variations. The body diode conduction loss can be approximated as follows:

$$P_{loss,BD} \approx V_{DS}.I_L.(t_{d1} + t_{d2}).f_{sw} \qquad (1)$$

where, V_{DS} is drain to source voltage or the diode forward voltage drop, I_L is inductor current during the dead times, t_{d1} and t_{d2} are turn-on and turn-off dead times respectively and f_{sw} is switching frequency. Referring to (1), it can be seen that excessive dead time increase the body diode conduction loss due to the diode forward voltage drop.

A number of research papers on PWM dead time optimization for various buck derived topologies have been published [1]-[10].

Switching current sensing was proposed in [1] for buck converters to measure the current flowing through the synchronous MOSFET and detect shoot-through current. Therefore, when it occurs, the dead times must be increased, otherwise they should be decreased. However, this method requires high bandwidth sensors for current measurement which makes its implementation difficult and adds cost.

Switching voltage sensing is sub-divided into the adaptive [2]-[4] and predictive [5]-[7] types. Adaptive control measures the switching voltage across the synchronous rectifier in order to adjust the dead times accordingly. However, predictive control uses the information from the previous switching cycle to adjust the dead times in the current cycle. Although the experimental results show a significant reduction in body diode losses, these control methods require additional circuit components which drive cost up and decrease reliability. Therefore, sensorless optimization methods have been proposed that do not require additional components.

Sensorless dead time optimization for dc-dc converters was proposed in [8] based on maximum power point tracking (MPPT). This method optimizes the dead time by indirectly measuring the efficiency of the converter. Another sensorless dead time optimization method proposed in [9] optimizes the efficiency by minimizing the duty cycle in dc-dc converters. This method is able to optimize both turn-on and turn-off dead times. A perturb and observe algorithm (P&O) was proposed in [10] to detect dead time resulting in maximum efficiency.

Several methods have been proposed to overcome the dead time issues in inverters. A correlation-based control technique was introduced in [11] to optimize dead time in motor drive inverters. The algorithm dynamically minimizes the input current resulting in minimum power loss. The technique proposed in [12] compensates the effect of dead time by adjusting the switching frequency in sinusoidal pulsewidth-modulated voltage-source inverters. Another dead time compensation method was presented in [13] based on harmonic analysis and filtering of the voltage distortion produced by dead time effects by adding a predetermined compensation signal to the voltage reference in ac motor drives. However, this method requires extra components which increases the cost and circuit complexity.

The method introduced in [14] improves THD and the magnitude of the inverter output voltage by forbidding the unnecessary dead time. Another switching strategy for PWM power converters was proposed in [15] on the basis of using independent on/off switching action according to the current polarity. Therefore, as long as the current polarity does not change, the dead time is not needed. These methods can be affected by detection circuit delay, harmonics and A/D converter errors.

In three-phase inverters and micro-inverters, it is advantageous to operate at or close to rated power where efficiency is highest. However, when power levels drop, the available power is distributed evenly among the three phases. This results in reduced efficiency because load independent loss will become a bigger portion of the overall losses. In single-phase inverters, techniques such as pulse skipping [16] are commonly employed in an effort to improve light load efficiency. This method is not predictable and can become problematic in large installations.

In this paper, two control techniques are introduced both of which improve efficiency in three-phase inverters and micro-inverters. The first technique is dynamic dead time optimization wherein PWM dead times are dynamically adjusted as a function of load current. The PWM dead times are calculated by sensing the grid voltage which is also used for duty cycle calculation. The second method is advanced

phase-skipping control which distributes power on individual phases depending on the available input power from PV source. Neither of the techniques require any additional components and both can be easily implemented in the digital controller firmware. The two techniques were designed and implemented in a 400W three-phase micro-inverter prototype and the experimental results confirm practical implementation of these techniques and demonstrate that significant efficiency improvement can be achieved using these digital control algorithms.

II. ZVS Boundary Conduction Mode Current Control of Three-Phase Grid-Tied Inverter

Fig. 2 depicts a standard three-phase half-bridge grid-tied inverter. MOSFET body diodes and parasitic capacitors play an important role when implementing ZVS in an inverter output stage. In order to achieve ZVS, the bidirectional current flow discharges the MOSFET parasitic capacitor and then the body diode conducts prior to each switching transition to apply zero voltage across the MOSFET.

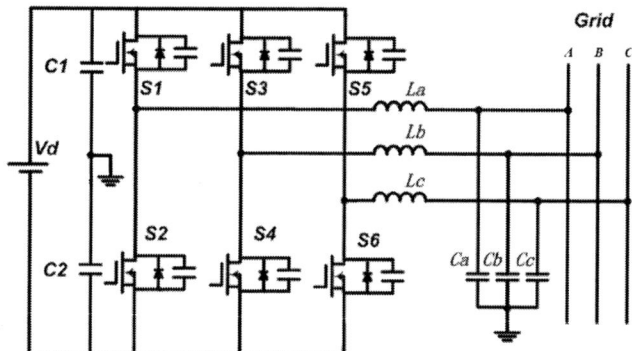

Fig. 2. Three-phase half-bridge inverter.

Although the three phases operate independently, it can be assumed that the system is balanced. Therefore, the analysis can be performed on one of the three phases. In order to simplify the analysis of the inverter stage, steady state operation is assumed with the output voltage equal to the grid voltage. The output voltage is considered to be constant during one switching period of the output stage due to the large difference between the inverter switching frequency and the grid frequency.

Stage 1 (t_1-t_3): In this stage, as shown in Fig. 2, the upper switch (S_1) is on while the lower switch (S_2) is off. Since voltage across the inductor is positive, the inductor current is linearly increasing. The inductor current can be calculated as follows:

$$i_L(t) = \frac{V_{dc}/2 - V_{ac}}{L}t + i_L(t_1) \qquad (2)$$

Stage 2 (t_3-t_4): In this stage, the inductor current reaches the upper limit and then the upper switch (S_1) is turned off. As shown in Fig. 3, both switches are off and the parasitic capacitors of the MOSFETs S_1 and S_2 are being respectively charged and discharged by the inductor current. This stage is

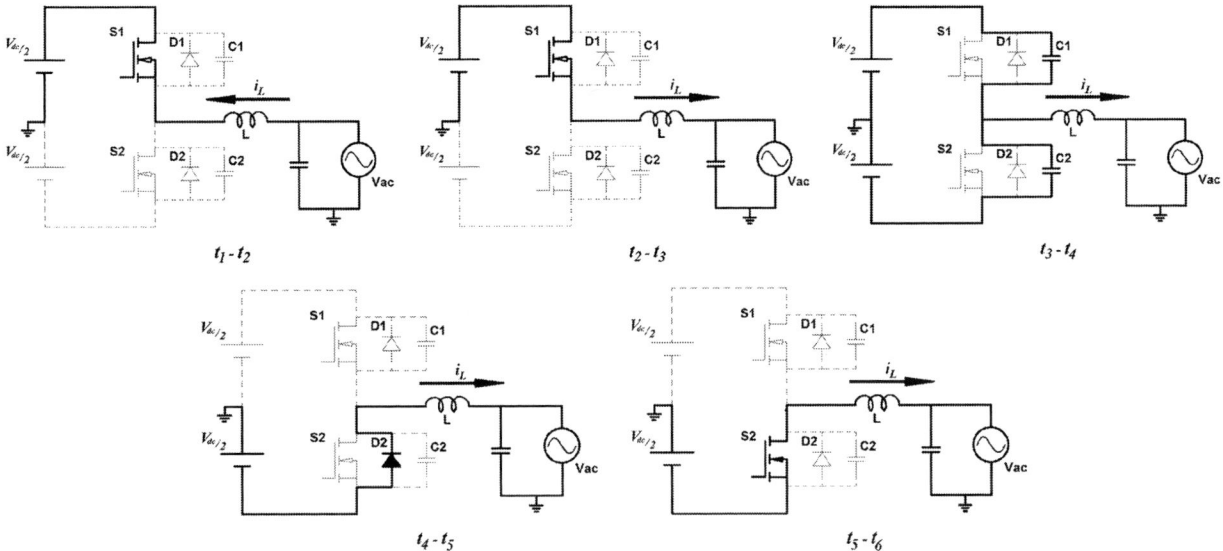

Fig. 3. Operating intervals of a single-phase half-bridge inverter.

resonant stage, and the inductor current and parasitic capacitors voltage can be calculated as follows:

$$
\begin{cases}
i_L(t) = \dfrac{V_{dc}/2 - V_{ac}}{Z_o} sin\omega_o(t - t_3) + i_L(t_3)cos\omega_o(t - t_3) \\[2mm]
V_{CS1}(t) = (V_{dc}/2 - V_{ac}) + i_L(t_3)Z_o sin\omega_o(t - t_3) \\
\qquad\qquad - (V_{dc}/2 - V_{ac})cos\omega_o(t - t_3) \\[2mm]
V_{CS2}(t) = (V_{dc}/2 + V_{ac}) - i_L(t_3)Z_o sin\omega_o(t - t_3) \\
\qquad\qquad + (V_{dc}/2 - V_{ac})cos\omega_o(t - t_3)
\end{cases}
\quad (3)
$$

where, ω_o, the natural resonant frequency, and Z_o, the characteristic impedance of the resonant tank are respectively defined as follows:

$$
\begin{cases}
\omega_0 = \dfrac{1}{\sqrt{L.2C}} \\[2mm]
Z_0 = \sqrt{\dfrac{L}{2C}} \\[2mm]
C_{s1} = C_{s2} = C
\end{cases}
\quad (4)
$$

Stage 3 (t_4-t_6): In this stage, as soon as the parasitic capacitors of the MOSFETs S_1 and S_2 are fully charged and discharged respectively and the voltage on C_2 is zero, the body diode of MOSFET S_2 (D_2) starts conducting at t_4. Therefore, zero voltage is applied across MOSFET S_2 and it is ready to be turned on under ZVS condition. At t_5, the lower switch (S_2) turns on and the inductor current flowing through the body diode is transferred to the switch (S_2).

During this stage, since voltage across the inductor is negative, the inductor current is linearly decreasing and can be approximated as follows:

$$
i_L(t) = \frac{-V_{dc}/2 - V_{ac}}{L}t + i_L(t_4)
\quad (5)
$$

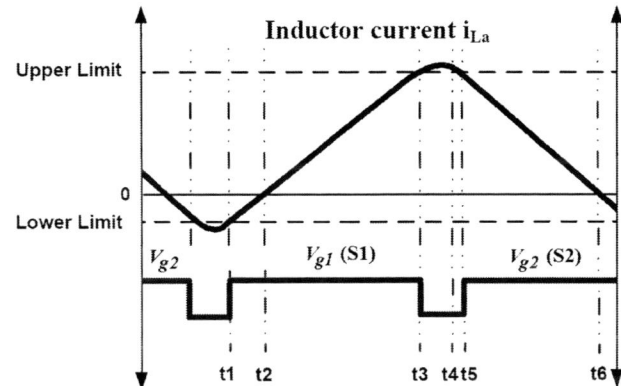

Fig. 4. Inductor current and driving signals for ZVS BCM current control.

For ZVS boundary conduction mode (BCM) control implementation, the upper and lower boundaries of the inductor current need to be determined so that the inverter injects only AC current to the grid. Therefore, the average output current must be equal to the reference current during each switching cycle.

There are three possible inductor current modulation methods satisfying this requirement: BCM with fixed reverse current, BCM with variable reverse current, and BCM with fixed bandwidth. Experimental results from the 400W three-phase micro-inverter prototype in [17] and [18] show that the maximum efficiency can be achieved in BCM with fixed reverse current modulation.

Fig. 5 demonstrates the upper and lower boundaries of BCM with fixed reverse current modulation along with the average inductor current. For the positive half cycle, the upper boundary is determined so that the average current is equal to the reference current during each switching cycle, while the lower boundary provides the reverse current to discharge the MOSFET parasitic capacitors for achieving ZVS during each switching frequency.

Similarly, for the negative half cycle, the upper limit provides ZVS for the MOSFETs and the lower limit produces the reference current. It can be seen that the average current is equal to the reference current. The upper and lower boundaries for this inductor current modulation can be obtained as follows:

$$\begin{cases} I_{upper} = 2I_{REF}\sin(\omega t) + I_o \\ I_{lower} = -I_o \end{cases} \text{, if } \sin(\omega t) \geq 0$$

$$\begin{cases} I_{upper} = I_o \\ I_{lower} = 2I_{REF}\sin(\omega t) - I_o \end{cases} \text{, if } \sin(\omega t) < 0$$

$$(6)$$

where, I_o is the reverse current for achieving ZVS and I_{REF} is the reference peak current.

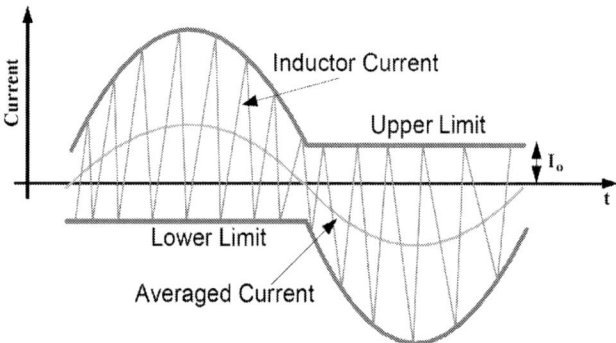

Fig. 5. Upper and lower boundaries for BCM ZVS current control with fixed reverse current.

Referring to the Fig. 5, during the time where the inductor current transitions from positive to negative for positive half cycle and from negative to positive for negative half cycle, ZVS is realized. The power dissipation during this interval is a function of the amount of time that the ZVS MOSFET's body diode is conducting. This time is determined by the amount of fixed dead times provided by the controller.

In a fixed dead time control, a constant period is applied for both turn-on and turn-off dead times throughout the entire operation. The dead times must be long enough to achieve ZVS by allowing the inductor current to fully charge and discharge the MOSFET parasitic capacitors. The required dead time can be calculated as follows:

$$t_d = \frac{2CV_{dc}}{I_o} \qquad (7)$$

where, C is the MOSFET parasitic capacitors, V_{dc} is the voltage on the MOSFET parasitic capacitors and I_o is the minimum worst-case reverse inductor current.

III. DYNAMIC DEAD TIME OPTIMIZATION

The optimum dead time required to minimize MOSFET body diode conduction time during ZVS is a function of load current and grid voltage. Referring to Fig. 6, if the voltage at point 'a' is known, the voltage on the MOSFET parasitic capacitors can be determined.

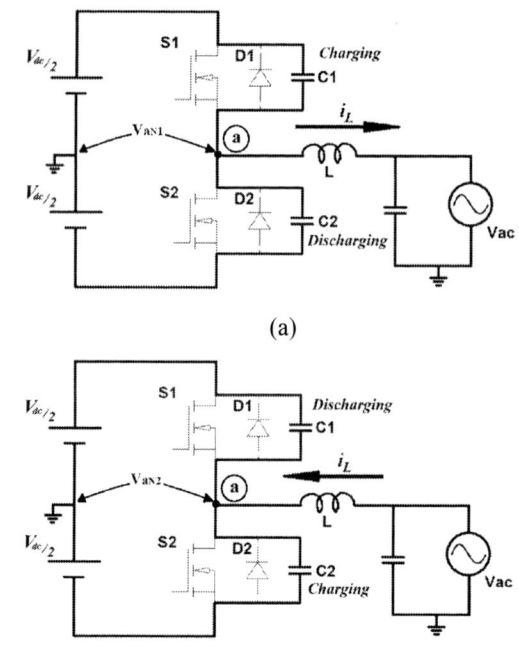

(a)

(b)

Fig. 6. ZVS dead time resonant intervals.

The voltage at point 'a' during the turn-on dead time (t_{d1}) and turn-off dead time (t_{d2}) can be calculated using the equations below, respectively.

$$V_{aN1}(t) = V_{ac} + \left(\frac{V_{dc}}{2} - V_{ac}\right)cos\omega_o(t_{d1})$$
$$- Z_o.I_{upper}sin\omega_o(t_{d1}) \qquad (8)$$

$$V_{aN2}(t) = V_{ac} - \left(\frac{V_{dc}}{2} + V_{ac}\right)cos\omega_o(t_{d2})$$
$$- Z_o.I_{lower}sin\omega_o(t_{d2}) \qquad (9)$$

From (8) and (9), the optimum dead times can now be calculated for each switching cycle as follows:

$$t_{d1} = \left| \frac{-i}{\omega_o}.log\frac{-2V_{ac} - V_{dc} + i2\sqrt{Z_o^2 I_{upper}^2 - 2V_{ac}V_{dc}}}{V_{dc} - 2V_{ac} + i2Z_o I_{upper}} \right| \quad (10)$$

$$t_{d2} = \left| \frac{-i}{\omega_o}.log\frac{2V_{ac} - V_{dc} + 2\sqrt{-Z_o^2 I_{lower}^2 - 2V_{ac}V_{dc}}}{V_{dc} + 2V_{ac} - i2Z_o I_{upper}} \right| \quad (11)$$

The optimum dead times for various power levels during positive and negative half cycles for a 400W three-phase micro-inverter prototype using BCM ZVS current control with fixed reverse current modulation are shown in Fig. 7. The fixed dead time for this prototype was set at 600 ns for both turn-on and turn-off dead times. The MOSFET used is FCB20N60F with V_{dc}=400 V and a 270 μH output inductor.

It should be noted that t_{d1} and t_{d2} represent turn-on and turn-off dead times respectively. As shown in Fig. 7, for positive half cycles, the optimum turn-on dead time (t_{d1}) varies with changes in power levels due to the fact that the upper boundary in BCM is sinusoidal, while the optimum turn-off dead time (t_{d2}) remains constant with load variations due to the fixed reverse current in BCM modulation. Similarly, during negative half cycles, t_{d1} remains constant and t_{d2} changes with power levels.

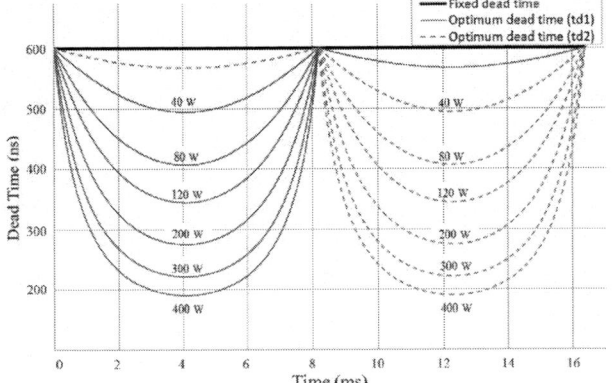

Fig. 7. Optimum dead time verses fixed dead time for different loads.

IV. ADVANCED PHASE-SKIPPING CONTROL

Advanced phase-skipping control maximizes the output stage efficiency of a three-phase micro-inverter by managing the amount of power on each phase based on the available power [19]. The available power is determined by the PV side MPPT controller.

In a three-phase system, there are three possible modes of operation: normal three-phase mode, two-phase mode and single-phase mode. Advanced phase-skipping control transitions between these three modes in order to maximize inverter efficiency. Based on the available power, the controller can disable one or two phases of the three-phase system. At power levels above two-thirds of rated power, all three phases are enabled. At one-third to two-thirds of rated power, the controller disables one phase. At power levels below one-third of rated power, the controller enables only one phase.

Available power may vary over a wide range during a typical day due to environmental factors such as clouds and shading as well as low irradiance during morning and afternoon. During times when available power is low, the three-phase micro-inverter will most likely be operating in single-phase mode. As available power increases during the day, the controller will enable two and eventually all three phases and distribute power evenly in each phase in order to maximize overall system efficiency.

V. EXPERIMENTAL RESULTS

The experimental results were obtained from a 400W three-phase half-bridge micro-inverter prototype using two different manufacturer's MOSFETs for switching devices. The MOSFETs specifications are included in Table I for reference.

The micro-inverter was tested in four different modes of operation; normal operation (with fixed dead time and no phase skipping), with dynamic Dead Time Optimization (DTO) only, with advanced phase-skipping control only, and with both dynamic DTO and advanced phase-skipping control.

TABLE I
Two Different Manufacturer's MOSFETs Specifications

Manufacturer Part Number	V_{dss} (V)	I_d (A)	R_{ds}-ON (Ω)	C_{iss} (pF)	C_{oss} (pF)	C_{rss} (pF)	V_{SD}^{*} (V)
FCB20N60F	600	20	0.150	2370	1280	95	1.4
STB45N65M5	650	35	0.067	3470	82	7	1.5

* Drain to Source Diode Forward Voltage

Figs. 8 and 9 show the efficiency measured with a Yokogawa PZ4000 power analyzer for each of the micro-inverter operating modes for FCB20N60F and STB45N65M5 MOSFETs respectively.

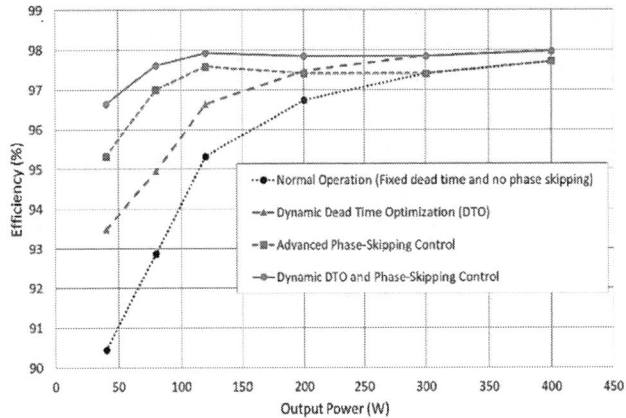

Fig. 8. Micro-inverter output stage efficiency with FCB20N60F MOSFET.

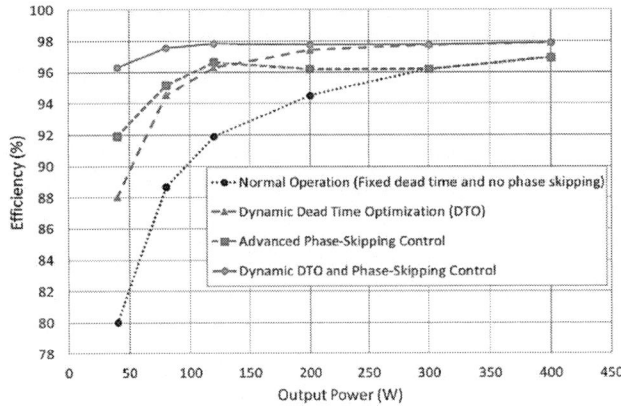

Fig. 9. Micro-inverter output stage efficiency with STB45N65M5 MOSFET.

Referring to the efficiency figures, it can be seen that both manufacturer's MOSFETs exhibited similar performance with respect to each of the four micro-inverter operating modes. Efficiency was the lowest over the entire load range when the micro-inverter was operated in normal mode for both manufacturer's MOSFETs. A noticeable improvement in efficiency was achieved by implementing dynamic DTO. Efficiency improvement was the highest at light loads where switching frequency is maximum. Implementing advanced phase-skipping control also resulted in significant efficiency improvement especially at light loads. The best efficiency was realized with a combination of dynamic DTO and advanced phase-skipping control. CEC weighted efficiency is shown in Table II for both manufacturer's MOSFETs and for each modes of operation.

TABLE II
CEC Weighted Efficiency for Different Modes of Operation

Manufacturer Part Number	Normal Operation (Fixed Dead Time)	Dynamic DTO	Phase Skipping	Dynamic DTO + Phase Skipping
FCB20N60F	96.52%	97.31%	97.33%	97.81%
STB45N65M5	94.34%	96.98%	96.07%	97.72%

Combining dynamic dead time optimization and advanced phase-skipping control produced an additional 1.3 percentage points in CEC efficiency for FCB20N60F MOSFET and 3.4 percentage points in CEC efficiency for STB45N65M5 MOSFET.

It is worth noting that neither of the two efficiency improvement techniques described in this paper require any additional circuit components. Therefore, reliability and cost will not be negatively impacted and in fact reliability will be improved due to lower operating junction temperatures. Although advanced phase-skipping control is applicable only to three-phase inverters and micro-inverters, dynamic dead time optimization will improve efficiency of both three-phase and single-phase inverters and micro-inverters that employ soft switching techniques such as BCM ZVS current control.

VI. CONCLUSION

In this paper, two efficiency improvement techniques have been introduced and implemented in a three-phase micro-inverter. Practical implementation of these two techniques has been verified by experimental results on a 400W three-phase micro-inverter prototype. It is very important to note that no additional circuit components are required and the techniques can be implemented entirely in the digital controller firmware.

Dynamic dead time optimization minimizes MOSFET body diode conduction time which reduces power dissipation. Therefore, reliability will be improved due to lower operating junction temperatures. This technique can be digitally implemented on three-phase and single-phase inverters and micro-inverters that employ soft switching techniques. Advanced phase-skipping control is applicable to three-phase inverters and micro-inverters. Depending on the available input power from the PV source, this control method features distributed power to individual phases so that the maximum DC/AC stage efficiency of the three-phase micro-inverter is achieved. The experimental results show that combining dynamic dead time optimization and advanced phase-skipping control can greatly improve the CEC efficiency of a 400W three-phase half-bridge micro-inverter prototype.

For future work, dynamic dead time optimization could be applied to other application areas such as power supplies and motor controllers. Advanced phase-skipping control implementation requires the development of a master controller in order to maintain a balanced three-phase system.

VII. REFERENCES

[1] H.W. Huang, C. Y. Hsieh, K. H. Chen, and S. Y. Kuo, "Load dependent dead-times controller based on minimized duty cycle technique for DC-DC buck converters," in Proc. PESC Record – IEEE Annual Power Electron. Spec. Conf., pp. 2037–2041, 2007.

[2] P. T. Krein and R. M. Bass, "Autonomous control technique for high-performance switches," IEEE Trans. Ind. Electron., vol. 39, no. 3, pp. 215–222, Jun. 1992.

[3] W. L. W. Lau and S. Sanders, "An integrated controller for a high frequency buck converter," in Proc. PESC '97 Record. 28th Annual IEEE Power Electron. Spec. Conf., vol. 1, 1997.

[4] B. Acker, C. Sullivan, and S. Sanders, "Synchronous rectification with adaptive timing control," in Proc. PESC '95 Record. 26th Annual IEEE Power Electron. Spec. Conf., vol. 1, 1995.

[5] A. Zhao, A. A. Fomani, and W. T. Ng, "One-step digital dead-time correction for DC-DC converters," in Proc. APEC '04. 19th Annual IEEE Appl. Power Electron. Conf. (APEC) 2010, pp. 132–137.

[6] L. Mei, D. Williams, and W. Eberle, "A Synchronous Buck Converter Using a New Predictive Analog Dead-Time Control Circuit to Improve Efficiency," Canadian Journal of Electrical and Computer Engineering, vol. 36, no. 4, pp. 181–187, 2013.

[7] O. Trescases, W. T. N. W. T. Ng, and S. C. S. Chen, "Precision gate drive timing in a zero-voltage-switching DC-DC converter," in Proc. The 16th International Symposium on Power Semiconductor Devices and ICs, ISPSD '04, 2004.

[8] J. Abu-Qahouq, H. M. H. Mao, H. Al-Atrash, and I. Batarseh, "Maximum Efficiency Point Tracking (MEPT) Method and Digital Dead Time Control Implementation," IEEE Trans. Power Electron., vol. 21, no. 5, pp. 1273–1281, Sept. 2006.

[9] V. Yousefzadeh and D. Maksimovi´c, "Sensorless optimization of dead times in dc-dc converters with synchronous rectifiers," IEEE Trans. Power Electron., vol. 21, no. 4, pp. 994–1002, Jul. 2006.

[10] A. V. Peterchev and S. R. Sanders, "Digital multimode buck converter control with loss-minimizing synchronous rectifier adaptation," IEEE Trans. Power Electron., vol. 21, no. 6, pp. 1588–1599, Nov. 2006.

[11] J. W. Kimball and P. T. Krein, "Real-time optimization of dead time for motor control inverters," in Proc. IEEE Power Electron. Specialists Conf. 1997, pp. 597–600.

[12] A. C. Oliveira, C. B. Jacobina, and A. M. N. Lima, "Improved dead-time compensation for sinusoidal PWM inverters operating at high switching frequencies," IEEE Trans. Ind. Electron., vol. 54, no. 4, pp. 2295–2304, Aug. 2007.

[13] S. Y. Kim and S.Y. Park, "Compensation of dead-time effects based on adaptive harmonic filtering in the vector-controlled AC motor drives," IEEE Trans. Ind. Electron., vol. 54, no. 3, pp. 1768–1777, Jun. 2007.

[14] J. S. Choi, J. Y. Yoo, S. W. Lim, and Y. S. Kim, "A novel dead-time minimization algorithm of the PWM inverter," in Proc. IEEE IAS Conf. Rec. 1999, vol. 4, pp. 2188–2193.

[15] K. M. Cho, W. S. Oh, Y. T. Kim, and H. J. Kim, "A new switching strategy for pulse width modulation (PWM) power converters," IEEE Trans. Ind. Electron., vol. 54, no. 1, pp. 330–337, Feb. 2007.

[16] H. Hu, W. Al-Hoor, N. H. Kutkut, I. Batarseh, and Z. J. Shen, "Efficiency Improvement of Grid-Tied Inverters at Low Input Power Using Pulse-Skipping Control Strategy," IEEE Trans. Power Electron., vol. 25, no. 12, pp. 3129-3138, Dec. 2010.

[17] A. Amirahmadi, H. Hu, A. Grishina, Q. Zhang, L. Chen, U. Somani, and I. Batarseh, "Hybrid ZVS BCM current controlled three-phase microinverter," IEEE Trans. Power Electron., vol. 29, no. 4, pp. 2124–2134, Apr. 2014.

[18] A. Amirahmadi, L. Chen, U. Somani, N. Kutkut, and I. Batarseh, "High Efficiency Dual Mode Current Modulation Method for Low Power DC/AC Inverters," IEEE Trans. Power Electron., vol. 29, no. 6, pp. 2638-2642, Jun. 2014.

[19] S. Milad Tayebi, Charles Jourdan, and Issa Batarseh, "Advanced phase-skipping control with improved efficiency of three-phase micro-inverters," IEEE Energy Convers. Congr. Expo. (ECCE) 2015, pp. 3802–3806.

Correcting Current Imbalances in Three-Phase Four-Wire Distribution Systems

Vinson Jones and Juan Carlos Balda
Department of Electrical Engineering
University of Arkansas at Fayetteville
Fayetteville, Arkansas, USA
email: vinson.j.jones@gmail.com, jbalda@uark.edu

Abstract—This paper presents the analysis and control of a current imbalance compensator designed to eliminate upstream negative- and zero- sequence current components caused by single-phase loads in three-phase four-wire systems. The proposed compensator significantly reduces the fundamental-frequency current component in the neutral conductor and can potentially be used to eliminate harmonic currents. The Matlab/Simulink™ simulations are based on a 3-MVA radial feeder case study and a scaled-down 10-kVA test setup used to validate the proposed ideas. Three TMS320F28335 DSPs are used to implement the control system and the results of both the simulations and the prototype testing are shown and discussed. Results from testing of the prototype demonstrate the ability of the compensator to reduce the current in the neutral conductor.

Keywords—*unbalanced compensation, power electronics, negative- and zero-sequence currents*

I. INTRODUCTION

The existence of single-phase loads and single-phase generation leads to inherent current imbalances in three-phase power distribution systems. These imbalances having negative- and zero-sequence components cause system adverse problems [1] □ [6]. Examples are: overheating of a generator, improper sizing of neutral conductors, MMF ripples inside the generator due to negative-sequence MMFs, and overheating of the holding tank for ungrounded wye-connected transformers.

Several methods have been proposed to eliminate negative- and zero-sequence components [7] □ [13]. A zig-zag transformer in parallel with the load is proposed in [7] in order to provide a low-impedance path for zero-sequence currents. In [8] a series inverter is placed on the neutral conductor in order to produce a high impedance path for zero-sequence currents. In [10] and [11] shunt active filters are used to compensate for the load harmonics. In [12] a 4-leg inverter topology is controlled using space-vector modulation in order to effectively compensate for neutral conductor currents. An instantaneous power alpha-beta reference frame controller in [13] is used to compensate for harmonics, power factor and current unbalances.

The solution proposed in this paper based upon the shunt active compensator concept is termed the Unbalanced Current Static Compensator (UCSC). This equipment uses three single-phase voltage source inverters connected to each phase of a three-phase four-wire system, sharing the same dc link capacitor bank as shown in Fig 1. The UCSC compensates for the negative- and zero-sequence components of unbalanced systems by compensating for the power factor of each phase, and then circulating the substation currents through the common dc-link capacitor in order to balance the upstream currents. This is accomplished through a d-q reference frame current controller. Harmonic filtering can also be potentially achieved if there is enough computational power and the switching frequency is high enough.

This paper is organized as follows: Section II presents an overview of the current compensator, Section III the control algorithm, Section IV simulation results, Section V experimental results and Section VI the conclusions.

II. OVERVIEW OF THE CURRENT COMPENSATOR

The connection of the proposed UCSC strategically located near the power transformer at a distribution substation and the selected case study are shown in Fig. 1. The compensator consists of three single-phase voltage-source inverters that are current controlled and coupled to the grid through three single-phase step-up transformers. These inverters are placed in parallel with the grid. The compensator is placed near to the power transformer in order to compensate for all the downstream loads. To balance the currents as seen by the transformer the inverter either draws or injects current onto the grid to bring the magnitude of each of the transformer currents to the same value, while also compensating for the reactive power consumption of the downstream loads.

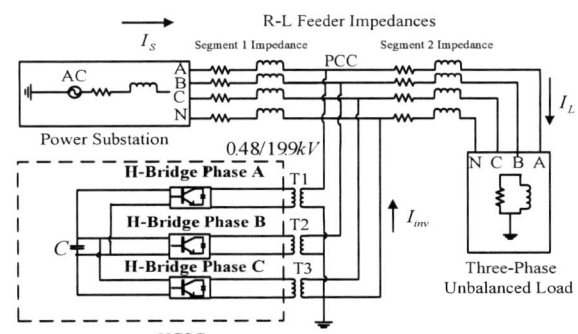

Fig. 1 Connection of the UCSC to the distribution system.

978-1-4673-9551-9/16 $31.00 © 2016 IEEE

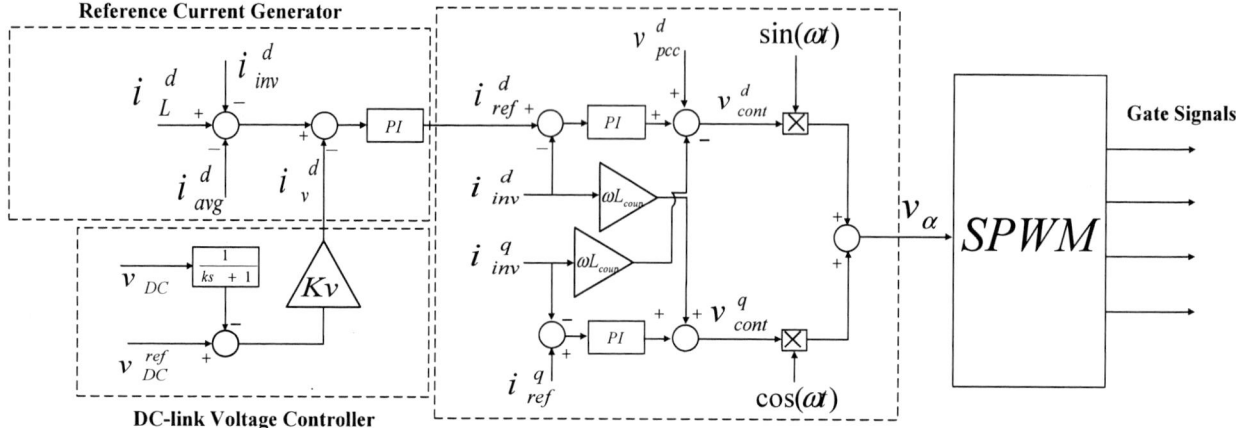

Fig. 2 Control block diagram for each UCSC inverter.

III. USCS CONTROL ALGORITHM

A. Current Controller Stage

The control block schematic for each inverter is shown in Fig. 2. At the point of common coupling (PCC) the UCSC injects current i_{inv} to compensate for the load current i_L such that the substation current i_s is balanced. The controller first transforms these currents along with phase voltages at the PCC into the d-q synchronously-rotating reference frame using a single-phase α-β to d-q transformation as shown in (1)-(2):

$$i_d = i_\alpha \cos(\omega t) + i_\beta \sin(\omega t) \tag{1}$$

$$i_q = -i_\alpha \sin(\omega t) + i_\beta \cos(\omega t) \tag{2}$$

where i_α is the present value of the current waveform, i_β equals i_a, but lagging by a quarter of the grid cycle, and ω is the grid frequency.

Next, an average d-axis load current for the three phases $i_{avg}{}^d$ is computed. Each of the substation currents ($i_a{}^d$, $i_b{}^d$, $i_c{}^d$) is compared to this average value through current summation at the PCC ($i_s{}^d = -i_{inv}{}^d - i_L{}^d$) to develop the d-axis current reference. A voltage controller is required to replenish lost energy in resistive components (e.g., semiconductor devices),

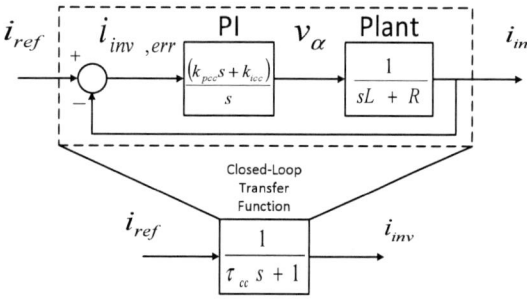

Fig. 3 Closed loop transfer function.

and thus, to maintain the required voltage for the dc bus. The voltage controller compares a reference voltage $v_{DC}{}^{ref}$ to the measured dc-link capacitor voltage v_{DC} and the developed error signal is used to modify the d-axis current reference error. This error signal is fed through a PI controller that determines the reference's response to changing loads and is sent to the current controller as $i_{ref}{}^d$.

D-axis reference $i_{ref}{}^d$ is then compared to the inverter d-axis current $i_{inv}{}^d$. The error between these two signals is then routed to a PI controller to establish the response characteristics of the current controller. The d-axis component of the grid voltage $v_{PCC}{}^d$ is added to the signal as a feed-forward element in order to bias the voltage output of the inverter. The same process is done for the q-axis current, however the current reference becomes the q-axis component of the load current, the feed-forward grid voltage is assumed to be 0. A decoupling term ωL_{coup} is added after the PI controllers for the d- and q-axis components in order for the two to act independently; namely:

$$\omega L_{coup} = \omega(L_{filter} + L_{leakage}) \tag{3}$$

In order to produce a control signal for the PWM generator the signals are converted back to the α-β reference frame, where the α component is the output voltage waveform.

B. Selection of the Current Controller PI Gains

Fig. 3 shows the simplification of the current controller to a single-pole transfer function. This is accomplished through pole cancellation; (4) shows the proportional gain, k_P and integral gain, k_I required for the simplification:

$$\frac{k_I}{k_P} = \frac{R_{coup}}{L_{coup}} \quad , \quad \frac{L_{coup}}{k_P} = \tau_{cc} \tag{4}$$

τ_{cc} is the inverse of the chosen controller bandwidth [15].

C. Selection of the Reference Generator PI Gains

(5) shows the relationship used to select the gains [16].

$$i_{cap} = C_{DC}\frac{dv_{DC}}{dt} = [k_{pref}(v_{DC}^{ref} - v_{DC}) + k_{iref}\int(v_{DC}^{ref} - v_{DC})] \qquad (5)$$

Transferring to the Laplace domain and finding v_{DC} / v_{DC}^{ref} results in:

$$\frac{v_{DC}}{v_{DC}^{ref}} = \frac{k_{pref}}{C_{DC}}\frac{(s+\frac{k_{iref}}{k_{pref}})}{(s^2+\frac{k_{pref}}{C_{DC}}s+\frac{k_{iref}}{C_{DC}})} \qquad (6)$$

This can be fitted to the coefficients of the general form of a two-pole second-order transfer function shown in (7); by matching the coefficients of the denominator, the polynomial (8) and (9) can be used to determine k_{pref} and k_{iref}:

$$T(s) = \frac{\omega_{cc}^2}{(s^2 + 2\zeta\omega_{cc}s + \omega_{cc}^2)} \qquad (7)$$

$$k_{pref} = 2\zeta\omega_{cr}(C_{DC}) \qquad (8)$$

$$k_{iref} = (C_{DC})\omega_{cr}^2 \qquad (9)$$

Where ζ is the damping coefficient and ω_{cr} is the required bandwidth of the reference controller.

D. d-q Trasformations and Phase-Lock Loop

To have a more robust system, the source impedance of the feeder where the UCSC is installed is assumed to have a non-negligible value. To deal with this issue the phase-lock loop and the d-q transformations required for operation were done on a single-phase basis. The single-phase d-q transformations require a quarter cycle delayed signal of the sampled voltage and current measurements. This was achieved with the use of a SOGI-QSG. The phase-lock loop uses a single-phase d-q transformation in which the predicted grid phase is used in the d-q transformation and the resulting value of the q-axis output is used as the phase error for the next prediction. The d-q transformations of i_{inv} and i_L use the predicted grid phase angle from the phase-lock loop as the reference phase angle.

IV. UCSC Simulations Results

A. Case Study

The 34.5 kV$_{\text{L-L}}$, 6 MVA system illustrated in Fig. 1 is simulated using Matlab/Simulink™ software. Each of the inverters in the UCSC is rated 480 V$_{\text{L-N}}$, 1 MVA. The substation current waveforms for the different modes of operation of the UCSC are shown in Fig. 4. From t = 0.00 s to t = 0.10 s the UCSC is inactive. From t = 0.10 s to t = 0.15 s the UCSC is compensating for the reactive power of each load. From t = 0.15 s to t = 0.35 s the UCSC is fully operational with a step change in load of 30% on Phase A at t = 0.25 s. The UCSC reaches steady state within 1 cycle at t = 0.15 s and attains an average UBF of 0.497% while fully operational.

$$\text{UBF} = \frac{|negative\ sequence|}{|postive\ sequence|}x100\% \qquad (10)$$

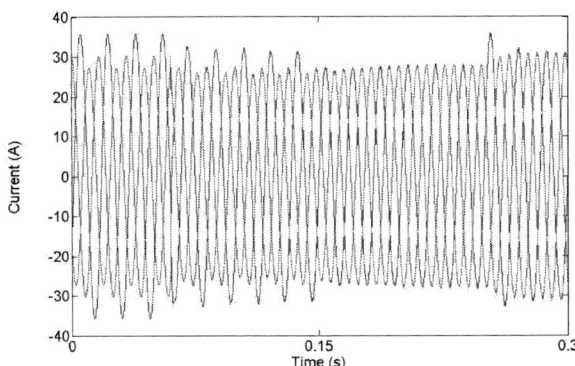

Fig. 4 Substation Currents.

B. UCSC Prototype

In order to reflect how the controller is run on the actual system at rated conditions and on the DSP, a simulation is constructed using the z-domain implementation of the control method which includes a switching dead-time of 0.7 µs. The sampling frequency for the system is 10 kHz and the switching frequency is 5 kHz. The load currents that are compensated by the UCSC are given in Fig. 5, the currents from the compensator in Fig. 6, and the currents at the substation in Fig. 7 with the black waveform being the neutral current; the neutral current has almost been eliminated at the substation.

V. Results from Prototype Testing

The prototype under testing shown in Fig. 8 is a 10-kVA scaled-down version of the case study simulated above. The

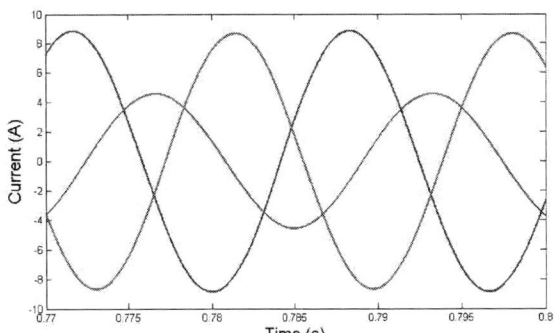

Fig. 5 Load currents from prototype simulations.

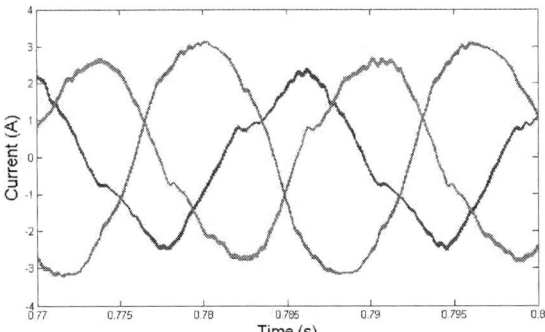

Fig. 6 UCSC currents from prototype simulations.

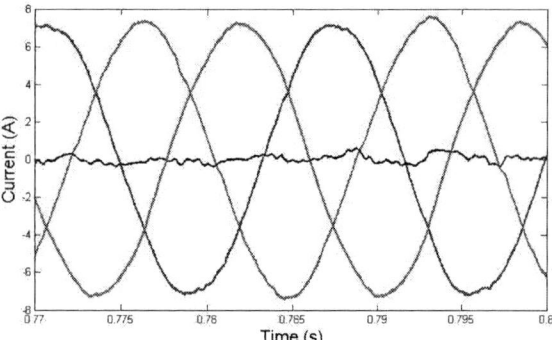

Fig. 7 Substation currents from prototype simulations.

Fig. 8 The prototype under testing.

UCSC is connected via a 1:1 isolation transformer to a variable voltage system. Each inverter is rated for 120 V, 1 kVA with a common dc bus voltage of 250 V and its components are given in Table I. The output LCL filter for each inverter is constructed based on the work done in [12] and its parameters are given in Table I. The dc-link capacitor was chosen using ripple current calculations from [17] and a desired dc bus voltage ripple of 1%.

TABLE I. PROTOTYPE COMPONENTS

Component	Part
IGBT	IRG4PC30FDPbF, 600 V, 31 A
DC bus capacitor	2200 uF, 450 V
Filter Inductor 1	2.2 mH, 10 A
Filter Capacitor	10 uF
Damping Resistor	3.2 Ω
Filter Inductor 2	0.5 mH, 10 A

The load currents that the UCSC is compensating for during lab testing are displayed in Fig. 9. The resulting substation currents during operation of the UCSC are captured in Fig. 10. In both figures, the neutral current is shown in black. The fundamental current component that is present in the neutral conductor of the load is no longer present in the upstream substation neutral conductor. Some lower-harmonic components still exist due to the effects of the dead-time in full-bridge inverter operation. Demonstrating this is Fig. 11 which displays the inverter current from the UCSC. An FFT of the neutral current is presented in Fig. 12.

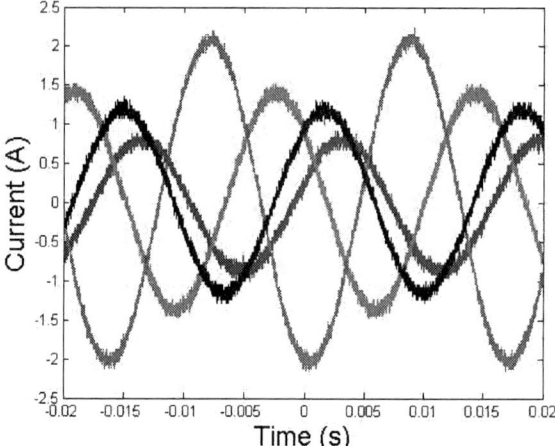

Fig. 9 Load currents from prototype testing.

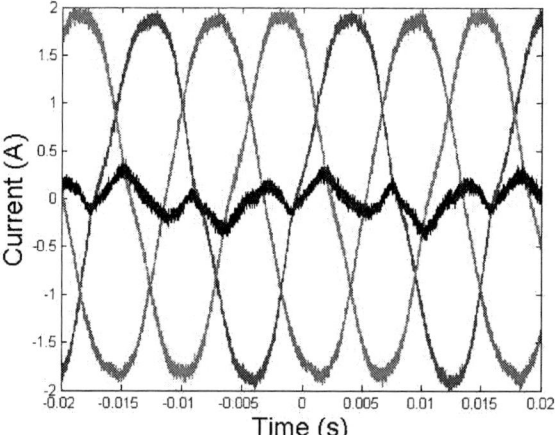

Fig. 10 Substation currents from prototype testing.

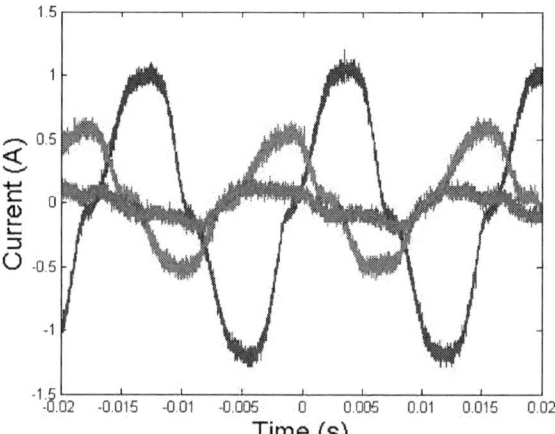

Fig. 11 Inverter currents from prototype testing.

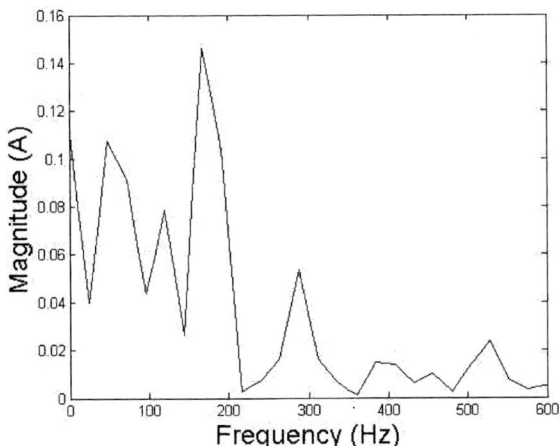

Fig. 12 FFT of the substation neutral current during UCSC operation

VI. CONCLUSIONS

A current controller based on the d-q rotating reference frame is used to control a three-phase active power filter in order to balance the fundamental currents in a four-wire electric power distribution system. After implementing a case-study consisting of a 6-MVA radial feeder system in Matlab/Simulink™, simulations were run to design the controller and analyze the effectiveness of the UCSC to eliminate the neutral currents in the upstream neutral conductor, and thus, achieving a negligible system UBF. After designing the controller in the z-domain, additional simulations were run to more closely emulate the behavior of a prototype rated 10-kVA and 120 V. The control algorithm was implemented using three TMS320F28335 DSPs. The UCSC was effective in removing the fundamental current component from the neutral conductor. However, some neutral current remained as a result of UCSC unipolar switching that causes a voltage distortion near the zero crossings. If harmonic voltages are present in the grid, the ability of the UCSC to suppress neutral current is severely reduced. This is because there is no mechanism to compensate for these harmonics in the current controller.

ACKNOWLEDGMENTS

The authors are grateful to the financial support from the NSF I/UCRC Grid-Connected Advanced Power Electronic Systems (GRAPES) under grant IIP-1439700.

REFERENCES

[1] A. P. S. Meliopoulos, J. Kennedy, C. A. Nucci, A. Borghetti, G. Contaxis, "Power distribution practices in USA and Europe: impact on power quality," in Proceedings of the 8th International Conference on Harmonics and Quality of Power Proceedings, vol. 1, pp. 24□29, October 1998.

[2] R.G. Harley, E.B. Makram, E.G. Duran. (1987); "The effects of unbalanced networks on synchronous and asynchronous machine transient stability," Electric Power System Research[Online].Available:http://www.sciencedirect.com/science/articl e/pii/0378779687900393

[3] W-J. Lee, T.-Y. Ho, J.-P. Liu, Y.-H. Liu, "Negative sequence current reduction for generator/turbine protection," Conference record of the 1993 IEEE Industry Applications Society Annual Meeting, vol. 2, pp. 1428□1433, October, 1993.

[4] R.H. Salim, R.A. Ramos, N.G. Bretas, "Analysis of the small signal dynamic performance of synchronous generators under unbalanced operating conditions," IEEE Power and Energy Society General Meeting, pp. 1□6, July, 2010.

[5] B. N. Gafford, W.C. Duesterhoeft, C.C. Mosher, "Heating of induction motors on unbalanced voltages," Transactions of the American Institute of Electrical Engineers Power Apparatus and Systems, Part III , vol. 78, no. 3, pp. 282□286, April 1959.

[6] T. A. Short. Electrical Power Distribution Handbook. Boca Raton, FL: CRC Press Taylor & Taylor Group, 2004.

[7] Hurng-Liahng Jou; Jinn-Chang Wu; Kuen-Der Wu; Wen-Jung Chiang; Yi-Hsun Chen, "Analysis of zig-zag transformer applying in the three-phase four-wire distribution power system," Power Delivery, IEEE Transactions on , vol.20, no.2, pp.1168-1173, April 2005

[8] Inoue, S.; Shimizu, T.; Wada, K., "Control Methods and Compensation Characteristics of a Series Active Filter for a Neutral Conductor," Industrial Electronics, IEEE Transactions on, vol.54, no.1, pp.433-440, Feb. 2007.

[9] Enjeti, P.N.; Shireen, W.; Packebush, P.; Pitel, I.J., "Analysis and Design of a New Active Power Filter to Cancel Neutral Current Harmonics in Three-Phase Four-Wire Electric Distribution Systems," Industry Applications, IEEE Transactions on , vol.30, no.6, pp.1565, Nov/Dec 1994.

[10] Sewan Choi; Minsoo Jang, "Analysis and Control of a Single-Phase-Inverter–Zigzag-Transformer Hybrid Neutral-Current Suppressor in Three-Phase Four-Wire Systems," Industrial Electronics, IEEE Transactions on , vol.54, no.4, pp.2201-2208, Aug. 2007.

[11] Chung-Chuan Hou; Yung-Fu Huang, "Design of single-phase shunt active filter for three-phase four-wire distribution systems," Energy Conversion Congress and Exposition (ECCE), 2010 IEEE , vol., no., pp.1525-1528, 12-16 Sept. 2010.

[12] Ning-Yi Dai; Man-Chung Wong; Ying-Duo Han, "Application of a three-level NPC inverter as a three-phase four-wire power quality compensator by generalized 3DSVM," Power Electronics, IEEE Transactions on , vol.21, no.2, pp.440,449, March 2006

[13] Singh, B.N.; Rastgoufard, P., "A new topology of active filter to correct power-factor, compensate harmonics, reactive power and unbalance of three-phase four-wire loads," in Applied Power Electronics Conference and Exposition, 2003. APEC '03. Eighteenth Annual IEEE , vol.1, no., pp.141-147 vol.1, 9-13 Feb. 2003

[14] Reznik, A.; Simoes, M.G.; Al-Durra, A.; Muyeen, S.M., " Filter Design and Performance Analysis for Grid-Interconnected Systems," Industry Applications, IEEE Transactions on , vol.50, no.2, pp.1225-1232, March-April 2014.

[15] A. Yazdani, R. Iravani, Voltage-Sourced Converters in Power Systems, John Wiley & Sons, Inc., 2010.

[16] M. S. Huertas, "The use of power electronics for emulating electric power distribution feeders," M.S. Thesis, Dept. of Elect. Eng., Univ. of Arkansas, Fayetteville, USA, 2008.

Salcone, M.; Bond, J., "Selecting film bus link capacitors for high performance inverter applications," in Electric Machines and Drives Conference, 2009. IEMDC '09. IEEE International , vol., no., pp.1692-1699, 3-6 May 2009.

New Design Methdology for Megahertz-Frequency Resonant dc-dc Converters using Impedance Control Network Architecture

Yushi Liu, Ashish Kumar, Jie Lu, Dragan Maksimovic and Khurram K. Afridi

Dept. of Electrical, Computer and Energy Engineering
University of Colorado
Boulder, USA
yushi.liu@colorado.edu

Abstract— **This paper introduces a new design methodology for high-frequency resonant dc-dc converters utilizing the recently proposed impedance control network (ICN) converter architecture. This design methodology guarantees zero voltage switching (ZVS) and near zero current switching (ZCS) of all transistors across the entire operating range of the converter. As compared to previous ICN converter design techniques, the new methodology does not require frequency adjustment to achieve ZVS. In addition, it uses a rigorous analytical approach to minimize the phase lag required for ZVS, thus ensuring near-ZCS operation. The advantages gained from the proposed design methodology are validated using a prototype 1-MHz 90-W ICN resonant converter designed to operate over an input voltage range of 36 V to 60 V and output voltage of 1.8 V. This prototype ICN resonant converter employs a stacked inverter structure to further enhance performance.**

Keywords— **dc/dc converter; resonant converter; converter for wide-range operation; converter for wide-range operation; impedance control network; ZVS and near ZCS.**

I. INTRODUCTION

The ever increasing demand on the compactness of power electronic converters necessitates high-frequency operation. Achieving high efficiency at high frequencies requires soft-switching in the form of zero voltage switching (ZVS) and/or zero current switching (ZCS) to reduce transistor switching losses. In addition, many applications such as power supplies and large step-down point-of-load converters also require large conversion ratios, isolation, and efficient operation under wide variations in input voltage and output power. Therefore, there is a need for isolated high-frequency converters that can provide large conversion ratios and achieve soft-switching across wide operating ranges. Unfortunately, most soft-switching converters can only maintain high efficiency under specific operating conditions, and lose ZVS or ZCS capability when input voltage or output power change. For example, in the case of resonant converters operated under frequency control [1]-[3], output voltage can be regulated in the face of wide variations in input voltage by substantially increasing the switching frequency (frequency is increased rather than decreased in order to maintain ZVS); however, this exacerbates switching losses. Furthermore, wide frequency range operation presents difficulties in optimizing magnetics and EMI filters

[4]. Several fixed-frequency techniques have also been proposed for resonant converter control. Conventional resonant converters using phase-shift control suffer from asymmetric current in the two inverter legs [4], [5]. Under partial loads, the leading inverter leg turns off at large current, while the lagging leg loses ZVS capability. Other fixed frequency control techniques also lose ZVS capability under low output power conditions [6]-[8]. Therefore, there is a requirement for converter architectures and associated design and control methods that can achieve soft-switching across wide input voltage and power ranges, and provide large voltage conversion ratios. One such architecture, the impedance control network (ICN) resonant converter, was recently proposed in [9].

Resonant dc-dc converters comprise an inverter stage, a transformation stage, and a rectifier stage, as shown in Fig. 1. The recently proposed impedance control network (ICN) converter architecture, shown in Fig. 2, incorporates multiple inverters and rectifiers, along with a transformation stage containing the impedance control network. When the inverters are operated under phase-shift control, the ICN enables soft-switching across widely varying operating conditions, thus mitigating the issues described earlier. The ICN architecture is inspired by lossless power combiners and resistance compression networks [13]-[16].

This paper introduces a new design methodology for high-

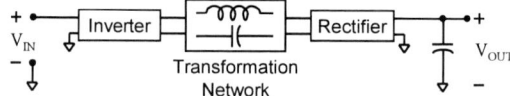

Figure 1: Block diagram for a conventional dc-dc resonant converter.

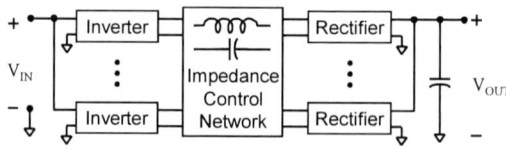

Figure 2: Architecture of the recently proposed impedance control network (ICN) resonant converter. Note that stacked inverter structures may also be employed, and can be advantageous for high step-down designs.

frequency resonant dc-dc converters utilizing the impedance control network (ICN) architecture. This design methodology guarantees zero voltage switching (ZVS) and near zero current switching (ZCS) of all transistors across the entire operating range of the converter. In contrast to previous ICN design techniques, the new methodology does not require frequency adjustment to achieve ZVS. Furthermore, it ensures near-ZCS operation through a rigorous analytical approach that minimizes the phase lag required for ZVS. The performance benefits obtained from the proposed design methodology are validated using a prototype 1-MHz, 90-W ICN resonant converter designed to operate over an input voltage range of 36 V to 60 V and output voltage of 1.8 V. This converter uses a stacked inverter structure to enhance performance.

The remainder of this paper is organized as follows: Section II describes the topology and principle of operation of the ICN dc-dc converter. Section III presents the new methodology used to design this converter for wide-range ZVS and near ZCS operation. Design details of a prototype of the proposed converter are provided in section IV, and experimental results obtained from this prototype are presented in section V. Finally, the conclusions of the paper are summarized in section VI.

II. IMPEDANCE CONTROL NETWORK RESONANT CONVERTER ARCHITECTURE

Several implementations of the ICN converter architecture have been recently reported [9]-[12]. A new implementation, with a stacked inverter structure suitable for step-down applications, is shown in Fig. 3. This converter is operated at a fixed switching frequency, and both inverter legs are operated at a fixed duty ratio (~50%). At the resonant frequency, the two branches of the impedance control network are designed to have equal and opposite reactances ($+jX$ and $-jX$). As the input and/or output voltage of the converter vary, the impedances seen at the two inverter switch-nodes can be made equal and both purely resistive, if the inverters are driven with a specific phase-shift given by:

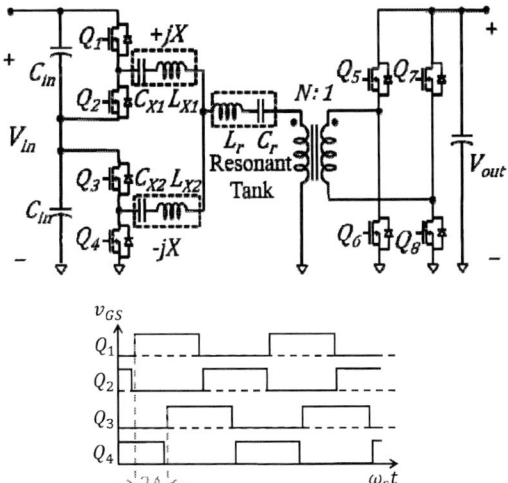

Figure 3: One implementation of impedance control network (ICN) resonant converter, appropriate for voltage step-down: (a) converter topology, and (b) inverter switch gating signals.

$$2\Delta = 2\cos^{-1}\left(\frac{V_{\text{IN}}}{4NV_{\text{OUT}}}\right). \quad (1)$$

Here V_{IN} is the input voltage, V_{OUT} is the output voltage, and N is the transformer turns ratio. This facilitates zero current switching (ZCS) of the inverter transistors under varying operating conditions.

In practice, owing to the non-negligible output capacitances of these transistors, they need to operate with zero voltage switching (ZVS) as well. In previous ICN converter designs, ZVS (and near-ZCS) operation of the inverter transistors was achieved through a small adjustment in switching frequency, so that the impedances seen by the inverters become slightly inductive. However, adjusting switching frequency to achieve ZVS is not robust as a soft-switching technique, particularly in absence of analytical expressions for the required frequency adjustment. Furthermore, frequency changes beyond certain limits may also significantly impact the steady-state operational characteristics and efficiency of the ICN converter.

To resolve this issue, we propose a new design methodology for the ICN converter architecture. This design methodology does not depend on frequency adjustment to achieve ZVS. Instead, it introduces a topological enhancement to the ICN converter, and rigorous analytical means to guarantee ZVS (and near-ZCS) operation across the entire operating range of the converter.

III. DESIGN METHODOLOGY FOR WIDE-RANGE ZVS AND NEAR-ZCS OPERATION

The proposed designed methodology introduces an additional positive reactance in series in each branch of the impedance control network. This is illustrated for the ICN converter described in the previous section in Fig. 4. The additional reactance, X_L, when optimally chosen, ensures that both the inverters see slightly inductive impedance at the design switching frequency across the full operating range, ensuring ZVS and near-ZCS. With this topological modification, it should also be ensured that the magnitudes of the two inverter impedances remain equal, so that the inverters share current equally. To maintain such slightly inductive inverter switch-node impedances of equal magnitude, the inverters are operated with a modified phase shift given by:

$$2\Delta_e = 2\cos^{-1}\sqrt{\frac{\frac{V_{in}^2}{16N^2V_{out}^2} - r^2}{1 - r^2}}. \quad (2)$$

Here r is the ratio of the additional reactance X_L to the differential reactance X. The imaginary and real parts of the inverter switch-node impedances, when operated with this phase shift for a particular design of this ICN converter, are shown in Fig. 5 (a) and (b). It may be observed that the imaginary part of the impedance is positive for all input voltages, indicating inductive loading of the inverters.

An additional advantage of this design methodology is that the output power of the ICN converter, when operating with the inverter phase shift given above, is a concave function of input voltage. This property may be deduced from Fig. 5 (b),

978-1-4673-9551-9/16 $31.00 © 2016 IEEE

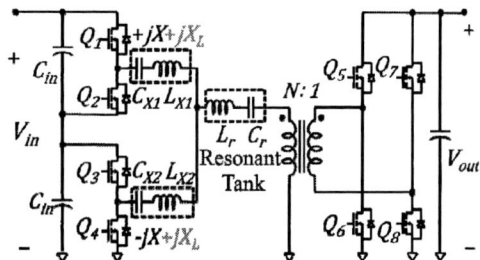

Figure 4: Modified implementation of the impedance control network (ICN) resonant converter, with additional reactance X_L in each inverter leg to ensure ZVS and near-ZCS at design switching frequency.

which shows that as the input voltage is increased, the resistance seen by the inverters also increases. The resultant output power curve of the ICN converter is illustrated in Fig. 6, and expressed mathematically as:

$$P_{out} = \frac{V_{in}}{\pi^2 X}\sqrt{16N^2 V_{out}{}^2 - V_{in}{}^2}\sqrt{\left(1 - r^2\frac{16N^2 V_{out}^2}{V_{in}^2}\right)\Big/(1 - r^2)} \quad (3)$$

This is a highly desirable characteristic as the output power is naturally limited over a wide input voltage range, enabling converter components to be optimally sized.

The next step in the proposed design methodology is to determine the transformer turns ratio, N, and the differential reactance of impedance control network, X. These are selected based on the design critera reported for ICN converters in the past [9]-[12], and modified to include the additional reactance X_L. Referring to Fig. 6, the converter is designed to deliver

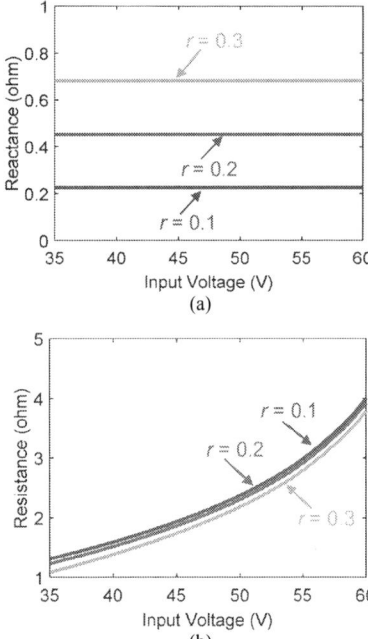

Figure 5: The impedance seen by the two half-bridge inverters of the ICN converter when operated with the proposed phase-shift, as input voltage is varied: (a) imaginary part, and (b) real part.

equal output power at the minimum input voltage (V_{IN_min}) and the maximum input voltage (V_{IN_max}). The resultant design equations are given by:

$$N = \frac{\sqrt{V_{IN,min}^2 + V_{IN,max}^2}}{4V_{out}}\frac{1}{\sqrt{1 + r^2}} \quad (4)$$

$$X = \frac{V_{IN,min} \cdot V_{IN,max}}{\pi^2 P_{out}(1 + r^2)}\sqrt{(1 - r^2\frac{V_{IN,min}^2}{V_{IN,max}^2})(1 - r^2\frac{V_{IN,max}^2}{V_{IN,min}^2})\Big/(1 - r^2)} \quad (5)$$

It was stated earlier that optimal choice of the reactance X_L results in wide-range ZVS and near-ZCS operation. This optimal value of X_L is obtained by evaluating the amount of phase lag required between the inverter output voltages and currents to achieve ZVS. The phase lag φ between the output voltage and current of each inverter can be expressed as:

$$\varphi = \tan^{-1}\left(\frac{-r}{\frac{V_{in,max}}{V_{in,min}} - r^2\frac{V_{in,min}}{V_{in,max}}}\sqrt{(1 - r^2\frac{V_{in,min}^2}{V_{in,max}^2})(1 - r^2\frac{V_{in,max}^2}{V_{in,min}^2})(1 - r^2)}\right) \quad (6)$$

The (equal-magnitude) output current of each inverter can be obtained from the input voltage of the converter, and the admittance seen at each inverter switch node. The magnitude of this admittance is given by:

$$|Y| = \frac{\pi^2 P_{out}}{V_{in\_max}^2}\sqrt{\frac{1 + r^2}{1 - r^2\frac{V_{in\_min}^2}{V_{in\_max}^2}}} \quad (7)$$

Using (6) and (7), it is possible to derive a condition on r such that the phase lag between the inverter output voltages and currents is large enough for the current to fully charge and discharge the output capacitances of the inverter transistors, at the worst-case operating point, where the input voltage is at its maximum value and inverter current at its minimum. Furthermore, it is ensured that while ZVS is achieved in the worst case, the phase lag φ is the minimum required for this purpose – this results in near-ZCS operation and produces the optimal, minimum value of r, and hence the additional inductive reactance X_L, for ZVS. This minimization of X_L ensures that the extra inductance minimally impacts conduction and core losses. The minimum phase-lag ZVS and near-ZCS condition can be mathematically represented as:

$$\int_{\frac{T_s}{2}}^{\frac{T_s}{2} + t_d} i_{inv}(t)dt = 2C_{oss}V_{IN,max} \quad (8)$$

Here, i_{inv} is the inverter output current, t_d is the dead-time duration during which the soft-switching transitions occur, and C_{oss} is the output capacitance of the inverter transistors. It may be noted that the value for C_{oss} used here is a linear charge-equivalent capacitance derived from datasheet curves. Simplifying (8) using (6) and (7), one obtains the following equivalent constraint:

$$|Y|[\cos(2\varphi) - \cos(\varphi)] = 8\pi C_{oss}f_s \quad (9)$$

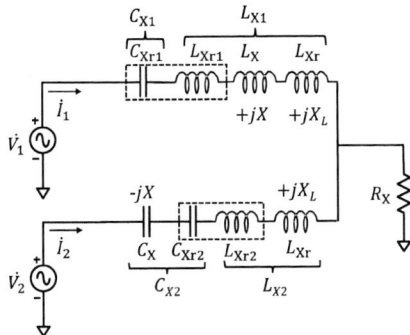

Figure 7. Primary side reactive elements broken up into their conceptual constituents: differential reactances L_X and C_X, series resonant tank elements (L_{Xr1}, C_{Xr1}, L_{Xr2} and C_{Xr2}), and additional reactance L_{Xr}. The $\dot{V}_1$ and $\dot{V}_2$ are the fundamental components of the output voltages of the inverters, and the $\dot{I}_1$ and $\dot{I}_2$ are the fundamental components of the output currents of the inverters.

With (9) as the final design criterion, the optimal value of X_L is arrived at, and the design procedure is complete.

IV. PROTOTYPE DESIGN

To validate the performance enhancements enabled by the proposed design methodology, a prototype ICN converter has been designed, built and tested. The topology of this converter is as shown in Fig. 4. It may be observed from Fig. 4 that this ICN converter utilizes a stacked inverter structure. Stacking the inverter halves the voltage stress on the inverter transistors, enabling utilization of low-voltage devices, which generally have lower on-state resistances. Furthermore, by halving inverter input voltage and doubling the inverter output current, the stacked inverters make it easier for inverter transistors to achieve ZVS in large step-down applications. A third advantage from inverter stacking is that the transformer turns ratio requirement is reduced by a factor of two.

The converter is operated at a fixed switching frequency of 1 MHz, and is designed for an input voltage range of 36-60 V, an output voltage of 1.8 V, and output power of 90 W. For the given specifications, using the design methodology described in the previous section, N is 9.6 and X is 2.8 Ω and r is 0.18. The resultant, optimal value of X_L is 0.504 Ω. A conceptual breakdown of the reactive components of the inverter is shown in Fig. 7. In the top branch L_{Xr1}, L_X, and L_{Xr} collectively form L_{X1}, and C_{Xr1} is simply the C_{X1} of Fig. 4. In the bottom branch, C_X and C_{Xr2} collectively form C_{X2}, and L_{Xr2} and L_{Xr} collectively form the L_{X2} shown in Fig. 4. It may be noted that the L_{Xr} components of the top and bottom inverter tanks are embodiments of the additional reactance X_L. The elements L_X and C_X provide the necessary differential reactance in the top and bottom branch, respectively. The values of these differential reactive elements are determined by:

$$L_X = \frac{X}{\omega_s}, \qquad (10)$$

$$C_X = \frac{1}{X\omega_s}, \qquad (11)$$

where ω_s is the angular switching frequency of the converter. Finally, the functional objective of L_{Xr1} and C_{Xr1} in the top

tank, and L_{Xr2} and C_{Xr2} in the bottom tank is to provide series resonant filtering. The values of these resonant tank elements are determined using:

$$L_{Xr1} = \frac{Z_{0X1}}{\omega_s}, \; L_{Xr2} = \frac{Z_{0X2}}{\omega_s}, \qquad (12)$$

$$C_{Xr1} = \frac{1}{Z_{0X1}\omega_s}, \; C_{Xr2} = \frac{1}{Z_{0X2}\omega_s}, \qquad (13)$$

where Z_{0X1} and Z_{0X2} are the desired characteristic impedances of the resonant tanks and their values are determined from $Z_{0X1} = Q_{0X1}R_X$ and $Z_{0X2} = Q_{0X2}R_X$, where Q_{0X1} and Q_{0X2} are the desired loaded quality factors of the resonant tanks, and $R_X = 8N^2 V_{OUT}^2/(\pi^2 P_{OUT})$ is the equivalent resistance of the rectifier referred to the primary side of the transformer. By selecting appropriate quality factors (approximately 2) for the resonant networks to obtain adequate filtering of the higher order harmonics, the value of primary side reactive components are determined as: $L_{X1} = (1+r)X/\omega_s + Q_{0X}R_X/\omega_s = 1.27 \; \mu H$, $C_{X1} = 1/Q_{0X}R_X\omega_s = 30 \; nF$, $L_{X2} = Q_{0X}R_X/\omega_s + rX/\omega_s = 0.91 \; \mu H$, and $C_{X2} = (1/Q_{0X}R_X\omega_s) \| (1/X\omega_s) = 21 \; nF$.

The core materials used to construct the magnetic elements (inductors and transformer) are L-material from Magnetics Inc. and N49 from EPCOS, chosen for their low core loss characteristics at 1 MHz. The resonant capacitors are 100-V low-ESR ceramic capacitors. For the inverters, EPC 40-V/10-A enhancement-mode gallium nitride (GaN) transistors (EPC2014) are used. These are driven by half-bridge bootstrap drivers designed for enhancement-mode GaN transistors (TI LM5113). For the rectifier, VICOR PI5101 transistors are used. The control signals for the converter are generated using a Microchip dsPIC33FJ64GS610, a 16-bit digital signal controller with high-speed PWM outputs. Fig. 9 shows the top and bottom views of the prototype ICN converter.

V. EXPERIMENTAL RESULTS

The 1-MHz, 90-W prototype ICN converter has been tested across its operating range. The converter is required to deliver

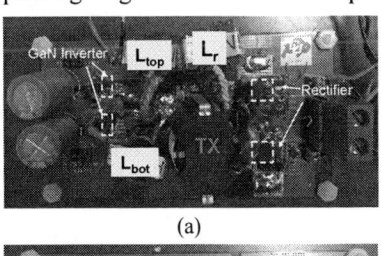

(a)

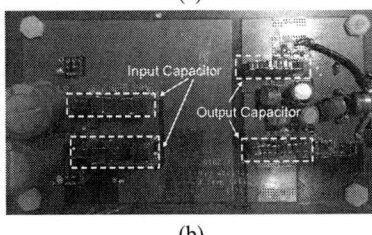

(b)

Figure 8: Photograph of the (a) top and (b) bottom of the prototype ICN resonant converter.

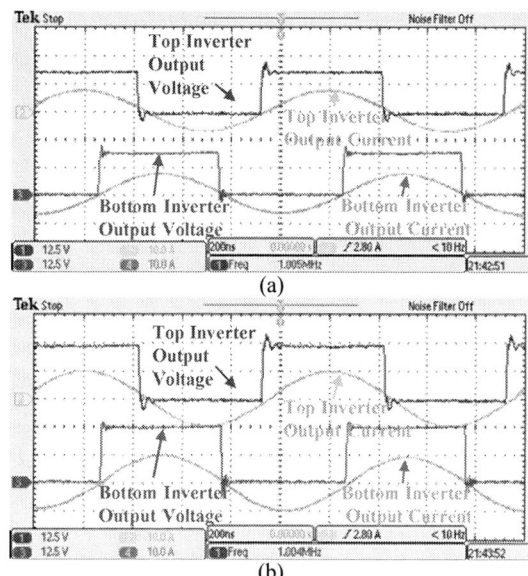

(a)

(b)

Figure 9: Measured waveforms of output voltage and output current of both (top and bottom) half-bridge inverters confirming ZVS and near ZCS operation of the ICN resonant converter. (a) $V_{IN} = 36$ V and (b) $V_{IN} = 48$ V.

an output power of 90 W at 1.8 V output voltage over an input voltage range of 36 V to 60 V (nominal value 48 V). Figure 9 shows the measured waveforms of the ICN converter at two input voltage levels, confirming ZVS and near-ZCS operation of all inverter switches as input voltage is varied. Magnified waveforms focusing on the inverter switch-node voltage transitions are presented in Fig. 10. Clearly, the switches of both the top and the bottom inverters achieve ZVS and near ZCS.

VI.   CONCLUSIONS

This paper introduces a new design methodology for high-frequency resonant dc-dc converters utilizing the recently proposed impedance control network (ICN) converter architecture. This design methodology guarantees zero voltage switching (ZVS) and near zero current switching (ZCS) of all transistors across the entire operating range of the converter. As compared to previous ICN converter design techniques, the new methodology does not require frequency adjustment to achieve ZVS. In addition, it also uses an analytical approach to minimize the phase lag required for ZVS, thus ensuring near-ZCS operation. A prototype 1-MHz, 90-W ICN resonant converter designed to operate over an input voltage range of 36 V to 60 V, and output voltage of 1.8 V is used to validate the advantages gained from the proposed design methodology. The performance of this converter is further enhanced using a stacked inverter structure, that reduces voltage stress on inverter transistors, making it easier to achieve ZVS, and reduces the transformer turns ratio requirement by a factor of two.

REFERENCES

[1]   R.L. Steigerwald, "High-Frequency Resonant Transistor DC-DC Converters," *IEEE Transactions on Industrial Electronics*, vol. IE-31, no. 2, pp. 181-191, May 1984.

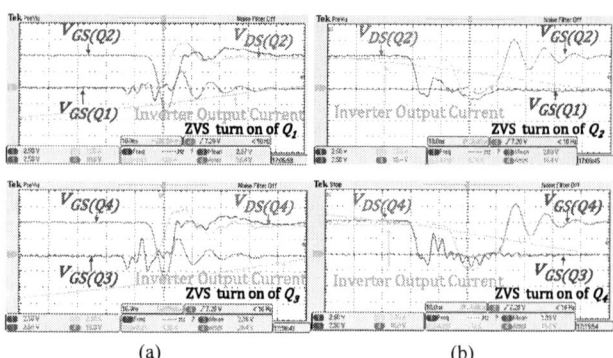

(a)   (b)

Figure 10: Measured waveforms confirming ZVS turn-on operation of the ICN resonant converter operating at full power 90 W: (a) ZVS turn-on of the transistors in the top inverter and (b) ZVS turn-on of the transistors in the bottom inverter.

[2]   R.L. Steigerwald, "A Comparison of Half-Bridge Resonant Converter Topologies," *IEEE Transactions on Power Electronics*, vol. 3, no. 2, pp. 174-182, April 1988.

[3]   A.K.S. Bhat, "Analysis and design of LCL-type series resonant converter," *IEEE Transactions on Industrial Electronics*, vol. 41, no. 1, pp. 118-124, February 1994.

[4]   J. Vandelac and P.D. Ziogas, "A DC to DC PWM Series Resonant Converter Operated at Resonant Frequency," *IEEE Transactions on Industrial Electronics*, vol. 35, no. 3, pp. 451-460, August 1988.

[5]   M.Z. Youssef and P.K. Jain, "A Review and Performance Evaluation of Control Techniques in Resonant Converters," *Proceedings of the IEEE Industrial Electronics Society*, pp. 215-221, Busan, Korea, November, 2004.

[6]   F.S. Tsai, P. Materu and F.C. Lee, "Constant-Frequency Clamped-Mode Resonant Converters," *IEEE Transactions on Power Electronics*, vol. 3, no. 4, pp. 460-473, October, 1988.

[7]   P. Jain, A. St-Martin and G. Edwards, "Asymmetrical Pulse Width Modulated Resonant DC/DC Converter Topologies," *Proceedings of the IEEE Power Electronics Specialists Conference (PESC)*, pp. 818-825, Seattle, WA, June 1993.

[8]   J.M. Burdio, F. Canales, P.M. Barbosa and F.C. Lee, "A Comparison Study of Fixed-Frequency Control Strategies for ZVS DC/DC Series Resonant Converters," *Proceedings of the IEEE Power Electronics Specialists Conference (PESC)*, pp. 427-432, Vancouver, Canada, June 2001.

[9]   J. Lu, D.J. Perreault and K.K. Afridi, "Impedance Control Network Resonant dc-dc Converter for Wide-Range High-Efficiency Operation," *Proceedings of the IEEE Applied Power Electronics Conference and Exposition (APEC)*, Charlotte, NC, March 2015.

[10]   J. Lu and K.K. Afridi, "High Efficiency Impedance Control Network Resonant DC-DC Converter with Optimized Startup Control," *Proceedings of the IEEE Energy Conversion Congress and Exposition (ECCE)*, Montreal, Canada, September, 2015.

[11]   S.J. Gunter, K.K. Afridi, D. Otten, R. Abramson and D.J. Perreault, "Impedance Control Network Resonant Step-Down DC-DC Converter Architecture," *Proceedings of the IEEE Energy Conversion Congress and Exposition (ECCE)*, Montreal, Canada, September, 2015.

[12]   A. Kumar and K.K. Afridi, "Megahertz-Frequency Isolated Resonant dc-dc Converter using Impedance Control Network for High-Efficiency Wide-Range Operation," Proceedings of the IEEE Energy Conversion Congress and Exposition (ECCE), Montreal, Canada, September, 2015.

[13]   H. Chireix, "High Power Outphasing Modulation," Proceedings of the IRE, vol. 23, no.11, pp. 1370-1392, November 1935.

[14]   Y. Han, O. Leitermann, D.A. Jackson, J.M. Rivas, D.J. Perreault, "Resistance Compression Networks for Radio-Frequency Power Conversion," *IEEE Transactions on Power Electronics*, pp. 41-53, January 2007.

978-1-4673-9551-9/16 $31.00 © 2016 IEEE   1396

[15] D.J. Perreault, "A New Power Combining and Outphasing Modulation System for High-Efficiency Power Amplification," *IEEE Transactions on Circuits and Systems – I,* vol. 58, no. 8, pp. 1713-1726, August 2011.

[16] A.S. Jurkov, L. Roslaniec, and D.J. Perreault, "Lossless Multi-Way Power Combining and Outphasing for High-Frequency Resonant Inverters," *2012 International Power Electronics and Motion Control Conference,* pp. 910-917, June 2012.

Dual Voltage Regulations of Single Switch Flyback Converter Using Variable Switching Frequency

Jin-Woong Kim and Jung-Ik Ha

Department of Electrical & Computer Engineering
Seoul National University
Seoul, Korea
E-mail: tndud37@snu.ac.kr, jungikha@snu.ac.kr

Abstract—**This paper proposes a dual voltage regulations of single switch flyback converter using variable frequency control. By adding an adjusted filter to secondary windings, each output shows different tendency according to the switching frequency. This means that even if the switching duty is same, the output voltage might be not equal at the different switching frequency. So the output voltage can be regulated by not only the switching duty control but also the switching frequency control. Though the adjusted filter is added to each secondary windings, the simplicity of circuit and control algorithm which are main features of the flyback converter is not disturbed. The dual voltage regulations of the proposed method is compared with that of the conventional method and the feasibility and performance of the proposed method is verified through experiment using a proto-type converter.**

Keywords—Dual output converter, Isolated dc/dc converter, Variable swithcin frequency.

I. INTRODUCTION

A power converter which has more than one output is used in most electrical equipment [1]-[3]. Due to a cost effective structure, simple control and galvanic isolation, a flyback converter is one of the most attractive topology in the area of dual output power converters [4]-[5]. In a common dual-output flyback converter, it is difficult to achieve tight cross regulation and load regulation because of leakage inductances in primary and secondary winding, a voltage drop of the rectifier, a clamp voltage of the snubber circuit and a load variation [6]-[8]. Usually, only main output is regulated by a single feedback and auxiliary output is cross regulated as shown in Fig. 1(a). The auxiliary output always has a regulation error according to the load variation of each output [9]-[10]. For example, when the load of the main output becomes heavy while the load of the auxiliary output is light, the auxiliary output voltage becomes higher than the reference. To reduce this biased error, weighted voltage control method is used as shown in Fig. 1(b). Both output voltages are sensed simultaneously and a sum of the weighted output voltage is fed back to the controller [11]-[12]. As the voltage of auxiliary output is used for the regulation, the output error of auxiliary output is reduced [13]. However, the weighting factors are fixed at the design process because the factors are determined by voltage dividing resistors. So the converter cannot satisfy a regulation specification under the different load status. Another method use time-multiplexing

control (TMC) for the flexible weighting factor [14]. As shown in Fig. 1(c), a control signal for output 1 is applied during T_{s1} and a control signal for output 2 is applied during T_{s2}. So, the weighting factor of each output can be adjusted by control the T_{s1} and T_{s2}. However, these methods only redistribute the error to the both outputs, the sum of the regulation error is not reduced.

In this paper, adjusted LCL filters and variable frequency

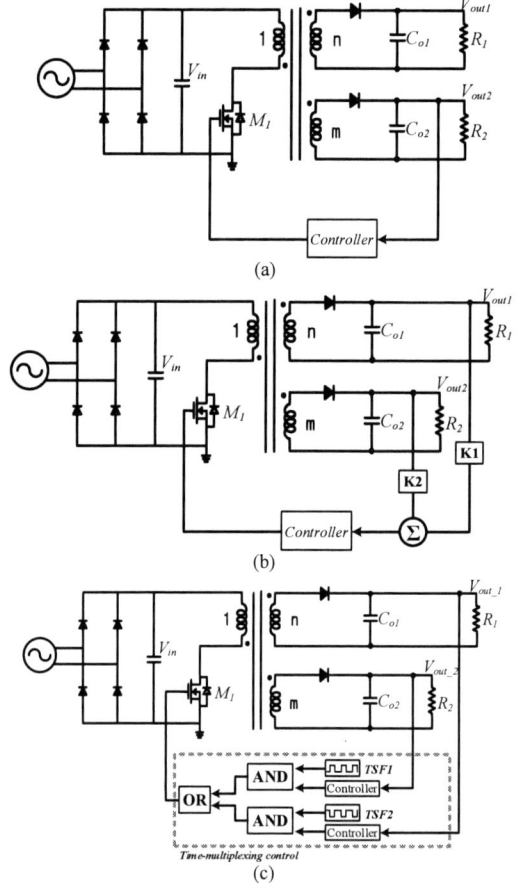

(a)

(b)

(c)

Fig. 1. The conventional converter. (a) With Single feedback. (b)With weighting feedback.

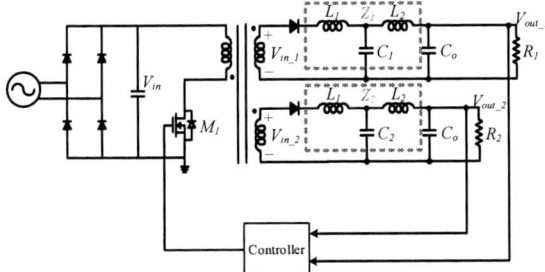

Fig. 2. The Proposed circuit configuration.

control are used to improve cross-regulation and load-regulation. The adjusted LCL filter is added to each secondary winding and it affects each output voltage differently according to the switching frequency. In order to perform the tight regulation under any load variation, both switching duty control and variable switching frequency control are applied. One feedback of output voltage is used for the switching duty control and another feedback is used for the variable switching frequency control. The proposed method does not need any additional active component and complicated control algorithm. So the simple structure and control algorithm which is the main feature of flyback converter are preserved. The enhanced regulation performance of the proposed converter has been validated using a prototype converter.

II.   PROPOESD METHOD

A. Filter analysis

The proposed dual output flyback converter is represented in Fig. 2. The basic circuit configuration is similar to the conventional circuit. The only difference is the filter. The adjusted LCL filter is added to each secondary windings and it makes different response to each output according to the switching frequency. So the output voltage can be controlled by switching frequency as well as switching duty. To show the relation between the output voltage and switching frequency, a filter impedance and transfer function of output voltage are analyzed. The transfer function from V_{in_k} to V_{out_k} is expressed as (1) and the impedance of the filter is expressed as (2).

$$\frac{V_{out\_k}}{V_{in\_k}} = \frac{R}{RAs^4 + L_2L_1C_ks^3 + RBs^2 + (L_1+L_2)s + R} \quad (1)$$

$$(A_k = C_oC_kL_2L_1, B_k = L_1C_k + L_1C_o + L_2C_o)$$

$$Z_k = \frac{C_kL_2s^3 + (L_1+L_2)s}{C_kL_2s^2 + 1} \quad (2)$$

where the subscript k means the order of output. V_{out_k} is a voltage of kth output and V_{in_k} is a voltage of kth secondary winding. L_1, L_2 and C_k mean inductance and capacitance of kth LCL filter. C_o represents the output capacitor of each winding and Z_k means the impedance of kth filter. If we assume that the value of the load R is much larger than other passive components, resonant frequencies of (1) and (2) can be derived as follows.

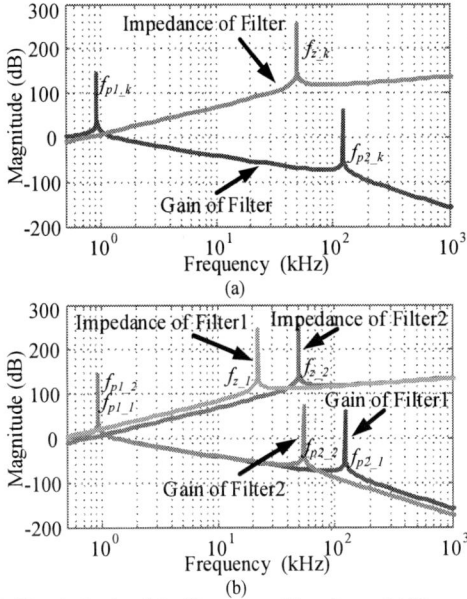

Fig. 3. Magnitude plot of the filter gain and impedance. (a) The proposed filter. (b) Two adjusted filters.

$$f_{p1\_k} = \frac{1}{2\pi}\sqrt{\frac{B_k - \sqrt{B_k^2 - 4A_k}}{2B_k}}$$

$$f_{p2\_k} = \frac{1}{2\pi}\sqrt{\frac{B_k + \sqrt{B_k^2 - 4A_k}}{2B_k}} \quad (3)$$

$$(A_k = C_oC_kL_2L_1, B_k = L_1C_k + L_1C_o + L_2C_o)$$

$$f_{z\_k} = \frac{1}{2\pi}\sqrt{\frac{1}{L_1C_k}} \quad (4)$$

A magnitude of the transfer function gain and impedance according to the switching frequency variation and their resonant frequencies are represented in Fig. 3. As can be seen in Fig. 3(a), the gain of the transfer function has two resonant frequencies and the magnitude of filter impedance has one resonant frequency. If each filter is designed to have different resonant frequencies, the gain or impedance of specific filter would be larger than that of another filter at the certain frequency range as shown in Fig. 3(b). This difference of the gain or impedance affects the output voltage. Even if the switching duty is same, the voltage of the output1 increases with switching frequency f_{p2_1} and it decreases with switching frequency f_{z1}. So if the specific output has the regulation error, the error can be adjusted by switching frequency control.

B. Filter design

The gain of the transfer function and resonant frequency of the output impedance can be designed by the filter capacitance. As the output capacitance is much larger than other parameters, second equation of (3) can be simplified as follows.

978-1-4673-9551-9/16 $31.00 © 2016 IEEE   1399

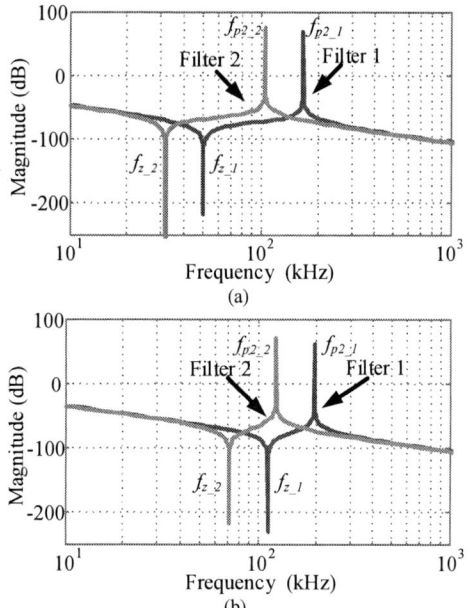

(a)

(b)

Fig. 4. Magnitude plot of the filter according to the design (a) Wide range of switching frequency (b) Narrow range of switching frequency.

$$f_{p2\_k} \approx \frac{1}{2\pi}\sqrt{\frac{L_1+L_2}{L_1 L_2 C_k}} = \sqrt{1+\frac{L_2}{L_1}}f_{z\_k} \quad (5)$$

From (4), f_{z_k} of each filter is determined by the filter capacitance and from (5), f_{p2_k} of each filter is determined by the filter inductances ratio. The ratio should be considered in design process. If the ratio is large value, a difference between f_{z_k} and f_{p2_k} becomes large as shown in Fig. 4(a). So, a rate of the gain variation according to the switching frequency is slow and a required switching frequency bandwidth becomes wide. It may cause EMI problem and affect to flyback core design. On the other hand, if the ratio is small value, a difference between f_{z_k} and f_{p2_k} becomes small as shown in Fig. 4(b). A rate of the gain variation according to the switching frequency is steep and a required switching frequency bandwidth becomes narrow. For the dual output voltage regulation, a precise switching frequency control is required. Therefore, the switching bandwidth and precision of switching frequency control is considered according to the application.

C. Control algorithm

Generally, an output voltage of a converter is controlled by a switching duty and the duty is determined by a feedback of the output voltage. As the adjusted filer makes the output

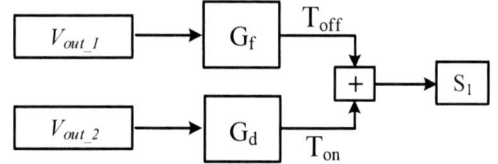

Fig. 5. Block diagram of proposed control algorithm.

voltage vary with switching frequency, the proposed controller uses not only a switching duty control but also a switching frequency control for the tight cross regulation and load regulation. But it is hard to change the switching frequency directly when the switching duty is already determined. In the proposed method, the switching frequency is controlled by a turn off time variation of the switching device. So even if the turn on time reference is same, the switching frequency can be a low frequency with the long turn off time reference and the switching frequency can be high frequency with the short turn off time reference. Fig. 5 shows a block diagram of the proposed control algorithm. All output voltages are sensed at the same time without any weighting factor. One of them is fed back to the duty controller to make a turn on time reference of the switching device and another is fed back to the frequency controller to make a turn off time reference of the switching device. In Fig. 5, G_f represents the switching frequency controller and G_d represents the switching duty controller. So the switching frequency is naturally determined according to the sum of the on and off time reference. Each controller is consisted of a proportional integral (PI) controller and the bandwidth of duty controller is wider than that of frequency controller due to a faster response.

III. SIMULATION RESULTS

To verify the performance of the proposed method, a 40W (15V/2A, 5V/2A) dual output flyback converter is designed for the simulation. L_1 and L_2 of both filters are designed to be $1\mu H$ and $5\mu H$ respectively. Capacitance of the first filter is designed to be $2\mu F$ and that of the second filter is $10\mu F$. A Feedback of the 5V output is used for the duty controller and a feedback of the 15V output is used for the switching frequency controller. With these adjusted filters and controller, the required switching frequency for the tight regulation under any load variation is represented in Fig 6.

The cross regulation and load regulation of the proposed method is compared with those of the conventional weighting factor control under the extreme load variation. All parameters of the conventional circuit is same as the proposed circuit and the weighting factor of each feedback is same. As shown in Fig. 7, the conventional method represents a regulation error at both output. Especially, the error becomes large when the load of 5V output is 100%. But the proposed method shows no regulation error at the steady state due to the switching frequency control.

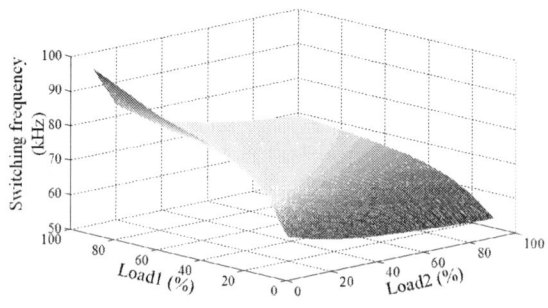

Fig. 6. Required switching frequency under the load variation.

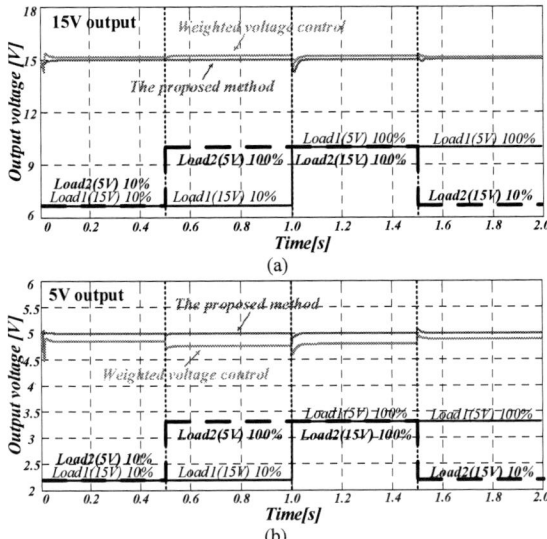

Fig. 7. Simulation results of regulation performance under the load variation. (a) Waveform of output 1 voltage (b) waveform of output2 voltage.

IV. EXPERIMENTAL RESULTS

For the experiment, a 10W (5V/1A, 3.3V/1.5A) prototype dual output flyback converter is designed. The output voltage V_1 is 5V and V_2 is 3.3V. The transformer turn ratio of the output 1 is 30:8 and that of the output2 is 30:5. Filter inductances, L_1 and L_2, are $1\mu H$ and $5\mu H$ respectively. And filter capacitance of the output 1 is $2\mu F$ and that of the output 2 is $10\mu F$. Other parameters are described in Table 1. Feedback of the 3.3V output is used for the duty controller and a feedback of the 5V output is used for the switching frequency controller. Fig. 8 shows a verification of the filter design. The transfer function gain of the designed filters are shown in Fig. 9(a) and voltage variation of output 1 according to the switching frequency is shown in Fig. 8(b). As the resonant frequency of the first filter is designed at 120 kHz, voltage of output 1 is increased according to the switching frequency and as the resonant frequency of the second filter is designed at 55 kHz, voltage of output1 is minimum at around 60 kHz. The difference between calculation value and real value comes from the component error.

Fig. 9 shows the regulation performance of the proposed method. A transition of output voltage and switching frequency is represented when one of two loads is changed.

Table I PARAMETERS OF PROTOTYPE CIRCUIT

Parameters	Values
Input voltage	30V
V_{out1}	5V / 1A
V_{out2}	3.3V / 1.5A
Filter 1	L1=$1\mu H$, L2=$5\mu H$, C2= $2\mu F$
Filter 2	L1=$1\mu H$, L2=$5\mu H$, C2= $10\mu F$
Output capacitor	3.3mF
N_p:N_{s1}	30:8
N_p:N_{s2}	30:5

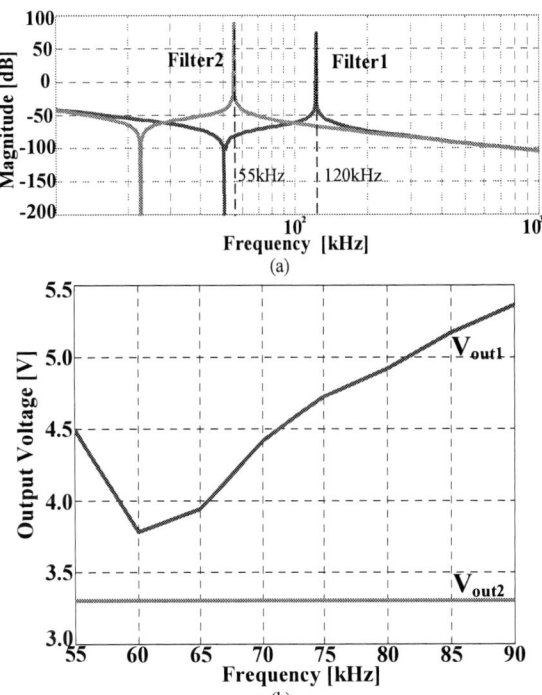

Fig. 8. Filter design verification (a) Magnitude plot of designed filter (b) Experimental plot of filter.

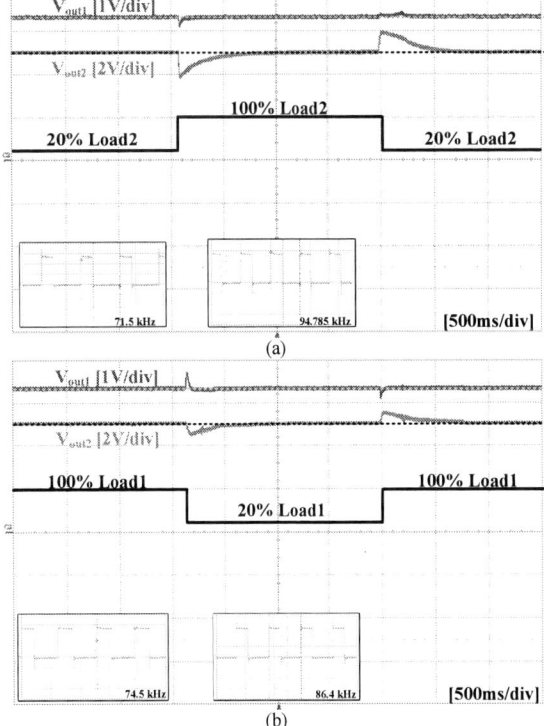

Fig. 9. Experimental results of regulation performance under the load variation. (a) Step variation of load 2. (b) Step variation of load 1.

The load of output 2 is changed from 20% to 100% and from 100% to 20% in Fig. 9(a) and the load of output 1 is changed from 100% to 20% and from 20% to 100% in Fig. 9(b). As shown in both figure, not only the switching duty but also the switching frequency of main switch is controlled for the dual voltage regulations. And it can be shown that both output is well-regulated under the load variation.

V. CONCLUSION

A dual-output flyback converter using a single switch and LCL filter has been proposed in this paper. Because of the load variation and leakage inductance of each windings, the cross-regulation is always a problem in the conventional dual-output converter topology. The proposed method adds an adjusted filter to each secondary winding for the enhanced cross-regulation and load-regulation. Due to the characteristic of the filter, the each output voltage varies with the switching frequency. So the cross regulation and load regulation can be performed using a switching frequency control as well as switching duty control. Both output voltages are sensed simultaneously and fed back to the controller. One of them is used for the duty controller to determine a turn on time of the switching device. The other one is used for the switching frequency controller to determine a turn off time of the switching device. From the experimental results using 10W (5V/1A, 3.3V/1.5A) prototype converter, the feasibility and performance have been confirmed. And the proposed method shows much better cross regulation and load regulation compared to the conventional method under any load variation. More detailed explanations about the filter design and additional experimental results will be included in the full paper.

ACKNOWLEDGMENT

This work was supported by the Robot industry fusion core technology development project of the Ministry of Trade, Industry & Energy of KOREA(10052980).

REFERENCES

[1] J. Choi and H. Cha, "A processor power management scheme for handheld systems considering off-chip contributions," *IEEE Trans Ind. Informat.*, vol. 6, no. 3, pp. 255–264, Aug. 2010.

[2] H. Tao, A. Kotsopulos, J. L. Duarte, and M. A. M. Hendrix, "Family of multiport bidirectional dc–dc converters," *Inst. Electr. Eng. Proc. Electr. Power Appl.*, vol. 153, no. 15, pp. 451–458, May 2006.

[3] G. Ma, W. Qu, G. Yu, Y. Liu, N. Liang, and W. Li, "A zero-voltage switching bidirectional dc–dc converter with state analysis and soft switching-oriented design consideration," *IEEE Trans. Ind. Electron.*, vol.56, no. 6, pp. 2174–2184, Jun. 2009.

[4] J.W. Shin, S.J. Choi and B.H. Cho, "High-Efficiency Bridgeless Flyback Rectifier With Bidirectional Switch and Dual Output Windings," *IEEE Trans. Power Electron.*, vol.29, no.9, pp.4752-4762, Sept. 2014

[5] V.L. Tran, W.J. Choi, "Novel Time Division Multiple Control Method for Multiple Output Battery Charger," *IEEE Trans. Power Electron.*, vol.29, no.10, pp.5102-5105, Oct. 2014

[6] H.S. Kim, J.H. Jung, J.W. Baek, and H.J. Kim, "Analysis and design of a multioutput converter using asymmetrical PWM half-bridge flyback

converter employing a parallel–series transformer," *IEEE Trans. Ind. Electron.*, vol. 60, no. 8, pp. 3115–3125, Aug. 2013.

[7] S., Kim, H. Todorovic., M. and Enjeti, P.N. "Three-phase active harmonic rectifier (Ahr) to improve utility input current Thd in telecommunication power distribution system," *IEEE Trans. Ind. Appl.*, vol. 39, no. 5, pp. 1414–1421, 2003.

[8] P. Patra, J. Ghosh, and A. Patra, "Control scheme for reduced cross-regulation in single-inductor multiple-output DC-DC converters," *IEEE Trans. Ind. Electron.*, vol. 60, no. 11, pp. 5095–5104, Nov. 2013.

[9] J. Chuanwen, M. Smith, K. M. Smedley, Jr., and K. King, "Cross regulation in flyback converters: Analytic model and solution," *IEEE Trans. Power Electron.*, vol. 16, no. 2, pp. 231–239, Mar. 2001.

[10] B. Su, H. Wen, J. Zhang, and Z. Lu, "A soft-switching post-regulator for multi-outputs dual forward DC/DC converter with tight output voltage regulation," *IET Power Electron.*, vol. 6, pp. 1069–1077, 2013

[11] Q. Chen, F. C. Lee, and M. M. Jovanovi´c, "Analysis and design of weighted voltage-mode-control for a multiple output forward converter," in *Proc. IEEE APEC*, Mar. 1993, pp. 449–455.

[12] Y. Xi and P. K. Jain, "A forward converter topology with independently and precisely regulated multiple outputs," *IEEE Trans. Power Electron.*, vol. 18, no. 2, pp. 648–658, Mar. 2003.

[13] Y. K. Lo, S.C. Yen and T.H. Song "Analysis and Design of a Double-Output Series-Resonant DC–DC Converter", *IEEE Trans. Power Electron.*, vol. 22, no.3, pp. 952 - 959, May 2007.

[14] R.Y.W. Chan, J.W. Hill, D.D.C. Lu, "Time Multiplexing Control for Power Converters," *Industrial Electronics and Applications*, May 2007, pp.126–131.

On-Chip PLL-Based Methods for Synchronizing Active Switches Across the Isolation Boundary in Dc-Dc Converters

Shahab Poshtkouhi, Miad Fard, Olivier Trescases

Department of Electrical and Computer Engineering, University of Toronto
10 King's College Road, Toronto, ON, M5S 3G4, Canada
E-mail: shahab.poshtkouhi@utoronto.ca

Abstract—Digital isolators are widely used in isolated power converters for transmitting high-frequency gating pulses. Opto-isolators generally suffer from low reliability, while emerging RF based isolators are expensive and power-hungry. In this work, two Phase-Locked-Loop (PLL) based synchronization schemes are introduced and demonstrated to operate a bidirectional Dual-Active-Bridge (DAB) dc-dc converter. The proposed Digital Isolator Sensing (DIS) scheme utilizes a miniature air-core pulse transformer for synchronization and data communication, while the Power Transformer Sensing (PTS) scheme eliminates the need for any digital isolator for synchronization by using the power transformer. The application of DIS scheme can be extended to any topology that requires active switches on both sides of the isolation boundary, whereas the PTS scheme is designed for the DAB converter. In both schemes, the reference clock is generated on the secondary-side controller, and sensed indirectly on the primary-side. A PLL with a 100 MHz delay line, precise 12-bit phase-shift control module, and high-frequency transceiver are implemented on-chip in a 0.18μm 80V BCD process, and are used to control an off-chip DAB power stage.

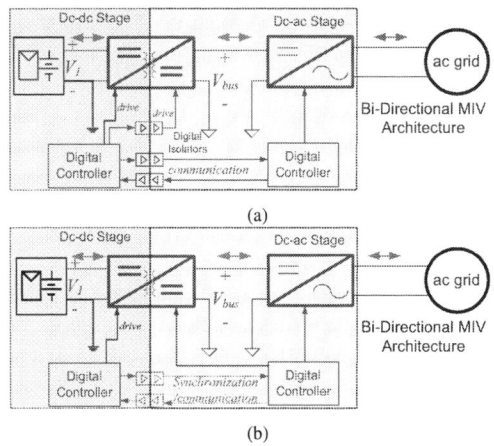

(a)

(b)

Fig. 1. (a) Architecture of a conventional isolated two-stage bi-directional dc-ac converter (one isolator per transistor). (b) Alternative architecture where the secondary-side controller drives the secondary-side switches, which requires a synchronization scheme.

I. INTRODUCTION

There is a growing demand for bidirectional power-flow capability and four-quadrant operation in Micro-Inverters (MIV) [3]. Multi-port MIVs with integrated storage, which is typically realized with a battery [4], [5] or super-capacitor [3], have been proposed for reactive power support and off-grid operation as well as active power smoothing [3]. Bidirectional EV charger topologies with Vehicle-to-Grid (V2G) capability can provide ac grid support using the on-board battery during peak load demand [6], [7].

A bidirectional MIV, as shown in Fig. 1(a) [7], requires active devices on both sides of the transformer in the dc-dc stage. Certain unidirectional dc-dc topologies, such as the *LLC* converter, may also be designed with active switches on both sides of the isolation for improved efficiency [8]. As the switching frequency is scaled up for improved power density in modern converters, driving multiple switches on both sides of the transformer with precise timing becomes a major challenge. Low-frequency communication is also required between the dc-dc and dc-ac stages for tasks such as start-up, fault-handling and feedforward control loops, hence the need for additional digital isolators. Opto-isolators are commonly used, however they suffer from poor matching, relatively long delays

(typically ≥15 ns) and short life-time at high temperature [9], [10]. More recently, digital isolators and isolated gate-drivers based on miniaturized magnetic components or capacitive coupling either on the PCB [11] or on-chip [12]–[16] have addressed some of these shortcomings, while offering a high level of integration. An isolated gate-driver architecture with off-chip magnetics is proposed in [15], [16]. Digital isolators and drivers typically modulate the PWM signal with a carrier in the range of tens of MHz to several GHz in order to reduce the size of the isolation passive components [15], [16]. They require precise mechanical alignment between primary and secondary-side coils. The cost of this technology is typically high, due to multi-die packages or high isolation ratings required for the process technology, as well as the cost of extra components and board space. These isolators are also susceptible to Electromagnetic Interference (EMI) issues and consume nearly as much power as the gate-driver itself. Moreover, the driver cost scales with the number of active switches driven from the opposite side of isolation.

This work presents the first phase in a long-term project

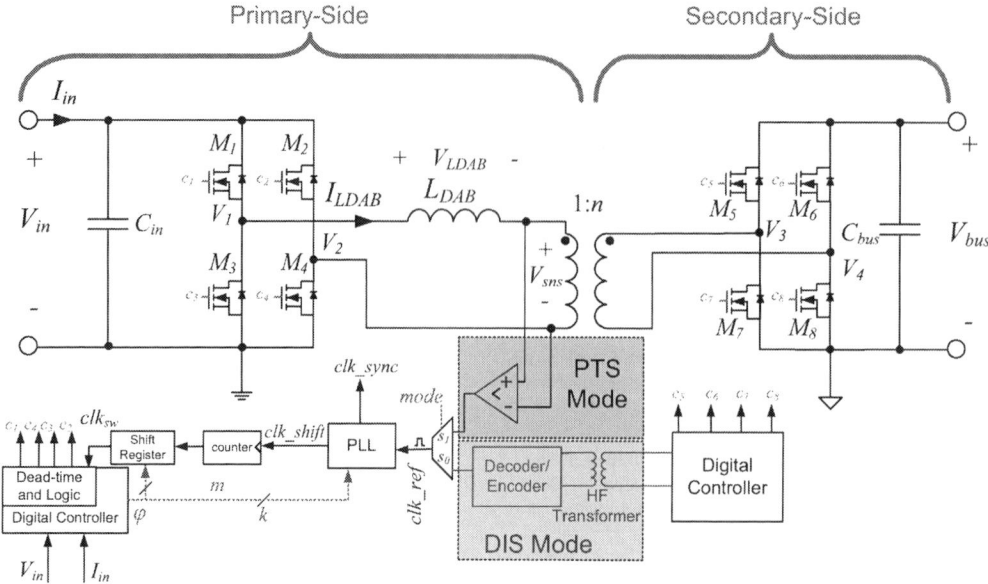

Fig. 2. Simplified architecture of the two proposed synchronization schemes (DIS and PTS) implemented on a DAB dc-dc converter.

to monolithically integrate the power transistors, gate-drivers, digital controller and the digital isolation capability in a BCD technology for optimized cost and performance. In this paper, two alternative synchronization schemes are demonstrated on-chip to eliminate the need for using one isolator per transistor:

1) The Power Transformer Sensing (PTS) scheme relies on extracting the switching signal on the secondary-side directly from the reflected voltage across the power transformer, and reconstructing the driving waveforms accordingly on the primary side. The PTS was first reported in [17] with an Field Programmable Gate Array (FPGA) and discrete components.

2) The Digital Isolator Sensing (DIS) scheme, where the switching information is encoded in the data packets sent through a miniature high-frequency Galvanically Isolated (GISO) transceiver.

In both schemes, the active devices on the secondary-side of the dc-dc stage are driven by the dc-ac stage controller, as shown in Fig. 1(b), and a Phase-Locked-Loop (PLL) is used for clock recovery. While both schemes are compatible with a variety of isolated topologies, this paper is focused on the soft-switching Dual Active Bridge (DAB) converter with simple Phase-Shift-Modulation (PSM), as shown in Fig. 2(a). Both bridges operate with a duty cycle, D, of 50% and the power flow is given by:

$$P_{DAB} = \frac{V_{PV} V_{bus}}{n \omega_s L_{DAB}} \phi \left(1 - \frac{|\phi|}{\pi}\right), \qquad (1)$$

where n is the transformer turns ratio, L_{DAB} is the DAB inductance, which is the sum of transformer's leakage inductance and the external inductance, ϕ is the phase-shift between

the two bridges, and $\omega_s = 2\pi f_s$, where f_s is the switching frequency.

This paper is organized as follows. The PTS and DIS schemes are discussed in detail in Section II. The wide-range PLL implementation is presented in Section III, while the performance of both of the synchronization schemes are demonstrated on an external DAB converter in Section IV. The concluding remarks are provided in Section V.

II. SYNCHRONIZATION SCHEMES

A custom IC was developed in a 80V 0.18μm BCD technology to demonstrate and compare the two synchronization schemes on a DAB converter. The high-level architecture is shown in Fig. 2. Synchronization can be achieved using either a high-frequency isolated communication channel (DIS mode, activated when $mode = 1$), or sensing through the power transformer (PTS mode, activated when $mode = 0$).

A. Digital Isolator Based Synchronization and Communication

The DIS scheme utilizes a single digital isolator to achieve two purposes: 1) synchronize both controllers on opposite sides of the isolation, and 2) to transmit data between the controllers, as needed for control and fault handling. A low-power, high-speed GISO transceiver was developed to demonstrate the DIS scheme. High-frequency modulation, up to 100 MHz, is adopted to reduce the air-core transformer size, which can also be implemented using PCB traces. The high-speed GISO transceiver is designed with 1.8V devices and transmits a differential low-amplitude signal (typically $\geq$50 mV) across an on-chip 20 Ω termination resistor, with a common-mode voltage of 0.9 V, as shown in Fig. 3. The critical GISO

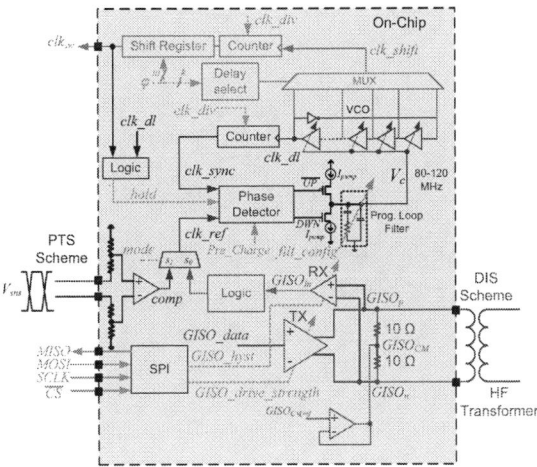

Fig. 3. Architecture of the on-chip PTS/DIS synchronization schemes and phase-shift generation.

parameters, including the driver signal-swing and the receiver hysteresis are digitally programmable via the SPI interface.

A typical GISO communication packet includes two main components, as shown in Fig. 4(a). The first period of the packet is a preamble and serves as the reference clock, clk_ref, for the PLL. The PLL is enabled during the preamble period. The bits following the preamble contain the data and can be recovered using a variety of well-known methods, such as Manchester coding [18]. The start-up sequence in DIS mode is shown in Fig. 4(b). The PLL is put into a *hold* state, where the phase detector is disabled in between consecutive preamble periods, as shown in Fig. 4(b). The Voltage-Controlled-Oscillator (VCO) is therefore free-running during this time. It is necessary to maintain regular communication in order to guarantee an acceptable PLL drift. Furthermore, the PLL is placed in the *hold* state for 3% of the switching period, T_s, during power-transistor switching, in order to minimize the disturbance on the synchronizing operation. The locked output of PLL, clk_sync, is used as the main system clock in order to generate the phase-shifted PWM signal, clk_sw, to control the primary-side bridge of the DAB converter. The phase-shift is achieved by using a combination of delay-line and a counter for fine and coarse delay adjustment, respectively, as shown in Fig. 3. The phase-shift resolution is $k + m$, where k and m are the number of delay elements and counter bits respectively; $k = 5$ and $m = 7$ are adopted in this work, which results in a phase-shift resolution of 300 ps.

B. Synchronization by Power Transformer Sensing and Communication

Using the PTS scheme, synchronization is achieved by sensing the reflected voltage across the power transformer, hence eliminating the need for any digital isolator. The PTS scheme was first reported in [17] using discrete components, where it was also shown that low-frequency unidirectional communication can be achieved through switching frequency

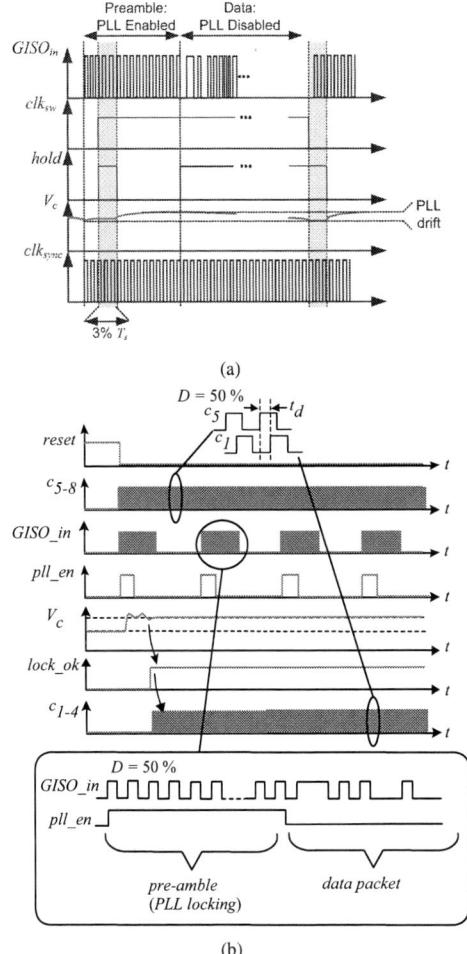

Fig. 4. (a) GISO packet including the preamble and data periods. The PLL is activated during preamble period. (b) Start-up sequence in DIS mode.

modulation. The reflected switching waveform on the secondary side is detected using a comparator, and the resulting waveform, $comp$, is fed as clk_ref to the PLL, as shown in Fig. 3. When locked, clk_sync is in phase with the secondary-side switching and can thus be phase-shifted by ϕ in order to generate four PWM signals, c_{1-4}. The resulting power flow is given by (1). The start-up sequence is illustrated in Fig. 5. Once the secondary controller activates the secondary-side bridge, the switching action is detected on the primary side and PLL is enabled. The charge-pump output voltage, V_c, is regulated in order to synchronize clk_sync to clk_ref. Once PLL locking is achieved, the converter can start-up by driving c_{1-4}.

III. PLL IMPLEMENTATION

In order to achieve synchronization across an isolated boundary, the system must be able to lock to an external clock source. The synchronization is achieved by using an on-chip PLL, as shown in Fig. 3. The PLL consists of a

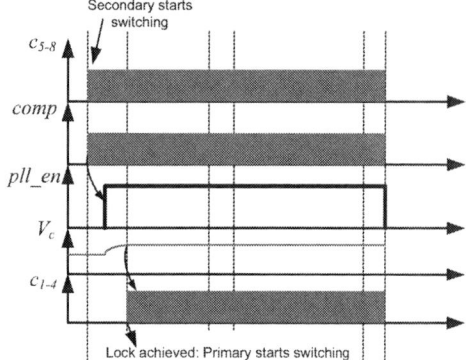

Fig. 5.   Start-up sequence in PTS mode.

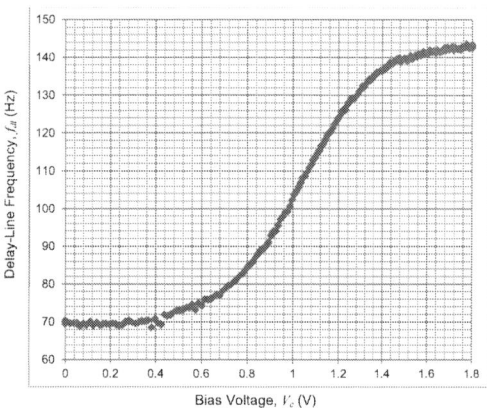

Fig. 6.   The simulated delay-line frequency, f_{dl}, vs. the bias voltage, V_c.

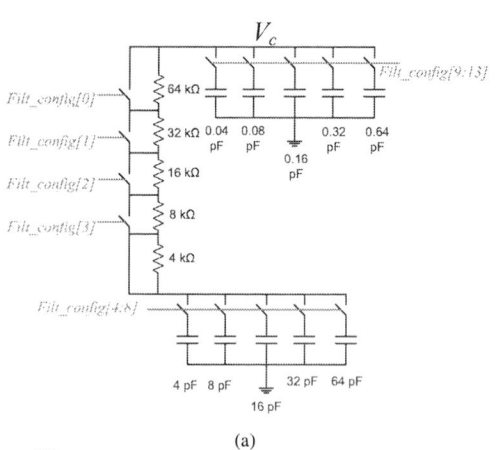

(a)

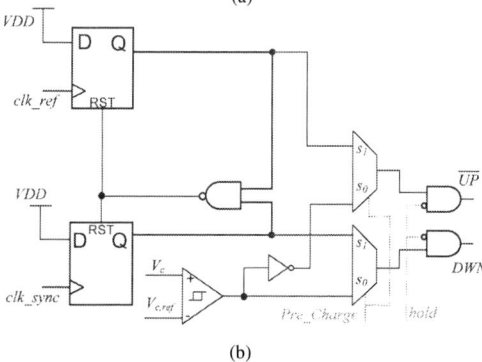

(b)

Fig. 7.   (a) PLL loop-filter, and (b) phase-detector block.

TABLE I
DAB CONVERTER PROTOTYPE SPECIFICATIONS WITH EXTERNAL POWER
STAGE

Parameter	Value	Unit
Switching Frequency, f_s	500	kHz
DAB Inductance, L_{DAB}	6.9	μH
Primary-Side Voltage, V_{in}	30	V
Bus Voltage, V_{bus}	90	V
Transformer Turns Ratio, n	3.5	
Nominal Power, P_{DAB}	50	W

VCO, clock divider, loop-filter, and phase detector blocks, as in a conventional architecture [19]. For this application, it is designed to lock over a very wide frequency range, from 500 kHz to 100 MHz, using digitally configurable clock divider and loop-filter in order to accommodate both PTS and DIS schemes.

The VCO is comprised of a ring of 32 current-starved inverters with voltage-controlled delay. The simulated frequency of the delay-line, f_{dl}, versus the inverter bias voltage, V_c, is shown in Fig. 6. A frequency range of 80-120 MHz is achieved for $V_c = 0.8$-1.2 V, in which the VCO is quite linear with a slope of 105 MHz/V. The maximum measured frequency range is lower due to the parasitics. The bias voltage V_c is generated from an analog loop-filter which consists of three on-chip passive elements, R_1, C_1, and C_2. All these elements are implemented as programmable passive banks with $2^{14} = 16384$ different combinations, as shown in Fig. 7(a), and occupy a total space of 260 μm $\times$ 400 μm on-chip.

The charge-pump current, I_{pump}, is programmable in the range of 20-200 μA. This current is either pumped in or out of the loop-filter, depending on the $\overline{UP}$ and DWN signals, generated by the phase-detector block. The phase-detector compares the rising edges of an external reference clock, clk_ref, and the internal synchronized clock of the delay-line, clk_sync, as shown in Fig. 7(b). It consists of two Flip-Flops and a NAND gate. A comparator is used in order to pre-charge the voltage V_c to $V_{c,ref}$, a programmable value which is set within the VCO's linear range.

IV. EXPERIMENTAL RESULTS

The fabricated DAB synchronization IC is shown in Fig. 8 and measures 2.5×4.5 mm². The chip was tested with a 50 W discrete DAB converter, with the parameters listed in Table I. The start-up process, as well as steady-state operation in the DIS and PTS modes, are shown in Figs. 9 and 10, respectively.

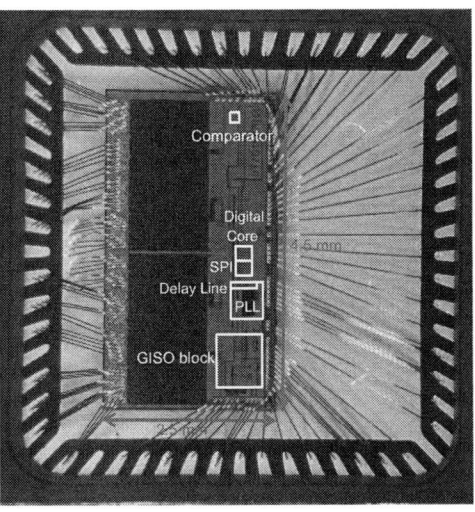

Fig. 8. The packaged chip micrograph (2.5×4.5 mm² in 0.18μm 80V BCD process).

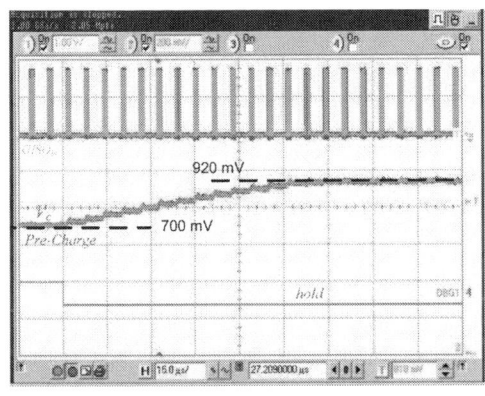

(a)

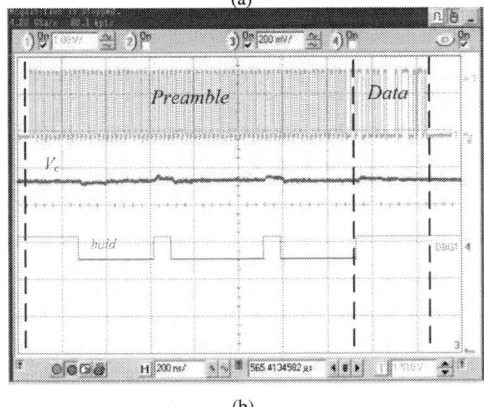

(b)

Fig. 9. (a) Start-up, and (b) steady state operation of the synchronization process in the DIS mode.

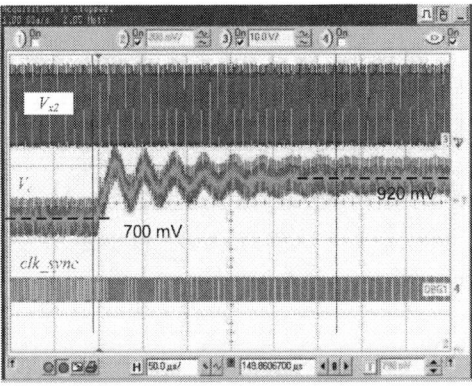

(a)

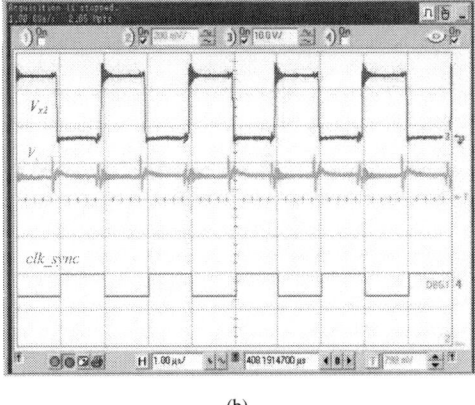

(b)

Fig. 10. (a) Start-up, and (b) steady state operation of the synchronization process in the PTS mode.

In both modes, V_c is well regulated to its final value in steady-state. With GISO frequency equal to 50 MHz (clk_ref = 50 MHz), it takes 18 GISO packets for the PLL to settle in DIS mode, while the settling time is 275 μs in PTS mode (clk_ref = 500 kHz). The PLL is constantly active in PTS mode, except for when $hold$ is high due to PWM edges. The PLL is put in the $hold$ state in between the data packets in the DIS mode. The DAB current waveform in the DIS and PTS modes are shown in Fig. 11(a), (b), respectively. The DAB current waveform in Fig. 11(a) is captured by activating the persistence mode of the oscilloscope, and shows very low PLL jitter in DAB operation in the DIS mode. The delay-line, which is the most power hungry part of the PLL, consumes 0.846 mW on average when oscillating at the maximum frequency of 100 MHz. The GISO transmission consumes 27 mW in the transmitter and 63 μW in the receiver when operating continuously at 100 MHz. The average power consumption of GISO block in DIS mode can be much lower based on the packet transmission frequency.

TABLE II

COMPARISON OF DRIVING SCHEMES IN ISOLATED CONVERTERS

Scheme	Conventional	PTS	DIS
Communication Capability	No	Yes (unidirectional)	Yes
Continuous Communication	No	Yes (required to avoid PLL drift)	Optional
PLL Operation	None	Low frequency	High frequency
Absolute Phase Reference	N/A	Available through *comp*	Communicated
Number of Digital Isolators Required for Driving (DAB)	4	0	1
Power Consumption	High	Lowest	Low
Power Converter Topology	No limitation	Limited by transformer leakage inductance	No limit
High Voltage Comparator	None	Yes	No

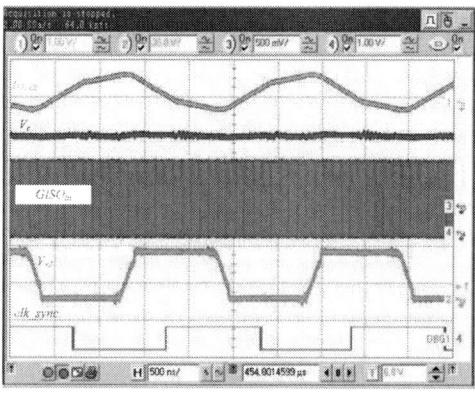

(a)

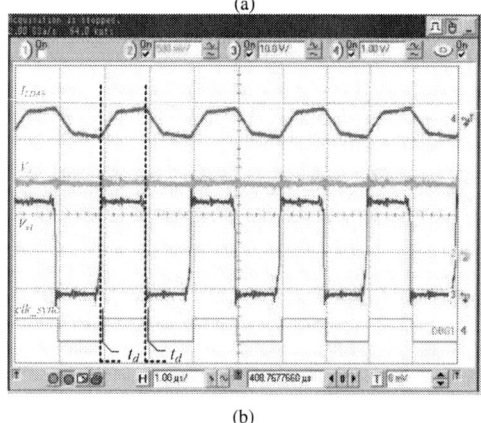

(b)

Fig. 11.   Synchronized DAB converter waveforms in the (a) DIS mode with continuous $GISO_{in}$ input clock, and (b) PTS mode (I_{LDAB}: 5A/div).

V. CONCLUSIONS

This work presents the first on-chip demonstration of both PTS and DIS synchronization schemes for isolated dc-dc converters. While both PLL-based schemes are superior to the conventional approach of using one opto-isolator per transistor, they have important differences: DIS is more expensive and consumes more power than PTS, since it requires an off-chip air-core transformer for the GISO, however it is much more flexible in terms of data communication and power topology.

A qualitative comparison of the PTS and DIS schemes is provided in Table II. The PTS scheme eliminates the need for any digital isolator, while the DIS scheme requires one digital isolator. However, the application of PTS is limited to certain topologies, while DIS scheme can be applied to a variety of isolated topologies. The PLL in PTS scheme is designed to lock to a low frequency compared to the DIS scheme. This shrinks the size of the on-chip PLL for the DIS scheme; however, the power consumption is higher. In general, both of these schemes are superior in cost and performance to the conventional approach of using one digital isolator per transistor in addition to the data channel. The developed IC in this work is promising for future use in isolated dc-dc converters with high switching frequencies beyond 1 MHz.

VI. ACKNOWLEDGEMENT

This project was supported by Solantro Semiconductor, Ontario Centres of Excellence and the Canada Foundation for Innovation. The authors thank Magnachip and Ikjoon Choi for fabrication support. The authors thank Ray Orr and Ben Bacque for their contributions as well.

REFERENCES

[1] Y.-C. Wang, Y.-C. Wu, and T.-L. Lee, "Design and implementation of a bidirectional isolated dual-active-bridge-based dc/dc converter with dual-phase-shift control for electric vehicle battery," in *2013 IEEE Energy Conversion Congress and Exposition (ECCE)*, Sept 2013, pp. 5468–5475.

[2] D. Costinett, K. Hathaway, M. Rehman, M. Evzelman, R. Zane, Y. Levron, and D. Maksimovic, "Active balancing system for electric vehicles with incorporated low voltage bus," in *2014 Twenty-Ninth Annual IEEE Applied Power Electronics Conference and Exposition (APEC)*, March 2014, pp. 3230–3236.

[3] S. Poshtkouhi, M. Fard, H. Hussein, L. Dos Santos, O. Trescases, M. Varlan, and T. Lipan, "A dual-active-bridge based bi-directional micro-inverter with integrated short-term li-ion ultra-capacitor storage and active power smoothing for modular pv systems," in *Applied Power Electronics Conference and Exposition (APEC), 2014 Twenty-Ninth Annual IEEE*, March 2014, pp. 643–649.

[4] W. Hu, H. Wu, Y. Xing, and K. Sun, "A full-bridge three-port converter for renewable energy application," in *Applied Power Electronics Conference and Exposition (APEC), 2014 Twenty-Ninth Annual IEEE*, March 2014, pp. 57–62.

[5] Q. Mei, X. Zhen-lin, and W.-Y. Wu, "A novel multi-port dc-dc converter for hybrid renewable energy distributed generation systems connected to power grid," in *Industrial Technology, 2008. ICIT 2008. IEEE International Conference on*, April 2008, pp. 1–5.

[6] C.-J. Shin and J.-Y. Lee, "An electrolytic capacitor-less bi-directional ev on-board charger using harmonic modulation technique," *Power Electronics, IEEE Transactions on*, vol. 29, no. 10, pp. 5195–5203, Oct 2014.

[7] B. Koushki, A. Safaee, P. Jain, and A. Bakhshai, "Review and comparison of bi-directional ac-dc converters with v2g capability for on-board ev and hev," in *Transportation Electrification Conference and Expo (ITEC), 2014 IEEE*, June 2014, pp. 1–6.

[8] S. Abe, T. Ninomiya, T. Zaitsu, J. Yamamoto, and S. Ueda, "Seamless operation of bi-directional llc resonant converter for pv system," in *Applied Power Electronics Conference and Exposition (APEC), 2014 Twenty-Ninth Annual IEEE*, March 2014, pp. 2011–2016.

[9] A. Thaduri, A. Verma, G. Vinod, and R. Gopalan, "Reliability prediction of optocouplers for the safety of digital instrumentation," in *2011 IEEE International Conference on Quality and Reliability (ICQR)*, Sept 2011, pp. 491–495.

[10] P. Jacob, G. Nicoletti, and M. Rutsch, "Reliability failures in small optocoupling and dc/dc converter devices," in *2006. 13th International Symposium on the Physical and Failure Analysis of Integrated Circuits*, July 2006, pp. 167–170.

[11] S. Hui, S. C. Tang, and H.-H. Chung, "Optimal operation of coreless pcb transformer-isolated gate drive circuits with wide switching frequency range," *IEEE Transactions on Power Electronics*, vol. 14, no. 3, pp. 506–514, May 1999.

[12] Y. Moghe, A. Terry, and D. Luzon, "Monolithic 2.5kv rms, 1.8v;3.3v dual-channel 640mbps digital isolator in 0.5 um sos," in *IEEE International SOI Conference (SOI)*, Oct 2012, pp. 1–2.

[13] T. V. Nguyen, J.-C. Crebier, and P.-O. Jeannin, "Design and investigation of an isolated gate driver using cmos integrated circuit and hf transformer for interleaved dc/dc converter," *IEEE Transactions on Industry Applications*, vol. 49, no. 1, pp. 189–197, Jan 2013.

[14] S. Nagai, T. Fukuda, N. Otsuka, D. Ueda, N. Negoro, H. Sakai, T. Ueda, and T. Tanaka, "A one-chip isolated gate driver with an electromagnetic resonant coupler using a spdt switch," in *2012 24th International Symposium on Power Semiconductor Devices and ICs (ISPSD)*, June 2012, pp. 73–76.

[15] K. Muhammad and D.-C. Lu, "Magnetically isolated gate driver with leakage inductance immunity," *IEEE Transactions on Power Electronics*, vol. 29, no. 4, pp. 1567–1572, April 2014.

[16] B. Chen, "icoupler products with isopower technology: Signal and power transfer across isolation barrier using microtransformers," Analog Devices Inc., 2006, available http://www.analog.com/static/imported-files/overviews/isoPower.pdf.

[17] S. Poshtkouhi, A. Eski, and O. Trescases, "Pll based bridge synchronization as an alternative to digital isolators for dual active bridge dc-dc converters," in *Applied Power Electronics Conference and Exposition (APEC), 2015 IEEE*, March 2015, pp. 9–14.

[18] G. Raghul, K. Sudhakar, and M. Devi, "Design and implementation of encoding techniques for wireless applications," in *Circuit, Power and Computing Technologies (ICCPCT), 2015 International Conference on*, March 2015, pp. 1–7.

[19] T. Chan Carusone, D. Johns, and K. Martin, *Analog Integrated Circuit Design, Second Ed.* Wiley, 2001.

An Isolated Soft-switching Buck-Boost Converter Utilizing Two Transformers and Embedded Bidirectional Switches on Secondary-side For Wide Voltage Applications

Tingting Liu, Hongfei Wu, Yan Xing,
Jiangsu Key Lab. of New Energy Generation and Power
Conversion
Nanjing University of Aeronautics and Astronautics
Nanjing, China
Email: liutingting@nuaa.edu.cn

Kai Sun
State Key Lab of Power Systems, Department of Electrical
Engineering
Tsinghua University
Beijing, China
Email: sun-kai@mail.tsinghua.edu.cn

Abstract—A dual-phase-shift controlled isolated buck-boost converter is presented for wide input or output voltage range applications. Two transformers and a voltage-doubler rectifier with embedded bidirectional switch are employed. The primary windings of the two transformers are in series and the secondary windings are in parallel. With optimized dual-phase-shift modulation strategies, zero voltage switching (ZVS) is achieved for both the primary- and secondary-side power MOSFETs in a wide load range. The reverse recovery problem of rectifying diodes is eliminated as well. Leakage inductances of the transformers are utilized for power transferring. The voltage stresses of primary-side and secondary-side power MOSFETs are clamped to the input voltage and half of the output voltage, respectively. Besides, the converter can operate in the buck, balance and boost modes to achieve a wide voltage range. The operational principles are analyzed and experimental results of a 1-kW 100-kHz prototype are provided to verify the effectiveness of the presented converter.

Keywords—DC-DC converter; isolated Buck-Boost converter; soft-switching; dual-phase-shift control

I. INTRODUCTION

Isolated DC-DC converters with wide input/output voltage ranges have been widely used in the applications of renewable energy, storage and electric vehicles' power systems. The isolated converters can be classified into three categories: Buck converters [1]-[2], Boost converters [3]-[4] and Buck-Boost converters. In comparison with Buck converters and Boost converters, an isolated Buck-Boost converter (IBB) [5], which can operate either as a Buck converter or as a Boost converter, is more flexible in terms of conversion efficiency and voltage range. Few works on IBB DC-DC converter topologies have been reported in literatures. The Flyback converter is the simplest IBB topology. However, efficiency of the Flyback converter is low because of the high voltage/current stresses on the components and the hard-switching of active switches and rectifying diodes. The LLC resonant converter [6]-[8] is very attractive due to its excellent soft-switching performance and

voltage step-up and step-down capability. However, it is difficult to realize a wide range of voltage regulation while maintaining high efficiency within the entire operating range, because the efficiency is poor once the operating frequency is far away from its resonant frequency of the resonant tank due to large circulating currents. The semi-dual-active-bridge (SDAB) converters presented in [8], which can be seen as simplified versions of dual-active-bridge (DAB) converters, can be good options for IBB power conversion. But the power/current stresses of these converters are still high, and the soft-switching range is limited.

A novel soft-switching IBB converter utilizing two transformers and embedded bidirectional switch on secondary side voltage-doubler rectifier is presented for wide voltage range applications. High efficiency and low stresses can be achieved with the proposed converter and control scheme.

II. PROPOSED CONVERTER AND OPERATIONAL PRICINPLE

A. Topology and analysis

The proposed soft-switching DC-DC converter with two transformers and a voltage-doubler rectifier is shown in Fig. 1, in which the D_1/D_2 and C_{o1}/C_{o2} construct as the output voltage-doubler rectifier while S_5/S_6 as the bidirectional switch. With optimized dual-phase-shift modulation, soft switching over wide output voltage and load range can be achieved. In order to simplify the analysis, the equivalent voltage gain G is defined

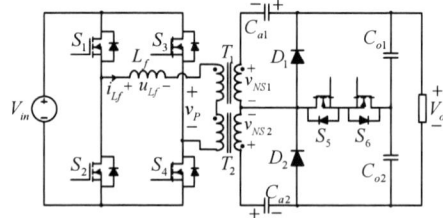

Fig.1. Topology of proposed converter

This work was supported by the National Natural Science Foundation of China (51407092, 51577102), the Natural Science Foundation of Jiangsu Province, China (BK20140812), and Foundation of State Key Lab of Power System(SKLD15KZ01, SKLD15M07).

as $G=nV_o/V_{in}$, where V_{in}, V_o and n are the input voltage, output voltage and transformers' turns ratio (n_P:n_S), respectively. The converter can work in either the Buck mode ($0<G<1$), or the Boost mode ($G>1$). In each mode, according to the waveform of inductor current i_{Lf}, the operation mode can be further divided into continuous-conduction mode (CCM) and discontinuous-conduction mode (DCM). Thus, the converter has four operation modes, namely Buck-CCM, Buck-DCM, Boost-CCM and Boost-DCM, respectively. For the sake of simplicity, the parasitic capacitances of MOSFETs are ignored and the transformers are assumed idea. The operational principle of proposed converter is analyzed as follows.

B. Operational Principle in Boost-CCM mode

The control strategy in Boost mode is the same as the traditional DAB converter. All the MOSFETs on the primary and secondary sides have a constant duty cycle of 0.5. S_1 and S_4 (S_2 and S_3) are always turned-on or off simultaneously while S_1 and S_2 (S_5 and S_6) are operated complementarily. A secondary-side phase-shift angle φ_S between S_1 and S_5 is employed to regulate the output power and voltage, as is shown in Fig. 2. As this phase shift angle serves the same function as duty cycle in PWM converter, the secondary-side phase-shift ratio D_S is defined as $D_S=\varphi_S/\pi$. It can be given that when $D_S>(G-1)/G$, the converter operates in Boost-CCM mode.

The key waveforms in Boost-CCM mode are shown in Fig. 2. Due to the symmetry of the circuit, five stages in half switching period are analyzed here and corresponding equivalent circuits for each operation stage are shown in Fig. 3.

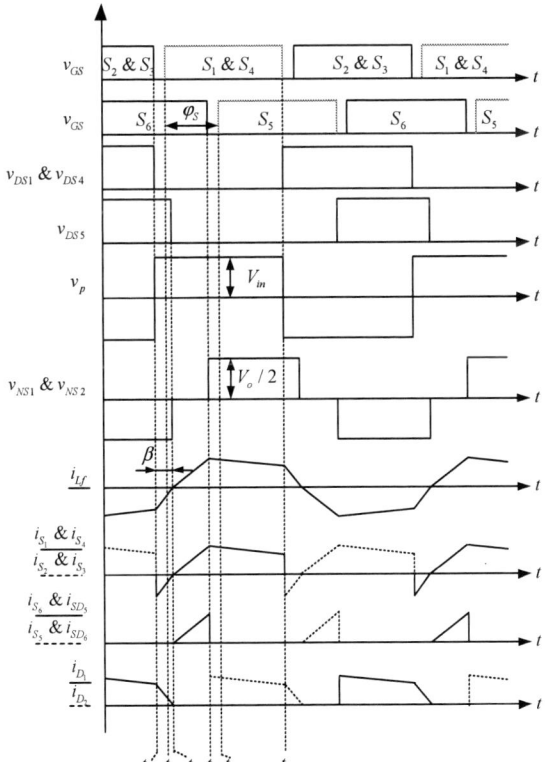

Fig. 2. Key waveforms of proposed converter in Boost-CCM mode

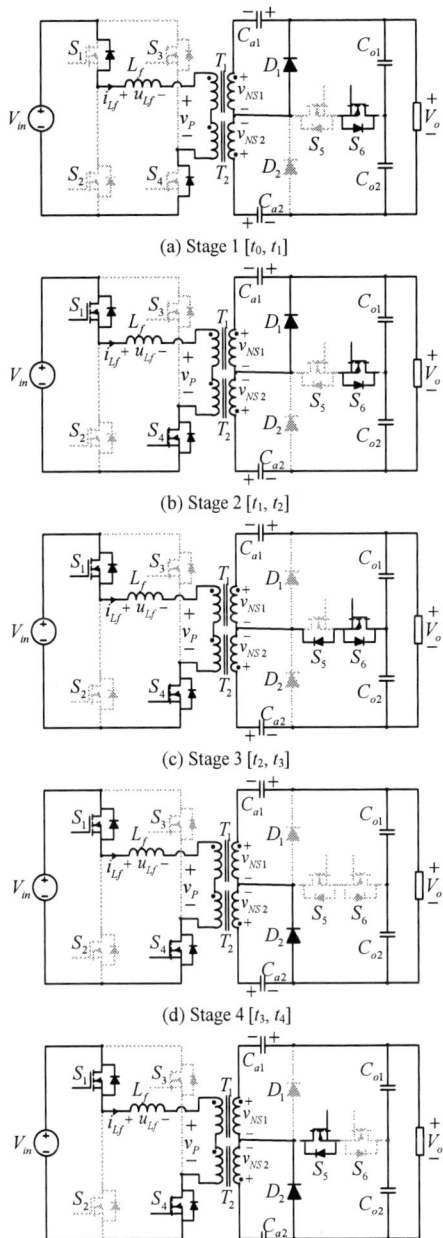

(a) Stage 1 $[t_0, t_1]$

(b) Stage 2 $[t_1, t_2]$

(c) Stage 3 $[t_2, t_3]$

(d) Stage 4 $[t_3, t_4]$

(e) Stage 5 $[t_4, t_5]$

Fig. 3. Equivalent circuits in Boost-CCM mode

Stage 1[t_0, t_1] [see Fig. 3(a)]: Before t_0, S_2, S_3, S_6 and D_1 are at on state and inductor current $i_{Lf}<0$. The energies from the input source and stored in the inductor L_f are delivered to the output and auxiliary capacitor C_{a1}. At t_0, S_2 and S_3 turn off and the body-diodes of S_1 and S_4 begin to conduct due to the energy stored in L_f. Inductor current i_{Lf} is defined as

$$i_{Lf}(t) = \frac{V_{in}+nV_o}{L_f}(t-t_0)+i_{Lf}(t_0) \qquad (1)$$

Stage 2[t_1, t_2] [see Fig. 3(b)]: At t_1, S_1 and S_4 turn on with ZVS.

Stage 3[t_2, t_3] [see Fig. 3(c)]: At t_2, i_{Lf} and the current through D_1 reach zero. D_1 shuts down without reverse recovery. Body diode of S_5 begins to conduct as i_{Lf} becomes positive. Secondary-side voltages of T_1 and T_2 are clamped to zero due to the conduction of bidirectional switch, thus L_f is charged by the input voltage and no energy is transferred to the output through the transformers. Inductor current i_{Lf} is defined as

$$i_{Lf}(t) = \frac{V_{in}}{L_f}(t - t_2) + i_{Lf}(t_2) \tag{2}$$

Stage 4[t_3, t_4] [see Fig. 3(d)]: At t_3, S_6 turns off. Since the current i_{Lf} is positive, D_2 begins to conduct. The energies from the input source and stored in the inductor L_f are delivered to the output and auxiliary capacitor C_{a2}. The inductor current i_{Lf} is given as

$$i_{Lf}(t) = \frac{V_{in} - nV_o}{L_f}(t - t_3) + i_{Lf}(t_3) \tag{3}$$

Stage 5[t_4, t_5] [see Fig. 3(e)]: At t_4, S_5 turns on with ZVS. Even so, the bidirectional switch is still off as the current i_{Lf} is positive and S_6 is already turned off. The current loop is no different. At t_5, S_1 and S_4 turn off and a similar operation works in the next half switching cycle.

C. Operational Principle in Buck-CCM mode

In Buck mode, optimized dual-phase-shift control strategy is adopted. Based on the phase-shift control strategy mentioned

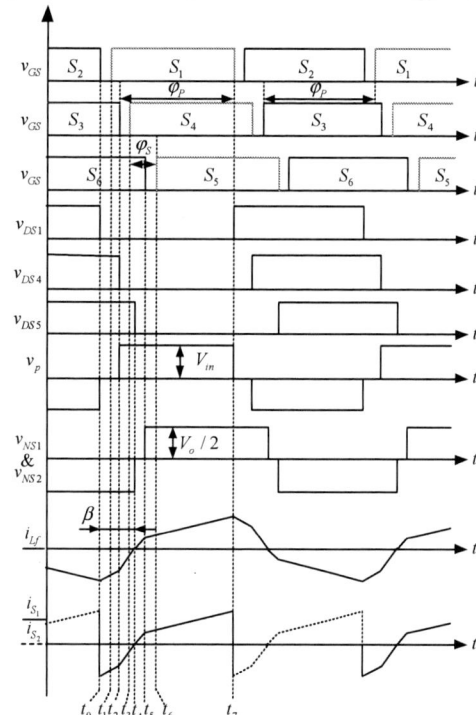

Fig. 4. Key waveforms of proposed converter in Buck-CCM mode

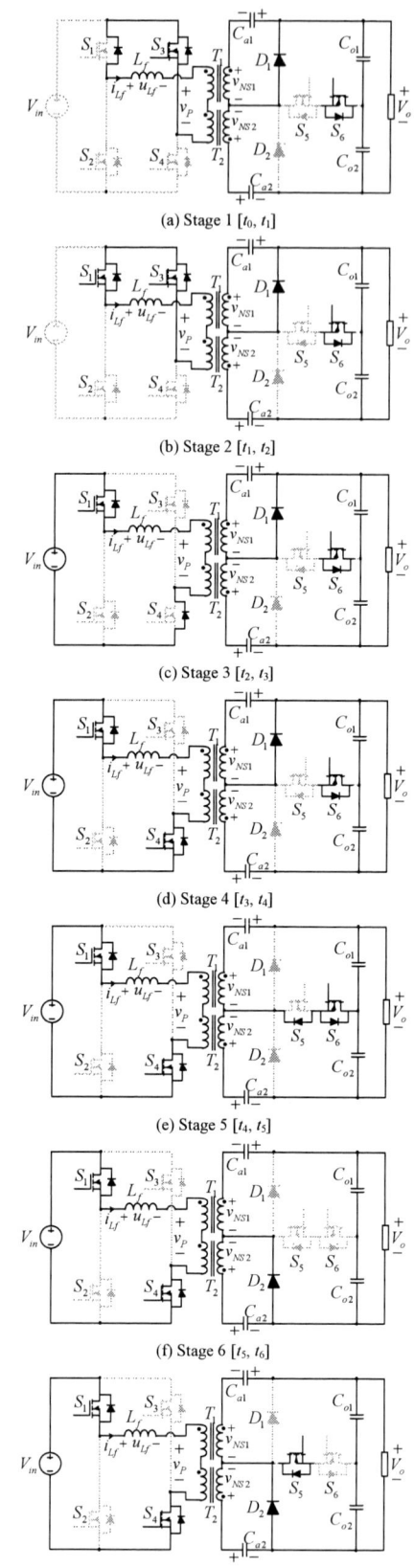

(a) Stage 1 [t_0, t_1]

(b) Stage 2 [t_1, t_2]

(c) Stage 3 [t_2, t_3]

(d) Stage 4 [t_3, t_4]

(e) Stage 5 [t_4, t_5]

(f) Stage 6 [t_5, t_6]

(g) Stage 7 [t_6, t_7]

Fig. 5. Equivalent circuits in Buck-CCM mode

above, an inner phase-shifted angle, φ_P, between S_1/S_2 arm and S_3/S_4 arm of primary-side full bridge circuit is employed to extend soft switching region of the proposed converter, as is shown in Fig. 5. D_P is the primary-side phase-shift ratio, defined as $D_P=\varphi_P/\pi$. In Buck-CCM mode, the primary-side phase-shift ratio is set to be constant, which equals to the equivalent voltage gain G. The secondary-side phase-shift ratio is adjusted to control the output voltage and power. While in Buck-DCM mode, D_S is set to be nearly zero and D_P is adjusted. The control strategy mentioned above can improve the soft-switching characteristics in a wide range, which will be analyzed in detail too. The key waveforms in Buck-CCM mode are shown in Fig. 4. Seven stages in half switching cycle are analyzed here and corresponding equivalent circuits for each operation stage are shown in Fig. 5.

Stage 1 $[t_0, t_1]$ [see Fig. 5(a)]: Before t_0, S_2, S_3, S_6 and D_1 are at on state and inductor current $i_{Lf}<0$. At t_0, S_2 turns off and the body-diode of S_1 conducts due to the energy stored in L_f. During this stage, the input source is cut off from the inductor and the energy stored in L_f is transferred to the secondary-side. Inductor current i_{Lf} is defined as

$$i_{Lf}(t) = \frac{nV_o}{L_f}(t-t_0) + i_{Lf}(t_0) \qquad (4)$$

Stage 2 $[t_1, t_2]$ [see Fig. 5(b)]: At t_1, S_1 turns on with ZVS.

Stage 3 $[t_2, t_3]$ [see Fig. 5(c)]: At t_2, S_3 turns off and body-diode of S_4 conducts due to the energy stored in L_f. During this stage, the energies from the input source and stored in the inductor L_f are delivered to the output and auxiliary capacitor C_{a1}. Inductor current i_{Lf} is defined as

$$i_{Lf}(t) = \frac{V_{in}+nV_o}{L_f}(t-t_2) + i_{Lf}(t_2) \qquad (5)$$

Stage 4 $[t_3, t_4]$ [see Fig. 5(d)]: At t_3, S_4 turns on with ZVS.

Stage 5 $[t_4, t_5]$ [see Fig. 5(e)]: At t_4, i_{Lf} and the current through D_1 reach zero. D_1 shuts down without reverse recovery. Body diode of S_5 begins to conduct as i_{Lf} becomes positive and S_6 is on. Secondary-side voltages of T_1 and T_2 are clamped to zero due to the conduction of bidirectional switch, thus L_f is charged by the input voltage and no energy is transferred to the output through the transformers. Inductor current i_{Lf} is defined as

$$i_{Lf}(t) = \frac{V_{in}}{L_f}(t-t_4) + i_{Lf}(t_4) \qquad (6)$$

Stage 6 $[t_5, t_6]$ [see Fig. 5(f)]: At t_5, S_6 turns off. Since i_{Lf} is positive, D_2 begins to conduct. The energies from the input source and stored in the inductor L_f are delivered to the output and auxiliary capacitor C_{a2}. The inductor current i_{Lf} is given as

$$i_{Lf}(t) = \frac{V_{in}-nV_o}{L_f}(t-t_5) + i_{Lf}(t_5) \qquad (7)$$

Stage 7 $[t_6, t_7]$ [see Fig. 5(g)]: At t_6, S_5 turns on with ZVS. At t_7, S_1 turns off and a similar operation works in the next half switching cycle.

D. Characterization

1) Soft-switching

By defining $\theta=2\pi f_s t$, $D=\theta/\pi$ and $I_b=V_{in}/2f_sL_f$, the dynamics of i_{Lf} in Boost-CCM mode can be further simplified as

$$\begin{cases} i_{Lf}(D) = (1+G)\cdot I_b \cdot (D-0) + i_{Lf}(0) & 0 \le D < \beta/\pi \\ i_{Lf}(D) = I_b \cdot (D-\beta/\pi) + i_{Lf}(\beta/\pi) & \beta/\pi \le D < D_s \\ i_{Lf}(D) = (1-G)\cdot I_b \cdot (D-D_s) + i_{Lf}(D_s) & D_s \le D < 1 \end{cases} \quad (8)$$

The average value of i_{Lf} must equal zero under steady state conditions, namely, $i_{Lf}(0)+i_{Lf}(1)=0$. With this constraint and $i_{Lf}(\beta/\pi)=0$, $i_{Lf}(0)$ and $i_{Lf}(D_s)$ can be derived as

$$\begin{cases} i_{Lf}(0) = -\dfrac{D_s \cdot G - G^2 + D_s \cdot G^2 + 1}{G+2} \cdot I_b \\ i_{Lf}(D_S) = \dfrac{2D_s + G - 1}{G+2} \cdot I_b \end{cases} \quad (9)$$

On the basis of principle analysis, it's easy to find the conclusion that ZVS of the semiconductor devices requires $i_{Lf}(0)<0$ (S_1-S_4) and $i_{Lf}(D_S)>0$ (S_5-S_6 and D_1-D_2). Thus, we can get the constraints as follows

$$\begin{cases} D_S > (G-1)/G & S_1 - S_4 \\ D_S > (1-G)/2 & S_5 - S_6, D_1 - D_2 \end{cases} \quad (10)$$

As $G>1$ in Boost mode, soft-switching for secondary-side MOSFETs S_5-S_6 and diodes D_1-D_2 can be easily achieved while D_S greater than $(G-1)/G$ is needed for ZVS of the primary-side MOSFETs. That is to say, ZVS of S_1-S_4 can't be achieved in Boost-DCM mode.

Similarly, in Buck-CCM mode, $i_{Lf}(0)$, $i_{Lf}(1-D_P)$ and $i_{Lf}(1-D_P+D_S)$ can be given as

$$\begin{cases} i_{Lf}(0) = -\dfrac{G\cdot[(1+G)\cdot(D_S-G)+2]}{G+2} \cdot I_b \\ i_{Lf}(1-D_P) = -\dfrac{G(G+1)D_S}{G+2} \cdot I_b \\ i_{Lf}(1-D_P+D_S) = \dfrac{2D_S}{G+2} \cdot I_b \end{cases} \quad (11)$$

Soft-switching of the switches requires $i_{Lf}(0)<0$ (S_1-S_2), $i_{Lf}(1-D_P)<0$ (S_3-S_4) and $i_{Lf}(1-D_P+D_S)>0$ (S_5-S_6 and D_1-D_2). Soft-switching requirements for all the switches can be derived as

$$\begin{cases} D_S > (G+2)(G-1)/(G+1) & S_1 - S_2 \\ D_S > 0 & S_3 - S_6, D_1 - D_2 \end{cases} \quad (12)$$

As $G<1$ in Buck-CCM mode, soft-switching of all the semiconductor devices can be easily achieved under the condition that $D_S>0$. In similar way, soft-switching for all the switches except for S_3/S_4 can be achieved once $D_P>0$ and S_3/S_4 can only achieve ZCS in Buck-DCM mode, theoretically. Actually, when the energy stored in L_f is too small to discharge the junction capacitors, soft-switching can hardly be achieved even if the phase-shift ratio meets the requirements mentioned above. The soft-switching characteristics associated with the operation modes are summarized in Table I.

TABLE I.   SOFT-SWITCHING CHARACTERISTICS

Operation Modes	Soft-Switching Characteristics			
	$S_1\&S_2$	$S_3\&S_4$	$S_5\&S_6$	$D_1\&D_2$
Boost-CCM	ZVS		ZVS&ZCS	ZC-OFF
Boost-DCM	ZCS		ZVS&ZCS	ZC-OFF
Buck-CCM	ZVS	ZVS	ZVS&ZCS	ZC-OFF
Buck-DCM	ZVS	ZCS	ZVS&ZCS	ZC-OFF

2) Power transmission capability

By defining $P_b=V_{in}^2/2f_sL_f$ and assuming that $\eta=1$ (ignoring loss of the converter), output power of the proposed converter in Boost-CCM mode ($G>1$, $D_S>(G-1)/G$) can be normalized as

$$\frac{P_o}{P_b}=\frac{-2(D_S-1)^2\cdot G^3+(4D_S-4D_S^2+1)\cdot G\cdot(G+1)}{2(G+2)^2} \quad (13)$$

Similarly, output power of the converter in Boost-DCM mode ($G>1$, $D_S<(G-1)/G$) can be derived as

$$\frac{P_o}{P_b}=\frac{D_S^2\cdot G}{2(G-1)} \quad (14)$$

While operating in Buck-CCM mode ($0<G<1$, $D_P=G$ and $D_S>0$), the normalized output power is as follow

$$\frac{P_o}{P_b}=\frac{1}{2(G+2)^2}\left\{\begin{array}{l}G^3\left[D_S(2+2G)-2D_S^2-(G^2+1)\right]-\\G(G+1)\left[4D_S^2-4G\cdot D_S+(3G^2-4G)\right]\end{array}\right\} \quad (15)$$

In Buck-DCM mode ($0<G<1$ and $D_S=0$), the primary-side phase-shift ratio is adjusted. P_o/P_b can be given as

$$\frac{P_o}{P_b}=\frac{(1-G)\cdot D_P^2}{2} \quad (16)$$

According to the above analysis, normalized output power versus phase-shift ratio D_P and D_S is illustrated in Fig. 6.

From these curves in Fig. 6, it can be seen that the converter can work either as a Buck converter ($0<G<1$) or a Boost

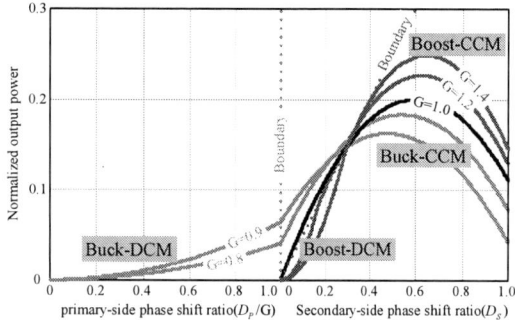

Fig. 6. Normalized output power curve vs. phase shift ratio D_P and D_S

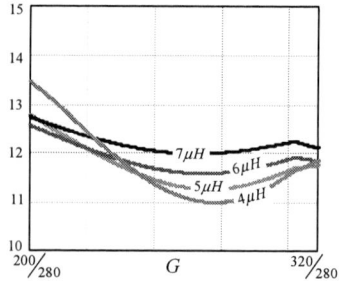

Fig. 6. RMS of i_{Lf} vs. voltage gain G with different inductance value

converter ($G>1$). Thus, a wide voltage range can be achieved. Besides, the output power can be adjusted from zero to full load continuously with dual-phase-shift control scheme.

3) Design considerations of magnetic components

The operating characteristics of the converter including its output capability and efficiency are closely related with the inductance value and turns-ratio of the transformers, which can be optimally designed to improve the efficiency with power loss analysis. As soft-switching can be achieved in a wide load range, only the conduction losses are taken into account and switching losses are ignored here. As the conduction losses are proportional to RMS of inductor current i_{Lf}, lower I_{Lf} over wide load and voltage range is the target of parameter design. What's more, inductance value should be checked as to ensure the output capability and efficiency at rated output voltage (310V) should be emphatically considered.

Generally speaking, conduction losses of Balance mode ($G=1$) is relatively less due to lower distortion factor. Taking this and the whole output voltage range into consideration, turns ratio of the transformers can be set as 6:17, under which condition the equivalent voltage gain of 310V output is around 1.09. With $V_{in}=100V$, $n=6:17$ and full load, curve of I_{Lf} versus voltage gain G can be drawn with fixed inductance value, as is shown in Fig. 7. To achieve lower conduction losses within the needed voltage range, 5uH is more suitable. With $n=6:17$ and $L_f=5uH$, output capability can meet the requirements.

III. EXPERIMENTAL VERIFICATION

A 1000W prototype of the proposed converter with 90-110V input, 200-320V output and 100kHz switching frequency is built and tested. The main parameters of the prototype are given as follows: $S_1\sim S_4$: IPP111N15N3G, $S_5\sim S_6$: IPP600N25N3G, $D_1\sim D_2$: MURF1660CTG, inductor L_f=4.97μH, transformers' turns ratio $n= n_P:n_S$ =6:17.

Steady state experimental waveforms in Boost-CCM mode are shown in Fig. 8 with 100V input and 310V output. v_{GS1} and v_{GS5} are the driving signals of S_1 and S_5 while v_{DS1} and v_{DS5} are the drain-to-source voltages of S_1 and S_5. v_P is midpoint voltage of the primary-side full bridge and v_{NS1} and v_{NS2} are the secondary-side voltages of T_1 and T_2. i_{Lf} represents the current of L_f. It can be seen that experimental results satisfy the analysis well and all the switches can achieve ZVS.

The steady state experimental results in Buck-CCM mode are given in Fig. 9 with 100V input and 260V output at 900W.

978-1-4673-9551-9/16 $31.00 © 2016 IEEE

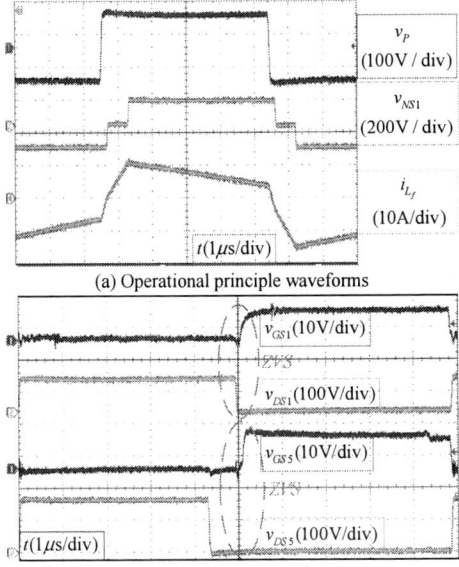

(a) Operational principle waveforms

(b) Soft switching waveforms

Fig. 8. Steady state experimental waveforms in Boost-CCM mode

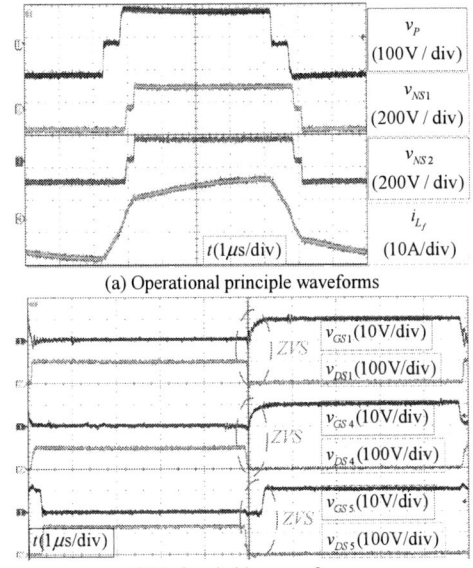

(a) Operational principle waveforms

(b) Soft switching waveforms

Fig. 9. Steady state experimental waveforms in Buck-CCM mode

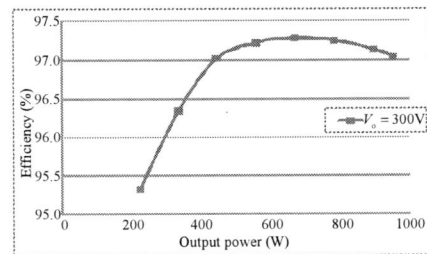

Fig. 10. Measured efficiency curves of proposed converter

It can be found that both the primary-side and secondary-side

switches have been realized ZVS and the experimental results satisfy previous theoretical analysis well.

Efficiency curve of the proposed converter with 100V input and 300V output is given in Fig. 10. When the output voltage is 300V, the maximum efficiency is 97.3% and the value at full load is about 97.0%. Full load efficiency of the converter is above 95% over whole output voltage range, with the minimum value being 95.3% at 200V output. It can be concluded that high efficiency over wide load and output voltage range has been achieved. The experimental results indicate that the proposed converter is suitable for high efficiency and wide voltage range applications.

IV. CONCLUSIONS

A dual-phase-shift controlled soft-switching DC/DC converter for high-efficiency, wide voltage range and galvanic isolation applications is proposed. Detailed operational principle, characterization and control strategy of proposed converter are analyzed and verified by experimental results. With optimized control scheme, soft switching is achieved for all MOSFETs and diodes within a wide load range. Leakage inductance of transformers is used to for power transferring and the voltage spike problem is overcome. A unique structure with two transformers and a voltage-doubler rectifier is employed in the proposed converter to decrease the voltage and current stresses of the transformers and rectifying devices. All the above-mentioned characterizations demonstrate that the proposed converter is a good candidate for the high efficiency and wide voltage range applications.

REFERENCES

[1] D. S. Gautam, F. Musavi, W. Eberle, and W. G. Dunford, "A zero-voltage switching full-bridge DC-DC converter with capacitive output filter for plug-in hybrid electric vehicle battery charging," IEEE Trans. Power Electronics, vol. 28, no. 12, pp. 5728-5735, Dec. 2013.

[2] H. Wu, Y. Xing, "Families of forward converters suitable for wide input voltage range applications," IEEE Trans. Power Electronics, vol. 29, no. 11, pp. 6006-6017, Nov. 2014.

[3] Y. Zhao, W. Li, Y. Deng, and X. He, "Analysis, design, and experimentation of an isolated ZVT boost converter with coupled inductors," IEEE Trans. Power Electronics, vol. 26, no. 2, pp. 541-550, Feb. 2011.

[4] Y. Wang, W. Liu, H. Ma, L. Chen, "Resonance analysis and soft-switching design of isolated boost converter with coupled inductors for vehicle inverter application," IEEE Trans. Power Electronics, vol. 30, no. 3, pp. 1383-1392, Mar. 2015.

[5] C. Yao, X. Ruan, X. Wang, and C. K. Tse, "Isolated Buck-Boost DC/DC converters suitable for wide input-voltage range," IEEE Trans. Power Electronics, vol. 26, no. 9, pp. 2599-2613, Sep. 2011.

[6] W. Feng, F. C. Lee, P. Mattavelli, "Optimal trajectory control of LLC resonant converters for LED PWM dimming," IEEE Trans. on Power Electronics, vol. 29, no. 2, pp. 979-987, Feb. 2014.

[7] I.-O. Lee and G.-W. Moon, "The k-Q analysis for an LLC series resonant converter," IEEE Trans. on Power Electronics, vol. 29, no. 1, pp. 13-16, Jan. 2014.

[8] X. Fang, H. Hu, F. Chen, U. Somani, E. Auadisian, J. Shen, I. Batarseh, "Efficiency-oriented optimal design of the LLC resonant converter based on peak gain placement," IEEE Trans. on Power Electronics, vol. 28, no.5, pp. 2285-2296, May 2013.

[9] H. Wu, Y. Lu, T. Mu, Y. Xing, "A family of soft-switching DC-DC converters based on a phase-shift-controlled active boost rectifier," IEEE Trans. Power Electronics, vol. 30, no. 2, pp. 657-667, Feb. 2015.

Effect of Transformer Design on Operation of Fundamental Duty Modulation for Dual-Active-Bridge Converter

Wooin Choi, Moonhyun Lee, and Bo-Hyung Cho
Department of Electrical and Computer Engineering
Seoul National University
Gwanak-Gu, Seoul 151-600, Korea

Abstract— In this paper, effects of transformer design on operational characteristics of the dual-active-bridge (DAB) converter are studied to further improve the fundamental duty modulation (FDM) scheme. As discussed in the previous work, the FDM was proved to exhibit improved light-load efficiency compared to the recent on-line modulation schemes. However, at a specific operating condition, it showed deteriorated performance due to the hard switching on the primary-side lagging-leg switches. To examine the operational variations originate from the practical implementation of the transformer, a generalized transformer model is defined with three parameters. It was explained from the quantitative analysis that rms current level and soft-switching range of the four legs of the DAB converter highly depends on the transformer design under the same series inductance. Following the systematic design method for medium-power DAB converter, experimental verification was carried out with multiple samples of the transformer stage. Efficiency improvements of the FDM according to the transformer design were observed especially at light- and medium-load operations.

I. INTRODUCTION

Since the demand for bidirectional conversion systems has gained increasing attention, the dual-active-bridge (DAB) converter has been a popular choice for bidirectional dc-dc conversion applications with galvanic isolation [1]-[3]. Numerous studies have reported utilizations of the DAB converter for medium-voltage applications, electric vehicles and energy storage systems [4]-[9]. One of the most highlighted features of the circuit is that the converter, in principle, requires a single passive element for power transfer leading to increased power density. Using the series inductance of a transformer, power can be delivered in bidirectional way through the symmetrical isolated circuit. Moreover, the DAB converter allows zero-voltage-switching (ZVS) turn-ons without any additional resonant component.

However, one of the practical limitations of the DAB converter is the low efficiency observed at light-load operation due to high circulating currents as well as hard switching [10]. There has been numerous literature on modulation schemes of the DAB converter for improved conversion efficiency. In [11]-[13], optimal strategies were suggested based on a thorough time-domain analysis for every operating point. However, these methods require look-up tables or off-line

calculations and have high parametric dependency. On the other hand, there have been proposed the modulation schemes based on the Fourier series analysis [14]-[17]. Although these schemes provide global solutions due to the universality of the Fourier series method, their limitations are the increased complexity [14], or sub-optimality from the approximated derivation [15], [16].

In the previous work [17], the fundamental duty modulation (FDM) scheme was presented which proposes an on-line modulation scheme for minimization of first harmonic rms currents. While the FDM was proved to exhibit improved light-load efficiency compared to the recent on-line modulation schemes in [15], [18], it has deteriorated efficiency at a specific operating condition. This is due to the fact that a non-ZVS area of the FDM scheme appears on the primary-side lagging-leg switches as discussed.

On the other hand, the most of the literature on the DAB converter describe the operation of the circuit using an ideal transformer model as in the previous study. The transformer stage in DAB converters practically has a magnetizing inductance value which is comparable to the series inductance. The effect of this practical transformer model which greatly alters the operational characteristics of the converter has rarely been explored in a quantitative way. While a large portion of the research have focused on integration and design of the transformer [5], [6], [19]-[22], the influence of the magnetizing inductance on the soft-switching region or conduction loss of the DAB converter was reported in [20]-[22].

In this paper, effect of transformer design on operational characteristics of the FDM is studied to seek the possibility for a further efficiency improvement of modulation schemes. A generalized model of the transformer was defined to precisely account for the effect of the practical transformer on converter losses. Critical operational characteristics which determine conduction and switching losses were analyzed based on the Fourier series method [23]. Also, a systematic design method for the transformer stage was presented for design of a medium-power few-ten-kilohertz DAB converter. With the method, the maximum magnetizing inductance of the transformer can be estimated in advance. To represent various case studies of the generalized model, multiple samples of transformer stage were built for verification. Efficiency

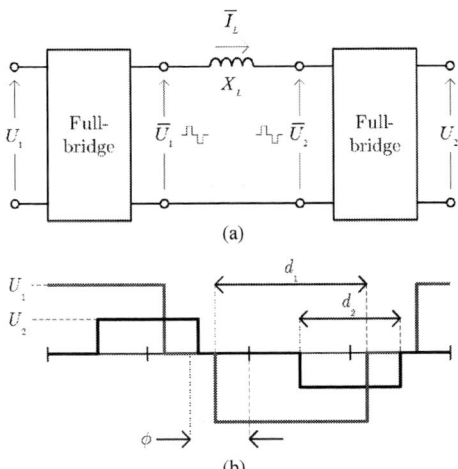

(a)

(b)

Fig. 1. (a) Circuit diagram of DAB converter and (b) voltage waveforms for d_1, d_2 and ϕ.

improvements of the FDM according to the transformer design were observed at light- and medium-load operations.

II. FUNDAMENTAL DUTY MODULATION SCHEME

A. DAB Converter Phasor Modeling

The DAB converter can be modeled by a simplified circuit diagram depicted in Fig. 1(a). The transformer is modeled by a series inductance assuming infinite magnetizing inductance and zero series resistance. Turn ratio is assumed 1 without loss of generality. The three-level voltages in Fig. 1(b) are induced through the full-bridges from the dc voltages U_1 and U_2, respectively. Regarding the fundamental component, the voltages induced at both sides of the inductor are expressed by their phasors:

$$\bar{U}_k = U_0 \bar{u}_k := U_0 \left(\alpha_k + j\beta_k \right) \tag{1}$$

where $\bar{U}_k$ is voltage phasor, $\bar{u}_k$ is normalized voltage phasor. U_0 is normalization factor defined by the output voltage U_2 as

$$U_0 = \frac{2\sqrt{2}}{\pi} U_2. \tag{2}$$

Then the normalized voltage phasor α_k's and β_k's in (1) are calculated using d_1, d_2, ϕ as

$$\begin{aligned} \alpha_1 &= M_1 \sin \pi d_1 \cos 2\pi \phi_1 \\ \beta_1 &= M_1 \sin \pi d_1 \sin 2\pi \phi_1 \\ \alpha_2 &= \sin \pi d_2 \\ \beta_2 &= 0 \end{aligned} \tag{3}$$

where M_1 is U_1/U_2. The current phasor of the inductor, $\bar{I}_L$, is

$$\bar{I}_L = \frac{1}{jX_L} \left(\bar{U}_1 - \bar{U}_2 \right) = \frac{U_0}{jX_L} \left(\bar{u}_1 - \bar{u}_2 \right) \tag{4}$$

where X_L is reactance of the series inductance. Substituting (1) into (4), the current is equated as

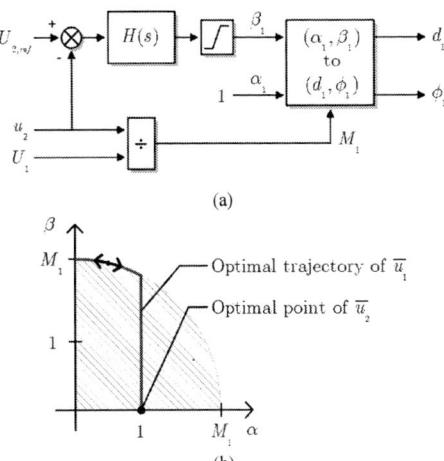

(a)

(b)

Fig. 2. (a) Controller block diagram and (b) operating trajectories of FDM.

$$\begin{aligned} \bar{I}_L &= \frac{U_0}{jX_L} \left\{ \left(\alpha_1 - \alpha_2 \right) + j \left(\beta_1 - 0 \right) \right\} \\ &= \frac{U_0}{X_L} \left(\beta_1 \right) - j \frac{U_0}{X_L} \left(\alpha_1 - \alpha_2 \right). \end{aligned} \tag{5}$$

The transferred power, P_1, is real part of product of the current and voltage phasor as follows:

$$\begin{aligned} P_1 &= \mathrm{Re} \left\{ \bar{U}_1 \bar{I}_L \right\} \\ &= \frac{U_0^2}{X_L} \left\{ \alpha_1 \left(\beta_1 \right) - \beta_1 \left(\alpha_1 - \alpha_2 \right) \right\} \\ &= \frac{U_0^2}{X_L} \beta_1 \alpha_2 \end{aligned} \tag{6}$$

B. Current Minimization Strategy

The objective of the FDM is on minimization of the current in (5) constrained by (3) and (6) as

$$\begin{aligned} \text{Minimize} \qquad & f = \beta_1^2 + \left(\alpha_1 - \alpha_2 \right)^2 \\ \text{subject to} \qquad & \beta_1 \alpha_2 = k \\ & 0 \le \alpha_1^2 + \beta_1^2 \le M_1^2 \\ & 0 \le \alpha_2 \le 1. \end{aligned} \tag{7}$$

The solution of (7) can be implemented by the controller diagram in Fig. 2(a) of which the operating trajectories are illustrated in phasor domain in Fig. 2(b). It is seen from Fig. 2(b) that $\bar{u}_2$ always stays on a single point, $\alpha_2 = 1$; that is, d_2 is 0.5 according to the relation in (5). On the other hand, $\bar{u}_1$ moves along the optimal trajectory with respect to the output power. When the output power is zero, $\bar{u}_1$ is on (1, 0). As power increases, α_1 is 1 while β_1 increases until it reaches the maximum operating range of $\bar{u}_1$ described by (7). Additional power increase is achieved along the circular boundary where d_1 is also 0.5. The controller of the FDM requires only the voltage information without the need for off-line calculation or mode selection.

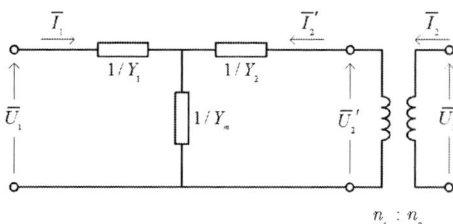

Fig. 3. Two-port circuit diagram of T-model transformer.

III. OPERATIONAL CHARACTERIZATION OF FDM

The aim of the FDM is minimization of the reactive power hence reducing conduction loss. However, limitations of the proposed scheme are expected to be: (1) that the optimal solution is derived dealing with only the fundamental components, and (2) that the soft-switching characteristics are not taken into consideration on the derivation of the scheme. Thus, to fully understand and estimate the effectiveness of the scheme, operational characteristics of the FDM are required to be analyzed.

A. Two-Port Modeling using Higher-Order Terms

As introduced in [23], time-domain characteristics of the DAB converter can be derived from the Fourier series analysis using higher-order terms. Considering the odd harmonics, m-th ($[2n-1]$-th) order normalized voltage phasors are defined as

$$\bar{u}_{k,m} = \alpha_{k,m} + \beta_{k,m} \tag{8}$$

where $k = 1$ and 2 for the primary and the secondary side, respectively. Each component in (8) is calculated as

$$\alpha_{k,m} = \frac{M_k}{m} \cdot (-1)^{n-1} \cdot \sin m\pi d_k \cdot \cos m2\pi\phi_k$$

$$\beta_{k,m} = \frac{M_k}{m} \cdot (-1)^{n-1} \cdot \sin m\pi d_k \cdot \sin m2\pi\phi_k. \tag{9}$$

Here, the practical transformer model in Fig. 3 is considered instead of the simplified inductor model. Y_1 and Y_2 are admittances of leakage inductances at each sides, and Y_m is admittance of magnetizing inductance. The two-port network is expressed as follows:

$$\begin{bmatrix} \bar{I}_1 \\ \bar{I}_2' \end{bmatrix} = \mathbf{Y} \begin{bmatrix} \bar{U}_1 \\ \bar{U}_2' \end{bmatrix} = \begin{bmatrix} Y_{11} & Y_{12} \\ Y_{21} & Y_{22} \end{bmatrix} \begin{bmatrix} \bar{U}_1 \\ \bar{U}_2' \end{bmatrix} \tag{10}$$

The elements of the admittances matrix $\mathbf{Y}$ are computed as

$$Y_{ii} = Y_i \cdot \frac{Y_j + Y_m}{Y_i + Y_j + Y_m}$$

$$Y_{ij} = Y_{ji} = -\frac{Y_i Y_j}{Y_i + Y_j + Y_m}. \tag{11}$$

The current phasor of m-th order harmonic is calculated according to (10) as,

$$\bar{I}_{1,m} = Y_{11,m} \cdot \bar{U}_{1,m} + Y_{12,m} \cdot \bar{U}_{2,m}'$$

$$\bar{I}_{2,m}' = Y_{21,m} \cdot \bar{U}_{1,m} + Y_{22,m} \cdot \bar{U}_{2,m}'. \tag{12}$$

The current in time-domain can be reconstructed from the following relationship:

TABLE I
ELECTRIC SPECIFICATIONS AND OPERATING RANGES

Parameter		Value
Input voltage	U_1	200 V
Output voltage	U_2	200 to 400 V
Turn ratio	$N\,(n_1{:}n_2)$	2 (11:22)
Maximum power	P_{max}	500 to 1000 W
Switching frequency	f_s	50 kHz

$$i_1(t) = \sum_n \left[\text{Re}\{\bar{I}_{1,m}\}\sin w_m t + \text{Im}\{\bar{I}_{1,m}\}\cos w_m t \right]$$

$$i_2(t) = \frac{n_1}{n_2}\sum_n \left[\text{Re}\{\bar{I}_{2,m}'\}\sin w_m t + \text{Im}\{\bar{I}_{2,m}'\}\cos w_m t \right] \tag{13}$$

The Fourier analysis above is applied to a DAB converter with the specifications given in Table I. The range of the output voltage is 200 to 400 V, which means that the voltage gain, $M = U_2/U_1 N$ is from 0.5 to 1.0. It should be noted that the maximum transferrable power also changes from 500 to 1000 W with respect to U_2. In the remaining part of this Section, two major characteristics, the rms current and the soft-switching property, of the converter are analyzed assuming an ideal transformer model.

B. Ideal Transformer Model: Conduction Loss

Fig. 4 depicts estimated conduction losses of the FDM compared to the other modulation schemes for $U_2 = 200$, 250, 300 and 350 V. The phase-shift modulation (PSM) is the most conventional scheme that utilizes only the phase-shifts between the bridges. The triangular modulation (TRM) scheme performs theoretically minimal rms current although the maximum power is limited with respect to the voltage gain condition. In Fig. 4(a), the light-load conduction of the PSM is

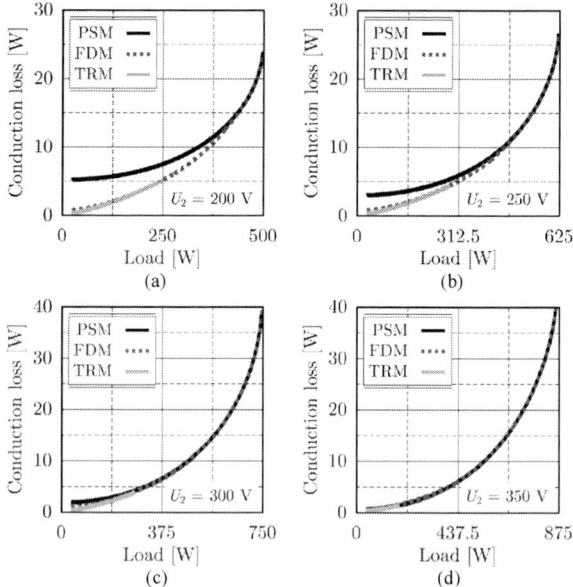

Fig. 4. Conduction losses of PSM, FDM and the reference method for (a) U_2 = 200 V, (b) 250 V, (c) 300 V, and (d) 350 V.

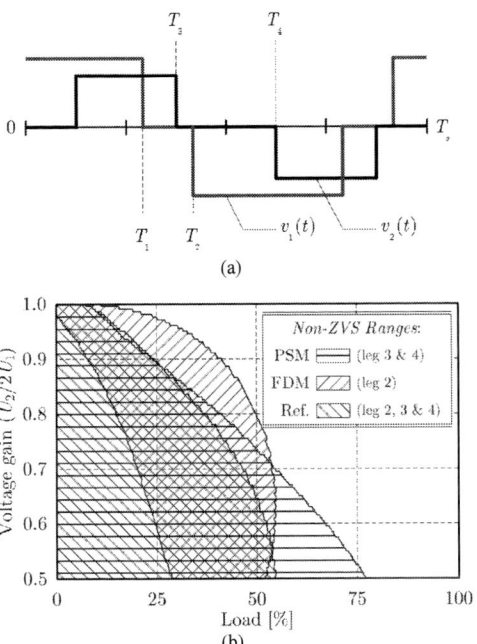

(a)

(b)

Fig. 5. (a) Switching instants of the switch legs and (b) non-ZVS area of the modulation schemes.

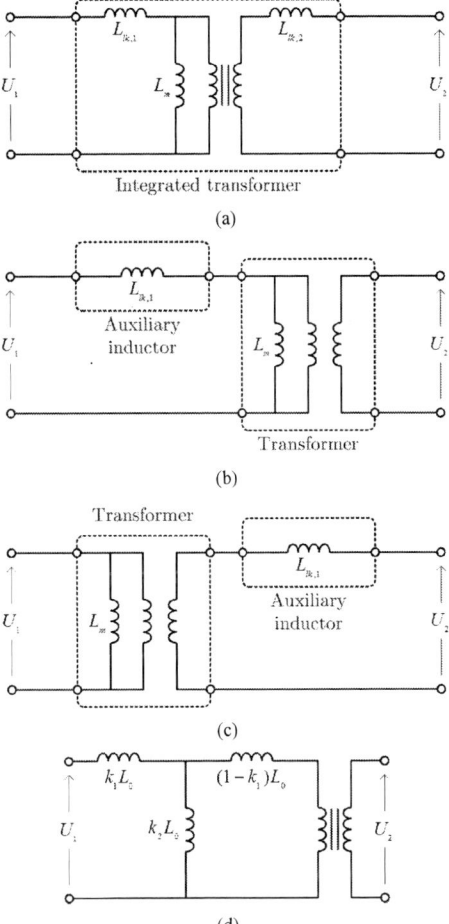

Fig. 6. Configurations of transformer stage: (a) equally divided series inductance, (b) an auxiliary inductor on primary side, and (c) on secondary side, and (d) generalized transformer model.

far greater than the TRM while that of the FDM is almost the same as the TRM for $U_2 = 200$ V ($M = 0.5$). However, as U_2 increases that M becomes closer to 1, differences of the conduction losses between the schemes becomes negligible.

C. Ideal Transformer Model: ZVS Range

Once the current waveforms are calculated utilizing (13), the soft-switching range can be determined by comparing the current magnitudes with the minimum current for ZVS at switching instants as

$$i(t = T_j) > i_{ZVS,\min} \qquad (14)$$

where T_j is switching instant of leg j. Leg 1 and 2 (3 and 4) are the leading and the lagging legs of primary (secondary) bridge, respectively. Exemplary switching instants according to arbitrary d_1, d_2, and ϕ are depicted in Fig. 5(a).

Fig. 5(b) shows non-ZVS ranges of the modulation schemes including the FDM. Switching properties of each leg were investigated for every operating conditions in the entire range where the voltage gain, M, is ranging from 0.5 to 1.0 and the load is from zero to full load. It should be remarked that comparing the ZVS ranges of the FDM, the PSM, and the reference method in [18], each method has non-ZVS range at leg 2, at leg 3 and 4, and at leg 2, 3 and 4, respectively.

The conventional PSM suffers from the hard switching of the secondary-side switches at lower M which leads to reduced efficiency. While the reference method exhibits improved soft-switching characteristics than that of the PSM, the strategy does not provide enough capacitive discharge currents, hence, still showing large non-ZVS range. Lastly, the FDM scheme performs ZVS turn-ons of all switches at light load which allows improved efficiency. However, there is a non-ZVS area

at medium-load range. In the following part, reduction of this non-ZVS area of the FDM are discussed considering the effect of the magnetic component configuration.

IV. GENERALIZED TRANSFORMER CONFIGURATION

The magnetizing inductance of the transformer is practically few to few ten times of the series inductance. An isolated magnetic component with this non-negligible series inductance can be practically implemented employing either a single transformer with enlarged leakage inductances (Fig. 6(a)), or a transformer in series with a primary-side auxiliary inductor (Fig. 6(b)), or with a secondary-side auxiliary inductor (Fig. 6(c)).

The generalized transformer stage illustrated in Fig. 6(d) is used to represent diversified configurations of the magnetic component. L_0 is a nominal series inductance value which is the sum of the series inductances at both sides. This series inductance is divided into the primary and the secondary part by a factor of k_1. The range of k_1 is from 0 (Fig.6(c)) to 1 (Fig. 6(b)). On the other hand, the magnetizing inductance is

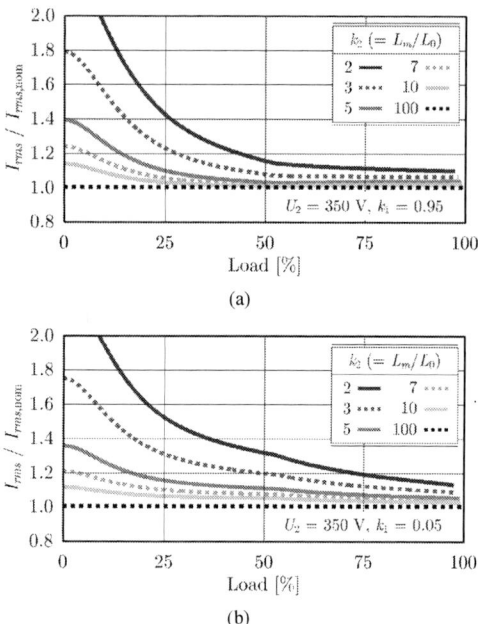

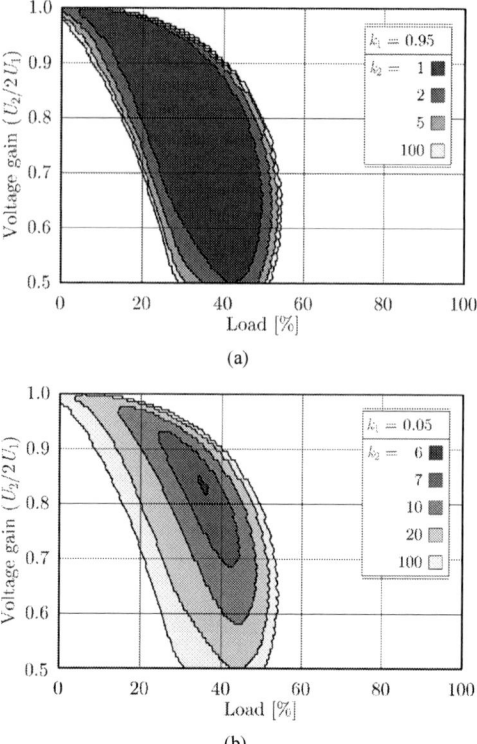

Fig. 7. RMS current level for various k_2 normalized by the ideal case ($L_m = \infty$) at $U_2 = 350$ V: (a) $k_1 = 0.95$, and (b) 0.05.

expressed by $k_2 L_0$ where k_2 is the ratio of magnetizing inductance and the series inductance. Practically, the maximum value of k_2 is limited by a given specifications regarding the maximum flux density, winding capacity, and, also, power density requirements.

Operating characteristics of a DAB converter with the generalized transformer model can be examined using the two-port model described previously in (13). The elements of the admittance matrix **Y** in (10) are replaced with that of the practical transformer model using the relation (11). Since the objective of the paper is efficiency improvements of the FDM, rms current and ZVS characteristics of the FDM were thoroughly investigated.

A. Practical Transformer Model: RMS Current

Fig. 7 depicts rms currents for various design conditions for $U_2 = 350$ V. Variations of the rms currents were normalized by that of the ideal case calculated in the previous analysis. Fig. 7(a) depicts the current variations for $k_1 = 0.95$ which implies the configuration of an auxiliary inductor at the primary side and a small leakage inductance of the transformer at the secondary side. The circulating current increases as k_2 decreases. For the opposite case of $k_2 = 0.05$, as seen from Fig. 7(d), the tendency is similar to Fig. 7(a) although showing increased rms currents for overall power range than $k_1 = 0.95$.

B. Practical Transformer Model: ZVS Range

As discussed previously, whereas the FDM has a remarkable non-ZVS range on the leg 2 (the red area in Fig. 5(b)), the switches on leg 1, 3 and 4 are always turned on with zero voltage. Since insertion of a reduced magnetizing inductance, in most cases, improves soft-switching properties,

Fig. 8. Non-ZVS area of leg 2 switches under FDM scheme for various k_2: (a) $k_1 = 0.95$, and (b) 0.05.

the ZVS areas of leg 1, 3 and 4 will not be altered from the ideal case.

Switching characteristics of leg 2 on the primary side were investigated with respect to k_1 and k_2. In Fig. 8(a), non-ZVS ranges of the FDM for various k_2 are illustrated for the case with the primary-side auxiliary inductor ($k_1 = 0.95$). When compared to $k_2 = 100$ which is almost the ideal case, the non-ZVS range does not decreases even for the small magnetizing inductance of $k_2 = 1$. In case of $k_1 = 0.05$ in Fig. 8(b), the non-ZVS area dramatically decreases according to k_2. $k_2 = 6$ is enough to turn on the switches of leg 2 with zero voltage in most of the operating conditions. This is because the circulating current induced by the magnetizing inductance could directly affect the switching properties of the primary-side switches. The placement of the series inductance at the secondary side ($k_1 = 0.05$) brings significant advantage for the switching loss reduction while keeping the conduction level at a moderate range.

V. PRACTICAL TRANSFORMER DESIGN

A. Estimation of Magnetizing Inductance

For a high-frequency transformer, two design criteria should be met: the minimum cross-sectional area, A_c, of a core for the flux density in (15); and the minimum window area, A_W, for the winding capacity in (16) [24].

TABLE II
GEOMETRIC COEFFICIENTS OF CORES

Core	$g_1 = A_c / A_p$		$g_2 = l_e / A_p^{1/2}$	
Type & No. of samples	mean	SD	mean	SD
PQ, 7	1.47	0.21	7.10	1.00
EE, 8	1.11	0.22	7.10	0.28
EER, 6	1.01	0.12	7.35	0.48
POT, 9	1.57	0.11	4.90	0.15

$$A_c \geq \frac{\lambda}{n_1 B_{max}} = \frac{1}{4} \frac{U_1 T_s}{n_1 B_{max}} \tag{15}$$

$$A_W \geq \frac{n_1 I_1 + n_2 I_2}{K_W J_W} = 2 \frac{n_1 I_1}{K_W J_W} \tag{16}$$

where I_1 and I_2 are rms currents at the primary and secondary windings, λ is the maximum flux, K_W is winding factor, J_W is current density of a copper wire, and B_{max} is the maximum flux density of a core. Multiplying (15) and (16), the key design figure, A_p, is defined independently to the core specifications:

$$A_p := (A_c W_A)^{1/2} \geq \sqrt{\frac{P_1 T_s}{2 K_W J_W B_{max}}}. \tag{17}$$

Note that the coefficients, such as K_W, J_W, and B_{max}, are given by characteristics of the materials. Therefore, from (17), a proper core selection can be determined from the maximum power and the frequency information.

Now, two dimension-less coefficients, g_1 and g_2, are defined which are given by geometries of individual core types:

$$\begin{aligned} g_1 &:= A_c / A_p \\ g_2 &:= l_e / A_p^{1/2} \end{aligned} \tag{18}$$

where l_e is effective core length of a core.

For medium power applications with operating frequency of few to few hundreds hertz, the ferrite magnetic core is a viable choice. Inspecting the geometric properties of popular core types including PQ, POT, EE and EER, there was found a tendency in the g_1 and g_2 for each core type. Table II shows averages and standard deviations of the coefficients. Once core type is decided, using g_1 and g_2, the transformer design can further proceed. With A_p calculated from (17), the estimated A_c for each core type can be obtained using g_1 as,

$$A_c = g_1 A_p. \tag{19}$$

Also, l_e is calculated as

$$l_e = g_2 A_p^{1/2}. \tag{20}$$

Then, the core reluctance is given as follows:

$$R_c = l_e / (\mu_c \mu_0 A_c). \tag{21}$$

Consequently, the maximum magnetizing inductance at primary side is estimated using n_1 from (15) to be

$$L_{m,1} = n_1^2 / R_c. \tag{22}$$

The total magnetizing inductance of the transformer is formed in parallel of $L_{m,1}$ and $L_{m,2}$, that is,

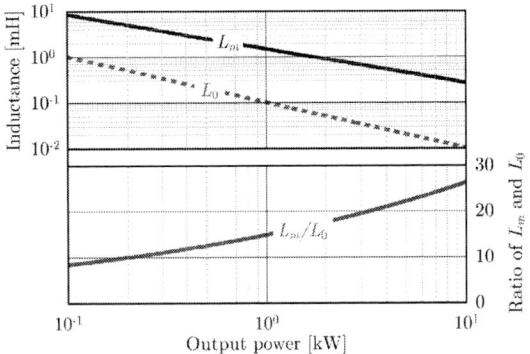

Fig. 9. Estimated magnetizing and series inductance values, and their ratio designed according to the maximum output power specification.

$$\begin{aligned} L_m &= L_{m,1} \| L_{m,2} = L_{m,1} \| (N^2 L_{m,1}) \\ &= \frac{N^2}{1+N^2} L_{m,1}. \end{aligned} \tag{23}$$

The estimated value of the maximum magnetizing inductance in (22) is obtained only from power, voltage, and frequency information. However, due to the limited number of available core types, core type selection cannot exactly match the design condition in (17). Also, it is worth to note that the inductance calculation in (20) is based on the assumption that the air-gap is zero. Hence, for a practical implementation, the magnetizing inductance can be smaller but not greater than the estimation.

B. Ratio of Inductances

The output power of the DAB converter is well-knowingly

$$P_1 = \frac{U_1 U_2}{N f_s L_0} \cdot \phi \cdot (1 - 2\phi). \tag{24}$$

The nominal series inductance value is obtained based on the assumptions: 1) turn-ratio of the transformer is $N = U_2/U_1$, and 2) the maximum power is achieved at $\phi = 0.25$. Then L_0 is given as

$$L_0 = \frac{U_1^2}{8 f_s P_{max}}. \tag{25}$$

Once we have a maximum output power and input voltage specification, a set of magnetizing and series inductance values can be estimated using (22) and (24). In Fig. 9, the magnetizing and the series inductances, and the ratio of them are depicted assuming a PQ-type core. In Fig. 9, as the output power varying from 0.1 to 10 kW, L_m/L_0 increases from 8 to 26.

VI. EXPERIMENTAL RESULTS

In the following experimental results, operational characteristics of the DAB converter are described for multiples of transformer-stage configurations. In Table III, two samples of each of the auxiliary inductor and the transformer used for the verification are listed. The auxiliary inductor I and II are placed at the primary and the secondary side, respectively. These inductors were elaborately designed that they have almost the same inductance value with 2% of difference. On the other hand, the two transformer samples were designed identically except for the air-gap lengths. By

TABLE III
MAGNETIC COMPONENTS DESIGNED FOR VERIFICATIONS

Parameter		Value	
Aux. inductor		I	II
	Placement	primary	secondary
	Series inductance, L	98.0 μH	97.5 μH*
Transformer		I	II
	Air-gap length, l_g	none	0.05 mm
	Mag. inductance, L_m	840 μH	480 μH
	Leak. Inductance, L_{lk}	3 μH	3 μH

*Reflected inductance onto the primary side by turn ratio:
$L' = L/N^2$ (97.5 μH = 390 μH / 2^2)

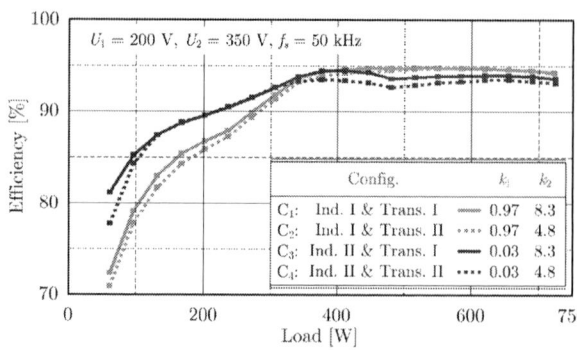

Fig. 10. Efficiency curves of FDM for various transformer configurations at $U_2 = 350$.

changing the air gaps, two different magnetizing inductances were obtained.

In Fig. 10, efficiency curves for four combinations of the inductor and the transformer were measured at $U_2 = 350$ V where the FDM had shown lower efficiency compared to the other methods [17]. The configuration and its coefficients of the transformer stage, C_{1-4}, are defined by the combinations of the magnetic components. Comparing configuration C_1 and C_2, whereas the auxiliary inductor is at the primary side for both, the magnetizing inductor of C_2 is reduced by the air-gap. The increased conduction loss of C_2 lowered the efficiency noting that the magnetizing inductance has little effect on the soft-switching characteristics for $k_1 = 0.97$.

When comparing configuration C_1 and C_3, the effects of the transformer configuration becomes clearer. At light-load range, C_3 shows much improved efficiency (up to 9%) than C_1 due to the extended ZVS capability of C_3 (Fig. 8). However, at the high-power region where conduction loss dominates, C_1 achieves about 1% higher conversion efficiency than C_3 owing to the increased rms current of C_3 than that of C_1. (Fig. 7).

On the other hand, efficiency differences between C_3 and C_4, also, can be explained by the conduction loss increment of C_4 due to the reduced magnetizing inductance. Nonetheless, the efficiency of C_4 is elevated at low-medium-power range that C_4 has the efficiency almost identical to C_3. This is due to the fact that the effect of reducing k_2 on the soft-switching range is larger for $k_1 = 0.03$ than $k_1 = 0.97$ so that the ZVS of C_4 boosted the efficiency in spite of the increased conduction losses.

VII. CONCLUSION

In this paper, effects of transformer design on operational characteristics was analyzed to improve the performance of the FDM scheme for the DAB converter. Following the description of the FDM discussed previously, the operational characteristics for conduction and switching loss estimation were obtained using the two-port Fourier series model. The generalized transformer model is proposed for analysis of the operation of the converter considering the practical transformer with non-infinite magnetizing inductance. Based on a systematic design procedure, four samples of the transformer stage were constructed from the combinations of two transformers with different magnetizing inductances and two auxiliary inductor. Experimental verification was carried out

with these multiple samples of the transformer stage. Efficiency improvements of the FDM according to the transformer design were observed at light- and medium-load operations.

REFERENCES

[1] R. W. A. A. De Doncker, D.M. Divan, and M.H. Kheraluwala, "A three-phase soft-switched high-power-density dc/dc converter for high-power applications," *IEEE Trans. Ind. Appl.*, vol. 27, no. 1, pp. 63-73, Jan./Feb. 1991.

[2] B. Zhao, Q. Song, W. Liu, and Y. Sun, "Overview of dual-active-bridge isolated bidirectional dc–dc converter for high-frequency-link power-conversion system," *IEEE Trans. Power Electron.*, vol. 29, no. 8, pp. 4091-4106, Aug. 2014.

[3] N. Schibli, "Symmetrical multilevel converters with two quadrant dc-dc feeding," Ph.D. dissertation, Ecole Polytechnique Federale de Lausanne, Lausanne, Switzerland, 2000.

[4] C. Gu, Z. Zheng, L. Xu, K. Wang, and Y. Li, "Modeling and control of a multiport power electronic transformer (PET) for electric traction applications," *IEEE Trans. Power Electron.*, vol. 31, no. 2, pp. 915-927, Feb. 2016.

[5] H. Fan and H. Li, "High-frequency transformer isolated bidirectional dc -dc converter modules with high efficiency over wide load range for 20 kVA solid-state transformer," *IEEE Trans. Power Electron.*, vol. 26, no. 12, pp. 3599-3608, Dec. 2011.

[6] M. Mu, L. Xue, D. Boroyevich, B. Hughes, and P. Mattavelli, "Design of integrated transformer and inductor for high frequency dual active bridge GaN Charger for PHEV," in *Proc. IEEE Applied Power Electron. Conf. and Expo.*, 2015, pp.579-585.

[7] D.S. Gautam, F. Musavi, W. Eberle, and W.G. Dunford, "A zero-voltage switching full-bridge dc-dc converter with capacitive output filter for plug-in hybrid electric vehicle battery charging," *IEEE Trans. Power Electron.*, vol. 28, no. 12, pp. 5728-5735, Dec. 2013.

[8] F. Jauch and J. Biela, J., "Novel isolated cascaded half-bridge converter for battery energy storage systems," in *Proc. IEEE Power Electron. and Appl.*, 2014, pp.1-10.

[9] L. Cao, K.H. Loo, and Y.M. Lai, "systematic derivation of a family of output-impedance shaping methods for power converters–a case study using fuel cell-battery-powered single-phase inverter system," *IEEE Trans. Power Electron.*, vol. 30, no. 10, pp. 5854-5869, Oct. 2015.

[10] F. Krismer, S. Round, and J. W. Kolar, "Performance optimization of a high current dual active bridge with a wide operating voltage range," in *Proc. IEEE Power Electron. Specialists Conf.*, 2006, pp. 1-7.

[11] F. Krismer and J. W. Kolar, "Efficiency-optimized high-current dual active bridge converter for automotive applications," *IEEE Trans. Ind. Electron.*, vol. 59, no. 7, pp. 2745-2760, July 2012.

[12] J. Everts, F. Krismer, J. Van den Keybus, J. Driesen, and J. W. Kolar, "Optimal ZVS modulation of single-phase single-stage bidirectional DAB ac–dc converters," *IEEE Trans. Power Electron.*, vol. 29, no. 8, pp. 3954-3970, Aug. 2014.

[13] J. Hiltunen, V. Vaisanen, R. Juntunen, and P. Silventoinen, P., "variable-frequency phase shift modulation of a dual active bridge converter," *IEEE Trans. Power Electron.*, vol. 30, no. 12, pp. 7138-7148, Dec. 2015.

[14] J. Everts, G.E. Sfakianakis, and E.A. Lomonova, "Using Fourier series to derive optimal soft-switching modulation schemes for dual active bridge converters," in *Proc. IEEE Energy Conversion Congress and Expo.* 2015, pp.4648-4655.

[15] B. Zhao, Q. Song, W. Liu, G. Liu, and Y. Zhao, "Universal high-frequency-link characterization and practical fundamental-optimal strategy for dual-active-bridge dc-dc converter under PWM plus phase-shift control," *IEEE Trans. Power Electron.*, vol. PP, no. 99, pp. 1, 1.

[16] J. Itoh, H. Higa, and T. Nagano, "A novel control method focusing on reactive power for a dual active bridge converter," in *Proc. IEEE Electron. and Appl. Conf. and Expo.*, 2014, pp. 1020-1025.

[17] W. Choi, K. Rho, and B. Cho, "fundamental duty modulation of dual-active-bridge converter for wide range operation," *IEEE Trans. Power Electron.* (in press).

[18] A.K. Jain and R. Ayyanar, "PWM control of dual active bridge: Comprehensive analysis and experimental verification," *IEEE Trans. Power Electron.*, vol. 26, no. 4, pp.1215-1227, April 2011.

[19] D. Rothmund, G. Ortiz, T. Guillod, and J.W. Kolar, "10kV SiC-based isolated dc-dc converter for medium voltage-connected solid-state transformers," in *Proc. IEEE Applied Power Electron. Conf. and Expo.*, 2015, pp.1096-1103.

[20] Y. Du, S. Baek, S. Bhattacharya, and A.Q. Huang, "High-voltage high-frequency transformer design for a 7.2kV to 120V/240V 20kVA solid state transformer," in *Proc. IEEE Annual Conf. on IEEE Ind. Electron. Society*, 2010, pp.493-498.

[21] M.N. Kheraluwala, R.W. Gascoigne, D.M. Divan, and E.D. Baumann, "Performance characterization of a high-power dual active bridge dc-to-dc converter," *IEEE Trans. Ind. Appl.*, vol. 28, no. 6, pp. 1294-1301, Nov./Dec. 1992.

[22] K. Mainali, A. Tripathi, D.C. Patel, S. Bhattacharya, and T. Challita, "Design, measurement and equivalent circuit synthesis of high power HF transformer for three-phase composite dual active bridge topology," in *Proc. IEEE Applied Power Electron. Conf. and Expo.*, 2014, pp.342-349.

[23] J. Riedel, D.G. Holmes, C. Teixeira, and B.P. McGrath, "Harmonic-based determination of soft switching boundaries for 3-level modulated single-phase dual active bridge converters," in *Proc. IEEE Energy Conversion Congress and Expo.*, 2015, pp. 1505-1512.

[24] R.W. Erickson, *Fundamentals of power electronics*, Springer Science & Business Media, 2007, pp. 539-583.

A High Step-up Bidirectional Isolated Dual-Active-Bridge Converter with Three-Level Voltage-Doubler Rectifier for Energy Storage Applications

Xiaohai Zhan, Hongfei Wu, Yan Xing, Hongjuan Ge
Jiangsu Key Lab. of New Energy Generation and Power
Conversion
Nanjing University of Aeronautics and Astronautics
Nanjing, China
Email: zhanxiaohai@nuaa.edu.cn

Xi Xiao
State Key Lab of Power Systems, Department of Electrical
Engineering
Tsinghua University
Beijing, China
Email: xiao_xi@tsinghua.edu.cn

Abstract—A bidirectional isolated dual-active-bridge (DAB) converter with three-level voltage-doubler rectifier (TL-VDR) is presented for energy storage applications. The voltage conversion ratio is enhanced and the voltage stresses on transformer windings are reduced with the help of TL-VDR. Optimized pulse-width-modulation (PWM) plus dual phase-shift-modulation (PSM) strategy is applied to the presented DAB converter. The circulating current is reduced and the soft-switching performance is achieved in wide voltage and load range. Therefore, high efficiency bidirectional step-up/step-down power conversion is achieved with the presented topology and control strategy. These features make the presented converter attractive for energy storage applications. The operation principles and characteristics are analyzed and verified with experimental results.

Keywords—bidirectional; dual-active-bridge (DAB); three-level voltage-doubler rectifier (TL-VDR); pulse-width-modulation (PWM) plus dual phase-shift-modulation (PSM)

I. INTRODUCTION

With the increasingly serious environmental pollution and energy shortage, renewable energy, like solar, wind, and tide energy, have attracted more and more attention and are considered as excellent alternatives for conventional electrical power generation which uses fossil fuels as the main source [1]-[2]. Since the renewable energy sources have the feature of intermittence and randomness [3]-[4], energy storage devices are always integrated with the renewable energy generation units to provide stable and continuous power [4]-[5]. In general, a high step-up bidirectional DC-DC converter is required to control bidirectional power transfer of the energy storage devices. Thus, high efficiency bidirectional DC-DC converters have attracted much interest in energy storage systems. A dual-active-bridge (DAB) converter is attractive for isolated bidirectional power conversion due to its symmetric structure and zero voltage switching performance [6]-[7]. However, this converter has a limited ZVS range and high circulating current under wide voltage variation. Many advanced modulation strategies for the full-bridge DAB converter have been presented to overcome these problems [8]-[12]. Since the

This work was supported by the National Natural Science Foundation of China (51407092, U1233127), the Natural Science Foundation of Jiangsu Province, China (BK20140812), and Foundation of State Key Lab of Power System(SKLD15KZ01, SKLD15M07).

voltages of energy storage elements, such as batteries and super-capacitors, are relatively low while the voltage of DC-bus is usually high to interface with AC-grid, high step-up ratio is required to interface the low voltage batteries with high voltage DC-bus. In this case, a half-bridge or voltage-doubler type rectifier is more suitable to be used on the high-voltage side of a DAB converter [13]. However, since the duty cycle of the switches in the half-bridge rectifier is constant to be 0.5, the advanced modulation strategies for the full-bridge DAB converter cannot be used. As a result, the problems of limited ZVS range and high circulating current of traditional DAB converters cannot be solved. If a zero-voltage level can be produced by the half-bridge circuit, then optimized modulations can be applied and problems can be solved.

To solve these problems and satisfy the requirement of high voltage ratio and high efficiency bidirectional power conversion, a novel bidirectional DAB converter utilizing a three-level voltage-doubler rectifier (TL-VDR) on the high-voltage side is presented in this paper. With PWM plus dual phase-shift-modulation (PSM) control, soft-switching for all switches can be realized over a wide range, which helps to reduce switching losses. The TL-VDR can be employed to reduce the circulating current by generating a trapezoidal three-level in the high-voltage side, which contributes to reduce conduction losses. These advantages make the proposed converter a good candidate for energy storage applications.

II. PROPOSED CONVERTER AND OPERATION PRINCIPLE

A. Proposed Topology and Analysis

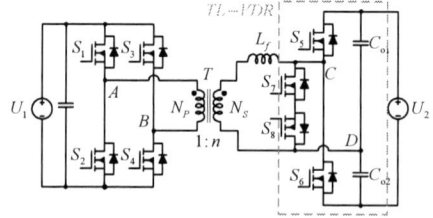

Fig. 1. Topology of proposed converter

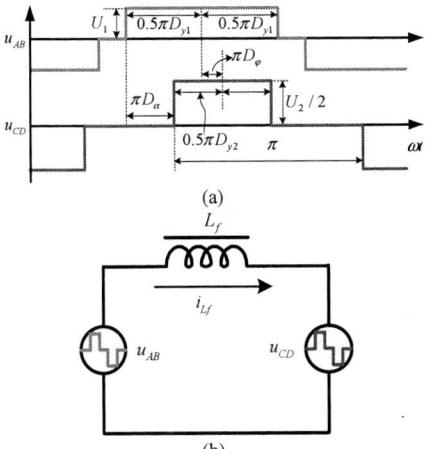

(a)

(b)

Fig. 2. Fundamental model of proposed converter with PWM plus dual PSM control: (a) simplified key waveforms; (b) equivalent model of proposed converter

The topology of proposed bidirectional DAB converter is shown in Fig. 1, where a full-bridge circuit is employed on the low-voltage side while a three-level voltage-doubler-rectifier (TL-VDR) is employed on the high-voltage side. The switches, S_7 and S_8, in the TL-VDR can provide a power flow path for the current of inductor L_f and generate a zero-voltage on the high-voltage side. Therefore, it is possible to generate three voltage-levels with the TL-VDR. As a result, optimized PWM plus PSM control strategies can be applied to both sides of the bidirectional converter, which is similar to the traditional full-bridge DAB converter.

B. Operation Principle

The phase shift angle φ is used to regulate the power flow of the DAB converter. G is defined to be the equivalent voltage gain, $G=U_2/(2nU_1)$. To reduce the circulating current and extend the ZVS range, PWM is applied to the full-bridge side if $G<1$, and PWM is applied to the TL-VDR side if $G>1$. With PWM plus PSM strategies, each side of the converter generates

a square wave voltage. Fig. 2(a) shows the simplified key waveforms of the proposed converter, while Fig. 2(b) shows the equivalent model of the converter, where u_{AB} is the voltage generated by the full-bridge side and u_{CD} is the voltage generated by the TL-VDR side. In Fig. 2(a), Dy_1 and Dy_2 are the duty ratio of u_{AB} and u_{CD} respectively and D_φ is the equivalent phase shift angle between u_{AB} and u_{CD}.

In the bidirectional converter, D_φ can either be greater than 0 or be less than 0. If $D_\varphi>0$, the power flows from the full-bridge side to the TL-VDR side and if $D_\varphi<0$, the power flows from the TL-VDR side to the full-bridge side.

According to the relationship of D_φ and G, the proposed converter has four working modes:

1) Forward-Boost mode with $D_\varphi>0$ and $G>1$, where $D_{y1}=1$, $D_{y2}<1$ and the key waveforms are given in Fig. 3(a).

2) Forward-Buck mode with $D_\varphi>0$ and $G<1$, where $D_{y1}<1$, $D_{y2}=1$ and the key waveforms are given in Fig. 3(b).

3) Back-Boost mode with $D_\varphi<0$ and $G<1$, where $D_{y1}<1$, $D_{y2}=1$ and the key waveforms are given in Fig. 3(c).

4) Back-Buck mode with $D_\varphi<0$ and $G>1$, where $D_{y1}=1$, $D_{y2}<1$ and the key waveforms are given in Fig. 3(d).

C. Operation Principle Analysis in Forward-Boost Mode

Since the operation principles of these working modes are similar, this paper takes the Forward-Boost mode as an example to analyze the operation principle.

In the Forward-Boost mode, since the PWM is applied to the TL-VDR side, so $D_{y1}=1$, $D_{y2}<1$. With the phase shift angle between u_{AB} and u_{CD} increasing, the Forward-Boost mode can be divided into three cases. The analysis of each case is given as follows.

1) CCM1: When the rising edge of u_{AB} (0 to U_1) lags that of u_{CD} (-$U_2/2$ to 0) and leads anther rising edge of u_{CD} (0 to $U_2/2$), ie: $0 \leq D_\varphi \leq 0.5(1-D_{y2})$, the key waveforms are shown in

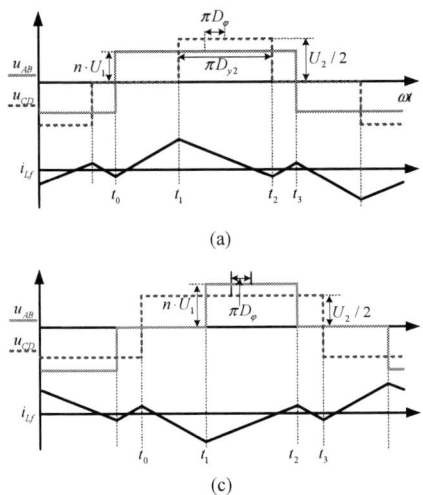

(a)

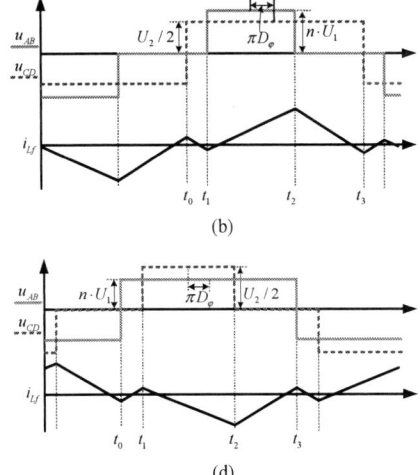

(b)

(c)

(d)

Fig. 3. Key waveforms of proposed converter with PWM plus dual PSM control in (a) Forward-Boost mode; (b) Forward-Buck mode; (c) Back-Boost mode and (d) Back-Buck mode.

978-1-4673-9551-9/16 $31.00 © 2016 IEEE

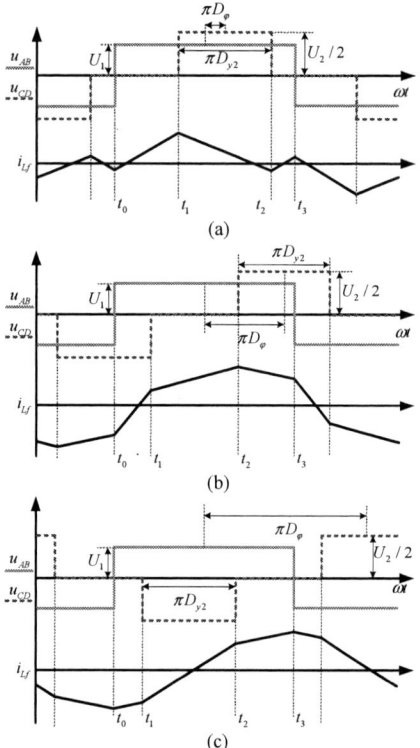

Fig. 4. Key waveforms of Forward-Boost mode in (a) CCM1; (b) CCM2 and (c) CCM3.
Fig. 4(a).

2) CCM2: When the rising edge of u_{AB} (0 to U_1) lags the falling edge of u_{CD} (0 to $-U_2/2$) and leads the rising edge of u_{CD} ($-U_2/2$ to 0), ie: $0.5(1-D_{y2}) \leq D_\varphi \leq 0.5(1+D_{y2})$, the key waveforms are shown in Fig. 4(b).

3) CCM3: When the falling edge of u_{AB} (U_1 to 0) lags the rising edge of u_{CD} ($-U_2/2$ to 0) and leads anther rising edge of u_{CD} (0 to $U_2/2$), ie: $0.5(1+D_{y2}) \leq D_\varphi \leq 1$, the key waveforms are shown in Fig. 4(c).

In order to simplify the analysis, CCM1 was taken as an example to analysis the operation principle of proposed converter working in Forward-Boost Mode.

Ignoring the dead time of MOSFET, the operation of a half switching period can be divided into three stages. The key waveforms are shown in Fig. 5 and the equivalent circuit in each switching stage is illustrated in Fig. 6. As given in Fig. 5, $u_{GS1} \sim u_{GS8}$ are the driving voltage of $S_1 \sim S_8$, u_{AB} and u_{CD} are the midpoint voltages of the full-bridge and TL-VDR sides, u_{Lf} and i_{Lf} are the voltage and current of L_f. And D_φ is the equivalent phase shift angle between u_{AB} and u_{CD}.

Stage I $[t_0 \sim t_1]$ [refer to Fig. 6(a)]: Before t_0, $S_2 \& S_3$ and $S_7 \& S_8$ were ON, the i_{Lf} flowed into the transformer from the dotted terminal. At t_0, $S_2 \& S_3$ turned OFF, in the dead time between $S_1 \& S_2$ ($S_3 \& S_4$), the body diode of $S_1 \& S_4$ were ON and $S_1 \& S_4$ turned ON with ZVS. Due to the forward voltage across the high-frequency inductor, i_{Lf} decreases linearly with the initial value $i_{Lf}(t_0)$, which is expressed as

$$i_{Lf}(t) = \frac{nU_1}{L_f}(t - t_0) + i_{Lf}(t_0) \tag{1}$$

This interval ends at t_1 when S_7 turned OFF.

Stage II $[t_1 \sim t_2]$ [refer to Fig. 6(b)]: At t_1, S_7 turned OFF. Firstly, i_{Lf} flowed through the body diode of S_5, thus, S_5 turned ON with ZVS. Due to the reverse voltage across the high-frequency inductor, i_{Lf} decreases linearly with the initial value $i_{Lf}(t_1)$, which is expressed as

$$i_{Lf}(t) = \frac{nU_1(1-G)}{L_f}(t - t_1) + i_{Lf}(t_1) \tag{2}$$

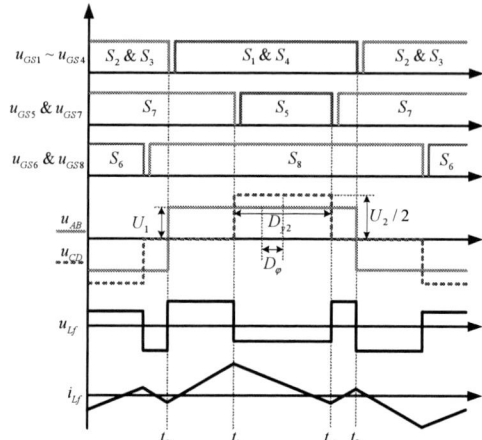

Fig. 5. Key waveforms in Forward-Boost CCM1 mode.

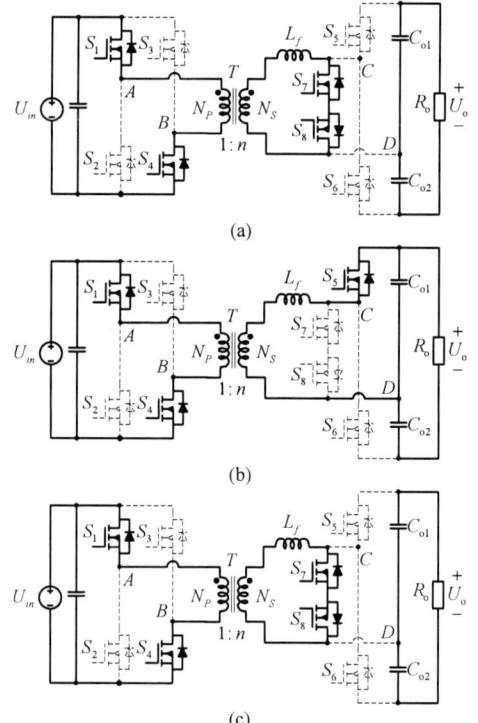

Fig. 6. Equivalent circuits of different switching stages in Forward-Boost CCM1 mode: (a) Stage I $[t_0 \sim t_1]$, (b) Stage II $[t_1 \sim t_2]$, (c) Stage III $[t_2 \sim t_3]$.

This interval ends at t_2 when S_5 turned OFF.

Stage III [$t_2\sim t_3$] [refer to Fig. 6(c)]: At t_2, S_5 turned OFF. Firstly, i_{Lf} flowed through the body diode of S_7, thus, S_7 turned ON with ZVS. Due to the forward voltage across the high-frequency inductor, i_{Lf} decreases linearly with the initial value $i_{Lf}(t_2)$, which is expressed as

$$i_{Lf}(t) = \frac{nU_1}{L_f}(t-t_2) + i_{Lf}(t_2) \tag{3}$$

At the end of stage III, i_{Lf} is positive but has the same absolute value as that at the beginning of stage I, which means

$$i_{Lf}(t_0) + i_{Lf}(t_3) = 0 \tag{4}$$

The next half switching period operates in the similar principle depicted above, with the roles of S_1 and S_2, S_3 and S_4, S_5 and S_6, S_7 and S_8 exchanged.

III. CHARACTERISTICS AND ANALYSIS

A. Power Transmission Capability

According to (1) ~ (4), the $i_{Lf}(t_0)$ can be expressed as

$$i_{Lf}(t_0) = nU_1(GD_{y2}-1)/4L_f f \tag{5}$$

Using (1) ~ (5), the transmission power of the proposed converter in CCM1 of Forward-Boost mode can be defined as follow:

$$P_{T1}(G, D_\varphi, D_{y2}) = n^2 U_1^2 GD_\varphi D_{y2} / 2L_f f \tag{6}$$

In order to simplify the analysis, the transmission power is normalized with P_B, which is expressed as

$$P_B = n^2 U_1^2 / 8L_f f \tag{7}$$

Thus, the normalized transmission power in CCM1 of Forward-Boost mode can be given as

$$P_{T1}^*(G, D_\varphi, D_{y2}) = 4GD_\varphi D_{y2} \tag{8}$$

Similarly, the normalized transmission power in CCM2 and CCM3 of Forward-Boost mode can be obtained as

$$P_{T2}^*(G, D_\varphi, D_{y2}) = G[1-(1-D_{y2})^2-(1-2D_\varphi)^2] \tag{9}$$

$$P_{T3}^*(G, D_\varphi, D_{y2}) = 4GD_{y2}(1-D_\varphi) \tag{10}$$

According to (8)-(10), the theoretical results of the

normalized transmission power curves versus normalized phase-shift angle D_φ in Forward-Boost mode, when the equivalent voltage gain $G = 1.2$, are depicted in Fig. 7, and the boundaries between different modes are plotted with dotted lines.

It can be seen that the transmission power increases along with D_{y2}, and all the transmission power curves are symmetric about D_φ=0.5. So, the maximum phase-shift angle of proposed converter can be limited to $\pi/2$. Because of the symmetry of the proposed topology, the transmission power curve verses phase-shift angle in Back-Buck mode can be acquired by rotating Fig. 7 180 degree in clockwise direction.

B. Soft-switching Analysis

Based on the analysis above, the converter can transfer the maximum power at D_φ=0.5 and the ripple of i_{Lf} while D_φ>0.5 is greater than that of i_{Lf} while D_φ<0.5 when the same power is transferred. So, the working range of the converter must be limited to 0<D_φ<0.5 and there is no need to analyze the characteristics of CCM3.

According to the characteristic of MOSFET, if the body diode of the MOSFET is in ON state before its driving signal is given, the ZVS turn ON is realized. The detailed constraints of ZVS for the Forward-Boost mode can be obtained from Table I.

From Table I, the specific constrains of ZVS for the Forward-Boost mode in CCM1 and CCM2 can be get as follow.

$$\begin{cases} CCM1: \dfrac{2D_\varphi}{G-1} \le D_{y2} \le \dfrac{1}{G} \\ CCM2: \dfrac{2(1-D_\varphi)}{G+1} \le D_{y2} \ \& \ D_\varphi \ge \dfrac{G-1}{2G} \end{cases} \tag{11}$$

Based on (8), (9) and (11), the ZVS region for all MOSFETs of the proposed converter in Forward-Boost mode can be given in Fig. 8. It can be shown that ZVS for all

TABLE I.   ZVS CONSTRAINS IN FORWARD-BOOST MODE

	$S_1\sim S_4$	$S_5\&S_6$	$S_7\&S_8$
CCM1	$i_{Lf}(t_0)$<0	$i_{Lf}(t_1)$>0	$i_{Lf}(t_2)$<0
CCM2	$i_{Lf}(t_0)$<0	$i_{Lf}(t_2)$>0	$i_{Lf}(t_1)$>0

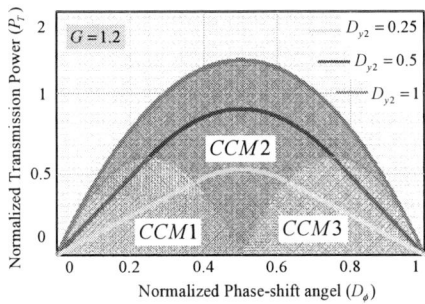

Fig. 7. Normalized transmission power curves versus phase-shift angle in Forward-Boost mode

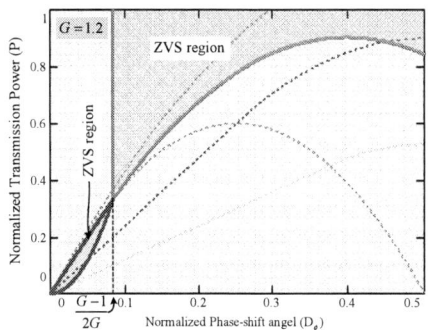

Fig. 8. ZVS region for all MOSFETs in Forward-Boost mode

MOSFETs can be achieved by coordinating control D_{y2} and D_φ. Because of the symmetry of the proposed topology, the ZVS region for all MOSFETs in Back-Buck mode can be acquired by rotating Fig. 8 180 degree in clockwise direction.

C. Voltage stress and output voltage gain analysis

Based on the previous analysis, it can be seen that the leakage inductance of transformer is operated as a part of L_f, which is used for power transmission. As a consequence, no voltage spike problem is caused and the voltage stress of the full-bridge side switches is clamped to input voltage. Meanwhile, thanks to the voltage-doubler, the voltage stress of S_7 and S_8 in the TL-VDR side is reduced to half of output voltage. Furthermore, with the implementation of active boost rectifier in the TL-VDR side, the proposed converter can achieve high voltage gain.

IV. EXPERIMENTAL VERIFICATION

In order to verify the validity of the analysis above, a prototype controlled by DSP MC56F8247 with the 60-80V/12A input and 380V output operating at 100 kHz switching frequency is built and tested. The main parameters of the prototype are given as follows: $S_1 \sim S_4$: IPP075N15N3G, $S_5 \sim S_6$: IXFH44N50P, $S_7 \sim S_8$: IXTQ82N25P, the inductor L_f=34.8μH, turns-ratio of transformer $n=N_S:N_P$=13:5.

The steady state experimental waveforms in Forward/Back and Buck/Boost modes are shown in Fig. 9, where u_{AB} and u_{CD} are the midpoint voltage of full-bridge side and TL-VDR side, respectively, and i_{Lf} is the currents of the energy transfer inductor, L_f. It can be observed that, the experimental results accord with theoretical analysis which is shown in Fig.3.

The experimental results of ZVS for all MOSFETs in Forward-Boost and Buck-Boost modes are given in Fig.10, where u_{GS} is the driving signal of MOSFET and u_{DS} is the drain-to-source voltage. It can be found that both the full-bridge side and TL-VDR side MOSFETs can achieve ZVS, which satisfied the previous analysis well.

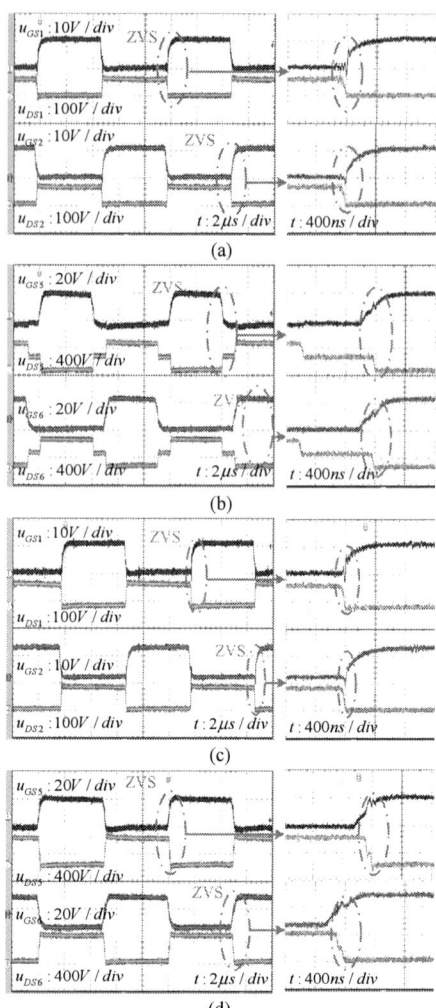

Fig. 10. Soft-switching waveforms for (a) & (c) full-bridge side and (b) & (d) TL-VDR side in Forward-Boost and Back-Boost mode.

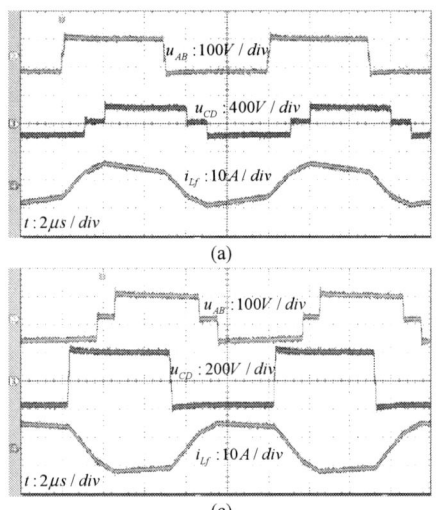

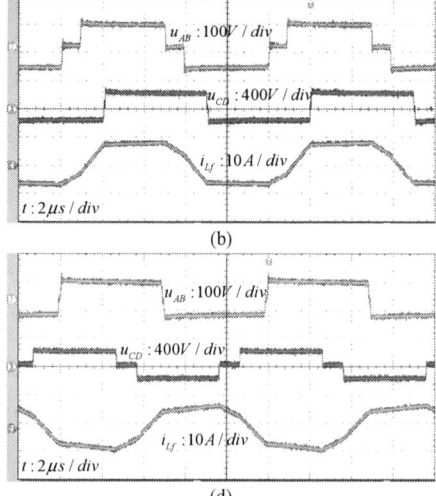

Fig. 9. Steady experimental waveforms in (a) Forward-Boost mode; (b) Forward-Buck mode; (c) Back-Boost mode and (d) Back-Buck mode.

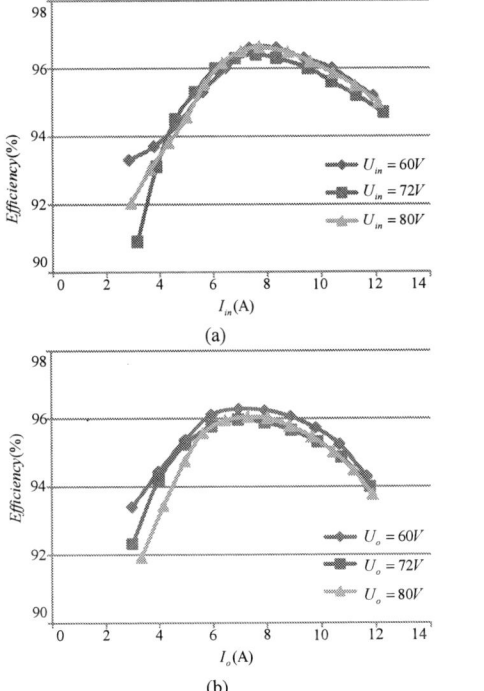

(a)

(b)

Fig. 11. Measured efficiency curve in (a) Forward mode and (b) Back mode.

The measured efficiency curve of the proposed converter in Forward and Back modes are given in Fig. 11. It can be seen that the efficiency are higher than 94% over wide load and input voltage range and the maximum value is about 96.6%. The experimental results indicate that the proposed converter suitable for high efficiency and high step-up application.

V. CONCLUSION

An isolated bidirectional dual-active-bridge converter is presented for energy storage application. A full-bridge circuit and a three-level voltage-doubler circuit are linked by a high-frequency transformer to realize bidirectional power conversion. The voltage stresses on high-voltage side transformer windings are reduced by using the three-level voltage-doubler-rectifier. PWM plus dual phase-shift control strategy can be applied to this converter to reduce the circulating current and extend the soft-switching range. Wide voltage gain can be achieved by operating the converter in Buck mode and Boost mode. All the above-mentioned characteristics indicate that the proposed converter is a good candidate for the high efficiency and high step-up applications in energy storage element.

REFERENCES

[1] D. Velasco de la Fuente, C. L. T. Rodriguez, E. Figueres, et.al, "Photovoltaic power system with battery backup with grid-connection and islanded operation capabilities," IEEE Trans. on Ind. Electron., vol. 60, no. 4, pp. 1571–1581, Apr. 2013.

[2] Zhe Zhang, Ziwei Ouyang, Ole C. Thomsen, Michael A. E. Andersen, "Analysis and Design of a Bidirectional Isolated DC–DC Converter for Fuel Cells and Supercapacitors Hybrid System," IEEE Trans. on Power. Electron., vol. 27, no. 2, pp. 848–859, Feb. 2012.

[3] S.N. Motapon, L.-A. Dessaint and K. Al-Haddad, "A comparative study of energy management schemes for a fuel-cell hybrid emergency power system of more-electric aircraft," IEEE Trans. on Ind. Electron., vol. 61, no. 3, pp. 1320–1334, Mar. 2014.

[4] Haihua Zhou, Ashwin M. Khambadkone, "Hybrid Modulation for Dual-Active-Bridge Bidirectional Converter With Extended Power Range for Ultracapacitor Application," IEEE Trans. on Ind. Electron., vol. 45, no. 4, pp. 1434–1442, Jul. 2009.

[5] Wuhua Li, Haimeng Wu, Hongbing Yu, and Xiangning He, "Isolated Winding-Coupled Bidirectional ZVS Converter With PWM Plus Phase-Shift (PPS) Control Strategy," IEEE Trans. Power Electron., vol. 26, no. 12, pp. 3077–3090, Dec. 2015

[6] Masatoshi Uno, Akio Kukita, "Bidirectional PWM Converter Integrating Cell Voltage Equalizer Using Series-Resonant Voltage Multiplier for Series-Connected Energy Storage Cells," IEEE Trans. Power Electron., vol. 30, no. 6, pp. 3560–3570, Jun. 2011.

[7] Biao Zhao, Qiang Song, Wenhua Liu, Yandong Sun, "Overview of Dual-Active-Bridge Isolated Bidirectional DC–DC Converter for High-Frequency-Link Power-Conversion System," IEEE Trans. Power Electron., vol. 29, no. 8, pp. 4091–4169, Aug. 2014.

[8] R. T. Naayagi, Andrew J. Forsyth, R. Shuttleworth, "High-Power Bidirectional DC–DC Converter for Aerospace Applications", IEEE Trans. Power Electron., vol. 21, no. 11, pp. 4366–4379, Nov. 2012.

[9] Germ´an G. Oggier, Guillermo O. Garc´ıa, Alejandro R. Oliva, "Modulation Strategy to Operate the Dual Active Bridge DC–DC Converter Under Soft Switching in the Whole Operating Range", IEEE Trans. Power Electron., vol. 26, no. 4, pp. 1228–1236, Apr. 2011.

[10] Biao Zhao, Qiang Song, Wenhua Liu, Guowei Liu and Yuming Zhao, "Universal High-Frequency-Link Characterization and Practical Fundamental-Optimal Strategy for Dual-Active-Bridge DC-DC Converter under PWM plus Phase-Shift Control", IEEE Trans. Power Electron., vol. 30, no. 12, pp. 6488–6494, Apr. 2015.

[11] Alberto Rodr´ıguez, Aitor V´azquez, Diego G. Lamar, Guowei Liu, Marta M. Hernando and Javier Sebasti´an, "Different Purpose Design Strategies and Techniques to Improve the Performance of a Dual Active Bridge With Phase-Shift Control", IEEE Trans. Power Electron., vol. 30, no. 2, pp. 790–804, Feb. 2015.

[12] Jordi Everts, Florian Krismer, Jeroen Van den Keybus, Guowei Liu, Johan Driesen and Johann W. Kolar, Fellow, "Optimal ZVS Modulation of Single-Phase Single-Stage Bidirectional DAB AC–DC Converters", IEEE Trans. Power Electron., vol. 29, no. 8, pp. 3954–3970, Aug. 2014.

[13] Zhe Zhang, Ziwei Ouyang, Ole C. Thomsen, Michael A. E. Andersen, "Analysis and Design of a Bidirectional Isolated DC–DC Converter for Fuel Cells and Supercapacitors Hybrid System," IEEE Trans. Power Electron., vol. 27, no. 2, pp. 848–859, Feb. 2012.

Digitized self-oscillating loop for piezoelectric transformer-based power converters

Marzieh Ekhtiari, Thomas Andersen, Zhe Zhang and Michael A. E. Andersen

Electronics Group, Department of Electrical Engineering
Technical University of Denmark
Richard Petersens Plads, bldg. 325, DK-2800, Kgs. Lyngby, Denmark
Web: http://www.ele.elektro.dtu.dk/
Email: maekh@elektro.dtu.dk

Abstract—A new method is implemented in designing of self-oscillating loop for driving piezoelectric transformers. The implemented method is based on combining both analog and digital control systems. Digitized delay, or digitized phase shift through the self-oscillating loop results in a very precise frequency control and ensures an optimum operation of the piezoelectric transformer in terms of voltage gain and efficiency. In this work, additional time delay is implemented digitally for the first time through 16 bit digital-to-analog converter to the self-oscillating loop. Delay control setpoints updates at a rate of 417 kHz. This allows the control loop to dynamically follow frequency changes of the transformer in each resonant cycle. The operation principle behind self-oscillating is discussed in this paper. Moreover, experimental results are reported.

Index Terms—Optimum delay line; self-oscillating loop; phase shift; switch-mode power supply; zero-voltage switching; piezoelectric transformer.

NOMENCLATURE

BPF	Band pass filter.
C	Resonant capacitance of the piezoelectric transformer.
C_A	Measurement capacitance.
C_{d1}	Input electrode capacitance of the piezoelectric transformer.
C_{d2}	Output electrode capacitance of the piezoelectric transformer.
CEZC	Current estimate zero crossing.
D	Duty cycle.
DAC	Digital-to-analog converter.
DDL	Digitized delay line.
DDL_{out}	Output of the DDL.
DT	Dead time.
$EDDL_{in}$	Edge detected of the input of the DDL.
FF	Flip-flop.
FPGA	Field-programmable gate array.

$i_{res}(t)$	Piezoelectric transformer's resonant current.
L	Internal inductance of the piezoelectric transformer.
LPF	Low pass filter.
MOSFET	Metal-oxide-semiconductor field-effect transistor.
PT	Piezoelectric transformer.
R	Dielectric losses inside the transformer.
R_A, R_B	Sensing resistors.
R_L	Load resistor.
$R_{matched}$	Matched load for the piezoelectric transformer.
$v_{FP}(t)$	Switching voltage in the primary side of the PT.
ZVS	Zero voltage switching.
ω	Angular frequency.
ϕ_I	Phase of the resonant current.

I. INTRODUCTION

Piezoelectric transformers' (PT) voltage gain, resonance frequency and efficiency change with variation of their load and temperature [1]–[4]. Therefore, in order to operate a PT in its optimum point, located slightly above the resonant frequency, it is necessary to follow changes in the resonance frequency [1]. It is required that the self-oscillating loop be able to adjust its phase shift to follow the PT's resonant frequency. Therefore, implemented adjustable phase shift compensates for the rest of the phase shift in the loop for frequency variations. For instance, when the energy transfered by the converter needs to be controlled to maintain DC output voltage at different voltage levels, the PT's load changes. Any changes in the PT's load causes a change in its operating point [4]. In order to keep the operation of the PT at its most efficient point, its operating frequency should

978-1-4673-9551-9/16 $31.00 © 2016 IEEE

be changed. This is performed by changing the converter's switching frequency. The switching frequency is controlled through a self-oscillating loop. Therefore, by changing the predesigned phase shift, the switching frequency follows variations in the PT's resonant current.

Phase shift compensation with high resolution becomes necessary especially when the load of the converter is variable. These variations directly translates to the PT's load variations. If the total phase shift of the loop is not properly adjusted to an integer factor of 360°, it causes a damping of the resonant current and, the closed-loop operation cannot be achieved, so the converter will not start working. Therefore, very fine resolution for the phase shift adjustment is required.

In previous research, an adjustable time delay block that controls the total phase of the loop has been implemented through an analog circuit. In the closed-loop operation, 360° phase difference in the desired frequency cannot be ensured for the load or temperature variations of the PT [5], [6]. To solve this problem, a digitalized phase shift compensation is applied in this paper. Changes in the PT's resonant frequency are compensated for the closed-loop by detecting and adding required phase shift in order to obtain a full loop phase shift of 360°. Compensation is performed by adding a finely-controlled time delay to the feedback chain. Resolution of the applied time delay is 1 ns. This ensures that the added time delay is finely controllable in order to precisely adapt the frequency of the self-oscillating loop and match changes in the PT's operating point. Furthermore, the delay can work in the range between 0° to 180°. This then, further ensures soft switching operation of the PT and efficiency increase.

This paper is organized as follows. Section II presents thorough explanation of the self-oscillating loop. Subsections II-A and II-B explain the principle behind self-oscillation in the prior art. Section II-C shows experimental results. Subsection III-A gives more explanation about the proposed method and subsection III-B shows experimental results for the proposed method.

II. SELF-OSCILLATING LOOP FOR PT-BASED CONVERTERS

A. Principle and design considerations

Essentially, two requirements need to be satisfied in order to produce sustained oscillation in a closed-loop. One is that phase angle of the entire loop should be an integer multiple of 360°; the other requirement is that the loop gain should be greater than unity to start-up the oscillation. The former condition is fulfilled

by adjusting phase shift through the loop. The latter condition is fulfilled by the use of a comparator since the comparator's gain can be considered infinite and therefore its input becomes saturated to the rail voltages. This results in generating square waves in the output of the comparator.

Fig. 1 shows the circuit for the self-oscillating loop together with the PT-based power stage with matched load, $R_{matched}$. Fig. 2 shows the block diagram for the self-oscillating loop [7]. A small perturbation in the loop starts the self-induced oscillation. This results in self-excitation of the resonant current inside the PT during the start-up. Self-excited square waves with a frequency lower than or close to the resonant frequency of the PT will excite resonant modes inside the transformer. This is achieved by the fundamental frequency of the square waves. Since the PT is operating as a high quality factor (Q) band pass filter (BPF), it filters higher order harmonics out and transfers the fundamental component to its output. The electrical quality factor of the PT is derived as:

$$Q = \frac{1}{\omega C_{d2} R_L} \qquad (1)$$

where $\omega = 1/\sqrt{LC}$ is the series-resonance angular frequency of the PT [8]. Therefore, the resonant current is considered as a sinusoidal waveform and is described as

$$i_{res}(t) = I_{pk}(t) \, sin(\omega t - \phi_I) \qquad (2)$$

where $\phi_I \in [0, \pi]$ is the current phase angle.

The amplitude of the fundamental resonant current grows with time until it reaches to a certain level that the self-oscillating loop becomes locked at the resonant frequency. Since the amplitude of the sinusoidal waveform at the output of low pass filter (LPF) shown in Fig. 2, becomes greater than the amplitude of the self-induced oscillation waveform, the oscillator operates as a comparator. This allows the comparator's behavior to change from that of an oscillator to that of a true comparator. Therefore, it compares the resonant current with a DC level in order to mark the zero crossing of the current. The loop is designed for the case in which the PT is connected to the resistive matched load [7].

$$R_{matched} = \frac{1}{\omega C_{d2}} \qquad (3)$$

The reasoning behind this design choice is that a matched load is considered to be the worst case scenario for a PT, in terms of achieving soft switching [9]. Zero-voltage switching (ZVS) is a form of soft switching considered in this work. At the matched load, the energy

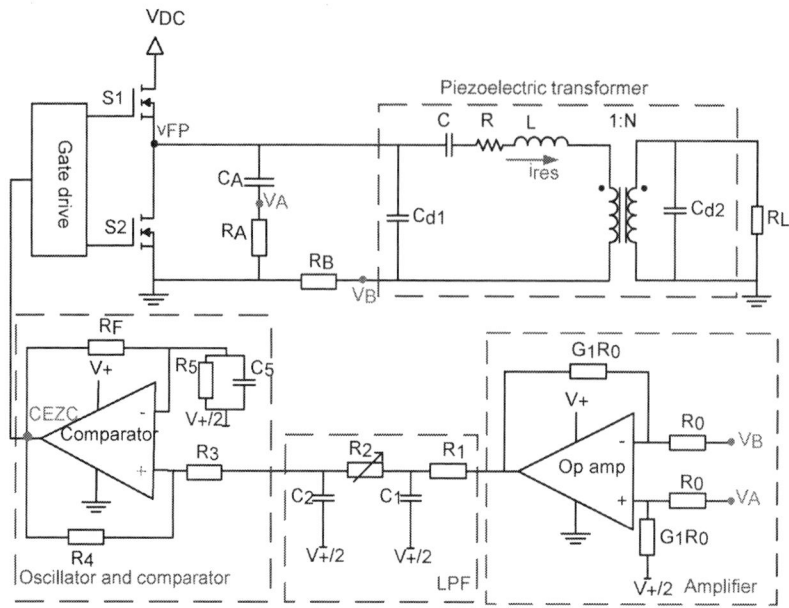

Fig. 1: Circuit design of self-oscillating loop in the PT-based SMPS.

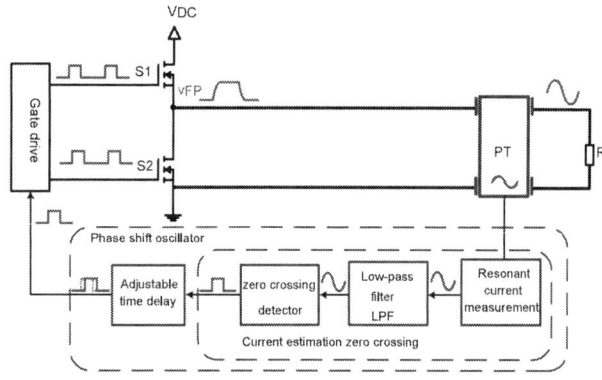

Fig. 2: Block diagram of the self-oscillating loop from the previous research.

transfer through the PT is maximum and therefore its efficiency is maximized as well. This results in a point of minima on the ZVS factor axis [4], [10]. The role of the ZVS factor is to provide the worst case scenario for analyzing PTs in terms of soft switching capability, which means that if the ZVS is achieved for the matched load, it will be obtained for other loads as well [9]. The worst case explanation of the approximated ZVS factor is expressed based on empiric analysis of the ZVS in the following equation [11]:

$$V'_p = (0.304 \; \frac{1}{n^2} \frac{C_{d2}}{C_{d1}} + 0.538) \cdot (0.585\eta + 0.414) \quad (4)$$

where η is the efficiency of the transformer. V'_p is the ZVS factor and should be above 1 in order to obtain soft switching. Therefore, the ZVS region is a narrow frequency range above the resonant frequency, where the PT behaves as an inductor and $V'_p > 1$. Furthermore, the ZVS bandwidth of the PT is a ratio of L/C for a constant resonant frequency [9]. Equation (4) is based on a primary analysis of ZVS in order to justify the necessity to design a driver and the self-oscillating loop.

B. Current estimation zero crossing (CEZC)

The resonant current in the PT is reconstructed within two time intervals. Voltage V_B across R_B shown in Fig. 1 measures this resonant current while the switches are on. Voltage V_A across R_A measures the resonant current during dead time (DT), while both switches are off. The resonant current in this period is supplied by the PT's input capacitor C_{d1}. Therefore, the current passing through C_{d1} follows the resonant current and can be measured by differentiating the voltage across C_{d1}. This is performed by using C_A and R_A as a differentiator. By proper subtraction of these two voltage waveforms through the op-amp, an approximate sinusoidal waveform representing the resonant current will be reconstructed in the output [7], [12], [13].

The estimated current has a 180° phase shift compared to the resonance current which results in the same zero crossing points. Thereafter, the estimated resonance current is transmitted to a second order LPF which

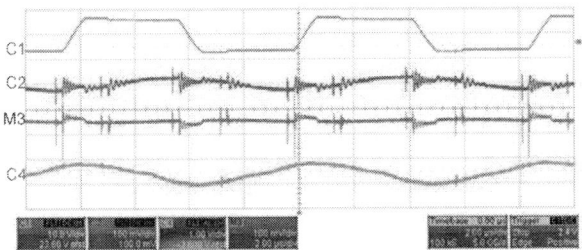

Fig. 3: Reconstruction of the resonant current. C1: switching voltage v_{FP}, C2: the voltage V_B, M3: the voltage V_B, C4: voltage output of the op-amp G1.(V_A - V_B)

Fig. 4: The CEZC board.

provides an additional phase shift through the feedback loop. The filtered signal is transmitted to the comparator and generates a square wave indicating the zero crossing of the input signal. The square wave signal is fed into the adjustable time delay in order to compensate for the rest of the phase shift to have a total phase shift of 360° in the entire loop from the input of the gate driver to the output of the feedback loop. In case the switching frequency needs to be decreased, the total phase shift should be increased. Additional phase shift in the loop is added through a digitally-controlled time delay. Since time delay is a positive value, the controlled delay adds an adequate phase shift in order to synchronize the frequency in the next cycle.

C. Experimental results

A radial mode PT with Mason's equivalent circuit, shown in Fig. 1, is used, driven by square wave signals with a switching frequency of 118.3 kHz, while driving a resistive load of 300 Ω. Moreover, reconstructed resonant current from voltages V_A and V_B is shown in Fig. 3. The designed board is shown in Fig. 4. Fig. 5 shows experimental waveforms for the self-oscillating loop. Phase shifts and corresponding time delays between stages are measured and shown in Table I. Furthermore, the equivalent parameter values of the PT are measured and shown in Table II.

TABLE I: Phase shifts during one switching cycle in the self-oscillating loop; switching frequency is 118.3 kHz with the time period of 8.45μs. HS: high-side control signal, HSGD: high-side gate voltage in the output of the gate driver, I_{est}: estimated current in the input of the LPF.

Delay	Time [μs]	Phase [°]	Duty cycle %
HS→HSGD	1.59	67.8	18.85
HSGD→I_{est}	0.27	11.5	3.19
I_{est}→LPF	6.09	259.4	72.04
LPF→CEZC	0.5	21.3	5.92
Total	8.45	360	100

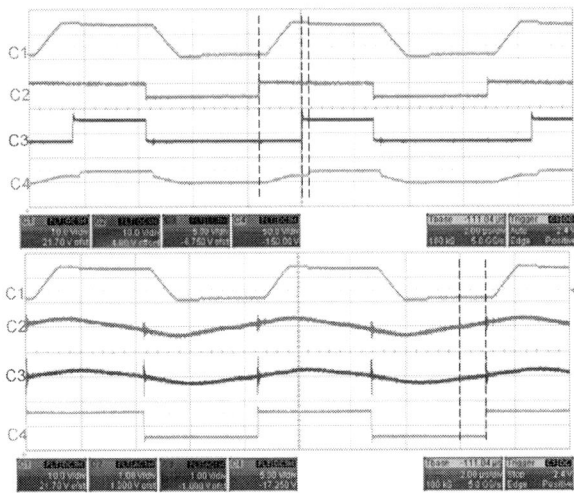

Fig. 5: The self-oscillating loop in the previous research; wave forms from top to bottom show phase shift during one switching cycle. On top: C1: the switching voltage (v_{FP}); C2: CEZC; C3: input signal to the high-side gate drive; C4: high-side gate drive output to the ground. On bottom: C1: switching voltage (v_{FP}); C2: estimated resonant current in the output of op-amp; C3: estimated resonant current in the output of the LPF; C4: the CEZC.

TABLE II: PT equivalent parameters

Parameter	Value	Parameter	Value
C_{d1}	3.8 nF	C_{d2}	626 pF
C	565 nF	R	5.6 Ω
L	3.5 mH	N	3.5

III. DIGITIZED DELAY LINE (DDL)

A. Design considerations

Initial investigation of the time step resolution shows need for a finely adjust of the total delay in the loop. The result of this investigation showed that there was a change in the amplitude of the PT's output voltage for every 10 Hz change in the switching frequency. For example, if the operating frequency of 100 kHz increases by 10 Hz, the output voltage shows a consid-

978-1-4673-9551-9/16 $31.00 © 2016 IEEE

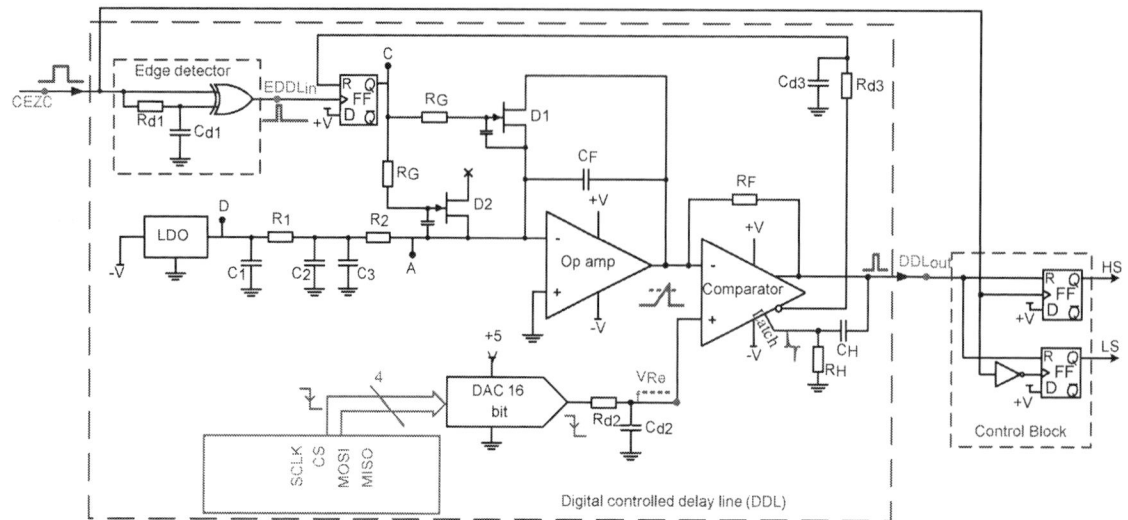

Fig. 6: The electrical circuit of the DDL block, with resolution of 1 ns.

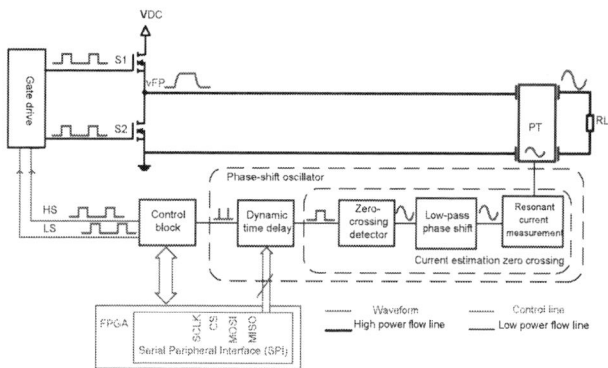

Fig. 7: Block digram of self-oscillating loop with dynamic time delay.

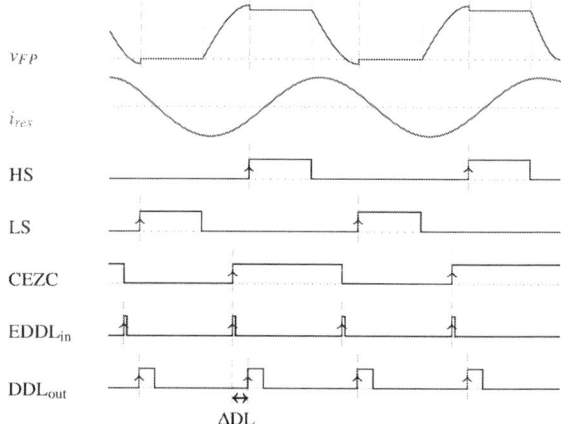

Fig. 8: Signal waveforms with negligible propagation delays in digital gates and the gate driver. ΔDL: applied delay with the resolution of 1ns.

erable change in the amplitude. Therefore, a precision of minimum 10 Hz in the frequency is required for a dynamic time delay which is equivalent to 1ns resolution. Fig. 6 shows a simplified version of the circuit used as a control block. Fig. 7 shows the block diagram of the self-oscillating closed-loop together with the contribution of the digital-to-analog converter (DAC) and field-programmable gate array (FPGA). These are used to implement a high-resolution time delay inside the DDL block.

Fig. 8 shows a detailed drawing of the main signal waveforms in the circuit. $v_{FP}(t)$ is the transformer's primary-side voltage while exhibiting soft switching. $i_{res}(t)$ shows the resonant current of the PT. This current is dependent on the characteristic parameters of the PT and it changes its polarity when either the switches are on or their body diodes conduct. Therefore, depending

on the operating frequency and temperature of the PT, there is a phase shift between the resonance frequency and the switching voltage, that is defined as ϕ_l in (2). Consequently, it is important that the circuit is capable of compensating for this phase shift. The estimated resonant current shown in Fig. 7 passes through a 180°phase shifter, then a LPF and a comparator. This is illustrated in Fig. 8. The output of the CEZC block which has the same frequency as the resonance is then tied to the DDL block. In DDL block, the input signal is first transformed into edge-detecting one-shot pulses (EDDL$_{in}$) which are then used as a clock source for the flip-flop (FF). The

Fig. 9: The DDL board

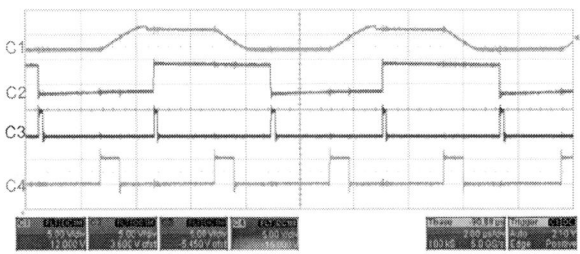

Fig. 10: Measurement: Signal waveforms where dynamic delay is applied; C1: switching waveform (v_{FP}); C2: CEZC as input of the DDL block; C3: one-shot pulse EDDL$_{in}$; C4: output of the DDL block (DDL$_{out}$).

TABLE III: One cycle phase shift measured in the self-oscillating loop, HS: high-side control signal, HSGD: high-side gate voltage in the output of the gate driver, I$_{est}$: estimated current in the input of the LPF.

Delay	Time [µs]	Phase [°]	Duty cycle %
HS→HSGD	0.28	11.94	3.3
HSGD→I$_{est}$	6.6	281.52	78.2
I$_{est}$→ LPF	1.56	66.5	18.5
Total	8.44	360	100
D:EDDL$_{in}$→DDL$_{out}$	2.5	106.6	29.6

signal from this FF is then used to reset the hardware integrator present in the circuit. Namely, when the input signal edge-triggers the FF, the feedback capacitor in the op-amp's feedback starts charging, thereby creating a fixed voltage slope. This is then compared to the variable reference voltage provided by the DAC. Thereafter, the complementary output of the comparator resets the FF which consequently turns the MOSFET, D1, on and discharges the feedback capacitor, thereby resetting the integrator. Since the input pulse triggers the start of the integration, the variable reference provided to the comparator by the DAC coupled with the voltage slope work together to create a time-delayed version of the input pulse which is proportional to the DAC output value. The output of the comparator is then latched for a short time by its own output through a high-passed signal to its latch pin, resulting in one-shot pulses at the output of the DDL.

The output of the DDL is capable of prolonging the switching period by changing the on time of the switches (T$_{on}$). The high-side and low-side switches are turned on by rising edges of CEZC and $\overline{\text{CEZC}}$ signal, respectively, which are used as clock inputs to the control block FFs. The output of the DDL block is then used to reset the FFs, thereby turning the switches off. Dead time (DT) is fixed for a specific design regarding a certain PT and switching frequency. Since this DT is fixed, by adjusting the time-delay for turning the switches off, the frequency of self-oscillation changes. DDL block delays its input (CEZC) to compensate the rest of phase shift through loop to reach 360°. The propagation delay in the control block is assumed negligible in the waveforms shown in Fig. 8.

B. Experimental results

A prototype has been built and tested. Fig. 9 shows designed modular board for DDL block. A radial mode PT with Mason's equivalent circuit is used, driven by square wave signals with a switching frequency of 118.3 kHz and a resistive load of 300 Ω. Furthermore, the equivalent parameter values of the PT are measured and shown in Table II. The input signal waveform of DDL circuit i.e. CEZC and one-shot input pulse together with the output of DDL are shown in Fig. 10. Table III shows measured phase shifts and corresponding time delays between stages.

IV. CONCLUSION

General operation of a self-oscillating loop for piezo-electric transformer-based power converters is explained. The designed circuit for operating the transformer together with experimental results are provided. The circuit is based on a new idea for compensating and controlling the phase shift for the self-oscillating in order to achieve and maintain soft switching. This is combined with a resonant current estimation that is able to start and maintain the circuit oscillation. The operation of and cooperation between the analog resonance current estimator and digital phase shift adjustment are presented in detail, together with some insight into the performance of the designed circuit. The concept is proven through the experimental results. 1 ns time step resolution is sufficient for adjusting the phase shift of the loop. The designed circuit is able to follow changes in the resonant frequency of the PT in every cycle.

978-1-4673-9551-9/16 $31.00 © 2016 IEEE

ACKNOWLEDGMENT

The authors would like to thank Noliac A/S for supplying the prototype of a piezoelectric transformer as well as "The Danish National Advanced Technology Foundation" for financial support.

REFERENCES

[1] M. Ekhtiari, Z. Zhang, and M. A. E. Andersen, "State-of-the-art piezoelectric transformer-based switch mode power supplies," *40th Annual Conference of the IEEE Industrial Electronics Society, IECON2014*, pp. 5072–5078, 2014.

[2] E. Horsley, M. Foster, and D. Stone, "State-of-the-art piezoelectric transformer technology," in *Power Electronics and Applications, 2007 European Conference on*, Sept 2007, pp. 1–10.

[3] A. V. Carazo, "50 years of piezoelectric transformers. trends in the technology," in *Materials and Devices for Smart Systems Symposium*, ser. MRS Proceedings, vol. 785, 2003.

[4] C. y. Lin, "Design and analysis of piezoelectric transformer converters," *Ph.D. Dissertation, Virginia Tech*, 1997.

[5] M. A. E. Andersen, K. Meyer, M. S. Rodgaard, and T. Andersen, "Piezoelectric power converter with bi-directional power transfer," *US patent 61/567,924 and 2013, WIPO, WO2013083679-A1*, 2011.

[6] M. Rodgaard, "Bi-directional piezoelectric transformer based converter for high-voltage capacitive applications," in *Applied Power Electronics Conference and Exposition (APEC), 2015 IEEE*, March 2015, pp. 1993–1998.

[7] M. Rodgaard, M. A. E. Andersen, T. Andersen, and K. Meyer, "Self-oscillating loop based piezoelectric power converter," *US patent 61/638,883 and WIPO, WO2013083678-A2*, 2011.

[8] G. Ivensky, I. Zafrany, and S. Ben-Yaakov, "Generic operational characteristics of piezoelectric transformers," *Power Electronics, IEEE Transactions on*, vol. 17, no. 6, pp. 1049–1057, Nov 2002.

[9] K. Meyer, M. Andersen, and F. Jensen, "Parameterized analysis of zero voltage switching in resonant converters for optimal electrode layout of piezoelectric transformers," pp. 2543–2548, June 2008.

[10] R. L. Lin, "Piezoelectric transformer characterization and application of electronic ballast," *Ph.D. Dissertation, Virginia Polytechnic Institute and State University*, 2001.

[11] M. Rodgaard, T. Andersen, and M. Andersen, "Empiric analysis of zero voltage switching in piezoelectric transformer based resonant converters," pp. 1–6, March 2012.

[12] T. Andersen, "Piezoelectric transformer based power supply for dielectric electro active polymers," *Ph.D. Thesis, Department of Electrical Engineering, Technical University of Denmark*, 2012.

[13] M. S. Rodgaard, "Piezoelectric transformer based power converters; design and control," *Ph.D. Thesis, Department of Electrical Engineering, Technical University of Denmark*, 2012.

An Isolated Topology for Reactive Power Compensation With a Modularized Dynamic-Current Building-Block

Hao Chen[1], Anish Prasai[2], and Deepak Divan[1, 2]

[1]School of Electrical and Computer Engineering
Georgia Institute of Technology
Atlanta, Georgia, USA
hchen95@gatech.edu
[2]Varentec, Inc.

Abstract—**This paper presents a novel topology for instantaneous reactive power compensation. The topology is derived from Dynamic-Current or Dyna-C, which is a patented power converter capable of transferring energy for two- or multi-terminal DC, single- and/or multi-phase AC systems. The proposed topology has a modularized low-voltage current source building block that can be stacked for medium-voltage (MV) applications to provide dynamic leading or lagging reactive power. In addition, the phases are coupled through a high-frequency transformer for inter-phase fault isolation among the three-phase. The converter functionality is validated through simulations and experimental results with a 480 V, 75 kVAr prototype.**

I. INTRODUCTION

Proliferation of non-resistive loads imposes power quality concerns over the existing grid network. The reactive power loads, which possess a low power factor, draw high amount of volt-ampere reactive power from the grid and thus restrict the active power transfer capability. The excessive reactive power flow over the network heavily burdens the transformer, distribution and transmission lines, increases the line losses, and impacts the voltage stability.

Reactive power compensators used for grid support with slow dynamic response include the mechanically-switched capacitor with damping network (MSCDN), the thyristor-switched capacitor (TSC), and the thyristor-controlled reactor (TCR) [1]-[6]. A static synchronous compensator (STATCOM), built using voltage source converter (VSC), can provide reactive power support with a less noisy sinusoidal current injection due to operation at higher effective frequency and generally take up less real estate compared to TSCs and TCRs for the same level of harmonics and power rating [7]. The Multi-level STATCOM architectures such as the diode- or neutral-clamped, flying capacitor, and cascaded H-bridges,

can be scaled to medium voltage with fractionally-rated components [8]-[10]. VSC-based VAr compensators require bulk energy storage, typically implemented using electrolytic capacitors, to support the DC link voltages. While the technology advances in electrolytic capacitors have come a long way, compared to film-based capacitors, they still have reliability and safety issues that contribute to a shorter life. In addition, fault management in a VSC-based architecture is made complex due to the presence of DC source(s) that can contribute to high fault currents, requiring fast detection and isolation [11]-[12]. As reliability and long life is of utmost concern with utility-grade equipment, a technology that can provide STATCOM functionality but with higher reliability and longer life is needed.

As an alternative to the VSC based VAr compensator, this paper proposes a novel modular current source inverter (CSI) based topology to provide STATCOM functionality. The topology is derived from Dynamic-Current or Dyna-C, a patented power converter capable of transferring energy for two- or multi-terminal DC, single- and/or multi-phase AC systems [13]-[14]. The topology, named Dynamic VAr Compensator (DVC), employs a high-frequency transformer rather than a DC capacitor as the means to store energy, which limits the maximum current that can contribute to a fault. The paper presents the topology and the operating principles, and validates the core functionality through simulation results and experimental measurements of a 480 V / 75 kVAr prototype.

I. TOPOLOGY

Figure 1 shows the proposed topology of the Dyna-C DVC. The topology, which is also a modular building block, consists of three single-phase CSI modules coupled together with a three-winding medium- to high-frequency transformer. Appropriate LC filters are used around the three low-frequency AC terminals for suppressing switching harmonics in the current. Unlike a CSI, the magnetizing inductance of the transformer itself is used to store energy. The modules are

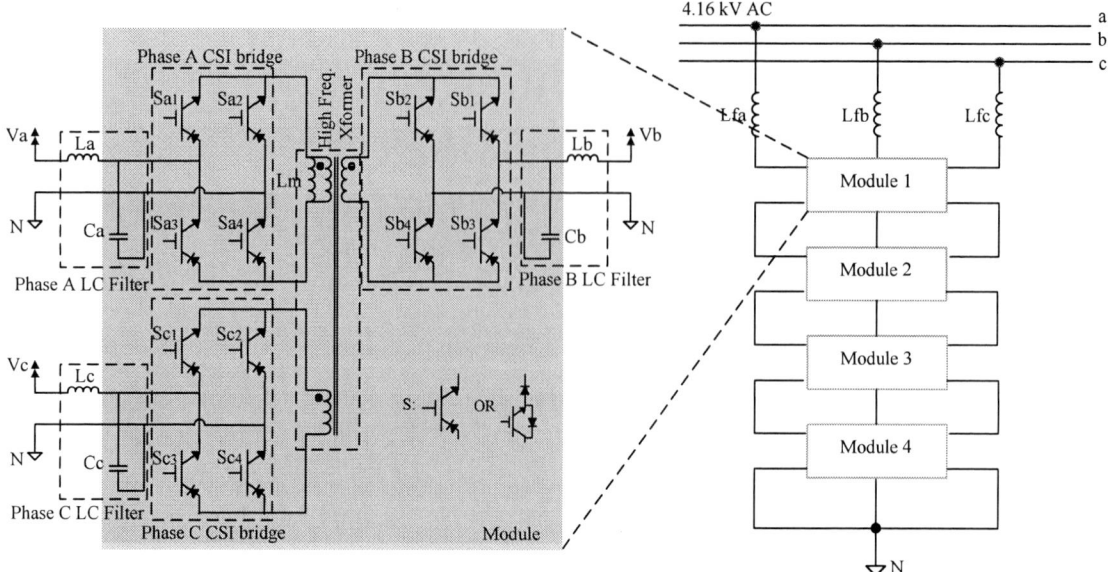

Figure 1. Dynamic VAr compensator (DVC) topologies: a) circuit architecture of one module; and b) multi-level configureation with modules connected in series.

stackable to scale to medium voltages. The transformer serves to isolate a fault on one phase from propagating to the other two phases, especially in medium-voltage application where the fault could quickly bridge the phases and propagate through all the modules, compromising safety and speed of fault isolation. However, in low-voltage applications in which alternate safety mechanisms are provides, the three-winding transformer can be replaced by an inductor with a single winding and a core. The switching devices of the CSI bridges conduct current in one direction but block voltages in both directions. This is typically implemented with either a MOSFET or IGBT in series with a diode, or a revere blocking IGBT (RB-IGBT).

Scaling to medium voltage is achieved through series stacking of the DVC modules as shown in Figure 1(b), in which four modules are series stacked to interface with a 4.16 kV three-phase AC system. The filter inductor, L_a, L_b, L_c, of all four modules can be lumped into one as of L_{fa}, L_{fb}, and L_{fc}. Actually, the series connected modules can have phase staggering on their switching operation such that the current ripple can be significantly reduced, leading to a much smaller filter design. Voltage sharing among the stacked modules is made simpler by series connection of the low-frequency AC terminals. Specific yet dynamic charge balance is maintained on all the series-stacked output filter capacitors at every switching period of the modules to regulate a certain AC voltage at the terminals.

II. PRINCIPLE OF OPERATION

DVC temporarily stores energy in the magnetizing inductance of the transformer. Depending on the instantaneous current and voltage polarities, the transformer is charged and discharged by three-phase voltages over one switching cycle. Volt-sec balance across the transformer is regulated such that a DC magnetizing current is maintained per a reference value. Each phase bridge is sequenced consecutively and a complete

switching period consists of cycling through all three phases. Without considering the transformer leakage effect, the converter operation over one switching cycle consists of four states, as shown in Figure 2. The transformer magnetizing inductance acts as a temporary energy transfer element and a DC constant current is regulated through the inductance.

Excluding the inter-bridge transition interval for leakage management, only one of the three bridges is operating and the active operation is sequenced consecutively through all three bridges within one switching cycle. Assuming an instant when the phase A reference current has the highest positive value, then phase B and C reference currents should be negative for a three-phase balanced system. The converter operation within one switching cycle starts from state 1 by turning on S_{a1} and S_{a4}, the transformer magnetizing current i_m flows through phase A, S_{a1}, L_m, S_{a4} and neutral line N. This state ends when the average current delivered to phase A over one switching cycle equals to its reference. Then, devices in bridge A are turned off and S_{b2} and S_{b3} of bridge B are turned on. The current i_m flows through phase B, S_{b2}, L_m, S_{b3} and N. After referenced current is delivered to phase B, state 3 starts by turning off devices of bridge B and turning on S_{c2} and S_{c3} of bridge C. Then, the current i_m flows through phase C, S_{c2}, L_m, S_{c3} and N. As soon as referenced current is delivered to phase C, the converter goes into state 4, the free-wheeling state, by turning on S_{c2} and S_{c4}. Converter operation starts over from state 1 again when a new cycle begins.

The cycle-by-cycle conceptualized waveforms of the magnetizing current and three-phase unfiltered currents are shown in Figure 3. At this particular instant, the voltage and current of phase A and B have the same polarity while phase C has a different one. As a result, phase A and B tend to charge L_m while phase C tends to discharge it. The voltages applied to L_m are shown as the top waveform of Figure 3. The highlighted areas indicate the volt-sec across L_m, which have

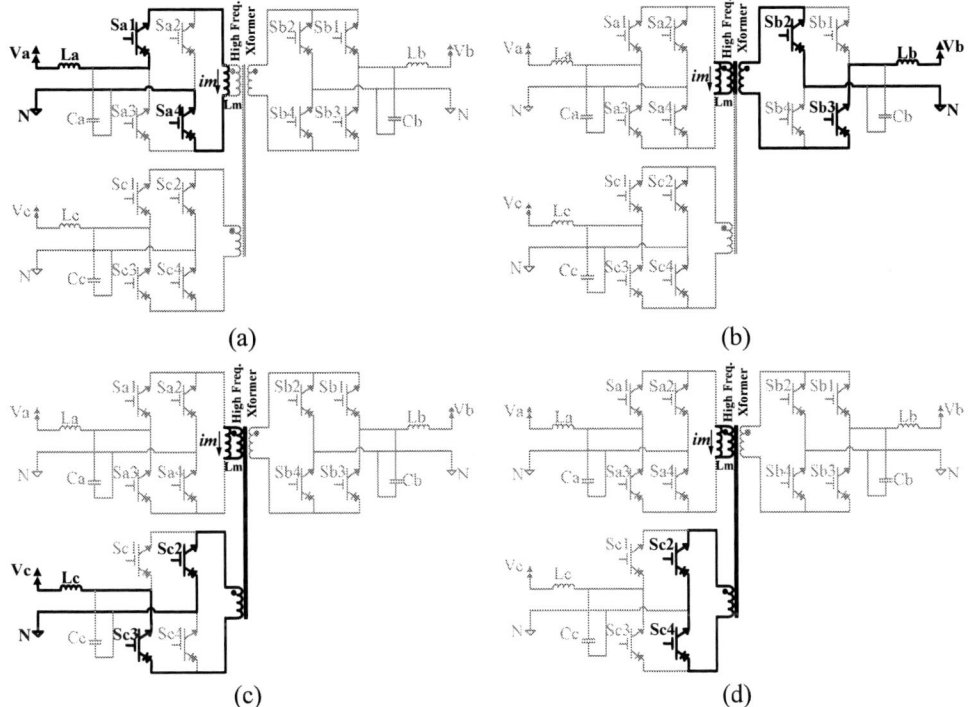

Figure 2. Operating states of DVC over one switching cycle: a) state 1: transformer delivers charge to the 1st L-N voltage; b) state 2: transformer delivers charge to the 2nd L-N voltage; c) state 3: transformer delivers charge to the 3rd L-N voltage; and d) state 4: free-wheeling.

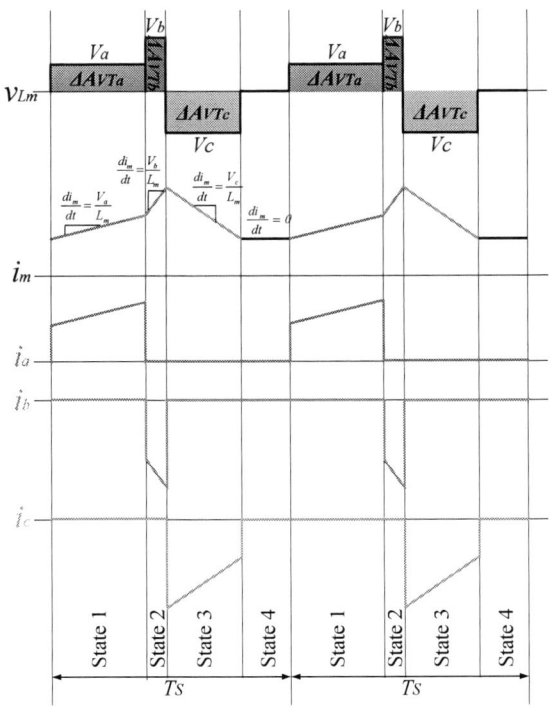

Figure 3. Cycle-by-cycle conceptualized waveforms of magnetizing current and three-phase unfiltered currents.

to be balanced to maintain a constant magnetizing current, as shown in (1) and (2), in which T_{M1}, T_{M2}, and T_{M3} are the time duration for state 1, 2, and 3, respectively.

$$\Delta V_{ATa} + \Delta V_{ATb} + \Delta V_{ATc} = 0 \qquad (1)$$

$$\begin{cases} \Delta V_{ATa} = V_a * T_{M1} \\ \Delta V_{ATb} = V_b * T_{M2} \\ \Delta V_{ATc} = V_c * T_{M3} \end{cases} \qquad (2)$$

III. LEAKAGE MANAGEMENT

The energy trapped in the transformer leakage has to be managed properly to avoid excessive stress over the devices. The transformer trapped leakage energy issue for Dyna-C DVC is similar as for the flyback converter, in which a RCD snubber is commonly used to deal with it. However, for high-power applications, this solution will neither be efficient nor cost effective. Instead, a methodology of managing leakage energy in a loss-less manner is proposed.

In addition to the four operation states shown in Figure 2, the Dyna-C DVC requires another three leakage management states, one for each of the bridge-to-bridge transitions. During the transition state, the devices of both incoming and outgoing bridges are switching to manage the leakage energy. The outgoing bridge, which has an initial winding current of the full magnetizing current, should be driven to zero. Simultaneously, the incoming bridge, which has an initial winding current of zero, should be driven to the full magnetizing current. This is achieved by applying a voltage across the transformer leakage in the direction of the current flow. Figure 4 shows one of the transition state required between state 1 and state 2. In this leakage management state, for the outgoing bridge, bridge A, the devices of S_{a2} and S_{a3} are turned on. While for the incoming bridge, bridge B, the devices of S_{b2} and S_{b3} are turned on. Therefore, the voltage applied to the transformer leakage is equal to $|V_a| + |V_b|$. The simplified equivalent circuit for this state is shown in Figure 5. Another two similar transition states are required between state 2 and state 3, state 4 and state 1 of Figure 2.

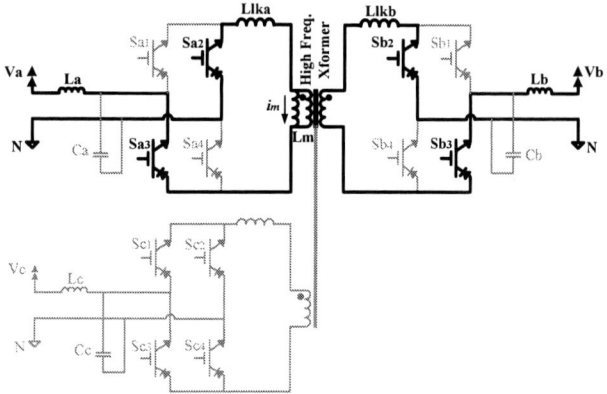

Figure 4.   Leakage management state for bridge A to bridge B transition.

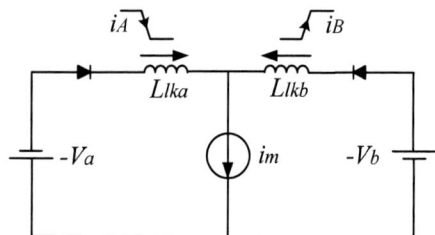

Figure 5.   Simplified equivalent circuit for the leakage management state.

IV.   CONTROL OF DYNA-C DVC

The Dyna-C DVC temporarily stores energy in the transformer. To maintain a compact and low cost converter design, the transformer magnetizing inductance, which carries a constant DC current, is designed to have a low inductance value. However, this leads to a large peak-to-peak current ripple on top of the DC component. Operating under large current ripple while still maintaining high quality injections with low THD makes the Dyna-C DVC control much more challenging.

To achieve high quality injections with a high DC current ripple, a charge-control based modulation scheme, initially introduced to control rectifiers [15], is selected for the DVC in this paper in which the charge balance across the transformer is satisfied within every switching cycle. Control of the Dyna-C DVC consists of two loops, as shown in Figure 6, in which the outer loop generates three-phase current references and the inner loop controls device duty cycles based on the charge modulation. When synchronized with the three-phase line voltage, the D component of the grid current is used to regulate the DC magnetizing current. The feedback loop generates I_D^* from the magnetizing current error Δi_m, and the compensator G_M, which can be a PI controller, is used to compensate the converter loss such that i_m is regulated at its reference I_m^*. The Q component I_Q^* is set based on the

required reactive power Q^*, which satisfies (5). The three-phase reference currents can be obtained through the DQ-ABC transformation. Figure 7 shows the inner control loop. The reference charge for each phase is calculated as the reference current multiplying with the switching period T_S, as shown in (3). The actual instantaneous charge, calculated as integration of magnetizing current, shown in (4), is continuously compared with the reference charge and it determines the duty cycle of each phase bridge. The bridge operation is sequenced consecutively such that a complete switching period consists of cycling through all three phases.

$$Q_{a,b,c}^* = I_{a,b,c}^* * T_S \qquad (3)$$

$$Q_m = \int_{\tau=t} i_m(\tau)d\tau \qquad (4)$$

$$Q^* = I_Q^* * V_D \qquad (5)$$

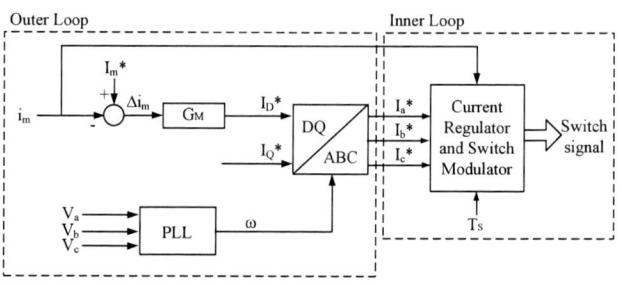

Figure 6.   Control architecture for the Dyna-C DVC.

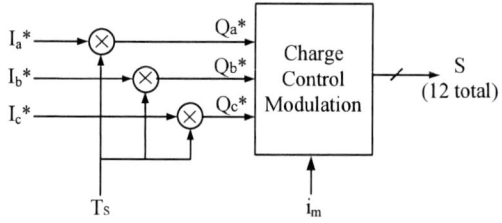

Figure 7.   Inner control loop.

To avoid current flow path interruption, there should be an additional overlap state between any of the switch transitions, in which the incoming-on devices and outgoing-off devices are both kept on. This state for the current source converter is in dual with the dead time for the voltage source converter. The free-wheeling state can be accomplished in any of the three phase legs. However, from the thermal balance point view, to distribute losses equally, the free-wheeling state should be performed on the leg that has the smallest duty cycle.

For the multi-level Dyna-C DVC, the charge-control strategy cannot be directly applied. The voltage balance among all series connected modules and the magnetizing current regulation for each module are critical for stable operation. Therefore, instead of using the charge-control modulation to directly regulate the grid current, the converter is operated with a dead-beat control strategy, in which the filter capacitor voltage of each module is tightly regulated to achieve voltage balance. The compensation of the magnetizing current is performed using the energy from filter capacitors on a switching cycle basis. In addition, the operation of all series connected modules can be in a phase-staggering manner such that the line current ripple is significantly reduced. With an *N*-level configuration, the phase delay of every switching cycle between two modules should be *2π/N* rad. In this way, the system equivalent switching frequency is *N* times of the actual switching frequency of each module. The detailed control strategy for the multi-level Dyna-C DVC was discussed in [16].

V. SIMULATION AND EXPERIMENTAL RESULTS

The Dyna-C DVC topology is simulated at 480 V, 75 kVar. Table II lists parameters of the converter to be simulated and fabricated. Figure 8 shows simulated waveforms of the three-phase grid voltage, the DVC injected current, and the transformer magnetizing current of the DVC. In spite of the high ripple presented in the transformer, grid side currents are still regulated with minimal harmonic distortion. Figure 9 presents cycle-by-cycle current waveforms of bridge A, B, and C, which shows that the leakage management strategy drives the outgoing bridge current to zero while simultaneously builds up the incoming bridge current.

Table II. Parameters of the Dyna-C DVC experimental prototype.

Power rating	Input/output voltage	Switching frequency	Inductor/ Magnetizing inductance
75 kVA	480 V	15 kHz	150 μH

Leakage inductance	Filter capacitor	Filter inductor
1.5 μH	40 μF	150 μH

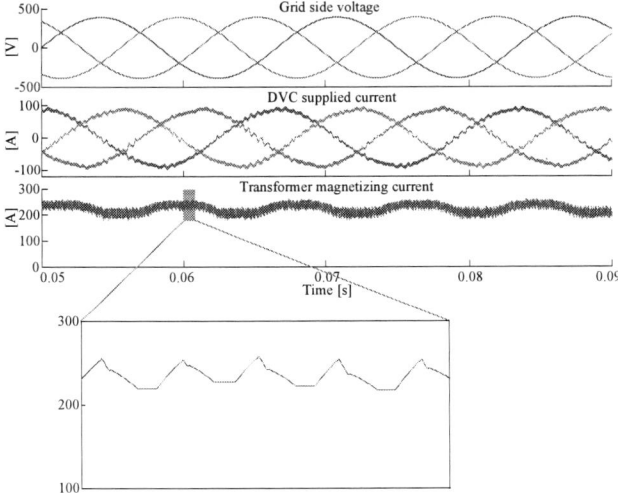

Figure 8. Simulated waveforms of the Dyna-C DVC.

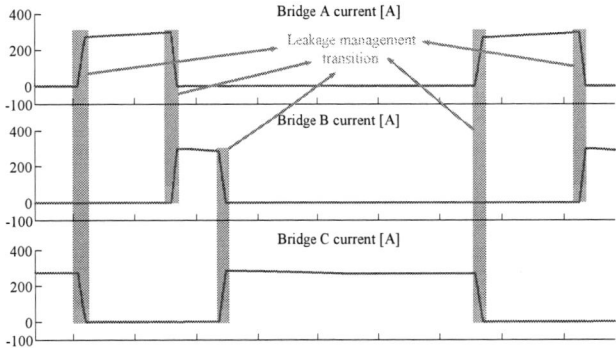

Figure 9. cycle-by-cycle current waveforms of bridge A, B, and C.

To experimentally verify the Dyna-C DVC functionality, a 480 V, 75 kVA Dyna-C DVC prototype is assembled in the lab, which is shown in Figure 10. The converter is built with 1200 V, 400 A IGBTs and diodes from Powerex. All the devices are connected with copper bus-planes, and filter capacitors are vertically integrated to each bridge module such that the parasitic inductances are minimized.

For this experiment, the Dyna-C DVC is connected to a three-phase 0~480 V AC source with a neutral line. The converter is controlled to inject reactive power that is specified by the controller. Figure 11 shows experimental waveforms of three-phase injected currents and phase A voltage. The voltage is 380 V and the current is 63 A, which correspond to a reactive power of 41 kVA. The phase of the current is controlled to lead the voltage by 90° such as to inject leading/capacitive VArs. A current waveform of one of the transformer windings over one switching period is shown in Figure 12. It can be seen that the transformer winding current is slowly driven up and down at the inter-bridge switching transitions, indicating an effective leakage management strategy that is proposed in this paper.

Figure 10. A 480 V/75 kVar prototype of isolated Dyna-C DVC.

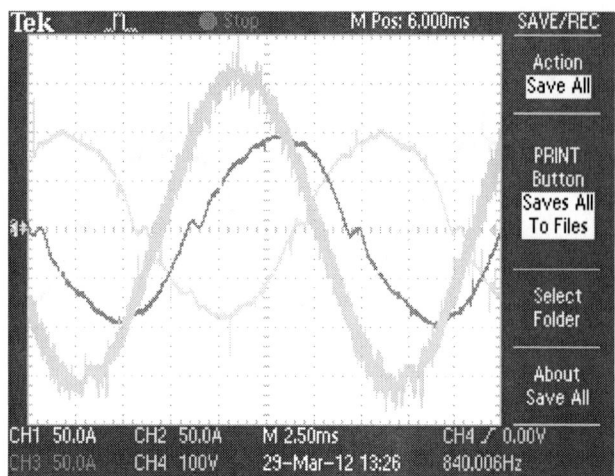

Figure 11. Experimental waveforms at 380 V, 63 A (three-phase line currents of CH1: phase A, CH2: phase B, CH3: Phase C, and CH4: phase A voltage).

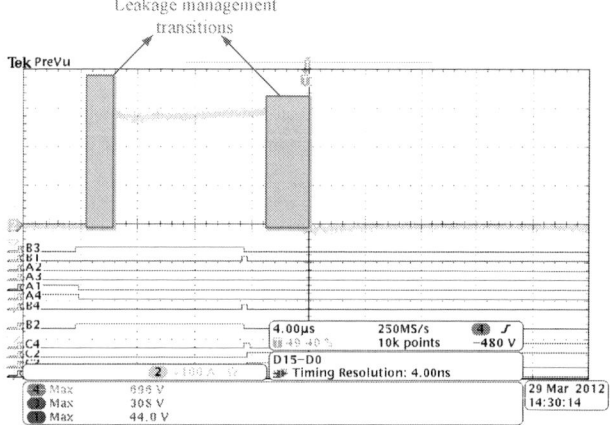

Figure 12. Cycle-by-cycle bridge current.

VI. CONCLUTIONS

This paper has presented a novel topology for reactive power compensation that can be scaled to medium voltage with simple series stacking of its AC terminals. Compared to the STATCOM which requires an additional outer control loop to control the line current, the proposed DVC provides much faster dynamic response since it directly regulates the line current on a cycle-to-cycle basis. The topology has a high-frequency transformer to provide DC storage as well as inter-phase fault isolation among the three-phase. The inherent nature of the topology limits the maximum fault current, contributing to qualities of long life, features that are important in utility-grade equipment. Cycle-by-cycle charge modulation control is adopted for the topology such that a compact converter design can be achieved. In addition, a key loss-less leakage management strategy is proposed for the bridge transitions. Finally, the modularity feature allows the converter to provide load compensation and voltage regulation for the distribution and transmission systems with a fractionally rated converter module design, retaining fast response for the medium-voltage system. The converter

functionality is validated through simulations as well as experimental results with a 75 kVAr prototype.

REFERENCES

[1] L. Gyugyi, and E.R. Taylor, "Characteristics of static, thyristor-controlled shunt compensators for power transmission system applications," *IEEE Transactions on Power Apparatus and Systems*, vol. PAS-99, no.5, pp. 1795-1804, Sep. 1980.

[2] L. Gyugyi, "Reactive power generation and control by thyristor circuits," *IEEE Transactions on Industry Applications*, vol. IA-15, no.5, pp. 521-532, Sep. 1979.

[3] N.G. Hingorani, "High power electronics and flexible AC transmission system," *IEEE Power Engineering Review*, vol. 8, no.7, pp. 3-4, Jul. 1988.

[4] Y. Sumi, Y. Harumoto, T. Hasegawa, M. Yano, K. Ikeda, and T. Matsuura, "New static var control using force-commutated inverters," *IEEE Transactions on Power Apparatus and Systems*, vol. PAS-100, no.9, pp. 4216-4224, Sep. 1981.

[5] J. Dixon, L. Moran, J. Rodriguez, and R. Domke, "Reactive power compensation technologies: state-of-the-art review," *Proceedings of the IEEE*, vol. 93, no.12, pp. 2144-2164, Dec. 2005.

[6] S. Mori, K. Matsuno, T. Hasegawa, S. Ohnishi, M. Takeda, M. Seto, S. Murakami, and F. Ishiguro, "Development of a large static VAR generator using self-commutated inverters for improving power system stability," *IEEE Transactions on Power Systems*, vol. 8, no.1, pp. 371-377, Feb. 1993.

[7] H. Akagi, Y. Kanazawa, and A. Nabae, "Instantaneous reactive power compensators comprising switching devices without energy storage components," *Proceedings of the IEEE*, vol. 93, no.12, pp. 2144-2164, Dec. 2005.

[8] N.S. Choi, G.C. Cho, and G.H. Cho, "Modeling and analysis of a static VAR compensator using multilevel voltage source inverter," in *IEEE Industry Applications Society Annual Meeting*, 1993, pp. 901-908.

[9] F.Z. Peng, J. Lai, J.W. McKeever, and J. VanCoevering, "A multilevel voltage-source inverter with separate DC sources for static VAR generation," *IEEE Transactions on Industry Applications*, vol. 32, no. 5, pp. 1130-1138, Sep. 1996.

[10] K. Fujii, U. Schwarzer, and R.W. DeDoncker, "Comparison of hard-switched multi-level inverter topologies for STATCOM by loss-implemented simulation and cost estimation," in *IEEE Power Electronics Specialists Conference*, 2005, pp. 340-346.

[11] K. Ilves, A. Antonopoulos, S. Norrga, and H.P. Nee, "Steady-state analysis of interaction between harmonic components of arm and line quantities of modular multilevel converters," *IEEE Transactions on Power Electronics*, vol. 27, no. 1, pp. 57-68, Jan . 2011.

[12] S. Shao, P.W. Wheeler, J.C. Clare, and A.J. Watson, "Fault detection for modular multilevel converters based on sliding mode observer," *IEEE Transactions on Power Electronics*, vol. 28, no. 11, pp. 4867-4872, Jan. 2013.

[13] D. Divan, A. Prasai, and H. Chen, "Isolated dynamic current converters," U. S. Patent US 9,065,321 B2, Aug. 8, 2013.

[14] A. Prasai, H. Chen, and D. Divan, "Dyna-C: A topology for a bi-directional solid-state transformer," in *IEEE Applied Power Electronics Conference and Exposition*, 2014, pp. 1219-1226.

[15] K. Wang. D. Boroyevich, and F.C. Lee, "Charge control of three-phase buck PWM rectifiers," in *IEEE Applied Power Electronics Conference and Exposition*, 2000, pp. 824-831.

[16] H. Chen, A. Prasai, and D. Divan, "Stacked modular isolated dynamic current source converters for medium voltage applications," in *IEEE Applied Power Electronics Conference and Exposition*, 2014, pp. 2278-2285.

Design and Control of a Compact MMC Submodule Structure with Reduced Capacitor Size Using the Stacked Switched Capacitor Architecture

Yuan Tang[†], Minjie Chen[‡] and Li Ran[†]

[†]the University of Warwick, Coventry, UK, CV4 7AL
[‡]Massachusetts Institute of Technology, Cambridge, MA 02139 USA
yuan.tang@warwick.ac.uk, minjie@mit.edu, l.ran@warwick.ac.uk

Abstract— **Modular multilevel converters (MMCs) are being developed for the grid connection of offshore wind or tidal farms. In order to reduce the construction and maintenance costs, it is desirable to reduce the weight and volume of the MMC systems. In most existing MMC submodule designs, the reservoir capacitor usually accounts for over 50% of the total volume and 80% of the total weight. This paper presents a new circuit topology and control principle for a submodule based on a stacked switched capacitor (SSC) architecture that can significantly reduce the capacitor size in an MMC. Practical considerations for a high-voltage high-power system implementation are presented in this paper through the design and simulation of a 21-level, 40 kV (pole-pole dc), 19.1 MW, grid-connected MMC system. The proposed MMC submodule design is further verified experimentally on a scaled down 400 V, 12.3 A$_{peak}$ laboratory prototype. It is shown that with the proposed SSC-architecture, the total volume of capacitors in each submodule can be reduced by more than 40% without significantly increasing power losses.**

Keywords—modular multilevel converter; submodule; switched capacitor circuits; ac-dc power conversion; energy storage

I. INTRODUCTION

High voltage dc (HVDC) with voltage source converters (VSCs), especially modular multilevel converters (MMCs) is being actively developed for offshore wind or tidal power collection and onwards transmission. Given the difficulties of constructing and maintaining an infrastructure in a hostile environment, it is important to keep the size as small as possible. In many of the existing MMC submodule (SM) designs for 50 or 60 Hz ac grid-interfaced systems, the reservoir capacitor needs to absorb low order harmonics and hence accounts for over 50% of the total size and 80% of weight of each SM. The SM capacitor needs to be large enough (capacitance) to constrain the voltage ripple, while having sufficient ripple current capability to avoid overheating. Metalized polypropylene film (MPPF) capacitors are commonly used in MMC SMs due to their stability and ripple current capability. It is found in [1] that when the capacitor is selected by the capacitance to satisfy the voltage ripple requirement, the ripple current capability of the MPPF capacitor is twice or three times higher than needed. As a result, there is an opportunity to improve the utilization of MPPF capacitor and reduce the capacitor size in the MMC SM.

Increasing the switching frequency to reduce the capacitor size is not effective in MMC SMs because a majority of the energy that needs to be buffered by the reservoir capacitor is caused by the imbalanced power between the ac and dc sides at low frequencies. To address the problem, various methods are proposed to reduce the capacitor voltage ripple, including harmonic current injection [2-6], high frequency injection [7], novel modulation technique [8,9], SM circuit and converter topology innovations [10-13].

This paper proposes a new SM structure utilizing the recently proposed stacked-switched capacitor (SSC) energy buffer architecture, which has been demonstrated to be effective in low-voltage, low-power dc-ac applications for energy buffer capacitor size reduction [14-16]. In an SSC based SM (SSC-SM), the SM dc bus voltage is synthesized by combining several series-connected capacitors. The overall volume of all capacitors is much smaller than the original capacitor. The voltage of each capacitor can fluctuate in a wider range towards specification limits. The increased voltage ripples in different capacitors are compensated through appropriate switching strategy, to control the SM dc bus voltage ripple. It is shown that with the SSC technique, the energy density of an SSC-SM can be almost doubled with only 7% higher power loss at the rated power. And under the normal operating point (when the loading power is substantially lower than the rated power), the SSC-SM behaves like a conventional HB-SM and negligible additional loss is introduced. As the SSC energy buffer functions the same as the original single capacitor seen from two input/output ports, on the system level the operation of the equivalent capacitor is indifference from the original capacitor but with a few higher order voltage ripples, which has negligible effect on system operation due to the operating mechanism of MMC (the SM ripple voltage is to be filtered with proper control). This topological innovation opens a new degree of freedom for reducing the capacitor, as a result, other capacitor reduction methods, which reduce the capacitor size by using other mechanisms, such as those proposed in [2-13] can be combined with the SSC-SM based MMC to enable better capacitor reduction performance.

The rest of the paper is organized as follows. The structure of the SSC based SM is described in Section II. Section III discusses the SM-level and system-level control. The design of a 21-level, 40 kV (pole-pole dc), 19.1 MW, grid-connected

978-1-4673-9551-9/16 $31.00 © 2016 IEEE

MMC system with SSC-SMs is presented in Section IV together with simulation results. The capacity of the system is then scaled down to 21-level, 8 kV (pole-pole dc), 92 kW for experimental verification. A 400 V, 12.3 A prototype SSC-SM is built and tested. Section V concludes the paper.

II. SSC BASED MMC SM

Fig.1 shows the proposed SSC-SM against the original half-bridge SM. In the SSC-SM, the lower switch S_L stays unchanged. S_{U1}, S_{U2_1} together with S_{U3_1} can be regarded as replacing the original upper switch S_U. The rest of the circuit is developed based on the *1-2 enhanced unipolar* SSC energy buffer presented in [16]. There are one backbone capacitor (C_0), two supporting capacitors (C_1 and C_2) and a capacitor-free branch (S_{U3_1} and S_{U3_2}). In the rest of the paper, the *SM dc bus voltage* (V_{bus}) refers to the total voltage of the backbone capacitor C_0 and the active supporting capacitor (C_1 or C_2 or none) that may be switched in. The two additional switches (S_{U2_2} and S_{U3_2}) are connected with S_{U2_1} and S_{U3_1} in a back-to-back manner to prevent parallel connection of C_1, C_2 and the capacitor-free branch. As will be discussed later, the switches only slightly increase power losses and some of them are of low voltage ratings. Through advanced packaging, it is possible to integrate all switches into a single customized module. With continuously improvement of semiconductor technologies, the volume and weight would be comparable with the HB power module [17-19]. Hence, the increased semiconductor chips are expected to be acceptable.

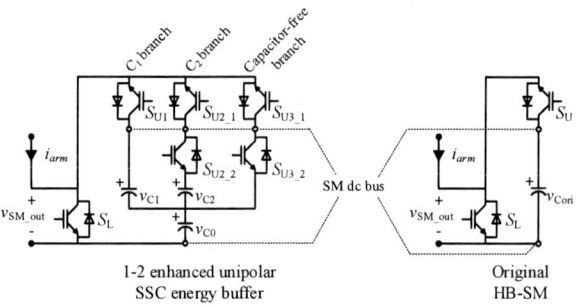

Fig. 1. Circuit diagrams of both SSC- and HB-SM

The backbone capacitor and the two supporting capacitors are chosen to have the same capacitance, but different voltage ratings. The capacitance of C_0, C_1 and C_2 are all selected to be half of C_{ori}. With identical charging and discharging currents as well as the same permitted SM dc bus voltage ripple, the voltage ripple on C_0 will be twice of that on C_{ori}. In general, C_0, C_1 and C_2 can be designed with different capacitance values. Ni *et al.* [20] offered a way to optimize the capacitance values for the enhanced unipolar SSC energy buffer aiming at further reducing the total capacitor size. When the allowed bus voltage ripple is small (as in our design, 15%), using equal capacitance and optimal capacitance leads to similar capacitor size reduction. As a result, all capacitance values in our prototype are selected to be equal.

Since the maximum voltage on C_0 is the same as C_{ori} when it is switched in alone, the rated voltage of C_0 is selected to be

the same as C_{ori}. Both C_1 and C_2 are used to compensate the voltage difference between the permitted SM dc bus voltage and the actual voltage of C_0. C_1 and C_2 have different voltage ratings. When the voltage difference is large, C_1 with higher voltage is switched in. When the voltage difference is small, C_2 with lower voltage will be switched in. In both cases, the rated voltages of C_1 and C_2 are defined by the maximum voltages they may experience depending on the permitted voltage ripple of the SM dc bus. The larger is the permitted SM dc bus voltage ripple, the higher rated voltages of C_1 and C_2 [14]. In the following analysis, the SM dc voltage ripple is assumed to be symmetrical about the average, and the p.u. value, $V_{ripplep.u}$ (p-p), is based on the SM rated dc voltage V_{SM_dc} (the MMC rated dc link voltage V_{dc} divided by the total number of SMs in one arm N). The rated voltages of C_0, C_1 and C_2 are chosen to be $(1+\frac{1}{2}V_{ripplep.u})V_{SM\_dc}$, $3/2V_{ripplep.u} \cdot V_{SM\_dc}$ and $V_{ripplep.u} V_{SM\_dc}$ respectively as to be detailed later in Fig. 2.

III. CONTROL AND OPERATION PRINCIPLES

A. SM-level control

Fig.2 sketches the waveforms of an SSC-SM at rated power when the SM is in the arm circuit. The current through the SSC energy buffer equals the arm current $i_{arm}(t)$. For simplicity, charging and discharging currents are both assumed to be constant with identical amplitudes. Hence, in each fundamental cycle there are symmetrical charging and discharging processes. Switching strategies for both processes can be divided into three phases similar to [14,16]:

Phase 1 – charging C_0 and C_1: when the energy buffer is to be charged, C_1 branch is activated (and all the other supporting branches are OFF). Detailed switching states are shown in Table I. C_0 and C_1 are charged in series, and C_2 is floating. When the SM dc bus voltage, V_{bus}, reaches its upper limit V_{SM_dcmax}, i.e. $(1+V_{ripplep.u}/2)V_{SM\_dc}$, v_{C1} becomes $(3/2)V_{ripplep.u} V_{SM\_dc}$. Then C_1 branch is turned OFF and C_2 branch is turned ON. v_{C2} adds to v_{C0} and the SM dc bus voltage drops back to the lower limit V_{SM_dcmin}, i.e. $[(1-V_{ripplep.u}/2)V_{SM\_dc}]$.

Phase 2 – charging C_0 and C_2: C_2 branch is ON and is charged in series with C_0 until the SM dc bus voltage reaches V_{SM_dcmax} again. At this point v_{C0} is $(1-V_{ripplep.u}/2)V_{SM\_dc}$, and v_{C2} increases to $V_{ripplep.u} V_{SM\_dc}$. Then C_2 branch is turned OFF, and the capacitor- free branch is turned ON. The SM dc bus voltage equals the voltage of C_0.

Phase 3 – charging C_0 only: during this period, only the capacitor-free branch is ON. C_0 is charged until the SM dc bus voltage reaches V_{SM_dcmax}.

At this time, all capacitors have reached their maximum and the discharging process will begin which is the reverse of the charging process. Fig.3 sketches operation waveforms when the SM capacitor is included or bypassed in the arm. The current passing through the SSC energy buffer (solid line) equals the arm current (dashed line) when the SM is switched in, which may contain a dc offset corresponding to the active power exchange between the ac and dc sides of the converter. The 2nd order harmonic component is usually eliminated to avoid the extra losses [21,22]. Hence, only the fundamental and the dc component are included in the arm current waveform.

978-1-4673-9551-9/16 $31.00 © 2016 IEEE

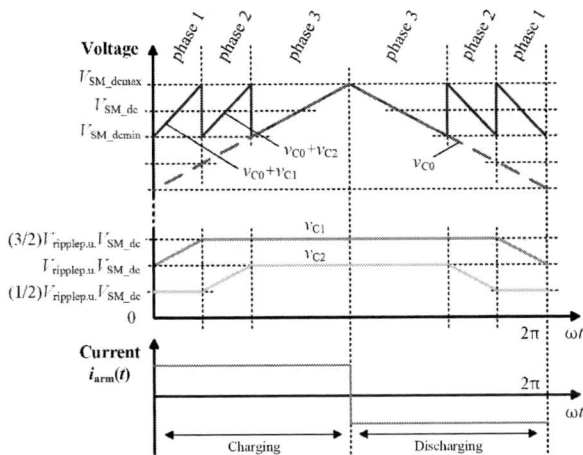

Fig. 2. Operating waveforms of a *1-2 enhanced unipolar* energy buffer based SSC-SM operates at rated power when the SM is always switched into the arm circuit with constant charging and discharging currents

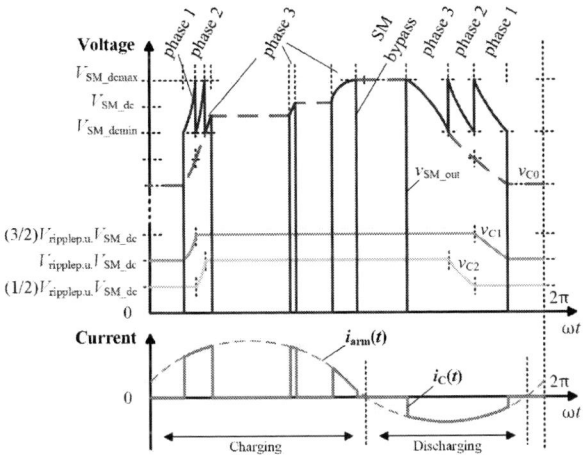

Fig. 3. Operating waveforms of a *1-2 enhanced unipolar* energy buffer based SSC-SM operates at rated power when the SM is being switched as installed in an MMC system with actual sinusoidal arm current containing a dc offset

Referring to Fig.3, the switching strategy between different supporting branches is the same as in Fig.2 where the next supporting branch will be turned ON every time when the upper limit of the SM dc bus voltage is met in the charging process or when the lower limit is met in the discharging process. The total voltage of the currently active capacitor(s), i.e. $(v_{C1}+v_{C0})$, $(v_{C2}+v_{C0})$ or v_{C0}, is used for control. The saw-tooth like waveform in Fig.3 sketches the SM output voltage v_{SM_out} that would drop to near 0 when the SM is bypassed, during which the current passing through the energy buffer is zero and all capacitor voltages remain constant. The SM-bypass mode is special in MMC applications. Although the SM may be bypassed for arbitrary periods of time in a line frequency cycle, the overall energy charging and discharging the capacitor is balanced by the MMC system-level controller.

As a result, the control strategy of the SSC-SM is still effective for all conditions because the capacitors are charged and discharged periodically.

An advantage of the SSC-SM approach is that the controller of the SSC energy buffer is embedded in each SM with minimum impact on the system-level MMC control. Each SSC-SM can be treated as an equivalent single capacitor with a certain level of charge from the viewpoint of the system-level (MMC) centralized controller. For each SM, the signal sent to the central controller is only the voltage of the backbone capacitor or a combined voltage signal indicating the state of charge of the energy buffer in order to implement the capacitor voltage balancing control at the system-level. The signal received from the centralized controller is still the SM switching signal indicating whether it is switched in or bypassed. Detailed gate switching signals are generated locally. For each SSC energy buffer controller at the SM-level, the required measurements are the voltages of the three capacitors and the arm current. Usually the current sensor for i_{arm} is already installed for the SM capacitor balancing control. Current-sensorless control is also possible [14]. As described in [14], all the capacitor charge information can be extracted from the SM dc bus voltage. Thus, only one voltage sensor would be necessary, which is the same as the HB-SM. Moreover, if the current sensor is already available, all capacitor voltage information can be estimated through the current measurement and switching signals. Voltage sensorless control for the HB-SM based MMC can be applied as well [23]. In the prototype, three voltage sensors are used for measurement in order to enhance design flexibility. In addition, the localized controller only needs to observe the SM dc bus voltage and ensure that the voltages of C_1 and C_2 will never exceed their limits. The local control within each SM can be implemented with a few discrete analog comparators, or a simple customized low-cost integrated circuit, as presented in [16]. All the control circuits and additional drivers can be powered by the backbone capacitor C_0 with no additional complexity to the overall architecture.

s1 to s6 in Table I are the gating signals for the switches. In normal operation when the SM is switched into the arm circuit, the switching strategies are shown in the third column marked as s1, s2 and s3 for phases 1, 2 and 3 respectively. In addition, the switching strategies to bypass the SM are listed in the second column marked as s4, s5 and s6 for the three operating phases. In additional to the forth column of Table I, Fig.4 sketches the switching strategies especially during dead-bands. In high-voltage high-power applications, a dead-band of a few microseconds is usually added before the turn-ON command in order to avoid shorting the dc capacitor.

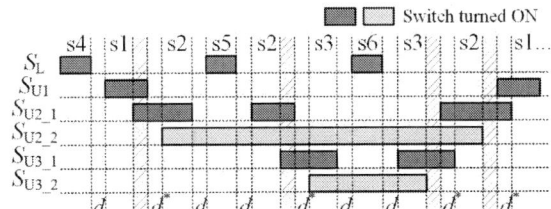

Fig. 4. Dead-band strategies for power switches.

978-1-4673-9551-9/16 $31.00 © 2016 IEEE   1445

TABLE I. SWITCHING STRATEGY AND SM OUTPUT VOLTAGE

States	SM bypassed						SM switched-in						Dead-band					
	S_L	S_{U1}	S_{U2_1}	S_{U2_2}	S_{U3_1}	S_{U3_2}	S_L	S_{U1}	S_{U2_1}	S_{U2_2}	S_{U3_1}	S_{U3_2}	S_L	S_{U1}	S_{U2_1}	S_{U2_2}	S_{U3_1}	S_{U3_2}
phase 1	1	0	0	0	0	0	0	1	0	0	0	0	0	0	0	0	0	0
$i_{arm} \geq 0$	$v_{SM\_out} = 0$ (s4)						$v_{SM\_out} = v_{C1} + v_{C0}$ (s1)						$v_{SM\_out} = v_{C1} + v_{C0}$					
$i_{arm} < 0$													$v_{SM\_out} = 0$					
phase 2	1	0	0	1	0	0	0	0	1	1	0	0	0	0	0	1	0	0
$i_{arm} \geq 0$	$v_{SM\_out} = 0$ (s5)						$v_{SM\_out} = v_{C2} + v_{C0}$ (s2)						$v_{SM\_out} = v_{C2} + v_{C0}$					
$i_{arm} < 0$													$v_{SM\_out} = 0$					
phase 3	1	0	0	1	0	1	0	0	0	1	1	1	0	0	0	1	0	1
$i_{arm} \geq 0$	$v_{SM\_out} = 0$ (s6)						$v_{SM\_out} = v_{C0}$ (s3)						$v_{SM\_out} = v_{C0}$					
$i_{arm} < 0$													$v_{SM\_out} = 0$					

d represents the dead-bands between the lower switch S_L and the upper main switches – S_{U1}, S_{U2_1} and S_{U3_1} – when the SM is switched in or out of the arm circuit (only the capacitor charging process is shown). When the SSC-SM is bypassed (s5 and s6 in Table I), other than S_L, S_{U2_2} is still ON in s5 and also both S_{U2_2} and S_{U3_2} are ON in s6. A similar case can be found in s3. The reason for this action is to ensure that the SM output voltage is always within the range $[V_{SM\_dcmin}, V_{SM\_dcmax}]$ during dead-bands d, especially when the current flows into the SM. As a result, the voltage that needs to be blocked by S_L would never exceed V_{SM_dcmax}, and thus no higher voltage blocking capability is required for S_L in the SSC-SM.

Dead-bands marked as $d*$ are added between the switching of supporting branches to prevent discharging the supporting capacitors. In addition, a short overlap time (a few micro-seconds, the hatched area in Fig.4) is added to the two upper main switches being switched to reduce their switching losses. In such a case, zero-voltage-switching (ZVS) or near ZVS is achieved. Since the switching losses of IGBT/diode modules are proportional to the blocking voltage [24], the switching losses are kept low. More specifically, in the discharging process, or when i_{arm} is negative, the voltage on the switch after being turned OFF is near zero, and the voltage on the switch to be switched ON is either $(v_{C1}-v_{C2})$ or v_{C2}. In the charging process, or when i_{arm} is positive, the voltage on the switch to be switched ON is always near zero, and the voltage on the switch after being turned OFF is either $(v_{C1}-v_{C2})$ or v_{C2}. The switching losses of the back-to-back connected switches S_{U2_2} and S_{U3_2} are always low due to the very low voltages they block.

B. System-level control

The system-level control of the SSC-SM based MMC is similar to the one with HB-SM [21,25]. The outer power controller converts the power references to current references (I_{d_ref} and I_{q_ref}) for the inner controller. The inner loop is based on the widely used d-q reference frame [25,26] that generates the required converter output voltages. In the SSC-SM based MMC, due to switching the supporting branches, additional harmonics, mainly fifth order, are found in the converter ac outputs. Thus, an additional closed-loop PI controller based on a d-q reference frame operating at five times of the line-frequency is added to suppress the fifth harmonic. Current signals share the ones used for the output current control. Reference values of the fifth-order PI controller are set to zero. Fourth order harmonic is also found in the circulating currents between converter arms. In addition to the circulating current suppressing controller (CCSC) working at twice line-frequency [21], another CCSC operating at four times of the line-frequency is added to eliminate the fourth harmonic.

Control signals generated by the inner current controllers and CCSCs are used to determine the required arm voltages. The acquired signals are then sent to the modulation [21] to decide the number of SMs to be switched-in for each arm. The SM capacitor voltage balancing control will then help to find the exact SMs. The signals sent to the voltage balancing controller can be only the backbone capacitor voltages, yielding a fully independent SM-level control. The supporting capacitor voltages are controlled and balanced locally with the SM-level controller embedded in each SM.

IV. SIMULATION AND EXPERIMENTAL VALIDATION OF THE SSC-SM BASED MMC

A. System Simulation

A grid-connected SSC-SM based MMC is modelled in Matlab/Simulink to assist the design of a SM. Effectiveness and validity of the system-level control, including the output current control, inner circulating current control and SM capacitor voltage balancing control are verified through computer simulation. Performance of the SSC-SM based system is evaluated in both steady state and transients. Power losses are compared with a HB-SM design. Finally, a scaled down prototype SSC-SM is built and tested with a model assisted single SM testing platform to verify the design of the localized controller as well as the low-voltage switching strategy.

Table II summarizes the system parameters. The converter is connected to a ±20 kV dc-link and a 23 kV (line-to-line rms), 50 Hz grid bus-bar through a three-phase transformer. The arm inductor is 0.2p.u. All p.u. values are based on the rated ac side voltage and current. There are 20 SMs in each arm with the SM rated dc voltage $V_{SM\_dc}=2000$ V. The converter is designed to invert or rectify 19.1 MW active power with a modulation index m of 0.9. The ac side power is measured at the converter output to exclude the coupling transformer effect. With the allowable 0.15p.u. p-p voltage ripple at rated power ($V_{ripplep.u.}$), the minimum required capacitance for one HB-SM is derived to be 2.6 mF [1] with 2150 V_{dc} rated voltage. For the SSC-SM, the capacitance of backbone or a supporting capacitor (C_0, C_1 or C_2) is half of this i.e. 1.3 mF.

TABLE II.	SYSTEM PARAMETERS FOR SIMULATION
Item	Parameters
dc-link voltage	40 kV (pole-pole)
Grid voltage	23 kV line-to-line (rms), 50Hz
$V_{nominal}$ at output point	18 kV phase to neutral
Rated line current	500 A_{rms}
Rated capacity	19.1 MW (PF = 1)
Transformer inductance	0.1p.u.
Number of submodules	20 per arm
Arm inductance	0.2p.u.
SM rated dc bus voltage	2000 V
$V_{ripple p.u.}$	0.15 p-p

Five 520 μF, 2200 V_{dc} capacitors from IXYS each with 120 A_{rms} ripple current capability [27] are selected for an HB-SM while two and half (assuming possible in a customized design in the future) of the same are chosen as the backbone capacitor in an SSC-SM. Both supporting capacitors use the 1000 μF, 450 V_{dc} capacitor from TDK each with 135 A_{rms} ripple current capability [28]. In order to offer the 1.3 mF capacitance, parameters including capacitance, volume and ripple current capability are scaled. The total volume of all capacitors in the HB-SM is 21.42 liters while that in the SSC-SM is 12.93 liters that is only 60% of the HB-SM. The rated energy stored in the HB-SM based MMC is derived to be 37.75 kJ/MW, while that in the SSC-SM based system is only 20.07 kJ/MW, which is 53% of the original.

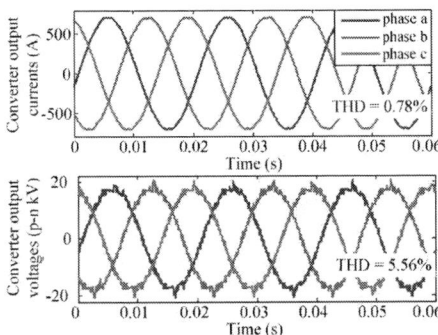

Fig. 5. Converter output ac current (top) and voltage (bottom) waveforms of SSC-SM based MMC in normal operation.

Fig.5 gives the converter ac output currents and voltages of the SSC-SM based MMC when inverting 19.1 MW active power. Harmonic analysis shows a slight increase of THD compared with the HB-SM based converter outputs. The THD of output current and voltage are increased from 0.6% and 4.28% to 0.78% and 5.56% respectively. Additional harmonics are caused by switching the supporting branches. Advanced MMC control strategy can be implemented to cancel the voltage ripple of different SSC-SMs if necessary.

Fig.6 shows the voltages and currents of the upper arm of phase a in the SSC-SM based MMC. Fig.6(a) is a SM backbone capacitor voltage. With half the capacitance of a HB-SM, the voltage ripple (approx. 600 V) is twice of the permitted voltage ripple of SM dc bus (300 V or 0.15 p.u. p-p). Through switching the supporting branches, the SM dc bus voltage is maintained within the allowable range.

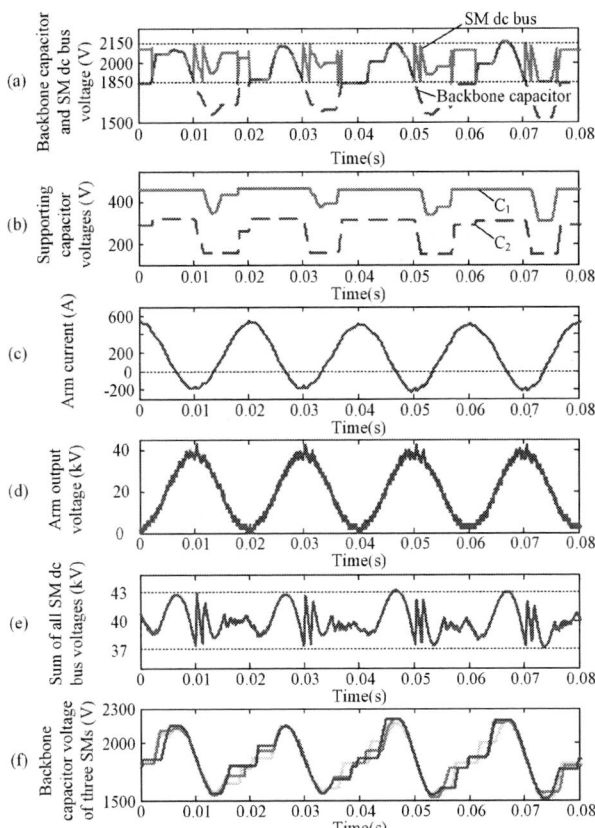

Fig. 6. (a) backbone capacitor and SM dc bus voltage of one SM in the upper arm in phase a; (b) supporting capacitor voltages in the same SM; (c) arm current; (d) arm output voltage; (e) sum of all SM dc bus voltages in the arm and (f) backbone capacitor voltage of three SMs in the arm.

Fig.6(b) sketches the supporting capacitor voltages in the same SM. v_{C1} varies between 300 V and 450 V while v_{C2} varies between 150 V and 300 V. All voltage waveforms in Fig.6(a) and (b) agree with the expected in Fig.3. Note that in both HB-SM and SSC-SM based MMCs, depending on the SM capacitor voltage balancing algorithm [21], the switching sequence of a SM may be different from cycle to cycle, i.e. the bypass time of a SM may vary. Also, due to low switching frequency, there are small derivations between the SM capacitor voltages in the same arm [Fig.6(f)]. The capacitor voltage in the same SM may vary from cycle to cycle as well. That is why the waveforms in Fig.6(a) and (b) are slightly different from the expected. Similar effects are found in a HB-SM MMC and this has no impact on the overall system performance because these variations are not reflected in the ac outputs. The arm current is shown in Fig.6(c). With 2nd and 4th order CCSCs, the harmonic circulating currents are negligible leaving only the 50 Hz and dc components. Fig.6(d) shows the arm output voltage. When the arm voltage is high, majority of the SMs are switched into the arm string, two voltage spikes can be found in each cycle due to the simultaneous switching of the supporting branches. The additional d-q rotational frame (4th and 5th order) based PI controllers in the inner current controller and the CCSC have helped to limit the impacts of the voltage spikes on the converter outputs. If higher quality is required, faster controllers such as predictive control can be applied. Fig.6(e) shows the sum of all SM dc bus voltages in the arm. The ripple of this voltage is within the permitted band (0.15p.u. p-p). Fig.6(f) gives three backbone capacitor voltages in the arm and they are balanced. The capacitor ripple currents in C_0, C_1 and C_2 are measured to be 138 A, 40 A and 75 A (rms) respectively, which are still much lower than their limits.

Fig.7 compares the power losses of one SM in HB-SM or SSC-SM based MMC system, at four operating points: 1) inverting, 2) rectifying 19.1 MW active power, 3) generating or 4) consuming 19.1 MVAr reactive power at the converter ac output. Fig.7 shows that power losses of one SSC-SM are slightly higher than one HB-SM in rated power operation with an increment of 7% to 15%. Power losses of the lower switch S_L are similar because of the similar modulation and voltage balancing controllers applied in both systems. The total loss of the three upper main switches (S_{U1}, S_{U2_1} and S_{U3_1}) in the SSC-SM is close to the upper switch in the HB-SM (S_U). It shows that the additional switching losses introduced by the switching of the supporting branches are insignificant owing to the low-voltage switching described in Section III.A. For all four operating points, power losses of the two low voltage switches (S_{U2_2} and S_{U3_2}) contribute to most of the additional losses. Note that conduction losses dominate in S_{U2_2} and S_{U3_2}.

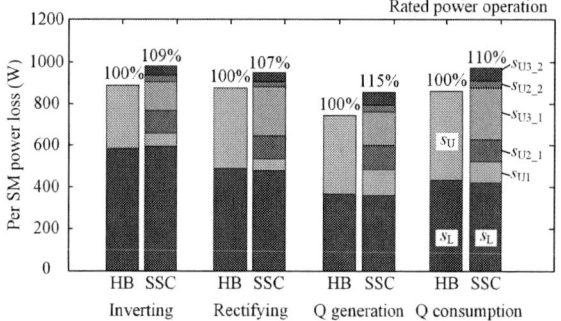

Fig. 7. Power losses comparison between one HB-SM and one SSC-SM at inverting, rectifying, reactive power (Q) generation or consumption.

B. Experiment – Single SSC-SM Testing

Computer simulation helped to show the effectiveness and validity of the system-level control. A prototype SSC-SM with reduced capacity is built and tested experimentally to verify the SM-level control. The rated capacity of the 21-level SSC-SM based MMC system is scaled down to 92 kW (PF=1) with 8 kV pole-to-pole dc and 4.4 kV ac side voltages. The rated ac current is 12 A_{rms}. With 20 SMs in each arm, the SM rated dc voltage is 400 V. Due to the lower rated current of the test platform, the SM dc bus voltage ripple drops down to 0.12p.u. p-p. The single SM test platform proposed in [29] is used to test the prototype SSC-SM. Signals of the arm current and switching sequence of one SM (only indicates whether the SM is switched-in or bypassed while the gate signals are generated locally in the test platform) are recorded in computer simulation for the scaled down MMC and used to run the test platform. The test platform uses a full bridge converter together

with a coupling inductor to synthesize the required arm current with current hysteresis control while sending the switching sequence to the prototype SSC-SM under test. To limit the SM dc bus voltage ripple to 0.12p.u. p-p, the required capacitance for one HB-SM is calculated to be 400 μF [1]. The capacitance for all backbone and supporting capacitors in the SSC-SM is 200 μF. A 200 μF with 450 V_{dc} rated voltage MPPF capacitor from VISHAY is selected as the backbone capacitor. The supporting capacitor C_1 consists of three parallel connected capacitors from AVX each with 68 μF and 100 V_{dc} rated voltage while C_2 is made up of two parallel connected capacitors each with 110 μF and 75 V_{dc} rated voltage, also from AVX. Six IGBT/diode devices are used that are driven separately.

Fig.8 shows the prototype SSC-SM. Fig.9 shows an oscilloscope snapshots in rated power operation, including SM output voltage, voltages of the backbone capacitor C_0 and supporting capacitors C_1 and C_2. All waveforms well agree with the theoretical analysis as depicted in Fig.3. The saw-tooth like waveform is the SM output voltage which is always controlled within the permitted range 0.12p.u. p-p (376 V to 424 V) when the SM is switched into the arm circuit. When the SM is bypassed, the SM output voltage will drop to near zero. The backbone capacitor voltage v_{C0} fluctuates at twice the range of the SM output voltage. When v_{C0} is less than the minimum SM dc bus voltage (376 V), one supporting capacitor C_1 or C_2 will be switched in to compensate the gap and satisfy the ripple requirement. The maximum voltage of C_1 is measured to be 72 V and that of C_2 is 48 V, both agreeing with design calculation.

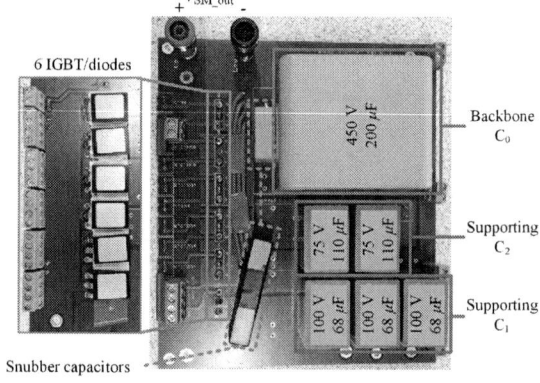

Fig. 8. A picture of the 400 V, 12.3 A_{peak} prototype SSC-SM.

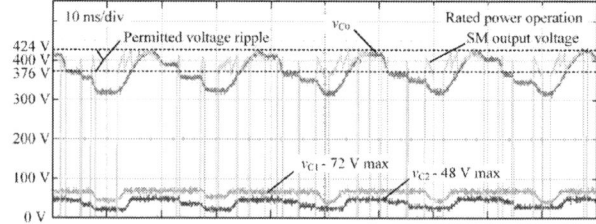

Fig. 9. Experiment results. Horizontal Axis: 10 ms/div. v_{C1}: 100 V/div, -2 div offset (yellow). v_{C2}: 100 V/div, -2 div offset (blue). v_{C0}: 100 V/div, -2 div offset (red). SM output voltage: 100 V/div, -2 div offset (green).

978-1-4673-9551-9/16 $31.00 © 2016 IEEE

V. CONCLUSION

This paper analyzes a *1-2 enhanced unipolar* SSC energy buffer architecture proposed for compact SMs in an MMC. The physical volume of all capacitors in one SSC-SM is 60% of the original HB-SM with 2000 V SM rated dc voltage and 0.15p.u. peak-to-peak voltage ripple. Computer simulation shows the performance of the SSC-SM based MMC system in steady state, which is close to the HB-SM based system. Modulation algorithm, capacitor voltage balancing and sorting controllers of the HB-SM based system can be used directly. The switching of the supporting branches is controlled locally within each SM. The main disadvantage of the SSC-SM based system is the increase of losses by 7% to 15% at rated power operation. Experiments on a prototype SSC-SM verified the validity of the design as well as the effectiveness of the SM-level controller.

ACKNOWLEDGMENT

The authors thank Professor Dave Perreault of Massachusetts Institute of Technology for his support.

REFERENCES

[1] Y. Tang, L. Ran, O. Alatise, and P. Mawby, "Capacitor Selection for Modular Multilevel Converter," in *Proc. IEEE ECCE*, Pittsburgh, PA, USA, Sep. 2014, pp. 2080-2087.

[2] K. Ilves, A. Antonopoulos, L. Harnefors, S. Norrga, L. Angquist, and H.-P. Nee, "Capacitor voltage ripple shaping in modular multilevel converters allowing for operating region extension," in *Proc. 37th Annu. Conf. IEEE Ind. Electron. Soc.*, Nov. 2011, pp. 4403–4408.

[3] R. Picas, J. Pou, S. Ceballos, V. G. Agelidis, and M. Saeedifard, "Minimization of the capacitor voltage fluctuations of a modular multilevel converter by circulating current control," in *Proc. IEEE IECON*, Montreal, QC, Canada, Oct. 25–28, 2012, pp. 4985–4991.

[4] J. Pou, S. Ceballos, G. Konstantinou, V. G. Agelidis, R. Picas, and J. Zaragoza, "Circulating current injection methods based on instantaneous information for the modular multilevel converter," *IEEE Trans. Ind. Electron.*, vol. 62, no. 2, pp. 777–788, Feb. 2015.

[5] S. P. Engel and R. W. De Doncker, "Control of the modular multi-level converter for minimized cell capacitance," in *Proc. 14th Eur. Conf. Power Electron. Appl.*, Aug./Sep. 2011, pp. 1-10.

[6] R. Picas, J. Pou, S. Ceballos, J. Zaragoza, G. Konstantinou, and V. G. Agelidis, "Optimal injection of harmonics in circulating currents of modular multilevel converters for capacitor voltage ripple minimization," in *Proc. IEEE ECCE Asia*, Melbourne, Australia, Jun. 3–6, 2013, pp. 318–324.

[7] A. J. Korn, M. Winkelnkemper, and P. Steimer, "Low output frequency operation of the modular multi-level converter," in *Proc. IEEE ECCE*, Atlanta, GA, USA, Sep. 12–16, 2010, pp. 3993–3997.

[8] R. Picas, S. Ceballos, J. Pou, J. Zaragoza, G. Konstantinou, and V. G. Agelidis, "Closed-loop discontinuous modulation technique for capacitor voltage ripples and switching losses reduction in modular multilevel converters," *IEEE Trans. Power Electron.*, vol. 30, no. 9, pp. 4714–4725, Sept. 2015.

[9] R. Picas, S. Ceballos, J. Pou, J. Zaragoza, G. Konstantinou, and V. G. Agelidis, "Improving capacitor voltage ripples and power losses of modular multilevel converter through discontinuous modulation," in *Proc. IEEE Ind. Electron. Conf.*, Vienna, Austria, Nov. 10–13, 2013, pp. 6233–6238.

[10] J. C. Clare, M. Tomasini, D. Trainer, and R. Whitehouse, Hybrid 2-level and multilevel HVDC converter, US patent, US8867242B2, 2014

[11] K. Wang, Y. Li, Z. Zheng, and L. Xu, "Voltage balancing and fluctuation suppression methods of floating capacitors in a new modular multilevel converter," *IEEE Trans. Ind. Electron.*, vol. 60, no. 5, pp. 1943–1954, May 2013.

[12] B. Li, Y. Zhang, G. Wang, W. Sun, D. Xu, and W. Wang, "A modified modular multilevel converter with reduced capacitor voltage fluctuation," *IEEE Trans. Ind. Electron.*, vol. 62, no. 10, pp. 6108–6119, Oct. 2015.

[13] K. Ilves, F. Taffner, S. Norrga, A. Antonopoulos, L. Harnefors, and H. P. Nee, "A submodule implementation for parallel connection of capacitors in modular multilevel converters," *IEEE Trans. Power Electron.*, vol. 30, no. 7, pp. 3518–3527, Jul. 2015.

[14] M. Chen, K. Afridi, and D. Perreault, "Stacked switched capacitor energy buffer architecture," *IEEE Trans. Power Electron.*, vol. 28, no. 11, pp. 5183–5195, Nov. 2013.

[15] K. Afridi, M. Chen, and D. Perreault, "Enhanced bipolar stacked switched capacitor energy buffers," *IEEE Trans. Ind. Appl.*, vol. 50, no. 2, pp. 1141-1149, Mar./Apr. 2014.

[16] M. Chen, Y. Ni, C. Serrano, B. Montgomery, D. Perreault, and K. Afridi, "An electrolytic-free offline LED driver with a ceramic-capacitor-based compact SSC energy buffer," in *Proc. IEEE ECCE*, Pittsburgh, PA, USA, Sep. 2014, pp. 2713-2718.

[17] A. B. Lostetter, F. Barlow and A. Elshabini, "An overview to integrated power module design for high power electronics packaging," *Microelectronics Reliability*, vol. 40, pp. 365-379, 2000.

[18] Z. Liang, J. D. van Wyk, and F. C. Lee, "Embedded power: a 3-D MCM integration technology for IPEM packaging application," *IEEE Trans. Adv. Packag.*, vol. 29, no. 3, pp. 504-512, Aug. 2006.

[19] C. Neeb, L. Boettcher, M. Conrad, and R. W. De Doncker, "Innovative and reliable power modules: a future trend and evolution of technologies," *IEEE Ind. Electron. Mag.*, vol. 8, no. 3, pp. 6-16, Sept. 2014.

[20] Y. Ni, S. Pervaiz, M. Chen and K.K. Afridi, "Energy Density Enhancement of Unipolar SSC Energy Buffers through Capacitance Ratio Optimization," *Proc. IEEE COMPEL*, Santander, Spain, June, 2014.

[21] Q. R. Tu, Z. Xu, and L. Xu, "Reduced switching-frequency modulation and circulating current suppression for modular multilevel converters," *IEEE Trans. Power Del.*, vol. 26, no. 3, pp. 2009-2017, Jul. 2011.

[22] K. Ilves, S. Norrga, L. Harnefors, and H. P. Nee, "On energy storage requirements in modular multilevel converters," *IEEE Trans. Power Electron.*, vol. 29, no. 1, pp. 77-88, Jan. 2014.

[23] A. Elserougi, M. I. Daoud, A. M. Massoud, A. S. Abdel-Khalik, and S. Ahmed, "Investigation of sensorless capacitor voltage balancing technique for modular multilevel converters," in *Proc. IEEE IECON*, Dallas, TX, USA, Oct.-Nov. 2014, pp. 1569-1574.

[24] S. Jahdi, O. Alatise, C. Fisher, L. Ran, and P. Mawby, "An evaluation of silicon carbide unipolar technologies for electric vehicle drive-trains," *IEEE J. Emerg. Sel. Topics in Power Electron.*, vol. 2, no. 3, pp. 517–528, Sept. 2014.

[25] S. Debnath, J. Qin, B. Bahrani, M. Saeedifard, and P. Barbosa, "Operation, control, and applications of the modular multilevel converter: a review," *IEEE Trans. Power Electron.*, vol. 30, no. 1, pp. 37–53, Jan. 2015.

[26] V. Blasko and V. Kaura, "A new mathematical model and control of a three-phase AC-DC voltage source converter," *IEEE Trans. Power Electron.*, vol. 12, no. 1, pp. 116–123, Jan. 1997.

[27] IXYS High Density DC Film Capacitors. (2013). [Online]. Available: http://www.westcode.com/publicity/prod_lit/flyer09.pdf

[28] TDK Film Capacitors. (2014). [Online]. Available: http://en.tdk.eu/blob/173582/download/10/pb-ppc-e-mobility.pdf

[29] Y. Tang, L. Ran, O. Alatise, and P. Mawby, "A model assisted testing scheme for modular multilevel converter," *IEEE Trans. Power Electron.*, vol. 31, no. 1, Jan. 2016.

Fundamental Frequency Sorting Strategy for Capacitor Voltage Balance of Modular Multilevel Converters with Phase Disposition PWM

Kun Wang, Yan Deng, Wenyu Li, Hao Peng, Guipeng Chen, Xiangning He

College of Electrical Engineering
Zhejiang University
Hangzhou, 310027, P.R. China

Abstract—**Modular multilevel converters (MMCs) become more and more attractive in high-voltage and high-power applications. The balance of the floating capacitor voltages is vital for MMCs. In order to balance the capacitor voltages, a fundamental frequency sorting strategy with Phase Disposition PWM is proposed in this paper. A dual sorting method, which sorts the capacitor voltage increments and the present capacitor voltages simultaneously, is implemented to change the relationship between the carriers and the sub-modules once in every fundamental period. Due to the low sorting frequency, the intensive computation of the controllers is avoided. Moreover, the switching commutations are evenly distributed among the power devices. Furthermore, the measurement of the arm current is eliminated, which further simplifies the communication among different controllers. Finally, the proposed voltage balance strategy is validated by simulation and experiment.**

Keywords—modular multilevel converters; fundamental frequency sorting strategy; dual sorting method; phase disposition PWM

I. INTRODUCTION

Modular multilevel converters (MMCs) have received increasing attention in the recent decade due to its outstanding merits, such as high modularity, redundancy, low switching frequency, and less need of output filter, etc. It is regarded as a state-of-the-art multilevel converter in the high-voltage and high-power applications. However, the control system becomes much more complex when the number of sub-modules (SMs) increases. Capacitor voltage balance is one of the most important challenges for MMCs.

Many methods have been proposed to balance the capacitor voltages. The sorting method in [1-4] is the most common-used method. This method selects the SMs with highest voltages or lowest voltages to be discharged or charged according to the direction of the arm current. However, this method has to sort the capacitor voltages in every switching period, which increases the calculation burden of the controller. And it causes excessive power devices' commutations, which increases the losses. Reference [5] proposed a diode-clamped balance method. This method utilizes extra diodes and isolation transformers to

simplify the balance control, but it increases the circuit complexity and cost at the same time. With the method presented in [6], each SM has an independent controller to modify its modulation signal. However, the design of each controller turns to be extremely difficult to keep the system stable in large-scale system. References [7-9] balance the capacitor voltages by alternating the pulse distribution of the drive signals. This method can improve the distribution of switching commutations. However, the mechanism analysis of this method is not clarified. A fundamental sorting frequency algorithm with Carrier Phase Shifted PWM (CPS-PWM) is proposed in [10]. It can achieve low sorting frequency and elimination of arm current measurement. However, CPS-PWM will induce excessive switching actions in large-scale multilevel converters, which leads to large switching losses.

In high-power applications, the switching frequency is expected to be low in order to minimize the switching losses. Among the common-used modulation schemes, nearest level modulation (NLM) is especially suitable for large-scale multilevel converters due to the fundamental switching frequency. In contrast, CPS-PWM is more suitable for small-scale multilevel converters due to the perfect harmonic performance. Phase Disposition PWM (PD-PWM) is an ideal compromised modulation scheme which offers low switching frequency and perfect harmonic performance in medium-scale or large-scale multilevel converters. In this paper, a voltage balance method based on fundamental frequency sorting strategy with PD-PWM scheme is proposed for the MMCs.

This paper is organized as follows. The structure and steady-state analysis are introduced in section II. In section III, the mechanism of the proposed strategy is explained. The Fourier transformation of PD-PWM and the capacitor voltage increments are analyzed in detail. And the operation process of the proposed strategy based on dual sorting method is also illustrated in this part. Finally, the effectiveness of the proposed strategy is validated by the simulation and experimental results in section IV. At last, the conclusions are summarized in section V.

This work is sponsored by the Major Program of National Natural Science Foundation of China (51490682).

II. STRUCTURE AND OPERATION PRINCIPLES OF MMC

A. Structure of MMC

The basic structure of a three-phase MMC and the configuration of a half-bridge sub-module are shown in Fig. 1. Each phase of the converter consists of two arms, namely the upper arm and lower arm. In the case of $N+1$ levels, each arm is composed of N half-bridge SMs and one arm inductor L.

The half-bridge SMs can generate two output voltages (v_c or zero) by inserting or bypassing the SM capacitors to the circuit, as shown in Table I. And the capacitor will be charged or discharged when the SM is inserted.

Table I Sub-module State

(S_1,S_2)	SM State	SM Output
(1,0)	inserted	v_c
(0,1)	bypassed	0

Define s_{pjk}, s_{njk} as the switching functions of the corresponding SMs ($j=a$, b, c). s_{pjk} and s_{njk} equal "1" when SM is inserted and equal "0" when SM is bypassed. Thus, the arm voltage can be calculated as,

$$\begin{cases} v_{pj} = \sum_{k=1}^{N} s_{pjk} \times v_{cpjk} \\ v_{nj} = \sum_{k=1}^{N} s_{njk} \times v_{cnjk} \end{cases} \quad (1)$$

where v_{pj}, v_{nj} are the upper arm voltage and the lower arm voltage of phase j, respectively.

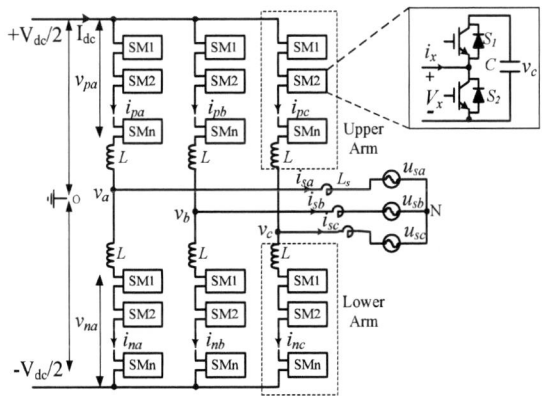

Fig. 1 Basic structure of three-phase MMC

B. Steady-state analysis of MMC

According to Kirchhoff's law, the output voltage and circulating current of phase j can be calculated as,

$$\begin{cases} v_j = \frac{1}{2}\left(v_{nj} - v_{pj}\right) - \frac{1}{2}L\frac{di_{sj}}{dt} \\ L\frac{di_{cirj}}{dt} = \frac{1}{2}V_{dc} - \frac{1}{2}\left(v_{nj} + v_{pj}\right) \end{cases} \quad (2)$$

where i_{sj} is the output current, and i_{cirj} is the circulating current, which flows both in the upper and lower arm and equals half of the sum of upper arm current and lower arm current.

From (1) and (2), it can be inferred that, the expected output voltage can be generated by controlling the number of SMs inserted into the upper arm and the lower arm. The difference between the lower arm voltage and the upper arm voltage should track the reference phase voltage [11]. And the summation of the upper and lower arm voltages controls the circulating current.

In steady state, the dc input current I_{dc} is supposed to be distributed equally among three phases. And if the second harmonics of the circulating current is neglected, the arm current i_{pj}, i_{nj} can be calculated as (3) [12],

$$\begin{cases} i_{pj} = \frac{I_{dc}}{3} + \frac{i_{sj}}{2} = \frac{I_{sm}}{2}[\frac{1}{2}m_V\cos\theta + \cos(\omega_o t + \theta_I)] \\ i_{nj} = \frac{I_{dc}}{3} - \frac{i_{sj}}{2} = \frac{I_{sm}}{2}[\frac{1}{2}m_V\cos\theta - \cos(\omega_o t + \theta_I)] \end{cases} \quad (3)$$

where I_{sm} is the amplitude of output current, m_v is the modulation index, ω_o is the output angle frequency, θ is the load impedance angle, $\theta=\theta_V-\theta_I$, and θ_V, θ_I are the phase angle of output voltage and output current respectively.

III. VOLTAGE BALANCE STRATEGY ANALYSIS

A. Analysis of PD-PWM

For $N+1$ level PD-PWM, N triangular carriers are adopted to generate the drive signals D_1, D_2…and D_N. A 5-level PD-PWM scheme is depicted in Fig. 2. The Fourier series expressions of the drive signals are necessary to analyze the charging abilities of carriers. Taking D_3 as an example, it can be divided into two parts D_{3D} and D_{3S}, where D_{3D} is related to the carrier frequency ω_c and output frequency ω_o. Using dual Fourier transformation presented in [13], D_{3D} can be calculated as

$$\begin{aligned} F_{3D}(\omega_c t, \omega_o t) = &\frac{A_{00}}{2} + \sum_{n=1}^{\infty}\left[A_{0n}\cos(n\omega_o t) + B_{0n}\sin(n\omega_o t)\right] \\ &+ \sum_{m=1}^{\infty}\left[A_{m0}\cos(m\omega_c t) + B_{m0}\sin(m\omega_c t)\right] \\ &+ \sum_{m=1}^{\infty}\sum_{\substack{n=-\infty \\ n\neq 0}}^{\infty}\left[A_{mn}\cos(m\omega_c t + n\omega_o t) + B_{mn}\sin(m\omega_c t + n\omega_o t)\right] \end{aligned} \quad (4)$$

$$\begin{aligned} \overline{C_{mn}} &= A_{mn} + jB_{mn} \\ &= \frac{1}{2\pi^2}\int_{-\pi}^{\pi}\int_{-\pi}^{\pi} f(x,y)e^{j(mx+ny)}dxdy \quad, x=\omega_c t, y=\omega_o t \end{aligned} \quad (5)$$

where m is the carrier index and n is the baseband index.

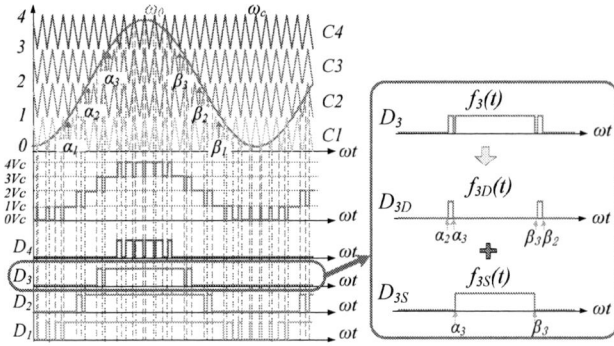

Fig. 2 PD-PWM for MMCs

In (5), the boundaries for the outer integral (α_K or β_K) and the inner integral (x_r and x_f) can be calculated as,

$$\alpha_K = -\varphi_K - \theta_V = \begin{cases} -\pi - \theta_V, & K \le \frac{N}{2}(1-m_v) \\ -\arccos\frac{2K-N}{Nm_v} - \theta_V, & \frac{N}{2}(1-m_v) < K < \frac{N}{2}(1+m_v) \\ -\theta_V. & K \ge \frac{N}{2}(1+m_v) \end{cases} \quad (6)$$

$$\begin{cases} x_r = -\frac{\pi}{2}\left(Nm_v\cos(\omega_o t + \theta_V) + N - 2K + 2\right) \\ x_f = \frac{\pi}{2}\left(Nm_v\cos(\omega_o t + \theta_V) + N - 2K + 2\right) \end{cases} \quad (7)$$

where K is the number of the carrier. The carrier ratio ω_c/ω_o is assumed to be an integer. Thus, α and β are symmetrical about the peak or valley ($-\theta_V \pm l\pi$, $l=0, \pm1, \pm2\ldots$) of the modulation waveform,

$$\beta_K = -2\theta_V - \alpha_K \quad (8)$$

Eq. (5) can be rewritten as (9) to calculate the left part of D_{3D}, $F_{3D\_left}(\omega_c t, \omega_o t)$.

$$\overline{C_{mn}} = \frac{1}{2\pi^2}\int_{\alpha_2}^{\alpha_3}\int_{x_r}^{x_f} f(x,y)e^{j(mx+ny)}dxdy \quad ,x=\omega_c t, y=\omega_o t \quad (9)$$

For $m=n=0$(the DC offset), (9) can be simplified to

$$C_{00} = \frac{2m_v}{\pi}(\sin(\alpha_3+\theta_V) - \sin(\alpha_2+\theta_V)) \quad (10)$$

For $m=0$, $n>0$(the baseband harmonics), (9) becomes

$$C_{0n} = \frac{m_v e^{j\theta_V}}{j\pi(n+1)}\left(e^{j(n+1)\alpha_3} - e^{j(n+1)\alpha_2}\right) + \frac{m_v e^{-j\theta_V}}{j\pi(n-1)}\left(e^{j(n-1)\alpha_3} - e^{j(n-1)\alpha_2}\right) \quad (11)$$

For $m>0$, $n=0$(the carrier harmonics), (9) becomes

$$C_{m0} = \frac{2}{m\pi^2}\sum_{k=1}^{\infty}\frac{1}{k}\sin\frac{k\pi}{2}J_k(m2\pi m_v) \\ \cdot\left[\sin(k(\alpha_3+\theta_V)) - \sin(k(\alpha_2+\theta_V))\right] \quad (12)$$

For $m>0$, $n\ne0$(the sideband harmonics), (9) can be integrated to

$$C_{mn} = \frac{1}{m\pi^2}\sum_{k=1}^{\infty}\sin\frac{k\pi}{2}J_k(m2\pi m_v) \\ \cdot\begin{bmatrix} \frac{1}{j(n+k)}e^{jk\theta_V}\left(e^{j(n+k)\alpha_3} - e^{j(n+k)\alpha_2}\right)\Big|n+k\ne0 \\ +\frac{1}{j(n-k)}e^{-jk\theta_V}\left(e^{j(n-k)\alpha_3} - e^{j(n-k)\alpha_2}\right)\Big|n-k\ne0 \\ +e^{jk\theta_V}\left(\alpha_3-\alpha_2\right)\Big|n=-k \\ +e^{-jk\theta_V}\left(\alpha_3-\alpha_2\right)\Big|n=k \end{bmatrix} \quad (13)$$

where $J_k(m2\pi m_v)$ is the Bessel function of order k and argument $m2\pi m_v$.

Then, $F_{3D\_left}(\omega_c t, \omega_o t)$ can be formed by substituting the results of (10), (11), (12), (13) back into (4). The exactly same method can be used to get the right part (β_3, β_2) of D_{3D}, $F_{3D\_right}(\omega_c t, \omega_o t)$. Then, $F_{3D}(\omega_c t, \omega_o t)$ can be calculated as,

$$F_{3D}\left(\omega_c t, \omega_o t\right) = F_{3D\_left}\left(\omega_c t, \omega_o t\right) + F_{3D\_right}\left(\omega_c t, \omega_o t\right) \quad (14)$$

In addition, the Fourier expression of D_{3S} can be calculated as

$$F_{3S}\left(\omega_c t, \omega_o t\right) = \frac{\varphi_3}{\pi} + \frac{2}{\pi}\sum_{m=1}^{\infty}\frac{1}{m}\sin(m\varphi_3)\cos(m(\omega_o t + \theta_V)) \quad (15)$$

Finally, the complete Fourier expression of the third drive signal $F_3(\omega_c t, \omega_o t)$ can be formed by adding $F_{3D}(\omega_c t, \omega_o t)$ and $F_{3S}(\omega_c t, \omega_o t)$ together.

B. Capacitor Voltage Increment Analysis

The capacitor voltage increment of the previous fundamental period is taken as a criterion to judge the charging ability of each carrier. Based on the above analysis and the assumption that the carrier ratio ω_c/ω_o equals to an integer, the capacitor voltage increment caused by the third carrier (lower arm for example) can be calculated as,

$$\Delta V_{cC3} = \frac{1}{C}\int_0^{T_o} F_3(\omega_c t, \omega_o t)\times i_{nj}dt \\ = \frac{\pi I_{sm}}{4\omega_o C}\left[\left(A_{00}+A'_{00}\right)\cos\theta - 2\left(A_{01}+A'_{01}\right)\cos\theta_I + 2\left(B_{01}+B'_{01}\right)\sin\theta_I\right] \\ + \frac{I_{sm}\cos\theta}{\omega_o C}\left(-\frac{\varphi_3}{2}m_v + \sin\varphi_3\right) \quad (16)$$

where the coefficients A_{00}, A_{01}, B_{01} relates to (α_2, α_3) and A'_{00}, A'_{01}, B'_{01} relates to (β_3, β_2) as shown in (17).

$$
\begin{cases}
A_{00} = \dfrac{2m_v}{\pi}\left[\sin(\alpha_3+\theta_V)-\sin(\alpha_2+\theta_V)\right], \\[2mm]
A'_{00} = \dfrac{2m_v}{\pi}\left[\sin(\beta_2+\theta_V)-\sin(\beta_3+\theta_V)\right], \\[2mm]
A_{01} = \dfrac{m_v}{\pi}\begin{bmatrix}\frac{1}{2}\left[\cos\theta_V(\sin2\alpha_3-\sin2\alpha_2)+\sin\theta_V(\cos2\alpha_3-\cos2\alpha_2)\right]\\ +(\alpha_3-\alpha_2)\cos\theta_V\end{bmatrix}, \\[4mm]
A'_{01} = \dfrac{m_v}{\pi}\begin{bmatrix}\frac{1}{2}\left[\cos\theta_V(\sin2\beta_2-\sin2\beta_3)+\sin\theta_V(\cos2\beta_2-\cos2\beta_3)\right]\\ +(\beta_2-\beta_3)\cos\theta_V\end{bmatrix}, \\[4mm]
B_{01} = \dfrac{m_v}{\pi}\begin{bmatrix}\frac{1}{2}\left[\sin\theta_V(\sin2\alpha_3-\sin2\alpha_2)-\cos\theta_V(\cos2\alpha_3-\cos2\alpha_2)\right]\\ -(\alpha_3-\alpha_2)\sin\theta_V\end{bmatrix}, \\[4mm]
B'_{01} = \dfrac{m_v}{\pi}\begin{bmatrix}\frac{1}{2}\left[\sin\theta_V(\sin2\beta_2-\sin2\beta_3)-\cos\theta_V(\cos2\beta_2-\cos2\beta_3)\right]\\ -(\beta_2-\beta_3)\sin\theta_V\end{bmatrix}.
\end{cases} \quad (17)
$$

It can be inferred from (16) and (17) that the voltage increment is determined by the carrier, output current, modulation index and the load power factor. The influence of the load power factor and the modulation index on the voltage increment are plotted in Fig. 3 and Fig. 4 respectively, where the voltage increments, ΔV_{cC3}^{*}, is normalized by $I_{sm}/\omega_o C$. The conclusion can be drawn that different carriers have different charging abilities, which is the basis of the proposed capacitor voltage balance strategy.

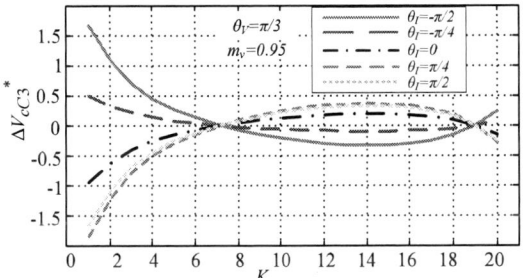

Fig. 3 The influence of the load power factor on voltage increment

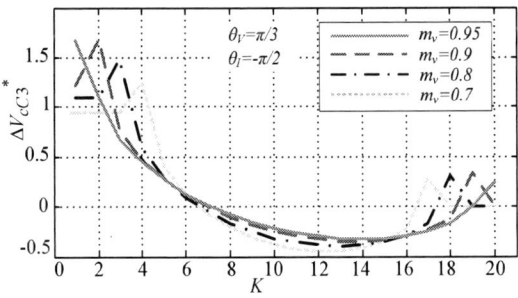

Fig. 4 The influence of modulation index on voltage increment

C. Operation Process of the Proposed Strategy

With the conclusion that different carriers have different charging abilities, and taking the voltage increment in the previous fundamental period as the criterion to judge the charging ability of the carrier, the proposed balance strategy is accomplished with the dual sorting method shown in Fig. 5.

The voltage increments are sorted in descending order and the present capacitor voltages are sorted in ascending order. Then the carrier with the largest voltage increment which charges the most will be assigned to the SM with the lowest capacitor voltage, while the carrier with the lowest voltage increment which discharges the most will be assigned to the SM with the highest capacitor voltage. The other carriers and sub-modules are handled in the same way.

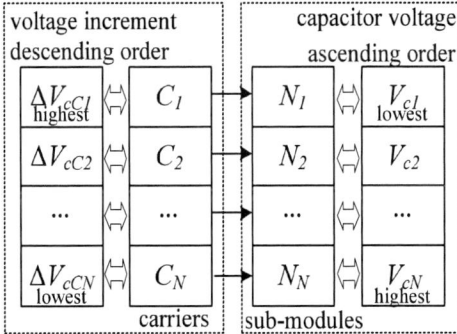

Fig. 5 Dual sorting method

IV. SIMULATION AND EXPERIMENT VERIFICATION

In order to verify the proposed balance strategy, a 21-level three-phase inverter model in MATLAB/Simulink and a downscaled 9-level three-phase prototype is built. The parameters of the simulation model and the prototype are shown in Table II.

Table II Parameters of the simulation model and prototype

Parameters	Simulation model	Prototype
DC bus voltage (V_{dc})	18kV	1kV
Modulation index (m_V)	0.9	0.9
Output frequency (f)	50Hz	50Hz
Number of SMs in each arm (N)	20	8
Arm inductance (L)	40mH	40mH
SM capacitance (C)	3mF	3mF
Load resistance (R)	100Ω	100Ω
Rated capacitor voltage (V_c)	900V	125V
Carrier frequency (f_c)	1050Hz	1050Hz

A. Simulation Verification

The AC output voltage and output current are displayed in Fig. 6. The simulation waveform of capacitor voltages in Fig. 7 demonstrates the dynamic situation where the proposed balance strategy is disabled at 0.3s, and re-enabled at 0.4s. Before 0.3s, with the help of the balance strategy, the sub-module capacitor voltages are maintained at the rated value. During the period when the balance strategy is disabled, the capacitor voltages diverge from the rated value. After the balance strategy returns at 0.4s, the capacitor voltages are able to converge to the balance state.

The waveforms of the capacitor voltages and output current, when the load increases, are demonstrated in Fig. 8. The capacitor voltages keep at rated value all the time, even though the voltage ripple becomes larger because of the increment of load current.

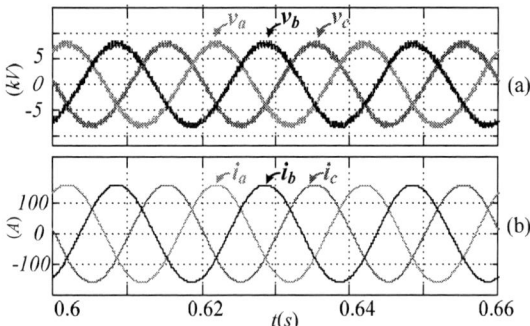

Fig. 6 The AC output voltage and output current

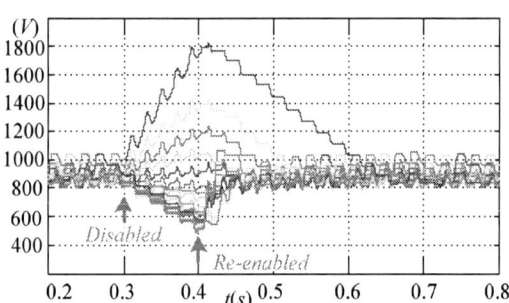

Fig. 7 Capacitor voltages with and without balance strategy

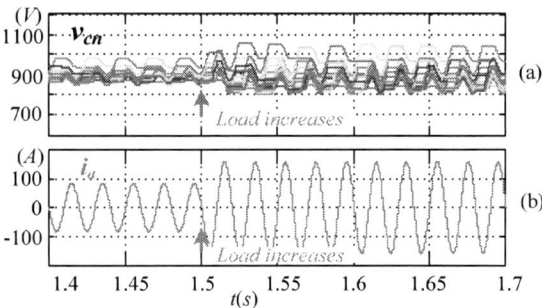

Fig. 8 Capacitor voltages when the load increases

B. Experimental Results

The experiment results of three-phase output voltage and output current in steady state are shown in Fig. 9. And the arm voltage and arm current of phase A are depicted Fig. 10.

The experimental waveforms of the capacitor voltages and output current, when the load increases, are demonstrated Fig. 11. It can be seen that the capacitor voltages can still keep balanced with a larger voltage ripple when the load increases.

The drive signals of the 1st, 3rd and 5th SMs of lower arm in Fig. 12 show that the switching commutation is distributed evenly with the proposed balance strategy.

All these simulation and experimental results verify the effectiveness of the proposed capacitor balance strategy both in steady state and dynamic state.

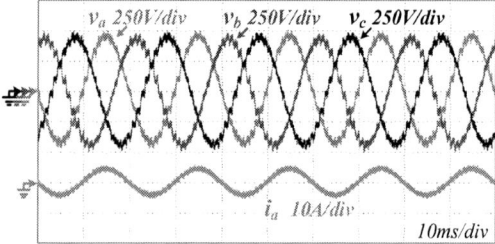

Fig. 9 Output voltage and current of phase A

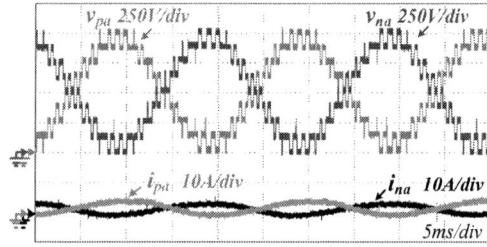

Fig. 10 Upper arm and lower arm voltage and current

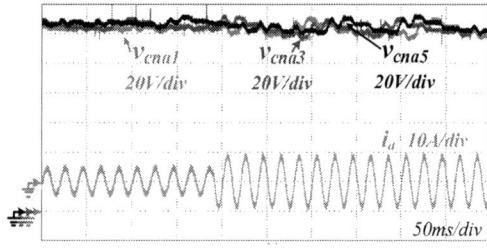

Fig. 11 Capacitor voltages when the load increases

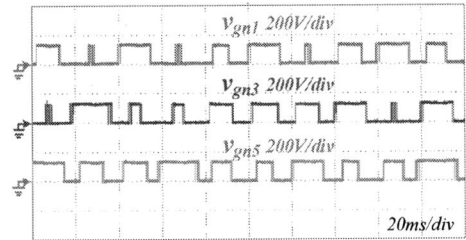

Fig. 12 Drive signals of the 1st, 3rd, 5th SM

V. CONCLUSION

In this paper, a fundamental frequency sorting strategy with PD-PWM is proposed to balance the capacitor voltage of MMCs. The corresponding relationship between the carriers and sub-modules is changed once in every fundamental period based on the conclusion that different carriers have different charging abilities. The low sorting frequency is achieved by dual sorting method, which takes the voltage increment in the previous fundamental period as the criterion to judge the charging ability of the driving signal.

The proposed balance strategy can achieve low sorting frequency which reduces the computational burden of the controller. And the switching commutations are evenly distributed among the power devices. Moreover, the measurement of arm current is eliminated. Furthermore, compared with NLM and CPS-PWM, the balance strategy with PD-PWM can achieve both low switching frequency and perfect harmonic performance in medium-scale or large-scale MMCs. The simulation and experimental results verify the effectiveness of the proposed strategy.

REFERENCES

[1] A. Lesnicar and R. Marquardt, "An innovative modular multilevel converter topology suitable for a wide power range," *Power Tech Conf. Proc.* 2003 IEEE Bologna, vol. 3, 2003, pp. 6

[2] P. M. Meshram and V. B. Borghate, "A Simplified Nearest Level Control (NLC) Voltage Balancing Method for Modular Multilevel Converter (MMC)," *IEEE Trans. Power Electron.*, vol. 30, pp. 450-462, 2015-01-01 2015.

[3] M. Saeedifard and R. Iravani, "Dynamic Performance of a Modular Multilevel Back-to-Back HVDC System," *IEEE Trans. Power Delivery*, vol. 25, 2010, pp. 2903-2912.

[4] K. Wang, Y. Li and Z. Zheng, "Voltage balancing control and experiments of a novel modular multilevel converter," in *Energy Conversion Congress and Exposition (ECCE)*, 2010 IEEE, 2010, pp. 3691-3696.

[5] C. Gao, X. Jiang, Y. Li, Z. Chen, and J. Liu, "A DC-Link Voltage Self-Balance Method for a Diode-Clamped Modular Multilevel Converter With Minimum Number of Voltage Sensors," *IEEE Trans. Power Electron.*, vol. 28, 2013, pp. 2125-2139.

[6] M. Hagiwara and H. Akagi, "Control and Experiment of Pulsewidth-Modulated Modular Multilevel Converters," *IEEE Trans. Power Electron.*, vol. 24, 2009, pp. 1737-1746.

[7] S. Fan, K. Zhang, J. Xiong, and Y. Xue, "An Improved Control System for Modular Multilevel Converters with New Modulation Strategy and Voltage Balancing Control," *IEEE Trans. Power Electron.*, vol. 30, 2015, pp. 358-371.

[8] J. Mei, K. Shen, B. Xiao, L. M. Tolbert, and J. Zheng, "A New Selective Loop Bias Mapping Phase Disposition PWM With Dynamic Voltage Balance Capability for Modular Multilevel Converter," *IEEE Trans. Ind. Electron.*, vol. 61, 2014, pp. 798-807

[9] Mei, J. and B. Xiao, et al. "Modular Multilevel Inverter with New Modulation Method and Its Application to Photovoltaic Grid-Connected Generator." *IEEE Trans. Power Electron.*, 2013, vol. 28, pp. 5063-5073

[10] H. Peng, Y. Wang, K. Wang, Y. Deng, X. He and R. Zhao, "A Simple Capacitor Voltage Balancing Method with a Fundamental Sorting Frequency for Modular Multilevel Converters," *Journal of Power Electronics*, vol. 14, 2014, No. 6, pp. 1109-1118

[11] Q. Tu, Z. Xu, and L. Xu, "Reduced switching-frequency modulation and circulating current suppression for modular multilevel converters," *IEEE Trans. Power Del.*, vol. 26, no. 3, pp. 2009–2017, Jul. 2011.

[12] Q. Tu, Z. Xu, H. Huang, and J. Zhang, "Parameter design principle of the arm inductor in modular multilevel converter based HVDC," *Power System Technology (POWERCON)*, 2010 International Conference, 2010, pp. 1-6

[13] D. G. Holmes and T. A. Lipo, Pulse Width Modulation for Power Converters Principles and Practice, NJ: IEEE Press, 2003.

Active Voltage Balancing Control for 10kV Three-level Converter Using Series-Connected HV-IGBTs

Shiqi Ji[1,2], Ting Lu[1], Zhengming Zhao[1], Hualong Yu[1], Fred Wang[2]

Email: sji6@utk.edu

[1]State Key Laboratory of Control and Simulation of Power System and Generation Equipment Department of Electrical Engineering, Tsinghua University, Beijing 100084, China

[2]Department of Electrical Engineering and Computer Science, University of Tennessee, Knoxville, TN, USA

Abstract—The integration of series connection of insulated gate bipolar transistors (IGBTs) and multi-level can achieve high voltage converters with low total harmonics distortion (THD). However, due to the transient voltage unbalance, the series connection technology has not been widely applied. Asynchronous gate drive signals, which cause series-connected IGBTs not to turn-on and turn-off at the same time, result in serious unbalanced voltage sharing. This paper presents an active voltage balancing control with its electromagnetic compatibility (EMC) design to solve the asynchronous gate signal problem. The effectiveness of the active voltage balancing control has been experimentally validated in a 10kV dc-link voltage three-level bridge using two 4.5kV HV-IGBTs in series-connection.

Keywords—series-connected HV-IGBTs, active voltage balancing control, EMC

I. INTRODUCTION

In recent years, medium/high voltage and high power converters have been widely applied in high voltage motor drive, the static synchronous compensator (STATCOM), new distributed power system and so on [1]-[3]. However, as the most commonly used semiconductor power devices, the maximum voltage blocking capability of insulated gate bipolar transistors (IGBT) is 6.5kV in commercial production. This is not sufficient for these applications in many cases with the common two-level and three-level topology.

Using series-connected HV-IGBTs enables a switching unit to operate at voltage levels higher than the rated voltage of one IGBT. It saves the device number compared to multi-level technology, but requires larger filters due to high THD. The integration of series connection and multi-level is considered to be a better choice for high voltage converters. However, due to the transient voltage unbalance, the series connection technology has not been widely applied.

Some control methods have been proposed for the voltage unbalance in series-connected IGBTs. The imbalance voltage can be decreased by limiting dv/dt during switching transient with snubber circuits, such as C, RC and RCD [4]-[6]. But it causes a large power loss in these snubber circuits. The gate active clamping circuit which is first proposed by Siemens and then improved by Concept, Alstom and ABB is also applied for unbalanced voltage suppression by limiting the peak voltage in

switching transient [7]-[10]. However, it causes a larger switching power loss of IGBT due to a longer operating time in active region. In recent years, researchers realized that the voltage unbalance is caused by different reasons, such as unequal collector-to-gate capacitance and stray capacitance of main circuit. The voltage balancing methods should be designed according to the specific cause. The main circuit was improved in order to achieve a same stray capacitance between collector and emitter for series-connected IGBTs [11]. An active voltage balance method was proposed to solve the voltage unbalance caused by different collector-to-gate capacitances [12][13]. However, the consistency of the parasitic capacitance can be ensured by IGBT manufactures. The asynchronous gate drive signals, which cause series-connected IGBTs not to turn-on and turn-off at the same time, result in serious unbalanced voltage sharing. This is mainly caused by different gate drive signal delays between series-connected IGBTs. A gate signal compensation method was proposed in [14] by measuring the voltage unbalance in tail period, but is greatly affected by the pulse width of off-state. A gate signal regulation strategy by sampling V_{ce} during turn off transient was discussed in [15] but only tested in a half bridge with several pulses.

A novel active voltage balancing control for series-connected HV-IGBTs is proposed in this paper to solve the asynchronous gate signal problem. The critical issues such as sample speed and EMC which highly influence the control reliability are designed such that the control method could be applied to industry products. A 10kV three-level bridge using two 4.5kV/600A HV-IGBTs in series connection is built to validate the control method.

II. ACTIVE VOLTAGE BALANCING CONTROL

The circuit for active voltage balancing control which consists of three parts is shown in Fig.1. A gate active clamping circuit is used to maintain the peak voltage of IGBT below its breakdown level. The status feedback circuit is integrated with gate active clamping circuit. The operation time of gate active clamping circuit is chosen as the feedback signal. The gate drive signal compensation is generated using the feedback status in the microcontroller and thus to achieve synchronous gate drive signals.

Supported by Major Program of the National Natural Science Foundation of China (51490683) and Program of State Key Laboratory of Power System in Tsinghua University (SKLD15Z01).

978-1-4673-9551-9/16 $31.00 © 2016 IEEE

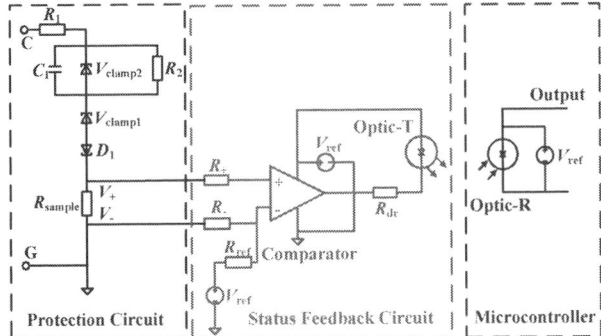

Fig. 1. Active voltage balancing control circuit

A. Protection circuit

A two-stage gate active clamping circuit using transient voltage suppressors (TVS) is applied to limit the maximum IGBT voltage in its safe operation area. In this technique, it requires several pulses to regulate the gate signals. In these pulses, a serious voltage unbalance still exists and the gate active clamping circuit is essential to limiting the peak voltage. The two-stage gate active clamping has been proposed in [9] and its principle is also given in this paper. By using this circuit, the maximum V_{ce} can be clamped to V_{clamp} effectively as in (1):

$$V_{clamp} = V_{clamp1} + V_{clamp2} \qquad (1)$$

V_{clamp} is determined by the safe operation area (SOA) of IGBT. The status feedback is integrated with the gate active clamping circuit here. When V_{ce} reaches V_{clamp1}, a current flows through the sample resistor (R_{sample}). The voltage unbalance can be presented by the current in the clamping circuit. Therefore, V_{clamp1} is peak voltage when IGBT operates at rated dc-link voltage and maximum load current (V_{bus}=10kV, I_L=600A) with perfect voltage balance and can be approximately represented as in (2):

$$V_{clamp1} = \frac{1}{4} V_{bus} + L_s \frac{I_L}{t_f} \qquad (2)$$

where L_s is the stray inductance in main circuit and t_f is the fall time during turn-off transient. If there is an asynchronous gate signal delay, one of IGBTs will be clamped and voltage unbalance can be detected by the current in the clamping circuit.

B. Status feedback circuit

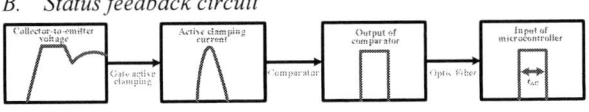

Fig. 2. Feedback signals of the active voltage balancing control

Clamping time instead of maximum V_{ce} is used as the feedback signal to describe the unbalanced voltage sharing. A longer clamping time means more serious voltage unbalance. The status feedback circuit transfers the current in gate active clamping circuit into 0V-5V pulse as shown in Fig.2. The pulse

width t_{AC} is the clamping time and can be recognized by microcontroller.

C. Microcontroller

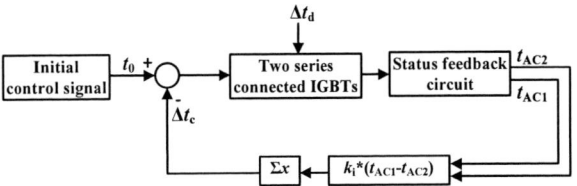

Fig. 3. Active voltage balancing control schematic

The control schematic is shown in Fig.3. If there is an asynchronous gate delay of Δt_d, one of series-connected IGBTs is clamped and the microcontroller can receive the feedback status t_{AC1} and t_{AC2} which represent the clamping time of Tx-1 and Tx-2 respectively. In the microcontroller, a compensation time Δt_c is generated before the control signal is sent to the gate driver. The voltage of Tx-1 and Tx-2 can be balanced only if $\Delta t_c = \Delta t_d$.

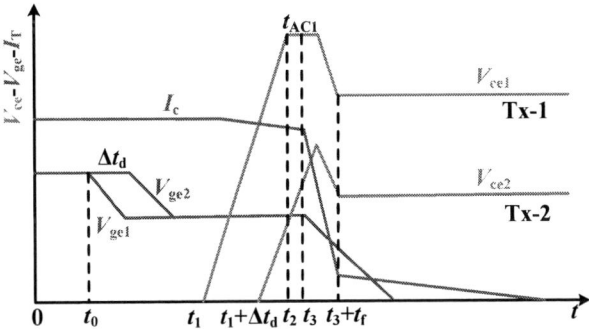

Fig. 4. Turn off waveforms with asynchronous gate drive signals

A schematic turn off waveform with asynchronous gate drive signals is shown in Fig.4. If the gate drive of Tx-2 has a Δt_d delay from Tx-1, then Tx-1 will withstand high voltage and has a feedback signal of t_{AC1}. At t_0, the gate voltage V_{ge} begins to fall. After a period of gate discharging, V_{ce1} begins to increase typically at t_1. The time between t_0 and t_1 is mainly decided by the gate driver resistance, gate-to-emitter capacitance and gate-to-collector oxide capacitance which do not have critical differences between Tx-1 and Tx-2. Therefore, V_{ce2} begins to rise typically at $t_1 + \Delta t_d$. When V_{ce1} gets to V_{clamp1} at t_2, the microcontroller can receive the feedback signal of T1. The change of V_{ce1} is very small and can be ignored after it reaches V_{clamp1}. At t_3, V_{ce2} is $V_{bus}/2 - V_{clamp1}$ and the current though the IGBT, I_c, begins to fall rapidly. After a fall time of t_f, I_c falls to the tail current. At one point between t_3 and $t_3 + t_f$, V_{ce1} is going to be smaller than V_{clamp1} and the feedback signal of T1 disappears. This point is difficult to calculate accurately but is much closer to t_3 especially with large asynchronous gate delay. Accordingly, the t_{AC1} can be approximately represented as in (3):

$$t_{AC1} = t_3 - t_2 \qquad (3)$$

The gate voltage of T1 and T2 remains constant in this period. Using the model in [16][17] to describe the voltage-dependent capacitance of the IGBT, the time can be solved in (4):

$$\Delta t_d = t_{AC1} + k\left(\sqrt{V_{clamp1}} - \sqrt{V_{bus}/2 - V_{clamp1}}\right) \tag{4}$$

where k is constant which is related to gate discharging current at miller stage and gate-to-collector capacitance.

It can be seen in equation (4) that there is a linear relationship between Δt_d and Δt_{AC} (t_{AC1}-t_{AC2}). The experimental result with 570V dc-link voltage and 100A load current is shown in Fig.5. Δt_{AC} shows linear change with Δt_d and the ratio approximately equals to 1. Therefore, an integrator is used to calculate Δt_c from Δt_{AC} in the microcontroller.

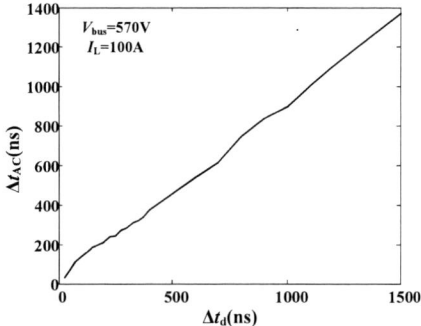

Fig. 5. Experimental result of relationship between Δt_d and Δt_{AC}

III. STATUS FEEDBACK CIRCUIT DESIGN

Some special requirements should be satisfied through status feedback circuit design in order to accurately measure the feedback status.

A. Electromagnetic interference caused by high dv/dt

High dv/dt causes large feedback interferences in the control. The TVS can be seen as an ideal TVS and a junction capacitor C_{TVS} in parallel connection as shown in Fig.6. Therefore, the clamping current I_{ac} is considerable with high dV_{cg}/dt, even though the transient voltages between series-connected IGBTs are well shared and the TVS does not reach its breakdown voltage. The interference current can be represented as in (5):

$$I_{ac} = C_{TVS}\frac{dV_{cg}}{dt} \tag{5}$$

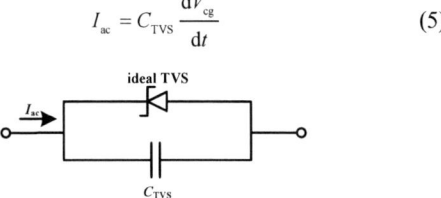

Fig. 6. Equivalent circuit of TVS

C_{TVS} is the junction voltage which is decided by V_{cg}:

$$C_{TVS} = \frac{C_{TVS,1V}}{\sqrt{V_{cg}}} \tag{6}$$

where $C_{TVS,1V}$ is the capacitance when V_{cg} equals 1V.

High positive dv/dt happens during turn-off transient and turn-on transient of its interlocking IGBT (cross-talk). The electromagnetic compatibility (EMC) methods for dv/dt in these two periods are designed separately.

For T1, during its turn-off transient, the rise of V_{ce1} mainly happens when V_{ge1} stays at miller stage and dv/dt can be described in (7):

$$\frac{dV_{cg1}}{dt} = \frac{\left(V_{miller} - V_{g(off)}\right)\sqrt{V_{cg1}}}{R_{g(off)}(C_{TVS,1V} + C_{gc,1V})} \tag{7}$$

And the interference current can be represented as in (8):

$$I_{ac} = \frac{\left(V_{miller} - V_{g(off)}\right)C_{TVS,1V}}{R_{g(off)}(C_{TVS,1V} + C_{gc,1V})} \tag{8}$$

During the turn-on transient of T3, V_{ce1} increases with the decreases of V_{ce3} and can be represented as in (9):

$$\frac{dV_{cg1}}{dt} = \frac{\left(V_{g(on)} - V_{miller}\right)\sqrt{V_{cg3}}}{R_{g(on)}(C_{TVS,1V} + C_{gc,1V})} \tag{9}$$

The interference current is:

$$I_{ac} = \sqrt{\frac{V_{cg3}}{V_{cg1}}}\frac{\left(V_{g(on)} - V_{miller}\right)C_{TVS,1V}}{R_{g(on)}(C_{TVS,1V} + C_{gc,1V})} \tag{10}$$

The active clamp circuit supplies all current when V_{ce} is clamped. The clamping current equals:

$$I_{ac} = \frac{V_{miller} - V_{g(off)}}{R_{g(off)}} \tag{11}$$

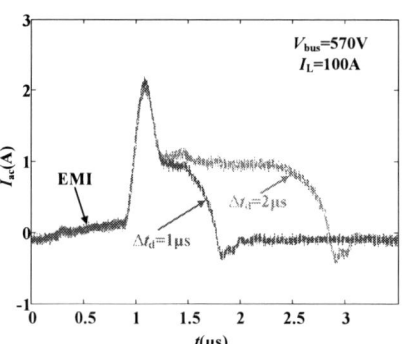

Fig. 7. Experimental results of I_{ac} with different asynchronous gate delay during turn-off transient

The clamping current when V_{ce} is clamped is always higher than the interference current caused by dv/dt during turn-off transient. The experimental result of clamping current with different asynchronous gate delay is shown in Fig.7. A method of classifying the useful feedback signal from EMI is to compare I_{ac} with a threshold current I_{TH}. When I_{ac} is higher than I_{TH}, it is classified into feedback signal. Therefore, I_{TH} should satisfy:

$$\frac{C_{TVS,1V}}{C_{TVS,1V}+C_{gc,1V}}\frac{V_{miller}-V_{g(off)}}{R_{g(off)}} < I_{TH} < \frac{V_{miller}-V_{g(off)}}{R_{g(off)}} \quad (12)$$

As shown in Fig.1, the voltage drop on R_{sample} equals $I_{ac}R_{sample}$ which is also anode voltage of comparator. Cathode voltage of comparator equals $V_{ref}R_-/(R_{ref}+R_-)$. When the anode voltage is higher than cathode voltage, output of comparator is at high level and there is a feedback signal. Therefore, I_{TH} is determined by V_{ref}, R_{ref} and R_- which can be represented as in (13):

$$I_{TH} = \frac{V_{ref}R_-}{R_{sample}(R_{ref}+R_-)} \quad (13)$$

However, the interference current caused by dv/dt during turn-on transient of its interlocking IGBT is much higher when V_{cg3} is larger than V_{cg1}. The experimental waveforms with and without the I_{TH} configuration when voltages are well shared and no active clamping happens are given in Fig.8. V_{out} is feedback signal. With the threshold configuration, the interference during T1's turn-off transient is eliminated, but the interference during T3's turn-on transient still exists.

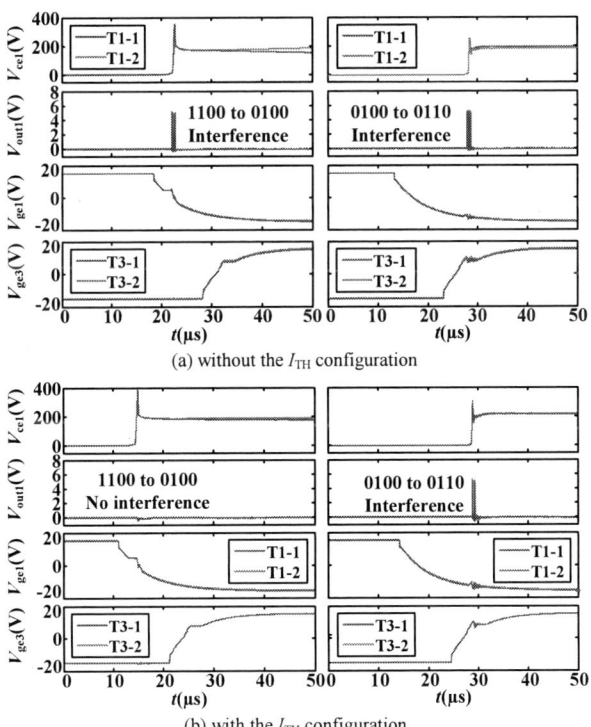

(a) without the I_{TH} configuration

(b) with the I_{TH} configuration

Fig. 8. Experimental waveforms when voltage is not clamped with and without the I_{TH} configuration

There is a dead time between T1's turn-off and T3's turn-on (commonly about 10μs for HV-IGBT). The interference

during turn-on transient of its interlocking IGBT is separated from real clamping feedback signal in time sequence. The interference in this period can be eliminated by ending the sample in a short period after its turn-off transient which should be shorter than the dead time.

B. High speed

The control effect goes worse with the increase of sample error including transmission and quantization error. If there is a sample error t_{err}, the feedback circuit may not detect a voltage unbalance if the difference of gate signal delay is smaller than t_{err}. Therefore, the voltage unbalance ratio $\Delta V_{ce}/V_{bus}$ caused by sample error t_{err} could reach:

$$\alpha = \frac{V_{miller}-V_{g(off)}}{R_{g(off)}C_{gd,1V}\sqrt{V_{bus}}}t_{err} \quad (14)$$

t_{err} should be smaller than 25ns if the voltage unbalance ratio α is lower than 1%.

The sampling error consists of transmission and quantization error. The transmission error is the pulse width difference between the current in R_{sample} and the output of optical fiber receiver. It is caused by the distortion through comparator, optical fiber transmitter and receiver and so on.

Therefore, the devices in the status feedback circuit should have fast response and high transfer speed. In order to transform and transfer the feedback signal timely, a high speed comparator (ADCMP600), 10MBd fiber optic transmitter (HFBR-1528) and receiver (HFBR-2528) are adopted. The fiber optic transmitter is driven directly by the comparator so as to avoid the signal transmission delay caused by additional optical drive circuit.

The output results under input signals with different pulse widths are shown in Table.I. The test waveforms with 26ns input signal is shown in Fig.9. It can be concluded from the test results that the transmission error is about 10ns.

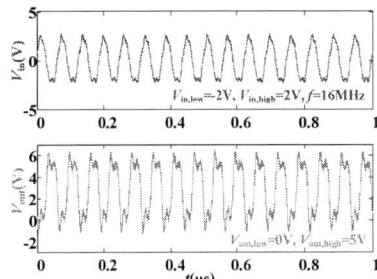

Fig. 9. Input and output signal with 26ns pulse width

TABLE I.   TEST DATA OF STATUS FEEDBACK SUB-CIRCUIT

T_{in}	Rise delay	Fall delay	T_{out}
491ns	80ns	95ns	506ns
93ns	80ns	93ns	106ns
43.6ns	81ns	92.4ns	55ns
26ns	82ns	90ns	34ns

The quantization error occurss when the microcontroller samples the output of optical fiber receiver with a clock frequency. Here, FPGA with 80MHz clock frequency is applied as the microcontroller. Therefore, the quantization error is smaller than 12.5ns and the total error of the control will not exceed 25ns.

IV. EXPERIMENTAL VERIFICATION

A 10kV three-level bridge using two 4.5kV/600A HV-IGBTs (CM600HG-90H_E made by Mitsubishi) in series connection is built. The experiment platform and the status feedback circuit are shown in Fig.10. The experimental results in 6kV dc-link and 10kV dc-link are given in Fig.11 and Fig.12. The voltage unbalances in Fig.11 and Fig.12 are caused by a nearly 800ns asynchronous gate signal delay. The performance is perfectly improved comparing the system without active voltage balancing control.

There are four kinds of commutation loops for three level converters and they work in different periods. The voltage balance for each of the four groups of series-connected IGBTs (T1-T4) is achieved separately in different periods. For example, the regulation for T1 happens when the load current flows from positive of dc-link to neutral point and the switch state transfers between 0100 and 1100. The regulations for four groups of IGBTs are shown in Table.II.

TABLE II.   VOLTAGE BALANCE REGULATION IN THREE-LEVEL

Period	Current direction	Switch state	Regulate group
1	Positive to neutral point	1100 to 0100	T1-1 and T1-2
2	Neutral point to negative	0110 to 0010	T2-1 and T2-2
3	Negative to neutral point	0011 to 0010	T4-1 and T4-2
4	Neutral point to positive	0110 to 0100	T3-1 and T3-2

(a) 10kV three-level test platform

(b) status feedback circuit

Fig. 10. The three-level test platform using two series-connected HVIGBTs

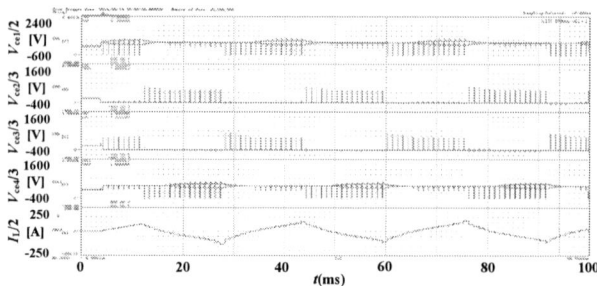

(a) without the control

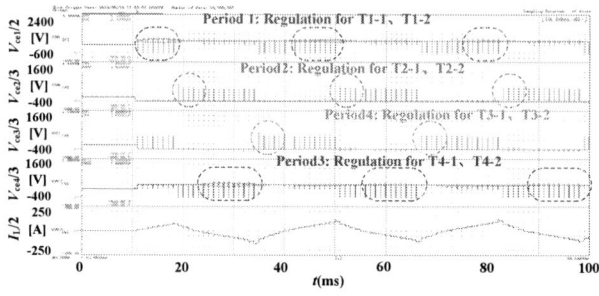

(b) with the control

Fig. 11. Experimental results without and with the active voltage balancing control in 6kV dc-link

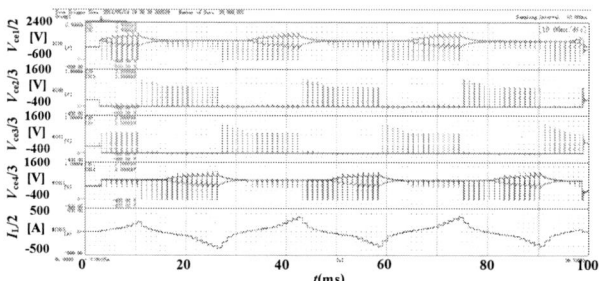

(a) without the control

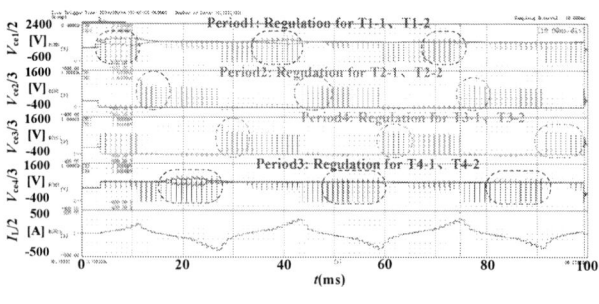

(b) with the control

Fig. 12. Experimental results without and with the active voltage balancing control in 10kV dc-link

I. CONCLUSION

An active voltage balancing control used in series-connected HV-IGBTs for the three-level converter is proposed in this paper. The unbalanced voltage sharing which is caused by asynchronous gate delay can be significantly restrained using the control. The EMI caused by dv/dt is analyzed and the EMC is designed in detail. A 10kV three-level bridge is built to validate the performance of the control.

[1] H. Kon, M. Tobita, H. Suzuki, J. Kanno, N. Nishizawa, T. Murao, S.Irokawa. "Development of a multiple series-connected IGBT converter for large-capacity STATCOM," in Proc.IPEC, pp.2024-2028, 2010.

[2] A. Huang, M. Crow, G. Heydt, J. Zheng, S. Dale. "The future renewable electric energy delivery and management (FREEDM) system: the energy internet," Proceedings of the IEEE, vol.99, no.1, pp.133-148, Jan.2011.

[3] Xiao Zhang, and Gabriela Hug. "Optimal regulation provision by aluminum smelters," in IEEE PES General Meeting, pp. 1-5, 2014.

[4] Chen J, Lin J, Ai T. "The techniques of the serial and paralleled IGBTs," Proceedings of IEEE Industrial Electronics, Control and Instrumentation (IECON), 1996: 999-1004.

[5] Nakajima T, Suzuki H, Tamura Y, et al. "Development of IEGT series and parallel connection technology for high power converters," International Power Electronics Conference (IPEC),2000: 670-675.

[6] Sung K, Akiyama H, Takao K, et al. "Switching characteristic of Si-IEGT and SiC-PiN diode pair connected in series," International Power Electronics Conference (IPEC), 2010: 170-175.

[7] Bruckmann M, Sommer R, Fasching M, et al. "Series connection of high voltage IGBT modules," Proceedings of IEEE Industry Applications, 1998: 1067-1072.

[8] Rüedi H, Köhli P. "SCALE driver for high voltage IGBTs," Proceedings of PCIM, 1999:357-364.

[9] Saiz J, Mermet M. Frey D, et al. "Optimisation and integration of an active clamping circuit for IGBT series association," Proceedings of IEEE Industry Applications, 2001: 1046-1051.

[10] Piazzesi A, Metsenc L. "Series connection of 3.3kV IGBTs with active voltage balancing," International Power Electronics Specialists Conference (PESC), 2004: 893-898.

[11] The Van Nguyen, Pierre-Olivier Jeannin, et al. "Series connection of IGBTs with self-powering technique and 3-D topology," IEEE Transaction on Industry Applications, vol. 47, no. 4, pp. 1844-1852, July/August 2011.

[12] T.C. Lim, B.W. Williams, S.J. Finney, P.R. Palmer. "Series-Connected IGBTs Using Active Voltage Control Technique," IEEE Transaction on Power Electronics, vol.28, no.8, pp.4083–4103, Aug.2013.

[13] I. Baraia, J. Barrena, G. Abad, J.M.C. Segade, U. Iraola. "An experimentally verified active gate control method for the series connection of IGBT/diodes," IEEE Transactions on Power Electronics, vol.27, no.2, pp.1025-1038, Feb.2012.

[14] Nakatake H, Iwata A. "Series connection of IGBTs used multilevel clamp circuit and turn off timing adjustment circuit," International Power Electronics Specialists Conference (PESC), 2003: 1910-1915.

[15] Ji Shiqi, Lu Ting, Zhao Zhengming, Yu Hualong, Yuan Liqiang. "Series-connected HV-IGBTs using active voltage balancing control with status feedback circuit," IEEE Transactions on Power Electronics, 2015, 30(8): 4165-4174.

[16] Lu Ting, Zhao Zhengming, Ji Shiqi, Yu Hualong. "Active clamping circuit with status feedback for series-connected HV-IGBTs," IEEE Transaction on Industry Applications, 2014, 50(5): 3579-3590.

[17] Ji shiqi, Zhao Zhengming, Lu Ting, Yuan Liqiang, Yu Hualong. "HVIGBT physical model analysis during transient," IEEE Transaction on Power Electronics, 2013, 28(5): 2616-2624.

Average-Value Model of Modular Multilevel Converters Considering Capacitor Voltage

Heya Yang, Yuxiang Chen, Wuhua Li, Xiangning He
College of Electrical Engineering
Zhejiang University
Hangzhou, China
Email: woohualee@zju.edu.cn

Wei Sun, Yongning Chi, Yan Li
Renewable Energy Department,
China Electric Power Research Institute
Beijing, China
Email: sunwei@epri.sgcc.com.cn

Abstract— The average-value model (AVM) of the modular multilevel converters (MMCs) is the simplified one for the average response of switching devices, converters and control. The effect of sub-module capacitor voltage ripple on the MMC dynamic behavior is explored and an accurate average-value model of MMCs is established. Based on the proposed AVM, the equivalent impedances of the DC and AC sides in the MMC converters can be easily extracted and evaluated. Furthermore, the detailed interaction of the AC side voltages/currents and arm voltages and the coupling relationship of AC output current i_d and i_q can be also precisely described by the introduced AVM. The effectiveness of the presented model on the MMC dynamic performance analysis is verified by the simulation.

Keywords—Modular multilevel converter (MMC), *capacitor voltage ripple, Average-value model* (AVM), *cross coupling.*

I. INTRODUCTION

The voltage-sourced converter based high-voltage direct-current (VSC-HVDC) system is a growing candidate of the HVDC technology with the development of the power devices [1]. The VSC-HVDC system has some clear advantages of independent reactive power control, black start capability, and smaller station footprint compared with the line-commutated converter based HVDC (LCC-HVDC) system [2]. Modular multilevel converters (MMCs) [3] are the promising topologies for VSC-HVDC transmission system due to their modularity, scalability and flexibility [4].

Compared with the two-level converters, the MMCs replace the direct-series-connected power devices in each arm by a stack of sub-modules, which consist of the storage capacitors and a set of power devices [5]. In terms of the practical projects and power capability requirements [6], where the voltage level numbers of MMCs can vary from tens to hundreds [7], the low device switching frequency can be employed to reduce the semiconductor switching losses and the equivalent high converter switching frequency offers high quality of the output voltages with small even no filters [8]. However, an account of MMC sub-modules bring serious difficulties for the appropriate control strategy design. Consequently, how to derive an accurate average-value model (AVM) for control performance analysis is a challengeable problem due to the complicated circuit structure with a large

number of identical sub-modules in series per arm.

In order to simplify the complicated structure of the MMC, the series-connected sub-modules in the arms are represented as controllable voltage sources [9] or variable capacitors [10]. Nevertheless, the capacitor features of the arms are not embodied by the controllable voltage source. Furthermore, the sub-module capacitor is regarded as infinite although the sub-module capacitor is variable in the practical applications [11]. And the capacitor voltage ripple is also ignored to simplify the analysis. Thus, it is not accurate to treat the arms of the MMCs as controllable voltage sources or variable capacitors. In order to expose the distinctive features of the MMCs with the distributed sub-module capacitors rather than the centralized bus capacitor, the accurate capacitor voltage ripple is estimated by the respective capacitive energy variation of the arms in [12]. However, the effect caused by the arm currents or the AC output currents on the capacitor voltage ripple is not revealed, and whether the capacitor voltage ripple impacts the performance of the converter is not investigated. Furthermore, it is necessary to develop a particular AVM of the MMC converters and design more optimized control strategy by considering the capacitor voltage ripple.

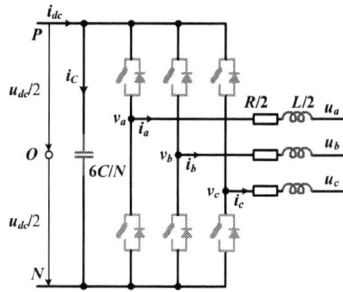

Fig. 1. Conventional MMC AVM in [13].

Usually, the AVM of MMCs is regarded as the same as that of the two-level converters by assuming infinite sub-module capacitors [13], as plotted in Fig. 1. As a result, the controller design of MMCs is similar to that of two-level converters [13]. However, controlling the MMCs as the two-level converters without considering the capacitive feature of the sub-module capacitors may lead to the coupled transient response between i_d and i_q during the active and reactive power change when the

---

This work is sponsored by the National Basic Research Program of China (973 Program 2014CB247400) and the National Nature Science Foundation of China (51490682).

sub-module capacitors are limited. In order to better design the controller of the MMCs, a switching-cycle state-space MMC model is proposed in [14], which can derive the equivalent circuit of the averaged PWM-switch model. However, the nature of this model considers that the MMC capacitors as centralized rather than the distributed capacitors in the real MMC systems. The MMCs are simplified to an equivalent boost-buck converter [15][16]. But the distributed sub-module capacitors are replaced by a single scaled bulk capacitor. Consequently, their influence on the AC side is not clearly investigated.

This paper derives that the total capacitor voltage ripple is controlled by the AC output currents and the sub-module capacitance. And the influence of the distributed capacitor voltage ripple on the interaction of AC output voltages and currents is also disclosed. Furthermore, an average-value model of MMC converters is developed with considering the total capacitor voltage ripple, which provides a more precise AC and DC side impedances of the MMCs. Moreover, the comparison of the AC current transfer functions between the proposed AVM and conventional one is made to reveal the cross-coupling relationship between the i_d and i_q in MMCs.

II. CAPACITOR VOLTAGE RIPPLE IN MMCS

A. Fundamental of MMC Topology

A three-phase MMC is plotted in Fig. 2, where each phase comprises two arms and each arm has N sub-modules and one arm inductor. The sub-module, which is the basic cell for MMCs, has various topologies to meet different requirements [17]. However, the sub-module configuration does not affect the dynamic behavior and control model. Consequently, the half-bridge sub-module is considered in this paper due to its simple structure.

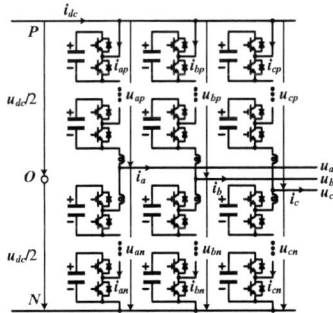

Fig. 2. MMC topology.

With the voltage and current variables and their direction defined in Fig. 2, the AC voltages and currents (u_j and i_j) are described by the following general expressions

$$\begin{cases} u_j = U_1 \sin(\omega t + \varphi_j) \\ i_j = I_1 \sin(\omega t + \varphi_j + \varphi) \end{cases}, (j = a, b, c) \quad (1)$$

where U_1 and I_1 are the peak values of AC voltages and currents, respectively, ω is the fundamental angular frequency, φ_j is the phase angle of phase j, and φ is the power factor angle. Neglecting the converter power loss, the steady-state

three-phase DC and AC powers (P_{dc} and P_{ac}) are

$$P_{dc} = u_{dc} \cdot i_{dc} = 3/2 U_1 I_1 \cos\varphi = P_{ac} \quad (2)$$

where u_{dc} and i_{dc} are DC voltages and currents, respectively.

On the other hand, the voltage modulation index k is defined as

$$k = \frac{U_1}{u_{dc}/2}. \quad (3)$$

Inserting (4) into (3) yields

$$i_{dc} = 3/4 \, k I_1 \cos\varphi. \quad (4)$$

Hence, the DC current can be expressed by the AC current peak value, voltage modulation index and power factor. Moreover, the arm voltages and currents are obtained in term of the Kirchhoff's law from Fig. 2

$$\begin{cases} u_{jp} = \frac{1}{2}u_{dc} - u_j \\ u_{jn} = \frac{1}{2}u_{dc} + u_j \end{cases}, (j = a, b, c) \quad (5)$$

$$\begin{cases} i_{jp} = \frac{1}{3}i_{dc} + \frac{1}{2}i_j \\ i_{jn} = \frac{1}{3}i_{dc} - \frac{1}{2}i_j \end{cases}, (j = a, b, c) \quad (6)$$

where u_{jp} and u_{jn} are the upper and lower arm voltages, i_{jp} and i_{jn} are upper and lower arm currents.

Unlike other multilevel converters, the MMCs depend on the inserting and bypassing operation of the sub-module capacitors to support the DC bus voltage and AC output voltage. Furthermore, the required numbers of sub-modules in upper and lower arms can be obtained as

$$\begin{cases} n_{jp} = \frac{v_{dc}/2 - v_j}{u_C} = \frac{1}{2}N[1 - k\sin(\omega t + \varphi_j)] \\ n_{jn} = \frac{v_{dc}/2 + v_j}{u_C} = \frac{1}{2}N[1 + k\sin(\omega t + \varphi_j)] \end{cases}, (j = a, b, c) \quad (7)$$

where n_{jp} and n_{jn} are the numbers of upper and lower inserted capacitors, v_{dc} is the inner DC alternating voltage, v_j is the inner AC alternating voltage, and u_C is the reference of the sub-module capacitor voltage, which is equal to v_{dc}/N. In the expressions of n_{jp} and n_{jn}, the harmonic components are ignored for the reason that the total sub-module number N is large even up to tens or hundreds.

B. Capacitor Voltage Ripple

As N is large, the equivalent converter switching frequency of MMCs is high enough to ignore the voltage difference among the sub-module capacitors in one arm. Therefore, the sub-module capacitor voltages in one arm are equal to the average value of the total capacitor voltage sum

$$\begin{cases} u_{Cjp} = \cdots = u_{Cjp,N} = \frac{1}{N}\sum_{h=1}^{N} u_{Cjp,h} \\ u_{Cjn} = \cdots = u_{Cjn,N} = \frac{1}{N}\sum_{h=1}^{N} u_{Cjn,h} \end{cases}. \quad (8)$$

where u_{Cjp} and u_{Cjn} are the average capacitor voltage in upper and lower arm, respectively. In terms of the sub-module currents, the average currents of the arm ($i_{Cjp,n}$) can be derived by the currents across the inserted capacitors

$$\begin{cases} i_{Cjp} = \dfrac{1}{N}\displaystyle\sum_{h=1}^{n_{jp}} i_{jp} = \dfrac{n_{jp}}{N} i_{jp} \\ i_{Cjn} = \dfrac{1}{N}\displaystyle\sum_{h=1}^{n_{jn}} i_{jn} = \dfrac{n_{jn}}{N} i_{jn} \end{cases}. \tag{9}$$

If the value of the sub-module capacitor is defined as C, the voltages on the sub-module capacitor in the upper and lower arms can be found by integrating the sub-module capacitor currents i_{Cjp} and i_{Cjn}, which is derived by

$$\begin{cases} u_{Cjp} = \dfrac{1}{C}\displaystyle\int i_{Cjp} dt + \dfrac{v_{dc}}{N} = \dfrac{1}{C}\int \dfrac{n_{jp}}{N} i_{jp} dt + \dfrac{v_{dc}}{N} \\ u_{Cjn} = \dfrac{1}{C}\displaystyle\int i_{Cjn} dt + \dfrac{v_{dc}}{N} = \dfrac{1}{C}\int \dfrac{n_{jn}}{N} i_{jn} dt + \dfrac{v_{dc}}{N} \end{cases},(j=a,b,c).\tag{10}$$

Once the capacitor voltage is balanced, v_{dc}/N in (10) means the DC component of the capacitor voltages. And the capacitor voltage ripples, which comes from the integrations in (10), are the periodic variation components of the sub-module voltages.

The total voltages of the inserted sub-modules ($u_{sjp,n}$) in the arms are found by multiplying the capacitor voltages in (10) with the numbers of inserted sub-modules in (7), which is derived by

$$\begin{cases} u_{sjp} = n_{jp} u_{Cjp} = u_{rjp} + v_{dc}/2 - v_j \\ u_{sjn} = n_{jn} u_{Cjn} = u_{rjn} + v_{dc}/2 + v_j \end{cases},(j=a,b,c)\tag{11}$$

where u_{rjp} and u_{rjn} are the total capacitor voltage ripples in the arms, which are derived as

$$\begin{cases} u_{rjp} = \dfrac{n_{jp}}{C}\displaystyle\int \dfrac{n_{jp}}{N} i_{jp} dt \\ u_{rjn} = \dfrac{n_{jn}}{C}\displaystyle\int \dfrac{n_{jn}}{N} i_{jn} dt \end{cases},(j=a,b,c).\tag{12}$$

On the other hand, the integral of the AC output currents can be obtained as

$$\int i_j dt = -\frac{1}{\omega} I_1 \cos(\omega t + \varphi_j + \varphi),(j=a,b,c).\tag{13}$$

By inserting (6), (7) and (13) into (12), and ignoring the small items for simplification, the total capacitor voltage ripples can be performed as

$$\begin{cases} u_{rjp} = n_{jp} \dfrac{1}{C}\displaystyle\int \dfrac{n_{jp}}{N} i_{jp} dt = \dfrac{N}{8C}(1-\dfrac{3}{8}k^2)\int i_j dt \\ u_{rjn} = n_{jn} \dfrac{1}{C}\displaystyle\int \dfrac{n_{jn}}{N} i_{jn} dt = -\dfrac{N}{8C}(1-\dfrac{3}{8}k^2)\int i_j dt \end{cases}.\tag{14}$$

Therefore, the total capacitor voltage ripples can be regarded as the voltage ripples on the equivalent capacitors charging by the AC output currents.

III. Proposed Average Value Model of MMC

In Section II, the capacitor voltage ripples are derived by the average sub-module capacitor voltages and average sub-module capacitor currents. Based on the capacitor voltage ripples, an average value model of MMC is proposed in this Section. From Fig. 2, the arm voltages can be obtained as

$$\begin{cases} u_{jp} = u_{sjp} + L\dfrac{di_{jp}}{dt} + Ri_{jp} \\ u_{jn} = u_{sjn} + L\dfrac{di_{jn}}{dt} + Ri_{jn} \end{cases},(j=a,b,c).\tag{15}$$

It can be seen that the arm voltages are the sums of the total voltages of the inserted sub-modules, the voltages of the arm inductors, and the voltages of the arm equivalent resistances. Given (5), (11), and (14), the AC output voltages can be expressed by the arm voltages, which is given as

$$u_j = \frac{1}{2}(u_{jn} - u_{jp}) = v_j - \frac{N}{8C}(1-\frac{3}{8}k^2)\int i_j dt - \frac{1}{2}L\frac{di_j}{dt} - \frac{1}{2}Ri_j.\tag{16}$$

Consequently, the AC side equivalent impedance is obtained as

$$Z_{ac} = \frac{N}{8C\omega}(1-\frac{3}{8}k^2) + \frac{1}{2}L\omega + \frac{1}{2}R.\tag{17}$$

On the other hand, the DC voltage is sustained by the sums of the upper and lower voltages. Inserting (6) and (15) into (5) yields

$$u_{dc} = u_{jp} + u_{jn} = v_{dc} + \frac{2}{3}L\frac{di_{dc}}{dt} + \frac{2}{3}Ri_{dc},(j=a,b,c).\tag{18}$$

Therefore, the DC bus voltage can be regarded as the total voltage of the inner DC alternating voltage and the voltage on the DC equivalent inductors and resistances. Moreover, the DC side equivalent capacitor can be expressed by the sub-module capacitors in terms of the conservation of energy

$$E_{dc} = \frac{1}{2}C_{dc}v_{dc}^2 = E_{tot} = \sum_{j=a,b,c}\frac{1}{2}CN(u_{Cjp}^2 + u_{Cjn}^2).\tag{19}$$

By inserting (10) into (19), the DC side equivalent capacitor can be obtained with ignoring the small items

$$C_{dc} = \frac{6C}{N}.\tag{20}$$

According to (18) and (20), the DC equivalent impedance can be expressed as

$$Z_{dc} = \frac{N}{6C\omega} + \frac{2}{3}L\omega + \frac{2}{3}RC_{dc}.\tag{21}$$

Consequently, the proposed average-value model of the MMC, which is illustrated in Fig. 3, is described by (17) and (21) in concert demonstrating the DC and AC output voltages and currents performance. As can be seen, the proposed AVM is illustrated as a two-level voltage source converter with equivalent DC and AC impedances of the MMCs. In Fig. 3, C_{ac} is equal to $8C/N(1-3k^2/8)$, and the DC equivalent inductors and resistances is divided into two impedances of $L/3$ and $R/3$ for symmetry. Furthermore, the proposed AVM can be used to develop the small signal model to design the control strategy.

978-1-4673-9551-9/16 $31.00 © 2016 IEEE

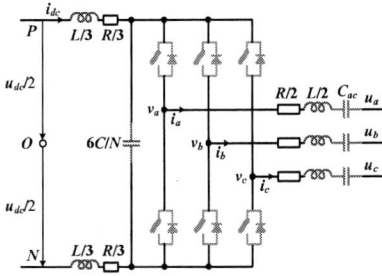

Fig. 3. Proposed MMC AVM.

IV. COMPARISON OF PROPOSED AND CONVENTIONAL AVMs

According to Fig. 1 and Fig. 3, the major difference between the proposed AVM and the conventional AVM, is the AC equivalent capacitor $8C/N(1-3k^2/8)$ appearing in the AC side of the proposed model, which may result in the difference in the AC output dynamic performance. (16) is rewritten to identify the expressions of the AC output voltages and currents

$$\begin{cases} u_j = v_j - u_{Cacj} - \dfrac{1}{2}L\dfrac{di_j}{dt} - \dfrac{1}{2}Ri_j \\ i_j = \dfrac{8C}{N(1-3k^2/8)}\dfrac{du_{Cacj}}{dt} \end{cases}, (j=a,b,c) \quad (22)$$

where u_{Cacj} is the voltage on the equivalent AC side capacitors. Furthermore, (22) under the Laplace transform, in terms of the synchronous dq-frame, is expressed as

$$\begin{cases} v_d = u_d + [\dfrac{1}{2}Ls + \dfrac{1}{2}R + \dfrac{N(1-3k^2/8)s}{8C(s^2+\omega^2)}]i_d - [\dfrac{1}{2}L\omega - \dfrac{N(1-3k^2/8)\omega}{8C(s^2+\omega^2)}]i_q \\ v_q = u_q + [\dfrac{1}{2}Ls + \dfrac{1}{2}R + \dfrac{N(1-3k^2/8)s}{8C(s^2+\omega^2)}]i_q + [\dfrac{1}{2}L\omega - \dfrac{N(1-3k^2/8)\omega}{8C(s^2+\omega^2)}]i_d \end{cases}. (23)$$

From (23), the cross-coupling items $\dfrac{1}{2}L\omega - \dfrac{N(1-3k^2/8)\omega}{8C(s^2+\omega^2)}$ in the AC output performance of the AVM are relative to the arm inductor, sub-module capacitor, sub-module number and voltage modulation index. Usually, the control strategy of the MMCs is the feed-forward decoupling control of the conventional AVM, the feed-forward decoupling control strategy is expressed as

$$\begin{cases} v_d = u_d + (k_{pi} + \dfrac{k_{ii}}{s})(i_{dr} - i_d) - \dfrac{1}{2}L\omega i_q \\ v_q = u_q + (k_{pi} + \dfrac{k_{ii}}{s})(i_{qr} - i_q) + \dfrac{1}{2}L\omega i_d \end{cases} \quad (24)$$

where i_{dr} and i_{qr} are the references of the d-axis current and q-axis current, respectively, and k_{pi} and k_{ii} are the PI parameters of the inner current control loop. Furthermore, when the conventional control strategy is used in the proposed AVM, the currents in d-axis and q-axis are not decoupled completely. The transfer functions of the AC currents are obtained by (23) and (24) in concert as

$$\begin{bmatrix} i_d \\ i_q \end{bmatrix} = \begin{bmatrix} G_{dd} & -G_{dq} \\ G_{dq} & G_{qq} \end{bmatrix}\begin{bmatrix} i_{dr} \\ i_{qr} \end{bmatrix} \quad (25)$$

where

$$\begin{cases} G_{dd} = G_{qq} = \dfrac{(a+b)b}{(a+b)^2 + c^2} \\ G_{dq} = \dfrac{bc}{(a+b)^2 + c^2} \end{cases} \quad (26)$$

and

$$\begin{cases} a = \dfrac{1}{2}Ls + \dfrac{1}{2}R + \dfrac{N(1-3k^2/8)s}{8C(s^2+\omega^2)} \\ b = k_{pi} + \dfrac{k_{ii}}{s} \\ c = \dfrac{N(1-3k^2/8)\omega}{8C(s^2+\omega^2)} \end{cases} \quad (27)$$

G_{dd} and G_{qq} in (25) are the same, which are the transfer functions of the closed inner current control loop, and G_{dq} is the cross-coupling transfer function between the currents in d-axis and q-axis. Therefore, the cross-coupling relationship of i_d and i_q is revealed by (25).

V. SIMULATION RESULTS

Verification of the theory contribution proposed in this paper is implemented in a three-phase MMC PSCAD/EMTDC simulation system, rated at 18MVA, with N=24 sub-modules per arm, AC and DC terminal voltages of 35kV and 60kV, fundamental frequency of 50Hz, with parameters L=0.14H, C=810μF, and R=0.5Ω. In order to confirm the voltages and currents performance, the MMC can be exhibited by the proposed AVM. And the parameters of the proposed AVM are relative to those of the MMC and SPWM strategy with the switching frequency of 12kHz to make the same inductor current ripples.

A. Simulation of Steady-State Operation

To verify the steady-state performance, the output voltages and currents of the MMC and the proposed AVM are shown respectively in Fig. 4(a) and Fig. 4(b) with active power P=18MW(1p.u.) and reactive power Q=0. It can be seen that the agreement between the output performances of the MMC circuit model and proposed AVM is good. The purpose of the simulation results in Fig. 4(c) is to verify (8) to confirm that the assumptions of the sub-module voltage and current are reasonable. The N capacitor voltages (solid) in the upper arm of phase a are shown in Fig. 4(c), synthesized as the average voltage (dashed) of them. As expected, the assumptions of the sub-module performance are confirmed to be valid in the derivation the proposed AVM.

B. Simulation of Dynamic Response

On the other hand, the output currents of the MMCs in d-axis and q-axis are not decoupled completely with the conventional control strategy revealed by the proposed AVM. In order to verify the dynamic response performances, two step response simulations are illustrated as follow.

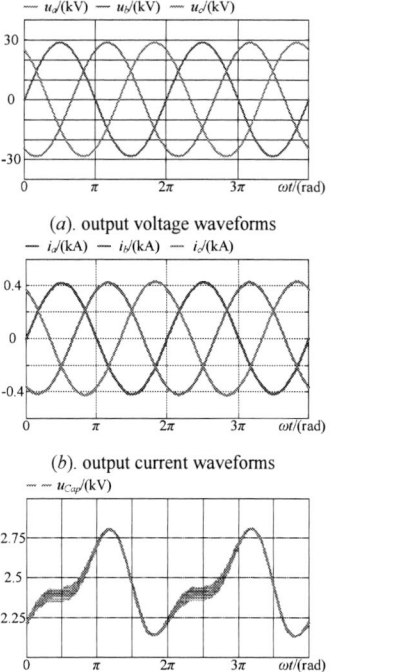

(a). output voltage waveforms

(b). output current waveforms

(c). capacitor voltages and average voltage (upper arm of phase a)

Fig. 4. MMC circuit steady-state waveforms.

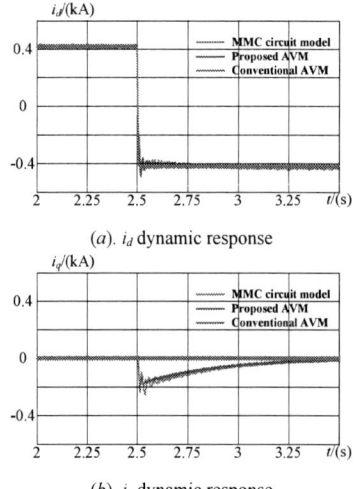

(a). i_d dynamic response

(b). i_q dynamic response

Fig. 5. dynamic response waveforms of MMC circuit model, proposed AVM and conventional AVM (k=1; i_d=0.4kA→-0.4kA;i_q=0).

In the first case, the active power is transmitted from the DC side to AC side before t=2.5s, and the active power is transmitted reversely after t=2.5s. i_d and i_q transient response of MMC system of the proposed and conventional AVMs are recorded in Fig. 5(a) and Fig. 5(b), respectively. As can be seen, the i_d dynamic responses of the MMC circuit model, proposed and conventional AVMs are similar. Furthermore, the proposed AVM with the little tail in the i_d dynamic response is more accurate than the conventional one, and the

difference between the i_q dynamic responses of the proposed AVM and conventional one is obvious. The tail current in the i_q dynamic response of the proposed AVM is approximate to the dynamic response of the MMC circuit model. Conversely, the conventional AVM failed to reflect the tail current. As a consequence, the simulation results verify the accurate dynamic performances of the proposed AVM in this case.

In the second case, the change of the power flowing is the same as that in the first case. However, the modulation index k reduces to 0.8 in the second case. The dynamic responses of the output currents are illustrated in Fig. 6. From Fig. 6, the proposed AVM has described the MMC performance that the dynamic responses of i_d and i_q have tail currents and approach the current commands slowly. As a result, the proposed AVM is verified to be effective even under the situations of the different voltage modulation index k.

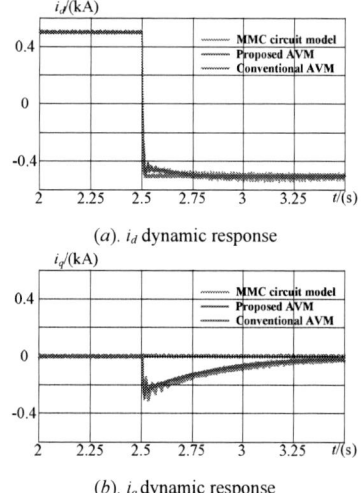

(a). i_d dynamic response

(b). i_q dynamic response

Fig. 6. dynamic response waveforms of MMC circuit model, proposed AVM and conventional AVM (k=0.8; i_d=0.5kA→-0.5kA; i_q=0).

VI. CONCLUSION

In conclusion, the inherent relationship between the AC output voltages and currents and system parameters RLC in the MMCs is derived by considering the distributed sub-module capacitor voltage ripple in this paper. Based the capacitor voltage ripple, a particular average-value model of the MMCs is presented to better show the real dynamic behavior of the MMCs. Moreover, the conspicuous difference between the proposed AVM and conventional AVM is the AC equivalent capacitor $8C/N(1-3k^2/8)$, which reveals that the coupled transient response between i_d and i_q in MMCs using the conventional control strategy. Furthermore, the cross-coupling relationship is explored by the transfer functions of the currents in the d-axis and q-axis and their commands. Finally, the simulation results verify the validity of the proposed AVM in the steady-state operation and dynamic response behavior in the paper.

978-1-4673-9551-9/16 $31.00 © 2016 IEEE

REFERENCES

[1] R. Marquardt, "Modular multilevel converter: An universal concept for HVDC-networks and extended DC-bus-applications," in *Proc. Int. Power Electron. Conf.*, Jun. 2010, pp. 502–507.

[2] N. Flourentzou, V. G. Agelidis, and G. D. Demetriades, "VSC-based HVDC power transmission systems: An overview," *IEEE Trans. Power Electron.*, vol. 24, no. 3, pp. 592–602, Mar. 2009.

[3] A. Lesnicar and R. Marquardt, "An innovative modular multilevel converter topology suitable for a wide power range," in *Proc. IEEE Bologna Power Tech Conf.*, 2003, pp. 3–6.

[4] K. Friedrich, "Modern HVDC PLUS application of VSC in modular multilevel converter topology," in *Proc. IEEE Int. Symp. Ind. Electron.*, 2010, pp. 3807–3810.

[5] N. Ahmed, S. Norrga, H. P. Nee, et al., "HVDC supergrids with modular multilevel converters— the power transmission backbone of the future," in *Proc. Int. Conf. Syst., Signals Devices*, 2012, pp. 1–7.

[6] F. Dijkhuizen, "Multilevel converters: Review, form, function and motivation," presented at the *EVER 2012*, Monte Carlo, Monaco, 2012.

[7] J. Peralta, H. Saad, S. Dennetière, et al., "Detailed and averaged models for a 401-level MMC-HVDC system," *IEEE Trans. Power Del.*, vol. 27, no. 3, pp. 1501–1508, Jul. 2012.

[8] T. Modeer, H. P. Nee, and S. Norrga, "Loss comparison of different sub-module implementations for modular multilevel converters in HVDC applications," in *Proc. 14th Eur. Conf. Power Electron. App.*, 2011, pp. 1–7.

[9] U. N. Gnanarathna, A. M. Gole, and R. P. Jayasinghe, "Efficient modeling of modular multilevel HVDC converters (MMC) on electromagnetic transient simulation programs," *IEEE Trans. Power Del.*, vol. 26, no. 1, pp. 316–324, Jan. 2011.

[10] S. P. Engel and R. W. De Doncker, "Control of the modular multi-level converter for minimized cell capacitance," in *Proc. 14th Eur. Conf. Power Electron. Appl.*, Aug. 30–Sep. 1, 2011, pp. 1–10.

[11] J. Xu, A. M. Gole and C. Zhao, "The Use of Averaged-Value Model of Modular Multilevel Converter in DC Grid," *IEEE Trans. Power Del.*, vol. 30, no. 2, pp. 519–528, April. 2015.

[12] M. Vasiladiotis, N. Cherix and A. Rufer, "Accurate Capacitor Voltage Ripple Estimation and Current Control Considerations for Grid-Connected Modular Multilevel Converters," *IEEE Trans. Power Electron.*, vol. 29, no. 9, pp. 4568–4579, Sept. 2014.

[13] A. Antonopoulos, L. Angquist, and H. P. Nee, "On dynamics and voltage control of the modular multilevel converter," in *Proc. 13th Eur. Conf. Power Electron. Appl.*, Sep. 8–10, 2009, pp. 1–10.

[14] J. Wang, R. Burgos and D. Boroyevich, "Switching-Cycle State-Space Modeling and Control of the Modular Multilevel Converter," *IEEE Journal of Emerging and Selected Topics in Power Electronics*, vol. 2, no. 4, pp. 2168-6777, Dec. 2014.

[15] D. C. Ludois and G. Venkataramanan, "Simplified Terminal Behavioral Model for a Modular Multilevel Converter," *IEEE Transactions on Power Electron.*, vol. 29, no. 4, pp. 1622-1631, April 2014.

[16] D. C. Ludois and G. Venkataramanan, "Simplified dynamics and control of modular multilevel converter based on a terminal behavioral model," in *Proc. IEEE Energy Convers. Congr. Expo.*, 2012, pp. 3520–3527.

[17] A. Nami, J. Liang, F. Dijkhuizen, and G. Demetriades, "Modular multilevel converters for HVDC applications: Review on converter cells and functionalities," *IEEE Trans. Power Electron.*, vol. 30, no. 1, pp. 18–36, Jan. 2015.

AUTHOR INDEX

A

Abdelmoaty, Ahmed	2437
Abdul Azeez, Najath	3140
Abe, Seiya	1640, 2422
Abedinpour, Siamak	3669
Abramov, Eli	111, 692
Abramson, Rose A.	1138
Abu Qahouq, Jaber A.	1868, 2114, 3611, 3684
Abu-Rub, Haitham	1214, 3663
Acero, J.	3020, 3026, 3566
Achanta, Prasanta K.	3273
Adhikari, Jeevan	9
Adragna, Claudio	564
Afridi, Khurram K.	1138, 1392, 1947, 2395
Afsharian, Jahangir	33, 899, 2312, 2320
Agamy, Mohammed	3403
Agelidis, Vassilios G.	236, 1702
Agostinelli, M.	339, 350
Agostini, Francesco	472
Agrawal, Neeraj	951
Aguilar, A.	2540
Ahmed, Emad M.	1505
Ahmed, Ibrahim	1882
Ahmed, Mahrous	1505
Ahmed, Shamim	1646
Ahmed, Shehab	936
Ahmed-Zaid, Said	2821
Ahn, Jung-Hoon	163, 1273, 2161
Ahsanuzzaman, S.M.	2497
Akagi, Hirofumi	1163
Akamatsu, Keiji	2607
Akin, Bilal	505, 1096, 1176, 2108, 2875
Alatise, Olayiwola	253, 2645
Al-Durra, Ahmed	1941
Alexandrov, Peter	2973
Al-Hallaj, Said	3128
Alharbi, Mahmood	3333
Ali, Kawsar	2491
Allard, Bruno	524
Allen, Scott	979
Allmeling, Jost	1108
Alonso, J. Marcos	1115
Alonso, R.	3020

Alonso, Rafael 3026
Alou, P. 2409
Al-Shyoukh, Mohammad 2437
Am, Sokchea 2401, 2700
Amaro, Mike 66
Ambacher, Oliver 2083
Amirabadi, Mahshid 3704
Amirahmadi, Ahmadreza 3333
Amon, Cristina 1350
Amouzandeh, Maryam S. 329
Andersen, Michael A.E. 1090, 1430, 1541, 1842, 2252, 2473
Andersen, Thomas 1430, 1842
Ando, Masato 2986
Aniruddhan, Sankaran 1878
Anthon, Alexander 1235, 2252
Antunes, Fernando L.M. 2592
Anwar, Saeed 424
Anwar, Usama 1947
Arafat, A.K.M. 1123
Arafat, Akm 2847
Arefifar, Ali 2561
Arias, Andrea 79
Arias, M. 1823
Arias, Manuel 822
Arnold, Cory 1597, 3273
Asa, Erdem 1323, 1756, 1767, 2587
Asensi, R. 1624
Ayers, Curtis 3529
Ayyanar, Raja 432, 3364
Azcondo, F.J. 2389
Azuma, Katsunori 283

B

Badawey, Mohammed 392
Baek, Jeihoon 3004
Bagawade, Snehal 544
Bahman, Amir Sajjad 261, 3012
Bahmani, M.A. 3043
Bai, Hua 529
Bai, Yongjiang 766, 3623
Baier, Thomas 2897
Bak, Claus Leth 3051
Bak, Yeongsu 2764, 3416
Baker, Michael W. 1597
Bakhshai, Alireza 460
Bakker, Cas 2457
Balasubramanian, Bharat 1868
Balda, Juan Carlos 143, 362, 651, 1387, 3712
Ball, Roy 2122, 3038

Bandyopadhyay, Santanu	3286
Banerjee, Arijit	2881
Baranwal, Rohit	2043
Barbosa, A.U.	3231
Bari, Syed	3259
Barlow, Matthew	1646
Barner, Alexander	106
Barrado, A.	2545, 3090
Barth, Christopher	1512
Barthelmebs, Clement	453
Batarseh, Issa	1381, 3333
Bawohl, Melanie	3069
Bayhan, Sertac	3663
Bazzi, Ali M.	2666
Bęczkowski, Szymon	704, 974, 3101
Bede, Lorand	1702, 2264
Beres, Remus	3051
Bergman, Joshua	79
Bermejo, M.	3090
Bermingham, Jack	2794
Berzoy, Alberto	928, 3200
Betz, Vaughn	1882
Bezdenezhnykh, Yevgeny	308
Bhalla, Anup	2973
Bhangu, Bicky	715
Bhardwaj, Manish	505
Bhattachaarjee, Parijat	1344
Bhattacharya, Subhashish	295, 601, 778, 886, 1497, 1632, 2076
Bhattacharya, Tanmoy	199
Bhowmik, Pankaj Kumar	2706
Biglarbegian, Mehrdad	2998
Biswas, Suvankar	1934
Bizjak, Luca	1663
Blaabjerg, Frede	221, 229, 261, 288, 370, 1253, 1872, 1941, 1995, 2011, 2154, 2207, 2215, 2264, 3012, 3051, 3416, 3431, 3500
Blalock, Benjamin J.	684, 893, 1569, 3255
Blasko, Vladimir	2167
Böcker, Joachim	1547
Bodano, Emanuele	1663
Bojarski, Mariusz	1756
Bonthu, Sai Sudheer Reddy	1131
Bonyadi, Roozbeh	253
Bonyadi, Yeganeh	253
Borges, Beatriz	3637
Born, Rachael	1148, 3243
Boroyevich, Dushan	177, 516, 524, 739, 1024, 1315, 1561
Botting, Chris	854
Boynuegri, Ali R.	207
Braga, A.P.S.	3231

Brandt, Tobias 3172
Brar, Berinder 79
Breaz, Elena 3476
Briggs, Roger 2927
Brohlin, Paul 838
Brothers, John A. 990, 1967
Brown, Alan 529
Burdío, J.M. 1040, 1762, 3020, 3026, 3566
Burgos, Rolando 177, 516, 524, 1024, 1561
Buticchi, Giampaolo 2449, 3493, 3629
Buttay, Cyril 524

C

Cai, Wen 1057, 1861, 2599
Campbell, S.L. 1307
Canacsinh, Hiren 3637
Canales, Francisco 472
Cao, Dong 3553
Cao, Jiankun 1371
Cao, Wenchao 2229
Cao, Yuan 1868, 3684
Carlos, Gregory A.A. 3641
Carretero, C. 663, 3020, 3026, 3566
Casady, Jeffrey 979
Castro, Ignacio 25, 822
Castro, Marcus R. 2592
Ceballos, Salvador 236
Cervera, Alon 111, 692, 2298
Cha, Hanju 511, 1708, 3598
Chae, Young-Ho 2801
Chakraborty, Shiladri 1954, 2652, 3389
Challingsworth, Mark 3069
Chang, C.-H. 951
Chatterjee, Urmimala 1183
Chattopadhyay, Ritwik 778
Chattopadhyay, Souvik 1954, 2652, 3389
Chawda, Pradeep 3266
Chee, Seung-Jun 1206, 2370
Chen, Alian 3453, 3465
Chen, Changdong 2981
Chen, Cheng-Po 3255
Chen, Chingchi 1554
Chen, Di 529
Chen, Fang 177
Chen, Guipeng 1450
Chen, Guodong 499
Chen, Guoliang 1227
Chen, Hao 1437
Chen, Hua 1947

Chen, Jie	3453
Chen, Min	138
Chen, Minjie	1138, 1443
Chen, Qianhong	2518
Chen, Runruo	1045
Chen, Weiqiang	2666
Chen, Wenjie	493, 766, 3115, 3623
Chen, Woei-Luen	3471
Chen, Xinwen	1358
Chen, Xuling	1788
Chen, Yang	899, 2304, 2312, 2320
Chen, Yang-Lin	558
Chen, Yaow-Ming	558
Chen, Ying	2071
Chen, Yuxiang	499, 1462
Chen, Zhe	3431
Cheng, Chun Sing	1795
Cheng, Kuang-Yao	118
Cheung, Chun	1616
Chi, Yongning	1462
Chinthavali, M.	1307
Cho, Bo-Hyung	487, 1416
Cho, Shin Young	3690
Choe, Songbaek	2051
Choi, Beomseok	1947
Choi, Byeung G.	1773
Choi, Hee-Su	3153
Choi, Seungdeog	631, 1123, 1131, 1748, 2847, 3004
Choi, Sewan	859
Choi, Sung-Jin	3153
Choi, Wooin	1416
Choi, Wooyoung	2679
Chou, Derek	1512
Chow, Jeff Po Wa	1795
Chowdhury, Md Asif Mahmood	207
Chub, Andrii	2533
Chun, Chang Yoon	3322
Chung, Henry Shu-Hung	1795, 1807, 2154
Chung, Steven	1350
Church, Ron	786
Ci, Song	3189
Ciobotaru, Mihai	1702
Cobos, J.A.	2409
Coelho, Ernane A.A.	3585
Colak, Kerim	1323, 1756, 1767, 2587
Colmenares, Juan	746, 1018
Comanescu, Mihai	2759, 2855
Connaughton, A.M.	355
Conway, Thomas	1670

Cook, M. 3537
Correa, Maurício B.R. 3641, 1032
Corzine, Keith 720, 1191, 1481, 2187, 2840
Cosetin, Marcelo 1115
Costa, Levy F. 2449
Costa, Louelson A. 1032
Costa, Paulo Junior Silva 2376
Costinett, Daniel 424, 872, 893, 1010, 1569, 2441, 3255, 3577
Craciun, Marian 854
Cui, Shenghui 2620
Cui, Yutian 893
Curuvija, Boris 3553
Cuzner, Robert 1577
Czarkowski, Dariusz 1323, 1756, 1767, 2587
Czwickla, Christoph 3069

D

Dahan, Nadav 802
Dai, Ke 1358
Dai, Zhiyong 3134
Dai, Ziwei 1358
Dally, William J. 86
Dang, Zhigang 2114, 3684
Daniel, Michael T. 1695
Dargahi, Vahid 720, 1191, 1481, 2187, 2840
Daryaei, Mohammad 2579
Das, Partha Pratim 2652
Das, Pritam 552, 2491
Dashmiz, Shadi 3297
Davari, Pooya 221, 229
Davletzhanova, Zarina 253
de Almeida Carlos, Gregory A. 2720
de Almeida, Bruno Ricardo 60
De Carne, Giovanni 3493
De Doncker, Rik W. 643
de Oliveira Pacheco, Juliano 3346
de Rooij, Michael 2292
de Souza Oliveira Jr., Demercil 60, 3231, 3346
De, Ankan 295, 1632
Debnath, Suman 1528
Deboy, Gerald 3570
Degner, Michael W. 241
Delhotal, J. 1926
Demerdash, Nabeel A.O. 1065, 2826
Deng, Cheng 143, 362, 651
Deng, Hao 816
Deng, Lu 3521
Deng, Yan 1450
Dias Jr., A.J.S. 3231

Diaz Reigosa, Paula	288
Diaz, Nelson	1227
Dimarino, Christina	516
Ding, Pengling	1371
Ding, Weisheng	440
Dinulovic, Dragan	3097
Ditze, Stefan	864
Divan, Deepak	1437
Dix, Jeffery	684
do Prado, Ricardo N.	1115
Dobmeier, Christian	1741
Domoto, Kazuhide	2422, 2465
Dong, Dong	3403
Dong, Zhou	73, 2518
Doolla, Suryanarayana	2245, 3376
Dorn-Gomba, Lea	453
dos Santos Jr., Euzeli C.	2720
Dos Santos, Gutemberg G.	1032
Dou, Manfeng	3134
Dou, Qingyun	2272
Dousoky, Gamal M.	2735
Driesen, Johan	1183
Driessen, Anton	2457
Drofenik, Uwe	472
Du, Weijing	1002
Du, Weijing	2334
Du, Xiong	2992
Duarte, J.L.	3158
Dujic, Drazen	156, 1108
Dumais, Alex	3219
Dusmez, Serkan	505, 1176, 2108
Dutta, Atanu	3012

E

Eberle, Wilson	1286
Ebrahimi, Mohammad	2579, 3207
Edpuganti, Amarendra	402, 943
Egan, Michael	2794
Ehrlich, Stefan	1741
Einspieler, Sascha	759
Ekhtiari, Marzieh	1430
Elrayyah, Ali	392, 2660, 2806
Elsayed, Ahmed T.	1267
El-Taweel, Nader A.	830
Emadi, Ali	453, 1300
Engelmann, Georges	643
Eni, Emanuel-Petre	974, 3101
Enjeti, Prasad	936, 1695, 2567, 3545
Enshaei, Hossein	2813

Enslin, Johan 2998
Erickson, Robert 1947
Ertl, H. 1
Escobar-Mejía, Andrés 362
Eskandari, Soheila 2127
Essakiappan, Somasundaram 2706
Eum, Hyunchul 2355
Evzelman, Michael 1603
Ezra, Ofer 308

F

Fabricio, Edgard L.L. 3641
Fan, Bo 334
Faraci, Eric 838
Fard, Miad 1403
Farhang, Peyman 733
Farley, Kathleen Blair 1737, 3526
Farnell, Chris 143
Faulkner, Bryan 54
Fayed, Ayman 2437
Fedison, J.B. 247
Fei, Chao 322
Feng, Junjie 1534, 2334
Ferdowsi, Mehdi 1962, 2687
Fernandes, B.G. 2245, 2673, 3376
Fernandes, Darlan A. 1032
Fernandez, A. 1823
Fernandez, C. 2545, 3090
Ferrieux, Jean-Paul 2700
Figge, Heiko 1547
Flankl, Michael 623
Foulkes, Thomas 1512
Francés, A. 1624
Francis, A. Matt 1646
Freitas, Antônio A.A. 2592
Freitas, Luiz C.G. 3585
Frey, David 2401, 2700
Friedrichs, Daniel 3577
Fröhleke, Norbert 1547
Fu, Lixing 1554, 1967
Fu, Shihang 1475
Furukawa, Keita 1336

G

Gaafar, Mahmoud A. 2735
Gafford, James 1577
Gajanayake, C.J. 2942, 3058
Gakhar, Vikram 1878
Galiano Zurbriggen, Ignacio 386

Gan, Yiliang	3560
Gandikota, Srikant	1051
Gao, Fei	3476
Gao, Feng	410, 536, 921, 2259, 2935
Gao, Mingzhi	138
Gao, Rui	3383
Gao, Sugu	2868
Gao, Xieping	3185
Gao, Yabiao	1737, 3526
Gao, Yikai	1861
Garces, Luis	3403
Garcia Rodriguez, Luciano Andres	651
Garcia, Jorge	3508
Garcia, O.	2409, 1624
Garcia, Pablo	3508
Garcia, Virginia	3069
Garcia-Rodriguez, Luciano A.	362, 3712
Gavagsaz-Ghoachani, Roghayeh	446, 3397
Ge, Baoming	1214
Ge, Hongjuan	1424
Ge, Ting	668
Ge, Xiongxuan	3080
Geng, Shengbao	3560
Georgious, Ramy	3508
Gerfer, Alexander	2553, 3097
Gerling, Dieter	215
Ghaffarzadeh, Hooman	3353
Ghandi, Reza	3255
Ghat, Mahendra B.	2342
Ghias, Amer M.Y.M.	236
Giezendanner, Florian	1018
Ginart, Antonio	1737, 3526
Glavanovics, Michael	759
Glover, S.	3537
Goetz, Stefan M.	2349
Gohil, Ghanshyamsinh	1702, 2264
Goktas, Taner	1096, 2875
Gong, Bing	33
Gonnet, Luc	2700
Gonzalez, S.	1926
Gonzalez-Llorente, Jesus	3712
Gorla, Naga Brahmendra Yadav	2491
Gotovac, Ante	1663
Gou, Ruifeng	2071
Gray, C. Thomas	86
Greer III, Thomas H.	86
Gritti, Giovanni	564
Grosse, Thorben	643
Gu, Bin	838

Gu, Dong-Jie 1243
Gu, Lei 2889
Guerrero, Josep M. 398, 1227, 1376, 3459, 3697
Gui, Han-Dong 1243
Guirguis, David 1350
Gulbudak, Ozan 3248
Gunasekaran, Deepak 1045, 2525
Gundel, Paul 3069
Gunter, Samantha J. 1138
Guo, Ben 1010
Guo, Feng 1682
Guo, Suxuan 2063
Guo, Xiaoqiang 398
Gupta, Amit K. 715, 2942, 3058
Gupta, Ankit 1344
Gupta, Mahima 2919
Gupta, Ranjan K. 1520
Gupta, Shalabh 2666
Gurusinghe, Nicoloy 2479

H

Ha, Jung-Ik 193, 487, 1398, 2801, 3717
Hadjidemetriou, Lenos 3500
Hafez, Bahaa 936
Halivni, Bar 111
Hameyer, Kay 643
Han, Di 2861, 2950
Han, Jung Kyu 3690
Han, Xiangyu 2957
Han, Yang 816
Hang, Lijun 3560
Hanrahan, Robert 3266
Hanson, Alex J. 98
Haque, Moinul Shahidul 3004
Hare, James 2666
Harfman-Todorovic, Maja 3403
Hariharan, K. 315
Hariya, Akinori 2430
Harris, Richard Kyle 3255
Harrison, M.J. 247
Hartmann, M. 1
Haryani, Nidhi 1024, 1561
Hasan, Iftekhar 638
Hasegawa, Kazunori 3032
Hata, Katsuhiro 1731
Hata, Yuki 468
Hatae, Shinji 468
Hattori, Yoshiyuki 3146
Haug, Martin 3097

Hayakawa, Seiichi	283
Hayes, John G.	2794
He, Dingyi	2599
He, Haibing	3185
He, Jiangbiao	1065, 1084, 2826
He, Jinwei	1249
He, Ruirui	766
He, Xiangning	499, 1450, 1462, 1475
He, Xiaobin	2182, 2194
Heckel, Thomas	864
Heldwein, Marcelo Lobo	2833
Henke, M.	700
Henkenius, Carsten	1547
Herbert, Joseph	1123, 3004
Hernando, Marta M.	822
Higaki, Yusuke	1713
Hilber, Patrik	746
Hinken, Reiner	303
Hitoshi, Ishii	676
Ho, Carl Ngai-Man	2905
Ho, Kwun Yuan Godwin	2328
Hofmann, Heath	1721, 1726
Hofmann, Wilfried	2175
Holmes, Grahame	2252
Hopkins, Douglas C.	295, 2141
Hori, Yoichi	1731
Hosseini, Rasoul	1577
Hou, Dongbin	657
Hou, Ruoyu	1300
Hsiehu, H.-C.	951
Hu, Haibing	2182, 2194
Hu, Ji	253
Hu, Sheng	3310
Hu, Xiaolei	1071, 3409, 3591
Hu, Zhiyuan	899, 2320
Huang, Alex Q.	132, 269, 983, 2063, 2365, 2727, 2927, 3383, 3648, 3677
Huang, J.-W.	1364
Huang, Kuohsien	2355
Huang, Qingjun	3521
Huang, Qingyun	2727, 3648
Huang, Xiucheng	1002, 1534, 1853, 2334
Huang, Yi	1616
Huang, Ying	1888, 1894
Huang, Yung-Ting	1900
Huang, Zhengrong	1847
Huber, Laszlo	38, 46
Huh, Sungjae	2370
Hui, S.Y. Ron	169, 913, 1169, 1888, 1894, 2328, 3302, 3481
Hui, Zhao	2019

Hull, Brett	979
Husain, Iqbal	2141, 2927, 3383
Hwu, K.I.	2415

I

Iannuzzo, Francesco	288
Ibrahim, Mahmoud	2401, 2700
Iijima, Ryuji	3722
Illa Font, Carlos Henrique	2376
Ilves, Kalle	276
Imura, Takehiro	1731
Inaba, Masamitsu	283
Inokuchi, Seiichiro	468
Inoue, Shuntaro	3146
Ishikuro, Hiroki	1802
Ishizuka, Yoichi	2422, 2430, 2465
Islam, Md. Zakirul	631
Islam, Rakib	3279
Isobe, Takanori	3722
Isurin, Alexander	880
Itagaki, Atsushi	2465
Itakura, Tetsuro	1907
Itoh, Jun-Ichi	1336, 1911
IV, Prasanna	9
Iyer, Vishnu Mahadeva	295
Izuka, Arata	468

J

Jacobina, Cursino B.	2720, 3641
Jahns, Thomas	2167
Jain, Parth	2553
Jain, Praveen	378, 460, 544
Jang, Yujin	3690
Jang, Yungtaek	595, 1292
Jayasinghe, Shantha Gamini	2813
Jedtberg, Holger	3629
Jensen, Scott	1947
Jerinic, Vladan	303
Ji, Junpeng	493, 3115
Ji, Lin	3446
Ji, Shiqi	1456
Jia, Xiaoyu	398
Jiang, Dan	1788
Jiang, Dong	3616
Jiang, Ling	872
Jiang, Qirong	1468
Jiang, W.Z.	2415
Jiang, Xinjian	907
Jiao, Ningfei	2776

Jin, Qian 2409
Jo, Hyunsik 511
Jo, Jongmin 1708
Joffe, Christopher 1741
John, Vinod 951, 2200, 3439
Johnson, J. 1926
Jones, David C. 3273
Jones, Edward A. 1010, 2441
Jones, Vinson 1387
Jourdan, Charlie 3333
Jovanović, Milan M. 38, 46, 1292
Jung, Jae-Jung 2620
Jung, Jee-Hoon 3213
Jung, Kyungsub 17

K

Kakitani, Hisao 2969
Kallfass, Ingmar 2969
Kang, Taeyong 3545
Kang, Yong 3521
Kapat, Santanu 315, 2504, 3224, 3237
Karbalaye Zadeh, Mehdi 446, 3397
Karimi-Ghartemani, Masoud 3165
Karki, Ujjwal 2525
Kashyap, Avinash 3255
Katzir, Liran 3655
Kawai, Yasufumi 2051
Kawajiri, Toru 1802
Kawase, Daisuke 283
Kazama, Taisuke 1585
Ke, Haotao 295
Ke, Xugang 94
Ke, Ziwei 241
Kelly, Anthony 1591, 1670
Kerekes, Tamas 974, 1702, 2264
Khajehoddin, S. Ali 2579, 3207
Khaligh, Alireza 54, 440
Kharezy, M. 3043
Khayat, Joseph 66
Khoshkbar Sadigh, Arash 720, 1191, 1481, 2187, 2840
Khurram, Adil 2782
Kieferndorf, Frederick 472
Kikuchi, Naoto 3146
Kim, Byeong-Heon 1220
Kim, Dong-Hee 1273
Kim, Hyeokjin 1947
Kim, Hyeon-Sik 1206
Kim, Jae-Gu 2161
Kim, Ji-Min 3690

Kim, Jin-Woong 1398
Kim, Jonghoon 1690, 3322
Kim, Minjae 859
Kim, Nari 1273
Kim, Youngjong 2355
Kim, Yun-Sung 163
Kirshenboim, Or 111, 802
Kirtley, James L. 2881
Knott, Arnold 1541, 1842
Ko, Youngjong 3629
Koga, Tomoya 2430
Kolar, Johann W. 615, 623, 1198
Kolluri, Sandeep 552, 2491
Koltsov, H. 339
Kondo, Ryota 1713
Kondo, Takeshi 2788
Konishi, Kyohei 1780
Konrad, Werner 3570
Kostov, Konstantin 1018
Kou, Lei 1489, 2278
Kouchaki, Alireza 2382
Krischan, K. 355, 759
Krishna Moorthy, Radha Sree 794
Krishnamurthy, Mahesh 880, 3128
Kshirsagar, Parag 2027, 3616
Kubendran, S. 663
Kubo, Hajime 2788
Kudva, Sudhir S. 86
Kularatna, Nihal 2479
Kulkarni, Abhijit 2200, 3439
Kulkarni, Onkar Vitthal 2245, 3376
Kulkarni, S. 663
Kulothungan, Gnana Sambandam 402, 943
Kumar, Ashish 1392, 1947, 2395
Kumar, Misha 38, 46, 1292
Kumar, Nikhil 1344
Kumar, V. Inder 3224, 3237
Kumar, V.V.S. Pradeep 2673
Kurokawa, Fujio 754
Kusaka, Keisuke 1336
Kusama, Fumito 2607
Kwak, Sangshin 1748
Kwon, Yong-Cheol 1206
Kyriakides, Elias 3500

L

Lago, Jackson 2833
Lai, Jih-Sheng 1148, 1974
Lai, Jih-Sheng Jason 3243

Lai, Wei-Han 1974
Lam, John 786, 830
Lamar, Diego G. 25, 822, 1823, 2545
Lamo, Paula 2389
Lamoureux, Carl 1882
Langham, Jeff 334
Langmaack, N. 700
Lashway, Christopher R. 1267
Lave, M. 1926
Lazaro, A. 2545, 3090
Lazaro, Orlando 66
Lazzarin, Telles Brunelli 2376
Le, Hoai Nam 1911
Lee, Albert T.L. 169
Lee, Byoung-Kuk 163, 1273, 2161
Lee, C.K. 913
Lee, Eun S. 1773
Lee, Fred C. 322, 343, 657, 1002, 1534, 1608, 1847, 1853, 2334, 3259
Lee, Hyun-jun 1690
Lee, Jaedo 3598
Lee, Jae-Hyun 3086
Lee, Jian-Hsing 1900
Lee, June-Seok 3416
Lee, Junwon 511
Lee, Kevin 2003
Lee, Kun Wang 2875
Lee, Kyo-Beum 2764, 3416
Lee, Kyu-Chan 487
Lee, Moonhyun 1416
Lee, Woongkul 2679, 2861
Lee, Yong-Duk 125
Lee, Yongjae 193
Lee, Younggi 2370
Leeb, Steven B. 2881
Lefranc, Pierre 2401, 2700
Lehman, Brad 417, 2122, 2286, 3038
Lei, Yang 2063
Lei, Yutian 1512
Lemmon, Andrew 1577
Leng, Mingzhi 1554
Leng, Siyu 1941
Lenz, Kevin 303
Leong, K.K. 355
Lequesne, Bruno 2927
Leubner, Martin 2175
Levy, Aron 138
Li, Chendan 3459
Li, Dan 595
Li, David K.W. 1350

Li, Guojie 3560
Li, He 990, 1554, 1967
Li, Helong 704, 3101
Li, Hongxu 1657
Li, Hui 1675, 2237
Li, Jie 2770
Li, Jun 2963
Li, Kai 2613
Li, Kaiyuan 3422
Li, Peide 3697
Li, Qiang 322, 343, 657, 1002, 1534, 1608, 1847, 1853, 2334, 3259
Li, River Tin-Ho 2905
Li, Rui 1675
Li, Sinan 169
Li, Tao 2573
Li, Tengfei 2992
Li, Virginia 343
Li, Wenyu 1450
Li, Wuhua 499, 1462
Li, Xing 728
Li, Xinlei 3623
Li, Xuan 2063
Li, Xueqing 2973
Li, Yalong 2637
Li, Yan 1462
Li, Yan-Cun 1853
Li, Yaohua 3080
Li, Yongdong 3317
Li, Yuan 417, 2525
Li, Yun Wei 1249
Li, Yungui 3560
Li, Yunwei 185
Li, Zhiqing 1329
Li, Zhongxi 2349
Li, Zhongyu 3697
Liang, Beihua 1249
Liang, Lin 2981
Liang, Tsorng-Juu 1900
Liang, Xinyu 2349, 2714
Liao, Yi-Hung 1831
Liao, Zitao 1512
Lidow, Alex 587
Lightbody, Gordon 2794
Liivik, Liisa 2533
Lim, Changjin 2370
Lim, Seungbum 98
Lima, Gustavo B. 3585
Lin, Hua 728, 1078
Lin, L.-C. 951, 1364

Lin, Ni	3189
Lin, P.-H.	1364
Liserre, Marco	2449, 3493, 3629
Lisi, G.	2540
Liu, Baojin	3328, 3370
Liu, Bing	843, 2754
Liu, Bo	966, 2441, 2637
Liu, Fuxin	1788
Liu, Gang	38, 46, 595, 1292
Liu, Haichun	1371
Liu, Haoyan	3180
Liu, Hongpeng	1253
Liu, Jingbo	2147
Liu, Jinjun	739, 2272, 3193, 3328, 3370
Liu, Liming	990
Liu, Pei-Hsin	343
Liu, Pengkun	132, 983
Liu, Sucheng	1489, 2278
Liu, Teng	2272
Liu, Tianshu	899
Liu, Tingting	1410
Liu, Weiguo	1726, 2748, 2776
Liu, Wenbo	899, 2095, 2320
Liu, Wen-Chuen	1512
Liu, Xianzhuo	3317
Liu, Xiaohu	3403
Liu, Xiaokang	2742
Liu, Yan-Fei	899, 1243, 1489, 2087, 2095, 2278, 2304, 2312, 2320
Liu, Yang	959
Liu, Yunting	1045
Liu, Yushan	1214
Liu, Yushi	1392
Liu, Yusi	143
Liu, Zeng	739, 2272, 3328, 3370
Liu, Zhengyang	1847, 1853
Liu, Zhichao	1155
Loh, Poh Chiang	229, 1253, 1872, 1995, 2011, 2207
Lomonova, E.A.	3158
Lope, I.	3020
López del Moral, D.	3090
López, Felipe	2389
Lopez, Ozzie	3065
Lorenz, Robert D.	215, 2055, 2167
Lotfi, Ashraf	1882
Lu, Daorong	2194
Lu, Fei	1721, 1726
Lu, Jie	1392, 1947
Lu, Juncheng	529
Lu, Minghui	1941

Lu, Sizhao	2613
Lu, Ting	1456
Lu, Yong	2215
Lu, Zhengang	2613
Lu, Zhengyu	2003
Lu, Zhigang	398
Lu, Zhouyu	1243
Lucia, Oscar	1040, 1762, 3566
Lukic, Srdjan M.	2349, 2714
Luna, Adriana	1227
Luo, Fang	709, 2981
Luo, Guangzhao	2748
Luo, Haoze	499
Luo, Min	1108
Luo, Tianyi	3065
Lynch, Brian	66

M

Ma, Cong	3279
Ma, Dongsheng	94
Ma, Hongbo	3243
Ma, Jun	499
Ma, Ke	261
Ma, Weizhong	816
Ma, Yingxian	2497
Ma, Yiwei	966, 1261, 2229, 3121
Madhusoodhanan, Sachin	886, 1497, 1632, 2076
Madsen, Mickey P.	1842
Magne, Pierre	453
Mahajan, Anirudh	601
Mahdavikhah, Behzad	329, 3297
Mahmoodzadeh, Zahra	3353
Mainali, Krishna	886, 1497, 1632, 2076
Maitra, Arindam	1974
Makhdoomi Kaviri, Sajjad	378
Makoschitz, M.	1
Maksimović, Dragan	580, 1392, 1947, 2292, 3273
Malcolm, Doug	2087, 2095
Mallik, Ayan	54
Manabe, Shinya	2465
Mandal, Arindam	3237
Mandi, Bipin Chandra	2504
Manjrekar, Madhav	2706
Mansour, Makram	3266
Mantooth, H. Alan	143, 1646, 3012, 3180
Mao, Shuai	2776
Marsili, S.	339
Martín, Kevin	25
Martineau, Donatien	524

Martínez, Gilberto	1115
März, Martin	864, 1741
Mátéfi-Tempfli, Stefan	733
Mathew, Dinto	3286
Matsumori, Hiroaki	676, 3051
Matsuura, Ken	2430
Mattavelli, Paolo	1315
Mauerer, M.	1198
Mawby, Philip	253, 2645
Mazhari, Iman	2998
Mazumder, Paromita	1344
Mazumder, Sudip K.	1344, 1989
Mazzola, Michael	1577
McAmmond, Matt	529
McCann, Roy A.	143
McCue, Benjamin M.	3255
McDonald, Brent	329, 334, 3297
McGrath, Brendan	2252
McHugh, Colin	1947
McIntrye, W.	2540
McKenzie, Craig	3038
McRae, T.	2540
Meder, Dirk	2083
Meere, Ronan	2629
Megyei, George	3255
Mehrizi-Sani, Ali	3353
Mehrotra, Vivek	79
Mekhilef, S.	1163
Méllo, João Paulo R.	2720
Meng, Jinhao	2748
Meng, Peipei	1102
Meng, Tao	2748, 2776
Meola, Marco	1591
Mertens, Axel	3172
Meyer, Jeffrey	1947
Mi, Chris	1721, 1726
Michihira, Masakazu	2607
Mikata, Atsushi	2969
Mikulla, Michael	2083
Miraoui, Abdellatif	3476
Mishima, Tomokazu	1780
Mishra, Richa	2342
Miskovic, Vlatko	2167
Mitra, Rakesh	3279
Miwa, Brett	1597, 3273
Miyazaki, Koutarou	1640
Miyazaki, Takayuki	1907
Modes, Christina	3069
Moeini, Amirhossein	2019

Mohamed, A.A.S. 928
Mohammad, Mostak 1748
Mohammadi, Danyal 2813, 2821
Mohammadi, Mehdi 848
Mohammadpour, Bahador 378
Mohammed, Osama 928, 1267, 3200
Mohan, Ned 1051, 1520, 1934, 1982, 2043
Mojab, Alireza 1989
Molinas, Marta 446, 3397
Mønster, Jakob D. 1842
Moon, Gun-Woo 3690
Moon, Intae 1512
Moon, Seung-Ryul 1974
Morgan, Adam 295, 2141
Moroto, Takahiro 1802
Morris, Casey 2861, 2950
Morsy, Ahmed 2567
Mosesian, Jerry 2122
Moss, Jim 668
Motto, Eric R. 468
Moury, Sanjida 786
Mu, Xianmin 1381
Muetze, A. 355, 759, 3570
Mukherjee, Subhajyoti 2687
Mukhopadhyay, Shayok 2782
Mukhopadhyay, Siddhartha 315
Munk-Nielsen, Stig 288, 704, 974, 1376, 3101
Murmann, Boris 1650
Musavi, Fariborz 772
Musumeci, Salvatore 3669
Muyeen, S.M. 1941

N

Na, Woonki 3322
Nadarajan, Sivakumar 715
Nademi, Hamed 3291
Nagai, Shuichi 2051
Nahid-Mobarakeh, Babak 446, 3397
Nakano, Toshiya 468
Nakao, Hiroshi 754
Nakaoka, Mutsuo 1780
Nakashima, Yoshiyasu 754
Nan, Chenhao 432
Narasimhan, Sneha 2043
Nasr, Miad 1350
Navarro, Angel 3508
Nawaz, Muhammad 276
Nee, Hans-Peter 746, 1018
Neely, J. 1926, 3537

Neft, Charles 79
Negoro, Noboru 2051
Ngo, Khai 668
Nguyen, Duy T. 1773
Ni, Tianheng 2754
Ni, Xijun 983
Niapour, S.A.Kh. Mozaffari 3704
Nikolaidis, Ilias 3069
Ning, Puqi 3080
Ninomiya, Tamotsu 2422, 2430
Nishizawa, Shin-Ichi 3032
Niu, He 2055
Niu, Ying 2770
Noh, Shinyoung 859
Nomura, Katsuya 3146
Nondahl, Thomas A. 2147
Noquil, Jonathan 3065
Norisada, Takaaki 2607
Norum, Lars Einar 3291
Nowak, Torsten 3069
Nymand, Morten 609, 2382

O

O'Donnell, Terence 2629
O'Donovan, Gerard 2794
Ogawa, Taichi 1907
Oh, Chang-Yeol 163
Oh, Jaeyoon 2370
Ojo, Olorunfemi 2035
Okubo, Hiizu 2465
Oliver, J.A. 2409
O'Loughlin, Cathal 2629
O'Mathuna, C. 663
Omura, Ichiro 1640, 3032
Oña, E. 2545
Onal, Yasemin 2693
Onar, O.C. 1307
Orabi, Mohamed 1505
Ordonez, Martin 386, 848, 854, 2561
Orikawa, Koji 1336, 1911
Orr, Ray 1350
Ortiz-Gonzalez, Jose 253
Ortiz-Rivera, Eduardo I. 3712
Otsuka, Masafumi 1350
Ouyang, Ziwei 2473
Ozimek, Patrick E. 1084
Ozpineci, Burak 3529

P

Padhee, Varsha ... 1982
Pagano, Rosario ... 3669
Pahlevani, Majid ... 378
Pala, Vipindas ... 979
Palaniappan, Vishal ... 1350
Palmour, John ... 979
Pam, Srikanth ... 3266
Pan, Bing ... 2613
Panda, S.K. ... 9, 552, 715, 2491
Park, Hwa-Pyeong ... 3213
Park, Joung-hu ... 1690
Park, Sung-Yeul ... 125
Parkhideh, Babak ... 2998
Parsa, Leila ... 2573
Parvez, M. ... 1163
Patel, Ankur ... 150
Pathan, Abrar Ahmed ... 2497
Patil, Devendra ... 2889
Patra, Amit ... 2504
Pavlick, Stephanie A. ... 1138
Pavlovic, Z. ... 663
Paz, Francisco ... 386
Pedersen, Jeppe A. ... 1541, 1842
Peixoto, Paulo P. ... 3346
Peng, Chang ... 132, 269, 983, 2927
Peng, Fang Z. ... 959, 1045, 2525
Peng, Hao ... 1450
Peng, Jichang ... 2748, 2776
Peng, Kang ... 2127
Peng, Li ... 1358
Perales, Mico ... 1967
Perdigão, Marina ... 1115
Peretz, Mor Mordechai ... 111, 308, 692, 802, 2298
Perez, Aday ... 215
Pérez-Tarragona, Mario ... 1762
Perreault, David J. ... 98, 1138
Perrin, Remi ... 524
Perry, Jeff ... 3266
Persons, Ryan ... 3069
Pervaiz, Saad ... 1947, 2395
Peterchev, Angel V. ... 2349
Pevere, Alessandro ... 1183
Phillips, Evan ... 3684
Piepenbreier, Bernhard ... 2897
Pierfederici, Serge ... 446, 3397
Pigazo, Alberto ... 2389
Pilawa-Podgurski, Robert C.N. ... 1512
Ping, Dinggang ... 38, 46

Piya, Prasanna 3165
Pong, M.H. Bryan 2328
Poshtkouhi, Shahab 1350, 1403
Pou, Josep 236
Praça, Paulo P. 60, 3231
Pramod, Prerit 3279
Prasai, Anish 1437
Preciat, Philippe 524
Prieto, R. 1624
Prodić, Aleksandar 329, 1597, 2497, 2540, 2553, 3297
Puukko, Joonas 990

Q

Qi, Feng 2912
Qian, Qiang 1919, 3446
Qiao, Wei 3514
Qin, Jiangchao 1528
Qin, Liang 2525
Qin, Shibin 1512
Qin, Xianhui 843
Qiu, Maohang 138
Qiu, Yajie 2320
Qu, Liyan 3279, 3514
Qu, Xiaohui 2154
Quan, Zhongyi 185
Quay, Rüdiger 2083
Quentin, Nicolas 524

R

Raciti, Angelo 3669
Rahnamaee, Arash 1989
Raizada, Shirish 1344
Rajashekara, Kaushik 2788
Ramachandran, Rakesh 609
Ramadass, Yogesh 838
Ramani, Ramanathan 66
Rambal-Vecino, Andres 3712
Ramezani, Medhi 2035
Ramos, Francisco 215
Ramu, Krishnan 2027
Ran, Li 253, 1443, 2645
Ranstad, Per 1018
Rao, Yuan 1585
Rashkin, L. 3537
Rathore, Akshay K. 402, 794, 943
Ravey, Alexandre 3476
Redondo, Luís M. 3637
Rehman, Habibur 2782
Reiner, Richard 2083

Reitz, Jessica 3069
Remus, Nico 2175
Ren, Hai-Peng 2770
Ren, Ren 2441
Ren, Xiaoyong 73, 2518, 3488
Ren, Xizhou 2912
Ren, Yu 2071, 2102
Rengifo, Johnny 3200
Renjit, Ajit A. 1682
Reusch, David 587
Riazmontazer, Hossein 1989
Rim, Chun T. 1773
Roasto, Indrek 2533
Robbins, William 1934
Roberts II, Charles 3255
Rodrigues, Danillo B. 3585
Rodriguez, A. 1823
Roehrs, Benjamin D. 3255
Rogers, Daniel J. 1650
Romero, David 1350
Rosahl, Thoralf 106
Roßkopf, Andreas 1741
Round, W. Howell 2479
Ruan, Xinbo 1788, 3488
Ruiz, Juan M. 1292

S

Sá Jr., Edilson M. 2592
Saasaa, Raed 1286
Sadik, Diane-Perle 746, 1018
Saeed, Sarah 3508
Saeedifard, Maryam 1528, 2136, 2957
Safaee, Alireza 460
Saha, Aparna 2806
Sahoo, Ashish Kumar 1982
Sahoo, Saroj Kumar 199
Saito, Katsuaki 283
Saito, Shoji 468
Saket, Mohammad Ali 854, 2561
Sakurai, Takayasu 1640
Salameh, Mohamad 3128
Salcines, Cristino 2969
Salem, Ahmed 1505
Sandoval, José Juan 3545
Sangwongwanich, Ariya 370
Sankman, Joseph 94
Santi, Enrico 2127, 3248
Santiago-González, Juan A. 98
Santos de Moura, Diogo Cesar 3553

Santos Guimarães, Jéssica 3346
Sanz, M. 2545, 3090
Sariri, Kouros 3255
Sarlioglu, Bulent 2679, 2861, 2950
Sarnago, Hector 1040, 1762, 3566
Satija, Yudhister 3266
Sato, Masaki 2465
Saublet, Louis-Marie 3397
Saur, Michael 215
Savaghebi, Mehdi 1227, 3697
Scandrett, Brad 417
Schmidt, Peter B. 2147
Schubert, Michael 643
Schweitzer, Ben 3128
Sebastián, Javier 25, 822, 1823
Seeman, Michael 838
Seltzer, Daniel 1947
Sen, Paresh C. 1489, 2278
Senanayake, Thilak 3722
Senol, Murat 643
Seo, Gab-Su 487
Sepahvand, Alihossein 580, 1947
Serrano, J. 3020
Setyawan, Leonardy 3409
Severson, Eric 2043
Shafiei, Navid 848, 854, 2561
Shagar, Viknash 2813
Shah, Neel 2998
Shahbazi, Caitlin 3069
Shamsi, Pourya 1962, 2687
Shang, Fei 880
Shao, Jianwen 3659
Sharkh, Suleiman M. 2223
Sharma, Ratnesh 1682
Shen, Ang 1962
Shen, Guangtong 2003
Shen, Zhiyu 516, 1561
Sheng, Su 417, 2286
Shenoy, Pradeep S. 66
Shi, Baoping 843
Shi, Jianjiang 1475
Shi, Xiaojie 2637
Shi, Yuxiang 1675
Shimizu, Toshihisa 676, 3051
Shiu, T.-H. 1364
Shmilovitz, Doron 3655
Shousha, Mahmoud 3097
Shoyama, Masahito 2735
Shrivastav, Ashish 601

Shu, Zhan	2223
Shukla, Anshuman	2342, 3286
Silva, J. Fernando	3637
Silva, Paulo R.	3585
Silva, Rafael V.	2592
Simanjorang, Rejeki	2942, 3058
Singh, Amandeep	3140
Singh, Shikhar	601
Singh, Surinder P.	1585
Sinha, Sreyam	1947
Siwakoti, Yam P.	1872
Sleik, Roland	759
Sohn, Hoon	3690
Soltani, Hamid	229
Somani, Apurva	1520
Somani, Utsav	3333
Son, Yeongrack	3717
Son, Young-Kwang	2370
Song, Xiaoqing	132, 269, 983
Soni, Harshit	1344
Sozer, Yilmaz	207, 392, 638, 2693, 2806
Srdic, Srdjan	2714
Srinivasan, Dipti	402, 943
Srivastava, Vineet	786
Stack, David	1670
Stamm, Thomas	3659
Steenis, Joel	3219
Steyn-Ross, D. Alistair	2479
Stillwell, Andrew	1512
Strydom, Johan	587, 2292
Stübig, Marc	2175
Styles, Julian	529
Su, Yipeng	118
Subotic, Ivan	623
Suh, Yongsug	17
Sul, Seung-Ki	1206, 1220, 2370, 2620
Sun, Bo	1376
Sun, Kai	1227, 1410, 2182, 2194
Sun, Lei	505
Sun, Lejia	810
Sun, Libing	1227
Sun, Pengju	2992
Sun, Wei	1462
Sung, Won-Yong	163
Sveum, Peter	3128

T

T T, Anandha Ruban	1878
Tabata, Osamu	2051

Tadano, Hiroshi 3722
Tadeparthy, Preetam 1878
Tai, Heng-Ming 2992
Takamiya, Makoto 1640
Takano, Koushi 676
Talebi Khanmiri, Dawood 2122, 3038
Tan, Kai 132, 983
Tan, Linlin 2102
Tan, Nadia M.L. 1163
Tan, Pingan 3185
Tan, Siew-Chong 169, 913, 1169, 1888, 1894, 3302, 3481
Tang, Yi 1071, 3591
Tang, Yichao 440
Tang, Yuan 1443, 2645
Tareilus, G. 700
Tayebi, S. Milad 1381
Teixeira, Carlos 2252
Tekgun, Burak 207
Teodorescu, Remus 974, 1702, 2264
Tewari, Saurabh 1520, 2043
Thiringer, T. 3043
Thone, Jef 3097
Tian, Shuilin 1608
Tian, Ye 499
Ting, Lo Pang-Yen 1900
Tkachov, Sergii 350, 1663
Tolbert, Leon M. 893, 966, 1261, 1307, 1569, 2637, 3121
Tomas-Manez, Kevin 1235
Tomioka, Satoshi 2430
Tong, C.F. 2942, 3058
Trabelsi, Mohamed 3663
Tran, Yan-Kim 156
Trento, Brad 3577
Trescases, Olivier 1350, 1403, 1882
Trintis, Ionut 1376
Tripathi, Awneesh 886, 1497, 1632, 2076
Tse, Zion Tsz Ho 1737, 3526
Tseng, King Jet 1071
Tseng, K.J. 2942, 3058, 3107, 3422, 3591
Tsukuda, Masanori 1640
Tung, Chung-Pui 1807
Tüysüz, Arda 615, 623, 1198

U

Uceda, J. 1624
Uddin, Md Wasi 638
Ueda, Tesuzo 2051
Ueno, Takeshi 1907
Ugur, Enes 1176

Urteaga, Miguel 79

V

Vasić, M. 2409
Vásquez, Juan C. 1227, 3459
Vázquez, A. 25, 2545
Vechalapu, Kasunaidu 295, 886, 1497, 1632, 2076
Vekslender, Timur 308
Venkataramanan, Giri 2919
Venkateswaran, Muthusubramanian 1878
Vermulst, B.J.D. 3158
Vermulst, Bas 2457
Vesti, S. 339
Vilathgamuwa, Mahinda 2813
Villarejo, J.A. 1823
Vinnikov, Dmitri 2533
Vitorino, Montiê A. 1032
Vrankovic, Zoran 1084
Vu, Trong Tue 1835, 2485
Vukadinović, Nenad 1597

W

Wada, Keiji 1640, 2986
Walsh, Ray 2794
Waltereit, Patrick 2083
Wang, Chengshan 1249
Wang, Fan 334
Wang, Feng 810, 2742
Wang, Fred 424, 893, 966, 1010, 1261, 1456, 1569, 2229, 2441, 2637, 3121
Wang, Gangyao 979
Wang, Guo-Xiang 1123
Wang, Haoyu 480, 1280, 1329
Wang, Hongliang 899, 1489, 2278, 2304, 2312, 2320
Wang, Huai 370, 2154
Wang, Jiangfeng 2182
Wang, Jin 990, 1554, 1967
Wang, Jun 516
Wang, Kui 3317
Wang, Kun 1450
Wang, Laili 899, 2087, 2095, 2320
Wang, Li 2365
Wang, Long 2754
Wang, Meilin 1078
Wang, Meng-Jie 3471
Wang, Mengqi 3648
Wang, Miao 709, 2912
Wang, Ming-Hao 3302
Wang, N. 663
Wang, Peng 1071, 3409, 3591

Wang, Qin 1657
Wang, Shike 3328, 3370
Wang, Shiliang 2889
Wang, Shuo 2019, 3603
Wang, Wei 1253
Wang, Xiaoping 572
Wang, Xingwei 1078
Wang, Xiongfei 704, 1253, 1941, 2011, 2207, 2215, 3051, 3101, 3431
Wang, Yanbo 3431
Wang, Yi 1815
Wang, Zhenxiong 3358
Wang, Zhiqiang 2637
Watanabe, Hiroki 1336
Watanabe, Yoshitoshi 3146
Wattes, J.L. 3231
Weber, Bastian 3172
Wei, Chun 3514
Wei, Jiadan 843, 2754
Wei, Lixiang 1065, 2826
Wei, Yingdong 1468
Weise, Nathan 1065
Weiss, Beatrix 2083
Wen, Changyun 3409
Wen, Lucheng 990
Wen, Xuhui 3080
Wens, Mike 3097
Wespel, Matthias 2083
Wicht, Bernhard 106
Wijnands, C.G.E. 3158
Wiktor, Wlodek 66
Williamson, Sheldon S. 3140
Wilson, D. 3537
Winterhalter, Craig 1084
Wittmann, Jürgen 106
Wu, Dalei 3189
Wu, Hongfei 1410, 1424
Wu, John 1967
Wu, Liyao 2136, 2957
Wu, Qunfang 1657
Wu, T.-F. 951, 1364
Wu, Teng 3328, 3370
Wu, Tong 3529

X

Xia, Yinglai 3364
Xiao, Guochun 2215
Xiao, Jianfang 3409
Xiao, Lan 1657
Xiao, Xi 1424, 3338

Xie, Shaojun 1371, 1919, 3446
Xie, Xiaogao 816
Xie, Yicong 2360
Xin, Zhen 1995, 2207, 3697
Xing, Xiangyang 3453, 3465
Xing, Yan 1410, 1424, 2182, 2194
Xiong, Liansong 2742
Xiong, Song 1888, 1894
Xu, Chen 1358
Xu, Dewei David 33
Xu, Dianguo 1253
Xu, Jialin 1657
Xu, Jing 990
Xu, Jinming 1919, 3446
Xu, Longya 709, 2912
Xu, Qianwen 3409
Xu, Tao 921
Xu, Wei 2868
Xu, Yang 2141
Xue, Fei 3677
Xue, Lingxiao 1315

Y

Yadav Gorla, Naga Brahmendra 552
Yadav, Akshat 2076
Yamada, Go 2607
Yamada, Masaki 1713
Yamaguchi, Koji 3075
Yamaguchi, Masahiro 2465
Yamamoto, Keiichi 283
Yamamoto, Yasuhiro 2788
Yan, Shuo 913
Yan, Xingda 2223
Yanagi, Hiroshige 2430
Yang, Ching-Chieh 558
Yang, Enxing 499
Yang, Heya 1462
Yang, Hongbin 1078
Yang, Jianwei 3134
Yang, Liu 1261, 3121
Yang, Pengzhi 1554
Yang, Qichen 2957
Yang, Shuitao 959
Yang, Shunfeng 1071, 3591
Yang, Tao 2629
Yang, Tianbo 913
Yang, Xu 493, 766, 2071, 2102, 3115, 3623
Yang, Yang 1243
Yang, Yong 1788

Yang, Yongheng	221, 370, 1253, 3500
Yang, Yuanyu	843
Yang, Yuchen	1534
Yang, Yun	1169, 3481
Yang, Zhihua	33, 899, 2312, 2320
Yao, Chengcheng	990, 1554
Yao, Jianhui	2182, 2194
Yao, Kai	572, 1815
Yao, Wenxi	2003
Yau, Y.T.	2415
Yazdanian, Mehrdad	3353
Ye, Qing	2237
Ye, Zichao	1512
Yeo, H.L.	3107
Yeong, Lee Meng	3409
Yi, Fan	1057, 1861, 2599
Yi, Hao	3358
Yin, Shan	2942, 3058
Yonezawa, Yu	754
Young, George	1835, 2485
Yu, Hualong	1456
Yu, Ruiyang	3677
Yu, Sheng-Yang	2511
Yu, Wensong	2063, 2141, 2365, 2714, 2727, 3648
Yu, Xinyu	1468
Yu, Yanqi	1286
Yuan, Huawei	907
Yuan, Liqiang	2613

Z

Zafarani, Mohsen	1096, 2875
Zagrodnik, Michael	1071, 3591
Zane, Regan	1603
Zare, Firuz	221, 229
Zefran, Milos	1989
Zeltser, Ilya	802
Zeng, Hulong	1045
Zeng, Xiangjun	2102
Zhak, Serhii M.	3273
Zhan, Xiaohai	1424
Zhan, Xiaoqing	2154
Zhang, Bin	1155
Zhang, Binfeng	1919
Zhang, Canhui	138
Zhang, Chengduo	2349
Zhang, Chenghui	3453, 3465
Zhang, Chi	2714
Zhang, Dan	766, 3623
Zhang, Fan	1947, 2071, 2102

Zhang, Hao	2973
Zhang, Haojiong	2167
Zhang, Hua	1721, 1726
Zhang, Hui	3338
Zhang, Jianqiu	595
Zhang, Julia	241
Zhang, Jun	2992
Zhang, Junfang	572
Zhang, Ke	3476
Zhang, Lanhua	1148, 1974, 3243
Zhang, Li	3488
Zhang, Li-Heng	2770
Zhang, Lin	1868
Zhang, Liqi	269, 2063, 2365
Zhang, Shuoting	966, 3121
Zhang, Weimin	424, 893
Zhang, Wenli	524, 1002
Zhang, Xiangming	1102
Zhang, Xuan	990, 1967, 2229
Zhang, Xuning	177, 1024, 1561
Zhang, Yongchang	2868
Zhang, Yuanzhe	580, 2292
Zhang, Yuzhi	3180
Zhang, Zhe	1090, 1235, 1430, 2252, 3514
Zhang, Zhen	3134
Zhang, Zheyu	684, 1010, 1569, 2441
Zhang, Zhigang	3358
Zhang, Zhiliang	73, 1243, 2518
Zhang, Zhiyu	1475
Zhang, Zicheng	3453, 3465
Zhao, Bo	2912
Zhao, Chongwen	3577
Zhao, Dongdong	3134
Zhao, Hengyang	1475
Zhao, Hui	3603
Zhao, Rende	1995, 2207, 3697
Zhao, Shuze	1882
Zhao, Tao	33
Zhao, Xiaonan	1148, 3243
Zhao, Xin	295
Zhao, Zhengming	1456, 2613
Zheng, Sheng	966
Zheng, Yue	417
Zheng, Zedong	3317
Zhi, Na	3338
Zhou, Bo	843, 2754
Zhou, Daming	3476
Zhou, Jinping	2360
Zhou, Keliang	1155

Zhou, Liwei	410, 2259, 2935
Zhou, Luowei	2992
Zhou, Min	2360
Zhou, Qi	536
Zhou, Sizhan	3193
Zhou, Yong	2567
Zhou, Yuan	73, 2518
Zhou, Zhe	2912
Zhu, Bohang	2788
Zhu, Donghai	3521
Zhu, Guorong	3310
Zhu, Ke	990
Zhu, Qianlai	2365
Zhu, Tianhua	810
Zhuo, Fang	810, 2742, 3358
Zojer, Bernhard	996
Zou, Juan	651
Zou, Ke	1554
Zou, Xudong	3521
Zou, Xuewen	73, 2518
Zou, Zhi-Xiang	3493
Zumel, P.	2545, 3090

IEEE
445 Hoes Lane
Piscataway, NJ 08854-4141

ISBN 978-1-4673-9551-9